Tropical Fruit Flies (Tephritidae: Dacinae) of South-East Asia

Indomalaya to North-West Australasia

Tropical Fruit Flies (Tephritidae: Dacinae) of South-East Asia

Indomalaya to North-West Australasia

Richard A.I. Drew and Meredith C. Romig

International Centre for the Management of Pest Fruit Flies, Griffith University, Australia

www.cabi.org

CABI is a trading name of CAB International

CABI
Nosworthy Way
Wallingford
Oxfordshire OX10 8DE
UK

Tel: +44 (0)1491 832111
Fax: +44 (0)1491 833508
E-mail: info@cabi.org
Website: www.cabi.org

CABI
38 Chauncey Street
Suite 1002
Boston MA 02111
USA

Tel: +1 800 552 3083 (toll free)
Tel: +1 617 395 4051
E-mail: cabi-nao@cabi.org

A catalogue record for this book is available from the British Library, London, UK.

Library of Congress Cataloguing-in-Publication Data

Drew, R. A. I.
Tropical fruit flies (Tephritidae: Dacinae) of South-East Asia: Indomalaya to North-West Australasia / by Richard A.I. Drew and Meredith C. Romig.
p. cm.
Includes bibliographical references and index.
ISBN 978-1-78064-035-8 (alk. paper)
1. Fruit-flies--Southeast Asia--Classification. I. Romig, Meredith C. II. Title.

QL537.M6D695 2013
595.77′4--dc23

2012043422

ISBN-13: 978 1 78064 035 8

Commissioning editor: Nigel Farrar
Editorial assistant: Alexandra Lainsbury
Production editor: Shankari Wilford

Typeset by SPi, Pondicherry, India
Printed and bound in the UK by CPI Group (UK) Ltd, Croydon, CR0 4YY

Contents

Acknowledgements vii

Abstract 1

Introduction 5

Methods and Abbreviations 7

Generic and Subgeneric Classification 11
Definitions of Genera 11
Definitions of Subgenera 12

Species Listed Under Genera and Subgenera 17

Species Complexes 23
Bactrocera (Bactrocera) dorsalis complex 23
Bactrocera (Bactrocera) nigrotibialis complex 24
Bactrocera (Zeugodacus) scutellaris complex 24
Bactrocera (Zeugodacus) tau complex 24

Identification of Cryptic Species 25

Description of Species Under Genera and Subgenera 27
Genus *Bactrocera* Macquart 29
Subgenus *Afrodacus* Bezzi 29
Asiadacus Perkins 29
Bactrocera Macquart 36
Bulladacus Drew & Hancock 195
Daculus Speiser 199
Gymnodacus Munro 201
Hemigymnodacus Hardy 205
Javadacus Hardy 208
Nesodacus Perkins 214
Papuodacus Drew 215
Paradacus Perkins 219
Parasinodacus new subgenus 225
Paratridacus Shiraki 237
Parazeugodacus Shiraki 243
Semicallantra Drew 249
Sinodacus Zia 251
Tetradacus Miyake 259
Zeugodacus Hendel 263

Genus *Dacus* Fabricius 373
Subgenus *Callantra* Walker 373
Didacus Collart 374
Leptoxyda Macquart 375
Mellesis Bezzi 376
Neodacus Perkins 401

Genus *Ichneumonopsis* Hardy 403

Genus *Monacrostichus* Bezzi 405

Pest Species in the Genera *Bactrocera* and *Dacus* 408

Figures 411

References 639

Taxonomic Index 647

Acknowledgements

Very large collections of specimens and biological data for this research have been obtained through fruit fly projects conducted in countries across Indomalaya, Wallacea and Australasia. These field activities have resulted in a world-class collection curated by M.C. Romig (one of the authors) and Sarah Romig, and now deposited in QDPC. Many institutions, noted in the text, provided specimens for study (particularly type specimens) and some, namely The Natural History Museum, London; the Bernice P. Bishop Museum, Honolulu; and Naturhistorisches Museum Wien, Vienna, made facilities available for on-site research. Bert Orr studied specimens and prepared artwork at BMNH and MZB. Many institutions provided staff and support for field collections and botanical identifications over many years and have been acknowledged in Allwood *et al*. (1999) and Drew & Hancock (1994a). In addition to these, field collections in Indonesia were conducted under the support of the Department of Horticulture Protection and Agricultural Quarantine Agency, Ministry of Agriculture, Jakarta; in Bhutan through the Ministry of Agriculture, Thimphu; and in Vietnam through the Ministry of Agriculture and Rural Development, Hanoi. The artwork was produced by Michelle Baker (MB), Clare Bremner (CB), Chris Lambkin (CLL), Amy Lawson (AL), Yanni Martin (YM), Bert Orr (AGO), Susan Phillips (SP), Meredith Romig (MCR) and Susan Sands (SS). Text and scanning of artwork were provided by Karen Drew and Sarah Romig. Dr Paul Ferrar, formerly of CSIRO and ACIAR, Australia, an internationally recognized Dipterist, provided support and encouragement for this research. Dr David Hancock, Australia provided advice on taxonomic nomenclature and the ICZN. All this assistance is gratefully acknowledged.

Abstract

The species within the Tribe Dacini, from Indomalaya to north-west Australasia, are described and illustrated. In all, 423 species are recorded from this region and treated under four genera, *Bactrocera* Macquart (with 18 subgenera), *Dacus* Fabricius (with five subgenera), *Ichneumonopsis* Hardy and *Monacrostichus* Bezzi. One new subgenus is described, *Parasinodacus*, to include eight previously known species and five new species.

123 new species are described –
Bactrocera (Asiadacus) careomacula; *Bactrocera (Bactrocera) affinibancroftii*, *B. (B.) apicopicta*, *B. (B.) avittata*, *B. (B.) bellisi*, *B. (B.) bhutaniae*, *B. (B.) binhduongiae*, *B. (B.) bitungiae*, *B. (B.) blairiae*, *B. (B.) bruneiae*, *B. (B.) careofascia*, *B. (B.) commensurata*, *B. (B.) curtivitta*, *B. (B.) curvosterna*, *B. (B.) dongnaiae*, *B. (B.) elongata*, *B. (B.) eurycosta*, *B. (B.) flavosterna*, *B. (B.) fuscoformosa*, *B. (B.) fuscolobata*, *B. (B.) fuscoptera*, *B. (B.) halmaherae*, *B. (B.) illusioscutellaris*, *B. (B.) inconspicua*, *B. (B.) jaceobancroftii*, *B. (B.) kalimantaniae*, *B. (B.) laithieuiae*, *B. (B.) mamaliae*, *B. (B.) mediorufula*, *B. (B.) nanoarcuata*, *B. (B.) nationigrotibialis*, *B. (B.) neofulvicauda*, *B. (B.) neoritsemai*, *B. (B.) nothaphoebe*, *B. (B.) obscurivitta*, *B. (B.) ochroma*, *B. (B.) ochroventer*, *B. (B.) opacovitta*, *B. (B.) paraarecae*, *B. (B.) paralatissima*, *B. (B.) paralimbifera*, *B. (B.) paranigrita*, *B. (B.) patula*, *B. (B.) peneallwoodi*, *B. (B.) penebeckerae*, *B. (B.) penecostalis*, *B. (B.) penephaea*, *B. (B.) pruniae*, *B. (B.) pseudobeckerae*, *B. (B.) quasiinfulata*, *B. (B.) quasineonigrita*, *B. (B.) ranganathi*, *B. (B.) sapaensis*, *B. (B.) suliae*, *B. (B.) tapahensis*, *B. (B.) ternatiae*, *B. (B.) truncata*, *B. (B.) waaiae*; *Bactrocera (Bulladacus) captiva*, *B. (B.) cinnabaria*; *Bactrocera (Daculus) rutengiae*; *Bactrocera (Gymnodacus) symplocos*; *Bactrocera (Hemigymnodacus) mukiae*; *Bactrocera (Javadacus) semisurstyli*; *Bactrocera (Papuodacus) fereuncinata*, *B. (P.) ochrosterna*; *Bactrocera (Paradacus) hancocki*; *Bactrocera (Parasinodacus) brevivitta*, *B. (P.) citrifusca*, *B. (P.) pahangiae*, *B. (P.) pantabanganiae*, *B. (P.) waimitaliae*; *Bactrocera (Parazeugodacus) fulvosterna*; *Bactrocera (Semicallantra) incompta*, *B. (S.) superba*; *Bactrocera (Sinodacus) speciosa*, *B. (S.) spectabilis*, *B. (S.) whitei*; *Bactrocera (Zeugodacus) aithonota*, *B. (Z.) apicofemoralis*, *B. (Z.) baliensis*, *B. (Z.) borongensis*, *B. (Z.) daclaciae*, *B. (Z.) dorsirufa*, *B. (Z.) flavolateralis*, *B. (Z.) flavoverticalis*, *B. (Z.) fuscoalata*, *B. (Z.) hatyaiensis*, *B. (Z.) havelockiae*, *B. (Z.) hengsawadae*, *B. (Z.) hoabinhiae*, *B. (Z.) hodgsoniae*, *B. (Z.) hululangatiae*, *B. (Z.) iriomotiae*, *B. (Z.) juxtuncinata*, *B. (Z.) khaoyaiae*, *B. (Z.) laguniensis*, *B. (Z.) laocaiae*, *B. (Z.) liquida*, *B. (Z.) macrophyllae*, *B. (Z.) melanofacies*, *B. (Z.) nakhonnayokiae*, *B. (Z.) namlingiae*, *B. (Z.) neoemittens*, *B. (Z.) neoflavipilosa*, *B. (Z.) neolipsana*, *B. (Z.) pemalangiae*, *B. (Z.) quasiinfesta*, *B. (Z.) sabahensis*, *B. (Z.) sasaotiae*, *B. (Z.) semongokensis*, *B. (Z.) sonlaiae*, *B. (Z.) tebeduiae*, *B. (Z.) trichosanthes*, *B. (Z.) tricuspidatae*, *B. (Z.) trivandrumensis*, *B. (Z.) ujungpandangiae*, *B. (Z.) uncinata*, *B. (Z.) yalaensis*; *Dacus (Mellesis) discretus*, *D. (M.) donggaliae*, *D. (M.) melanopectus*, *D. (M.) triquetrus*.

One new name is proposed –
Bactrocera (Bactrocera) linduensis for the preoccupied name *enigmatica* Hardy.

Fifty-six species are newly placed in synonymy –
Asiadacus ***absonus*** Hering of *Bactrocera (Gymnodacus) tillyardi* (Perkins)
Dacus (Strumeta) ***aculeus*** Hardy of *B. (Zeugodacus) rubella* (Hardy)
Dacus (Zeugodacus) ***adustus*** Wang & Zhao of *B. (Zeugodacus) biguttata* (Bezzi)
Bacterodacus [sic] *(Zeugodacus)* ***apicinigra*** Yu, Bai & Yang of *B. (Zeugodacus) maculata* (Perkins)
Dacus (Zeugodacus) ***aptatus*** Hardy of *B. (Paratridacus) melanius* (Hardy & Adachi)
Dacus (Zeugodacus) ***ascitus*** Hardy of *B. (Sinodacus) bogorensis* (Hardy)
Dacus (Callantra) ***bimaculatus*** Lin, Wang & Zeng of *Dacus (Mellesis) conopsoides* de Meijere
Bactrocera (Bactrocera) ***cheni*** Zhang, Ji & Chen of *B. (Bactrocera) lombokensis* Drew & Hancock

Sinodacus ***chonglui*** Chao & Lin of *D. (Neodacus) sphaeroidalis* (Bezzi)
Dacus (Zeugodacus) ***deceptus*** Hardy of *B. (Parazeugodacus) tetrachaeta* (Bezzi)
Dacus (Asiadacus) ***dianensis*** Wang & Zhao of *B. (Asiadacus) apicalis* (de Meijere)
Dacus (Pacifodacus) ***drewi*** Hardy of *B. (Parasinodacus) vinnulus* (Hardy)
Dacus (Zeugodacus) ***elegantis*** Tseng, Chen & Chu of *B. (Zeugodacus) tau* (Walker)
Dacus (Zeugodacus) ***flavus*** Tseng, Chen & Chu of *B. (Zeugodacus) tau* (Walker)
Bactrocera (Zeugodacus) ***fraserensis*** Chua of *B. (Zeugodacus) scutellaris* (Bezzi)
Bactrocera (Zeugodacus) ***gansuica*** Chen, Han & Zhou of *B. (Zeugodacus) assamensis* White
Dacus (Zeugodacus) ***guangxianus*** Chao & Lin of *B. (Zeugodacus) diaphora* (Hendel)
Dacus (Callantra) ***haikouensis*** Wang & Chen of *D. (Mellesis) pictus* Hardy
Bactrocera (Bactrocera) ***hardyi*** Zhang, Ji & Chen of *B. (Bactrocera) thailandica* Drew & Hancock
Bactrocera (Bactrocera) ***heppneri*** White of *B. (Bactrocera) obscurata* (de Meijere)
Dacus (Bactrocera) ***hsui*** Tseng, Chen & Chu of *B. (Bactrocera) hyalina* (Shiraki)
Sinodacus ***jiannanus*** Chao & Lin of *B. (Sinodacus) hochii* (Zia)
Bactrocera (Bactrocera) ***jinghongensis*** Zhang, Ji & Chen of *B. (Bactrocera) citima* (Hardy)
Sinodacus ***jinreni*** Chao & Lin of *B. (Sinodacus) hochii* (Zia)
Dacus (Zeugodacus) ***katoi*** Hardy of *B. (Paratridacus) yayeyamanus* (Matsumura)
Bactrocera (Bactrocera) ***lacerata*** White & Evenhuis of *B. (Bactrocera) umbrosa* (Fabricius)
Sinodacus ***laterum*** Wang of *B. (Sinodacus) hochii* (Zia)
Dacus ***latifae*** Anwar Cheema of *B. (Hemigymnodacus) diversa* (Coquillett)
Dacus (Zeugodacus) ***lunulatus*** Tseng, Chen & Chu of *B. (Zeugodacus) diaphora* (Hendel)
Dacus (Strumeta) ***luzonae*** Hardy & Adachi of *B. (Bactrocera) nigrotibialis* (Perkins)
Dacus (Asiadacus) ***modicus*** Hardy of *B. (Asiadacus) apicalis* (de Meijere)
Dacus (Bactrocera) ***mulyonoi*** Hardy of *B. (Bactrocera) nigrotibialis* (Perkins)
Dacus (Asiadacus) ***nadanus*** Chao & Lin of *B. (Asiadacus) apicalis* (de Meijere)
Dacus (Bactrocera) ***paratappanus*** Tseng, Chen & Chu of *B. (Zeugodacus) tappanus* (Shiraki)
Dacus (Zeugodacus) ***parater*** Chao & Lin of *B. (Zeugodacus) atrifacies* (Perkins)
Dacus (Zeugodacus) ***parvifoliaceus*** Tseng, Chen & Chu of *B. (Parazeugodacus) arisanica* (Shiraki)
Dacus (Zeugodacus) ***pauleiensis*** Tseng, Chen & Chu of *B. (Zeugodacus) ambigua* (Shiraki)
Dacus (Zeugodacus) ***personatus*** Hardy of *B. (Zeugodacus) vulta* (Hardy)
Bactrocera (Bactrocera) ***philippinensis*** Drew & Hancock of *B. (Bactrocera) papayae* Drew & Hancock
Bactrocera (Zeugodacus) ***proprediaphora*** Wang, Xiao & Chen of *B. (Zeugodacus) diaphora* (Hendel)
Dacus (Zeugodacus) ***pseudoscutellatus*** Tseng, Chen & Chu of *B. (Zeugodacus) scutellata* (Hendel)
Sinodacus ***qionganus*** Chao & Lin of *B. (Sinodacus) hochii* (Zia)
Sinodacus ***quaternum*** Wang of *B. (Sinodacus) hochii* (Zia)
Sinodacus ***rubzovi*** Chao & Lin of *B. (Sinodacus) hochii* (Zia)
Dacus (Bactrocera) limbifer ***rufulus*** Hardy of *B. (Bactrocera) ritsemai* (Weyenbergh)
Bactrocera (Bactrocera) ***ruiliensis*** Wang, Long & Zhang of *B. (Bactrocera) thailandica* Drew & Hancock
Dacus (Bactrocera) ***semifemoralis*** Tseng, Chen & Chu of *B. (Bactrocera) dorsalis* (Hendel)
Dacus (Bactrocera) ***silvaticus*** Hardy of *B. (Bactrocera) lata* (Perkins)
Dacus (Zeugodacus) ***stenomus*** Wang & Zhao of *B. (Zeugodacus) synnephes* (Hendel)
Dacus (Bactrocera) ***sumatranus*** Hardy of *B. (Bactrocera) nigrotibialis* (Perkins)
Dacus (Zeugodacus) ***tenuifinis*** Hardy of *B. (Parasinodacus) cilifera* (Hendel)
Dacus (Bactrocera) ***tenuivittatus*** Tseng, Chen & Chu of *B. (Bactrocera) pernigra* (Ito)
Strumeta ***transtillum*** Hering of *B. (Javadacus) javanensis* (Perkins)
Dacus (Zeugodacus) ***ubiquitus*** Hardy of *B. (Zeugodacus) synnephes* (Hendel)
Bactrocera (Bactrocera) ***variabilis*** Lin & Wang of *B. (Bactrocera) dorsalis* (Hendel)
Dacus (Bactrocera) ***yilanensis*** Tseng, Chen & Chu of *B. (Bactrocera) dorsalis* (Hendel)

Seven species have been removed from synonymy –
Dacus ***melanius*** Hardy & Adachi from *B. (Paratridacus) garciniae* Bezzi
Dacus ***parvulus*** Hendel from *B. (Bactrocera) latifrons* (Hendel)
Dacus ***speculifer*** Walker from *B. (Bactrocera) curvifera* Walker
Dacus ***yayeyamanus*** Matsumura from *B. (Paratridacus) garciniae* Bezzi
Zeugodacus ***bezzianus*** Hering from *B. (Zeugodacus) tau* Walker
Zeugodacus ***heinrichi*** Hering from *B. (Zeugodacus) tau* Walker
Zeugodacus ***signata*** Hering from *B. (Zeugodacus) tau* Walker

Two species listed by Norrbom *et al.* (1998) as *Bactrocera Incertae Sedis* (of uncertain placement) are included and described – *Dacus ritsemae* Weyenbergh as *B. (Bactrocera) ritsemai* (Weyenbergh) and *Dacus terminifer* Walker as *B. (Paradacus) terminifera* (Walker).

Lectotypes are designated for six species –
B. (Bactrocera) costalis (Shiraki), *B. (Bactrocera) nigrotibialis* (Perkins), *B. (Zeugodacus) depressa* (Shiraki), *B. (Zeugodacus) nigrifacies* (Shiraki), *B. (Zeugodacus) okunii* (Shiraki), *B. (Zeugodacus) scutellaris* (Bezzi).

One species, *Dacus (Mellesis) discipennis* Walker is transferred to the genus *Dacus* from genus *Bactrocera.* Two species, *Dacus expandens* ssp. *melanius* Hardy & Adachi and *Dacus yayeyamanus* Matsumura are transferred to the genus *Bactrocera* from the genus *Dacus.* Their most recent listing, in Norrbom *et al.* (1998), was in the genus *Dacus* as synonyms of *Bactrocera garciniae.*

Introduction

The research presented herein covers studies undertaken on the dacine fauna of Indomalaya, Wallacea and north-west Australasia.

Taxonomic studies in the tephritid subfamily Dacinae have their origins in the early research of Fabricius (1794; 1805), followed by Wiedemann (1819; 1830) and Macquart (1835; 1843). The voyage of Alfred Russel Wallace to the Malay Archipelago (1854–1862), the area now known as Maluku (formerly Moluccas), and New Guinea resulted in pioneering biological collections, including some valuable dacines, that are still preserved in The Natural History Museum (BMNH). This fruit fly material was studied by Francis Walker who published a large number of new species (Walker, 1856a, b; 1859; 1860a, b; 1861a, b, c; 1864), the types of which are still in good condition. Hardy (1959) designated Lectotypes for some of these species and this new publication records and describes, for the first time, some of the Walker species collected in recent surveys within the region now called Wallacea. Other early researchers were Saunders (1842), Doleschall (1856; 1858), Schiner (1868) and Weyenbergh (1869).

The realisation that the Tephritidae, particularly the subfamily Dacinae, included a number of pest species that damaged important food crops, led to a large number of taxonomists collecting and describing species during the 20th century. Across the region from the Indian subcontinent to South-East Asia, a large number of species were described by some 50 workers. Some researchers have been working on fauna within confined geographic areas and have thus published a number of synonyms. During the first decade of the 21st century, at least ten workers have published taxonomic papers on the Dacinae of the Indomalayan region.

Some workers have concentrated on fauna within their regions, notably China, Japan and the Indian subcontinent. In China, Zia (1936; 1937; 1939; 1955; 1963), Chen (1940) and Wang (1988; 1990; 1996) have described most of the known Chinese dacine fauna. In the Indian subcontinent, Senior-White (1922), Kapoor (1971; 1993), Hardy (1964), Munro (1935), Kapoor *et al.* (1980), Tsuruta & White (2001), Agarwal *et al.* (1989) and Agarwal & Sueyoshi (2005) described and listed the bulk of species. The fauna of the Andaman Islands was recorded by Ranganath & Veenakumari (1995; 1996). In the Japanese islands, Matsumura (1916), Miyake (1919), Shiraki (1933; 1968) and Ito (1983) made the major contributions to our knowledge of species in that region. For the remainder of the region covered in this publication, the majority of known species have been described by Bezzi (1908a; 1909; 1913; 1914a, b; 1916; 1919), de Meijere (1911; 1914), Hendel (1912; 1927a, b; 1928), Enderlein (1920), Hering (1938; 1939; 1941; 1942; 1952a, b; 1953; 1956), Perkins (1938a, b; 1939), Hardy (1954; 1955a, b; 1959; 1964; 1968; 1969a, b; 1970; 1973; 1974; 1981; 1982; 1983; 1984), Hardy & Adachi (1954), Drew (1979), Drew & Hancock (1994a, b), Drew & Raghu (2002), Drew *et al.* (1998; 2007) and White & Evenhuis (1999).

Some large tephritid monographic works have been produced by Hardy (1973; 1974), Shiraki (1933; 1968) and Wang (1996) but these have had only small sections on the Dacinae. The largest overview of the dacines is in the White & Hancock (1997) CABIKEY which has been extremely valuable in listing species and their main morphological features and in the collation of taxonomic literature.

Some additional studies have been published that enhance our understanding of the taxonomy of the Dacinae. Drew (2004) discussed aspects of biogeography and speciation of the subfamily, Clarke *et al.* (2005) produced a comprehensive overview of the *dorsalis* complex with additional discussion of this group by Drew *et al.* (2008). Hancock & Drew (2006) presented a revised classification of the genus *Dacus* and Allwood *et al.* (1999) published a comprehensive list of host records.

In carrying out this comprehensive revision of the Asian and South-East Asian Dacinae, the species listings of Hardy (1977), White & Hancock (1997) in CABIKEY and Norrbom *et al.* (1998) have been valuable.

The history of taxonomic research on the Tephritidae of the Australasian and Oceanian regions was discussed by Drew (1989).

In discussions on the biogeography and speciation of the Dacini, Drew & Hancock (2000) and Drew (2004) proposed that the parental stock originated in the Indian block of Gondwana as it drifted northwards. At this point the genera *Ichneumonopsis* and *Monacrostichus* evolved, the *Dacus* dispersed to Africa while the *Bactrocera* entered a long period of speciation in the rainforests of southern and eastern Asia. Prolific speciation in these rainforest habitats, across the many islands, has resulted in the largest number of endemic dacine species occurring in South-East Asia and Papua New Guinea (Drew, 2004), many of which occur within complexes based on morphological similarity. These complexes present great challenges for taxonomists, and accurate diagnosis of species will depend on data collated from research into the behaviour, ecology and molecular aspects in addition to morphological features.

There are four identifiable species complexes within the genus *Bactrocera*, i.e. *B. (Bactrocera) dorsalis* complex, *B. (Bactrocera) nigrotibialis* complex, *B. (Zeugodacus) scutellaris* complex and the *B. (Zeugodacus) tau* complex. The speciation processes within the genus *Bactrocera*, in particular, have led to reasonably rapid morphological changes (Clarke *et al.*, 2005) resulting in complex taxonomic difficulties for researchers attempting to define species. As a consequence, a large number of synonyms have been described by taxonomists working in isolation, attempting to diagnose and describe species without studying the more widely distributed regional fauna.

The supraspecific classification within the Dacinae has always been in a state of flux and open to change as a result of differing opinions on the value of morphological characters at this level. Early taxonomists, namely Perkins (1937; 1939), Hering (1941a, b, c, d, e) and May (1963), chose to place all species within genera and not use subgenera. However, Hardy (1951) and subsequent workers placed species within subgenera, particularly within the genera *Bactrocera* and *Dacus*, based on their opinion that the available character states were not worthy of generic level status. The generic classification of Drew (1989) where *Bactrocera* and *Dacus* were divided on the basis of fused abdominal terga (genus *Dacus*) and free terga (genus *Bactrocera)* has been widely accepted, as this morphological feature is supported with biogeographical distribution data and host plant specificity. That is, the *Dacus* appear to have originated and speciated in the African continent where they utilize host plants in the dry savannah environment, and *Bactrocera* in Indomalaya and Australasia where their endemic host plants are associated with the subtropical and tropical rainforests. The African fauna has been revised by White (2006).

The subgeneric classifications of Drew (1972a; 1989) have generally been found satisfactory in the studies on South-East Asian fauna by Hardy (1973; 1974) and the South Pacific fauna by Drew (1989) but not consistent with available molecular data. While in this publication we have further refined the subgeneric classification of the genus *Bactrocera*, more improvements can be made, especially in combining data from in-depth molecular research and host plant associations. For the genus *Dacus*, we have followed the subgeneric classification of Hancock & Drew (2006). For the region, we recognize only five subgenera within *Dacus*.

The research presented in this publication has resulted from extensive field surveys of fruit flies in Bhutan, India, Sri Lanka, Thailand, Vietnam, Peninsular Malaysia, East Malaysia, the Philippines and all provinces of Indonesia. All surveys were conducted by teams organized by, and funded by projects from, the International Centre for Management of Pest Fruit Flies, Griffith University, Brisbane, except for that in Sri Lanka which was conducted by Mr Kenji Tsuruta, Quarantine Service, Japan. Within each country, field sites representing a range of ecosystems were chosen. At each site, male lure traps (cue lure and methyl eugenol) were set and serviced for 1–2 years on a 2–3 week cycle. Also, host fruits, commercial/edible and wild, were sampled and fruit flies reared from those infested. This programme resulted in extensive collections of flies, now curated and held within each country and the Queensland Primary Industries Insect Collection (QDPC), Brisbane. Also, the surveys were so comprehensive that the host records have been published in Allwood *et al.* (1999). In this publication, we are presenting full descriptions of all known species, previously undescribed species, updated distribution records, male lure records and an evaluation of the economic pest status of those species reared from commercial/edible host fruits. New host records since Allwood *et al.* (1999) are also presented.

The research presented herein is the first holistic taxonomic review of the Dacinae covering the region from Indomalaya to north-west Australasia. Seven species, considered as belonging to the Australasian region, are included because they were collected in the Indonesian provinces of Papua and West Papua (originally part of Irian Jaya). These species are *Bactrocera (Bactrocera) bryoniae* (Tryon), *Bactrocera (Bactrocera) fulvicauda* (Perkins), *Bactrocera (Bactrocera) musae* (Tryon), *Bactrocera (Bactrocera) neocheesmanae* Drew, *Bactrocera (Bactrocera) paramusae* Drew, *Bactrocera (Bactrocera) speculifera* (Walker) and *Bactrocera (Zeugodacus) abdoangusta* (Drew).

White and Evenhuis (1999) published complete known records of dacines from the Indonesian territory of Papua.

Methods and Abbreviations

The research in this publication is based on major collections of dacine specimens from extensive field surveys in Bhutan, India, Palau, Sri Lanka, Thailand, Malaysia (Peninsular and East Malaysia), Vietnam, Indonesia and parts of the Philippines. In addition to these surveys, specimens were received from China, Japan, Nepal and Taiwan. These surveys were carried out between 1983 and 2009 and were based on cue lure and methyl eugenol trapping and host fruit collections. The male lure traps were set for a minimum of 1 year at sites selected to represent a wide range of ecosystems. The traps were serviced on a 2–3 week cycle. The host fruits were collected from rainforest areas, plantations and village gardens, and flies were reared from these under laboratory conditions. All specimens were forwarded to the International Centre for the Management of Pest Fruit Flies, Griffith University, Brisbane, for identification.

To establish a foundation for the taxonomic research, type specimens of most known species were received on loan from the institutions where they are held, or studied during visits to some museums. Types of species described by Shiraki (1933), previously recorded as lost (Norrbom *et al.*, 1998), have been found in the Laboratory of Insect Systematics, National Institute of Agro-Environmental Sciences (NIAS), Japan, and studied. Upon this foundation, we were able to identify undescribed species and synonyms and to withdraw some species from previously declared synonymies.

To our knowledge, we have reviewed all taxonomic works on Dacinae published before early 2012. Since 2008, ten papers have been published describing new species of Dacinae from China. Approximately 50% of these species are synonyms and some use preoccupied names (homonyms). We forecast that this process will continue into the future and have decided to not review any taxonomic papers on Chinese fauna published after January 2012.

From the field surveys, host records were published by Allwood *et al.* (1999) with some more recent updates included in this publication. Further, we have been able to present male lure records and geographic distribution records for all species, and obtain an accurate diagnosis of all economic pest species.

Generally, we have conformed with McAlpine (1981) and White *et al.* (2000) for the morphological terminology. We have only departed from this system in a few instances where we believe it is essential to have conformity with our previous works used by Agriculture and Quarantine services. In particular, we have used:

- Mesopleural stripe for the anepisternal stripe
- Postnotum for the mediotergite
- *r-m* for the R-M crossvein; *dm-cu* for the DM-Cu crossvein
- Shining spots for the ceromae on abdominal tergum V.

See Figs 1–5 for morphological terminologies and Fig. 6 for diagrammatic colours used in descriptions and illustrations.

Measurements for new species have been taken from the Holotype. For revised species, primary types or Syntypes have been used where available.

The acronyms for institutions in which specimens are deposited are listed as follows:

AMNH	American Museum of Natural History, Department of Entomology, Central Park West at 79th St., New York, New York 10024, USA
ANIC	Australian National Insect Collection, CSIRO, GPO Box 1700, Canberra, Australian Capital Territory 2601, Australia
APQS	Animal and Plant Quarantine Service, Guangzhou, Guangdong 510060, China
BBM	Bernice P. Bishop Museum, Department of Entomology Collection, P.O. Box 19000A, 1525 Bernice St, Honolulu, Hawaii 96819, USA
BMNH	The Natural History Museum, Department of Entomology, Cromwell Road, London SW7 5BD, England, UK
BPIM	Bureau of Plant Industry, Department of Agriculture, 692 San Andres St., Malate, Manila, Philippines
CAIQ	Chinese Academy of Inspection and Quarantine, No A3 Gaobeidian Bei Lu, Chaoyang District, Beijing 100025, China
DEI	Deutsches Entomologisches Institut, Deutschen Akademie der Landwirtswissenschaften zu Berlin, Schicklerstrasse 5, 13 Eberswalde D-1300, Germany
DOAC	Department of Agriculture, Chiang Mai 50002, Thailand
FAFU	Fujian Agriculture and Forestry University, Fuzhou 350002, China
FSCA	Florida State Collection of Arthropods, Division of Plant Industry, 1911 34th St., SW, P.O. Box 147100, Gainesville, Florida 32614, USA
HEIQ	Insect Museum, Center for Tropical Plant Quarantine, Hainan Entry-Exit Inspection and Quarantine Bureau, Haikou 570311, Hainan, China
HORDI	Entomology Division, Horticulture and Research and Development Institute, Gannoruwa, Sri Lanka
HUS	Entomological Institute, Faculty of Agriculture, Hokkaido University, Sapporo, Hokkaido 060, Japan
INPC	National Pusa Collections, Division of Entomology, Indian Agriculture Research Institute, New Delhi, Delhi 110012, India
IZAS	Institute of Zoology, Academia Sinica, Insect Collection, 19 Zhongguancun Lu, Haidian, Beijing 10080, China
KUB	Kasetsart University, 50 Ngam Wong Wan Rd, Ladyaow Chatuchak, Bangkok 10900, Thailand
MCSNM	Museo Civico di Storia Naturale, Corso Venezia 55, Milan 20121, Italy
MEA	Plant Quarantine Laboratory, Tainan Branch Office, Bureau of Commodity Inspection and Quarantine, Ministry of Economic Affairs, Taiwan
MFN	Museum fur Naturkunde der Humboldt Universitat zu Berlin, Invalidenstrasse 43, Berlin 10115, Germany
MHNG	Museum d'histoire naturelle, 1 Route de Malagnou, CH-1208, Geneve, Switzerland
MLUH	Wissenschaftsbereich Zoologie, Sektion Biowissenschaften Martin-Luther-Universitata Halle, WB Zoologie, Domplatz 4, Halle/Salle D-4020, Germany
MNHNP	Museum National d'Histoire Naturelle, National Collection of Insects, 45, rue Buffon, Paris 75005, France
MNM	Magyar Termeszettudomanyi Muzeum Allattara (Hungarian Natural History Museum), Baross u. 13, 1088 Budapest, Hungary
MUA	Zoology Museum, Muslim University, Aligarh, India
MVMA	National Museum of Victoria, Department of Entomology, 71 Victoria Crescent, Abbotsford, Victoria 3067, Australia
MZB	Museum Zoologicum Bogoriense, P.O. Box 110, Jalan. Juanda 3, Bogor, Java, Indonesia
NIAS	Laboratory of Insect Systematics, National Institute of Agro-Environmental Sciences, Kannondai, Tsukuba, Ibaraki Pref. 305, Japan
NMB	Naturhistorisches Museum, Entomology Department, Augustinergasse 2, Basel CH-4001, Switzerland
NMKE	National Museum of Kenya, Section of Entomology, P.O. Box 40658, Nairobi 00100, Kenya
NMW	Naturhistorisches Museum Wien, Postfach 417, Burgring 7, Vienna A-1040, Austria
NMWC	National Museum of Wales, Cathays Park, Subdepartment of Entomology, Department of Zoology, Cardiff, South Glamorgan CF1 3NP, Wales, UK
NPPC	National Plant Protection Centre, Thimphu, Bhutan
NRS	Naturhistoriska Riksmuseet, Sektionen fur entomologi, Stockholm S-10405, Sweden
NTU	National Taiwan University, No 1, Sec 4, Roosevelt Rd, Taipei 10617, Taiwan, ROC
NUS	National University of Singapore, 21 Lower Kent Ridge Road, Singapore 119077, Republic of Singapore
PAU	Punjab Agricultural University, Ludhiana, Punjab, India
PQMAB	Plant Quarantine Institute, Ministry of Agriculture, Beijing 100029, China
QDPC	Queensland Primary Industries Insect Collection, Department of Agriculture, Fisheries and Forestry, Queensland, Ecosciences Precinct, GPO Box 267, Brisbane, Queensland 4001, Australia

Continued

(continued)

QMIC	Queensland Museum, P.O. Box 3300, Brisbane, Queensland 4101, Australia (UQIC Insect Collection, Dept of Entomology, University of Queensland, St Lucia, Brisbane, Queensland 4067, Australia, transferred to QMIC)
RNH	Nationaal Natuurhistorische Museum, Raamsteeg 2, Leiden 2311 PL, Netherlands
SANC	South African National Collection of Insects, Private X134, Pretoria, Transvaal 0001, South Africa
SMKM	Selangor Museum, Kuala Lumpur, Malaysia
TAUI	Insect Collection, Zoological Museum, Tel Aviv University, Tel Aviv 69978, Israel
UOPJ	Entomological Laboratory, University of Osaka Prefecture, Mosu, Umemachi Sakai, Osaka 593, Japan
USNM	United States National Museum of Natural History, United States National Entomological Collection, Washington, DC 20560, USA
UZMC	University of Copenhagen, Zoological Museums, Department of Entomology, Universitetsparken, Copenhagen DK-2100, Denmark
UZMH	Zoological Museum, Finnish Museum of Natural History, University of Helsinki, P. Rautatiek 13, Helsinski, SF-00100, Finland
ZMAN	Zoologisch Museum, Instituut voor Taxonomische Zoologie, Universiteit van Amsterdam, Plantage Middenlaan 64, Amsterdam 1018 DH, Netherlands
ZRC	Zoological Reference Collection of the Raffles Museum of Biodiversity Research, National University of Singapore, 21 Lower Kent Ridge Road, Singapore 119077
ZSI	Zoological Survey of India, National Zoological Collection, 34, Chittaranjan Avenue, Calcutta, West Bengal 700 012, India

Some localities in Peninsular Malaysia are abbreviated as follows:

TN	Transect North (Peninsular Malaysia), from Kuala Kangsar (west coast, TN 1) to Kota Bharu (east coast, TN 17), sites approximately 20 km apart
TS	Transect South (Peninsular Malaysia), from Jalan Kebun (west coast, TS 1) to Kuantan (east coast, TS 20) through Kuala Lumpur, sites approximately 20 km apart

Other acronyms used:

ACIAR	Australian Centre for International Agricultural Research
AQIS	Australian Quarantine and Inspection Service
AusAID	Australian Agency for International Development
ICZN	International Code of Zoological Nomenclature, Fourth Edition (Anon., 1999)
SPC	Secretariat of the Pacific Community
UNDP	United Nations Development Programme

Generic and Subgeneric Classification

The Dacinae from Indomalaya to Australasia fall primarily into the Tribe Dacini. This tribe is concentrated in two areas of the world, the Afrotropical region on one hand and the Indomalayan, Australasian and Oceanian regions on the other, with significantly smaller numbers of species occurring away from these centres. There is a small amount of overlap between these regions, particularly at the genus level where some species of *Dacus* have evolved away from the Afrotropical region, into the Indomalayan, Australasian and Oceanian regions.

In the research published herein, we have followed the basic principle of defining genera, subgenera and species on morphological characters, biological features (host plant associations and male lure responses) and geographic distributions. For some species in the *dorsalis* complex, we have drawn on available molecular data.

Within the Dacini, we recognize four genera, *Bactrocera* Macquart, *Dacus* Fabricius, *Ichneumonopsis* Hardy and *Monacrostichus* Bezzi. For the regions covered by this publication, we have defined 18 subgenera within the genus *Bactrocera* and five within the genus *Dacus*. Herein we have followed the supraspecific classification of Drew (1989) where the genus *Bactrocera* is divided into two major groups of subgenera, the *Bactrocera* group and the *Zeugodacus* group, and two smaller groups, the *Melanodacus* group and the *Queenslandacus* group. Whether each of these groups of subgenera should be incorporated within separate genera remains a subject for future research.

DEFINITIONS OF GENERA

Genus *Bactrocera* Macquart

Bactrocera Macquart, 1835: 452. Type species *Bactrocera longicornis* Macquart, 1835, by monotypy.

Definition: Abdominal terga not fused; abdomen generally oval in shape; scutellum triangular in shape; distribution primarily across the Indomalayan, Australasian and Oceanian regions; host plants tropical and subtropical rainforest species.

Genus *Dacus* Fabricius

Dacus Fabricius, 1805: 272. Type species *Dacus armatus* Fabricius, 1805, by designation by Speiser (1924: 140) confirmed by Hendel (1927a: 25).

Definition: Abdominal terga fused; abdomen generally elongate-oval and often petiolate; scutellum short and broad; distribution primarily African with some incursion into the Indomalayan, Australasian and Oceanian regions; host plants primarily in the families *Asclepiadaceae*, *Cucurbitaceae* and *Passifloraceae*.

Genus *Ichneumonopsis* Hardy

Ichneumonopsis Hardy, 1973: 132. Type species *Ichneumonopsis burmensis* Hardy, 1973, by original designation.

Definition: Abdominal terga not fused; abdomen elongate-oval; pecten on abdominal tergum III of male absent; two rounded spermathecae; wing cell bm narrow at base; wing vein R_{4+5} setose; *i.or.* setae on head reduced or absent; antennae with arista plumose; distribution South-East Asia.

Genus *Monacrostichus* Bezzi

Monacrostichus Bezzi, 1914a: 322. Type species *Monacrostichus citricola* Bezzi, 1913, by original designation.

Definition: Abdominal terga not fused; abdomen elongate-oval; pecten on abdominal tergum III of male absent; wing with cell sc elongate and broad, cell bm narrow and basally broadened, cell dm basally narrow, veins R_{2+3} and M with pronounced curvatures, vein A_1+CuA_2 ending well before wing margin; scutum and face with transverse furrows; frontal setae on head absent; distribution South-East Asia; host plants *Citrus* spp. (family *Rutaceae*).

Placing *Ichneumonopsis* and *Monacrostichus* as separate genera within the Dacinae appears valid on the basis of the characters detailed above. Both genera possess a number of characters that are considered primitive: narrow cell bm; free abdominal terga; absence of shining spots on abdominal tergum V; a moderately deep concavity on posterior margin of abdominal sternum V in males; absence of a pecten of cilia on abdominal tergum III of males; tergum VI not disassociated from tergum V in females.

Ichneumonopsis differs from *Monacrostichus* in details of the wing venation; the lack of facial and scutal furrows; the shorter antenna with a plumose arista; the long surstylus lobe in males and the longer ovipositor and rounded spermathecae in females. Both of these genera can be separated from the other dacine genera, *Dacus* Fabricius and *Bactrocera* Macquart, by the narrow cell bm; the lack of abdominal shining spots on tergum V and the presence of a distinct tergum VI in females.

DEFINITIONS OF SUBGENERA

Subgenera in genus *Bactrocera* Macquart

Subgenus *Afrodacus* Bezzi

Afrodacus Bezzi, 1924: 469. Type species *Chaetodacus biguttulus* Bezzi, 1922, by original designation.

Definition: *Bactrocera* species with posterior lobe of male surstylus short; abdominal sternum V of male deeply concave on posterior margin; pecten of cilia present on abdominal tergum III of male; postpronotal setae absent; *a.sa.* setae absent; *prsc.* setae present; one pair of *sc.* setae present.

Subgenus *Asiadacus* Perkins

Asiadacus Perkins, 1937: 57. Type species *Chaetodacus bakeri* Bezzi, 1919, by original designation.

Definition: *Bactrocera* species with posterior lobe of male surstylus long; abdominal sternum V of male slightly concave on posterior margin; pecten of cilia present on abdominal tergum III of male; postpronotal setae absent; *a.sa.* setae absent; *prsc.* setae absent; one pair of *sc.* setae present.

Subgenus *Bactrocera* Macquart

Bactrocera Macquart, 1835: 452. Type species *Bactrocera longicornis* Macquart, 1835, by monotypy.

Definition: Species of the genus *Bactrocera* with posterior lobe of male surstylus short; abdominal sternum V of male deeply concave on posterior margin; pecten of cilia present on abdominal tergum III of male; postpronotal setae absent; *a.sa.* setae present; *prsc.* setae present; one pair of *sc.* setae present; bulla in male wing absent.
Included species: A major subgenus with over 100 species from Indomalaya to north-west Australasia. A small number of species fit imperfectly into this subgenus in having the *prsc.* setae absent, e.g. *B. curvifera* (Walker).

Subgenus *Bulladacus* Drew & Hancock

Bulladacus Drew & Hancock, 1995: 9. Type species *Bactrocera (Bulladacus) gnetum* Drew & Hancock, 1995, by original designation.

Definition: *Bactrocera* species with posterior lobe of male surstylus short; abdominal sternum V of male deeply concave on posterior margin; pecten of cilia present on abdominal tergum III of male; postpronotal setae absent; *a.sa.* setae present or absent; *prsc.* setae present; one pair of *sc.* setae present; bulla present on cell cup extension in male wing; shining spots absent on abdominal tergum V of male.

Subgenus *Daculus* Speiser

Daculus Speiser, 1924: 140. Type species *Musca oleae* Gmelin, 1790, by original designation.

Definition: *Bactrocera* species with an oval-shaped abdomen; posterior lobe of male surstylus short; abdominal sternum V of male deeply concave on posterior margin; pecten of cilia present on abdominal tergum III of male; postpronotal setae absent; *a.sa.* and *prsc.* setae absent; one pair of *sc.* setae present; bulla in male wing absent.

Comments: Drew (1989) synonymised the subgenera *Daculus* Speiser and *Heterodaculus* Hardy with subgenus *Polistomimetes* Enderlein. Subsequently, White & Wang (1992) placed *Polistomimetes* as a synonym of subgenus *Tetradacus* Miyake. Species placed in *Tetradacus* by White & Wang (1992) included *B. minax* (Enderlein) and the type species *B. tsuneonis* (Miyake). The main characters that set *Tetradacus* apart from other close subgenera of *Bactrocera* are the elongate-oval-shaped abdomen which is sometimes petiolate, and the bulbous-shaped female oviscape which is rounded in cross section. Some South Pacific species placed by Drew (1989) in *Polistomimetes* also possess these characters, namely *B. neopagdeni* Drew and *B. pagdeni* (Malloch). Other species placed in *Polistomimetes* by Drew (1989) possess the standard oval-shaped abdomen of the subgenus *Bactrocera* and may well be aberrant species of this subgenus. Here we have decided to maintain the subgenus *Daculus* for those species that are similar to subgenus *Bactrocera* but lack *a.sa.* and *prsc.* setae. However, as White & Wang (1992) point out, *Bactrocera oleae* is the only species which we can be sure fits into *Daculus*.

Subgenus *Gymnodacus* Munro

Gymnodacus Munro, 1938: 117. Type species *Dacus mesomelas* Bezzi, 1908a, by original designation.

Definition: Species of *Bactrocera* with posterior lobe of male surstylus short; abdominal sternum V of male deeply concave on posterior margin; pecten of cilia absent from abdominal tergum III of male; postpronotal setae absent; *a.sa.* setae present; *prsc.* setae present; one pair of *sc.* setae present; bulla in male wing absent.

Subgenus *Hemigymnodacus* Hardy

Hemigymnodacus Hardy, 1973: 19. Type species *Dacus diversus* Coquillett, 1904, by original designation.

Definition: *Bactrocera* species with posterior lobe of male surstylus long; abdominal sternum V of male slightly concave on posterior margin; pecten of cilia absent from abdominal tergum III of male; postpronotal setae absent; *a.sa.* setae present; *prsc.* setae present; one pair of *sc.* setae present; bulla in male wing absent.

Subgenus *Javadacus* Hardy

Javadacus Hardy, 1983: 26. Type species *Dacus montanus* Hardy, 1983, by original designation.

Definition: *Bactrocera* species with posterior lobe of male surstylus long; abdominal sternum V of male slightly concave on posterior margin; pecten of cilia present on abdominal tergum III of male; postpronotal setae absent; *a.sa.* setae absent; *prsc.* setae present; one pair of *sc.* setae present; bulla in male wing absent.

Subgenus *Nesodacus* Perkins

Nesodacus Perkins, 1937: 57. Type species *Chaetodacus atrichus* Bezzi, 1919, by original designation.

Definition: *Bactrocera* species with posterior lobe of male surstylus long; abdominal sternum V of male slightly concave on posterior margin; pecten of cilia absent from abdominal tergum III of male; postpronotal setae absent; *a.sa.* setae present; *prsc.* setae absent; one pair of *sc.* setae present; bulla in male wing absent.

Subgenus *Papuodacus* Drew

Papuodacus Drew, 1972a: 13. Type species *Dacus pallescentis* Drew, 1971, by original designation [= *neopallescentis* Drew].

Definition: *Bactrocera* species with posterior lobe of male surstylus long; abdominal sternum V of male slightly concave on posterior margin; pecten of cilia present on abdominal tergum III of male; postpronotal setae absent; *a.sa.* setae absent; *prsc.* setae generally present; two pairs of *sc.* setae present; bulla in male wing absent.

Subgenus *Paradacus* Perkins

Paradacus Perkins, 1938a: 143. Type species *Paradacus fulvipes* Perkins, 1938a, by original designation.

Definition: *Bactrocera* species with posterior lobe of male surstylus long; abdominal sternum V of male slightly concave on posterior margin; pecten of cilia present on abdominal tergum III of male; postpronotal setae absent; *a.sa.* setae present; *prsc.* setae present or absent; two pairs of *sc.* setae present; bulla in male wing absent.

Subgenus *Parasinodacus* new subgenus

Type species: *Bactrocera incisa* (Walker)

This new subgenus has been established to include a group of species that previously were placed in either subgenus *Paradacus* or subgenus *Sinodacus*. It differs from *Paradacus* in having one pair of *sc.* setae and from *Sinodacus* in having the medial post-sutural yellow vitta absent.

Definition: *Bactrocera* species with posterior lobe of male surstylus long; abdominal sternum V of male slightly concave on posterior margin; pecten of cilia present on abdominal tergum III of male; postpronotal setae absent; *a.sa.* setae present; *prsc.* setae present or absent; one pair of *sc.* setae present; bulla in male wing absent; medial longitudinal postsutural yellow vitta on scutum absent; generally with a yellow spot anterior to the mesonotal suture in front of the lateral postsutural yellow vittae.

Subgenus *Paratridacus* Shiraki

Paratridacus Shiraki, 1933: 109. Type species *Dacus yayeyamanus* Matsumura, 1916, by original designation.

Definition: *Bactrocera* species with posterior lobe of male surstylus long; abdominal sternum V of male slightly concave on posterior margin; pecten of cilia absent from abdominal tergum III of male; postpronotal setae absent; *a.sa.* setae present; *prsc.* setae present; two pairs of *sc.* setae present.

Subgenus *Parazeugodacus* Shiraki

Parazeugodacus Shiraki, 1933: 107. Type species *Parazeugodacus matsumurai* Shiraki, 1933, by original designation.

Definition: *Bactrocera* species with posterior lobe of male surstylus short; abdominal sternum V of male slightly concave on posterior margin; pecten of cilia present on abdominal tergum III of male; humeral setae absent; *a.sa.* setae present; *prsc.* setae present; two pairs of *sc.* setae present.

Subgenus *Semicallantra* Drew

Dacus (Semicallantra) Drew, 1989: 17. Type species *Dacus aquilus* Drew, 1989, by original designation.

Definition: *Bactrocera* species with posterior lobe of male surstylus short; pecten of cilia present on abdominal tergum III of male; postpronotal setae absent; *prsc.* setae present; one pair of *sc.* setae present; abdomen shape elongate-oval and petiolate; abdominal terga not fused; antennal segments long (*Callantra*-like).
Comments: Drew (1989) established subgenus *Semicallantra* to include three South Pacific species but incorrectly placed it in the genus *Dacus*. The type species, *aquilus* Drew, has free (not fused) abdominal terga and thus must be placed in the genus *Bactrocera*. The *B. (Semicallantra)* species appear to form a link between the genera *Bactrocera* and *Dacus* as they exhibit characters of both.

Subgenus *Sinodacus* Zia

Sinodacus Zia, 1936: 157. Type species *Sinodacus hochii* Zia, 1936, by original designation.

Definition: *Bactrocera* species with posterior lobe of male surstylus long; abdominal sternum V of male slightly concave on posterior margin; pecten of cilia present on abdominal tergum III of male; postpronotal setae absent; *a.sa.* setae present; *prsc.* setae absent; one pair of *sc.* setae present; bulla in male wing absent; medial longitudinal postsutural yellow vitta present on scutum.

Subgenus *Tetradacus* Miyake

Dacus (Tetradacus) Miyake, 1919: 92. Type species *Dacus (Tetradacus) tsuneonis* Miyake, 1919, by original designation.

Definition: *Bactrocera* species with posterior lobe of male surstylus short; abdominal sternum V of male slightly concave on posterior margin; postpronotal setae absent; one pair of *sc.* setae present; dense microtrichia around A_1+CuA_2 absent; abdomen shape elongate-oval; oviscape with a bulbous base and rounded in transverse cross section; a medial longitudinal postsutural yellow vitta present; antennal segments short.
Comments: Drew (1989) placed a group of South Pacific species in the *Bactrocera* subgenus *Polistomimetes* Enderlein. However, White & Wang (1992) established *Polistomimetes* as a synonym of *B. (Tetradacus)* and this appears to be valid. As for *Bactrocera (Semicallantra)*, species of *B. (Tetradacus)* form a link between the genera *Bactrocera* and *Dacus*.

Subgenus *Zeugodacus* Hendel

Dacus (Zeugodacus) Hendel, 1927a: 26. Type species *Dacus caudatus* Fabricius, 1805, by original designation.

Definition: *Bactrocera* species with posterior lobe of male surstylus long; abdominal sternum V of male slightly concave on posterior margin; pecten of cilia present on abdominal tergum III of male; postpronotal setae absent; *a.sa.* setae present; *prsc.* setae present; two pairs of *sc.* setae present.
Included species: This is a major subgenus with approximately 100 species in the region from Indomalaya to north-west Australasia.

Subgenera in Genus *Dacus* Fabricius

Subgenus *Callantra* Walker

Callantra Walker, 1860a: 153. Type species *Callantra smieroides* Walker, 1860a, by monotypy. (Note: *Callantra smieroides* Walker is a synonym of *Dacus (Callantra) longicornis* Wiedemann, see Drew *et al.* (1998).)

Definition: *Dacus* species with abdomen elongate and strongly petiolate with tergum I at least as long as wide, abdominal sternum V of male slightly concave on posterior margin; *prsc.* setae absent; *a.sa.* setae present; combined lengths of antennal segments greater than vertical length of head; anal streak present in wing. Host plants in family *Cucurbitaceae.*

Subgenus *Didacus* Collart

Dacus (Didacus) Collart, 1935: 33. Type species *Dacus (Didacus) ciliatus* Loew, 1862, by original designation.

Definition: *Dacus* species with abdominal sternum V of male slightly concave on posterior margin; combined lengths of antennal segments equal to or less than vertical length of head; *a.sa.* setae absent; *prsc.* setae absent; abdomen shape elongate-oval with tergum I wider than long. Host plants primarily in family *Cucurbitaceae*.

Subgenus *Leptoxyda* Macquart

Leptoxyda Macquart, 1835: 452. Type species *Leptoxyda testacea* Macquart, 1835, by monotypy. (Note: *Leptoxyda testacea* is a synonym of *Dacus longistylus* Wiedemann.)

Definition: *Dacus* species with abdominal sternum V of male with a moderately deep concavity on posterior margin; combined lengths of antennal segments equal to or less than vertical length of head; *prsc.* setae absent; *a.sa.* setae absent; abdomen shape elongate-oval with tergum I wider than long; basal segment of ovipositor (oviscape) elongate and narrow, at least 1.75 times longer than tergum V.

Subgenus *Mellesis* Bezzi

Mellesis Bezzi, 1916: 114. Type species *Monacrostichus crabroniformis* Bezzi, 1916, by monotypy.

Definition: *Dacus* species with abdomen petiolate with tergum I at least as long as wide; abdominal sternum V of male weakly concave or straight; *prsc.* setae absent; *a.sa.* setae usually present; combined lengths of antennal segments greater than vertical length of head; anal streak in wing often absent. Host plants in family *Asclepiadaceae.*
Included species: Subgenus *Mellesis* is the largest subgenus of *Dacus* in the region from Indomalaya to north-west Australasia, with 43 species.

Subgenus *Neodacus* Perkins

Neodacus Perkins, 1937: 58. Type species *Neodacus newmani* Perkins, 1937, by original designation.

Definition: *Dacus* species with abdomen elongate-oval to oval; abdominal sternum V of male with posterior margin weakly concave to straight; *prsc.* setae absent; *a.sa.* setae present or absent; combined lengths of antennal segments less than vertical length of head; anal streak in wing present; pecten of cilia on abdominal tergum III of male present. Host plants in family *Asclepiadaceae*.
Included species: *ortholomatus* Hardy, *santongae* Drew & Hancock, *sphaeroidalis* Bezzi.

Species Listed Under Genera and Subgenera

Genus *Bactrocera* Macquart

Afrodacus **Bezzi**

fastigata Tsuruta & White

Asiadacus **Perkins**

absoluta (Walker)
apicalis (de Meijere)
apiciflava Yu, He & Chen
bakeri (Bezzi)
careomacula new species
maculifacies (Hardy)
melanopsis (Hardy)

Bactrocera **Macquart**

abdonigella (Drew)
aemula Drew
aethriobasis (Hardy)
affinibancroftii new species
affinidorsalis (Hardy)
affinis (Hardy)
albistrigata (de Meijere)
amarambalensis Drew
andamanensis (Kapoor)
apicofuscans White & Tsuruta
apiconigroscutella Drew
apicopicta new species
arecae (Hardy & Adachi)
armillata (Hering)
atrifemur Drew & Hancock
avittata new species
bancroftii (Tryon)
beckerae (Hardy)
bellisi new species
bhutaniae new species
bifasciata (Hardy)
bimaculata Drew & Hancock
binhduongiae new species
bitungiae new species
bivittata Lin & Wang
blairiae new species
bruneiae new species
brunneola White & Tsuruta
bryoniae (Tryon)
carambolae Drew & Hancock
careofascia new species
caryeae (Kapoor)
ceylanica Tsuruta & White
cibodasae Drew & Hancock
citima (Hardy)
cognata (Hardy & Adachi)
collita Drew & Hancock
commensurata new species
contigua Drew
correcta (Bezzi)
costalis (Shiraki)
curtivitta new species
curvifera (Walker)
curvosterna new species
dispar (Hardy)
dongnaiae new species
dorsalis (Hendel)
dorsaloides (Hardy & Adachi)
elongata new species
epicharis (Hardy)
eurycosta new species
fernandoi Tsuruta & White
flavipennis (Hardy)
flavoscutellata Lin & Wang
flavosterna new species
floresiae Drew & Hancock
frauenfeldi (Schiner)
fulvicauda (Perkins)
fulvifemur Drew & Hancock
fuscitibia Drew & Hancock
fuscoformosa new species

fuscolobata new species
fuscoptera new species
gombokensis Drew & Hancock
halmaherae new species
hantanae Tsuruta & White
holtmanni (Hardy)
hyalina (Shiraki)
illusioscutellaris new species
impunctata (de Meijere)
inconspicua new species
indonesiae Drew & Hancock
infulata Drew & Hancock
invadens Drew, Tsuruta & White
involuta (Hardy)
irvingiae Drew & Hancock
jaceobancroftii new species
kalimantaniae new species
kanchanaburi Drew & Hancock
kandiensis Drew & Hancock
kinabalu Drew & Hancock
laithieuiae new species
lata (Perkins)
lateritaenia Drew & Hancock
latifrons (Hendel)
latilineola Drew & Hancock
limbifera (Bezzi)
linduensis new name
lombokensis Drew & Hancock
maculigera Doleschall
makilingensis Drew & Hancock
malaysiensis Drew & Hancock
mamaliae new species
mediorufula new species
megaspilus (Hardy)
melastomatos Drew & Hancock
merapiensis Drew & Hancock
minuscula Drew & Hancock
moluccensis (Perkins)
muiri (Hardy & Adachi)
musae (Tryon)
nanoarcuata new species
nationigrotibialis new species
neoarecae Drew
neocheesmanae Drew
neocognata Drew & Hancock
neofulvicauda new species
neonigrotibialis Drew
neopropinqua Drew & Hancock
neoritsemai new species
nigrifacia Zhang, Ji & Chen
nigrifemorata Lin & Wang
nigrofemoralis White & Tsuruta
nigrotibialis (Perkins)
nothaphoebe new species
obscurata (de Meijere)
obscurivitta new species
occipitalis (Bezzi)
ochroma new species
ochroventer new species
opacovitta new species
osbeckiae Drew & Hancock
papayae Drew & Hancock
paraarecae new species
paradiospyri Chen, Zhou & Li
paralatissima new species
paralimbifera new species
paramusae Drew
paranigrita new species
paraosbeckiae Drew
paraverbascifoliae Drew
parvula (Hendel)
patula new species
pectoralis (Walker)
pedestris (Bezzi)
peneallwoodi new species
penebeckerae new species
penecognata Drew & Hancock
penecorrecta Drew
penecostalis new species
penephaea new species
perigrapha White & Tsuruta
pernigra (Ito)
pictipennis Lin & Zeng
profunda Tsuruta & White
propinqua (Hardy & Adachi)
pruniae new species
pseudobeckerae new species
pseudodistincta (Drew)
pseudoversicolor Drew
pusilla (Hardy)
pyrifoliae Drew & Hancock
quasiinfulata new species
quasineonigrita new species
quasipropinqua Drew & Hancock
raiensis Drew & Hancock
ranganathi new species
recurrens (Hering)
resima (Drew)
ritsemai (Weyenbergh)
rubigina (Wang & Zhao)
sapaensis new species
selenophora Tsuruta & White
sembaliensis Drew & Hancock
speculifera (Walker)
sulawesiae Drew & Hancock
suliae new species
sumbawaensis Drew & Hancock
syzygii White & Tsuruta
tapahensis new species
ternatiae new species
thailandica Drew & Hancock
thistletoni Drew
trifasciata (Hardy)
truncata new species
tuberculata (Bezzi)
umbrosa (Fabricius)
unimacula Drew & Hancock
usitata Drew & Hancock

venefica (Hering)
verbascifoliae Drew & Hancock
versicolor (Bezzi)
vishnu Drew & Hancock
waaiae new species
wuzhishana Lin & Yang
zonata (Saunders)

***Bulladacus* Drew & Hancock**

bullifera (Hardy)
captiva new species
cinnabaria new species
mcgregori (Bezzi)
peterseni (Hardy)

***Daculus* Speiser**

digressa Radhakrishnan
rutengiae new species

***Gymnodacus* Munro**

calophylli (Perkins & May)
continua (Bezzi)
kuniyoshii (Shiraki)
symplocos new species
tillyardi (Perkins)

***Hemigymnodacus* Hardy**

diversa (Coquillett)
mukiae new species

***Javadacus* Hardy**

javanensis (Perkins)
montana (Hardy)
nigrita (Hardy)
pallescentis (Hardy)
scutellaria (Bezzi)
semisurstyli new species
trilineata (Hardy)

***Nesodacus* Perkins**

atrichus (Bezzi)

***Papuodacus* Drew**

complicata White
fereuncinata new species
maculifemur (Hering)
ochrosterna new species
sinensis Yu, Bai & Chen

***Paradacus* Perkins**

angustifinis (Hardy)
areolata (Walker)
duplicata (Bezzi)
fulvipes (Perkins)
hancocki new species
terminifera (Walker)
urens White

***Parasinodacus* new subgenus**

ablepharus (Bezzi)
binoyi Drew
brevivitta new species
cilifera (Hendel)
citrifusca new species
eurylomata (Hardy)
incisa (Walker)
longicaudata (Perkins)
pahangiae new species
pantabanganiae new species
pseudocucurbitae White
vinnulus (Hardy)
waimitaliae new species

***Paratridacus* Shiraki**

banneri White
expandens (Walker)
garciniae (Bezzi)
icelus (Hardy)
melanius (Hardy & Adachi), new combination
yayeyamanus (Matsumura), new combination

***Parazeugodacus* Shiraki**

abbreviata (Hardy)
arisanica (Shiraki)
bipustulata (Bezzi)
fulvosterna new species
matsumurai (Shiraki)
pendleburyi (Perkins)
tetrachaeta (Bezzi)

***Semicallantra* Drew**

incompta new species
superba new species

***Sinodacus* Zia**

bogorensis (Hardy)
hochii (Zia)
infesta (Enderlein)
longivittata Chua & Ooi
speciosa new species
spectabilis new species
transversa (Hardy)
whitei new species

***Tetradacus* Miyake**

brachycera (Bezzi)
minax (Enderlein)
splendida (Perkins)
tsuneonis (Miyake)

***Zeugodacus* Hendel**

abdoangusta (Drew)
abnormis (Hardy)
aithonota new species
ambigua (Shiraki)
apicofemoralis new species
assamensis White
atrifacies (Perkins)
baliensis new species
bezziana (Hering)
biguttata (Bezzi)
borongensis new species
buruensis White
calumniata (Hardy)
caudata (Fabricius)
connexa (Hardy)
cucurbitae (Coquillett)
daclaciae new species
depressa (Shiraki)
diaphora (Hendel)
diaphoropsis (Hering)
dorsirufa new species
dubiosa (Hardy)
elegantula (Hardy)
emittens (Walker)
exornata (Hering)
flavipilosa (Hardy)
flavolateralis new species
flavopectoralis (Hering)
flavoverticalis new species
freidbergi White
fuscoalata new species
gavisa (Munro)
hatyaiensis new species
havelockiae new species
heinrichi (Hering)
hekouana Yu, He & Yang
hengsawadae new species
hoabinhiae new species
hodgsoniae new species
hululangatiae new species
indentus (Hardy)
iriomotiae new species
ishigakiensis (Shiraki)
isolata (Hardy)
javadica Mahmood
juxtuncinata new species
kaghanae Mahmood
khaoyaiae new species
laguniensis new species
laocaiae new species
lipsana (Hendel)
liquida new species
macrophyllae new species
maculata (Perkins)
melanofacies new species
menglana Yu, Liu & Yang
minima (Hering)
munda (Bezzi)
nakhonnayokiae new species
namlingiae new species
neoelegantula White
neoemittens new species
neoflavipilosa new species
neolipsana new species
nigrifacies (Shiraki)
okunii (Shiraki)
pemalangiae new species
perplexa (Walker)
persignata (Hering)
platamus (Hardy)
proprescutellata Zhang, Chen & Gao
pubescens (Bezzi)
quasiinfesta new species
rubella (Hardy)
sabahensis new species
sasaotiae new species
scutellaris (Bezzi)
scutellata (Hendel)
scutellina (Bezzi)
semongokensis new species
signata (Hering)
sonlaiae new species
sumbensis (Hering)
synnephes (Hendel)
tapervitta Mahmood
tappanus (Shiraki)
tau (Walker)
tebeduiae new species
timorensis (Perkins)
trichosanthes new species
tricuspidatae new species
trimaculata (Hardy & Adachi)
trivandrumensis new species
ujungpandangiae new species
uncinata new species
vargus (Hardy)
vulta (Hardy)
watersi (Hardy)
yalaensis new species
yoshimotoi (Hardy)
zahadi Mahmood

Genus *Dacus* Fabricius

Callantra Walker

longicornis Wiedemann

Didacus Collart

ciliatus Loew
hainanus Wang & Zhao
keiseri (Hering)

Leptoxyda Macquart

persicus Hensdel

Mellesis Bezzi

ambonensis Drew & Hancock
atrimarginatus White
bannatus (Wang)
bispinosus (Wang)
califrayae Drew & Hancock
conopsoides de Meijere
crabroniformis (Bezzi)
discipennis Walker, new combination
discophorus (Hering)
discretus new species
donggaliae new species
dorjii Drew & Romig
esakii (Shiraki)
feijeni White
fletcheri Drew
formosanus (Tseng & Chu)
icariiformis (Enderlein)
indecorus (Hardy)
infernus (Hardy)
insulosus Drew & Hancock
lagunae Drew & Hancock
leongi Drew & Hancock
maculipterus White
melanopectus new species
murphyi White
nanggalae Drew & Hancock
nummularius (Bezzi)
ooii Drew & Hancock
pedunculatus (Bezzi)
pictus (Hardy)
polistiformis (Senior-White)
pullus (Hardy)
ramanii Drew & Hancock
satanas (Hering)
siamensis Drew & Hancock
sinensis (Wang)
subsessilis (Bezzi)
tenebrosus Drew & Hancock
trimacula (Wang)
triquetrus new species
vijaysegarani Drew & Hancock
vittatus (Hardy)
wallacei White

Neodacus Perkins

ortholomatus Hardy
santongae Drew & Hancock
sphaeroidalis (Bezzi)

Genus *Ichneumonopsis* Hardy

burmensis Hardy

Genus *Monacrostichus* Bezzi

citricola Bezzi
malaysiae Drew & Hancock

Species Complexes

Within the genus *Bactrocera*, we recognize four complexes that contain species with close morphological similarities. These are defined as follows:

BACTROCERA (BACTROCERA) DORSALIS COMPLEX

Definition: Species of *B. (Bactrocera)* with distinct dark facial spots; scutum black, lateral postsutural yellow vittae present, medial postsutural yellow vitta absent; wing colourless except for a narrow costal band (never confluent with R_{4+5}) and anal streak, cells bc and c colourless or, at most, with an extremely pale tint, without dense microtrichia covering cells bc and c; femora entirely or mostly fulvous but may possess dark patterns particularly on or around apices; scutellum yellow with a dark basal band and never with other dark patterns; abdominal terga III-V with a dark 'T' pattern and with, in some species, variable lateral dark patterns.

Included species: *affinidorsalis* (Hardy), *amarambalensis* Drew, *arecae* (Hardy & Adachi), *atrifemur* Drew & Hancock, *bimaculata* Drew & Hancock, *binhduongiae* new species, *bitungiae* new species, *bivittata* Lin & Wang, *carambolae* Drew & Hancock, *caryeae* (Kapoor), *cibodasae* Drew & Hancock, *cognata* (Hardy & Adachi), *collita* Drew & Hancock, *dongnaiae* new species, *dorsalis* (Hendel), *dorsaloides* (Hardy & Adachi), *fernandoi* Tsuruta & White, *flavosterna* new species, *floresiae* Drew & Hancock, *fulvifemur* Drew & Hancock, *fuscitibia* Drew & Hancock, *gombokensis* Drew & Hancock, *hantanae* Tsuruta & White, *holtmanni* (Hardy), *indonesiae* Drew & Hancock, *infulata* Drew & Hancock, *involuta* (Hardy), *irvingiae* Drew & Hancock, *kalimantaniae* new species, *kanchanaburi* Drew & Hancock, *kandiensis* Drew & Hancock, *kinabalu* Drew & Hancock, *laithieuiae* new species, *lateritaenia* Drew & Hancock, *latilineola* Drew & Hancock, *lombokensis* Drew & Hancock, *makilingensis* Drew & Hancock, *malaysiensis* Drew & Hancock, *melastomatos* Drew & Hancock, *merapiensis* Drew & Hancock, *minuscula* Drew & Hancock, *muiri* (Hardy & Adachi), *neoarecae* Drew, *neocognata* Drew & Hancock, *neopropinqua* Drew & Hancock, *occipitalis* (Bezzi), *ochroma* new species, *osbeckiae* Drew & Hancock, *papayae* Drew & Hancock, *paraarecae* new species, *paraverbascifoliae* Drew, *pedestris* (Bezzi), *penecognata* Drew & Hancock, *profunda* Tsuruta & White, *propinqua* (Hardy & Adachi), *pyrifoliae* Drew & Hancock, *quasiinfulata* new species, *quasipropinqua* Drew & Hancock, *raiensis* Drew & Hancock, *ranganathii* new species, *sapaensis* new species, *sembaliensis* Drew & Hancock, *sulawesiae* Drew & Hancock, *sumbawaensis* Drew & Hancock, *syzygii* White & Tsuruta, *thailandica* Drew & Hancock, *unimacula* Drew & Hancock, *usitata* Drew & Hancock, *verbascifoliae* Drew & Hancock, *vishnu* Drew & Hancock.

Because some species in the *dorsalis* complex are of major economic importance, they have attracted considerable research attention since the publication of Drew & Hancock (1994a). Doubt has been raised over the specific status of *Bactrocera carambolae* Drew & Hancock, *Bactrocera dorsalis* (Hendel), *Bactrocera papayae* Drew & Hancock and *Bactrocera philippinensis* Drew & Hancock. In addition, *Bactrocera invadens* Drew, Tsuruta & White is referred to as *B. dorsalis* by some workers.

In this publication we have synonymised *B. papayae* and *B. philippinensis* (see explanation in Remarks under *B. papayae*). However, recent analyses of whole body transcriptomes have identified polymorphisms specific to *B. philippinensis* (Armstrong, Lincoln University, New Zealand, pers. comm. 2013). This genetic differentiation in addition to previously recorded differences in external morphology, aedeagus and ovipositor piercer lengths and wing shapes, will probably lead to the specific status of these two species being reviewed, particularly if further genetic data become available.

BACTROCERA (BACTROCERA) NIGROTIBIALIS COMPLEX

Definition: Species of *B. (Bactrocera)* with face varying from fulvous with dark spots to entirely black; scutum entirely black (never with pale patterns), lateral postsutural yellow vittae generally present, medial postsutural yellow vitta absent; wing colourless except for a narrow costal band (may be almost confluent with R_{4+5}) and anal streak, cells bc and c colourless or, at most, with an extremely pale tint, without dense microtrichia covering cells bc and c; fore and mid femora mostly or entirely black and hind femora with apical ⅓ to ½ black; scutellum yellow with or without dark patterns; abdominal terga mostly or entirely black with some species possessing pale colour centrally over terga III–V.
Included species: *apiconigroscutella* Drew, *bellisi* new species, *lata* (Perkins), *nationigrotibialis* new species, *neonigrotibialis* Drew, *nigrifacia* Zhang, Ji & Chen, *nigrifemorata* Lin & Wang, *nigrofemoralis* White & Tsuruta, *nigrotibialis* (Perkins), *paradiospyri* Chen, Zhou & Li, *perigrapha* White & Tsuruta, *pernigra* Ito, *wuzhishana* Lin & Yang.

BACTROCERA (ZEUGODACUS) SCUTELLARIS COMPLEX

Definition: Species of *B. (Zeugodacus)* with face varying from fulvous with dark spots to entirely black; scutum entirely black, lateral and medial postsutural yellow vittae generally short and narrow; wing colourless except for a narrow costal band confluent with R_{2+3} before becoming very narrow beyond apex of R_{2+3} and then widening across apex of R_{4+5}, anal streak present or absent, cells bc and c colourless, dense microtrichia in outer corner of cell c only; femora varying from entirely fulvous to fulvous with dark patterns; scutellum varying from entirely yellow to yellow with dark patterns, scutellum with 2 or 4 *sc.* setae; abdominal terga III–V varying from entirely dark fuscous to black to red-brown with a variety of dark patterns.
Included species: *assamensis* White, *atrifacies* (Perkins), *biguttata* (Bezzi), *diaphora* (Hendel), *dorsirufa* new species, *freidbergi* White, *hekouana* Yu, He & Yang, *hoabinhiae* new species, *kaghanae* Mahmood, *laocaiae* new species, *liquida* new species, *menglana* Yu, Liu & Yang, *nakhonnayokiae* new species, *namlingiae* new species, *proprescutellata* Zhang, Chen & Gao, *scutellaris* (Bezzi), *scutellata* (Hendel), *sonlaiae* new species, *tappanus* (Shiraki), *yoshimotoi* (Hardy).

BACTROCERA (ZEUGODACUS) TAU COMPLEX

Definition: Species of *B. (Zeugodacus)* with scutum black with or without large areas of red-brown, lateral and medial postsutural yellow vittae present; wing with a narrow costal band generally overlapping R_{2+3} and expanding into a spot at apex and a distinct anal streak, occasionally with infuscation on *dm-cu* crossvein, cells bc and c colourless or with a pale tint, dense microtrichia in outer corner of cell c only; femora fulvous with or without dark apical spots; scutellum yellow with or without dark patterns, scutellum with 4 *sc.* setae (rarely with 2 *sc.* setae); abdominal terga III–V with a dark 'T' pattern and variable lateral dark patterns.
Included species: *bezziana* (Hering), *borongensis* new species, *flavopectoralis* (Hering), *heinrichi* (Hering), *hodgsoniae* new species, *isolata* (Hardy), *javadica* Mahmood, *khaoyaiae* new species, *macrophyllae* new species, *okunii* (Shiraki), *pubescens* (Bezzi), *signata* (Hering), *synnephes* (Hendel), *tapervitta* Mahmood, *tau* (Walker), *tebeduiae* new species, *trichosanthes* new species, *tricuspidatae* new species, *trivandrumensis* new species, *yalaensis* new species, *zahadi* Mahmood.

Identification of Cryptic Species

For several decades taxonomists have been defining species of Dacinae, particularly within the genus *Bactrocera,* on criteria additional to external morphology. For example, Drew & Hardy (1981) using cytology and host records and Drew & Hancock (1994a) using differing combinations of male pheromone chemistry, enzyme electrophoresis, DNA data, host plant records, larval morphology and adult fly morphometrics as supporting evidence, defined species within the *dorsalis* complex. Some of these criteria were revisited by Clarke *et al.* (2005) and Drew *et al.* (2008) for some species in the *dorsalis* complex. While congruence between these various criteria is difficult to achieve for some cryptic species, it is essential that all avenues of research are explored in depth and all resulting data used in order to make decisions on species identity within species complexes. In this section, we review some of the criteria that past and present research has covered to assist in defining cryptic species, in addition to the normal external morphological analyses.

Morphometrics: Significant differences in male aedeagus lengths have been found between *B. carambolae*, *B. dorsalis*, *B. papayae* and *B. philippinensis* (Drew *et al.*, 2008). In that study, correlation between the length of the aedeagus and the female aculeus within species was also recorded. These data could be relevant as part of the answer if these morphological characters contribute to species isolation. However, sound morphometric data are dependent on a correct and systematic field sampling strategy. It is essential that host reared specimens be used, in addition to male lure trapped material.
Wing shape: A study of the wing shapes by Schutze *et al.* (2012) has shown that *B. carambolae*, *B. dorsalis*, *B. papayae* and *B. philippinensis* can all be separated with a high degree of accuracy. In addition, there was a high level of similarity in wing shape between *B. carambolae* from Malaysia and the introduced population in Suriname. Given that there has been clear documentation of the use of wings in courtship premating within *Bactrocera* species, these wing shape analyses appear highly significant to the elucidation of species. For example, Arakaki *et al.* (1984), Kuba and Sokei (1988) and Drew *et al.* (2008) have described male flies taking up positions on the lower leaf surfaces of host plants, wiping the hind legs over the anal area of the abdomen to collect pheromone droplets, then transferring the droplets to the anal area of the wing before vibrating or fanning the wing over the pecten of cilia on abdominal tergum III. These actions atomize the male pheromones into the surrounding atmosphere and produce an accompanying high frequency sound, the combination of which attract females. Under laboratory conditions, Arakaki *et al.* (1984) found that not all males exhibited these functions over the period of experimentation, but those that did achieved a higher rate of mating success. Given that we have observed large differences in wing shape between *Bactrocera* species that do not fit within complexes, we conclude that differences in wing shape could become very important in diagnosing cryptic species.
Chromosome studies: Research on mitotic chromosomes by Baimai *et al.* (1999) has shown that, at least 10 species in the *dorsalis* complex can be successfully separated. In particular, they have shown significant differences between *B. carambolae* and *B. dorsalis* in the patterns of heterochromatin in the mitotic chromosomes and have concluded that 'differences in mitotic chromosomes can be used as species-specific diagnostic characters.' Clearly, further cytological research into populations within the *dorsalis* complex should be undertaken as such data must be used as part of any final conclusion. Similar studies by Baimai *et al.* (2000) resulted in the diagnosis of species in the *Bactrocera tau* (Walker) complex.
Molecular data: Comparison of species based on mitochondrial DNA has been carried out over many years in laboratories around the world. To date, little significant data have been published on the *dorsalis* complex. A summary of evidence was discussed by Drew *et al.* (2008). Further, the relevance of molecular research to defining species remains unconvincing in the absence of the elucidation of entire genomes. Also, this technique cannot be applied to allopatric populations.
Hybridization experiments: The standard methodology for laboratory cross-breeding and choice experiments, utilizes the development of cultures isolined from individual field collected females. This approach guarantees that experiments are conducted on cultures that represent a single species. Laboratory based experiments where two species are interbred (hybridized)

have been performed in Drosophilidae for over half a century. Applying these procedures to the Tephritidae is less convincing as explained in the following examples: *Bactrocera (Bactrocera) neohumeralis* (Hardy) and *Bactrocera (Bactrocera) tryoni* (Froggatt) are considered to be distinct species with separate daily mating periods but interbreed in the laboratory to produce fertile progeny. Another pair, *Bactrocera (Afrodacus) jarvisi* (Tryon) and *Bactrocera (Bactrocera) tryoni* (Froggatt), each in different subgenera, have been interbred under laboratory conditions to produce fertile F_1 and F_2 generations (Cruickshank *et al.*, 2001), while *B. dorsalis* and *B. invadens* (African material) similarly interbreed in the laboratory to produce fertile progeny (Jessup, Department Primary Industries, New South Wales, Australia, pers. comm. 2012). Further, laboratory hybridization is diametrically opposed to the standard biological definition of species which specifically excludes interbreeding in captivity. In defining species, little assistance can be provided from interbreeding experiments as they are unrelated to the behaviour of populations under field conditions. While cross-breeding populations in captivity to define post-zygotic isolation is not helpful, male and female choice experiments can be of assistance. However, in the final analysis, it is essential to define the Specific Mate Recognition Systems that, in nature, confine species populations.

Pheromone chemistry: The chemistry of male pheromones has been thoroughly researched and reviewed by Fletcher & Kitching (1995), for some 20 *Bactrocera* species. On the basis of this work, Drew & Hancock (1994a) used major differences in the chemical components of the male pheromones of *B. carambolae* and *B. dorsalis* as one character upon which to describe *B. carambolae* as a good biological species. Fletcher & Kitching (1995) recorded 'profound differences in the chemistry of the rectal gland secretions' between *B. carambolae* and *B. papayae* and 'minor consistent differences between *B. papayae*, *B. philippinensis* and *B. dorsalis*'. They concluded that 'the nature of the rectal glandular components may be a powerful taxonomic criterion'. As for morphometric studies, it is essential to obtain specimens for this research from a correct field collecting programme using host reared specimens.

Host records: When analysing host plant records, it is important to concentrate on those deemed to be endemic, specifically those associated with rainforests. Many species in the *dorsalis* complex occupy specific host plants. For example, *B. arecae*, *B. irvingiae*, *B. melastomatos* and *B. pyrifoliae* to name a few (Allwood *et al.*, 1999). For the polyphagous species such as *B. carambolae* and *B. papayae*, while there is considerable overlap in host records, each exhibits some specificity in their major hosts. Extensive field surveys have been conducted throughout the Indomalayan and Pacific regions providing a rich resource of data (Allwood *et al.*, 1999; Leblanc *et al.*, 2012).

When undertaking research in the categories listed above, accurate comparisons can only be achieved if all research groups use the same field populations. The best method of reaching this standard is to collect infested host fruits and rear out the flies. Because many species can be collected in the same male lure trap, we do not recommend using specimens from this source unless their identity is confirmed by a specialist diagnostician or another research strategy.

In conclusion, we emphasize the need to undertake as many research avenues as possible and compare all relevant data in order to define cryptic species. Also, all such approaches must be based on carefully constructed field collecting programmes that are consistent over all areas of research.

Description of Species Under Genera and Subgenera

Comprehensive descriptions of species are presented together with hand-drawn figures that illustrate essential diagnostic characters. These figures provide maximum assistance in the identification of species and are grouped together at the end of the publication to facilitate comparing species within specific groups or subgenera.

Data listed for all specimens studied have been transcribed exactly from the specimen labels, which may result in occasional differences in date presentation, spelling of names of geographic locations and collectors. Not all acronyms on specimen labels could be defined.

GENUS *BACTROCERA* Macquart

SUBGENUS *AFRODACUS* Bezzi

Bactrocera (Afrodacus) fastigata Tsuruta & White

(Fig. 7)

Bactrocera (Afrodacus) fastigata Tsuruta & White, 2001: 72; Drew & Raghu, 2002: 330. Holotype male in HORDI.

Non-type material examined: INDIA: 1 male, S. India, Hort Farm Coonoor, 1450 m, 29.v.1992, coll. D.L. Hancock, attracted to cue lure; 1 male, New Amarambalam Forest, Kerala, 15.ii.1999-25.ii.1999, coll. S. Raghu, attracted to cue lure. **SRI LANKA (formerly CEYLON)**: 1 male, Padiya Pelella, c6, 23.ix.1993; 1 male, Sri Lanka, 21.x.1993, Gannoruwa; 1 male, Sri Lanka, 8.xi.1993, Gannoruwa. Specimens in QDPC.

Diagnosis: Face fulvous with a pair of large subquadrate black spots; postpronotal lobes and notopleura yellow; scutum black with brown lateral to postsutural vittae; very narrow and short lateral postsutural yellow vittae ending before level of *p.sa.* seta; medial postsutural yellow vitta absent; mesopleural stripe reaching midway between anterior margin of notopleuron and anterior *npl.* seta dorsally; scutellum yellow; *a.sa.* setae absent; legs with all femora fulvous with elongate oval dark fuscous to black preapical spots; fore and hind tibiae fuscous, mid tibiae fulvous with base fuscous; wing with cells bc and c with a pale fuscous tint, microtrichia in outer corner of cell c only; a narrow fuscous costal band confluent with R_{2+3} and remaining narrow apically to end between extremities of R_{4+5} and M; a narrow fuscous anal streak; supernumerary lobe of medium development; abdominal terga III–V red-brown with a 'T' pattern consisting of a broad transverse dark fuscous band across anterior margin of tergum III which expands laterally to cover the entire lateral margins and a narrow medial longitudinal dull black band over all three terga, a pair of oval red-brown shining spots on tergum V, abdominal sterna dark coloured; posterior lobe of surstylus short, sternum V with a deep concavity on posterior margin.

Attractant: Cue lure.
Female: No known record.
Distribution:
Type locality: Sri Lanka (Kandy District, Padiya Pellella).
Previously recorded: Type locality and other district locations in Sri Lanka (Tsuruta & White, 2001); Southern India (Kerala) (Drew & Raghu, 2002).
Hosts: No known record.
Remarks: *Bactrocera (Afrodacus) fastigata* Tsuruta & White has the general appearance of the *dorsalis* complex species but in lacking the *a.sa.* setae fits best into the subgenus *Afrodacus*. It can be best diagnosed by the black scutum, very narrow and short lateral postsutural vittae, the possession of an elongate oval dark-coloured preapical spot on each femur, the wing with a narrow fuscous costal band and abdominal terga III–V red-brown with a dark-coloured 'T' pattern. Species of subgenus *Javadacus*, particularly *B. (Javadacus) scutellaria* (Bezzi) and *B. (Javadacus) trilineata* (Hardy), are similar but these species differ in possessing a medial longitudinal postsutural vitta and a long posterior surstylus lobe.

SUBGENUS *ASIADACUS* Perkins

Bactrocera (Asiadacus) absoluta (Walker)

(Fig. 8)

Dacus absolutus Walker, 1861c: 22. Lectotype female in BMNH.
Dacus (Neodacus) absolutus – Hardy, 1959: 162. Lectotype designation by inference of Holotype.
Bactrocera (Asiadacus) absoluta – Norrbom *et al.*, 1998: 87.

Type material examined: Lectotype female, **INDONESIA**: with the following eight labels – 'Dacus absolutus Wlk., type'; 'Ceram, A.R. Wallace, 68–4' (in Walker's handwriting); 'Cer.' (in Walker's handwriting); 'T.89'; 'absolutus' (in Walker's handwriting); 'BMNH (E) No. 53276';

'Lectotype♀ of Dacus absolutus Walker 1861 des. D.E. Hardy 1959, Indonesia: Maluku, Ceram, A.R. Wallace'; 'Dacus (Neodacus) absolutus Walker, det. 1959, D.E. Hardy'. Studied at BMNH.

Diagnosis: A large species; face fulvous with a pair of medium-sized circular black spots; postpronotal lobes and notopleura yellow; scutum uniformly black; broad parallel-sided lateral postsutural yellow vittae present; medial postsutural yellow vitta absent; no yellow spot anterior to notopleural suture; mesopleural stripe reaching to anterior *npl.* seta dorsally; scutellum yellow; wing with cells bc and c colourless, dense microtrichia covering all of cell c and outer corner of cell bc; a broad fuscous costal band confluent with R_{4+5} and remaining narrow around margin of wing; a broad fuscous anal streak; abdominal terga III–V pale orange-brown with a narrow transverse dark fuscous band across anterior margin of tergum III becoming black and gradually widening to cover lateral margins, anterolateral corners of tergum IV and V fuscous to dark fuscous, a narrow medial longitudinal black band on tergum V.

Description: female.

Head: Frons – pale fuscous with fulvous around lateral and dorsal margins; orbital setae black: 1 *s.or.*, 2 *i.or.*; lunule pale fuscous. Ocellar triangle black. Vertex fulvous. Face fulvous with a pair of medium-sized circular black spots. Occiput mostly fuscous, fulvous along eye margins. Antennae with segments 1 and 2 pale fuscous, segment 3 mostly fuscous.

Thorax: Scutum uniformly black except for red-brown below and behind lateral postsutural vittae. Pleural areas black except fulvous below postpronotal lobes. Yellow markings as follows: postpronotal lobes; notopleura; mesopleural stripe reaching to anterior *npl.* seta dorsally, extending to katepisternum as a narrow transverse spot, anterior margin straight; two broad parallel-sided lateral postsutural vittae extending to *ia.* seta. Medial longitudinal postsutural yellow vitta absent. Postnotum black. Scutellum yellow except for narrow black basal band. Setae: *sc.* 2; *prsc.* absent; *ia.* 1; *p.sa.* 1; *a.sa.* 1; *mpl.* 1; *npl.* 2; *scp.* 4.

Legs: All segments entirely fulvous except fuscous to dark fuscous on apical ventral surfaces of hind femora; fore and hind tibiae pale fuscous; mid tibiae each with an apical black spur.

Wings: Length 7.7 mm; cells bc and c colourless (possibly would be pale fuscous in fully matured specimens); microtrichia covering all of cell c and outer corner of cell bc; remainder of wings colourless except pale fuscous cell sc, fuscous costal band confluent with R_{4+5} and not widening around margin of wing, to end beyond extremity of R_{4+5}; a moderately broad pale fuscous anal streak; no dense aggregation of microtrichia around A_1+CuA_2; supernumerary lobe weak. This specimen appears post-teneral as recorded by Hardy (1959) and consequently the exact colour of the wing patterns cannot be determined.

Abdomen: Oval; terga free; no pecten on tergum III. Tergum I and sterna I and II wider than long. Tergum I dark fuscous to black laterally tending to pale fuscous medially and with a narrow transverse fulvous band across posterior margin but not reaching lateral margins; tergum II mostly fulvous with a narrow subanterior transverse dark fuscous band; terga III–V pale orange-brown with a narrow transverse fuscous band across anterior margin of tergum III which becomes black as it gradually widens to cover lateral margins, fuscous to dark fuscous anterolateral corners on terga IV and V, a narrow medial longitudinal black band on tergum V. A pair of oval orange-brown shining spots on tergum V. Abdominal sterna dark. Ovipositor basal segment orange-brown, dorsoventrally flattened and tapering slightly posteriorly in dorsal view.

Attractant: No known record.

Male: No known record.

Distribution:

Type locality: Indonesia (Moluccas = Maluku – Seram I. = Ceram I.).

Hosts: No known record.

Remarks: *Bactrocera (Asiadacus) absoluta* (Walker) fits imperfectly into the subgenus *Asiadacus* in that it possesses *a.sa.* setae. In the absence of male specimens which carry important subgeneric characters, we are leaving this species in *Asiadacus* where it has been placed by previous workers. It differs from all other species placed in the subgenus *Asiadacus* in lacking a medial longitudinal yellow vitta on the scutum, from *Bactrocera (Asiadacus) apicalis* (de Meijere) and *Bactrocera (Asiadacus) bakeri* (Bezzi) in lacking a large fuscous spot in the apex of the wing and from *Bactrocera (Asiadacus) maculifacies* (Hardy) and *Bactrocera (Asiadacus) melanopsis* (Hardy) in having entirely fulvous femora without dark markings and in lacking a black 'T'-shaped pattern on abdominal terga III–V.

The female type in BMNH appears to be the only specimen that Walker studied for the original description. It bears a Lectotype label in Hardy's handwriting. Although Hardy (1959) did not refer to his Lectotype designation, this specimen can be recognized as a Lectotype by inference of Holotype under Article 74.6 in the International Code of Zoological Nomenclature (ICZN).

Bactrocera (Asiadacus) apicalis (de Meijere)

(Fig. 9)

Dacus apicalis de Meijere, 1911: 376. Holotype male in RNH.

Dacus (Strumeta) apicalis – Hardy & Adachi, 1954: 161.

Dacus (Asiadacus) modicus Hardy, 1973: 17. **New Synonym.**

Dacus (Asiadacus) dianensis Wang & Zhao, 1989: 215. **New Synonym** (previously synonymised with *Dacus modicus* Hardy, see Norrbom *et al.*, 1998: 87).

Dacus (Asiadacus) nadanus Chao & Lin, 1993b: 139. **New Synonym.**

Bactrocera (Asiadacus) apicalis – Norrbom *et al.*, 1998: 87.

Non-type material examined: INDONESIA: JAVA – 1 male, West Java, Cirebon, Panguragan, 24.ix.2005, coll. Naim S., attracted to cue lure in guava; 1 male (14.vii.2005), 1 male (21.vii.2005), West Java, Majalengka, Ampel 25 dpl, coll. Yaya Senjaya, attracted to cue lure in mango; 1 male, East Java, Sidorajo, Gelang, 12.vii.2005, coll. Agus Riadi, attracted to cue lure in village; 1 male (28.viii.2005), 1 male (4.ix.2005), 19 males (13.xi.2005), 32 males (4.xii.2005),

Central Java, Pemalang, Pegongsoran, coll. Sapto Darmono, attracted to cue lure in forest; **KALIMANTAN** – 1 male (19.vii.2005), 1 male (16.vii.2005), 2 males (23.vii.2005), West Kalimantan, Pontianak, Sei Duri I, coll. A. Irom, attracted to cue lure in bananas; 1 male (1.viii.2005), 1 male (8.viii.2005), 1 male (29.viii.2005), 1 male (5.ix.2005), 1 male (12.ix.2005), East Kalimantan, Kutai Kartanegara, Batuah, coll. Ertha DP, attracted to cue lure in bananas; **SULAWESI** – 1 male, Sulawesi Selatan, Takalar, Topejawa, 2.ix.2006, coll. Sirajuddin SP, attracted to cue lure in village; 2 males, Sulawesi Barat, Mamuju, Tambayako, 30.x.2006, coll. Yan P Fachrie, attracted to cue lure in village. **EAST MALAYSIA: SABAH** – 1 male, ACIAR Site Ht4, Sabah orchard, 9.x.1991, attracted to cue lure in fruit orchard; 1 male, Ulu Dusun, Sandakan fruit orchard, 20.xi.1991, coll. Jinius *et al.*, attracted to cue lure; **SARAWAK**: 1 male, ARC Semongok, nr Kuching, 18.vii.1992, coll. S. Leong, attracted to cue lure; 1 male (16.iv.1992), 1 male (24.vi.1992), Sg Sui, Coastal area nr Kuching, coll. S. Leong, attracted to cue lure. **PENINSULAR MALAYSIA**: A large series of specimens collected over a wide geographic range during field surveys. **THAILAND**: A large series of specimens collected over a wide geographic range during field surveys. **VIETNAM: Southern Vietnam** – 1 male, Vinh Long Province, Phuan An Village, 29.x.1999, attracted to cue lure; 1 male, Vinh Long Province, Long Ho District, Phuoc Hau Village, 25.ii.2002, coll. SOFRI Staff, attracted to cue lure; 1 male, Dong Thap Province, Lap Vo District, Tan Khanh Trung, 18.viii.2002, attracted to cue lure; 1 male, Cantho Province, Chau Thanh District, Tan Phu Village, 28.vii.2002, attracted to cue lure; 1 male, Can Tho Province, Binh Thuy Ward, 29.x.1999, attracted to cue lure; 1 male, Dong Thap Province, Lap Vo District, Tan Khanh Trung, 11.viii..2002, attracted to cue lure; 1 male (17.xii.2001), 3 males (24.xii.2001), 1 male (5.i.2002), 7 males (12.i.2002), 2 males (19.i.2002), 1 male (1.vi.2002), 1 male (8.vi.2002), 2 males (14.vii.2002), 1 male (21.vii. 2002), 1 male (27.vii.2002), 1 male (3.viii.2002), 3 males (7.ix.2002), Ben Tre Province, Chau Thanh District, Tien Thuy Village, Site VS04, coll. SOFRI Staff, attracted to cue lure; 2 males, Dong Nai Province, Long Khanh District, Nhan Nghia Village, 6.vii.2002, attracted to cue lure; 1 male, Tien Giang Province, Cho Gao Village, 7.viii.2000, attracted to cue lure; 1 male, Tien Giang Province, Cai Lay District, Long Khanh Village, 4.i.2000; 1 male, Lam Dong Province, 3 District, Da Lat Town, 8.x.2002, Site VS08, attracted to cue lure; 1 male (30.xi.1999), 1 male (12.i.2000), Tien Giang Province, Cai Be District, My Thirn Village, Site S008, attracted to cue lure; 3 males (8.vii.1999), 2 males (10.v.2000), Dong Nai Province, Vinh Cuu District, Tan Binh Village, Site S011, attracted to cue lure; 1 male, Dong Nai Province, Long Nhat District, Long Nhat Village, 7.i.2000, Site S012, attracted to cue lure; 1 male, Tien Giang Province, Cai Be District, My Thirn Village, 17.ii.2000, Site S013, attracted to cue lure; 1 male, Dong Nai Province, Long Khanh District, Tan Phuong Village, Site S014, 20.i.2000, attracted to cue lure; 1 male, Dong Nai Province, Zuan Phu Village, Site S015, 7.viii.1999, attracted to cue lure. Specimens in QDPC.

Diagnosis: A small species; face fulvous with a pair of small- to medium-sized circular to pear-shaped black spots; postpronotal lobes and notopleura yellow; scutum generally red-brown; lateral and medial postsutural yellow vittae present; a yellow spot present anterior to notopleural suture; mesopleural stripe reaching midway between anterior margin of notopleuron and anterior *npl.* seta dorsally; scutellum yellow; wing with cells bc and c with a pale tint, microtrichia in outer corner of cell c only; a narrow fuscous costal band confluent with R_{2+3} and ending at apex of this vein, a large dark fuscous spot in apex of wing; a broad fuscous anal streak; abdominal terga III–V fulvous with a black 'T' pattern consisting of a narrow transverse band across anterior margin of tergum III and a narrow- to medium-width medial longitudinal band over all three terga, dark fuscous to black anterolateral corners on terga IV and V.

Description: male.

Head: Height 1.2 mm. Frons – length 1.1 times breadth; red-brown with fulvous around lateral and ventral margins; orbital setae small and black: 1 *s.or.*, 2 *i.or.*; lunule red-brown. Ocellar triangle black. Vertex black. Face fulvous with a pair of small- to medium-sized circular to pear-shaped black spots; length 0.43 mm. Genae fulvous, small fuscous subocular spot present; weak black seta present. Occiput red-brown, fulvous along eye margins; occipital row with a small number of weak pale setae. Antennae with segments 1 and 2 red-brown, segment 3 red-brown with fuscous on apex and outer surface; length of segments: 0.12 mm; 0.24 mm; 0.53 mm.

Thorax: Scutum generally red-brown with variable areas of dark fuscous to black along margins of lateral and medial postsutural vittae, between postpronotal lobes and notopleura, and a narrow medial longitudinal line running from anterior margin of medial postsutural vitta to anterior margin of scutum. Pleural areas red-brown with diffuse areas of dark fuscous anterior and posterior to mesopleural stripe and on katepisternum. Yellow markings as follows: postpronotal lobes; notopleura; mesopleural stripe reaching midway between anterior margin of notopleuron and anterior *npl.* seta dorsally, continuing to katepisternum as a large transverse spot, anterior margin convex; anatergite (posterior apex red-brown to fuscous); anterior ¾ katatergite (remainder red-brown to fuscous); two narrow parallel-sided lateral postsutural vittae beginning with a triangular spot anterior to notopleural suture and ending before *ia.* seta (in occasional specimens these vittae appear to end at *ia.* seta); a broad triangular medial longitudinal postsutural vitta. Postnotum red-brown. Scutellum yellow except for narrow red-brown basal band. Setae: *sc.* 2; *prsc.* absent; *ia.* 1; *p.sa.* 1; *a.sa.* absent; *mpl.* 1; *npl.* 2; *scp.* 4.

Legs: All femora entirely fulvous; fore tibiae pale fuscous, mid tibiae fulvous with pale fuscous basally, hind tibiae fuscous; tarsi with basal two segments fulvous and apical three segments fuscous; mid tibiae each with an apical black spur.

Wings: Length 5.1 mm; cells bc and c with a pale tint (cell bc slightly darker); microtrichia in outer corner of cell c only; remainder of wings colourless except fuscous cell sc, narrow fuscous costal band confluent with R_{2+3} and ending at apex of this vein, a large isolated dark fuscous spot in apex of wing, a broad fuscous anal streak; a dense aggregation of microtrichia around A_1+CuA_2; supernumerary lobe of medium development.

Abdomen: Elongate oval; terga free; pecten present on tergum III. Tergum I and sterna I and II wider than long. Tergum I red-brown with broad dark fuscous to black lateral margins; tergum II fulvous with a narrow transverse black band across anterior margin but not reaching narrow fuscous lateral margins; terga III–V fulvous with a black 'T' pattern consisting of a narrow transverse band across anterior margin of tergum III which covers narrow lateral margins of this tergum and a narrow- to medium-width medial longitudinal band over all three terga, black anterolateral corners on tergum IV consisting of a narrow transverse band running from lateral margins towards centre of tergum and then recurving back towards posterior margin of the tergum, dark fuscous to black anterolateral corners on tergum V. A pair of oval fulvous shining spots on tergum V. Posterior lobe of surstylus long, sternum V with a slight concavity on posterior margin. Abdominal sterna I and V dark fuscous to black, remainder fulvous.
Attractant: Cue lure.
Female: No known record.
Distribution:
Type locality: Indonesia (Java).
Previously recorded: Type locality and the following locations – Brunei; China; Malaysia; Thailand; Vietnam (White & Hancock, 1997). The record from Philippines (Mindanao) in CABIKEY (White & Hancock, 1977) is incorrect as it refers to *B. (Asiadacus) bakeri* (Bezzi).
New records: Indonesia (Kalimantan; Sulawesi).
Hosts: *Trichosanthes wawraei* flowers (family *Cucurbitaceae*) (Allwood *et al.*, 1999).
Remarks: *Bactrocera (Asiadacus) apicalis* (de Meijere) is similar to *Bactrocera (Asiadacus) bakeri* (Bezzi) in the general colour patterns of the thorax, wing and abdomen; however, it differs from this species in having the costal band discontinuous after apex of R_{2+3}, a weak supernumerary lobe in the male wing and narrow lateral postsutural vittae that end before the *ia.* seta. *Bactrocera apicalis* is a unique species in possessing a break in the costal band beyond apex of R_{2+3}, narrow lateral postsutural vittae ending before the *ia.* seta and the lateral dark pattern on abdominal tergum IV 'L' shaped and enclosing an irregularly circular-shaped orange-brown area. These characters are common to *Dacus (Asiadacus) dianensis* Wang & Zhao, *Dacus (Asiadacus) modicus* Hardy and *Dacus (Asiadacus) nadanus* Chao & Lin, and thus we regard these as **New Synonyms** of *B. apicalis*.

Bactrocera (Asiadacus) apiciflava Yu, He & Chen

(Fig. 10)

Bactrocera (Javadacus) apiciflava Yu, He & Chen, in Yu *et al.*, 2011: 605. Holotype male in CAIQ.

Type material studied: CHINA: A study of the colour photographs and description of *Bactrocera (Javadacus) apiciflava* Yu, He & Chen in the original publication.
Non-type material examined: THAILAND: 1 male, Pha Yao, Chiang Rai, 27.xi.1991, coll. Charuwan, Site CR257T, attracted to cue lure, in QDPC.

Diagnosis: A medium-sized species; face fulvous with a pair of large angulate black spots; postpronotal lobes and notopleura yellow; scutum black; lateral and medial postsutural yellow vittae present; a yellow spot present anterior to notopleural suture; mesopleural stripe reaching almost to anterior *npl.* seta dorsally; scutellum yellow; wing with cells bc and c colourless, microtrichia in outer corner of cell c only; a narrow, dark fuscous costal band confluent with R_{2+3} and extremely narrow beyond apex of this vein before expanding into a spot across apex of R_{4+5} (*scutellaris*-like); a broad fuscous anal streak; abdominal terga III–V orange-brown with broad medial and lateral longitudinal black bands joined over anterior ¾ of tergum III.
Description: male.
Head: Height 1.48 mm. Frons – length 1.3 times breadth; red-brown with dark fuscous around orbital setae and on anteromedial hump; orbital setae black: 1 *s.or.*, 2 *i.or.*; lunule dark fuscous. Ocellar triangle black. Vertex black. Face fulvous with a pair of large elongate and angular black spots extending to oral margin (Fig. 10); length 0.5 mm. Genae fulvous, large dark fuscous subocular spot present; black seta present. Occiput dark fuscous, tending black dorsally and fulvous along eye margins; occipital row with 3–4 dark setae. Antennae with segments 1 and 2 red-brown, segment 3 absent; length of segments: 0.12 mm; 0.24 mm.
Thorax: Scutum black with fuscous below lateral postsutural vittae and around notopleural suture. Pleural areas entirely black. Yellow markings as follows: postpronotal lobes; notopleura; mesopleural stripe reaching almost to anterior *npl.* seta dorsally, continuing to katepisternum as a small transverse spot, anterior margin convex; anatergite (posterior apex black); anterior ⅔ katatergite (remainder black); two medium-width parallel-sided lateral postsutural vittae ending at *ia.* seta; a narrow medial longitudinal postsutural vitta; a small spot anterior to notopleural suture in front of each lateral postsutural vitta. Postnotum black. Scutellum yellow except for narrow black basal band. Setae: *sc.* 2; *prsc.* absent; *ia.* 1; *p.sa.* 1; *a.sa.* absent; *mpl.* 1; *npl.* 2; *scp.* 4.
Legs: Fore femora fulvous with a large dark fuscous to black spot over most of outer surface, mid femora fulvous with a dark fuscous to black subapical spot on outer surface, hind femora fulvous with dark fuscous to black around apical ⅓; fore tibiae dark fuscous basally to fulvous apically, mid tibiae fuscous to dark fuscous basally to pale fuscous apically, hind tibiae black basally to dark fuscous apically; tarsi with all segments entirely fulvous; mid tibiae each with an apical black spur.
Wings: Length 5.3 mm; cells bc and c colourless; microtrichia in outer corner of cell c only; remainder of wings colourless except dark fuscous cell sc, narrow dark fuscous costal band confluent with R_{2+3} and extremely narrow beyond apex of this vein before expanding into a spot around apex of R_{4+5}, a broad fuscous anal streak; a dense aggregation of microtrichia around A_1+CuA_2; supernumerary lobe weak.
Abdomen: Oval; terga free; pecten present on tergum III. Tergum I and sterna I and II wider than long. Tergum I dark fuscous to black with a narrow transverse orange-brown band across posterior margin but not reaching lateral margins;

tergum II fulvous with a narrow transverse subanterior black band which just reaches the narrow black anterolateral corners; terga III–V orange-brown with broad medial and lateral longitudinal black bands joined across anterior ¾ of tergum III (the lateral bands may also be joined across the anterior ½ of tergum IV). A pair of oval fuscous shining spots on tergum V. Posterior lobe of surstylus long, sternum V with a shallow concavity on posterior margin. Abdominal sterna dark.
Attractant: Cue lure.
Female: No known record.
Distribution:
Type locality: China (Yunnan Province, Hekou).
New records: Thailand (Chiang Rai).
Hosts: No known record.
Remarks: In the specimen studied from Thailand, the *prsc.* setae and the *a.sa.* setae are absent. Consequently, we are placing *B. apiciflava* in the subgenus *Asiadacus. Bactrocera apiciflava* Yu, He & Chen is similar to *Bactrocera (Asiadacus) absoluta* (Walker) in possessing a fulvous face with a pair of black spots and the wing with the costal band not expanded into a large apical spot but differs from this species in having the costal band extremely narrow beyond apex of R_{2+3} before widening into a small spot around apex of R_{4+5} (*scutellaris*-like), *a.sa.* setae absent, cell c of wing bare of microtrichia and abdominal terga III and IV each with a dark 'T' pattern.

Bactrocera (Asiadacus) bakeri (Bezzi)

(Fig. 11)

Chaetodacus bakeri Bezzi, 1919: 426. Holotype male in MCSNM.
Dacus (Neodacus) bakeri – Hardy, 1954: 8; Hardy & Adachi, 1954: 154.
Bactrocera (Asiadacus) bakeri – Norrbom *et al.*, 1998: 87.

Type material examined: Holotype male, **PHILIPPINES**: with the following four labels – 'Davao, Mindanao, Baker'; '6432'; 'bakeri n.' in Bezzi's handwriting; 'Type male'. Received on loan from MCSNM.
Diagnosis: A medium-sized species; face fulvous with medium-sized oval black spots; postpronotal lobes and notopleura yellow; scutum black with extensive red-brown markings; lateral and medial postsutural yellow vittae present; a large triangle along anterior margin of notopleural suture; mesopleural stripe reaching midway between anterior margin of notopleuron and anterior *npl.* seta dorsally; scutellum yellow with a narrow black basal band; wing with cells bc and c with a very pale tint, microtrichia in outer corner of cell c only; a narrow dark fuscous costal band confluent with R_{2+3} and expanding into a large spot across apex of wing; a broad dark fuscous anal streak; abdominal terga III–V orange-brown with a black 'T' pattern consisting of a narrow black basal band across anterior margin of tergum III and expanding over lateral margins and a medium-width medial longitudinal black band over all three terga, a large inverted 'U'-shaped black marking on anterolateral corners of tergum IV and a large black triangle on anterolateral margins of tergum V.

Description: male.
Head: Height 1.55 mm. Frons – length 1.1 times breadth; red-brown without dark markings; orbital setae small and weak: 1 *s.or.*, 1 *i.or.*; lunule red-brown. Ocellar triangle black. Vertex dark fuscous. Face fulvous with a pair of medium-sized oval black spots; length 0.5 mm. Genae fulvous. Occiput red-brown, dark fuscous dorsolaterally, fulvous along eye margins; occipital row consisting of very short and weak pale setae. Antennae with segments 1 and 2 red-brown, segment 3 red-brown with pale fuscous on apex and outer surface; length of segments: 0.14 mm; 0.28 mm; 0.58 mm.
Thorax: Scutum black with the following red-brown markings: below and behind lateral postsutural vittae, a triangular marking around posterior of medial postsutural yellow vitta, extensive areas between notopleural suture and anterior margin of scutum. Pleural areas black except red-brown below postpronotal lobes. Yellow markings as follows: postpronotal lobes; notopleura; mesopleural stripe reaching midway between anterior margin of notopleuron and anterior *npl.* seta dorsally, continuing to katepisternum as a large transverse spot, anterior margin slightly convex; anatergite (posterior apex black); anterior ⅗ katatergite (remainder black); two broad parallel-sided lateral postsutural vittae ending at *ia.* seta; a short broad medial longitudinal postsutural vitta; a large triangle along anterior margin of notopleural suture with base at notopleuron. Postnotum black laterally, red-brown centrally. Scutellum yellow except for narrow black basal band. Setae: *sc.* 2; *prsc.* absent; *ia.* 1; *p.sa.* 1; *a.sa.* absent; *mpl.* 1; *npl.* 2; *scp.* 4.
Legs: All femora entirely fulvous; fore tibiae fuscous, mid tibiae fuscous basally tending pale fuscous apically, hind tibiae dark fuscous; all tarsi with basal segment fulvous and apical four segments pale fuscous; mid tibiae each with an apical black spur.
Wings: Length 6.0 mm; cells bc and c with a very pale tint; microtrichia in outer corner of cell c only; remainder of wings colourless except dark fuscous cell sc, narrow dark fuscous costal band confluent with R_{2+3} and becoming extremely narrow beyond apex of this vein before expanding into a large spot across apex of wing to apex of M, broad dark fuscous anal streak; dense aggregation of microtrichia around A_1+CuA_2; supernumerary lobe strong.
Abdomen: Oval; terga free; pecten present on tergum III. Tergum I orange-brown with narrow dark fuscous to black lateral margins; tergum II orange-brown with a narrow transverse dark fuscous to black band across anterior margin but not reaching narrow dark fuscous to black anterolateral corners; terga III–V orange-brown with a 'T' pattern consisting of a narrow transverse black band across anterior margin of tergum III and expanding over lateral margins, a medium-width medial longitudinal black band over all three terga, a large inverted 'U'-shaped black marking on anterolateral corners of tergum IV, broad black anterolateral corners of tergum V. A pair of oval orange-brown shining spots on tergum V. Posterior lobe of surstylus long, sternum V with a slight concavity on posterior margin.
Attractant: No known record.
Female: No known record.

Distribution:
Type locality: Philippines (Mindanao I., Davao).
Hosts: No known record.
Remarks: *Bactrocera (Asiadacus) bakeri* (Bezzi) is similar to *Bactrocera (Asiadacus) apicalis, Bactrocera (Asiadacus) maculifacies* (Hardy) and *Bactrocera (Asiadacus) melanopsis* (Hardy) in the general colour patterns of the thorax, abdomen and wing. It differs from *B. apicalis* in having black over the scutum and a strong supernumerary lobe in the male wing, from *B. maculifacies* in having the costal band expanded into a large spot in the apex of the wing and from *B. melanopsis* in having a strong supernumerary lobe in the male wing and in lacking infuscation along the *dm-cu* crossvein and dark-coloured femora.

Bactrocera (Asiadacus) careomacula Drew & Romig, new species

(Fig. 12)

Material examined: Holotype male, **PHILIPPINES: LUZON ISLAND** – Mt Makiling, 600 m, Philippines, 5.ix.1983, coll. R.A.I. Drew, attracted to cue lure, deposited in BMNH.
Paratypes: **PHILIPPINES: LUZON ISLAND** – 22 males, same data as holotype; 11 males, Mt Makiling, 500 m, Philippines, 9.v.1983, coll. R. Drew & B. Ramos, attracted to cue lure; 6 males, Mt Makiling, 500 m, Philippines, 15. vi.1983, coll. B. Ramos, attracted to cue lure; 10 males, Mt Makiling, 500 m, Philippines, 5.ix.1983, coll. R.A.I. Drew, attracted to cue lure; 8 males (8.vi.1983), 2 males (15.vi. 1983), Mt Makiling (400 m), Philippines, coll. B. Ramos, attracted to cue lure; 1 male (8.vi.1983), 1 male (15.vi.1983), Mt Makiling, 300 m, Philippines, coll. B. Ramos, attracted to cue lure; 3 males, Mt Makiling, Philippines, 8.vi.1983, coll. B. Ramos, attracted to cue lure. 10 paratypes in ANIC, 10 paratypes in BMNH, 34 paratypes in QDPC, 10 paratypes in QMIC.
Diagnosis: A medium-sized species; face entirely fulvous without dark markings; postpronotal lobes and notopleura yellow; scutum black; lateral and medial postsutural yellow vittae present; a yellow spot present anterior to notopleural suture; mesopleural stripe reaching postpronotal lobe dorsally; scutellum yellow; wing with cells bc and c colourless, microtrichia in outer corner of cell c only; a narrow dark fuscous costal band widening across apex of wing; a broad dark fuscous anal streak; a narrow dark fuscous band enclosing *dm-cu* crossvein; abdominal terga III–V fulvous with a broad medial and a moderately broad lateral longitudinal dark fuscous to black bands over all three terga and joined along anterior margin of tergum III.
Description: male.
Head: Height 1.47 mm. Frons – length 1.3 times breadth; fulvous with dark fuscous to black around bases of *s.or.* setae and dark fuscous on anteromedial hump; orbital setae black: 1 *s.or.*, 2 *i.or.*; lunule dark fuscous. Ocellar triangle black. Vertex black. Face entirely fulvous without dark markings; length 0.6 mm. Genae fulvous, large dark fuscous subocular spot present; black seta present. Occiput fuscous, fulvous along eye margins; occipital row with a large number of short pale setae. Antennae with segments 1 and 2 red-brown, segment 3 entirely dark fuscous; length of segments: 0.1 mm; 0.26 mm; 0.64 mm.
Thorax: Scutum black with dark red-brown below and behind lateral postsutural vittae and around notopleural suture. Pleural areas black except for red-brown below postpronotal lobes. Yellow markings as follows: postpronotal lobes; notopleura; broad mesopleural stripe reaching postpronotal lobe dorsally, continuing to katepisternum as a large transverse spot, anterior margin straight; anatergite (posterior apex black); anterior ⅔ katatergite (remainder black); two broad parallel-sided lateral postsutural vittae ending at *ia.* seta; a broad medial longitudinal postsutural vitta; a large spot anterior to notopleural suture in front of each lateral postsutural vitta. Postnotum black. Scutellum yellow except for narrow black basal band. Setae: *sc.* 2; *prsc.* absent; *ia.* 1; *p.sa.* 1; *a.sa.* absent; *mpl.* 1; *npl.* 2; *scp.* 2.
Legs: Fore femora fulvous with dark fuscous to black around apical ½ (generally paler on inner surface), mid femora fulvous with dark fuscous to black around apical ½, hind femora fulvous with dark fuscous to black around apical ⅓; fore and mid tibiae fulvous, hind tibiae dark fuscous; all tarsal segments entirely fulvous; mid tibiae each with an apical black spur.
Wings: Length 5.5 mm; cells bc and c colourless; microtrichia in outer corner of cell c only; remainder of wings colourless except dark fuscous cell sc, a narrow dark fuscous costal band confluent with R_{2+3} and widening across apex of R_{4+5}, a broad dark fuscous anal streak, a narrow dark fuscous band enclosing the *dm-cu* crossvein; a dense aggregation of microtrichia around A_1+CuA_2; supernumerary lobe weak.
Abdomen: Oval; terga free; pecten present on tergum III. Tergum I and sterna I and II wider than long. Tergum I dark fuscous to black with fulvous across posterior ½ but not reaching lateral margins; tergum II fulvous with long transverse anterolateral black corners; terga III–V fulvous with a broad medial and moderately broad lateral longitudinal dark fuscous to black bands over all three terga and joined along anterior margin of tergum III. A pair of oval fulvous shining spots on tergum V. Posterior lobe of surstylus long, sternum V with a shallow concavity on posterior margin. Abdominal sterna red-brown with dark fuscous lateral and posterior margins.
Attractant: Cue lure.
Female: No known record.
Distribution: Philippines (Luzon I., Mt Makiling).
Hosts: No known record.
Remarks: *Bactrocera (Asiadacus) careomacula* n.sp. is similar to *Bactrocera (Asiadacus) melanopsis* (Hardy) in possessing infuscation along the *dm-cu* crossvein in the wing and to *Bactrocera (Asiadacus) apiciflava* Yu, He & Chen in possessing a black scutum, dark markings on all femora and dark fuscous to black 'T'-shaped patterning on abdominal terga III–V. It differs from both species in having the face entirely fulvous and from *B. apiciflava* in having infuscation along the *dm-cu* crossvein in the wing.

Bactrocera (Asiadacus) maculifacies (Hardy)

(Fig. 13)

Dacus maculifacies Hardy, 1973: 15. Holotype male in BBM.
Bactrocera (Asiadacus) maculifacies – Norrbom *et al.*, 1998: 87.

Type material examined: Holotype male, **THAILAND**: with the following five labels – 'Town of Chantaburi 18-20. vi.1969'; 'Steiner's trap with cue lure +1% naled'; 'J.J.S. Burton Collector'; 'Holotype male, *Dacus (Asiadacus) maculifacies*, D.E. Hardy'; 'Bishop Museum Holotype No. 9946'. Received on loan from BBM.

Non-type material examined: THAILAND: 4 males and 4 females, Doi Luong N.P., Chiang Rai, 14.vi.1991, coll. Charuwan Supasatien, bred from wild fruit sample No. CR60; 1 male, Phan, Chiang Rai, 13.xi.1991, coll. Charuwan, trap No. CR232T, attracted to cue lure; 2 males and 3 females, Pang Chang, Chiang Mai, 18.v.1983, coll. Chantanee *et al.*, bred ex *Siphonodon celastrineus*; 2 males and 1 female, 23 km Muang, Prachinburi, 4.vi.1993, coll. Montree; 1 male, NRDC, Chiang Mai, 23.vi.1991, coll. Chantanee, attracted to cue lure. **VIETNAM: Southern Vietnam** – 1 male, Dong Nai Province, Thong Nhat District, Cay Gao Village, Site S013, 9.vii.1999, attracted to cue lure; 10 males (20.vii.2002), 8 males (3.viii.2002), 1 male (31.viii.2002), Dong Nai Province, Long Khanh District, Nhan Nghia Village, Site VS06, attracted to cue lure. Specimens in QDPC.

Diagnosis: A medium-sized species; face fulvous with a broad transverse black band; postpronotal lobes and notopleura yellow; scutum black with a broad medial longitudinal red-brown band anterior to notopleural suture; lateral and medial postsutural yellow vittae present; a yellow spot anterior to notopleural suture; mesopleural stripe reaching midway between anterior margin of notopleuron and anterior *npl.* seta dorsally; scutellum yellow; wing with cells bc and c colourless, microtrichia in outer corner of cell c only; a narrow fuscous costal band and broad fuscous anal streak; abdominal terga III–V fulvous with a black 'T' pattern consisting of a narrow transverse black band across anterior margin of tergum III, a narrow medial longitudinal black band over all three terga, broad dark fuscous lateral margins on tergum IV and dark fuscous anterolateral corners on tergum V.

Description: male.

Head: Height 1.6 mm. Frons – length 1.46 times breadth; fulvous with dark fuscous around orbital setae and fuscous on anteromedial hump; orbital setae black: 1 *s.or.*, 3 *i.or.*; lunule dark fuscous. Ocellar triangle black. Vertex dark fuscous. Face fulvous with a broad black transverse band; length 0.55 mm. Genae fulvous, large fuscous to dark fuscous subocular spot present; black seta present. Occiput fuscous, fulvous along eye margins; occipital row with a large number of short pale setae. Antennae with segments 1 and 2 red-brown, segment 3 red-brown with dark fuscous on apex and outer surface; length of segments: 0.16 mm; 0.28 mm; 0.78 mm.

Thorax: Scutum black except for red-brown below and behind lateral postsutural vittae, on lateral margin between postpronotal lobe and notopleuron, a broad medial longitudinal band running from line of notopleural suture to anterior margin (broken by a narrow medial longitudinal black line through centre). Pleural areas black except red-brown below postpronotal lobes. Yellow markings as follows: postpronotal lobes; notopleura; mesopleural stripe reaching midway between anterior margin of notopleuron and anterior *npl.* seta dorsally, continuing to katepisternum as a large transverse spot, anterior margin convex; anatergite (posterior apex black); anterior $^{3}/_{5}$ katatergite (remainder black); two broad parallel-sided lateral postsutural vittae beginning with a spot anterior to notopleural suture and ending well behind *ia.* seta; a very broad pear-shaped medial longitudinal postsutural vitta. Postnotum black laterally, red-brown medially. Scutellum yellow except for a narrow black basal band. Setae: *sc.* 2; *prsc.* absent; *ia.* 1; *p.sa.* 1; *a.sa.* absent; *mpl.* 1; *npl.* 2; *scp.* 4.

Legs: Fore and mid femora fulvous with a small dark fuscous subapical spot on outer surfaces, hind femora fulvous with a small dark fuscous subapical spot on inner surfaces; fore tibiae fulvous with a small dark fuscous line basally, mid tibiae fulvous, hind tibiae pale fuscous; tarsi with all segments fulvous; mid tibiae each with an apical black spur.

Wings: Length 6.2 mm; cells bc and c colourless; microtrichia in outer corner of cell c only; remainder of wings colourless except fuscous cell sc, narrow fuscous costal band confluent with R_{2+3} and remaining narrow around margin of wing to end just beyond extremity of R_{4+5}, a moderately broad fuscous anal streak; a dense aggregation of microtrichia around A_1+CuA_2; supernumerary lobe large and rounded.

Abdomen: Oval; terga free; pecten present on tergum III. Tergum I fulvous with narrow lateral black margins; tergum II fulvous with a narrow subanterior transverse band which joins with narrow lateral black margins; terga III–V fulvous with a black 'T' pattern consisting of a narrow transverse band across anterior margin of tergum III expanding over lateral margins, a narrow medial longitudinal black band over all three terga, broad lateral dark fuscous margins on tergum IV and dark fuscous anterolateral corners on tergum V. A pair of oval fulvous shining spots on tergum V. Posterior lobe of surstylus long, sternum V with a slight concavity on posterior margin.

Attractant: Cue lure.

Female: As for male except dark fuscous to black around apices of all femora (not preapical spots); no dense aggregation of microtrichia around A_1+CuA_2; supernumerary lobe weak; pecten of cilia on abdominal tergum III absent; ovipositor basal segment red-brown, dorsoventrally compressed and tapering posteriorly in dorsal view; ratio of length of oviscape to length of tergum V, 2:1; apex of piercer needle-shaped.

Distribution:

Type locality: Thailand (Chantaburi).

New records: Southern Vietnam; Thailand (Chiang Mai).

Hosts: *Siphonodon celastrineus* (family *Celastraceae*) (Allwood *et al.*, 1999).

Remarks: *Bactrocera (Asiadacus) maculifacies* (Hardy) is similar to *Bactrocera (Asiadacus) apicalis*, *Bactrocera (Asiadacus) bakeri* and *Bactrocera (Asiadacus) melanopsis*

(Hardy) in the general colour patterns of the thorax, abdomen and wings. It differs from *B. apicalis* and *B. bakeri* in lacking a large spot in the apex of the wing and from *B. melanopsis* in lacking infuscation on the *dm-cu* crossvein and in possessing a large supernumerary lobe in the male wing and mostly fulvous femora. *Bactrocera maculifacies* is superficially similar to *Bactrocera (Zeugodacus) hengsawadae* n.sp. but differs in having a broad transverse black band across the face, a large rounded supernumerary lobe in the male wing, two *sc.* setae, *prsc.* and *a. sa.* setae absent.

Bactrocera (Asiadacus) melanopsis (Hardy)

(Fig. 14)

Dacus melanopsis Hardy, 1982: 187. Holotype male in MZB.
Bactrocera (Asiadacus) melanopsis – Norrbom *et al.*, 1998: 87.

Type material examined: Holotype male, **INDONESIA: SULAWESI** – Lindu Valley, Central Sulawesi, 960 m, in forest at cue lure, Aug. 1975, coll. D.E. Hardy. Studied in MZB.
Non-type material examined: INDONESIA: SULAWESI – 18 males, 'Bententingi, 5 km W Malino, Sulawesi, 27.vii.1981, 850–1000 m, coll. J. Turner'; 'Residual secondary forest'; 'cue lure'. 1 male, 'Bantimurung, 40 km NW Ujung Pandang, Sulawesi, 50 m, 27.vii.1981, coll. J. Turner'. Specimens in QDPC.
Diagnosis: A medium-sized species; face entirely glossy black; postpronotal lobes and notopleura yellow; scutum entirely glossy black; lateral and medial postsutural yellow vittae present; a yellow spot anterior to notopleural suture; mesopleural stripe reaching to postpronotal lobe dorsally; scutellum yellow; wing with cells bc and c colourless, microtrichia in outer corner of cell c only; a narrow fuscous costal band confluent with R_{2+3} and expanding slightly across apex of R_{4+5}, distinct infuscation along *dm-cu* crossvein; a broad fuscous anal streak; abdominal terga III–V fulvous with a black 'T' pattern consisting of a narrow transverse band across anterior margin of tergum III and a medium-width medial longitudinal band over all three terga but broken at posterior area of tergum III, moderately broad lateral longitudinal black bands over terga IV and V.
Description: male.
Head: Frons – fulvous with small black markings around bases of *s.or.* setae and a narrow transverse black mark across anteromedial hump; orbital setae black: 1 *s.or.*, 1 *i.or.*; lunule fulvous. Ocellar triangle black. Vertex fulvous with small lateral black markings. Face entirely black. Genae fulvous with a small black subocular spot. Occiput red-brown with fulvous along eye margins. Antennae with all segments mostly pale fuscous.
Thorax: Scutum entirely glossy black. Yellow markings as follows: postpronotal lobes; notopleura; mesopleural stripe extending to postpronotal lobe dorsally; two broad parallel-sided lateral postsutural vittae beginning with a spot anterior to notopleural suture and ending just behind *ia.* seta; a medial longitudinal postsutural vitta. Postnotum black. Scutellum yellow with a broad black basal band. Setae: *sc.* 2; *prsc.* absent; *ia.* 1; *p.sa.* 1; *a.sa.* absent; *mpl.* 1; *npl.* 2; *scp.* 4.
Legs: Fore femora dark fuscous to black except for small area of fulvous basally, mid femora dark fuscous on apical ⅔ and fulvous on basal ⅓, hind femora dark fuscous to black on apical ½ and fulvous over basal ½; fore and mid tibiae fulvous, hind tibiae dark fuscous basally to fuscous apically; all tarsi with basal segment fulvous and apical four segments pale fuscous.
Wings: Length 6.0 mm; cells bc and c colourless; microtrichia in outer corner of cell c only; remainder of wings colourless except fuscous cell sc, narrow fuscous costal band confluent with R_{2+3} and expanding slightly across apex of R_{4+5}, distinct infuscation along *dm-cu* crossvein, a broad fuscous anal streak; a dense aggregation of microtrichia around A_1+CuA_2; supernumerary lobe of medium development.
Abdomen: Oval; terga free; pecten present on tergum III. Tergum I and sterna I and II wider than long. Tergum I fulvous with small black anterolateral corners; tergum II fulvous with black anterolateral corners connected to two small transverse black markings subanteriorly; terga III–V fulvous with a black 'T' pattern consisting of a narrow transverse band across anterior margin of tergum III and a medium-width medial longitudinal black band over all three terga but broken on posterior area of tergum III, moderately broad lateral longitudinal black bands over terga IV and V. A pair of oval fulvous shining spots on tergum V. Posterior lobe of surstylus long, sternum V with a slight concavity on posterior margin.
Attractant: Cue lure.
Female: No known record.
Distribution:
Type locality: Indonesia (Sulawesi – Central Sulawesi, Lindu Valley).
New records: Indonesia – Sulawesi (Bantimurung and Bententingi).
Hosts: No known record.
Remarks: *Bactrocera (Asiadacus) melanopsis* (Hardy) is similar to *Bactrocera (Asiadacus) apicalis* (de Meijere), *Bactrocera (Asiadacus) bakeri* (Bezzi) and *Bactrocera (Asiadacus) maculifacies* (Hardy) in the general colour patterns of the thorax, abdomen and wings. It differs from *B. apicalis* and *B. bakeri* in lacking a large dark fuscous spot in the apex of the wing and from *B. maculifacies* in possessing a weak supernumerary lobe in the male wing, distinct infuscation along *dm-cu* crossvein and large areas of dark fuscous on the femora.

SUBGENUS *BACTROCERA* Macquart

Bactrocera (Bactrocera) abdonigella (Drew)

(Fig. 15)

Dacus (Strumeta) abdonigellus Drew, 1971: 52. Holotype male in QMIC.
Bactrocera (Bactrocera) abdonigella – Drew, 1989: 117; White & Evenhuis, 1999: 492; Norrbom *et al.*, 1998: 87.

Dacus maculosus Walker, 1866, nomen nudum (see White & Evenhuis, 1999: 492).

Non-type material examined: INDONESIA: MOLUCCAS (MALUKU) – 1 male, Western Seram, Waimital, Moluccas (Maluku), 19.vii.2007, coll. A. Haris Tualepe, attracted to cue lure in orange. Specimen in QDPC.

Diagnosis: Face fulvous with a pair of small circular to triangular black spots; postpronotal lobes and notopleura yellow (the specimen studied herein from Moluccas (Maluku) appears to have the postpronotal lobe darkened but this is probably due to damage after collection); scutum red-brown with irregular fuscous markings; broad parallel-sided lateral postsutural yellow vittae ending behind *ia.* seta; medial postsutural yellow vitta absent; mesopleural stripe slightly wider than notopleuron dorsally; scutellum yellow; legs with femora entirely fulvous; fore and mid tibiae fulvous, hind tibiae fulvous with fuscous on outer apical surfaces; wing with cells bc and c fulvous, microtrichia in outer corner of cell c only; a narrow fuscous costal band overlapping R_{2+3} and remaining of uniform width to end between extremities of R_{4+5} and M; a broad fuscous anal streak; supernumerary lobe of medium development; abdominal terga III–V mostly black, a pair of oval black shining spots on tergum V.

Attractant: Cue lure.

Female: No known record.

Distribution:

Type locality: Papua New Guinea (Bubia).

Previously recorded: Type locality and the following locations – Indonesia (Moluccas = Maluku–Bacan I.); Indonesia (West Papua, formerly Irian Jaya), White & Evenhuis, 1999.

New records: Indonesia (Moluccas = Maluku – Seram I.).

Hosts: No known record.

Remarks: *Bactrocera (Bactrocera) abdonigella* (Drew) is similar to *Bactrocera (Bactrocera) abdofuscata* (Drew) and *Bactrocera (Bactrocera) dyscrita* (Drew) in possessing a red-brown scutum, wings with a narrow fuscous costal band and anal streak and cells bc and c with microtrichia in outer corner of cell c only. It is unique and different from these species in having abdominal terga III–V entirely black. As discussed by White & Evenhuis (1999), *B. abdonigella* is probably conspecific with *Dacus maculosus* Walker which has to be regarded as a nomen nudum as it is only a label name in the BMNH and not a manuscript name. The species *Dacus maculosus* was collected on Bacan Island in the Moluccas and, on the basis of conspecifity, this record is noted under distribution above.

Bactrocera (Bactrocera) aemula Drew

(Fig. 16)

Bactrocera (Bactrocera) aemula Drew, 1989: 26; Norrbom *et al.*, 1998: 87. Holotype male in QMIC.

Non-type material examined: INDONESIA: MOLUCCAS (MALUKU) – 1 male, Western Seram, Waimital, Moluccas (Maluku), 12.vii.2007, coll. A. Haris Tualepe, attracted to cue lure in orange. Specimen in QDPC.

Diagnosis: Face fulvous with a pair of small to medium oval black spots; postpronotal lobes and notopleura yellow; scutum uniformly black without pale markings; moderately broad lateral postsutural yellow vittae narrowing sharply posteriorly to end well before *ia.* seta; medial postsutural yellow vitta absent; mesopleural stripe reaching midway between anterior margin of notopleuron and anterior *npl.* seta dorsally; scutellum yellow; legs with femora fulvous except pale fuscous around apices; fore and mid tibiae pale fuscous, hind tibiae fuscous; wing length 5.0 mm, wing with cells bc and c with a very pale tint, microtrichia in outer ⅓ of cell c only; a narrow fuscous costal band just overlapping R_{2+3} and remaining narrow before ending beyond extremity of R_{4+5}; a narrow fuscous anal streak; supernumerary lobe of medium development; abdominal tergum III black, tergum IV dark fuscous to black except for small posterocentral red-brown areas each side of a broad medial longitudinal black band, tergum V red-brown with dark fuscous to black anterolateral corners and a broad medial longitudinal black band, a pair of oval dark fuscous shining spots on tergum V; abdominal sterna dark coloured; posterior lobe of surstylus short, sternum V with a deep concavity on posterior margin.

Attractant: Cue lure.

Female: No known record.

Distribution:

Type locality: Papua New Guinea (Mt Hagen).

New records: Indonesia (Moluccas = Maluku – Seram I.).

Hosts: No known record.

Remarks: *Bactrocera (Bactrocera) aemula* Drew is similar to *Bactrocera (Bactrocera) beckerae* (Hardy) and *Bactrocera (Bactrocera) mimulus* Drew in the general colour patterns of the thorax, wings, legs and abdomen and particularly in having a black scutum with the lateral postsutural vittae narrowing posteriorly to end before the *ia.* setae, but differs from these species in possessing a narrower costal band (not extending to or nearly to R_{4+5}) and extensive areas of dark fuscous to black over abdominal terga III–V.

Bactrocera (Bactrocera) aethriobasis (Hardy)

(Fig. 17)

Dacus aethriobasis Hardy, 1973: 30. Holotype male in KUB.
Bactrocera (Bactrocera) aethriobasis – Norrbom *et al.*, 1998: 87; Drew *et al.*, 2007: 2.

Non-type material examined: BHUTAN: 2 males (13.v.2000), 1 male (20.v.2000), 1 male (27.v.2000), 2 males (2.vi.2000), Phuentsholing, coll. S. Thapa, attracted to methyl eugenol. **PENINSULAR MALAYSIA**: 67 males, Gajah Mati, April–August 1986, coll. Sulimandesa, attracted to methyl eugenol. **THAILAND**: 1 male (7.vii.1991), 1 male (24.vii.1991), Phan, Chiang Rai, coll. Charuwan, attracted to methyl eugenol; 1 male, Pha Yao, Chiang Rai, 14.viii.1991, coll. Charuwan, attracted to methyl eugenol; 3 males, Mae Salong, Chiang Rai, 13.viii.1991, coll. Charuwan, attracted to methyl eugenol; 1 male, Mae Salong, Chiang Rai, 29.vii.1991, coll. Charuwan, attracted to methyl eugenol; 1 male

(12.ii.1992), 1 male (4.iii.1992) Pha Yao, Chiang Rai, coll. Charuwan, attracted to methyl eugenol; 2 males, 7k T. Nemhom, A. Muang, Prachinburi, 27.iii.1992, coll. Yuthana, bred from *Azadirachta indica*; 1 female, A. Thamuang, Kanchanaburi, 22.iv.1992, coll. Yuthana, bred from *Azadirachta indica*. **VIETNAM: Southern Vietnam** – 3 males (23.i.2002), 1 male (23.ii.2002), 1 male (8.iii.2002), 2 males (8.iv.2002), Dac Lac (P), Ku Mi Nga (D), Quang Phu (V), coll. SOFRI staff, attracted to methyl eugenol; 2 males, Dong Nai (P), Long Khanh (D), Nhan Nghia (V), 15.ii.2002, coll. SOFRI staff, attracted to methyl eugenol; 1 male, Binh Dinh (P), Tay Son District, Binh Thanh Village, South Vietnam, 30.vi.2002, attracted to methyl eugenol. Specimens in QDPC.

Diagnosis: A medium-sized species; face fulvous with a pair of medium-sized circular to oval black spots; postpronotal lobes and notopleura yellow; scutum mostly red-brown with irregular dark fuscous markings; lateral postsutural yellow vittae present; medial postsutural yellow vitta absent; no yellow spot anterior to notopleural suture; mesopleural stripe reaching to anterior *npl.* seta dorsally; scutellum yellow with a narrow dark basal band; wing with cells bc and c colourless and totally devoid of microtrichia; a very narrow fuscous costal band confluent with R_{2+3} and remaining extremely narrow around apex of wing; anal streak reduced to a pale fuscous tint within cell cup; abdominal terga III–V mostly dark fuscous to black with central areas of pale to dark orange-brown either side of a medial longitudinal black band which runs from centre of tergum III to posterior area of tergum V.

Description: male.

Head: Height 1.6 mm. Frons – length 1.48 times breadth; fulvous with fuscous to dark fuscous around orbital setae and occasionally on anteromedial hump; orbital setae dark red-brown: 1 *s.or.*, 2 *i.or.*; lunule dark fuscous. Ocellar triangle black. Vertex dark fuscous. Face fulvous with a pair of medium-sized circular to oval black spots; length 0.45 mm. Genae fulvous, small pale fuscous subocular spot present; red-brown seta present. Occiput red-brown, fulvous along eye margins; occipital row with short pale setae. Antennae with segments 1 and 2 red-brown, segment 3 red-brown with fuscous on apex and outer surface; length of segments: 0.26 mm; 0.34 mm; 0.84 mm.

Thorax: Scutum red-brown with pale brown below lateral postsutural vittae and irregular fuscous to dark fuscous patterns across posterior margin and occasional dark fuscous spots across anterior area. Pleural areas black except red-brown below postpronotal lobes. Yellow markings as follows: postpronotal lobes; notopleura; mesopleural stripe reaching to anterior *npl.* seta dorsally, continuing to katepisternum as a transverse spot, anterior margin slightly convex; anatergite (posterior apex black); anterior $^3/_5$ katatergite (remainder black); two broad parallel-sided lateral postsutural vittae beginning at notopleural suture and broadening around *ia.* seta posteriorly to end well behind this seta. Medial longitudinal postsutural yellow vitta absent. Postnotum black, tending dark red-brown medially. Scutellum yellow except for narrow red-brown basal band. Setae: *sc.* 2; *prsc.* 2; *ia.* 1; *p.sa.* 1; *a.sa.* 1; *mpl.* 1; *npl.* 2; *scp.* 4.

Legs: All femora entirely fulvous; all tibiae fulvous with small areas of dark fuscous basally; tarsi with all segments entirely fulvous; mid tibiae each with an apical pale red-brown spur.

Wings: Length 6.2 mm; cells bc and c colourless; no microtrichia in either cell; remainder of wings colourless except fuscous cell sc, very narrow fuscous costal band confluent with R_{2+3} and remaining extremely narrow beyond apex of R_{2+3} to apex of wing; anal streak absent but with a pale fuscous tint within cell cup; dense aggregation of microtrichia around A_1+CuA_2; supernumerary lobe of medium development.

Abdomen: Oval; terga free; pecten present on tergum III. Tergum I and sterna I and II wider than long. Terga I and II mostly orange-brown but may have dark fuscous to black narrow lateral margins on tergum I and anterolaterally on tergum II and a small black marking anterocentrally on tergum II; terga III–V dark fuscous to black except for pale to dark orange-brown areas either side of a medium-width medial longitudinal black band which runs from the centre of tergum III to the posterior area of tergum V (these orange-brown areas broaden across posterior margin of tergum V). A pair of fuscous to black shining spots on tergum V. Posterior lobe of surstylus short, sternum V with a deep concavity on posterior margin.

Attractant: Methyl eugenol.

Female: As for male except no dense aggregation of microtrichia around A_1+CuA_2; supernumerary lobe weak; pecten of cilia on abdominal tergum III absent; ovipositor basal segment red-brown, dorsoventrally compressed and tapering posteriorly in dorsal view; ratio of length of oviscape to length of tergum V, 0.8:1; apex of piercer needle-shaped.

Distribution:

Type locality: Thailand (Trang).

Previously recorded: Type locality and Peninsular Malaysia (Norrbom *et al.*, 1998).

New records: Bhutan; Southern Vietnam; Thailand (Chiang Rai).

Hosts: New records: *Azadirachta indica* (family *Meliaceae*).

Remarks: *Bactrocera (Bactrocera) aethriobasis* (Hardy) is similar to *Bactrocera (Bactrocera) invadens* Drew, Tsuruta & White, *Bactrocera (Bactrocera) pseudoversicolor* Drew, *Bactrocera (Bactrocera) rubigina* (Wang & Zhao) and *Bactrocera (Bactrocera) versicolor* (Bezzi) in possessing a basically red-brown scutum, the costal band confluent with R_{2+3}, cells bc and c colourless and the lateral postsutural vittae ending at or behind the *ia.* seta. It differs from these species in having the lateral postsutural vittae ending well behind the *ia.* seta and the mesopleural stripe reaching to the anterior *npl.* seta dorsally.

Bactrocera (Bactrocera) affinibancroftii Drew & Romig, new species

(Fig. 18)

Material examined: Holotype male, **INDONESIA: SULAWESI** – Sulawesi Utara, Bitung, Manembonembo, 11.ix.2006, coll. Karel F Lala, attracted to methyl eugenol in citrus, deposited in BMNH.

Paratypes: **INDONESIA: SULAWESI** – 15 males, same data as holotype; 9 males, Sulawesi Utara, Bitung, Danowudu, 11.ix.2006, coll. Karel F. Lala, attracted to methyl eugenol in forest; 3 males (12.viii.2006), 4 males (10.ix.2006), Sulawesi Tengah, Donggala, Nupabomba, coll. Rusdin, attracted to methyl eugenol in forest; 3 males (14.viii.2006), 1 male (11.ix.2006), 1 male (26.ix.2006), Sulawesi Utara, Minahasa Utara, Sarongsong, coll. Karel F. Lala, attracted to methyl eugenol in village; 1 male, Sulawesi, Gorontalo, Limboto, Biyonga, 5.ix.2006, coll. Yosias Y Huru, attracted to methyl eugenol in village jambu; 1 male, Sulawesi Selatan, Maros, Limampoccoe, 27.ix.2006, coll. Sirajuddin, attracted to methyl eugenol in forest; 1 male, Sulawesi Selatan, Takalar, Limampoccoe, 29.viii.2006, coll. Sirajuddin, SP.; 1 male (12.viii.2006), 1 male (15.x.2006), Sulawesi Tengah, Donggala, Nupabomba, coll. Lyar, attracted to methyl eugenol in forest; 1 male (12.ix.2006), 1 male (9.xi.2006), Sulawesi, Gorontalo, Bone Bolango, Lombongo, coll. Yusbar Ismail, attracted to methyl eugenol in forest; 1 male, Sulawesi, Gorontalo, Bone Bolango, Suwawa, 26.ix.2006, coll. Yusbar Ismail, attracted to cue lure in village nangka. 10 paratypes in ANIC, 5 paratypes in BMNH, 19 paratypes in QDPC, 10 paratypes in QMIC.

Diagnosis: A medium-sized species; face fulvous with a pair of medium-sized circular black spots; postpronotal lobes and notopleura yellow; scutum red-brown with variable dark fuscous to black patterns from a narrow lanceolate shape to almost covering the entire scutum surface; lateral postsutural yellow vittae present; medial postsutural yellow vitta absent; no yellow spot anterior to notopleural suture; mesopleural stripe just wider than notopleuron dorsally; scutellum yellow; wing with cells bc and c colourless, microtrichia in outer corner of cell c only; a narrow dark fuscous costal band overlapping R_{2+3} where it is pale fuscous and widening slightly across apex of R_{4+5}; a narrow pale fuscous anal streak which is almost absent in some specimens; abdominal terga entirely orange-brown and with the overall shape of *Bactrocera (Bactrocera) bancroftii* (Tryon) and in some specimens with dark fuscous transverse spots on anterolateral corners of tergum III and narrow fuscous to black medial longitudinal bands over part of terga IV and V.

Description: male.

Head: Height 1.8 mm. Frons – length 1.48 times breadth; red-brown with dark fuscous around orbital setae and on anteromedial hump; orbital setae black: 1 *s.or.*, 2 *i.or.*; lunule dark fuscous. Ocellar triangle black. Vertex dark fuscous. Face fulvous with a pair of medium-sized circular black spots; length 0.52 mm. Genae red-brown, large fuscous subocular spot present; red-brown seta present. Occiput red-brown, fulvous along eye margins; occipital row with 3–6 black setae. Antennae with segments 1 and 2 red-brown, segment 3 red-brown with fuscous on apex and outer surface; length of segments: 0.18 mm; 0.36 mm; 0.76 mm.

Thorax: Scutum dark red-brown with dark fuscous to black patterns varying from a narrow lanceolate shape to almost covering the entire scutum. Pleural areas entirely red-brown. Yellow markings as follows: postpronotal lobes; notopleura; mesopleural stripe slightly wider than notopleuron dorsally, continuing to katepisternum as a small transverse spot, anterior margin slightly convex; anatergite (posterior apex black); anterior ⅔ katatergite (remainder black); two moderately broad lateral postsutural vittae either subparallel or narrowing slightly posteriorly to end at or just before *ia.* seta. Medial longitudinal postsutural yellow vitta absent. Postnotum black laterally, red-brown centrally. Scutellum yellow with a narrow red-brown basal band. Setae: *sc.* 2; *prsc.* 2; *ia.* 1; *p.sa.* 1; *a.sa.* 1; *mpl.* 1; *npl.* 2; *scp.* 4.

Legs: All segments entirely fulvous except basal area of mid tibiae fuscous and hind tibiae varying from fuscous basally to pale fuscous apically; mid tibiae each with an apical black spur.

Wings: Length 5.9 mm; cells bc and c colourless; microtrichia in outer corner of cell c only; remainder of wings colourless except dark fuscous cell sc, narrow dark fuscous costal band slightly overlapping R_{2+3} where it is pale fuscous and widening apically across apex of R_{4+5}, a narrow pale fuscous anal streak almost absent in some specimens; a dense aggregation of microtrichia around A_1+CuA_2; supernumerary lobe of medium development.

Abdomen: Oval; terga free; pecten present on tergum III. Tergum I and sterna I and II wider than long. All abdominal terga entirely orange-brown with the overall shape of *Bactrocera (Bactrocera) bancroftii* (Tryon) with, in some specimens, transverse dark fuscous spots on anterolateral corners of tergum III and narrow fuscous to black medial longitudinal bands over part of tergum IV and tergum V. A pair of oval orange-brown to fuscous shining spots on tergum V. Posterior lobe of surstylus short, sternum V with a deep concavity on posterior margin. Abdominal sterna pale coloured, generally fulvous.

Attractant: Methyl eugenol.

Female: No known record.

Distribution: Indonesia (Sulawesi).

Hosts: No known record.

Remarks: *Bactrocera (Bactrocera) affinibancroftii* n.sp. is similar to *Bactrocera (Bactrocera) bhutaniae* n.sp., *Bactrocera (Bactrocera) brunneola* White and Tsuruta, *Bactrocera (Bactrocera) paraosbeckiae* Drew and *Bactrocera (Bactrocera) suliae* n.sp. in possessing an entire costal band overlapping R_{2+3} and cells bc and c colourless but differs from these species in having the abdomen shape similar to *Bactrocera (Bactrocera) bancroftii* (Tryon), terga III–V orange-brown and from all of these species except *B. suliae* n.sp. in having abdominal sterna pale coloured.

Bactrocera (Bactrocera) affinidorsalis (Hardy)

(Fig. 19)

Dacus (Bactrocera) affinidorsalis Hardy, 1982: 215. Holotype male in MZB.

Bactrocera (Bactrocera) affinidorsalis – Drew & Hancock, 1994a: 7; Norrbom *et al.*, 1998: 87.

Type material examined: Holotype male, **INDONESIA: SULAWESI** – Sadaunta, 65 km S.E. Palu, 600 m, Aug. 1975, coll. D.E. & A.T. Hardy, attracted to cue lure in primary forest. Studied at MZB.

Non-type material examined: INDONESIA: SULAWESI – 1 male (19.ix.2006), 1 male (26.ix.2006), Gorontalo, Bone Bolango, Lombongo, coll. Yusbar Ismail, attracted to cue lure in forest; 1 male (26.ix.2006), 3 males (3.x.2006), 1 male (17.x.2006), 1 male (31.x.2006), Gorontalo, Bone Bolango, Suwawa, coll. Yusbar Ismail, attracted to cue lure in forest; 1 male (2.ix.2006), 1 male (23.ix.2006), 1 male (15.x.2006), 2 males (22.x.2006), 4 males (29.x.2006), Sulawesi Tengah, Donggala, Nupabomba, coll. Rusdin, attracted to cue lure in forest. Specimens in QDPC.

Diagnosis: Face fulvous laterally, fuscous centrally with a pair of medium-sized black spots; postpronotal lobes and notopleura yellow; scutum black with brown below and behind lateral postsutural vittae, around notopleural suture and inside postpronotal lobes; broad parallel-sided or subparallel lateral postsutural yellow vittae ending before *ia.* seta; medial postsutural yellow vitta absent; mesopleural stripe reaching midway between anterior margin of notopleuron and anterior *npl.* seta dorsally; scutellum yellow; legs with femora fulvous with large fuscous to black markings around apices; fore and mid tibiae fulvous and hind tibiae pale fuscous to fuscous; wing with cells bc and c colourless, microtrichia in outer corner of cell c only; a narrow dark fuscous costal band confluent with R_{2+3} and widening gradually across apex of R_{4+5}; a narrow fuscous anal streak; supernumerary lobe of medium development; abdominal terga III–V orange-brown with a 'T' pattern consisting of a narrow transverse black band across anterior margin of tergum III which becomes dark fuscous laterally and a medium-width medial longitudinal black band over all three terga, narrow fuscous anterolateral corners on tergum IV, a pair of oval fuscous shining spots on tergum V; abdominal sterna dark coloured.

Attractant: Cue lure.

Female: No known record.

Distribution:

Type locality: Indonesia (Sulawesi, Sadaunta).

Previously recorded: Type locality and Philippines (Luzon I.) (Drew & Hancock, 1994a).

New records: Indonesia (Sulawesi – Gorontalo; Central Sulawesi, Donggala).

Hosts: No known record.

Remarks: *Bactrocera (Bactrocera) affinidorsalis* (Hardy), within the *dorsalis* complex, is similar to *Bactrocera (Bactrocera) latilineola* Drew & Hancock, *Bactrocera (Bactrocera) malaysiensis* Drew & Hancock and *Bactrocera (Bactrocera) unimacula* Drew & Hancock in possessing broad parallel-sided or subparallel lateral postsutural vittae, the costal band confluent with R_{2+3} and expanding across apex of wing. It differs from *B. malaysiensis* in possessing large fuscous to black spots on apices of all femora and the lateral postsutural vittae ending before the *ia.* seta, from *B. latilineola* in having the dark patterns on apices of all femora (not just on fore femora), the medial longitudinal black band on terga III–V (not on terga II–V), the shining spots on tergum V pale coloured (not dark fuscous to black) and in having the males attracted to cue lure (not methyl eugenol). *Bactrocera unimacula* is distinct in possessing a large distinct spot in the apex of the costal band and in having the males attracted to methyl eugenol.

Bactrocera (Bactrocera) affinis (Hardy)

(Fig. 199B)

Dacus (Neodacus) affinis Hardy, 1954: 7. Holotype male in USNM.
Dacus (Polistomimetes) oscinae Agarwal & Kapoor, 1983: 171; Norrbom *et al.*, 1998: 87 (Synonym).
Bactrocera (Bactrocera) affinis – Norrbom *et al.*, 1998: 87.

Type material examined: INDIA – Paratype male with the following five labels – 'Paratype'; 'Yercaud, India, vi.1950'; 'ex Methyl eugenol bait'; 'G. Beavor collector'; 'Paratype, *Dacus (Neodacus) affinis* Hardy'. Studied at BMNH.

Non-type material examined: INDIA: 6 males, India, Karnataka, Bangalore, 916m, 6.vi.1987, coll. G. Bhat, attracted to methyl eugenol; 2 males (21.v.1992), 1 male (30-31.v.1992), S. India, Bangalore, coll. D.L. Hancock and I.M. White, attracted to methyl eugenol; 2 males, S. India, 15km from Yercaud, 750m, 3.vi.1992, coll. D.L. Hancock and I.M. White, attracted to methyl eugenol. Specimens in QDPC.

Diagnosis: Face fulvous with a pair of medium-sized oval black spots; postpronotal lobes and notopleura yellow; scutum red-brown with pale fuscous markings as three longitudinal bands joined along posterior margin; narrow, parallel-sided lateral postsutural yellow vittae ending before or at the *ia.* seta; medial postsutural yellow vitta absent; mesopleural stripe equal in width to notopleuron dorsally; scutellum yellow; legs with all segments fulvous; wing with cells *bc* and *c* colourless and devoid of microtrichia; a narrow fuscous costal band confluent with R_{2+3} and ending at apex of this vein, a small fuscous spot around apex of R_{4+5}; anal streak absent; supernumerary lobe of medium development; all abdominal terga orange-brown with a narrow fuscous transverse band across anterior margin of tergum III and a small medial longitudinal dark fuscous spot on tergum V, a pair of oval orange-brown shining spots on tergum V. Posterior lobe of surstylus short; sternum V with a deep concavity on posterior margin. Abdominal sterna pale.

Attractant: Methyl eugenol.

Female: As for male except no dense aggregation of microtrichia around A_1+CuA_2; supernumerary lobe weak; pecten of cilia on abdominal tergum III absent; apex of piercer trilobed (see Mahmood & Hasan, 2005, for photography of piercer).

Distribution:

Type locality: India (Yercaud).

Previously recorded: Type locality and other locations in Southern India.

Hosts: No known record.

Remarks: *Bactrocera (Bactrocera) affinis* (Hardy) is similar to *Bactrocera (Bactrocera) zonata* (Saunders) in possessing a red-brown scutum, orange-brown abdominal terga, abdominal sterna pale, the costal band broken beyond apex of R_{2+3} and a small oval fuscous spot across apex of R_{4+5}. It differs from *B. zonata* in lacking *prsc.* setae and in having the apex of the female piercer trilobed, not needle-shaped. Generally, in *B. affinis*, the lateral postsutural yellow vittae end before the *ia.* seta whereas in *B. zonata* they end at or behind this seta.

Bactrocera (Bactrocera) albistrigata (de Meijere)

(Figs 20A, 20B)

Dacus albistrigatus de Meijere, 1911: 377; Hardy & Adachi, 1954: 168 as synonym of *B. frauenfeldi* (Schiner). Holotype female in ZMAN.

Bactrocera (Bactrocera) albistrigata – Norrbom *et al.*, 1998: 88.

Non-type material examined: AUSTRALIAN TERRITORY: CHRISTMAS ISLAND – 12 males, 'Xmas Island, 27.i.1983, coll. A.N. Sproul'; 'Indian Ocean'; 'Insecoll 0-131500 to 0-131511'; attracted to cue lure; 112 males, Christmas Island, September 2002, coll. R. Kitching, attracted to cue lure. **INDIA: ANDAMAN ISLANDS** – 4 females, G. Nicobar, C. bay, 22.vi.1997, coll. Ranganath, bred ex *Calophyllum*; 1 female and 3 males, G. Nicobar, Katchal, 26.vi.1997, coll. Ranganath, bred ex *Syzygium*; 3 females, Havelock Is., 1.vi.1998, coll. Ranganath, bred ex *Ochrosia*; 1 female and 3 males, L. Andaman, 22.ii.1996, coll. Ranganath, bred ex *Ochrosia*; 1 female and 3 males, 'India, Great Nicobar, 10.viii.1993, coll. Veena Kumar'; 'sl No. 37, Terminalia sp.'; 4 males, 'India, Great Nicobar, 10.viii.1993, coll. Veena Kumar'; sl No. 38, Neisosperma sp. **INDONESIA: JAVA** – 7 males, West Java, Cirebon, Pelangon, 30.xi.2005, coll. Herty, attracted to cue lure in forest; 6 males, West Java, Cirebon, Panguragan, 30.xii.2005, coll. Nam S., attracted to cue lure in Guava; 5 males, Cirebon, Pelangon, 19.vii.2005, coll. Herty, cue lure; 1 male, Majalengka, Ampel, West Java, 21.vii.2005, coll. Yaya Senjaya, cue lure; 15 males, DKI Jakarta, Jakarta Selaton, Tanjung Barat, 23.vii.2005, coll. Mulyana, cue lure; 14 males, Sidorajo, Gelang, 12.vii.2005, coll. Agus Riadi, cue lure; 4 males, Demak, Betokan, Central Java, 16.viii.2005, coll. Gunawan & Suharsono, cue lure; **KALIMANTAN** – 12 males (19.vii.2005), 1 male (2.viii.2005), West Kalimantan, Pontianak, Sei Duri II, coll. A. Irom, attracted to cue lure in citrus; 11 males (26.vii.2005), 3 males (2.vii.2005), West Kalimantan, Pontianak, Sei Pinyuh, coll. A. Irom, attracted to cue lure in Village (mango); 2 males (21.vii.2005), 1 male (28.vii.2005), 10 males (2.viii.2005), South Kalimantan, Banjarbaru, Guntung Payung, coll. Sidrotul & Sutini, attracted to cue lure in village; 2 males (14.xi.2005), 2 males (28.xi.2005), East Kalimantan, Kutai Kartanegara, Batuah, coll. Ertha DP, attracted to cue lure in banana; 18 males (19.vii.2005), 9 males (26.vii.2005), 6 males (2.viii.2005), Pontianak, West Kalimantan, coll. A. Irom, cue lure; 5 males (21.vii.2005), 1 male (28.vii.2005), Banjarbaru, Guntung Payung, South Kalimantan, coll. Sidrotul & Sutini, cue lure; **LOMBOK** – 4 males, Mataram, Indonesia, iv.1988, coll. J. Turner, attracted to cue lure; **SULAWESI** – 11 males and 16 females (6.viii.2006), 19 males (1.x.2006), Central Sulawesi, Donggala, Bulu Pountu, coll. Rusdin, bred ex sample GF14(9), Lampijo; 12 males and 13 females, Tengah, Donggala, Lolu, no date, coll. Rusdin, bred ex sample GF14, Jambu Biji; 2 males and 4 females, Tengah, Donggala, Guntarano, 27.viii.2006, coll. Rusdin, bred ex sample GF14, Mangga; 5 males and 2 females, Selatan, Maros, Turikale, 19.viii.2006, coll. Sirajuddin SP, bred ex sample GF15, *Syzygium samarangenae* (jambu air); 27 males, Gorontalo, Bone Bolango, Ayma Ultara, 7.xii.2006, coll. Yusbar Ismail, attracted to cue lure in forest; 9 males and 14 females, Tengah, Donggala, Bulu Pountu, 6.viii.2006, coll. Rusdin, bred ex sample GF14, Lampijo; 8 males and 11 females, Central Sulawesi, Donggala, Bulu Pountu, 1.x.2006, coll. Rusdin, bred ex sample GF14(9), Lampijo; 9 males and 17 females, Sulawesi Selatan, Maros, Turikale, 18.x.2006, coll. Sirajuddin SP, bred ex sample GF15, *Eugenia javanica* (Fam. Myrtaceae), jambu putik; 6 males (8.viii.2006), 3 males (22.viii.2006), West Sulawesi, Mamuju, Belang-Belang, coll. Sri Hastuti, Yan PF, Sukri, attracted to cue lure in Forest Jeruk. **PENINSULAR MALAYSIA**: 1 male, RS Sungai Baging, Malaysia, 16.vi.1986, coll. A. Samat Wahid, attracted to cue lure; 2 males, TS19 Malaysia, 16.vi.1986, coll. Mahdzir Ahmad, attracted to cue lure; 1 male, TS20 Malaysia, 16.vi.1986, coll. Mahdzir Ahmad, attracted to cue lure; 2 males, Kuantan, Malaysia, 2.vi.1986, coll. Vignes, attracted to cue lure in rainforest; 1 male, Puchong, Malaysia, 26.i.1987, coll. Bahari, attracted to cue lure. **THAILAND**: 1 male, Songkhla, Tone-Nga, Chang, 8.v.1991, coll. R. Drew *et al.*, bred from *Artocarpus heterophyllus*; 1 male, Songkhla, Sikao, Trang, 24.v.1991, coll. Sieng *et al.*, bred from *Mangifera indica*; 2 males (28.v.1991), 1 male (31.v.1991), 1 female (2.vi.1991), PSU, Songkhla, coll. Anuchit *et al.*, bred from *Syzygium samarangense*; 1 male, DPM, PSU Songkhla, 2.vi.1991, coll. Anuchit *et al.*, bred from *Syzygium malaccense*; 1 female, Songkhla, Pikul Thong, Narathiwat, 14.vi.1991, coll. Sieng *et al.*, bred from *Psidium guajava*; 3 males and 2 females, Songkhla, Amphoe Muang, Narathiwat, 6.vii.1991, coll. Anuchit *et al.*, bred from *Terminalia catappa*. Specimens in QDPC.

Diagnosis: A medium-sized species; face fulvous with a pair of circular to oval black spots; postpronotal lobes yellow (anteromedial corners black); notopleura yellow; scutum mostly black; lateral postsutural yellow vittae present; medial postsutural yellow vitta absent; no yellow spot anterior to notopleural suture; mesopleural stripe reaching to anterior *npl.* seta dorsally; scutellum yellow with a broad black basal band; wing with cells bc and c pale fuscous, microtrichia in outer corner of cell c only; a narrow fuscous costal band which is extremely pale beyond extremity of cell sc to apex of wing, a narrow dark fuscous transverse band across wing enclosing *r-m* and *dm-cu* crossveins; a broad fuscous to dark fuscous anal streak; abdominal terga III–V orange-brown with a narrow- to medium-width medial longitudinal dark fuscous to black band over all three terga and lateral dark markings which vary from narrow anterolateral dark fuscous to black corners on all three terga to broad lateral longitudinal dark fuscous to black bands over all three terga.

Description: male.

Head: Height 1.46 mm. Frons – length 1.6 times breadth; fuscous, fulvous along lateral margins and with dark fuscous around orbital setae and on anteromedial hump; orbital setae black: 1 *s.or.*, 2 *i.or.*; lunule dark fuscous. Ocellar triangle black. Vertex dark fuscous. Face fulvous with a pair of large oval black spots; length 0.46 mm. Genae fulvous, dark fuscous subocular spot present; black seta present. Occiput red-brown to fuscous, black laterally and fulvous along eye

margins; occipital row with a large number of strong black setae. Antennae with segments 1 and 2 fuscous, segment 3 fuscous with dark fuscous of apex and outer surface; length of segments: 0.2 mm; 0.3 mm; 0.8 mm.
Thorax: Scutum entirely shining black except dark red-brown behind lateral postsutural vittae and inside notopleura along notopleural suture in some specimens. Pleural areas entirely black. Yellow markings as follows: postpronotal lobes (anterodorsal corners black); notopleura; mesopleural stripe reaching anterior *npl.* seta dorsally, continuing to katepisternum as a large transverse spot, anterior margin straight; anatergite (posterior apex black); anterior ⅔ katatergite (remainder black); two moderately broad lateral postsutural vittae tapering gradually posteriorly to end just before *ia.* seta. Medial longitudinal postsutural yellow vitta absent. Postnotum black. Scutellum yellow except for broad black basal band. Setae: *sc.* 2; *prsc.* 2; *ia.* 1; *p.sa.* 1; *a.sa.* 1; *mpl.* 1; *npl.* 2; *scp.* 4.
Legs: All femora generally fulvous tending red-brown apically (in the hind femora there can be diffuse pale fuscous to fuscous apically); fore and mid tibiae fulvous, hind tibiae fuscous to dark fuscous; tarsi with all segments fulvous; mid tibiae each with an apical black spur.
Wings: Length 5.0 mm; cells bc and c pale fuscous; microtrichia in outer corner of cell c only; remainder of wings colourless except fuscous cell sc, diffuse pale fuscous costal band beyond apex of cell sc to end between apices of R_{4+5} and M and generally not confluent with R_{4+5} but distinctly overlapping R_{2+3}, a narrow dark fuscous transverse band across wing enclosing both *r-m* and *dm-cu* crossveins, a broad dark fuscous anal streak; dense aggregation of microtrichia around A_1+CuA_2; supernumerary lobe of medium development.
Abdomen: Oval; terga free; pecten present on tergum III. Tergum I and sterna I and II wider than long. Tergum I dark fuscous to black with a narrow transverse orange-brown to dark orange-brown band across posterior margin but not reaching lateral margins; tergum II orange-brown with a broad transverse dark fuscous to black band across anterior margin and broadening over lateral margins but not reaching posterolateral corner; terga III–V orange-brown with a narrow- to medium-width medial longitudinal dark fuscous to black band over all three terga and darkened lateral margins which vary from narrow elongate fuscous to black anterolateral corners on all three terga to broader anterolateral dark fuscous to black corners over all three terga to broad lateral longitudinal dark fuscous to black bands over all three terga (Fig. 20B). A pair of oval fuscous shining spots on tergum V. Posterior lobe of surstylus short, sternum V with a deep concavity on posterior margin.
Attractant: Cue lure.
Female: As for male except no dense aggregation of microtrichia around A_1+CuA_2; supernumerary lobe weak; pecten of cilia on abdominal tergum III absent; ovipositor basal segment dark fuscous, dorsoventrally compressed and tapering posteriorly in dorsal view; ratio of length of oviscape to length of tergum V, 0.5:1; apex of piercer needle-shaped.
Distribution:
Type locality: Indonesia (Java, Jakarta – formerly Batavia).
Previously recorded: Type locality, widespread across South-East Asia (White & Hancock, 1997; Norrbom *et al.*, 1998); India (Andaman Is), David & Ramani, 2011.
New records: Australian Territory (Christmas I.); Indonesia (Kalimantan; Lombok; Sulawesi).
Hosts: This species has been recorded from nine families with a wide range of edible and wild hosts (see Allwood *et al.*, 1999 for host records).
Remarks: *Bactrocera (Bactrocera) albistrigata* (de Meijere) belongs to the *frauenfeldi* complex described by Drew (1989). The other species in the complex, *Bactrocera (Bactrocera) caledoniensis* Drew, *Bactrocera (Bactrocera) frauenfeldi* (Schiner), *Bactrocera (Bactrocera) parafrauenfeldi* Drew and *Bactrocera (Bactrocera) trilineola* Drew all possess the same basic body and wing colour patterns; however, *B. albistrigata* is the only species that occurs in South-East Asia and is distinguished by having a combination of moderately broad and elongate lateral postsutural vittae, face with a pair of black spots and abdominal terga III–V fulvous with dark colour patterns (not entirely black).

Hardy (1954) made *B. albistrigata* a synonym of *B. frauenfeldi* and continued on with this position (see Hardy, 1973; Hardy & Foote, 1989). We believe that these are separate species.

Bactrocera (Bactrocera) amarambalensis Drew

(Fig. 21)

Bactrocera (Bactrocera) amarambalensis Drew, in Drew & Raghu 2002: 331. Holotype male in BMNH.

Type material examined: Holotype male, **INDIA**: New Amarambalam Forest, Kerala, 15.ii.1999–25.ii.1999, coll. S. Raghu, methyl eugenol; one male paratype, same data as holotype. Received on loan from BMNH.
Diagnosis: A medium-sized species; face fulvous with a pair of moderately large oval black spots; postpronotal lobes yellow (anteromedial corners red-brown); notopleura yellow; scutum black except red-brown below lateral postsutural vittae, anterior to notopleural suture and inside postpronotal lobes; narrow parallel-sided lateral postsutural yellow vittae ending at *ia.* seta; medial postsutural yellow vitta absent; mesopleural stripe reaching midway between anterior margin of notopleuron and anterior *npl.* seta dorsally; scutellum yellow with a narrow black basal band; legs with femora entirely fulvous except for a small elongate black spot on subapical surfaces of fore femora; wing with cells bc and c colourless, microtrichia in outer corner of cell c only; a narrow fuscous costal band slightly overlapping R_{2+3} where it is a pale tint and with a slight swelling around apex of R_{4+5}; a narrow pale fuscous anal streak contained mostly within boundaries of cell cup; abdominal terga III–V red-brown except for a 'T' pattern consisting of a transverse band across anterior margin of tergum III which is dark fuscous across lateral margins and fuscous and narrower towards midline and a narrow black medial longitudinal band over all three terga, narrow lateral fuscous to dark fuscous margins on terga IV and V; a pair of oval fuscous shining spots on tergum V;

posterior lobe of surstylus short, sternum V with a deep concavity on posterior margin.
Attractant: Methyl eugenol.
Female: No known record.
Distribution:
Type locality: Southern India (Kerala).
Hosts: No known record.
Remarks: *Bactrocera (Bactrocera) amarambalensis* Drew belongs to the *dorsalis* complex and is similar to *Bactrocera (Bactrocera) arecae* (Hardy & Adachi), *Bactrocera (Bactrocera) paraverbascifoliae* Drew and *Bactrocera (Bactrocera) verbascifoliae* Drew & Hancock in possessing narrow parallel-sided lateral postsutural vittae and femora entirely fulvous or with, at most, a dark subapical spot on the outer surfaces of the fore femora. It differs from *B. verbascifoliae* and *B. paraverbascifoliae* in having a dark subapical spot on the fore femora and from *B. arecae* in having the scutum entirely black, not with large areas of red-brown anterocentrally, and in possessing red-brown anteromedial corners on the postpronotal lobes.

Bactrocera (Bactrocera) andamanensis (Kapoor)

(Fig. 22)

Dacus andamanensis Kapoor, 1971: 477. Holotype male in INPC.
Bactrocera (Bactrocera) andamanensis – Norrbom *et al.*, 1998: 88.

Non-type material examined: INDIA: ANDAMAN ISLANDS – 1 male, 'S. Andaman, Port Blair, 6.vi.1992, coll. Ranganath'; 'Sl No. 14, cue lure'; 'B. andamanensis'; 'BMNH (E), 533096', studied at BMNH. 4 males, India, South Andaman, Port Blair, 6.vi.1992, coll. Ranganath, Sl No. 14, attracted to cue lure; 1 male, Andaman Is., 14. vi.1991, attracted to cue lure. 3 males, S. Andaman, Havelock Is., 25.vi.1998, coll. Ranganath, attracted to cue lure, Sl. No. 66. Specimens in QDPC.
Diagnosis: A large species; face with a pair of medium-sized oval to sub-quadrate black spots; postpronotal lobes and notopleura yellow; scutum primarily dark red-brown with irregular black markings between lateral postsutural vittae and between notopleural suture and postpronotal lobes; lateral postsutural yellow vittae present; medial postsutural yellow vitta absent; no yellow spot anterior to notopleural suture; mesopleural stripe reaching midway between anterior margin of notopleuron and anterior *npl.* seta dorsally; scutellum yellow with narrow red-brown basal band; wing with cells bc and c with a pale tint, microtrichia in outer corner of cell c only; a dark fuscous costal band reaching almost to R_{4+5} but distinctly paler between R_{2+3} and R_{4+5} and expanding into a distinct spot in apex of wing; a narrow fuscous to dark fuscous anal streak; abdominal terga III–V either entirely orange-brown or orange-brown and tending darker orange-brown laterally, with a pair of dark fuscous spots along anterior margin of tergum III but not meeting in midline and a medium-width medial longitudinal dark fuscous to black band over all three terga or terga IV and V.
Description: male.
Head: Frons – fulvous with dark fuscous around orbital setae and occasionally on anteromedial hump; orbital setae black: 1 *s.or.*, 2 *i.or.*; lunule fuscous. Ocellar triangle black. Vertex dark fuscous. Face fulvous with a pair of medium-sized oval to sub-quadrate black spots. Occiput red-brown, fulvous along eye margins. Antennae with segments 1 and 2 fulvous, segment 3 fulvous with dark fuscous on apex and outer surface.
Thorax: Scutum dark red-brown with irregular dull black patterns between lateral postsutural vittae and anteriorly between notopleural suture and postpronotal lobes. Pleural areas primarily dark red-brown. Yellow markings as follows: postpronotal lobes; notopleura; mesopleural stripe reaching midway between anterior margin of notopleuron and anterior *npl.* seta dorsally, continuing to katepisternum as a small transverse spot, anterior margin convex; two narrow parallel-sided lateral postsutural vittae ending at or just before *ia.* seta. Medial longitudinal postsutural yellow vitta absent. Postnotum dark red-brown. Scutellum yellow except for narrow dark red-brown basal band. Setae: *sc.* 2; *prsc.* 2 (absent in some specimens); *ia.* 1; *p.sa.* 1; *a.sa.* 1; *mpl.* 1; *npl.* 2; *scp.* 4.
Legs: All femora fulvous with large preapical fuscous to dark fuscous spots (on the hind femora the fuscous to dark fuscous can surround the entire apical ½), all tibiae entirely dark fuscous; tarsi with all segments fulvous; mid tibiae each with an apical black spur.
Wings: Length 7.6 mm; cells bc and c with a pale tint; microtrichia in outer corner of cell c only; remainder of wings colourless except dark fuscous cell sc, dark fuscous costal band extending almost to R_{4+5} (distinctly paler between R_{2+3} and R_{4+5}) and expanding into a distinct spot at apex of wing, narrow fuscous to dark fuscous anal streak becoming paler at margins; dense aggregation of microtrichia around A_1+CuA_2; supernumerary lobe weak.
Abdomen: Elongate oval; terga free; pecten present on tergum III. Tergum I red-brown with narrow fuscous lateral margins; tergum II orange-brown tending fulvous along posterior margin, with narrow dark fuscous anterolateral corners; terga III–V either entirely orange-brown or orange-brown tending to dark orange-brown laterally and with dark fuscous spots along anterior margin of tergum III but not joining on midline, a medium-width medial longitudinal dark fuscous to black band over all three terga (this band may extend only from the centre of tergum IV to posterior margin of tergum V) and anterolateral corners of tergum V tending pale fuscous to fuscous. A pair of oval dark red-brown to fuscous shining spots on tergum V. Abdominal sterna mostly pale red-brown. Posterior lobe of surstylus short, sternum V with a deep concavity on posterior margin.
Attractant: Cue lure.
Female: No known record.
Distribution:
Type locality: India (Andaman Is, Stawartsound).
New records: India – Andaman Is (South Andaman I., Port Blair; Havelock I.).
Hosts: No known record.
Remarks: *Bactrocera (Bactrocera) andamanensis* (Kapoor) is similar to *Bactrocera (Bactrocera) megaspilus* (Hardy) in possessing the costal band confluent with R_{4+5} and expanded into a large spot in the apex of the wing, but differs from this

species in having the scutum basically black, narrow lateral postsutural yellow vittae and a medial longitudinal black band on abdominal terga III–V.

Bactrocera (Bactrocera) apicofuscans White & Tsuruta

(Fig. 23)

Bactrocera (Bactrocera) apicofuscans White & Tsuruta, in Tsuruta & White, 2001: 73; Drew & Raghu, 2002: 332. Holotype male in HORDI.

Non-type material examined: INDIA: 1 male, New Amarambalam Forest, Kerala, 15.ii.1999-25.ii.1999, coll. S. Raghu, attracted to methyl eugenol; 1 male, Karnataka, Chethalli, 970m, 3.iv.1986, coll. Ramani, attracted to methyl eugenol; 1 male, India, Karnataka, Chethalli, 970m, 3.iv. 1986, coll. Ramani, attracted to methyl eugenol. Specimens in QDPC. **SRI LANKA (formerly CEYLON)**: 1 male, Gannoruwa, 4 Aug.1987, coll. K. Tsuruta; 1 male, Hunasigiriya, 11.v.1994, coll. K. Tsuruta. Specimens donated by Kenji Tsuruta and in QDPC.

Diagnosis: A large species; face fulvous with a pair of large elongate oval black spots; postpronotal lobes yellow (anteromedial corners red-brown); notopleura yellow; scutum red-brown either without dark markings or with narrow elongate dull black along inside of lateral postsutural vittae and circular dull black spots between notopleural suture and posterior margin of postpronotal lobes; narrow lateral postsutural yellow vittae ending before *ia.* seta; medial postsutural yellow vitta absent; no yellow spot anterior to notopleural suture; mesopleural stripe slightly wider than postpronotal lobe dorsally; scutellum yellow except for narrow red-brown basal band; legs with fore femora fulvous basally, red-brown apically and with a large elongate oval dark fuscous spot on outer surfaces, mid femora fulvous basally and red-brown apically with a large dark fuscous spot on outer apical surfaces in some specimens, hind femora fulvous basally and red-brown apically with a large dark fuscous spot on outer apical surfaces in some specimens; wing with cell bc fuscous, cell c pale fuscous, microtrichia in outer corner of cell c only; a narrow fuscous costal band almost confluent with R_{4+5} and ending in a dark fuscous spot at apex; a narrow dark fuscous anal streak; abdominal terga III–V red-brown with a black 'T' pattern consisting of a narrow transverse band across anterior margin of tergum III and a narrow- to medium-width medial longitudinal band over all three terga; posterior lobe of surstylus short, sternum V with a deep concavity on posterior margin; abdominal sterna dark fuscous to black.

Attractant: Methyl eugenol.

Female: No known record.

Distribution:

Type locality: Sri Lanka (Kandy District, Gannoruwa).

Previously recorded: Type locality and other district locations in Sri Lanka; Southern India (Karnataka) (Tsuruta & White, 2001); Southern India (Kerala) (Drew & Raghu, 2002).

Hosts: No known record.

Remarks: *Bactrocera (Bactrocera) apicofuscans* White & Tsuruta has been adequately described in Tsuruta & White (2001) and by Drew & Raghu (2002). It is similar to *Bactrocera (Bactrocera) abdonigella* (Drew) and *Bactrocera (Bactrocera) moluccensis* (Perkins) in possessing a complete costal band overlapping R_{2+3}, cells bc and c distinctly coloured and the basic colour of the scutum red-brown, but differs from these species in having narrow lateral postsutural vittae tapering posteriorly to end well before the *ia.* setae and the *prsc.* setae absent. Other species with a red-brown scutum that occur in the Indian subcontinental area are *Bactrocera (Bactrocera) brunneola* White & Tsuruta, *Bactrocera (Bactrocera) paraosbeckiae* Drew, *Bactrocera (Bactrocera) pseudoversicolor* Drew and *Bactrocera (Bactrocera) versicolor* (Bezzi). The first two species can be differentiated by having colourless cells bc and c and the latter two in having the costal band confluent with R_{2+3}.

Bactrocera (Bactrocera) apiconigroscutella Drew

(Fig. 24)

Bactrocera (Bactrocera) apiconigroscutella Drew, in Drew & Raghu, 2002: 333. Holotype male in BMNH.

Type material examined: Holotype male, **INDIA**: Bababudangiri near Kemmannugundi, 1600 m, 21.v.1992, coll. D.L. Hancock, cue lure. Studied at BMNH.

Diagnosis: A small species; face fulvous with a pair of large oval black spots reaching oral margin; postpronotal lobes yellow (anteromedial corners black and extending into centre of lobe); notopleura yellow (black encroaching on anterior and posterior margins); scutum entirely black without pale markings; lateral postsutural yellow vittae very short and narrow reaching posteriorly to level of *a.sa.* seta; medial postsutural yellow vitta absent; mesopleural stripe reaching midway between anterior margin of notopleuron and anterior *npl.* seta dorsally; scutellum yellow except for broad black basal band and an apical black spot enclosing both setae; legs with fore femora black except fulvous around base and apex and along entire inner surface, mid femora entirely black except fulvous around base and apex, hind femora fulvous on basal ½ and black on apical ½; wing with cells bc and c colourless, microtrichia in outer corner of cell c only; narrow fuscous costal band confluent with R_{2+3} and remaining very narrow beyond apex of this vein to end just beyond apex of R_{4+5}; narrow fuscous anal streak; abdominal terga III–V dark fuscous to black except for small areas of dark red-brown posterocentrally on terga IV and V either side of a medium-width medial longitudinal dark fuscous to black band; posterior lobe of surstylus short, sternum V with a deep concavity on posterior margin; abdominal sterna dark fuscous to black.

Attractant: Cue lure.

Female: No known record.

Distribution:

Type locality: India (Bababudangiri).

Hosts: No known record.

Remarks: *Bactrocera (Bactrocera) apiconigroscutella* Drew has been adequately described in Drew & Raghu (2002). It belongs to the *nigrotibialis* complex of species in possessing a predominantly black thorax, wings colourless except for the costal band and anal streak and legs, particularly the femora, with extensive areas of black. It differs from *Bactrocera (Bactrocera) nigrofemoralis* White & Tsuruta and *Bactrocera (Bactrocera) nigrotibialis* (Perkins) in possessing a fulvous face with a pair of black spots (not entirely black) and from all species in the complex by having a distinct black spot on the apex of the scutellum.

Bactrocera (Bactrocera) apicopicta Drew & Romig, new species

(Fig. 25)

Material examined: Holotype male, **INDONESIA: FLORES** – Borong, Flores, Indonesia (200–400 m), 25. vi.1981, coll. J. Turner, attracted to cue lure, primary sclerophyll forest dry, deposited in BMNH.

Paratypes: **INDONESIA: FLORES** – 1 male, same data as holotype; **LOMBOK** – 1 male, Bentek, 20 km N. Mataram, Lombok, Indonesia (100–250 m), 2.vi.1981, coll. J. Turner. 1 paratype in ANIC, 1 paratype in QDPC.

Diagnosis: A medium-sized species; face fulvous with a pair of medium-sized oval black spots; postpronotal lobes and notopleura yellow; scutum with a variegated colour pattern varying from fuscous posteriorly and laterally to red-brown centrally and lateral and medial longitudinal irregularly shaped black bands; lateral postsutural yellow vittae present; medial postsutural yellow vitta absent; no yellow spot anterior to notopleural suture; mesopleural stripe reaching midway between anterior margin of notopleuron and anterior *npl.* seta dorsally; scutellum yellow; wing with cells bc and c dark fuscous, dense microtrichia covering all of both cells; a broad dark fuscous costal band confluent with R_{4+5} and expanding across apex of wing and then continuing as a pale fuscous tint around hind margin of wing; a narrow dark fuscous anal streak; abdominal terga III–V orange-brown with a dark 'T' pattern consisting of a narrow transverse dark fuscous band across anterior margin of tergum III which widens over the lateral margins and a broad medial longitudinal black to dark fuscous band over all three terga and narrow fuscous to dark fuscous lateral margins on terga IV and V.

Description: male.

Head: Height 1.9 mm. Frons – length 1.62 times breadth; red-brown with fuscous around orbital setae; orbital setae black: 1 *s.or.*, 2 *i.or.*; lunule fuscous. Ocellar triangle black. Vertex red-brown. Face fulvous with a pair of medium-sized oval black spots; length 0.52 mm. Genae fulvous, pale fuscous subocular spot present; black seta present. Occiput red-brown, fulvous along eye margins; occipital row with 6–8 black setae. Antennae with segments 1 and 2 red-brown, segment 3 red-brown with fuscous on apex and outer surface; length of segments: 0.19 mm; 0.36 mm; 0.96 mm.

Thorax: Scutum with a variegated pattern consisting of fuscous below and behind lateral postsutural vittae and across posterior margin, red-brown centrally and anteriorly with lateral and medial irregularly shaped longitudinal black bands and fuscous to dark fuscous laterally anterior to notopleural suture. Pleural areas dark fuscous to black with dark red-brown below postpronotal lobes. Yellow markings as follows: postpronotal lobes; notopleura; mesopleural stripe reaching midway between anterior margin of notopleuron and anterior *npl.* seta dorsally, continuing to katepisternum as a small transverse spot, anterior margin slightly convex; anatergite (posterior apex black); anterior ¾ katatergite (remainder black); two medium-width lateral postsutural vittae narrowing posteriorly to end before *ia.* seta. Medial longitudinal postsutural yellow vitta absent. Postnotum black, tending red-brown centrally. Scutellum yellow with a moderately broad black basal band. Setae: *sc.* 2; *prsc.* 2; *ia.* 1; *p.sa.* 1; *a.sa.* 1; *mpl.* 1; *npl.* 2; *scp.* 4.

Legs: Fore and mid femora fulvous with dark fuscous to black over most of the outer surfaces, hind femora fulvous with a subapical dark fuscous to black spot over ⅓ of the inner surfaces; all tibiae dark fuscous; fore tarsi with basal segment fulvous and apical four segments pale fuscous, mid and hind tarsi with basal segment fulvous and apical four segments red-brown; mid tibiae each with an apical black spur.

Wings: Length 6.85 mm; cells bc and c dark fuscous; dense microtrichia covering all of both cells; remainder of wings colourless except dark fuscous cell sc, broad dark fuscous costal band confluent with R_{4+5} and expanding across apex of wing and continuing then as a pale fuscous tint around hind margin of wing, narrow dark fuscous anal streak; a dense aggregation of microtrichia around A_1+CuA_2; supernumerary lobe of medium development.

Abdomen: Oval; terga free; pecten present on tergum III. Tergum I and sterna I and II wider than long. Tergum I dark fuscous; tergum II orange-brown with narrow dark fuscous lateral margins; terga III–V orange-brown with a dark 'T' pattern consisting of a narrow transverse dark fuscous band across anterior margin of tergum III which widens across lateral margins, a broad medial longitudinal black band over terga III–V (this band tends to dark fuscous on tergum V), narrow fuscous to dark fuscous lateral margins on terga IV and V. A pair of oval fuscous to dark fuscous shining spots on tergum V. Posterior lobe of surstylus short, sternum V with a deep concavity on posterior margin. Abdominal sterna red-brown to dark red-brown.

Attractant: Cue lure.

Female: No known record.

Distribution: Indonesia (Flores, Borong; Lombok).

Hosts: No known record.

Remarks: *Bactrocera (Bactrocera) apicopicta* n.sp. is similar to *Bactrocera (Zeugodacus) emittens* (Walker) in wing colour pattern but is a unique species of subgenus *Bactrocera* in possessing infuscation around the apex of the wing, dark fuscous to black femora (except basal ¾ of hind femora) and a red-brown scutum with dark patterns.

Bactrocera (Bactrocera) arecae (Hardy & Adachi)

(Fig. 26)

Dacus (Strumeta) arecae Hardy & Adachi, 1954: 161; Hardy, 1969a: 405; Hardy, 1973: 32. Holotype male in USNM.
Dacus (Bactrocera) arecae – Hardy, 1977: 48.
Bactrocera (Bactrocera) arecae – White & Elson-Harris, 1992: 274; Drew & Hancock, 1994a: 8; Norrbom *et al.*, 1998: 88.

Non-type material examined: No new specimens available for study since the publication of Drew & Hancock (1994a).
Diagnosis: Face fulvous with a pair of large oval black spots; postpronotal lobes and notopleura yellow; scutum black with dark red-brown below and behind lateral postsutural vittae, around notopleural suture, inside postpronotal lobes, a broad medial longitudinal area anterocentrally; narrow- to medium-width parallel-sided lateral postsutural yellow vittae ending at *ia.* seta; medial postsutural yellow vitta absent; mesopleural stripe reaching midway between anterior margin of notopleuron and anterior *npl.* seta dorsally; scutellum yellow; legs with femora fulvous with a large dark fuscous to black preapical spot on outer surfaces of fore femora; tarsi dark fuscous to black; wing with cells bc and c colourless or with a very pale tint, microtrichia in outer corner of cell c only; a narrow dark fuscous costal band confluent with R_{2+3} and widening slightly across apex of R_{4+5}; a narrow dark fuscous anal streak; supernumerary lobe of medium development; abdominal terga III–V orange-brown with a broad black medial longitudinal band and two broad lateral longitudinal dark fuscous to black bands over all three terga and joined along anterior margin of tergum III, a pair of oval orange-brown shining spots on tergum V; abdominal sterna dark coloured.
Attractant: No known record.
Female: As for male in the general body colour patterns. See Drew & Hancock (1994a) for a detailed description of the female characters.
Distribution:
Type locality: Singapore.
Previously recorded: Type locality and the following locations – Peninsular Malaysia; Southern Thailand (Drew and Hancock, 1994a).
Hosts: This species breeds primarily in *Areca catechu* (betel nut palm) (family *Arecaceae*) (see Allwood *et al.*, 1999 for host records).
Remarks: *Bactrocera (Bactrocera) arecae* (Hardy & Adachi), within the *dorsalis* complex, is similar to *Bactrocera (Bactrocera) amarambalensis* Drew, *Bactrocera (Bactrocera) caryeae* (Kapoor), *Bactrocera (Bactrocera) kandiensis* Drew & Hancock and *Bactrocera (Bactrocera) neoarecae* Drew in possessing narrow parallel-sided lateral postsutural vittae, dark apical patterns on at least one pair of femora in addition to the general *dorsalis* complex characteristics. It differs from *B. caryeae*, *B. kandiensis* and *B. neoarecae* in possessing dark preapical spots on the fore femora only and from *B. amarambalensis* in possessing large areas of dark red-brown on the scutum, particularly anterocentrally (i.e. not entirely black).

Bactrocera (Bactrocera) armillata (Hering)

(Fig. 27)

Strumeta armillata Hering, 1938: 6. Holotype female in NRS.
Bactrocera (Bactrocera) armillata – Norrbom *et al.*, 1998: 88.

Type material examined: Paratype female, **MYANMAR (formerly BURMA)**: with the following six labels – 'Paratype'; 'Paratype (red label)'; 'N.E. Burma, Kambaiti, 2000 m, 9.6.1934, Malaise'; 'Strumeta armillatum n, PT female, det. M. Hering 1938'; 'Purchased from E.M. Hering, B.M. 1965-270'; 'BMNH (E) No. 532764'. Studied at BMNH.
Diagnosis: A medium-sized species; face fulvous with a pair of large oval black spots; postpronotal lobes yellow with anterodorsal corners black; notopleura dark fuscous to black; scutum entirely shining black; very short and narrow lateral postsutural yellow vittae present; medial postsutural yellow vitta absent; no yellow spot anterior to notopleural suture; mesopleural stripe reaching midway between anterior margin of notopleuron and anterior *npl.* seta dorsally; scutellum entirely black except for small yellow anterolateral corners; wing with cells bc and c colourless and totally devoid of microtrichia; a narrow fuscous costal band confluent with R_{2+3} and remaining extremely narrow beyond apex of this vein before expanding across apex of R_{4+5}, a pale fuscous spot on hind wing margin around CuA_1; a narrow pale fuscous anal streak consisting of colour within margins of cell cup; all abdominal terga black except for narrow dark red-brown bands along posterior margins of terga I and II.
Description: female.
Head: Frons – fulvous with dark fuscous around bases of orbital setae; orbital setae black: 1 *s.or.*, 2 *i.or.*; lunule fulvous. Ocellar triangle black. Vertex black. Face fulvous with a pair of large oval black spots. Occiput dark fuscous to black reaching to eye margins. Antennae with segments 1 and 2 red-brown, segment 3 missing.
Thorax: Scutum entirely shining black without pale markings. Pleural areas entirely black without pale markings. Yellow markings as follows: postpronotal lobes (anterodorsal corners black); mesopleural stripe reaching midway between anterior margin of notopleuron and anterior *npl.* seta dorsally, not continuing to katepisternum, anterior margin convex; two very short and narrow lateral postsutural vittae ending just posterior of *a.sa.* seta. Medial longitudinal postsutural yellow vitta absent. Notopleura dark fuscous to black. Postnotum black. Scutellum with dorsal surface entirely black except for small yellow anterolateral corners. Setae: *sc.* 2; *prsc.* 2; *ia.* 1; *p.sa.* 1; *a.sa.* 1; *mpl.* 1; *npl.* 2; *scp.* 4.
Legs: Fore femora black except basal ⅓ fulvous, mid femora apical ⅖ black and basal ⅗ fulvous, hind femora fulvous except apical ⅓ black; fore and mid tibiae dark fuscous, hind tibiae black; all tarsal segments fulvous; mid tibiae each with an apical black spur.
Wings: Length 6.6 mm; cells bc and c colourless and totally devoid of microtrichia; remainder of wings colourless except dark fuscous cell sc, narrow paler fuscous costal band

confluent with R_{2+3} and remaining very narrow beyond apex of R_{2+3} before expanding across apex of R_{4+5} (from apex of R_{2+3} to apex of wing the band is dark fuscous), a pale fuscous spot on hind wing margin around CuA_1, a pale fuscous anal streak consisting of colour within the margins of cell cup; basal area of cell br colourless and devoid of microtrichia; no dense aggregation of microtrichia around A_1+CuA_2; supernumerary lobe weak.

Abdomen: Oval; terga free; no pecten on tergum III. Tergum I and sterna I and II wider than long. All abdominal terga entirely black except for narrow dark red-brown bands along posterior margins of terga I and II. A pair of oval black shining spots on tergum V. Abdominal sterna dark fuscous to black. Ovipositor basal segment dark fuscous to black, dorsoventrally flattened and tapering posteriorly in dorsal view; ratio of length of oviscape to length of tergum V, 0.3:1; apex of piercer needle-shaped.

Attractant: No known record.

Male: No known record.

Distribution:

Type locality: Myanmar (Kambaiti).

Hosts: No known record.

Remarks: *Bactrocera (Bactrocera) armillata* (Hering) is almost certainly a synonym of *Bactrocera (Zeugodacus) biguttata* (Bezzi) in the *scutellaris* complex. As it is only known from the female types, the male genital characters cannot be compared with those of *B. biguttata*. The two species are similar in the unusual characters of dark fuscous to black notopleura and almost black scutellum in addition to the general characters of the *scutellaris* complex. They differ in minor characters, namely the slightly broader mesopleural stripe and presence of a narrow pale anal streak in *B. armillata*. Until male specimens, preferably collected at the type locality, are available for study, *B. armillata* should be retained.

Bactrocera (Bactrocera) atrifemur Drew & Hancock

(Fig. 28)

Bactrocera (Bactrocera) atrifemur Drew & Hancock, 1994a: 9; Norrbom *et al.*, 1998: 88. Holotype male in BMNH.

Non-type material examined: Since the publication of Drew & Hancock (1994a), the following specimens have been studied – **INDONESIA: BANGKA ISLAND** – 1 male, Indonesia, Bangka, Mendo Barat, Kace, 29.viii.2006, coll. Hendra K., attracted to methyl eugenol in Cempedak. **PENINSULAR MALAYSIA**: 6 males (4.iii.1986), 20 males (11.iii.1986), 9 males (18.iii.1986), 1 male (22.iv.1986), Sungai Baging, Kuantan, Malaysia, coll. A. Samat Wahid, attracted to methyl eugenol. Specimens in QDPC.

Diagnosis: Face fulvous with a pair of very large oval black spots; postpronotal lobes and notopleura yellow; scutum black without pale areas; broad lateral postsutural yellow vittae narrowing slightly posteriorly to end at or just before *ia.* seta; medial postsutural yellow vitta absent; mesopleural stripe reaching midway between anterior margin of notopleuron and anterior *npl.* seta dorsally; scutellum yellow; legs with apical ¾ fore, apical ⅓ mid and apical ⅖ hind femora shining black (remainder fulvous); fore and mid tibiae fuscous and hind tibiae black; wing with cells bc and c colourless, microtrichia in outer corner of cell c only; a narrow fuscous costal band just overlapping R_{2+3} and with a very slight expansion across apex of R_{4+5} before ending between extremities of R_{4+5} and M; a narrow fuscous anal streak; supernumerary lobe of medium development; abdominal terga III–V orange-brown with a broad transverse black band across anterior ½ of tergum III and a broad medial and two broad lateral longitudinal black bands over all three terga, a pair of oval fuscous shining spots on tergum V; abdominal sterna dark coloured.

Attractant: Methyl eugenol.

Female: No known record.

Distribution:

Type locality: Peninsular Malaysia (Sungai Baging).

New records: Indonesia (Bangka I.).

Hosts: *Myristica fragrans* (family *Myristicaceae*) (Allwood *et al.*, 1999).

Remarks: *Bactrocera (Bactrocera) atrifemur* Drew & Hancock, within the *dorsalis* complex, is similar to *Bactrocera (Bactrocera) arecae* (Hardy & Adachi) and *Bactrocera (Bactrocera) occipitalis* (Bezzi) in possessing broad parallel or subparallel-sided lateral postsutural vittae, costal band overlapping R_{2+3}, abdominal terga III–V with broad lateral longitudinal dark markings in addition to the general characteristics of the *dorsalis* complex. It differs from these species in possessing shining black markings on the apices of all femora, tibiae dark fuscous to black and fuscous to black shining spots on abdominal tergum V.

Bactrocera (Bactrocera) avittata Drew & Romig, new species

(Fig. 29)

Material examined: Holotype male, **INDONESIA: MOLUCCAS (MALUKU): AMBON ISLAND** – Moluccas, Maluku, Middle Maluku, Suli, 18.vi.2007, coll. La Naibu, attracted to cue lure in Jeruk Nipis, deposited in BMNH.

Paratypes: **INDONESIA: MOLUCCAS (MALUKU)** – 1 male, same data as holotype; 41 males (2.vi.2007), 5 males (4.vi.2007), 39 males (9.vi.2007), 3 males (16.vi.2007), 4 males (1.vii.2007), 1 male (12.vii.2007), 1 male (26.vii. 2007), Moluccas, Maluku, Middle Maluku, Suli, coll. La Naibu, attracted to cue lure; 1 male (24.v.2007), 1 male (29.vi.2007), 6 males (12.vii.2007), 3 males (19.vii.2007), Moluccas, Maluku, Seram (Western part), Waimital, coll. Haris Tualepe, attracted to cue lure; 1 male, Mamala, Ambon, Indonesia, 50–150 m, 7.viii.1981, coll. J. Turner, residual forest with cloves, nutmeg and banana groves, attracted to cue lure. 10 paratypes in ANIC, 25 paratypes in BMNH, 57 paratypes in QDPC, 15 paratypes in QMIC.

Diagnosis: A medium-sized species; face fulvous with a pair of medium-sized oval black spots; postpronotal lobes dark fuscous; notopleura yellow; scutum entirely dull black without pale markings; lateral and medial postsutural yellow vittae absent; no yellow spot anterior to notopleural suture; mesopleural stripe reaching midway between anterior margin of notopleuron and anterior *npl.* seta dorsally; scutellum yellow with a broad black basal band; wing with cells bc and c colourless, microtrichia in outer corner of cell c only; a broad dark fuscous costal band overlapping R_{2+3} where it becomes paler but not confluent with R_{4+5}, a broad diffuse fuscous band across wing from costal band to hind margin and enclosing both crossveins; a broad fuscous to dark fuscous anal streak; all abdominal terga entirely fuscous to dark fuscous with dark red-brown posterocentrally on tergum IV and centrally on tergum V in some specimens.

Description: male.

Head: Height 1.7 mm. Frons – length 1.6 times breadth; red-brown with pale fuscous around bases of *s.or.* setae; orbital setae black: 1 *s.or.*, 2 *i.or.*; lunule fuscous. Ocellar triangle black. Vertex red-brown. Face fulvous with a pair of medium-sized oval black spots; length 0.52 mm. Genae red-brown, small pale fuscous subocular spot present; black seta present. Occiput red-brown, fulvous along eye margins; occipital row with a large number of short black setae. Antennae with segments 1 and 2 red-brown, segment 3 red-brown with fuscous on apex and outer surface; length of segments: 0.22 mm; 0.34 mm; 0.84 mm.

Thorax: Scutum entirely dull black without pale markings. Pleural areas black except dark red-brown to fuscous below postpronotal lobes. Yellow markings as follows: notopleura; mesopleural stripe reaching midway between anterior margin of notopleuron and anterior *npl.* seta dorsally, continuing to katepisternum as a small transverse spot, anterior margin straight; anatergite (posterior apex black); a yellow patch across centre of katatergite with anterior and posterior areas and lower margin black. Lateral and medial postsutural yellow vittae absent. Postpronotal lobes dark fuscous. Postnotum black. Scutellum yellow except for broad black basal band. Setae: *sc.* 2; *prsc.* 2; *ia.* 1; *p.sa.* 1; *a.sa.* 1; *mpl.* 1; *npl.* 2; *scp.* 2.

Legs: Fore femora fuscous tending fulvous basally, mid femora fulvous on basal ¼ with remainder fuscous, hind femora fulvous on basal ⅔ with remaining ⅓ fuscous; fore tibiae fuscous, mid tibiae fulvous, hind tibiae dark fuscous; fore tarsi with basal segment fulvous and apical four segments fuscous, mid tarsi with basal segment fulvous and apical four segments pale fuscous, hind tarsi with all segments fulvous; mid tibiae each with an apical black spur.

Wings: Length 6.7 mm; cells bc and c colourless; microtrichia in outer corner of cell c only; remainder of wings colourless except dark fuscous cell sc, broad dark fuscous costal band overlapping R_{2+3} where it becomes paler but not confluent with R_{4+5}, a broad diffuse fuscous band across wing from costal band to hind margin and enclosing both crossveins, a broad fuscous to dark fuscous anal streak; a dense aggregation of microtrichia around A_1+CuA_2; supernumerary lobe of medium development.

Abdomen: Oval; terga free; pecten present on tergum III. Tergum I and sterna I and II wider than long. Terga I and II dark fuscous except for a narrow dark red-brown band across posterior margin of tergum II; tergum III entirely dark fuscous; tergum IV dark fuscous tending fuscous medially; tergum V entirely fuscous. A pair of oval dull black shining spots on tergum V. Posterior lobe of surstylus short, sternum V with a deep concavity on posterior margin. Abdominal sterna dark coloured.

Attractant: Cue lure.

Female: No known record.

Distribution: Indonesia (Moluccas = Maluku – Ambon I.; Seram I.).

Hosts: No known record.

Remarks: *Bactrocera (Bactrocera) avittata* n.sp. is similar to *Bactrocera (Bactrocera) neofulvicauda* n.sp. in possessing similar wing colour patterns, a black scutum and dark patterns on the abdominal terga. It differs from *B. neofulvicauda* in lacking postsutural yellow vittae and in having dark fuscous postpronotal lobes, a broad black basal band on the scutellum, colourless cells bc and c and all abdominal terga mostly fuscous to dark fuscous.

Bactrocera (Bactrocera) bancroftii (Tryon)

(Fig. 30)

Chaetodacus bancroftii Tryon, 1927: 199. Lectotype male in QMIC.
Strumeta bancroftii – May, 1963: 43, 50.
Bactrocera (Bactrocera) bancroftii – Drew, 1989: 91, Lectotype designation; Norrbom *et al.*, 1998: 88; Drew *et al.*, 2011: 149.

Non-type material examined: INDONESIA: WEST TIMOR – 10 males, Indonesia, West Timor, 15km S.E. Kupang, 30.x.1986, coll. E.S.C. Smith, attracted to methyl eugenol trap; 7 males, Indonesia, West Timor, Kupang ADAB office, 31.x.1986, coll. E.S.C. Smith, attracted to methyl eugenol trap. Specimens in QDPC.

Diagnosis: A medium-sized species; face fulvous with a pair of medium-sized circular black spots; postpronotal lobes and notopleura yellow; scutum black; lateral postsutural yellow vittae present; medial postsutural yellow vitta absent; no yellow spot anterior to notopleural suture; mesopleural stripe reaching midway between anterior margin of notopleuron and anterior *npl.* seta dorsally; scutellum yellow with a large apical fuscous spot; wing with cells bc and c colourless to a very pale tint, microtrichia in outer corner of cell c only; a narrow fuscous costal band; a narrow fuscous anal streak; abdominal terga generally entirely orange-brown but some specimens may have fuscous on tergum I and narrow transverse dark fuscous to black spots on anterolateral corners of tergum III.

Description: male.

Head: Height 1.65 mm. Frons – length 1.45 times breadth; red-brown, fulvous along lateral and ventral margins, dark fuscous around bases of orbital setae; orbital setae black: 1 *s.or.*, 2 *i.or.*; lunule fuscous. Ocellar triangle black. Vertex fuscous. Face fulvous with a pair of medium-sized circular black spots; length 0.48 mm. Genae fulvous, red-brown seta present. Occiput fuscous, fulvous along eye margins; occipital row with 6 black setae. Antennae with segments 1 and 2 red-brown, segment 3 red-brown with fuscous on apex and outer surface; length of segments: 0.18 mm; 0.3 mm; 0.76 mm.

Thorax: Scutum black with red-brown below and behind lateral postsutural vittae, around notopleural suture, inside postpronotal lobes. Pleural areas black tending red-brown below postpronotal lobes. Yellow markings as follows: postpronotal lobes; notopleura; mesopleural stripe reaching midway between anterior margin of notopleuron and anterior *npl.* seta dorsally, continuing to katepisternum as a transverse spot, anterior margin straight; anatergite (posterior apex black); anterior ½ katatergite (remainder black); two broad parallel-sided lateral postsutural vittae ending at or just behind *ia.* seta. Medial longitudinal postsutural yellow vitta absent. Postnotum black. Scutellum yellow except for narrow black basal band and a large fuscous spot around apex. Setae: *sc.* 2; *prsc.* 2; *ia.* 1; *p.sa.* 1; *a.sa.* 1; *mpl.* 1; *npl.* 2; *scp.* 4.

Legs: All leg segments entirely fulvous; mid tibiae each with an apical black spur.

Wings: Length 5.5 mm; cells bc and c colourless to having a very pale tint; microtrichia in outer corner of cell c only; remainder of wings colourless except fuscous cell sc, narrow fuscous costal band confluent with or just overlapping R_{2+3} and remaining narrow around apex of wing, a narrow fuscous anal streak; a dense aggregation of microtrichia around A_1+CuA_2; supernumerary lobe of medium development.

Abdomen: Oval; terga free; pecten present on tergum III. Tergum I and sterna I and II wider than long. All abdominal terga entirely orange-brown with some specimens showing variable patterns where tergum I may be fuscous and some specimens with small transverse dark fuscous to black spots on anterolateral corners of tergum III. A pair of oval orange-brown to red-brown shining spots on tergum V. Posterior lobe of stylus short, sternum V with a deep concavity on posterior margin. Abdominal sterna orange-brown.

Attractant: Methyl eugenol (a weak response).

Female: As for male except no dense aggregation of microtrichia around A_1+CuA_2; supernumerary lobe weak; pecten of cilia on abdominal tergum III absent; ovipositor basal segment orange-brown, dorsoventrally compressed and tapering posteriorly in dorsal view; ratio of length of oviscape to length of tergum V, 0.45:1; apex of piercer needle-shaped.

Distribution:

Type locality: Australia, Queensland, Gympie District.

Previously recorded: Type locality and the following locations – Australia (coastal areas of Queensland; Torres Strait Is.); Papua New Guinea (Morobe Province, Lae) (Drew *et al.*, 2011).

New records: Indonesia (West Timor).

Hosts: *Maclura cochinchinensis* and *Morus nigra* (family *Moraceae*) (Hancock *et al.*, 2000).

Remarks: *Bactrocera (Bactrocera) bancroftii* (Tryon) belongs to the *musae* complex of species recently revised by Drew *et al.* (2011). It is distinct in having abdominal terga III–V entirely orange-brown, a large fuscous spot on the apex of the scutellum and the costal band narrow, at most, just overlapping R_{2+3} and remaining very narrow around the apex of the wing. In South-East Asia, *Bactrocera (Bactrocera) latifrons* (Hendel) is similar to *B. bancroftii*; however, the apex of the piercer is trilobed compared with needle-shaped in *B. bancroftii*.

Bactrocera (Bactrocera) beckerae (Hardy)

(Fig. 31)

Dacus (Bactrocera) beckerae Hardy, 1982: 217. Holotype male in MZB.

Bactrocera (Bactrocera) beckerae – Norrbom *et al.*, 1998: 88.

Type material examined: Holotype male, **INDONESIA: SULAWESI** – Lindu Valley, Central Sulawesi, 960 m, Aug. 1975, coll. D.E. & A.T. Hardy, cue lure. Studied at MZB.

Paratypes: **INDONESIA: SULAWESI** – 4 males, 'Sadaunta River, Central Sulawesi, March 1976, 3000′, coll. Margareta Becker'; blue label 'Paratype male, *Dacus (Bactrocera) beckerae*, D.E. Hardy'. Specimens in QDPC.

Non-type material examined: INDONESIA: SULAWESI – 2 males (12.ix.2006), 5 males (3.x.2006), 1 male (10.x.2006), 1 male (17.x.2006), 1 male (31.x.2006), 1 male (16.xi.2006), Gorontalo, Bone Bolango, Lombongo, coll. Yusbar Ismail, attracted to cue lure in forest and village; 1 male (4.ix.2006), 1 male (30.x.2006), Sulawesi Barat, Mamuju, Belang-Belang, coll. Sri Hastuti, Yan PF, Sukri, attracted to cue lure in forest and orange (jeruk); 1 male (12.viii.2006), 1 male (20.viii.2006), 1 male (2.ix.2006), 1 male (27.ix.2006), Sulawesi Tengah, Donggala, Nupabomba, coll. Rusdin, attracted to cue lure in forest; 1 male, Gorontalo, Bone Bolango, Lombongo, 26.ix.2006, coll. Yosis, attracted to cue lure in forest. Specimens in QDPC.

Diagnosis: A large species; face fulvous with a pair of medium-sized circular to slightly oval black spots; postpronotal lobes and notopleura yellow; scutum mostly black; lateral postsutural yellow vittae present; medial postsutural yellow vitta absent; no yellow spot anterior to notopleural suture; mesopleural stripe reaching midway between anterior margin of notopleuron and anterior *npl.* seta dorsally; scutellum yellow except for narrow black basal band; wing with cells bc and c fulvous, microtrichia in outer ⅓ of cell c only; a broad dark fuscous costal band confluent with R_{4+5} and overlapping this vein towards apex; a broad dark fuscous anal streak; abdominal terga III–V orange-brown with a moderately broad black band across anterior margin of tergum III expanding to cover lateral margins and a very narrow medial longitudinal dark fuscous to black band over terga IV and V.

Description: male.

Head: Height 1.8 mm. Frons – length 1.5 times breadth; red-brown, fulvous along eye margins and fuscous around orbital setae and on anteromedial hump; orbital setae black: 1 *s.or.*, 2 *i.or.*; lunule fuscous. Ocellar triangle black. Vertex fuscous. Face fulvous with a pair of medium-sized circular to slightly oval black spots; length 0.55 mm. Genae fulvous, large fuscous subocular spot present; black seta present. Occiput fuscous, fulvous along eye margins; occipital row with 5–8 strong black setae. Antennae with segments 1 and

2 red-brown, segment 3 red-brown with fuscous on apex and outer surface; length of segments: 0.18 mm; 0.39 mm; 0.81 mm.

Thorax: Scutum black except dark red-brown below and behind lateral postsutural vittae, around notopleural suture and base of notopleuron, inside postpronotal lobes. Pleural areas black except dark red-brown below postpronotal lobes. Yellow markings as follows: postpronotal lobes; notopleura; mesopleural stripe reaching midway between anterior margin of notopleuron and anterior *npl.* seta dorsally, continuing to katepisternum as a transverse spot, anterior margin straight to slightly convex; anatergite (posterior apex black); anterior $^3/_5$ katatergite (remainder black); two moderately broad lateral postsutural vittae narrowing posteriorly to end at or just before *ia.* seta. Medial longitudinal postsutural yellow vitta absent. Postnotum black. Scutellum yellow except for narrow black basal band. Setae: *sc.* 2; *prsc.* 2; *ia.* 1; *p.sa.* 1; *a.sa.* 1; *mpl.* 1; *npl.* 2; *scp.* 4.

Legs: All femora entirely fulvous (in some specimens there may be a very small diffuse pale fuscous spot on outer subapical surfaces of fore femora); fore and mid tibiae pale fuscous, hind tibiae fuscous; tarsi with all segments fulvous (in some specimens the apical four segments of all tarsi are pale fuscous); mid tibiae each with an apical black spur.

Wings: Length 7.2 mm; cells bc and c fulvous; microtrichia in outer $^1/_3$ of cell c only; remainder of wings colourless except dark fuscous cell sc, broad dark fuscous costal band confluent with R_{4+5} and overlapping this vein at apex, broad dark fuscous anal streak; dense aggregation of microtrichia around A_1+CuA_2; supernumerary lobe of medium development.

Abdomen: Oval; terga free; pecten present on tergum III. Tergum I and sterna I and II wider than long. Tergum I black except for a narrow orange-brown transverse band across posterior margin but not reaching lateral margins; tergum II orange-brown except for a narrow transverse subanterior black band across anterior margin which may or may not reach black anterolateral corners; terga III–V orange-brown except for a moderately broad transverse black band across anterior margin of tergum III expanding over lateral margins of tergum and a very narrow medial longitudinal dark fuscous to black band over terga IV and V. A pair of red-brown to fuscous shining spots on tergum V. Posterior lobe of surstylus short, sternum V with a moderately deep concavity on posterior margin.

Attractant: Cue lure.

Female: No known record.

Distribution:

Type locality: Indonesia (Sulawesi, Lindu Valley).

Previously recorded: Type locality and Indonesia (Sulawesi – Central Sulawesi, Sadaunta River).

New records: Indonesia (Sulawesi – Gorontalo; Central Sulawesi, Donggala; West Sulawesi, Mamuju).

Hosts: No known record.

Remarks: *Bactrocera (Bactrocera) beckerae* (Hardy) fits into the *bryoniae* complex represented in the South Pacific region by seven species (Drew, 1989). It is close to *Bactrocera (Bactrocera) bryoniae* (Tryon) in this complex but differs in having a much narrower medial longitudinal dark line (band) on abdominal terga IV–V and lateral postsutural vittae that taper posteriorly and sometimes end before the *ia.* seta. In South-East Asia, *Bactrocera (Bactrocera) limbifera* (Bezzi) and *Bactrocera (Bactrocera) ritsemai* (Weyenbergh) all possess broad costal bands confluent with R_{4+5} but differ in having colourless cells bc and c. Species with broad costal bands and coloured cells bc and c include *Bactrocera (Bactrocera) costalis* (Shiraki), *Bactrocera (Bactrocera) epicharis* (Hardy), *Bactrocera (Bactrocera) linduensis* new name and *Bactrocera (Bactrocera) paramusae* Drew; however, *B. linduensis* has the costal band overlapping R_{4+5}, *Bactrocera (Bactrocera) costalis* (Shiraki) and *Bactrocera (Bactrocera) epicharis* (Hardy) have parallel-sided lateral postsutural vittae and *B. paramusae* abdominal terga III–V mostly orange-brown.

Bactrocera (Bactrocera) bellisi Drew & Romig, new species

(Fig. 32)

Material examined: Holotype male, **INDONESIA: WEST TIMOR** – Nusa Tenggara Timur, Belo, Kt Kupang, 2.vi.2007, coll. Ir Hendri Wihanto, attracted to cue lure in Kebun-belimbing, deposited in BMNH.

Paratypes: **INDONESIA: WEST TIMOR** – 1 male, same data as holotype; 4 males (1.vi.2007), 1 male (31.v.2007), Nusa Tenggara Timur, Oebufu, Kt Kupang, coll. Ir Hendri Wihanto, attracted to cue lure in Village Jambu; 4 males, Indonesia, West Timor, Kupang, Wisma Maleita, 31.x.1986, coll. E.S.C. Smith, attracted to cue lure; 1 male, Indonesia, West Timor, 15 km S E Kupang, 30.x.1986, coll. E.S.C. Smith. 1 paratype in ANIC, 1 paratype in BMNH, 8 paratypes in QDPC, 1 paratype in QMIC.

Diagnosis: A small to medium-sized species; face dark fuscous with a pair of large elongate oval black spots; postpronotal lobes yellow (anterodorsal corners black); notopleura yellow; scutum entirely shining black without pale markings; very short and narrow lateral postsutural yellow vittae present; medial postsutural yellow vitta absent; no yellow spot anterior to notopleural suture; mesopleural stripe reaching midway between anterior margin of notopleuron and anterior *npl.* seta dorsally; scutellum yellow with a broad black basal band which is sometimes triangular in shape; wing with cells bc and c colourless, microtrichia in outer corner of cell c only; a narrow dark fuscous costal band; a narrow dark fuscous anal streak; abdominal terga III–V dark fuscous to black except for narrow red-brown margins along a medial longitudinal dark fuscous to black band on tergum V.

Description: male.

Head: Height 1.2 mm. Frons – length 1.67 times breadth; dark fuscous with fulvous along lateral margins between bases of *i.or.* setae; orbital setae black: 1 *s.or.*, 2 *i.or.*; lunule dark fuscous. Ocellar triangle black. Vertex dark fuscous. Face dark fuscous with very large elongate oval black spots in the antennal furrows; length 0.36 mm. Genae dark red-brown, large dark fuscous subocular spot present; black seta present. Occiput black, fulvous along eye margins; occipital row with 4–6 black setae. Antennae with segments

1 and 2 fuscous, segment 3 dark fuscous; length of segments: 0.18 mm; 0.23 mm; 0.62 mm.

Thorax: Scutum entirely shining black without pale markings. Pleural areas entirely black without pale markings. Yellow markings as follows: postpronotal lobes (anterodorsal corners black); notopleura; mesopleural stripe reaching midway between anterior margin of notopleuron and anterior *npl.* seta dorsally, continuing to katepisternum as a very small dark red-brown spot, anterior margin straight; anatergite (posterior apex black); anterior $^{3}/_{5}$ katatergite (remainder black); two very short and narrow lateral postsutural vittae ending before level of *a.sa.* seta (in some specimens these vittae are barely visible as a red-brown line). Medial longitudinal postsutural yellow vitta absent. Postnotum black. Scutellum yellow except for a very broad black basal band (in some specimens this band is a triangular shape). Setae: *sc.* 2; *prsc.* 2; *ia.* 1; *p.sa.* 1; *a.sa.* 1; *mpl.* 1; *npl.* 2; *scp.* 4.

Legs: Fore and mid femora entirely black, hind femora with basal ½ fulvous and apical ½ black; fore tibiae dark fuscous tending fuscous apically, mid tibiae dark fuscous basally to fuscous apically, hind tibiae dark fuscous; all tarsal segments entirely fulvous; mid tibiae each with an apical black spur.

Wings: Length 4.4 mm; cells bc and c colourless; microtrichia in outer corner of cell c only; remainder of wings colourless except fuscous cell sc, narrow dark fuscous costal band confluent with R_{2+3} and remaining very narrow around apex of wing, a narrow dark fuscous anal streak; a dense aggregation of microtrichia around A_1+CuA_2; supernumerary lobe of medium development.

Abdomen: Oval; terga free; pecten present on tergum III. Tergum I and sterna I and II wider than long. Terga I and II black except for a narrow dark red-brown band along intersegmental line between both terga but not reaching lateral margins and fulvous along posterior margins of tergum II; terga III and IV entirely black except tergum IV tending dark fuscous centrally, tergum V black except for red-brown either side of a medial longitudinal dark fuscous to black band and red-brown along posterior margin. A pair of oval dark fuscous shining spots on tergum V. Posterior lobe of surstylus short, sternum V with a deep concavity on posterior margin. Abdominal sterna dark fuscous to black.

Attractant: Cue lure.

Female: No known record.

Distribution: Indonesia (West Timor).

Hosts: No known record.

Etymology: This species is named after Mr Glen Bellis, Northern Australia Quarantine Strategy, Darwin, Australia, in recognition of his contributions to fruit fly survey work in Indonesia and Northern Australia.

Remarks: *Bactrocera (Bactrocera) bellisi* n.sp. belongs to the *nigrotibialis* complex of species. It is similar to *Bactrocera (Bactrocera) nigrofemoralis* White & Tsuruta and *Bactrocera (Bactrocera) nigrotibialis* (Perkins) in possessing a dark fuscous face with large oval black spots in the antennal furrows, lateral postsutural yellow vittae present and the costal band confluent with or just overlapping R_{2+3}. It differs from *B. nigrofemoralis* in having a very broad basal band on the scutellum, the fore femora entirely black and the mesopleural stripe narrower, generally midway between the anterior margin of the notopleuron and the anterior *npl.* seta and from *B. nigrotibialis* in having the lateral postsutural yellow vittae very short and narrow, the costal band confluent with R_{2+3} and abdominal terga I–V entirely black except for small areas of dark red-brown centrally on tergum V. *Bactrocera (Bactrocera) paradiospyri* Chen, Zhou & Li from Bhutan is also similar but differs from *B. bellisi* in having the lateral postsutural vittae entirely absent, a narrow black basal band on the scutellum, the costal band widening beyond apex of R_{2+3} and all abdominal terga entirely black.

Bactrocera (Bactrocera) bhutaniae Drew & Romig, new species

(Fig. 33)

Material examined: Holotype male, **BHUTAN**: Rimchu 1, 10.viii.2000, coll. C. Dorji *et al.*, attracted to cue lure, deposited in BMNH.

Paratypes: **BHUTAN**: 43 males, same data as holotype; 4 males (28.vi.2000), 5 males (5.vii.2000), 65 males (19.vii. 2000), 13 males (23.vii.2000), 76 males (26.vii.2000), 25 males (2.viii.2000), Rimchu 1, coll. C. Dorji *et al.*, attracted to cue lure; 1 male (24.v.2000), 1 male (31.v.2000), 10 males (16.viii.2000), 4 males (30.viii.2000), 8 males (6.ix.2000), 1 male (12.ix.2000), 7 males (20.ix.2000), 1 male (27.ix. 2000), Rimchu 1, coll. C. Dorji, attracted to cue lure; 2 males, Rimchu 1, 27.v.2000, coll. Thapa, attracted to cue lure; 1 male, Rimchu 1, 7.vi.2000, coll. C. Dorji, Nado & Tshomo, attracted to cue lure; 2 males, Rimchu 1, 14.vi.2000, coll. Nado & Tshomo, attracted to cue lure; 2 males, Rimchu 1, 21.vi.2000, coll. Nado *et al.*, attracted to cue lure; 1 male, Rimchu, 17.v.2001, coll. R. Drew & M. Romig; 3 males (7.vi.2000), 3 males (14.vi.2000), 2 males (21.vi.2000), 5 males (28.vi.2000), 1 male (5.vii.2000), 23 males (12.vii.2000), 17 males (19.vii.2000), 18 males (26.vii.2000), 11 males (2.viii.2000), 34 males (10.viii.2000), 18 males (16.viii.2000), 9 males (6.ix.2000), 2 males (5.x.2000), Rimchu 2, coll. C. Dorji *et al.*, attracted to cue lure; 2 males (6.v.2000), 2 males (13.v.2000), 3 males (20.v.2000), Phuentsholing, coll. S. Thapa, attracted to cue lure; 4 males, Tsirang, 17. vii.2000, coll. S.D., attracted to cue lure; 1 male, Lumitshawa, 7.vi.2000, coll. C. Dorji, Nado & Tshomo; 2 males, Lumitshawa, 18.x.2000, coll. C. Dorji; 4 males (15.xi.2004), 5 males (15.xii.2004), 1 male (29.xii.2004), East Bhutan, Sandrup Jongkhu District, coll. Wangdi, attracted to cue lure; 13 males, Ganglephug, Tatapani, 9-13.ix.1990, coll. H.R. Fiejen; 6 males, Bhur, Gaylephug 200 m, 13.ix.1990, coll. H.R. Fiejen. **INDIA: ANDAMAN ISLANDS** – 1 male and 1 female, India, S. Andaman, Port Blair, 20.x.1994, coll. Ranganath, Sl No. 42, Icacinaceae. **THAILAND**: 1 male (11.xi.1991), 1 male (26.xi.1991), Philu Hort. Res. Sta., A. Laemsingh, Chanthaburi, BK 138T, coll. Montree, attracted to cue lure; 1 male (27.ix.1991), 1 male (4.x.1991), 2 males (25.x.1991), 1 male (1.xi.1991), 1 male (8.xi.1991), 1 male (20.xii.1991), Wang Jum Pee, Pakchong, Nakhonratsima, coll. Montree, attracted to cue lure; 2 males (28.vi.1991), 1 male

(30.iii.1991), Khao Yai, Moosi, Pakchong, Nakhonratsima, coll. Montree, attracted to cue lure; 8 males (14.vi.1991), 3 males (21.vi.1991), Nr. Bangkok, Khao Yai, Nakhonnayok, coll. Montree, attracted to cue lure; 1 male (27.ix.1991), 1 male (18.x.1991), 1 male (25.x.1991), 1 male (8.xi. 1991), 1 male (15.xi.1991), 2 males (29.xi.1991), 1 male (6.iii.1992), 1 male (13.iii.1992), 2 males (29.v.1992), Khao Kheaw, Pakphli, Nakhonnayok, coll. Montree, attracted to cue lure; 3 males (25.vi.1991), 1 male (11.xi.1991), 1 male (25.xi.1991), 3 males (24.xii.1991), Yang Pong Rat, Chanthaburi, coll. Montree, attracted to cue lure; 1 male (30.x.1991), 2 males (6.xi.1991), 2 males (27.xi.1991), 2 males (4.xii.1991), 1 male (11.xii.1991), 2 males (18.xii. 1991), 3 males (22.vii.1992), 4 males (26.vi.1991), Pha Yao, Chiang Rai, coll. Charuwan, attracted to cue lure; 6 males (19.vi.1991), 3 males (26.vi.1991), 1 male (24.x.1991), 1 male (13.xi.1991), 1 male (4.xii.1991), Phan, Chiang Rai, coll. Charuwan, attracted to cue lure; 1 male (9.i.1992), 1 male (5.iii.1992), 2 males (28.v.1992), 1 male (25.vi.1992), 3 males (9.vii.1992), 1 male (23.vii.1992), Khun Khon, Chiang Rai, coll. Charuwan, attracted to cue lure; 11 males, Mae Sai, Chiang Rai, 10.vi.1991, coll. Charuwan, attracted to cue lure; 3 males (10.vi.1991), 1 male (15.vii.1991), Chiang Sang, Chiang Rai, coll. Charuwan, attracted to cue lure; 2 males, Mae Salong, Chiang Rai, 22.vii.1991, coll. Charuwan, attracted to cue lure; 3 females, Doi Luong N.P., Phayoa, Chiang Rai, 14.vi.1991, coll. Charuwan, bred from wild fruit Sample No. CR67; 4 males, N.R.D.C., Chiang Mai, 4.vi.1991, coll. Chantanee Hengsawad, attracted to cue lure; male (29.v.1991), 7 males (18.vi.1991), 1 male (2.i.1992), Doi Pui N.P., Chiang Mai, coll. Chantanee *et al.*, attracted to cue lure; 12 males (11.vi.1991), 4 males (18. vi.1991), 1 male (27.ix.1991), 1 male (20.vii.1992), N.R.D.C., Chiang Mai, coll. Chantanee *et al.*, attracted to cue lure; 5 males, Chiang Mai, 13.vi.1991, coll. Chantanee *et al.*, attracted to cue lure in guava orchard; 10 males, Doi Pui Nat. Park, Chiang Mai, 29.v.1991, coll. Chantanee *et al.*, attracted to cue lure; 2 males, Tum Rusee, Chiang Mai, 1.x.1992, coll. Chantanee, bred from *Xylosma brachystachys* Craib; 4 males and 3 females, Doi Pui, Chiang Mai, coll. Chantanee, bred from *Xylosma brachystachys* Craib. **VIETNAM: Northern Vietnam**: 1 male, Hanoi Province, Tu Liem District, Con Hue, 31.xii.1999, attracted to cue lure; 3 males, Son La Province, Moc Chau District, Long Luong Village, 15.iv.2002, coll. NIPP staff, attracted to cue lure; 2 males (26.ix.1999), 1 male (1.xi.1999), 3 males (26.vi.2000), 1 male (15.ix.2000), Son La Province, Moc Chau District, Co Do, attracted to cue lure; 2 males (12.vi.1999), 1 male (26.vi.1999), 6 males (9.viii.1999), 1 male (23.viii.1999), 1 male (6.ix.1999), 3 males (18.x.1999), Son La Province, Moc Chau District, Tra Long, attracted to cue lure; 1 male, Son La Province, Moc Chau District, Moc Chau House Garden, 6.ix.1999, attracted to cue lure; 2 males, Lao Cai Province, Sapa 6 Village, 9.vi.1999, attracted to cue lure; 1 male, Lao Cai Province, Sapa Oquy Ho, 30.vii.1999, attracted to cue lure; **Southern Vietnam**: 1 male, Can Tho Province, Binh Thuy Village, 10.xii.1999, attracted to cue lure; 1 male, Lam Dong Province, 3 District, Da Lat Town, 23.v.2002, coll. SOFRI staff, attracted to cue lure; 1 male, Dac Lac Province, Ku Mi Nga District, Quang Phu Village, 24.vi. 2002. 50 paratypes in ANIC, 30 paratypes in BMNH, 506 paratypes in QDPC, 70 paratypes in QMIC.

Diagnosis: A medium-sized species; face fulvous with a pair of moderately large oval black spots; postpronotal lobes and notopleura yellow; scutum red-brown with a lanceolate dark pattern which varies in intensity from fuscous through to dark fuscous and black; lateral postsutural yellow vittae present; medial postsutural yellow vitta absent; no yellow spot anterior to notopleural suture; mesopleural stripe reaching midway between anterior margin of notopleuron and anterior *npl.* seta dorsally; scutellum yellow; wing with cells bc and c colourless, microtrichia in outer corner of cell c only; a narrow dark fuscous costal band; a narrow dark fuscous anal streak; abdominal terga III–V orange-brown with a 'T' pattern consisting of a broad medial longitudinal black band over all three terga and a narrow transverse black band across anterior margin of tergum III which widens to cover lateral margins of tergum where it becomes paler, narrow dark fuscous to black anterolateral corners on terga IV and V.

Description: male.

Head: Height 1.3 mm. Frons – length 1.4 times breadth; fulvous with pale fuscous around anteromedial hump and continuing around upper pair of *i.or.* setae; orbital setae black: 1 *s.or.*, 2 *i.or.*; lunule fuscous. Ocellar triangle black. Vertex fuscous. Face fulvous with a pair of moderately large oval black spots; length 0.4 mm. Genae fulvous, large dark fuscous subocular spot present; black seta present. Occiput fuscous, fulvous along eye margins; occipital row with 4–6 black setae. Antennae with segments 1 and 2 red-brown, segment 3 red-brown with dark fuscous on apex and outer surface; length of segments: 0.18 mm; 0.28 mm; 0.72 mm.

Thorax: Scutum red-brown with a dark lanceolate pattern which varies in colour intensity from fuscous through to dark fuscous and black, dark fuscous between postpronotal lobes and notopleura on lateral margins, a black spot inside postpronotal lobes (the lanceolate pattern is similar to that in *Bactrocera cacuminata* (Hering) (see Drew, 1989, Fig. 91) and *Bactrocera rubigina* (Wang)) (the dark patterns on the scutum vary in size as well as intensity of colour). Pleural areas dark fuscous to black except red-brown below postpronotal lobes and above hind coxae. Yellow markings as follows: postpronotal lobes; notopleura; mesopleural stripe reaching midway between anterior margin of notopleuron and anterior *npl.* seta dorsally, continuing to katepisternum as a transverse spot, anterior margin convex; anatergite (posterior apex dark fuscous to black); anterior $^{3}/_{5}$ katatergite (remainder dark fuscous to black); two moderately broad parallel-sided lateral postsutural vittae ending at or just behind *ia.* seta. Medial longitudinal postsutural yellow vitta absent. Postnotum red-brown with black lateral margins. Scutellum yellow except for narrow black basal band. Setae: *sc.* 2; *prsc.* 2; *ia.* 1; *p.sa.* 1; *a.sa.* 1; *mpl.* 1; *npl.* 2; *scp.* 4.

Legs: All femora entirely fulvous; fore tibiae dark fuscous basally to fuscous apically, mid tibiae dark fuscous basally to fulvous apically, hind tibiae dark fuscous on basal ¾ to fuscous on apical ¼; all tarsi with basal segment fulvous and apical segments red-brown; mid tibiae each with an apical black spur.

Wings: Length 5.1 mm; cells bc and c colourless; microtrichia in outer corner of cell c only; remainder of wings colourless except dark fuscous cell sc, narrow dark fuscous costal band just overlapping R_{2+3} where it is a pale tint and remaining narrow around margin of wing to end just beyond extremity of R_{4+5} (the costal band may have a very slight swelling around apex of R_{4+5}), a narrow dark fuscous anal streak; a dense aggregation of microtrichia around A_1+CuA_2; supernumerary lobe of medium development.

Abdomen: Oval; terga free; pecten present on tergum III. Tergum I and sterna I and II wider than long. Tergum I fuscous centrally, dark fuscous laterally and with a narrow transverse orange-brown band across posterior margin but not reaching lateral margins; tergum II orange-brown with fulvous posterolaterally, a narrow transverse subanterior dark fuscous to black band but not reaching narrow dark fuscous anterolateral corners; terga III–V orange-brown with a black 'T' pattern consisting of a broad medial longitudinal band over all three terga and a narrow transverse band across anterior margin of tergum III which expands to cover lateral margins where it becomes paler, narrow dark fuscous to black anterolateral corners on terga IV and V. A pair of oval orange-brown shining spots on tergum V. Posterior lobe of surstylus short, sternum V with a deep concavity on posterior margin. Abdominal sterna dark fuscous to black.

Attractant: Cue lure.

Female: As for male except no dense aggregation of microtrichia around A_1+CuA_2, supernumerary lobe weak, pecten of cilia on abdominal tergum III absent; ovipositor basal segment red-brown, dorsoventrally compressed and tapering posteriorly in dorsal view; ratio of length of oviscape to length of tergum V, 0.8:1; apex of piercer needle-shaped.

Distribution: Bhutan; India (Andaman Is); Vietnam; Central to Northern Thailand.

Hosts: *Xylosma brachystachys* (family *Flacourtiaceae*).

Remarks: *Bactrocera (Bactrocera) bhutaniae* n.sp. is a general red-brown species similar to *Bactrocera (Bactrocera) bivittata* Lin & Wang, *Bactrocera (Bactrocera) invadens* Drew, Tsuruta & White and *Bactrocera (Bactrocera) rubigina* (Wang & Zhao) in possessing a red-brown scutum with variable dark colour patterns, lateral postsutural yellow vittae, wings with a narrow fuscous costal band and anal streak, femora entirely fulvous and abdominal terga III–V with a dark 'T' pattern. It differs from *B. bivittata* in having broad lateral postsutural vittae, the costal band slightly overlapping R_{2+3} and males attracted to cue lure, from *B. invadens* in having two broad longitudinal red-brown bands on the scutum either side of the mid line and running from approximately the centre to the anterior margin, broad lateral postsutural vittae, the costal band slightly overlapping R_{2+3} and males attracted to cue lure and from *B. rubigina* in having extensive dark colour patterns on the scutum, broad parallel-sided lateral postsutural vittae, the costal band slightly overlapping R_{2+3} and abdominal terga III–V without lateral dark colour patterns.

Bactrocera (Bactrocera) bifasciata (Hardy)

(Fig. 34)

Dacus (Bactrocera) bifasciatus Hardy, 1982: 219. Holotype male in MZB.
Bactrocera (Bactrocera) bifasciata – Norrbom *et al.*, 1998: 88.

Type material examined: Holotype male, **INDONESIA: SULAWESI** – Sadaunta, 65 km S.E. Palu, Central Sulawesi, 650 m, Aug. 1975, coll. D.E. Hardy. Studied at MZB.
Paratypes: **INDONESIA: SULAWESI** – 1 male, 'Kuala Navusa, Malakosa, Sulawesi, Indonesia, Aug.–Nov. 1975, 100′, Margareta Becker'; blue label 'Paratype male, Dacus (Bactrocera) bifasciatus, D.E. Hardy'; white label 'Dacus (Bactrocera) bifasciatus Hardy'. 1 male, 'Tolai, Sungai Tolewonu, C. Sulawesi, Indonesia, Feb. 11–26, 1976, 450′, Margareta Becker'; blue label 'Paratype male, Dacus (Bactrocera) bifasciatus, D.E. Hardy'. Specimens in QDPC.

Non-type material examined: INDONESIA: NORTH MOLUCCAS (NORTH MALUKU) – 1 male, Moluccas (North Maluku), Kodya Ternate, Foramadiahi, 3.vi.2007, coll. La Ruti, attracted to cue lure in forest; **CENTRAL SULAWESI** – 1 male, Sulawesi, Sulawesi Tengah, Donggala, Nupabomba, 2.ix.2006, coll. Rusdin, attracted to cue lure in forest. Specimens in QPDC.

Diagnosis: A medium-sized species; face fulvous with a pair of medium-sized sub-quadrate black spots; postpronotal lobes and notopleura yellow; scutum entirely dark fuscous to black without pale markings; lateral postsutural yellow vittae present; medial postsutural yellow vitta absent; no yellow spot anterior to notopleural suture; mesopleural stripe reaching to postpronotal lobe dorsally; scutellum yellow except for narrow black basal band; wing with cell c pale fuscous, cell bc colourless, microtrichia covering most of cell c; a broad fuscous costal band confluent with R_{4+5} and two transverse fuscous bands across wing running from costal band to hind margin; a broad fuscous anal streak; abdominal terga III–V mostly dark fuscous with a broad medial longitudinal pale fuscous band running from centre of tergum III to posterior margin of tergum V.

Description: male.

Head: Frons – fulvous with fuscous around orbital setae; orbital setae black: 1 *s.or.*, 2 *i.or.*; lunule fuscous. Ocellar triangle black. Vertex fuscous. Face fulvous with a pair of medium-sized sub-quadrate black spots. Occiput red-brown, fulvous along eye margins. Antennae with segments 1 and 2 red-brown, segment 3 red-brown with fuscous on apex and outer surface.

Thorax: Scutum entirely dark fuscous to black without pale markings. Pleural areas black without pale markings. Yellow markings as follows: postpronotal lobes; notopleura; mesopleural stripe reaching postpronotal lobe dorsally, continuing to katepisternum as a small transverse spot, anterior margin straight; two broad parallel-sided lateral postsutural vittae ending at *ia.* seta. Medial longitudinal postsutural yellow vitta absent. Postnotum black. Scutellum yellow except for narrow black basal band. Setae: *sc.* 2; *prsc.* 2; *ia.* 1; *p.sa.* 1; *a.sa.* 1; *mpl.* 1; *npl.* 2; *scp.* 4.

Legs: All segments fulvous except apical two to three segments of tarsi darker red-brown; mid tibiae each with an apical black spur.
Wings: Cell c pale fuscous, cell bc colourless; microtrichia covering most of cell c; remainder of wings colourless except broad fuscous costal band confluent with R_{4+5} and ending between apices of R_{4+5} and M, a broad transverse fuscous band across centre of wing enclosing both crossveins, a narrow transverse fuscous band across wing between the broad central band and the apex of the wing, a broad fuscous anal streak; dense aggregation of microtrichia around A_1+CuA_2; supernumerary lobe of medium development.
Abdomen: Oval; terga free; pecten present on tergum III. Tergum I and sterna I and II wider than long. Terga I and II dark fuscous to black except fuscous posteromedially on tergum II; terga III–V dark fuscous except for a broad fuscous medial longitudinal band over posterior ½ of tergum III and all of terga IV and V. A pair of dark fuscous shining spots on tergum V. Posterior lobe of surstylus short, sternum V with a moderately deep concavity on posterior margin.
Attractant: Cue lure (see 'Non-type material examined' above). The Holotype was obviously attracted to a male lure (Hardy, 1982) but not therein recorded.
Female: No known record.
Distribution:
Type locality: Indonesia (Central Sulawesi, Sadaunta).
Previously recorded: Type locality and Indonesia (Sulawesi – Kuala Navusa and Tolai).
New records: Indonesia (North Moluccas = North Maluku – Ternate).
Hosts: No known record.
Remarks: *Bactrocera (Bactrocera) bifasciata* (Hardy) is similar to *Bactrocera (Bactrocera) umbrosa* (Fabricius) in possessing broad dark-coloured transverse bands across the wing but differs from *B. umbrosa* in having two transverse bands (not three) which are fuscous (not red-brown), cell bc colourless and cell c pale fuscous, in lacking a medial longitudinal black band on abdominal terga III–V and being attracted to cue lure, not methyl eugenol.

Bactrocera (Bactrocera) bimaculata Drew & Hancock

(Fig. 35)

Bactrocera (Bactrocera) bimaculata Drew & Hancock, 1994a: 10; Norrbom *et al.*, 1998: 88. Holotype male in BBM.

Non-type material examined: Since the publication of Drew & Hancock (1994a), the following specimens have been studied – **INDONESIA: FLORES** – 3 males, Borong, Flores, Indonesia, iii.1988, coll. J. Turner, attracted to cue lure; **JAVA** – 1 male, East Java, Lumajang, Kadang Tepus, 19.vii.2005, coll. Utomo, attracted to cue lure in banana. **VIETNAM: Northern Vietnam** – 1 male, Son La Province, Moc Chau District, Long Luong Village, Site VN01, 15.v.2002, coll. NIPP Staff, attracted to cue lure. Specimens in QDPC.

Diagnosis: Face fulvous with a pair of medium-sized oval black spots; postpronotal lobes and notopleura yellow; scutum black with dark brown below lateral postsutural vittae, around notopleural suture, between postpronotal lobe and notopleuron and inside postpronotal lobes; broad parallel-sided lateral postsutural yellow vittae ending behind *ia.* seta; medial postsutural yellow vitta absent; mesopleural stripe reaching midway between anterior margin of notopleuron and anterior *npl.* seta dorsally; scutellum yellow; legs with femora fulvous with small fuscous to dark fuscous preapical spots on outer surfaces of fore and inner surfaces of mid and hind femora; tibiae pale fuscous to fuscous; wing with cells bc and c colourless, microtrichia in outer corner of cell c only; a narrow dark fuscous costal band just overlapping R_{2+3} where it is very pale and expanding slightly across apex of R_{4+5}; a narrow fuscous anal streak; supernumerary lobe of medium development; abdominal terga III–V orange-brown with a narrow transverse black band across anterior margin of tergum III and expanding over lateral margins, anterolateral corners of terga IV and V dark fuscous, a medium-width medial longitudinal black band over all three terga, a pair of oval dark fuscous shining spots on tergum V; abdominal sterna dark coloured.
Attractant: Cue lure.
Female: No known record.
Distribution:
Type locality: Indonesia (Lombok I., Sasaot).
New records: Indonesia (East Java; Flores); Northern Vietnam (Moc Chau District).
Hosts: No known record.
Remarks: *Bactrocera (Bactrocera) bimaculata* Drew & Hancock, within the *dorsalis* complex, is similar to some specimens of *Bactrocera (Bactrocera) fuscitibia* Drew & Hancock and *Bactrocera (Bactrocera) dongnaiae* n.sp. in possessing broad parallel-sided or subparallel lateral postsutural vittae, costal band overlapping R_{2+3}, abdominal terga III–V with narrow- to medium-width dark lateral margins, shining spots on abdominal tergum V dark fuscous to black in addition to the general characteristics of the *dorsalis* complex. It differs from these species in having dark preapical spots on or markings around apices of all femora and in being a large species (wing length 6.7 mm).

Bactrocera (Bactrocera) binhduongiae Drew & Romig, new species

(Fig. 36)

Material examined: Holotype male, **VIETNAM: Southern Vietnam** – Dac Lac Province, Ku Mi Nga District, Quang Phu Village, 8.vi.2002, attracted to methyl eugenol, deposited in BMNH.
Paratypes: **VIETNAM: Southern Vietnam** – 3 males (8.v.2002), 6 males (8.vi.2002), 2 males (24.vii.2002), same locality data as holotype, coll. SOFRI staff, attracted to methyl eugenol; 1 male (4.x.2002), 1 male (15.xii.2002), Binh Duong Province, Thuan An District, An Thanh Village, attracted to methyl eugenol; 4 males (20.vii.2002),

6 males (28.ix.2002), Dong Nai Province, Long Khanh District, Nhan Nghia Village, attracted to methyl eugenol; 1 male, Lam Dong Province, 3 district, Da Lat town, 8.vii. 2002, attracted to methyl eugenol; 3 males (6.vii.2002), 1 male (20.vii.2002), 2 males (3.viii.2002), 3 males (8.ix. 2002), Khanh Hoa Province, Dien Khanh District, Suoi Tan Village, attracted to methyl eugenol. 6 paratypes in ANIC; 6 paratypes in BMNH; 15 paratypes in QDPC; 6 paratypes in QMIC.

Diagnosis: A small species; face fulvous with a pair of medium-sized oval black spots; postpronotal lobes and notopleura yellow; scutum black; lateral postsutural yellow vittae present; medial postsutural yellow vitta absent; no yellow spot anterior to notopleural suture; mesopleural stripe slightly wider than notopleuron dorsally; scutellum yellow; wing with cells bc and c colourless, microtrichia in outer corner of cell c only; a narrow dark fuscous costal band overlapping R_{2+3} for its entire length; a narrow fuscous anal streak; abdominal terga III–V orange-brown with a broad medial and two broad lateral longitudinal dark fuscous to black bands joined along anterior margin of tergum III.

Description: male.

Head: Height 1.29 mm. Frons – length 1.46 times breadth; fulvous with fuscous around bases of orbital setae and on anteromedial hump; orbital setae black: 1 *s.or.*, 2 *i.or.*; lunule dark fuscous. Ocellar triangle black. Vertex fuscous. Face fulvous with a pair of medium-sized oval black spots; length 0.39 mm. Genae fulvous, small dark fuscous subocular spot present; black seta present. Occiput black, fulvous along eye margins; occipital row with 2–6 small black setae. Antennae with segments 1 and 2 red-brown, segment 3 red-brown with dark fuscous on apex and outer surface; length of segments: 0.16 mm; 0.3 mm; 0.72 mm.

Thorax: Scutum black with dark red-brown below and behind lateral postsutural vittae. Pleural areas black with red-brown below postpronotal lobes. Yellow markings as follows: postpronotal lobes; notopleura; mesopleural stripe slightly wider than notopleuron dorsally, continuing to katepisternum as a small transverse spot, anterior margin straight; anatergite (posterior apex black); anterior ⅔ katatergite (remainder black); two moderately broad parallel-sided lateral postsutural vittae ending at *ia.* seta. Medial longitudinal postsutural yellow vitta absent. Postnotum black. Scutellum yellow except for narrow black basal band. Setae: *sc.* 2; *prsc.* 2; *ia.* 1; *p.sa.* 1; *a.sa.* 1; *mpl.* 1; *npl.* 2; *scp.* 4.

Legs: Fore femora fulvous with a large dark fuscous spot on outer apical surfaces, mid and hind femora entirely fulvous; fore tibiae dark fuscous, mid tibiae fuscous basally to fulvous apically, hind tibiae dark fuscous; tarsi with all segments fulvous; mid tibiae each with an apical black spur.

Wings: Length 4.8 mm; cells bc and c colourless; microtrichia in outer corner of cell c only; remainder of wings colourless except dark fuscous cell sc, narrow dark fuscous costal band overlapping R_{2+3} and remaining of uniform width to end beyond extremity of R_{4+5}, a narrow fuscous anal streak; a dense aggregation of microtrichia around A_1+CuA_2; supernumerary lobe of medium development.

Abdomen: Oval; terga free; pecten present on tergum III. Tergum I and sterna I and II wider than long. Tergum I dark fuscous with a narrow transverse orange-brown band along posterior margin but not reaching lateral margins; tergum II fulvous with a narrow transverse subanterior black band but not reaching narrow dark fuscous anterolateral corners; terga III–V orange-brown with a broad medial longitudinal black band and broad lateral longitudinal dark fuscous to black bands over all three terga and joined along anterior margin of tergum III. A pair of oval orange-brown shining spots on tergum V. Posterior lobe of surstylus short, sternum V with a deep concavity on posterior margin.

Attractant: Methyl eugenol.

Female: No known record.

Distribution: Southern Vietnam.

Hosts: No known record.

Remarks: *Bactrocera (Bactrocera) binhduongiae* n.sp., within the *dorsalis* complex, is similar to *Bactrocera (Bactrocera) kanchanaburi* Drew & Hancock in general body characters but differs in having the costal band distinctly overlapping R_{2+3} and a broader medial longitudinal black band over terga III–V.

Bactrocera (Bactrocera) bitungiae Drew & Romig, new species

(Fig. 37)

Material examined: Holotype male, **INDONESIA: SULAWESI** – North Sulawesi, Bitung, Danowudu, 14.viii.2006, coll. Karel F. Lala, attracted to cue lure in forest, deposited in BMNH.

Paratypes: **INDONESIA: SULAWESI** – 4 males, same data as holotype; 14 males, same locality data as holotype, 26.ix.2006, coll. Karel F. Lala, attracted to cue lure in forest. 3 paratypes in ANIC, 3 paratypes in BMNH, 9 paratypes in QDPC, 3 paratypes in QMIC.

Diagnosis: A small species; face fulvous with a pair of large oval black spots; postpronotal lobes and notopleura yellow; scutum black; lateral postsutural yellow vittae present; medial postsutural yellow vitta absent; no yellow spot anterior to notopleural suture; mesopleural stripe reaching midway between anterior margin of notopleuron and anterior *npl.* seta dorsally; scutellum yellow; wing with cells bc and c with a pale tint, microtrichia in outer corner of cell c only; a narrow very dark fuscous costal band overlapping R_{2+3}; a broad very dark fuscous anal streak; abdominal terga III–V orange-brown with a 'T' pattern consisting of a broad transverse black band across anterior margin of tergum III and widening over lateral margins, a broad medial longitudinal black band over all three terga, moderately wide dark fuscous anterolateral corners on terga IV and V.

Description: male.

Head: Height 1.47 mm. Frons – length 1.75 times breadth; red-brown with fuscous around orbital setae and on anteromedial hump; orbital setae black: 1 *s.or.*, 2 *i.or.*; lunule fuscous. Ocellar triangle black. Vertex dark fuscous. Face fulvous with a pair of large elongate oval black spots; length 0.49 mm. Genae fulvous, large dark fuscous subocular spot present; black seta present. Occiput black, dark fuscous dorsocentrally, fulvous along eye margins; occipital row with 4–6 strong black setae. Antennae with segments 1 and 2 red-brown, segment 3 red-brown with dark fuscous on apex and outer surface; length of segments: 0.18 mm; 0.32 mm; 0.86 mm.

Thorax: Scutum black except for a small area of dark fuscous anterior to notopleural suture. Pleural areas entirely black. Yellow markings as follows: postpronotal lobes; notopleura; mesopleural stripe reaching midway between anterior margin of notopleuron and anterior *npl.* seta dorsally, continuing to katepisternum as a small transverse spot, anterior margin straight; anatergite (posterior apex black); anterior $^3/_5$ katatergite (remainder black); two broad parallel-sided to sub-parallel lateral postsutural vittae ending at or just before *ia.* seta. Medial longitudinal postsutural yellow vitta absent. Postnotum black. Scutellum yellow except for narrow black basal band. Setae: *sc.* 2; *prsc.* 2; *ia.* 1; *p.sa.* 1; *a.sa.* 1; *mpl.* 1; *npl.* 2; *scp.* 4.

Legs: Fore femora fulvous with a moderately large dark fuscous subapical spot on outer surfaces, mid and hind femora entirely fulvous; fore tibiae dark fuscous, mid tibiae dark fuscous basally to fuscous apically, hind tibiae dark fuscous to black; fore tarsi with basal segment fulvous and apical four segments fuscous; mid and hind tarsi with basal two segments fulvous and apical three segments red-brown; mid tibiae each with an apical black spur.

Wings: Length 4.9 mm; cells bc and c with a pale tint; microtrichia in outer corner of cell c only; remainder of wings colourless except very dark fuscous cell sc, narrow very dark fuscous costal band overlapping R_{2+3} and widening slightly around apex of wing, a broad very dark fuscous anal streak; a dense aggregation of microtrichia around A_1+CuA_2; supernumerary lobe of medium development.

Abdomen: Oval; terga free; pecten present on tergum III. Tergum I and sterna I and II wider than long. Tergum I black with a narrow transverse orange-brown band across posterior margin but not reaching lateral margins; tergum II orange-brown anteriorly, fulvous posteriorly and with a narrow transverse subanterior black band which reaches black anterolateral corners; terga III–V orange-brown with a black 'T' pattern consisting of a broad transverse band across anterior margin of tergum III which expands across lateral margins and a broad medial longitudinal black band over all three terga, moderately wide dark fuscous anterolateral corners on terga IV and V. A pair of oval orange-brown to fuscous shining spots on tergum V. Posterior lobe of surstylus short, sternum V with a deep concavity on posterior margin. Abdominal sterna dark.

Attractant: Cue lure.

Female: No known record.

Distribution: Indonesia (Sulawesi – North Sulawesi, Bitung).

Hosts: No known record.

Remarks: *Bactrocera (Bactrocera) bitungiae* n.sp., within the *dorsalis* complex, is similar to some specimens of *Bactrocera fuscitibia* Drew & Hancock, *Bactrocera laithieuiae* n.sp. and some specimens of *Bactrocera occipitalis* (Bezzi) in the general colour patterns of the body and wings. It differs from *B. laithieuiae* in having broadly oval facial spots, from *B. fuscitibia* in having a broad tomentum-free bare area on the anterocentral region of the scutum and wider anterolateral dark margins on abdominal terga IV and V and from *B. occipitalis* in having a broad anal streak, the scutum entirely black, dark preapical spots on the outer surfaces of the fore femora and the wing markings very dark fuscous.

Bactrocera (Bactrocera) bivittata Lin & Wang

(Fig. 38)

Bactrocera (Bactrocera) bivittata Lin & Wang, in Lin *et al.*, 2005: 843. Holotype male in HEIQ.

Material examined: No specimens available for study. The following description has been taken from literature.

Diagnosis: Face fulvous with a pair of oval black spots; postpronotal lobes and notopleura yellow; scutum red-brown with black markings; broad parallel-sided lateral postsutural yellow vittae ending at *ia.* seta; medial postsutural yellow vitta absent; mesopleural stripe reaching midway between anterior margin of notopleuron and anterior *npl.* seta dorsally; scutellum yellow; wing with cells bc and c colourless; a narrow fuscous costal band confluent with R_{2+3} and widening across apex of wing from apex of R_{2+3}; a narrow fuscous anal streak; supernumerary lobe of medium development; abdominal terga III–V red-brown with a black 'T' pattern consisting of a narrow transverse band across anterior margin of tergum III which expands to cover lateral margins and a broad medial longitudinal band over all three terga, a pair of oval fuscous shining spots on tergum V.

Description: male.

Head: Frons – red-brown with dark fuscous on anteromedial hump; orbital setae black: 1 *s.or.*, 2 *i.or.*; lunule red-brown. Ocellar triangle black. Vertex fuscous. Face fulvous with a pair of oval black spots. Genae red-brown. Occiput dark fuscous, fulvous along eye margins. Antennae with segments 1 and 2 red-brown, segment 3 red-brown with fuscous on apex and outer surface.

Thorax: Scutum red-brown with a central black pattern consisting of black across posterior area and from this area three narrow longitudinal black lines extending to anterior margin. Pleural areas red-brown to pale fuscous. Yellow markings as follows: postpronotal lobes; notopleura; mesopleural stripe reaching midway between anterior margin of notopleuron and anterior *npl.* seta dorsally, continuing to katepisternum as a transverse spot; anatergite (posterior apex black); anterior $^2/_3$ katatergite (remainder black); two moderately broad parallel-sided lateral postsutural vittae ending at *ia.* seta. Medial longitudinal postsutural yellow vitta absent. Postnotum black laterally, red-brown centrally. Scutellum yellow except for narrow red-brown basal band. Setae: *sc.* 2; *prsc.* 2; *ia.* 1; *p.sa.* 1; *a.sa.* 1; *mpl.* 1; *npl.* 2; *scp.* 4.

Legs: All segments entirely fulvous except hind tibiae fuscous; mid tibiae each with an apical black spur.

Wings: Cells bc and c colourless; microtrichia in outer corner of cell c only; remainder of wings colourless except dark fuscous cell sc, narrow dark fuscous costal band confluent with R_{2+3} and widening slightly across apex of wing, a narrow dark fuscous anal streak; a dense aggregation of microtrichia around A_1+CuA_2; supernumerary lobe of medium development.

Abdomen: Oval; terga free; pecten present on tergum III. Tergum I and sterna I and II wider than long. Tergum I black with a narrow transverse red-brown band across posterior margin but not reaching lateral margins; tergum II fulvous with narrow black anterolateral corners; terga III–V red-brown with a dark 'T' pattern consisting of a narrow transverse black band across anterior margin of tergum III which widens along lateral margins and a medium-width medial longitudinal black band over all three terga. A pair of oval fuscous shining spots on tergum V. Posterior lobe of surstylus short, sternum V with a deep concavity on posterior margin.

Attractant: Methyl eugenol.

Female: No known record.

Distribution:

Type locality: China (Hainan I.).

Hosts: No known record.

Remarks: In the original description, in Lin *et al.* (2005) record 4 *sc.* setae present under the thorax description and 2 *sc.* in the remarks. It appears that 2 *sc.* setae are correct and that this species belongs to the *Bactrocera (Bactrocera).*

Bactrocera (Bactrocera) bivittata Lin & Wang is similar to *Bactrocera (Bactrocera) affinidorsalis* (Hardy), *Bactrocera (Bactrocera) latilineola* Drew & Hancock and *Bactrocera (Bactrocera) malaysiensis* Drew & Hancock in possessing broad parallel-sided lateral postsutural vittae and the costal band confluent with R_{2+3} and expanding across the apex of the wing. It differs from these species in having all femora entirely fulvous, without dark subapical or apical spots.

Bactrocera bivittata is also similar to *Bactrocera (Bactrocera) invadens* Drew, Tsuruta & White and *Bactrocera (Bactrocera) rubigina* (Wang & Zhao) in possessing a red-brown scutum, costal band confluent with R_{2+3}, cells bc and c colourless, lateral postsutural vittae reaching the *ia.* setae and abdominal terga III–V with a general black 'T' pattern. It differs from *B. rubigina* in being attracted to methyl eugenol and from *B. invadens* in having broader parallel-sided lateral postsutural vittae. Another species, *Bactrocera (Bactrocera) bhutaniae* n.sp. is similar in the overall body colour patterns but is attracted to cue lure. See 'Remarks' under *B. bhutaniae* for a full comparison of these species.

Bactrocera (Bactrocera) blairiae Drew & Romig, new species

(Fig. 39)

Material examined: Holotype male, **INDIA: ANDAMAN ISLANDS** – S. Andaman, Port Blair, 29.vi.1992, coll. Ranganath, Sl No. 8, attracted to methyl eugenol, deposited in BMNH.

Paratypes: **INDIA: ANDAMAN ISLANDS** – 4 males, same data as holotype; 2 males (15.ii.1996), 2 males (22.ii.1996), L. Andaman, coll. Ranganath, Sl No. 54, attracted to methyl eugenol; 2 males, S. Andaman, Havelock I., 25.vi.1998, coll. Ranganath, Sl No. 65, attracted to methyl eugenol. 2 paratypes in ANIC, 1 paratype in BMNH, 4 paratypes in QDPC, 3 paratypes in QMIC.

Diagnosis: A medium-sized species; face fulvous with a pair of medium-sized oval black spots; postpronotal lobes and notopleura yellow; scutum entirely red-brown without dark markings; lateral postsutural yellow vittae present; medial postsutural yellow vitta absent; no yellow spot anterior to notopleural suture; mesopleural stripe reaching midway between anterior margin of notopleuron and anterior *npl.* seta dorsally; scutellum yellow; wing with cells bc and c with a pale fuscous tint, microtrichia in outer corner of cell c only; a narrow dark fuscous costal band confluent with R_{2+3}; a narrow dark fuscous anal streak; abdominal terga III–V orange-brown with a dark 'T' pattern consisting of a narrow transverse black band across anterior margin of tergum III which expands to cover lateral margins, a moderately broad medial longitudinal black band over all three terga, narrow dark fuscous anterolateral corners on terga IV and V.

Description: male.

Head: Height 1.6 mm. Frons – length 1.72 times breadth; red-brown without dark markings; orbital setae dark red-brown: 1 *s.or.*, 2 *i.or.*; lunule red-brown. Ocellar triangle black. Vertex red-brown. Face fulvous with a pair of medium-sized oval black spots; length 0.49 mm. Genae fulvous, large fuscous subocular spot present; red-brown seta present. Occiput red-brown, fulvous along eye margins; occipital row with 4–7 dark setae. Antennae with segments 1 and 2 red-brown, segment 3 red-brown with fuscous on apex and outer surface; length of segments: 0.15 mm; 0.3 mm; 0.8 mm.

Thorax: Scutum entirely red-brown without dark markings. Pleural areas red-brown with small areas of pale fuscous on katepisternum. Yellow markings as follows: postpronotal lobes; notopleura; mesopleural stripe reaching midway between anterior margin of notopleuron and anterior *npl.* seta dorsally, continuing to katepisternum as a large transverse spot, anterior margin slightly convex; anatergite (posterior apex red-brown); anterior ⅔ katatergite (remainder red-brown); two medium-width parallel-sided lateral postsutural vittae ending at *ia.* seta. Medial longitudinal postsutural yellow vitta absent. Postnotum red-brown centrally, dark fuscous laterally. Scutellum yellow except for narrow red-brown basal band. Setae: *sc.* 2; *prsc.* 2; *ia.* 1; *p.sa.* 1; *a.sa.* 1; *mpl.* 1; *npl.* 2; *scp.* 4.

Legs: All femora entirely fulvous; fore and mid tibiae dark fuscous basally to fulvous apically, hind tibiae dark fuscous tending fuscous around apex; all tarsal segments fulvous except apical four segments of fore tarsi fuscous; mid tibiae each with an apical black spur.

Wings: Length 5.5 mm; cells bc and c with a pale fuscous tint; microtrichia in outer corner of cell c only; remainder of wings colourless except dark fuscous cell sc, narrow dark fuscous costal band confluent with R_{2+3} and widening slightly across apex of R_{4+5}, a narrow dark fuscous anal streak; a dense aggregation of microtrichia around A_1+CuA_2; supernumerary lobe of medium development.

Abdomen: Oval; terga free; pecten present on tergum III. Tergum I and sterna I and II wider than long. Tergum I orange-brown with narrow dark fuscous lateral margins;

tergum II orange-brown with a narrow transverse black band subanteriorly and narrow lateral black margins; terga III–V orange-brown with a dark 'T' pattern consisting of a narrow transverse black band across anterior margin of tergum III which expands to cover lateral margins and a moderately broad medial longitudinal black band over all three terga, narrow dark fuscous anterolateral corners on terga IV and V. A pair of oval red-brown shining spots on tergum V. Posterior lobe of surstylus short, sternum V with a deep concavity on posterior margin. Abdominal sterna pale orange-brown.
Attractant: Methyl eugenol.
Female: No known record.
Distribution: India (Andaman Is).
Hosts: No known record.
Remarks: *Bactrocera (Bactrocera) blairiae* n.sp. belongs to a complex of red-brown species in South-East Asia and is similar to *Bactrocera (Bactrocera) commensurata* n.sp., *Bactrocera (Bactrocera) invadens* Drew, Tsuruta & White and *Bactrocera (Bactrocera) rubigina* (Wang & Zhao) in possessing a red-brown scutum, costal band confluent with R_{2+3}, cells bc and c colourless (or with a pale tint), lateral postsutural vittae ending at or just behind *ia.* seta and mesopleural stripe extending, at most, midway between anterior margin of notopleuron and anterior *npl.* seta dorsally. It differs from *B. invadens* and *B. rubigina* in having pale-coloured abdominal sterna and from *B. commensurata* in having the costal band widening across the wing apex and in lacking dark patterns on the scutum.

Bactrocera (Bactrocera) bruneiae Drew & Romig, new species

(Fig. 40)

Material examined: Holotype male, **BRUNEI**: Brunei Darassalam, KBFC, Bato Apou Nat. Pk, Forest, 29.i.2002, coll. K. Hurley, attracted to methyl eugenol, deposited in BMNH.
Paratypes: **BRUNEI**: 6 males, same data as holotype; 1 male (23.i.2002), 4 males (25.i.2002), 2 males (27.i.2002), Brunei Darassalam, KBFC, Bato Apou Nat. Pk, Forest, coll. K. Hurley, attracted to methyl eugenol; 1 male (23.i.2002), 1 male (24.i.2002), Brunei Darassalam, KBFC, Bato Apou Nat. Pk, Riparian, coll. K. Hurley, attracted to methyl eugenol; 1 male (23.i.2002), 4 males (25.i.2002), 3 males (27.i.2002), 3 males (29.i.2002), Brunei Darassalam, KBFC, Bato Apou Nat. Pk, Upland, coll. K. Hurley, attracted to methyl eugenol. **PENINSULAR MALAYSIA**: 1 male, TS Mentakab, Malaysia, 25.iii.1986, coll. Bahari, attracted to methyl eugenol; 4 males, TS Jalan Kebun, Malaysia, 28.i.1986, coll. Bahari, attracted to methyl eugenol; 2 males, TS15, Malaysia, 28.vii.1986, coll. Bahari, attracted to methyl eugenol; 1 male, TS18, Malaysia, 14.vii.1986, coll. Bahari, attracted to methyl eugenol; 5 males, TS15 Temerloh, Malaysia, 25.viii.1986, coll. Bahari, attracted to methyl eugenol; 1 male, TS Maran, Malaysia, 22.iv.1986, coll. Anthony, attracted to methyl eugenol; 2 males, TS 5 and 7, Malaysia, 11.viii.1986, coll. Anthony, attracted to methyl eugenol; 1 male, TS Temerloh, Malaysia, 16.vi.1986, coll. Vijay, attracted to methyl eugenol; 1 male, RS Kluang, Malaysia, 29.iv.1986, coll. Jasni Zamman, attracted to methyl eugenol; 1 male, RS Jalan Bekun, Malaysia, 1.iv.1986, coll. Abdul Khalib, attracted to methyl eugenol; 1 male, Malaysia, Cameron Highlands, 5500 ft., Trap 6, 22.v.1983, coll. R. Drew, attracted to methyl eugenol. 10 paratypes in ANIC, 10 paratypes in BMNH, 16 paratypes in QDPC, 10 paratypes in QMIC.

Non-type material examined: INDONESIA: KALIMANTAN – 1 male with the following label data, 'Kalimantan, Kerang, Bengkirai'; '2°19′62″S, 113°54′53″E, 13.viii.1996'; 'methyl eugenol trap, mixed swamp forest, Nottingham Univ. Exp. 1996, IIE23602'; 'Bactrocera ritsemai (WY), det. I.M. White, 1997'; 'shown to DLH, ix. 98 – agree to assume as *B. ritsemai* (IMW 29.ix.1998)'; 'BMNH (E) No. 533126'. Specimen studied at BMNH.
Diagnosis: A medium-sized species; face fulvous with a pair of large elongate oval black spots filling most of the antennal furrow; postpronotal lobes and notopleura yellow; scutum black; lateral postsutural yellow vittae present; medial postsutural yellow vitta absent; no yellow spot anterior to notopleural suture; mesopleural stripe reaching midway between anterior margin of notopleuron and anterior *npl.* seta dorsally; scutellum yellow; wing with cells bc and c with a pale tint, microtrichia in outer ⅓ of cell c only; a broad dark fuscous costal band; a moderately broad dark fuscous anal streak; abdominal terga III–V orange-brown with a 'T' pattern consisting of a narrow transverse black band across anterior margin of tergum III and a medium-width medial longitudinal black band over all three terga, narrow dark fuscous lateral margins on terga IV and V.
Description: male.
Head: Height 1.26 mm. Frons – length 1.4 times breadth; red-brown with dark fuscous on anteromedial hump and around bases of orbital setae; orbital setae black: 1 *s.or.*, 2 *i.or.*; lunule dark fuscous. Ocellar triangle black. Vertex dark fuscous. Face fulvous with a pair of large elongate oval black spots filling most of the antennal furrow; length 0.45 mm. Genae fulvous, large dark fuscous subocular spot present; black seta present. Occiput dark fuscous centrally, black laterally, fulvous along eye margins; occipital row with 4–5 strong black setae. Antennae with segments 1 and 2 red-brown, segment 3 red-brown with dark fuscous on apex and outer surface; length of segments: 0.15 mm; 0.24 mm; 0.66 mm.
Thorax: Scutum black except for dark fuscous along anterior margin of notopleural suture and inside postpronotal lobes. Pleural areas entirely black, tending very dark fuscous below postpronotal lobes. Yellow markings as follows: postpronotal lobes; notopleura; mesopleural stripe reaching midway between anterior margin of notopleuron and anterior *npl.* seta dorsally, continuing to katepisternum as a large transverse spot, anterior margin straight; anatergite (posterior apex black); anterior ⅔ katatergite (remainder black); two broad parallel-sided to sub-parallel lateral postsutural vittae ending at *ia.* seta. Medial longitudinal postsutural yellow vitta absent. Postnotum black. Scutellum yellow except for

narrow black basal band. Setae: *sc.* 2; *prsc.* 2; *ia.* 1; *p.sa.* 1; *a.sa.* 1; *mpl.* 1; *npl.* 2; *scp.* 4.
Legs: All femora entirely fulvous except for an elongate dark fuscous subapical spot on outer surfaces of fore femora; all tibiae dark fuscous; all tarsi with basal segment fulvous and apical four segments fuscous; mid tibiae each with an apical black spur.
Wings: Length 5.0 mm; cells bc and c with a pale tint; microtrichia in outer $^1/_3$ of cell c only; remainder of wings colourless except dark fuscous cell sc, broad dark fuscous costal band confluent with R_{4+5} (a fine pale fuscous line adjacent to R_{4+5}) and remaining of uniform width throughout, a moderately broad dark fuscous anal streak; a dense aggregation of microtrichia around A_1+CuA_2; supernumerary lobe of medium development.
Abdomen: Oval; terga free; pecten present on tergum III. Tergum I and sterna I and II wider than long. Tergum I dark fuscous except for a narrow transverse fulvous band across posterior margin but not reaching lateral margins; tergum II fulvous except for a narrow transverse dark fuscous band subanteriorly and reaching narrow dark fuscous anterolateral corners; terga III–V orange-brown with a dark 'T' pattern consisting of a narrow transverse black band across anterior margin of tergum III and a medium-width medial longitudinal black band over all three terga, narrow dark fuscous margins on terga IV and V. A pair of oval pale fuscous shining spots on tergum V. Posterior lobe of surstylus short, sternum V with a deep concavity on posterior margin. Abdominal sterna dark.
Attractant: Methyl eugenol.
Female: No known record.
Distribution: Brunei; Indonesia (Kalimantan); Peninsular Malaysia.
Hosts: No known record.
Remarks: *Bactrocera (Bactrocera) bruneiae* n.sp. is similar to *Bactrocera (Bactrocera) costalis* (Shiraki), *Bactrocera (Bactrocera) eurycosta* n.sp. and *Bactrocera (Bactrocera) ritsemai* (Weyenbergh) in possessing a black scutum, the costal band confluent with R_{4+5} and of uniform width throughout, cells bc and c colourless or with a pale tint and all femora entirely fulvous or with, at most, a dark subapical spot on the fore femora only. It differs from these species in having consistently large, narrow elongate oval black facial spots in the antennal furrows.

Bactrocera (Bactrocera) brunneola White & Tsuruta

(Fig. 41)

Bactrocera (Bactrocera) brunneola White & Tsuruta, in Tsuruta & White, 2001: 75. Holotype male in HORDI.

Non-type material examined: SRI LANKA (formerly CEYLON): 1 male, Akressa, 21.ix.1994; 2 males, Labuduwa, 19.x.1994. Specimens donated by Kenji Tsuruta and in QDPC.
Diagnosis: A medium-sized species; face fulvous with a pair of large elongate oval black spots; postpronotal lobes and notopleura yellow; scutum red-brown; lateral postsutural yellow vittae present; medial postsutural yellow vitta absent; no yellow spot anterior to notopleural suture; mesopleural stripe equal in width to notopleuron dorsally; scutellum yellow with a narrow red-brown basal band; wing with cells bc and c colourless, microtrichia in outer corner of cell c only; a narrow fuscous costal band slightly overlapping R_{2+3} and remaining narrow around apex of wing to end just beyond apex of R_{4+5}; a narrow fuscous anal streak; abdominal terga III–V orange-brown with a 'T' pattern consisting of a narrow black band across anterior margin of tergum III and a narrow medial longitudinal dark fuscous to black band over all three terga, narrow fuscous anterolateral corners on terga IV and V.
Description: male.
Head: Height 1.5 mm. Frons – length 1.4 times breadth; red-brown, fulvous along lateral and ventral margins, pale fuscous around orbital setae and on anteromedial hump; orbital setae dark red-brown: 1 *s.or.*, 2 *i.or.*; lunule red-brown. Ocellar triangle black. Vertex red-brown. Face fulvous with large elongate oval black spots reaching to oral margin; length 0.5 mm. Genae fulvous, dark fuscous subocular spot present; black seta present. Occiput red-brown, fulvous along eye margins; occipital row with a few short dark setae. Antennae with segments 1 and 2 red-brown, segment 3 red-brown with fuscous on apex and outer surface; length of segments: 0.19 mm; 0.31 mm; 0.78 mm.
Thorax: Scutum entirely red-brown without distinct dark markings (in some specimens there may be an irregular pattern of fuscous markings across posterior region with narrow lines running anteriorly). Pleural areas pale fuscous except red-brown below postpronotal lobes. Yellow markings as follows: postpronotal lobes; notopleura; mesopleural stripe equal in width to notopleuron dorsally, continuing to katepisternum as a small transverse spot, anterior margin slightly convex; anatergite (posterior apex dark fuscous); anterior $^3/_5$ katatergite (remainder dark fuscous); two narrow lateral postsutural vittae tapering posteriorly to end well before *ia.* seta. Medial longitudinal postsutural yellow vitta absent. Postnotum red-brown. Scutellum yellow except for narrow red-brown basal band. Setae: *sc.* 2; *prsc.* 2; *ia.* 1; *p.sa.* 1; *a.sa.* 1; *mpl.* 1; *npl.* 2; *scp.* 4.
Legs: All femora entirely fulvous; fore tibiae dark fuscous, mid tibiae fuscous, hind tibiae dark fuscous; all tarsal segments entirely fulvous; mid tibiae each with an apical dark fuscous spur.
Wings: Length 5.0 mm; cells bc and c colourless; microtrichia in outer corner of cell c only; remainder of wings colourless except fuscous cell sc, narrow fuscous costal band slightly overlapping R_{2+3} where it is pale and ending beyond extremity of R_{4+5}, a narrow fuscous anal streak; dense aggregation of microtrichia around A_1+CuA_2; supernumerary lobe of medium development.
Abdomen: Oval; terga free; pecten present on tergum III. Tergum I and sterna I and II wider than long. Tergum I orange-brown with narrow fuscous lateral margins; tergum II orange-brown tending fulvous posterolaterally and with narrow dark fuscous anterolateral corners; terga III–V orange-brown with a narrow transverse black band across anterior margin of tergum III, a very narrow medial longitudinal fuscous to dark

fuscous band over all three terga and narrow fuscous anterolateral corners on terga IV and V. A pair of oval red-brown to fuscous shining spots on tergum V. Posterior lobe of surstylus short, sternum V with a deep concavity on posterior margin. Abdominal sterna dark fuscous to black.

Attractant: Cue lure.

Female: No known record.

Distribution:

Type locality: Sri Lanka (Matara District, Kanake).

Previously recorded: Type locality and other district locations in Sri Lanka (Tsuruta & White, 2001).

Hosts: No known record.

Remarks: *Bactrocera (Bactrocera) brunneola* White & Tsuruta is similar to *Bactrocera (Bactrocera) abdonigella* Drew, *Bactrocera (Bactrocera) apicofuscans* White & Tsuruta, *Bactrocera (Bactrocera) moluccensis* (Perkins) and *Bactrocera (Bactrocera) paraosbeckiae* Drew in possessing a red-brown scutum, the costal band complete and just overlapping R_{2+3} and remaining narrow around wing apex. It differs from *B. abdonigella*, *B. apicofuscans* and *B. moluccensis* in having colourless cells bc and c and from *B. paraosbeckiae* in having the black 'T' pattern on abdominal terga III–V consisting of narrow bands and the lateral dark markings on terga IV and V narrow. In addition, *B. brunneola* appears to be inseparable morphologically from *Bactrocera (Bactrocera) rubigina* (Wang & Zhao) and both respond to cue lure. These species may be conspecific.

Bactrocera (Bactrocera) bryoniae (Tryon)

(Fig. 42)

Chaetodacus bryoniae Tryon, 1927: 192. Lectotype female in QMIC.

Strumeta bryoniae – Perkins, 1938a: 127; May, 1963: 50.

Dacus (Strumeta) bryoniae – Hardy, 1951: 148; Drew, 1974: 21.

Bactrocera (Bactrocera) bryoniae – Drew, 1989: 46, Lectotype designation; Norrbom *et al.*, 1998: 88.

Material examined: INDONESIA: PAPUA (formerly part of IRIAN JAYA) – 1 male, Papua, Keerom, Intaimilyan, 16.vi.2007, coll. Eikanus Karma, attracted to cue lure in forest; 1 male, Papua, Keerom, Wiantre, 23.vi.2007, coll. Eikanus Karma. Specimens in QDPC.

Diagnosis: A medium-sized species; face fulvous with a pair of medium-sized subquadrate to oval black spots; postpronotal lobes and notopleura yellow; scutum dull black; lateral postsutural yellow vittae present; medial postsutural yellow vitta absent; no yellow spot anterior to notopleural suture; mesopleural stripe reaching midway between anterior margin of notopleuron and anterior *npl.* seta dorsally; scutellum yellow; wing with cells bc and c fuscous, microtrichia in outer ½ of cell c only; a broad dark fuscous costal band confluent with R_{4+5}; a moderately broad dark fuscous anal streak; abdominal terga III–V red-brown with a 'T' pattern consisting of a narrow- to medium-width transverse black band across anterior margin of tergum III which widens over the lateral margins, a narrow medial longitudinal black band over all three terga, narrow dark fuscous to black lateral margins on terga IV and V.

Description: male.

Head: Height 1.6 mm. Frons – length 1.57 times breadth; red-brown without distinct dark markings; orbital setae black: 1 *s.or.*, 2 *i.or.*; lunule pale fuscous. Ocellar triangle black. Vertex pale fuscous. Face fulvous with a pair of medium-sized pear-shaped to oval black spots; length 0.5 mm. Genae red-brown, large fuscous subocular spot present; black seta present. Occiput red-brown, fulvous along eye margins; occipital row with 4–7 strong black setae. Antennae with segments 1 and 2 red-brown, segment 3 red-brown with fuscous on apex and outer surface; length of segments: 0.18 mm; 0.3 mm; 0.76 mm.

Thorax: Scutum dull black with red-brown below lateral postsutural vittae, around notopleural suture, along lateral margin between postpronotal lobe and notopleuron, inside postpronotal lobe. Pleural areas dark fuscous to dull black with red-brown below postpronotal lobe. Yellow markings as follows: postpronotal lobes; notopleura; mesopleural stripe reaching midway between anterior margin of notopleuron and anterior *npl.* seta dorsally, continuing to katepisternum as a transverse spot, anterior margin straight; anatergite (posterior apex black); anterior ⅔ katatergite (remainder black); two moderately broad parallel-sided lateral postsutural vittae ending at *ia.* seta. Medial longitudinal postsutural yellow vitta absent. Postnotum black, tending dark red-brown centrally. Scutellum yellow except for narrow black basal band. Setae: *sc.* 2; *prsc.* 2; *ia.* 1; *p.sa.* 1; *a.sa.* 1; *mpl.* 1; *npl.* 2; *scp.* 4.

Legs: All femora entirely fulvous; fore and mid tibiae pale fuscous, hind tibiae fuscous; tarsi with all segments fulvous; mid tibiae each with an apical black spur.

Wings: Length 5.7 mm; cells bc and c fuscous; microtrichia in outer ½ of cell c only; remainder of wings colourless except dark fuscous costal band confluent with R_{4+5} and remaining of uniform width throughout, a moderately broad dark fuscous anal streak; a dense aggregation of microtrichia around A_1+CuA_2; supernumerary lobe of medium development.

Abdomen: Oval; terga free; pecten present on tergum III. Tergum I and sterna I and II wider than long. Tergum I dark fuscous with a narrow transverse red-brown band across posterior margin but not reaching lateral margins; tergum II red-brown anteriorly to fuscous posteriorly and with a narrow transverse dark fuscous band across centre of tergum but not reaching the narrow dark fuscous anterolateral corners; terga III–V red-brown with a dark 'T' pattern consisting of a narrow- to medium-width transverse black band across anterior margin of tergum III which widens over lateral margins and a narrow medial longitudinal black band over all three terga, narrow dark fuscous to black lateral margins on terga IV and V. A pair of oval red-brown shining spots on tergum V. Posterior lobe of surstylus short, sternum V with a deep concavity on posterior margin. Abdominal sterna dark fuscous to black.

Attractant: Cue lure.

Female: As for male except no dense aggregation of microtrichia around A_1+CuA_2, supernumerary lobe weak, pecten of cilia on abdominal tergum III absent; ovipositor basal segment red-brown, dorsoventrally compressed and tapering posteriorly in dorsal view; ratio of length of oviscape to length of tergum V, 0.6:1; apex of piercer needle-shaped.

Distribution:

Type locality: Australia (Queensland, Eidsvold).

Previously recorded: Type locality and the following locations – Australia (Queensland, Northwestern Australia, Torres Strait Is.); Papua New Guinea; New Britain; New Ireland; Lihir I.

New records: Indonesia (Papua, formerly part of Irian Jaya – Keerom).

Hosts: Recorded from five plant families – *Cucurbitaceae*; *Loganiaceae*; *Musaceae*; *Passifloraceae*; and *Solanaceae* (see Drew, 1989; Hancock *et al.*, 2000).

Remarks: *Bactrocera (Bactrocera) bryoniae* (Tryon) is similar to *Bactrocera (Bactrocera) epicharis* (Hardy), *Bactrocera (Bactrocera) neoritsemai* n.sp. and *Bactrocera (Bactrocera) penecostalis* n.sp. in possessing a broad costal band confluent with R_{4+5}, cells bc and c coloured and lateral postsutural vittae parallel-sided or subparallel. It differs from *B. epicharis* in having abdominal terga III–V not entirely black, from *B. neoritsemai* in having dark lateral margins on abdominal terga IV and V and the costal band of uniform width and colour throughout, and from *B. penecostalis* in having cells bc and c paler (generally fulvous) and a broad black 'T' pattern over abdominal terga III–V.

Bactrocera (Bactrocera) carambolae Drew & Hancock

(Fig. 43)

Bactrocera (Bactrocera) carambolae Drew & Hancock, 1994a: 11; Norrbom *et al.*, 1998: 89. Holotype male in BMNH.

Non-type material examined: Since the publication of Drew & Hancock (1994a), the following specimens have been studied – **INDIA: ANDAMAN ISLANDS** – 2 males and 2 females, South Andaman, Port Blair, 20.ii.1996, coll. Ranganath, bred ex Papaya, Sample No. 55; South Andaman, Port Blair, 15.iv.1998, coll. Verna, bred ex *Trichilia* sp., Sample No. 60; 2 males and 2 females, South Andaman, Port Blair, 16.vi.1998, coll. Suryanarayan, bred ex. *Artocarpus heterophyllus*, Sample No. 61; 1 male and 2 females, S. Andaman, Port Blair, 2.vii.1995, coll. Ranganath, bred ex *Polyalthia* sp., Sl No. 46; 1 male, S. Andaman, Port Blair, 2.vii.1995, coll. Ranganath, bred ex *Fragrea* sp., Sl No. 45; 1 male and 3 females, S. Andaman, Port Blair, 12.vi.1995, Ranganath, bred ex *Fragrea* sp. Sl No. 45; 2 males and 1 female, South Andaman, Havelock Is., 10.vi.1998, coll. Ranganath, bred ex *Calophyllum* sp., Sample No. 64; 3 males and 1 female, South Andaman, Havelock Is., 10.vi.1998, coll. Ranganath, bred ex *Syzygium* sp., Sample No. 68. **INDONESIA: JAVA** – 60 males, West Java, Majalengka, Ampel, 14.vii.2005, coll. Yaya Senjaya, attracted to methyl eugenol in mango. **PENINSULAR MALAYSIA**: 16 males and 79 females, MARDI Carambola orchard, Selangor, vi.2002, coll. Vijaysegaran, attracted to protein bait; 36 males, MARDI Carambola orchard, 21.ix.2000, coll. R. Drew, attracted to methyl eugenol; 18 males and 14 females, MARDI, ix.2000, coll. R. Drew, bred ex *Carambola*. **VIETNAM: Southern Vietnam** – 1 male, Vinhlong Province, Long Ho District, Phuoc Hau Village, Site VS01, 1.iv.2002, coll. SOFRI Staff, attracted to methyl eugenol; 1 male, Can Tho Province, Binh Thuy Village, Site S003, 31.xii.1999; 3 males, Ben Tre Province, Chau Thanh District, Tien Thuy Village, Site VS04, 21.iv.2002, coll. SOFRI Staff; 4 males (26.ii.2002), 3 males (12.iii.2002), 6 males (9.iv.2002), 7 males (23.iv.2002), 1 male (4.vi.2002), 1 male (2.vii.2002), 1 male (16.vii.2002), 1 male (27.viii.2002), 1 male (4.x.2002), Binh Duong Province, Thuan An District, An Thanh Village, Site VS05, attracted to methyl eugenol; 1 male (15.ii.2002), 10 males (2.iii.2002), 5 males (16.iii.2002), 5 males (30.iii.2002), 1 male (27.iv.2002), 4 males (22.vi.2002), 2 males (20.vii.2002), 2 males (31.vii.2002), 2 males (3.viii.2002), 1 male (9.xi.2002), 1 male (12.x.2002), Dong Nai Province, Long Khanh District, Nham Nghia Village, Site VS06, coll. SOFRI Staff, attracted to methyl eugenol; 1 male (1.vii.2002), 3 males (15.vii.2002), 1 male (29.vii.2002), 1 male (12.viii.2002), Binh Thuan Province, Ham Thuan Bac District, No. 1 Village, Site VS07, attracted to methyl eugenol; 1 male (31.v.2002), 1 male (14.vi.2002), Tien Giang Province, Cai Lay District, Long Khanh Village, Site S007, attracted to methyl eugenol; 1 male, Lam Dong Province, 3 District, Da Lat Town, 8.vii.2002, Site VS08, attracted to methyl eugenol; 3 males, Dac Lac Province, Ku Mi Nga District, Quang Phu Village, Site VS09, 8.iv.2002, coll. SOFRI Staff, attracted to methyl eugenol; 1 male (24.viii.2002), 1 male (8.ix.2002), Khanh Hoa Province, Dien Khanh District, Suoi Tan Village, Site VS10, attracted to methyl eugenol; 1 male, Dong Nai Province, Long Nhat District, Long Nhat Village, 7.vii.2000, Site S012, attracted to methyl eugenol; 1 male, Dong Nai Province, Thong Nhat District, Cay Gao Village, 10.v.2000, Site S013, attracted to methyl eugenol; 1 male (17.ii.2000), 2 males (28.iv.2000), 1 male (26.v.2000), 2 males (23.vi.2000), 2 males (7.vii.2000), 4 males (11.xi.1999), Dong Nai Province, Long Khanh District, Tan Phong Village, Site S014, attracted to methyl eugenol; 1 male, Dong Nai Province, Xuan Loc District, Xuan Phu Village, Site S015, 23.vi.2000, attracted to methyl eugenol; 1 male (12.xi.1999), 1 male (26.xi.1999), 1 male (21.i.2000), 2 males (26.v.2000), Binh Duong Province, Lai Thieu Village, Site S020, attracted to methyl eugenol. Specimens in QDPC.

Diagnosis: Face fulvous with a pair of medium-sized oval black spots; postpronotal lobes and notopleura yellow; scutum dull black with brown behind lateral postsutural vittae, around notopleural suture and inside postpronotal lobes; broad parallel-sided lateral postsutural yellow vittae ending at or behind *ia.* seta; medial postsutural yellow vitta absent; mesopleural stripe reaching midway between anterior margin of notopleuron and anterior *npl.* seta dorsally; scutellum yellow; legs with femora fulvous with a large elongate oval dark fuscous to black preapical spot on outer surfaces of fore femora in some specimens; tibiae dark fuscous (except mid tibiae paler apically); wing with cells bc and c colourless, microtrichia in outer corner of cell c only; a narrow fuscous costal band slightly overlapping R_{2+3} and expanding slightly beyond apex of R_{2+3} across apex of R_{4+5}; a narrow fuscous anal streak; supernumerary lobe of medium development; abdominal terga III–V orange-brown with a 'T' pattern consisting of a narrow transverse black band across anterior margin of tergum III and widening to cover lateral margins, a medium-width medial longitudinal black band over all three terga, anterolateral corners of terga IV dark fuscous to black and rectangular in shape and anterolateral corners of tergum V dark fuscous, a pair of oval orange-brown shining spots on tergum V; abdominal sterna dark coloured.

Attractant: Methyl eugenol.

Female: As for male in the general body colour patterns. See Drew & Hancock (1994a) for a detailed description of female characters.

Distribution:

Type locality: Peninsular Malaysia (Kuala Kangsar).

Previously recorded: Type locality and the following locations – East Malaysia; French Guyana; Guyana; India (Andaman Is); Indonesia (widespread); Southern Thailand; Surinam (Drew & Hancock, 1994a; Norrbom *et al.*, 1998).

New records: Southern Vietnam.

Hosts: This species is a major economic pest throughout the region where it occurs (see Allwood *et al.*, 1999 for host records).

Remarks: *Bactrocera (Bactrocera) carambolae* Drew & Hancock, within the *dorsalis* complex, is similar to *Bactrocera (Bactrocera) propinqua* (Hardy & Adachi) and some specimens of *Bactrocera (Bactrocera) papayae* Drew & Hancock in possessing broad parallel-sided or subparallel lateral postsutural vittae, costal band slightly overlapping R_{2+3}, abdominal terga III–V with narrow- to medium-width dark lateral margins, shining spots on abdominal tergum V pale (orange-brown to fuscous), femora entirely fulvous or with, at most, subapical dark spots on fore femora only, in addition to the general characteristics of the *dorsalis* complex. It differs from *B. papayae* in having a broad medial longitudinal black band on abdominal terga III–V, a broader costal band apically, and shorter male aedeagus and female ovipositor and from *B. propinqua* in having a narrower medial longitudinal black band on abdominal terga III–V (in *B. propinqua* this band is very broad) and apex of the piercer needle-shaped (in *B. propinqua* the apex of the piercer is trilobed).

In our molecular analyses using the mitochondrial genes, COI and ND5, *B. carambolae* is separate from *B. dorsalis* (Taiwan specimens), *B. invadens* (Sri Lanka specimens), *B. occipitalis* (Kalimantan specimens) and *B. papayae* (specimens from Christmas I., Java, Kalimantan, Peninsular Malaysia; Drew, unpublished data). In addition, Drew & Hancock (1994a) published male pheromone analyses that indicated *B. carambolae* possessed entirely different chemicals in its glandular extracts. In combination with the morphological features of a broad costal band and wider medial longitudinal dark band over abdominal terga III–V, we maintain that *B. carambolae* is a good species. See discussion under section titled 'Identification of Cryptic Species'.

Bactrocera (Bactrocera) careofascia Drew & Romig, new species

(Fig. 44)

Material examined: Holotype male, **INDONESIA**: **CENTRAL SULAWESI** – Sulawesi Tengah, Donggala, Nupabomba, 23.ix.2006, coll. Rusdin, attracted to cue lure in forest, deposited in BMNH.

Diagnosis: A medium-sized species; face fulvous with a pair of medium-sized oval black spots; postpronotal lobes and notopleura yellow; scutum black; lateral postsutural yellow vittae present; medial postsutural yellow vitta absent; no yellow spot anterior to notopleural suture; mesopleural stripe reaching midway between anterior margin of notopleuron and anterior *npl.* seta dorsally; scutellum yellow; wing with cells bc and c fuscous, dense microtrichia covering all of cell c and outer ½ of cell bc; a broad dark fuscous costal band; a broad dark fuscous anal streak; abdominal terga III–V fulvous with broad lateral longitudinal fuscous to dark fuscous bands over all three terga which join along anterior margin of tergum III.

Description: male.

Head: Height 1.47 mm. Frons – length 1.5 times breadth; red-brown with fuscous around orbital setae; orbital setae black: 1 *s.or.*, 2 *i.or.*; lunule fuscous. Ocellar triangle black. Vertex fuscous. Face fulvous with a pair of medium-sized oval black spots; length 0.44 mm. Genae fulvous, large fuscous subocular spot present; black seta present. Occiput fuscous, fulvous along eye margins; occipital row with 8–9 strong black setae. Antennae with segments 1 and 2 red-brown, segment 3 missing; length of segments: 0.22 mm; 0.34 mm.

Thorax: Scutum entirely black without pale markings. Pleural areas black except red-brown above mid and hind coxae and below postpronotal lobes. Yellow markings as follows: postpronotal lobes; notopleura; mesopleural stripe reaching midway between anterior margin of notopleuron and anterior *npl.* seta dorsally, continuing to katepisternum as a small transverse spot, anterior margin straight; anatergite (posterior apex black); anterior ⅔ katatergite (remainder black); two broad parallel-sided lateral postsutural vittae narrowing posteriorly to end just before *ia.* seta. Medial longitudinal postsutural yellow vitta absent. Postnotum black. Scutellum yellow except for narrow black basal band. Setae: *sc.* 2; *prsc.* 2; *ia.* 1; *p.sa.* 1; *a.sa.* 1; *mpl.* 1; *npl.* 2; *scp.* 4.

Legs: Fore femora fulvous, mid femora fulvous on basal ¼ and pale fuscous on apical ¾, hind femora fulvous with fuscous around apical ¼; fore tibiae pale fuscous, mid tibiae fulvous, hind tibiae fuscous; fore tarsi with basal segment fulvous and apical four segments fuscous, mid tarsi with all segments entirely fulvous, hind tarsi with basal segment

fulvous and apical four segments red-brown; mid tibiae each with an apical black spur.
Wings: Length 6.0 mm; cells bc and c fuscous; dense microtrichia covering all of cell c and outer ½ of cell bc; remainder of wings colourless except dark fuscous cell sc, broad dark fuscous costal band confluent with R_{4+5} but slightly paler adjacent to that vein and remaining of uniform width to apex of wing, broad dark fuscous anal streak; a dense aggregation of microtrichia around A_1+CuA_2; supernumerary lobe of medium development.
Abdomen: Oval; terga free; pecten present on tergum III. Tergum I and sterna I and II wider than long. Tergum I fuscous to dark fuscous with a narrow transverse fulvous band across posterior margin but not reaching lateral margins; tergum II fulvous with a narrow transverse dark fuscous band subanteriorly and reaching narrow fuscous anterolateral corners; terga III–V fulvous with broad fuscous to dark fuscous lateral longitudinal bands over all three terga and joining along anterior margin of tergum III. A pair of oval fuscous to dark fuscous shining spots on tergum V. Posterior lobe of surstylus short, sternum V with a deep concavity on posterior margin. Abdominal sterna dark fuscous.
Attractant: Cue lure.
Female: No known record.
Distribution: Indonesia (Sulawesi – Central Sulawesi, Donggala).
Hosts: No known record.
Remarks: *Bactrocera (Bactrocera) careofascia* n.sp. is similar to *Bactrocera (Bactrocera) curvosterna* n.sp., *Bactrocera (Bactrocera) nanoarcuata* n.sp. and *Bactrocera (Bactrocera) paralatissima* n.sp., in possessing a broad costal band confluent with R_{4+5}, fuscous cells bc and c, lateral postsutural yellow vittae tapering posteriorly and abdominal terga III–V with dark colour patterns, but differs from these species in having abdominal terga III–V with broad lateral dark margins joined along anterior margin of tergum III and lacking a medial longitudinal dark band over these terga.

Bactrocera (Bactrocera) caryeae (Kapoor)

(Fig. 45)

Chaetodacus ferrugineus incisus Bezzi, 1916: 105 (partim), not Walker, 1861 [Misidentification].
Dacus (Strumeta) caryeae Kapoor, 1971: 479. Holotype male in INPC.
Dacus (Bactrocera) caryeae – Hardy, 1977: 49.
Bactrocera (Bactrocera) caryeae – White & Elson-Harris, 1992: 186; Drew & Hancock,1994a:13; Norrbom *et al.*, 1998: 89; Drew & Raghu, 2002: 334.

Non-type material examined: Since the publication of Drew & Raghu (2002), the following specimens have been studied – **INDIA**: 29 males and 26 females, S. India, Chethalli, 24.v.1992, coll. D.L. Hancock, bred from Rose Apple *Syzygium jambos*; 11 males and 8 females, S. India, Chethalli, 24.v.1992, coll. D.L. Hancock, bred from West Indian Cherry *Malpighia glabra* [= *emarginata*]; 9 males and 5 females, S. India, Kotigehar, 20.v.1992, coll. D.L. Hancock, bred from Sapota *Calocarpum sapote*; 1 female, S. India, Mudigere, 21.v.1992, coll. D.L. Hancock, bred from Pomelo *Citrus grandis*. Specimens in QDPC.
Diagnosis: Face fulvous with a pair of large elongate oval black spots; postpronotal lobes yellow (except anterodorsal corners fuscous); notopleura yellow; scutum black with a small area of dark brown posterolateral to lateral postsutural vittae; narrow lateral postsutural yellow vittae which are either parallel-sided or narrowing slightly posteriorly to end at or just before *ia.* seta; medial postsutural yellow vitta absent; mesopleural stripe reaching midway between anterior margin of notopleuron and anterior *npl.* seta dorsally; scutellum yellow with a broad black basal band; legs with femora fulvous with large dark fuscous to black preapical spots on outer surfaces of fore femora and inner surfaces of mid and hind femora; fore tibiae fuscous, mid tibiae fulvous, hind tibiae dark fuscous; wing with cells bc and c colourless, sparse microtrichia in outer corner of cell c only; a very narrow fuscous costal band confluent with R_{2+3} and remaining very narrow around apex of wing; a narrow fuscous anal streak contained within cell cup; supernumerary lobe of medium development; abdominal terga III–V orange-brown with dark fuscous to black across anterior ⅓ to ½ of tergum III, two broad lateral longitudinal dark fuscous to black bands and a narrow medial longitudinal black band over all three terga, a pair of oval orange-brown shining spots on tergum V; abdominal sterna dark coloured.
Attractant: Methyl eugenol.
Female: As for male in the general body colour patterns. See Drew & Hancock (1994a) for a detailed description of female characters.
Distribution:
Type locality: Southern India (Mysore).
Previously recorded: Southern India (widespread) (Drew & Raghu, 2002).
The records from Sri Lanka (in Norrbom *et al.*, 1998; Drew & Raghu, 2002) are incorrect. This species has not been collected in Sri Lanka (Tsuruta & White, 2001, p. 86).
Hosts: This species is a serious economic pest and has been recorded from seven plant families – *Anacardiaceae*, *Lecythidaceae*, *Malpighiaceae*, *Moraceae*, *Myrtaceae*, *Rutaceae*, *Sapotaceae* (see Allwood *et al.*, 1999 and White & Elson-Harris, 1992, for host records).
New records: *Syzygium jambos* (family *Myrtaceae*).
Remarks: *Bactrocera (Bactrocera) caryeae* (Kapoor), within the *dorsalis* complex, is similar to *Bactrocera (Bactrocera) kandiensis* Drew & Hancock and *Bactrocera (Bactrocera) arecae* (Hardy & Adachi) in possessing narrow parallel-sided lateral postsutural vittae, preapical dark markings on at least one pair of femora in addition to the general characteristics of the *dorsalis* complex. It differs from *B. arecae* in possessing preapical dark markings on all femora (in *B. arecae* the preapical dark markings are on fore femora only) and from *B. kandiensis* in possessing a broad medial longitudinal dark band and broad lateral longitudinal dark bands over abdominal terga III–V.

Tsuruta & White (2001) page 86 recorded Drew & Hancock incorrectly as the authors of *B. caryeae*.

Bactrocera (Bactrocera) ceylanica Tsuruta & White

(Fig. 46)

Bactrocera (Bactrocera) ceylanica Tsuruta & White, 2001: 77. Holotype male in HORDI.

Non-type material examined: SRI LANKA (formerly CEYLON): 1 male, 'Pitawala, 26.x.1994, cue lure 53'. This specimen was donated by Kenji Tsuruta and is listed under other non-type series in Tsuruta & White (2001), where Pitawala is incorrectly spelt as Pitawara. Specimen in QDPC.

Diagnosis: A small species; face fulvous with a pair of very large elongate oval black spots; postpronotal lobes and notopleura yellow; scutum mostly red-brown to fuscous with patterns of small black markings; lateral postsutural yellow vittae present; medial postsutural yellow vitta absent; no yellow spot anterior to notopleural suture; mesopleural stripe slightly wider than notopleuron dorsally; scutellum yellow with a broad black basal band; wing with cells bc and c colourless, microtrichia in outer corner of cell c only; a narrow fuscous to dark fuscous costal band; a narrow fuscous to dark fuscous anal streak; abdominal terga III–V dark red-brown with a broad transverse black band across most of tergum III, a broad medial longitudinal black band over all three terga, broad dark fuscous to black anterolateral corners on terga IV and V.

Description: male.

Head: Height 1.44 mm. Frons – length 1.67 times breadth; red-brown with fuscous on anteromedial hump and dark fuscous around bases of orbital setae; orbital setae black: 1 *s.or.*, 2 *i.or.*; lunule fuscous. Ocellar triangle black. Vertex dark fuscous. Face fulvous with a pair of very large elongate oval black spots; length 0.4 mm. Genae fulvous, fuscous subocular spot present; black seta present. Occiput fuscous, fulvous along eye margins; occipital row with 6–7 strong black setae. Antennae with segments 1 and 2 red-brown, segment 3 red-brown with dark fuscous on apex and outer surface; length of segments: 0.16 mm; 0.28 mm; 0.7 mm.

Thorax: Scutum mostly red-brown to dark fuscous with a pair of longitudinal black markings inside lateral postsutural vittae and large lateral black areas between postpronotal lobes and notopleura and a narrow medial longitudinal black line from level of *a.sa.* setae to anterior margin. Pleural areas black except dark fuscous below postpronotal lobes. Yellow markings as follows: postpronotal lobes; notopleura; mesopleural stripe slightly wider than notopleuron dorsally, continuing to katepisternum as a small transverse spot, anterior margin straight; anatergite (posterior apex black); anterior $^{3}/_{5}$ katatergite (remainder black); two narrow parallel-sided lateral postsutural vittae ending just before *ia.* seta. Medial postsutural yellow vitta absent. Postnotum black laterally, red-brown centrally. Scutellum yellow except for broad black basal band. Setae: *sc.* 2; *prsc.* 2; *ia.* 1; *p.sa.* 1; *a.sa.* 1; *mpl.* 1; *npl.* 2; *scp.* 4.

Legs: All femora entirely fulvous except dark fuscous around apical $^{1}/_{5}$ of hind femora; fore tibiae fulvous except fuscous basally, mid tibiae fulvous, hind tibiae dark fuscous; tarsi with all segments entirely fulvous; mid tibiae each with an apical black spur.

Wings: Length 4.8 mm; cells bc and c colourless; microtrichia in outer corner of cell c only; remainder of wings colourless except dark fuscous cell sc, narrow fuscous to dark fuscous costal band confluent with R_{2+3} and remaining very narrow as it continues around apex of wing to end just beyond apex of R_{4+5}, a narrow fuscous to dark fuscous anal streak; a dense aggregation of microtrichia around A_1+CuA_2; supernumerary lobe of medium development.

Abdomen: Oval; terga free; pecten present on tergum III. Tergum I and sterna I and II wider than long. Tergum I dark fuscous centrally, black laterally and with a narrow transverse fuscous band across posterior margin but not reaching lateral margins; tergum II fuscous with a narrow transverse subanterior black band reaching narrow anterolateral black corners; terga III–V dark red-brown with a broad transverse black band across most of tergum III and a broad medial longitudinal black band over all three terga, broad dark fuscous to black anterolateral corners on terga IV and V. A pair of oval dark fuscous shining spots on tergum V. Posterior lobe of surstylus short, sternum V with a deep concavity on posterior margin. Abdominal sterna dark fuscous to black.

Attractant: Cue lure.

Female: No known record.

Distribution:

Type locality: Sri Lanka (Matale District, Pitawala).

Previously recorded: Type locality and other district locations in Sri Lanka (Tsuruta & White, 2001).

Hosts: No known record.

Remarks: *Bactrocera (Bactrocera) ceylanica* Tsuruta & White is similar to *Bactrocera (Bactrocera) brunneola* White & Tsuruta, *Bactrocera (Bactrocera) paraosbeckiae* Drew and *Bactrocera (Bactrocera) rubigina* (Wang & Zhao) in possessing a basic red-brown scutum, narrow costal band and colourless cells bc and c, and in having the males attracted to cue lure. It differs from *B. brunneola* and *B. paraosbeckiae* in having the costal band confluent with R_{2+3} (not overlapping this vein) and from *B. rubigina* in having the lateral postsutural vittae ending before the *ia.* seta. *Bactrocera ceylanica* was placed in the *dorsalis* complex by Tsuruta & White (2001); however, with the red-brown scutum, it does not belong to this complex of species.

Bactrocera (Bactrocera) cibodasae Drew & Hancock

(Fig. 47)

Bactrocera (Bactrocera) cibodasae Drew & Hancock, 1994a: 15; Norrbom *et al.*, 1998: 89. Holotype male in BBM.

Non-type material examined: Since the publication of Drew & Hancock (1994a), the following specimens have been studied – **INDONESIA: JAVA** – 1 male, East Java, Lumajang, Kadang Tepua, 19.vii.2005, coll. Utomo, attracted to cue lure in banana; 1 male, Java, Banten, Lebak, Pasir Nini Nini, 8.viii.2005, coll. Aden Haer, attracted to cue lure. **VIETNAM: Northern Vietnam** – 1 male, Northern Vietnam, Son La Province, Moc Chau District, Co Do, Site

N006, 29.v.2000, attracted to cue lure; **Southern Vietnam** – 1 male, Southern Vietnam, Dac Lac Province, Ku Mi Nga District, Quang Phu Village, Site VS09, coll. SOFRI Staff, attracted to cue lure; 1 male, Southern Vietnam, Lam Dong Province, 3 District, Da Lat town, Site VS08, 8.x.2002, attracted to cue lure; 1 male, Southern Vietnam, Dong Nai Province, Long Khanh District, Nhan Nghia Village, 3.viii. 2002, attracted to cue lure. Specimens in QDPC.

Diagnosis: Face fulvous with a pair of small to medium-sized circular to oval black spots; postpronotal lobes and notopleura yellow; scutum black with dark red-brown between *prsc.* setae, a small area in centre of scutum, behind lateral postsutural vittae, around notopleural suture and inside postpronotal lobes; narrow lateral postsutural yellow vittae tapering posteriorly to end before *ia.* seta; medial postsutural yellow vitta absent; mesopleural stripe slightly wider than notopleuron dorsally; scutellum yellow; legs with femora entirely fulvous; fore tibiae pale fuscous, mid tibiae fulvous, hind tibiae fuscous; wing with cells bc and c colourless, microtrichia in outer corner of cell c only; narrow dark fuscous costal band confluent with R_{2+3} and widening across apex of wing; a narrow fuscous anal streak; supernumerary lobe of medium development; abdominal terga III–V orange-brown with a narrow transverse dark fuscous to black band across anterior margin of tergum III which widens across lateral margins, anterolateral corners of terga IV and V dark fuscous to black, a narrow medial longitudinal dark fuscous to black band over all three terga, a pair of oval fuscous shining spots on tergum V; abdominal sterna dark coloured.

Attractant: Cue lure.

Female: No known record.

Distribution:

Type locality: Indonesia (Java, Cibodas).

New records: Indonesia (East Java – Lumajang; West Java – Banten Province, Lebak); Vietnam.

Hosts: No known record.

Remarks: *Bactrocera (Bactrocera) cibodasae* Drew & Hancock, within the *dorsalis* complex, is similar to *Bactrocera (Bactrocera) merapiensis* Drew & Hancock, *Bactrocera (Bactrocera) osbeckiae* Drew & Hancock, *Bactrocera (Bactrocera) sumbawaensis* Drew & Hancock and *Bactrocera (Bactrocera) thailandica* Drew & Hancock in possessing narrow lateral postsutural vittae that taper posteriorly, legs with femora entirely fulvous or with dark preapical spots on apices of fore femora only, costal band confluent with R_{2+3}, abdominal terga III–V with distinct dark colour patterns consisting of a black 'T' pattern and narrow to broad lateral markings. It differs from these species in the unique shape of the costal band which is extremely narrow beyond apex of R_{2+3} then widening gradually across the apex of the wing.

Bactrocera (Bactrocera) citima (Hardy)

(Fig. 48)

Dacus (Strumeta) citimus Hardy, 1973: 36. Holotype male in BBM.

Bactrocera (Bactrocera) citima – Norrbom *et al.*, 1998: 89.

Bactrocera (Bactrocera) jinghongensis Zhang, Ji & Chen, in Zhang *et al.*, 2011: 600. Holotype male in FAFU. **New Synonym**.

Type material examined: Holotype male, **THAILAND**: with the following three labels – 'NW Chiangmai Prov., Chiangdao, 450 m, 5–11.iv.1958'; 'J.L. Gressitt Collector'; 'Holotype male, Dacus (Strumeta) citimus, D.E. Hardy'. Received on loan from BBM.

Other type material studied: CHINA: A study of the colour photographs and description of *Bactrocera (Bactrocera) jinghongensis* Zhang, Ji & Chen in the original publication.

Diagnosis: A large species; face fulvous with a pair of large circular black spots; postpronotal lobes and notopleura yellow; scutum entirely black; lateral postsutural yellow vittae present; medial postsutural yellow vitta absent; no yellow spot anterior to notopleural suture; mesopleural stripe reaching midway between anterior margin of notopleuron and anterior *npl.* seta dorsally; scutellum yellow with a narrow black basal band; wing with cells bc and c colourless or with a pale tint, microtrichia in outer corner of cell c only; a narrow dark fuscous costal band overlapping R_{2+3} and widening slightly across apex of R_{4+5}; a broad dark fuscous anal streak; abdominal terga III–V dark orange-brown with a narrow medial longitudinal fuscous band and two broad lateral longitudinal dark fuscous to black bands over all three terga and joining along anterior margin of tergum III.

Description: male.

Head: Height 1.74 mm. Frons – length 1.36 times breadth; red-brown centrally, fulvous around margins and dark fuscous around orbital setae and on anteromedial hump; orbital setae black: 1 *s.or.*, 2 *i.or.*; lunule dark fuscous. Ocellar triangle black. Vertex dark fuscous. Face fulvous with a pair of large circular black spots reaching oral margins; length 0.5 mm. Genae red-brown, fuscous subocular spot present. Occiput dark fuscous, fulvous along eye margins; occipital row with 4–6 strong black setae. Antennae with segments 1 and 2 pale fuscous, segment 3 missing; length of segments: 0.19 mm; 0.34 mm.

Thorax: Scutum entirely black without pale markings. Pleural areas entirely black, tending dark fuscous below postpronotal lobes. Yellow markings as follows: postpronotal lobes (anterodorsal corners fuscous); notopleura; mesopleural stripe reaching midway between anterior margin of notopleuron and anterior *npl.* seta dorsally, continuing to katepisternum as a small transverse spot, anterior margin straight; anatergite (posterior apex black); anterior $^{3}/_{5}$ katatergite (remainder black); two moderately broad lateral postsutural vittae narrowing posteriorly to end just before *ia.* seta. Medial longitudinal postsutural yellow vitta absent. Postnotum black. Scutellum yellow except for narrow black basal band. Setae: *sc.* 2; *prsc.* 2; *ia.* 1; *p.sa.* 1; *a.sa.* 1; *mpl.* 1; *npl.* 2; *scp.* 4.

Legs: Fore femora fulvous with a large dark fuscous subapical spot on outer surfaces, mid femora fulvous with dark fuscous around apical ⅓ hind femora fulvous with dark fuscous around apex; fore tibiae dark fuscous basally to fuscous apically, mid femora fuscous basally to fulvous apically, hind tibiae entirely dark fuscous; fore tarsi with basal segment

fulvous and apical four segments fuscous, mid and hind tarsi entirely fulvous; mid tibiae each with an apical black spur.
Wings: Length 6.9 mm; cells bc and c colourless or with a pale tint; microtrichia in outer corner of cell c only; remainder of wings colourless except fuscous to dark fuscous cell sc, a narrow fuscous to dark fuscous costal band overlapping R_{2+3} and widening slightly across apex of R_{4+5}, broad fuscous to dark fuscous anal streak; dense aggregation of microtrichia around A_1+CuA_2; supernumerary lobe of medium development.
Abdomen: Oval; terga free; pecten present on tergum III. Tergum I and sterna I and II wider than long. Tergum I dark fuscous with a narrow transverse orange-brown band across posterior margin but not reaching lateral margins; tergum II orange-brown anteriorly to fulvous posteriorly and with a broad transverse dark fuscous to black band across anterior margin which connects with narrow dark fuscous to black anterolateral corners; terga III–V dark orange-brown with a narrow medial longitudinal fuscous band and two broad lateral longitudinal dark fuscous to black bands over all three terga and joining along anterior margin of tergum III. A pair of oval dark fuscous shining spots on tergum V. Posterior lobe of surstylus short, sternum V with a deep concavity on posterior margin.
Attractant: Cue lure.
Female: No known record.
Distribution:
Type locality: Thailand (Chiang Mai, Chiangdao).
New records: China (Yunnan Province) due to the **New Synonymy** of *B. jinghongensis*.
Hosts: No known record.
Remarks: *Bactrocera (Bactrocera) citima* (Hardy) is similar to *Bactrocera (Bactrocera) aemula* Drew, *Bactrocera (Bactrocera) latifrons* (Hendel), *Bactrocera (Bactrocera) musae* (Tryon), *Bactrocera (Bactrocera) parvula* (Hendel) and *Bactrocera (Bactrocera) pectoralis* (Walker) in possessing an entire costal band overlapping R_{2+3}, a black scutum and abdominal tergum III without a distinct 'T' pattern that is common in the *dorsalis* complex. It differs from *B. latifrons* and *B. musae* in having broad lateral dark fuscous to black margins on abdominal terga III–V, from *B. aemula* in having subparallel lateral postsutural vittae narrowing only slightly posteriorly, while *B. pectoralis* differs in possessing very short and narrow lateral postsutural vittae that do not extend posterior to the *a.sa.* setae. It is very close to *B. parvula* but differs in having the mesopleural stripe narrower ending midway between the anterior margin of the notopleuron and the anterior *npl.* seta and the apex of the piercer needle-shaped. We can find no good species level characters to separate *Bactrocera (Bactrocera) jinghongensis* Zhang, Ji & Chen and we hereby make this species a **New Synonym** of *B. citima*.

Bactrocera (Bactrocera) cognata (Hardy & Adachi)

(Fig. 49)

Dacus (Strumeta) cognatus Hardy & Adachi, 1954: 162. Holotype female in USNM.
Dacus (Bactrocera) cognatus – Hardy, 1977: 49.
Bactrocera (Bactrocera) cognata – Drew & Hancock, 1994a: 16; Norrbom *et al.*, 1998: 89.

Material examined: No new specimens available for study since the publication of Drew & Hancock (1994a).
Diagnosis: Face fulvous with a pair of medium-sized oval black spots; postpronotal lobes and notopleura yellow; scutum black except brown below and behind lateral postsutural vittae, around notopleural suture, inside postpronotal lobes; broad triangular lateral postsutural yellow vittae narrowing sharply posteriorly to end before *ia.* seta; medial postsutural yellow vitta absent; mesopleural stripe reaching midway between anterior margin of notopleuron and anterior *npl.* seta dorsally; scutellum yellow; legs with femora entirely fulvous except for a fuscous subapical spot on outer surfaces of fore femora; fore and mid tibiae fuscous, hind tibiae dark fuscous; wing with cells bc and c colourless, microtrichia in outer corner of cell c only; a narrow fuscous costal band just overlapping R_{2+3} and remaining of uniform width around apex of wing; a narrow fuscous anal streak; supernumerary lobe of medium development; abdominal terga III–V orange-brown with broad lateral longitudinal dark fuscous to black bands and a broad medium longitudinal black band over all three terga and joined along anterior margin of tergum III; a pair of oval orange-brown shining spots on tergum V that tend towards fuscous laterally; abdominal sterna dark coloured.
Attractant: No known record.
Male: As for female in the general body colour patterns. See Drew & Hancock (1994a) for a detailed description of male characters.
Distribution:
Type locality: Philippines (Luzon I., Los Banos).
Previously recorded: Hardy (1983) recorded this species from Java, Indonesia, but this record is in error.
Hosts: *Eugenia* sp. (family *Myrtaceae*) (Hardy & Adachi, 1954).
Remarks: *Bactrocera (Bactrocera) cognata* (Hardy & Adachi), within the *dorsalis* complex, is similar to *Bactrocera (Bactrocera) fulvifemur* Drew & Hancock, *Bactrocera (Bactrocera) lateritaenia* Drew & Hancock, *Bactrocera (Bactrocera) neocognata* Drew & Hancock, *Bactrocera (Bactrocera) penecognata* Drew & Hancock and *Bactrocera (Bactrocera) usitata* Drew & Hancock in possessing tapering lateral postsutural vittae, femora either entirely fulvous or with dark spots on outer apical surfaces of fore femora only and costal band just overlapping R_{2+3}. It differs from *B. neocognata* in possessing a broad medial longitudinal black band over abdominal terga III–V and from the other four species in possessing a costal band of uniform width that does not expand or widen at the apex.

Bactrocera (Bactrocera) collita Drew & Hancock

(Fig. 50)

Bactrocera (Bactrocera) collita Drew & Hancock, 1994a: 16; Norrbom *et al.*, 1998: 89. Holotype male in BBM.

Material examined: No new specimens available for study since the publication of Drew & Hancock (1994a).
Diagnosis: Face fulvous with a pair of large elongate oval black spots; postpronotal lobes and notopleura yellow; scutum black except brown around notopleural suture and inside postpronotal lobes; broad triangular lateral postsutural yellow vittae narrowing sharply posteriorly to end just before *ia.* seta; medial postsutural yellow vitta absent; mesopleural stripe reaching anterior *npl.* seta dorsally; scutellum yellow; legs with femora fulvous with a large black subapical spot on outer ½ of fore femora, a black spot on inner subapical surfaces of mid femora and dark fuscous to black on both lateral surfaces of apex of hind femora; all tibiae dark fuscous; wing with cells bc and c colourless, microtrichia in outer corner of cell c only; a narrow fuscous costal band just overlapping R_{2+3} and widening markedly across apex of R_{4+5}; a narrow fuscous anal streak; supernumerary lobe of medium development; abdominal terga III–V orange-brown with broad lateral and medial longitudinal black bands over all three terga which are joined along anterior margin of tergum III, a pair of oval orange-brown shining spots on tergum V; abdominal sterna dark coloured.
Attractant: Methyl eugenol.
Female: No known record.
Distribution:
Type locality: Philippines (Luzon I., Mt Makiling).
Hosts: No known record.
Remarks: *Bactrocera (Bactrocera) collita* Drew & Hancock, within the *dorsalis* complex, is similar to *Bactrocera (Bactrocera) dorsaloides* (Hardy & Adachi), *Bactrocera (Bactrocera) makilingensis* Drew & Hancock and *Bactrocera (Bactrocera) thailandica* Drew & Hancock in possessing triangular-shaped lateral postsutural vittae that taper distinctly posteriorly, dark markings on apices of all femora, a narrow black basal band on the scutellum and abdominal terga III–V with a black 'T' pattern and dark lateral colour patterns over all three terga. It differs from *B. thailandica* in having narrow- to medium-width dark lateral margins on abdominal terga III–V (in *B. thailandica* these markings are broad), from *B. dorsaloides* in having distinct dark spots on apices of all femora (in *B. dorsaloides* the dark patterns encircle the apices of all femora), and from *B. makilingensis* in having a broad expansion of the costal band in the apex of the wing, lateral postsutural yellow vittae ending before *ia.* setae and lateral dark margins on abdominal terga IV and V broader (in *B. makilingensis* these markings are very narrow).

Bactrocera (Bactrocera) commensurata Drew & Romig, new species

(Fig. 51)

Material examined: Holotype male, **PHILIPPINES: LUZON ISLAND** – Mt Makiling, 500 m, Philippines, 9.vi.1986, coll. B. Ramos, attracted to methyl eugenol, deposited in BMNH.
Paratypes: **PHILIPPINES: LUZON ISLAND** – 1 male (8.vi.1983), 7 males (9.vi.1983), 2 males (15.vi.1983), same data as holotype; 1 male (15.vi.1983), 1 male (12.vii.1983), Mt Makiling, 400 m, Philippines, coll. B. Ramos, attracted to methyl eugenol; 2 males, Mt Makiling, Philippines, 5.ix.1983, coll. R. Drew & B. Ramos, attracted to methyl eugenol; 5 males, Mt Makiling, 400 m, Philippines, 5.ix.1983, coll. R. Drew, attracted to cue lure; 8 males, Mt Makiling, 400 m, Philippines, 5.ix.1983, coll. R. Drew & Nette, attracted to methyl eugenol; 4 males, Mt Makiling, 300 m, Philippines, 8.vi.1983, coll. B. Ramos, attracted to methyl eugenol; 1 male, Mt Makiling, 200 m, Philippines, 8.vi.1983, coll. B. Ramos, 3 males (8.vi.1983), 2 males (15.vi.1983), Mt Makiling, 100 m, Philippines, coll. B. Ramos, attracted to methyl eugenol; 26 males, Mt Makiling, 100 m, 5.ix.1983, coll. R. Drew; 18 males, Mt Makiling, 100 m, 5.ix.1983, coll. Drew, Nette, Albert, attracted to methyl eugenol; 1 female and 10 males, Dasmarinas, Cavite, Philippines, 13.v.1983, coll. B. Ramos, attracted to methyl eugenol; 1 male (31.v.1983), 1 male (8.vi.1983), Dasmarinas, Cavite, Philippines, coll. A.M. Arida & N.E. Rivera; 8 males (17.vi.1983), 4 males (25.vi.1983), Hamtik, Antique, coll. A.S. Nogodula, attracted to methyl eugenol; 6 males, Hamtik, Antique, Philippines, 17.vi.1983, coll. A.M. Tenorio, attracted to methyl eugenol; 1 male (19.vi.1983), 12 males (10.vii.1983), Sta Cruz, Zambales, Philippines, coll. L. Montevirgen, attracted to methyl eugenol; 4 males, Pantabangan, Nueva Ecija, Philippines, 30.vi.1983, RCPC, Maligaya, attracted to methyl eugenol; 2 males, Guimaras Exp St, Jordan, Philippines, 24.vi.1983, coll. S.N. Iligau, attracted to methyl eugenol; 5 males, Bgy San Rogue, Philippines, 17.vi.1983, coll. M. de La Cruz, attracted to methyl eugenol; 3 males, Viejo, Iloilo, Philippines, 10.vi.1983, coll. A.S. Nogodula, attracted to methyl eugenol; 3 males, Guimbal, Iloilo, Philippines, 17.vi.1983, coll. A.S. Nogodula, attracted to methyl eugenol; 2 males (8.vi.1983), 1 male (14.vi.1983), Nueva Ecija, Philippines, coll. A.M. Tenorio, attracted to methyl eugenol; 8 males, Salawag, Dasmarinas, Cavite, 6.x.1983, coll. Bert, attracted to methyl eugenol; 1 male, Sepmin, Concepcion, Tarlae, Philippines, 7.vi.1983, coll. A.M. Tenorio, attracted to methyl eugenol; 1 male, Hugo Perez, Trece Martires City, Philippines, 8.vi.1983, coll. A.M. Arida & E. Rivera; 1 male, UPLB, Los Banos, Philippines, 30.vii.1983, coll. B. Ramos, attracted to methyl eugenol. 20 paratypes in ANIC, 20 paratypes in BMNH, 96 paratypes in QDPC, 20 paratypes in QMIC.
Diagnosis: A medium-sized species; face fulvous with a pair of small circular black spots; postpronotal lobes and notopleura yellow; scutum generally red-brown with a pair of irregularly shaped longitudinal black bands; lateral postsutural yellow vittae present; medial postsutural yellow vitta absent; no yellow spot anterior to notopleural suture; mesopleural stripe reaching midway between anterior margin of notopleuron and anterior *npl.* seta dorsally; scutellum yellow; wing with cells bc and c with a pale fuscous tint, microtrichia in outer corner of cell c only; a narrow fuscous costal band confluent with R_{2+3}; a narrow fuscous anal streak; abdominal terga III–V orange-brown with a 'T' pattern consisting of a narrow transverse dark fuscous band across anterior margin of tergum III and a narrow medial longitudinal black band over all three terga, narrow dark fuscous spot on anterolateral corners of tergum IV.

Description: male.
Head: Height 1.8 mm. Frons – length 1.38 times breadth; fulvous with fuscous on anteromedial hump; orbital setae black: 1 *s.or.*, 2 *i.or.*; lunule red-brown. Ocellar triangle black. Vertex fuscous. Face fulvous with a pair of small circular black spots; length 0.46 mm. Genae fulvous, pale fuscous subocular spot present; red-brown seta present. Occiput red-brown, fulvous along eye margins; occipital row with 4–7 black setae. Antennae with all segments entirely red-brown; length of segments: 0.15 mm; 0.3 mm; 0.72 mm.
Thorax: Scutum dark red-brown posteriorly to red-brown laterally and anteriorly with two irregularly shaped longitudinal black bands inside the lateral postsutural vittae. Pleural areas dark fuscous to black with red-brown below postpronotal lobes and above mid and hind coxae. Yellow markings as follows: postpronotal lobes; notopleura; mesopleural stripe reaching midway between anterior margin of notopleuron and anterior *npl.* seta dorsally, continuing to katepisternum as a large transverse spot, anterior margin convex; anatergite (posterior apex black); anterior ⅔ katatergite (remainder black); two broad parallel-sided lateral postsutural vittae ending just behind *ia.* seta. Medial longitudinal postsutural yellow vitta absent. Postnotum black, red-brown centrally. Scutellum yellow except for narrow black basal band. Setae: *sc.* 2; *prsc.* 2; *ia.* 1; *p.sa.* 1; *a.sa.* 1; *mpl.* 1; *npl.* 2; *scp.* 4.
Legs: All segments entirely fulvous except for all femora tending red-brown apically; mid tibiae each with an apical black spur.
Wings: Length 5.7 mm; cells bc and c with a pale fuscous tint; microtrichia in outer corner of cell c only; remainder of wings colourless except fuscous cell sc, narrow fuscous costal band confluent with R_{2+3} and remaining narrow around apex of wing, a narrow fuscous anal streak; a dense aggregation of microtrichia around A_1+CuA_2; supernumerary lobe of medium development.
Abdomen: Oval; terga free; pecten present on tergum III. Tergum I and sterna I and II wider than long. Tergum I orange-brown with narrow fuscous lateral margins; tergum II orange-brown anteriorly, fulvous posteriorly and with a narrow transverse dark fuscous band subanteriorly but not reaching the narrow fuscous anterolateral corners; terga III–V orange-brown with a dark 'T' pattern consisting of a narrow transverse dark fuscous band across anterior margin of tergum III, a narrow medial longitudinal black band over all three terga and small dark fuscous anterolateral spots on tergum IV. A pair of oval red-brown shining spots on tergum V. Posterior lobe of surstylus short, sternum V with a deep concavity on posterior margin. Abdominal sterna pale orange-brown.
Attractant: Methyl eugenol.
Female: As for male except no dense aggregation of microtrichia around A_1+CuA_2, pecten of cilia on abdominal tergum III absent; ovipositor basal segment red-brown, dorsoventrally compressed and tapering posteriorly in dorsal view; ratio of length of oviscape to length of tergum V, 0.75:1; apex of piercer needle-shaped.
Distribution: Philippines (Luzon I.).
Hosts: No known record.

Remarks: *Bactrocera (Bactrocera) commensurata* n.sp. is similar to *Bactrocera (Bactrocera) papayae* Drew & Hancock in possessing a narrow costal band and anal streak on the wing, femora entirely fulvous, abdominal terga III–V with a narrow black 'T' pattern. It differs from *B. papayae* in having a red-brown scutum with large areas of black, coloured cells bc and c and abdominal sterna pale coloured (generally orange-brown to pale red-brown). *Bactrocera commensurata* n.sp. is also similar to *Bactrocera (Bactrocera) blairiae* n.sp. in being a general red-brown species with a narrow costal band and pale-coloured abdominal sterna but differs from this species in having a narrow costal band that remains narrow at the apex and the scutum with large areas of dark patterns.

Bactrocera (Bactrocera) contigua Drew
(Fig. 52)

Bactrocera (Bactrocera) contigua Drew, 1989: 40; Norrbom *et al.*, 1998: 89. Holotype female in QMIC.

Non-type material examined: INDONESIA: MOLUCCAS (MALUKU): AMBON ISLAND – 2 males, Hila/Mamua, Ambon, Indonesia, 25-100m, 5.viii.1981, coll. J. Turner, residual forest with clove, nutmeg and banana groves, attracted to methyl eugenol; **NORTH MOLUCCAS (NORTH MALUKU)** – 23 males, North Maluku, West Halmahera, Goal, 18.vi.2006, coll. Raais Abdullah, attracted to methyl eugenol in forest; 40 males, Moluccas (North Maluku), West Halmahera, Gamlamo, 29.v.2007, coll. Raais Abdullah, attracted to methyl eugenol in mango and forest. Specimens in QDPC.
Diagnosis: Face fulvous with a pair of small circular black spots; postpronotal lobes and notopleura yellow; scutum black with brown to dark brown below and behind lateral postsutural vittae, along notopleural suture, between postpronotal lobes and notopleura, inside postpronotal lobes; lateral postsutural yellow vittae of medium-width and narrowing slightly posteriorly to end at *ia.* seta; medial postsutural yellow vitta absent; mesopleural stripe reaching midway between anterior margin of notopleuron and anterior *npl.* seta dorsally; scutellum yellow; legs with femora entirely fulvous; all tibiae pale fuscous to fuscous; wing length 6.0 mm, wing with cells bc and c with a very pale tint, microtrichia in outer corner of cell c only; broad dark fuscous costal band almost confluent with R_{4+5} and remaining of uniform width around apex of wing; a broad fuscous anal streak; supernumerary lobe of medium development; abdominal terga III–V orange-brown narrow dark fuscous anterolateral corners on tergum III and a very narrow pale fuscous medial longitudinal band over all three terga and usually broken at intersegmental lines, a pair of oval orange-brown shining spots on tergum V; abdominal sterna fuscous to dark fuscous.
Attractant: Methyl eugenol.
Female: As for male in the general body colour patterns. See Drew (1989) for a detailed description of female characters.
Distribution:
Type locality: Papua New Guinea (Bulolo, Robbies Creek).
New records: Indonesia (Moluccas = Maluku – Ambon I.); (North Moluccas = North Maluku – Halmahera I.).

Hosts: No known record.
Remarks: *Bactrocera (Bactrocera) contigua* Drew was adequately described and illustrated by Drew (1989) and belongs to the *musae* complex as defined by Drew *et al.* (2011), having a black scutum, wings colourless except for costal band and anal streak, abdominal terga III–V mostly orange-brown and males attracted to methyl eugenol. It is similar to *Bactrocera (Bactrocera) musae* (Tryon) in the above characters but differs in possessing a broad dark fuscous costal band almost confluent with R_{4+5} and a pale fulvous tint in cells bc and c.

Bactrocera (Bactrocera) correcta (Bezzi)

(Fig. 53)

Chaetodacus correctus Bezzi, 1916: 107. Syntypes in ZSI.
Bactrocera zonata – Bezzi, 1913: 94, [Misidentification] see Bezzi, 1916: 106.
Dacus dutti Kapoor, 1971: 480; Kapoor 1993: 74 (Synonym).
Strumeta paratuberculatus Philip, 1950: 31; Norrbom *et al.*, 1998: 89 (Synonym).
Dacus bangaloriensis Agarwal & Kapoor, 1983: 169; Norrbom *et al.*, 1998: 89 (Synonym).
Dacus (Strumeta) correctus – Hardy, 1973: 38.
Bactrocera (Bactrocera) correcta – Norrbom *et al.*, 1998: 89; Drew & Raghu, 2002: 335; Drew *et al.*, 2007: 2.

Non-type material examined: Since the publication of Drew & Raghu (2002), the following specimens have been studied – **PENINSULAR MALAYSIA**: 1 male (8.ix.1986), 1 male (17.xi.1986), TN1, Malaysia, coll. Yusoff Salleh, attracted to methyl eugenol. **SRI LANKA (formerly CEYLON)**: 1 male, Gannoruwa, 14.x.1993, attracted to methyl eugenol; 1 male, Looloowatta (Knuckle Bay), 1150m, 6.viii.1997, coll. K. Tsuruta. **VIETNAM: Northern Vietnam** – 3 males, Northern Vietnam, Tu Liem, Hanoi, 25.vii.2002, coll. NIPP staff, bred ex *Eugenia javanica* (rose apple), sample No. FN0985; 10 males (1.x.1999), 12 males (7.x.1999), 10 males (15.x.1999), 7 males (22.x.1999), Hanoi Province, Tu Liem District, Con Hue, Site N001, attracted to methyl eugenol; 2 males (14.x.1999), 4 males (21.x.1999), Hanoi Province, Tu Liem District, Lien Mac, Site N003, attracted to methyl eugenol; 1 male, Hanoi Province, Tu Liem District, Lien Mac Village, Site VN03, 13.vi.2002, attracted to methyl eugenol; 2 males, Hoa Binh Province, Ky Son District, Cao Phong, Site N014, attracted to methyl eugenol; **Southern Vietnam** – 6 females and 6 males, 21.iv.2000, coll. SOFRI, bred ex *Mangifera indica* (Fam. Anacardiaceae, mango); a number of specimens bred ex guava, mango, *Terminalia* sp. and water apple; 1 male (14.i.2002), 1 male (20.i.2002), 4 males (18.iii.2002), 4 males (1.iv.2002), Vinh Long Province, Long Ho District, Phuoc Hau Village, Site VS01, coll. SOFRI Staff, attracted to methyl eugenol; 2 males, Vinh Long Province, Long Ho District, Phuoc Hau Village, Site VS01, 2.vi.2002, attracted to methyl eugenol; 11 males (22.xii.2001), 8 males (25.ii.2002), 11 males (25.iii.2002), 3 males (5.v.2002), Dong Thap Province, Lap Vo District, Tan Khanh Trung Village, Site VS03, coll. SOFRI Staff, attracted to methyl eugenol; 2 males, Tien Giang Province, My Tho Village, Site S005, attracted to methyl eugenol; 6 males (26.ii.2002), 1 male (12.iii.2002), Binh Duong Province, Thuan An District, An Thanh Village, Site VS05, coll. SOFRI Staff, attracted to methyl eugenol; 1 male (28.ii.2002), 2 males (25.iii.2002), Binh Thuan Province, Ha Thuan Bac District, No. 1 Village, Site VS07, coll. SOFRI Staff, attracted to methyl eugenol; 1 female, Tien Giang Province, Cai Be District, An Huu Village, Site S009, in cue lure trap; 37 males, Hoi An, 21.v.1997, coll. G. Pattemore, attracted to methyl eugenol; 4 females, South Vietnam, Long Tri, Long An, 25.vii.2004, coll. Chris Clark, bred from Dragonfruit, (white flesh), *Hylocereus undatus*, cv Binh Thuan. Specimens in QDPC.
Diagnosis: Face fulvous with a pair of transverse elongate black spots almost meeting in centre; postpronotal lobes and notopleura yellow; scutum black with dark red-brown along lateral and posterior margins; broad parallel-sided lateral postsutural yellow vittae ending behind *ia.* seta; medial postsutural yellow vitta absent; mesopleural stripe reaching almost to anterior *npl.* seta dorsally; scutellum yellow with narrow black basal band; legs with all segments entirely fulvous except hind tibiae pale fuscous; wing with cells bc and c colourless, both cells entirely devoid of microtrichia; a narrow pale fuscous costal band confluent with R_{2+3} and ending at apex of this vein, a small oval fuscous spot across apex of R_{4+5}; anal streak absent but with a pale fuscous tint within cell cup; supernumerary lobe of medium development; abdominal terga III–V red-brown with a 'T' pattern consisting of a narrow transverse black band across anterior margin of tergum III and a narrow medial longitudinal black band over all three terga, narrow black anterolateral corners on terga IV and V, a pair of oval red-brown shining spots on tergum V.
Attractant: Methyl eugenol.
Female: As for male except no dense aggregation of microtrichia around A_1+CuA_2; supernumerary lobe weak; pecten of cilia on abdominal tergum III absent; ovipositor basal segment orange-brown, dorsoventrally compressed and tapering posteriorly in dorsal view; ratio of length of oviscape to length of tergum V, 0.9: 1; apex of piercer needle-shaped.
Distribution:
Type locality: India: Pusa, Bihar; Coimbatore; Guindy, Madras; Hagari, Bellary District, Madras. In the original description, Bezzi (1916) listed specimens studied from these four locations but did not designate a Holotype.
Previously recorded: Type locality and the following locations – Bhutan; Myanmar; Nepal; Pakistan; Southern China; Sri Lanka; Thailand.
New records: Peninsular Malaysia (northern area); Vietnam.
Hosts: Recorded from a wide range of commercial/edible host fruits and wild forest fruits. See Allwood *et al.* (1999) for detailed host records and Tsuruta *et al.* (1997) for records in Sri Lanka.
New records: *Hylocereus undatus* (family *Cactaceae*) (dragonfruit).
Remarks: *Bactrocera (Bactrocera) correcta* (Bezzi) is similar to *Bactrocera (Bactrocera) dorsalis* (Hendel) in the general colour patterns of the body, wings and legs but

differs from *B. dorsalis* in possessing transverse facial spots and an incomplete costal band. It is also similar to *Bactrocera (Bactrocera) penecorrecta* Drew in the general colour patterns of the body and wings but differs from this species in having abdominal terga III–V mostly pale coloured (not mostly black as in *B. penecorrecta*) and the scutellum with a narrow black basal band.

Bactrocera (Bactrocera) costalis (Shiraki)

(Fig. 54)

Chaetodacus costalis Shiraki, 1933: 66. Lectotype male and Paralectotype female, here designated, in NIAS.
Bactrocera (Bactrocera) costalis – Norrbom *et al.*, 1998: 89.

Type material examined: TAIWAN (formerly FORMOSA): Lectotype male here designated with the following five labels – 'Kuranu, 25.v.1921, Col. T. Shiraki'; 'Japanese writing, Col. T. Shiraki'; 'Caetodacus costalis n sp det. T. Shiraki'; 'Type'; 'Kőrper'. One female Paralectotype here designated with the following four labels – 'Pinan, 2.vi.1921, Col. T. Shiraki'; 'Japanese writing, Col. T. Shiraki'; 'Caetodacus costalis n sp T. Shiraki'; 'Type'. Note the Misspelling of *Chaetodacus* in the label data. Received on loan from NIAS.

Non-type material examined: PHILIPPINES: LUZON ISLAND – 15 males, Mt Makiling, 600m, Philippines, 9.v.1983, coll. R. Drew, B. Ramos, attracted to cue lure; 7 males, Mt Makiling, 500m, Philippines, 9.v.1983, coll. R. Drew, B. Ramos, attracted to cue lure; 19 males, Mt Makiling, 400m, 9.v.1983, coll. Drew, Nette, Bert, attracted to cue lure; 9 males, Mt Makiling, 200m, 5.ix.1983, coll. Drew, Nette, Bert, attracted to cue lure; 1 male, Mt Makiling, 100m, 5.ix.1983, coll. Drew, Nette, Bert; 1 male, Dasmarinas, Cavite, Philippines, 13.v.1983, coll. B. Ramos, R. Drew, attracted to cue lure; 1 male, Pantabangan, Nueva Ecija, Philippines, 30.vi.1983, R.C.P.C. Maligaya, attracted to cue lure; 1 male, Bagumbayan, Batangas, Philippines, 17.v.1983, coll. A.M. Arida, attracted to cue lure; **MINDANAO ISLAND** – 1 male, Malagos, Davao, Philippines, 15.v.1983, coll. R. de la Cruz, attracted to cue lure; 1 male, D.E.S. Bago Oshiro, Philippines, 13.v.1983, coll. D.A. Nicolas, attracted to cue lure; 1 male, in cultivation, D.E.S. Davao City, Philippines, 16.v.1983, coll. D.A. Nicolas, attracted to cue lure. Specimens in QDPC.

Diagnosis: A medium-sized species; face fulvous with a pair of medium-sized circular black spots; postpronotal lobes and notopleura yellow; scutum black; lateral postsutural yellow vittae present; medial postsutural yellow vitta absent; no yellow spot anterior to notopleural suture; mesopleural stripe reaching midway between anterior margin of notopleuron and anterior *npl.* seta dorsally; scutellum yellow with a narrow black basal band; wing with cells bc and c pale fuscous, microtrichia in outer ½ of cell c only; a broad fuscous costal band confluent with R_{4+5}; a broad fuscous anal streak; abdominal terga III–V orange-brown with a 'T' pattern consisting of a narrow black band across anterior margin on tergum III and a narrow medial longitudinal fuscous band over all three terga.

Description: male.

Head: Height 1.8 mm. Frons – length 1.4 times breadth; red-brown with fuscous around orbital setae and on anteromedial hump; orbital setae black: 1 *s.or.*, 2 *i.or.*; lunule fuscous. Ocellar triangle black. Vertex fuscous. Face fulvous with a pair of medium-sized circular black spots; length 0.5 mm. Genae fulvous, small dark fuscous subocular spot present; red-brown seta present. Occiput red-brown, fulvous along eye margins; occipital row with 3–5 strong black setae. Antennae with segments 1 and 2 red-brown, segment 3 red-brown with fuscous on apex and outer surface; length of segments: 0.22 mm; 0.34 mm; 0.78 mm.

Thorax: Scutum black with dark red-brown across posterior margin and red-brown below and behind lateral postsutural vittae, around notopleural suture, and along lateral margins and inside postpronotal lobes anterior to notopleural suture. Pleural areas black except red-brown below postpronotal lobes. Yellow markings as follows: postpronotal lobes; notopleura; mesopleural stripe reaching midway between anterior margin of notopleuron and anterior *npl.* seta dorsally, continuing to katepisternum as a small transverse spot, anterior margin straight; anatergite (posterior apex black); anterior ¾ katatergite (remainder black); two moderately broad lateral postsutural vittae narrowing slightly posteriorly to end at *ia.* seta. Medial longitudinal postsutural yellow vitta absent. Postnotum black laterally, dark red-brown centrally. Scutellum yellow except for narrow black basal band. Setae: *sc.* 2; *prsc.* 2; *ia.* 1; *p.sa.* 1; *a.sa.* 1; *mpl.* 1; *npl.* 2; *scp.* 4.

Legs: Femora fulvous except for a small oval-shaped pale fuscous spot subapically on fore femora; fore tibiae pale fuscous basally to red-brown apically, mid tibiae fuscous basally to pale fuscous apically, hind tibiae dark fuscous; tarsal segments with basal segment fulvous and apical four segments red-brown; mid tibiae each with an apical black spur.

Wings: Length 6.3 mm; cells bc and c pale fuscous; microtrichia in outer ½ of cell c only; remainder of wings colourless except fuscous cell sc, broad fuscous costal band confluent with R_{4+5} and ending at apex of wing between R_{4+5} and M, broad fuscous anal streak; dense aggregation of microtrichia around A_1+CuA_2; supernumerary lobe weak.

Abdomen: Oval; terga free; pecten present on tergum III. Tergum I and sterna I and II wider than long. Tergum I fuscous centrally, dark fuscous laterally and with a narrow orange-brown transverse band across posterior margin but not reaching lateral margins; tergum II orange-brown anteriorly, fulvous posteriorly and with a small transverse black spot anteriorly on mid-line; terga III–V orange-brown with a 'T' pattern consisting of a narrow transverse black band across anterior margin of tergum III and a narrow medial longitudinal fuscous band over all three terga (this band may be broken at the intersegmental line between terga III–IV). A pair of oval orange-brown shining spots on tergum V. Posterior lobe of surstylus short, sternum V with a slight concavity of posterior margin.

Attractant: Cue lure.

Female: As for male except no dense aggregation of microtrichia around A_1+CuA_2; pecten of cilia on abdominal tergum III absent; ovipositor basal segment orange-brown, dorsoventrally compressed and tapering posteriorly in dorsal view;

ratio of length of oviscape to length of tergum V, 0.3:1; apex of piercer broad and needle-shaped.

Distribution:

Type locality: Taiwan (Kuranu).

New records: Philippines (Luzon I.; Mindanao I.).

Hosts: *Bryonopsis laciniosa* (family *Cucurbitaceae*), *Solanum suratense* (family *Solanaceae*), recorded by Shiraki (1933).

Remarks: *Bactrocera (Bactrocera) costalis* (Shiraki) is similar to *Bactrocera (Bactrocera) bruneiae* n.sp., *Bactrocera (Bactrocera) eurycosta* n.sp. and *Bactrocera (Bactrocera) ritsemai* (Weyenbergh) in possessing a black scutum, the costal band of uniform width throughout and confluent with R_{4+5}, cells bc and c colourless or with a pale fuscous tint and femora entirely fulvous with, at most, a dark preapical spot on the fore femora. It differs from *B. bruneiae* in having oval facial spots, from *B. eurycosta* in having a black scutum with red-brown margins and from *B. ritsemai* in having the costal band of uniform colour throughout, a broad anal streak and a narrow- to medium-width fuscous medial longitudinal band on abdominal terga III–V.

The subgeneric position of *B. costalis* is uncertain, having the posterior lobe of the surstyli short and a slight concavity on posterior margin of abdominal sternum V. With this combination, it fits into the *Melanodacus* group of subgenera but we have decided to leave this species in *B. (Bactrocera)* until more specimens are available for study. In the original description, Shiraki (1933) did not designate a holotype, listing all specimens studied as types. Consequently, the male listed above in 'Type material examined' is hereby designated as the Lectotype and the female the Paralectotype.

Bactrocera (Bactrocera) curtivitta Drew & Romig, new species

(Fig. 55)

Material examined: Holotype male, **INDIA: ANDAMAN ISLANDS** – India, S. Andaman, Port Blair, 15.iv.1994, coll. Ranganath, Sl. No. 41, probably hand collected on Meliaceae, deposited in BMNH.

Paratypes: **INDIA: ANDAMAN ISLANDS** – 1 male and 2 females, same data as holotype. 1 paratype in ANIC, 1 paratype in QDPC, 1 paratype in QMIC.

Diagnosis: A medium-sized species; face fulvous with a pair of moderately large oval black spots; postpronotal lobes and notopleura yellow; scutum red-brown; very short and narrow lateral postsutural yellow vittae present; medial postsutural yellow vitta absent; no yellow spot anterior to notopleural suture; mesopleural stripe reaching midway between anterior margin of notopleuron and anterior *npl.* seta dorsally; scutellum yellow; wing with cells bc and c colourless, microtrichia in outer corner of cell c only; a narrow dark fuscous costal band slightly overlapping R_{2+3}; a narrow dark fuscous anal streak; abdominal terga III–V red-brown with a 'T' pattern consisting of a narrow transverse black band across anterior margin of tergum III and a moderately broad medial longitudinal black band over all three terga, narrow dark fuscous lateral margins on terga IV and V.

Description: male.

Head: Height 1.47 mm. Frons – length 1.57 times breadth; red-brown without dark markings; orbital setae dark red-brown: 1 *s.or.*, 2 *i.or.*; lunule red-brown. Ocellar triangle black. Vertex red-brown. Face fulvous with a pair of moderately large oval black spots; length 0.42 mm. Genae fulvous, small dark fuscous subocular spot present; black seta present. Occiput red-brown, fulvous along eye margins; occipital row with 3–5 red-brown setae. Antennae with segments 1 and 2 red-brown, segment 3 red-brown with fuscous on apex; length of segments: 0.16 mm; 0.3 mm; 0.74 mm.

Thorax: Scutum red-brown with narrow fuscous longitudinal bands either side of midline. Pleural areas red-brown with fuscous on anterior and posterior margins of mesopleural stripe and on katepisternum. Yellow markings as follows: postpronotal lobes; notopleura; mesopleural stripe reaching midway between anterior margin of notopleuron and anterior *npl.* seta dorsally, continuing to katepisternum as a transverse spot, anterior margin convex; anatergite (posterior apex black); anterior ⅔ katatergite (remainder black); two very short and narrow lateral postsutural vittae barely reaching the line of the *a.sa.* seta. Medial longitudinal postsutural yellow vitta absent. Postnotum red-brown with black on lateral margins. Scutellum yellow with narrow red-brown basal band. Setae: *sc.* 2; *prsc.* 2; *ia.* 1; *p.sa.* 1; *a.sa.* 1; *mpl.* 1; *npl.* 2; *scp.* 4.

Legs: All femora entirely fulvous; fore and hind tibiae dark fuscous, mid tibiae dark fuscous basally to fulvous apically; fore tarsi with basal segment fulvous and apical four segments fuscous, mid and hind tarsi with all segments entirely fulvous; mid tibiae each with an apical black spur.

Wings: Length 5.0 mm; cells bc and c colourless; microtrichia in outer corner of cell c only; remainder of wings colourless except dark fuscous cell sc, narrow dark fuscous costal band overlapping R_{2+3} where it is pale fuscous and widening slightly around apex of wing, narrow dark fuscous anal streak; a dense aggregation of microtrichia around A_1+CuA_2; supernumerary lobe of medium development.

Abdomen: Oval; terga free; pecten present on tergum III. Tergum I and sterna I and II wider than long. Tergum I fuscous medially, dark fuscous on lateral margins and red-brown along posterior margin; tergum II red-brown anteriorly, fulvous posteriorly with a narrow transverse dark fuscous band subanteriorly and narrow dark fuscous anterolateral corners; terga III–V red-brown with a dark 'T' pattern consisting of a narrow transverse black band across anterior margin of tergum III and a moderately broad medial longitudinal black band over all three terga, narrow dark fuscous lateral margins on terga IV and V. A pair of oval fuscous shining spots on tergum V. Posterior lobe of surstylus short, sternum V with a deep concavity on posterior margin. Abdominal sterna dark coloured.

Attractant: No known record.

Female: As for male except no dense aggregation of microtrichia around A_1+CuA_2, supernumerary lobe weak, pecten of cilia on abdominal tergum III absent; ovipositor basal segment red-brown, dorsoventrally compressed and tapering posteriorly in dorsal view; ratio of length of oviscape to length of tergum V, 0.55:1; apex of piercer needle-shaped.

Distribution: India (South Andaman I., Port Blair).
Hosts: No known record.
Remarks: *Bactrocera (Bactrocera) curtivitta* n.sp. is similar to *Bactrocera (Bactrocera) bhutaniae* n.sp., *Bactrocera (Bactrocera) brunneola* White & Tsuruta and *Bactrocera (Bactrocera) suliae* n.sp. in possessing a red-brown scutum, the costal band overlapping R_{2+3}, cells bc and c colourless, abdominal terga III–V with a black 'T' pattern and narrow lateral dark margins. It differs from *B. suliae* in having dark-coloured abdominal sterna, from *B. bhutaniae* in having narrow lateral postsutural vittae ending before the *ia.* seta and from *B. brunneola* in having the costal band clearly overlapping R_{2+3} and widening at apex of wing, facial spots circular and medium-sized and lateral postsutural vittae very short and ending at level of *ia.* seta.

Bactrocera (Bactrocera) curvifera (Walker)

(Figs 56A, 56B, 56C)

Dacus curvifer Walker, 1864: 229. Lectotype male in BMNH.
Neodacus curvifer – Perkins, 1939: 21.
Dacus (Neodacus) curvifer – Hardy, 1954: 8; Hardy, 1959: 168, Lectotype designation by inference of Holotype.
Dacus speculifer Walker, 1865: 122; Hardy, 1959: 168, 182, as synonym of *B. curvifer* and Lectotype designation by inference of Holotype.
Dacus (Strumeta) curvifer – Drew 1974: 27.
Bactrocera (Bactrocera) curvifera – Drew, 1989: 128; Norrbom *et al.*, 1998: 89.

Type material examined: Lectotype male, **INDONESIA: WEST PAPUA (formerly part of IRIAN JAYA)** – with the following seven labels – 'Type, *Dacus curvifer* Wlk'; 'Waigiou, A.R. Wallace, 68-4'; 'Wag'; '68.4'; 'curvifer'; 'Lectotype male of *Dacus curvifer* Walker 1864, Des D.E. Hardy 1959, INDONESIA, Irian Jaya, Waigiou, A.R. Wallace'; '*Dacus (Neodacus) curvifer* Walker, Det 1954, D.E. Hardy'. Received on loan from BMNH.
Non-type material examined: INDONESIA: MOLUCCAS (MALUKU) – 1 male, Seram (Western part), Waimital, 2.viii.2007, coll. A. Haris Tualepe, attracted to methyl eugenol in jeruk; 1 male, Maluku, Western part, Kairatu, Waimital, 19.vii.2006, coll. A. Haris Tualepe, attracted to methyl eugenol in jeruk; 4 males, Maluku, western part, Kairatu, Nuruwe, 6.vii.2007, coll. A. Haris Tualepe, attracted to methyl eugenol in forest; 1 male (26.vii.2007), 1 male (2.viii.2007), Maluku, Seram (western part), Waimital, coll. A. Haris Tualepe, attracted to methyl eugenol in jeruk; 16 males, Ambon, Waai, <20m, 6.vii.1981, coll. J. Turner, attracted to methyl eugenol in residual forest with cloves, nutmeg and banana groves; 5 males, Ambon, Paso, 20-50m, 7.vii.1981, coll. J. Turner in residual rainforest; 6 males, Ambon, Soya, 300m, 4.viii.1981, coll. J. Turner, attracted to methyl eugenol; 12 males, Ambon, Hila/Mamua, 25-100m, 5.viii.1981, coll. J. Turner, attracted to methyl eugenol in residual rainforest; 2 males, Ambon, Mamala, 50-150m, 7.viii.1981, coll. J. Turner, attracted to methyl eugenol in residual forest with cloves, nutmeg and banana groves; **NORTH MOLUCCAS (NORTH MALUKU)** – 2 males (3.vi.2007), 1 male (10.vi.2007), 6 males (17.vi.2007), North Maluku, Kodya Ternate, Gambesi, coll. La Ruti, attracted to methyl eugenol in jeruk; 4 males, North Maluku, Kodya Ternate, Kalumata, 10.vi.2007, coll. La Ruti, attracted to methyl eugenol in mangga; 54 males (29.v.2007), 50 males (12.vi.2007), 36 males (18.vi.2007), North Maluku, West Halmahera Gaol, coll. Raais Abdullah, attracted to methyl eugenol in forest. Specimens in QDPC.
Diagnosis: A large species; face fulvous with a pair of medium-sized irregularly circular black spots; postpronotal lobes and notopleura yellow; scutum entirely dull black except dark fuscous below lateral postsutural vittae; lateral postsutural yellow vittae present; medial postsutural yellow vitta absent; no yellow spot anterior to notopleural suture; mesopleural stripe reaching to anterior *npl.* seta dorsally; scutellum yellow; *prsc.* setae absent; wing with cells bc and c fuscous, microtrichia in outer ⅓ of cell c only; a broad dark fuscous costal band confluent with R_{4+5}, a narrow transverse dark fuscous band connecting costal band to anal streak; a broad transverse dark fuscous band across centre of wing which re-curves back towards vein M, a narrow transverse dark fuscous to fuscous band at apex of wing connecting costal band to hind margin; a broad dark fuscous anal streak; abdominal terga III–V entirely dull black.
Description: male.
Head: Height 2.1 mm. Frons – length 1.5 times breadth; red-brown with dark fuscous around bases of orbital setae; orbital setae black: 1 *s.or.*, 2 *i.or.*; lunule pale fuscous. Ocellar triangle black. Vertex pale fuscous. Face fulvous with a pair of medium-sized irregularly circular black spots; length 0.74 mm. Genae fulvous, pale fuscous subocular spot present; black seta present. Occiput red-brown, fulvous along eye margins; occipital row with a large number of small and large black setae. Antennae with segments 1 and 2 red-brown, segment 3 red-brown with dark fuscous on apex and outer surface; length of segments: 0.27 mm; 0.47 mm; 1.2 mm.
Thorax: Scutum entirely dull black except dark fuscous below lateral postsutural vittae. Pleural areas entirely black. Yellow markings as follows: postpronotal lobes; notopleura; mesopleural stripe reaching anterior *npl.* seta dorsally, continuing to katepisternum as a transverse spot, anterior margin convex; anatergite (posterior apex black); anterior ⅔ katatergite (remainder black); two broad parallel-sided lateral postsutural vittae ending at or just behind *ia.* seta. Medial longitudinal postsutural yellow vitta absent. Postnotum black. Scutellum yellow except for narrow black basal band. Setae: *sc.* 2; *prsc.* absent; *ia.* 1; *p.sa.* 1; *a.sa.* 1; *mpl.* 1; *npl.* 2; *scp.* 4.
Legs: All femora fulvous, tending red-brown on apical ⅓ to ¼; all tibiae and tarsal segments fulvous; mid tibiae each with an apical black spur.
Wings: Length 7.8 mm; cells bc and c fuscous (except for a paler fuscous area in centre of cell c); dense microtrichia covering outer ⅓ of cell c only; remainder of wings colourless except for a dark fuscous cell sc, a broad dark fuscous costal band confluent with R_{4+5} and remaining of uniform width throughout, a narrow transverse dark fuscous band connecting costal band to anal streak, a broad dark fuscous transverse

band across centre of wing enclosing both crossveins and covering outer ½ of cell dm and with a narrow re-curved arm turning back towards base of wing and anterior to vein M, a narrow dark fuscous to fuscous transverse band at apex of wing from costal band to hind margin, a broad dark fuscous anal streak; a dense aggregation of microtrichia around A_1+CuA_2; supernumerary lobe of medium development.
Abdomen: Oval; terga free; pecten present on tergum III. Tergum I and sterna I and II wider than long. Tergum I entirely black; tergum II black except for a moderately broad transverse fulvous band across posterior margin and reaching posterolateral corners; abdominal terga III–V entirely dull black. A pair of oval black shining spots on tergum V. Posterior lobe of surstylus short, sternum V with a deep concavity on posterior margin. Abdominal sterna dark fuscous to black.
Attractant: Methyl eugenol.
Female: No known record.
Distribution:
Type locality: Indonesia (West Papua – Waigiou I., formerly part of Irian Jaya).
New records: Indonesia (Moluccas = Maluku – Ambon I.; Seram I.); (North Moluccas = North Maluku – West Halmahera; Ternate).
Hosts: No known record.
Remarks: Hardy (1959) synonymised *Dacus curvifer* Walker from Waigiou and *Dacus speculifer* Walker from Papua New Guinea. We have now had the opportunity to study Lectotypes and large series of specimens of both populations and are satisfied that the synonymy of Hardy is incorrect and that these are both good species. The specimens of *Bactrocera (Bactrocera) curvifera* (Walker) studied for this description were collected from the Moluccas (Maluku). The type locality of *B. curvifera* is Waigiou I., formerly part of the Moluccas (now part of West Papua). All specimens of *B. curvifera*, and the male Lectotype, have the same wing pattern with very little intraspecific variation (Fig. 56B). Similarly, in a large series of specimens of *Bactrocera (Bactrocera) speculifera* (Walker) from Papua New Guinea, and the male Lectotype, all have a distinctly different wing pattern to *B. curvifera* with little intraspecific variation (Fig. 56C).

Bactrocera curvifera is clearly diagnosed by the wing pattern which has a re-curved arm on the central dark fuscous broad transverse band. Clearly, this re-curved pattern was the reason for which Walker (1864) gave the specific name *curvifer* (Fig. 56B). In addition, this is a large species with a black scutum, broad parallel-sided lateral postsutural vittae and abdominal terga III–V entirely black.

Bactrocera (Bactrocera) curvosterna Drew & Romig, new species

(Figs 57A, 57B)

Material examined: Holotype male, **INDONESIA: SULAWESI** – Sulawesi Tengah, Donggala, Nupabomba, 15.x.2006, coll. Rusdin, attracted to cue lure in forest, deposited in BMNH.

Paratypes: **INDONESIA: SULAWESI** – 1 male, Sulawesi Tengah, Donggala, Nupabomba, 27.ix.2006, coll. Rusdin, attracted to cue lure in forest; 2 males (12.ix.2006), 13 males (17.x.2006), 12 males (31.x.2006), Sulawesi, Gorontalo, Bone Bolango, Lombongo, coll. Yusbar Ismail, attracted to cue lure in forest; 1 male, Sulawesi, Gorontalo, Bone Bolango, Suwawa, 26.ix.2006, coll. Yusbar Ismail, attracted to cue lure in forest; 1 male, Sulawesi, Sulawesi Barat, Mamuju, Belang-Belang, 30.x.2006, coll. Sri Hastuti, Yan P.F., Sukri, attracted to cue lure in forest, orange (jeruk). 5 paratypes in ANIC, 5 paratypes in BMNH, 15 paratypes in QDPC, 5 paratypes in QMIC.
Diagnosis: A medium-sized species; face fulvous with a pair of large circular to oval black spots; postpronotal lobes and notopleura yellow; scutum black; lateral postsutural yellow vittae present; medial postsutural yellow vitta absent; no yellow spot anterior to notopleural suture; mesopleural stripe reaching midway between anterior margin of notopleuron and anterior *npl.* seta dorsally; scutellum yellow; wing with cell bc pale fuscous, cell c paler fuscous, microtrichia in outer corner of cell c only; a broad dark fuscous costal band confluent with R_{4+5} and becoming paler between R_{2+3} and R_{4+5} and widening slightly across apex of wing; a broad dark fuscous anal streak; abdominal terga III–V dark orange-brown with a dark 'T' pattern consisting of a narrow dark fuscous transverse band across anterior margin of tergum III which expands to fuscous across lateral margins of that tergum and a moderately broad medial longitudinal black band over all three terga, fuscous shining spots on tergum V.
Description: male (Fig. 57A).
Head: Height 1.57 mm. Frons – length 1.54 times breadth; red-brown with dark fuscous on anteromedial hump and around bases of orbital setae; orbital setae black: 1 *s.or.*, 2 *i.or.*; lunule dark fuscous. Ocellar triangle black. Vertex dark fuscous. Face fulvous with a pair of large circular to oval black spots; length 0.44 mm. Genae fulvous, large dark fuscous subocular spot present; black seta present. Occiput fuscous, fulvous along eye margins; occipital row with 4–6 black setae. Antennae with segments 1 and 2 red-brown, segment 3 red-brown with fuscous on apex and outer surface; length of segments: 0.19 mm; 0.31 mm; 0.74 mm.
Thorax: Scutum black with dark red-brown below lateral postsutural vittae, around notopleural suture, along lateral margins between postpronotal lobe and notopleuron, inside postpronotal lobes. Pleural areas black except dark fuscous below postpronotal lobes. Yellow markings as follows: postpronotal lobes; notopleura; mesopleural stripe reaching midway between anterior margin of notopleuron and anterior *npl.* seta dorsally, continuing to katepisternum as a transverse spot, anterior margin straight; anatergite (posterior apex black); anterior ¾ katatergite (remainder black); two broad lateral postsutural vittae narrowing posteriorly to end at *ia.* seta. Medial longitudinal postsutural yellow vitta absent. Postnotum black. Scutellum yellow except for narrow black basal band. Setae: *sc.* 2; *prsc.* 2; *ia.* 1; *p.sa.* 1; *a.sa.* 1; *mpl.* 1; *npl.* 2; *scp.* 4.
Legs: All femora entirely fulvous; fore tibiae pale fuscous, mid tibiae pale fuscous basally to paler fuscous apically,

hind tibiae dark fuscous; all tarsi with basal segment fulvous and apical four segments red-brown; mid tibiae each with an apical black spur.

Wings: Length 6.0 mm; cell bc pale fuscous, cell c paler fuscous; microtrichia in outer corner of cell c only; remainder of wings colourless except dark fuscous cell sc, broad dark fuscous costal band confluent with R_{4+5} but paling to fuscous between R_{2+3} and R_{4+5} and slightly wider in shape across apex of wing, a broad dark fuscous anal streak; a dense aggregation of microtrichia around A_1+CuA_2; supernumerary lobe of medium development.

Abdomen: Oval; terga free; pecten present on tergum III. Tergum I and sterna I and II wider than long. Tergum I dark orange-brown with narrow dark fuscous lateral margins; tergum II dark orange-brown anteriorly and fulvous posteriorly with a narrow fuscous transverse band subanteriorly and reaching narrow fuscous lateral margins; terga III–V dark orange-brown with a dark 'T' pattern consisting of a narrow dark fuscous transverse band across anterior margin of tergum III which expands across lateral margins of this tergum where it becomes fuscous, a moderately broad medial longitudinal black band over all three terga. A pair of oval fuscous shining spots on tergum V. Posterior lobe of surstylus short, sternum V with an unusual horseshoe-shape (see Fig. 57B for shape of abdominal sterna). Abdominal sterna orange-brown to red-brown.

Attractant: Cue lure.

Female: No known record.

Distribution: Indonesia (Sulawesi).

Hosts: No known record.

Remarks: *Bactrocera (Bactrocera) curvosterna* n.sp. is similar to *Bactrocera (Bactrocera) nanoarcuata* n.sp. and *Bactrocera (Bactrocera) paralatissima* n.sp. in possessing an entire costal band confluent with R_{4+5}, a black scutum, cells bc and c coloured and tapering lateral postsutural vittae. It differs from *B. nanoarcuata* in having a horseshoe-shaped abdominal sternum V and from *B. paralatissima* in having distinct facial spots and pale abdominal sterna that are generally orange-brown to red-brown.

Bactrocera (Bactrocera) dispar (Hardy)

(Fig. 58)

Dacus (Bactrocera) dispar Hardy, 1982: 222. Holotype male in BBM.

Bactrocera (Bactrocera) dispar – Norrbom *et al.*, 1998: 90.

Type material examined: Holotype male, **INDONESIA: SULAWESI** – Central Sulawesi, Wotu, Aug. 1975, coll. N. Gough, collected in lowland forest. Studied at MZB.

Paratypes: **INDONESIA: SULAWESI** – 1 male, 'paredean via Palopo, 600 m, South Sulawesi'; 'Aug. 1975, N. Gough'; 'Paratype male, Dacus (Bactrocera) dispar, D.E. Hardy'. 1 male, 'Wotu, Central Sulawesi, Aug. 1975, N. Gough'; 'Paratype male, Dacus (Bactrocera) dispar, D.E. Hardy'. Received on loan from BBM.

Diagnosis: A large species; face fulvous with a pair of medium-sized circular black spots; postpronotal lobes and notopleura yellow; scutum basically red-brown with irregularly shaped lateral longitudinal black markings; broad parallel-sided lateral postsutural yellow vittae present; medial postsutural yellow vitta absent; a yellow spot anterior to notopleural suture; mesopleural stripe reaching just anterior to anterior *npl.* seta dorsally; scutellum yellow with a narrow red-brown basal band; wing with cells bc and c pale fuscous, microtrichia in outer ½ of cell c only; a broad fuscous costal band confluent with R_{4+5} and expanding slightly across apex of wing where it becomes dark fuscous; a broad pale fuscous anal streak; abdominal terga III–V orange-brown with a black 'T' pattern consisting of a narrow black band across anterior margin of tergum III and widening to cover lateral margins of this tergum and a medium-width medial longitudinal band over all three terga which is black on terga III and IV and tending dark fuscous on tergum V.

Description: male.

Head: Height 1.92 mm. Frons – length 1.36 times breadth; red-brown with dark fuscous around orbital setae; orbital setae black: 1 *s.or.*, 2 *i.or.*; lunule fuscous. Ocellar triangle black. Vertex dark fuscous. Face fulvous with medium-sized circular black spots; length 0.62 mm. Genae red-brown, small pale fuscous subocular spot present; red-brown seta present. Occiput red-brown, fulvous along eye margins; occipital row with 2–5 strong black setae. Antennae with segments 1 and 2 red-brown, segment 3 red-brown with fuscous on apex and outer surface; length of segments: 0.19 mm; 0.4 mm; 0.87 mm.

Thorax: Scutum with base colour red-brown with two irregularly shaped lateral longitudinal black markings running from inside lateral postsutural vittae to anterior margin of scutum. Pleural areas red-brown with irregular black markings on katepisternum, anterior and posterior to mesopleural stripe and above hind coxae. Yellow markings as follows: postpronotal lobes; notopleura; broad mesopleural stripe reaching anterior to anterior *npl.* seta dorsally, continuing to katepisternum as a transverse spot, anterior margin straight; anatergite (posterior apex black); anterior ⅔ katatergite (remainder black); two broad parallel-sided lateral postsutural vittae beginning with a spot anterior to notopleural suture anteriorly and ending posteriorly just behind *ia.* seta. Medial longitudinal postsutural yellow vitta absent. Postnotum red-brown centrally, black laterally. Scutellum yellow with a narrow red-brown basal band. Setae: *sc.* 2; *prsc.* absent; *ia.* 1; *p.sa.* 1; *a.sa.* 1; *mpl.* 1; *npl.* 2; *scp.* 4.

Legs: All segments entirely fulvous; mid tibiae each with an apical black spur.

Wings: Length 7.65 mm; cells bc and c pale fuscous; microtrichia in outer ½ of cell c only; remainder of wings colourless except fuscous cell sc, broad fuscous costal band confluent with R_{4+5} before widening across apex of wing where it becomes darker fuscous, broad pale fuscous anal streak; dense aggregation of microtrichia around A_1+CuA_2; supernumerary lobe of medium development.

Abdomen: Elongate oval; terga free; pecten present on tergum III. Tergum I and sterna I and II wider than long. Tergum I fuscous to dark fuscous with a narrow orange-brown band across posterior margin but not reaching lateral margins; tergum II orange-brown with a narrow transverse black band subanteriorly but not reaching dark fuscous narrow lateral margins; terga III–V orange-brown with a 'T' pattern

consisting of a narrow transverse black band across anterior margin of tergum III and expanding over lateral margins of this tergum and a medium-width medial longitudinal black band over all three terga (the band tends to dark fuscous on tergum V). A pair of oval dark fuscous shining spots on tergum V. Posterior lobe of surstylus short, sternum V with a deep concavity on posterior margin. All abdominal sterna appear pale coloured, generally red-brown.
Attractant: No known record.
Female: No known record.
Distribution:
Type locality: Indonesia (Sulawesi – Central Sulawesi, Wotu).
Previously recorded: Type locality and Indonesia (Sulawesi – South Sulawesi, Palopo).
Hosts: No known record.
Remarks: *Bactrocera (Bactrocera) dispar* (Hardy) is unique in possessing yellow spots anterior to notopleural suture in line with the lateral postsutural vittae, an elongate oval-shaped abdomen and *prsc.* setae absent. It is similar to *Bactrocera (Bactrocera) flavipennis* (Hardy) in possessing a broad costal band confluent with R_{4+5}, the base colour of the scutum red-brown and the costal band lacking a large spot in the apex of the wing. It differs from *B. flavipennis* in having the costal band of uniform width throughout, abdominal terga III–V orange-brown without a dark 'T' pattern and in lacking a yellow spot anterior to the notopleural suture in front of the lateral postsutural vittae.

Bactrocera (Bactrocera) dongnaiae Drew & Romig, new species

(Fig. 59)

Material examined: Holotype male, **VIETNAM: Southern Vietnam** – Dong Nai Province, Long Khanh District, Nhan Nghia Village, 22.vi.2002, attracted to cue lure, deposited in BMNH.
Paratypes: **VIETNAM: Southern Vietnam** – 1 male, same data as holotype; 3 males, same locality data as holotype, 20.vii.2002, attracted to cue lure; 1 male, Binh Duong Province, Thuan An District, An Thanh Village, 15.i.2002, coll. SOFRI staff, attracted to cue lure; 1 male, Tien Giang Province, Cai Be District, My Thirn Village, 9.xi.1999, attracted to cue lure. 1 paratype in ANIC; 1 paratype in BMNH, 3 paratypes in QDPC, 1 paratype in QMIC.
Diagnosis: A medium-sized species; face fulvous with a pair of medium-sized circular black spots; postpronotal lobes and notopleura yellow; scutum black; lateral postsutural yellow vittae present; medial postsutural yellow vitta absent; no yellow spot anterior to notopleural suture; mesopleural stripe of medium-width reaching midway between anterior margin of notopleuron and anterior *npl.* seta dorsally; scutellum yellow; wing with cells bc and c colourless, microtrichia in outer corner of cell c only; a moderately broad dark fuscous costal band; a narrow fuscous anal streak; abdominal terga III–V orange-brown with a medium-width medial longitudinal black band over all three terga, a narrow transverse black band across anterior margin of tergum III and expanding to broad dark fuscous to black lateral margins, narrow fuscous to dark fuscous anterolateral corners on terga IV and V.
Description: male.
Head: Height 1.72 mm. Frons – length 1.55 times breadth; red-brown with dark fuscous around bases of orbital setae and on anteromedial hump; orbital setae black: 1 *s.or.*, 2 *i.or.*; lunule fuscous. Ocellar triangle black. Vertex fuscous. Face fulvous with a pair of medium-sized circular black spots; length 0.49 mm. Genae fulvous, small dark fuscous subocular spot present; black seta present. Occiput fuscous, fulvous along eye margins; occipital row with 3–5 strong black setae. Antennae with segments 1 and 2 red-brown, segment 3 red-brown with fuscous on apex and outer surface; length of segments: 0.2 mm; 0.36 mm; 0.8 mm.
Thorax: Scutum black with red-brown to dark red-brown below and behind lateral postsutural vittae and around notopleural suture. Pleural areas black with red-brown below postpronotal lobes. Yellow markings as follows: postpronotal lobes; notopleura; mesopleural stripe reaching midway between anterior margin of notopleuron and anterior *npl.* seta dorsally, continuing to katepisternum as a small transverse spot, anterior margin slightly convex; anatergite (posterior apex black); anterior $^{3}/_{5}$ katatergite (remainder black); two moderately broad parallel-sided lateral postsutural vittae ending at or just behind *ia.* seta. Medial longitudinal postsutural yellow vitta absent. Postnotum black. Scutellum yellow except for narrow black basal band. Setae: *sc.* 2; *prsc.* 2; *ia.* 1; *p.sa.* 1; *a.sa.* 1; *mpl.* 1; *npl.* 2; *scp.* 4.
Legs: Fore femora fulvous with a moderately large subapical black spot on outer surfaces, mid and hind femora entirely fulvous; fore tibiae dark fuscous basally to fuscous apically, mid tibiae dark fuscous basally to fulvous apically, hind tibiae dark fuscous; all tarsi with basal segment fulvous and apical four segments fuscous; mid tibiae each with an apical black spur.
Wings: Length 5.9 mm; cells bc and c colourless; microtrichia in outer corner of cell c only; remainder of wings colourless except dark fuscous cell sc, moderately broad dark fuscous costal band overlapping R_{2+3} and almost confluent with R_{4+5} and expanding slightly across apex of R_{4+5}, a narrow fuscous anal streak; a dense aggregation of microtrichia around A_1+CuA_2; supernumerary lobe of medium development.
Abdomen: Oval; terga free; pecten present on tergum III. Tergum I and sterna I and II wider than long. Tergum I dark fuscous centrally, black laterally and with a narrow orange-brown transverse band across posterior margin but not reaching lateral margins; tergum II orange-brown anteriorly, fulvous laterally, with a narrow transverse subanterior fuscous to dark fuscous band reaching narrow anterolateral black corners; terga III–V orange-brown with a medium-width medial longitudinal black band over all three terga, a narrow transverse black band across anterior margin of tergum III and expanding to broad dark fuscous to black lateral margins on this tergum, narrow fuscous to dark fuscous anterolateral corners on terga IV and V. A pair of dark fuscous shining spots on tergum V. Posterior lobe of surstylus short, sternum V with a deep concavity on posterior margin. Abdominal sterna mostly dark.

Attractant: Cue lure.
Female: No known record.
Distribution: Southern Vietnam (Binh Duong, Dong Nai, Tien Giang Provinces).
Hosts: No known record.
Remarks: *Bactrocera (Bactrocera) dongnaiae* n.sp. is similar to *Bactrocera (Bactrocera) arecae* (Hardy & Adachi) and *Bactrocera (Bactrocera) occipitalis* (Bezzi) in general body colour patterns. It differs from these species in having the males attracted to cue lure, a broader costal band almost confluent with R_{4+5} and from *B. arecae* in having a uniformly black scutum.

Bactrocera (Bactrocera) dorsalis (Hendel)

(Figs 60A, 60B, 60C)

Musca ferruginea Fabricius, 1794: 342. Preoccupied by *Musca ferruginea* Scopoli, 1763 (see Hardy, 1969a: 396).
Dacus ferrugineus – Fabricius, 1805: 274.
Dacus dorsalis Hendel, 1912: 18. Lectotype female in BMNH.
Bactrocera ferruginea – Bezzi, 1913: 95.
Chaetodacus ferrugineus var. *dorsalis* – Hendel, 1915: 426.
Chaetodacus ferrugineus – Bezzi, 1916: 104.
Chaetodacus ferrugineus dorsalis – Bezzi, 1916: 104.
Chaetodacus ferrugineus var. *okinawanus* Shiraki, 1933: 62; Hardy & Adachi, 1956: 8; Hardy, 1969a: 402 (Synonym).
Dacus (Strumeta) dorsalis – Hardy & Adachi, 1956: 7; Hardy, 1969a: 395; 1973: 41; 1974: 29.
Strumeta dorsalis – Hering, 1956: 63.
Strumeta ferruginea – Hering, 1956: 63.
Strumeta dorsalis okinawana – Shiraki, 1968: 23.
Dacus (Bactrocera) dorsalis – Hardy, 1977: 49; Drew, 1982: 60; Drew, 1989: 63.
Dacus (Bactrocera) semifemoralis Tseng, Chen & Chu, 1992: 46. **New Synonym.**
Dacus (Bactrocera) yilanensis Tseng, Chen & Chu, 1992: 52. **New Synonym.**
Bactrocera (Bactrocera) dorsalis – Drew & Hancock, 1994a: 17, Lectotype designation; Norrbom *et al.*, 1998: 90; Drew *et al.*, 2007: 3.
Bactrocera (Bactrocera) variabilis Lin & Wang, in Lin *et al.*, 2011b: 896. Holotype in HEIQ. **New Synonym**.

Other type material studied: CHINA: A study of the colour photographs and description of *Bactrocera (Bactrocera) variabilis* Lin & Wang in the original publication.
Non-type material examined: Since the publication of Drew & Hancock (1994a), the following specimens have been studied – **BHUTAN**: 42 males (15 Nov.2004), 21 males (5.xii.2004), 12 males (29.xii.2004), East Bhutan, Sandrup Jongkhar District, coll. Wangdi, attracted to methyl eugenol; 9 males (2.xi.2004), 1 male (9.xi.2004), 7 males (16.xi.2004), 5 males (23.xi.2004), 4 males (26.ii.2005), Gelephu Purana Bastey, coll. Maha Prasad & Karma Namgyel, attracted to methyl eugenol; 1 male, Gelephu Purana Bastey, 21.ii. 2005, coll. M.P. Homagal & Karma Namgyel, attracted to methyl eugenol; 7 males (16.xii.2004), 1 male (1.ii.2005), 3 males (5.iii.2005), 4 males (12.iii.2005), 2 males (20.iii.2005), 2 males (3.iv.2005), Gelephu Purana Bastey, coll. Xangyal Nooh & Karma Namgyel, attracted to methyl eugenol; 5 males, Purana Bastey, 30.xi. 2004, coll. Karma Namgyel, attracted to methyl eugenol; 29 males (10.v.2002), 9 males (17.v.2002), 47 males (24.v.2002), Rimchu, coll. C. Dorji, attracted to methyl eugenol; 5 males (21.iv.2000), 2 males (26.iv.2000), Rimchu, coll. Clarke & Fletcher, attracted to methyl eugenol; 40 males, Rimchu, 17.v.2001, coll. R. Drew & M. Romig, attracted to methyl eugenol; 30 males, Rimchu bagging site No. 8, 17.v.2001, coll. R. Drew & M. Romig, attracted to methyl eugenol; 1 male, Rimchu 2, 21.v.2000, coll. C. Dorji; 4 males (4.v.2000), 13 Males (10.v.2000), Rimchu 2, coll. B.S. Fletcher, attracted to methyl eugenol; 5 males (4.v.2000), 7 males (10.v.2000), Rimchu 3, coll. B.S. Fletcher, attracted to methyl eugenol; 54 males (6.v.2000), 1 male (10.vi.2000), Phuentsholing, coll. Thapa, attracted to methyl eugenol; 1 male (11.vi.2000), 1 male (9.xii.2004), Tsirang, Suntalay, coll. Karma Yanezon & Karma Dukpa, attracted to methyl eugenol; 4 males, Lumitshawa, 5.vii.2000, coll. C. Dorji; 7 males, Namling 16km, altitude 1600m, subtropical rainforest, 9.vi.2001, coll. Brian Fletcher, attracted to methyl eugenol; 16 males, Lingmethang, altitude 700m, mango tree, 9.vi.2001, coll. Brian Fletcher, attracted to methyl eugenol; 18 males, Lingmethang 11km, altitude 1100m, subtropical rainforest, 9.vi.2001, coll. Brian Fletcher, attracted to methyl eugenol; 24 males, Lingmethang, Manchungan, 3.x.1991, coll. H.R. Fiejen, attracted to methyl eugenol. **INDIA: ANDAMAN ISLANDS** – 1 male and 1 female (19.v.1998), 1 male and 1 female (1.vi.1998), South Andamans, Havelock Island, coll. Ranganath, bred ex *Antiaris* sp. sample No. 62. **TAIWAN (formerly FORMOSA)**: 1 male and 1 female, Kaonsiung, 17.x.1990, coll. Y.H. Tseng, attracted to methyl eugenol. **VIETNAM: Southern Vietnam** – 5 males and 2 females, South Vietnam, Long Tri, Long An, 25.vii.2004, coll. Chris Clark, bred from Dragonfruit, (white flesh), *Hylocereus undatus*, cv Binh Thuan; 3 males (14.i.2002), 8 males (1.iv.2002), 1 female (20.iv.2002), 4 females (26.v.2002), 6 males and 1 female (2.vi.2002), 3 males and 4 females (9.vi.2002), 2 females (30.vi.2002), 3 females (14.vii.2002), 1 female (28.vii.2002), 1 female (1.ix.2002), Vinhlong Province, Long Ho District, Phuoc Hau Village, Site VS01, coll. SOFRI Staff, attracted to methyl eugenol; 2 males (15.xii.2001), 1 male (23.iii. 2002), 1 female (17.iv.2002), 5 males and 1 female (20.iv.2002), 1 male and 1 female (5.v.2002), 2 males (12.v.2002), 2 females (2.vi.2002), Cantho Province, Chau Thanh District, Tan Phu Village, Site VS02, coll. SOFRI Staff, attracted to methyl eugenol; 1 male (18.iv.2002), 2 males (23.v.2000), Dong Thap Province, Lap Vo District, Tan Khanh Trung Village, Site VS03, coll. SOFRI Staff, attracted to methyl eugenol; 2 males, Can Tho Province, O Mon District, Tan Thoi Village, Site S004, 17.iii.2000, attracted to methyl eugenol; 1 female (19.i.2002), 1 female (2.iii.2002), 1 female (24.iii.2002), 2 females (31.iii.2002), 4 females (7.iv.2002), 7 females (15.iv.2002), 1 female (28.iv.2002), 2 females (1.vi.2002), 1 male and 1 female (8.vi.2002), 1 male (12.x.2002), Ben Tre Province, Chau Thanh District, Tien Thuy Village, Site VS04, coll. SOFRI Staff, attracted to methyl eugenol; 1 male (25.vii.2000), 1 female (15.i.2002),

1 female (26.ii.2002), 1 female (21.v.2002), 1 male (25.v.2002), 2 males (30.vii.2002), Binh Duong Province, Thuan An District, An Thanh Village, Site VS05, coll. SOFRI Staff, attracted to methyl eugenol; 3 males (26.iv.2002), 1 male (26.v.2002), 2 males (3.viii.2002), Dong Nai Province, Long Khanh District, Nhan Nghia Village, Site VS06, coll. SOFRI Staff; 1 male, Tien Giang Province, Cai Be District, An Huu Village, Suite S009, 18.v.2000, attracted to methyl eugenol; 5 males, Dac Lac Province, Ku Mi Nga District, Quang Phu Village, Site VS09, 8.vi.2002, attracted to methyl eugenol; 1 male, Dong Nai Province, Vinh Cuu District, Tan Binh Village, Site S011, 21.i.2000, attracted to methyl eugenol; 2 males (15.vi.2002), 1 male (15.vii.2002), Binh Dinh Province, Tay Son District, Bin Thanh Village, Site VS11, attracted to methyl eugenol; 2 males (14.x.1999), 1 male (26.v.2000), Dong Nai Province, Long Nhat District, Long Nhat Village, Site S012, attracted to methyl eugenol; 1 male (1.x.1999), 1 male (28.iv.2000), Dong Nai Province, Thong Nhat District, Cay Gao Village, Site S013, attracted to methyl eugenol; 1 male (6.viii.1999), 2 females (21.i.2000), 1 male (2.ii.2000), 1 female (18.ii.2000), Binh Duong Province, Lai Thieu Village, Site S020, attracted to methyl eugenol; **Northern Vietnam** – 3 males, Son La Province, Moc Chau District, Long Luong Village, Site VN01, attracted to methyl eugenol; 2 males (14.viii.1999), 2 males (28.viii.1999), 2 males (30.vi.2000), 1 female (7.vii.2000), Hanoi Province, Tu Liem District, Con Hue, Site N001, attracted to methyl eugenol; 3 males (16.v.2002), 1 female (13.vi.2002), Hanoi Province, Tu Liem District, Lien Mac Village, Site VN03, coll. NIPP Staff, attracted to methyl eugenol; 2 males (14.viii.1999), 1 male (21.viii.1999), 1 female (30.xi.2000), 1 female (13.ix.2001), 4 males (4.x.2001), 1 male (8.xi.2001), Hanoi Province, Tu Liem District, Lien Mac, Site N003, attracted to methyl eugenol; 1 female, Hai Phong Province, An Nai District, Nam Son Village, Site VN04, attracted to methyl eugenol; 1 male (19.vi.1999), 1 female (7.viii.1999), 1 male and 1 female (9.vi.2000), 2 males (23.vi.2000), 1 male (14.vii.2000), Hanoi Province, Tu Liem District, Xuan Dinh, Site N004, attracted to methyl eugenol; 1 male, Son La Province, Moc Chau District, Co Do State Farm, FAO Site VN06, coll. NIPP Staff, attracted to methyl eugenol; 1 male (3.v.1999), 3 females (15.vi.2002), 1 male and 2 females (30.vi.2002), Son La Province, Moc Chau District, Co Do State Farm, Site VN06, attracted to methyl eugenol; 1 male (6.ix.1999), 1 male (4.x.1999), 1 male (12.vi.2000), Son La Province, Moc Chau District, Tra Long, Site N007, attracted to methyl eugenol; 1 male, Phu Tho Province, Phu Ninh District, Phu Tho Fruit Res Centre, Site VN07, 9.v.2002, coll. NIPP Staff, attracted to methyl eugenol; 2 females, Son La Province, Moc Chau District, Moc Chau house garden, Site N008, 12.vi.2000, attracted to methyl eugenol; 2 males, Hoa Binh Province, Moc Chau District, Pa Co Village, 15.vi.2002, FAO Site VN10, coll. NIPP Staff, attracted to methyl eugenol; 1 male, Bac Giang Province, Luc Ngan District, Site VN11, 30.v.2002, coll. NIPP Staff, attracted to methyl eugenol; 1 female (1.xii.1999), 1 female (27.vi.2000), Hoa Binh Province, Ky Son District, Cao Phong, Site N012, attracted to methyl eugenol; 1 male (15.v.1999), 1 male (28.ix.1999), 1 female (15.xi.1999), Hoa Binh Province, Ky Son District, Cao Phong, Site N013, attracted to methyl eugenol; 1 male (10.ix.1999), 1 male (8.x.1999), Lao Cai Province, Sapa 2 Village, Site N015, attracted to methyl eugenol; 3 males, Lao Cai Province, Sapa Town, 7.vii.2000, Site N016, attracted to methyl eugenol; 1 male, Lao Cai Province, Sapa 6 Village, Site N017, 12.xii.1999, attracted to methyl eugenol. A large series of specimens reared from carambola, gourd, guava, longan, lychee, mango, orange, peach, persimmon, plum, pomelo, rambutan, rose apple, sapodilla, squash, terminalia and water apple in northern and southern Vietnam. Specimens in QDPC.

Diagnosis: Face fulvous with a pair of medium-sized circular black spots; postpronotal lobes and notopleura yellow; scutum black with extensive areas of red-brown to brown below and behind lateral postsutural vittae, around notopleural suture, between postpronotal lobes and notopleura, inside postpronotal lobes; broad parallel-sided lateral postsutural yellow vittae ending behind *ia.* seta; medial postsutural yellow vitta absent; mesopleural stripe reaching midway between anterior margin of notopleuron and anterior *npl.* seta dorsally; scutellum yellow; legs with femora entirely fulvous; fore tibiae pale fuscous and hind tibiae fuscous, mid tibiae fulvous with a small area of dark fuscous basally; wing with cells bc and c colourless, microtrichia in outer corner of cell c only; a narrow fuscous costal band confluent with R_{2+3} and remaining very narrow around apex of wing (occasionally there can be a very slight swelling around apex of R_{4+5}); a narrow pale fuscous anal streak; supernumerary lobe of medium development; abdominal terga III–V exhibits a wide range of colour patterns (Figs 60A, 60B, also see Drew & Hancock, 1994a) but normally possesses the basic pattern of a black ‘T’ consisting of a narrow transverse black band across anterior margin of tergum III, a narrow medial longitudinal black band over all three terga, narrow anterolateral fuscous to dark fuscous corners on terga IV and V; a pair of oval orange-brown to pale fuscous shining spots on tergum V; abdominal sterna dark coloured.

Attractant: Methyl eugenol.

Female: As for male in the general body colour patterns. See Drew & Hancock (1994a) for a detailed description of female characters (Fig. 60C).

Distribution:

Type locality: Taiwan (Koshun).

Previously recorded: Type locality and the following locations – Bhutan, Cambodia, Hong Kong, India, Laos, Myanmar, Nepal, northern Thailand, southern China, Sri Lanka, Taiwan, Vietnam (Drew & Hancock, 1994a; Norrbom *et al.*, 1998). We cannot confirm that *B. dorsalis* occurs in India and Sri Lanka. Introduced into Hawaii and Mariana Islands. Recently eradicated from Ryukyu Is., Japan (including Okinawa). The record from Bhutan is incorrect (see ‘Remarks’ below).

Hosts: This species is a major economic pest and utilizes a wide range of commercial/edible and rainforest fruits (see Allwood *et al.*, 1999 for host records).

New records: *Hylocereus undatus* (family *Cactaceae*) (dragonfruit).

Remarks: *Bactrocera (Bactrocera) dorsalis* (Hendel), within the *dorsalis* complex, is similar to *Bactrocera (Bactrocera) carambolae* Drew & Hancock, *Bactrocera (Bactrocera) papayae* Drew & Hancock and *Bactrocera (Bactrocera) verbascifoliae* Drew & Hancock in possessing broad parallel-sided lateral postsutural vittae, costal band confluent with or very slightly overlapping R_{2+3}, and to

B. papayae and *B. verbascifoliae* in having the costal band remaining very narrow beyond apex of R_{2+3}, femora entirely fulvous and abdominal terga III–V with a narrow medial longitudinal dark band. It differs from *B. carambolae* in possessing a very narrow apical section of the costal band, a narrow medial longitudinal dark band on abdominal terga III–V and triangular-shaped anterolateral dark corners on abdominal terga IV and V (these markings are rectangular in *B. carambolae*). It differs from *B. verbascifoliae* in possessing narrow lateral dark margins on abdominal terga IV and V and from *B. papayae* in having a short male aedeagus and female ovipositor. Other *dorsalis* complex species that are similar to *B. dorsalis* are *Bactrocera (Bactrocera) hantanae* Tsuruta & White, *Bactrocera (Bactrocera) irvingiae* Drew & Hancock, *Bactrocera (Bactrocera) raiensis* Drew & Hancock and *Bactrocera (Bactrocera) syzygii* White & Tsuruta; however, all of these species possess a broad medial longitudinal dark band on abdominal terga III–V and have not been recorded as having males responding to methyl eugenol. See Drew & Hancock (1994a) for a full discussion of type specimens, relationships and synonymies.

Following the publication on the *dorsalis* complex by Drew & Hancock (1994a), there has been considerable research to investigate the integrity of many of the morphologically close species. The review of Clarke *et al.* (2005) summarized the bulk of this research and has demonstrated that most taxa within the complex can be satisfactorily resolved and that species in the complex are undergoing rapid morphological change. *Dacus semifemoralis* and *D. yilanensis* were each described from a single lure-trapped male specimen. There appear to be no valid characters to separate these species from *B. dorsalis* and consequently we are making them **New Synonyms** of the latter. In a large number of specimens collected in Yunnan Province in 2012, the colour patterns on abdominal terga III–V varied from pale red-brown with an incomplete 'T' to a complete 'T' with broad lateral dark fuscous to black bands. We believe that *Bactrocera (Bactrocera) variabilis* Lin & Wang fits within these variations studied and thus we hereby make it a **New Synonym** of *B. dorsalis*.

In our molecular studies, using the mitochondrial genes COI and ND5, *B. dorsalis* (Taiwan specimens) was significantly different from *B. carambolae* (specimens from Bali and Peninsular Malaysia), *B. invadens* (specimens from Sri Lanka), *B. occipitalis* (specimens from Kalimantan), *B. ochroma* n.sp. and specimens previously recorded as *B. dorsalis* from Bhutan in Drew *et al.* (2007) (Drew, unpublished data). Some of the specimens listed above, under 'Non-type material examined' from Bhutan, may eventually be proven on molecular grounds to be not *B. dorsalis*. See discussion under section titled 'Identification of Cryptic Species'.

Bactrocera (Bactrocera) dorsaloides (Hardy & Adachi)

(Fig. 61)

Dacus (Strumeta) dorsaloides Hardy & Adachi, 1954: 167; Hardy, 1969a: 412; 1973: 42; 1974: 32. Holotype female in USNM.

Dacus (Bactrocera) dorsaloides – Hardy, 1977: 50.
Bactrocera (Bactrocera) dorsaloides – Drew & Hancock, 1994a: 20; Norrbom *et al.*, 1998: 90.

Non-type material examined: No new specimens available for study since the publication of Drew & Hancock (1994a).
Diagnosis: Face fulvous with a pair of medium-sized irregularly oval black spots; postpronotal lobes and notopleura yellow; scutum black with pale brown below and behind lateral postsutural vittae, around notopleural suture, between postpronotal lobes and notopleura and inside postpronotal lobes; short narrow lateral postsutural yellow vittae narrowing sharply posteriorly to end at or just behind *a.sa.* seta; medial postsutural yellow vitta absent; mesopleural stripe reaching almost to anterior *npl.* seta dorsally; scutellum yellow; legs with femora fulvous except apical ⅓ of each fuscous; all tibiae fuscous; wing with cells bc and c colourless, microtrichia in outer corner of cell c only; a narrow fuscous costal band overlapping R_{2+3} and ending between extremities of R_{4+5} and M (in some specimens there is a slight expansion around apex of R_{4+5}); a broad fuscous anal streak; supernumerary lobe of medium development; abdominal terga III–V orange-brown with a narrow- to medium-width medial longitudinal black band and moderately broad lateral longitudinal dark fuscous to black bands over all three terga and joined along anterior margin of tergum III, a pair of oval orange-brown shining spots on tergum V; abdominal sterna dark coloured.
Attractant: No known record.
Female: As for male in the general body colour patterns. See Drew & Hancock (1994a) for a detailed description of female characters.
Distribution:
Type locality: Philippines (Luzon I., Mt Makiling).
Hosts: Recorded only from rainforest plant species in the family *Sapotaceae* (Allwood *et al.*, 1999).
Remarks: *Bactrocera (Bactrocera) dorsaloides* (Hardy & Adachi), within the *dorsalis* complex, is similar to *Bactrocera (Bactrocera) collita* Drew & Hancock, *Bactrocera (Bactrocera) makilingensis* Drew & Hancock and *Bactrocera (Bactrocera) thailandica* Drew & Hancock in possessing lateral postsutural vittae that narrow posteriorly, dark markings on the apices of all femora and broad lateral dark markings on abdominal terga III–V. *B. thailandica* differs in having broad lateral longitudinal dark bands on terga III–V and tergum III often entirely dark fuscous to black, while *B. dorsaloides* differs from *B. collita* and *B. makilingensis* in having fuscous colouration around the apical ⅓ of all femora (not distinct dark preapical spots on the outer surfaces).

Bactrocera (Bactrocera) elongata Drew & Romig, new species

(Fig. 62)

Material examined: Holotype male, **INDONESIA: SULAWESI** – Gorontalo, Bone Bolango, Suwawa, 30.xi.2006, coll. Yusbar Ismail, attracted to cue lure in forest, deposited in BMNH.

Diagnosis: A large species; face fulvous with a pair of small oval black spots; postpronotal lobes and notopleura yellow; scutum base colour black; lateral postsutural yellow vittae present; medial postsutural yellow vitta absent; no yellow spot anterior to notopleural suture; mesopleural stripe reaching midway between anterior margin of notopleuron and anterior *npl.* seta dorsally; scutellum yellow; wing with cells bc and c colourless, microtrichia in outer corner of cell c only; a narrow dark fuscous costal band; a narrow fuscous anal streak; abdomen with a pronounced elongate oval shape; abdominal terga III–V orange-brown with a dark 'T' pattern consisting of a narrow transverse black band across anterior margin of tergum III and a moderately broad medial longitudinal black band over all three terga.

Description: male.

Head: Height 1.9 mm. Frons – length 1.57 times breadth; red-brown with fuscous on anteromedial hump and around bases of orbital setae: orbital setae black: 1 *s.or.*, 2 *i.or.*; lunule fuscous. Ocellar triangle black. Vertex fuscous. Face fulvous with a pair of small oval black spots; length 0.48 mm. Genae red-brown, large dark fuscous subocular spot present; large black seta present. Occiput red-brown, fulvous along eye margins; occipital row with a large number of strong black setae. Antennae with segments 1 and 2 red-brown, segment 3 red-brown with fuscous on apex and outer surface; length of segments: 0.26 mm; 0.31 mm; 0.89 mm.

Thorax: Scutum entirely black except red-brown across posterior margin, below and behind lateral postsutural vittae and around notopleural suture and dark red-brown inside postpronotal lobes. Pleural areas entirely black except red-brown below postpronotal lobes. Yellow markings as follows: postpronotal lobes; notopleura; mesopleural stripe reaching midway between anterior margin of notopleuron and anterior *npl.* seta dorsally, continuing to katepisternum as a transverse spot, anterior margin straight; anatergite (posterior apex black); anterior ⅔ katatergite (remainder black); two moderately broad lateral postsutural vittae narrowing slightly posteriorly to end just before *ia.* seta. Medial longitudinal postsutural yellow vitta absent. Postnotum black. Scutellum yellow except for narrow black basal band. Setae: *sc.* 2; *prsc.* 2; *ia.* 1; *p.sa.* 1; *a.sa.* 1; *mpl.* 1; *npl.* 2; *scp.* 4.

Legs: All femora entirely fulvous; fore and mid tibiae dark fuscous basally to fuscous apically, hind tibiae dark fuscous; all tarsal segments with basal segment fulvous and apical four segments red-brown; mid tibiae each with an apical black spur.

Wings: Length 6.5 mm; cells bc and c colourless; microtrichia in outer corner of cell c only; remainder of wings colourless except dark fuscous cell sc, narrow dark fuscous costal band overlapping R_{2+3} with a slight tint in the apical area of R_{2+3} and remaining narrow around apex of wing, a narrow fuscous anal streak; a dense aggregation of microtrichia around A_1+CuA_2; supernumerary lobe weak.

Abdomen: Elongate oval; terga free; pecten present on tergum III. Tergum I and sterna I and II wider than long. Tergum I orange-brown with a narrow medial longitudinal fuscous line and narrow black lateral margins; tergum II orange-brown anteriorly to fulvous posteriorly and with a narrow transverse black band across centre joining narrow black anterolateral corners; terga III–V orange-brown with a dark 'T' pattern consisting of a narrow transverse black band across anterior margin of tergum III and widening over lateral margins of that tergum and a medium-width medial longitudinal black band over all three terga (the transverse black band across anterior margin of tergum III may be diluted to a fuscous colour either side of the join with the medial longitudinal black band). A pair of oval dark fuscous shining spots on tergum V. Posterior lobe of surstylus short, sternum V with a deep concavity on posterior margin. Abdominal sterna dark fuscous to black.

Attractant: Cue lure.

Female: No known record.

Distribution: Indonesia (Sulawesi – Gorontalo).

Hosts: No known record.

Remarks: *Bactrocera (Bactrocera) elongata* n.sp. possesses the primary characters of the *dorsalis* complex of species, particularly an entire costal band confluent with or just overlapping R_{2+3}, a black scutum, cells bc and c colourless and abdominal terga III–V with a black 'T' pattern. It differs from this group of species in having a distinctly elongate oval-shaped abdomen.

Bactrocera (Bactrocera) epicharis (Hardy)

(Fig. 63)

Dacus (Strumeta) epicharis Hardy, 1970: 119; Drew, 1974: 37. Holotype male in UZMC.

Bactrocera (Bactrocera) epicharis – Drew, 1989: 47; Norrbom *et al.*, 1998: 90.

Non-type material examined: INDONESIA: NORTH MOLUCCAS (NORTH MALUKU) – 1 male (29.v.2007), 2 males (5.vi.2007), Maluku, North Maluku, West Halmahera, Goal, coll. Raais Abdullah, attracted to cue lure in forest; 1 male (3.vi.2007), 1 male (10.vi.2007), North Maluku, Kodya Ternate, Foramadiahi, coll. La Ruti, attracted to cue lure in forest. Specimens in QDPC.

Diagnosis: A large species; face fulvous with a pair of medium-sized to large circular black spots; postpronotal lobes and notopleura yellow; scutum black; lateral postsutural yellow vittae present; medial postsutural yellow vitta absent; no yellow spot anterior to notopleural suture; mesopleural stripe reaching midway between anterior margin of notopleuron and anterior *npl.* seta dorsally; scutellum yellow; wing with cells bc and c fuscous, microtrichia in outer corner of cell c only; a broad dark fuscous costal band becoming paler between R_{2+3} and R_{4+5} and remaining of uniform width throughout; a broad dark fuscous anal streak; abdominal terga III–V entirely black.

Description: male.

Head: Height 2.0 mm. Frons – length 1.57 times breadth; fuscous with fulvous along lateral and ventral margins and dark fuscous on anteromedial hump; orbital setae black: 1 *s.or.*, 2 *i.or.*; lunule dark fuscous. Ocellar triangle black. Vertex fuscous. Face fulvous with a pair of medium-sized to large circular black spots; length 0.58 mm. Genae red-brown, dark fuscous subocular spot present; black seta present. Occiput fuscous, red-brown along eye margins; occipital row with 5–8 strong black setae. Antennae with segments 1 and 2 red-brown, segment 3 red-brown with fuscous on apex and outer surface; length of segments: 0.24 mm; 0.44 mm; 1.06 mm.

Thorax: Scutum black with dark fuscous below and behind lateral postsutural vittae, around notopleural suture, along lateral margins between postpronotal lobe and notopleuron, inside postpronotal lobe. Pleural areas entirely black. Yellow markings as follows: postpronotal lobes; notopleura; mesopleural stripe reaching midway between anterior margin of notopleuron and anterior *npl.* seta dorsally, continuing to katepisternum as a small transverse spot, anterior margin slightly convex; anatergite (posterior apex black); anterior ⅔ katatergite (remainder black); two broad parallel-sided lateral postsutural vittae ending just behind *ia.* seta. Medial longitudinal postsutural yellow vitta absent. Postnotum black. Scutellum yellow except for narrow black basal band. Setae: *sc.* 2; *prsc.* 2; *ia.* 1; *p.sa.* 1; *a.sa.* 1; *mpl.* 1; *npl.* 2; *scp.* 4.

Legs: All femora entirely fulvous; fore and hind tibiae dark fuscous, mid tibiae fuscous; fore tarsi with basal segment fulvous and apical four segments red-brown, mid and hind tarsi with all segments entirely fulvous; mid tibiae each with an apical black spur.

Wings: Length 7.5 mm; cells bc and c fuscous; microtrichia in outer corner of cell c only; remainder of wings colourless except dark fuscous cell sc, dark fuscous costal band confluent with R_{4+5} but tending fuscous between R_{2+3} and R_{4+5} and remaining of uniform width throughout, a broad dark fuscous anal streak; a dense aggregation of microtrichia around A_1+CuA_2; supernumerary lobe of medium development.

Abdomen: Oval; terga free; pecten present on tergum III. Tergum I and sterna I and II wider than long. Tergum I black with a narrow transverse red-brown band across posterior margin but not reaching lateral margins; tergum II black with a narrow transverse fulvous band across posterior margin but not reaching lateral margins; terga III–V entirely black. A pair of oval black shining spots on tergum V. Posterior lobe of surstylus short, sternum V with a deep concavity on posterior margin. Abdominal sterna dark fuscous to black.

Attractant: Cue lure.

Female: No known record.

Distribution:

Type locality: Papua New Guinea (Mussau I., Malakata).

Previously recorded: Type locality and the Solomon Is. (Shortland I.).

New records: Indonesia (North Moluccas = North Maluku – West Halmahera; Ternate).

Hosts: No known record.

Remarks: *Bactrocera (Bactrocera) epicharis* (Hardy) is similar to *Bactrocera (Bactrocera) bryoniae* (Tryon), *Bactrocera (Bactrocera) neoritsemai* n.sp. and *Bactrocera (Bactrocera) penecostalis* n.sp. in possessing coloured cells bc and c, a broad costal band confluent with R_{4+5} and parallel-sided or subparallel lateral postsutural vittae. It differs from these species in having abdominal terga III–V entirely black.

Bactrocera (Bactrocera) eurycosta Drew & Romig, new species

(Fig. 64)

Material examined: Holotype male, **VIETNAM: Southern Vietnam** – Dong Nai Province, Vinh Cuu District, Tan Binh Village, 8.vii.1999, Site S011, attracted to cue lure, deposited in BMNH.

Paratypes: **VIETNAM: Southern Vietnam** – 1 male, same locality data as holotype, 17.ix.1999; 1 male, South Vietnam, Ben Tre Province, Chau Thanh District, Tien Thuy Village, 23.iv.2002, Site VS03, attracted to cue lure; 1 male, South Vietnam, Tien Giang Province, Cai Lay District, Long Khanh Village, 13.x.1999, Site S007, attracted to cue lure; 1 male, South Vietnam, Tien Giang Province, An Huu Village, 13.x.1999, Site S009, attracted to cue lure; 2 males (9.vii.1999), 1 male (26.v.2000), South Vietnam, Dong Nai Province, Long Khanh District, Tan Phong Village, Site S014, attracted to cue lure; 1 male (8.vii.1999), 1 male (14.x.1999), 1 male (14.iv.2000), South Vietnam, Binh Duong Province, Lai Thieu Village, Site S020, attracted to cue lure. 2 paratypes in ANIC, 2 paratypes in BMNH, 4 paratypes in QDPC, 2 paratypes in QMIC.

Non-type material examined: BRUNEI: 1 male, Brunei Darassalam, KBFC, Bato Apou Nat Pk, Forest, 23.i.2002, coll. K. Hurley, attracted to cue lure; 1 male (23.i.2002), 1 male (27.i.2002), Brunei Darassalam, KBFC, Bato Apou Nat Pk, Riparian, coll. K. Hurley, attracted to cue lure; 1 male (25.i.2002), 1 male (27.i.2002), Brunei Darassalam, KBFC, Bato Apou Nat Pk, Upland Ridge, coll. K. Hurley, attracted to cue lure. **EAST MALAYSIA: SABAH** – Luasong Camp Via Tawau, North Borneo Timber, Sabah, Malaysia, ix.1975, coll. N. Gough. **PENINSULAR MALAYSIA**: 1 male, TS Gombak, Malaysia, 8.iv.1986, coll. Mahdzir, attracted to cue lure; 3 males, TS 7 Malaysia, 25.viii.1986, coll. Mahdzir, attracted to cue lure; 1 male, TS 8 Malaysia, 1.vii.1986, coll. Mahdzir, attracted to cue lure; 1 male TS 11 Malaysia, 1.vii.1986, coll. Mahdzir, attracted to cue lure; 1 male, TS 17 Malaysia, 1.vii.1986, coll. Mahdzir, attracted to cue lure; 1 male, TS 18 Malaysia, 1.vii.1986, coll. Mahdzir, attracted to cue lure; 3 males, TS 19 Malaysia, 16.vi.1986, coll. Mahdzir Ahmad, attracted to cue lure; 1 male, Butong, Malaysia, 19.v.1986, coll. S. Vijaysegaran, attracted to cue lure; 3 males, Gambang/Kuantan, Malaysia, 19.v.1986, coll. S. Vijaysegaran, attracted to cure lure; 2 males, Jeram Pasu I, Malaysia, 23.vi.1986, coll. Yusoff Salleh, attracted to cue lure; 1 male, TN 9 Malaysia, 25.viii.1986, coll. Yusoff Salleh, attracted to cue lure; 1 male TS 7 Malaysia, 20.x.1986, coll. Vignes, attracted to cue lure. Specimens in QDPC.

Diagnosis: A medium-sized species; face fulvous with a pair of large oval black spots; postpronotal lobes and notopleura yellow; scutum black; lateral postsutural yellow vittae present; medial postsutural yellow vitta absent; no yellow spot anterior to notopleural suture; mesopleural stripe reaching midway between anterior margin of notopleuron and anterior *npl.* seta dorsally; scutellum yellow; wing with cells bc and c with a pale fuscous tint, microtrichia in outer corner of cell c only; a broad dark fuscous costal band confluent with R_{4+5} but pale fuscous adjacent to this vein; a narrow fuscous anal streak; abdominal terga III–V orange-brown with a dark 'T' pattern consisting of a broad black transverse band covering most of tergum III and a medium-width to narrow medial longitudinal black to dark fuscous band from centre of tergum III to posterior margin of tergum V and narrow dark fuscous anterolateral corners on terga IV and V.

Description: male.

Head: Height 1.98 mm. Frons – length 1.47 times breadth; red-brown with fuscous on anteromedial hump and around bases of orbital setae; orbital setae black: 1 *s.or.*, 2 *i.or.*; lunule fuscous. Ocellar triangle black. Vertex fuscous. Face fulvous with a pair of large oval black spots; length 0.54 mm. Genae fulvous, large dark fuscous subocular spot present; black seta present. Occiput fuscous centrally, dark fuscous laterally, fulvous along eye margins; occipital row with 5–7 strong black setae. Antennae with segments 1 and 2 red-brown, segment 3 red-brown with fuscous on apex and outer surface; length of segments: 0.24 mm; 0.4 mm; 1.0 mm.

Thorax: Scutum entirely black without pale markings. Pleural areas black with dark fuscous below postpronotal lobes. Yellow markings as follows: postpronotal lobes; notopleura; mesopleural stripe reaching midway between anterior margin of notopleuron and anterior *npl.* seta dorsally, continuing to katepisternum as a large transverse spot, anterior margin slightly convex; anatergite (posterior apex black); anterior $^{3}/_{5}$ katatergite (remainder black); two broad parallel-sided lateral postsutural vittae ending just behind *ia.* seta. Medial longitudinal postsutural yellow vitta absent. Postnotum black. Scutellum yellow except for narrow black basal band. Setae: *sc.* 2; *prsc.* 2; *ia.* 1; *p.sa.* 1; *a.sa.* 1; *mpl.* 1; *npl.* 2; *scp.* 4.

Legs: Fore femora fulvous with a distinct elongate dark fuscous subapical spot on outer surfaces, mid and hind femora fulvous; fore tibiae dark fuscous, mid tibiae fuscous basally to fulvous apically, hind tibiae fuscous; all tarsi with basal segment fulvous and apical four segments fuscous; mid tibiae each with an apical black spur.

Wings: Length 6.9 mm; cells bc and c with a pale fuscous tint; microtrichia in outer corner of cell c only; remainder of wings colourless except dark fuscous cell sc, broad dark fuscous costal band confluent with R_{4+5} but pale fuscous adjacent to this vein and remaining of uniform width around apex of wing, a narrow fuscous anal streak; a dense aggregation of microtrichia around A_1+CuA_2; supernumerary lobe of medium development.

Abdomen: Oval; terga free; pecten present on tergum III. Tergum I and sterna I and II wider than long. Tergum I dark fuscous to black with narrow orange-brown band across posterior margin but not reaching lateral margins; tergum II with a dark fuscous to black transverse band across anterior ½ interrupted by a narrow transverse orange-brown band across anterior margin and posterior ½ fulvous; terga III–V orange-brown with a broad transverse black band covering most of tergum III and narrow dark fuscous to fuscous anterolateral corners on terga IV and V and a moderately broad medial longitudinal black band from centre of tergum III to posterior margin of tergum V (this band is generally narrower and dark fuscous on tergum V). A pair of oval fuscous to dark fuscous shining spots on tergum V. Posterior lobe of surstylus short, sternum V with a deep concavity on posterior margin. Abdominal sterna dark fuscous to black.

Attractant: Cue lure.

Female: No known record.

Distribution: Brunei; East Malaysia (Sabah); Peninsular Malaysia; Southern Vietnam (Ben Tre, Dong Nai, Binh Duong, Tien Giang Provinces).

Hosts: No known record.

Remarks: *Bactrocera (Bactrocera) eurycosta* n.sp. is similar to *Bactrocera (Bactrocera) bruneiae* n.sp., *Bactrocera (Bactrocera) costalis* (Shiraki) and *Bactrocera (Bactrocera) ritsemai* (Weyenbergh) in possessing a black scutum, a costal band of uniform width and confluent with R_{4+5}, cells bc and c colourless or with a pale fuscous tint and femora entirely fulvous with, at most, a dark subapical spot on the apices of the fore femora. It differs from *B. bruneiae* in having oval facial spots and from *B. costalis* and *B. ritsemai* in having the scutum entirely black without red-brown margins.

Bactrocera (Bactrocera) fernandoi Tsuruta & White

(Fig. 65)

Bactrocera (Bactrocera) fernandoi Tsuruta & White, 2001: 78. Holotype male in HORDI.

Non-type material examined: SRI LANKA (formerly CEYLON): 1 male, 'Kadugannawa, 21.x.1993, cue lure'; 1 male, 'Gannoruwa Mountain, Kandy, 25.x.1993, cue lure'. Specimens donated by Kenji Tsuruta and in QDPC.

Diagnosis: A large species; face fulvous with a pair of medium-sized oval black spots; postpronotal lobes yellow (anterodorsal corners red-brown); notopleura yellow; scutum black; lateral postsutural yellow vittae present; medial postsutural yellow vitta absent; no yellow spot anterior to notopleural suture; mesopleural stripe reaching midway between anterior margin of notopleuron and anterior *npl.* seta dorsally; scutellum yellow; wing with cells bc and c colourless, microtrichia in outer corner of cell c only; a narrow dark fuscous costal band; a narrow dark fuscous anal streak; abdominal terga III–V orange-brown with a black 'T' pattern consisting of a narrow transverse band across anterior margin of tergum III which widens slightly towards lateral margins and a medium-width medial longitudinal band over all three terga. Abdominal terga IV and V may have very narrow pale fuscous anterolateral corners.

Description: male.

Head: Height 2.0 mm. Frons – length 1.58 times breadth; red-brown with fuscous to dark fuscous on anteromedial hump and around bases of orbital setae; orbital setae black:

1 *s.or.*, 2 *i.or.*; lunule dark fuscous. Ocellar triangle black. Vertex fuscous. Face fulvous with a pair of medium-sized oval black spots; length 0.65 mm. Genae fulvous, fuscous subocular spot present; black seta present. Occiput red-brown, fulvous along eye margins; occipital row with 5–10 strong black setae. Antennae with segments 1 and 2 red-brown, segment 3 red-brown with dark fuscous on apex and outer surface; length of segments: 0.3 mm; 0.4 mm; 1.1 mm.

Thorax: Scutum black with red-brown below and behind lateral postsutural vittae, around notopleural suture, along lateral margins between notopleuron and anterior margin of scutum. Pleural areas red-brown except dark fuscous to black along anterior and posterior margins of mesopleural stripe, on katepisternum, above hind coxae. Yellow markings as follows: postpronotal lobes (anterodorsal corners red-brown); notopleura; mesopleural stripe reaching midway between anterior margin of notopleuron and anterior *npl.* seta dorsally, continuing to katepisternum as a small transverse spot, anterior margin straight; anatergite (posterior apex black); anterior ⅔ katatergite (remainder black); two medium-width parallel-sided lateral postsutural vittae ending at *ia.* seta. Medial longitudinal postsutural yellow vitta absent. Postnotum red-brown centrally, black laterally. Scutellum yellow except for narrow black basal band. Setae: *sc.* 2; *prsc.* 2; *ia.* 1; *p.sa.* 1; *a.sa.* 1; *mpl.* 1; *npl.* 2; *scp.* 4.

Legs: Fore femora fulvous with a large fuscous to black elongate subapical spot on outer surface, mid femora fulvous tending red-brown apically and with a large black elongate marking on apical ⅔ of outer surface, hind femora fulvous with fuscous to dark fuscous around apical ¼; fore tibiae fuscous to dark fuscous, mid tibiae fulvous tending fuscous basally, hind tibiae dark fuscous; tarsi with all segments fulvous; mid tibiae each with an apical black spur.

Wings: Length 7.2 mm; cells bc and c colourless; microtrichia in outer corner of cell c only; remainder of wings colourless except fuscous cell sc, narrow dark fuscous costal band confluent with R_{2+3} and remaining narrow around apex of wing to end just beyond apex of R_{4+5}, a narrow dark fuscous anal streak; a dense aggregation of microtrichia around A_1+CuA_2; supernumerary lobe of medium development.

Abdomen: Oval; terga free; pecten present on tergum III. Tergum I and sterna I and II wider than long. Tergum I entirely orange-brown; tergum II orange-brown with narrow dark fuscous lateral margins; terga III–V orange-brown with a black 'T' pattern consisting of a narrow transverse band across anterior margin of tergum III which widens slightly towards lateral margins and a medium-width medial longitudinal band over all three terga. A pair of orange-brown oval shining spots on tergum V. Posterior lobe of surstylus short, sternum V with a deep concavity on posterior margin. Abdominal sterna dark fuscous to black. Abdominal terga IV and V may have very narrow pale fuscous anterolateral corners.

Attractant: Cue lure.

Female: No known record.

Distribution:

Type locality: Sri Lanka (Kandy District, Gannoruwa).

Previously recorded: Type locality and other district locations in Sri Lanka (Tsuruta and White, 2001).

Hosts: No known record.

Remarks: *Bactrocera (Bactrocera) fernandoi* Tsuruta & White, within the *dorsalis* complex, is similar to *Bactrocera (Bactrocera) gombokensis* Drew & Hancock, *Bactrocera (Bactrocera) lombokensis* Drew & Hancock and *Bactrocera (Bactrocera) pedestris* (Bezzi) in possessing a narrow costal band confluent with R_{2+3} and distinct dark markings on the apices of all femora. It differs from *B. gombokensis* and *B. pedestris* in having a slight expansion of the costal band beyond the apex of R_{2+3} and from *B. lombokensis* in having large dark fuscous to black markings on the apices of the mid and hind femora and in lacking dark patterns on the lateral margins of abdominal terga III–V.

Bactrocera (Bactrocera) flavipennis (Hardy)

(Fig. 66)

Dacus flavipennis Hardy, 1982: 226. Holotype male in MZB.

Bactrocera (Bactrocera) flavipennis – Norrbom *et al.*, 1998: 90.

Type material examined: Holotype male, **INDONESIA: SULAWESI** – Central Sulawesi, Lindu Valley, 960 m, Aug. 1975, coll. D.E. Hardy. Studied at MZB.

Other type material examined: INDONESIA: SULAWESI – 1 male, 'Lindu Valley, C. Sulawesi, 960 m, Aug. 1975, D.E. Hardy'; blue label – 'Paratype ♂ *Dacus (Zeugodacus) flavipilosus* D.E. Hardy'; 'Acc. No. 1983-454'. Received on loan from BBM (see 'Remarks' below).

Non-type material examined: INDONESIA: SULAWESI – 1 male, Sulawesi, 6km E Bantimurang, 15.v.1997, coll. G. Pattemore, attracted to cue lure; 2 males, Sulawesi Barat, Mamuju, Belang-Belang, 4.ix.2006, coll. Sri Hastuti, Yan P.F., Sukri, attracted to cue lure in Forest, orange (jeruk); 3 males, Sulawesi Selatan, Takalar, Limampoccoe, 16. viii.2006, coll. Sirajuddin SP, attracted to cue lure in forest; 3 males, West Sulawesi, Mamuju, Polo Pangale, 29.viii.2006, coll. Sri Hastuti, Yan P.F., Sukri, attracted to cue lure in jeruk; 16 males, Sulawesi Barat, Mamuju, Binanga, 30.x.2006, coll. Sri Hastuti, Yan P.F., Sukri, attracted to cue lure in orange (jeruk); 25 males (10.ix.2006), 11 males (9.x.2006), Sulawesi Tenggara, Konawe, Laomea, coll. Syamsul Ma'arif, attracted to cue lure in village; 8 males, Sulawesi Tenggara, Kendari, Wua-Wua, 10.ix.2006, coll. Syamsul M., attracted to cue lure in village; 3 males (6.viii.2006), 5 males (22.viii.2006), Sulawesi Barat, Mamuju, Belang-Belang, coll. Sri Hastuti, Sukri, attracted to cue lure in forest, citrus; 1 male (20.viii.2006), 1 male (2.ix.2006), 1 male (17.ix.2006), 1 male (23.ix.2006), Sulawesi Tengah, Donggala, Nupabomba, coll. Rusdin, attracted to cue lure in forest; 1 male, Gorontalo, Bone Bolango, Suwawa, 31.x.2006, coll. Yusbar Ismail, attracted to cue lure in forest; 1 male, Sulawesi Barat, Mamuju, Sese, 30.x.2006, coll. Yan P. Fachrie, attracted to cue lure in village; 1 male, Sulawesi Barat, Mamuju, Bulutakkang, 30.x.2006, coll. Yan P. Fachrie, attracted to cue lure in forest; 2 males, Sulawesi Utara,

Bitung, Manembonembo, 14.viii.2006, coll. Karel F. Lola, attracted to cue lure; 1 male, Sulawesi Utara, Minahasa, Utara, Sarongsong, 14.viii.2006, coll. Karel F. Lola, attracted to cue lure; 1 male (21.viii.2006), 3 males (28.viii.2006), Sulawesi Utara, Minahasa, Utara, Suwoon, coll. Karel F. Lola, attracted to cue lure. Specimens in QDPC.

Diagnosis: A large species; face fulvous with a pair of small oval black spots; postpronotal lobes and notopleura yellow; scutum red-brown with large black markings; lateral postsutural yellow vittae present; medial postsutural yellow vitta absent; no yellow spot anterior to notopleural suture; mesopleural stripe equal in width to notopleuron dorsally; scutellum yellow except for a moderately broad black basal band; wing with cells bc and c pale fuscous, microtrichia in outer corner of cell c only; a general pale fulvous tint, a broad fulvous costal band confluent with R_{4+5}; a moderately broad fulvous anal streak; abdominal terga III–V red-brown with a narrow transverse black band across anterior margin of tergum III.

Description: male.

Head: Height 1.68 mm. Frons – length 1.4 times breadth; fulvous to pale fuscous; orbital setae black: 1 *s.or.*, 3 *i.or.*; lunule pale fuscous. Ocellar triangle black. Vertex pale fuscous. Face fulvous with a pair of small oval black spots; length 0.48 mm. Occiput red-brown, pale fuscous laterally and fulvous along eye margins. Antennae with segments 1 and 2 red-brown, segment 3 pale fuscous; length of segments: 0.15 mm; 0.3 mm; 0.8 mm.

Thorax: Scutum red-brown tending fulvous along lateral margins, with large broad longitudinal black bands inside lateral postsutural vittae and large triangular black spots inside postpronotal lobes. Yellow markings as follows: postpronotal lobes; notopleura; mesopleural stripe equal in width to notopleuron dorsally; two medium-width parallel-sided lateral postsutural vittae ending before *ia.* seta. Medial longitudinal postsutural yellow vitta absent. Scutellum yellow except for moderately broad black basal band. Setae: *sc.* 2; *prsc.* absent; *ia.* 1; *p.sa.* 1; *a.sa.* 1; *mpl.* 1; *npl.* 2; *scp.* 4.

Legs: All segments fulvous except tibiae tending pale fuscous; mid tibiae each with an apical black spur.

Wings: Length 5.5 mm; cells bc and c pale fuscous (cell c slightly paler than bc); microtrichia in outer corner of cell c only; remainder of wings with a faint tint across membrane and with a fulvous coloured cell sc, a broad fulvous costal band confluent with R_{4+5} and ending with a small pale fuscous to fuscous spot in apex, a moderately broad fulvous anal streak; a dense aggregation of microtrichia around A_1+CuA_2; supernumerary lobe of medium development.

Abdomen: Oval; terga free; pecten present on tergum III. Tergum I and sterna I and II wider than long. Tergum I red-brown with narrow black lateral margins; tergum II red-brown with a narrow transverse black band across anterior margin but not reaching lateral margins; terga III–V red-brown with a narrow transverse black band across anterior margin of tergum III. A pair of oval red-brown shining spots on tergum V. Posterior lobe of surstylus short, sternum V with a deep concavity on posterior margin.

Attractant: Cue lure.

Female: No known record.

Distribution:

Type locality: Indonesia (Sulawesi – Central Sulawesi, Lindu Valley).

New records: Indonesia (many other locations in Sulawesi listed under 'Non-type material examined').

Hosts: No known record.

Remarks: *Bactrocera (Bactrocera) flavipennis* (Hardy) is similar to *Bactrocera (Bactrocera) dispar* (Hardy) and *Bactrocera (Bactrocera) megaspilus* (Hardy), all recorded from Sulawesi, in possessing a broad costal band confluent with R_{4+5} and the base colour of the scutum red-brown. It differs from *B. megaspilus* in lacking a large spot in the apex of the costal band and from *B. dispar* in having the costal band of uniform width (lacking apical expansion), lacking a yellow spot anterior to the notopleural suture in front of the lateral postsutural vittae and abdominal terga III–V orange-brown without a black 'T' pattern (only a transverse black band across anterior margin of tergum III).

In the type series of *Dacus (Zeugodacus) flavipilosus* Hardy, three species were represented. One of these is a specimen of *Bactrocera (Bactrocera) flavipennis* (Hardy) which is listed above under 'Other type material examined'. The other two species are *Bactrocera (Zeugodacus) flavipilosa* (Hardy) and *Bactrocera (Zeugodacus) neoflavipilosa* n.sp. and are discussed under subgenus *Zeugodacus*.

Bactrocera (Bactrocera) flavoscutellata Lin & Wang

(Fig. 67)

Bactrocera (Bactrocera) flavoscutellata Lin & Wang, in Lin *et al.*, 2005: 844. Holotype male in HEIQ.

Material examined: THAILAND: 2 males, Chiang Mai, N19 54.649, E99 02.679, Doi Ang, Khang Agr station, 15–17.ii.2011, coll. M. San Jose and D. Rubinoff, FF061 and FF065, cue lure trap, specimens in QDPC.

Diagnosis: A medium-sized species; face fulvous with a pair of large oval black spots; postpronotal lobes and notopleura yellow; scutum entirely black; lateral postsutural yellow vittae present; medial postsutural yellow vitta absent; no yellow spot anterior to notopleural suture; mesopleural stripe equal in width to notopleuron dorsally; scutellum yellow; wing with cells bc and c colourless, microtrichia in outer corner of cell c only; a narrow dark fuscous costal band; narrow dark fuscous anal streak; abdominal terga III–V red-brown with a narrow medial and two broad lateral longitudinal black bands joined over anterior margin of tergum III.

Description: male.

Head: Height 1.4 mm. Frons – length 1.4 times breadth; red-brown with dark fuscous on anteromedial hump and around bases of orbital setae; orbital setae black: 1 *s.or.*, 2 *i.or.*; lunule dark fuscous. Ocellar triangle black. Vertex dark fuscous. Face fulvous with a pair of large oval black spots; length 0.46 mm. Genae red-brown, dark fuscous subocular spot present; black

seta present. Occiput black, fulvous along eye margins; occipital row with 2–3 large black setae. Antennae with segments 1 and 2 fuscous, segment 3 red-brown with dark fuscous on apex and outer surface; length of segments: 0.16 mm; 0.28 mm; 0.72 mm.
Thorax: Scutum entirely black without pale markings. Pleural areas entirely black. Yellow markings as follows: postpronotal lobes; notopleura; mesopleural stripe equal in width to notopleuron dorsally, continuing to katepisternum as a small transverse spot, anterior margin convex; anatergite (posterior apex black); anterior $^{3}/_{5}$ katatergite (remainder black); two short and narrow lateral postsutural vittae narrowing sharply posteriorly to end at level of *a.sa.* seta. Medial longitudinal postsutural yellow vitta absent. Postnotum black. Scutellum yellow except for narrow black basal band. Setae: *sc.* 2; *prsc.* 2; *ia.* 1; *p.sa.* 1; *a.sa.* 1; *mpl.* 1; *npl.* 2; *scp.* 4.
Legs: Fore and mid femora fulvous with small black subapical spots on outer surfaces, hind femora fulvous with black around apical $^{1}/_{5}$; fore tibiae fuscous, mid tibiae fulvous with fuscous basally, hind tibiae dark fuscous; all tarsal segments entirely fulvous; mid tibiae each with an apical black spur.
Wings: Length 6.3 mm; cells bc and c colourless; microtrichia in outer corner of cell c only; remainder of wings colourless except dark fuscous cell sc, narrow dark fuscous costal band confluent with R_{2+3} and remaining narrow around apex of wing, a narrow dark fuscous anal streak; a dense aggregation of microtrichia around A_1+CuA_2; supernumerary lobe of medium development.
Abdomen: Oval; terga free; pecten present on tergum III. Tergum I and sterna I and II wider than long. Tergum I dark fuscous to black with a narrow transverse fulvous band along posterior margin but not reaching lateral margins; tergum II black with a broad transverse fulvous band across anterior ½ but not reaching lateral margins; terga III–V red-brown centrally with a narrow medial and two broad lateral longitudinal black bands over all three terga and joining along anterior ⅓ to ½ of tergum III and, in some specimens, joining along anterior margin of tergum IV. A pair of oval dark fuscous shining spots on tergum V. Posterior lobe of surstylus short, sternum V with a deep concavity on posterior margin. Abdominal sterna dark fuscous to black.
Attractant: Cue lure.
Female: No known record.
Distribution:
Type locality: China (Hainan I., Wuzhishan).
New records: Thailand (Chiang Mai).
Hosts: No known record.
Remarks: *Bactrocera (Bactrocera) flavoscutellata* Lin & Wang is similar to *Bactrocera (Bactrocera) perigrapha* White & Tsuruta and *Bactrocera (Bactrocera) pernigra* (Ito) in possessing a uniform black scutum, a narrow costal band confluent with R_{2+3}, very short and narrow lateral postsutural vittae and abdominal terga III–V with large areas of dark colouration. It differs from *B. perigrapha* in having entirely yellow postpronotal lobes and notopleura and the scutellum without a black triangular pattern on the dorsal surface and from *B. pernigra* in having the scutellum with a narrow black basal band. However, there is a strong possibility that *B. flavoscutellata* is a synonym of *B. pernigra* as the width of the basal band on the scutellum can vary between specimens of *B. (Bactrocera)* species and is therefore not a sound character to use alone to separate species. More specimens from Hainan I., the type locality of *B. flavoscutellata*, should be studied before this synonymy can be confirmed.

Bactrocera (Bactrocera) flavosterna Drew & Romig, new species

(Fig. 68)

Material examined: Holotype male, **INDONESIA: SULAWESI** – Sulawesi Tengah, Donggala, Nupabomba, 29.x.2006, coll. Rusdin, attracted to cue lure in forest, deposited in BMNH.
Paratypes: **INDONESIA: SULAWESI** – 5 males, same data as holotype; 5 males (20.viii.2006), 3 males (27.viii.2006), 5 males (2.ix.2006), 3 males (10.ix.2006), 6 males (17.ix.2006), 19 males (15.x.2006), 4 males (22.x.2006), Sulawesi Tengah, Donggala, Nupabomba, coll. Rusdin, attracted to cue lure in forest; 1 male, Sulawesi Tengah, Donggala, Nupabomba, 23.ix.2006, coll. Lyar, attracted to cue lure in forest; 1 male, Sulawesi, Gorontalo, Bone Bolango, Lombongo, 19.ix.2006, coll. Yusbar Ismail, attracted to cue lure in forest; 1 male, Sulawesi, Gorontalo, Bone Bolango, Suwawa, 26.ix.2006, coll. Yusbar Ismail, attracted to cue lure in forest. 10 paratypes in ANIC, 10 paratypes in BMNH, 23 paratypes in QDPC, 10 paratypes in QMIC.
Diagnosis: A medium-sized species; face fulvous with a pair of small oval to circular black spots; postpronotal lobes and notopleura yellow; scutum black with dark red-brown posterocentrally, below and behind lateral postsutural vittae and around notopleural suture; lateral postsutural yellow vittae present; medial postsutural yellow vitta absent; no yellow spot anterior to notopleural suture; mesopleural stripe reaching midway between anterior margin of notopleuron and anterior *npl.* seta dorsally; scutellum yellow; wing with cells bc and c with a pale tint, microtrichia in outer corner of cell c only; a narrow dark fuscous costal band distinctly overlapping R_{2+3} and expanding across apex of wing; a narrow fuscous anal streak; abdominal terga III–V orange-brown with a dark 'T' pattern consisting of a narrow transverse black band across anterior margin of tergum III and expanding to fuscous across lateral margins of that tergum and a narrow medial longitudinal dark fuscous to black band over terga III and IV.
Description: male.
Head: Height 1.6 mm. Frons – length 1.5 times breadth; fulvous with fuscous on anteromedial hump and dark fuscous around bases of orbital setae; orbital setae black: 1 *s.or.*, 2 *i.or.*; lunule fuscous. Ocellar triangle black. Vertex fuscous. Face fulvous with a pair of small circular to oval black spots; length 0.48 mm. Genae fulvous, large dark fuscous subocular spot present; dark seta present. Occiput fuscous, fulvous along eye margins; occipital row with 5–7 strong black setae. Antennae with segments 1 and 2 red-brown, segment 3 red-brown with fuscous on apex and outer surface; length of segments: 0.16 mm; 0.3 mm; 0.8 mm.
Thorax: Scutum black with dark red-brown posterocentrally which often extends along the midline, below and behind

lateral postsutural vittae and around notopleural suture. Pleural areas black except red-brown above mid coxae and below postpronotal lobes. Yellow markings as follows: postpronotal lobes; notopleura; mesopleural stripe reaching midway between anterior margin of notopleuron and anterior *npl.* seta dorsally, continuing to katepisternum as a transverse spot, anterior margin slightly convex; anatergite (posterior apex black); anterior $^3/_5$ katatergite (remainder black); two very broad parallel to subparallel-sided lateral postsutural vittae ending at *ia.* seta. Medial longitudinal postsutural yellow vitta absent. Postnotum black, tending dark red-brown centrally. Scutellum yellow except for narrow black basal band. Setae: *sc.* 2; *prsc.* 2; *ia.* 1; *p.sa.* 1; *a.sa.* 1; *mpl.* 1; *npl.* 2; *scp.* 4.

Legs: All segments entirely fulvous except for small subapical fuscous spot on inner surfaces of hind femora and fuscous on bases of hind tibiae; mid tibiae each with an apical black spur.

Wings: Length 5.4 mm; cells bc and c with a pale tint; microtrichia in outer corner of cell c only; remainder of wings colourless except dark fuscous cell sc, narrow dark fuscous costal band distinctly overlapping R_{2+3} where it is paler and expanding across apex of wing, a narrow fuscous anal streak; a dense aggregation of microtrichia around A_1+CuA_2; supernumerary lobe of medium development.

Abdomen: Oval; terga free; pecten present on tergum III. Tergum I and sterna I and II wider than long. Tergum I orange-brown with a fuscous medial longitudinal spot and narrow fuscous lateral margins; tergum II fulvous with a narrow transverse black band across anterior margin but not reaching the narrow dark fuscous anterolateral corners; terga III–V orange-brown with a dark 'T' pattern consisting of a narrow transverse black band across anterior margin of tergum III and becoming fuscous as it expands across lateral margins and a narrow medial longitudinal dark fuscous to black band over terga III and IV. A pair of oval red-brown shining spots on tergum V. Posterior lobe of surstylus short, sternum V with a deep concavity on posterior margin. Abdominal sterna pale coloured, generally fulvous to orange-brown.

Attractant: Cue lure.

Female: No known record.

Distribution: Indonesia (Sulawesi – Central Sulawesi, Donggala; Gorontalo).

Hosts: No known record.

Remarks: *Bactrocera (Bactrocera) flavosterna* n.sp. belongs to the *dorsalis* complex and is similar to *Bactrocera (Bactrocera) muiri* (Hardy & Adachi), some specimens of *Bactrocera (Bactrocera) papayae* Drew & Hancock, *Bactrocera (Bactrocera) profunda* Tsuruta & White and *Bactrocera (Bactrocera) vishnu* Drew & Hancock in possessing broad parallel-sided lateral postsutural vittae, the costal band overlapping R_{2+3}, reduced lateral dark markings and a narrow medial longitudinal dark band on abdominal terga III–V. It differs from *B. muiri* and *B. vishnu* in having the lateral postsutural vittae ending at the *ia.* seta, from *B. papayae* and *B. profunda* in having a small fuscous subapical spot on the hind femora and from all of these species in having the costal band expanded across the wing apex similar to that in *Bactrocera (Bactrocera) occipitalis* (Bezzi).

Bactrocera (Bactrocera) floresiae Drew & Hancock

(Fig. 69)

Bactrocera (Bactrocera) floresiae Drew & Hancock, 1994a: 22; Norrbom *et al.*, 1998: 90. Holotype male in BBM.

Non-type material examined: Since the publication of Drew & Hancock (1994a), the following specimens have been studied – **INDONESIA: BALI** – 6 males, Buleleng, Singaraja, Kaliuntu, 27.viii.2006, coll. Dewa den Eny, attracted to methyl eugenol in village; 1 male, Bali Buleleng, Kubutambahan, Depeha, 22.x.2006, coll. Dewa den Eny, attracted to methyl eugenol in mangga; **FLORES** – 1 male, Borong, iv.1989, coll. J. Turner; **JAVA** – 1 male (21.viii.2005), 1 male (28.viii.2005), Central Java, Pemalang, Surajaya, coll. Sapto Darmono, attracted to methyl eugenol in forest; **SULAWESI** – 4 males, Tengah, Donggala, Nupabomba, 15.x.2006, coll. Rusdin, attracted to methyl in forest. Specimens in QDPC.

Diagnosis: Face fulvous with a pair of medium-sized circular black spots; postpronotal lobes and notopleura yellow; scutum black with dark brown below and behind lateral postsutural vittae, around notopleural suture and inside postpronotal lobes; broad triangular lateral postsutural yellow vittae narrowing sharply posteriorly to end before *ia.* seta; medial postsutural yellow vitta absent; mesopleural stripe equal in width to, or slightly wider than, notopleuron dorsally; scutellum yellow; legs with all femora entirely fulvous; fore and mid tibiae fulvous and hind tibiae fulvous to fuscous; wing with cells bc and c colourless, microtrichia in outer corner of cell c only; a narrow fuscous costal band confluent with R_{2+3}, remaining narrow except for a slight expansion around apex of R_{4+5}; a narrow fuscous anal streak; supernumerary lobe of medium development; abdominal terga III–V orange-brown with a 'T' pattern consisting of a narrow transverse dark fuscous band across anterior margin of tergum III and expanding to broad lateral margins on this tergum, a narrow medial longitudinal dark fuscous band over all three terga (generally the lateral margins of terga IV and V are orange-brown but in some specimens there are narrow pale fuscous lateral margins), a pair of oval orange-brown to fuscous shining spots on tergum V; abdominal sterna dark coloured.

Attractant: Methyl eugenol.

Female: No known record.

Distribution:

Type locality: Indonesia (Flores, Borong).

Previously recorded: Type locality and Indonesia (Sumbawa) (Drew & Hancock, 1994a).

New records: Indonesia (Bali; Java; Sulawesi).

Hosts: No known record.

Remarks: *Bactrocera (Bactrocera) floresiae* Drew & Hancock, within the *dorsalis* complex, is similar to *Bactrocera (Bactrocera) cibodasae* Drew & Hancock, *Bactrocera (Bactrocera) osbeckiae* Drew & Hancock, *Bactrocera (Bactrocera) sumbawaensis* Drew & Hancock and *Bactrocera (Bactrocera) thailandica* Drew & Hancock in possessing lateral postsutural vittae that narrow sharply posteriorly, femora entirely fulvous, costal band confluent with R_{2+3} and abdominal terga III–V with a general dark 'T' pattern. It differs from these four species in lacking lateral dark colour patterns on abdominal terga IV and V and in possessing a narrow medial longitudinal dark band over terga III–V and having the males attracted to methyl eugenol.

Bactrocera (Bactrocera) frauenfeldi (Schiner)

(Fig. 70)

Dacus frauenfeldi Schiner, 1868: 262; Froggatt, 1909: 96; Malloch, 1939a: 232. Lectotype female and Paralectotype female in NMW.
Strumeta frauenfeldi – Perkins, 1938a: 128.
Dacus (Strumeta) frauenfeldi – Hardy & Adachi, 1954: 168; Hardy, 1968: 138, Lectotype designation; Drew, 1974: 41.
Dacus (Bactrocera) frauenfeldi – Drew, 1982: 27.
Bactrocera (Bactrocera) frauenfeldi – Drew, 1989: 76; Norrbom *et al*., 1998: 90.

Non-type material examined: INDONESIA: MOLUCCAS (MALUKU) – 3 males, 'Waai, Ambon, (<20m), 6.vii.1981, coll. J. Turner'; 'Residual forest with clove, nutmeg & banana groves'; 4 males, 'Paso, Ambon, (20-50m), 7.vii.1981, coll. J. Turner'; 'Residual forest'; 1 male, 'Hila/Mamua, Ambon, (25-100m), 5.viii.1981, coll. J. Turner'; 'Residual forest'; 13 males, 'Mamala, Ambon 50-150m, 7.viii.1981, coll. J. Turner'; 'Residual forest with clove, nutmeg and banana groves'. All attracted to cue lure. 18 males, Middle Maluku, Suli, 9.vii.2007, coll. La Naibu, attracted to cue lure in Jambu; **NORTH MOLUCCAS (NORTH MALUKU)** – 18 males, Moluccas, North Maluku, West Halmahera, Goal, 29.v.2007, coll. Raais Abdullah, attracted to cue lure in forest; 6 males, Maluku, North Maluku, West Halmahera, Goal, 5.vi.2007, coll. Raais Abdullah, attracted to cue lure in forest. Specimens in QDPC.
Diagnosis: Face fulvous with a pair of medium-sized circular to oval black spots; postpronotal lobes black; notopleura yellow; scutum shining black; medium-width lateral postsutural yellow vittae narrowing posteriorly to end before *ia.* seta; medial postsutural yellow vitta absent; mesopleural stripe reaching midway between anterior margin of notopleuron and anterior *npl.* seta or, in some specimens, almost to anterior *npl.* seta dorsally; scutellum yellow with a black triangle on dorsal surface; legs with fore femora fulvous, mid femora fulvous basally with apical ⅔ dark fuscous to black, hind femora fulvous with apical ⅓ dark fuscous to black; fore and mid tibiae fuscous, hind tibiae dark fuscous; wing length 4.6 mm, wing with cells bc and c fuscous, dense microtrichia covering all of cell c; cell sc dark fuscous; costal band extremely pale fuscous and overlapping R_{2+3} and not expanding apically; a narrow dark fuscous transverse band across wing and enclosing both crossveins; a broad dark fuscous anal streak; supernumerary lobe of medium development; abdominal terga III–V orange-brown with a broad medial and two broad lateral longitudinal black bands over all three terga and which are sometimes joined across anterior margin of tergum III, a pair of oval dark fuscous to black shining spots on tergum V; abdominal sterna dark coloured.
Attractant: Cue lure.
Female: As for male in the general body colour patterns. See Drew (1989) for a detailed description of female characters.
Distribution:
Type locality: Solomon Islands (Stuart I.).
Previously recorded: Type locality and the following locations – Australia (North Queensland; Torres Strait Is.); Bismarck Archipelago; Bougainville I., Papua New Guinea (Drew, 1989); Indonesia (West Papua, formerly part of Irian Jaya).
New records: Indonesia (Moluccas = Maluku – Ambon I.); (North Moluccas = North Maluku – Halmahera I.).
Hosts: A wide range of commercial/edible and wild host fruits (see Drew, 1989; Hancock *et al*., 2000).
Remarks: *Bactrocera (Bactrocera) frauenfeldi* Schiner belongs to the *frauenfeldi* complex of five species (see 'Remarks' under *B. albistrigata*, for discussions on the complex). In South-East Asia, *B. frauenfeldi* has been confused with *B. albistrigata* but can be distinguished from this species in having black postpronotal lobes, dark fuscous to black on apical areas of mid and hind femora and a very broad medial longitudinal black band on abdominal terga III–V. See 'Remarks' under *B. albistrigata* where Hardy treated these species incorrectly as synonymous.

Bactrocera (Bactrocera) fulvicauda (Perkins)

(Figs 71A, 71B)

Strumeta fulvicauda Perkins, 1939: 13. Holotype female in QMIC.
Dacus (Strumeta) fulvicaudus – Drew, 1974: 47.
Bactrocera (Bactrocera) fulvicauda – Drew, 1989: 79; Norrbom *et al*., 1998: 90.

Material examined: INDONESIA: WEST PAPUA (formerly part of IRIAN JAYA) – 1 male, West Papua, Manokwari Utara, Wosi, 21.vi.2007, coll. Caroline, attracted to methyl eugenol in forest; **PAPUA (formerly part of IRIAN JAYA)** – 2 males, Papua, Keerom, Wiantre, 16. vi.2007, coll. Eikanus Karma, attracted to methyl eugenol in orange and jambu; 1 male, Papua, Keerom, Jaifuri, 16. vi.2007, coll. Eikanus Karma, attracted to methyl eugenol in jambu. Specimens in QDPC.
Diagnosis: A medium-sized species; face fulvous with a pair of very small elongate oval black spots; postpronotal lobes and notopleura yellow; scutum black; lateral postsutural yellow vittae present; medial postsutural yellow vitta absent; no yellow spot anterior to notopleural suture; mesopleural stripe reaching to postpronotal lobe dorsally; scutellum yellow; wing with cells bc and c fuscous, dense microtrichia covering most of cell c and outer corner of cell bc; a broad dark fuscous costal band confluent with R_{4+5}, a broad dark fuscous band across centre of wing and enclosing both crossveins, a narrow transverse dark fuscous band across base of cell dm; a broad dark fuscous anal streak; abdominal terga III–V with variation from entirely black to red-brown with broad medial and lateral longitudinal black bands joined along anterior margin of tergum III.

Description: male (Fig. 71A).

Head: Height 1.86 mm. Frons – length 1.4 times breadth; red-brown with pale fuscous on anteromedial hump and around bases of orbital setae; orbital setae black: 1 *s.or.*, 2 *i.or.*; lunule red-brown. Ocellar triangle black. Vertex red-brown. Face fulvous with a pair of very small elongate oval black spots; length 0.53 mm. Genae red-brown, fuscous subocular spot present; black seta present. Occiput red-brown, fulvous along eye margins; occipital row with 6–8 strong black setae. Antennae with segments 1 and 2 red-brown, segment 3 red-brown with fuscous on apex and outer surface; length of segments: 0.16 mm; 0.38 mm; 0.88 mm.

Thorax: Scutum entirely black with dark fuscous below lateral postsutural vittae. Pleural areas entirely black. Yellow markings as follows: postpronotal lobes; notopleura; mesopleural stripe reaching to postpronotal lobe dorsally, continuing to katepisternum as a transverse spot, anterior margin straight; anatergite (posterior apex black); anterior $^3/_5$ katatergite (remainder black); two broad parallel-sided to subparallel lateral postsutural vittae ending at *ia.* seta. Medial longitudinal postsutural yellow vitta absent. Postnotum black. Scutellum yellow except for narrow black basal band. Setae: *sc.* 2; *prsc.* 2; *ia.* 1; *p.sa.* 1; *a.sa.* 1; *mpl.* 1; *npl.* 2; *scp.* 4.

Legs: All segments entirely red-brown; mid tibiae each with an apical black spur.

Wings: Length 6.3 mm; cells bc and c fuscous; dense microtrichia over most of cell c and outer corner of cell bc; remainder of wings colourless except dark fuscous cell sc, broad dark fuscous costal band confluent with R_{4+5} and remaining of uniform width throughout, a broad transverse dark fuscous band across centre of wing enclosing both crossveins and outer ½ of cell dm, a narrow transverse band across base of cell dm, a broad dark fuscous anal streak; a dense aggregation of microtrichia around A_1+CuA_2; supernumerary lobe of medium development.

Abdomen: Oval; terga free; pecten present on tergum III. Tergum I and sterna I and II wider than long. Tergum I black with a narrow transverse dark red-brown band across posterior margin but not reaching lateral margins; tergum II dark red-brown anteriorly, fulvous posteriorly with a narrow transverse black band across centre but not reaching lateral margins; terga III–V variable from entirely black to red-brown with a broad medial and two broad lateral longitudinal black bands over all three terga and joined along anterior margin of tergum III (see Drew (1989) for full description and Fig. 71B for artwork of abdomen). A pair of oval fuscous to black shining spots on tergum V. Posterior lobe of surstylus short, sternum V with a deep concavity on posterior margin. Abdominal sterna black.

Attractant: Methyl eugenol.

Female: As for male except no dense aggregation of microtrichia around A_1+CuA_2, supernumerary lobe weak, pecten of cilia on abdominal tergum III absent; ovipositor basal segment red-brown, conical in transverse cross section and tapering posteriorly in dorsal view; ratio of length of oviscape to length of tergum V, 2.8:1; apex of piercer needle-shaped.

Distribution:

Type locality: Papua New Guinea (Mt Lamington).

Previously recorded: Type locality and lowland rainforests of Papua New Guinea (Drew, 1989); Indonesia (West Papua, formerly part of Irian Jaya, White & Evenhuis, 1999).

New records: Indonesia (Papua, formerly part of Irian Jaya – Keerom).

Hosts: No known record.

Remarks: *Bactrocera (Bactrocera) fulvicauda* (Perkins) is similar to *Bactrocera (Bactrocera) neofulvicauda* n.sp. and *Bactrocera (Bactrocera) pseudodistincta* (Drew) in possessing a single distinct fuscous band across the wing and a broad costal band confluent with R_{4+5}. It differs from these species in the shape of the transverse band across the wing and in having the mesopleural stripe reaching to the postpronotal lobe dorsally.

Bactrocera (Bactrocera) fulvifemur Drew & Hancock

(Fig. 72)

Bactrocera (Bactrocera) fulvifemur Drew & Hancock, 1994a: 23; Norrbom *et al.*, 1998: 90. Holotype male in BBM.

Non-type material examined: Since the publication of Drew & Hancock (1994a), the following specimen has been studied – **VIETNAM: Southern Vietnam** – 1 male, Dong Nai Province, Long Khanh District, Nhan Nghia Village, Site VS06, 12.x.2002, attracted to cue lure. Specimen in QDPC.

Diagnosis: Face fulvous with a pair of medium-sized oval black spots; postpronotal lobes and notopleura yellow; scutum black with red-brown to dark brown below and behind lateral postsutural vittae, around notopleural suture and inside postpronotal lobes; broad triangular lateral postsutural yellow vittae narrowing sharply posteriorly to end at or before level of *p.sa.* seta; medial postsutural yellow vitta absent; mesopleural stripe reaching midway between anterior margin of notopleuron and anterior *npl.* seta dorsally; scutellum yellow; legs with femora entirely fulvous; fore tibiae pale fuscous, mid tibiae fulvous, hind tibiae dark fuscous; wing with cells bc and c pale fulvous to fulvous, microtrichia in outer corner of cell c only; a narrow dark fuscous costal band overlapping R_{2+3} and sometimes reaching R_{4+5} as a pale fuscous pattern and expanding across apex of wing; a broad dark fuscous anal streak; supernumerary lobe of medium development; abdominal terga III–V orange-brown with broad lateral and medial longitudinal black bands over all three terga and joining along anterior margin of tergum III, a pair of oval orange-brown shining spots on tergum V; abdominal sterna dark coloured.

Attractant: Cue lure.

Female: No known record.

Distribution:

Type locality: Philippines (Luzon I., Mt Makiling).

New records: Southern Vietnam (Dong Nai Province).

Hosts: No known record.

Remarks: *Bactrocera (Bactrocera) fulvifemur* Drew & Hancock, within the *dorsalis* complex, is similar to *Bactrocera (Bactrocera) melastomatos* Drew & Hancock and *Bactrocera (Bactrocera) usitata* Drew & Hancock in possessing lateral postsutural vittae that narrow sharply posteriorly, femora entirely or mostly fulvous and costal band overlapping R_{2+3} and almost confluent with R_{4+5}. It differs from *B. melastomatos* in possessing clearly defined dark patterns on abdominal terga III–V and the lateral postsutural vittae ending well before *ia.* setae and from *B. usitata* in possessing short triangular-shaped lateral postsutural vittae ending well before *ia.* setae and having the wings with cells bc and c pale fulvous to fulvous, not colourless.

Bactrocera (Bactrocera) fuscitibia Drew & Hancock

(Fig. 73)

Bactrocera (Bactrocera) fuscitibia Drew & Hancock, 1994a: 24; Norrbom *et al.*, 1998: 91. Holotype male in BBM.

Non-type material examined: Since the publication of Drew & Hancock (1994a), the following specimens have been studied – **INDONESIA: BALI** – 10 males (4.viii.2006), 9 males (11.viii.2006), 4 males (18.viii.2006), Bali, Buleleng, Kubutambahan, Tajun, coll. Kebut Rawa W, attracted to cue lure in forest; **JAVA** – 2 males (8.viii.2005), 1 male (16.viii.2005), 1 male (6.x.2005), Java, Banten, Lebok, Pasir Nini Nini, coll. Aden Haer, attracted to cue lure in forest; 1 male, East Java, Sidoarjo, Gelang, 12.vii.2005, coll. Agus Riadi, attracted to cue lure in village; 1 male, East Java, Probolinggo, Betektanam 150m, 14.xii.2005, coll. Tjipto Waluyo, attracted to cue lure in mango; **KALIMANTAN** – 1 male, East Kalimantan, Kutai Kalimantan, Batuah, 1.viii.2005, coll. Heru G, attracted to cue lure in forest; 1 male (8.viii.2005), 1 male (15.viii.2005), East Kalimantan, Kutai Kartanegara, Batuah, coll. Ertha D.P., attracted to cue lure in banana; 6 males (22.viii.2005), 3 males (29.viii.2005), East Kalimantan, Kutai Kartanegara, Batuah, coll. Ertha & Keru G, attracted to cue lure in forest; 2 males (27.vii.2005), 2 males (3.viii.2005), 1 male (17.viii.2005), 1 male (24.viii.2005), 2 males (31.viii.2005), West Kalimantan, Pontianak, Pancaroba, coll. A. Irom, attracted to cue lure in forest; 1 male, South Kalimantan, Banjar, Mandiangin, 8.viii.2005, coll. Bilal & Sidrotul, attracted to cue lure in forest; **SULAWESI** – 1 male, Sulawesi, Gorontalo, Bone Bolango, Lombongo, 26.ix.2006, coll. Yusbar Ismail, attracted to cue lure in forest; **SUMATRA** – 10 males (9.viii.2006), 12 males (4.x.2006), Sumatra, Jambi, Muaro Jambi, Sebapo, coll. Hendri Y, attracted to cue lure in forest; 6 males, Sumatra, Bengkulu, Bengkulu Selatan, Kebun Jali, 12.x.2006, coll. M. Nelson, attracted to cue lure in forest. **PENINSULAR MALAYSIA**: 1 male, TS8 Malaysia, 25.viii.1986, coll. Mahdzir, attracted to cue lure; 1 male, TS18 Malaysia, 1.vii.1986, coll. Mahdzir, attracted to cue lure. **VIETNAM: Northern Vietnam** – 5 males, northern Vietnam, Hoa Binh Province, Moc Chau District, Pa Co, Site N010, 28.ix.1999, attracted to cue lure; 1 male, northern Vietnam, Hoa Binh Province, Ky Son District, Cao Phong, Site N012, 31.viii.1999, attracted to cue lure; 1 male, northern Vietnam, Hoa Binh Province, Ky Son District, Cao Phong, Site N014, 17.x.1999, attracted to cue lure; **Southern Vietnam** – 1 male, southern Vietnam, Can Tho Province, Binh Thuy Village, Site S003, 12.xi.1999, attracted to cue lure; 1 male, southern Vietnam, Tien Giang Province, My Tho Village, Site S005, 6.vii.1999, attracted to cue lure; 3 males, southern Vietnam, Binh Duong Province, Thuan An District, An Thanh Village, Site VS05, 18.vii.2002, attracted to cue lure; 2 males (22.vi.2002), 2 males (20.vii.2002), 2 males (3.viii.2002), southern Vietnam, Dong Nai Province, Long Khanh District, Nhan Nghia Village, Site VS06, attracted to cue lure; 1 male, southern Vietnam, Dong Nai Province, Xuan Loc District, Xuan Phu Village, Site S015, 18.ix.1999, attracted to cue lure; 2 males, southern Vietnam, Binh Duong Province, Lai Thieu Village, Site S020, 8.vii.1999, attracted to cue lure. Specimens in QDPC.

Diagnosis: Face fulvous with a pair of large elongate oval black spots; postpronotal lobes and notopleura yellow; scutum black without pale markings; broad parallel-sided or subparallel lateral postsutural yellow vittae ending at *ia.* seta; medial postsutural yellow vitta absent; mesopleural stripe equal in width to notopleuron dorsally; scutellum yellow; legs with femora entirely fulvous or, at most, a dark fuscous subapical spot on outer surfaces of fore femora; fore and hind tibiae dark fuscous and mid tibiae fuscous; wing with cells bc and c colourless, microtrichia in outer corner of cell c only; a narrow dark fuscous costal band slightly overlapping R_{2+3} (particularly in apical ⅓ of this vein) and expanding slightly across apex of wing; a narrow fuscous anal streak; supernumerary lobe of medium development; abdominal terga III–V orange-brown with a narrow transverse black band across anterior margin of tergum III which expands to cover outer ½ of this tergum, a moderately broad medial longitudinal black band over all three terga, narrow anterolateral corners of terga IV and V fuscous to dark fuscous, a pair of oval shining spots on tergum V which vary from fuscous through to black; abdominal sterna dark coloured.

Attractant: Cue lure.

Female: No known record.

Distribution:

Type locality: Indonesia (Java).

Previously recorded: Type locality and the following locations – Indonesia (Sulawesi); East Malaysia (Sabah); Peninsular Malaysia (Drew & Hancock, 1994a).

New records: Indonesia (Bali; Kalimantan; Sumatra); Vietnam.

Hosts: No known record.

Remarks: *Bactrocera (Bactrocera) fuscitibia* Drew & Hancock, within the *dorsalis* complex, is similar to *Bactrocera (Bactrocera) bimaculata* Drew & Hancock, *Bactrocera (Bactrocera) bitungiae* n.sp., *Bactrocera (Bactrocera) dongnaiae* n.sp., *Bactrocera (Bactrocera) laithieuiae* n.sp. and *Bactrocera (Bactrocera) occipitalis* (Bezzi) in possessing broad parallel-sided or subparallel lateral postsutural vittae, costal band overlapping R_{2+3} and narrow dark lateral patterns on abdominal terga IV and V. It differs from *B. bitungiae*,

B. laithieuiae and *B. occipitalis* in possessing a narrow anteromedial bare area on the scutum and very narrow anterolateral dark corners on terga IV and V, from *B. bimaculata* in possessing all femora fulvous or at most with a dark preapical spot on outer surfaces on fore femora and from *B. dongnaiae* in possessing a broad medial longitudinal dark band on abdominal terga III–V.

Bactrocera (Bactrocera) fuscoformosa Drew & Romig, new species

(Fig. 74)

Material examined: Holotype male, **INDONESIA: FLORES** – Borong, Flores, Indonesia, iii.1988, coll. J. Turner, attracted to cue lure, deposited in BMNH.

Paratypes: **INDONESIA: FLORES** – 2 males, same data as holotype. 1 paratype in ANIC, 1 paratype in QDPC.

Diagnosis: A medium-sized species; face fulvous with a pair of large oval black spots; postpronotal lobes and notopleura yellow; scutum dark red-brown with a fuscous to dark fuscous pattern across posterior margin extending anteriorly into three narrow longitudinal fuscous to dark fuscous lines which meet across anterior margin; lateral postsutural yellow vittae present; medial postsutural yellow vitta absent; no yellow spot anterior to notopleural suture; mesopleural stripe reaching midway between anterior margin of notopleuron and anterior *npl.* seta dorsally; scutellum yellow; wing with cells bc and c fuscous, dense microtrichia covering all of cell c and outer ½ of cell bc; a broad dark fuscous costal band; a broad dark fuscous anal streak; abdominal terga II–V entirely red-brown.

Description: male.

Head: Height 1.5 mm. Frons – length 1.6 times breadth; fulvous with dark fuscous on anteromedial hump and around bases of orbital setae; orbital setae black: 1 *s.or.*, 2 *i.or.*; lunule dark fuscous. Ocellar triangle black. Vertex dark fuscous. Face fulvous with a pair of moderately large oval black spots; length 0.39 mm. Genae fulvous, small dark fuscous subocular spot present; black seta present. Occiput fuscous, fulvous along eye margins; occipital row with 5–7 black setae. Antennae with segments 1 and 2 red-brown, segment 3 red-brown with fuscous on apex and outer surface; length of segments: 0.18 mm; 0.3 mm; 0.72 mm.

Thorax: Scutum dark red-brown with a fuscous to dark fuscous pattern across posterior margin leading to three narrow longitudinal dark fuscous to black bands joining along anterior margin. Pleural areas dark fuscous to black with dark red-brown below postpronotal lobes and above mid coxae. Yellow markings as follows: postpronotal lobes; notopleura; mesopleural stripe reaching midway between anterior margin of notopleuron and anterior *npl.* seta dorsally, continuing to katepisternum as a transverse spot, anterior margin slightly convex; anatergite (posterior apex dark fuscous); anterior ⅔ katatergite (remainder dark fuscous); two narrow parallel-sided lateral postsutural vittae ending just before *ia.* seta. Medial longitudinal postsutural yellow vitta absent. Postnotum black, tending dark red-brown centrally. Scutellum yellow except for narrow black basal band. Setae: *sc.* 2; *prsc.* 2; *ia.* 1; *p.sa.* 1; *a.sa.* 1; *mpl.* 1; *npl.* 2; *scp.* 4.

Legs: All femora fulvous with fuscous around apical ⅓ to ¼; fore tibiae fuscous, mid tibiae fuscous basally to fulvous apically, hind tibiae dark fuscous; fore tarsi with basal segment fulvous and apical four segments fuscous, mid and hind tarsi with all segments entirely fulvous; mid tibiae each with an apical black spur.

Wings: Length 5.5 mm; cells bc and c fuscous; dense microtrichia covering all of cell c and outer ½ of cell bc; remainder of wings colourless except dark fuscous cell sc, broad dark fuscous costal band confluent with R_{4+5} but paler adjacent to this vein and of uniform width throughout, a broad dark fuscous anal streak; a dense aggregation of microtrichia around A_1+CuA_2; supernumerary lobe of medium development.

Abdomen: Oval; terga free; pecten present on tergum III. Tergum I and sterna I and II wider than long. Tergum I fuscous with a narrow transverse red-brown band across posterior margin but not reaching lateral margins; terga II–V entirely red-brown with fulvous posterolaterally on tergum II. A pair of oval dark fuscous shining spots on tergum V. Posterior lobe of surstylus short, sternum V with a moderately deep concavity on posterior margin. Abdominal sterna dark fuscous to black.

Attractant: Cue lure.

Female: No known record.

Distribution: Indonesia (Flores, Borong).

Hosts: No known record.

Remarks: *Bactrocera (Bactrocera) fuscoformosa* n.sp. is similar to *Bactrocera (Bactrocera) dispar* (Hardy), *Bactrocera (Bactrocera) flavipennis* (Hardy), *Bactrocera (Bactrocera) ochroventer* n.sp. and *Bactrocera (Bactrocera) patula* n.sp. in possessing a basic red-brown scutum, the costal band confluent with R_{4+5} (but sometimes pale adjacent to this vein) and of uniform width throughout. It differs from *B. dispar* in having *prsc.* setae, abdominal terga III–V mostly orange-brown and in lacking a yellow spot anterior to the notopleural suture, from *B. flavipennis* and *B. patula* in having abdominal terga III–V entirely orange-brown and *prsc.* setae present, and from *B. ochroventer* in having the scutum with fuscous to dark fuscous patterns and longer lateral postsutural vittae reaching almost to the *ia.* setae.

Bactrocera (Bactrocera) fuscolobata Drew & Romig, new species

(Fig. 75)

Material examined: Holotype male, **INDONESIA: CENTRAL SULAWESI** – Tolai, Sungai Tolewonu, C. Sulawesi, Indonesia, 11–26.ii.1976, 450 ft., coll. Margareta Becker, deposited in BMNH.

Paratypes: **INDONESIA: SOUTH SULAWESI** – 1 male, Palopo-Rantepao Rd., 16–23 km Palopo, Sulawesi, Indonesia, 250–500 m, 24.vii.1981, coll. J. Turner, attracted to cue lure, Nanggala III forest, deposited in QDPC.

Diagnosis: A medium-sized species; face fulvous with a pair of medium-sized oval black spots; postpronotal lobes fuscous; notopleura yellow; scutum black; lateral postsutural yellow vittae present; medial postsutural yellow vitta absent; no yellow spot anterior to notopleural suture; mesopleural stripe reaching midway between anterior margin of notopleuron and anterior *npl.* seta dorsally; scutellum yellow; wing with cells bc and c fuscous, dense microtrichia covering all of cell c and outer ½ of cell bc; a broad dark fuscous costal band overlapping R_{4+5}; a broad dark fuscous anal streak; abdominal terga III–V entirely orange-brown with narrow transverse dark fuscous to black spots on anterolateral corners of tergum III.

Description: male.

Head: Height 1.68 mm. Frons – length 1.44 times breadth; red-brown with fulvous along lateral margins; orbital setae black: 1 *s.or.*, 2 *i.or.*; lunule fulvous. Ocellar triangle black. Vertex red-brown. Face fulvous with a pair of medium-sized oval black spots; length 0.48 mm. Genae red-brown; black seta present. Occiput fuscous, fulvous along eye margins; occipital row with 6–8 strong black setae. Antennae with segments 1 and 2 red-brown, segment 3 entirely fuscous; length of segments: 0.18 mm; 0.3 mm; 0.78 mm.

Thorax: Scutum black with dark red-brown below and behind lateral postsutural vittae and around notopleural suture. Pleural areas black with fuscous below postpronotal lobes and above mid coxae. Yellow markings as follows: notopleura; mesopleural stripe reaching midway between anterior margin of notopleuron and anterior *npl.* seta dorsally, continuing to katepisternum as a small transverse spot, anterior margin straight; anatergite (posterior apex black); anterior ⅔ katatergite (remainder black); two medium-width lateral postsutural vittae narrowing slightly posteriorly to end before *ia.* seta. Medial longitudinal postsutural yellow vitta absent. Postpronotal lobes fuscous. Postnotum black. Scutellum yellow except for narrow black basal band. Setae: *sc.* 2; *prsc.* 2; *ia.* 1; *p.sa.* 1; *a.sa.* 1; *mpl.* 1; *npl.* 2; *scp.* 4.

Legs: Fore femora red-brown to fuscous, mid and hind femora fulvous basally to red-brown to fuscous apically; fore and mid tibiae red-brown, hind tibiae red-brown to fuscous; all tarsi with basal segment fulvous and apical four segments red-brown; mid tibiae each with an apical black spur.

Wings: Length 6.4 mm; cells bc and c fuscous; dense microtrichia covering all of cell c and outer ½ of cell bc; remainder of wings colourless except dark fuscous cell sc, broad dark fuscous costal band overlapping R_{4+5} and of uniform width throughout, a broad dark fuscous anal streak; a dense aggregation of microtrichia around A_1+CuA_2; supernumerary lobe of medium development.

Abdomen: Oval; terga free; pecten present on tergum III. Tergum I and sterna I and II wider than long. Tergum I dark fuscous with a narrow orange-brown transverse band across posterior margin but not reaching lateral margins; tergum II orange-brown with narrow fulvous posterolateral margins; terga III–V entirely orange-brown with narrow transverse dark fuscous to black markings on anterolateral corners of tergum III. A pair of oval dark fuscous shining spots on tergum V. Posterior lobe of surstylus short, sternum V with a deep concavity on posterior margin. Abdominal sterna red-brown.

Attractant: Cue lure.

Female: No known record.

Distribution: Indonesia (Sulawesi – Central Sulawesi; South Sulawesi).

Hosts: No known record.

Remarks: *Bactrocera (Bactrocera) fuscolobata* n.sp. is similar to *Bactrocera (Bactrocera) linduensis* new name and *Bactrocera (Bactrocera) paralimbifera* n.sp. in possessing a black scutum, a broad fuscous costal band of uniform width and overlapping R_{4+5} and cells bc and c fuscous. It differs from *B. paralimbifera* in having abdominal terga III–V almost entirely orange-brown and from *B. linduensis* in having the postpronotal lobes dark fuscous and cells bc and c dark fuscous.

Bactrocera (Bactrocera) fuscoptera Drew & Romig, new species

(Fig. 76)

Material examined: Holotype male, **INDONESIA: SOUTH SULAWESI** – Palopo – Rantepao Rd., 15–22 km Rantepao, Sulawesi, Indonesia (1000–1100 m), 24.vii.1981, coll. J. Turner, attracted to methyl eugenol in primary rainforest (some disturbance), deposited in BMNH.

Diagnosis: A large species; face fulvous with a pair of medium-sized circular black spots; postpronotal lobes yellow (anterodorsal corners fuscous); notopleura yellow; scutum red-brown with fuscous across posterior margin and irregularly shaped black patterns; lateral postsutural yellow vittae present; medial postsutural yellow vitta absent; no yellow spot anterior to notopleural suture; mesopleural stripe reaching midway between anterior margin of notopleuron and anterior *npl.* seta dorsally; scutellum yellow; wing with cells bc and c fuscous, microtrichia in outer corner of cell c only; an overall pale fuscous colouration throughout the membrane and a broad fuscous costal band confluent with R_{4+5} and a broad fuscous anal streak each merging into the pattern across the wing membrane; abdominal terga III–V orange-brown with a dark 'T' pattern consisting of a narrow transverse black band across anterior margin of tergum III and a narrow- to medium-width medial longitudinal dark fuscous to black band over terga III and IV.

Description: male.

Head: Height 2.1 mm. Frons – length 1.32 times breadth; red-brown with small areas of pale fuscous around bases of orbital setae; orbital setae red-brown: 1 *s.or.*, 2 *i.or.*; lunule fuscous. Ocellar triangle black. Vertex red-brown. Face fulvous with a pair of medium-sized circular black spots; length 0.59 mm. Genae fulvous to red-brown, pale red-brown seta present. Occiput red-brown, fulvous along eye margins; occipital row with a small number of short pale setae. Antennae with all segments red-brown; length of segments: 0.22 mm; 0.36 mm; 0.94 mm.

Thorax: Scutum red-brown, tending fuscous across posterior margin and with elongate black markings inside lateral postsutural vittae and black between notopleural suture and postpronotal lobe. Pleural areas red-brown, tending dark fuscous to black along both margins of the mesopleural stripe, above hind coxae and on katepisternum. Yellow markings as follows: postpronotal lobes (fuscous on anterodorsal corners); notopleura; mesopleural stripe reaching midway between anterior margin of notopleuron and anterior *npl.* seta dorsally, continuing to katepisternum as a transverse spot, anterior margin straight; anatergite (posterior apex dark fuscous); anterior ¾ katatergite (remainder dark fuscous); two moderately broad parallel-sided lateral postsutural vittae ending at *ia.* seta. Medial longitudinal postsutural yellow vitta absent. Postnotum black laterally, dark red-brown centrally. Scutellum yellow except for narrow black basal band. Setae: *sc.* 2; *prsc.* 2; *ia.* 1; *p.sa.* 1; *a.sa.* 1; *mpl.* 1; *npl.* 2; *scp.* 4.

Legs: All segments entirely fulvous; mid tibiae each with an apical black spur.

Wings: Length 9.1 mm; cells bc and c fuscous; microtrichia in outer corner of cell c only; remainder of wing membrane entirely coloured with a pale fuscous tint except fuscous cell sc, broad fuscous costal band confluent with R_{4+5} and merging into the colour across the wing membrane and with a dark fuscous spot in apex of wing, a broad fuscous anal streak merging into colour across the wing membrane; a dense aggregation of microtrichia around A_1+CuA_2; supernumerary lobe of medium development.

Abdomen: Oval; terga free; pecten present on tergum III. Tergum I and sterna I and II wider than long. Tergum I orange-brown with narrow dark fuscous lateral margins; tergum II orange-brown with a narrow transverse fuscous to black band subanteriorly but not reaching narrow fuscous lateral margins; terga III–V orange-brown with a dark 'T' pattern consisting of a narrow transverse black band across anterior margin of tergum III and a narrow- to medium-width medial longitudinal dark fuscous to black band over terga III and IV. A pair of oval dark red-brown shining spots on tergum V. Posterior lobe of surstylus short, sternum V with a deep concavity on posterior margin. Abdominal sterna pale orange-brown.

Attractant: Methyl eugenol.

Female: No known record.

Distribution: Indonesia (South Sulawesi).

Hosts: No known record.

Remarks: *Bactrocera (Bactrocera) fuscoptera* n.sp. is similar to *Bactrocera (Polistomimetes) fuscalata* Drew in possessing infuscation over almost the entire wing membrane. However, it is a unique species of subgenus *Bactrocera* in having a red-brown scutum with dark patterns, abdominal terga III–V orange-brown with a narrow black 'T' pattern and cells bc and c fuscous in addition to the infuscated wing membrane.

Bactrocera (Bactrocera) gombokensis Drew & Hancock

(Fig. 77)

Bactrocera (Bactrocera) gombokensis Drew & Hancock, 1994a: 24; Norrbom *et al*., 1998: 91; Drew *et al*., 2007: 3. Holotype male in BMNH.

Non-type material examined: Since the publication of Drew & Hancock (1994a), the following specimens have been studied – **BHUTAN**: 1 male, Phuentsholing, 6.v.2000, coll. S. Thapa, attracted to cue lure; 1 male, Tsirang, 19. vi.2000, coll. T.D. & S.D., attracted to cue lure; 2 males, Namling 16km, 9.vi.2001, coll. Brian Fletcher, attracted to cue lure in subtropical rainforest, altitude 1600m; 1 male, Lingmethang, 9.vi.2001, coll. Brian Fletcher, attracted to cue lure in mango tree, altitude 700m; 4 males (15.xi.2004), 3 males (5.xii.2004), East Bhutan, Sandrup Jongkhar District, coll. Wangdi, attracted to cue lure; 6 males, Ganglephug, Tatapani, 9-13.ix.1990, coll. H.R. Fiejen; 5 males, Manchungang, Tashi, 4.x.1991, coll. H.R. Fiejen, attracted to cue lure. **VIETNAM: Northern Vietnam** – 1 male, Hoi An, 21.v.1997, coll. G. Pattemore, attracted to cue lure; 1 male, Northern Vietnam, Bac Giang Province, Luc Ngan District, Cao Cao, Quy Son, Site VN05, 14.vi.2002, attracted to cue lure; 1 male, Northern Vietnam, Son La Province, Moc Chau District, Co Do State Farm, site VN06, 15. vii.2002, attracted to cue lure; 1 male, Northern Vietnam, Son La Province, Moc Chau District, Co Do, site N006, 1.v.2000, attracted to cue lure; 1 male, Northern Vietnam, Hoa Binh Province, Moc Chau District, Pa Co Village, Site VN10 (FAO site), 30.iv.2002, coll. NIPP Staff, attracted to cue lure; northern Vietnam, Hoa Binh Province, Ky Son District, Cao Phong, site N012, 15.v.2000, attracted to cue lure; 1 male (29.iv.2000), 1 male (30.v.2000), Northern Vietnam, Hoa Binh Province, Ky Son District, Cao Phong, site N013, attracted to cue lure; 2 males, Northern Vietnam, Lao Cai Province, Sapa 2 Village, Site N015, 30.vii.1999, attracted to cue lure; 5 males (30.vii.1999), 3 males (13.viii.1999), Northern Vietnam, Lao Cai Province, Sapa 6 Village, Site N017, attracted to cue lure; 1 male, Northern Vietnam, Lao Cai Province, Sapa Oquy Ho, site N019, 30. vii.1999, attracted to cue lure; **Southern Vietnam** – 1 female, Southern Vietnam, Cantho Province, Chau Thanh District, Tan Phu Village, site VS02, 6.vii.2002, attracted to cue lure; 1 male, Southern Vietnam, Binh Duong Province, Thuan An District, An Thanh Village, site VS05, 4 Oct.2002; 1 male (31.viii.1999), 1 male (16.viii.2000), Southern Vietnam, Tien Giang Province, Cai Lay District, Long Khanh Village, Site S007, attracted to cue lure; 1 male Southern Vietnam, Tien Giang Province, Cai Be District, An Huu Village, site S009, 9.xi.1999, attracted to cue lure; 1 male (22.vii.1999), 1 male (3.iii.2000), 2 males (28.iv.2000), 1 male (10.v.2000), Southern Vietnam, Dong Nai Province, Vinh Cuu District, Tan Binh Village, site S011, attracted to cue lure; 1 male, Southern Vietnam, Tien Giang Province, Cai Be District, My Thirn Village, site S013, 28.iv.2000, attracted to cue lure; 1 male, Southern Vietnam, Dong Nai Province, Long Khanh District, Tan Phong Village, site S014, 9.vii.1999, attracted to cue lure. Specimens in QDPC.

Diagnosis: Face fulvous with a pair of large circular black spots; postpronotal lobes and notopleura yellow; scutum black without pale markings; medium-width parallel-sided lateral postsutural yellow vittae ending at *ia.* seta; medial postsutural yellow vitta absent; mesopleural stripe reaching midway between anterior margin of notopleuron and anterior *npl.* seta dorsally; scutellum yellow; legs with femora fulvous with distinct black spots on outer subapical surfaces of fore femora and inner subapical surfaces of mid and hind

femora; fore tibiae fuscous, mid tibiae fulvous tending fuscous basally, hind tibiae fuscous; wing with cells bc and c colourless, microtrichia in outer corner of cell c only; narrow fuscous costal band confluent with R_{2+3} and remaining very narrow around apex of wing; anal streak absent; supernumerary lobe of medium development; abdominal terga III–V orange-brown with a narrow transverse black band across anterior margin of tergum III and widening to cover outer ⅓ of lateral margins of this tergum, a narrow medial longitudinal black band over all three terga, narrow anterolateral corners of terga IV and V dark fuscous to black, a pair of oval fuscous shining spots on tergum V; abdominal sterna dark coloured.
Attractant: Cue lure.
Female: No known record.
Distribution:
Type locality: Peninsular Malaysia (Gombak).
Previously recorded: Type locality and Bhutan.
New records: Vietnam.
Hosts: No known record.
Remarks: *Bactrocera (Bactrocera) gombokensis* Drew & Hancock, within the *dorsalis* complex, is similar to *Bactrocera (Bactrocera) fernandoi* Tsuruta & White, *Bactrocera (Bactrocera) lombokensis* Drew & Hancock and *Bactrocera (Bactrocera) pedestris* (Bezzi) in possessing moderately broad parallel-sided lateral postsutural vittae, costal band confluent with R_{2+3} and dark apical markings on all femora. It differs from *B. fernandoi* and *B. lombokensis* in having the costal band extremely uniform and narrow around apex of wing beyond apex of R_{2+3} and from *B. pedestris* in having the apical dark markings on the femora as distinct spots, lacking black rings around bases of antennae, lateral postsutural vittae ending at the *ia.* seta (not behind the seta) and the anal streak absent.

The species name *gombokensis* is a misspelling of the district name Gombak in Malaysia. However, under the rules of the ICZN, the original spelling *gombokensis* must be retained.

Bactrocera (Bactrocera) halmaherae Drew & Romig, new species

(Fig. 78)

Material examined: Holotype male, **INDONESIA: NORTH MOLUCCAS (NORTH MALUKU)** – Moluccas, North Maluku, West Halmahera Gaol, 12.vi.2007, coll. Raais Abdullah, attracted to cue lure in citrus, deposited in BMNH. Paratypes: **INDONESIA: NORTH MOLUCCAS (NORTH MALUKU)** – 31 males, same data as holotype; 18 males (29.v.2007), 33 males (5.vi.2007), Maluku, North Maluku, West Halmahera Gaol, coll. Raais Abdullah, attracted to cue lure; 1 male, Moluccas, North Maluku, Kodya, Ternate, Foramadiahi, 10.vi.2007, coll. La Ruti, attracted to cue lure in citrus; 3 males, Moluccas, North Maluku, West Halmahera, Gamlamo, 19.vi.2007, coll. Raais Abdullah, attracted to cue lure in mango. 16 paratypes in ANIC, 20 paratypes in BMNH, 40 paratypes in QDPC, 10 paratypes in QMIC.

Diagnosis: A medium-sized species; face fulvous with a pair of very small irregularly circular black spots; postpronotal lobes and notopleura yellow; scutum black with narrow red-brown to dark red-brown lateral margins; lateral postsutural yellow vittae present; medial postsutural yellow vitta absent; no yellow spot anterior to notopleural suture; mesopleural stripe reaching anterior *npl.* seta dorsally; scutellum yellow; wing with cells bc and c dark fuscous, microtrichia in outer ⅓ of cell c only; an irregularly shaped broad dark fuscous costal band overlapping R_{4+5} and extending along *r-m* cross-vein, a broad irregularly shaped dark fuscous 'U' pattern posterocentrally; a broad dark fuscous anal streak; abdominal terga III–V orange-brown with a black 'T' pattern consisting of a narrow transverse black band across anterior margin and extending in as narrow dark fuscous lateral margins on this tergum and a medium-width medial longitudinal black band over all three terga.
Description: male.
Head: Height 1.76 mm. Frons – length 1.65 times breadth; red-brown with dark fuscous around bases of orbital setae; orbital setae black: 1 *s.or.*, 2 *i.or.*; lunule dark fuscous. Ocellar triangle black. Vertex dark fuscous. Face fulvous with a pair of very small irregularly circular black spots; length 0.49 mm. Genae fulvous, large fuscous subocular spot present; red-brown seta present. Occiput fuscous, fulvous along eye margins; occipital row with a large number of short black setae. Antennae with segments 1 and 2 dark fuscous, segment 3 fuscous; length of segments: 0.17 mm; 0.31 mm; 0.84 mm.
Thorax: Scutum black with red-brown lateral to lateral postsutural vittae and around notopleural suture and dark red-brown along lateral margins between postpronotal lobe and notopleuron. Pleural areas dark fuscous to black except red-brown below postpronotal lobes. Yellow markings as follows: postpronotal lobes; notopleura; mesopleural stripe reaching anterior *npl.* seta dorsally, continuing to katepisternum as a small transverse spot, anterior margin straight; anatergite (posterior apex black); anterior ⅗ katatergite (remainder black); two broad lateral postsutural vittae narrowing slightly posteriorly to end at or just before *ia.* seta. Medial longitudinal postsutural yellow vitta absent. Postnotum black. Scutellum yellow except for narrow black basal band. Setae: *sc.* 2; *prsc.* 2; *ia.* 1; *p.sa.* 1; *a.sa.* 1; *mpl.* 1; *npl.* 2; *scp.* 4.
Legs: Fore femora fulvous with diffuse fuscous around apical ⅓, mid tibiae fulvous with a diffuse fuscous subapical spot on outer surfaces, hind femora fulvous with a diffuse fuscous line along apical ⅓ all tibiae fuscous; mid and hind tarsi with basal segment fulvous and apical four segments red-brown; mid tibiae each with an apical black spur.
Wings: Length 6.0 mm; cells bc and c dark fuscous; microtrichia in outer ⅓ of cell c only; remainder of wings colourless except dark fuscous cell sc, a broad dark fuscous costal band overlapping R_{4+5} and extending along cross-vein *r-m* and apically extending in a point to vein M, a broad dark fuscous 'U' pattern posterocentrally, a broad dark fuscous anal streak; a dense aggregation of microtrichia around A_1+CuA_2; supernumerary lobe of medium development.

Abdomen: Oval; terga free; pecten present on tergum III. Tergum I and sterna I and II wider than long. Tergum I dark fuscous to black with a narrow transverse orange-brown band posterocentrally; tergum II orange-brown with a narrow transverse dark fuscous to black band across anterior margin but not reaching lateral margins; terga III–V orange-brown with a 'T' pattern consisting of a narrow transverse black band across anterior margin of tergum III and continuing around lateral margins of that tergum as a dark fuscous pattern, a medium-width medial longitudinal black band over all three terga. A pair of oval red-brown to dark fuscous shining spots on tergum V. Posterior lobe of surstylus short, sternum V with a deep concavity on posterior margin. Abdominal sterna red-brown to fuscous.
Attractant: Cue lure.
Female: No known record.
Distribution: Indonesia (North Moluccas = North Maluku – Halmahera I.; Ternate).
Hosts: No known record.
Remarks: *Bactrocera (Bactrocera) halmaherae* n.sp. is a unique species with an unusual wing pattern unlike any other *Bactrocera* species. The broad dark fuscous costal band overlapping R_{4+5} and extending along *r-m* crossvein and to M at apex together with the broad 'U'-shaped dark fuscous pattern posterocentrally, sets this species apart.

Bactrocera (Bactrocera) hantanae Tsuruta & White

(Fig. 79)

Bactrocera (Bactrocera) hantanae Tsuruta & White, 2001: 79. Holotype male in HORDI.

Non-type material examined: SRI LANKA (formerly CEYLON): 3 males, 'Erminigolla, 31.x.1994, cue lure 29'. This locality is spelt Eraminigolla in Tsuruta & White, 2001. Specimens donated by Kenji Tsuruta and in QDPC.
Diagnosis: A medium-sized species; face fulvous with a pair of medium-sized circular to oval black spots; postpronotal lobes and notopleura yellow; scutum black; lateral postsutural yellow vittae present; medial postsutural yellow vitta absent; no yellow spot anterior to notopleural suture; mesopleural stripe reaching midway between anterior margin of notopleuron and anterior *npl.* seta dorsally; scutellum yellow; wing with cells bc and c colourless, microtrichia in outer corner of cell c only; a narrow dark fuscous costal band; anal streak reduced to a pale fuscous tint within cell cup; abdominal terga III–V orange-brown with a 'T' pattern consisting of a narrow transverse fuscous to black band across anterior margin of tergum III but not reaching lateral margins and a very broad medial longitudinal black band over all three terga, very narrow pale fuscous anterolateral corners on terga IV and V.
Description: male.
Head: Height 1.75 mm. Frons – length 1.6 times breadth; red-brown with fuscous on anteromedial hump and around orbital setae; orbital setae black: 1 *s.or.*, 2 *i.or.*; lunule dark fuscous. Ocellar triangle black. Vertex fuscous. Face fulvous with a pair of medium-sized circular to oval black spots; length 0.55 mm. Genae fulvous, fuscous subocular spot present; black seta present. Occiput red-brown, fulvous along eye margins; occipital row with 5–7 strong black setae. Antennae with segments 1 and 2 red-brown, segment 3 red-brown with fuscous on apex and outer surface; length of segments: 0.16 mm; 0.28 mm; 0.8 mm.
Thorax: Scutum black with red-brown below and behind lateral postsutural vittae, around notopleural suture, along lateral margins from notopleuron to anterior margin of scutum. Pleural areas black except red-brown below postpronotal lobes. Yellow markings as follows: postpronotal lobes; notopleura; mesopleural stripe reaching midway between anterior margin of notopleuron and anterior *npl.* seta dorsally, continuing to katepisternum as a small transverse spot, anterior margin slightly convex; anatergite (posterior apex black); anterior $^{3}/_{5}$ katatergite (remainder black); two broad parallel-sided lateral postsutural vittae ending slightly behind *ia.* seta. Medial longitudinal postsutural yellow vitta absent. Postnotum black. Scutellum yellow except for narrow black basal band. Setae: *sc.* 2; *prsc.* 2; *ia.* 1; *p.sa.* 1; *a.sa.* 1; *mpl.* 1; *npl.* 2; *scp.* 4.
Legs: All femora entirely fulvous; fore and mid tibiae pale fuscous, hind tibiae fuscous; tarsi with all segments fulvous; mid tibiae each with an apical black spur.
Wings: Length 6.0 mm; cells bc and c colourless; microtrichia in outer corner of cell c only; remainder of wings colourless except fuscous cell sc, narrow dark fuscous costal band confluent with R_{2+3} and remaining narrow around margin of wing to end just beyond extremity of R_{4+5}, anal streak reduced to pale fuscous tint within cell cup; a dense aggregation of microtrichia around A_1+CuA_2; supernumerary lobe of medium development.
Abdomen: Oval; terga free; pecten present on tergum III. Tergum I and sterna I and II wider than long. Tergum I fuscous with dark fuscous lateral margins; tergum II fulvous with a narrow subanterior transverse black band but not reaching lateral margins; terga III–V orange-brown with a 'T' pattern consisting of a narrow transverse fuscous to black band across anterior margin but not reaching lateral margins and a very broad medial longitudinal black band over all three terga and narrow anterolateral pale fuscous margins on terga IV and V. A pair of oval dark fuscous shining spots on tergum V. Posterior lobe of surstylus short, sternum V with a deep concavity on posterior margin. Abdominal sterna I and V dark fuscous to black, remainder pale.
Attractant: Cue lure.
Female: No known record.
Distribution:
Type locality: Sri Lanka (Kegalle District, Eraminigolla).
Previously recorded: Type locality and other district locations in Sri Lanka (Tsuruta & White, 2001).
Hosts: No known record.
Remarks: *Bactrocera (Bactrocera) hantanae* Tsuruta & White, within the *dorsalis* complex, is similar to *Bactrocera (Bactrocera) irvingiae* Drew & Hancock, *Bactrocera (Bactrocera) raiensis* Drew & Hancock and *Bactrocera (Bactrocera) syzygii* White & Tsuruta in possessing broad

parallel-sided lateral postsutural vittae, costal band confluent with R_{2+3}, femora entirely fulvous, a broad medial longitudinal black band on abdominal terga III–V. It differs from *B. raiensis* and *B. syzygii* in lacking dark lateral markings on abdominal terga III–V and from *B. irvingiae* in having dark fuscous shining spots on abdominal tergum V.

Bactrocera (Bactrocera) holtmanni (Hardy)

(Fig. 80)

Dacus (Strumeta) holtmanni Hardy, 1974: 34. Holotype female in BBM.
Dacus (Bactrocera) holtmanni – Hardy, 1977: 50.
Bactrocera (Bactrocera) holtmanni – Drew & Hancock, 1994a: 25; Norrbom *et al.*, 1998: 91.

Non-type material examined: Since the publication of Drew & Hancock (1994a), the following specimens have been studied – **EAST MALAYSIA**: 1 male, Ranchang Pool, Serian, near Kuching, 30.ix.1991, coll. S. Leong, attracted to cue lure; 1 male, Kpg Tema Mawang, Tebedu, near Kuching, 23.vi.1992, coll. S. Leong, attracted to cue lure. **VIETNAM: Northern Vietnam** – 2 males (15.vi.2001), 3 males (26.vi.2000), Northern Vietnam, Son La Province, Moc Chau District, Co Do, site N006, attracted to cue lure; 2 males (26.vi.1999), 1 male (12.vi.2000), Northern Vietnam, Son La Province, Moc Chau District, Tra Long, site N007, attracted to cue lure; 2 males, Northern Vietnam, Hoa Binh Province, Moc Chau District, Pa Co, site N010, 30.vi.2001, attracted to cue lure; 1 male, Northern Vietnam, Lao Cai Province, Sapa Oquy Ho, site N019, 13.viii.1999, attracted to cue lure. Specimens in QDPC.

Diagnosis: Face fulvous with a pair of medium-sized oval black spots; postpronotal lobes and notopleura yellow; scutum black with brown to dark fuscous below lateral postsutural vittae, brown around notopleural suture and inside postpronotal lobes; broad parallel-sided lateral postsutural yellow vittae ending just behind *ia.* seta; medial postsutural yellow vitta absent; mesopleural stripe reaching midway between anterior margin of notopleuron and anterior *npl.* seta dorsally; scutellum yellow; legs with femora fulvous with dark fuscous preapical spots on outer surfaces of fore and inner surfaces of mid and hind femora; fore tibiae fuscous, mid tibiae fulvous tending fuscous basally, hind tibiae dark fuscous; wing with cells bc and c colourless, microtrichia in outer corner of cell c only; a narrow fuscous costal band overlapping R_{2+3} and expanding apically across apex of R_{4+5}; a narrow fuscous anal streak; supernumerary lobe of medium development; abdominal terga III–V orange-brown with a broad dark fuscous to black transverse band across anterior margin of tergum III but narrowing more in centre of tergum, a medium-width medial longitudinal black band over all three terga, narrow anterolateral corners of terga IV and V dark fuscous, a pair of oval orange-brown shining spots on tergum V; abdominal sterna dark coloured.
Attractant: Cue lure.
Female: No known record.
Distribution:
Type locality: Philippines (Busuanga I.).
Previously recorded: Type locality and Peninsular Malaysia (Drew & Hancock, 1994a).
New records: East Malaysia; Northern Vietnam (Hoa Binh, Lao Cai, Son La Provinces).
Hosts: No known record.
Remarks: *Bactrocera (Bactrocera) holtmanni* (Hardy), within the *dorsalis* complex, is similar to *Bactrocera (Bactrocera) affinidorsalis* (Hardy) and *Bactrocera (Bactrocera) quasipropinqua* Drew & Hancock in possessing broad parallel-sided lateral postsutural vittae, costal band overlapping R_{2+3} and expanding apically, pale coloured shining spots on abdominal tergum V, lateral markings on abdominal terga IV and V narrow and dark patterns on apices of all femora or at least the fore and mid femora. It differs from *B. quasipropinqua* in having preapical dark spots on all femora (not only on the fore and mid femora) and the apex of the female piercer needle-shaped and from *B. affinidorsalis* in having the lateral postsutural vittae ending behind the *ia.* seta, not ending before the *ia.* seta.

Bactrocera (Bactrocera) hyalina (Shiraki)

(Fig. 81)

Chaetodacus hyalinus Shiraki, 1933: 62. Holotype female in NIAS.
Strumeta asatoi Shiraki, 1968: 26; Norrbom *et al.*, 1998: 91.
Bactrocera (Bactrocera) hyalina – Norrbom *et al.*, 1998: 91.
Dacus (Bactrocera) hsui Tseng, Chen & Chu, 1992: 34. **New Synonym**.

Type material examined: Holotype female, **JAPAN**: with the following five labels – 'Kagoshima, 10.vii, Col. T. Shiraki'; 'Japanese writing, 7/10'; '611'; 'Chaetod. hyalinus n sp, Det. T. Shiraki'; 'Type'. Received on loan from NIAS.
Non-type material examined: JAPAN: 1 male, Kogoshima, Japan, 2.viii.1990, D. Hyalinus; 1 male, Amami Is., Kagoshima Pref., 7.iv.1975, coll. T. Imamura, bred ex *Osmanthus insularis*. **THAILAND**: 4 males and 2 females, Doi Pui, Chiang Mai, 9.ix.1993, coll. Chantanee *et al.*, bred from *Phoebe* sp. (Family Lauraceae), Sample No. CM2026. Specimens in QDPC.
Diagnosis: A medium-sized species; face fulvous with a pair of small circular black spots; postpronotal lobes and notopleura yellow; scutum black; very short and narrow lateral postsutural yellow vittae present; medial postsutural yellow vitta absent; no yellow spot anterior to notopleural suture; mesopleural stripe reaching midway between anterior margin of notopleuron and anterior *npl.* seta dorsally; scutellum yellow with a large dark fuscous spot enclosing apical *sc.* seta; wing with cells bc and c colourless, a narrow line of microtrichia along anterior margin of cell c only; a narrow fuscous costal band confluent with R_{2+3} and ending at apex of this vein, a small narrow fuscous spot around apex of R_{4+5}; a narrow fuscous anal streak; abdominal terga III–V red-brown with a 'T' pattern consisting of a broad black transverse band across anterior margin of tergum III, a narrow medial longitudinal black band over all three terga, broad

black lateral margins on tergum IV, narrow black anterolateral corners on tergum V.

Description: female.

Head: Height 1.55 mm. Frons – length 1.36 times breadth; fulvous with dark fuscous around orbital setae and fuscous on anteromedial hump; orbital setae black: 1 *s.or.*, 2 *i.or.*; lunule dark fuscous. Ocellar triangle black. Vertex dark fuscous. Face fulvous with a pair of small circular black spots; length 0.56 mm. Genae fulvous, dark fuscous subocular spot present; black seta present. Occiput black, fulvous along eye margins; occipital row with a large number of strong black setae. Antennae with segments 1 and 2 red-brown, segment 3 red-brown with fuscous on apex and outer surface; length of segments: 0.19 mm; 0.31 mm; 0.87 mm.

Thorax: Scutum almost entirely black except red-brown below and behind lateral postsutural vittae and inside notopleura. Pleural areas black except red-brown below postpronotal lobes. Yellow markings as follows: postpronotal lobes (fuscous on anterodorsal margins); notopleura; mesopleural stripe reaching midway between anterior margin of notopleuron and anterior *npl.* seta dorsally, continuing to katepisternum as a small transverse spot, anterior margin straight; anatergite (posterior apex black); anterior ⅔ katatergite (remainder black); two very short and narrow lateral postsutural vittae ending at a point just behind level of *a.sa.* seta. Medial longitudinal postsutural yellow vitta absent. Postnotum black. Scutellum yellow except for a narrow black basal band and a large black apical spot enclosing apical *sc.* setae. Setae: *sc.* 2; *prsc.* 2; *ia.* 1; *p.sa.* 1; *a.sa.* 1; *mpl.* 1; *npl.* 2; *scp.* 4.

Legs: All segments entirely fulvous except hind tibiae pale fuscous; mid tibiae each with an apical black spur.

Wings: Length 6.9 mm; cells bc and c colourless; microtrichia along fore margin of cell c only; remainder of wings colourless except fuscous cell sc, narrow pale fuscous costal band confluent with R_{2+3} and ending at apex of this vein, a small narrow fuscous spot around apex of R_{4+5}, a narrow fuscous anal streak; no dense aggregation of microtrichia around A_1+CuA_2; supernumerary lobe weak.

Abdomen: Oval; terga free; no pecten on tergum III. Tergum I and sterna I and II wider than long. Tergum I black except for a narrow red-brown band along posterior margin but not reaching lateral margins; tergum II red-brown except for a narrow transverse black band across tergum just posterior of anterior margin and reaching narrow anterolateral black corners; terga III–V red-brown with a 'T' pattern consisting of a broad transverse black band across anterior margin of tergum III and a narrow medial longitudinal black band over all three terga, broad black lateral margins on tergum IV, narrow black anterolateral corners on tergum V. A pair of oval red-brown shining spots on tergum V. Ovipositor with basal segment red-brown, dorsoventrally compressed and tapering posteriorly in dorsal view; ratio of length of oviscape to length of tergum V, 0.8:1.

Attractant: No known record.

Male: As for female except for dense aggregation of microtrichia around A_1+CuA_2 present; supernumerary lobe of medium development; pecten of cilia on abdominal tergum III present; posterior lobe of surstylus short; abdominal sternum V with a deep concavity on posterior margin; abdominal sterna pale.

Distribution:

Type locality: Japan (Kyushu I., Kagoshima).

Previously recorded: Type locality and China (Norrbom *et al.*, 1998).

New records: Thailand (Chiang Mai, Doi Pui). Taiwan, due to **New Synonymy** of *D. hsui*.

Hosts: Recorded from a range of hosts in the family *Lauraceae* (Allwood *et al.*, 1999).

New records: *Osmanthus insularis* (family *Oleaceae*).

Remarks: *Bactrocera (Bactrocera) hyalina* (Shiraki) could belong to the *Bactrocera (Zeugodacus) scutellaris* complex in that it possesses wing markings, scutum and abdominal terga colour patterns similar to species in that group. However, in the absence of male specimens, it is here regarded as a species of *Bactrocera (Bactrocera)*. It is similar to *Bactrocera (Bactrocera) venefica* (Hering) in possessing a broken costal band resulting in a small fuscous spot around apex of R_{4+5}, a black scutum, short narrow lateral postsutural vittae and a yellow scutellum with a distinct black apical spot. It differs from *B. venefica* in having the costal band entirely absent beyond apex of R_{4+5}, femora entirely fulvous and a broader mesopleural stripe reaching midway between the anterior margin of the notopleuron and the anterior *npl.* seta dorsally.

Strumeta asatoi Shiraki was listed as a synonym of *B. hyalina* by Norrbom *et al.* (1998), however the original designation of this synonymy is unclear. Further, there are no significant characters that can be used to separate *Dacus (Bactrocera) hsui* Tseng, Chen & Chu from *B. hyalina* and thus we are making *D. hsui* a **New Synonym** of the latter species.

Bactrocera (Bactrocera) illusioscutellaris Drew & Romig, new species

(Fig. 82)

Material examined: Holotype male, **VIETNAM: Northern Vietnam** – North Vietnam, Hanoi Province, Tu Liem District, Lien Mac, 22.xi.2001, Site N003, attracted to cue lure, deposited in BMNH.

Paratypes: **BHUTAN**: 2 males, Rimchu 2, 10.v.2000, coll. B.S. Fletcher, attracted to cue lure. **VIETNAM: Northern Vietnam** – 3 males, same data as holotype; 2 males (23.xi. 2000), 1 male (6.xii.2001), North Vietnam, Hanoi Province, Tu Liem District, Lien Mac, Site N003, attracted to cue lure; 1 male, North Vietnam, Hanoi Province, Tu Liem District, Con Hue, 29.xii.1999, Site N001, attracted to cue lure; 1 male (24.xii.1999), 1 male (31.xii.1999), North Vietnam, Hanoi Province, Tu Liem District, Xuan Dinh, Site N004, attracted to cue lure; 1 male, North Vietnam, Son La Province, Moc Chau District, Moc Chau house garden, 15.xi.1999, Site N008, attracted to cue lure; 1 male, North Vietnam, Hanoi Province, Dong Anh District, Bae Ha, 24.xii.1999, Site N005, attracted to cue lure. 3 paratypes in ANIC, 3 paratypes in BMNH, 4 paratypes in QDPC, 3 paratypes in QMIC.

Diagnosis: A small species; face fulvous with a pair of moderately large oval black spots; postpronotal lobes and notopleura yellow; scutum entirely black; lateral postsutural yellow vittae present; medial postsutural yellow vitta absent; no yellow spot anterior to notopleural suture; mesopleural stripe moderately broad but curved inwards dorsally to be equal in width to notopleuron; scutellum yellow; wing with cells bc and c colourless, microtrichia in outer corner of cell c only; a narrow fuscous costal band (*scutellaris*-like); a narrow fuscous anal streak; abdominal terga III–V mostly black with red-brown longitudinal bands along margin of a moderately broad medial longitudinal black band running from centre of tergum III to posterior margin of tergum V.

Description: male.

Head: Height 1.2 mm. Frons – length 1.42 times breadth; fulvous with dark fuscous on anteromedial hump and around bases of orbital setae; orbital setae black: 1 *s.or.*, 2 *i.or.*; lunule dark fuscous. Ocellar triangle black. Vertex fuscous. Face fulvous with a pair of moderately large oval black spots; length 0.33 mm. Genae fulvous, small dark fuscous subocular spot present; black seta present. Occiput dark fuscous, fulvous along eye margins; occipital row with a large number of small pale setae. Antennae with segments 1 and 2 red-brown, segment 3 red-brown with fuscous on apex and outer surface; length of segments: 0.1 mm; 0.2 mm; 0.58 mm.

Thorax: Scutum entirely black without pale markings. Pleural areas dark fuscous to black. Yellow markings as follows: postpronotal lobes; notopleura; mesopleural stripe moderately broad but curved inwards on dorsal surface to be equal in width to notopleuron, continuing to katepisternum as a transverse spot, anterior margin convex; anatergite (posterior apex black); anterior $^{3}/_{5}$ katatergite (remainder black); two narrow lateral postsutural vittae either parallel-sided or narrowing slightly posteriorly to end before *ia.* seta. Medial longitudinal postsutural yellow vitta absent. Postnotum black. Scutellum yellow except for narrow black basal band. Setae: *sc.* 2; *prsc.* 2; *ia.* 1; *p.sa.* 1; *a.sa.* 1; *mpl.* 1; *npl.* 2; *scp.* 4.

Legs: All segments entirely fulvous except hind tibiae dark fuscous; mid tibiae each with an apical black spur.

Wings: Length 4.4 mm; cells bc and c colourless; microtrichia in outer corner of cell c only; remainder of wings colourless except dark fuscous cell sc, narrow fuscous costal band confluent with R_{2+3} and extremely narrow beyond apex of this vein before expanding into an elongate oval spot across apex of wing, narrow fuscous anal streak; a dense aggregation of microtrichia around A_1+CuA_2; supernumerary lobe weak.

Abdomen: Oval; terga free; pecten present on tergum III. Tergum I and sterna I and II wider than long. Tergum I black with a narrow transverse red-brown band across posterior margin but not reaching lateral margins; tergum II black with fulvous across posterior ½; terga III–V black except for narrow red-brown margins either side of a broad medial longitudinal black band extending from centre of tergum III to posterior margin of tergum V. A pair of oval dark fuscous shining spots on tergum V. Posterior lobe of surstylus short, sternum V with a deep concavity on posterior margin. Abdominal sterna dark fuscous to black.

Attractant: Cue lure.

Female: No known record.

Distribution: Bhutan (Rimchu); Northern Vietnam (Hanoi Province; Son La Province).

Hosts: No known record.

Remarks: *Bactrocera (Bactrocera) illusioscutellaris* n.sp., on general appearance seems close to species in *Bactrocera (Zeugodacus) scutellaris* complex in possessing a black scutum, costal band in the wing extremely narrow after extremity of R_{2+3} before expanding into a small spot around apex of R_{4+5} and abdominal terga III–V almost entirely black. However, *B. illusioscutellaris* belongs to the subgenus *Bactrocera*, not subgenus *Zeugodacus*.

Bactrocera (Bactrocera) impunctata (de Meijere)

(Fig. 83)

Dacus impunctatus de Meijere, 1914: 188. Holotype male in ZMAN.

Dacus (Strumeta) impunctatus – Hardy & Adachi, 1954: 171.

Dacus (Bactrocera) impunctata – Hardy, 1983: 14.

Bactrocera (Bactrocera) impunctata – Norrbom *et al.*, 1998: 91.

Non-type material examined: INDONESIA: JAVA – 8 males (23.vii.2005), 10 males (13.viii.2005), 16 males (20.viii.2005), 14 males (2.ix.2005), Java, DKI Jakarta Depok, Kukusan (U I Kampus), coll. Endang, attracted to methyl eugenol in forest; 1 male, Java, DKI Jakarta, Jakarta Selatan, Tanjung Barat, 27.viii.2005, coll. Mulyana, attracted to methyl eugenol in forest; **LOMBOK** – 3 males, Sapit, 675-850 m, 1.vi.1981, coll. J. Turner, attracted to methyl eugenol; 3 males, Sembalia & Belanting, 10-50m, 1.vi.1981, coll. J. Turner, attracted to methyl eugenol. Specimens in QDPC. 1 male, Lombok, Sapit (675-850m), 2.vii.1981, coll. J. Turner. Specimen studied in MZB.

Diagnosis: A small species; face entirely fulvous; postpronotal lobes and notopleura yellow; scutum entirely shining black; very broad and short lateral postsutural yellow vittae present; medial postsutural yellow vitta absent; no yellow spot anterior to notopleural suture; mesopleural stripe reaching midway between anterior margin of notopleuron and anterior *npl.* seta dorsally; scutellum yellow; wing with cells bc and c pale fuscous, microtrichia in outer corner of cell c only; a narrow dark fuscous costal band; a moderately broad fuscous anal streak; abdominal terga generally orange-brown but with a unique black marking on tergum III consisting of broad lateral black bands joined along anterior margin of tergum III but not reaching extreme lateral margins of this tergum and a moderately broad medial longitudinal black band over terga IV and V which may extend on to posterior margin of tergum III.

Description: male.

Head: Height 0.95 mm. Frons – length 1.4 times breadth; red-brown with fuscous on anteromedial hump; orbital setae black: 1 *s.or.*, 2 *i.or.*; lunule red-brown. Ocellar triangle black. Vertex red-brown. Face entirely fulvous without dark markings; length 0.3 mm. Genae fulvous, small fuscous subocular spot present; black seta present. Occiput red-brown centrally, black laterally, fulvous along eye margins; occipital row with a large number of very small pale setae. Antennae with segments 1 and 2 red-brown, segment 3 red-brown with fuscous on apex and outer surface; length of segments: 0.08 mm; 0.2 mm; 0.5 mm.

Thorax: Scutum entirely shining black without pale markings. Pleural areas entirely shining black without pale markings. Yellow markings as follows: postpronotal lobes; notopleura; mesopleural stripe reaching midway between anterior margin of notopleuron and anterior *npl.* seta dorsally, continuing to katepisternum as a small pale spot, anterior margin strongly convex; anatergite (posterior apex black); anterior ¾ katatergite (remainder black); two very broad and short triangular lateral postsutural vittae tapering strongly posteriorly to end well before level of *p.sa.* setae. Medial longitudinal postsutural yellow vitta absent. Postnotum entirely shining black. Scutellum yellow except for narrow black basal band. Setae: *sc.* 2; *prsc.* 2; *ia.* 1; *p.sa.* 1; *a.sa.* 1; *mpl.* 1; *npl.* 2; *scp.* 4.

Legs: All femora entirely fulvous; fore and mid tibiae fulvous, hind tibiae fuscous; tarsi with basal two segments fulvous and apical three segments red-brown; mid tibiae each with an apical black spur.

Wings: Length 3.6 mm; cells bc and c with pale fuscous tint; microtrichia in outer corner of cell c only; remainder of wings colourless except dark fuscous cell sc, narrow dark fuscous costal band confluent with R_{2+3} and widening slightly across apex of R_{4+5} before ending between extremities of R_{4+5} and M, a moderately broad fuscous anal streak; a dense aggregation of microtrichia around A_1+CuA_2; supernumerary lobe of medium development.

Abdomen: Oval; terga free; pecten present on tergum III. Tergum I and sterna I and II wider than long. Terga I and II entirely orange-brown; tergum III orange-brown with unique black markings consisting of very broad lateral black bands which join along anterior margin of tergum III (these bands do not reach the extreme lateral margins of the tergum); terga IV–V orange-brown with a moderately broad medial longitudinal black band that may reach forward on to the posterior margin of tergum III. A pair of oval orange-brown shining spots on tergum V. Posterior lobe of surstylus short, sternum V with a deep concavity on posterior margin. Abdominal sterna entirely pale, generally orange-brown.

Attractant: Methyl eugenol.

Female: No known record.

Distribution:

Type locality: Indonesia (Java, Semarang).

New records: Indonesia (Java, Jakarta; Lombok).

Hosts: No known record.

Remarks: *Bactrocera (Bactrocera) impunctata* (de Meijere) is a unique, extremely small species, similar to *Bactrocera (Bactrocera) selenophora* Tsuruta & White in possessing a narrow costal band confluent with R_{2+3}, a black scutum, pale cells bc and c and lateral postsutural vittae tapering posteriorly to end before *ia.* setae. It differs from *B. selenophora* in having the scutum entirely black, lateral postsutural vittae broad and triangular, femora entirely fulvous and a unique black pattern on abdominal terga III–V.

Bactrocera (Bactrocera) inconspicua Drew & Romig, new species

(Fig. 84)

Material examined: Holotype male, **INDONESIA: NORTH MOLUCCAS (NORTH MALUKU)** – Moluccas (Maluku), North Maluku, West Halmahera Gaol, 29.v.2007, coll. Raais Abdullah, attracted to methyl eugenol in forest, deposited in BMNH.

Paratypes: **INDONESIA: NORTH MOLUCCAS (NORTH MALUKU)** – 15 males, same data as holotype; 16 males (5.vi.2007), 33 males (12.vi.2007), 11 males (18.vi.2007), 1 male (19.vi.2007), North Maluku, West Halmahera Gaol, coll. Raais Abdullah, attracted to methyl eugenol in forest or jeruk; 2 males, North Maluku, Kodya, Ternate, Foramadiahi, 27.v.2007, coll. La Ruti, attracted to methyl eugenol in forest; 1 male (3.vi.2007), 3 males (17.vi.2007), North Maluku, Kodya, Ternate, Gambesi, coll. La Ruti, attracted to methyl eugenol in jeruk; 10 males, North Maluku, Kodya, Ternate, Kalumata, 10.vi.2007, coll. La Ruti, attracted to methyl eugenol in mangga. 15 paratypes in ANIC, 15 paratypes in BMNH, 50 paratypes in QDPC, 12 paratypes in QMIC.

Diagnosis: A medium-sized species; face fulvous with a pair of very small circular black spots; postpronotal lobes and notopleura yellow; scutum dull red-brown with narrow dark red-brown borders along posterior margin and lateral margins; lateral postsutural yellow vittae present; medial postsutural yellow vitta absent; no yellow spot anterior to notopleural suture; mesopleural stripe reaching to anterior *npl.* seta dorsally; scutellum yellow; wing with cells bc and c fuscous, microtrichia in outer corner of cell c only; a broad dark fuscous costal band overlapping R_{4+5} basal to *r-m* crossvein and confluent with R_{4+5} apical to this vein and paler between R_{2+3} and R_{4+5}, a fuscous to very pale fuscous band across wing which is narrow and fuscous along *r-m* crossvein and broader and diffuse from this vein to posterior margin of wing; a narrow dark fuscous anal streak; abdominal terga III–V mostly orange-brown except for narrow dark fuscous anterolateral corners on tergum III and a very narrow and pale fuscous medial longitudinal line from posterior area of tergum IV to posterior margin of tergum V.

Description: male.

Head: Height 1.52 mm. Frons – length 1.57 times breadth; red-brown with pale fuscous on anteromedial hump and around bases of orbital setae; orbital setae black: 1 *s.or.*, 2 *i.or.*; lunule red-brown. Ocellar triangle black. Vertex pale fuscous. Face fulvous with a pair of small circular black spots; length 0.39 mm. Genae red-brown, small dark fuscous subocular spot present; black seta present. Occiput red-brown, fulvous along eye margins; occipital row with 3–4 black setae. Antennae with segments 1 and 2 red-brown, segment 3 red-brown with dark fuscous on apex and outer surface; length of segments: 0.18 mm; 0.32 mm; 0.78 mm.

Thorax: Scutum dull black with dark red-brown along posterior margin, below lateral postsutural vittae, around notopleural suture and on lateral margins from notopleural suture to anterior margin of scutum. Pleural areas black except fuscous below postpronotal lobes. Yellow markings as follows: postpronotal lobes; notopleura; mesopleural stripe reaching anterior *npl.* seta dorsally, continuing to katepisternum as a transverse spot, anterior margin straight; anatergite (posterior apex black); anterior ¾ katatergite (remainder black); two moderately broad parallel to sub-parallel lateral postsutural vittae ending at *ia.* seta. Medial longitudinal postsutural yellow vitta absent. Postnotum black. Scutellum yellow except for narrow black basal band. Setae: *sc.* 2; *prsc.* 2; *ia.* 1; *p.sa.* 1; *a.sa.* 1; *mpl.* 1; *npl.* 2; *scp.* 4.

Legs: All segments entirely fulvous except hind tibiae pale fuscous basally to red-brown apically and apical four segments of all tarsi pale fuscous; mid tibiae each with an apical black spur.

Wings: Length 5.6 mm; cells bc and c fuscous; microtrichia in outer corner of cell c only; remainder of wings colourless except dark fuscous cell sc, broad dark fuscous costal band overlapping R_{4+5} basal to *r-m* crossvein and confluent with R_{4+5} apical to *r-m* crossvein and tending fuscous between R_{2+3} and R_{4+5}, a pale fuscous narrow band enclosing *r-m* crossvein and broadening and becoming diffuse from *r-m* crossvein to posterior margin of wing, a narrow dark fuscous anal streak; a dense aggregation of microtrichia around A_1+CuA_2; supernumerary lobe of medium development.

Abdomen: Oval; terga free; pecten present on tergum III. Tergum I and sterna I and II wider than long. Tergum I dark fuscous with a narrow transverse orange-brown band across posterior margin but not reaching lateral margins; tergum II fulvous with a narrow transverse dark fuscous band subanteriorly but not reaching narrow fuscous anterolateral corners; terga III–V orange-brown except for narrow transverse dark fuscous bands anterolaterally on tergum III and a very narrow medial longitudinal pale fuscous line from posterior ¼ of tergum IV to posterior margin of tergum V. A pair of oval dark fuscous shining spots on tergum V. Posterior lobe of surstylus short, sternum V with a deep concavity on posterior margin. Abdominal sterna fulvous to pale red-brown.

Attractant: Methyl eugenol.

Female: No known record.

Distribution: Indonesia (North Moluccas = North Maluku – Halmahera I.; Ternate).

Hosts: No known record.

Remarks: *Bactrocera (Bactrocera) inconspicua* n.sp. is similar to *Bactrocera (Bactrocera) unistriata* (Drew) in possessing a black scutum, broad parallel-sided lateral postsutural vittae, a broad costal band confluent with R_{4+5} and transverse infuscation across wing enclosing both crossveins but differs from *B. unistriata* in having darker cells bc and c, the transverse band across the wing pale and diffuse and abdominal terga III–V mostly orange-brown, not with broad lateral longitudinal dark bands.

Bactrocera (Bactrocera) indonesiae Drew & Hancock

(Fig. 85)

Bactrocera (Bactrocera) indonesiae Drew & Hancock, 1994a: 26; Norrbom *et al.*, 1998: 91. Holotype male in BBM.

Non-type material examined: Since the publication of Drew & Hancock (1994a), the following specimens have been studied – **INDONESIA: JAVA** – 1 male, Java, Banten, Lebak, Pasir Nini Nini, 8.viii.2005, coll. Aden Haer; **KALIMANTAN** – 2 males, South Kalimantan, Banjar, Karang Intan, Mandiangin, 13.xi.2005, coll. Bilal. Specimens in QDPC.

Diagnosis: Face fulvous with a pair of large elongate oval black spots; postpronotal lobes and notopleura yellow; scutum black except brown below and behind lateral postsutural vittae, around notopleural suture and inside postpronotal lobes; triangular-shaped lateral postsutural yellow vittae narrowing sharply posteriorly to end before *ia.* seta; medial postsutural yellow vitta absent; mesopleural stripe slightly wider than notopleuron dorsally; scutellum yellow; legs with femora entirely fulvous; fore tibiae fuscous to dark fuscous, mid tibiae fulvous tending fuscous basally, hind tibiae dark fuscous; wing with cells bc and c colourless, microtrichia in outer corner of cell c only; a narrow fuscous costal band confluent with R_{2+3} and remaining narrow around apex of wing to end between apices of R_{4+5} and M; a narrow fuscous anal streak; supernumerary lobe of medium development; abdominal terga III–V orange-brown except dark fuscous across most of tergum III, diffuse fuscous markings across outer ⅔ of tergum IV and narrow lateral fuscous markings on tergum V, a medium-width medial longitudinal dark fuscous band over all three terga, a pair of oval fuscous shining spots on tergum V; abdominal sterna dark coloured.

Attractant: Methyl eugenol.

Female: No known record.

Distribution:

Type locality: Indonesia (Java, Bogor).

New records: Indonesia (Kalimantan).

Hosts: No known record.

Remarks: *Bactrocera (Bactrocera) indonesiae* Drew & Hancock, within the *dorsalis* complex, is similar to *Bactrocera (Bactrocera) cibodasae* Drew & Hancock, *Bactrocera (Bactrocera) floresiae* Drew & Hancock, *Bactrocera (Bactrocera) minuscula* Drew & Hancock, *Bactrocera (Bactrocera) osbeckiae* Drew & Hancock, *Bactrocera (Bactrocera) sumbawaensis* Drew & Hancock and *Bactrocera (Bactrocera) thailandica* Drew & Hancock in possessing triangular-shaped lateral postsutural vittae that narrow sharply posteriorly, femora entirely fulvous or at most with dark preapical spots on fore femora and costal band confluent with R_{2+3}. It differs from all of these species in possessing broad diffuse lateral dark markings on abdominal terga III–V, not markings that are distinctly defined.

Bactrocera (Bactrocera) infulata Drew & Hancock

(Fig. 86)

Bactrocera (Bactrocera) infulata Drew & Hancock, 1994a: 27; Norrbom *et al.*, 1998: 91. Holotype male in BBM.

Non-type material examined: Since the publication of Drew & Hancock (1994a) the following specimens have been studied – **INDONESIA: SOUTH SULAWESI** – 3 males, South Sulawesi, Maros, Limampoccoe, 4.x.2006, coll.

Sirajuddin SP., attracted to methyl eugenol in forest. Specimens in QDPC.

Diagnosis: Face fulvous with a pair of large oval black spots; postpronotal lobes and notopleura yellow; scutum and pleural areas entirely black without pale markings; narrow lateral postsutural yellow vittae that narrow sharply posteriorly to end before level of *p.sa.* seta; medial postsutural yellow vitta absent; mesopleural stripe slightly wider than notopleuron dorsally; scutellum yellow with a broad black basal band; legs with femora fulvous except dark fuscous to black on outer apical ½ of fore femora, around apical ⅓ of mid and hind femora; all tibiae fuscous to dark fuscous; wing with cells bc and c colourless, microtrichia in outer ⅓ of cell c only; a narrow fuscous costal band almost confluent with R_{4+5} and remaining of uniform width around apex of wing; a broad fuscous anal streak; supernumerary lobe of medium development; abdominal terga III–V orange-brown with a broad transverse dark fuscous band across anterior ½ of tergum III, two broad lateral and a broad medial longitudinal dark fuscous to black bands over all three terga and joined by the broad transverse dark fuscous band across tergum III, a pair of oval fuscous shining spots on tergum V; abdominal sterna dark coloured.

Attractant: Methyl eugenol.

Female: No known record.

Distribution:

Type locality: Indonesia (Sulawesi–South Sulawesi, Bententingi).

New records: Indonesia (Sulawesi–South Sulawesi, Maros).

Hosts: No known record.

Remarks: *Bactrocera (Bactrocera) infulata* Drew & Hancock, within the *dorsalis* complex, is similar to *Bactrocera (Bactrocera) kinabalu* Drew & Hancock in possessing short lateral postsutural vittae that narrow sharply posteriorly to end well before *ia.* seta and dark markings on apices of all femora. It differs from *B. kinabalu* in possessing a broad black basal band on the scutellum and being attracted to methyl eugenol.

Bactrocera (Bactrocera) invadens Drew, Tsuruta & White

(Figs 87A, 87B, 87C)

Bactrocera (Bactrocera) invadens Drew, Tsuruta & White, 2005: 149; Drew *et al.*, 2007: 4. Holotype male in NMKE.

Non-type material examined: AFRICA: **BURUNDI** – 47 males, Isabu, ix.2010-i.2011, coll. M. De Meyer; **MOZAMBIQUE** – 5 males, Dombe, 19°58′33.6″S 33°23′42.7″E, coll. M. De Meyer and M. Mansell; **TANZANIA** – 11 males, Hingawali, 5.xii.2010, coll. Maulid Mwatawala, attracted to methyl eugenol; 17 males, Jaribu-Mpakani, 5.xii.2010, coll. Maulid Mwatawala, attracted to methyl eugenol; 20 males, Kilambo/Mpakani, 5.xii.2010, coll. Maulid Mwatawala, attracted to methyl eugenol; 16 males, Kilangara, 5.xii.2010, coll. Maulid Mwatawala; 10 males, Mavuji, 5.xii.2010, coll. Maulid Mwatawala, attracted to methyl eugenol; 15 males, Mbonde, 5.xii.2010, coll. Maulid Mwatawala, attracted to methyl eugenol; 17 males, Nari, 5.xii.2010, coll. Maulid Mwatawala, attracted to methyl eugenol; 76 males, Sokoine University of Agriculture, Horticulture Unit, Morogoro, 1.xi.2010, attracted to methyl eugenol. **GRAND COMORE ISLAND, near Madagascar**: 38 males, 15.x.2005, coll. A. Chadhouliati, attracted to methyl eugenol lure. **SRI LANKA**: 1 male, Kandy, Piachaud Gardens, 17.xi.1993; 1 male (24.xii.1993), 1 male (27.xii.1993), Kandy, Udawattekele, attracted to methyl eugenol; 1 male, Kandy market, 16.ix.1993; 3 males (22.ix.1993), 2 males (4.x.1993), 1 male (6.x.1993), 2 males (11.x.1993), 2 males (15.x.1993), 2 males (18.x.1993), 1 male (20.x.1993), 2 males (25.x.1993), Kandy, Gannoruwa Mountain; 2 males, Hingurakgoda, 26.x.1994; 11 males, Pelwehera, 25.x.1994; 2 males (18.xii.1994), 4 males (21.xii.1994), 4 males (22.xii.1994), Sri Lanka, coll. Thalduma. Specimens in QDPC. **SRI LANKA**: 1 female with the following label data – 'Ceylon Horn'; 'coll. Lichtwardt'; 'coll. Hendel'; '*Chaetodacus ferrugineus* F.', studied at NMW. DEI Museum records indicate that this specimen was collected by Walther Horn in 1899.

Diagnosis: Face fulvous with a pair of medium-sized to large oval black spots; postpronotal lobes and notopleura yellow; scutum with basic colour mostly dark orange-brown, occasionally black, when dark orange-brown it possesses the basic lanceolate fuscous to black pattern which varies in size and extent and often large dark markings between the postpronotal lobe and notopleuron; narrow- to medium-width parallel-sided lateral postsutural yellow vittae ending at or just behind *ia.* seta; medial postsutural yellow vitta absent; mesopleural stripe reaching midway between anterior margin of notopleuron and anterior *npl.* seta dorsally; scutellum yellow; legs with femora entirely fulvous; fore tibiae pale fuscous to fuscous, mid tibiae fuscous basally to pale fuscous apically, hind tibiae dark fuscous; wing with cells bc and c colourless, microtrichia in outer corner of cell c only; a narrow fuscous costal band confluent with R_{2+3} and remaining very narrow around apex of wing to end just beyond extremity of R_{4+5}; a narrow pale fuscous anal streak; supernumerary lobe of medium development; abdominal terga III–V orange-brown with a 'T' pattern consisting of a transverse black band across anterior margin of tergum III which varies from narrow to broad and almost covering entire tergum and a narrow- to medium-width medial longitudinal dark fuscous to black band over all three terga, narrow lateral dark fuscous margins on terga IV and V, a pair of oval orange-brown to dark orange-brown shining spots on tergum V.

Attractant: Methyl eugenol.

Female: As for male in the general body colour patterns. See Drew *et al.* (2005) for a detailed description of female characters.

Distribution:

Type locality: Kenya (Matuga).

Previously recorded: Type locality and the following locations – Sri Lanka; tropical regions of Central Africa to which it has been introduced (see Drew *et al.*, 2005 for distribution).

Current distribution: Uncertainty about the ability to diagnose this species accurately and separate it from other South-East Asian red-brown scutum species, based on morphological features, throws doubt on some published records. One example is the record from Bhutan (Drew *et al.*, 2007). The only distribution records that we can confirm are those listed above (Drew *et al.*, 2005) plus Grand Camore Island, near Madagascar.

Hosts: Recorded from a wide range of edible/commercial host fruits in Africa (see Drew *et al.*, 2005 for host records).

Remarks: The original definition of *Bactrocera (Bactrocera) invadens* Drew, Tsuruta & White was based on a population of flies collected in Sri Lanka by Kenji Tsuruta. Before publication, additional specimens were added to the type series from localities in Africa with the Holotype from Kenya. As described in the original publication, *B. invadens* possesses a pale-coloured scutum mostly dark orange-brown to red-brown with or without a range of dark patterns (Figs 87A, 87B) which can expand to where it is entirely black (Fig. 87C; also see Fig. 4 in Drew *et al.*, 2005). In recent studies of new material from Africa (Burkina Faso, Burundi and Tanzania), 62–68% of specimens possessed a pale-coloured scutum with or without dark patterns, while the remainder were black and similar to *Bactrocera dorsalis* (Hendel). Because of these dark forms, Drew *et al.* (2008) included *B. invadens* in their study of the *dorsalis* complex. However, the majority of specimens, those with a pale scutum, do not fit the *dorsalis* complex.

In 1794, Fabricius described *Musca ferruginea* from a location listed as India Orientali. This species in now considered to belong to the subgenus *Bactrocera* and was placed as a synonym of *B. dorsalis* by Hardy (1969a) purely as a matter of convenience, as he gave no justification for this. Although the type locality of *B. ferruginea* is unknown, it can be considered to be within the Indian subcontinent. Fabricius described the thorax as red-brown and it is likely that he studied specimens that we are now calling *B. invadens*. The specimen listed above labelled *Chaetodacus ferrugineus* F. by Hendel is in the Hendel collection in NMW. It was collected in 1899 in Sri Lanka by Walther Horn and is a specimen of *B. invadens*. This specimen in NMW is probably the oldest specimen available for study of what Fabricius called *M. ferruginea*.

In Sri Lanka, there are four other red-brown species of *Bactrocera (Bactrocera)*, i.e. *B. apicofuscans* White & Tsuruta, *B. brunneola* White & Tsuruta, *B. versicolor* (Bezzi) and *B. zonata* (Saunders). *Bactrocera apicofuscans* differs from *B. invadens* in possessing coloured cells bc and c and in lacking *prsc.* setae, *B. brunneola* differs in possessing narrow lateral postsutural vittae that end before the *ia.* setae and in being attracted to cue lure, *B. versicolor* differs in possessing a large dark fuscous spot on the apex of the scutellum and *B. zonata* differs in having the costal band broken beyond the apex of R_{2+3}.

Other general red-brown species in South-East Asia are *Bactrocera (Bactrocera) bhutaniae* n.sp., *Bactrocera (Bactrocera) bivittata* Lin & Wang and *Bactrocera (Bactrocera) rubigina* (Wang & Zhao). Two of these, *B. bhutaniae* and *B. rubigina* are attracted to cue lure while *B. bivittata* is attracted to methyl eugenol. See 'Remarks' under *B. bhutaniae* for a full comparison of these species.

Those specimens of *B. invadens* with a black scutum are similar to *Bactrocera (Bactrocera) dorsalis* (Hendel) and *Bactrocera (Bactrocera) kandiensis* Drew & Hancock in possessing a very narrow costal band and anal streak in the wing, parallel-sided lateral postsutural vittae and abdominal terga III–V with a general dark 'T' pattern. It differs from these species in having the scutum base colour red-brown with or without dark fuscous to black lanceolate patterns, from *B. dorsalis* in the males having a longer aedeagus and narrower lateral postsutural vittae and from *B. kandiensis* in having the femora entirely fulvous. It is noted that some specimens of *B. dorsalis* possess large areas of red-brown on the scutum. The pale forms of *B. dorsalis* occur in less than 20% of the population and are distinctly different from the pale forms of *B. invadens* which possess definitive dark patterns when they do occur. To call *B. invadens* a synonym of *B. dorsalis* would require a major change to our current morphological definition of *B. dorsalis*. See Drew *et al.* (2005) for a detailed discussion of these species.

Based on morphological comparisons, the populations in Africa and Sri Lanka appear to be conspecific. However, some researchers believe that there have been two separate invasions of Africa involving two *Bactrocera* species. The situation in Africa is complex and requires resolution, particularly as the fly has invaded that continent and is a major pest of fruit and vegetable crops. Such resolution will probably only be achieved through advanced genetic research where entire genomes of cryptic species can be defined.

In our molecular studies, based on the mitochondrial genes COI and ND5, *B. invadens* (Sri Lanka specimens) is markedly different from *B. carambolae* (specimens from Bali and Peninsular Malaysia), *B. dorsalis* (Taiwan specimens), *B. ochroma* n.sp., *B. occipitalis* (Kalimantan specimens) and *B. papayae* (specimens from Christmas I. (Australian Territory), Java, Kalimantan and Peninsular Malaysia) (J. Hughes, Griffith University Genetics Laboratory, unpublished data). Based on these data and the significant morphological differences described above, we conclude that *B. invadens* is a valid species. See discussion under section titled 'Identification of Cryptic Species'.

Bactrocera (Bactrocera) involuta (Hardy)

(Fig. 88)

Dacus (Bactrocera) involutus Hardy, 1982: 229. Holotype male in MZB.

Bactrocera (Bactrocera) involuta – Drew & Hancock, 1994a: 28; Norrbom *et al.*, 1998: 91.

Type material examined: Holotype male, **INDONESIA: SOUTH SULAWESI** – Sulawesi, Bantimurung, via Maros, 50 m, July 1975, coll. N. Gough. Studied at MZB.

Non-type material examined: Since the publication of Drew & Hancock (1994a), the following specimens have been studied – **INDONESIA: NORTH SULAWESI** – 3 males, Indonesia, Dumoga, Taraut, N. Sulawesi, 0′30′N 124′E, 22.xii.1985, coll. L. Santianawati, attracted to cue lure; 8 males, Indonesia, Dumoga, Taraut, 225m, N. Sulawesi,

21.xii.1985, coll. L. Santianawati, attracted to cue lure. Specimens in QDPC.

Diagnosis: Face fulvous with a pair of medium-sized circular black spots; postpronotal lobes and notopleura yellow; scutum entirely black without pale markings; broad parallel-sided lateral postsutural yellow vittae ending behind *ia*. seta; medial postsutural yellow vitta absent; mesopleural stripe slightly wider than notopleuron dorsally; scutellum yellow with a slightly expanded black basal band; legs with all femora entirely fulvous; fore and mid tibiae entirely fulvous, hind tibiae fuscous; wing with cells bc and c colourless, microtrichia in outer corner of cell c only; a narrow fuscous costal band almost confluent with R_{4+5} and widening across apex of wing; a narrow fuscous anal streak; supernumerary lobe of medium development; abdominal terga III–V orange-brown with a narrow transverse black band across anterior margin of tergum III and a narrow medial longitudinal black band over all three terga (in some specimens there are narrow fuscous to dark fuscous anterolateral corners on terga IV and V), a pair of oval dark fuscous to black shining spots on tergum V; abdominal sterna dark coloured.

Attractant: Cue lure.

Female: No known record.

Distribution:

Type locality: Indonesia (Sulawesi – South Sulawesi, Bantimurung).

New records: Indonesia (Sulawesi – North Sulawesi, Dumoga).

Hosts: No known record.

Remarks: *Bactrocera (Bactrocera) involuta* (Hardy), within the *dorsalis* complex, is similar to *Bactrocera (Bactrocera) bimaculata* Drew & Hancock and *Bactrocera (Bactrocera) fuscitibia* Drew & Hancock in possessing broad parallel-sided lateral postsutural vittae, costal band overlapping R_{2+3}, narrow dark lateral margins on abdominal terga IV and V, and dark fuscous to black shining spots on abdominal tergum V. It differs from *B. bimaculata* in possessing entirely fulvous femora (in *B. bimaculata* there are subapical dark spots on all femora) and from *B. fuscitibia* in having entirely fulvous fore and mid tibiae, a broad black basal band on the scutellum, a broader costal band in the wing and narrower dark lateral margins on abdominal terga III–V.

Bactrocera (Bactrocera) irvingiae Drew & Hancock

(Fig. 89)

Bactrocera (Bactrocera) irvingiae Drew & Hancock, 1994a: 29; Norrbom *et al*., 1998: 91. Holotype female in BMNH.

Non-type material examined: No new specimens available for study since the publication of Drew & Hancock (1994a).

Diagnosis: Face fulvous with a pair of medium-sized oval black spots; postpronotal lobes and notopleura yellow; scutum black except dark brown below and behind lateral postsutural vittae and dark brown around notopleural suture and inside postpronotal lobes; broad parallel-sided lateral postsutural yellow vittae ending at *ia*. seta; medial postsutural yellow vitta absent; mesopleural stripe reaching midway between anterior margin of notopleuron and anterior *npl*. seta dorsally; scutellum yellow; legs with femora entirely fulvous in males and with a small dark preapical spot on outer surfaces of fore femora in females; fore and hind tibiae dark fuscous, mid tibiae fuscous basally tending fulvous apically; wing with cells bc and c colourless, microtrichia in outer corner of cell c only; a narrow fuscous costal band confluent with R_{2+3} and with a very slight expansion around apex of R_{4+5}; a narrow pale fuscous anal streak; supernumerary lobe of medium development; abdominal terga III–V orange-brown with a narrow transverse black band across anterior margin of tergum III which expands to cover narrow lateral margins of this tergum, a moderately broad medial longitudinal black band over all three terga, narrow anterolateral black corners on terga IV and V, a pair of oval dark orange-brown shining spots on tergum V; abdominal sterna dark coloured.

Attractant: No known record.

Female: As for male in the general body colour patterns. See Drew & Hancock (1994a) for a detailed description of female characters.

Distribution:

Type locality: Central Thailand (Khao Yai).

Hosts: Recorded from three plant families – *Meliaceae*, *Moraceae*, *Simaroubaceae* (see Allwood *et al*., 1999 for host records).

Remarks: *Bactrocera (Bactrocera) irvingiae* Drew & Hancock, within the *dorsalis* complex, is similar to *Bactrocera (Bactrocera) hantanae* Tsuruta & White, *Bactrocera (Bactrocera) raiensis* Drew & Hancock and *Bactrocera (Bactrocera) syzygii* White & Tsuruta in possessing broad parallel-sided lateral postsutural vittae, costal band confluent with R_{2+3}, femora entirely fulvous and a broad medial longitudinal black band over abdominal terga III–V. It differs from *B. raiensis* and *B. syzygii* in having narrow lateral dark markings on abdominal terga III–V and from *B. hantanae* in being a small species with pale, usually orange-brown, shining spots on abdominal tergum V.

Bactrocera (Bactrocera) jaceobancroftii Drew & Romig, new species

(Fig. 90)

Material examined: Holotype male, **VIETNAM: Northern Vietnam** – Northern Vietnam, Hanoi Province, Tu Liem District, Con Hue, 4.ix.1999, Site N001, attracted to methyl eugenol, deposited in BMNH.

Paratypes: **VIETNAM: Northern Vietnam** – 1 male, same data as holotype; 1 male, Northern Vietnam, Son La Province, Moc Chau District, Co Do, 1.i.2000, Site N006, attracted to methyl eugenol. 1 paratype in ANIC, 1 paratype in QDPC.

Diagnosis: A medium-sized species; face fulvous with a pair of large oval black spots; postpronotal lobes and notopleura yellow; scutum red-brown with a dark lanceolate pattern varying from fuscous posteriorly to dark fuscous to black on the arms that protrude forwards; lateral postsutural yellow vittae present; medial postsutural yellow vitta absent;

no yellow spot anterior to notopleural suture; mesopleural stripe reaching midway between anterior margin of notopleuron and anterior *npl.* seta dorsally; scutellum yellow; wing with cells bc and c colourless, microtrichia in outer corner of cell c only; a narrow fuscous costal band which remains narrow around apex of wing; a narrow fuscous anal streak; abdominal terga III–V orange-brown with a fuscous to dark fuscous transverse marking anterolaterally on tergum III and narrow pale fuscous anterolateral corners on terga IV and V.

Description: male.

Head: Height 1.86 mm. Frons – length 1.45 times breadth; red-brown with fuscous to dark fuscous on anteromedial hump and around bases of orbital setae; orbital setae black: 1 *s.or.*, 2 *i.or.*; lunule fuscous to dark fuscous. Ocellar triangle black. Vertex fuscous to dark fuscous. Face fulvous with a pair of large oval black spots; length 0.48 mm. Genae fulvous, large fuscous subocular spot present; dark red-brown seta present. Occiput fuscous, fulvous along eye margins; occipital row with 3–4 black setae. Antennae with segments 1 and 2 fuscous, segment 3 red-brown with fuscous on apex and outer surface; length of segments: 0.22 mm; 0.34 mm; 0.79 mm.

Thorax: Scutum red-brown with a dark lanceolate pattern varying from fuscous posteriorly to dark fuscous to black on the arms that protrude forward, dark fuscous to black spots between postpronotal lobe and notopleuron. Pleural areas dark fuscous to black except red-brown above mid and hind coxae and below postpronotal lobes. Yellow markings as follows: postpronotal lobes; notopleura; mesopleural stripe reaching midway between anterior margin of notopleuron and anterior *npl.* seta dorsally, continuing to katepisternum as a large transverse spot, anterior margin convex; anatergite (posterior apex dark fuscous); anterior ⅔ katatergite (remainder dark fuscous); two broad parallel-sided lateral postsutural vittae ending just behind *ia.* seta. Medial longitudinal postsutural yellow vitta absent. Postnotum red-brown. Scutellum yellow except for narrow black basal band. Setae: *sc.* 2; *prsc.* 2; *ia.* 1; *p.sa.* 1; *a.sa.* 1; *mpl.* 1; *npl.* 2; *scp.* 4.

Legs: All femora entirely fulvous; fore and hind tibiae pale fuscous, mid tibiae pale fuscous basally to fulvous apically; all tarsi with segments entirely fulvous; mid tibiae each with an apical black spur.

Wings: Length 6.4 mm; cells bc and c colourless; microtrichia in outer corner of cell c only; remainder of wings colourless except fuscous cell sc, narrow fuscous costal band confluent with R_{2+3} and remaining narrow around apex of wing, a narrow fuscous anal streak; a dense aggregation of microtrichia around A_1+CuA_2; supernumerary lobe of medium development.

Abdomen: Oval; terga free; pecten present on tergum III. Tergum I and sterna I and II wider than long. Tergum I orange-brown with narrow lateral fuscous margins; tergum II orange-brown with narrow fuscous anterolateral corners and fulvous areas posterolaterally; terga III–V orange-brown with a large transverse fuscous to dark fuscous marking on anterolateral corners of tergum III and narrow pale fuscous anterolateral corners on terga IV and V. A pair of oval dark orange-brown shining spots on tergum V. Posterior lobe of surstylus short, sternum V with a deep concavity on posterior margin. Abdominal sterna pale, generally orange-brown.

Attractant: Methyl eugenol.

Female: No known record.

Distribution: Northern Vietnam (Hanoi Province; Son La Province).

Hosts: No known record.

Remarks: *Bactrocera (Bactrocera) jaceobancroftii* n.sp. is similar to *Bactrocera (Bactrocera) invadens* Drew, Tsuruta & White and *Bactrocera (Bactrocera) rubigina* (Wang & Zhao) in possessing a basic red-brown scutum, costal band confluent with R_{2+3}, cells bc and c colourless, mesopleural stripe ending midway between anterior margin of notopleuron and anterior *npl.* seta dorsally, lateral postsutural vittae ending just behind *ia.* seta and scutellum entirely yellow. It differs from these species in having all abdominal terga mostly orange-brown without a black 'T' pattern and an abdomen shape broadly oval similar to *Bactrocera (Bactrocera) bancroftii* (Tryon) and from *B. rubigina* in the males being attracted to methyl eugenol.

Bactrocera (Bactrocera) kalimantaniae Drew & Romig, new species

(Fig. 91)

Material examined: Holotype male, **INDONESIA: KALIMANTAN** – Kalimantan, South Kalimantan, Banjar, Karang Intan, Mandiangin, 13.xi.2005, coll. Bilal, attracted to cue lure in forest, deposited in BMNH.

Paratypes: **INDONESIA: KALIMANTAN** – 3 males, same data as holotype; 1 male (20.xi.2005), 1 male (4.xii.2005), 2 males (11.xii.2005), same locality data as holotype, attracted to cue lure in forest. 1 paratype in ANIC, 1 paratype in BMNH, 3 paratypes in QDPC, 2 paratypes in QMIC.

Diagnosis: A small species; face fulvous with a pair of large elongate oval black spots; postpronotal lobes and notopleura yellow; scutum black; lateral postsutural yellow vittae present; medial postsutural yellow vitta absent; no yellow spot anterior to notopleural suture; mesopleural stripe equal in width to notopleuron dorsally; scutellum yellow; wing with cells bc and c with a very pale tint, microtrichia in outer ½ of cell c only; a moderately broad dark fuscous costal band almost confluent with R_{4+5}; a narrow dark fuscous anal streak; abdominal terga III–V orange-brown with a narrow- to medium-width medial and two broad lateral longitudinal dark bands over all three terga and joined by a moderately broad transverse black band across anterior margin of tergum III.

Description: male.

Head: Height 1.1 mm. Frons – length 1.43 times breadth; fulvous with fuscous to dark fuscous centrally and around orbital setae; orbital setae black: 1 *s.or.*, 2 *i.or.*; lunule dark fuscous. Ocellar triangle black. Vertex dark fuscous. Face fulvous with a pair of large elongate oval black spots; length 0.28 mm. Genae fulvous, small dark fuscous subocular spot present; black seta present. Occiput dark fuscous to black, fulvous along eye margins; occipital row with 2–5 small black setae. Antennae with segments 1 and 2 fuscous, segment 3 red-brown with dark fuscous on apex and outer surface; length of segments: 0.12 mm; 0.2 mm; 0.5 mm.

Thorax: Scutum entirely black except, at most, dark red-brown behind lateral postsutural vittae and anterior to notopleural suture. Pleural areas entirely black. Yellow markings as follows: postpronotal lobes; notopleura; mesopleural stripe equal in width to notopleuron dorsally, continuing to katepisternum as a small transverse spot, anterior margin slightly convex; anatergite (posterior apex black); anterior ¾ katatergite (remainder black); two broad parallel-sided to sub-parallel lateral postsutural vittae ending at *ia.* seta. Medial longitudinal postsutural yellow vitta absent. Postnotum black. Scutellum yellow except for narrow black basal band. Setae: *sc.* 2; *prsc.* 2; *ia.* 1; *p.sa.* 1; *a.sa.* 1; *mpl.* 1; *npl.* 2; *scp.* 4.

Legs: Fore femora fulvous with a large dark fuscous subapical spot on outer surfaces, mid and hind femora entirely fulvous; all tibiae dark fuscous to black; fore tarsi with basal segment fulvous and apical four segments fuscous, mid and hind tarsi with basal two segments fulvous and apical three segments red-brown; mid tibiae each with an apical black spur.

Wings: Length 3.7 mm; cells bc and c with a very pale tint; microtrichia in outer ½ of cell c only; remainder of wings colourless except dark fuscous cell sc, moderately broad dark fuscous costal band almost confluent with R_{4+5} and remaining of uniform width before ending between apices of R_{4+5} and M, a narrow dark fuscous anal streak; a dense aggregation of microtrichia around A_1+CuA_2; supernumerary lobe of medium development.

Abdomen: Oval; terga free; pecten present on tergum III. Tergum I and sterna I and II wider than long. Tergum I dark fuscous to black with a narrow transverse fulvous band across posterior margin but not reaching lateral margins; tergum II fulvous with a moderately broad transverse black band across anterior margin and reaching narrow black lateral margins; terga III–V orange-brown with two broad lateral longitudinal dark bands which vary between fuscous to dark fuscous and black and a narrow- to medium-width medial longitudinal dark fuscous to black band over all three terga and joined along anterior margin of tergum III by a moderately broad transverse black band. A pair of oval fuscous to dark fuscous shining spots on tergum V. Posterior lobe of surstylus short, sternum V with a deep concavity on posterior margin. Abdominal sterna dark.

Attractant: Cue lure.

Female: No known record.

Distribution: Indonesia (South Kalimantan).

Hosts: No known record.

Remarks: *Bactrocera kalimantaniae* n.sp. is similar to *Bactrocera arecae* (Hardy & Adachi), *Bactrocera atrifemur* Drew & Hancock, *Bactrocera occipitalis* (Bezzi), *Bactrocera verbascifoliae* Drew & Hancock and *Bactrocera binhduongiae* n.sp. in possessing broad parallel-sided lateral postsutural vittae, costal band overlapping R_{2+3} and broad lateral dark bands on abdominal terga III–V. It differs from *B. arecae*, *B. atrifemur*, *B. occipitalis* and *B. verbascifoliae* in having dark preapical spots on outer surfaces of fore femora only, and from all of these species in having the costal band almost confluent with R_{4+5}.

Bactrocera (Bactrocera) kanchanaburi Drew & Hancock

(Fig. 92)

Bactrocera (Bactrocera) kanchanaburi Drew & Hancock, 1994a: 30; Norrbom *et al.*, 1998: 91. Holotype male in BMNH.

Non-type material examined: Since the publication of Drew & Hancock (1994a), the following specimen has been studied – **VIETNAM: Southern Vietnam** – 1 male, Southern Vietnam, Khanh Hoa Province, Dien Khanh District, Suoi Tan, Site VS10, 21.vi.2002, attracted to methyl eugenol. Specimen in QDPC.

Diagnosis: Face fulvous with a pair of large oval black spots; postpronotal lobes and notopleura yellow; scutum black except brown below and behind lateral postsutural vittae and red-brown around notopleural suture and inside postpronotal lobes; broad parallel-sided lateral postsutural yellow vittae ending behind *ia.* seta; medial postsutural yellow vitta absent, mesopleural stripe reaching midway between anterior margin of notopleuron and anterior *npl.* seta dorsally; scutellum yellow; legs with femora fulvous with a large subapical dark fuscous to black spot on outer surfaces of fore femora; fore tibiae dark fuscous, mid tibiae fuscous to dark fuscous basally and fulvous apically, hind tibiae dark fuscous to black; wing with cells bc and c colourless, microtrichia in outer corner of cell c only; a narrow fuscous costal band confluent with R_{2+3} and expanding slightly beyond apex of R_{2+3} to end between extremities of R_{4+5} and M; a narrow fuscous anal streak; supernumerary lobe of medium development; abdominal terga III–V dark orange-brown with a medial longitudinal black band and two broad lateral longitudinal dark fuscous to black bands over all three terga and which are joined across anterior ½ of tergum III, a pair of oval dark fuscous shining spots on tergum V; abdominal sterna dark coloured.

Attractant: Methyl eugenol.

Female: As for male in the general body colour patterns. See Drew & Hancock (1994a) for a detailed description of female characters.

Distribution:

Type locality: Thailand (Kanchanaburi, Phutonnam).

New records: Southern Vietnam (Khanh Hoa Province).

Hosts: *Artabotrys siamensis* and *Goniothalamus gigantifolius* (family *Annonaceae*) (Allwood *et al.*, 1999).

Remarks: *Bactrocera (Bactrocera) kanchanaburi* Drew & Hancock, within the *dorsalis* complex, is similar to *Bactrocera (Bactrocera) neopropinqua* Drew & Hancock and *Bactrocera (Bactrocera) raiensis* Drew & Hancock in possessing broad parallel-sided lateral postsutural vittae, costal band confluent with R_{2+3} and dark spots on outer subapical surfaces of fore femora only. It differs from *B. neopropinqua* in possessing broad lateral longitudinal dark fuscous to black bands over abdominal terga III–V and from *B. raiensis* in having a narrow medium longitudinal black band over abdominal terga III–V and dark fuscous shining spots on abdominal tergum V.

Bactrocera (Bactrocera) kandiensis Drew & Hancock

(Fig. 93)

Bactrocera (Bactrocera) kandiensis Drew & Hancock, 1994a: 31; Norrbom *et al.*, 1998: 91. Holotype female in BMNH.
Strumeta pedestris – Hering, 1956: 63 [Misidentification], see Drew & Hancock, 1994a: 31.
Bactrocera sp. nr *B. dorsalis* (D), White & Elson-Harris, 1992: 198.

Non-type material examined: Since the publication of Drew & Hancock (1994a), the following specimens have been studied – **SRI LANKA (formerly CEYLON)**: 4 males, Pasyala (Site M1), 3.x.1994; 2 males, Horana (Site M3), 8.x.1994; 1 male, Horana (Site M4), 19.x.1994; 1 male, Bataota (Site M10), 14.x.1994; 1 male, Looloowatta (Knuckle Reg), 1150m, 6.vii.1997, coll. K. Tsuruta; 2 males, Kundrasale, 17.xi.1993, attracted to methyl eugenol; 2 males, Maha, Illuppallama, 26.x.1993, attracted to methyl eugenol; 1 male, Sigiriya (Site M2), 19.xi.1993; 2 males, Polwohera, 25.x.1994; 2 males (18.xi.1994), 1 male (7.iv.1995), Rahangala (Site 35); 4 males, Bibile (Site M41), 28.x.1994; 1 male (15.ix.1993), 2 males (16.ix.1993), 5 males (17.ix.1993), Piachaud Gardens, Kandy; 2 males (22.ix.1993), 9 males (4.x.1993), 4 males (6.x.1993), 3 males (8.x.1993), 2 males (11.x.1993), 1 male (15.x.1993), 4 males (18.x.1993), 2 males (20.x.1993), 5 males (25.x.1993), Gannoruwa Mountain, Kandy. Specimens in QDPC.
Diagnosis: Face fulvous with a pair of large oval black spots; postpronotal lobes yellow (anteromedial corners red-brown); notopleura yellow; scutum black except brown below and behind lateral postsutural vittae, around notopleural suture, inside postpronotal lobes, around *prsc.* setae and on anterocentral margin; narrow parallel-sided lateral postsutural yellow vittae ending at *ia.* seta; medial postsutural yellow vitta absent; mesopleural stripe slightly wider than notopleuron dorsally; scutellum yellow with a moderately broad black basal band; legs with femora fulvous with dark fuscous on outer apical ⅔ of fore femora, inner apical ½ of mid and inner apical ⅓ of hind femora; fore tibiae fuscous, mid tibiae fulvous and hind tibiae dark fuscous; wing with cells bc and c colourless, microtrichia in outer corner of cell c only; a narrow fuscous costal band confluent with R_{2+3} and remaining narrow around margin of wing to end between extremities of R_{4+5} and M; a narrow fuscous anal streak; supernumerary lobe of medium development; abdominal terga III–V orange-brown with a narrow transverse black band across anterior margin of tergum III but not covering lateral margins, a very narrow medial longitudinal fuscous to dark fuscous band over all three terga (occasionally interrupted at intersegmental lines) and very narrow fuscous to dark fuscous anterolateral corners on terga IV and V, a pair of oval orange-brown shining spots on tergum V; abdominal sterna dark coloured.
Attractant: Methyl eugenol.
Female: As for male in the general body colour patterns. See Drew & Hancock (1994a) for a detailed description of female characters.
Distribution:
Type locality: Sri Lanka (Peradeniya).
New records: Sri Lanka (many other locations listed under 'Non-type material examined').
Hosts: This species is an economic pest in Sri Lanka and has been recorded from *Mangifera indica* (family *Anacardiaceae*) and *Garcinia* sp. (family *Clusiaceae*) (Drew & Hancock, 1994a).
Remarks: *Bactrocera (Bactrocera) kandiensis* Drew & Hancock, within the *dorsalis* complex, is similar to *Bactrocera (Bactrocera) caryeae* (Kapoor) and *Bactrocera (Bactrocera) neoarecae* Drew in possessing narrow parallel-sided lateral postsutural vittae and dark patterns on the apices of all femora or, at least, on fore and mid femora. It differs from *B. neoarecae* in possessing a single 'T' pattern over abdominal terga III–V (not on each of the three separate terga), a narrow black basal band on the scutellum and dark markings on the apices of all femora and from *B. caryeae* in possessing a very narrow medial longitudinal dark band on abdominal terga III–V and narrow dark anterolateral corners on terga IV and V.

Bactrocera (Bactrocera) kinabalu Drew & Hancock

(Fig. 94)

Bactrocera (Bactrocera) kinabalu Drew & Hancock, 1994a: 33; Norrbom *et al.*, 1998: 91. Holotype female in BMNH.

Non-type material examined: Since the publication of Drew & Hancock (1994a), the following specimens have been studied – **EAST MALAYSIA: SABAH** – 5 males (9.i.1992), 1 male (12.vi.1992), 1 male (24.vi.1992), Mt Kinabalu, 1500m, coll. Jinius *et al.*, attracted to cue lure. **PENINSULAR MALAYSIA**: 1 male (6.v.1986), 1 male (13.v.1986), 1 male (19.v.1986), 1 male (19.vi.1986), 1 male (30.vi.1986), R S Cameron Highlands, Malaysia, coll. Zan Darus, attracted to cue lure; 3 males, Cameron Highlands, Malaysia, 5.i.1987, coll. Azizan Darus, attracted to cue lure. Specimens in QDPC.
Diagnosis: Face fulvous with a pair of medium-sized oval black spots; postpronotal lobes and notopleura yellow; scutum and pleural areas entirely black without pale markings; narrow to broad triangular lateral postsutural yellow vittae narrowing sharply posteriorly to end before *ia.* seta; medial postsutural yellow vitta absent; mesopleural stripe equal in width to notopleuron dorsally; scutellum yellow; legs with femora fulvous except black on outer apical ½ to entire outer surfaces of fore femora and around apices of mid and hind femora; all tibiae dark fuscous to black; wing with cells bc and c colourless, microtrichia in outer corner of cell c only;

a narrow dark fuscous costal band overlapping R_{2+3} and expanding around apex of R_{4+5} (the costal band is occasionally almost confluent with R_{4+5} and expanding at apex); a narrow dark fuscous anal streak; supernumerary lobe of medium development; abdominal tergum III entirely dark fuscous to black, terga IV and V orange-brown except for a moderately broad medial longitudinal black band over both terga, broad dark fuscous lateral margins on tergum IV and dark fuscous anterolateral corners on tergum V, a pair of oval dark fuscous shining spots on tergum V; abdominal sterna dark coloured.
Attractant: Cue lure.
Female: As for male in the general body colour patterns. See Drew & Hancock (1994a) for a detailed description of female characters.
Distribution:
Type locality: East Malaysia (Sabah, Mt Kinabalu).
New records: Peninsular Malaysia (Cameron Highlands).
Hosts: *Melastoma* sp. (family *Melastomataceae*) (Allwood *et al*., 1999).
Remarks: *Bactrocera (Bactrocera) kinabalu* Drew & Hancock, within the *dorsalis* complex, is similar to *Bactrocera (Bactrocera) collita* Drew & Hancock, *Bactrocera (Bactrocera) dorsaloides* (Hardy & Adachi), *Bactrocera (Bactrocera) infulata* Drew & Hancock, *Bactrocera (Bactrocera) makilingensis* Drew & Hancock, *Bactrocera (Bactrocera) pyrifoliae* Drew & Hancock, *Bactrocera (Bactrocera) sulawesiae* Drew & Hancock and *Bactrocera (Bactrocera) thailandica* Drew & Hancock in possessing triangular-shaped lateral postsutural vittae that narrow distinctly posteriorly and dark markings on the apices of all femora. It differs from these species in having the outer apical ½ to entire outer surfaces of fore femora black and distinct black markings encircling the apices of the mid and hind femora (not with dark preapical spots on outer surfaces of the femora). *Bactrocera kinabalu* is also similar to *Bactrocera melastomatos* Drew & Hancock in the general body and wing colour patterns but differs from this species in having dark spots on the apices of all femora (the dark patterns on the fore femora covering the outer apical ½ to the entire outer surfaces), the lateral postsutural vittae more distinctly triangular and tapering to a point posteriorly and abdominal tergum III entirely dark fuscous to black.

Bactrocera (Bactrocera) laithieuiae Drew & Romig, new species

(Fig. 95)

Material examined: Holotype male, **VIETNAM: Southern Vietnam** – Southern Vietnam, Binh Duong Province, Lai Thieu Village, 26.v.2000, attracted to cue lure, site S020, deposited in BMNH.
Paratypes: **VIETNAM: Southern Vietnam** – 1 male, 26.xi.1999, same locality data as holotype, attracted to cue lure; 3 males, Dong Nai Province, Long Khanh District, Tan Phong Village, 9.vi.2000, site S014, attracted to cue lure; 1 male, Dong Nai Province, Long Khanh District, Tan Phong Village, 11.xi.1999; 1 male, Ben Tre Province, Chau Thanh District, Tien Thuy Village, 28.i.2002, coll. SOFRI Staff, Site VS 04, attracted to cue lure; 1 male, Dong Nai Province, Long Khanh District, Nhan Nghia Village, 2.iii.2002, coll. SOFRI Staff, Site VS 06, attracted to cue lure; 1 male, Dac Lac Province, Ku Mi Nga District, Quang Phu Village, 23.iii.2002, coll. SOFRI Staff, Site VS 09, attracted to cue lure. **THAILAND**: 5 males (25.vi.1991), 2 males (24.xii.1991), 1 male (21.i.1992), Thailand, Yang Pong Rat, A. Muang, Chanthaburi, coll. Montree, attracted to cue lure; 1 male, Thailand, Site SK91TC, Tone Nga Chaeng Waterfall, 2.iii.1993, coll. Anuchit & Sieng, attracted to cue lure. 3 paratypes in ANIC, 3 paratypes in BMNH, 7 paratypes in QDPC, 4 paratypes in QMIC.
Diagnosis: A small species; face fulvous with a pair of large elongate oval black spots; postpronotal lobes and notopleura yellow; scutum black; lateral postsutural yellow vittae present; medial postsutural yellow vitta absent; no yellow spot anterior to notopleural suture; mesopleural stripe reaching midway between anterior margin of notopleuron and anterior *npl.* seta dorsally; scutellum yellow; wing with cells bc and c colourless, microtrichia in outer ½ of cell c only; a narrow dark fuscous costal band overlapping R_{2+3}; a narrow dark fuscous anal streak; abdominal terga III–V orange-brown with black across most of tergum III, broad anterolateral corners of tergum IV black, anterolateral corners of tergum V fuscous to dark fuscous, a broad medial longitudinal black band over all three terga.
Description: male.
Head: Height 1.35 mm. Frons – length 1.6 times breadth; fulvous with dark fuscous around bases of orbital setae and on anteromedial hump; orbital setae black: 1 *s.or.*, 2 *i.or.*; lunule dark fuscous. Ocellar triangle black. Vertex dark fuscous. Face fulvous with a pair of large elongate oval black spots; length 0.36 mm. Genae fulvous, small fuscous subocular spot present; black seta present. Occiput dark fuscous to black, fulvous along eye margins; occipital row with 2–6 small black setae. Antennae with segments 1 and 2 red-brown, segment 3 red-brown with fuscous on apex and outer surface; length of segments: 0.18 mm; 0.28 mm; 0.74 mm.
Thorax: Scutum black without pale markings. Pleural areas entirely dark fuscous to black. Yellow markings as follows: postpronotal lobes; notopleura; mesopleural stripe reaching midway between anterior margin of notopleuron and anterior *npl.* seta dorsally, continuing to katepisternum as a small transverse spot, anterior margin straight; anatergite (posterior apex black); anterior ⅔ katatergite (remainder black); two broad parallel-sided lateral postsutural vittae ending just behind *ia.* seta. Medial longitudinal postsutural yellow vitta absent. Postnotum black. Scutellum yellow except for narrow black basal band. Setae: *sc.* 2; *prsc.* 2; *ia.* 1; *p.sa.* 1; *a.sa.* 1; *mpl.* 1; *npl.* 2; *scp.* 4. The tomentum-free bare area anterocentrally is narrow.
Legs: Fore femora fulvous with a small elongate fuscous to dark fuscous preapical spot on outer surfaces in some specimens, mid and hind femora entirely fulvous; fore and hind tibiae fuscous to dark fuscous, mid tibiae dark fuscous basally to fuscous apically; tarsi with basal segment fulvous and apical four segments red-brown; mid tibiae each with an apical black spur.

Wings: Length 4.5 mm; cells bc and c colourless; microtrichia over outer ½ of cell c only; remainder of wings colourless except dark fuscous cell sc, narrow dark fuscous costal band overlapping R_{2+3} and widening only slightly around apex of wing, a narrow dark fuscous anal streak; a dense aggregation of microtrichia around A_1+CuA_2; supernumerary lobe of medium development.
Abdomen: Oval; terga free; pecten present on tergum III. Tergum I and sterna I and II wider than long. Tergum I mostly black with a narrow transverse orange-brown band along posterior margin but not reaching lateral margins; tergum II orange-brown anteriorly, fulvous posteriorly, with a narrow transverse subanterior black band reaching narrow black lateral margins; terga III–V orange-brown with a transverse black band across anterior of tergum III and widening to cover most of the lateral areas of the tergum, anterolateral corners of tergum IV black, anterolateral corners of tergum V fuscous to dark fuscous, a broad medial longitudinal black band over all three terga. A pair of oval red-brown shining spots on tergum V. Posterior lobe of surstylus short, sternum V with a deep concavity on posterior margin. Abdominal sterna dark.
Attractant: Cue lure.
Female: No known record.
Distribution: Southern Vietnam (Ben Tre, Dac Lac, Dong Nai Provinces); Thailand.
Hosts: No known record.
Remarks: *Bactrocera laithieuiae* n.sp. is similar to *Bactrocera fuscitibia* Drew & Hancock and *Bactrocera occipitalis* (Bezzi) in the general body and wing colour patterns. It differs from *B. fuscitibia* in possessing elongate pear-shaped facial spots, a very broad medial longitudinal black band over abdominal terga III–V and pale-coloured shining spots on abdominal tergum V. *B. laithieuiae* differs from *B. occipitalis* in having a narrow, tomentum-free bare area anterocentrally on the scutum, facial spots narrow and pear-shaped, a dark preapical spot on the outer surfaces of the fore femora and being attracted to cue lure.

Bactrocera (Bactrocera) lata (Perkins)

(Figs 96A, 96B)

Strumeta nigrotibialis var. *lata* Perkins, 1938: 130. Holotype female in QMIC.
Bactrocera (Bactrocera) lata – Norrbom *et al*., 1998: 91.
Dacus (Bactrocera) silvaticus Hardy, 1983: 20. **New Synonym**.

Type material examined: Holotype female, **PENINSULAR MALAYSIA**: with the following two labels – 'Malay Penin., Kedah Peak, 3300–3950 ft., 14.iii.1928'; 'Holotype, Strumeta nigrotibialis var. lata, F.A. Perk'. Received on loan from QMIC.
Paratype: **PENINSULAR MALAYSIA**: 1 female, 'Malay Penin., Kedah Peak, 3000–3300 ft., 12.iii.1928'; 'Paratype, Strumeta nigrotibialis var. lata, F.A. Perk'. Received on loan from QMIC.
Other type material examined: Holotype male of *Dacus (Bactrocera) silvaticus* Hardy, **INDONESIA: SUMATRA** – North Sumatra, 12 km N.W. Bohorok, 200 m, 9 May 1975, coll. D.E. & A.T. Hardy, attracted to cue lure in rainforest. Studied at MZB. Paratype male of *Dacus (Bactrocera) silvaticus* Hardy, same data as Holotype (Fig. 96B). Received on loan from BBM.
Non-type material examined: BRUNEI: 1 male (25.i.2002), 2 males (27.i.2002), KBFC, Bato Apou Nat Pk, Riparian, coll. K. Hurley, attracted to cue lure; 1 male (25.i.2002), 2 males (27.i.2002), 1 male (29.i.2002), KBFC, Bato Apou Nat Pk, Upland, coll. K. Hurley, attracted to cue lure; 2 males (25.i.2002), 1 male (27.i.2002), KBFC, Bato Apou Nat Pk, Forest, coll. K. Hurley, attracted to cue lure. **EAST MALAYSIA: SARAWAK** – 2 males (9.xii.1991), 2 males (8.i.1992), 1 male (16.iii.1992), 1 male (23.iii.1992), Tarat Agriculture Station, Nr Kuching, coll. S. Leong, attracted to cue lure; 1 male (21.x.1991), 1 male (22.iv.1992), Kpg, Tema Mawang, Tebedu, Nr Kuching, coll. S. Leong, attracted to cue lure; 1 male, Forest Arboretum, Semongok, Nr Kuching, 28.x.1991, coll. S. Leong, attracted to cue lure; 1 male, Sarawak, Ramchang Pool, Serian, nr. Kuching, 23.iii.1992, coll. S. Leong, attracted to cue lure; 1 male, Sarawak, Sg. Sui coastal area, 22.iv.1993, coll. S. Leong, attracted to cue lure. 7 males and 2 females, Sarawak, Kpg. Kapayong, 7.i.1992, coll. S. Leong, bred from buah daba, *Canarium odontophyllum* (Family Burseraceae). **INDONESIA: KAMIMANTAN** – 1 male, West Kalimantan, Pontianak, Pancaroba, 14.xii.2005, coll. A. Irom, attracted to cue lure; **SUMATRA** – 2 males (14.xi.2006), 2 males (21.xi.2006), 1 male (28.xi.2006), Sumatra, Nanggroe Aceh, Aceh Besar, Tumbo baro, coll. Kasman, attracted to cue lure in forest; 2 males, Sumatra, Nanggroe Aceh, Aceh Besar, Lambaro Samahoni, 28.xi.2006, coll. Kasman, attracted to cue lure in village; 1 male, Sumatra, Lampung Province NP, Bukit Barisan Selatan, 37km W of Kota Agung, Sukaraya 520m, 11.iii.2000, coll. P. Schwendinger, attracted to cue lure. **PENINSULAR MALAYSIA**: 1 male (12.iii.1986), 1 male (15.iii.1986), coll. Zaidi b. Salleh, attracted to cue lure; 1 male (12.iii.1986), 1 male (15.iv.1986), 1 male (29.iv.1986), RS Jalan Kebun, coll. Mahdzir, Abdul Khalib, Mat Saod and Abdul Kebun; 1 male (20.v.1986), 1 male (2.vi.1986), 2 males (16.vi.1986), 3 males (23.vi.1986), 1 male (21.vii.1986), RS Jaram Pasu I and RS Jeram Pasu II, coll. Yusoff Salleh, attracted to cue lure; 1 male, RS Jeram Pasu II, 12.vi.1986, coll. Ismail Mat, attracted to cue lure; 1 male, R.S. Jerangan, 19.v.1986, coll. Md Ismail, attracted to cue lure; 2 males, RS Kemaman, 13.v.1986, coll. Raja Azid & Raja Samat, attracted to cue lure; 1 male (11.ii.1986), 1 male (4.iii.1986), R S Kluang, coll. Jasni Zaman, attracted to cue lure; 1 male, RS Pontian, 25.ii.1986, coll. Salleh Masri, attracted to cue lure; 1 male, TS Bentong, 28.i.1986, coll. Mahdzir, attracted to cue lure; 2 males (28.i.1986), 1 male (12.ii.1986), 1 male (11.iii.1986), 1 male (25.iii.1986), TS Gombak, coll. Mahdzir, attracted to cue lure; 1 male, Gombak rainforest, 9.ix.1987, coll. R. Drew & C.S. Ooi, attracted to cue lure; 1 male, TS Kuantan, 12.ii.1986, coll. Mahdzir, attracted to cue lure; 4 males, TS Mentakab, 12.ii.1986, coll. Mahdzir, attracted to cue lure; 1 male (12.ii.1986), 1 male (22.iv.1986) TS Puchong/Serdang, coll. Mahdzir, attracted to cue lure; 1 male, TS Temerloh, 28.i.1986, coll. Mahdzir, attracted to cue lure; 1 male, TS17, 2.vi.1986, coll.

Vignes, attracted to cue lure; 1 male, TS20, 16.vi.1986, coll. Mahdzir Ahmad, attracted to cue lure; 6 males, TN10, 25. viii.1986, coll. Ahmad Radzuan & Panjang Abd, attracted to cue lure; 2 males, TN11, 25.viii.1986, coll. Ahmad Radzuan & Panjang Abd, attracted to cue lure; 2 males (25.viii.1986), 1 male (28.viii.1986), TN13, coll. Ahmad Radzuan & Panjang Abd, attracted to cue lure; 1 male, TN20, 16.vi.1986, coll. Mahdzir Ahmad, attracted to cue lure; 1 male, TN5, 25.viii.1986, coll. Yusoff Salleh, attracted to cue lure; 1 male, TN6, 25.viii.1986, coll. Yusoff Salleh, attracted to cue lure; 4 males, TN7, 25.viii.1986, coll. Yusoff Salleh, attracted to cue lure; 1 male, Butong, 19.v.1986, coll. S. Vijaysegaran, attracted to cue lure; 2 males, Gombak/Kuantan, 19.v.1986, coll. S. Vijaysegaran, attracted to cue lure; 1 male, Jeram Pasu II, Malaysia, 21.vii.1986, coll. Y. Salleh, attracted to cue lure; 1 male, R.S. Jeram Pasu I, Malaysia, 2.vi.1986, coll. Yusoff Salleh, attracted to cue lure; 1 male, Malaysia, site TS8 (Transect South 8), 15.i.1986, coll. R. Drew, attracted to cue lure; 4 males and 2 females, Malaysia, 15.i.1986, coll. Allwood, Vijay, Gombak Old Road, 233km to Kuantan from KL. **THAILAND**: 1 male, Khun Khon, Chiang Rai, 23.vii.1992, coll. Charuwan, attracted to cue lure. Specimens in QDPC.

Diagnosis: A medium-sized species; face red-brown with large oval black spots in antennal furrows; postpronotal lobes and notopleura yellow; scutum entirely black; lateral postsutural yellow vittae present; medial postsutural yellow vitta absent; no yellow spot anterior to notopleural suture; mesopleural stripe reaching midway between anterior margin of notopleuron and anterior *npl.* seta dorsally; scutellum yellow with a moderately broad black basal band; wing with cells bc and c pale fuscous, microtrichia in outer ⅔ of cell c only; a broad dark fuscous costal band almost confluent with R_{4+5} and broad dark fuscous anal streak; abdominal terga III–V mostly black except for narrow longitudinal red-brown to dark red-brown or fuscous longitudinal bands either side of a medial longitudinal fuscous to dark fuscous band which is broken on posterior margins of terga IV and V.

Description: female.

Head: Height 1.74 mm. Frons – length 1.4 times breadth; red-brown centrally, fulvous on lateral margins, dark fuscous around orbital setae and on anteromedial hump; orbital setae black: 1 *s.or.*, 2 *i.or.*; lunule fuscous. Ocellar triangle black. Vertex dark fuscous. Face varies from fulvous to fuscous and dark fuscous with a pair of large oval black spots in antennal furrows; length 0.55 mm. Genae fulvous, fuscous subocular spot present; black seta present. Occiput dark fuscous, fulvous along eye margins; occipital row with 8–9 strong black setae. Antennae with segments 1 and 2 red-brown, segment 3 mostly fuscous; length of segments: 0.2 mm; 0.32 mm; 0.96 mm.

Thorax: Scutum entirely black except for a small area of dark red-brown above notopleuron along anterior margin of notopleural suture. Pleural areas entirely black, tending dark fuscous below postpronotal lobes. Yellow markings as follows: postpronotal lobes; notopleura; mesopleural stripe reaching midway between anterior margin of notopleuron and anterior *npl.* seta dorsally, continuing to katepisternum as a small transverse spot, anterior margin slightly convex; anatergite (posterior apex black); anterior ⅔ katatergite (remainder black); two medium-width sub-parallel lateral postsutural vittae ending at *ia.* seta (in some specimens these vittae taper posteriorly to end just before the *ia.* seta). Medial longitudinal postsutural yellow vitta absent. Postnotum black. Scutellum yellow except for a moderately broad black basal band. Setae: *sc.* 2; *prsc.* 2; *ia.* 1; *p.sa.* 1; *a.sa.* 1; *mpl.* 1; *npl.* 2; *scp.* 4.

Legs: Fore and mid femora entirely black, hind femora with basal $^3/_5$ fulvous and apical $^2/_5$ black; fore and hind tibiae dark fuscous, mid tibiae fuscous; fore and hind tarsi with basal segment fulvous and apical four segments fuscous, mid tarsi with all segments fulvous; mid tibiae each with an apical black spur.

Wings: Length 6.5 mm; cells bc and c pale fuscous to fuscous; microtrichia covering outer ⅔ to all of cell c only; remainder of wings colourless except dark fuscous cell sc, broad dark fuscous costal band almost confluent with R_{4+5} and ending beyond apex of this vein, a broad dark fuscous anal streak; no dense aggregation of microtrichia around A_1+CuA_2; supernumerary lobe weak.

Abdomen: Oval; terga free; no pecten on tergum III. Tergum I and sterna I and II wider than long. Terga I and II black except for narrow transverse dark red-brown bands posterolaterally; terga III and IV mostly black except for narrow dark fuscous longitudinal markings over posterior ½ of tergum III either side of a moderately broad medial longitudinal black band and narrow dark red-brown markings on tergum IV either side of a medium-width medial longitudinal dark fuscous band, tergum V red-brown except for broad dark fuscous to black anterolateral markings and a narrow medial longitudinal fuscous band over anterior ¾ of tergum (in some specimens, particularly from East Malaysia (Sarawak), the central regions have larger areas of orange-brown to red-brown colouration with reduced areas of dark patterning). A pair of oval shining spots on tergum V that vary from fulvous through to black. Ovipositor with basal segment red-brown, dorsoventrally flattened and tapering posteriorly in dorsal view; ratio of length of oviscape to length of tergum V, 1:1; apex of piercer needle-shaped.

Attractant: Cue lure.

Male: As for female except for dense aggregation of microtrichia around A_1+CuA_2 present; supernumerary lobe of medium development; pecten of cilia on abdominal tergum III present; posterior lobe of surstylus short; abdominal sternum V with a deep concavity on posterior margin.

Distribution:

Type locality: Peninsular Malaysia (Kedah Peak).

Previously recorded: Type locality and Peninsular Malaysia.

New records: Brunei; East Malaysia (Sarawak); Indonesia (Sumatra – Aceh and Lampung; West Kalimantan); Thailand (Chaing Rai).

Hosts: Recorded from one wild host, *Canarium odontophyllum* (family *Burseraceae*) (Allwood *et al.*, 1999).

Remarks: *Bactrocera (Bactrocera) lata* (Perkins) cannot be separated from *Dacus (Bactrocera) silvaticus* Hardy. Only slight differences occur, mainly in the shape and length of the lateral postsutural vittae. Both species exhibit the same small variations in colour of fore femora and width of the costal band. Consequently, herein *D. silvaticus* is made a **New Synonym** of *B. lata*. *Bactrocera lata* belongs to the *nigrotibialis* complex in possessing a generally black thorax and abdomen and in having the legs (especially femora) with large areas of glossy black but it is unique within this complex in having a broad costal band almost confluent with R_{4+5}.

In the original description, Perkins (1938) had two females and one male before him. These were presumed to be in SMKM or BMNH by Hardy (1973) and Norrbom *et al.* (1998). However, the two females, one with a Holotype label and the other with a Paratype label have been found in UQIC (now transferred to QMIC). The male is presumed lost.

Bactrocera (Bactrocera) lateritaenia Drew & Hancock

(Fig. 97)

Bactrocera (Bactrocera) lateritaenia Drew & Hancock, 1994a: 34; Norrbom *et al.*, 1998: 91. Holotype male in BMNH.

Non-type material examined: Since the publication of Drew & Hancock (1994a), the following specimens have been studied – **BRUNEI**: 1 male, KBFC, Bato Apou Nat Pk, Riparian, 23.i.2002, coll. K. Hurley, attracted to cue lure; 3 males, KBFC, Bato Apou Nat Pk, Upland Ridge, 23.i.2002, coll. K. Hurley, attracted to cue lure; 2 males, KBFC, Bato Apou Nat Pk, Forest, 23.i.2002, coll. K. Hurley, attracted to cue lure; 1 male (25.i.2002), 6 males (27.i.2002), 4 males (29.i.2002), Brunei Darassalam, KBFC, Bato Apou Nat Pk, coll. K. Hurley, attracted to cue lure in forest. **VIETNAM: Southern Vietnam** – 1 male, southern Vietnam, Dong Nai Province, Long Khanh District, Nhan Nghia Village, Site VS06, 16.iii.2002, coll. SOFRI staff, attracted to cue lure. Specimens in QDPC.

Diagnosis: Face fulvous with a pair of medium-sized elongate oval black spots; postpronotal lobes and notopleura yellow; scutum black with dark brown below *a.sa.* seta and in some specimens two narrow longitudinal dark red-brown bands each side of the midline over anterior ½; broad parallel-sided lateral postsutural yellow vittae that narrow sharply posteriorly to end at or occasionally before *ia.* seta; medial postsutural yellow vitta absent; mesopleural stripe slightly wider than notopleuron dorsally; scutellum yellow; legs with femora entirely fulvous; fore tibiae dark fuscous, mid tibiae fuscous (darker basally), hind tibiae dark fuscous to black; wing with cells bc and c colourless, microtrichia in outer corner of cell c only; a narrow fuscous costal band slightly overlapping R_{2+3} and widening slightly across apex of R_{4+5} to end between extremities of this vein and M; a narrow fuscous anal streak; supernumerary lobe of medium development; abdominal terga III–V orange-brown except for a broad medial and two broad lateral longitudinal black bands over all three terga and joined along anterior margin of tergum III, a pair of oval orange-brown shining spots on tergum V; abdominal sterna dark coloured.

Attractant: Cue lure.

Female: No known record.

Distribution:

Type locality: Peninsular Malaysia (Cameron Highlands).

Previously recorded: Type locality and East Malaysia (Sabah) (Drew & Hancock, 1994a).

New records: Brunei; Southern Vietnam (Dong Nai Province).

Hosts: No known record.

Remarks: *Bactrocera (Bactrocera) lateritaenia* Drew & Hancock, within the *dorsalis* complex, is similar to *Bactrocera (Bactrocera) cognata* (Hardy & Adachi), *Bactrocera (Bactrocera) fulvifemur* Drew & Hancock, *Bactrocera (Bactrocera) penecognata* Drew & Hancock and *Bactrocera (Bactrocera) usitata* Drew & Hancock in possessing lateral postsutural vittae that narrow sharply posteriorly, femora entirely fulvous or with, at most, a dark spot on fore femora only, costal band slightly overlapping R_{2+3} and broad lateral dark markings on abdominal terga III–V. It differs from *B. cognata* in possessing the costal band widening slightly across apex of wing and from the other three species in having the lateral postsutural vittae ending at the *ia.* seta.

Bactrocera (Bactrocera) latifrons (Hendel)

(Fig. 98)

Chaetodacus latifrons Hendel, 1915: 425. Lectotype male in BMNH.

Dacus amoyensis Froggatt, 1909: 36, nomen nudum.

Dacus parvulus Hendel, 1912: 21; White & Liquido, 1995: 250–252 (Synonym).

Chaetodacus antennalis Shiraki, 1933: 56; Hardy, 1973: 49 as synonym of *D. parvulus* Hendel; White & Liquido, 1995: 250–252.

Bactrocera (Bactrocera) latifrons – White & Liquido, 1995: 251, Lectotype designation; Norrbom *et al.*, 1998: 91.

Type material examined: Lectotype male, **TAIWAN (formerly FORMOSA)**, with the following seven labels – 'Paratype'; 'Paratype'; 'Tainan, iii.1912'; 'Formosa, Sauter'; 'latifrons Hend, typus, det. Hendel'; 'Purch. from E.M. Hering, B.M. 1965–270'; 'Lectotype'. Received on loan from BMNH. 1 male with the following six labels – 'Formosa, Sauter'; 'Tainan, iii.1912'; 'Dacus latifrons H, det. Hendel'; 'Cotype male'; 'coll. Hendel'; 'cotype male marked by D.E. Hardy 1961'. Studied at NMW.

Other type material examined: TAIWAN (formerly FORMOSA): 1 male cotype of *Dacus parvulus* Hendel with the following four labels – 'Cotypus'; 'Kanshirei, Formosa, iv.09, H. Sauter'; '*Dacus parvulus* H., det. Hendel'; '*Chaetodacus parvulus* Hend. det. H.K. Munro, 1935'. Received on loan from UQIC (now transferred to QMIC).

Non-type material examined: INDONESIA: KALIMANTAN – 4 males and 11 females, South Kalimantan, Banjarbaru Landasan Ulin, 29.x.2005, coll. Bahriman, bred from sample No. GF10(27); **SULAWESI** – 2 males and 6 females, Sulawesi, Central Sulawesi, Donggala, Nupabomba, 21.viii.2006, coll. Rusdin, bred from Sample No. GF14, *Solanum melongena* (eggplant). **SRI LANKA (formerly CEYLON)** – 3 males and 3 females, Sri Lanka, Eppawala,

14.vi.1996, coll. J.P. Rajapakshe & H. Rajapakshe, hand collected on *Solanum melongena*. **THAILAND**: 5 males and 3 females, Prachinburi, Jan. 1992, coll. Montree, bred from *Solanum* sp.; a large series reared from host fruits in a range of localities. **VIETNAM: Northern Vietnam** – 1 female, North Vietnam, Hoa Binh Province, Ky Son District, Cao Phong, 1.xii.1999, Site N013, in cue lure trap. Specimens in QDPC. **TAIWAN**: 1 female, 'Tainan, Formosa H. Sauter, v.12'; 'coll. Hendel'. Studied at NMW.

Diagnosis: A medium-sized species; face fulvous with a pair of large oval black spots; postpronotal lobes and notopleura yellow; scutum dull black; lateral postsutural yellow vittae present; medial postsutural yellow vitta absent; no yellow spot anterior to notopleural suture; mesopleural stripe extending to anterior *npl.* seta dorsally; scutellum yellow; wing with cells bc and c colourless, microtrichia in outer corner of cell c only; a narrow fuscous costal band overlapping R_{2+3} and expanding into a small spot around apex of R_{4+5}; a medium-width fuscous anal streak; all abdominal terga entirely dark orange-brown.

Description: male.

Head: Height 1.5 mm. Frons – length 1.25 times breadth; red-brown centrally, fulvous along lateral margins, dark fuscous around bases of orbital setae and on anteromedial hump; orbital setae black: 1 *s.or.*, 2 *i.or.*; lunule fuscous to dark fuscous. Ocellar triangle black. Vertex fuscous to dark fuscous. Face fulvous with a pair of large oval black spots; length 0.4 mm. Genae fulvous, large fuscous subocular spot present; black seta present. Occiput fuscous, fulvous along eye margins; occipital row with a large number of short pale setae. Antennae with segments 1 and 2 red-brown, segment 3 red-brown with fuscous on apex and outer surface; length of segments: 0.2 mm; 0.3 mm; 0.68 mm.

Thorax: Scutum dull black with red-brown below and behind lateral postsutural vittae, around notopleural suture, along lateral margins anterior to suture and inside postpronotal lobes, and in some specimens across posterior margin of scutum including *prsc.* setae. Pleural areas red-brown with black on katepisternum, along posterior and anterior margins of mesopleural stripe and above hind coxae. Yellow markings as follows: postpronotal lobes; notopleura; mesopleural stripe reaching anterior *npl.* seta dorsally, continuing to katepisternum as a large transverse spot, anterior margin straight; anatergite (posterior apex black); anterior ⅔ katatergite (remainder black); two moderately broad parallel-sided lateral postsutural vittae ending at *ia.* seta. Medial longitudinal postsutural yellow vitta absent. Postnotum black. Scutellum yellow except for narrow red-brown basal band. Setae: *sc.* 2; *prsc.* 2; *ia.* 1; *p.sa.* 1; *a.sa.* 1; *mpl.* 1; *npl.* 2; *scp.* 4.

Legs: All segments fulvous except hind tibiae fuscous; mid tibiae each with an apical black spur.

Wings: Length 5.0 mm; cells bc and c colourless; microtrichia in outer corner of cell c only; remainder of wings colourless except dark fuscous cell sc, narrow fuscous to dark fuscous costal band overlapping R_{2+3} where it becomes paler and ending with a slight expansion around apex of R_{4+5}, a medium-width fuscous anal streak; a dense aggregation of microtrichia around A_1+CuA_2; supernumerary lobe of medium development.

Abdomen: Oval; terga free; pecten present on tergum III. Tergum I and sterna I and II wider than long. All abdominal terga dark orange-brown without distinct dark colour patterns and with transverse fulvous markings posterolaterally on tergum II. A pair of oval fuscous to dark fuscous shining spots on tergum V. Posterior lobe of surstylus short, sternum V with a deep concavity on posterior margin. Abdominal sterna pale, generally red-brown.

Attractant: No known record.

Female: As for male except no dense aggregation of microtrichia around A_1+CuA_2; supernumerary lobe weak; pecten of cilia on abdominal tergum III absent; ovipositor basal segment dark fuscous to black, dorsoventrally compressed and tapering posteriorly in dorsal view; ratio of length of oviscape to length of tergum V, 0.5:1; apex of piercer trilobed (Fig. 98).

Distribution:

Type locality: Taiwan.

Previously recorded: Widespread from the Indian subcontinent across South-East Asia (White & Hancock, 1997 (CABIKEY); Norrbom *et al.*, 1998). Introduced into Hawaii.

New records: Indonesia (South Kalimantan; Sulawesi–Central Sulawesi); Northern Vietnam (Hoa Binh Province).

Hosts: Recorded from ten plant families (see Allwood *et al.*, 1999 for host records). The family *Solanaceae* contains the major host species and *B. latifrons* is a major pest of *Capsicum* and *Solanum* species.

New records: *Syzygium samarangense* (family *Myrtaceae*).

Remarks: *Bactrocera (Bactrocera) latifrons* (Hendel) can be confused with species in the *Bactrocera musae* complex and the *Bactrocera dorsalis* complex in possessing similar body colour patterns. However, it is distinct in having a trilobed apex on the piercer and uniformly dark orange-brown abdominal terga. It is similar to *Bactrocera (Bactrocera) citima* (Hardy) in possessing a generally black scutum, costal band overlapping R_{2+3}, cells bc and c colourless and parallel-sided lateral postsutural vittae but differs from this species in having red-brown around lateral and posterior margins of the scutum, femora entirely fulvous and abdominal terga III–V entirely red-brown. We have withdrawn *B. latifrons* from synonymy with *B. parvula* (Hendel) and for our reasons see 'Remarks' under *B. parvula*.

Bactrocera (Bactrocera) latilineola Drew & Hancock

(Fig. 99)

Bactrocera (Bactrocera) latilineola Drew & Hancock, 1994a: 35; Norrbom *et al.*, 1998: 92. Holotype male in BMNH.

Non-type material examined: Since the publication of Drew & Hancock (1994a), no new specimens have been studied.

Diagnosis: Face fulvous with a pair of medium-sized elongate oval black spots; postpronotal lobes and notopleura yellow; scutum and pleural areas entirely black without pale markings; broad parallel-sided lateral postsutural yellow vittae ending at *ia.* seta; medial postsutural yellow vitta absent; mesopleural stripe equal in width to notopleuron dorsally; scutellum yellow; legs with femora fulvous except for a large oval subapical black spot on outer surfaces of fore femora; fore tibiae fuscous tending darker basally, mid tibiae fulvous tending fuscous basally, hind tibiae dark fuscous; wing with cells bc and c colourless, microtrichia in outer corner of cell c only; a narrow fuscous costal band confluent with R_{2+3} and expanding across apex of wing; a very narrow fuscous anal streak; supernumerary lobe of medium development; abdominal terga III–V orange-brown with broad lateral longitudinal black bands over all three terga and a very broad medial longitudinal black band over terga II–V and joined by a broad transverse black band across anterior ⅓ of tergum III, a pair of oval dark fuscous to black shining spots on tergum V; abdominal sterna dark coloured.

Attractant: Methyl eugenol.

Female: No known record.

Distribution:

Type locality: Peninsular Malaysia (Cameron Highlands).

Hosts: No known record.

Remarks: *Bactrocera (Bactrocera) latilineola* Drew & Hancock, within the *dorsalis* complex, is similar to *Bactrocera (Bactrocera) affinidorsalis* (Hardy), *Bactrocera (Bactrocera) malaysiensis* Drew & Hancock and *Bactrocera (Bactrocera) unimacula* Drew & Hancock in possessing broad parallel-sided lateral postsutural vittae and the costal band confluent with R_{2+3} and expanding across apex of wing. It differs from *B. unimacula* in having the costal band gradually widening across apex of wing (not in the distinct spot as in *B. unimacula*) and from the other two species in having the broad medial longitudinal black band on the abdomen over terga II–V (not terga III–V), the mid and hind femora entirely fulvous and the shining spots on abdominal tergum V dark fuscous to black.

Bactrocera (Bactrocera) limbifera (Bezzi)

(Fig. 100)

Chaetodacus ferrugineus var. *limbiferus* Bezzi, 1919: 424. Lectotype male in MCSNM.

Strumeta pedestris var. *limbiferus* – Perkins, 1938a: 126.

Dacus (Strumeta) limbiferus – Hardy & Adachi, 1954: 172; Hardy, 1969b: 479, Lectotype designation.

Dacus (Strumeta) limbifer – Hardy, 1974: 35.

Bactrocera (Bactrocera) limbifera – Norrbom *et al.*, 1998: 92.

Type material examined: Lectotype male, **PHILIPPINES**: with the following three labels – 'Batbatan Isl, R.C. McGregor'; 'limbiferus n.' in Bezzi's handwriting; 'Lectotype male, Chaetodacus limbiferus Bez. desig. by Hardy – 68'. Six female Paralectotypes, 'Batbatan Isl, R.C. McGregor'. Received on loan from MCSNM.

Other type material examined: Holotype male of *Dacus (Bactrocera) limbifer rufulus* Hardy, **INDONESIA: SULAWESI** – Sadaunta, 65 km S.E. Palu, Central Sulawesi, August 1975, coll. D.E. & A.T. Hardy, attracted to cue lure. Studied at MZB. One male paratype of *Dacus (Bactrocera) limbifer rufulus* Hardy, Lindu Valley, C. Sulawesi, 960 m, August 1975, D.E. Hardy, in QDPC (see 'Remarks' below).

Non-type material examined: BRUNEI: 1 male, KBFC, Bato Apou Nat Pk, Forest, 29.i.2002, coll. K. Hurley, attracted to cue lure. **EAST MALAYSIA: SARAWAK** – 1 male (19.ix.1991), 2 males (30.ix.1991), 1 male (25.xi.1991), 1 male (10.vi.1992), Ranchang Pool, Serian, Nr Kuching, coll. S. Leong, attracted to cue lure; 1 male, Tarat Agriculture Station, Nr Kuching, 30.ix.1991, coll. S. Leong, attracted to cue lure; 1 male, ARC Semongok, Nr Kuching, 13.vi.1992, coll. S. Leong, attracted to cue lure; 15 males, 13 females, Sarawak, Mulu National Park, 28.vii.1992, coll. Leong, bred from *Sterculia* sp., K463. **INDIA: ANDAMAN ISLANDS** – 1 male and 2 females, India, S. Andaman, Port Blair, 4. vi.1992, coll. Ranganath, Sl No. 6, collected on *Excoecaria agallocha* (Family Euphorbiaceae). **INDONESIA: FLORES** – 65 males, Borong, iv.1989, coll. J. Turner; **JAVA** – 1 male, East Java, Sidoarjo, Tulangan, Gelang, 4.x.2005, coll. Agus Riadi, attracted to cue lure in village; 1 male (8.xi.2005), 1 male (27.xii.2005), East Java, Lumajang, Kadang Tepus, coll. Utomo, attracted to cue lure in banana; 2 males (1 xi.2005), 1 male (22.xi.2005), D.I. Yogyakarta, Sieman, Kaliurang, coll. Agus W, attracted to cue lure in forest; 1 male (30.xi.2005), 2 males (14.xii.2005), East Java, Probolinggo, Betektanam 150m, coll. Tjipto Waluyo, attracted to cue lure in mango; 1 female (24.x.2005), 1 male (2.i.2006), Banten, Lebak, Pasir Nini-Nini, coll. T B Aden Haer, attracted to cue lure in forest; **KAMIMANTAN** – 1 male (8.viii.2005), 1 male (22.viii.2005), 1 male (29.viii.2005), 2 males (5.ix.2005), East Kalimantan, Kutal Kartanegara Batuah, coll. Ertha & Heru G, attracted to cue lure in forest; 4 males, South Kalimantan, Banjar, Mandiangan, 31.vii.2005, coll. Bilal & Sidtotul; **SULAWESI** – 36 males, Sulawesi Barat, Mamuju, De Bulutakang, 11.ix.2006, coll. Yan P. Fachrie, attracted to cue lure in Hutan (Jeruk); 15 males, Sulawesi Barat, Mamuju, DS Tambayako, 28 Aug. 2006, coll. Yan P. Fachrie, attracted to cue lure in Hutan (Jeruk); 6 males (4.ix.2006), 24 males (30.x.2006), Sulawesi Barat, Mamuju, Belang-Belang, coll. Sri Hastuti, Yan P.F. Sukri, attracted to cue lure in forest, orange (jeruk); 4 males (8.viii.2006), 3 males (11.ix.2006), 7 males (30.x.2006), Sulawesi Barat, Mamuju, Binanga, coll. Sri Hastuti, Yan P.F., Sukri, attracted to cue lure in orange (jeruk); 12 males (12.ix.2006), 26 males (26.ix.2006), Gorontalo, Bone Bolango, Lombongo, coll. Yusbar Ismail, attracted to cue lure in forest; 13 males, Gorontalo, Bone Bolango, Lombongo, 26.ix.2006, coll. Yosis, attracted to cue lure in forest; 2 males (16.viii.2006), 3 males (27.viii.2006), Sulawesi Selatan, Maros, Limampoccoe, coll. Sirajuddin, attracted to cue lure in forest; 3 males (27.viii.2006), 4 males (2.ix.2006), 3 males (10.ix.2006), 5 males (17.ix.2006), 2 males (23.ix.2006), 5 males (27.ix.2006), 7 males (15.x.2006), 9 males (22.x.2006), Sulawesi Tengah, Donggala, Nupabomba, coll. Rusdin, attracted to cue lure in forest; **SUMATRA** – 2 males, Nanggroe Aceh, Aceh Besar, Lam Ara Tunong, 3.x.2006, coll. Kasman, attracted to cue lure in Kebun, Mangga; 2 males, Nanggroe Aceh, Aceh Besar,

Tumbo Baro, 14.xi.2006, coll. Kasman, attracted to cue lure in forest; **SUMBAWA** – 6 males, 'Semongkat, 250-350m, 12.vi.1981, coll. J. Turner'; 'Primary sclerophyll forest dry'; attracted to cue lure. **VIETNAM: Northern Vietnam** – 1 male, Vietnam North, Hoa Binh Province, Ky Son District, Cao Phong, Site N014, 17.x.1999, attracted to cue lure; **Southern Vietnam** – 1 male, Vietnam South, Tien Giang Province, My Tho Village, Site S005, 20.xii.1999, attracted to cue lure; 1 male (23.i.2002), 3 males (24.ii.2002), 2 males (8.iv.2002), 1 male (8.v.2002), Dac Lac Province, Ku Mi Nga District, Quang Phu Village, Site VS09, coll. SOFRI Staff, attracted to cue lure; 1 male (18.x.2002), 3 males (16.xi.2002), Vietnam South, Binh Duong Province, Thuan An District, An Thanh Village, Site VS05, coll. SOFRI Staff, attracted to cue lure. Specimens in QDPC.

Diagnosis: A large species; face fulvous with a pair of medium-sized oval black spots; postpronotal lobes and notopleura yellow; scutum mostly black; lateral postsutural yellow vittae present; medial postsutural yellow vitta absent; no yellow spot anterior to notopleural suture; mesopleural stripe reaching midway between anterior margin of notopleuron and anterior *npl.* seta dorsally; scutellum yellow; wing with cells bc and c colourless, microtrichia in outer corner of cell c only; a broad dark fuscous costal band usually confluent with R_{4+5} and expanding slightly across apex of this vein; a moderately broad fuscous anal streak; abdominal terga III–V dark orange-brown with a dark fuscous to black 'T' pattern consisting of a narrow transverse band across anterior margin of tergum III and broadening to cover lateral margins, a broad medial longitudinal black band over all three terga and anterolateral corners of terga IV and V varying from red-brown to dark fuscous to black.

Description: male.

Head: Height 1.8 mm. Frons – length 1.59 times breadth; red-brown with fuscous around orbital setae and on anteromedial hump; orbital setae black: 1 *s.or.*, 2 *i.or.*; lunule dark fuscous. Ocellar triangle black. Vertex fuscous. Face fulvous with a pair of medium-sized oval black spots; length 0.53 mm. Genae fulvous, fuscous subocular spot present; black seta present. Occiput fuscous, fulvous along eye margins; occipital row with 5–6 strong black setae. Antennae with segments 1 and 2 red-brown, segment 3 red-brown with dark fuscous on apex and outer surface; length of segments: 0.2 mm; 0.33 mm; 0.88 mm.

Thorax: Scutum black except dark red-brown behind lateral postsutural vittae, around notopleural suture and inside postpronotal lobes. Pleural areas entirely black. Yellow markings as follows: postpronotal lobes; notopleura; mesopleural stripe of medium-width reaching midway between anterior margin of notopleuron and anterior *npl.* seta dorsally, continuing to katepisternum as a transverse spot, anterior margin convex; anatergite (posterior apex black); anterior ⅔ katatergite (remainder black); two broad subparallel lateral postsutural vittae ending at *ia.* seta. Medial longitudinal postsutural yellow vitta absent. Postnotum black. Scutellum yellow except for narrow black basal band. Setae: *sc.* 2; *prsc.* 2; *ia.* 1; *p.sa.* 1; *a.sa.* 1; *mpl.* 1; *npl.* 2; *scp.* 4.

Legs: Fore femora fulvous with dark fuscous to black along outer apical ⅔, mid femora fulvous except fuscous to black around apical ⅓, hind femora fulvous except dark fuscous to black around apical ⅓; fore and mid tibiae fulvous, hind tibiae pale fuscous; fore and hind tarsi with basal segment fulvous and apical four segments pale fuscous, mid tarsi with basal two segments fulvous and apical three segments red-brown; mid tibiae each with an apical black spur; fore femora with rows of strong black setae over apical ½ of dorsal and ventral surfaces.

Wings: Length 6.5 mm; cells bc and c colourless; microtrichia in outer corner of cell c only; remainder of wings colourless except for a dark fuscous cell sc, broad dark fuscous costal band confluent with R_{4+5} and widening slightly around apex of this vein (in some specimens the costal band does not quite reach R_{4+5}), a moderately broad fuscous anal streak; dense aggregation of microtrichia around A_1+CuA_2; supernumerary lobe of medium development.

Abdomen: Oval; terga free; pecten present on tergum III. Tergum I and sterna I and II wider than long. Tergum I dark fuscous to black except for a narrow transverse orange-brown band along posterior margin but not reaching lateral margins; tergum II orange-brown to dark orange-brown with a narrow dark fuscous to black transverse band across anterior margin but not reaching dark fuscous to black narrow anterolateral corners; terga III–V dark orange-brown except for a 'T' pattern consisting of a narrow dark fuscous to black transverse band across anterior margin of tergum III and broadening across lateral margins and a broad medium longitudinal dark fuscous to black band over all three terga, dark fuscous to black anterolateral corners on terga IV and V. In some specimens the lateral margins on terga IV and V are paler, often red-brown to dark red-brown. A pair of oval dark orange-brown shining spots on tergum V. Posterior lobe of surstylus short, sternum V with a deep concavity on posterior margin.

Attractant: Cue lure.

Female: As for male except no dense aggregation of microtrichia around A_1+CuA_2; supernumerary lobe weak; pecten of cilia on abdominal tergum III absent; ovipositor basal segment red-brown, long and narrow, conical in transverse cross section and tapering posteriorly in dorsal view; ratio of length of oviscape to length of tergum V, 1.8:1; apex of piercer needle-shaped.

Distribution:

Type locality: Philippines (Batbatan I.).

Previously recorded: Type locality and the following locations – India (Andaman Is); Indonesia (Java; Lombok; Sumatra) (Norrbom *et al.*, 1998).

New records: Brunei; East Malaysia (Sarawak); Indonesia (Flores; Kalimantan; Sulawesi; Sumbawa); Vietnam.

Hosts: *Dracontomelon dao* (family *Anacardiaceae*) and *Aglaia* sp. (family *Meliaceae*) (Allwood *et al.*, 1999).

New records: *Sterculia* sp. (family *Sterculiaceae*).

Remarks: *Bactrocera (Bactrocera) limbifera* (Bezzi) is similar to *Bactrocera (Bactrocera) bryoniae* (Tryon) in possessing a broad costal band, black scutum and abdominal terga III–V with a general dark fuscous to black 'T' pattern. It differs from *B. bryoniae* in having dark patterns on the femora and colourless cells bc and c. There appears to be extensive variation in the dark colour patterns on the femora and on the lateral margins of abdominal terga IV and V. The Holotype and one Paratype of *Bactrocera (Bactrocera) rufula* (Hardy) have been studied and this species has been synonymised with *B. ritsemai* (see 'Remarks' under *B. ritsemai*). *Bactrocera (Bactrocera) involuta* (Hardy) also appears similar to *B. limbifera* in most significant specific character states. However, *B. involuta* possesses an elongate oval-shaped abdomen, all femora entirely fulvous and dark fuscous to black shining spots on abdominal tergum V.

Bactrocera (Bactrocera) linduensis Drew & Romig, new name

(Fig. 101)

Dacus (Bactrocera) enigmaticus Hardy, 1982: 224. Holotype male in MZB.
Bactrocera (Bactrocera) enigmatica – Norrbom *et al.*, 1998: 90.

Type material examined: Holotype male, **INDONESIA: SULAWESI** – Lindu Valley, Central Sulawesi, 960 m, Aug. 1975, coll. D.E. & A.T. Hardy. Studied at MZB.
Paratypes: **INDONESIA: SULAWESI** – 2 males, 'Lindu Valley, C. Sulawesi, 960 m, Aug. 1975, coll. D.E. Hardy'; 'Paratype male, Dacus (Bactrocera) enigmaticus, D.E. Hardy'. Specimens in QDPC.
Non-type material examined: INDONESIA: FLORES – 2 males, Borong, iv.1989, coll. J. Turner; **NORTH MOLUCCAS: (NORTH MALUKU)** – 10 males (27.v.2007), 2 males (3.vi.2007), 1 male (10.vi.2007), North Maluku, Kodya Ternate, Foramadiahi, coll. La Ruti, attracted to cue lure; 4 males, North Maluku, Kodya Ternate, Kalumata, 30.vi.2007, coll. La Ruti, attracted to cue lure; 1 male, North Maluku, Kodya Ternate, Kalumata, 17.vi.2007, coll. Raais Abdullah, attracted to cue lure in mango; 3 males, North Maluku, West Halmahera, Goal, 12.vi.2007, coll. Raais Abdullah, attracted to cue lure in orange (jeruk); **MOLUCCAS (MALUKU)** – 1 male, Maluku, Middle Maluku, Suli, 9.vi.2007, coll. La Naibu, attracted to cue lure in Hutan amalatuei; 1 male, Maluku, Seram (Western part), Nuruwe, 29.vi.2007, A. Haris Tualepe, attracted to cue lure in forest (200 mdpl); 1 male (24.v.2007), 1 male (12.vii.2007), 1 male (19.vii.2007), 1 male (2.viii.2007), Maluku, Western Seram, Waimital, coll. A. Haris Tualepe, attracted to cue lure in orange (jeruk); **SULAWESI** – 1 male (19.ix.2006), 3 males (17.x.2006), 14 males (31.x.2006), Gorontalo, Bone Bolango, Suwawa, coll. Yusbar Ismail, attracted to cue lure in forest; 1 male, Gorontalo, Bone Bolango, Ayma Utara, 12.ix.2006, coll. Yusbar Ismail, attracted to cue lure in Cabe; 1 male, Gorontalo, Bone Bolango, Lombongo, 26.ix.2006, coll. Yusbar Ismail, attracted to cue lure in forest; 1 male, Sulawesi Tengah, Donggala, Nupabomba, 12.viii.2006, coll. Rusdin, attracted to cue lure; 1 male, Sulawesi Selatan, Maros, Limampoccoe, 30.viii.2006, coll. Sirajuddin, attracted to cue lure in forest; 1 male, Sulawesi Utara, Minahasa Utara, Sarongsong, 11.ix.2006, coll. Karel F. Lala, attracted to cue lure; 2 males (16.viii.2006), 1 male (27.viii.2006), 1 male (13.ix.2006), Sulawesi Selatan, Takalar, Limampoccoe, coll. Sirajuddin SP, attracted to cue lure in forest; **SUMBAWA** – 1 male, Kanar, 12.vi.1981, coll. J. Turner, teak forest with sclerophyll remnants and gallery forest, attracted to cue lure; **WEST TIMOR** – 2 males, Indonesia, West Timor, Besi Pae, 80km SE Kupang, 28.x.1986, coll. E.S.C. Smith, attracted to cue lure; 1 male, Indonesia, West Timor, 15km S.E. Kupang, 30.x.1986, coll. E.S.C. Smith. Specimens in QDPC collection.
Diagnosis: A large species; face fulvous with a pair of medium-sized oval black spots; postpronotal lobes and notopleura yellow; scutum black; lateral postsutural yellow vittae present; medial postsutural yellow vitta absent; no yellow spot anterior to notopleural suture; mesopleural stripe reaching midway between anterior margin of notopleuron and anterior *npl.* seta dorsally; scutellum yellow; wing with cells bc and c pale fuscous, microtrichia over cells bc and c variable in extent; a broad fuscous costal band overlapping R_{4+5}; a broad fuscous anal streak; abdominal terga III–V orange-brown with a narrow transverse black band across anterior margin of tergum III.
Description: male.
Head: Height 1.96 mm. Frons – length 1.56 times breadth; fulvous with pale fuscous around orbital setae and on anteromedial hump; orbital setae black: 1 *s.or.*, 2 *i.or.*; lunule fuscous. Ocellar triangle black. Vertex fulvous. Face fulvous with a pair of medium-sized oval black spots; length 0.59 mm. Occiput very pale fuscous, fulvous along eye margins. Antennae with segments 1 and 2 red-brown, segment 3 red-brown with fuscous on apex and outer surface; length of segments: 0.21 mm; 0.42 mm; 0.9 mm.
Thorax: Scutum black except red-brown below and behind lateral postsutural vittae, around notopleural suture, around anterolateral margins and inside postpronotal lobes. Yellow markings as follows: postpronotal lobes; notopleura; mesopleural stripe reaching midway between anterior margin of notopleuron and anterior *npl.* seta dorsally; two medium-width parallel-sided lateral postsutural vittae ending at *ia.* seta. Medial longitudinal postsutural yellow vitta absent. Scutellum yellow except for narrow black basal band. Setae: *sc.* 2; *prsc.* 2; *ia.* 1; *p.sa.* 1; *a.sa.* 1; *mpl.* 1; *npl.* 2; *scp.* 4.
Legs: All femora entirely fulvous; fore tibiae pale fuscous, mid tibiae fuscous, hind tibiae darker fuscous; tarsi with all segments entirely fulvous; mid tibiae each with an apical black spur.
Wings: Length 7.1 mm; cells bc and c pale fuscous; microtrichia covering outer ½ of cell c (Hardy (1982) recorded variable covering of microtrichia over cells bc and c); remainder of wings colourless except fuscous to dark fuscous cell sc, broad fuscous to dark fuscous costal band overlapping R_{4+5} for its entire length, a broad fuscous anal streak; a dense aggregation of microtrichia around A_1+CuA_2; supernumerary lobe of medium development.
Abdomen: Oval; terga free; pecten present on tergum III. Tergum I and sterna I and II wider than long. Tergum I fuscous to black except for a narrow orange-brown transverse band across posterior margin but not reaching lateral margins; tergum II orange-brown with a narrow transverse subanterior black band but not reaching lateral margins; terga III–V orange-brown with a narrow transverse black band

across anterior margin of tergum III. A pair of fuscous to dark fuscous shining spots on tergum V. Posterior lobe of surstylus short, sternum V with a deep concavity on posterior margin.
Attractant: Cue lure.
Female: No known record.
Distribution:
Type locality: Indonesia (Central Sulawesi, Lindu Valley).
New records: Indonesia (Flores; Moluccas = Maluku – Ambon I.; Seram I.; North Moluccas = North Maluku – Halmahera I.; Ternate; Sumbawa; West Timor).
Hosts: No known record.
Remarks: Under Article 58 of the ICZN, the species name *enigmaticus* Hardy (1982) is preoccupied by *aenigmaticus* Malloch (1931). As both names are now placed together in the genus *Bactrocera*, *enigmaticus* Hardy becomes a secondary homonym of *aenigmaticus* Malloch and is hereby replaced with the new name, *linduensis*. *Bactrocera linduensis*, new name, is similar to *Bactrocera (Bactrocera) beckerae* (Hardy), *Bactrocera (Bactrocera) bryoniae* (Tryon), *Bactrocera (Bactrocera) costalis* (Shiraki), *Bactrocera (Bactrocera) epicharis* (Hardy) and *Bactrocera (Bactrocera) paramusae* Drew in possessing a broad costal band of uniform width and confluent with or overlapping R_{4+5} and cells bc and c coloured. It differs from these species in having the costal band distinctly overlapping R_{4+5}.

Bactrocera (Bactrocera) lombokensis Drew & Hancock
(Fig. 102)

Bactrocera (Bactrocera) lombokensis Drew & Hancock, 1994a: 36; Norrbom *et al*., 1998: 92. Holotype male in BBM.
Bactrocera (Bactrocera) cheni Zhang, Ji & Chen, in Zhang *et al*., 2012: 206. Holotype male in FAFU. **New Synonym.**

Other type material studied: CHINA: A study of the colour photographs and description of *Bactrocera (Bactrocera) cheni* Zhang, Ji & Chen in the original publication.
Non-type material examined: Since the publication of Drew & Hancock (1994a), the following specimens have been studied – **THAILAND**: 1 male, Khao Kheaw, Pakphli, Nakhonnayok, 7.ii.1992, coll. Montree, attracted to methyl eugenol (probable error due to lure contamination). **VIETNAM: Northern Vietnam** – 1 male (15.x.2001), 1 male (30.x.2001), northern Vietnam, Hoa Binh province, Mai Chau District, Pa Co, Site N010, attracted to cue lure. Specimens in QDPC.
Diagnosis: Face fulvous with a pair of large oval black spots; postpronotal lobes and notopleura yellow; scutum entirely black without pale markings; moderately broad parallel-sided lateral postsutural yellow vittae ending at *ia*. seta; medial postsutural yellow vitta absent; mesopleural stripe reaching midway between anterior margin of notopleuron and anterior *npl*. seta dorsally; scutellum yellow; legs with femora fulvous with a black spot on outer apical ½ of fore femora and smaller black spots on inner apical surfaces of mid and hind femora; all tibiae dark fuscous; wing with cells bc and c colourless, microtrichia in outer corner of cell c only; a narrow fuscous costal band confluent with R_{2+3} and widening slightly across apex of wing; a narrow fuscous anal streak; supernumerary lobe of medium development; abdominal terga III–V orange-brown with a black band across anterior margin of tergum III which expands laterally to cover outer ⅔ of the tergum, antero-lateral corners of terga IV and V dark fuscous, a moderately broad medial longitudinal black band over all three terga, a pair of oval fuscous shining spots on tergum V; abdominal sterna dark coloured.
Attractant: Cue lure.
Female: No known record.
Distribution:
Type locality: Indonesia (Lombok).
New records: Northern Vietnam (Hoa Binh Province); Thailand. China (Yunnan Province, Jinghong) due to **New Synonymy** of *B. cheni*.
Hosts: No known record.
Remarks: *Bactrocera (Bactrocera) lombokensis* Drew & Hancock, within the *dorsalis* complex, is similar to *Bactrocera (Bactrocera) fernandoi* Tsuruta & White, *Bactrocera (Bactrocera) gombokensis* Drew & Hancock and *Bactrocera (Bactrocera) pedestris* (Bezzi) in possessing broad parallel-sided lateral postsutural vittae, the costal band confluent with R_{2+3} and dark markings on the apices of all femora. It differs from *B. gombokensis* and *B. pedestris* in having the costal band widening slightly across apex of wing (not remaining narrow) and from *B. pedestris* in having the apical dark markings on some femora as small spots (not as dark markings encircling the apices of the femora) and from *B. fernandoi* in having broad lateral dark markings over abdominal terga III–V and small dark preapical spots on the mid and hind femora (*B. fernandoi* possesses a large preapical dark spot on the mid femora and dark encircling the apices of the hind femora). *Bactrocera (Bactrocera) cheni* Zhang, Ji & Chen was separated from *B. lombokensis* on the size of the black spots on the apices of the mid and hind femora. However, these spots are variable within *B. lombokensis* and thus we hereby make *B. cheni* a **New Synonym** of this species. Further, the name *Bactrocera cheni* is preoccupied by *cheni* Chao (1987) which is a synonym of *Bactrocera tsuneonis* (Miyake). As both names are now placed in *Bactrocera*, *cheni* Zhang, Ji & Chen is a **Secondary Homonym** of *cheni* Chao.

Bactrocera (Bactrocera) maculigera Doleschall
(Fig. 103)

Bactrocera maculigera Doleschall, 1858: 122. Syntype male in MFN.
Dacus maculiger – Froggatt, 1910: 30.
Bactrocera maculigera – Hardy, 1973: 54; [Unconfirmed Synonym of *Bactrocera zonata* (Saunders)]; Norrbom *et al*., 1998: 96.

Non-type material examined: INDONESIA: MOLUCCAS (MALUKU) – 10 males, Ambon, Waai,

<20m, 6.vii.1981, coll. J. Turner, attracted to methyl eugenol, residual forest with clove, nutmeg and banana groves; 3 males, Ambon, Paso, 20-50m, 7.vii.1981, coll. J. Turner, attracted to methyl eugenol, residual forest with clove, nutmeg and banana groves; 6 males, Ambon, Hila/Mamua, 25–100m, 5.viii.1981, coll. J. Turner, attracted to methyl eugenol, residual forest with clove, nutmeg and banana groves; 3 males, Ambon, Mamala, 50-150m, 7.viii.1981, coll. J. Turner, attracted to methyl eugenol, residual forest with clove, nutmeg and banana groves; 1 male, Maluku, Middle Maluku, Ambon, Suli, 29.vii.2007, coll. La Naibu, attracted to methyl eugenol in forest; 2 males, Seram, Western Part, Kairatu, Kamarian, 29.vi.2007, coll. A. Haris Tualepe, attracted to methyl eugenol in village; 1 male, Maluku, Western Part, Kairatu, Nuruwe, 6.vii.2007, coll. A. Haris Tualepe, attracted to methyl eugenol in forest; 4 males, Maluku, Western Part, Kairatu, Waimital, 19. vii.2006, coll. A. Harris Tualepe, attracted to methyl eugenol in jeruk; 3 males, Maluku, Western Part, Waimital, 26. vii.2007, coll. A. Haris Tualepe, attracted to methyl eugenol in jeruk; 2 males, Maluku, Seram (Western Part), Waimital, 2.viii.2007, coll. A. Haris Tualepe, attracted to methyl eugenol in jeruk. Specimens in QDPC.

Diagnosis: A large species; face fulvous with a pair of medium-sized circular black spots; postpronotal lobes fuscous; notopleura yellow; scutum uniformly dull black without pale markings and covered with a fine dense golden pubescence; very short and narrow lateral postsutural yellow vittae present; medial postsutural yellow vitta absent; no yellow spot anterior to notopleural suture; mesopleural stripe reaching anterior *npl.* seta dorsally; scutellum yellow (in some specimens there is a small fuscous apical spot between the apical setae); wing with cell c fuscous, cell bc pale fuscous, dense microtrichia covering most of both cells; a broad dark fuscous costal band confluent with R_{4+5}, a broad fuscous band across centre of wing from hind margin but not reaching to vein R_{4+5}; a broad dark fuscous anal streak; abdominal terga III–V orange-brown with broad lateral fuscous to dark fuscous margins on tergum III and broad fuscous lateral margins on tergum IV and a narrow medial longitudinal dark fuscous line on tergum V.

Description: male.

Head: Height 2.2 mm. Frons – length 1.6 times breadth; red-brown with dark fuscous on anteromedial hump and around bases of orbital setae; orbital setae black: 1 *s.or.*, 2 *i.or.*; lunule fuscous. Ocellar triangle black. Vertex red-brown. Face red-brown with a pair of medium-sized circular black spots; length 0.6 mm. Genae fulvous, small dark fuscous subocular spot present; no seta present. Occiput red-brown, fulvous along eye margins; occipital row with 6–9 strong black setae. Antennae with segments 1 and 2 fuscous, segment 3 red-brown with dark fuscous on apex and outer surface; length of segments: 0.24 mm; 0.4 mm; 1.06 mm.

Thorax: Scutum entirely dull black without pale markings and covered with a short golden pubescence. Pleural areas black, tending dark fuscous below postpronotal lobes. Yellow markings as follows: notopleura; mesopleural stripe reaching anterior *npl.* seta dorsally, continuing to katepisternum as a small transverse spot, anterior margin straight; anatergite (posterior apex black); anterior $^3/_5$ katatergite (remainder black); two very short and narrow lateral postsutural vittae ending before the level of the *a.sa.* seta and in most cases these vittae are almost not visible. Medial longitudinal postsutural yellow vitta absent. Postpronotal lobes pale fuscous. Postnotum black. Scutellum yellow except for a narrow- to medium-width black basal band (in some specimens there is a small fuscous spot around the apical setae). Setae: *sc.* 2; *prsc.* absent; *ia.* 1; *p.sa.* 1; *a.sa.* 1; *mpl.* 1; *npl.* 2; *scp.* 4.

Legs: Fore femora fulvous except red-brown around apical $^1/_3$, mid and hind femora fulvous except pale fuscous around apical $^1/_3$ to $^1/_4$; all tibiae and tarsal segments entirely fulvous; mid tibiae each with an apical black spur.

Wings: Length 7.5 mm; cell c fuscous, cell bc pale fuscous; dense microtrichia covering almost all of both cells; remainder of wings colourless except dark fuscous cell sc, broad dark fuscous costal band confluent with R_{4+5} and remaining of uniform width around apex of wing, a broad fuscous band across centre of wing reaching hind margin but not extending forward to R_{4+5}, a broad dark fuscous anal streak; a dense aggregation of microtrichia around A_1+CuA_2; supernumerary lobe of medium development.

Abdomen: Oval; terga free; pecten present on tergum III. Tergum I and sterna I and II wider than long. Tergum I dark fuscous to black except for a narrow transverse red-brown band across posterior margin but not reaching lateral margins; tergum II dark fuscous to black with fulvous across posterior $^1/_2$; terga III–V orange-brown except for broad dark fuscous to fuscous lateral margins on tergum III and broad fuscous lateral margins on tergum IV and a narrow medial longitudinal dark fuscous line on tergum V. A pair of oval dark fuscous shining spots on tergum V. Posterior lobe of surstylus short, sternum V with a deep concavity on posterior margin. Abdominal sterna dark fuscous to black.

Attractant: Methyl eugenol.

Female: No known record.

Distribution:

Type locality: Indonesia (Moluccas = Maluku – Ambon I.).

New records: Indonesia (Moluccas = Maluku, Seram I.).

Hosts: No known record.

Remarks: *Bactrocera (Bactrocera) maculigera* Doleschall is a unique species and similar to *Bactrocera (Bactrocera) beckerae* (Hardy) in possessing a broad costal band confluent with R_{4+5}, cells bc and c coloured, a black scutum and abdominal terga III–V orange-brown without a black ‘T’ pattern. It is also similar to *Bactrocera (Bactrocera) fulvicauda* (Perkins) from Papua New Guinea in these characters and in possessing a fuscous pattern across the centre of the wing. It differs from both of these species in having fuscous postpronotal lobes, very narrow and short lateral postsutural vittae that are almost not discernable and a broad fuscous band across wing that does not connect with the costal band. Abdominal terga III–V in *B. fulvicauda* are black, not orange-brown with dark lateral margins, as in *B. maculigera*.

Doleschall (1858) did not specify the number of specimens he had before him to do the description of

B. maculigera. Consequently, we regard the specimen in MFN as a syntype.

In CABIKEY, White & Hancock (1997) reported characters from this syntype which has the head missing. We are confident that the specimens studied herein are conspecific with the type.

Bactrocera (Bactrocera) makilingensis Drew & Hancock

(Fig. 104)

Bactrocera (Bactrocera) makilingensis Drew & Hancock, 1994a: 36; Norrbom *et al.*, 1998: 92. Holotype male in BBM.

Non-type material examined: Since the publication of Drew & Hancock (1994a), no new specimens have been studied.

Diagnosis: Face fulvous with a pair of medium-sized circular black spots; postpronotal lobes and notopleura yellow; scutum black except dark brown below lateral postsutural vittae; broad lateral postsutural yellow vittae narrowing slightly posteriorly to end at *ia.* seta; medial postsutural yellow vitta absent; mesopleural stripe equal in width to anterior *npl.* seta before cutting back dorsally to the width of the notopleuron; scutellum yellow; legs with femora fulvous with small- to medium-sized preapical dark spots on outer surfaces of fore femora and inner surfaces of mid and hind femora; fore tibiae fuscous, mid tibiae fulvous tending fuscous basally, hind tibiae dark fuscous; wing with cells bc and c colourless, microtrichia in outer corner of cell c only; a narrow fuscous costal band overlapping R_{2+3} where it becomes paler and expanding into a small spot around apex of R_{4+5}; a narrow fuscous anal streak; supernumerary lobe of medium development; abdominal terga III–V orange-brown except for a narrow transverse black band across anterior margin of tergum III which expands into broad lateral margins, anterolateral corners of tergum IV black and of tergum V fuscous to dark fuscous, a broad medial longitudinal black band over all three terga, a pair of oval pale fuscous to fuscous shining spots on tergum V; abdominal sterna dark coloured.

Attractant: Cue lure.

Female: No known record.

Distribution:

Type locality: Philippines (Luzon I., Mt Makiling).

Hosts: No known record.

Remarks: *Bactrocera (Bactrocera) makilingensis* Drew & Hancock, within the *dorsalis* complex, is similar to *Bactrocera (Bactrocera) collita* Drew & Hancock, *Bactrocera (Bactrocera) dorsaloides* (Hardy & Adachi) and *Bactrocera (Bactrocera) thailandica* Drew & Hancock in possessing broad lateral postsutural vittae that narrow slightly posteriorly, dark markings on the apices of all femora, scutellum yellow with a narrow basal band and distinct dark lateral colour patterns on abdominal terga III–V. It differs from *B. thailandica* in possessing narrow- to medium-width lateral dark colour patterns on abdominal terga III–V, from *B. dorsaloides* in having distinct dark preapical spots on the apices of all femora (not dark patterns encircling the apices of the femora) and from *B. collita* in having the lateral postsutural vittae ending at the *ia.* seta (not before the seta), abdominal terga IV and V with relatively narrow dark lateral colour patterns and the costal band with a slight expansion across the apex of R_{4+5}.

Bactrocera (Bactrocera) malaysiensis Drew & Hancock

(Fig. 105)

Bactrocera (Bactrocera) malaysiensis Drew & Hancock, 1994a: 37; Norrbom *et al.*, 1998: 92. Holotype male in BMNH.

Non-type material examined: No new specimens available for study since the publication of Drew & Hancock (1994a).

Diagnosis: Face fulvous with a pair of small- to medium-sized elongate oval black spots; postpronotal lobes and notopleura yellow; scutum black except dark brown behind lateral postsutural vittae; broad parallel-sided lateral postsutural yellow vittae ending at or behind *ia.* seta; medial postsutural yellow vitta absent; mesopleural stripe reaching midway between anterior margin of notopleuron and anterior *npl.* seta dorsally; scutellum yellow; legs with femora fulvous with a moderately large preapical oval black spot on outer surfaces of fore femora and small elongate preapical black spots on inner surfaces of mid and hind femora; fore and hind tibiae dark fuscous, mid tibiae fulvous tending darker basally; wing with cells bc and c colourless, microtrichia in outer corner of cell c only; a narrow dark fuscous costal band confluent with R_{2+3} and widening uniformly across apex of wing; a narrow fuscous anal streak; supernumerary lobe of medium development; abdominal terga III–V orange-brown except for a narrow transverse black band across anterior margin of tergum III which widens laterally to form narrow lateral margins, a moderately broad medial longitudinal black band over all three terga, anterolateral corners of terga IV and V black, a pair of oval fuscous to black shining spots on tergum V; abdominal sterna dark coloured.

Attractant: Cue lure.

Female: No known record.

Distribution:

Type locality: Peninsular Malaysia.

Hosts: No known record.

Remarks: *Bactrocera (Bactrocera) malaysiensis* Drew & Hancock, within the *dorsalis* complex, is similar to *Bactrocera (Bactrocera) affinidorsalis* (Hardy), *Bactrocera (Bactrocera) latilineola* Drew & Hancock and *Bactrocera (Bactrocera) unimacula* Drew & Hancock in possessing broad parallel-sided lateral postsutural vittae and the costal band confluent with R_{2+3} and widening markedly across apex of wing. It differs from *B. unimacula* in having a gradual widening of the costal band across of apex of wing (not as a distinct large spot) and from the other species in having small preapical dark spots on all femora and from *B. affinidorsalis* in having the lateral postsutural vittae ending at or behind the *ia.* seta.

Bactrocera (Bactrocera) mamaliae Drew & Romig, new species

(Fig. 106)

Material examined: Holotype male, **INDONESIA: MOLUCCAS (MALUKU): AMBON ISLAND** – Mamala, Ambon, Indonesia, 50–150 m, 7.viii.1981, coll. J. Turner, attracted to cue lure, in residual forest with clove, nutmeg and banana groves, deposited in BMNH.
Paratype: **INDONESIA: MOLUCCAS (MALUKU): AMBON ISLAND** – 1 male, same data as holotype, deposited in QDPC.
Diagnosis: A large species; face fulvous with a pair of medium-sized subquadrate black spots; postpronotal lobes fuscous; notopleura yellow; scutum entirely red-brown to pale fuscous; narrow and short lateral postsutural yellow vittae present; medial postsutural yellow vitta absent; no yellow spot anterior to notopleural suture; mesopleural stripe reaching midway between anterior margin of notopleuron and anterior *npl.* seta dorsally; scutellum yellow; wing with cells bc and c fuscous, microtrichia in outer corner of cell c only; a broad dark fuscous costal band confluent with R_{4+5}; a broad fuscous anal streak; abdominal terga III–V red-brown with a 'T' pattern consisting of a narrow transverse dark fuscous to black band across anterior margin of tergum III which widens over lateral margins and a medium-width medial longitudinal dark fuscous to black band over all three terga.
Description: male.
Head: Height 1.9 mm. Frons – length 1.68 times breadth; red-brown, fulvous along lateral margins and with fuscous on anteromedial hump and around bases of orbital setae; orbital setae black: 1 *s.or.*, 2 *i.or.*; lunule fuscous. Ocellar triangle black. Vertex fuscous. Face red-brown with a pair of medium-sized subquadrate black spots; length 0.58 mm. Genae red-brown, large dark fuscous subocular spot present; black seta present. Occiput red-brown, fulvous along eye margins; occipital row with 8–10 strong black setae. Antennae with segments 1 and 2 red-brown, segment 3 red-brown with fuscous on apex and outer surface; length of segments: 0.24 mm; 0.42 mm; 0.93 mm.
Thorax: Scutum entirely red-brown to pale fuscous without dark markings. Pleural areas entirely red-brown with fuscous along anterior margin of mesopleural stripe. Yellow markings as follows: notopleura; mesopleural stripe reaching midway between anterior margin of notopleuron and anterior *npl.* seta dorsally, not continuing to katepisternum, anterior margin straight; anatergite (posterior apex red-brown); anterior ⅔ katatergite (remainder red-brown); two very narrow lateral postsutural vittae ending before *ia.* seta (these vittae might be short and pale and almost indistinguishable). Medial longitudinal postsutural yellow vitta absent. Postpronotal lobes fuscous. Postnotum red-brown. Scutellum yellow except for narrow red-brown basal band. Setae: *sc.* 2; *prsc.* 2; *ia.* 1; *p.sa.* 1; *a.sa.* 1; *mpl.* 1; *npl.* 2; *scp.* 4.
Legs: All femora red-brown, tending fulvous basally; all tibiae red-brown; all tarsal segments fulvous; mid tibiae each with an apical black spur.
Wings: Length 7.3 mm; cells bc and c fuscous; microtrichia in outer corner of cell c only; remainder of wings with a very pale tint, dark fuscous cell sc, broad dark fuscous costal band confluent with R_{4+5} and becoming pale adjacent to that vein, a broad fuscous anal streak; a dense aggregation of microtrichia around A_1+CuA_2; supernumerary lobe of medium development.
Abdomen: Oval; terga free; pecten present on tergum III. Tergum I and sterna I and II wider than long. Tergum I entirely dark fuscous; tergum II red-brown with a narrow transverse dark fuscous band subanteriorly; terga III–V red-brown with a dark 'T' pattern consisting of a narrow transverse dark fuscous to black band across anterior margin of tergum III and widening over the lateral margins of that tergum and a narrow- to medium-width medial longitudinal dark fuscous to black band over all three terga. A pair of oval red-brown shining spots on tergum V. Posterior lobe of surstylus short, sternum V with a deep concavity on posterior margin. Abdominal sterna red-brown.
Attractant: Cue lure.
Female: No known record.
Distribution: Indonesia (Moluccas = Maluku – Ambon I., Mamala).
Hosts: No known record.
Remarks: *Bactrocera (Bactrocera) mamaliae* n.sp. is similar to *Bactrocera (Bactrocera) flavipennis* (Hardy), *Bactrocera (Bactrocera) patula* n.sp., *Bactrocera (Bactrocera) fuscoformosa* n.sp. and *Bactrocera (Bactrocera) ochroventer* n.sp. in possessing a broad costal band confluent with R_{4+5}, a red-brown scutum and femora mostly fulvous. It differs from these species in having fuscous postpronotal lobes.

Bactrocera (Bactrocera) mediorufula Drew & Romig, new species

(Fig. 107)

Material examined: Holotype male, **INDONESIA: KALIMANTAN** – East Kalimantan, Kutai Kertanegara, Batuah, 15.viii.2005, coll. Heru G & Ertha, attracted to methyl eugenol in forest, deposited in BMNH.
Paratypes: **INDONESIA: KALIMANTAN** – 3 males, same data as holotype; 1 male (1.viii.2005), 12 males (8.viii.2005), East Kalimantan, Kutai Kertanegara, Batuah, coll. Heru G & Ertha, attracted to methyl eugenol in forest; 1 male, Kalimantan Selatan, Banjar Baru, Mandiangin, 12.ii.2006, coll. Bilal, attracted to methyl eugenol in forest. **EAST MALAYSIA: SARAWAK** – 1 male, Kpg Tema Mawang, Tebedu, KT5, 18.viii.1992, coll. Leong, attracted to methyl eugenol; 1 male (28.v.1992), 3 males (5.vi.1992), 1 male (23.vi.1992), Kpg Tema Mawang, Tebedu, Nr Kuching, coll. S. Leong, attracted to methyl eugenol; 2 males, E. Malaysia, Tarat Agriculture Station, Nr Kuching, 19.v.1992, coll. S. Leong, attracted to methyl eugenol; 1 male (22.iv.1992), 1 male (10.vi.1992), E. Malaysia, Ranchang Pool, Serian,

Nr Kuching, coll. S. Leong, attracted to methyl eugenol. **PHILIPPINES**: 1 male, Sta Maria, Tarlac, Philippines, 18. vi.1983, coll. B. Ramos, attracted to cue lure; 2 males, Hantik, Antique, Philippines, 17.vi.1983, coll. A.M. Tenorio, attracted to methyl eugenol. 4 paratypes in ANIC, 4 paratypes in BMNH, 15 paratypes in QDPC, 7 paratypes in QMIC.

Non-type material examined: PENINSULAR MALAYSIA: 3 males and 3 females, Paso Stn., Prien Jempol, Malaysia, 12.xi.1986, coll. S. Vijaysegaran *et al.*, bred ex Sample No. 0885, no details supplied. **THAILAND**: 5 males and 13 females, Bansut Tha, Songkhla, 13.viii.1992, coll. Anuchit *et al.*, bred from *Brucea javanica*; 2 males and 4 females, Khlongtaklaeng, Thataklab, Chachoengsao, 31.x.1992, coll. Prasert; 2 females, Hat Yai, Songkhla, 24.x.1992, coll. Surakrai *et al.*, bred from *Brucea javanica*. Specimens in QDPC.

Diagnosis: A medium-sized species; face red-brown with a pair of small elongate oval black spots; postpronotal lobes and notopleura yellow; scutum black; lateral postsutural yellow vittae present; medial postsutural yellow vitta absent; no yellow spot anterior to notopleural suture; mesopleural stripe slightly wider than notopleuron dorsally; scutellum yellow; wing with cells bc and c fuscous, dense microtrichia covering all of cell c; a narrow dark fuscous costal band slightly overlapping R_{2+3} where it becomes paler; a broad dark fuscous anal streak; abdominal terga III–V orange-brown with a broad medial and two moderately broad lateral longitudinal black bands joined along anterior margin of tergum III.

Description: male.

Head: Height 1.17 mm. Frons – length 1.59 times breadth; red-brown without dark markings; orbital setae black: 1 *s.or.*, 2 *i.or.*; lunule red-brown. Ocellar triangle black. Vertex red-brown. Face red-brown with a pair of small elongate oval black spots; length 0.3 mm. Genae fulvous, large fuscous subocular spot present; black seta present. Occiput black, fulvous along eye margins; occipital row with 3–5 black setae. Antennae with segments 1 and 2 red-brown, segment 3 red-brown with dark fuscous on apex and outer surface; length of segments: 0.12 mm; 0.26 mm; 0.6 mm.

Thorax: Scutum black with red-brown on extreme lateral margins below lateral postsutural vittae, behind lateral postsutural vittae and around notopleural suture and a broad medial longitudinal red-brown band over centre. Pleural areas black without pale markings. Yellow markings as follows: postpronotal lobes; notopleura; mesopleural stripe slightly wider than notopleuron dorsally, continuing to katepisternum as a small transverse spot, anterior margin slightly convex; anatergite (posterior apex black); anterior $^3/_5$ katatergite (remainder black); two moderately broad parallel-sided lateral postsutural vittae ending at *ia.* seta. Medial longitudinal postsutural yellow vitta absent. Postnotum black. Scutellum yellow except for narrow black basal band. Setae: *sc.* 2; *prsc.* 2; *ia.* 1; *p.sa.* 1; *a.sa.* 1; *mpl.* 1; *npl.* 2; *scp.* 4.

Legs: All femora entirely fulvous; fore tibiae fuscous to dark fuscous, mid tibiae fuscous basally to fulvous apically, hind tibiae dark fuscous; fore and hind tarsi with basal segment fulvous and apical four segments red-brown, mid tarsal segments entirely fulvous; mid tibiae each with an apical black spur.

Wings: Length 4.2 mm; cells bc and c fuscous; dense microtrichia covering all of cell c; remainder of wings colourless except dark fuscous cell sc, narrow dark fuscous costal band slightly overlapping R_{2+3} where it is pale fuscous and widening slightly across apex of R_{4+5}, a broad dark fuscous anal streak; dense aggregation of microtrichia around A_1+CuA_2; supernumerary lobe of medium development.

Abdomen: Oval; terga free; pecten present on tergum III. Tergum I and sterna I and II wider than long. Tergum I entirely black; tergum II orange-brown with a narrow transverse black band subanteriorly and not reaching narrow lateral black margins; terga III–V orange-brown with a broad medial and moderately broad lateral longitudinal black bands over all three terga and joined along anterior margin of tergum III. A pair of oval orange-brown shining spots on tergum V. Posterior lobe of surstylus short, sternum V with a deep concavity on posterior margin. Abdominal sterna dark fuscous to black.

Attractant: Methyl eugenol.

Female: No known record.

Distribution: Indonesia (East Kalimantan); East Malaysia (Sarawak); Peninsular Malaysia; Philippines; Thailand.

Hosts: Recorded from two hosts – *Brucea javanica* (family *Simaroubaceae*) and unknown host – 0885 (Malaysia).

Remarks: *Bactrocera (Bactrocera) mediorufula* n.sp. is similar to *Bactrocera (Bactrocera) pusilla* (Hardy) in possessing a continuous costal band confluent with or just overlapping R_{2+3}, a black scutum, cells bc and c fuscous and covered with dense microtrichia but differs from this species in lacking a lateral postsutural yellow band connecting the postpronotal lobes and notopleuron.

Bactrocera (Bactrocera) megaspilus (Hardy)

(Fig. 108)

Dacus megaspilus Hardy, 1982: 232. Holotype male in MZB.

Bactrocera (Bactrocera) megaspilus – Norrbom *et al.*, 1998: 92.

Type material examined: Holotype male, **INDONESIA: SULAWESI** – Central Sulawesi, Wotu, Aug. 1975, coll. N. Gough. Studied at MZB.

Non-type material examined: INDONESIA: SOUTH SULAWESI – 1 male (27.viii.2006), 1 male (13.ix.2006), South Sulawesi, Maros, Limampoccoe, coll. Sirajuddin, attracted to cue lure in forest; 5 males, Palopo-Rantepao Rd., 16-23km Palopo, 250-500m, 24.vii.1981, coll. J. Turner, Nanggala III forest, attracted to cue lure; 1 male, Lamona, 12km S Palopo, 20m, 23.vii.1981, coll. J. Turner, Primary Rattan forest, attracted to cue lure. Specimens in QDPC.

Diagnosis: A large species; face fulvous with a pair of medium-sized sub-quadrate black spots; postpronotal lobes and notopleura yellow; scutum red-brown to fuscous with large black markings that diffuse into base colour; lateral postsutural yellow vittae present; medial postsutural yellow vitta absent; no yellow spot anterior to notopleural suture; mesopleural stripe equal in width to notopleuron dorsally; scutellum yellow; wing with cells bc and c with a pale fulvous tint, microtrichia in outer corner of cell c only; a broad fulvous costal band confluent with R_{4+5} and expanding into a large fuscous apical spot which overlaps M; a moderately broad pale fuscous anal streak; abdominal terga III–V fulvous with a narrow transverse black band across anterior margin of tergum III.

Description: male.

Head: Height 2.1 mm. Frons – length 1.4 times breadth; fulvous with dark fuscous around orbital setae; orbital setae black: 1 *s.or.*, 3 *i.or.*; lunule pale fuscous. Ocellar triangle black. Vertex dark fuscous. Face fulvous with a pair of medium-sized sub-quadrate-shaped black spots; length 0.65 mm. Occiput pale red-brown, fulvous along eye margins. Antennae with segments 1 and 2 red-brown, segment 3 red-brown with pale fuscous on apex and outer surface (taken from MZB specimen); length of segments: 0.25 mm; 0.37 mm; third segment missing on non-type material.

Thorax: Scutum red-brown to fuscous with moderately broad longitudinal black bands inside lateral postsutural vittae and large black spots inside postpronotal lobes. Yellow markings as follows: postpronotal lobes; notopleura; mesopleural stripe equal in width to notopleuron dorsally; two moderately broad lateral postsutural vittae narrowing posteriorly to end at *ia.* seta. Medial longitudinal postsutural yellow vitta absent. Scutellum yellow except for narrow black basal band. Setae: *sc.* 2; *prsc.* 2; *ia.* 1; *p.sa.* 1; *a.sa.* 1; *mpl.* 1; *npl.* 2; *scp.* 4.

Legs: All segments entirely fulvous; mid tibiae each with an apical black spur.

Wings: Length 8.2 mm; cells bc and c with a pale fulvous tint; microtrichia in outer corner of cell c only; remainder of wings colourless except fulvous cell sc, broad fulvous costal band confluent with R_{4+5} (paler between R_{2+3} and R_{4+5}) and expanding into a large fuscous apical spot which overlaps M, a moderately broad pale fuscous anal streak; a dense aggregation of microtrichia around A_1+CuA_2; supernumerary lobe of medium development.

Abdomen: Elongate oval; terga free; pecten present on tergum III. Tergum I and sterna I and II wider than long. Tergum I red-brown with dark fuscous anterolateral corners; tergum II red-brown anteriorly, fulvous posteriorly, with a narrow transverse black band across anterior margin but not reaching lateral margins; terga III–V fulvous with a narrow transverse black band across anterior margin of tergum III. A pair of oval red-brown shining spots on tergum V. Posterior lobe of surstylus short, sternum V with a deep concavity on posterior margin.

Attractant: Cue lure.

Female: No known record.

Distribution:

Type locality: Indonesia (Sulawesi – Central Sulawesi, Wotu).

New records: Indonesia (other locations in South Sulawesi listed under 'Non-type material examined').

Hosts: No known record.

Remarks: *Bactrocera (Bactrocera) megaspilus* (Hardy) is similar to *Bactrocera (Bactrocera) dispar* (Hardy) and *Bactrocera (Bactrocera) flavipennis* (Hardy) in possessing a broad costal band confluent with R_{4+5} and a generally red-brown scutum but differs from these species in having a large fuscous spot across the apex of the wing and overlapping M.

Bactrocera (Bactrocera) melastomatos Drew & Hancock

(Fig. 109)

Bactrocera (Bactrocera) melastomatos Drew & Hancock, 1994a: 38; Norrbom *et al.*, 1998: 92. Holotype female in BMNH.

Non-type material examined: Since the publication of Drew & Hancock (1994a), the following specimens have been studied – **INDONESIA: JAVA** – 2 males (14.viii.2005), 2 males (21.viii.2005), Central Java, Pemalang, Surajaya, coll. Sapto Darmono, attracted to cue lure in forest; 12 males (5.vii.2005), 21 males (20.vii.2005), 11 males (2.viii.2005), D.I. Yogyakarta, Sleman, Kaliurang, coll. Agus W, attracted to cue lure in forest; 6 males (8.viii.2005), 4 males (15.viii.2005), 3 males (5.ix.2005), Banten, Lebak, Pasir Nini Nini, coll. Aden Haer, attracted to cue lure in forest; **KALIMANTAN** – 1 male (15.viii.2005), 11 males (22.viii.2005), 13 males (29.viii.2005), 1 male (24.x.2005), 8 males (28.xi.2005), 4 males (5.xii.2005), East Kalimantan, Kutai Kartanegara, Batuah, coll. Ertha D.P., attracted to cue lure in banana; 8 males (1.viii.2005), 4 males (5.ix.2005), East Kalimantan, Kutai Kartanegara, Batuah, coll. Heru G., attracted to cue lure in forest; 2 males (21.vii.2005), 1 male (2.xii.2005), South Kalimantan, Banjarbaru, Guntung Payung, coll. Sidrotul & Sutimi, attracted to cue lure in village; 11 males (13.xi.2005), 6 males (20.xi.2005), 15 males (4.xii.2005), 16 males (11.xii.2005), South Kalimantan, Banjar Karang Intan, Mandiangin, coll. Bilal, attracted to cue lure in forest; 1 male (19.vii.2005), 1 male (9.viii.2005), 1 male (23.viii.2005), West Kalimantan, Pontianak, Sei Pinyuh, coll. A. Irom, attracted to cue lure in village mango; 2 males (26.vii.2005), 3 males (2.viii.2005), West Kalimantan, Pontianak, Sei Duri II, coll. A. Irom, attracted to cue lure in citrus; 1 male (3.viii.2005), 1 male (27.vii.2005), West Kalimantan, Pontianak, Pancaroba, coll. A. Irom, attracted to cue lure in forest; 2 males (26.vii.2005), 4 males (9.viii.2005), 2 males (16.viii.2005), Central Kalimantan, Kota Waringin Timur, Kota Besi, coll. Utik D & Kusyono; **SUMATRA** – 11 males (16.viii.2006), 6 males (23.viii.2006), Riau, Kampar, Kuala, coll. Nasrun, attracted to cue lure; 41 males, South Sumatra, Ogan Komering Llir, Celikal, 4.ix.2006, coll. Imam Muhayani, attracted to cue lure in Duku; 2 females, Lampung, Lampung Utara, Bangun Sari, 6.ix.2006, coll. Suhenda Arsol, attracted to cue lure in forest. Specimens in QDPC.

Diagnosis: Face fulvous with a pair of moderately large elongate oval black spots; postpronotal lobes and notopleura yellow; scutum black with, at most, dark fuscous behind lateral postsutural vittae and around notopleural suture; narrow- to medium-width lateral postsutural yellow vittae narrowing posteriorly to end before *ia.* seta; medial postsutural yellow vitta absent; mesopleural stripe equal in width to notopleuron dorsally; scutellum yellow; legs with femora fulvous with a large fuscous to black spot on outer apical surfaces of fore femora (in some specimens this spot is paler and less distinct); fore and hind tibiae dark fuscous to black, mid tibiae dark fuscous tending paler apically; wing with cells bc and c colourless, microtrichia in outer corner of cell c only; a narrow dark fuscous costal band distinctly overlapping R_{2+3} and expanding around apex of R_{4+5}; a narrow dark fuscous anal streak; supernumerary lobe of medium development; abdominal terga III–V dark orange-brown with a medium-width medial longitudinal dark fuscous band and broad lateral longitudinal fuscous to dark fuscous bands over all three terga and joined along anterior ⅓ to ½ of tergum III (the broad lateral longitudinal dark fuscous bands tend to be diffuse and not distinctly defined on the inner margins), a pair of oval fuscous to dark fuscous shining spots on tergum V; abdominal sterna dark coloured.

Attractant: Cue lure.

Female: As for male in general body colour patterns. See Drew & Hancock (1994a) for a detailed description of female characters.

Distribution:

Type locality: Peninsular Malaysia (Slim River).

Previously recorded: Type locality and the following locations – India (Andaman Is.); Singapore; Thailand (Norrbom *et al.*, 1998).

New records: Indonesia (Java; Kalimantan; Sumatra).

Hosts: *Melastoma malabathrica* and *Melastoma sanguineum* (family *Melastomataceae*) (Allwood *et al.*, 1999).

Remarks: *Bactrocera (Bactrocera) melastomatos* Drew & Hancock, within the *dorsalis* complex, is similar to *Bactrocera (Bactrocera) fulvifemur* Drew & Hancock and *Bactrocera (Bactrocera) usitata* Drew & Hancock in possessing lateral postsutural vittae that narrow posteriorly, femora with, at most, dark spots on outer apical surfaces of fore femora and costal band extending well over R_{2+3}. It differs from these species in possessing diffuse broad lateral dark margins on abdominal terga III–V and longer more elongate lateral postsutural vittae (not short and triangular). Hardy (1973) studied specimens reared from *Melastoma malabathrica* collected at Kuala Lumpur, Malaysia, in 1958 by N.L.H. Krauss. He misidentified these specimens which we now call *B. melastomatos* as *Bactrocera occipitalis* (Bezzi) which is a major pest species and only occurs in the Philippines, Kalimantan, Brunei and East Malaysia.

Bactrocera (Bactrocera) merapiensis Drew & Hancock

(Fig. 110)

Bactrocera (Bactrocera) merapiensis Drew & Hancock, 1994a: 40; Norrbom *et al.*, 1998: 92. Holotype male in BBM.

Non-type material examined: No new specimens available for study since the publication of Drew & Hancock (1994a).

Diagnosis: Face fulvous with a pair of small oval black spots; postpronotal lobes and notopleura yellow; scutum entirely black without pale markings; lateral postsutural yellow vittae of medium-width and narrowing posteriorly to end before *ia.* seta; medial postsutural yellow vitta absent; mesopleural stripe slightly wider than notopleuron dorsally; scutellum yellow; legs with femora entirely fulvous; fore tibiae fuscous, mid tibiae fulvous (tending darker basally), hind tibiae dark fuscous; wing with cells bc and c colourless, microtrichia in outer corner of cell c only; a narrow fuscous costal band confluent with R_{2+3} and widening slightly around apex of wing; a narrow fuscous anal streak; supernumerary lobe of medium development; abdominal tergum III entirely dark fuscous to black, terga IV and V pale orange-brown except for broad lateral dark fuscous to black margins on tergum IV joined along anterior margin of this tergum, anterolateral corners of tergum V dark fuscous to black, a moderately broad medial longitudinal dark fuscous to black band over both terga, a pair of oval fuscous to dark fuscous shining spots on tergum V; abdominal sterna dark coloured.

Attractant: Cue lure.

Female: No known record.

Distribution:

Type locality: Indonesia (Java, Mt Merapi).

Previously recorded: Type locality and the following locations – Southern India (Kerala) (Drew & Raghu, 2002); Indonesia (Sumatra) (Norrbom *et al.*, 1998); India (Andaman Is.) (David & Ramani, 2011).

Hosts: No known record.

Remarks: *Bactrocera (Bactrocera) merapiensis* Drew & Hancock, within the *dorsalis* complex, is similar to *Bactrocera (Bactrocera) cibodasae* Drew & Hancock, *Bactrocera (Bactrocera) osbeckiae* Drew & Hancock, *Bactrocera (Bactrocera) sumbawaensis* Drew & Hancock and *Bactrocera (Bactrocera) thailandica* Drew & Hancock in possessing lateral postsutural vittae that narrow posteriorly, femora entirely fulvous or with, at most, a dark spot subapically on fore femora, costal band confluent with R_{2+3} and distinct dark lateral markings on abdominal terga III–V. It differs from *B. cibodasae* in having a more uniform costal band and broader medial and lateral longitudinal dark bands on abdominal terga III–V, from *B. sumbawaensis* in possessing elongate lateral postsutural vittae (not short and triangular), from *B. osbeckiae* in having the scutum entirely black without pale markings and from *B. thailandica* in having abdominal tergum III entirely dark fuscous to black and broad lateral dark margins on terga IV and V.

Bactrocera (Bactrocera) minuscula Drew & Hancock

(Fig. 111)

Bactrocera (Bactrocera) minuscula Drew & Hancock, 1994a: 41. Holotype male in BBM.

Bactrocera (Bactrocera) miniscula – Norrbom *et al.*, 1998: 92 [Misspelt].

Non-type material examined: Since the publication of Drew & Hancock (1994a), the following specimens have been studied – **INDONESIA: BALI** – 1 male, Buleleng,

Depeha, 4.viii.2006, coll. Ketut Rawa W, attracted to methyl eugenol in mango; 1 male (13.viii.2006), 1 male (8.x.2006), 3 males (22.x.2006), Buleleng, Singaraja, Kaliuntu, coll. Dewa dan Eny, attracted to methyl eugenol in village. Specimens in QDPC.

Diagnosis: Face fulvous with a pair of small circular black spots; postpronotal lobes and notopleura yellow; scutum black except dark brown behind lateral postsutural vittae and, in some specimens, below these vittae, around notopleural suture and inside postpronotal lobes; narrow lateral postsutural yellow vittae tapering posteriorly to end before *ia.* seta; medial postsutural yellow vitta absent; mesopleural stripe slightly wider than notopleuron dorsally; scutellum yellow; legs with femora entirely fulvous; fore and mid tibiae fulvous and hind tibiae fuscous; wing with cells bc and c colourless, microtrichia in outer corner of cell c only; a narrow fuscous costal band confluent with R_{2+3} and very narrow beyond apex of this vein to end between extremities of R_{4+5} and M; a very narrow fuscous to pale fuscous anal streak; supernumerary lobe of medium development; abdominal terga III–V orange-brown except lateral ⅓ of each side of tergum III black and sometimes joined by a narrow band along anterior margin of this tergum, a narrow medial longitudinal black band over all three terga (in some specimens the anterolateral corners of terga IV and V are dark fuscous to black), a pair of oval red-brown to pale fuscous shining spots on tergum V; abdominal sterna dark coloured.

Attractant: Methyl eugenol.

Female: No known record.

Distribution:

Type locality: Indonesia (West Timor, Kupang).

New records: Indonesia (Bali).

Hosts: No known record.

Remarks: *Bactrocera (Bactrocera) minuscula* Drew & Hancock, within the *dorsalis* complex, is similar to *Bactrocera (Bactrocera) cibodasae* Drew & Hancock, *Bactrocera (Bactrocera) indonesiae* Drew & Hancock and *Bactrocera (Bactrocera) floresiae* Drew & Hancock in possessing lateral postsutural vittae that narrow posteriorly, femora entirely fulvous or with, at most, a dark preapical spot on outer surfaces of fore femora and costal band confluent with R_{2+3}. It differs from *B. indonesiae* in having distinct dark patterns on abdominal terga III–V and from *B. cibodasae* and from *B. floresiae* in the unique abdominal colour patterns with dark lateral margins on abdominal tergum III and dark patterns absent on abdominal terga IV and V.

Bactrocera (Bactrocera) moluccensis (Perkins)

(Fig. 112)

Strumeta moluccensis Perkins, 1939: 17. Holotype female in BMNH.

Dacus (Strumeta) moluccensis – Hardy, 1970: 120; Drew, 1974: 60.

Bactrocera (Bactrocera) moluccensis – Drew, 1989: 145; Norrbom *et al.*, 1998: 92.

Non-type material examined: Since the publication of Drew (1989), the following specimens have been studied – **INDONESIA: BALI** – 1 male, Sanur Beach, 18-20.vii.1977, coll. J.T. Medler, attracted to cue lure; 1 male, Legian Beach (rural), S.W. Coast, 12-14.viii.1978, coll. J. Turner, attracted to cue lure; 1 male, Sanur Beach (urban), S.E. Coast, 15-18. viii.1978, coll. J. Turner, attracted to cue lure; **JAVA** – 8 males (12.vii.2005), 3 males (2.viii.2005), 1 male (16. viii.2005), East Java, Sidorajo, Gelang, coll. Agus Riadi, attracted to cue lure in village; 6 males, East Java, Sidoarjo, Gelang, 19.vii.2005, coll. Trigiyanto, attracted to cue lure in village; 2 males (16.vii.2005), 2 males (7.viii.2005), West Java, Indramayu, Dermayu, coll. Rifki K, attracted to cue lure in village; 1 male, West Java, Cirebon, Pelangon, 4.viii.2005, coll. Herty, attracted to cue lure in forest; 3 males, Kebun Raya, Bogor, 300m, 17.v.1981, coll. J. Turner, attracted to cue lure; **LOMBOK** – Sembolia & Belanting, 10-50m, 1.vi.1981, coll. J. Turner, Primary sclerophyll forest, attracted to cue lure; 1 male, Suranadi, W. Lombok, 4.vi.1978, coll. J.T. Medler, attracted to cue lure; 1 male, Mataram, April 1988, coll. J. Turner, attracted to cue lure; **MOLUCCAS (MALUKU): AMBON ISLAND** – 1 female with the following label data – 'Dr Doleschal, 1859, Amboina'; '*Strumeta (Chaetodacus) ferruginea* F. det. M. Hering, 1938', studied at NMW: This specimen was misidentified as *S. ferruginea*; 4 males, Soya, 300m, 4.viii.1981, coll. J. Turner, attracted to cue lure; 7 males, Hila/Mamua, 25–100m, 5.viii.1981, coll. J. Turner, Residual forest with clove, nutmeg & banana groves, attracted to cue lure; 9 males, Waai, <20m, 6.vii.1981, coll. J. Turner, Residual forest with clove, nutmeg & banana groves, attracted to cue lure; 5 males, Paso, 20-50m, 7.vii.1981, coll. J. Turner, Residual rainforest, attracted to cue lure; 9 males, Mamala, 50-150m, 7.viii.1981, coll. J. Turner, Residual forest with clove, nutmeg & banana groves, attracted to cue lure; 13 males, Maluku, Middle Maluku, Suli, 9.vii.2007, coll. La Naibu, attracted to cue lure in jambu; **NORTH MOLUCCAS (NORTH MALUKU)** – 3 males (29.v.2007), 3 males (5. vi.2007), 2 males (12.vi.2007), North Maluku, West Halmahera Goal, coll. Raais Abdullah, attracted to cue lure in citrus, chilli (G31a2); **CENTRAL SULAWESI** – 2 males (10.ix.2006), 2 males (23.ix.2006), Sulawesi Tengah, Donggala, Nupabomba, coll. Rusdin, attracted to cue lure in forest. Specimens in QDPC.

Diagnosis: Face fulvous with a pair of medium-sized oval black spots; postpronotal lobes and notopleura yellow; scutum red-brown with fuscous spots between postpronotal lobes and notopleura and narrow longitudinal fuscous lines laterally and medially; broad parallel-sided lateral postsutural yellow vittae ending at or just behind *ia.* seta; medial postsutural yellow vitta absent; mesopleural stripe reaching midway between anterior margin of notopleuron and anterior *npl.* seta dorsally; scutellum yellow; legs with all segments entirely fulvous; wing with cells bc and c fulvous, microtrichia in outer corner of cell c only; a narrow fuscous costal band overlapping R_{2+3} and remaining narrow around apex of wing; a broad fuscous anal streak; supernumerary lobe of medium development; abdominal terga III–V fulvous with a 'T' pattern consisting of a

narrow transverse black band across anterior margin of tergum III and a narrow medial longitudinal fuscous to black band over all three terga, a pair of oval fulvous shining spots on tergum V.
Attractant: Cue lure.
Female: As for male in general body colour patterns. See Drew (1989) for a detailed description of female characters.
Distribution:
Type locality: Indonesia (Moluccas = Maluku – Buru I.).
Previously recorded: Type locality and the following locations – Indonesia (Java); Papua New Guinea; Bougainville I.; Lihir I.; New Britain; New Ireland; Solomon Is. (Drew, 1989; Norrbom *et al.*, 1998).
New records: Indonesia – Bali; Lombok; Moluccas = Maluku (Ambon I.); North Moluccas = North Maluku (Halmahera I.); Sulawesi (Central Sulawesi, Donggala).
Hosts: Recorded from one wild host, *Inocarpus fagifer* (family *Fabaceae*) (Drew, 1989).
Remarks: *Bactrocera (Bactrocera) moluccensis* (Perkins) is similar to *Bactrocera (Bactrocera) robiginosa* (May) and *Bactrocera (Bactrocera) bidentata* complex species in possessing fulvous cells bc and c without dense microtrichia, abdominal terga III–V fulvous and lacking lateral dark colour patterns but possessing a medial longitudinal dark band over some or all three terga and scutum pale (orange-brown to red-brown). It differs from these species in having abdominal terga III–V with a narrow transverse black band across anterior margin of tergum III and a medial longitudinal fuscous to black band over all three terga. *Bactrocera (Bactrocera) turneri* Drew is similar to *B. moluccensis* in the general body and wing colour patterns but differs in possessing a needle-shaped apex on the female piercer (in *B. moluccensis* the apex of the piercer is trilobed).

Bactrocera (Bactrocera) muiri (Hardy & Adachi)

(Fig. 113)

Dacus (Strumeta) muiri Hardy & Adachi, 1954: 177; Hardy, 1969b: 419. Holotype female in BBM.
Dacus (Bactrocera) muiri – Hardy, 1977: 51.
Bactrocera (Bactrocera) muiri – Drew & Hancock, 1994a: 42; Norrbom *et al.*,1998: 92.

Non-type material examined: No new specimens available for study since the publication of Drew & Hancock (1994a).
Diagnosis: Face fulvous with a pair of medium-sized oval black spots; postpronotal lobes and notopleura yellow; scutum black except brown around notopleural suture and inside postpronotal lobes; broad parallel-sided lateral postsutural yellow vittae ending just before *ia.* seta; medial postsutural yellow vitta absent; mesopleural stripe slightly wider than notopleuron dorsally; scutellum yellow; legs with femora entirely fulvous; fore and mid tibiae fuscous, hind tibiae dark fuscous; wing with cells bc and c colourless, microtrichia in outer corner of cell c only; a narrow fuscous costal band overlapping R_{2+3} and ending between extremities of R_{4+5} and M and with a slight swelling around apex of R_{4+5}; a narrow fuscous anal streak; supernumerary lobe weak; abdominal tergum III with a narrow dark fuscous transverse band across anterior margin and expanding as it becomes paler over lateral margins, abdominal terga IV and V missing in type which is the only specimen available for study (Hardy & Adachi (1954) described the abdomen as similar to that in *Bactrocera dorsalis* (Hendel)).
Attractant: No known record.
Female: No known record.
Distribution:
Type locality: Indonesia (Kalimantan, Pontianak).
Hosts: No known record.
Remarks: *Bactrocera (Bactrocera) muiri* (Hardy & Adachi) is similar to some specimens of *Bactrocera (Bactrocera) papayae* Drew & Hancock, *Bactrocera (Bactrocera) profunda* Tsuruta & White and *Bactrocera (Bactrocera) vishnu* Drew & Hancock in possessing broad parallel-sided lateral postsutural vittae, costal band overlapping R_{2+3}, pale coloured shining spots on abdominal tergum V and femora entirely fulvous or with, at most, a preapical dark spot on outer surfaces of fore femora only. It differs from *B. vishnu* in possessing narrow dark lateral margins on abdominal terga IV and V and from *B. papayae* and *B. profunda* in having the lateral postsutural vittae ending before the *ia.* seta.

Bactrocera (Bactrocera) musae (Tryon)

(Fig. 114)

Chaetodacus musae Tryon, 1927: 197. Lectotype female in QMIC.
Dacus ornatissimus Froggatt, 1909: 93 (partim, Paralectotypes only); Drew1974: 64.
Chaetodacus tryoni var. *Musa* Tryon, 1927: 187 (nomen nudum).
Chaetodacus musae var. *dorsopicta* Tryon, 1927: 198.
Dacus (Strumeta) musae – Hardy, 1951: 162; Drew, 1974: 63.
Strumeta musae – May, 1953: 58, 1963: 50, 66.
Dacus (Bactrocera) musae – Drew, 1982: 35.
Bactrocera (Bactrocera) musae – Drew, 1989: 92, Lectotype designation; Norrbom *et al.*, 1998: 93; Drew *et al.*, 2011: 151.

Material examined: INDONESIA: WEST PAPUA (formerly part of IRIAN JAYA) – 46 males, West Papua, Manokwari Utara, Wosi, 21.vi.2007, coll. Caroline, attracted to methyl eugenol in forest. Specimens in QDPC.
Diagnosis: A medium-sized species; face fulvous with a pair of small- to medium-sized circular black spots; postpronotal lobes and notopleura yellow; scutum dull black; lateral postsutural yellow vittae present; medial postsutural yellow vitta absent; no yellow spot anterior to notopleural suture; mesopleural stripe slightly wider than notopleuron dorsally; scutellum yellow; wing with cells bc and c colourless, microtrichia in outer corner of cell c only; a narrow dark fuscous costal band distinctly overlapping R_{2+3}; a moderately broad dark fuscous anal streak; abdominal terga III–V variable from entirely orange-brown with narrow transverse dark fuscous to black markings on anterolateral corners of tergum III to orange-brown with a narrow dark 'T' pattern consisting of

a narrow dark fuscous to black transverse band across anterior margin of tergum III and a narrow medial longitudinal dark fuscous to black band over all three terga.

Description: male.

Head: Height 1.75 mm. Frons – length 1.44 times breadth; red-brown with pale fuscous on anteromedial hump and around bases of orbital setae; orbital setae black: 1 *s.or.*, 2 *i.or.*; lunule pale fuscous. Ocellar triangle black. Vertex pale fuscous. Face fulvous with a pair of small- to medium-sized circular black spots; length 0.5 mm. Genae fulvous, dark fuscous subocular spot present; black seta present. Occiput red-brown, fulvous along eye margins; occipital row with a large number of short black setae. Antennae with segments 1 and 2 red-brown, segment 3 red-brown with dark fuscous on apex and outer surface; length of segments: 0.19 mm; 0.33 mm; 0.84 mm.

Thorax: Scutum dull black with dark red-brown below and behind lateral postsutural vittae, around notopleural suture, inside postpronotal lobes. Pleural areas black with dark red-brown below postpronotal lobes. Yellow markings as follows: postpronotal lobes; notopleura; mesopleural stripe slightly wider than notopleuron dorsally, continuing to katepisternum as a transverse spot, anterior margin slightly convex; anatergite (posterior apex black); anterior $^{3}/_{5}$ katatergite (remainder black); two medium-width lateral postsutural vittae narrowing slightly posteriorly to end just before *ia.* seta. Medial longitudinal postsutural yellow vitta absent. Postnotum black, dark red-brown centrally. Scutellum yellow except for narrow black basal band. Setae: *sc.* 2; *prsc.* 2; *ia.* 1; *p.sa.* 1; *a.sa.* 1; *mpl.* 1; *npl.* 2; *scp.* 4.

Legs: All femora entirely red-brown; fore tibiae fuscous to dark fuscous, mid tibiae fuscous basally to red-brown apically, hind tibiae dark fuscous; tarsi with all segments fulvous; mid tibiae each with an apical black spur.

Wings: Length 6.2 mm; cells bc and c colourless; microtrichia in outer corner of cell c only; remainder of wings colourless except dark fuscous cell sc, narrow dark fuscous costal band distinctly overlapping R_{2+3} and remaining of uniform width around apex of wing, a moderately broad dark fuscous anal streak; a dense aggregation of microtrichia around A_1+CuA_2; supernumerary lobe of medium development.

Abdomen: Oval; terga free; pecten present on tergum III. Tergum I and sterna I and II wider than long. Tergum I dark fuscous with a narrow transverse red-brown band across posterior margin; tergum II red-brown anteriorly to fulvous posteriorly with a broad transverse dark fuscous to black band subanteriorly but not reaching lateral margins; terga III–V variable from entirely orange-brown with narrow transverse dark fuscous to black markings on anterolateral corners of tergum III to orange-brown with a narrow black 'T' pattern (see Drew, 1989, for full description of abdominal colour patterns). A pair of oval orange-brown to red-brown shining spots on tergum V. Posterior lobe surstylus short, sternum V with a deep concavity on posterior margin. Abdominal sterna red-brown.

Attractant: Methyl eugenol.

Female: As for male except no dense aggregation of microtrichia around A_1+CuA_2, supernumerary lobe weak, pecten of cilia on abdominal tergum III absent; ovipositor basal segment orange-brown, dorsoventrally compressed and tapering posteriorly in dorsal view; ratio of length of oviscape to length of tergum V, 0.9: 1; apex of piercer needle-shaped.

Distribution:

Type locality: Australia (Queensland, Meringa).

Previously recorded: Type locality and the following locations – Australia (Queensland – East Coast, north from Mt Spec); Torres Strait Is.; Papua New Guinea and nearby islands; Bismarck Archipelago; Solomon Is.

New records: Indonesia (West Papua, formerly part of Irian Jaya – North Manokwari, Wosi).

Hosts: Recorded from nine plant families (see Drew, 1989; Hancock *et al.*, 2000). The statement that *B. musae* is 'not known from hard green' bananas (Hancock *et al.*, 2000) is incorrect.

Remarks: *Bactrocera (Bactrocera) musae* (Tryon) belongs to the *Bactrocera musae* complex recently reviewed by Drew *et al.* (2011). It is the only species within the *B. musae* complex that utilizes bananas as its host fruit and is best distinguished by having the lateral postsutural vittae narrowing posteriorly to end at or just before the *ia.* seta, the costal band dark fuscous and remaining this colour as it overlaps R_{2+3} and abdominal terga III–V with variable dark colour patterns (see Drew, 1989).

Bactrocera (Bactrocera) nanoarcuata Drew & Romig, new species

(Figs 115A, 115B)

Material examined: Holotype male, **INDONESIA: SULAWESI** – Sulawesi Barat, Mamuju, Belang-Belang, 22.viii.2006, coll. Sri Hastuti, Yan PF, Sukri, attracted to cue lure in forest, jeruk, deposited in BMNH.

Paratypes: **INDONESIA: SULAWESI** – 10 males (22.viii.2006), 1 male (4.ix.2006), 26 males (30.x.2006), same data as holotype, attracted to cue lure in forest, orange (jeruk); 2 males (3.x.2006), 1 male (17.x.2006), 1 male (31.x.2006), Sulawesi, Gorontalo, Bone Bolango, Lombongo, coll. Yusbar Ismail, attracted to cue lure in forest; 1 male, Sulawesi, Gorontalo, Limboto, Bionga, 31.x.2006, coll. Yusbar Ismail, attracted to cue lure in forest; 1 male (12.viii.2006), 1 male (15.x.2006), Sulawesi Tengah, Donggala, Nupabomba, coll. Rusdin, attracted to cue lure in forest; 1 male, Sulawesi Selatan, Maros, Limampoccoe, 13.ix.2006, coll. Sirajuddin, attracted to cue lure in forest; 1 male, North Sulawesi, Bitung, Danowudu, 26.ix.2006, coll. Karel F. Lala, attracted to cue lure in forest; 2 males, Sulawesi Tenggara, Konawe, Laomea, 10.ix.2006, coll. Syamsul Ma'arif, attracted to cue lure in mangga, jeruk; 1 male, Sulawesi Tenggara, Kendari, Wua-Wua, 10.ix.2006, coll. Syamsul M., attracted to cue lure in village. 10 paratypes in ANIC, 9 paratypes in BMNH, 20 paratypes in QDPC, 10 paratypes in QMIC.

Diagnosis: A medium-sized species; face fulvous with a pair of medium-sized circular black spots; postpronotal lobes and notopleura yellow; scutum black with red-brown along entire lateral margins and around the notopleural suture; lateral postsutural yellow vittae present; medial postsutural yellow vitta absent; no yellow spot anterior to notopleural suture; mesopleural stripe slightly wider than notopleuron dorsally; scutellum yellow; wing with cells bc and c pale fuscous, microtrichia in outer corner of cell c only; a broad fuscous costal band confluent with R_{4+5} and expanding across apex of wing; a moderately broad fuscous anal streak; abdominal terga III–V orange-brown with a dark 'T' pattern consisting of a narrow transverse black band across anterior margin of tergum III which expands across lateral margins of that tergum, a medium-width medial longitudinal black band over all three terga, narrow fuscous to dark fuscous anterolateral corners on terga IV and V, fuscous shining spots on tergum V.

Description: male (Fig. 115A).

Head: Height 1.57 mm. Frons – length 1.48 times breadth; fulvous with fuscous on anteromedial hump and around bases of orbital setae; orbital setae black: 1 *s.or.*, 2 *i.or.*; lunule red-brown. Ocellar triangle black. Vertex red-brown. Face fulvous with a pair of medium-sized circular black spots; length 0.49 mm. Genae fulvous, pale fuscous subocular spot present; small dark seta present. Occiput red-brown, fulvous along eye margins; occipital row with 4–6 strong black setae. Antennae with segments 1 and 2 red-brown, segment 3 red-brown with pale fuscous around apex; length of segments: 0.22 mm; 0.31 mm; 0.84 mm.

Thorax: Scutum black with red-brown below and behind lateral postsutural vittae, around notopleural suture, along lateral margins between postpronotal lobes and notopleura and inside postpronotal lobes. Pleural areas dark fuscous to black except red-brown above mid coxae and below postpronotal lobes. Yellow markings as follows: postpronotal lobes; notopleura; mesopleural stripe slightly wider than notopleuron dorsally, continuing to katepisternum as a transverse spot, anterior margin straight; anatergite (posterior apex black); anterior ⅔ katatergite (remainder black); two broad lateral postsutural vittae narrowing posteriorly to end just before *ia.* seta. Medial longitudinal postsutural yellow vitta absent. Postnotum black with a red-brown midline. Scutellum yellow except for narrow black basal band. Setae: *sc.* 2; *prsc.* 2; *ia.* 1; *p.sa.* 1; *a.sa.* 1; *mpl.* 1; *npl.* 2; *scp.* 4.

Legs: All femora fulvous with red-brown on apical ¼ to ⅓ fore and hind tibiae pale fuscous, mid tibiae fulvous; all tarsi with basal segment fulvous and apical four segments red-brown; mid tibiae each with an apical black spur.

Wings: Length 5.3 mm; cells bc and c pale fuscous; microtrichia in outer corner of cell c only; remainder of wings colourless except fuscous cell sc, broad fuscous costal band confluent with R_{4+5} but becoming pale fuscous adjacent to that vein and becoming dark fuscous around apex of wing where it expands across apex of R_{4+5}, a moderately broad fuscous anal streak; a dense aggregation of microtrichia around A_1+CuA_2; supernumerary lobe of medium development.

Abdomen: Oval; terga free; pecten present on tergum III. Tergum I and sterna I and II wider than long. Tergum I red-brown centrally with narrow dark fuscous lateral margins and a narrow fulvous band across posterior margin that reaches lateral margins; tergum II fulvous with a narrow transverse dark fuscous band subanteriorly but not reaching narrow dark fuscous anterolateral corners; terga III–V orange-brown with a dark 'T' pattern consisting of a narrow transverse black band across anterior margin of tergum III which expands over lateral margins, a medium-width medial longitudinal black band over all three terga, narrow fuscous to dark fuscous anterolateral corners on terga IV and V. A pair of oval fuscous shining spots on tergum V. Posterior lobe of surstylus short, sternum V with a moderately deep concavity on posterior margin (see Fig. 115B for shape of abdominal sterna). Abdominal sterna pale, tending orange-brown to red-brown.

Attractant: Cue lure.

Female: No known record.

Distribution: Indonesia (Sulawesi).

Hosts: No known record.

Remarks: *Bactrocera (Bactrocera) nanoarcuata* n.sp. is similar to *Bactrocera (Bactrocera) curvosterna* n.sp. and *Bactrocera (Bactrocera) paralatissima* n.sp. in possessing an entire costal band confluent with R_{4+5}, a black scutum, cells bc and c coloured and lateral postsutural vittae tapering posteriorly. It differs from *B. curvosterna* in having abdominal sternum V with a deep concavity on posterior margin but not horseshoe shaped and from *B. paralatissima* in having distinct facial spots and the abdominal sterna pale coloured, generally orange-brown to red-brown.

Bactrocera (Bactrocera) nationigrotibialis Drew & Romig, new species

(Fig. 116)

Material examined: Holotype male, **INDONESIA: SULAWESI** – Sulawesi Tengah, Donggala, Lolu, 29.x.2006, coll. Lyar, attracted to methyl eugenol in mangga, deposited in BMNH.

Paratypes: **INDONESIA: SULAWESI** – 1 male, same data as holotype; 2 males (12.viii.2006), 6 males (20.viii.2006), 5 males (27.viii.2006), 4 males (2.ix.2006), 3 males (10.ix.2006), 1 male (17.ix.2006), Sulawesi Tengah, Donggala, Lolu, coll. Rusdin, attracted to methyl eugenol in mangga; 3 males (17.ix.2006), 2 males (23.ix.2006), Sulawesi Tengah, Donggala, Lolu, coll. Lyar, attracted to methyl eugenol in mangga; 4 males, Sulawesi Tengah, Donggala, Nupabomba, 12.viii.2006, coll. Rusdin, attracted to methyl eugenol in forest; 3 males (20.viii.2006), 1 male (8.x.2006), 3 males (15.x.2006), Sulawesi Tengah, Donggala, Nupabomba, coll. Lyar, attracted to methyl eugenol in forest; 1 male (20.viii.2006), 1 male (23.ix.2006), 2 males (15.x.2006), Sulawesi Tengah, Donggala, Jono Oge, coll. Lyar, attracted to methyl eugenol in jackfruit; 1 male (27.viii.2006), 3 males (10.ix.2006), 1 male (23.ix.2006), Sulawesi Tengah, Donggala, Jono Oge, coll. Rusdin, attracted to methyl eugenol in jackfruit (nangka); 1 male (2.viii.2006), 1 male (15.x.2006), Sulawesi Tengah, Donggala, Pombewe, coll. Lyar, attracted to methyl eugenol in village; 1 male (10.ix.2006), 1 male (23.ix.2006), 1 male (1.x.2006), 2 males (29.x.2006), Sulawesi Tengah, Donggala, Pombewe, coll. Rusdin, attracted to methyl eugenol in village; 2 males, Gorontalo, Limboto,

Polohungo, 15.viii.2006, coll. Yusbar Ismail, attracted to methyl eugenol in forest; 3 males, Gorontalo, Limboto, Bionga, 22.viii.2006, coll. Yusbar Ismail, attracted to methyl eugenol in forest; 1 male, Gorontalo, Bone Bolango, Lombongo, 9.xi.2006, coll. Yusbar Ismail, attracted to methyl eugenol in village. 10 paratypes in ANIC, 10 paratypes in BMNH, 30 paratypes in QDPC, 10 paratypes in QMIC.

Diagnosis: A small species; face fulvous with a pair of large circular black spots; postpronotal lobes yellow (anterior dorsal corners black); notopleura yellow; scutum entirely shining black; lateral postsutural yellow vittae present; medial postsutural yellow vitta absent; no yellow spot anterior to notopleural suture; mesopleural stripe reaching midway between anterior margin of notopleuron and anterior *npl.* seta dorsally; scutellum yellow with a broad black basal band; wing with cells bc and c colourless, microtrichia in outer corner of cell c only; a narrow dark fuscous costal band; a narrow fuscous anal streak; abdominal terga III–V entirely black except for small areas of red-brown on posterior margin of tergum V.

Description: male.

Head: Height 1.08 mm. Frons – length 1.3 times breadth; generally fuscous to dark fuscous with narrow lateral fulvous margins; orbital setae black: 1 *s.or.*, 2 *i.or.*; lunule dark fuscous. Ocellar triangle black. Vertex fuscous. Face fulvous with a pair of large circular black spots; length 0.3 mm. Genae fulvous, large dark fuscous subocular spot present; black seta present. Occiput black, fulvous along eye margins; occipital row with 4–6 black setae. Antennae with segments 1 and 2 red-brown, segment 3 red-brown with dark fuscous on apex and outer surface; length of segments: 0.16 mm; 0.24 mm; 0.6 mm.

Thorax: Scutum entirely black without pale markings. Pleural areas black, tending very dark fuscous below postpronotal lobes. Yellow markings as follows: postpronotal lobes (anterior dorsal corners black); notopleura; mesopleural stripe reaching midway between anterior margin of notopleuron and anterior *npl.* seta dorsally, continuing to katepisternum as a transverse spot, anterior margin convex; anatergite (posterior apex black); anterior ¾ katatergite (remainder black); two moderately broad lateral postsutural vittae narrowing posteriorly to end well before *ia.* seta (generally ending at level of *p.sa.* seta). Medial longitudinal postsutural yellow vitta absent. Postnotum black. Scutellum yellow except for broad black basal band. Setae: *sc.* 2; *prsc.* 2; *ia.* 1; *p.sa.* 1; *a.sa.* 1; *mpl.* 1; *npl.* 2; *scp.* 4.

Legs: Fore and mid femora black with small areas of fulvous at the base and apex, hind femora fulvous with apical ½ to ⅓ black; fore tibiae dark fuscous, mid and hind tibiae dark fuscous to black; all tarsi with segments entirely fulvous; mid tibiae each with an apical black spur.

Wings: Length 3.5 mm; cells bc and c colourless; microtrichia in outer corner of cell c only; remainder of wings colourless except dark fuscous cell sc, narrow dark fuscous costal band confluent with R_{2+3} and remaining narrow around apex of wing, a narrow fuscous anal streak; a dense aggregation of microtrichia around A_1+CuA_2; supernumerary lobe of medium development.

Abdomen: Oval; terga free; pecten present on tergum III. Tergum I and sterna I and II wider than long. Tergum I black except for a narrow transverse red-brown band along posterior margin but not reaching lateral margins; tergum II black except for a broad fulvous transverse band across posterior margin but not reaching lateral margins; terga III–V black except for small areas of red-brown on posterior margins of tergum V. A pair of oval black shining spots on tergum V. Posterior lobe of surstylus short, sternum V with a deep concavity on posterior margin. Abdominal sterna dark fuscous to black.

Attractant: Methyl eugenol.

Female: No known record.

Distribution: Indonesia (Sulawesi – Central Sulawesi; Gorontalo).

Hosts: No known record.

Remarks: *Bactrocera (Bactrocera) nationigrotibialis* n.sp. belongs to the *nigrotibialis* complex and is similar to *Bactrocera (Bactrocera) neonigrotibialis* Drew and *Bactrocera (Bactrocera) wuzhishana* Lin & Yang in possessing a fulvous face with distinct black facial spots, costal band in wing confluent with R_{2+3}, cells bc and c colourless, a narrow anal streak in wing and scutellum without black markings in addition to the basal band. It differs from *B. neonigrotibialis* in having abdominal terga III–V entirely black and from *B. wuzhishana* in having broad tapering lateral postsutural vittae, circular facial spots and the costal band uniformly narrow around wing apex.

Bactrocera (Bactrocera) neoarecae Drew

(Fig. 117)

Bactrocera (Bactrocera) neoarecae Drew, in Drew & Raghu, 2002: 337. Holotype male in BMNH.

Non-type material examined: No new specimens available for study since the publication of Drew & Raghu (2002).

Diagnosis: Face fulvous with a pair of small oval transverse black spots not meeting in midline; postpronotal lobes and notopleura yellow; scutum black without pale markings except for red-brown anterior to notopleural suture; medium-width parallel-sided lateral postsutural yellow vittae ending behind *ia.* seta; medial postsutural yellow vitta absent; mesopleural stripe reaching midway between anterior margin of notopleuron and anterior *npl.* seta dorsally; scutellum yellow with a broad black basal band; legs with femora entirely fulvous except for a small dark fuscous to black preapical spot on outer surfaces of fore and hind femora; fore and mid tibiae fulvous, hind tibiae dark fuscous; wing with cells bc and c colourless, microtrichia in outer corner of cell c only; a narrow fuscous costal band confluent with R_{2+3} and widening gradually across apex of wing; a narrow very pale fuscous anal streak; supernumerary lobe of medium development; abdominal terga III–V each with a 'T' pattern consisting of a broad dark fuscous to black transverse band across anterior margin of the tergum and a broad medial longitudinal dark fuscous to black band, remainder of tergum III fuscous and terga IV and V red-brown, a pair of oval red-brown shining spots on tergum V; abdominal sterna dark coloured.

Attractant: Methyl eugenol.

Female: No known record.
Distribution:
Type locality: Southern India (Kerala).
Hosts: No known record.
Remarks: *Bactrocera (Bactrocera) neoarecae* Drew is similar to *Bactrocera (Bactrocera) amarambalensis* Drew, *Bactrocera (Bactrocera) arecae* (Hardy & Adachi), *Bactrocera (Bactrocera) caryeae* (Kapoor) and *Bactrocera (Bactrocera) kandiensis* Drew & Hancock in possessing narrow- to medium-width parallel-sided lateral postsutural vittae and preapical dark spots on at least the fore femora. It differs from *B. amarambalensis* and *B. arecae* in possessing preapical dark spots on the fore and hind femora (not only the fore femora), and from *B. caryeae* and from *B. kandiensis* in having a dark 'T' pattern on each of abdominal terga III–V.

Bactrocera (Bactrocera) neocheesmanae Drew

(Fig. 118)

Bactrocera (Bactrocera) neocheesmanae Drew, 1989: 147; Norrbom *et al.*, 1998: 93. Holotype male in QMIC.

Material examined: INDONESIA: PAPUA (formerly part of IRIAN JAYA) – 1 male, Indonesia, Papua, Keerom, Wiantre, 2.vi.2007, coll. Eikanus Karma, attracted to methyl eugenol in orange (kebun jeruk). Specimen in QDPC.
Diagnosis: A small species; face fulvous with a pair of small narrow elongate oval black spots; postpronotal lobes and notopleura yellow; scutum black; lateral and medial postsutural yellow vittae present; no yellow spot anterior to notopleural suture; mesopleural stripe reaching to postpronotal lobe dorsally; scutellum yellow; wing with cells bc and c pale fuscous, dense microtrichia covering all of cell c and outer ½ of cell bc; a broad fuscous costal band confluent with R_{4+5}, a moderately broad transverse fuscous band across wing and enclosing both crossveins; a broad fuscous to dark fuscous anal streak; abdominal terga III–V orange-brown with medium-width to broad lateral longitudinal fuscous to dark fuscous bands over all three terga.
Description: male.
Head: Height 1.2 mm. Frons – length 1.36 times breadth; fulvous with pale fuscous on anteromedial hump and around bases of *s.or.* setae; orbital setae black: 1 *s.or.*, 2 *i.or.*; lunule red-brown. Ocellar triangle black. Vertex red-brown. Face fulvous with a pair of small narrow elongate oval black spots; length 0.39 mm. Genae fulvous, small fuscous subocular spot present; black seta present. Occiput fuscous, fulvous along eye margins; occipital row with a large number of small black setae. Antennae with segments 1 and 2 red-brown, segment 3 red-brown with fuscous on apex and outer surface; length of segments: 0.14 mm; 0.24 mm; 0.56 mm.
Thorax: Scutum black with variable dark red-brown to fuscous markings below lateral postsutural vittae, on lateral margin between postpronotal lobe and notopleuron, over anteromedial portion of scutum. Pleural areas entirely dark red-brown to fuscous. Yellow markings as follows: postpronotal lobes; notopleura; mesopleural stripe reaching to postpronotal lobe dorsally, continuing to katepisternum as a small transverse spot, anterior margin strongly concave; anatergite (posterior apex black); anterior ⅔ katatergite (remainder dark red-brown to fuscous); two narrow lateral postsutural vittae narrowing to a point posteriorly to end before *ia.* seta; a broad medial longitudinal postsutural vitta. Postnotum black laterally to dark red-brown centrally. Scutellum yellow except for narrow black basal band. Setae: *sc.* 2; *prsc.* 2; *ia.* 1; *p.sa.* 1; *a.sa.* 1; *mpl.* 1; *npl.* 2; *scp.* 4.
Legs: All femora entirely fulvous; fore and mid tibiae dark fuscous basally to pale fuscous apically, hind tibiae dark fuscous; tarsi with all segments fulvous; mid tibiae each with an apical black spur.
Wings: Length 4.6 mm; cells bc and c pale fuscous; dense microtrichia covering all of cell c and outer ½ of cell bc; remainder of wings colourless except fuscous cell sc, broad fuscous costal band confluent with R_{4+5} and of uniform width throughout, a broad transverse fuscous band across centre of wing and enclosing both crossveins, a broad fuscous to dark fuscous anal streak; a dense aggregation of microtrichia around A_1+CuA_2; supernumerary lobe weak.
Abdomen: Elongate oval; terga free; pecten present on tergum III. Tergum I and sterna I and II wider than long. Tergum I entirely dark fuscous; tergum II fulvous with a broad transverse dark fuscous band across anterior margin which widens to cover lateral margins; terga III–V orange-brown with medium-width to broad lateral longitudinal fuscous to dark fuscous bands over all three terga. A pair of oval orange-brown shining spots on tergum V. Posterior lobe of surstylus short, sternum V with a moderately deep concavity on posterior margin. Abdominal sterna red-brown.
Attractant: Methyl eugenol.
Female: No known record.
Distribution:
Type locality: Papua New Guinea (Mt Hagen).
Previously recorded: Type locality and Papua New Guinea (Central District) (Drew, 1989).
New records: Indonesia (Papua, formerly part of Irian Jaya – Keerom).
Hosts: No known record.
Remarks: *Bactrocera (Bactrocera) neocheesmanae* Drew is similar to *Bactrocera (Bactrocera) cheesmanae* (Perkins) in possessing lateral and medial postsutural yellow vittae on the scutum, an elongate oval-shaped abdomen and a broad transverse fuscous band across the wing. It differs from *B. cheesmanae* in having the scutum black and abdominal terga III–V mostly orange-brown. In possessing a medial postsutural yellow vitta on the scutum, these two species probably are not genuine members of the subgenus *Bactrocera* and are here placed in this subgenus for convenience.

Bactrocera (Bactrocera) neocognata Drew & Hancock

(Fig. 119)

Bactrocera (Bactrocera) neocognata Drew & Hancock, 1994a: 43. Holotype male in BBM.
Bactrocera (Bactrocera) neocongnata – Norrbom *et al.*, 1998: 93 [Misspelt].

Non-type material examined: Since the publication of Drew & Hancock (1994a), the following specimens have been studied – **BRUNEI**: 5 males (23.i.2002), 5 males (25.i.2002), 14 males (27.i.2002), 14 males (29.i.2002), KBFC, Bato Apou Nat. Park, coll. K. Hurley, attracted to cue lure. **INDONESIA: KALIMANTAN** – 7 males (22.viii.2005), 7 males (29. viii.2005), 12 males (28.xi.2005), East Kalimantan, Kutai Kartanegara, Batuah, coll. Ertha D.P., attracted to cue lure in banana; 1 male, East Kalimantan, Kutai Kartanegara, Batuah, 1.viii.2005, coll. Heru G, attracted to cue lure in forest; 1 male, East Kalimantan, Kota Samarinda, Lempake, 1.viii.2005, coll. Arofah, attracted to cue lure in village; 16 males (31.vii.2005), 12 males (13.xi.2005), 19 males (20.xi.2005), South Kalimantan, Banjar, Karang Intan, Mandiangin, coll. Bilal, attracted to cue lure in forest; 2 males, South Kalimantan, Banjarbaru, Guntung Payung, 21.vii.2005, coll. Sidrotul & Sutini, attracted to cue lure in village; 8 males, South Kalimantan, Banjar, Mandiangin, 7.viii.2005, coll. Bilal & Sidrotul, attracted to cue lure in forest; 7 males, South Kalimantan, Banjar Karang Intan, Mandiangin, 21.viii.2005, coll. Bilal & Sidrotul, attracted to cue lure in forest; 1 male, West Kalimantan, Pontianak, Sei Pinyuh, 19. vii.2005, coll. A. Irom, attracted to cue lure in Village mango; 2 males, West Kalimantan, Pontianak, Pancaroba, 10.viii.2005, coll. A. Irom, attracted to cue lure in forest; 3 males, West Kalimantan, Pontianak, Sei Duri II, 26.vii.2005, coll. A. Irom, attracted to cue lure in citrus; 1 male, Central Kalimantan, Kota Waringin Timur, Kota Besai, 26.vii.2005, coll. Utik D & Kusyono, attracted to cue lure in citrus. **PENINSULAR MALAYSIA**: 48 females and 32 males, Jeram Pasu, Kelantan, Peninsular Malaysia, 8.ix.1986, coll. Mahdzir & Tony, bred from *Eugenia* sp. (fruit sample No. 0414). **VIETNAM: Northern Vietnam** – 1 male, northern Vietnam, Hoa Binh Province, Ky Son District, Cao Phong, Site N014, 17.x.1999, attracted to cue lure; 1 male, northern Vietnam, Hoa Binh Province, Mai Chau District, Pa Co, Site N010, 1.x.2001, attracted to cue lure; 1 male, northern Vietnam, Bac Giang Province, Luc Ngan District, Cao Cao, Quy Son, Site VN05, 25.vi.2002, attracted to cue lure. Specimens in QDPC.

Diagnosis: Face fulvous with a pair of small- to medium-sized black facial spots; postpronotal lobes and notopleura yellow; scutum black except dark brown behind lateral postsutural vittae and around notopleural suture; moderately broad lateral postsutural yellow vittae that taper posteriorly to end just before *ia.* seta; medial postsutural yellow vitta absent; mesopleural stripe slightly wider than notopleuron dorsally; scutellum yellow; legs with femora entirely fulvous (in some specimens there is a small pale preapical dark spot on outer surfaces of fore femora); fore and hind tibiae dark fuscous and mid tibiae fuscous tending darker basally; wing with cells bc and c colourless, microtrichia in outer corner of cell c only; a narrow fuscous costal band slightly overlapping R_{2+3} and with a slight widening around apex of R_{4+5}; a narrow fuscous anal streak; supernumerary lobe of medium development; abdominal terga III–V orange-brown with a narrow transverse dark fuscous to black band across anterior margin of tergum III, fuscous laterally on tergum III, anterolateral corners of terga IV and V fuscous, a narrow medial longitudinal fuscous to dark fuscous band over all three terga, a pair of oval pale fuscous shining spots on tergum V; abdominal sterna dark coloured.

Attractant: Cue lure.

Female: As for male in the general body colour patterns. See Drew & Hancock (1994a) for a detailed description of female characters.

Distribution:

Type locality: Indonesia (Lombok, Suranadi).

Previously recorded: Type locality and the following locations – Indonesia (Java; Kalimantan); East Malaysia (Sabah) (Drew & Hancock, 1994a; Norrbom *et al.*, 1998).

New records: Brunei; Peninsular Malaysia; Northern Vietnam (Bac Giang, Hoa Binh Provinces).

Hosts:

New records: *Eugenia* sp. (family *Myrtaceae*).

Remarks: *Bactrocera (Bactrocera) neocognata* Drew & Hancock is similar to *Bactrocera (Bactrocera) cognata* (Hardy & Adachi), some specimens of *Bactrocera (Bactrocera) fulvifemur* Drew & Hancock, *Bactrocera (Bactrocera) lateritaenia* Drew & Hancock, *Bactrocera (Bactrocera) penecognata* Drew & Hancock and some specimens of *Bactrocera (Bactrocera) usitata* Drew & Hancock in possessing lateral postsutural vittae that narrow posteriorly, femora entirely fulvous or with, at most, dark preapical spots on fore femora only and costal band just overlapping R_{2+3}. It differs from these species in possessing a very narrow medial longitudinal dark band on abdominal terga III–V.

Bactrocera (Bactrocera) neofulvicauda Drew & Romig, new species

(Fig. 120)

Material examined: Holotype male, **INDONESIA: NORTH MOLUCCAS (NORTH MALUKU)** – Moluccas, North Maluku, Kodya Ternate, Foramadiahi, 3.vi.2007, coll. La Ruti, attracted to cue lure in forest, deposited in BMNH. Paratypes: **INDONESIA: NORTH MOLUCCAS (NORTH MALUKU)** – 7 males (5.vi.2007), 1 male (12.vi.2007), Maluku, North Maluku, West Halmahera Gaol, coll. Raais Abdullah, attracted to cue lure in forest and orange (jeruk). 1 paratype in ANIC, 1 paratype in BMNH, 4 paratypes in QDPC, 1 paratype in QMIC.

Diagnosis: A medium-sized species; face fulvous with a pair of large circular black spots; postpronotal lobes and notopleura yellow; scutum entirely dull black; lateral postsutural yellow vittae present; medial postsutural yellow vitta absent; no yellow spot anterior to notopleural suture; mesopleural stripe reaching to or almost to anterior *npl.* seta dorsally; scutellum yellow; wing with cells bc and c dark fuscous, dense microtrichia covering all of cell c and most of cell bc; a broad dark fuscous costal band confluent with R_{4+5}, a broad transverse band across wing from costal band to hind margin and enclosing both crossveins; a broad dark fuscous anal streak; abdominal tergum III entirely black, tergum IV black with small posterocentral orange-brown spots, tergum V fuscous with dark fuscous anterolateral corners.

Description: male.

Head: Height 1.3 mm. Frons – length 1.55 times breadth; fulvous with an overall fuscous colouration; orbital setae black: 1 *s.or.*, 2 *i.or.*; lunule red-brown. Ocellar triangle black. Vertex fuscous. Face dark red-brown with a pair of

large oval black spots; length 0.36 mm. Genae red-brown, small fuscous subocular spot present; black seta present. Occiput dark fuscous, fulvous along eye margins; occipital row with 4–6 black setae. Antennae with segments 1 and 2 fuscous, segment 3 dark fuscous; length of segments: 0.18 mm; 0.28 mm; 0.66 mm.
Thorax: Scutum dull black without pale markings. Pleural areas entirely black, tending dark fuscous below postpronotal lobes. Yellow markings as follows: postpronotal lobes; notopleura; mesopleural stripe reaching to or almost to anterior *npl.* seta dorsally, continuing to katepisternum as a transverse spot, anterior margin straight; anatergite (posterior apex black); anterior $^{3}/_{5}$ katatergite (remainder black); two broad lateral postsutural vittae narrowing slightly posteriorly to end at *ia.* seta. Medial longitudinal postsutural yellow vitta absent. Postnotum black. Scutellum yellow with a narrow black basal band. Setae: *sc.* 2; *prsc.* 2; *ia.* 1; *p.sa.* 1; *a.sa.* 1; *mpl.* 1; *npl.* 2; *scp.* 4.
Legs: Fore femora fulvous to red-brown with diffuse fuscous on apical ¼, mid and hind femora fulvous with diffuse fuscous around apical ¼; fore tibiae red-brown, mid tibiae fuscous, hind tibiae fuscous to dark fuscous; fore tarsi with basal segment fulvous and apical four segments red-brown, mid and hind tarsi with basal segment fulvous and apical four segments fuscous; mid tibiae each with an apical black spur.
Wings: Length 5.5 mm; cells bc and c dark fuscous; dense microtrichia covering all of cell c and most of cell bc; remainder of wings colourless except dark fuscous cell sc, broad dark fuscous costal band confluent with R_{4+5} but slightly paler adjacent to this vein in centre of wing, a broad fuscous transverse band across wing from costal band to hind margin and enclosing both crossveins and becoming diffuse along posterior margin of wing towards supernumerary lobe, a broad dark fuscous anal streak; a dense aggregation of microtrichia around A_1+CuA_2; supernumerary lobe weak.
Abdomen: Oval; terga free; pecten present on tergum III. Tergum I and sterna I and II wider than long. Tergum I entirely black; tergum II dark fuscous with a moderately broad transverse orange-brown band across posterior margin; tergum III entirely black; tergum IV black with small orange-brown spots posterocentrally; tergum V fuscous with dark fuscous anterolateral corners. A pair of oval black shining spots on tergum V. Posterior lobe of surstylus short, sternum V with a deep concavity on posterior margin. Abdominal sterna dark fuscous.
Attractant: Cue lure.
Female: No known record.
Distribution: Indonesia (North Moluccas = North Maluku – Halmahera I.; Ternate).
Hosts: No known record.
Remarks: *Bactrocera (Bactrocera) neofulvicauda* n.sp. is similar to *Bactrocera (Bactrocera) fulvicauda* (Perkins) in possessing a black scutum, abdominal terga III–V mostly black, wing with a broad costal band confluent with R_{4+5} and a broad transverse band from costal band to hind margin that encloses both crossveins. It differs from *B. fulvicauda* in having the costal band paler between R_{2+3} and R_{4+5}, cells bc and c dark fuscous and the mesopleural stripe narrower, reaching the anterior *npl.* seta, not extending to the postpronotal lobe dorsally.

Bactrocera (Bactrocera) neonigrotibialis Drew

(Fig. 121)

Bactrocera (Bactrocera) neonigrotibialis Drew, in Drew & Raghu, 2002: 338. Holotype male in BMNH.

Non-type material examined: Since the publication of Drew & Raghu (2002), the following specimens have been studied – **BHUTAN**: 4 males, Namling 16km, 1600m, 9.vi.2001, coll. Brian Fletcher, attracted to cue lure in subtropical forest; 2 males, Tsirang, Darachu-La, 30.v.2001, coll. Brian Fletcher, attracted to cue lure; 4 males (15.xi.2004), 8 males (5.xii.2004), 1 male (29.xii.2004), East Bhutan, Samdrup Jongkhar District, coll. Wangdi, attracted to cue lure; 1 male, Purana Bastey, 3.iv.2005, coll. Yangyal Nooh, Karma Namgyel. **INDIA**: 1 male, S. India, Talakaveri, Nr Madikeri, 1100m, 23.v.1992, coll. D.L. Hancock, attracted to cue lure. Specimens in QDPC.
Diagnosis: Face fulvous with a pair of large elongate narrow to broad oval black spots; postpronotal lobes yellow (anteromedial corners black); notopleura yellow; scutum black without pale markings; lateral postsutural yellow vittae narrow, short and narrowing sharply posteriorly to end at or just behind level of *a.sa.* seta; medial postsutural yellow vitta absent; mesopleural stripe equal in width to notopleuron dorsally; scutellum yellow except for narrow to broad black basal band; legs with fore femora fulvous except apical ⅔ of outer surfaces shining black, mid femora fulvous except apical ⅓ to ½ dark fuscous to black, hind femora fulvous except apical ⅓ dark fuscous to black; fore tibiae fuscous to dark fuscous, mid tibiae fuscous, hind tibiae dark fuscous; wing with cells bc and c colourless, microtrichia in outer corner of cell c only; a narrow fuscous costal band confluent with R_{2+3} and widening slightly between apex of R_{2+3} and where it ends just beyond extremity of R_{4+5}; a narrow fuscous anal streak; supernumerary lobe of medium development; abdominal tergum III dark fuscous to black with a medium-width medial longitudinal black band, terga IV–V dark red-brown centrally with a medium-width medial longitudinal black band and broad lateral longitudinal dark fuscous bands over both terga, a pair of oval dark fuscous shining spots on tergum V (the specimens from Bhutan have terga III–V almost entirely black).
Attractant: Cue lure.
Female: No known record.
Distribution:
Type locality: Southern India (Kerala).
New records: Bhutan.
Hosts: No known record.
Remarks: *Bactrocera (Bactrocera) neonigrotibialis* Drew belongs to the *nigrotibialis* complex which consists of species with a mostly black thorax, wings with a narrow costal band and anal streak and without additional dark patterning, legs with extensive black markings particularly over the femora. It differs from the other species in this complex in having a combination of a fulvous face with a pair of black spots, the lateral postsutural vittae very short, narrow and ending approximately at the level of the *a.sa.* seta, the femora fulvous with extensive areas of black on or around the apices.

Bactrocera (Bactrocera) neopropinqua Drew & Hancock

(Fig. 122)

Bactrocera (Bactrocera) neopropinqua Drew & Hancock, 1994a: 44; Norrbom *et al.*, 1998: 93. Holotype female in MCSNM.

Non-type material examined: No new specimens available for study since the publication of Drew & Hancock (1994a).

Diagnosis: Face fulvous with a pair of small- to medium-sized elongate oval black spots; postpronotal lobes and notopleura yellow; scutum black except red-brown below and behind lateral postsutural vittae, around notopleural suture and inside postpronotal lobes; broad parallel-sided lateral postsutural yellow vittae ending behind *ia.* seta; medial postsutural yellow vitta absent; mesopleural stripe almost reaching anterior *npl.* seta dorsally; scutellum yellow; legs with femora entirely fulvous except for small oval preapical dark fuscous spot on outer surfaces of fore femora; fore and mid tibiae fuscous, hind tibiae dark fuscous; wing with cells bc and c colourless, microtrichia in outer corner of cell c only; a narrow fuscous costal band confluent with R_{2+3} and remaining narrow and of uniform width around apex of wing; a narrow fuscous anal streak; supernumerary lobe of medium development in male, weak in female; abdominal terga III–V orange-brown except for a narrow transverse black band across anterior margin of tergum III which widens to cover lateral margins, anterolateral corners of terga IV and V dark fuscous to black, a very broad medial longitudinal black band over all three terga, a pair of oval fuscous shining spots on tergum V; abdominal sterna dark coloured.

Attractant: No known record.

Female: As for male in the general body colour patterns. See Drew & Hancock (1994a) for a detailed description of female characters.

Distribution:

Type locality: Philippines (Batbatan I.).

Hosts: No known record.

Remarks: *Bactrocera (Bactrocera) neopropinqua* Drew & Hancock is similar to *Bactrocera (Bactrocera) irvingiae* Drew & Hancock, *Bactrocera (Bactrocera) kanchanaburi* Drew & Hancock, *Bactrocera (Bactrocera) raiensis* Drew & Hancock and *Bactrocera (Bactrocera) sapaensis* n.sp. in possessing broad parallel-sided lateral postsutural vittae, costal band confluent with R_{2+3} and not expanding apically and dark preapical spots on outer surfaces of fore femora only. It differs from *B. kanchanaburi* and *B. raiensis* in possessing dark anterolateral corners on terga IV and V (not broad lateral longitudinal dark bands), from *B. sapaensis* n.sp. in having the anal streak present and from *B. irvingiae* in having the mesopleural stripe reaching almost to the anterior *npl.* seta dorsally and the apex of the piercer trilobed.

Bactrocera (Bactrocera) neoritsemai Drew & Romig, new species

(Fig. 123)

Material examined: Holotype male, **INDONESIA: SULAWESI** – Sulawesi, Sulawesi Tengah, Donggala, Nupabomba, 27.ix.2006, coll. Rusdin, attracted to cue lure in forest, deposited in BMNH.

Paratypes: **INDONESIA: SULAWESI** – 1 male, Sulawesi, Sulawesi Tenggara, Kendari, Wua-Wua, 10.ix.2006, coll. Syamsul M, attracted to cue lure in village; 1 male (20.viii.2006), 2 males (1.x.2006), 3 males (2.ix.2006), 1 male (22.x.2006), 1 male (29.x.2006), Sulawesi, Sulawesi Tengah, Donggala, Nupabomba, coll. Rusdin, attracted to cue lure in forest. 1 paratype in ANIC, 1 paratypes in BMNH, 5 paratypes in QDPC, 2 paratypes in QMIC.

Diagnosis: A large species; face fulvous with a pair of medium-sized circular to oval black spots; postpronotal lobes and notopleura yellow; scutum dull black; lateral postsutural yellow vittae present; medial postsutural yellow vitta absent; no yellow spot anterior to notopleural suture; mesopleural stripe reaching midway between anterior margin of notopleuron and anterior *npl.* seta dorsally; scutellum yellow; wing with cells bc and c pale fuscous to fuscous, microtrichia in outer corner of cell c only; a broad fuscous costal band confluent with R_{4+5} and becoming dark fuscous at apex; a medium-width pale fuscous anal streak; abdominal terga III–V orange-brown with a 'T' pattern consisting of a narrow transverse black band across anterior margin of tergum III and a medium-width medial longitudinal dark fuscous band over all three terga.

Description: male.

Head: Height 2.1 mm. Frons – length 1.35 times breadth; fulvous with black on anteromedial hump and around bases of orbital setae; orbital setae black: 1 *s.or.*, 2 *i.or.*; lunule red-brown. Ocellar triangle black. Vertex black. Face fulvous with a pair of medium-sized circular to oval black spots; length 0.6 mm. Genae fulvous, black seta present. Occiput red-brown, fulvous along eye margins; occipital row with 5–8 strong black setae. Antennae with segments 1 and 2 red-brown, segment 3 red-brown with fuscous on apex and outer surface; length of segments: 0.18 mm; 0.42 mm; 0.9 mm.

Thorax: Scutum dull black with red-brown below and behind lateral postsutural vittae, around notopleural suture and inside postpronotal lobes. Pleural areas black with red-brown below postpronotal lobes and a dark red-brown to fuscous behind mesopleural stripe. Yellow markings as follows: postpronotal lobes; notopleura; mesopleural stripe reaching midway between anterior margin of notopleuron and anterior *npl.* seta dorsally, continuing to katepisternum as a transverse spot, anterior margin slightly convex; anatergite (posterior apex black); anterior $^{3}/_{5}$ katatergite (remainder black); two broad parallel-sided lateral postsutural vittae ending behind *ia.* seta. Medial longitudinal postsutural yellow vitta absent. Postnotum black tending red-brown centrally. Scutellum yellow with moderately broad black basal band. Setae: *sc.* 2; *prsc.* 2; *ia.* 1; *p.sa.* 1; *a.sa.* 1; *mpl.* 1; *npl.* 2; *scp.* 4.

Legs: All femora entirely fulvous; fore and mid tibiae pale fuscous, hind tibiae fuscous basally to pale fuscous apically; all tarsi with segments entirely fulvous; mid tibiae each with an apical black spur.

Wings: Length 7.7 mm; cells bc and c pale fuscous to fuscous; microtrichia in outer corner of cell c only; remainder of wings colourless except fuscous cell sc, broad fuscous costal band confluent with R_{4+5} and becoming dark fuscous at apex, a medium-width pale fuscous anal streak; a dense aggregation of microtrichia around A_1+CuA_2; supernumerary lobe of medium development.

Abdomen: Oval; terga free; pecten present on tergum III. Tergum I and sterna I and II wider than long. Tergum I orange-brown with narrow fuscous lateral margins; tergum II fulvous with a narrow transverse dark fuscous band across anterior margin and narrow dark fuscous anterolateral corners; terga III–V orange-brown with a 'T' pattern consisting of a narrow transverse black band across anterior margin of tergum III and a medium-width medial longitudinal dark fuscous band over all three terga. A pair of oval black shining spots on tergum V. Posterior lobe of surstylus short, sternum V with a deep cavity on posterior margin. Abdominal sterna pale coloured (generally red-brown).

Attractant: Cue lure.

Female: No known record.

Distribution: Indonesia (Central Sulawesi, Donggala; South-East Sulawesi, Kendari).

Hosts: No known record.

Remarks: *Bactrocera (Bactrocera) neoritsemai* n.sp. is similar to *Bactrocera (Bactrocera) costalis* (Shiraki) and *Bactrocera (Bactrocera) ritsemai* (Weyenbergh) in possessing a black scutum, a broad costal band confluent with R_{4+5}, parallel-sided lateral postsutural vittae and abdominal terga III–V with a black 'T' pattern. It differs from *B. costalis* in having the lateral postsutural vittae ending behind the *ia.* setae, all femora entirely fulvous and a dark fuscous spot within the apex of the costal band and from *B. ritsemai* in having cells bc and c fuscous, not colourless.

Bactrocera (Bactrocera) nigrifacia Zhang, Ji & Chen

(Fig. 124)

Bactrocera (Bactrocera) nigrifacia Zhang, Ji & Chen, in Zhang *et al.*, 2011: 598. Holotype male in FAFU.

Other type material studied: CHINA: A study of the colour photographs and description of *Bactrocera (Bactrocera) nigrifacia* Zhang, Ji & Chen in the original publication.

Non-type material examined: THAILAND: 2 males and 5 females, Huai Muang, Chiang Mai, 2.ix.1993, coll. Chantanee *et al.*, bred from *Capparis seblaria* L., Sample No. CM2011; 2 males and 2 females, Tet Yoa, Chiang Mai, 8.ix.1993, coll. Chantanee, bred from *Melothria wallichii* cl., Sample No. CM2017; 2 males and 1 female, Mae Sapat, Chiang Mai, 2.ix.1993, coll. Chantanee *et al.*, bred from *Capparis seblaria* L., Sample No. CM2004; 8 males and 7 females, Huai Som, T Mae Sai, Ping Pro, Chiang Mai, 10.ix.1992, coll. Chantanee, bred from *Securinega virosa* (Roxb. ex Wild) Baill; 1 male and 3 females, Denpakun, T Maesay, Ping Proj, Chiang Mai, 19.viii.1992, coll. Chantanee, bred from *Securinega virosa* (Roxb. ex Wild) Baill; 6 males and 9 females, Huai Som, T Mae Sai, Ping Pro, Chiang Mai, 10.ix.1992, coll. Chantanee, bred from *Securinega virosa* (Roxb. ex Wild) Baill; 7 males and 17 females, Huai Som, T Mae Sai, Ping Pro, Chiang Mai, 10.ix.1992, coll. Chantanee, bred from *Securinega virosa* (Roxb. ex Wild) Baill.; 1 male, N. Thailand, 330, Chiang Mai Prov., Chiang Mai (Suandok), 18.79N/98.97E, 19.x.2000, coll. B. Merz, attracted to cue lure; 1 male (24.vi.1991), 2 males (7.vii.1991), Mae Sai, Chiang Rai, coll. Charuwan, attracted to cue lure; 1 male (3.vii.1991), 2 males (22.vii.1992), Pa Yao, Chiang Rai, coll. Charuwan, attracted to cue lure; 1 male (17.vi.1991), 2 males (8. vii.1991), Chiang San, Chiang Rai, coll. Charuwan, attracted to cue lure; 2 males (3.vii.1991), 1 male (13.xi.1991), 1 male (27.xi.1991), Phan, Chiang Rai, coll. Charuwan, attracted to cue lure; 1 male, Khun Khon, Chiang Rai, 23.i.1992, coll. Charuwan, CR320T, attracted to cue lure; 7 males, N.R.D.C., Chiang Mai, 11.vi.1991, coll. Chantanee Hengsawad, attracted to cue lure; 2 males (16.i.1992), 3 males (30.i.1992), 1 male (5.ii.1992), 1 male (12.ii.1992), 2 males (25.vi.1991), 2 males (27.ix.1991), N.R.D.C., Chiang Mai, coll. Chantanee, attracted to cue lure; 27 males, N.R.D.C., Chiang Mai, 11.vi.1991, coll. Chantanee *et al.*, attracted to cue lure; 1 male (7.vii.1991), 1 male (8.vii.1991), Mae Sai, Chiang Rai, coll. Charuwan, attracted to cue lure; 1 male (22. vii.1992), 1 male (18.viii.1992), 1 male (27.xi.1991), 2 males (26.ii.1992), Pha Yao, Chiang Rai, coll. Charuwan, attracted to cue lure; 1 male (3.vii.1991), 1 male (8.i.1992), 2 males (15.i.1992), Phan, Chiang Rai, coll. Charuwan, attracted to cue lure; 1 male, Phan, Chiang Rai, 3.vi.1991, coll. Charuwan Supasatien; 1 male (9.i.1992), 1 male (5.iii.1992), Khun Khon, Chiang Rai, coll. Charuwan, attracted to cue lure; 1 male (4.ii.1992), 1 male (3.iii.1992), Pa Ha, Chiang Rai, coll. Charuwan, attracted to cue lure; 1 male, Mae Hae, Chiang Mai, 18.ii.1992, coll. Chantanee, attracted to cue lure; 1 male, Doi Suthep – Pui Nat Park, 1425m, 17.v.1991, coll. D.L. Hancock, attracted to cue lure; 1 male, Doi Luong Nat Park, S.E. Chiang Mai, 24.v.1991, coll. D.L. Hancock, attracted to cue lure; 9 females and 7 males, Mae Hae, Chiang Mai, 20.vii.1993, coll. Chantanee *et al.*, bred from *Callicarpa arborea* Roxb, CM1975, 6 females and 5 males, Mon Yoa, Chiang Mai, 20.vii.1993, coll. Chantanee *et al.*, bred from *Securinega virosa*, CM1978; 7 females and 2 males, Pang Chang, Chiang Mai, 20.vii.1993, coll. Chantanee *et al.*, bred from *Securinega virosa*, CM1972. Specimens in QDPC.

Diagnosis: A small species; face fuscous to dark fuscous with large elongate oval black spots in antennal furrows (in some specimens the face appears entirely black); postpronotal lobes and notopleura yellow; scutum black; medium-width parallel-sided lateral postsutural yellow vittae present; medial postsutural yellow vitta absent; no yellow spot anterior to notopleural suture; mesopleural stripe reaching anterior *npl.* seta dorsally; scutellum yellow; wing with cells bc and c colourless, microtrichia in outer ⅓ of cell c only; a narrow dark fuscous costal band confluent with R_{2+3}; a narrow dark fuscous anal streak; abdominal terga III–V orange-brown with broad dark fuscous to black lateral longitudinal bands and a medium-width medial longitudinal black band over all three terga joined over anterior ½ of tergum III (in some specimens tergum III is entirely dark fuscous to black).

Description: male.

Head: Height 1.5 mm. Frons – length 1.43 times breadth; fuscous with fulvous lateral margins, dark fuscous on anteromedial hump and around bases of orbital setae; orbital setae black: 1 *s.or.*, 2 *i.or.*; lunule dark fuscous. Ocellar triangle black. Vertex dark fuscous. Face fuscous to dark fuscous with large elongate oval black spots in antennal grooves (in some specimens the face appears entirely black); length 0.45 mm. Genae fulvous, large fuscous to dark fuscous subocular spot present; black seta present. Occiput fuscous centrally, black laterally, fulvous along eye margins; occipital row with 2–6 black setae. Antennae with segments 1 and 2 red-brown, segment 3 red-brown with fuscous on apex and outer surface; length of segments: 0.2 mm; 0.32 mm; 0.74 mm.

Thorax: Scutum black with dark red-brown below and behind lateral postsutural vittae, around notopleural suture and inside postpronotal lobes. Pleural areas entirely black. Yellow markings as follows: postpronotal lobes; notopleura; mesopleural stripe reaching anterior *npl.* seta dorsally; continuing to katepisternum as a large transverse spot, anterior margin straight; anatergite (posterior apex black); anterior $^3/_5$ katatergite (remainder black); two medium-width parallel-sided lateral postsutural vittae ending at or just behind *ia.* seta. Medial longitudinal postsutural yellow vitta absent. Postnotum black. Scutellum yellow except for narrow black basal band. Setae: *sc.* 2; *prsc.* 2; *ia.* 1; *p.sa.* 1; *a.sa.* 1; *mpl.* 1; *npl.* 2; *scp.* 4.

Legs: Fore femora fulvous with dark fuscous covering entire outer surfaces and merging to inner surfaces, mid femora entirely black centrally with dark fuscous basally and apically, hind femora fulvous with dark fuscous around apical ⅓ to ¼; fore tibiae dark fuscous basally to fuscous apically, mid tibiae fuscous basally to fulvous apically, hind tibiae dark fuscous; tarsi with all segments entirely fulvous; mid tibiae each with an apical black spur.

Wings: Length 4.7 mm; cells bc and c colourless; microtrichia in outer ⅓ of cell c only; remainder of wings colourless except dark fuscous cell sc, narrow dark fuscous costal band confluent with R_{2+3} and remaining narrow around apex of wing, a narrow dark fuscous anal streak; a dense aggregation of microtrichia around A_1+CuA_2; supernumerary lobe of medium development.

Abdomen: Oval; terga free; pecten present on tergum III. Tergum I and sterna I and II wider than long. Tergum I orange-brown with narrow black lateral margins; tergum II black anteriorly with a narrow transverse orange-brown band across anterior margin but not reaching lateral margins and fulvous across posterior ½; terga III–V orange-brown with broad dark fuscous to black lateral longitudinal bands and a medium-width medial longitudinal black band joined along anterior ½ of tergum III (in some specimens tergum III is entirely dark fuscous to black). A pair of oval black shining spots on tergum V. Posterior lobe of surstylus short, sternum V with a deep concavity on posterior margin. Abdominal sterna dark fuscous to black.

Attractant: Cue lure.

Female: As for male except no dense aggregation of microtrichia around A_1+CuA_2, supernumerary lobe weak, pecten of cilia on abdominal tergum III absent; ovipositor basal segment dark fuscous, dorsoventrally compressed and tapering posteriorly in dorsal view; ratio of length of oviscape to length of segment V, 0.54:1; apex of piercer needle-shaped.

Distribution:

Type locality: China (Yunnan Province, Jinghong).

New records: Thailand (Chiang Mai).

Hosts: *Callicarpa arborea* (family *Verbenaceae*); *Capparis seblaria* L. (family *Capparaceae*); *Melothria wallichii* cl. (family *Cucurbitaceae*); *Securinega virosa* (Roxb. ex Wild.) Baill. (family *Euphorbiaceae*).

Remarks: *Bactrocera (Bactrocera) nigrifacia* Zhang, Ji & Chen is a member of the *nigrotibialis* complex with extensive dark fuscous to black markings over all femora and is similar to *Bactrocera (Bactrocera) nigrofemoralis* White & Tsuruta and *Bactrocera (Bactrocera) nigrotibialis* (Perkins) in possessing a dark fuscous face with large oval black spots and lateral postsutural vittae, but differs from these species in having the lateral postsutural vittae parallel-sided and reaching to the *ia.* seta (not tapering posteriorly to end before this seta). This species was noted by Hardy (1973: 47) as *Dacus (Strumeta)* new species near *nigrotibialis* (Perkins) but not formally described until further material could be collected for study.

Bactrocera (*Bactrocera*) *nigrifemorata* Lin & Wang

(Fig. 125)

Bactrocera (Bactrocera) nigrifemorata Lin & Wang, in Lin *et al.*, 2011b: 898. Holotype male in HEIQ.

Type material studied: CHINA: A study of the colour photographs and description of *Bactrocera (Bactrocera) nigrifemorata* Lin & Wang in the original description.

Diagnosis: A medium-sized species; face fulvous with a broad transverse black band across oral margin; postpronotal lobes and notopleura yellow; scutum black; broad parallel-sided lateral postsutural yellow vittae present; medial postsutural yellow vitta absent; no yellow spot anterior to notopleural suture; mesopleural stripe reaching midway between anterior margin of notopleuron and anterior *npl.* seta dorsally; scutellum yellow; wing with cells bc and c colourless, microtrichia in outer corner of cell c only; a narrow dark fuscous costal band confluent with R_{2+3}; a narrow dark fuscous anal streak; all abdominal terga entirely black except red-brown along posterior margin of tergum V.

Description: male.

Head: Frons fulvous with fuscous on anteromedial hump; orbital setae black: 1 *s.or.*, 2 *i.or.*; lunule fulvous. Ocellar triangle black. Vertex dark fuscous to black. Face fulvous with a broad transverse black band across oral margin. Genae fulvous. Occiput dark fuscous to black, fulvous along eye margins, occipital row with 4–6 black setae. Antennae with segments 1 and 2 red-brown, segment 3 red-brown with fuscous on outer surfaces.

Thorax: Scutum black without pale markings. Pleural areas entirely black. Yellow markings as follows: postpronotal lobes; notopleura; mesopleural stripe reaching midway between anterior margin of notopleuron and anterior *npl.* seta dorsally, continuing to katepisternum as a small transverse spot; anatergite (posterior apex black); katatergite (area covered by yellow not visible); two broad parallel-sided lateral postsutural vittae ending behind *ia.* seta. Medial longitudinal postsutural yellow vitta absent. Postnotum black. Scutellum yellow except for narrow black basal band. Setae: *sc.* 2; *prsc.* 2; *ia.* 1; *p.sa.* 1; *a.sa.* 1; *mpl.* 1; *npl.* 2; *scp.* 4.

Legs: All femora entirely black except for fulvous around basal ½ of hind femora; fore and mid tibiae fuscous, hind tibiae dark fuscous; all tarsal segments fulvous; mid tibiae each with an apical black spur.

Wings: Length 5.0 mm; cells bc and c colourless; microtrichia in outer corner of cell c only; remainder of wings colourless except dark fuscous cell sc, narrow dark fuscous costal band confluent with R_{2+3} and remaining very narrow around wing margin before expanding slightly around apex of R_{4+5}, a narrow dark fuscous anal streak; a dense aggregation of microtrichia around A_1+CuA_2; supernumerary lobe of medium development.

Abdomen: Oval; terga free; pecten present on tergum III. All terga entirely black except red-brown across posterior margin of tergum V. Posterior lobe of surstylus short, sternum V with a deep concavity on posterior margin.

Attractant: No known record.

Female: No known record.

Distribution:

Type locality: China (Hainan Province, Haikou).

Hosts: No known record.

Remarks: *Bactrocera (Bactrocera) nigrifemorata* Lin & Wang belongs to the *nigrotibialis* complex and is similar to *Bactrocera (Bactrocera) nigrifacia* Zhang, Ji & Chen in possessing dark facial patterns and parallel-sided lateral postsutural vittae, but differs from this species in having the face with a black band across the oral margin and abdominal terga I–V entirely black.

Bactrocera (Bactrocera) nigrofemoralis White & Tsuruta

(Fig. 126)

Bactrocera (Bactrocera) nigrofemoralis White & Tsuruta, in Tsuruta & White, 2001: 79; Drew & Raghu, 2002: 339; Drew *et al.*, 2007: 4. Holotype male in BMNH.

Non-type material examined: Since the publication of Drew & Raghu (2002), the following specimens have been studied – **BHUTAN**: 2 males (6.v.2000), 2 males (20.v.2000), Phuentsholing, coll. S. Thapa, attracted to cue lure. **INDIA**: 3 females (10.x.1987), 1 female and 2 males (21.x.1987), Karnataka, Bangalore, 916m, coll. G. Bhat, bred ex *Santalum album*; 1 male (4.x.1987), 1 male (10.x.1987), 1 male (13.x.1987), 1 male (16.x.1987), Karnataka, Bangalore, 916m, coll. G. Bhat, attracted to cue lure; 1 female and 2 males, S. India, Chethalli, 24.v.1992, coll. D.L. Hancock, bred from West Indian cherry, *Malpighia glabra* (L.); 1 female, S. India, Mudigere, 21.v.1992, coll. D.L. Hancock, bred from Pomelo, *Citrus grandis* Osbeck; 1 male, S. India, Kotigehar, Nr Mudigere, 900m, 20.v.1992, coll. D.L. Hancock, ex fallen fruit of *Callocarpum sapote* [= *Pouteria sapota*]; 11 males, S. India, 15km from Yercaud, 750m, 3.vi.1992, coll. D.L. Hancock, attracted to cue lure; 1 male, S. India, Kotigehar, Nr Mudigere, 900m, 20.v.1992, coll. D.L. Hancock; 1 male, S. India, Appangala, Nr Madikeri, 950m, 23.v.1992, coll. D.L. Hancock, attracted to cue lure; 1 male (20.v.1992), 1 male (21.v.1992), 1 male (22.v.1992), S. India, Mudigere, 980m, coll. D.L. Hancock, attracted to cue lure; 15 males, S. India, 37km E. Kodaikanal, 850m, 2.vi.1992, coll. D.L. Hancock, attracted to cue lure; 13 males, S. India, Talakaveri, Nr. Madikeri, 1100m, 23.v.1992, coll. D.L. Hancock, attracted to cue lure; 5 males, S. India, Burliar, Nr Coonoor, 760m, 29.v.1992, coll. D.L. Hancock; 1 male, Kotigehar, Nr Mudigere, 900m, 20.v.1992, coll. D.L. Hancock; 2 males, S. India, Thadiyankudisai, Nr Kodaikanal, 1000m, 1.vi.1992, coll. D.L. Hancock, attracted to cue lure; 1 male, S. India, Kodaikanal, 1900m, 31.v.1992, coll. D.L. Hancock, attracted to cue lure; 1 male, S. India, Kodaikanal, 1680m, 31.v.1992, coll. D.L. Hancock, attracted to cue lure; 1 male, S. India, Chethalli, Nr. Madikeri, 950m, 22-23.v.1992, coll. D.L. Hancock, attracted to cue lure; 7 males, S. India, Hort Farm, Coonoor, 1450m, 29.v.1992, coll. D.L. Hancock, attracted to cue lure; 2 males, S. India, Hort Farm, 1450m, Coonoor, 30.v.1992, coll. D.L. Hancock, attracted to cue lure; 4 males, Karadiputhur, Gumudipundi, Chinglepet, India, 4.viii.1988, coll. S. Vijay, attracted to cue lure in guava tree. **SRI LANKA (formerly CEYLON)**: 1 male, Gannoruwa, 3.v.1994, trap C6; 1 male, Gannoruwa, 26.xii.1994, trap C27; 1 male, Horana, 7.iii.1995, trap C4. Specimens donated by Kenji Tsuruta. Specimens in QDPC.

Diagnosis: Face shining black; postpronotal lobes yellow (anteromedial corners dark fuscous to black); notopleura yellow (except black around base forming a small yellow apex); scutum shining black without pale markings; narrow lateral postsutural yellow vittae, narrowing posteriorly to end well before *ia.* seta; medial postsutural yellow vitta absent; mesopleural stripe reaching to anterior *npl.* seta dorsally; scutellum yellow except for a moderately broad black basal band; legs with fore femora mostly shining black but may be dark fulvous on some or all of the inner surfaces, mid femora entirely shining black, hind femora fulvous with apical ⅓ shining black; fore tibiae fuscous (tending darker basally), mid tibiae fulvous (tending fuscous basally), hind tibiae dark fuscous; wing with cells bc and c colourless, microtrichia in outer corner of cell c only; a narrow fuscous costal band confluent with R_{2+3} and remaining very narrow around wing margin beyond extremity of R_{2+3} to end just beyond extremity of R_{4+5}; a narrow fuscous anal streak; supernumerary lobe of medium development; abdominal terga III–V red-brown except for a broad black band across anterior margin of tergum III which may expand to cover the entire tergum, broad black lateral longitudinal bands over all three terga or terga IV–V if abdominal tergum III is entirely black and a narrow medial longitudinal black band from base of tergum III to apex of tergum V, a pair of oval black shining spots on tergum V.

Attractant: Cue lure.

Female: No known record.
Distribution:
Type locality: Southern India (Karnataka).
Previously recorded: Type locality and many district locations in Southern India and Sri Lanka; Pakistan (Rawalpindi) (Tsuruta & White, 2001; Drew & Raghu, 2002); Bhutan (Drew *et al.*, 2007).
New records: Southern India (Appangala; Bangalore; Chethalli; Kotigehar; Nandi Hills).
Hosts: Recorded from one host, *Terminalia catappa* (family *Combretaceae*) (Tsuruta & White, 2001).
New records: *Pouteria sapota* (family *Sapotaceae*); *Citrus grandis* (family *Rutaceae*); *Malpighia glabra* (family *Malpighiaceae*); *Santalum album* (family *Santalaceae*).
Remarks: *Bactrocera (Bactrocera) nigrofemoralis* White & Tsuruta belongs to the *nigrotibialis* complex of species as discussed under *Bactrocera (Bactrocera) neonigrotibialis*. It is distinct in this complex in possessing a black face, the mesopleural stripe reaching to the anterior *npl.* seta dorsally and a broad black basal band on the scutellum.

Bactrocera (Bactrocera) nigrotibialis (Perkins)

(Figs 127A, 127B, 127C, 127D)

Strumeta nigrotibialis Perkins, 1938: 129. Lectotype male and 7 Paralectotypes here designated in QMIC.
Dacus (Strumeta) nigrotibialis – Hardy & Adachi, 1954: 178; Hardy, 1973: 47; Hardy, 1974: 38.
Bactrocera (Bactrocera) nigrotibialis – Norrbom *et al.*, 1998: 93.
Dacus (Strumeta) luzonae Hardy & Adachi, 1954: 174. **New Synonym**.
Dacus (Bactrocera) mulyonoi Hardy, 1983: 15. **New Synonym**.
Dacus (Bactrocera) sumatranus Hardy, 1983: 22. **New Synonym**.

Type material examined: PENINSULAR MALAYSIA: Lectotype male here designated with the following two labels – 'Malay Penin: Selangor, Bukit Kutu, 3500 ft., 13.9.1929, H.M. Pendlebury'; 'Holotype, Strumeta nigrotibialis, F.A. Perk'. Paralectotypes here designated: 2 females (one marked Holotype), same data as Lectotype male; 1 male and 1 female, 'Malay Penin: West Coast, Pulau Angoa, light house at light, Oct 26.1926, E. Seimund'; 1 female, 'Malay Penin, Kedah Peak, 3950 ft, 21 March 1928'; 1 male, 'Perak F.M.S., Larut Hills, 3700–4000 ft, 9 Feb. 1932, H.M. Pendlebury'; 1 male, 'Selangor F.M.S., Kuala Lumpur, Bukit Kutu, Dec 15, 1915, H.M. Pendlebury'. All Paralectotypes, except one marked holotype with the following label – 'Paratype, Strumeta nigrotibialis F.A. Perk.'. Received on loan from QMIC.
Other type material examined: Holotype male of *Dacus (Strumeta) luzonae* Hardy, with the following two labels – 'Luzon Is., P.I., Oct., 1947'; 'Collector Q.C. Chock', received on loan from USNM; paratype female with the following five labels – 'paratype'; 'Luzon P.I. Sept.-Oct. 1947'; 'paratype female *Dacus (Strumeta) luzonae* Hardy'; 'Pres. by D.E. Hardy BM1954-320'; 'BMNH(E) 533140'. Studied at BMNH. (Fig. 127B).
Holotype male of *Dacus (Bactrocera) mulyonoi* Hardy with the following three labels – 'Lombok, S.E. Slope of Mt. Rindjani, 400 m, 3 km N. Pesugulan', 'in primary forest at cue lure, June 12, 1975'; 'coll. D.E. & A.T. Hardy', studied at MZB; two paratypes of *Dacus (Bactrocera) mulyonoi* Hardy with the following label – 'Batu Belik, Lombok, 450 m, Dec. 1975, E. Malyono', received on loan from USNM. (Fig. 127C).
Holotype male of *Dacus (Bactrocera) sumatranus* Hardy with the following label – '12 km N.W. Bohorok, N. Sumatra, 200 m, 4 May 1975, coll. D.E. & A.T. Hardy, at cue lure', studied at MZB; two paratypes of *Dacus (Bactrocera) sumatranus* Hardy with the following label – 'Panti Forest Res, Pasaman, W. Sumatra, 280 m, May 7–8, 1975, D.E. Hardy'; one paratype of *Dacus (Bactrocera) sumatranus* Hardy with the following label – '12 km N.W. Bohorok, N. Sumatra 200 m, 4 May 1975, D.E. Hardy', 3 paratypes received on loan from USNM. One paratype of *Dacus (Bactrocera) sumatranus* Hardy, same data as previous paratype on loan from USNM from Bohorok, N. Sumatra, in QDPC (Fig. 127D).
Non-type material examined: BRUNEI: 5 males (23.i.2002), 4 males (25.i.2002), 7 males (27.i.2002), 6 males (29.i.2002), KBFC, Bato Apou Nat Pk, Riparian, coll. K. Hurley, attracted to cue lure; 6 males (23.i.2002), 6 males (25.i.2002), 7 males (27.i.2002), 7 males (29.i.2002), KBFC, Bato Apou Nat Pk, Upland, coll. K. Hurley, attracted to cue lure; 5 males (23.i.2002), 6 males (25.i.2002), 5 males (27.i.2002), 1 male (29.i.2002), KBFC, Bato Apou Nat. Pk, Forest, coll. K. Hurley, attracted to cue lure. **INDONESIA: KALIMANTAN** – 2 males (1.viii.2005), 2 males (8.viii.2005), 11 males (15.viii.2005), 6 males (22.viii.2005), 11 males (29. viii.2005), 9 males (5.ix.2005), East Kalimantan, Kutai Kartanegara, Batuah, coll. Ertha & Heru G., attracted to cue lure in forest; 1 male, East Kalimantan, Kutai Kartanegara, Batuah, coll. Ertha DP, attracted to cue lure in banana; 1 male, East Kalimantan, Kota Samarinda, Lempake, 15.viii.2005, coll. Arofah, attracted to cue lure in village; 2 males, West Kalimantan, Pontianak Pancaroba, 7.ix.2005, coll. Ertha & Heru G, attracted to cue lure in forest; 5 males (27.vii.2005), 2 males (3.viii.2005), 4 males (10.viii.2005), 2 males (17. viii.2005), West Kalimantan, Pontianak, Pancaroba, coll. A. Irom, attracted to cue lure in forest; 1 male, West Kalimantan, Pontianak, Sei Duri II, 26.vii.2005, coll. A. Irom, attracted to cue lure in citrus; 18 males, South Kalimantan, Banjar, Karang Intan, Mandiangin, 13.xi.2005, coll. Bilal, attracted to cue lure in forest; 4 males (31.vii.2005), 26 males (21. viii.2005), 15 males (2.x.2005), 32 males (16.x.2005), South Kalimantan, Banjar, Mandiangin, coll. Bilal & Sidrotul, attracted to cue lure in forest; **KRAKATOA** – 1 male, Rakata I., 1.ix.1990, coll. E. Schmidt, attracted to cue lure; **LOMBOK** – 10 males, Sapit, Lombok, 675-850m, 1.vi.1981, coll. J. Turner, attracted to cue lure; 3 males, Santong, Lombok, 1200-1300m, 2.vi.1981, coll. J. Turner, attracted to cue lure; 3 males, Sembalia & Belanting, Lombok, 10-50m, 1.vi.1981, coll. J. Turner; 2 males, Sasaot, Lombok, 400-550m, 31.v.1981, coll. J. Turner, attracted to cue lure; 1 male and 1 female, Mataram, Lombok, 19.xi.1988, coll. Slamet Ry, bred ex Katapang; 1 female, Mataram, Lombok, xi.1988, coll. J. Turner, bred from *Syzygium jambos*; 1 female, Sweta,

Lombok, 4.xii.1988, coll. J. Turner, bred ex jambu; **SINGKEP ISLAND** – 1 male, 80m, Singkep Isl., 10km NW Dabo, 17-18.vi.2001, leg. P. Schwendinger; **SUMBAWA** – 9 males, Tiumera & Brangrea, Sumbawa, 100-150m, 13.vi.1981, coll. J. Turner, Sclerophyll forest, wet & dry; 4 males, Semongkat, Sumbawa, 250-350m, 12.vi.1981, coll. J. Turner, attracted to cue lure; 1 male, Pulau Moyo, Sumbawa, 11.vi.1981, coll. J. Turner, Savannah & Sclerophyll forest; 1 male, Semamung & Kerang, Sumbawa, 100m, 13.vi.1981, coll. J. Turner, Teak forest with Sclerophyll remnants and Gallery forest, attracted to cue lure; **SUMATRA** – 1 male (9.viii.2006), 1 male (13.ix.2006), 1 male (20.ix.2006), 2 males (4.x.2006), Jambi, Muaro Jambi, Sebapo, coll. Hendri Y, attracted to cue lure in orange (Jeruk); 1 male (30.viii.2006), 1 male (6.ix.2006), Lampung, Lampung Utara, Bangun Sari, coll. Suhenda Arsol, attracted to cue lure in forest. **EAST MALAYSIA: SABAH** – 1 male, ACIAR Site Ht 4, Sabah Orchard, 25.ix.1991, attracted to cue lure in fruit orchard; 1 male, Sabah, Kg. Tabiu, nr. Tapang, W. of Sandakan, 23.vii.1992, coll. D.L. Hancock, attracted to cue lure; 1 male, Sabah, Mt. Kinabalu, 1500m, 17.x.1991, coll. P. Anthony, attracted to cue lure; 1 male, Sabah, Poring Hot Springs, nr. Mt. Kinabalu, 24.vii.1992, coll. D.L. Hancock, attracted to cue lure; **SARAWAK** – 2 males (3.ii.1992), 4 males (18.ii.1992), 3 males (3.iii.1992), Sarawak, Forest Arboretum, Semongok, Nr. Kuching, coll. S. Leong, attracted to cue lure; 2 males, Sarawak, Forest Arboretum, Semongok KT2, 17.x.1992, coll. Leong, attracted to cue lure; 2 males (10.viii.1992), 1 male (15.viii.1992), Sarawak, Tarat Agric Station, coll. Leong, KT3, attracted to cue lure; 2 males, Sarawak, ARC Semongok KT1, 8.viii.1992, coll. Leong, attracted to cue lure; 1 male, Sarawak, Ranchan Pool, Serian, 10.viii.1992, coll. Leong, KT4, attracted to cue lure; 1 male, Sarawak, Kpg Tema Mawang, Tebedu, Nr. Kuching, 22.iv.1992, coll. S. Leong, attracted to cue lure; 1 male, Sarawak, Gunung Mulu Nat Park, 28-30.vii.1992, coll. D.L. Hancock, attracted to cue lure. **PENINSULAR MALAYSIA**: 4 males, Pulau, Perhentian, August 1986, coll. C.S. Ooi, attracted to cue lure; 1 male, Pangsoon, Hulu Langat, 26.ix.1990, coll. D.L. Hancock, attracted to cue lure; 4 males, Hulu Langat, Kuala Lumpur, 15.v.1992, coll. D.L. Hancock, attracted to cue lure; 10 males, Telok Chempedak, nr. Kuantan, Malaya, 23-24.vii.1977, coll. D. Kistner, attracted to cue lure; 1 male, Dungan, Malaya, Malaysia, 28-30.vii.1977, coll. D. Kistner; 2 males, Mersing, Malaya, Malaysia, 2-3.viii.1977, coll. D. Kistner, attracted to cue lure; 3 males, Kuantan, Malaysia, 5-6.vi.1976, coll. G.R. Young, attracted to cue lure. **PHILIPPINES**: **LUZON ISLAND** – 1 male, Hugo Perez, Trece Martires City, 14.vi.1983, coll. A.M. Arida and N.E. Rivera; 1 male, Mt. Makiling (100m), 8.vi.1983, coll. B. Ramos, attracted to cue lure; 2 males, Nueva Ecija, 14.vi.1983, attracted to cue lure; 20 males, Pantabangan, Nueva Ecija, 30.vi.1983, R.C.P.C., Maligaya, attracted to cue lure; 3 males (26.v.1983), 8 males (7.vi.1983), Sepmin, Concepcion, Tarlae, coll. A.M. Tenorio, attracted to cue lure; 8 males, Sta Maria, Concepcion, Tarlae, coll. R. Mangan, attracted to cue lure; 4 males, Nueva Ecija, 8.vi.1983, coll. A.M. Tenorio; 1 male, Viejo, Iloilo, 10.vi.1983, coll. A.S. Nogodula, attracted to cue lure; 4 males, Santa Cruz, Zambales, 10.vii.1983, coll. Montevirgen, attracted to cue lure. **THAILAND**: 3 males (29.xi.1991), 1 male (6.xii.1991), 3 males (27.xii.1991), 2 males (10.i.1992), 1 male (17.i.1992), 10 males (31.i.1992), 2 males (4.ii.1992), 1 male (6.iii.1992), 2 males (27.iii.1992), 6 males (10.iv.1992), 1 male (7.vii.1992), Khao Kheaw, A. Pakphli, Nakhonnayok, coll. Montree, attracted to cue lure; 8 males (25.vi.1991), 2 males (8.vii.1991), 1 male (11.xi.1991), 5 males (23.vii.1991), 4 males (26.xi.1991), 1 male (10.xii.1991), 3 males (24.xii.1991), 1 male (27.xii.1991), 2 males (7.i.1992), 2 males (17.iii.1992), 8 males (1.iv.1992), Yang Pong Rat, A. Muang, Chanthaburi, coll. Montree, attracted to cue lure; 1 male (13.xii.1991), 1 male (20.xii.1991), 4 males (27.xii.1991), 1 male (3.i.1992), 1 male (10.i.1992), 4 males (31.i.1992), 6 males (7.ii.1992), 6 males (14.ii.1992), 1 male (6.iii.1992), 1 male (10.iv.1992), Wang Jum Pee, Pakchong, Nakhonratsima, coll. Montree, attracted to cue lure; 7 males, Khao Yai, Nakhonnayok, 14.vi.1991, coll. Montree, attracted to cue lure; 2 males, Khao Yai, T. Moosi, A. Pakchong, Nakhonratsima, 30.viii.1991, coll. Montree, BK70T, attracted to cue lure; 2 males, Khao Yai, Khao Kheaw, A. Pakphli, Nakhonnayok, 28.vi.1991, coll. Montree, BK20T, attracted to cue lure; 8 males, Khao Yai, Nr. Bangkok, Nakhonnayok, 21.vi.1991, coll. Montree, attracted to cue lure; 2 males (25.vi.1991), 3 males (23.vii.1991), 2 males (1.iv.1992), Philu Hort Res Stn, Laemsingh, Chanthaburi, coll. Montree, BK40T, attracted to cue lure; 5 females and 3 males, Tepatule, Pakphli, Nakhonnayok, 18.x.1992, coll. Prasert, bred from guava, *Psidium guajava* L.; 3 females and 3 males, Songkhla, Amphoe Sikao, Trang, 23.vii.1993, coll. Sieng *et al.*, bred from fruit sample No. SK11575; 2 males, Songkhla, Ko Hong Hill, 21.vi.1993, coll. Anuchit & Sieng, attracted to cue lure; 2 males, Songkhla, Tone Nga Chaeng Waterfall, 14.vi.1993, coll. Anuchit & Sieng, attracted to cue lure; 1 male, Pa Yao, Chiang Rai, 3.vii.1991, coll. Charuwan, attracted to cue lure; 1 male, Pa Yao, Chiang Rai, 26.vi.1991, coll. Charuwan Supasatien, attracted to cue lure; 1 male, Rayong, 80km W of Chanthaburi, 27.v.1991, coll. D.L. Hancock, coll. On *Occimum* sp.; 1 male, Tone-Nga Chang waterfall, Hat Yai, 8.v.1991, coll. D.L. Hancock, attracted to cue lure; 6 males, Philu Hort Res Stn, Chanthaburi, 28.v.1991, coll. D.L. Hancock, attracted to cue lure; 1 male, 200m, Trang, Khao Banthat Wild Isc, 2.viii.2000, leg. P. Schwendinger; 1 male, 650m, N. Thailand, Lamphun Prov., Doi Khuntan NP, 45km SE Chiang Mai, 18.50N/99.27E, 2.xi.2000, Merz & Schwendinger leg, attracted to cue lure; 1 male, Tone Nga, Chaeng Waterfall, 16.iii.1993, coll. Anuchit & Sieng, attracted to cue lure. **VIETNAM: Northern Vietnam** – 1 male, Son La Province, Moc Chau District, Tra Long, Site N007, 6.ix.1999, attracted to cue lure; 1 male, Hoa Binh Province, Ky Son District, Cao Phong, Site N011, 31.viii.1999, attracted to cue lure; 1 male, Lao Cai Province, Sapa 6 Village, Site N017, 30.vii.1999, attracted to cue lure; **Southern Vietnam** – 11 males, Dac Lac Province, Ku Mu Nga District, Quang Phu Village, Site VS09, 24.x.2002, attracted to cue lure; 1 male (30.iii.2002), 1 male (13.iv.2002), 1 male (11.v.2002), 4 males (3.viii.2002), Dong Nai Province, Long Khanh District, Nhan Nghia Village, Site VS06, coll. SOFRI staff, attracted to cue lure; 1 male (16.vii.2002), 2 males (23.vii.2002), Ben Tre Province, Chau Thanh District, Tien Thuy Village, Site VS04, attracted to cue lure; 4 males, Dong Nai Province, Long Khanh District, Tan Phong Village, Site S014, 9.vii.2000, attracted to cue lure; 4 males, Binh Duong Province, Lai Thieu Village, Site S020, 9.viii.1999, attracted to cue lure; Specimens in QDPC.

Diagnosis: A small species; face red-brown to dark fuscous with large oval black spots filling antennal furrows; postpronotal lobes and notopleura yellow; scutum entirely black; lateral postsutural yellow vittae present; medial postsutural yellow vitta absent; no yellow spot anterior to notopleural suture; mesopleural stripe reaching midway between anterior margin of notopleuron and anterior *npl.* seta dorsally; scutellum yellow with a moderately broad black basal band; wing with cells bc and c colourless, microtrichia in outer corner of cell c only; a narrow dark fuscous costal band slightly overlapping R_{2+3} on apical ½; a narrow fuscous anal streak; abdominal terga III–V generally dark fuscous to black and tending paler posterocentrally on each tergum.

Description: male (Fig. 127A).

Head: Height 1.2 mm. Frons – length 1.6 times breadth; red-brown with dark fuscous around orbital setae and on anteromedial hump; orbital setae black: 1 *s.or.*, 2 *i.or.*; lunule dark fuscous. Ocellar triangle black. Vertex dark fuscous. Face red-brown to dark fuscous with very large oval black spots filling most of the antennal furrow; length 0.4 mm. Genae fulvous, fuscous subocular spot present; black seta present. Occiput black, fulvous along eye margins; occipital row with 7–8 black setae. Antennae with segments 1 and 2 red-brown, segment 3 mostly dark fuscous; length of segments: 0.16 mm; 0.25 mm; 0.68 mm.

Thorax: Scutum entirely black except for a small area of dark red-brown above notopleura along anterior margin of notopleural suture. Pleural areas entirely black. Yellow markings as follows: postpronotal lobes; notopleura; mesopleural stripe of medium-width reaching midway between anterior margin of notopleuron and anterior *npl.* seta dorsally, continuing to katepisternum as a small transverse spot, anterior margin straight; anatergite (posterior apex black); anterior ⅔ katatergite (remainder black); two medium-width lateral postsutural vittae narrowing sharply posteriorly to end well before level of *ia.* seta. Medial longitudinal postsutural yellow vitta absent. Postnotum black. Scutellum yellow with a moderately broad black basal band. Setae: *sc.* 2; *prsc.* 2; *ia.* 1; *p.sa.* 1; *a.sa.* 1; *mpl.* 1; *npl.* 2; *scp.* 4.

Legs: Fore and mid femora entirely black, hind femora with basal $^3/_5$ fulvous and apical $^2/_5$ black; fore and hind tibiae entirely dark fuscous, mid tibiae dark fuscous tending fuscous apically; tarsi with all segments entirely fulvous; mid tibiae each with an apical black spur.

Wings: Length 4.7 mm; cells bc and c colourless; microtrichia in outer corner of cell c only; remainder of wings colourless except for dark fuscous cell sc, narrow dark fuscous costal band slightly overlapping R_{2+3} on apical ½ and remaining narrow before ending just beyond apex of R_{4+5}, a narrow fuscous anal streak; dense aggregation of microtrichia around A_1+CuA_2; supernumerary lobe of medium development.

Abdomen: Oval; terga free; pecten present on tergum III. Tergum I and sterna I and II wider than long. Tergum I entirely black; tergum II black with narrow transverse red-brown markings posterolaterally; tergum III black, merging to dark fuscous posterocentrally; tergum IV fuscous to dark fuscous with orange-brown posterocentrally either side of a medium-width fuscous medial longitudinal band; tergum V orange-brown with broad fuscous anterolateral margins and a narrow medial longitudinal fuscous band; a pair of oval dark fuscous shining spots on tergum V. Posterior lobe of surstylus short, sternum V with a deep concavity on posterior margin. Abdominal sterna dark.

Attractant: Cue lure.

Female: As for male except no dense aggregation of microtrichia around A_1+CuA_2; pecten of cilia on abdominal tergum III absent; ovipositor basal segment dark red-brown, dorsoventrally compressed and tapering posteriorly in dorsal view; ratio of length of oviscape to length of tergum V, 1:1; apex of piercer needle-shaped.

Distribution:

Type locality: Peninsular Malaysia (Selangor, Bukit Kutu).

Previously recorded: Type locality and the following locations – India; Laos; Sri Lanka; Thailand (Norrbom *et al.*, 1998).

New records: Brunei; East Malaysia (Sabah; Sarawak); Indonesia (Kalimantan; Krakatoa; Lombok; Singkep I.; Sumbawa; Sumatra); Philippines (Luzon I.); Vietnam.

Hosts: Recorded from one host, *Psidium guajava* (family *Myrtaceae*) (Allwood *et al.*, 1999).

New records: *Occimum* sp. (family *Lamiaceae*); *Syzygium jambos* (family *Myrtaceae*).

Remarks: *Bactrocera (Bactrocera) nigrotibialis* (Perkins) is not only a widespread species across South-East Asia but also occurs in high population numbers in rainforests. Consequently, large numbers of specimens are often collected in cue lure traps in these habitats. Hardy (1983) recorded large numbers of paratypes of *Dacus (Bactrocera) mulyonoi* (Hardy) and *Dacus (Bactrocera) sumatranus* (Hardy) collected at cue lure in rainforests. For this study, large numbers of *B. nigrotibialis* have been studied from localities across South-East Asia. Variations in face colour patterns, from fuscous with large oval black spots to dark fuscous with the spots barely visible to almost entirely black, are exhibited. Similarly, the abdominal terga III–V have variations in the amount of pale colour that occurs over the central areas of these terga either side of a medial longitudinal dark band. Hardy & Adachi (1954) for *Dacus (Strumeta) luzonae* Hardy & Adachi, and Hardy (1983) for *D. mulyonoi* and *D. sumatranus*, used these variations, together with slight variations in fore femora and abdominal tergum II colour patterns, to establish these species. Consequently, *D. luzonae*, *D. mulyonoi* and *D. sumatranus* are here made **New Synonyms** of *B. nigrotibialis* as there are no good species level characters that can be used to separate them. It is also noteable that guava has been recorded as the host for both *B. nigrotibialis* (Allwood *et al.*, 1999) and *D. luzonae* (Hardy, 1974).

In the original description, Perkins (1938) noted that 10 specimens were available for study. We have studied eight of these, two which have Holotype labels and the remainder have Paratype labels. Also their other label data agree with that listed in the original description. Because Perkins (1938) did not specify which particular specimen should be the Holotype, we hereby designate the male listed above with the Holotype label as the Lectotype and the remaining specimens as Paralectotypes.

Bactrocera (Bactrocera) nothaphoebe Drew & Romig, new species

(Fig. 128)

Material examined: Holotype female, **PENINSULAR MALAYSIA**: 10 km Gap Rest House, Pahang, Malaysia, 22.x.1986, coll. S. Vijaysegaran *et al.*, bred ex *Litsea* sp., sample No. 0774, deposited in BMNH.

Paratypes: **PENINSULAR MALAYSIA**: 2 males, same data as holotype. **THAILAND**: 4 males and 8 females, Thailand, Songkhla, Khao Nui Park, Rattaphum, 13.ii.1992, coll. Sieng *et al.*, bred from *Nothaphoebe ringiana* Gamble; 1 male and 2 females, Thailand, Songkhla, Tam Khoa Nui, Rattaphum, 7.v.1992, coll. Dusadee *et al.*, bred from *Nothaphoebe ringiana* Gamble; 1 male and 3 females, Thailand, Than To waterfall, Yala, Songkhla, 10.vii.1992, coll. Anuchit *et al.*, bred from *Eugenia* sp. 3 paratypes in ANIC, 2 paratypes in BMNH, 12 paratypes in QDPC, 4 paratypes in QMIC.

Diagnosis: A medium-sized species; face fulvous with a pair of medium-sized oval black spots; postpronotal lobes and notopleura yellow; scutum black; short lateral postsutural yellow vittae present; medial postsutural yellow vitta absent; no yellow spot anterior to notopleural suture; mesopleural stripe equal in width to notopleuron dorsally; scutellum yellow; wing with cells bc and c colourless, microtrichia in outer corner of cell c only; a broad dark fuscous costal band; a broad dark fuscous anal streak; abdominal terga III–V red-brown with a broad medial longitudinal black band and two broad lateral longitudinal dark fuscous bands over all three terga and joined along anterior margin of tergum III.

Description: female.

Head: Height 1.48 mm. Frons – length 1.45 times breadth; fulvous with dark fuscous on anteromedial hump and around bases of orbital setae; orbital setae black: 1 *s.or.*, 2 *i.or.*; lunule dark fuscous. Ocellar triangle black. Vertex dark fuscous. Face fulvous with a pair of medium-sized oval black spots; length 0.42 mm. Genae red-brown, large dark fuscous subocular spot present; black seta present. Occiput fuscous, fulvous along eye margins; occipital row with 4–6 black setae. Antennae with segments 1 and 2 red-brown, segment 3 red-brown with fuscous on apex and outer surface; length of segments: 0.16 mm; 0.28 mm; 0.69 mm.

Thorax: Scutum black with dark red-brown along lateral margins, around notopleural suture and inside postpronotal lobes. Pleural areas entirely black, tending dark fuscous below postpronotal lobes. Yellow markings as follows: postpronotal lobes; notopleura; mesopleural stripe equal in width to notopleuron dorsally, continuing to katepisternum as a transverse spot, anterior margin straight; anatergite (posterior apex black); anterior ⅔ katatergite (remainder black); two moderately broad and short lateral postsutural vittae ending at level of *a.sa.* seta. Medial longitudinal postsutural yellow vitta absent. Postnotum black. Scutellum yellow except for broad black basal band. Setae: *sc.* 2; *prsc.* 2; *ia.* 1; *p.sa.* 1; *a.sa.* 1; *mpl.* 1; *npl.* 2; *scp.* 4.

Legs: Fore and mid femora entirely dark fuscous except for fulvous around basal ends, hind femora entirely fulvous; all tibiae entirely dark fuscous; fore tarsi with basal segment fulvous and apical four segments dark fuscous, mid and hind tarsi with all segments entirely fulvous; mid tibiae each with an apical black spur.

Wings: Length 5.1 mm; cells bc and c colourless; microtrichia in outer corner of cell c only; remainder of wings colourless except dark fuscous cell sc, broad dark fuscous costal band confluent with R_{4+5} and of uniform width throughout, a broad dark fuscous anal streak; no dense aggregation of microtrichia around A_1+CuA_2; supernumerary lobe weak.

Abdomen: Oval; terga free; no pecten present on tergum III. Tergum I and sterna I and II wider than long. Tergum I dark fuscous with a narrow transverse red-brown band across posterior margin but not reaching lateral margins; tergum II fulvous with a broad transverse dark fuscous band subanteriorly and joining broad dark fuscous anterolateral corners; terga III–V red-brown with a broad medial longitudinal black band and two broad lateral longitudinal dark fuscous bands joined along anterior margin of tergum III. Basal segment of ovipositor red-brown, dorsoventrally compressed and tapering posteriorly in dorsal view; length of oviscape to length of tergum V, 0.65:1. Apex of piercer needle-shaped. A pair of oval red-brown shining spots on tergum V. Sterna dark coloured.

Attractant: No known record.

Male: As for female except with a dense aggregation of microtrichia around A_1+CuA_2 present, pecten of cilia on abdominal tergum III present, posterior lobe of surstylus short; abdominal sternum V with a deep concavity on posterior margin.

Distribution: Peninsular Malaysia (Pahang); Thailand (Songkhla).

Hosts: *Eugenia* sp. (family *Myrtaceae*); *Nothaphoebe ringiana* (family *Lauraceae*).

Remarks: *Bactrocera (Bactrocera) nothaphoebe* n.sp. is similar to *Bactrocera (Bactrocera) limbifera* (Bezzi) in possessing a black scutum, the costal band confluent with R_{4+5} and not expanding into a spot at apex and femora with large areas of dark fuscous to black. It differs from *B. limbifera* in having dark patterns on the fore and mid femora with the hind femora entirely fulvous and the lateral postsutural vittae very short and triangular.

Bactrocera (Bactrocera) obscurata (de Meijere)

(Figs 129A, 129B)

Dacus ferrugineus var *obscurata* de Meijere, 1911: 373. Holotype female in ZMAN.

Dacus (Bactrocera) obscuratus – Hardy, 1983: 17.

Bactrocera (Bactrocera) obscurata – Norrbom *et al.*, 1998: 93.
Bactrocera (Bactrocera) heppneri White, in White & Evenhuis, 1999: 496. **New Synonym**.

Type material examined: Holotype female, **INDONESIA**: with the following five labels – 'E. Jacobson, Enkluizen, Poeloe Njamoek Ketjil, Java, Sept. 1907'; '32'; 'Dacus ferrugineus var obscuratus, det. de Meijere, Type'; 'Examined det. I.M. White, 1995'; 'Dacus ferrugineus obscurata de Meijere, 1911, ZMAN type DIPT.1105.1'. Reported lost by Hardy (1983) and Norrbom *et al.* (1998), but received on loan from ZMAN.
Other type material examined: Holotype male of *Bactrocera (Bactrocera) heppneri* White, **TAIWAN**: with the following three labels – 'holotype'; 'TAIWAN, Pingtung Co., Kenting Park 50 m, 31.x.–1.xi.1984, Coll. J.B. Heppner & H. Wang'; 'Holotype, Bactrocera heppneri White, det. I.M. White 1999' (Fig. 129B); two female paratypes of *B. heppneri* with the same label data as Holotype. Received on loan from FSCA for study (see 'Remarks' below).
Non-type material examined: JAPAN: IRIOMOTE ISLAND – 5 males and 8 females, Iriomote I., Okinawa, 4.vii.1984, coll. T. Teruya, bred ex *Diospyros maritima*. Specimens in QDPC.
Diagnosis: A medium-sized species; face fulvous with a pair of medium-sized oval black spots; postpronotal lobes and notopleura yellow; scutum entirely black; lateral and medial postsutural yellow vittae absent; no yellow spot anterior to notopleural suture; mesopleural stripe reaching to anterior *npl.* seta dorsally; scutellum yellow with narrow black basal band; wing with cells bc and c colourless, microtrichia in outer ½ of cell c only; a narrow fuscous costal band slightly overlapping R_{2+3} on apical ½ of this vein and remaining of uniform width to end between apices of R_{4+5} and M; a broad fuscous anal streak; abdominal terga III–V mostly black with small fulvous markings posterocentrally either side of medial longitudinal dark bands.
Description: female.
Head: Missing in type specimen. This description of the head is taken from the Holotype of *Bactrocera heppneri* White. Height 1.75 mm; Frons – length 1.67 times breadth; fulvous with dark fuscous around orbital setae and on anteromedial hump; orbital setae large and black: 1 *s.or.*, 2 *i.or.*; lunule dark fuscous. Ocellar triangle black. Vertex dark fuscous. Face fulvous with a pair of medium-sized oval black spots; length 0.6 mm. Genae fulvous, fuscous subocular spot present; black seta present. Occiput dark fuscous to black, fulvous dorsocentrally and along eye margins; occipital row with 5 strong black setae. Antennae with segments 1 and 2 red-brown, segment 3 red-brown with dark fuscous on apex and outer surface; length of segments: 0.12 mm; 0.34 mm; 0.93 mm.
Thorax: Scutum entirely black without pale markings. Pleural areas entirely black without pale markings. Yellow markings as follows: postpronotal lobes; notopleura (yellow extending slightly above notopleuron onto scutum); mesopleural stripe reaching to anterior *npl.* seta dorsally, continuing to katepisternum as a transverse spot, anterior margin slightly convex; anatergite (posterior apex black); anterior $^3/_5$ katatergite (remainder black). Lateral and medial longitudinal postsutural yellow vittae absent. Postnotum black. Scutellum yellow except for narrow black basal band. Setae: *sc.* 2; *prsc.* 2; *ia.* 1; *p.sa.* 1; *a.sa.* 1; *mpl.* 1; *npl.* 2; *scp.* 4.
Legs: Fore femora fuscous with dark fuscous on outer surfaces, mid femora dark fuscous except fulvous basally, hind femora fulvous tending red-brown apically; fore tibiae pale fuscous, mid tibiae fuscous on basal ½ and pale fuscous on apical ½, hind tibiae dark fuscous; tarsi with all segments entirely fulvous; mid tibiae each with an apical black spur. In some specimens the femora are entirely fulvous.
Wings: Length 5.25 mm; cells bc and c colourless; microtrichia in outer ½ of cell c only; remainder of wings colourless except fuscous cell sc, narrow fuscous costal band slightly overlapping apical ½ of R_{2+3} and remaining of even width around apex of wing to end between apices of R_{4+5} and M, broad fuscous anal streak; no dense aggregation of microtrichia around A_1+CuA_2; supernumerary lobe weak.
Abdomen: Oval; terga free; no pecten on tergum III. Tergum I and sterna I and II wider than long. Tergum I black except for a narrow transverse fulvous band across posterior margin interrupted medially; tergum II black except for a narrow transverse fulvous band across posterior margin interrupted by a narrow medial longitudinal black band; terga III and IV black except for narrow transverse fulvous spots posterocentrally; tergum V black except for a transverse fulvous area across posterior margin which extends anteriorly to anterior margin either side of a medial longitudinal black band. A pair of dark fuscous shining spots on tergum V. Ovipositor basal segment dark fuscous, dorsoventrally flattened and tapering posteriorly in dorsal view; ratio of length of oviscape to length of tergum V, 1:1; apex of piercer needle-shaped.
Attractant: No known record.
Male: The following description is taken from the Holotype of *Bactrocera heppneri* White and specimens listed under 'Non-type material examined'. As for female except for dense aggregation of microtrichia around A_1+CuA_2 present; supernumerary lobe of medium development; pecten of cilia on abdominal tergum III present; posterior lobe of surstylus short; abdominal sternum V with a slight concavity on posterior margin.
Distribution:
Type locality: Indonesia (Java).
New records: Japan (Iriomote I.). Taiwan, due to **New Synonymy** of *B. heppneri* listed under 'Other type material examined'.
Hosts:
New records: *Diospyros maritima* (family *Ebenaceae*).
Remarks: For this study the female Holotype of *Bactrocera (Bactrocera) obscurata* (de Meijere), and the male Holotype and two female Paratypes of *Bactrocera (Bactrocera) heppneri* White have been studied and compared. No differences can be observed between these species and so *B. heppneri* is considered a **New Synonym** of *B. obscurata*. The short surstylus lobe and shallow posterior margin on sternum V are present on the types of *B. obscurata* and *B. heppneri* and also the male specimens from Japan which are listed above under 'Non-type material examined'. This species is provisionally placed in subgenus *Bactrocera*. However, in possessing short

surstylus lobes and the male abdominal sternum V with a shallow posterior margin it could be placed in subgenus *Hemizeugodacus*.

B. obscurata is similar to *Bactrocera (Hemizeugodacus) abdomininigra* Drew and *Bactrocera (Bactrocera) thistletoni* Drew, in general body colour patterns, particularly the black scutum without postsutural vittae and the black abdomen; however, it differs from *B. abdomininigra* in having yellow postpronotal lobes, entirely yellow notopleura and a narrow basal band on the scutellum and from *B. thistletoni* in having yellow postpronotal lobes and the narrow basal band on the scutellum.

Bactrocera (Bactrocera) obscurivitta Drew & Romig, new species

(Fig. 130)

Material examined: Holotype male, **INDONESIA: MOLUCCAS (MALUKU): AMBON ISLAND** – Middle Maluku, Suli, 2.vi.2007, coll. La Naibu, attracted to cue lure in Hutan amalatuei, deposited in BMNH.

Paratypes: **INDONESIA: MOLUCCAS (MALUKU): AMBON ISLAND** – 1 male (9.vi.2007), 1 male (1.vii.2007), Middle Maluku, Suli, coll. La Naibu, attracted to cue lure in Hutan amalatuei. 1 paratype in QDPC, 1 paratype in QMIC.

Diagnosis: A large species; face fulvous with a pair of small irregularly circular black spots; postpronotal lobes fuscous; notopleura yellow; scutum generally fuscous; lateral postsutural yellow vittae absent (in one specimen there are very short and pale indistinct vittae present); medial postsutural yellow vitta absent; no yellow spot anterior to notopleural suture; mesopleural stripe reaching midway between anterior margin of notopleuron and anterior *npl.* seta dorsally; scutellum yellow; wing with cells bc and c fuscous, microtrichia in outer corner of cell c only; a broad dark fuscous costal band confluent with R_{4+5}; a broad dark fuscous anal streak; abdominal terga III–V with an overall dark appearance and generally with a variable pattern from dark red-brown through fuscous to dark fuscous.

Description: male.

Head: Height 2.1 mm. Frons – length 1.58 times breadth; red-brown with fuscous on anteromedial hump and around bases of orbital setae; orbital setae black: 1 *s.or.*, 2 *i.or.*; lunule red-brown. Ocellar triangle black. Vertex fuscous. Face fulvous with a pair of small irregularly circular black spots; length 0.6 mm. Genae fulvous, pale fuscous subocular spot present; red-brown seta present. Occiput fuscous, fulvous along eye margins; occipital row with six black setae. Antennae with segments 1 and 2 red-brown, segment 3 red-brown with fuscous on outer surface; length of segments: 0.2 mm; 0.4 mm; 1.04 mm.

Thorax: Scutum entirely fuscous with indistinct narrow lateral longitudinal dark fuscous lines. Pleural areas entirely fuscous to dark fuscous. Yellow markings as follows: notopleura; mesopleural stripe reaching midway between anterior margin of notopleuron and anterior *npl.* seta dorsally, not continuing to katepisternum, anterior margin straight; anatergite (posterior apex dark fuscous); anterior ⅔ katatergite (remainder dark fuscous); two very short and narrow indistinct lateral postsutural vittae in one specimen only. Medial longitudinal postsutural yellow vitta absent. Postpronotal lobes fuscous. Postnotum fuscous to dark fuscous. Scutellum yellow with a narrow dark basal band. Setae: *sc.* 2; *prsc.* 2; *ia.* 1; *p.sa.* 1; *a.sa.* 1; *mpl.* 1; *npl.* 2; *scp.* 4.

Legs: All femora entirely fulvous; fore tibiae pale fuscous basally to fulvous apically, mid tibiae fulvous, hind tibiae fuscous; all tarsal segments entirely fulvous; mid tibiae each with an apical black spur.

Wings: Length 7.8 mm; cells bc and c fuscous; microtrichia in outer corner of cell c only; remainder of wings with a pale fuscous tint across membrane except for a dark fuscous cell sc, broad dark fuscous costal band confluent with R_{2+3} and becoming paler adjacent to this vein and remaining of uniform width around apex of wing, a broad dark fuscous anal streak; a dense aggregation of microtrichia around A_1+CuA_2; supernumerary lobe of medium development.

Abdomen: Oval; terga free; pecten present on tergum III. Tergum I and sterna I and II wider than long. Tergum I fuscous to dark fuscous; tergum II red-brown to dark fuscous anteriorly and fulvous to dark red-brown posteriorly; terga III–V with an overall dark appearance with dark fuscous across anterior margin of tergum III and a medial longitudinal dull black band over all three terga and very broad lateral longitudinal dark fuscous to dark red-brown bands over all three terga and tending dark red-brown in medial sections of these terga either side of the medial longitudinal dull black band. A pair of oval dark fuscous shining spots on tergum V. Posterior lobe of surstylus short, sternum V with a deep concavity on posterior margin. Abdominal sterna dark fuscous to black.

Attractant: Cue lure.

Female: No known record.

Distribution: Indonesia (Moluccas = Maluku, Ambon I., Suli).

Hosts: No known record.

Remarks: *Bactrocera (Bactrocera) obscurivitta* n.sp. is unique within the subgenus *Bactrocera* in having postsutural yellow vittae either entirely absent or, at most, with very short, pale, almost indistinct lateral postsutural vittae. It is similar to *Bactrocera (Bactrocera) obscurata* (de Meijere) and *Bactrocera (Bactrocera) thistletoni* Drew which lack lateral and medial postsutural yellow vittae but differs from *B. obscurata* in having fuscous postpronotal lobes and from *B. thistletoni* in being an overall fuscous colouration (not black) and in having a broad costal band confluent with R_{4+5} and fuscous cells bc and c.

Bactrocera (Bactrocera) occipitalis (Bezzi)

(Fig. 131)

Chaetodacus ferrugineus var. *occipitalis* Bezzi, 1919: 423. Lectotype male in MCSNM.

Dacus (Strumeta) dorsalis var. *occipitalis* – Hardy & Adachi, 1954: 166 (partim); Hardy, 1969a: 405; Hardy, 1969b: 479, Lectotype designation.

Dacus (Strumeta) occipitalis – Hardy, 1974: 39.
Dacus (Bactrocera) occipitalis – Hardy, 1977: 51; Drew, 1982: 81.
Bactrocera (Bactrocera) occipitalis – White & Elson-Harris, 1992: 192; Drew & Hancock, 1994a: 45; Norrbom *et al.*, 1998: 93.

Non-type material examined: Since the publication of Drew & Hancock (1994a), the following specimens have been studied – **BRUNEI**: 6 males (23.i.2002), 10 males (25.i.2002), 24 males (27.i.2002), 28 males (29.i.2002), KBFC, Bato Apou Nat. Park, coll. K. Hurley, attracted to methyl eugenol. **INDONESIA: KALIMANTAN** – 27 males, Central Kalimantan, Kota Waringin Timur, Kota Besi, 13.ix.2005, coll. Utik Darmanto & Kusyono, attracted to methyl eugenol in citrus; 14 males, West Kalimantan, Pontianak, Pasir, 16.viii.2005, coll. A. Irom, attracted to methyl eugenol in village carambola; 1 male, West Kalimantan, Pontianak, Pancaroba, 3.viii.2005, coll. A. Irom, attracted to methyl eugenol in forest; 5 males, South Kalimantan, Banjar, Mandiangin, 24.vii.2005, coll. Bilal Sidrotul, attracted to methyl eugenol in forest. **PHILIPPINES**: 26 males (12.xii.1999), 4 males (12.i.2000), Mt. Makiling botanic gardens, alt. 150m, coll. A. O'Toole, attracted to Zorge 95 (methyl eugenol); 9 males, Sirang Lupa, Calamba, Luzon, 7.i.2000, coll. G. Quimio, attracted to Zorgen 95 (methyl eugenol) in guava (mixed) orchard; 66 males, Sirang Lupa, Calamba, Laguna, 7.i.2000, coll. G. Quimio, attracted to Zorgen 95 (methyl eugenol) in citrus orchard; 31 males, Sirang Lupa, Calamba, Laguna, 7.i.2000, coll. G. Quimio, attracted to Zorgen 95 (methyl eugenol) in carambola/guava orchard; 32 males, Sirang Lupa, calamba, Laguna, xii.1999, coll. A. O'Toole & G. Quimio, attracted to Zorgen 95 (methyl eugenol) in citrus orchard; 1 male, Sirang Lupa, Calamba, Laguna, 20.xii.1999, coll. A. O'Toole & G. Quimio, attracted to Zorgen 95 (methyl eugenol) in mixed orchard; 11 males, Sirang Lupa, Calamba, Luzon, 20.xii.1999, coll. A. O'Toole & G. Quimio, attracted to Zorgen 95 (methyl eugenol) in guava (big) orchard; 7 males, Sirang Lupa, Calamba, Laguna, 20.xii.1999, coll. A. O'Toole & G. Quimio, attracted to Zorgen 95 (methyl eugenol) in Apple guava orchard; 15 males, UPLB Campus, 8.i.2000, coll. G. Quimio, attracted to Zorgen 95 (methyl eugenol) in rainforest area mixed vegetation. Specimens in QDPC.

Diagnosis: Face fulvous with a pair of large oval black spots; postpronotal lobes and notopleura yellow; scutum black except dark red-brown along posterior margin and enclosing *prsc.* setae, below and behind lateral postsutural vittae, around notopleural suture, around anterior margin of notopleura and inside postpronotal lobes; broad parallel-sided or subparallel lateral postsutural yellow vittae ending at *ia.* seta (in some specimens the vittae end behind the *ia.* seta); medial postsutural yellow vitta absent; mesopleural stripe reaching midway between anterior margin of notopleuron and anterior *npl.* seta dorsally; scutellum yellow; legs with femora entirely fulvous; fore tibiae pale fuscous to fuscous, mid tibiae pale fuscous to fuscous basally tending paler apically, hind tibiae fuscous; wing with cells bc and c colourless, microtrichia in outer corner of cell c only; a narrow fuscous costal band distinctly overlapping R_{2+3} and widening markedly across apex of wing; a narrow fuscous anal streak; supernumerary lobe of medium development; abdominal terga III–V with a narrow transverse black band across anterior margin of tergum III and expanding to cover lateral margins, dark fuscous to black rectangular markings anterolaterally on tergum IV which sometimes continue to cover posterolateral margins of this tergum, dark fuscous to black anterolateral corners on tergum V, a very broad medial longitudinal black band over all three terga, a pair of oval orange-brown shining spots on tergum V; abdominal sterna dark coloured.

Attractant: Methyl eugenol.

Female: As for male in the general body colour patterns. See Drew & Hancock (1994a) for a detailed description of female characters.

Distribution:

Type locality: Philippines (Luzon I.).

Previously recorded: Type locality and East Malaysia (Sabah); Philippines.

New records: Brunei; Indonesia (Kalimantan).

Hosts: Recorded from three plant families – *Anacardiaceae*; *Myrtaceae*; *Rutaceae* (see Allwood *et al.*, 1999 for host records). These host records are incomplete due to a lack of field host survey work throughout the area of distribution of the species. *Bactrocera occipitalis* is a major pest species within the *dorsalis* complex of South-East Asia.

Remarks: *Bactrocera (Bactrocera) occipitalis* (Bezzi), within the *dorsalis* complex, is similar to *Bactrocera (Bactrocera) bitungiae* n.sp., some specimens of *Bactrocera (Bactrocera) fuscitibia* Drew & Hancock and *Bactrocera (Bactrocera) laithieuiae* n.sp. in possessing broad parallel-sided lateral postsutural vittae, costal band overlapping R_{2+3}, narrow- to medium-width dark patterns on abdominal terga III–V, shining spots on abdominal tergum V pale coloured, femora entirely fulvous or with a dark spot on outer apical surfaces of fore femora only and a broad medial longitudinal dark band on abdominal terga III–V. It differs from *B. laithieuiae* n.sp. in having broad oval facial spots (not narrow elongate oval spots), from *B. fuscitibia* in having the anteromedial bare area on the scutum broad and lateral dark markings on abdominal terga IV and V of medium-width (not narrow) and from *B. bitungiae* n.sp. in having the scutum with large areas of red-brown (not entirely black), the anal streak narrow, the femora entirely fulvous, cells bc and c colourless and the costal band and anal streak fuscous to dark fuscous (not very dark fuscous). Some populations of fruit flies throughout South-East Asia have been misidentified as *B. occipitalis* in previous literature. See Drew & Hancock (1994a) for a complete discussion on this species and previous misidentifications.

Norrbom *et al.* (1998: 93) stated that the Lectotype designation by Hardy (1969b) was invalid. However, this specimen was received on loan from MCSNM and studied by Drew & Hancock (1994a). It has the following label data – Luzon, Manila, October 1918, McGregor – which agrees with some of the type series listed by Bezzi (1919) in the original description. Further, Norrbom (pers. comm.) advised that the types of Bezzi species are in the MCSNM, not the USNM where some non-type specimens are lodged.

Consequently, we accept that the Lectotype designation by Hardy (1969b) is valid.

Within the *dorsalis* complex there is debate over the validity of some pest species. However, we are of the opinion that *B. occipitalis* is a good species and can be easily distinguished from the other pest species in the complex. Based on an analysis of the mitochondrial genes, COI and ND5, *B. occipitalis* is significantly different from *B. carambolae* (Bali, Peninsular Malaysia specimens), *B. dorsalis* (Taiwan specimens), *B. invadens* (Sri Lanka specimens) and *B. papayae* (specimens from Christmas Island, Java, Kalimantan, Peninsular Malaysia) (Drew, unpublished data).

Bactrocera (Bactrocera) ochroma Drew & Romig, new species

(Fig. 132)

Material examined: Holotype male, **INDONESIA: JAVA** – East Java, Probolinggo, Betektaman, 20.vii.2005, coll. Tjipto Waluyo, attracted to methyl eugenol in mango, deposited in BMNH.

Paratypes: **INDONESIA: JAVA** – 16 males, same data as holotype; 2 males, East Java, Megetan, Delotan, 19.vii.2005, coll. Trigi Yanto, attracted to methyl eugenol in pomelo; 2 males, East Java, Lumajang, Kadang Tepus, 2.viii.2005, coll. Utomo, attracted to methyl eugenol in banana; 69 males (14.vii.2005), 18 males (21.vii.2005), 9 males (18.viii.2005), West Java, Majalengka, Ampel (25 dpl), coll. Yaya Senjaya, attracted to methyl eugenol in mango; 1 male, Java, Banten, Lebok, Pasir Ninl–nini, 1.viii.2005, coll. TB. Aden Haer, attracted to methyl eugenol in forest; 2 males, Java, Banten, Tangerang, Tanah Tinggi, 23.viii.2005, coll. Komarudin, attracted to methyl eugenol in village mango. 20 paratypes in ANIC, 20 paratypes in BMNH, 59 paratypes in QDPC, 20 paratypes in QMIC.

Non-type material examined: INDONESIA: BALI – 11 males, Bali, Buleleng, Tajun, 8.ix.06, coll. Ketut Rawa W, attracted to methyl eugenol in forest; **SULAWESI** – 3 males, Sulawesi Selatan, Maros, Turikale, 19.ix.06, coll. Sirajuddin SP, bred ex *Mangifera indica*, family Anacardiaceae (mango), sample No. GF15(3); **SUMATRA** – 2 males, Sumatera Selatan, Ogan, Komering Llir, Srinanti, 21.viii.06, coll. Imam Muhayani, attracted to methyl eugenol in Duku, village; 1 male, Sumatera Selatan, Komering Llir, Celikah, 11.ix.06, coll. Imam Muhayani, attracted to methyl eugenol in Duku.

Diagnosis: A medium-sized species; face fulvous with a pair of medium-sized circular black spots; postpronotal lobes and notopleura yellow; scutum dull black; lateral postsutural yellow vittae present; medial postsutural yellow vitta absent; no yellow spot anterior to notopleural suture; mesopleural stripe reaching midway between anterior margin of notopleuron and anterior *npl.* seta dorsally; scutellum yellow; wing with cells bc and c colourless, microtrichia in outer corner of cell c only; a very narrow fuscous to dark fuscous costal band confluent with R_{2+3}; a narrow pale fuscous anal streak; abdominal terga III–V orange-brown with a 'T' pattern consisting of a narrow transverse black band across anterior margin of tergum III and a narrow- to medium-width medial longitudinal black band over all three terga; abdominal sterna II–V pale orange-brown.

Description: male.

Head: Height 1.6 mm. Frons – length 1.36 times breadth; red-brown centrally, fulvous along lateral margins and fuscous around orbital setae and on anteromedial hump; orbital setae black: 1 *s.or.*, 2 *i.or.*; lunule fuscous. Ocellar triangle black. Vertex fuscous. Face fulvous with a pair of medium-sized circular black spots; length 0.44 mm. Genae fulvous; black seta present. Occiput red-brown, fulvous along eye margins; occipital row with 6–8 small black setae. Antennae with segments 1 and 2 red-brown, segment 3 red-brown with fuscous on apex and outer surface; length of segments: 0.19 mm; 0.26 mm; 0.67 mm.

Thorax: Scutum dull black with red-brown below and behind lateral postsutural vittae, around notopleural suture, inside postpronotal lobes and a triangular marking of dark red-brown enclosing *prsc.* setae (in some specimens the red-brown areas expand over the central area of the scutum). Pleural areas dark fuscous to black except red-brown below postpronotal lobes. Yellow markings as follows: postpronotal lobes; notopleura; mesopleural stripe reaching midway between anterior margin of notopleuron and anterior *npl.* seta dorsally, continuing to katepisternum as a small transverse spot, anterior margin slightly convex; anatergite (posterior apex black); anterior ⅔ katatergite (remainder black); two moderately broad subparallel lateral postsutural vittae ending at or before *ia.* seta (in some specimens the lateral postsutural vittae tend to narrow a little more sharply posteriorly). Medial longitudinal postsutural yellow vitta absent. Postnotum black. Scutellum yellow except for narrow black basal band. Setae: *sc.* 2; *prsc.* 2; *ia.* 1; *p.sa.* 1; *a.sa.* 1; *mpl.* 1; *npl.* 2; *scp.* 4.

Legs: All femora entirely fulvous; fore and mid tibiae pale fuscous basally to fulvous apically, hind tibiae fuscous basally to pale fuscous apically; tarsi with basal two segments fulvous and apical three segments red-brown; mid tibiae each with an apical black spur.

Wings: Length 5.5 mm; cells bc and c colourless; microtrichia in outer corner of cell c only; remainder of wings colourless except fuscous to dark fuscous cell sc, narrow fuscous to dark fuscous costal band confluent with R_{2+3} and remaining narrow around apex of wing or with a slight fish-hook pattern around apex of R_{4+5}, a narrow very pale fuscous anal streak; a dense aggregation of microtrichia around A_1+CuA_2; supernumerary lobe of medium development.

Abdomen: Oval; terga free; pecten present on tergum III. Tergum I and sterna I and II wider than long. Tergum I dark fuscous except for a narrow transverse orange-brown band across posterior margin but not reaching lateral margins; tergum II orange-brown with fulvous across posterior margin and, at most, very narrow fuscous anterolateral corners; terga III–V orange-brown with a 'T' pattern consisting of a narrow transverse black band across anterior margin of tergum III and not reaching lateral margins and a narrow- to medium-width medial longitudinal black band over all three terga. In most specimens there are no dark patterns on the lateral

margins of terga III–V. A pair of oval orange-brown shining spots on tergum V. Posterior lobe of surstylus short, sternum V with a deep concavity on posterior margin. Abdominal sterna pale orange-brown.
Attractant: Methyl eugenol.
Female: No known record.
Distribution: Indonesia (Bali; Java; Sulawesi; Sumatra).
Hosts: *Mangifera indica* (family *Anacardiaceae*) (mango).
Remarks: *Bactrocera ochroma* n.sp. is similar to *Bactrocera dorsalis* (Hendel) in the general colour patterns of the body, wings and legs. It differs from *B. dorsalis* in having pale orange-brown abdominal sterna, lateral postsutural vittae tapering posteriorly to end at or before *ia.* seta, legs with femora and tibiae pale (generally fulvous), and the absence of dark lateral margins on abdominal terga III to V. Based on analyses of the mitochondrial genes, COI and ND5, *B. ochroma* is significantly different from *B. dorsalis* from Taiwan (Drew, unpublished data).

Bactrocera (Bactrocera) ochroventer Drew & Romig, new species

(Fig. 133)

Material examined: Holotype male, **INDONESIA: MOLUCCAS (MALUKU)** – Seram (western part), Waimital, 2.viii.2007, coll. A. Haris Tualepe, attracted to methyl eugenol in jeruk, deposited in BMNH.
Paratype: **INDONESIA: MOLUCCAS (MALUKU): AMBON ISLAND** – 1 male, Moluccas (Maluku), Middle Maluku, Suli, 9.vi.2007, coll. La Naibu, attracted to cue lure in Hutan amalatuei; 1 male, Hila/Mamua, Ambon, Indonesia, 25–100 m, 5.viii.1981, coll. J. Turner, residual rainforest, attracted to cue lure. 1 paratype in ANIC, 1 paratype in QDPC.
Diagnosis: A medium-sized species; face fulvous with a pair of small circular to oval black spots; postpronotal lobes and notopleura yellow; scutum entirely red-brown; lateral postsutural yellow vittae very short and narrow ending before level of *a.sa.* setae; medial postsutural yellow vitta absent; no yellow spot anterior to notopleural suture; mesopleural stripe reaching anterior *npl.* seta dorsally; scutellum yellow; wing with cells bc and c pale fuscous, microtrichia in outer ⅓ of cell c only; a broad dark fuscous costal band confluent with R_{4+5}; a broad dark fuscous anal streak; all abdominal terga entirely orange-brown.
Description: male.
Head: Height 1.8 mm. Frons – length 1.36 times breadth; red-brown with fuscous around orbital setae; orbital setae black: 1 *s.or.*, 2 *i.or.*; lunule fuscous. Ocellar triangle black. Vertex fuscous. Face fulvous with a pair of small circular to oval black spots; length 0.48 mm. Genae fulvous, large fuscous to dark fuscous subocular spot present; black seta present. Occiput red-brown, fulvous along eye margins; occipital row with 4–6 black setae. Antennae with segments 1 and 2 red-brown, segment 3 red-brown with fuscous on apex and outer surface; length of segments: 0.18 mm; 0.33 mm; 0.81 mm.
Thorax: Scutum entirely red-brown with a small fuscous triangular marking anterocentrally. Pleural areas entirely red-brown except dark fuscous to black on the katepisternum. Yellow markings as follows: postpronotal lobes; notopleura; mesopleural stripe reaching anterior *npl.* seta dorsally, continuing to katepisternum as a small transverse spot, anterior margin straight; anatergite (posterior apex black); anterior ¾ katatergite (remainder black); two narrow and very short lateral postsutural vittae ending before level of *a.sa.* seta. Medial longitudinal postsutural yellow vitta absent. Postnotum black laterally, red-brown centrally. Scutellum yellow except for narrow red-brown basal band. Setae: *sc.* 2; *prsc.* 2; *ia.* 1; *p.sa.* 1; *a.sa.* 1; *mpl.* 1; *npl.* 2; *scp.* 4.
Legs: Fore femora fulvous, mid and hind femora fulvous with red-brown around apical ¼; fore and mid tibiae fulvous, hind tibiae red-brown; fore and mid tarsi with all segments fulvous, hind tarsi with basal segment fulvous and apical four segments red-brown; mid tibiae each with an apical black spur.
Wings: Length 6.08 mm; cells bc and c pale fuscous; microtrichia in outer ⅓ of cell c only; remainder of wings colourless except dark fuscous cell sc, broad dark fuscous costal band confluent with R_{4+5} but becoming paler adjacent to this vein and remaining of uniform width around apex of wing, a broad dark fuscous anal streak; a dense aggregation of microtrichia around A_1+CuA_2; supernumerary lobe of medium development.
Abdomen: Oval; terga free; pecten present on tergum III. Tergum I and sterna I and II wider than long. All abdominal terga entirely orange-brown except fulvous posterolaterally on tergum II. A pair of oval orange-brown shining spots on tergum V. Posterior lobe of surstylus short, sternum V with a deep concavity on posterior margin. Abdominal sterna orange-brown.
Attractant: Uncertain.
Female: No known record.
Distribution: Indonesia (Moluccas = Maluku – Ambon I.; Seram I.).
Hosts: No known record.
Remarks: *Bactrocera (Bactrocera) ochroventer* n.sp. is similar to *Bactrocera (Bactrocera) dispar* (Hardy), *Bactrocera (Bactrocera) flavipennis* (Hardy) and *Bactrocera (Bactrocera) megaspilus* (Hardy) in possessing a broad costal band confluent with R_{4+5} and the scutum red-brown. It differs from *B. dispar* and *B. megaspilus* in having the costal band of uniform width (not expanded apically) and from *B. flavipennis* in having *prsc.* setae present, abdominal terga III–V entirely orange-brown and the lateral postsutural vittae very short, ending before the level of the *a.sa.* setae.

Bactrocera (Bactrocera) opacovitta Drew & Romig, new species

(Fig. 134)

Material examined: Holotype male, **INDONESIA: MOLUCCAS (MALUKU): AMBON ISLAND** – Mamala, Ambon, Indonesia, 50–150 m, 7.viii.1981, coll. J. Turner, attracted to methyl eugenol, in residual forest with cloves, nutmeg and banana groves, deposited in BMNH.
Paratypes: **INDONESIA: MOLUCCAS (MALUKU): AMBON ISLAND** – 1 male, same data as holotype; 2 males,

Paso, Ambon, Indonesia, 20–50 m, 7.vii.1981, coll. J. Turner, attracted to methyl eugenol, in residual forest with cloves, nutmeg and banana groves; 5 males, Hila/Mamua, Ambon, Indonesia, 25–100 m, 5.viii.1981, coll. J. Turner, attracted to methyl eugenol, in residual rainforest. 1 paratype in ANIC, 1 paratype in BMNH, 4 paratypes in QDPC, 2 paratypes in QMIC.

Diagnosis: A medium-sized species; face fulvous with a pair of medium-sized circular black spots; postpronotal lobes fuscous; notopleura yellow; scutum black; very narrow and short lateral postsutural yellow vittae present but sometimes not easy to distinguish; medial postsutural yellow vitta absent; no yellow spot anterior to notopleural suture; mesopleural stripe reaching midway between anterior margin of notopleuron and anterior *npl.* seta dorsally; scutellum yellow; wing with cells bc and c colourless to having a pale fuscous tint, microtrichia in outer corner of cell c only; a broad fuscous costal band confluent with R_{4+5} but paler between R_{2+3} and R_{4+5}; a broad fuscous anal streak; all abdominal terga entirely orange-brown except tergum I which may be pale fuscous.

Description: male.

Head: Height 1.48 mm. Frons – length 1.67 times breadth; fulvous with fuscous on anteromedial hump and around bases of orbital setae; orbital setae black: 1 *s.or.*, 2 *i.or.*; lunule fuscous. Ocellar triangle black. Vertex fuscous. Face fulvous with a pair of medium-sized circular black spots; length 0.39 mm. Genae red-brown, small dark fuscous subocular spot present; red-brown seta present. Occiput red-brown, fulvous along eye margins; occipital row with a large number of strong black setae. Antennae with all segments entirely red-brown; length of segments: 0.18 mm; 0.34 mm; 0.8 mm.

Thorax: Scutum black with dark fuscous along posterior lateral and anterior margins, around notopleural suture and between postpronotal lobe and notopleuron. Pleural areas fuscous with black on katepisternum. Yellow markings as follows: notopleura; mesopleural stripe reaching midway between anterior margin of notopleuron and anterior *npl.* seta dorsally, continuing to katepisternum as a small transverse spot, anterior margin straight; anatergite (posterior apex black); anterior ½ katatergite (remainder red-brown to dark fuscous); two very narrow and short lateral postsutural vittae ending well before *ia.* seta (these lateral postsutural vittae are almost indistinguishable in some specimens). Medial longitudinal postsutural yellow vitta absent. Postpronotal lobes fuscous. Postnotum black. Scutellum yellow except for narrow fuscous basal band. Setae: *sc.* 2; *prsc.* 2; *ia.* 1; *p.sa.* 1; *a.sa.* 1; *mpl.* 1; *npl.* 2; *scp.* 4.

Legs: All femora and tibiae entirely red-brown; all tarsal segments fulvous; mid tibiae each with an apical black spur.

Wings: Length 5.4 mm; cells bc and c colourless to a very pale tint; microtrichia in outer corner of cell c only; remainder of wings colourless except fuscous cell sc, broad fuscous costal band confluent with R_{4+5} but paler between R_{2+3} and R_{4+5} and remaining of uniform width throughout, a broad fuscous anal streak; a dense aggregation of microtrichia around A_1+CuA_2; supernumerary lobe of medium development.

Abdomen: Oval; terga free; pecten present on tergum III. Tergum I and sterna I and II wider than long. All abdominal terga entirely orange-brown except tergum I which may be pale fuscous. A pair of oval orange-brown shining spots on tergum V. Posterior lobe of surstylus short, sternum V with a deep concavity on posterior margin. Abdominal sterna orange-brown.

Attractant: Methyl eugenol.

Female: No known record.

Distribution: Indonesia (Moluccas = Maluku – Ambon I.).

Hosts: No known record.

Remarks: *Bactrocera (Bactrocera) opacovitta* n.sp. belongs to a group of *Bactrocera* species that possess broad costal bands confluent with R_{4+5}. It is similar to *Bactrocera (Bactrocera) flavipennis* (Hardy) and *Bactrocera (Bactrocera) paramusae* Drew in possessing lateral postsutural vittae that taper posteriorly and abdominal terga mostly orange-brown but differs from these species in having fuscous postpronotal lobes.

Bactrocera (Bactrocera) osbeckiae Drew & Hancock

(Fig. 135)

Bactrocera (Bactrocera) osbeckiae Drew & Hancock, 1994a: 46; Norrbom *et al.*, 1998: 93. Holotype female in BMNH.

Non-type material examined: No new specimens available for study since the publication of Drew & Hancock (1994a).

Diagnosis: Face fulvous with a pair of large oval black spots; postpronotal lobes and notopleura yellow; scutum black except brown to dark brown below lateral postsutural vittae and red-brown around notopleural suture, inside postpronotal lobes, two broad longitudinal bands running anteriorly from notopleural suture each side of a narrow medial longitudinal black band; narrow parallel-sided lateral postsutural yellow vittae (occasionally narrowing slightly posteriorly) and ending at or just before *ia.* seta; medial postsutural yellow vitta absent; mesopleural stripe reaching midway between anterior margin of notopleuron and anterior *npl.* seta dorsally; scutellum yellow; legs with femora entirely fulvous; fore tibiae dark fuscous, mid tibiae dark fuscous basally to fuscous apically, hind tibiae dark fuscous to black; wing with cells bc and c colourless, microtrichia in outer corner of cell c only; a narrow fuscous costal band confluent with R_{2+3} (occasionally just overlapping this vein) and remaining narrow around apex of wing; a narrow fuscous anal streak; supernumerary lobe of medium development in males and weak in females; abdominal terga III–V orange-brown except for a moderately narrow dark fuscous to black transverse band across anterior margin of tergum III which expands to cover lateral margins as a broad diffuse pattern, broad diffuse fuscous patterns on lateral margins of tergum IV and anterolateral corners of tergum V, a narrow medial longitudinal black band over all three terga, a pair of oval fuscous shining spots on tergum V; abdominal sterna dark coloured.

Attractant: No known record.

Female: As for male in the general body colour patterns. See Drew & Hancock (1994a) for a detailed description of female characters.

Distribution:
Type locality: Thailand (Nakhonratsima).
Previously recorded: Type locality and Vietnam.
Hosts: This species breeds in the flowers of species in the family *Melastomataceae* (Drew & Hancock, 1994a; Allwood *et al*., 1999).
Remarks: *Bactrocera (Bactrocera) osbeckiae* Drew & Hancock, within the *dorsalis* complex, is similar to *Bactrocera (Bactrocera) cibodasae* Drew & Hancock, *Bactrocera (Bactrocera) merapiensis* Drew & Hancock, *Bactrocera (Bactrocera) sumbawaensis* Drew & Hancock and *Bactrocera (Bactrocera) thailandica* Drew & Hancock in possessing lateral postsutural vittae that narrow slightly posteriorly, femora entirely fulvous or with, at most, preapical dark spots on outer surfaces of fore femora only, costal band confluent with R_{2+3} and distinct dark lateral markings on abdominal terga III–V. It differs from *B. cibodasae* in having the costal band of uniform width around apex of wing, from *B. sumbawaensis* in having the lateral postsutural vittae elongate (not broad and triangular) and the shining spots on abdominal tergum V fuscous to dark fuscous (not black), and from *B. merapiensis* and *B. thailandica* in having large areas of red-brown anterocentrally on the scutum.

Bactrocera (*Bactrocera*) *papayae* Drew & Hancock

(Fig. 136)

Bactrocera (Bactrocera) papayae Drew & Hancock, 1994a: 48; Norrbom *et al*., 1998: 93. Holotype female in BMNH.
Bactrocera conformis Doleschall, 1858: 122 (see Norrbom *et al*., 1998: 93).
Bactrocera (Bactrocera) philippinensis Drew & Hancock, 1994a: 52. **New Synonym**.

Non-type material examined: Since the publication of Drew & Hancock (1994a), the following specimens have been studied – **AUSTRALIAN TERRITORY: CHRISTMAS ISLAND** – 10 males, ix.2002, coll. R. Kitching, attracted to methyl eugenol. **BRUNEI**: 122 males (23.i.2002), 4 males (24.i.2002), 100 males (25.i.2002), 150 males (27.i.2002), 141 males (29.i.2002), KBFC Bato Apou Nat. Park, coll. K. Hurley, attracted to methyl eugenol. **INDONESIA: JAVA** – 2 males, D.I. Yogyakarta, Sleman, Kaliurang, 12.vii.2005, coll. Agus W, attracted to methyl eugenol in forest; 50 males, West Java, Majalengka, Ampel (25 dpl), 14.vii.2005, coll. Yaya Senjaya, attracted to methyl eugenol in mango; 6 males, West Java, Cirebon, Panguragan, 21.vii.2005, coll. Naim S.; **KALIMANTAN** – 5 males, East Kalimantan, Kutai Kertanegara, Batual, 12.ix.2005, coll. Heru G., attracted to methyl eugenol in forest; 1 male (13.viii.2005), 9 males (16.viii.2005), 30 males (13.ix.2005), Central Kalimantan, Kota Waringin Timur, Kota Besi, coll. Utik Darmanto & Kusyono, attracted to methyl eugenol in citrus. **PENINSULAR MALAYSIA**: 16 females, MARDI, 21.ix.2000, coll. R. Drew, hand collected on bananas; 16 males, MARDI, 21.ix.2000, coll. R. Drew, attracted to methyl eugenol in carambola orchard; 86 males, MARDI, 21.ix.2000, coll. R. Drew, attracted to methyl eugenol in banana plantation; 5 males and 3 females, MARDI, Selangor, vi.2002, coll. Vijaysegaran, protein bait trapped in carambola orchard. **PHILIPPINES**: 99 males, Sirang Lupa, Calamba, Laguna, xii.1999, coll. A. O'Toole & G. Quimio, citrus orchard; 13 males, Mt Makiling botanic gardens, alt 150m, 12.xii.1999, coll. A. O'Toole, attracted to Zorgen 95 (methyl eugenol); 73 males, Sirang Lupa, Calamba, Laguna, 20.xii.1999, coll. A. O'Toole & G. Quimio, in apple guava mixed orchard; 43 males, Sirang Lupa, Calamba, Laguna, 20.xii.1999, coll. A. O'Toole & G. Quimio, attracted to Zorgen 95 (methyl eugenol) in mixed orchard; 69 males, Sirang Lupa, Calamba, Laguna, 20.xii.1999, coll. A. O'Toole & G. Quimio, attracted to Zorgen 95 (methyl eugenol) in Guava (big) orchard; 92 males, Sirang Lupa, Calamba, Laguna, 7.i.2000, coll. G. Quimio, attracted to Zorgen 95 (methyl eugenol) in carambola/guava orchard; 132 males and 1 female, Sirang Lupa, Calamba, Laguna, 7.i.2000, coll. G. Quimio, attracted to Zorgen 95 (methyl eugenol) in guava (mixed) orchard; 223 males, Sirang Lupa, Calamba, Laguna, 7.i.2000, coll. G. Quimio, in citrus orchard. Specimens in QDPC.
Diagnosis: Face fulvous with a pair of large oval black spots; postpronotal lobes and notopleura yellow; scutum black with dark brown below and behind lateral postsutural vittae, around notopleural suture and inside postpronotal lobes; broad parallel-sided lateral postsutural yellow vittae ending at or behind *ia*. seta; medial postsutural yellow vitta absent; mesopleural stripe reaching midway between anterior margin of notopleuron and anterior *npl*. seta dorsally; scutellum yellow; legs with femora entirely fulvous; fore and hind tibiae dark fuscous, mid tibiae fuscous basally and fulvous apically; wing with cells bc and c colourless, microtrichia in outer corner of cell c only; a narrow fuscous costal band confluent with R_{2+3} or just overlapping this vein where it becomes paler and remaining narrow around wing apex (in some specimens there is a slight expansion or a small fish-hook shape around apex of R_{4+5}); a narrow fuscous anal streak; supernumerary lobe of medium development in males and weak in females; abdominal terga III–V orange-brown with a 'T' pattern consisting of a narrow transverse black band across anterior margin of tergum III which expands laterally into narrow margins and a medium-width medial longitudinal black band over all three terga, antero-lateral corners of terga IV and V dark fuscous to black (in occasional specimens the transverse black band across anterior margin of tergum III is broken in the midline), a pair of oval orange-brown shining spots on tergum V; abdominal sterna dark coloured.
Attractant: Methyl eugenol.
Female: As for male in the general body colour patterns. See Drew & Hancock (1994a) for a detailed description of female characters.
Distribution:
Type locality: Peninsular Malaysia (Kuala Kangsar).
Previously recorded: Type locality and the following locations – Australian Territory (Christmas Island); East Malaysia (Sabah); Indonesia (Bali; Flores; Kalimantan; Lombok; Sulawesi; Sumbawa); Peninsular Malaysia; Singapore; Thailand; West Papua (formerly part of Irian Jaya); West Timor.

New records: Brunei; Indonesia (Java; North Moluccas = North Maluku; Sumatra); Palau. Philippines, due to **New Synonymy** of *B. philippinensis*.

Hosts: Recorded from a wide range of commercial/edible and wild forest fruits (see Allwood *et al.*, 1999 for host records of *B. papayae* and *B. philippinensis*). *Bactrocera papayae* is a major pest species within the *dorsalis* complex of South-East Asia.

Remarks: *Bactrocera (Bactrocera) papayae* Drew & Hancock, within the *dorsalis* complex, is similar to *Bactrocera (Bactrocera) dorsalis* (Hendel), *Bactrocera (Bactrocera) muiri* (Hardy & Adachi), *Bactrocera (Bactrocera) profunda* Tsuruta & White, *Bactrocera (Bactrocera) verbascifoliae* Drew & Hancock and *Bactrocera (Bactrocera) vishnu* Drew & Hancock in possessing broad parallel-sided lateral postsutural vittae, costal band confluent with or slightly overlapping R_{2+3}, femora entirely fulvous, abdominal terga III–V with a narrow- to medium-width medial longitudinal dark band and shining spots on tergum V pale. *Bactrocera verbascifoliae* and *B. vishnu* differ in having moderately broad lateral dark markings on abdominal terga III to V, *B. muiri* in having the lateral postsutural vittae ending before the *ia.* setae, *B. profunda* in having subapical dark spots on the outer surfaces of the fore femora and *B. dorsalis* in having a short male aedeagus and female piercer (see Drew *et al.*, 2008).

The original definition of the Philippine population as *Bactrocera philippinensis* Drew & Hancock (1994a) was based on some minor morphological differences, especially the length of the female piercer, electron microscopy of the ovipositor and shape of the costal band. Recent studies by Drew *et al.* (2008) supported this differentiation, particularly on the basis of morphometric analyses of various body parts, as did the studies of Schutze *et al.* (2012) where they separated *B. dorsalis*, *B. carambolae*, *B. papayae* and *B. philippinensis* on wing shape.

In a recent major collection of Dacinae across all Indonesian provinces, we have compared a large number of specimens from Java and Kalimantan with those from Brunei and the Philippines. The land mass formerly known as Borneo, a large portion of which Brunei and Kalimantan now occupy, contains a number of dacine species previously only known from the Philippines. We have found that the minor morphological characters used to identify *B. papayae* and *B. philippinensis* are insufficient to positively diagnose specimens from Brunei and Kalimantan.

In summary, *B. papayae* and *B. philippinensis* exhibit some genetic differentiation and possess minor external differences in morphology, distinct differences in aedeagus and piercer lengths, and in wing shapes. Given that it is impossible to separate these species morphologically when attempting to identify specimens microscopically, and that the Philippines and 'Borneo' possess species in common, our current assessment is that these allopatric populations, particularly that in the Philippines versus other locations across South-East Asia, represent diverging populations in evolutionary time and are probably not distinct species, at present. Consequently, we feel justified in making *B. philippinensis* a **New Synonym** of *B. papayae*, the latter name being preferred on the basis of publication page priority. See discussion under section titled 'Identification of Cryptic Species'.

Bactrocera (Bactrocera) paraarecae Drew & Romig, new species

(Fig. 137)

Material examined: Holotype male, **BHUTAN**: Tsirang, 10.vii.2000, attracted to methyl eugenol, deposited in BMNH.

Paratypes: **BHUTAN**: 4 males, same data as holotype; 3 males, Suntalay, Tsirang, 26.vi.2000, attracted to methyl eugenol; 7 males, Suntalay, Tsirang, 3.vii.2000, attracted to methyl eugenol. 2 paratypes in ANIC, 2 paratypes in BMNH, 6 paratypes in QDPC, 4 paratypes in QMIC.

Diagnosis: A medium-sized species; face fulvous with a pair of large oval black spots; postpronotal lobes and notopleura yellow; scutum black; lateral postsutural yellow vittae present; medial postsutural yellow vitta absent; no yellow spot anterior to notopleural suture; mesopleural stripe reaching midway between anterior margin of notopleuron and anterior *npl.* seta dorsally; scutellum yellow; wing with cells bc and c colourless, microtrichia in outer corner of cell c only; a narrow dark fuscous costal band confluent with R_{2+3}; a fuscous anal streak; abdominal terga III–V orange-brown with a broad medial longitudinal black band and two broad lateral longitudinal dark fuscous to black bands over all three terga and joined over anterior ½ of tergum III.

Description: male.

Head: Height 1.72 mm. Frons – length 1.44 times breadth; red-brown with fulvous along lateral margins, fuscous on anteromedial hump and around bases of orbital setae; orbital setae black: 1 *s.or.*, 2 *i.or.*; lunule fuscous. Ocellar triangle black. Vertex fuscous. Face red-brown with a pair of large oval black spots; length 0.49 mm. Genae red-brown, large fuscous subocular spot present; black seta present. Occiput red-brown to fuscous, fulvous along eye margins; occipital row with 4–6 black setae. Antennae with segments 1 and 2 fuscous, segment 3 dark fuscous; length of segments: 0.19 mm; 0.31 mm; 0.84 mm.

Thorax: Scutum black except red-brown below and behind lateral postsutural vittae, around notopleural suture, inside postpronotal lobes. Pleural areas entirely black, tending dark fuscous below postpronotal lobes. Yellow markings as follows: postpronotal lobes; notopleura; mesopleural stripe reaching midway between anterior margin of notopleuron and anterior *npl.* seta dorsally, continuing to katepisternum as a small transverse spot, anterior margin straight; anatergite (posterior apex black); anterior ½ katatergite (remainder black); two medium-width parallel-sided lateral postsutural vittae ending at *ia.* seta. Medial longitudinal postsutural yellow vitta absent. Postnotum black. Scutellum yellow except for narrow black basal band. Setae: *sc.* 2; *prsc.* 2; *ia.* 1; *p.sa.* 1; *a.sa.* 1; *mpl.* 1; *npl.* 2; *scp.* 4.

Legs: Fore femora fulvous with a large dark fuscous to black elongate spot covering most of outer surfaces, mid femora fulvous with a large dark fuscous apical spot on inner surfaces, hind femora fulvous with dark fuscous around apical surfaces; fore and hind tibiae dark fuscous, mid tibiae dark fuscous basally to fulvous apically; tarsi with all segments fulvous; mid tibiae each with an apical black spur.

Wings: Length 5.9 mm; cells bc and c colourless; microtrichia in outer corner of cell c only; remainder of wings colourless except dark fuscous cell sc, narrow dark fuscous costal band confluent with R_{2+3} and remaining narrow around apex of wing, a narrow fuscous anal streak; a dense aggregation of microtrichia around A_1+CuA_2; supernumerary lobe of medium development.

Abdomen: Oval; terga free; pecten present on tergum III. Tergum I and sterna I and II wider than long. Tergum I black with a narrow transverse orange-brown band along posterior margin but not reaching lateral margins; tergum II orange-brown anteriorly, fulvous posteriorly with a moderately broad transverse black band subanteriorly and reaching black anterolateral corners; terga III–V with a dark 'T' pattern consisting of a moderately broad transverse black band across anterior margin of tergum III expanding broadly over lateral margins and a broad medial longitudinal black band over all three terga, broad dark fuscous to black anterolateral corners on terga IV and V, behind the anterolateral corners on tergum IV the colouration tends to fuscous. A pair of oval orange-brown shining spots on tergum V. Posterior lobe of surstylus short, sternum V with a deep concavity on posterior margin. Abdominal sterna dark fuscous to black.

Attractant: Methyl eugenol.

Female: No known record.

Distribution: Bhutan (Tsirang).

Hosts: No known record.

Remarks: *Bactrocera (Bactrocera) paraarecae* n.sp. is similar to *Bactrocera (Bactrocera) caryeae* (Kapoor) and *Bactrocera (Bactrocera) kandiensis* Drew & Hancock in the *dorsalis* complex in possessing narrow parallel-sided lateral postsutural vittae, dark patterns on the apices of all femora, terga III–V with a black 'T' pattern and the scutellum with a narrow black basal band. It differs from *B. kandiensis* in having a broad medial and two broad lateral longitudinal dark bands on abdominal terga III–V and from *B. caryeae* in having the dark patterns on the mid and hind femora around the apices and red-brown colouration below the lateral postsutural yellow vittae on the scutum.

Bactrocera (Bactrocera) paradiospyri Chen, Zhou & Li

(Fig. 138)

Bactrocera (Bactrocera) paradiospyri Chen, Zhou & Li, in Chen *et al.*, 2011: 60. Holotype male in IZAS.

Non-type material examined: BHUTAN: 2 males, Tsirang, Darachu-La, 30.v.2001, coll. Brian Fletcher, altitude 1500m, subtropical forest, attracted to methyl eugenol; 1 male, Namling 16km, 9.vi.2001, coll. Brian Fletcher, altitude 1600m, subtropical forest, attracted to methyl eugenol; 1 male, Rontong, 30.ix.1991, coll. H.R. Feijen, attracted to methyl eugenol. Specimens in QDPC.

Diagnosis: A medium-sized species; face dark fuscous with a pair of large elongate oval black spots in antennal furrows; postpronotal lobes and notopleura yellow; scutum entirely black; lateral and medial postsutural yellow vittae absent; no yellow spot anterior to notopleural suture; mesopleural stripe reaching midway between anterior margin of notopleuron and anterior *npl.* seta dorsally; scutellum yellow; wing with cells bc and c colourless, microtrichia in extreme outer corner of cell c only; a narrow dark fuscous costal band confluent with R_{2+3} and widening slightly around apex of wing; a narrow dark fuscous anal streak; a general pale infuscation across entire wing membrane; all abdominal terga entirely black.

Description: male.

Head: Height 1.72 mm. Frons – length 1.4 times breadth; dark fuscous to black across centre and around bases of orbital setae with fulvous along lateral and ventral margins; orbital setae black: 1 *s.or.*, 2 *i.or.*; lunule dark fuscous to black. Ocellar triangle black. Vertex dark fuscous to black. Face dark fuscous with very large elongate oval black spots in antennal furrows; length 0.49 mm. Genae dark fuscous, black subocular spot present; black seta present. Occiput black; occipital row with a small number of black setae. Antennae with segments 1 and 2 dark fuscous to black, segment 3 dark fuscous with dark fuscous to black around apex and outer surface; length of segments: 0.19 mm; 0.31 mm; 0.84 mm.

Thorax: Scutum entirely black without pale markings. Pleural areas entirely black. Yellow markings as follows: postpronotal lobes (anterodorsal corners dark fuscous to black); notopleura; mesopleural stripe reaching midway between anterior margin of notopleuron and anterior *npl.* seta dorsally, not continuing to katepisternum as a yellow spot, anterior margin convex; anatergite (posterior apex black); anterior ⅔ katatergite (remainder and ventral surfaces black); lateral and medial postsutural yellow vittae absent. Postnotum black. Scutellum yellow except for narrow black basal band. Setae: *sc.* 2; *prsc.* 2; *ia.* 1; *p.sa.* 1; *a.sa.* 1; *mpl.* 1; *npl.* 2; *scp.* 4.

Legs: All femora shining black with slight indications of fulvous on basal ends of mid and hind femora; fore and mid tibiae dark fuscous to black, hind tibiae black; all tarsi with basal segment fulvous and apical four segments dark fuscous to black; mid tibiae each with an apical black spur.

Wings: Length 6.8 mm; cells bc and c colourless; microtrichia in extreme outer corner of cell c only; remainder of wings with fuscous colouration across entire wing membrane and with a dark fuscous cell sc, narrow dark fuscous costal band confluent with R_{2+3} and widening slightly from the apex of R_{2+3} to apex of wing, a narrow dark fuscous anal streak; a dense aggregation of microtrichia around A_1+CuA_2; supernumerary lobe of medium development.

Abdomen: Oval; terga free; pecten present on tergum III. Tergum I and sterna I and II wider than long. All abdominal terga entirely black. A pair of oval black shining spots on tergum V. Posterior lobe of surstylus short, sternum V with a deep concavity on posterior margin. Abdominal sterna black.

Attractant: Methyl eugenol.

Female: No known record.
Distribution:
Type locality: China (Yunnan Province, Nujiang Lushui, Pianma).
New records: Bhutan (Namling; Rontong; Tsirang).
Hosts: No known record.
Remarks: *Bactrocera (Bactrocera) paradiospyri* Chen, Zhou & Li is similar to *Bactrocera (Bactrocera) nigrofemoralis* White & Tsuruta and *Bactrocera (Bactrocera) nigrotibialis* (Perkins) in the *nigrotibialis* complex in possessing the face dark fuscous with large oval black spots in the antennal furrows, a black scutum and dark fuscous to black on all or most of the femora. It differs from these species in having the lateral postsutural yellow vittae absent, femora entirely black and wings with a light infuscation over the entire membrane in some specimens.

Bactrocera (Bactrocera) paralatissima Drew & Romig, new species

(Fig. 139)

Material examined: Holotype male, **INDONESIA: NORTH MOLUCCAS (NORTH MALUKU)** – Moluccas, North Maluku, West Halmahera, Gamlamo, 12.vi.2007, coll. Raais Abdullah, attracted to cue lure in mango, deposited in BMNH.
Paratypes: **INDONESIA: NORTH MOLUCCAS (NORTH MALUKU)** – 2 males, Maluku, North Maluku, West Halmahera Gaol, 5.vi.2007, coll. Raais Abdullah, attracted to cue lure in forest. 1 paratype in BMNH, 1 paratype in QDPC.
Diagnosis: A medium-sized species; face either entirely fulvous or with a pair of very small circular dark spots; postpronotal lobes and notopleura yellow; scutum black with dark fuscous below lateral postsutural vittae; lateral postsutural yellow vittae present; medial postsutural yellow vitta absent; no yellow spot anterior to notopleural suture; mesopleural stripe reaching to anterior *npl.* seta dorsally; scutellum yellow; wing with cell bc fulvous, cell c paler fulvous, microtrichia in outer corner of cell c only; a broad dark fuscous costal band confluent with R_{4+5}; a broad dark fuscous anal streak; abdominal terga III–V orange-brown with a 'T' pattern consisting of a narrow transverse black band across anterior margin of tergum III and broadening over lateral margins and a medium-width medial longitudinal black band over all three terga, narrow dark fuscous anterolateral corners on tergum IV and narrow fuscous anterolateral corners on tergum V.
Description: male.
Head: Height 1.8 mm. Frons – length 1.65 times breadth; red-brown with dark fuscous along anteromedial hump and pale fuscous around bases of *s.or.* setae; orbital setae black: 1 *s.or.*, 2 *i.or.*; lunule dark fuscous. Ocellar triangle black. Vertex pale fuscous. Face either entirely fulvous or with a pair of very small dark spots; length 0.44 mm. Genae red-brown; black seta present. Occiput red-brown ventrally, fuscous dorsally, fulvous along eye margins; occipital row with 5–8 black setae. Antennae with segments 1 and 2 fuscous, segment 3 dark fuscous; length of segments: 0.17 mm; 0.24 mm; 0.69 mm.
Thorax: Scutum black except dark fuscous below lateral postsutural vittae. Pleural areas black except red-brown below postpronotal lobes. Yellow markings as follows: postpronotal lobes; notopleura; mesopleural stripe reaching anterior *npl.* seta dorsally, continuing to katepisternum as a transverse spot, anterior margin slightly convex; anatergite (posterior apex black); anterior $^3/_5$ katatergite (remainder black); two very broad lateral postsutural vittae narrowing sharply posteriorly to end before *ia.* seta. Medial longitudinal postsutural yellow vitta absent. Postnotum black. Scutellum yellow except for narrow black basal band. Setae: *sc.* 2; *prsc.* 2; *ia.* 1; *p.sa.* 1; *a.sa.* 1; *mpl.* 1; *npl.* 2; *scp.* 4.
Legs: All femora entirely fulvous; fore and mid tibiae pale fuscous, hind tibiae dark fuscous; fore and mid tarsi with basal segment fulvous and apical four segments fuscous, hind tarsi with basal segment fulvous and apical four segments pale fuscous; mid tibiae each with an apical black spur.
Wings: Length 6.3 mm; cell c pale fulvous, cell bc fulvous; microtrichia in outer corner of cell c only; remainder of wings colourless except dark fuscous cell sc, broad dark fuscous costal band confluent with R_{4+5} and slightly paler adjacent to this vein in centre of wing, broad dark fuscous anal streak; a dense aggregation of microtrichia around A_1+CuA_2; supernumerary lobe of medium development.
Abdomen: Oval; terga free; pecten present on tergum III. Tergum I and sterna I and II wider than long. Terga I and II black except for a narrow transverse orange-brown band across intersegmental line between these two terga and not reaching lateral margins and fulvous across posterior half of tergum II; terga III–V orange-brown with a black 'T' pattern consisting of a narrow transverse band across anterior margin of tergum III and widening over lateral margins and a medium-width medial longitudinal band across all three terga, narrow dark fuscous anterolateral corners on tergum IV and narrow fuscous anterolateral corners on tergum V. A pair of oval pale fuscous shining spots on tergum V. Posterior lobe of surstylus short, sternum V with a deep concavity on posterior margin. Abdominal sterna dark coloured.
Attractant: Cue lure.
Female: No known record.
Distribution: Indonesia (North Moluccas = North Maluku – Halmahera I.).
Hosts: No known record.
Remarks: *Bactrocera (Bactrocera) paralatissima* n.sp. is similar to *Bactrocera (Bactrocera) latissima* Drew in possessing a black scutum, lateral postsutural vittae ending before *ia.* setae, coloured cells bc and c and a broad costal band confluent with R_{4+5}. It differs from *B. latissima* in having very broad lateral postsutural vittae, mesopleural stripe wider dorsally reaching to anterior *npl.* seta and abdominal terga III–V with a distinct black 'T' pattern.

Bactrocera (Bactrocera) paralimbifera Drew & Romig, new species

(Fig. 140)

Material examined: Holotype male, **INDONESIA: NORTH MOLUCCAS (NORTH MALUKU)** – Moluccas,

North Maluku, Kodya Ternate, Foramadiahi, 3.vi.2007, coll. La Ruti, attracted to cue lure in forest, deposited in BMNH. Paratypes: **INDONESIA: NORTH MOLUCCAS (NORTH MALUKU)** – 2 males, Maluku, North Maluku, West Halmahera Gaol, 5.vi.2007, coll. Raais Abdullah, attracted to cue lure in forest. 1 paratype in QDPC, 1 paratype in QMIC.

Diagnosis: A medium-sized species; face fulvous with a pair of medium-sized irregularly circular black spots; postpronotal lobes and notopleura yellow; scutum black; lateral postsutural yellow vittae present; medial postsutural yellow vitta absent; no yellow spot anterior to notopleural suture; mesopleural stripe reaching midway between anterior margin of notopleuron and anterior *npl.* seta dorsally; scutellum yellow with a broad black basal band; wing with cells bc and c dark fuscous, microtrichia over outer ⅓ of cell c only; a broad very dark fuscous costal band overlapping R_{4+5} throughout; a broad very dark fuscous anal streak; abdominal tergum III black except for small red-brown spots posterocentrally, terga IV and V red-brown with dark fuscous margins on tergum IV and dark fuscous anterolateral corners on tergum V, a broad medial longitudinal black band over all three terga; fuscous shining spots on tergum V.

Description: male.

Head: Height 1.96 mm. Frons – length 1.38 times breadth; fulvous with dark fuscous on anteromedial hump and around bases of orbital setae; orbital setae black: 1 *s.or.*, 2 *i.or.*; lunule dark fuscous. Ocellar triangle black. Vertex fuscous. Face fulvous with a pair of medium-sized irregularly circular black spots; length 0.54 mm. Genae fulvous, large dark fuscous subocular spot present; black seta present. Occiput red-brown ventrally, fuscous dorsally, fulvous along eye margins; occipital row with a large number of small black setae. Antennae with segments 1 and 2 fuscous, segment 3 red-brown with dark fuscous on apex and outer surface; length of segments: 0.22 mm; 0.38 mm; 0.84 mm.

Thorax: Scutum black except red-brown around notopleural suture. Pleural areas black, tending dark red-brown below postpronotal lobes. Yellow markings as follows: postpronotal lobes; notopleura; mesopleural stripe reaching midway between anterior margin of notopleuron and anterior *npl.* seta dorsally, continuing to katepisternum as a transverse spot, anterior margin straight; anatergite (posterior apex black); anterior ⅔ katatergite (remainder black); two broad lateral postsutural vittae tapering slightly posteriorly to end at *ia.* seta. Medial longitudinal postsutural yellow vitta absent. Postnotum black. Scutellum yellow except for broad black basal band. Setae: *sc.* 2; *prsc.* 2; *ia.* 1; *p.sa.* 1; *a.sa.* 1; *mpl.* 1; *npl.* 2; *scp.* 4.

Legs: Fore femora red-brown with a small fuscous subapical spot on outer surfaces, mid and hind femora fulvous basally to red-brown apically; fore tibiae fuscous, mid tibiae dark fuscous basally to fuscous apically, hind tibiae dark fuscous; all tarsi with basal segment fulvous and apical four segments fuscous; mid tibiae each with an apical black spur.

Wings: Length 6.5 mm; cells bc and c dark fuscous; microtrichia covering outer ⅓ of cell c only; remainder of wings colourless except very dark fuscous cell sc, broad very dark fuscous costal band overlapping R_{4+5} throughout, a broad very dark fuscous anal streak; a dense aggregation of microtrichia around A_1+CuA_2; supernumerary lobe of medium development.

Abdomen: Oval; terga free; pecten present on tergum III. Tergum I and sterna I and II wider than long. Terga I and II black except for a narrow orange-brown transverse band across intersegmental line between these terga but not reaching lateral margins and a narrow fulvous band across posterior margin of tergum II; tergum III black except for a pair of small red-brown spots posterocentrally; terga IV and V red-brown except for medium-width dark fuscous margins on tergum IV and narrow dark fuscous anterolateral corners on tergum V, a broad medial longitudinal black band over all three terga. A pair of oval fuscous shining spots on tergum V. Posterior lobe of surstylus short, sternum V with a deep concavity on posterior margin. Abdominal sterna dark coloured.

Attractant: Cue lure.

Female: No known record.

Distribution: Indonesia (North Moluccas = North Maluku – Halmahera I.; Ternate).

Hosts: No known record.

Remarks: *Bactrocera (Bactrocera) paralimbifera* n.sp. is similar to *Bactrocera (Bactrocera) linduensis* new name in possessing a black scutum, a broad costal band overlapping R_{4+5} and cells bc and c coloured. It differs from *B. linduensis* new name in having a subapical fuscous spot on fore femora, cells bc and c dark fuscous and abdominal terga III–V with a black 'T' pattern and dark fuscous margins on terga IV and V.

Bactrocera (Bactrocera) paramusae Drew

(Fig. 141)

Bactrocera (Bactrocera) paramusae Drew, 1989: 48; Norrbom *et al.*, 1998: 94. Holotype male in QMIC.

Non-type material examined: INDONESIA: WEST PAPUA (formerly part of IRIAN JAYA) – 1 male, Indonesia, West Papua, Manokwari, Ambon, 24.vi.2007, coll. Caroline, attracted to cue lure in Jambu Air. Specimen in QDPC.

Diagnosis: A medium-sized species; face fulvous with a pair of small- to medium-sized triangular black spots; postpronotal lobes and notopleura yellow; scutum black; lateral postsutural yellow vittae present; medial postsutural yellow vitta absent; no yellow spot anterior to notopleural suture; mesopleural stripe reaching midway between anterior margin of notopleuron and anterior *npl.* seta dorsally; scutellum yellow; wing with cells bc and c fuscous, microtrichia in outer ⅓ of cell c only; a broad dark fuscous costal band confluent with R_{4+5}; a broad dark fuscous anal streak; abdominal terga III–V uniformly orange-brown with narrow dark fuscous anterolateral corners on tergum III.

Description: male.

Head: Height 1.67 mm. Frons – length 1.5 times breadth; red-brown with fuscous on anteromedial hump and around bases of orbital setae; orbital setae black: 1 *s.or.*, 2 *i.or.*; lunule fuscous. Ocellar triangle black. Vertex fuscous. Face fulvous with a pair of medium-sized triangular black spots; length 0.46 mm. Genae fulvous, dark fuscous subocular spot present; black seta present. Occiput fuscous, fulvous along eye margins; occipital row with 5–7 black setae. Antennae with segments 1 and 2 fuscous, segment 3 red-brown with dark fuscous on apex and outer surface; length of segments: 0.18 mm; 0.33 mm; 0.81 mm.

Thorax: Scutum entirely black with dark fuscous below lateral postsutural vittae. Pleural areas entirely black with dark fuscous below postpronotal lobe. Yellow markings as follows: postpronotal lobes; notopleura; mesopleural stripe reaching midway between anterior margin of notopleuron and anterior *npl.* seta dorsally, continuing to katepisternum as a transverse spot, anterior margin straight; anatergite (posterior apex black); anterior $^3/_5$ katatergite (remainder black); two narrow- to medium-width lateral postsutural vittae narrowing slightly posteriorly to end just before *ia.* seta. Medial longitudinal postsutural yellow vitta absent. Postnotum black. Scutellum yellow except for narrow black basal band. Setae: *sc.* 2; *prsc.* 2; *ia.* 1; *p.sa.* 1; *a.sa.* 1; *mpl.* 1; *npl.* 2; *scp.* 4.

Legs: All segments entirely red-brown except basal segment of all tarsi fulvous; mid tibiae each with an apical black spur.

Wings: Length 6.16 mm; cells bc and c fuscous; microtrichia in outer ⅓ of cell c only; remainder of wings colourless except dark fuscous cell sc, broad dark fuscous costal band confluent with R_{4+5} and remaining of uniform width throughout, a broad dark fuscous anal streak; a dense aggregation of microtrichia around A_1+CuA_2; supernumerary lobe of medium development.

Abdomen: Oval; terga free; pecten present on tergum III. Tergum I and sterna I and II wider than long. Tergum I dark fuscous with a narrow transverse dark red-brown band across posterior margin but not reaching lateral margins; tergum II dark red-brown with a narrow transverse dark fuscous band subanteriorly but not reaching lateral margins and tending fulvous along posterior margin; terga III–V orange-brown with dark fuscous on anterolateral corners of tergum III. A pair of oval dark fuscous shining spots on tergum. Posterior lobe of surstylus short, sternum V with a deep concavity on posterior margin. Abdominal sterna red-brown.

Attractant: Cue lure.

Female: No known record.

Distribution:

Type locality: Papua New Guinea (Morobe Province, Bulolo).

Previously recorded: Type locality and Papua New Guinea (Central District; Northern District; Western Highlands District; Western District; Milne Bay District).

New records: Indonesia (West Papua, formerly part of Irian Jaya – Manokwari).

Hosts: No known record.

Remarks: *Bactrocera (Bactrocera) paramusae* Drew is similar to *Bactrocera (Bactrocera) flavipennis* (Hardy) in possessing a black scutum, mostly orange-brown abdominal terga III–V and a broad dark fuscous costal band confluent with R_{4+5}. It differs from *B. flavipennis* in having fuscous to dark fuscous cells bc and c with microtrichia in outer corner of cell c only. *Bactrocera paramusae* differs from all other species in the *Bactrocera musae* complex in having the costal band dark fuscous and confluent with R_{4+5}.

Bactrocera (Bactrocera) paranigrita Drew & Romig, new species

(Fig. 142)

Material examined: Holotype male, **INDONESIA: NORTH MOLUCCAS (NORTH MALUKU)** – North Maluku, West Halmahera Gaol, 18.vi.2007, coll. Raais Abdullah, attracted to methyl eugenol in forest, deposited in BMNH.

Paratypes: **INDONESIA: NORTH MOLUCCAS (NORTH MALUKU)** – 3 males, same data as holotype; 3 males (5.vi.2007), 3 males (12.vi.2007), 3 males (29.v.2007), North Maluku, West Halmahera Gaol, coll. Raais Abdullah, attracted to methyl eugenol in forest; 1 male, North Maluku, West Halmahera Gaol, Gamlamo, 29.v.2007, coll. Raais Abdullah, attracted to methyl eugenol in mangga; 1 male, North Maluku, Kodya Ternate, Kalumata, 10.vi.2007, coll. La Ruti, attracted to methyl eugenol in jeruk. 3 paratypes in ANIC, 3 paratypes in BMNH, 5 paratypes in QDPC, 3 paratypes in QMIC.

Diagnosis: A medium-sized species; face fulvous with a pair of medium-sized circular black spots; postpronotal lobes and notopleura yellow; scutum dull black; lateral postsutural yellow vittae present; medial postsutural yellow vitta absent; no yellow spot anterior to notopleural suture; mesopleural stripe reaching postpronotal lobe dorsally; scutellum yellow; wing with cells bc and c with a pale tint, microtrichia in outer corner of cell c only; a moderately broad fuscous costal band overlapping R_{2+3}, infuscation along *r-m* and *dm-cu* crossveins which are connected by a pale fuscous diffuse pattern; a broad fuscous anal streak; abdominal terga III–V orange-brown with narrow dark fuscous to black transverse lines on anterolateral margins of tergum III and a narrow medial longitudinal black band on terga IV and V.

Description: male.

Head: Height 1.64 mm. Frons – length 1.57 times breadth; red-brown with pale fuscous around orbital setae; orbital setae black: 1 *s.or.*, 2 *i.or.*; lunule red-brown. Ocellar triangle black. Vertex pale fuscous. Face fulvous with a pair of medium-sized circular black spots; length 0.48 mm. Genae fulvous, large fuscous to dark fuscous subocular spot present; black seta present. Occiput red-brown, fulvous along eye margins; occipital row with 4–5 black setae. Antennae with segments 1 and 2 red-brown, segment 3 red-brown with dark fuscous on apex and outer surface; length of segments: 0.22 mm; 0.31 mm; 0.77 mm.

Thorax: Scutum dull black with dark red-brown below and behind lateral postsutural vittae, around notopleural suture and along lateral margins between postpronotal lobe and notopleural suture. Pleural areas black except red-brown below postpronotal lobes. Yellow markings as follows: postpronotal lobes; notopleura; mesopleural stripe reaching postpronotal lobe dorsally, continuing to katepisternum as a small transverse spot, anterior margin straight; anatergite

(posterior apex black); anterior ⅔ katatergite (remainder black); two moderately broad parallel to sub-parallel-sided lateral postsutural vittae ending at *ia.* seta. Medial longitudinal postsutural yellow vitta absent. Postnotum black. Scutellum yellow except for narrow black basal band. Setae: *sc.* 2; *prsc.* 2; *ia.* 1; *p.sa.* 1; *a.sa.* 1; *mpl.* 1; *npl.* 2; *scp.* 4.
Legs: All segments entirely fulvous; mid tibiae each with an apical black spur.
Wings: Length 5.5 mm; cells bc and c with a pale tint; microtrichia in outer corner of cell c only; remainder of wings colourless except fuscous cell sc, a moderately broad fuscous costal band overlapping R_{2+3} and remaining of uniform width around apex of wing, fuscous around *r-m* and *dm-cu* crossveins which are connected by a broad diffuse pale fuscous area, a broad fuscous anal streak; a dense aggregation of microtrichia around A_1+CuA_2; supernumerary lobe of medium development.
Abdomen: Oval; terga free; pecten present on tergum III. Tergum I and sterna I and II wider than long. Tergum I dark fuscous to black; tergum II orange-brown; terga III–V orange-brown with narrow dark fuscous to black transverse lines on anterolateral margins of tergum III and a narrow medial longitudinal black band on terga IV and V. A pair of oval fuscous shining spots on tergum V. Posterior lobe of surstylus short, sternum V with a deep concavity on posterior margin. Abdominal sterna pale, generally fulvous.
Attractant: Methyl eugenol.
Female: No known record.
Distribution: Indonesia (North Moluccas = North Maluku – Halmahera I.; Ternate).
Hosts: No known record.
Remarks: *Bactrocera (Bactrocera) paranigrita* n.sp. is similar to *Bactrocera (Bactrocera) neonigrita* Drew in possessing a similar wing colour pattern of infuscation around the *r-m* crossvein becoming diffuse as it extends across the wing. However, *B. neonigrita* differs in possessing abdominal terga III–V mostly black and dark fuscous to black on the apices of all femora.

Bactrocera (Bactrocera) paraosbeckiae Drew
(Fig. 143)

Bactrocera (Bactrocera) paraosbeckiae Drew, in Drew & Raghu, 2002: 340. Holotype male in BMNH.

Non-type material examined: No new specimens available for study since the publication of Drew & Raghu (2002).
Diagnosis: Face fulvous with a pair of large oval black spots; postpronotal lobes and notopleura yellow; scutum dark red-brown with irregular-shaped patterns of fuscous markings; narrow lateral postsutural yellow vittae narrowing sharply posteriorly to end before *ia.* seta; medial postsutural yellow vitta absent; mesopleural stripe reaching midway between anterior margin of notopleuron and anterior *npl.* seta dorsally; scutellum yellow; legs with femora entirely fulvous (there may be a small diffuse dark fuscous subapical spot on outer surfaces of fore femora); fore and hind tibiae dark fuscous, mid tibiae dark fuscous basally to fuscous apically; wing with cells bc and c colourless, microtrichia in outer corner of cell c only; a narrow fuscous costal band slightly overlapping R_{2+3} and remaining pale and narrow around costal margin to end just beyond extremity of R_{4+5}; a narrow pale fuscous to fuscous anal streak; supernumerary lobe of medium development; abdominal terga III–V red-brown with a dark 'T' pattern consisting of a transverse black band across anterior margin of tergum III and a narrow medial longitudinal black band over all three terga, broad fuscous to dark fuscous lateral longitudinal bands over all three terga, a pair of oval fuscous shining spots on tergum V.
Attractant: Cue lure.
Female: No known record.
Distribution:
Type locality: Southern India (Kerala).
Hosts: No known record.
Remarks: *Bactrocera (Bactrocera) paraosbeckiae* Drew is similar to *Bactrocera (Bactrocera) osbeckiae* Drew & Hancock in the general colour patterns of the body, wings and legs. It differs from *B. osbeckiae* in having a dark red-brown scutum with irregular fuscous patterns (not mostly black).

Bactrocera (Bactrocera) paraverbascifoliae Drew
(Fig. 144)

Bactrocera (Bactrocera) paraverbascifoliae Drew, in Drew & Raghu, 2002: 341. Holotype male in BMNH.

Non-type material examined: Since the publication of Drew & Raghu (2002), the following specimens have been studied – **SRI LANKA (formerly CEYLON)**: 9 males (14.x.1994), 5 males (1.xii.1994), 1 male (15.xii.1994), 2 males (29.xii.1994), 2 males (8.ii.1995), Bindunuwewa, attracted to methyl eugenol; 14 males, Hakgala, 14.x.1994; 2 males, Rendapola, 14.x.1994; 2 males, Rahangala, 14.x.1994. Specimens in QDPC.
Diagnosis: Face fulvous with medium-sized circular to oval black spots; postpronotal lobes and notopleura yellow; scutum black with red-brown to dark red-brown below and behind lateral postsutural vittae, around notopleural suture and inside postpronotal lobes; lateral postsutural yellow vittae parallel-sided and of medium-width and ending just before *ia.* seta; medial postsutural yellow vitta absent; mesopleural stripe reaching midway between anterior margin of notopleuron and anterior *npl.* seta dorsally; scutellum yellow; legs with all segments fulvous except for a small area of dark fuscous on base of fore tibiae and fuscous on base of mid tibiae and hind tibiae dark fuscous basally to fuscous apically; wing with cells bc and c colourless, microtrichia in outer corner of cell c only; a narrow fuscous to dark fuscous costal band confluent with R_{2+3} and widening gradually across apex of wing to end between extremities of R_{4+5} and M; a narrow fuscous anal streak; supernumerary lobe of medium development; abdominal terga III–V orange-brown except for a 'T' pattern consisting of a transverse band across anterior margin of tergum III which widens to cover lateral margins, a narrow- to medium-width medial longitudinal black band over all three terga and broad lateral longitudinal dark fuscous bands over all three terga which are interrupted by orange-brown on posterolateral areas of tergum IV, a pair of

oval red-brown shining spots on tergum V; abdominal sterna dark coloured.
Attractant: Methyl eugenol.
Female: No known record.
Distribution:
Type locality: Southern India (Kerala, New Amarambalam Forest).
New records: Sri Lanka.
Hosts: No known record.
Remarks: *Bactrocera (Bactrocera) paraverbascifoliae* Drew, within the *dorsalis* complex, is similar to *Bactrocera (Bactrocera) verbascifoliae* Drew & Hancock in possessing narrow- to medium-width parallel-sided lateral postsutural vittae and femora entirely fulvous. It differs from *B. verbascifoliae* in having the lateral postsutural vittae ending before the *ia.* setae and the costal band confluent with R_{2+3} and widening across the apex of the wing.

Bactrocera (Bactrocera) parvula (Hendel)

(Fig. 145)

Dacus parvulus Hendel, 1912: 21; White & Liquido, 1995: 250 (synonymised with *B. (Bactrocera) latifrons* (Hendel)). Syntypes in DEI and NMW.
Chaetodacus parvulus – Shiraki, 1933: 54.
Dacus (Strumeta) parvulus – Hardy, 1968: 113; 1973: 49.

Type material examined: TAIWAN (formerly FORMOSA): 1 male with the following three labels – 'Kanshirei, Formosa, Sauter, 07–09'; '*Dacus parvulus* H., det. Hendel'; 'TYPUS'. 2 females with the following three labels – 'Kanshirei, Formosa, Sauter, IV.09'; '*Dacus parvulus* H., det Hendel'; 'COTYPUS'. Received on loan from DEI. 1 male with the following five labels – 'Kanshirei, Formosa, Sauter, IV.09'; '*Dacus parvulus* H. det. Hendel'; 'coll. Hendel'; 'COTYPE'; '*Bactrocera latifrons* (H.D.) det I. M. White, 1997'. 1 male with the following five labels – 'Kanshirei, Formosa, Sauter, IV.09'; '*Dacus parvulus* H. det. Hendel'; 'coll. Hendel'; 'COTYPE'; 'cotype by D.E. Hardy 1961'. Studied at NMW. The above five specimens represent the type series of Hendel.
Diagnosis: A small species; face fulvous with a pair of very large oval black spots; postpronotal lobes and notopleura yellow; scutum black; lateral postsutural yellow vittae present; medial postsutural yellow vitta absent; no yellow spot anterior to notopleural suture; mesopleural stripe reaching anterior *npl.* seta dorsally; scutellum yellow; wing with cells bc and c colourless, microtrichia in outer corner of cell c only; a narrow dark fuscous costal band slightly overlapping R_{2+3}; a narrow dark fuscous anal streak; abdominal tergum III mostly dark fuscous to black, tergum IV with a dark fuscous transverse band across anterior ⅓ and widening to cover lateral margins and with a narrow medial longitudinal dark fuscous band, tergum V red-brown with dark fuscous anterolateral corners and a narrow medial longitudinal dark fuscous band.
Description: male.
Head: Height 1.1 mm. Frons – length 1.4 times breadth; red-brown with dark fuscous on anteromedial hump and around bases of orbital setae; orbital setae black: 1 *s.or.*, 2 *i.or.*; lunule dark fuscous. Ocellar triangle black. Vertex dark fuscous. Face fulvous with a pair of very large oval black spots; length 0.46 mm. Genae fulvous, pale fuscous subocular spot present; red-brown seta present. Occiput dark fuscous, fulvous along eye margins; occipital row with a large number of small pale setae. Antennae with segments 1 and 2 red-brown, segment 3 red-brown with fuscous on apex and outer surface; length of segments: 0.14 mm; 0.3 mm; 0.57 mm.
Thorax: Scutum entirely black except dark fuscous behind lateral postsutural vittae. Pleural areas dark fuscous to black except red-brown below postpronotal lobes and above mid coxae. Yellow markings as follows: postpronotal lobes; notopleura; mesopleural stripe reaching anterior *npl.* seta dorsally, continuing to katepisternum as a large transverse spot, anterior margin concave; anatergite (posterior apex black); anterior $^3/_5$ katatergite (remainder black); two narrow parallel-sided lateral postsutural vittae ending at *ia.* seta. Medial longitudinal postsutural yellow vitta absent. Postnotum black. Scutellum yellow except for narrow black basal band. Setae: *sc.* 2; *prsc.* 2; *ia.* 1; *p.sa.* 1; *a.sa.* 1; *mpl.* 1; *npl.* 2; *scp.* 4.
Legs: Fore femora fulvous basally with dark fuscous around apical ½ to ¾, mid femora fulvous with dark fuscous around apical ⅓, hind femora fulvous with dark fuscous around apical ¼; fore and hind tibiae dark fuscous, mid tibiae dark fuscous basally to fuscous apically; all tarsi with basal segment fulvous and apical four segments fuscous; mid tibiae each with an apical black spur.
Wings: Length 4.2 mm; cells bc and c colourless; microtrichia in outer corner of cell c only; remainder of wings colourless except dark fuscous cell sc, narrow dark fuscous costal band slightly overlapping R_{2+3} where it becomes fuscous and widening over apex of R_{4+5}, a narrow dark fuscous anal streak; a dense aggregation of microtrichia around A_1+CuA_2; supernumerary lobe of medium development.
Abdomen: Oval; terga free; pecten present on tergum III. Tergum I and sterna I and II wider than long. Tergum I dark fuscous to black with a narrow transverse red-brown band across posterior margin but not reaching lateral margins; tergum II with a broad transverse dark fuscous to black band across anterior margin but not reaching the narrow dark fuscous to black anterolateral corners and with fulvous across posterior ½; tergum III mostly dark fuscous to black (may have a narrow medial longitudinal dark fuscous to black band with small areas of red-brown either side of this band); tergum IV with a dark fuscous band across anterior ⅓ and widening to cover lateral margins and with a narrow medial longitudinal dark fuscous band which does not reach posterior margin; tergum V red-brown with dark fuscous anterolateral corners and a narrow dark fuscous medial longitudinal band. A pair of oval red-brown shining spots on tergum V which tend towards dark fuscous laterally. Posterior lobe of surstylus short, sternum V with a deep concavity on posterior margin. Abdominal sterna dark fuscous to black.
Attractant: No known record.
Female: As for male except no dense aggregation of microtrichia around A_1+CuA_2; supernumerary lobe weak; pecten of cilia on abdominal tergum III absent; ovipositor basal

segment fuscous, dorsoventrally compressed and tapering posteriorly in dorsal view; ratio of length of oviscape to length of tergum V, 0.6: 1; apex of piercer trilobed.
Distribution:
Type locality: Taiwan (Kanshirei).
Hosts: No known record.
Remarks: *Bactrocera (Bactrocera) parvula* (Hendel) is similar to *Bactrocera (Bactrocera) latifrons* (Hendel) in possessing a black scutum, the costal band narrow but widening at apex and the apex of the ovipositor piercer trilobed. It differs from *B. latifrons* in having distinct dark fuscous patterns around the apices of all femora, dark fuscous to black over abdominal tergum III, dark fuscous medial and lateral longitudinal bands over terga IV and V joined over the anterior margin of tergum IV and dark fuscous to black abdominal sterna. It is also significant to note that in the original description of Hendel (1912) and the revised description of Shiraki (1933), they both recognized the dark colour patterns on the femora and abdominal terga. These characters contrast with those for *B. latifrons. Bactrocera (Bactrocera) citima* (Hardy) is very close to *B. parvula* and differs only in having a narrower mesopleural stripe and a needle-shaped apex of the ovipositor piercer. Because there are good characters that separate *B. latifrons* and *B. parvula*, we are here withdrawing them from the synonymy in which they were placed by White & Liquido (1995). Hardy (1968) stated that he was 'quite certain that *B. parvula* is a synonym of *Dacus pectoralis* Walker'. However, we have now studied the types of both species and do not agree with this assessment. We believe that both *B. parvula* and *B. pectoralis* are good species.

Bactrocera (Bactrocera) patula Drew & Romig, new species

(Fig. 146)

Material examined: Holotype male, **INDIA: ANDAMAN ISLANDS** – India, S. Andaman, Port Blair, 18.vii.1992, coll. Ranganath, Sl. No. 13, attracted to cue lure, deposited in BMNH. Paratypes: **INDIA: ANDAMAN ISLANDS** – 4 males, same data as holotype. 1 paratype in ANIC, 2 paratypes in QDPC, 1 paratype in QMIC.
Diagnosis: A medium-sized species; face fulvous with a pair of moderately large circular black spots; postpronotal lobes and notopleura yellow; scutum entirely red-brown; lateral postsutural yellow vittae present; medial postsutural yellow vitta absent; no yellow spot anterior to notopleural suture; mesopleural stripe reaching midway between anterior margin of notopleuron and anterior *npl.* seta dorsally; scutellum yellow; wing with cells bc and c with a pale fuscous tint, microtrichia in outer corner of cell c only; a broad dark fuscous costal band; a moderately broad dark fuscous anal streak; abdominal terga III–V orange-brown with a narrow medial longitudinal dark fuscous band over terga III and IV, narrow fuscous to dark fuscous anterolateral corners on all three terga.
Description: male.
Head: Height 1.66 mm. Frons – length 1.57 times breadth; red-brown with pale fuscous on anteromedial hump and around bases of orbital setae; orbital setae black: 1 *s.or.*, 2 *i.or.*; lunule fuscous. Ocellar triangle black. Vertex dark fuscous. Face fulvous with a pair of moderately large circular black spots; length 0.49 mm. Genae fulvous, dark fuscous subocular spot present; black seta present. Occiput red-brown, fulvous along eye margins; occipital row with 4–6 black setae. Antennae with segments 1 and 2 red-brown, segment 3 red-brown with fuscous on apex and outer surface; length of segments: 0.2 mm; 0.33 mm; 0.84 mm.
Thorax: Scutum entirely red-brown without dark markings. Pleural areas entirely red-brown without dark markings. Yellow markings as follows: postpronotal lobes; notopleura; mesopleural stripe reaching midway between anterior margin of notopleuron and anterior *npl.* seta dorsally, continuing to katepisternum as a transverse spot, anterior margin straight; anatergite (posterior apex black); anterior ⅔ katatergite (remainder red-brown); two moderately broad parallel-sided lateral postsutural vittae ending at *ia.* seta. Medial longitudinal postsutural yellow vitta absent. Postnotum red-brown. Scutellum yellow except for narrow red-brown basal band. Setae: *sc.* 2; *prsc.* absent in some specimens with one or two present in other specimens; *ia.* 1; *p.sa.* 1; *a.sa.* 1; *mpl.* 1; *npl.* 2; *scp.* 4.
Legs: All femora fulvous with subapical dark fuscous spots on outer surfaces; fore tibiae fuscous, mid tibiae dark fuscous basally to red-brown apically, hind tibiae dark fuscous; all tarsi with basal segment fulvous and apical four segments pale fuscous to fuscous; mid tibiae each with an apical black spur.
Wings: Length 6.0 mm; cells bc and c with a pale fuscous tint; microtrichia in outer corner of cell c only; remainder of wings colourless except dark fuscous cell sc, broad dark fuscous costal band confluent with R_{4+5} and remaining of uniform width throughout, a moderately broad dark fuscous anal streak; a dense aggregation of microtrichia around A_1+CuA_2; supernumerary lobe of medium development.
Abdomen: Oval; terga free; pecten present on tergum III. Tergum I and sterna I and II wider than long. Tergum I orange-brown with narrow dark fuscous lateral margins; tergum II orange-brown with fulvous on posterolateral margins, narrow dark fuscous lateral margins, a narrow medial longitudinal dark fuscous band over anterior ½ which may extend onto posterior area of tergum I; terga III–V orange-brown with a narrow medial longitudinal dark fuscous band over terga III and IV and narrow fuscous to dark fuscous anterolateral corners on all three terga. A pair of oval red-brown to dark red-brown shining spots on tergum V. Posterior lobe of surstylus short, sternum V with a deep concavity on posterior margin. Abdominal sterna dark coloured.
Attractant: Cue lure.
Female: No known record.
Distribution: India (South Andaman I., Port Blair).
Hosts: No known record.
Remarks: *Bactrocera (Bactrocera) patula* n.sp. is similar to *Bactrocera (Bactrocera) flavipennis* (Hardy) in possessing a broad costal band confluent with R_{4+5} and of uniform width throughout, *prsc.* setae absent (in some specimens) and abdominal terga III–V orange-brown with dark colour patterns. It differs from *B. flavipennis* in having the scutum entirely red-brown, cells bc and colourless (or with, at most, a pale tint), all femora with a dark subapical spot on outer

surfaces and abdominal terga III–V with a medial longitudinal dark band over terga III and IV and narrow anterolateral dark corners on all three terga.

Bactrocera (Bactrocera) pectoralis (Walker)

(Fig. 147)

Dacus pectoralis Walker, 1859: 114. Lectotype female in BMNH.
Dacus (Strumeta) pectoralis – Hardy & Adachi, 1954: 179; Hardy, 1959: 179. Lectotype designation by inference of Holotype.
Bactrocera (Bactrocera) pectoralis – Norrbom *et al*., 1998: 94.

Type material examined: Lectotype female, **INDONESIA: MOLUCCAS (MALUKU)** – with the following six labels – 'Type'; 'pectoralis'; 'Aru'; 'Borneo, Aru, W.W. Saunders, BM 1868-4'; 'BMNH (E) No. 533122'; 'Lectotype female of *Dacus pectoralis* Walker, 1859, des. D.E. Hardy 1959, Indonesia: Java, Maluku, A.R. Wallace'. Studied at BMNH.
Diagnosis: A large species; face fulvous with a pair of medium-sized circular black spots; postpronotal lobes and notopleura yellow; scutum black; very short and narrow lateral postsutural yellow vittae present; medial postsutural yellow vitta absent; no yellow spot anterior to notopleural suture; mesopleural stripe reaching midway between anterior margin of notopleuron and anterior *npl.* seta dorsally; scutellum yellow; wing with cells bc and c fulvous, dense microtrichia covering cell bc and most of cell c; a narrow fuscous costal band; a broad fuscous anal streak; abdominal terga mostly dark fuscous to dull black with a broad medial longitudinal dark red-brown band running from posterior margin of tergum III to posterior margin of tergum V.
Description: female.
Head: Frons – fulvous with dark fuscous on anteromedial hump and around bases of lower pair of *i.or.* setae; orbital setae black: 1 *s.or.*, 2 *i.or.*; lunule fulvous. Ocellar triangle black. Vertex red-brown. Face fulvous with a pair of medium-sized circular black spots. Occiput pale red-brown, fulvous along eye margins. Antennae with segments 1 and 2 red-brown, segment 3 missing.
Thorax: Scutum black except red-brown anterior to notopleural suture and around lateral margins. Pleural areas generally black with dark fuscous below postpronotal lobes. Yellow markings as follows: postpronotal lobes; notopleura; mesopleural stripe reaching midway between anterior margin of notopleuron and anterior *npl.* seta dorsally, continuing to katepisternum as a transverse spot, anterior margin slightly convex; two very narrow and short lateral postsutural vittae tapering posteriorly and ending just posterior to level of *a.sa.* seta. Medial longitudinal postsutural yellow vitta absent. Scutellum yellow except for a moderately broad black basal band. Setae: *sc.* 2; *prsc.* 2; *ia.* 1; *p.sa.* 1; *a.sa.* 1; *mpl.* 1; *npl.* 2; *scp.* 2.
Legs: Fore and mid femora fuscous, hind femora changing from fulvous on basal ⅔ to fuscous on apical ⅓ fore and mid tibiae fuscous, hind tibiae dark fuscous; tarsi with all segments fulvous; mid tibiae each with an apical black spur.
Wings: Length 7.8 mm; cells bc and c fulvous; dense microtrichia covering all of cell c and most of cell bc; remainder of wings colourless except fuscous cell sc, narrow fuscous costal band overlapping R_{2+3} but not reaching R_{4+5} except in basal portion of R_{4+5} and remaining narrow around apex of wing to end just beyond extremity of R_{4+5}, a broad fuscous anal streak; no dense aggregation of microtrichia around A_1+CuA_2; supernumerary lobe weak.
Abdomen: Oval; terga free; pecten absent on tergum III. Tergum I and sterna I and II wider than long. All terga dark fuscous to dull black except for dark red-brown as follows: a narrow transverse band across posterior margins of terga I and II, on lateral margins of terga II–V and posterior margin of tergum V, a broad medial longitudinal band from posterior margin of tergum III to posterior margin of tergum V. A pair of dark fuscous shining spots on tergum V. Abdominal sterna dark fuscous. Ovipositor with basal segment dark red-brown, dorsoventrally compressed and tapering posteriorly in dorsal view.
Attractant: No known record.
Male: No known record.
Distribution:
Type locality: Indonesia (Moluccas = Maluku – Aru I.). The record of Java listed by D.E. Hardy on the Lectotype label in 1959 is incorrect as the type specimen was collected on Aru I.
Hosts: No known record.
Remarks: *Bactrocera (Bactrocera) pectoralis* (Walker) is similar to *Bactrocera (Bactrocera) citima* (Hardy) and *Bactrocera (Bactrocera) latifrons* (Hendel) in possessing the costal band overlapping R_{2+3}, black scutum and abdominal terga III–V without a distinct 'T' pattern, but differs from these species in having very short and narrow lateral postsutural vittae not extending posterior to level of *a.sa.* setae and coloured cells bc and c which are covered with dense microtrichia.

Hardy (1968) stated that *Dacus (Strumeta) parvulus* Hendel was a synonym of *B. pectoralis* although he did not formally declare this. Hardy (1968) was incorrect as *B. pectoralis* is clearly a distinct species.

Hardy (1959) stated that *Dacus ferrugineus* var. *obscurata* de Meijere is probably a synonym of *B. pectoralis*. However, we have studied the Holotype female of *Bactrocera (Bactrocera) obscurata* (de Meijere) and Lectotype female of *B. pectoralis* and confirm that they are separate species. In particular, *B. pectoralis* has cells bc and c fulvous and covered with dense microtrichia, a broader costal band almost confluent with R_{4+5} and abdominal terga III–V without a medial longitudinal dark band.

Bactrocera (Bactrocera) pedestris (Bezzi)

(Fig. 148)

Chaetodacus ferrugineus var. *pedestris* Bezzi, 1914a: 322. Lectotype female in MCSNM.
Chaetodacus ferrugineus pedestris – Bezzi, 1919: 423.

Dacus (Strumeta) pedestris – Hardy & Adachi, 1954: 180 (partim); Hardy, 1969a: 421 (partim); Hardy, 1969b: 481, Lectotype designation invalid; 1974: 40 (partim).
Bactrocera (Bactrocera) pedestris – White & Elson-Harris, 1992: 194; Drew & Hancock, 1994a: 50, Lectotype designation; Norrbom *et al*., 1998: 94.

Non-type material examined: Since the publication of Drew & Hancock (1994a) the following specimens have been studied – **PHILIPPINES**: 1 male, Mt Makiling, Botanic Gardens, Alt 150m, 12.i.2000, coll. A. O'Toole, attracted to cue lure. **THAILAND**: 7 males and 14 females, Mae Sapox, Chiang Mai, 22.vi.1993, coll. Chantanee, bred from *Walsura intermedia*, Sample No. CM1951; 1 male and 1 female, Jae Chon Nat Park, Chiang Rai, 28.vii.1992, coll. Charuwan, bred from *Chukrasia venlutina* (Misspelt on label data). Specimens in QDPC.
Diagnosis: Face fulvous with a pair of large oval black spots and a dark ring around bases of antennae; postpronotal lobes and notopleura yellow; scutum black except brown below and behind lateral postsutural vittae (only below these vittae in some specimens), around notopleural suture and inside postpronotal lobes; broad parallel-sided or subparallel lateral postsutural yellow vittae ending at or behind *ia.* seta; medial postsutural yellow vitta absent; mesopleural stripe reaching almost to anterior *npl.* seta dorsally; scutellum yellow; legs with femora fulvous with dark fuscous to black around apices as follows: around apical ⅔ of fore femora and apical ⅓ of mid and hind femora (in some specimens these apical dark markings are reduced to subapical spots on outer apical surfaces); tibiae fuscous to dark fuscous; wing with cells bc and c colourless, microtrichia in outer corner of cell c only; a narrow fuscous costal band confluent with or very slightly overlapping R_{2+3} and remaining narrow around apex of wing; a moderately broad anal streak; supernumerary lobe weak in females and of medium development in males; abdominal terga III–V orange-brown except for a 'T' pattern consisting of a transverse black band across anterior ⅓ of tergum III which expands to cover lateral margins and a broad medial longitudinal black band over all three terga, lateral ¼ of tergum IV and anterolateral corners of tergum V dark fuscous to black; a pair of oval orange-brown shining spots on tergum V; abdominal sterna dark coloured.
Attractant: Cue lure.
Female: As for male in the general body colour patterns. See Drew & Hancock (1994a) for a detailed description of female characters.
Distribution:
Type locality: Philippines (Luzon I., Los Banos).
New records: Thailand (Chiang Mai; Chiang Rai).
Hosts:
New records: *Chukrasia velutina* and *Walsura intermedia* (family *Meliaceae*).
Remarks: *Bactrocera (Bactrocera) pedestris* (Bezzi), within the *dorsalis* complex, is similar to *Bactrocera (Bactrocera) fernandoi* Tsuruta & White, *Bactrocera (Bactrocera) gombokensis* Drew & Hancock and *Bactrocera (Bactrocera) lombokensis* Drew & Hancock in possessing broad parallel-sided or subparallel lateral postsutural vittae, costal band confluent with or very slightly overlapping R_{2+3} and dark apical markings on all femora. Most specimens of *B. pedestris* differ from these species in possessing the apical dark markings on all femora surrounding the apices (not as subapical spots) and in having a ring of dark colour around bases of antennae. See Drew & Hancock (1994a) for a detailed discussion on the status of the type specimens and why the Lectotype designation of Hardy (1969b) is invalid.

Bactrocera (Bactrocera) peneallwoodi Drew & Romig, new species

(Fig. 149)

Material examined: Holotype male, **INDONESIA: WEST TIMOR** – Indonesia, West Timor, Besi Pae, 80 km SE Kupang, 28.x.1986, coll. E.S.C. Smith, ex cue lure trap, deposited in BMNH.
Paratypes: **INDONESIA**: **WEST TIMOR** – 2 males, same data as holotype; 7 males, Indonesia, West Timor, Kupang ADAB office, 31.x.1986, coll. E.S.C. Smith, ex methyl eugenol trap. 2 paratypes in ANIC, 1 paratype in BMNH, 4 paratypes in QDPC, 2 paratypes in QMIC.

Diagnosis: A small species; face fulvous with a pair of large circular black spots; postpronotal lobes and notopleura yellow; scutum black; very short and narrow lateral postsutural yellow vittae present; medial postsutural yellow vitta absent; no yellow spot anterior to notopleural suture; mesopleural stripe reaching midway between anterior margin of notopleuron and anterior *npl.* seta dorsally; scutellum yellow; wing with cells bc and c pale fuscous, microtrichia in outer corner of cell c only; a narrow dark fuscous costal band; a narrow dark fuscous anal streak; abdominal terga III–V orange-brown with narrow transverse black spots on anterolateral corners of tergum III and a narrow medial longitudinal dark fuscous to black band which varies from occurring on only tergum V to covering all three terga.

Description: male.
Head: Height 1.56 mm. Frons – length 1.7 times breadth; red-brown with fulvous along lateral margins and dark fuscous around bases of orbital setae; orbital setae black: 1 *s.or.*, 2 *i.or.*; lunule dark fuscous. Ocellar triangle black. Vertex fuscous. Face fulvous with a pair of large circular black spots; length 0.46 mm. Genae fulvous, large fuscous subocular spot present; red-brown seta present. Occiput red-brown, fulvous along eye margins; occipital row with 5–6 black setae. Antennae with segments 1 and 2 red-brown, segment 3 red-brown with fuscous on apex and outer surface; length of segments: 0.16 mm; 0.26 mm; 0.72 mm.
Thorax: Scutum black with red-brown along the entire lateral margins. Pleural areas dark fuscous to black with red-brown below postpronotal lobes and above mid and hind coxae. Yellow markings as follows: postpronotal lobes; notopleura; mesopleural stripe reaching midway between anterior margin of notopleuron and anterior *npl.* seta dorsally, continuing to katepisternum as a transverse spot, anterior margin straight; anatergite (posterior apex black); anterior ⅔ katatergite (remainder black); two very short and narrow lateral postsutural vittae narrowing posteriorly and

generally ending around or just posterior to line of *a.sa.* seta. Medial longitudinal postsutural yellow vitta absent. Postnotum black. Scutellum yellow except for narrow black basal band. Setae: *sc.* 2; *prsc.* 2; *ia.* 1; *p.sa.* 1; *a.sa.* 1; *mpl.* 1; *npl.* 2; *scp.* 4.
Legs: Fore and mid femora entirely red-brown, hind femora entirely fulvous; fore tibiae red-brown, mid tibiae fuscous basally to red-brown apically, hind tibiae entirely fuscous to dark fuscous; fore tarsi with basal segment fulvous and apical four segments red-brown, mid and hind tarsi with all segments fulvous; mid tibiae each with an apical black spur.
Wings: Length 4.9 mm; cell c pale fuscous, cell bc slightly darker fuscous; microtrichia in outer corner of cell c only; remainder of wings colourless except dark fuscous cell sc, narrow dark fuscous costal band confluent with R_{2+3} and remaining of uniform width around apex of wing, a narrow dark fuscous anal streak; a dense aggregation of microtrichia around A_1+CuA_2; supernumerary lobe of medium development.
Abdomen: Oval; terga free; pecten present on tergum III. Tergum I and sterna I and II wider than long. Tergum I dark fuscous with a narrow transverse orange-brown band across posterior margin but not reaching lateral margins; tergum II orange-brown with fulvous across posterior margin; terga III–V orange-brown with narrow transverse black spots on anterolateral corners of tergum III and a narrow medial longitudinal dark fuscous to black band which varies from being only on tergum V to covering all three terga. A pair of oval dark red-brown shining spots on tergum V. Posterior lobe of surstylus short, sternum V with a deep concavity on posterior margin. Abdominal sterna orange-brown.
Attractant: Uncertain.
Female: No known record.
Distribution: Indonesia (West Timor, Kupang).
Hosts: No known record.
Remarks: *Bactrocera (Bactrocera) peneallwoodi* n.sp. is similar to *Bactrocera (Bactrocera) flavoscutellata* Lin & Wang, *Bactrocera (Bactrocera) perigrapha* White & Tsuruta and *Bactrocera (Bactrocera) pernigra* (Ito) in possessing a black scutum, a narrow fuscous costal band confluent with R_{2+3}, abdominal terga III–V without a distinct dark 'T' pattern and short narrow lateral postsutural vittae not extending posterior to *a.sa.* setae. It differs from *B. perigrapha* in having the scutellum and postpronotal lobes entirely yellow and from *B. flavoscutellata* and *B. pernigra* in having all femora entirely fulvous.

Bactrocera (Bactrocera) penebeckerae Drew & Romig, new species

(Fig. 150)

Material examined: Holotype male, **INDONESIA: FLORES** – Borong, iv.1989, coll. J. Turner, deposited in BMNH.
Paratypes: **INDONESIA: FLORES** – 8 males, same data as holotype; **SULAWESI** – 1 male, Sulawesi Tengah, Donggala, Nupabomba, 12.viii.2006, coll. Rusdin, attracted to methyl eugenol in forest. 2 paratypes in ANIC, 1 paratype in BMNH, 4 paratypes in QDPC, 2 paratypes in QMIC.

Diagnosis: A medium-sized species; face fulvous with a pair of medium-sized oval black spots; postpronotal lobes and notopleura yellow; scutum black with dark fuscous behind lateral postsutural vittae and dark red-brown around notopleural suture and inside postpronotal lobes; lateral postsutural yellow vittae present; medial postsutural yellow vitta absent; no yellow spot anterior to notopleural suture; mesopleural stripe reaching midway between anterior margin of notopleuron and anterior *npl.* seta dorsally; scutellum yellow; wing with cells bc and c colourless, microtrichia in outer corner of cell c only; a narrow dark fuscous costal band; a narrow dark fuscous anal streak; abdominal terga III–V similar to *Bactrocera (Bactrocera) beckerae* (Hardy) being mostly orange-brown with a narrow black band across anterior margin of tergum III, narrow fuscous anterolateral corners on terga IV and V and a narrow medial longitudinal fuscous line on tergum V.
Description: male.
Head: Height 1.8 mm. Frons – length 1.45 times breadth; red-brown with dark fuscous on anteromedial hump and around bases of orbital setae; orbital setae black: 1 *s.or.*, 2 *i.or.*; lunule dark fuscous. Ocellar triangle black. Vertex red-brown. Face fulvous with a pair of medium-sized oval black spots; length 0.54 mm. Genae red-brown, large dark fuscous subocular spot present; dark red-brown seta present. Occiput pale fuscous, fulvous along eye margins; occipital row with 4–6 black setae. Antennae with segments 1 and 2 red-brown, segment 3 red-brown with fuscous on apex and outer surface; length of segments: 0.21 mm; 0.39 mm; 0.84 mm.
Thorax: Scutum black with dark fuscous behind lateral postsutural vittae and red-brown around notopleural suture and inside postpronotal lobes. Pleural areas entirely black, tending dark fuscous to black below postpronotal lobes. Yellow markings as follows: postpronotal lobes; notopleura; mesopleural stripe reaching midway between anterior margin of notopleuron and anterior *npl.* seta dorsally, continuing to katepisternum as a small transverse spot, anterior margin straight; anatergite (posterior apex black); anterior ⅔ katatergite (remainder black); two medium-width lateral postsutural vittae narrowing slightly posteriorly to end at *ia.* seta. Medial longitudinal postsutural yellow vitta absent. Postnotum black. Scutellum yellow except for narrow black basal band. Setae: *sc.* 2; *prsc.* 2; *ia.* 1; *p.sa.* 1; *a.sa.* 1; *mpl.* 1; *npl.* 2; *scp.* 4.
Legs: All femora entirely fulvous; fore tibiae fuscous, mid tibiae dark fuscous basally to fuscous apically, hind tibiae dark fuscous; all tarsi with basal segment fulvous and apical four segments red-brown; mid tibiae each with an apical black spur.
Wings: Length 6.4 mm; cells bc and c colourless; microtrichia in outer corner of cell c only; remainder of wings colourless except dark fuscous cell sc, narrow dark fuscous costal band slightly overlapping R_{2+3} where it becomes fuscous and continuing around apex of wing where it expands slightly across apex of R_{4+5}, a narrow dark fuscous anal streak; a dense aggregation of microtrichia around A_1+CuA_2; supernumerary lobe of medium development.
Abdomen: Oval; terga free; pecten present on tergum III. Tergum I and sterna I and II wider than long. Tergum I

fuscous centrally with broad black lateral margins and a narrow orange-brown band across posterior margin but not reaching lateral margins; tergum II orange-brown anteriorly and fulvous posteriorly with a narrow transverse black band across centre of tergum and reaching narrow black lateral margins; terga III–V orange-brown with a narrow transverse black band across anterior margin of tergum III which in some specimens changes from fuscous to dark fuscous in midline and narrow anterolateral fuscous corners on terga IV and V and a narrow medial longitudinal fuscous line on tergum V. A pair of oval fuscous shining spots on tergum V. Posterior lobe of surstylus short, sternum V with a deep concavity on posterior margin. Abdominal sterna dark fuscous.
Attractant: No known record.
Female: No known record.
Distribution: Indonesia (Flores; Sulawesi – Central Sulawesi).
Hosts: No known record.
Remarks: *Bactrocera (Bactrocera) penebeckerae* n.sp. is similar to *Bactrocera (Bactrocera) aemula* Drew, *Bactrocera (Bactrocera) contigua* Drew, *Bactrocera (Bactrocera) musae* (Tryon) and *Bactrocera (Bactrocera) pseudobeckerae* n.sp., in possessing an entire costal band overlapping R_{2+3}, abdominal terga III–V without a distinct dark 'T' pattern and elongate lateral postsutural vittae reaching to or almost to *ia.* setae. It differs from *B. aemula* in having abdominal terga III–V almost entirely orange-brown, from *B. contigua* and *B. pseudobeckerae* in having the costal band just overlapping R_{2+3} and from *B. musae* in having *beckerae*-like abdominal patterns with a dark transverse band across anterior margin of tergum III and with fuscous to dark fuscous ceromae.

Bactrocera (Bactrocera) penecognata Drew & Hancock
(Fig. 151)

Bactrocera (Bactrocera) penecognata Drew & Hancock, 1994a: 51; Norrbom *et al.*, 1998: 94. Holotype male in BBM.

Non-type material examined: Since the publication of Drew & Hancock (1994a), the following specimens have been studied – **INDONESIA: JAVA** – 2 males, East Java, Lumajang, Kadang Tepus, 12.vii.2005, coll. Utomo, attracted to methyl eugenol in banana; 2 males, East Java, Probolinggo, Betektaman, 20.vii.2005, coll. Tjipto Waluyo, attracted to methyl eugenol in mango (the methyl eugenol records are probably in error). Specimens in QDPC.
Diagnosis: Face fulvous with a pair of medium-sized circular black spots; postpronotal lobes and notopleura yellow; scutum black except dark brown behind lateral postsutural vittae (occasionally below these vittae also), around notopleural suture and inside postpronotal lobes; broad triangular lateral postsutural yellow vittae narrowing sharply posteriorly to end before level of *p.sa.* seta; medial postsutural yellow vitta absent; mesopleural stripe equal to or slightly wider than notopleuron dorsally; scutellum yellow; legs with femora entirely fulvous (except for a large dark fuscous spot on outer apical surfaces of fore femora in some specimens); fore and mid tibiae fuscous to dark fuscous (mid tibiae sometimes paler apically) and hind tibiae dark fuscous; wing with cells bc and c colourless (or with a very pale tint), microtrichia in outer corner of cell c only; a narrow dark fuscous costal band slightly overlapping R_{2+3} and widening slightly across apex of R_{4+5}; a moderately broad dark fuscous anal streak; supernumerary lobe of medium development; abdominal terga III–V orange-brown except for a narrow transverse black band across anterior margin of tergum III which widens to cover lateral ¼ of this tergum, a moderately broad medial longitudinal black band over all three terga, narrow black anterolateral margins on terga IV and V, a pair of oval orange-brown shining spots on tergum V; abdominal sterna dark coloured.
Attractant: Cue lure.
Female: No known record.
Distribution:
Type locality: Indonesia (Lombok I., Sasaot).
Previously recorded: Type locality and Indonesia (Sumbawa).
New records: Indonesia (East Java).
Hosts: No known record.
Remarks: *Bactrocera (Bactrocera) penecognata* Drew & Hancock, within the *dorsalis* complex, is similar to *Bactrocera (Bactrocera) fulvifemur* Drew & Hancock, *Bactrocera (Bactrocera) lateritaenia* Drew & Hancock and *Bactrocera (Bactrocera) usitata* Drew & Hancock in possessing lateral postsutural vittae that narrow posteriorly, femora entirely fulvous or with, at most, a preapical dark spot on fore femora, costal band slightly overlapping R_{2+3} and a broad medial longitudinal black band over abdominal terga III–V. It differs from *B. lateritaenia* and *B. usitata* in possessing short triangular lateral postsutural vittae and from *B. fulvifemur* in having cells bc and c colourless and abdominal terga IV and V with narrow dark lateral margins.

Bactrocera (Bactrocera) penecorrecta Drew
(Fig. 152)

Bactrocera (Bactrocera) penecorrecta Drew, in Drew & Raghu, 2002: 342. Holotype male in BMNH.

Non-type material examined: No new specimens available for study since the publication of Drew & Raghu (2002).
Diagnosis: Face fulvous with a pair of transverse oval black spots almost meeting in midline; postpronotal lobes and notopleura yellow; scutum black with red-brown posterolateral to lateral postsutural vittae and anterior to notopleural suture; moderately broad parallel-sided lateral postsutural yellow vittae ending behind *ia.* seta; medial postsutural yellow vitta absent; mesopleural stripe reaching almost to anterior *npl.* seta dorsally; scutellum yellow except for a moderately broad black basal band; legs with all femora entirely fulvous; fore tibiae fuscous, hind tibiae dark fuscous; wing with cells bc and c colourless, microtrichia in outer corner of cell c only; narrow fuscous costal band confluent with R_{2+3} and ending at apex of this vein before continuing as a

large oval fuscous spot across apex of R_{4+5}; a narrow pale fuscous anal streak confined within cell cup; supernumerary lobe of medium development; abdominal terga III–V black except dark fuscous spots posterocentrally on tergum III and dark red-brown spots posterocentrally on terga IV and V either side of a medial longitudinal black band, a pair of oval dark red-brown shining spots on tergum V.
Attractant: Methyl eugenol.
Female: No known record.
Distribution:
Type locality: Southern India (Kerala, New Amarambalam Forest).
Hosts: No known record.
Remarks: *Bactrocera (Bactrocera) penecorrecta* Drew is similar to *Bactrocera (Bactrocera) correcta* (Bezzi) in possessing a fulvous face with transverse oval black spots, a black scutum, parallel-sided lateral postsutural vittae, femora entirely fulvous and costal band interrupted beyond apex of R_{2+3} but differs from this species in having a moderately broad black band on the scutellum and abdominal terga III–V mostly black with small dark red-brown spots posterocentrally on either side of a medial longitudinal black band on terga IV and V.

Bactrocera (Bactrocera) penecostalis Drew & Romig, new species

(Fig. 153)

Material examined: Holotype male, **INDONESIA: SULAWESI** – Sulawesi Utara, Bitung, Danowudu, 14. viii.2006, coll. Karel F. Lala, attracted to cue lure in forest, deposited in BMNH.
Paratypes: **INDONESIA: SULAWESI** – 1 male, Sulawesi, Sulawesi Barat, Mamuju, Binanga, 4.ix.2006, coll. Sri Hastuti, Yan P.F., Sukri, attracted to cue lure in orange (jeruk); 1 male (12.viii.2006), 2 males (2.ix.2006), 1 male (17. ix.2006), 1 male (27.ix.2006), 1 male (15.x.2006), Sulawesi, Sulawesi Tengah, Donggala, Nupabomba, coll. Rusdin, attracted to cue lure in forest. 1 paratype in ANIC, 1 paratype in BMNH, 4 paratypes in QDPC, 1 paratype in QMIC.
Diagnosis: A large species; face fulvous with a pair of medium-sized oval black spots; postpronotal lobes and notopleura yellow; scutum dull black; lateral postsutural yellow vittae present; medial postsutural yellow vitta absent; no yellow spot anterior to notopleural suture; mesopleural stripe reaching midway between anterior margin of notopleuron and anterior *npl.* seta dorsally; scutellum yellow; wing with cells bc and c fuscous, microtrichia in outer corner of cell c only; a broad fuscous to dark fuscous costal band confluent with R_{4+5}; a broad fuscous anal streak; abdominal terga III–V orange-brown with a 'T' pattern consisting of a narrow transverse black band across anterior margin of tergum III expanding into dark fuscous on lateral margins, a narrow fuscous to black medial longitudinal band over terga III and IV, narrow fuscous anterolateral corners on tergum IV, dark fuscous to black oval shining spots on tergum V.
Description: male.
Head: Height 1.98 mm. Frons – length 1.54 times breadth; red-brown with dark fuscous on anteromedial hump and around bases of orbital setae; orbital setae black: 1 *s.or.*, 2 *i.or.*; lunule dark fuscous. Ocellar triangle black. Vertex fuscous. Face fulvous with a pair of medium-sized oval black spots; length 0.54 mm. Genae fulvous, small fuscous subocular spot present; black seta present. Occiput red-brown, fulvous along eye margins; occipital row with a large number of strong black setae. Antennae with segments 1 and 2 red-brown, segment 3 red-brown with fuscous on apex and outer surface; length of segments: 0.19 mm; 0.36 mm; 0.96 mm.
Thorax: Scutum dull black with dark red-brown below and behind lateral postsutural vittae, around notopleural suture, inside postpronotal lobes and along anterior margin (in some specimens, the black is reduced to two broad longitudinal bands either side of midline, Fig. 153). Pleural areas black except red-brown below postpronotal lobe. Yellow markings as follows: postpronotal lobes; notopleura; mesopleural stripe reaching midway between anterior margin of notopleuron and anterior *npl.* seta dorsally, continuing to katepisternum as a transverse spot, anterior margin slightly convex; anatergite (posterior apex black); anterior $^{3}/_{5}$ katatergite (remainder black); two moderately broad parallel-sided lateral postsutural vittae ending at *ia.* seta. Medial longitudinal postsutural yellow vitta absent. Postnotum black. Scutellum yellow except for narrow black basal band. Setae: *sc.* 2; *prsc.* 2; *ia.* 1; *p.sa.* 1; *a.sa.* 1; *mpl.* 1; *npl.* 2; *scp.* 4.
Legs: All segments entirely fulvous; mid tibiae each with an apical black spur.
Wings: Length 7.3 mm; cells bc and c fuscous; microtrichia in outer corner of cell c only; remainder of wings colourless except dark fuscous cell sc, broad costal band that is dark fuscous to R_{2+3} and around margin of wing and fuscous between R_{2+3} and R_{4+5}, a broad fuscous anal streak; a dense aggregation of microtrichia around A_1+CuA_2; supernumerary lobe of medium development.
Abdomen: Oval; terga free; pecten present on tergum III. Tergum I and sterna I and II wider than long. Tergum I fuscous centrally with narrow black lateral margins and a narrow transverse orange-brown band across posterior margin; tergum II fulvous with a narrow transverse black band across anterior margin and narrow dark fuscous anterolateral corners; terga III–V orange-brown with a 'T' pattern consisting of a narrow transverse black band across anterior margin of tergum III which expands into dark fuscous around lateral margins and a narrow medial longitudinal band over terga III and IV which varies from black to fuscous, narrow fuscous anterolateral corners on tergum IV. A pair of oval dark fuscous to black shining spots on tergum V. Posterior lobe of surstylus short, sternum V with a deep concavity on posterior margin. Abdominal sterna pale.
Attractant: Cue lure.
Female: No known record.
Distribution: Indonesia (Sulawesi – North Sulawesi (Bitung); Central Sulawesi (Donggala); West Sulawesi (Mamuju)).
Hosts: No known record.

Remarks: *Bactrocera (Bactrocera) penecostalis* n.sp. is similar to *Bactrocera (Bactrocera) costalis* (Shiraki) in possessing a black scutum, a uniform width costal band confluent with R_{4+5}, parallel-sided lateral postsutural vittae and coloured cells bc and c but differs from *B. costalis* in having abdominal terga III–V with dark lateral margins and all femora entirely fulvous.

Bactrocera (Bactrocera) penephaea Drew & Romig, new species

(Fig. 154)

Material examined: Holotype male, **INDONESIA: NORTH MOLUCCAS (NORTH MALUKU)** – Moluccas, North Maluku, Kodya Ternate, Foramadiahi, 3.vi.2007, coll. La Ruti, attracted to cue lure in forest, deposited in BMNH. Paratypes: **INDONESIA: NORTH MOLUCCAS (NORTH MALUKU)** – 2 males, same data as holotype; 5 males, Moluccas, North Maluku, Kodya Ternate, Foramadiahi, 10.vi.2007, coll. La Ruti, attracted to cue lure in forest; 1 male (5.vi.2007), 2 males (12.vi.2007), 2 males (19.vi.2007), 4 males (29.v.2007), Moluccas, North Maluku, West Halmahera Gaol, coll. Raais Abdullah, attracted to cue lure in forest, jeruk, mangga, citrus and chilli. 2 paratypes in ANIC, 4 paratypes in BMNH, 6 paratypes in QDPC, 4 paratypes in QMIC.

Non-type material examined: INDONESIA: NORTH MOLUCCAS (NORTH MALUKU) – 1 male (29.v.2007), 3 males (5.vi.2007), 9 males (12.vi.2007), 2 males (18.vi.2007), North Maluku, West Halmahera Gaol, coll. Raais Abdullah, attracted to methyl eugenol in forest; 7 males, North Maluku, Kodya, Ternate, Foramadiahi, 27.v.2007, coll. La Ruti, attracted to methyl eugenol in forest; 2 males, North Maluku, Kodya, Ternate, Kalumata, 10.vi.2007, coll. La Ruti, attracted to methyl eugenol in Jeruk. Specimens in QDPC.

Diagnosis: A medium-sized species; face fulvous with a pair of medium-sized circular black spots; postpronotal lobes and notopleura yellow; scutum generally dull black; lateral postsutural yellow vittae present; medial postsutural yellow vitta absent; no yellow spot anterior to notopleural suture; mesopleural stripe reaching almost to anterior *npl.* seta dorsally; scutellum yellow; wing with cells bc and c with a pale fuscous tint, microtrichia in outer corner of cell c only; a broad dark fuscous costal band which is slightly paler between R_{2+3} and R_{4+5} and confluent with R_{4+5}, narrow fuscous margins enclosing *r-m* and *dm-cu* crossveins; a broad dark fuscous anal streak; abdominal terga III–V orange-brown with a dark 'T' pattern consisting of a narrow transverse black band across anterior margin of tergum III which widens across lateral margins into dark fuscous, a moderately broad medial longitudinal black band over three terga, narrow fuscous to dark fuscous anterolateral corners on terga IV and V.

Description: male.

Head: Height 1.67 mm. Frons – length 1.67 times breadth; red-brown with fuscous on anteromedial hump and around orbital setae; orbital setae black: 1 *s.or.*, 2 *i.or.*; lunule fuscous. Ocellar triangle black. Vertex fuscous. Face fulvous with a pair of medium-sized circular black spots; length 0.44 mm. Genae fulvous, small fuscous subocular spot present; black seta present. Occiput dark fuscous, fulvous along eye margins; occipital row with 7–8 strong black setae. Antennae with segments 1 and 2 red-brown, segment 3 red-brown with fuscous on apex and outer surface; length of segments: 0.14 mm; 0.31 mm; 0.79 mm.

Thorax: Scutum dull black with dark fuscous below and behind lateral postsutural vittae and dark red-brown around notopleural suture and inside postpronotal lobes. Pleural areas dull black except dark fuscous below postpronotal lobes. Yellow markings as follows: postpronotal lobes; notopleura; mesopleural stripe reaching almost to anterior *npl.* seta dorsally, continuing to katepisternum as a transverse spot, anterior margin slightly convex; anatergite (posterior apex black); anterior $^4/_5$ katatergite (remainder black); two broad lateral postsutural vittae narrowing slightly posteriorly to end at or just before *ia.* seta. Medial longitudinal postsutural yellow vitta absent. Postnotum black. Scutellum yellow with a narrow black basal band. Setae: *sc.* 2; *prsc.* 2; *ia.* 1; *p.sa.* 1; *a.sa.* 1; *mpl.* 1; *npl.* 2; *scp.* 2.

Legs: All femora entirely fulvous; fore and mid tibiae fuscous, hind tibiae dark fuscous; all tarsi with basal segment fulvous and apical four segments red-brown; mid tibiae each with an apical black spur.

Wings: Length 5.5 mm; cells bc and c with a pale fuscous tint; microtrichia in outer corner of cell c only; remainder of wings colourless except dark fuscous cell sc, broad dark fuscous costal band which tends paler between R_{2+3} and R_{4+5} and confluent with R_{4+5} before ending between apices of R_{4+5} and M, pale fuscous enclosing *r-m* and *dm-cu* crossveins, a broad dark fuscous anal streak; a dense aggregation of microtrichia around A_1+CuA_2; supernumerary lobe weak.

Abdomen: Oval; terga free; pecten present on tergum III. Tergum I and sterna I and II wider than long. Tergum I dark fuscous except for a narrow orange-brown band along posterior margin but not reaching lateral margins; tergum II orange-brown to fulvous with a narrow transverse black band across anterior margin and reaching dark fuscous anterolateral corners; terga III–V orange-brown with a 'T' pattern consisting of a narrow transverse black band across anterior margin of tergum III which widens to cover lateral margins as a dark fuscous area, a moderately broad medial longitudinal black band over all three terga, dark fuscous to fuscous narrow anterolateral corners on terga IV and V. A pair of oval fuscous to dark fuscous shining spots on tergum V. Posterior lobe of surstylus short, sternum V with a deep concavity on posterior margin. Abdominal sterna dark coloured.

Attractant: Cue lure.

Female: No known record.

Distribution: Indonesia (North Moluccas = North Maluku – Halmahera I.; Ternate).

Hosts: No known record.

Remarks: *Bactrocera (Bactrocera) penephaea* n.sp. is similar to *Bactrocera (Bactrocera) phaea* (Drew) from New Britain and Lihir Island in possessing a black scutum, broad lateral postsutural vittae, coloured cells bc and c, a broad costal band confluent with R_{4+5}, infuscation enclosing *r-m* crossvein and abdominal terga III–V with a black 'T' pattern.

It differs from *B. phaea* in having infuscation enclosing both crossveins and the mesopleural stripe not reaching the postpronotal lobe dorsally.

Bactrocera (Bactrocera) perigrapha White & Tsuruta

(Fig. 155)

Bactrocera (Bactrocera) perigrapha White & Tsuruta, in Tsuruta & White, 2001: 82. Holotype male in BMNH.

Type material examined: Holotype male, **SRI LANKA (formerly CEYLON)**: with the following three labels – 'Holotype'; 'Ceylon, Punduluoya, E.E. Green, 1890.11.5'; 'Holotype male of Bactrocera (Bactrocera) perigrapha White & Tsuruta, 2001'. Received on loan from BMNH.
Non-type material examined: BHUTAN: 1 male, Namling 11km, 9.vi.2001, coll. B. Fletcher, altitude 2000m, broad leafed forest, attracted to cue lure. Specimen in QDPC.
Diagnosis: A medium-sized species; face fulvous with a pair of very large elongate oval black spots; postpronotal lobes yellow (anterodorsal corners black); notopleura yellow; scutum entirely black; lateral postsutural yellow vittae present; medial postsutural yellow vitta absent; no yellow spot anterior to notopleural suture; mesopleural stripe slightly wider than notopleuron dorsally; scutellum yellow with a large black triangular marking covering most of dorsal surface; wing with cells bc and c colourless, microtrichia in outer corner of cell c only; a narrow dark fuscous costal band; a narrow dark fuscous anal streak; all abdominal terga entirely black except for narrow dark red-brown transverse bands posterolaterally on tergum II.
Description: male.
Head: Height 1.5 mm. Frons – length 1.5 times breadth; red-brown without dark markings; orbital setae dark fuscous: 1 *s.or.*, 2 *i.or.*; lunule red-brown. Ocellar triangle black. Vertex black. Face fulvous with a pair of large elongate oval black spots reaching oral margin; length 0.5 mm. Genae fulvous, small pale fuscous subocular spot present; black seta present. Occiput black, fulvous along eye margins; occipital row with 4–5 strong black setae. Antennae with segments 1 and 2 red-brown, segment 3 red-brown with dark fuscous on apex and outer surface; length of segments: 0.16 mm; 0.28 mm; 0.8 mm.
Thorax: Scutum entirely black without pale markings. Pleural areas entirely black without pale markings. Yellow markings as follows: postpronotal lobes (anterodorsal corners black); notopleura; mesopleural stripe slightly wider than notopleuron dorsally, continuing to katepisternum as a very small transverse spot, anterior margin straight; anatergite (posterior apex black); anterior ½ katatergite (ventral margin and posterior ½ black); two very short and narrow lateral postsutural vittae ending well before level of *a.sa.* seta. Medial longitudinal postsutural yellow vitta absent. Postnotum black. Scutellum yellow with almost entire dorsal surface covered by a black triangular marking. Setae: *sc.* 2; *prsc.* 2; *ia.* 1; *p.sa.* 1; *a.sa.* 1; *mpl.* 1; *npl.* 2; *scp.* 4.
Legs: Fore femora fulvous with apical ⅔ of outer surface black, mid and hind femora fulvous with apical ½ black; fore and mid tibiae fulvous, hind tibiae dark fuscous; tarsi with all segments fulvous; mid tibiae each with an apical black spur.
Wings: Length 5.8 mm; cells bc and c colourless; microtrichia in outer corner of cell c only; remainder of wings colourless except dark fuscous cell sc, narrow dark fuscous costal band confluent with R_{2+3} and widening slightly across apex of wing to end just beyond apex of R_{4+5}, a narrow dark fuscous anal streak; a dense aggregation of microtrichia around A_1+CuA_2; supernumerary lobe of medium development.
Abdomen: Oval; terga free; pecten present on tergum III. Abdominal terga entirely black except for narrow dark red-brown transverse bands posterolaterally on tergum II. Posterior lobe of surstylus short, sternum V with a deep concavity on posterior margin.
Attractant: Cue lure.
Female: No known record.
Distribution:
Type locality: Sri Lanka (Punduluoya).
Previously recorded: Type locality and Sri Lanka (Matale District, Pitawala [Misspelt in publication]) (Tsuruta & White, 2001).
New records: Bhutan (Namling).
Hosts: No known record.
Remarks: *Bactrocera (Bactrocera) perigrapha* White & Tsuruta fits within the *nigrotibialis* complex and is similar to *Bactrocera (Bactrocera) apiconigroscutella* Drew in possessing a pair of large elongate oval black facial spots, the costal band confluent with R_{2+3}, anal streak narrow, cells bc and c colourless and scutellum not entirely yellow, but differs in having a large black triangular pattern over the dorsal surface of the scutellum, mid femora with apical ½ to ⅓ black (not entirely black) and abdominal terga III–V entirely black. *Bactrocera perigrapha* is also similar to *Bactrocera (Bactrocera) pernigra* (Ito) but differs in having the anterodorsal corners of the postpronotal lobes black and a large black elongate spot on the outer surfaces of the fore femora (not black around the apices of the fore femora).

Bactrocera (Bactrocera) pernigra Ito

(Fig. 156)

Bactrocera pernigra Ito, 1983: 23. Holotype male in UOPJ.
Bactrocera (Bactrocera) pernigra – Norrbom *et al.*, 1998: 94.
Dacus (Bactrocera) tenuivittatus Tseng, Chen & Chu, 1992: 47. **New Synonym**.

Non-type material examined: JAPAN: SHIKOKU ISLAND – 2 males, Komatsujima, Tokushima Prefecture, 22.xi.1977, coll. Hiura. Studied at BMNH. 1 male, Kaifu, Tokushima, 22.xi.1978. Specimen in QDPC. **THAILAND**: 2 males, N. Thailand, 1200m, Chiang Mai Prov., Doi Suthep – Pui N.P., above What Phrathat, 18.80N/98.92E, 1.xi.2000, coll. Merz and Schwendinger, attracted to cue lure. Specimens in QDPC.

Diagnosis: A medium-sized species; face fulvous with a pair of large oval black spots; postpronotal lobes yellow with anterodorsal corners black; notopleura yellow; scutum entirely black; very short and narrow lateral postsutural yellow vittae present; medial postsutural yellow vitta absent; no yellow spot anterior to notopleural suture; mesopleural stripe equal in width to notopleuron dorsally; scutellum yellow with a broad black basal band; wing with cells bc and c colourless, microtrichia in outer corner of cell c only; a narrow pale fuscous to fuscous costal band (*scutellaris*-like in shape); a narrow fuscous anal streak; all abdominal terga entirely black except for narrow transverse fulvous bands on posterolateral margins of tergum II and dark red-brown areas posterocentrally on tergum V either side of a narrow medial longitudinal black band.
Description: male.
Head: Frons – fulvous with fuscous around orbital setae, on anteromedial hump and on lateral margins between *s.or.* setae and vertex; orbital setae black: 1 *s.or.*, 2 *i.or.*; lunule fulvous. Ocellar triangle black. Vertex fuscous. Face fulvous with a pair of large oval black spots. Occiput black, fulvous along eye margins. Antennae with segments 1 and 2 red-brown, segment 3 mostly fuscous to dark fuscous.
Thorax: Scutum entirely black. Pleural areas entirely black. Yellow markings as follows: postpronotal lobes (anterodorsal corners black); notopleura; mesopleural stripe equal in width to notopleuron dorsally, continuing to katepisternum as a small transverse spot, anterior margin straight; two very short and narrow lateral postsutural vittae ending before *a.sa.* seta. Medial longitudinal postsutural yellow vitta absent. Scutellum yellow with a broad black basal band. Setae: *sc.* 2; *prsc.* 2; *ia.* 1; *p.sa.* 1; *a.sa.* 1; *mpl.* 1; *npl.* 2; *scp.* 2.
Legs: All femora fulvous with black around apical ⅓ to ¼; fore and mid tibiae fuscous basally to pale fuscous apically, hind tibiae dark fuscous; tarsi with all segments entirely fulvous; mid tibiae each with an apical black spur.
Wings: Length 6.1 mm; cells bc and c colourless; microtrichia in outer corner of cell c only; remainder of wings colourless except fuscous cell sc, narrow fuscous costal band confluent with R_{2+3} and becoming very narrow beyond apex of this vein before widening slightly and becoming fuscous before ending beyond extremity of R_{4+5} (*scutellaris*-like), a narrow fuscous anal streak; a dense aggregation of microtrichia around A_1+CuA_2; supernumerary lobe of medium development.
Abdomen: Oval; terga free; pecten present on tergum III. Tergum I and sterna I and II wider than long. Abdominal terga entirely black except pale fulvous areas on posterolateral margins of tergum II and dark red-brown posterocentrally on tergum V either side of a narrow medial longitudinal black band. A pair of oval black shining spots on tergum V. Posterior lobe of surstylus short, sternum V with a deep concavity on posterior margin. Abdominal sterna dark.
Attractant: Cue lure.
Female: No known record.
Distribution:
Type locality: Japan (Kyushu I.).
Previously recorded: Type locality and Japan (Honshu I.; Shikoku I.) (Norrbom *et al.*, 1998).
New records: Thailand (Chiang Mai).
Hosts: No known record.
Remarks: *Bactrocera (Bactrocera) pernigra* Ito fits within the *nigrotibialis* complex and is similar to *Bactrocera (Bactrocera) perigrapha* White & Tsuruta in possessing large elongate oval black facial spots, costal band confluent with R_{2+3}, cells bc and c colourless, scutellum not entirely yellow and with, at least, a broad black basal band. It differs from this species in having black on apical ⅓ to ¼ of all femora, scutellum with a broad black basal band and postpronotal lobes entirely yellow.

There are no suitable characters that can be used to separate *Dacus (Bactrocera) tenuivittatus* Tseng, Chen & Chu from *B. pernigra* and consequently we are making *D. tenuivittatus* a **New Synonym** of the latter species.

Bactrocera (Bactrocera) pictipennis Lin & Zeng

(Fig. 157)

Bactrocera (Bactrocera) pictipennis Lin & Zeng, in Lin *et al.*, 2011b: 897. Holotype male in HEIQ.

Type material studied: CHINA: A study of the colour photographs and description of *Bactrocera (Bactrocera) pictipennis* Lin & Zeng in the original description.
Diagnosis: A medium-sized species; face fulvous with a pair of medium-sized oval black spots; postpronotal lobes and notopleura yellow; scutum black; broad parallel-sided lateral postsutural yellow vittae present; medial postsutural yellow vitta absent; no yellow spot anterior to notopleural suture; mesopleural stripe reaching anterior *npl.* seta dorsally; scutellum yellow; wing with cells bc and c colourless, microtrichia in outer corner of cell c only; a narrow dark fuscous costal band confluent with R_{2+3} and widening at apex; a narrow dark fuscous anal streak; all abdominal terga entirely pale red-brown except for a transverse fulvous marking across posterior margin of tergum II.
Description: male.
Head: Frons fulvous with fuscous on anteromedial hump and around bases of orbital setae; orbital setae black: 1 *s.or.*, 2 *i.or.*; lunule fuscous. Ocellar triangle black. Vertex fuscous. Face fulvous with a pair of medium-sized oval black spots. Genae fulvous. Occiput red-brown, fulvous along eye margins; occipital row with 5–7 black setae. Antennae with segments 1 and 2 red-brown, segment 3 red-brown with fuscous on outer surface.
Thorax: Scutum black with red-brown below and behind lateral postsutural vittae, around notopleural suture and inside postpronotal lobes. Pleural areas black except red-brown below postpronotal lobes. Yellow markings as follows: postpronotal lobes; notopleura; mesopleural stripe reaching anterior *npl.* seta dorsally and continuing to katepisternum as a small transverse spot; anatergite (posterior apex black); anterior ⅔ katatergite (remainder black); two broad parallel-sided lateral postsutural yellow vittae ending behind *ia.* seta. Medial longitudinal postsutural

yellow vitta absent. Postnotum black. Scutellum yellow except for narrow black basal band. Setae: *sc.* 2; *prsc.* 2; *ia.* 1; *p.sa.* 1; *a.sa.* 1; *mpl.* 1; *npl.* 2; *scp.* 4.

Legs: All femora fulvous with small fuscous subapical spots on outer surfaces; all tibiae fulvous; all tarsi with basal segment fulvous and apical four segments fuscous; mid tibiae each with an apical black spur.

Wings: Length 5.0 mm; cells bc and c colourless; microtrichia in outer corner of cell c only; remainder of wings colourless except dark fuscous cell sc, narrow dark fuscous costal band confluent with R_{2+3} and widening markedly just beyond apex of this vein, a narrow dark fuscous anal streak; a dense aggregation of microtrichia around A_1+CuA_2; supernumerary lobe of medium development.

Abdomen: Oval; terga free; pecten present on tergum III. All abdominal terga pale red-brown except for a fulvous transverse marking across posterior margin of tergum II. A pair of oval red-brown shining spots on tergum V. Posterior lobe of surstylus short, sternum V with a deep concavity on posterior margin.

Attractant: Methyl eugenol.

Female: No known record.

Distribution:

Type locality: China (Hainan Province, Qionghai).

Hosts: No known record.

Remarks: *Bactrocera (Bactrocera) pictipennis* Lin & Zeng is similar to *Bactrocera (Bactrocera) bancroftii* (Tryon), *Bactrocera (Bactrocera) latifrons* (Hendel) and *Bactrocera (Bactrocera) pruniae* n.sp. in possessing a narrow costal band, parallel-sided lateral postsutural vittae, a black scutum and abdominal terga mostly or entirely orange-brown to red-brown. It differs from *B. bancroftii* in having the costal band confluent with R_{2+3} and widening markedly at the apex, from *B. latifrons* in having spots on the apices of all femora, the costal band confluent with R_{2+3} and males attracted to methyl eugenol and from *B. pruniae* in having the costal band widening at the apex, the lateral postsutural vittae ending behind the *ia.* seta and fuscous spots on the apices of all femora.

Bactrocera (Bactrocera) profunda Tsuruta & White

(Fig. 158)

Bactrocera (Bactrocera) profunda Tsuruta & White, 2001: 83. Holotype male in HORDI.

Non-type material examined: SRI LANKA: (formerly CEYLON) – 1 male with the following label data – 'Uraniya, 11.v.1994, cue lure 17'; 1 male – 'Gannoruwa, 26.xii.1994, cue lure 27'; 1 male – 'Horana, 3.v.1995, cue lure 3'. All specimens donated by Kenji Tsuruta and in QDPC. The first two specimens were listed under other non-type series material examined and the last specimen under Paratypes, by Tsuruta & White, 2001. The last specimen, however, does not have a Paratype label and thus cannot be considered as one.

Diagnosis: A medium-sized species; face fulvous with a pair of large oval black spots; postpronotal lobes and notopleura yellow; scutum black; lateral postsutural yellow vittae present; medial postsutural yellow vitta absent; no yellow spot anterior to notopleural suture; mesopleural stripe reaching midway between anterior margin of notopleuron and anterior *npl.* seta dorsally; scutellum yellow; wing with cells bc and c colourless, microtrichia in outer corner of cell c only; a narrow dark fuscous costal band; a narrow fuscous anal streak; abdominal terga III–V orange-brown with a 'T' pattern consisting of a narrow transverse fuscous to darker fuscous band across anterior margin of tergum III and expanding over lateral margins, a narrow medial longitudinal black band over all three terga, narrow fuscous anterolateral corners on terga IV and V.

Description: male.

Head: Height 1.67 mm. Frons – length 1.53 times breadth; red-brown with pale fuscous on anteromedial hump and around bases of orbital setae; orbital setae black: 1 *s.or.*, 2 *i.or.*; lunule dark fuscous. Ocellar triangle black. Vertex pale fuscous. Face fulvous with a pair of large oval black spots; length 0.5 mm. Genae fulvous, fuscous subocular spot present; black seta present. Occiput fuscous, fulvous along eye margins; occipital row with 5–7 strong black setae. Antennae with segments 1 and 2 red-brown, segment 3 red-brown with dark fuscous on apex and outer surface; length of segments: 0.19 mm; 0.31 mm; 0.9 mm.

Thorax: Scutum black except dark fuscous posterior to lateral postsutural vittae, around notopleural suture and inside postpronotal lobes. Pleural areas black except red-brown below postpronotal lobes. Yellow markings as follows: postpronotal lobes; notopleura; mesopleural stripe reaching midway between anterior margin of notopleuron and anterior *npl.* seta dorsally, continuing to katepisternum as a transverse spot, anterior margin straight; anatergite (posterior apex black); anterior $^4/_7$ katatergite (remainder black); two moderately broad parallel-sided lateral postsutural vittae ending at *ia.* seta. Medial longitudinal postsutural yellow vitta absent. Postnotum black. Scutellum yellow except for narrow black basal band. Setae: *sc.* 2; *prsc.* 2; *ia.* 1; *p.sa.* 1; *a.sa.* 1; *mpl.* 1; *npl.* 2; *scp.* 4.

Legs: All femora fulvous except for a large elongate black spot on outer surfaces of fore femora; fore tibiae fuscous tending dark fuscous basally, mid tibiae fulvous tending dark fuscous basally, hind tibiae fuscous to dark fuscous; fore tarsi with basal two segments fulvous and apical three segments red-brown, mid and hind tarsi entirely fulvous; mid tibiae each with an apical black spur.

Wings: Length 5.9 mm; cells bc and c colourless; microtrichia in outer corner of cell c only; remainder of wings colourless except dark fuscous cell sc, narrow fuscous costal band slightly overlapping R_{2+3} where it is paler and remaining narrow around apex of wing to end just beyond apex of R_{4+5}, a narrow fuscous anal streak; a dense aggregation of microtrichia around A_1+CuA_2; supernumerary lobe of medium development.

Abdomen: Oval; terga free; pecten present on tergum III. Tergum I and sterna I and II wider than long. Tergum I dark fuscous to black except for a narrow transverse fulvous band across posterior margin but not reaching lateral margin; tergum II fulvous except for a narrow transverse subanterior black

band reaching dark fuscous to black lateral margins; terga III–V orange-brown except for a 'T' pattern consisting of a narrow transverse fuscous to darker fuscous band across anterior margin and widening over lateral margins and a narrow medial longitudinal black band over all three terga, narrow fuscous anterolateral margins on terga IV and V. A pair of oval fuscous shining spots on tergum V. Posterior lobe of surstylus short, sternum V with a deep concavity on posterior margin. Abdominal sterna II–V pale, sternum I dark fuscous to black.
Attractant: Cue lure.
Female: No known record.
Distribution:
Type locality: Sri Lanka (Anuradhapura District, Mahaillupallama).
Previously recorded: Type locality and many other district locations in Sri Lanka (Tsuruta & White, 2001).
Hosts: No known record.
Remarks: *Bactrocera (Bactrocera) profunda* Tsuruta & White, within the *dorsalis* complex, is similar to *Bactrocera (Bactrocera) muiri* (Hardy & Adachi) and some *Bactrocera (Bactrocera) papayae* Drew & Hancock in possessing broad parallel-sided lateral postsutural vittae, cells bc and c colourless, costal band slightly overlapping R_{2+3}, shining spots pale and abdominal terga III–V with narrow lateral dark margins and a narrow- to medium-width medial longitudinal dark band. It differs from *B. muiri* in having the lateral postsutural vittae reaching the *ia.* setae and from *B. papayae* in having a black preapical spot on outer surfaces of fore femora.

Bactrocera (Bactrocera) propinqua (Hardy & Adachi)

(Fig. 159)

Dacus (Strumeta) propinquus Hardy & Adachi, 1954: 182; Hardy, 1969a: 423; 1973: 50; 1974: 40. Holotype female in USNM.
Dacus (Bactrocera) propinquus – Hardy, 1977: 52.
Bactrocera (Bactrocera) propinqua – Drew & Hancock, 1994a: 54; Norrbom *et al*., 1998: 94.

Non-type material examined: Since the publication of Drew & Hancock (1994a), the following specimens have been studied – **VIETNAM: Northern Vietnam** – 3 males, Son La Province, Moc Chau District, Long Luong Village, Site VN01, 15.iv.2002, coll. NIPP Staff, attracted to cue lure; 1 male (29.x.1999), 1 male (5.xi.1999), Hanoi Province, Tu Liem District, Con Hue, Site N001, attracted to cue lure; 1 male (12.xi.1999), 1 male (10.v.2001), 1 male (26.vii.2001), 1 male (13.ix.2001), 1 male (26.x.2001), 1 male (22.xi.2001), Hanoi Province, Tu Liem District, Lien Mac, Site N003, attracted to cue lure; 1 male, Hanoi City Province, Tu Liem District, Lien Mac Village, Site VN03, 12.ix.2002, attracted to cue lure; 1 male (12.xi.1999), 1 male (19.xi.1999), 1 male (26.xi.1999), Hanoi Province, Dong Anh District, Bae Ha, Site N005, attracted to cue lure; 1 male, Son La Province, Moc Chau District, Tra Long, Site N007, 20.ix.1999, attracted to cue lure; 7 males (30.vi.2001), 1 male (1.x.2001), Hoa Binh Province, Mai Chau District, Pa Co, Site N010, attracted to cue lure; 1 male, Lao Cai Province, Sapa town, Site N016, 30.vii.1999, attracted to cue lure; 2 males (13.viii.1999), 1 male (10.ix.1999), Lao Cai Province, Sapa 6 Village, Site N017, attracted to cue lure; 1 male, Lao Cai Province, Sapa Oquy Ho, 30.vii.1999, Site N019, attracted to cue lure; **Southern Vietnam** – 4 males (22.vi.2002), 1 male (15.vii.2002), 1 male (3.viii.2002), 1 male (14.ix.2002), Dong Nai Province, Long Khanh District, Nhan Nghia Village, Site VS06, attracted to cue lure; 1 male, Dong Nai Province, Long Khanh District, Tan Phong Village, Site S014, attracted to cue lure; 1 male (29.x.1999), 2 males (24.xii.1999), Binh Duong Province, Lai Thieu Village, Site S020, attracted to cue lure. Specimens in QDPC.
Diagnosis: Face fulvous with a pair of large oval black spots; postpronotal lobes and notopleura yellow; scutum black except dark brown below and behind lateral postsutural vittae, around notopleural suture and inside postpronotal lobes; broad parallel-sided lateral postsutural yellow vittae ending just behind *ia.* seta; medial postsutural yellow vitta absent; mesopleural stripe reaching just anterior of notopleuron dorsally; scutellum yellow; legs with femora fulvous except for an oval dark fuscous spot on outer apical surfaces of fore femora; fore tibiae fuscous, mid tibiae fulvous apically and fuscous basally, hind tibiae dark fuscous; wing with cells bc and c colourless, microtrichia in outer corner of cell c only; a narrow fuscous costal band overlapping R_{2+3} where it becomes paler and expanding slightly across apex of R_{4+5}; a narrow fuscous anal streak; supernumerary lobe weak in females and of medium development in males; abdominal terga III–V orange-brown except for a narrow transverse black band across anterior margin of tergum III which widens to cover narrow lateral margins, a moderately broad medial longitudinal black band over all three terga, narrow dark fuscous to black anterolateral corners on terga IV and V, a pair of oval orange-brown shining spots on tergum V; abdominal sterna dark coloured.
Attractant: Cue lure.
Female: As for male in the general body colour patterns. See Drew & Hancock (1994a) for a detailed description of female characters.
Distribution:
Type locality: East Malaysia (Sabah, Tagap).
Previously recorded: Type locality and the following locations – Peninsular Malaysia; Singapore; Southern Thailand; Cambodia (Drew & Hancock, 1994a).
New records: Vietnam.
Hosts: *Garcinia* species (family *Clusiaceae*). See Drew & Hancock (1994a) for host records.
Remarks: *Bactrocera (Bactrocera) propinqua* (Hardy & Adachi), within the *dorsalis* complex, is similar to *Bactrocera (Bactrocera) bitungiae* n.sp., *Bactrocera (Bactrocera) carambolae* Drew & Hancock, *Bactrocera (Bactrocera) fuscitibia* Drew & Hancock, *Bactrocera (Bactrocera) laithieuiae* n.sp. and *Bactrocera (Bactrocera) occipitalis* (Bezzi) in possessing broad parallel-sided lateral postsutural vittae, costal band slightly overlapping R_{2+3} and not expanding markedly apically, abdominal terga III–V with narrow- to medium-width lateral dark markings, pale coloured shining spots on abdominal tergum V, femora entirely fulvous or with dark spots on outer apical surfaces of fore femora only, costal band overlapping R_{2+3} but not extending to or

nearly to R_{4+5} and abdominal terga III–V with a moderately broad to broad medial longitudinal black band over all three terga. It differs from *B. bitungiae* n.sp., *B. fuscitibia*, *B. laithieuiae* n.sp. and *B. occipitalis* in having the costal band only slightly overlapping R_{2+3} where it becomes paler and only widening slightly at apex and from *B. carambolae* in possessing a very broad medial longitudinal black band over all three terga and apex of female piercer trilobed (not needle-shaped).

Bactrocera (Bactrocera) pruniae Drew & Romig, new species

(Fig. 160)

Material examined: Holotype male, **VIETNAM: Northern Vietnam** – North Vietnam, coll. NIPP, bred *Prunus persica* Rosaceae, peach-untreated, deposited in BMNH.

Paratypes: **VIETNAM: Northern Vietnam** – 6 males and 7 females, same data as holotype; 3 males and 2 females, North Vietnam, coll. NIPP, Bred ex FUN4, Prunus persica Rosaceae, peach-treated; 2 males and 6 females, North Vietnam, Sapa, Lao Cai, 1.viii.2000, coll. NIPP, Bred ex FNO539, *Prunus persica* Rosaceae (peach). 4 paratypes in ANIC, 4 paratypes in BMNH, 12 paratypes in QDPC, 6 paratypes in QMIC.

Diagnosis: A medium-sized species; face fulvous with a pair of medium-sized circular black spots; postpronotal lobes and notopleura yellow; scutum black; lateral postsutural yellow vittae present; medial postsutural yellow vitta absent; no yellow spot anterior to notopleural suture; mesopleural stripe equal in width to notopleuron dorsally; scutellum yellow; wing with cells bc and c colourless, microtrichia in outer corner of cell c only; a narrow fuscous costal band; a narrow fuscous anal streak; all abdominal terga entirely orange-brown except for narrow anterolateral fuscous margins on terga I, IV and V and narrow transverse dark fuscous markings on anterolateral areas of tergum III.

Description: male.

Head: Height 1.68 mm. Frons – length 1.48 times breadth; red-brown with fuscous around orbital setae; orbital setae black: 1 *s.or.*, 2 *i.or.*; lunule fuscous. Ocellar triangle black. Vertex fuscous. Face fulvous with a pair of medium-sized circular black spots; length 0.42 mm. Genae fulvous, large dark fuscous subocular spot present; dark seta present. Occiput red-brown with fuscous around lateral and dorsal margins and fulvous along eye margins; occipital row with 4–7 strong black setae. Antennae with segments 1 and 2 red-brown, segment 3 red-brown with fuscous on apex and outer surface; length of segments: 0.24 mm; 0.33 mm; 0.78 mm.

Thorax: Scutum black with red-brown along entire lateral and posterior margins. Pleural areas dark fuscous to black with red-brown above mid coxae and below postpronotal lobes. Yellow markings as follows: postpronotal lobes; notopleura; mesopleural stripe equal in width to notopleuron dorsally, continuing to katepisternum as a transverse spot, anterior margin straight; anatergite (posterior apex dark fuscous); anterior $^{3}/_{5}$ katatergite (remainder dark fuscous); two narrow lateral postsutural vittae narrowing slightly posteriorly to end just before *ia.* seta. Medial longitudinal postsutural yellow vitta absent. Postnotum black, red-brown centrally. Scutellum yellow except for narrow red-brown basal band. Setae: *sc.* 2; *prsc.* 2; *ia.* 1; *p.sa.* 1; *a.sa.* 1; *mpl.* 1; *npl.* 2; *scp.* 4.

Legs: Fore and mid femora red-brown, hind femora fulvous with apical $^{1}/_{3}$ red-brown; fore tibiae pale fuscous, mid tibiae fuscous basally to fulvous apically, hind tibiae fuscous; all tarsi with basal segment fulvous and apical four segments red-brown; mid tibiae each with an apical black spur.

Wings: Length 5.9 mm; cells bc and c colourless; microtrichia in outer corner of cell c only; remainder of wings colourless except fuscous cell sc, narrow fuscous costal band confluent with R_{2+3} and remaining uniformly narrow to apex of wing, a narrow fuscous anal streak; a dense aggregation of microtrichia around A_1+CuA_2; supernumerary lobe of medium development.

Abdomen: Oval; terga free; pecten present on tergum III. Tergum I and sterna I and II wider than long. All abdominal terga entirely orange-brown except for narrow fuscous lateral margins on tergum I, fulvous along posterior area of tergum II, narrow transverse dark fuscous anterolateral corners on tergum III and narrow fuscous anterolateral corners on terga IV and V. A pair of oval fuscous to dark fuscous shining spots on tergum V. Posterior lobe of surstylus short, sternum V with a deep concavity on posterior margin. Abdominal sterna dark fuscous.

Attractant: No known record.

Female: As for male except no dense aggregation of microtrichia around A_1+CuA_2, supernumerary lobe weak, pecten of cilia on abdominal tergum III absent; ovipositor basal segment red-brown, dorsoventrally compressed and tapering posteriorly in dorsal view; ratio of length of oviscape to length of segment V, 0.7:1; apex of piercer needle-shaped.

Distribution: Northern Vietnam.

Hosts: *Prunus persica* (peach), family *Rosaceae*.

Remarks: *Bactrocera (Bactrocera) pruniae* n.sp. is similar to species in the *Bactrocera (Bactrocera) musae* complex of Australia and Papua New Guinea. It is similar to *Bactrocera (Bactrocera) musae* (Tryon) in possessing a black scutum, lateral postsutural vittae tapering slightly posteriorly to end just before *ia.* seta, femora pale without dark markings, wings colourless except for costal band and anal streak, abdominal terga mostly orange-brown. It differs from *B. musae* in having the costal band confluent with R_{2+3} (not overlapping) and all abdominal terga uniformly orange-brown with fuscous to dark fuscous shining spots on abdominal tergum V. It is also similar to *Bactrocera (Bactrocera) latifrons* (Hendel) in the characters listed above for *B. musae* but *B. latifrons* possesses a trilobed apex on the piercer of the female ovipositor and a slight expansion of the costal band around the apex of R_{4+5}.

Bactrocera (Bactrocera) pseudobeckerae Drew & Romig, new species

(Fig. 161)

Material examined: Holotype male, **INDONESIA: SULAWESI** – Sulawesi Tengah, Donggala, Nupabomba,

15.x.2006, coll. Rusdin, attracted to cue lure in forest, deposited in BMNH.
Paratypes: **INDONESIA: SULAWESI** – 3 males, Sulawesi Tengah, Donggala, Nupabomba, 22.x.2006, coll. Rusdin, attracted to cue lure in forest. 1 paratype in ANIC, 1 paratype in QDPC, 1 paratype in QMIC.
Diagnosis: A medium-sized species; face fulvous with a pair of medium-sized circular black spots; postpronotal lobes yellow (anteromedial corners dark fuscous); notopleura yellow; scutum uniformly black except dark fuscous below lateral postsutural vittae; broad and short lateral postsutural yellow vittae present; medial postsutural yellow vitta absent; no yellow spot anterior to notopleural suture; mesopleural stripe reaching midway between anterior margin of notopleuron and anterior *npl.* seta dorsally; scutellum yellow with a broad black basal band; wing with cells bc and c colourless, microtrichia in outer corner of cell c only; a broad costal band almost confluent with R_{4+5}; a narrow fuscous anal streak; abdominal terga III–V dark orange-brown with narrow transverse dark fuscous markings anterolaterally on tergum III and connected in midline by a fuscous pattern and black shining spots on tergum V.
Description: male.
Head: Height 1.98 mm. Frons – length 1.43 times breadth; red-brown with fuscous around orbital setae and on anteromedial hump; orbital setae black: 1 *s.or.*, 2 *i.or.*; lunule fuscous. Ocellar triangle black. Vertex fuscous. Face fulvous with a pair of medium-sized circular black spots; length 0.48 mm. Genae red-brown, large dark fuscous subocular spot present; strong black seta present. Occiput fuscous, fulvous along eye margins; occipital row with a large number of strong black setae. Antennae with segments 1 and 2 red-brown, segment 3 red-brown with fuscous on apex and outer surface; length of segments: 0.14 mm; 0.24 mm; 0.65 mm.
Thorax: Scutum uniformly black except dark fuscous below lateral postsutural vittae. Pleural areas entirely black except dark red-brown below postpronotal lobes. Yellow markings as follows: postpronotal lobes (anteromedial corners dark fuscous); notopleura; mesopleural stripe reaching midway between anterior margin of notopleuron and anterior *npl.* seta dorsally, continuing to katepisternum as a small transverse spot, anterior margin straight; anatergite (posterior apex black); anterior $^3/_5$ katatergite (remainder black); two broad and short lateral postsutural vittae tapering sharply posteriorly to end well before *ia.* seta. Medial longitudinal postsutural yellow vitta absent. Postnotum black. Scutellum yellow except for a broad black basal band. Setae: *sc.* 2; *prsc.* 2; *ia.* 1; *p.sa.* 1; *a.sa.* 1; *mpl.* 1; *npl.* 2; *scp.* 4.
Legs: All femora with large black elongate spots on ventral surfaces, fore femora overall fuscous with fulvous on basal ¼, mid femora fulvous on basal ⅔ and dark fuscous on apical ⅓, hind femora fulvous on basal ⅔ to ¾ and dark fuscous on the remainder; fore tibiae fuscous, mid tibiae dark fuscous basally to fuscous apically, hind tibiae dark fuscous; all tarsal segments entirely fulvous; mid tibiae each with an apical black spur.
Wings: Length 6.2 mm; cells bc and c colourless; microtrichia in outer corner of cell c only; remainder of wings colourless except dark fuscous cell sc, a broad costal band almost confluent with R_{4+5} (but not absolutely confluent with this vein) and fuscous between costal margin and R_{2+3}, pale fuscous overlapping R_{2+3} and dark fuscous around apex of wing, a narrow fuscous anal streak; a dense aggregation of microtrichia around A_1+CuA_2; supernumerary lobe of medium development.
Abdomen: Oval; terga free; pecten present on tergum III. Tergum I and sterna I and II wider than long. Tergum I fuscous centrally with broad dark fuscous lateral margins; tergum II dark orange-brown with narrow transverse fuscous lines centrally; terga III–V almost entirely orange-brown with transverse dark fuscous markings anterolaterally on tergum III and joined in the midline by a fuscous pattern. A pair of oval black shining spots on tergum V. Posterior lobe of surstylus short, sternum V with a deep concavity on posterior margin. Abdominal sterna dark fuscous to black.
Attractant: Cue lure.
Female: No known record.
Distribution: Indonesia (Sulawesi – Central Sulawesi, Donggala).
Hosts: No known record.
Remarks: *Bactrocera (Bactrocera) pseudobeckerae* n.sp. is similar to *Bactrocera (Bactrocera) aemula* Drew, *Bactrocera (Bactrocera) contigua* Drew, *Bactrocera (Bactrocera) musae* (Tryon) and *Bactrocera (Bactrocera) penebeckerae* n.sp. in possessing an entire costal band overlapping R_{2+3}, abdominal terga III–V without a distinct dark 'T' pattern and elongate lateral postsutural vittae reaching to or almost to the *ia.* seta. It differs from *B. aemula* in having abdominal terga III–V almost entirely orange-brown, from *B. musae* and *B. penebeckerae* in having the costal band almost confluent with R_{4+5} and from *B. contigua* in having fuscous to dark fuscous markings on apices of all femora and a broad basal black band on the scutellum.

Bactrocera (Bactrocera) pseudodistincta (Drew)

(Fig. 162)

Dacus (Strumeta) pseudodistinctus Drew, 1971: 79. Holotype male in QMIC.
Bactrocera (Bactrocera) pseudodistincta – Drew, 1989: 61; Norrbom *et al.*, 1998: 94.

Non-type material examined: INDONESIA: MOLUCCAS (MALUKU) – 2 males, Middle Maluku, Suli, 9.vi.2007, coll. La Naibu, attracted to cue lure; 1 male, Seram (Western Part), Waimital, 12.vii.2007, coll. A. Haris Tualepe, attracted to cue lure in Jeruk; 1 male, Seram (Western Part), Kairatu, Waimital, 19.vii.2007, coll. A. Haris Tualepe. Specimens in QDPC.
Diagnosis: A medium-sized species; face fulvous with a pair of medium-sized oval black spots; postpronotal lobes and notopleura yellow; scutum entirely dull black without pale markings; lateral postsutural yellow vittae present; medial postsutural yellow vitta absent; no yellow spot anterior to notopleural suture; mesopleural stripe reaching

anterior *npl.* seta dorsally; scutellum yellow; wing with cells bc and c dark fuscous, microtrichia in outer ⅓ of cell c only; a broad dark fuscous costal band confluent with R_{4+5}, a moderately broad parallel-sided dark fuscous band across wing and enclosing both crossveins; a broad dark fuscous anal streak; abdominal terga II–V orange-brown with a 'T' pattern consisting of a narrow transverse black band across anterior margin of tergum II but not reaching lateral margins and a medium-width to broad medial longitudinal black band over all four terga, moderately broad lateral longitudinal pale fuscous bands over terga III–V, a pair of oval pale fuscous spots on tergum V.

Description: male.

Head: Height 1.8 mm. Frons – length 1.6 times breadth; red-brown with pale fuscous around orbital setae; orbital setae black: 1 *s.or.*, 2 *i.or.*; lunule red-brown. Ocellar triangle black. Vertex red-brown. Face fulvous with a pair of medium-sized oval black spots; length 0.48 mm. Genae red-brown, fuscous subocular spot present; black seta present. Occiput red-brown, fulvous along eye margins; occipital row with a large number of strong black setae. Antennae with segments 1 and 2 red-brown, segment 3 red-brown with fuscous on apex and outer surface; length of segments: 0.17 mm; 0.22 mm; 0.86 mm.

Thorax: Scutum dull black without pale markings. Pleural areas black except red-brown to fuscous below postpronotal lobes. Yellow markings as follows: postpronotal lobes; notopleura; mesopleural stripe reaching anterior *npl.* seta dorsally, continuing to katepisternum as a transverse spot, anterior margin straight; anatergite (posterior apex black); anterior $^{3}/_{5}$ katatergite (remainder black); two broad parallel-sided lateral postsutural vittae ending at *ia.* seta. Medial longitudinal postsutural yellow vitta absent. Postnotum black. Scutellum yellow except for narrow black basal band. Setae: *sc.* 2; *prsc.* 2; *ia.* 1; *p.sa.* 1; *a.sa.* 1; *mpl.* 1; *npl.* 2; *scp.* 4.

Legs: All segments entirely fulvous except hind tibiae pale fuscous; mid tibiae each with an apical black spur.

Wings: Length 6.1 mm; cells bc and c dark fuscous; microtrichia in outer ⅓ of cell c only; remainder of wings colourless except dark fuscous cell sc, broad dark fuscous costal band confluent with R_{4+5} and slightly overlapping this vein at its extremity, a moderately broad parallel-sided dark fuscous transverse band across wing and enclosing both crossveins, a broad dark fuscous anal streak; supernumerary lobe of medium development.

Abdomen: Oval; terga free; pecten present on tergum III. Tergum I and sterna I and II wider than long. Tergum I black except for a narrow transverse orange-brown band across posterior margin but not reaching lateral margins. Terga II–V orange-brown except for a 'T'-shaped pattern consisting of a narrow transverse black band across anterior margin of tergum II that does not reach lateral margins and a medium-width to broad medial longitudinal black band over all four terga but usually not reaching posterior margin of tergum V, moderately broad lateral longitudinal pale fuscous bands over terga III–V. A pair of oval pale fuscous shining spots on tergum V. Posterior lobe of surstylus short, sternum V with a deep concavity on posterior margin. Abdominal sterna dark red-brown to fuscous.

Attractant: Cue lure.

Female: No known record.

Distribution:

Type locality: New Britain (Keravat).

Previously recorded: Type locality plus New Ireland; Lihir Island; Papua New Guinea (Central District, Morobe Province); Torres Strait Islands (Boigu; Darnley; Saibai; Yam).

New records: Indonesia (Moluccas = Maluku – Ambon I.; Seram I.).

Hosts: No known record.

Remarks: *Bactrocera (Bactrocera) pseudodistincta* (Drew) is similar to *Bactrocera (Bactrocera) fergussoniensis* Drew and *Bactrocera (Bactrocera) propedistincta* Drew in the *distincta* complex of the South Pacific region in possessing coloured cells bc and c, costal band confluent with R_{4+5}, lateral postsutural vittae ending at *ia.* setae, mesopleural stripe not reaching postpronotal lobe and a transverse fuscous band across wing enclosing both *r-m* and *dm-cu* crossveins. It is unique in having the combination of a medial longitudinal dark band over abdominal terga II–V and lateral longitudinal dark bands over terga III–V, cell bc paler fuscous and dense microtrichia covering outer ½ of cell c only.

Bactrocera (Bactrocera) pseudoversicolor Drew

(Fig. 163)

Bactrocera (Bactrocera) pseudoversicolor Drew, in Drew & Raghu, 2002: 343. Holotype male in BMNH.

Non-type material examined: Since the publication of Drew & Raghu (2002), the following specimens have been studied – **INDIA**: 1 male, S. India, Nandi Hills, nr Bangalore, 1450m, 26.v.1992, coll. D.L. Hancock & I.M. White, attracted to methyl eugenol; 1 male, India, Karnataka, Chethalli, 970m, 1.iv.1986, coll. Ramani, attracted to methyl eugenol; 1 male, India, Karnataka, Bangalore, 916m, 19. vi.1987, coll. G. Bhat, attracted to methyl eugenol. Specimens in QDPC.

Diagnosis: Face fulvous with a pair of medium-sized oval black spots; postpronotal lobes and notopleura yellow; scutum red-brown without dark markings; narrow- to medium-width parallel-sided lateral postsutural yellow vittae ending just behind *ia.* seta; medial postsutural yellow vitta absent; mesopleural stripe equal in width to notopleuron dorsally; scutellum yellow except for narrow dark red-brown basal band; legs with all femora fulvous except for an oval fuscous subapical spot on outer surfaces of mid femora; fore tibiae pale fuscous, mid tibiae fulvous except for a small fuscous area basally, hind tibiae fuscous; wing with cells bc and c colourless, microtrichia in outer corner of cell c only; a narrow fuscous costal band confluent with R_{2+3} and remaining narrow around apex of wing to end between extremities of R_{4+5} and M; a narrow fuscous anal streak; supernumerary lobe of medium development; abdominal terga III–V red-brown with a 'T' pattern consisting of a narrow transverse black band across anterior margin of tergum III and a narrow medial longitudinal fuscous band over all three terga, narrow

fuscous to dark fuscous margins on terga IV and V, a pair of oval fuscous to dark fuscous shining spots on tergum V.
Attractant: Methyl eugenol.
Female: No known record.
Distribution:
Type locality: Southern India (Kerala, New Amarambalam Forest).
New records: Southern India (Karnataka – Bangalore; Chethalli; Nandi Hills).
Hosts: No known record.
Remarks: *Bactrocera (Bactrocera) pseudoversicolor* Drew is similar to *Bactrocera (Bactrocera) versicolor* (Bezzi) and *Bactrocera (Bactrocera) apicofuscans* White & Tsuruta in the general red-brown colouration of the thorax and abdomen but differs from *B. apicofuscans* in having a narrower costal band confluent with R_{2+3}, colourless cells bc and c, *prsc.* setae present and dark lateral margins on abdominal terga III–V and from *B. versicolor* in possessing a narrow mesopleural stripe equal in width to notopleuron dorsally, lacking a dark apical spot on the scutellum and dark preapical spots on the fore and hind femora.

Bactrocera (Bactrocera) pusilla (Hardy)

(Fig. 164)

Dacus pusillus Hardy, 1983: 18. Holotype male in MZB.
Bactrocera (Bactrocera) pusilla – Norrbom *et al.*, 1998: 94.

Type material examined: Holotype male, **INDONESIA: JAVA** – West Java, 120 km S.E. Bogor, Lengkong Forest 600 m, 24 Apr. 1975, coll. A.T. Hardy, collected in dense jungle at cue lure. Studied at MZB.
Diagnosis: A small species; face fulvous with a pair of large oval black spots; postpronotal lobes and notopleura yellow; scutum black with narrow longitudinal dark fulvous bands either side of midline; lateral postsutural yellow vittae present; medial postsutural yellow vitta absent; a broad yellow band connecting postpronotal lobes and notopleural suture; mesopleural stripe equal in width to notopleuron dorsally; scutellum yellow; wing with cells bc and c dark fuscous, dense microtrichia covering all of both cells; a broad dark fuscous costal band broadening around apex of wing to end at apex of M; a broad dark fuscous anal streak; abdominal terga III–V fulvous with a black 'T' pattern consisting of a broad transverse black band across anterior ⅓ of tergum III and a moderately broad medial longitudinal black band over all three terga, moderately broad lateral black bands on terga IV and V.
Description: male.
Head: Frons – mostly fulvous without dark markings; orbital setae black: 1 *s.or.*, 2 *i.or.*; lunule red-brown. Ocellar triangle black. Vertex dark fuscous to black. Face fulvous with a pair of large oval black spots. Occiput fuscous, fulvous along eye margins. Antennae with segments 1 and 2 red-brown, segment 3 red-brown with fuscous on apex and outer surface.
Thorax: Scutum black with narrow dark fulvous longitudinal bands either side of midline. Yellow markings as follows: postpronotal lobes; notopleura; broad lateral yellow band connecting postpronotal lobes and notopleural suture; mesopleural stripe equal in width to notopleuron dorsally; two moderately broad triangular-shaped lateral postsutural vittae narrowing posteriorly to end well before *ia.* seta. Medial longitudinal postsutural yellow vitta absent. Scutellum yellow except for narrow black basal band. Setae: *sc.* 2; *prsc.* 2; *ia.* 1; *p.sa.* 1; *a.sa.* 1; *mpl.* 1; *npl.* 2; *scp.* 4.
Legs: Femora fulvous except for a small fuscous area around apex of mid femora and dark fuscous around apex of hind femora; fore tibiae pale fuscous, mid tibiae fuscous, hind tibiae dark fuscous; all tarsi with basal segment fulvous and apical four segments pale fuscous; mid tibiae each with an apical black spur.
Wings: Cells bc and c dark fuscous; dense microtrichia covering all of both cells; remainder of wings colourless except broad dark fuscous costal band confluent with R_{4+5} and widening slightly around apex of wing to end at apex of M (there is a narrow paler fuscous area above R_{4+5}), a broad dark fuscous anal streak; a dense aggregation of microtrichia around A_1+CuA_2; supernumerary lobe of medium development.
Abdomen: Oval; terga free; pecten present on tergum III. Tergum I and sterna I and II wider than long. Tergum I fulvous with narrow anterolateral black margins; tergum II fulvous with a narrow subanterior transverse black band not reaching lateral margins; terga III–V fulvous with a black 'T' pattern consisting of a broad transverse band across anterior ⅓ of tergum III and a moderately broad medial longitudinal band over all three terga, moderately broad lateral black margins on terga IV and V. A pair of oval fulvous shining spots on tergum V. Posterior lobe of surstylus short, sternum V with a deep concavity on posterior margin.
Attractant: Cue lure.
Female: No known record.
Distribution:
Type locality: Indonesia (West Java, Lengkong Forest).
Hosts: No known record.
Remarks: *Bactrocera (Bactrocera) pusilla* (Hardy) is unique without any genuine close species in the South-East Asian region. It is easily identified by the dark fuscous costal band expanded across the wing apex, dark fuscous cells bc and c, broad tapering lateral postsutural vittae ending well before the *ia.* setae and lateral yellow band connecting the postpronotal lobe and notopleuron.

Bactrocera (Bactrocera) pyrifoliae Drew & Hancock

(Fig. 165)

Bactrocera (Bactrocera) pyrifoliae Drew & Hancock, 1994a: 55; Norrbom *et al.*, 1998: 94. Holotype female in BMNH.

Non-type material examined: Since the publication of Drew & Hancock (1994a), the following specimens have been studied – **VIETNAM: Northern Vietnam** – 1 male, Lao Cai Province, Sapa Oquy Ho, Site N019, 27.viii.1999, attracted to cue lure; 1 male, Lao Cai Province, Sapa

6 Village, Site N017, 12.v.2000, attracted to cue lure; 1 female, Dao TN Kong Phun Mao, vi.2004, coll. Dept of Agric, bred ex peach. Specimens in QDPC.

Diagnosis: Face fulvous with a pair of medium-sized circular black spots; postpronotal lobes and notopleura yellow; scutum black except dark brown posterolateral to lateral postsutural vittae and dark red-brown anterior to notopleural suture and inside postpronotal lobes; narrow lateral postsutural yellow vittae tapering posteriorly to end before *ia.* seta; medial postsutural yellow vitta absent; mesopleural stripe equal in width to notopleuron dorsally; scutellum yellow; legs with femora fulvous except for a small subapical oval black spot on outer surfaces of fore femora and dark fuscous around apices of mid and hind femora; fore and mid tibiae dark fuscous, hind tibiae black; wings with cells bc and c colourless, microtrichia in outer corner of cell c only; a narrow fuscous costal band confluent with R_{2+3} and with a slight swelling around apex of R_{4+5}; a narrow fuscous anal streak; supernumerary lobe weak in females, of medium development in males; abdominal terga III and IV orange-brown and generally each with its own 'T' pattern consisting of a narrow- to medium-width transverse dark fuscous to black band across anterior margin which expands to cover outer $\frac{1}{3}$ of lateral margins, a narrow- to medium-width medial longitudinal dark fuscous to black band, tergum V with a narrow- to medium-width medial longitudinal dark fuscous to black band and dark fuscous to black anterolateral corners which may also meet along anterior margin, a pair of oval dark fuscous shining spots on tergum V; abdominal sterna dark coloured.

Attractant: May be a weak attraction to cue lure.

Female: As for male in the general body colour patterns. See Drew & Hancock (1994a) for a detailed description of female characters.

Distribution:

Type locality: Thailand (Chiang Rai, Wawi).

Previously recorded: Type locality and Thailand (Chiang Mai).

New records: Northern Vietnam.

Hosts: Recorded from five plant families – *Araliaceae*; *Euphorbiaceae*; *Myrtaceae*; *Polygalaceae*; *Rosaceae* (see Allwood *et al.*, 1999, for host records).

Remarks: *Bactrocera (Bactrocera) pyrifoliae* Drew & Hancock, within the *dorsalis* complex, is similar to *Bactrocera (Bactrocera) infulata* Drew & Hancock and *Bactrocera (Bactrocera) kinabalu* Drew & Hancock in possessing lateral postsutural vittae that narrow posteriorly and all femora with apical dark markings. It differs from *B. kinabalu* in possessing smaller dark markings on apices of femora and distinct 'T' patterns on abdominal terga III and IV and from *B. infulata* in having a narrow dark basal band on the scutellum and the unique 'T' patterns on abdominal terga III and IV.

Bactrocera (Bactrocera) quasiinfulata Drew & Romig, new species

(Fig. 166)

Material examined: Holotype male, **VIETNAM: Northern Vietnam** – Lao Cai Province, Sapa 6 Village, 10.ix.1999, Site N017, attracted to cue lure, deposited in BMNH.

Paratypes: **VIETNAM: Northern Vietnam** – 1 male, Northern Vietnam, Hoa Binh Province, Mai Chau District, Pa Co, 15.vi.2001, Site N010; 1 male, Northern Vietnam, Son La Province, Moc Chau District, Long Luong Village, 15.v.2002, Site VN01, coll. NIPP Staff, attracted to cue lure. Paratypes in QDPC.

Diagnosis: A large species; face fulvous with a pair of large oval black spots; postpronotal lobes and notopleura yellow; scutum black; lateral postsutural yellow vittae present; medial postsutural yellow vitta absent; no yellow spot anterior to notopleural suture; mesopleural stripe reaching midway between anterior margin of notopleuron and anterior *npl.* seta dorsally; scutellum yellow with a broad black basal band; wing with cells bc and c with a pale fuscous tint, microtrichia in outer corner of cell c only; a broad dark fuscous costal band almost confluent with R_{4+5} and remaining of uniform width around apex of wing; a broad dark fuscous anal streak; abdominal terga III–V orange-brown with a narrow- to medium-width medial and two broad lateral longitudinal dark fuscous to black bands over all three terga and joined along anterior margin of tergum III.

Description: male.

Head: Height 2.04 mm. Frons – length 1.47 times breadth; fulvous with fuscous on anteromedial hump and dark fuscous around bases of orbital setae; orbital setae black: 1 *s.or.*, 2 *i.or.*; lunule dark fuscous. Ocellar triangle black. Vertex dark fuscous. Face fulvous with a pair of large oval black spots; length 0.48 mm. Genae fulvous, large dark fuscous subocular spot present; black seta present. Occiput dark fuscous, fulvous along eye margins; occipital row with 7–10 black setae. Antennae with segments 1 and 2 red-brown, segment 3 red-brown with dark fuscous on apex and outer surface; length of segments: 0.26 mm; 0.34 mm; 0.91 mm.

Thorax: Scutum entirely black without pale markings. Pleural areas entirely black. Yellow markings as follows: postpronotal lobes; notopleura; mesopleural stripe reaching midway between anterior margin of notopleuron and anterior *npl.* seta dorsally, continuing to katepisternum as a small transverse spot, anterior margin straight; anatergite (posterior apex black); anterior $\frac{3}{5}$ katatergite (remainder black); two narrow- to medium-width lateral postsutural vittae narrowing sharply posteriorly to end before *ia.* seta. Medial longitudinal postsutural yellow vitta absent. Postnotum black. Scutellum yellow except for broad black basal band. Setae: *sc.* 2; *prsc.* 2; *ia.* 1; *p.sa.* 1; *a.sa.* 1; *mpl.* 1; *npl.* 2; *scp.* 4.

Legs: Fore femora fulvous with a large elongate subapical black spot on outer surfaces, mid femora dark fuscous to black with basal $\frac{1}{3}$ fulvous, hind femora fulvous with dark fuscous around apical $\frac{1}{4}$; fore tibiae dark fuscous, mid tibiae fuscous basally to fulvous apically, hind tibiae dark fuscous; fore and hind tarsi with basal segment fulvous and apical four segments fuscous, mid tarsi with all segments fulvous; mid tibiae each with an apical black spur.

Wings: Length 6.9 mm; cells bc and c with a pale fuscous tint; microtrichia in outer corner of cell c only; remainder of wings colourless except dark fuscous cell sc, broad dark fuscous costal band almost confluent with R_{4+5} and of uniform width around apex of wing, a broad dark fuscous anal streak;

a dense aggregation of microtrichia around A_1+CuA_2; supernumerary lobe of medium development.
Abdomen: Oval; terga free; pecten present on tergum III. Tergum I and sterna I and II wider than long. Tergum I black except for a narrow transverse orange-brown band across posterior margin but not reaching lateral margins; tergum II black with fulvous across posterior ½; terga III–V orange-brown with a narrow- to medium-width medial longitudinal black band over all three terga and two broad lateral longitudinal dark fuscous to black bands over all three terga and joined along anterior margin of tergum III. A pair of oval dark fuscous shining spots on tergum V. Posterior lobe of surstylus short, sternum V with a deep concavity on posterior margin. Abdominal sterna dark fuscous to black.
Attractant: Cue lure.
Female: No known record.
Distribution: Northern Vietnam (Hoa Binh, Lao Cai, Son La Provinces).
Hosts: No known record.
Remarks: *Bactrocera (Bactrocera) quasiinfulata* n.sp. is similar to *Bactrocera (Bactrocera) infulata* Drew & Hancock in the *Bactrocera (Bactrocera) dorsalis* complex. Its prominent characters are a black scutum without pale markings, a broad black basal band on the scutellum, narrowing lateral postsutural vittae ending before the *ia.* setae, wings with a broad fuscous costal band almost confluent with R_{4+5}, a broad anal streak and abdominal terga III–V with broad medial and lateral longitudinal dark bands joined along anterior margin of tergum III. It differs from *B. infulata* in having cells bc and c with a pale infuscation (not clear) and mid femora with apical ⅔–¾ dark fuscous to black.

Bactrocera (Bactrocera) quasineonigrita Drew & Romig, new species

(Fig. 167)

Material examined: Holotype male, **INDONESIA: NORTH MOLUCCAS (NORTH MALUKU)** – Moluccas (Maluku), North Maluku, West Halmahera Gaol, 29.v.2007, coll. Raais Abdullah, attracted to methyl eugenol in forest, deposited in BMNH.
Paratypes: **INDONESIA: NORTH MOLUCCAS (NORTH MALUKU)** – 2 males, same data as holotype; 2 males (12.vi.2007), 1 male (18.vi.2007), Moluccas, North Maluku, West Halmahera Gaol, coll. Raais Abdullah, attracted to methyl eugenol in forest. 1 paratype in ANIC, 1 paratype in BMNH, 2 paratypes in QDPC, 1 paratype in QMIC.
Diagnosis: A medium-sized species; face fulvous with a pair of very large oval black spots; postpronotal lobes and notopleura yellow; scutum entirely black; broad parallel-sided lateral postsutural yellow vittae present; medial postsutural yellow vitta absent; no yellow spot anterior to notopleural suture; mesopleural stripe reaching postpronotal lobes dorsally; scutellum yellow; wing with cells bc and c colourless, microtrichia in outer corner of cell c only; a broad fuscous costal band almost confluent with R_{4+5}, a transverse band across wing which is fuscous and narrow along *r-m* crossvein then becoming diffuse as its expands across posterior area of wing; a broad fuscous anal streak; abdominal terga III–V entirely black except for dark fuscous posterior to shining spots on tergum V, black shining spots on tergum V.
Description: male.
Head: Height 1.72 mm. Frons – length 1.58 times breadth; red-brown with fuscous on anteromedial hump; orbital setae black: 1 *s.or.*, 2 *i.or.*; lunule red-brown. Ocellar triangle black. Vertex red-brown. Face fulvous with a pair of very large oval black spots; length 0.49 mm. Genae red-brown, large dark fuscous subocular spot present; black seta present. Occiput black, fulvous along eye margins; occipital row with a large number of strong black setae. Antennae with segments 1 and 2 red-brown, segment 3 red-brown with dark fuscous on apex and outer surface; length of segments: 0.19 mm; 0.34 mm; 0.94 mm.
Thorax: Scutum entirely black without pale markings. Pleural areas black without pale markings. Yellow markings as follows: postpronotal lobes; notopleura; mesopleural stripe reaching to postpronotal lobe dorsally, continuing to katepisternum as a small transverse spot, anterior margin straight; anatergite (posterior apex black); anterior ³/₅ katatergite (remainder black); two broad parallel-sided lateral postsutural vittae ending behind *ia.* seta. Medial longitudinal postsutural yellow vitta absent. Postnotum black. Scutellum yellow except for narrow black basal band. Setae: *sc.* 2; *prsc.* 2; *ia.* 1; *p.sa.* 1; *a.sa.* 1; *mpl.* 1; *npl.* 2; *scp.* 4.
Legs: Fore femora red-brown except for a small dark fuscous subocular spot on outer surfaces, mid and hind femora fulvous except dark fuscous around apical ¼; fore tibiae red-brown, mid tibiae fulvous, hind tibiae fuscous; all tarsi with basal segment fulvous and apical four segments red-brown; mid tibiae each with an apical black spur.
Wings: Length 5.2 mm; cells bc and c colourless; microtrichia in outer corner of cell c only; remainder of wings colourless except fuscous cell sc, broad fuscous costal band almost confluent with R_{4+5}, a narrow fuscous cross band enclosing *r-m* crossvein and expanding or broadening into a diffuse fuscous pattern over posterior area of wing, a broad fuscous anal streak; a dense aggregation of microtrichia around A_1+CuA_2; supernumerary lobe of medium development.
Abdomen: Oval; terga free; pecten present on tergum III. Tergum I and sterna I and II wider than long. Tergum I entirely black; tergum II black with small red-brown spots anterocentrally and a broad dark red-brown band across posterior margin; terga III–V entirely black except for dark fuscous areas on tergum V posterior to shining spots; a pair of oval black shining spots on tergum V. Posterior lobe of surstylus short, sternum V with a deep concavity on posterior margin. Abdominal sterna dark coloured.
Attractant: Methyl eugenol.
Female: No known record.
Distribution: Indonesia (North Moluccas = North Maluku – Halmahera I.).
Hosts: No known record.

Remarks: *Bactrocera (Bactrocera) quasineonigrita*, n.sp. is similar to *Bactrocera (Bactrocera) fulvicauda* (Perkins), *Bactrocera (Bactrocera) neonigrita* Drew, and *Bactrocera (Bactrocera) unistriata* (Drew) in possessing a black scutum, entirely yellow scutellum, abdominal terga III–V with an extensive amount of dark colouration and the wing with a broad costal band and anal streak and transverse band which encloses both crossveins. It differs from *B. fulvicauda* in having colourless cells bc and c, a diffuse transverse band across the wing, the costal band not confluent with R_{4+5} and dense microtrichia confined to outer corner of cell c only, from *B. neonigrita* in having the mesopleural stripe reaching the postpronotal lobe dorsally and the lateral postsutural vittae extending posterior to the *ia.* setae and from *B. unistriata* in having the mesopleural stripe reaching the postpronotal lobe dorsally, lateral postsutural vittae extending posterior to the *ia.* seta, cells bc and c colourless, the costal band not confluent with R_{4+5} and abdominal terga III–V almost entirely black.

Bactrocera (Bactrocera) quasipropinqua Drew & Hancock

(Fig. 168)

Bactrocera (Bactrocera) quasipropinqua Drew & Hancock, 1994a: 56; Norrbom *et al.*, 1998: 94. Holotype female in BBM.

Non-type material examined: No new specimens available for study since the publication of Drew & Hancock (1994a).
Diagnosis: Face fulvous with a pair of large oval black spots; postpronotal lobes and notopleura yellow; scutum black except brown behind lateral postsutural vittae, around notopleural suture and inside postpronotal lobes; broad parallel-sided lateral postsutural yellow vittae ending at *ia.* seta; medial postsutural yellow vitta absent; mesopleural stripe reaching midway between anterior margin of notopleuron and anterior *npl.* seta dorsally; scutellum yellow; legs with femora fulvous with a medium-sized dark fuscous subapical spot on outer surfaces of fore femora and a large dark fuscous spot around apical $^4/_9$ of mid femora; tibiae dark fuscous; wing with cells bc and c colourless, microtrichia in outer corner of cell c only; a narrow fuscous costal band almost confluent with R_{4+5} (paler between R_{2+3} and R_{4+5}) and widening across apex of R_{4+5}; a narrow fuscous anal streak; supernumerary lobe weak in female, of medium development in male; abdominal terga III–V orange-brown with a 'T' pattern consisting of a narrow transverse black band across anterior margin of tergum III and a moderately narrow medial longitudinal black band over all three terga, a narrow rectangular black marking on anterolateral corners of tergum IV and a triangular dark fuscous to black marking on anterolateral corners of tergum V, a pair of oval orange-brown shining spots on tergum V; abdominal sterna dark coloured.
Attractant: No known record.
Female: As for male in the general body colour patterns. See Drew & Hancock (1994a) for a detailed description of female characters.
Distribution:
Type locality: Philippines (Luzon I., Mt Makiling).
Hosts: *Pouteria macranthum* (family *Sapotaceae*) (Drew & Hancock, 1994a).
Remarks: *Bactrocera (Bactrocera) quasipropinqua* Drew & Hancock, within the *dorsalis* complex, is similar to *Bactrocera (Bactrocera) affinidorsalis* (Hardy) and *Bactrocera (Bactrocera) holtmanni* (Hardy) in possessing broad parallel-sided lateral postsutural vittae, costal band overlapping R_{2+3}, dark lateral margins on abdominal terga III–V narrow- to medium-width, pale coloured shining spots on abdominal tergum V pale and dark apical spots on all femora or at least on fore and mid femora. It differs from these species in possessing apical dark spots on fore and mid femora only and the apex of the female piercer trilobed.

Bactrocera (Bactrocera) raiensis Drew & Hancock

(Fig. 169)

Bactrocera (Bactrocera) raiensis Drew & Hancock, 1994a: 57; Norrbom *et al.*, 1998: 94. Holotype female in BMNH.

Non-type material examined: Since the publication of Drew & Hancock (1994a), the following specimens have been studied – **VIETNAM: Southern Vietnam** – 1 female, Tien Giang Province, My Tho Village, Site S005, 19.v.2000, attracted to methyl eugenol; 1 male, Dong Nai Province, Long Khanh District, Nhan Nghia Village, Site VS06, 12.x.2002, attracted to methyl eugenol; 1 male, Binh Dinh Province, Tay Son District, Binh Thanh Village, Site VS11, 15.vi.2002, attracted to methyl eugenol; 1 male, Dong Nai Province, Vinh Cuu District, Tan Binh Village, Site S011, 8. vii.1999. Specimens in QDPC.
Diagnosis: Face fulvous with a pair of large oval black spots; postpronotal lobes and notopleura yellow; scutum black except brown below and behind lateral postsutural vittae and red-brown around notopleural suture and inside postpronotal lobes; broad parallel-sided lateral postsutural yellow vittae ending just behind *ia.* seta; medial postsutural yellow vitta absent; mesopleural stripe reaching midway between anterior margin of notopleuron and anterior *npl.* seta dorsally; scutellum yellow; legs with femora entirely fulvous or with, at most, a small subapical fuscous spot on outer surfaces of fore femora; fore and hind tibiae dark fuscous, mid tibiae dark fuscous basally to fulvous apically; wing with cells bc and c colourless or with a very pale tint, microtrichia in outer corner of cell c only; a narrow fuscous costal band confluent with R_{2+3} and with a very slight swelling around apex of R_{4+5}; a narrow fuscous anal streak; supernumerary lobe weak in female, of medium development in male; abdominal terga III–V orange-brown with a moderately broad medial longitudinal black band and two broad lateral longitudinal dark fuscous to black bands over all three terga and which are joined along anterior margin of tergum III, a pair of oval orange-brown shining spots on tergum V; abdominal sterna dark coloured.

Attractant: Methyl eugenol (**New record**).
Female: As for male in the general body colour patterns. See Drew & Hancock (1994a) for a detailed description of female characters.
Distribution:
Type locality: Thailand (Chiang Rai, Doi Luong).
New records: Southern Vietnam (Binh Dinh, Dong Nai, Tien Giang Provinces).
Hosts: Recorded from four plant families – *Asclepiadaceae; Dilleniaceae*; *Meliaceae*; *Moraceae* (see Allwood *et al.*, 1999 for host records).
Remarks: *Bactrocera (Bactrocera) raiensis* Drew & Hancock, within the *dorsalis* complex, is similar to *Bactrocera (Bactrocera) hantanae* Tsuruta & White, *Bactrocera (Bactrocera) irvingiae* Drew & Hancock and *Bactrocera (Bactrocera) syzygii* White & Tsuruta in possessing broad parallel-sided lateral postsutural vittae, costal band confluent with R_{2+3}, femora entirely fulvous and abdominal terga III–V with a broad medial longitudinal black band. It differs from *B. hantanae* and *B. irvingiae* in possessing broad lateral dark markings on abdominal terga III–V and from *B. syzygii* in having pale shining spots on abdominal tergum V. Specimens of *B. raiensis* that possess a small dark subapical spot on the outer surfaces of the fore femora are also similar to *Bactrocera (Bactrocera) kanchanaburi* Drew & Hancock and *Bactrocera (Bactrocera) neopropinqua* Drew & Hancock. However, *B. raiensis* differs from *B. neopropinqua* in possessing broad lateral longitudinal dark fuscous to black bands on abdominal terga III–V and from *B. kanchanaburi* in having a broad medial longitudinal black band on abdominal terga III–V and pale shining spots on abdominal tergum V.

Bactrocera (Bactrocera) ranganathi Drew & Romig, new species
(Fig. 170)

Material examined: Holotype male, **INDIA: ANDAMAN ISLANDS** – India, S. Nicobar, C. Bay, 16.vi.1997, coll. Ranganath, attracted to methyl eugenol, No. 58, deposited in BMNH.
Paratypes: **INDIA: ANDAMAN ISLANDS** – 3 males, same data as holotype; 4 males, Arong, Car Nicobar, 16.vi.1995, coll. Ranganath, attracted to methyl eugenol, Sl No. 41. 1 paratype in ANIC, 1 paratype in BMNH, 3 paratypes in QDPC, 2 paratypes in QMIC.
Diagnosis: A medium-sized species; face fulvous with a pair of large elongate oval black spots; postpronotal lobes and notopleura yellow; scutum black; lateral postsutural yellow vittae present; medial postsutural yellow vitta absent; no yellow spot anterior to notopleural suture; mesopleural stripe slightly wider than notopleuron dorsally; scutellum yellow; wing with cells bc and c colourless, microtrichia in outer corner of cell c only; a narrow dark fuscous costal band which widens slightly across apex of R_{4+5}; a narrow fuscous anal streak; abdominal terga III–V orange-brown with a 'T' pattern consisting of a narrow transverse black band across anterior margin of tergum III which widens across lateral margins and a narrow- to medium-width longitudinal black band over all three terga, narrow anterolateral black corners on tergum IV.
Description: male.
Head: Height 1.3 mm. Frons – length 1.7 times breadth; red-brown with fuscous around bases of orbital setae; orbital setae black: 1 *s.or.*, 2 *i.or.*; lunule fuscous. Ocellar triangle black. Vertex fuscous. Face fulvous with a pair of large elongate oval black spots; length 0.4 mm. Genae fulvous, small dark fuscous subocular spot present; black seta present. Occiput red-brown, fulvous along eye margins; occipital row with 3–6 black setae. Antennae with segments 1 and 2 red-brown, segment 3 red-brown with fuscous on apex and outer surface; length of segments: 0.14 mm; 0.24 mm; 0.67 mm.
Thorax: Scutum entirely black except red-brown around notopleural suture. Pleural areas entirely dark fuscous to black. Yellow markings as follows: postpronotal lobes; notopleura; mesopleural stripe slightly wider than notopleuron dorsally, continuing to katepisternum as a transverse spot, anterior margin convex; anatergite (posterior apex black); anterior ¾ katatergite (remainder black); two broad parallel to sub-parallel lateral postsutural vittae ending at *ia.* seta. Medial longitudinal postsutural yellow vitta absent. Postnotum black. Scutellum yellow except for narrow black basal band. Setae: *sc.* 2; *prsc.* 2; *ia.* 1; *p.sa.* 1; *a.sa.* 1; *mpl.* 1; *npl.* 2; *scp.* 4.
Legs: All femora entirely red-brown; fore tibiae fuscous basally to pale fuscous apically, mid tibiae red-brown, hind tibiae fuscous; all tarsi entirely fulvous; mid tibiae each with an apical black spur.
Wings: Length 5.0 mm; cells bc and c colourless; microtrichia in outer corner of cell c only; remainder of wings colourless except fuscous cell sc, narrow dark fuscous costal band confluent with R_{2+3} and widening slightly across apex of R_{4+5}, a narrow fuscous anal streak; a dense aggregation of microtrichia around A_1+CuA_2; supernumerary lobe of medium development.
Abdomen: Oval; terga free; pecten present on tergum III. Tergum I and sterna I and II wider than long. Tergum I orange-brown with narrow lateral black margins; tergum II orange-brown with a narrow transverse subanterior black band which almost reaches the narrow black anterolateral corners; terga III–V orange-brown with a 'T' pattern consisting of a narrow transverse black band across anterior margin of tergum III and which widens over the lateral margins of that tergum and a narrow- to medium-width medial longitudinal black band over all three terga, narrow anterolateral black corners on tergum IV. A pair of oval black shining spots on tergum V. Posterior lobe of surstylus short, sternum V with a deep concavity on posterior margin. Abdominal sterna dark fuscous to black.
Attractant: Methyl eugenol.
Female: No known record.
Distribution: India: Andaman Islands (South Nicobar).
Hosts: No known record.
Etymology: This species is named after the collector, Dr Ranganath.
Remarks: *Bactrocera (Bactrocera) ranganathi* n.sp. belongs to the *dorsalis* complex of species but appears

to have no economic significance. It is similar to *Bactrocera (Bactrocera) dorsalis* (Hendel), *Bactrocera (Bactrocera) papayae* Drew & Hancock and *Bactrocera (Bactrocera) verbascifoliae* Drew & Hancock in possessing broad parallel-sided lateral postsutural vittae, the costal band confluent with R_{2+3}, all femora entirely fulvous and abdominal terga III–V with a narrow medial longitudinal dark band, but differs in having distinct shining black ceromae on abdominal tergum V.

Bactrocera (Bactrocera) recurrens (Hering)

(Fig. 171)

Strumeta (Bactrocera) recurrens Hering, 1941a: 47. Holotype male in MNM.
Strumeta recurrens – May, 1963:50.
Dacus (Strumeta) recurrens – Drew, 1974: 78 (partim, material from Papua New Guinea).
Bactrocera (Bactrocera) recurrens – Drew, 1989: 104; Norrbom *et al*., 1998: 94.

Non-type material examined: INDONESIA: NORTH MOLUCCAS (NORTH MALUKU) – 1 male, North Maluku, West Halmahera Goal, 5.vi.2007, coll. Raais Abdullah, attracted to cue lure in forest; **MOLUCCAS (MALUKU)** – 1 male, Moluccas (Maluku), Seram (Western part), Nuruwe, 9.vi.2007, coll. A. Haris Tualepe, attracted to cue lure in forest; 1 male, Moluccas (Maluku), Western Seram, Waimital, 19. vii.2007, coll. A. Haris Tualepe, attracted to cue lure in orange (jeruk); 1 male, Mamala, Ambon, Indonesia, 50-150m, 7. viii.1981, coll. J. Turner, residual forest with clove, nutmeg and banana groves, attracted to cue lure; 1 male, Moluccas (Maluku), Middle Maluku, Suli, 9.viii.2007, coll. La Naibu, attracted to cue lure in orange (jeruk). Specimens in QDPC.
Diagnosis: Face fulvous with a pair of medium-sized circular black spots; postpronotal lobes and notopleura yellow; scutum dull black tending dark red-brown to fuscous below lateral postsutural vittae; lateral postsutural yellow vittae of medium-width and tapering sharply posteriorly to end just before *ia*. seta; medial postsutural yellow vitta absent; mesopleural stripe reaching postpronotal lobes dorsally; scutellum yellow except for narrow black basal band; legs with femora entirely fulvous; fore and mid tibiae fulvous, hind tibiae dark red-brown; wing with cells bc and c fuscous, dense microtrichia covering all of cell c and outer ½ of cell bc; a broad fuscous costal band confluent with R_{4+5} and broad anal streak, an 'S'-shaped fuscous band across wing; supernumerary lobe of medium development; abdominal terga III–V variable from mostly orange-brown with fuscous anterolateral corners on tergum III to orange-brown with a 'T'-shaped pattern consisting of a dark fuscous to black narrow transverse band across anterior margin of tergum III, a moderately broad medial longitudinal dark fuscous to black band over all three terga and narrow dark fuscous to black lateral margins on all three terga, a pair of oval fuscous shining spots on tergum V; abdominal sterna pale coloured.
Attractant: Cue lure.
Female: No known record.
Distribution:
Type locality: Papua New Guinea (Madang, Friedrich Wilhelms-Hafen).
Previously recorded: Type locality and Papua New Guinea (Milne Bay District, Samari I.) (Drew, 1989); Indonesia (West Papua, formerly part of Irian Jaya) (White & Evenhuis, 1999). The published record of *B. recurrens* from Queensland, Australia by Norrbom *et al*. (1998) is incorrect.
New records: Indonesia (Moluccas = Maluku – Ambon I., Mamala and Suli; Seram I.; North Moluccas = North Maluku – West Halmahera).
Hosts: No known record.
Remarks: *Bactrocera (Bactrocera) recurrens* (Hering) is unlike any other species in South-East Asia in possessing a distinct 'S'-shaped fuscous band across the wing. In the South Pacific area it is similar to *Bactrocera (Bactrocera) manskii* (Perkins and May) in possessing an 'S'-shaped fuscous band across the wing, but differs from this species in having the scutellum entirely black, lateral postsutural vittae broad and narrowing posteriorly to end at or just before the *ia*. seta, abdominal terga III–V mostly orange-brown. See Drew (1989) for a full discussion of *B. recurrens* and *B. manskii* and the distribution records.

Bactrocera (Bactrocera) resima (Drew)

(Fig. 172)

Dacus (Strumeta) resimus Drew, 1971: 85. Holotype male in QMIC.
Bactrocera (Bactrocera) resima – Drew, 1989: 105; Norrbom *et al*., 1998: 95.

Non-type material examined: INDONESIA: SUMATRA – 1 male, Sumatera Selatan, Ogan Komering Ilir, Celikah, 2.x.2006, coll. Imam Muhayani, attracted to cue lure in Duku. Specimen in QDPC.
Diagnosis: Face fulvous with a pair of small circular black spots; postpronotal lobes and notopleura yellow; scutum black with dark red-brown to fuscous below lateral postsutural vittae; broad parallel-sided lateral postsutural yellow vittae; medial postsutural yellow vitta absent; mesopleural stripe reaching postpronotal lobes dorsally; scutellum yellow with a narrow black basal band; legs with all segments entirely fulvous; wing with cell c colourless to having a very pale tint, cell bc fulvous, dense microtrichia in outer corner of cell c only; a narrow fuscous costal band overlapping R_{2+3} but not confluent with R_{4+5}; a broad fuscous anal streak, a 'U'-shaped fuscous pattern across wing that is often ill-defined at posterior margin of wing; supernumerary lobe of medium development; abdominal terga generally orange-brown with a narrow medial longitudinal dark fuscous to black band over all five terga and narrow dark fuscous to black lateral margins on tergum III, shining spots on tergum V orange-brown; abdominal sterna orange-brown.
Attractant: Cue lure.
Female: No known record.
Distribution:
Type locality: Papua New Guinea (Sepik, Ambunti).

Previously recorded: Type locality and Australia (Torres Strait Islands – Saibai I.).
New records: Indonesia (Sumatra).
Hosts: No known record.
Remarks: *Bactrocera (Bactrocera) resima* (Drew) is unlike any other species in South-East Asia in that it possesses a 'U'-shaped fuscous band across the centre of the wing. In this character it is similar to *Bactrocera (Bactrocera) absidata* Drew that occurs in Papua New Guinea, but differs from this species in having a medial longitudinal dark fuscous to black band over all five terga and narrow lateral dark margins on tergum III (not with the abdominal terga entirely orange-brown).

Bactrocera (Bactrocera) ritsemai (Weyenbergh)

(Figs 173A, 173B)

Dacus ritsemae Weyenbergh, 1869: 360. Holotype lost; Hardy, 1983: 20.
Bactrocera ritsemae, incertae sedis – Norrbom *et al.*, 1998: 104.
Dacus (Bactrocera) limbifer rufulus Hardy, 1982: 231. Holotype male in MZB. **New Synonym.**
Bactrocera (Bactrocera) rufula – Norrbom *et al.*, 1998: 95.

Other type material examined: Holotype male of *Dacus (Bactrocera) limbifer rufulus* Hardy, **INDONESIA: SULAWESI**: with the following label data – Sadaunta, 65 km S.E. Palu, Central Sulawesi 650 m, Aug. 1975, coll. D.E. & A.T. Hardy, attracted to cue lure. Studied at MZB (Fig. 173B). Paratypes of *Dacus (Bactrocera) limbifer rufulus* Hardy: **INDONESIA: SULAWESI** – 3 males each with the following three labels – 'Kuala Navusa, Malakosa, Sulawesi, Indonesia, Aug–Nov 1975, 100', Margareta Becker'; 'Paratype male, Dacus (Bactrocera) limbifer rufulus d.e. Hardy'; 'Dacus (Bactrocera) limbifer rufulus Hardy det. 1983, d.e. Hardy'(last two labels in Hardy's handwriting). 2 males each with the following three labels – 'Lindu Valley, Cent. Sulawesi, 960 m, Aug 1975, d.e. Hardy'; 'Paratype male, Dacus (Bactrocera) limbifer rufulus, d.e. Hardy'; 'BMNH (E) No. 532962'. Studied at BMNH. 1 male, same data as last two paratypes. Specimens in QDPC.
Non-type material examined: INDONESIA: JAVA – 2 males (19.vii.2005), 1 male (26.vii.2005), East Java, Lumajang, Kadang Tepus, coll. Utomo, attracted to cue lure in banana plants; **LOMBOK** – 1 male, Bentek, 20km N. Mataram, Lombok, Indonesia, 100-250m, 2.vi.1981, coll. J. Turner, attracted to cue lure; **SULAWESI** – 3 males, 6km E. Bantimurang, 15.v.1997, coll. G. Pattemore, attracted to cue lure; 4 males (27.vii.2006), 2 males (16.viii.2006), 3 males (2. ix.2006), 1 male (10.ix.2006), 4 males (17.ix.2006), 2 males (22.x.2006), 1 male (29.x.2006), Sulawesi Tengah, Donggala, Nupabomba, coll. Rusdin, attracted to cue lure in forest; 24 males, Sulawesi, Gorontalo, Limboto, Bionga, 31.x.2006, coll. Yusbar Ismail, attracted to cue lure in forest; 1 male, Sulawesi Selatan, Takalar, Topejawa, 2. ix.2006, coll. Sirajuddin, attracted to cue lure in village; 1 male (27. vii.2006), 14 males (16.viii.2006), 1 male (27.viii.2006), 34 males (30.viii.2006), 80 males (13.ix.2006), Sulawesi Selatan, Maros, Limampoccoe, coll. Sirajuddin, attracted to cue lure in forest; 15 males, 'Bantimurung, 40km N.W. Ujung Pandang, Sulawesi, 50m, 27.vii.1981, coll. J. Turner', 'Primary Rainforest', attracted to cue lure; 1 male, Gorontalo, Bone Bolango, Lombongo, 19. ix.2006, coll. Yusbar Ismail, attracted to cue lure in forest; 13 males, Sulawesi Selatan, Takalar, Limampoccoe, 13.ix.2006, coll. Sirajuddin SP., attracted to cue lure in forest. Specimens in QDPC.
Diagnosis: A large species; face fulvous with a pair of small- to medium-sized circular to subquadrate black spots; postpronotal lobes and notopleura yellow; scutum black; lateral postsutural yellow vittae present; medial postsutural yellow vitta absent; no yellow spot anterior to notopleural suture; mesopleural stripe equal in width or slightly wider than notopleuron dorsally; scutellum yellow; wing with cells bc and c colourless, microtrichia in outer corner of cell c only; a broad costal band which is dark fuscous to R_{2+3}, pale fuscous overlapping R_{2+3} to be almost confluent with R_{4+5} and a very dark fuscous apical spot from apex of R_{2+3} across apex of R_{4+5}; a medium-width fuscous anal streak; abdominal terga III–V orange-brown with a black 'T' pattern consisting of a narrow transverse band across anterior margin of tergum III and widening to moderately broad dark fuscous lateral margins, a moderately broad medial longitudinal black band over all three terga, narrow fuscous to dark fuscous anterolateral corners on tergum IV.
Description: male.
Head: Height 1.95 mm. Frons – length 1.4 times breadth; red-brown with dark fuscous around orbital setae and on anteromedial hump; orbital setae black: 1 *s.or.*, 2 *i.or.*; lunule dark fuscous. Ocellar triangle black. Vertex red-brown. Face fulvous with a pair of small- to medium-sized circular to sub-quadrate black spots; length 0.55 mm. Genae fulvous, dark fuscous subocular spot present; black seta present. Occiput fuscous, fulvous along eye margins; occipital row with 5–10 strong black setae. Antennae with segments 1 and 2 red-brown, segment 3 red-brown with fuscous on apex; length of segments: 0.25 mm; 0.37 mm; 0.9 mm.
Thorax: Scutum black with fuscous below lateral postsutural vittae. Pleural areas dark fuscous to black except fuscous below postpronotal lobes. Yellow markings as follows: postpronotal lobes; notopleura; mesopleural stripe equal in width or slightly wider than notopleuron dorsally, continuing to katepisternum as a transverse spot, anterior margin slightly convex; anatergite (posterior apex black); anterior ⅔ katatergite (remainder black); two broad parallel-sided or sub-parallel lateral postsutural vittae ending at or just behind *ia.* seta. Medial longitudinal postsutural yellow vitta absent. Postnotum black. Scutellum yellow except for narrow black basal band. Setae: *sc.* 2; *prsc.* 2; *ia.* 1; *p.sa.* 1; *a.sa.* 1; *mpl.* 1; *npl.* 2; *scp.* 4.
Legs: Fore femora fulvous basally, red-brown apically and with a small fuscous to dark fuscous subapical spot on outer surfaces of some specimens only, mid femora fulvous basally, red-brown apically and with a small fuscous to dark fuscous

spot on outer apical surfaces of some specimens only, hind femora fulvous basally, red-brown apically and with a pale fuscous to fuscous spot on inner apical ¼ in some specimens only; fore tibiae fulvous to pale fuscous, mid tibiae fulvous, hind tibiae fuscous; tarsi with all segments fulvous; mid tibiae each with an apical black spur.

Wings: Length 7.0 mm; cells bc and c colourless; microtrichia in outer corner of cell c only; remainder of wings colourless except dark fuscous cell sc, broad costal band which is dark fuscous to R_{2+3} and pale fuscous overlapping this vein to be almost confluent with R_{4+5} and with a very dark fuscous apical spot from apex of R_{2+3} to end beyond apex of R_{4+5}, a moderately broad fuscous anal streak; a dense aggregation of microtrichia around A_1+CuA_2; supernumerary lobe of medium development.

Abdomen: Oval; terga free; pecten present on tergum III. Tergum I and sterna I and II wider than long. Tergum I fuscous with narrow dark fuscous lateral margins; tergum II orange-brown anteriorly, fulvous posteriorly, with a narrow dark fuscous subanterior transverse band reaching narrow dark fuscous anterolateral corners; terga III–V orange-brown with a black 'T' pattern consisting of a narrow transverse band across anterior margin of tergum III which expands to form moderately broad dark fuscous lateral margins and a moderately broad medial longitudinal black band over all three terga, fuscous to dark fuscous narrow anterolateral corners on tergum IV. A pair of oval dark fuscous to black shining spots on tergum V. Posterior lobe of surstylus short, sternum V with a deep concavity on posterior margin. Abdominal sterna dark fuscous to black.

Attractant: Cue lure.

Female: No known record.

Distribution:

Type locality: Indonesia (Java, Surabaya).

New records: Indonesia (Lombok; Sulawesi).

Hosts: No known record.

Remarks: *Bactrocera (Bactrocera) ritsemai* (Weyenbergh) was named after Mr M.L. Ritsema and thus must have the masculine ending to the species name. The original description is adequate to define this species and after a study of many specimens collected over Java, Lombok and Sulawesi together with the Holotype and some Paratypes of *Dacus (Bactrocera) limbifer rufulus* Hardy, we believe that these are conspecific and here place *D. rufulus* a **New Synonym** of *B. ritsemai*.

Bactrocera (Bactrocera) ritsemai (Weyenbergh) is similar to *Bactrocera (Bactrocera) bruneiae* n.sp., *Bactrocera (Bactrocera) costalis* (Shiraki) and *Bactrocera (Bactrocera) eurycosta* n.sp. in possessing the costal band confluent with R_{4+5} and of uniform width throughout, cells bc and c colourless or with a pale fuscous tint, a black scutum and femora entirely fulvous with, at most, small dark subapical markings on fore femora. It differs from *B. bruneiae* in having oval facial spots, from *B. eurycosta* in having the scutum black with red-brown margins, from all three species in having the costal band dark fuscous to R_{2+3}, paler between R_{2+3} and R_{4+5} and with a darker fuscous spot within the apex of the band, and from *B. costalis* in having the anal streak narrow and a broad black medial longitudinal band over abdominal terga III–V.

Bactrocera (Bactrocera) rubigina (Wang & Zhao)

(Fig. 174)

Dacus rubiginus Wang & Zhao, 1989: 211. Holotype male in IZAS.

Bactrocera (Bactrocera) rubigina – Norrbom *et al.*, 1998: 95; Drew *et al.*, 2007: 5.

Non-type material examined: BHUTAN: with the following label data – 1 male (13.v.2000), 2 males (20.v.2000), Phuentsholing, coll. S. Thapa, attracted to cue lure; 1 male, Phuentsholing, 27.v.2000, coll. MS RAI, attracted to cue lure; 1 male, Rimchu, 17.v.2002, coll. C. Dorji, attracted to cue lure; 1 male, Rimchu 1, 10.v.2000, coll. B.S. Fletcher, attracted to cue lure; 1 male (17.v.2000), 1 male (31.v.2000), Rimchu 1, coll. C. Dorji, attracted to cue lure; 1 male, Rimchu 1, 21.vi.2000, coll. Nado *et al.*, attracted to cue lure; 1 male, Rimchu 2, 17.v.2000, coll. C. Dorji, attracted to cue lure; 2 males, Rimchu 2, 7.vi.2000, coll. Dorji and Nado, attracted to cue lure; 1 male, Tsirang, 10.vii.2000, coll. T.D. & S.D., attracted to cue lure; 1 male, Tsirang, 31.vii.2000, attracted to cue lure; 1 male, Suntalay, 15.v.2000, coll. S. Wangdi *et al.*, attracted to cue lure; 1 male, Lingmethang 11km, 9.vi.2001, coll. Brian Fletcher, attracted to cue lure in Subtropical rainforest 1100m; 1 male, Gelephu Purana Bastey, 1.ii.2005, coll. Maha Prasad & Karma Nangyel, attracted to cue lure; 10 males (15.xi.2004), 7 males (5. xii.2004), East Bhutan, Samdrup Jongkhar District, coll. Wangdi, attracted to cue lure. **THAILAND**: A large series of specimens collected from many localities in Thailand. **VIETNAM: Northern Vietnam** – 1 male, Son La Province, Moc Chau District, Long Luong Village, 15.iv.2002, coll. NIPP staff, attracted to cue lure; 1 female (17.i.2002), 1 male (11.vii.2002), 1 male (22.viii.2002), Hanoi Province, Tu Liem District, Lien Mac Village, attracted to cue lure; 1 male, Phu Tho Province, Phu Ninh District, Phu Tho fruit research centre, 9.v.2002, coll. NIPP staff, attracted to cue lure; 1 male, Son La Province, Moc Chau District, Moc Chau house garden, 9.viii.1999, attracted to cue lure; 1 male, Hoa Binh Province, Ky Son District, Cao Phong, 15.iv.2000, attracted to cue lure; 12 males, Lao Cai Province, Sapa 2 Village, 13.viii.1999, attracted to cue lure; 14 males, Lao Cai Province, Sapa 2 Village, 30.vii.1999, attracted to cue lure; 1 male, Lao Cai Province, Sapa 6 Village, 13.viii.1999, attracted to cue lure. Specimens in QDPC.

Diagnosis: A medium-sized species; face fulvous with a pair of medium-sized oval black spots; postpronotal lobes and notopleura yellow; scutum red-brown with lanceolate black patterns; lateral postsutural yellow vittae present; medial postsutural yellow vitta absent; no yellow spot anterior to notopleural suture; mesopleural stripe reaching midway between anterior margin of notopleuron and anterior *npl.* seta dorsally; scutellum yellow; wing with cells bc and c colourless, microtrichia in outer corner of cell c only; a narrow fuscous costal band; a narrow fuscous anal streak; abdominal terga III–V orange-brown with a 'T' pattern consisting of a narrow transverse black band across anterior margin of tergum III which widens to a broad dark fuscous to black lateral margin, a narrow- to medium-width medial longitudinal black band over all three terga and narrow dark fuscous anterolateral corners on terga IV and V.

Description: male.

Head: Height 1.6 mm. Frons – length 1.35 times breadth; red-brown with dark fuscous around bases of orbital setae and on anteromedial hump; orbital setae black: 1 *s.or.*, 2 *i.or.*; lunule dark fuscous. Ocellar triangle black. Vertex fuscous. Face fulvous with a pair of medium-sized oval black spots; length 0.4 mm. Genae fulvous, large dark fuscous subocular spot present; black seta present. Occiput fuscous, fulvous along eye margins; occipital row with 1–3 short black setae. Antennae with segments 1 and 2 fuscous, segment 3 red-brown with dark fuscous on apex and outer surface; length of segments: 0.12 mm; 0.31 mm; 0.78 mm.

Thorax: Scutum red-brown with a dark fuscous to black lanceolate pattern (the extent of this pattern varies, in some specimens it is confined to the area posterior to the notopleural suture and in others in extends forward from the posterior area of the scutum to the anterior margin). Pleural areas black except red-brown below postpronotal lobes and above mid coxae. Yellow markings as follows: postpronotal lobes; notopleura; mesopleural stripe reaching midway between anterior margin of notopleuron and anterior *npl.* seta dorsally, continuing to katepisternum as a transverse spot, anterior margin straight; anatergite (posterior apex black); anterior ⅔ katatergite (remainder black); two narrow- to medium-width sub-parallel lateral postsutural vittae ending at or just before *ia.* seta. Medial longitudinal postsutural yellow vitta absent. Postnotum black. Scutellum yellow except for narrow black basal band. Setae: *sc.* 2; *prsc.* 2; *ia.* 1; *p.sa.* 1; *a.sa.* 1; *mpl.* 1; *npl.* 2; *scp.* 4.

Legs: All femora entirely fulvous; fore tibiae fuscous to dark fuscous, mid tibiae dark fuscous basally to fulvous apically, hind tibiae dark fuscous; fore and mid tarsi with all segments fulvous, hind tarsi with basal segment fulvous and apical four segments red-brown; mid tibiae each with an apical black spur.

Wings: Length 5.3 mm; cells bc and c colourless; microtrichia in outer corner of cell c only; remainder of wings colourless except dark fuscous cell sc, narrow fuscous costal band confluent with R_{2+3} (or in some specimens just overlapping R_{2+3} where it becomes very pale) and remaining narrow around apex of wing to end just beyond apex of R_{4+5} (the costal band is constricted in width at apex of R_{2+3}), a narrow fuscous anal streak; a dense aggregation of microtrichia around A_1+CuA_2; supernumerary lobe of medium development.

Abdomen: Oval; terga free; pecten present on tergum III. Tergum I and sterna I and II wider than long. Tergum I dark fuscous except for a narrow orange-brown transverse band across posterior margin but not reaching lateral margins; tergum II orange-brown with fulvous areas posterolaterally, a narrow transverse black band subanteriorly but not reaching narrow black anterolateral corners; terga III–V orange-brown with a 'T' pattern consisting of a narrow transverse black band across anterior margin of tergum III which expands to a broad dark fuscous to black lateral margin and a narrow- to medium-width medial longitudinal black band over all three terga, narrow anterolateral dark fuscous margins on terga IV and V (the abdominal colour patterns vary from pale fuscous to fuscous lateral margins that are often extremely narrow to broad dark fuscous to black lateral margins). A pair of oval red-brown shining spots on tergum V (in some specimens these tend to fuscous on lateral margins). Posterior lobe of surstylus short, sternum V with a deep concavity on posterior margin. Abdominal sterna dark fuscous to black.

Attractant: Cue lure.

Female: As for male except no dense aggregation of microtrichia around A_1+CuA_2; supernumerary lobe weak; pecten of cilia on abdominal tergum III absent; ovipositor basal segment red-brown, dorsoventrally compressed and tapering posteriorly in dorsal view; ratio of length of oviscape to length of tergum V, 1.4:1; apex of piercer needle-shaped.

Distribution:

Type locality: China (Hainan I.).

Previously recorded: Type locality and Bhutan.

New records: Thailand; Northern Vietnam.

Hosts: *Litsea verticillata* (family *Lauraceae*) (Liang *et al*., 1993).

Remarks: *Bactrocera (Bactrocera) rubigina* Wang & Zhao is similar to *Bactrocera (Bactrocera) invadens* Drew, Tsuruta & White in the general colour patterns of the wing and body. It differs from *B. invadens* in having the males attracted to cue lure whereas those of *B. invadens* are attracted to methyl eugenol. Other similar general red-brown species are *Bactrocera (Bactrocera) bhutaniae* n.sp. and *Bactrocera (Bactrocera) bivittata* Lin & Wang. See 'Remarks' under *B. bhutaniae* for a full comparison of these species.

There is a large series of specimens in the QDPC collection from Southern Vietnam with a broad costal band in the wing, some almost confluent with R_{4+5}. In all other characters, these specimens are identical with *B. rubigina* from Bhutan, China, Northern Vietnam and Thailand. These two populations are also similar on the length of the male aedeagus, the mean length of the narrow costal band form being 2.58 mm and that for the broad costal band form being 2.48 mm. We have decided to classify all specimens from Northern and Southern Vietnam as *B. rubigina*.

Bactrocera (Bactrocera) sapaensis Drew & Romig, new species

(Fig. 175)

Material examined: Holotype male, **VIETNAM: Northern Vietnam** – Lao Cai Province, Sapa 6 Village, 9.vi.2000, attracted to cue lure, deposited in BMNH.

Paratypes: **Northern Vietnam** – 2 males, same data as holotype; 2 males (23.vi.2000), 1 male (7.vii.2000), 1 male (4.viii.2000), same locality and lure data as holotype; 1 male, Lao Cai Province, Sapa Oquy Ho, 12.v.2000, attracted to cue lure; 1 male, Son La Province, Moc Chau District, Tra Long, 29.v.2000, attracted to cue lure; 1 male, Hoa Binh Province, Mai Chau District, Pa Co Village, 30.iv.2002, coll. NIPP staff, attracted to cue lure; 1 male, Hanoi province, Tu Liem District, Con Hue, 7.iv.2000, attracted to cue lure; Southern Vietnam: 1 male (24.ii.2002), 2 males (8.iii.2002), 1 male (8.v.2002), Dac Lac Province, Ku Mi Nga District, Quang

Phu Village, coll. SOFRI staff, attracted to cue lure; 1 male (13.iv.2002), 1 male (11.v.2002), Dong Nai Province, Long Khanh District, Nhan Nghia Village, coll. SOFRI staff, attracted to cue lure. 2 paratypes in ANIC; 2 paratypes in BMNH; 10 paratypes in QDPC; 2 paratypes in QMIC.

Diagnosis: A medium-sized species; face fulvous with a pair of medium-sized oval black spots; postpronotal lobes and notopleura yellow; scutum black; lateral postsutural yellow vittae present; medial postsutural yellow vitta absent; no yellow spot anterior to notopleural suture; mesopleural stripe reaching midway between anterior margin of notopleuron and anterior *npl.* seta dorsally; scutellum yellow; wing with cells bc and c colourless, microtrichia in outer corner of cell c only; a narrow dark fuscous costal band confluent with R_{2+3}; anal streak absent (there is a pale fuscous to fuscous tint within cell cup); abdominal terga III–V dark orange-brown with a 'T' pattern consisting of a narrow transverse black band across anterior margin of tergum III which widens to form broad dark fuscous to black margins, a broad medial longitudinal black band over all three terga, dark fuscous anterolateral corners on terga IV and V.

Description: male.

Head: Height 1.76 mm. Frons – length 1.61 times breadth; red-brown with dark fuscous around orbital setae and on anteromedial hump; orbital setae black: 1 *s.or.*, 2 *i.or.*; lunule dark fuscous. Ocellar triangle black. Vertex dark fuscous. Face fulvous with a pair of medium-sized oval black spots; length 0.39 mm. Genae fulvous, dark fuscous subocular spot present; black seta present. Occiput dark red-brown, fulvous along eye margins; occipital row with 4–6 strong black setae. Antennae with segments 1 and 2 red-brown, segment 3 red-brown with dark fuscous on apex and outer surface; length of segments: 0.17 mm; 0.31 mm; 0.82 mm.

Thorax: Scutum black with red-brown below lateral postsutural vittae and inside postpronotal lobes. Pleural areas black except red-brown below postpronotal lobes. Yellow markings as follows: postpronotal lobes; notopleura; mesopleural stripe reaching midway between anterior margin of notopleuron and anterior *npl.* seta dorsally, continuing to katepisternum as a transverse spot, anterior margin convex; anatergite (posterior apex black); anterior $^{3}/_{5}$ katatergite (remainder black); two medium-width to broad parallel-sided lateral postsutural vittae ending just behind *ia.* seta. Medial longitudinal postsutural yellow vitta absent. Postnotum black. Scutellum yellow except for narrow black basal band. Setae: *sc.* 2; *prsc.* 2; *ia.* 1; *p.sa.* 1; *a.sa.* 1; *mpl.* 1; *npl.* 2; *scp.* 4.

Legs: Fore femora fulvous with a small subapical fuscous spot on outer surfaces, mid and hind femora entirely fulvous; fore tibiae dark fuscous, mid tibiae dark fuscous basally to fulvous apically, hind tibiae dark fuscous; tarsi with basal segment fulvous and apical four segments red-brown to fuscous; mid tibiae each with an apical black spur.

Wings: Length 6.5 mm; cells bc and c colourless; microtrichia in outer corner of cell c only; remainder of wings colourless except dark fuscous cell sc, narrow dark fuscous costal band confluent with R_{2+3} and remaining narrow around apex of wing to end just beyond apex of R_{4+5}, anal streak absent (there is a pale fuscous to fuscous tint within cell cup); a dense aggregation of microtrichia around A_1+CuA_2; supernumerary lobe of medium development.

Abdomen: Oval; terga free; pecten present on tergum III. Tergum I and sterna I and II wider than long. Tergum I black with a narrow transverse orange-brown band across posterior margin but not reaching lateral margins; tergum II orange-brown with a dark fuscous to black narrow transverse band subanteriorly and just reaching narrow black anterolateral corners; terga III–V dark orange-brown with a 'T' pattern consisting of a narrow transverse black band across anterior margin of tergum III which expands to form broad fuscous to black lateral margins and a broad medial longitudinal black band over all three terga, narrow dark fuscous anterolateral corners on terga IV and V. A pair of oval fuscous shining spots on tergum V. Posterior lobe of surstylus short, sternum V with a deep concavity on posterior margin. Abdominal sterna dark fuscous to black.

Attractant: Cue lure.

Female: No known record.

Distribution: Northern Vietnam.

Hosts: No known record.

Remarks: *Bactrocera (Bactrocera) sapaensis* n.sp. is similar to *Bactrocera (Bactrocera) neopropinqua* Drew & Hancock in general body and wing colour patterns but differs from this species in having the anal streak absent, facial spots broadly oval, mesopleural stripe narrow, dark lateral margins on terga IV and V narrow and males attracted to cue lure.

Bactrocera (Bactrocera) selenophora Tsuruta & White

(Fig. 176)

Bactrocera (Bactrocera) selenophora Tsuruta & White, 2001: 84. Holotype male in HORDI.

Type material examined: Paratype male, **SRI LANKA (formerly CEYLON)**: with the following four labels – 'Paratype'; 'Sri Lanka, 21.ix.1994, Sinharaja, C7'; 'Sri Lanka NPQS Project, Sp. R. No. 8'; 'Paratype male of *Bactrocera (Bactrocera) selenophora* Tsuruta & White 2001'. Received on loan from BMNH.

Diagnosis: A medium-sized species; face fulvous with a pair of large elongate oval black spots; postpronotal lobes yellow (anterodorsal corners red-brown); notopleura yellow; scutum dull black with large areas of red-brown to fuscous markings; lateral postsutural yellow vittae present; medial postsutural yellow vitta absent; no yellow spot anterior to notopleural suture; mesopleural stripe slightly wider than notopleuron dorsally; scutellum yellow; wing with cells bc and c colourless, microtrichia in outer corner of cell c only; a narrow fuscous costal band; a narrow fuscous anal streak; abdominal terga III–V orange-brown to dark orange-brown with a narrow black transverse band across anterior margin of tergum III joining with narrow fuscous to dark fuscous lateral margins, fuscous to black anterolateral corners on tergum IV and fuscous to dark fuscous anterolateral corners on tergum V and with a narrow fuscous medial longitudinal line on tergum V.

Description: male.

Head: Height 1.4 mm. Frons – length 1.6 times breadth; red-brown with fuscous around *s.or.* setae; orbital setae dark fuscous: 1 *s.or.*, 2 *i.or.*; lunule red-brown. Ocellar triangle black. Vertex red-brown. Face fulvous with a pair of large oval black spots reaching oral margin; length 0.45 mm. Genae fulvous, small pale fuscous subocular spot present; black seta present. Occiput dark fuscous laterally, red-brown centrally, fulvous along eye margins; occipital row with 3–4 fuscous setae. Antennae with segments 1 and 2 red-brown, segment 3 missing; length of segments: 0.19 mm; 0.28 mm.

Thorax: Scutum dull black with red-brown below and behind lateral postsutural vittae, around notopleural suture, around anterolateral margins to end inside postpronotal lobes and a dark red-brown to fuscous broad medial longitudinal band over most of scutum. Pleural areas black except red-brown below postpronotal lobes. Yellow markings as follows: postpronotal lobes (anterodorsal corners red-brown); notopleura; narrow mesopleural stripe slightly wider than notopleuron dorsally, continuing to katepisternum as transverse spot, anterior margin straight; anatergite (posterior apex black); anterior $^{3}/_{5}$ katatergite (remainder black); two short narrow lateral postsutural vittae tapering posteriorly to end in a point just posterior of *a.sa.* setae. Medial longitudinal postsutural yellow vitta absent. Postnotum dark fuscous. Scutellum yellow except for narrow black basal band. Setae: *sc.* 2; *prsc.* 2; *ia.* 1; *p.sa.* 1; *a.sa.* 1; *mpl.* 1; *npl.* 2; *scp.* 4.

Legs: Fore femora fulvous with a large dark fuscous spot on outer surfaces, mid femora fulvous with dark fuscous around lateral and ventral margins towards apex, hind femora fulvous tending red-brown around apical ¼; fore tibiae pale fuscous, mid tibiae fuscous, hind tibiae dark fuscous; tarsi with all segments fulvous; mid tibiae each with an apical black spur.

Wings: Length 5.2 mm; cells bc and c colourless; microtrichia in outer corner of cell c only; remainder of wings colourless except fuscous cell sc, narrow fuscous costal band confluent with R_{2+3} and remaining narrow around apex of wing to end just beyond apex of R_{4+5}, a narrow fuscous anal streak; a dense aggregation of microtrichia around A_1+CuA_2; supernumerary lobe of medium development.

Abdomen: Oval; terga free; pecten present on tergum III. Tergum I dark orange-brown with dark fuscous across anterior and lateral margins; tergum II fulvous across posterior ½, dark orange-brown anterolaterally and with a narrow black transverse band across anterior margin but not reaching lateral margins; tergum III dark orange-brown with a narrow transverse black band across anterior margin joining fuscous to dark fuscous narrow lateral margins; tergum IV dark orange-brown with black to fuscous anterolateral corners; tergum V orange-brown with fuscous to dark fuscous anterolateral corners; a very narrow fuscous medial longitudinal line on tergum V. A pair of oval black shining spots on tergum V. Posterior lobe of surstylus short, sternum V with a deep concavity on posterior margin.

Attractant: Cue lure (Tsuruta & White, 2001).

Female: No known record.

Distribution:

Type locality: Sri Lanka (Matara District, Enselwatta).

Previously recorded: Type locality and Sri Lanka (Matale District, Pitawala [Misspelt in publication]) (Tsuruta & White, 2001).

Hosts: No known record.

Remarks: *Bactrocera (Bactrocera) selenophora* Tsuruta & White is similar to *Bactrocera (Bactrocera) impunctata* (de Meijere) in possessing the costal band confluent with R_{2+3}, scutum with base colour black, abdominal terga III–V without a black 'T' pattern and lateral postsutural vittae tapering posteriorly to end between *a.sa.* and *ia.* setae. It differs from *B. impunctata* in having large areas of the scutum red-brown, lateral postsutural vittae narrow, fore and mid femora with dark preapical markings and abdominal terga III–V with a dark transverse band across anterior margin of tergum III, narrow dark anterolateral corners on terga IV and V and dark shining spots.

Bactrocera selenophora was placed in the *dorsalis* complex by Tsuruta & White (2001); however, this species does not possess the complete set of characters that are necessary for inclusion in this complex (it does not possess a dark 'T' pattern over abdominal terga III–V).

Bactrocera (Bactrocera) sembaliensis Drew & Hancock

(Fig. 177)

Bactrocera (Bactrocera) sembaliensis Drew & Hancock, 1994a: 58; Norrbom *et al.*, 1998: 95. Holotype male in BBM.

Non-type material examined: Since the publication of Drew & Hancock (1994a), the following specimens have been studied – **INDONESIA: SUMATRA** – 2 males, Sumatra, Riau, Kampar, Kualu, 16.viii.2006, coll. Nasrun, attracted to cue lure in forest. Specimens in QDPC.

Diagnosis: Face fulvous with a pair of large circular to oval black spots; postpronotal lobes and notopleura yellow; scutum black except brown below lateral postsutural vittae; lateral postsutural yellow vittae of medium-width and ending at *ia.* seta; medial postsutural yellow vitta absent; mesopleural stripe reaching midway between anterior margin of notopleuron and anterior *npl.* seta dorsally; scutellum yellow; legs with femora fulvous except for a dark fuscous subapical spot on outer surfaces of fore femora (this spot is occasionally absent); all tibiae dark fuscous (mid tibiae may be paler apically); wing with cells bc and c colourless, microtrichia in outer corner of cell c only; a fuscous costal band almost confluent with R_{4+5} and widening slightly across apex of R_{4+5}; a narrow pale fuscous anal streak; supernumerary lobe of medium development; abdominal terga III–V orange-brown except for a transverse black band across anterior ⅓ of tergum III which widens to cover outer ¼ of lateral margins, a narrow medial longitudinal dark fuscous band over all three terga, anterolateral corners of terga IV and V dark fuscous (these lateral corners are occasionally absent), a pair of oval orange-brown shining spots on tergum V; abdominal sterna dark coloured.

Attractant: Cue lure.
Female: No known record.
Distribution:
Type locality: Indonesia (Lombok I., Sembalia and Belanting).
Previously recorded: Type locality and Indonesia (Sumbawa) (Drew & Hancock, 1994a).
New records: Indonesia (Sumatra – Riau, Kampar).
Hosts: No known record.
Remarks: *Bactrocera (Bactrocera) sembaliensis* Drew & Hancock, within the *dorsalis* complex, is similar to *Bactrocera (Bactrocera) affinidorsalis* (Hardy), *Bactrocera (Bactrocera) holtmanni* (Hardy) and *Bactrocera (Bactrocera) quasipropinqua* Drew & Hancock in possessing moderately broad parallel-sided lateral postsutural vittae, the costal band overlapping R_{2+3}, the lateral margins of abdominal terga III–V narrow- to medium-width, shining spots on abdominal tergum V pale and with dark markings on at least the apices of the fore femora. It differs from these species in having the costal band almost confluent with R_{4+5} and dark subapical spots on the fore femora only.

Bactrocera (Bactrocera) speculifera (Walker)

(Fig. 178)

Dacus speculifer Walker, 1865: 122; Norrbom *et al.*, 1998: 89, as synonym of *Bactrocera curvifera*. Lectotype male in BMNH.
Dacus (Neodacus) speculifer – Hardy, 1959: 168, 182, as synonym of *B. curvifera* and Lectotype designation by inference of Holotype.

Type material examined: OLD NEW GUINEA: Lectotype male with the following five labels – '62.91'; 'male type'; 'New Guinea, A.R. Wallace, Bt of Stevens, B.M. 1862-91'; 'Lectotype male of *Dacus speculifer* Walker, 1865, Des. D.E. Hardy 1959, New Guinea'; '*Dacus (Neodacus)* = *curvifer* Walker, det 1954, D.E. Hardy'. Paralectotype: 1 male with the following two labels – 'New Guinea, W.W. Saunders, B.M. 1868-4'; 'Paralectotype male of *Dacus speculifer* Walker, 1865, Des D.E. Hardy 1959, New Guinea'. Received on loan from BMNH.
Non-type material examined: INDONESIA: WEST PAPUA (formerly part of IRIAN JAYA) – 5 males, Papua, West Papua, Manokwari, Wosi, 21.vi.2007, coll. Caroline, attracted to methyl eugenol in forest. **PAPUA NEW GUINEA**: 2 males (17.v.1999), 21 males (29.x.1998), PNG, Central Province, Galley Reach Ranch, Hiri Tano Highway, coll. D. Tenakanai, Site P020, attracted to methyl eugenol; 1 female, PNG, Central Province, Botanical Gardens, 29.x.1998, coll. D. Tenakanai, attracted to methyl eugenol, Site P011; 1 male (30.vi.1999), 2 males (4.viii.1999), PNG, E New Britain Province, Bainings Mts, coll. L. Leblanc *et al.*, Site P209, attracted to methyl eugenol; 3 males, PNG, East New Britain, Keravat L.A.E.S., 24.vi.1999, coll. L. Leblanc *et al.*, Site P202, attracted to methyl eugenol; 1 male, PNG, East New Britain Prov, LAES Keravat, Fidelis Hela house, 22.i.1999, Site P202, coll. L. Leblanc & A. Mararuai, attracted to methyl eugenol; 1 male, PNG, E New Britain Prov, Keravat CIS, Prison Guava Orchard, 14.ii.2001, Site P213, Coll. A. Mararuai, attracted to methyl eugenol; 1 female, PNG, Madang Province, Ramu Sugar Residential Area, 6.i.1999, Site P411, attracted to methyl eugenol; 3 males, PNG, Morobe Province, Bubia Ag Exp Centre, 9.vii.1999, coll. S. Sar & S. Balagawi, Site P401, attracted to methyl eugenol; 8 males, PNG, Morobe Province, Bulolo Forestry Uni, 31.viii.1999, coll. S. Sar & S. Balagawi, Site P404, attracted to methyl eugenol; 1 male (20.vi.2000), 1 male (20.vii.2000), PNG, E. New Britain Prov, Vudal Settlement, Site P230, coll. Kalu Namun, attracted to methyl eugenol; 1 female, PNG, Madang Province, Brahman High School, 6.vi.1999, Site P412, attracted to methyl eugenol. Specimens in QDPC. 1 male, 'Dacus speculifer Wlk, Det. G.A.K. Marshall'; 'D.210'; 'New Britain, Rabaul, 1.iv.1934, J.L. Froggatt'; 'Pres by Imp Inst Ent, B.M. 1937-55'; 1 male, 'Terr. Papua & New Guinea, Maprik, 17.x.1957, J. Smart'; 'Brit. Mus., 1957-693'; 3 males, 'Terr. Papua & New Guinea, Maprik, 28.x.1957, J. Smart'; 'Brit. Mus., 1957-693', one of these specimens has the following label – '*Dacus (Neodacus) curvifer* Walk, Det. 1969, D.E. Hardy'. Received on loan from BMNH.
Diagnosis: A large species; face fulvous with a pair of small irregularly circular black spots; postpronotal lobes and notopleura yellow; scutum entirely black; lateral postsutural yellow vittae present; medial postsutural yellow vitta absent; no yellow spot anterior to notopleural suture; mesopleural stripe reaching anterior *npl.* seta dorsally; scutellum yellow; wing with cells bc and c fuscous (cell c slightly paler), microtrichia in outer corner of cell c only; almost the entire membrane covered in an irregularly shaped dark fuscous pattern; a broad dark fuscous anal streak; abdominal tergum III dull black with dark fuscous posterocentrally; abdominal terga IV– V predominantly dark fuscous with a dull black medial longitudinal band over both terga (in some specimens abdominal terga III–V are entirely black).
Description: male.
Head: Height 2.1 mm. Frons – length 1.5 times breadth; fulvous with fuscous on anteromedial hump and around bases of orbital setae; orbital setae black: 1 *s.or.*, 2 *i.or.*; lunule fuscous. Ocellar triangle black. Vertex fuscous. Face fulvous with a pair of small irregularly circular black spots; length 0.74 mm. Genae fulvous, small fuscous to dark fuscous subocular spot present; black seta present. Occiput red-brown, fulvous along eye margins; occipital row with a large number of small to dark setae. Antennae with segments 1 and 2 fuscous, segment 3 red-brown with fuscous on apex and outer surface; length of segments: 0.39 mm; 0.59 mm; 1.4 mm.
Thorax: Scutum entirely black with, at most, dark red-brown behind lateral postsutural vittae. Pleural areas entirely black. Yellow markings as follows: postpronotal lobes; notopleura; mesopleural stripe reaching anterior *npl.* seta dorsally, continuing to katepisternum as a small transverse spot, anterior margin straight; anatergite (posterior apex black); anterior ½ katatergite (remainder black); two broad parallel-sided lateral postsutural vittae ending at *ia.* scta. Medial longitudinal postsutural yellow vitta absent. Postnotum black. Scutellum yellow except for narrow black basal band. Setae: *sc.* 2; *prsc.* absent; *ia.* 1; *p.sa.* 1; *a.sa.* 1; *mpl.* 1; *npl.* 2; *scp.* 4.
Legs: All leg segments entirely fulvous; mid tibiae each with an apical black spur.

Wings: Length 8.3 mm; cells bc and c fuscous (cell c slightly paler in centre); microtrichia in outer corner of cell c only; remainder of wings with a dark fuscous cell sc, a broad irregularly shaped dark fuscous pattern across centre of wing and covering almost the entire wing membrane, a broad dark fuscous anal streak; a dense aggregation of microtrichia around A_1+CuA_2; supernumerary lobe of medium development.

Abdomen: Oval; terga free; pecten present on tergum III. Tergum I and sterna I and II wider than long. Tergum I dull black centrally to dark fuscous laterally and with a narrow transverse fulvous band across posterior margin but not reaching lateral margins; tergum II fulvous with a narrow transverse dull black band subanteriorly but not reaching narrow dark fuscous anterolateral corners; terga III–V variable in colour but generally tergum III dull black with dark fuscous posterocentrally, terga IV–V dark fuscous with a medium-width medial longitudinal dull black band over both terga and tending to fuscous on tergum V in some specimens (in some specimens terga III–V are entirely black). A pair of oval dark fuscous to black shining spots on tergum V. Posterior lobe of surstylus short, sternum V with a deep concavity on posterior margin. Abdominal sterna dark.

Attractant: Methyl eugenol.

Female: As for male except no dense aggregation of microtrichia around A_1+CuA_2; supernumerary lobe weak; pecten of cilia on abdominal tergum III absent; ovipositor basal segment dark fuscous, conical in transverse cross section and tapering posteriorly in dorsal view; ratio of length of oviscape to length of tergum V, 1.5:1; apex of piercer needle-shaped.

Distribution:

Type locality: New Guinea (it is uncertain where this locality is within Old New Guinea).

Previously recorded: Type locality and the following islands of Papua New Guinea (Samarai I., Goodenough I., New Britain, New Ireland, Lihir I.) (listed under *B. curvifera* Drew, 1989).

New records: Indonesia (West Papua, formerly part of Irian Jaya).

Hosts: *Artocarpus altilis* (family *Moraceae*). This record is due to the withdrawal of *B. speculifera* from synonymy with *B. curvifera*. Record in SPC (2001).

Remarks: *Bactrocera (Bactrocera) speculifera* (Walker) is here deemed to be a good species and withdrawn from synonymy with *Bactrocera (Bactrocera) curvifera* (Walker). See 'Remarks' under *B. curvifera* for a detailed discussion on this decision. While both species are similar, they each possess a distinct wing colour pattern upon which they can be separated (Figs 56B, 56C). Further, while abdominal terga III–V are uniformly black in *B. curvifera*, in *B. speculifera* they vary from dull black on tergum III to dark fuscous with a hint of a dull black medial longitudinal band on terga IV and V. In Drew (1989), the description of *B. curvifera* now refers to *B. speculifera*. These species are clearly allopatric with *B. curvifera* restricted to Wallacea (Ambon I. and Maluku) and *B. speculifera* to West Papua (formerly part of Irian Jaya), Papua New Guinea and eastern islands close to Papua New Guinea.

Bactrocera (Bactrocera) sulawesiae Drew & Hancock

(Fig. 179)

Bactrocera (Bactrocera) sulawesiae Drew & Hancock, 1994a: 59; Norrbom *et al*., 1998: 95. Holotype male in BBM.

Non-type material examined: No new specimens available for study since the publication of Drew & Hancock (1994a).

Diagnosis: Face fulvous with a pair of large oval black spots; postpronotal lobes and notopleura yellow; scutum entirely black without pale markings; narrow triangular lateral postsutural yellow vittae narrowing sharply posteriorly to end before level of *p.sa*. setae; medial postsutural yellow vitta absent; mesopleural stripe reaching midway between anterior margin of notopleuron and anterior *npl*. seta dorsally; scutellum yellow; legs with femora fulvous except for a large dark fuscous spot on outer apical ½ of fore femora and dark fuscous around apical ¼ of mid and hind femora; tibiae dark fuscous; wing with cells bc and c colourless, microtrichia in outer corner of cell c only; a narrow fuscous costal band overlapping R_{2+3} where it is paler and widening beyond apex of R_{2+3} to end beyond extremity of R_{4+5}; a broad fuscous anal streak; supernumerary lobe of medium development; abdominal terga III–V orange-brown except for a narrow transverse dark fuscous band across anterior margin of tergum III which widens to cover lateral ¼, a narrow medial longitudinal fuscous band over all three terga, a pair of oval fuscous shining spots on tergum V; abdominal sterna dark coloured.

Attractant: Methyl eugenol.

Female: No known record.

Distribution:

Type locality: Indonesia (Sulawesi, Bententingi).

Hosts: No known record.

Remarks: *Bactrocera (Bactrocera) sulawesiae* Drew & Hancock, within the *dorsalis* complex, is similar to *Bactrocera (Bactrocera) collita* Drew & Hancock, *Bactrocera (Bactrocera) dorsaloides* (Hardy & Adachi), *Bactrocera (Bactrocera) makilingensis* Drew & Hancock and *Bactrocera (Bactrocera) thailandica* Drew & Hancock in possessing lateral postsutural vittae that narrow posteriorly, dark markings on apices of all femora and abdominal terga III–V with a general 'T' pattern with or without dark lateral markings. It differs from these species in having a broader costal band extending well over R_{2+3} and lacking dark lateral margins on abdominal terga IV and V.

Bactrocera (Bactrocera) suliae Drew & Romig, new species

(Fig. 180)

Material examined: Holotype male, **INDONESIA: MOLUCCAS (MALUKU): AMBON ISLAND** – Moluccas (=Maluku), middle Maluku, Suli, 13.viii.2007, coll. La Naibu, attracted to methyl eugenol in Jambu, deposited in BMNH.

Paratypes: **INDONESIA: MOLUCCAS (MALUKU): AMBON ISLAND** – 15 males (11.vi.2007), 10 males (25. vi.2007), 19 males (13.viii.2007), 9 males (20.viii.2007), same data as holotype; 2 males (10.vi.2007), 6 males (5. viii.2007), Middle Maluku, Tulehu, coll. La Naibu, attracted to methyl eugenol in Cabai; **SERAM ISLAND** – 12 males (26.vii.2007), 11 males (2.viii.2007), Maluku, Seram (Western Part), Waimital, coll. A. Harris Tualepe, attracted to methyl eugenol in Jeruk; **SULAWESI** – 2 males (11.ix.2006), 5 males (26.ix.2006), Sulawesi Utara, Bitung, Manembonembo, coll. Karel F Lala, attracted to methyl eugenol in citrus; 1 male, Sulawesi Utara, Minahasa Utara, Sarongsong, 11.ix.2006, coll. Karel F Lala, attracted to methyl eugenol in village; 5 males, Sulawesi Selatan, Takalar, Topejawa, 2.ix.2006, coll. Karel F Lala, attracted to methyl eugenol in village; 18 males, Sulawesi Selatan, Maros, Topejawa, 27.ix.2006, coll. Karel F Lala, attracted to methyl eugenol in forest; 12 males (19.ix.2006), 7 males (26.ix.2006), Gorontalo, Bone Bolango, Suwawa, Lambongo, coll. Yosias Y Huru, attracted to methyl eugenol in jackfruit; 7 males, Sulawesi Barat, Mamuju, Kaluku, Kubuloang, 30.x.2006, coll. Yan P Fachrie, attracted to methyl eugenol in forest, citrus; 4 males, Sulawesi Barat, Mamuju, Topoyo, Salugatta, 30.x.2006, coll. Yan P Fachrie, attracted to methyl eugenol in village citrus; 1 male, Sulawesi Barat, Mamuju, Rongas, Ds Sese, 11.ix.2006, coll. Yan P Fachrie, attracted to methyl eugenol in Desa (Jeruk); 3 males, Gorontalo, Bone Bolango, Suwawa, 19.ix.2006, coll. Yusbar Ismail, attracted to methyl eugenol in village Nangka; 3 males, Gorontalo, Bone Bolango, Lombongo, 9.xi.2006, coll. Yusbar Ismail, attracted to methyl eugenol in forest; 2 males, Gorontalo, Limboto, Biyonga, 5.ix.2006, coll. Yosis, attracted to methyl eugenol in village mangga; 6 males, Bone Bolango, Suwawa, Lambongo, 5.ix.2006, coll. Yosis, attracted to methyl eugenol in forest; 1 male (2.ix.2006), 2 males (17.ix.2006), Sulawesi Tengah, Donggala, Lolu, coll. Lyar, attracted to methyl eugenol in forest; 1 male, Sulawesi Tengah, Donggala, Nupabomba, 8.x.2006, coll. Lyar, attracted to methyl eugenol in forest; 1 male, Sulawesi Tengah, Donggala, Nupabomba, 10.ix.2006, coll. Rusdin, attracted to methyl eugenol in forest; 1 male (20. viii.2006), 2 males (23.ix.2006), Sulawesi Tengah, Donggala, Lolu, coll. Rusdin, attracted to methyl eugenol in mangga; 2 males, Sulawesi Tenggara, Konsel, Puosu Jaya, 30.ix.2006, coll. Tasrif, attracted to methyl eugenol in citrus; 1 male, Sulawesi Tenggara, Kota Kendari, Banggoea, 10.x.2006, coll. Tasrif, attracted to methyl eugenol in village; 1 male, Sulawesi Selatan, BKT Makasar, 26. viii.2006, coll. Sirajuddin, attracted to methyl eugenol in village; 1 male, Rantepao, 13.v.2007, coll. G. Pattemore, attracted to methyl eugenol; 2 males, 6 km E Bantimurang, coll. G. Pattemore, attracted to methyl eugenol. 40 paratypes in ANIC, 20 paratypes in BMNH, 65 paratypes in QDPC, 50 paratypes in QMIC.

Non-type material examined: INDONESIA: MOLUCCAS (MALUKU): AMBON ISLAND – 3 males, Hila/Mamua, Ambon, Indonesia, 25–100m, 5.viii.1981, coll. J. Turner, residual forest with clove, nutmeg & banana groves, attracted to methyl eugenol; 7 males, Mamala, Ambon, Indonesia, 50-150m, 7.viii.1981, coll. J. Turner, residual forest with clove, nutmeg & banana groves, attracted to methyl eugenol; 3 males, Paso, Ambon, Indonesia, 20-50m, 7.vii.1981, coll. J. Turner, residual forest with clove, nutmeg and banana groves, attracted to methyl eugenol; 2 males, Waai, Ambon, Indonesia, <20m, 6.vii.1981, coll. J. Turner, residual forest with clove, nutmeg and banana groves, attracted to methyl eugenol; **SULAWESI** – 4 males, Ujung Pandang, 4.v.1982, coll. Andi. Specimens in QDPC.

Diagnosis: A medium-sized species; face fulvous with a pair of medium-sized circular black spots; postpronotal lobes and notopleura yellow; scutum red-brown with irregularly shaped dark fuscous to black patterns on midline over anterior ⅔ to ¾; lateral postsutural yellow vittae present; medial postsutural yellow vitta absent; no yellow spot anterior to notopleural suture; mesopleural stripe reaching midway between anterior margin of notopleuron and anterior *npl.* seta dorsally; scutellum yellow; wing with cells bc and c colourless, microtrichia in outer corner of cell c only; a narrow fuscous to dark fuscous costal band very slightly overlapping R_{2+3} and remaining narrow around apex of wing; a narrow fuscous anal streak; abdominal terga III–V orange-brown with at most a dark 'T' pattern consisting of a narrow transverse black band across anterior margin of tergum III and a very narrow medial longitudinal dark fuscous to black band over all three terga but generally broken on intersegmental lines, very narrow fuscous to dark fuscous anterolateral corners on terga IV and V.

Description: male.

Head: Height 1.76 mm. Frons – length 1.41 times breadth; red-brown without dark markings; orbital setae black: 1 *s.or.*, 2 *i.or.*; lunule pale fuscous. Ocellar triangle black. Vertex red-brown. Face fulvous with a pair of medium-sized circular black spots; length 0.48 mm. Genae red-brown, no dark subocular spot; dark red-brown seta present. Occiput red-brown, fulvous along eye margins; occipital row with 3–6 strong black setae. Antennae with segments 1 and 2 red-brown, segment 3 red-brown with fuscous on apex and outer surface; length of segments: 0.17 mm; 0.29 mm; 0.79 mm.

Thorax: Scutum red-brown with broad dark fuscous to black irregularly shaped markings centrally over anterior ⅔ to ¾. Pleural areas dark fuscous to black except red-brown below postpronotal lobes, above hind coxae and posterior to mesopleural stripe. Yellow markings as follows: postpronotal lobes; notopleura; mesopleural stripe reaching midway between anterior margin of notopleuron and anterior *npl.* seta dorsally, continuing to katepisternum as a transverse spot, anterior margin convex; anatergite (posterior apex black); anterior ¾ katatergite (remainder black); two narrow- to medium-width lateral postsutural vittae narrowing slightly posteriorly to end before *ia.* seta (on occasional specimens these vittae are more sub-parallel and reaching to *ia.* seta). Medial longitudinal postsutural yellow vitta absent. Postnotum black. Scutellum yellow except for narrow red-brown basal band. Setae: *sc.* 2; *prsc.* 2; *ia.* 1; *p.sa.* 1; *a.sa.* 1; *mpl.* 1; *npl.* 2; *scp.* 4.

Legs: All segments entirely fulvous except hind tibiae dark fuscous; mid tibiae each with an apical black spur.

Wings: Length 6.1 mm; cells bc and c colourless; microtrichia in outer corner of cell c only; remainder of wings colourless except fuscous cell sc, narrow fuscous costal band slightly overlapping R_{2+3} where it is paler and remaining narrow around apex of wing to end beyond extremity of R_{4+5} (the costal band tends to dark fuscous from apex of R_{2+3} to apex of wing), a narrow fuscous anal streak; a dense aggregation of microtrichia around A_1+CuA_2; supernumerary lobe of medium development.

Abdomen: Oval; terga free; pecten present on tergum III. Tergum I and sterna I and II wider than long. Tergum I dark fuscous with two narrow transverse orange-brown spots on posterior margin; tergum II entirely orange-brown; terga III–V orange-brown with a dark 'T' pattern consisting of a narrow transverse black band across anterior margin and widening across lateral margins where it becomes dark fuscous, a narrow medial longitudinal dark fuscous to black band over all three terga but generally broken on intersegmental lines, narrow dark fuscous anterolateral corners on tergum IV and very narrow fuscous anterolateral corners on tergum V (in most specimens abdominal terga III–V are entirely orange-brown with virtually no dark patterns). A pair of oval orange-brown shining spots on tergum V. Posterior lobe of surstylus short, sternum V with a deep concavity on posterior margin. Abdominal sterna orange-brown.

Attractant: Methyl eugenol.

Female: No known record.

Distribution: Indonesia – Moluccas = Maluku (Ambon I.; Seram I.); Sulawesi.

Hosts: No known record.

Remarks: *Bactrocera (Bactrocera) suliae* n.sp. is similar to *Bactrocera (Bactrocera) affinibancroftii* n.sp., *Bactrocera (Bactrocera) brunneola* White and Tsuruta and *Bactrocera (Bactrocera) paraosbeckiae* Drew in possessing an entire costal band overlapping R_{2+3}, a scutum with base colour red-brown and cells bc and c colourless. It differs from *B. affinibancroftii* in having a general dark 'T' pattern on abdominal terga III–V, from *B. paraosbeckiae* in having narrow lateral dark margins on abdominal terga III–V and from *B. brunneola* in having dark fuscous to black patterns on the scutum and orange-brown abdominal sterna.

Bactrocera (Bactrocera) sumbawaensis Drew & Hancock

(Fig. 181)

Bactrocera (Bactrocera) sumbawaensis Drew & Hancock, 1994a: 60; Norrbom *et al.*, 1998: 95. Holotype male in BBM.

Non-type material examined: Since the publication of Drew & Hancock (1994a), the following specimens have been studied – **INDONESIA: SUMATRA** – 27 males, Aceh, Aceh Besar, Lam Ara Tunong, 9.viii.2006, coll. Kasman, attracted to cue lure in mangga; 10 males (14.xi.2006), 3 males (21.xi.2006), 10 males (28.xi.2006), Aceh, Aceh Besar, Tumbo Baro, coll. Kasman, attracted to cue lure in forest. Specimens in QDPC.

Diagnosis: Face fulvous with a pair of large oval black spots; postpronotal lobes and notopleura yellow; scutum entirely black without pale markings; narrow triangular lateral postsutural yellow vittae narrowing sharply posteriorly to end before level of *p.sa.* setae; medial postsutural yellow vitta absent; mesopleural stripe slightly wider than notopleuron dorsally; scutellum yellow; legs with femora entirely fulvous; fore tibiae fuscous, mid tibiae fuscous basally to fulvous apically, hind tibiae dark fuscous; wing with cells bc and c colourless, microtrichia in outer corner of cell c only; a very narrow fuscous costal band confluent with R_{2+3} and remaining narrow around apex of wing to end between extremities of R_{4+5} and M; a narrow fuscous anal streak; supernumerary lobe of medium development; abdominal terga III–V orange-brown with a moderately broad medial and two broad lateral longitudinal black bands over all three terga and joining across tergum III to cover most of this tergum, a pair of oval dark fuscous to black shining spots on tergum V; abdominal sterna dark coloured.

Attractant: Cue lure.

Female: No known record.

Distribution:

Type locality: Indonesia (Sumbawa, Surading).

Previously recorded: Type locality and other locations in Sumbawa, Indonesia (Drew & Hancock, 1994a).

New records: Indonesia (Sumatra–Aceh).

Hosts: No known record.

Remarks: *Bactrocera (Bactrocera) sumbawaensis* Drew & Hancock, within the *dorsalis* complex, is similar to *Bactrocera (Bactrocera) merapiensis* Drew & Hancock, *Bactrocera (Bactrocera) osbeckiae* Drew & Hancock and *Bactrocera (Bactrocera) thailandica* Drew & Hancock in possessing lateral postsutural vittae that narrow posteriorly, femora entirely fulvous or with, at most, a dark preapical spot on fore femora only, costal band confluent with R_{2+3} and abdominal terga III–V with a distinct 'T' pattern with narrow to broad lateral dark margins. It differs from these species in having the lateral postsutural vittae short and triangular and the shining spots on abdominal tergum V black.

Bactrocera (Bactrocera) syzygii White & Tsuruta

(Fig. 182)

Bactrocera (Bactrocera) syzygii White & Tsuruta, in Tsuruta & White, 2001: 85. Holotype female in HORDI.

Non-type material examined: SRI LANKA: one female, with the following label – 'Penideniya, 21.vi.1994, ex *Syzygium jambos*'. This specimen bears the same label data as the type series in Tsuruta & White (2001) but does not have a type label so cannot be considered as a type. Specimen donated by Kenji Tsuruta and in QDPC.

Diagnosis: A small species; face fulvous with a pair of large elongate oval black spots; postpronotal lobes and notopleura yellow; scutum black without pale markings; lateral postsutural yellow vittae present; medial postsutural yellow vitta absent; no yellow spot anterior to notopleural suture; mesopleural stripe equal in width to notopleuron dorsally; scutellum yellow; wing with cells bc and c colourless, microtrichia in outer corner of cell c only; a narrow fuscous costal band; a narrow fuscous anal streak; abdominal terga III–V red-brown with a medium-width medial and two broad lateral longitudinal dark fuscous to black bands joined along anterior ⅓ of tergum III.

Description: female.

Head: Height 1.36 mm. Frons – length 1.65 times breadth; red-brown with fuscous on anteromedial hump and around bases of orbital setae; orbital setae black: 1 *s.or.*, 2 *i.or.*; lunule fuscous. Ocellar triangle black. Vertex dark fuscous. Face fulvous with a pair of large elongate oval black spots; length 0.37 mm. Genae fulvous, dark fuscous subocular spot present; black seta present. Occiput black, fulvous along eye margins; occipital row with seven large black setae. Antennae with segments 1 and 2 red-brown, segment 3 red-brown with dark fuscous on apex and outer surface; length of segments: 0.16 mm; 0.22 mm; 0.71 mm.

Thorax: Scutum black without pale markings. Pleural areas black except dark fuscous below postpronotal lobes. Yellow markings as follows: postpronotal lobes; notopleura; mesopleural stripe equal in width to notopleuron dorsally, continuing to katepisternum as a transverse spot, anterior margin slightly convex; anatergite (posterior apex black); anterior ⅔ katatergite (remainder black); two medium-width parallel-sided lateral postsutural vittae ending just before *ia.* seta. Medial longitudinal postsutural yellow vitta absent. Postnotum black. Scutellum yellow except for narrow black basal band. Setae: *sc.* 2; *prsc.* 2; *ia.* 1; *p.sa.* 1; *a.sa.* 1; *mpl.* 1; *npl.* 2; *scp.* 4.

Legs: All femora entirely fulvous; fore tibiae dark fuscous, mid tibiae dark fuscous on basal ¾ tending fuscous apically, hind tibiae dark fuscous; tarsi with all segments fulvous; mid tibiae each with an apical black spur.

Wings: Length 4.8 mm; cells bc and c colourless; microtrichia in outer corner of cell c only; remainder of wings colourless except dark fuscous cell sc, narrow fuscous costal band confluent with R_{2+3} and remaining narrow around apex of wing to end beyond apex of R_{4+5}, a narrow fuscous anal streak; no dense aggregation of microtrichia around A_1+CuA_2; supernumerary lobe weak.

Abdomen: Oval; terga free; no pecten on tergum III. Tergum I and sterna I and II wider than long. Tergum I dark fuscous to black; tergum II red-brown with a narrow transverse dark fuscous to black band subanteriorly and reaching narrow dark fuscous to black lateral margins; terga III–V red-brown with a moderately broad medial and two broad lateral longitudinal dark fuscous to black bands joined along anterior ⅓ of tergum III. A pair of oval dark fuscous to black shining spots on tergum V. Ovipositor basal segment black, dorsoventrally compressed and tapering posteriorly in dorsal view. Ratio of length of oviscape to length of tergum V, 0.6:1. Apex of piercer needle-shaped. Abdominal sterna black.

Attractant: No known record.

Male: As for female except a dense aggregation of microtrichia around A_1+CuA_2 present; supernumerary lobe of medium development; pecten of cilia on abdominal tergum III present; posterior lobe of surstylus short; abdominal sternum V with a deep concavity on posterior margin.

Distribution:

Type locality: Sri Lanka (Kandy District, Penideniya).

Hosts: *Syzygium jambos* (family *Myrtaceae*) (Tsuruta & White, 2001).

Remarks: *Bactrocera (Bactrocera) syzygii* White & Tsuruta, within the *dorsalis* complex, is similar to *Bactrocera (Bactrocera) hantanae* Tsuruta & White, *Bactrocera (Bactrocera) irvingiae* Drew & Hancock and some *Bactrocera (Bactrocera) raiensis* Drew & Hancock possessing broad parallel-sided lateral postsutural vittae, costal band confluent with R_{2+3} and not expanded markedly at apex, femora entirely fulvous and abdominal terga III–V with a broad medial longitudinal dark band. It differs from *B. hantanae* and *B. irvingiae* in having broad lateral dark markings on abdominal terga III–V and from *B. raiensis* in having dark fuscous to black shining spots on abdominal tergum V.

Bactrocera (Bactrocera) tapahensis Drew & Romig, new species

(Fig. 183)

Material examined: Holotype male, **PENINSULAR MALAYSIA**: TS Gombak, Malaysia, 12.ii.1986, coll. Bahari, attracted to methyl eugenol, deposited in BMNH.

Paratypes: **PENINSULAR MALAYSIA**: 2 males, TS Gombak, Malaysia, 8.iv.1986, coll. Bahari, attracted to methyl eugenol; 5 males (1.vii.1986), 2 males (14.vii.1986), 3 males (28.vii.1986), 1 male (17.xi.1986), TS7 Malaysia, coll. Bahari, attracted to methyl eugenol; 1 male (1.vii.1986), 1 male (17.xi.1986), TS8 Malaysia, coll. Bahari, attracted to methyl eugenol; 1 male (30.vi.1986), 2 males (12.i.1987), 1 male (26.i.1987), 1 male (1.vi.1987), TN8 Malaysia, coll. Yusoff Saleh, attracted to methyl eugenol; 1 male (30.vi.1986), 3 males (11.viii.1986), 3 males (25.viii.1986), 3 males (20.x.1986), 1 male (29.vi.1987), 3 males (13.vii.1987), 1 male (27.vii.1987), TN9 Malaysia, coll. Yusoff Salleh, attracted to methyl eugenol; 1 male, TS Gombak, Malaysia, 19.vi.1986, coll. S. Vijaysegaran, attracted to methyl eugenol; 1 male, Tapah, 18th mile, Malaysia, trap 3, 24.vi.1983, coll. S. Vijaysegaran, attracted to methyl eugenol; 2 males, Cameron Hlds, Malaysia, Tapah, 18th mile, trap 3, 7.vii.1983, coll. S. Vijaysegaran, attracted to methyl eugenol; 2 males, TN9 Malaysia, 28.vii.1986, coll. Ahmad Radzuan & Panjang Abd, attracted to methyl eugenol; 1 male, TS Gombak, Malaysia, 28.i.1986, coll. Antony, attracted to methyl eugenol; 2 males, TS10 Bentong, Malaysia, 23.iii.1987, coll. Antony, attracted to methyl eugenol; 1 male, TS8 Gombak, Malaysia, 26.i.1987, coll. Mahdzir, attracted to methyl eugenol; 1 male, F.E.S. Gombak, Selangor, Malaysia, ii.1987, coll. C.S. Ooi, attracted to methyl eugenol; 1 male, RS Cameron Hlds 2, Malaysia, 30.vi.1986, coll. Z. Davies; 1 male, Malaysia, Sel., Hulu Langat, Pangsun, 2.ii.1989, coll. YHS, attracted to methyl eugenol. 10 paratypes in ANIC, 8 paratypes in BMNH, 20 paratypes in QDPC, 10 paratypes in QMIC.

Diagnosis: A medium-sized species; face red-brown to fuscous centrally, fulvous along lateral margins and with a pair of medium-sized oval black spots; postpronotal lobes and notopleura yellow; scutum black with red-brown patterns; lateral postsutural yellow vittae present; medial postsutural yellow vitta absent; no yellow spot anterior to notopleural suture; mesopleural stripe reaching midway between anterior margin of notopleuron and anterior *npl.* seta dorsally; scutellum yellow; wing with cells bc and c fuscous to dark fuscous, dense microtrichia covering all of cell c and anterior ½ of cell bc; a narrow dark fuscous costal band expanding into a spot apically; a narrow fuscous anal streak; abdominal terga III–V fulvous with a broad medial longitudinal black band and two moderately broad lateral longitudinal dark fuscous to black bands over all three terga and joined along anterior margin of tergum III.

Description: male.

Head: Height 1.5 mm. Frons – length 1.5 times breadth; red-brown with fuscous on anteromedial hump; orbital setae black: 1 *s.or.*, 2 *i.or.*; lunule fuscous. Ocellar triangle black. Vertex fuscous. Face red-brown centrally, fulvous along lateral margins, with a pair of medium-sized oval black spots; length 0.49 mm. Genae red-brown, large dark fuscous subocular spot present; black seta present. Occiput fuscous, fulvous along eye margins; occipital row with 4–6 black setae. Antennae with segments 1 and 2 dark fuscous, segment 3 red-brown with dark fuscous on apex and outer surface; length of segments: 0.18 mm; 0.3 mm; 0.78 mm.

Thorax: Scutum black with red-brown along lateral margins of lateral postsutural vittae, around notopleural suture, inside postpronotal lobes, a broad medial longitudinal band from centre of scutum to anterior margin. Pleural areas entirely black except dark fuscous below postpronotal lobes. Yellow markings as follows: postpronotal lobes; notopleura; mesopleural stripe reaching midway between anterior margin of notopleuron and anterior *npl.* seta dorsally, continuing to katepisternum as a large transverse spot, anterior margin straight; anatergite (posterior apex black); anterior ⅔ katatergite (remainder black); two very broad lateral postsutural vittae narrowing slightly posteriorly to end before *ia.* seta. Medial longitudinal postsutural yellow vitta absent. Postnotum black, tending red-brown centrally. Scutellum yellow except for narrow black basal band. Setae: *sc.* 2; *prsc.* 2; *ia.* 1; *p.sa.* 1; *a.sa.* 1; *mpl.* 1; *npl.* 2; *scp.* 4.

Legs: All femora entirely fulvous except for a small dark fuscous subapical spot on outer surfaces of fore femora; fore tibiae fuscous, mid tibiae dark fuscous basally to fuscous apically, hind tibiae dark fuscous; fore tarsi with basal segment fulvous and apical four segments fuscous, mid and hind tarsi with basal two segments fulvous and apical three segments fuscous; mid tibiae each with an apical black spur.

Wings: Length 5.8 mm; cells bc and c fuscous to dark fuscous; dense microtrichia covering all of cell c and anterior ½ of cell bc; remainder of wings colourless except dark fuscous cell sc, narrow dark fuscous costal band confluent with R_{2+3} and expanding into a large spot across apex of R_{4+5}, a narrow fuscous anal streak; a dense aggregation of microtrichia around A_1+CuA_2; supernumerary lobe of medium development.

Abdomen: Oval; terga free; pecten present on tergum III. Tergum I and sterna I and II wider than long. Tergum I dark fuscous with a narrow fulvous band along posterior margin but not reaching lateral margins; tergum II fulvous with a narrow transverse dark fuscous band subanteriorly but not reaching narrow dark fuscous lateral margins; terga III–V fulvous with a broad medial longitudinal black band and moderately broad lateral longitudinal black to dark fuscous bands over all three terga and joined along anterior margin of tergum III. A pair of oval red-brown shining spots on tergum V. Posterior lobe of surstylus short, sternum V with a deep concavity on posterior margin. Abdominal sterna dark coloured.

Attractant: Methyl eugenol.

Female: No known record.

Distribution: Peninsular Malaysia.

Hosts: No known record.

Remarks: *Bactrocera (Bactrocera) tapahensis* n.sp. is similar to *Bactrocera (Bactrocera) mediorufula* n.sp. and *Bactrocera (Bactrocera) pusilla* (Hardy) in possessing a narrow costal band confluent with or just overlapping R_{2+3}, a black scutum and cells bc and c fuscous to dark fuscous. It differs from *B. pusilla* in lacking a lateral yellow band between the postpronotal lobe and notopleuron and from *B. mediorufula* in having the costal band expanded into a distinct spot at the apex of the wing and a dark fuscous spot on the outer apical surfaces of the fore femora.

Bactrocera (Bactrocera) ternatiae Drew & Romig, new species

(Fig. 184)

Material examined: Holotype male, **INDONESIA: NORTH MOLUCCAS (NORTH MALUKU)** – North Maluku, Kodya Ternate, Foramadiahi, 27.v.2007, coll. Raais Abdullah, attracted to methyl eugenol in forest, deposited in BMNH.

Paratypes: **INDONESIA: NORTH MOLUCCAS (NORTH MALUKU)** – 3 males (29.v.2007), 3 males (12.vi.2007), 1 male (19.vi.2007), North Maluku, West Halmahera Gaol, coll. Raais Abdullah, attracted to methyl eugenol in forest; 2 males (27.v.2007), 2 males (10.vi.2007), 2 males (17. vi.2007), North Maluku, Kodya Ternate, Gambesi, coll. La Ruti, attracted to methyl eugenol in jeruk; 2 males, North Maluku, Kodya Ternate, Kalumata, 10.vi.2007, coll. La Ruti, attracted to methyl eugenol in mangga. 3 paratypes in ANIC, 3 paratypes in BMNH, 7 paratypes in QDPC, 2 paratypes in QMIC.

Diagnosis: A large species; face fulvous with a pair of medium-sized circular black spots; postpronotal lobes and notopleura yellow; scutum entirely black except dark fuscous below lateral postsutural vittae; lateral postsutural yellow vittae present; medial postsutural yellow vitta absent; no yellow spot anterior to notopleural suture; broad mesopleural stripe reaching postpronotal lobe dorsally; scutellum yellow; wing with cells bc and c fuscous, microtrichia covering all of cell c and most of cell bc; a broad dark fuscous costal band confluent with R_{4+5} and three transverse dark fuscous bands across wing from costal band to hind margin; a broad dark fuscous anal streak; abdominal terga III–V entirely dark fuscous to black with a broad medial longitudinal black band over all three terga.

Description: male.

Head: Height 1.98 mm. Frons – length 1.67 times breadth; red-brown with fuscous on anteromedial hump and around bases of orbital setae; orbital setae black: 1 *s.or.*, 2 *i.or.*; lunule red-brown. Ocellar triangle black. Vertex red-brown. Face fulvous with a pair of medium-sized circular black spots; length 0.48 mm. Genae red-brown, small dark fuscous subocular spot present; black seta present. Occiput fuscous, fulvous along eye margins; occipital row with 6–9 strong black setae. Antennae with segments 1 and 2 red-brown, segment 3 red-brown with fuscous on apex and outer surface; length of segments: 0.24 mm; 0.45 mm; 0.96 mm.

Thorax: Scutum black except dark fuscous below lateral postsutural vittae. Pleural areas entirely black except dark fuscous below postpronotal lobes. Yellow markings as follows: postpronotal lobes; notopleura; broad mesopleural stripe reaching postpronotal lobe dorsally, continuing to katepisternum as a small transverse spot, anterior margin straight; anatergite (posterior apex black); anterior ⅔ katatergite (remainder black); two moderately broad lateral postsutural vittae narrowing slightly posteriorly to end just before *ia.* seta. Medial longitudinal postsutural yellow vitta absent. Postnotum black. Scutellum yellow except for narrow black basal band. Setae: *sc.* 2; *prsc.* 2; *ia.* 1; *p.sa.* 1; *a.sa.* 1; *mpl.* 1; *npl.* 2; *scp.* 4.

Legs: All femora fulvous except red-brown on apical ⅓ to ½ of fore and mid femora and apical ¼ of hind femora; fore and mid tibiae fulvous, hind tibiae red-brown; fore tarsi with basal segment fulvous and apical four segments red-brown, mid and hind tarsi with all segments fulvous; mid tibiae each with an apical black spur.

Wings: Length 7.5 mm; cells bc and c fuscous; dense microtrichia covering all of cell c and most of cell bc; remainder of wings colourless except for dark fuscous cell sc, a broad dark fuscous costal band confluent with R_{4+5}, three transverse dark fuscous bands across wing as follows: a narrow band at base running from costal band to anal streak, a broad band across centre of wing and expanding over apex of cell m and running from costal band to hind margin and enclosing both crossveins, a narrow transverse dark fuscous to fuscous band in apex of wing running from costal band to hind margin, a broad dark fuscous anal streak; a dense aggregation of microtrichia around A_1+CuA_2; supernumerary lobe of medium development.

Abdomen: Oval; terga free; pecten present on tergum III. Tergum I and sterna I and II wider than long. Tergum I black with a narrow transverse red-brown band across posterior margin but not reaching lateral margins; tergum II black anteriorly and over lateral margins, fulvous posteriorly; terga III–V entirely dark fuscous to black with a broad medial longitudinal black band over all three terga. A pair of dark red-brown shining spots on tergum V. Posterior lobe of surstylus short, sternum V with a deep concavity on posterior margin. Abdominal sterna dark fuscous to black.

Attractant: Methyl eugenol.

Female: No known record.

Distribution: Indonesia (North Moluccas = North Maluku – Ternate; West Halmahera).

Hosts: No known record.

Remarks: *Bactrocera (Bactrocera) ternatiae* n.sp. is similar to *Bactrocera (Bactrocera) bifasciata* (Hardy) and *Bactrocera (Bactrocera) umbrosa* (Fabricius) in possessing transverse dark bands across the wing and colour in cells bc and c. It differs from *B. bifasciata* in having three transverse dark bands across the wing and from *B. umbrosa* in having the dark patterns on the wing dark fuscous, not red-brown, and abdominal terga III–V dark fuscous to black with a broad medial longitudinal black band over all three terga. Also, the middle transverse band across the wing of *B. ternatiae* is not parallel-sided and widens across the distal ½ of cell dm. *Bactrocera (Bactrocera) fulvicauda* (Perkins) in Papua New Guinea is similar to *B. ternatiae* in the colour patterns of the thorax and abdomen but possesses only one transverse fuscous band across the wing.

Bactrocera (Bactrocera) thailandica Drew & Hancock

(Fig. 185)

Bactrocera (Bactrocera) thailandica Drew & Hancock, 1994a: 61; Norrbom *et al.*, 1998: 95; Liu *et al.*, 2009: 661. Holotype female in BMNH.

Bactrocera (Bactrocera) ruiliensis Wang, Long & Zhang, in Wang *et al.*, 2008: 74. **New Synonym.**

Bactrocera (Bactrocera) hardyi Zhang, Ji & Chen, in Zhang *et al.*, 2011: 600. Holotype male in FAFU. **New Synonym.**

Other type material studied: CHINA: A study of the colour photographs and description of *Bactrocera (Bactrocera) hardyi* Zhang, Ji & Chen in the original publication.

Non-type material examined: Since the publication of Drew & Hancock (1994a), the following specimens have been studied – **BHUTAN**: 6 males, Ganglephug, Tatapani, 9-13.ix.1990, coll. H.R. Fiejen; 1 male, Rontong, 2.x.1991, coll. H.R. Fiejen; 3 males, Rimchu 1, 7.vi.2000, coll. Dorji *et al.*; 8 males, Rimchu 2, 7.vi.2000, coll. Dorji, Nado, attracted to cue lure; 1 male (17.v.2000), 1 male (28.vi.2000), 1 male (12.vii.2000), Rimchu 2, coll. Dorji; 1 male, Purana Bastey, 2.xi.2004, coll. Maha Prasard, Karma Nangyel; 4 males, Lumitshawa, 7.vi.2000, coll. Dorji *et al.*; 4 males, Phuentsholing, 10km, 6.v.2000, coll. S. Thapa; 1 male (13.v.2000), 10 males (20.v.2000), Phuentsholing, coll. S. Thapa; 1 male, Namling 11km, 9.vi.2001, coll. Brian Fletcher, altitude 2000m, broad leafed forest; 1 male, Namling 14km, 7.vi.2001, coll. Brian Fletcher, altitude 1800m, subtropical forest; 1 male, Lingmethang 11km, 9. vi.2001, coll. Brian Fletcher, altitude 1100m, subtropical forest; 5 males, Lingmethang, 9.vi.2001, coll. Brian Fletcher, altitude 700m, mango tree; 3 males (15.xi.2004), 1 male (5.xii.2004), Sandrup Jongkhar District, coll. Wangdi; 6 males, Tsirang, Darachu-La, 30.v.2001, coll. Brian Fletcher; 6 males (19.vi.2000), 1 male (10.vii.2000), Tsirang, coll. T.D. & S.D.; 2 males, Tsirang, 17.vii.2000, coll. S.D.; 1 male, Tsirang, 7.viii.2000; 4 males, Suntalay, 5.vi.2000, coll.

S. Wangdi *et al.*; 2 males, Tsirang, Suntalay, 11.vi.2000. All above specimens attracted to cue lure. **BRUNEI**: 2 males, KBFC, Bato Apou Nat Pk, Upland Ridge, 23.i.2002, coll. K. Hurley, attracted to cue lure. **VIETNAM: Northern Vietnam** – 5 males (30.iii.2002), 4 males (15.v.2002), Son La Province, Moc Chau District, Long Luong Village, Site VN01, coll. NIPP Staff, attracted to cue lure; 1 male, Hanoi Province, Tu Liem District, Xuan Dinh, Site N004, 31. iii.2000, attracted to cue lure; 2 males, Hanoi Province, Dong Anh District, Bae Ha, Site N005, 22.x.1999, attracted to cue lure; 1 male (31.v.1999), 1 male (28.vi.1999), 1 male (20 ix.1999), Son La Province, Moc Chau District, Co Do, Site N006; 3 males, Son La Province, Moc Chau District, Co Do State Farm, FAO Site VN06, 15.iv.2002, coll. NIPP Staff, attracted to cue lure; 1 male (12.vii.1999), 1 male (9. viii.1999), 2 males (23.viii.1999), Son La Province, Moc Chau District, Tra Long, Site N007, attracted to cue lure; 1 male, Son La Province, Moc Chau District, Moc Chau house garden, Site N008, 9.viii.1999, attracted to cue lure; 15 males, Hoa Binh Province, Mai Chau District, Pa Co, Site N010, 28.ix.1999, attracted to cue lure; 1 male, Hoa Binh Province, Ky Son District, Cao Phong, Site N014, 17.x.1999, attracted to cue lure; 2 males (30.vii.1999), 6 males (27. viii.2000), Lao Cai Province, Sapa 2 Village, Site N015, attracted to cue lure; 3 males, Lao Cai Province, Sapa town, Site N016, 13.viii.1999, attracted to cue lure; 2 males (30. vii.1999), 15 males (13.viii.1999), 3 males (27.viii.1999), 5 males (10.ix.1999), Lao Cai Province, Sapa 6 village, Site N017, attracted to cue lure; 18 males (30.vii.1999), 8 males (27.viii.1999), 1 male (10.ix.1999), Lao Cai Province, Sapa 2 Village, Site N018, attracted to cue lure; 5 males (30. vii.1999), 24 males (13.viii.1999), 3 males (27.viii.1999), 9 males (10.ix.1999), Lao Cai Province, Sapa Oquy Ho, Site N019, attracted to cue lure; **Southern Vietnam** – 1 male, Cantho Province, Chau Thanh District, Tan Phu Village, Site VS02, 1.iv.2002, coll. SOFRI staff, attracted to cue lure; 1 male (16.iii.2002), 1 male (30.iii.2002), 1 male (11.v.2002), Dong Nai Province, Long Khanh District, Nhan Nghia Village, Site VS06, coll. SOFRI staff, attracted to cue lure; 1 male, Lam Dong Province, 3 District, Da Lat town, Site VS08, coll. SOFRI staff, attracted to cue lure; 1 male, Dac Lac Province, Ku Mi Nga District, Quang Phu Village, Site VS09, 8.iv.2002, coll. SOFRI staff, attracted to cue lure; 1 male, Dong Nai Province, Xuan Loc District, Xuan Phu Village, Site S015, 25.xi.1999, attracted to cue lure. Specimens in QDPC.

Diagnosis: Face fulvous with a pair of medium-sized oval black spots; postpronotal lobes and notopleura yellow; scutum entirely black without pale markings; narrow lateral postsutural yellow vittae tapering posteriorly to end before *ia.* seta; medial postsutural yellow vitta absent; mesopleural stripe equal in width to notopleuron dorsally; scutellum yellow; legs with femora fulvous with a dark fuscous subapical spot on outer surfaces of fore femora and narrow black bands around apices of mid and hind femora; fore and mid tibiae fuscous to dark fuscous, hind tibiae dark fuscous to black; wing with cells bc and c colourless, microtrichia in outer corner of cell c only; a narrow fuscous costal band confluent with R_{2+3} or just overlapping this vein where it becomes very pale and remaining narrow around apex of wing before expanding slightly across apex of R_{4+5}; a narrow pale fuscous anal streak; supernumerary lobe weak in females, of medium development in males; abdominal terga III–V dark red-brown with very broad lateral longitudinal dark fuscous to black bands and a medium-width medial longitudinal dark fuscous to black band over all three terga and joined over anterior ½ of tergum III (the colour patterns on the abdomen are quite variable, the medial longitudinal band varies in width and, in some specimens the dark fuscous to black covers most or all of tergum III and the bands may be joined along anterior margin of tergum IV), a pair of oval dark fuscous shining spots on tergum V; abdominal sterna dark coloured.

Attractant: Cue lure.

Female: As for male in the general body colour patterns. See Drew & Hancock (1994a) for a detailed description of female characters.

Distribution:

Type locality: Thailand (Haewnarok).

Previously recorded: Type locality and China (Yunnan Province) (Liu *et al.*, 2009).

New records: Bhutan; Brunei; Vietnam.

Hosts: *Elaeocarpus lanceafolius* (family *Elaeocarpaceae*) (Drew & Hancock, 1994a; Allwood *et al.*, 1999).

Remarks: *Bactrocera (Bactrocera) thailandica* Drew & Hancock, within the *dorsalis* complex, is similar to *Bactrocera (Bactrocera) collita* Drew & Hancock, *Bactrocera (Bactrocera) dorsaloides* (Hardy & Adachi) and *Bactrocera (Bactrocera) makilingensis* Drew & Hancock in possessing lateral postsutural vittae that narrow posteriorly, dark markings on apices of all femora, a narrow black basal band on the scutellum and lateral dark markings on abdominal terga III–V. It differs from these species in having abdominal tergum III entirely or almost entirely dark fuscous to black, abdominal terga IV and V with broad lateral dark bands and a narrow- to medium-width medial longitudinal dark band over abdominal terga III–V or IV–V and the dark patterns on mid and hind femora surrounding the apices (not as subapical spots). *Bactrocera thailandica* is also similar to *Bactrocera (Bactrocera) pyrifoliae* Drew & Hancock in the general colour patterns of the body, wings and legs. However, *B. pyrifoliae* has the abdomen with a unique elongate oval shape and distinct dark 'T' patterns on each of abdominal terga III–V. Also, *B. pyrifoliae* is rarely attracted to the male lure, cue lure, whereas *B. thailandica* readily responds to this lure. Each species also has different host records. Molecular studies will be required to deduce whether or not these two populations are conspecific.

Bactrocera (Bactrocera) ruiliensis Wang, Long & Zhang was originally separated from *B. thailandica* on having longer lateral postsutural yellow vittae. However, these vittae in *B. thailandica* vary in length and this character is not adequate to separate these species. Consequently, we are making *B. ruiliensis* a **New Synonym** of *B. thailandica*. Further, there are no species level differences that set *Bactrocera (Bactrocera) hardyi* Zhang, Ji & Chen aside and we are making this species a **New Synonym** of *B. thailandica*.

Bactrocera (Bactrocera) thistletoni Drew
(Fig. 186)

Bactrocera (Bactrocera) thistletoni Drew, 1989: 163; Norrbom *et al.*, 1998: 95. Holotype male in QMIC.

Non-type material examined: INDONESIA: MOLUCCAS (MALUKU) – 1 male, Western Seram, Waimital, 19.vii.2007, coll. A. Haris Tualepe, attracted to cue lure in orange (jeruk). Specimen in QDPC.

Diagnosis: Face fuscous with a pair of large elongate oval black spots; postpronotal lobes dark fuscous to black; notopleura yellow; scutum entirely black without pale markings; lateral and medial postsutural yellow vittae absent; mesopleural stripe reaching midway between anterior margin of notopleuron and anterior *npl.* seta dorsally; scutellum yellow; legs with fore and mid femora entirely black, hind femora with apical ½ black and basal ½ fulvous; fore and mid tibiae fuscous, hind tibiae dark fuscous; wing with cells bc and c colourless, microtrichia in outer corner of cell c only; a narrow dark fuscous costal band confluent with or slightly overlapping R_{2+3} and remaining narrow around apex of wing; a broad fuscous anal streak; supernumerary lobe of medium development; abdominal terga III–V black except for a narrow transverse orange-brown band across posterior margin of tergum V which extends as a narrow margin either side of a narrow medial longitudinal dark fuscous to black band over this tergum and in some specimens onto tergum IV, a pair of oval black shining spots on tergum V; abdominal sterna dark coloured.

Attractant: Cue lure.

Female: No known record.

Distribution:

Type locality: Papua New Guinea (Western Highlands District, Mt Hagen).

Previously recorded: Type locality and Papua New Guinea (Central Province and Western District).

New records: Indonesia (Moluccas = Maluku – Seram I.).

Hosts: No known record.

Remarks: *Bactrocera (Bactrocera) thistletoni* Drew has previously only been recorded from Papua New Guinea where it is similar to *Bactrocera (Bactrocera) lineata* (Perkins) in general body colour patterns but differs from this species in having dense microtrichia in only the outer corner of cell c, face dark fuscous with large elongate oval black spots, lateral postsutural yellow vittae absent and mesopleural stripes broader.

Bactrocera (Bactrocera) trifasciata (Hardy)
(Fig. 187)

Dacus trifasciatus Hardy, 1982: 237. Holotype male in MZB.

Bactrocera (Bactrocera) trifasciata – Norrbom *et al.*, 1998: 96.

Type material examined: Holotype male, **INDONESIA: SULAWESI** – Central Sulawesi 650 m, Sadaunta, 65 km S.E. of Palu, Aug. 1975, coll. D.E. & A.T. Hardy, attracted to cue lure. Studied at MZB.

Paratype: **INDONESIA: SULAWESI** – 1 male, 'Lindu Valley, C. Sulawesi, 960 m, Aug. 1975, coll. D.E. Hardy'; 'Paratype male, Dacus (Bactrocera) trifasciatus, D.E. Hardy'. Specimen in QDPC.

Non-type material examined: INDONESIA: SULAWESI – 1 male, Sulawesi, Sulawesi Tengah, Donggala Nupabomba, 17.ix.2006, coll. Rusdin, attracted to cue lure in forest. Specimen in QDPC.

Diagnosis: A medium-sized species; face fulvous with a pair of medium-sized oval black spots; postpronotal lobes and notopleura yellow; scutum orange-brown with broad lateral longitudinal black bands running from inside lateral postsutural vittae to anterior margin adjacent to inner margins of postpronotal lobes; lateral postsutural yellow vittae present; medial postsutural yellow vitta absent; a yellow spot anterior to notopleural suture; mesopleural stripe reaching almost to anterior *npl.* seta dorsally; scutellum yellow; wing with cells bc and c pale fuscous, microtrichia in outer corner to outer ½ of cell c only; a narrow pale fuscous costal band; a moderately broad pale fuscous anal streak; abdominal terga II–V orange-brown with a broad medial longitudinal black band over all four terga and two narrow sub-marginal lateral longitudinal black bands running from posterior ½ of tergum II to posterior margin of tergum V.

Description: male.

Head: Frons – entirely fulvous without dark markings; orbital setae black: 1 *s.or.*, 2 *i.or.*; lunule fulvous. Ocellar triangle black. Vertex fuscous. Face fulvous with a pair of medium-sized oval black spots. Occiput fulvous without dark markings. Antennae with all segments entirely fulvous.

Thorax: Scutum orange-brown with two broad lateral longitudinal black bands running from inside lateral postsutural vittae to anterior margin adjacent to inner surface of postpronotal lobes. Yellow markings as follows: postpronotal lobes; notopleura; mesopleural stripe reaching almost to anterior *npl.* seta dorsally; two broad parallel-sided lateral postsutural vittae beginning with a spot anterior to notopleural suture and ending at *ia.* seta. Medial longitudinal postsutural yellow vitta absent. Scutellum yellow except for narrow orange-brown basal band. Setae: *sc.* 2; *prsc.* 2; *ia.* 1; *p.sa.* 1; *a.sa.* 1; *mpl.* 1; *npl.* 2; *scp.* 4.

Legs: All segments entirely fulvous; mid tibiae each with an apical black spur.

Wings: Length 6.4 mm; cells bc and c pale fuscous; microtrichia in outer corner to outer ½ of cell c only; remainder of wings colourless except pale fuscous cell sc, narrow pale fuscous costal band confluent with R_{2+3} and remaining narrow as it extends around apex of wing (there is a slight widening of the costal band around apex of R_{4+5}), a moderately broad pale fuscous anal streak; a dense aggregation of microtrichia around A_1+CuA_2; supernumerary lobe of medium development.

Abdomen: Oval; terga free; pecten present on tergum III. Tergum I and sterna I and II wider than long. Tergum I dark fuscous to black with a narrow transverse orange-brown band across posterior margin but not reaching lateral margins; terga II–V orange-brown with a broad medial longitudinal

black band over all four terga and two narrow sub-marginal lateral longitudinal black bands running from posterior ½ of tergum II to posterior margin of tergum V. A pair of oval orange-brown shining spots on tergum V. Posterior lobe of surstylus short, sternum V with a moderately deep concavity on posterior margin.
Attractant: Cue lure.
Female: No known record.
Distribution:
Type locality: Indonesia (Central Sulawesi, Sadaunta).
Previously recorded: Type locality and Indonesia: Sulawesi – Central Sulawesi (Lindu Valley; Kuala Navusa; Tolai) (Hardy, 1982).
New records: Indonesia (Sulawesi – Central Sulawesi, Donggala).
Hosts: No known record.
Remarks: *Bactrocera (Bactrocera) trifasciata* (Hardy) is unique in South-East Asia and similar to *Bactrocera (Bactrocera) enochra* (Drew) that has been recorded from the Solomon Is and Bougainville I., in possessing three longitudinal black stripes on the abdomen, a triangular red-brown marking on the dorsal surface of the scutellum and a narrow fuscous costal band confluent with R_{2+3}. It differs from *B. enochra* in having yellow postpronotal lobes and a yellow spot anterior to the notopleural suture in front of the lateral postsutural vittae.

Bactrocera (Bactrocera) truncata Drew & Romig, new species

(Fig. 188)

Material examined: Holotype male, **INDONESIA: MOLUCCAS (MAKUKU)**: **AMBON ISLAND** – Waai, Ambon, Indonesia, <20 m, 6.vii.1981, coll. J. Turner, residual forest with clove, nutmeg and banana groves, attracted to cue lure, deposited in BMNH.
Paratypes: **INDONESIA: MOLUCCAS (MALUKU): AMBON ISLAND** – 6 males, same data as holotype; 1 male, Hila/Mamua, Ambon, Indonesia, 25–100 m, 5. viii.1981, coll. J. Turner, residual forest, attracted to cue lure; 1 male, Paso, Ambon, Indonesia, 20–50 m, 7.vii.1981, coll. J. Turner, residual forest, attracted to cue lure; 1 male, Indonesia, Moluccas, Maluku, middle Maluku, Suli, 2. vi.2007, coll. La Naibu, attracted to cue lure. 2 paratypes in ANIC; 1 paratype in BMNH; 4 paratypes in QDPC; 2 paratypes in QMIC.
Diagnosis: A medium-sized species; face fulvous with a pair of medium-sized oval black spots; postpronotal lobes fuscous; notopleura yellow; scutum black with dark red-brown to dark fuscous patterns; very short and broad lateral postsutural yellow vittae present; medial postsutural yellow vitta absent; no yellow spot anterior to notopleural suture; mesopleural stripe reaching almost to anterior *npl.* seta dorsally; scutellum yellow; wing with cells bc and c fuscous, dense microtrichia covering outer ⅓ of cell c only; a broad dark fuscous costal band confluent with R_{4+5} and becoming paler adjacent to that vein; a broad dark fuscous anal streak, a broad transverse band across wing enclosing both crossveins and running from costal band to hind margin; abdominal tergum III pale fuscous; abdominal terga IV–V red-brown with very narrow lateral dark fuscous margins; a medium-width medial longitudinal black band over terga III–V.
Description: male.
Head: Height 1.6 mm. Frons – length 1.64 times breadth; red-brown with fuscous around orbital setae; orbital setae black: 1 *s.or.*, 2 *i.or.*; lunule fuscous. Ocellar triangle black. Vertex fuscous. Face fulvous with a pair of medium-sized oval black spots; length 0.44 mm. Genae fulvous, large dark fuscous subocular spot present; black seta present. Occiput pale fuscous, fulvous along eye margins; occipital row with six black setae. Antennae with segments 1 and 2 red-brown, segment 3 red-brown with fuscous on apex and outer surface; length of segments: 0.14 mm; 0.28 mm; 0.82 mm.
Thorax: Scutum black with dark fuscous across posterior margin, dark fuscous lateral margins over entire length of scutum, a broad medial longitudinal dark red-brown band running from centre of scutum to anterior margin. Pleural areas dark fuscous except dark red-brown below postpronotal lobes. Yellow markings as follows: notopleura; mesopleural stripe reaching almost to anterior *npl.* seta dorsally, continuing to katepisternum as a small transverse spot, anterior margin straight; anatergite (posterior apex black); anterior ⅔ katatergite (remainder black); two broad and very short lateral postsutural vittae narrowing sharply posteriorly to end at or before level of *a.sa.* seta. Medial longitudinal postsutural yellow vitta absent. Postpronotal lobes fuscous. Postnotum black laterally and dark red-brown centrally. Scutellum yellow except for narrow black basal band. Setae: *sc.* 2; *prsc.* 2; *ia.* 1; *p.sa.* 1; *a.sa.* 1; *mpl.* 1; *npl.* 2; *scp.* 4.
Legs: Fore femora fulvous, mid and hind femora fulvous with red-brown around apical ¼; fore tibiae pale fuscous, mid tibiae fulvous, hind tibiae fuscous; fore tarsi with basal two segments fulvous and apical three segments pale fuscous, mid and hind tarsi with all segments fulvous; mid tibiae each with an apical black spur.
Wings: Length 5.8 mm; cells bc and c fuscous; microtrichia in outer ⅓ of cell c only; remainder of wings colourless except dark fuscous cell sc, broad dark fuscous costal band confluent with R_{4+5} but paler towards this vein and remaining of uniform width around apex of wing, a broad dark fuscous anal streak, a dark fuscous band across centre of wing enclosing both crossveins and extending from costal band to hind margin of wing; a dense aggregation of microtrichia around A_1+CuA_2; supernumerary lobe weak.
Abdomen: Oval; terga free; pecten present on tergum III. Tergum I and sterna I and II wider than long. Tergum I dark fuscous with a narrow red-brown band along posterior margin but not reaching lateral margins; tergum II pale fuscous with fulvous across posterior ½ but not reaching lateral margins; tergum III entirely pale fuscous; terga IV–V red-brown with extremely narrow dark fuscous lateral margins, a medium-width medial longitudinal black band over terga III–V. A pair of oval red-brown shining spots on tergum V. Posterior lobe of surstylus short, sternum V with a deep concavity on posterior margin. Abdominal sterna dark fuscous.

Attractant: Cue lure.
Female: No known record.
Distribution: Indonesia (Moluccas = Maluku – Ambon I.).
Hosts: No known record.
Remarks: *Bactrocera (Bactrocera) truncata* n.sp. is similar to *Bactrocera (Bactrocera) fulvicauda* (Perkins), *Bactrocera (Bactrocera) inconspicua* n.sp., *Bactrocera (Bactrocera) neofulvicauda* n.sp. and *Bactrocera (Bactrocera) pseudodistincta* (Drew) in possessing a single fuscous band across the wing in addition to the costal band and anal streak and the costal band confluent with R_{4+5} but differs from *B. fulvicauda* in having the mesopleural stripe not reaching the postpronotal lobe, from *B. inconspicua* and *B. pseudodistincta* in having the lateral postsutural vittae tapering posteriorly (not parallel-sided) and from *B. neofulvicauda* in having the postpronotal lobe fuscous, lateral postsutural vittae very short (at most, to level of *a.sa.* seta) and abdominal terga III–V red-brown with fuscous over tergum III, very narrow dark fuscous lateral margins on terga IV and V and a medial longitudinal black band over all three terga.

Bactrocera (Bactrocera) tuberculata (Bezzi)

(Fig. 189)

Chaetodacus tuberculatus Bezzi, 1916: 106. Holotype male in BMNH.
Dacus (Strumeta) tuberculatus (Bezzi) – Hardy, 1973: 50.
Bactrocera (Bactrocera) tuberculata – Norrbom *et al.*, 1998: 96; Drew *et al.*, 2007: 5.

Type material examined: Holotype male, **MYANMAR (formerly BURMA)**: with the following five labels – 'Type, Chaetodacus tuberculatus Bezzi'; 'Pres by J.B. Fletcher, 1915.414'; 'Upper Burma, Southern Shan State, Taunggyi, about 4000 feet elevation'; 'Taunggyi, 24.v.1912, on peach, K.D.S.'; 'Chaetodacus tuberculatus male n.sp. Bezzi, T. B. Fletcher'. Studied at BMNH.
Non-type material examined: THAILAND: A series of specimens from Chiang Mai, Chiang Rai, host reared and attracted to methyl eugenol traps. **VIETNAM: Northern Vietnam** – 1 male, Son La Province, Moc Chau District, Co Do, 6.ix.1999, attracted to methyl eugenol; 2 males, Hoa Binh Province, Mai Chau District, Paco Village, 15.ix.2002, attracted to methyl eugenol; 1 male, Son La Province, Moc Chau District, Co Do State Farm, 30.vi.2002, attracted to methyl eugenol; 5 males, Son La Province, Moc Chau District, Long Luong Village, 1.x.2002, attracted to methyl eugenol. Specimens in QDPC.
Diagnosis: A medium-sized species; face fulvous with a pair of large circular black spots; postpronotal lobes and notopleura yellow, scutum mostly black; lateral postsutural yellow vittae present; medial postsutural yellow vitta absent; no yellow spot anterior to notopleural suture; mesopleural stripe reaching midway between anterior margin of notopleuron and anterior *npl.* seta dorsally; scutellum yellow; wing with cells bc and c colourless, microtrichia in extreme outer corner of cell c only; a narrow pale fuscous to fuscous costal band confluent with R_{2+3} and broken beyond apex of this vein and with a small fuscous to dark fuscous spot around apex of R_{4+5}; a narrow pale fuscous anal streak confined within cell cup; abdominal terga III–V entirely black.
Description: male.
Head: Height 1.48 mm. Frons – length 1.2 times breadth; fulvous with fuscous around orbital setae and on anteromedial hump; orbital setae black: 1 *s.or.*, 2 *i.or.*; lunule dark red-brown. Ocellar triangle black. Vertex dark red-brown. Face fulvous with a pair of large circular black spots; length 0.48 mm. Genae fulvous; black seta present. Occiput black, fulvous along eye margins; occipital row with 1–4 strong black setae. Antennae with segments 1 and 2 red-brown, segment 3 fulvous with fuscous on apex and outer surface; length of segments: 0.19 mm; 0.31 mm; 0.84 mm.
Thorax: Scutum uniformly black with narrow dark red-brown below lateral postsutural vittae and around notopleural suture. Pleural areas entirely black without pale markings. Yellow markings as follows: postpronotal lobes; notopleura; mesopleural stripe reaching midway between anterior margin of notopleuron and anterior *npl.* seta dorsally, continuing to katepisternum as a small transverse spot, anterior margin convex; anatergite (posterior apex black); anterior ⅔ katatergite (remainder black); two broad parallel-sided lateral postsutural vittae, sometimes widening slightly posteriorly and ending just behind *ia.* seta. Medial longitudinal postsutural yellow vitta absent. Postnotum black. Scutellum yellow except for narrow black basal band. Setae: *sc.* 2; *prsc.* 2; *ia.* 1; *p.sa.* 1; *a.sa.* 1; *mpl.* 1; *npl.* 2; *scp.* 4.
Legs: All segments entirely fulvous; mid tibiae each with an apical black spur.
Wings: Length 6.0 mm; cells bc and c colourless; microtrichia in extreme outer corner of cell c only; remainder of wings colourless except fuscous cell sc, narrow pale fuscous to fuscous costal band confluent with R_{2+3} and ending at apex of this vein, a small elongate fuscous to dark fuscous spot around apex of R_{4+5}, a narrow pale fuscous anal streak mostly confined to cell cup; a dense aggregation of microtrichia around A_1+CuA_2; supernumerary lobe of medium development.
Abdomen: Oval; terga free; pecten present on tergum III. Tergum I and sterna I and II wider than long. Tergum I mostly black (in some specimens there is a narrow red-brown transverse band across posterior margin but not reaching lateral margins); tergum II pale orange-brown basally to fulvous adjacent to tergum III, with a narrow transverse black band across anterior margin and reaching narrow black lateral margins; terga III–V entirely black. A pair of oval black shining spots on tergum V. Posterior lobe of surstylus short, sternum V with a deep concavity on posterior margin. Abdominal sterna dark fuscous to black.
Attractant: Methyl eugenol.
Female: As for male except no dense aggregation of microtrichia around A_1+CuA_2; supernumerary lobe weak; pecten of cilia on abdominal tergum III absent; ovipositor red-brown, dorsoventrally compressed and tapering in dorsal view; ratio of length of oviscape to length of tergum V, 0.65:1; apex of piercer needle-shaped.
Distribution:
Type locality: Myanmar (Taunggyi).

Previously recorded: Type locality and the following locations – Bhutan (Drew *et al.*, 2007); China, Yunnan Province (Norrbom *et al.*, 1998); Thailand and Southern Vietnam (Hardy, 1973).
New records: Northern Vietnam (Son La and Hoa Binh Provinces).
Hosts: Recorded from eight plant families including some commercial/edible species – *Anacardiaceae*; *Caricaceae*; *Euphorbiaceae*; *Lecythidaceae*; *Myrtaceae*; *Polygalaceae*; *Rosaceae*; *Sapotaceae* (see Allwood *et al.*, 1999 for host records).
Remarks: *Bactrocera (Bactrocera) tuberculata* (Bezzi) is similar to *Bactrocera (Bactrocera) correcta* (Bezzi), *Bactrocera (Bactrocera) penecorrecta* Drew and *Bactrocera (Bactrocera) zonata* (Saunders) in possessing a broken costal band that ends at apex of R_{2+3} and then has a fuscous spot around apex of R_{4+5}. It differs from *B. zonata* in having a black scutum, from *B. correcta* in having the abdominal terga III–V mostly black and from *B. penecorrecta* in having very broad lateral postsutural vittae and abdominal terga III–V entirely black.

Bactrocera (Bactrocera) umbrosa (Fabricius)

(Figs 190A, 190B)

Dacus umbrosus Fabricius, 1805: 274. Lectotype in UZMC.
Dacus fascipennis Wiedemann, 1819: 28; Bezzi, 1913: 72 (Synonym); Hardy, 1968: 146.
Bactrocera fasciatipennis Doleschall, 1856: 412; Osten-Sacken, 1881 (fide Bezzi, 1913: 71) (Synonym).
Dacus conformis Walker, 1856a: 34; Osten-Sacken, 1881 (fide Bezzi, 1913: 69) (Synonym).
Dacus diffusus Walker, 1860: 153; Hardy, 1959: 169–70 (Synonym and invalid Lectotype designation).
Dacus frenchi Froggatt, 1909: 92–3; Bezzi, 1913: 72; Bezzi, 1919: 417 (Synonym).
Dacus (Bactrocera) umbrosus – Malloch, 1939b: 412; Drew, 1982: 75–8; Hardy, 1982: 240, 1983: 25.
Strumeta umbrosa – Perkins, 1939: 12.
Strumeta frenchi – Perkins, 1939: 13.
Dacus (Strumeta) umbrosus – Hardy and Adachi, 1954: 184; Hardy, 1973: 52, Lectotype designation by inference of holotype; Drew, 1974a: 88–92.
Bactrocera (Bactrocera) umbrosa – Drew, 1989: 165; Norrbom *et al.*, 1998: 96.
Bactrocera (Bactrocera) lacerata White & Evenhuis, 1999: 497. **New Synonym**.

Other type material examined: INDONESIA: Holotype female of *Bactrocera (Bactrocera) lacerata* White & Evenhuis with the following five labels – 'Indonesia: Timor I., Balical 200–300 m, 14–24 Dec. 1963'; 'J. Sedlacek, Collector, Bishop'; 'Bactrocera sp. No. 10, nr. *umbrosa* (Fabr), det. N.L. Evenhuis 1989'; 'Holotype *Bactrocera lacerata* n.sp., det I.M. White, 1998 and N.L. Evenhuis'; 'Bishop Museum holotype No. 16004'. Received on loan from BBM (Fig. 190B).
Non-type material examined: Since the publication of Drew (1989), the following specimens have been studied – **AUSTRALIAN TERRITORY: CHRISTMAS ISLAND** – 7 males, 'Xmas Island, 27.i.1983, coll. A.N. Sproul, M.E.'; 'Dacus umbrosus'; 'Indian Ocean'. **INDONESIA: JAVA** – 1 male, West Java, Subang, Kali Jati, 24.xi.2005, coll. I.N. Raga & Tri, bred from unidentified host fruit; **KALIMANTAN** – 23 males, Central Kalimantan, Kota Waringin Timur, Kota Besi, 13.ix.2005, coll. Utik Darmanto & Kusyono, attracted to methyl eugenol in citrus; 13 males, West Kalimantan, Pontianak, Pasir, 16.viii.2005, coll. A. Irom, attracted to methyl eugenol in village carambola; 15 males, East Kalimantan, Kutai Kertanegara, Batuah, 12.ix.2005, coll. Heru G, attracted to methyl eugenol in forest; 22 males, South Kalimantan, Banjarbaru, Guntung Payung, 21. vii.2005, coll. Sutini, attracted to methyl eugenol in village. **PHILIPPINES**: 8 males, Sirang Lupa, Calamba, Luzon, 7.i.2000, coll. G. Quimio, attracted to Zorgen 95 (methyl eugenol) in guava mixed orchard; 10 males, Sirang Lupa, Calamba, Laguna, 7.i.2000, coll. G. Quimio, attracted to Zorgen 95 (methyl eugenol) in citrus orchard; 7 males, Sirang Lupa, Calamba, Laguna, 7.i.2000, coll. G. Quimio, attracted to Zorgen 95 (methyl eugenol) in Carambola/guava orchard; 25 males, UPLB Campus, 8.i.2000, coll. G. Quimio, attracted to Zorgen 95 (methyl eugenol), in Rainforest area mixed vegetation; 3 males and 1 female, Pinaglapatan, Infonta, Quezon, 10.xii.1999, coll. G. Quimio, reared from Jackfruit; 1 male, Sirang Lupa, Calamba, Laguna, 20. xii.1999, coll. A. O'Toole & G. Quimio, attracted to Zorgen 95 (methyl eugenol) in apple guava orchard; 2 males, Sirang Lupa, Calamba, Laguna, 20.xii.1999, coll. A. O'Toole & G. Quimio, attracted to Zorgen 95 (methyl eugenol) in guava (big) orchard; 2 males, Sirang Lupa, Calamba, Laguna, 20.xii.1999, coll. A. O'Toole & G. Quimio, attracted to Zorgen 95 (methyl eugenol) in mixed orchard; 4 males, Sirang Lupa, Calamba, Laguna, xii.1999, coll. A. O'Toole & G. Quimio, attracted to Zorgen 95 (methyl eugenol) in citrus orchard; 3 males, Mt. Makiling botanic gardens, alt 150m, 12.xii.1999, coll. A. O'Toole, attracted to Zorgen 95 (methyl eugenol). **THAILAND**: 2 males, Songkhla, Ka Chong, Trang, 24.v.1991, coll. Sieng *et al.*, bred from *Artocarpus heterophyllus*. Specimens in QDPC.
Diagnosis: Face fulvous with a pair of medium-sized circular black spots; postpronotal lobes and notopleura yellow; scutum black except brown along lateral margins and around notopleural suture; broad lateral postsutural yellow vittae almost parallel-sided and ending at or just behind *ia.* seta; medial postsutural yellow vitta absent; mesopleural stripe reaching postpronotal lobe dorsally; scutellum yellow; legs with all segments entirely fulvous; wing with cells bc and c fulvous, microtrichia in outer ½ of cell c only; the patterning on the wings consists of a unique reddish fuscous colour and comprises a broad costal band confluent with R_{4+5} and overlapping this vein at its apex, a broad transverse band across centre of wing and enclosing both crossveins, a narrow transverse band across apex of wing, a broad transverse band at base of wing uniting with a broad anal streak; supernumerary lobe of medium development; abdominal terga III–V varying from orange-brown with a medial longitudinal black stripe over terga IV and V to orange-brown with broad medial and lateral

longitudinal black bands over all three terga, a pair of oval orange-brown shining spots on tergum V.
Attractant: Methyl eugenol.
Female: See Drew (1989) for a detailed description of female characters.
Distribution:
Type locality: Indonesia (Sumatra).
Previously recorded: Widespread across South-East Asia and the South Pacific region to New Caledonia. This species has not been recorded east of New Caledonia or in the Indian subcontinent (Norrbom *et al*., 1998).
New records: Australian Territory (Christmas I.). Indonesia (Timor I.) due to **New Synonymy** of *B. lacerata*.
Hosts: This species only breeds in species of the genus *Artocarpus* (family *Moraceae*). It is classified as an economic pest as these hosts are edible/commercial (see Allwood *et al*., 1999 for host records).
Remarks: *Bactrocera (Bactrocera) umbrosa* (Fabricius) is a unique species, particularly in the reddish fuscous colour of the wing patterns, and has no strong similarities to other species (Drew, 1989). Some variations in the wing patterns do occur, and we have studied the type of *Bactrocera (Bactrocera) lacerata* White & Evenhuis (Fig. 190B), and can see no characters that are convincing in setting *B. lacerata* aside as a distinct species; therefore we are making it a **New Synonym** of *B. umbrosa*.

Bactrocera (Bactrocera) unimacula Drew & Hancock

(Fig. 191)

Bactrocera (Bactrocera) unimacula Drew & Hancock, 1994a: 62; Norrbom *et al*., 1998: 96. Holotype male in BMNH.

Non-type material examined: Since the publication of Drew & Hancock (1994a), the following specimen has been studied – **PENINSULAR MALAYSIA**: 1 male, Malaysia, x.1986, coll. C.S. Ooi, attracted to methyl eugenol. Specimen in QDPC.
Diagnosis: Face fulvous with a pair of small elongate black spots; postpronotal lobes and notopleura yellow; scutum black except dark brown around notopleural suture and occasionally inside postpronotal lobes and a distinctive broad medial longitudinal brown band from centre of scutum to anterior margin and extending to the posterior margin of the scutum in some specimens; broad parallel-sided lateral postsutural yellow vittae ending at or just behind *ia*. seta; medial postsutural yellow vitta absent; mesopleural stripe reaching midway between anterior margin of notopleuron and anterior *npl*. seta dorsally; scutellum yellow; legs with femora entirely fulvous; fore tibiae dark fuscous basally to fulvous apically, mid tibiae fuscous basally to fulvous apically, hind tibiae dark fuscous; wing with cells bc and c colourless, microtrichia in outer ⅓ of cell c only; a narrow dark fuscous costal band confluent with R_{2+3} and expanding into a distinct large spot around apex of R_{4+5}; a narrow fuscous anal streak; supernumerary lobe of medium development; abdominal terga III–V orange-brown except for a distinct 'T' pattern consisting of a narrow transverse black band across anterior margin of tergum III and gradually widening to cover lateral margins, a broad medial longitudinal black band over all three terga, anterolateral corners of terga IV and V black, a pair of oval orange-brown shining spots on tergum V; abdominal sterna dark coloured.
Attractant: Methyl eugenol.
Female: No known record.
Distribution:
Type locality: Peninsular Malaysia (Transect South, trap 7).
Previously recorded: Type locality and the following locations – East Malaysia (Sabah); Indonesia (Kalimantan) (Drew & Hancock, 1994a; Norrbom *et al*., 1998).
Hosts: No known record.
Remarks: *Bactrocera (Bactrocera) unimacula* Drew & Hancock, within the *dorsalis* complex, is similar to *Bactrocera (Bactrocera) affinidorsalis* (Hardy), *Bactrocera (Bactrocera) latilineola* Drew & Hancock and *Bactrocera (Bactrocera) malaysiensis* Drew & Hancock in possessing broad parallel-sided lateral postsutural vittae and the costal band confluent with R_{2+3} and expanding into a distinct large spot in apex of wing. It differs from these species in possessing a unique dark spot in apex of costal band that surrounds the apex of R_{4+5}.

Bactrocera (Bactrocera) usitata Drew & Hancock

(Fig. 192)

Bactrocera (Bactrocera) usitata Drew & Hancock, 1994a: 63; Norrbom *et al*., 1998: 96. Holotype male in BMNH.

Non-type material examined: Since the publication of Drew & Hancock (1994a), the following specimens have been studied – **BRUNEI**: 2 males, KBFC Bato Apou Nat Park, 29.i.2002, coll. K. Hurley, attracted to cue lure. **INDONESIA: KALIMANTAN** – 1 male, East Kalimantan, Kutai Kartanegara, Batuah, 29.i.2006, coll. Ertha D.P., attracted to cue lure in banana. **PHILIPPINES**: 4 males, Mt. Makiling Botanic Gardens, alt 150m, 12.i.2000, coll. A. O'Toole, attracted to cue lure. **THAILAND**: 1 male (25.x.1991), 1 male (27.xii.1991), 1 male (6.iii.1992), 1 male (13.iii.1992), 1 male (27.iii.1992), 1 male (3.iv.1992), 1 male (24.vii.1992), Khao Kheaw, Pakphli, Nakhonnayok, coll. Montree, attracted to cue lure; 2 males, Wang Jum Pee, Pakchong, Nakhonratsima, 10.iv.1992, coll. Montree, attracted to cue lure; 5 males (25.vi.1991), 1 male and 1 female (23.vii.1991), Yang Pong Rat, Plubpla, A. Muang, Chanthaburi, Site BK38T, coll. Montree, attracted to cue lure; 2 males (17.iii.1992), 1 male (1.iv.1992), Yang Pong Rat, A. Muang, Chanthaburi, coll. Montree, attracted to cue lure. **VIETNAM: Southern Vietnam** – 1 male, Dong Nai Province, Vinh Cuu District, Tan Binh Village, Site S011, 28.iv.2000, attracted to cue lure; 1 male (18.vi.2002), 1 male (4.x.2002), Binh Duong Province, Thuan An

District, An Thanh Village, Site VS05, attracted to cue lure. Specimens in QDPC.

Diagnosis: Face fulvous with a pair of large oval black spots; postpronotal lobes and notopleura yellow; scutum black except in some specimens, dark brown below lateral postsutural vittae and inside postpronotal lobes (some specimens from the Philippines have two broad red-brown longitudinal bands down the scutum each side of the midline); broad triangular lateral postsutural yellow vittae tapering sharply posteriorly to end before *ia.* seta; medial postsutural yellow vitta absent; mesopleural stripe reaching midway between anterior margin of notopleuron and anterior *npl.* seta dorsally; scutellum yellow; legs with femora entirely fulvous (some specimens from the Philippines have a dark fuscous preapical spot on outer surfaces of fore femora); fore and mid tibiae fuscous, hind tibiae dark fuscous; wing with cells bc and c colourless or with a pale fulvous tint, microtrichia in outer corner of cell c only; a narrow dark fuscous costal band almost confluent with R_{4+5} and widening across apex of wing (similar to that in *B. occipitalis*); a broad fuscous anal streak; supernumerary lobe of medium development; abdominal terga III–V orange-brown except for a transverse black band across anterior margin of tergum III which widens laterally to form broad lateral margins, a very broad medial longitudinal black band over all three terga, distinct narrow lateral longitudinal black bands over terga IV and V, a pair of oval orange-brown shining spots on tergum V; abdominal sterna dark coloured.

Attractant: Cue lure.

Female: As for male except no dense aggregation of microtrichia around A_1+CuA_2; supernumerary lobe weak; pecten of cilia on abdominal tergum III absent; ovipositor basal segment red-brown, conical in transverse cross section and tapering posteriorly in dorsal view; ratio of length of oviscape to length of tergum V, 2:1; apex of piercer needle-shaped.

Distribution:

Type locality: Peninsular Malaysia (Transect South, trap 7).

Previously recorded: Type locality and the following locations – East Malaysia (Sabah); Philippines; Singapore (Drew & Hancock, 1994a).

New records: Brunei; Indonesia (Kalimantan); Thailand; Southern Vietnam (Binh Duong, Dong Nai Provinces).

Hosts: No known record.

Remarks: *Bactrocera (Bactrocera) usitata* Drew & Hancock, within the *dorsalis* complex, is similar to *Bactrocera (Bactrocera) fulvifemur* Drew & Hancock and *Bactrocera (Bactrocera) melastomatos* Drew & Hancock in possessing lateral postsutural vittae that narrow posteriorly, femora entirely fulvous or with, at most, dark spots on apices of fore femora and costal band extending well over R_{2+3}. It differs from *B. melastomatos* in possessing clearly defined black patterns on abdominal terga III–V (not diffuse as in *B. melastomatos*), and more triangular-shaped lateral postsutural vittae that end well before *ia.* seta, and from *B. fulvifemur* in having more elongate lateral postsutural vittae that end just before *ia.* seta and cells bc and c in the wing colourless.

Bactrocera (Bactrocera) venefica (Hering)

(Fig. 193)

Strumeta venefica Hering, 1938: 5. Holotype male in NRS.
Bactrocera (Bactrocera) venefica – Norrbom *et al.*, 1998: 96.

Type material examined: Holotype male, **MYANMAR (formerly BURMA)**: with the following three labels – 'Type'; 'N.E. Burma, Kambaiti, 2000 m, 12/4/1934, Malaise'; 'Strumeta venefica Type, det. M. Hering, 1938'. Received on loan from NRS.

Diagnosis: A small species; face pale fuscous with a pair of medium-sized circular black spots; postpronotal lobes and notopleura yellow; scutum entirely black; very short and narrow lateral postsutural yellow vittae present; medial postsutural yellow vitta absent; no yellow spot anterior to notopleural suture; mesopleural stripe equal in width to notopleuron dorsally; scutellum yellow with a broad black basal band and a large black apical spot enclosing apical setae; wing with cells bc and c colourless, a narrow margin of microtrichia in cell c adjacent to costal vein; a narrow fuscous to pale fuscous costal band confluent with R_{2+3} and extremely narrow beyond apex of this vein before widening across apex of R_{4+5}; a narrow fuscous anal streak; abdominal terga III–V orange-brown with a medium-width medial and two broad lateral longitudinal dark fuscous to black bands joined along anterior margin of tergum III.

Description: male.

Head: Height 1.32 mm. Frons – length 1.29 times breadth; red-brown with dark fuscous around orbital setae and on anteromedial hump; orbital setae large and black: 1 *s.or.*, 2 *i.or.*; lunule dark fuscous. Ocellar triangle black. Vertex dark fuscous. Face fulvous with a pair of medium-sized circular black spots; length 0.44 mm. Genae fulvous, dark fuscous subocular spot present; black seta present. Occiput black, fulvous along eye margins; occipital row with two short dark setae each side. Antennae with segments 1 and 2 red-brown, segment 3 red-brown with dark fuscous on apex and outer surface; length of segments: 0.13 mm; 0.25 mm; 0.65 mm.

Thorax: Scutum entirely black except for a very small area of red-brown inside notopleura. Pleural areas entirely black. Yellow markings as follows: postpronotal lobes (black on anterodorsal corners); notopleura; mesopleural stripe equal in width to notopleuron dorsally, not continuing to katepisternum, anterior margin straight; anatergite (posterior apex black); anterior $^{3}/_{5}$ katatergite (remainder black); two very narrow and short parallel-sided lateral postsutural vittae ending just posterior of line of *a.sa.* setae. Medial longitudinal postsutural yellow vittae absent. Postnotum black. Scutellum yellow except for a broad black basal band and a large black apical spot enclosing apical setae. Setae: *sc.* 2; *prsc.* 2; *ia.* 1; *p.sa.* 1; *a.sa.* 1; *mpl.* 1; *npl.* 2; *scp.* 2.

Legs: All femora entirely fulvous except for a small area of dark fuscous on upper apical surfaces of mid femora and around apices of hind femora; fore tibiae entirely fulvous except for a dark fuscous line on basal part of outer surfaces, mid tibiae entirely fulvous except dark fuscous basally, hind tibiae entirely dark fuscous; fore tarsi with basal segment fulvous and apical four segments pale fuscous, mid tarsi with basal segment fulvous and apical four segments red-brown, hind tarsi with basal segment fulvous and apical four segments pale fuscous; mid tibiae each with an apical black spur.

Wings: Length 4.8 mm; cells bc and c colourless; a narrow margin of microtrichia adjacent to costal vein; remainder of wings colourless except fuscous cell sc, narrow pale fuscous costal band confluent with R_{2+3} then extremely narrow and fuscous beyond apex of this vein before widening apically across apex of R_{4+5}, a narrow fuscous anal streak; area between cell bm and vein R mostly colourless and devoid of microtrichia; dense aggregation of microtrichia around A_1+CuA_2; supernumerary lobe weak.

Abdomen: Oval; terga free; pecten present on tergum III. Tergum I and sterna I and II wider than long. Tergum I red-brown with broad black lateral margins; tergum II fulvous with a broad black transverse band across anterior margin tending to dark fuscous on lateral margins; terga III–V orange-brown with a medium-width medial longitudinal black band and two broad lateral longitudinal bands which change from black on tergum III to dark fuscous to black on tergum IV and dark fuscous on tergum V (these bands are joined across anterior margin of tergum III). A pair of oval orange-brown shining spots which tend to fuscous laterally. Posterior lobe of surstylus short, sternum V with a deep concavity on posterior margin.

Attractant: No known record.

Female: No known record.

Distribution:

Type locality: Myanmar (Kambaiti).

Hosts: No known record.

Remarks: *Bactrocera (Bactrocera) venefica* (Hering) is similar to *Bactrocera (Bactrocera) armillata* (Hering) and *Bactrocera (Bactrocera) hyalina* (Shiraki) in possessing a costal band that is discontinued beyond apex of R_{2+3} (or extremely narrow and pale in this area), lateral postsutural vittae very short and narrow and a large dark spot over dorsal surface or on apex of scutellum. It differs from *B. armillata* in having a large black spot on apex of scutellum (not over dorsal surface) and femora mostly or entirely fulvous and from *B. hyalina* in having the mesopleural stripe equal in width to the notopleuron dorsally, costal band very narrow and pale beyond apex of R_{2+3} (not discontinuous) and small preapical dark fuscous spots on mid and hind femora (not entirely fulvous).

Bactrocera (Bactrocera) verbascifoliae Drew & Hancock

(Fig. 194)

Bactrocera (Bactrocera) verbascifoliae Drew & Hancock, 1994a: 64; Norrbom *et al.*, 1998: 96; Drew *et al.*, 2007: 5. Holotype female in BMNH.

Non-type material examined: Since the publication of Drew & Hancock (1994a), the following specimens have been studied – **BHUTAN**: 3 males, Phuentsholing, 10km, 10.vi.2000, coll. S. Thapa, attracted to methyl eugenol; 1 male, East Bhutan, Sandrup Jongkhar District, 5.xii.2004, coll. Wangdi, attracted to methyl eugenol. **INDONESIA: JAVA** – 1 male (2.viii.2005), 3 males (9.viii.2005), East Java, Lumajang, Kadang, Tepus, coll. Utomo, attracted to methyl eugenol in banana; 2 males (21.vii.2005), 4 males (11.viii.2005), 1 male (18.viii.2005), West Java, Majalengka Ampel (25 dpl), coll. Yaya Senjaya, attracted to methyl eugenol in mango (cengkir); 1 male, Banten, Tangerang, Tanah Tinggi, 23.viii.2005, coll. Komarudin, attracted to methyl eugenol in village mango; 1 male, Banten, Lebak, Pasir Nini–nini, 22.viii.2005, coll. TB Aden Haer, attracted to methyl eugenol in forest; 1 male, D.I. Yogyakarta, Sieman, Kaliurang, 12.vii.2005, coll. Agus W, attracted to methyl eugenol in forest. **VIETNAM: Northern Vietnam** – 21 males, Son La Province, Moc Chau District, Long Luong Village, Site VN01, 15.ix.2002, attracted to methyl eugenol; 15 males, Son La Province, Moc Chau District, Long Luong Village, Site VN01, coll. NIPP Staff, attracted to methyl eugenol; 1 male (1.x.1999), 1 male (16.vi.1999), 1 male (30. vi.1999), Hanoi Province, Tu Liem District, Con Hue, Site N001, attracted to methyl eugenol; 1 male (15.v.1999), 1 male (7.viii.1999), 3 males (14.viii.1999), 3 males (21.viii.1999), 1 male (28.viii.1999), 1 male (14.vii.2000), 1 male (8.xi.2001), Hanoi Province, Tu Liem District, Lien Mac, Site N003, attracted to methyl eugenol; 2 males, Hanoi Province, Tu Liem District, Xuan Dinh, Site N004, 15.iv.1999, attracted to methyl eugenol; 1 male, Hanoi Province, Dong Anh District, Bae Ha, Site N005, attracted to methyl eugenol; 18 males (30.iii.2002), 9 males (15.iv.2002), 17 males (1.ix.2002), Son La Province, Moc Chau District, Co Do State Farm, FAO Site VN06, coll. NIPP Staff, attracted to methyl eugenol; 3 males (30.vi.2002), 8 males (30.vii.2002), 2 males (15.viii.2002), 12 males (15.ix.2002), Son La Province, Moc Chau District, Co Do State Farm, FAO Site VN06, attracted to methyl eugenol; 6 males, Son La Province, Moc Chau District, Co Do, Site N006, attracted to methyl eugenol; 2 males (14.vi.1999), 1 male (6.ix.1999), 1 male (20.xi.1999), Son La Province, Moc Chau District, Tra Long, Site N007, attracted to methyl eugenol; 1 male (23.viii.1999), 1 female (21.viii.2000), Son La Province, Moc Chau District, Moc Chai Town, Site N009, attracted to methyl eugenol; 2 males, Hoa Binh Province, Mai Chau District, Pa Co Village, FAO Site VN10, 15.iv.2002, coll. NIPP Staff, attracted to methyl eugenol; 15 males, Hoa Binh Province, Mai Chau District, Pa Co Village, Site VN10, 15.ix.2002, attracted to methyl eugenol; 10 males (14.xii.2000), 2 males (30.xi.2001), Hoa Binh Province, Mai Chau District, Pa Co, Site N010, attracted to methyl eugenol; 7 males (2.viii.1999), 1 male (1.xii.1999), Hoa Binh Province, Ky Son District, Cao Phong, Site N011, attracted to methyl eugenol; 2 males, Hoa Binh Province, Ky Son District, Cao Phong, Site N012, attracted to methyl eugenol; 1 male, Hoa Binh Province, Ky Son District, Cao Phong, Site N013, attracted to methyl eugenol; 1 male (13.viii.1999), 1 male (8.x.1999), Lao Cai Province, Sapa Town, Site N016, attracted to methyl eugenol; **Southern Vietnam** – 1 male, Vinh Long Province, Thuan An Village, Site S001, 25.viii.2000, attracted to methyl eugenol; 3 males (2.vi.2002), 1 male,

(22.vi.2002), 1 male (30.vi.2002), 1 male (28.vii.2002), Vinh Long Province, Long Ho District, Phuoc Hau Village, Site VS01, attracted to methyl eugenol; 10 males, Cantho Province, Chau Thanh District, Tan Phu Village, Site VS02, 23.iii.2002, coll. SOFRI Staff, attracted to methyl eugenol; 1 male (2.vi.2002), 1 male (28.vii.2002), Cantho Province, Chau Thanh District, Tan Phu Village, Site VS02, attracted to methyl eugenol; 1 male (4.iv.2002), 1 male (9.v.2002), 1 male (30.v.2002), Dong Thap Province, Lap Vo District, Tan Khanh Trung Village, Site VS03, coll. SOFRI Staff, attracted to methyl eugenol; 1 male (12.i.2002), 1 male (19.i.2002), 2 males (28.i.2002), 2 males (2.v.2002), 1 male (25.iii.2002), Ben Tre Province, Chau Thanh District, Tien Thuy Village, Site VS04, coll. SOFRI Staff, attracted to methyl eugenol; 1 male (1.vi.2002), 1 male (23.vi.2002), 2 males (7.vii.2002), Ben Tre Province, Chau Thanh District, Tien Thuy Village, Site VS04, attracted to methyl eugenol; 2 males, Binh Duong Province, Thuan An District, An Thanh Village, Site VS05, 21.v.2002, coll. SOFRI Staff, attracted to methyl eugenol; 1 male, Binh Duong Province, Thuan An District, An Thanh Village, Site VS05, 30.vii.2002, attracted to methyl eugenol; 1 male, Tien Giang Province, My Tho Village, Site S005, 1.iii.2000, attracted to methyl eugenol; 7 males (13.iv.2002), 14 males (27.iv.2002), Dong Nai Province, Long Khanh District, Nhan Nghia Village, Site VS06, coll. SOFRI Staff, attracted to methyl eugenol; 2 males (28.ix.2002), 1 male (23.xi.2002), Dong Nai Province, Long Khanh District, Nhan Nghia Village, Site VS06, attracted to methyl eugenol; 1 male, Lam Dong Province, 3 District, Da Lat town, Site VS08, 8.x.2002, attracted to methyl eugenol; 3 males, Dac Lac Province, Ku Mi Nga District, Quang Phu Village, Site VS09, coll. SOFRI Staff, attracted to methyl eugenol; 1 male (8.i.2002), 5 males (8.v.2002), 15 males (8.vi.2002), 1 male (24.vi.2002), 1 male (6.vii.2002), 1 male (24.vii.2002), Dac Lac Province, Ku Mi Nga District, Quang Phu Village, Site VS09, attracted to methyl eugenol; 3 males (22.xii.2001), 1 male (21.i.2002), Khanh Hoa Province, Dien Khanh District, Suoi Tan Village, Site VS10, coll. SOFRI Staff, attracted to methyl eugenol; 2 males, Khanh Hoa Province, Dien Khanh District, Suoi Tan Village, Site VS10, attracted to methyl eugenol; 1 male, Dong Nai Province, Vinh Cuu District, Tan Vinh Village, Site S011, 3.iii.2000, attracted to methyl eugenol; 2 males (7.vii.2000), 1 male (3.viii.2000), Dong Nai Province, Long Nhat District, Long Nhat Village, Site S012, attracted to methyl eugenol; 18 males (21.i.2000), 2 males (18.ii.2000), 2 males (3.iii.2000), 1 male (31.iii.2000), 2 males (14.iv.2000), 2 males (26.v.2000), 3 males (3.viii.2000), Binh Duong Province, Lai Thieu Village, Site S020, attracted to methyl eugenol. Specimens in QDPC.

Diagnosis: Face fulvous with a pair of medium-sized circular black spots; postpronotal lobes and notopleura yellow; scutum black except red-brown below and behind lateral postsutural vittae, around notopleural suture and inside postpronotal lobes; narrow parallel-sided lateral postsutural yellow vittae ending at *ia.* seta; medial postsutural yellow vitta absent; mesopleural stripe reaching midway between anterior margin of notopleuron and anterior *npl.* seta dorsally; scutellum yellow; legs with femora entirely fulvous; fore tibiae pale fuscous, mid tibiae pale fuscous basally tending fulvous apically, hind tibiae pale fuscous; wing with cells bc and c colourless, microtrichia in outer corner of cell c only; a narrow fuscous costal band confluent with R_{2+3} or slightly overlapping this vein where it becomes paler and remaining narrow and of uniform width around wing margin to end between the extremities of R_{4+5} and M (it may have a slight swelling around apex of R_{4+5}); a narrow fuscous anal streak; supernumerary lobe of medium development in males and weak in females; abdominal terga III–V orange-brown with a narrow transverse black band across anterior margin which expands to cover lateral margins, a narrow medial longitudinal black band over all three terga, anterolateral corners of terga IV and V black, in some specimens the anterolateral corners on terga IV and V broaden to cover the entire lateral margins giving the image of broad lateral longitudinal dark bands over terga III–V, a pair of oval orange-brown shining spots on tergum V; abdominal sterna dark coloured.

Attractant: Methyl eugenol.

Female: As for male in the general body colour patterns. See Drew & Hancock (1994a) for a detailed description of female characters.

Distribution:

Type locality: Thailand (Chiang Mai, Doi Pui).

Previously recorded: Type locality and the following locations – Bhutan (Drew *et al.*, 2007); India (Drew & Hancock, 1994a).

New records: Indonesia (Java); Vietnam.

Hosts: *Solanum* species, particularly *Solanum donianum* (previously *S. verbascifolium*) (family *Solanaceae*).

Remarks: *Bactrocera (Bactrocera) verbascifoliae* Drew & Hancock, within the *dorsalis* complex, is similar to *Bactrocera (Bactrocera) dorsalis* (Hendel) and *Bactrocera (Bactrocera) papayae* Drew & Hancock in possessing parallel-sided lateral postsutural vittae, costal band confluent with or just overlapping R_{2+3} and not expanded across apex of wing, femora entirely fulvous and abdominal terga III–V with a narrow- to medium-width medial longitudinal dark band. It differs from these species in having broad lateral longitudinal dark markings over abdominal terga III–V. *Bactrocera verbascifoliae* is also similar to *Bactrocera (Bactrocera) paraverbascifoliae* Drew in possessing narrow parallel-sided lateral longitudinal postsutural vittae and the femora entirely fulvous. It differs from this species in having the lateral postsutural vittae ending at *ia.* seta (not before the seta) and the costal band not widening markedly across the apex of the wing.

Bactrocera (Bactrocera) versicolor (Bezzi)

(Fig. 195)

Chaetodacus ferrugineus var. *versicolor* Bezzi, 1916: 105. Syntypes in BMNH.
Bactrocera (Bactrocera) versicolor – Norrbom *et al.*, 1998: 96; Drew & Raghu, 2002: 344.

Type material examined: Syntype male, **INDIA**: with the following five labels – 'Type, Chaetodacus ferrugineus var versicolor Bezzi'; 'Pres by T.B. Fletcher 1915.4.14'; 'On Sapoto fruit, Pusa, Bihar 14.vi.1914, D.P.S.'; 'Chaetodacus ferrugineus versicolor n. var. Bezzi, type male, T.B. Fletcher'; 'Syntype of *Chaetodacus ferrugineus* var *versicolor* Bezzi 1916'. Studied at BMNH.

Non-type material examined: Since the publication of Drew & Raghu (2002), the following specimens have been studied – **INDIA**: 1 male, 3 females, S. India, Mudigere, 980 m, 21.v.1992, coll. D.L. Hancock; 1 female, Mysore St., Bangalore, 7.vi.1974, coll. Viraktanath C.A.; 1 male, Karnataka, Chethalli, 970m, 3.iv.1986, coll. Ramani, attracted to methyl eugenol. **SRI LANKA (formerly CEYLON)**: 2 males, Hantana Hill, 75°w, coll. K. Tsuruta. Specimens in QDPC.

Diagnosis: Face fulvous with a pair of medium-sized oval black spots; postpronotal lobes and notopleura yellow; scutum red-brown with narrow longitudinal fuscous lines and narrow fuscous lateral margins; medium-width parallel-sided lateral postsutural yellow vittae ending behind *ia.* seta; medial postsutural yellow vitta absent; mesopleural stripe reaching midway between anterior margin of notopleuron and anterior *npl.* seta dorsally; scutellum yellow with a dark fuscous apical spot; legs with all femora entirely fulvous except for pale fuscous subapical spots on outer surfaces of fore and mid femora and pale fuscous around apices of hind femora; hind tibiae entirely fuscous; wing with cells bc and c colourless, microtrichia in outer corner of cell c only; a very narrow fuscous costal band confluent with R_{2+3} and remaining extremely narrow around apex of wing to end beyond extremity of R_{4+5}; a very narrow pale fuscous anal streak confined within cell cup; supernumerary lobe of medium development; abdominal terga III–V red-brown with a 'T' pattern consisting of a narrow transverse black band across anterior margin of tergum III and a narrow medial longitudinal black band over all three terga, narrow dark fuscous anterolateral corners on terga IV and V, dark red-brown to fuscous oval shining spots on tergum V, abdominal sterna dark.

Attractant: Methyl eugenol.

Female: As for male except no dense aggregation of microtrichia around A_1+CuA_2; supernumerary lobe weak; pecten of cilia on abdominal tergum III absent; ovipositor basal segment red-brown, conical in transverse cross section and tapering posteriorly in dorsal view; ratio of length of oviscape to length of tergum V, 1.4: 1; apex of piercer needle-shaped.

Distribution:

Type locality: India (Bihar, Pusa; Tamil Nadu, Coimbatore); Sri Lanka (Peradeniya). No one location can be regarded as the type locality as no Holotype or Lectotype has been designated.

Previously recorded: Type locality and other locations in Southern India – Kerala (Drew & Raghu, 2002); Sri Lanka (Norrbom *et al.*, 1998). The record from Bhutan in Drew & Raghu, 2002 has not been confirmed through extensive field survey work.

New records: Southern India (Karnataka).

Hosts: *Manilkara zapota* (family *Sapotaceae*) (White & Elson-Harris, 1992).

Remarks: *Bactrocera (Bactrocera) versicolor* (Bezzi) is similar to *Bactrocera (Bactrocera) paraosbeckiae* Drew and *Bactrocera (Bactrocera) pseudoversicolor* Drew in being a general red-brown species and having the wings with a narrow costal band and anal streak. It differs from *B. paraosbeckiae* in lacking the widespread dark patterns on the scutum and in possessing a dark apical spot on the scutellum and in lacking lateral dark patterns on abdominal terga III–V. It differs from *B. pseudoversicolor* in having a broader mesopleural stripe, a dark apical spot on the scutellum and dark preapical spots on the fore and hind femora.

In the original description, Bezzi (1916) had a series of specimens before him for study. Because we cannot be sure that the specimen in BMNH was set aside as the type, we are regarding it as a Syntype.

Bactrocera (Bactrocera) vishnu Drew & Hancock

(Fig. 196)

Bactrocera (Bactrocera) vishnu Drew & Hancock, 1994a: 65; Norrbom *et al.*, 1998: 96; Drew & Raghu, 2002: 345; Drew *et al.*, 2007: 6. Holotype male in BMNH.

Non-type material examined: Since the publication of Drew & Hancock (1994a), the following specimens have been studied – **BHUTAN**: 1 male, Rimchu trap 1, 24.v.2000, coll. C. Dorji, attracted to cue lure; 1 male, Rimchu trap 1, 14.vi.2000, coll. Nado, attracted to cue lure; 1 male, Rimchu trap 1, 21.vi.2000, coll. Nado *et al.*, attracted to cue lure. Specimens in QDPC.

Diagnosis: Face fulvous with a pair of large oval black spots; postpronotal lobes and notopleura yellow; scutum black with a small area of dark brown posterolateral to lateral postsutural vittae and inside postpronotal lobes; broad parallel-sided lateral postsutural yellow vittae ending at or just behind *ia.* seta; medial postsutural yellow vitta absent; mesopleural stripe reaching midway between anterior margin of notopleuron and anterior *npl.* seta dorsally; scutellum yellow; legs with femora fulvous except for a large oval subapical black spot on outer surfaces of fore femora; fore and mid tibiae fuscous to dark fuscous basally and fulvous apically, hind tibiae dark fuscous; wing with cells bc and c colourless, microtrichia in outer corner of cell c only; a narrow fuscous costal band just overlapping R_{2+3} where it becomes very pale and remaining very narrow around apex of wing to end just beyond extremity of R_{4+5}; a narrow fuscous anal streak; supernumerary lobe of medium development; abdominal terga III–V orange-brown except for black across anterior ½ of tergum III which expands to cover lateral margins, dark fuscous to black anterolateral corners on terga IV and V, a narrow medial longitudinal black band over all three terga, a pair of oval fuscous shining spots on tergum V; abdominal sterna dark coloured.

Attractant: Cue lure.

Female: No known record.

Distribution:

Type locality: Southern India (Palani Hills).

Previously recorded: Type locality and Southern India (Kerala) and Bhutan (Drew *et al.*, 2007).

Hosts: No known record.
Remarks: *Bactrocera (Bactrocera) vishnu* Drew & Hancock, within the *dorsalis* complex, is similar to *Bactrocera (Bactrocera) muiri* (Hardy & Adachi), *Bactrocera (Bactrocera) papayae* Drew & Hancock and *Bactrocera (Bactrocera) profunda* Tsuruta & White in possessing moderately broad parallel-sided lateral postsutural vittae, costal band confluent with or just overlapping R_{2+3} and not widening across apex of wing, abdominal terga IV and V with moderately narrow lateral dark markings, shining spots on abdominal tergum V pale and femora entirely fulvous or with, at most, subapical dark spots on fore femora. It differs from these species in having the lateral dark margins on abdominal terga IV and V of medium-width (not narrow).

Bactrocera (Bactrocera) waaiae Drew & Romig, new species

(Fig. 197)

Material examined: Holotype male, **INDONESIA: MOLUCCAS (MALUKU): AMBON ISLAND** – Waai, Ambon, Indonesia, <20 m, 6.vii.1981, coll. J. Turner, attracted to methyl eugenol in residual forest with cloves, nutmeg and banana groves, deposited in BMNH.
Diagnosis: A medium-sized species; face fulvous with a pair of medium-sized circular black spots; postpronotal lobes and notopleura yellow; scutum entirely red-brown; lateral and medial postsutural yellow vittae absent; no yellow spot anterior to notopleural suture; mesopleural stripe equal in width to notopleuron dorsally; scutellum yellow; wing with cells bc and c pale fuscous, microtrichia in outer corner of cell c only; a broad dark fuscous costal band confluent with R_{4+5} and paler adjacent to that vein; a broad pale fuscous anal streak; all abdominal terga entirely orange-brown.
Description: male.
Head: Height 1.56 mm. Frons – length 1.48 times breadth; entirely red-brown without dark markings; orbital setae black: 1 *s.or.*, 2 *i.or.*; lunule red-brown. Ocellar triangle black. Vertex red-brown. Face fulvous with a pair of medium-sized circular black spots; length 0.5 mm. Genae red-brown, small fuscous subocular spot present; red-brown seta present. Occiput red-brown, fulvous along eye margins; occipital row with 5–7 black setae. Antennae with segments 1 and 2 red-brown, segment 3 red-brown with fuscous on apex and outer surface; length of segments: 0.16 mm; 0.3 mm; 0.79 mm.
Thorax: Scutum entirely red-brown without dark markings. Pleural areas red-brown with dark fuscous on katepisternum. Yellow markings as follows: postpronotal lobes; notopleura; mesopleural stripe equal in width to notopleuron dorsally, continuing to katepisternum as a transverse spot, anterior margin straight; anatergite (posterior apex red-brown); anterior $^3/_5$ katatergite (remainder including ventral margin red-brown). Lateral and medial postsutural yellow vittae absent. Postnotum red-brown. Scutellum yellow except for narrow red-brown basal band. Setae: *sc.* 2; *prsc.* 2; *ia.* 1; *p.sa.* 1; *a. sa.* 1; *mpl.* 1; *npl.* 2; *scp.* 4.
Legs: All segments entirely fulvous except fore and hind tibiae tending pale fuscous; mid tibiae each with an apical black spur.
Wings: Length 5.6 mm; cells bc and c pale fuscous; microtrichia in outer corner of cell c only; remainder of wings colourless except dark fuscous costal band confluent with R_{4+5} and paler adjacent to that vein, broad pale fuscous anal streak; a dense aggregation of microtrichia around A_1+CuA_2; supernumerary lobe of medium development.
Abdomen: Oval; terga free; pecten present on tergum III. Tergum I and sterna I and II wider than long. All abdominal terga entirely orange-brown. A pair of oval orange-brown shining spots on tergum V. Posterior lobe of surstylus short, sternum V with a deep concavity on posterior margin. Abdominal sterna orange-brown.
Attractant: Methyl eugenol.
Female: No known record.
Distribution: Indonesia (Moluccas = Maluku – Ambon I., Waai).
Hosts: No known record.
Remarks: *Bactrocera (Bactrocera) waaiae* n.sp. is similar to *Bactrocera (Bactrocera) obscurata* (de Meijere), *Bactrocera (Bactrocera) thistletoni* Drew and *Bactrocera (Bactrocera) obscurivitta* n.sp. in having the wings colourless except for the costal band and anal streak and in lacking lateral and medial postsutural yellow vittae on the scutum. It differs from *B. thistletoni* and *B. obscurivitta* in having yellow postpronotal lobes and from *B. obscurata* in having a red-brown scutum and broad costal band confluent with R_{4+5}.

Bactrocera (Bactrocera) wuzhishana Lin & Yang

(Fig. 198)

Bactrocera (Bactrocera) wuzhishana Lin & Yang, in Lin *et al.*, 2006: 312. Holotype male in HEIQ.

Non-type material examined: HONG KONG: 1 male, Tai Po Kau Orchard, 13.iii.1981, coll. R. Winney, attracted to methyl eugenol; 2 males, Tai Po Kau, TPK5, 25.iii.1981, coll. R. Winney, attracted to methyl eugenol; 1 male, Tai Po Kau Orchard, 13.v.1981, coll. R. Winney, attracted to methyl eugenol. **INDONESIA: SULAWESI** – 2 males, Burung, 10km E. Malino (1350m), 27.vii.1981, coll. J. Turner, attracted to methyl eugenol in primary forest. **THAILAND**: 2 males, Nong Hoi, Chiang Mai, 15.iii.1989, trap; 2 males, Doi Suthep-Pui Nat Park, 1100m, Chiang Mai, 12.v.1991, coll. D.L. Hancock, attracted to methyl eugenol; 1 male (13. vi.1991), 3 males (31.x.1991), 4 males (15.xi.1991), 9 males (13.xii.1991), 12 males (2.i.1992), 1 male (8.i.1992), 8 males (10.i.1992), 19 males (23.i.1992), 6 males (8.ii.1992), Doi Pui, Chiang Mai, coll. Chantanee, attracted to methyl eugenol; 1 male (17.xii.1991), 2 males (30.i.1992), Mae Hae, Chiang Mai, coll. Chantanee, attracted to methyl eugenol; 1 male, NRDC, Chiang Mai, 30.i.1992, coll. Chantanee, attracted to methyl eugenol; 1 male (8.i.1992), 3 males (22.i.1992), 1 male (29.i.1992), Pha Yao, Chiang Rai, coll. Charuwan, attracted to methyl eugenol; 1 male (8.i.1992), 3 males (22.i.1992), 1 male (29.i.1992), Pha Yao, Chiang Rai, coll. Charuwan, attracted to methyl eugenol; 1 male (4. ii.1992), 1 male (27.ii.1992), Khun Khon, Chiang Rai, coll. Charuwan, attracted to methyl eugenol; 1 male, Phan, Chiang Rai, 29.i.1992, coll. Charuwan, attracted to methyl eugenol; 1 male (9.viii.1991), 1 male (18.x.1991), 1 male (1.xi.1991), 1 male (15.xi.1991), 3 males (29.xi.1991), 1 male (13.xii.1991), 1 male (20.xii.1991), 2 males (3.i.1992), 3 males (7.ii.1992), Khao Khaew, Pakphli, Nakhonnayok, coll. Montree, attracted to methyl eugenol; 3 males, 14. vi.1991, Khao Yai, Nr Bangkok, Nakhonnayok, coll. Montree, attracted to methyl eugenol; 1 male (4.x.1991), 1 male (13. xii.1991), 1 male (24.i.1992), 1 male (27.iii.1992), 1 male (3.iv.1992), Wang Jum Pee, Nakhonratsima, coll. Montree, attracted to methyl eugenol. Specimens in QDPC. **VIETNAM: Northern Vietnam** – 1 male (30.xi.2000), 4 males (1.ii.2001), 2 males (15.iv.2001), Son La Province, Moc Chau District, Co Do, attracted to methyl eugenol; 1 male, Son La Province, Moc Chau District, Long Luong Village, 30.iii.2002, attracted to methyl eugenol; 1 male, Hoa Binh Province, Mai Chau District, Pa Co Village, 1.xii.2002; 3 males (15.iii.2002), 2 males (30.iii.2002), 1 male (30. iv.2002), 5 males (1.x.2002), Son La Province, Moc Chau District, Long Luong Village, Site VN01, coll. NIPP Staff, attracted to methyl eugenol; 1 male, Vietnam North, Site N002, 20.xi.1999, attracted to methyl eugenol; 1 male, Vietnam North, Hanoi Province, Tu Liem District, Xuan Dinh, 24.xii.1999, Site N004, attracted to methyl eugenol; 1 male (3.v.1999), 1 male (29.xi.1999), 1 male (10.i.2000), 1 male (21.ii.2000), 1 male (1.iv.2000), 6 males (1.iv.2001), Son La Province, Moc Chau District, Co Do, Site N006, attracted to methyl eugenol; 3 males (30.iii.2002), 2 males (15.iv.2002), Son La Province, Moc Chau District, Co Do State Farm, Site VN06 (FAO Site), coll. NIPP Staff, attracted to methyl eugenol; 2 males, Son La Province, Moc Chau District, Co Do, Site N008, 1.iv.2001, attracted to methyl eugenol; 1 male, Son La Province, Moc Chau District, Moc Chau house garden, Site N008, 10.i.2000, attracted to methyl eugenol; 1 male, Son La Province, Moc Chau District, Moc Chai town, Site N009, 10.i.2000, attracted to methyl eugenol; 3 males, Hoa Binh Province, Mai Chau District, Pa Co Village, Site VN10 (FAO Site), 15.iv.2002, coll. NIPP Staff, attracted to methyl eugenol; 1 male (17.x.1999), 1 male (1.xii.1999), 2 males (15.iii.2000), 5 males (24.xi.2000), 4 males (30.xi.2001), 4 males (15.xii.2001), 4 males (30. xii.2001), 7 males (1.xii.2002), 26 males (1.i.2003), Hoa Binh Province, Mai Chau District, Pa Co Village, Site VN10, attracted to methyl eugenol; 1 male (12.xii.1999), 2 males (28.iv.2000), Lao Cai Province, Sapa 6 village, Site N017, attracted to methyl eugenol; 1 male, Lao Cai Province, Sapa Oquy Ho, 23.vi.2000, Site N019, attracted to methyl eugenol. Specimens in QDPC.

Diagnosis: A medium-sized species; face dark fuscous with large elongate oval black spots; postpronotal lobes and notopleura yellow; scutum black without pale markings; very short and narrow lateral postsutural yellow vittae present; medial postsutural yellow vitta absent; mesopleural stripe reaching midway between anterior margin of notopleuron and anterior *npl.* seta dorsally; scutellum yellow with a narrow black basal band; wing with cells bc and c colourless, microtrichia in outer corner of cell c only; a narrow dark fuscous costal band confluent with R_{2+3} and widening markedly around apex of wing; a narrow dark fuscous anal streak; abdominal terga III–V almost entirely black except for a narrow transverse orange-brown band across posterior margin of tergum V which extends anteriorly along a narrow medial longitudinal black band on this tergum.

Description: male.

Head: Height 1.54 mm. Frons – length 1.5 times breadth; generally fuscous, small dark fuscous areas around bases of orbital setae, red-brown along ventral margin; orbital setae black: 1 *s.or.*, 2 *i.or.*; lunule fuscous. Ocellar triangle black. Vertex fuscous. Face red-brown to fuscous with a pair of large elongate oval black spots; length 0.46 mm. Genae fulvous, large dark fuscous subocular spot present; black seta present. Occiput shining black, fulvous along eye margins; occipital row with four strong black setae. Antennae with segments 1 and 2 fuscous, segment 3 red-brown with dark fuscous on apex and outer surface; length of segments: 0.18 mm; 0.3 mm; 0.78 mm.

Thorax: Scutum entirely black without pale markings. Pleural areas entirely black without pale markings. Yellow markings as follows: postpronotal lobes; notopleura; mesopleural stripe reaching midway between anterior margin of notopleuron and anterior *npl.* seta dorsally, continuing to katepisternum as a small transverse spot, anterior margin straight; anatergite (posterior apex black); anterior ⅔ katatergite (remainder black); two very short and narrow lateral postsutural vittae narrowing sharply posteriorly to end at or just before level of *a.sa.* seta. Medial longitudinal postsutural yellow vitta absent. Postnotum black. Scutellum yellow except for narrow black basal band. Setae: *sc.* 2; *prsc.* 2; *ia.* 1; *p.sa.* 1; *a.sa.* 1; *mpl.* 1; *npl.* 2; *scp.* 4.

Legs: Fore femora black tending fulvous on inner surfaces, mid femora with apical ½ black and basal ½ fulvous, hind femora with apical ⅓ black and basal ⅔ fulvous; all tibiae dark fuscous to black; tarsi with basal segment fulvous and apical four segments fuscous; mid tibiae each with an apical black spur.

Wings: Length 5.8 mm; cells bc and c colourless; microtrichia in outer corner of cell c only; remainder of wings colourless except dark fuscous cell sc, narrow dark fuscous costal band confluent with R_{2+3} and widening markedly beyond apex of this vein across apical margin of wing, a narrow dark fuscous anal streak; a dense aggregation of microtrichia around A_1+CuA_2; supernumerary lobe of medium development.

Abdomen: Oval; terga free; pecten present on tergum III. Tergum I and sterna I and II wider than long. Terga I and II entirely black except for a narrow dark red-brown band along intersegmental line between terga I and II and narrow transverse red-brown bands posterolaterally on tergum II; terga III–V entirely black except for a narrow transverse orange-brown band across posterior margin of tergum V which extends medially to form narrow orange-brown margins

along a narrow medial longitudinal black band on this tergum. A pair of oval dark fuscous to black shining spots on tergum V. Posterior lobe of surstylus short, sternum V with a deep concavity on posterior margin. Abdominal sterna black.
Attractant: Methyl eugenol.
Female: No known record.
Distribution:
Type locality: China (Hainan I., Mt Wuzhishan).
New records: Hong Kong; Indonesia (Sulawesi); Thailand; Northern Vietnam.
Hosts: No known record.
Remarks: *Bactrocera (Bactrocera) wuzhishana* Lin & Yang belongs to the *nigrotibialis* complex of species, having the base colour of the thorax and abdomen mostly black, wings colourless except for a narrow costal band and anal streak and distinct, usually large, areas of black on the femora. It is similar to *Bactrocera (Bactrocera) neonigrotibialis* Drew, *Bactrocera (Bactrocera) nationigrotibialis* n.sp. and *Bactrocera (Bactrocera) nigrifacia* Zhang, Ji & Chen in possessing a fulvous face with distinct black spots, costal band confluent with R_{2+3}, cells bc and c colourless and scutellum yellow without large black markings. It differs from *B. neonigrotibialis* and *B. nigrifacia* in having the abdominal terga III–V entirely black and from *B. nationigrotibialis* in having the facial spots elongate oval, lateral postsutural vittae very short and narrow and the costal band widening around apex of wing distal to extremity of R_{2+3}.

Bactrocera (Bactrocera) zonata (Saunders)

(Figs 199A, 199B)

Dasyneura zonatus Saunders, 1842: 61. Location of types unknown.
Bactrocera maculigera Doleschall, 1858: 122; Norrbom *et al.*, 1998: 96 (Synonym).
Rivellia persicae Bigot, 1890: 192.
Dacus ferrugineus var. *mangiferae* Cotes, 1893: 17.
Dacus (Strumeta) zonatus – Hardy, 1973: 54.
Bactrocera (Bactrocera) zonata – Norrbom *et al.*, 1998: 96; Drew & Raghu, 2002: 345; Drew *et al.*, 2007: 6.

Non-type material examined: Since the publication of Drew & Raghu (2002), the following specimens have been studied – A large series of specimens collected in India, Mauritius and Thailand in the QDPC collection plus the following – **BHUTAN**: 26 males, Phuentsholing, 6.v.2000, coll. Thapa, attracted to methyl eugenol; 5 males, Rimchu trap 5, 17.v.2001, coll. R. Drew & M. Romig, attracted to methyl eugenol; 10 males, Rimchu Bagging Site 8, 17.v.2001, coll. R. Drew & M. Romig, attracted to methyl eugenol; 38 males, Rimchu traps 1, 3, 4, 6, 7, 17.v.2001, coll. R. Drew & M. Romig, attracted to methyl eugenol; 1 male, Rimchu, 21.iv.2000, coll. Clarke & Fletcher, attracted to methyl eugenol; 1 male, Rimchu 2, 10.v.2000, coll. B.S. Fletcher, attracted to methyl eugenol; 1 male, Rimchu 3, 10.v.2000, coll. B.S. Fletcher, attracted to methyl eugenol; 9 males (1.ii.2005), 4 males (26.ii.2005), Gelephu Purana Bastey, 1.ii.2005, coll. Maha Prasad & Karma Namgyel, attracted to methyl eugenol; 24 males, Gelephu Purana Bastey, 5.iii.2005, coll. Yangyal Noah & Karma Namgyel, attracted to methyl eugenol; 52 males, Purana Bastey, 3.iv.2005, coll. Yangyal Noah & Karma Namgyel; 39 males (12.iii.2005), 43 males (20.iii.2005), Purana Bastey, coll. Yangnal & Karma Namgyel, attracted to methyl eugenol. **INDIA**: 8 specimens listed by Drew & Raghu (2002, p. 345). **SRI LANKA (formerly CEYLON)**: 1 male, Illukumbra, 5.viii.1997, coll. K. Tsuruta. **VIETNAM: Northern Vietnam** – 1 male (22.v.1999), 1 male (29.v.1999), Hanoi Province, Tu Liem District, Xuan Dinh, Site N004, attracted to methyl eugenol; 1 male, Hanoi Province, Dong Anh District, Site N005, 23. vi.2000, attracted to methyl eugenol; **Southern Vietnam** – 1 male (13.iv.2002), 1 male (27.iv.2002), 3 males (11.v.2002), Dong Nai Province, Long Khanh District, Nhan Nghia Village, Site VS06, coll. SOFRI Staff, attracted to methyl eugenol; 7 males (8.iv.2002), 2 males (8.v.2002), Dac Lac Province, Ku Mi Nga District, Quang Phu Village, Site VS09, coll. SOFRI Staff, attracted to methyl eugenol. Specimens in QDPC.
Diagnosis: Face fulvous with a pair of medium-sized oval black spots; postpronotal lobes and notopleura yellow; scutum red-brown with pale fuscous patterning posteriorly; medium-width parallel-sided lateral postsutural yellow vittae ending at or just behind *ia.* seta; medial postsutural yellow vitta absent; mesopleural stripe reaching to or almost to anterior *npl.* seta dorsally; scutellum yellow; legs with all segments entirely fulvous except apices of femora red-brown and hind tibiae pale fuscous to fuscous; wing with cells bc and c colourless and entirely devoid of microtrichia; a narrow fuscous costal band confluent with R_{2+3} and ending at apex of this vein, a small oval fuscous spot across apex of R_{4+5}; anal streak reduced to a pale tint within cell cup; supernumerary lobe of medium development; abdominal terga III–V red-brown with a 'T' pattern consisting of a narrow transverse black band across anterior margin of tergum III (this band is often broken in the central region) and a narrow medial longitudinal black band over all three terga (this band is often reduced to a stripe over parts of terga IV and V), narrow anterolateral fuscous corners on terga IV and V, a pair of oval red-brown shining spots on tergum V.
Attractant: Methyl eugenol.
Female: As for male except no dense aggregation of microtrichia around A_1+CuA_2; supernumerary lobe weak; pecten of cilia on abdominal tergum III absent; ovipositor basal segment red-brown, dorsoventrally compressed and tapering posteriorly in dorsal view.
Distribution:
Type locality: Unknown.
Previously recorded: Widespread from the Indian subcontinent across South-East Asia. Introduced to Mauritius and Egypt (Drew & Raghu, 2002). After extensive field surveys throughout Indonesia this species has not been recorded and

thus throws in doubt the record from Maluku recorded by Norrbom *et al.* (1998). United Arab Emirates (Merz, 2011).
New records: Bhutan.
Hosts: A wide range of commercial/edible host fruits. See Allwood *et al.* (1999) for host records and Tsuruta *et al.* (1997) for hosts recorded in Sri Lanka.
Remarks: *Bactrocera (Bactrocera) zonata* (Saunders) is similar to *Bactrocera (Bactrocera) affinis* (Hardy) in possessing a red-brown scutum, orange-brown abdominal terga, pale abdominal sternites, the costal band broken beyond apex of R_{2+3} and a small fuscous spot across apex of R_{4+5}. It differs from *B. affinis* in having *prsc.* setae present and the apex of the female piercer needle-shaped. Because of the very close similarities, we are including the illustrations of both *B. zonata* (Fig. 199A) and *B. affinis* (Fig. 199B) for comparison. *Bactrocera zonata* is also similar to *Bactrocera correcta* (Bezzi) in possessing similar wing patterns, but differs in having a red-brown scutum, not black as in *B. correcta.* The specimens listed above in 'Non-type material examined', from Vietnam, exhibit slight differences from *B. zonata* in the shape and size of the facial spots, mesopleural stripes and lateral postsutural yellow vittae. These may eventually be proven to be a distinct species after more extensive field collections are obtained.

SUBGENUS *BULLADACUS* Drew & Hancock

Bactrocera (Bulladacus) bullifera (Hardy)

(Fig. 200)

Dacus bulliferus Hardy, 1973: 32. Holotype male in BBM.
Bactrocera (Bulladacus) bullifera – Norrbom *et al.*, 1998: 97.

Type material examined: Holotype male, **THAILAND**: with the following five labels – 'Songkhla, Thai, 19.vii.1963, Col R Kawasaki'; 'on the tree of Ficus maltissima'; 'No. 13'; 'Holotype male, *Dacus (Strumeta) bulliferus*, D.E. Hardy'; 'Bishop Museum holotype 9947'. Received on loan from BBM.
Non-type material examined: EAST MALAYSIA: SARAWAK – 12 males and 3 females, 'Sg Kapit, 3.viii.1993, coll. S. Leong'; 'bred from Buah ma (rambutan family), sample No. K983'. **INDONESIA: JAVA** – 10 males and 3 females, Ciledug, Tangerang, 1994, coll. Supandi; 'ex larva collected in ripe fruit of Melinjo (*Gnetum gnemon*)'; **SUMATRA** – 1 female, Sumatra, Bangka Belitung, Bangka, Tengah, Air Mesu, 5.xii.2006, coll. Hendra K., in cue lure trap in Rambutan. **PENINSULAR MALAYSIA**: 2 males and 5 females, Kuala Lumpur, Univ. Malaya, Feb. 1994, coll. H.S. Yong; 'bred ex *Gnetum gnemon* (L.) Bulso'. Specimens in QDPC.
Diagnosis: A medium-sized species; face fulvous with a pair of black spots in female and entirely fulvous in male; postpronotal lobes yellow; notopleura entirely yellow in female and with posterior ½ yellow and anterior ½ black in male; scutum black with red-brown patterns on anterior ½; lateral postsutural yellow vittae present; medial postsutural yellow vitta absent; a yellow spot anterior to notopleural suture; mesopleural stripe reaching almost to anterior *npl.* seta dorsally; scutellum yellow; wing with cells bc and c dark fuscous, dense microtrichia covering all of cell c and most of cell bc; a narrow dark fuscous costal band expanding into an irregular-shaped spot in apex of wing, a distinct large dark fuscous bulla across extension of cell cup; no distinct anal streak present; abdominal terga III–V orange-brown with a broad medial and two broad lateral longitudinal black bands over all three terga and joining along anterior margin of tergum III.
Description: male.
Head: Height 1.3 mm. Frons – length 1.67 times breadth; red-brown without dark markings; orbital setae black: 1 *s.or.*, 2 *i.or.*; lunule red-brown. Ocellar triangle black. Vertex red-brown. Face entirely fulvous without dark markings; length 0.47 mm. Genae fulvous without dark markings; red-brown seta present. Occiput fuscous, fulvous along eye margins; occipital row with 1–3 short black setae. Antennae with segments 1 and 2 red-brown; segment 3 missing; length of segments: 0.12 mm; 0.25 mm.
Thorax: Scutum black with longitudinal red-brown patterns over anterior ⅔. Pleural areas entirely black. Yellow markings as follows: postpronotal lobes; posterior ½ notopleura (anterior ½ black); mesopleural stripe reaching almost to anterior *npl.* seta dorsally, continuing to katepisternum as a small transverse spot, anterior margin straight; anatergite (posterior apex black); anterior ⅔ katatergite (remainder black); two broad parallel-sided lateral postsutural vittae beginning with a spot anterior to notopleural suture and ending at *ia.* seta. Medial longitudinal postsutural yellow vitta absent. Postnotum black. Scutellum yellow except for narrow black basal band. Setae: *sc.* 2; *prsc.* 2; *ia.* 1; *p.sa.* 1; *a.sa.* 1; *mpl.* 1; *npl.* 2; *scp.* 4.
Legs: Fore and hind femora entirely fulvous, mid femora fulvous with a distinct black subapical spot on outer surfaces; fore tibiae fulvous, mid tibiae dark fuscous over basal ⅓ and remainder fulvous, hind tibiae dark fuscous; tarsi with all segments entirely fulvous; mid tibiae each with an apical black spur.
Wings: Length 5.1 mm; cells bc and c dark fuscous; dense microtrichia covering all of cell c and most of cell bc; remainder of wings colourless except dark fuscous cell sc, narrow dark fuscous costal band overlapping R_{2+3} and expanding into an irregularly shaped spot across apex of wing, a distinct large dark fuscous bulla across extension of cell cup and a faint fuscous tint in area normally covered by anal streak; a dense aggregation of microtrichia around A_1+CuA_2; supernumerary lobe of medium development.
Abdomen: Oval; terga free; pecten present on tergum III. Tergum I and sterna I and II wider than long. Tergum I orange-brown with broad lateral black margins; tergum II orange-brown with a narrow transverse black band across anterior margin but not reaching narrow black anterolateral corners; terga III–V orange-brown with a moderately broad medial and two broad lateral longitudinal black bands over all three terga and joining across anterior margin of tergum III. No shining spots visible on tergum V. Posterior lobe of surstylus short, sternum V with a slight concavity on posterior margin.
Attractant: No known record. The female from Sumatra was collected in a cue lure trap but this is not regarded as a genuine male lure record.
Female: As for male, except face fulvous with a pair of medium-sized elongate oval black spots; notopleura entirely

yellow; cells bc and c pale fuscous and cell bc bare of microtrichia, costal band narrow and confluent with R_{2+3} and widening slightly across wing apex; abdominal terga IV and V with anterolateral corners black; ovipositor basal segment dark orange-brown, dorsoventrally compressed and tapering posteriorly in dorsal view; ratio of length of oviscape to length of tergum V, 0.6:1; apex of piercer needle-shaped.

Distribution:

Type locality: Thailand (Songkhla).

New records: East Malaysia (Sarawak); Indonesia (Java; Sumatra); Peninsular Malaysia (Kuala Lumpur).

Hosts: This species has been recorded from *Gnetum gnemon* (family *Gnetaceae*). The record Buah ma (family *Sapindaceae*) from label data on specimens collected in Sarawak, East Malaysia, may be incorrect as all confirmed host records for species in the subgenus *Bulladacus* are from *Gnetum gnemon*.

Remarks: *Bactrocera (Bulladacus) bullifera* (Hardy) belongs to a small subgenus in which males have a darkened swelling (bulla) in the wing across the cell cup extension. Other South-East Asian species are *Bactrocera (Bulladacus) mcgregori* (Bezzi) and *Bactrocera (Bulladacus) peterseni* (Hardy) but *B. bullifera* differs from these in possessing a broader costal band overlapping R_{2+3} and expanding markedly across apex of the wing of the male and in lacking a lateral yellow band connecting the postpronotal lobe and notopleuron.

Bactrocera (Bulladacus) captiva Drew & Romig, new species

(Figs 201A, 201B)

Material examined: Holotype male, **PHILIPPINES**: intercepted at Nagoya airport, Japan, from Philippines, bred from host fruit *Eugenia javanica*, coll. K. Funabashi, deposited in BMNH.

Paratype: **PHILIPPINES**: 1 female, same data as holotype. Paratype in QDPC.

Diagnosis: A small species; face entirely fulvous without dark markings; postpronotal lobes yellow; notopleura with anterior ½ black and posterior ½ yellow; scutum black; moderately parallel-sided lateral postsutural yellow vittae present; a broad medial postsutural yellow vitta present; no yellow spot anterior to notopleural suture; mesopleural stripe reaching to postpronotal lobe dorsally; scutellum yellow with a narrow black basal band which expands into distinct spots laterally; wing with cells bc and c dark fuscous, dense microtrichia covering all of both cells bc and c; a broad costal band of variable colour patterns and width, fuscous and confluent with vein M basally before contracting to very pale fuscous and confluent with R_{4+5} centrally then expanding into a very large fuscous spot across apex of wing, a large red-brown bulla across extension of cell cup; anal streak absent; abdominal terga III–V fulvous with narrow black lateral margins on terga IV and V which extend anteriorly across the lateral and anterior margins of tergum III but not meeting in midline, a narrow transverse black band across posterior margin of tergum IV which extends anteriorly on the midline of that tergum, a narrow medial longitudinal black band on tergum V.

Description: male (Fig. 201A).

Head: Height 1.03 mm. Frons – length 1.57 times breadth; dark fuscous with fulvous along lateral margins; orbital setae black: 1 *s.or.*, 2 *i.or.*; lunule fulvous. Ocellar triangle black. Vertex fulvous. Face entirely fulvous without dark markings; length 0.36 mm. Genae red-brown without dark markings; red-brown seta present. Occiput dark fuscous centrally, black laterally, fulvous along eye margins; occipital row with a large number of very small pale setae. Antennae with all segments red-brown; length of segments: 0.08 mm; 0.14 mm; 0.46 mm.

Thorax: Scutum entirely black with small areas of red-brown behind lateral postsutural vittae. Pleural areas entirely black. Yellow markings as follows: postpronotal lobes; posterior ½ notopleura (anterior ½ black); mesopleural stripe reaching postpronotal lobe dorsally, continuing to katepisternum as a large transverse spot, anterior margin slightly concave; anatergite (posterior apex black); anterior $^{3}/_{5}$ katatergite (remainder black); two broad parallel-sided lateral postsutural vittae ending at *ia.* seta; a broad medial longitudinal postsutural yellow vitta present. Postnotum black. Scutellum yellow except for narrow black basal band which expands into distinct spots laterally. Setae: *sc.* 2; *prsc.* 2; *ia.* 1; *p.sa.* 1; *a.sa.* 1; *mpl.* 1; *npl.* 2; *scp.* 2.

Legs: All segments entirely fulvous except apical ½ of hind femora black and hind tibiae dark fuscous to black; mid tibiae each with an apical black spur.

Wings: Length 3.8 mm; cells bc and c dark fuscous; dense microtrichia covering both cells bc and c; remainder of wings colourless except dark fuscous cell sc, a broad costal band confluent with vein M and fuscous to level of *r-m* crossvein, confluent with R_{4+5} from level of *r-m* crossvein and very pale fuscous before expanding into a large fuscous spot across apex of wing and extending over vein M, a large red-brown bulla over extension of cell cup; anal streak absent; a small area of dense microtrichia around A_1+CuA_2; supernumerary lobe very large and keel shaped.

Abdomen: Oval; terga free; pecten present on tergum III. Tergum I and sterna I and II wider than long. Tergum I fulvous with narrow black lateral margins; tergum II fulvous with large black spots on anterolateral corners; terga III–V fulvous with distinct black markings as follows – narrow lateral margins on terga IV and V which extend forward across lateral and anterior margins of tergum III but not meeting in the midline of that tergum, a narrow transverse band across posterior margin of tergum IV which joins the lateral margins and extends forward in the midline to the anterior margin of that tergum, a narrow medial longitudinal black band on tergum V. Shining spots on tergum V absent. Posterior lobe of surstylus short, sternum V with a deep concavity on posterior margin. Abdominal sterna pale.

Attractant: No known record.

Female: (Fig. 201B) – As for male except no dense aggregation of microtrichia around A_1+CuA_2; supernumerary lobe weak; bulla on wing absent, a narrow dark fuscous costal band overlapping R_{2+3} and widening slightly as it continues around apex of wing to M; pecten of cilia on abdominal

tergum III absent; abdominal terga III–V fulvous with a narrow transverse black band across anterior margin of tergum III, narrow lateral black margins over all three terga and a broad medial longitudinal black band over terga IV and V; ovipositor basal segment red-brown, dorsoventrally compressed and tapering posteriorly in dorsal view; ratio of length of oviscape to length of tergum V, 0.85:1; apex of piercer needle-shaped (Fig. 201B).
Distribution: Philippines.
Hosts: *Eugenia javanica* (family *Myrtaceae*) – this record may be incorrect (See 'Hosts' under *B. bullifera*).
Remarks: *Bactrocera (Bulladacus) captiva* n.sp. is similar to *Bactrocera (Bulladacus) bullifera* (Hardy) in possessing a black scutum without a lateral yellow band between the postpronotal lobe and the notopleuron but differs from this species in having a medial longitudinal yellow vitta on the scutum, the mesopleural stripe reaching to the postpronotal lobe dorsally and the scutum being entirely black.

Bactrocera (Bulladacus) cinnabaria Drew & Romig, new species

(Fig. 202)

Material examined: Holotype male, **INDIA: NICOBAR ISLANDS**–Arong, Carnicobar, 10.vii.1995, coll. Ranganath, bred from *Gnetum* sp., Sl No. 42, deposited in BMNH.
Paratypes: **INDIA: NICOBAR ISLANDS** – 3 females, same data as holotype. 1 paratype in ANIC, 1 paratype in QDPC, 1 paratype in QMIC.
Diagnosis: A small species; face entirely fulvous without dark markings; postpronotal lobes yellow; notopleura yellow on posterior ½ and red-brown on anterior ½; scutum uniformly red-brown; lateral postsutural yellow vittae present; medial postsutural yellow vitta absent; no yellow spot anterior to notopleural suture; mesopleural stripe reaching to postpronotal lobe dorsally; scutellum yellow; wing with cells bc and c pale fuscous, microtrichia covering most of cell c; a narrow dark fuscous costal band widening slightly as it continues around apex of wing to cross the vein M; a moderately broad pale fuscous anal streak, a large bulla over extension of cell cup; abdominal terga III–V red-brown with narrow transverse black bands on anterolateral corners of tergum III, narrow lateral black bands on terga IV and V, a broad medial longitudinal black band over tergum V.
Description: male.
Head: Height 1.2 mm. Frons – length 1.8 times breadth; red-brown without dark markings; orbital setae large and black: 1 *s.or.*, 2 *i.or.*; lunule red-brown. Ocellar triangle black. Vertex red-brown. Face entirely fulvous without dark markings; length 0.42 mm. Genae entirely fulvous, black seta present. Occiput red-brown, small fuscous to dark fuscous areas laterally and fulvous along eye margins; occipital row with 5–7 strong black setae and a large number of short dark setae. Antennae with all segments red-brown; length of segments: 0.08 mm; 0.2 mm; 0.44 mm.
Thorax: Scutum uniformly red-brown with a narrow transverse black posterior margin adjacent to the scutellum. Pleural areas red-brown with black along anterior and posterior margins of mesopleural stripe, above hind coxae and on posterior margin of katepisternum. Yellow markings as follows: postpronotal lobes; posterior ½ notopleura (anterior ½ red-brown); a broad mesopleural stripe reaching to postpronotal lobe dorsally, continuing to katepisternum as a large transverse spot, anterior margin straight; anatergite (posterior apex black); anterior $^{3}/_{5}$ katatergite (remainder black); two medium-width parallel-sided lateral postsutural vittae ending before *ia.* seta. Medial longitudinal postsutural yellow vitta absent. Postnotum black tending dark red-brown centrally. Scutellum yellow except for narrow black basal band. Setae: *sc.* 2; *prsc.* 2; *ia.* 1; *p.sa.* 1; *a.sa.* 1; *mpl.* 1; *npl.* 2; *scp.* 4.
Legs: Fore and mid femora fulvous, hind femora fulvous with red-brown around apical ⅓; fore and mid tibiae fulvous, hind tibiae fuscous to dark fuscous; fore and mid tarsi with all segments fulvous, hind tarsi with basal segment fulvous and apical four segments red-brown; mid tibiae each with an apical black spur.
Wings: Length 4.0 mm; cells bc and c pale fuscous; microtrichia over most of cell c; remainder of wings colourless except dark fuscous cell sc, a narrow dark fuscous costal band confluent with R_{2+3} and widening slightly as it continues around apex of wing to cross the vein M, a moderately broad pale fuscous anal streak, a large pale fuscous bulla over extension of cell cup; a dense aggregation of microtrichia around A_1+CuA_2; supernumerary lobe weak.
Abdomen: Oval; terga free; pecten present on tergum III. Tergum I and sterna I and II wider than long. Tergum I entirely red-brown; tergum II red-brown anteriorly to fulvous posteriorly; terga III–V red-brown with black markings as follows – narrow transverse bands on anterolateral corners of tergum III, narrow lateral margins on terga IV and V, a broad medial longitudinal band on tergum V. There are no visible shining spots on tergum V. Posterior lobe of surstylus short, sternum V with a deep concavity on posterior margin. Abdominal sterna pale.
Attractant: No known record.
Female: As for male except no dense aggregation of microtrichia around A_1+CuA_2, bulla on wing absent, pecten of cilia on abdominal tergum III absent; ovipositor basal segment orange-brown, dorsoventrally compressed and tapering slightly posteriorly in dorsal view; ratio of length of oviscape to length of tergum V, 0.3:1; apex of piercer needle-shaped.
Distribution: India – Nicobar Is (Car Nicobar I.).
Hosts: *Gnetum* sp. (family *Gnetaceae*).
Remarks: *Bactrocera (Bulladacus) cinnabaria* n.sp. is a unique species of the subgenus *Bulladacus* in South-East Asia in being primarily red-brown with, at most, small areas of black on abdominal terga III–V.

Bactrocera (Bulladacus) mcgregori (Bezzi)

(Fig. 203)

Chaetodacus mcgregori Bezzi, 1919: 426. Lectotype male and Paralectotype female in MCSNM.

Dacus (Strumeta) mcgregori – Hardy & Adachi, 1954: 176; Hardy, 1969b: 479, Lectotype designation; Hardy, 1974: 37.
Bactrocera (Bulladacus) mcgregori – Norrbom *et al*., 1998: 97.

Type material examined: Lectotype male, **PHILIPPINES**: with the following three labels – 'Batbatan Isl, R.C. McGregor'; 'macgregori n.' (Bezzi's handwriting); 'Lectotype male, Chaetodacus macgregori Bez., desig. by Hardy 68'. One Paralectotype female, 'Batbatan Isl, R.C. McGregor'; 'Chaetodacus macgregori Bezz., Typ.' (Bezzi's handwriting). Received on loan from MCSNM.
Diagnosis: A small species; face entirely fulvous; postpronotal lobes and notopleura yellow; scutum red-brown with dark markings; lateral postsutural yellow vittae present; medial postsutural yellow vittae absent; a yellow spot anterior to notopleural suture; a lateral yellow band running from hind margin of postpronotal lobes almost to notopleuron; mesopleural stripe reaching postpronotal lobe dorsally; scutellum yellow; wing with cells bc and c with very pale tint, microtrichia over most of cell c; a narrow dark fuscous costal band confluent with R_{2+3}, a faint tint in cell cup in male and very pale anal streak in female; a large bulla in wing of male over extension of cell cup; abdominal terga III–V orange-brown with a narrow black band across anterior margin of tergum III but not reaching lateral margins, a broad medial longitudinal black band over terga IV and V and black anterolateral corners on terga IV and V.
Description: male.
Head: Height 1.43 mm. Frons – length 1.56 times breadth; red-brown without dark markings; orbital setae large and dark fuscous: 1 *s.or.*, 2 *i.or.*; lunule red-brown. Ocellar triangle black. Vertex red-brown. Face fulvous without dark markings; length 0.47 mm. Genae fulvous; red-brown seta present. Occiput red-brown, pale fuscous dorsally and fulvous along eye margins; occipital row with four strong dark setae. Antennae with all segments entirely red-brown; segment 3 short and truncated apically; length of segments: 0.12 mm; 0.24 mm; 0.56 mm.
Thorax: Scutum red-brown with dark markings as follows: a narrow black band along outer margin of lateral postsutural vittae, a narrow black 'U'-shaped pattern between lateral postsutural vittae, dark fuscous around anterior *npl*. seta and dark fuscous markings on anterior margin of scutum. Pleural areas black except red-brown below postpronotal lobes. Yellow markings as follows: postpronotal lobes; notopleura; a yellow vitta running from posterior margin of postpronotal lobe almost reaching notopleuron; a broad mesopleural stripe reaching postpronotal lobe dorsally, continuing to katepisternum as a transverse spot, anterior margin straight; anatergite (posterior apex black); anterior ⅔ katatergite (remainder black); two broad parallel-sided lateral postsutural vittae beginning in a spot anterior to notopleural suture anteriorly and ending at the *ia*. seta. Medial longitudinal postsutural yellow vitta absent. Postnotum black. Scutellum yellow except for narrow black basal band. Setae: *sc*. 2; *prsc*. 2; *ia*. 1; *p.sa*. 1; *a.sa*. 1; *mpl*. 1; *npl*. 2; *scp*. 4.
Legs: All segments entirely fulvous except hind tibiae pale fuscous; mid tibiae each with an apical black spur.
Wings: Length 4.65 mm; cells bc and c with a very pale tint; microtrichia over most of cell c; remainder of wings colourless except for a dark fuscous cell sc, narrow dark fuscous costal band confluent with R_{2+3} and widening very slightly around apex of wing to end at apex of M, an indistinct anal streak represented by a pale tint in cell cup and around a fuscous bulla on cell cup extension; no dense aggregation of microtrichia around A_1+CuA_2; supernumerary lobe weak.
Abdomen: Oval; terga free; pecten present on tergum III. Tergum I and sterna I and II wider than long. Tergum I orange-brown with narrow dark fuscous anterolateral margins; tergum II orange-brown with a narrow transverse black band across anterior margin but not reaching lateral margins; terga III–V orange-brown except for a narrow transverse black band across anterior margin of tergum III but not reaching lateral margins, a broad medial longitudinal black band over terga IV and V, black anterolateral corners on terga IV and V. Oval shining spots on tergum V absent. Posterior lobe of surstylus short, sternum V with a deep concavity on posterior margin.
Attractant: No known record.
Female: As for male except no pecten present on abdominal tergum III; no bulla present on wing; a pale anal streak present in wing; abdominal tergum III with a narrow transverse black band across anterior margin and reaching lateral margins; abdominal tergum IV with a distinct black 'T' pattern; ovipositor basal segment orange-brown, dorsoventrally compressed and tapering posteriorly in dorsal view; ratio of length of oviscape to length of tergum V, 0.5:1; apex of piercer needle-shaped.
Distribution:
Type locality: Philippines (Batbatan I., Antique Province).
Previously recorded: Type locality and Singapore (Norrbom *et al*., 1998); India (Andaman Is) (David & Ramani, 2011).
Hosts: *Gnetum gnemon* (family *Gnetaceae*) (Allwood *et al*., 1999).
Remarks: *Bactrocera (Bulladacus) mcgregori* (Bezzi) is very similar to *Bactrocera (Bulladacus) peterseni* (Hardy), also from the Philippines. The latter species differs in having fuscous to dark fuscous on apical ⅓ of mid femora, a narrower mesopleural stripe reaching midway between anterior margin of notopleuron and anterior *npl*. seta dorsally, notopleura with anterior ½ black and costal cells dark fuscous. Information on the biology and host plant behaviour was presented by Yong (1994).

Bactrocera (Bulladacus) peterseni (Hardy)

(Fig. 204)

Dacus peterseni Hardy, 1970: 75. Holotype male in UZMC.
Bactrocera petersoni – Hardy, 1974: 41 [Misspelling for *peterseni*].
Bactrocera (Bulladacus) peterseni – Norrbom *et al*., 1998: 97.

Type material examined: Holotype male, **PHILIPPINES**: with the following four labels – 'Philippines, Tawi Tawi, Tarawakan, north of Batu Batu, 6. Nov 1961, Noona Dan Exp. 61-62'; 'Caught in Malaise traps'; '*Dacus (Strumeta)* *petersoni* [Misspelling], Det. 1968, Hardy, D.E. Hardy'; 'Holotype male, Dacus (Strumeta) petersoni [Misspelling], D.E. Hardy'. Received on loan from UZMC.

Diagnosis: A small species; face entirely fulvous without dark markings; postpronotal lobes yellow; notopleura with anterior ½ black and posterior ½ yellow; scutum black with a large broad medial longitudinal dark red-brown area; lateral postsutural yellow vittae present; medial longitudinal postsutural yellow vitta absent; a large yellow spot anterior to notopleural suture in line with lateral postsutural vittae; lateral yellow bands connecting postpronotal lobe to spot anterior to notopleural suture; mesopleural stripe reaching almost to anterior *npl.* seta dorsally; scutellum yellow with a narrow black basal band which broadens laterally; wing with cells bc and c dark fuscous, dense microtrichia covering all of both cells; a narrow dark fuscous costal band confluent with R_{2+3} and broadening across apex of wing; anal streak absent but with a large dark fuscous bulla over extension of cell cup and a pale fuscous tint between bulla and basal area of cell cup, ill-defined areas of dark fuscous along posterior margins of R_{2+3} and R_{4+5}; abdominal terga III–V fulvous with a black 'T' pattern consisting of a broad transverse black band across anterior margin of tergum III which widens to cover lateral margins and a broad medial longitudinal black band over all three terga, narrow lateral black margins on terga IV and V.

Description: male.

Head: Height 1.27 mm. Frons – length 1.8 times breadth; red-brown, fulvous along lateral margins and pale fuscous on anteromedial hump; orbital setae large and black: 1 *s.or.*, 2 *i.or.*; lunule red-brown. Ocellar triangle black. Vertex red-brown to fulvous. Face entirely fulvous without dark markings; length 0.43 mm. Genae fulvous, large pale fuscous to fuscous subocular spot present; black seta present. Occiput dark fuscous, fulvous along eye margins; occipital row with two strong black setae. Antennae with segments 1 and 2 fulvous, segment 3 fulvous with pale fuscous on apex and outer surface; length of segments: 0.08 mm; 0.2 mm; 0.54 mm.

Thorax: Scutum black with a broad medial longitudinal dark red-brown area within which is a narrow medial longitudinal black band, dark red-brown posterolateral to lateral postsutural vittae. Pleural areas entirely black. Yellow markings as follows: postpronotal lobes; posterior ½ notopleura (anterior ½ black); mesopleural stripe reaching almost to anterior *npl.* seta dorsally, continuing to katepisternum as a broad transverse spot, anterior margin slightly convex; anatergite (posterior apex black); anterior ¾ katatergite (remainder black); two broad parallel-sided lateral postsutural vittae beginning with a large spot anterior to notopleural suture and ending posteriorly at level of *ia.* seta; a moderately broad yellow band on lateral margins connecting postpronotal lobe to yellow spot anterior to notopleural suture. Medial longitudinal postsutural yellow vitta absent. Postnotum black. Scutellum yellow except for narrow black basal band which broadens laterally (this is typical of species in subgenus *Bulladacus*). Setae: *sc.* 2; *prsc.* 2; *ia.* 1; *p.sa.* 1; *a.sa.* 1; *mpl.* 1; *npl.* 2; *scp.* 4.

Legs: Fore and hind femora entirely fulvous, mid femora fulvous with apical ⅓ fuscous to dark fuscous; fore tibiae entirely fulvous, mid tibiae dark fuscous tending fulvous apically, hind tibiae dark fuscous to black; all tarsal segments fulvous; mid tibiae each with an apical black spur.

Wings: Length 4.8 mm; cells bc and c dark fuscous; dense microtrichia over all of both cells; remainder of wings colourless except for a dark fuscous cell sc, narrow dark fuscous costal band confluent with R_{2+3} and broadening across apex of wing to end at apex of M; anal streak absent but with a large dark fuscous bulla on cell cup extension and a pale fuscous tint between bulla and base of cell cup; small ill-defined areas of dark fuscous tint along posterior margins of R_{2+3} and R_{4+5}; dense aggregation of microtrichia around A_1+CuA_2; supernumerary lobe of medium development.

Abdomen: Oval; terga free; pecten present on tergum III. Tergum I and sterna I and II wider than long. Tergum I dark fuscous with a broad transverse fulvous band across anterior margin but not reaching lateral margins; tergum II fulvous with a broad inverted 'U'-shaped black band across anterior margin but not reaching the narrow black anterolateral corners; terga III–V fulvous with a 'T' pattern consisting of a moderately broad transverse black band across anterior margin of tergum III which widens to cover lateral margins and a broad medial longitudinal black band over all three terga, narrow lateral black margins on terga IV and V. Oval shining spots on tergum V absent. Posterior lobe of surstylus short, sternum V with a deep concavity on posterior margin. Abdominal sterna dark coloured.

Attractant: No known record.

Female: No known record.

Distribution:

Type locality: Philippines (Tawitawi I.).

Hosts: No known record.

Remarks: *Bactrocera (Bulladacus) peterseni* (Hardy) is similar to *Bactrocera (Bulladacus) bullifera* (Hardy) in the general colour patterns of the scutum and abdomen but differs in possessing a narrow costal band confluent with R_{2+3}, a lateral yellow band connecting the postpronotal lobe and notopleuron, black over the anterior half of the notopleuron and dark fuscous around the apices of the mid femora.

SUBGENUS *DACULUS* Speiser

Bactrocera (Daculus) digressa Radhakrishnan

(Fig. 205)

Bactrocera (Bactrocera) digressa Radhakrishnan, 1999: 1; David and Ramani, 2011:1. Holotype female in ZSI.
Bactrocera (Daculus) yercaudiae Drew, in Drew & Raghu, 2002: 346; David and Ramani, 2011: 16 (Synonym).

Other type material examined: Holotype male of *Bactrocera (Daculus) yercaudiae*, **INDIA**: 15 km Yercaud, 750 m, 3.vi.1992, coll. D. Hancock, attracted to cue lure. Studied in BMNH.

Diagnosis: Face fulvous with a pair of medium-sized elongate oval black spots; postpronotal lobes yellow (anterodorsal corners orange-brown); notopleura yellow (orange-brown on anterior and posterior margins); scutum orange-brown with red-brown margins along either side of lateral postsutural vittae and irregularly shaped patterns occasionally anterolaterally; narrow parallel-sided lateral postsutural yellow vittae ending at *ia.* setae; medial postsutural yellow vitta absent; mesopleural stripe equal in width to notopleuron dorsally; scutellum yellow with a broad red-brown basal band; legs with fore femora red-brown with fuscous to dark fuscous patterns on outer surfaces, mid femora red-brown with pale fulvous basally and dark markings as for fore femora, hind femora with basal ½ pale fulvous and apical ½ red-brown; tibiae generally pale fuscous to fuscous; wing with cells bc and c colourless, microtrichia in outer corner of cell c only; a narrow fuscous costal band just overlapping R_{2+3} and remaining narrow around apex of wing; a narrow fuscous anal streak; abdominal terga III–V red-brown with a narrow black 'T' pattern; posterior lobe of surstylus short; abdominal sternum V with a deep concavity on posterior margin; *a.sa.* and *prsc.* setae absent.

Attractant: Cue lure.

Female: As for male except no dense aggregation of microtrichia around A_1+CuA_2, supernumerary lobe weak, pecten of cilia on abdominal tergum III absent, basal segment of ovipositor red-brown, apex of piercer trilobed.

Distribution:

Type locality: Southern India (Tamil Nadu, Shevroy Hills).

Previously recorded: Type locality and Southern India (other localities in Karnataka; Tamil Nadu, Yercaud) (David & Ramani, 2011).

Hosts: No known record.

Remarks: *Bactrocera (Daculus) digressa* Radhakrishnan is similar to *Bactrocera (Javadacus) pallescentis* (Hardy) in possessing a red-brown scutum and *a.sa.* setae absent, but differs from this species in having the *prsc.* setae absent and short posterior surstylus lobes. It is also similar to *Bactrocera (Bactrocera) zonata* (Saunders) in possessing a red-brown scutum and a narrow mesopleural stripe but differs in having a complete costal band and *a.sa.* setae absent.

Bactrocera (Daculus) rutengiae Drew & Romig, new species

(Fig. 206)

Material examined: Holotype male, **INDONESIA: FLORES** – Ruteng forest, S. Ruteng, Flores, Indonesia, 1000–1350 m, 24.vi.1981, coll. J. Turner, Primary Wet Montane forest, attracted to methyl eugenol, deposited in BMNH.

Paratype: **INDONESIA**: one male, same data as holotype, in QDPC.

Diagnosis: A medium-sized species; face fulvous with a pair of medium-sized oval black spots; postpronotal lobes and notopleura yellow; scutum black; moderately broad and very short lateral postsutural yellow vittae present; medial postsutural yellow vitta absent; no yellow spot anterior to notopleural suture; mesopleural stripe equal in width to notopleuron dorsally; scutellum yellow with a dark fuscous apical spot; wing with cell c pale fuscous, cell bc fuscous, microtrichia over outer ½ of cell c only; a narrow dark fuscous costal band; a moderately broad dark fuscous anal streak; abdominal terga III–V fulvous with a moderately broad medial longitudinal black band and moderately broad lateral longitudinal bands which vary from black on tergum III to fuscous on tergum V and joined along anterior margin of tergum III.

Description: male.

Head: Height 1.4 mm. Frons – length 1.65 times breadth; red-brown without dark markings; orbital setae black: 1 *s. or.*, 2 *i.or.*; lunule red-brown. Ocellar triangle black. Vertex red-brown. Face fulvous with a pair of medium-sized oval black spots; length 0.46 mm. Genae red-brown; red-brown seta present. Occiput red-brown, fulvous along eye margins; occipital row with a large number of strong black setae. Antennae with segments 1 and 2 red-brown, segment 3 red-brown with dark fuscous on apex and outer surface; length of segments: 0.16 mm; 0.28 mm; 0.74 mm.

Thorax: Scutum entirely black except red-brown below and behind lateral postsutural vittae. Pleural areas entirely black except dark fuscous below postpronotal lobes. Yellow markings as follows: postpronotal lobes; notopleura; mesopleural stripe equal in width to notopleuron dorsally, not continuing to katepisternum, anterior margin slightly convex; anatergite (posterior apex black); anterior $^3/_5$ katatergite (remainder black); two moderately broad and very short lateral postsutural vittae ending at or just behind level of *ia.* seta. Medial longitudinal postsutural yellow vitta absent. Postnotum black. Scutellum yellow except for a moderately broad black basal band and a large dark fuscous apical spot enclosing both apical setae. Setae: *sc.* 2; *prsc.* absent; *ia.* 1; *p.sa.* 1; *a.sa.* absent; *mpl.* 1; *npl.* 2; *scp.* 4.

Legs: Fore femora dark fuscous to black on outer surfaces and fulvous on inner surfaces, mid femora entirely dark fuscous to black, hind femora fulvous with dark fuscous to black on apical ⅓ fore tibiae fuscous, mid tibiae fuscous basally to red-brown apically, hind tibiae dark fuscous; all tarsi with basal segment fulvous and apical four segments red-brown to fuscous; mid tibiae each with an apical black spur.

Wings: Length 5.7 mm; cell c pale fuscous, cell bc fuscous; microtrichia over outer ½ of cell c only; remainder of wings colourless except dark fuscous cell sc, narrow dark fuscous costal band confluent with R_{2+3} and remaining of uniform width around apex of wing, a moderately broad dark fuscous anal streak; a dense aggregation of microtrichia around A_1+CuA_2; supernumerary lobe of medium development.

Abdomen: Oval; terga free; pecten present on tergum III. Tergum I and sterna I and II wider than long. Tergum I black with a narrow red-brown band along posterior margin; tergum II red-brown anteriorly, fulvous posteriorly and with narrow dark fuscous anterolateral corners; terga III–V fulvous with a moderately broad medial longitudinal black band and medium-width lateral longitudinal bands which vary from black on tergum III to fuscous on tergum V and all joined along anterior margin of tergum III. A pair of oval red-brown shining spots on tergum V. Posterior lobe of surstylus short, sternum V with a deep concavity on posterior margin. Abdominal sterna dark fuscous to black.

Attractant: Methyl eugenol.

Female: No known record.

Distribution: Indonesia (Flores).

Hosts: No known record.
Remarks: *Bactrocera (Daculus) rutengiae* n.sp. is one of only two South-East Asian species which we have placed in the subgenus *Daculus*, the other being *Bactrocera (Daculus) digressa* Radhakrishnan. *Bactrocera rutengiae* differs from *B. digressa* in having a black scutum, cells bc and c pale fuscous to fuscous, all femora with large areas of fuscous to black, and abdominal terga III–V with broad medial and lateral longitudinal dark fuscous to black bands joined over the anterior margin of tergum III.

SUBGENUS *GYMNODACUS* Munro

Bactrocera (Gymnodacus) calophylli (Perkins & May)

(Fig. 207)

Asiadacus calophylli Perkins & May, 1949: 16. Lectotype female in QMIC.
Dacus (Gymnodacus) calophylli – Hardy, 1951: 130.
Gymnodacus calophylli – May, 1963: 42, 49.
Bactrocera (Gymnodacus) calophylli – Drew, 1989: 167. Lectotype designation.

Non-type material examined: INDIA: ANDAMAN ISLANDS – 1 female, S. Andaman, Havelock Is., 10.vi.1998, coll. Ranganath, Sl. No. 64, *Calophyllum*. 1 male and 3 females, C. Lapathi, Carnicobar, 22.vi.1995, coll. Ranganath, bred ex *Calophyllum* sp., Sl No. 43. **THAILAND**: a large series of specimens, bred ex *Calophyllum inophyllum* at Prince of Songkhla University, Songkhla. Specimens in QDPC.
Diagnosis: Face fulvous with a pair of small elongate fuscous spots; postpronotal lobes and notopleura yellow; scutum black with brown around lateral margins; lateral postsutural yellow vittae of medium-width and narrowing posteriorly to end well before *ia.* seta; medial postsutural yellow vitta absent; mesopleural stripe reaching anterior *npl.* seta dorsally; scutellum yellow; legs with all segments fulvous except fore and hind tibiae pale fuscous; wing with cells bc and c with a pale tint, microtrichia in outer corner of cell c only; a broad fuscous costal band confluent with R_{4+5} and not widening at apex, pale fuscous within cell cup, supernumerary lobe weak in males and females; no dense aggregation of microtrichia around A_1+CuA_2; abdominal terga III–V orange-brown except for a black 'T' pattern and narrow fuscous anterolateral corners on terga IV and V; posterior lobe of surstylus short; abdominal sternum V with a deep concavity on posterior margin; pecten of setae absent on tergum III.
Attractant: No known record.
Female: As for male in the general body colour patterns. See Drew (1989) for a detailed description of female characters.
Distribution:
Type locality: Australia (Queensland, Cairns).
Previously recorded: Type locality and the following locations – Peninsular Malaysia; Singapore (Norrbom *et al.*, 1998); India (Andaman Is) (David & Ramani, 2011).
New records: Thailand (Songkhla).
Hosts: *Calophyllum inophyllum* (family *Clusiaceae*) (Hancock *et al.*, 2000).
Remarks: *Bactrocera (Gymnodacus) calophylli* (Perkins & May) is an unusual species in that it is distributed from Australia across South-East Asia. It is similar to *Bactrocera (Gymnodacus) kuniyoshii* (Shiraki) in possessing a black scutum, wing with a broad costal band confluent with R_{4+5} and abdominal terga III–V with a black 'T' pattern, but differs from this species in having the lateral postsutural vittae ending before the *ia.* seta and femora entirely fulvous.

Bactrocera (Gymnodacus) continua (Bezzi)

(Fig. 208)

Chaetodacus continuus Bezzi, 1919: 424. Holotype female in MCSNM.
Dacus (Strumeta) continuus – Hardy & Adachi, 1954: 164; Hardy, 1974: 27.
Bactrocera (Bactrocera) continua – Norrbom *et al.*, 1998: 89.

Non-type material examined: PHILIPPINES: LUZON ISLAND – 1 male, Mt. Makiling, Laguna, P.I., 9.ix.1930, coll. F.C. Hadden; 1 female, same data as male plus a white label added by Hardy – '*Dacus (Bactrocera) continuus* (Bez), det. D.E. Hardy 1977'. Received on loan from USNM.
Diagnosis: A medium-sized species; face entirely fulvous in male and with a pair of large elongate black spots in female; postpronotal lobes yellow; notopleura with posterior ½ yellow, anterior ½ dark fuscous; scutum red-brown with narrow medial and lateral longitudinal dark fuscous lines; two short broad triangular lateral postsutural yellow vittae; medial postsutural yellow vitta absent; a lateral yellow band joining the postpronotal lobe and the notopleural suture; mesopleural stripe reaching to the postpronotal lobe dorsally; scutellum yellow; legs with femora mostly fulvous in males and with large dark fuscous pattern in females; wing with cells bc and c colourless, dense microtrichia covering all of both cells; a narrow fuscous costal band confluent with R_{2+3} and widening slightly across apex; a broad fuscous anal streak; no dense aggregation of microtrichia around A_1+CuA_2; abdominal terga III–V orange-brown with a dark fuscous 'T' pattern; *a.sa.* seta absent.
Description: male.
Head: Height 1.3 mm. Frons – length 1.7 times breadth; red-brown with fuscous on anteromedial hump; orbital setae red-brown: 1 *s.or.*, 2 *i.or.*; lunule red-brown. Ocellar triangle black. Vertex red-brown. Face entirely fulvous in male; length 0.7 mm. Genae entirely fulvous, red-brown seta present. Occiput red-brown, fulvous along eye margins; no distinct setae in occipital row. Antennae with segments 1 and 2 red-brown, segment 3 red-brown with fuscous on outer surface; length of segments: 0.14 mm; 0.2 mm; 0.7 mm.
Thorax: Scutum uniformly red-brown with a narrow medial and two narrow lateral longitudinal dark fuscous lines. Pleural areas red-brown with fuscous along posterior and anterior margins of mesopleural stripe. Yellow markings as follows: postpronotal lobes; posterior ½ of notopleura (anterior ½ dark fuscous); broad mesopleural stripe reaching to

postpronotal lobes dorsally, continuing to katepisternum as a large transverse spot, anterior margin convex; anatergite (posterior apex red-brown); anterior ⅔ katatergite (remainder red-brown); two very broad triangular lateral postsutural vittae narrowing sharply posteriorly to end before level of *p.sa.* seta; a distinct yellow band running from the posterior margin of the postpronotal lobe to the notopleural suture on each side of the scutum. Medial longitudinal postsutural yellow vitta absent. Postnotum red-brown. Scutellum yellow except for narrow red-brown basal band. Setae: *sc.* 2; *prsc.* 2; *ia.* 1; *p.sa.* 1; *a.sa.* absent; *mpl.* 1; *npl.* 2; *scp.* 4.
Legs: All segments fulvous except hind tibiae dark fuscous; note female leg colours below.
Wings: Length 5.1 mm; cells bc and c colourless; dense microtrichia covering all of both cells; remainder of wings colourless except fuscous cell sc, narrow fuscous costal band confluent with R_{2+3} and remaining narrow around wing margin before widening across apex of wing, a moderately broad fuscous anal streak; no dense aggregation of microtrichia around A_1+CuA_2; supernumerary lobe weak.
Abdomen: Oval; terga free; no pecten on tergum III. Tergum I and sterna I and II wider than long. Tergum I dark fuscous tending paler centrally and with a narrow transverse fulvous band across posterior margin but not reaching lateral margins; tergum II fulvous with a narrow transverse subanterior dark fuscous line reaching dark fuscous anterolateral corners; terga III–V orange-brown with a dark fuscous 'T' pattern consisting of a narrow transverse band across anterior margin of tergum III widening to cover lateral margins and a narrow medial longitudinal band over all three terga. A pair of oval orange-brown shining spots on tergum V. Posterior lobe of surstylus short, sternum V with a deep concavity on posterior margin. Abdominal sterna pale.
Attractant: No known record.
Female: As for male except face fulvous with a pair of large elongate oval black spots filling the bottom ½ of the antennal sockets and reaching the oral margin; legs with the fore femora fulvous with a very large dark fuscous spot covering outer apical ⅔, mid femora fulvous with a dark fuscous outer apical spot, hind femora fulvous with apical ⅔ dark fuscous; anterolateral corners of abdominal terga IV and V with narrow fuscous patterns; ovipositor basal segment orange-brown, dorsoventrally compressed and tapering posteriorly in dorsal view; ratio of length of oviscape to length of tergum V, 0.65:1; apex of piercer needle-shaped.
Distribution:
Type locality: Philippines (Batbatan I., Antique Province).
New records: Philippines (Luzon I., Mt Makiling).
Hosts: No known record.
Remarks: *Bactrocera (Gymnodacus) continua* (Bezzi) is one of only four species of the subgenus *Gymnodacus* recorded in South-East Asia. It is most similar to *Bactrocera (Gymnodacus) calophylli* (Perkins & May) in the general colour patterns of the wing, legs and abdomen but differs from this species in having an entirely red-brown scutum, broad triangular lateral postsutural vittae, a yellow band running from the postpronotal lobe to the notopleural suture, a broad mesopleural stripe reaching the postpronotal lobe, notopleura not entirely yellow, cells bc and c covered with dense microtrichia and a narrow fuscous costal band.

Bactrocera (Gymnodacus) kuniyoshii (Shiraki)
(Fig. 209)

Gymnodacus kuniyoshii Shiraki, 1968: 9. Holotype male in NIAS.
Bactrocera (Gymnodacus) kuniyoshii – Norrbom *et al.*, 1998: 97.

Type material examined: Holotype male, **JAPAN**: with the following four labels – '1-8-1953, Ryukyu Is., T. Shiraki'; 'Kőrper'; 'Gymnodacus kuniyoshii sp.n., Det. T. Shiraki'; 'Holotype'. Received on loan from NIAS.
Non-type material examined: JAPAN: 1 male and 1 female, Ishigaki, Shiraho, Teriha-Boku, 30.i.1989, coll. H. Nakaima. Specimens in QDPC.
Diagnosis: A small species; face fulvous with a pair of very small elongate oval dark fuscous spots; postpronotal lobes and notopleura yellow; scutum black; lateral postsutural yellow vittae present; medial postsutural yellow vitta absent; no yellow spot anterior to notopleural suture; mesopleural stripe reaching anterior *npl.* seta dorsally; scutellum yellow with a narrow black basal band; wing with cells bc and c with very pale tint, microtrichia in outer ⅓ of cell c only; a broad fuscous costal band confluent with R_{4+5}; anal streak reduced to a pale fuscous tint in cell cup; no dense aggregation of microtrichia around A_1+CuA_2; abdominal terga III–V orange-brown with a black 'T' pattern consisting of a narrow black band across anterior margin of tergum III and a very broad medial longitudinal black band over all three terga, broad dark fuscous to black margins on tergum III and narrow fuscous margins on terga IV and V.
Description: male.
Head: Height 1.48 mm. Frons – length 1.9 times breadth; fulvous with dark fuscous on anteromedial hump; orbital setae black: 1 *s.or.*, 2 *i.or.*; lunule fuscous. Ocellar triangle black. Vertex fuscous. Face fulvous with a pair of very small elongate oval spots; length 0.44 mm. Genae fulvous, fuscous subocular spot present; red-brown seta present. Occiput dark fuscous, fulvous along eye margin; occipital row with 3–5 small dark setae. Antennae with segments 1 and 2 red-brown, segment 3 red-brown with fuscous on apex and outer surface; length of segments: 0.12 mm; 0.26 mm; 0.64 mm.
Thorax: Scutum black with red-brown behind lateral postsutural vittae and inside notopleura. Pleural areas dark fuscous to black without pale markings. Yellow markings as follows: postpronotal lobes; notopleura; mesopleural stripe reaching anterior *npl.* seta dorsally, continuing to katepisternum as a transverse spot, anterior margin convex; anatergite (posterior apex black); anterior ⅔ katatergite (remainder black); two broad sub-parallel lateral postsutural vittae ending at *ia.* seta. Medial longitudinal postsutural yellow vitta absent. Postnotum black. Scutellum yellow except for narrow black basal band. Setae: *sc.* 2; *prsc.* 2; *ia.* 1; *p.sa.* 1; *a.sa.* 1; *mpl.* 1; *npl.* 2; *scp.* 2.
Legs: Fore femora fulvous with fuscous on apical ⅓, mid femora fulvous with fuscous to dark fuscous on apical ⅓, hind femora fulvous with dark fuscous on apical ⅓; all tibiae dark fuscous; all tarsi with basal segment fulvous

and apical four segments pale fuscous; mid tibiae each with an apical black spur.

Wings: Length 4.3 mm; cells bc and c with a very pale tint; microtrichia in outer ⅓ of cell c only; remainder of wings colourless except fuscous cell sc, broad fuscous costal band confluent with R_{4+5} but paler fuscous between R_{2+3} and R_{4+5} and ending midway between apices of R_{4+5} and M, anal streak reduced to a pale fuscous tint within cell cup; no dense aggregation of microtrichia around A_1+CuA_2; supernumerary lobe weak.

Abdomen: Oval; terga free; no pecten on tergum III. Tergum I and sterna I and II wider than long. Tergum I black laterally, dark fuscous centrally with a narrow orange-brown band along posterior margin but not reaching lateral margins; tergum II fulvous with a narrow dark fuscous to black transverse band across anterior margin but not reaching narrow black anterolateral corners; terga III–V orange-brown with a black 'T' pattern consisting of a narrow transverse band across anterior margin of tergum III and a very broad medial longitudinal band over all three terga, broad fuscous to black lateral margins on tergum III, narrow fuscous lateral margins on terga IV and V. A pair of oval orange-brown shining spots on tergum V. Posterior lobe of surstylus short, sternum V with a deep concavity on posterior margin.

Attractant: No known record.

Female: As for male except basal segment of ovipositor red-brown, dorsoventrally compressed and tapering posteriorly in dorsal view, ratio of length of oviscape to length of tergum V, 0.6:1; apex of piercer needle-shaped.

Distribution:

Type locality: Japan (Ryukyu Is – Iriomote I., Funauki).

New records: Japan (Ryukyu Is – Ishigaki I.).

Hosts: No known record.

Remarks: *Bactrocera (Gymnodacus) kuniyoshii* (Shiraki) is very similar to *Bactrocera (Gymnodacus) calophylli* (Perkins & May) in the facial spot markings, thorax, wing and abdominal colour patterns. It differs from *B. calophylli* in having the lateral postsutural vittae ending at the *ia.* setae (not before) and in possessing dark colour patterns on the apices of all femora (not with femora entirely fulvous).

Bactrocera (Gymnodacus) symplocos Drew & Romig, new species

(Fig. 210)

Material examined: Holotype female, **THAILAND**: Doi Pui, Chiang Mai, 27.vi.1991, coll. Chantanee, bred from *Symplocos cochinchinensis*, sample No. CM129, deposited in BMNH.

Paratypes: **THAILAND**: 2 males and 7 females, same data as holotype; 5 males and 5 females, Doi Pui N.P., Chiang Mai, 13.vi.1991, coll. Chantanee, bred from *Symplocos cochinchinensis*, Sample No. CM94; 3 males and 1 female, Doi Pui, Chiang Mai, 25.vii.1991, coll. Chantanee, bred from *Symplocos cochinchinensis*, Sample No. CM207; 4 males and 9 females, Doi Pui, Chiang Mai, 20.vi.1991, coll. Chantanee, bred from *Symplocos cochinchinensis*, Sample No. CM115; 10 males and 8 females, Doi Pui, Chiang Mai, 27.vi.1991, coll. Chantanee, bred from *Symplocos cochinchinensis*, Sample No. CM129; 10 males and 12 females, Doi Pui, Chiang Mai, 4.vii.1991, coll. Chantanee, bred from *Symplocos cochinchinensis*, Sample No. CM148; 8 males and 11 females, Doi Pui, Chiang Mai, 11.vii.1991, coll. Chantanee, bred from *Symplocos cochinchinensis*, Sample No. CM185; 3 males and 2 females, Doi Pui, Chiang Mai, 19.vii.1991, coll. Chantanee, bred from *Symplocos cochinchinensis*, Sample No. CM198; 3 males and 10 females, Doi Pui N.P., Chiang Mai, 13.vi.1991, coll. Chantanee, bred from *Symplocos cochinchinensis*, Sample No. CM090; 1 male and 1 female, Doi Pui, Chiang Mai, 25.vii.1991, coll. Chantanee, bred from *Symplocos cochinchinensis*, Sample No. CM207; 5 males and 2 females, Doi Pui, Chiang Mai, 1.viii.1991, coll. Chantanee, bred from *Symplocos cochinchinensis*, Sample No. CM218; 5 males and 3 females, Doi Pui, Chiang Mai, 20.ix.1991, coll. Chantanee, bred from *Symplocos cochinchinensis*, Sample No. CM258; 5 males and 7 females, Doi Pui, Chiang Mai, 4.x.1991, coll. Chantanee, bred from *Symplocos cochinchinensis*, Sample No. CM304; 5 males and 9 females, Doi Pui, Chiang Mai, 25.vi.1992, coll. Chantanee *et al.*, bred from *Symplocos cochinchinensis*; 4 males and 9 females, Doi Pui, Chiang Mai, 9.vii.1992, coll. Chantanee *et al.*, bred from *Symplocos cochinchinensis*; 1 male and 1 female, Doi Pui, Chiang Mai, 17.x.1992, coll. Chantanee, bred from *Symplocos cochinchinensis*; 4 females, Tum Rusee, Chiang Mai, 1.x.1992, coll. Chantanee, bred from *Symplocos cochinchinensis*; 2 females, Tum Rusee, Chiang Mai, 15.vi.1993, coll. Chantanee *et al.*, bred from *Symplocos cochinchinensis*, Sample No. CM1938; 2 females, Doi Suthep, Chiang Mai, 13.vii.1993, coll. Chantanee *et al.*, bred from *Sapium baccatum*, Sample No. CM1964; 3 males and 4 females, Pa Yao, Chiang Rai, 24.iii.1993, Coll. Charuwan, bred from *Spondias pinnata*, Sample No. CR1346; 4 males and 3 females, Songkhla, Nang Nual Waterfall, Thung Wa, Satun, 23.i.1992, coll. Sieng *et al.*, bred from *Spondias pinnata*, Sample No. SK2912. 20 paratypes in ANIC, 20 paratypes in BMNH, 133 paratypes in QDPC, 20 paratypes in QMIC.

Diagnosis: A medium-sized species; face fulvous with a pair of large circular black spots; postpronotal lobes and notopleura yellow; scutum black; lateral postsutural yellow vittae present; medial postsutural yellow vitta absent; no yellow spot anterior to notopleural suture; mesopleural stripe reaching almost to anterior *npl.* seta dorsally; scutellum yellow; wing with cells bc and c with a pale tint, microtrichia in outer corner of cell c only; a broad fuscous costal band confluent with R_{4+5}; a narrow fuscous anal streak; no dense aggregation of microtrichia around A_1+CuA_2; abdominal terga III–V red-brown with broad medial and lateral longitudinal black bands joined along anterior margin of tergum III.

Description: female.

Head: Height 1.24 mm. Frons – length 1.5 times breadth; red-brown with a large area of fulvous centrally over anteromedial hump; orbital setae black: 1 *s.or.*, 2 *i.or.*; lunule dark fuscous. Ocellar triangle black. Vertex fuscous. Face fulvous with a pair of large circular black spots; length 0.4 mm. Genae red-brown, large pale fuscous subocular spot present;

dark red-brown seta present. Occiput fuscous, fulvous along eye margins; occipital row with 5–6 black setae. Antennae with segments 1 and 2 red-brown, segment 3 dark red-brown with dark fuscous on apex and outer surface; length of segments: 0.16 mm; 0.24 mm; 0.6 mm.

Thorax: Scutum black with red-brown below and behind lateral postsutural vittae, around notopleural suture, inside postpronotal lobes. Pleural areas black except red-brown below postpronotal lobes. Yellow markings as follows: postpronotal lobes; notopleura; mesopleural stripe reaching almost to anterior *npl.* seta dorsally, continuing to katepisternum as a transverse spot, anterior margin straight; anatergite (posterior apex black); anterior ⅔ katatergite (remainder black); two broad parallel-sided lateral postsutural vittae ending at *ia.* seta. Medial longitudinal postsutural yellow vitta absent. Postnotum black. Scutellum yellow except for narrow black basal band. Setae: *sc.* 2; *prsc.* 2; *ia.* 1; *p.sa.* 1; *a.sa.* 1; *mpl.* 1; *npl.* 2; *scp.* 4.

Legs: Fore femora fulvous with a large dark fuscous spot over outer apical ½, mid and hind femora fulvous with small areas of dark fuscous around apices; fore and hind tibiae dark fuscous, mid tibiae dark fuscous basally to fuscous apically; fore tarsi with basal two segments fulvous and apical three segments red-brown, mid and hind tarsi with basal segment fulvous and apical four segments fuscous; mid tibiae each with an apical black spur.

Wings: Length 4.7 mm; cells bc and c with a pale tint; microtrichia in outer corner of cell c only; remainder of wings colourless except dark fuscous cell sc, broad fuscous costal band confluent with R_{4+5} and remaining of uniform width to apex of wing, a narrow short fuscous anal streak; no dense aggregation of microtrichia around A_1+CuA_2; supernumerary lobe weak.

Abdomen: Oval; terga free; no pecten present on tergum III. Tergum I and sterna I and II wider than long. Tergum I black with a narrow transverse red-brown band across posterior margin but not reaching lateral margins; tergum II red-brown anteriorly, fulvous posteriorly, with a narrow transverse black band subanteriorly but not reaching narrow black lateral margins; terga III–V red-brown with broad medial and lateral longitudinal black bands joined over anterior ¾ of tergum III. A pair of oval red-brown shining spots on tergum V. Abdominal sterna dark. Basal segment of ovipositor red-brown, dorsoventrally compressed and tapering posteriorly in dorsal view; ratio of length of oviscape to length of tergum V, 1.5:1; apex of piercer needle-shaped.

Attractant: No known record.

Male: As for female except posterior lobe of surstylus short; abdominal sternum V with a deep concavity on posterior margin.

Distribution: Thailand (Chiang Mai; Chiang Rai; Songkhla).

Hosts: *Symplocos cochinchinensis* (family *Symplocaceae*); *Sapium baccatum* (family *Euphorbiaceae*); *Spondias pinnata* (family *Anacardiaceae*).

Remarks: *Bactrocera (Gymnodacus) symplocos* n.sp. is similar to *Bactrocera (Gymnodacus) kuniyoshii* (Shiraki) in possessing a black scutum, lateral postsutural vittae ending at *ia.* seta and femora fulvous with dark markings but differs from this species in having a pair of large circular spots on the face, fore femora with a large dark fuscous spot on outer apical ½, dark fuscous around extreme apices of mid and hind femora and abdominal terga III–V with broad lateral longitudinal black bands.

Bactrocera (Gymnodacus) tillyardi (Perkins)

(Figs 211A, 211B)

Strumeta tillyardi Perkins, 1938a: 131. Holotype female in QMIC.
Dacus (Strumeta) tillyardi – Hardy & Adachi, 1954: 183.
Bactrocera (Bactrocera) tillyardi – Norrbom *et al.*, 1998: 95.
Asiadacus absonus Hering, 1941c: 1. **New Synonym**.

Type material examined: Holotype female, **PENINSULAR MALAYSIA**: with the following two labels – 'Malay Penin: Selangor Bukit Kutu, 3500 ft., 11.ix.1929, H.M. Pendlebury'; 'Holotype, Strumeta tillyardi, F.A. Perk'. Received on loan from QMIC.

Paratype: **PENINSULAR MALAYSIA**: 1 female, 'Malay Penin., Pahang F.M.S., Frasers Hill, 4000 ft., 28.i.1929'; 'Paratype, Strumeta tillyardi F.A. Perk'. Received on loan from QMIC.

Other type material examined: Holotype male of *Asiadacus absonus* Hering, **MYANMAR (formerly BURMA)**: with the following three labels – 'Type'; 'N.E. Burma, Kambaiti, 7000 ft, 18.v.1934, R. Malaise'; 'Asiadacus absonus Type, det Mart Hering'. Received on loan from NRS (Fig. 211B) (See 'Remarks' below).

Non-type material examined: PENINSULAR MALAYSIA: 1 female, with the following label data – 'Perak, Larut Hills, 4000–4500 ft., Capt Feb-March 1905, pres 1905 by R. Shelford'; '1905'; '*Bactrocera tillyardi* (PK) det IM White 1994'; 'BMNH(E) No. 532992'. Studied at BMNH. 1 female, Ringlet, Cameron Highlands, 25.xii.1990, coll. C.S. Ooi, hand collected on guava fruit. Specimen in QDPC.

Diagnosis: A medium-sized species; face entirely fulvous without dark markings; postpronotal lobes yellow (anterodorsal corners black); notopleura yellow; scutum entirely black; lateral and medial postsutural yellow vittae absent; no yellow spot anterior to notopleural suture; mesopleural stripe equal in width to notopleuron dorsally; scutellum yellow with a large black triangular dorsal marking; wing with cells bc and c colourless, a very small area of microtrichia in outer corner of cell c only; a broken or incomplete costal band consisting of a dark fuscous cell sc and narrow fuscous apical band from apex of R_{2+3} to midway between apices of R_{4+5} and M, dark fuscous bands enclosing *r-m* and *dm-cu* crossveins; a narrow pale fuscous to fuscous anal streak confined mostly within cell cup; no dense aggregation of microtrichia around A_1+CuA_2; abdominal terga III–V orange-brown with a broad transverse black band across anterior margin of tergum III and widening to lateral margins, broad anterolateral black corners on tergum IV and narrow anterolateral black corners on tergum V.

Description: female (Fig. 211A).

Head: Height 1.2 mm. Frons – length 1.4 times breadth; fuscous, fulvous around lateral and ventral margins, lateral black markings extending from vertex to *s.or.* setae; lunule fuscous. Ocellar triangle black. Vertex black, as a continuation from occiput. Face entirely fulvous without dark markings; length 0.34 mm. Genae fulvous, pale fuscous subocular spot present; weak black seta present. Occiput entirely black and confluent with eye margins; occipital row with 2–3 black setae. Antennae with segments 1 and 2 fuscous, segment 3 red-brown with dark fuscous on apex and outer surface; length of segments: 0.09 mm; 0.22 mm; 0.56 mm.

Thorax: Scutum entirely shining black without pale markings. Pleural areas entirely shining black without pale markings. Yellow markings as follows: postpronotal lobes (anterodorsal corners black); notopleura; narrow mesopleural stripe equal in width to notopleuron dorsally, not continuing to katepisternum, anterior margin straight; anatergite (posterior apex black); anterior ½ katatergite (remainder black). Lateral and medial postsutural yellow vittae absent. Postnotum black. Scutellum yellow with basal band extended into a broad triangular black marking extending over ½ of dorsal surface. Setae: *sc.* 2; *prsc.* 2; *ia.* 1; *p.sa.* 1; *a.sa.* 1; *mpl.* 1; *npl.* 2; *scp.* 4.

Legs: All segments entirely fulvous except small areas of dark fuscous on apices of mid and hind femora and hind tibiae entirely dark fuscous; mid tibiae each with an apical black spur.

Wings: Length 5.3 mm; cells bc and c colourless; extremely small area of microtrichia in outer corner of cell c only; remainder of wings colourless except for dark fuscous cell sc, a pale tint on wing margin at apex of cell r, a narrow apical fuscous costal band from apex of R_{2+3} to end midway between apices of R_{4+5} and M, a narrow pale fuscous to fuscous anal streak mostly confined within cell cup, narrow transverse dark fuscous bands enclosing *r-m* and *dm-cu* crossveins; no dense aggregation of microtrichia around A_1+CuA_2; supernumerary lobe weak.

Abdomen: Oval; terga free; no pecten on tergum III. Tergum I and sterna I and II wider than long. Tergum I dark fuscous centrally to black laterally; tergum II with a transverse black band across anterior ½ and fulvous posteriorly; tergum III orange-brown with a broad transverse black band across anterior ½ and expanding towards lateral margins; terga IV and V orange-brown with broad black anterolateral transverse markings on tergum IV and narrow black anterolateral corners on tergum V. A pair of oval fuscous shining spots on tergum V. Basal segment of ovipositor red-brown, dorsoventrally compressed and tapering posteriorly in dorsal view; ratio of length of oviscape to length of tergum V, 1:1; apex of piercer needle-shaped.

Attractant: No known record.

Male: As for female except posterior lobe of surstylus short, sternum V with a deep concavity on posterior margin.

Distribution:

Type locality: Peninsular Malaysia (Selangor, Bukit Kutu).

Previously recorded: Type locality and Peninsular Malaysia (Fraser's Hill).

New records: Myanmar (Kambaiti) due to **New Synonymy** of *A. absonus*.

Hosts: No known record.

Remarks: *Bactrocera (Gymnodacus) tillyardi* (Perkins) is a unique species of subgenus *Gymnodacus* and easily separated from the other species in this subgenus in having black anterodorsal corners on the postpronotal lobes, no lateral and medial postsutural yellow vittae, a distinct black triangle on the dorsal surface of the scutellum, an incomplete costal band interrupted after apex of R_1, infuscation along *r-m* and *dm-cu* crossveins and abdominal terga III–V with black across anterior margin of tergum III and on anterolateral corners of terga IV and V (the black patterns may join along lateral margins to form a distinct lateral longitudinal band).

Norrbom *et al.* (1998) listed the type series as Syntypes and possibly lost. The original description, Perkins (1938a), recorded that the type series consisted of two females. These have been found in the QMIC and were labelled separately by Perkins, one as the Holotype and the other the Paratype.

After studying the Holotypes of *B. tillyardi* and *Asiadacus absonus* Hering, we can find no characters to separate them and so herein are making *A. absonus* a **New Synonym** of *B. tillyardi.*

SUBGENUS *HEMIGYMNODACUS* Hardy

Bactrocera (Hemigymnodacus) diversa (Coquillett)

(Fig. 212)

Dacus diversus Coquillett, 1904: 139. Syntypes in USNM.
Asiadacus diversa – Perkins, 1937: 57; 1938a: 134 (as *diversus*).
Dacus (Gymnodacus) diversus – Hardy, 1954: 18.
Dacus (Hemigymnodacus) diversus – Hardy, 1973: 19.
Dacus quadrifidus Hendel, 1928: 343; Norrbom *et al.*, 1998: 99 (Synonym).
Dacus citronellae Kapoor & Katiyar, 1969: 123; Norrbom *et al.*, 1998: 99 (Synonym).
Dacus latifae Anwar Cheema, 1964: 302. **New Synonym.**
Bactrocera (Paratridacus) diversa – Norrbom *et al.*, 1998: 99.

Non-type material examined: INDIA: 2 females with the following label data – 'Calcutta'; 'Indian Museum'; '*Strumeta (Chaetodacus) diversa* (Coquillett) det M. Hering, 1938', studied at NMW. 1 male, 1 female, Mudigere, 980 m, 20.v.1992, coll. D.L. Hancock; 1 male, Karnataka, Chethalli, 970 m, 3.vii.1998, coll. Ramani, attracted to methyl eugenol. 3 males, 3 females, Rawalpindi, 16–20.x.1961, 'CIBC'; 'Maggots in flowers of *L. cylindrica*'. **THAILAND**: 2 males and 2 females, Nong Chang, Kuun L.P., Chiang Mai, 15.vi.1993, coll. Chantanee *et al.*, bred from flowers of *Coccinia grandis* (L.); 3 males, Klongnumken, A. Muang, Chanthaburi, 25.vi.1991, coll. Montree, attracted to methyl eugenol; 2 males, Pang Da, Samoeng, Chiang Mai, 11.iii.1992, coll. Chantanee, bred from *Cucurbita moschata*, Sample No. CM768; 2 males and 1 female, Pong Khu, Sarapu, Chiang Mai, 6.viii.1992, coll. Chantanee *et al.*, bred from flower of *Coccinia grandis*; 1 female, SPT Chiang Mai, 23.vii.1992, coll. Chantanee, bred from flower of *Coccinia grandis*; 1 male, Nons Pla, Sawai, Chiang Mai, 9.xi.1993, coll.

Chantanee, bred from flower of *Luffa cylindrica*, sample No. CM2047; 3 males and 2 females, Huai Koa Leep, Chiang Mai, 18.viii.1992, coll. Chantanee, bred from *Cucurbita moschata*; 3 males, Khun Khon, Chiang Rai, 29.iv.1993, coll. Charuwan, bred from flower of *Cucurbita moschata*, sample No. CR1368. A series of specimens collected at Chiang Mai and Chiang Rai, bred from flowers of *Benincasa hispida*, *Coccinia grandis*, *Cucurbita moschata*, *Luffa acutangula* and *Luffa cylindrica* (family *Cucurbitaceae*). **VIETNAM: Southern Vietnam** – 1 male, Southern Vietnam, Binh Duong Province, Lai Thieu Village, 18.ii.2000, attracted to methyl eugenol; 1 male, Southern Vietnam, Dong Nai Province, Long Khanh District, Nhan nghia Village, 12.x.2002, attracted to methyl eugenol; 1 female (23.iii.2002), 6 males (8.viii.2002), 1 male and 1 female (24.viii.2002), 2 males (8.x.2002), 5 males (24.xi.2002), Southern Vietnam, Dac Lac Province, Ku Mi Nga District, Quang Phu Village, attracted to methyl eugenol. Specimens in QDPC.

Diagnosis: A medium-sized species; face in males entirely fulvous, in females fulvous with a narrow transverse dark fuscous to black band across oral margin; postpronotal lobes and notopleura yellow; scutum black; lateral and medial postsutural yellow vittae present; a yellow spot present anterior to notopleural suture; mesopleural stripe reaching to or almost to anterior *npl.* seta dorsally; scutellum yellow with a narrow black basal band; wing with cells bc and c colourless, microtrichia in outer corner of cell c only; a narrow dark fuscous costal band expanding slightly into an elongate spot across apex of R_{4+5}; a moderately broad dark fuscous anal streak; a pale tint on posterior margin around apex of CuA_1; abdominal terga III–V orange-brown with each segment having a black 'T' pattern consisting of a broad medial longitudinal black band and a narrow transverse black band across anterior margin expanding over lateral margins (in some specimens tergum IV and V have the dark markings reduced to anterolateral dark fuscous corners and incomplete medial longitudinal black bands, and in some specimens tergum V is entirely orange-brown).

Description: male.

Head: Height 1.23 mm. Frons – length 1.3 times breadth; red-brown centrally, fulvous along lateral margins and dark fuscous around bases of orbital setae and on anteromedial hump; orbital setae black: 1 *s.or.*, 2 *i.or.*; lunule red-brown. Ocellar triangle black. Vertex dark fuscous. Face entirely fulvous without dark markings; length 0.39 mm. Genae fulvous, small dark fuscous subocular spot present; pale seta present. Occiput red-brown centrally, dark fuscous laterally, fulvous along eye margins; occipital row with a large number of short pale setae. Antennae with segments 1 and 2 red-brown, segment 3 red-brown with pale fuscous on apex and outer surface; length of segments: 0.1 mm; 0.26 mm; 0.64 mm.

Thorax: Scutum black except red-brown below and behind lateral postsutural vittae and anterior to notopleural suture. Pleural areas black except red-brown below postpronotal lobes. Yellow markings as follows: postpronotal lobes; notopleura; mesopleural stripe reaching almost to anterior *npl.* seta dorsally, continuing to katepisternum as a transverse spot, anterior margin convex; anatergite (posterior apex black); anterior ¾ katatergite (remainder black); two broad parallel-sided lateral postsutural vittae beginning with a spot anterior to notopleural suture and ending just behind *ia.* seta; a moderately broad medial longitudinal postsutural vitta beginning at level of *prsc.* setae and narrowing anteriorly to end at level of notopleural suture. Postnotum black. Scutellum yellow except for narrow black basal band. Setae: *sc.* 2 (occasionally 4); *prsc.* 2; *ia.* 1; *p.sa.* 1; *a.sa.* 1; *mpl.* 1; *npl.* 2; *scp.* 2.

Legs: Fore femora pale fulvous, mid and hind femora pale fulvous basally tending red-brown apically; fore and mid tibiae dark fuscous basally to fulvous apically, hind tibiae dark fuscous basally to fuscous apically; tarsi with all segments entirely fulvous; mid tibiae each with an apical black spur.

Wings: Length 4.9 mm; cells bc and c colourless; microtrichia in outer corner of cell c only; remainder of wings colourless except dark fuscous cell sc, narrow dark fuscous costal band confluent with R_{2+3} and remaining narrow beyond apex of this vein before expanding into an elongate spot across apex of R_{4+5}, a moderately broad dark fuscous anal streak, a pale tint on hind margin of wing around apex of CuA_1; no dense aggregation of microtrichia around A_1+CuA_2; supernumerary lobe strong and rounded.

Abdomen: Oval; terga free; no pecten on tergum III. Tergum I and sterna I and II wider than long. Tergum I fuscous centrally, black laterally and orange-brown across posterior margin; tergum II orange-brown anteriorly, fulvous posteriorly and with a narrow transverse subanterior black band joining narrow anterolateral black margins; tergum III orange-brown with a black 'T' pattern consisting of a broad medial longitudinal black band and a narrow transverse band across anterior margin widening to cover lateral margins; tergum IV orange-brown with a broad medial longitudinal black band that may not cover the entire length of the tergum and moderately broad dark fuscous anterolateral corners; tergum V orange-brown with narrow dark fuscous anterolateral corners (in some specimens tergum IV possesses a narrow transverse black band across the entire anterior margin, and tergum V occasionally is entirely orange-brown without dark markings). A pair of oval orange-brown shining spots on tergum V. Posterior lobe of surstylus long, sternum V with a slight concavity on posterior margin. Abdominal sterna dark fuscous to black.

Attractant: This species appears to have a weak attraction to methyl eugenol.

Female: As for male except face with a narrow transverse dark fuscous to black band across oral margin; supernumerary lobe weak; ovipositor basal segment orange-brown on basal ⅓ and black on apical ⅔, tapering strongly posteriorly, conical in transverse cross section; ratio of length of oviscape to length of tergum V, 3:1; apex of piercer needle-shaped.

Distribution:

Type locality: India (Bangalore); Sri Lanka (Colombo). The original type series was collected at both localities and no primary type specimen has been designated.

Previously recorded: Type locality and the following locations – China (Sichuan; Yunnan); India (Bihar); Nepal; Sri Lanka; Thailand (Norrbom *et al.*, 1998).

New records: Southern Vietnam (Binh Duong, Dac Lac, Dong Nai Provinces). Pakistan (Lyallpur), due to **New Synonymy** of *D. latifae*.
Hosts: Recorded from flowers of a wide range of species in the family *Cucurbitaceae* (this species does not appear to breed in fruits) (see Allwood *et al.*, 1999 for host records).
Remarks: *Bactrocera (Hemigymnodacus) diversa* (Coquillett) is superficially similar to *Bactrocera (Zeugodacus) caudata* (Fabricius) and both species are similar biologically in that they breed in flowers of *Cucurbitaceae*. However, *B. diversa* can be separated in the absence of the pecten of cilia on abdominal tergum III of the male, the presence of entirely fulvous femora and a large supernumerary lobe in the male wing. There are no characters that separate *B. diversa* from *Dacus latifae* Anwar Cheema and so the latter is herein made a **New Synonym** of *B. diversa*.

Bactrocera (Hemigymnodacus) mukiae Drew & Romig, new species

(Figs 213A, 213B)

Material examined: Holotype male, **THAILAND**: Doi Lung Nat. Park, Chiang Rai, 27.i.1993, coll. Charuwan, bred from *Mukia melothria wallichi* (= *Mukia wallichii*), deposited in BMNH.
Paratypes: **THAILAND**: 3 males and 2 females, same data as holotype; 1 female, Mae Hae, Chiang Mai, 9.vii.1991, coll. Chantanee, bred from *Zehneria maysorensis*, Sample No. CM164. **VIETNAM: Northern Vietnam** – 1 female, Son La (P), Moc Chau (D), Long Luong (V), 15.v.2002, Site VN01, coll. NIPP Staff. 2 paratypes in ANIC; 4 paratypes in QDPC; 1 paratype in QMIC.
Non-type material examined: VIETNAM: Northern Vietnam – 1 male, Lao Cai Province, Sapa 6 village, trap N017, urban forest, 23.iv.1999, coll. Le Duc Khanh *et al.*, attracted to cue lure. **BHUTAN**: 1 male, Tsirang, Darachu-la, 30.v.2001, coll. Brian Fletcher, altitude 1500m, subtropical forest, attracted to cue lure. These two specimens differ from the types only in the lack of an apical dark spot on the scutellum and in possessing a pecten on tergum III. Although they appear conspecific, they are not being included in the type series. Specimens in QDPC.
Diagnosis: A medium-sized species; face fulvous with a pair of large oval black spots; postpronotal lobes and notopleura yellow; scutum shining black; lateral and medial postsutural yellow vittae present; a yellow spot present anterior to notopleural suture; mesopleural stripe reaching almost to anterior *npl.* seta dorsally; scutellum yellow with a small dark fuscous to black apical spot; wing with cells bc and c colourless, microtrichia in outer corner of cell c only; a narrow fuscous costal band expanding into a small spot around apex of R_{4+5} (*scutellaris*-like) and narrow pale fuscous anal streak; abdominal terga III–V black with a pair of small red-brown spots on posterior margins of terga III and IV either side of the medial longitudinal black band and a very narrow red-brown band along posterior margin of tergum V.
Description: male (Fig. 213A).
Head: Height 1.44 mm. Frons – length 1.24 times breadth; red-brown with fulvous along lateral margins, dark fuscous around orbital setae and fuscous on anteromedial hump; orbital setae black:1 *s.or.*, 2 *i.or.*; lunule dark fuscous. Ocellar triangle black. Vertex black. Face fulvous with a pair of large oval black spots reaching oral margin; length 0.55 mm. Genae fulvous, fuscous subocular spot present; black seta present. Occiput dark fuscous to black, fulvous along eye margins; occipital row with a small number of weak pale setae. Antennae with segments 1 and 2 red-brown, segment 3 red-brown with fuscous on apex and outer surface; arista black (red-brown basally); length of segments: 0.1 mm; 0.29 mm; 0.74 mm.
Thorax: Scutum shining black with red-brown to dark red-brown behind lateral postsutural vittae. Pleural areas shining black tending dark fuscous below postpronotal lobe. Yellow markings as follows: postpronotal lobes; notopleura; mesopleural stripe reaching almost to anterior *npl.* seta dorsally, continuing to katepisternum as a small transverse spot, anterior margin straight; anatergite (posterior apex black); anterior ⅔ katatergite (remainder black); two medium-width parallel-sided lateral postsutural vittae beginning with a spot anterior to notopleural suture and ending at *ia.* seta; a short narrow medial longitudinal postsutural vitta. Postnotum black. Scutellum yellow except for narrow black basal band and a dark fuscous to black apical spot between apical setae. Setae: *sc.* 2; *prsc.* present or absent; *ia.* 1; *p.sa.* 1; *a.sa.* absent; *mpl.* 1; *npl.* 2; *scp.* 2.
Legs: Fore femora almost entirely shining black except fulvous apically and basally and tending fulvous to fuscous on inner surfaces, mid and hind femora with apical ⅓ shining black remainder fulvous; fore tibiae fuscous to dark fuscous, mid tibiae dark fuscous basally to fuscous apically, hind tibiae dark fuscous to black; tarsi with basal segment fulvous and apical four segments fuscous; mid tibiae each with an apical black spur.
Wings: Length 5.8 mm; cells bc and c colourless; microtrichia in outer corner of cell c only; remainder of wings colourless except dark fuscous cell sc, narrow fuscous costal band confluent with R_{2+3} and remaining very narrow beyond apex of this vein before expanding into a spot around apex of R_{4+5} (*scutellaris*-like), a narrow pale fuscous anal streak; no dense aggregation of microtrichia around A_1+CuA_2; supernumerary lobe weak.
Abdomen: Oval; terga free; pecten on tergum III absent. Tergum I and sterna I and II wider than long. Tergum I black with a narrow transverse fulvous band across posterior margin but not reaching lateral margins; tergum II fulvous with a broad transverse black band across anterior ½ but not expanding to cover lateral margins; terga III–V black except for red-brown spots posterocentrally on terga III and IV either side of a medial longitudinal black band and a very narrow red-brown band along posterior margin of tergum V. A pair of oval black shining spots on tergum V. Posterior lobe of surstylus long, sternum V with a slight concavity on posterior margin. Abdominal sterna black, except sternum II fulvous.
Attractant: Cue lure.
Female: (Fig. 213B) – As for male except ovipositor basal segment black, dorsoventrally flattened and tapering

posteriorly in dorsal view; ratio of length of oviscape to length of tergum V, 0.75:1.
Distribution: Thailand (Chiang Rai); Northern Vietnam (San La Province); Bhutan.
Hosts: *Mukia wallichii* (current botanical databases infer that this plant species is *Zehneria wallichii*); *Zehneria maysorensis* (family *Cucurbitaceae*).
Remarks: *Bactrocera (Hemigymnodacus) mukiae* n.sp. is similar to *Bactrocera (Zeugodacus) hoabinhiae* n.sp., *Bactrocera (Zeugodacus) laocaiae* n.sp. and *Bactrocera (Zeugodacus) yoshimotoi* (Hardy) in possessing two *sc.* setae, yellow notopleura and face not entirely black. It differs from these species in having a weak supernumerary lobe in the male wing, a narrow pale anal streak, *a.sa.* seta absent and the pecten of cilia on male abdominal tergum III absent. It also differs from *B. hoabinhiae* in having facial spots, not a dark band across the oral margin.

SUBGENUS *JAVADACUS* Hardy

Bactrocera (Javadacus) javanensis (Perkins)

(Fig. 214)

Afrodacus javanensis Perkins, 1938a: 132. Holotype female in BMNH.
Dacus (Afrodacus) javanensis – Hardy, 1955a: 9; 1983: 9.
Bactrocera (Javadacus) javanensis – Norrbom *et al.*, 1998: 98.
Strumeta transtillum Hering, 1952b: 265. **New Synonym**.

Type material examined: Holotype female, **INDONESIA: JAVA** – with the following five labels – 'Type'; 'Holotype, Afrodacus javanensis F.A. Perk'; 'Purch. from E. Le Moult, B.M. 1937-32'; 'E. Java, Mt Ardjoeno, 6000 ft, i.1936'; 'BMNH (E), No. 532993'. Studied at BMNH.
Diagnosis: A medium-sized species; face fulvous with a narrow transverse black band across oral margin but not reaching antennal grooves; postpronotal lobes and notopleura yellow; scutum black; broad lateral postsutural yellow vittae present; moderately broad medial postsutural yellow vitta present; a yellow spot anterior to notopleural suture; broad mesopleural stripe reaching anterior *npl.* seta dorsally; scutellum yellow; wing with cells bc and c colourless, microtrichia in outer corner of cell c only; a narrow fuscous costal band confluent with R_{2+3} and widening across apex of R_{4+5}, a narrow fuscous band enclosing *dm-cu* crossvein, a narrow fuscous anal streak; abdominal terga III–V fulvous with a black 'T' pattern and black anterolateral corners on terga IV and V.
Description: Female.
Head: Frons – fulvous with pale red-brown on anteromedial hump; orbital setae black: 1 *s.or.*, 3 *i.or.*; lunule pale fuscous. Ocellar triangle black. Vertex fulvous with small dark fuscous spots laterally adjacent to eye margins but not joining with the ocellar triangle. Face fulvous with a narrow transverse black band across oral margin but not reaching antennal grooves. Occiput red-brown with fuscous around dorsal margins. Antennae with segments 1 and 2 red-brown, segment 3 red-brown with fuscous on apex and outer surface.
Thorax: Scutum black except for narrow red-brown lateral margins posterior to notopleural suture and lateral to yellow spot anterior to the notopleural suture. Pleural areas dark fuscous to black. Yellow markings as follows: postpronotal lobes; notopleura; mesopleural stripe reaching to anterior *npl.* seta dorsally, continuing to katepisternum as a small transverse spot, anterior margin mostly convex; two broad parallel-sided lateral postsutural vittae beginning as a yellow spot anterior to notopleural suture and ending behind *ia.* seta; moderately broad medial postsutural vitta present. Scutellum yellow except for narrow black basal band. Setae: *sc.* 2; *prsc.* 2; *ia.* 1; *p.sa.* 1; *a.sa.* absent; *mpl.* 1; *npl.* 2; *scp.* 4.
Legs: All femora fulvous with small dark fuscous apical to subapical spots on outer surfaces; fore tibiae fulvous tending fuscous basally, mid tibiae dark fuscous basally to fulvous apically, hind tibiae dark fuscous; tarsi with basal segment fulvous and apical four segments fuscous.
Wings: Length 6.1 mm; cells bc and c colourless; microtrichia in outer corner of cell c only; remainder of wings colourless except dark fuscous cell sc, narrow fuscous costal band confluent with R_{2+3} and widening across apex of R_{4+5}, a narrow fuscous band enclosing *dm-cu* crossvein, a narrow fuscous anal streak; no dense aggregation of microtrichia around A_1+CuA_2; supernumerary lobe weak.
Abdomen: Oval; terga free; no pecten present on tergum III. Tergum I and sterna I and II wider than long. Tergum I fulvous with narrow black lateral margins; tergum II fulvous with a narrow transverse black band across anterior margin; terga III–V fulvous with a black 'T' pattern consisting of a broad transverse band across anterior margin of tergum III which expands to cover lateral margins and a moderately broad medial longitudinal band over all three terga, anterolateral corners of terga IV and V black. A pair of fulvous shining spots on tergum V. Basal segment of ovipositor dark fuscous to black, dorsoventrally compressed and tapering posteriorly in dorsal view; apex of piercer needle-shaped.
Attractant: No known record.
Male: No known record.
Distribution:
Type locality: Indonesia (Java, Mt Ardjoeno).
Hosts: No known record.
Remarks: *Bactrocera (Javadacus) javanensis* (Perkins) is similar to *Bactrocera (Javadacus) montana* (Hardy), *Bactrocera (Javadacus) scutellaria* (Bezzi) and *Bactrocera (Javadacus) trilineata* (Hardy) in possessing a black scutum, lateral and medial postsutural yellow vittae, a yellow spot anterior to notopleural suture, a broad mesopleural stripe reaching the anterior *npl.* seta dorsally, a narrow costal band and anal streak and dark patterns on abdominal terga III–V. It differs from these species in having the face fulvous with a narrow transverse black band across oral margin and a fuscous band enclosing *dm-cu* crossvein.

After comparing *B. javanensis* with *Strumeta transtillum* Hering we can find no sound characters to separate them and thus declare *S. transtillum* a **New Synonym** of the former species.

Bactrocera (Javadacus) montana (Hardy)

(Fig. 215)

Dacus montanus Hardy, 1983: 27. Holotype male in MZB.
Bactrocera (Javadacus) montana – Norrbom *et al.*, 1998: 98.

Type material examined: Holotype male, **INDONESIA: JAVA** – Cibodas, West Java, 1400 m, 24–25.vi.1975, coll. D.E. & A.T. Hardy, attracted to methyl eugenol. Studied at MZB.
Paratype: **INDONESIA**: 1 male, same label data as holotype. Studied at MZB.
Non-type material examined: INDONESIA: JAVA – Cibodas, West Java, 1300-1600m, 16.v.1981, coll. J. Turner, attracted to cue lure. Specimen in QDPC.
Diagnosis: A medium-sized species; face entirely dark fuscous to black; postpronotal lobes and notopleura yellow; scutum black; lateral and medial postsutural yellow vittae present; a yellow spot anterior to notopleural suture; mesopleural stripe reaching to postpronotal lobe dorsally; scutellum yellow; wing with cells bc and c colourless, microtrichia in outer corner of cell c only; a narrow fuscous costal band expanding to a small spot around apex of R_{4+5} and a moderately broad anal streak; abdominal terga III–V fuscous with a black 'T' pattern over all three terga and broad lateral black markings on terga IV and V.
Description: male.
Head: Frons – fulvous with pale fuscous on anteromedial hump; orbital setae black: 1 *s.or.*, 2–3 *i.or.*; lunule pale fuscous. Ocellar triangle black. Vertex fulvous. Face entirely dark fuscous to black. Occiput red-brown, black laterally and fulvous along eye margins. Antennae with segments 1 and 2 red-brown, segment 3 fuscous to dark fuscous.
Thorax: Scutum entirely black without pale markings. Pleural areas dark fuscous to black. Yellow markings as follows: postpronotal lobes; notopleura; broad mesopleural stripe reaching to postpronotal lobe dorsally; two moderately broad parallel-sided lateral postsutural vittae beginning with a spot anterior to notopleural suture and ending at *ia.* seta; a moderately broad medial longitudinal postsutural vitta present. Scutellum yellow with a narrow black basal band. Setae: *sc.* 2; *prsc.* 2; *ia.* 1; *p.sa.* 1; *a.sa.* absent; *mpl.* 1; *npl.* 2; *scp.* 4.
Legs: All femora fulvous basally with glossy dark fuscous to black around apical ⅓; fore and mid tibiae fulvous tending pale fuscous basally, hind tibiae fuscous to dark fuscous; tarsi with all segments fulvous except apical four segments of hind tarsi pale fuscous.
Wings: Length 5.7 mm; cells bc and c colourless; microtrichia in outer corner of cell c only; remainder of wings colourless except fuscous cell sc, narrow fuscous costal band confluent with R_{2+3} and then very narrow beyond extremity of R_{2+3} before expanding into a distinct but small fuscous spot around apex of R_{4+5}, a moderately broad fuscous anal streak; a dense aggregation of microtrichia around A_1+CuA_2; supernumerary lobe of medium development.
Abdomen: Oval; terga free; pecten present on tergum III. Tergum I and sterna I and II wider than long. Tergum I black except for a narrow transverse fulvous band across posterior margin but not reaching lateral margins; tergum II fuscous except for a narrow transverse black band across anterior margin; terga III–V fuscous with a black 'T' pattern consisting of a broad transverse band across anterior margin of tergum III and a narrow medial longitudinal band over all three terga, broad lateral black markings on terga IV and V. A pair of oval fuscous shining spots on tergum V. Posterior lobe of surstylus long, sternum V with a shallow concavity on posterior margin.
Attractant: Cue lure (the record of methyl eugenol by Hardy, 1983, is probably incorrect).
Female: No known record.
Distribution:
Type locality: Indonesia (West Java, Cibodas).
Hosts: No known record.
Remarks: *Bactrocera (Javadacus) montana* (Hardy) is similar to *Bactrocera (Javadacus) javanensis* (Perkins), *Bactrocera (Javadacus) scutellaria* (Bezzi) and *Bactrocera (Javadacus) trilineata* (Hardy) in possessing a black scutum, lateral and medial postsutural yellow vittae with a yellow spot anterior to notopleural suture, a broad mesopleural stripe reaching to or almost to postpronotal lobe dorsally, wing with a narrow fuscous costal band and abdominal terga III–V with dark colour patterns. It differs from all three species in having the face entirely dark fuscous to black and glossy black patterns around apical ⅓ of all femora.

Bactrocera (Javadacus) nigrita (Hardy)

(Fig. 216)

Dacus aberrans ssp. *nigritus* Hardy, 1955a: 5. Holotype male in USNM.
Bactrocera (Javadacus) nigrita – Norrbom *et al.*, 1998: 98.

Type material examined: Holotype male, **SINGAPORE**: with the following three labels – 'Singapore, Malaya, May 1949'; 'ex *Cinnamomum iners* fruit, N.L.H. Krauss'; 'Holotype male, *Dacus (Afrodacus) aberrans nigritus* D.E. Hardy'. Received on loan from USNM.
Paratype: **SINGAPORE**: 1 female paratype labelled Allotype female with same data labels as holotype. Received on loan from USNM.
Non-type material examined: VIETNAM: Northern Vietnam – 1 male, Northern Vietnam, Lao Cai Province, Sapa town, Site N016, 30.vii.1999, attracted to methyl eugenol. Specimen in QDPC.
Diagnosis: A small species; face fulvous with a pair of small circular to sub-quadrate black spots; postpronotal lobes and notopleura yellow; scutum dull black; lateral postsutural yellow vittae present; medial postsutural yellow vitta absent; no yellow spot anterior to notopleural suture; mesopleural stripe reaching midway between anterior margin of notopleuron and anterior *npl.* seta dorsally; scutellum yellow; wing with cells bc and c colourless, microtrichia in outer corner of cell c only; a narrow fuscous costal band; a narrow fuscous anal streak; abdominal terga III–V orange-brown with a black 'T' pattern consisting of a moderately broad transverse band across anterior margin of tergum III and a broad medial

longitudinal black band over all three terga, anterolateral corners of terga IV and V black.

Description: male.

Head: Height 1.32 mm. Frons – length 1.38 times breadth; red-brown with fuscous between *i.or.* setae; orbital setae black: 1 *s.or.*, 2 *i.or.*; lunule red-brown. Ocellar triangle black. Vertex red-brown. Face fulvous with a pair of small circular to sub-quadrate black spots; length 0.44 mm. Genae fulvous, small dark fuscous subocular spot present; black seta present. Occiput black laterally, red-brown centrally, fulvous along eye margins; occipital row with 2–4 short dark setae. Antennae with segments 1 and 2 red-brown, segment 3 red-brown with fuscous on apex and outer surface; length of segments: 0.13 mm; 0.23 mm; 0.6 mm.

Thorax: Scutum dull black with red-brown below and behind lateral postsutural vittae, around notopleural suture, inside postpronotal lobes. Pleural areas black except red-brown below postpronotal lobes. Yellow markings as follows: postpronotal lobes; notopleura; mesopleural stripe reaching midway between anterior margin of notopleuron and anterior *npl.* seta dorsally, continuing to katepisternum as a small transverse spot, anterior margin straight; anatergite (posterior apex black); anterior ⅔ katatergite (remainder red-brown); two moderately broad lateral postsutural vittae tapering posteriorly to end well before *ia.* seta. Medial longitudinal postsutural yellow vitta absent. Postnotum black. Scutellum yellow except for narrow black basal band. Setae: *sc.* 2; *prsc.* 2; *ia.* 1; *p.sa.* 1; *a.sa.* absent; *mpl.* 1; *npl.* 2; *scp.* 4.

Legs: All segments entirely fulvous except hind tibiae fuscous; mid tibiae each with an apical black spur.

Wings: Length 4.8 mm; cells bc and c colourless; microtrichia in outer corner of cell c only; remainder of wings colourless except fuscous cell sc, narrow fuscous costal band confluent with R_{2+3} and widening slightly across apex of wing to end just beyond extremity of R_{4+5}, narrow fuscous anal streak; a dense aggregation of microtrichia around A_1+CuA_2; supernumerary lobe of medium development.

Abdomen: Oval; terga free; pecten present on tergum III. Tergum I and sterna I and II wider than long. Tergum I fulvous with black across anterior and lateral margins; tergum II fulvous with a narrow black subanterior transverse band reaching black anterolateral corners; terga III–V orange-brown with a black 'T' pattern consisting of a moderately broad transverse band across anterior margin of tergum III and a broad medial longitudinal band over all three terga, black anterolateral corners on terga IV and V. A pair of oval orange-brown shining spots on tergum V. Posterior lobe of surstylus long, sternum V with a slight concavity on posterior margin.

Attractant: Methyl eugenol.

Female: As for male except no dense aggregation of microtrichia around A_1+CuA_2; supernumerary lobe weak; pecten of cilia on abdominal tergum III absent. The posterior part of the abdomen including the ovipositor has been dissected from the female Paratype and apparently lost as it has not been available for study.

Distribution:

Type locality: Singapore.

New records: Northern Vietnam (Lao Cai Province, Sapa town).

Hosts: *Cinnamomum iners* (family *Lauraceae*) (Hardy, 1955a).

Remarks: *Bactrocera (Javadacus) nigrita* (Hardy) appears more closely related to subgenus *Bactrocera* species and similar to the *Bactrocera (Javadacus)* species in the South Pacific region, especially *Bactrocera (Javadacus) aberrans* (Hardy) and *Bactrocera (Javadacus) melanothoracica* Drew in possessing a dull black scutum, lateral postsutural vittae ending before the *ia.* setae, a narrow costal band, pale legs and abdominal terga III–V with a dark 'T' pattern. It differs from *B. melanothoracica* in having the costal band confluent with R_{2+3} (not overlapping), cells bc and c colourless, the mesopleural stripe narrower (reaching midway between the anterior margin of the notopleuron and the anterior *npl.* seta dorsally) and distinct black anterolateral corners on abdominal terga IV and V. *Bactrocera nigrita* is very difficult to separate from *B. aberrans* and they may be conspecific, *B. nigrita* differing only in having broader bands in the black 'T' pattern on abdominal terga III–V and black anterolateral corners on terga IV and V.

Bactrocera (Javadacus) pallescentis (Hardy)

(Fig. 217)

Dacus aberrans ssp. *pallescentis* Hardy, 1955a: 5. Holotype male in USNM.
Bactrocera (Javadacus) pallescentis – Norrbom *et al*., 1998: 98.

Type material examined: Holotype male, **INDIA**: with the following four labels – 'Ranikhet, U.P. India, June 1949'; 'ex Mimusops sp.'; 'I.M. Newell Collector'; 'holotype male, *Dacus (Afrodacus) aberrans pallescentis*, D.E. Hardy'. Received on loan from USNM.

Paratypes: **INDIA**: 1 female paratype labelled Allotype female with same data labels as holotype. Received on loan from USNM. 1 male and 1 female paratype, same data as holotype and each with a blue label – 'Paratype, *Dacus (Afrodacus) aberrans pallescentis*, D.E. Hardy'. Specimens in QDPC.

Diagnosis: A medium-sized species; face fulvous with a pair of medium-sized circular black spots; postpronotal lobes and notopleura yellow; scutum entirely red-brown without dark markings; lateral postsutural yellow vittae present; medial postsutural yellow vitta absent; no yellow spot anterior to notopleural suture; mesopleural stripe reaching midway between anterior margin of notopleuron and anterior *npl.* seta dorsally; scutellum yellow; wing with cells bc and c colourless, microtrichia in outer corner of cell c only; a narrow fuscous costal band; a narrow pale fuscous anal streak; abdominal terga III–V orange-brown with a black 'T' pattern consisting of a transverse band across anterior margin of tergum III and a broad medial longitudinal band over all three terga, anterolateral corners of terga IV and V black.

Description: male.

Head: Height 1.4 mm. Frons – length 1.44 times breadth; fulvous with red-brown anterocentrally and dark fuscous around orbital setae; orbital setae dark red-brown: 1 *s.or.*, 2 *i.or.*; lunule dark fuscous. Ocellar triangle black. Vertex dark fuscous. Face fulvous with a pair of medium-sized circular black spots; length 0.45 mm. Genae fulvous; red-brown seta present. Occiput red-brown, fulvous along eye margins; no distinct setae in occipital row. Antennae with segments 1 and 2 red-brown, segment 3 red-brown with fuscous on apex and outer surface; length of segments: 0.12 mm; 0.25 mm; 0.84 mm.

Thorax: Scutum red-brown without dark markings. Pleural areas red-brown without dark markings. Yellow markings as follows: postpronotal lobes; notopleura; mesopleural stripe of medium-width reaching midway between anterior margin of notopleuron and anterior *npl.* seta dorsally, continuing to katepisternum as a small transverse spot, anterior margin slightly convex; anatergite (posterior apex red-brown); anterior ¾ katatergite (remainder red-brown); two medium-width lateral postsutural vittae tapering posteriorly to end at level of *p.sa.* seta. Medial longitudinal postsutural yellow vitta absent. Postnotum red-brown. Scutellum yellow except for narrow red-brown basal band. Setae: *sc.* 2; *prsc.* 2; *ia.* 1; *p.sa.* 1; *a.sa.* absent; *mpl.* 1; *npl.* 2; *scp.* 4.

Legs: All segments fulvous except hind tibiae pale fuscous; mid tibiae each with an apical black spur.

Wings: Length 5.0 mm; cells bc and c colourless; microtrichia in outer corner of cell c only; remainder of wings colourless except fuscous cell sc, narrow fuscous costal band confluent with R_{2+3} and remaining narrow as it continues around apex of wing to end beyond extremity of R_{4+5}, a narrow pale fuscous anal streak; a dense aggregation of microtrichia around A_1+CuA_2; supernumerary lobe of medium development.

Abdomen: Oval; terga free; pecten present on tergum III. Tergum I and sterna I and II wider than long. Tergum I orange-brown with narrow dark fuscous lateral margins; tergum II orange-brown, fulvous posteriorly and with narrow dark fuscous anterolateral corners; terga III–V orange-brown with a black 'T' pattern consisting of a narrow transverse band across anterior margin of tergum III but not reaching lateral margins and a broad medial longitudinal band over all three terga, anterolateral corners of terga IV and V black. A pair of oval orange-brown shining spots on tergum V. Posterior lobe of surstylus long, sternum V with a slight concavity on posterior margin.

Attractant: No known record.

Female: As for male except no dense aggregation of microtrichia around A_1+CuA_2; supernumerary lobe weak; pecten of cilia on abdominal tergum III absent; ovipositor basal segment orange-brown, dorsoventrally compressed and tapering posteriorly in dorsal view; ratio of length of oviscape to length of tergum V, 0.8:1; apex of piercer needle-shaped.

Distribution:

Type locality: India (Uttar Pradesh – 22 miles S. of Ranikhet).

Hosts: *Mimusops* sp. (family *Sapotaceae*) (Hardy, 1955a).

Remarks: *Bactrocera (Javadacus) pallescentis* (Hardy), like *Bactrocera (Javadacus) nigrita* (Hardy), appears closely related to species in the subgenus *Bactrocera*. It is similar to *Bactrocera (Daculus) digressa* Radhakrishnan in possessing an overall red-brown basic colouration, a narrow costal band and black 'T' pattern on abdominal terga III–V. It differs from *B. digressa* in lacking yellow spots anterior to the notopleural suture, and in having a broader mesopleural stripe, lateral postsutural vittae ending before the *ia.* setae, broader bands in the black 'T' pattern on abdominal terga III–V and *prsc.* setae. *Bactrocera pallescentis* is readily separated from *B. nigrita* in the possession of a red-brown scutum. Although Hardy (1955a) made these subspecies of *Bactrocera (Javadacus) aberrans* (Hardy), there is no valid reason to believe that they are closely related, being readily separated on morphological and biogeographic grounds.

Bactrocera (Javadacus) scutellaria (Bezzi)

(Fig. 218)

Chaetodacus scutellarius Bezzi, 1916: 110. Holotype female in BMNH.
Dacus (Bactrocera) scutellarius – Hardy, 1977: 52.
Bactrocera (Bactrocera) scutellaria – Norrbom *et al.*, 1998: 95.
Bactrocera (Javadacus) scutellaria – Drew & Raghu, 2002: 328.

Type material examined: Holotype female, **INDIA**: with the following five labels – 'Pres. by T.B. Fletcher, 1915. 414'; 'Chaetodacus scutellarius Bezzi Type'; 'S. Mysore, 3300 ft, Goorghalli Estate, 14: 24 MAR 13, P.S. Coll'; 'Chaetodacus scutellarius n.sp. Bezzi type female, T.B. Fletcher'; 'a det. label by T.B. Fletcher that is not entirely legible'. Specimen studied in BMNH.

Non-type material examined: INDIA: 1 male, Hort Farm, 1450m, Coonoor, S. India, 30.v.1992, coll. D.L. Hancock, attracted to cue lure. Specimen in QDPC.

Diagnosis: A medium-sized species; face fulvous with a pair of transverse oval black spots in outer corners adjacent to the oral margin; postpronotal lobes yellow; notopleura black; scutum entirely black; lateral and medial postsutural yellow vittae present; a small yellow spot anterior to notopleural suture; mesopleural stripe reaching to postpronotal lobe dorsally; scutellum yellow with a large shining black apical spot surrounding apical setae; wing with cells bc and c colourless, microtrichia in outer corner of cell c only; a narrow fuscous costal band confluent with R_{2+3} and widening slightly around apex of wing, a small area of fuscous on hind margin around CuA_2; a broad fuscous anal streak; abdominal terga III–V black with small transverse oval fulvous spots posterocentrally either side of a black midline on terga III and IV.

Description: female.

Head: Frons – fulvous with fuscous around orbital setae and on anteromedial hump; orbital setae black: 1 *s.or.*, 3 *i.or.*; lunule fuscous. Ocellar triangle black. Vertex with two black spots adjacent to the eye margins. Face fulvous with a pair of large transverse oval black spots adjacent to the outer corners of the oral margin. Occiput fuscous to dark fuscous, fulvous along eye margins. Antennae with segments 1 and 2 red-brown, segment 3 red-brown with fuscous on apex.

Thorax: Scutum entirely black without pale markings. Pleural areas dark fuscous to black. Yellow markings as follows: postpronotal lobes; broad mesopleural stripe reaching to postpronotal lobe dorsally, not extending to katepisternum ventrally; two moderately broad parallel-sided lateral postsutural vittae beginning with a spot anterior to notopleural suture and ending before *ia.* seta; a narrow medial postsutural vitta. Notopleuron black. Scutellum yellow except for narrow black basal band and large glossy black apical spots surrounding the apical *sc.* setae. Setae: *sc.* 2; *prsc.* 2; *ia.* 1; *p.sa.* 1; *a.sa.* absent; *mpl.* 1; *npl.* 2; *scp.* 4.

Legs: All femora black except for small basal areas of fore and mid femora fulvous and basal ⅔ of hind femora fulvous; all tibiae dark fuscous (fore and mid tibiae with central areas of paler fuscous); all tarsal segments fulvous.

Wings: Length 5.2 mm; cells bc and c colourless; microtrichia in outer corner of cell c only; remainder of wings colourless except fuscous cell sc, narrow fuscous costal band confluent with R_{2+3} and continuing around apex of wing with a very slight widening, a small area of pale fuscous on hind margin of wing around CuA_2, a moderately broad fuscous anal streak; no dense aggregation of microtrichia around A_1+CuA_2; supernumerary lobe weak.

Abdomen: Oval; terga free; no pecten present on tergum III. Tergum I and sterna I and II wider than long. Tergum I black with a narrow transverse fulvous band across posterior margin but not reaching lateral margins; tergum II black with a narrow transverse fulvous band across posterior margin; terga III–V black except for small transverse oval fulvous spots posterocentrally either side of a black midline on terga III and IV. A pair of oval pale fuscous shining spots on tergum V. Basal segment of ovipositor black, dorsoventrally compressed and tapering posteriorly in dorsal view.

Attractant: Cue lure.

Male: As for female except for dense aggregation of microtrichia around A_1+CuA_2 present, supernumerary lobe strong and rounded, pecten of cilia on abdominal tergum III present; posterior lobe of surstylus long, posterior margin of tergum V with a slight concavity; abdominal sterna dark coloured.

Distribution:

Type locality: Southern India (South Mysore, Goorghalli Estate).

New records: Southern India (Coonoor).

Hosts: No known record.

Remarks: *Bactrocera (Javadacus) scutellaria* (Bezzi) is similar to *Bactrocera (Javadacus) javanensis* (Perkins), *Bactrocera (Javadacus) montana* (Hardy) and *Bactrocera (Javadacus) trilineata* (Hardy) in possessing a black scutum, a broad mesopleural stripe reaching to or almost to postpronotal lobe dorsally, lateral and medial postsutural yellow vittae, a yellow spot anterior to notopleural suture, a narrow fuscous costal band confluent with R_{2+3} and black patterns on abdominal terga III–V. It differs from these species in having the face with a pair of large transverse oval black spots, black notopleura, a black spot on the apex of the scutellum and large areas of black on all femora.

Bactrocera (Javadacus) semisurstyli Drew & Romig, new species

(Fig. 219)

Material examined: Holotype male, **INDONESIA: MOLUCCAS (MALUKU): AMBON ISLAND** – Middle Maluku, Suli, 1.vii.2007, coll. La Naibu, attracted to cue lure, deposited in BMNH.

Paratype: **INDONESIA: MOLUCCAS (MALUKU): AMBON ISLAND** – 1 male, same data as holotype, deposited in QDPC.

Diagnosis: A medium-sized species; face fulvous with a pair of medium-sized oval black spots; postpronotal lobes and notopleura yellow; scutum dark red-brown with large irregularly shaped black patterns; very short and narrow lateral postsutural yellow vittae present; a large broad medial postsutural yellow vitta present; a yellow spot anterior to notopleural suture; mesopleural stripe reaching anterior *npl.* seta dorsally; scutellum yellow; wing with cells bc and c with a pale tint, microtrichia in outer corner of cell c only; a narrow dark fuscous costal band widening across apex of wing, infuscation along *r-m* and *dm-cu* crossveins and around hind margin of wing, a broad fuscous anal streak; abdominal terga III–V dark orange-brown with moderately broad black lateral margins on tergum III and narrow dark fuscous lateral margins on terga IV and V.

Description: male.

Head: Height 1.64 mm. Frons – length 1.67 times breadth; red-brown without dark markings; orbital setae black: 1 *s.or.*, 2 *i.or.*; lunule fuscous. Ocellar triangle black. Vertex red-brown. Face fulvous with a pair of medium-sized oval black spots; length 0.48 mm. Genae red-brown; black seta present. Occiput red-brown, fulvous along eye margins; occipital row with a large number of short pale setae. Antennae with segments 1 and 2 red-brown, segment 3 red-brown with fuscous on outer surface; length of segments: 0.14 mm; 0.22 mm; 0.67 mm.

Thorax: Scutum dark red-brown with two longitudinal irregularly shaped dull black bands running from inside lateral postsutural vittae to inside postpronotal lobes. Pleural areas dark fuscous, red-brown below postpronotal lobes. Yellow markings as follows: postpronotal lobes; notopleura; a lateral yellow band running from posterior margin of postpronotal lobe to notopleural suture; mesopleural stripe reaching anterior *npl.* seta dorsally, continuing to katepisternum as a large transverse spot, anterior margin convex; anatergite (posterior apex black); anterior ⅔ katatergite (remainder black); two moderately broad but very short lateral postsutural vittae beginning with a spot anterior to notopleural suture and ending well before level of *p.sa.* seta; a broad medial postsutural yellow vitta. Postnotum black, tending red-brown centrally. Scutellum yellow except for narrow black basal band. Setae: *sc.* 2; *prsc.* 2; *ia.* 1; *p.sa.* 1; *a.sa.* absent; *mpl.* 1; *npl.* 2; *scp.* 4.

Legs: All femora entirely red-brown; fore and mid tibiae red-brown, hind tibiae fuscous; all tarsi with basal segment fulvous and apical four segments fuscous; mid tibiae each with an apical black spur.

Wings: Length 6.5 mm; cells bc and c with a pale tint; microtrichia in outer corner of cell c only; remainder of wings colourless except dark fuscous cell sc, a narrow dark fuscous costal band overlapping R_{2+3} and widening across apex of wing, pale infuscation around *r-m* and *dm-cu* crossveins and across hind margin of wing, a moderately broad fuscous anal streak; a dense aggregation of microtrichia around A_1+CuA_2; supernumerary lobe of medium development.

Abdomen: Elongate oval; terga free; pecten present on tergum III. Tergum I and sterna I and II wider than long. Tergum I dark fuscous with a narrow transverse orange-brown band across posterior margin but not reaching lateral margins; tergum II orange-brown with moderately broad dark fuscous lateral margins; terga III–V dark orange-brown with moderately broad lateral black markings on tergum III, narrow dark fuscous lateral margins on terga IV and V and a medium-width medial longitudinal black band over terga IV and V. A pair of oval dark orange-brown shining spots on tergum V. Posterior lobe of surstylus long, sternum V with a slight concavity on posterior margin. Abdominal sterna dark fuscous to black.
Attractant: Cue lure.
Female: No known record.
Distribution: Indonesia (Moluccas = Maluku – Ambon I., Suli).
Hosts: No known record.
Remarks: *Bactrocera (Javadacus) semisurstyli* n.sp. fits into the subgenus *Javadacus* as it possesses long surstyli in the male. This species is best identified through the combination of the lateral yellow band connecting the postpronotal lobe and notopleuron, very short and narrow lateral postsutural yellow vittae, a broad medial postsutural yellow vitta, the costal band widened across apex of wing and pale infuscation enclosing *r-m* and *dm-cu* crossveins. *Bactrocera (Javadacus) javanensis* (Perkins) is similar to *B. semisurstyli* but differs in having the scutum entirely black, infuscation only on *dm-cu* crossvein, a complete black 'T' pattern over abdominal terga III–V and in lacking a lateral yellow band joining the postpronotal lobe and the notopleuron.

Bactrocera (Javadacus) trilineata (Hardy)
(Fig. 220)

Dacus (Afrodacus) trilineatus Hardy, 1955a: 12; Hardy, 1977: 47. Holotype male in BMNH.
Bactrocera (Javadacus) trilineata – Norrbom *et al.*, 1998: 98; Drew & Raghu, 2002: 328.

Non-type material examined: INDIA: 1 male (19.i.1988), 1 male (ii.1988), Karnataka, Bangalore, 916m, coll. G. Bhat, attracted to cue lure. Specimens in QDPC. **SRI LANKA (formerly CEYLON)**: 1 male, Galkulama, C8, 27.v.1994 and 1 male, Ralangala, 1.xii.1994. Specimens in QDPC. 1 male (7.x.1994), 1 male (22.x.1994), 1 male (1.xi.1994), 'Sri Lanka, Tabbowa, C16'; 'Sri Lanka, NPQS Project, Sp.R.No. 21'. 1 male (14.x.1994), 1 male (18.xi.1994), 'Sri Lanka, Rendapola, C34'; 'Sri Lanka, NPQS Project, Sp. R. No. 21'. 3 males, 'Sri Lanka, Peradeniya, CARI – trapped, viii-xi.1992, IIE 22664'; 'Bactrocera trilineata (Hardy), det. I.M. White, 1993'. Specimens in BMNH. **THAILAND**: 1 male, 'Chiang Mai Prov., San Sai, ix-x.1992, coll. P. Kaewchoung'. Specimen in BMNH. 6 males and 5 females, Sarapee, Chiang Mai, 6.vi.1991, coll. Chantanee Hengsawad, bred from *Coccinia grandis* (L.) Voigt, Sample No. CM61; 7 males and 3 females, Hang Chat, Chiang Mai, 10.x.1991, coll. Chantanee, bred from *Coccinia grandis* (L.) Voigt, Sample No. CM340; 4 males and 9 females, P. Kanchang, Potharam, Ratchaburi, 7.x.1992, coll. Prasert, bred from *Coccinia grandis* (L.) Voigt; 5 males and 2 females, M004, T. Samphraya, A. Cha-Am, Petchaburi, 25.ii.1992, coll. Montree, bred from *Coccinia grandis* (L.) Voigt, Sample No. BK1511; 1 male, Thailand, Guava orchard, Chiang Mai, 28.xi.1991, coll. Chantanee, Sample No. CM249T, attracted to cue lure; 1 male, Thailand, Guava orchard, Chiang Mai, 2.i.1992, coll. Chantanee, Sample No. CM281T, attracted to cue lure; 6 males and 6 females, Doi Saket, 3.i.1991, coll. Chant, bred from *Coccinia grandis*, Sample No. TCM0808. Specimens in QDPC. **VIETNAM: Northern Vietnam** – 1 male, Vietnam North, Son La Province, Moc Chau District, Co Do, Site N006, 29.xi.1999; 1 male, Vietnam North, Son La Province, Moc Chau District, Moc Chau town, Site N009, 29.xi.1999, attracted to cue lure; **Southern Vietnam** – 1 male (25.ii.2002), 2 males (25.iii.2002), 1 male (8.iv.2002), 3 males (6.v.2002), 1 male (20.v.2002), Binh Thuan Province, Ham thuan District, No. 1 village, Site VS07, Coll. SOFRI Staff, attracted to cue lure. Specimens in QDPC.
Diagnosis: A small species; face entirely fulvous; postpronotal lobes yellow; notopleura with posterior area yellow and anterior black; scutum entirely black; lateral and medial postsutural yellow vittae present; a small yellow spot anterior to notopleural suture; mesopleural stripe reaching to postpronotal lobe dorsally; scutellum yellow with a narrow black basal band; wing with cells bc and c colourless, microtrichia in outer corner of cell c only; a narrow dark fuscous costal band confluent with R_{2+3}, remaining very narrow beyond apex of this vein before expanding across apex of R_{4+5}; anal streak reduced to fuscous within cell cup; abdominal terga III–V variable in colour patterns from orange-brown with a 'T' pattern and broad lateral margins on terga III–V to almost entirely black with orange-brown patterns posterocentrally either side of a medial longitudinal black band on terga IV and V or terga III, IV and V.
Description: male.
Head: Height 1.23 mm. Frons – length 1.25 times breadth; fulvous with dark fuscous around orbital setae and on anteromedial hump; orbital setae black: 1 *s.or.*, 2–3 *i.or.*; lunule dark fuscous. Ocellar triangle black. Vertex red-brown. Face entirely fulvous without dark markings; length 0.4 mm. Genae fulvous without dark markings; red-brown seta present. Occiput fuscous to dark fuscous, fulvous along eye margins; occipital row with a large number of small pale setae. Antennae with segments 1 and 2 red-brown, segment 3 red-brown with fuscous on apex and outer surface; length of segments: 0.12 mm; 0.21 mm; 0.6 mm.
Thorax: Scutum entirely black without pale markings. Pleural areas entirely black. Yellow markings as follows: postpronotal lobes; notopleura (anterior ¼ to ½ black); broad mesopleural stripe reaching to postpronotal lobe dorsally, continuing to katepisternum as a small transverse spot, anterior margin strongly convex; anatergite (posterior apex black); anterior $^{3}/_{5}$ katatergite (remainder black); two narrow to broad parallel-sided lateral postsutural vittae widening slightly posteriorly and ending well behind *ia.* seta; a small yellow spot anterior to notopleural suture; a short narrow medial postsutural vitta. Postnotum black. Scutellum yellow

except for narrow black basal band. Setae: *sc.* 2; *prsc.* 2; *ia.* 1; *p.sa.* 1; *a.sa.* absent; *mpl.* 1; *npl.* 2; *scp.* 4.
Legs: All segments entirely fulvous or with dark fuscous around apical ¼ of hind femora in some specimens (this mark varies in size); mid tibiae each with an apical black spur.
Wings: Length 4.8 mm; cells bc and c colourless; microtrichia in outer corner of cell c only; remainder of wings colourless except dark fuscous cell sc, narrow dark fuscous costal band confluent with R_{2+3} and becoming extremely narrow beyond apex of this vein before expanding across apex of R_{4+5}, anal streak reduced to pale fuscous within cell cup; a dense aggregation of microtrichia around A_1+CuA_2; supernumerary lobe of medium development.
Abdomen: Oval; terga free; pecten present on tergum III. Tergum I and sterna I and II wider than long. Tergum I orange-brown with broad black lateral margins; tergum II fulvous with a broad transverse black band across anterior margin; terga III–V variable in colour patterns from orange-brown with a 'T' pattern consisting of a broad transverse black band across tergum III covering almost the entire tergum and a narrow medium longitudinal black band over all three terga and with broad black anterolateral corners on terga IV and V to abdominal terga III–V entirely black with small orange-brown markings posterocentrally either side of a medial longitudinal black band on terga IV and V or terga III, IV and V. A pair of oval orange-brown shining spots on tergum V. Posterior lobe of surstylus long, sternum V with a slight concavity on posterior margin.
Attractant: Cue lure.
Female: No known record.
Distribution:
Type locality: India (Karnataka – Bangalore, Sarakki Village).
Previously recorded: Type locality and India (Tamil Nadu); Sri Lanka (Norrbom *et al.*, 1998).
New records: Thailand; Vietnam.
Hosts: *Coccinia grandis* (family *Cucurbitaceae*) equals *Cephalandra indica* as recorded by Hardy (1955a).
Remarks: *Bactrocera (Javadacus) trilineata* (Hardy) is similar to *Bactrocera (Javadacus) javanensis* (Perkins), *Bactrocera (Javadacus) montana* (Hardy) and *Bactrocera (Javadacus) scutellaria* (Bezzi) in possessing a black scutum with lateral and medial postsutural yellow vittae, a small yellow spot anterior to the notopleural suture, a broad mesopleural stripe reaching to or almost to the postpronotal lobe dorsally, a narrow costal band confluent with R_{2+3} and black patterns on abdominal terga III–V. It is distinct from these species in having an entirely fulvous face and all leg segments entirely fulvous in some specimens.

The populations of flies sampled from Thailand and Vietnam show slight differences from those collected in India and Sri Lanka. The former possess a smaller area of black on the anterior margin of the notopleuron, broader lateral postsutural yellow vittae and entirely fulvous femora. It is possible that the population from South-East Asia represents a distinct species separate from *B. trilineata* in the Indian subcontinent. However, at present we are not making this separation.

SUBGENUS *NESODACUS* Perkins

Bactrocera (Nesodacus) atrichus (Bezzi)

(Figs 221A, 221B)

Chaetodacus atrichus Bezzi, 1919: 420. Lectotype male and 2 Paralectotypes (1 male and 1 female) in MCSNM.
Chaetodacus atrichus var. *davaoanus* Bezzi, 1919: 421; Norrbom *et al.*, 1998: 98. Holotype female in MCSNM.
Dacus (Nesodacus) atrichus – Hardy, 1969b: 479, Lectotype designation; Hardy, 1974: 19.
Dacus (Nesodacus) atrichus davaoanus – Hardy, 1974: 19.
Bactrocera (Nesodacus) atrichus – Norrbom *et al.*, 1998: 98.

Type material examined: PHILIPPINES: Lectotype male with the following three labels – 'Batbatan Isl., R.C. McGregor'; '*atrichus* n.' (Bezzi's handwriting); 'Lectotype male, *Chaetodacus atrichus* Bez., desig by Hardy – 68'. Paralectotypes, 1 male and 1 female each labelled – 'Batbatan Isl., R.C. McGregor'. Received on loan from MCSNM.
Other type material examined: Holotype female of *Chaetodacus atrichus* var *davaoanus* Bezzi with the following three labels – 'Davao, Mindanao, Baker'; 'davaoanus n.' (Bezzi's handwriting); 'Type female'. Received on loan from MCSNM (Fig. 221B) (see 'Remarks' below).
Diagnosis: A small species; males with face dark fuscous and with a pair of medium-sized oval black spots, females with face fulvous with a pair of small oval dark fuscous to black spots; postpronotal lobes and notopleura yellow; scutum shining black; lateral postsutural yellow vittae present; medial postsutural yellow vitta absent; a yellow spot anterior to notopleural suture; mesopleural stripe reaching to anterior *npl.* seta dorsally; scutellum yellow with a narrow black basal band; wing with cells bc and c colourless, microtrichia in outer corner of cell c only; a narrow dark fuscous costal band slightly overlapping R_{2+3} and widening slightly apically; a broad dark fuscous anal streak; abdominal terga III–V with a broad transverse black band across anterior margin of tergum III and a broad dark fuscous band across anterior margin of tergum IV and broad anterolateral dark fuscous corners on tergum V, a pair of black oval shining spots on tergum V.
Description: male (Fig. 221A).
Head: Height 1.25 mm. Frons – length 1.25 times breadth; fulvous with dark fuscous around *s.or.* and upper pair of *i.or.* setae and on anteromedial hump; orbital setae black: 1 *s.or.*, 2 *i.or.*; lunule red-brown. Ocellar triangle black. Vertex dark fuscous. Face dark fuscous with a pair of medium-sized oval black spots; length 0.45 mm. Genae fulvous, fuscous subocular spot present; black seta present. Occiput dark fuscous, fulvous along eye margins; occipital row indistinct and with a large number of very small pale setae. Antennae with segments 1 and 2 red-brown, segment 3 red-brown with dark fuscous on apex and outer surface; length of segments: 0.125 mm; 0.225 mm; 0.6 mm.
Thorax: Scutum shining black with a small area of dark red-brown around notopleural suture. Pleural areas shining black without distinct pale markings. Yellow markings as follows: postpronotal lobes; notopleura; mesopleural stripe reaching

to anterior *npl.* seta dorsally, continuing to katepisternum as a small transverse spot, anterior margin slightly convex; anatergite (posterior apex black); anterior $^{3}/_{5}$ katatergite (remainder black); two broad sub-parallel lateral postsutural vittae beginning in a spot anterior to notopleural suture and ending at or just behind *ia.* seta. Medial longitudinal postsutural yellow vitta absent. Postnotum black. Scutellum yellow except for narrow black basal band. Setae: *sc.* 2; *prsc.* absent; *ia.* 1; *p.sa.* 1; *a.sa.* 1; *mpl.* 1; *npl.* 2; *scp.* 4.

Legs: Fore femora with fuscous to dark fuscous on apical ¾ and with basal ¼ fulvous, mid femora with apical ⅔ dark fuscous and basal ⅓ fulvous, hind femora with apical ½ dark fuscous and basal ½ fulvous; fore and mid tibiae fulvous, hind tibiae pale fuscous; tarsi with all segments fulvous; mid tibiae each with an apical black spur.

Wings: Length 4.7 mm; cells bc and c colourless; microtrichia in outer corner of cell c only; remainder of wings colourless except dark fuscous cell sc, narrow dark fuscous costal band slightly overlapping R_{2+3} where it is paler and widening slightly around apex of wing, broad dark fuscous anal streak; no dense aggregation of microtrichia around A_1+CuA_2; supernumerary lobe weak.

Abdomen: Oval; terga free; no pecten on tergum III. Tergum I black with a very narrow fulvous band across posterior margin but not reaching lateral margins; tergum II fulvous with a narrow transverse black band anteriorly and reaching dark fuscous anterolateral corners; tergum III fulvous with a broad transverse black band across anterior margin and covering lateral margins, tergum IV orange-brown with a broad transverse dark fuscous band across anterior margin and covering lateral margins, tergum V orange-brown with broad anterolateral dark fuscous corners. A pair of oval black shining spots on tergum V. Posterior lobe of surstylus long, sternum V with a slight concavity on posterior margin.

Attractant: No known record.

Female: As for male except face fulvous with a pair of small oval dark fuscous to black spots; ovipositor basal segment red-brown to fuscous, conical in transverse cross section and tapering posteriorly in dorsal view; ratio of length of oviscape to length of tergum V, 1:1; apex of piercer needle-shaped.

Distribution:

Type locality: Philippines (Batbatan I., Antique Province).

Previously recorded: Type locality and Philippines (Mindanao I., Davao) (Norrbom *et al.*, 1998).

Hosts: No known record.

Remarks: *Bactrocera (Nesodacus) atrichus* (Bezzi) is a unique species based on the combination of dark femora markings and dark patterns on abdominal terga III–V. After studying the Holotype of *Chaetodacus atrichus* var. *davaoanus* Bezzi, the placing of this in synonymy with *B. atrichus* appears correct. In the revisions of Philippine Tephritidae, Hardy (1954) treated *B. atrichus* and *C. davaoanus* as separate species while in Hardy (1974) he stated that they were probably not separate species. Also, in CABIKEY (White and Hancock, 1997) the authors stated that these were inseparable, while Norrbom *et al.* (1998) listed *C. davaoanus* as a synonym of *B. atrichus*. In these treatments, there does not appear to be any declaration of a formal synonymy. *Bactrocera atrichus* possesses the subgeneric characters of *Sinodacus* except that it has the cilia on abdominal terga III of male absent. Illustrations of the Lectotype of *B. atrichus* and Holotype of *B. davaoanus* are presented for comparison.

SUBGENUS *PAPUODACUS* Drew

Bactrocera (Papuodacus) complicata White

(Fig. 222)

Bactrocera (Papuodacus) complicata White, in White & Evenhuis, 1999: 514. Holotype male in RNH.

Non-type material examined: EAST TIMOR: with the following label data – 3 males, Suai, 5.v.2002, coll. G. Bellis, ex cue lure trap; 4 males, Dili – Gleno Road, 8.v.2002, coll. G. Bellis, cue lure trap; 1 male, Brian Palmer's, Dili, 10.v.2002, coll. G. Bellis; 1 male, Santa Cruz, Dili, 18/19.xi.2003, coll. G. Bellis, ex cue lure trap; 1 male, Heru, 12.ii.2004, coll. G. Bellis, cue lure trap. Specimens in AQIS, Darwin. **INDONESIA: WEST TIMOR** – 1 male, Nusa Tenggara Timur, Oebufu, Kt Kupang, 31.v.2007, coll. Ir Hendri Wihanto, attracted to cue lure in Village Jambu; 1 male, Soe, 800m elev, 30.x.1986, coll. E.S.C. Smith, attracted to cue lure; 1 male, Kupang, ADAB office, 31.x.1986, coll. E.S.C. Smith, attracted to cue lure. Specimens in QDPC.

Diagnosis: A medium-sized species; face fulvous with a pair of medium-sized circular black spots; postpronotal lobes and notopleura yellow; scutum black with areas of red-brown to dark red-brown; lateral and medial postsutural yellow vittae present; a yellow spot present anterior to notopleural suture; mesopleural stripe reaching midway between anterior margin of notopleuron and anterior *npl.* seta dorsally; scutellum yellow; wing with cell bc colourless, cell c colourless except dark fuscous over apical end, microtrichia in outer corner of cell c only; a broad dark fuscous costal band confluent with R_{4+5} and expanding across apex of wing; a moderately broad dark fuscous anal streak, broad transverse dark fuscous bands enclosing both crossveins; abdominal terga III–V orange-brown with moderately broad dark fuscous to black lateral margins over all three terga and a narrow fuscous to dark fuscous medial longitudinal band on tergum V.

Description: male.

Head: Height 1.7 mm. Frons – length 1.36 times breadth; red-brown with very small areas of fuscous around orbital setae and on anteromedial hump; orbital setae black: 1 *s.or.*, 3 *i.or.*; lunule fuscous. Ocellar triangle black. Vertex red-brown. Face fulvous with a pair of medium-sized circular black spots; length 0.59 mm. Genae fulvous; black seta present. Occiput red-brown, fulvous along eye margins; occipital row with a large number of very small pale setae. Antennae with all segments red-brown; length of segments: 0.12 mm; 0.29 mm; 0.79 mm.

Thorax: Scutum black with areas of dark red-brown below and behind lateral postsutural vittae, posterior to medial longitudinal yellow vitta, anterior to notopleural suture, inside postpronotal lobes and a broad medial longitudinal red-brown band from anterior point of medial postsutural vitta to anterior margin of scutum. Pleural areas black with red-brown below postpronotal lobes. Yellow markings as follows: postpronotal lobes; notopleura; mesopleural stripe reaching midway between anterior margin of notopleuron and anterior *npl.* seta dorsally, continuing to katepisternum as a transverse spot, anterior margin slightly convex; anatergite (posterior apex black); anterior ⅔ katatergite (remainder black); two narrow lateral postsutural vittae beginning with a yellow spot anterior to notopleural suture and narrowing sharply posteriorly to end before *ia.* seta. Medial longitudinal postsutural yellow vitta present. Postnotum black laterally, red-brown centrally. Scutellum yellow except for narrow black basal band. Setae: *sc.* 2 (in most specimens there are also 2 very weak pale basal setae); *prsc.* 2; *ia.* 1; *p.sa.* 1; *a.sa.* absent; *mpl.* 1; *npl.* 2; *scp.* 4.

Legs: All femora fulvous with red-brown around apices; fore tibiae fulvous, mid tibiae pale fuscous basally to fulvous apically, hind tibiae fuscous; all tarsi with basal two segments fulvous and apical three segments red-brown; mid tibiae each with an apical black spur.

Wings: Length 7.2 mm; cell bc colourless, cell c colourless with dark fuscous across apical end; microtrichia in outer corner of cell c only; remainder of wings colourless except dark fuscous cell sc, broad dark fuscous costal band confluent with R_{4+5} and expanding across apex of wing to end at apex of M, broad dark fuscous transverse bands enclosing *r-m* and *dm-cu* crossveins, a moderately broad dark fuscous anal streak; a dense aggregation of microtrichia around A_1+CuA_2; supernumerary lobe very strong.

Abdomen: Oval; terga free; pecten present on tergum III. Tergum I and sterna I and II wider than long. Tergum I orange-brown with broad black lateral margins; tergum II orange-brown with moderately broad black lateral margins; terga III–V orange-brown with a narrow fuscous to dark fuscous medial longitudinal band on tergum V and moderately broad lateral dark fuscous to black margins over all three terga. A pair of oval orange-brown shining spots on tergum V. Posterior lobe of surstylus long, sternum V with a slight concavity on posterior margin.

Attractant: Cue lure.

Female: No known record.

Distribution:

Type locality: Indonesia (Timor).

New records: Indonesia (West Timor, Kupang); East Timor.

Hosts: No known record.

Remarks: *Bactrocera (Papuodacus) complicata* White is a unique species that fits imperfectly into the subgenus *Papuodacus* Drew. Specimens with 4 *sc.* setae fit *Papuodacus* while those with 2 *sc.* setae fit into subgenus *Javadacus* Hardy. The known species in subgenus *Papuodacus* possess a medial longitudinal postsutural yellow vitta on the scutum similar to *B. complicata*, whilst in species of *Javadacus* this medial vitta is absent. This species appears to be restricted in geographic distribution to the island of Timor.

Bactrocera (Papuodacus) fereuncinata Drew & Romig, new species

(Fig. 223)

Type material examined: Holotype male, **THAILAND**: Khao Kheaw, A. Pakphli, Nakhonnayok, 24.vii.1992, coll. Montree, attracted to cue lure, deposited in BMNH.

Paratypes: **THAILAND**: 3 males, Nakhonratsima, Pakdong, Wang Jum Pee, Thailand, 18.x.1991, 15.xi.1991, 14.viii.1992, coll. Montree, attracted to cue lure. 1 paratype in ANIC, 2 paratypes in QDPC.

Diagnosis: A medium-sized species; large circular facial spots present; postpronotal lobes and notopleura yellow; scutum red-brown with large black patterns; lateral and medial postsutural yellow vittae present; a yellow triangle along anterior margin of notopleural suture; mesopleural stripe equal in width to notopleuron dorsally; scutellum yellow; wing with cells bc and c colourless, microtrichia in outer corner of cell c only; a narrow dark fuscous costal band and narrow fuscous anal streak; abdominal terga III–V fulvous with a distinct black 'T' pattern, narrow lateral black margins on terga IV and V and a recurved black band on tergum IV running from posterior margin almost to anterior margin.

Description: male.

Head: Height 1.62 mm. Frons – length 1.52 times breadth; fulvous with dark fuscous on anteromedial hump; orbital setae black: 1 *s.or.*, 2 *i.or.*; lunule dark fuscous. Ocellar triangle black. Vertex dark fuscous to black. Face fulvous with large circular black spots, length 0.54 mm. Genae fulvous, large fuscous to dark fuscous subocular spot present; black seta present. Occiput fuscous, fulvous along eye margins; occipital row with a large number of small setae. Antennae with segments 1 and 2 red-brown, segment 3 red-brown with fuscous on apex and outer surface; arista black (red-brown basally); length of segments: 0.20 mm; 0.27 mm; 0.72 mm.

Thorax: Scutum red-brown with large areas of black between lateral and medial postsutural vittae, between postpronotal lobes and notopleura and a narrow longitudinal line from inner point of notopleural suture to midline on postpronotal lobe. Pleural areas black except red-brown below postpronotal lobes and on ventral surfaces of katepisternum. Yellow markings as follows: postpronotal lobes; notopleura; mesopleural stripe equal in width to notopleuron dorsally, extending to katepisternum as a large spot, anterior margin strongly convex; anatergite (posterior apex black); anterior $^{3}/_{5}$ katatergite (remainder black); a yellow triangle with base on notopleuron and extending along anterior margin of notopleural suture; two parallel-sided lateral postsutural vittae of medium-width and ending at *ia.* setae; a broad medial longitudinal postsutural vitta beginning anterior to *prsc.* setae and ending in a point anterior to notopleural suture. Postnotum red-brown centrally, black laterally. Scutellum yellow except for narrow black basal band. Setae: *sc.* 4; *prsc.* 2; *ia.* 1; *p.sa.* 1; *a.sa.* absent; *mpl.* 1; *npl.* 2; *scp.* 4.

Legs: Femora entirely fulvous; fore tibiae dark fuscous, mid tibiae fuscous basally to fulvous apically, hind tibiae dark

fuscous; tarsi with basal segment fulvous and apical four segments pale fuscous; mid tibiae each with an apical black spur.
Wings: Length 6.4 mm; cells bc and c colourless; microtrichia in outer corner of cell c only; remainder of wings colourless except dark fuscous cell sc, narrow dark fuscous costal band slightly overlapping R_{2+3} and expanding into a small spot around apex of R_{4+5}, narrow pale fuscous anal streak; dense aggregation of microtrichia around A_1+CuA_2; supernumerary lobe weak.
Abdomen: Elongate oval; terga free; pecten present on tergum III. Tergum I and sterna I and II wider than long. Tergum I fulvous with narrow dark fuscous lateral margins; tergum II fulvous with a narrow subanterior transverse dark fuscous band which joins into narrow lateral dark fuscous margins; terga III–V fulvous with a distinct 'T' consisting of a narrow transverse black band across anterior margin of tergum III that expands across lateral margins and a broad medial longitudinal black band over all three terga, narrow lateral black margins over terga IV and V and a recurved black marking on tergum IV from hind margin to almost reach anterior margin leaving a fulvous area between this marking and the narrow black lateral margin. A pair of oval fulvous shining spots on tergum V. Posterior lobe of surstylus long, sternum V with a slight concavity on posterior margin.
Attractant: Cue lure.
Female: No known record.
Distribution: Thailand (Nakhonnayok; Nakhonratsima).
Hosts: No known record.
Remarks: *Bactrocera (Papuodacus) fereuncinata* n.sp. belongs to the *Bactrocera* subgenus *Papuodacus* Drew, separating from the subgenus *Zeugodacus* by the absence of *a.sa.* setae. It is similar to *Bactrocera (Zeugodacus) uncinata* n.sp. in the colour patterns of the scutum and the unique black markings on abdominal terga III–V. It differs from *B. uncinata* in having parallel-sided lateral postsutural vittae, a broader and shorter medial postsutural vitta, narrower mesopleural stripe dorsally (equal in width to notopleuron), a yellow triangle along anterior margin of notopleural suture, large circular facial spots and costal band with a *tau*-like spot in apex. It is also similar to *Bactrocera (Zeugodacus) tau* (Walker) in the general colour patterns of the scutum, wings and abdomen. However, *B. tau* possesses a long narrow transverse yellow spot on the upper margin of the katepisternum, lateral postsutural yellow vittae extending well posterior to the *ia.* seta, generally more black on the scutum and a strong keel-shaped supernumerary lobe in the male wing.

Bactrocera (Papuodacus) maculifemur (Hering)
(Fig. 224)

Zeugodacus maculifemur Hering, 1938: 3. Holotype male in NRS.
Bactrocera (Zeugodacus) maculifemur – Norrbom *et al.*, 1998: 103.

Type material examined: Holotype male, **MYANMAR (formerly BURMA)**: with the following three labels – 'Type'; 'N.E. Burma, Kambaiti, 2000 m, 14.5.1934, Malaise'; 'Zeugodacus maculifemur, Type male, det. M. Hering, 1938'. Received on loan from NRS.
Diagnosis: A medium-sized species; face entirely fulvous; postpronotal lobes and notopleura yellow; scutum black; lateral and medial postsutural yellow vittae present; a small yellow spot anterior to notopleural suture anterior to lateral postsutural vittae; mesopleural stripe equal in width to notopleuron dorsally; scutellum yellow with broad black basal band and a large black apical spot; wing with cells bc and c colourless, microtrichia in extreme outer corner of cell c only; a narrow fuscous costal band confluent with R_{2+3} and extremely narrow around apex of wing before expanding into a fishhook-shaped spot around apex of R_{4+5}; anal streak absent; abdominal terga I–V entirely black except for small orange-brown markings on posterior margins of terga II–V.
Description: male.
Head: Height 1.3 mm. Frons – length 1.16 times breadth; red-brown with fuscous around orbital setae and on anteromedial hump; orbital setae black: 1 *s.or.*, 2 *i.or.*; lunule dark fuscous. Ocellar triangle black. Vertex dark fuscous. Face entirely fulvous without dark markings; length 0.37 mm. Genae fulvous, pale fuscous subocular spot present. Occiput black, fulvous along eye margins; occipital row with a large number of short pale setae. Antennae with segments 1 and 2 dark fuscous, segment 3 absent; length of segments: 0.09 mm; 0.25 mm.
Thorax: Scutum black with red-brown behind lateral postsutural vittae. Pleural areas entirely black without pale markings. Yellow markings as follows: postpronotal lobes (inner margins black); notopleura; mesopleural stripe equal in width to notopleuron dorsally, continuing to katepisternum as a small transverse spot, anterior margin straight; anatergite (posterior apex black); anterior ⅔ katatergite (remainder black); two narrow parallel-sided lateral postsutural vittae beginning with a small spot anterior to notopleural suture and ending just before *ia.* seta; a very short and narrow medial longitudinal postsutural vitta. Postnotum black. Scutellum yellow except for a broad black basal band and a large black apical spot enclosing apical setae. Setae: *sc.* 4; *prsc.* 2; *ia.* 1; *p.sa.* 1; *a.sa.* absent; *mpl.* 1; *npl.* 2; *scp.* 4.
Legs: Fore femora fulvous except for an elongate oval black spot on outer apical surfaces, mid femora fulvous except for elongate black spot on apical ½ of outer surfaces, hind femora fulvous except for an elongate black spot on apical ⅓ of outer surfaces; fore tibiae fulvous except dark fuscous on basal ½ of outer surfaces, mid tibiae entirely fulvous, hind tibiae fulvous except dark fuscous basally; tarsi with all segments entirely fulvous; mid tibiae each with an apical black spur.
Wings: Length 6.04 mm; cells bc and c colourless; microtrichia in extreme outer corner of cell c only; remainder of wings colourless except dark fuscous cell sc, narrow fuscous costal band confluent with R_{2+3} and remaining extremely narrow beyond apex of this vein around margin of wing and ending in a fishhook-shaped enlarged spot around apex of R_{4+5}; anal streak absent; dense aggregation of microtrichia around A_1+CuA_2; supernumerary lobe of medium development.

Abdomen: Elongate oval; terga free; pecten present on tergum III. Tergum I and sterna I and II wider than long. All terga entirely black except for narrow transverse dark orange-brown bands posterolaterally on tergum II, a narrow transverse orange-brown band posteromedially on terga III and IV and orange-brown across posterior margin of tergum V. A pair of oval black shining spots on tergum V. Posterior lobe of surstylus long, sternum V with a slight concavity on posterior margin.
Attractant: No known record.
Female: No known record.
Distribution:
Type locality: Myanmar (Kambaiti).
Hosts: No known record.
Remarks: *Bactrocera (Papuodacus) maculifemur* (Hering) is placed in the subgenus *Papuodacus* on the basis of lacking *a.sa.* setae. It is very similar to species in the *Bactrocera (Zeugodacus) scutellaris* complex in the thoracic and wing colour patterns, especially *Bactrocera (Zeugodacus) namlingiae* n.sp. from which it differs in possessing parallel-sided lateral postsutural vittae, a yellow spot anterior to the notopleural suture and subapical dark spots on the apices of all femora (*B. namlingiae* has dark colouration surrounding the apices of the mid and hind femora).

Bactrocera (Papuodacus) ochrosterna Drew & Romig, new species

(Fig. 225)

Material examined: Holotype male, **VIETNAM: Northern Vietnam** – North Vietnam, Hoa Binh Province, Mai Chau District, Pa Co, 30.vi.2001, Site N010, attracted to cue lure, deposited in BMNH.
Paratype: **VIETNAM: Northern Vietnam** – 1 male, North Vietnam, Hoa Binh Province, Mai Chau District, Pa Co, 30.i.2002, Site N010, attracted to cue lure. Specimen in QDPC.
Diagnosis: A large species; face red-brown with a pair of medium-sized oval black spots; postpronotal lobes and notopleura yellow; scutum black; lateral and medial postsutural yellow vittae present; yellow spot present anterior to notopleural suture; mesopleural stripe equal in width to notopleuron; scutellum yellow with a large black apical spot; wing with cells bc and c entirely colourless and devoid of microtrichia; a narrow fuscous costal band; a narrow fuscous anal streak; all abdominal terga entirely orange-brown.
Description: male.
Head: Height 1.62 mm. Frons – length 1.43 times breadth; red-brown with dark fuscous on anteromedial hump and around bases of *s.or.*; orbital setae black: 1 *s.or.*, 3 *i.or.*; lunule red-brown. Ocellar triangle black. Vertex dark fuscous. Face red-brown with a pair of medium-sized oval black spots; length 0.44 mm. Genae red-brown, small dark fuscous subocular spot present; red-brown seta present. Occiput black, fulvous along eye margins; occipital row with 3–5 weak pale setae. Antennae with segments 1 and 2 fuscous, segment 3 red-brown with fuscous on apex and outer surface; length of segments: 0.12 mm; 0.26 mm; 0.77 mm.
Thorax: Scutum black except fuscous across posterior margin and dark red-brown below and behind lateral postsutural vittae. Pleural areas dark fuscous except red-brown below postpronotal lobes. Yellow markings as follows: postpronotal lobes (anterodorsal corners dark fuscous); notopleura; mesopleural stripe equal in width to notopleuron dorsally, extending to katepisternum as a large transverse spot, anterior margin straight; anatergite (posterior apex black); anterior $^3/_5$ katatergite (remainder black); two narrow parallel-sided lateral postsutural vittae extending to *ia.* seta; a narrow medial longitudinal postsutural vitta; a small yellow spot anterior to notopleural suture. Postnotum black tending red-brown centrally. Scutellum yellow with a narrow black basal band and a large black apical spot enclosing bases of all four setae. Setae: *sc.* 4; *prsc.* 2; *ia.* 1; *p.sa.* 1; *a.sa.* absent; *mpl.* 1; *npl.* 2; *scp.* 4.
Legs: Fore femora fulvous with a large black subapical spot on outer surfaces, mid and hind femora fulvous with black around apical ⅓ to ¼; fore tibiae fulvous with dark fuscous along outer surfaces, mid tibiae fulvous tending dark fuscous basally, hind tibiae dark fuscous; tarsi with all segments entirely fulvous; mid tibiae each with an apical black spur.
Wings: Length 7.2 mm; cells bc and c colourless and entirely devoid of microtrichia; remainder of wings colourless except dark fuscous cell sc, narrow fuscous costal band confluent with R_{2+3} and extremely narrow beyond apex of this vein before expanding slightly around apex of R_{4+5}, a narrow fuscous anal streak; a dense aggregation of microtrichia around A_1+CuA_2; supernumerary lobe of medium development.
Abdomen: Oval; terga free; pecten present on tergum III. Tergum I and sterna I and II wider than long. All abdominal terga entirely orange-brown except for narrow dark fuscous lateral margins on tergum I and narrow dark fuscous anterolateral corners on tergum II. A pair of oval orange-brown shining spots on tergum V. Posterior lobe of surstylus long, sternum V with a slight concavity on posterior margin. Abdominal sterna pale coloured, generally orange-brown.
Attractant: Cue lure.
Female: No known record.
Distribution: Vietnam (Northern Vietnam – Hoa Binh Province).
Hosts: No known record.
Remarks: *Bactrocera (Papuodacus) ochrosterna* n.sp. is similar to *Bactrocera (Papuodacus) maculifemur* (Hering) in possessing a black scutum, an apical black spot on the scutellum, wing with a narrow costal band confluent with R_{2+3} and all femora with black apical markings. It differs from *B. maculifemur* in having all abdominal terga almost entirely orange-brown and abdominal sterna orange-brown.

Bactrocera (Papuodacus) sinensis Yu, Bai & Chen

(Fig. 226)

Bactrocera (Papuodacus) sinensis Yu, Bai & Chen, in Yu *et al.*, 2011: 607. Holotype male in CAIQ.

Type material studied: CHINA: A study of the colour photographs and description of *Bactrocera (Papuodacus) sinensis* Yu, Bai & Chen in the original publication.
Non-type material examined: PENINSULAR MALAYSIA: 2 males, TN10, Malaysia, 30.vi.1986, coll. Yusoff Salleh, attracted to cue lure, in QDPC.
Diagnosis: A medium-sized species; face fulvous with a pair of small circular black spots; postpronotal lobes and notopleura yellow; scutum black; lateral and medial postsutural yellow vittae present; a yellow spot present anterior to notopleural suture; mesopleural stripe equal in width to notopleuron dorsally; scutellum yellow with a small dark fuscous spot between the apical setae; wing with cells bc and c colourless, microtrichia in outer corner of cell c only; a narrow dark fuscous costal band confluent with R_{2+3} and *scutellaris*-like; a moderately broad dark fuscous anal streak; abdominal terga III–V orange-brown with a broad medial and two broad lateral longitudinal black bands joined along anterior margin of tergum III.
Description: male.
Head: Height 1.3 mm. Frons – length 1.47 times breadth; red-brown with fuscous on anteromedial hump; orbital setae black: 1 *s.or.*, 3 *i.or.*; lunule dark fuscous. Ocellar triangle black. Vertex black. Face fulvous with a pair of small circular black spots; length 0.4 mm. Genae fulvous, dark fuscous subocular spot present; black seta present. Occiput dark fuscous, fulvous along eye margins; occipital row with 3–4 black setae. Antennae with segments 1 and 2 red-brown, segment 3 red-brown with fuscous on apex and outer surface; length of segments: 0.09 mm; 0.21 mm; 0.67 mm.
Thorax: Scutum black except red-brown below and behind lateral postsutural vittae and inside notopleura. Pleural areas black except dark fuscous below postpronotal lobes. Yellow markings as follows: postpronotal lobes; notopleura; mesopleural stripe equal in width to notopleuron dorsally, continuing to katepisternum as a large transverse spot, anterior margin straight; anatergite (posterior apex black); anterior ⅔ katatergite (remainder black); two parallel-sided lateral postsutural vittae of medium-width, beginning with a spot anterior to notopleural suture and ending at *ia.* seta; a narrow short medial longitudinal postsutural vitta. Postnotum black. Scutellum yellow except for narrow black basal band and a small dark fuscous spot between the apical setae. Setae: *sc.* 4; *prsc.* 2; *ia.* 1; *p.sa.* 1; *a.sa.* absent; *mpl.* 1; *npl.* 2; *scp.* 4.
Legs: Fore femora fulvous with a large black subapical spot on outer surfaces, mid femora fulvous with dark fuscous to black around apical ¼, hind femora fulvous with black around apical ¼; fore and hind tibiae dark fuscous, mid tibiae dark fuscous basally to fulvous apically; fore and mid tarsi with all segments fulvous, hind tarsi with basal segment fulvous and apical four segments fuscous; mid tibiae each with an apical black spur.
Wings: Length 5.5 mm; cells bc and c colourless; microtrichia in outer corner of cell c only; remainder of wings colourless except dark fuscous cell sc, narrow dark fuscous costal band confluent with R_{2+3} and becoming very narrow beyond apex of this vein before widening slightly across apex of R_{4+5}, a moderately broad dark fuscous anal streak; a dense aggregation of microtrichia around A_1+CuA_2; supernumerary lobe weak.
Abdomen: Oval; terga free; pecten present on tergum III. Tergum I and sterna I and II wider than long. Tergum I black with fulvous across posterior margin; tergum II fulvous with a narrow transverse band across anterior margin which meets the narrow black anterolateral corners; terga III–V orange-brown with a broad medial and two broad lateral longitudinal black bands which join across anterior margin of tergum III. A pair of oval orange-brown shining spots on tergum V. Posterior lobe of surstylus long, sternum V with a slight concavity on posterior margin. Abdominal sterna dark.
Attractant: Cue lure.
Female: No known record.
Distribution:
Type locality: China (Yunnan Province, Xishuangbanna).
New records: Peninsular Malaysia (Transect North Site 10).
Hosts: No known record.
Remarks: *Bactrocera (Papuodacus) sinensis* Yu, Bai & Chen belongs to a small subgenus with only five known species in South-East Asia, the others being *Bactrocera (Papuodacus) complicata* White, *Bactrocera (Papuodacus) fereuncinata* n.sp., *Bactrocera (Papuodacus) maculifemur* (Hering) and *Bactrocera (Papuodacus) ochrosterna* n.sp. It differs from *B. complicata* in having a narrow costal band confluent with R_{2+3} and lacking infuscation on the *r-m* and *dm-cu* crossveins, from *B. fereuncinata* in having a black scutum, a black apical spot on the scutellum and black markings on all femora, from *B. ochrosterna* in having dark abdominal sterna and from *B. maculifemur* in having a pair of black facial spots and abdominal terga III–V not entirely black.

SUBGENUS *PARADACUS* Perkins

Bactrocera (Paradacus) angustifinis (Hardy)

(Fig. 227)

Dacus (Paradacus) angustifinis Hardy, 1982: 197–199. Holotype male in MZB.
Bactrocera (Paradacus) angustifinis – Norrbom *et al.*, 1998: 99.

Type material examined: Holotype male, **INDONESIA: SULAWESI** – Lindu Valley, Central Sulawesi, 960 m, August 1975, coll. D.E. Hardy, attracted to cue lure in forest. Studied at MZB.
Paratypes: **INDONESIA: SULAWESI** – 1 male, 'Lindu Valley, C. Sulawesi, 960 m, August 1975. coll. D.E. Hardy'; 'Paratype, Dacus (Paradacus) angustifinis, D.E. Hardy'. 1 male, 'Kuala Navusa, Malakosa, Sulawesi, viii–xi.1975, 100 ft, coll. Margareta Becker'; 'attracted to cue lure'; 'paratype male Dacus (Paradacus) angustifinis, D.E. Hardy'; 'Dacus (Paradacus) angustifinis Hardy, 1980 det. D.E. Hardy'. Specimens in QDPC.
Non-type material examined: INDONESIA: SULAWESI – 4 males (20.xii.1985), 1 male (21.xii.1985), Dumoga Taraut

1350m, N. Sulawesi, coll. Santianawati, attracted to cue lure; 1 male, Dumoga Taraut, N. Sulawesi, (0°30′N, 124° E), 22.xii.1985, coll. L. Santianawati, attracted to cue lure; 2 males, Palopo-Rantepao Rd, 16-23km Palopo, Sulawesi, Indonesia (250–500m), 24.vii.1981, coll. J. Turner, Nanggala III forest, attracted to cue lure. Specimens in QDPC.

Diagnosis: A medium-sized species; face pale fuscous with a pair of large oval black spots; postpronotal lobes and notopleura yellow; scutum shining black without pale markings; lateral postsutural yellow vittae present; medial postsutural yellow vitta absent; yellow spot anterior to notopleural suture present; mesopleural stripe equal in width to notopleuron dorsally; scutellum yellow with a broad black basal band; wing with cells bc and c colourless, microtrichia in outer corner of cell c only; a narrow pale fuscous costal band confluent with R_{2+3} and remaining narrow before widening very slightly across apex of R_{4+5}; a broad pale fuscous anal streak; abdominal terga III–V dark fuscous to black except for narrow fulvous areas either side of a narrow medial longitudinal black band on tergum V.

Description: male.

Head: Frons – fulvous to red-brown with pale fuscous around orbital setae and on anteromedial hump; orbital setae black: 1 *s.or.*, 3 *i.or.*; lunule red-brown. Ocellar triangle black. Vertex black. Face pale fuscous with a pair of large oval black spots. Occiput fuscous, fulvous along eye margins. Antennae with all segments pale fuscous.

Thorax: Scutum entirely black without pale markings. Pleural areas black. Yellow markings as follows: postpronotal lobes; notopleura; mesopleural stripe equal in width to notopleuron dorsally; two moderately broad parallel-sided lateral postsutural vittae beginning with a small spot anterior to notopleural suture and ending at *ia.* seta. Medial longitudinal postsutural yellow vitta absent. Scutellum yellow except for a broad black basal band. Setae: *sc.* 4; *prsc.* absent; *ia.* 1; *p.sa.* 1; *a.sa.* 1; *mpl.* 1; *npl.* 2; *scp.* 4.

Legs: Fore and mid femora fuscous to black with small areas of fulvous basally, hind femora with apical ½ fuscous to black and basal ½ fulvous; fore and mid tibiae very pale fuscous, hind tibiae dark fuscous; tarsi with all segments fulvous; mid tibiae each with an apical black spur.

Wings: Length 5.5 mm; cells bc and c colourless; microtrichia in outer corner of cell c only; remainder of wings colourless except fuscous cell sc, narrow pale fuscous costal band confluent with R_{2+3} and remaining narrow around apex of wing except for a very slight expansion across apex of R_{4+5}, a broad pale fuscous anal streak; a dense aggregation of microtrichia around A_1+CuA_2; supernumerary lobe of medium development.

Abdomen: Oval; terga free; pecten present on tergum III. Tergum I and sterna I and II wider than long. Tergum I dark fuscous to black except for a narrow transverse fulvous band across posterior margin but not reaching lateral margins; tergum II dark fuscous to black except for a broad transverse fulvous band across posterior margin; terga III–V dark fuscous to black except for narrow fulvous areas either side of a narrow medial longitudinal black band on tergum V. A pair of oval fulvous shining spots on tergum V which tend to dark fuscous to black laterally. Posterior lobe of surstylus long, sternum V with a slight concavity on posterior margin.

Attractant: Cue lure.

Female: No known record.

Distribution:

Type locality: Indonesia (Sulawesi – Central Sulawesi, Lindu Valley).

Previously recorded: Type locality and several locations in Central Sulawesi (Hardy, 1982).

New records: Indonesia – Sulawesi (North Sulawesi – Dumoga Taraut; South Sulawesi – Palopo-Rantepao Road).

Hosts: No known record.

Remarks: *Bactrocera (Paradacus) angustifinis* (Hardy) is similar to *Bactrocera (Parasinodacus) incisa* (Walker) in the colour patterns of the thorax, wing, abdomen and legs and in possessing parallel-sided lateral postsutural vittae that begin with a yellow spot anterior to the notopleural suture. It differs from *B. incisa* in having distinct black facial spots (not a black line across oral margin), four *sc.* setae, a narrow mesopleural stripe equal in width to notopleuron dorsally and a broad black basal band on the scutellum. Other known species that have been placed in subgenus *Paradacus* have been recorded in Papua New Guinea, East New Britain and Maluku, and all have a distinct medial longitudinal yellow vitta on the scutum and entirely pale abdominal terga.

Bactrocera (Paradacus) areolata (Walker)

(Fig. 228)

Dacus areolatus Walker, 1861b: 295. Lectotype female in BMNH.

Dacus (Paradacus) areolatus – Hardy, 1959: 163. Lectotype designation by inference of Holotype.

Bactrocera (Paradacus) areolata – Norrbom *et al.*, 1998: 99.

Type material examined: Lectotype female, **INDONESIA: NORTH MOLUCCAS (NORTH MALUKU)** – with the following seven labels – 'Type, Dacus areolatus Wlk'; 'Batchian, A.R. Wallace, 68-4' (Wallace's handwriting); '684'; 'Bae.'; 'areolatus'; 'BMNH(E) No. 532999'; 'Dacus (Paradacus) areolatus Walker det. 1954 D.E. Hardy'. Studied at BMNH.

Diagnosis: A medium-sized species; face fulvous with a pair of medium-sized oval black spots; postpronotal lobes and notopleura yellow; scutum entirely red-brown; lateral postsutural yellow vittae present; medial postsutural yellow vitta absent; a yellow spot anterior to notopleural suture; mesopleural stripe reaching midway between anterior margin of notopleuron and anterior *npl.* seta dorsally; scutellum yellow; wings with cells bc and c colourless, microtrichia covering outer ½ of cell c only; almost entirely fuscous with small colourless areas along hind margin and in cells bm, dm, br, dr; a broad fuscous anal streak; abdominal terga entirely orange-brown.

Description: female.

Head: Frons – red-brown with fuscous on anteromedial hump and around bases of orbital setae; orbital setae black: 1 *s.or.*, 2 *i.or.*; lunule fuscous. Ocellar triangle black. Vertex red-brown. Face fulvous with a pair of medium-sized oval black spots. Occiput red-brown with fulvous along eye margins. Antennae with all segments fulvous.

Thorax: Scutum entirely red-brown. Pleural areas entirely red-brown. Yellow markings as follows: postpronotal lobes; notopleura; mesopleural stripe reaching midway between anterior margin of notopleuron and anterior *npl.* seta dorsally, anterior margin straight; two broad parallel-sided lateral postsutural vittae beginning with a spot anterior to notopleural suture and ending just behind *ia.* seta. Medial longitudinal postsutural yellow vitta absent. Scutellum yellow except for narrow red-brown basal band. Setae: *sc.* 4; *prsc.* absent; *ia.* 1; *p.sa.* 1; *a.sa.* 1; *mpl.* 1; *npl.* 2; *scp.* 4.

Legs: All femora fulvous tending red-brown apically; all tibiae red-brown; tarsi with all segments fulvous.

Wings: Length 6.1 mm; cells bc and c colourless; microtrichia over outer ½ of cell c only; remainder of wings almost entirely fuscous except for small colourless areas around hind margin and in cells bm and dm, br and dr, a broad fuscous anal streak; no dense aggregation of microtrichia around A_1+CuA_2; supernumerary lobe weak.

Abdomen: Oval; terga free; no pecten present on tergum III. Tergum I and sterna I and II wider than long. All terga entirely orange-brown. Basal segment of ovipositor orange-brown, extremely long and cylindrical and longer than the entire abdomen.

Attractant: No known record.

Male: No known record.

Distribution:

Type locality: Indonesia (North Moluccas = North Maluku – Bacan I., Batchian).

Hosts: No known record.

Remarks: *Bactrocera (Paradacus) areolata* (Walker) is similar to *Bactrocera (Paradacus) fulvipes* (Perkins), *Bactrocera (Parasinodacus) longicaudata* (Perkins) and *Bactrocera (Paradacus) urens* White in possessing lateral postsutural vittae with a yellow spot anterior to notopleural suture, abdominal terga almost entirely pale coloured, legs with femora pale with or without darker apical markings and the basal segment of the female ovipositor extremely long, narrow and cylindrical. It differs from these species in having the wings almost entirely fuscous.

Bactrocera (Paradacus) duplicata (Bezzi)

(Fig. 229)

Chaetodacus duplicatus Bezzi, 1916: 107. Syntypes in BMNH.

Dacus (Zeugodacus) duplicatus – Hardy, 1977: 57.

Bactrocera (Zeugodacus) duplicata – Norrbom *et al.*, 1998: 102.

Type material examined: Syntypes, **INDIA**: 1 male with the following four labels – 'Chaetodacus duplicatus Bezzi Type male'; 'Larva in peaches'; 'Central India, Pachmarhi, 3500 ft, 30.v.1909, Ratiram coll., Pres by T.B. Fletcher, 1915.414'; 'Chaetodacus duplicatus n.sp. Bezzi, Type male and female, T.B. Fletcher'. One female with the following three labels – 'Chaetodacus duplicatus Bezzi Type female'; 'Larva in peaches'; 'Central India, Pachmarhi, 3500 ft, 30.v.1909, Ratiram coll., Pres by T.B. Fletcher, 1915.414'. Studied at BMNH.

Non-type material examined: INDIA: 1 male, 'INDIA (KA): Mudigere 980m UASB Res.stn. 20-22 v.1992', 'I.M. White, D.L. Hancock & S. Ramani, IIE 22353'. Studied at BMNH. **SRI LANKA (formerly CEYLON)**: 1 male, Sri Lanka, 10.x.1994, Ambatenna, C18; 1 male, Sri Lanka, Laggala C26, 2.ii.1985. Specimens in QDPC.

Diagnosis: A medium-sized species; face fulvous with two narrow transverse black bands, one across oral margin and one just below antennal sockets; postpronotal lobes and notopleura yellow; scutum almost entirely black except dark fuscous below lateral postsutural vittae; lateral postsutural yellow vittae present; medial postsutural yellow vitta absent; a small yellow spot anterior to notopleural suture; mesopleural stripe reaching postpronotal lobe dorsally; scutellum entirely yellow; legs with femora fulvous except for a small subapical dark fuscous to black spot on fore femora and dark fuscous to black around apical ¼ of mid and hind femora; wing with cells bc and c colourless, microtrichia in outer corner of cell c only; a narrow fuscous costal band confluent with R_{2+3}, generally absent beyond apex of R_{2+3} and with a small darker fuscous spot around apex of R_{4+5}; a narrow fuscous anal streak; abdominal terga III–V entirely black.

Description: male.

Head: Frons – fulvous with small dark fuscous areas around orbital setae and on anteromedial hump; orbital setae black: 1 *s.or.*, 2 *i.or.*; lunule fuscous. Ocellar triangle black. Vertex black. Face fulvous with two transverse black bands, one across oral margin and one dorsally below antennal sockets. Occiput black, fulvous along eye margins. Antennae with segments 1 and 2 red-brown, segment 3 red-brown with fuscous on apex and outer surface.

Thorax: Scutum entirely black except dark fuscous below lateral postsutural vittae. Pleural areas black without pale markings. Yellow markings as follows: postpronotal lobes; notopleura; mesopleural stripe reaching postpronotal lobe dorsally, continuing to katepisternum as a narrow transverse spot, anterior margin convex; two broad parallel-sided lateral postsutural vittae beginning with a small spot anterior to notopleural suture and ending behind *ia.* seta. Medial longitudinal postsutural yellow vitta absent. Scutellum yellow except for narrow black basal band. Setae: *sc.* 4; *prsc.* 2; *ia.* 1; *p.sa.* 1; *a.sa.* 1; *mpl.* 1; *npl.* 2; *scp.* 4.

Legs: Fore femora fulvous with a small dark fuscous to black subapical spot on outer surfaces, mid and hind femora fulvous with dark fuscous to black around apical ¼; fore tibiae fuscous, mid and hind tibiae fuscous basally to pale fuscous apically; tarsi with basal segments fuscous and apical four segments pale fuscous.

Wings: Length 5.5 mm; cells bc and c colourless; microtrichia in outer corner of cell c only; remainder of wings colourless except dark fuscous cell sc, fuscous costal band confluent with R_{2+3} and ending at apex of this vein and then with a small dark fuscous apical spot around apex of R_{4+5}, a narrow pale fuscous anal streak; dense aggregation of microtrichia around A_1+CuA_2; supernumerary lobe strong and rounded.

Abdomen: Oval; terga free; pecten present on tergum III. Tergum I fulvous with a dark fuscous band across anterior margin and narrow lateral black margins; tergum II fulvous with a narrow transverse subanterior black band reaching narrow anterolateral black corners; terga III–V entirely black. A pair of oval orange-brown shining spots on tergum V. Posterior lobe of surstylus long, sternum V with a slight concavity on posterior margin. All abdominal sterna pale coloured.
Attractant: No known record.
Female: As for male except no dense aggregation of microtrichia around A_1+CuA_2; pecten of cilia on abdominal tergum III absent; supernumerary lobe weak; ovipositor basal segment red-brown, dorsoventrally compressed and tapering posteriorly in dorsal view.
Distribution:
Type locality: Central India (Madhya Pradesh – Pachmarhi).
New records: Southern India (Karnataka – Mudigere); Sri Lanka.
Hosts: No known record.
Remarks: *Bactrocera (Paradacus) duplicata* (Bezzi) is very close to *Bactrocera (Parasinodacus) cilifera* (Hendel) and differs only in possessing four *sc.* setae, a small yellow spot anterior to notopleural suture, a slightly wider mesopleural stripe and smaller areas of dark markings on apices of femora. It is similar to *Bactrocera (Paradacus) hancocki* n.sp. in possessing four *sc.* setae and a black scutum and in lacking a medial postsutural yellow vitta, but differs in having lateral postsutural yellow vittae.

Bactrocera (Paradacus) fulvipes (Perkins)

(Fig. 230)

Paradacus fulvipes Perkins, 1938a: 143. Holotype male in QMIC.
Dacus (Paradacus) fulvipes – Hardy & Adachi, 1954: 155; Hardy, 1974: 20.
Bactrocera (Paradacus) fulvipes – Norrbom *et al.*, 1998: 99.

Type material examined: Holotype male, **EAST MALAYSIA: SABAH (formerly part of North Borneo)** – with the following three labels – "N. Borneo, Bettotan, nr. Sandakan, Aug 5th 1927"; "Holotype, *Paradacus fulvipes* F.A. Perk"; "UQIC Reg. No. 45244". Received on loan from QMIC.

Non-type material examined: INDONESIA: SULAWESI – 1 female with the following label data – 'N.E. Sulawesi, 47km WSW Kotamobagu, Dumoga – Bone, N. Pk, Toraut (forest edge), 211m, June 1985'; 'Project Wallace, G.R. Else, BM 1985-10'; 'Bactrocera fulvipes (Perkins), det. I.M. White, 1981' [probable year error]; 'BMNH(E) No. 533000'. Studied at BMNH.
Diagnosis: A medium-sized species; face fulvous with a pair of medium-sized angulate oval black spots; postpronotal lobes and notopleura yellow; scutum entirely black; broad parallel-sided lateral postsutural yellow vittae present; medial postsutural yellow vitta absent; yellow spot anterior to notopleural suture; mesopleural stripe equal in width to notopleuron dorsally; scutellum yellow; wing with cells bc and c colourless, microtrichia in outer ⅓ of cell c only; a broad fuscous costal band confluent with R_{4+5} and expanding into a distinct spot across apex of wing; a moderately broad fuscous anal streak; abdominal terga III–V orange-brown with a broad transverse black band across anterior margin of tergum III and narrow diffuse dark fuscous lateral margins over all three terga.
Description: female.
Head: Frons – fulvous with pale fuscous on anteromedial hump; orbital setae black: 1 *s.or.*, 2 *i.or.*; lunule fuscous. Ocellar triangle black. Vertex fuscous. Face fulvous with a pair of medium-sized angulate oval black spots. Occiput red-brown, fulvous along eye margins. Antennae with segments 1 and 2 red-brown, segment 3 red-brown tending pale fuscous along outer surface.
Thorax: Scutum entirely black without pale markings. Pleural areas black without pale markings. Yellow markings as follows: postpronotal lobes; notopleura; mesopleural stripe equal in width to notopleuron dorsally, continuing to katepisternum as a small transverse spot, anterior margin straight; two broad parallel-sided lateral postsutural vittae beginning with a spot anterior to notopleural suture and ending behind *ia.* seta. Medial longitudinal postsutural yellow vitta absent. Scutellum yellow except for narrow black basal band. Setae: *sc.* 4; *prsc.* absent; *ia.* 1; *p.sa.* 1; *a.sa.* 1; *mpl.* 1; *npl.* 2; *scp.* 4.
Legs: Fore femora fulvous basally to pale fuscous apically, mid femora fulvous basally with apical ½ dark fuscous, hind femora fulvous with apical ⅓ dark fuscous; fore tibiae pale fuscous, mid and hind tibiae dark fuscous; tarsi with all segments fulvous; mid tibiae each with an apical black spur.
Wings: Length 6.3 mm; cells bc and c colourless; microtrichia in outer ⅓ of cell c only; remainder of wings colourless except fuscous cell sc, broad fuscous costal band confluent with R_{4+5} (paler adjacent to R_{4+5}) and expanding into a distinct spot across apex of wing but becoming paler fuscous across apex of M, a moderately broad fuscous anal streak; no dense aggregation of microtrichia around A_1+CuA_2; supernumerary lobe weak.
Abdomen: Oval; terga free; no pecten on tergum III. Tergum I and sterna I and II wider than long. Tergum I orange-brown with narrow lateral black margins; tergum II with a broad black transverse band across anterior half and expanding to cover lateral margins, remainder fulvous; terga III–V orange-brown with a broad transverse black band across anterior margin of tergum III and narrow diffuse dark fuscous lateral margins over all there terga. A pair of oval orange-brown shining spots on tergum V. Basal segment of ovipositor orange-brown, extremely long and cylindrical, longer than the entire abdomen.
Attractant: No known record.
Male: No known record.
Distribution:
Type locality: East Malaysia (Sabah, Bettotan near Sandakan).
Previously recorded: The record from the Philippines (Luzon I.) (Hardy, 1974; Norrbom *et al.*, 1998) is probably incorrect as Hardy noted that the female specimen he studied 'may belong to a distinct species'.
New records: Indonesia (Sulawesi – North Sulawesi, Dumoga Bone National Park).
Hosts: No known record.
Remarks: *Bactrocera (Paradacus) fulvipes* (Perkins) is similar to *Bactrocera (Parasinodacus) longicaudata* (Perkins) and *Bactrocera (Paradacus) urens* White in possessing a black scutum, broad parallel-sided lateral postsutural vittae

beginning with a yellow spot anterior to notopleural suture, abdominal terga III–V mostly pale and female ovipositor basal segment long and cylindrical. It differs from *B. longicaudata* in having four *sc.* setae and the costal band confluent with R_{4+5} and from *B. urens* in possessing black facial spots. It is possible that *B. urens* is a synonym of *B. fulvipes* but more specimens are needed for study before this can be evaluated.

Bactrocera (Paradacus) hancocki Drew & Romig, new species

(Fig. 231)

Type material examined: Holotype male, **INDONESIA: SOUTH SULAWESI** – Latupa, 9 km W. Palopo (100 m), Sulawesi, 23.vii.1981, coll. J. Turner, Primary Rainforest (some disturbance), attracted to cue lure, deposited in BMNH. Paratypes: **INDONESIA: SOUTH SULAWESI** – 1 male, Indonesia, Bententingi, 5 km W. Malino (850–1000 m), Sulawesi, 27.vii.1981, coll. J. Turner, Residual Secondary rainforest, attracted to cue lure; 2 males (16.viii.2006), 1 male (27.viii.2006), 4 males (13.xi.2006), South Sulawesi, Maros, Lamimpoccoe, coll. Sirajuddin. 1 paratype in ANIC, 1 paratype in BMNH, 4 paratypes in QDPC, 2 paratypes in QMIC.
Diagnosis: A unique large species with postpronotal setae (*ppn.* setae) present; face dark fuscous centrally with a pair of medium-sized circular black spots; postpronotal lobes and notopleura yellow; scutum dull black; lateral and medial postsutural yellow vittae absent; a large yellow triangle along anterior margin of notopleural suture; narrow mesopleural stripe equal in width to notopleuron dorsally; scutellum yellow with a narrow transverse dark fuscous subapical band; wing with cells bc and c fuscous, dense microtrichia covering all of both cells; a broad dark fuscous costal band overlapping R_{4+5} and expanding to form a large dark fuscous apical spot that includes almost the apical ⅓ of wing; broad fuscous to dark fuscous anal streak; abdominal terga III–V red-brown with a dark fuscous transverse band across anterior margin of tergum III and widening to cover lateral margins.
Description: male.
Head: Height 1.96 mm. Frons – length 1.25 times breadth; red-brown with fuscous on anteromedial hump and dark fuscous around bases of orbital setae; orbital setae black: 1 *s.or.*, 3 *i.or.*; lunule dark fuscous. Ocellar triangle black. Vertex dark red-brown. Face red-brown with a large area of dark fuscous centrally and a pair of medium-sized circular black spots; length 0.69 mm. Genae red-brown, large fuscous subocular spot present; black seta present. Occiput fuscous, fulvous along eye margins; occipital row with a large number of strong black setae. Antennae with all segments red-brown; arista black (red-brown basally); length of segments: 0.15 mm; 0.24 mm; 0.96 mm.
Thorax: Scutum dull black with dark red-brown along lateral margins behind the notopleural suture and posterocentrally between *prsc.* setae, and dark red-brown inside postpronotal lobes and anterocentrally around inner *scp.* setae. Pleural areas dull black except dark red-brown below postpronotal lobes. Yellow markings as follows: postpronotal lobes; notopleura; narrow mesopleural stripe equal in width to notopleuron dorsally, continuing to katepisternum as a small transverse spot, anterior margin straight; anatergite (posterior apex black); anterior ⅔ katatergite (remainder black); a large triangle along anterior margin of notopleural suture with a base at notopleuron. Postnotum black with a small area of dark red-brown centrally. Scutellum yellow except for narrow black basal band and a narrow transverse subapical dark fuscous to black band. Lateral and medial postsutural yellow vittae absent. Setae: *sc.* 4; *prsc.* 2; *ia.* 1; *p.sa.* 1; *a.sa.* 1; *mpl.* 1; *npl.* 2; *scp.* 4; *ppn* 1.
Legs: Fore femora entirely fuscous to dark fuscous, mid femora fuscous to dark fuscous except for a small area of fulvous basally, hind femora with apical ⅓ fuscous to dark fuscous and basal ⅔ fulvous; fore and hind tibiae fuscous to dark fuscous, mid tibiae fuscous basally to fulvous apically; tarsi with all segments fulvous.
Wings: Length 6.98 mm; cells bc and c fuscous; dense microtrichia covering all of both cells; remainder of wings colourless except fuscous cell sc, broad fuscous to dark fuscous costal band overlapping R_{4+5} and widening markedly apically to overlap M, broad dark fuscous anal streak; dense aggregation of microtrichia around A_1+CuA_2; supernumerary lobe weak.
Abdomen: Elongate oval; terga free; pecten present on tergum III. Tergum I and sterna I and II wider than long. Tergum I red-brown with a dark fuscous band across anterior margin and widening laterally to form broad dark fuscous lateral margins; tergum II red-brown with a narrow transverse dull black band across anterior margin which becomes dark fuscous before joining with narrow dark fuscous anterolateral corners; terga III–V dark orange-brown with a narrow transverse dark fuscous band across anterior margin of tergum III and widening to include lateral margins and narrow fuscous lateral margins on tergum V. A pair of pale dark orange-brown shining spots on tergum V. Posterior lobe of surstylus long, sternum V with a slight concavity on posterior margin.
Attractant: Cue lure.
Female: No known record.
Distribution: Indonesia (Sulawesi – South Sulawesi).
Hosts: No known record.
Etymology: This species is named after Dr David Hancock in recognition of his extensive contributions to tephritid taxonomic research.
Remarks: *Bactrocera (Paradacus) hancocki* n.sp. is a unique species that fits into subgenus *Paradacus* in having long surstylus lobes, abdominal sternum V with a slight concavity on posterior margin, four *sc.* setae, pecten present on abdominal sternum V, *a.sa.* present and medial postsutural yellow vittae absent from the scutum. The only exception is that the *ppn.* seta is present. In some species where this seta is present, it varies from weak to strong (for example, *Bactrocera (Zeugodacus) flavolateralis* n.sp.) and consequently this character is variable intraspecifically and not considered sound enough to separate off another subgenus. *Bactrocera hancocki* differs from all known *Zeugodacus* species in having the lateral and medial postsutural yellow

vittae absent and from the other South-East Asian *Paradacus* in lacking the lateral postsutural yellow vittae and in possessing a distinct costal band that overlaps R_{4+5} throughout and expands to overlap M at apex of wing.

Bactrocera (Paradacus) terminifera (Walker)

Dacus terminifer Walker, 1860a: 152. Lectotype female in BMNH.
Zeugodacus terminifer – Perkins, 1939: 28. Lectotype designation by inference of Holotype.
Dacus (Zeugodacus) terminifer – Hardy & Adachi, 1954: 195; Hardy, 1959: 183; 1982: 212; Drew, 1973: 33.
Bactrocera (Zeugodacus) terminifera – Drew, 1989: 219.
Bactrocera terminifera, incertae sedis – Norrbom *et al.*, 1998: 104.

Type material examined: Lectotype female, **INDONESIA: SULAWESI (formerly CELEBES)** – with the following four labels – 'Celebes, Macassar, W.W. Saunders, B.M. 1868-4'; 'Terminifer'; 'Mak'; '68-4'. The latter three labels are in Walker's handwriting. Studied by Drew (1989), the Lectotype is too damaged to provide complete data and thus has not been illustrated.
Diagnosis: Face entirely black; postpronotal lobes yellow with anterodorsal corners glossy black; notopleura yellow; scutum glossy black; very short and narrow lateral postsutural vittae present; medial postsutural vitta absent; no yellow spot anterior to notopleural suture; mesopleural stripe reaching to postpronotal lobe dorsally; scutellum yellow with a narrow black basal band; legs with fore and mid femora and apical ½ of hind femora black (basal ½ hind femora fulvous); wing with cells bc and c colourless, microtrichia in outer corner of cell c only; a narrow pale fuscous costal band confluent with R_{2+3} and ending at apex of this vein, a small pale fuscous spot at apex of R_{4+5}; a broad fuscous anal streak.
Attractant: No known record.
Male: No known record.
Distribution:
Type locality: Indonesia (Sulawesi – South Sulawesi, Macassar = Makassar).
Hosts: No known record.
Remarks: *Bactrocera (Paradacus) terminifera* (Walker) is a relatively unknown species from Sulawesi. The Lectotype is badly damaged with the abdomen missing. Drew (1989) provided comprehensive data from the Lectotype but more specimens will be needed in order to obtain a complete species definition.

Bactrocera (Paradacus) urens White
(Fig. 232)

Bactrocera (Paradacus) urens White, in White & Evenhuis, 1999: 518. Holotype female in ZMAN.

Type material examined: Holotype female, **INDONESIA: MOLUCCAS (MALUKU)** – with the following two labels – 'Buru 1921, Station 9, coll. L.J. Toxopeus'; 'Holotype, Bactrocera urens White, det. I.M. White, 1999'. Studied at BMNH.
Diagnosis: A small species; face entirely fulvous without dark markings; postpronotal lobes and notopleura yellow; scutum dark fuscous centrally to black laterally; lateral postsutural yellow vittae present; medial postsutural yellow vitta absent; a yellow spot anterior to notopleural suture; mesopleural stripe reaching midway between anterior margin of notopleuron and anterior *npl.* seta dorsally; scutellum yellow; wing with cells bc and c colourless, microtrichia in outer corner of cell c only; a broad pale fuscous costal band almost confluent with R_{4+5} and expanding across apex of wing; a narrow pale fuscous anal streak; all abdominal terga pale red-brown.
Description: Female.
Head: Frons – fulvous with pale fuscous spots around bases of orbital setae; orbital setae black: 1 *s.or.*, 2 *i.or.*; lunule fulvous. Ocellar triangle black. Vertex fulvous. Face entirely fulvous without dark spots. Occiput mostly fulvous. Antennae with segments I and II red-brown, segment III pale fuscous.
Thorax: Scutum dark fuscous centrally tending black laterally and red-brown around notopleural suture and lateral to lateral postsutural vittae. Yellow markings as follows: postpronotal lobes; notopleura; mesopleural stripe reaching midway between anterior margin of notopleuron and anterior *npl.* seta dorsally, extending to katepisternum as a small yellow spot, anterior margin straight; two broad parallel-sided lateral postsutural vittae beginning with a spot anterior to notopleural suture and ending at *ia.* seta. Medial longitudinal postsutural yellow vitta absent. Scutellum yellow except for narrow black basal band. Setae: *sc.* 4; *prsc.* 2; *ia.* 1; *p.sa.* 1; *a.sa.* 1; *mpl.* 1; *npl.* 2; *scp.* 4.
Legs: Fore femora fulvous, mid femora fulvous with pale fuscous around apical ⅓, hind femora fulvous with a small area of pale fuscous around apices; all tibiae pale fuscous; tarsi with basal segment fulvous and apical four segments pale fuscous.
Wings: Length 4.6 mm; cells bc and c colourless; microtrichia in outer corner of cell c only; remainder of wings colourless except pale fuscous cell sc, broad pale fuscous costal band almost confluent with R_{4+5} and expanding across apex of wing, a narrow pale fuscous anal streak; no dense aggregation of microtrichia around A_1+CuA_2; supernumerary lobe weak.
Abdomen: Oval; terga free; no pecten on tergum III. All terga pale red-brown except for a narrow transverse pale fulvous band across posterior margin of tergum II. Ovipositor basal segment red-brown, extremely long and narrow (at least as long as the entire abdomen) and of uniform width throughout.
Attractant: No known record.
Male: No known record.
Distribution:
Type locality: Indonesia (Moluccas = Maluku – Buru I.).
Hosts: No known record.
Remarks: *Bactrocera (Paradacus) urens* White is similar to *Bactrocera (Paradacus) fulvipes* (Perkins) and *Bactrocera (Parasinodacus) longicaudata* (Perkins) in

possessing a black scutum, broad parallel-sided lateral postsutural vittae with a yellow spot anterior to the notopleural suture, medial postsutural yellow vitta absent, femora with apical dark patterns, wings with costal band expanded apically, abdominal terga III–V mostly red-brown and the basal segment of the ovipositor long and narrow. It differs from *B. fulvipes* in having an entirely fulvous face and from *B. longicaudata* in having four *sc.* setae and paler colour patterns on the femora.

SUBGENUS *PARASINODACUS* new subgenus

Bactrocera (Parasinodacus) ablepharus (Bezzi)

(Figs 233A, 233B)

Chaetodacus ablepharus Bezzi, 1919: 422. Lectotype male and Paralectotype female in MCSNM.
Chaetodacus ablepharus var. *mindanaus* Bezzi, 1919: 422; Norrbom *et al.*, 1998: 98. Holotype female in MCSNM.
Nesodacus ablepharus – Perkins, 1938a: 134.
Dacus (Nesodacus) ablepharus – Hardy, 1969b: 479, Lectotype designation, Hardy, 1974: 18.
Bactrocera (Nesodacus) ablepharus – Norrbom *et al.*, 1998: 98.

Type material examined: Lectotype male, **PHILIPPINES**: with the following three labels – 'Malinao, Tayabas, Baker'; 'ablepharus n.' (Bezzi's handwriting); 'Lectotype [male], Chaetodacus ablepharus Bez., desig. by Hardy – 68'. Paralectotype female of *Chaetodacus ablepharus* Bezzi labelled – 'Mt Banahao, P.I. Baker'. Types on loan from MCSNM.
Other type material examined: PHILIPPINES: Holotype female of *Chaetodacus mindanaus* with the following three labels – 'Davao, Mindanao, Baker'; 'mindanaus n.' (Bezzi's handwriting); 'Type female'. Type on loan from MCSNM (Fig. 233B) (see 'Remarks' below).
Non-type material examined: VIETNAM: Southern Vietnam – 1 female, South Vietnam, Dac Lac Province, Ku Mi Nga District, Quang Phu Village, 24.viii.2002, Site VS09, collected in methyl eugenol trap. Specimen in QDPC.
Diagnosis: A small species; face in male and one female with large oval black spots filling antennal furrows and connected across centre of face (in the female of *mindanaus* the face is fulvous with a pair of large oval black spots); postpronotal lobes and notopleura yellow; scutum entirely black without pale markings; lateral postsutural yellow vittae present; medial postsutural vitta absent; no yellow spot anterior to notopleural suture; mesopleural stripe reaching almost to anterior *npl.* seta dorsally; scutellum yellow with a very narrow black basal band; wing with cells bc and c colourless, microtrichia in outer corner of cell c only; a narrow fuscous costal band confluent with R_{2+3} and widening slightly across apex of wing; a broad fuscous anal streak; abdomen with all terga entirely black except for a narrow transverse dark red-brown band across posterior margin of tergum V.
Description: male (Fig. 233A).
Head: Height 1.13 mm. Frons – length 1.33 times breadth; red-brown without dark markings; anteromedial hump with a small number of short pale hairs; orbital setae black: 1 *s.or.*, 2 *i.or.*; lunule red-brown. Ocellar triangle black. Vertex black. Face fulvous with two very large oval black spots filling antennal furrows and connected with a broad transverse band across centre; length 0.38 mm. Genae fulvous, small dark fuscous subocular spot. Occiput black, fulvous along eye margins; no distinct setae in occipital row. Antennae with segments 1 and 2 red-brown, segment 3 red-brown with fuscous on apex and outer surface; length of segments: 0.1 mm; 0.19 mm; 0.53 mm.
Thorax: Scutum uniformly black without pale markings. Pleural areas entirely black. Yellow markings as follows: postpronotal lobes; notopleura; mesopleural stripe reaching almost to anterior *npl.* seta dorsally, continuing to katepisternum as a small transverse spot, anterior margin convex; anatergite (posterior apex black); anterior ⅔ katatergite (remainder black); two narrow lateral postsutural vittae narrowing further posteriorly to end at *ia.* seta. Medial longitudinal postsutural yellow vitta absent. Postnotum black. Scutellum yellow except for a narrow black basal band. Setae: *sc.* 2; *prsc.* absent; *ia.* 1; *p.sa.* 1; *a.sa.* 1; *mpl.* 1; *npl.* 2; *scp.* 4.
Legs: Fore femora entirely black, mid femora black on apical ¾ and fulvous on basal ¼, hind femora black on apical ½ and fulvous on basal ½; fore tibiae fuscous basally to fulvous apically, mid tibiae fulvous, hind tibiae fuscous; tarsi with all segments fulvous; mid tibiae each with an apical black spur.
Wings: Length 4.4 mm; cells bc and c colourless; microtrichia in outer corner of cell c only; remainder of wings colourless except dark fuscous cell sc, narrow fuscous costal band confluent with R_{2+3} and widening slightly across apex of wing to end midway between apex of R_{4+5} and M, a broad fuscous anal streak; dense aggregation of microtrichia around A_1+CuA_2; supernumerary lobe strong and rounded.
Abdomen: Oval; terga free; pecten present on tergum III. Tergum I and sterna I and II wider than long. All segments entirely black except for a narrow transverse dark red-brown marking across posterior margin of tergum V. A pair of oval black shining spots on tergum V. Posterior lobe of surstylus long, sternum V with a slight concavity on posterior margin.
Attractant: No known record. The female specimen from Vietnam was collected in a methyl eugenol trap but this cannot be used to confirm the male lure response.
Female: As for male except no dense aggregation of microtrichia around A_1+CuA_2; supernumerary lobe weak; pecten of cilia on abdominal tergum III absent; ovipositor basal segment dark fuscous to black, dorsoventrally compressed and tapering posteriorly in dorsal view; ratio of length of oviscape to length of tergum V, 0.74:1. In the female specimen of *mindanaus* the face is fulvous with two large oval black spots filling the antennal furrows and not connected across centre. The female specimen of *ablepharus* has a narrow black band across centre connecting the large oval black spots in the antennal furrows. These probably represent some intraspecific variation in the facial colour patterns.

Distribution:
Type locality: Philippines (Luzon I., Tayabas, Malinao).
Previously recorded: Type locality and the Philippines (Mt Banahao, now spelt Banahaw). Philippines (Mindanao I., Davao) due to synonymy of *Chaetodacus mindanaus* (Norrbom *et al.*, 1998) listed under 'Other type material examined'. Perkins (1938a) recorded one female specimen collected at Bettotan near Sandakan (now East Malaysia, Sabah). The description of the specimen presented by Perkins is consistent with *B. ablepharus* and a number of dacine species are shared between the Philippines, East Malaysia and Kalimantan. Consequently, this record is accepted as correct.
New records: Southern Vietnam (Dac Lac Province).
Hosts: No known record.
Remarks: *Bactrocera (Parasinodacus) ablepharus* (Bezzi) fits into the subgenus *Parasinodacus* in having the posterior lobe of the surstylus long, posterior margin of male abdominal sternum V with a slight concavity and the pecten of cilia present on abdominal tergum III, in contrast to the type species of subgenus *Nesodacus*, *B. atrichus* (Bezzi), which has the cilia on abdominal tergum III absent. It is similar to *Bactrocera (Parasinodacus) incisa* (Walker), *Bactrocera (Parasinodacus) eurylomata* (Hardy), *Bactrocera (Parasinodacus) longicaudata* (Perkins) and *Bactrocera (Parasinodacus) vinnulus* (Hardy) in possessing a black scutum and lateral postsutural yellow vittae and lacking a medial postsutural yellow vitta. It differs from all of these species in lacking a yellow spot anterior to the notopleural suture in front of the lateral postsutural vittae and in having the abdominal terga entirely black.

Hardy (1954, 1974) and White & Hancock (1997) in CABIKEY treated *B. ablepharus* and *C. mindanaus* as separate species. However, Norrbom *et al.* (1998) listed *Chaetodacus ablepharus var. mindanaus* Bezzi as a synonym of *B. ablepharus* (Bezzi). This appears to be the first record of this synonymy in literature; however, there was no formal declaration of this synonymy. We have studied the types of both species and herein are able to confirm this synonymy.

Bactrocera (Parasinodacus) binoyi Drew

(Fig. 234)

Bactrocera (Sinodacus) binoyi Drew, in Drew & Raghu, 2002: 347. Holotype male in BMNH.

Type material examined: Holotype male, **INDIA**: New Amarambalam Forest, Kerala, 5.iii.1999 – 22.iii. 1999, coll. C.F. Binoy, attracted to cue lure. Studied at BMNH.
Non-type material examined: INDIA: 2 males, S. India, Hort Farm, Coonoor, 1450m, 29.v.1992, coll. D.L. Hancock, attracted to cue lure. Specimens in QDPC.
Diagnosis: A medium-sized species; face fulvous with a pair of medium-sized circular black spots; postpronotal lobes and notopleura yellow; a yellow triangular marking anterior to the notopleural suture; scutum dull black with a large area of dark fuscous across posterior margin and dark red-brown spots anterocentrally and inside postpronotal lobes; lateral and medial postsutural yellow vittae absent; mesopleural stripe reaching midway between anterior margin of notopleuron and anterior *npl.* seta dorsally; scutellum yellow; legs with femora fulvous with dark fuscous to black on apical ⅔ of fore femora, apical ⅓ mid femora and apical ¼ hind femora; fore and mid tibiae fulvous and hind tibiae fuscous to dark fuscous; wing with cells bc and c colourless, microtrichia in outer corner of cell c only; a narrow fuscous costal band just overlapping R_{2+3} as a pale tint and expanding around apex of R_{4+5}; a broad fuscous anal streak; supernumerary strong; abdominal terga III–V orange-brown with a broad black 'T' pattern and broad lateral longitudinal dark fuscous bands over all three terga, a pair of oval orange-brown shining spots on tergum V; posterior lobe of surstylus long; sternum V with a shallow concavity on posterior margin; Setae: *sc.* 2; *prsc.* absent; *ia.* 1; *p.sa.* 1; *a.sa.* 1; *mpl.* 1; *npl.* 2; *scp.* 2.
Attractant: Cue lure.
Female: No known record.
Distribution:
Type locality: Southern India (Kerala, New Amarambalam Forest).
Previously recorded: Type locality and Southern India – Tamil Nadu, Coonoor.
Hosts: No known record.
Remarks: *Bactrocera (Parasinodacus) binoyi* Drew is similar to *Bactrocera (Parasinodacus) eurylomata* (Hardy), *Bactrocera (Parasinodacus) incisa* (Walker), *Bactrocera (Parasinodacus) longicaudata* (Perkins) and *Bactrocera (Parasinodacus) vinnulus* (Hardy) in possessing a black scutum with a yellow spot anterior to the notopleural suture and an elongate oval abdomen with dark colour patterns and in lacking a medial postsutural yellow vitta. It differs from these species in lacking lateral postsutural yellow vittae.

Bactrocera (Parasinodacus) brevivitta Drew & Romig, new species

(Fig. 235)

Material examined: Holotype male, **PENINSULAR MALAYSIA**: Jln Kf Mimaland, Malaysia, 7.v.1986, coll. Mahdzir and Bahari, deposited in BMNH.
Paratypes: **PENINSULAR MALAYSIA**: 2 males and 1 female, Kg Bukit Cempedak, Kl-Kuantan, Malaysia, 29.i.1986, coll. S. Vijaysegaran, bred ex *Melastoma malabathricum*, sample No. 0026. 1 paratype in ANIC, 2 paratypes in QDPC.
Diagnosis: A small species; face fulvous with a pair of medium-sized oval black spots; postpronotal lobes and notopleura yellow; scutum entirely black; lateral postsutural yellow vittae present; medial postsutural yellow vitta absent; no yellow spot anterior to notopleural suture; mesopleural stripe reaching anterior *npl.* seta dorsally; scutellum yellow; wing with cells bc and c with a pale tint, microtrichia in outer ⅓ of cell c only; a narrow dark fuscous costal band slightly overlapping R_{2+3}; a broad dark fuscous anal streak; abdominal terga III–V orange-brown with a broad medial and two broad

lateral longitudinal black bands over all three terga and joining along anterior margin of tergum III.

Description: male.

Head: Height 1.3 mm. Frons – length 1.4 times breadth; fulvous with dark fuscous around bases of *s.or.* setae; orbital setae black: 1 *s.or.*, 2 *i.or.*; lunule dark fuscous. Ocellar triangle black. Vertex dark fuscous. Face fulvous with a pair of large oval black spots; length 0.39 mm. Genae red-brown, small fuscous subocular spot present; black seta present. Occiput dark fuscous, fulvous along eye margins; occipital row with 4–7 strong black setae. Antennae with segments 1 and 2 red-brown, segment 3 red-brown with fuscous on apex and outer surface; length of segments: 0.14 mm; 0.28 mm; 0.69 mm.

Thorax: Scutum entirely black without pale markings. Pleural areas black tending dark fuscous below postpronotal lobes. Yellow markings as follows: postpronotal lobes; notopleura; mesopleural stripe reaching anterior *npl.* seta dorsally, continuing to katepisternum as a small transverse spot, anterior margin straight; anatergite (posterior apex black); anterior ⅔ katatergite (remainder black); two medium-width very short lateral postsutural yellow vittae ending just posterior to level of *a.sa.* setae. Medial longitudinal postsutural yellow vitta absent. Postnotum black. Scutellum yellow except for narrow black basal band. Setae. *sc.* 2; *prsc.* 2; *ia.* 1; *p.sa.* 1; *a.sa.* 1; *mpl.* 1; *npl.* 2; *scp.* 4.

Legs: Fore and mid femora dark fuscous to black with small areas of fulvous at basal extremity, hind femora fulvous with dark fuscous to black around apical ⅓ fore tibiae entirely fuscous, mid tibiae fuscous basally to fulvous apically, hind tibiae entirely dark fuscous; fore and hind tarsi with basal segment fulvous and apical four segments fuscous, mid tarsi with all segments entirely fulvous; mid tibiae each with an apical black spur.

Wings: Length 4.5 mm; cells bc and c with a pale tint; microtrichia in outer ⅓ of cell c only; remainder of wings colourless except dark fuscous cell sc, narrow dark fuscous costal band slightly overlapping R_{2+3} at extremity and remaining of uniform width around apex of wing, a broad dark fuscous anal streak; a dense aggregation of microtrichia around A_1+CuA_2; supernumerary lobe of medium development.

Abdomen: Oval; terga free; pecten present on tergum III. Tergum I and sterna I and II wider than long. Tergum I black with a narrow transverse red-brown band across posterior margin but not reaching lateral margins; tergum II fulvous with a narrow transverse black band subanteriorly and reaching narrow black lateral margins; terga III–V orange-brown with a moderately broad medial and two broad lateral longitudinal black bands joined along anterior margin of tergum III. A pair of oval orange-brown shining spots on tergum V. Posterior lobe of surstylus long, sternum V with a slight concavity on posterior margin. Abdominal sterna dark.

Attractant: No known record.

Female: As for male except no dense aggregation of microtrichia around A_1+CuA_2, supernumerary lobe weak; pecten of cilia on abdominal tergum III absent; ovipositor basal segment dark red-brown, dorsoventrally compressed and tapering posteriorly in dorsal view; ratio of length of oviscape to length of tergum V, 0.4:1; apex of piercer needle-shaped.

Distribution: Peninsular Malaysia.

Hosts: *Melastoma malabathricum* (family *Melastomataceae*).

Remarks: *Bactrocera (Parasinodacus) brevivitta* n.sp. is similar to *Bactrocera (Parasinodacus) cilifera* (Hendel) and *Bactrocera (Parasinodacus) pahangiae* n.sp. in possessing a black scutum with lateral postsutural vittae, wings colourless except for the costal band and anal streak, the costal band mostly confluent with R_{2+3} and in lacking a lateral yellow band between the postpronotal lobe and notopleuron. It differs from *B. cilifera* in having the costal band entire and from *B. pahangiae* in having the scutellum entirely yellow, the fore and mid femora dark fuscous to black and hind femora with the apical ⅓ black and abdominal terga III–V with a black 'T' pattern and broad lateral dark fuscous to black margins.

Bactrocera (Parasinodacus) cilifera (Hendel)

(Fig. 236)

Dacus cilifer Hendel, 1912: 15. Syntypes in DEI and NMW.

Dacus (Strumeta) cilifera – Hardy & Adachi, 1954: 162.

Dacus (Zeugodacus) tenuifinis Hardy, 1983: 42. **New Synonym**.

Bactrocera (Zeugodacus) cilifera – Norrbom *et al.*, 1998: 102.

Type material examined: TAIWAN (formerly FORMOSA): 4 Syntypes with the following label data – 1 female, white label 'Koshun Formosa Sauter ix.08'; white label 'Dacus cilifer, H. det. Hendel'; red label 'TYPUS'; white label 'Chaetodacus cilifer Hendel'; white label 'Coll. DEI Eberswalde'. 2 females, white label 'Koshun Formosa Sauter ix.08'; white label 'Dacus cilifer, H. det. Hendel'; red label 'COTYPUS'; white label 'Coll. DEI Eberswalde' (one specimen has one extra white label 'Chaetodacus cilifer Hend. Det. HK Munro 1935.'). 1 male, white label 'Koshun Formosa Sauter ix.08'; white label 'Dacus cilifer H. det. Hendel'; red label 'COTYPUS'; white label 'Chaetodacus cilifer Hendel'; white label 'Coll. DEI Eberswalde'. These 4 specimens received on loan from DEI. 1 specimen (abdomen missing) with the following label data – white circular label with green outer circle 'Co-type'; white label 'Dacus cilifer H. det Hendel'; originally red label 'COTYPUS'; white label 'Koshun Formosa Sauter IX.08'; white label 'Formosa H. Sauter Brit. Mus. 1923-61'; white label '*Dacus cilifer* Hendel'; white label 'BMNH(E) #533022'. Studied at BMNH.

Non-type material examined: PENINSULAR MALAYSIA: 1 male (11.iii.1986), 1 male (8.iv.1986), T.S. Gombak, Malaysia, coll. Mahdzir, attracted to cue lure; 1 male, West Malaysia [Peninsular Malaysia], Perak, 2000 feet, 17.v.1983, coll. Dr Drew, MARDI M, caught off bamboo shoot. **THAILAND**: 8 males, Mae Hae, 1300m. 50km. West of Chiang Mai, 13-14.v.1991, coll. D.L. Hancock, attracted to cue lure; 1 male, Doi Suthep-Pui Nat Park, 1425m., Chiang Mai, 17.v.1991, coll. D.L. Hancock, attracted to cue lure; 1 male, Doi Luong Nat. Park, S.E. Chiang Mai, 24.v.1991, coll. D.L. Hancock, attracted to cue lure; 4 males,

Wawi Exp. Sta. 1400. 45km. South of Chiang Rai, 22.v.1991, coll. D.L. Hancock, attracted to cue lure; 1 male, Phkchong, Nakhon Ratchasima, 5.vi.05, attracted to cue lure. **TAIWAN**: 1 male, Chuchi (300m), 3.vi.78, coll. Y.H. Tseng, attracted to cue lure. **VIETNAM: Northern Vietnam**: 1 male, Hanoi (P), Tu Lien (D), Xuan Dinh (V), 26.xi.1999, attracted to cue lure, northern Vietnam, N004; 1 male, Lao Cai (P), Sapa 2 (V), 27.viii.1999, attracted to cue lure, northern Vietnam, N018; 1 male, Hoa Binh (P), Mai Chau (D), Pa Co (V), 15.xii.2001, attracted to cue lure, northern Vietnam, N010; 1 male, Son la (P), Moc chau (D), Co do State Farm, 15.ii.2002, (FAO site), attracted to cue lure, northern Vietnam, VN06; 1 male, Son La (P), Moc chau (D), Long luong (V), 15.iv.2002, attracted to cue lure, northern Vietnam, VN01. Specimens in QDPC.

Diagnosis: A small species; face fulvous with a broad transverse black band across oral margin which is occasionally broken in centre and a transverse black band across dorsal area just below antennal sockets; postpronotal lobes and notopleura yellow; scutum black; lateral postsutural yellow vittae present; medial postsutural yellow vitta absent; no yellow spot anterior to notopleural suture in front of lateral postsutural vittae; mesopleural stripe reaching midway between anterior margin of notopleuron and anterior *npl.* seta dorsally; scutellum yellow; wing with cells bc and c colourless, microtrichia in outer corner of cell c only; a narrow fuscous costal band confluent with R_{2+3} and ending at apex of this vein and a small fuscous spot around apex of R_{4+5}, a broad fuscous anal streak; abdominal terga almost entirely black.

Description: male.

Head: Height 1.28 mm. Frons – length 1.29 times breadth; fulvous with dark fuscous around orbital setae and on anteromedial hump; orbital setae black: 1 *s.or.*, 2 *i.or.*; lunule dark fuscous. Ocellar triangle black. Vertex dark fuscous. Face fulvous with a broad transverse black band across oral margin which is occasionally broken in the centre and a transverse black band across dorsal area just below antennal sockets; length 0.43 mm. Genae fulvous, large dark fuscous subocular spot; black seta present. Occiput dark fuscous, fulvous along eye margins; occipital row with no distinct setae. Antennae with segments 1 and 2 red-brown, segment 3 red-brown with fuscous on apex and outer surface; length of segments: 0.1 mm; 0.19 mm; 0.6 mm.

Thorax: Scutum black without pale markings. Pleural areas black tending dark fuscous below postpronotal lobes. Yellow markings as follows: postpronotal lobes (anteromedial corners dark fuscous); notopleura; mesopleural stripe of medium-width reaching midway between anterior margin of notopleuron and anterior *npl.* seta dorsally, continuing to katepisternum as a transverse spot, anterior margin strongly convex; anatergite (posterior apex black); anterior $^3/_5$ katatergite (remainder black); two narrow lateral postsutural vittae parallel-sided and widening slightly posteriorly to end behind level of *ia.* setae. Medial longitudinal postsutural yellow vitta absent. Postnotum black. Scutellum yellow with a narrow black basal band. Setae: *sc.* 2; *prsc.* 2; *ia.* 1; *p.sa.* 1; *a.sa.* 1; *mpl.* 1; *npl.* 2; *scp.* 4.

Legs: Fore femora black, mid femora with basal ⅓ fulvous and apical ⅔ black, hind femora with basal ⅔ fulvous, apical ⅓ black; all tibiae and tarsi entirely fulvous; mid tibiae each with an apical black spur.

Wings: Length 4.85 mm; cells bc and c colourless; microtrichia in outer corner of cell c only; remainder of wings colourless except fuscous cell sc, narrow fuscous costal band confluent with R_{2+3} and ending at apex of this vein, a small fuscous spot around apex of R_{4+5}, a broad fuscous anal streak; a dense aggregation of microtrichia around A_1+CuA_2; supernumerary lobe moderately strong.

Abdomen: Oval; terga free; pecten present on tergum III. Tergum I and sterna I and II wider than long. Tergum I black centrally merging to dark fuscous laterally; tergum II dark fuscous tending black along anterior margin; terga III–V dark fuscous tending black along anterior margin of tergum III and medially on tergum V. A pair of oval black shining spots on tergum V. Posterior lobe of surstylus long, sternum V with a slight concavity on posterior margin.

Attractant: Cue lure.

Female: As for male except no dense aggregation of microtrichia around A_1+CuA_2; supernumerary lobe weak; pecten of cilia on abdominal tergum III absent; ovipositor basal segment fuscous, dorsoventrally compressed and tapering posteriorly in dorsal view; ratio of length of oviscape to tergum V, 1.2:1.

Distribution:

Type locality: Taiwan (Koshun).

Previously recorded: Type locality and the following locations – China; Laos; Thailand; Vietnam (Norrbom *et al.*, 1998).

New records: Peninsular Malaysia (Gombak and Perak). Indonesia, Sumatra, Pasaman due to **New Synonymy** of *D. tenuifinis*.

Hosts: Recorded from flowers of *Thladiantha hookeri* (family *Cucurbitaceae*) (Allwood *et al.*, 1999).

Remarks: *Bactrocera (Parasinodacus) cilifera* (Hendel) belongs to a group of species previously placed in subgenus *Zeugodacus* with two *sc.* setae, medial postsutural vitta absent and no yellow spot anterior to notopleural suture. It is similar to *Bactrocera (Parasinodacus) pseudocucurbitae* White but differs in having an incomplete costal band and narrower mesopleural stripe reaching midway between anterior margin of notopleuron and anterior *npl.* seta dorsally, and in lacking infuscation on *r-m* and *dm-cu* crossveins. The measurements included are from specimens collected in Thailand, not the type. It is similar to *Bactrocera (Parasinodacus) ablepharus* (Bezzi) in the general colour patterns of the thorax, wing and abdomen but differs in possessing *prsc.* setae and entirely fulvous fore femora.

Dacus (Zeugodacus) tenuifinis Hardy cannot be separated from *B. cilifera* on any sound specific characters and is here made a **New Synonym** of *B. cilifera*.

Bactrocera (Parasinodacus) citrifusca Drew & Romig, new species

(Fig. 237)

Material examined: Holotype male, **THAILAND**: Thailand, 74/63 M.6 Bangkhuntak, Muang, Samutsongeram, 22.x.1992, coll. Prasert, deposited in BMNH.

Paratypes: **THAILAND**: 15 males and 17 females, same data as holotype. 5 paratypes in ANIC, 5 paratypes in BMNH, 17 paratypes in QDPC, 5 paratypes in QMIC.

Diagnosis: A medium-sized species; face entirely fulvous without dark markings; postpronotal lobes and notopleura yellow; scutum entirely red-brown without dark markings; lateral postsutural yellow vittae present; medial postsutural yellow vitta absent; no yellow spot anterior to notopleural suture; mesopleural stripe reaching almost to anterior *npl.* seta dorsally; scutellum yellow; wing with cells bc and c with a pale tint, microtrichia in outer corner of cell c only; a narrow fuscous costal band widening into a small spot around apex of R_{4+5} (*scutellaris*-like); pale fuscous within cell cup; abdominal terga III–V entirely orange-brown except for narrow transverse black spots on anterolateral corners of tergum III.

Description: male.

Head: Height 1.47 mm. Frons – length 1.5 times breadth; red-brown without dark markings; orbital setae black: 1 *s.or.*, 2 *i.or.*; lunule red-brown. Ocellar triangle black. Vertex red-brown. Face entirely fulvous without dark markings; length 0.49 mm. Genae fulvous, pale fuscous subocular spot present; pale red-brown seta present. Occiput red-brown, fulvous along eye margins; occipital row with 4–6 pale setae. Antennae with segments 1 and 2 red-brown, segment 3 red-brown with fuscous on apex and outer surface; length of segments: 0.18 mm; 0.24 mm; 0.63 mm.

Thorax: Scutum entirely red-brown without dark markings. Pleural areas entirely red-brown. Yellow markings as follows: postpronotal lobes; notopleura; mesopleural stripe reaching almost to anterior *npl.* seta dorsally, continuing to katepisternum as a large transverse spot, anterior margin slightly convex; anatergite (posterior apex red-brown); anterior ⅘ katatergite (remainder red-brown); two moderately broad parallel-sided lateral postsutural vittae ending at *ia.* seta. Medial longitudinal postsutural yellow vitta absent. Postnotum red-brown. Scutellum yellow except for narrow red-brown basal band. Setae: *sc.* 2; *prsc.* 2; *ia.* 1; *p.sa.* 1; *a. sa.* 1; *mpl.* 1; *npl.* 2; *scp.* 4.

Legs: All segments entirely fulvous except dark fuscous on basal end of hind tibiae; mid tibiae each with an apical black spur.

Wings: Length 5.4 mm; cells bc and c with a pale tint; microtrichia in outer corner of cell c only; remainder of wings colourless except dark fuscous cell sc, narrow fuscous costal band confluent with R_{2+3} and becoming very narrow beyond apex of R_{2+3} before widening into a small spot around apex of R_{4+5}, pale fuscous within cell cup; a dense aggregation of microtrichia around A_1+CuA_2; supernumerary lobe of medium development.

Abdomen: Oval; terga free; pecten present on tergum III. Tergum I and sterna I and II wider than long. Tergum I entirely orange-brown; tergum II orange-brown with fulvous across posterior margin; terga III–V entirely orange-brown with narrow transverse black spots on anterolateral corners of tergum III. A pair of oval orange-brown shining spots on tergum V. Posterior lobe of surstylus long, sternum V with a shallow concavity on posterior margin. Abdominal sterna orange-brown.

Attractant: No known record.

Female: As for male except no dense aggregation of microtrichia around A_1+CuA_2, supernumerary lobe weak, pecten of cilia on abdominal tergum III absent; ovipositor basal segment red-brown, dorsoventrally compressed and tapering posteriorly in dorsal view; ratio of length of oviscape to length of tergum V, 1:1; apex of piercer needle-shaped.

Distribution: Central Thailand (Samutsongeram – Misspelt on label = Samut Song Khram).

Hosts: No known record.

Remarks: *Bactrocera (Parasinodacus) citrifusca* n.sp. is similar to *Bactrocera (Parasinodacus) brevivitta* n.sp. and *Bactrocera (Parasinodacus) pahangiae* n.sp. in possessing lateral postsutural vittae, wings colourless except for a narrow costal band confluent with R_{2+3} and anal streak and *prsc.* setae and in lacking a lateral yellow band between the postpronotal lobe and notopleuron. It differs from these species in having a red-brown scutum and the face entirely fulvous without dark markings.

Bactrocera (Parasinodacus) eurylomata (Hardy)

(Fig. 238)

Dacus eurylomatus Hardy, 1982: 191. Holotype male in MZB.
Bactrocera (Sinodacus) eurylomata – Norrbom *et al.*, 1998: 100.

Type material examined: Holotype male, **INDONESIA: SULAWESI** – Lindu Valley, Central Sulawesi, 960 m, August 1975, coll. D.E. Hardy. Studied at MZB.

Diagnosis: A large species; face entirely fuscous; postpronotal lobes and notopleura yellow; a broad yellow band connecting postpronotal lobe and notopleura; scutum dark fuscous to dull black and tending paler in midline; lateral postsutural yellow vittae present; medial postsutural yellow vitta absent; yellow spot present anterior to notopleural suture; mesopleural stripe in centre of mesopleural area but not reaching either the dorsal margin or the katepisternum; scutellum yellow; wing with cells bc and c colourless, microtrichia in outer corner of cell c only; a very broad fuscous costal band overlapping R_{4+5} in centre of wing and vein M in apex of wing; a moderately broad pale fuscous anal streak; abdominal terga III–V dark fuscous with a large fulvous area in centre from posterior margin of tergum III to posterior margin of tergum V.

Description: male.

Head: Frons – fulvous tending red-brown adjacent to lunule; orbital setae black: 1 *s.or.*, 2 *i.or.*; lunule dark red-brown. Ocellar triangle black. Vertex red-brown. Face entirely fuscous. Occiput pale red-brown, fulvous along eye margins. Antennae with all segments fuscous to dark fuscous.

Thorax: Scutum dark fuscous to dull black tending slightly paler along midline and fulvous below and behind lateral postsutural vittae. Pleural areas dark fuscous to dull black. Yellow markings as follows: postpronotal lobes; notopleura; an unusual mesopleural stripe that does not reach the dorsal suture that runs between the postpronotal lobe and notopleuron; broad lateral postsutural vittae that taper posteriorly to end at *ia.*

seta; medial postsutural vitta absent; a broad yellow band joining the postpronotal lobe and notopleuron; a yellow spot anterior to notopleural suture in front of lateral postsutural vittae. Scutellum yellow except for narrow fuscous basal band. Setae: *sc.* 2; *prsc.* absent; *ia.* 1; *p.sa.* 1; *a.sa.* 1; *mpl.* 1; *npl.* 2; *scp.* 4.
Legs: Legs with all segments fulvous except pale fuscous on inner basal surfaces of mid femora and outer basal surfaces of hind femora.
Wings: Cells bc and c colourless; microtrichia in outer corner of cell c only; remainder of wings with a fuscous cell sc, a very broad fuscous costal band that overlaps R_{4+5} in centre of wing and M at apex of wing and merges as a pale tint across cell m, a moderately broad pale fuscous anal streak; supernumerary lobe weak.
Abdomen: Elongate oval; terga free; pecten present on tergum III. Tergum I and sterna I and II wider than long. Terga I and II dark fuscous except for a narrow pale fulvous line across posterior margin of tergum II; terga III–V entirely dark fuscous except for a broad pale fulvous area centrally from posterior margin of tergum III to posterior margin of tergum V. A pair of oval pale fulvous shining spots on tergum V. Posterior lobe of surstylus long, sternum V with a slight concavity on posterior margin.
Attractant: No known record.
Female: No known record.
Distribution:
Type locality: Indonesia (Sulawesi – Central Sulawesi, Lindu Valley).
Hosts: No known record.
Remarks: *Bactrocera (Parasinodacus) eurylomata* (Hardy) is similar to *Bactrocera (Parasinodacus) incisa* (Walker), *Bactrocera (Parasinodacus) longicaudata* (Perkins) and *Bactrocera (Parasinodacus) vinnulus* (Hardy) in possessing lateral postsutural yellow vittae with a yellow spot anterior to the notopleural suture, a black scutum and abdominal terga III–V with dark colour patterns. It differs from these species in having a lateral yellow band connecting the postpronotal lobe and notopleuron, a unique mesopleural stripe that does not reach the notopleuron dorsally or the katepisternum ventrally and a very broad costal band that overlaps veins R_{4+5} centrally and M apically.

Bactrocera (Parasinodacus) incisa (Walker)

(Fig. 239)

Dacus incisus Walker, 1861d: 323. Lectotype male in BMNH.
Dacus (Strumeta) incisus – Hardy, 1959: 175. Lectotype designation by inference of Holotype.
Bactrocera (Bactrocera) incisa – Norrbom *et al.*, 1998: 91.
Dacus poonensis Kapoor, 1971: 478; Norrbom *et al.*, 1998: 91 (Synonym).
Bactrocera (Zeugodacus) incisa – Chen *et al.*, 2011: 60.

Type material examined: Lectotype male, **MYANMAR (formerly BURMA)**: with the following six labels – 'Type, Dacus incisus Wlk'; 'Burma: ex Coll. W.W. Saunders 68-4'; 'incisus Wlk' (Walker's handwriting); 'Burma'; 'BMNH(E) No 533033'; 'Dacus (Stumeta) incisus Walker det. 1954 D.E. Hardy'. Studied at BMNH.
Non-type material examined: INDIA: ANDAMAN ISLANDS – 2 males, India, S. Andaman, Port Blair, 21.vii.1992, coll. Ranganath, Sl No. 12, attracted to cue lure; 3 males, India, S. Andaman, Port Blair, 31.vii.1992, coll. Ranganath, Sl No. 12, attracted to cue lure; 5 males, India, S. Andaman, Port Blair, 5.viii.1992, coll. Ranganath, Sl No. 11, attracted to cue lure; 1 male, S. Andaman, Port Blair, 11.v.1992, coll. H.R.R., 4 males, India, S. Andaman, 21.v.1997, coll. Ranganath, No 56, attracted to cue lure. **PENINSULAR MALAYSIA**: 1 male (11.iii.1986), 1 male (8.iv.1986), TS Gombak, coll. Mahdzir, attracted to cue lure; 1 male (15.iv.1986), 5 males (22.iv.1986), 4 males (29.iv.1986), 4 males (6.v.1986), 2 males (20.v.1986), 2 males (25.v.1986), 4 males (16.vi.1986), R.S. Jeram Pasu II, Malaysia, coll. Yusoff Salleh, attracted to cue lure; 1 male (12.v.1986), 5 males (12.vi.1986), R.S. Jeram Pasu II, Malaysia, coll. Ismail Mat, attracted to cue lure; 4 males, R.S. Jeram Pasu II, Malaysia, 13.v.1986, coll. Fauzi, attracted to cue lure; 1 male, TN9, Malaysia, 8.ix.1986, coll. Yusoff Salleh, attracted to cue lure. **THAILAND**: 1 male, Nakornratchasrima Province, north-east Thailand, 5.vi.2005, coll. Yuvarin Boontop. **VIETNAM: Northern Vietnam** – 1 male, Son La Province, Moc Chau District, Moc Chau house garden (trap site N008), 6.ix.1999, attracted to cue lure. Specimens in QDPC.
Diagnosis: A medium-sized species; face fulvous with a transverse black band expanding into oval black spots in antennal grooves and a transverse dark fuscous to black band below antennal sockets; postpronotal lobes and notopleura yellow; scutum black with small areas of red-brown; lateral postsutural yellow vittae present; medial postsutural yellow vitta absent; a yellow spot present anterior to notopleural suture; mesopleural stripe reaching almost to anterior *npl.* seta dorsally; scutellum yellow; wing with cells bc and c colourless, microtrichia in outer corner of cell c only; a narrow dark fuscous costal band confluent with R_{2+3} and remaining narrow around apex of wing; a moderately broad dark fuscous anal streak; abdominal tergum III dark fuscous to dull black, tergum IV dark fuscous to dull black with fuscous centrally, tergum V dark fuscous to black with dark red-brown shining spots.
Description: male.
Head: Frons – red-brown without dark markings; orbital setae black: 1 *s.or.*, 2 *i.or.*; lunule red-brown. Ocellar triangle black. Vertex black. Face fulvous with a transverse black line across oral margin which expands into large oval black spots in the antennal grooves and a transverse dark fuscous to black band below antennal sockets. Occiput black, fulvous along eye margins. Antennae with segments 1 and 2 red-brown, segment 3 red-brown with fuscous on apex and outer surface.
Thorax: Scutum black with red-brown as follows: a narrow band below lateral postsutural vittae, inside notopleura, inside postpronotal lobes. Pleural areas black. Yellow markings as follows: postpronotal lobes; notopleura; mesopleural stripe reaching almost to anterior *npl.* seta dorsally, extending to katepisternum as a large yellow spot, anterior margin

slightly convex; two moderately broad parallel-sided lateral postsutural vittae beginning with a spot anterior to notopleural suture and ending well behind *ia.* seta. Medial longitudinal postsutural yellow vitta absent. Scutellum yellow except for a narrow black basal band. Setae: *sc.* 2; *prsc.* absent; *ia.* 1; *p.sa.* 1; *a.sa.* 1; *mpl.* 1; *npl.* 2; *scp.* 4.
Legs: Fore femora entirely dark fuscous, mid femora with basal ⅕ fulvous (remainder dark fuscous), hind femora with basal ½ fulvous and apical ½ dark fuscous; fore tibiae fulvous, hind tibiae dark fuscous; fore and hind tarsal segments entirely fulvous; middle leg entirely missing.
Wings: Length 5.96 mm; cells bc and c colourless; microtrichia in outer corner of cell c only; remainder of wings colourless except dark fuscous cell sc, narrow dark fuscous costal band confluent with R_{2+3} and remaining narrow around apex of wing to end just beyond apex of R_{4+5}, a moderately broad dark fuscous anal streak; a dense aggregation of microtrichia around A_1+CuA_2; supernumerary lobe large.
Abdomen: Oval; terga free; pecten present on tergum III. Tergum I and sterna I and II wider than long. Tergum I black with a narrow transverse orange-brown band across posterior margin but not reaching lateral margins; tergum II mostly fulvous with a narrow transverse black band across anterior margin extending into narrow black lateral margins; tergum III uniformly dark fuscous to dull black, tergum IV dark fuscous to dull black over anterior and lateral margins and fuscous in centre, tergum V dark fuscous anterolaterally and with a broad black medial longitudinal band and dark red-brown along posterior margin posterior to shining spots. A pair of oval dark red-brown shining spots on tergum V. Posterior lobe of surstylus long, sternum V with a slight concavity on posterior margin. Abdominal sterna dark.
Attractant: Cue lure.
Female: No known record.
Distribution:
Type locality: Myanmar.
Previously recorded: Type locality and the following locations – Andaman Is (David & Ramani, 2011); China (Yunnan Province) (Chen *et al.*, 2011); India; Thailand (Norrbom *et al.*, 1998).
New records: Peninsular Malaysia; Northern Vietnam (Son La Province, Moc Chau).
Hosts: No known record.
Remarks: *Bactrocera (Parasinodacus) incisa* (Walker) is here placed in the new subgenus *Parasinodacus* in having the posterior lobe of the surstylus long, a slight concavity on the posterior margin of abdominal sternum V, two *sc.* setae, pecten present on abdominal tergum III, *a.sa.* seta present and medial postsutural yellow vitta absent. It differs from *Bactrocera (Parasinodacus) ablepharus* (Bezzi), *Bactrocera (Parasinodacus) cilifera* (Hendel) and *Bactrocera (Parasinodacus) pseudocucurbitae* White in possessing a yellow spot anterior to the notopleural suture in front of the lateral postsutural vittae, and from *Bactrocera (Parasinodacus) longicaudata* (Perkins) and *Bactrocera (Parasinodacus) vinnulus* (Hardy) in having the costal band narrow (not expanding) around apex of the wing, fore femora entirely dark fuscous and a black band across the face.

Bactrocera (Parasinodacus) longicaudata (Perkins)

(Figs 240A, 240B)

Nesodacus longicaudatus Perkins, 1938a: 134. Holotype female in QMIC.
Bactrocera (Paradacus) longicaudata – Norrbom *et al.*, 1998: 99.

Type material examined: Holotype female, **EAST MALAYSIA**: with the following two labels – 'N. Borneo, Bettotan, nr Sandakan, Aug 4th 1927'; 'Holotype, Nesodacus longicaudatus, F.A. Perk'. Received on loan from QMIC.
Non-type material examined: EAST MALAYSIA: SABAH – 2 males (25.ix.1991), 1 male (9.x.1991), 1 male (23.x.1991), ACIAR Site Ht4, Sabah orchard, attracted to cue lure in fruit orchard; 1 male, East Malaysia, Ulu Dusun, Sandakan forest, 22.iv.1992, coll. Jinius *et al.*, attracted to cue lure; **SARAWAK** – 1 male, Sarawak, E. Malaysia, Tarat Agriculture Station, Nr. Kuching, 15.vii.1992, coll. S. Leong, attracted to cue lure; 2 males, Sarawak, East Malaysia, Gunung Mulu Nat. Park, 28-30.vii.1992, coll. D.L. Hancock, attracted to cue lure; 1 male, Sarawak, E. Malaysia, Ranchang Pool, Serian, Nr Kuching, 23.vi.1992, coll. S. Leong, attracted to cue lure. **THAILAND**: 1 male, Nakornratchasrima Province, north-east Thailand, 5.vi.2005, coll. Yuvarin Boontop. Specimens in QDPC.
Diagnosis: A medium-sized species; face fulvous with a pair of large elongate oval black spots; postpronotal lobes and notopleura yellow; scutum entirely black; lateral postsutural yellow vittae present; medial postsutural yellow vitta absent; a yellow spot anterior to notopleural suture; mesopleural stripe slightly wider than notopleuron dorsally; scutellum yellow with a narrow black basal band; wing with cells bc and c with a pale tint, microtrichia in outer corner of cell c only; a narrow fuscous to dark fuscous costal band slightly overlapping R_{2+3} and expanding into a distinct spot in apex of wing; a moderately broad fuscous to dark fuscous anal streak; abdominal terga III–V red-brown with a narrow fuscous band across anterior margin of tergum III which expands to cover lateral margins of this tergum.
Description: female (Fig. 240A).
Head: Height 1.55 mm. Frons – length 1.3 times breadth; red-brown without dark markings; orbital setae black: 1 *s.or.*, 2 *i.or.*; lunule fuscous. Ocellar triangle black. Vertex black. Face fulvous with a pair of large elongate oval black spots reaching to oral surface; length 0.6 mm. Genae fulvous, dark fuscous subocular spot present; black seta present. Occiput dark fuscous, fulvous along eye margins; no distinct setae in occipital row. Antennae with segments 1 and 2 fuscous, segment 3 entirely dark fuscous; length of segments: 0.12 mm; 0.28 mm; 0.8 mm.
Thorax: Scutum entirely black except for a small area of dark fuscous behind lateral postsutural vittae and a small area of red-brown inside notopleura. Pleural areas entirely dark fuscous to black. Yellow markings as follows: postpronotal lobes; notopleura; mesopleural stripe slightly wider than notopleuron dorsally, continuing to katepisternum as a transverse spot, anterior margin straight; anatergite

(posterior apex black); anterior $^3/_5$ katatergite (remainder black); two broad parallel-sided lateral postsutural vittae beginning in a spot anterior to notopleural suture and ending posteriorly at *ia.* seta. Medial longitudinal postsutural yellow vitta absent. Postnotum black. Scutellum yellow except for narrow black basal band. Setae: *sc.* 2; *prsc.* absent; *ia.* 1; *p.sa.* 1; *a.sa.* 1; *mpl.* 1; *npl.* 2; *scp.* 4.
Legs: Fore femora entirely dark fuscous tending red-brown basally, mid femora with basal $^2/_5$ red-brown and apical $^3/_5$ dark fuscous, hind femora with basal $^2/_3$ red-brown and apical $^1/_3$ dark fuscous; all tibiae dark fuscous; tarsi with all segments red-brown; mid tibiae each with an apical black spur.
Wings: Length 6.2 mm; cells bc and c with a pale tint; microtrichia in outer corner of cell c only; remainder of wings colourless except fuscous to dark fuscous cell sc, narrow fuscous to dark fuscous costal band slightly overlapping R_{2+3} and expanding into a distinct spot in apex of wing, moderately broad fuscous to dark fuscous anal streak; no dense aggregation of microtrichia around A_1+CuA_2; supernumerary lobe weak.
Abdomen: Oval; terga free; no pecten on tergum III. Tergum I dark fuscous with a narrow red-brown transverse band across posterior margin but not reaching lateral margins; tergum II fulvous posteriorly with a broad transverse dark fuscous band across anterior margin; terga III–V red-brown with a narrow transverse fuscous band across anterior margin of tergum III which expands across lateral margins of this tergum. A pair of oval red-brown shining spots on tergum V. Ovipositor basal segment red-brown, extremely long and conical in transverse cross-section; ratio of length of oviscape to length of tergum V, 2.86:1.
Attractant: Cue lure.
Male: (Fig. 240B) – As for female except for dense aggregation of microtrichia around A_1+CuA_2 present; supernumerary lobe strong and rounded; pecten of cilia on abdominal tergum III present; posterior lobe of surstylus long, posterior margin of abdominal sternum V with a slight concavity.
Distribution:
Type locality: East Malaysia (Sabah, Bettotan).
New records: East Malaysia (Sarawak); Thailand (Nakornratchasrima Province).
Hosts: No known record.
Remarks: *Bactrocera (Parasinodacus) longicaudata* (Perkins) is here placed in the subgenus *Parasinodacus* in that it possesses the combination of two *sc.* setae and the medial postsutural yellow vitta absent from the scutum. It is similar to *Bactrocera (Paradacus) fulvipes* (Perkins) and *Bactrocera (Paradacus) urens* White in possessing a black scutum with broad parallel-sided lateral postsutural vittae and a yellow spot anterior to the notopleural suture in front of each lateral vitta, medial postsutural yellow vitta absent, femora with apical dark patterns, wings with costal band expanded apically, abdominal terga III–V mostly orange-brown to red-brown and the basal segment of the ovipositor long and narrow (approximately as long as the entire abdomen). It differs from both species in having two *sc.* setae, from *B. urens* in having a pair of dark spots on the face and darker patterns on the femora.

Bactrocera (Parasinodacus) pahangiae Drew & Romig, new species

(Fig. 241)

Material examined: Holotype male, **PENINSULAR MALAYSIA**: 10 km Gap Rest House, Pahang, Malaysia, 22.x.1986, coll. S. Vijaysegaran *et al.*, bred ex *Litsea* sp. (fam. Lauraceae), deposited in BMNH.
Paratypes: **PENINSULAR MALAYSIA**: 6 males and 7 females, same data as holotype. 1 paratype in ANIC, 1 paratype in BMNH, 9 paratypes in QDPC, 2 paratypes in QMIC.
Diagnosis: A small species; face fulvous with a pair of medium-sized subquadrate black spots; postpronotal lobes and notopleura yellow; scutum entirely black; short, medium-width lateral postsutural yellow vittae present; medial postsutural yellow vitta absent; no yellow spot anterior to notopleural suture; mesopleural stripe equal in width to notopleuron dorsally; scutellum yellow with a large black apical spot; wing with cells bc and c with a pale tint, microtrichia in outer corner of cell c only; a narrow fuscous costal band confluent with R_{2+3} and remaining narrow around apex of wing; a narrow fuscous anal streak; abdominal tergum III entirely black; abdominal terga IV–V red-brown with a broad medial and two broad lateral longitudinal black bands joined along anterior margin of tergum IV.
Description: male.
Head: Height 1.4 mm. Frons – length 1.47 times breadth; red-brown with pale fuscous extending from vertex along outer margins to level of *s.or.* setae; orbital setae black: 1 *s.or.*, 2 *i.or.*; lunule fuscous. Ocellar triangle black. Vertex pale fuscous. Face fulvous with a pair of medium-sized subquadrate black spots; length 0.46 mm. Genae fulvous, small fuscous subocular spot present; black seta present. Occiput black, fulvous along eye margins; occipital row with a large number of strong black setae. Antennae with segments 1 and 2 red-brown, segment 3 red-brown with fuscous on apex and outer surface; length of segments: 0.12 mm; 0.24 mm; 0.67 mm.
Thorax: Scutum entirely black except for a small area of red-brown inside notopleura. Pleural areas entirely black. Yellow markings as follows: postpronotal lobes; notopleura; mesopleural stripe equal in width to notopleuron dorsally, continuing to katepisternum as a small spot, anterior margin straight; anatergite (posterior apex black); anterior $^3/_5$ katatergite (remainder black); two medium-width short lateral postsutural vittae ending just posterior of level of *a.sa.* setae. Medial longitudinal postsutural yellow vitta absent. Postnotum black. Scutellum yellow with a moderately broad black basal band and a large black apical spot which encloses the apical setae. Setae: *sc.* 2; *prsc.* 2; *ia.* 1; *p.sa.* 1; *a.sa.* 1; *mpl.* 1; *npl.* 2; *scp.* 4.
Legs: All femora fulvous with small areas of fuscous around the apices; fore tibiae pale fuscous, mid tibiae dark fuscous basally to fuscous apically, hind tibiae black basally to dark fuscous apically; fore and hind tarsi with basal segment fulvous and apical four segments fuscous, mid tarsi with all segments entirely fulvous; mid tibiae each with an apical black spur.
Wings: Length 4.9 mm; cells bc and c with a pale tint; microtrichia in outer corner of cell c only; remainder of

wings colourless except dark fuscous cell sc, narrow fuscous costal band confluent with R_{2+3} and remaining narrow around apex of wing, a narrow fuscous anal streak; a dense aggregation of microtrichia around A_1+CuA_2; supernumerary lobe of medium development.

Abdomen: Oval; terga free; pecten present on tergum III. Tergum I and sterna I and II wider than long. Tergum I black with red-brown along posterior margin but not reaching lateral margins; tergum II red-brown with a narrow transverse black band subanteriorly and reaching narrow black lateral margins; tergum III entirely black; terga IV–V red-brown with a moderately broad medial and two broad lateral longitudinal black bands which join along anterior margin of tergum IV. A pair of oval red-brown shining spots on tergum V. Posterior lobe of surstylus long, sternum V with a slight concavity on posterior margin. Abdominal sterna dark.

Attractant: No known record.

Female: As for male except no dense aggregation of microtrichia around A_1+CuA_2, supernumerary lobe weak, pecten of cilia on abdominal tergum III absent; ovipositor basal segment red-brown, dorsoventrally compressed and tapering posteriorly in dorsal view; ratio of length of oviscape to length of tergum V, 1:1; apex of piercer needle-shaped.

Distribution: Peninsular Malaysia (Pahang).

Hosts: *Litsea* sp. (family Lauraceae).

Remarks: *Bactrocera (Parasinodacus) pahangiae* n.sp. is similar to *Bactrocera (Parasinodacus) brevivitta* n.sp. and *Bactrocera (Parasinodacus) cilifera* (Hendel) in possessing a black scutum with lateral postsutural vittae, wings colourless except for costal band and anal streak, the costal band confluent with or just overlapping R_{2+3} and in lacking a lateral yellow band between the postpronotal lobe and notopleuron. It differs from *B. cilifera* in having the costal band entire and from *B. brevivitta* in having the scutellum with a large black apical spot, the femora entirely fulvous with very small areas of fuscous on the apices and abdominal terga III–V mostly black.

Bactrocera (Parasinodacus) pantabanganiae Drew & Romig, new species

(Fig. 242)

Material examined: Holotype male, **PHILIPPINES: LUZON ISLAND** – Pantabangan, Nueva Ecija, Philippines, 30.vi.1983, coll. RCPC, Maligaya, attracted to cue lure, deposited in BMNH.

Paratypes: **PHILIPPINES: LUZON ISLAND** – 3 males, same data as holotype; 1 male, Mt Makiling, 500 m, 8.vi.1983, coll. B. Ramos; 1 male, Mt Makiling, 400 m, Philippines, 8.vi.1983, coll. B. Ramos, attracted to cue lure. 1 paratype in ANIC, 3 paratypes in QDPC, 1 paratype in QMIC.

Diagnosis: A medium-sized species; face fulvous with a pair of medium-sized transverse oval black spots pointing towards midline and a narrow transverse black line below antennal sockets; postpronotal lobes and notopleura yellow; scutum black; lateral postsutural yellow vittae present; medial postsutural yellow vitta absent; a large yellow spot anterior to notopleural suture; mesopleural stripe reaching midway between anterior margin of notopleuron and anterior *npl.* seta dorsally; scutellum yellow; wing with cells bc and c with a pale tint, microtrichia in outer ½ of cell c only; a narrow fuscous costal band; a broad fuscous anal streak; abdominal tergum III orange-brown with a narrow transverse black band across anterior margin widening to cover lateral margins, abdominal terga IV–V orange-brown with a broad medial and two moderately broad lateral longitudinal dark fuscous to black bands joined over anterior margin of tergum IV.

Description: male.

Head: Height 1.4 mm. Frons – length 1.4 times breadth; red-brown with fuscous around bases of orbital setae and on anteromedial hump; orbital setae black: 1 *s.or.*, 2 *i.or.*; lunule dark fuscous. Ocellar triangle black. Vertex dark fuscous. Face fulvous with a pair of medium-sized transverse oval black spots pointing towards the midline and a narrow transverse black band beneath the antennal sockets; length 0.5 mm. Genae fulvous, large dark fuscous subocular spot present; black seta present. Occiput dark fuscous, fulvous along eye margins; occipital row with a large number of small pale setae. Antennae with segments 1 and 2 fuscous, segment 3 fulvous with dark fuscous on apex and outer surface; length of segments: 0.12 mm; 0.24 mm; 0.67 mm.

Thorax: Scutum black except dark red-brown below and behind lateral postsutural vittae and inside postpronotal lobes. Pleural areas black, tending dark fuscous below postpronotal lobes. Yellow markings as follows: postpronotal lobes; notopleura; mesopleural stripe reaching midway between anterior margin of notopleuron and anterior *npl.* seta dorsally, continuing to katepisternum as a large transverse spot, anterior margin straight; anatergite (posterior apex black); anterior ⅔ katatergite (remainder black); two broad parallel to subparallel lateral postsutural vittae ending behind *ia.* seta; a large spot anterior to notopleural suture in front of each lateral postsutural vitta. Medial longitudinal postsutural yellow vitta absent. Postnotum black. Scutellum yellow except for narrow black basal band. Setae: *sc.* 2; *prsc.* absent; *ia.* 1; *p.sa.* 1; *a.sa.* 1; *mpl.* 1; *npl.* 2; *scp.* 4.

Legs: Fore femora fulvous with a large dark fuscous subapical spot on outer surface, mid and hind femora fulvous with fuscous to dark fuscous around apical ¼; fore tarsi fuscous basally to fulvous apically, mid tarsi entirely fulvous, hind tarsi entirely fuscous; all tarsi with segments entirely fulvous; mid tibiae each with an apical black spur.

Wings: Length 5.3 mm; cells bc and c with a pale tint; microtrichia covering outer ½ of cell c only; remainder of wings colourless except fuscous cell sc, narrow fuscous costal band overlapping R_{2+3} and widening slightly across apex of R_{4+5}, a broad fuscous anal streak; a dense aggregation of microtrichia around A_1+CuA_2; supernumerary lobe large and rounded.

Abdomen: Oval; terga free; pecten present on tergum III. Tergum I and sterna I and II wider than long. Tergum I fulvous with narrow black lateral margins; tergum II fulvous with a narrow transverse dark fuscous to black band subanteriorly and joining narrow dark fuscous to black lateral margins; tergum III orange-brown with a narrow transverse black band across anterior margin and continuing around lateral margins and extending posteriorly along midline but not reaching posterior margin; terga IV–V orange-brown with a broad medial and a moderately broad lateral longitudinal dark fuscous to black bands joined along anterior margin of tergum IV. A pair of oval orange-brown shining spots on tergum V. Posterior lobe of surstylus long, sternum V with a slight concavity on posterior margin. Abdominal sterna I, IV and V dark.

Attractant: Cue lure.

Female: No known record.

Distribution: Philippines (Luzon I. – Nueva Ecija; Mt Makiling).

Hosts: No known record.

Remarks: *Bactrocera (Parasinodacus) pantabanganiae* n. sp. is similar to *Bactrocera (Parasinodacus) incisa* (Walker), *Bactrocera (Parasinodacus) longicaudata* (Perkins) and *Bactrocera (Parasinodacus) vinnulus* (Hardy) in possessing lateral postsutural yellow vittae with a yellow spot anterior to the notopleural suture, *prsc.* setae absent, the costal band complete and in lacking a lateral yellow band between the postpronotal lobe and the notopleuron. It differs from *B. incisa* in having the costal band overlapping R_{2+3} and widening slightly at the apex, from *B. longicaudata* in having dark patterns on abdominal terga III–V and from *B. vinnulus* in having dark 'T' patterns on tergum III and over terga IV–V.

Bactrocera (Parasinodacus) pseudocucurbitae White

(Fig. 243)

Bactrocera (Bactrocera) pseudocucurbitae White, in White & Evenhuis, 1999: 502. Holotype male in NMWC.

Type material examined: Holotype male, **EAST MALAYSIA**: with the following six labels – 'BORNEO: Sabah, Danum Valley, 5.01'N 117.47'E. 19–24.ix.1987 150 m A.H. Kirk-Spriggs'; 'lowland mixed dipterocarp forest Grid No.'; 'Grid No WON3 tephritid trap No. 6 baited cue lure'; 'DANUM 3 NAT MUS WALES pseudocucurbitae White MS'; 'NMW Sabah (Borneo) Expedition NMW. Z.1987.094'; 'Bactrocera (B.) pseudocucurbitae HOLOTYPE male White in White & Evenhuis 1999 Raffles Bull. of Zool. 47(2):502'. Received on loan from NMWC.

Non-type material examined: EAST MALAYSIA: SARAWAK – 1 male, Kpg. Tema Mawang, Tebedu, nr. Kuching, 25.xi.1991, coll. S. Leong, attracted to cue lure. **PENINSULAR MALAYSIA**: 2 males, Cameron Highlands, 15.xii.1986, coll. Azizan Darus, attracted to cue lure; 1 male, RS Cameron Highlands, 13.v.1986, coll. Zan Davies, attracted to cue lure; 1 male, Cameron Highlands, 1.i.1987, coll. C.S. Ooi, attracted to cue lure; 1 male, TS7 Malaysia, 15.xii.1986, coll. Mahdzir, attracted to cue lure; 1 male, TS8 Malaysia, 3.xi.1986, coll. Fauzi, attracted to cue lure; 1 male (12.i.1987), 1 male (1.xii.1986), TN9 Malaysia, coll. Yusoff Salleh, attracted to cue lure; 1 male, TN10 Malaysia, 23.ix.1986, coll. Ahmad Radzuan & Panjang Abdullah, attracted to cue lure; 1 male, TN14 Malaysia, 9.ii.1987, coll. Y. Salleh, attracted to cue lure; 1 male, TN14 Malaysia, 23.ii.1987, coll. A.R. & P.A., attracted to cue lure. **INDONESIA: BALI** – 1 male, Ds. Depha, Kubutambanan, Buleleng, 6.viii.2006, coll. Ketut Rawa W., attracted to cue lure in mangoes; 3 males (4.viii.2006), 3 males (11.viii.2006), 6 males (18.viii.2006), Bali, Buleleng, Tajun, coll. Ketut Rawa W., attracted to cue lure in forest; 7 males, Pura Batakaru, Bali (800m), 11.vii.1981, coll. J. Turner, attracted to cue lure; 2 males, Sari Kuning nr Negara, Bali (<200m), 7.vii.1981, coll. J. Turner, attracted to cue lure; 2 males, Teluk Terima, Bali (0m), 9.vii.1981, coll. J. Turner, primary coastal scrub; 1 male, Medewi, nr Pulakan, Bali (400m), 10.vii.1981, coll. J. Turner, attracted to cue lure; 2 males, Bedugul, Bali (1300m), 11.vii.1981, coll. J. Turner; attracted to cue lure; **FLORES** – 1 male, Borong, Flores (0-50m), 24.vi.1981, coll. J. Turner, attracted to cue lure; 16 males, Borong, Flores (200-400m), 25.vi.1981, coll. J. Turner, attracted to cue lure; 4 males, Ruteng Forest, S. Ruteng, Flores (1000-1350m), 24.vi.1981, coll. J. Turner, attracted to cue lure in primary wet montane forest; 7 males, Sita, S. Ruteng, Flores (500-800m), 24.vi.1981, coll. J. Turner, attracted to cue lure; 1 male, Waelaku, Ruteng, Flores (100-150m), 24.vi.1981, coll. J. Turner, attracted to cue lure; 101 males, Borong, Flores, iv.1989, coll. J. Turner; 2 males, Borong, Flores, Indonesia, iii.1988, coll. J. Turner, attracted to cue lure; **KALIMANTAN**: 1 male, East Kalimantan, Kutai Kartanegara, Batuah, 29.viii.2005, coll. Ertha & Heru G, attracted to cue lure in forest; **LOMBOK**: 67 males, Sasaot (400-550m), Lombok, 31.v.1981, coll. J. Turner, old plantation-Mahogany, attracted to cue lure; 15 males, Suranadi, Lombok (200m), 31.v.1981, coll. J. Turner, attracted to cue lure, Monkey forest – variety of small trees; 4 males, Suranadi, W. Lombok, 4.vi.1978, coll. J.T. Medler, attracted to cue lure; 10 males, Sapit, Lombok (675-850m), 1.vi.1981, coll. J. Turner, attracted to cue lure; 20 males, Santong, Lombok (1200-1300m), 2.vi.1981, coll. J. Turner, attracted to cue lure; 35 males, Bentek, 20km N. Mataram, Lombok (100-250m), 2.vi.1981, attracted to cue lure; **SUMBAWA**: 81 males, Semongkot, 250-350m, 12.vi.1981, coll. J. Turner, attracted to cue lure; 33 males, Pulau Moyo, 0-75m, 10.vi.1981, coll. J. Turner, attracted to cue lure; 10 males, Kanar, 12.vi.1981, coll. J. Turner, attracted to cue lure in Teak forest with Sclerophyll remnants and Gallery forest; 7 males, Tiumera and Brangrea, 100-150m, 13.vi.1981, coll. J. Turner, attracted to cue lure in Primary sclerophyll wet forest. **THAILAND**: 16 males (25.vi.1991), 2 males (23.vii.1991), 2 males (5.viii.1991), 1 male (9.vi.1992), 2 males (21.vii.1992), Yang Pong Rat, Plubpla, A. Muang, Chanthaburi, coll. Montree, attracted to cue lure; 1 male (23.vii.1991), 1 male (20.viii.1991), 1 male (7.i.1992), Philu Hort Res Stn, Laensingh, Chanthaburi, coll. Montree, attracted to cue lure. Specimens in QDPC.

Diagnosis: A medium-sized species; face with a pair of large circular black spots; postpronotal lobes and notopleura yellow; scutum black; lateral postsutural yellow vittae present; medial postsutural yellow vitta absent; mesopleural stripe almost reaching postpronotal lobe dorsally; scutellum yellow; wing with cells bc and c colourless, microtrichia in outer corner of cell c only; a narrow fuscous costal band slightly widening apically and broad fuscous anal streak; abdominal tergum II fulvous and terga III–V red-brown with a 'T' pattern which consists of a narrow black transverse band across anterior margin of tergum II and a broad medial longitudinal black band over all four terga and medium-width lateral longitudinal black bands over terga III–V.

Description: male.

Head: Height 1.52 mm. Frons – length 1.43 times breadth; red-brown with fulvous along lateral margins, fuscous on anteromedial hump and dark fuscous around bases of orbital setae; orbital setae black: 1 *s.or.*, 2 *i.or.*; lunule red-brown. Ocellar triangle black. Vertex red-brown to fuscous. Face fulvous with a pair of large circular black spots, length 0.49 mm. Genae fulvous, small dark fuscous subocular spot present; black seta present. Occiput fuscous, fulvous along eye margins; occipital row with 4–8 strong black setae. Antennae with segments 1 and 2 red-brown, segment 3 red-brown with dark fuscous on apex and outer surface; arista black (red-brown basally); length of segments: 0.14 mm; 0.26 mm; 0.74 mm.

Thorax: Scutum entirely black with no pale markings. Pleural areas entirely black with at most a small area of dark red-brown below postpronotal lobes. Yellow markings as follows: postpronotal lobes; notopleura; broad mesopleural stripe, almost reaching postpronotal lobe dorsally, extending to katepisternum as a small transverse spot, anterior margin straight; anatergite (posterior apex black); anterior ⅔ katatergite (remainder black); two broad parallel-sided lateral postsutural vittae ending at *ia.* setae. Medial longitudinal postsutural yellow vitta absent. Postnotum black. Scutellum yellow except for narrow black basal band. Setae: *sc.* 2; *prsc.* 2; *ia.* 1; *p.sa.* 1; *a.sa.* 1; *mpl.* 1; *npl.* 2; *scp.* 4.

Legs: Femora entirely fulvous; fore tibiae pale fuscous, mid tibiae fuscous basally to pale fuscous apically, hind tibiae dark fuscous; tarsi with basal segment fulvous and apical four segments pale fuscous; mid tibiae each with an apical black spur.

Wings: Length 5.62 mm; cells bc and c colourless; microtrichia in outer corner of cell c only; remainder of wings colourless except fuscous cell sc, narrow fuscous costal band distinctly overlapping R_{2+3} and widening slightly as it continues to end between apices of R_{4+5} and M, broad infuscation along *r-m* and *dm-cu* crossveins, broad fuscous anal streak ending at wing margin; dense aggregation of microtrichia around A_1+CuA_2; supernumerary lobe weak.

Abdomen: Oval; terga free; pecten present on tergum III. Tergum I and sterna I and II wider than long. Tergum I dark fuscous to black with dark red-brown posterocentrally; tergum II fulvous and terga III–V red-brown with a distinct 'T' pattern consisting of a narrow transverse black band subanteriorly across tergum II but not reaching lateral margins and a broad medial longitudinal black band over all four terga, anterolateral corners of tergum II narrowly dark fuscous, medium-width lateral longitudinal black bands over terga III–V. A pair of oval red-brown to fuscous shining spots on tergum V. Posterior lobe of surstylus long and stout, sternum V with a slight concavity on posterior margin.

Attractant: Cue lure.

Female: No known record.

Distribution:

Type locality: East Malaysia (Sabah, Danum Valley).

New records: Indonesia (Bali; Flores; Kalimantan; Lombok; Sumbawa); East Malaysia (Sarawak); Peninsular Malaysia; Thailand (Chanthaburi).

Hosts: No known record.

Remarks: *Bactrocera (Parasinodacus) pseudocucurbitae* White is placed in the new subgenus *Parasinodacus* in possessing two *sc.* setae and the medial longitudinal yellow vittae absent. It is similar to *Bactrocera (Parasinodacus) ablepharus* (Bezzi) and *Bactrocera (Parasinodacus) cilifera* (Hendel) in lacking a yellow spot anterior to the notopleural suture and in possessing yellow notopleura. It differs from *B. ablepharus* in having *prsc.* setae and from *B. cilifera* in possessing a complete costal band and a broader mesopleural stripe dorsally (reaching almost to postpronotal lobe).

Bactrocera (Parasinodacus) vinnulus (Hardy)

(Fig. 244)

Dacus (Pacifodacus) vinnulus Hardy, 1973: 23. Holotype male in KUB.
Bactrocera (Sinodacus) vinnula – Norrbom *et al.*, 1998: 101.
Dacus (Pacifodacus) drewi Hardy, 1983: 29. **New Synonym.**

Other type material examined: Holotype male of *Dacus drewi* Hardy, **INDONESIA**: N. Sumatra, 12 km NW Bohorok, 200 m, 4.v.1975, coll. D.E. & A.T. Hardy, attracted to cue lure, studied at MZB; 1 male paratype of *Dacus drewi* Hardy, '12 km W.W. Bohorok, N. Sumatra 200 m, May 4 1975, D.E. Hardy'; 'Paratype male, *Dacus (Pacifodacus) drewi*, D.E. Hardy'. Specimen in QDPC. (See 'Remarks' below.)

Non-type material examined: PENINSULAR MALAYSIA: 1 male (28.i.1986), 10 males (12.ii.1986), 3 males (25.ii.1986), 2 males (11.iii.1986), 3 males (25.iii.1986), 4 males (8.iv.1986), 5 males (22.iv.1986), 2 males (6.v.1986), T.S. Gombak, coll. Mahdzir, attracted to cue lure; 1 male, Gombak, 18.iv.1986, coll. Mahdzir, attracted to cue lure; 9 males, Gombak, 19.v.1986, coll. S. Vijaysegaran, attracted to cue lure; 7 males, Gombak, 2.vi.1986, coll. Vignes, attracted to cue lure in rainforest; 1 male, T.S. 6 Malaysia, 15.i.1986, coll. R. Drew, attracted to cue lure; 10 males, T.S. 7 Malaysia, 16.vi.1986, coll. Mahdzir Ahmad, attracted to cue lure; 1 male, T.S. 7 Malaysia, 1.xii.1986, coll. Antony, attracted to cue lure; 1 male, T.S. 8 Malaysia, 16.vi.1986, coll. Mahdzir Ahmad, attracted to cue lure; 6 males, Malaysia T.S. 8, coll. Yusoff Salleh, attracted to cue lure; 2 males (11.viii.1986), 1 male (23.ix.1986), T.N. 13 Malaysia, coll. Ahmad Radzuan and Panjang Abdullah, attracted to cue lure; 1 male, T.S. Gombak, 25.ii.1986, coll. Antony, attracted to cue lure; 1 male T.S. Puchong Malaysia, 25.ii.1986, coll. Antony, attracted to cue lure; 2 males, Gombak, Malaysia, rainforest, 9.x.1987, coll. R. Drew & C.S. Ooi, attracted to cue lure; 1 male, Pangsoon, Hulu Langat, 26.x.1990, coll. D.L. Hancock, attracted to cue lure; 1 male, Malaysia, Hulu Langat, Punsun, 9.x.1988, coll. YHS, attracted to cue lure; 1 male, Pulau, Perhentian, viii.1986, coll. C.S. Ooi, attracted to cue lure. Specimens in QDPC.

Diagnosis: A medium-sized species; face fuscous to dark fuscous with a pair of very large elongate oval black spots; postpronotal lobes and notopleura yellow; scutum black; lateral postsutural yellow vittae present; medial postsutural yellow vitta absent; a yellow spot present anterior to notopleural suture; mesopleural stripe reaching midway between anterior margin of notopleuron and anterior *npl.* seta dorsally; scutellum yellow; wing with cells bc and c colourless, microtrichia in outer corner of cell c only; a dark fuscous costal band overlapping R_{2+3} and sometimes almost confluent with R_{4+5} and expanding distinctly across apex of wing; a broad dark fuscous anal streak; abdominal tergum III dark fuscous centrally to black anteriorly and laterally, tergum IV fuscous centrally to dark fuscous laterally, tergum V orange-brown with fuscous anterolateral corners; in some specimens there is a narrow medial longitudinal dark fuscous to black band over terga IV and V or terga III–V.

Description: male.

Head: Height 1.47 mm. Frons – length 1.25 times breadth; fulvous with dark fuscous around orbital setae and on anteromedial hump (in some specimens the patterns around the base of the setae are absent); orbital setae black: 1 *s.or.*, 2 *i.or.*; lunule fuscous to dark fuscous. Ocellar triangle black. Vertex black. Face fuscous to dark fuscous with a pair of very large elongate oval black spots; length 0.49 mm. Genae fulvous with large dark fuscous subocular spots; black seta present. Occiput dark fuscous to black; occipital row with a large number of small pale setae. Antennae with segments 1 and 2 fuscous, segment 3 dark fuscous; length of segments: 0.12 mm; 0.22 mm; 0.74 mm.

Thorax: Scutum black with at most, dark red-brown inside postpronotal lobes. Pleural areas entirely black, tending dark fuscous below postpronotal lobes. Yellow markings as follows: postpronotal lobes; notopleura; mesopleural stripe reaching midway between anterior margin of notopleuron and anterior *npl.* seta dorsally, continuing to katepisternum as a transverse spot, anterior margin convex; anatergite (posterior apex black); anterior ⅔ katatergite (remainder black); two broad parallel-sided lateral postsutural vittae beginning with a yellow spot anterior to notopleural suture and ending behind *ia.* seta. Medial longitudinal postsutural yellow vitta absent. Postnotum black. Scutellum yellow except for narrow black basal band. Setae: *sc.* 2; *prsc.* absent; *ia.* 1; *p.sa.* 1; *a.sa.* 1; *mpl.* 1; *npl.* 2; *scp.* 4.

Legs: Fore femora fulvous basally, tending dark fuscous to black on outer apical ⅔ and fuscous on inner apical ⅔, mid femora fulvous on basal ½ and fuscous to black on apical ½, hind femora fulvous with dark fuscous to black around apical ⅓ fore and hind tibiae dark fuscous basally to fuscous apically, mid tibiae fuscous basally to fulvous apically; tarsi with all segments fulvous; mid tibiae each with an apical black spur.

Wings: Length 5.6 mm; cells bc and c colourless; microtrichia in outer corner of cell c only; remainder of wings colourless except dark fuscous cell sc, narrow dark fuscous costal band overlapping R_{2+3} (sometimes almost confluent with R_{4+5}) and widening markedly across apex of wing before ending midway between extremities of R_{4+5} and M, broad dark fuscous anal streak; a dense aggregation of microtrichia around A_1+CuA_2; supernumerary lobe strong and rounded.

Abdomen: Oval; terga free; pecten present on tergum III. Tergum I and sterna I and II wider than long. Tergum I dark fuscous centrally to black laterally and with a narrow orange-brown transverse band across posterior margin but not reaching lateral margins; tergum II fulvous with a narrow transverse black band across anterior margin and widening to cover lateral margins; tergum III dark fuscous centrally with a narrow transverse black band across anterior margin and widening to cover lateral margins, tergum IV fuscous centrally to dark fuscous laterally, tergum V orange-brown with fuscous anterolateral corners, a narrow medial longitudinal dark fuscous to black band over all three terga. A pair of oval orange-brown shining spots on tergum V. Posterior lobe of surstylus long, sternum V with a slight concavity on posterior margin. Abdominal sterna red-brown.

Attractant: Cue lure.

Female: No known record.

Distribution:

Type locality: Thailand (Yala).

New records: Peninsular Malaysia. Indonesia (Sumatra) due to **New Synonymy** of *D. drewi*.

Hosts: No known record.

Remarks: *Bactrocera (Parasinodacus) vinnulus* (Hardy) is similar to *Bactrocera (Parasinodacus) incisa* (Walker) and *Bactrocera (Parasinodacus) longicaudata* (Perkins) in possessing a yellow spot anterior to the notopleural suture in front of the lateral postsutural vittae, a black scutum, parallel-sided lateral postsutural vittae and legs with dark patterns on the femora. It differs from *B. incisa* in having large black facial spots (not a band across face), the costal band overlapping R_{2+3} and expanding over wing apex and pale abdominal sterna and from *B. longicaudata* in having the costal band overlapping R_{2+3} and abdominal terga III–V darker coloured (not mostly orange-brown to red-brown).

Following a thorough study of the Holotype and one male Paratype of *Dacus (Pacifodacus) drewi* Hardy, it is clear that this species is a **New Synonym** of *B. vinnulus* Hardy.

Bactrocera (Parasinodacus) waimitaliae Drew & Romig, new species

(Fig. 245)

Material examined: Holotype male, **INDONESIA: MOLUCCAS (MALUKU)** – Seram (Western part), Waimital, 12.vii.2007, coll. A. Haris Tualepe, attracted to cue lure in jeruk, deposited in BMNH.

Paratypes: **INDONESIA: MOLUCCAS (MALUKU)** – 1 male, same data as holotype; 1 male, Maluku, Seram (Western part), Waimital, 2.viii.2007, coll. A. Haris Tualepe, attracted to cue lure in jeruk; 1 male, Maluku, Seram (Western part), Nuruwe, 6.vii.2007, coll. A. Haris Tualepe, attracted to cue lure in forest; **AMBON ISLAND**: 1 male, Paso, Ambon, Indonesia, 20–50 m, 7.vii.1981, coll. J. Turner, residual rainforest, attracted to cue lure; 1 male, Hila/Mamua, Ambon, Indonesia, 25–100 m, 5.viii.1981, coll. J. Turner, residual rainforest, attracted to cue lure. 1 paratype in ANIC, 2 paratypes in QDPC, 2 paratypes in QMIC.

Diagnosis: A medium-sized species; face fuscous with a pair of narrow elongate dark fuscous spots; postpronotal lobes fuscous; notopleura yellow; scutum entirely uniformly red-brown; lateral and medial postsutural yellow vittae absent; no yellow spot anterior to notopleural suture; mesopleural stripe reaching almost to anterior *npl.* seta dorsally; scutellum yellow; wing with cells bc and c fuscous, microtrichia in outer ½ of cell c only; a broad dark fuscous costal band slightly overlapping R_{4+5} and remaining uniform around apex of wing, a transverse band across centre of wing which is dark fuscous around *r-m* crossvein and broad and diffuse to pale fuscous across remainder of wing; a broad dark fuscous anal streak; abdominal terga entirely orange-brown except for narrow fuscous lateral margins on tergum I and narrow transverse fuscous lines on anterolateral corners of tergum III.

Description: male.

Head: Height 1.67 mm. Frons – length 1.6 times breadth; red-brown with fuscous on anteromedial hump and around bases of orbital setae; orbital setae black: 1 *s.or.*, 2 *i.or.*; lunule fuscous. Ocellar triangle black. Vertex red-brown. Face fuscous with a pair of narrow elongate dark fuscous spots; length 0.49 mm. Genae red-brown; dark seta present. Occiput red-brown, fulvous along eye margins; occipital row with a large number of short pale setae. Antennae with segments 1 and 2 red-brown, segment 3 red-brown with fuscous on apex and outer surface; length of segments: 0.14 mm; 0.24 mm; 0.84 mm.

Thorax: Scutum entirely uniformly dark red-brown. Pleural areas dark red-brown. Yellow markings as follows: notopleura; mesopleural stripe reaching almost to anterior *npl.* seta dorsally, continuing to katepisternum as a small transverse spot, anterior margin straight; anatergite (posterior apex red-brown); anterior $^{3}/_{5}$ katatergite (remainder plus ventral margin red-brown). Lateral and medial longitudinal postsutural yellow vittae absent. Postpronotal lobes fuscous. Postnotum red-brown. Scutellum yellow with a moderately broad dark red-brown basal band. Setae: *sc.* 2; *prsc.* absent; *ia.* 1; *p.sa.* 1; *a.sa.* 1; *mpl.* 1; *npl.* 2; *scp.* 4.

Legs: All segments entirely fulvous; mid tibiae each with an apical black spur.

Wings: Length 6.5 mm; cells bc and c fuscous; microtrichia in outer ½ of cell c only; remainder of wings colourless except dark fuscous cell sc, broad dark fuscous costal band slightly overlapping R_{4+5} and remaining of uniform width around apex of wing, a transverse band across centre of wing where it is dark fuscous enclosing *r-m* crossvein and then broad and diffuse across remainder of wing to hind margin, a broad dark fuscous anal streak; a dense aggregation of microtrichia around A_1+CuA_2; supernumerary lobe of medium development.

Abdomen: Oval to elongate oval in shape; terga free; pecten present on tergum III. Tergum I and sterna I and II wider than long. All terga entirely orange-brown except for narrow lateral fuscous margins on tergum I and narrow transverse fuscous lines on anterolateral margins of tergum III. A pair of oval orange-brown shining spots on tergum V. Posterior lobe of surstylus long, sternum V with a slight concavity on posterior margin. Abdominal sterna orange-brown.

Attractant: Cue lure.

Female: No known record.

Distribution: Indonesia (Moluccas = Maluku – Ambon I.; Seram I.).

Hosts: No known record.

Remarks: *Bactrocera (Parasinodacus) waimitaliae* n.sp. is similar to *Bactrocera (Parasinodacus) binoyi* Drew in possessing the basic characters of the subgenus *Parasinodacus* and lacking medial and lateral postsutural vittae. It differs from *B. binoyi* in having fuscous postpronotal lobes, a dark red-brown scutum, costal band broad and just overlapping R_{4+5} and a transverse band of fuscous colouration across the centre of the wing.

SUBGENUS *PARATRIDACUS* Shiraki

Bactrocera (Paratridacus) banneri White

(Fig. 246)

Bactrocera (Paratridacus) banneri White, in White & Evenhuis, 1999: 520. Holotype male in AMNH.

Type material examined: Holotype male, **INDONESIA: NORTH MOLUCCAS (NORTH MALUKU)** – with the following two labels – 'Morotai, Moluccas, iii.1945, Gilbert Banner'; 'Bactrocera banneri, Ian White, not published'. Received on loan from AMNH.

Diagnosis: A small species; face fulvous without dark markings; postpronotal lobes and notopleura yellow; scutum entirely black; lateral postsutural yellow vittae present; medial postsutural yellow vitta absent; no yellow spot anterior to notopleural suture; broad mesopleural stripe reaching to postpronotal lobe dorsally; scutellum yellow with a narrow black basal band; wing with cells bc and c colourless, microtrichia in outer corner of cell c only; wing colourless without a costal band but with cell sc fuscous and pale fuscous between cell sc and base of vein R_{4+5}, pale fuscous surrounding *r-m* crossvein; broad dark fuscous anal streak with colour around margins of cell bm and into base of cell br; abdominal terga I–III entirely black, tergum IV dark fuscous tending fuscous along posterior margin, tergum V entirely fuscous.

Description: male.

Head: Height 1.05 mm. Frons – length 1.6 times breadth; red-brown without dark markings; orbital setae black: 1 *s.or.*, 2 *i.or.*; lunule red-brown. Ocellar triangle black. Vertex dark fuscous. Face fulvous without dark markings; length 0.45 mm. Genae fulvous, dark fuscous subocular spot present; black seta present. Occiput dark fuscous, fulvous along eye margins; occipital row without distinct setae. Antennae with all segments entirely red-brown; length of segments: 0.08 mm; 0.2 mm; 0.6 mm.

Thorax: Scutum entirely black without pale markings. Pleural areas entirely dark fuscous to black. Yellow markings as follows: postpronotal lobes; notopleura; broad mesopleural stripe reaching to postpronotal lobe dorsally, continuing to katepisternum as a small transverse spot, anterior margin

convex; anatergite (posterior apex black); anterior ⅔ katatergite (remainder black); two broad parallel-sided lateral postsutural vittae ending posteriorly at *ia.* seta. Medial longitudinal postsutural yellow vitta absent. Postnotum black. Scutellum yellow except for narrow black basal band. Setae: *sc.* 4; *prsc.* 2; *ia.* 1; *p.sa.* 1; *a.sa.* 1; *mpl.* 1; *npl.* 2; *scp.* 4.
Legs: Fore and mid femora entirely dark fuscous to black, hind femora with basal ½ fulvous and apical ½ dark fuscous to black; fore tibiae entirely fulvous, mid tibiae fulvous tending fuscous basally, hind tibiae dark fuscous; tarsi with all segments entirely fulvous; mid tibiae each with an apical black spur.
Wings: Length 3.7 mm; cells bc and c colourless; microtrichia in outer corner of cell c only; remainder of wings colourless and without a costal band but with cell sc fuscous, pale fuscous between cell sc and basal area of R_{4+5}, a narrow pale infuscation along *r-m* crossvein, a broad dark fuscous anal streak with colour surrounding margins of cell bm and covering basal area of cell br; no dense aggregation of microtrichia around A_1+CuA_2; supernumerary lobe of medium development.
Abdomen: Oval; terga free; no pecten on tergum III. Tergum I and sterna I and II wider than long. Terga I–III entirely black; tergum IV dark fuscous with a narrow fuscous band along posterior margin but not reaching lateral margins; tergum V entirely fuscous. A pair of oval red-brown shining spots on tergum V. Posterior lobe of surstylus long, sternum V with a slight concavity on posterior margin.
Attractant: No known record.
Female: No known record.
Distribution:
Type locality: Indonesia (North Moluccas = North Maluku – Morotai I.).
Hosts: No known record.
Remarks: *Bactrocera (Paratridacus) banneri* White is a unique species of subgenus *Paratridacus* in possessing a black scutum and abdominal terga mostly dark fuscous to black. The other known South-East Asian species are mostly pale coloured. Also, the wing lacks a costal band which is present in the other species of this subgenus.

Bactrocera (Paratridacus) expandens (Walker)

(Fig. 247)

Dacus expandens Walker, 1859: 114. Lectotype male in BMNH.
Paratridacus expandens – Perkins, 1939: 33.
Dacus (Paratridacus) expandens – Hardy, 1951: 140; Hardy, 1959: 171, Lectotype designation by inference of Holotype; Drew, 1973: 18.
Bactrocera (Paratridacus) expandens – Drew, 1989: 198; Norrbom *et al.*, 1998: 99.

Non-type material examined: AUSTRALIA and **PAPUA NEW GUINEA**: a large series of specimens collected in eastern coastal areas of Queensland north from Cairns and throughout Papua New Guinea.
Diagnosis: Face fulvous with a pair of medium-sized oval black spots; postpronotal lobes and notopleura yellow; scutum red-brown without dark markings; broad lateral postsutural yellow vittae narrowing slightly posteriorly to end at *ia.* seta; medial postsutural yellow vitta absent; mesopleural stripe reaching midway between anterior margin of notopleuron and anterior *npl.* seta dorsally; scutellum yellow with a narrow red-brown basal band; legs with all segments entirely fulvous; wing with cells bc and c dark fulvous, microtrichia covering cell c and outer corner of cell bc; a broad dark fulvous costal band confluent with R_{4+5} and of uniform width throughout; a broad dark fulvous anal streak; supernumerary lobe of medium development; abdominal terga III–V orange-brown with a narrow transverse fuscous to dark fuscous spot on anterolateral corners of tergum III and a narrow medial longitudinal fuscous to dark fuscous band over posterior part of tergum IV and all of tergum V; apex of female ovipositor trilobed (see Drew, 1989, Fig. 361).
Attractant: No known record.
Distribution:
Type locality: Indonesia (Moluccas = Maluku – Aru Is).
Previously recorded: Type locality and Northeastern Australia and Papua New Guinea. Records published by Hardy (1951), India, Japan, Philippines and Sri Lanka are incorrect as they refer to either *Bactrocera garciniae* (Bezzi) (India, Sri Lanka) or *Bactrocera yayeyamanus* (Matsumura) (Japan, Philippines). The record of *B. expandens* from the Andaman Islands (David and Ramani, 2011) is incorrect and possibly refers to *B. garciniae*. The record of *B. garciniae* from India cannot be confirmed.
Hosts: Recorded from many species of *Garcinia* (family *Clusiaceae*) (Drew, 1989; Hancock *et al.*, 2000).
Remarks: *Bactrocera (Paratridacus) expandens* (Walker) was described adequately by Hardy (1951) and Drew (1973). In South-East Asia it is very close to *Bactrocera (Paratridacus) garciniae* Bezzi in the colour patterns of the scutum and abdomen, including the yellow patterns on the thorax. It differs from *B. garciniae* in lacking dark fuscous patterns on the femora and in having the costal band of uniform width throughout (not expanding across apex of R_{4+5}). Hardy (1951) made *B. garciniae* and *B. yayeyamanus* synonyms of *B. expandens* but there appears to be adequate character differences to establish these as separate species and thus these synonymies cannot be justified. This early declaration by Hardy (1951) led to future taxonomists listing further new synonymies (see Norrbom *et al.*, 1998). We have undertaken extensive studies of types and a number of specimens collected across South-East Asia and believe that the following are good species – *Bactrocera (Paratridacus) expandens* (Walker), *Bactrocera (Paratridacus) garciniae* Bezzi, *Bactrocera (Paratridacus) melanius* (Hardy & Adachi) and *Bactrocera (Paratridacus) yayeyamanus* (Matsumura).

Drew (1989) questioned the validity of the type status of the male specimen in BMNH as Walker (1859) based his description on a female. However, Norrbom *et al.* (1998) accepted the male specimen as the type on the basis that Walker misstated the sex in his description. Consequently, herein we accept that the male with the following three labels – ‘expandens’; ‘Borneo, Aru Is, W. Saunders, BM 1868-4’; ‘Aru 28’, is the Lectotype by inference of Holotype by Hardy (1959).

Bactrocera (Paratridacus) garciniae Bezzi

(Fig. 248)

Bactrocera garciniae Bezzi, 1913: 97. Syntypes in ZSI and MCSNM.
Bactrocera (Paratridacus) garciniae – Norrbom *et al.*, 1998: 99.

Type material examined: 1 male and 2 female Syntypes, **SRI LANKA (formerly CEYLON)**: Peradeniya, 8-09 (1 female with a label – '*garciniae* n.sp.' in Bezzi's handwriting). Received on loan from MCSNM.
Non-type material examined: SRI LANKA: 1 female, Sri Lanka, Peradeniya, 25.iii.1996, ex *Garcinia tinctoria*; 1 female, Sri Lanka, Peradeniya, 15.iv.1996, ex *Garcinia*. Specimens in QDPC.
Diagnosis: A large species; face fulvous with a pair of large oval black spots; postpronotal lobes and notopleura yellow; scutum uniformly red-brown; lateral postsutural yellow vittae present; medial postsutural yellow vitta absent; no yellow spot anterior to notopleural suture; mesopleural stripe reaching midway between anterior margin of notopleuron and anterior *npl.* seta dorsally; scutellum yellow with a broad fuscous basal band; wing with cells bc and c with a pale fuscous tint, microtrichia in outer corner of cell c only; a broad fuscous costal band confluent with R_{4+5} and expanding over apex of R_{4+5}; a narrow fuscous anal streak; abdominal terga III–V orange-brown with a broken 'T' pattern consisting of dark fuscous transverse spots on anterolateral margins of tergum III and a narrow medial longitudinal dark band over posterior portion of tergum IV and most of tergum V (in one female type specimen studied the band was over all of terga IV and V and posterior part of tergum III).
Description: male.
Head: Height 1.86 mm. Frons – length 1.36 times breadth; red-brown with fulvous along lateral margins; orbital setae large and dark fuscous: 1 *s.or.*, 2 *i.or.*; lunule red-brown. Ocellar triangle black. Vertex red-brown. Face fulvous with a pair of large oval black spots; length 0.62 mm. Genae fulvous, pale fuscous subocular spot present; dark fuscous seta present. Occiput red-brown, fulvous along eye margins; occipital row with 7–8 strong black setae. Antennae with segments 1 and 2 red-brown, segment 3 mostly dark fuscous; length of segments: 0.25 mm; 0.45 mm; 1.1 mm.
Thorax: Scutum uniformly red-brown except a pale fuscous tint along inner margins of lateral postsutural vittae. Pleural areas red-brown except fuscous to dark fuscous along anterior margin of mesopleural stripe, posterior to mesopleural stripe and on katepisternum. Yellow markings as follows: postpronotal lobes; notopleura; mesopleural stripe reaching midway between anterior margin of notopleuron and anterior *npl.* seta dorsally, continuing to katepisternum as a transverse spot, anterior margin straight; anatergite (posterior apex red-brown); anterior $^3/_5$ katatergite (remainder red-brown); two medium-width parallel-sided lateral postsutural vittae ending at *ia.* seta. Medial longitudinal postsutural yellow vitta absent. Postnotum red-brown. Scutellum yellow except for broad fuscous basal band. Setae: *sc.* 4; *prsc.* 2; *ia.* 1; *p.sa.* 1; *a.sa.* 1; *mpl.* 1; *npl.* 2; *scp.* 4.
Legs: Fore femora fulvous except dark fuscous along outer surface, mid femora pale fulvous basally to dark fulvous apically and dark fuscous along outer surface, hind femora pale fulvous with fuscous to dark fuscous on apical $^2/_5$; fore and mid tibiae pale fuscous, hind tibiae fuscous; fore tarsi with basal segment fulvous and apical four segments pale fuscous, mid and hind tarsi with basal segment pale fulvous and apical four segments fulvous; mid tibiae each with an apical black spur.
Wings: Length 7.3 mm; cells bc and c with a pale fuscous tint; microtrichia in outer corner of cell c only; remainder of wings colourless except for fuscous cell sc, broad fuscous costal band confluent with R_{4+5} and broadening across apex of R_{4+5}, a narrow fuscous anal streak; no dense aggregation of microtrichia around A_1+CuA_2; supernumerary lobe weak.
Abdomen: Elongate oval; terga free; no pecten on tergum III. Tergum I and sterna I and II wider than long. Tergum I orange-brown with fuscous narrow lateral margins; tergum II orange-brown tending fulvous along posterolateral margins and with narrow pale fuscous lateral margins; terga III–V orange-brown with dark fuscous spots anterolaterally on tergum III and a narrow medial longitudinal dark fuscous to black band on posterior ⅓ of tergum IV and over most of tergum V. A pair of oval orange-brown shining spots on tergum V. Posterior lobe of surstylus long and broad, sternum V with a slight concavity on posterior margin.
Attractant: No known record.
Female: As for male except one specimen showing the narrow medial longitudinal dark band on the abdomen over most of tergum III and all of terga IV and V; ovipositor basal segment orange-brown, dorsoventrally flattened and tapering posteriorly in dorsal view; ratio of length of oviscape to length of tergum V, 0.5:1; apex of piercer trilobed.
Distribution:
Type locality: Sri Lanka (Peradeniya). Records published in Norrbom *et al.* (1998) are incorrect due to confusion over the taxonomy of *B. garciniae* and related species. See 'Remarks' under *B. expandens* (Walker).
Hosts: *Garcinia* sp. (family *Clusiaceae*). Records published in Allwood *et al.* (1999) for *B. garciniae* are in error due to confusion over the taxonomy of this and other related species. See 'Remarks' for *B. expandens* (Walker).
Remarks: *Bactrocera (Paratridacus) garciniae* Bezzi was synonymised with *Bactrocera (Paratridacus) expandens* (Walker) by Hardy (1951) and maintained in that position by Drew (1989). While these two species are very similar, *B. garciniae* differs in having the costal band broadly expanded over the apex of R_{4+5}, extensive dark fuscous patterns on the femora and parallel-sided lateral postsutural vittae. Both species have been reared from fruit of *Garcinia* and have a similar unique trilobed apex on the female piercer.

Bactrocera (Paratridacus) icelus (Hardy)

Dacus icelus Hardy, 1974: 47. Holotype female in BPIM.
Bactrocera (Zeugodacus) icelus – Norrbom *et al.*, 1998: 103.

Type material examined: This species is known only from the female Holotype which was not available for study.

Diagnosis: A medium-sized species (wing length 6.0 mm); face fulvous with large oval black spots; postpronotal lobes and notopleura yellow; scutum red-brown with large medial and lateral longitudinal black bands; two broad parallel-sided lateral postsutural yellow vittae ending at *ia.* seta; medial postsutural yellow vitta absent; mesopleural stripe equal in width to notopleuron dorsally; scutellum yellow with a narrow black basal band; legs with all femora entirely fulvous; fore and mid tibiae pale fuscous, hind tibiae fuscous; wing with a narrow fuscous costal band just overlapping R_{2+3} and ending beyond apex of R_{4+5}; a distinct fuscous anal streak; abdominal terga III–V orange-brown with fuscous lateral margins, a black 'T' pattern consisting of a transverse band across base of tergum III and a medial longitudinal band over all three terga; ovipositor with basal segment fulvous, overall length short, apex of piercer needle-shaped.
Attractant: No known record.
Distribution:
Type locality: Philippines (Luzon I., Muntinlupa).
Hosts: No known record.
Remarks: *Bactrocera (Paratridacus) icelus* (Hardy) is here placed in the subgenus *Paratridacus* although more specimens, including males, need to be collected and studied before definitive subgeneric placement. Having been described from only one female specimen, which is now not available for study, it is not possible to define this species further.

Bactrocera (Paratridacus) melanius (Hardy & Adachi), new combination

(Fig. 249)

Dacus expandens ssp. *melanius* Hardy & Adachi, 1954: 157; Norrbom *et al.*, 1998: 100 as Synonym of *B. garciniae* Bezzi. Holotype male in USNM.
Dacus (Zeugodacus) aptatus Hardy, 1973: 57. **New Synonym.**

Type material examined: SINGAPORE: Holotype male, with the following four labels – 'Singapore, Malaya, v.1949'; 'ex *Garcinia dulcis*'; 'N.L.H. Krauss, Collector'; 'Holotype male, *Dacus (Paratridacus) expandens melanius* D.E. Hardy'. Received on loan from USNM.
Paratypes: Allotype female, 2 male and 4 female paratypes, same locality label data as holotype. Received on loan from USNM.
Non-type material examined: CHINA: 1 female with two labels in Chinese which include the data – 23.iv.1958 and 1600 m; 1 male with two labels in Chinese which include the date 17.iv.1958 and another label '*Dacus (Paratridacus)* sp., X.J. Wang'. **INDONESIA: JAVA** – 1 male and 3 females, 'Indonesia, Bogor, x.1953, E. Dresner'. **PENINSULAR MALAYSIA**: 21 males and 12 females, Kaki Bukit, Perlis, North Malaysia, vii.1991, coll. C.S. Ooi. **THAILAND: CHIANG MAI** – 6 males and 1 female (20.ii.1992), 2 females (5.iii.1992), Doi Pui, Chiang Mai, coll. Chantanee, bred from *Garcinia xanthochymus*; 6 males and 4 females (6.ii.1992), 2 females (17.ix.1992), 7 males and 6 females (1.x.1992), 1 male (15.x.1992), 4 males and 6 females (29.x.1992), 3 males and 1 female (10.xi.1992), 6 males and 6 females (8.xii.1992), 2 males (14.xii.1994) and 1 male and 1 female (28.xii.1994), Tum Rusee, Chiang Mai, coll. Chantanee, bred from *Garcinia xanthochymus*; **CHIANG RAI** – 2 males and 5 females, Doi Luang Nat. Park, Pha Yao, Chiang Rai, 29.i.1992, coll. Charuwan, bred from *Planchonella* sp., 3 males and 4 females (14.vi.1991), 2 males and 9 females (29.xi.1991), Doi Luang Nat Park, Phayao, Chiang Rai, coll. Charuwan, bred from *Garcinia xanthochymus*; 6 females, Doi Luang N.P., Chiang Rai, 24.v.1991, coll. Charuwan Supasatien, bred from wild-fruit; 1 female (2.vi.1992), 1 male and 4 females (25.ix.1992), Jae Chon Nat. Park, Lampong, Chiang Rai, coll. Charuwan, bred from *Garcinia xanthochymus*; **SONGKHLA** – 2 males and 2 females, Trang H.R.S., Trang, Songkhla, 4.i.1992, coll. Sieng *et al.*, bred from *Garcinia dulcis*; 1 male, Songkhla, Si Kao Hort Cultural Res. Station, Trang, 4.iii.1992, coll. Sieng *et al.*, bred from *Garcinia dulcis*; 1 male and 1 female, Songkhla, Sikao Hort. Res. Centre, Trang, 19.iii.1993, coll. Sieng *et al.*, bred from *Garcinia dulcis*; 2 males and 2 females, Songkhla, Sikao Horticultural Res Station, Trang, 31.v.1992, coll. Sieng *et al.*, bred from *Garcinia dulcis*; 3 males, Sikao Hort. Res. Stn., Trang, Songkhla, 11.vi.1993, coll. Surakrai *et al.*, bred from *Garcinia dulcis*; 1 female, Songkhla, Horticultural Research Station, Trang, 8.v.1992, coll. Dusadee *et al.*, bred from *Garcinia dulcis*; 1 female, Songkhla, Khuan Khanum, Phatthalung, 10.viii.1992, coll. Supanee, bred from *Garcinia dulcis*; 2 males and 2 females, Trang H.R.S., Sikao, Trang, Songkhla, 13.xii.1991, coll. Anuchit *et al.*, bred from *Garcinia dulcis*; 4 females, Songkhla, Sikao, Trang, 28. vi.1991, coll. Anuchit *et al.*, bred from *Garcinia dulcis*; 3 males and 3 females, Auy Yer Wang, Yala, Songkhla, 18.x.1991, coll. Anuchit *et al.*, bred from *Solanum stramonifolium* [Misspelt]. Specimens in QDPC.
Diagnosis: A large species; face fulvous with a pair of medium-sized to large subquadrate black spots; postpronotal lobes and notopleura yellow; scutum black; lateral postsutural vittae present; medial postsutural yellow vitta absent; no yellow spot anterior to notopleural suture; mesopleural stripe reaching midway between anterior margin of notopleuron and anterior *npl.* seta dorsally; scutellum yellow; wing with cells bc and c with a pale fuscous tint and becoming darker at area covered by microtrichia in cell c, microtrichia in outer ½ of cell c only; a broad dark fuscous costal band overlapping R_{4+5} and widening at apex of wing; a narrow fuscous to dark fuscous anal streak; abdominal terga III–V orange-brown with a 'T' pattern consisting of a broad transverse black band covering almost the entire area of tergum III and a broad medial longitudinal black band over all three terga, narrow dark fuscous to black anterolateral margins on terga IV and V.
Description: male.
Head: Height 1.7 mm. Frons – length 1.68 times breadth; red-brown centrally, fulvous along lateral margins, dark fuscous on anteromedial hump and around bases of orbital setae; orbital setae black: 1 *s.or.*, 2 *i.or.*; lunule fuscous. Ocellar triangle black. Vertex dark fuscous. Face fulvous with a pair of medium-sized to large subquadrate black spots; length 0.48 mm. Genae fulvous, fuscous subocular spot present; black seta present. Occiput red-brown, fulvous along eye margins; occipital row with a large number of strong black

setae. Antennae with segments 1 and 2 fuscous, segment 3 dark fuscous; length of segments: 0.24 mm; 0.41 mm; 0.89 mm.
Thorax: Scutum black except red-brown below and behind lateral postsutural vittae, around notopleural suture and inside postpronotal lobes. Pleural areas black, tending fuscous to dark fuscous below postpronotal lobes. Yellow markings as follows: postpronotal lobes; notopleura; mesopleural stripe reaching midway between anterior margin of notopleuron and anterior *npl.* seta dorsally, continuing to katepisternum as a large transverse spot, anterior margin convex; anatergite (posterior apex black); anterior ¾ katatergite (remainder black); two broad parallel-sided lateral postsutural vittae reaching to *ia.* seta. Medial longitudinal postsutural yellow vitta absent. Postnotum black. Scutellum yellow except for narrow black basal band. Setae: *sc.* 4; *prsc.* 2; *ia.* 1; *p.sa.* 1; *a.sa.* 1; *mpl.* 1; *npl.* 2; *scp.* 4.
Legs: Fore femora fulvous with a small black subapical spot on outer surfaces, mid and hind femora fulvous with a dark pattern around apical ⅓ to ¼ represented as black around lateral and ventral surfaces and fuscous on dorsal surface; fore and hind tibiae dark fuscous, mid tibiae dark fuscous basally to fuscous apically; all tarsi with basal segment fulvous and apical four segments dark fuscous; mid tibiae each with an apical black spur.
Wings: Length 6.7 mm; cells bc and c with a pale fuscous tint tending darker on area covered by microtrichia in cell c; microtrichia covering outer ½ of cell c only; remainder of wings colourless except dark fuscous cell sc, broad dark fuscous costal band overlapping R_{4+5} and widening over apical ⅓ of this vein, a narrow fuscous to dark fuscous anal streak; no dense aggregation of microtrichia around A_1+CuA_2; supernumerary lobe weak.
Abdomen: Oval; terga free; no pecten on tergum III. Tergum I and sterna I and II wider than long. Tergum I black with two narrow transverse orange-brown spots on posterior margin; tergum II orange-brown anteriorly, fulvous posteriorly with a narrow irregularly shaped transverse black band across centre which reaches narrow black anterolateral corners; terga III–V orange-brown with a dark 'T' pattern consisting of a broad transverse black band covering almost entire tergum III and a broad medial longitudinal black band over all three terga, narrow dark fuscous to black anterolateral margins on terga IV and V. A pair of oval orange-brown shining spots on tergum V. Posterior lobe of surstylus long and stout, sternum V with a slight concavity on posterior margin. Abdominal sterna dark fuscous to black.
Attractant: No known record.
Female: As for male except ovipositor basal segment red-brown, dorsoventrally compressed and tapering posteriorly in dorsal view, ratio of length of oviscape to length of tergum V, 0.5:1, apex of piercer trilobed (Fig. 249) and appears more stout in stem basal to lateral lobes than in *B. yayeyamanus* (Matsumura).
Distribution:
Type locality: Singapore.
Previously recorded: Type locality and China (Yunnan Province). Records in Norrbom *et al.* (1998) under *B. garciniae* (Bezzi) are in error as *B. melanius* is a distinct species and not a synonym of that species.
New records: Indonesia (Java, Bogor); Peninsular Malaysia (Perlis); Thailand (Chiang Mai; Chiang Rai; Songkhla). Thailand (Phu Kae), due to **New Synonymy** of *D. aptatus*.
Hosts: *Garcinia dulcis* and *Garcinia xanthochymus* (family *Clusiaceae*). Under non-type material examined, there are records of specimens bred from *Planchonella* sp. and *Solanum stramoniifolium* which should be regarded as doubtful unless they can be confirmed through field collecting of fruits as *Bactrocera (Paratridacus)* species have not previously been recorded from these plant genera.
Remarks: *Bactrocera (Paratridacus) melanius* (Hardy & Adachi) was originally described as a subspecies of *Bactrocera (Paratridacus) expandens* (Walker) (Hardy & Adachi, 1954) and later listed as a synonym of *Bactrocera (Paratridacus) garciniae* (Bezzi) by Norrbom *et al.* (1998). However, it is a good species, most similar to *Bactrocera (Paratridacus) yayeyamanus* (Matsumura) in possessing a broad costal band that expands over R_{4+5} at apical ⅓ of this vein and abdominal terga III–V orange-brown with a broad medial and two broad lateral longitudinal black bands over all three terga and joined across anterior margin of tergum III. It differs from *B. yayeyamanus* in having the scutum entirely black, the mesopleural stripe reaching midway between the anterior margin of the notopleuron and the anterior *npl.* seta dorsally, broad parallel-sided lateral postsutural vittae and the costal band slightly overlapping R_{4+5} in the centre of wing before widening apically.

See 'Remarks' under *B. expandens* for further comments regarding previous synonymies. We can find no good characters to define *Dacus (Zeugodacus) aptatus* Hardy as a separate species and are therefore making it a **New Synonym** of *B. melanius*.

Bactrocera (Paratridacus) yayeyamanus (Matsumura), new combination

(Figs 250A, 250B)

Dacus yayeyamanus Matsumura, 1916: 412; Hardy, 1951: 140 (as synonym of *B. expandens* (Walker)); Norrbom *et al.*, 1998: 100 (as synonym of *B. garciniae* Bezzi). Holotype female in HUS.
Paratridacus yayeyamanus – Shiraki, 1933: 110; Shiraki, 1968: 14 (as synonym of *B. expandens* (Walker)).
Dacus (Zeugodacus) katoi Hardy, 1974: 50. **New Synonym.**
Dacus yayemanus – Hendel, 1927a: 221. [Misspelling of *yayeyamanus* (Matsumura)].

Other type material examined: PHILIPPINES: LUZON ISLAND – Allotype female of *Dacus (Zeugodacus) katoi* Hardy, with the following three labels – 'No. 296-E, Bataan, P.I., Date 20 Jan 1948, coll. L.T. Kargonillo'; '*Dacus (Zeugodacus) katoi* Hardy, Det 1970 D.E. Hardy'; 'Allotype female, *Dacus (Zeugodacus) katoi* Hardy'. 1 female paratype, same locality label data as Allotype; 1 female paratype with the following two labels – 'Lamao, Bataan, 11 Nov 1947, 1003, Q.C. Chock'; 'Paratype, *Dacus (Zeugodacus) katoi*, D.E. Hardy'. Received on loan from BBM (Fig. 250B). (See 'Remarks' below.)

Non-type material examined: JAPAN: (RYUKYU ISLANDS) – 1 male and 1 female, 'Ryukyu Is, 1.i.1953, T. Shiraki'; '*Paratridacus yayeyamanus* Mats, det. T. Shiraki'; 1 male, 'Ishigaki Is, Okinawa, 7.x.1973'; 'Ex. *Garcinia subelliptica*'; 'Collected by T. Imamura'; '*Dacus expandens* Walker'; 2 males and 2 females, Taketomi I, Okinawa, 29.vii.1990; 1 male, 'Ishigaki I, *D. expandens*, M. Kaneda'; 1 female, 29.vii.1989, M. Kaneda. Specimens in QDPC. **PHILIPPINES**: 1 male and 1 female, 'Luzon, P.I., ix-x.1947'; 'Collector Q.C. Chock'. Received on loan from USNM.

Diagnosis: A large species; face fulvous with a pair of medium-sized circular black spots; postpronotal lobes and notopleura yellow; scutum with a variegated pattern of dark fuscous to fuscous covering most areas, narrow longitudinal black lines from centre of scutum to anterior margin either side of midline and red-brown in medial area from centre of scutum to anterior margin, below lateral postsutural vittae and around notopleural suture; lateral postsutural vittae present; medial postsutural yellow vitta absent; no yellow spot anterior to notopleural suture; mesopleural stripe reaching anterior *npl.* seta dorsally; scutellum yellow; wing with cell bc pale fuscous, cell c fuscous tending dark fuscous in area covered by dense microtrichia, microtrichia covering outer ½ of cell c only; a broad dark fuscous costal band confluent with R_{4+5} and widening over apical ⅓ of this vein; a narrow dark fuscous anal streak; abdominal terga III–V orange-brown with a dark 'T' pattern consisting of a narrow transverse black band across anterior margin of tergum III and widening over lateral margins and a broad medial longitudinal black band over all three terga, narrow dark fuscous to black anterolateral corners on terga IV and V.

Description: male (Fig. 250A).

Head: Height 1.76 mm. Frons – length 1.75 times breadth; red-brown with fuscous on anteromedial hump and dark fuscous around bases of *s.or.* and lower pair of *i.or.* setae; orbital setae black: 1 *s.or.*; 2 *i.or.*; lunule dark fuscous. Ocellar triangle black. Vertex red-brown. Face fulvous with a pair of medium-sized circular black spots; length 0.49 mm. Genae fulvous, fuscous subocular spot present; black seta present. Occiput red-brown, fulvous along eye margins; occipital row with a large number of strong black setae. Antennae with segments 1 and 2 fuscous, segment 3 red-brown with dark fuscous on apex and outer surface; length of segments: 0.24 mm; 0.4 mm; 0.94 mm.

Thorax: Scutum with a variegated colour pattern basically composed of dark fuscous posteriorly and fuscous to dark fuscous anterolaterally with narrow black longitudinal lines running from inside of lateral postsutural vittae to anterior margin of scutum, a black spot between postpronotal lobe and notopleural suture, a black spot posteriorly on inner margin of lateral postsutural vittae and red-brown below lateral postsutural vittae, around notopleural suture and medially from centre of scutum to anterior margin. Pleural areas black except red-brown below postpronotal lobes. Yellow markings as follows: postpronotal lobes; notopleura; mesopleural stripe reaching anterior *npl.* seta dorsally, continuing to katepisternum as a large transverse spot, anterior margin slightly convex; anatergite (posterior apex black); anterior ¾ katatergite (remainder black); two moderately broad lateral postsutural vittae narrowing slightly posteriorly to end before *ia.* seta. Medial longitudinal postsutural yellow vitta absent. Postnotum black, red-brown centrally. Scutellum yellow except for narrow black basal band. Setae: *sc.* 4; *prsc.* 2; *ia.* 1; *p.sa.* 1; *a.sa.* 1; *mpl.* 1; *npl.* 2; *scp.* 4.

Legs: Fore femora fulvous with a small fuscous subapical spot on outer surfaces in some specimens, mid and hind femora fulvous with an elongate oval subapical black spot on ventral surfaces; fore and mid tibiae fulvous, hind tibiae fuscous to dark fuscous; fore tarsi with basal segment fulvous and apical four segments red-brown, mid and hind tarsi entirely fulvous; mid tibiae each with an apical black spur.

Wings: Length 7.0 mm; cell bc pale fuscous, cell c fuscous tending dark fuscous on area covered by dense microtrichia; microtrichia covering outer ½ of cell c only; remainder of wings colourless except dark fuscous cell sc, broad dark fuscous costal band confluent with R_{4+5} above which it becomes slightly paler and widening across apical ⅓ of this vein, a narrow dark fuscous anal streak; no dense aggregation of microtrichia around A_1+CuA_2; supernumerary lobe weak.

Abdomen: Oval; terga free; no pecten on tergum III. Tergum I and sterna I and II wider than long. Tergum I orange-brown with narrow dark fuscous lateral margins; tergum II orange-brown anteriorly, fulvous posteriorly and with narrow dark fuscous anterolateral corners; terga III–V orange-brown with a dark 'T' pattern consisting of a narrow transverse black band across anterior margin of tergum III which widens across lateral margins and a broad medial longitudinal black band over all three terga, narrow dark fuscous to black anterolateral corners on terga IV and V. A pair of oval orange-brown shining spots on tergum V. Posterior lobe of surstylus long and stout, sternum V with a slight concavity on posterior margin. Abdominal sterna red-brown in males and dark fuscous to black in females.

Attractant: No known record.

Female: As for male except ovipositor basal segment fulvous, dorsoventrally compressed and taping posteriorly in dorsal view, ratio of length of oviscape to length of tergum V, 0.5:1, apex of piercer trilobed (Fig. 250A) and appears more slender in the stem, basal to the lateral lobes, than that in *B. melanius* (Hardy & Adachi).

Distribution:

Type locality: Japan, Ryukyu Is (Yayeyama Is [probably Misspelling for Yaeyama Is]).

Previously recorded: Type locality; Japan, Ryukyu Is. (Iriomote I., Ishigaki I.; Miyako I.; Okinawa I.) (Shiraki, 1968: 17). Records in Norrbom *et al.* (1998) under *B. garciniae* are in error, as *B. yayeyamanus* is a distinct species and not a synonym of that species.

New records: Philippines (Negros I., La Granja; Luzon I., Bataan) Hardy, 1974: 52, due to **New Synonymy** of *D. katoi* listed under 'Other type material examined'.

Hosts: *Garcinia subelliptica* (family *Clusiaceae*).

Remarks: *Bactrocera (Paratridacus) yayeyamanus* (Matsumura) is similar to *Bactrocera (Paratridacus) melanius* (Hardy & Adachi) in possessing a broad costal band that expands over R_{4+5} at apical ⅓ of this vein and abdominal terga III–V orange-brown with broad medial and lateral longitudinal black bands over all three terga and joining along anterior margin of tergum III. It differs from *B. melanius* in having the scutum fuscous to dark fuscous with areas of red-brown and black, the lateral postsutural vittae narrowing slightly posteriorly to end just before the *ia.* seta, the mesopleural stripe reaching to the anterior *npl.* seta dorsally and the costal band confluent with R_{4+5} (not overlapping this vein). See 'Remarks' under *B. expandens* for comments regarding previous synonymies.

Bactrocera (Paratridacus) yayeyamanus (Matsumura) was placed as a synonym of *Bactrocera (Paratridacus) expandens* (Walker) by Hardy (1951) and Shiraki (1968) and a synonym of *Bactrocera (Paratridacus) garciniae* Bezzi by Norrbom *et al.* (1998). However, *B. yayeyamanus* is a good species and these synonymies are incorrect. We have studied three female Paratypes of *Dacus (Zeugodacus) katoi* Hardy, one marked Allotype and the other two marked Paratypes. We cannot find sound characters to establish this species and are making it a **New Synonym** of *B. yayeyamanus*.

SUBGENUS *PARAZEUGODACUS* Shiraki

Bactrocera (Parazeugodacus) abbreviata (Hardy)
(Fig. 251)

Dacus (Zeugodacus) abbreviatus Hardy, 1974: 44. Holotype male in BBM.
Bactrocera (Zeugodacus) abbreviata – Norrbom *et al.*, 1998: 101.

Type material examined: Holotype male, **PHILIPPINES: LUZON ISLAND** – with the following four labels – 'Los Banos, Laguna, PI, 7 July 1930'; 'Collected by F.C. Hadden'; 'Holotype male, Dacus (Zeugodacus) abbreviatus, D.E. Hardy'; 'Dacus (Zeugodacus) abbreviatus Hardy, Det. 1969, D.E. Hardy'. Received on loan from BBM.
Non-type material examined: CHINA: HONG KONG – 1 male, 'Hong Kong, N.T., Sai Kung Farm, 18.ii.1981, coll. R. Winney, ex methyl eugenol trap, MEYE 3'; 'C.I.E. coll. A.I. 3101'. **THAILAND**: 6 males, 8 females, Mae Hong Son, Chiang Rai, 26.iv.1993, coll. Charuwan, bred from *Olea salicifolia*, sample no. CR1361. Specimens in QDPC.
Diagnosis: A small species; face fulvous with a pair of medium-sized elongate oval black spots; postpronotal lobes and notopleura yellow; scutum almost entirely black; lateral postsutural yellow vittae present and short and triangular; medial postsutural yellow vitta absent; no yellow spot anterior to notopleural suture; mesopleural stripe slightly wider than notopleuron dorsally; scutellum with a broad black basal band extending as a broad medial longitudinal black band over dorsal surface; wing with cells bc and c colourless and totally devoid of microtrichia; entirely colourless except for a fuscous cell sc and a pale fuscous tint between R_1 and R_{2+3}; anal streak absent; abdominal terga III–V entirely black except for narrow longitudinal orange-brown bands running either side of medial longitudinal black band from posterior area of tergum III to posterior margin of tergum V.
Description: male.
Head: Height 1.55 mm. Frons – length 1.54 times breadth; red-brown with dark fuscous around *s.or.* setae only; lunule dark fuscous. Ocellar triangle black. Vertex dark fuscous. Face fulvous with a pair of medium-sized elongate oval black spots; length 0.4 mm. Genae red-brown, pale fuscous subocular spot present; black seta present. Occiput black, fulvous along eye margins; occipital row with 4–5 black setae. Antennae with segments 1 and 2 red-brown, segment 3 red-brown with fuscous on apex and outer surface; length of segments: 0.16 mm; 0.2 mm; 0.78 mm.
Thorax: Scutum entirely black except for a triangular-shaped red-brown marking anterior to notopleural suture. Pleural areas entirely black, tending dark fuscous below postpronotal lobes. Yellow markings as follows: postpronotal lobes (anterodorsal corners dark fuscous); notopleura; mesopleural stripe slightly wider than notopleuron dorsally, continuing to katepisternum as a transverse spot, anterior margin slightly convex; anatergite (posterior apex black); anterior $^3/_5$ katatergite (remainder black); two very broad, short and triangular lateral postsutural vittae ending at level of *a.sa.* seta. Medial longitudinal postsutural yellow vitta absent. Postnotum black. Scutellum yellow except for a broad black basal band which continues as a broad medial longitudinal black band over entire dorsal surface. Setae: *sc.* 4; *prsc.* 2; *ia.* 1; *p.sa.* 1; *a.sa.* 1; *mpl.* 1; *npl.* 2; *scp.* 4.
Legs: All femora entirely fulvous except for very small dark fuscous apices on mid and hind femora; fore and hind tibiae dark fuscous, mid tibiae dark fuscous basally tending to fuscous apically; all tarsi with basal segment fulvous and apical four segments dark fuscous; mid tibiae each with an apical black spur.
Wings: Length 4.8 mm; cells bc and c colourless and totally devoid of microtrichia; remainder of wings entirely colourless except for a fuscous cell sc and a pale fuscous tint between R_1 and the basal ½ of R_{2+3}; dense aggregation of microtrichia around A_1+CuA_2; supernumerary lobe weak.
Abdomen: Oval; terga free; pecten present on tergum III. Tergum I and sterna I and II wider than long. Tergum I entirely black; tergum II black except for small transverse fulvous areas posterolaterally; terga III–V entirely black except for narrow longitudinal orange-brown bands either side of medial longitudinal black band which runs from posterior area of tergum III to posterior margin of tergum V. A pair of oval orange-brown shining spots on tergum V. Posterior lobe of surstylus short, sternum V with a slight concavity on posterior margin.
Attractant: No known record.
Female: As for male except basal segment of ovipositor red-brown to fuscous, dorsoventrally compressed and tapering posteriorly in dorsal view, ratio of length of oviscape to length of tergum V, 0.5:1; apex of piercer needle-shaped.
Distribution:
Type locality: Philippines (Luzon I., Laguna).
New records: China (Hong Kong); Thailand (Chaing Rai).
Hosts: *Linociera ramiflora* and *Olea salicifolia* (family *Oleaceae*) (Allwood *et al.*, 1999).
Remarks: *Bactrocera (Parazeugodacus) abbreviata* (Hardy) is similar to *Bactrocera (Parazeugodacus) bipustulata* Bezzi in possessing an entirely black scutum, yellow postpronotal

lobes with anterodorsal corners dark, yellow notopleura, very short tapering lateral postsutural vittae, a broad medial black band on the scutellum, wings without a costal band and anal streak, abdominal terga III–V with broad medial and lateral longitudinal black bands joined over tergum III. It differs from *B. bipustulata* in having a pair of black spots on the face and all femora entirely fulvous.

Bactrocera (Parazeugodacus) arisanica (Shiraki)
(Fig. 252)

Zeugodacus arisanicus Shiraki, 1933: 81. Holotype male in NIAS.
Bactrocera (Zeugodacus) arisanica – Norrbom *et al.*, 1998: 102.
Dacus (Zeugodacus) parvifoliaceus Tseng, Chen & Chu, 1992: 71. **New Synonym**.

Type material examined: Holotype male, **TAIWAN (formerly FORMOSA)**: with the following five labels – 'Formosa, Arisan, 1918.X.2-23, J. Sonan, M. Yoshino', 'near scutellaris Bezzi Det. T. Shiraki', 'Zeugodacus arisanicus, Det. T. Shiraki', 'Type', 'Korper'. Received on loan from NIAS.
Non-type material examined: TAIWAN: 12 males, Jiauliping, Chiayi, Itsien, 1.ix-i.x.1977, coll. Y.H. Tseng, attracted to cue lure; 1 male, Alisnan, Chiayi, 1800m, 4. iv.1978, coll. Y.H. Tseng, attracted to cue lure; 1 male, Tashashan, Taichung, 5.x.1990, Sugar. **THAILAND**: Wang Jum Pee, Pakchong, Nakhonratsima, trap BK112T, 25.x.1991, coll. Montree, attracted to cue lure; 1 male, Doi Pui, Chiang Mai, 15.xi.1991, trap CM231T, coll. Chantanee, attracted to cue lure. Specimens in QDPC.
Diagnosis: A medium-sized species; face fulvous with a pair of medium-sized oval black spots; postpronotal lobes and notopleura yellow; scutum entirely black; narrow lateral and medial postsutural yellow vittae present; a yellow spot anterior to notopleural suture anterior to lateral postsutural vittae; mesopleural stripe reaching midway between anterior margin of notopleuron and anterior *npl.* seta dorsally; scutellum yellow with a large black apical spot; wing with cells bc and c colourless, microtrichia in outer corner of cell c only; a narrow fuscous costal band confluent with R_{2+3} and remaining very narrow before expanding into a spot around apex of R_{4+5}, a broad fuscous band along *dm-cu* crossvein; a broad fuscous anal streak; abdominal terga III–V black except for small fulvous markings posterocentrally on tergum III and fulvous longitudinal bands either side of a medial longitudinal black band over terga IV and V.
Description: male.
Head: Height 1.35 mm. Frons – length 1.29 times breadth; red-brown with dark fuscous around orbital setae; orbital setae large and black: 1 *s.or.*, 3 *i.or.*; lunule fuscous. Ocellar triangle black. Vertex black. Face fulvous with a pair of medium-sized oval black spots; length 0.5 mm. Genae fulvous, small fuscous subocular spot present; black seta present. Occiput dark fuscous, fulvous along eye margins; occipital row with 3–5 dark setae. Antennae with segments 1 and 2 red-brown, segment 3 red-brown with fuscous on apex and outer surface; length of segments: 0.13 mm; 0.25 mm; 0.63 mm.
Thorax: Scutum black without pale markings. Pleural areas black without pale markings. Yellow markings as follows: postpronotal lobes (narrow black margins on inner surfaces); notopleura; mesopleural stripe reaching midway between anterior margin of notopleuron and anterior *npl.* seta dorsally, continuing to katepisternum as a small transverse spot, anterior margin slightly convex; anatergite (posterior apex black); anterior ⅔ katatergite (remainder black); two narrow lateral postsutural vittae beginning in a small spot anterior to notopleural suture and narrowing posteriorly to end before *ia.* seta; a very short and narrow medial longitudinal postsutural vitta. Postnotum black. Scutellum yellow except for narrow black basal band and a large black apical spot covering ½ of dorsal surface. Setae: *sc.* 4; *prsc.* 2; *ia.* 1; *p.sa.* 1; *a.sa.* 1; *mpl.* 1; *npl.* 2; *hm.* 1 (very weak); *scp.* 4.
Legs: Fore femora fulvous (apical ⅓ dark fuscous), mid femora fulvous (apical ¼ dark fuscous), hind femora fulvous (apical ⅓ dark fuscous); fore and mid tibiae dark fuscous, hind tibiae dark fuscous to black; tarsi with all segments fulvous; mid tibiae each with an apical black spur.
Wings: Length 6.4 mm; cells bc and c colourless; microtrichia in outer corner of cell c only; remainder of wings colourless except dark fuscous cell sc, very narrow fuscous costal band confluent with R_{2+3} and remaining very narrow beyond apex of this vein before widening into a dark fuscous spot around apex of R_{4+5}, a broad fuscous marking along *dm-cu* crossvein, broad fuscous anal streak; dense aggregation of microtrichia around A_1+CuA_2; supernumerary lobe weak.
Abdomen: Oval; terga free; pecten present on tergum III. Tergum I and sterna I and II wider than long. Tergum I black with a narrow orange-brown band along posterior margin but not reaching lateral margins; tergum II black across anterior ½ and orange-brown across posterior ½; terga III–V mostly black except for small fulvous markings posterocentrally on tergum III and fulvous longitudinal bands either side of a medial longitudinal black band over terga IV and V. In some specimens, terga III–V have small fulvous to orange-brown spots posterocentrally on terga III and IV and a narrow pale band along posterior margin of tergum V. A pair of oval dark fuscous to fulvous shining spots on tergum V. Posterior lobe of surstylus short, sternum V with a slight concavity on posterior margin.
Attractant: Cue lure.
Female: No known record.
Distribution:
Type locality: Taiwan (Arisan).
New records: Taiwan (Jiauliping; Alisnan; Taichung); Thailand (Chiang Mai; Nakhonratsima). Taiwan, Nantou, due to **New Synonymy** of *D. parvifoliaceus*.
Hosts: No known record.
Remarks: *Bactrocera (Parazeugodacus) arisanica* (Shiraki) is similar in general body colour patterns to species in the *scutellaris* complex. However, species within this complex are all in the subgenus *Zeugodacus* which have long surstylus lobes in contrast to the short lobes of species in the subgenus *Parazeugodacus*. It differs from other species of *Parazeugodacus* in having lateral and medial postsutural yellow vittae, yellow spots anterior to the notopleural suture, apical half of

scutellum black, apices of all femora dark fuscous, a narrow dark fuscous costal band and dark fuscous over *dm-cu* crossvein.

Dacus (Zeugodacus) parvifoliaceus Tseng, Chen & Chu was separated from *B. arisanica* on minor differences in colour patterns which are not valid characters and thus we are making *B. parvifoliaceus* a **New Synonym** of *B. arisanica*.

Bactrocera (Parazeugodacus) bipustulata Bezzi

(Fig. 253)

Bactrocera bipustulata Bezzi, 1914b: 153. Holotype male in BMNH.
Bactrocera (Zeugodacus) bipustulata – Norrbom *et al.*, 1998: 102.

Type material examined: Holotype male, **INDIA**: with the following six labels – '*Bractocera* [Misspelling in handwriting] *bipustulata* Bezzi, Type'; '*Bactrocera bipunctulata* [Misspelling in handwriting] Bezzi, Prof M. Bezzi det.'; 'Mysore 4100 ft, Arabidacool Est, DAMAR, 29.iv.1913, P.S. Coll'; 'Pres by Imp. Bur. Ent, Brit. Mus, 1924-291'; 'Bactrocera bipustulata, n.sp. [male]'; Bezzi's handwriting 'Holotype [male] of Bactrocera (Chaetodacus) bipustulata Bezzi, 1914'. Studied at BMNH.
Non-type material examined: INDIA: 19 males, S. India, *Appangala*, Nr. Madikeri, 950m, 23.v.1992, coll. D.L. Hancock, attracted to cue lure. 2 males, S. India, *Talakaveri*, Nr. Madikeri, 1100m, 23.v.1992, coll. D.L. Hancock, attracted to cue lure. 1 male, S. India, *Bababudangiri*, Nr. Kemmannugundi, 1600m, 21.v.1992, coll. D.L. Hancock, attracted to cue lure. Specimens in QDPC.
Diagnosis: A small species; face fulvous on dorsal ½ and dark fuscous to black on ventral ½; postpronotal lobes yellow with black anterodorsal corners; notopleura yellow; scutum entirely black; very short and narrow lateral postsutural yellow vittae present; medial postsutural yellow vitta absent; no yellow spot anterior to notopleural suture; mesopleural stripe reaching midway between anterior margin of notopleuron and anterior *npl.* seta dorsally; scutellum yellow with a broad medial longitudinal black band; wing with cells bc and c colourless and entirely devoid of microtrichia; entirely colourless except for fuscous cell sc; abdominal terga III–V black except for small areas of dark red-brown posterocentrally on tergum IV either side of a medial longitudinal black band.
Description: male.
Head: Frons – red-brown with fuscous to dark fuscous around orbital setae and on anteromedial hump; orbital setae black: 1 *s.or.*, 2 *i.or.*; lunule dark fuscous. Ocellar triangle black. Vertex dark fuscous. Face with upper ½ fulvous and lower ½ dark fuscous tending to black in antennal grooves. Occiput dark fuscous to black, fulvous along eye margins. Antennae with segments 1 and 2 red-brown, segment 3 pale fuscous.
Thorax: Scutum entirely black without pale markings. Pleural areas entirely black without pale markings. Yellow markings as follows: postpronotal lobes (anterodorsal corners black); notopleura; mesopleural stripe reaching midway between anterior margin of notopleuron and anterior *npl.* seta dorsally, extending to katepisternum as a small transverse spot, anterior margin straight; two very short and narrow lateral postsutural vittae not reaching to level of *a.sa.* seta. Medial longitudinal postsutural yellow vitta absent. Scutellum with a broad medial longitudinal black band. Setae: *sc.* 4; *prsc.* 2; *ia.* 1; *p.sa.* 1; *a.sa.* 1; *mpl.* 1; *npl.* 2; *scp.* 4.
Legs: Fore femora fuscous, mid femora fulvous on basal ⅓ and fuscous on apical ⅔, hind femora fulvous on basal ⅔ and fuscous on apical ⅓ all tibiae fuscous with hind tibiae tending darker fuscous; tarsi with all segments entirely fulvous.
Wings: Length 4.8 mm; cells bc and c colourless and without microtrichia; remainder of wings entirely colourless except for fuscous cell sc; supernumerary lobe of medium development.
Abdomen: Oval; terga free; pecten present on tergum III. Tergum I and sterna I and II wider than long. Tergum I dark fuscous to black; tergum II dark fuscous to black with a transverse fulvous band across posterior margin reaching posterolateral corners; terga III–V black except for small red-brown areas posterocentrally on tergum IV either side of a medial black band. A pair of oval dark red-brown shining spots on tergum V. Posterior lobe of surstylus short, sternum V with a slight concavity on posterior margin. Abdominal sterna dark coloured.
Attractant: Cue lure.
Female: No known record.
Distribution:
Type locality: Southern India (Karnataka State, Mysore).
Previously recorded: Type locality and Sri Lanka.
New records: Southern India (other locations in Karnataka State listed under Non-type material examined).
Hosts: No known record.
Remarks: *Bactrocera (Parazeugodacus) bipustulata* Bezzi is similar to *Bactrocera (Parazeugodacus) abbreviata* (Hardy) in possessing an entirely black scutum, postpronotal lobes yellow with dark anterodorsal corners, yellow notopleura, very short lateral postsutural vittae, a broad medial black band on the scutellum, wings without a costal band and anal streak, abdominal terga III–V with broad medial and lateral longitudinal black bands joined over tergum III. It differs from *B. abbreviata* in having the lower half of the face dark fuscous to black and femora with extensive areas of fuscous.

Bactrocera (Parazeugodacus) fulvosterna Drew & Romig, new species

(Fig. 254)

Material examined: Holotype male, **EAST MALAYSIA**: Sarawak, Gunung Mulu National Park, 28–30.vii.1992, coll. D.L. Hancock, attracted to cue lure, deposited in BMNH. Paratypes: **EAST MALAYSIA**: 6 males, same data as holotype. 1 paratype in ANIC; 2 paratypes in QDPC; 3 paratypes in QMIC.

Diagnosis: A medium-sized species; face fulvous with very small oval dark fuscous spots; postpronotal lobes and notopleura yellow; scutum mostly black; lateral and medial postsutural yellow vittae present; mesopleural stripe of medium-width reaching midway between anterior margin of notopleuron and anterior *npl.* seta; scutellum yellow with a dark fuscous to black apical spot; wing with cells bc and c colourless, microtrichia in outer corner of cell c only; a narrow fuscous costal band and broad fuscous anal streak and infuscation along *dm-cu* crossvein; abdominal terga III–V fulvous with a narrow black transverse band across anterior margin of tergum III, dark anterolateral corners on terga IV and V, medium-width medial longitudinal dark band on terga IV and V that is interrupted at intersegmental lines; pecten present on tergum III.

Description: male.

Head: Height 1.37 mm. Frons – length 1.33 times breadth; fulvous with pale fuscous on anteromedial hump; bristles black: 1 *s.or.*, 3 *i.or.*; lunule red-brown. Ocellar triangle black. Vertex red-brown. Face fulvous with very small oval dark fuscous spots; length 0.54 mm. Genae fulvous with a small fuscous subocular spot. Occiput red-brown with fulvous along eye margins; occipital row with a large number of small pale setae. Antennae with segments 1 and 2 red-brown, segment 3 red-brown with fuscous on apex and outer surface; arista black (red-brown basally); length of segments: 0.07 mm; 0.24 mm; 0.58 mm.

Thorax: Scutum black with red-brown posterolateral to lateral postsutural vittae, around notopleural suture and yellow spot anterior to notopleural suture and centrally on anterior margin. Pleural areas shining black except red-brown below postpronotal lobe. Yellow markings as follows: postpronotal lobes; notopleura; mesopleural stripe of medium-width, reaching midway between anterior margin of notopleuron and anterior notopleural seta dorsally, continuing to katepisternum as a large transverse spot, anterior margin strongly convex; broad parallel-sided lateral postsutural vittae beginning in a spot anterior to notopleural suture and ending just posterior to *ia.* setae; a moderately broad medial longitudinal postsutural vitta rounded posteriorly at level of *prsc.* setae and narrowing to a point just anterior to level of notopleural suture; anatergite (posterior apex black); anterior ⅔ katatergite (remainder black). Postnotum black. Scutellum yellow with a narrow black basal band and a dark fuscous to black apical spot. Setae: *sc.* 4; *prsc.* 2; *ia.* 1; *p.sa.* 1; *a.sa.* 1; *npl.* 2; *scp.* 4.

Legs: Femora entirely fulvous; fore tibiae pale fuscous, mid tibiae pale fuscous basally and fulvous apically, hind tibiae dark fuscous; tarsal segments with basal segment fulvous and apical four segments fuscous; mid tibiae each with an apical black spur.

Wings: Length 5.52 mm; cells bc and c colourless; microtrichia in outer corner of cell c only; remainder of wings colourless except fuscous cell sc, narrow fuscous costal band confluent with R_{2+3} and expanding into a small spot around apex of R_{4+5}, fuscous around *dm-cu* crossvein that extends to hind margin of wing, broad fuscous anal streak; dense aggregation of microtrichia around A_1+CuA_2; supernumerary lobe weak.

Abdomen: Oval; terga free; pecten present on tergum III. Tergum I and sterna I and II wider than long. Tergum I orange-brown with dark fuscous on lateral margins. Tergum II orange-brown, fulvous posteriorly and with small fuscous spots just posterior to anterior margin; terga III–V fulvous with a narrow black transverse band across anterior margin of tergum III, broad black anterolateral corners of tergum IV, dark fuscous anterolateral corners of tergum V, a medium-width black medial longitudinal band over terga IV and V that is interrupted at intersegmental lines. A pair of fulvous oval shining spots on tergum V. Abdominal sterna entirely fulvous. Posterior lobe of surstylus short, sternum V with a slight concavity on posterior margin.

Attractant: Cue lure.

Female: No known record.

Distribution: East Malaysia (Sarawak).

Hosts: No known record.

Remarks: *Bactrocera (Parazeugodacus) fulvosterna* n.sp. is similar to some species of subgenus *Zeugodacus* but differs in having short posterior surstylus lobes. It is similar to *Bactrocera (Parazeugodacus) matsumurai* (Shiraki) in the colour patterns of the abdominal terga but differs from this species in having broad parallel-sided lateral postsutural vittae, a large medial longitudinal postsutural yellow vitta, an entirely yellow scutellum and wings with *dm-cu* crossvein infuscated.

Bactrocera (Parazeugodacus) matsumurai (Shiraki)

(Fig. 255)

Parazeugodacus matsumurai Shiraki, 1933: 107. Syntypes male and female in NIAS.

Dacus boninensis Hardy & Adachi, 1956: 12; Norrbom *et al.*, 1998:100 (Synonym).

Bactrocera (Parazeugodacus) matsumurai – Norrbom *et al.*, 1998: 100.

Type material examined: Syntype male, **JAPAN**: with the following three labels – 'Ogasawara, 20.viii.1905, Col. T. Shiraki'; 'Zeugodacus matsumurai, Det. T. Shiraki'; 'Type'; Syntype female with same labels as male. Received on loan from NIAS.

Non-type material examined: JAPAN: 2 males and 2 females, Chichi Jima, Ogasawara Islands, Tokyo Pref., coll. M. Dozono, bred ex Litsea japonica Juss; 1 female, 'Ogasawara Is., 2.v.1981, Dacus matsumuri'. Specimens in QDPC.

Diagnosis: A medium-sized species; face fulvous with a pair of small circular black spots; postpronotal lobes and notopleura yellow; scutum almost entirely black; lateral postsutural yellow vittae present; medial postsutural yellow vitta absent; no yellow spot anterior to notopleural suture; mesopleural stripe equal in width to notopleuron dorsally; scutellum yellow with a broad medial longitudinal black band over dorsal surface; wing with cells bc and c colourless and devoid of microtrichia; a narrow pale fuscous to fuscous costal band confluent with R_{2+3} and widening slightly around apex of wing; anal streak absent; abdominal terga III–V orange-brown with a narrow transverse black band across anterior margin of tergum III, narrow transverse black spots anterolaterally on tergum IV, a medial longitudinal black spot anteriorly on tergum IV (the female has dark fuscous anterolateral corners on tergum V and a medial longitudinal black spot anteriorly on tergum V).

Description: male.

Head: Height 1.68 mm. Frons – length 1.5 times breadth; red-brown with fuscous around orbital setae and on antero-medial hump; orbital setae black: 1 *s.or.*, 2 *i.or.*; lunule pale fuscous. Ocellar triangle black. Vertex fuscous. Face fulvous with a pair of small circular black spots; length 0.52 mm. Genae fulvous, pale fuscous subocular spot present; black seta present. Occiput dark fuscous centrally, black laterally, fulvous along eye margins; occipital row with 3–5 strong black setae. Antennae with segments 1 and 2 red-brown, segment 3 red-brown with pale fuscous on apex and outer surface; length of segments: 0.16 mm; 0.34 mm; 0.78 mm.

Thorax: Scutum entirely black except for a small area of red-brown anterior to notopleural suture. Pleural areas black except dark fuscous below postpronotal lobes. Yellow markings as follows: postpronotal lobes; notopleura; mesopleural stripe equal in width to notopleuron dorsally, continuing to katepisternum as a small transverse spot, anterior margin straight; anatergite (posterior apex black); anterior ⅔ katatergite (remainder black); two very short lateral postsutural vittae beginning at notopleural suture and ending just posterior to level of *a.sa.* seta. Medial longitudinal postsutural yellow vitta absent. Postnotum black. Scutellum yellow except for a broad medial longitudinal black band over dorsal surface. Setae: *sc.* 4; *prsc.* 2; *ia.* 1; *p.sa.* 1; *a.sa.* 1; *mpl.* 1; *npl.* 2; *scp.* 4.

Legs: All femora entirely fulvous except for a small fuscous spot on dorsal apical surfaces of hind femora; fore tibiae pale fuscous, mid tibiae pale fuscous basally to fulvous apically, hind tibiae dark fuscous; tarsi with all segments entirely fulvous; mid tibiae each with an apical black spur.

Wings: Length 6.1 mm; cells bc and c colourless; both cells devoid of microtrichia; remainder of wings colourless with areas of pale fuscous tint except for a fuscous cell sc, narrow pale fuscous costal band confluent with R_{2+3} and widening slightly as it becomes fuscous around apex of wing; anal streak absent; dense aggregation of microtrichia around A_1+CuA_2; supernumerary lobe weak.

Abdomen: Oval; terga free; pecten present on tergum III. Tergum I and sterna I and II wider than long. Tergum I entirely red-brown; tergum II red-brown with areas of fulvous posteriorly and a narrow black transverse band across anterior margin but not reaching lateral margins; terga III–V red-brown except for a narrow transverse black band across anterior margin of tergum III but not reaching lateral margins, narrow transverse spots on anterolateral margins of tergum IV and a moderately broad medial longitudinal black spot on anterior margin of tergum IV. A pair of oval red-brown shining spots on tergum V. Posterior lobe of surstylus short, sternum V with a slight concavity on posterior margin.

Attractant: No known record.

Female: As for male except mesopleural stripe slightly wider; no dense aggregation of microtrichia around A_1+CuA_2; pecten of cilia on abdominal tergum III absent; ovipositor basal segment red-brown, conical in transverse cross section and tapering posteriorly in dorsal view; ratio of length of oviscape to length of tergum V, 0.62:1; apex of piercer needle-shaped; tergum V with dark fuscous anterolateral corners and a medial longitudinal black spot on anterior margin.

Distribution:

Type locality: Japan (Bonin Is – Ogasawara Is).

Previously recorded: Type locality and Japan (Bonin Is – Chichi Jima & Haha Jima) due to synonymy of *D. boninensis* in Norrbom *et al.* (1998).

Hosts: *Osmanthus insularis* (family *Oleaceae*) (Ito, 1983; Allwood *et al.*, 1999). The record of *Osmanthus insularis* (Allwood *et al.*, 1999) is probably incorrect as this is a host of *B. (Bactrocera) hyalina* (Shiraki), a fly species that looks similar to *B. matsumurai* in external colour patterns.

New record: *Litsea japonica* Juss.

Remarks: *Bactrocera (Parazeugodacus) matsumurai* (Shiraki) is a unique species in possessing the *Bactrocera (Zeugodacus)* chaetotaxy, short posterior lobes on the surstyli and slight concavity on the posterior margin of the male abdominal sternum V. It is readily identified by the very short lateral postsutural vittae, the broad medial longitudinal band on the dorsal surface of the scutellum and the absence of a fuscous anal streak in the wing. It is similar to *Bactrocera (Parazeugodacus) abbreviata* (Hardy) and *Bactrocera (Parazeugodacus) bipustulata* (Bezzi) in possessing a medial longitudinal black band on the scutellum and in lacking a medial longitudinal postsutural yellow vitta on the scutum but differs from these species in having a continuous costal band around the fore margin of the wing.

Bactrocera (Parazeugodacus) pendleburyi (Perkins)

(Figs 256A, 256B)

Zeugodacus pendleburyi Perkins, 1938a: 141. Holotype female in QMIC.

Dacus (Zeugodacus) pendleburyi – Hardy & Adachi, 1954: 194.

Bactrocera (Zeugodacus) pendleburyi – Norrbom *et al.*, 1998: 103.

Type material examined: Holotype female, **PENINSULAR MALAYSIA**: with the following two labels – 'Malay Penin: Selangor, Bukit Kutu, 1500 ft., 14.ix.1929, H.M. Pendlebury'; 'Holotype, Zeugodacus pendleburyi, F.A. Perk'. Received on loan from QMIC.

Paratypes: **PENINSULAR MALAYSIA**: 1 female, 'Malay Penin: Selangor, Bukit Kutu, 3500 ft., 7.ix.1929, H.M. Pendlebury'; 1 female, 'Perak F.M.S., Latut Hills, 3700–4000 ft., 9.ii.1932, H.M. Pendlebury'; both paratypes with the label – 'Paratype, Zeugodacus pendleburyi F.A. Perk'. Received on loan from QMIC.

Non-type material examined: PENINSULAR MALAYSIA: 2 males 'Malaysia, 9 Aug.1986, Allwood, Vijay, Collins'; 'Old Bentong Road, 29km after Genting – Krak

Junction'; 'bred out of *Gmelina arborea*'; 'BMNH(E) No. 533007'. Studied at BMNH. 4 males and 5 females, 'Malaysia, 18.viii.1986, coll. Allwood *et al.*'; 'Kg Belimbing, Pekan Pahang'; 'Bred out of *Symplocos cochinchinensis*'. **THAILAND**: 1 female, Khun Khon, Nat. Park, Chiang Rai, 16.vii.1992, coll. Charuwan, bred from *Symplocos racemosa*. Specimens in QDPC.

Diagnosis: A medium-sized species; face fulvous with a pair of large oval black spots; postpronotal lobes and notopleura yellow; scutum entirely black; short tapering lateral postsutural yellow vittae present; medial postsutural yellow vitta absent; no yellow spot anterior to notopleural suture; mesopleural stripe equal in width to notopleuron dorsally; scutellum yellow with a moderately broad black basal band; wing with cells bc and c colourless, microtrichia in outer corner of cell c only; a narrow fuscous costal band confluent with R_{2+3}; anal streak absent; abdominal terga III–V red-brown with a broad medial and two broad lateral longitudinal black bands joined over tergum III which is almost entirely black.

Description: female (Fig. 256A).

Head: Height 1.36 mm. Frons – length 1.5 times breadth; fuscous with fulvous around lateral and ventral margins and dark fuscous around orbital setae; orbital setae black: 1 *s.or.*, 2 *i.or.*; lunule dark fuscous. Ocellar triangle black. Vertex fuscous. Face fulvous with a pair of large oval black spots; length 0.43 mm. Genae fulvous, fuscous subocular spot present; black seta present. Occiput black, fulvous along eye margins; occipital row with 4–6 black setae. Antennae with segments 1 and 2 fuscous, segment 3 red-brown with dark fuscous on apex and outer surface; length of segments: 0.13 mm; 0.35 mm; 0.68 mm.

Thorax: Scutum entirely black without pale markings. Pleural areas entirely black without pale markings. Yellow markings as follows: postpronotal lobes; notopleura; mesopleural stripe equal in width to notopleuron dorsally, continuing to katepisternum as a small transverse spot, anterior margin straight; anatergite (posterior apex black); anterior $^3/_5$ katatergite (remainder black); two medium-width very short lateral postsutural vittae narrowing posteriorly to end at or just posterior to level of *a.sa.* seta. Medial longitudinal postsutural yellow vitta absent. Postnotum black. Scutellum yellow except for moderately broad black basal band. Setae: *sc.* 4; *prsc.* 2; *ia.* 1; *p.sa.* 1; *a.sa.* 1; *mpl.* 1; *npl.* 2; *scp.* 4.

Legs: All femora fulvous with small areas of dark fuscous around apices in Holotype (in the paratypes these areas of dark fuscous are considerably larger); fore and hind tibiae dark fuscous, mid tibiae dark fuscous basally tending fuscous apically; fore and hind tarsi with basal two segments fulvous and apical three segments red-brown, mid tarsi with all segments entirely fulvous; mid tibiae each with an apical black spur.

Wings: Length 5.3 mm; cells bc and c colourless; microtrichia in outer corner of cell c only; remainder of wings colourless except fuscous cell sc, narrow fuscous costal band confluent with R_{2+3} and becoming very narrow apically to end just beyond apex of R_{4+5}; anal streak absent but with a pale tint within base of cell cup; no dense aggregation of microtrichia around A_1+CuA_2; supernumerary lobe weak.

Abdomen: Oval; terga free; no pecten on tergum III. Tergum I black; tergum II fulvous except for narrow dark red-brown bands anterolaterally along intersegmental line between terga I and II and a broad transverse black band anteriorly which broadens laterally to cover lateral margins; terga III–V red-brown with a broad medial and two broad lateral longitudinal black bands which join over tergum III (tergum III being almost entirely black). A pair of oval red-brown shining spots on tergum V. Basal segment of ovipositor red-brown, dorsoventrally compressed and tapering posteriorly in dorsal view; ratio of length of oviscape to length of tergum V, 1:1; apex of piercer needle-shaped. Abdominal sterna dark.

Attractant: No known record.

Male: (Fig. 256B) – As for female except dense aggregation of microtrichia around A_1+CuA_2 present; pecten of cilia on tergum III present; posterior lobe of surstylus short; posterior margin of abdominal sternum V with a slight concavity (Fig. 256B).

Distribution:

Type locality: Peninsular Malaysia (Selangor, Bukit Kutu).

Previously recorded: Type locality and Peninsular Malaysia, Perak.

New records: Thailand (Chiang Rai).

Hosts: Recorded from two plant families – *Symplocaceae* and *Verbenaceae* (see Allwood *et al.*, 1999).

Remarks: *Bactrocera (Parazeugodacus) pendleburyi* (Perkins) is similar to *Bactrocera (Parazeugodacus) abbreviata* (Hardy), *Bactrocera (Parazeugodacus) bipustulata* Bezzi and *Bactrocera (Parazeugodacus) matsumurai* (Shiraki) in having a black scutum with short lateral postsutural yellow vittae, four *sc.* setae, *prsc.* setae, *a.sa.* setae, and in the medial postsutural yellow vitta and anal streak being absent. It differs from these species in having an entirely yellow scutellum (not with a medial longitudinal black band), from *B. abbreviata* and *B. bipustulata* in possessing a complete costal band in the wing and from *B. matsumurai* in having abdominal terga III–V with broad medial and lateral longitudinal black bands over all three terga and joined across tergum III.

Bactrocera (Parazeugodacus) tetrachaeta (Bezzi)

(Fig. 257)

Chaetodacus tetrachaetus Bezzi, 1919: 431. Lectotype male in MCSNM.

Dacus (Zeugodacus) tetrachaetus – Hardy & Adachi, 1954: 196; Hardy, 1969b: 479, Lectotype designation; Hardy, 1974: 57.

Bactrocera (Zeugodacus) tetrachaeta – Norrbom *et al.*, 1998: 104.

Dacus (Zeugodacus) deceptus Hardy, 1974: 46. **New Synonym**.

Type material examined: Lectotype male, **PHILIPPINES**: with the following three labels – 'Batbatan Isl., R.C. McGregor'; 'tetrachaetus n.' (Bezzi's handwriting); 'Lectotype male, Chaetodacus tetrachaetus Bez., desig. by Hardy – 68'. 1 male Paralectotype, 'Batbatan Isl, R.C. McGregor'. Received on loan from MCSNM.

Diagnosis: A medium-sized species; face fulvous with a pair of large elongate oval black spots; postpronotal lobes and notopleura yellow and joined by a lateral yellow band;

scutum black with red-brown markings; lateral and medial postsutural yellow vittae present; no yellow spot anterior to notopleural suture; mesopleural stripe reaching to anterior *npl.* seta dorsally; scutellum yellow with a narrow black basal band; wing with cells bc and c colourless, microtrichia in outer corner of cell c only; a narrow fuscous costal band confluent with R_{2+3} and remaining narrow around apex of wing; anal streak absent; abdominal terga III–V orange-brown with a broad medial and broad to narrow lateral longitudinal black bands joined along anterior margin of tergum III.

Description: male.

Head: Height 1.61 mm. Frons – length 1.24 times breadth; red-brown with dark fuscous on anteromedial hump; orbital setae black: 1 *s.or.*, 3 *i.or.*; lunule dark fuscous. Ocellar triangle black. Vertex red-brown. Face fulvous with a pair of large elongate oval black spots; length 0.62 mm. Genae fulvous, fuscous subocular spot present; black seta present. Occiput red-brown, fulvous along eye margin; occipital row with 2–4 setae. Antennae with segments 1 and 2 red-brown, segment 3 red-brown with fuscous on apex and outer surface; length of segments: 0.13 mm; 0.3 mm; 0.8 mm.

Thorax: Scutum black with red-brown below and behind lateral postsutural vittae, around notopleural suture, around posterior end of medial postsutural yellow vitta, as a narrow medial longitudinal band running anteriorly from anterior point of medial longitudinal yellow vitta, across anterior margin of scutum. Pleural areas black except red-brown below postpronotal lobes. Yellow markings as follows: postpronotal lobes; notopleura; yellow band connecting postpronotal lobe and notopleural suture; mesopleural stripe reaching anterior *npl.* seta dorsally, continuing to katepisternum as a large transverse spot, anterior margin convex; anatergite (posterior apex black); anterior $^3/_5$ katatergite (remainder black); two medium-width lateral postsutural vittae narrowing over posterior ½ to end at *ia.* seta; a medium-width medial postsutural vitta broadly rounded posteriorly at level of *prsc.* setae and narrowing to a point approximately level with notopleural suture. Postnotum black. Scutellum yellow except for narrow black basal band. Setae: *sc.* 4; *prsc.* 2; *ia.* 1; *p.sa.* 1; *a.sa.* 1; *mpl.* 1; *npl.* 2; *scp.* 4.

Legs: All femora fulvous; fore tibiae fuscous, mid tibiae fuscous basally to pale fuscous apically, hind tibiae dark fuscous; all tarsi with basal segment fulvous and apical four segments fuscous; mid tibiae each with an apical black spur.

Wings: Length 6.4 mm; cells bc and c colourless; microtrichia in outer corner of cell c only; remainder of wings colourless except narrow fuscous costal band confluent with R_{2+3} and remaining narrow to end just beyond apex of R_{4+5}; anal streak absent; dense aggregation of microtrichia around A_1+CuA_2; supernumerary lobe weak.

Abdomen: Elongate oval; terga free; pecten present on tergum III. Tergum I and sterna I and II wider than long. Tergum I orange-brown with narrow black lateral margins; tergum II orange-brown with a narrow transverse black band anteriorly and reaching narrow anterolateral black corners; terga III–V orange-brown with a broad medial longitudinal black band and two lateral longitudinal black bands which are broad on tergum III and narrowing towards posterior margin of tergum V and are joined along anterior margin of tergum III. A pair of oval orange-brown shining spots on tergum V. Posterior lobe of surstylus short, sternum V with a slight concavity on posterior margin.

Attractant: No known record.

Female: No known record.

Distribution:

Type locality: Philippines (Batbatan I.).

New records: Philippines (Mindanao I.) due to **New Synonymy** of *D. deceptus*.

Hosts: No known record.

Remarks: *Bactrocera (Parazeugodacus) tetrachaeta* (Bezzi) is similar to *Bactrocera (Parazeugodacus) fulvosterna* n.sp. in possessing broad lateral and medial longitudinal yellow vittae, a yellow scutellum and abdominal terga III–V orange-brown with black patterns. It differs from *B. fulvosterna* in having lateral yellow bands joining the postpronotal lobe and notopleural suture, broad medial and lateral longitudinal black bands over abdominal terga III–V which are joined along anterior margin of tergum III and in the wing lacking an anal streak and infuscation on the *dm-cu* crossvein.

We have found no sound characters that can separate *B. tetrachaeta* and *Dacus (Zeugodacus) deceptus* Hardy and are therefore making the latter a **New Synonym** of *B. tetrachaeta*.

SUBGENUS *SEMICALLANTRA* Drew

Bactrocera (Semicallantra) incompta Drew & Romig, new species

(Fig. 258)

Material examined: Holotype male, **INDONESIA: MOLUCCAS (MALUKU): AMBON ISLAND** – Waai, Ambon, Indonesia, <20 m, 6.vii.1981, coll. J. Turner, attracted to cue lure, residual forest with clove, nutmeg & banana groves, deposited in BMNH.

Diagnosis: A medium-sized species; elongate antennae as in species of *Dacus (Callantra)* but with free abdominal terga; face red-brown with a pair of large elongate black spots in antennal furrows; postpronotal lobes and notopleura dark fuscous; scutum dull black; lateral and medial postsutural yellow vittae absent; no yellow spot anterior to notopleural suture; mesopleural stripe reaching midway between anterior margin of notopleuron and anterior *npl.* seta dorsally; scutellum yellow; wing with cells bc and c pale fuscous, microtrichia covering all of cell c; a narrow pale fuscous costal band; a moderately broad pale fuscous anal streak; abdominal terga entirely red-brown with fuscous lateral margins on tergum III.

Description: male.

Head: Height 1.7 mm. Frons – length 1.3 times breadth; red-brown with fuscous on anteromedial hump; orbital setae absent; lunule fuscous. Ocellar triangle black. Vertex red-brown. Face red-brown with a pair of large elongate black spots in antennal furrows; length 0.6 mm. Genae red-brown, pale seta present. Occiput red-brown, fulvous along eye margins; occipital row with 4–6 black setae. Antennae with segments 1 and 2 red-brown, segment 3 fuscous; length of segments: 0.34 mm; 0.39 mm; 0.93 mm.

Thorax: Scutum entirely dull black with dark fuscous across posterior margin. Pleural areas entirely dark fuscous. Yellow markings as follows: mesopleural stripe reaching midway between anterior margin of notopleuron and anterior *npl.* seta dorsally, not continuing to katepisternum, anterior margin straight; anatergite (posterior apex dark fuscous), anterior ¾ katatergite (remainder dark fuscous). Postpronotal lobes and notopleura dark fuscous. Lateral and medial longitudinal postsutural yellow vittae absent. Postnotum dark fuscous. Scutellum yellow except for narrow black basal band. Setae: *sc.* 2; *prsc.* 2; *ia.* 1; *p.sa.* 1; *a.sa.* absent; *mpl.* 1; *npl.* 2; *scp.* 4.

Legs: All femora entirely red-brown; fore tibiae fuscous basally to red-brown apically, mid and hind tibiae entirely fuscous; all tarsi entirely fulvous; mid tibiae each with an apical black spur.

Wings: Length 6.0 mm; cells bc and c pale fuscous; dense microtrichia covering all of cell c; remainder of wings colourless except dark fuscous cell sc, narrow pale fuscous costal band remaining very narrow around apex of wing and overlapping R_{2+3} as a very pale tint, medium-width pale fuscous anal streak; a dense aggregation of microtrichia around A_1+CuA_2; supernumerary lobe weak.

Abdomen: Elongate oval and petiolate; terga free; pecten present on tergum III. Tergum I and sterna I and II longer than wide. All abdominal terga entirely red-brown except for fuscous lateral margins on tergum III. A pair of oval red-brown shining spots on tergum V. Posterior lobe of surstylus long, sternum V with a deep concavity on posterior margin. Abdominal sterna red-brown.

Attractant: Cue lure.

Female: No known record.

Distribution: Indonesia (Moluccas = Maluku – Ambon I., Waai).

Hosts: No known record.

Remarks: *Bactrocera (Semicallantra) incompta* n.sp. is one of only two species of subgenus *Semicallantra* that have been discovered in South-East Asia. It is distinguishable in having a black scutum without yellow markings, postpronotal lobes and notopleura dark fuscous, wings colourless except for costal band and anal streak and all abdominal terga mostly or entirely orange-brown.

Bactrocera (Semicallantra) superba Drew & Romig, new species

(Fig. 259)

Material examined: Holotype male, **INDONESIA: NORTH MOLUCCAS (NORTH MALUKU)** – Moluccas (Maluku), North Maluku, West Halmahera Gaol, 5.vi.2007, coll. Raais Abdullah, attracted to cue lure in forest, Site G31b2, deposited in BMNH.

Paratypes: **INDONESIA: NORTH MOLUCCAS (NORTH MALUKU)** – 4 males, same data as holotype. 1 paratype in BMNH, 2 paratypes in QDPC, 1 paratype in QMIC.

Diagnosis: A large species; face dark red-brown with a pair of very large oval black spots in antennal furrows and a narrow black line running from the top of each spot to the base of each antenna; postpronotal lobes dark fuscous; notopleura yellow; scutum entirely black; lateral postsutural yellow vittae present; medial postsutural yellow vitta absent; no yellow spot anterior to notopleural suture; mesopleural stripe reaching midway between anterior margin of notopleuron and anterior *npl.* seta dorsally; scutellum yellow; wing with cells bc and c dark fuscous, dense microtrichia covering all of cell c and outer corner of cell bc; almost entirely dark fuscous with clear areas in cell bm, cell dm, a narrow line running from base of cell dm to hind margin of wing, the entire anal lobes; abdominal terga III–V orange-brown with fuscous to dark fuscous broad lateral longitudinal bands over all three terga and joined on the anterior midline of tergum III.

Description: male.

Head: Height 2.2 mm. Frons – length 1.35 times breadth; fuscous with dark fuscous on anteromedial hump; orbital setae black: 1 *s.or.*, 2 *i.or.*; lunule fuscous. Ocellar triangle black. Vertex dark fuscous. Face dark red-brown with a pair of very large oval black spots in antennal furrows and a narrow black line running vertically from the top of each black spot to the base of each antenna; length 0.72 mm. Genae dark red-brown, large fuscous subocular spot present; black seta present. Occiput dark fuscous, red-brown along eye margins; occipital row with 5–7 strong black setae. Antennae with segments 1 and 2 dark red-brown, segment 3 dark red-brown with fuscous on apex and outer surface; length of segments: 0.45 mm; 0.6 mm; 1.17 mm.

Thorax: Scutum entirely dull black without pale markings. Pleural areas entirely dark fuscous to black. Yellow markings as follows: notopleura; mesopleural stripe reaching midway between anterior margin of notopleuron and anterior *npl.* seta dorsally, not continuing to katepisternum, anterior margin straight; anatergite (posterior apex dark red-brown); anterior $^3/_5$ katatergite (remainder dark red-brown); two moderately broad parallel-sided lateral postsutural vittae ending at or just before *ia.* seta. Medial longitudinal postsutural yellow vitta absent. Postpronotal lobes dark fuscous. Postnotum dark fuscous. Scutellum yellow except for narrow black basal band. Setae: *sc.* 2; *prsc.* 2; *ia.* 1; *p.sa.* 1; *a.sa.* 1; *mpl.* 1; *npl.* 2; *scp.* 4. Occasionally one or both *prsc.* setae may be absent.

Legs: All femora entirely dark fuscous except fuscous at base and apex of fore femora; all tibiae fuscous; fore and hind tarsi with basal segment red-brown and apical four segments fuscous, mid tarsi with all segments fuscous; mid tibiae each with an apical black spur.

Wings: Length 7.8 mm; cells bc and c dark fuscous; dense microtrichia covering all of cell c and outer corner of cell bc; remainder of wings entirely covered with dark fuscous except for clear areas as follows – cell bm, anterior of cell dm, a line running from base of cell dm to hind margin of wing and anal lobes; a dense aggregation of microtrichia around A_1+CuA_2; supernumerary lobe weak.

Abdomen: Elongate oval and slightly petiolate; terga free; pecten present on tergum III. Tergum I and sterna I and II longer than wide. Terga I and II entirely dark fuscous; terga III–V

orange-brown with broad fuscous to dark fuscous lateral longitudinal bands over all three terga and joined on the anterior midline of tergum III. A pair of oval fuscous to black shining spots on tergum V. Posterior lobe of surstylus short, sternum V with a slight concavity on posterior margin. Abdominal sterna dark red-brown.
Attractant: Cue lure.
Female: No known record.
Distribution: Indonesia (North Moluccas = North Maluku – Halmahera I.).
Hosts: No known record.
Remarks: *Bactrocera (Semicallantra) superba* n.sp. is here placed in the subgenus *Semicallantra* in possessing free (non-fused) abdominal terga, elongate antennal segments (*Callantra*-like) and the posterior lobe of the male surstyli short. It is distinct from other species in this subgenus, namely *Bactrocera (Semicallantra) aquila* (Drew) and *Bactrocera (Semicallantra) nigricula* (Drew) in having fuscous postpronotal lobes, abdominal terga fuscous to dark fuscous with orange-brown medially from anterior margin of tergum III to posterior margin of tergum V.

SUBGENUS *SINODACUS* Zia

Bactrocera (Sinodacus) bogorensis (Hardy)

(Fig. 260)

Dacus (Zeugodacus) bogorensis Hardy, 1983: 34. Holotype male in MZB.
Dacus (Zeugodacus) ascitus Hardy, 1983: 32. **New Synonym**.
Bactrocera (Zeugodacus) bogorensis – Norrbom *et al.*, 1998: 102.

Type material examined: Holotype male of *Dacus bogorensis*, **INDONESIA: JAVA** – with the following label – 'Kebun Raya (botanical gardens), Bogor, West Java, Apr.1975, coll. D.E. Hardy, attracted to cue lure'. Studied at MZB.
Paratype: **INDONESIA: JAVA** – 1 male paratype of *Dacus bogorensis*, with following two labels – 'Cimanggu, Bogor, Java, 200 m, Dec.76, S. Isnadi'; '*Dacus (Zeugodacus) bogorensis* Hardy' (this specimen does not have a paratype label but fits the collection data for the paratypes listed by Hardy (1983)). Received on loan from BBM.
Other type material examined: Holotype male of *Dacus ascitus*, **INDONESIA: SUMATRA** – Lembak Anai, 63 km north Padang, W. Sumatra, 9.v.1975, coll. D.E. Hardy, attracted to cue lure. Studied at MZB.
Non-type material examined: INDONESIA: NORTH SULAWESI – 1 male, Dumoga, Taraut, 22.xii.1985, coll. L. Santianawati; **SUMATRA** – 1 male, Sumatera Barat, Tanah Datar, Sawah Tangah, 13.ix.2006, coll. Ir Janimar, attracted to cue lure in village. Specimens in QDPC.
Diagnosis: A medium-sized species; face black; postpronotal lobes and notopleura yellow; scutum black; lateral and medial postsutural yellow vittae present; a yellow spot anterior to notopleural suture in front of lateral postsutural vittae; mesopleural stripe reaching midway between anterior margin of notopleuron and anterior *npl.* seta dorsally; scutellum yellow; wing with cells bc and c colourless, microtrichia in outer corner of cell c only; a narrow dark fuscous costal band confluent with R_{2+3} and widening across apex of R_{4+5} and broad dark fuscous anal streak; abdominal terga III–V red-brown with a broad very dark band across anterior margin of tergum III and a narrow medial longitudinal and two broad lateral longitudinal very dark fuscous to black bands over all three terga.
Description: male.
Head: Height 1.34 mm. Frons – length 1.16 times breadth; red-brown with pale fuscous around *s.or.* setae and on anteromedial hump; orbital setae black: 1 *s.or.*, 2 *i.or.*; lunule red-brown. Ocellar triangle black. Vertex black. Face entirely black; length 0.72 mm. Genae fulvous, large fuscous to dark fuscous subocular spot present; black seta present. Occiput dark fuscous with dorsolateral black spots and fulvous along eye margins; occipital row with a large number of short pale setae. Antennae with all segments fuscous to dark fuscous; length of segments: 0.13 mm; 0.26 mm; 0.6 mm.
Thorax: Scutum entirely black with, at most, dark fuscous behind lateral postsutural vittae. Pleural areas entirely black. Yellow markings as follows: postpronotal lobes; notopleura; mesopleural stripe of medium-width reaching midway between anterior margin of notopleuron and anterior *npl.* seta dorsally, continuing to katepisternum as a transverse spot, anterior margin slightly convex; anatergite (posterior apex black); anterior ⅔ katatergite (remainder black); two moderately broad sub-parallel lateral postsutural vittae beginning with a spot anterior to notopleural suture and ending at *ia.* seta; a broad medial longitudinal postsutural vitta broadly rounded posteriorly at level of *prsc.* setae and ending at a point anteriorly at level of notopleural suture. Postnotum black. Scutellum yellow except for narrow black basal band. Setae: *sc.* 2; *prsc.* 2 (absent in Holotype); *ia.* 1; *p.sa.* 1; *a.sa.* 1; *mpl.* 1; *npl.* 2; *scp.* 4.
Legs: Fore femora with basal ¼ fulvous and remainder dark fuscous to black, mid femora with basal $^2/_5$ fulvous and remainder dark fuscous to black, hind femora with basal $^3/_5$ fulvous and remainder dark fuscous to black; fore tibiae fulvous except light fuscous basally, mid tibiae fulvous with pale fuscous basally, hind tibiae fuscous basally tending pale fuscous apically; tarsal segments entirely fulvous; mid tibiae each with an apical black spur.
Wings: Length 5.78 mm; cells bc and c colourless; microtrichia in outer corner of cell c only; remainder of wings colourless except dark fuscous cell sc, narrow dark fuscous costal band confluent with R_{2+3} and widening slightly across apex of R_{4+5}, broad dark fuscous anal streak; dense aggregation of microtrichia around A_1+CuA_2; supernumerary lobe well developed.
Abdomen: Oval; terga free; pecten present on tergum III. Tergum I and sterna I and II wider than long. Tergum I fulvous with dark fuscous across anterior margin and along lateral margins; tergum II fulvous with dark fuscous to black band across anterior margin; terga III–V red-brown with a broad very dark fuscous band across anterior margin of tergum III, a narrow medial longitudinal and two broad lateral longitudinal very dark fuscous to black bands over all three terga. A pair of oval red-brown shining spots on tergum V.

Posterior lobe of surstylus long, sternum V with a slight concavity on posterior margin. Abdominal sterna dark.
Attractant: Cue lure.
Female: No known record.
Distribution:
Type locality: Indonesia (Java, Bogor).
New records: Indonesia (Sulawesi – North Sulawesi, Dumoga); (Sumatra – West Sumatra, Tanah Datar). Sumatra (West Sumatra – Lembak Anai, north of Padang), due to **New Synonymy** of *D. ascitus* listed under 'Other type material examined'.
Hosts: No known record.
Remarks: The Holotype of *Bactrocera (Sinodacus) bogorensis* (Hardy) fits the subgenus *Sinodacus* Zia in having two *sc.* setae, *prsc.* setae absent, posterior lobe of surstylus long, a weak concavity on posterior margin of abdominal sternum V of male and a medial postsutural yellow vitta present. The Paratype that we studied has the *prsc.* setae present. We prefer to place this species in *Sinodacus* at present, pending more specimens becoming available for study. It is similar to *Bactrocera (Zeugodacus) hengsawadae* n.sp. in general body and wing colour patterns but differs in having the face entirely black, mesopleural stripe broader and reaching midway between anterior margin of notopleuron and anterior *npl.* seta dorsally, all femora with apices surrounded by large areas of dark fuscous to black and male supernumerary lobe strong and keel-shaped.

Bactrocera (Sinodacus) longivittata Chua & Ooi is similar in possessing a black scutum with lateral and medial postsutural yellow vittae, a yellow spot anterior to the notopleural suture, and abdominal terga III–V with a general black 'T' pattern, but differs in having a yellow band connecting the postpronotal lobe and notopleuron, a broad costal band overlapping R_{4+5} and a very long medial postsutural yellow vitta and in lacking broad dark patterns on abdominal terga III–V.

In comparing the Holotypes of *Dacus (Zeugodacus) ascita* (Hardy) and *Dacus (Zeugodacus) bogorensis* (Hardy) the facial colour patterns are identical and there are no other characters useful in separating them. Although *D. ascita* has page priority over *D. bogorensis*, it was described from one specimen (the Holotype) which is mouldy and damaged whereas *D. bogorensis* was described from a Holotype and two Paratypes that are in good condition. Consequently, *D. ascita* is here declared a **New Synonym** of *D. bogorensis*.

Bactrocera (Sinodacus) hochii (Zia)

(Fig. 261)

Sinodacus hochii Zia, 1936: 159. Holotype male in IZAS.
Dacus (Strumeta) hochii – Hardy & Adachi, 1954: 171.
Sinodacus hainanus Chao & Lin, 1996: 126; Norrbom *et al.*, 1998: 100 (Synonym).
Sinodacus laterum Wang, 1988: 294. **New Synonym**.
Sinodacus quaternum Wang, 1988: 293. **New Synonym**.
Sinodacus jiannanus Chao & Lin, 1996: 130. **New Synonym**.
Sinodacus jinreni Chao & Lin, 1996: 128. **New Synonym**.
Sinodacus qionganus Chao & Lin, 1996: 127. **New Synonym**.
Sinodacus rubzovi Chao & Lin, 1996: 129. **New Synonym**.
Bactrocera (Sinodacus) hochii – Norrbom *et al.*, 1998: 100.

Other type material examined: Holotype female of *Sinodacus laterum* Wang with the following two labels – '*Sinodacus laterum* Wang, 1986', 'HOLOTYPE' (red label); Holotype male of *Sinodacus quaternum* Wang with the following two labels – '*Sinodacus quaternum* Wang', 'HOLOTYPE' (red label); Holotype male of *Sinodacus jiannanus* Chao & Lin; Holotype male of *Sinodacus qionganus* Chao & Lin with the following three labels – '*Sinodacus qionganus* Chao', 'Sinodacus 6', '1985.7.16, cue'. All studied and photographed by Dr Simon Song at IZAS (*laterum* Wang and *quaternum* Wang) and PQMAB (*jiannanus* Chao & Lin, *qionganus* Chao & Lin).
Paratypes: 1 female paratype of *Sinodacus quaternum* Wang with the following five labels – '17 APR 1958' (Chinese writing); '650 m' (Chinese writing); 'PARATYPE', '*Bactrocera (Sinodacus) quaternum* (WANG) det X.J. Wang', 'CHINA: Pres. by X.-J. Wang, BMNH(E) No. 1999-233'. Studied at BMNH.
Non-type material examined: CHINA: 1 male, Unnan Prov., 5.viii.1983. **VIETNAM: Southern Vietnam** – 1 male (17.vii.1999), 2 males (25.vii.1999), 9 males (14.viii.1999), Vinh Long Province, Thuan An Village, Site S001, attracted to cue lure; 2 males (18.ix.1999), 1 male (8.x.1999), Can Tho Province, Binh Thuy Ward, Site S003, attracted to cue lure; 2 males (22.xii.2001), 1 male (14.i.2002), 1 male (20.i.2002), 1 male (25.ii.2002), 1 male (5.v.2002), 1 male (2.vi.2002), 1 male (30.vi.2002), 2 males (6.vii.2002), 3 males (14.vii.2002), 2 males (28.vii.2002), 1 male (4.viii.2002), Southern Vietnam, Cantho Province, Chau Thanh District, Tan Phu Village, attracted to cue lure; 2 males (24.xii.2001), 1 male (27.vii.2002), 5 males (16.xi.2002), 1 male (23.xi.2002), Southern Vietnam, Ben Tre Province, Chau Thanh District, Tien Thuy Village, attracted to cue lure; 1 male (30.xii.2001), 1 male (6.vii.2002), Southern Vietnam, Vinh Long Province, Long Ho District, Phuoc Village, attracted to cue lure; 1 male (11.v.2002), 1 male (9.xi.2002), Southern Vietnam, Dong Nai Province, Long Khanh District, Nhan Nghia Village, coll. SOFRI staff, attracted to cue lure; 1 male, Southern Vietnam, Tien Giang Province, Cai Ben District, My Thirn Village, 16.viii.2000, attracted to cue lure; 1 male, Southern Vietnam, Tien Giang Province, Cai Be District, An Huu Village, 8.viii.2000, attracted to cue lure; 1 male, Southern Vietnam, Dac Lac Province, Ku Mi Nga District, Quang Phu Village, 8.ix.2002, attracted to cue lure; 1 male, Southern Vietnam, Tien Giang Province, Cai Be District, My Thirn Village, 17.ii.2000, attracted to cue lure; 2 males, Southern Vietnam, Dong Nai Province, Thong Nhat District, Cay Gao Village, 28.iv.2000, attracted to cue lure; **Northern Vietnam** – 1 male, Northern Vietnam, Son La Province, Moc Chau District, Tra Long Village, 26.vi.1999, attracted to cue lure; 1 male, Northern Vietnam, Son La Province, Moc Chau District, Moc Chai Town, 29.xi.1999, attracted to cue lure. **INDONESIA: JAVA** – 1 male (12.vii.2005), 3 males (23.viii.2005), 2 males (11.x.2005),

East Java, Sidoarjo, Gelang, coll. Agus Riadi, attracted to cue lure in village; 1 male, East Java, Sidoarjo, 19.vii.2005, coll. Trigiyanto, attracted to cue lure in village; 1 male, East Java, Probolinggo, Betektanam 150m asl, 21.x.2005, coll. Tjipto Waluyo; **SUMATRA** – 1 male, Riau, Kampar, Kualu, 16.viii.2006, coll. Nasrum, attracted to cue lure in forest; 1 male, Sumatera Selatan, Ogan Komering Ilir, Srinanti, 21.viii.2006, coll. Iman Muhajani, attracted to cue lure in village. A large series of specimens widespread across Peninsular Malaysia and Thailand. Specimens in QDPC.

Diagnosis: A large *Callantra*-like species; face fulvous with a broad transverse black band just above oral margin; postpronotal lobes and notopleura yellow; scutum uniformly red-brown; lateral postsutural yellow vittae absent; medial postsutural yellow vitta present; a large yellow triangular marking anterior to notopleural suture; mesopleural stripe reaching midway between anterior margin of notopleuron and anterior *npl.* seta to almost reaching anterior *npl.* seta dorsally; scutellum yellow; wing with cells bc and c pale fuscous, microtrichia in outer ½ of cell c only; a broad pale fuscous to paler fuscous costal band confluent with R_{4+5} and expanding into a large circular dark fuscous spot across apex of wing; a broad pale fuscous to fuscous anal streak, a general pale fulvous tint across wing membrane; abdominal terga III–V orange-brown with a black 'T' pattern consisting of a narrow- to medium-width medial longitudinal band over all three terga and a narrow transverse black band across anterior margin of tergum III expanding to cover lateral margins, large anterolateral black markings on terga IV and V (on tergum IV the lateral black marking can extend to posterior margin of this tergum enclosing a circular orange-brown marking).

Description: male.

Head: Height 1.7 mm. Frons – length 1.1 times breadth; red-brown with fuscous to dark fuscous around *i.or.* setae and on anteromedial hump; orbital setae black: 1 *s.or.*, 2 *i.or.*; lunule fuscous. Ocellar triangle black. Vertex fuscous. Face fulvous with a broad transverse black band just above oral margin; length 0.6 mm. Genae fulvous, pale fuscous subocular spot present; black seta present. Occiput red-brown, fulvous along eye margins; occipital row with a large number of short black setae. Antennae with segments 1 and 2 fuscous, segment 3 red-brown with dark fuscous on apex and outer surface; length of segments: 0.2 mm; 0.3 mm; 0.86 mm.

Thorax: Scutum red-brown with at most a narrow fuscous medial longitudinal band from anterior point of medial postsutural yellow vitta to anterior margin (in some specimens there is an area of dark fuscous to black on the lateral margins posterior to the postpronotal lobes). Pleural areas red-brown with narrow dark fuscous margins along anterior and posterior margins of notopleural suture and on katepisternum. Yellow markings as follows: postpronotal lobes; notopleura; a large triangular marking anterior to notopleural suture; mesopleural stripe varying from midway between anterior margin of notopleuron and anterior *npl.* seta to almost reaching anterior *npl.* seta above, continuing to katepisternum as a large transverse spot, anterior margin convex; anatergite (posterior apex red-brown); anterior ⅔ katatergite (remainder red-brown); a very short and narrow medial longitudinal postsutural vitta; lateral postsutural vittae absent (occasional specimens have reduced narrowing lateral vittae). Postnotum red-brown with narrow black lateral margins. Scutellum yellow except for narrow black basal band. Setae: *sc.* 2 (weak basal setae present in occasional specimens); *prsc.* absent; *ia.* 1; *p.sa.* 1; *a.sa.* 1; *mpl.* 1; *npl.* 2; *scp.* 4.

Legs: Fore and mid femora red-brown, hind femora fulvous on basal ½ to red-brown on apical ½; fore tibiae pale fuscous basally to fulvous apically, mid tibiae fuscous basally to fulvous apically, hind tibiae fuscous; tarsi with all segments fulvous; mid tibiae each with an apical black spur.

Wings: Length 7.5 mm; cells bc and c pale fuscous; microtrichia in outer ½ of cell c only; remainder of wings with a pale fulvous tint except for a fuscous cell sc, broad costal band confluent with R_{4+5}, pale fuscous to a position overlapping R_{2+3} and paler fuscous to R_{4+5}, a large circular dark fuscous spot across apex of wing, a broad fuscous to pale fuscous anal streak; a dense aggregation of microtrichia around A_1+CuA_2; supernumerary lobe strong and rounded.

Abdomen: Elongate oval and petiolate; terga free; pecten present on tergum III. Tergum I and sterna I and II longer than wide. Tergum I entirely orange-brown; tergum II orange-brown anteriorly to fulvous posteriorly and with a small boomerang-shaped black marking anteriorly and narrow black anterolateral corners; terga III–V orange-brown with a black 'T' pattern consisting of a narrow- to medium-width medial longitudinal band over all three terga and a narrow transverse band across anterior margin of tergum III widening to cover lateral margins, large anterolateral black markings on terga IV and V (the marking on tergum IV can continue from its inner margin to the posterior margin of tergum IV enclosing a circular orange-brown pattern). A pair of oval orange-brown shining spots on tergum V. Posterior lobe of surstylus long, sternum V with a slight concavity on posterior margin. Abdominal sterna mostly dark fuscous to black.

Attractant: Cue lure.

Female: As for male except no dense aggregation of microtrichia around A_1+CuA_2, supernumerary lobe weak; pecten of cilia on abdominal tergum III absent; ovipositor basal segment red-brown, conical in transverse cross section and tapering posteriorly in dorsal view; ratio of length of oviscape to length of tergum V, 1:1.

Distribution:

Type locality: China (Hainan I.).

Previously recorded: Type locality and the following locations – China (Yunnan Province); Indonesia (Nusa Tenggara); Peninsular Malaysia; Thailand (Norrbom *et al.*, 1998).

New records: Indonesia (Java; Sumatra); Vietnam. China (Guangxi and Guizhou Provinces), due to **New Synonymies** listed under 'Other type material examined'.

Hosts: Recorded from three species in the family *Cucurbitaceae* (Allwood *et al.*, 1999).

Remarks: *Bactrocera (Sinodacus) hochii* (Zia) is a unique species and superficially similar to *Bactrocera (Asiadacus) apicalis* (de Meijere) and *Bactrocera (Sinodacus) infesta* (Enderlein) in possessing a large dark fuscous spot in the apex of the wing; however, *B. apicalis* differs in having a pair of black spots on the face, the costal band broken after apex of R_{2+3} and the *a.sa.* setae absent and *B. infesta* differs in having distinct broad parallel-sided lateral postsutural vittae, a pair of black spots on the face and large areas of dark fuscous to black on all femora.

We have studied the available types of *Sinodacus laterum* Wang, *Sinodacus quaternum* Wang, *Sinodacus jiannanus* Chao & Lin and *Sinodacus qionganus* Chao & Lin. In addition we have studied the descriptions, translated into English, of these species plus *Sinodacus hainanus* Chao & Lin, *Sinodacus jinreni* Chao & Lin and *Sinodacus rubzovi* Chao & Lin. Each of these species are identical to *B. hochii* in the face colour pattern of a black band across the oral margin, the wing colour patterns and a large rounded supernumerary lobe in the male wing, the femora entirely pale without fuscous to dark fuscous patterns, the basic red-brown scutum and abdominal terga III–V with a black 'T' pattern and dark anterolateral corners on terga IV and V. The only characters that have been used to separate these species from *B. hochii* and from each other are the number of *sc.* setae, lateral and medial postsutural yellow vittae present or absent and slight variations in the width of the costal band. In our examination of large series of specimens of *B. hochii* from Indonesia, Peninsular Malaysia, Thailand and Vietnam, we have found irregular variations in the number of *sc.* setae (usually two), the appearance of narrow and short lateral postsutural vittae that, if present, are extremely pale and red-brown, and changes in the width of the costal band due to variations in the intensity of colour. The short medial postsutural vitta is always present and although it was recorded as absent in *S. jiannanus*, it has probably faded with ageing of the specimen. Consequently, we cannot find sound characters to separate these species and agree with Norrbom *et al.* (1998) placing *S. hainanus* a synonym of *B. hochii* and we consider the other six species as **New Synonyms** of *B. hochii*.

Bactrocera (Sinodacus) infesta (Enderlein)

(Fig. 262)

Polistomimetes infestus Enderlein, 1920: 359. Holotype female in MFN.
Callantra (Polistomimetes) infestus – Hardy & Adachi, 1954: 153.
Dacus (Pacifodacus) infestus – Hardy, 1973: 21.
Bactrocera (Sinodacus) infesta – Norrbom *et al.*, 1998: 100.

Non-type material examined: A large series of specimens widespread across Peninsular Malaysia and Thailand. **VIETNAM: Northern Vietnam** – 1 male (1.xi.1999), 1 male (15.xi.1999), Northern Vietnam, Son La Province, Moc Chau District, Tra Long, attracted to cue lure; 1 male, Hoa Binh Province, Moc Chau District, Pa Co Village, 15.ix.2002. Specimens in QDPC.

Diagnosis: A large *Callantra*-like species; face fulvous with a pair of moderately large circular to oval black spots; postpronotal lobes and notopleura yellow; scutum uniformly red-brown; lateral and medial postsutural yellow vittae present; a yellow spot anterior to notopleural suture; mesopleural stripe reaching midway between anterior margin of notopleuron and anterior *npl.* seta dorsally; scutellum yellow; wing with cell c fulvous, cell bc pale fulvous, microtrichia in outer corner of cell c only; a broad pale fuscous to fuscous costal band confluent with R_{4+5} and expanding into a large circular dark fuscous spot across apex of wing; a very narrow fulvous anal streak; abdominal terga III–V red-brown with a black 'T' pattern consisting of a narrow- to medium-width longitudinal black band over all three terga and a broad transverse black band across anterior ½ of tergum III expanding to cover lateral margins, broad fuscous to dark fuscous anterolateral corners on terga IV and V which may be joined along anterior margin of tergum IV.

Description: male.

Head: Height 1.86 mm. Frons – length 1.36 times breadth; red-brown with dark fuscous around bases of *i.or.* setae and on anteromedial hump; orbital setae red-brown: 1 *s.or.*, 2 *i.or.*; lunule dark fuscous. Ocellar triangle black. Vertex red-brown. Face fulvous with a pair of moderately large circular to oval black spots; length 0.66 mm. Genae fulvous, small dark fuscous subocular spot present; black seta present. Occiput red-brown centrally, fuscous to dark fuscous laterally, fulvous along eye margins; occipital row with 5–6 strong dark setae. Antennae with segments 1 and 2 fuscous, segment 3 fuscous with dark fuscous on apex and outer surface; length of segments: 0.12 mm; 0.31 mm; 0.89 mm.

Thorax: Scutum uniformly red-brown with at most a small area of dark fuscous to black posterior to postpronotal lobes. Pleural areas red-brown with dark fuscous to black along anterior and posterior margins of mesopleural stripe and on katepisternum. Yellow markings as follows: postpronotal lobes; notopleura; mesopleural stripe reaching midway between anterior margin of notopleuron and anterior *npl.* seta dorsally, continuing to katepisternum as a very large transverse spot, anterior margin convex; anatergite (posterior apex black); anterior ⅔ katatergite (remainder black); two narrow curved lateral postsutural vittae beginning with a spot anterior to notopleural suture and ending well behind *ia.* seta; a very short and narrow medial longitudinal postsutural vitta. Postnotum red-brown with narrow black lateral margins. Scutellum yellow with a narrow black basal band. Setae: *sc.* 2; *prsc.* absent; *ia.* 1; *p.sa.* 1; *a.sa.* 1; *mpl.* 1; *npl.* 2; *scp.* 4–6.

Legs: Fore femora fulvous basally to fuscous apically and with most of outer surface black, mid and hind femora mostly black with small areas of fulvous basally and fuscous apically; fore tibiae dark fuscous, mid tibiae fuscous to dark fuscous, hind tibiae dark fuscous; fore tarsi dark fuscous except apical segment fuscous, mid and hind tarsi with all segments fuscous; mid tibiae each with an apical black spur.

Wings: Length 8.5 mm; cell c fulvous, cell bc pale fulvous; microtrichia in outer corner of cell c only; remainder of wings colourless except fuscous cell sc, broad fuscous costal band confluent with R_{4+5} (pale fuscous between R_{2+3} and R_{4+5}) and expanding into a large round dark fuscous spot across apex of wing and enclosing apices of R_{4+5} and M, a narrow fulvous anal streak mostly contained

within cell cup; a dense aggregation of microtrichia around A_1+CuA_2; supernumerary lobe weak.
Abdomen: Elongate oval and petiolate (*Callantra*-like); terga free; pecten present on tergum III. Tergum I and sterna I and II longer than wide. Tergum I entirely red-brown; tergum II red-brown with a boomerang-shaped black pattern anteriorly and narrow black lateral margins; terga III–V red-brown with a black 'T' pattern consisting of a narrow- to medium-width medial longitudinal black band over all three terga and a broad transverse black band across anterior ½ of tergum III expanding to cover lateral margins, broad fuscous to dark fuscous antero-lateral corners on terga IV and V (on tergum IV the anterolateral dark patterns may be joined along anterior margin). A pair of oval pale fuscous shining spots on tergum V. Posterior lobe of surstylus slightly elongate, sternum V with a slight concavity on posterior margin. Abdominal sterna dark fuscous to black.
Attractant: Cue lure.
Female: No known record.
Distribution:
Type locality: Indonesia (Sumatra, Deli).
Previously recorded: Type locality and the following locations – Indonesia (Java); Thailand (Norrbom *et al.*, 1998; Hardy, 1973).
New records: Northern Vietnam (Hoa Binh and Son La Provinces); Peninsular Malaysia.
Hosts: No known record.
Remarks: *Bactrocera (Sinodacus) infesta* (Enderlein) is similar to *Bactrocera (Asiadacus) apicalis* (de Meijere) and *Bactrocera (Sinodacus) hochii* (Zia) in possessing a large dark fuscous spot in the apex of the wing. It differs from *B. apicalis* in having *a.sa.* setae present, a complete costal band (not broken after apex of R_{2+3}) and extensive areas of dark fuscous to black on all femora and from *B. hochii* in having the extensive areas of dark fuscous to black on the femora, broad parallel-sided lateral postsutural vittae and a pair of black spots on the face (not a transverse band across the face).

Bactrocera (Sinodacus) longivittata Chua & Ooi

(Fig. 263)

Bactrocera (Sinodacus) longivittata Chua & Ooi, 1998: 487. Holotype male in NUS.

Type material examined: No specimens available for study. The following description has been taken from literature.
Diagnosis: A medium-sized species; face fulvous with a pair of small elongate oval black spots; postpronotal lobes and notopleura yellow; a broad lateral yellow band connecting postpronotal lobe and notopleuron; scutum black; lateral and medial postsutural yellow vittae present; a yellow spot anterior to notopleural suture; mesopleural stripe reaching to anterior *npl.* seta dorsally; scutellum yellow; wing with cells bc and c colourless, microtrichia in outer corner of cell c only; a very broad dark fuscous costal band overlapping R_{4+5} throughout; anal streak absent but with pale fuscous tint within cell cup; abdominal terga III–V fulvous with a black 'T' pattern consisting of a broad transverse band across anterior margin of tergum III and a broad medial longitudinal band over all three terga but narrower on tergum III.
Description: male.
Head: Frons – red-brown, fulvous along eye margins and with fuscous on anteromedial hump; orbital setae black: 1 *s.or.*, 2 *i.or.*; lunule red-brown. Ocellar triangle black. Vertex fuscous. Face fulvous with a pair of small elongate oval black spots. Genae entirely red-brown; black seta present. Occiput fuscous, fulvous along eye margins. Antennae with segments 1 and 2 red-brown, segment 3 red-brown with fuscous on apex and outer surface; length of segments: 0.2 mm; 0.3 mm; 0.9 mm.
Thorax: Scutum black with fuscous below and behind lateral postsutural vittae. Pleural areas black. Yellow markings as follows: postpronotal lobes; notopleura; a broad lateral band connecting postpronotal lobe and notopleuron; mesopleural stripe reaching to anterior *npl.* seta dorsally, continuing to katepisternum as a transverse spot, anterior margin convex; two broad parallel to sub-parallel lateral postsutural vittae beginning with a spot anterior to notopleural suture and ending just behind *ia.* seta; an extremely long medial postsutural vitta broadly rounded posteriorly at level of *ia.* seta and narrowing anteriorly to end just before anterior margin of scutum. Scutellum yellow except for narrow black basal band. Setae: *sc.* 2; *prsc.* absent; *ia.* 1; *p.sa.* 1; *a. a.* 1; *mpl.* 1; *npl.* 2; *scp.* 2.
Legs: All segments entirely fulvous; mid tibiae each with an apical black spur.
Wings: Length 5.6 mm; cells bc and c colourless; microtrichia in outer corner of cell c only; remainder of wings colourless except dark fuscous cell sc, a very broad dark fuscous costal band overlapping R_{4+5} for its entire length, anal streak absent but with a pale fuscous tint within cell cup; a dense aggregation of microtrichia around A_1+CuA_2; supernumerary lobe of medium development.
Abdomen: Elongate oval; terga free; pecten present on tergum III. Tergum I and sterna I and II wider than long. Tergum I black with a narrow transverse fulvous band across posterior margin but not reaching lateral margins; tergum II fulvous with a narrow transverse subanterior black band not reaching lateral margins; terga III–V fulvous with a black 'T' pattern consisting of a broad transverse band across anterior margin of tergum III and a broad medial longitudinal band over all three terga (this band is narrower on tergum III), narrow black anterolateral corners on terga IV and V. A pair of oval fulvous shining spots on tergum V. Posterior lobe of surstylus long, sternum V with a slight concavity on posterior margin. Abdominal sterna black.
Attractant: Methyl eugenol (Chua & Ooi, 1998).
Female: No known record.
Distribution:
Type locality: Peninsular Malaysia (Hulu Langat).
Hosts: No known record.
Remarks: *Bactrocera (Sinodacus) longivittata* Chua & Ooi is a unique species of subgenus *Sinodacus*. It is similar to *Bactrocera (Sinodacus) spectabilis* n.sp. and *Bactrocera (Sinodacus) speciosa* n.sp. in possessing a lateral yellow band joining the postpronotal lobe and notopleuron, the extremely long medial postsutural yellow vitta on the scutum and the very broad dark fuscous costal band overlapping R_{4+5} throughout. It differs from

these species in having the face fulvous with a pair of black spots, the femora entirely fulvous and the anal streak in the wing absent.

Bactrocera (Sinodacus) speciosa Drew & Romig, new species

(Fig. 264)

Material examined: Holotype male, **EAST MALAYSIA**: Sabah, E. Malaysia, Mt Kinabalu, 2000 m, 19.iii.1992, coll. Jinius *et al*., attracted to cue lure, deposited in BMNH.

Diagnosis: A medium-sized species; face glossy black; postpronotal lobes and notopleura yellow; a lateral yellow band joining the postpronotal lobe and notopleuron; scutum entirely glossy black; lateral and medial postsutural yellow vittae present; a yellow spot anterior to the notopleural suture; an unusual mesopleural stripe which does not reach the line of the postpronotal lobe and notopleuron dorsally and extends to the katepisternum as a large yellow spot; scutellum yellow; wing with cells bc and c fuscous, microtrichia in outer ⅓ of cell c only; a broad dark fuscous costal band overlapping vein R_{4+5} throughout; a moderately broad dark fuscous anal streak; abdominal tergum III entirely black, abdominal terga IV–V fulvous with a broad medial and two broad lateral longitudinal black bands over both terga.

Description: male.

Head: Height 1.5 mm. Frons – length 1.4 times breadth; red-brown without dark markings; orbital setae black: 1 *s.or*., 2 *i.or*.; lunule red-brown. Ocellar triangle black. Vertex red-brown. Face glossy black with red-brown lateral margins and a black line along each ptilinal suture; length 0.6 mm. Genae red-brown, a large number of strong black setae present in the form of a comb. Occiput fuscous, tending dark fuscous dorsally and fulvous along eye margins; occipital row with a large number of short pale setae. Antennae with segments 1 and 2 dark fuscous, segment 3 missing; length of segments: 0.12 mm; 0.28 mm.

Thorax: Scutum entirely glossy black without pale markings. Pleural areas entirely glossy black without pale markings. Yellow markings as follows: postpronotal lobes; notopleura; an unusual mesopleural stripe which does not extend dorsally to the margin with the postpronotal lobes and notopleura, it extends anteriorly to the level of the anterior *npl*. seta and ventrally to the katepisternum as a large transverse spot and has the anterior margin slightly concave; a lateral yellow band connecting the postpronotal lobe and notopleuron; two moderately broad lateral postsutural vittae narrowing posteriorly to end in a point at the *ia*. seta; a very long narrow medial longitudinal postsutural vitta extending over most of the scutum; a small spot anterior to the notopleural suture in front of each lateral postsutural vitta. Postnotum black. Scutellum yellow except for a broad black basal band. Setae: *sc*. 2; *prsc*. absent; *ia*. 1; *p.sa*. 1; *a.sa*. 1; *mpl*. 1; *npl*. 2; *scp*. 4.

Legs: Fore femora fulvous with black along the entire outer surface, mid femora fulvous with dark fuscous over the apical ⅔ of the dorsal surface, hind femora entirely black; fore tibiae entirely fulvous, mid tibiae dark fuscous basally to fulvous apically, hind tibiae dark fuscous tending fulvous in the central area; all tarsi with basal segment fulvous and apical four segments dark fuscous; mid tibiae each with an apical black spur.

Wings: Length 6.3 mm; cells bc and c fuscous; microtrichia in outer ⅓ of cell c only; remainder of wings colourless except dark fuscous cell sc, a broad dark fuscous costal band overlapping vein R_{4+5} throughout, a moderately broad dark fuscous anal streak; a dense aggregation of microtrichia around A_1+CuA_2; supernumerary lobe large and rounded.

Abdomen: Elongate oval; terga free; pecten present on tergum III. Tergum I and sterna I and II wider than long. Tergum I black with an irregularly shaped transverse fulvous band across posterior margin but not reaching lateral margins; tergum II fulvous with a broad transverse black band across anterior margin but not reaching the narrow black lateral margins; tergum III entirely black; terga IV–V fulvous with a broad medial and two broad lateral longitudinal black bands over both terga. A pair of oval fulvous shining spots on tergum V. Posterior lobe of surstylus long and narrow, sternum V with a slight concavity on posterior margin. Abdominal sterna dark fuscous to black.

Attractant: Cue lure.

Female: No known record.

Distribution: East Malaysia (Sabah – Mt Kinabalu).

Hosts: No known record.

Remarks: *Bactrocera (Sinodacus) speciosa* n.sp. is similar to *Bactrocera (Sinodacus) bogorensis* (Hardy), *Bactrocera (Sinodacus) longivittata* Chua & Ooi and *Bactrocera (Sinodacus) spectabilis* n.sp. in possessing the costal band of uniform width with, at most, only a slight expansion at the apex and lateral and medial longitudinal postsutural yellow vittae present. It differs from *B. bogorensis* in having a lateral yellow band joining the postpronotal lobe and notopleuron and a very long medial longitudinal postsutural yellow vitta, from *B. longivittata* in having an anal streak and dark patterns on all femora, and from *B. spectabilis* in having a glossy black face, the scutum entirely glossy black, the costal band just overlapping R_{4+5} at centre of wing, the pleural areas entirely black and the lateral postsutural yellow vittae narrowing posteriorly to end before the *ia*. setae.

Bactrocera (Sinodacus) spectabilis Drew & Romig, new species

(Fig. 265)

Material examined: Holotype male, **PHILIPPINES: LUZON ISLAND** – Mt Makiling, Botanic Gardens, 12.i.2000, coll. A. O'Toole, attracted to cue lure, deposited in BMNH. Paratypes: **PHILIPPINES: LUZON ISLAND** – 1 male, same data as holotype; 1 male, Forestry near UPLB, Philippines, 24.ix.2004, attracted to cue lure, coll. Greg Quimio; 2 males, Mt Makiling, 500 m alt., 8.vi.1983, coll. B. Ramos, attracted to cue lure. 1 paratype in BMNH, 2 paratypes in QDPC, 1 paratype in QMIC.

Diagnosis: A large species; face variable from fuscous to dark fuscous with a pair of small black spots; postpronotal lobes and notopleura yellow; a broad lateral yellow band connecting the postpronotal lobe and notopleuron; scutum red-brown with a pattern of longitudinal black lines; lateral postsutural yellow vittae present; a very long and narrow medial longitudinal postsutural yellow vitta present; a yellow spot present anterior to notopleural suture; mesopleural stripe of an unusual 'L' shape and not reaching the dorsal margin of the pleural areas but continuing on to katepisternum as a large transverse yellow spot; scutellum yellow; wing with cells bc and c fuscous, microtrichia covering outer ½ of cell c only; a broad dark fuscous costal band overlapping R_{4+5} and of uniform width throughout; a narrow dark fuscous anal streak; abdominal terga III–V fulvous with a black 'T' pattern and broad lateral dark fuscous to black margins on terga IV and V.

Description: male.

Head: Height 1.56 mm. Frons – length 1.36 times breadth; fulvous ventrally to red-brown dorsally; orbital setae very small and pale: 1 *s.or.*, 2 *i.or.*; lunule dark fuscous to black. Ocellar triangle black. Vertex red-brown. Face variable from fuscous to dark fuscous with a pair of small oval black spots from which narrow black lines run to the base of each antenna; length 0.48 mm. Genae fulvous with a row of very strong large black setae. Occiput red-brown, fulvous along eye margins; occipital row with a large number of very small pale setae. Antennae with segments 1 and 2 fuscous, segment 3 fulvous with black around outer surfaces; length of segments: 0.14 mm; 0.36 mm; 0.74 mm.

Thorax: Scutum red-brown with four narrow longitudinal black lines joined along posterior margin, the lateral pair interrupted at the notopleural suture and with a large black spot on each side anterior to the notopleural suture. Pleural areas black with fulvous above mid coxae and below postpronotal lobes. Yellow markings as follows: postpronotal lobes; notopleura; a broad lateral yellow band connecting the postpronotal lobe and notopleuron on each side; a unique mesopleural stripe which is 'L'-shaped and not reaching the dorsal margin of the pleural areas and extending ventrally as a large spot on the katepisternum; anatergite (posterior apex black); anterior ⅘ katatergite (remainder black); two medium-width parallel-sided lateral postsutural vittae beginning as a spot anterior to notopleural suture and ending posteriorly at *ia.* seta; a very long and narrow medial longitudinal postsutural vitta. Postnotum black. Scutellum yellow except for narrow black basal band. Setae: *sc.* 2; *prsc.* absent; *ia.* 1; *p.sa.* 1; *a.sa.* 1; *mpl.* 1; *npl.* 2; *scp.* 2.

Legs: Fore femora fulvous with an elongate dark fuscous to black band on outer surfaces, mid femora fulvous with elongate dark fuscous to black bands on inner surfaces and black sub-apically on ventral surfaces, hind femora entirely black except fulvous on basal and apical margins; fore tibiae fulvous, mid tibiae pale fuscous, hind tibiae fulvous centrally with dark fuscous on apex and base; fore and mid tarsi with basal segment fulvous and apical four segments fuscous, hind tarsi with all segments fulvous; mid tibiae each with an apical black spur.

Wings: Length 6.4 mm; cells bc and c fuscous; microtrichia covering outer ½ of cell c only; remainder of wings colourless except dark fuscous cell sc, broad dark fuscous costal band overlapping R_{4+5} and of uniform width throughout, a narrow dark fuscous anal streak; a dense aggregation of microtrichia around A_1+CuA_2; supernumerary lobe strong and rounded.

Abdomen: Elongate oval and slightly petiolate; terga free; pecten present on tergum III. Tergum I and sterna I and II longer than wide. Tergum I fulvous with narrow black lateral margins; tergum II fulvous with a transverse black band subanteriorly but not reaching narrow black anterolateral corners; terga III–V fulvous with a dark 'T' pattern consisting of an irregularly shaped broad black band across anterior margin of tergum III which continues around the margin into dark fuscous colouration and a medium-width medial longitudinal black band over all three terga and broad dark fuscous to black lateral margins on terga IV and V. A pair of oval fulvous shining spots on tergum V. Posterior lobe of surstylus long, sternum V with a slight concavity on posterior margin. Abdominal sterna dark fuscous to black.

Attractant: Cue lure.

Female: No known record.

Distribution: Philippines (Luzon I. – Los Banos; Mt Makiling).

Hosts: No known record.

Remarks: *Bactrocera (Sinodacus) spectabilis* n.sp. is a unique species of this subgenus with a black face, a mesopleural stripe that does not reach the dorsal margin of the pleural area, a lateral yellow band joining the postpronotal lobe and notopleuron, lateral and medial longitudinal postsutural yellow vittae (the medial vitta extremely long), black markings on all femora, a broad fuscous costal band overlapping R_{4+5} and abdominal terga III–V with a black 'T' pattern and broad black lateral margins.

Bactrocera (Sinodacus) transversa (Hardy)

(Fig. 266)

Dacus (Pacifodacus) transversus Hardy, 1982: 192. Holotype male in MZB.

Bactrocera (Sinodacus) transversa – Norrbom *et al.*, 1998: 101.

Type material examined: Holotype male, **INDONESIA: SULAWESI** – Lindu Valley, Central Sulawesi, 960 m, August 1975, coll. D.E. Hardy, attracted to cue lure. Studied at MZB.

Non-type material examined: INDONESIA: SULAWESI – 1 male, 'Sadaunta, 65km S.E. Palu, Sulawesi, 650m, August 1975, coll. D.E. Hardy'; 'BMNH(E) No. 533139'. Studied in BMNH. 2 males, Palopo-Rantepao Rd., 15-22km Rantepao, Sulawesi, Indonesia (1000-1100m), 24.vii.1981, coll. J. Turner, primary rainforest some disturbance; 3 males, Palopo-Rantepao Rd., 16-23km Rantepao, Sulawesi, Indonesia (250-500m), 24.vii.1981, coll. J. Turner, Nanggala III forest. Specimens in QDPC.

Diagnosis: A large species; face fulvous with a pair of medium-sized circular to oval black spots; postpronotal lobes fuscous; notopleura yellow; scutum entirely black; lateral postsutural yellow vittae reduced to very short narrow stripes; medial postsutural yellow vitta as a narrow transverse band across posterior margin of scutum; a large yellow triangular marking along anterior margin of notopleural suture; mesopleural stripe equal in width to notopleuron dorsally; scutellum yellow with a broad black basal band; wing with cells bc and c fuscous, dense microtrichia covering all of both cells; a very broad dark fuscous costal band confluent with R_{4+5} and then expanding markedly across apex of wing to form a large spot; a moderately broad pale fuscous anal streak; abdominal terga III–V pale red-brown with dark fuscous to black bands across anterior margins of terga III and IV which expand laterally to cover lateral margins and dark fuscous to black anterolateral corners on tergum V.

Description: male.

Head: Frons – pale fuscous tending fulvous anteriorly; orbital setae black: 1 *s.or.*, 2 *i.or.*; lunule fulvous. Ocellar triangle black. Vertex pale fuscous. Face fulvous with a pair of medium-sized circular to oval black spots. Occiput pale red-brown, fulvous along eye margins. Antennae with segments 1 and 2 red-brown, segment 3 pale fuscous.

Thorax: Scutum entirely black. Pleural areas black. Yellow markings as follows: notopleura; mesopleural stripe equal in width to notopleuron dorsally; two very short and narrow lateral postsutural vittae ending well before line of *a.sa.* setae; medial postsutural vitta as a narrow transverse band across posterior area of scutum just in front of level of *ia.* setae. Postpronotal lobes fuscous. Scutellum yellow with a broad black basal band. Setae: *sc.* 2; *prsc.* absent; *ia.* 1; *p.sa.* 1; *a.sa.* 1; *mpl.* 1; *npl.* 2; *scp.* 4.

Legs: Femora fulvous with irregular red-brown patches apically; all tibiae pale red-brown; all tarsal segments fulvous; mid tibiae each with an apical black spur.

Wings: Cells bc and c fuscous; dense microtrichia covering all of both cells; remainder of wings colourless except dark fuscous cell sc, broad dark fuscous costal band confluent with R_{4+5} and expanding into a large spot across apex of wing (there is a pale fuscous tint between R_{4+5} and M in centre of wing), a moderately broad pale fuscous anal streak; a dense aggregation of microtrichia around A_1+CuA_2; supernumerary lobe of medium development.

Abdomen: Elongate oval; terga free; pecten present on tergum III. Tergum I and sterna I and II wider than long. Tergum I dark fuscous to black with a transverse red-brown band across posterior margin but not reaching lateral margins; tergum II fulvous with a broad transverse dark fuscous to black band across anterior margin and widening to cover most of lateral margins; terga III–V pale red-brown with a broad transverse dark fuscous to black band across anterior margin of tergum III and widening over lateral margins, a narrow transverse dark fuscous to black band across anterior margin of tergum IV and broadly expanding over lateral margins, broad dark fuscous to black anterolateral margins on tergum V, a narrow black medial longitudinal black band on tergum V. In some specimens the dark patterns on terga IV and V are reduced to narrow dark fuscous anterolateral corners. A pair of oval pale red-brown shining spots on tergum V. Posterior lobe of surstylus long, sternum V with a slight concavity on posterior margin.

Attractant: Cue lure.

Female: No known record.

Distribution:

Type locality: Indonesia (Sulawesi – Central Sulawesi, Lindu Valley).

Previously recorded: Type locality and other locations in Sulawesi – Kuala Navusa; Paredean via Palopo; Sadaunta River; Tolai (Hardy, 1982).

Hosts: No known record.

Remarks: *Bactrocera (Sinodacus) transversa* (Hardy) is a unique species of subgenus *Sinodacus* in having fuscous postpronotal lobes, very short and narrow lateral postsutural vittae, a transverse yellow band across posterior margin of scutum and a very broad fuscous costal band confluent with R_{4+5} and expanding across apex of wing.

Bactrocera (Sinodacus) whitei Drew & Romig, new species

(Fig. 267)

Material examined: Holotype male, **PHILIPPINES: LUZON ISLAND** – Mt Makiling, 400 m, 9.v.1983, coll. Drew, Nette, Bert, attracted to cue lure, deposited in BMNH.

Paratypes: **PHILIPPINES: LUZON ISLAND** – 1 male, Mt Makiling, 500 m, Philippines, 9.v.1983, coll. R. Drew, B. Ramos, attracted to cue lure. **EAST MALAYSIA**: 1 male, Sabah, Mt Kinabalu, 1500 m, 5.ix.1991, coll. P. Anthony, attracted to cue lure. 1 paratype in QDPC, 1 paratype in QMIC.

Diagnosis: A large species; face fulvous with a pair of medium-sized oval black spots pointed towards the oral margin and a broad dark fuscous transverse band across oral margin; postpronotal lobes and notopleura yellow; scutum black with red-brown to fuscous patterns; short lateral postsutural yellow vittae present; a large triangular medial postsutural yellow vitta present; large yellow spots anterior to notopleural suture; mesopleural stripe reaching almost to postpronotal lobe dorsally; scutellum yellow; wing with cells bc and c fuscous, dense microtrichia covering all of both cells bc and c; a broad dark fuscous costal band confluent with R_{4+5} and expanding into a large spot across apex; a moderately broad fuscous anal streak; abdominal terga III–V with a 'T' pattern consisting of a broad transverse black band across anterior margin of tergum III and a broad medial longitudinal black band over all three terga and broad lateral longitudinal dark fuscous bands over terga IV and V.

Description: male.

Head: Height 1.6 mm. Frons – length 1.4 times breadth; fuscous to dark fuscous with fulvous on the lower lateral and ventral margins; orbital setae black: 1 *s.or.*, 2 *i.or.*; lunule dark fuscous. Ocellar triangle black. Vertex dark fuscous. Face fulvous with a pair of medium-sized oval black spots pointed towards the oral margin and, in some specimens, a transverse dark fuscous band across the oral margin; length 0.53 mm. Genae fulvous, large dark fuscous subocular spot present; black seta present. Occiput fuscous, fulvous along eye margins; occipital row with 6–8 black setae. Antennae with segments 1 and 2 red-brown, segment 3 red-brown with fuscous on apex and outer surface; length of segments: 0.16 mm; 0.33 mm; 0.88 mm.

Thorax: Scutum black except dark fuscous across posterior margin, red-brown below and behind lateral postsutural vittae and running anteriorly off the medial postsutural yellow vitta, red-brown to dark fuscous across anterior margin between postpronotal lobes. Pleural areas dark fuscous to black except red-brown above hind coxae and below postpronotal lobes. Yellow markings as follows: postpronotal lobes; notopleura; mesopleural stripe reaching almost to postpronotal lobe dorsally, continuing to katepisternum as a small transverse spot, anterior margin straight; anatergite (posterior apex black); anterior ⅔ katatergite (remainder black); two very short and narrow lateral postsutural vittae ending at level of *a.sa.* setae; a large triangular medial longitudinal postsutural vitta; a large spot anterior to notopleural suture in front of each lateral postsutural vitta. Postnotum black. Scutellum yellow except for narrow black basal band. Setae: *sc.* 2; *prsc.* absent; *ia.* 1; *p.sa.* 1; *a.sa.* 1; *mpl.* 1; *npl.* 2; *scp.* 4.

Legs: Fore femora fulvous with a large dark fuscous spot along outer surface, mid femora fulvous with fuscous to dark fuscous around apical ⅔, hind femora fulvous with fuscous to dark fuscous around apical ⅓; all tibiae dark fuscous; all tarsi with basal segment fulvous and apical four segments fuscous; mid tibiae each with an apical black spur.

Wings: Length 6.6 mm; cells bc and c fuscous; dense microtrichia covering all of both cells bc and c; remainder of wings colourless except dark fuscous cell sc, a broad dark fuscous costal band confluent with R_{4+5} and broadening to a large spot across apex of wing and covering vein M, a moderately broad fuscous anal streak; a dense aggregation of microtrichia around A_1+CuA_2; supernumerary lobe weak.

Abdomen: Elongate oval; terga free; pecten present on tergum III. Tergum I and sterna I and II wider than long. Tergum I dark fuscous with narrow transverse fulvous band across posterior margin but not reaching lateral margins; tergum II fulvous with a narrow transverse dark fuscous band subanteriorly and joining the narrow dark fuscous lateral margins; terga III–V with a 'T' pattern consisting of a broad transverse black band across anterior margin of tergum III and a broad medial longitudinal black band over all three terga and two broad lateral longitudinal dark fuscous bands over terga IV and V. A pair of oval orange-brown shining spots on tergum V. Posterior lobe of surstylus moderately long and stout, sternum V with a slight concavity on posterior margin. Abdominal sterna dark.

Attractant: Cue lure.

Female: No known record.

Distribution: East Malaysia (Sabah, Mt Kinabalu); Philippines (Luzon I., Mt Makiling).

Hosts: No known record.

Etymology: This species is named after Dr Ian White, CABI and BMNH, in recognition of his major contributions to tephritid taxonomic research.

Remarks: *Bactrocera (Sinodacus) whitei* n.sp. is similar to *Bactrocera (Sinodacus) transversa* (Hardy), *Bactrocera (Sinodacus) hochii* (Zia) and *Bactrocera (Sinodacus) infesta* (Enderlein) in having the costal band expanded into a large spot over the wing apex and to *B. transversa* in possessing a black scutum and a fuscous anal streak in the wing. It differs from *B. hochii* and *B. infesta* in having a black scutum, from *B. hochii* in having dark patterns on all femora and from *B. transversa* in having the medial postsutural yellow vitta on the scutum as a distinct triangle.

SUBGENUS *TETRADACUS* Miyake

Bactrocera (Tetradacus) brachycera (Bezzi)

(Fig. 268)

Mellesis brachycera Bezzi, 1916: 116. Holotype female in BMNH.
Bactrocera (Tetradacus) brachycera – Norrbom *et al.*, 1998: 101.
Sinodacus fuscans Wang, 1988: 292; Drew *et al.*, 1998: 569 (Synonym).

Type material examined: Holotype female, **INDIA**: with the following four labels – 'Type, Mellesis brachycera Bezzi'; 'India, Bhimtal Kumtal, 20.vi.1912, pres. by Prof A. D. Imms, 1915.414'; 'For Zool Coll, Bhimtal Kumtal, 20.vi.1912'; 'Mellesis n.gen. brachycera n.sp. Bezzi, type [female], A.D. I mms'. Studied at BMNH.

Non-type material examined: BHUTAN: 1 male, Tashi, Malathion bait spray, 26.x.1991, coll. H.R. Fiejen. **INDIA**: 1 male, on mulberry ripe fruit, Jolikote, VP, 18.v.1915, Revd Exch for ident. Fletcher; Pres by E. Brunetti, B.M. 1927-184. Studied at BMNH.

Diagnosis: A large species; face fulvous with a pair of large elongate black spots covering ventral ½ of antennal sockets; postpronotal lobes and notopleura yellow; scutum red-brown with dark fuscous to black irregular patterns posterocentrally; lateral postsutural yellow vittae present; medial postsutural yellow vitta absent; yellow spot present anterior to notopleural suture; mesopleural stripe reaching midway between anterior margin of notopleuron and anterior *npl.* seta dorsally; scutellum yellow; wing with cells bc and c with a pale tint, microtrichia in outer corner of cell c only; a broad fuscous costal band confluent with R_{4+5} and expanding markedly over apex of wing to overlap M; a narrow fuscous anal streak; abdominal terga III–V orange-brown with a black 'T' pattern and narrow black anterolateral corners on terga IV and V.

Description: female.

Head: Frons – fulvous with dark fuscous around orbital setae; orbital setae black: 1 *s.or.*, 3 *i.or.*; lunule fulvous. Ocellar triangle black. Vertex fulvous. Face fulvous with a pair of large elongate black spots filling bottom ½ of antennal furrow. Occiput dark fulvous, fulvous along eye margins. Antennae with segments 1 and 2 red-brown, segment 3 red-brown with fuscous on apex and outer surface.

Thorax: Scutum red-brown with dark fuscous to black irregular markings posteriorly between lateral postsutural vittae. Mesopleural areas red-brown with large areas of black anterior to mesopleural stripe and on katepisternum and fuscous posterior to mesopleural stripe and below postpronotal lobes. Yellow markings as follows: postpronotal lobes; notopleura; mesopleural stripe reaching midway between anterior margin of notopleuron and anterior *npl.* seta dorsally, extending on to katepisternum as a large transverse spot, anterior margin straight; two moderately broad parallel- sided lateral postsutural vittae beginning with a small yellow spot anterior to notopleural suture and ending just behind *ia.* seta. Medial longitudinal postsutural yellow vitta absent. Scutellum yellow with narrow red-brown basal band. Setae: *sc.* 2; *prsc.* absent; *ia.* 1; *p.sa.* 1; *a.sa.* absent; *mpl.* 1; *npl.* 2; *scp.* 4.

Legs: Fore femora fuscous except fulvous on basal ½ of inner surfaces, mid and hind femora fulvous on basal ½ and fuscous on apical ½; fore and hind tibiae fuscous, mid tibiae fulvous; all tarsal segments fulvous; mid tibiae each with an apical black spur.

Wings: Length 8.8 mm; cells bc and c with a pale tint; microtrichia in outer corner of cell c only; remainder of wings colourless except fuscous cell sc, broad fuscous costal band confluent with R_{4+5} (paler between R_{2+3} and R_{4+5}) and expanding into a very large spot across apex of wing and overlapping M; a narrow fuscous anal streak; no dense aggregation of microtrichia around A_1+CuA_2; supernumerary lobe weak.

Abdomen: Elongate oval and petiolate; terga free; no pecten on tergum III. Tergum I and sterna I and II wider than long. Tergum I entirely orange-brown; tergum II orange-brown tending fulvous posteriorly and with a large central dark fuscous to black spot anteriorly; terga III–V orange-brown with a black 'T' pattern consisting of a moderately broad black band across anterior margin of tergum III and a narrow medial longitudinal black band over all three terga but absent on posterior area of terga III and V, narrow anterolateral black corners on terga IV and V. A pair of oval orange-brown shining spots on tergum V. Abdominal sterna dark. Basal segment of ovipositor orange-brown, dorsoventrally compressed and tapering posteriorly in dorsal view.

Attractant: No known record.

Male: As for female except pecten of cilia on abdominal tergum III present; posterior lobe of surstylus short; abdominal sternum V with a slight concavity on posterior margin.

Distribution:

Type locality: Northern India (Uttar Pradesh State, Dehra Dun, Bhimtal Kumtal).

Previously recorded: Type locality and China (Yunnan Province) due to synonymy (Drew *et al.*, 1998: 569).

New records: Bhutan (Tashi); India (Jolikote).

Hosts: No known record.

Remarks: *Bactrocera (Tetradacus) brachycera* (Bezzi) is similar to *Bactrocera (Tetradacus) minax* (Enderlein) and *Bactrocera (Tetradacus) tsuneonis* (Miyake) in the general shape of the abdomen and in possessing a red-brown scutum and a broad costal band. It differs from both species in lacking a medial longitudinal yellow vitta and a lateral yellow band connecting the postpronotal lobe and notopleuron and in having a yellow spot anterior to the notopleural suture and the costal band expanding into a large spot in apex of wing.

Bactrocera (Tetradacus) minax (Enderlein)

(Fig. 269)

Polistomimetes minax Enderlein, 1920: 358. Lectotype male in BMNH.
Mellesis citri Chen, 1940: 133; White & Wang, 1992: 276 (Synonym).
Dacus (Polistomimetes) minax – Drew, 1979: 76.
Bactrocera (Polistomimetes) minax – Drew, 1989: 13.
Bactrocera (Tetradacus) minax – White & Wang, 1992: 276, Lectotype designation; Norrbom *et al.*, 1998: 101.

Type material examined: Lectotype male, **INDIA: (SIKKIM)** with the following four labels – 'Cotype'; 'Polistomimetes minax type Endere, male, Dr. Enderlein det. 1920'; 'Sikkim coll. Bingham rec. from Dr. Enderlein 1920. 375'; 'lectotype Polistomimetes minax Enderlein det. I.M. White 1991'. Studied at BMNH.

Non-type material examined: BHUTAN: 2 males and 1 female (17.v.2000), 2 males and 2 females (24.v.2000), 2 females (31.v.2000), Rimchu 1, coll. C. Dorji, attracted to methyl eugenol; 1 male and 1 female, Rimchu 1, 7.vi.2000, coll. C. Dorji, Nado & Tshomo, attracted to methyl eugenol; 1 female, Rimchu 1, new trap, 12.vii.2000, coll. C. Dorji, P. Codey, attracted to methyl eugenol; 1 male, Rimchu 2, 21.vi.2000, attracted to methyl eugenol; 3 males and 3 females, Western Bhutan, Punakha, Phuntsho Pelri, coll. Chencho Dorji, bred ex *Citrus reticulata*, emerged iv.1998. Specimens in QDPC. **NEPAL**: 1 female (1985.i), 1 male (1.i.1986), ENTO. Division, Helumbu (Central Nepal), Junar (Orange), coll. P.D. Shreota. Specimens in QDPC. **INDIA**: 1 female, 'Kahimpong, Darjeeling, Host orange, Aug-Sept 1965'; 'ZSI, Lot No. 27, 1966'; 'CIE Coll. No. A1414'; 'Pres by Com Inst Ent, BM 1967-3'. Studied at BMNH.

Diagnosis: A large *Callantra*-like species; face fulvous with a pair of large elongate black spots filling most of antennal furrow; postpronotal lobes and notopleura yellow; a lateral yellow band running from posterior margin of postpronotal lobe towards notopleuron but not reaching notopleuron; scutum red-brown with a mottled appearance from dorsoventral flight muscles and occasional irregular dark fuscous markings; lateral and medial postsutural yellow vittae present; no yellow marking anterior to notopleural suture; mesopleural stripe reaching anterior *npl.* seta dorsally; scutellum yellow; wing with cells bc and c fuscous, microtrichia covering outer ½ of cell c only; a broad fuscous costal band overlapping R_{4+5} and with a dark fuscous spot within apex of this band; anal streak absent but with pale fuscous within cell cup; abdominal terga III–V fulvous to red-brown and generally with a dark fuscous to black 'T' pattern and a dark fuscous to black anterolateral corners on terga IV and V.

Description: male.

Head: Height 2.4 mm. Frons – length 1.2 times breadth; red-brown, fulvous along eye margins and with small dark fuscous areas around bases of orbital setae and on anteromedial hump; orbital setae black: 1 *s.or.*, 2 *i.or.*; lunule dark fuscous. Ocellar triangle black. Vertex red-brown. Face fulvous with a pair of large elongate black spots filling most of antennal furrow; length 0.98 mm. Genae red-brown, large dark

fuscous subocular spot present; red-brown seta present. Occiput red-brown, fulvous along eye margins; occipital row with 1–4 weak black setae. Antennae with all segments red-brown; length of segments: 0.26 mm; 0.5 mm; 1.25 mm.
Thorax: Scutum red-brown with a mottled appearance from dorsoventral flight muscles and irregular dark fuscous patterns and particularly with a narrow medial longitudinal dark fuscous line running from anterior point of medial postsutural yellow vitta to anterior margin of scutum. Pleural areas red-brown with fuscous along anterior and posterior margins of mesopleural stripe. Yellow markings as follows: postpronotal lobes; notopleura; a lateral yellow band running from posterior margin of postpronotal lobes towards notopleuron but finishing before the notopleuron; a broad mesopleural stripe reaching the anterior *npl.* seta dorsally, extending to the katepisternum as a large transverse spot, anterior margin slightly convex; anatergite (posterior apex red-brown); anterior ⅔ katatergite (remainder red-brown); two medium-width parallel-sided lateral postsutural vittae ending at *ia.* seta and turned slightly inwards along notopleural suture; a broad medial postsutural vitta broadly rounded posteriorly at level of *p.sa.* seta and narrowing to a point to end anteriorly before level of notopleural suture. Postnotum red-brown. Scutellum yellow except for narrow red-brown basal band. Setae: *sc.* 2; *prsc.* absent; *ia.* 1; *p.sa.* 1; *a.sa.* absent; *mpl.* 1; *npl.* 2; *scp.* 2.
Legs: Femora fulvous except for small areas of fuscous around bases of all femora and larger areas of fuscous to dark fuscous around apices of mid and hind femora; fore tibiae fuscous to dark fuscous basally tending paler apically, mid tibiae fuscous basally to fulvous apically, hind tibiae fuscous to dark fuscous; all tarsal segments fulvous; mid tibiae each with an apical black spur.
Wings: Length 10.8 mm; cells bc and c fuscous; microtrichia in outer ½ of cell c only; remainder of wings colourless except fuscous cell sc, broad fuscous costal band overlapping R_{4+5} and having a distinct dark fuscous spot within apex, a distinct anal streak absent but with pale fuscous within cell cup; no dense aggregation of microtrichia around A_1+CuA_2; supernumerary lobe weak.
Abdomen: Elongate oval and petiolate; terga free; pecten present on tergum III. Tergum I and sterna I and II longer than wide. Terga I and II fulvous with narrow fuscous to dark fuscous lateral margins and a broad medial longitudinal dark fuscous band that is diffuse on the lateral margins; terga III–V variable from fulvous to red-brown and with a dark fuscous to black 'T' pattern consisting of a broad transverse band across anterior margin of tergum III which widens to cover lateral margins and a narrow medial longitudinal band over all three terga but ending before posterior margin of tergum III, fuscous to dark fuscous anterolateral corners on terga IV and V (in some specimens the dark patterns on terga III–V are paler and reduced to a fuscous medial longitudinal band over all three terga and pale fuscous anterolateral corners on terga IV and V). A pair of oval fulvous to red-brown shining spots on tergum V. Posterior lobe of surstylus short, sternum V with a slight concavity on posterior margin. Abdominal sterna variable from red-brown to fuscous.
Attractant: Methyl eugenol (see Drew *et al.*, 2007).
Female: As for male except pecten of cilia on abdominal tergum III absent, ovipositor basal segment red-brown, broadly conical in transverse cross section at base and narrowing and circular in transverse cross section towards apex and extremely long; ratio of length of oviscape to length of tergum V, 2.7:1; apex of piercer needle-shaped.
Distribution:
Type locality: India (Sikkim).
Previously recorded: Type locality and the following locations – Bhutan; southern China; India (West Bengal) (White & Wang, 1992; Norrbom *et al.*, 1998).
New records: Nepal, Helumbu.
Hosts: A range of *Citrus* species and *Fortunella crassifolia* (family *Rutaceae*) (see Allwood *et al.*, 1999).
Remarks: *Bactrocera (Tetradacus) minax* (Enderlein) is similar to *Bactrocera (Tetradacus) tsuneonis* (Miyake) in the general shape of the thorax and abdomen and in the basic body colour patterns to a point where these species could be confused in diagnosis. However, *B. minax* differs in lacking *a.sa.* setae and in having an incomplete lateral yellow band between the postpronotal lobe and notopleuron, a broader costal band overlapping R_{4+5} with a distinct dark fuscous spot within its apex and a needle-shaped apex on the piercer (not trilobed).

Bactrocera (Tetradacus) splendida (Perkins)

(Fig. 270)

Callantra splendida Perkins, 1938a: 136. Syntypes male and female in BMNH.
Bactrocera splendida – White & Wang, 1992: 278.
Bactrocera (Tetradacus) splendida – Norrbom *et al.*, 1998: 101.

Type material examined: INDONESIA: WEST JAVA – Syntype male, with the following five labels – 'Type'; 'Holotype, Callantra splendida, F.A. Perkins'; 'Purch from E. Le Moult, BM 1937-32'; 'W. Java, Preanger Djampang Tenggah, FEB 1935'; 'BMNH(E) No. 533013'. Syntype female, with the following five labels – 'Type'; 'Holotype Callantra splendida F.A. Perkins'; 'Purch from E. Le Moult, BM 1937-32'; 'W. Java, Preanger Djampang Tenggah, April 1935'; 'BMNH(E) No 533012'. Studied at BMNH.
Diagnosis: A large species; face fulvous with a pair of very large elongate black spots; postpronotal lobes and notopleura yellow; a lateral yellow band connecting postpronotal lobe and notopleuron; scutum black with red-brown markings; lateral and medial postsutural yellow vittae present; no yellow spot anterior to notopleural suture; mesopleural stripe reaching midway between anterior margin of notopleuron and anterior *npl.* seta dorsally; scutellum yellow; wing with cells bc and c dark fuscous, dense microtrichia covering both cells; a broad dark fuscous costal band confluent with R_{4+5} and overlapping this vein at apex; anal streak absent; abdominal terga III–V orange-brown with broad black lateral margins that change to diffuse dark fuscous on posterolateral corners of terga III and IV.
Description: male.

Head: Frons – red-brown with dark fuscous around bases of orbital setae and on anteromedial hump; orbital setae black: 1 *s.or.*, 2 *i.or.*; lunule fuscous. Ocellar triangle black. Vertex red-brown. Face fulvous with a pair of very large elongate black spots filling the entire antennal furrow. Occiput red-brown, fulvous along eye margins. Antennae with segments 1 and 2 elongate (similar to *Dacus (Mellesis)*) and, red-brown, segment 3 fuscous.
Thorax: Scutum black with red-brown markings as follows: below and behind lateral postsutural vittae, around notopleural suture, inside postpronotal lobes, a narrow medial longitudinal band from anterior point of lateral postsutural yellow vitta to anterior margin of scutum. Pleural areas black except for a vertical red-brown band below posterior area of postpronotal lobes. Yellow markings as follows: postpronotal lobes; notopleura; a narrow lateral yellow band joining the postpronotal lobe to the notopleuron; mesopleural stripe reaching midway between anterior margin of notopleuron and anterior *npl.* seta dorsally, extending to katepisternum as a small spot, anterior margin straight; two moderately broad parallel-sided lateral postsutural vittae ending at level of *ia.* seta; a moderately broad medial postsutural vitta rounded posteriorly at level of *ia.* seta and ending anteriorly at approximately the level of the notopleural suture. Scutellum yellow except for narrow black basal band. Setae: *sc.* 2; *prsc.* absent; *ia.* 1; *p.sa.* 1; *a.sa.* absent; *mpl.* 1; *npl.* 2; *scp.* 4.
Legs: Fore and mid femora fulvous with large black spots over apical ⅔ of outer surfaces, hind femora entirely fulvous; all tibiae dark fuscous; all tarsal segments fulvous; mid tibiae each with an apical black spur.
Wings: Length 6.7 mm; cells bc and c dark fuscous; dense microtrichia covering all of both cells; remainder of wings colourless except dark fuscous cell sc, broad dark fuscous costal band confluent with R_{4+5} and overlapping this vein at apex (the costal band fades to pale fuscous adjacent to R_{4+5}), anal streak absent; a dense aggregation of microtrichia around A_1+CuA_2; supernumerary lobe weak.
Abdomen: Elongate oval; terga free; pecten present on tergum III. Tergum I orange-brown with broad black lateral margins; tergum II orange-brown anteriorly and fulvous posteriorly with a narrow transverse black band across centre; terga III–V orange-brown with broad black lateral margins that do not meet in centre and change to diffuse dark fuscous on posterolateral corners of terga III and IV. A pair of oval dark fuscous shining spots on tergum V. Posterior lobe of surstylus short, sternum V with a slight concavity on posterior margin. Abdominal sterna dark.
Attractant: No known record.
Female: As for male except pecten of cilia on abdominal tergum III absent; no dense aggregation of microtrichia around A_1+CuA_2; basal segment of ovipositor red-brown, cylindrical in transverse cross section, ratio of length of oviscape to length of tergum V, 1.5:1, apex of piercer needle-shaped.
Distribution:
Type locality: Indonesia (West Java).
Hosts: No known record.
Remarks: *Bactrocera (Tetradacus) splendida* (Perkins) is a unique species of this subgenus and is readily distinguished in possessing a black scutum, an elongate oval abdomen that is not petiolate, abdominal terga III–V orange-brown with broad black lateral margins and cells bc and c dark fuscous and covered with dense microtrichia. The subgeneric placement of *B. splendida* was recorded as uncertain by White & Wang (1992). They noted that it possessed all the characters of subgenus *Tetradacus* except that it possessed a dense patch of microtrichia around A_1+CuA_2 in the wing. We have decided to keep *B. splendida* in subgenus *Tetradacus* at present.

Bactrocera (Tetradacus) tsuneonis (Miyake)

(Fig. 271)

Dacus tsuneonis Miyake, 1919: 92. Types lost.
Dacus cheni Chao, 1987: 4; White & Wang, 1992: 277 (Synonym).
Bactrocera (Tetradacus) tsuneonis – Norrbom *et al.*, 1998: 101.

Non-type material examined: JAPAN: 1 male and 1 female, 'Japonia, Kyushu, Tsukumi (Oita), vi.1948, coll. T. Tachu Kawa'; 'Tetradacus tsuneonis Miyake, Det. Syusiro Ito, 1950'; 'Purch from E.M. Hering, B.M. 1965-270'; 1 male, 'Tsukumi, 1.viii.1957, R. Matsuda'; 'Purch. from E.M. Hering, B.M. 1965-270'; 1 female, 'Uhasiro, Prope Tsukumi, 27.vii.1957, R. Matsuda'; 'Purch from E.M. Hering, B.M. 1965-270'. Studied at BMNH. 1 male and 1 female, Nakano Shima, Tokara Islands, Kagoshima Pref., 7.xi.1975, coll. S. Takagi, bred 'ex Citrus kinokuni Hort ex Tanaka'. Specimens in QDPC. **CHINA**: 1 male, 'Tetradacus tsuneonis Miyake, identified by X.J. Wang', 'Yangshuo, Guangxi, 1959, Chinese Academy of Science'; 1 female, ' Hexian, Guangxi, 1959, Chinese Academy of Science'; 1 male, 'Dacus tsuneonis, Japan'. Specimens in QDPC.
Diagnosis: A large species; face fulvous with a pair of small- to medium-sized oval black spots; postpronotal lobes and notopleura yellow; a lateral yellow band connecting postpronotal lobe and notopleuron; scutum red-brown; lateral and medial postsutural yellow vittae present; no yellow spot anterior to notopleural suture; mesopleural stripe reaching anterior *npl.* seta dorsally; scutellum yellow; wing with cells bc and c with a strong pale fuscous tint, microtrichia in outer corner of cell c only; a broad fuscous to pale fuscous costal band confluent with R_{4+5} at base and overlapping this vein towards apex of wing; anal streak reduced to pale fuscous within cell cup; abdominal terga III–V orange-brown to fulvous with a dark 'T' pattern consisting of a narrow transverse dark fuscous band across anterior margin of tergum III and a narrow medial longitudinal black band over all three terga and in some specimens fuscous to dark fuscous in anterolateral corners of tergum IV.
Description: male.
Head: Height 2.16 mm. Frons – length 1.2 times breadth; red-brown with fuscous on anteromedial hump; orbital setae black: 1 *s.or.*, 2 *i.or.*; lunule red-brown. Ocellar triangle black. Vertex red-brown. Face fulvous with a pair of small- to medium-sized oval black spots; length 0.77 mm. Genae fulvous with a very small dark fuscous subocular spot; black

seta present. Occiput red-brown, fulvous along eye margins; occipital row with a large number of small dark setae. Antennae with all segments red-brown; length of segments: 0.28 mm; 0.48 mm; 1.16 mm.

Thorax: Scutum red-brown with a mottled appearance from dorsoventral flight muscles and occasionally with irregularly shaped small areas of fuscous. Pleural areas red-brown without dark markings. Yellow markings as follows: postpronotal lobes; notopleura; a moderately broad lateral yellow band joining the postpronotal lobe and notopleuron; mesopleural stripe reaching anterior *npl.* seta dorsally, continuing to katepisternum as a large yellow spot, anterior margin convex; almost entire surfaces of anatergite and katatergite; a pair of moderately broad parallel-sided lateral postsutural vittae ending at *ia.* seta and turned inwards along notopleural suture; a short narrow medial postsutural vitta. Yellow spot anterior to notopleural suture absent. Postnotum red-brown. Scutellum yellow except for narrow red-brown basal band. Setae: *sc.* 2; *prsc.* absent; *ia.* 1; *p.sa.* 1; *a.sa.* 1; *mpl.* 1; *npl.* 2; *scp.* 2.

Legs: All femora fulvous except fuscous on extreme apical ventral surfaces of mid femora; fore and mid tibiae entirely fulvous, hind tibiae red-brown; fore and hind tarsal segments with basal segment fulvous and apical four segments red-brown, mid tarsi with all segments red-brown, mid tibiae each with an apical black spur.

Wings: Length 9.2 mm; cells bc and c with a strong pale fulvous tint; microtrichia in outer corner of cell c only; remainder of wings colourless except dark fuscous cell sc, broad fuscous costal band confluent with R_{4+5} basally and overlapping this vein just beyond *r-m* crossvein to apex of wing (the band is paler between R_{2+3} and R_{4+5}), anal streak reduced to pale fuscous within cell cup; no dense aggregation of microtrichia around A_1+CuA_2; supernumerary lobe weak.

Abdomen: Elongate oval (not as petiolate as in *Bactrocera minax*); terga free; pecten present on tergum III. Tergum I and sterna I and II wider than long. Terga I and II vary from red-brown to fulvous; terga III–V red-brown to fulvous with a dark 'T' pattern consisting of a narrow transverse dark fuscous to black band across anterior margin of tergum III and a narrow medial longitudinal black band over all three terga (in some specimens the anterolateral corners of tergum IV are fuscous to dark fuscous); a pair of oval red-brown shining spots on tergum V. Posterior lobe of surstylus short, sternum V with a slight concavity on posterior margin. Abdominal sterna pale.

Attractant: No known record.

Female: As for male except pecten of cilia on abdominal tergum III absent; basal segment of ovipositor red-brown and strongly bulbous and rounded basally and narrowing to a cylinder apically, ratio of length of oviscape to length of tergum V, 1.8:1; apex of piercer trilobed.

Distribution:

Type locality: Japan (Ryukyu Is, Kyushu I.).

Previously recorded: Type locality and the following provinces in China (Guangxi, Hunan, Jiangsu, Sichuan) (White & Wang, 1992).

New records: Japan – Kyushu I., Tsukumi; Ryukyu Is (Tokara Is Group, Nakano-shima I.).

Hosts: A range of *Citrus* species and *Fortunella* species (family *Rutaceae*) (see Allwood *et al.*, 1999).

Remarks: *Bactrocera (Tetradacus) tsuneonis* (Miyake) is very close to *Bactrocera (Tetradacus) minax* (Enderlein) in the general body shape and colour patterns but differs from this species in having a complete yellow band joining the postpronotal lobe and notopleuron, *a.sa.* setae, paler cells bc and c, costal band without a distinct dark fuscous spot at apex and the overall abdomen shape elongate oval but not strongly petiolate.

SUBGENUS *ZEUGODACUS* Hendel

Bactrocera (Zeugodacus) abdoangusta (Drew)

(Fig. 272)

Dacus (Zeugodacus) abdoangustus Drew, 1972b: 191; Hardy, 1982: 199. Holotype male in QMIC.

Bactrocera (Zeugodacus) abdoangusta – Drew 1989: 210; Norrbom *et al.*, 1998: 101.

Type material examined: Holotype male, **BOUGAINVILLE ISLAND**: Daru Village, v.1970, attracted to cue lure, coll. R.M. Berena. Specimen in QMIC.

Diagnosis: A large species; face fulvous with medium-sized black facial spots; postpronotal lobes and notopleura yellow; scutum red-brown with black markings between lateral and medial postsutural vittae and between notopleuron and postpronotal lobe; lateral and medial postsutural yellow vittae present; a yellow spot anterior to notopleural suture; mesopleural stripe reaching to anterior *npl.* seta dorsally; scutellum entirely yellow; all leg segments entirely fulvous; wing with cells bc and c fulvous, microtrichia in outer ½ of cell c only; a general pale fulvous tint throughout the membrane and a broad fuscous costal band confluent with R_{4+5} and widening across apex of wing, infuscation along *r-m* and *dm-cu* crossveins; a broad fuscous anal streak; supernumerary lobe strong and rounded; abdomen with pecten present on tergum III, abdominal terga III–V fulvous with a narrow transverse black band across anterior margin of tergum III and a moderately broad medial longitudinal black band over all three terga, narrow black lateral margins on all three terga. Posterior lobe of surstylus long, sternum V with a slight concavity on posterior margin.

Attractant: Cue lure.

Female: No known record.

Distribution:

Type locality: Papua New Guinea (Bougainville I., Daru Village).

Previously recorded: Type locality. The recording by Hardy (1982) from Sulawesi, Indonesia is probably a misidentification. (See 'Remarks' below.)

Hosts: No known record.

Remarks: *Bactrocera (Zeugodacus) abdoangusta* (Drew) has been adequately described and revised by Drew (1972b, 1989). The recording by Hardy (1982) from Sulawesi is probably in error based on misidentification of specimens of *Bactrocera (Zeugodacus) synnephes* (Hendel). *Bactrocera* *abdoangusta* has only been recorded from Bougainville Island in the South Pacific and is unlikely to occur in South-East Asia. It differs from *B. synnephes* in possessing a broader costal band almost confluent with R_{4+5}, a wider mesopleural stripe reaching anterior *npl.* seta dorsally, coloured cells bc and c and infuscation on *r-m* crossvein in addition to that on *dm-cu* crossvein.

Bactrocera (Zeugodacus) abnormis (Hardy)

(Fig. 273)

Dacus (Zeugodacus) abnormis Hardy, 1982: 201. Holotype male in MZB.
Bactrocera (Zeugodacus) abnormis – Norrbom *et al.*, 1998: 101.

Type material examined: Holotype male, **INDONESIA: CENTRAL SULAWESI** – Sulawesi, 65 km SE Palu, Sadaunata 650 m, viii.1975, cue lure, primary forest, coll. D.E. Hardy. Studied at MZB.
Non-type material examined: INDONESIA: SULAWESI – 1 male, Dumoga, Taraut, N. Sulawesi, 22.xii.1985, coll. L. Santianawati. 1 male, Dumoga, Taraut, 350m, N. Sulawesi, 21.xii.1985, coll. Santianawati; 4 males (12.viii.2006), 16 males (27.viii.2006), 23 males (2.ix.2006), 2 males (17.ix.2006), 12 males (23.ix.2006), 4 males (27.ix.2006), 2 males (8.x.2006), 1 male (15.x.2006), Sulawesi, Sulawesi Tengah, Donggala, Nupabomba, coll. Rusdin, attracted to cue lure in forest; 1 male, Sulawesi, Gorontalo, Limboto, Bionga, 29.viii.2006, coll. Yusbar Ismail, attracted to cue lure in forest; 1 male (4.ix.2006), 2 males (11.ix.2006), Sulawesi, Sulawesi Barat, Simboro, Ds Tambayako, coll. Yan P. Fachrie, attracted to cue lure; 1 male, Sulawesi, Gorontalo, Bone Bolango, Suwawa, 31.x.2006, coll. Yusbar Ismail, attracted to cue lure in forest; **NORTH MOLUCCAS (NORTH MALUKU)** – 2 males, Moluccas, North Maluku, Kodya Ternate, Foramadiahi, 27.v.2007, coll. La Ruti, attracted to cue lure in forest; 1 male, Moluccas, North Maluku, Kodya Ternate, Kalumata, 3.vi.2007, attracted to cue lure in mango; 1 male, Maluku, North Maluku, West Halmahera, Goal, 29.v.2007, coll. Raais Abdullah, attracted to cue lure in citrus/chilli. **EAST MALAYSIA: SABAH** – 1 male, Poring Hot Springs, nr. Mt. Kinabalu, Sabah, 24.vii.1992, D.L. Hancock, attracted to cue lure. Specimens in QDPC.
Diagnosis: A medium-sized species; face entirely fulvous; postpronotal lobes and notopleura yellow; scutum mostly black; lateral and medial postsutural yellow vittae present; no yellow spot anterior to notopleural suture; mesopleural stripe reaching to postpronotal lobe dorsally; scutellum yellow; legs entirely fulvous without dark markings; wing with cells bc and c colourless, microtrichia in outer corner of cell c only; a narrow fuscous costal band broadening across apex of wing; a broad fuscous anal streak; supernumerary lobe weak; abdominal terga entirely pale red-brown except for a pair of small dark fuscous spots on anterolateral corners of tergum III; abdominal sterna pale.

Description: male.
Head: Height 1.29 mm. Frons – length 1.38 times breadth; fulvous without dark markings; orbital setae black: 1 *s.or.*, 2 *i. or.*; lunule fulvous. Ocellar triangle dark fuscous. Vertex fulvous. Face entirely fulvous without dark markings; length 0.43 mm. Genae fulvous. Occiput entirely fulvous. Antennae with segments 1 and 2 red-brown, segment 3 red-brown tending fuscous on apex; length of segments: 0.1 mm; 0.2 mm; 0.56 mm.
Thorax: Scutum black except dark red-brown on extreme lateral margins below lateral postsutural yellow vittae, on posterolateral margins of medial postsutural yellow vitta, anterior to notopleural suture and as a medial longitudinal band from anterior point of medial postsutural yellow vitta to anterior margin of scutum. Pleural areas dark fuscous to black except red-brown below postpronotal lobes. Yellow markings as follows: postpronotal lobes; notopleura; broad mesopleural stripe reaching to postpronotal lobe dorsally, continuing to katepisternum as a transverse spot, anterior margin straight; two broad parallel-sided lateral postsutural vittae beginning at notopleural suture and ending behind *ia.* seta; a broad medial longitudinal postsutural vitta broadly rounded posteriorly at level of *prsc.* setae and narrowing to a point approximately level with notopleural suture; no yellow spot anterior to notopleural suture. Scutellum yellow except for narrow black basal band. Setae: *sc.* 4; *prsc.* 2; *ia.* 1; *p.sa.* 1; *a.sa.* 1; *mpl.* 1; *npl.* 2; *scp.* 4.
Legs: All segments fulvous to dark fulvous; mid tibiae each with an apical black spur; two long setae on lower surfaces of basal segment of hind tarsi.
Wings: Length 4.9 mm; cells bc and c colourless; microtrichia in outer corner of cell c only; remainder of wings colourless except pale fuscous cell sc, narrow fuscous costal band confluent with R_{2+3} and widening beyond apex of R_{2+3} to end midway between apices of R_{4+5} and M, a broad fuscous anal streak; dense aggregation of microtrichia around A_1+CuA_2; supernumerary lobe weak.
Abdomen: All terga pale red-brown except for a small dark fuscous spot on anterolateral corners of tergum III. In the specimen from Sabah, East Malaysia, there is a narrow transverse black band across anterior margin of tergum III and a narrow medial longitudinal black band over tergum V. A pair of oval pale red-brown shining spots on tergum V. Posterior lobe of surstylus long, sternum V with a slight concavity on posterior margin. All abdominal sterna pale.
Attractant: Cue lure.
Female: No known record.
Distribution:
Type locality: Indonesia (Sulawesi – Central Sulawesi, Sadaunta).
New records: East Malaysia (Sabah); Indonesia (North Moluccas = North Maluku – Kodya Ternate, West Halmahera; Sulawesi – North, Gorontalo, Central, West Provinces).
Hosts: No known record.
Remarks: *Bactrocera (Zeugodacus) abnormis* (Hardy) is similar to *Bactrocera (Zeugodacus) connexa* (Hardy) in possessing pale abdominal sterna, scutum mostly black and wings colourless except for the costal band and anal streak. It differs from this species in having the face

entirely fulvous, lacking a lateral yellow band connecting the notopleuron and postpronotal lobe and abdominal terga III–V without a black 'T' pattern.

Bactrocera (Zeugodacus) aithonota Drew & Romig, new species

(Fig. 274)

Material examined: Holotype male, **VIETNAM: Northern Vietnam** – Hoabinh Province, Mai Chau-Paco, 6.v.1999, coll. Le Duc Khanh *et al.*, attracted to cue lure, deposited in BMNH.
Paratype: **VIETNAM: Northern Vietnam** – 1 male, Pa Co, Mai Chau District, Hoa Binh Province, North Vietnam, 1. iv.2001, attracted to cue lure. Deposited in QDPC.
Diagnosis: A large species; face with a distinct inverted 'U'-shaped black pattern; postpronotal lobes yellow; notopleura dark red-brown to fuscous; scutum black; lateral and medial postsutural yellow vittae present; a yellow spot anterior to notopleural suture; mesopleural stripe broad reaching almost to postpronotal lobe dorsally; scutellum yellow; wing with cells bc and c colourless, microtrichia in outer corner of cell c only; a narrow fuscous costal band, a broad dark fuscous band along *dm-cu* crossvein; a broad fuscous anal streak; abdominal terga III–V red-brown with a 'T' pattern consisting of a broad black band across anterior margin of tergum III which becomes narrow and fuscous in centre where it meets the moderately broad medial longitudinal black band which extends over all three terga, broad black anterolateral corners on terga IV and V.
Description: male.
Head: Height 1.93 mm. Frons – length 1.18 times breadth; red-brown with fulvous laterally and pale fuscous on anteromedial hump and around bases of orbital setae; orbital bristles black: 1 *s.or.*, 3 *i.or.*; lunule dark fuscous. Ocellar triangle black. Vertex red-brown with dark fuscous laterally not meeting at Ocellar triangle. Face fulvous with a pair of large oval black spots joined in centre to form a 'U'-shaped black pattern; length 0.66 mm. Genae fulvous, large fuscous to dark fuscous subocular spot present; black seta present. Occiput fuscous, fulvous along eye margins; occipital row with a large number of short pale setae. Antennae with segments 1 and 2 dark red-brown, segment 3 red-brown with dark fuscous on apex and outer surface; arista black (red-brown basally); length of segments: 0.16 mm; 0.36 mm; 0.93 mm.
Thorax: Scutum uniformly black with red-brown below and behind lateral postsutural vittae and around notopleural suture. Pleural areas black except red-brown below postpronotal lobes. Yellow markings as follows: postpronotal lobes; broad mesopleural stripe reaching just anterior to anterior *npl.* seta dorsally, continuing to katepisternum as a transverse spot, anterior margin straight; anatergite (posterior apex black); anterior ⅔ katatergite (remainder black); two parallel-sided lateral postsutural vittae of medium-width beginning with a spot anterior to notopleural suture and ending at *ia.* seta; a short narrow medial longitudinal postsutural vitta. Notopleuron dark red-brown to fuscous. Postnotum black. Scutellum yellow except for narrow black basal band (black spot on apex in Paratype). Setae: *sc.* 2; *prsc.* 2; *ia.* 1; *p.sa.* 1; *a.sa.* 1; *mpl.* 1; *npl.* 2; *scp.* 2.
Legs: Fore femora dark fuscous to black except fulvous on basal ½ of inner surface, mid femora fulvous with apical ½ dark fuscous to black, hind femora fulvous with apical ⅓ dark fuscous to black; fore and mid tibiae fuscous tending darker basally, hind tibiae dark fuscous; tarsi with basal segment fulvous and apical four segments fuscous; mid tibiae each with an apical black spur.
Wings: Length 8.57 mm; cells bc and c colourless; microtrichia in outer corner of cell c only; remainder of wings colourless except dark fuscous cell sc, narrow fuscous costal band confluent with R_{2+3} and ending in a 'fishhook' pattern at apex of R_{4+5}, a broad dark fuscous band around *dm-cu* crossvein, a broad fuscous anal streak; a dense aggregation of microtrichia around A_1+CuA_2; supernumerary lobe moderately strong. In the Paratype, the costal band overlaps R_{2+3}.
Abdomen: Oval; terga free; pecten present on tergum III. Tergum I and sterna I and II wider than long. Tergum I fuscous with broad black lateral margins; tergum II red-brown with a broad black transverse band across anterior margin but not reaching the broad black anterolateral corners; terga III–V red-brown with a broad black transverse band across anterior margin of tergum III which is narrow and fuscous where it joins in the centre, a moderately broad medial longitudinal black band over all three terga, broad black anterolateral corners on terga IV and V. A pair of oval red-brown shining spots on tergum V. Posterior lobe of surstylus long, sternum V with a slight concavity on posterior margin.
Attractant: Cue lure.
Female: No known record.
Distribution: Northern Vietnam (Hoa Binh Province).
Hosts: No known record.
Remarks: *Bactrocera (Zeugodacus) aithonata* n.sp. is similar to *Bactrocera (Zeugodacus) elegantula* (Hardy), *Bactrocera (Zeugodacus) hululangatiae* n.sp., *Bactroera (Zeugodacus) neoelegantula* White and some specimens of *Bactrocera (Zeugodacus) synnephes* (Hendel) with two *sc.* setae in possessing two scutellar setae, a medial longitudinal postsutural yellow vitta in addition to the lateral postsutural yellow vittae, prescutellar setae, a black scutum and infuscation around *dm-cu* crossvein. It is unique in having dark red-brown to fuscous notopleura, large dark markings on the femora and broad black lateral markings on abdominal terga III–V.

Bactrocera (Zeugodacus) ambigua (Shiraki)

(Fig. 275)

Zeugodacus ambiguus Shiraki, 1933: 85. Holotype male in NIAS.
Bactrocera (Zeugodacus) ambigua – Norrbom *et al.*, 1998: 101.
Dacus (Zeugodacus) pauleiensis Tseng, Chen & Chu, 1992: 72. **New Synonym**.

Type material examined: Holotype male, **TAIWAN (formerly FORMOSA)**: with the following three labels – 'Formosa, Koshun, 1918.IV.25-V.25, J. Sonan, K. Miyake, M. Yoshino'; 'Zeugodacus ambiguus sp. nov. Det. by T. Shiraki, Taihoku, viii, 1931'; 'Type'. Received on loan from NIAS.
Non-type material examined: TAIWAN: 13 males, Band Hsie, Nantour, 21.ix.1978, coll. Lena Lee Wen, attracted to cue lure. Specimens in QDPC.
Diagnosis: A large species; face fulvous with medium-sized circular black spots; scutum black with large areas of red-brown anteriorly; broad lateral and medial postsutural yellow vittae present; a large yellow spot anterior to notopleural suture; mesopleural stripe reaching midway between anterior margin of notopleuron and anterior *npl.* seta dorsally; scutellum yellow with a large dark fuscous apical spot; wing with cells bc and c colourless, microtrichia in outer corner of cell c only; a narrow fuscous costal band slightly overlapping R_{2+3} and expanding slightly across apex of R_{4+5}, a narrow pale fuscous margin along *dm-cu* crossvein; broad fuscous anal streak; abdominal terga III–V red-brown with a narrow transverse black band across anterior margin of tergum III, a narrow medial longitudinal black band over terga IV and V, rectangular-shaped dark fuscous to black anterolateral corners on tergum IV and large triangular dark fuscous anterolateral corners on tergum V.
Description: male.
Head: Missing in Holotype. Frons fulvous with dark fuscous around bases of orbital setae and fuscous on anteromedial hump; orbital setae black: 1 *s.or.*, 3 *i.or.*; lunule dark fuscous. Ocellar triangle black. Vertex black. Face fulvous with a pair of medium-sized circular black spots. Genae red-brown, dark fuscous subocular spot present; black seta present. Occiput red-brown, fulvous along eye margins. Antennae with segments 1 and 2 red-brown, segment 3 fuscous.
Thorax: Scutum black with large areas of red-brown anteriorly and below and behind lateral postsutural vittae. Pleural areas red-brown with fuscous along anterior and posterior margins of mesopleural stripe and black on katepisternum. Yellow markings as follows: postpronotal lobes; notopleura; mesopleural stripe of medium-width reaching midway between anterior margin of notopleuron and anterior *npl.* seta dorsally, continuing to katepisternum as a large transverse spot, anterior margin straight; anatergite (posterior apex black); anterior ¾ katatergite (remainder black); two moderately broad lateral postsutural vittae beginning in a large spot anterior to notopleural suture anteriorly and narrowing slightly posteriorly to end behind level of *ia.* setae at which point they turn inwards slightly; a broad medial longitudinal postsutural vitta broadly rounded posteriorly and ending at a point level with notopleural suture. Postnotum black tending dark red-brown centrally. Scutellum yellow except for narrow black basal band and a large dark fuscous apical spot which encloses the apical setae. Setae: *sc.* 4; *prsc.* 2; *ia.* 1; *p.sa.* 1; *a.sa.* 1; *mpl.* 1; *npl.* 2; *scp.* 4.
Legs: Missing in Holotype. All femora entirely fulvous; fore tibiae fuscous, mid tibiae fulvous, hind tibiae dark fuscous; all tarsi with basal segment fulvous and apical four segments fuscous; mid tibiae each with an apical black spur.
Wings: Length 7.03 mm; cells bc and c colourless; microtrichia in outer corner of cell c only; remainder of wings colourless except fuscous cell sc, narrow fuscous costal band slightly overlapping R_{2+3} as a pale tint and expanding slightly across apex of R_{4+5}, a pale fuscous marking around *dm-cu* crossvein, a broad fuscous anal streak; dense aggregation of microtrichia around A_1+CuA_2; supernumerary lobe strong and rounded.
Abdomen: Oval; terga free; pecten present on tergum III. Tergum I and sterna I and II wider than long. Tergum I red-brown with narrow lateral black margins; tergum II red-brown with a narrow transverse black band across anterior margin but not reaching the narrow black anterolateral corners; terga III–V red-brown with a narrow transverse black band across anterior margin of tergum III which extends along the midline into a narrow black medial longitudinal band which does not reach posterior margin, rectangular-shaped dark fuscous to black anterolateral corners on tergum IV and a narrow medial longitudinal black band over the entire tergum, large dark fuscous anterolateral corners on tergum V and a narrow medial longitudinal black band over most of this tergum. A pair of oval red-brown shining spots on tergum V. Posterior lobe of surstylus long, sternum V with a slight concavity on posterior margin. Abdominal sterna mostly dark coloured.
Attractant: Cue lure.
Female: No known record.
Distribution:
Type locality: Taiwan (Koshun).
New records: Type locality and Taiwan (Band Hsie, Nantour). Taiwan (Paulei, Kaohsiung Hsien), due to **New Synonymy** of *D. pauleiensis*.
Hosts: No known record.
Remarks: *Bactrocera (Zeugodacus) ambigua* (Shiraki) is similar to *Bactrocera (Zeugodacus) calumniata* (Hardy), *Bactrocera (Zeugodacus) ishigakiensis* (Shiraki), *Bactrocera (Zeugodacus) neoemittens* n.sp. and *Bactrocera (Zeugodacus) synnephes* (Hendel) in possessing four *sc.* setae, a black scutum, wing with infuscation on either one or both crossveins and the costal band with a slight expansion across wing apex. It differs from *B. calumniata*, *B. neoemittens* and *B. synnephes* in having a black spot on the apex of the scutellum and from *B. ishigakiensis* in having infuscation on only the *dm-cu* crossvein, cells bc and c colourless and the lateral postsutural yellow vittae extending posterior to the *ia.* setae and turned inwards. *Bactrocera ambigua* is similar to *Bactrocera (Zeugodacus) scutellata* (Hendel) but differs in having broader medial and lateral postsutural yellow vittae and a dark 'T' pattern over abdominal terga III–V. We can find no good characters to separate *B. ambigua* and *Dacus (Zeugodacus) pauleiensis* Tseng, Chen & Chu and are thus making the latter a **New Synonym** of *B. ambigua*.

Bactrocera (Zeugodacus) apicofemoralis Drew & Romig, new species

(Fig. 276)

Type material examined: Holotype male, **INDONESIA: JAVA** – Pelabuhan Ratu (0–100 m), Java, 12.v.1981, coll. J. Turner, attracted to cue lure, deposited in BMNH.

Paratypes: **INDONESIA: FLORES** – 1 male, Borong, Flores, iii.1988, coll. J. Turner, attracted to cue lure; 86 males, Borong, Flores, iv.1989, coll. J. Turner; **JAVA** – 3 males, Gunung Salak, Bogor (800 m), Java, Indonesia, 17.v.1981, coll. J. Turner, secondary rainforest; 3 males, Pasir Nini Nini, Lebak, Banten, Java, 20.vii.2005, coll. Aden Haer, attracted to cue lure in forest. 22 paratypes in ANIC; 10 paratypes in BMNH; 50 paratypes in QDPC; 11 paratypes in QMIC.

Non-type material examined: INDONESIA: JAVA – 1 male, Banten, Lebak, Pasir Nini Nini, 22.viii.2005, coll. TB Aden Haer, attracted to cue lure in forest; 1 male, East Java, Probolinggo, Betektanan, 20.vii.2005, coll. Tijipto Waluyo, attracted to cue lure in mango; **KALIMANTAN** – 2 males, West Kalimantan, Pontianak, Sei Duri II, 19.vii.2005, coll. A. Irom, attracted to cue lure in citrus; 1 male, West Kalimantan, Pontianak, Sei Pinyuh, 26.vii.2005, coll. A. Irom; **SUMATRA** – 1 female, Bengkulu Selatan, Kebon Jati, 28.ix.2006, coll. M. Nelson, attracted to cue lure in citrus. Specimens in QDPC.

Diagnosis: A large species; face fulvous with a broad black band across oral margin; postpronotal lobes and notopleura yellow; scutum black; lateral and medial postsutural yellow vittae present; a yellow spot anterior to notopleural suture; mesopleural stripe reaching midway between anterior margin of notopleuron and anterior *npl.* seta; scutellum yellow; wing with cells bc and c colourless, microtrichia in outer corner of cell c only; a narrow fuscous costal band expanding into a small spot around apex of R_{4+5} and a broad fuscous anal streak; abdominal terga III–V orange-brown with a general dark fuscous to black 'T' pattern which may have the medial longitudinal band broken on posterior areas of terga III and IV and broad anterolateral dark fuscous corners on terga IV and V (on tergum IV the corners may be joined across anterior margin of the tergum).

Description: male.

Head: Height 1.64 mm. Frons – length 1.41 times breadth; red-brown with fulvous on lateral margins, fuscous around bases of orbital setae and on anteromedial hump; orbital setae black: 1 *s.or.*, 3 *i.or.*; lunule dark fuscous. Ocellar triangle black. Vertex black. Face fulvous with a broad black band across oral margin; length 0.59 mm. Genae fulvous, a large dark fuscous subocular spot present; black seta present. Occiput fuscous with dark fuscous dorsolaterally, fulvous along eye margins; occipital row with a small number of short dark setae. Antennae with segments 1 and 2 red-brown, segment 3 red-brown with fuscous on apex and outer surface; arista black (red-brown basally); length of segments: 0.12 mm; 0.31 mm; 0.82 mm.

Thorax: Scutum entirely black except dark red-brown to fuscous around notopleural suture and posterolateral to lateral postsutural vittae. Pleural areas entirely black. Yellow markings as follows: postpronotal lobes; notopleura; mesopleural stripe reaching midway between anterior margin of notopleuron and anterior *npl.* seta dorsally, continuing to katepisternum as a small transverse spot, anterior margin convex; anatergite (posterior apex black); anterior ⅔ katatergite (remainder black); two broad parallel-sided lateral postsutural vittae beginning with a spot anterior to notopleural suture and ending at or just behind *ia.* seta; a broad medial longitudinal postsutural vitta beginning posterior to *prsc.* setae and ending in a point at level of notopleural suture. Postnotum black. Scutellum yellow except for narrow black basal band. Setae: *sc.* 4; *prsc.* 2; *ia.* 1; *p.sa.* 1; *a.sa.* 1; *mpl.* 1; *npl.* 2; *scp.* 4.

Legs: Fore femora with apical ⅔ dark fuscous to black and basal ⅓ fulvous, mid femora with apical ⅓ to ½ dark fuscous and remainder fulvous, hind femora with apical ⅓ dark fuscous and basal ⅔ fulvous; fore tibiae fuscous tending darker basally, mid tibiae fulvous with fuscous basally, hind tibiae dark fuscous; tarsi with basal segment fulvous and apical four segments fuscous; mid tibiae each with an apical black spur.

Wings: Length 6.7 mm; cells bc and c colourless; microtrichia in outer corner of cell c only; remainder of wings colourless except fuscous cell sc, narrow fuscous costal band confluent with R_{2+3} and widening into a spot around apex of R_{4+5}, a moderately broad fuscous anal streak; dense aggregation of microtrichia around A_1+CuA_2; supernumerary lobe strong and rounded.

Abdomen: Oval; terga free; pecten present on tergum III. Tergum I and sterna I and II wider than long. Tergum I fuscous centrally and dark fuscous laterally except for a small transverse orange-brown band on posterocentral margin; tergum II fulvous with a narrow transverse dark fuscous band subanteriorly which may be broken in the centre and joining dark fuscous anterolateral margins; tergum III with a moderately broad transverse dark fuscous to black band across anterior margin and expanding to include lateral margins, tergum IV with broad dark fuscous anterolateral corners which sometimes join along anterior margin as a fuscous band in centre of tergum, tergum V with broad anterolateral dark fuscous corners, a narrow- to medium-medial longitudinal band which varies from fuscous to dark fuscous and can continue over all three terga or be broken on posterior margins of terga III and IV. A pair of oval orange-brown shining spots on tergum V. Posterior lobe of surstylus long, sternum V with a slight concavity on posterior margin. Abdominal sterna dark except sternum II pale.

Attractant: Cue lure.

Female: No known record.

Distribution: Indonesia (Flores; Java; Kalimantan; Sumatra).

Hosts: No known record.

Remarks: *Bactrocera (Zeugodacus) apicofemoralis* n.sp. is similar to *Bactrocera (Zeugodacus) laguniensis* n.sp. in general body colour patterns and in possessing a black band across the oral margin of the face. It differs from *B. laguniensis* in having parallel-sided lateral postsutural vittae, a strong supernumerary lobe in the male wing, an apical spot in the costal band and broad anterolateral dark corners on terga IV and V. It is also similar to *Bactrocera (Zeugodacus) caudata* (Fabricius) but differs in possessing dark fuscous to black over fore femora and around apices of mid and hind femora, rather than small fuscous spots on outer apical surfaces of all femora, and abdominal sternum V broad and oval shaped rather than a narrow transverse shape.

Bactrocera (Zeugodacus) assamensis White

(Fig. 277)

Bactrocera (Zeugodacus) assamensis White, in White & Evenhuis, 1999: 522. Holotype female in BMNH.
Bactrocera (Zeugodacus) gansuica Chen, Han & Zhou, in Chen *et al*., 2012: 352. **New Synonym**.

Type material examined: Holotype female, **INDIA: ASSAM** – with the following five labels – 'holotype'; 'BURMA: Mishmi Hills, Lohit R, 17.iii.1935, M. Steele'; 'Brit Mus 1935-312'; 'HOLOTYPE, *Bactrocera assamensis* White, det I.M. White, 1999'; 'BMNH(E), #533014'. Studied at BMNH.
Non-type material examined: BHUTAN: 1 male, Zingman, 4.x.1991, coll. H. Fiejen, attracted to cue lure (poss. contaminated by methyl eugenol); 1 male, Rimchu trap No. 1, 7.vi.2000, coll. C. Dorji *et al*., attracted to cue lure; 1 male, Rimchu trap No. 1, 14.vi.2000, coll. Nado T. Shomo, attracted to cue lure; 1 male, Rimchu trap No. 1, 5.x.2000, coll. C. Dorji, attracted to cue lure; 1 male, Rimchu trap No. 2, 10.v.2000, coll. B.S. Fletcher, attracted to cue lure; 1 male, Rimchu trap No. 3, 4.v.2000, coll. B.S. Fletcher, attracted to cue lure; 1 male, Tsirang, 17.vi.2000, coll. T.D. & S.D., attracted to cue lure; 1 male (10.v.2000), 1 male (17.v.2000), Lumitshawa, C. Dorji, attracted to cue lure; 2 males, Namling 14 km, 7.vi.2001, coll. Brian Fletcher, attracted to cue lure. **VIETNAM: Northern Vietnam** – 1 male (23.iv.1999), 1 male (13.viii.1999), 1 male (27.viii.1999), Lao Cai Province, Sapa town, trap No. 16, coll. Le Duc Khanh *et al*., attracted to cue lure; 1 male, Son La (P), Moc Chau (D), Long Luong (V), 15.iii.2002, trap VN01, coll. NIPP staff, attracted to cue lure. Specimens in QDPC.
Diagnosis: A large species; face fulvous with a pair of large elongate oval black spots; postpronotal lobes yellow with inner margins narrowly black; notopleura either entirely dark fuscous to black or dark fuscous to black with apical yellow spots; scutum shining black; narrow lateral postsutural yellow vittae present; very short and narrow medial postsutural yellow vitta present; mesopleural stripe reaching midway between anterior margin of notopleuron and anterior *npl.* seta dorsally; scutellum yellow with a broad medial longitudinal dark fuscous to black band or in some specimens a broad apical dark fuscous to black spot; wing with cells bc and c colourless, microtrichia in outer corner of cell c only; a narrow dark fuscous costal band extremely narrow or absent beyond apex of R_{2+3} and widening into a distinct spot across apex of R_{4+5} and a narrow fuscous anal streak; a small diffuse fuscous spot around CuA_1 at hind margin of wing; abdominal terga III–V mostly dark fuscous to black with small red-brown spots posterocentrally on terga III and IV either side of a narrow medial longitudinal black band and red-brown across posterior margin of tergum V.
Description: male.
Head: Height 1.7 mm. Frons – length 1.35 times breadth; mostly fuscous with narrow red-brown lateral margins, red-brown across anterior margin and dark fuscous around bases of orbital setae; orbital setae black: 1 *s.or.*, 3 *i.or.*; lunule dark fuscous. Ocellar triangle black. Vertex black. Face fulvous with a pair of large elongate oval black spots; length 0.51 mm. Genae fuscous; black seta present. Occiput black, fulvous along eye margins; occipital row with a large number of small pale setae. Antennae with segments 1 and 2 fuscous, segment 3 red-brown with dark fuscous on apex and outer surface; arista black (red-brown basally); length of segments: 0.12 mm; 0.33 mm; 0.69 mm.
Thorax: Scutum entirely shining black with, at most, dark red-brown posterior to lateral postsutural vittae in some specimens. Pleural areas entirely black. Yellow markings as follows: postpronotal lobes (with a narrow black margin along inner surfaces); mesopleural stripe of medium-width, reaching midway between anterior margin of notopleuron and anterior *npl.* seta dorsally, anterior margin slightly convex (the stripe does not extend onto katepisternum); anatergite (posterior apex black); anterior ⅔ katatergite (remainder black); two narrow parallel-sided lateral postsutural vittae ending just before or at *ia.* seta; a very short and narrow medial longitudinal postsutural vitta; there is no yellow spot anterior to notopleural suture. Notopleuron either entirely dark fuscous to black or dark fuscous to black with variable amounts of yellow around apex. Postnotum black. Scutellum yellow with a narrow black basal band and a broad medial longitudinal band which widens slightly apically to enclose the apical scutellar setae (in some specimens, there is only a broad apical dark fuscous to black spot enclosing the apical scutellar setae). Setae: *sc.* 2; *prsc.* 2; *ia.* 1; *p.sa.* 1; *a.sa.* 1; *mpl.* 1; *npl.* 2; *scp.* 4.
Legs: Fore femora almost entirely shining black (may have small areas of fulvous apically and basally), mid femora with apical ⅓ to ½ shining black and remainder fulvous, hind femora with apical ⅓ shining black with remainder fulvous; fore tibiae dark fuscous to black tending paler apically, mid tibiae dark fuscous on basal ½ tending fulvous apically, hind tibiae dark fuscous to black; tarsi with basal segment fulvous and apical four segments fuscous to dark fuscous; mid tibiae each with an apical black spur.
Wings: Length 7.28 mm; cells bc and c colourless; microtrichia in outer corner of cell c only; remainder of wings colourless except dark fuscous cell sc, narrow dark fuscous costal band confluent with R_{2+3} and extremely narrow (sometimes absent) beyond apex of R_{2+3} before expanding into a distinct spot around apex of R_{4+5}, a small diffuse fuscous spot on hind margin around CuA_1, a narrow fuscous anal streak; a dense aggregation of microtrichia around A_1+CuA_2; supernumerary lobe weak. The area of cell br normally covered with microtrichia is variable in density of these fine hairs.
Abdomen: Oval; terga free; pecten present on tergum III. Tergum I and sterna I and II wider than long. Tergum I dark fuscous to black except for a narrow- to medium-width transverse fulvous band across posterior margin that usually does not reach lateral margins; tergum II black except for a moderately broad transverse fulvous band across posterior margin (sometimes this band is divided by a narrow medial longitudinal black band); terga III–V dark fuscous to black except for small red-brown spots posterocentrally on terga III and IV either side of a medial longitudinal black band and red-brown across posterior margin of tergum V. A pair of oval dark fuscous shining spots on tergum V. Posterior lobe of surstylus

long, sternum V with a slight concavity on posterior margin. Abdominal sterna black, except sternum II fulvous.
Attractant: Cue lure.
Female: As for male except no dense aggregation of microtrichia around A_1+CuA_2; pecten of cilia on abdominal tergum III absent; basal segment of ovipositor dark fuscous to black, dorsoventrally flattened and tapering posteriorly in dorsal view; ratio of length of oviscape to length of tergum V, 2.1:1.
Distribution:
Type locality: India (Assam, Mishmi Hills).
New records: Bhutan; Northern Vietnam (Lao Cai and Son La Provinces). China, Gansu Province, Wenxian County, due to **New Synonymy** of *B. gansuica*.
Hosts: No known record.
Remarks: *Bactrocera (Zeugodacus) assamensis* White, within the *scutellaris* complex, is similar to *Bactrocera (Zeugodacus) biguttata* (Bezzi), *Bactrocera (Zeugodacus) freidbergi* White and *Bactrocera (Zeugodacus) tappanus* (Shiraki) in having two *sc.* setae and dark colour on or around notopleura. It differs from *B. freidbergi* in having dark patterns on the scutellum, from *B. tappanus* in having abdominal terga III–V mostly dark fuscous to black and femora uniformly black around apices, and from *B. biguttata* in having a medial longitudinal dark band on the scutellum, microtrichia on cell br of wing and anal streak present. Measurements included above are from specimens collected in Moc Chau, Northern Vietnam, not the type. In the specimens of *B. assamensis* from Bhutan and Vietnam, there is variation on the amount of dark colouration and yellow on the notopleuron. Both *B. assamensis* and *B. gansuica* fit within this range of variation and these species are identical in all other characters. Consequently, we are making *B. gansuica* a **New Synonym** of *B. assamensis*.

Bactrocera (Zeugodacus) atrifacies (Perkins)

(Fig. 278)

Zeugodacus atrifacies Perkins, 1938a: 140. Holotype male in QMIC.
Dacus (Zeugodacus) atrifacies – Hardy, 1973: 58; Hardy, 1977: 57.
Bactrocera (Zeugodacus) atrifacies – Norrbom *et al.*, 1998: 102.
Dacus (Zeugodacus) parater Chao & Lin, 1993b: 138. **New Synonym**.

Type material examined: Holotype male, **PENINSULAR MALAYSIA**: with the following two labels – 'Malay Penin: Selangor, Bukit Kutu, 3550 ft, 14.ix.1929, H.M. Pendlebury'; 'Zeugodacus limbipennis Macquart, det. F.A. Perkins'. Received on loan from QMIC.
Non-type material examined: BHUTAN: 7 males (10.v.2002), 9 males (17.v.2002), 15 males (24.v.2002), Rimchu, coll. C. Dorji, attracted to cue lure; 4 males (17.v.2000), 14 males (24.v.2000), 9 males (31.v.2000), 1 male (30.viii.2000), Rimchu trap No. 1, coll. C. Dorji, attracted to cue lure; 11 males (7.vi.2000), 2 males (28.vi.2000), 4 males (5.vii.2000), 4 males (19.vii.2000), 2 males (26.vii.2000), 3 males (2.viii.2000), 1 male (16.viii.2000), Rimchu trap No. 1, coll. C. Dorji *et al.*, attracted to cue lure; 2 males, Rimchu trap No. 1, 14.vi.2000, coll. Nado, attracted to cue lure; 3 males, Rimchu trap No. 1, 21.vi.2000, coll. Nado *et al.*, attracted to cue lure; 1 male, Rimchu trap No. 1, 4.v.2000, coll. B.S. Fletcher, attracted to cue lure; 4 males (10.v.2000), 2 males (14.vi.2000), Rimchu trap No. 1, attracted to cue lure; 1 male, Rimchu trap No. 2, 4.v.2000, coll. B.S. Fletcher, attracted to cue lure; 5 males (17.v.2000), 4 males (21.v.2000), 5 males (24.v.2000), 1 male (10.viii.2000), Rimchu trap No. 2, coll. C. Dorji, attracted to cue lure; 1 male, Rimchu trap No. 3, 17.v.2000, coll. C. Dorji, attracted to cue lure; 3 males, 5.vi.2000, Suntalay, attracted to cue lure; 1 male, Suntalay, 22.v.2000, coll. S.P. & C. Dorji, attracted to cue lure; 4 males (11.vi.2000), 4 males (3.vii.2000), 1 male (31.vii.2000), Tsirang, Suntalay, attracted to cue lure; 1 male, Tsirang, 19.viii.2000, attracted to cue lure; 1 male (19.vi.2000), 2 males (10.vii.2000), Tsirang, coll. T.D. & S.D, attracted to cue lure; 1 male (6.v.2000), 1 male (13.v.2000), Phuentsholing, coll. S. Thapa, attracted to cue lure; 2 males, Namling 16km, 9.vi.2001, coll. Brian Fletcher, attracted to cue lure; 1 male, Lumitshawa, 24.v.2000, coll. C. Dorji, attracted to cue lure. **EAST MALAYSIA: SARAWAK** – 2 males, East Malaysia, Sarawak, Gunung Mulu, Nat. Park, 28-30.vii.1992, coll. D.L. Hancock, attracted to cue lure. 4 males, Tarat, Sarawak, East Malaysia, 16.vii.1977, coll. B.H. Voon, attracted to cue lure. 1 male, Sarawak, Ranchan Pool, Serian, 8.iii.1993, coll. S. Leong, attracted to cue lure; 1 male, Sarawak, Tarat Agric. Station, 19.iii.1993, coll. S. Leong, attracted to cue lure; 1 male, Sarawak, Tarat Agriculture Station, Nr Kuching, 10.vi.1992, coll. S. Leong, attracted to cue lure; 4 males, Sarawak, Tarat Agric. Station, 2.iv.1993, coll. S. Leong, attracted to cue lure. **THAILAND**: 3 males, 'Chumphon Prov, Khao Kai Jae, 80m, 5-8.v.2003/2'; '9.55.07N, 98.56.34E, semi-evergreen rainforest/2'; 'attracted to cue lure'; 1 male, 'Sunat Thanai Prov., Khao Sok N.P., 12–14.v.2003/7'; '8.54.55N/98.31.39E, P. Schwendinger leg, secondary forest/7'; 'attracted to cue lure'. Received on loan from MHNG. 1 male, Tone-Nga Chang Waterfall, near Hat Yai, 8.v.1991, D.L. Hancock, cue lure; 1 male, Wawi Exp. Sta., 1400m, 45km South of Chiang Rai, 22.v.1991, D.L. Hancock, cue lure. 1 male (20.viii.1991), 2 males (17.viii.1992), A. Laemsingh, Chanthaburi, trap BK64T, coll. Montree, attracted to cue lure; 2 males (20.viii.1991), 2 males (21.vii.1992), Yang Pong Rat, A. Muang, Chanthaburi, coll. Montree, attracted to cue lure; 1 male (20.viii.1991), 2 males (3.ix.1991), Yang Pong Rat, Plubpla, A. Muang, Chanthaburi, coll. Montree, attracted to cue lure; 1 male (6.ix.1991), 1 male (10.ix.1991), 1 male (20.ix.1991), Khao Yai, T. Moosi, A. Pakchong, Nakhonratsima, coll. Montree, attracted to cue lure; 1 male, Wang Jum Pee, Pakchong, Nakhonratsima, 1.xi.1991, coll. Montree, attracted to cue lure; 1 male (11.x.1991), 1 male (22.v.1992), 1 male (29.v.1992), 1 male (24.vii.1992), Khao Kheaw, A. Pakphli, Nakhonnayok, coll. Montree, attracted to cue lure; 2 males, Wang Jum Pee, A. Pakchong, Nakhonratsima, 14.viii.1992, coll. Montree, attracted to cue lure; 1 male, Wang Jum Pee,

A. Pakchong, Nakhonratsima, 21.viii.1992, coll. Montree, attracted to cue lure; 1 male (23.i.1992), 1 male (30.i.1992), 1 male (6.viii.1992), Khun Khon, Chiang Rai, coll. Charuwan, attracted to cue lure; 2 males, Pha Yao, Chiang Rai, 16.x.1991, coll. Charuwan, attracted to cue lure. 2 males, Nakkon Naoyk Prov., Khao Yai Nat. Park, ca. 700m, 29.ix.-6.x.1984, coll. Karsholt, Lomholdt & Nielsen leg. (these 2 males in UZMC). **VIETNAM: Northern Vietnam** – 1 male (15.v.2002), 1 male (30.iv.2002), Son La (P.), Moc Chau (D), Long Luong (V), coll. NIPP Staff, VN01, attracted to cue lure; 1 male, Son La Province, Moc Chau Dist., Co Do, 12.vi.2000, N006, attracted to cue lure; 1 male (29.v.2000), 1 male (12.vi.2000), 1 male (26.vi.2000), Son La Province, Moc Chau Dist, Tra Long, N007, attracted to cue lure; 5 males (20.iv.1999), 1 male (15.v.2000), Hoabinh Prov, Mai Chau, PaCo, N010, coll. Le Duc Khanh *et al.*, attracted to cue lure; 1 male, Hoa Binh Province, Ky Son Dist, Cao Phong, 29.iv.2000, N013, attracted to cue lure; 1 male, Lao Cai Province, Sapa 2 Village, 4.viii.2000, N015, attracted to cue lure; 1 male (23.iv.1999), 1 male (13.viii.1999), 1 male (26.v.2000), 1 male (23.vi.2000), Lao Cai Prov, Sapa Town, N016, attracted to cue lure; 1 male (13.viii.1999), 1 male (9.vi.2000), 5 males (23.vi.2000), 4 males (7.vii.2000), Lao Cai Province, Sapa 6 Village, N017, attracted to cue lure; 1 male (23.iv.1999), 5 males (9.vi.2000), 3 males (23.vi.2000), Lao Cai Province, Sapa 2 Village, N018, attracted to cue lure; 1 male (23.iv.1999), 1 male (23.vi.2000), 6 males (30.vii.1999), 2 males (12.v.2000), 4 males (23.vi.2000), Lao Cai Province, Sapa, Oquy Ho, N019, attracted to cue lure. All specimens in QDPC except where specified.

Diagnosis: A medium-sized species; face entirely shining black; postpronotal lobes and notopleura yellow; scutum shining black; medium-width lateral and narrow medial postsutural yellow vittae present; a yellow spot anterior to notopleural suture; mesopleural stripe reaching midway between anterior margin of notopleuron and anterior *npl.* seta dorsally; scutellum yellow; wing with cells bc and c colourless, microtrichia in outer corner of cell c only; a narrow dark fuscous costal band expanding slightly across apex of R_{4+5} and broad dark fuscous anal streak; abdominal terga almost entirely dark fuscous to black with, in some specimens, small red-brown to dark red-brown spots posterocentrally on terga III and IV and with bright orange-brown to fulvous shining spots on tergum V.

Description: male.

Head: Height 1.27 mm. Frons – length 1.36 times breadth; fulvous, sometimes red-brown centrally, dark fuscous to black around bases of orbital setae and dark fuscous on anteromedial hump; orbital setae black: 1 *s.or.*, 3 *i.or.*; lunule dark fuscous. Ocellar triangle black. Vertex black. Face entirely shining black with fulvous along lateral margins in antennal grooves; length 0.45 mm. Genae dark fuscous with black subocular spot present; small pale seta present. Occiput black, fulvous along eye margins; occipital row with a large number of small dark setae. Antennae with all segments dark fuscous; arista black (dark fuscous basally); length of segments: 0.12 mm; 0.33 mm; 0.6 mm.

Thorax: Scutum entirely shining black with dark red-brown to dark fuscous posterior to lateral postsutural vittae in some specimens (also in some specimens dark red-brown below lateral postsutural vittae). Pleural areas entirely shining black. Yellow markings as follows: postpronotal lobes; notopleura; mesopleural stripe of medium-width, reaching midway between anterior margin of notopleuron and anterior *npl.* seta dorsally, continuing to katepisternum as a small transverse spot, anterior margin straight; anatergite (posterior apex black); anterior ⅔ katatergite (remainder black); two medium-width lateral postsutural vittae beginning with a spot anterior to notopleural suture and either mostly parallel-sided and narrowing very slightly posteriorly or narrowing gradually posteriorly to end at or just behind *ia.* seta; a very short and narrow medial longitudinal postsutural vitta. Postnotum black. Scutellum yellow except for narrow black basal band. Setae: *sc.* 4; *prsc.* 2; *ia.* 1; *p.sa.* 1; *a.sa.* 1; *mpl.* 1; *npl.* 2; *scp.* 4.

Legs: Fore femora entirely shining black, mid femora with apical ½ to ⅔ shining black and remainder fulvous, hind femora with apical ⅓ to ½ shining black and remainder fulvous; fore tibiae fulvous except basal ⅓ dark fuscous, mid tibiae fulvous tending fuscous basally, hind tibiae fuscous tending dark fuscous on basal ½; tarsi with all segments entirely fulvous; mid tibiae each with an apical black spur.

Wings: Length 5.34 mm; cells bc and c colourless; microtrichia in outer corner of cell c only; remainder of wings colourless except dark fuscous cell sc, narrow dark fuscous costal band confluent with R_{2+3} and remaining narrow beyond apex of R_{2+3} before expanding slightly across apex of R_{4+5} to end between apices of R_{4+5} and M; a broad dark fuscous anal streak; dense aggregation of microtrichia around A_1+CuA_2; supernumerary lobe strong and rounded.

Abdomen: Oval; terga free; pecten present on tergum III. Tergum I and sterna I and II wider than long. Tergum I fulvous to red-brown with dark fuscous to black anterolateral corners which extend along lateral margins and which are occasionally joined along anterior margin of tergum by a narrow dark fulvous transverse band; tergum II fulvous with a narrow black transverse band across anterior margin expanding into black anterolateral corners (in some specimens the transverse band is not joined to anterolateral corners); terga III–V vary from (a) entirely dark fuscous to black with a narrow medial longitudinal black band over tergum V and red-brown across posterior margin of tergum V, to (b) dark fuscous to black with very small dark red-brown spots on posterocentral areas of terga III and IV either side of a medial longitudinal dark fuscous to black band and with tergum V as for '(a)', to (c) terga III–V dark fuscous to black with larger and clearer red-brown spots posterocentrally on terga III and IV either side of a medial longitudinal dark fuscous to black band. A pair of bright orange-brown to fulvous shining spots on tergum V on all colour forms of abdomen. Posterior lobe of surstylus long, sternum V with a slight concavity on posterior margin. Abdominal sterna shining black, except sternum II fulvous.

Attractant: Cue lure.

Female: No known record.

Distribution:

Type locality: Peninsular Malaysia (Selangor, Bukit Kutu).

Previously recorded: Type locality and Thailand (CABIKEY – White & Hancock, 1997). The record from India (Norrbom *et al*., 1998) cannot be confirmed.
New records: Bhutan; East Malaysia (Sarawak); Northern Vietnam (Hoa Binh, Lao Cai, Son La Provinces). China, due to **New Synonymy** of *D. parater*.
Hosts: No known record.
Remarks: *Bactrocera (Zeugodacus) atrifacies* (Perkins) belongs to the *scutellaris* complex and particularly the group with four *sc*. setae. It differs from other species in this group in possessing an entirely black face. In facial colour, general leg colour patterns and with abdominal terga III–V mostly dark fuscous to black, it is similar to *Bactrocera (Zeugodacus) diaphora* (Hendel) but the latter possesses two *sc*. setae. Measurements included above are from specimens collected in Bhutan, not the type. Perkins (1938a), page 140, described this species with the name *atrifacies* under the heading *Zeugodacus limbipennis* Macq. 1843. The single specimen he studied is in QMIC and its label data are in agreement with those listed in his description and can thus be considered the Holotype of *B. atrifacies*. The description on page 140 can be accepted as the original description of *B. atrifacies*. Another specimen in QMIC with the following labels is not the type of *B. atrifacies* – 'Pahang FMS, Kuala Teku 500', Dec 6th 1921, H.M. Pendlebury', 'TYPE', '*Zeugodacus maculipennis* Dol. det. F.A. Perkins', '*Dacus (Zeugodacus) atrifacies* (Perkins), poss. as *limbipennis* Macq. From Bukit Kuta, Selangor'. This is a specimen of *Bactrocera (Zeugodacus) caudata* (Fabricius) and the label data are not in agreement with those recorded by Perkins (1938a) for *B. atrifacies*. There are no sound characters that separate *B. atrifacies* and *Dacus (Zeugodacus) parater* Chao & Lin and thus we are making *D. parater* a **New Synonym** of *B. atrifacies*.

Bactrocera (Zeugodacus) baliensis Drew & Romig, new species

(Fig. 279)

Material examined: Holotype male, **INDONESIA: BALI** – Pulaki, Bali (50 m), 9.vii.1981, coll. J. Turner, Primary Rainforest (some disturbance), attracted to cue lure, deposited in BMNH.
Paratypes: **INDONESIA: BALI** – 3 males, same data as holotype; **SUMBAWA** – 4 males, Pulau Moyo, Sumbawa (0–75 m), 10.vi.1981, coll. J. Turner, Sclerophyll Forest, attracted to cue lure; 2 males, Semongkat, Sumbawa (250–350 m), 12.vi.1981, coll. J. Turner, Primary Sclerophyll Forest, attracted to cue lure; 1 male, Kanar, Sumbawa, 12.vi.1981, coll. J. Turner, Teak Forest with Sclerophyll remnants and Gallery Forest, attracted to cue lure. 2 paratypes in ANIC; 2 paratypes in BMNH; 4 paratypes in QDPC; 2 paratypes in QMIC.
Diagnosis: A medium-sized species; face fulvous with medium-sized oval black spots; postpronotal lobes and notopleura yellow; scutum black; lateral and medial postsutural yellow vittae present; a large triangular yellow marking along anterior margin of notopleural suture; mesopleural stripe reaching midway between anterior margin of notopleuron and anterior *npl*. seta; scutellum yellow; wing with cells bc and c colourless, microtrichia in outer corner of cell c only; a narrow dark fuscous costal band expanding into a large apical spot and a medium-width fuscous anal streak; abdominal terga III–V with a broad transverse black band along anterior margin of tergum III and a broad medial and two broad lateral longitudinal black bands over all three terga.
Description: male.
Head: Height 1.72 mm. Frons – length 1.33 times breadth; red-brown, fulvous along lateral margins; orbital setae black: 1 *s.or*., 2 *i.or*.; lunule dark fuscous. Ocellar triangle black. Vertex black. Face fulvous with a pair of medium-sized oval black spots; length 0.59 mm. Genae fulvous, large dark fuscous subocular spot present; black seta present. Occiput fuscous, fulvous along eye margins; occipital row with a large number of small pale setae. Antennae with segments 1 and 2 red-brown, segment 3 red-brown with fuscous on apex and outer surface; arista black (red-brown basally); length of segments: 0.12 mm; 0.29 mm; 0.79 mm.
Thorax: Scutum black with dark red-brown below lateral postsutural vittae, as a narrow band around inner margins of postpronotal lobes, along anterior margin but divided by a medial longitudinal black vitta. Pleural areas black except dark red-brown below postpronotal lobes. Yellow markings as follows: postpronotal lobes; notopleura; mesopleural stripe of medium-width, reaching midway between anterior margin of notopleuron and anterior *npl*. seta dorsally, broken by a narrow black band before continuing to katepisternum as a large narrow transverse spot, anterior margin convex; anatergite (posterior apex black); anterior $^{3}/_{5}$ katatergite (remainder black); two medium-width parallel-sided lateral postsutural vittae ending behind *ia*. seta; a short, broad medial longitudinal postsutural vitta beginning anterior to *prsc*. setae and ending in a point just posterior of line of notopleural suture; a large triangle running along anterior margin of notopleural suture with base adjacent to notopleuron. Postnotum black with dark red-brown centrally. Scutellum yellow except for narrow black basal band. Setae: *sc*. 4; *prsc*. 2; *ia*. 1; *p.sa*. 1; *a.sa*. 1; *mpl*. 1; *npl*. 2; *scp*. 4.
Legs: Femora entirely fulvous; fore tibiae fuscous, mid tibiae fuscous basally to pale fuscous apically, hind tibiae dark fuscous; tarsal segments entirely fulvous.
Wings: Length 6.4 mm; cells bc and c colourless; microtrichia in outer corner of cell c only; remainder of wings colourless except dark fuscous cell sc, narrow dark fuscous costal band overlapping R_{2+3} (almost confluent with R_{4+5}) and expanding into a large apical spot that reaches to vein M, a medium-width fuscous anal streak; dense aggregation of microtrichia around A_1+CuA_2; supernumerary lobe of medium development.
Abdomen: Elongate oval; terga free; pecten present on tergum III. Tergum I and sterna I and II wider than long. Tergum I dark fuscous centrally, black laterally and with a narrow red-brown transverse band along posterior margin but not reaching lateral margins; tergum II fulvous with a broad dark fuscous to black transverse band across anterior margin which joins narrow dark fuscous to black lateral margins; terga III–V red-brown with a broad transverse black band across anterior margin of tergum III, a broad medial and two broad

lateral longitudinal black bands over all three terga. A pair of oval red-brown shining spots on tergum V. Posterior lobe of surstylus long, sternum V with a slight concavity on posterior margin.
Attractant: Cue lure.
Female: No known record.
Distribution: Indonesia (Bali; Sumbawa).
Hosts: No known record.
Remarks: *Bactrocera (Zeugodacus) baliensis* n.sp. is unique in possessing a large dark fuscous apical spot in the wing which is similar to *Bactrocera (Sinodacus) infesta* (Enderlein) and *Bactrocera (Asiadacus) apicalis* (de Meijere). However, these two species are in different subgenera. It is also close to species in the *tau* complex but is differentiated by having a larger apical spot on the apex of the costal band.

Bactrocera (Zeugodacus) bezziana (Hering)
(Fig. 280)

Zeugodacus bezzianus Hering, 1941e: 26; Hardy & Adachi, 1954: 188 as Synonym of *Dacus (Zeugodacus) hageni* de Meijere. Holotype female in BMNH.
Dacus (Zeugodacus) bezzianus – Hardy, 1973: 70 as Synonym of *Dacus (Zeugodacus) tau* Walker.
Bactrocera (Zeugodacus) bezzianus – Norrbom *et al.*, 1998: 104 as Synonym of *Bactrocera (Zeugodacus) tau* Walker; Mahmood, 1999: 220.

Type material examined: Holotype female, **CHINA**: with the following eight labels – '♀'; 'Paratype'; 'Type (red label)'; 'Museum Paris, Mou-Pin, A. David 1870'; 'Museum Paris, Thibet (Mou Pin), A. David 1689-70'; 'Tausch V Mus. Paris'; 'Zeugodacus bezzianus Type n det. Hering 1940'; 'Purch. from E.M. Hering B.M. 1965-270'. This specimen appears to be the type Hering used in his brief description of *Zeugodacus bezzianus* (1941e: 26) and thus should be considered a Holotype, not Paratype. Studied at BMNH.
Diagnosis: A large species; face fulvous with a pair of medium-sized oval black spots; postpronotal lobes and notopleura yellow; scutum black with large areas of red-brown; lateral and medial postsutural yellow vittae present; a yellow spot anterior to notopleural suture; mesopleural stripe almost reaching midway between anterior margin of notopleuron and anterior *npl.* seta dorsally; scutellum yellow; legs with all segments fulvous except fore tibiae dark fuscous to pale fuscous; wing with cells bc and c colourless, microtrichia in outer corner of cell c only; a narrow fuscous costal band overlapping R_{2+3} and widening into a spot around apex of R_{4+5}; anal streak present; abdominal terga III–V fulvous with separate black 'T' patterns on each tergum; all abdominal sterna dark fuscous to black; apex of piercer trilobed.
Description: female.
Head: Frons – fulvous with dark fuscous on anteromedial hump; orbital setae black: 1 *s.or.*, 3 *i.or.*; lunule fuscous. Ocellar triangle black. Vertex dark fuscous. Face fulvous with a pair of medium-sized oval black spots. Occiput red-brown. Antennae with segments mostly red-brown.
Thorax: Scutum black except red-brown below and behind lateral postsutural vittae, around notopleural suture, across anterior and posterior margins, these latter areas connected by a narrow medial longitudinal red-brown band (within this medial band, there is an area of dark fuscous anterior to the medial longitudinal yellow vitta). Pleural areas mostly dark fuscous except red-brown below postpronotal lobes. Yellow markings as follows: postpronotal lobes; notopleura; mesopleural stripe of medium-width reaching midway between anterior margin of notopleura and anterior *npl.* seta dorsally, continuing to katepisternum as a small transverse spot, anterior margin slightly convex; two broad parallel-sided lateral postsutural vittae beginning with a spot anterior to notopleural suture and ending just behind *ia.* seta; a medium-width medial longitudinal postsutural vitta beginning anterior to *prsc.* setae and narrowing to a point just before level of notopleural suture. Scutellum yellow except for narrow black basal band. Setae: *sc.* 4; *prsc.* 2; *ia.* 1; *p.sa.* 1; *a.sa.* 1; *mpl.* 1; *npl.* 2; *scp.* 4.
Legs: All segments fulvous except fore tibiae dark fuscous basally to pale fuscous apically; mid tibiae each with an apical black spur.
Wings: Length 7.9 mm; cells bc and c colourless; microtrichia in outer corner of cell c only; remainder of wings colourless except fuscous cell sc, narrow fuscous costal band overlapping R_{2+3} where it is pale fuscous and extending to apex of wing where it expands into a small fuscous spot around apex of R_{4+5}, a fuscous anal streak; no dense aggregation of microtrichia around A_1+CuA_2; supernumerary lobe weak.
Abdomen: Tergum I fulvous with narrow lateral black margins; tergum II pale orange-brown with a narrow transverse black band across anterior margin but not reaching narrow black anterolateral corners; terga III–V pale orange-brown and each with a separate 'T' pattern consisting of a narrow transverse black band across anterior margin and expanding laterally to cover anterolateral corners and a moderately broad medial longitudinal black band. A pair of oval pale orange-brown shining spots on tergum V. Abdominal sterna dark fuscous to black. Basal segment of ovipositor fuscous, apex of piercer trilobed.
Attractant: No known record.
Male: No known record.
Distribution:
Type locality: China, Sichuan Province (Mou Pin).
Hosts: No known record.
Remarks: *Bactrocera (Zeugodacus) bezziana* (Hering) belongs to the *tau* complex. Hardy & Adachi (1954) placed *B. bezziana* as a synonym of *Dacus (Zeugodacus) hageni* de Meijere and later Hardy (1973) made *B. bezziana* and *D. hageni* synonyms of *Dacus (Zeugodacus) tau* (Walker). This synonymy was maintained by Norrbom *et al.* (1998). However, *B. bezziana* can be separated from *B. tau* by the distinct black 'T' patterns on each of abdominal terga III, IV and V, and the possession of a trilobed apex on the piercer. Consequently, we regard *B. bezziana* as a good species and are withdrawing it from Synonymy with *B. tau*. We still regard *D. hageni* as a Synonym of *B. tau*.

Bactrocera (Zeugodacus) biguttata (Bezzi)

(Fig. 281)

Chaetodacus biguttatus Bezzi, 1916: 111. Holotype male in ZSI.
Bactrocera (Bactrocera) biguttatus – Hardy, 1977: 49.
Dacus (Zeugodacus) adustus Wang & Zhao, 1989: 218. **New Synonym**.
Bactrocera (Zeugodacus) biguttata – Norrbom *et al.*, 1998: 102.

Type material examined: Holotype male, **INDIA**: with the following four labels – '$^{490}/_{20}$'; 'B'; 'IND. MUS. Darjiling, Alt. 7000 ft., 15–22.vii.16, Brunetti'; 'Chaetodacus biguttatus Type M Bezzi'. Studied at ZSI.
Non-type material examined: BHUTAN: 1 male, Tsirang, Darachu-La, 30.v.2001, coll. Brian Fletcher, attracted to cue lure; 4 males, Namling 7km, 9.vi.2001, coll. Brian Fletcher, attracted to cue lure.
Diagnosis: A unique medium- to large-sized species; face fulvous with a pair of medium-sized oval black spots; postpronotal lobes yellow with crescent-shaped black markings on inner surfaces; notopleura shining black; scutum entirely shining black; very narrow lateral and medial postsutural yellow vittae present; mesopleural stripe reaching midway between anterior margin of notopleuron and anterior *npl.* seta dorsally; scutellum shining black with small anterolateral yellow corners; wing with cells bc and c colourless, cells bc and c entirely devoid of microtrichia; a narrow fuscous costal band absent or almost absent beyond apex of R_{2+3} and a dark fuscous spot across apex of R_{4+5}; anal streak reduced to pale fuscous within cell cup; abdominal terga III–V shining black with, at most, small red-brown spots on posterior margins of terga III and IV either side of a medial longitudinal black band and narrow red-brown band across posterior margin of tergum V.
Description: male.
Head: Height 1.62 mm. Frons – length 1.22 times breadth; generally fuscous with narrow lateral fulvous margins and dark fuscous around orbital setae; orbital setae black: 1 *s.or.*, 3 *i.or.*; lunule dark fuscous. Ocellar triangle black. Vertex black. Face fulvous with a pair of medium-sized oval black spots; length 0.57 mm. Genae fulvous, small fuscous subocular spot present; black seta present. Occiput shining black, fulvous along eye margins; occipital row with a large number of pale setae. Antennae with segments 1 and 2 fuscous, segment 3 red-brown with dark fuscous on apex and outer surface; arista black (red-brown basally); length of segments: 0.9 mm; 0.27 mm; 0.66 mm.
Thorax: Scutum entirely shining black without pale markings. Pleural areas entirely shining black without pale markings. Yellow markings as follows: postpronotal lobes (with a broad crescent-shaped shining black marking on inner margin); mesopleural stripe of medium-width, reaching midway between anterior margin of notopleuron and anterior *npl.* seta dorsally, continuing to katepisternum as a very narrow transverse spot, anterior margin straight; anatergite (posterior apex black); anterior ½ katatergite (remainder black); two narrow parallel-sided lateral postsutural vittae ending just before *ia.* seta; a very narrow and short medial longitudinal postsutural vitta. Notopleura shining black. No yellow spot anterior to notopleural suture. Postnotum shining black. Scutellum shining black except for small yellow anterolateral corners. Setae: *sc.* 2; *prsc.* 2; *ia.* 1; *p.sa.* 1; *a.sa.* 1; *mpl.* 1; *npl.* 2; *scp.* 2.
Legs: Fore femora with apical ⅔ shining black and basal ⅓ fulvous (usually with a narrow apical fulvous area), mid and hind femora with apical ⅓ shining black and basal ⅔ fulvous; fore tibiae fuscous to dark fuscous, mid tibiae dark fuscous basally and fuscous apically, hind tibiae shining black; tarsi with basal segment fulvous and apical four segments fuscous; mid tibiae each with an apical black spur.
Wings: Length 7.0 mm; cells bc and c colourless and entirely devoid of microtrichia; remainder of wings colourless except dark fuscous cell sc, narrow fuscous costal band confluent with R_{2+3} and extremely narrow or entirely absent beyond apex of R_{2+3} and with a dark fuscous spot across apex of R_{4+5}, a diffuse pale fuscous spot on posterior margin around CuA_1, very pale fuscous within cell cup; dense aggregation of microtrichia around A_1+CuA_2; supernumerary lobe weak. Cell br is colourless and entirely devoid of microtrichia. The anal streak is either absent or reduced to pale infuscation within cell cup.
Abdomen: Oval; terga free; pecten present on tergum III. Tergum I and sterna I and II wider than long. Tergum I shining black with a very narrow transverse fulvous band across posterior margin but not reaching lateral margins; tergum II shining black with broad posterolateral transverse fulvous bands; terga III–V shining black with small red-brown posterocentral spots on posterior margins of terga III and IV either side of a medial longitudinal black band and red-brown across posterior margin of tergum V (in one specimen, terga III–V are entirely shining black). A pair of oval dark fuscous to black shining spots on tergum V. Posterior lobe of surstylus long, sternum V with a slight concavity on posterior margin. Abdominal sterna shining black, except sternum II fulvous.
Attractant: Cue lure.
Female: No known record.
Distribution:
Type locality: India (Darjiling = Darjeeling).
New records: Bhutan (Tsirang; Namling). China (Sichuan), due to **New Synonymy** of *D. adustus*.
Hosts: No known record.
Remarks: *Bactrocera (Zeugodacus) biguttata* (Bezzi), within the *scutellaris* complex, is similar to *Bactrocera (Zeugodacus) assamensis* White in possessing dark-coloured notopleura, dark markings on the scutellum and femora, abdominal terga III–V mostly black. It differs from *B. assamensis* in having the scutellum almost entirely shining black, cell br colourless and devoid of microtrichia, a crescent-shaped black marking on inner margins of postpronotal lobes and the anal streak reduced to pale fuscous within cell cup. Measurements included above are from specimens collected in Bhutan, not the type.

Bactrocera (Bactrocera) armillata (Hering) is probably a synonym of *B. biguttata* as there are no good characters that can be used to separate them. However, until male specimens are found that can be attributed to *B. armillata*, the subgeneric position of this species and the synonymy cannot be confirmed. We can find no sound characters to separate *Dacus (Zeugodacus) adustus* Wang & Zhao and *B. biguttata* and thus are making *D. adustus* a **New Synonym** of *B. biguttata*.

Bactrocera (Zeugodacus) borongensis Drew & Romig, new species

(Fig. 282)

Material examined: Holotype male, **INDONESIA: FLORES** – Borong, Flores, iii.1988, coll. J. Turner, attracted to cue lure, deposited in BMNH.

Diagnosis: A large species; face fulvous with a pair of small oval black spots; postpronotal lobes and notopleura yellow; scutum mostly black; lateral and medial postsutural yellow vittae present; a triangular yellow spot anterior to notopleural suture; mesopleural stripe just wider than notopleuron dorsally; scutellum yellow; wing with cells bc and c colourless, microtrichia in outer corner of cell c only; narrow dark fuscous costal band broadening into a large rounded apical spot; a moderately broad anal streak; abdominal terga III–V fulvous with a narrow transverse black band across anterior margin of tergum III and a broad medial and two broad lateral longitudinal black bands over all three terga.

Description: male.

Head: Height 1.7 mm. Frons – length 1.28 times breadth; red-brown with small areas of pale fuscous around *i.or.* setae; orbital setae black: 1 *s.or.*, 3 *i.or.*; lunule dark fuscous. Ocellar triangle black. Vertex red-brown. Face fulvous with a pair of small oval black spots; length 0.54 mm. Genae red-brown, large fuscous subocular spot present; black seta present. Occiput red-brown, fulvous along eye margins; occipital row with a large number of small pale setae. Antennae with segments 1 and 2 pale fuscous, segment 3 missing; length of segments: 0.17 mm; 0.32 mm.

Thorax: Scutum black with dark red-brown around bases of *prsc.* setae and red-brown below and behind lateral postsutural vittae, around notopleural suture and yellow spot anterior to notopleural suture, between notopleura and postpronotal lobes, inside postpronotal lobes, a broad medial longitudinal area from anterior point of medial postsutural vitta to anterior margin of scutum (this area has a narrow medial longitudinal black line). Pleural areas dark fuscous to black except red-brown below postpronotal lobes and above mid and hind coxae. Yellow markings as follows: postpronotal lobes; notopleura; mesopleural stripe reaching just anterior to *npl.* seta dorsally, continuing to katepisternum as a large elongate transverse spot, anterior margin strongly convex; anatergite (posterior apex black); anterior $^5/_8$ katatergite (remainder black); two broad parallel-sided lateral postsutural vittae ending behind *ia.* seta; a large triangular marking anterior to notopleural suture with base at notopleuron; a broad medial longitudinal postsutural vitta rounded posteriorly where it begins at level of *prsc.* setae and narrowing to a point anteriorly to end at level of notopleural suture. Postnotum black laterally and dark red-brown medially. Scutellum yellow except for narrow black basal band. Setae: *sc.* 4; *prsc.* 2; *ia.* 1; *p.sa.* 1; *a.sa.* 1; *mpl.* 1; *npl.* 2; *scp.* 4.

Legs: All femora entirely fulvous; fore tibiae dark fuscous basally tending fuscous apically, mid tibiae fuscous basally tending fulvous apically, hind tibiae entirely dark fuscous; tarsi with all segments fulvous; mid tibiae each with an apical black spur.

Wings: Length 6.8 mm; cells bc and c colourless; microtrichia in outer corner of cell c only; remainder of wings colourless except dark fuscous cell sc, narrow dark fuscous costal band just overlapping R_{2+3} and ending with a large spot around apex of R_{4+5}, a moderately broad anal streak (not broadening towards wing margin as in many *Zeugodacus*); a dense aggregation of microtrichia around A_1+CuA_2; supernumerary lobe weak.

Abdomen: Oval; terga free; pecten present on tergum III. Tergum I and sterna I and II wider than long. Tergum I dark fuscous to black with a narrow transverse fulvous band along posterior margin but not reaching lateral margins; tergum II fulvous with a narrow transverse dark fuscous band across anterior margin which reaches the narrow dark fuscous lateral margins; terga III–V fulvous with a narrow transverse black band across anterior margin of tergum III and a broad medial and two broad lateral longitudinal black bands over all three terga. A pair of fulvous shining spots on tergum V. Posterior lobe of surstylus long, sternum V with a slight concavity on posterior margin. Abdominal sterna dark except sternum II fulvous.

Attractant: Cue lure.

Female: No known record.

Distribution: Indonesia (Flores, Borong).

Hosts: No known record.

Remarks: *Bactrocera (Zeugodacus) borongensis* n.sp., within the *tau* complex, is similar to *Bactrocera (Zeugodacus) javadica* Mahmood in possessing dark sterna, an entirely yellow scutellum, broad parallel-sided lateral postsutural vittae, femora entirely fulvous and lateral longitudinal dark bands on abdominal terga III–V. It differs from *B. javadica* in having *prsc.* setae, supernumerary lobe weak and broad lateral longitudinal dark bands on terga III–V.

Bactrocera (Zeugodacus) buruensis White

(Fig. 283)

Bactrocera (Zeugodacus) buruensis White, in White & Evenhuis, 1999: 524. Holotype female in ZMAN.

Type material examined: INDONESIA: MOLUCCAS (MALUKU) – Holotype female, Buru, Station 77, 1921, Leg. L.J. Toxopeus. Studied at BMNH.

Non-type material examined: INDONESIA: SULAWESI – 1 male, Manado, Nth. Sulawesi, 10.ii.79, coll. J.T. Medler, attracted to cue lure. Specimen in QDPC.

Diagnosis: A large species; face fulvous with a pair of medium-sized circular black spots; postpronotal lobes and notopleura yellow; scutum red-brown to fuscous with large areas of dull black between lateral postsutural vittae and between notopleural suture and postpronotal lobes; lateral and medial postsutural yellow vittae present; a small yellow spot anterior to notopleural suture; mesopleural stripe reaching almost to anterior *npl.* seta; scutellum yellow; wing with cells bc and c pale fuscous, microtrichia in outer corner of cell c only; a broad fuscous to dark fuscous costal band expanding across apex of wing to M, very narrow pale infuscation around *r-m* crossvein and dark fuscous infuscation around *dm-cu* crossvein and narrow fuscous anal streak; abdominal terga III–V red-brown with a narrow transverse dull black band across anterior margin of tergum III, a diffuse narrow pale fuscous medial longitudinal line on terga IV and V and fuscous anterolateral corners on terga IV and V.

Description: male.

Head: Height 1.77 mm. Frons – length 1.42 times breadth; red-brown with fuscous on anteromedial hump; orbital setae black: 1 *s.or.*, 3 *i.or.*; lunule dark fuscous. Ocellar triangle black. Vertex red-brown. Face fulvous with a pair of medium-sized circular black spots; length 0.59 mm. Genae fulvous, fuscous subocular spot present; black seta present. Occiput fuscous, fulvous along eye margins; occipital row with a large number of strong pale setae. Antennae with segments 1 and 2 red-brown, segment 3 red-brown with fuscous on outer surface; arista black (red-brown basally); length of segments: 0.16 mm; 0.27 mm; 0.82 mm.

Thorax: Scutum red-brown to pale fuscous with large black areas between postsutural vittae and between notopleural suture and postpronotal lobe each side. Pleural areas generally dark fuscous to black except red-brown above mid coxae and below postpronotal lobes. Yellow markings as follows: postpronotal lobes; notopleura; mesopleural stripe reaching almost to anterior *npl.* seta dorsally, continuing to katepisternum as a large transverse spot, anterior margin slightly convex; anatergite (posterior apex black); anterior ⅔ katatergite (remainder black); two broad lateral postsutural vittae beginning with a small spot anterior to notopleural suture and parallel-sided for the anterior ½ then narrowing sharply posteriorly to end just before *ia.* seta; a moderately broad medial longitudinal postsutural vitta beginning just anterior to *prsc.* setae and narrowing to end in a point anterior to level of notopleural suture. Postnotum black with red-brown centrally. Scutellum yellow except for narrow dark basal band. Setae: *sc.* 4; *prsc.* 2; *ia.* 1; *p.sa.* 1; *a.sa.* absent; *mpl.* 1; *npl.* 2; *scp.* 4.

Legs: Femora fulvous with a diffuse fuscous spot on outer apical surfaces of fore femora, around apices of mid femora and inner surfaces of hind femora; fore tibiae pale fuscous, mid tibiae fulvous tending red-brown basally, hind tibiae fuscous; tarsi with all segments fulvous; mid tibiae each with an apical black spur.

Wings: Length 7.9 mm; cells bc and c pale fuscous; microtrichia in outer corner of cell c only; remainder of wings colourless except fuscous cell sc, fuscous to dark fuscous costal band confluent with R_{4+5} and expanding into a spot across apex of wing but becoming pale as it reaches M (the costal band is paler between R_{2+3} and R_{4+5}), very narrow and pale infuscation around *r-m* crossvein, dark fuscous infuscation around *dm-cu* crossvein and becoming very pale fuscous as it extends around hind margin of wing to reach the moderately narrow fuscous anal streak (the infuscation around *dm-cu* crossvein begins anterior to vein M); dense aggregation of microtrichia around A_1+CuA_2; supernumerary lobe strong and pronounced.

Abdomen: Elongate oval; terga free; pecten present on tergum III. Tergum I and sterna I and II wider than long. Tergum I red-brown with broad dark fuscous to black lateral margins; tergum II red-brown with a broad transverse dark fuscous band across anterior margin but not reaching the narrow dark fuscous lateral margins and a narrow fulvous band across posterior margin; terga III–V red-brown with a narrow transverse dull black band across anterior margin and partly covering lateral margins of tergum III, a narrow diffuse pale fuscous to fuscous medial longitudinal band over terga IV and V and very narrow fuscous anterolateral corners on terga IV and V. A pair of oval fuscous shining spots on tergum V. Posterior lobe of surstylus long, sternum V with a slight concavity on posterior margin. All abdominal sterna pale coloured (fulvous).

Attractant: Cue lure.

Female: As for male except no dense aggregation of microtrichia around A_1+CuA_2; supernumerary lobe weak; pecten of cilia on abdominal tergum III absent; basal segment of ovipositor fuscous and tapering posteriorly in dorsal view.

Distribution:

Type locality: Indonesia (Moluccas = Maluku, Buru I.).

New records: Indonesia (Sulawesi – North Sulawesi, Manado).

Hosts: No known record.

Remarks: The specimen described here appears to be *Bactrocera (Zeugodacus) buruensis* White but does differ in having dark patterns on the apices of all femora, very narrow and pale infuscation around *r-m* crossvein, a longer medial longitudinal postsutural yellow vitta on the scutum and a slightly wider mesopleural stripe dorsally. *Bactrocera buruensis* is very similar to *Bactrocera (Zeugodacus) flavipilosa* (Hardy) and can only be separated on some minor wing colour patterns. It has some pale infuscation around the *dm-cu* crossvein whereas *B. flavipilosa* has a pale fuscous tint around the wing hind margin.

Bactrocera (Zeugodacus) calumniata (Hardy)

(Fig. 284)

Dacus calumniatus Hardy, 1970: 77. Holotype male in UZMC.

Bactrocera (Zeugodacus) calumniata – Norrbom *et al.*, 1998: 102.

Type material examined: Holotype male, **PHILIPPINES**: with the following label data – Tawi Tawi, Tarawakan, north of Batu Batu, 7.xi.1961, Noona Dan Exp 61–62 caught in Malaise traps. Received on loan from UZMC.

Non-type material examined: INDONESIA: BALI – 1 male, Pulaki, Bali (50 m), 9.vii.1981, coll. J. Turner, primary rainforest (some disturbance), attracted to cue lure; 2 males, Bali, Buleleng, Depeha, 28.viii.2006, coll. Ketut Rawa W., attracted to cue lure in mango; **JAVA** – 4 males, Tirto Argo

Ungaran, Java, 7.ix.1976, coll. Miss Santianawati, attracted to cue lure; 1 male, Kebun Raya, Bogor, Java, 12-19.vi.1977, coll. D. Kistner; 1 male, West Java, viii.1978, coll. S. Isnadi; 3 males (19.vii.2005), 4 males (26.vii.2005), 1 male (9. viii.2005), East Java, Lumajang, Kadang Tepus, coll. Utomo, attracted to cue lure in banana; 4 males, East Java, Sidoajo, Tulangan, Gelang, 4.x.2005, coll. Agus Riadi, attracted to cue lure in village; 1 male, East Java, Probolinggo, Betektanan, 20.vii.2005, coll. Tjipto Waluyo, attracted to cue lure in mango; 1 male (31.vii.2005), 1 male (13.xi.2005), 1 male (4.xii.2005), Central Java, Pemalang, Pegongsoran, coll. Sapto Darmono, attracted to cue lure in forest; 5 males, D.I. Yogyakarta, Sieman, Parwodadi, 19.vii.2005, coll. Agus W., attracted to cue lure in chilli; 1 male (5.vii.2005), 5 males (26.vii.2005), D.I. Yogyakarta, Sieman, Kaliurang, coll. Agus W., attracted to cue lure in forest; 5 males (20.vii.2005), 3 males (8.viii.2005), 2 males (22.viii.2005), Banten, Lebak, Pasir Nini Nini, coll. Aden Haer, attracted to cue lure in forest. Specimens in QDPC.

Three additional specimens, on loan from UZMC, are also most likely to be this species – 1 female, Mus. Western.; 1 male, Mus. Western, *D. caudatus* Fab., Batavia, Aug.1815; 1 female, Java, Louendal. These all have parallel-sided lateral postsutural vittae as for *B. calumniata*.

Diagnosis: A medium-sized species; face fulvous with a pair of medium-sized oval black spots; postpronotal lobes and notopleura yellow; scutum black; lateral and medial postsutural yellow vittae present; a yellow spot anterior to notopleural suture in front of lateral postsutural vittae; mesopleural stripe reaching midway between anterior margin of notopleuron and anterior *npl.* seta dorsally; scutellum yellow; wing with cells bc and c colourless, microtrichia in outer corner of cell c only; a narrow dark fuscous costal band overlapping R_{2+3} and widening across apex of R_{4+5}, a broad dark fuscous band along *dm-cu* crossvein; a broad fuscous anal streak; abdominal terga III–V orange-brown with a black 'T' pattern consisting of a narrow transverse black band across anterior margin of tergum III and a medium-width medial longitudinal black band over all three terga, anterolateral corners of terga IV and V black.

Description: male.

Head: Height 1.4 mm. Frons – length 1.39 times breadth; fulvous with fuscous on anteromedial hump; orbital setae black: 1 *s.or.*, 3 *i.or.*; lunule fuscous. Ocellar triangle black. Vertex black. Face fulvous with a pair of medium-sized oval black spots; length 0.45 mm. Genae fulvous, pale fuscous subocular spot present; black seta present. Occiput fuscous with a pair of small black dorsocentral spots and fulvous along eye margins; occipital row with a few weak pale setae. Antennae with segments 1 and 2 red-brown, segment 3 red-brown with pale fuscous on apex and outer surface; length of segments: 0.12 mm; 0.24 mm; 0.65 mm.

Thorax: Scutum black with pale markings as follows: fuscous below lateral postsutural yellow vittae, dark red-brown around notopleural suture and as a medial longitudinal band running from anterior point of medial postsutural yellow vitta to anterior margin of scutum. Pleural areas dark fuscous to black except dark red-brown behind postpronotal lobes. Yellow markings as follows: postpronotal lobes; notopleura; mesopleural stripe of medium-width reaching midway between anterior margin of notopleuron and anterior *npl.* seta dorsally, continuing to katepisternum as a large transverse spot, anterior margin slightly convex; anatergite (posterior apex black); anterior $^3/_5$ katatergite (remainder black); two broad lateral postsutural vittae beginning in a spot anterior to notopleural suture and ending posteriorly behind *ia.* seta; a broad medial longitudinal postsutural vitta beginning posteriorly at level of *prsc.* setae and ending at a point just anterior of notopleural suture. Postnotum black. Scutellum yellow with a narrow black basal band. Setae: *sc.* 4; *prsc.* 2; *ia.* 1; *p.sa.* 1; *a.sa.* 1; *mpl.* 1; *npl.* 2; *scp.* 4.

Legs: Legs entirely fulvous except basal segment of fore and hind tarsi pale fulvous and hind tibiae fuscous basally tending to pale fuscous apically; in some specimens, there are small fuscous to dark fuscous preapical spots on some or all femora; mid tibiae each with an apical black spur.

Wings: Length 6.0 mm; cells bc and c colourless; microtrichia in outer corner of cell c only; remainder of wings colourless except dark fuscous cell sc, narrow dark fuscous costal band overlapping R_{2+3} and widening slightly across apex of R_{4+5}, a broad dark fuscous band along *dm-cu* crossvein which begins just anterior of vein M and ends at hind wing margin, a broad fuscous anal streak; a dense aggregation of microtrichia around A_1+CuA_2; supernumerary lobe strong and keel shaped.

Abdomen: Elongate oval; terga free; pecten present on tergum III. Tergum I and sterna I and II wider than long. Tergum I orange-brown with narrow lateral black margins; tergum II orange-brown with a narrow transverse black band across anterior margin and widening into anterolateral corners; terga III–V orange-brown with a black 'T' consisting of a narrow transverse black band across anterior margin of tergum III and widening around lateral margins and a medium-width medial longitudinal black band over all three terga, anterolateral corners of terga IV and V black. A pair of oval orange-brown shining spots on tergum V. Posterior lobe of surstylus long, sternum V with a slight concavity on posterior margin.

Attractant: Cue lure.

Female: No known record.

Distribution:

Type locality: Philippines (Tawi Tawi).

Previously recorded: Type locality and Indonesia (Java – Jakarta = Batavia and Louendal); Philippines (White & Hancock, 1997).

New records: Indonesia (Bali; several other locations in Java listed under 'Non-type material examined').

Hosts: No known record.

Remarks: *Bactrocera (Zeugodacus) calumniata* (Hardy) is similar to *Bactrocera (Zeugodacus) emittens* (Walker), *Bactrocera (Zeugodacus) ishigakiensis* (Shiraki) and *Bactrocera (Zeugodacus) synnephes* (Hendel) in possessing four *sc.* setae, lateral and medial postsutural yellow vittae, black scutum, wing with infuscation in addition to costal band and anal streak and costal band with some expansion across apex of wing. It differs from *B. ishigakiensis* in having an entirely yellow scutellum, from *B. synnephes* in having parallel-sided lateral postsutural yellow vittae, cells bc and c colourless and supernumerary lobe large, and from *B. emittens* in having cells bc and c colourless and apex of costal band not expanded to M. The measurements included are from specimens collected in Java, Indonesia, not the type.

Bactrocera (Zeugodacus) caudata (Fabricius)

(Fig. 285)

Dacus caudatus Fabricius, 1805: 276. Lectotype female in UZMC.
Bactrocera maculipennis Doleschall, 1856: 412; Hardy, 1973: 60 (Synonym).
Dacus (Zeugodacus) caudatus – Hardy & Adachi, 1954: 186; Hardy, 1973: 60, Lectotype designation by inference of Holotype; Hardy, 1983: 35.
Bactrocera (Zeugodacus) caudata – Norrbom *et al.*, 1998: 102.

Type material examined: INDONESIA: JAVA – Lectotype female with the following two labels – red label 'TYPE', discoloured white label 'D.caudatus e Java', the latter in original handwriting. Received on loan from UZMC.
Non-type material examined: INDONESIA: BALI – 1 male, Pura Batakaru, 800m, 11.vii.1981, coll. J. Turner, attracted to cue lure. 4 males, Buleleng, Depeha, 28.viii.2006, coll. Ketut Rawa W., attracted to cue lure in mango; **FLORES** - 78 males, Borong, Flores, iv.1989, coll. J. Turner; **JAVA** – 1 male, DKI Jakarta, Jakarta Selatan, Tanjung Barat, 23.vii.2005, coll. Mulyana, attracted to cue lure in carambola; 4 males, Banten, Lebak, Pasir Nini Nini, 20.vii.2005, coll. Aden Haer, attracted to cue lure in forest; 1 male, Banten, Tangerang, Tanah Tinggi, 20.vii.2005, coll. B. Guritno, attracted to cue lure in village (mango and guava); 1 male, West Java, Cirebon, Pelangon, 28.vii.2005, coll. Herty, attracted to cue lure in forest; 10 males (14.vii.2005), 13 males (27.vii.2005), 7 males (28.vii.2005), West Java, Majalengka, Ampel 25 dpl, coll. Yaya Senjaya, attracted to cue lure in mango; 4 males, East Java, Sidorajo, Gelang, 12.vii.2005, coll. Agus Riadi, attracted to cue lure in village; 6 males, East Java, Probolinggo, Betektanan, 20.vii.2005, coll. Tjipto Waluyo, attracted to cue lure in mango; 8 males, East Java, Lumajang, Kadang Tepus, 19.vii.2005, coll. Utomo, attracted to cue lure in banana; 1 male (13.xi.2005), 6 males (4.xii.2005), Central Java, Pemalang, Pegongsoran, coll. Sapto Darmono, attracted to cue lure in forest; **KALIMANTAN** – 11 males, West Kalimantan, Pontianak, Sei Pinyuh, 26.vii.2005, coll. A. Irom, attracted to cue lure in village mango; 16 males, West Kalimantan, Pontianak, Sei Duri II, 19.vii.2005, coll. A. Irom, attracted to cue lure in citrus; 12 males, South Kalimantan, Pontianak, Guntung Payung, 21.vii.2005, coll. Sidratul, attracted to cue lure in village; 3 males, South Kalimantan, Banjarbaru, Guntung Payung, 28.vii.2005, coll. Sidrotul & Sutini, attracted to cue lure in village; **KRAKATOA**: 1 male, Rakata I., 1.ix.1990, coll. E. Schmidt, attracted to cue lure; **SUMATRA** – 1 female, Sumatra, P. Koogua; **SUMBAWA** – 1 male, Pulau Moyo, Sumbawa (0-75m), 10.vi.1981, coll. J. Turner, Sclerophyll Forest, attracted to cue lure; **WEST TIMOR** – 13 males, Nusa Tenggara Timur, Oebufu, Kt. Kupang, 31.v.2007, Ir. Hendri Wihanto, attracted to cue lure in village jambu; 31 males, Nusa Tenggara Timur, Belu, Kupang, 2.vii.2007, coll. Ir. Hendri Wihanto, attracted to cue lure in starfruit; 2 females, Nusa Tenggara Timur, Alak, Kt. Kupang, 7.v.2007, coll. Ir. Hendri Wihanto, attracted to cue lure in village jambu. 2 males, Indonesia, West Timor, 15km S.E. Kupang, 30.x.1986, coll. E.S.C. Smith, attracted to cue lure; 1 male, Indonesia, West Timor, Kupang, Wisma Maleita, 31.x.1986, coll. E.S.C. Smith, attracted to cue lure; 2 males, Indonesia, West Timor, Kupang, ADAB Office, 31.x.1986, coll. E.S.C. Smith, attracted to cue lure. **EAST MALAYSIA: SARAWAK** – 2 males, Sarawak, Forest Arboretum, Semongok, nr Kuching, 11.i.1992, coll. S. Leong, attracted to cue lure; 6 males (16.iv.1992), 1 male (23.iv.1992), 1 male (30.iv.1992), 2 males (7.v.1992), 7 males (14.v.1992), 1 male (3.vi.1992), 2 males (24.vi.1992), 1 male (12.viii.1992), Sarawak, Sg Sui, Coastal area, coll. S. Leong, attracted to cue lure; 2 males (18.i.1992), 1 male (18.ii.1992), Sarawak, ARC Semongok, nr. Kuching, coll. S. Leong, attracted to cue lure; 1 female, Sarawak, ARC Semongok KT1, 1.viii.1992, coll. Leong, attracted to cue lure; 4 males (8.i.1992), 1 male (5.vi.1992), 1 male (10.vi.1992), Sarawak, Tarat Agriculture Station, Nr. Kuching, coll. S. Leong, attracted to cue lure; 1 male, Sarawak, Tarat Agric. Station, 19.iii.1993, coll. S. Leong, attracted to cue lure. **SRI LANKA (formerly CEYLON)**: 2 males, Sri Lanka, Maha, Illuppallama, 26.x.1993, attracted to cue lure. **THAILAND**: 7 males, 'Chumphon Prov., Lang Suan, 40m, 4-8.v.2003/3'; '9.57.01N, 99.04.23E, P. Schwendinger leg, garden/3'; 'attracted to cue lure'. Received on loan from MHNG. 15 males, Kunthuli, Surat Thani, 22.x.1991, coll. Pichate, attracted to cue lure; 2 males, Tha-u-Tac, Kamchanadit, Surat Thani, 26.xi.1991, coll. Pichate, attracted to cue lure; 1 male, Yang Pong Rat, Plubpla, A. Muong, Chanthaburi, 3.ix.1991, coll. Montree, attracted to cue lure; 1 male, NRDC, Chiang Mai, 3.vii.1992, coll. Chantanee, attracted to cue lure; 1 male, Songkhla, Ko Hong Hill, 7.vi.1993, coll. Anuchit & Sieng, attracted to cue lure; 2 males, Ya Ha, Yala, Songkhla, 18.i.1993, coll. Sieng *et al*, bred from *Cucurbita moschata* (SK10397); 7 females and 2 males, Rattaphum, Songkhla, 20.iii.1993, coll. Sieng *et al.*, bred from flower of *Cucurbita moschata* (SK10941). **VIETNAM: Southern Vietnam** – 1 male (28.x.1999), 2 males (9.vii.2000), Dong Nai (P), Long Khanh (D), Tan Phong (V), attracted to cue lure, southern Vietnam, S014; 1 male, Dong Nai (P), Long Nhat (D), Long Nhat (V), 26.v.2000, attracted to cue lure, southern Vietnam, S012; 1 male, Dong Nai (P), Thong Nhat (D), Cay Gao (V), 9.vi.2000, attracted to cue lure, southern Vietnam, S013; 1 male, Binh duong (P), Thuan An (D), An Thanh (V), 4.vi.2002, attracted to cue lure, southern Vietnam, VS05; 1 male, Vinhlong (P), Long ho (D), Phuoc (V), 29.ix.2002, attracted to cue lure, southern Vietnam, VS01; 1 male, Ben Tre (P), Chau thanh (D), Tien Thuy (V), 7.vii.2002, attracted to cue lure, southern Vietnam, VS04; 1 male (2.iii.2002), 1 male (11.v.2002), 1 male (6.vi.2002), 1 male (20.vii.2002), 1 male (3.viii.2002), Dong Nai (P), Long Khanh (D), Nhan nghia (V), attracted to cue lure, southern Vietnam, VS06. Specimens in QDPC except where specified.

Diagnosis: A medium-sized species; face fulvous with a narrow transverse black line across oral margin; postpronotal lobes and notopleura yellow; scutum black; lateral and medial postsutural yellow vittae present; a yellow spot anterior to notopleural suture in front of lateral postsutural vittae; mesopleural stripe reaching almost to anterior *npl.* seta dorsally; scutellum yellow; wing with cells bc and c colourless, microtrichia in outer corner of cell c only; a narrow fuscous costal band confluent with R_{2+3} and widening around apex of R_{4+5} and broad fuscous anal streak; abdominal terga III–V orange-brown except for a 'T' pattern consisting of a dark fuscous to black transverse band across anterior margin of tergum III and a narrow medial longitudinal fuscous to dark fuscous band over all three terga, anterolateral corners of terga IV and V fuscous to black.

Description: female.

Head: Height 1.47 mm. Frons – length 1.4 times breadth; fulvous with dark fuscous to black around orbital setae and on anteromedial hump; orbital setae black: 1 *s.or.*, 3 *i.or.*; lunule dark fuscous to black. Ocellar triangle black. Vertex red-brown. Face fulvous with a transverse black line just above oral margin; length 0.45 mm. Genae fulvous, large fuscous subocular spot present; black bristle present. Occiput red-brown, fulvous along eye margins; occipital row with weak pale setae. Antennae with all segments red-brown except fuscous on apex of segment 3; length of segments: 0.12 mm; 0.27 mm; 0.66 mm.

Thorax: Scutum black except areas of red-brown and fuscous below and behind lateral postsutural yellow vittae. Pleural areas black except red-brown below postpronotal lobes. Yellow markings as follows: postpronotal lobes; notopleura; mesopleural stripe reaching almost to anterior *npl.* seta dorsally, continuing to katepisternum as a transverse spot, anterior margins slightly convex; anatergite (posterior apex black); anterior $^{3}/_{5}$ katatergite (remainder black); two moderately broad parallel-sided lateral postsutural vittae beginning as a spot anterior to notopleural suture and ending posterior to *ia.* seta; a moderately broad medial longitudinal postsutural vitta rounded posteriorly at level of *prsc.* setae and narrowing to a point anteriorly at level of notopleural suture. Postnotum black. Scutellum yellow except for narrow black basal band. Setae: *sc.* 4; *prsc.* 2; *ia.* 1; *p.sa.* 1; *a.sa.* 1; *mpl.* 1; *npl.* 2; *scp.* 4.

Legs: Femora fulvous except for a fuscous to dark fuscous spot on outer apical surfaces of fore femora and fuscous around apices of mid and hind femora; fore tibiae fuscous to dark fuscous, mid tibiae fuscous tending dark fuscous basally, hind tibiae fuscous; all tarsi with basal segment fulvous and apical four segments fuscous; mid tibiae each with an apical black spur.

Wings: Length 6.0 mm; cells bc and c colourless; microtrichia in outer corner of cell c only; remainder of wings colourless except fuscous cell sc, narrow fuscous costal band confluent with R_{2+3} and widening around apex of R_{4+5}, a broad fuscous anal streak; no dense aggregation of microtrichia around A_1+CuA_2; supernumerary lobe weak.

Abdomen: Oval; terga free; no pecten on tergum III. Tergum I and sterna I and II wider than long. Tergum I orange-brown except dark fuscous anteriorly and as narrow lateral margins; tergum II fulvous with a narrow orange-brown anterior margin and a narrow transverse black band just posterior of anterior margin and filling anterolateral corners; terga III–V orange-brown except for a 'T' pattern consisting of a dark fuscous to black narrow transverse band across anterior margin of tergum III which becomes black in anterolateral corners of this tergum and a narrow medial longitudinal fuscous to dark fuscous band over all three terga and anterolateral corners of terga IV and V changing from fuscous to black. A pair of oval orange-brown shining spots on tergum V. Abdominal sterna dark except sternum II pale. Basal segment of ovipositor red-brown, tapering posteriorly in dorsal view and dorsoventrally compressed; ratio of length of oviscape to length of tergum V, 2.2:1.

Attractant: Cue lure.

Male: As for female except for a dense aggregation of microtrichia around A_1+CuA_2; supernumerary lobe of medium development; pecten present on abdominal tergum III; posterior lobe of surstylus long; abdominal sternum V with a slight concavity on posterior margin; abdominal sterna dark except sterna II and III pale.

Distribution:

Type locality: Indonesia (Java).

Previously recorded: Type locality and the following locations – Brunei; China; India; Indonesia (Sumatra); Peninsular Malaysia; Myanmar; Sri Lanka; Taiwan; Thailand; Vietnam.

New records: Indonesia (Bali; Flores; Kalimantan; Krakatoa; Sumbawa; West Timor); East Malaysia (Sarawak).

Hosts: This species has been recorded from male flowers of *Cucurbita moschata* (family *Cucurbitaceae*).

Remarks: *Bactrocera (Zeugodacus) caudata* (Fabricius) is a widely distributed species across South-East Asia. It can be confused, superficially, with species in the *tau* complex but is best differentiated by the transverse black line across the oral margin. It is similar to *Bactrocera (Zeugodacus) apicofemoralis* n.sp. in possessing four *sc.* setae, a black scutum with lateral and medial postsutural yellow vittae, wings colourless except for costal band and anal streak, cells bc and c colourless, costal band confluent with R_{2+3}, abdominal sterna dark coloured, abdominal terga III–V with a black 'T' pattern and black line across oral margin of face. It differs from *B. apicofemoralis* in having a small preapical fuscous spot on fore femora and small fuscous apices on mid and hind femora, not extensive dark fuscous to black areas and abdominal sternum V with a narrow transverse shape. The measurements included are from specimens collected in Indonesia, not the type.

Bactrocera (Zeugodacus) connexa (Hardy)

(Fig. 286)

Dacus (Zeugodacus) connexus Hardy, 1982: 203. Holotype female in BMNH.

Bactrocera (Zeugodacus) connexa – Norrbom *et al.*, 1998: 102.

Type material examined: Holotype female, **INDONESIA: SULAWESI (formerly CELEBES)** – Celebes (Sulawesi), Minahassa, Tomohon, vI–vii.1954, coll. A.H.G. Alston, B.M. 1954-414. Studied at BMNH.

Diagnosis: A large species; face fulvous with a pair of small- to medium-sized oval black spots; postpronotal lobes and notopleura yellow; a narrow lateral yellow band connecting postpronotal lobe and notopleuron; scutum mostly black; lateral and medial postsutural yellow vittae present; no yellow spot anterior to notopleural suture; mesopleural stripe reaching postpronotal lobe dorsally; scutellum yellow; fore and middle legs absent, hind legs with femora and tibiae entirely fulvous; wing with cells bc and c colourless, microtrichia in outer corner of cell c only; a narrow fuscous costal band confluent with R_{2+3} and widening slightly around apex of wing to end at apex of M; anal streak absent; abdominal terga III–V fulvous with a black 'T' pattern consisting of a narrow transverse black band across anterior margin of tergum III and a moderately broad medial longitudinal black band over all three terga, narrow black lateral margins on terga IV and V.
Description: female.
Head: Frons – fulvous without dark markings; orbital setae black: 1 *s.or.*, 2 *i.or.*; lunule fulvous. Ocellar triangle black. Vertex fulvous. Face fulvous with a pair of small to medium oval black spots. Occiput fulvous. Antennae absent.
Thorax: Scutum black except red-brown below and behind lateral postsutural vittae and inside postpronotal lobes. Pleural areas black except red-brown below postpronotal lobes. Yellow markings as follows: postpronotal lobes; notopleura; broad mesopleural stripe reaching postpronotal lobe dorsally, continuing to katepisternum as a transverse spot, anterior margin straight; anatergite and katatergite mostly yellow; a narrow lateral yellow band connecting postpronotal lobe and notopleuron; two broad parallel-sided lateral postsutural vittae ending well behind *ia.* seta; a broad medial longitudinal postsutural vitta beginning posteriorly at level of *prsc.* setae and narrowing to a point at level of notopleural suture. No yellow spot anterior to notopleural suture. Postnotum black. Scutellum yellow except for narrow black basal band. Setae: *sc.* 4; *prsc.* 2; *ia.* 1; *p.sa.* 1; *a.sa.* 1; *mpl.* 1; *npl.* 2; *scp.* 4.
Legs: Fore and middle legs broken off; hind legs with all segments entirely fulvous.
Wings: Cells bc and c colourless; microtrichia in outer corner of cell c only; remainder of wings colourless except fuscous cell sc, narrow fuscous costal band confluent with R_{2+3} and widening slightly around apex of wing to end at apex of M, anal streak absent (a pale fuscous tint in basal area of cell cup); no dense aggregation of microtrichia around A_1+CuA_2; supernumerary lobe weak.
Abdomen: Tergum I fulvous with narrow black lateral margins; tergum II fulvous with a narrow transverse black band across anterior margin extending into anterolateral corners; terga III–V fulvous with a 'T' pattern consisting of a narrow transverse black band across anterior margin widening to cover lateral margins of tergum III and a moderately broad medial longitudinal black band over all three terga, narrow black lateral margins on terga IV and V. A pair of oval fulvous shining spots on tergum V. Abdominal sterna fulvous. Basal segment of ovipositor red-brown and elongate; ratio of length of oviscape to length of tergum V, 1.5:1.
Attractant: No known record.
Male: No known record.
Distribution:
Type locality: Indonesia (Sulawesi – North Sulawesi, Minahassa).
Hosts: No known record.
Remarks: *Bactrocera (Zeugodacus) connexa* (Hardy) is similar to *Bactrocera (Zeugodacus) abnormis* (Hardy) in possessing a black scutum, wings colourless except for the costal band and anal streak, cells bc and c colourless, costal band confluent with R_{2+3} and abdominal sterna pale. It differs from this species in having facial spots, a yellow band connecting the notopleuron and postpronotal lobe and abdominal terga III–V with a black 'T' pattern and narrow black lateral margins.

Bactrocera (Zeugodacus) cucurbitae (Coquillett)

(Fig. 287)

Dacus cucurbitae Coquillett, 1899: 129; Froggatt, 1909: 84. Lectotype female and two Paralectotypes (1 male and 1 female) in USNM.
Bactrocera cucurbitae – Bezzi, 1913: 96.
Chaetodacus cucurbitae – Bezzi, 1916: 109; Hendel, 1915: 426, 1927a: 28; Shiraki, 1933: 73.
Strumeta cucurbitae – Perkins, 1938a: 127.
Dacus (Strumeta) cucurbitae – Hardy & Adachi, 1954: 164; Hardy, 1973: 38, 1974: 27.
Dacus (Zeugodacus) cucurbitae – Drew, 1973: 23; Hardy, 1977: 57, 1982: 203, 1983: 36.
Dacus aureus Tseng & Chu, 1982: 85.
Bactrocera (Zeugodacus) cucurbitae – Drew, 1989: 212, Lectotype designation.
Dacus yuiliensis Tseng & Chu, 1992: 84 (new name for *aureus* preoccupied by May, 1951: 10). Record of original synonymy unknown, see Norrbom *et al.*, 1998: 102.

Type material examined: HAWAII: Lectotype female, 2 Paralectotypes (male and female), Honolulu, coll. G. Compere. Received on loan from USNM.
Non-type material examined: A large series of specimens collected from Christmas I., Nepal, Peninsular Malaysia, Thailand, Papua New Guinea, Solomon Islands. Also, the following – **BHUTAN**: 1 male, Rimchu 1, 20.ix.2000, coll. C. Dorji; 1 male (17.vii.2000), 1 male (2.x.2000), Tsirang, Coll. S.D.; 1 male, Gelephu, Purana Bastey, 1.ii.2005, coll. Maha Prasad & Karma Namgyel, all attracted to cue lure. **INDONESIA: JAVA** – 4 males, DKI Jakarta Selatan, Tanjung Barat, 23.vii.2005, coll. Mulyana, attracted to cue lure in carambola; 5 males, West Java, Cirebon, Pelangon, 14.vii.2005, coll. Herty, attracted to cue lure in forest; 1 male, Banten, Tangerang, Tanah Tinggi, 13.vii.2005, coll. B. Guritno, attracted to cue lure in village (mango, guava); 1 male, West Java, Indramayu, Lobener, 16.vii.2005, coll. Rifki K, attracted to cue lure in village; 1 male (21.vii.2005), 1 male (28.vii.2005), West Java, Majalengka, Ampel 25 dpl, coll. Yaya Senjaya, attracted to cue lure in mango; 4 males (7.viii.2005), 1 male (14.viii.2005), 2 males (21.viii.2005),

1 male (28.viii.2005), 1 male (4.xii.2005), Central Java, Pemalang, Pegongsoran, coll. Sapto Darmono, attracted to cue lure in forest; 8 females and 2 males, Central Java, Sukoharjo, Mrangger, coll. Duei Hariyanto, reared from GF4a(8); 4 males, East Java, Magetan, Belatan, 12.vii.2005, coll. Trigiyanto, attracted to cue lure in citrus/pomelo; 4 males, East Java, Sidoarjo, Gelang, 12.vii.2005, coll. Agus Riadi, attracted to cue lure in village; 2 males, East Java, Lumajang, Kadang Tepus, 19.vii.2005, coll. Utomo, attracted to cue lure in banana; **KALIMANTAN** – 2 males, South Kalimantan, Banjarbaru, Guntung Payung, 21.vii.2005, coll. Sidratul, attracted to cue lure in village; 10 males (1.viii.2005), 4 males (8.viii.2005), East Kalimantan, Kota Samarinda, Lempake, coll. Arofah, attracted to cue lure in village; 10 males, East Kalimantan, Kutai Katanegara, Batuah, 12.ix.2005, coll. Ertha, attracted to cue lure in bananas; 9 males, Central Kalimantan, Kota Waringin Timur, Kota Besi, 9.viii.2005, coll. Utik D & Kusyono, attracted to cue lure in citrus; 3 males (19.vii.2005), 3 males (26.vii.2005), 1 male (16.viii.2005), West Kalimantan, Pontianak, Sei Duri II, coll. A. Irom, attracted to cue lure in citrus; **NORTH MOLUCCAS (NORTH MALUKU)** – 15 males (29.v.2007), 8 males (12.vi.2007), Moluccas, North Maluku, West Halmahera, Goal, coll. Raais Abdullah, attracted to cue lure in orange; 8 males, Maluku, North Maluku, West Halmahera, Goal, 5.vi.2007, coll. Raais Abdullah, attracted to cue lure in forest; **SULAWESI** – 1 female and 2 males, Central Sulawesi, Donggala, Jono onge, 7.viii.2006, coll. Rusdin, bred ex *Momordica charantia* (balsam pear), sample no. GF14(16); 5 females and 5 males, Central Sulawesi, Donggala, Jono onge, 23.viii.2006, coll. Rusdin, bred ex cucumber, sample no. GF14(17); 13 females and 2 males, Central Sulawesi, Donggala, Nupabomba, 17.ix.2006, coll. Rusdin, bred ex Lampijo, sample no. GF14; 2 females and 4 males, Central Sulawesi, Donggala, Bula pountu, 1.x.2006, coll. Rusdin, bred ex Lampijo, sample no. GF14(8); 1 male, Sulawesi Selatan, Takalar, Limampoccoe, 16.viii.2006, coll. Sirajuddin SP., attracted to cue lure in forest; 2 males, Sulawesi South, BKT Makasar, 9.ix.2006, coll. Sirajuddin SP., attracted to cue lure in village; **SUMATRA** – 6 females and 1 male, Jambi, Muara Jambi, Baru, 30.viii.2006, coll. Hendri Y, bred ex *Luffa acutangula* (sample GF21). **PAPUA (formerly IRIAN JAYA)**: 32 males, Papua, Keerom, Wiantre, 16.vi.2007, coll. Elkanus Karma, attracted to cue lure in Kebun Jeruk; 10 males, Papua, Keerom, Wiantre, 16.vi.2007, coll. Elkanus Karma, attracted to cue lure in citrus. **PHILIPPINES**: 75 males, UPLB Campus, rainforest area, mixed vegetation, 8.i.2000, coll. G. Quimio, attracted to cue lure; 16 males, Sirong Lupa, calamba Languna, carambola/guava orchard, 7.i.2000, coll. G. Quimio, attracted to cue lure; 1 male (15.xii.1999), 1 male (20.xii.1999), Sirang Lupa, calamba, Laguna, mixed orchard, coll. A. O'Toole & G. Quimio, attracted to cue lure; 1 male, Sirang Lupa, calamba laguna, citrus orchard, 7.i.2000, coll. G. Quimio, attracted to cue lure; 1 male, Sirang Lupa, calamba, Laguna, citrus orchard, 7.i.2000, coll. A. O'Toole & G. Quimio, attracted to cue lure; 1 female, Sirang Lupa, apple guava orchard, 20.xii.1999, coll. A. O'Toole & G. Quimio, attracted to cue lure; 6 males, Sirang Lupa, calamba, laguna, citrus orchard, 20.xii.2000, coll. A. O'Toole, attracted to cue lure; 1 male, 5 females, Darigayos, luna, la union, 15.xi.1999, coll. G. Quimio, reared from *Luffa cylindrica*; 2 female, Pinaglasatan, infanta, quezan, 10.xii.1999, coll. G. Quimio, reared from *Luffa cylindrica*. **SRI LANKA (formerly CEYLON)**: 1 male, Sri Lanka, Eraminigolla, 8.viii.1997, attracted to cue lure. **THAILAND**: 4 males, 'Chumphon Prov, Khao Kai Jae, Watf, 80m, 5-8.v.2003/2'; '9.55.07N, 98.56.34E, P. Schwendinger leg, semi-evergreen rainforest/2'; 'attracted to cue lure'; 1 male, 'Ban Lak Wo, Surin, 160m, 3km S Surin City, 3-5.xii.2002/12'; '14.54.06N, 103.27.14E, P. Schwendinger/12'; 'attracted to cue lure'. Received on loan from MHNG. **VIETNAM: Northern Vietnam** –1 male, Lai Chau (P), Dien Bien (D), Sam Mun (V), 23.iii.2002, coll. NIPP staff, attracted to cue lure, northern Vietnam trap, VN02; **Southern Vietnam** – 1 male (15.xii.2001), 1 male (10.iii.2002), 1 male (5.v.2002), 1 male (6.vii.2002), Vinhlong (P), Long ho (D), Phuoc hau (V), attracted to cue lure, southern Vietnam VS01; 1 male (23.iii.2002), 9 males (14.vii.2002), 7 males (28.vii.2002), 1 male (4.viii.2002), Cantho (P), Chau thanh (D), Tan phu (V), attracted to cue lure, southern Vietnam VS02; 2 males (24.xii.2001), 1 male (12.i.2002), Ben tre (P), Chau thanh (D), Tien thuy (V), attracted to cue lure, southern Vietnam, VS04; 1 male, Binh thuan (P), Ham thuan (D), No 1 (V), 20.v.2002, attracted to cue lure, southern Vietnam VS07; 1 male (24.viii.2002), 1 male (24.ix.2002), 2 males (8.x.2002), Dac lac (P), Hu mi nga (D), Quang phu (V), attracted to cue lure, southern Vietnam VS09; 1 male, Vinh long (P), Thuan an (V), 25.viii.2000, attracted to cue lure, southern Vietnam S001; 1 male, Tien giang (P), Cho gao (V), 25.vii.2000, attracted to cue lure, southern Vietnam S006; 1 male, Tien giang (P), Cai lee (D), An huu (V), 8.viii.2000, attracted to cue lure, southern Vietnam S009. A large series of specimens bred from *Momordica charantia*, *Momordica cochinchinensis*, *Sechium edule*, Gourd and Luffa, in Northern and Southern Vietnam. Specimens in QDPC except where specified.

Diagnosis: A medium-sized species; face fulvous with large black oval spots; postpronotal lobes and notopleura yellow; scutum red-brown with or without fuscous markings; lateral and medial postsutural yellow vittae present; a yellow spot anterior to notopleural suture; mesopleural stripe reaching midway between anterior margin of notopleuron and anterior *npl.* seta dorsally; scutellum entirely yellow; legs with all segments fulvous with apical ⅓ of all femora, fore and hind tibiae dark fulvous; wing with cells bc and c colourless, microtrichia in outer corner of cell c only; a broad fuscous costal band paler between R_{2+3} and R_{4+5} and expanding into a large spot in apex of wing, pale infuscation along *r-m* crossvein and dark fuscous along *dm-cu* crossvein; a broad fuscous anal streak; abdominal terga III–V orange-brown with a 'T' pattern consisting of a narrow black transverse band across anterior margin of tergum III and a medium to broad medial longitudinal band over all three terga, anterolateral corners of terga IV and V fuscous to dark fuscous.

Attractant: Cue lure.

Distribution:

Type locality: Hawaii (Honolulu).

Previously recorded: Widespread around the world (Drew, 1982, 1989; Norrbom *et al.*, 1998). Introduced into many countries including Hawaii, Papua New Guinea, Solomon Islands (Drew, 1982, 1989; Drew & Romig, 2001).
Hosts: This species has been recorded from 12 host plant families (Allwood *et al.*, 1999).
Remarks: *Bactrocera (Zeugodacus) cucurbitae* (Coquillett) is one species of Dacinae that has been extensively studied, both ecologically and taxonomically. It is regarded as one of the major pest species within the Dacinae and has a worldwide distribution in the tropical and subtropical zones. During the late 1990s, it spread across the Solomon Islands. It is distinguished by the possession of a red-brown scutum, a large spot in the apex of the costal band, infuscation over both the *r-m* and *dm-cu* crossveins, cells bc and c colourless, abdominal tergum III with a narrow transverse black band across base and a medial longitudinal black band over terga III–V.

Bactrocera (Zeugodacus) daclaciae Drew & Romig, new species

(Fig. 288)

Material examined: Holotype male, **VIETNAM: Southern Vietnam** – Dac Lac (P), Ku Mi Nga (D), Quang Phu (V), Site VS09, 23.iii.2002, coll. SOFRI Staff, attracted to cue lure, deposited in BMNH.
Paratypes: **THAILAND**: 4 males, 'Phuket Prov, Siray Isl, E Phuket City, 30 m, 22–26.vii.2002'; '7.53N, 98.26E, P. Schwendinger Leg, secondary forest, at cue lure'; 1 male (21.ii.1992), 1 male (6.iii.1992), 1 male (13.iii.1992), Khao Kheaw, Pakphli, Nakhonnayok, coll. Montree, attracted to cure lure. 1 paratype in ANIC; 3 paratypes in MHNG; 2 paratypes in QDPC; 1 paratype in QMIC.
Diagnosis: A small species; face fulvous with a broad inverted 'U'-shaped black band; postpronotal lobes and notopleura yellow; scutum shining black; broad lateral and medial postsutural yellow vittae present; a distinct broad yellow triangle along anterior margin of notopleural suture with base on notopleuron; mesopleural stripe equal in width to yellow surrounding notopleuron dorsally; scutellum yellow; wing with cell bc fuscous, cell c colourless, microtrichia in outer corner of cell c only; a narrow dark fuscous costal band expanding slightly across apex of R_{4+5}; abdominal terga III–V fulvous with a general black 'T' pattern over terga III–V and broad lateral longitudinal black bands over terga IV and V which join as a narrow black band around posterior margin of tergum V.
Description: male.
Head: Height 1.13 mm. Frons – length 1.33 times breadth; fulvous without dark markings; orbital setae black:1 *s.or.*, 2 *i.or.*; lunule fulvous. Ocellar triangle black. Vertex black. Face fulvous with a broad inverted 'U'-shaped black band consisting of two very large elongate-oval black spots meeting or almost meeting in midline; length 0.32 mm. Genae fulvous, small pale fuscous subocular spot present; dark seta present. Occiput black laterally, fuscous centrally, fulvous along eye margins; occipital row with a large number of small pale setae. Antennae with segments 1 and 2 fulvous, segment 3 fulvous with fuscous on apex and outer surface; length of segments: 0.09 mm; 0.18 mm; 0.54 mm.
Thorax: Scutum black with dark red-brown below and behind lateral postsutural yellow vittae (some specimens have a medial longitudinal red-brown band anterior to the notopleural suture). Pleural areas black except for a small area of red-brown above fore coxae. Yellow markings as follows: postpronotal lobes; notopleura; mesopleural stripe equal in width to yellow surrounding notopleuron dorsally, continuing to katepisternum as a small transverse spot, anterior margin convex; anatergite (posterior apex black); anterior ¾ katatergite (remainder black); two broad parallel-sided lateral postsutural vittae ending behind *ia.* seta; a broad medial longitudinal postsutural vitta beginning at level of *prsc.* setae posteriorly and narrowing slightly anteriorly to end at level of notopleural suture; a distinct large yellow triangle along anterior margin of notopleural suture with base at notopleuron. Postnotum black. Scutellum yellow except for a narrow black basal band. Setae: *sc.* 4; *prsc.* 2; *ia.* 1; *p.sa.* 1; *a.sa.* 1; *mpl.* 1; *npl.* 2; *scp.* 4.
Legs: Femora pale fulvous tending red-brown apically and with small areas of dark fuscous to black around apices of mid and hind femora; fore and mid tibiae fulvous, hind tibiae dark fuscous to black on basal ⅔ and fulvous apically; tarsi with basal two segments fulvous and apical three segments fuscous.
Wings: Length 4.2 mm; cell c colourless, cell bc fuscous; microtrichia in outer corner of cell c only; remainder of wings colourless except dark fuscous cell sc, narrow dark fuscous costal band confluent with R_{2+3} and gradually widening across apex of R_{4+5} to end midway between apices of R_{4+5} and M, a broad dark fuscous anal streak; dense aggregation of microtrichia around A_1+CuA_2; supernumerary lobe strong and rounded.
Abdomen: Oval; terga free; pecten present on tergum III. Tergum I and sterna I and II wider than long. Tergum I red-brown with a narrow transverse black band across anterior margin turning around to include lateral margins; tergum II fulvous with a narrow transverse black band across anterior margin and widening laterally to cover lateral margins; terga III–V fulvous with a distinct black 'T' pattern formed by a broad black transverse band across most of tergum III and a moderately broad medial longitudinal black band over all three terga and broad lateral longitudinal black bands over terga IV and V which join across posterior margin of tergum V as a narrow black band. A pair of oval fulvous shining spots on tergum V. Posterior lobe of surstylus long, sternum V with a slight concavity on posterior margin. Abdominal sterna black, except sterna II and III pale.
Attractant: Cue lure.
Female: No known record.
Distribution: Southern Vietnam (Dac Lac Province); Thailand (Phuket and Nakhonnayok).
Hosts: No known record.
Remarks: *Bactrocera (Zeugodacus) daclaciae* n.sp. is similar to *Bactrocera (Zeugodacus) flavipilosa* (Hardy) in possessing fuscous colouration on at least one costal cell in addition to four *sc.* setae, a medial longitudinal postsutural

yellow vitta, scutum black and wings colourless except for costal band and anal streak. It differs from *B. flavipilosa* in having a broad inverted 'U'-shaped black band across the face and abdominal terga III–V with a black 'T' pattern and broad lateral longitudinal black bands over terga IV and V which join along posterior margin of tergum V. It is also similar to *Bactrocera (Zeugodacus) laguniensis* n.sp. from the Philippines but is distinct in having the inverted 'U'-shaped black band across the face, the costal band not overlapping R_{2+3} and a distinct yellow triangle on anterior margin of notopleural suture (not yellow spots).

Bactrocera (Zeugodacus) depressa (Shiraki)

(Fig. 289)

Zeugodacus depressus Shiraki, 1933: 90. Lectotype male and Paralectotype female, here designated in NIAS.
Paradacus depressus – Shiraki, 1968: 12.
Bactrocera (Paradacus) depressa – Norrbom *et al*., 1998: 99.

Type material examined: TAIWAN (formerly FORMOSA): Lectotype male here designated, with the following three labels – white label '14.ii.1925 29.i emerged Col. T. Shiraki' on one side and 'Kisofutussima Nagano 25.x.1924' on opposite side; white label 'Zeugodacus deprimerus Det. T. Shiraki'; white circular label with red circle 'Type' in red handwriting. Paralectotype female here designated, with the following three labels – white label 'Kiso Shinano Col. T. Shiraki' on one side and 'from Pumpkin' on opposite side; white label 'Zeugodacus deprimerus Det. T. Shiraki'; white circular label with red circle 'Type' in red handwriting. Received on loan from NIAS.

Non-type material examined: JAPAN: 1 female and 16 males, Honshu, Minamiminowa, Ina, Nagano, 1.vi.1952. 1 male, 'Japan – Kyushu, Fukuoka: Hikosan, 20.vi.1946, coll. Syusiro Ito'; 'Paradacus depressus Shiraki det. Syusiro Ito, 1950'; 'Purch from E.M. Hering, B.M. 1965-270'. 1 female, 'Oosumi, Satamisaki, 21.v.1952, coll. Syusiro Ito'; 'Purch from E.M. Hering, B.M. 1965-270'; 'See slide, IMW-1483'. Specimens in BMNH. 1 male, 'Oosumi, Sata, 17.v.1952, coll. Syusiro Ito'; '6'; 'Purch. from E.M. Hering, B.M. 1965-270'; 1 female, 'x.1988, 16.ii.1989, TKY'; 'x.1988, Yokohama, Japan, D. depressus'. Specimens in QDPC. **TAIWAN**: 1 female, 'Toyenmongai bei, Taiwan, Formosa, Rolle, v.1910'; 'F. Bearbeit, erf. v. Mus.'; 'Purch from E.M. Hering, B.M. 1965-270'; 'In coll. Hering as D. depressus'. Specimens in BMNH.

Diagnosis: A large species; face fulvous with a pair of small to medium ill-defined fuscous to dark fuscous spots; postpronotal lobes and notopleura yellow; scutum basically red-brown with dark markings; lateral and medial postsutural yellow vittae present; a yellow spot anterior to notopleural suture in front of lateral postsutural yellow vittae; mesopleural stripe reaching midway between anterior margin of notopleuron and anterior *npl.* seta dorsally; scutellum yellow; wing with cells bc and c with a pale fuscous tint, microtrichia in outer corner of cell c only; a narrow fuscous costal band slightly overlapping R_{2+3}; a broad fuscous anal streak; a pale tint around medial posterior area of wing margin; abdominal terga III–V fulvous with a black 'T' pattern consisting of a narrow transverse black band across anterior margin of tergum III and a medium-width medial longitudinal black band over all three terga that is broken on posterior margin of tergum III, anterolateral corners of terga IV and V dark fuscous.

Description: male.

Head: Height 2.2 mm. Frons – length 1.5 times breadth; fulvous with dark fuscous around orbital setae and on anteromedial hump; orbital setae black: 1 *s.or.*, 3 *i.or.*; lunule dark fuscous. Ocellar triangle black. Vertex fulvous. Face fulvous with a pair of small to medium ill-defined fuscous to dark fuscous spots; length 0.7 mm. Genae fulvous, ill-defined fuscous subocular spot; black seta present. Occiput red-brown, fulvous along eye margins; occipital row with six well-developed red-brown setae. Antennae with segments 1 and 2 red-brown, segment 3 red-brown with pale fuscous on apex and outer surface; length of segments: 0.2 mm; 0.36 mm; 1.04 mm.

Thorax: Scutum basically red-brown with variable dark markings as follows: black inside lateral postsutural yellow vittae merging to dark fuscous as it reaches medial postsutural yellow vitta, a large area varying from pale fuscous to black between postpronotal lobe and notopleural suture. Pleural areas red-brown except dark fuscous anterior to mesopleural stripe and on katepisternum. Yellow markings as follows: postpronotal lobes; notopleura; mesopleural stripe of medium-width reaching midway between anterior margin of notopleuron and anterior *npl.* seta dorsally, continuing to katepisternum as a large transverse spot, anterior margin slightly convex; anatergite (posterior apex black); anterior ⅔ katatergite (remainder black); two moderately broad lateral postsutural vittae beginning anterior to notopleural suture and narrowing only slightly posteriorly to end at or just before *ia.* seta; a broad and short medial longitudinal postsutural vitta beginning anterior to *prsc.* setae and ending at a point at level of notopleural suture. Postnotum black laterally and red-brown medially. Scutellum yellow with a narrow red-brown basal band (a small fuscous to dark fuscous apical spot present in some specimens). Setae: *sc.* 4; *prsc.* pale and weak in male Lectotype, absent in female Paralectotype (it appears that these can be absent or present with 1 or 2 setae); *ia.* 1; *p.sa.* 1; *a.sa.* 1; *mpl.* 1; *npl.* 2; *scp.* 4–6.

Legs: Femora pale fulvous tending fulvous apically; tibiae and all tarsal segments fulvous; mid tibiae each with an apical black spur.

Wings: Length 9.2 mm; cells bc and c with a pale fuscous tint; microtrichia in outer corner of cell c only; remainder of wings colourless except for fuscous cell sc, narrow fuscous costal band slightly overlapping R_{2+3} where it becomes pale fuscous and widening across apex of vein R_{4+5}, a broad fuscous anal streak, a pale tint around medial posterior area of wing margin; a dense aggregation of microtrichia around A_1+CuA_2; supernumerary lobe of medium development.

Abdomen: Elongate oval; terga free; pecten present on tergum III. Tergum I and sterna I and II wider than long. Tergum I fulvous with narrow dark fuscous lateral margins; tergum II fulvous with a dark fuscous transverse band along anterior margin but not reaching lateral margins and anterolateral corners dark fuscous; tergum III fulvous with a black 'T' consisting of a narrow transverse black band across anterior margin which expands to cover lateral margins and a medium-width medial longitudinal black band which does not reach posterior margin of this tergum; terga IV and V fulvous with anterolateral corners dark fuscous and extending inwards along anterior margin of each tergum and a medium-width medial longitudinal dark fuscous band extending over each tergum. A pair of oval fulvous shining spots on tergum V. Posterior lobe of surstylus long, sternum V with a slight concavity on posterior margin.

Attractant: No known record.

Female: As for male except no dense aggregation of microtrichia around A_1+CuA_2; supernumerary lobe weak; pecten of cilia on abdominal tergum III absent; basal segment of ovipositor dark red-brown, dorsoventrally flattened and tapering posteriorly in dorsal view; apex of piercer trilobed; ratio of length of oviscape to length of tergum V, 1.2:1.

Distribution:

Type locality: – Taiwan (Kisofutussima Nagano).

Previously recorded: Type locality and the following locations – Japan (all major islands); Ryukyu Is; Korea (White & Elson-Harris, 1992; Norrbom *et al*., 1998).

Hosts: Recorded from three species in the family *Cucurbitaceae* (Allwood *et al*., 1999).

Remarks: The types of *Bactrocera (Zeugodacus) depressa* (Shiraki), previously presumed lost, have now been found in NIAS. This has made possible the full definition of this species presented above. It belongs to a group of species with four *sc.* setae, scutum base colour red-brown, wings colourless except for costal band and anal streak, costal band expanded into a spot apically, lateral postsutural vittae ending before *ia.* setae and costal band overlapping R_{2+3}. It is similar to *Bactrocera (Zeugodacus) okunii* (Shiraki) and *Bactrocera (Zeugodacus) uncinata* n.sp. in these characters, but differs from *B. uncinata* in not having a recurved black barb on abdominal tergum IV and from *B. okunii* in having cells bc and c with a pale tint (not colourless) and a small spot in costal band around apex of R_{4+5} (not large and almost to vein M).

The two specimens labelled 'type', and received on loan from NIAS, were the specimens studied by Shiraki (1933) for the original description and are thus Syntypes as Shiraki (1933) did not designate a Holotype. Consequently, we are here designating the male as the Lectotype and female the Paralectotype.

Bactrocera (Zeugodacus) diaphora (Hendel)

(Fig. 290)

Chaetodacus diaphorus Hendel, 1915: 425; Shiraki, 1933:69. Syntypes male and female in BMNH and NMW.
Chaetodacus ater Chen, 1940: 131 – Preoccupied by *atra* Malloch, 1938: 113. Liang *et al*., 1993: 139 (Synonym).
Dacus (Strumeta) diaphorus – Hardy, 1973: 40.
Dacus (Bactrocera) diaphorus – Hardy, 1977: 49; Tseng, Chen & Chu, 1992: 28.
Dacus (Zeugodacus) lunulatus Tseng, Chen & Chu, 1992: 66. **New Synonym**.
Dacus (Zeugodacus) guangxianus Chao & Lin, 1993b: 137. **New Synonym**.
Dacus (Zeugodacus) sicieni Chao & Lin, 1993a: 77, new name for *ater* Chen, 1940. Record of original synonymy unknown, see Norrbom *et al*., 1998: 102.
Bactrocera (Zeugodacus) diaphora – Wang, 1996: 68; Norrbom *et al*., 1998: 102.
Bactrocera (Zeugodacus) proprediaphora Wang, Xiao & Chen, in Wang *et al*., 2008: 73. **New Synonym**.

Type material examined: TAIWAN (formerly FORMOSA): 2 male and 2 female Syntypes with the following four labels – 'Formosa Sauter'; 'Tapani 1911.iii'; 'diaphorus Hend. typus det. Hendel' and a white label with red margins and one word in red 'typus'; 1 male syntype with the following four labels – 'Formosa Sauter'; 'Suisharyo 1911.x'; 'Dacus diaphorus H. typus det. Hendel' and a white label with red margins and one word in red 'typus'. Received on loan from MNM.

Paratypes: **TAIWAN**: one male labelled paratype with the following seven labels – white circular label with yellow margins 'Para-type'; red label 'Paratype'; 'Tapani 1911.iii.'; 'Formosa Sauter'; 'diaphorus Typus Hendel det Hendel'; 'Purch. from E.M. Hering B.M. 1965-270'; 'BMNH(E) #533024'. Studied at BMNH.

Non-type material examined: TAIWAN: 1 female in poor condition in QMIC with the following labels is probably not a type of *B. diaphora* – 'Koshun, Formosa, Sauter, 11 SEPT '08'; 'Dacus sp. B det. Hendel'; 'Type, *Chaetodacus diaphorus* Hend. det. H.K. Munro 1934'; handwritten in German 'probably observed by Munro as Type in error'. **BHUTAN**: 2 males, Rontong, 30.ix.1991, coll. H.R. Fiejen, attracted to cue lure (poss. contaminated by methyl eugenol); 1 male, Hurongpong, 7.x.1991, coll. H.R. Fiejen, attracted to cue lure (poss. contaminated by methyl eugenol). 1 male (17.v.2000), 1 male (26.vii.2000), 2 males (16.viii.2000), 2 males (6.ix.2000), 1 male (12.ix.2000), 1 male (20.ix.2000), 1 male (27.ix.2000), Rimchu trap No. 1, coll. C. Dorji, all attracted to cue lure; 1 male (7.vi.2000), 3 males (28.vi.2000), 5 males (5.vii.2000), 1 male (2.viii.2000), 1 male (10. viii.2000), Rimchu trap No. 1, coll. C. Dorji *et al*., all attracted to cue lure; 5 males, Rimchu trap No. 1, 14.vi.2000, coll. Nado, attracted to cue lure; 1 male, Rimchu trap No. 1, 21.vi.2000, coll. Nado *et al*., attracted to cue lure; 2 males (21.v.2000), 1 male (24.v.2000), 2 males (28.vi.2000), 1 male (26.vii.2000), 3 males (10.viii.2000), 2 males (16.viii.2000), 1 male (30.viii.2000), 1 male (6.ix.2000), 1 male (27. ix.2000), Rimchu trap No. 2, coll. C. Dorji, all attracted to cue lure; 1 male, Tsirang, 17.vii.2000, coll. S.D., attracted to cue lure; 1 male, Tsirang, 11.ix.2000, attracted to cue lure; 1 male, Lumitshawa, 24.v.2000, coll. C. Dorji, attracted to cue lure; 1 male, Lumitshawa, 10.v.2000, coll. B.S. Fletcher, attracted to cue lure; 1 male, Namling 16km, 9. vi.2001, coll. Brian Fletcher, attracted to cue lure; 1 male, Lingmethang, 9.vi.2001, coll. Brian Fletcher, attracted to cue lure. **CHINA**: 1 male (2.ix.1940) and 1 female (28.vi.1940), Beipei, Sichuan Province. **INDONESIA: JAVA** – 1 male, East Java, Lumajang, Kadang Tepus, 30.viii.2005, coll. Utomo, attracted to cue lure in banana. **PENINSULAR MALAYSIA**: 1 male, Pangsoon, Hulu Langat, 26.ix.1990, coll. D.L. Hancock, attracted to cue lure; 2 males, Pulau Perhentian, viii.1986, coll. C.S. Ooi, attracted to cue lure; a series of specimens collected in Peninsular Malaysia. **THAILAND**: 1 male, N. Thailand, Chiang Mai Prov., Doi Suthep – Pui N.P., What Phrathat, attracted to cue lure, 18.80N/98.92E, 19.x.2000, coll. Merz & Schwendinger; 1 male, Yang Pong Rat, Plubpla, A. Muang, Chanthaburi, 3.ix.1991, coll. Montree, BK74T, attracted to cue lure; 1 male (6.iii.1992), 1 male (17.iv.1992), 2 males (22.v.1992), 1 male (29.v.1992), 1 male (21.viii.1992), Wang Jum Pee, A. Pakchong, Nakhonratsima, coll. Montree, all attracted to cue lure; 2 males, Yang Pong Rat, A. Muang, Chanthaburi, 9.vi.1992, coll. Montree, attracted to cue lure; 4 males (22.v.1992), 1 male (12.vi.1992), 1 male (10.vii.1992), Khao Kheaw, A. Pakphli, Nakhonnayok, coll. Montree, attracted to cue lure. **VIETNAM: Northern Vietnam** – 1 male, Lao Cai Province, Sapa (Town), trap No. 16, 13.viii.1999, attracted to cue lure; 1 male, Lao Cai Province, Sapa 6 Village, trap No. 17, 13.viii.1999, attracted to cue lure; 1 male (23.vi.2000), 1 male (7.vii.2000), Lao Cai Province, Sapa 2 Village, trap No. 18, attracted to cue lure; 2 males (30.iv.2002), 3 males (15.v.2002), Son La Province, Moc Chau (D), Long Luong (V), coll. NIPP Staff, attracted to cue lure; 1 male (15.iii.2000), 1 male (15.v.2000), Hoa Binh Province, Mai Chau District, Pa Co, trap No. N010, attracted to cue lure; **Southern Vietnam** – 1 male, Southern Vietnam, Cantho (P), Chau Thanh (D), Tan Phu (V), 23.iii.2002, trap No. VS02, coll. SOFRI staff, attracted to cue lure; Southern Vietnam, Dac Lac Province, Ku Mi Nga District, Quang Phu Village, trap No. VS09, 8.vi.2002. Specimens in QDPC.

Diagnosis: A medium-sized species; face shining black; postpronotal lobes and notopleura yellow; scutum shining black; narrow lateral and medial postsutural yellow vittae; a yellow spot anterior to notopleural suture; mesopleural stripe reaching almost to anterior *npl.* seta dorsally; scutellum yellow with a dark fuscous to black apical spot enclosing apical setae; wing with cells bc and c colourless, microtrichia in outer corner of cell c only; a narrow dark fuscous costal band widening slightly across apex of R_{4+5}; a narrow fuscous anal streak; abdominal tergum III entirely dark fuscous to black, terga IV and V dark fuscous except red-brown on posterior apex of tergum V.

Description: male.

Head: Height 1.27 mm. Frons – length 1.23 times breadth; fulvous with dark fuscous around orbital setae and on anteromedial hump; latter with a small number of short pale hairs; orbital setae black: 1 *s.or.*, 3 *i.or.*; lunule dark fuscous. Ocellar triangle black. Vertex black. Face shining black except for a small fulvous area below antennal sockets; length 0.41 mm. Fuscous along inside eye margins which extends to a large dark fuscous area on genae; pale seta present. Occiput dark fuscous, fulvous along eye margins; occipital row with a small number of short pale setae. Antennae with segments 1 and 2 red-brown, segment 3 dark fuscous except for a narrow inner lateral red-brown margin; length of segments: 0.09 mm; 0.21 mm; 0.68 mm.

Thorax: Scutum black except dark red-brown behind lateral postsutural vittae and red-brown anterior to notopleural suture. Pleural areas entirely dark fuscous to black. Yellow markings as follows: postpronotal lobes; notopleura; mesopleural stripe reaching almost to anterior *npl.* seta dorsally, continuing to katepisternum as a transverse spot, anterior margin convex; anatergite (posterior apex dark fuscous); anterior ¾ katatergite (remainder dark fuscous); two narrow lateral postsutural vittae, subparallel to tapering posteriorly, beginning anterior to notopleural suture and ending slightly before or at *ia.* setae; a narrow medial longitudinal postsutural vitta. Postnotum black. Scutellum yellow except for narrow black basal band and dark fuscous to black apical spot enclosing apical pair of *sc.* setae, in some specimens. Setae: *sc.* 2; *prsc.* 2; *ia.* 1; *p.sa.* 1; *a.sa.* 1; *mpl.* 1; *npl.* 2; *scp.* 4.

Legs: Fore femora dark fuscous, mid femora with basal ⅓ fulvous and apical ⅔ dark fuscous, hind femora with basal ½ fulvous and apical ½ dark fuscous; all tibiae fulvous except dark fuscous on basal ⅓ to ¼; all tarsal segments fulvous; mid tibiae each with an apical black spur.

Wings: Length 5.5 mm; cells bc and c colourless; microtrichia in outer corner of cell c only; remainder of wings colourless except dark fuscous cell sc, narrow dark fuscous costal band confluent with R_{2+3} and widening slightly across apex of R_{4+5}, a dark fuscous anal streak of medium-width; dense aggregation of microtrichia around A_1+CuA_2; supernumerary lobe strong.

Abdomen: Oval; terga free; pecten present on tergum III. Tergum I and sterna I and II wider than long. Tergum I fuscous centrally, dark fuscous laterally and with a narrow red-brown band along posterior margin but not reaching lateral margins; tergum II red-brown except for a dark fuscous to black narrow transverse band across anterior margin and expanding to anterolateral corners of that tergum; tergum III entirely dark fuscous to black; terga IV and V dark fuscous except for red-brown on posterior apex of tergum V (in some specimens, there are red-brown areas either side of a dark fuscous to black medial longitudinal vitta on posterior of tergum III and on terga IV and V). A pair of oval dark fuscous shining spots on tergum V. Posterior lobe of surstylus long, sternum V with a slight concavity on posterior margin.

Attractant: Cue lure.

Female: As for male except mid femora with apical ½ dark fuscous and hind femora with apical ⅓ dark fuscous, all tibiae entirely dark fuscous; no dense aggregation of microtrichia around A_1+CuA_2; supernumerary lobe weak; pecten of cilia on abdominal tergum III absent; basal segment of ovipositor red-brown to fuscous, dorsoventrally compressed and tapering posteriorly in dorsal view, extremely long; ratio of length of oviscape to length of tergum V, 2.9:1.

Distribution:

Type locality: Taiwan (Tapani; Suisharyo).

Previously recorded: Type locality and the following locations – India and Sri Lanka to China (Sichuan Province); Indonesia (Sumatra); Peninsular Malaysia; Thailand; Vietnam (White & Hancock, 1997; Norrbom *et al.*, 1998).

New records: Bhutan; Indonesia (Java). China (Yunnan Province), due to **New Synonymy** of *D. proprediaphora*. China (Guangxi Province), due to **New Synonymy** of *D. guangxianus*.
Hosts: No known record.
Remarks: *Bactrocera (Zeugodacus) diaphora* (Hendel), within the *scutellaris* complex, is similar to *Bactrocera (Zeugodacus) hoabinhiae* n.sp., *Bactrocera (Zeugodacus) laocaiae* n.sp., *Bactrocera (Zeugodacus) yoshimotoi* (Hardy) and *Bactrocera (Hemigymnodacus) mukiae* n.sp. in possessing two *sc.* setae and yellow notopleura. It differs from these species in having the face entirely black and all femora with large areas of dark fuscous to black.

There are no good characters that can be used to separate *B. diaphora* from *Dacus (Zeugodacus) guangxianus* Chao & Lin, *Dacus (Zeugodacus) lunulatus* Tseng, Chen & Chu and *Bactrocera (Zeugodacus) proprediaphora* Wang, Xiao & Chen, and we are therefore making these three species **New Synonyms** of *B. diaphora*. Further, we confirm that *Dacus (Zeugodacus) sicieni* Chao & Lin is a synonym of *B. diaphora* as recorded by Norrbom *et al. (*1998).

Bactrocera (Zeugodacus) diaphoropsis (Hering)

(Fig. 291)

Zeugodacus diaphoropsis Hering, 1952b: 268. Holotype male in RNH.
Dacus (Zeugodacus) diaphoropsis – Hardy & Adachi, 1954: 187.
Bactrocera (Zeugodacus) diaphoropsis – Norrbom *et al.*, 1998: 102.

Type material examined: Holotype male, **INDONESIA: KALIMANTAN (formerly BORNEO)** – with the following four labels - white label 'E. Borneo – 50 m, Samarinda xi.50, Muara Kaman, A.M.R. Wegner'; red square label 'Type'; white label 'Zeugodacus diaphoropsis n. det. Hering. Type'; red rectangular label 'Holotype'. Received on loan from RNH.
Diagnosis: A medium-sized species; face fulvous with a pair of very small pale fuscous spots on lower margin; postpronotal lobes and notopleura yellow; scutum black; lateral and medial postsutural yellow vittae present; a yellow spot anterior to notopleural suture in front of lateral postsutural yellow vittae; mesopleural stripe reaching to anterior *npl.* seta dorsally; scutellum yellow; wing with cells bc and c colourless, microtrichia in outer corner of cell c only; a narrow dark fuscous costal band confluent with R_{2+3}, a pale tint across centre of wing, pale fuscous around CuA_1; broad pale fuscous to fuscous anal streak; abdominal terga III–V orange-brown with a black 'T' pattern consisting of a transverse black band across anterior margin of tergum III and a medium-width medial longitudinal band over all three terga and anterolateral corners of terga IV and V dark fuscous.
Description: male.
Head: Frons – fulvous with fuscous around orbital setae and pale fuscous on anteromedial hump; orbital setae strong and black: 1 *s.or.*, 3 *i.or.*; lunule fuscous. Ocellar triangle black. Vertex fuscous. Face fulvous with a pair of very small pale fuscous spots on lower margin. Genae fulvous, a large fuscous subocular spot present; black seta present. Occiput red-brown centrally, pale fuscous laterally, fulvous along eye margins; occipital row with a few short pale setae. Antennae with segments 1 and 2 red-brown, segment 3 red-brown with fuscous on apex and outer surface.
Thorax: Scutum black except fuscous posterior to lateral postsutural yellow vittae and red-brown lateral to these vittae, anterior to notopleural suture and inside postpronotal lobes. Pleural areas black except fuscous below postpronotal lobes. Yellow markings as follows: postpronotal lobes; notopleura; mesopleural stripes reaching anterior *npl.* seta dorsally, extending to katepisternum as a transverse spot, anterior margin convex; anatergite (posterior apex black); anterior $^5\!/_7$ katatergite (remainder dark fuscous); two broad lateral postsutural vittae beginning as a spot anterior to notopleural suture and narrowing slightly posteriorly to end at *ia.* seta; a broad medial longitudinal postsutural vitta broadly rounded posteriorly and ending in a rounded point anteriorly at level of notopleural suture. Postnotum black. Scutellum yellow with a narrow black basal band. Setae: *sc.* 4; *prsc.* 2; *ia.* 1; *p.sa.* 1; *a.sa.* 1; *mpl.* 1; *npl.* 2; *scp.* 4.
Legs: Fore femora dark fuscous with a small area of fulvous basally, mid femora fulvous except for a large dark fuscous spot on outer apical surface, hind femora fulvous with dark fuscous spots on outer apical surfaces; fore tibiae fulvous on apical ½ and dark fuscous on basal ½, mid tibiae fulvous with a small area of fuscous basally, hind tibiae changing from dark fuscous basally to pale fuscous apically; fore and mid tarsi entirely fulvous, hind tarsi fulvous on basal two segments and pale fuscous on apical three segments; mid tibiae each with an apical black spur.
Wings: Cells bc and c colourless; microtrichia in outer corner of cell c only; remainder of wings colourless except dark fuscous cell sc, a narrow dark fuscous costal band confluent with R_{2+3} and continuing around wing margin to end between apices of R_{4+5} and M, a pale tint across centre of wing, pale fuscous around CuA_1, a broad pale fuscous to fuscous anal streak; a dense aggregation of microtrichia around A_1+CuA_2; supernumerary lobe strong and rounded.
Abdomen: Oval; terga free; pecten present on Tergum III. Tergum I and sterna I and II wider than long. Tergum I orange-brown with narrow black lateral margins; tergum II orange-brown with fulvous across posterior margin and with a narrow transverse black band anteriorly but not reaching lateral margins and anterolateral corners narrowly black; terga III to V orange-brown with a black 'T' pattern consisting of a narrow transverse band across anterior margin of tergum III expanding to cover lateral margins and a medium-width medial longitudinal band which is black on tergum III and dark fuscous on terga IV and V, anterolateral corners of terga IV and V with broad dark fuscous patterns. A pair of oval orange-brown shining spots on tergum V. Posterior lobe of surstylus long, sternum V with a slight concavity on posterior margin.
Attractant: No known record.
Female: No known record.
Distribution:
Type locality: Indonesia (Kalimantan – Samarinda, Muara Kama).
Hosts: No known record.

Remarks: *Bactrocera (Zeugodacus) diaphoropsis* (Hering) is similar to *Bactrocera (Zeugodacus) munda* (Bezzi) in possessing four *sc.* setae, a black scutum with medial and lateral postsutural yellow vittae, wings with infuscation in addition to costal band and anal streak, costal band without a distinct apical spot, scutellum entirely yellow and face not entirely black. It differs from *B. munda* in having the lateral postsutural yellow vittae ending at the *ia.* setae, femora with reduced dark colour patterns (particularly the mid and hind femora) and small areas of infuscation along posterior portion of *dm-cu* crossvein.

Bactrocera (Zeugodacus) dorsirufa Drew & Romig, new species

(Fig. 292)

Material examined: Holotype male, **BHUTAN**: Namling 14 km, 7.vi.2001, coll. Brian Fletcher, attracted to cue lure, deposited in BMNH.
Paratypes: **BHUTAN**: 3 males, same data as holotype. 1 paratype in ANIC; 1 paratype in QDPC; 1 paratype in QMIC.
Diagnosis: A large species; face fulvous with a pair of medium-sized circular black spots; postpronotal lobes and notopleura yellow; scutum shining black; lateral and medial postsutural yellow vittae present; a small yellow spot present anterior to notopleural suture; mesopleural stripe equal to or slightly wider than notopleuron dorsally; scutellum yellow with a large black apical spot on dorsal surface; wing with cells bc and c colourless, microtrichia in outer corner of cell c only; a narrow pale fuscous costal band not extending beyond apex of R_{2+3} and with a small fuscous to dark fuscous spot around apex of R_{4+5}; anal streak absent; cell br with a very pale tint and devoid of microtrichia; abdominal terga III–V red-brown with broad lateral longitudinal black bands over all three terga.
Description: male.
Head: Height 1.62 mm. Frons – length 1.35 times breadth; red-brown centrally with fulvous along lateral margins, dark fuscous around orbital setae and on anteromedial hump; orbital setae black: 1 *s.or.*, 3 *i.or.*; lunule dark fuscous. Ocellar triangle black. Vertex dark fuscous to black. Face fulvous with a pair of medium-sized circular black spots; length 0.54 mm. Genae fulvous, fuscous subocular spot present; weak pale seta present. Occiput shining black laterally tending dark fuscous centrally and fulvous along eye margins; occipital row with 2–4 black setae. Antennae with segments 1 and 2 red-brown, segment 3 red-brown with fuscous on apex and outer surface; arista black (red-brown basally); length of segments: 0.12 mm; 0.24 mm; 0.69 mm.
Thorax: Scutum shining black with red-brown below and behind lateral postsutural vittae and anterior to notopleural suture between notopleuron and yellow spot. Pleural areas shining black without pale markings. Yellow markings as follows: postpronotal lobes (anterodorsal corners fuscous); notopleura; mesopleural stripe equal to or slightly wider than notopleuron dorsally, continuing to katepisternum as a small transverse spot, anterior margin convex; anatergite (posterior apex black); anterior ½ katatergite remainder including ventral margin black); two narrow parallel-sided lateral postsutural vittae beginning with a small spot anterior to notopleural suture and ending at *ia.* seta; a very narrow and short medial longitudinal postsutural vitta. Postnotum black. Scutellum yellow except for narrow black basal band and large black apical spot on dorsal surface. Setae: *sc.* 4; *prsc.* 2; *ia.* 1; *p.sa.* 1; *a.sa.* 1; *mpl.* 1; *npl.* 2; *scp.* 2.
Legs: Fore femora fulvous with a small black preapical spot on outer surfaces, mid femora fulvous with a small black preapical spot on inner surfaces, hind femora fulvous with dark fuscous to black spot around apical ¼; fore tibiae fuscous to dark fuscous, mid tibiae fuscous to dark fuscous basally to fulvous apically, hind tibiae dark fuscous on basal ½ to pale fuscous to fulvous apically; tarsi with basal segment fulvous and apical four segments pale fuscous; mid tibiae each with an apical black spur.
Wings: Length 7.85 mm; cells bc and c colourless; microtrichia in outer corner of cell c only; remainder of wings colourless except dark fuscous cell sc, narrow pale fuscous costal band confluent with R_{2+3} and absent beyond apex of this vein and with a small fuscous to dark fuscous spot around apex of R_{4+5}; dense aggregation of microtrichia around A_1+CuA_2; supernumerary lobe weak; anal streak absent; cell br with a very pale tint and devoid of microtrichia.
Abdomen: Elongate oval; terga free; pecten present on tergum III. Tergum I and sterna I and II wider than long. Tergum I red-brown with broad lateral black margins that may meet along anterior margin; tergum II fulvous with a broad transverse black band across anterior margin and widening over lateral margins; terga III–V red-brown with broad lateral longitudinal black bands over all three terga and a small circular black spot anterocentrally on tergum III (in some specimens there is a very short and very narrow medial longitudinal fuscous line on tergum V but not reaching anterior or posterior margins). A pair of oval orange-brown shining spots on tergum V. Posterior lobe of surstylus long, sternum V with a slight concavity on posterior margin. Abdominal sterna dark fuscous to black, except sternum V red-brown.
Attractant: Cue lure.
Female: No known record.
Distribution: Bhutan (Namling).
Hosts: No known record.
Remarks: *Bactrocera (Zeugodacus) dorsirufa* n.sp. is a unique species with typical *scutellaris* complex characters on the thorax and wings. It differs from the species in the complex with four *sc.* setae, in having abdominal terga III–V red-brown with broad lateral longitudinal black bands, cell br with a very pale tint and devoid of microtrichia and the anal streak absent.

Bactrocera (Zeugodacus) dubiosa (Hardy)

(Fig. 293)

Dacus (Zeugodacus) dubiosus Hardy, 1982: 205. Holotype male in MZB.
Bactrocera (Zeugodacus) dubiosa – Norrbom *et al.*, 1998: 102.

Type material examined: Holotype male, **INDONESIA: SULAWESI** –Sadaunta, Sulawesi, 650 m, in primary forest, cue lure, viii.1975, coll. D.E. Hardy. Studied at MZB.

Diagnosis: Face fulvous with a pair of medium-sized circular black spots; postpronotal lobes and notopleura yellow; scutum pale red-brown with dark fuscous along inner margins of lateral postsutural yellow vittae and between notopleural suture and postpronotal lobe; lateral and medial postsutural yellow vittae present; yellow spot anterior to notopleural suture; mesopleural stripe reaching midway between anterior margin of notopleuron and anterior *npl.* seta dorsally; scutellum entirely yellow; legs with all segments entirely fulvous; wing with cells bc and c colourless, microtrichia in outer one half of cell c only; a broad fuscous costal band confluent with R_{4+5} and expanding into a spot across apex of wing to reach M, a very pale tint along *r-m* crossvein, pale fuscous along *dm-cu* crossvein which may connect with large spot in apex of wing; a moderately broad fuscous anal streak; abdominal terga III–V fulvous with a black 'T' pattern consisting of a narrow transverse band across anterior margin of tergum III and a narrow medial longitudinal dark fuscous to black band over all three terga, narrow anterolateral black corners on terga IV and V.

Description: male.

Head: Frons – fulvous without dark markings; orbital setae black: 1 *s.or.*, 2 *i.or.*; lunule fuscous. Ocellar triangle black. Vertex fulvous. Face fulvous with a pair of medium-sized circular black spots. Occiput fulvous without dark markings. Antennae with segments 1 and 2 fulvous, segment 3 red-brown with pale fuscous on apex and outer surface.

Thorax: Scutum pale red-brown with dark fuscous along the inner margins of the lateral postsutural yellow vittae and between the notopleural suture and postpronotal lobe. Yellow markings as follows: postpronotal lobes; notopleura; mesopleural stripe reaching midway between anterior margin of notopleuron and anterior *npl.* seta dorsally; broad almost parallel-sided lateral postsutural vittae beginning with a spot anterior to notopleural suture and ending just behind *ia.* seta; a broad medial longitudinal postsutural vitta beginning posteriorly at level of *prsc.* setae and ending in a point anterior to notopleural suture. Scutellum yellow except for narrow red-brown basal band. Setae: *sc.* 4; *prsc.* 2; *ia.* 1; *p.sa.* 1; *a. sa.* 1; *mpl.* 1; *npl.* 2; *scp.* 4.

Legs: All segments fulvous except with some pale infuscation on the tarsi.

Wings: Cells bc and c colourless; microtrichia in outer ½ of cell c only; remainder of wings colourless except fuscous cell sc, broad fuscous costal band distinct as it overlaps R_{2+3} and then fading to be confluent with R_{4+5} and expanding into a large fuscous to pale fuscous spot across apex of wing, a very pale tint along *r-m* crossvein, pale fuscous surrounding *dm-cu* crossvein and tending to connect with the large spot across apex of wing, a moderately broad fuscous anal streak; dense aggregation of microtrichia around A_1+CuA_2; supernumerary lobe strong and rounded.

Abdomen: Tergum I red-brown with narrow dark fuscous lateral margins; tergum II fulvous with a narrow transverse dark fuscous band across anterior margin but not reaching the narrow dark fuscous lateral margins; terga III–V fulvous with a 'T' pattern consisting of a narrow transverse black band across anterior margin of tergum III and a narrow medial longitudinal dark fuscous to black band over all three terga, small black anterolateral corners on terga IV and V. A pair of oval fulvous shining spots on tergum V. Posterior lobe of surstylus long, sternum V with a slight concavity on posterior margin. Abdominal sterna dark fuscous to black, except sternum II pale.

Attractant: Cue lure.

Female: No known record.

Distribution:

Type locality: Indonesia (Sulawesi, Sadaunta).

Hosts: No known record.

Remarks: *Bactrocera (Zeugodacus) dubiosa* (Hardy) appears to be a synonym of *Bactrocera (Zeugodacus) emittens* (Walker). Hardy (1982) based this species on a description of one specimen and, until more specimens are available for study, we cannot confirm this synonymy. There appear to be only minor differences in the dark colour patterns on abdominal terga III–V and in the infuscation on the apex of the wing. In particular, *B. dubiosa* has a complete black 'T' pattern on abdominal terga III–V whereas *B. emittens* has an incomplete 'T'. It is also close to *Bactrocera (Zeugodacus) cucurbitae* (Coquillett) but differs in having broad lateral and medial postsutural yellow vittae and a broad mesopleural stripe.

Bactrocera (Zeugodacus) elegantula (Hardy)

(Fig. 294)

Dacus (Strumeta) elegantulus Hardy, 1974: 32. Holotype male in UZMH.
Bactrocera (Bactrocera) elegantula – Norrbom *et al.*, 1998: 90.

Type material examined: Holotype male, **PHILIPPINES**: with the following label data – Kalambuge, 12.i.15. Received on loan from UZMH.

Non-type material examined: EAST MALAYSIA: SARAWAK – 1 male (25.xi.1991), 1 male (7.iii.1992), 1 male (25.iv.1992), 2 males (2.v.1992), 1 male (5.vi.1992), Sarawak, ARC Semongok Nr Kuching, coll. S. Leong, attracted to cue lure; 1 male (1.viii.1992), 1 male (2.i.1993), Sarawak, ARC Semongok KT1, coll. S. Leong, attracted to cue lure. **PHILIPPINES**: 10 males, Mt. Makiling (600m), 2 males (500m), 5.ix.1983, coll. R.A.I. Drew, attracted to cue lure; 1 male, Sirang Lupa, Calamba, Laguna, Apple Guava mixed orchard, 20.xii.1999, attracted to cue lure. **INDONESIA: KALIMANTAN** – 3 males, East Kalimantan, Kutai Kartanegara, Batuah, 1.vii.2005, coll. Heru G., attracted to cue lure in forest; 1 male, West Kalimantan, Pontianak, Sei Duri I, 13.xii.2005, coll. A. Irom, attracted to cue lure in citrus; 1 male, West Kalimantan, Sei Kunyit, Sei Duri I, 15.xi.2005, coll. A. Irom, attracted to cue lure in citrus; 1 male, East Kalimantan, Kutai Kartanegara, Batuah, 9.i.2005, coll. Ertha & Heru G., attracted to cue lure in forest; 1 male, South Kalimantan, Banjarbaru, Guntung Payung, 28.vii.2005, coll. Sidrotul & Sutini, attracted to cue lure in village. Specimens in QDPC.

Diagnosis: A medium-sized species; face fulvous with a pair of medium-sized oval black spots; postpronotal lobes and notopleura yellow; scutum black; lateral and medial postsutural yellow vittae present; a yellow spot anterior to notopleural suture in front of lateral postsutural yellow vittae; mesopleural stripe almost reaching anterior *npl.* seta dorsally; scutellum yellow; wing with cells bc and c with a pale fuscous tint, microtrichia in outer ⅓ of cell c only; a narrow fuscous costal band overlapping R_{2+3} and widening across apex of wing to M, a fuscous band enclosing *dm-cu* crossvein; a broad fuscous anal streak; abdominal terga III–V orange-brown with a narrow transverse black band across anterior margin of tergum III which continues around lateral margins of all three terga and a medial longitudinal black band which is narrow on anterior ½ of tergum III then widening over the remainder of terga III–V.

Description: male.

Head: Height 1.32 mm. Frons – length 1.5 times breadth; fulvous with pale fuscous on anteromedial hump; orbital setae black: 1 *s.or.*, 2 *i.or.*; lunule fulvous. Ocellar triangle black. Vertex with two lateral black spots. Face fulvous with a pair of medium-sized oval black spots; length 0.44 mm. Genae fulvous, pale fuscous subocular spot; black bristle present. Occiput fuscous with red-brown centrally and fulvous along eye margins; occipital row with a few small pale setae. Antennae with segments I and II red-brown, segment 3 red-brown with fuscous on apex and outer surface; length of segments: 0.12 mm; 0.24 mm; 0.57 mm.

Thorax: Scutum black with red-brown below lateral postsutural yellow vittae and dark red-brown inside postpronotal lobes and on midline anterior to medial postsutural yellow vitta. Pleural areas dark fuscous to black except red-brown below postpronotal lobes. Yellow markings as follows: postpronotal lobes; notopleura; mesopleural stripe almost reaching anterior *npl.* seta dorsally, continuing to katepisternum as a transverse spot, anterior margin slightly convex; anatergite (posterior apex black); anterior ⅔ katatergite (remainder black); a distinct triangle along anterior margin of notopleural suture; two broad lateral postsutural vittae narrowing sharply posteriorly to end before *ia.* seta; a broad medial longitudinal postsutural vitta broadly rounded with lateral flanges posteriorly and narrowing anteriorly to end well anterior to notopleural suture. Postnotum black with a narrow dark red-brown area centrally. Scutellum yellow with a narrow black basal band. Setae: *sc.* 2; *prsc.* 2; *ia.* 1; *p.sa.* 1; *a.sa.* 1; *mpl.* 1; *npl.* 2; *scp.* 2.

Legs: All femora entirely fulvous; fore tibiae pale fuscous, mid tibiae pale fuscous basally and fulvous apically, hind tibiae dark fuscous; all tarsal segments fulvous; mid tibiae each with an apical black spur.

Wings: Length 5.2 mm; cells bc and c with a pale fuscous tint; microtrichia in outer ⅓ of cell c only; remainder of wings colourless except for a narrow fuscous costal band overlapping R_{2+3} and broadening across apex of wing, a fuscous band enclosing *dm-cu* crossvein, a broad fuscous anal streak; a dense aggregation of microtrichia around A_1+CuA_2; supernumerary lobe strong.

Abdomen: Elongate oval; terga free; pecten present on tergum III. Terga I and sterna I and II wider than long. Tergum I orange-brown except black across anterior and lateral margins; tergum II orange-brown except for a narrow transverse black band across anterior area that does not reach lateral margins and narrow black lateral margins; terga III to V orange-brown except for a narrow transverse black band across anterior margin of tergum III which continues around lateral margins of all three terga, medial longitudinal black band which is narrow on anterior ½ on tergum III then broadens markedly over the remainder of terga III, IV and V. A pair of oval orange-brown shining spots on tergum V. Posterior lobe of surstylus long, sternum V with a slight concavity on posterior margin.

Attractant: Cue lure.

Female: No known record.

Distribution:

Type locality: Philippines (Mindano I., Kalambuge).

New records: East Malaysia (Sarawak); Indonesia (Kalimantan); Philippines (Luzon I., Mt Makiling).

Hosts: No known record.

Remarks: *Bactrocera (Zeugodacus) elegantula* (Hardy), previously placed in subgenus *Bactrocera*, belongs to a group of *Zeugodacus* species with two *sc.* setae, medial and lateral postsutural yellow vittae present, *a.sa.* and *prsc.* setae present, a black scutum and *dm-cu* crossvein with distinct infuscation. It is similar to *Bactrocera (Zeugodacus) hululangatiae* n.sp., *Bactrocera (Zeugodacus) neoelegantula* White and some specimens of *Bactrocera (Zeugodacus) synnephes* (Hendel) that have two *sc.* setae in the above noted characters but differs from these in having the femora entirely fulvous and a yellow triangle above the notopleuron along the anterior margin of the notopleural suture. The measurements included above are from specimens collected in the Philippines, not the type.

Bactrocera (Zeugodacus) emittens (Walker)

(Fig. 295)

Dacus emittens Walker, 1860a: 152; Froggatt, 1909: 96. Syntypes male and female in BMNH and MVMA.

Dacus chrysotoxus Hendel, 1912: 24; Perkins, 1939: 29 (Synonym).

Zeugodacus emittens – Perkins, 1939: 29.

Dacus (Zeugodacus) emittens – Hardy & Adachi, 1954: 187–8; Hardy, 1959: 171, 1977: 58; Drew, 1973: 27–30.

Bactrocera (Zeugodacus) emittens – Drew, 1989: 215; Norrbom *et al.*, 1998: 102.

Type material examined: INDONESIA: SULAWESI (formerly CELEBES) – 1 syntype female, Sulawesi, in BMNH; 4 Syntypes referred to as cotypes by Hardy & Adachi (1954) and labelled '*Dacus emittens*, Celebes' in Walker's handwriting, collected at Makassar, Celebes. Syntypes received on loan from BMNH and MVMA.

Non-type material examined: INDONESIA: SULAWESI – 2 males, Sulawesi Selatan, Takalar, Limampoccoe, 16.viii.2006, coll. Sirajuddin SP, attracted to cue lure in forest; 2 males, South Sulawesi, Maros,

Limampoccoe, 13.ix.2006, coll. Sirajuddin, attracted to cue lure in forest. Specimens in QDPC.

Diagnosis: A large species; face fulvous with medium-sized oval black spots; postpronotal lobes and notopleura yellow; scutum red-brown with small irregular dark markings; lateral and medial postsutural yellow vittae present; a yellow spot anterior to notopleural suture; mesopleural stripe reaching midway between anterior margin of notopleuron and anterior *npl.* seta dorsally; scutellum entirely yellow; legs entirely fulvous; wing with cells bc and c fuscous, microtrichia in outer ⅓ of cell c only; a broad fuscous costal band confluent with R_{4+5} and expanding into a large spot in apex of wing which recurves back along M to connect with broad infuscation along *dm-cu* crossvein; a moderately broad fuscous anal streak; abdominal terga III–V orange-brown with a narrow dark fuscous to black band across anterior margin on tergum III and a broad medial longitudinal dark fuscous to black band over terga IV and V.

Attractant: Cue lure.

Distribution:

Type locality: Indonesia, Sulawesi (formerly Celebes) – South Sulawesi, Makassar. Published records from the Lesser Sunda Is are incorrect, referring to the closely related species, *Bactrocera (Zeugodacus) neoemittens* n.sp.

Previously recorded: Type locality and the Moluccas = Maluku (Kai I. = Key I.).

Hosts: No known records.

Remarks: *Bactrocera (Zeugodacus) emittens* (Walker) is a relatively common species restricted to Sulawesi. It is similar to a closely related species *Bactrocera (Zeugodacus) neoemittens* n.sp. in possessing a broad costal band on the wing confluent with R_{4+5} and expanding into a large spot across the wing apex and then recurving back to include the *dm-cu* crossvein, but differs in having a predominantly red-brown scutum, no dark subapical spot on the outer surfaces of the fore femora, no distinct black 'T' over abdominal terga III–V, a medium-sized rounded supernumerary lobe in the male wing and fuscous cells bc and c. The description of the abdomen and wing photograph of *D. chrysotoxus* Hendel (1912) are consistent with *B. emittens*, not *B. neoemittens* n.sp.

Bactrocera (Zeugodacus) exornata (Hering)

(Fig. 296)

Zeugodacus exornatus Hering, 1941b: 55. Holotype male in MNM.

Dacus (Zeugodacus) exornatus – Drew, 1973: 30; Hardy, 1983: 37.

Bactrocera (Zeugodacus) exornata – Drew, 1989: 215; Norrbom *et al.*, 1998: 102.

Type material examined: Holotype male, **INDONESIA**: Dammer I. (probably Damar Island). Studied by Drew (1989).

Non-type material examined: INDONESIA: BALI – 1 male, Pulaki, Bali, 50m, 9.vii.1981, coll. J. Turner, attracted to cue lure, Primary Rainforest (some disturbance); 1 male, Bali, Buleleng, Tajun, 18.viii.2006, coll. Hendri Y, attracted to cue lure in forest; **FLORES** – 4 males, Borong, Flores, iv.1989, coll. J. Turner; **JAVA** – 1 male, D. I. Yogyakarta, Pandak-Bantul, Gesikan, 30.viii.2005, coll. Agus W., attracted to cue lure in village; 1 male (18.ix.2005), 1 male (11.xii.2005), Central Java, Pemalang, Pegongsoran, coll. Sapto Darmono, attracted to cue lure in forest; **SULAWESI** – 1 male, Latupa, 9km W. Palopo, 100m, 23.vii.1981, coll. J. Turner; 1 male, Bengo Forest N. Ujung Pandang, 250m, 26.vii.1981, coll. J. Turner, primary rainforest, attracted to cue lure; 3 males, Padukku Forest, 5km S. Enrekang, 150m, 25.vii.1981, coll. J. Turner, primary rainforest (some disturbance), attracted to cue lure; 1 male, Sulawesi Utara, Bitung, Danowudu, 26.ix.2006, coll. Karel F. Lala, attracted to cue lure in forest; 3 males, Sulawesi Tengah, Donggala, Nupabomba, 23.iv.2006, coll. Lyar; 1 male (20.viii.2006), 6 males (27.viii.2006), 1 male (8.x.2006), 2 males (15.x.2006), 2 males (22.x.2006), 1 male (29.x.2006), Sulawesi Tengah, Donggala, Nupabomba, coll. Rusdin, attracted to cue lure in forest; 2 males (19.ix.2006), 2 males (26.ix.2006), 3 males (3.x.2006), Gorontalo, Bone Bolango, Lombongo, coll. Yusbar Ismail, attracted to cue lure in forest; 4 males (25.vii.2006), 5 males (10.x.2006), Gorontalo, Bone Bolango, Suwawa, coll. Yusbar Ismail, attracted to cue lure in village; 1 male, Gorontalo, Bone Bolango, Buonga, 3.x.2006, coll. Yusbar Ismail, attracted to cue lure in village; 2 males, Gorontalo, Ayma Utara, 12.ix.2006, coll. Yusbar Ismail, attracted to cue lure in Cabe; 3 males (27.viii.2006), 10 males (13.ix.2006), Sulawesi Selatan, Takalar, Limampoccoe, coll. Sirajuddin SP, attracted to cue lure in forest. **PENINSULAR MALAYSIA**: 1 male (11.viii.1986), 5 males (25.viii.1986), 1 male (20.iv.1987), 1 male (18.v.1987), TN3, coll. Yusoff Salleh, attracted to cue lure; 1 male, TN3, 3.xi.1986, coll. Fanzi, attracted to cue lure; 1 male, TN3, 4.v.1987, coll. Ahmad Radzuan, bin Panjang Abd, attracted to cue lure; 4 males (11.viii.1986), 3 males (29.vi.1987), TN15, coll. Yusoff Salleh, attracted to cue lure; 5 males (28.vii.1986), 3 males (8.ix.1986), 1 male (6.x.1986), 6 males (20.x.1986), TN15/TN16, coll. Ahmad Radzuan, Panjang Abdullah, attracted to cue lure; 7 males (6.iv.1987), 5 males (20.iv.1987), 13 males (18.v.1987), 4 males (4.vi.1987), 4 males (15.vi.1987), TN15, coll. Ahmad Radzuan, bin Panjang Abd, attracted to cue lure; 4 males, TN15, 25.viii.1986, coll. Ahmad Radzuan, Panjang Abd, attracted to cue lure; 2 males (26.i.1987), 3 males (9.ii.1987), 1 male (23.ii.1987), TN15, coll. A.R. & P.A., attracted to cue lure. Specimens in QDPC.

Diagnosis: A small species; face entirely black; postpronotal lobes and notopleura yellow; scutum entirely black; broad lateral and medial postsutural yellow vittae present; no yellow spot anterior to notopleural suture; mesopleural stripe reaching to anterior *npl.* seta dorsally; scutellum entirely yellow; wing with cells bc and c colourless, microtrichia in outer corner of cell c only; a narrow fuscous costal band confluent with R_{2+3} and widening slightly around apex of wing; a broad fuscous anal streak narrowing towards wing margin; abdominal terga III–V fulvous with a broad medial and two broad lateral longitudinal dark fuscous to black bands joined along anterior margin of tergum III.

Description: male.

Head: Height 1.38 mm. Frons – length 1.47 times breadth; fulvous with fuscous to dark fuscous around orbital setae and on anteromedial hump; orbital setae black: 1 *s.or.*, 2 *i.or.*; lunule fuscous to dark fuscous. Ocellar triangle black. Vertex dark fuscous to black. Face entirely black; length 0.48 mm. Genae fulvous, large fuscous to dark fuscous subocular spot present; black seta present. Occiput black, fulvous along eye margins; occipital row with a number of short pale setae. Antennae with segments 1 and 2 red-brown, segment 3 red-brown with fuscous on apex and outer surface; arista black (red-brown basally); length of segments: 0.09 mm; 0.21 mm; 0.57 mm.

Thorax: Scutum entirely black without pale markings. Pleural areas entirely black without pale markings. Yellow markings as follows: postpronotal lobes; notopleura; mesopleural stripe reaching to anterior *npl.* seta dorsally, continuing to katepisternum as a transverse spot, anterior margin straight; anatergite (posterior apex black); anterior ⅔ katatergite (remainder black); two broad parallel-sided lateral postsutural vittae beginning anteriorly at notopleural suture and ending just behind *ia.* seta; a broad medial longitudinal postsutural vitta almost parallel-sided and beginning posteriorly at level of *prsc.* setae and ending anteriorly at level of notopleural suture. Postnotum black. Scutellum yellow except for narrow black basal band. Setae: *sc.* 4; *prsc.* 2; *ia.* 1; *p.sa.* 1; *a.sa.* 1; *mpl.* 1; *npl.* 2; *scp.* 2.

Legs: Fore femora entirely black (there may be small areas of dark fulvous on the apices and bases), mid femora with basal ½ fulvous and apical ½ dark fuscous to black, hind femora fulvous except apical ⅓ dark fuscous to black; fore and mid tibiae fulvous, hind tibiae dark fuscous; tarsi with all segments fulvous; mid tibiae each with an apical black spur.

Wings: Length 4.56 mm; cells bc and c colourless; microtrichia in outer corner of cell c only; remainder of wings colourless except fuscous to dark fuscous cell sc, narrow fuscous costal band confluent with R_{2+3} and widening slightly apically to end between apices of R_{4+5} and M, a broad fuscous anal streak which tapers towards wing margin; a dense aggregation of microtrichia around A_1+CuA_2; supernumerary lobe weak.

Abdomen: Oval; terga free; pecten present on tergum III. Tergum I and sterna I and II wider than long. Tergum I fulvous with dark fuscous along anterior margin and dark fuscous to black lateral margins; tergum II fulvous with a narrow transverse black band across anterior margin and reaching narrow dark fuscous anterolateral corners; terga III–V fulvous with a broad medial and two broad lateral longitudinal dark fuscous to black bands joined along anterior margin of tergum III. A pair of oval fulvous shining spots on tergum V. Posterior lobe of surstylus long, sternum V with a slight concavity on posterior margin. Abdominal sterna I, IV and V dark, sterna II and III pale.

Attractant: Cue lure.

Female: No known record.

Distribution:

Type locality: Indonesia (Moluccas = Maluku, Damar I.).

Previously recorded: Type locality and the following locations in Indonesia (Sulawesi; Sumatra (Norrbom *et al.*, 1998)).

New records: Peninsular Malaysia; Indonesia (Bali; Flores; Java).

Hosts: No known record.

Remarks: The badly damaged Holotype was previously studied by Drew (1973) with some additional notes provided by Hardy (1983). The collection of a large series of specimens in Peninsular Malaysia has now provided the opportunity to do a complete revisionary description. *Bactrocera exornata* (Hering) is similar to *Bactrocera (Zeugodacus) trimaculata* (Hardy & Adachi) in the general colour patterns of the scutum, wings and abdominal terga. It differs from *B. trimaculata* in having the face black across the lower ½, femora mostly black except for the fulvous bases of mid and hind femora and the scutellum entirely yellow.

Bactrocera (Zeugodacus) flavipilosa (Hardy)

(Fig. 297)

Dacus (Zeugodacus) flavipilosa Hardy, 1982: 208, partim type series. Holotype male in MZB.

Bactrocera (Zeugodacus) flavipilosa – Norrbom *et al.*, 1998: 102.

Type material examined: Holotype male, **INDONESIA: SULAWESI** – Tolai, Tolewonu River, Central Sulawesi, 133 m, ii.1976, coll. M. Becker. Studied at MZB.

Paratype: one male with the following four labels – 'Kuala Navusa, Malakosa, Sulawesi, Aug.-Nov. 1975, 100 ft., coll. Margareta Becker'; blue label – 'Paratype ♂, Dacus (Zeugodacus) flavipilosus, D.E. Hardy'; 'Note: 7 Scut. br.'; 'Acc No. 1983-454'. Received on loan from BBM.

Non-type material examined: INDONESIA: SULAWESI – 1 male, Palopo – Rantepao Rd., 15-22km Rantepao, Primary Rainforest, 1000-1100m, Sulawesi, 24 vii.1981, coll. J. Turner, attracted to cue lure. Specimen in QDPC.

Diagnosis: A large species; face fulvous with a pair of moderately large oval black spots; postpronotal lobes and notopleura yellow; scutum fuscous posteriorly to red-brown anteriorly and with a pair of large irregular black patterns inside lateral postsutural vittae and running anteriorly to almost meet postpronotal lobes; lateral and medial postsutural yellow vittae present; yellow spot anterior to notopleural suture; mesopleural stripe reaching almost to anterior *npl.* seta dorsally; scutellum yellow; legs entirely fulvous without any dark patterns; wing with cells bc and c fulvous, microtrichia in outer corner of cell c only; a broad fuscous costal band confluent with R_{4+5} and expanding into a large dark fuscous spot across apex of wing which reaches M, a pale fuscous tint around posterior margin of wing from apical spot to anal streak; a broad fuscous anal streak present; abdominal terga III–V light red-brown with a broad black transverse band across anterior margin of tergum III, narrow dark fuscous anterolateral corners on tergum IV, a narrow medial longitudinal black band over terga IV and V.

Description: male.

Head: Height 1.92 mm. Frons – length 1.53 times breadth; fulvous with a small area of fuscous on anteromedial hump; orbital setae black: 1 *s.or.*, 3 *i.or.*; lunule dark fuscous. Ocellar triangle black. Vertex fuscous. Face fulvous with a pair of medium to large oval black spots; length 0.6 mm.

Genae fulvous, fuscous subocular spot present; dark seta present. Occiput red-brown, fulvous along eye margins; occipital row with four black setae. Antennae with segments 1 and 2 fuscous, segment 3 entirely red-brown; arista black (red-brown basally); length of segments: 0.18 mm; 0.33 mm; 0.9 mm.

Thorax: Scutum fuscous posteriorly tending red-brown anteriorly with a pair of large irregularly shaped black patterns inside lateral postsutural vittae and running anteriorly to almost reach postpronotal lobes. Pleural areas black except red-brown above mid coxae and below postpronotal lobes. Yellow markings as follows: postpronotal lobes; notopleura; mesopleural stripe reaching almost to anterior *npl.* seta dorsally, continuing to katepisternum as a large transverse spot, anterior margin slightly convex; anatergite (posterior apex black); anterior $\frac{5}{8}$ katatergite (remainder black); two broad lateral postsutural vittae beginning with a spot anterior to notopleural suture and narrowing posteriorly to end before *ia.* seta; a broad medial longitudinal postsutural vitta beginning just anterior of *prsc.* setae and narrowing to a point anteriorly to end anterior to notopleural suture. Postnotum black tending dark red-brown centrally. Scutellum yellow except for narrow black basal band. Setae: *sc.* 4; *prsc.* 2; *ia.* 1; *p.sa.* 1; *a.sa.* 1; *mpl.* 1; *npl.* 2; *scp.* 4.

Legs: All segments entirely fulvous; mid tibiae each with an apical black spur.

Wings: Length 9.5 mm; cells bc and c fulvous; microtrichia in outer corner of cell c only; remainder of wings colourless except fuscous cell sc, broad fuscous costal band confluent with R_{4+5} (slightly paler between R_{2+3} and R_{4+5}) and expanding into a dark fuscous spot across apex of wing and reaching M, a broad pale fuscous tint around posterior margin of wing from apical spot to anal streak, a broad fuscous anal streak; dense aggregation of microtrichia around A_1+CuA_2; supernumerary lobe weak.

Abdomen: Elongate oval; terga free; pecten present on tergum III. Tergum I and sterna I and II wider than long. Tergum I pale red-brown with dark fuscous on lateral margins; tergum II pale red-brown with a narrow transverse black band across anterior margin but not reaching lateral margins and narrow anterolateral black margins; terga III–V pale red-brown with a broad transverse black band across anterior margin of tergum III and expanding over lateral margins of this tergum, dark fuscous anterolateral corners on tergum IV and a narrow black medial longitudinal band over terga IV and V. A pair of oval pale red-brown shining spots on tergum V. Posterior lobe of surstylus long, sternum V with a slight concavity on posterior margin. All abdominal sterna pale, tending red-brown. A dense golden pubescence covers all abdominal terga.

Attractant: Cue lure.

Female: No known record.

Distribution:

Type locality: Indonesia (Sulawesi – Central Sulawesi, Tolewonu River).

Previously recorded: Type locality and Indonesia (Sulawesi – Central Sulawesi, Malakosa, Kuala Navusa).

New records: Indonesia (Sulawesi – South Sulawesi, 15–22 km from Rantepao).

Hosts: No known record.

Remarks: The type series of *Bactrocera (Zeugodacus) flavipilosa* (Hardy) (Hardy, 1982) included three different species, two species of *Bactrocera (Zeugodacus)* and one *Bactrocera (Bactrocera)*. The latter is *Bactrocera (Bactrocera) flavipennis* (Hardy). This description, based on the Holotype and one Paratype from Hardy's type series plus one non-type specimen (in QDPC), provides a clear definition of *Bactrocera (Zeugodacus) flavipilosa* (Hardy). The other *Bactrocera (Zeugodacus)* species in the type series is treated in this paper as *B. (Zeugodacus) neoflavipilosa* n.sp. *Bactrocera flavipilosa* belongs to the group of *B. (Zeugodacus)* with four *sc.* setae, medial postsutural yellow vitta present and scutum red-brown with large dark patterns. It is similar to *Bactrocera (Zeugodacus) buruensis* White and *Bactrocera (Zeugodacus) neoflavipilosa* n.sp. in possessing pale abdominal sterna and abdominal terga III–V with a black transverse band across anterior margin of tergum III and a medial longitudinal black band over terga IV and V. *Bactrocera neoflavipilosa* n.sp. differs in having a strong keel-shaped supernumerary lobe in the male wing and a 'U'-shaped black pattern over the red-brown scutum. The only character that can be found to separate *B. buruensis* and *B. flavipilosa* is the presence of a pale tint around the hind margin of the wing in *B. flavipilosa* and a pale tint around the *dm-cu* crossvein in *B. buruensis*.

Bactrocera (Zeugodacus) flavolateralis Drew & Romig, new species

(Fig. 298)

Material examined: Holotype female, **THAILAND**: Songkhla, Kao Luang, Lan Saka, Naklon Si, Thammarat, 23.v.1991, coll. Sieng *et al.*, bred from *Drypetes* sp., Sample No. SK087, deposited in BMNH.

Diagnosis: A medium-sized species; face fulvous with a pair of large oval black spots; postpronotal lobes and notopleura yellow and connected with a narrow lateral yellow band; scutum black; lateral and medial postsutural yellow vittae present; mesopleural stripe reaching almost to anterior *npl.* seta; scutellum yellow; wing with cells bc and c with a pale tint, microtrichia in outer corner of cell c only; a narrow fuscous costal band; anal streak absent; abdominal terga III–VI red-brown with a general black 'T' pattern and narrow black lateral margins on all four terga; in addition to the standard subgenus *Zeugodacus* chaetotaxy, this species has a humeral seta present on one postpronotal lobe.

Description: female.

Head: Height 1.48 mm. Frons – length 1.39 times breadth; red-brown with fulvous laterally and pale fuscous on anteromedial hump; orbital setae black: 1 *s.or.*, 3 *i.or.* (all three are evenly separated); lunule fuscous. Ocellar triangle black. Vertex red-brown. Face fulvous with a pair of large oval black spots reaching to oral margin; length 0.55 mm. Genae fulvous, small dark fuscous subocular spot present; black seta present. Occiput fuscous, fulvous along eye margins; occipital row with five black setae. Antennae with segments 1 and 2 fuscous, segment 3 red-brown with fuscous on outer

surface; arista black (red-brown basally); length of segments: 0.12 mm; 0.19 mm; 0.70 mm.
Thorax: Scutum black with red-brown below and behind lateral postsutural yellow vittae, forming a narrow margin around medial postsutural vitta and along inner margin of lateral postsutural vittae, around notopleural suture, inside postpronotal lobes and along anterior margin of scutum. Pleural areas black with red-brown below postpronotal lobes and above mid coxae. Yellow markings as follows: postpronotal lobes; notopleura; mesopleural stripe reaching almost to anterior *npl.* seta dorsally, continuing to katepisternum as a large transverse spot, anterior margin slightly convex; anatergite (posterior apex black); anterior ⅔ katatergite (remainder black); two parallel-sided lateral postsutural vittae of medium-width and ending at *ia.* seta; a broad medial longitudinal postsutural vitta beginning at level of *prsc.* setae and ending just anterior of level of notopleural suture; a narrow lateral yellow band connecting postpronotal lobe and notopleuron. Postnotum black. Scutellum yellow except for very narrow black basal band. Setae: *sc.* 4; *prsc.* 2; *ia.* 1; *p.sa.* 1; *a.sa.* 1; *mpl.* 1; *npl.* 2; *scp.* 4; *hm.* 1 (on one postpronotal lobe only).
Legs: Femora entirely fulvous; fore and hind tibiae dark fuscous, mid tibiae fuscous basally to pale fuscous apically; tarsi with basal segment fulvous and apical four segments pale fuscous; mid tibiae each with an apical black spur.
Wings: Length 5.5 mm; cells bc and c with pale tint; microtrichia in outer corner of cell c only; remainder of wings colourless except fuscous cell sc, narrow fuscous costal band confluent with R_{2+3} and widening into a *tau* spot around apex of R_{4+5}; anal streak absent; no dense aggregation of microtrichia around A_1+CuA_2; supernumerary lobe weak to absent.
Abdomen: Elongate oval; terga free; pecten absent on tergum III. Tergum I and sterna I and II wider than long. Abdominal sterna I and V dark fuscous to black; sterna II, III and IV fulvous. Tergum I red-brown with fuscous anterocentrally and distinct black lateral narrow lateral margins; tergum II fulvous with a narrow transverse black band across anterior margin connecting with narrow black lateral margins; terga III–VI red-brown with a distinct black 'T' consisting of a broad transverse black band across anterior margin of tergum III and a broad medial longitudinal black band over all four terga, distinct narrow black lateral margins over all four terga. A pair of oval red-brown shining spots on tergum V. Abdominal sterna I, V and VI dark, remainder pale. Basal segment of ovipositor red-brown, dorsoventrally flattened and tapering posteriorly in dorsal view; ratio of length of oviscape to length of tergum V, 1.5:1; apex of piercer needle-shaped.
Attractant: No known record.
Male: No known record.
Distribution: Southern Thailand (Songkhla).
Hosts: Recorded from one host, *Drypetes* sp. (family *Euphorbiaceae*).
Remarks: *Bactrocera (Zeugodacus) flavolateralis* n.sp. is provisionally placed in the subgenus *Zeugodacus* but possesses a humeral seta on one postpronotal lobe (it is absent on the other lobe). It is similar to *Bactrocera (Zeugodacus) flavopectoralis* (Hering) and *Bactrocera (Zeugodacus) zahadi* Mahmood in general body and wing colour patterns and in having the face with a pair of black spots. It is distinct in possessing a lateral yellow band connecting the postpronotal lobe and notopleuron, a humeral seta and the anal streak absent.

Bactrocera (Zeugodacus) flavopectoralis (Hering)

(Fig. 299)

Zeugodacus flavopectoralis Hering, 1953: 77. Holotype female in NMB.
Bactrocera (Zeugodacus) flavopectoralis – Norrbom *et al.*, 1998: 102.

Type material examined: Holotype female, **INDONESIA: SUMBA ISLAND** – with the following three labels – one white label 'O-SUMBA, Prai Jawang, 12.6.1949, Expedition Buhler-Sutter'; one red label 'TYPUS'; one white label '*Zeugodacus flavopectoralis*, det. Hering, Holotypus'. Received on loan from NMB.
Diagnosis: A medium-sized species; face fulvous with a pair of medium-sized subquadrate black spots; postpronotal lobes and notopleura yellow; scutum black; lateral and medial postsutural yellow vittae present; a yellow spot anterior to notopleural suture in front of lateral postsutural yellow vittae; mesopleural stripe reaching midway between anterior margin of notopleuron and anterior *npl.* seta dorsally; scutellum yellow; wing with cells bc and c colourless, microtrichia in outer corner of cell c only; a narrow fuscous costal band confluent with R_{2+3} and forming a fish-hook pattern across apex of R_{4+5}; a narrow pale fuscous to fuscous anal streak; abdominal terga III–V fulvous with a 'T' pattern consisting of a narrow transverse black band across anterior margin of tergum III, a medium-width dark fuscous medial longitudinal band over terga III and IV and a narrow dark fuscous medial longitudinal spot on tergum V, dark fuscous anterolateral corners on terga IV and V.
Description: female.
Head: Height 1.52 mm. Frons – length 1.29 times breadth; fulvous with black around orbital setae and dark fuscous on anteromedial hump; latter with a small number of short pale hairs; orbital setae black: 1 *s.or.*, 3 i.*or.*; lunule black. Ocellar triangle black. Vertex dark fuscous. Face fulvous (darker over lower ½) with a pair of medium-sized subquadrate black spots; length 0.5 mm. Genae fulvous, fuscous subocular spot; black seta present. Occiput fuscous with dark fuscous spots dorsally and fulvous along eye margins; occipital row with two to three pale setae. Antennal segments 1 and 2 fulvous, segment 3 fulvous with fuscous on apex and outer surface; length of segments: 0.12 mm; 0.27 mm; 0.71 mm.
Thorax: Scutum black with fuscous behind lateral postsutural vittae and red-brown below lateral postsutural vittae and inside notopleura. Pleural areas black, tending fuscous below postpronotal lobes. Yellow markings as follows: postpronotal lobes; notopleura; mesopleural stripe reaching midway between anterior margin of notopleuron and anterior *npl.* seta dorsally, continuing to katepisternum as a

moderately large transverse spot, anterior margin convex; anatergite (posterior apex black); anterior ¾ katatergite (remainder black); two broad parallel-sided lateral postsutural vittae, beginning as a spot anterior to notopleural suture and ending behind *ia.* seta; a broad medial longitudinal postsutural vitta broadly rounded posteriorly and narrowing to a point anterior to notopleural suture. Postnotum black. Scutellum yellow except for a narrow black basal band. Setae: *sc.* 4; *prsc.* 2; *ia.* 1; *p.sa.* 1; *a.sa.* 1; *mpl.* 1; *npl.* 2; *scp.* 4.

Legs: Fore femora with basal ½ fulvous, apical ½ fuscous on inner margin and dark fuscous on outer margin, mid femora fulvous on basal ⅔ and fuscous to dark fuscous on apical ⅓, hind femora fulvous with fuscous to dark fuscous around apical ¼; fore tibiae fuscous on inner margins and dark fuscous on outer margins, mid tibiae fuscous basally tending to pale fuscous apically, hind tibiae dark fuscous; tarsal segments with fulvous on basal segment and fuscous on apical four segments; mid tibiae each with an apical black spur.

Wings: Length 6.2 mm; cells bc and c colourless; microtrichia in outer corner of cell c only; remainder of wings colourless except dark fuscous cell sc, narrow fuscous costal band confluent with R_{2+3} and widening with a fish hook pattern across apex of R_{4+5} where it is also dark fuscous, narrow pale fuscous to fuscous anal streak; no dense aggregation of microtrichia around A_1+CuA_2; supernumerary lobe weak.

Abdomen: Oval; terga free; pecten present on tergum III. Tergum I and sterna I and II wider than long. Tergum I fulvous except dark fuscous across anterior margin and narrow black along lateral margins; tergum II fulvous with a narrow black band across anterior margin which expands into black anterolateral corners; terga III–V fulvous with a narrow black band across anterior margin of tergum III which expands across lateral margins and a medium-width dark fuscous medial longitudinal band over terga III and IV and a narrow dark fuscous medial longitudinal spot on tergum V, dark fuscous anterolateral corners on terga IV and V. A pair of oval fulvous shining spots on tergum V. Basal segment of ovipositor red-brown, dorsoventrally flattened and tapering posteriorly in dorsal view; ratio of length of oviscape to length of tergum V, 2.3:1.

Attractant: No known record.

Male: No known record.

Distribution:

Type locality: Indonesia (Sumba I.).

Hosts: No known record.

Remarks: *Bactrocera (Zeugodacus) flavopectoralis* (Hering) is similar to some species in the *tau* complex that have the costal band confluent with R_{2+3}, the costal band expanded into a spot apically and abdominal terga III–V with a black 'T' pattern. It is close to *Bactrocera (Zeugodacus) flavolateralis* n.sp., *Bactrocera (Zeugodacus) hatyaiensis* n. sp. and *Bactrocera (Zeugodacus) zahadi* Mahmood in possessing a pair of black facial spots. It differs from *B. flavolateralis* n.sp. and *B. zahadi* in having broad lateral postsutural vittae and a small fish-hook patterned apical spot on the costal band and from *B. hatyaiensis* n.sp. in lacking *hm.* setae and in having an entirely yellow scutellum and anterolateral corners of terga IV and V dark coloured.

Bactrocera (Zeugodacus) flavoverticalis Drew & Romig, new species

(Fig. 300)

Material examined: Holotype male, **THAILAND**: Khun Tan National Park, Chiang Rai, 4.vi.1992, coll. Charuwan, bred from *Strychnos nux-vomica* L, deposited in BMNH. Paratypes: **THAILAND**: 7 females, same data as holotype. 1 paratype in ANIC; 1 paratype in BMNH; 4 paratypes in QDPC; 1 paratype in QMIC.

Diagnosis: A medium-sized species; face fulvous with a pair of large oval black spots; postpronotal lobes and notopleura yellow; scutum black with large areas of red-brown; lateral and medial postsutural yellow vittae present; yellow spot along anterior margin of notopleural suture; mesopleural stripe reaching almost to anterior *npl.* seta dorsally; scutellum yellow; wing with cells bc and c colourless, microtrichia in outer corner of cell c only; a very narrow fuscous costal band and broad fuscous anal streak; abdominal terga III–V fulvous with a 'T' pattern consisting of a narrow transverse black band across anterior margin of tergum III and a medium-width medial longitudinal black band over all three terga and anterolateral corners of terga IV and V black.

Description: male.

Head: Height 1.62 mm. Frons – length 1.36 times breadth; fuscous centrally, fulvous along lateral margins and dark fuscous on anteromedial hump; orbital setae black: 1 *s.or.*, 2 or 3 *i.or.* (if 3 then lower pair weak); lunule dark fuscous. Ocellar triangle black. Vertex dark fuscous. Face fulvous with a pair of large oval black spots; length 0.59 mm. Genae fulvous, fuscous to dark fuscous subocular spot present; black seta present. Occiput fuscous, dark fuscous dorsolaterally, fulvous along eye margins; occipital row with a large number of small pale setae. Antennae with segments 1 and 2 red-brown, segment 3 red-brown with fuscous on apex and outer surface; arista black (red-brown basally); length of segments: 0.14 mm; 0.29 mm; 0.79 mm.

Thorax: Scutum black with red-brown below and behind lateral postsutural vittae and over a large area of the anterior ½ of scutum interspersed with black patterns. Pleural areas dark fuscous to black except for a unique vertical yellow band running from posterior corner of postpronotal lobe to fore coxae, with the dorsal width from postpronotal lobe almost to anterior margin of mesopleural stripe. Yellow markings as follows: postpronotal lobes; notopleura; mesopleural stripe reaching almost to anterior *npl.* seta dorsally, continuing to katepisternum as a large elongate transverse spot, anterior margin straight; anatergite (posterior apex fuscous to black); anterior ⅔ katatergite (remainder black); two lateral postsutural vittae of medium-width beginning with a spot anterior to notopleural suture and widening slightly posteriorly to end well behind *ia.* seta; a medial longitudinal postsutural vitta broadly rounded posteriorly to begin at posterior margin of scutum and enclose *prsc.* setae and narrowing anteriorly to end at level of notopleural suture. Postnotum black, red-brown centrally. Scutellum yellow except for narrow black basal band. Setae: overall the setae are weaker than in most dacine species: *sc.* 4

(basal pair often very weak); *prsc.* 2; *ia.* 1; *p.sa.* 1; *a.sa.* 1; *mpl.* 1; *npl.* 2; *scp.* 4.

Legs: Femora fulvous tending pale red-brown apically; fore tibiae fuscous to dark fuscous, mid tibiae pale fuscous tending dark fuscous basally, hind tibiae fuscous to dark fuscous; tarsi with basal segment fulvous and apical four segments fuscous; mid tibiae each with an apical black spur.

Wings: Length 6.2 mm; cells bc and c colourless; microtrichia in outer corner of cell c only; remainder of wings colourless except fuscous cell sc, very narrow fuscous costal band slightly overlapping R_{2+3} and remaining very narrow before ending just beyond apex of R_{4+5}, broad fuscous anal streak; dense aggregation of microtrichia around A_1+CuA_2; supernumerary lobe strong.

Abdomen: Oval; terga free; pecten present on tergum III. Tergum I and sterna I and II wider than long. Tergum I fulvous with fuscous to dark fuscous across anterior margin and reaching narrow black lateral margins; tergum II fulvous with a narrow black transverse band across anterior margin but not reaching lateral margins and narrow black lateral margins; terga III–V fulvous with a 'T' pattern consisting of a narrow black transverse band across anterior margin of tergum III and a medium-width medial longitudinal black band over all three terga and anterolateral corners of terga IV and V black. A pair of oval fulvous shining spots on tergum V. Posterior lobe of surstylus long, sternum V with a slight concavity on posterior margin. Abdominal sterna dark except sterna II and III paler.

Attractant: No known record.

Female: As for male except no dense aggregation of microtrichia around A_1+CuA_2; supernumerary lobe weak; pecten of cilia on abdominal tergum III absent; basal segment of ovipositor red-brown, dorsoventrally flattened and tapering posteriorly in dorsal view; ratio of length of oviscape to length of tergum V, 2.3:1; apex of piercer needle-shaped.

Distribution: Thailand (Chiang Rai, Khun Tan National Park).

Hosts: Bred from one host *Strychnos nux-vomica* (family *Loganiaceae*).

Remarks: *Bactrocera (Zeugodacus) flavoverticalis* n.sp. is similar to *Bactrocera (Zeugodacus) vultus* (Hardy), *Bactrocera (Zeugodacus) melanofacies* n.sp. and *Bactrocera (Zeugodacus) sasaotiae* n.sp. in general body colour patterns and in possessing a strong supernumerary lobe in the male wing. It differs from *B. vultus* and *B. melanofacies* in having distinct facial spots (not entirely black) and from *B. sasaotiae* in having the anterolateral corners of terga IV and V dark coloured. It is also distinct from all known *Bactrocera* species in possessing a unique vertical yellow band on the pleural area from the posterior corner of the postpronotal lobe to the fore coxa.

Bactrocera (Zeugodacus) freidbergi White

(Fig. 301)

Bactrocera (Zeugodacus) freidbergi White, in White & Evenhuis, 1999: 526. Holotype male in TAUI.

Type material examined: Holotype male, **INDIA: ASSAM** – with the following three labels – 'INDIA: Meghalaya, Shillong, Bot. Gardens, 23 Apr 1980, A. Freidberg'; 'Holotype, *Bactrocera freidbergi* White, det. I.M. White, 1999'; 'Tel Aviv University'. Received on loan from TAUI. Paratypes: **INDIA: ASSAM** – 2 female paratypes, same locality label data as holotype; 1 male paratype, 'INDIA: Meghalaya, Shillong, Botan Gardens, 20 Apr 1980, A. Freidbergi'. Received on loan from TAUI.

Diagnosis: A medium-sized species; face fulvous with a pair of medium-sized circular black spots; postpronotal lobes yellow; notopleura yellow with either a narrow dark band around base or a broader dark fuscous pattern leaving only a yellow apex; scutum shining black; narrow lateral and medial postsutural yellow vittae present; mesopleural stripe reaching slightly beyond midway between anterior margin of notopleuron and anterior *npl.* seta dorsally; scutellum yellow; wing with cells bc and c colourless, microtrichia in outer corner of cell c only; a narrow pale fuscous to fuscous costal band confluent with R_{2+3} and extremely narrow to absent beyond apex of this vein and widening into a dark fuscous spot around apex of R_{4+5}; a pale infuscation around CuA_1 at posterior margin of wing and very narrow pale fuscous anal streak; cell br colourless or with a very pale tint and with variable sparse setae; abdominal terga III–V black with dark red-brown to fuscous spots posterocentrally on terga III and IV either side of a medial longitudinal black band and red-brown along posterior margin of tergum V (tergum V may have an overall paler appearance in some specimens).

Description: male.

Head: Height 1.33 mm. Frons – length 1.2 times breadth; red-brown, tending fulvous anteriorly, dark fuscous around orbital setae and fuscous on anteromedial hump; orbital setae black: 1 *s.or.*, 3 *i.or.*; lunule dark fuscous. Ocellar triangle black. Vertex dark fuscous. Face fulvous with a pair of medium-sized circular black spots; length 0.43 mm. Genae fulvous, fuscous subocular spot present; black seta present. Occiput fuscous ventrally, dark fuscous dorsally, fulvous along eye margins; occipital row with a large number of strong black setae. Antennae with segments 1 and 2 red-brown, segment 3 red-brown with dark fuscous on apex and outer surface; arista black (red-brown basally); length of segments: 0.1 mm; 0.22 mm; 0.55 mm.

Thorax: Scutum shining black with dark red-brown to fuscous below and behind lateral postsutural vittae and along anterior margin of notopleural suture. Pleural areas shining black without pale markings. Yellow markings as follows: postpronotal lobes; notopleura (in the two male Paratypes and one female Paratype there is a narrow band of dark fuscous around base and in one female Paratype most of the notopleuron is dark fuscous with a small apical yellow spot); mesopleural stripe reaching midway between anterior margin of notopleuron and anterior *npl.* seta dorsally, continuing to katepisternum as a dark red-brown spot, anterior margin convex; anatergite (posterior apex black); anterior $\frac{5}{8}$ katatergite (remainder black); two narrow parallel-sided lateral postsutural vittae ending at *ia.* seta (in the Holotype there is a small yellow spot anterior to notopleural suture); a very short and narrow medial longitudinal postsutural vitta. Postnotum black. Scutellum yellow except for narrow black basal band (in the Holotype there is a very small dark fuscous line on the apex between the apical setae). Setae: *sc.* 2; *prsc.* 2; *ia.* 1; *p.sa.* 1; *a.sa.* 1; *mpl.* 1; *npl.* 2; *scp.* 2.

Legs: Femora fulvous with fuscous to dark fuscous preapical spot on outer ½ of fore femora, fuscous to dark fuscous around apical ⅓ of mid and hind femora (in some specimens the fuscous to dark fuscous on the mid femora is in the form of a preapical spot on the inner surfaces); fore tibiae pale fuscous, mid tibiae pale fuscous tending darker basally, hind tibiae dark fuscous; tarsi with basal two segments fulvous and apical three segments pale fuscous; mid tibiae each with an apical black spur.

Wings: Length 5.5 mm; cells bc and c colourless; microtrichia in outer corner of cell c only; remainder of wings colourless except dark fuscous cell sc, narrow pale fuscous to fuscous costal band confluent with R_{2+3} and extremely narrow to absent beyond apex of this vein and expanding into a dark fuscous spot around apex of R_{4+5} (*scutellaris*-like) a pale fuscous tint on hind wing margin around CuA_1, a narrow pale fuscous anal streak almost contained within cell cup; dense aggregation of microtrichia around A_1+CuA_2; supernumerary lobe weak. Cell br colourless to having a very pale tint and with microtrichia variable but sparse.

Abdomen: Oval; terga free; pecten present on tergum III. Tergum I and sterna I and II wider than long. Tergum I black with a narrow transverse orange-brown band across intersegmental line between terga I and II but not reaching lateral margins; tergum II fulvous with a broad transverse black band across anterior ½ and reaching but not covering lateral margins; terga III–V black except dark red-brown to fuscous spots posterocentrally on terga III and IV either side of a medial longitudinal broad black vitta and red-brown to fuscous along posterior margin of tergum V (in some specimens tergum V has an overall slightly paler appearance). A pair of oval fuscous to dark fuscous shining spots on tergum V. Posterior lobe of surstylus long, sternum V with a slight concavity on posterior margin. Abdominal sterna dark fuscous.

Attractant: No known record.

Female: As for male except no dense aggregation of microtrichia around A_1+CuA_2; pecten of cilia on abdominal tergum III absent; basal segment of ovipositor red-brown, conical in transverse cross section and tapering posteriorly in dorsal view; ratio of length of oviscape to length of tergum V, 1.3:1; apex of piercer needle-shaped.

Distribution:

Type locality: India [Assam] (Shillong, Botanical Gardens).

Hosts: No known record.

Remarks: *Bactrocera (Zeugodacus) freidbergi* White, within the *scutellaris* complex, is similar to *Bactrocera (Zeugodacus) assamensis* White and *Bactrocera (Zeugodacus) tappanus* (Shiraki) in possessing two *sc.* setae, femora with dark patterns and notopleura not entirely yellow. It differs from *B. assamensis* in having an entirely yellow scutellum and larger areas of dark patterning on fore femora, and from *B. tappanus* in having an entirely yellow scutellum. It is possible that *B. freidbergi* is a synonym of *B. tappanus*, as in the Holotype of the latter there is a black line on the apex of the scutellum between the setae, indicating that *B. freidbergi* could possess dark apical patterns similar to that in *B. tappanus*, if large collections of both species could be studied.

Bactrocera (Zeugodacus) fuscoalata Drew & Romig, new species

(Fig. 302)

Material examined: Holotype female, **INDIA: SOUTH ANDAMAN ISLANDS** – Port Blair, 13.iv.1995, coll. Ranganath, SI No. 47, collected on *Ficus* sp., deposited in BMNH.

Paratype: **INDIA: NICOBAR ISLANDS** – 1 female, Nicobar, coll. Roepstorff, deposited in UZMC.

Diagnosis: A large species; face fulvous with a pair of small pear-shaped black spots; postpronotal lobes and notopleura yellow; scutum red-brown; lateral and medial postsutural yellow vittae present; yellow spot anterior to notopleural suture; mesopleural stripe reaching midway between anterior margin of notopleuron and anterior *npl.* seta dorsally; scutellum yellow with a broad dark fuscous apical spot; wing with cells bc and c fuscous, microtrichia in outer corner of cell c only; a broad fuscous to dark fuscous costal band confluent with R_{4+5} and expanding into a broad fuscous to dark fuscous spot across apex of wing and recurving back along hind margin to meet a broad fuscous anal streak; abdominal terga III–V red-brown with a general 'T' pattern consisting of a narrow transverse dark fuscous band across anterior margin of tergum III and a narrow- to medium-width medial longitudinal dark fuscous band over all three terga and narrow lateral fuscous margins on all three terga.

Description: female.

Head: Height 1.72 mm. Frons – length 1.29 times breadth; dark fuscous anteriorly, red-brown on remainder centrally and fulvous on lateral margins; orbital setae black: 1 *s.or.*, 3 *i.or.*; lunule dark fuscous. Ocellar triangle black. Vertex red-brown. Face fulvous with a pair of small pear-shaped black spots; length 0.69 mm. Genae fulvous, dark fuscous subocular spot present; black seta present. Occiput red-brown, fulvous along eye margins; occipital row with a large number of small and large black setae. Antennae with segments 1 and 2 red-brown, segment 3 red-brown with pale fuscous to fuscous on apex and outer surface; arista black (red-brown basally); length of segments: 0.14 mm; 0.26 mm; 0.84 mm.

Thorax: Scutum uniformly red-brown with a very narrow medial longitudinal dark fuscous line running from anterior point of medial postsutural yellow vitta to anterior margin of scutum. Pleural areas fuscous except red-brown above mid coxae and below postpronotal lobes. Yellow markings as follows: postpronotal lobes; notopleura; mesopleural stripe of medium-width, reaching midway between anterior margin of notopleuron and anterior *npl.* seta dorsally, continuing to katepisternum as a large elongate transverse spot, anterior margin convex; anatergite (posterior apex red-brown); anterior ³/₅ katatergite (remainder red-brown); two medium-width parallel-sided lateral postsutural vittae beginning with a yellow spot anterior to notopleural suture and ending at *ia.* seta; a broad medial longitudinal postsutural vitta beginning anterior to *prsc.* setae and ending in a point level with notopleural suture. Postnotum red-brown, tending fuscous laterally. Scutellum yellow except for a broad dark fuscous apical spot. Setae: *sc.* 4; *prsc.* 2; *ia.* 1; *p.sa.* 1; *a.sa.* 1; *mpl.* 1; *npl.* 2; *scp.* 4.

Legs: Fore femora red-brown, mid and hind femora fulvous tending red-brown apically; fore and hind tibiae dark fuscous, mid tibiae fuscous tending dark fuscous basally; tarsi with all segments red-brown; mid tibiae each with an apical black spur.

Wings: Length 7.8 mm; cells bc and c fuscous; microtrichia in outer corner of cell c only; remainder of wings colourless except dark fuscous cell sc, broad fuscous to dark fuscous costal band confluent with R_{4+5} expanding across apex of wing and around hind margin to join the broad fuscous anal streak; the costal band is fuscous between R_{2+3} and R_{4+5} and possesses a large dark fuscous spot around apex of R_{4+5}; no dense aggregation of microtrichia around A_1+CuA_2; supernumerary lobe weak.

Abdomen: Elongate oval; terga free; pecten absent on tergum III. Tergum I and sterna I and II wider than long. Tergum I dark red-brown with a narrow fulvous lateral band across posterior margin but not reaching lateral margins; tergum II red-brown with a diffuse dark fuscous band across anterior margin reaching to dark fuscous anterolateral corners and narrow fulvous posterolateral margins; terga III–V red-brown with a 'T' pattern consisting of a narrow dark fuscous transverse band across anterior margin of tergum III and a narrow- to medium-width medial longitudinal dark fuscous band over all three terga and narrow lateral fuscous margins over all three terga. A pair of oval fuscous shining spots on tergum V. Basal segment of ovipositor dark red-brown, oval in transverse cross section and tapering posteriorly in dorsal view; ratio of length of oviscape to length of tergum V, 2.5:1; apex of piercer trilobed.

Attractant: No known record.

Male: No known record.

Distribution: India (South Andaman Is, Port Blair; Nicobar Is).

Hosts: No known record.

Remarks: *Bactrocera (Zeugodacus) fuscoalata* n.sp. is generally similar to *Bactrocera (Zeugodacus) watersi* (Hardy) but differs in possessing extensive fuscous wing colour patterns and *a.sa.* setae. It is also similar to *Bactrocera (Zeugodacus) emittens* (Walker) in having fuscous colour patterns across the wing but differs in the specific shape of the wing colour patterns, especially the broad fuscous band around the hind margin and in possessing fuscous cells bc and c.

Bactrocera (Zeugodacus) gavisa (Munro)

(Fig. 303)

Dacus gavisus Munro, 1935: 15. Holotype male in ZSI.
Bactrocera (Zeugodacus) gavisa – Norrbom *et al.*, 1998: 102; Tsuruta & White, 2001: 70; Drew & Raghu, 2002: 349.

Type material examined: Holotype male, **INDIA**: with the following five labels – 'Zool. S. Ind, Barkuda Id., Chilka Lake, Ganjam dist., Madras Pres., 15–22.vii.16, F.H. Gravely'; 'Type, *Dacus (Zeugodacus) gavisus* Mro., Det H.K. Munro, 1934'; 'Bactrocera near caudatus F.'; 'TYPE'; '$^{1034}/_{H6}$'. Studied by Drew & Raghu (2002).

Non-type material examined: INDIA: 1 male, Karnataka, Bangalore, 916m, ii.1988, coll. G. Bhat, attracted to cue lure. 1 male, S. India, Hort. Farm, Coonoor, 1450m, 29.v.1992, coll. D.L. Hancock, attracted to cue lure. Specimens in QDPC collection. **SRI LANKA (formerly CEYLON)**: 1 male, Sri Lanka, 20.x.1994, Bataatha, C9; 1 male, Sri Lanka, Pelwehera, 15.xi.1994, C20. Specimens in QDPC.

Diagnosis: A medium-sized species; face fulvous with a pair of large oval black spots; postpronotal lobes and notopleura yellow; a broad triangular yellow marking along the anterior margin of notopleural suture; scutum shining black except for a broad area of red-brown anteromedially; lateral and medial postsutural yellow vittae present; a unique inverted 'L'-shaped mesopleural stripe reaching to postpronotal lobe dorsally; scutellum entirely yellow; legs with femora entirely fulvous; wing with cells bc and c colourless, microtrichia in outer corner of cell c only; a narrow fuscous costal band confluent with R_{2+3} and remaining narrow before widening slightly across apex of R_{4+5}; a narrow pale fuscous anal streak; abdominal terga III–V fulvous with a 'T' pattern consisting of a narrow transverse black band across anterior margin of tergum III and a moderately broad medial longitudinal black band over all three terga, broad anterolateral black markings on terga IV and V; abdominal sterna orange-brown.

Attractant: Cue lure.

Female: No known record.

Distribution:

Type locality: India (Orissa State, Ganjam district, Chilka Lake).

Previously recorded: Type locality and the following locations – India – Kerala State (Vaikom, Ernakulum; Coimbatore District, Marudamalai) (Drew & Raghu, 2002); Sri Lanka (Tsuruta & White, 2001).

Hosts: No known record.

Remarks: *Bactrocera (Zeugodacus) gavisa* (Munro) has been adequately described and illustrated by Drew and Raghu (2002). It appears to be restricted in distribution to the southern Indian region. It is a distinct species easily diagnosed in having a broad inverted 'L'-shaped mesopleural stripe and lateral postsutural vittae that widen around and behind the *ia.* setae.

Bactrocera (Zeugodacus) hatyaiensis Drew & Romig, new species

(Fig. 304)

Material examined: Holotype male, **THAILAND**: Tone Nga Chang Waterfall, near Hat Yai, 8.v.1991, coll. D.L. Hancock, attracted to cue lure, deposited in BMNH.

Paratypes: 5 males, same data as holotype. 1 paratype in ANIC; 1 paratype in BMNH; 2 paratypes in QDPC; 1 paratype in QMIC.

Diagnosis: A medium-sized species; face fulvous with a pair of medium-sized oval black spots; postpronotal lobes and notopleura yellow; scutum black; lateral and medial postsutural yellow vittae present; yellow spot present anterior to notopleural suture; mesopleural stripe reaching midway between anterior margin of notopleuron and anterior *npl.* seta dorsally; scutellum yellow with a very pale fuscous apical spot; wing with cells bc and c colourless, microtrichia in outer corner of cell c only; a narrow dark fuscous costal band confluent with R_{2+3} and expanding into a small spot around apex of R_{4+5} and broad dark fuscous anal streak; abdominal terga III–V fulvous with a black 'T' pattern consisting of a transverse black band across anterior margin of tergum III and a medium-width medial longitudinal black band over all three terga and broad lateral longitudinal dark fuscous to black bands over terga IV and V; in addition to the standard subgenus *Zeugodacus* chaetotaxy, this species has a humeral seta (*hm.*) present on each postpronotal lobe.

Description: male.

Head: Height 1.37 mm. Frons – length 1.39 times breadth; red-brown, fulvous on lateral margins and at most pale fuscous around bases of orbital setae and on anteromedial hump; orbital setae black: 1 *s.or.*, 3 *i.or.*; lunule red-brown. Ocellar triangle black. Vertex black (sometimes incomplete). Face fulvous with a pair of medium-sized oval black spots; length 0.41 mm. Genae fulvous, small dark fuscous subocular spot present; black seta present. Occiput red-brown, fulvous along eye margins; occipital row with a large number of small and larger black setae. Antennae with segments 1 and 2 red-brown, segment 3 fuscous except red-brown on basal area of inner surface; arista black (red-brown basally); length of segments: 0.10 mm; 0.22 mm; 0.65 mm.

Thorax: Scutum almost entirely black except dark red-brown to fuscous below and behind lateral postsutural vittae and around yellow spot anterior to notopleural suture. Pleural areas dark fuscous to black except dark red-brown to fuscous below postpronotal lobes. Yellow markings as follows: postpronotal lobes; notopleura; mesopleural stripe of medium-width, reaching midway between anterior margin of notopleuron and anterior *npl.* seta dorsally, continuing to katepisternum as a transverse spot, anterior margin straight; anatergite (posterior apex black); anterior ¾ katatergite (remainder black); two broad parallel-sided lateral postsutural vittae beginning with a spot anterior to notopleural suture and ending behind *ia.* seta; a broad medial longitudinal postsutural vitta beginning at posterior margin of scutum and ending at a point level with notopleural suture (in some specimens there is an indication of a narrow dark red-brown medial longitudinal band running anteriorly off the anterior point of the medial postsutural yellow vitta). Postnotum black. Scutellum yellow except for narrow black basal band and with a pale fuscous apical spot between apical setae. Setae: *sc.* 4; *prsc.* 2; *ia.* 1; *p.sa.* 1; *a.sa.* 1; *mpl.* 1; *npl.* 2; *scp.* 4; *hm.* 1 (one weak pale to dark seta on each postpronotal lobe).

Legs: Fore femora fulvous on basal ⅓ and dark fuscous to black on apical ⅔, mid femora dark fuscous to black on approximately apical ½ and fulvous basally, hind femora dark fuscous to black on apical ¼ and remainder fulvous; fore tibiae dark fuscous basally and fuscous apically, mid tibiae fuscous basally and fulvous apically, hind tibiae dark fuscous; tarsi with basal segment fulvous and apical four segments fuscous; mid tibiae each with an apical black spur.

Wings: Length 5.6 mm; cells bc and c colourless; microtrichia in outer corner of cell c only; remainder of wings colourless except dark fuscous cell sc, narrow dark fuscous costal band confluent with R_{2+3} and widening into a small spot around apex of R_{4+5}, broad dark fuscous anal streak; dense aggregation of microtrichia around A_1+CuA_2; supernumerary lobe strong.

Abdomen: Oval; terga free; pecten present on tergum III. Tergum I and sterna I and II wider than long. Tergum I fulvous with narrow dark fuscous to black lateral margins; tergum II fulvous with a narrow broken subanterior transverse dark fuscous to black band that does not reach lateral margins and dark fuscous to black anterolateral corners; terga III–V fulvous with a 'T' pattern consisting of a narrow transverse black band across anterior margin of tergum III that expands over lateral margins and a medium-width medial longitudinal black band over all three terga (sometimes this band is broken on tergum III) and broad lateral longitudinal black bands over terga IV and V (sometimes there is a pale line along posterior margin of tergum IV interrupting this band). A pair of oval fulvous shining spots on tergum V. Posterior lobe of surstylus long, sternum V with a slight concavity on posterior margin. Abdominal sterna dark.

Attractant: Cue lure.

Female: No known record.

Distribution: Southern Thailand (Hat Yai).

Hosts: No known record.

Remarks: *Bactrocera (Zeugodacus) hatyaiensis* n.sp. is close to *Bactrocera (Zeugodacus) flavopectoralis* (Hering) in general colour patterns of the body, wings and legs. It differs from *B. flavopectoralis* in having humeral setae on the postpronotal lobe, a small fuscous spot on the apex of the scutellum and broad lateral longitudinal dark fuscous to black bands over terga IV and V.

Bactrocera (Zeugodacus) havelockiae Drew & Romig, new species

(Figs 305A, 305B)

Material examined: Holotype male, **INDIA: ANDAMAN ISLANDS** – S. Andaman, Havelock Is., 26.vi.1998, coll. Ranganath, attracted to cue lure, deposited in BMNH.

Paratypes: **INDIA: ANDAMAN ISLANDS** – 2 males, same data as holotype; 1 male, S. Andaman, Havelock I., 25.vi.1998, coll. Ranganath, attracted to cue lure; 2 males and 2 females, L. Andaman, 22.ii.1996, coll. Ranganath, bred from *Trichosanthes* sp., Sl No. 52; 2 males and 2 females, India, Port Blair, coll. Ranganath, 14.vi.1993, bred from cucurbits, Sl No. 39; 2 males and 2 females, India, S. Andaman, Port Blair, 26.viii.1994, coll. Veena Kumari, bred from *M. dioica*, Sl No. 43; 1 male, India, S. Andaman, Port Blair, 18.v.1992, coll. H.R.R. 3 paratypes in ANIC, 3 paratypes in BMNH, 6 paratypes in QDPC, 4 paratypes in QMIC.

Diagnosis: A medium-sized species; face fulvous with a pair of medium-sized oval black spots; postpronotal lobes and notopleura yellow; scutum uniformly red-brown without dark patterns; lateral postsutural yellow vittae short and narrow; medial postsutural yellow vitta present; a yellow spot anterior to notopleural suture; mesopleural stripe reaching midway between anterior margin of notopleuron and anterior *npl.* seta dorsally; scutellum yellow; wing with cell c pale fuscous, cell bc fuscous, microtrichia in outer corner of cell c only; an overall pale fuscous tint, a narrow costal band which varies from pale fuscous to dark fuscous and expands into a large fuscous to dark fuscous spot across apex of wing; a broad dark fuscous anal streak; abdominal terga III–V orange-brown with a 'T' pattern consisting of a narrow transverse black band across anterior margin of tergum III and a broad medial longitudinal black band over all three terga, narrow dark fuscous anterolateral corners on terga IV and V.

Description: male.

Head: Height 1.56 mm. Frons – length 1.45 times breadth; red-brown with fuscous on anteromedial hump; orbital setae black: 1 *s.or.*, 3 *i.or.*; lunule fuscous. Ocellar triangle black. Vertex dark red-brown. Face fulvous with a pair of medium-sized oval black spots; length 0.58 mm. Genae red-brown, small fuscous subocular spot present; black seta present. Occiput red-brown, fulvous along eye margins; occipital row with 5–8 black setae. Antennae with all segments red-brown; length of segments: 0.12 mm; 0.26 mm; 0.76 mm.

Thorax: Scutum entirely red-brown without dark markings. Pleural areas entirely red-brown without dark markings. Yellow markings as follows: postpronotal lobes; notopleura; mesopleural stripe reaching midway between anterior margin of notopleuron and anterior *npl.* seta dorsally, continuing to katepisternum as a large transverse spot, anterior margin convex; anatergite (posterior apex red-brown); anterior ⅔ katatergite (remainder red-brown); two medium-width very short lateral postsutural vittae ending at level of *a.sa.* seta (variable in shape and size, Figs 305A, 305B); a narrow medial longitudinal postsutural vitta present; a small yellow spot anterior to notopleural suture in front of each lateral postsutural vitta. Postnotum red-brown. Scutellum yellow except for narrow red-brown basal band. Setae: *sc.* 4; *prsc.* absent; *ia.* 1; *p.sa.* 1; *a.sa.* 1; *mpl.* 1; *npl.* 2; *scp.* 4.

Legs: Fore femora fulvous basally to red-brown apically, mid and hind femora fulvous basally to red-brown apically and with dark fuscous on the ventral apical surfaces; fore tibiae fuscous, mid tibiae dark fuscous basally to fuscous apically, hind tibiae dark fuscous; fore tarsi with all segments red-brown, mid and hind tarsi with basal segment fulvous and apical four segments red-brown; mid tibiae each with an apical black spur.

Wings: Length 6.2 mm; cell c pale fuscous and cell bc fuscous; microtrichia in outer corner of cell c only; remainder of wings with a pale fuscous tint, a dark fuscous cell sc, a narrow costal band which is dark fuscous from anterior margin of wing to vein R_{2+3} and then overlapping this vein where it is pale fuscous and expanding into a large spot across apex of wing which varies from dark fuscous to fuscous, a broad dark fuscous anal streak; a dense aggregation of microtrichia around A_1+CuA_2; supernumerary lobe very large and keel-shaped. The wing patterns vary between extremes shown in Figs 305A, 305B.

Abdomen: Elongate oval; terga free; pecten present on tergum III. Tergum I and sterna I and II wider than long. Tergum I fuscous centrally to dark fuscous on lateral margins and with a narrow red-brown transverse band across posterior margin; tergum II red-brown anteriorly and fulvous posteriorly with a narrow transverse dark fuscous band subanteriorly but not reaching the narrow dark fuscous lateral margins; terga III–V orange-brown with a 'T' pattern consisting of a narrow transverse black band across anterior margin of tergum III and a moderately broad medial longitudinal black band over all three terga, narrow dark fuscous anterolateral corners on terga IV and V. A pair of oval orange-brown shining spots on tergum V. Posterior lobe of surstylus long, sternum V with a slight concavity on posterior margin. Abdominal sterna dark.

Attractant: Cue lure.

Female: As for male except no dense aggregation of microtrichia around A_1+CuA_2, supernumerary lobe weak, pecten of cilia on abdominal tergum III absent; basal segment of ovipositor red-brown, dorsoventrally compressed and tapering posteriorly in dorsal view; ratio of length of oviscape to length of tergum V, 1.0:1; apex of piercer needle-shaped.

Distribution: India: Andaman Is (South Andaman I.; Little Andaman I.).

Hosts: *Trichosanthes* sp.; *Momordica dioica* (family *Cucurbitaceae*).

Remarks: *Bactrocera (Zeugodacus) havelockiae* n.sp. is similar to *Bactrocera (Zeugodacus) quasiinfesta* n.sp., in possessing a red-brown scutum with lateral and medial postsutural yellow vittae, four *sc.* setae and *prsc.* setae absent. It differs from *B. quasiinfesta* in having the lateral postsutural yellow vittae short and tapering posteriorly and the supernumerary lobe large and keel shaped. Although the *prsc.* setae are absent, we are placing *B. havelockiae* in the subgenus *Zeugodacus* as it fits all other criteria in the definition for this subgenus.

Bactrocera (Zeugodacus) heinrichi (Hering)

(Figs 306A, 306B)

Zeugodacus nubilus ssp. *heinrichi* Hering, 1941d: 11; Hardy & Adachi 1954: 188 as Synonym of *Dacus (Zeugodacus) hageni* de Meijere. Hardy, 1973: 70 Synonym of *B. tau* Walker. Holotype female in MFN.

Bactrocera (Zeugodacus) nubilus ssp. *heinrichi* – Norrbom *et al.*, 1998: 104 Synonym of *B. tau* Walker.

Bactrocera (Zeugodacus) heinrichi – Mahmood, 1999: 222.

Type material examined: Holotype female, **INDONESIA: SULAWESI (formerly CELEBES)** – with the following four labels – red label 'Type'; cream label 'Celebes, Bantimoeroeng, 8.1931, G. Heinrich'; white label 'Zeugodacus nubilus heinrichi, Type, det. Hering 1941'; white label 'Zool. Mus. Berlin'. Received on loan from MFN.

Female type with the following six labels – white label 'female'; white circular label with yellow outer circle

'Para-type'; red label 'Paratype'; white label 'Celebes Bantimoeroeng 8.1931 G. Heinrich'; white label 'Zeugodacus heinrichi W.P.T. det. Hering 1941'; white label 'I. Tausch erh. v. Mus. Berlin 27.III.1941'. Studied at BMNH.

Non-type material examined: INDONESIA: JAVA – 3 males, Pelabuhawatu, w. Java, 13.iv.1975, coll. D.E. Hardy; 2 males (iii.1975), 6 males (vii.1975), Kebrun Raya, Bog or W. Java, coll. D.E. Hardy; 19 males, Java, Bogor, Kebun Raya (300m), 17.v.1981, coll. J. Turner, attracted to cue lure; 1 male, Java, Banten, Lebak, Pisir Nini-nini, 9.i.2006, coll. TB Aden Haer, attracted to cue lure in forest; 1 male, Java, DKI Jakarta, Jakarta, Selatan, Tanjung Barat, 23.vii.2005, Mulyana, Carambola, attracted to cue lure; **LOMBOK** – 5 males, Lombok, Sasaot (400-550m), 31.v.1981, coll. J. Turner, old plantation Mahogany, attracted to cue lure; **SULAWESI** – 3 males, Dumoga, Taraut, 225m N. Sulawesi, 21.xii.1985, coll. Santianawati; **SUMATRA** – 1 male, 12km N.W. Bohorok, Sumatra 200m, 4.v.1975, coll. D.E. Hardy; 6 males, Bukit, Lawang Reserve, Sumatra, Indonesia, 2-12.vii.1977, coll. D. Kistner, attracted to cue lure. **EAST MALAYSIA: SARAWAK** – 1 male, Tarat Agriculture Station, Nr Kutching, Sarawak, 27.iv.1992, coll. S. Leong, attracted to cue lure; 1 male, Ranchang Pool, Serian, Nr. Kutching, 11.v.1992, coll. S. Leong, attracted to cue lure; 1 male, Tarat Agricultural Station, 5.iv.1993, coll. S. Leong, KT3, attracted to cue lure. **PENINSULAR MALAYSIA**: 2 males, Pangsoon, Hulu, Langat, 26.ix.1990, coll. D.L. Hancock, attracted to cue lure; 2 males, Malaysia, Hulu Langat, Kuala Lumpur, 15.v.1992, coll. D.L. Hancock, attracted to cue lure; 1 male, Malaysia, Hulu Langat, Pansun, Sel, 9.x.1988, coll. VHS; 1 male, Malaysia, Selangor, Hulu Langat, 9.iv.1994, coll. VHS; 1 male, T.S. Maran, Malaysia, 25.ii.1986, coll. Antony, attracted to cue lure; 1 male, T.S. Maran, Malaysia, 12.ii.1986, coll. Mahdzir, attracted to cue lure; 2 males, T.S.6, Malaysia, 15.i.1986, coll. R. Drew, attracted to cue lure; 1 male (11.ii.1986), 1 male (25.iii.1986), R.S. Kluang, coll. Jasni Zaman, attracted to cue lure; 1 male, Cameron Highlands, Malaysia, Tapah, 6th mile, Trap 1, 24.vi.1983, coll. S. Vijaysegaran, attracted to cue lure; 1 male, Cameron H'lds, Malaysia, Tapah, 6th mile, site 1, 7.vii.1983, coll. S. Vijaysegaran, attracted to cue lure; 1 male, Cameron H'lds, Malaysia, Tanah Ratal Ringlet, site 5, 7.vii.1983, coll. S. Vijaysegaran, attracted to cue lure; 1 male, Cameron Highlands, Malaysia, Tapah, 18th mile, trap 3, 24.vi.1983, coll. S. Vijaysegaran, attracted to cue lure. **THAILAND**: 1 male, 'Phang Nga Prov., Hin Lad Waterfall, 60 m, 13.v.2003/8'; '8.39.26N, 98.27.44E, P. Schwendinger leg, secondary forest/8'; 'attracted to cue lure'; 1 male, 'Chumphon Prov, Khao Kai Jae Watf, 5-8.v.2003/2'; '9.55.07N, 98.56.34E, P. Schwendinger leg, semi-evergreen rainforest/2'; 'attracted to cue lure'. Received on loan from MHNG. 3 males, Trang Province, Khao Chong (7 32 N 99 47 E), 24.ii-2.iii.2009, McPhail tr. MP-M18, ii.2009-D, coll. P. Kongnoo & T. Tongrod, KHC 1384, KHC 1466, KHC 1553. 1 male, Thailand, 50m, Phuket pr, Ko Siray, 15.viii.2000, coll. P. Schwendinger; 1 male, S. Thailand, NP, Khao Phanom Bench, Kralei Prov & Dist., 11.x.1999, coll. Schwendinger. **VIETNAM: Northern Vietnam** – 1 male, Pa Co Village, Mai Chau District, Hoa Binh Province, North Vietnam, 1.x.2002, coll. NIPP Staff, attracted to cue lure. Specimens in QDPC except where specified.

Diagnosis: A medium-sized species; face fulvous with a pair of small oval black spots; postpronotal lobes and notopleura yellow; scutum base colour black; lateral and medial postsutural yellow vittae present; a yellow spot anterior to notopleural suture in front of lateral postsutural vittae; mesopleural stripe reaching midway between anterior margin of notopleuron and anterior *npl.* seta dorsally; scutellum yellow; wing with cells bc and c colourless, microtrichia in outer corner of cell c only; a narrow fuscous costal band overlapping R_{2+3} and broadening across apex of wing, a small fuscous area on posterior margin around vein CuA_1; a broad fuscous to pale fuscous anal streak; abdominal terga III–V fulvous with a black 'T' pattern consisting of a transverse band across anterior margin of tergum III, a medium-width medial longitudinal band over all three terga, anterolateral corners of terga IV and V black.

Description: female (Fig. 306A).

Head: Height 1.52 mm. Frons – length 1.25 times breadth; red-brown with fuscous dorsolaterally; orbital setae black: 1 *s.or.*, 3 *i.or.*; lunule red-brown. Ocellar triangle black. Vertex dark fuscous to black. Face fulvous with a pair of small oval black spots; length 0.48 mm. Genae fulvous, pale fuscous subocular spot present; black seta present. Occiput red-brown, fulvous along eye margins; occipital row with a few small pale setae. Antennae with segment 1 red-brown, segments 2 and 3 fuscous; length of segments: 0.72 mm; 0.24 mm; 0.14 mm.

Thorax: Scutum black except dark red-brown below and behind lateral postsutural yellow vittae and posterior to medial postsutural yellow vitta, red-brown anterior to notopleural suture and across anterior margin of scutum which joins with anterior end of medial longitudinal yellow vitta. Pleural areas black except red-brown below postpronotal lobes. Yellow markings as follows: postpronotal lobes; notopleura; mesopleural stripe of medium-width reaching midway between anterior margin of notopleuron and anterior *npl.* seta dorsally, extending to katepisternum as a large transverse spot, anterior margin slightly convex; anatergite (posterior apex black); anterior $^5/_7$ katatergite (remainder black); two broad parallel-sided lateral postsutural vittae beginning anteriorly as a spot in front of notopleural suture and ending posteriorly behind *ia.* seta; a broad medial longitudinal postsutural vitta broadly rounded posteriorly at level of *prsc.* setae and ending in a point anteriorly just anterior of notopleural suture. Postnotum black except dark red-brown centrally. Scutellum yellow except for a narrow black basal band. Setae: *sc.* 4; *prsc.* 2; *ia.* 1; *p.sa.* 1; *a.sa.* 1; *mpl.* 1; *npl.* 2; *scp.* 4.

Legs: All femora entirely fulvous; fore and hind tibiae fuscous, mid tibiae fulvous; all tarsal segments entirely fulvous; mid tibiae each with an apical black spur.

Wings: Length 6.24 mm; cells bc and c colourless; microtrichia in outer corner of cell c only; remainder of wings colourless except fuscous cell sc, narrow fuscous costal band overlapping R_{2+3} and broadening across apex of wing, a small pale fuscous area on posterior margin around vein

CuA_1, a broad fuscous to pale fuscous anal streak; supernumerary lobe weak.
Abdomen: Oval; terga free; no pecten on tergum III. Tergum I and sterna I and II wider than long. Tergum I fulvous with narrow black lateral margins; tergum II fulvous with a narrow black transverse band anteriorly but not reaching lateral margins and anterolateral corners black; terga III to V fulvous with a black 'T' consisting of a transverse black band across anterior margin of tergum III which widens around to form narrow black lateral margins, a medium-width medial longitudinal black band over all three terga and anterolateral corners of terga IV and V broadly black. A pair of oval fulvous shining spots on tergum V. All abdominal sterna pale. Basal segment of ovipositor red-brown, dorsoventrally compressed and tapering posteriorly in dorsal view; ratio of length of oviscape to length of tergum V, 1.95:1; apex of piercer needle-shaped.
Attractant: Cue lure.
Male: (Fig. 306B) – As for female except for a dense aggregation of microtrichia around A_1+CuA_2; supernumerary lobe strong and rounded; pecten present on abdominal tergum III; posterior lobe of surstylus long; abdominal sternum V with a slight concavity on posterior margin.
Distribution:
Type locality: Indonesia (Sulawesi – South Sulawesi, Bantimoeroeng).
Previously recorded: Type locality and the following locations – Brunei; Indonesia (Java) (White & Hancock, 1997).
New records: Indonesia (Lombok; Sumatra; Sulawesi – North Sulawesi); East Malaysia (Sarawak); Peninsular Malaysia; Thailand; Northern Vietnam (Hoa Binh Province). See locations listed under 'Non-type material examined'.
Hosts: No known record.
Remarks: *Bactrocera (Zeugodacus) heinrichi* (Hering) belongs to the *tau* complex of species that possess four *sc.* setae, a black scutum with lateral and medial postsutural yellow vittae, wings with cells bc and c colourless or with a pale tint, costal band overlapping R_{2+3} and expanded into a spot apically and face with a pair of black spots. The entire fulvous abdominal sterna sets *B. heinrichi* apart from the other species in the *tau* complex. The measurements included above are from specimens collected in West Java, not the type.

Bactrocera heinrichi was placed in synonymy with *Dacus (Zeugodacus) hageni* de Meijere by Hardy & Adachi (1954) and with *Bactrocera (Zeugodacus) tau* (Walker) by Hardy (1973) and Norrbom *et al.* (1998). However, we regard *B. heinrichi* as a good species and are thus withdrawing it from these previous synonymies. We still regard *D. hageni* a synonym of *B. tau*.

Bactrocera (Zeugodacus) hekouana Yu, He & Yang

(Fig. 307)

Bacterodacus (Zeugodacus) hekouana Yu, He & Yang, in Yu *et al.*, 2011: 317. [*Bactrocera* Misspelt as *Bacterodacus*]. Holotype male in CAIQ.

Material examined: No specimens available for study. The following has been developed from the description and photographs in Yu *et al.* (2011).
Diagnosis: A medium-sized species; face red-brown with a pair of large oval black spots; postpronotal lobes and notopleura yellow; scutum shining black without pale colour patterns; narrow short lateral and medial postsutural yellow vittae; yellow spot present anterior to notopleural suture; mesopleural stripe reaching almost to level of anterior *npl.* seta before curving back to midway between anterior margin of notopleuron and anterior *npl.* seta; scutellum yellow; wing with cells bc and c colourless, microtrichia in outer corner of cell c only; a narrow dark fuscous costal band confluent with R_{2+3} and becoming very narrow beyond apex of this vein before expanding into a distinct spot around apex of R_{4+5} (*scutellaris*-like); a broad dark fuscous anal streak; abdominal terga III–V mostly black.
Description: male.
Head: Frons fulvous with dark fuscous around orbital setae and on anteromedial hump; orbital setae black: 1 *s.or.*, 3 *i.or.*; lunule fuscous. Ocellar triangle black. Vertex dark fuscous. Face red-brown with a pair of large oval black spots. Genae fulvous, fuscous subocular spot present; black seta present. Occiput dark fuscous, fulvous along eye margins. Antennae with segments 1 and 2 fulvous, segment 3 missing.
Thorax: Scutum uniformly shining black. Pleural areas black without pale markings. Yellow markings as follows: postpronotal lobes; notopleura; mesopleural stripe reaching almost to level of anterior *npl.* seta before curving back to midway between anterior margin of notopleuron and anterior *npl.* seta dorsally, not extending to katepisternum, anterior margin straight; anatergite (posterior apex black); most of katatergite (proportion of black not visible); two narrow lateral postsutural vittae narrowing posteriorly to end before *ia.* seta; a short, narrow medial longitudinal postsutural vitta; a yellow spot anterior to notopleural suture. Postnotum black. Scutellum yellow with a narrow black basal band. Setae: *sc.* 2; *prsc.* 2; *ia.* 1; *p.sa.* 1; *a.sa.* 1; *mpl.* 1; *npl.* 2; *scp.* 2.
Legs: Fore femora dark fuscous to black with small areas of fulvous on apex and around base, mid femora fulvous with black around apical ½, hind femora fulvous with black around apical ⅓ fore and mid tibiae dark fuscous basally to fulvous apically, hind tibiae dark fuscous; tarsi with basal segment fulvous and apical four segments fuscous; mid tibiae each with an apical red-brown spur.
Wings: Cells bc and c colourless; microtrichia in outer corner of cell c only; remainder of wings colourless except dark fuscous cell sc, narrow dark fuscous costal band confluent with R_{2+3} and becoming very narrow beyond apex of this vein before expanding into a distinct spot around apex of R_{4+5} (*scutellaris*-like), a broad dark fuscous anal streak; dense aggregation of microtrichia around A_1+CuA_2; supernumerary lobe strong; cell br pale fuscous with dense microtrichia.
Abdomen: Oval; terga free; pecten present on tergum III. Tergum I wider than long. Tergum I black; tergum II black with fulvous along posterior margin; terga III–V black tending paler posterocentrally on each tergum. A pair of oval fulvous shining spots on tergum V. Posterior lobe of surstylus long,

sternum V with a slight concavity on posterior margin. Abdominal sterna dark.
Attractant: No known record.
Female: No known record.
Distribution:
Type locality: China (Yunnan Province, Hekou).
Hosts: No known record.
Remarks: *Bactrocera (Zeugodacus) hekouana* Yu, He & Yang belongs to the *scutellaris* complex and is similar to *Bactrocera (Zeugodacus) yoshimotoi* (Hardy) in possessing two *sc.* setae, yellow notopleura, face with a pair of black spots and the supernumerary lobe in the wing well developed. It differs from this species in having large oval facial spots vertically placed in the antennal groves, distinct dark fuscous to black patterns on all femora and abdominal terga III–V mostly black (not with a black 'T' pattern). Other similar species are *Bactrocera (Zeugodacus) hoabinhiae* n.sp. and *Bactrocera (Zeugodacus) laocaiae* n.sp., but *B. hoabinhiae* possesses a black band across the oral margin of the face, a black spot on the apex of the scutellum and abdominal terga III–V with a black 'T' pattern, and *B. laocaiae* has small facial spots pointed towards the midline of the face which can form into a transverse black band.

Bactrocera (Zeugodacus) hengsawadae Drew & Romig, new species

(Fig. 308)

Material examined: Holotype male, **THAILAND**: Rayong, 80 km W of Chanthaburi, 27.v.1991, coll. D.L. Hancock, on *Ocimum* sp., deposited in BMNH.
Paratypes: **THAILAND**: 5 males and 6 females, 23 km from Muang, Prachinburi, 4.vi.1993, coll. Montree, bred ex *Siphonodon* sp. (family *Celastraceae*); 2 males and 7 females, Jundish, Pakchong, Nakhonratsima, 4.vi.1993, coll. Montree, bred ex *Siphonodon* sp. (Family Celastraceae). 3 paratypes in ANIC; 2 paratypes in BMNH; 12 paratypes in QDPC; 3 paratypes in QMIC.
Diagnosis: A medium-sized species; face fulvous with a pair of large oval black spots; postpronotal lobes and notopleura yellow; scutum black with a large medial longitudinal dark red-brown area from anterior point of medial postsutural yellow vitta to anterior margin of scutum; lateral and medial postsutural yellow vittae present; a yellow spot on anterior margin of notopleural suture; mesopleural stripe equal in width to notopleuron dorsally; scutellum yellow; wing with cells bc and c colourless, microtrichia in outer corner of cell c only; a narrow dark fuscous costal band and medium-width anal streak; abdominal terga III–V fulvous with a black 'T' pattern consisting of a narrow transverse black band across anterior margin of tergum III and a narrow- to medium-width medial longitudinal black band over all three terga and broad lateral longitudinal black bands over terga IV and V.
Description: male.
Head: Height 1.44 mm. Frons – length 1.4 times breadth; fulvous with dark fuscous on anteromedial hump and reaching to lower *i.or.* setae; orbital setae black: 1 *s.or.*, 3 *i.or.*; lunule dark fuscous. Ocellar triangle black. Vertex black. Face fulvous with a pair of large oval black spots reaching to oral margin; length 1.37 mm. Genae fulvous, large fuscous to dark fuscous subocular spot present; black seta present. Occiput fuscous, dark fuscous dorsolaterally, fulvous along eye margins; occipital row with a large number of short pale setae. Antennae with segments 1 and 2 fuscous, segment 3 red-brown with fuscous on apex and outer surface; arista black (red-brown basally); length of segments: 0.12 mm; 0.24 mm; 0.68 mm.
Thorax: Scutum black except for dark red-brown below and behind lateral postsutural vittae, between notopleuron and yellow spot anterior to notopleural suture, a broad medial longitudinal band from anterior point of medial postsutural yellow vitta to anterior margin of scutum (this area has a narrow medial longitudinal black band over almost the entire length). Pleural areas black except dark red-brown below postpronotal lobe. Yellow markings as follows: postpronotal lobes; notopleura; mesopleural stripe equal in width to notopleuron dorsally, continuing to katepisternum as a transverse spot, anterior margin convex; anatergite (posterior apex black); anterior ¾ katatergite (remainder black); two broad parallel-sided lateral postsutural vittae beginning with a spot anterior to notopleural suture and ending behind *ia.* seta; a broad medial longitudinal postsutural vitta beginning behind *prsc.* setae and ending anterior to notopleural suture. Postnotum black. Scutellum yellow except for narrow black basal band. Setae: *sc.* 2 (occasionally one or two weak basal setae are present); *prsc.* 2; *ia.* 1; *p.sa.* 1; *a.sa.* 1; *mpl.* 1; *npl.* 2; *scp.* 4.
Legs: Femora entirely fulvous except for a small fuscous to dark fuscous preapical spot on outer surfaces of fore femora and inner surfaces of mid and hind femora; fore tibiae fuscous, mid tibiae pale fuscous tending darker basally, hind tibiae dark fuscous; tarsi with basal segment fulvous and apical four segments pale fuscous; mid tibiae each with an apical black spur.
Wings: Length 6.2 mm; cells bc and c colourless; microtrichia in outer corner of cell c only; remainder of wings colourless except dark fuscous cell sc, narrow dark fuscous costal band confluent with R_{2+3} and remaining of uniform width until a slight expansion beyond apex of R_{4+5}, a medium-width fuscous anal streak; dense aggregation of microtrichia around A_1+CuA_2; supernumerary lobe of medium development.
Abdomen: Oval; terga free; pecten present on tergum III. Tergum I and sterna I and II wider than long. Tergum I orange-brown with narrow black lateral margins; tergum II fulvous with a narrow black transverse band across anterior margin reaching to narrow anterolateral black corners; terga III–V fulvous with a 'T' pattern consisting of a narrow transverse black band across anterior margin and expanding narrowly across lateral margins and a narrow- to medium-width medial longitudinal black band over all three terga, broad lateral longitudinal black bands over terga IV and V. A pair of oval fulvous shining spots on tergum V. Posterior lobe of surstylus long, sternum V with a slight concavity on posterior margin. Abdominal sterna dark, except sternum II pale.
Attractant: No known record.

Female: As for male except no dense aggregation of microtrichia around A_1+CuA_2; supernumerary lobe weak; pecten of cilia on abdominal tergum III absent; basal segment of ovipositor red-brown, dorsoventrally compressed and tapering posteriorly in dorsal view; ratio of length of oviscape to length of tergum V, 1.5:1; apex of piercer narrow and needle-shaped.
Distribution: Thailand.
Hosts: Recorded from *Ocimum* sp. (family *Lamiaceae*); *Siphonodon* sp. (family *Celastraceae*).
Etymology: This species is named after Dr Chantanee Hengsawad, Department of Agriculture, Chiang Mai, who undertook extensive field collections in Thailand.
Remarks: *Bactrocera (Zeugodacus) hengsawadae* n.sp. is similar to *Bactrocera (Zeugodacus) yoshimotoi* (Hardy) in possessing two *sc.* setae, lateral and medial longitudinal postsutural yellow vittae, scutum mostly black, wings colourless except for costal band and anal streak, costal band narrow, scutellum entirely yellow and face with a pair of black spots. It differs from *B. yoshimotoi* in having vertical oval facial spots, parallel-sided lateral postsutural vittae and all femora with small dark preapical spots. It is distinguished from the *scutellaris* complex species, especially, in possessing broad lateral and medial longitudinal postsutural vittae. *Bactrocera hengsawadae* is also superficially similar to *Bactrocera (Asiadacus) maculifacies* (Hardy) but differs in having distinct facial spots, a smaller supernumerary lobe in the male wing and *prsc.* and *a.sa.* setae present.

Bactrocera (Zeugodacus) hoabinhiae Drew & Romig, new species

(Fig. 309)

Material examined: Holotype male, **VIETNAM: Northern Vietnam** – Hoabinh Prov, Mai Chau, Paco, trap No. N010 rainforest, 20.iv.1999, coll. Le Duc Khanh *et al.*, attracted to cue lure, deposited in BMNH.
Paratypes: **VIETNAM: Northern Vietnam** – 1 male (15.vii.2000), 1 male (30.xi.2001), Hoa Binh Province, Mai Chau Dist, Paco, trap N010, attracted to cue lure. **THAILAND**: 1 male, Khun Khon, Chiang Rai CR303T, 9.i.1992, coll. Charuwan, attracted to cue lure; 1 male, Khun Khon, Chiang Rai, CR320T, 23.i.1992, coll. Charuwan, attracted to cue lure. 1 paratype in ANIC; 2 paratypes in QDPC; 1 paratype in QMIC.
Non-type material examined: INDIA: ASSAM – 1 female, Assam, Shillong, 6.x.1920, coll. R. Senior White, B.M. 1924-100, TR. 36; 1 female, Assam, Shillong, 7.x.1920, coll. Senior White, B.M. 1924-100, TR. 36. Specimens in BMNH.
Diagnosis: A large species; face fulvous with a broad black band across oral margin; postpronotal lobes yellow (anterodorsal corners fuscous); notopleura yellow; scutum shining black; lateral and medial postsutural yellow vittae present; a yellow spot present anterior to notopleural suture; mesopleural stripe slightly wider than notopleuron dorsally; scutellum yellow with a dark fuscous to black apical spot; wing with cells bc and c colourless, microtrichia in outer corner of cell c only; a narrow dark fuscous costal band expanding across apex of R_{4+5} (*scutellata*-like) and broad dark fuscous anal streak; abdominal terga III–V red-brown with a 'T' pattern consisting of a broad black transverse band across anterior margin of tergum III and a narrow medial longitudinal dark fuscous to black band over all three terga and broad lateral longitudinal dark fuscous to black bands over terga IV and V.
Description: male.
Head: Height 1.6 mm. Frons – length 1.33 times breadth; red-brown with fulvous along lateral margins and fuscous to dark fuscous around orbital setae and on anteromedial hump; orbital setae black: 1 *s.or.*, 3 *i.or.*; lunule dark fuscous. Ocellar triangle black. Vertex black. Face fulvous with a broad black band across oral margin (this consists of two elongate oval black spots connected with a broad black band); length 0.54 mm. Genae fuscous, dark fuscous subocular spot present; black seta present. Occiput red-brown with fuscous to dark fuscous along dorsal and lateral margins in some specimens, fulvous along eye margins; occipital row with a large number of short dark setae. Antennae with segments 1 and 2 red-brown, segment 3 dark fuscous; arista black (red-brown basally); length of segments: 0.15 mm; 0.3 mm; 0.75 mm.
Thorax: Scutum shining black with red-brown below and behind lateral postsutural vittae and inside notopleura. Pleural areas shining black except dark red-brown below postpronotal lobes. Yellow markings as follows: postpronotal lobes (anterodorsal corners fuscous); notopleura; mesopleural stripe slightly wider than notopleuron dorsally, continuing to katepisternum as a transverse spot, anterior margin slightly convex; anatergite (posterior apex black); anterior ⅔ katatergite (remainder black); two narrow- to medium-width lateral postsutural vittae beginning with a spot anterior to notopleural suture and narrowing posteriorly to end before *ia.* seta; a narrow and short medial longitudinal postsutural vitta. Postnotum black. Scutellum yellow except for narrow black basal band and a dark fuscous to black apical spot enclosing the apical setae. Setae: *sc.* 2; *prsc.* 2; *ia.* 1; *p.sa.* 1; *a.sa.* 1; *mpl.* 1; *npl.* 2; *scp.* 4.
Legs: Fore femora shining black except basal ⅓ fulvous, apical ½ mid femora shining black and remainder fulvous, apical ⅓ hind femora shining black and remainder fulvous; fore tibiae dark fuscous basally tending fuscous apically, mid tibiae fulvous tending fuscous basally, hind tibiae dark fuscous; tarsal segments with basal segment fulvous and apical four segments pale fuscous to fuscous; mid tibiae each with an apical black spur.
Wings: Length 7.2 mm; cells bc and c colourless; microtrichia in outer corner of cell c only; remainder of wings colourless except dark fuscous cell sc, narrow dark fuscous costal band confluent with R_{2+3} and remaining narrow beyond apex of R_{2+3} before widening across apex of R_{4+5} (*scutellata*-like), broad dark fuscous anal streak; dense aggregation of microtrichia around A_1+CuA_2; supernumerary lobe very strong and keel-shaped.
Abdomen: Oval; terga free; pecten present on tergum III. Tergum I and sterna I and II wider than long. Tergum I red-brown with broad black anterolateral margins; tergum II fulvous with a broad black transverse band across anterior

margin but not enclosing lateral margins; terga III–V red-brown with a 'T' pattern consisting of a broad transverse black band across anterior margin of tergum III which encloses lateral margins and a narrow medial longitudinal dark fuscous to black band over all three terga and broad dark fuscous to black lateral longitudinal bands over terga IV and V. A pair of oval red-brown shining spots on tergum V. Posterior lobe of surstylus long, sternum V with a slight concavity on posterior margin. Abdominal sterna shining dark fuscous to black, except sternum II fulvous.

Attractant: Cue lure.

Female: No known record.

Distribution: India (Assam – Shillong); Thailand (Chiang Rai); Northern Vietnam (Hoa Binh Province).

Hosts: No known record.

Remarks: *Bactrocera (Zeugodacus) hoabinhiae* n.sp., within the *scutellaris* complex, is similar to *Bactrocera (Zeugodacus) laocaiae* n.sp., *Bactrocera (Hemigymnodacus) mukiae* n.sp. and *Bactrocera (Zeugodacus) yoshimotoi* (Hardy) in possessing two *sc.* setae, yellow notopleura and face with either a pair of black spots or a transverse black band across oral margin. It differs from *B. yoshimotoi* and *B. mukiae* in having a black band across the face, from *B. yoshimotoi* in having black patterns on the femora, from *B. mukiae* in having *a.sa.* setae and a strong supernumerary lobe in the male wing and from *B. laocaiae* in having abdominal terga III–V with a dark 'T' pattern and broad lateral longitudinal dark bands on terga IV and V.

Bactrocera (Zeugodacus) hodgsoniae Drew & Romig, new species

(Fig. 310)

Material examined: Holotype male, **PENINSULAR MALAYSIA**: Kuala Woh, Tapeh, Perak, 29.ix.1986, coll. Mahdzir and Bahari, bred from *Hodgsonia capniocarpa*, deposited in BMNH.

Paratypes: **INDONESIA: JAVA** – 1 female, Gunung Salak, Bogor, Java, Indonesia, 800 m, 17.v.1981, coll. J. Turner, secondary rainforest, hand collected on red cucurbit. **PENINSULAR MALAYSIA**: 12 males and 23 females, same data as holotype; 1 male and 2 females, Old Gombak Road, 16.ix.1986, coll. Anthony *et al.*, bred from host fruit. 2 males, Km 87 KL, nr Felds Trolak Jen, Malaysia, 29. ix.1986, coll. Mahdzir, Bahari, bred ex *Trichosanthes tricuspidata*, sample No. 0642. 4 paratypes in ANIC; 4 paratypes in BMNH; 29 paratypes in QDPC; 4 paratypes in QMIC.

Diagnosis: A medium-sized species; face with a pair of medium-sized circular black spots; postpronotal lobes and notopleura yellow; scutum shining black with a small area of red-brown anteriorly; lateral and medial postsutural yellow vittae present (medial vitta very long and narrow); a large yellow spot anterior to notopleural suture; mesopleural stripe reaching midway between anterior margin of notopleuron and anterior *npl.* seta dorsally; scutellum yellow; wing with cells bc and c with a very pale tint, microtrichia in outer corner of cell c only; a narrow dark fuscous costal band overlapping R_{2+3} and ending in a small *tau*-like spot across apex of R_{4+5}; anal streak absent; supernumerary lobe weak and no aggregation of microtrichia around A_1+CuA_2 in male; abdominal terga III–V fulvous with a 'T' pattern consisting of a narrow transverse black band across anterior margin of tergum III and a medium-width medial longitudinal black band over all three terga, broad black anterolateral corners on terga IV and V.

Description: male.

Head: Height 1.26 mm. Frons – length 1.35 times breadth; fulvous with pale fuscous on anteromedial hump; orbital setae weak and black: 1 *s.or.*, 3 *i.or.*; lunule red-brown. Ocellar triangle black. Vertex black. Face fulvous with a pair of medium-sized circular black spots; length 0.5 mm. Genae fulvous, large fuscous subocular spot present; black seta present. Occiput fuscous, yellow along eye margins; occipital row with a large number of small weak setae. Antennae with segments 1 and 2 red-brown, segment 3 mostly fuscous; arista black (red-brown basally); length of segments: 0.08 mm; 0.2 mm; 0.62 mm.

Thorax: Scutum shining black with dark red-brown behind lateral postsutural vittae and red-brown across anterior margin and joining anterior point of medial postsutural yellow vitta. Pleural areas dark fuscous to shining black. Yellow markings as follows: postpronotal lobes; notopleura; mesopleural stripe reaching midway between anterior margin of notopleuron and anterior *npl.* seta dorsally, continuing to katepisternum as a very large transverse spot, anterior margin strongly convex; anatergite (posterior apex black); anterior ¾ katatergite (remainder black); two broad parallel-sided lateral postsutural vittae beginning with a large spot anterior to notopleural suture and ending behind *ia.* seta; a very long and narrow medial longitudinal postsutural vitta beginning posteriorly at level of *prsc.* setae and ending anteriorly at level of centre of postpronotal lobes. Postnotum black, tending dark red-brown centrally. Scutellum yellow except for narrow black basal band. Setae: *sc.* 4; *prsc.* 2; *ia.* 1; *p.sa.* 1; *a.sa.* 1; *mpl.* 1; *npl.* 2; *scp.* 4.

Legs: Femora entirely fulvous except for a large subapical dark fuscous spot on outer surfaces of fore femora; fore tibiae dark fuscous, mid tibiae fuscous tending dark fuscous basally, hind tibiae dark fuscous to black; tarsi with basal segment fulvous and apical four segments fuscous; no distinct apical black spur on mid tibiae.

Wings: Length 5.2 mm; cells bc and c with a very pale tint; microtrichia in outer corner of cell c only; remainder of wings colourless except dark fuscous cell sc, narrow dark fuscous costal band overlapping R_{2+3} and ending in a small *tau*-like spot across apex of R_{4+5} (this pattern is not clear in many of the types because they were reared from host fruit); anal streak absent; no dense aggregation of microtrichia around A_1+CuA_2; supernumerary lobe weak.

Abdomen: Elongate oval; terga free; pecten present on tergum III. Tergum I and sterna I and II wider than long. Tergum I fulvous with moderately broad black lateral margins; tergum II fulvous with a narrow black transverse band across anterior margin joining narrow black lateral margins; terga III–V fulvous with a 'T' pattern consisting of a narrow transverse black band across anterior margin of tergum III and a medium-width medial longitudinal black band over all three terga, broad black anterolateral corners on terga IV and V. A pair of oval fulvous shining spots on tergum V. Posterior lobe of surstylus long, sternum V with a slight concavity on posterior margin. Abdominal sterna black, except sternum II fulvous.

Attractant: No known record.

Female: As for male except pecten of cilia on abdominal tergum III absent. Basal segment of ovipositor red-brown, dorsoventrally flattened and tapering posteriorly in dorsal view; ratio of length of oviscape to length of tergum V, 1:1; apex of piercer broad and needle-shaped.

Distribution: Indonesia (Java); Peninsular Malaysia.

Hosts: Bred from *Hodgsonia macrocarpa* var. *capniocarpa* and *Trichosanthes tricuspidata* (family *Cucurbitaceae*).

Remarks: *Bactrocera (Zeugodacus) hodgsoniae* n.sp., within the *tau* complex, is a distinct species in possessing a long, narrow medial postsutural yellow vitta on the scutum and in having the anal streak in the wing absent. It differs from *Bactrocera (Zeugodacus) tau* (Walker) in having the supernumerary lobe of the male wing very weak.

Bactrocera (Zeugodacus) hululangatiae Drew & Romig, new species

(Fig. 311)

Material examined: Holotype male, **PENINSULAR MALAYSIA**: Pangsoon, Hulu Langat, 26.ix.1990, coll. D.L. Hancock, attracted to cue lure, deposited in BMNH.

Paratypes: **PENINSULAR MALAYSIA**: 1 male, same data as holotype; 1 male, Hulu Langat, Kuala Lumpur, 15.v.1992, coll. D.L. Hancock, attracted to cue lure; 1 male (8.ix.1986), 1 male (3.xi.1986), 1 male (12.i.1987), Transect South Site 8 (TS8), Malaysia, coll. Mahdzir, attracted to cue lure; 1 male, TS8, Malaysia, 20.x.1986, coll. Vignes, attracted to cue lure; 1 male, TS8, Gombak, Malaysia, 9.ii.1987, coll. Bahari, attracted to cue lure; 1 male, Malaysia, sel [Selangor], Hulu Langat, Pangsun, Dec 1988, coll. YHS [Yong Hoi Seng]. 1 paratype in ANIC; 1 paratype in BMNH; 5 paratypes in QDPC; 1 paratype in QMIC.

Diagnosis: A large species; face fulvous with large elongate oval black spots; postpronotal lobes and notopleura yellow; scutum almost entirely black; lateral and medial postsutural yellow vittae present; yellow spot anterior to notopleural suture; mesopleural stripe broad reaching to or almost to postpronotal lobe; scutellum yellow; wing with cells bc and c colourless, microtrichia in outer corner of cell c only; a narrow dark fuscous costal band, broad dark fuscous band along *dm-cu* crossvein, broad dark fuscous anal streak; abdominal terga III–V fulvous with a general black 'T' pattern, broad black anterolateral corners on tergum IV and broad dark fuscous anterolateral corners on tergum V.

Description: male.

Head: Height 1.8 mm. Frons – length 1.19 times breadth; red-brown with fulvous along lateral margins; orbital setae black: 1 *s.or.*, 3 *i.or.* (upper pair reduced and pale); lunule fuscous. Ocellar triangle black. Vertex varies from red-brown to fuscous. Face fulvous with a pair of large elongate oval black spots reaching to oral margin; length 0.63 mm. Genae fulvous, small pale subocular spot present; black seta present. Occiput fuscous, fulvous along eye margins; occipital row with a large number of small pale setae. Antennae with segments 1 and 2 red-brown, segment 3 red-brown with fuscous on apex and outer surface; arista black (red-brown basally); length of segments: 0.12 mm; 0.31 mm; 0.86 mm.

Thorax: Scutum entirely black except dark red-brown below lateral postsutural vittae and inside notopleura. Pleural areas black except red-brown below postpronotal lobes and a small area above mid coxae. Yellow markings as follows: postpronotal lobes; notopleura; broad mesopleural stripe almost reaching to or reaching to postpronotal lobe dorsally, continuing to katepisternum as a large transverse spot, anterior margin straight; anatergite (posterior apex black); anterior $^{3}/_{5}$ katatergite (remainder black); two broad parallel-sided lateral postsutural vittae beginning with a spot anterior to notopleural suture and ending behind *ia.* seta; a medium-width medial longitudinal postsutural vitta beginning just anterior of *prsc.* setae and ending at a point just anterior of notopleural suture. Postnotum black. Scutellum yellow except for narrow black basal band. Setae: *sc.* 2; *prsc.* 2; *ia.* 1; *p.sa.* 1; *a.sa.* 1; *mpl.* 1; *npl.* 2; *scp.* 4.

Legs: Femora fulvous with dark fuscous to black patterns on outer apical $^{2}/_{3}$ of fore femora and around apical $^{1}/_{3}$ of mid and hind femora; fore and mid tibiae fuscous basally to pale fuscous apically, hind tibiae dark fuscous; tarsi with basal segment fulvous and apical four segments fuscous; mid tibiae each with an apical black spur.

Wings: Length 7.7 mm; cells bc and c colourless; microtrichia in outer corner of cell c only; remainder of wings colourless except dark fuscous cell sc, narrow dark fuscous costal band very slightly overlapping R_{2+3} and widening slightly across apex of R_{4+5}, broad dark fuscous band enclosing *dm-cu* crossvein and extending around hind margin as a very pale tint, broad dark fuscous anal streak; dense aggregation of microtrichia around A_1+CuA_2; supernumerary lobe moderately strong and rounded.

Abdomen: Oval; terga free; pecten present on tergum III. Tergum I and sterna I and II wider than long. Tergum I red-brown with fuscous to dark fuscous centrally and moderately broad black lateral margins; tergum II fulvous with a moderately broad transverse black band subanteriorly which may or may not reach black anterolateral corners; terga III–V fulvous with a 'T' pattern consisting of a moderately broad black transverse band across anterior margin of tergum III and extending around lateral margins and a broad medial longitudinal black band over all three terga, broad black anterolateral corners on tergum IV and broad dark fuscous to black anterolateral corners on tergum V. A pair of oval fulvous shining spots on tergum V. Posterior lobe of surstylus long, sternum V with a slight concavity on posterior margin.

Attractant: Cue lure.

Female: No known record.

Distribution: Peninsular Malaysia (Hulu Langat; Gombak).

Hosts: No known record.

Remarks: *Bactrocera (Zeugodacus) hululangatiae* n.sp. belongs to the group of *Bactrocera* placed in subgenus

Zeugodacus that possess two scutellar setae. In all other characters, they are consistent with *Zeugodacus*. It is similar to *Bactrocera (Zeugodacus) elegantula* (Hardy) in possessing two scutellar setae, lateral and medial postsutural yellow vittae, a black scutum and infuscation along *dm-cu* crossvein. It differs from *B. elegantula* in having a broader mesopleural stripe, a yellow spot anterior to notopleural suture (not a yellow triangle) and dark fuscous to black apices on all femora. *Bactrocera hululangatiae* is also similar to *Bactrocera (Zeugodacus) neoelegantula* White and some specimens of *Bactrocera (Zeugodacus) synnephes* (Hendel) with two *sc.* setae, but differs from the latter in having the mesopleural stripe reaching the postpronotal lobe dorsally and the costal band only slightly widened across apex of R_{4+5} and from *B. neoelegantula* in having a narrow basal band on the scutellum, fore femora with large apical dark spots (not entirely dark), abdominal terga IV and V with anterolateral corners dark and abdominal sternum V of male with a shallow concavity on posterior margin. *B. hululangatiae* n.sp. is also similar to *B. aithonata* n.sp. but differs in possessing yellow notopleura (not red-brown as in *B. aithonata*).

Bactrocera (Zeugodacus) indentus (Hardy)

(Fig. 312)

Dacus (Zeugodacus) indentus Hardy, 1974: 49–50. Holotype female in BBM.
Bactrocera (Zeugodacus) indentus – Norrbom *et al.*, 1998: 103.

Type material examined: Holotype female, **PHILIPPINES**: **LUZON ISLAND** – with the following four labels – 'Los Banos, Philippine Ids, ix.1916'; 'F.X. Williams Collector'; 'Holotype female, *Dacus (Zeugodacus) indentus*, D.E. Hardy'; '*Dacus (Zeugodacus) indentus* Hardy, Det. 1969, D.E. Hardy'. Received on loan from BBM.

Diagnosis: A large species; face with a pair of medium-sized oval black spots; postpronotal lobes and notopleura yellow; scutum base colour red-brown; lateral and medial postsutural yellow vittae present; a yellow spot anterior to notopleural suture in front of lateral postsutural vittae; mesopleural stripe reaching to or almost to postpronotal lobe dorsally; scutellum yellow; wing with cells bc and c colourless, microtrichia in outer corner of cell c only; a narrow fuscous costal band confluent with R_{2+3} and a narrow pale fuscous to fuscous anal streak; abdominal terga III–V fulvous with a dark 'T' pattern consisting of a narrow transverse dark fuscous band across anterior margin of tergum III, a moderately broad dark fuscous medial longitudinal band over all three terga, moderately broad dark fuscous lateral margins on terga IV and V.

Description: female.

Head: Height 1.7 mm. Frons – length 1.37 times breadth; red-brown with fuscous around lower *i.or.* setae; anteromedial hump with a number of short red-brown hairs; orbital setae: 1 *s.or.* (strong and black), 2 *i.or.* (weak and red-brown); lunule red-brown. Ocellar triangle black. Vertex red-brown. Face fulvous with a pair of medium-sized oval black spots; length 0.98 mm. Genae fulvous; dark seta present. Occiput red-brown, fulvous along eye margins; occipital row with 5–6 dark setae. Antennae with segments 1 and 2 fuscous, segment 3 red-brown with fuscous on apex and outer surface; length of segments: 0.18 mm; 0.31 mm; 0.84 mm.

Thorax: Scutum red-brown with a complex pattern of pale fuscous, fuscous and black markings. Pleural areas pale fuscous to fuscous except fulvous below postpronotal lobes. Yellow markings as follows: postpronotal lobes; notopleura; mesopleural stripe reaching to or almost to postpronotal lobe dorsally, continuing to katepisternum as a very large transverse spot, anterior margin slightly convex; anatergite (posterior apex dark fuscous); anterior ⅔ katatergite (remainder dark fuscous); two broad lateral postsutural vittae beginning with a spot anterior to notopleural suture and narrowing slightly posteriorly to end before *ia.* seta; a broad medial longitudinal postsutural vitta narrowing to a point anteriorly to end anterior to notopleural suture. Postnotum dark fuscous. Scutellum yellow except for narrow red-brown basal band. Setae: *sc.* 4; *prsc.* 2; *ia.* 1; *p.sa.* 1; *a.sa.* 1; *mpl.* 1; *npl.* 2; *scp.* 4.

Legs: Fore femora fulvous tending red-brown on apical ⅓, mid femora entirely fulvous, hind femora fulvous tending red-brown on apical ¼; fore tibiae fuscous tending pale fuscous on apical ½, mid tibiae pale fuscous, hind tibiae fuscous; fore and hind tarsi with basal segment fulvous and apical four segments pale fuscous, mid tarsi with basal two segments fulvous and apical three segments pale fuscous; mid tibiae each with an apical black spur.

Wings: Length 7.0 mm; cells bc and c colourless; microtrichia in outer corner of cell c only; remainder of wings colourless except fuscous cell sc, narrow fuscous costal band confluent with R_{2+3} and remaining very narrow around margin of wing to end just beyond extremity of R_{4+5}, narrow pale fuscous to fuscous anal streak; no dense aggregation of microtrichia around A_1+CuA_2; supernumerary lobe weak.

Abdomen: Elongate oval; terga free; no pecten on tergum III. Tergum I and sterna I and II wider than long. Tergum I fulvous with narrow dark fuscous lateral margins; tergum II fulvous with a narrow dark fuscous band across anterior margin turning around to lateral margins; terga III–V fulvous with a dark 'T' pattern consisting of a narrow transverse dark fuscous band across anterior margin widening over lateral margins and a moderately broad medial longitudinal dark fuscous band over all three terga and moderately broad dark fuscous lateral margins on terga IV and V. A pair of oval fulvous shining spots on tergum V. Basal segment of ovipositor red-brown, dorsoventrally compressed and tapering posteriorly in dorsal view; ratio of length of oviscape to length of tergum V, 1.5:1; apex of piercer trilobed.

Attractant: No known record.

Male: No known record.

Distribution:

Type locality: Philippines (Luzon I., Los Banos).

Hosts: No known record.

Remarks: *Bactrocera (Zeugodacus) indentus* (Hardy) is a distinct species which belongs to a group that have four *sc.* setae, scutum base colour red-brown with medial and lateral postsutural yellow vittae and wings colourless except for costal band and anal streak. It is distinct from the other species in this group in having the mesopleural stripe reaching to the postpronotal lobe dorsally, the costal band confluent with R_{2+3} and not widening beyond apex of this vein.

Bactrocera (Zeugodacus) iriomotiae Drew & Romig, new species

(Fig. 313)

Material examined: Holotype male, **JAPAN**: Iriomote I., Okinawa, Japan, 4.viii.1984, coll. T. Teruya, attracted to methyl eugenol, deposited in BMNH.

Paratypes: **JAPAN**: 7 males, same data as holotype; 3 males, Japan, Hirara, Miyako I., 1.viii.1990, coll. Tsuguo Kohama, attracted to methyl eugenol plus cue lure mixture; 1 male, Japan, Kuroshima I., 23.vii.1991, coll. Tsuguo Kohama, attracted to methyl eugenol plus cue lure mixture; 1 male, Japan, Hira Kubo, Ishigaki I., 25.vii.1990, coll. Tsuguo Kohama, attracted to methyl eugenol; 1 male, Japan Gusukube, Miyako I., 17.viii.1990, coll. Tsuguo Kohama, attracted to methyl eugenol plus cue lure mixture. 2 paratypes in ANIC, 2 paratypes in BMNH, 7 paratypes in QDPC, 2 paratypes in QMIC.

Diagnosis: A medium-sized species; face varies from entirely fulvous to fuscous to dark fuscous centrally with red-brown laterally and black along the ptilinal sutures; postpronotal lobes and notopleura yellow; scutum black with red-brown markings of variable size and shape; lateral and medial postsutural yellow vittae present; a yellow spot present anterior to notopleural suture; mesopleural stripe reaching anterior *npl.* seta dorsally; scutellum yellow with a small black spot between apical setae; wing with cells bc and c with a pale tint, microtrichia in outer corner of cell c only; a narrow dark fuscous costal band, a broad dark fuscous anal streak; abdominal terga III–V orange-brown with narrow transverse black markings on anterolateral corners of terga III and IV and a narrow- to medium-width medial longitudinal black band over all three terga.

Description: male.

Head: Height 1.4 mm. Frons – length 1.6 times breadth; red-brown to black centrally with fulvous lateral margins; orbital setae black: 1 *s.or.*, 2 *i.or.*; lunule fuscous. Ocellar triangle black. Vertex fulvous. Face varies from entirely fulvous to fuscous to dark fuscous centrally, red-brown laterally and with distinct narrow black lines along the ptilinal sutures; length 0.42 mm. Genae entirely dark fuscous; small black seta present. Occiput red-brown, fulvous along eye margins; occipital row with a large number of small pale setae. Antennae with segments 1 and 2 red-brown, segment 3 red-brown with pale fuscous on apex and outer surface; length of segments: 0.1 mm; 0.24 mm; 0.64 mm.

Thorax: Scutum black with red-brown below and behind lateral postsutural vittae, along posterior margin, around notopleural suture, across the anteromedial area. Pleural areas entirely red-brown except for a narrow black margin along anterior of the mesopleural stripe and a small area on the katepisternum. Yellow markings as follows: postpronotal lobes; notopleura; mesopleural stripe reaching to anterior *npl.* seta dorsally, continuing to katepisternum as a large transverse spot, anterior margin convex; anatergite (posterior apex black); anterior ¾ katatergite (remainder black); two broad parallel-sided lateral postsutural vittae beginning with a spot anterior to notopleural suture and ending behind *ia.* seta; a medial postsutural vitta which is very broad at level of *prsc.* setae and narrowing to a point at level of notopleural suture; a narrow lateral band joining the postpronotal lobe and the notopleuron. Postnotum black tending red-brown centrally. Scutellum yellow except for narrow red-brown basal band and a small black spot between apical setae. Setae: *sc.* 4; *prsc.* 2; *ia.* 1; *p.sa.* 1; *a.sa.* 1; *mpl.* 1; *npl.* 2; *scp.* 4.

Legs: All segments entirely fulvous except for a small fuscous spot on inner apical surfaces of hind tibiae; mid tibiae each with an apical black spur.

Wings: Length 5.3 mm; cells bc and c with a pale tint; microtrichia in outer corner of cell c only; remainder of wings colourless except dark fuscous cell sc, narrow dark fuscous costal band confluent with R_{2+3} and widening slightly around apex of wing, a broad dark fuscous anal streak; a dense aggregation of microtrichia around A_1+CuA_2; supernumerary lobe weak.

Abdomen: Oval; terga free; pecten present on tergum III. Tergum I and sterna I and II wider than long. Tergum I entirely orange-brown; tergum II orange-brown anteriorly to fulvous posteriorly; terga III–V orange-brown except for small black transverse spots on anterolateral corners of terga IV and V and a narrow- to medium-width medial longitudinal black band over all three terga. A pair of oval orange-brown shining spots on tergum V. Posterior lobe of surstylus long, sternum V with a slight concavity on posterior margin. Abdominal sterna entirely pale orange-brown.

Attractant: Methyl eugenol.

Female: No known record.

Distribution: Japan (Iriomote I., Ishigaki I., Kuroshima I., Miyako I.).

Hosts: No known record.

Remarks: *Bactrocera (Zeugodacus) iriomotiae* n.sp. is similar to *Bactrocera (Zeugodacus) abnormis* (Hardy), *Bactrocera (Zeugodacus) connexa* (Hardy) and *Bactrocera (Zeugodacus) pemalangiae* n.sp. in possessing four *sc.* setae, lateral and medial postsutural yellow vittae, a black scutum, cells bc and c colourless or with a very pale tint, costal band confluent with R_{2+3} and abdominal sterna pale (generally fulvous to orange-brown). It differs from *B. abnormis* and *B. pemalangiae* in having a lateral yellow band joining the postpronotal lobe and notopleuron and from *B. connexa* in having the face fuscous to dark fuscous with black margins and abdominal terga III–V orange-brown with an incomplete 'T' pattern.

Bactrocera (Zeugodacus) ishigakiensis (Shiraki)

(Fig. 314)

Zeugodacus ishigakiensis Shiraki, 1968: 20. Holotype male in USNM.

Bactrocera (Zeugodacus) ishigakiensis – Norrbom *et al.*, 1998: 103.

Type material examined: Holotype male, **JAPAN: RYUKYU ISLANDS** – with the following two labels - 'RYUKYU IS., 11.xii.1953, T. Shiraki, Ishigaki Is., Kabira'; '*Zeugodacus ishigakiensis* sp. nov., Det. T. Shiraki'. Received on loan from USNM.

Non-type material examined: JAPAN: RYUKYU ISLANDS – 1 male, Ishigaki I., Okinawa, Japan, 29.vii.1987, coll. H. Nakaima, cue lure trap; 1 male, Ishigaki I., Japan, 29.vii.1989, coll. M. Kaneda. Specimens in QDPC.

Diagnosis: A large species; face with a pair of medium-sized circular black spots; postpronotal lobes and notopleura yellow; scutum black; lateral and medial postsutural yellow vittae present; a yellow spot anterior to notopleural suture in front of lateral postsutural vittae; mesopleural stripe reaching midway between anterior margin of notopleuron and anterior *npl.* seta dorsally; scutellum yellow with an apical black spot; wing with cells bc and c with a pale tint, microtrichia in outer corner of cell c only; a narrow dark fuscous costal band slightly overlapping R_{2+3} and expanding across wing apex to M, dark fuscous surrounding *r-m* and *dm-cu* crossveins; a broad dark fuscous anal streak; abdominal tergum III red-brown with a distinct black 'T' pattern, terga IV and V red-brown with a 'T' pattern consisting of a narrow transverse dark fuscous band across anterior margin of tergum IV and a medial longitudinal dark fuscous to black band over both terga, dark fuscous to black anterolateral corners on terga IV and V.

Description: male.

Head: Height 1.51 mm. Frons – length 1.06 times breadth; red-brown with fuscous around orbital setae and on anteromedial hump; latter with a small number of short dark hairs; orbital setae black: 1 *s.or.*, 3 *i.or.*; lunule dark fuscous. Ocellar triangle black. Vertex black. Face red-brown with a pair of medium-sized circular black spots; length 0.82 mm. Genae red-brown, small pale fuscous subocular spot present; black seta present. Occiput fuscous with dorsolateral dark fuscous spots; occipital row with a large number of weak setae. Antennae with segments 1 and 2 red-brown; segment 3 absent; length of segments: 0.19 mm; 0.29 mm.

Thorax: Scutum black with dark fuscous posterolateral to lateral postsutural vittae and inside notopleura. Pleural areas black except dark red-brown below postpronotal lobes. Yellow markings as follows: postpronotal lobes; notopleura; medium-sized mesopleural stripe reaching midway between anterior margin of notopleuron and anterior *npl.* seta dorsally, continuing to katepisternum as a small transverse spot, anterior margin slightly convex; anatergite (posterior apex black); anterior ⅔ katatergite (remainder black); two lateral postsutural vittae of medium-width, beginning with a spot anterior to notopleural suture, subparallel and ending at *ia.* seta; a narrow short medial longitudinal postsutural vitta ending anteriorly at a point behind notopleural suture. Postnotum black. Scutellum yellow except for narrow black basal band and an apical black spot surrounding apical setae. Setae: *sc.* 4; *prsc.* 2; *ia.* 1; *p.sa.* 1; *a.sa.* 1; *mpl.* 1; *npl.* 2; *scp.* 4.

Legs: Femora entirely fulvous except for a small elongate fuscous subapical spot on outer surface of fore femora and a small area of dark fuscous around apex of hind femora; fore and mid tibiae fulvous, hind tibiae dark fuscous basally tending fuscous apically; tarsi with basal segment fulvous and apical four segments pale fuscous; mid tibiae each with an apical black spur.

Wings: Length 7.08 mm; cells bc and c with pale tint; microtrichia in outer corner of cell c only; remainder of wings colourless except dark fuscous cell sc, narrow dark fuscous costal band slightly overlapping R_{2+3} and expanding across apex of wing where it fades from dark fuscous to pale fuscous, dark fuscous bands surrounding *r-m* and *dm-cu* crossveins, a broad dark fuscous anal streak; a dense aggregation of microtrichia around A_1+CuA_2; supernumerary lobe strong and rounded.

Abdomen: Oval; terga free; pecten present on tergum III. Tergum I and sterna I and II wider than long. Tergum I red-brown with dark fuscous across anterior margin and along lateral margins; tergum II red-brown except for a narrow transverse black band across anterior margin but not reaching lateral margins and small dark fuscous anterolateral corners; tergum III red-brown with a narrow transverse black band across anterior margin and widening over lateral margins and a medial longitudinal black band over anterior ½; terga IV–V red-brown with a 'T' pattern consisting of a narrow transverse dark fuscous band across anterior margin of tergum IV widening to cover lateral margins and a medium-width medial longitudinal band which is dark fuscous on tergum IV and black on tergum V, dark fuscous to black anterolateral corners of tergum V. A pair of oval fuscous shining spots on tergum V. Posterior lobe of surstylus long, sternum V with a slight concavity on posterior margin.

Attractant: Cue lure.

Female: No known record.

Distribution:

Type locality: Japan (Ryukyu Is, Ishigaki I.).

Hosts: No known record.

Remarks: *Bactrocera (Zeugodacus) ishigakiensis* (Shiraki) is similar to *Bactrocera (Zeugodacus) calumniata* (Hardy) and *Bactrocera (Zeugodacus) synnephes* (Hendel) in possessing four *sc.* setae, a black scutum with medial and lateral postsutural yellow vittae, wings with infuscation in addition to costal band and anal streak and costal band with a broad expansion across the wing apex. *Bactrocera ishigakiensis* is distinct from these species in having a black apical spot on the scutellum and no fuscous colouration around the hind margin of the wing.

Bactrocera (Zeugodacus) isolata (Hardy)

(Fig. 315)

Dacus (Zeugodacus) isolatus Hardy, 1973: 61. Holotype male in KUB.

Bactrocera (Zeugodacus) isolata – Norrbom *et al.*, 1998: 103.

Type material examined: Holotype male, **THAILAND**: with the following label data – Nan, collected on *Citrus nobilis*, 10.vii.1963, coll. A. Vattanatungum. Received on loan from KUB.

Non-type material examined: THAILAND: 1 male, 'Phang Nga Prov., Raman Waterfall, 50 m, 16.v.2003/9'; '8.27.05N, 98.26.52E, P. Schwendinger leg, evergreen forest/9'; 'attracted to cue lure'. Received on loan from MHNG. 1 male, Bangkok, Khao Kheaw, A. Pakphli, Nakhonnayok, 10.iv.1992, coll. Montree, attracted to cue lure; 1 male, Songkhla, Ko Hong Hill, 21.vi.1993, coll. Anuchit and Sieng, attracted to cue lure. **VIETNAM: Southern Vietnam** – 1 male, Tien Giang (P), Cho Gao (V), 21.iii.2000, attracted to cue lure, southern Vietnam, S006; 1 male, Tien Giang (P), Cai Be (D), My Thim (V), 3.ii.2000, attracted to cue lure, southern Vietnam, S013; 1 male, Dac lac (P), Ku mi nga (D), Quang phu (V), 23.iii.2002, attracted to cue lure, VS09; 1 male (2.iii.2002), 2 males (30.iii.2002), 1 male (13.iv.2002), 1 male (11.v.2002), 2 males (22.vi.2002), 1 male (6.vii.2002), 4 males (20.vii.2002), 3 males (3.viii.2002), Dong nai (P), Long khanh (D), Nhan nghia (V), attracted to cue lure, southern Vietnam, VS06. Specimens in QDPC except where specified.

Diagnosis: A medium-sized species; face with a pair of large oval black spots; postpronotal lobes and notopleura yellow; scutum base colour black; lateral and medial postsutural yellow vittae present; a yellow triangle along anterior margin of notopleural suture with base on notopleuron; mesopleural stripe reaching to anterior *npl.* seta dorsally; scutellum yellow; wing with cells bc and c colourless, microtrichia in outer corner of cell c only; a narrow fuscous costal band slightly overlapping R_{2+3} and forming a small spot around apex of R_{4+5}; a moderately broad fuscous anal streak; abdominal terga III–V fulvous with a 'T' pattern consisting of a narrow transverse black band across anterior margin of tergum III, a narrow medial longitudinal black band over all three terga, a broad lateral black pattern on tergum IV that encloses a circular fulvous area, anterolateral corners of tergum V black.

Description: male.

Head: Height 1.5 mm. Frons – length 1.35 times breadth; red-brown with fuscous around bases of lower *i.or.* setae; orbital setae black: 1 *s.or.*, 2 *i.or.*; lunule red-brown. Ocellar triangle black. Vertex with two lateral black spots. Face fulvous with a pair of large oval black spots; length 0.6 mm. Genae fulvous, pale fuscous subocular spot present; black seta present. Occiput red-brown, fulvous along eye margins; occipital row with a few weak pale setae. Antennae with segments 1 and 2 red-brown, segment 3 red-brown with fuscous on apex and outer surface; length of segments: 0.12 mm; 0.31 mm; 0.74 mm.

Thorax: Scutum black with dark red-brown posterior to medial longitudinal yellow vitta, and red-brown below lateral postsutural vittae, laterally between postpronotal lobe and notopleuron and across anterior margin leading back to anterior point of medial longitudinal postsutural yellow vitta. Pleural areas dark fuscous to black except red-brown below postpronotal lobes. Yellow markings as follows: postpronotal lobes; notopleura; mesopleural stripe reaching to anterior *npl.* seta dorsally, continuing to katepisternum as a large elongate transverse spot, anterior margin slightly convex; anatergite (posterior apex black); anterior ⅔ katatergite (remainder black); two moderately broad parallel-sided lateral postsutural vittae ending behind *ia.* seta; a large triangle along anterior margin of notopleural suture; a broad medial longitudinal postsutural vitta broadly rounded posteriorly at level of *prsc.* setae and ending in a point anteriorly at level of notopleural suture. Postnotum black. Scutellum yellow with a narrow black basal band. Setae: *sc.* 4; *prsc.* 2; *ia.* 1; *p.sa.* 1; *a.sa.* 1; *mpl.* 1; *npl.* 2; *scp.* 2.

Legs: All femora pale fulvous tending fulvous on apical ¼; fore tibiae pale fuscous, mid tibiae fulvous tending pale fuscous basally, hind tibiae fuscous; all tarsi with basal segment fulvous and apical four segments pale fuscous; mid tibiae with an apical black spur.

Wings: Length 5.7 mm; cells bc and c colourless; microtrichia in outer corner of cell c only; remainder of wings colourless except for fuscous cell sc, narrow fuscous costal band slightly overlapping R_{2+3} as a pale tint and forming a small spot around apex of R_{4+5}, a moderately broad fuscous anal streak; a dense aggregation of microtrichia around A_1+CuA_2; supernumerary lobe of medium development.

Abdomen: Elongate oval; terga free; pecten present on tergum III. Tergum I and sterna I and II wider than long. Tergum I fulvous with narrow dark fuscous lateral margins; tergum II fulvous with a narrow transverse black band anteriorly that merges into black anterolateral corners; terga III to V fulvous with a 'T' pattern consisting of a narrow transverse black band across anterior margin which widens to form narrow lateral black margins and a narrow medial longitudinal black band over all three terga; a broad lateral black pattern on tergum IV that encloses a circular fulvous area and anterolateral corners of tergum V black. A pair of oval fulvous shining spots on tergum V. Posterior lobe of surstylus short, sternum V with a slight concavity on posterior margin.

Attractant: Cue lure.

Female: As for male except no dense aggregation of microtrichia around A_1+CuA_2; supernumerary lobe weak; pecten of cilia on abdominal tergum III absent; basal segment of ovipositor red-brown, dorsoventrally compressed and tapering posteriorly in dorsal view; ratio of length of oviscape to length of tergum V, 1:1; apex of piercer needle-shaped.

Distribution:

Type locality: Thailand (Nan).

Previously recorded: Type locality and the following locations – China; Laos; Thailand (other locations listed under 'Non-type material examined').

New records: Southern Vietnam (Dac Lac, Dong Nai, Tien Giang Provinces).

Hosts: Recorded from species of the genera *Capparis* and *Maerua* (family *Capparaceae*) (Allwood *et al.*, 1999).

Remarks: *Bactrocera (Zeugodacus) isolata* (Hardy) belongs to the *tau* complex of species with four *sc.* setae, a black scutum with medial and lateral postsutural yellow vittae, wings colourless except for costal band and anal streak, costal cells colourless or with a pale tint, costal band overlapping R_{2+3} and expanded into a spot apically. It is unique in having a fulvous spot on each side of abdominal tergum IV formed by a recurved black band. *Bactrocera (Zeugodacus) juxtuncinata* n.sp. and *Bactrocera (Zeugodacus) uncinata* n.sp. have similar abdominal colour patterns but *B. juxtuncinata* does not have a spot in the apex of the costal band and *B. uncinata* has a basic red-brown scutum. The measurements included above are from specimens collected in Thailand, not the type.

Bactrocera (Zeugodacus) javadica Mahmood
(Fig. 316)

Bactrocera (Zeugodacus) javadica Mahmood, 1999: 224. Holotype male in BMNH.

Type material examined: Holotype male, **INDONESIA: JAVA** – with the following six labels – 'Larva attacks pods and seeds of *Phaseolus lunulatus*'; 'Lalerkatjang, Katak $^{31}/_{10}$.22'; 'Java 6, Buitenzorg, P.V. de Goot'; 'Pres by Imp. Bur. Ent, Brit. Mus., 1924-291'; 'Type'; 'Holotype, *Bactrocera (Z.) javadica*, K. Mahmood 1999'. Studied at BMNH.
Paratype: **INDONESIA: JAVA** – One male with the first five labels as for the Holotype above and the last label as follows – 'Paratype, *Bactrocera (Z.) javadica*, K. Mahmood 1999.' Specimens studied at BMNH.
Diagnosis: A medium-sized species; face fulvous with a pair of medium-sized circular black spots; postpronotal lobes and notopleura yellow; scutum black with extensive red-brown patterns; lateral and medial postsutural yellow vittae present; yellow spot anterior to notopleural suture; mesopleural stripe reaching midway between anterior margin of notopleuron and anterior *npl.* seta dorsally; scutellum yellow; legs with all femora entirely fulvous; wing with cells bc and c colourless, microtrichia in outer corner of cell c only; a narrow fuscous costal band overlapping R_{2+3} and expanding into a *tau*-like spot in wing apex; a moderately broad fuscous anal streak; abdominal terga III–V fulvous with a black 'T' pattern consisting of a narrow transverse band across anterior margin and a medium to broad medial longitudinal band over all three terga and narrow dark fuscous margins on all three terga.
Description: male.
Head: Frons – fulvous with fuscous on anteromedial hump; orbital setae black: 1 *s.or.*, 2 *i.or.*; lunule fuscous. Ocellar triangle black. Vertex fulvous. Face fulvous with a pair of medium-sized circular black spots. Occiput fulvous centrally with red-brown laterally. Antennae with segments 1 and 2 red-brown, segment 3 red-brown with pale fuscous on outer surfaces.
Thorax: Scutum black with red-brown below and behind lateral postsutural vittae, inside notopleura, across full width of anterior margin from anterior point of medial postsutural vitta, and dark fuscous across posterior margin. Pleural areas black except red-brown below postpronotal lobes. Yellow markings as follows: postpronotal lobes; notopleura; mesopleural stripe reaching midway between anterior margin of notopleuron and anterior *npl.* seta dorsally, continuing to katepisternum as a transverse spot, anterior margin slightly convex; two broad parallel-sided lateral postsutural vittae beginning with a spot anterior to notopleural suture and ending at *ia.* seta; a broad medial longitudinal postsutural vitta narrowing slightly anteriorly to end at approximately the level of the anterior *npl.* setae. Scutellum entirely yellow with a narrow black basal band. Setae: *sc.* 4; *prsc.* 2 (absent in Holotype); *ia.* 1; *p.sa.* 1; *a.sa.* 1; *mpl.* 1; *npl.* 2; *scp.* 4.
Legs: All femora entirely fulvous without dark markings; fore and mid tibiae with pale fuscous basally to fulvous apically, hind tibiae dark fuscous; all tarsi with basal segment fulvous and apical four segments pale fuscous.
Wings: Length 5.6 mm; cells bc and c colourless; microtrichia in outer corner of cell c only; remainder of wings colourless except dark fuscous cell sc, narrow fuscous costal band overlapping R_{2+3} and expanding into a *tau*-like spot in apex of wing, a moderately broad fuscous anal streak; dense aggregation of microtrichia around A_1+CuA_2; supernumerary lobe of medium development.
Abdomen: Oval; terga free; pecten present on tergum III. Tergum I fulvous with dark red-brown across anterior margin and covering anterolateral corners; tergum II fulvous with a narrow transverse red-brown to dark red-brown band across anterior margin; terga III–V fulvous with a 'T' pattern consisting of a narrow transverse black band across anterior margin which widens over lateral margins, a medium to broad medial longitudinal black band over all three terga, narrow fuscous to dark fuscous lateral margins on terga IV and V. Posterior lobe of surstylus long, sternum V with a slight concavity on posterior margin. A pair of oval fulvous shining spots on tergum V. Abdominal sterna dark, except sternum II pale.
Attractant: No known record.
Female: No known record.
Distribution:
Type locality: Indonesia (Java, Bogor, formerly Buitenzorg).
Hosts: No known record.
Remarks: *Bactrocera (Zeugodacus) javadica* Mahmood appears to be very similar to *Bactrocera (Zeugodacus) tau* (Walker) and belongs to the *tau* complex, but differs in possessing a weak supernumerary lobe in the male wing. The abdominal colour patterns used by Mahmood to distinguish this species, i.e. lateral margins of terga III and IV with black, also occur in *B. tau*. From a study of large numbers of *B. tau* collected across South-East Asia, it is clear that the lateral dark markings on abdominal terga IV and V are extremely variable.

Bactrocera (Zeugodacus) juxtuncinata Drew & Romig, new species
(Fig. 317)

Material examined: Holotype male, **PHILIPPINES: MINDANAO ISLAND** – Polomolok, S. Cotabato, 20. vii.1983, coll. E.T. Figueroa, attracted to cue lure, deposited in BMNH.
Diagnosis: A medium-sized species; face with a pair of large elongate oval black spots; postpronotal lobes and notopleura yellow; scutum mostly black with areas of red-brown; lateral and medial postsutural yellow vittae present; a yellow triangle along anterior margin of notopleural suture; mesopleural stripe of medium-width reaching midway between anterior margin of notopleuron and anterior *npl.* seta; scutellum yellow; wing with cells bc and c colourless, microtrichia in outer corner of cell c only; a narrow fuscous costal band widening slightly across apex of wing and a narrow fuscous anal streak; abdominal terga III–V fulvous with a distinct black

'T' pattern, narrow black margins on terga IV and V and a recurved black marking on tergum IV from hind margin to anterior margin and leaving a fulvous area between this band and the narrow lateral black margin.

Description: male.

Head: Height 1.52 mm. Frons – length 1.48 times breadth; fulvous with fuscous centrally from anteromedial hump to anterior margin; orbital setae black: 1 *s.or.*, 2 *i.or.*; lunule dark fuscous. Ocellar triangle black. Vertex dark fuscous to black. Face fulvous with a pair of large elongate oval black spots, length 0.49 mm. Genae fulvous, large dark fuscous subocular spot present. Occiput fuscous, fulvous along eye margins; occipital row with a large number of small pale setae. Antennae with segments 1 and 2 red-brown; segment 3 missing; length of segments: 0.10 mm; 0.24 mm.

Thorax: Scutum mostly black with red-brown below and behind lateral postsutural vittae, around notopleural suture, centrally on posterior and anterior margins and inside postpronotal lobes. Pleural areas dark fuscous to black except red-brown below postpronotal lobes and on ventral area of katepisternum. Yellow markings as follows: postpronotal lobes; notopleura; medium-sized mesopleural stripe, reaching midway between anterior margin of notopleuron and anterior *npl.* seta dorsally, extending to katepisternum as a large spot, anterior margin slightly convex; anatergite (posterior apex black); anterior ⅔ katatergite (remainder black); a yellow triangle along anterior margin of notopleural suture with base on notopleuron; two broad parallel-sided lateral postsutural yellow vittae ending at *ia.* seta; a long and broad medial postsutural yellow vitta beginning just anterior of *prsc.* setae and ending in a point level with the centre of the postpronotal lobes. Postnotum mostly black with a small area of dark red-brown centrally. Scutellum yellow except for narrow black basal band. Setae: *sc.* 4; *prsc.* 2; *ia.* 1; *p.sa.* 1; *a.sa.* 1; *mpl.* 1; *npl.* 2; *scp.* 4.

Legs: Femora entirely fulvous; fore and hind tibiae dark fuscous, mid tibiae dark fuscous basally to fuscous apically; tarsi with all segments fulvous; mid tibiae without a distinct apical black spur.

Wings: Length 5.62 mm; cells bc and c colourless; microtrichia in outer corner of cell c only; remainder of wings colourless except fuscous cell sc, narrow fuscous costal band just overlapping R_{2+3} and widening across apex of wing to apex of vein M, a narrow fuscous anal streak; a dense aggregation of microtrichia around A_1+CuA_2; supernumerary lobe weak.

Abdomen: Oval; terga free; pecten present on tergum III. Tergum I and sterna I and II wider than long. Tergum I dark fuscous centrally and black laterally; tergum II fulvous with dark fuscous anterolateral corners and a narrow black transverse band across anterior margin but not reaching lateral margins; terga III–V fulvous with a 'T' pattern consisting of a narrow transverse black band across anterior margin of tergum III and extending over lateral margins and a broad medial longitudinal black band over all three terga, narrow lateral black margins on terga IV and V and a recurved black marking on tergum IV extending from hind margin to anterior margin and leaving a fulvous area between this band and the lateral black margin. A pair of oval fulvous shining spots on tergum V. Posterior lobe of surstylus long, sternum V with a slight concavity on posterior margin. Abdominal sterna dark.

Attractant: Cue lure.

Female: No known record.

Distribution: Philippines (Mindanao I., South Cotabato).

Hosts: No known record.

Remarks: *Bactrocera (Zeugodacus) juxtuncinata* n.sp. is close to *Bactrocera (Zeugodacus) isolata* (Hardy), *Bactrocera (Zeugodacus) fereuncinata* n.sp. and *Bactrocera (Zeugodacus) uncinata* n.sp. in the abdominal and general wing colour patterns. It differs from these species in having broad parallel-sided lateral postsutural vittae ending at *ia.* seta, a very long medial postsutural yellow vitta extending anteriorly to a level approximately the centre of the postpronotal lobes and abdomen shape oval (not elongate oval). In addition, it differs from *B. isolata* in having the mesopleural stripe narrower, a short wide yellow spot on the katepisternum, the apex of the costal band widening across wing apex and the recurved black band on tergum IV not enclosing a fulvous spot, from *B. fereuncinata* and *B. uncinata* in possessing the scutum mostly black, from *B. fereuncinata* in having *a.sa.* setae, elongate oval facial spots and expansion of costal band across apex of wing and from *B. uncinata* in having a triangular yellow marking along anterior margin of notopleural suture.

Bactrocera (Zeugodacus) kaghanae Mahmood

Bactrocera (Zeugodacus) kaghanae Mahmood, 2005: 161. Holotype male in BMNH.

Material examined: No specimens available for study. The following diagnosis has been taken from literature.

Diagnosis: Face fulvous with a pair of black spots; postpronotal lobes yellow; notopleura vary from yellow to black; scutum black without pale markings; parallel sided lateral postsutural yellow vittae ending just before *ia.* seta; medial postsutural yellow vitta present; a small yellow spot anterior to notopleural suture; mesopleural stripe reaching anterior *npl.* seta dorsally; scutellum yellow with a black spot on apex. Setae: *sc.* 2; *prsc.* 2; *ia.* 1; *p.sa.* 1; *a.sa.* 1; *mpl.* 1; *npl.* 2; *scp* 4. Wing with cells bc and c colourless; a narrow fuscous costal band confluent with R_{2+3} and ending at apex of this vein, a fuscous spot across apex of R_{4+5}; a moderately broad anal streak; legs with femora fulvous with preapical black spots on outer surfaces; all abdominal terga dark fuscous to black with narrow transverse fulvous bands across posterior margins of terga I and II, a pair of black shining spots on tergum V. Posterior lobe of surstylus long, sternum V with a shallow concavity on posterior margin.

Attractant: Cue lure.

Female: No known record.

Distribution:

Type locality: Pakistan (Kaghan).

Hosts: No known record.

Remarks: *Bactrocera (Zeugodacus) kaghanae* Mahmood, within the *scutellaris* complex, possesses a black scutum, narrow lateral and medial postsutural yellow vittae, costal band confluent with R_{2+3} and absent or very narrow beyond apex of this vein before widening slightly around wing apex. It is similar to *Bactrocera (Zeugodacus) assamensis* White in possessing two *sc.* setae, femora with dark apical markings, notopleura variable from yellow to black and scutellum with an apical black spot, but differs in having the femora fulvous with subapical black spots on outer surfaces and a small yellow spot anterior to the notopleural suture. *Bactrocera kaghanae* is also inseparable from *Bactrocera (Zeugodacus) scutellaris* (Bezzi) except that it possesses two *sc.* setae, not four as in *B. scutellaris.*

Bactrocera (Zeugodacus) khaoyaiae Drew & Romig, new species

(Fig. 318)

Material examined: Holotype male, **THAILAND**: Khao Yai, Nakhonnayok, 21.vi.1991, coll. Montree, attracted to cue lure, deposited in BMNH.
Paratypes: **THAILAND**: 3 males, Chumphon Prov., 80 m, Khao Kai Jae waterfall, 9.55.07N, 98.56.34E, semi-evergreen rainforest, 5–8.v.2003, coll. P. Schwendinger, attracted to cue lure; 1 male, Chiang Mai prov, 570 m, Tham Klaeb Yai, 30 Km N Chiang Dao, 19.34.34N, 99.03.40E, 23–24.xii.2002, coll. P. Schwendinger. **PENINSULAR MALAYSIA**: 2 males, RS Cameron Highlands, 6.v.1986, coll. Zan Davies, attracted to cue lure. 1 paratype in ANIC; 1 paratype in BMNH; 2 paratypes in MHNG; 1 paratype in QDPC; 1 paratype in QMIC.
Diagnosis: A medium-sized species; face fulvous with a pair of large circular black spots; postpronotal lobes and notopleura yellow; scutum mostly black; lateral and medial postsutural yellow vittae present; a large yellow spot anterior to notopleural suture; mesopleural stripe equal in width to notopleuron dorsally; scutellum yellow; wing with cells bc and c colourless, microtrichia in outer corner of cell c only; a narrow dark fuscous costal band clearly overlapping R_{2+3}, a large dark fuscous spot around apex of R_{4+5}; a broad dark fuscous anal streak; abdominal terga III–V fulvous with a black ‘T’ pattern consisting of a narrow transverse band across anterior margin of tergum III and a medium-width medial longitudinal black band over all three terga and broad lateral longitudinal dark fuscous to black bands over terga IV and V.
Description: male.
Head: Height 1.52 mm. Frons – length 1.45 times breadth; fulvous with dark fuscous on anteromedial hump; orbital setae black: 1 *s.or.*, 3 *i.or.*; lunule fuscous. Ocellar triangle black. Vertex dark fuscous to black. Face fulvous with a pair of large circular black spots; length 0.49 mm. Genae fulvous, large fuscous subocular spot present; dark seta present. Occiput fuscous to dark fuscous, fulvous along eye margins; occipital row with a large number of pale setae. Antennae with segments 1 and 2 red-brown, segment 3 red-brown with fuscous on apex and outer surface; arista black (red-brown basally); length of segments: 0.14 mm; 0.26 mm; 0.74 mm.
Thorax: Scutum black except fuscous below and behind lateral postsutural vittae, surrounding yellow spot anterior to notopleural suture and red-brown anterocentrally from anterior point of medial postsutural yellow vitta to anterior margin of scutum. Pleural areas black except red-brown above mid coxae and red-brown to fuscous below postpronotal lobes. Yellow markings as follows: postpronotal lobes; notopleura; mesopleural stripe equal in width to notopleuron dorsally, extending to katepisternum as a large transverse spot, anterior margin slightly convex; anatergite (posterior apex black); anterior $^{5}/_{8}$ katatergite (remainder black); two broad parallel-sided lateral postsutural vittae beginning with a large spot anterior to notopleural suture and ending well behind *ia.* seta. A broad medial longitudinal postsutural vitta beginning posteriorly at level of *prsc.* setae and ending anteriorly in a point anterior to notopleural suture. Postnotum black. Scutellum yellow except for narrow black basal band. Setae: *sc.* 4; *prsc.* 2; *ia.* 1; *p.sa.* 1; *a.sa.* 1; *mpl.* 1; *npl.* 2; *scp.* 4.
Legs: Femora fulvous except for a large elongate black spot on outer surfaces of fore femora and distinct black subapical spots on inner surfaces of mid and hind femora; fore tibiae dark fuscous, mid tibiae fuscous basally to fulvous apically, hind tibiae dark fuscous; tarsi with all segments fulvous; mid tibiae each with an apical black spur.
Wings: Length 6.2 mm; cells bc and c colourless; microtrichia in outer corner of cell c only; remainder of wings colourless except dark fuscous cell sc, narrow dark fuscous costal band clearly overlapping R_{2+3} and ending in a spot around apex of R_{4+5}, a broad dark fuscous anal streak; a dense aggregation of microtrichia around A_1+CuA_2; supernumerary lobe large and rounded.
Abdomen: Oval; terga free; pecten present on tergum III. Tergum I and sterna I and II wider than long. Tergum I fulvous with fuscous along anterior margin and black on lateral margins; tergum II fulvous except for a narrow black transverse band across anterior margin but not reaching the narrow black lateral margins; terga III–V fulvous except for a black ‘T’ pattern consisting of a narrow transverse band across anterior margin of tergum III which covers lateral margins and a medium-width medial longitudinal black band over all three terga, broad lateral longitudinal black bands over terga IV and V (these lateral bands may be fuscous to dark fuscous on posterior section of tergum IV and in one specimen on tergum IV there is a large anterolateral black marking not covering the entire lateral margin). A pair of oval fulvous shining spots on tergum V. Posterior lobe of surstylus long, sternum V with a slight concavity on posterior margin. Abdominal sterna black, except sternum II fulvous.
Attractant: Cue lure.
Female: No known record.
Distribution: Peninsular Malaysia (Cameron Highlands); Thailand.
Hosts: No known record.
Remarks: *Bactrocera (Zeugodacus) khaoyaiae* n.sp., within the *tau* complex, is similar to *Bactrocera (Zeugodacus)*

yalaensis n.sp. in possessing dark abdominal sterna, an entirely yellow scutellum, broad lateral dark bands over terga IV and V, broad parallel-sided lateral postsutural yellow vittae, facial spots present and dark markings on all femora. It differs from *B. yalaensis* in having the dark markings on the fore femora large and elongate, facial spots large and the supernumerary lobe in the male wing large and rounded.

Bactrocera (Zeugodacus) laguniensis Drew & Romig, new species

(Fig. 319)

Material examined: Holotype male, **PHILIPPINES: LUZON ISLAND** – Sirang Lupa, Calamba, Laguna, Apple and Guava mixed orchard, 20.xii.1999, coll. G. Quimio, attracted to cue lure, deposited in BMNH.

Paratypes: **PHILIPPINES: LUZON ISLAND** – 2 males, sama data as holotype; 1 male, Sirang Lupa, Calamba, Laguna, Guava and Carambola mixed orchard, 7.i.2000, coll. G. Quimio, attracted to cue lure; 5 males, UPLB campus rainforest area mixed vegetation (near NCPC dorms), 8.i.2000, coll. G. Quimio, attracted to cue lure; 5 males, Mt. Makiling Botanic Gardens, 12.i.2000, coll. A. O'Toole, attracted to cue lure; 1 male, Mt. Makiling (600 m), 5.ix.1983, coll. R.A.I. Drew; 13 males, Mt. Makiling (500 m), 5.ix.1983, coll. R.A.I. Drew; 5 males, Mt. Makiling (500 m), 15.vi.1983, coll. B. Ramos; 1 male, Mt. Makiling (500 m), 9.v.1983, coll. R. Drew & B. Ramos; 3 males (8.vi.1983), 10 males (15.vi.1983), Mt. Makiling (400 m), coll. B. Ramos; 3 males, Mt. Makiling, 8.vi.1983, coll. B. Ramos; all attracted to cue lure. 6 paratypes in ANIC; 6 paratypes in BMNH; 27 paratypes in QDPC; 10 paratypes in QMIC.

Diagnosis: A medium-sized species; face fulvous with a broad black band across oral margin; postpronotal lobes and notopleura yellow; scutum black; lateral and medial postsutural yellow vittae present; a yellow spot anterior to notopleural suture; mesopleural stripe almost reaching anterior *npl.* seta; scutellum yellow; wing with cells bc and c colourless, microtrichia in outer corner of cell c only; a narrow dark fuscous costal band confluent with R_{2+3} and widening across apex of R_{4+5} and a broad dark fuscous anal streak; abdominal terga III–V orange-brown with a general black 'T' pattern consisting of a broad black band across anterior margin of tergum III, a broad medial longitudinal black band over all three terga and broad lateral longitudinal black bands on terga IV and V.

Description: male.

Head: Height 1.47 mm. Frons – length 1.32 times breadth; red-brown with fuscous to dark fuscous around orbital setae and on anteromedial hump (sometimes spreading from anteromedial hump to *s.or.* and lower *i.or.* setae); orbital setae black: 1 *s.or.*, 3 *i.or.*; lunule dark fuscous. Ocellar triangle black. Vertex fuscous to dark fuscous. Face fulvous with a distinct broad black band around oral margin; length 0.54 mm. Genae fulvous, fuscous to dark fuscous subocular spot present; black seta present. Occiput fuscous, fulvous along eye margins; occipital row with 4–6 black setae. Antennae with segments 1 and 2 red-brown, segment 3 red-brown with fuscous on apex and outer surface; arista black (red-brown basally); length of segments: 0.12 mm; 0.24 mm; 0.6 mm.

Thorax: Scutum uniformly black with red-brown below and behind lateral postsutural vittae, between notopleuron and yellow spot anterior to notopleural suture and inside postpronotal lobes. Pleural areas black except red-brown above fore coxae. Yellow markings as follows: postpronotal lobes; notopleura; mesopleural stripe almost reaching anterior *npl.* seta dorsally, continuing to katepisternum as a transverse spot, anterior margin convex; anatergite (posterior apex black); anterior ⅔ katatergite (remainder black); two moderately broad lateral postsutural vittae narrowing posteriorly to end at or just before *i.or.* seta; a moderately broad medial longitudinal postsutural vitta beginning just anterior to *prsc.* setae and ending in a point at level of notopleural suture; a yellow spot anterior to notopleural suture. Postnotum black. Scutellum yellow except for narrow black basal band. Setae: *sc.* 4; *prsc.* 2; *ia.* 1; *p.sa.* 1; *a.sa.* 1; *mpl.* 1; *npl.* 2; *scp.* 4.

Legs: Fore femora fulvous with dark fuscous to black around apical ½ to ⅔, mid femora fulvous with fuscous to dark fuscous around apical ½ and hind femora fulvous with fuscous to dark fuscous around apical ⅓; fore and mid tibiae fuscous tending darker basally, hind tibiae dark fuscous; tarsi with basal segment fulvous and apical four segments dark fuscous; mid tibiae each with an apical black spur.

Wings: Length 5.78 mm; cells bc and c colourless; microtrichia in outer corner of cell c only; remainder of wings colourless except dark fuscous cell sc, narrow dark fuscous costal band confluent with R_{2+3} and widening across apex of R_{4+5}, broad dark fuscous anal streak; dense aggregation of microtrichia around A_1+CuA_2; supernumerary lobe weak.

Abdomen: Oval; terga free; pecten present on tergum III. Tergum I and sterna I and II wider than long. Tergum I orange-brown with narrow fuscous band across anterior margin and black lateral margins; tergum II orange-brown to fulvous with a narrow transverse subanterior band across anterior margin sometimes reaching the narrow black lateral margins; terga III–V orange-brown with a 'T' pattern consisting of a broad transverse black band across anterior margin of tergum III and widening over lateral margins and a broad medial longitudinal black band over all three terga and broad lateral longitudinal bands over terga IV and V (usually there is an orange-brown band along posterolateral margin of tergum III which gives the appearance of the broad bands on terga IV and V). A pair of oval orange-brown shining spots on tergum V. Posterior lobe of surstylus long, sternum V with a slight concavity on posterior margin. Abdominal sterna dark, except sternum II paler.

Attractant: Cue lure.

Female: No known record.

Distribution: Philippines (Luzon I.).

Hosts: No known record.

Remarks: *Bactrocera (Zeugodacus) laguniensis* n.sp. is similar to *Bactrocera (Zeugodacus) sabahensis* n.sp. and specimens of *Bactrocera (Zeugodacus) hengsawadae* n.sp. with four *sc.* setae in possessing a black scutum, wings colourless except for costal band and anal streak, colourless costal cells, costal band confluent with R_{2+3} and dark abdominal sterna. It differs from these species in having a broad black band across the oral margin of the face.

Bactrocera (Zeugodacus) laocaiae Drew & Romig, new species

(Fig. 320)

Material examined: Holotype male, **VIETNAM: Northern Vietnam** – Lao Cai Province, Sapa town, trap No. 16 urban, 23.iv.1999, coll. Le Duc Khanh *et al*., attracted to cue lure, deposited in BMNH.

Paratypes: **VIETNAM: Northern Vietnam** – 17 males, same data as holotype; 2 males (12.v.1999), 5 males (30.vii.1999), 3 males (13.viii.1999), 5 males (28.iv.2000), 1 male (7.vii.2000), 1 male (4.viii.2000), Lao Cai Province, Sapa town, trap No. 16, attracted to cue lure; 8 males (23.iv.1999), 1 male (13.viii.1999), 1 male (10.ix.1999), Lao Cai Province, Sapa, Oquy Ho, trap No. 19, coll. Le Duc Khanh *et al*., attracted to cue lure; 1 male (30.vii.1999), 5 males (13.viii.1999), 1 male (27.viii.1999), 2 males (21.vii.2000), 1 male (27.viii.2000), Lao Cai Province, Sapa 2 Village, trap No. 15/18, attracted to cue lure; 2 males (30.vii.1999), 4 males (13.viii.1999), 1 male (7.vii.2000), Lao Cai Province, Sapa 6 Village, trap No. 17, attracted to cue lure. **THAILAND**: 1 female, Wawi Exp Sta, 1400 m, 45 km South of Chiang Rai, 22.v.1991, coll. D.L. Hancock, caught sweeping. 10 paratypes in ANIC; 10 paratypes in BMNH; 32 paratypes in QDPC; 10 paratypes in QMIC.

Diagnosis: A medium-sized species; face fulvous with a black band across oral margin; postpronotal lobes and notopleura yellow; scutum shining black; lateral and medial postsutural yellow vittae present; a yellow spot present anterior to notopleural suture; mesopleural stripe reaching midway between anterior margin of notopleuron and anterior *npl.* seta dorsally; scutellum yellow without any dark markings; wing with cells bc and c colourless, microtrichia in outer corner of cell c only; a narrow dark fuscous costal band expanding into a spot around apex of R_{4+5} (*scutellaris*-like) and broad dark fuscous anal streak; abdominal terga III–V mostly dark fuscous to black with red-brown to dark red-brown posterocentrally on terga III and IV and bordering the medial longitudinal dark fuscous to black band on tergum V.

Description: male.

Head: Height 1.2 mm. Frons – length 1.4 times breadth; fulvous with fuscous to dark fuscous around orbital setae and on anteromedial hump; orbital setae black: 1 *s.or.*, 3 *i.or.*; lunule dark fuscous. Ocellar triangle black. Vertex dark fuscous. Face fulvous with a broad black transverse band across oral margin (occasionally broken in centre); length 0.38 mm. Genae fulvous; black seta present. Occiput dark fuscous to black, fulvous along eye margins; occipital row with a large number of small pale setae. Antennae with segments 1 and 2 red-brown, segment 3 red-brown with dark fuscous on apex and outer surface; arista black (red-brown basally); length of segments: 0.1 mm; 0.2 mm; 0.54 mm.

Thorax: Scutum shining black with dark red-brown below and behind lateral postsutural vittae in some specimens only. Pleural areas shining black tending dark fuscous below postpronotal lobes. Yellow markings as follows: postpronotal lobes; notopleura; mesopleural stripe reaching midway between anterior margin of notopleuron and anterior *npl.* seta dorsally, continuing to katepisternum as a small transverse spot, anterior margin convex; anatergite (posterior apex black); anterior ⅔ katatergite (remainder black); two lateral postsutural vittae of medium-width, beginning with a spot anterior to notopleural suture and narrowing sharply posteriorly to end before *ia.* seta; a narrow medial longitudinal postsutural vitta beginning at level of *prsc.* setae and narrowing to a point anteriorly to end at level of notopleural suture. Postnotum black. Scutellum yellow except for narrow black basal band. Setae: *sc.* 2; *prsc.* 2; *ia.* 1; *p.sa.* 1; *a.sa.* 1; *mpl.* 1; *npl.* 2; *scp.* 4.

Legs: Femora fulvous with dark fuscous to shining black around apical ½ of fore femora and apical ⅓ of mid and hind femora (in some specimens the dark fuscous to black marking is on the outer apical surfaces of the fore femora and the inner apical surfaces of the hind femora); all tibiae fulvous tending pale fuscous basally (in some specimens the hind tibiae are pale fuscous); all tarsi with basal segment fulvous and apical four segments pale fuscous; mid tibiae each with an apical black spur.

Wings: Length 5.0 mm; cells bc and c colourless; microtrichia in outer corner of cell c only; remainder of wings colourless except dark fuscous cell sc, narrow dark fuscous costal band confluent with R_{2+3} and remaining very narrow beyond apex of this vein before expanding into a distinct spot around apex of R_{4+5} (*scutellaris*-like), a broad dark fuscous anal streak; dense aggregation of microtrichia around A_1+CuA_2; supernumerary lobe very strong.

Abdomen: Oval; terga free; pecten present on tergum III. Tergum I and sterna I and II wider than long. Tergum I fulvous with black anterolateral margins; tergum II fulvous with a dark fuscous to black transverse band across anterior margin sometimes reaching the dark fuscous to black anterolateral corners; terga III–V mostly dark fuscous to black with red-brown to dark red-brown posterocentrally on terga III and IV either side of a medial longitudinal black band and red-brown bordering a medial longitudinal dark fuscous to black band on tergum V. A pair of oval red-brown shining spots on tergum V. Posterior lobe of surstylus long, sternum V with a slight concavity on posterior margin. Abdominal sterna shining dark fuscous to black, except sternum II fulvous.

Attractant: Cue lure.

Female: As for male except face fulvous with a pair of transverse oval black spots not meeting in centre; supernumerary lobe weak; no dense aggregation of microtrichia around A_1+CuA_2; pecten of cilia on abdominal tergum III absent; basal segment of ovipositor black, conical in transverse cross section and tapering posteriorly in dorsal view; ratio of length of oviscape to length of tergum V, 1.4:1; apex of piercer blunt with three small sharp lobes.

Distribution: Northern Vietnam (Lao Cai Province); Thailand (Chiang Rai).

Hosts: No known record.
Remarks: *Bactrocera (Zeugodacus) laocaiae* n.sp., within the *scutellaris* complex, is a distinct, medium-sized species that has only been collected in far northern Vietnam. It is similar to *Bactrocera (Zeugodacus) hoabinhiae* n.sp. and *Bactrocera (Zeugodacus) yoshimotoi* (Hardy) in possessing two *sc.* setae, yellow notopleura and the face with a pair of black spots or a black band across oral margin. It differs from *B. yoshimotoi* in having black patterns on the femora and a black band across the oral margin of the face and from *B. hoabinhiae* in having an entirely yellow scutellum and abdominal terga III–V mostly dark fuscous to black.

Bactrocera (Zeugodacus) lipsana (Hendel)

(Fig. 321)

Chaetodacus lipsanus Hendel, 1915: 427. Syntypes male and female in MNM.
Bactrocera (Zeugodacus) lipsanus – Norrbom *et al.*, 1998: 103.

Type material examined: TAIWAN: (formerly FORMOSA) – Syntype male, with the following four labels – 'Formosa Saunter', 'Koshun [1]908,VIII', 'lipsanus Hend. typus det. Hendel' and a white label with red margins and one word in red 'typus'. 1 female and 1 male syntype, with the following four labels – 'Formosa Sauter'; 'Suisharyo, 1911.X'; 'lipsanus Hend typus det Hendel' and a white label with red margins and one word in red 'typus'. 2 male Syntypes, with the following four labels – 'Formosa Sauter'; 'Tainan 1912.iv'; 'lipsanus Hend. typus det. Hendel' and a white label with red margins and one word in red 'typus'. Received on loan from MNM.
Diagnosis: A medium-sized species; face mostly dark fuscous; postpronotal lobes and notopleura yellow; scutum black; lateral and medial postsutural yellow vittae present; no yellow spot anterior to notopleural suture; mesopleural stripe reaching almost to anterior *npl.* seta dorsally; scutellum yellow with a large dark fuscous apical spot; wing with cells bc and c colourless or with a very pale tint, microtrichia in outer corner of cell c only; a narrow fuscous costal band confluent with R_{2+3} and widening slightly across apex of R_{4+5}, a broad fuscous anal streak; abdominal terga III–V fulvous with a black 'T' pattern consisting of a narrow transverse band across anterior margin of tergum III and a moderately broad medial longitudinal band over all three terga, broad lateral black margins on terga IV and V.
Description: male.
Head: Height 1.37 mm. Frons – length 1.43 times breadth; fulvous with dark fuscous anteriorly and around orbital setae and fuscous on anteromedial hump; latter with a small number of short pale hairs; orbital setae black: 1 *s.or.*, 2 i.*or.*; lunule fuscous. Ocellar triangle black. Vertex red-brown. Face mostly dark fuscous except for fulvous lateral longitudinal sub-marginal lines in antennal furrows and sometimes with fulvous along oral margin; length 0.41 mm. Genae fulvous, large fuscous subocular spot present; black seta present. Occiput dark red-brown with small areas of fuscous dorsocentrally and fulvous along eye margins; occipital row with a number of small pale setae. Antennae with segment 1 red-brown except dark fuscous laterally and segment 2 red-brown, segment 3 mostly dark fuscous (red-brown basally); length of segments: 0.09 mm; 0.24 mm; 0.62 mm.
Thorax: Scutum black with fuscous behind lateral postsutural vittae and dark red-brown below lateral postsutural vittae and around notopleural suture. Pleural areas black except red-brown below postpronotal lobes. Yellow markings as follows: postpronotal lobes; notopleura; mesopleural stripe reaching almost to anterior *npl.* seta dorsally, continuing to katepisternum as a transverse spot, anterior margin slightly convex; anatergite (posterior apex black); anterior ¾ katatergite (remainder black); two broad parallel-sided lateral postsutural vittae ending behind *ia.* setae; a moderately broad medial longitudinal postsutural vitta. No yellow spot anterior to notopleural suture. Postnotum black. Scutellum yellow except for a narrow black basal band and a large dark fuscous spot on apex. Setae: *sc.* 4; *prsc.* 2; *ia.* 1; *p.sa.* 1; *a.sa.* 1; *mpl.* 1; *npl.* 2; *scp.* 2.
Legs: Fore femora fulvous tending dark fuscous apically and with a dark fuscous spot on outer apical surface, mid femora fulvous tending fuscous apically and with a dark fuscous spot on outer apical surface, hind femora fulvous with a dark fuscous spot around apical extremity; fore and mid tibiae fulvous, hind tibiae dark fuscous; tarsal segments fulvous except apical four segments of fore and hind tarsi and apical three segments of mid tarsi pale fuscous to fuscous; mid tibiae each with an apical black spur.
Wings: Length 5.2 mm; cells bc and c colourless to having a pale fuscous tint; microtrichia in outer corner of cell c only; remainder of wings colourless except fuscous cell sc, narrow fuscous costal band confluent with R_{2+3} and widening slightly across apex of R_{4+5}, a broad fuscous anal streak; a dense aggregation of microtrichia around A_1+CuA_2; supernumerary lobe weak.
Abdomen: Oval; terga free; pecten present on tergum III. Tergum I and sterna I and II wider than long. Tergum I fulvous except for narrow black lateral margins; tergum II fulvous except for a narrow transverse black band across anterior margin which extends to cover narrow lateral margins; terga III–V fulvous except for a black 'T' pattern which consists of a transverse band across anterior ⅓ of tergum III and a moderately broad medial longitudinal band across all terga, broad lateral black margins on terga IV and V. A pair of oval fulvous shining spots on tergum V. Posterior lobe of surstylus long, sternum V almost straight across posterior margin.
Attractant: No known record.
Female: As for male except no dense aggregation of microtrichia around A_1+CuA_2; pecten of cilia on abdominal tergum III absent; ovipositor basal segment red-brown, dorsoventrally compressed and tapering posteriorly in dorsal view; ratio of length of oviscape to length of tergum V, 1.4:1.
Distribution:
Type locality: Taiwan.
Hosts: No known record.
Remarks: *Bactrocera (Zeugodacus) lipsana* (Hendel) is similar to *Bactrocera (Zeugodacus) neolipsana* n.sp. and

Bactrocera (Zeugodacus) scutellina (Bezzi) in possessing four *sc.* setae, a black scutum with lateral and medial longitudinal postsutural vittae, wings colourless except for costal band and anal streak, costal cells pale or colourless, costal band confluent with R_{2+3} and widening only slightly beyond apex of this vein and having no yellow spot or mark on anterior margin of notopleural suture in front of lateral postsutural vittae. It is separate from *B. scutellina* in having the face mostly dark fuscous to black and from *B. neolipsana* in having a dark fuscous spot on apex of the scutellum, the fore femora with a preapical dark fuscous spot on outer surface (not predominantly black), abdominal terga IV and V with lateral longitudinal dark bands (not over III–V) and red-brown patterns around margins of scutum.

Bactrocera (Zeugodacus) liquida Drew & Romig, new species

(Fig. 322)

Material examined: Holotype male, **THAILAND**: Chiang Mai, N19 54 998, E 99 02 393, Doi Ang Khang Agr station, 15–17.ii.2011, coll. M. San Jose & D. Rubinoff, FF063, cue lure trap, deposited in BBM.

Diagnosis: A medium-sized species; face fulvous with a pair of small circular black spots; postpronotal lobes and notopleura yellow; scutum shining black; narrow lateral postsutural yellow vittae present; a short narrow medial postsutural yellow vitta present; a small yellow spot anterior to notopleural suture; mesopleural stripe equal in width to notopleuron dorsally; scutellum yellow with a large black apical spot surrounding apical setae; wing with cells bc and c colourless, a small number of microtrichia in apex of cell c; entirely colourless except for a dark fuscous cell sc and dark fuscous spot around apex of R_{4+5} and a pale tint within cell cup; abdominal terga III–V orange-brown with a narrow medial and two broad lateral longitudinal black bands joined along anterior margin of tergum III.

Description: male.

Head: Height 1.38 mm. Frons – length 1.3 times breadth; mostly fuscous with fulvous along ventral margin; orbital setae black: 1 *s.or.*, 3 *i.or.*; lunule dark fuscous. Ocellar triangle black. Vertex fuscous. Face fulvous with a pair of small circular black spots; length 0.45 mm. Genae fulvous, pale fuscous subocular spot present; seta absent. Occiput red-brown with small areas of black laterally, fulvous along eye margins; occipital row with a large number of small pale setae. Antennae with segments 1 and 2 red-brown, segment 3 red-brown with dark fuscous on apex and outer surface; length of segments: 0.09 mm; 0.24 mm; 0.54 mm.

Thorax: Scutum shining black with red-brown below and behind lateral postsutural vittae and around notopleural suture. Pleural areas entirely shining black. Yellow markings as follows: postpronotal lobes; notopleura; mesopleural stripe equal in width to notopleuron dorsally, continuing to katepisternum as a very small spot, anterior margin convex; anatergite (posterior apex black); anterior ½ katatergite (remainder black); two narrow lateral postsutural vittae ending at *ia.* seta; a yellow spot anterior to notopleural suture; a very short and narrow medial postsutural vitta. Postnotum black. Scutellum yellow except for a narrow black basal band and a large black apical spot surrounding apical setae. Setae: *sc.* 4; *prsc.* 2; *ia.* 1; *p.sa.* 1; *a.sa.* 1; *mpl.* 1; *npl.* 2; *scp.* 2.

Legs: Fore femora fulvous with a black subapical spot on outer surfaces, mid femora fulvous with apical ⅕ with fuscous on outer surfaces and black on inner surfaces, hind femora fulvous with black around apical ¼; fore tibiae dark fuscous, mid tibiae fulvous with dark fuscous basally extending to a narrow line along dorsal surface, hind tibiae dark fuscous to black; all tarsi with basal segment fulvous and apical four segments dark fuscous; mid tibiae each with an apical black spur.

Wings: Length 6.1 mm; cells bc and c colourless; a small number of microtrichia in apex of cell c; remainder of wings entirely colourless except for a dark fuscous cell sc, a dark fuscous spot around apex of R_{4+5}, a pale tint within cell cup; a dense aggregation of microtrichia around A_1+CuA_2; supernumerary lobe weak.

Abdomen: Oval; terga free; pecten present on tergum III. Tergum I and sterna I and II wider than long. Tergum I black with orange-brown across posterior margin but not reaching lateral margins; tergum II with a broad transverse black band subanteriorly with orange-brown along anterior margin and fulvous across posterior ½; terga III–V orange-brown with a narrow medial and two broad lateral longitudinal black bands joined along anterior margin of tergum III. A pair of oval orange-brown shining spots on tergum V. Posterior lobe of surstylus long, sternum V with a slight concavity on posterior margin. Abdominal sterna dark fuscous to black.

Attractant: Cue lure.

Female: No known record.

Distribution: Thailand (Chiang Mai).

Hosts: No known record.

Remarks: *Bactrocera (Zeugodacus) liquida* n.sp. belongs to the *scutellaris* complex within the subgenus *Zeugodacus*. It is similar to *Bactrocera dorsirufa* n.sp. in possessing a shining black scutum, narrow lateral and medial postsutural yellow vittae and costal band reduced or absent beyond extremity of R_{2+3}. It differs from *B. dorsirufa* in having the costal band absent between the apex of cell sc and the dark fuscous spot at apex of R_{4+5} and abdominal terga III–V with a narrow medial and broad lateral longitudinal black bands joined over anterior margin of tergum III.

Bactrocera (Zeugodacus) macrophyllae Drew & Romig, new species

(Fig. 323)

Material examined: Holotype female, **THAILAND**: Hoy Ekong, Chiang Mai, 17.xii.1991, coll. Chantanee, bred from *Momordica macrophylla*, deposited in BMNH.

Paratypes: **MALAYSIA**: 1 male, Site TN10, 30.vi.1986, coll. Yusoff Salleh, attracted to cue lure. **THAILAND**: 4 females and 3 males, Ban Huai, Muang, Chiang Mai, 30.

vi.1992, coll. Chantanee *et al.*, bred from *Momordica macrophylla*. 1 paratype in ANIC; 1 paratype in BMNH; 5 paratypes in QDPC; 1 paratype in QMIC.

Diagnosis: A large species; face with a pair of large elongate oval black spots and a dorsocentral fuscous spot; postpronotal lobes and notopleura yellow; scutum black; lateral and medial postsutural yellow vittae present; a yellow spot anterior to notopleural suture; mesopleural stripe reaching midway between anterior margin of notopleuron and anterior *npl.* seta dorsally; scutellum yellow (with a dark fuscous apical spot in Holotype); wing with cells bc and c colourless, microtrichia in outer corner of cell c only; a narrow fuscous costal band overlapping R_{2+3} and ending in a spot around apex of R_{4+5}; a broad fuscous anal streak; abdominal terga III–V fulvous with a black 'T' pattern consisting of a narrow transverse black band across anterior margin of tergum III and a moderately broad medial longitudinal black band over all three terga, broad dark fuscous to black lateral longitudinal bands over terga III to V or terga IV and V.

Description: female.

Head: Height 1.74 mm. Frons – length 1.15 times breadth; fulvous with dark fuscous on anteromedial hump and around bases of orbital setae; orbital setae black: 1 *s.or.*, 3 *i.or.*; lunule dark fuscous. Ocellar triangle black. Vertex dark fuscous to black. Face fulvous with a pair of large elongate oval black spots and a dark fuscous to black spot centrally just below antennal sockets; length 0.59 mm. Genae fulvous, large fuscous subocular spot present; dark seta present. Occiput red-brown, fulvous along eye margins; occipital row with 5–9 well-developed black setae. Antennae with segments 1 and 2 red-brown, segment 3 red-brown with dark fuscous on apex and outer surface; arista black (red-brown basally); length of segments: 0.15 mm; 0.28 mm; 0.84 mm.

Thorax: Scutum black with red-brown patterns as follows: across posterior margin and surrounding posterior base of medial postsutural yellow vitta, below and behind lateral postsutural yellow vittae, around notopleural suture including area between yellow spot anterior to suture and notopleuron, across anterior margin between postpronotal lobes continuing back to anterior point of medial postsutural yellow vitta. Pleural areas dark fuscous to black except red-brown below postpronotal lobes. Yellow markings as follows: postpronotal lobes; notopleura; mesopleural stripe reaching midway between anterior margin of notopleuron and anterior *npl.* seta dorsally, continuing to katepisternum as a large transverse yellow spot, anterior margin convex; anatergite (posterior apex black); anterior ⅔ katatergite (remainder black); two broad parallel-sided lateral postsutural vittae beginning with a spot anterior to notopleural suture and ending just behind *ia.* seta; a medial longitudinal postsutural vitta broadly rounded posteriorly and narrowing to a point at level of notopleural suture. Postnotum black laterally, red-brown centrally. Scutellum yellow except for narrow black basal band (in the Holotype there is a small dark fuscous apical spot surrounding apical setae). Setae: *sc.* 4; *prsc.* 2; *ia.* 1; *p.sa.* 1; *a.sa.* 1; *mpl.* 1; *npl.* 2; *scp.* 2.

Legs: Fore femora fulvous except for a large elongate dark fuscous spot along outer surfaces, mid femora entirely fulvous, hind femora fulvous with a very small area of dark fuscous around apex; fore tibiae dark fuscous, mid tibiae fulvous tending fuscous basally, hind tibiae dark fuscous; all tarsi with basal segment fulvous and apical four segments fuscous; mid tibiae each with an apical black spur.

Wings: Length 7.65 mm; cells bc and c colourless; microtrichia in outer corner of cell c only; remainder of wings colourless except fuscous cell sc, narrow fuscous costal band overlapping R_{2+3} and ending in a *tau*-like spot around apex of R_{4+5}, a broad fuscous anal streak; no dense aggregation of microtrichia around A_1+CuA_2; supernumerary lobe very weak.

Abdomen: Oval; terga free; no pecten present on tergum III. Tergum I and sterna I and II wider than long. Tergum I fulvous except fuscous across anterior margin and dark fuscous to black broad lateral margins; tergum II fulvous with a narrow transverse black band across anterior margin but not reaching lateral margins and narrow black lateral margins; terga III–V fulvous with a distinct 'T' pattern consisting of a narrow transverse black band across anterior margin of tergum III and a moderately broad medial longitudinal black band over all three terga and either broad lateral longitudinal dark bands that are fuscous to dark fuscous on tergum III and dark fuscous to black on terga IV and V or broad lateral longitudinal dark fuscous to black bands over terga IV and V only. A pair of oval fulvous shining spots on tergum V. Abdominal sterna black, except sternum II fulvous. Basal segment of ovipositor red-brown, dorsoventrally flattened and tapering posteriorly in dorsal view; ratio of length of oviscape to length of tergum V, 2.2:1; apex of piercer trilobed.

Attractant: No known record.

Male: As for female except dense aggregation of microtrichia around A_1+CuA_2; supernumerary lobe of medium development; pecten of cilia on abdominal tergum III present; posterior lobe of surstylus long, sternum V with a slight concavity on posterior margin.

Distribution: Malaysia (Site TN10); Thailand (Chiang Mai).

Hosts: Bred from one host *Momordica macrophylla* (family *Cucurbitaceae*).

Remarks: *Bactrocera (Zeugodacus) macrophyllae* n.sp., within the *tau* complex, is similar to *Bactrocera (Zeugodacus) trichosanthes* n.sp. in being a large species with a reduced supernumerary lobe in the male wing and apex of piercer trilobed. It differs from *B. trichosanthes* in having a dark spot dorsocentrally on the face, fore and hind femora with dark apical spots, larger areas of red-brown on the scutum and abdominal terga III–V with broad lateral dark margins.

Bactrocera (Zeugodacus) maculata (Perkins)

(Fig. 324)

Zeugodacus caudatus var. *maculatus* Perkins, 1938a: 139. Holotype female in QMIC.

Dacus (Zeugodacus) maculatus – Hardy & Adachi, 1954: 192; Hardy, 1973: 63.

Bactrocera (Zeugodacus) maculata – Norrbom *et al.*, 1998: 103.

Bacterodacus (Zeugodacus) apicinigra Yu, Bai & Yang, in Yu *et al.*, 2011: 315. [*Bactrocera* Misspelt as *Bacterodacus*]. Holotype male in CAIQ. **New Synonym**.

Type material examined: Holotype female, **PENINSULAR MALAYSIA**: with the following 3 labels – "Malay Penin, Selangor, Bukit Kutu, 3300ft, 26.ix.1932, H.M. Pendlebury"; "Holotype, *Zeugodacus caudatus* var. *maculatus* F.A. Perk"; "UQIC Reg. No. 45728". Received on loan from QMIC.
Other type material studied: CHINA: A study of the colour photographs and description of *Bactrocera (Zeugodacus) apicinigra* Yu, Bai & Yang in the original publication.
Non-type material examined: PENINSULAR MALAYSIA: 3 males, TN17, 6.iv.1987, coll. Ahmad Radzuan & bin Panjang Abd, attracted to cue lure; 1 male, Hulu Langat, Kuala Lumpur, 15.v.1992, coll. D.L. Hancock, attracted to cue lure; 2 males, Cameron Highlands trap 5, 24.vi.1983, coll. S. Vijaysegaran, attracted to cure lure; 1 male, Cameron Highlands, Malaysia, 10.vi.1983, coll. S. Vijaysegaran, Tapah, 6th Mile, trap M1, attracted to cue lure; 1 male, trap TN14, 9.iii.1987, coll. A.R. & P.A., attracted to cue lure; 1 male, Gombak, 19.v.1986, coll. S. Vijaysegaran, attracted to cue lure. **THAILAND**: 1 male (8.vii.1991), 2 males (25.vi.1991), Yang Pong Rat, Phubpla, A. Muang, Chanthaburi, coll. Montree, attracted to cue lure; 1 male, Philu HRS, Laemsing, Chanthaburi, 25.vi.1991, coll. Montree, attracted to cue lure; 1 male, Pha Yao, Chiang Rai, 11.viii.1992, coll. Charuwan, attracted to cue lure. Specimens in QDPC.
Diagnosis: A medium-sized species; face fulvous with a pair of large triangular black spots with apex on oral margin and the bases almost joining in the midline; postpronotal lobes and notopleura yellow; scutum predominantly black; lateral and medial postsutural yellow vittae present; yellow spot present anterior to notopleural suture; mesopleural stripe varies from just wider than notopleuron to almost reaching anterior *npl.* seta dorsally; scutellum yellow with a black spot on apex; wing with cells bc and c colourless, microtrichia in outer corner of cell c only; a narrow dark fuscous costal band slightly overlapping R_{2+3} and widening across apex of wing; a moderately broad dark fuscous anal streak; abdominal terga III–V fulvous with a 'T' pattern consisting of a narrow transverse black band across anterior margin of tergum III and a moderately broad medial longitudinal black band over all three terga, broad dark fuscous lateral longitudinal bands over terga IV and V.
Description: male.
Head: Height 1.47 mm. Frons – length 1.4 times breadth; red-brown with dark fuscous around bases of orbital setae; orbital setae black: 1 *s.or.*, 3 *i.or.*; lunule dark fuscous. Ocellar triangle black. Vertex dark fuscous. Face fulvous with a pair of large triangular black spots which have their apices at the oral margin and their bases almost joining in the midline; length 0.44 mm. Genae fulvous, large dark fuscous subocular spot present; black seta present. Occiput red-brown, fulvous along eye margins; occipital row with a large number of short dark setae. Antennae with segments 1 and 2 red-brown, segment 3 red-brown with fuscous on apex and outer surface; length of segments: 0.1 mm; 0.24 mm; 0.72 mm.
Thorax: Scutum black except red-brown below lateral postsutural vittae, inside notopleuron and on anteromedial surface. Pleural areas dark fuscous to black except red-brown below postpronotal lobes. Yellow markings as follows: postpronotal lobes; notopleura; mesopleural stripe varies from just wider than notopleuron to almost reaching anterior *npl.* seta dorsally, continuing to katepisternum as a large transverse spot, anterior margin straight; anatergite (posterior apex black); anterior $^{3}/_{5}$ katatergite (remainder black); two broad parallel-sided lateral postsutural vittae beginning with a spot anterior to notopleural suture and ending behind *ia.* seta; a large, broad medial postsutural vitta. Postnotum black, red-brown centrally. Scutellum yellow except for narrow black basal band and a black spot on apex between apical setae. Setae: *sc.* 4; *prsc.* 2; *ia.* 1; *p.sa.* 1; *a.sa.* 1; *mpl.* 1; *npl.* 2; *scp.* 4.
Legs: All femora fulvous except for a black pattern around apical ⅓ to ¼ that tends to dark fuscous on dorsal surfaces; fore tibiae fulvous with dark fuscous along outer surfaces, mid tibiae pale fuscous basally to fulvous apically, hind tibiae dark fuscous; fore and hind tarsi with basal segment fulvous and apical four segments fuscous, mid tarsi with all segments fulvous; mid tibiae each with an apical black spur.
Wings: Length 5.9 mm; cells bc and c colourless; microtrichia in outer corner of cell c only; remainder of wings colourless except dark fuscous cell sc, narrow dark fuscous costal band slightly overlapping R_{2+3} and widening across apex of wing, a moderately broad dark fuscous anal streak; a dense aggregation of microtrichia around A_1+CuA_2; supernumerary lobe large and rounded.
Abdomen: Oval; terga free; pecten present on tergum III. Tergum I and sterna I and II wider than long. Tergum I fulvous, dark fuscous anterocentrally and with narrow black lateral margins; tergum II fulvous with a narrow transverse black band subanteriorly which reaches narrow dark fuscous anterolateral corners; terga III–V fulvous with a black 'T' pattern consisting of a narrow transverse band across anterior margin of tergum III which widens across lateral margins and a moderately broad medial longitudinal band over all three terga, broad dark fuscous lateral longitudinal bands over terga IV and V (in some specimens this band is broken on posterior area of tergum IV). A pair of oval fulvous shining spots on tergum V. Posterior lobe of surstylus long, sternum V with a slight concavity on posterior margin. Abdominal sterna dark fuscous to black.
Attractant: Cue lure.
Female: No known record.
Distribution:
Type locality: Peninsular Malaysia (Selangor, Bukit Kutu).
New records: Peninsular Malaysia (Cameron Highlands; Gombak; Hulu Langat; Transect North adjacent to Thailand border); Thailand (Chanthaburi; Chiang Rai). China (Yunnan Province) due to **New Synonymy** of *B. apicinigra.*
Hosts: No known record.
Remarks: *Bactrocera (Zeugodacus) maculata* (Perkins) is similar to *Bactrocera (Zeugodacus) caudata* (Fabricius) and *Bactrocera (Zeugodacus) platamus* (Hardy) in possessing a black scutum, broad lateral and medial postsutural yellow vittae on the scutum and abdominal terga III–V with a black 'T' pattern. It differs from both species in having the face with large triangular black spots with apices of triangle at oral margin and bases almost meeting in centre of face, supernumerary lobe in male wing large and rounded and lateral margins of abdominal terga IV

and V with broad dark fuscous lateral longitudinal bands and from *B. caudata* in having the apex of the costal band gradually widening, not narrow then suddenly developed into a distinct spot. There are no good species level characters to separate *B. apicinigra* and thus we are making it a **New Synonym** of *B. maculata*.

Bactrocera (Zeugodacus) melanofacies Drew & Romig, new species

(Fig. 325)

Material examined: Holotype male, **EAST MALAYSIA: SARAWAK** – Gunung Mulu National Park, Sarawak, 28–30. vii.1992, coll. D.L. Hancock, attracted to cue lure, deposited in BMNH.
Paratypes: **EAST MALAYSIA: SARAWAK** – 1 male, same data as holotype. 1 male, 19.iii.1993 and 1 male, 2.iv.1993, East Malaysia, Sarawak, Tarat Agriculture Station, coll. S. Leong, attracted to cue lure. **PENINSULAR MALAYSIA**: 2 males, TS7 Gombak, Malaysia, 6.iv.1987, coll. Antony, attracted to cue lure. **THAILAND**: 4 males, 80 m, Chumphon Prov., Khao Kai Jae Watf., 5–8.v.2003/2, 9.55.07 N, 98.56.34 E, P. Schwendinger leg., semI–evergreen rainforest/2, attracted to cue lure. 2 paratypes in ANIC; 5 paratypes in QDPC, 2 paratypes in QMIC.
Diagnosis: A medium-sized species; face entirely shining black; postpronotal lobes and notopleura yellow; scutum black; lateral and medial postsutural yellow vittae present; yellow spot anterior to notopleural suture; mesopleural stripe just wider than notopleuron dorsally; scutellum yellow; wing with cells bc and c colourless, microtrichia in outer corner of cell c only; a narrow dark fuscous costal band expanding slightly across apex of wing and a broad dark fuscous anal streak; supernumerary lobe very strong; abdominal terga III–V red-brown with a broad medial and two broad lateral longitudinal black bands joined along anterior margin of tergum III.
Description: male.
Head: Height 1.44 mm. Frons – length 1.28 times breadth; fulvous with pale fuscous around *i.or.* setae and on anteromedial hump; orbital setae black: 1 *s.or.*, 2 *i.or.*; lunule red-brown. Ocellar triangle black. Vertex black. Face entirely shining black; length 0.54 mm. Genae fuscous with darker fuscous subocular spot; black seta present. Occiput dark fuscous (paler centrally), fulvous along eye margins; occipital row with a large number of small pale setae. Antennae with segments 1 and 2 red-brown, segment 3 entirely fuscous; arista black (red-brown basally); length of segments: 0.1 mm; 0.24 mm; 0.68 mm.
Thorax: Scutum entirely black except dark fuscous below and behind lateral postsutural vittae and between notopleura and yellow spot anterior to notopleural suture. Pleural areas entirely black. Yellow markings as follows: postpronotal lobes; notopleura; mesopleural stripe just wider than notopleuron dorsally, continuing to katepisternum as a small transverse spot, anterior margin slightly convex; anatergite (posterior apex black); anterior ¾ katatergite (remainder black); two broad lateral postsutural vittae beginning with a yellow spot anterior to notopleural suture and narrowing slightly posteriorly to end at or just behind *ia.* seta; a broad medial longitudinal postsutural vitta beginning in a point posterior to level of *prsc.* setae and narrowing anteriorly to end at level of notopleural suture. Postnotum black. Scutellum yellow except for narrow black basal band. Setae: *sc.* 4; *prsc.* 2; *ia.* 1; *p.sa.* 1; *a.sa.* 1; *mpl.* 1; *npl.* 2; *scp.* 4.
Legs: All segments entirely fulvous except for a small pale fuscous spot on preapical outer surface of fore femora, fuscous basally on hind tibiae and pale fuscous on apical four segments of all tarsi; mid tibiae each with an apical black spur.
Wings: Length 5.1 mm; cells bc and c colourless; microtrichia in outer corner of cell c only; remainder of wings colourless except a dark fuscous cell sc, narrow dark fuscous costal band slightly overlapping R_{2+3} (especially apical ½ of R_{2+3}) and expanding slightly across apex of wing, a broad dark fuscous anal streak; dense aggregation of microtrichia around A_1+CuA_2; supernumerary lobe very strong and keel shaped.
Abdomen: Oval; terga free; pecten present on tergum III. Tergum I and sterna I and II wider than long. Tergum I red-brown with dark fuscous anterocentrally and moderately broad lateral black margins; tergum II orange-brown to fulvous with a narrow black transverse band subanteriorly across anterior margin joining with narrow dark fuscous to black lateral margins; terga III–V red-brown with a broad medial and two broad lateral longitudinal black bands over all three terga and joined along anterior margin of tergum III. A pair of oval red-brown shining spots on tergum V. Posterior lobe of surstylus long, sternum V with a slight concavity on posterior margin. Abdominal sterna dark fuscous to black except sternum II fulvous.
Attractant: Cue lure.
Female: No known record.
Distribution: East Malaysia (Sarawak); Peninsular Malaysia (Gombak); Thailand (Chumphon Province).
Hosts: No known record.
Remarks: *Bactrocera (Zeugodacus) melanofacies* n.sp. is similar to *Bactrocera (Zeugodacus) vulta* (Hardy) in possessing four *sc.* setae, broad well-developed lateral and medial longitudinal postsutural vittae, a black scutum, wings colourless except for costal band and anal streak, cells bc and c colourless and costal band overlapping R_{2+3} and of uniform width throughout or widening only slightly towards apex of wing and face dark fuscous to black. It differs from *B. vulta* in having a distinct widening of the costal band at apex, the male supernumerary lobe large and with a strong keel, the male anal streak broad and widening distinctly towards apex and small fuscous preapical spots on the outer surfaces of the fore femora in the specimens from Sarawak.

Bactrocera (Zeugodacus) menglana Yu, Liu & Yang

(Fig. 326)

Bacterodacus (Zeugodacus) menglana Yu, Liu & Yang, in Yu *et al.*, 2011: 317. [*Bactrocera* Misspelt as *Bacterodacus*]. Holotype male in CAIQ.

Type material studied: CHINA: A study of the colour photographs and description of *Bactrocera (Zeugodacus) menglana* Yu, Liu & Yang in the original publication.

Non-type material examined: THAILAND: 1 male, Doi Pui, Chiang Mai, 25.vi.1992, coll. Chantanee, attracted to cue lure; **BHUTAN**: 1 male, Phuentsholing, 1.v.2000, coll. Thapa, attracted to cue lure; 1 male, Lumitshawa, 10.v.2000, coll. C. Dorji, attracted to cue lure; 1 male, Namling 16km, 9.vi.2001, coll. Brian Fletcher, attracted to cue lure. Specimens in QDPC.

Diagnosis: A medium-sized species; face fulvous with a narrow black band across oral margin; postpronotal lobes yellow (inner anterior corner dark fuscous to black); notopleura yellow; scutum entirely black; very narrow lateral and medial postsutural yellow vittae present; a very small yellow spot on anterior margin of notopleural suture; mesopleural stripe reaching midway between anterior margin of notopleuron and anterior *npl.* seta dorsally; scutellum yellow with a distinct black apical spot; wing with cells bc and c colourless, microtrichia in outer corner of cell c only; a very narrow dark fuscous costal band expanding into a spot across apex of R_{4+5} (*scutellaris*-like) and broad dark fuscous anal streak; abdominal terga III–V entirely black.

Description: male.

Head: Height 1.37 mm. Frons – length 1.28 times breadth; dark red-brown, fulvous along lateral margins and pale fuscous around orbital setae; orbital setae black: 1 *s.or.*, 3 *i.or.*; lunule fuscous. Ocellar triangle black. Vertex black. Face fulvous with a narrow black band across oral margin; length 0.43 mm. Genae fulvous, large dark fuscous subocular spot present. Occiput black, fulvous along eye margins; occipital row with 5–6 small black setae. Antennae with segments 1 and 2 pale fuscous, segment 3 missing; length of segments: 0.1 mm; 0.22 mm.

Thorax: Scutum shining black without pale markings. Pleural areas entirely shining black. Yellow markings as follows: postpronotal lobes (inner anterior corners dark fuscous to black); notopleura; mesopleural stripe reaching midway between anterior margin of notopleuron and anterior *npl.* seta dorsally, narrowing ventrally to end on katepisternum as a very small spot, anterior margin convex; anatergite (posterior apex black); anterior ⅝ katatergite (remainder black); two narrow lateral postsutural vittae beginning with a very small spot anterior to notopleural suture and ending well before *ia.* seta (at level of *p.sa.* seta); a very narrow and short medial longitudinal postsutural vitta. Postnotum black. Scutellum yellow except for narrow black basal band and distinct black apical spot that surrounds apical setae. Setae: *sc.* 4; *prsc.* 2; *ia.* 1; *p.sa.* 1; *a.sa.* 1; *mpl.* 1; *npl.* 2; *scp.* 4.

Legs: Femora entirely shining black except small basal fulvous areas on mid and hind femora (up to ⅓ of femora); fore tibiae fulvous tending dark fuscous on basal ½; mid tibiae fulvous with black basally, hind tibiae dark fuscous to black with small fulvous area apically; tarsi entirely fulvous; mid tibiae each with an apical black spur.

Wings: Length 5.6 mm; cells bc and c colourless; microtrichia in outer corner of cell c only; remainder of wings colourless except dark fuscous cell sc, very narrow dark fuscous costal band confluent with R_{2+3} and remaining very narrow beyond apex of R_{2+3} before expanding into a spot around apex of R_{4+5} (*scutellaris*-like), a broad dark fuscous anal streak; dense aggregation of microtrichia around A_1+CuA_2; supernumerary lobe very strong.

Abdomen: Oval; terga free; pecten present on tergum III. Tergum I and sterna I and II wider than long. Terga I and II dark fuscous to black except for a narrow transverse orange-brown band across intersegmental line between these terga but not reaching lateral margins and a narrow transverse fulvous band across posterior margin of tergum II; terga III–V entirely black except for a very narrow transverse dark red-brown band across posterior margin of tergum V. A pair of oval black shining spots on tergum V. Posterior lobe of surstylus long, sternum V with a slight concavity on posterior margin. Abdominal sterna dark fuscous to black.

Attractant: Cue lure.

Female: No known record.

Distribution:

Type locality: China (Yunnan Province, Mengla).

New records: Bhutan, Thailand (Chiang Mai).

Hosts: No known record.

Remarks: *Bactrocera (Zeugodacus) menglana* Yu, Liu & Yang, within the *scutellaris* complex, is similar to *Bactrocera (Zeugodacus) scutellaris* (Bezzi) in possessing four *sc.* setae and similar general body and wing colour patterns but differs in having a distinct black band across the oral margin of the face and femora almost entirely black. It is also similar to *Bactrocera (Zeugodacus) atrifacies* (Perkins) but differs from this species in having a black spot on the apex of the scutellum and a black band across the oral margin of the face.

Bactrocera (Zeugodacus) minima (Hering)

(Fig. 327)

Paradacus minimus Hering, 1952a: 42. Holotype female in NMB.

Dacus (Paradacus) minimus – Hardy & Adachi, 1954: 155; Hardy, 1982: 196.

Bactrocera (Paradacus) minima – Norrbom *et al.*, 1998: 99.

Type material examined: Holotype female, **INDONESIA: FLORES** – with the following three labels – 'TYPUS'; 'W. Flores, Reo, 7.xi.49, Expedition Bűhler – Sutter'; 'Paradacus minimus det. Hering, 1951, female holotype'. Received on loan from NMB.

Diagnosis: A small species; face fulvous with a pair of very small pale fuscous spots just above oral margin; postpronotal lobes and notopleura yellow; scutum almost entirely shining black; lateral and medial postsutural yellow vittae present; a triangular yellow spot anterior to notopleural suture and attached to notopleuron; mesopleural stripe reaching midway between anterior margin of notopleuron and anterior *npl.* seta dorsally; scutellum yellow with a narrow black basal band; wing with cell bc fuscous, cell c with a pale fuscous tint, microtrichia in outer corner of cell c only; a narrow dark fuscous costal band expanded across apex of wing where it becomes paler; a narrow fuscous anal streak; abdominal terga III–VI fulvous with a broad medial and two broad lateral longitudinal black bands joined along anterior ⅔ of tergum III.

Description: female.

Head: Height 1.21 mm. Frons – length 1.25 times breadth; fulvous with pale fuscous across anteromedial hump and enclosing upper pair of *i.or.* setae; orbital setae black: 1 *s.or.*, 2 *i.or.*; lunule red-brown. Ocellar triangle black. Vertex dark fuscous. Face fulvous with a pair of very small pale fuscous oval spots just above oral margin; length 0.37 mm. Genae fulvous, pale fuscous subocular spot present; black seta present. Occiput dark fuscous, fulvous along eye margins; occipital row with 3–4 pale setae. Antennae with segments 1 and 2 red-brown, segment 3 red-brown with fuscous on apex and outer surface; length of segments: 0.1 mm; 0.18 mm; 0.58 mm.

Thorax: Scutum entirely shining black except red-brown inside postpronotal lobes and a small area anterocentrally. Pleural areas entirely shining black. Yellow markings as follows: postpronotal lobes; notopleura; mesopleural stripe extending midway between anterior margin of notopleuron and anterior *npl.* seta dorsally before recurving back towards notopleuron, continuing to katepisternum as a transverse spot, anterior margin convex; anatergite (posterior apex black); anterior ¾ katatergite (remainder black); two medium-width lateral postsutural vittae beginning with a spot anterior to notopleural suture which connects with the notopleuron and narrowing posteriorly before ending at or just before *ia.* seta; a narrow medial longitudinal postsutural vitta running almost entire length of scutum. Postnotum black. Scutellum yellow except for narrow black basal band. Setae: *sc.* 4; *prsc.* absent; *ia.* 1; *p.sa.* 1; *a.sa.* 1; *mpl.* 1; *npl.* 2; *scp.* 4. A weak *hm.* seta is also present.

Legs: Fore femora fulvous with a narrow elongate dark fuscous spot on outer surface, mid and hind femora fulvous tending dark fuscous on apex; fore tibiae fulvous, mid tibiae dark fuscous tending fuscous apically, hind tibiae dark fuscous; tarsi with basal two segments fulvous and apical three segments red-brown; mid tibiae each with an apical black spur.

Wings: Length 4.4 mm; cell bc fuscous, cell c with a pale fuscous tint; microtrichia in outer corner of cell c only; remainder of wings colourless except dark fuscous cell sc, narrow dark fuscous costal band slightly overlapping R_{2+3} and expanding across apex of wing where it becomes paler before ending at apex of M, a narrow fuscous anal streak; no dense aggregation of microtrichia around A_1+CuA_2; supernumerary lobe weak.

Abdomen: Oval; terga free; no pecten on tergum III. Tergum I and sterna I and II wider than long. Tergum I fulvous with narrow black lateral margins; tergum II black with a transverse fulvous band across anterior margin but not reaching lateral margins; terga III–VI fulvous with a broad medial and two broad lateral longitudinal black bands joined across-anterior ⅔ of tergum III. A pair of oval fulvous shining spots on tergum V. Basal segment of ovipositor red-brown, circular in transverse cross section and tapering posteriorly in dorsal view; ratio of length of oviscape to length of tergum V, 2:1.

Attractant: No known record.
Male: No known record.
Distribution:
Type locality: Indonesia (Flores, Reo).
Hosts: No known record.
Remarks: *Bactrocera (Zeugodacus) minima* (Hering) is provisionally placed in subgenus *Zeugodacus* until male specimens become available for study. The thoracic chaetotaxy is that for *Zeugodacus* except that the *prsc.* setae are absent. However, the presence of *a.sa.* setae, four *sc.* setae and a medial postsutural yellow vitta are now recognized as important characters of subgenus *Zeugodacus*. It is a unique species in possessing an extremely long medial longitudinal yellow vitta on the scutum in addition to the lateral postsutural yellow vittae, a narrow costal band expanded across apex of wing, cell bc of wing fuscous and abdominal terga III–VI with broad medial and lateral longitudinal black bands joined across tergum III.

Bactrocera (Zeugodacus) munda (Bezzi)

(Fig. 328)

Chaetodacus mundus Bezzi, 1919: 429. Syntypes male and female in MCSNM.
Zeugodacus tibialis Shiraki, 1933: 96; Hardy & Adachi, 1954: 193 (Synonym).
Dacus (Zeugodacus) mundus – Hardy, 1969b: 481, Lectotype designation invalid; Hardy, 1974: 52.
Bactrocera (Zeugodacus) munda – Norrbom *et al.*, 1998: 103.

Type material examined: PHILIPPINES: Syntype male with the following three labels – 'Batbatan Isl. R.C. McGregor'; 'mundus n.' (in handwriting probably Bezzi's); 'Lectotype male, *Chaetodacus mundus* Bez. desig. by Hardy 68'. Syntype male, 'Los Banos, P.I. Baker'. Syntype female, 'Paete, Lag. McGregor' (in handwriting). Received on loan from MCSNM.

Diagnosis: A medium-sized species; face entirely fulvous; postpronotal lobes and notopleura yellow; scutum black; lateral and medial postsutural yellow vittae present; a yellow spot anterior to notopleural suture in front of lateral postsutural vittae; mesopleural stripe reaching almost to anterior *npl.* seta dorsally; scutellum yellow; wing with cells bc and c colourless, microtrichia in outer corner of cell c only; a narrow dark fuscous costal band just overlapping R_{2+3}; a broad fuscous anal streak; fuscous enclosing *dm-cu* crossvein and continuing around hind margin to meet anal streak; abdominal terga III–V fulvous with a dark 'T' pattern consisting of a transverse black band across anterior margin of tergum III and a medial longitudinal dark fuscous band over all three terga, anterolateral corners of terga IV and V dark fuscous.

Description: male.

Head: Height 1.37 mm. Frons – length 1.28 times breadth; red-brown with dark fuscous around orbital setae and on anteromedial hump; orbital setae black: 1 *s.or.*, 3 *i.or.*; lunule dark fuscous. Ocellar triangle black. Vertex fuscous. Face fulvous

without any dark markings; length 0.48 mm. Genae fulvous, pale fuscous subocular spot present; black seta present. Occiput fuscous centrally, dark fuscous laterally; occipital row with a large number of small pale setae. Antennae with segments 1 and 2 red-brown, segment 3 red-brown with fuscous on apex and outer surface; length of segments: 0.1 mm; 0.24 mm; 0.62 mm.
Thorax: Scutum black with dark red-brown below and behind lateral postsutural vittae and red-brown around notopleural suture. Pleural areas black except red-brown below postpronotal lobes. Yellow markings as follows: postpronotal lobes; notopleura; mesopleural stripe reaching almost to anterior *npl.* seta dorsally, continuing to katepisternum as a transverse spot, anterior margin convex; anatergite (posterior apex black); anterior ¾ katatergite (remainder black); two broad lateral postsutural vittae beginning in a spot anterior to notopleural suture and narrowing sharply posteriorly to end before *ia.* seta; a broad medial longitudinal postsutural vitta beginning posteriorly at line of *prsc.* setae and ending in a point just anterior of notopleural suture. Postnotum black. Scutellum yellow with a narrow black basal band. Setae: *sc.* 4; *prsc.* 2; *ia.* 1; *p.sa.* 1; *a.sa.* 1; *mpl.* 1; *npl.* 2; *scp.* 4.
Legs: Fore femora dark fuscous except fuscous basally and apically, mid femora dark fuscous to fuscous on apical ½ and fulvous on basal ½, hind femora fuscous to dark fuscous on apical ⅓ and fulvous on the remainder; fore tibiae dark fuscous basally tending pale fuscous apically, mid tibiae fuscous basally tending fulvous apically, hind tibiae dark fuscous basally tending fuscous apically; all tarsal segments fulvous; mid tibiae each with an apical black spur.
Wings: Length 6.0 mm; cells bc and c colourless; microtrichia in outer corner of cell c only; remainder of wings colourless except dark fuscous cell sc, narrow dark fuscous costal band just overlapping R_{2+3} over apical ½ of the vein and remaining of uniform width to end midway between the apices of R_{4+5} and M, broad fuscous anal streak with pale fuscous to fuscous extending from anal streak around hind margin of wing and along *dm-cu* crossvein; supernumerary lobe strong and rounded.
Abdomen: Oval; terga free; pecten present on tergum III. Tergum I and sterna I and II wider than long. Tergum I fulvous with fuscous across anterior margin and narrow black lateral margins; tergum II fulvous with a narrow transverse black band anteriorly but not reaching lateral margins and dark fuscous anterolateral corners; terga III–V fulvous with a dark fuscous to black 'T' consisting of a narrow transverse band across anterior margin of tergum III which widens around lateral margins and a medial longitudinal dark fuscous band of medium-width over all three terga, dark fuscous anterolateral corners of terga IV and V. A pair of oval fulvous shining spots on tergum V. Posterior lobe of surstylus long, sternum V with posterior margin straight.
Attractant: No known record.
Female: As for male except no dense aggregation of microtrichia around A_1+CuA_2; supernumerary lobe weak; pecten of cilia on abdominal tergum III absent; basal segment of ovipositor dark fuscous, dorsoventrally flattened and tapering posteriorly in dorsal view; ratio of length of oviscape to length of tergum V, 2.1:1.

Distribution:
Type locality: Philippines (Luzon I. – Los Banos, Paete; Panay I. – Batbatan I.; Mindanao I. – Davao). All locations constitute the type locality as all type specimens are regarded as Syntypes.
Previously recorded: Type locality and Taiwan, due to synonymy of *Zeugodacus tibialis*.
Hosts: No known record.
Remarks: *Bactrocera (Zeugodacus) munda* (Bezzi) is similar to *Bactrocera (Zeugodacus) diaphoropsis* (Hering) in possessing four *sc.* setae, a black scutum with medial and lateral longitudinal postsutural yellow vittae, wing with infuscation in addition to costal band and anal streak, costal band of uniform width throughout and scutellum entirely yellow. It differs from *B. diaphoropsis* in having the lateral postsutural vittae ending before *ia.* seta, infuscation enclosing *dm-cu* crossvein and continuing around hind margin of wing to anal streak, fore femora entirely dark fuscous to black and apical ½ of mid femora and apical ⅓ hind femora dark fuscous to black.

Hardy (1969) designated a female Lectotype in the USNM. However, this designation is invalid (Norrbom *et al.*, 1998: 103) because there is no confirmation that Bezzi (1919) studied these USNM specimens when describing *B. munda*. Also, in MCSNM, there is a male specimen with a Lectotype label placed on it by Hardy in 1968. However, this specimen cannot be regarded as a Lectotype as there is no formal designation in literature by Hardy. Consequently, the specimens in MCSNM must be regarded as Syntypes and those in USNM non-types.

Bactrocera (Zeugodacus) nakhonnayokiae Drew & Romig, new species
(Fig. 329)

Material examined: Holotype male, **THAILAND**: Khao Kheaw, A. Pakphli, Nakhonnayok, 22.v.1992, coll. Montree, attracted to cue lure, deposited in BMNH.
Paratypes: **PENINSULAR MALAYSIA**: 1 male, Gombak, Malaysia, 6.v.1986, coll. Mahdzir, attracted to cue lure; 1 male, TS Gombak, Malaysia, 25.iii.1986, coll. Mahdzir, attracted to cue lure; 2 males, TS8 Gombak, Malaysia, 28.vii.1986, coll. Mahdzir, attracted to cue lure; 1 male, TS8 Malaysia, 1.xii.1986, coll. Anthony, attracted to cue lure; 1 male (26.i.1987), 1 male (23.ii.1987), TS8 Gombak, Malaysia, coll. Bahari, attracted to cue lure; 1 male, Kemaman, Malaysia, 7.vii.1986, coll. Raja Azid, attracted to cue lure. **THAILAND**: 2 males (25.x.1991), 2 males (1.xi.1991), 1 male (8.xi.1991), same data as holotype; 2 males (18.x.1991), 2 males (25.x.1991), 1 male (10.i.1992), 1 male (15.v.1992), Wang Jum Pee, Pakchong, Nakhonratsima, coll. Montree, attracted to cue lure; 1 male, Yang Pong Rat, Plubpla, A. Muang, Chanthaburi, trap No. BK38T, 23.vii.1991, coll. Montree; 1 male, Pha Yao, Chiang Rai, 6.xi.1991, coll. Charuwan, Site CR220T, attracted to cue lure; 2 males (4.ii.1992), 2 males (13.ii.1992), Khun Khon, Chiang Rai, coll. Charuwan, Sites CR309T and CR349T, attracted to cue lure; 1 female,

Nakhon Naoyk Prov, Khao Yai Nat. Park, c. 700 m, 29.ix-6.x.1984, coll. Karsholt, Lomholdt & Nielsen leg., Zool. Mus., Copenhagen. **VIETNAM: Northern Vietnam** – 1 male, Son La Province, Moc Chau Dist, Co Do, 20.ix.1999, trap No. N006, attracted to cue lure; **Southern Vietnam** – 1 male, Dong Nai Province, Long Khanh District, Nhan Nghia Village, 3.viii.2002, trap No. VS06, attracted to cue lure. 4 paratypes in ANIC; 4 paratypes in BMNH; 16 paratypes in QDPC; 4 paratypes in QMIC.

Diagnosis: A large species; face fulvous with a broad black arched transverse band; postpronotal lobes and notopleura yellow; scutum shining black; lateral and medial postsutural yellow vittae present; a yellow spot present anterior to notopleural suture; mesopleural stripe reaching almost to anterior *npl.* seta dorsally; scutellum yellow; wing with cells bc and c colourless, microtrichia in outer corner of cell c only; a narrow fuscous costal band widening across apex of R_{4+5} (*scutellata*-like) and broad fuscous anal streak; abdominal terga III–V mostly dark fuscous with narrow pale margins posterocentrally on terga III and IV and across hind margin of tergum V.

Description: male.

Head: Height 1.62 mm. Frons – length 1.38 times breadth; red-brown centrally with fulvous along lateral margins and dark fuscous around orbital setae and on anteromedial hump; orbital setae black:1 *s.or.*, 3 *i.or.*; lunule dark fuscous. Ocellar triangle black. Vertex black. Face fulvous with a broad black band which begins on oral margins at positions of antennal furrows and arches around centre of face; length 0.53 mm. Genae fulvous, large fuscous subocular spot present; black seta present. Occiput fuscous tending dark fuscous on dorsolateral margins, fulvous along eye margins; occipital row with 2–6 strong black setae. Antennae with segments 1 and 2 red-brown, segment 3 red-brown with fuscous on apex and outer surface; arista black (red-brown basally); length of segments: 0.12 mm; 0.26 mm; 0.77 mm.

Thorax: Scutum shining black with dark fuscous behind lateral postsutural vittae. Pleural areas entirely shining black. Yellow markings as follows: postpronotal lobes (small fuscous anterodorsal corners); notopleura; mesopleural stripe reaching almost to anterior *npl.* seta dorsally, continuing to katepisternum as a transverse spot, anterior margin straight to slightly convex; anatergite (posterior apex black); anterior $^{3}/_{5}$ katatergite (remainder black); two medium-width parallel-sided lateral postsutural vittae beginning with a small spot anterior to notopleural suture and ending at *ia.* seta; a short narrow medial longitudinal postsutural vitta. Postnotum black. Scutellum yellow except for narrow black basal band. Setae: *sc.* 4; *prsc.* 2; *ia.* 1; *p.sa.* 1; *a.sa.* 1; *mpl.* 1; *npl.* 2; *scp.* 4.

Legs: Fore femora shining black except up to ¼ of basal area fulvous, mid femora with apical ⅓ dark fuscous to black and remainder fulvous, hind femora with apical ¼ dark fuscous to black and remainder fulvous; fore tibiae dark fuscous tending paler apically, mid tibiae fuscous to dark fuscous tending darker basally, hind tibiae dark fuscous; tarsi with basal segment fulvous and apical four segments pale fuscous; mid tibiae each with an apical black spur.

Wings: Length 6.6 mm; cells bc and c colourless; microtrichia in outer corner of cell c only; remainder of wings colourless except fuscous cell sc, narrow fuscous costal band confluent with R_{2+3} and remaining narrow beyond apex of this vein before widening across apex of R_{4+5} (*scutellata*-like), broad fuscous anal streak; dense aggregation of microtrichia around A_1+CuA_2; supernumerary lobe strong and rounded.

Abdomen: Oval; terga free; pecten present on tergum III. Tergum I and sterna I and II wider than long. Tergum I dark fuscous tending darker laterally and with a broad transverse fulvous band across intersegmental line between terga I and II but not reaching lateral margins; tergum II fulvous with a broad transverse dark fuscous band across anterior margin and reaching dark fuscous anterolateral corners; terga III–V fuscous to dark fuscous tending paler (dark red-brown to fuscous) posterocentrally on terga III and IV and along posterior margin of tergum V (in some specimens the overall basic colour of terga III–V can be slightly paler). A pair of oval red-brown to dark red-brown shining spots on tergum V. Posterior lobe of surstylus long, sternum V with a slight concavity on posterior margin. Abdominal sterna dark fuscous, except sternum II paler (fulvous to fuscous).

Attractant: Cue lure.

Female: As for male except no dense aggregation of microtrichia around A_1+CuA_2; supernumerary lobe weak; pecten of cilia on abdominal tergum III absent. Basal segment of ovipositor red-brown, dorsoventrally flattened and tapering posteriorly in dorsal view; ratio of length of oviscape to length of tergum V, 2.5:1.

Distribution: Peninsular Malaysia; Thailand; Vietnam (Dong Nai and Son La Provinces).

Hosts: No known record.

Remarks: *Bactrocera (Zeugodacus) nakhonnayokiae* n.sp., within the *scutellaris* complex, is similar to *Bactrocera (Zeugodacus) menglana* Yu, Liu & Yang and *Bactrocera (Zeugodacus) scutellaris* (Bezzi) in possessing four *sc.* setae, abdominal terga III–V mostly dark fuscous to black and a strong supernumerary lobe. It differs from *B. scutellaris* in having a black transverse band across the face and a scutellata-like costal band on the wing and from *B. menglana* in having the scutellum entirely yellow, lateral postsutural vittae basically parallel-sided and extending to *ia.* seta, costal band *scutellata*-like and the mid and hind femora fulvous with dark fuscous to black apices, not entirely black.

Bactrocera (Zeugodacus) namlingiae Drew & Romig, new species

(Fig. 330)

Material examined: Holotype male, **BHUTAN**: Namling 16 km, 9.vi.2001, coll. Brian Fletcher, attracted to cue lure, deposited in BMNH.

Paratype: **BHUTAN**: 1 male, Namling 14 km, 7.vi.2001, coll. Brian Fletcher, attracted to cue lure, deposited in QDPC.

Diagnosis: A medium-sized species; face entirely pale fulvous; postpronotal lobes yellow with broad black bands along dorsal and ventral margins; notopleura yellow; scutum shining black; lateral and medial postsutural yellow vittae present but reduced to very short narrow lines; mesopleural stripe equal in width to notopleuron dorsally; scutellum yellow with apical ½ of dorsal surface shining black; wing with cells bc and c colourless, microtrichia in outer corner of cell c only; a narrow fuscous costal band remaining extremely narrow, almost absent, beyond apex of R_{2+3} and widening across apex of R_{4+5} (the expansion is more *scutellata*-like); a narrow fuscous anal streak; abdominal terga III and IV shining black with small dark red-brown spots on posterior margin either side of the medial longitudinal black band on terga III and IV, and tergum V dark fuscous to shining black with narrow longitudinal red-brown bands either side of a medial longitudinal black band.

Description: male.

Head: Height 1.08 mm. Frons – length 1.32 times breadth; fulvous with fuscous to dark fuscous around orbital setae and on anteromedial hump; orbital setae black: 1 *s.or.*, 3 *i.or.*; lunule dark fuscous. Ocellar triangle black. Vertex black. Face entirely pale fulvous without dark markings; length 0.45 mm. Genae fulvous, dark fuscous subocular spot present; black seta present. Occiput shining black, fulvous along eye margins; occipital row with 3–5 strong black setae. Antennae with segments 1 and 2 red-brown, segment 3 red-brown with fuscous on apex and outer surface; arista black (red-brown basally); length of segments: 0.09 mm; 0.21 mm; 0.57 mm.

Thorax: Scutum and pleural areas shining black without pale markings. Yellow markings as follows: postpronotal lobes (broad black bands along dorsal and ventral margins); notopleura (in one specimen black encroaches on the yellow around the base of the notopleuron); narrow mesopleural stripe equal in width to notopleuron dorsally, continuing to katepisternum as a small transverse spot, anterior margin straight; anatergite (posterior apex black); anterior $^{3}/_{5}$ katatergite (remainder black); two very short and narrow lateral postsutural vittae (these are extremely narrow and short with the appearance of fine lines); a very short and narrow medial longitudinal postsutural vitta (with similar appearance to lateral vittae). No yellow spot anterior to notopleural suture. Postnotum black. Scutellum yellow with a narrow black basal band and a large black spot over apical ½ of dorsal surface. Setae: *sc.* 4; *prsc.* 2; *ia.* 1; *p.sa.* 1; *a.sa.* 1; *mpl.* 1; *npl.* 2; *scp.* 4.

Legs: Fore femora fulvous with a large shining black spot on apical ½ of outer surface, mid and hind femora fulvous with shining black around apical ½; fore tibiae dark fuscous tending paler apically, mid tibiae fuscous basally tending pale fuscous apically, hind tibiae dark fuscous to black; tarsi with all segments fulvous; mid tibiae each with an apical black spur.

Wings: Length 5.4 mm; cells bc and c colourless; microtrichia in outer corner of cell c only; remainder of wings colourless except dark fuscous cell sc, narrow fuscous costal band confluent with R_{2+3} and remaining very narrow, almost absent, beyond apex of this vein before widening across apex of R_{4+5}, a narrow fuscous anal streak; dense aggregation of microtrichia around A_1+CuA_2; supernumerary lobe weak.

Abdomen: Elongate oval; terga free; pecten present on tergum III. Tergum I and sterna I and II wider than long. Tergum I shining black except for a narrow transverse dark red-brown band along posterior margin but not reaching lateral margins; tergum II shining black except for transverse fulvous posterolateral bands across posterior margin either side of a medial longitudinal black band; terga III and IV shining black except for dark red-brown spots posterocentrally on posterior margins either side of a medial longitudinal black band on terga III and IV and tergum V dark fuscous to black except for red-brown longitudinal bands either side of the broad medial longitudinal black band. A pair of oval shining spots on tergum V that are red-brown adjacent to medial longitudinal black band and dark fuscous laterally. Posterior lobe of surstylus long, sternum V with a slight concavity on posterior margin. All abdominal sterna dark fuscous to black.

Attractant: Cue lure.

Female: No known record.

Distribution: Bhutan (Namling).

Hosts: No known record.

Remarks: *Bactrocera (Zeugodacus) namlingiae* n.sp., within the *scutellaris* complex, is similar to *Bactrocera (Zeugodacus) atrifacies* (Perkins) and *Bactrocera (Zeugodacus) scutellaris* (Bezzi) in possessing four *sc.* setae and abdominal terga III–V mostly dark fuscous to black (*scutellaris*-like). It is distinct in having an entirely pale fulvous face without dark markings, broad black bands along dorsal and ventral margins of postpronotal lobes and anal streak narrow and pale fuscous.

Bactrocera (Zeugodacus) neoelegantula White

(Fig. 331)

Bactrocera (Zeugodacus) neoelegantula White, in White & Evenhuis, 1999: 532. Holotype male in ZMAN.

Type material examined: Holotype male, **INDONESIA: SUMATRA** – Fort de Kock, Sumatra, 920 m, 1925, coll. E. Jacobson. Studied at BMNH before it was deposited in ZMAN.

Diagnosis: A large species; face entirely fuscous to dark fuscous; postpronotal lobes and notopleura yellow; scutum entirely black without pale markings; lateral and medial postsutural yellow vittae present; a yellow spot anterior to notopleural suture; mesopleural stripe reaching almost to postpronotal lobe; scutellum entirely yellow; legs with fore femora entirely fuscous to dark fuscous, mid and hind femora fulvous with apical ⅓ fuscous to dark fuscous; wing with cells bc and c colourless, microtrichia in outer corner of cell c only; a narrow fuscous costal band just overlapping R_{2+3} and remaining narrow around apex of wing, infuscation along *dm-cu* crossvein; a broad pale fuscous anal streak present; abdominal terga III–V fulvous with broad lateral dark fuscous to black markings on lateral margins of terga III and IV and narrow dark fuscous to black margins on tergum V and a broad medial longitudinal black band over all three terga.

Description: male.
Head: Frons – fulvous with fuscous spots around *s.or.* setae; orbital setae black: 1 *s.or.*, 2 *i.or.*; lunule fuscous. Ocellar triangle black. Vertex with fuscous spots laterally. Face entirely fuscous to dark fuscous, paler centrally on oral margin and below antennae. Occiput fuscous, fulvous along eye margins. Antennae with all segments red-brown.
Thorax: Scutum entirely black without pale markings. Pleural areas dark fuscous to black. Yellow markings as follows: postpronotal lobes; notopleura; mesopleural stripe reaching almost to postpronotal lobe dorsally, extending to katepisternum as a large spot, anterior margin straight; two broad parallel-sided lateral postsutural vittae beginning with a spot anterior to notopleural suture and ending behind *ia.* seta; a broad medial longitudinal postsutural vitta beginning at level of *prsc.* setae and ending at a point just anterior to notopleural suture. Scutellum yellow with a broad black basal band. Setae: *sc.* 2; *prsc.* 2; *ia.* 1; *p.sa.* 1; *a.sa.* 1; *mpl.* 1; *npl.* 2; *scp.* 4.
Legs: Fore femora entirely fuscous to dark fuscous, mid and hind femora fulvous with fuscous to dark fuscous around apical ⅓; fore tibiae fulvous with a small fuscous spot basally, mid tibiae fuscous basally to fulvous apically, hind tibiae fuscous to dark fuscous; tarsi with basal segment fulvous and apical four segments pale fuscous.
Wings: Length 7.05 mm; cells bc and c colourless; microtrichia in outer corner of cell c only; remainder of wings colourless except fuscous cell sc, narrow pale fuscous costal band just overlapping R_{2+3} and remaining narrow as it extends to apex of wing, infuscation along *dm-cu* crossvein, broad pale fuscous anal streak; supernumerary lobe of medium development.
Abdomen: Oval; pecten present on tergum III. Tergum I fulvous with dark fuscous across anterior margin and around anterolateral corners; tergum II fulvous with a broad dark fuscous to black band across anterior margin and continuing along lateral margins; terga III–V fulvous with broad dark fuscous to black markings laterally on terga III and IV and narrowly on tergum V and a broad medial longitudinal black band over all three terga. Posterior lobe of surstylus long, sternum V with a 'V'-shaped indentation of moderate size in posterior margin. Abdominal sterna dark, except sternum II pale.
Attractant: No known record.
Female: No known record.
Distribution:
Type locality: Indonesia (Sumatra).
Hosts: No known record.
Remarks: *Bactrocera (Zeugodacus) neoelegantula* White belongs to the group of species with two *sc.* setae, medial postsutural yellow vitta and *prsc.* setae present, scutum basically black, notopleura yellow and *dm-cu* crossvein infuscated. It is similar to *B. hululangatiae* n.sp. and *B. synnephes* (Hendel) in possessing a yellow spot anterior to the notopleural suture and dark apices on the femora. It differs from *B. synnephes* in having the mesopleural stripe reaching to the postpronotal lobe dorsally and the costal band not expanded across apex of R_{4+5}, and from *B. hululangatiae* in having a broad black basal band on the scutellum, fore femora entirely fuscous to black, abdominal terga III–V with broad lateral dark margins and abdominal sternum V with a deep 'V'-shaped concavity on posterior margin.

Bactrocera (Zeugodacus) neoemittens Drew & Romig, new species

(Fig. 332)

Material examined: Holotype male, **INDONESIA: FLORES** – Borong, Flores, Apr.1989, coll. J. Turner, attracted to cue lure, deposited in BMNH.
Paratypes: **INDONESIA: FLORES** – 106 males, same data as holotype; **LOMBOK** – 3 males, Lombok, W. Lombok, Suranadi, 4.vi.78, coll. J.T. Medler; 1 male, C. Lombok, Aik Bukak, 2.vii.78, coll. J. T. Medler. 10 paratypes in ANIC; 10 paratypes in BMNH; 80 paratypes in QDPC; 10 paratypes in QMIC.
Diagnosis: A large species; face with a pair of small oval black spots; postpronotal lobes and notopleura yellow; scutum black; lateral and medial postsutural yellow vittae present; a yellow spot anterior to notopleural suture; mesopleural stripe reaching midway between anterior margin of notopleuron and anterior *npl.* seta dorsally; scutellum yellow; wing with cells bc and c with a pale tint, microtrichia in outer corner of cell c only; a broad dark fuscous costal band expanding into a large spot across apex of wing and recurved back along *dm-cu* crossvein and a moderately broad dark fuscous anal streak; abdominal terga III–V orange-brown with a dark fuscous to black 'T' pattern and broad fuscous to dark fuscous lateral margins over all three terga (in some specimens these margins are narrower and paler).
Description: male.
Head: Height 1.56 mm. Frons – length 1.18 times breadth; red-brown with pale fuscous on anteromedial hump; orbital setae black: 1 *s.or.*, 3 *i.or.*; lunule dark fuscous. Ocellar triangle black. Vertex red-brown. Face fulvous with a pair of small oval black spots; length 0.59 mm. Genae fulvous, pale fuscous subocular spot present; black seta present. Occiput red-brown to pale fuscous, fulvous along eye margins; occipital row with a large number of small to large black setae. Antennae with segments 1 and 2 red-brown, segment 3 red-brown with fuscous on apex and outer surface; arista black (red-brown basally); length of segments: 0.14 mm; 0.26 mm; 0.7 mm.
Thorax: Scutum mostly black with red-brown below and behind lateral postsutural vittae, around notopleural suture and yellow spot anterior to suture, posterocentrally around *prsc.* setae and anteriorly between postpronotal lobes and back to anterior point of medial postsutural yellow vitta (there is a narrow medial longitudinal black line in centre of the anteromedial red-brown area). Pleural areas dark fuscous to black except red-brown above mid coxae and below postpronotal lobes. Yellow markings as follows: postpronotal lobes; notopleura; mesopleural stripe of medium-width, reaching midway between anterior margin of notopleuron and anterior *npl.* seta dorsally, continuing to katepisternum as a large transverse spot, anterior margin slightly convex; anatergite (posterior apex black); anterior ¾ katatergite

(remainder black); two broad parallel-sided lateral postsutural vittae beginning with a spot anterior to notopleural suture and ending at or just behind *ia.* seta; a broad medial longitudinal postsutural vitta beginning at level of *prsc.* setae and ending at a point just anterior of notopleural suture. Postnotum black with a medial red-brown area. Scutellum yellow except for narrow black basal band. Setae: *sc.* 4 (occasional specimens with 2); *prsc.* 2; *ia.* 1; *p.sa.* 1; *a.sa.* 1; *mpl.* 1; *npl.* 2; *scp.* 4.
Legs: Femora fulvous except for a small to large fuscous to dark fuscous spot on outer apical surfaces of fore femora and mid and hind femora tending red-brown apically; fore and hind tibiae fuscous to dark fuscous, mid tibiae fuscous basally and fulvous apically; tarsi with basal segments fulvous and apical four segments fuscous; mid tibiae each with an apical black spur.
Wings: Length 6.6 mm; cells bc and c colourless; microtrichia in outer corner of cell c only; remainder of wings colourless except dark fuscous cell sc, broad dark fuscous costal band overlapping R_{2+3} and becoming paler as it becomes confluent with R_{4+5} and expanding into a large spot across apex of wing to vein M and then recurving back along *dm-cu* crossvein, a moderately broad dark fuscous anal streak (this pattern is very similar to *Bactrocera emittens* and in some specimens there is a gap between the apical spot and the infuscation along the *dm-cu* crossvein); a dense aggregation of microtrichia around A_1+CuA_2; supernumerary lobe strong.
Abdomen: Elongate oval; terga free; pecten present on tergum III. Tergum I and sterna I and II wider than long. Tergum I orange-brown with fuscous across anterior margin and dark fuscous lateral margins; tergum II fulvous with a narrow transverse dark fuscous band across anterior margin and reaching with dark fuscous lateral margins; terga III–V orange-brown with a 'T' pattern consisting of a narrow transverse dark fuscous to black band across anterior margin of tergum III and a medium-width medial longitudinal dark fuscous to black band over all three terga, broad lateral longitudinal dark fuscous bands over all three terga (in some specimens the lateral bands are considerably narrower and paler fuscous). A pair of oval pale fuscous shining spots on tergum V. Posterior lobe of surstylus long, sternum V with a slight concavity on posterior margin. Abdominal sterna dark.
Attractant: Cue lure.
Female: No known record.
Distribution: Indonesia (Flores; Lombok).
Hosts: No known record.
Remarks: *Bactrocera (Zeugodacus) neoemittens* n.sp. is almost identical to *Bactrocera (Zeugodacus) emittens* (Walker) in the wing colour pattern but differs in having a predominantly black scutum, fuscous to dark fuscous preapical spots on outer surfaces of fore femora, abdominal terga III–V with a dark fuscous to black 'T' pattern and broad lateral longitudinal fuscous to dark fuscous margins over all three terga, and a large keel-shaped supernumerary lobe in the male wing and colourless cells bc and c. It is also similar to *Bactrocera (Zeugodacus) calumniata* (Hardy) and *Bactrocera (Zeugodacus) ishigakiensis* (Shiraki) in the general colour patterns of the thorax and abdomen but differs in the distinctive *emittens*-like wing colour pattern and the abdominal patterns, particularly the continuous lateral longitudinal dark bands.

Bactrocera (Zeugodacus) neoflavipilosa Drew & Romig, new species

(Fig. 333)

Material examined: Holotype male, **INDONESIA: SULAWESI** – Kuala Navusa, Malakosa, Sulawesi, 100 ft, Aug–Nov. 1975, coll. Margareta Becker, with blue paratype label 'Paratype ♂, *Dacus (Zeugodacus) flavipilosus*, D.E. Hardy', deposited in BBM.
Paratypes: **INDONESIA: SULAWESI** – 1 male with the following four labels – 'Sadaunta, 65 km S.E. Palu, Sulawesi, 650 m, Aug 1975, D.E. Hardy'; blue paratype label 'Paratype ♂, *Dacus (Zeugodacus) flavipilosus*, D.E. Hardy'; white label '*Dacus (Zeugodacus) flavipilosus*, Det. 1976 Hardy, D.E. Hardy'; 'Bishop Museum Acc # 1997.256'. 2 males, 'Sungai, Sadaunta, Cent. Sulawesi, 3000 ft, March '76, M. Becker' blue paratype labels 'Paratype ♂, *Dacus (Zeugodacus) flavipilosus*, D.E. Hardy' – one specimen with 'Acc # 1983-454' and white label '*Dacus (Zeugodacus) flavipilosus* Hardy, Det 1983, D.E. Hardy' and the other specimen with 'Bishop Museum Acc #1997.256'. These 3 paratypes in BBM. 1 male, 'Sadaunta, 65km S.E. Palu, Sulawesi, 650 m, Aug 1975, coll. D.E. Hardy'; blue paratype label 'Paratype ♂, *Dacus (Zeugodacus) flavipilosus* D.E. Hardy'; white label 'Dacus (Zeugodacus) flavipilosus Hardy, Det D.E. Hardy, 1980'; paratype in QDPC.

Non-type material examined: INDONESIA: SULAWESI – 1 male, Sulawesi, Gorontalo, Bone Bolango, Suwawa, 26.ix.2006, Yusbar Ismail, attracted to cue lure in forest; 4 males, North Sulawesi, Bitung, Danowudu, 26.ix.2006, coll. Karel F. Lala, attracted to cue lure in forest; 2 males, Central Sulawesi, Donggala, Nupabomba, 2.ix.2006, coll. Rusdin, attracted to cue lure in forest; 1 male, Sulawesi Utara, Suwoon, 14.viii.2006, coll. Karel F. Lala, attracted to cue lure in citrus. Specimens in QDPC.
Diagnosis: A large species; face fulvous with a pair of small circular black spots; postpronotal lobes and notopleura yellow; scutum predominantly red-brown with small black patterns; lateral and medial postsutural yellow vittae present; a large triangle along anterior margin of notopleural suture; mesopleural stripe equal in width to notopleuron dorsally; scutellum yellow; wing with cells bc and c fulvous, microtrichia in outer ⅓ of cell c only; a broad fuscous costal band confluent with R_{4+5} and expanding into a fuscous to dark fuscous spot across apex of wing; a broad fuscous anal streak, a fuscous tint around posterior margin of wing from apical spot to anal streak; abdominal terga III–V light red-brown with a narrow dark fuscous to black transverse band across anterior margin of tergum III and a narrow dark fuscous to black medial longitudinal band over terga IV and V (this band is sometimes narrower and shorter or even absent).

Description: male.
Head: Height 1.77 mm. Frons – length 1.35 times breadth; entirely red-brown without dark markings; orbital setae pale: 1 *s.or.*, 3 *i.or.*; lunule red-brown. Ocellar triangle black. Vertex red-brown. Face fulvous with a pair of small circular black spots; length 0.64 mm. Genae red-brown; pale seta present. Occiput red-brown, fulvous along eye margins; occipital row with a large number of small pale setae. Antennae with segments 1 and 2 red-brown, segment 3 pale fuscous to fuscous; arista black (red-brown basally); length of segments: 0.12 mm; 0.27 mm; 0.84 mm.
Thorax: Scutum mostly red-brown with black patterns as follows: an irregular 'U'-shaped pattern between lateral and medial postsutural yellow vittae, between postpronotal lobes and yellow triangle at notopleural suture, a short narrow medial longitudinal stripe on anterior area. Pleural areas dark fuscous to black except red-brown above mid coxae and below postpronotal lobes. Yellow markings as follows: postpronotal lobes; notopleura; mesopleural stripe equal in width to notopleuron dorsally, extending to katepisternum as a large yellow spot, anterior margin straight; anatergite (posterior apex black); anterior ⅔ katatergite (remainder black); two broad subparallel lateral postsutural vittae ending just behind *ia.* seta; a broad medial longitudinal postsutural vitta beginning at level of *prsc.* setae and narrowing anteriorly to end anterior to notopleural suture; a broad triangle along anterior margin of notopleural suture with base at notopleuron. Postnotum black tending dark red-brown centrally. Scutellum yellow except for narrow black basal band. Setae pale: *sc.* 4; *prsc.* 2; *ia.* 1; *p.sa.* 1; *a.sa.* 1; *mpl.* 1; *npl.* 2; *scp.* 4.
Legs: All segments entirely fulvous; mid tibiae each with an apical black spur.
Wings: Length 7.8 mm; cells bc and c fulvous; microtrichia in outer ⅓ of cell c only; remainder of wings colourless except fuscous cell sc, broad pale fuscous costal band confluent with R_{4+5} and expanding into a spot across apex of wing reaching M (in some specimens this spot is fuscous with a dark fuscous apical spot within), a pale fuscous tint around posterior margin of wing from apical spot to anal streak, a broad fuscous anal streak; dense aggregation of microtrichia around A_1+CuA_2; supernumerary lobe strong and keel shaped.
Abdomen: Elongate oval; terga free; pecten present on tergum III. Tergum I and sterna I and II wider than long. Tergum I red-brown with narrow dark fuscous lateral margins in some specimens; tergum II fulvous with the following dark patterns in some specimens: a narrow dark fuscous transverse band across anterior margin but not reaching lateral margins, narrow dark fuscous anterolateral corners; terga III–V light red-brown with a narrow dark fuscous to black band across anterior margin of tergum III, a narrow dark fuscous to black medial longitudinal band over terga IV and V and narrow pale fuscous to fuscous anterolateral corners on terga IV and V. In some specimens the medial longitudinal band on terga IV and V is very narrow and pale and almost absent and the anterolateral dark corners on terga IV and V absent. A pair of light red-brown oval shining spots on tergum V. Posterior lobe of surstylus long, sternum V with a slight concavity on posterior margin. Abdominal sterna entirely pale (fulvous to light red-brown).
Attractant: Cue lure.
Female: No known record.
Distribution: Indonesia (Sulawesi).
Hosts: No known record.
Remarks: We have studied the Holotype (in MZB), 6 Paratypes (in BBM) and 1 Paratype (in QDPC) of Hardy's type series of *Bactrocera (Zeugodacus) flavipilosa* (Hardy). *Bactrocera (Zeugodacus) neoflavipilosa* n.sp. is described from five specimens from the type series of *B. flavipilosa*. These five specimens now become the type series of *B. neoflavipilosa*. When Hardy (1982) described *B. flavipilosa*, he inadvertently included three species in the type series. In the part of this series that we have studied, the Holotype and one Paratype represent *B. flavipilosa*, there is one specimen of *Bactrocera (Bactrocera) flavipennis* (Hardy) that we have treated under that subgenus, and specimens of *B. neoflavipilosa*. *Bactrocera neoflavipilosa* is similar to *Bactrocera (Zeugodacus) buruensis* White and *B. flavipilosa* in possessing four *sc.* setae, lateral and medial postsutural vittae, a basic red-brown scutum, wings with a pale fuscous tint around hind margin in addition to the costal band and anal streak, face with a pair of black spots, abdominal sterna pale and terga III–V with an incomplete 'T' pattern. It differs from these species in having a strong keel-shaped supernumerary lobe in the male wing (large and rounded in the other species) and an irregular 'U'-shaped black pattern on the scutum between the lateral and medial postsutural yellow vittae.

Bactrocera (Zeugodacus) neolipsana Drew & Romig, new species

(Fig. 334)

Material examined: Holotype male, **INDONESIA: LOMBOK** – Lombok, Karang Bayan, 25.x.1976, 380 m, coll. Eko Mulyono, deposited in BMNH.
Paratypes: **INDONESIA: LOMBOK** – 4 males, same data as holotype; **SULAWESI** – 1 male, Palopo – Rantepao Rd, 16–23 km from Palopo, Sulawesi, Indonesia (250–500 m), 24.vii.1981, coll. J. Turner, Nanggala III Forest, attracted to cue lure. 1 paratype in ANIC; 1 paratype in BMNH; 2 paratypes in QDPC; 1 paratype in QMIC.
Diagnosis: A small species; face black with dorsal and lateral fulvous margins; postpronotal lobes and notopleura yellow; scutum entirely shining black; lateral and medial postsutural yellow vittae present; no yellow spot or triangle anterior to notopleural suture; mesopleural stripe reaching to anterior *npl.* seta dorsally; scutellum yellow; wing with cells bc and c colourless, microtrichia in outer corner of cell c only; a narrow fuscous costal band confluent with R_{2+3} and widening slightly towards apex of wing and a narrow fuscous anal streak; abdominal terga III–V red-brown with a medium-width medial and two broad lateral longitudinal dark fuscous to black bands joined along anterior margin of tergum III.
Description: male.

Head: Height 1.13 mm. Frons – length 1.63 times breadth; fulvous with fuscous to dark fuscous around bases of orbital setae; orbital setae black: 1 *s.or.*, 2 *i.or.*; lunule fulvous to fuscous. Ocellar triangle black. Vertex dark fuscous to black. Face black with fulvous around lateral and dorsal margins; length 0.41 mm. Genae fulvous, dark fuscous subocular present; black seta present. Occiput dark fuscous to black, fulvous along eye margins; occipital row with a large number of small pale setae. Antennae with segments 1 and 2 red-brown, segment 3 fuscous except red-brown on basal area of inner surface; arista black (red-brown basally); length of segments: 0.08 mm; 0.2 mm; 0.52 mm.
Thorax: Scutum entirely shining black without pale markings. Pleural areas shining black without pale markings. Yellow markings as follows: postpronotal lobes; notopleura; mesopleural stripe reaching to anterior *npl.* seta dorsally, continuing to katepisternum as a transverse spot, anterior margin straight to very slightly convex; anatergite (posterior apex black); anterior ¾ katatergite (remainder black); two broad parallel-sided lateral postsutural vittae ending just behind *ia.* seta (no spot or triangle in front of notopleural suture); a broad medial longitudinal postsutural vitta beginning posteriorly at level of *prsc.* setae and narrowing slightly anteriorly to end at level of notopleural suture. Postnotum black. Scutellum yellow except for narrow black basal band. Setae: *sc.* 4; *prsc.* 2; *ia.* 1; *p.sa.* 1; *a.sa.* 1; *mpl.* 1; *npl.* 2; *scp.* 4.
Legs: Fore femora fulvous basally and apically with dark fuscous to black around centre of variable size, mid femora fulvous with large dark fuscous to black spot subapically around apical ½, hind femora fulvous with dark fuscous to black around apical ⅓; tibiae and tarsi entirely fulvous; mid tibiae each with an apical black spur.
Wings: Length 4.2 mm; cells bc and c colourless; microtrichia in outer corner of cell c only; remainder of wings colourless except fuscous cell sc, narrow fuscous costal band confluent with R_{2+3} and widening slightly between apex of R_{2+3} and apex of wing where it ends between apices of R_{4+5} and M, fuscous anal streak that narrows markedly towards wing margin; dense aggregation of microtrichia around A_1+CuA_2; supernumerary lobe weak.
Abdomen: Oval; terga free; pecten present on tergum III. Tergum I and sterna I and II wider than long. Tergum I fulvous with broad lateral black margins; tergum II fulvous with a transverse black band across anterior margin and meeting broad anterolateral black corners; terga III–V red-brown with a transverse black band across anterior margin of tergum III, a medium-width medial longitudinal dark fuscous to black band and broad lateral longitudinal dark fuscous to black bands over all three terga; in some specimens the transverse band across anterior margin of tergum III is incomplete in centre of tergum and there is a pale transverse narrow red-brown band on posterolateral margins of tergum III in line of pecten; in one specimen the medial longitudinal band is broken on tergum III. A pair of oval pale fuscous shining spots on tergum V. Posterior lobe of surstylus long, sternum V with a slight concavity on posterior margin. Abdominal sterna dark.
Attractant: Cue lure.
Female: No known record.
Distribution: Indonesia (Lombok; Sulawesi – South Sulawesi).
Hosts: No known record.
Remarks: *Bactrocera (Zeugodacus) neolipsana* n.sp. is similar to *Bactrocera (Zeugodacus) ascita* (Hardy) and *Bactrocera (Zeugodacus) bogorensis* Hardy in the colour patterns of the thorax, wings and abdomen but differs in having no yellow spot anterior to notopleural suture, four *sc.* setae, *prsc.* setae present and a weak supernumerary lobe in the wing. It is also similar to *Bactrocera (Zeugodacus) lipsanus* (Hendel) and *Bactrocera (Zeugodacus) scutellina* (Bezzi) but differs in having a black face, entirely yellow scutellum, no pale markings on the scutum and abdominal terga III–V with lateral longitudinal black bands.

Bactrocera (Zeugodacus) nigrifacies (Shiraki)

(Fig. 335)

Zeugodacus nigrifacies Shiraki, 1933: 99. Lectotype male and Paralectotype female, here designated, in NIAS.
Bactrocera (Zeugodacus) nigrifacies – Norrbom *et al.*, 1998: 103.

Type material examined: TAIWAN (formerly FORMOSA): Lectotype male here designated with the following three labels – white label 'Tamaru (Rota), 2.ix.1923, Col. T. Shiraki'; white label 'Zeugodacus nigrifacies sp. nov. Det. by T. Shiraki, TAIHOKU .ix.1931'; white circular label with red outer circle 'Type'. Paralectotype female here designated with the following four labels – white label 'Arisan, 25.iv.1917, Col. T. Shiraki'; white label 'near biguttata Bezzi, Det. T. Shiraki'; white label 'Zeugodacus nigrifacies sp. n., Det. by T. Shiraki, TAIHOKU .ix.1931'; white circular label with red outer circle 'Type'. Received on loan from NIAS.
Diagnosis: A medium-sized species; face entirely black; postpronotal lobes and notopleura yellow; scutum black; lateral and medial postsutural yellow vittae present; a small yellow spot anterior to notopleural suture in front of lateral postsutural vittae; mesopleural stripe equal in width to notopleuron dorsally; scutellum yellow with a large black apical spot; wing with cells bc and c colourless, microtrichia in outer corner of cell c only; a narrow fuscous costal band confluent with R_{2+3} and widening slightly across apex of R_{4+5}, an irregular-shaped transverse pale fuscous to fuscous band across centre of wing and enclosing both crossveins; a moderately broad fuscous anal streak; abdominal terga predominately black.
Description: male.
Head: Frons – fuscous centrally, fulvous along lateral and ventral margins, dark fuscous around lower *i.or.* setae and around *s.or.* setae; orbital setae black: 1 *s.or.*, 3 *i.or.*; lunule fuscous. Ocellar triangle black. Vertex black. Face entirely black. Genae fuscous; black seta present. Occiput dark fuscous except fuscous centrally and fulvous along eye margins; occipital row with a number of weak pale setae. Antennae with all segments uniformly dark fuscous.
Thorax: Scutum black except fuscous below lateral postsutural vittae. Pleural areas entirely dark fuscous to black.

Yellow markings as follows: postpronotal lobes; notopleura; mesopleural stripe equal in width to notopleuron dorsally, extending to katepisternum as a small transverse spot, anterior margin convex; anatergite (posterior apex black); anterior ¾ katatergite (remainder black); two parallel-sided lateral postsutural vittae of medium-width, beginning as a small spot anterior to notopleural suture and ending before *ia.* setae; a short narrow medial longitudinal postsutural vitta. Postnotum black. Scutellum yellow except for a broad black basal band and a large black apical spot that encloses both apical setae. Setae: *sc.* 4; *prsc.* 2; *ia.* 1; *p.sa.* 1; *a.sa.* 1; *mpl.* 1; *npl.* 2; *scp.* 4.
Legs: All femora dark fuscous to black with very small areas of fulvous basally on mid and hind femora; fore and mid tibiae fulvous except small areas of pale fuscous to fuscous basally, hind tibiae dark fuscous tending fuscous apically; all tarsal segments fulvous; mid tibiae each with an apical black spur.
Wings: Cells bc and c colourless; microtrichia in outer corner of cell c only; remainder of wings colourless except fuscous cell sc, narrow fuscous costal band confluent with R_{2+3} and widening slightly across apex of R_{4+5}, an irregularly shaped transverse band across wing which encloses both crossveins and varies from very pale fuscous to fuscous, a moderately broad fuscous anal streak; a dense aggregation of microtrichia around A_1+CuA_2; supernumerary lobe strong.
Abdomen: Oval; terga free; pecten present on tergum III. Tergum I and sterna I and II wider than long. Tergum I black except for a narrow dark red-brown band across posterior margin which does not reach lateral margins; tergum II black except for a dark red-brown band across posterior ½; terga III to V black except for dark fuscous posterocentrally on terga III and IV and dark red-brown posterocentrally on tergum V. A pair of oval black shining spots on tergum V. Posterior lobe of surstylus long, sternum V with a slight concavity on posterior margin.
Attractant: No known record.
Female: As for male except no dense aggregation of microtrichia around A_1+CuA_2; supernumerary lobe weak; pecten of cilia on abdominal tergum III absent; basal segment of ovipositor black, conical in transverse cross section and tapering posteriorly in dorsal view; ratio of length of basal segment to length of tergum V, 1.2:1.
Distribution:
Type locality: Taiwan (Tamaru).
Hosts: No known record.
Remarks: The types of *Bactrocera (Zeugodacus) nigrifacies* (Shiraki) were presumed lost but have now been found at NIAS and obtained for study. It is similar to *Bactrocera (Zeugodacus) diaphoropsis* (Hering) and *Bactrocera (Zeugodacus) munda* (Bezzi) in possessing four *sc.* setae, a black scutum with medial and lateral postsutural yellow vittae, wings with infuscation in addition to costal band and anal streak and costal band only slightly widening apically. It differs from both of these species in having a black face, infuscation across wing as a band enclosing both crossveins and abdominal terga I–V mostly black.

In the original description, Shiraki (1933) did not designate a Holotype. Consequently, for taxonomic stability and confirmation of the subgeneric position of this species, we are here designating the male specimen as the Lectotype and the female the Paralectotype, received on loan from NIAS for study.

Bactrocera (Zeugodacus) okunii (Shiraki)

(Fig. 336)

Zeugodacus okunii Shiraki, 1933: 104. Lectotype male and Paralectotype female, here designated, in NIAS.
Bactrocera (Zeugodacus) okunii – Norrbom *et al.*, 1998: 103.

Type material examined: TAIWAN (formerly FORMOSA): Lectotype male here designated with the following three labels – white label 'Kotosho, 10.iiI–15.iv.1920, Col. T. Okuni'; white label 'Parazeug. Okunii Det. T. Shiraki'; white circular label with red outer circle 'Type'. Paralectotype female here designated with the following four labels – white label 'Kotosho, 10.iiI–15.iv.192, Col. T. Okuni'; white label, 'Col. J. Sonan'; white label, 'Parazeug. Okunii, Det. T. Shiraki'; white circular label with red outer circle 'Type'. Received on loan from NIAS.
Non-type material examined: INDONESIA: JAVA – 1 male, Java, 1884, Klein; 1 female, Batavia, viii.1815, Mus. Westerm. Specimens in QDPC.
Diagnosis: A medium-sized species; face entirely fulvous; postpronotal lobes and notopleura yellow; scutum base colour red-brown; lateral and medial postsutural yellow vittae present; a yellow spot anterior to notopleural suture in front of lateral postsutural vittae; mesopleural stripe reaching midway between anterior margin of notopleuron and anterior *npl.* seta dorsally; scutellum yellow; wing with cells bc and c with a pale fuscous tint, microtrichia in outer corner of cell c only; a narrow fuscous costal band just overlapping R_{2+3} and widening into a spot across apex of wing almost to M, a small fuscous tint on hind margin around CuA_1; a broad pale fuscous anal streak; abdominal terga III–V orange-brown with a black 'T' pattern consisting of a narrow transverse band across anterior margin of tergum III and a narrow- to medium-medial longitudinal band over all three terga, anterolateral corners of terga IV and V dark fuscous.
Description: male.
Head: Frons – fulvous with pale fuscous on anteromedial hump; orbital setae black: 1 *s.or.*, 2 *i.or.*; lunule red-brown. Ocellar triangle black. Vertex fulvous. Face fulvous without dark markings. Genae fulvous, no dark subocular spot; weak pale seta present. Occiput red-brown, fulvous along eye margins; occipital row with weak pale setae. Antennae with all segments red-brown except fuscous on apex and outer surface of segment 3.
Thorax: Scutum red-brown except fuscous across posterior margin and a dark fuscous to black 'U'-shaped pattern between lateral and medial postsutural yellow vittae and fuscous to black areas anterior to notopleural suture. Pleural areas red-brown except for a narrow dark fuscous band along anterior margin of mesopleural stripe, a dark fuscous spot posterior to notopleural stripe and dark fuscous on katepisternum.

Yellow markings as follows: postpronotal lobes; notopleura; mesopleural stripe of medium-width reaching midway between anterior margin of notopleuron and anterior *npl.* seta dorsally, continuing to katepisternum as a large transverse spot, anterior margin convex; anatergite (posterior apex black); anterior ⅔ katatergite (remainder black); two broad lateral postsutural vittae beginning anterior to notopleural suture and narrowing sharply posteriorly to end well before *ia.* seta; a broad medial longitudinal postsutural yellow vitta beginning just anterior to *prsc.* setae where it is broadly rounded and narrowing to a point anteriorly to end at level of notopleural suture. Postnotum black laterally and red-brown centrally. Scutellum yellow with a narrow red-brown to black basal band. Setae: *sc.* 4; *prsc.* 2; *ia.* 1; *p.sa.* 1; *a.sa.* 1; *mpl.* 1; *npl.* 2; *scp.* 4.
Legs: All segments entirely fulvous without dark markings; mid tibiae each with an apical black spur.
Wings: Cells bc and c with a pale fuscous tint; microtrichia in outer corner of cell c only; remainder of wings colourless except fuscous cell sc, narrow fuscous costal band just overlapping R_{2+3} where it is a pale tint and expanding into an egg-shaped spot across apex of R_{4+5}, a small pale fuscous tint on hind wing margin around vein CuA_1, broad pale fuscous anal streak; dense aggregation of microtrichia around A_1+CuA_2; supernumerary lobe well developed.
Abdomen: Elongate oval; terga free; pecten present on tergum III. Tergum I and sterna I and II wider than long. Terga I and II orange-brown except for a narrow transverse black band across anterior margin of tergum II but not reaching lateral margins and anterolateral corners of tergum II dark fuscous; terga III to V orange-brown except for a 'T' pattern consisting of a narrow black band across anterior margin of tergum III and a narrow- to medium-medial longitudinal black band over all three terga but broken at intersegmental line between terga IV and V, anterolateral corners of terga IV and V dark fuscous. A pair of oval orange-brown shining spots on tergum V. Posterior lobe of surstylus long, sternum V with a slight concavity on posterior margin.
Attractant: No known record.
Female: As for male except no dense aggregation of microtrichia around A_1+CuA_2; supernumerary lobe weak; pecten of cilia on abdominal tergum III absent; basal segment of ovipositor red-brown, dorsoventrally flattened and tapering posteriorly in dorsal view; ratio of length of basal segment to length of tergum V, 0.9:1; apex of piercer needle-shaped.
Distribution:
Type locality: Taiwan (Kotosho).
New record: Indonesia (Java).
Hosts: No known record.
Remarks: The types of *Bactrocera (Zeugodacus) okunii* (Shiraki) were presumed lost but have now been found at NIAS and obtained for study. It is here placed within the *tau* complex and also similar to *Bactrocera (Zeugodacus) depressa* (Shiraki) in possessing four *sc.* setae, medial and lateral postsutural yellow vittae, scutum base colour red-brown, wings colourless except for costal band and anal streak, costal band overlapping R_{2+3} and expanded into an apical spot and lateral postsutural vittae ending before *ia.* seta. It differs from *B. depressa* in having a large apical spot in the costal band almost to M and cells bc and c colourless.

In the original description, Shiraki (1933) did not designate a holotype. Consequently, for taxonomic stability and confirmation of the subgeneric position of this species, we are here designating the male specimen as the Lectotype and the female the Paralectotype, received on loan from NIAS for study.

Bactrocera (Zeugodacus) pemalangiae Drew & Romig, new species

(Fig. 337)

Material examined: Holotype male, **INDONESIA: JAVA** – Central Java, Pemalang, Pegongsoran, 14.viii.2005, coll. Sapto Dormono, attracted to cue lure, deposited in BMNH. Paratypes: **INDONESIA: JAVA** – 1 male, same data as holotype; 1 male (7.viii.2005), 1 male (21.viii.2005), same data as holotype. 1 paratype in ANIC, 2 paratypes in QDPC.
Diagnosis: A small species; face black with broad lateral and dorsal fulvous margins; postpronotal lobes and notopleura yellow; scutum black; lateral and medial postsutural yellow vittae present; no yellow spot anterior to notopleural suture; mesopleural stripe reaching almost to anterior to *npl.* seta dorsally; scutellum yellow with a narrow black basal band; wing with cells bc and c colourless, microtrichia in outer corner of cell c only; a narrow fuscous to dark fuscous costal band confluent with R_{2+3} and a broad fuscous to dark fuscous anal streak; abdominal terga III–V orange-brown with a black 'T' pattern consisting of a narrow transverse black band across anterior margin of tergum III which is broader towards lateral margins and a narrow- to medium-width medial longitudinal fuscous to black band over all three terga, broad anterolateral black corners on tergum IV.
Description: male.
Head: Height 1.28 mm. Frons – length 1.5 times breadth; fulvous with dark fuscous around orbital setae and on anteromedial hump; orbital setae black: 1 *s.or.*, 2 *i.or.*; lunule fuscous. Ocellar triangle black. Vertex dark fuscous. Face black with broad lateral and dorsal fulvous margins; length 0.48 mm. Genae fulvous, dark fuscous subocular spot present; black seta present. Occiput fuscous to dark fuscous, fulvous along eye margins; occipital row with a large number of small pale setae. Antennae with segments 1 and 2 red-brown, segment 3 missing; length of segments: 0.08 mm; 0.24 mm.
Thorax: Scutum entirely black with dark fuscous behind lateral postsutural vittae. Pleural areas black except red-brown below postpronotal lobes. Yellow markings as follows: postpronotal lobes; notopleura; mesopleural stripe reaching almost to anterior *npl.* seta dorsally, continuing to katepisternum as a transverse spot, anterior margin straight; anatergite (posterior apex black); anterior ⅔ katatergite (remainder black); two broad parallel-sided lateral postsutural vittae ending behind *ia.* seta; a broad medial longitudinal postsutural vitta beginning just anterior of *prsc.* setae and ending at level of notopleural suture. Postnotum black. Scutellum

yellow except for narrow black basal band. Setae: *sc.* 4; *prsc.* 2; *ia.* 1; *p.sa.* 1; *a.sa.* 1; *mpl.* 1; *npl.* 2; *scp.* 4.
Legs: Fore femora with basal ⅖ fulvous and apical ⅗ red-brown to fuscous with a large dark fuscous spot on outer surface, mid femora with basal ½ fulvous and apical ½ red-brown to fuscous with a large dark fuscous spot on outer surface, hind femora with basal ⅔ fulvous and apical ⅓ red-brown to fuscous with a dark fuscous spot on outer surface; fore and mid tibiae entirely fulvous, hind tibiae red-brown to fuscous; tarsi with all segments entirely fulvous; mid tibiae each with an apical black spur.
Wings: Length 4.5 mm; cells bc and c colourless; microtrichia in outer corner of cell c only; remainder of wings colourless except fuscous to dark fuscous cell sc, narrow fuscous to dark fuscous costal band confluent with R_{2+3} and remaining narrow before ending between extremities of R_{4+5} and M, a broad fuscous to dark fuscous anal streak; dense aggregation of microtrichia around A_1+CuA_2; supernumerary lobe of medium development.
Abdomen: Oval; terga free; pecten present on tergum III. Tergum I and sterna I and II wider than long. Tergum I orange-brown with dark fuscous to black across anterior and lateral margins; tergum II orange-brown tending fulvous posteriorly and with a narrow transverse dark fuscous to black band across anterior margin; terga III–V orange-brown with a 'T' pattern consisting of a transverse black band across anterior margin of tergum III which is narrow in centre and broadens laterally, a narrow- to medium-width medial longitudinal fuscous to black band over all three terga, broad anterolateral black corners on tergum IV. A pair of oval orange-brown shining spots on tergum V. Posterior lobe of surstylus long, sternum V with a slight concavity on posterior margin. Abdominal sterna entirely fulvous.
Attractant: Cue lure.
Female: No known record.
Distribution: Indonesia (Java – Central Java, Pemalang).
Hosts: No known record.
Remarks: *Bactrocera (Zeugodacus) pemalangiae* n.sp. is similar to *Bactrocera (Zeugodacus) abnormis* (Hardy) and *Bactrocera (Zeugodacus) connexa* (Hardy) in possessing four *sc.* setae, a medial longitudinal postsutural yellow vitta, a black scutum, wings colourless except for the costal band and anal streak, the costal band confluent with R_{2+3}, cells bc and c colourless and abdominal sterna entirely fulvous. It differs from *B. abnormis* and *B. connexa* in possessing a black face. *Bactrocera (Zeugodacus) vulta* (Hardy) also possesses a black face but differs in having a large rounded supernumerary lobe in the male wing and dark abdominal sterna.

Bactrocera (Zeugodacus) perplexa (Walker)

(Fig. 338)

Dacus perplexus Walker, 1861c: 14. Lectotype female in MVMA.
Paradacus perplexus – Perkins, 1939: 33.
Dacus implexus Perkins, 1939: 34, nomen nudum.
Dacus (Paradacus) perplexus – Hardy & Adachi, 1954: 156; Hardy, 1959: 180, Lectotype designation by inference of Holotype; Drew, 1973: 13.
Bactrocera (Paradacus) perplexa – Drew, 1989: 196; Norrbom *et al.*, 1998: 99.

Type material examined: INDONESIA: MOLUCCAS (MALUKU) – Lectotype female, Gilolo, Moluccas, with the following label – 'Dacus implexus' (Walker's handwriting), received on loan from MVMA.
Non-type material examined: INDONESIA: MOLUCCAS (MALUKU) – 1 male with the following six labels – 'Gil. S.' (a round label in Walker's handwriting); '68.4'; 'perplexus' (probably Walker's handwriting); 'East Indies, Gilolo, W.W. Saunders, BM1868-4'; 'BMNH(E) # 533003'; 'Dacus (Paradacus) perplexus Walker, det. 1954, D.E. Hardy'. Studied at BMNH. There is no evidence that Walker (1862) had this male specimen before him when he described this species. Consequently, this specimen cannot be regarded as a type.
Diagnosis: A large species; face fulvous with a pair of medium-sized transverse oval black spots; postpronotal lobes and notopleura yellow; scutum dark fuscous with red-brown around lateral margins and a broad medial longitudinal red-brown band from anterior margin of medial postsutural yellow vitta to anterior margin of scutum; lateral postsutural yellow vittae present and tapering to end before *ia.* seta; medial postsutural yellow vitta present; yellow spot present anterior to notopleural suture; mesopleural stripe reaching to postpronotal lobe dorsally; scutellum yellow; wing with cells bc and c pale fuscous, microtrichia in outer ½ of cell c only; a broad fuscous costal band mostly overlapping R_{4+5} and ending between extremities of R_{4+5} and M; a broad fuscous anal streak, a broad 'V'-shaped band across centre of wing; abdominal terga III–V orange-brown with broad lateral longitudinal dark bands which are dark fuscous on terga III and IV and black on tergum V and narrow medial longitudinal black lines on terga IV and V.
Description: male.
Head: Frons – fulvous with fuscous around orbital setae and on anteromedial hump; orbital setae black: 1 *s.or.*, 2 *i.or.*; lunule fuscous. Ocellar triangle black. Vertex fulvous. Face fulvous with a pair of medium-sized transverse oval black spots. Occiput red-brown, fulvous along eye margins. Antennae with all segments fulvous.
Thorax: Scutum dark fuscous with red-brown as follows: below and behind lateral postsutural vittae, around notopleural suture, along lateral margins between postpronotal lobe and notopleuron, inside postpronotal lobes, a broad medial longitudinal band from anterior region of medial postsutural yellow vitta to anterior margin of scutum. Pleural areas fuscous. Yellow markings as follows: postpronotal lobes; notopleura; mesopleural stripe reaching to postpronotal lobe dorsally, extending to katepisternum as a large oval spot, anterior margin straight; two broad lateral postsutural vittae beginning as a spot anterior to notopleural suture and tapering posteriorly to end just before *ia.* seta; a short broad

medial postsutural vitta. Scutellum yellow except for narrow dark fuscous basal band. Setae: *sc.* 4; *prsc.* absent; *ia.* 1; *p.sa.* 1; *a.sa.* 1; *mpl.* 1; *npl.* 2; *scp.* 2.

Legs: Fore and mid femora fulvous with red-brown around apical ¼, hind femora fulvous with fuscous pre-apically around apical ¼; all tibiae and tarsi entirely fulvous.

Wings: Length 8.4 mm; cells bc and c pale fuscous; microtrichia in outer ½ of cell c only; remainder of wings colourless except fuscous cell sc, broad fuscous costal band overlapping R_{4+5} basally to extend along *r-m* crossvein and then almost confluent with R_{4+5} beyond *r-m* crossvein before overlapping this vein to end midway between extremities of R_{4+5} and M, a broad 'V'-shaped fuscous band across centre of wing, a broad fuscous anal streak; a dense aggregation of microtrichia around A_1+CuA_2; supernumerary lobe medium to strong development.

Abdomen: Oval; terga free; pecten present on tergum III. Tergum I and sterna I and II wider than long. Tergum I dark fuscous except for a narrow transverse orange-brown band across posterior margin but not reaching lateral margins; tergum II orange-brown except for a narrow transverse dark fuscous band across anterior margin and expanding to cover lateral margins and a narrow orange-brown band across anterior margin at intersegmental line but not reaching lateral margins; terga III–V orange-brown except for moderately broad lateral longitudinal dark bands over all three terga which are dark fuscous on terga III and IV and black on tergum V, narrow medial longitudinal black bands over posterior ¾ of tergum IV and most of tergum V. A pair of oval orange-brown shining spots on tergum V. Posterior lobe of surstylus long, sternum V with a slight concavity on posterior margin. Abdominal sterna dark.

Attractant: No known record.

Female: As for male except no dense aggregation of microtrichia around A_1+CuA_2; supernumerary lobe weak; pecten of cilia on abdominal tergum III absent; basal segment of ovipositor red-brown, oval in transverse cross-section and tapering posteriorly in dorsal view; ratio of length of oviscape to length of tergum V, 0.77:1; apex of piercer needle-shaped.

Distribution:

Type locality: Indonesia (Moluccas = Maluku, Gilolo).

Hosts: No known record.

Remarks: *Bactrocera (Zeugodacus) perplexa* (Walker) has previously been placed in subgenus *Paradacus* by most authors but is here provisionally placed in subgenus *Zeugodacus*, subject to the availability of more male specimens for study. It possesses four *sc.* setae, *a.sa.* setae and a medial longitudinal yellow vitta on the scutum, all important characters of subgenus *Zeugodacus. Bactrocera perplexa* is similar to *Bactrocera (Paradacus) decipiens* Drew in possessing a unique fuscous pattern across the wing but differs in this pattern being 'V' shaped, and in having dense microtrichia over the outer ½ of cell c, the mesopleural stripe reaching to the postpronotal lobe dorsally, abdominal terga III–V orange-brown with a narrow medial longitudinal fuscous to black band over all three terga, or at least terga IV and V, and apex of female piercer needle-shaped.

Bactrocera (Zeugodacus) persignata (Hering)

(Fig. 339)

Strumeta persignata Hering, 1941e: 27. Syntypes in DEI and MLUH.

Dacus (Zeugodacus) persignatus (Hering) – Hardy, 1981: 355.

Dacus goughi Hardy, 1982: 210; Hardy, 1984: 81 (Synonym).

Bactrocera (Zeugodacus) persignata – Norrbom *et al.*, 1998: 103.

Type material examined: INDONESIA: FLORES – 6 male Syntypes and 2 female Syntypes, with the following label data – 1 female red label 'TYPE', white label 'Sunda Exp. Rensch Endo Flores 12.vi.1927', white label 'Strumeta persignata Type female det. Hering 1940', white label 'Strumeta persignata Hrg.', white label 'Coll. DEI Eberswalde'; 1 female, white label 'Sunda Exp. Rensch Endo Flores 12.vi.1927', white label 'Strumeta persignata Hrg.', red label 'Paratypus', brown label 'PARATYPUS', white label 'Coll. DEI Eberswalde'; 2 males, white label 'Sunda Exp. Rensch Endo Flores 15.vi.1927', white label 'Strumeta persignata Hrg', red label 'Paratypus', brown label 'PARATYPUS', white label 'Coll. DEI Eberswalde'; 1 male, white label 'Sunda-Exp. Rensch Poeloe Endeh 11.6.27', red label 'Paratypus', brown label 'PARATYPUS', white label 'Coll. DEI Eberswalde'; 2 males, white label 'Sunda-Exp. Rensch Endeh 10–16.6.27', red label 'PARATYPUS', red label '1927', white label 'Coll. DEI Eberswalde'. Received on loan from DEI. One male paratype with the following eight labels – white label 'male'; white circular label with yellow outer circle 'Para-type'; red label 'Paratype'; white label 'Mittel Flores Endeh'; white label '10–16.6.27 B. Rensch S.'; white label 'F. Baarb erh. r. Mus. Dahlen'; white label 'Strumeta persignata m. det. Hering 1940'; white label 'Purch, from E.M. Hering B.M. 1965-270'. Studied at BMNH.

Hering (1941) listed a male and female type and a number of Paratypes as the specimens before him when he described *B. persignata*. As no particular specimen was designated a Holotype, the entire series is here regarded as Syntypes.

Non-type material examined: INDONESIA: MOLUCCAS (MALUKU) – 1 male (4.vi.2007), 2 males (18.vi.2007), 1 male (1.vii.2007), 1 male (26.vii.2007), Maluku, Middle Maluku, Suli, coll. La Naibu, attracted to cue lure in Jeruk; 4 males, Hila/Mamua, Ambon, Indonesia, 25-100m, 5.viii.1981, coll. J. Turner, residual forest with clove, nutmeg & banana groves; attracted to cue lure; 1 male (12.vii.2007), 2 males (19.vii.2007), 1 male (2.viii.2007), Maluku, Seram (Western part), Waimital, coll. A. Haris Tualepe; 2 males, Maluku, Seram (Western part), Nuruwe, 6.vii.2007, coll. A. Haris Tualepe, attracted to cue lure in forest; 1 male (24.vi.2007), 1 male (8.vii.2007), 1 male (12.viii.2007), Maluku, Middle Maluku, Tulefu, coll. La Naibu, attracted to cue lure in Cabal; **SULAWESI** – 4 males (12.viii.2006), 2 males (20.viii.2006), 4 males (2.ix.2006), 1 male (8.x.2006), 2 males (15.x.2006), Sulawesi, Sulawesi Tengah, Donggla Lolu, coll. Rusdin, attracted to cue lure in mango; 1 male, Sulawesi

Tengah, Donggala, Jono Oge, 27.viii.2006, coll. Rusdin, attracted to cue lure in Jackfruit; 5 males, Sulawesi Tengah, Donggala, Pombewe, 27.viii.2006, coll. Rusdin, attracted to cue lure in village; 4 males, Sulawesi Tengah, Donggala, Nupobomba, 2.ix.2006, coll. Rusdin, attracted to cue lure in forest; 2 males, Sulawesi Utara Bitung, Manembonembo, 11.ix.2006, coll. Karel F. Lala, attracted to cue lure in chilli; 4 males, Sulawesi Selatan, Takala, Topejawa, 2.ix.2006, coll. Sirajuddin SP, attracted to cue lure in village; 1 male, South Sulawesi, Maros, Laminpoccoe, 13.ix.2006, coll. Sirajuddin, attracted to cue lure in forest; 1 male, Sulawesi, Gorontalo, Bone Belango, Lombongo, 26.ix.2006, coll. Yusbar Ismail, attracted to cue lure in forest; 2 males, Sulawesi, Gorontalo, Hulawa, Teloga, 26.ix.2006, coll. Yusbar Ismail, attracted to cue lure in chilli; 2 males, Sulawesi, Utara, Minalasa Utara, Sarongsong, 26.ix.2006, coll. Karel F. Lala, attracted to cue lure in village. 1 male, Palopo-Rantepao Rd., 16-23km Palopo, Sulawesi, Indonesia, 250-500m, 24.vii.1981, coll. J. Turner, Nanggala III forest; attracted to cue lure; 1 male, Latupa, 9km W. Palopo, Sulawesi, Indonesia, 100m, 23. vii.1981, coll. J. Turner, Primary rainforest (some disturbance), attracted to cue lure. Specimens in QDPC.

Diagnosis: A medium-sized species; face fulvous with a pair of circular black spots; postpronotal lobes and notopleura yellow; scutum base colour red-brown; lateral and medial postsutural yellow vittae present; a yellow triangle along anterior margin of notopleural suture with base on notopleuron; mesopleural stripe reaching midway between anterior margin of notopleuron and anterior *npl.* seta dorsally; scutellum yellow; wing with cells bc and c pale fuscous, microtrichia in outer corner of cell c only; a broad fuscous costal band confluent with R_{4+5} and broadening across apex of wing to M, a pale fuscous to fuscous band along *r-m* crossvein, a broad fuscous band along *dm-cu* crossvein; a broad fuscous anal streak; abdominal terga III–V orange-brown with a small black spot on each anterolateral corner of tergum III and a small medial longitudinal black spot on tergum V.

Description: male.

Head: Height 1.48 mm. Frons – length 1.59 times breadth; fulvous without dark markings; orbital setae black; 1 *s.or.*, 2 *i.or.*; luule fuscous. Face fulvous with medium-sized circular black spots; length 0.48 mm. Genae fulvous, small pale fuscous subocular spot; black seta present. Occiput red-brown, fulvous along eye margins; occipital row with no distinct setae. Antennae with all segments red-brown; length of segments: 0.15 mm; 0.21 mm; 0.66 mm.

Thorax: Scutum basically red-brown with regularly shaped lateral longitudinal dark fuscous to black patterns between medial and lateral postsutural vittae and extending towards anterior margin of scutum. Pleural areas red-brown except dark fuscous to black on katepisternum below yellow spot and a small dark fuscous to black spot anterodorsal to mesopleural stripe. Yellow markings as follows: postpronotal lobes; notopleura; mesopleural stripe of medium-width reaching midway between anterior margin of notopleuron and anterior *npl.* seta dorsally, continuing to katepisternum as a large transverse spot, anterior margin slightly convex; anatergite (posterior apex black); anterior ¾ katatergite (remainder dark fuscous to black); two broad lateral postsutural vittae narrowing slightly posteriorly to end behind *ia.* seta; a large triangular marking anterior to each lateral postsutural vitta and anterior to notopleural suture; a broad medial longitudinal vitta broadly rounded posteriorly at level of *prsc.* setae and ending in a point anteriorly just anterior of notopleural suture. Postnotum dark fuscous to black laterally and red-brown centrally. Scutellum yellow with a narrow red-brown basal band. Setae: *sc.* 2; *prsc.* 2; *ia.* 1; *p.sa.* 1; *a. sa.* 1; *mpl.* 1; *npl.* 2; *scp.* 4.

Legs: All segments fulvous without any dark markings; mid tibiae each with an apical black spur.

Wings: Length 5.9 mm; cell bc pale fuscous and cell c paler fuscous; microtrichia in outer corner of cell c only; remainder of wings with a broad fuscous costal band confluent with R_{4+5} and broadening across apex of wing (it is narrowly pale fuscous along R_{4+5}), a fuscous to pale fuscous band along *r-m* crossvein, a broad fuscous band along *dm-cu* crossvein beginning anterior to vein M and extending around hind margin of wing but not reaching anal streak, a broad fuscous anal streak; a dense aggregation of microtrichia around A_1+CuA_2; supernumerary lobe strong.

Abdomen: Oval; terga free; pecten present on tergum III. Terga I and sterna I and II wider than long. Terga I and II red-brown except for a fulvous band along posterior margin of tergum II but not reaching lateral margins; terga III–V orange-brown with small anterolateral black spots on tergum III and a small medial longitudinal black spot on tergum V (the anterolateral spots on tergum III may extend along anterior margin to form a black to fuscous band along anterior margin). A pair of oval orange-brown shining spots on tergum V. Posterior lobe of surstylus long, sternum V with a deep concavity on posterior margin.

Attractant: Cue lure.

Female: As for male except no dense aggregation of microtrichia around A_1+CuA_2; supernumerary lobe weak; pecten of cilia on abdominal tergum III absent; basal segment of ovipositor red-brown elongate and conical in transverse cross section and tapering posteriorly in dorsal view; ratio of length of oviscape to length of tergum V, 1.73:1; apex of piercer needle-shaped.

Distribution:

Type locality: Indonesia (Flores).

Previously recorded: Type locality and Indonesia (Central Sulawesi), due to synonymy of *D. goughi*.

New records: Indonesia (Moluccas = Maluku – Ambon I.; Seram I. and several new locations in Sulawesi listed under 'Non-type material examined').

Hosts: No known record.

Remarks: *Bactrocera (Zeugodacus) persignata* (Hering) possesses a long posterior surstylus lobe and sternum V with a deep concavity on the posterior margin. With these characters, it fits imperfectly into the subgenus *Zeugodacus*. Hardy (1981) recorded that the posterior margin of sternum V possessed a slight concavity on the posterior margin but this is not evident on the male Paratypes which have a deep concavity.

This species is distinct within the subgenus *Zeugodacus* in possessing two *sc.* setae, scutum red-brown, cells bc and c pale fuscous, infuscation on *r-m* and *dm-cu* crossveins and abdominal terga III–V orange-brown with black spots on anterolateral corners of tergum III and a medial black spot on tergum V.

The female labelled 'TYPE' in DEI was treated as the Holotype by Hardy (1981). However, we are regarding all specimens in the type series as Syntypes.

Bactrocera (Zeugodacus) platamus (Hardy)

(Fig. 340)

Dacus (Zeugodacus) platamus Hardy, 1973: 65. Holotype male in KUB.
Bactrocera (Zeugodacus) platamus – Norrbom *et al.*, 1998: 103.

Type material examined: Holotype male, **THAILAND**: with the following label data – Yala, 15.viii.1965, coll. Anand, CL (= cue lure). Received on loan from KUB. Paratypes: **THAILAND**: 2 males, 'Yala, CL, 15 AUG 1965'; 'Paratype male, *Dacus (Zeugodacus) platamus*, D.E. Hardy'. Received on loan from BBM.

Non-type material examined: EAST MALAYSIA: SABAH – 1 male, Mt Kinabalu, 2000m, 19.iii.1992, coll. Jinius *et al.*, attracted to cue lure. **PENINSULAR MALAYSIA**: 1 male, TS7 (Transect South), 14.vii.1986, coll. Tony, attracted to cue lure; 1 male, TS7, 3.xi.1986, coll. Mahdzir, attracted to cue lure; 2 males, TS7, 29.xii.1986, attracted to cue lure; 2 males, TS7 Gombak, 15.vi.1987, coll. Bahari, attracted to cue lure; 1 male (12.i.1986), 3 males (11.viii.1986), 1 male (12.i.1987), TS8, coll. Mahdzir, attracted to cue lure; 2 males, TS8, 29.xii.1986, attracted to cue lure; 1 male, TS8, 12.i.1987, attracted to cue lure; 2 males, TS9, 3.xi.1986, coll. Mahdzir, attracted to cue lure; 1 male, TS9, 29.xii.1986, attracted to cue lure; 1 male, TN7 (Transect North), 26.i.1987, coll. Bahari, attracted to cue lure; 1 male, TN9, 26.i.1987, coll. Bahari, attracted to cue lure; 1 male, Malaysia, Sel [Selangor], Hulu Langat, Pangsun, 12 Aug 1989, coll. YHS [Yong Hoi Seng]. **THAILAND**: 11 males, 'Chumphon Prov., Khao Kai Jae, Watf, 5-8.v.2003/2'; '9.55.07N, 98.56.34E, P. Schwendinger leg, semi-evergreen rainforest/2'; 'attracted to cue lure'. Received on loan from MHNG. 1 male, Khun Khon, Chiang Rai, CR303T, 9.i.1992, coll. Charuwan, attracted to cue lure. Specimens in QDPC except where specified.

Diagnosis: A large species; face with a pair of large elongate oval black spots; postpronotal lobes and notopleura yellow; scutum black; lateral and medial postsutural yellow vittae present; a yellow spot on anterior margin of notopleural suture anterior to lateral postsutural vittae; mesopleural stripe reaching midway between anterior margin of notopleuron and anterior *npl.* seta dorsally; scutellum yellow with a small apical black spot; wing with cells bc and c colourless, microtrichia in outer corner of cell c only; a narrow dark fuscous costal band overlapping R_{2+3} and widening slightly across apex of wing; a broad fuscous anal streak; abdominal terga III–V fulvous with a black 'T' pattern consisting of a narrow black transverse band across anterior margin of tergum III and a moderately broad medial longitudinal band over all three terga, anterolateral corners of terga IV and V dark fuscous to black.

Description: male.

Head: Height 1.5 mm. Frons length 1.4 times breadth; fulvous with fuscous around *s.or.* setae and on anteromedial hump; orbital setae strong and black: 1 *s.or.*, 3 *i.or.*; lunule fuscous. Ocellar triangle black. Vertex black. Face fulvous with a pair of large elongate oval black spots; length 0.5 mm. Genae fulvous, pale fuscous subocular spot present; black seta present. Occiput fuscous, dark fuscous spots laterally and fulvous along eye margins; occipital row with a few weak pale setae. Antennae with segments 1 and 2 red-brown, segment 3 red-brown with fuscous on apex and outer surface; length of segments: 0.14 mm; 0.26 mm; 0.7 mm.

Thorax: Scutum black except red-brown behind lateral postsutural vittae, around notopleural suture and anteromedially. Pleural areas black except red-brown below postpronotal lobes. Yellow markings as follows: postpronotal lobes; notopleura; mesopleural stripe reaching midway between anterior margin of notopleuron and anterior *npl.* seta dorsally, continuing to katepisternum as a large transverse spot, anterior margin convex; anatergite (posterior apex black); anterior ⅘ katatergite (remainder black); two broad parallel-sided lateral postsutural vittae beginning with a spot anterior to notopleural suture and ending posteriorly at *ia.* seta; a broad medial longitudinal postsutural vitta broadly rounded posteriorly at level of *prsc.* setae and ending in a point anterior to notopleural suture. Postnotum black. Scutellum yellow except for a narrow black basal band and a small apical black spot between apical setae. Setae: *sc.* 4; *prsc.* 2; *ia.* 1; *p.sa.* 1; *a.sa.* 1; *mpl.* 1; *npl.* 2; *scp.* 2.

Legs: Fore femora fulvous except for a dark fuscous apical spot on outer surfaces, mid and hind femora fulvous except pale fuscous around apices of mid femora and dark fuscous around apices of hind femora; fore tibiae dark fuscous, mid tibiae fulvous except for pale fuscous basally, hind tibiae dark fuscous; all tarsal segments fulvous; mid tibiae each with an apical black spur.

Wings: Length 6.9 mm; cells bc and c colourless; microtrichia in outer corner of cell c only; remainder of wings colourless except dark fuscous cell sc, narrow dark fuscous costal band overlapping R_{2+3} and almost confluent with R_{4+5} and widening slightly across apex of wing, a broad fuscous anal streak; dense aggregation of microtrichia around A_1+CuA_2; supernumerary lobe of medium development.

Abdomen: Elongate oval; terga free; pecten present on tergum III. Tergum I and sterna I and II wider than long. Tergum I fulvous with narrow black lateral margins; tergum II fulvous with a narrow black transverse band anteriorly which connects with black anterolateral corners; terga III to V fulvous except for a 'T' pattern consisting of a narrow transverse black band across anterior margin of tergum III and a moderately broad medial longitudinal black band over all three terga, anterolateral corners of tergum IV broadly black and anterolateral corners of tergum V dark fuscous. A pair of oval fulvous shining spots on tergum V. Posterior lobe of surstylus long, sternum V with a deep concavity on posterior margin in types but not evident in Non-type material examined.

Attractant: Cue lure.
Female: No known record.
Distribution:
Type locality: Southern Thailand (Yala).
Previously recorded: Type locality and Indonesia (Java; Sumatra) (Hardy, 1983).
New records: East Malaysia (Sabah); Peninsular Malaysia; Thailand (Chiang Rai; Chumphon Province).
Hosts: No known record.
Remarks: *Bactrocera (Zeugodacus) platamus* (Hardy) belongs to a group of species that possess four *sc.* setae, a black scutum with medial and lateral longitudinal postsutural vittae, wings colourless except for costal band and anal streak, cells bc and c colourless or with a pale tint, costal band overlapping R_{2+3} and not expanded into a spot apically. It is close to *Bactrocera (Zeugodacus) juxtuncinata* n.sp. and *Bactrocera (Zeugodacus) ujungpandangiae* n.sp. in possessing a fulvous face with a pair of black spots and a weak supernumerary lobe. It differs from *B. ujungpandangiae* in having dark abdominal sterna and from *B. juxtuncinata* in having the scutellum with a dark apical spot, a short broad medial yellow vitta on the scutum, femora with dark apices and terga IV and V with dark anterolateral corners.

Bactrocera (Zeugodacus) proprescutellata Zhang, Chen & Gao

(Fig. 341)

Bactrocera (Zeugodacus) proprescutellata Zhang, Chen & Gao, in Zhang *et al.*, 2011: 321. Holotype male in FAFU.

Type material studied: CHINA: A study of the colour photographs and description of *Bactrocera (Zeugodacus) proprescutellata* Zhang, Chen & Gao in the original publication.
Non-type material examined: BHUTAN: one male, Namling 16km, 9.vi.2001, coll. Brian Fletcher, attracted to cue lure, in QDPC.
Diagnosis: A medium-sized species; face fulvous with a pair of very large triangular black spots with apex on oral margin; postpronotal lobes and notopleura yellow; scutum shining black; narrow lateral and medial postsutural yellow vittae present; yellow spot present anterior to notopleural suture; mesopleural stripe equal in width to notopleuron dorsally; scutellum yellow with a large apical black spot on dorsal surface; wing with cells bc and c colourless, microtrichia in outer corner of cell c only; a narrow dark fuscous costal band remaining narrow beyond apex of R_{2+3} and then widening across apex of R_{4+5} (*scutellata*-like); a broad fuscous anal streak; abdominal terga III–V fulvous with a general black 'T' pattern and broad black lateral longitudinal bands over terga IV and V.
Description: male.
Head: Height 1.25 mm. Frons – length 1.31 times breadth; fulvous with fuscous around orbital setae and on antero-medial hump; orbital setae black: 1 *s.or.*, 3 *i.or.*; lunule fuscous. Ocellar triangle black. Vertex black. Face fulvous with a pair of very large triangular black spots with apex of triangle at oral margin; length 0.4 mm. Genae fulvous, fuscous subocular spot present; black seta present. Occiput dark fuscous tending black along dorsolateral margins, fulvous along eye margins; occipital row with 4–5 black setae. Antennae with segments 1 and 2 red-brown, segment 3 red-brown with fuscous on apex and outer surface; arista black (red-brown basally); length of segments: 0.08 mm; 0.18 mm; 0.56 mm.
Thorax: Scutum shining black with red-brown below and behind lateral postsutural vittae. Pleural areas shining black with red-brown below postpronotal lobes. Yellow markings as follows: postpronotal lobes; notopleura; mesopleural stripe equal in width to notopleuron dorsally, extending to katepisternum as a large transverse spot, anterior margin straight; anatergite (posterior apex black); anterior ⅔ katatergite (remainder black); two narrow parallel-sided lateral postsutural vittae beginning with a spot anterior to notopleural suture and ending at *ia.* seta; a narrow medial longitudinal postsutural vitta beginning posteriorly approximately at level of *prsc.* setae and narrowing to a point anteriorly to end at level of notopleural suture. Postnotum black. Scutellum yellow with a narrow black basal band and a large apical black spot over apical ½ of dorsal surface. Setae: *sc.* 4; *prsc.* 2; *ia.* 1; *p.sa.* 1; *a.sa.* 1; *mpl.* 1; *npl.* 2; *scp.* 4.
Legs: Fore femora fulvous with apical ½ dark fuscous to shining black, mid femora fulvous with apical ½ shining black, hind femora fulvous with apical ⅓ shining black; fore tibiae fulvous with basal ½ dark fuscous, mid tibiae fulvous tending fuscous basally, hind tibiae dark fuscous; tarsi with basal segment fulvous and apical four segments pale fuscous; mid tibiae each with an apical black spur.
Wings: Length 6.0 mm; cells bc and c colourless; microtrichia in outer corner of cell c only; remainder of wings colourless except dark fuscous cell sc, narrow dark fuscous costal band confluent with R_{2+3} remaining narrow beyond extremity of R_{2+3} and then widening across apex of R_{4+5} (*scutellata*-like), a broad fuscous anal streak; dense aggregation of microtrichia around A_1+CuA_2; supernumerary lobe strong and rounded. Cell br fuscous and covered with dense microtrichia.
Abdomen: Oval; terga free; pecten present on tergum III. Tergum I and sterna I and II wider than long. Tergum I fulvous with a narrow dark fuscous transverse band across anterior margin joining with broad anterolateral shining black margins; tergum II fulvous with a narrow transverse black band across anterior margin broadening to cover anterolateral corners; terga III–V fulvous with a 'T' pattern consisting of a broad transverse black band across anterior margin and a moderately broad medial longitudinal black band over all three terga and broad lateral longitudinal black bands over terga IV and V (the lateral longitudinal band may join with the black transverse band across anterior margin of tergum III with a fuscous to dark fuscous pattern). A pair of oval fulvous shining spots on tergum V which tend to black on lateral margins. Posterior lobe of surstylus long, sternum V with a slight concavity on posterior margin. Abdominal sterna shining black, except sternum II fulvous.
Attractant: Cue lure.
Female: No known record.

Distribution:
Type locality: China (Yunnan Province, Jinghong).
New records: Bhutan (Namling).
Hosts: No known record.
Remarks: *Bactrocera (Zeugodacus) proprescutellata* Zhang, Chen & Gao, within the *scutellaris* complex, is similar to *Bactrocera (Zeugodacus) scutellata* (Hendel) in possessing four *sc.* setae, terga III–V with, at least, a black 'T' pattern on at least one tergum and a pair of facial spots. It differs from *B. scutellata* in having the 'T' pattern over terga III–V (not on each tergum), black apices on all femora, very large triangular black facial spots and a narrower mesopleural stripe. Also, in *B. scutellata*, the *dm-cu* crossvein is infuscated in most specimens.

Bactrocera (Zeugodacus) pubescens (Bezzi)

(Figs 342A, 342B)

Chaetodacus pubescens Bezzi, 1919: 434. Lectotype male in MCSNM.
Chaetodacus scutellatus Bezzi (nec Hendel) 1916: 114 [Misidentification].
Dacus (Zeugodacus) pubescens – Hardy & Adachi, 1954: 194.
Dacus (Zeugodacus) pubescens – Hardy, 1969b: 479, Lectotype designation; 1974: 54.
Bactrocera (Zeugodacus) pubescens – Norrbom *et al.*, 1998: 103.

Type material examined: PHILIPPINES: LUZON ISLAND – Lectotype male, with the following three labels – 'Los Banos, P.I., Baker'; 'pubescens n.' (in handwriting, probably Bezzi's) and 'scutellatus Hend.' (on opposite side); 'Lectotype male, Chaetodacus pubescens Bez., desig. by Hardy 68'. Received on loan from MCSNM.
Non-type material examined: PHILIPPINES: LUZON ISLAND – 1 male, Trece Martires, Cavite, 22.vi.1983, coll. N.E. Rivera, attracted to methyl eugenol; 1 female, Trece Martires City, 17.v.1983, coll. N.E. Rivera and A.M. Arida, attracted to methyl eugenol; 1 female, Sirang Lupa, Calamba, Laguna, 15.i.2000, coll. A. O'Toole. Specimens in QDPC. 1 female, 'Acc. No. 9941 Lot. Bu.g Sci. P.I.'; 'Collected by C.S. Banks'; 'Manilla Phillipine Is. ?.I.1909, C.S. Banks, 1909-222'; 'Dacus sp (apparently new)'; 'BMNH(E) # 533132'. Studied at BMNH.
Diagnosis: A large species; face fulvous with fuscous along oral margin between a pair of medium-sized oval black spots; postpronotal lobes and notopleura yellow; scutum black; lateral and medial postsutural yellow vittae present; a yellow spot anterior to notopleural suture anterior to lateral postsutural vittae; mesopleural stripe reaching midway between anterior margin of notopleuron and anterior *npl.* seta dorsally; scutellum yellow with a large black apical spot; wing with cells bc and c with a very pale tint, microtrichia in outer corner of cell c only; a narrow fuscous costal band slightly overlapping R_{2+3} and expanding into a fuscous to paler fuscous spot in apex; a broad fuscous anal streak; abdominal terga III–V fulvous with a black 'T' pattern consisting of a narrow black band across anterior margin of tergum III and a broad medium longitudinal black band over all three terga, broad lateral dark margins on terga IV and V.
Description: male (Fig. 342A).
Head: Height 1.91 mm. Frons – length 1.35 times breadth; fulvous with pale fuscous around orbital setae and on anteromedial hump; orbital setae large and dark fuscous to black: 1 *s.or.*, 3 *i.or.*; lunule fuscous. Ocellar triangle black. Vertex dark fuscous laterally tending pale fuscous adjacent to ocellar triangle. Face fulvous except fuscous along oral margin between a pair of medium-sized oval black spots; length 0.6 mm. Genae fulvous, fuscous subocular spot present; dark fuscous seta present. Occiput pale fuscous, fulvous along eye margins; occipital row with a few short pale setae. Antennae with segments 1 and 2 fulvous, segment 3 fulvous with fuscous on apex and outer surface; length of segments: 0.17 mm; 0.31 mm; 0.86 mm.
Thorax: Scutum black with the following pale markings: dark fuscous posterior to lateral postsutural vittae and around notopleural suture, red-brown below lateral postsutural vittae, along lateral margin between postpronotal lobe and notopleuron, inside postpronotal lobes and a broad medium longitudinal band extending from anterior point of medial postsutural yellow vitta to anterior margin of scutum (there is a medial longitudinal black line within this red-brown band). Pleural areas dark fuscous except red-brown below postpronotal lobes. Yellow markings as follows: postpronotal lobes; notopleura; mesopleural stripe of medium-width reaching midway between anterior margin of notopleuron and anterior *npl.* seta dorsally, continuing to katepisternum as a large transverse spot, anterior margin convex. Anatergite (posterior apex black); anterior ⅔ katatergite (remainder black); two broad parallel-sided lateral postsutural vittae beginning in a spot anterior to notopleural suture anteriorly and ending posteriorly at level of *ia.* seta; a broad medial longitudinal postsutural vitta broadly rounded posteriorly and ending at a point just anterior of notopleural suture. Postnotum black. Scutellum yellow except for a narrow black basal band and a black apical spot. Setae: *sc.* 4; *prsc.* 2; *ia.* 1; *p.sa.* 1; *a.sa.* 1; *mpl.* 1; *npl.* 2; *scp.* 2.
Legs: All femora entirely fulvous; fore tibiae fuscous with a small area of pale fuscous apically, mid tibiae pale fuscous basally tending fulvous apically, hind tibiae entirely fuscous; all tarsal segments entirely fulvous; mid tibiae each with an apical black spur.
Wings: Length 7.9 mm; cells bc and c with a very pale tint; microtrichia in outer corner of cell c only; remainder of wings colourless except for a fuscous cell sc, narrow fuscous costal band slightly overlapping R_{2+3} where it is pale fuscous and expanding into a fuscous to paler fuscous spot in apex of wing, broad fuscous anal streak; dense aggregation of microtrichia around A_1+CuA_2; supernumerary lobe of medium development.
Abdomen: Oval; terga free; pecten present on tergum III. Tergum I and sterna I and II wider than long. Tergum I red-brown with dark fuscous anteriorly and narrow black lateral margins; tergum II fulvous with a small 'T' pattern consisting

of a narrow transverse black band across anterior margin but not reaching lateral margins and a narrow medial longitudinal black band over most of tergum and with narrow black lateral margins; terga III–V fulvous with a black 'T' pattern consisting of a narrow black band across anterior margin of tergum III and a broad medial longitudinal black band over all three terga, broad lateral dark fuscous to black margins on tergum IV and broad fuscous to dark fuscous lateral margins on tergum V. A pair of oval fuscous shining spots on tergum V. Posterior lobe of surstylus long, sternum V with a slight concavity on posterior margin.

Attractant: No known record.

Female: (Fig. 342B) – As for male except no dense aggregation of microtrichia around A_1+CuA_2; supernumerary lobe weak; pecten of cilia on abdominal tergum III absent; basal segment of ovipositor red-brown, conical in transverse cross section and tapering posteriorly in dorsal view; ratio of length of oviscape to length of tergum V, 1:1; apex of piercer trilobed.

Distribution:

Type locality: Philippines (Luzon I., Los Banos).

New records: Philippines, Luzon I. (Cavite; Trece Martires City; Calamba; Manila).

Hosts: No known record.

Remarks: *Bactrocera (Zeugodacus) pubescens* (Bezzi) belongs to the *tau* complex of species possessing four *sc.* setae, a black scutum with lateral and medial postsutural yellow vittae, wings colourless except for costal band and anal streak, cells bc and c with a pale tint, costal band overlapping R_{2+3} and expanded into a spot apically and abdominal terga III–V with a distinct black 'T' pattern. It is very close to *Bactrocera (Zeugodacus) bezziana* (Hering) which has a slightly larger apical spot on the costal band.

Bactrocera (Zeugodacus) quasiinfesta Drew & Romig, new species

(Fig. 343)

Material examined: Holotype male, **THAILAND**: Khun Khon, Chiang Rai, 9.iv.1992, coll. Charuwan, attracted to cue lure, deposited in BMNH.

Paratypes: **THAILAND**: 1 male, Phan, Chiang Rai, 15. iv.1992, coll. Charuwan, attracted to cue lure; 1 male, C.H.R.C., Chiang Rai, 12.v.1992, coll. Charuwan, attracted to cue lure; 1 male, Pa Ha, Chiang Rai, 30.vi.1992, coll. Charuwan, 1 male, Yang Pong Rat, A. Muang, Chanthaburi, 21.vii.1992, coll. Montree; 1 male (23.vii.1991), 1 male (20.viii.1991), 1 male (27.ix.1991), 1 male (12.vi.1992), 1 male (30.vii.1992), 1 male (7.viii.1992), 1 male (28.viii.1992), NRDC, Chiang Mai, coll. Chantanee, attracted to cue lure. 3 paratypes in ANIC, 3 paratypes in BMNH, 4 paratypes in QDPC, 1 paratype in QMIC.

Diagnosis: A large species; face fulvous with a pair of medium-sized oval black spots joined across the face by a narrow black line; postpronotal lobes and notopleura yellow; scutum uniformly red-brown with small black patterns; lateral and medial postsutural yellow vittae present; a yellow spot anterior to notopleural suture; mesopleural stripe reaching midway between anterior margin of notopleuron and anterior *npl.* seta dorsally; scutellum yellow; wing with cell c colourless, cell bc with a pale tint, microtrichia in outer corner of cell c only; a broad pale fuscous to fuscous costal band almost confluent with R_{4+5} and expanding into a large dark fuscous spot across apex; a broad dark fuscous anal streak; abdominal terga III–V orange-brown with a 'T' pattern consisting of a narrow transverse black band across anterior margin of tergum III and a medium-width medial longitudinal dark fuscous to black band over all three terga, narrow dark fuscous lateral margins on tergum IV and narrow dark fuscous anterolateral corners on tergum V.

Description: male.

Head: Height 1.8 mm. Frons – length 1.2 times breadth; red-brown with dark fuscous around bases of orbital setae and on anteromedial hump; orbital setae black: 1 *s.or.*, 2 *i.or.*; lunule dark fuscous. Ocellar triangle black. Vertex dark fuscous. Face fulvous with a pair of medium-sized oval black spots which are joined across the centre of the face by a narrow black line; length 0.7 mm. Genae fulvous, small dark fuscous subocular spot present; black seta present. Occiput red-brown, fulvous along eye margins; occipital row with 3–5 black setae. Antennae with segments 1 and 2 red-brown, segment 3 red-brown with fuscous on apex and outer surface; length of segments: 0.2 mm; 0.36 mm; 0.96 mm.

Thorax: Scutum red-brown with black markings as follows: narrow black margins inside lateral postsutural yellow vittae, below anterior end of lateral postsutural vittae, extending from anterior end of medial postsutural yellow vitta, between postpronotal lobe and notopleuron. Pleural areas dark fuscous except red-brown below postpronotal lobes. Yellow markings as follows: postpronotal lobes; notopleura; mesopleural stripe reaching midway between anterior margin of notopleuron and anterior *npl.* seta dorsally, continuing to katepisternum as a large transverse spot, anterior margin strongly convex; anatergite (posterior apex dark fuscous); anterior $\frac{5}{7}$ katatergite (remainder dark fuscous); two narrow parallel-sided lateral postsutural vittae ending well behind *ia.* seta; a short, broad medial longitudinal postsutural vitta present; a small spot anterior to notopleural suture in front of lateral postsutural vitta. Postnotum red-brown tending dull black laterally. Scutellum yellow except for narrow red-brown basal band. Setae: *sc.* 4; *prsc.* absent; *ia.* 1; *p.sa.* 1; *a.sa.* 1; *mpl.* 1; *npl.* 2; *scp.* 4.

Legs: Fore femora fulvous with a narrow elongate dark fuscous spot along outer surface, mid and hind femora entirely fulvous; fore and hind tibiae dark fuscous basally to fuscous apically, mid tibiae dark fuscous basally to fulvous apically; tarsi with all segments entirely fulvous; mid tibiae each with an apical black spur.

Wings: Length 8.4 mm; cell c colourless, cell bc with a pale tint; microtrichia in outer corner of cell c only; remainder of wings colourless except fuscous cell sc, broad pale fuscous costal band almost confluent with R_{4+5} and expanding into a large dark fuscous spot across apex of wing, a broad dark fuscous anal streak; a dense aggregation of microtrichia around A_1+CuA_2; supernumerary lobe large and rounded.

Abdomen: Elongate oval and petiolate; terga free; pecten present on tergum III. Tergum I and sterna I and II longer than wide. Tergum I fulvous with narrow dark fuscous lateral margins; tergum II fulvous with a transverse dark fuscous band subanteriorly but not reaching the narrow dark fuscous lateral margins; terga III–V orange-brown with a 'T' pattern consisting of a narrow transverse black band across anterior margin of tergum III which extends into narrow black lateral margins on this tergum and a narrow- to medium-width medial longitudinal dark fuscous to black band over all three terga, narrow dark fuscous lateral margins on tergum IV and narrow anterolateral dark fuscous corners on tergum V. A pair of oval orange-brown shining spots on tergum V. Posterior lobe of surstylus long, sternum V with a slight concavity on posterior margin. Abdominal sterna dark.
Attractant: Cue lure.
Female: No known record.
Distribution: Thailand (Chiang Rai; Chiang Mai; Chanthaburi).
Hosts: No known record.
Remarks: *Bactrocera (Zeugodacus) quasiinfesta* n.sp. is similar to *Bactrocera (Zeugodacus) havelockiae* n.sp., in possessing a red-brown scutum, lateral and medial postsutural yellow vittae, four *sc.* setae and *prsc.* setae absent. It differs from *B. havelockiae* in having parallel-sided lateral postsutural vittae extending posterior to the *ia.* setae and the supernumerary lobe large and rounded. Both *B. havelockiae* and *B. quasiinfesta* are placed in the subgenus *Zeugodacus* although the *prsc.* setae are absent (see 'Remarks' under *B. havelockiae*).

Bactrocera (Zeugodacus) rubella (Hardy)

(Fig. 344)

Dacus (Zeugodacus) rubellus Hardy, 1973: 66. Holotype male in KUB.
Dacus (Strumeta) aculeus Hardy, 1973: 28. **New Synonym**.
Bactrocera (Zeugodacus) rubella – Norrbom *et al.*, 1998: 103.

Type material examined: Holotype male, **THAILAND**: with the following label data – Bangkok, 3.x.1962. Received on loan from KUB.
Non-type material examined: THAILAND: 12 males and 10 females, P.T.T. Station, T. Payoobnai, A. Wangchan, Rayong, 13.xi.1991, coll. Yuthana, bred from *Trichosanthes wawraei*; 1 male and 3 females, P.T.T. Station, A. Wangchan, Rayong, 25.xi.1991, coll. Montree, bred from *Trichosanthes wawraei*; 14 males and 13 females, 55/7, M002, Rayong, A. Wangchan, 16.x.1991, coll. Montree, bred from *Cucurbitaceae*. Specimens in QDPC.
Diagnosis: A large species; face with a pair of medium-sized oval black spots; postpronotal lobes and notopleura yellow; scutum base colour red-brown; lateral and medial postsutural yellow vittae present; a yellow spot anterior to notopleural suture in front of lateral postsutural vittae; mesopleural stripe reaching midway between anterior margin of notopleuron and anterior *npl.* seta dorsally; scutellum yellow; wing with cells bc and c colourless, microtrichia in outer corner of cell c only; a narrow fuscous costal band overlapping R_{2+3} and expanding into a large fuscous spot in apex and reaching to M; a broad fuscous anal streak; abdominal terga III–V with an incomplete 'T' pattern consisting of a narrow transverse dark fuscous band across anterior margin of tergum III and a narrow medial longitudinal fuscous to dark fuscous band over terga IV and V, anterolateral corners of terga IV and V fuscous.
Description: male.
Head: Height 1.8 mm. Frons length 1.35 times breadth; red-brown with dark fuscous around all orbital setae; orbital setae black: 1 *s.or.*, 3 *i.or.*; lunule red-brown. Ocellar triangle black. Vertex with lateral dark fuscous spots. Face fulvous with a pair of medium-sized oval black spots; length 0.62 mm. Genae fulvous, large fuscous subocular spot present; black bristle present. Occiput red-brown, fulvous along eye margins; occipital row with four or five weak dark setae. Antennae with all segments entirely red-brown; length of segments: 0.17 mm; 0.34 mm; 0.89 mm.
Thorax: Scutum red-brown with dark fuscous along inner margins of lateral postsutural yellow vittae and irregularly shaped fuscous spots across anterior area and adjacent to posterolateral margins of medial longitudinal postsutural yellow vitta. Pleural areas red-brown except for fuscous along anterior margins of mesopleural stripe, posterior to mesopleural stripe and on katepisternum. Yellow markings as follows: postpronotal lobes; notopleura; mesopleural stripe of medium-width extending midway between anterior margin of notopleuron and anterior *npl.* seta dorsally, continuing to katepisternum as a large transverse spot, anterior margin convex; anatergite (posterior apex dark fuscous); anterior ¾ katatergite (remainder dark fuscous); two parallel-sided lateral postsutural yellow vittae of medium-width, beginning as a spot anterior to notopleural suture anteriorly and ending posterior to *ia.* seta; a broad medial longitudinal postsutural yellow vitta rounded posteriorly where it begins anterior to *prsc.* setae and ending anteriorly in a point at level of notopleural suture. Postnotum dark fuscous. Scutellum yellow with a narrow fuscous basal band. Setae: *sc.* 4; *prsc.* 2; *ia.* 1; *p.sa.* 1; *a.sa.* 1; *mpl.* 1; *npl.* 2; *scp.* 2.
Legs: All femora entirely fulvous; fore and hind tibiae fuscous, mid tibiae fulvous with pale fuscous basally; tarsi with basal segments fulvous and apical four segments pale fuscous; mid tibiae each with an apical black spur.
Wings: Length 8.3 mm; cells bc and c colourless; microtrichia in outer corner of cell c only; remainder of wings colourless except fuscous cell sc, narrow fuscous costal band overlapping R_{2+3} where it becomes paler and almost reaches R_{4+5} and expanding into a large pale fuscous to fuscous spot in apex of wing, a broad fuscous anal streak; a dense aggregation of microtrichia around A_1+CuA_2; supernumerary lobe of medium development.
Abdomen: Oval; terga free; pecten present on tergum III. Tergum I and sterna I and II wider than long. Tergum I orange-brown with narrow dark fuscous lateral margins; tergum II orange-brown with fulvous areas posteriorly and a narrow dark fuscous band across anterior margin which does not meet lateral margins and narrow dark fuscous

anterolateral corners; terga III–V orange-brown except for a narrow transverse band across anterior margin of tergum III which is dark fuscous centrally and black laterally and widening to form narrow black lateral margins, a medial longitudinal fuscous to dark fuscous band which is narrow on tergum IV and of medium-width on tergum V, anterolateral corners of terga IV and V fuscous. A pair of oval orange-brown shining spots on tergum V. Posterior lobe of surstylus short, sternum V with a slight concavity on posterior margin.
Attractant: No known record.
Female: As for male except no dense aggregation of microtrichia around A_1+CuA_2; supernumerary lobe weak; pecten of cilia on abdominal tergum III absent; basal segment of ovipositor orange-brown, conical in transverse cross section and tapering posteriorly in dorsal view; ratio of length of oviscape to length of tergum V, 1.8:1; apex of piercer trilobed.
Distribution:
Type locality: Thailand (Bangkok).
New records: Central Thailand (Rayong).
Hosts: Bred from one host *Trichosanthes wawraei* (family *Cucurbitaceae*).
Remarks: *Bactrocera (Zeugodacus) rubella* (Hardy) belongs to a group of species with four *sc.* setae, scutum base colour red-brown, wings colourless except for costal band and anal streak, costal band expanded into an apical spot and lateral postsutural yellow vittae reaching to or behind *ia.* seta. It is close to *Bactrocera (Zeugodacus) signata* (Hering) but differs in having a reduced supernumerary lobe in the male wing, costal band almost reaching R_{4+5} and large spot in apex of costal band. The measurements included above are from specimens of host reared material from Thailand, not the type. *Bactrocera rubella* is similar to *Bactrocera (Zeugodacus) trichosanthes* n. sp. in the leg and wing colour patterns but differs in having a red-brown scutum and abdominal terga III–V without a complete black 'T' pattern.

We can find no sound specific characters for separating *Dacus (Strumeta) aculeus* Hardy and *B. rubella* and are consequently making *D. aculeus* a **New Synonym** of *B. rubella*.

Bactrocera (Zeugodacus) sabahensis Drew & Romig, new species

(Fig. 345)

Material examined: Holotype male, **EAST MALAYSIA: SABAH** – ACIAR Site Ht 4, Sabah Orchard, 9.x.1991, Sabah, attracted to cue lure in fruit orchard, deposited in BMNH.
Diagnosis: A medium-sized species; face entirely shining black; postpronotal lobes and notopleura yellow; scutum black; lateral and medial postsutural vittae present; a yellow spot anterior to notopleural suture; mesopleural stripe reaching anterior *npl.* seta dorsally; scutellum yellow; wing with cells bc and c colourless, microtrichia in outer corner of cell c only; a narrow fuscous costal band expanding slightly across apex of R_{4+5} and medium-width fuscous anal streak; abdominal terga III–V orange-brown with a medium-width medial and broad lateral longitudinal black bands joined along anterior margin of tergum III and also joined with a narrow fuscous to dark fuscous line across anterior margin of tergum IV (posterolateral corners of terga III and IV fuscous to dark fuscous).
Description: male.
Head: Height 1.33 mm. Frons – length 1.27 times breadth; red-brown centrally, fulvous on lateral margins, dark fuscous around bases of orbital setae and pale fuscous on antero-medial hump; orbital setae black: 1 *s.or.*, 3 *i.or.*; lunule red-brown. Ocellar triangle black. Vertex black. Face shining black with lateral red-brown to pale fuscous margins; length 0.42 mm. Genae pale fuscous with dark fuscous subocular spot present. Occiput dark fuscous, fulvous along eye margins; occipital row with 3–4 weak red-brown setae. Antennae with segments 1 and 2 fuscous; segment 3 missing; length of segments: 0.1 mm; 0.22 mm.
Thorax: Scutum entirely black except red-brown between postpronotal lobe and yellow spot anterior to notopleural suture. Pleural areas entirely black except for a small area of red-brown above fore coxae. Yellow markings as follows: postpronotal lobes; notopleura; mesopleural stripe reaching to anterior *npl.* seta dorsally, continuing to katepisternum as a small transverse spot, anterior margin straight; anatergite (posterior apex black); anterior ¾ katatergite (remainder black); two moderately broad parallel-sided lateral postsutural vittae beginning with a spot anterior to notopleural suture and ending just behind *ia.* seta; a moderately broad medial longitudinal postsutural vitta beginning at level of *prsc.* setae and narrowing to a point at level of notopleural suture. Postnotum black. Scutellum yellow except for narrow black basal band. Setae: *sc.* 4; *prsc.* 2; *ia.* 1; *p.sa.* 1; *a.sa.* 1; *mpl.* 1; *npl.* 2; *scp.* 2.
Legs: Fore femora entirely shining black, mid femora with apical ⅔ black and basal ⅓ fulvous, hind femora with apical ½ black and basal ½ fulvous (the extreme apices of mid and hind femora are slightly paler, probably dark fuscous); fore tibiae fulvous with dark fuscous basally, mid and hind tibiae pale fuscous with fuscous to dark fuscous basally; all tarsal segments entirely fulvous.
Wings: Length 5.3 mm; cells bc and c colourless; microtrichia in outer corner of cell c only; remainder of wings colourless except fuscous cell sc, narrow fuscous costal band confluent with R_{2+3} and remaining narrow but expanding slightly across apex of R_{4+5}, a medium-width fuscous anal streak; dense aggregation of microtrichia around A_1+CuA_2; supernumerary lobe strong.
Abdomen: Elongate oval; terga free; pecten present on tergum III. Tergum I and sterna I and II wider than long. Tergum I orange-brown with fuscous to dark fuscous across anterior margin and black anterolateral margins; tergum II orange-brown to fulvous with a broken narrow transverse subanterior band and narrow anterolateral black corners; terga III–V orange-brown with a medium-width medial and two broad lateral longitudinal black bands joined along anterior margin of tergum III (the posterolateral areas of terga III and IV tend to fuscous and the bands are joined along anterior margin of tergum IV with a narrow transverse fuscous line). A pair of oval orange-brown shining spots on tergum V.

Posterior lobe of surstylus long and stout, sternum V with a slight concavity on posterior margin. Abdominal sterna dark, except sternum I which is pale.
Attractant: Cue lure.
Female: No known record.
Distribution: East Malaysia (Sabah).
Hosts: No known record.
Remarks: *Bactrocera (Zeugodacus) sabahensis* n.sp. belongs to a group of *Zeugodacus* that possess a black scutum, no infuscation on wing other than costal band and anal streak, cells bc and c colourless, costal band confluent with R_{2+3} and dark abdominal sterna. It is unique in having a yellow spot anterior to notopleural suture, an entirely black face, femora almost entirely black and mesopleural stripe reaching anterior *npl.* seta. It is very close to *Bactrocera (Zeugodacus) hengsawadae* n.sp. and *Bactrocera (Zeugodacus) laguniensis* n.sp. in the characters stated above but differs from these in possessing a black face.

Bactrocera (Zeugodacus) sasaotiae Drew & Romig, new species

(Fig. 346)

Material examined: Holotype male, **INDONESIA: LOMBOK** – Sasaot, Lombok (400–550 m), 31.v.1981, coll. J. Turner, Old Plantation – Mahogany, attracted to cue lure, deposited in BMNH.
Diagnosis: A medium-sized species; face fuscous with a pair of large elongate oval black spots; postpronotal lobes and notopleura yellow; scutum almost entirely black; lateral and medial postsutural yellow vittae present; a yellow spot anterior to notopleural suture; mesopleural stripe reaching midway between anterior margin of notopleuron and anterior *npl.* seta dorsally; scutellum entirely yellow; wing with cells bc and c colourless, microtrichia in outer corner of cell c only; a moderately broad dark fuscous costal band and broad dark fuscous anal streak; abdominal terga III–V fulvous with a ‘T’ pattern consisting of a narrow transverse dark fuscous to black band across anterior margin of tergum III and a narrow- to medium-width medial longitudinal dark fuscous to black band over all three terga and anterolateral corners of terga IV and V dark fuscous.
Description: male.
Head: Height 1.29 mm. Frons – length 1.36 times breadth; red-brown with fulvous lateral margins and fuscous on anteromedial hump; orbital setae black: 1 *s.or.*, 2 *i.or.*; lunule fuscous. Ocellar triangle black. Vertex black. Face fuscous with a pair of large elongate oval black spots reaching to oral margin; length 0.45 mm. Genae fulvous, pale fuscous subocular spot present; red-brown seta present. Occiput fuscous, fulvous along eye margin; occipital row with a large number of short pale setae. Antennae with segments 1 and 2 red-brown, segment 3 entirely fuscous; arista black (red-brown basally); length of segments: 0.12 mm; 0.24 mm; 0.66 mm.
Thorax: Scutum entirely black except dark red-brown below and behind lateral postsutural vittae. Pleural areas entirely dark fuscous to black. Yellow markings as follows: postpronotal lobes; notopleura; mesopleural stripe of medium-width, reaching midway between anterior margin of notopleuron and anterior *npl.* seta dorsally, continuing to katepisternum as transverse spot, anterior margin convex; anatergite (posterior apex black); anterior $^3/_5$ katatergite (remainder black); two broad parallel-sided lateral postsutural vittae beginning with a large spot anterior to notopleural suture and ending behind *ia.* seta; a broad medial longitudinal postsutural vitta beginning at level of *prsc.* setae and narrowing anteriorly to end just anterior of level of notopleural suture. Postnotum dark fuscous. Scutellum yellow except for narrow black basal band. Setae: *sc.* 4; *prsc.* 2; *ia.* 1; *p.sa.* 1; *a.sa.* 1; *mpl.* 1; *npl.* 2; *scp.* 4.
Legs: All segments entirely fulvous; mid tibiae with a short apical black spur.
Wings: Length 5.24 mm; cells bc and c colourless; microtrichia in outer corner of cell c only; remainder of wings colourless except dark fuscous cell sc, moderately broad dark fuscous costal band almost confluent with R_{4+5} and remaining broad around apex of wing to end between apices of R_{4+5} and M, a broad dark fuscous anal streak; dense aggregation of microtrichia around A_1+CuA_2; supernumerary lobe strong.
Abdomen: Oval; terga free; pecten present on tergum III. Tergum I and sterna I and II wider than long. Tergum I fulvous with narrow dark fuscous lateral margins; tergum II fulvous with a short and narrow subanterior transverse dark fuscous band and narrow lateral dark fuscous bands; terga III–V fulvous with a ‘T’ pattern consisting of a narrow transverse dark fuscous to black band across anterior margin of tergum III and covering lateral margins and a narrow- to medium-width medial longitudinal dark fuscous to black band over all three terga, dark fuscous anterolateral corners on terga IV and V. A pair of oval fulvous shining spots on tergum V. Posterior lobe of surstylus long, sternum V with a slight concavity on posterior margin. Abdominal sterna dark fuscous to black.
Attractant: Cue lure.
Female: No known record.
Distribution: Indonesia (Lombok, Sasaot).
Hosts: No known record.
Remarks: *Bactrocera (Zeugodacus) sasaotiae* n.sp. is a smaller species similar to *Bactrocera (Zeugodacus) vultus* (Hardy) and *Bactrocera (Zeugodacus) platamus* (Hardy) in general body and wing colour patterns. It differs from *B. platamus* in having a strong, well-developed, supernumerary lobe in the male wing and fulvous femora without dark apical spots and from *B. vultus* in having a pair of black facial spots (not entirely black) and anterolateral corners of terga IV and V dark (not lateral longitudinal dark bands on these terga). It is also similar to *Bactrocera (Zeugodacus) ujungpandangiae* n.sp. but differs in having dark abdominal sterna.

Bactrocera (Zeugodacus) scutellaris (Bezzi)

(Fig. 347)

Bactrocera scutellaris Bezzi, 1913: 98. Lectotype male, here designated, in ZSI.
Zeugodacus malaisei Hering, 1938: 4; Norrbom *et al.*, 1998: 103 (Synonym).

Dacus pusaensis Kapoor and Katiyar, 1970: 252; Norrbom *et al.*, 1998: 103 (Synonym).
Dacus (Zeugodacus) scutellaris – Hardy, 1973: 68; Hardy, 1977: 59; Wang, 1996: 70.
Bactrocera (Zeugodacus) scutellaris – Norrbom *et al.*, 1998: 103.
Bactrocera (Zeugodacus) fraserensis Chua, 2010: 483. **New Synonym.**

Type material examined: INDIA: ASSAM – Lectotype male here designated, with the following eight labels – 'Shillong Assam La Touche'; '118'; 'B'; 'TYPE'; 'Dacus (Bactrocera) scutellaris M Bezzi'; 'TYPE'; 'B'; '$^{5910}/_{15}$'. Studied at ZSI.

Non-type material examined: BHUTAN: 13 males, Chirang, Lamidara, 4-7.ix.1990, coll. H.R. Fiejen; 2 males (10.v.2002), 1 male (24.v.2002), Rimchu, coll. C. Dorji, attracted to cue lure; 5 males (17.v.2000), 1 male (24.v.2000), 2 males (31.v.2000), 4 males (26.vii.2000), 3 males (10.viii.2000), 3 males (16.viii.2000), 13 males (6.ix.2000), 7 males (12.ix.2000), 21 males (20.ix.2000), 7 males (27.ix.2000), 12 males (5.x.2000), Rimchu trap No. 1, coll. C. Dorji, attracted to cue lure; 2 males (7.vi.2000), 7 males (28.vi.2000), 2 males (5.vii.2000), 5 males (19.vii.2000), 12 males (23.vii.2000), 1 male (2.viii.2000), 6 males (30.viii.2000), Rimchu trap No. 1, coll. C. Dorji *et al.*, attracted to cue lure; 2 males, Rimchu trap No. 1, 4.v.2000, coll. B.S. Fletcher, attracted to cue lure; 3 males, Rimchu trap No. 1, 14.vi.2000, coll. Nado T. Shomo, attracted to cue lure; 1 male, Rimchu trap No. 1, 21.vi.2000, coll. Nado *et al.*, attracted to cue lure; 3 males, Rimchu trap No. 1, 10.v.2000, attracted to cue lure; 3 males (28.vi.2000), 2 males (12.vii.2000), 3 males (16.vii.2000), 5 males (26.vii.2000), 2 males (5.x.2000), 1 male (16.xi.2000), Rimchu trap No. 2, coll. C. Dorji, attracted to cue lure; 3 males (4.v.2000), 2 males (10.v.2000), Rimchu trap No. 2, coll. B.S. Fletcher, attracted to cue lure; 3 males (4.v.2000), 1 male (10.v.2000), Rimchu trap No. 3, coll. B.S. Fletcher, attracted to cue lure; 2 males, Tsirang, 11.vi.2000, Suntalay, attracted to cue lure; 1 male, Tsirang, 28.vi.2000, coll. S. Wangdi, attracted to cue lure; 1 male, Tsirang, 12.vii.2000, coll. C. Dorji, attracted to cue lure; 1 male (19.vi.2000), 2 males (10.vii.2000), Tsirang, coll. T.D. & S.D., attracted to cue lure; 4 males (3.vii.2000), 4 males (17.vii.2000), 7 males (31.vii.2000), 4 males (21.viii.2000), Tsirang, attracted to cue lure; 3 males (31.v.2000), 1 male (13.ix.2000), Lumitshawa, coll. C. Dorji, attracted to cue lure; 1 male, Lumitshawa, 17.v.2000, coll. C. Dorji *et al.*, attracted to cue lure; 1 male, Lumitshawa, 3.v.2000, coll. B.S. Fletcher, attracted to cue lure; 2 males, Lumitshawa, 21.vi.2000, coll. Nado *et al.*, attracted to cue lure; 3 males, Namling 7km, 9.vi.2001, coll. Brian Fletcher, attracted to cue lure; 3 males, Namling 11km, 9.vi.2001, coll. Brian Fletcher, attracted to cue lure; 26 males, Namling 14km, 7.vi.2001, coll. Brian Fletcher, attracted to cue lure; 50 males, Namling 16km, 9.vi.2001, coll. Brian Fletcher, attracted to cue lure; 2 males, Lingmethang 11km, 9.vi.2001, coll. Brian Fletcher, attracted to cue lure. **CHINA**: 1 male, Xisuang Banna, Yunnan province, 1200 metres, 20.viii.1938, Fuji Pu, ident. X.J. Wang. **INDIA**: 1 male and 1 female, 'Murree, 2.vii.64', 'C.I.B.C. FF9 7/64-26', 'maggot in flowers of C. Maxima'; 2 females, 'Murree, 5.ix.64', 'C.I.B.C. FF9 9/64-35', 'maggot in flowers of C. Maxima'; 2 males 'Murree, 11.ix.63', 'C.I.B.C. FF9 9/63-21', 'maggot in flowers of C. Maxima'. **NEPAL**: 2 males, Kathmandu, PAPB, 20.v.1988, Shashi Adhikari, ENTO. Division, Dip 124. **PENINSULAR MALAYSIA**: A large series of specimens collected throughout Peninsular Malaysia held in QDPC. **THAILAND**: 1 male, 'Chiang Mai Prov, Tham Klaeb Yai/24, 570m, 30km N Chiang Dao'; '23-24.xii.2002, 19.34.34N, 99.03.40E, P. Schwendinger/24'; 'attracted to methyl eugenol'. Received on loan from MHNG. 3 males, Wawi Exp. Sta. 1400m, 45km South of Chiang Rai, 22.v.1991, coll. D.L. Hancock, attracted to cue lure; 2 males, Mae Salong, CR110T, Chiang Rai, 26.viii.1991, coll. Charuwan; 1 male, Khun Khon, Chiang Rai, 5.iii.1992, coll. Charuwan, attracted to cue lure; 2 males, Mae Hae, 1300m, 50km west of Chiang Mai, 13-14.v.1991, coll. D.L. Hancock; 1 male, Doi Suthep-Pui Nat. Park, 1425m, Chiang Mai, 17.v.1991, coll. D.L. Hancock; 1 male, Doi Inthanon, S of Chiang Mai, 1700m, 23.ix.1991, coll. D.L. Hancock, attracted to cue lure in rainforest; 1 male, Mae Hae, Chiang Mai, 26.xi.1991, Coll. Chantanee, attracted to cue lure; 6 males, Mae Hae, CM200T, Chiang Mai, 16.x.1991, coll. Chantanee, attracted to cue lure; 4 males, Mae Hae, CM243T, Chiang Mai, 26.xii.1991, coll. Chantanee, attracted to cue lure; 3 females and 6 males, Wavee, Chiang Rai, 17.vi.1992, coll. Charuwan, bred from flowers of pumpkin (*Cucurbita moschata*); 1 male, Wa Vee, Chiang Rai, 13.x.1992, coll. Charuwan, bred from flowers of pumpkin (*Cucurbita moschata*); 5 females and 3 males, Mae Hae Sta., Chiang Mai, 17.viii.1991, coll. Chantanee *et al.*, bred from growing shoot of pumpkin (*Cucurbita moschata*), sample No. CM2001; 1 female and 2 males (18.viii.1992), 1 female and 2 males (8.ix.1992), Huai Koa Leep, Chiang Mai, 18.viii.1992, coll. Chantanee, bred from flowers of pumpkin (*Cucurbita moschata*); 2 females and 3 males, Ban Khun, Boa Luang, 1010m, Chiang Mai, 7.x.1992, coll. Chantanee, bred from flowers of pumpkin (*Cucurbita moschata*); 1 male, Huai Kar Leep, Chiang Mai, 17.xi.1992, coll. Chantanee, bred from flowers of pumpkin (*Cucurbita moschata*); 4 females and 1 male, Mae Hae Stn, 1080m, Chiang Mai, 16.xii.1992, coll. Chantanee, bred from flowers of pumpkin (*Cucurbita moschata*); 2 females and 2 males, Mae Hae Sta., Chiang Mai, 12.i.1993, coll. Chantanee, bred from flowers of pumpkin (*Cucurbita moschata*). **VIETNAM: Northern Vietnam** – 1 male (6.ix.1999), 1 male (15.vi.2002), 1 male (30.vii.2002), Son La Province, Moc Chau District, Co Do State Farm, trap No. VN06; 1 male (6.ix.1999), 1 male (29.xi.1999), 1 male (16.iv.2000), Son La Province, Moc Chau District, trap Nos N007, N008, N009, attracted to cue lure; 4 males (30.iii.2002), 2 males (30.iv.2002), 2 males (15.v.2002), Son La (P), Moc Chau (D), Long Luong (V), trap No. VN01, coll. NIPP staff, attracted to cue lure; 12 males (28.ix.1999), 1 male (15.iii.2000), Hoa Binh Province, Mai Chau District, Pa Co, trap No. N010, attracted to cue lure; 1 male (30.vii.1999), 1 male (8.x.1999), 1 male (18.viii.2000), Lao Cai Province, Sapa 2 Village, trap N015, attracted to cue lure; 1 male, Lao Cai Province, Sapa Town, trap N016, 13.viii.1999, attracted to cue lure; 1 male, Lao Cai Province, Sapa 6 Village, trap N017, attracted to cue lure. Specimens in QDPC except where specified.

Diagnosis: A small- to medium-sized species; face fulvous with a pair of transverse oval black spots pointed towards centre; postpronotal lobes and notopleura yellow; scutum shining black; narrow lateral and medial postsutural yellow vittae present; a very small yellow spot anterior to notopleural suture; mesopleural stripe equal in width to notopleuron dorsally; scutellum yellow with a distinct black apical spot; wing with cells bc and c colourless, microtrichia in outer corner of cell c only; a narrow dark fuscous costal band which is extremely narrow beyond apex of R_{2+3} and widening into a distinct spot around apex of R_{4+5}; broad dark fuscous anal streak present; abdominal terga III–V mostly dark fuscous to black.

Description: male.

Head: Height 1.42 mm. Frons – length 1.45 times breadth; fulvous around margins with fuscous to dark fuscous over central area and around bases of orbital setae; orbital setae black:1 *s.or.*, 3 *i.or.*; lunule fuscous to dark fuscous. Ocellar triangle black. Vertex black. Face fulvous with a pair of medium-sized transverse oval black spots pointed towards centre but not joining to form a transverse band; length 0.41 mm. Genae fulvous, small dark fuscous subocular spot present; black seta present. Occiput black (sometimes fulvous in centre), fulvous along eye margin; occipital row with a large number of strong black setae. Antennae with segments 1 and 2 fuscous, segment 3 red-brown with fuscous on apex and outer surface; arista black (red-brown basally); length of segments: 0.1 mm; 0.22 mm; 0.62 mm.

Thorax: Scutum shining black without pale markings. Pleural areas shining black without pale markings. Yellow markings as follows: postpronotal lobes (in some specimens there is black on the anterodorsal corners and/or along inner margins); notopleura; mesopleural stripe equal in width to notopleuron dorsally, continuing to katepisternum as a small transverse spot, anterior margin straight; anatergite (posterior apex black); anterior ⅔ katatergite (remainder black); two narrow lateral postsutural vittae beginning with a very small spot anterior to notopleural suture and tapering to a point posteriorly to end well before *ia.* seta; a short and narrow medial longitudinal postsutural vitta. Postnotum black. Scutellum yellow except for narrow black basal band and distinct black apical spot enclosing apical setae. Setae: *sc.* 4; *prsc.* 2; *ia.* 1; *p.sa.* 1; *a.sa.* 1; *mpl.* 1; *npl.* 2; *scp.* 4.

Legs: Fore femora shining black on apical ½ to ⅔ with fulvous on remainder, mid femora with shining black on apical ⅓ to ½ and fulvous on remainder, hind femora with apical ⅓ shining black and remainder fulvous; fore tibiae dark fuscous to black, mid tibiae fuscous basally to pale fuscous apically, hind tibiae shining black; tarsi with basal segments fulvous and apical four segments fuscous; mid tibiae each with an apical black spur.

Wings: Length 6.1 mm; cells bc and c colourless; microtrichia in outer corner of cell c only; remainder of wings colourless except dark fuscous cell sc, narrow dark fuscous costal band confluent with R_{2+3} and very narrow beyond apex of R_{2+3} before expanding into a spot around apex of R_{4+5}, broad dark fuscous anal streak; dense aggregation of microtrichia around A_1+CuA_2; supernumerary lobe strong.

Abdomen: Oval; terga free; pecten present on tergum III. Tergum I and sterna I and II wider than long. Tergum I shining black with a narrow transverse fulvous band across posterior margin but not reaching lateral margins; tergum II shining black with a broad transverse fulvous band across posterior margin; terga III–V dark fuscous to black with a narrow red-brown band across posterior margin of tergum V (in some specimens there are areas of red-brown posterocentrally on terga III and IV). A pair of oval dark fuscous to black shining spots on tergum V. Posterior lobe of surstylus long, sternum V with a slight concavity on posterior margin. Abdominal sterna shining black, except sternum II dark fuscous.

Attractant: Cue lure.

Female: As for male except no dense aggregation of microtrichia around A_1+CuA_2; supernumerary lobe weak; pecten of cilia on abdominal tergum III absent; basal segment of ovipositor dark fuscous to black, dorsoventrally flattened and tapering posteriorly in dorsal view; ratio of length of oviscape to length of tergum V, 1.5:1; apex of piercer trilobed.

Distribution:

Type locality: India (Assam, Shillong).

Previously recorded: Type locality and the following locations – China; India; Myanmar; Nepal; Thailand.

New records: Bhutan; Northern Vietnam (Hoa Binh, Lao Cai, Son La Provinces); Peninsular Malaysia.

Hosts: Bred from flowers of species of *Cucurbitaceae* (Allwood *et al.*, 1999).

Remarks: *Bactrocera (Zeugodacus) scutellaris* Bezzi appears to be the most widespread of all *scutellaris* complex species. It is similar to *Bactrocera (Zeugodacus) dorsirufa* n.sp., *Bactrocera (Zeugodacus) menglana* Yu, Liu & Yang and *Bactrocera (Zeugodacus) sonlaiae* n.sp. in possessing four *sc.* setae, face with a pair of spots or a black band across oral margin, lateral postsutural vittae ending before *ia.* seta and the costal band *scutellaris*-like (i.e. extremely narrow, almost absent beyond extremity of R_{2+3} and expanding into a distinct spot across apex of R_{4+5}). It differs from *B. menglana* in having distinct facial spots and some areas of fulvous on bases of femora and from *B. sonlaiae* and *B. dorsirufa* in having lateral postsutural vittae tapering posteriorly (not parallel-sided), *dm-cu* crossvein not infuscated and a strong supernumerary lobe. Measurements included above are from specimens collected in Northern Vietnam, not the type.

Bezzi (1913) described *B. scutellaris* from a type series of four males and six females from the following locations – 'Shillong, Assam ($^{5907}/_{15}$ – $^{5910}/_{15}$); Siliguri, N. Bengal, in July ($^{5931}/_{15}$); Kurseong, 5000 ft., E. Himalayas, ix.09; Bhowali, Kumaon, 5700 ft., 15-vi-10 (A. D. Imms)'. Bezzi did not designate a Holotype and Paratypes and only one specimen could be found for study. This male specimen, described under material examined in this paper, is in ZSI and the label data fit with that listed by Bezzi. We hereby designate this male specimen as the Lectotype.

Bactrocera (Zeugodacus) fraserensis Chua is a **New Synonym** of *B. scutellaris*. We have studied very large numbers of specimens of *B. scutellaris* from across South-East Asia and have found significant intraspecific variation in colour patterns, especially the characters used to define *B. fraserensis*. In particular, the facial spot size, shape of the lateral and medial postsutural yellow vittae on the scutum, the size of the spot in the apex of the costal band, the size of the dark patterns on the apices of the femora and the infusion of paler colour patterns into abdominal terga III–V, all show considerable variation within *B. scutellaris*. Other characters used to separate *B. fraserensis* by Chua (2010) are never used to define species within the genus *Bactrocera* and are not considered species level characters, namely, shape of the dark spot on the apex of the scutellum and wing length.

Bactrocera (Zeugodacus) scutellata (Hendel)

(Figs 348A, 348B, 348C)

Dacus scutellatus Hendel, 1912: 20. Holotype male in DEI.
Zeugodacus scutellatus – Shiraki, 1933: 82.
Dacus (Zeugodacus) scutellatus – Hardy, 1977: 59; Tseng *et al*., 1992:74; Wang, 1996: 71.
Dacus (Chaetodacus) bezzii Miyake, 1919: 146. Shiraki, 1933: 85 (Synonym).
Dacus trivittatus Matsumura, 1916: 411. Miyake, 1919: 146 (Synonym).
Dacus (Zeugodacus) pseudoscutellatus Tseng, Chen & Chu, 1992: 74. **New Synonym**.
Bactrocera (Zeugodacus) scutellata – Norrbom *et al*., 1998: 103.

Type material examined: Holotype male, **TAIWAN (formerly FORMOSA)**: with the following four labels – white label 'Koshun Formosa Sauter X 08'; white label 'Dacus scutellatus, H. det. Hendel'; red label 'Holotypus'; white label 'Coll. DEI Eberswalde'. Received on loan from DEI.
Non-type material examined: BHUTAN: 1 male (19.vii.2000), 1 male (16.viii.2000), Rimchu trap No. 1, coll. C. Dorji, attracted to cue lure; 2 males, Rimchu trap No. 1, 23.vii.2000, coll. C. Dorji *et al*., attracted to cue lure. **CHINA**: 1 male (13.ix.1924), 1 male (19.ix. 1924), 1 female (25.x.1929), Shanghai, coll. E. Svenson (in ZMC); 2 males and 2 females, 20.ii.1984, Guangzhou; 1 male, Yangsuo, Guangi province, 150 metres, 18.vii.1963, coll. Suyong Wang; 1 female, Simao, Yunnan province, 28.viii.1957, coll. Suyong Wang. **JAPAN**: 24 males, Shuri-Kinjo-Cho, Okinawa, 30-31.vii.1977, T. Teruya. 1 male and 1 female, Yokohama City, Kanagawa Pref., 5.ix.1981, coll. F. Ichinoke; 1 male, Zushi, Kanagawa, vi.1989; 1 male, Yokohama, 30.iv.1990, coll. M. Kaneda. **TAIWAN**: 1 female with the following four labels – 'Formosa, Sauter'; 'Suisharyo, 1917 or 1911.x'; 'coll. Hendel'; 'Dacus scutellatus H. det. Hendel', studied at NMW. 1 male, Yilan, x.1978, coll. Y.H. Tseng, attracted to cue lure; 1 male, Sinlo, Pingtung, 11.xi.1990, coll. Y.H. Tseng, 18 males, Kaohsiung, Hsien, Liukuei, Shan-Ping LTER Site, 1-4.iv.2003, leg. M. Foldvari, on paper with cue lure; 5 males, Ilan Hsien, Fu-Shan LTER Site, 26-27.iii.2003, on paper with methyl eugenol, (incorrect lure record), leg. M. Foldvari. **THAILAND**: 2 males (28.viii.1991), 2 males (18.ix.1991), 1 male (9.x.1991), Phan, Chiang Rai, coll. Charuwan, attracted to cue lure; 1 male (29.vii.1991), 1 male (26.viii.1991), 1 male (5.viii.1991), Mae Salong, Chiang Rai, coll. Charuwan, attracted to cue lure; 1 male (16.x.1991), 1 male (30.x.1991), Pha Yao, Chiang Rai, coll. Charuwan, attracted to cue lure; 1 male (24.v.1991), 1 male (26.ix.1991), Doi Luong National Park, S.E. Chiang Mai, coll. D.L. Hancock, attracted to cue lure; 1 male, N.R.D.C., Chiang Mai, 3.ix.1991, coll. Chantanee, attracted to cue lure; 3 females, Huai Koa Leep, Chiang Mai, 18.viii.1992, coll. Chantanee, bred ex *Cucurbita moschata*; 2 females, Mae Mood, Mae Win, Chiang Mai, 8.ix.1992, coll. Chantanee, bred ex *Cucurbita moschata*; 2 females, Had Care Mae Sin, Srisatchanalai, Sukhothai 80m, 23.ix.1992, coll. Chantanee, bred ex *Cucurbita moschata*; 1 male, 2 females, Mae Na T.Mae Na, A. Mae Tang, Chiang Mai, 29.ix.1992, coll. Chantanee, bred ex *Cucurbita moschata*; 1 male, Ban Dong, Suk Ngarm, T. Pa Pai, Chiang Mai, 14.x.1992, coll. Chantanee, bred ex *Cucurbita moschata*; 1 male (1.v.1992), 1 male (8.v.1992), 1 male (15.v.1992), 4 males (22.v.1992), 5 males (29.v.1992), Khao Kheaw, A. Pakphli, Nakhonnayok, coll. Montree, attracted to cue lure. **VIETNAM: Northern Vietnam** – 2 males (30.iii.2002), 3 males (30.iv.2002), 3 males (15.v.2002), Son La (P), Moc Chau (D), Long Luong (V), trap No. VN01, coll. NIPP staff, attracted to cue lure; 1 male (6.ix.1999), 2 males (4.x.1999), 2 males (3.iv.2000), Son La Province, Moc Chau District, Co Do, trap No. N006, attracted to cue lure; 1 male (12.iv.1999), 1 male (19.iv.1999), 4 males (3.iv.2000), 1 male (9.v.2002), Son La Province, Moc Chau District, Tra Long, coll. Le Duc Khanh *et al*., trap No. N007, attracted to cue lure; 1 male (5.iv.1999), 4 males (20.iv.1999), 2 males (28.ix.1999), Hoabinh Province, Mai Chau, Paco, trap No. 10 (rainforest), coll. Le Duc Khanh *et al*., attracted to cue lure; 36 males, Hoa Binh Province, Ky Son District, Cao Phong, 2.viii.1999, trap No. N011, attracted to cue lure; 3 males, Hoabinh Province, Kyson, Caophong, 20.iv.1999, coll. Le Duc Khanh *et al*., trap No. N012 (State Farm), attracted to cue lure; 2 males, Lao Cai Province, Sapa 2 village, 23.iv.1999, trap No. N015 (urban), coll. Le Duc Khanh *et al*., attracted to cue lure; 1 male, Lao Cai Province, Sapa 6 village, 23.iv.1999, trap No. 17 (urban forest), coll. Le Duc Khanh *et al*., attracted to cue lure; 1 male, Lao Cai Province, Sapa, 23.iv.1999, trap No. N019 (urban forest), coll. Le Duc Khanh *et al*., attracted to cue lure; **Southern Vietnam** – 1 male, Southern Vietnam, Dac Lac Province, Ku mi nga District, Quong Phu Village, 8.x.2002, trap No. VS03, attracted to cue lure; 1 male (15.ii.2002), 1 male (14.iv.2002), Southern Vietnam, Binh Dinh (P), Tay Son (D), Binh Khanh (V), trap No. VS11, coll. SOFRI staff, attracted to cue lure. Specimens in QDPC.

Diagnosis: A large species; face fulvous with a pair of medium-sized circular black spots; postpronotal lobes and notopleura yellow; scutum shining black; lateral and medial postsutural yellow vittae present; mesopleural stripe reaching midway between anterior margin of notopleuron and anterior *npl.* seta; scutellum yellow with a black apical spot; wing with cells bc and c colourless, microtrichia in outer corner of cell c only; a narrow fuscous costal band confluent with R_{2+3} and remaining narrow beyond apex of this vein before expanding across apex of R_{4+5}; a broad fuscous anal streak, pale fuscous along *dm-cu* crossvein; abdominal terga III–V each with a dark fuscous to black 'T' pattern.

Description: male (Fig. 348A).

Head: Height 1.96 mm. Frons – length 1.3 times breadth; fulvous with dark fuscous around bases of frontal setae and on anteromedial hump; setae black: 1 *s.or.*, 3 *i.or.*; lunule dark fuscous. Ocellar triangle black. Vertex dark fuscous. Face fulvous with a pair of medium-sized circular black spots; length 0.64 mm. Genae fulvous with a large dark fuscous subocular spot; black bristle present. Occiput red-brown with fulvous along eye margins; occipital row with a large number of weak pale setae. Antennae with segments 1 and 2 red-brown, segment 3 red-brown with fuscous on apex and outer surface; arista black (pale basally); length of segments: 0.14 mm; 0.41 mm; 0.86 mm.

Thorax: Scutum black except red-brown below and behind lateral postsutural vittae. Pleural areas dark fuscous to black except red-brown below postpronotal lobes. Yellow markings as follows: postpronotal lobes; notopleura; mesopleural stripe of medium-width, reaching midway between anterior margin of notopleuron and anterior notopleural seta dorsally, continuing on to katepisternum as a transverse spot, anterior margin straight; lateral postsutural vittae of medium-width, parallel-sided and ending at or just behind *ia.* setae; a medium-width medial postsutural vitta beginning posteriorly at level of *prsc.* setae and ending anteriorly just anterior of notopleural suture; anatergite (posterior apex black); anterior ¾ katatergite (remainder black). Postnotum black. Scutellum yellow with a narrow black basal band and a black apical spot that includes the apical setae (small or absent in some specimens). Setae: *sc.* 4; *prsc.* 2; *ia.* 1; *p.sa.* 1; *sa.* 1; *mpl.* 1; *npl.* 2; *scp.* 4.

Legs: All femora pale fulvous tending fulvous apically (in some specimens there is a small subapical fuscous spot on outer surfaces of fore femora and small fuscous areas around apices of mid and hind femora); fore and mid tibiae fulvous with fuscous basally, hind tibiae fuscous to dark fuscous; all tarsi with basal segment fulvous and apical four segments darker fulvous; mid tibiae each with an apical black spur.

Wings: Length 8.5 mm; cells bc and c colourless; microtrichia in outer corner of cell c only; remainder of wings colourless except fuscous cell sc, a narrow fuscous costal band confluent with R_{2+3} and remaining narrow until it expands into a small spot around apex of R_{4+5}, pale fuscous along *dm-cu* crossvein and ending at hind wing margin (pale and almost absent in some specimens), a broad fuscous anal streak; a dense aggregation of microtrichia around A_1+CuA_2; supernumerary lobe large and rounded.

Abdomen: Oval; terga free; pecten present on tergum III. Tergum I and sterna I and II wider than long. Tergum I dark fuscous with a narrow transverse orange-brown band along posterior margin but not reaching lateral margins; tergum II orange-brown with a narrow transverse black band across anterior margin and spreading around lateral margins; tergum III orange-brown with a dark fuscous to black 'T' pattern, tergum IV orange-brown with a dark fuscous to black 'T' pattern and narrow dark fuscous lateral margins, tergum V orange-brown with a dark fuscous to black 'T' pattern which expands laterally around lateral margins (there is considerable variation in the amount of dark colour patterning on terga III–V (Fig. 348C)). A pair of orange-brown oval shining spots on tergum V. Posterior lobe of surstylus long; posterior margin of sternum V without a concavity.

Attractant: Cue lure.

Female: (Fig. 348B) – As for male except no dense aggregation of microtrichia around A_1+CuA_2; supernumerary lobe weak; pecten of cilia on abdominal tergum III absent; basal segment of ovipositor dark red-brown, dorsoventrally compressed and tapering posteriorly in dorsal view; ratio of length of oviscape to length of tergum V, 1:1; apex of piercer trilobed.

Distribution:

Type locality: Taiwan (Koshun).

Previously recorded: Type locality and the following locations – China; Japan; Taiwan (Norrbom *et al.*, 1998).

New records: Bhutan; Thailand; Vietnam.

Hosts: Bred from flowers of species of *Cucurbitaceae* (Allwood *et al.*, 1999).

Remarks: *Bactrocera (Zeugodacus) scutellata* (Hendel), within the *scutellaris* complex, is a large species and similar to *Bactrocera (Zeugodacus) proprescutellata* Zhang, Chen & Gao in possessing four *sc.* setae, wing with a strong supernumerary lobe and broad anal streak, costal band confluent with R_{2+3} and remaining narrow beyond apex of R_{2+3} before widening gradually across apex of R_{4+5}, abdominal terga III–V with a general 'T' pattern. It differs from *B. proprescutellata* in having a dark 'T' pattern on each of terga III, IV and V, circular facial spots and femora entirely fulvous. *Bactrocera scutellata* is also similar to *Bactrocera (Zeugodacus) ambigua* (Shiraki) but differs in having lateral and medial postsutural yellow vittae narrower and separate dark 'T' patterns on abdominal terga III, IV and V. The measurements included above are from specimens collected in Vietnam, not the type.

Dacus (Zeugodacus) pseudoscutellatus Tseng, Chen & Chu, was separated from *B. scutellata* on minor characters that vary within the species and consequently we are making *D. pseudoscutellatus* a **New Synonym** of *B. scutellata*.

Bactrocera (Zeugodacus) scutellina (Bezzi)

(Fig. 349)

Chaetodacus scutellinus Bezzi, 1916: 113; 1919: 432. Syntype male in MCSNM.

Dacus (Zeugodacus) scutellinus – Hardy & Adachi, 1954: 195; Hardy, 1969b: 481, Lectotype designation invalid; Hardy, 1974: 55.

Bactrocera (Zeugodacus) scutellina – Norrbom *et al.*, 1998: 103.

Type material examined: PHILIPPINES: **LUZON ISLAND** – Syntype male, with the following four labels – 'Mt. Makiling, Luzon, Baker'; '1550'; 'scutellinus Bezzi'; 'Lectotype ♂, Chaetodacus scutellinus Bez., desig. by Hardy 68'. Received on loan from MCSNM.

Non-type material examined: PHILIPPINES: **LUZON ISLAND** – 4 males (13.v.1983), 2 males (27.v.1983), Dasmarinas, Cavite, coll. B. Ramos, attracted to cue lure; 1 male, Salawag, Cavite, 10.vi.1983, coll. B. Ramos, attracted to cue lure; 1 male, Hugo Perez, Trece Martires City, 14.vi.1983, coll. A.M. Arida & N.E. Rivera. Specimens in QDPC.

Diagnosis: A medium-sized species; face fulvous with fuscous along oral margin and two small circular dark fuscous spots; postpronotal lobes and notopleura yellow; scutum black; lateral and medial postsutural yellow vittae present; no yellow spot on anterior margin of notopleural suture; mesopleural stripe reaching to anterior *npl.* seta dorsally; scutellum yellow with a fuscous apical spot; wing with cells bc and c colourless, microtrichia in outer corner of cell c only; a narrow fuscous costal band confluent with R_{2+3} and widening slightly across apex of wing; a broad fuscous anal streak; abdominal terga III–V red-brown with a 'T' pattern consisting of a transverse black band across anterior margin of tergum III and a broad medial longitudinal black band over all three terga, a moderately broad lateral longitudinal black band over terga IV and V.

Description: male.

Head: Height 1.27 mm. Frons – length 1.38 times breadth; fulvous with pale fuscous around upper orbital setae, dark fuscous on anteromedial hump and fuscous to red-brown from anteromedial hump to lunule; orbital setae black: 1 *s.or.*, 2 *i.or.*; lunule dark fuscous. Ocellar triangle black. Vertex dark fuscous. Face fulvous, fuscous along oral margin and two small circular dark fuscous spots; length 0.43 mm. Genae fulvous, dark fuscous subocular spot present; black seta present. Occiput fuscous with dark fuscous spots dorsally; occipital row with a few short pale setae. Antennae with segments 1 and 2 red-brown, segment 3 red-brown with fuscous on apex and outer surface; length of segments: 0.1 mm; 0.26 mm; 0.55 mm.

Thorax: Scutum black with red-brown below and behind lateral postsutural vittae and dark red-brown around notopleural suture, inside postpronotal lobes and anterocentrally. Pleural areas black except red-brown to fuscous below postpronotal lobes. Yellow markings as follows: postpronotal lobes; notopleura; mesopleural stripe reaching to anterior *npl.* seta dorsally, continuing to katepisternum as a large transverse spot, anterior margin straight; anatergite (posterior apex black); anterior ⅔ katatergite (remainder black); two broad parallel-sided lateral postsutural vittae, ending just posterior to *ia.* seta; a broad medial longitudinal postsutural vitta beginning just posterior to level of *prsc.* setae and ending anterior to notopleural suture. Postnotum black. Scutellum yellow except for a narrow black basal band and a fuscous apical spot. Setae: *sc.* 4; *prsc.* 2; *ia.* 1; *p.sa.* 1; *a.sa.* 1; *mpl.* 1; *npl.* 2; *scp.* 2.

Legs: Fore femora with basal ½ fulvous and apical ½ fuscous to black, mid femora fulvous except dark fuscous on apical ⅓, hind femora fulvous except dark fuscous on apical ¼; fore tibiae pale fuscous, mid tibiae fulvous, hind tibiae dark fuscous; tarsi with basal segment fulvous and apical four segments fuscous; mid tibiae each with an apical black spur.

Wings: Length 5.1 mm; cells bc and c colourless; microtrichia in outer corner of cell c only; remainder of wings colourless except dark fuscous cell sc, narrow fuscous costal band confluent with R_{2+3} and widening slightly across apex of wing, broad fuscous anal streak; dense aggregation of microtrichia around A_1+CuA_2; supernumerary lobe of medium development.

Abdomen: Oval; terga free; pecten present on tergum III. Tergum I and sternum I and II wider than long. Tergum I red-brown with narrow black lateral margins; tergum II red-brown with a narrow black band across anterior margin and turning to cover lateral margins; terga III–V red-brown with a 'T' pattern consisting of a narrow transverse black band across anterior margin of tergum III and a broad medial longitudinal black band over all three terga, a moderately broad lateral longitudinal black band over terga IV and V. A pair of oval fulvous shining spots on tergum V. Posterior lobe of surstylus long, sternum V with a slight concavity of posterior margin.

Attractant: No known record.

Female: No known record.

Distribution:

Type locality: Philippines (Luzon I. – Mt Makiling; Palawan I. – Puerto Princesa). All locations constitute the type locality as all type specimens are regarded as Syntypes.

New records: Philippines (Luzon I. – Cavite and Trece Martires City).

Hosts: No known record.

Remarks: *Bactrocera (Zeugodacus) scutellina* (Bezzi) is similar to *Bactrocera (Zeugodacus) lipsana* (Hendel) and *Bactrocera (Zeugodacus) neolipsana* n.sp. in possessing four *sc.* setae, a black scutum with medial and lateral postsutural yellow vittae, wings colourless except for costal band and anal streak, cells bc and c colourless, costal band confluent with R_{2+3} and not expanding into a spot at apex, abdominal terga III–V with a distinct black 'T' pattern and no yellow spot anterior to notopleural suture in front of lateral postsutural vittae. It differs from these species in having the face fulvous with one pair of small dark fuscous spots and a fuscous line across the oral margin.

Hardy (1969b) designated a female Lectotype in the USNM. However, this designation is invalid (Norrbom *et al.*, 1998: 103) because there is no confirmation that Bezzi (1916, 1919) studied these USNM specimens when describing *B. scutellina*. Also, in MCSNM, there is a male specimen with a Lectotype label placed on it by Hardy in 1968. However, this specimen cannot be regarded as a Lectotype as there is no formal designation in literature by Hardy. Consequently, specimens in MCSNM must be regarded as Syntypes and those in USNM non-types.

Bactrocera (Zeugodacus) semongokensis Drew & Romig, new species

(Fig. 350)

Material examined: Holotype male, **EAST MALAYSIA**: Sarawak, Forest Arboretum Semongok, nr. Kuching, 3.ii.1992, coll. S. Leong, attracted to cue lure, deposited in BMNH.

Paratypes: **EAST MALAYSIA**: 1 male, Sabah, E. Malaysia, Mt. Kinabalu, 1500 m, 9.i.1992, coll. Jinius *et al*., attracted to cue lure; 1 male, Sarawak, E. Malaysia, Kpg. Tema Mawang, Tebedu (Rain Forest), 1.iii.1993, coll. S. Leong, attracted to cue lure; 1 male, Sarawak, E. Malaysia, Kpg. Tema Mawang, Tebedu Nr. Kuching, 8.i.1992, coll. S. Leong, attracted to cue lure; 1 male, Sarawak, E. Malaysia, Ranchang Pool, Serian, Nr. Kuching, 23.iii.1992, coll. S. Leong, attracted to cue lure; 3 males (8.i.1992), 1 male (27.iv.1992), 1 male (15.vii.1992), 2 males (19.iii.1993), Sarawak, E. Malaysia, Tarat Agric Station, coll. S. Leong, attracted to cue lure. 2 paratypes in ANIC; 2 paratypes in BMNH; 5 paratypes in QDPC; 2 paratypes in QMIC.

Diagnosis: A large species; face fulvous with a broad black band across oral margin that is sometimes broken in centre; postpronotal lobes and notopleura yellow; scutum mostly black; lateral and medial postsutural yellow vittae present; mesopleural stripe reaching almost to anterior *npl*. seta dorsally; scutellum yellow; a yellow spot anterior to notopleural suture; wing with cells bc and c colourless, microtrichia in outer corner of cell c only; a narrow dark fuscous costal band expanding into a small spot around apex of R_{4+5} and broad dark fuscous anal streak; abdominal terga III–V mostly orange-brown with a narrow dark fuscous to black transverse band across anterior margin of tergum III and dark fuscous anterolateral corners on terga IV and V.

Description: male.

Head: Height 1.68 mm. Frons – length 1.45 times breadth; red-brown with fulvous along lateral margins, dark fuscous on anteromedial hump and around bases of orbital setae; orbital setae black: 1 *s.or*., 3 *i.or*.; lunule dark fuscous. Ocellar triangle black. Vertex black. Face fulvous with a broad black band around oral margin which is sometimes broken in centre leaving two large elongate black spots; length 0.59 mm. Genae fulvous, large dark fuscous subocular spot present; dark fuscous seta present. Occiput fuscous with dark fuscous laterally and fulvous along eye margins; occipital row with a large number of small and larger black setae. Antennae with segments 1 and 2 red-brown, segment 3 red-brown with fuscous on apex and outer surface; arista black (red-brown basally); length of segments: 0.14 mm; 0.29 mm; 0.8 mm.

Thorax: Scutum black except dark red-brown below lateral postsutural vittae and around yellow spot anterior to notopleural suture. Pleural areas black except dark fuscous below postpronotal lobes. Yellow markings as follows: postpronotal lobes; notopleura; mesopleural stripe reaching almost to anterior *npl*. seta dorsally, continuing to katepisternum as a transverse spot, anterior margin slightly convex; anatergite (posterior apex black); anterior $^3/_5$ katatergite (remainder dark fuscous); two broad sub-parallel lateral postsutural vittae beginning with a spot anterior to notopleural suture and ending at or just behind *ia*. setae; a medium-width medial longitudinal postsutural vitta beginning at level of *prsc*. setae, remaining almost parallel and ending at level of notopleural suture. Postnotum black. Scutellum yellow except for narrow black basal band. Setae: *sc*. 4; *prsc*. 2; *ia*. 1; *p.sa*. 1; *a.sa*. 1; *mpl*. 1; *npl*. 2; *scp*. 4.

Legs: Fore femora dark fuscous to black with fulvous on basal $^1/_3$, mid femora dark fuscous to black with fulvous on basal $^1/_2$, hind femora fulvous with fuscous to dark fuscous on apical $^1/_3$; fore and hind tibiae dark fuscous, mid tibiae dark fuscous basally and fuscous apically; tarsi with basal segment fulvous and apical four segments dark fuscous; mid tibiae each with an apical black spur.

Wings: Length 6.8 mm; cells bc and c colourless; microtrichia in outer corner of cell c only; remainder of wings colourless except dark fuscous cell sc, narrow dark fuscous costal band confluent with R_{2+3} and expanding into a small spot around apex of R_{4+5}, broad dark fuscous anal streak; dense aggregation of microtrichia around A_1+CuA_2; supernumerary lobe strong.

Abdomen: Elongate oval; terga free; pecten present on tergum III. Tergum I and sterna I and II wider than long. Tergum I fuscous to dark fuscous except for a narrow orange-brown band along posterior margin but not reaching lateral margins; tergum II fulvous except for a narrow transverse dark fuscous to black band across anterior margin and reaching dark fuscous anterolateral corners; terga III–V orange-brown except for a narrow dark fuscous to black band across anterior margin of tergum III and dark fuscous anterolateral corners on terga IV and V. A pair of oval orange-brown shining spots on tergum V. Posterior lobe of surstylus long, sternum V with a slight concavity on posterior margin. Abdominal sterna fuscous to dark fuscous.

Attractant: Cue lure.

Female: No known record.

Distribution: East Malaysia (Sabah and Sarawak).

Hosts: No known record.

Remarks: *Bactrocera (Zeugodacus) semongokensis* n.sp. is a unique species of the subgenus *Zeugodacus* with a broad black band across oral margin of face and abdominal terga mostly orange-brown. It is similar to *Bactrocera (Zeugodacus) timorensis* (Perkins) in possessing four *sc*. setae, a black scutum, lateral and medial longitudinal postsutural yellow vittae, wings colourless except for costal band and anal streak, costal band confluent with R_{2+3}, cells bc and c colourless, abdominal sterna dark coloured and no distinct 'T' pattern on terga III–V. It differs from *B. timorensis* in having a broad black band across the face and abdominal terga III–V mostly orange-brown.

Bactrocera (Zeugodacus) signata (Hering)

(Figs 351A, 351B)

Zeugodacus bezzianus f. *signata* Hering, 1941d: 10; Norrbom *et al*., 1998: 104 as synonym of *B. tau* (Walker). Syntypes male and female in BMNH and MFN.

Bactrocera (Zeugodacus) signata – Mahmood, 1999: 225.

Type material examined: INDIA: SIKKIM – Syntype female, with the following six labels – 'Paratype, circular yellow outer and white inner'; 'Paratype, rectangular pink'; 'Sikhim, Coll. Bingham'; 'Z. bezzianus P. signatum P.T., det. Hering 1940'; 'F. Bearbeit. erh. v. Mus';

'Purch. from E.M. Hering B.M. 1965-270'. Received on loan and studied at BMNH.

Hering (1941) described *B. signata* from male and female types but he did not designate a Holotype. Consequently, we recognize the specimen studied above in BMNH as a Syntype.

Non-type material examined: BHUTAN: 5 males, Rimchu 2, 10.v.2000, coll. B.S. Fletcher, attracted to cue lure; 1 male, Tsirang, 10.vii.2000, coll. T.D. & S.D, attracted to cue lure; 1 male, Suntalay, 28.vi.2000, coll. S. Wangdi, attracted to cue lure; 1 male, Lingmethang 11km, 9.vi.2001, coll. Brian Fletcher, attracted to cue lure; 1 male, Rimchu 1, 4.v.2000, coll. B.S. Fletcher, attracted to cue lure; 1 male, Rimchu 1, 7.vi.2000, coll. C. Dorji et. al., attracted to cue lure; 1 male, Rimchu 1, 24.v.2000, coll. C. Dorji, attracted to cue lure; 1 male, Rimchu 1, 21.vi.2000, coll. Mado et. al., attracted to cue lure; **THAILAND**: 1 male, Khun Khon, Chiang Rai, 6.viii.1992, coll. Charuwan, attracted to cue lure. Specimens in QDPC.

Diagnosis: A large species; face fulvous with a pair of medium-sized oval black spots; postpronotal lobes and notopleura yellow; scutum orange-brown with extensive areas of black; lateral and medial postsutural yellow vittae present; a yellow spot anterior to notopleural suture; mesopleural stripe reaching midway between anterior margin of notopleuron and anterior *npl.* seta dorsally; scutellum yellow with a dark fuscous apical spot; wing with cells bc and c colourless, microtrichia in outer corner of cell c only; a narrow fuscous costal band overlapping R_{2+3} and ending in a *tau*-like spot in apex of wing; a moderately broad fuscous anal streak; abdominal terga III–V fulvous except for a 'T' pattern consisting of a broad black transverse band across anterior margin of tergum III and a moderately broad medial longitudinal black band over all three terga, broad anterolateral black corners on terga IV and V.

Description: female (Fig. 351A).

Head: Height 1.65 mm. Frons – length 1.24 times breadth; fulvous with fuscous to dark fuscous around bases of the pair of *s.or.* setae and on anteromedial hump; orbital setae black: 1 *s.or.*, 3 *i.or.*; lunule fuscous to dark fuscous. Ocellar triangle black. Vertex fuscous to dark fuscous. Face fulvous with a pair of medium-sized oval black spots; length 0.59 mm. Genae fulvous with a large dark subocular spot. Occiput red-brown with a pair of lateral fuscous marks and fulvous along eye margins; occipital row with 6–8 dark setae. Antennae with segments 1 and 2 fulvous, segment 3 fulvous with fuscous on apex and outer surface; arista black (pale basally); length of segments: 0.18 mm; 0.28 mm; 0.94 mm.

Thorax: Scutum orange-brown with broad black lateral longitudinal bands and a short, narrow, medial longitudinal black line anterior to medial yellow vitta. Pleural areas orange-brown below postpronotal lobes, broadly black anterior to mesopleural stripe and on katepisternum and dark fuscous posterior to stripe. Yellow markings as follows: postpronotal lobes; notopleura; mesopleural stripe of medium-width, reaching midway between anterior margin of notopleuron and anterior notopleural seta dorsally, continuing to katepisternum as a large transverse spot, anterior margin strongly convex; medium-width parallel-sided lateral postsutural vittae beginning in a spot anterior to notopleural suture and ending just posterior to *ia.* seta; a broad medial longitudinal postsutural vitta rounded posteriorly and narrowing to a point anteriorly at line of notopleural suture; anatergite (posterior apex black); anterior ¾ katatergite (remainder black). Postnotum black laterally, red-brown centrally. Scutellum yellow with a narrow black basal band and a dark fuscous apical spot that includes the apical setae. Setae: *sc.* 4; *prsc.* 2; *ia.* 1; *p.sa.* 1; *a.sa.* 1; *mpl.* 1; *npl.* 2; *scp.* 4.

Legs: Fore femora fulvous with a small subapical dark fuscous spot, mid and hind femora fulvous with apical ¼ dark fuscous (paler on inner surfaces of apex of mid-femora); fore and hind tibiae dark fuscous, mid tibiae dark fuscous basally and fulvous apically; tarsi with basal segment fulvous and apical four segments pale fuscous; mid tibiae each with an apical black spur.

Wings: Length 7.9 mm; cells bc and c colourless; microtrichia in outer corner of cell c only; remainder of wings colourless except dark fuscous cell sc, dark fuscous costal band becoming paler after it overlaps R_{2+3} and expanding into a large apical spot that ends midway between apices of R_{4+5} and M, a medium-width dark fuscous anal streak ending well before wing margin; no dense aggregation of microtrichia around A_1+CuA_2; supernumerary lobe weak.

Abdomen: Oval; terga free; pecten absent on tergum III. Tergum I and sterna I and II wider than long. Tergum I fulvous with narrow black anterior and lateral margins; tergum II fulvous with a narrow transverse black band anteriorly which does not reach lateral margins and narrow black anterolateral corners; terga III–V fulvous except for a black 'T' pattern and broad black anterolateral markings on terga IV and V. A pair of fulvous oval shining spots on tergum V. Abdominal sterna dark except sternum II fulvous. Basal segment of ovipositor orange-brown, dorsoventrally flattened and tapering posteriorly in dorsal view; ratio of length of oviscape to length of tergum V, 1.4:1; apex of piercer needle-shaped.

Attractant: Cue lure.

Male: (Fig. 351B) – As for female except for a dense aggregation of microtrichia around A_1+CuA_2 present; supernumerary lobe strong and keel-shaped; pecten of cilia on abdominal tergum III present; posterior lobe of surstylus long; abdominal sternum V with a slight concavity on posterior margin.

Distribution:

Type locality: India (Sikkim).

New records: Bhutan; Thailand (Chiang Rai).

Hosts: No known record.

Remarks: *Bactrocera (Zeugodacus) signata* (Hering) in many characters conforms with *tau* complex species. It is similar to *Bactrocera (Zeugodacus) rubella* (Hardy) in possessing extensive areas of red-brown on the scutum, four *sc.* setae, medial and lateral postsutural yellow vittae, wings colourless except for costal band and anal streak, costal band expanded into a spot at apex and lateral postsutural vittae reaching to or behind *ia.* setae. It differs from *B. rubella* in having a well-developed supernumerary lobe in the male wing, costal band just overlapping R_{2+3} (not almost to R_{4+5}) and apical spot in costal band around apex of R_{4+5} (not reaching M). Within the *tau* complex, it is similar to *Bactrocera*

(Zeugodacus) pubescens (Bezzi) in having a dark fuscous spot on the apex of the scutellum but differs in possessing dark fuscous spots on the apices of all femora and the apex of the piercer needle-shaped. In *B. pubescens* the femora are entirely fulvous and the apex of the piercer trilobed. In studying large numbers of *B. tau* specimens from across South-East Asia, none has been observed with a dark spot on the apex of the scutellum.

Bactrocera signata was placed in synonymy with *Bactrocera (Zeugodacus) tau* (Walker) by Norrbom *et al.* (1998); however, we recognize it as a good species and are thus withdrawing it from this previous synonymy.

Bactrocera (Zeugodacus) sonlaiae Drew & Romig, new species

(Fig. 352)

Material examined: Holotype male, **VIETNAM: Northern Vietnam** – Son La Province, Moc Chau District, Co Do, trap No. 006, attracted to cue lure, deposited in BMNH.
Paratypes: **BHUTAN**: 1 male (27.ix.2000), 1 male (5.x.2000), Rimchu trap No. 2, coll. C. Dorji, attracted to cue lure. **VIETNAM: Northern Vietnam** – 1 male (15.vi.2001), 1 male (15.x.2001), same data as holotype; 1 male (30.vi.2001), 1 male (15.x.2001), Hoa Binh Province, Mai Chau District, Pa Co, trap N010, attracted to cue lure. 2 paratypes in ANIC; 3 paratypes in QDPC; 1 paratype in QMIC.
Diagnosis: A medium-sized species; face fulvous with a pair of large elongate oval black spots; postpronotal lobes and notopleura yellow; scutum shining black; lateral and medial postsutural yellow vittae present; a yellow spot present anterior to notopleural suture; mesopleural stripe slightly wider than notopleuron dorsally; scutellum yellow with a shining black apical spot over apical ½ which sometimes joins with the narrow black basal band; wing with cells bc and c colourless, microtrichia in outer corner of cell c only; a narrow dark fuscous costal band expanding into a small spot apically (*scutellaris*-like), a broad dark fuscous band along *dm-cu* crossvein and a broad dark fuscous anal streak; abdominal terga III–V mostly black with small red-brown spots posterocentrally on terga III and IV and a narrow dark red-brown band along posterior margin of tergum V.
Description: male.
Head: Height 1.32 mm. Frons – length 1.47 times breadth; red-brown centrally with fulvous lateral margins and dark fuscous around orbital setae and on anteromedial hump; orbital setae black: 1 *s.or.*, 3 *i.or.*; lunule dark fuscous. Ocellar triangle black. Vertex black. Face fulvous with a pair of large elongate oval black spots; length 0.48 mm. Genae fulvous, large dark fuscous subocular spot present; black seta present. Occiput black, fulvous along eye margins; occipital row with 4–6 black setae. Antennae with segments 1 and 2 fuscous, segment 3 red-brown with dark fuscous on apex and outer surface; arista black (red-brown basally); length of segments: 0.1 mm; 0.22 mm; 0.6 mm.
Thorax: Scutum shining black without pale markings. Pleural areas shining black without pale markings. Yellow markings as follows: postpronotal lobes; notopleura; mesopleural stripe slightly wider than notopleuron dorsally, continuing to katepisternum as a small transverse spot, anterior margin convex; anatergite (posterior apex black); anterior ⅔ katatergite (remainder black); two narrow parallel-sided lateral postsutural vittae beginning with a very small spot anterior to notopleural suture and ending just before *ia.* seta (in one specimen the postsutural vittae narrow posteriorly to end at *ia.* seta); a very short and narrow medial longitudinal postsutural vitta. Postnotum black. Scutellum yellow with a narrow black basal band and either a shining black spot over apical ½ or with the shining black apical spot joined to the narrow black basal band to form a medial longitudinal black band. Setae: *sc.* 4; *prsc.* 2; *ia.* 1; *p.sa.* 1; *a.sa.* 1; *mpl.* 1; *npl.* 2; *scp.* 4.
Legs: Femora fulvous with apical ⅓ to ¼ shining black (inner apical surfaces of fore femora tending dark fuscous); all tibiae dark fuscous to black; tarsi with basal segment fulvous and apical four segments dark fuscous; mid tibiae each with an apical black spur.
Wings: Length 5.8 mm; cells bc and c colourless; microtrichia in outer corner of cell c only; remainder of wings colourless except dark fuscous cell sc, narrow dark fuscous costal band confluent with R_{2+3} and very narrow beyond apex of this vein before expanding into a spot around apex of R_{4+5}, broad dark fuscous band along *dm-cu* crossvein (broader at hind wing margin), broad dark fuscous anal streak; dense aggregation of microtrichia around A_1+CuA_2; supernumerary lobe weak.
Abdomen: Oval; terga free; pecten present on tergum III. Tergum I and sterna I and II wider than long. Tergum I dark fuscous to shining black with a broad transverse fulvous band across posterior margin but not reaching lateral margins; tergum II fulvous with a broad transverse shining black band across anterior half but not covering entire lateral margins; terga III–V shining black with dark red-brown spots posterocentrally on terga III and IV either side of a medial longitudinal shining black band and dark red-brown along posterior margin of tergum V. A pair of oval dark fuscous to black shining spots on tergum V. Posterior lobe of surstylus elongate but shorter and more stout that in most subgenus *Zeugodacus*, sternum V with a slight concavity on posterior margin. Abdominal sterna shining black, except sternum II fulvous.
Attractant: Cue lure.
Female: No known record.
Distribution: Bhutan; Northern Vietnam (Hoa Binh and Son La Provinces).
Hosts: No known record.
Remarks: *Bactrocera (Zeugodacus) sonlaiae* n.sp., within the *scutellaris* complex, is similar to *Bactrocera (Zeugodacus) proprescutellata* Zhang, Chen & Gao and *Bactrocera (Zeugodacus) scutellaris* (Bezzi) in possessing four *sc.* setae, abdominal terga mostly dark fuscous to black (*scutellaris*-like) and face with a pair of black spots. It differs from *B. scutellaris* in having a weak supernumerary lobe in the male wing and parallel-sided lateral postsutural yellow vittae and from *B. proprescutellata* in having the *dm-cu* crossvein infuscated, anal streak present and male surstyli short and stout.

Bactrocera (Zeugodacus) sumbensis (Hering)

(Fig. 353)

Zeugodacus nubilus ssp. *sumbensis* Hering, 1953: 75. Holotype female in NMB.
Bactrocera (Zeugodacus) sumbensis – Norrbom *et al.*, 1998: 103.

Type material examined: Holotype female, **INDONESIA**: with the following three labels – one red label 'TYPUS'; one white label 'O-SUMBA, Baing, 1.7.1949, Expedition Buhler-Sutter'; one white label '*Zeugodacus nubilus sumbensis* n., Holotype, det Hering 1952'. Received on loan from NMB.
Diagnosis: A large species; face with a pair of small circular black spots; postpronotal lobes and notopleura yellow; scutum base colour red-brown; lateral and medial postsutural yellow vittae present; a yellow spot anterior to notopleural suture in front of lateral postsutural vittae; mesopleural stripe reaching midway between anterior margin of notopleuron and anterior *npl.* seta dorsally; scutellum yellow; wing with cells bc and c with a very pale tint, microtrichia in outer ⅓ of cell c only; a broad pale fuscous to fuscous costal band expanding into a spot across apex to M; a broad fuscous to pale fuscous anal streak; abdominal terga III–V red-brown with a pale 'T' pattern consisting of a narrow transverse pale fuscous band across anterior margin of tergum III and a narrow- to medium-width medial longitudinal pale fuscous to black band over all three terga, narrow dark fuscous lateral margins over all three terga.
Description: female.
Head: Height 1.81 mm. Frons – length 1.4 times breadth; fulvous with fuscous on anteromedial hump; latter with a large number of short pale hairs; orbital setae weak and dark fuscous: 1 *s.or.*, 3 *i.or.*; lunule fuscous. Ocellar triangle black. Vertex red-brown. Face fulvous with a pair of small circular black spots; length 0.66 mm. Genae fulvous, pale fuscous subocular spot; dark fuscous seta present. Occiput red-brown, narrow transverse fuscous band dorsally, fulvous along eye margins; occipital row with 5–7 dark setae. Antennae with segments 1 and 2 red-brown, segment 3 red-brown with fuscous on apex and outer surface; length of segments: 0.14 mm; 0.32 mm; 0.87 mm.
Thorax: Scutum red-brown with large irregularly shaped dark fuscous to black markings between medial and lateral postsutural yellow vittae and dark fuscous to black spots between the postpronotal lobe and notopleuron. Pleural areas red-brown except fuscous anterior and posterior to mesopleural stripe and below yellow spot on katepisternum. Yellow markings as follows: postpronotal lobes; notopleura; mesopleural stripe of medium-width reaching midway between anterior margin of notopleuron and anterior *npl.* seta dorsally, continuing to katepisternum as a large transverse spot, anterior margin slightly convex; anatergite (posterior apex black); anterior ¾ katatergite (remainder black); two narrow lateral postsutural vittae narrowing to a point posteriorly to end before level of *p.sa.* setae; a moderately broad medial longitudinal postsutural vitta rounded posteriorly and narrowing to a point anteriorly just anterior of notopleural suture. Postnotum black laterally and red-brown centrally. Scutellum yellow except for a narrow dark fuscous basal band. Setae: *sc.* 4; *prsc.* 2; *ia.* 1; *p.sa.* 1; *a.sa.* 1; *mpl.* 1; *npl.* 2; *scp.* 4.
Legs: Fore femora fulvous with a large fuscous spot on outer apical surface, mid femora fulvous on basal ½ and dark fulvous on apical ½, hind femora fulvous with dark fulvous on apical ⅓ fore tibiae pale fuscous on inner surface and fuscous on outer surface, mid tibiae pale fuscous basally and fulvous apically, hind tibiae fuscous; all tarsal segments entirely fulvous; mid tibiae each with an apical black spur.
Wings: Length 8.2 mm; cells bc and c with a very pale tint; microtrichia in outer ⅓ of cell c only; remainder of wings colourless except fuscous cell sc, broad fuscous costal band remaining fuscous narrowly over R_{2+3} and then becoming pale fuscous to R_{4+5} and with a large pale fuscous to fuscous apical spot reaching to vein M, a broad fuscous to pale fuscous anal streak; no dense aggregation of microtrichia around A_1+CuA_2; supernumerary lobe weak.
Abdomen: Elongate oval; terga free; no pecten present on tergum III. Tergum I and sterna I and II wider than long. Tergum I fuscous with dark fuscous narrow lateral margins and a narrow fulvous transverse line on posterior margin but not reaching lateral margins; tergum II red-brown tending fulvous posterolaterally, pale fuscous to fuscous spots anterolaterally and with dark fuscous narrow lateral margins; terga III–V red-brown with a 'T' pattern consisting of a narrow transverse pale fuscous band across anterior margin of tergum III and a narrow- to medium-width medial longitudinal band over all three terga which varies from pale fuscous to black, narrow dark fuscous lateral margins on all three terga; a pair of oval fulvous shining spots on tergum V. Basal segment of ovipositor red-brown, dorsoventrally flattened and tapering posteriorly in dorsal view; ratio of length of oviscape to length of tergum V, 1.2:1.
Attractant: No known record.
Male: No known record.
Distribution:
Type locality: Indonesia (Sumba I., Baing).
Hosts: No known record.
Remarks: *Bactrocera (Zeugodacus) sumbensis* (Hering) belongs to a group of species with four *sc.* setae, scutum base colour red-brown with medial and lateral postsutural yellow vittae, wings colourless except for costal band and anal streak, costal band expanded into an apical spot and lateral postsutural vittae ending before *ia.* seta. It is similar to *Bactrocera (Zeugodacus) depressa* (Shiraki), *Bactrocera (Zeugodacus) okunii* (Shiraki) and *Bactrocera (Zeugodacus) uncinata* n.sp. but differs from these in having the costal band broad and confluent with R_{4+5} and a subapical fuscous spot on outer surfaces of fore femora.

Bactrocera (Zeugodacus) synnephes (Hendel)

(Figs 354A, 354B)

Dacus synnephes Hendel, 1913: 40. Holotype female in DEI.
Zeugodacus synnephes var. *dobaensis* Shiraki, 1933: 95.
Dacus synnethes – Hardy, 1968: 113 [Misspelling].

Dacus (Zeugodacus) ubiquitus Hardy, 1973: 71. **New Synonym**.
Dacus (Zeugodacus) stenomus Wang and Zhao, 1989: 213. **New Synonym**.
Bactrocera (Zeugodacus) synnephes – Norrbom *et al*., 1998: 103.

Type material examined: Holotype female, **TAIWAN (formerly FORMOSA)**: with the following four labels – white label 'III.Fuhosho Formosa H. Sauter 09'; white label 'Dacus synnephes, H. det Hendel'; red label 'Holotypus'; white label 'Coll. DEI Eberswalde'. Received on loan from DEI.
Other type material examined: PHILIPPINES: Holotype male of *Bactrocera (Zeugodacus) ubiquitus* Hardy with the following label data – 'P.I.: Luzon, Comarines Sur, Mt. Isarog, PI11, 800–900 m, 22.iv.1965', 'H.M. Torrevillas, Collector, Bishop', red label 'Holotype male, *Dacus (Zeugodacus) ubiquitus*, D.E. Hardy'. The following paratypes of *B. ubiquitus* – Female, 'P.I., Mindanao, Cotabato Prov., Polo, nr base of Mt Matutum, 600 m, 14.viii.1958', 'H.E. Milliron Collector'; blue label 'Paratype female, *Dacus (Zeugodacus) ubiquitus*, D.E. Hardy'. Paratype female 'Polo, nr base of Mt Matutum, Cotabato, Mindanao, 14.viii.1958, H.E. Milliron'; blue label 'paratype female, *Dacus (Zeugodacus) ubiquitus*, D.E. Hardy'. **THAILAND**: Paratype male 'Thailand, NW Chiangmai Prov, Chiangdao 450 m, 5–11.iv.1958'; 'T.C. Mao Collector, No. 319'; blue label 'Paratype male, *Dacus (Zeugodacus) ubiquitus* D.E. Hardy'. Paratype female 'Thailand, Phu Kae, 7 SEPT 1965'; blue label 'paratype female, *Dacus (Zeugodacus) ubiquitus*, D.E. Hardy'. Received on loan from BBM.
Non-type material examined: INDONESIA: BALI – 1 male, Bali Buleleng, Tajun, 18.viii.2006, coll. Hendri Y, attracted to cue lure in forest; **JAVA** – 1 male, East Java, Pasuruan, Tretes, 26.vii.2005, coll. Ketut, attracted to cue lure in forest; 1 male (16.viii.2005), 2 males (23.viii.2005), 2 males (15.xii.2005), East Java, Lumajang, Kadang Tepus, coll. Utomo, attracted to cue lure in banana; 1 male (1.xi.2005), 1 male (22.xi.2005), D.I. Yogyakarta, Sieman, Kaliurang, coll. Agus W, attracted to cue lure in forest; 1male, Gunung Salak, Bogor, Java, Indonesia, 800m, 17.v.1981, coll. J. Turner, secondary rainforest; **LOMBOK** – 1 male, Aik Bukak, C. Lombok, 2.vii.78, J.T.; 1 female, Aik Bukak, c810m, nr Mataram, Mal trap 4, 10-28.iii.1991, coll. C.V. Achterberg, RMNH '91; 1 male, Suranadi, W. Lombok, 4.vi.1978, coll. J.T. Medler; 1 male, Aik Bukak, C. Lombok, 2.vii.1978, coll. J.T. Medler, attracted to cue lure; **RAKATA ISLAND** – 1 male, Krakatoa, Rakata, 2.ix.1990, coll. E. Schmidt; **SULAWESI** – 3 males (19.ix.2006), 1 male (3.x.2006), 1 male (10.x.2006), Gorontalo, Bone Bolango, Lombongo, coll. Yusbar Ismail, attracted to cue lure in forest. A large series of specimens from Flores and Sumbawa. **PENINSULAR MALAYSIA**: 2 males, transect North Trap No. 1 (TN1), 17.xi.1986, coll. Yusoff Salleh, attracted to cue lure; 1 male (25.viii.1986), 1 male (8.ix.1986), 1 male (6.x.1986), 2 males (20.x.1986), 1 male (29.xii.1986), 1 male (12.i.1987), 3 males (26.i.1987), 4 males (9.ii.1987), 6 males (23.ii.1987), 1 male (27.vii.1987), 1 male (15.xii.1987), Transect North Trap No. 9 (TN9), coll. Yusoff Salleh, attracted to cue lure; 1 male, TN9, 3 Nov.1986, coll. Fauzi, attracted to cue lure; 1 male, TN10, 26.i.1987, coll. Y. Salleh, attracted to cue lure. **PHILIPPINES**: 29 males, Mt Makiling (600m), 5.ix.1983, coll. R.A.I. Drew, attracted to cue lure; 9 males (9.v.1983), 4 males (15.vi.1983), 13 males (5.ix.1983), Mt Makiling (500m), coll. R.A.I. Drew, attracted to cue lure; 5 males (8.vi.1983), 8 males (15.vi.1983), Mt Makiling (400m), coll. B. Ramos, attracted to cue lure; 1 male, Mt Makiling (400m), 9.v.1983, coll. Drew, Nette and Bert, attracted to cue lure; 1 male (8.vi.1983), 2 males (15.vi.1983), Mt Makiling (300m), coll. B. Ramos, attracted to cue lure; 1 male (8.vi.1983), 2 males (15.vi.1983), Mt Makiling (200m), coll. B. Ramos, attracted to cue lure; 1 male, Mt Makiling (200m), 5.ix.1983, coll. Drew, Nette and Bert, attracted to cue lure; 3 males, 8.vi.1983, coll. B. Ramos, attracted to cue lure; 1 male, Malandag, 21.vii.1983, coll. A. Yasana; 3 males, D.E.S. Bago Oshiro, 13.v.1983, coll. D.A. Nicolas; 1 male, Guimarae Exp. St., 24.vi.1983, coll. S.N. Iligau; 4 males, in cultivation, D.E.S. Davao City, 16.v.1983, coll. D.A. Nicolas, attracted to cue lure; 5 males (11.v.1983), 2 males (13.v.1983), 1 male (17.v.1983), 1 male (23.v.1983), 2 males (15.vi.1983), D.E.S. Davao City, coll. D.A. Nicolas, attracted to cue lure; 2 males (12.xii.1999), 5 males (12.i.2000), Mt Makiling Botanic Gardens, alt 150m, coll. A. O'Toole, attracted to cue lure; 2 males, Sirang Lupa, Calamba, Laguna, Apple Guava orchard, 20.xii.2000, coll. A. O'Toole and G. Quimio, attracted to cue lure. **TAIWAN**: 39 males, Jiauliping Chiayi, Itsein, 1.ix – 1.x.1977, coll. Y.H. Tseng, attracted to cue lure; 23 males, Frenchifu, 30.xi.1978, coll. Lena Lee Wen, attracted to cue lure; 3 males, Chuchi, 30.xi.1978, coll. Lena Lee Wen, attracted to cue lure; 3 males, Yangtlen, Taichung Hs, 16.xi.1990, coll. Y.S. Tseng, attracted to cue lure. **THAILAND**: 1 male, Khao Yai, Moosi, Pakchong, Nakhonratsima, 28.vi.1991, coll. Montree, attracted to cue lure; 4 males (24.x.1991), 1 male (13.xi.1991), 2 males (27.xi.1991), Phan, Chiang Rai, coll. Charuwan, attracted to cue lure; 4 males (30.x.1991), 2 males (27.xi.1991), Pha Yao, Chiang Rai, coll. Charuwan, attracted to cue lure; 1 male, Wang Jum Pee, Pakchong, Nakhonratsima, 27.xii.1991, coll. Montree, attracted to cue lure; 1 male (10.i.1992), 1 male (15.v.1992), 1 male (22.v.1992), 1 male (4.ix.1992), Khao Kheaw, A. Pakphli, Nakhonnayok, coll. Montree, attracted to cue lure; 1 male (10.i.1992), 2 males (29.v.1992), 1 male (12.vi.1992), Wang Jum Pee, Pakchong, Nakhonratsima, coll. Montree, attracted to cue lure; all specimens in QDPC except where specified. 3 males, 570m, Chiang Mai Prov, Tham Klaeb Yai/16, 30km N Chiang Dao, 14.xii.2002, 19.34.34N, 99.03.40E, coll. P. Schwendinger/16, attracted to cue lure; 1 male, 660m, Chiang Mai Prov, Doi Suthep/14, Monthathan waterfall, 7.xii.2002, 18.48.58N, 98.55.39E, coll. P. Schwendinger/14, attracted to cue lure; 1 male, 7km NW of Fong, Horticultural Experimental Station, 30.x - 2.xi.1979, Zool. Mus. Copenhagen Exped; specimens in MHNG.
Diagnosis: A large species; face fulvous with a pair of circular to oval black spots; postpronotal lobes and notopleura yellow; scutum black with large areas of red-brown; lateral and medial postsutural yellow vittae present; yellow spot anterior to notopleural suture in front of lateral postsutural

vittae; mesopleural stripe reaching midway between anterior margin of notopleuron and anterior *npl.* seta dorsally; scutellum yellow; wing with cell bc pale fuscous and cell c slightly paler fuscous, microtrichia in outer corner of cell c only; a dark fuscous costal band overlapping R_{2+3} and widening across apex; a broad dark fuscous anal streak; dark fuscous enclosing *dm-cu* crossvein, a pale tint around *r-m* crossvein and around hind margin of wing; abdominal terga III–V fulvous with a general black 'T' pattern and broad dark fuscous to black lateral longitudinal bands over terga IV and V (in some specimens these bands are broken along posterior margin of tergum IV by a narrow transverse fulvous band).

Description: female (Fig. 354A).

Head: Height 1.67 mm. Frons – length 1.13 times breadth; fulvous with dark fuscous on anteromedial hump; orbital setae black: 1 *s.or.*, 3 *i.or.*; lunule dark fuscous. Ocellar triangle black. Vertex fuscous. Face fulvous with medium-sized circular to oval black spots; length 0.55 mm. Genae fulvous, large fuscous subocular spot; black seta present. Occiput red-brown, fulvous along eye margins; occipital row with 4–6 weak red-brown setae each side. Antennae with all segments red-brown; length of segments: 0.17 mm; 0.36 mm; 0.79 mm.

Thorax: Scutum red-brown with large irregularly shaped lateral longitudinal black markings or black with large red-brown areas as follows: below and behind lateral postsutural vittae, around yellow spot anterior to notopleural suture, on lateral margin between postpronotal lobe and notopleuron, and a large area anteriorly from anterior point of medial postsutural yellow vitta to anterior margin of scutum and across full width of scutum (there is a narrow medial longitudinal black line within this red-brown area from anterior point of medial postsutural yellow vitta to anterior margin of scutum); in the Holotype there is a large dark red-brown area on posterior of scutum reaching from posterior margin to posterior point of medial postsutural yellow vitta. Pleural areas red-brown except for dark fuscous to black along anterior margin of mesopleural stripe and below yellow spot on katepisternum and fuscous along posterior margin of mesopleural stripe. Yellow markings as follows: postpronotal lobes; notopleura; mesopleural stripe reaching midway between anterior margin of notopleuron and anterior *npl.* seta dorsally, continuing to katepisternum as a large transverse spot, anterior margin slightly convex; anatergite (posterior apex black); anterior $^{3}/_{5}$ katatergite (remainder black); two moderately broad lateral postsutural vittae beginning with a spot anterior to notopleural suture and narrowing slightly posteriorly to end before *ia.* seta; a moderately broad medial longitudinal postsutural vitta beginning slightly anterior of *prsc.* setae and ending at a point anterior to notopleural suture. Postnotum black laterally, red-brown centrally. Scutellum yellow with a narrow red-brown basal band. Setae: *sc.* 4; *prsc.* 2; *ia.* 1; *p.sa.* 1; *a.sa.* 1; *mpl.* 1; *npl.* 2; *scp.* 4. The *sc.* setae vary from 2 to 4 or sometimes with 2 weak basal setae.

Legs: Femora fulvous except for a dark fuscous spot on outer apical surfaces of fore femora and dark fulvous around apices of mid femora; fore and hind tibiae dark fuscous and mid tibiae fuscous; all tarsi with basal segment fulvous and apical four segments dark fulvous; mid tibiae each with an apical black spur. In some specimens there are dark fuscous areas around apices of hind femora and as a subapical spot on outer surfaces of mid femora.

Wings: Length 7.3 mm; cell bc pale fuscous and cell c slightly paler fuscous; microtrichia in outer corner of cell c only; wings with cell sc dark fuscous, a narrow dark fuscous costal band overlapping R_{2+3} and widening across apex of wing to M but more as a distinct spot in some specimens, a narrow fuscous tint along *r-m* crossvein and dark fuscous along *dm-cu* crossvein which begins anterior to vein M and continues around hind margin of wing as a pale tint, a broad fuscous anal streak; no dense aggregation of microtrichia around A_1+CuA_2; supernumerary lobe weak.

Abdomen: Elongate oval; terga free; no pecten on tergum III. Tergum I and sterna I and II wider than long. Tergum I fulvous except narrow fuscous margin anteriorly and narrow dark fuscous to black lateral margins; tergum II fulvous except for a narrow dark fuscous to black band along anterior margin which does not reach lateral margins and narrow dark fuscous to black lateral margins; terga III to V fulvous except for a black 'T' made up of a narrow dark fuscous to black band across anterior margin of tergum III and turning around lateral margins and a medium-width medial longitudinal dark fuscous to black band over all three terga, anterolateral corners of terga IV and V broadly dark fuscous to black. A pair of oval fulvous shining spots on tergum V. Abdominal sterna dark fuscous to black, except sternum II pale. Basal segment of ovipositor red-brown, dorsoventrally flattened; ratio of length of oviscape to length of tergum V, 0.9:1; apex of piercer needle-shaped.

Attractant: Cue lure.

Male: (Fig. 354B) – As for female except for a dense aggregation of microtrichia around A_1+CuA_2; supernumerary lobe strong and keel shaped; pecten present on abdominal tergum III; posterior lobe of surstylus long; abdominal sternum V with a slight concavity on posterior margin.

Distribution:

Type locality: Taiwan (Fuhosho).

New records: Indonesia (the following islands – Bali; Flores; Java; Lombok; Rakata; Sulawesi; Sumbawa); Peninsular Malaysia. Philippines and Thailand (due to **New Synonymy** of *D. ubiquitus*). China (due to **New Synonymy** of *D. stenomus*).

Hosts: No known record.

Remarks: *Bactrocera (Zeugodacus) synnephes* (Hendel) is similar to *Bactrocera (Zeugodacus) hululangatiae* n.sp. and *Bactrocera (Zeugodacus) neoelegantula* White where specimens possess two *sc.* setae, a medial postsutural yellow vitta, *prsc.* setae, *dm-cu* crossvein infuscated, notopleura yellow, a yellow spot anterior to notopleural suture and dark patterns on the apices of all femora. It differs from these species in having the mesopleural stripe reaching midway between anterior margin of notopleuron and anterior *npl.* seta dorsally and the costal band expanding into a distinct large spot at apex.

The specimens with four *sc.* setae are similar to *Bactrocera (Zeugodacus) abdoangusta* (Drew), *Bactrocera (Zeugodacus) calumniata* (Hardy) and *Bactrocera (Zeugodacus) ishigakiensis* (Shiraki). They differ from *B. abdoangusta* in having a narrower mesopleural stripe (not reaching anterior *npl.* seta), dark patterns on some or all femora and the costal band overlapping R_{2+3} but not confluent with R_{4+5}. It differs from *B. ishigakiensis* in having the scutellum entirely yellow and from *B. calumniata* in having the lateral postsutural vittae tapering to end before the *ia.* setae.

The type series of *Dacus ubiquitus* Hardy actually contains three species. The Holotype is a specimen of *Bactrocera (Zeugodacus) synnephes* (Hendel), a Paratype from Mindanao, Philippines is very similar to *Bactrocera (Zeugodacus) baliensis* n.sp. described in this volume, and three Paratypes from Mindanao, Philippines and Chiang Mai and Phu Kae in Thailand are what appears to be another undescribed species. All Paratypes were dissected at the time of original study by Hardy (1973) and are now too damaged to use for description of the undescribed species. Small tubes were used to hold the abdomens after dissection. In one tube for one paratype the abdomen is missing and has been replaced by a large ichneumonid wasp. The measurements included are from specimens collected in Thailand, not the type.

Dacus (Zeugodacus) stenomus Wang & Zhao and *Dacus (Zeugodacus) ubiquitus* Hardy cannot be separated from *Bactrocera synnephes* on any sound specific characters and consequently are here made **New Synonyms** of *B. synnephes*.

Bactrocera (Zeugodacus) tapervitta Mahmood

(Fig. 355)

Bactrocera (Zeugodacus) tapervitta Mahmood, 1999: 227. Holotype male in BMNH.

Type material examined: Holotype male, **PHILIPPINES**: with the following nine labels – 'Penal Colony, Davao, Mindanao, PI8-'50'; 'Ex *Momordica cochinchinensis*'; 'F. Skinner Collector'; 'Dacus (Zeugodacus) hageni de Meij det 1958, D.E. Hardy'; 'Purch from E.M. Hering, B.M. 1965-270'; '6'; 'Type'; 'Holotype, *Bactrocera (Z.) tapervitta* K. Mahmood 1999'; 'BMNH (E), #533050'. Studied at BMNH.

Non-type material examined: PHILIPPINES: 1 female with the following five labels – 'Penal Colony, Davao, Mindanao, PI8-'50'; 'Ex *Momordica cochinchinensis*'; 'F. Skinner Collector'; 'Purch from E.M. Hering, B.M. 1965-270'; '5'; studied at BMNH. 4 males, 1 female, Pinaglapatan, Infanta, Quezon, reared from *Luffa cylindrica*, 10.xii.1999, coll. G. Quimio; in QDPC.

Diagnosis: A medium-sized species; face fulvous with a pair of small- to medium-sized circular black spots; postpronotal lobes and notopleura yellow; scutum black with large areas of red-brown laterally and centrally; lateral and medial postsutural yellow vittae present; yellow spot anterior to notopleural suture; mesopleural stripe reaching midway between anterior margin of notopleuron and anterior *npl.* seta dorsally; scutellum entirely yellow; wing with cells bc and c with a fulvous tint, microtrichia in outer corner of cell c only; a narrow fuscous costal band overlapping R_{2+3} and ending in a *tau*-like spot in apex of wing; a medium-width fuscous anal streak; abdominal terga III–V fulvous with a black 'T' pattern and either broad anterolateral black marks on terga IV and V or a broad black lateral longitudinal band over terga IV and V.

Description: male.

Head: Height 1.44 mm. Frons – length 1.3 times breadth; red-brown, fulvous on lateral margins and dark fuscous on anteromedial hump; orbital setae black: 1 *s.or.*, 2 *i.or.*; lunule dark fuscous. Ocellar triangle black. Vertex fuscous. Face fulvous with a pair of small- to medium-sized circular black spots; length 0.53 mm. Genae fulvous, fuscous subocular spot present; dark seta present. Occiput dark fuscous, fulvous along eye margins; occipital row with a large number of strong dark setae. Antennae with segments 1 and 2 red-brown, segment 3 red-brown with fuscous on apex and outer surface; arista black (red-brown basally); length of segments: 0.1 mm; 0.26 mm; 0.76 mm.

Thorax: Typically, the scutum is black with large red-brown areas as follows: below and behind lateral postsutural vittae, around yellow spot anterior to notopleural suture and notopleuron, a broad medial longitudinal band from posterior margin to anterior margin (anteriorly it includes the entire area between the postpronotal lobes). In some specimens the black is more extensive, reducing the area of red-brown. Pleural areas dark fuscous to black except red-brown below postpronotal lobes, above mid and hind coxae and on ventral area of katepisternum. Yellow markings as follows: postpronotal lobes; notopleura; mesopleural stripe reaching midway between anterior margin of notopleuron and anterior *npl.* seta dorsally, continuing to katepisternum as a large transverse spot, anterior margin slightly convex; anatergite (posterior apex black); anterior $^{5}/_{7}$ katatergite (remainder black); two broad lateral postsutural vittae beginning with a large spot anterior to notopleural suture and tapering posteriorly to end at level of *p.sa.* seta; a broad medial longitudinal postsutural vitta broadly rounded posteriorly and tapering to end just anterior to notopleural suture. Postnotum black, red-brown centrally. Scutellum yellow except for narrow black basal band. Setae: *sc.* 4; *prsc.* 2; *ia.* 1; *p.sa.* 1; *a.sa.* 1; *mpl.* 1; *npl.* 2; *scp.* 4.

Legs: All femora fulvous with black laterally and ventrally on apical $^{1}/_{3}$; fore tibiae fuscous, mid tibiae fuscous basally to fulvous apically, hind tibiae dark fuscous; tarsi with basal segment fulvous and apical four segments pale fuscous.

Wings: Length 6.4 mm; cells bc and c with a strong fulvous tint; microtrichia in outer corner of cell c only; remainder of wings colourless except dark fuscous cell sc, narrow fuscous costal band overlapping R_{2+3} and ending with a *tau*-like spot in apex of wing, a medium-width fuscous anal streak; dense aggregation of microtrichia around A_1+CuA_2; supernumerary lobe large and rounded.

Abdomen: Oval; terga free; pecten present on tergum III. Tergum I and sterna I and II wider than long. Tergum I black with a narrow pale orange-brown band across posterior margin but not reaching lateral margins; tergum II pale orange-brown with a narrow transverse black band across anterior margin but not reaching the narrow black anterolateral corners; terga III–V pale orange-brown with a 'T' pattern consisting of a narrow transverse black band across anterior margin which includes lateral margins and a medium-width to broad medial longitudinal black band over all three terga, broad black anterolateral corners on terga IV and V. In the female specimen the medial longitudinal black band is broader on terga IV and V and the lateral margins on terga IV and V form continuous lateral longitudinal black bands. A pair of oval fulvous shining spots on tergum V. Posterior lobe of surstylus long, sternum V with a slight concavity on posterior margin. All abdominal sterna dark, except sternum II pale.

Attractant: No known record.

Female: As for male except no dense aggregation of microtrichia around A_1+CuA_2; supernumerary lobe weak; pecten of cilia on abdominal tergum III absent; basal segment of ovipositor red-brown, dorsoventrally compressed and tapering posteriorly in dorsal view; ratio of length of oviscape to length of tergum V, 1:1; apex of piercer needle-shaped.

Distribution:

Type locality: Philippines (Mindanao I., Davao).

New records: Philippines (Luzon I., Quezon, Infanta).

Hosts: Bred from *Luffa cylindrica* and *Momordica cochinchinensis* (family *Cucurbitaceae*).

Remarks: *Bactrocera (Zeugodacus) tapervitta* Mahmood is similar to *Bactrocera (Zeugodacus) trivandrumensis* n.sp. and *Bactrocera (Zeugodacus) tau* (Walker) in the general colour patterns of the thorax, wings and abdomen. It differs in having broad tapering lateral postsutural yellow vittae on the scutum, small circular facial spots and a large rounded supernumerary lobe in the male wing.

Bactrocera (Zeugodacus) tappanus (Shiraki)

(Fig. 356)

Chaetodacus tappanus Shiraki, 1933: 76. Holotype male in NIAS.

Strumeta tappanus – Perkins, 1938a: 122.

Dacus (Bactrocera) tappanus – Hardy, 1977: 52.

Decus (Bactrocera) tappanus [Misspelling for *Dacus*] – Tseng, Chen & Chu, 1992: 46.

Dacus (Bactrocera) paratappanus Tseng, Chen & Chu, 1992: 38. **New Synonym.**

Bactrocera (Bactrocera) tappanus – Wang, 1996: 56; Norrbom *et al.*, 1998: 95.

Type material examined: Holotype male, **TAIWAN (formerly FORMOSA)**: with the following four labels – white label 'Tappan, 18.vi.1917, Col. T. Shiraki'; white label 'Dacus near scutellarius Bezzi, Det. T. Shiraki'; white label 'Caetodacus tappanensis n sp, det. T. Shiraki'; white circular label with red circle 'Type' in red handwriting. Received on loan from NIAS.

Non-type material examined: TAIWAN: 1 male, 'uling Farm, Taichung HS, 22.ix.1990, C.C. Chen', '*Dacus (Bactrocera) tappanus* Shiraki, det. Y.H. Tseng'. Specimen in QDPC.

Diagnosis: A large species; face fulvous with a pair of medium-sized oval black spots; postpronotal lobes yellow; notopleura black basally with yellow on apex; scutum shining black; broad lateral postsutural yellow vittae present; a short narrow medial postsutural yellow vitta present; mesopleural stripe reaching midway between anterior margin of notopleuron and anterior *npl.* seta dorsally; scutellum yellow except for a medium-width black basal band and a black spot on apex; wing with cells bc and c colourless, microtrichia in outer corner of cell c only; a narrow fuscous costal band confluent with R_{2+3}, very narrow beyond apex of this vein and expanding into a distinct spot around apex of R_{4+5}, a small fuscous spot on posterior margin around CuA_1; a narrow fuscous anal streak; abdominal terga III and IV orange-brown each with a black 'T' pattern, tergum V mostly dark fuscous with a broad medial longitudinal black band and small areas of orange-brown on posterior margin.

Description: male.

Head: Height 1.56 mm. Frons – length 1.36 times breadth; fuscous centrally with fulvous around lateral and ventral margins; orbital setae well developed and black: 1 *s.or.*, 3 *i.or.*; lunule fuscous. Ocellar triangle black. Vertex black. Face fulvous with a pair of medium-sized oval black spots; length 0.56 mm. Genae fulvous with no distinct subocular spot; black seta present. Occiput black with a small red-brown area centrally and fulvous along eye margins; occipital row with a few weak pale setae each side. Antennae with segments 1 and 2 red-brown, segment 3 red-brown with fuscous on apex and outer surface; length of segments: 0.16 mm; 0.3 mm; 0.74 mm.

Thorax: Scutum shining black with a small area of brown posterior to *ia.* setae. Pleural areas shining black without any distinct pale markings. Yellow markings as follows: postpronotal lobes; apices of notopleura (black around base); mesopleural stripe of medium-width, reaching midway between anterior margin of notopleuron and anterior *npl.* seta dorsally, continuing to katepisternum as a small transverse spot, anterior margin strongly convex; anatergite (posterior apex black); anterior $^3/_5$ katatergite (remainder black); two moderately broad parallel-sided lateral postsutural vittae beginning as a small spot anterior to notopleural suture and ending at *ia.* seta; a short narrow medial longitudinal postsutural vitta. Scutellum yellow except for a medium-width black basal band and a black spot on apex that encloses both apical setae. Setae: *sc.* 2; *prsc.* 2; *ia.* 1; *p.sa.* 1; *a.sa.* 1; *mpl.* 1; *npl.* 2; *scp.* 2.

Legs: Femora fulvous except for black spots on outer apical surfaces of fore and mid femora and black around apical ¼ of hind femora; fore tibiae fulvous with a narrow fuscous line on outer surfaces, mid tibiae fulvous and hind tibiae dark fuscous; all tarsal segments fulvous; mid tibiae each with an apical black spur.

Wings: Length 8.0 mm; cells bc and c colourless; microtrichia in outer corner of cell c only; remainder of wings colourless except dark fuscous cell sc, narrow fuscous costal band confluent with R_{2+3} and remaining very narrow beyond apex of this vein until it expands into a distinct spot around apex of R_{4+5}, a small fuscous spot on posterior margin around vein CuA_1, a narrow fuscous anal streak; a dense aggregation of microtrichia around A_1+CuA_2; supernumerary lobe weak.

Abdomen: Oval; terga free; pecten present on tergum III. Tergum I black except for a fulvous band across posterior margin but not reaching lateral margins; tergum II black

except for an orange-brown transverse band across posterior margin; terga III and IV each orange-brown except for a distinct black 'T' which consists of a broad transverse black band across anterior ½ and widening to cover lateral margins and a broad medial longitudinal black band; tergum V mostly dark fuscous except for a broad medial longitudinal black band and small areas of orange-brown posteriorly either side of medial longitudinal black band. A pair of oval dark fuscous shining spots on tergum V. Posterior lobe of surstylus long, sternum V with a slight concavity on posterior margin.
Attractant: No known record.
Female: No known record.
Distribution:
Type locality: Taiwan (Tappan).
Hosts: No known record.
Remarks: *Bactrocera (Zeugodacus) tappanus* (Shiraki), within the *scutellaris* complex, is similar to *Bactrocera (Zeugodacus) assamensis* White, *Bactrocera (Zeugodacus) biguttata* (Bezzi) and *Bactrocera (Zeugodacus) freidbergi* White in possessing two *sc.* setae, notopleura entirely or partly dark coloured and femora with dark fuscous to black patterns. It differs from *B. assamensis* and *B. biguttata* in having abdominal terga III–V *scutellata*-like with dark 'T' patterns on terga III and IV and legs with preapical dark fuscous to black spots on outer surfaces of fore and mid femora and entire apical ¼ of hind femora. It differs from *B. freidbergi* in having a large elongate dark spot on the apex of the scutellum. The measurements included above are from a specimen collected in Taiwan, not the type.

Dacus (Bactrocera) paratappanus Tseng, Chen & Chu was separated from *B. tappanus* on a minor variation of the dark colour on the apex of the scutellum and the shape of the median cell in the wing. These characters are not sufficient to separate these species and thus we are making *D. paratappanus* a **New Synonym** of *B. tappanus*.

Bactrocera (Zeugodacus) tau (Walker)

(Figs 357A, 357B)

Dasyneura tau Walker, 1849: 1074. Syntype male in BMNH.
Dacus hageni de Meijere, 1911: 375, type locality, Sumatra.
Dacus caudatus var. *nubilus* Hendel, 1912: 16, type locality Taiwan; White & Wang, 1992: 278 (Synonym).
Dacus nubilus ssp. *femoralis* Hendel, 1934: 11, type locality China.
Dacus (Zeugodacus) tau – Hardy, 1977: 60.
Bactrocera (Zeugodacus) tau – Wang, 1996: 72.
Bactrocera (Zeugodacus) tau – Norrbom *et al*., 1998: 104.
Dacus (Zeugodacus) elegantis Tseng, Chen & Chu, 1992: 61. **New Synonym**.
Dacus (Zeugodacus) flavus Tseng, Chen & Chu, 1992: 63. **New Synonym**.

Type material examined: CHINA: Syntype male, with the following four labels – 'Type'; 'Dasyneura tau Wlk'; 'China Foochow G.T. Lay B.M. 1845-65'; 'Dacus (Zeugodacus) tau (Wlk.) (nubilus Hendel is a synonym, Det 1954 D.E. Hardy)'. Studied at BMNH.
Non-type material examined: BHUTAN: 11 males, Manchungang, Tashi, 4.x.1991, coll. H.R. Feijen, attracted to cue lure (poss. contaminated by methyl eugenol); 1 male, Lingmethang, Manchungang, 3.x.1991, coll. H.R. Feijen (methyl eugenol poss. contaminated by cue lure); 2 males, Zingman, 4.x.1991, coll. H.R. Feijen (cue lure poss. contaminated by methyl eugenol); 1 male, Rontong, 2.x.1991, coll. H.R. Feijen (cue lure poss. contaminated by methyl eugenol); 3 males, Songasa, 1045-1215, 13.ix.1991, coll. H.R. Feijen (cue lure poss. contaminated by methyl eugenol); 1 male, Chirang, Suntalay, 4-7.ix.1990, coll. H.R. Feijen. 2 males, Rimchu, Punakha, 1400m, citrus, cue lure, 24.x.1990, coll. Feijen & Dorji, det. I.M. White; 1 male, Chirang, Lamidara, 1200m, malaise, citrus, 4/7.ix.1990, Feijen & Dorji. 9 males (10.v.2000), 8 males (17.v.2000), 4 males (24.v.2000), 5 males (31.v.2000), 4 males (21.vi.2000), 8 males (23.vii.2000), 7 males (10.viii.2000), 6 males (16. viii.2000), 17 males (30.viii.2000), 23 males (6.ix.2000), 18 males (12.ix.2000), 25 males (20.ix.2000), 21 males (27.ix.2000), 18 males (5.x.2000), 8 males (18.x.2000), 7 males (16.xi.2000), Rimchu trap No. 1, coll. C. Dorji, attracted to cue lure; 3 males (7.vi.2000), 1 male (28.vi.2000), 6 males (19.vii.2000), Rimchu trap No. 1, coll. C. Dorji *et al*., attracted to cue lure; 5 males, Rimchu trap No. 1, 4.v.2000, coll. B.S. Fletcher, attracted to cue lure; 4 males, Rimchu trap No. 1, 14.vi.2000, coll. Nado, attracted to cue lure; 1 male, Rimchu trap No. 1, 21.vi.2000, coll. Nado *et al*., attracted to cue lure; 2 males, Rimchu trap No. 2, 10.v.2000, coll. B.S. Fletcher, attracted to cue lure; 1 male, Rimchu trap No. 3, 10.v.2000, coll. B.S. Fletcher, attracted to cue lure; 7 males (10.v.2002), 4 males (17.v.2002), 1 male (24.v.2002), Rimchu, coll. C. Dorji, attracted to cue lure; 2 males, Rimchu, 26.iv.2000, coll. Clarke/Fletcher, attracted to cue lure; 6 males, Phuentsholing, 6.v.2000, attracted to cue lure; 6 males (17.vii.2000), 6 males (31.vii.2000), 6 males (7.viii.2000), 2 males (2.x.2000), 2 males (23.x.2000), Tsirang, attracted to cue lure; 5 males (19.vi.2000), 11 males (10.vii.2000), Tsirang, coll. T.D. & S.D., attracted to cue lure; 4 males, Tsirang (Damphu orchard), 20.iv.2000, attracted to cue lure; 6 males, Suntalay, 28.vi.2000, coll. S. Wangdi, attracted to cue lure; 9 males, Lingmethang 11km, 9.vi.2001, coll. Brian Fletcher, attracted to cue lure; 10 males, Gelephu Purana Bastey, 1.ii.2005, coll. Maha Prasad & Karma Nangyel, attracted to cue lure. **BRUNEI**: 3 males and 3 females, H.S. Semabat, vi.1972, coll. D.J. McCrae, bred ex *Luffa* sp. **INDIA**: 2 males, Dewal, 21.x.64, coll. C.I.B.C., maggot in flower of *C. maxima*; 3 females, Ghari Habidullah, 24.x.64, coll. C.I.B.C., maggot in flowers of *C. maxima*. **INDONESIA: BALI** – 1 male, Bali, Buleleng, Tajun, 18.viii.2006, coll. Hendri Y, attracted to cue lure in forest; 3 males, Sumbar, Solak, Alahan Panjang, 13.ix.2006, coll. Katut Rawa W, bred ex Sample No. GF11, *Passiflora edulis* (passionfruit); **JAVA** – 6 males (20.vii.2005), 1 male (1.viii.2005), Banten, Lebak, Pasir Nini Nini, coll. Aden Haer, attracted to cue lure in forest; 4 males, DKI Jakarta, Selatan, Tanjung Barat, 23.vii.2005, coll. Mulyana, attracted to cue lure in carambola; **SUMATRA** – 1 male, Bukit Larwang Reserve, Sumatra,

2-12.vii.1977, coll. D. Kistner, attracted to cue lure; 3 males, South Sumatra, viii.1978, coll. S. Isnadi, attracted to cue lure; 2 females and 2 males, Jambi, Muara Jambi, Baru, 30.viii.2006, coll. Hendri Y, bred ex Sample No. GF21, *Luffa acutangula*; 2 males, Bengkulu, Tebat Kubu, Manna, 21. ix.2006, coll. M. Nelson; 2 males, Bengkulu, Bengkulu Selatan, Kebon Jati, 29.ix.2006, coll. M. Nelson, attracted to cue lure in citrus. **EAST MALAYSIA: SABAH** – 5 males, ACIAR Site Ht 5, Sabah Forest, 27.x.1991, attracted to cue lure; **SARAWAK** – 7 males, Tarat, Sarawak, 16.vii.1977, coll. B.H. Voon, attracted to cue lure; 1 male, Ranchang Poo, Serian, nr Kuching, Sarawak, 7.v.1992, coll. S. Leong, attracted to cue lure; 1 male (15.vii.1992), 1 male (28. vii.1992), Tarat Agriculture Station, nr Kuching, Sarawak, coll. S. Leong, attracted to cue lure. **PENINSULAR MALAYSIA**: 5 males, Kuala Brang, 14.vi. 1976, coll. G.H. Young, attracted to cue lure; 2 males, Mersing, 2-3.viii.1977, coll. D. Kistner, attracted to cue lure; 1 male, R S Jeram Pasu I, 2.vi.1986, coll. Yusoff Salleh, attracted to cue lure; 3 males, R S Jeram Pasu I, 12.v.1986, coll. Ismail Mat, attracted to cue lure; 1 male, T S Karak, 12.ii.1986, coll. Mahdzir, attracted to cue lure; 1 male, T S Maran, 12.ii.1986, coll. Mahdzir, attracted to cue lure; 1 male, R S Pontian, 29. iv.1986, coll. Salleh, attracted to cue lure; 1 male, R S Pontian, 8.vi.1986, coll. Moktar, attracted to cue lure; 1 male, Temerloh, 25.ii.1986, coll. Antony, attracted to cue lure; 1 male R S Jalan Kebun, 13.v.1986, coll. Abdul Khalib, attracted to cue lure; 1 male, Km 87-K (nr Felds), Trolak Jn, 29.ix.1986, coll. Mahadzir & Bahari, bred from *Trichosanthes tricuspidata*; 18 females and 13 males, Matan Lipur, Kanching, 15.ix.1986, coll. Anthony *et al.*, bred from *Muntingia calabura*; 2 males and 4 females, Kuala Wah, Tapah, 29.ix.1986, coll. Mahadzir *et al.*, bred from *Syzygium malaccense*; 27 males and 35 females, Kuala Wah, Tapah, Rebusi, Perak, 29.ix.1986, coll. Mahadzir & Bahari, bred from unknown host; 16 males and 14 females, Km 87-K (Nr Felds Trolak Jn), 29.ix.1986, coll. Mahadzir & Bahari, bred from *Trichosanthes tricuspidata*; 2 males and 1 female, Old Gombak Rd., 16.ix.1986, coll. Anthony *et al.*, bred from unknown host; 1 female, Tanah Rata, Tapah, Cameron Highlands, 1.x.1986, coll. S. Vijaysegaran *et al.*, bred from *Ficus tinctoria*. **SRI LANKA (formerly CEYLON)**: 1 male, Sri Lanka, Rahangala, 1.xii.1994, C35; 1 male, Sri Lanka, 1.xii.1994, Rahangala. **TAIWAN (formerly FORMOSA)**: 1 female with the following label data – 'Fuhosho, Formosa, H. Sauter, iii.09'; 'coll. Hendel'; 'Cotype female'; '*Bactrocera tau* (Walker) det. I.M. White, 1993', studied at NMW (this specimen was not part of the original type series of Walker); 105 males, Jiauliping Chiayi, Itsein, 1.ix – 1.x.1977, coll. Y.H. Tseng, attracted to cue lure; 1 male (25.ii.1990), 1 male (5.x.1990), Taichung H S, coll. Y.H. Tseng, attracted to cue lure; 1 male, Tsaochiao, Miao Li H S, 17.x.1990, coll. Y.H. Tseng, attracted to cue lure; 10 males, Fun Chi Hu, 25.x.1978, coll. Lena Lee Wen, attracted to cue lure; 8 males, Nanfeng, 27.xi.1978, coll. Lena Lee Wen, attracted to cue lure; 7 males, Wusha, 27.xi.1978, coll. Lena Lee Wen, attracted to cue lure; 2 males, Chuchi, 30.xi.1978, coll. Lena Lee Wen, attracted to cue lure. **THAILAND**: 6 males, Yang Pong Rat, Plubpla, A. Muang, Chanthaburi, 25.vi.1991, coll. Montree, attracted to cue lure; 1 male, Phila H.R.S., Laemsing, Chanthaburi, 25.vi.1991, coll. Montree, attracted to cue lure; 1 male (14.ii.1992), 1 male (6.iii.1992), Wang Jum Pee, Pakchong, Nakhonratsima, coll. Montree, attracted to cue lure; 1 male, Kunthuli, Surat Thani, 3.xii.1991, coll. Pichate, attracted to cue lure; 1 male, Khun Khon, Chiang Rai, 6.viii.1992, coll. Charuwan, attracted to cue lure; 4 males, Chiang Mai, 13.vi.1991, coll. Chantanee *et al.*, cue lure in guava orchard; 3 males, Mae Sai, Chiang Rai, 10.vi.1991, coll. Charuwan, attracted to cue lure; 1 male, Doi Suang N P, Chiang Rai, 26.ix.1991, coll. D.L. Hancock; 1 male, Khun Korn N P, Doi Chang, 20 km SW Chiang Rai, 27.ix.1991, coll. D.L. Hancock; 9 males, Wawi Exp. Sta. 1400m, 45km South of Chiang Rai, 22.v.1991, coll. D.L. Hancock, attracted to cue lure; 2 males, Chiang Rai Horticultural Stn, 19.v.1991, coll. D.L. Hancock, attracted to cue lure; 1 male, Mae Salong, 900m, NW of Chiang Rai, 19.v.1999, coll. D.L. Hancock, attracted to cue lure; 1 male, Tone-Nga-Chang waterfall, near Hat Yai, 8.v.1991, coll. D.L. Hancock, attracted to cue lure; 1 male (17.v.1991), 1 male and 1 female (31.v.1991), 1 male and 2 females (4.vi.1991), PSU, Songkhla, coll. Anuchit *et al.*, bred from *Cucumis sativus*; 4 females, Chiang Mai, 9.x.1991, coll. Chantanee, bred from *Momordica macrophylla*; 1 male and 5 females, Chiang Mai, 9.x.1991, coll. Chantanee, bred from *Trichosanthes rubriflos*; 11 males and 20 females, Hat Yai, Songkhla, 24.vi.1991, coll. Anuchit, bred from *Gymnopetalum integrifolium*; 3 males and 3 females, Ma Rue Bo, Narathiwat, 25.iv.1992, coll. Anuchit *et al.*, bred from *Gymnopetalum integrifolium*; 2 males and 4 females, Km 11 Pluang, Makham, Chanthaburi, 6.vii.1992, coll. Prasert, bred from *Coccinia grandis*; 1 male and 1 female, Wat Koo Ha Phi Moor, Yala, Songkhla, 20.ii.1992, coll. Surakrai *et al.*, bred from *Luffa cylindrica*; 1 male and 4 females, Thon NGA, Chaeng, Hat Yai, Songkhla, 28.i.1992, coll. Anuchit *et al.*, bred from *Gymnopetalum integrifolium*; 1 female, Jae Chon Nat Park, Chiang Rai, 8.i.1992, coll. Charuwan, bred from *Fagraea ceilanica*; 4 females, Maharaj Hospital, Nakhon Si, Thammarat, 23.x.1992, coll. Surakrai *et al.*, bred from *Tetrastigma lanceolarium*; 2 males and 3 females, Songkhla, Chawang, Nakhon Si, Thammarat, 23.x.1992, coll. Surakrai *et al.*, bred from *Gymnopetalum integrifolium*; 1 male and 6 females, Ma Rue Bo, Narathiwat, 21.ii.1992, coll. Surakrai *et al.*, bred from *Gymnopetalum integrifolium*; 3 females, Be Tong, Yala, Songkhla, 24.vii.1992, coll. Sieng *et al.*, bred from *Luffa cylindrica*; 1 male, Amphur, Muang, Chiang Rai, 25.viii.1992, coll. Charuwan, bred from *Coccinia grandis*; 1 male and 3 females, Ru So, Narathiwat, 3.iv.1992, coll. Anuchit *et al.*, bred from *Melastoma malabathricum*; 1 male, Chong Chang, Bannasan, Surat Thani, 18.vii.1991, coll. Pichate, bred from *Luffa acutangula*; 1 male and 1 female, Surat Thani HRC, Surat Thani, 2.viii.1991, coll. Pichate, bred from *Benincasa hispida*; 1 male and 2 females, Chumporn HRC, Surat Thani, 11.vii.1991, coll. Pichate, bred from *Cucumis sativus*; 1 female, MPO Angkhiri Makham, Chanthaburi, 8.xii.1992, coll. Montree, bred from *Strychnos nux-vomica*; 2 males and 4 females, Noenham, A. Muang, Prachinburi, 14.vi.1991, coll. Montree, bred from *Cucumis sativus*; 1 female, Songkhla, Pikut Thong,

Narathiwat, 14.vi.1991, coll. Sieng *et al.*, bred from *Artocarpus heterophyllus*; 1 male and 3 females, Songkhla, Bankuan, Trang, Amphoe Muang, 28.vi.1991, coll. Anuchit *et al.*, bred from *Luffa acutangula*; 1 male and 2 females, Rattaphum, Songkhla, 23.vi.1991, coll. Vanpen, bred from *Benincasa hispida*; 2 females, Rattaphum, Songkhla, 23.vi.1991, coll. Vanpen, bred from *Cucumis sativus*; 2 males and 2 females, Rattaphum, Songkhla, 23.vi.1991, coll. Vanpen, bred from *Cucurbita moschata*; 1 male and 1 female, Ban Nabin HLA, Amphoe Muang, Trang, Songkhla, 28.vi.1991, coll. Anuchit *et al.*, bred from *Benincasa hispida*; 2 females, Khok Pho, Pattari, 24.vii.1991, coll. Anuchit *et al.*, bred from *Benincasa hispida*; 2 males and 6 females, Khuan Ka Long, Satrum, Songkhla, 8.viii.1991, coll. Sieng *et al.*, bred from *Benincasa hispida*; 2 males and 2 females, T. Tungkrang, A. Khaosokiao, Chanthaburi, 23.vii.1991, coll. Montree, bred from *Cucumis sativus*; 3 males, Rattaphum, Songkhla, 23.vi.1991, coll. Vanpen, bred from *Cucumis sativus*; 1 male and 1 female bred from *Luffa acutangula* and 2 females bred from *Citrullus lonatus*, Tha-U-Tae, Kanchanadit, Surat Thani, 23.vii.1991, coll. Pichate; 2 females, T. Pangland, A. Klang, Rayong, 24.vii.1991, coll. Montree, bred from *Cucumis melo*; 1 female, Had Nai Yang Nat Park, Phuket, 18.iii.1992, coll. Sieng *et al.*, bred from *Gymnopetalum integrifolium*; 1 female, Km10-11, T. Puong, A. Makham, Chanthaburi, 8.vi.1992, coll. Prasert, bred from Ivy gourd *Coccinia grandis*; 3 males and 3 females, Hat Yai, Songkhla, 24.vi.1991, coll. Anuchit *et al.*, bred from *Morinda citrifolia*; 3 males and 3 females, Kungone Nat. Park, Chiang Rai, 4.x.1991, coll. Charuwan, bred from wild fruit; 1 female, Nong Pla, Sawai, Chiang Mai, 9.x.1993, coll. Chantanee, bred from *Luffa cylindrica*; 2 males and 3 females, Ko Yo, Songkhla, 10.vi.1992, coll. Vanpen *et al.*, bred from *Gymnopetalum integrifolium*; 1 male and 1 female, Muang, Songkhla, Narathiwat, 14.vi.1991, coll. Sieng *et al.*, bred from *Momordica charantia*; 2 males and 2 females, T. Banpra, A. Muang, Prachinburi, 26.vii.1991, coll. Montree, bred from *Luffa cylindrica*; 2 males and 3 females, Mae Tien, Mae Tam, Chiang Mai, 8.i.1992, coll. Chantanee, bred from *Trichosanthes rubriflos*; 2 females, SPT, Chiang Mai, 14.xi.1991, coll. Chantanee, bred from *Bambusa pallida*. **VIETNAM: Southern Vietnam** – 1 male (17.vii.1999), 1 male (1.x.1999), Can Tho Province, Binh They Village, southern Vietnam, trap S003, attracted to cue lure; 2 males (12.xi.1999), 1 male (23.vi.2000), Dong Nai Province, Long Nhat District, Long Nhat Village, southern Vietnam trap S012, attracted to cue lure; 2 males (25.xi.1999), 1 male (20.i.2000), 1 male (3.ii.2000), Dong Nai Province, Long Khanh District, Tan Phong Village, southern Vietnam trap S014, attracted to cue lure; 1 male (12.xi.1999), 1 male (14.xii.1999), Tien Giang Province, Cai Lay District, Long Khanh Village, southern Vietnam trap S007, attracted to cue lure; 1 male (20.i.2000), 2 males (17.ii.2000), 2 males (18.viii.2000), Dong Nai Province, Thong Nhat District, Cay Gao Village, southern Vietnam trap S013, attracted to cue lure; 2 males (21.i.2000), 1 male (18.ii.2001), Binh Duong Prov., Lai Thieu Village, southern Vietnam trap S020, attracted to cue lure; 1 male (17.xii.2001), 1 male (24.xii.2001), 1 male (29.xii.2001), 2 males (12.i.2002), Ben Tre (P), Chau Thanh (D), Tien Thuy (V), southern Vietnam trap VS04, coll. SOFRI staff, attracted to cue lure; 3 males (25.ii.2002), 2 males (23.iii.2002), 2 males (1.iv.2002), 1 male (5.v.2002), 1 male (4.viii.2002), Cantho (P), Chau Thanh (D), Tan Phu (V), southern Vietnam trap VS02, coll. SOFRI staff, attracted to cue lure; 1 male (8.ix.2002), 10 males (8.x.2002), Dac Lac Province, Ku Mi Nga District, Quang Phu Village, southern Vietnam trap VS09, attracted to cue lure; 2 males (23.ix.2002), 6 males (8.x.2002), 6 males (8.xii.2002), Lam Dong Province, 3 district, Da Lat town, southern Vietnam trap VS08, attracted to cue lure; **Northern Vietnam** – A series of specimens bred from *Momordica charantia*, *Momordica cochinchinensis*, Gourd, Luffa, pumpkin, squash and Rose Apple in Northern Vietnam. 3 males (6.ix.1999), 1 male (15.xi.1999), Son La Province, Moc Chau Dist., Tra Long, northern Vietnam trap N007, attracted to cue lure; 2 females, Son La Province, Moc Chau Dist., Moc Chau house garden, 6.ix.1999 and 7.viii.2000, northern Vietnam trap N008; 1 male (6.ix.1999), 4 males (18.x.1999), 1 male (29.xi.1999), 2 females (15.i.2000), Son La Province, Moc Chau Dist., Co Do, northern Vietnam trap N006, attracted to cue lure; 1 male (30.iii.2002), 2 males (15.ix.2002), Son La Province, Moc Chau District, Long Luong Village, northern Vietnam trap VN01, attracted to cue lure; 1 female, Lao Cai Province, Sapa town, 5.xi.1999, northern Vietnam trap N016, attracted to cue lure. Specimens in QDPC except where specified.

Diagnosis: A medium-sized species; face fulvous with a pair of medium-sized circular to oval black spots; postpronotal lobes and notopleura yellow; scutum black with large areas of red-brown centrally and anterocentrally; lateral and medial postsutural yellow vittae present; yellow spot anterior to notopleural suture in front of lateral postsutural vittae; mesopleural stripe reaching midway between anterior margin of notopleuron and anterior *npl.* seta; scutellum entirely yellow; wing with cells bc and c colourless, microtrichia in outer corner of cell c only; a narrow dark fuscous costal band overlapping R_{2+3} and expanding into a distinct apical spot and broad dark fuscous anal streak; abdominal terga III–V fulvous with a black 'T' pattern and anterolateral corners of terga IV and V with broad black markings.

Description: male (Fig. 357A).

Head: Height 1.6 mm. Frons – length 1.5 times breadth; fulvous to red-brown with fuscous to dark fuscous on anteromedial hump; orbital setae black: 1 *s.or.*, 3 *i.or.*; lunule fuscous. Ocellar triangle black. Vertex fuscous to dark fuscous. Face fulvous with a pair of medium-sized circular to oval black spots; length 0.44 mm. Genae fulvous, fuscous subocular spot present; dark seta present. Occiput red-brown with or without fuscous to dark fuscous laterodorsally, fulvous along eye margins; occipital row with 2–5 pale setae. Antennae with segments 1 and 2 red-brown, segment 3 red-brown with fuscous to dark fuscous on apex and outer surface; arista black (red-brown basally); length of segments: 0.15 mm; 0.21 mm; 0.69 mm.

Thorax: Scutum black with red-brown below and behind lateral postsutural vittae, around notopleural suture and yellow spot, around borders of medial postsutural yellow vitta and anterocentrally from anterior point of medial vitta to anterior margin. Pleural areas red-brown with black along anterior margin of mesopleural stripe and on centre of katepisternum and dark fuscous along posterior margin of mesopleural stripe. Yellow markings as follows: postpronotal lobes; notopleura; mesopleural stripe of medium-width, reaching midway between anterior margin of notopleuron and anterior *npl.* seta dorsally, continuing to katepisternum as a large transverse spot, anterior margin slightly convex; anatergite (posterior apex black); anterior ⅔ katatergite (remainder black); two moderately broad parallel-sided lateral postsutural vittae beginning with a spot anterior to notopleural suture and ending just behind *ia.* seta; a moderately broad medial longitudinal postsutural vitta beginning posteriorly approximately at level of *prsc.* setae and narrowing anteriorly to end just anterior of notopleural suture. Postnotum black laterally, red-brown centrally. Scutellum yellow except for narrow black basal band. Setae: *sc.* 4; *prsc.* 2; *ia.* 1; *p.sa.* 1; *a.sa.* 1; *mpl.* 1; *npl.* 2; *scp.* 4.

Legs: Femora mostly fulvous (in some specimens there are small preapical to apical, pale fuscous to dark fuscous spots or surrounding marks). In females, the most common femora pattern is a dark preapical spot on the fore femora, with the mid and hind femora entirely fulvous. However, apical dark patterns on all femora are relatively common also. In males, the most common femora pattern is for all femora to be entirely fulvous but a small pale fuscous preapical spot on the fore femora is also relatively common. Fore tibiae fuscous to dark fuscous, mid tibiae fuscous basally and fulvous apically, hind tibiae dark fuscous; tarsi with basal segments fulvous and apical four segments pale fuscous to fuscous; mid tibiae each with an apical black spur.

Wings: Length 6.4 mm; cells bc and c colourless; microtrichia in outer corner of cell c only; remainder of wings colourless except dark fuscous cell sc, narrow dark fuscous costal band overlapping R_{2+3} where it is sometimes paler (very occasionally the costal band is confluent with R_{2+3}) and expanding into a distinct apical spot which surrounds the apex of R_{4+5} and ends midway between the apices of R_{4+5} and M, a small pale fuscous tint at wing margin around CuA_1, a broad dark fuscous anal streak broadly rounded apically; dense aggregation of microtrichia around A_1+CuA_2; supernumerary lobe strong and keel-shaped.

Abdomen: Elongate oval; terga free; pecten present on tergum III. Tergum I and sterna I and II wider than long. Tergum I fulvous with narrow dark fuscous anterior and lateral margins; tergum II fulvous with a narrow transverse subanterior black band not reaching lateral margins and narrow dark fuscous to black anterolateral margins; terga III–V fulvous with a black 'T' pattern consisting of a narrow transverse band across anterior margin of tergum III and a narrow- to medium-width medial longitudinal band over all three terga and broad dark fuscous to black anterolateral corners on terga IV and V (these usually extended in transverse direction along anterior margin of the terga). A pair of oval fulvous shining spots on tergum V. Posterior lobe of surstylus long, sternum V with a slight concavity on posterior margin. Abdominal sterna dark, except sternum II fulvous.

Attractant: Cue lure.

Female: (Fig. 357B) – As for male except no dense aggregation of microtrichia around A_1+CuA_2; supernumerary lobe weak; pecten of cilia on abdominal tergum III absent; basal segment of ovipositor red-brown, dorsoventrally flattened and tapering posteriorly in dorsal view; ratio of length of oviscape to length of tergum V, 1.4 to 1.6:1; apex of piercer needle-shaped.

Distribution:

Type locality: China (Fujian Province, Foochow).

Previously recorded: Type locality and the following locations – China (Southern to Central) (Liang *et al.*, 1993); India; Sikkim; Sri Lanka; Taiwan; Indonesia (Java; Sulawesi) (Norrbom *et al.*, 1998). India (Andaman Is) (David & Ramani, 2011). Indonesia (Sumatra), due to synonymy of *D. hageni*.

New records: Bhutan; Brunei; East Malaysia (Sabah & Sarawak); Indonesia (Bali); Peninsular Malaysia; Thailand; Vietnam. (See specific locations listed under 'Non-type material examined'.)

Hosts: Recorded from nine plant families but primarily species of family *Cucurbitaceae* (see Allwood *et al.*, 1999 for host records) and four additional families by White & Elson-Harris (1992).

New records: *Artocarpus heterophyllus* and *Ficus tinctoria* (family *Moraceae*); *Melastoma malabathrica* (family *Melastomataceae*); *Passiflora edulis* (family *Passifloraceae*); *Morinda citrifolia* (family *Rubiaceae*); *Syzygium jambos* (family *Myrtaceae*). The record of two female specimens reared from *Bambusa pallida* (family *Poaceae*) listed above under 'Non-type material examined', Thailand, is probably incorrect.

Remarks: *Bactrocera (Zeugodacus) tau* (Walker) is a very common species throughout South-East Asia. It is an economic pest species, mainly in cucurbit crops, but can be misidentified as it belongs to a complex of closely related species. The *tau* complex includes *Zeugodacus* species with a black scutum, wings colourless except for a costal band and anal streak, cells bc and c colourless or with an extremely pale tint, costal band overlapping R_{2+3} and expanding into a distinct spot at apex. *Bactrocera tau* is distinct in having an entirely yellow scutellum, dark abdominal sterna, abdominal terga III–V with a distinct dark 'T' pattern and all femora usually with dark preapical spots.

There appears to be only one type specimen of *B. tau* in existence, the syntype male in BMNH. As Walker (1849) did not say how many specimens he had before him when he described *B. tau*, we are treating the specimen listed above under 'Type material examined' as a syntype.

Hardy & Adachi (1954) made *Dacus caudatus* var. *nubilus* Hendel a synonym of *Dacus (Zeugodacus) hageni* de Meijere. Later, Hardy (1973) listed *D. hageni* as a synonym of *B. tau* and *D. nubilus* as a good species. White & Wang (1992) studied the types of *D. nubilus* and placed it back into synonymy with *B. tau. Dacus (Zeugodacus) elegantis* Tseng, Chen & Chu was separated from *B. tau* on minor differences in the colour patterns on the frons, face and abdominal tergum III. These differences are not sufficient to establish *D. elegantis* as a separate species and we are therefore making it a **New Synonym** of *B. tau*. Similarly,

Dacus (Zeugodacus) flavus Tseng, Chen & Chu was established on minor differences in the chaetotaxy of the frons and microtrichia in cell c of the wing, and we are thus making it a **New Synonym** of *B. tau*.

Bactrocera (Zeugodacus) tebeduiae Drew & Romig, new species

(Fig. 358)

Material examined: Holotype male, **EAST MALAYSIA**: Sarawak, Kpg. Tema Mawang, Tebedu, Nr Kuching, 13. iii.1992, coll. S. Leong, attracted to cue lure, deposited in BMNH.

Paratypes: **BRUNEI**: 4 males (23.i.2002), 4 males (25.i.2002), 6 males (27.i.2002), 4 males (29.i.2002), KBFC, Bato Apou Nat. Pk., Riparian, coll. K. Hurley, attracted to cue lure; 6 males (23.i.2002), 11 males (25.i.2002), KBFC, Bato Apou Nat. Pk, Upland Ridge, coll. K. Hurley, attracted to cue lure; 5 males (23.i.2002), 7 males (27.i.2002), 5 males (29.i.2002), KBFC, Bato Apou Nat. Pk, Forest, coll. K. Hurley, attracted to cue lure. **EAST MALAYSIA: SARAWAK** – 1 male, Ranchang Pool, Serian, Nr Kuching, 25.ix.1991, coll. S. Leong, attracted to cue lure; 2 males (30. ix.1991), 2 males (8.i.1992), Kpg Tema Mawang, Tebedu, Nr Kuching, coll. S. Leong, attracted to cue lure. 1 male, same data as holotype; 2 males, 28.x.1991, same data as holotype; 1 male (28.x.1991), 1 male (27.i.1992), Sarawak, Forest Arboretum, Semongok, Nr Kuching, coll. S. Leong, attracted to cue lure; 1 male, Sarawak, Ranchang Pool, Serian, Nr Kuching, 13.iv.1992, coll. S. Leong, attracted to cue lure; 3 males, Sarawak, Semongok Forest Reserve, 16.vii.1977, coll. B.H. Voon, attracted to cue lure; 2 males, Sarawak, Tarat, 16.vii.1977, coll. B.H. Voon; **SABAH** – 1 male, Sabah, Mt Kinabalu, 1500 m, 17.x.1991, coll. P. Anthony, attracted to cue lure. **PENINSULAR MALAYSIA**: 2 males, Cameron Highlands, 3000', trap 4, 22.v.1983, coll. R. Drew, attracted to cue lure; 2 males, Cameron Highlands, Tapah 12th mile, trap M2, 10.vi.1983, coll. S. Vijaysegaran, attracted to cue lure; 1 male (10.vi.1983), 1 male (24.vi.1983), Cameron Highlands, Tapah 6th mile, trap 1, coll. S. Vijaysegaran, attracted to cue lure; 2 males (6.v.1986), 1 male (9.vi.1986), R S Cameron Highlands, coll. Zan Davies, attracted to cue lure; 1 male, TN13, 30.vi.1986, coll. Ahmad Radzuan Panjang Abd, attracted to cue lure; 1 male, Pangsoon, Hulu Langat, 26.ix.1990, coll. D.L. Hancock, attracted to cue lure; 1 male, T.S. Temerloh, 25.ii.1986, coll. Antony, attracted to cue lure. **INDONESIA: SUMATRA** – 1 male, W. Sumatra, 12 km N.W. Bokorok, 200 m, 4.v.1975, coll. D.E. Hardy. **THAILAND**: 1 male, Kunthuli, Surat Thani, 3.xii.1991, coll. Pichate, attracted to cue lure. 8 paratypes in ANIC; 8 paratypes in BMNH; 59 paratypes in QDPC; 8 paratypes in QMIC.

Non-type material examined: INDONESIA: SULAWESI – 3 males (21.xii.1985), 4 males (22.xii.1985), N. Sulawesi, Dumoga, Taraut, coll. L. Santianawati, attracted to cue lure; **SUMATRA** – 7 males, Sumatra, Bukit Lawong Reserve, 2-12.vii.1977, coll. D. Kistner, attracted to cue lure; **JAVA** – 1 male, Java, Banten, Tangerong, Tanah Tinggi, 23.viii.2005, Komarudin. Specimens in QDPC. **THAILAND**: 1 male, 'Chumphon Prov., Khao Kai Jae Watf, 80m, 5-8.v.2003/2'; '9.55.07N, 98.56.34E, P. Schwendinger leg, semi-evergreen rainforest/2'; 'attracted to cue lure'; 1 male, 'Phang Nga Prov., Raman Waterfall, 50m, 16.v.2003/9'; '8.27.05N, 98.26.52E, P. Schwendinger leg, evergreen forest/9'; 'attracted to cue lure'; 1 male, 'Sunat Thanai Prov., Khao Sok N.P., 100m, 12-14.v.2003/7'; '8.54.55N, 98.31.39E, P. Schwendinger leg, secondary forest/7'; 'attracted to cue lure'. Received on loan from MHNG.

Diagnosis: A large species; face with a pair of narrow elongate black spots; postpronotal lobes and notopleura yellow; scutum mostly black; lateral and medial postsutural yellow vittae present; a yellow spot anterior to notopleural suture; mesopleural stripe reaching midway between anterior margin of notopleuron and anterior *npl.* seta dorsally; scutellum yellow; wing with cells bc and c colourless, microtrichia in outer corner of cell c only; a narrow fuscous costal band expanding into a small *tau*-like spot around apex of R_{4+5}; a medium-width fuscous anal streak; abdominal terga III–V fulvous with a 'T' pattern consisting of a narrow transverse black band across anterior margin of tergum III and a broad medial longitudinal black band over all three terga, broad anterolateral dark fuscous to black markings on terga IV and V.

Description: male.

Head: Height 1.92 mm. Frons – length 1.38 times breadth; red-brown with fulvous along lateral margins; orbital setae black: 1 *s.or.*, 3 *i.or.*; lunule fuscous. Ocellar triangle black. Vertex black. Face fulvous with a pair of narrow elongate black spots; length 0.63 mm. Genae fulvous, small fuscous subocular spot present; black seta present. Occiput red-brown, fulvous along eye margins; occipital row with 4–8 well-developed dark setae. Antennae with segments 1 and 2 red-brown, segment 3 fulvous with dark fuscous on apex and outer surface; arista black (red-brown basally); length of segments: 0.18 mm; 0.24 mm; 0.93 mm.

Thorax: Scutum black with dark red-brown below lateral postsutural vittae posterior to *a.sa.* seta, around notopleural suture and yellow spot, inside postpronotal lobes and as a broad medial longitudinal band from anterior point of medial postsutural yellow vitta to anterior margin of scutum (there is a narrow medial longitudinal black line in the centre of this band). Pleural areas black except red-brown below postpronotal lobes. Yellow markings as follows: postpronotal lobes; notopleura; mesopleural stripe reaching midway between anterior margin of notopleuron and anterior *npl.* set dorsally, continuing to katepisternum as a large transverse spot, anterior margin convex; anatergite (posterior apex black); anterior ⅔ katatergite (remainder black); two broad parallel-sided lateral postsutural vittae beginning with a spot anterior to notopleural suture and ending behind *ia* seta; a narrow medial longitudinal postsutural vitta beginning posteriorly between *prsc.* setae and ending at a point anteriorly at level of notopleural suture. Postnotum black with dark red-brown centrally. Scutellum yellow except for narrow black basal band. Setae: *sc.* 4; *prsc.* 2; *ia.* 1; *p.sa.* 1; *a.sa.* 1; *mpl.* 1; *npl.* 2; *scp.* 4.

Legs: All femora entirely fulvous; fore tibiae fuscous, mid tibiae fulvous tending pale fuscous to fuscous basally, hind tibiae fuscous to dark fuscous; tarsi with basal segment fulvous and apical four segments pale fuscous; mid tibiae each with an apical black spur.
Wings: Length 8.0 mm; cells bc and c colourless; microtrichia in outer corner of cell c only; remainder of wings colourless except fuscous cell sc, narrow fuscous costal band confluent with or just overlapping R_{2+3} and remaining narrow before ending in a *tau*-like spot around apex of R_{4+5} (this spot is smaller than in *tau*), a medium-width fuscous anal streak; a dense aggregation of microtrichia around A_1+CuA_2; supernumerary lobe of medium development.
Abdomen: Oval; terga free; pecten present on tergum III. Tergum I and sterna I and II wider than long. Tergum I fulvous with narrow lateral dark fuscous to black margins; tergum II fulvous with a narrow transverse black band across anterior margin reaching narrow dark fuscous to black lateral margins; terga III–V fulvous with a 'T' pattern consisting of a narrow transverse black band across anterior margin of tergum III and a moderately broad medial longitudinal black band over all three terga, broad anterolateral dark fuscous to black corners on terga IV and V. A pair of fulvous shining spots on tergum V. Posterior lobe of surstylus long, sternum V with a slight concavity on posterior margin. Abdominal sterna dark fuscous to black except sternum II and posterocentral areas of sternum V fulvous.
Attractant: Cue lure.
Female: No known record.
Distribution: This species appears widespread across South-East Asia having been collected in Brunei; East Malaysia (Sabah and Sarawak); Peninsular Malaysia; Indonesia (Java; Sulawesi; Sumatra); Thailand.
Hosts: No known record.
Remarks: *Bactrocera (Zeugodacus) tebeduiae* n.sp., within the *tau* complex, is similar to *Bactrocera (Zeugodacus) okunii* (Shiraki), *Bactrocera (Zeugodacus) tau* (Walker) and *Bactrocera (Zeugodacus) trichosanthes* n.sp. in possessing an entirely yellow scutellum, terga III–V with a black 'T' pattern, terga IV and V with broad dark anterolateral corners and femora mostly fulvous. It differs from *B. tau* in having a weaker supernumerary lobe in the male wing and from all these species in having the facial spots narrow and elongate. In *B. okunii* the face is entirely fulvous without dark markings.

Bactrocera (Zeugodacus) timorensis (Perkins)

(Fig. 359)

Zeugodacus timorensis Perkins, 1939: 30. Holotype female in ANIC.
Dacus (Zeugodacus) timorensis – Hardy & Adachi, 1954: 196; Drew 1973: 33.
Bactrocera (Zeugodacus) timorensis – Drew, 1989: 220; Norrbom *et al*., 1998: 104.

Type material examined: Holotype female, **INDONESIA: WEST TIMOR (Kupang, formerly Koepang, Timor)** – Koepang, vi.1929, coll. I.M. Mackerras. Received on loan from ANIC and studied by Drew (1989).
Non-type material examined: INDONESIA: WEST TIMOR: 1 male, Besi Pae, 80km SE Kupang, 28.x.1986, coll. ESC Smith, Ex Cue lure trap; 1 male (31.v.2007), 3 males (1.vi.2007), Nusa Tenggara Timur, Oebufu, Kupang, coll. Ir Hendri Wihanto, attracted to cue lure in village jambu; 3 males, Nusa Tenggara Timur, Belo, Kt Kupang, 2.vi.2007, Ir Hendri Wihanto, attracted to cue lure in Kebun-belimbing. Specimens in QDPC.
Diagnosis: A small species; face fulvous with a pair of very small fuscous spots; postpronotal lobes and notopleura yellow; scutum glossy black; lateral and medial postsutural yellow vittae present; a small yellow spot anterior to notopleural suture; mesopleural stripe reaching midway between anterior margin of notopleuron and anterior *npl.* seta dorsally; scutellum yellow with a broad black basal band; legs with fore and mid femora almost entirely shining black, hind femora fulvous with apical $^2/_5$ shining black; wing with cells bc and c colourless, microtrichia in outer corner of cell c only; a narrow fuscous costal band confluent with R_{2+3} and widening slightly as it continues around the apex of the wing; a broad fuscous anal streak; abdominal terga III–V fulvous with two broad lateral longitudinal dark fuscous bands over all three terga and which meet in the anterocentral point of tergum III; abdominal sterna dark; pecten of cilia on abdominal tergum III of male present; posterior lobe of male surstylus long; male abdominal sternum V with a slight concavity; supernumerary lobe in male wing moderately large and rounded.
Attractant: Cue lure.
Distribution:
Type locality: Timor, Koepang, now known as Kupang, West Timor, Indonesia.
Previously recorded: Type locality and Indonesia (Flores) (Drew, 1989).
Hosts: No known record.
Remarks: *Bactrocera (Zeugodacus) timorensis* (Perkins) is a morphologically unique species restricted in distribution to Timor and Flores, within Wallacea. It has been adequately described by Drew (1989). It possesses a fulvous face with small fuscous spots, a black scutum with narrow tapering lateral postsutural yellow vittae and a broad medial postsutural yellow vitta, a broad black basal band on the scutellum, fore femora and apical ½ of mid and hind femora dark fuscous and abdominal terga III–V with broad dark fuscous lateral bands joined along anterior margin of tergum III.

Bactrocera (Zeugodacus) trichosanthes Drew & Romig, new species

(Fig. 360)

Material examined: Holotype male, **PENINSULAR MALAYSIA**: Serdang, viii.1987, coll. S. Vijaysegaran, bred from *Trichosanthes tricuspidata*, deposited in BMNH.
Paratypes: **PENINSULAR MALAYSIA**: 27 males, 4 females, same data as holotype; 1 male (18.viii.1986), 1 male (8.ix.1986), 1 male (27.x.1986), 1 male (3.xi.1986), Jalan Kebun, coll. Abdul Khalib bin Mat Saad; 2 males

and 1 female, MARDI, Jalan Kebun, 3.x.1986, coll. S. Vijaysegaran, bred from *Momordica charantia*; 1 male, Kuala Kangsaar, 4.vi.1987, coll. Ahmad Radzuan bin Panjang Abd; 1 male (7.vii.1986), 1 male (28.vii.1986), Teluk Intan, Abd Razak; 1 male, Transect North trap No. 17, 1.vi.1987, coll. A.R. and P.A. **EAST MALAYSIA: SARAWAK** – 1 male (28.x.1991), 1 male (26.x.1992), Sarawak, Tarat Agric. Station, coll. S. Leong; 2 males, Sarawak, Kpg Tema Mawang, KTS, 18.viii.1992, coll. Leong; 1 male (8.ix.1992), 1 male (29.ix.1992), Sarawak, Ranchan Pool, Serian, KT4, coll. Leong; 9 males and 13 females, Sarawak, ARI Semongok, 6.x.1993, coll. S. Leong, bred from fruit Sample No. K1084. **THAILAND**: 5 males and 9 females, Km 19 Khlongtakoh, Tahtakiah, Chachoeng Sao, 29.ix.1992, coll. Prasert, bred from *Trichosanthes tricuspidata*; 4 males and 5 females, Thatakiab, Sanamchai, Chachoeng, 3.viii.1992, coll. Prasert, bred from *Trichosanthes tricuspidata*; 2 males and 2 females, Km 5 Thatakiab, Chachoengsao, 30.ix.1992, coll. Prasert, bred from host fruit; 3 males and 5 females, Km 21 Pakphli, Nakhonnayok, 4.x.1992, coll. Prasert, bred from host fruit; 5 males and 1 female (2.viii.1991), 13 males and 9 females (23.viii.1991), Elephant Salt Lick, Moosi, A. Pakchong, Nakhonratsima, coll. Montree, bred from *Trichosanthes tricuspidata*; 3 males and 5 females, M002, Changtoon, Borrai Trad, 22.xii.1992, coll. Montree, bred from host fruit; 3 males, Km 37 Moosi, Pakchong, Nakhonratsima, 14.ix.1992, coll. Yuthana, bred from host fruit; 2 males and 4 females, Thone Nga, Chang, Hat Yai, Songkhla, 28.vii.1992, coll. Surin, bred from *Gymnopetalum integrifolium*; 3 males and 7 females, Songkhla, Ya Rong, Pattani, 30.i.1992, coll. Surakrai *et al.*, bred from *Gymnopetalum integrifolium*; 3 males and 2 females, Songkhla, Betong, Yala, 31.i.1992, coll. Surakrai, bred from *Gymnopetalum integrifolium*; 1 male and 1 female, Hat Yai, Songkhla, 25.i.1992, coll. Sieng *et al.*, bred from *Gymnopetalum integrifolium*; 6 males and 15 females, Amphoe Muang, Nakhon Si, Thammarat, 14.ii.1992, coll. Sieng *et al.*, bred from *Terminalia catappa*; 5 males and 6 females, Ban Nang Sata, Yala, Songkhla, 23.vii.1992, coll. Sieng *et al.*, bred from *Gymnopetalum integrifolium*; 4 males and 3 females, Songkhla, Na Pradoo, Pattani, 23.vii.1992, coll. Sieng *et al.*, bred from *Gymnopetalum integrifolium*; 3 males and 3 females, Be Tong, Yala, Songkhla, 24.vii.1992, coll. Sieng *et al.*, bred from *Gymnopetalum integrifolium*; 4 males and 9 females, Songkhla, Ra Nga, Narathiwat, 19.ii.1993, coll. Sieng *et al.*, bred from *Gymnopetalum integrifolium*; 6 males and 14 females, Betong, Yala, Songkhla, 5.ix.1991, coll. Anuchit, bred from *Gymnopetalum integrifolium*; 19 males and 23 females, Bung Nung Star, Yala, Songkhla, 27.ix.1991, coll. Anuchit *et al.*, bred from *Gymnopetalum integrifolium*; 15 males and 15 females, Bung Nung Star, Yala, Songkhla, 8.xi.1991, coll. Anuchit *et al.*, bred from *Gymnopetalum integrifolium*; 5 males and 5 females, Songkhla, Betong, Yala, 20.xii.1991, coll. Anuchit *et al.*, bred from *Gymnopetalum integrifolium*; 3 males and 5 females, Songkhla, Betong, Yala, 29.xi.1991, coll. Anuchit *et al.*, bred from *Gymnopetalum integrifolium*; 5 males and 3 females, Songkhla, Ya Rung, Patthani, 20.ii.1992, coll. Anuchit *et al.*, bred from *Gymnopetalum integrifolium*; 2 males and 4 females, Ya Rung, Patthani, 12.iii.1992, coll. Anuchit *et al.*, bred from *Gymnopetalum integrifolium*; 7 males and 12 females, Songkhla, Ya Rung, Patthani, 9.i.1992, bred from *Gymnopetalum integrifolium*; 2 males and 2 females, Kok Po, Patthani, Songkhla, 3.ix.1992, coll. Anuchit *et al.*, bred from *Gymnopetalum integrifolium*; 1 male and 1 female, Yala, Bun Nung, Sata, Songkhla, 3.ix.1992, coll. Anuchit *et al.*, bred from *Gymnopetalum integrifolium*; 2 males and 1 female, Be Tong, Yala, Songkhla, 3.ix.1992, coll. Anuchit *et al.*, bred from *Gymnopetalum integrifolium*; 2 females, Be Tong, Yala, Songkhla, 16.x.1992, coll. Anuchit *et al.*, bred from *Gymnopetalum integrifolium*. **INDONESIA: JAVA** – 2 males and 2 females, Kebun Raya, Bogor, Java, vii.1975, coll. D.E. Hardy, bred from *Trichosanthes* sp.; 4 males, Kebun Raya, Bogor, W. java, vii.1975, Coll D.E. Hardy; 1 female, W. Java, Bandoeng, 700 m, 12–15.vi.1941, coll. J. Olthof; **RAKATA ISLAND** – 1 male, Rakata, Krakatoa, 2.ix.1990, coll. E. Schmidt, attracted to methyl eugenol. 45 paratypes in ANIC; 30 paratypes in BMNH; 244 paratypes in QDPC; 70 paratypes in QMIC.

Non-type material examined: INDONESIA: JAVA – 2 males and 2 females, Gunung Salak, Bogor, Java (800m), 17.v.1981, coll. J. Turner, bred ex Red cucurbit, coll. Gunung Salak, Bogor, Java, v.1981, coll. J. Turner. Specimens in QDPC.

Diagnosis: A large species; face with a pair of moderately large oval black spots; postpronotal lobes and notopleura yellow; scutum mostly black; lateral and medial postsutural yellow vittae present; a yellow spot anterior to notopleural suture; mesopleural stripe of medium-width, reaching midway between anterior margin of notopleuron and anterior *npl.* seta; scutellum yellow; wing with cells bc and c colourless or with a very pale tint, microtrichia in outer corner of cell c only; a broad dark fuscous costal band almost confluent with R_{4+5} and expanding into a large spot around apex of R_{4+5}; a broad dark fuscous anal streak, a pale tint on hind margin around CuA_1; abdominal terga III–V fulvous with a 'T' pattern consisting of a narrow black band across anterior margin of tergum III and a narrow medial longitudinal black band over all three terga, broad dark fuscous to black anterolateral corners on terga IV and V.

Description: male.

Head: Height 1.8 mm. Frons – length 1.19 times breadth; red-brown with small pale fuscous markings around orbital setae and on anteromedial hump; orbital setae black: 1 *s.or.*, 3 *i.or.*; lunule pale fuscous. Ocellar triangle black. Vertex black. Face fulvous with a pair of moderately large oval black spots; length 0.69 mm. Genae fulvous, large fuscous to dark fuscous subocular spot present; black seta present. Occiput pale fuscous, fulvous along eye margins; occipital row with 4–8 strong black setae each side. Antennae with segments 1 and 2 pale fuscous, segment 3 red-brown with fuscous on apex and outer surface; arista black (red-brown basally); length of segments: 0.12 mm; 0.31 mm; 0.9 mm.

Thorax: Scutum black except dark red-brown around *prsc.* setae (in some specimens) and red-brown below and behind lateral postsutural yellow vittae (in some specimens), inside postpronotal lobes, between notopleuron and yellow spot anterior to notopleural suture, and a broad medial longitudinal band from anterior point of medial postsutural yellow vitta to anterior margin of scutum (there is a fine medial longitudinal black line in the centre of this band). Pleural areas black tending red-brown to fuscous below postpronotal lobes. Yellow markings as follows: postpronotal lobes; notopleura; medium-sized mesopleural stripe reaching midway between anterior margin of notopleuron and anterior *npl.* seta dorsally, continuing to katepisternum as a large transverse spot, anterior margin convex; anatergite (posterior apex black); anterior $^{3}/_{5}$ katatergite (remainder black); two broad parallel-sided lateral postsutural vittae beginning with a spot anterior to notopleural suture and ending behind *ia.* seta; a broad medial longitudinal postsutural vitta. Postnotum black tending red-brown medially. Scutellum yellow except for narrow black basal band. Setae: *sc.* 4; *prsc.* 2; *ia.* 1; *p.sa.* 1; *a.sa.* 1; *mpl.* 1; *npl.* 2; *scp.* 4; in most specimens there are a pair of weak dark humeral setae present.

Legs: All femora entirely fulvous; fore tibiae fuscous, mid tibiae fulvous tending fuscous basally, hind tibiae dark fuscous; fore tarsi with all segments fuscous, mid and hind tarsi with basal segment fulvous and apical four segments fuscous; mid tibiae each with an apical black spur.

Wings: Length 7.7 mm; cells bc and c colourless or with a very pale tint; microtrichia in outer corner of cell c only; remainder of wings colourless except dark fuscous cell sc, broad fuscous costal band almost confluent with R_{4+5} and ending in a large spot around apex of R_{4+5}, a broad dark fuscous anal streak, a pale fuscous tint on hind wing margin around CuA_1; a dense aggregation of microtrichia around A_1+CuA_2; supernumerary lobe weak.

Abdomen: Oval; terga free; pecten present on tergum III. Tergum I and sterna I and II wider than long. Tergum I fulvous with a narrow dark fuscous band across anterior margin joining broad lateral black margins; tergum II fulvous with a narrow transverse black band across anterior margin joining narrow lateral longitudinal black bands; terga III–V fulvous with a 'T' pattern consisting of a narrow transverse black band across anterior margin of tergum III and a medium-width medial longitudinal black band over all three terga and broad dark fuscous to black anterolateral corners on terga IV and V. A pair of oval fulvous shining spots on tergum V. Posterior lobe of surstylus long, sternum V with a slight concavity on posterior margin. Abdominal sterna dark fuscous to black, except sternum II fulvous.

Attractant: Cue lure.

Female: As for male except no dense aggregation of microtrichia around A_1+CuA_2; supernumerary lobe weaker than in male; pecten of cilia on abdominal tergum III absent; basal segment of ovipositor red-brown, conical in transverse cross section and tapering posteriorly in dorsal view; ratio of length of oviscape to length of tergum V, 2.6:1; apex of piercer trilobed (Fig. 360).

Distribution: This species is probably widespread across South-East Asia, having been recorded from Indonesia (Java and Rakata); East Malaysia (Sarawak); Peninsular Malaysia; Thailand.

Hosts: Recorded from *Trichosanthes tricuspidata*; *Momordica charantia*; *Gymnopetalum integrifolium* (family *Cucurbitaceae*); *Terminalia catappa* (family *Combretaceae*).

Remarks: *Bactrocera (Zeugodacus) trichosanthes* n.sp., within the *tau* complex, is similar to *Bactrocera (Zeugodacus) tau* (Walker) and *Bactrocera (Zeugodacus) okunii* (Shiraki). It differs from *B. tau* in being a large species with entirely fulvous femora and in having a weak supernumerary lobe in the male wing and a trilobed apex on the piercer. It differs from *B. okunii* in being a large species and in possessing a broader costal band (almost confluent with R_{4+5}) and an almost uniformly black scutum. *Bactrocera trichosanthes* is similar to *Bactrocera (Zeugodacus) rubella* (Hardy) and *Bactrocera (Bactrocera) tricuspidatae* n.sp. in the wing and leg colour patterns but differs from *B. rubella* in having a black scutum and abdominal terga III–V with a complete black 'T' pattern and from *B. tricuspidatae* in having anterolateral corners of terga IV and V dark and an extremely long female ovipositor with the ratio of the basal segment to the length of abdominal tergum V, 2.6:1. Further, the trilobed apex of the piercer of *B. trichosanthes* (Fig. 360) is a different shape to that of *B. tricuspidatae* (Fig. 361).

Bactrocera (Zeugodacus) tricuspidatae Drew & Romig, new species

(Fig. 361)

Material examined: Holotype male, **THAILAND**: Elephant Salt Lick, Moosi, A. Pakchong, Nakhonratsima, 23.viii.1991, coll. Montree, bred from *Trichosanthes tricuspidata*, deposited in BMNH.

Paratypes: **THAILAND**: 5 males and 12 females, same data as holotype. 2 paratypes in ANIC; 2 paratypes in BMNH; 11 paratypes in QDPC; 2 paratypes in QMIC.

Diagnosis: A large species; face with a pair of large oval black spots; postpronotal lobes and notopleura yellow; scutum mostly black; lateral and medial postsutural yellow vittae present; a small yellow spot anterior to notopleural suture; mesopleural stripe reaching midway between anterior margin of notopleuron and anterior *npl.* seta dorsally; scutellum yellow; wing with bc and c colourless, microtrichia in outer corner of cell c only; a broad dark fuscous costal band confluent with R_{4+5}; a broad dark fuscous anal streak; cells abdominal terga III–V fulvous with a 'T' pattern consisting of a narrow transverse black band across anterior margin of tergum III and a narrow- to medium-width medial longitudinal black band over all three terga, broad lateral longitudinal black bands over terga IV and V.

Description: male.

Head: Height 1.76 mm. Frons – length 1.29 times breadth; red-brown with small areas of pale fuscous around bases of orbital setae; orbital setae black: 1 *s.or.*, 3 *i.or.*; lunule fuscous. Ocellar triangle black. Vertex pale fuscous. Face fulvous with a pair of large oval black spots; length 0.64 mm. Genae fulvous, pale fuscous subocular spot present; black seta present. Occiput pale fuscous, red-brown along eye margins; occipital row with a large number of short dark setae. Antennae with segments 1 and 2 fuscous, segment 3 red-brown with pale fuscous on apex and outer surface; arista black (red-brown basally); length of segments: 0.18 mm; 0.27 mm; 0.79 mm.

Thorax: Scutum uniformly black with red-brown behind lateral postsutural yellow vittae, around yellow spot anterior to notopleural suture and across anterior margin between postpronotal lobes (this area is interspersed with medial and lateral longitudinal narrow black bands). Pleural areas black except red-brown below postpronotal lobes. Yellow markings as follows: postpronotal lobes; notopleura; mesopleural stripe reaching midway between anterior margin of notopleuron and anterior *npl.* seta dorsally, continuing to katepisternum as a large transverse spot, anterior margin convex; anatergite (posterior apex black); anterior ¾ katatergite (remainder including ventral margin black); two moderately broad parallel-sided lateral postsutural vittae beginning with a small spot anterior to notopleural suture and expanding to end behind *ia.* seta; a short, medium-width medial longitudinal postsutural vitta beginning anterior to *prsc.* setae and ending at level of notopleural suture. Postnotum black. Scutellum yellow except for narrow black basal band. Setae: *sc.* 4; *prsc.* 2; *ia.* 1; *p.sa.* 1; *a.sa.* 1; *mpl.* 1; *npl.* 2; *scp.* 4.
Legs: Femora entirely fulvous except for a small, diffuse pale fuscous spot on outer subapical surfaces of fore femora; fore and hind tibiae dark fuscous, mid tibiae dark fuscous basally to pale fuscous apically; tarsi with all segments fulvous; mid tibiae each with an apical black spur.
Wings: Length 6.9 mm; cells bc and c colourless; microtrichia in outer corner of cell c only; remainder of wings colourless except dark fuscous cell sc, a broad dark fuscous costal band tending paler fuscous as it becomes confluent with R_{4+5} and ending in a distinct spot around apex of R_{4+5}, a broad dark fuscous anal streak, an area of diffuse pale fuscous on hind margin of wing; a dense aggregation of microtrichia around A_1+CuA_2; supernumerary lobe weak.
Abdomen: Elongate oval; terga free; pecten present on tergum III. Tergum I and sterna I and II wider than long. Tergum I mostly black; tergum II fulvous with a narrow transverse black band across anterior margin joining moderately broad black lateral margins; terga III–V fulvous with a 'T' pattern consisting of a narrow transverse black band across anterior margin of tergum III and a narrow- to medium-width medial longitudinal black band over all three terga and broad lateral longitudinal black bands over terga IV and V. A pair of oval fulvous shining spots on tergum V. Posterior lobe of surstylus long, sternum V with a slight concavity on posterior margin. Abdominal sterna black except sternum II red-brown.
Attractant: No known record.
Female: As for male except no dense aggregation of microtrichia around A_1+CuA_2; pecten of cilia on abdominal tergum III absent; basal segment of ovipositor red-brown, conical in transverse cross section and tapering posteriorly in dorsal view; ratio of length of oviscape to length of tergum V, 1:1; apex of piercer trilobed (Fig. 361).
Distribution: Thailand (Nakhonratsima).
Hosts: Bred from one host *Trichosanthes tricuspidata* (family *Cucurbitaceae*).
Remarks: *Bactrocera (Zeugodacus) tricuspidatae* n.sp., within the *tau* complex, is a unique species similar to *Bactrocera (Zeugodacus) trichosanthes* n.sp. in general body colour patterns, femora mostly fulvous and supernumerary lobe weak. It differs from *B. trichosanthes* in possessing a short ovipositor with the ratio of the basal segment to length of tergum V, 1:1, compared with 2.6:1 in *B. trichosanthes*. Also, *B. tricuspidatae* possesses broad lateral longitudinal black bands over terga IV and V whereas *B. trichosanthes* has dark anterolateral corners on these terga. *Bactrocera tricuspidatae* is also similar to *Bactrocera (Zeugodacus) macrophyllae* n.sp. in general body colour patterns and in having the supernumerary lobe weak but differs in having the costal band confluent with R_{4+5} and a shorter ovipositor.

Bactrocera (Zeugodacus) trimaculata (Hardy & Adachi)

(Fig. 362)

Dacus (Zeugodacus) trimaculatus Hardy & Adachi, 1954: 196; Hardy: 1974: 58. Holotype female in USNM.
Bactrocera (Zeugodacus) trimaculata – Norrbom *et al.*, 1998: 104.

Type material examined: Holotype female and the Allotype (now Paratype) male, **PHILIPPINES**: with the following four labels – 'Penal Colony, Davao, Mindanao, P.I. 8-''50'; 'F. Skinner Collector'; 'Ex *Momordica cochinchinensis*'; 'Allotype male, *Dacus (Zeugodacus) trimaculatus*, D.E. Hardy'. Received on loan from USNM.
Diagnosis: A large species; face with fuscous along oral margin connecting a pair of medium-sized oval black spots; postpronotal lobes and notopleura yellow; scutum base colour black; lateral and medial postsutural yellow vittae present; a yellow spot anterior to notopleural suture in front of lateral postsutural vittae; mesopleural stripe reaching almost to anterior *npl.* seta dorsally; scutellum yellow with a small dark fuscous apical spot; wing with cells bc and c with a very pale tint, microtrichia in outer corner of cell c only; a narrow fuscous costal band confluent with R_{2+3} and broad pale fuscous anal streak; abdominal terga III–V red-brown with a black 'T' pattern consisting of a narrow transverse band across anterior margin of tergum III and a medial longitudinal band over all three terga, narrow black lateral margins on terga IV and V.
Description: male (Fig. 362).
Head: Height 1.61 mm. Frons – length 1.16 times breadth; fulvous with dark fuscous on anteromedial hump and fuscous between this hump and the lunule; orbital setae black: 1 *s.or.*, 2 *i.or.*; lunule dark fuscous. Ocellar triangle black. Vertex dark fuscous. Face fulvous with a pair of medium-sized oval black spots connected by a fuscous band along oral margin; length 0.86 mm. Genae fulvous, large fuscous subocular spot present; black seta present. Occiput fuscous with dark fuscous dorsolaterally and fulvous along eye margins; occipital row with 5–7 weak setae each side. Antennae with segments 1 and 2 fuscous, segment 3 red-brown with fuscous along apex and outer surface; length of segments: 0.14 mm; 0.29 mm; 0.77 mm.
Thorax: Scutum black with dark red-brown posterolateral to lateral postsutural vittae, dark fuscous below lateral postsutural vittae, dark red-brown inside notopleura and a broad fuscous medial longitudinal band from anterior point of

medial postsutural yellow vitta to anterior margin of scutum and fuscous across anterior margin of scutum to inside postpronotal lobes. Pleural areas black except dark fuscous below postpronotal lobes. Yellow markings as follows: postpronotal lobes; notopleura; mesopleural stripe reaching almost to anterior *npl.* seta dorsally, continuing to katepisternum as a large transverse spot, anterior margin straight; anatergite (posterior apex black); anterior $^3/_5$ katatergite (remainder black); two moderately broad parallel-sided lateral postsutural vittae beginning with a spot anterior to notopleural suture and ending at *ia.* seta; a broad medial longitudinal postsutural vitta broadly rounded posteriorly at level of *prsc.* setae and narrowing to a point anteriorly just anterior of notopleural suture. Postnotum black, tending dark fuscous centrally. Scutellum yellow except for a narrow black basal band and a small dark fuscous spot between apical setae. Setae: *sc.* 4; *prsc.* 2; *ia.* 1; *p.sa.* 1; *a.sa.* 1; *mpl.* 1; *npl.* 2; *scp.* 4.

Legs: Fore femora with basal ½ pale fulvous and apical ½ fulvous with a narrow fuscous posterolateral elongate spot, mid femora with basal ⅓ pale fulvous and remainder fulvous, hind femora with basal ½ pale fulvous and remainder fulvous; fore tibiae dark fuscous on basal ½ and fulvous apically, mid tibiae fuscous on basal ¼ and remainder fulvous, hind tibiae entirely fuscous; all tarsal segments entirely fulvous.

Wings: Length 7.6 mm; cells bc and c with a very pale tint (cell c paler); microtrichia in outer corner of cell c only; remainder of wings colourless except fuscous cell sc, narrow fuscous costal band confluent with R_{2+3} and remaining very narrow around margin of wing to end just beyond extremity of R_{4+5}, broad pale fuscous anal streak; dense aggregation of microtrichia around A_1+CuA_2; supernumerary lobe well developed.

Abdomen: Elongate oval; terga free; pecten present on tergum III. Tergum I and sterna I and II wider than long. Tergum I fuscous centrally with broad black lateral margins; tergum II red-brown with a narrow transverse black band anteriorly but not reaching lateral margins and narrow black lateral margins; terga III–V red-brown with a 'T' pattern consisting of a narrow transverse black band across anterior margin of tergum III and a medial longitudinal black band of medium-width over all three terga and narrow black lateral margins on tergum IV and V. A pair of oval fuscous shining spots on tergum V. Posterior lobe of surstylus long, sternum V with a slight concavity on posterior margin.

Attractant: No known record.

Female: As for male except no dense aggregation of microtrichia around A_1+CuA_2; supernumerary lobe weak; pecten of cilia on abdominal tergum III absent; basal segment of ovipositor red-brown, dorsoventrally flattened and tapering posteriorly in dorsal view; ratio of length of oviscape to length of tergum V, 2:1; apex of piercer trilobed.

Distribution:

Type locality: Philippines (Mindanao I., Davao).

Hosts: *Momordica cochinchinensis* and *M. charantia* (family *Cucurbitaceae*) (Allwood *et al.*, 1999).

Remarks: *Bactrocera (Zeugodacus) trimaculata* (Hardy and Adachi) is similar to *Bactrocera (Zeugodacus) exornata* (Hering) in possessing four *sc.* setae, a black scutum with medial and lateral postsutural yellow vittae, wings colourless except for costal band and anal streak, cells bc and c with a very pale tint, costal band confluent with R_{2+3} and not expanded apically and abdominal terga III–V with a basic black 'T' pattern. It differs from *B. exornata* in having the face fulvous with a pair of dark spots and a transverse dark line across the oral margin, femora almost entirely fulvous and the scutellum with a small dark apical spot.

Bactrocera (Zeugodacus) trivandrumensis Drew & Romig, new species

(Fig. 363)

Material examined: Holotype male, **INDIA**: Trivandrum District, Kerala State, Poonmudi Range 3000 ft, v.1972, coll. TRSN, deposited in MHNG.

Paratypes: **INDIA**: 1 female, same data as holotype; 1 female, Coimbatore, Madras State 1400 ft, xi.1972, coll. TRSN. Paratypes deposited in MHNG.

Diagnosis: A medium to large species; face fulvous with a pair of large oval black spots; postpronotal lobes and notopleura yellow; scutum mostly black; lateral and medial postsutural yellow vittae present; a small yellow spot anterior to notopleural suture; mesopleural stripe narrow and equal in width to notopleuron dorsally; scutellum yellow; wing with cells bc and c colourless, microtrichia in outer corner of cell c only; a narrow fuscous costal band, a moderately large elongate fuscous spot across apex of wing; a medium-width fuscous anal streak; abdominal terga III–V fulvous with a 'T' pattern consisting of a moderately broad transverse black band across anterior margin of tergum III and a moderately broad medial and broad lateral longitudinal dark fuscous to black bands over all three terga.

Description: male.

Head: Height 1.4 mm. Frons – length 1.27 times breadth; fulvous with dark fuscous around bases of orbital setae and on anteromedial hump; orbital setae black: 1 *s.or.*, 3 *i.or.*; lunule dark fuscous. Ocellar triangle black. Vertex dark fuscous. Face fulvous with a pair of large oval black spots; length 0.72 mm. Genae fulvous, dark fuscous subocular spot present; black seta present. Occiput fuscous centrally, dark fuscous laterally, fulvous along eye margins; occipital row with 8–10 dark setae. Antennae with segments 1 and 2 fuscous, segment 3 red-brown with fuscous on apex and outer surface; arista black (red-brown basally); length of segments: 0.12 mm; 0.26 mm; 0.76 mm.

Thorax: Scutum mostly dark fuscous to black except red-brown inside postpronotal lobes, along posterior and lateral margins and as a broad medial longitudinal band from anterior point of medial postsutural yellow vitta to anterior margin of scutum. Pleural areas dark fuscous to black except red-brown below postpronotal lobes. Yellow markings as follows: postpronotal lobes; notopleura; narrow mesopleural stripe equal in width to notopleuron dorsally, continuing to katepisternum as a large transverse spot, anterior margin straight; anatergite (posterior apex black); anterior ⅔ katatergite (remainder black); two narrow lateral postsutural vittae beginning with a small spot anterior to notopleural suture and narrowing posteriorly to end in a point before *ia.* seta; a narrow medial longitudinal postsutural vitta beginning at level of *prsc.* setae and ending anteriorly at level of notopleural suture. Postnotum red-brown centrally, black laterally. Scutellum yellow except for narrow black basal band. Setae: *sc.* 4; *prsc.* 2; *ia.* 1; *p.sa.* 1; *a.sa.* 1; *mpl.* 1; *npl.* 2; *scp.* 2.

Legs: Fore femora fulvous except for a large black spot along apical ¾ of outer surfaces, mid and hind femora fulvous except black on apical ¼ of inner surfaces; fore and hind tibiae dark fuscous basally to fuscous apically, mid tibiae dark fuscous basally to fulvous apically; tarsi with basal segment fulvous and apical fore segments fuscous; mid tibiae each with an apical black spur.

Wings: Length 6.5 mm; cells bc and c colourless; microtrichia in outer corner of cell c only; remainder of wings colourless except fuscous cell sc, narrow fuscous costal band confluent with R_{2+3} and remaining narrow before expanding into an elongate spot across apex of wing, a medium-width fuscous anal streak; a dense aggregation of microtrichia around A_1+CuA_2; supernumerary lobe of medium development and rounded.

Abdomen: Oval; terga free; pecten present on tergum III. Tergum I and sterna I and II wider than long. Tergum I dark fuscous to black except for a narrow transverse fulvous band across posterior margin but not reaching lateral margins; tergum II fulvous except for a narrow transverse dark fuscous band across anterior margin but not reaching the narrow dark fuscous lateral margins; terga III–V fulvous except for a 'T' pattern consisting of a moderately broad transverse dark fuscous to black band across anterior margin of tergum III and a moderately broad medial and broad lateral longitudinal dark fuscous to black bands over all three terga (the lateral bands may be paler on the posterior areas of tergum III). A pair of oval fulvous shining spots on tergum V. Posterior lobe of surstylus long, sternum V with a slight concavity on posterior margin. Abdominal sterna dark fuscous except sternum II fulvous.

Attractant: No known record.

Female: As for male except no dense aggregation of microtrichia around A_1+CuA_2; supernumerary lobe weak; pecten of cilia on abdominal tergum III absent; basal segment of ovipositor fuscous, conical in transverse cross section and tapering posteriorly in dorsal view; ratio of length of oviscape to length of tergum V, 2.6:1; apex of piercer needle-shaped.

Distribution: India (Kerala State; Tamil Nadu – formerly Madras State).

Hosts: No known record.

Remarks: *Bactrocera (Zeugodacus) trivandrumensis* n.sp., within the *tau* complex, is similar to *Bactrocera (Zeugodacus) tapervitta* Mahmood in possessing dark abdominal sterna, an entirely yellow scutellum, broad dark lateral bands over terga III–V or IV–V and tapering lateral postsutural yellow vittae. It differs from *B. tapervitta* in having the lateral postsutural vittae narrow, a reduced supernumerary lobe in the wing (medium-sized) and a long female ovipositor (basal segment 2.6 times length of abdominal tergum V).

Bactrocera (Zeugodacus) ujungpandangiae Drew & Romig, new species

(Fig. 364)

Material examined: Holotype male, **INDONESIA: SULAWESI** – Bengo Forest, nr. Ujung Pandang (220 m), Sulawesi, 26.vii.1981, coll. J. Turner, Residual Secondary Forest, attracted to cue lure, deposited in BMNH.

Diagnosis: A medium-sized species; face with a pair of medium-sized circular black spots; postpronotal lobes and notopleura yellow; scutum mostly black; lateral and medial postsutural yellow vittae present; a large yellow triangle along anterior margin of notopleural suture; mesopleural stripe reaching almost to anterior *npl.* seta dorsally; scutellum yellow; wing with cells bc and c with a pale fulvous tint, microtrichia in outer corner of cell c only; a narrow fuscous costal band slightly overlapping R_{2+3} and medium-width fuscous anal streak; abdominal terga orange-brown except for a narrow subanterior black band across tergum III but not reaching lateral margins and a medium-width medial longitudinal black band on terga IV and V.

Description: male.

Head: Height 1.62 mm. Frons – length 1.53 times breadth; red-brown with fuscous around upper pair of *i.or.* setae; orbital setae black: 1 *s.or.*, 2 *i.or.*; lunule fuscous. Ocellar triangle black. Vertex red-brown. Face fulvous with a pair of medium-sized circular black spots; length 0.55 mm. Genae fulvous, very pale fuscous subocular spot present; black seta present. Occiput red-brown, fulvous along eye margins; occipital row with a large number of small pale setae. Antennae with all segments mostly fuscous; arista black (red-brown basally); length of segments: 0.12 mm; 0.28 mm; 0.68 mm.

Thorax: Scutum almost entirely black except for a small area of dark red-brown anteromedially running from anterior point of medial longitudinal postsutural yellow vitta but ending before anterior margin of scutum (in some specimens, this red-brown is below and behind lateral postsutural yellow vittae and across hind margin of scutum). Pleural areas dark fuscous posterior to mesopleural stripe, black anterior to mesopleural stripe except red-brown below postpronotal lobes and on ventral area of katepisternum. Yellow markings as follows: postpronotal lobes; notopleura; mesopleural stripe reaching almost to anterior *npl.* seta dorsally, continuing to katepisternum as a large elongate transverse spot, anterior margin strongly convex; anatergite (posterior apex black); anterior ⅔ katatergite (remainder black); two broad lateral postsutural vittae widening slightly posteriorly to end behind *ia.* seta; a broad medial longitudinal postsutural vitta beginning just anterior of *prsc.* setae and narrowing to a point to end at level of anterior *npl.* setae; a large triangular marking along anterior margin of notopleural suture with base at notopleuron. Postnotum black. Scutellum yellow except for narrow black basal band. Setae: *sc.* 4; *prsc.* 2; *ia.* 1; *p.sa.* 1; *a.sa.* 1; *mpl.* 1; *npl.* 2; *scp.* 4.

Legs: All leg segments entirely fulvous; no apical black spur on mid tibiae.
Wings: Length 5.8 mm; cells bc and c with a pale fulvous tint; microtrichia in outer corner of cell c only; remainder of wings colourless except fuscous cell sc, narrow fuscous costal band slightly overlapping R_{2+3} and widening slightly across apex of R_{4+5}, a medium-width fuscous anal streak narrowing towards wing margin; dense aggregation of microtrichia around A_1+CuA_2; supernumerary lobe weak.
Abdomen: Elongate oval; terga free; pecten present on tergum III. Tergum I and sterna I and II wider than long. Abdominal terga entirely orange-brown except for a narrow transverse black band subanteriorly across tergum III but not reaching lateral margins and a narrow- to medium-width medial longitudinal black band on terga IV and V (these patterns vary with the transverse band across tergum III incomplete and the medial longitudinal band over terga III–V, in some specimens). A pair of oval pale fuscous shining spots on tergum V. Posterior lobe of surstylus long, sternum V with a slight concavity on posterior margin. Abdominal sterna entirely fulvous.
Attractant: Cue lure.
Female: No known record.
Distribution: Indonesia (Sulawesi – South Sulawesi, near Ujung Pandang = Makassar).
Hosts: No known record.
Remarks: *Bactrocera (Zeugodacus) ujungpandangiae* n. p. is close to *Bactrocera (Zeugodacus) abnormis* (Hardy) and *Bactrocera (Zeugodacus) connexa* (Hardy) in possessing pale-coloured abdominal sterna, a black scutum, wings without infuscation except for costal band and anal streak and to *B. abnormis* in having pale-coloured abdominal terga. It differs from *B. abnormis* in having a large yellow triangle along anterior margin of notopleural suture, a narrower mesopleural stripe (not reaching to postpronotal lobe), costal band overlapping R_{2+3}, a narrower anal streak and a medial longitudinal black band on abdominal terga IV and V. It differs from *B. connexa* in having the costal band overlapping R_{2+3}, facial spots present, and in lacking the yellow band connecting the postpronotal lobe and notopleuron and having abdominal terga mostly pale coloured.

Bactrocera (Zeugodacus) uncinata Drew & Romig, new species

(Fig. 365)

Material examined: Holotype male, **PHILIPPINES: LUZON ISLAND** – Mt Makiling (400 m), 9.v.1983, coll. R.A.I. Drew, Nette, Bert, attracted to cue lure, deposited in BMNH.
Paratypes: **PHILIPPINES: LUZON ISLAND** – 2 males, Mt Makiling (100 m), 8.vi.1983, coll. B. Ramos, attracted to cue lure; 7 males, same locality, 5.ix.1983, coll. Drew, Nette, Albert; 2 males, Mt Makiling (200 m), 5.ix.1983, coll. Drew, Nette, Bert, attracted to cue lure; 2 males, same locality, 8.vi.1983, coll. B. Ramos, attracted to cue lure; 3 males, 8.vi.1983 and 1 male, 15.vi.1983, Mt Makiling (300 m), coll. B. Ramos, attracted to cue lure; 2 males, Mt Makiling (400 m), 9.v.1983, coll. Drew, Nette, Bert, attracted to cue lure; 1 male, Mt Makiling (600 m), 9.v.1983, coll. Drew, Nette, Bert, attracted to cue lure. 2 paratypes in ANIC; 2 paratypes in BMNH; 10 paratypes in QDPC; 6 paratypes in QMIC.
Non-type material examined: PHILIPPINES: LUZON ISLAND – 2 males, Batangas City, 24.v.1983, coll. N.E. Rivera, attracted to cue lure; 3 males, Batangas City, 31.v.1983, coll. A.M. Arida & N.E. Rivera, attracted to cue lure; 1 male, Mt Makiling (300m), 15.vi.1983, coll. B. Ramos, attracted to cue lure; 1 male, Dasmarinas, Cavite, 13.v.1983, coll. B. Ramos & R. Drew, attracted to cue lure; 1 male, Dasmarinas, Cavite, 27.v.1980, coll. B. Ramos, attracted to cue lure; 1 male, Obando, Bulacan, 24.vi.1983, coll. B. Ramos, attracted to cue lure. Specimens in QDPC.
Diagnosis: A medium-sized species; face fulvous with a pair of medium-sized elongate oval black spots; postpronotal lobes and notopleura yellow; scutum red-brown with large black markings; lateral and medial postsutural yellow vittae present; a yellow spot on anterior margin of notopleural suture; mesopleural stripe of medium-width reaching midway between anterior margin of notopleuron and anterior *npl.* seta dorsally; scutellum yellow; wing with cells bc and c with very pale tint, microtrichia in outer corner of cell c only; a narrow fuscous costal band and medium-width anal streak; abdominal terga III–V fulvous with a distinct black 'T', narrow black lateral margins over terga IV and V and a recurved black band from posterior to anterior margins of tergum IV.
Description: male.
Head: Height 1.72 mm. Frons – length 1.44 times breadth; fulvous with dark fuscous on anteromedial hump; orbital setae black: 1 *s.or.*, 2 *i.or.*; lunule dark fuscous. Ocellar triangle black. Vertex dark fuscous. Face fulvous with medium-sized elongate oval black spots reaching to or almost to oral margin, length 0.59 mm. Genae fulvous, large dark fuscous subocular spot present; black seta present. Occiput fuscous, fulvous along eye margins; occipital row with a large number of short pale setae. Antennae with segments 1 and 2 red-brown, segment 3 red-brown with fuscous on apex and outer surface; arista black (red-brown basally); length of segments: 0.12 mm; 0.24 mm; 0.77 mm.
Thorax: Scutum red-brown with black between medial and lateral postsutural yellow vittae and extending anteriorly into an irregular pattern between postpronotal lobe and notopleural suture. Pleural areas black except red-brown below postpronotal lobe and on ventral area of katepisternum. Yellow markings as follows: postpronotal lobes; notopleura; mesopleural stripe of medium-width, extending midway between anterior margin of notopleuron and anterior *npl.* seta dorsally, continuing to katepisternum as a large broad spot, anterior margin almost straight; anatergite (posterior apex black); anterior $^5/_7$ katatergite (remainder black); two moderately broad triangular lateral postsutural vittae beginning with a distinct large spot anterior to notopleural suture and ending well before *ia.* seta (at level of *p.sa.* seta); a narrow medial longitudinal postsutural vitta beginning anterior to *prsc.*

setae and narrowing to a point to end well anterior of notopleural suture at level of anterior *npl.* setae. Postnotum black laterally, red-brown centrally. Scutellum yellow except for narrow black basal band. Setae: *sc.* 4; *prsc.* 2; *ia.* 1; *p.sa.* 1; *a.sa.* 1; *mpl.* 1; *npl.* 2; *scp.* 4.

Legs: Femora entirely fulvous; fore tibiae fuscous, mid tibiae fuscous basally and fulvous apically, hind tibiae dark fuscous; tarsi with basal segment fulvous and apical four segments fuscous; mid tibiae with no distinct apical dark spur.

Wings: Length 6.4 mm; cells bc and c with a very pale tint; microtrichia in outer corner of cell c only; remainder of wings colourless except for fuscous cell sc, narrow fuscous costal band just overlapping R_{2+3} and expanding across apex of wing to apex of vein M, a medium-width pale fuscous anal streak ending before wing margin; dense aggregation of microtrichia around A_1+CuA_2; supernumerary lobe weak.

Abdomen: Elongate oval; terga free; pecten present on tergum III. Tergum I and sterna I and II wider than long. Sternum I and sterna III, IV and V dark fuscous to black, sternum II fulvous. Tergum I fulvous with broad dark fuscous lateral margins; tergum II fulvous with narrow dark fuscous lateral margins and a broad transverse black band across anterior margin which does not reach to lateral margins; terga III–V fulvous with a distinct black 'T' pattern consisting of a broad transverse band across anterior margin of tergum III which expands around lateral margins, a broad medial longitudinal band over all three terga, narrow lateral black margins along terga IV and V which occasionally extend around posterior margin of tergum V, a recurved black band extending from posterior margin of tergum IV to anterior margin of this tergum and leaving a distinct area of fulvous between this band and the narrow black lateral margins. A pair of oval fulvous shining spots on tergum V. Posterior lobe of surstylus long, sternum V with a slight concavity on posterior margin.

Attractant: Cue lure.

Female: No known record.

Distribution: Philippines (Luzon I. – Batangas City; Bulacan; Cavite; Mt Makiling).

Hosts: No known record.

Remarks: *Bactrocera (Zeugodacus) uncinata* n.sp. is similar to *Bactrocera (Zeugodacus) isolata* (Hardy) in possessing a recurved black marking on abdominal tergum V in addition to lateral and medial postsutural yellow vittae, entirely fulvous femora and general colour patterns on the abdomen. *Bactrocera isolata* differs in having the scutum mostly black, broad parallel-sided lateral postsutural yellow vittae, a shorter medial longitudinal postsutural yellow vitta (ending anteriorly at line of notopleural suture), mesopleural stripe extending dorsally to anterior *npl.* seta, the yellow spot on the katepisternum long and narrow, facial spots circular, an enclosed fulvous spot on abdominal tergum IV and apex of costal band with a *tau*-like swelling or spot. It is also similar to *Bactrocera (Zeugodacus) juxtuncinata* n.sp. in having the recurved black band on abdominal tergum IV but differs in having the base colour of the scutum red-brown, not black.

Nine specimens in QDPC from Batangas City, Dasmarinas (Cavite) and Obando (Bulacan), Luzon I., Philippines, are similar to *B. uncinata* but differ in having broad lateral postsutural yellow vittae that narrow posteriorly to end at the *ia.* seta. These are probably conspecific and are set aside as Non-type material studied.

Bactrocera (Zeugodacus) vargus (Hardy)

(Fig. 366)

Dacus (Zeugodacus) vargus Hardy, 1982: 213. Holotype male in MZB.

Bactrocera (Zeugodacus) vargus – Norrbom *et al.*, 1998: 104.

Type material examined: Holotype male, **INDONESIA: SULAWESI** – Central Sulawesi, Sadaunta, 65 km S.E. Palu, 650 m, Aug. 1975, coll. D.E. Hardy, studied in MZB.

Non-type material examined: INDONESIA: SULAWESI – 23 males, Sulawesi Utara, Bitung, Danowudu, 26.ix.2006, coll. Karel F Lala, attracted to cue lure in forest; 18 males, Gorontalo, Bone Bolango, Ayma Utara, 12.ix.2006, coll. Yusbar Ismail, attracted to cue lure in Cabe; 11 males (19.ix.2006), 8 males (26.ix.2006), 3 males (3.x.2006), 1 male (10.x.2006), 2 males (17.x.2006), Gorontalo, Bone Bolango, Lombongo, coll. Yusbar Ismail, attracted to cue lure in forest; 2 males, South Sulawesi, Maros, Limampoccoe, 13.ix.2006, coll. Sirajuddin, attracted to cue lure in forest; 1 male, Central Sulawesi, Donggala, Nupabomba, 2.ix.2006, coll. Rusdin, attracted to cue lure in forest; 1 male (16.viii.2006), 1 male (13.ix.2006), Sulawesi Selatan, Takala, Limampoccoe, coll. Sirajuddin SP, attracted to cue lure in forest; 1 male, Sulawesi Barat, Mamuju, Belang-belang, 22.viii.2006, coll. Sri Hastuti, Sukri, attracted to cue lure in forest jeruk; 1 male, Sulawesi Tengah, Donggala, Nupabomba, 23.ix.2006, coll. Rusdin, attracted to cue lure in forest. Specimens in QDPC.

Diagnosis: A large species; face fulvous with a broad black band across oral margin; postpronotal lobes and notopleura yellow; scutum red-brown; lateral and medial postsutural yellow vittae present; a yellow spot anterior to notopleural suture; mesopleural stripe reaching to anterior *npl.* seta dorsally; scutellum entirely yellow; wing with cells bc and c with a pale tint, microtrichia in outer corner of cell c only; a broad costal band that is pale fuscous to R_{2+3} and very pale between R_{2+3} and R_{4+5} and ending in a large spot in apex of wing which varies from pale fuscous to a fulvous type tint, pale fuscous along *dm-cu* crossvein; a broad pale fuscous anal streak; abdominal terga III–V fulvous with a narrow red-brown transverse band across anterior margin and including lateral margins, narrow lateral red-brown shading on terga IV and V, narrow medial longitudinal red-brown bands on each of terga III, IV and V.

Description: male.

Head: Height 1.56 mm. Frons – length 1.34 times breadth; fulvous with red-brown on anteromedial hump; orbital setae black: 1 *s.or.*, 2 *i.or.*; lunule pale fuscous. Ocellar triangle black. Vertex red-brown. Face fulvous with a broad transverse black band across oral margin. Occiput red-brown, fulvous along eye margins; length 0.56 mm. Antennae with segments 1 and 2 red-brown, segment 3 fulvous; length of segments: 0.12 mm; 0.24 mm; 0.74 mm.

Thorax: Scutum black with red-brown across posterior margin, below and behind lateral postsutural vittae, around notopleural suture, along lateral margins between postpronotal lobe and notopleuron, across anterior section anterior to medial yellow vitta; yellow markings as follows: postpronotal lobes; notopleura; mesopleural stripe reaching to anterior *npl.* seta dorsally; two broad parallel-sided lateral postsutural vittae beginning with a spot anterior to notopleural suture and ending well behind *ia.* seta; a moderately broad medial longitudinal postsutural vitta beginning posteriorly at level of *prsc.* setae and narrowing to a point anteriorly approximately at level of anterior *npl.* seta. Scutellum yellow except for narrow dark basal band. Setae: *sc.* 4; *prsc.* 2; *ia.* 1; *p.sa.* 1; *a.sa.* 1; *mpl.* 1; *npl.* 2; *scp.* 4.

Legs: Fore femora fulvous except apical ½ red-brown dorsally and dark fuscous ventrally, mid femora fulvous except apical ¼ red-brown dorsally and dark fuscous ventrally, hind femora fulvous with apical ⅓ red-brown; all tibiae and tarsal segments fulvous.

Wings: Length 7.0 mm; cells bc and c dark fulvous; microtrichia in outer corner of cell c only; remainder of wings with a pale tint except dark fulvous cell sc, broad dark fulvous costal band confluent with R_{4+5} and paler between R_{2+3} to R_{4+5} and ending in apex of wing with a large spot which merges from pale fuscous to a yellow tint, pale fuscous along *dm-cu* crossvein, a broad pale fuscous anal streak; dense aggregation of microtrichia around A_1+CuA_2; supernumerary lobe of medium development.

Abdomen: Elongate oval; pecten present on tergum III. Tergum I fuscous to dark fuscous with orange-brown across posterior ½; tergum II fulvous with a narrow transverse black band across anterior margin but not reaching the narrow black lateral margins; terga III–V orange-brown with a narrow transverse black band across anterior margin of tergum III and widening to include lateral margins, narrow red-brown shading on lateral margins of terga IV and V, a narrow medial longitudinal black band over terga III–V which may be broken at the intersegmental lines. A pair of oval orange-brown shining spots on tergum V. Posterior lobe of surstylus long, sternum V with a slight concavity on posterior margin. Abdominal sterna orange-brown.

Attractant: Cue lure.

Female: No known record.

Distribution:

Type locality: Indonesia (Sulawesi – Central Sulawesi, Sadaunta).

New records: Indonesia (Sulawesi – North, Gorontalo, West and South Provinces).

Hosts: No known record.

Remarks: *Bactrocera (Zeugodacus) vargus* (Hardy) is a distinct species with the wings entirely coloured. It is close to *Bactrocera (Zeugodacus) emittens* (Walker) but differs in possessing a black band across the face oral margin, a broader mesopleural stripe, a tint across the entire wing membrane and infuscation only along *dm-cu* crossvein which does not connect with the spot in the apex of the wing.

Bactrocera (Zeugodacus) vulta (Hardy)

(Fig. 367)

Dacus (Zeugodacus) vultus Hardy, 1973: 74. Holotype male in KUB.

Dacus (Zeugodacus) personatus Hardy, 1983: 38. **New Synonym.**

Bactrocera (Zeugodacus) vulta – Norrbom *et al.*, 1998: 104.

Type material examined: Holotype male, **THAILAND**: with the following label data – Yala, 15.viii.1965, coll. Anand, cL (= cue lure). Received on loan from KUB.

Other type material examined: INDONESIA: JAVA – Holotype male of *Dacus (Zeugodacus) personatus* Hardy, Pelabuhan Ratu, W. Java, 13.iv.1975, D.E. Hardy; 4 paratypes, same data as holotype; 4 paratypes, Pelabuhan Ratu, West Java, iv.1975, D.E. Hardy. Studied at MZB.

Non-type material examined: BHUTAN: 1 male (26. vii.2000), 1 male (16.viii.2000), Rimchu trap No. 1, coll. C. Dorji. **INDONESIA: BALI** – Medewi, nr. Pulakon, 400m, 10.vii.1981, coll. J. Turner, attracted to cue lure; 11 males, Pura Batakaru, 800m, 11.vii.1981, coll. J. Turner, attracted to cue lure; **JAVA** – 5 males (1.viii.2005), 2 males (8.viii.2005), 1 male (15.viii.2005), 1 male (29.viii.2005), 8 males (5.ix.2005), 7 males (6.ix.2005), 7 males (19.ix.2005), 4 males (27.ix.2005), 3 males (6.x.2005), 1 male (15.xi.2005), Banten, Lebak, Pasir Nininini, coll. T.B. Aden Haer, attracted to cue lure in forest. 7 males, Pelabuhan Ratu, 0-100m, 12.v.1981, coll. J. Turner, attracted to cue lure; **LOMBOK** – 3 males, Suranadi, W. Lombok, 4.vi.78, coll. J. T. Medler, attracted to cue lure. **EAST MALAYSIA: SARAWAK** – 1 male, Malaysia, Sarawak, Santubong, 25.v.1991, coll. YHS [Yong Hoi Seng]; 1 male (8.i.1992), 1 male (3.ii.1992), Tarat Agriculture Station, Nr. Kuching, coll. S. Leong, attracted to cue lure; 1 male, Sarawak, Kpg Tema Mawang, Tebedu, 5.iv.1993, coll. S. Leong, attracted to cue lure; **SABAH** – 1 male, Sabah, Poring, Kinabalu, 20.i.1995, coll. H.S. Yong. **PENINSULAR MALAYSIA**: 1 male, Cameron Highlands, Malaysia, 24.vi.1983, Tapah 6th mile, coll. S. Vijaysegaran, attracted to cue lure; 1 male, TN3 Malaysia, 8.ix.1986, coll. Yusoff Salleh, attracted to cue lure; 2 males, TN4 Malaysia, 23.ix.1986, coll. Ahmad Radzuan & Panjang Abdullah, attracted to cue lure; 1 male (28.vii.1986), 3 males (15.vi.1987), 1 male (29.vi.1987), 2 males (27.vii.1987), TN5 Malaysia, coll. Yusoff Salleh, attracted to cue lure; 1 male, TN6 Malaysia, 28.vii.1986, coll. Yusoff Salleh, attracted to cue lure; 2 males, TN7 Malaysia, 18.v.1987, coll. Yusoff Salleh, attracted to cue lure; 1 male, TN8 Malaysia, 9.iii.1987, coll. Y. Salleh, attracted to cue lure; 3 males (25.viii.1986), 1 male (23.ii.1987), TN9 Malaysia, coll. Yusoff Salleh, attracted to cue lure; 1 male, TN11 Malaysia, 29.xii.1986, coll. A.R. & P.A., attracted to cue lure; 1 male (30.vi.1986), 1 male (8.ix.1986), TN12 Malaysia, coll.

Ahmad Radzuan & Panjang Abdullah, attracted to cue lure; 1 male (29.xii.1986), 2 males (20.iv.1987), TN14 Malaysia, coll. Ahmad Radzuan & Panjang Abdullah, attracted to cue lure; 1 male (20.x.1986), 1 male (4.v.1987), 1 male (18.v.1987), TN16 Malaysia, coll. Ahmad Radzuan & Panjang Abdullah, attracted to cue lure; 3 males, TS10 Malaysia, 20.x.1986, coll. Vignes, attracted to cue lure; 1 male, TS11 Malaysia, 20.x.1986, coll. Vignes, attracted to cue lure; 1 male (22.iv.1986), 1 male (30.vi.1986), Jerangau, Malaysia, coll. Md Ismail, attracted to cue lure; 1 male, RS Jeram Pasu I, Malaysia, 12.vi.1986, coll. Ismail Mat, attracted to cue lure; 1 male, Kuala Kangsar, Malaysia, 17.iii.1987, coll. Ahmad Radzuan & Panjang Abd., 1 male, Malaysia, 14.vii.1986, coll. Ahmad Radzuan & Panjang Abd., attracted to cue lure; 1 male, Malaysia, 14.vii.1986, coll. Yusoff Salleh, attracted to cue lure; 1 male, University of Malaya, Kuala Lumpur, 8.iii.1990, coll. C.S. Ooi. **PHILIPPINES**: **LUZON ISLAND** – 3 males, Mt Makiling (300m), Luzon, 21.x.1991, coll. D. Hancock; 2 males, Mt Makiling (300m), 8.vi.1983, coll. B. Ramos, attracted to cue lure; 3 males, Mt Makiling (300m), 21.x.1991, coll. D.L. Hancock; 2 males, Mt Makiling (300m), 8.vi.1983, coll. B. Ramos, attracted to cue lure. **THAILAND**: 2 males, Pha Yao, Chiang Rai, 14. viii.1991, coll. Charuwan, attracted to cue lure; 1 male (23.i.1992), 3 males (30.i.1992), 1 male (5.iii.1992), 1 male (9.iv.1992), Khun Khon, Chiang Rai, coll. Charuwan, attracted to cue lure; 1 male, Phliu Hort. Station, A. Laemsingh, Chanthaburi, 17.viii.1992, coll. Montree, attracted to cue lure. **VIETNAM: Southern Vietnam** – South Vietnam, 1 male, Dong Nai Province, Long Khanh District, Nhan Nghia Village, 20.vii.2002, attracted to cue lure; **Northern Vietnam** – 1 male, Phu tho (P), Phu ninh (D), Phu tho Fruit Res Centre, Site VN07, 14.vi.2002, coll. NIPP staff, attracted to cue lure. Specimens in QDPC collection.

Diagnosis: A medium-sized species; face with two extremely large oval black spots; postpronotal lobes and notopleura yellow; scutum black; lateral and medial postsutural yellow vittae present; a yellow spot anterior to notopleural suture in front of lateral postsutural yellow vittae; mesopleural stripe equal in width to notopleuron dorsally; scutellum yellow; wing with cells bc and c colourless, microtrichia in outer corner of cell c only; a narrow fuscous to dark fuscous costal band almost confluent with R_{4+5} and a broad fuscous to dark fuscous anal streak; abdominal terga III–V fulvous with a black 'T' pattern consisting of a narrow transverse band across anterior margin of tergum III, a broad medial longitudinal band over all three terga, a medium-width lateral dark fuscous band on tergum IV and narrow lateral dark fuscous band on tergum V.

Description: male.

Head: Height 1.4 mm. Frons – length 1.41 times breadth; red-brown with pale fuscous around lower orbital setae and on anteromedial hump; orbital setae black: 1 *s.or.*, 2 *i.or.*; lunule red-brown. Ocellar triangle black. Vertex black. Face with two extremely large oval black spots surrounded by narrow margins of fuscous to dark fuscous (the pattern gives the impression that the face is entirely black); length 0.44 mm. Genae fuscous, small dark fuscous subocular spot present; black seta present. Occiput dark fuscous, fulvous along eye margins; occipital row with a few weak pale setae. Antennae with segments 1 and 2 red-brown, segment 3 fuscous; length of segments: 0.1 mm; 0.32 mm; 0.8 mm.

Thorax: Scutum black with or without small areas of red-brown posterolateral to posterior end of lateral postsutural vittae, red-brown just anterior of notopleural suture and dark red-brown as two short narrow longitudinal bands inside postpronotal lobes. Pleural areas entirely dark fuscous to black. Yellow markings as follows: postpronotal lobes; notopleura; mesopleural stripe equal in width to notopleura dorsally, extending to katepisternum as a transverse spot, anterior margin strongly convex; anatergite (posterior apex black); anterior ⅔ katatergite (remainder dark fuscous to black); two moderately broad parallel-sided lateral postsutural vittae beginning with a spot anterior to notopleural suture and ending just behind *ia.* seta; a broad medial longitudinal postsutural vitta broadly rounded posteriorly at level of *prsc.* setae and ending in a point at level of notopleural suture. Postnotum black. Scutellum yellow except for a narrow black basal band. Setae: *sc.* 4; *prsc.* 2; *ia.* 1; *p.sa.* 1; *a.sa.* 1; *mpl.* 1; *npl.* 2; *scp.* 4.

Legs: All segments entirely fulvous except a small area of fuscous basally on fore tibiae and dark fuscous basally on hind tibiae; mid tibiae each with an apical black spur.

Wings: Length 5.2 mm; cells bc and c colourless; microtrichia in outer corner of cell c only; remainder of wings colourless except fuscous to dark fuscous cell sc, narrow fuscous to dark fuscous costal band almost confluent with R_{4+5} and ending between extremities of R_{4+5} and M, a broad fuscous to dark fuscous anal streak; a dense aggregation of microtrichia around A_1+CuA_2; supernumerary lobe strong and rounded.

Abdomen: Elongate oval; terga free; pecten present on tergum III. Tergum I and sterna I and II wider than long. Tergum I fulvous with narrow dark fuscous pattern across anterior margin and narrow black lateral margins; tergum II fulvous except for a narrow transverse dark fuscous band anteriorly which connects with the narrow lateral black margins; terga III to V fulvous with a 'T' pattern consisting of a narrow transverse black band across anterior margin of tergum III that extends around lateral margins and a moderately broad medial longitudinal band which is dark fuscous on tergum III and black on terga IV and V, a medium-width lateral dark fuscous band on tergum IV and a narrow lateral dark fuscous band on tergum V. A pair of oval fulvous shining spots on tergum V. Posterior lobe of surstylus long, sternum V with a slight concavity on posterior margin.

Attractant: Cue lure (also recorded for *Dacus personatus* Hardy (1983)).

Female: No known record.

Distribution:

Type locality: Thailand (Yala).

Previously recorded: Type locality and China (Yunnan Province); Laos (Menglong).

New records: Bhutan; East Malaysia (Sabah; Sarawak); Peninsular Malaysia; Indonesia (Bali; Lombok); Philippines (Luzon I., Mt Makiling); Vietnam. In addition Indonesia (Java), due to **New Synonymy** of *D. personatus*.

Hosts: No known record.

Remarks: *Bactrocera (Zeugodacus) vulta* (Hardy) belongs to a group of species that possess four *sc.* setae, broad well-developed lateral and medial longitudinal postsutural yellow vittae, a black scutum, wings colourless except for costal band and anal streak, cells bc and c colourless, costal band overlapping R_{2+3} and not expanded into a distinct spot in wing apex. It is closest to *Bactrocera (Zeugodacus) melanofacies* n.sp. in possessing a strong supernumerary lobe in the male wing and face entirely dark fuscous to black. *Bactrocera vulta* differs from *B. melanofacies* in having the costal band of uniform width throughout or only widening slightly towards the apex of wing, the male supernumerary lobe strong and rounded and the male anal streak broad but not widening apically. The measurements included above are from specimens collected in Thailand, not the type.

We can find no sound characters to separate *Dacus (Zeugodacus) personatus* Hardy from *B. vulta* and are making *D. personatus* a **New Synonym** of *B. vulta*.

Bactrocera (Zeugodacus) watersi (Hardy)

(Fig. 368)

Dacus (Neodacus) watersi Hardy, 1954:12. Holotype male in USNM.
Bactrocera (Paradacus) watersi – Norrbom *et al.*, 1998: 99.
Bactrocera (Sinodacus) watersi – Drew & Raghu, 2002: 328.

Type material examined: Holotype male, **INDIA**: with the following four labels – 'Kodaikanal, S. India, i.1951'; 'ex *Bryonia* sp.'; 'N.D. Waters Collector'; 'Holotype male, Dacus (Neodacus) watersi, D.E. Hardy'. Received on loan from USNM.
Paratypes: **INDIA**: 1 male specimen with the following six labels – 'Paratype'; 'Kodaikanal, S. India, i.1951'; 'ex *Trichosanthes primata*'; 'N.D. Waters collector'; 'Paratype male, Dacus (Neodacus) watersi Hardy'; 'Pres. by D.E. Hardy, BM1954-320'. Studied at BMNH. 1 male with the following five labels – 'Kodaikanal, S. India, i.1951'; 'Ex *Bryonia* sp.'; 'N.D. Waters Collector'; 'Paratype, Dacus (Neodacus) watersi, D.E. Hardy'; 'Dacus (Paradacus) watersi Hardy, 1980, det. D.E. Hardy'. Specimen in QDPC.
Non-type material examined: INDIA: 1 female, Karnataka, Chethalli, 970 m, 31.x.1987, coll. Ramani, attracted to light; 3 males and 3 females (20.v.1992), 1 female (21.v.1992), S. India, Mudigere, coll. D. Hancock. Specimens in QDPC.
Diagnosis: A large species; face fulvous with a pair of narrow elongate black spots reaching oral margin; postpronotal lobes pale fuscous; notopleura yellow; scutum red-brown without dark markings; lateral and medial postsutural yellow vittae present; a yellow spot anterior to notopleural suture; mesopleural stripe equal in width to notopleuron dorsally; scutellum yellow with a large dark fuscous to black apical spot; wing with cell c pale fulvous, cell bc fulvous, microtrichia in outer corner of cell c only; a broad fuscous costal band confluent with R_{4+5} and expanding into a large fuscous spot around apex of this vein; a broad fuscous anal streak; abdominal terga III–V red-brown with a 'T' pattern consisting of a narrow transverse black band across anterior margin of tergum III and a medial longitudinal black band which is narrow on tergum III and becoming wider as it continues to tergum V (in some specimens the medial longitudinal black band is interrupted at the intersegmental lines and over the anterior ½ of tergum III), narrow fuscous anterolateral corners on terga IV and V.
Description: male.
Head: Height 1.48 mm. Frons – length 1.3 times breadth; red-brown with fuscous between *i.or.* setae; orbital setae black: 1 *s.or.*, 3 *i.or.*; lunule red-brown. Ocellar triangle black. Vertex red-brown. Face fulvous with a pair of narrow elongate black spots reaching oral margin; length 0.64 mm. Genae fulvous, large fuscous subocular spot present; black seta present. Occiput red-brown, fulvous along eye margins; occipital row with a large number of short pale setae. Antennae with segments 1 and 2 red-brown, segment 3 red-brown with fuscous on outer surface; length of segments: 0.2 mm; 0.36 mm; 1.04 mm.
Thorax: Scutum red-brown without dark markings. Pleural areas red-brown with fuscous along anterior margin of mesopleural stripe. Yellow markings as follows: notopleura; mesopleural stripe equal in width to notopleuron dorsally, continuing to katepisternum as a large transverse spot, anterior margin straight; anatergite (posterior apex red-brown); anterior $^{3}/_{5}$ katatergite (remainder red-brown); two narrow parallel-sided lateral postsutural vittae beginning with a spot anterior to notopleural suture and ending at *ia.* seta; a short narrow medial longitudinal postsutural vitta. Postpronotal lobes pale fuscous. Postnotum red-brown. Scutellum yellow with a large apical dark fuscous to black spot. Setae: *sc.* 2 or 4; *prsc.* absent; *ia.* 1; *p.sa.* 1; *a.sa.* 1; *mpl.* 1; *npl.* 2; *scp.* 2.
Legs: Fore femora fuscous to dark fuscous tending fulvous basally, mid femora fulvous with fuscous to dark fuscous around apical $^{2}/_{5}$, hind femora fulvous with fuscous to dark fuscous around apical ⅓; all tibiae fuscous to dark fuscous; tarsi with basal segment fulvous and apical four segments red-brown; mid tibiae each with an apical black spur.
Wings: Length 7.5 mm; cell c pale fulvous, cell bc fulvous; microtrichia in outer corner of cell c only; remainder of wings colourless except fuscous cell sc, broad fuscous costal band confluent with R_{4+5} and slightly paler between R_{2+3} and R_{4+5} and ending in a large fuscous spot around apex of R_{4+5}, a broad fuscous anal streak; a dense aggregation of microtrichia around A_1+CuA_2; supernumerary lobe of medium development.
Abdomen: Oval; terga free; pecten present on tergum III. Tergum I and sterna I and II wider than long. Terga I and II red-brown with narrow fuscous anterolateral corners on both terga and narrow fulvous transverse band across posterior margin of tergum II; terga III–V red-brown with a black 'T' pattern consisting of a narrow transverse band across anterior margin of tergum III and a medial longitudinal vitta over all three terga which generally is narrow on tergum III and widens over terga IV and V (in some specimens the medial longitudinal black vitta is broken on the intersegmental lines and does not join the transverse band across anterior margin

of tergum III), narrow fuscous anterolateral corners on terga IV and V. A pair of oval red-brown shining spots on tergum V. Posterior lobe of surstylus long, sternum V with a slight concavity on posterior margin. Abdominal sterna mostly dark fuscous to black.

Attractant: No known record.

Female: As for male except no dense aggregation of microtrichia around A_1+CuA_2; supernumerary lobe weak; pecten of cilia on abdominal tergum III absent; basal segment of ovipositor red-brown, dorsoventrally compressed and tapering posteriorly in dorsal view; ratio of length of oviscape to length of tergum V, 2.8:1; apex of piercer trilobed.

Distribution:

Type locality: India, Tamil Nadu (Kodaikanal).

New records: India, Karnataka (Chethalli).

Hosts: The Holotype bears the label bred ex *Bryonia* sp. (family *Cucurbitaceae*) while the paratype in the BMNH bears the label bred ex *Trichosanthes primata* (family *Cucurbitaceae*).

Remarks: *Bactrocera (Zeugodacus) watersi* Hardy has variable numbers of *sc.* setae as recorded by Hardy (1954). Here it is provisionally placed in subgenus *Zeugodacus*, although those specimens with 2 *sc.* setae could indicate that the subgenus *Sinodacus* would be valid. Some specimens also have small humeral setae. Although the *prsc.* setae are absent, the presence of 4 *sc.* setae, *a.sa.* setae and a medial postsutural yellow vitta on the scutum are characters now recognized for the subgenus *Zeugodacus*.

Bactrocera (Zeugodacus) yalaensis Drew & Romig, new species

(Fig. 369)

Material examined: Holotype male, **THAILAND**: Thanto Waterfall, Yala, Songkhla, 29.xi.1991, coll. Anuchit *et al.*, bred from *Fagraea ceilanica*, deposited in BMNH.

Paratypes: 6 males and 3 females, same data as holotype. 1 paratype in ANIC; 1 paratype in BMNH; 6 paratypes in QDPC; 1 paratype in QMIC.

Diagnosis: A large species; face with a pair of small to medium circular black spots; postpronotal lobes and notopleura yellow; scutum black with large areas of red-brown; lateral and medial postsutural yellow vittae present; a yellow spot anterior to notopleural suture; mesopleural stripe equal in width to notopleuron dorsally; scutellum yellow; wing with cells bc and c colourless, microtrichia in outer corner of cell c only; a narrow fuscous costal band just overlapping R_{2+3} and expanding into a *tau*-like spot around apex of R_{4+5}; a broad fuscous anal streak; abdominal terga III–V fulvous with a 'T' pattern consisting of a narrow transverse black band across anterior margin of tergum III and a narrow- to medium-width medial longitudinal black band over all three terga, broad lateral longitudinal black bands over terga IV and V.

Description: male.

Head: Height 1.52 mm. Frons – length 1.27 times breadth; fulvous with pale fuscous on anteromedial hump; orbital setae black: 1 *s.or.*, 3 *i.or.*; lunule fuscous. Ocellar triangle black. Vertex dark fuscous to black. Face fulvous with a pair of small- to medium-sized circular black spots; length 0.51 mm. Genae fulvous, fuscous subocular spot present; black seta present. Occiput red-brown, fulvous along eye margins; occipital row with 4–5 dark setae. Antennae with segments 1 and 2 red-brown, segment 3 mostly fuscous; arista black (red-brown basally); length of segments: 0.13 mm; 0.26 mm; 0.67 mm.

Thorax: Scutum black with large areas of red-brown as follows – below and behind lateral postsutural vittae, around notopleural suture, across posterior margin and surrounding posterior end of medial postsutural yellow vitta, narrow lateral margins along medial postsutural vitta and expanding anteriorly to cover area between postpronotal lobes (there is a narrow medial longitudinal black line from anterior point of medial postsutural vitta to anterior margin of scutum). Pleural areas dark fuscous to black except red-brown below postpronotal lobes. Yellow markings as follows: postpronotal lobes; notopleura; mesopleural stripe equal in width to notopleuron dorsally, continuing to katepisternum as a large transverse spot, anterior margin slightly convex; anatergite (posterior apex black); anterior ¾ katatergite (remainder black); two moderately broad parallel-sided lateral postsutural vittae beginning with a spot anterior to notopleural suture and ending well behind *ia.* seta; a broad medial longitudinal postsutural vitta beginning at level of *prsc.* setae and narrowing to a point anteriorly to end anterior to notopleural suture. Postnotum black laterally, dark red-brown medially. Scutellum yellow except for narrow black basal band. Setae: *sc.* 4; *prsc.* 2; *ia.* 1; *p.sa.* 1; *a.sa.* 1; *mpl.* 1; *npl.* 2; *scp.* 2.

Legs: Femora fulvous except distinct dark fuscous subapical spots on outer surfaces of fore and mid femora and inner surfaces of hind femora; fore tibiae dark fuscous, mid tibiae fulvous tending fuscous basally, hind tibiae dark fuscous to black; tarsi with basal segments fulvous and apical four segments pale fuscous; mid tibiae each with an apical black spur.

Wings: Length 6.7 mm; cells bc and c colourless; microtrichia in outer corner of cell c only; remainder of wings colourless except fuscous cell sc, narrow fuscous costal band overlapping R_{2+3} and ending in a *tau*-like spot around apex of R_{4+5}, a broad fuscous anal streak; a dense aggregation of microtrichia around A_1+CuA_2; supernumerary lobe weak.

Abdomen: Oval; terga free; pecten present on tergum III. Tergum I and sterna I and II wider than long. Tergum I fulvous with dark fuscous across anterior margin and along lateral margins; tergum II fulvous with a narrow transverse black band across anterior margin but not reaching the narrow black lateral margins; terga III–V fulvous with a 'T' pattern consisting of a narrow transverse black band across anterior margin of tergum III and a narrow- to medium-width medial longitudinal black band over all three terga, a broad lateral longitudinal black band over terga IV and V. A pair of oval fulvous shining spots on tergum V. Posterior lobe of surstylus long, sternum V with a slight concavity on posterior margin. Abdominal sterna dark fuscous to black.

Attractant: No known record.

Female: As for male except no dense aggregation of microtrichia around A_1+CuA_2; pecten of cilia on abdominal tergum

III absent; basal segment of ovipositor red-brown, dorsoventrally flattened and tapering posteriorly in dorsal view; ratio of length of oviscape to length of tergum V, 1.7:1; apex of piercer needle-shaped.
Distribution: Thailand (Songkhla, Yala).
Hosts: Bred from one host *Fagraea ceilanica* (family *Loganiaceae*).
Remarks: *Bactrocera (Zeugodacus) yalaensis* n.sp., within the *tau* complex, is similar to *Bactrocera (Zeugodacus) macrophyllae* n.sp. and *Bactrocera (Zeugodacus) khaoyaiae* n.sp. in possessing a yellow scutellum, abdominal terga III–V with a black 'T' pattern and broad lateral longitudinal dark bands on terga IV and V, broad parallel-sided lateral postsutural vittae and femora not entirely fulvous. It differs from *B. macrophyllae* in having dark spots on the apices of all femora and from *B. khaoyaiae* in having smaller facial spots, supernumerary lobe of male wing weak and fore femora with a smaller subapical dark spot.

Bactrocera (Zeugodacus) yoshimotoi (Hardy)

(Fig. 370)

Dacus (Strumeta) yoshimotoi Hardy, 1973: 53. Holotype male in BBM.
Bactrocera (Bactrocera) yoshimotoi – Hardy, 1977: 52; Norrbom *et al*., 1998: 96.

Type material examined: Holotype male, **VIETNAM**: with the following three labels – 'Viet Nam, 22 km S of Nha Trang, 20–26.xi.1960'; 'C.M. Yoshimoto, Collector'; 'HOLOTYPE male, *Dacus (Strumeta) yoshimotoi*, D.E. Hardy'. Received on loan from BBM.
Non-type material examined: BHUTAN: 5 males, Chirang, Lamidara, 4-7.ix.1990, coll. H.R. Feijen. 12 males (10.v.2002), 11 males (17.v.2002), 7 males (24.v.2002), Rimchu, coll. C. Dorji, attracted to cue lure; 2 males (10.v.2000), 29 males (17.v.2000), 24 males (24.v.2000), 3 males (31.v.2000), 11 males (7.vi.2000), 18 males (28.vi.2000), 13 males (5.vii.2000), 12 males (19.vii.2000), 27 males (23.vii.2000), 4 males (26.vii.2000), 4 males (2.viii.2000), 12 males (10.viii.2000), 11 males (16.viii.2000), 33 males (30.viii.2000), 50 males (6.ix.2000), 37 males (12.ix.2000), 47 males (20.ix.2000), 17 males (27.ix.2000), 1 male (5.x.2000), Rimchu trap No. 1, coll. C. Dorji, attracted to cue lure; 25 males (4.v.2000), 26 males (10.v.2000), Rimchu trap No. 1, coll. B.S. Fletcher, attracted to cue lure; 22 males, Rimchu trap No. 1, 21.vi.2000, coll. Nado *et al*., attracted to cue lure; 4 males, Rimchu trap No. 1, 14.vi.2000, coll. Nado, Tshomo, attracted to cue lure; 3 males, Rimchu trap No. 2, 4.v.2002, coll. B.S. Fletcher, attracted to cue lure; 1 male, Rimchu trap No. 2, 28.vi.2000, coll. C. Dorji, attracted to cue lure; 1 male (10.v.2000), 1 male (17.v.2000), Lumitshawa, coll. C. Dorji, attracted to cue lure; 3 males (3.vii.2000), 4 males (21.viii.2000), Tsirang, Suntalay, attracted to cue lure; 1 male, Tsirang, 7.viii.2000, attracted to cue lure; 1 male, Phuentsholing, 6.v.2000, coll. Thapa, attracted to cue lure. **NEPAL**: 3 males, 'Kathmandu, PA PB, 4.v.1988, coll. Noor Kumar, ENTO. Division'; 'Dip 123'. **THAILAND**: 1 male (28.v.1992), 1 male (6.viii.1992), Khun Khon, Chiang Rai, coll. Charuwan, attracted to cue lure. **VIETNAM: Northern Vietnam** – 3 males (30.iv.2002), 2 males (15.v.2002), Son La (P), Moc Chau (D), Long Luong (V), trap No. VN01, coll. NIPP staff, attracted to cue lure; 1 male (30.vi.2002), 2 males (10.vii.2000), 2 males (24.vii.2000), 3 males (30.vii.2002), 1 male (7.viii.2000), Son La Province, Moc Chau District, Co Do, trap No. N006, attracted to cue lure; 2 males, Son La Province, Moc Chau District, Tra Long, 26.vi.2000, trap No. N007, attracted to cue lure; 1 male, Sonla Province, Moc Chau town, 12.iv.1999, trap No. N009 (urban), coll. Le Duc Khanh *et al*., attracted to cue lure; 1 male, Son La Province, Moc Chau District, Moc Chau town, 21.viii.2000, trap No. N009, attracted to cue lure. Specimens in QDPC.
Diagnosis: A small- to medium-sized species; face fulvous with a pair of small to large transverse oval black spots; postpronotal lobes and notopleura yellow; two narrow lateral postsutural yellow vittae; a moderately broad medial postsutural yellow vitta; mesopleural stripe slightly wider than notopleuron dorsally; scutellum yellow; wing with cells bc and c colourless, microtrichia in outer corner of cell c only; a narrow fuscous costal band confluent with R_{2+3} and remaining narrow beyond apex of this vein before widening across apex of R_{4+5}, a broad fuscous anal streak; abdominal terga III and IV red-brown with a general black 'T' pattern on each tergum, tergum V red-brown with a broad medial longitudinal black band and moderately broad black lateral margins.
Description: male.
Head: Height 1.1 mm. Frons – length 1.17 times breadth; fulvous with dark fuscous around orbital setae and on anteromedial hump; latter with a small number of short pale hairs; orbital setae black: 1 *s.or.*, 3 *i.or.* (one of these pair is reduced); lunule fuscous. Ocellar triangle black. Vertex black. Face fulvous with a pair of small to large transverse oval black spots but not meeting in midline; length 0.67 mm. Genae fulvous, dark fuscous subocular spot present. Occiput dark fuscous, fulvous along eye margins; occipital with a large number of short pale setae. Antennae with segments 1 and 2 red-brown, segment 3 red-brown with pale fuscous on apex and outer surface; length of segments: 0.14 mm; 0.22 mm; 0.58 mm.
Thorax: Scutum black except dark fuscous behind lateral postsutural vittae and red-brown inside notopleura. Pleural areas black except dark red-brown below postpronotal lobes. Yellow markings as follows: postpronotal lobes; notopleura; mesopleural stripe slightly wider than notopleuron dorsally, continuing to katepisternum as a small transverse spot, anterior margin strongly convex; anatergite (posterior apex dark fuscous); anterior $^3/_5$ katatergite (remainder dark fuscous); two narrow lateral postsutural vittae beginning with a spot anterior to notopleural suture and narrowing posteriorly to end before *ia.* seta (over the large number of specimens studied, these vittae may end just before, at or just behind *ia.* seta); a moderately broad medial longitudinal postsutural vitta, broadly rounded posteriorly and ending in a point just

posterior to notopleural suture. Postnotum black. Scutellum yellow except for a narrow black basal band. Setae: *sc.* 2; *prsc.* 2; *ia.* 1; *p.sa.* 1; *a.sa.* 1; *mpl.* 1; *npl.* 2; *scp.* 4.

Legs: All femora entirely red-brown; fore and mid tibiae fuscous basally tending pale fuscous apically, hind tibiae entirely dark fuscous. All tarsal segments entirely fulvous; mid tibiae each with an apical black spur.

Wings: Length 5.0 mm; cells bc and c colourless; microtrichia in outer corner of cell c only; remainder of wings colourless except dark fuscous cell sc, narrow fuscous costal band confluent with R_{2+3}, continuing narrow beyond apex of R_{2+3} and then widening across apex of R_{4+5}, broad fuscous anal streak; dense aggregation of microtrichia around A_1+CuA_2; supernumerary lobe strong and rounded.

Abdomen: Oval; terga free; pecten present on tergum III. Tergum I and sterna I and II wider than long. Tergum I fuscous with narrow black lateral margins; tergum II red-brown except for an irregularly shaped narrow transverse black band across anterior margin but not reaching lateral margins and narrow black lateral margins; tergum III red-brown with a 'T' pattern consisting of a narrow transverse black band across anterior margin turning to cover lateral margins and a medium-width medial longitudinal black band; tergum IV with a moderately broad medial longitudinal black band and very broad lateral black margins which meet the medial longitudinal black band at the anterocentral point of the tergum (the remainder of this tergum is red-brown); tergum V red-brown except for a moderately broad medial longitudinal black band and moderately broad lateral black margins. Over the large number of specimens studied, the pattern on terga III–V varies from (a) a black 'T' pattern over all three terga and with broad black anterolateral corners on terga IV and V to (b) a black 'T' pattern on terga III and IV and with a black medial longitudinal vitta and black anterolateral corners on tergum V to (c) that described above for the Holotype. A pair of red-brown oval shining spots on tergum V. Posterior lobe of surstylus long, sternum V with a slight concavity on posterior margin. Abdominal sterna dark fuscous, except sterna II and III fulvous.

Attractant: Cue lure.

Female: No known record.

Distribution:

Type locality: Southern Vietnam (Khanh Hoa Province, Nha Trang).

New records: Bhutan; Nepal; Northern Vietnam (Son La Province); Thailand.

Hosts: No known record.

Remarks: *Bactrocera (Zeugodacus) yoshimotoi* (Hardy), within the *scutellaris* complex, is similar to *Bactrocera (Zeugodacus) hoabinhiae* n.sp., *Bactrocera (Zeugodacus) laocaiae* n.sp. and *Bactrocera (Hemigymnodacus) mukiae* n.sp. in possessing two *sc.* setae, yellow notopleura and face with either a band across lower margin or a pair of black spots. It differs from *B. hoabinhiae* in having a pair of dark facial spots, from *B. mukiae* in having a strong supernumerary lobe in the male wing and *a.sa.* setae present and from *B. laocaiae* in having the femora entirely fulvous, large transverse oval dark facial spots and abdominal terga III–V with a black 'T' pattern and broad lateral longitudinal black bands over terga IV and V.

Bactrocera (Zeugodacus) zahadi Mahmood

(Fig. 371)

Bactrocera (Zeugodacus) zahadi Mahmood, 1999: 232; Drew & Raghu, 2002: 350. Holotype male in BMNH.

Type material examined: Holotype male, **INDIA**: with the following four labels – 'India (KA) Mudigere, 980 m, UASB Res. Sta., 20–22.v.1992', 'I.M. White, D.L. Hancock & S. Ramani, IIE 22353', 'Type', 'Holotype, *Bactrocera (Z.) zahadi*, K. Mahmood, 1999'. Studied at BMNH.

Non-type material examined: BHUTAN: 1 male, Tsirang, Damphu orchard, 20.iv.2000, attracted to cue lure; 1 male, Rimchu 1, 5.x.2000, coll. C. Dorji, attracted to cue lure. **INDIA**: 2 females, 'South India, Mysore State, T.R.S. Nathan'; 'Coorg Dist., Mercara, 4000 ft., v.1973'. Received on loan from MHNG. 1 female, 10 males, Mudigere, 980m, 20.v.1992, coll. D.L. Hancock; 1 female, 6 males, Mudigere, 980m, 22.v.1992, coll. D.L. Hancock; 2 females, 4 males, Mudigere, 980m, 21.v.1992, coll. D.L. Hancock; 1 male, Talakaveri, Nr. Madikeri, 1100m, 23.v.1992, coll. D.L. Hancock, attracted to cue lure; 3 males, Kemmannugundi, Nr. Tari Kere, 1400m, 21.v.1992, coll. D.L. Hancock, attracted to cue lure; 2 males, Hort. Farm, Coonoor, 1450m, 29.v.1992, coll. D.L. Hancock, attracted to cue lure; 3 males, Hort. Farm, 1450m, Coonoor, 30.v.1992, coll. D.L. Hancock, attracted to cue lure; 4 males, Pothiyl, Alwaye, 11.viii.1988, coll. S. Vijay, attracted to cue lure. **SRI LANKA (formerly CEYLON)**: 1 female, Mus. Westerm. Specimens in QDPC except where specified.

Diagnosis: A large species; face fulvous with a pair of large irregularly oval black spots; postpronotal lobes and notopleura yellow; scutum black with areas of red-brown; lateral and medial postsutural yellow vittae present; small yellow spot anterior to notopleural suture; mesopleural stripe just slightly wider than notopleuron dorsally; scutellum entirely yellow; legs with femora fulvous except for subapical dark fuscous spots on fore and mid femora and dark fuscous around apical ¼ of hind femora; wing with cells bc and c colourless, microtrichia in outer corner of cell c only; a narrow fuscous costal band expanding into a *tau*-like spot in apex of wing; a broad fuscous anal streak; abdominal terga III–V fulvous with a 'T' pattern consisting of a narrow transverse black band across anterior margin of tergum III and a moderately broad medial longitudinal black band over all three terga, large anterolateral black corners on terga IV and V.

Description: male.

Head: Frons – fulvous without dark markings; orbital setae black: 1 *s.or.*, 3 *i.or.*; lunule dark fuscous. Ocellar triangle black. Vertex dark fuscous. Face fulvous with a pair of large irregularly oval dark fuscous to black spots. Occiput red-brown, dark fuscous dorsally and laterally, fulvous along eye margins. Antennae with segments 1 and 2 red-brown, segment 3 red-brown with fuscous on apex and outer surface.

Thorax: Scutum black except red-brown below and behind lateral postsutural vittae, across posterior margin surrounding *prsc.* setae, around notopleural suture and yellow spot anterior to notopleural suture, as a large medial longitudinal band from anterior point of medial postsutural yellow vitta to anterior margin of scutum and including the entire area between the postpronotal lobes (there is a medial longitudinal black line within this latter area). Pleural areas dark fuscous to black except red-brown below postpronotal lobes. Yellow markings as follows: postpronotal lobes; notopleura; mesopleural stripe just slightly wider than notopleuron dorsally, continuing to katepisternum as a transverse spot, anterior margin slightly convex; two parallel-sided lateral postsutural vittae of medium-width and beginning with a small spot anterior to notopleural suture and ending just behind *ia*. seta; a narrow medial longitudinal postsutural vitta. Scutellum entirely yellow except for narrow black basal band. Setae: *sc*. 4; *prsc*. 2; *ia*. 1; *p.sa*. 1; *a.sa*. 1; *mpl*. 1; *npl*. 2; *scp*. 4.

Legs: Fore and mid femora fulvous with small dark fuscous subapical spots on outer surfaces, hind femora fulvous with fuscous around apical ¼; fore and hind tibiae dark fuscous, mid tibiae fulvous except for fuscous basally; tarsi with basal segment fulvous and apical four segments fuscous; mid tibiae each with an apical black spur.

Wings: Length 6.7 mm; cells bc and c colourless; microtrichia in outer corner of cell c only; remainder of wings colourless except dark fuscous cell sc, narrow fuscous costal band confluent with R_{2+3} and widening into a *tau*-like spot in apex of wing, a broad fuscous anal streak; dense aggregation of microtrichia around A_1+CuA_2; supernumerary lobe strong and keel-shaped.

Abdomen: Oval; terga free; pecten present on tergum III. Tergum I fulvous with dark fuscous to black across anterior and along lateral margins; tergum II fulvous except for a narrow black transverse band across anterior margin but not reaching narrow lateral black margins; terga III–V pale orange-brown with a 'T' pattern consisting of a narrow black transverse band across anterior margin and including lateral margins and a medium-width medial longitudinal black band over all three terga, large anterolateral black corners on terga IV and V. Posterior lobe of surstylus long, sternum V with a slight concavity on posterior margin. A pair of pale orange-brown oval shining spots on tergum V. All abdominal sterna dark, except sternum II pale.

Attractant: Cue lure.

Female: As for male except no dense aggregation of microtrichia around A_1+CuA_2; supernumerary lobe weak; pecten of cilia on abdominal tergum III absent; basal segment of ovipositor red-brown, dorsoventrally compressed and tapering posteriorly in dorsal view; ratio of length of oviscape to tergum V, 1.3:1; apex of piercer needle-shaped.

Distribution:

Type locality: Southern India (Mudigere).

Previously recorded: Type locality and the following locations – Bhutan; Myanmar; Nepal; Pakistan; Southern India (Kerala); Sri Lanka; Vietnam (Mahmood, 1999). The records from Nepal and Vietnam must be regarded as doubtful as no specimens were listed under material examined (Mahmood, 1999: 232–233).

New records: Southern India (Karnataka – Madikeri, Mysore, Tari Kere; Tamil Nadu).

Hosts: No known record.

Remarks: *Bactrocera (Zeugodacus) zahadi* Mahmood belongs to the *tau* complex and is extremely close to *Bactrocera (Zeugodacus) tau* (Walker) and may be a synonym of the latter. After studying large numbers of *B. tau* from across South-East Asia, it is clear that the lateral dark colour patterns on abdominal terga IV and V vary from small dark anterolateral corners to broad lateral longitudinal dark bands over both terga. Also, the femora colour varies from entirely fulvous to fulvous with preapical dark spots on some or all femora. One sound characteristic of *B. tau* is that it possesses a large keel-shaped supernumerary lobe in the male wing which also is present in the holotype of *B. zahadi*. No good characters can be found to separate *B. tau* and *B. zahadi*.

GENUS *DACUS* Fabricius

SUBGENUS *CALLANTRA* Walker

Dacus (Callantra) longicornis Wiedemann

(Fig. 372)

Dacus longicornis Wiedemann, 1830: 524 – de Meijere, 1911: 380; Bezzi, 1909: 292. Holotype female in UZMC.
Bactrocera vespoides Doleschall, 1858: 123. Holotype male in MFN.
Callantra smieroides Walker, 1860a: 154 – Bezzi, 1916: 120 (as *smicroides*, lapsus); Hardy & Adachi, 1954: 151; Hardy, 1959: 162; Drew, 1973: 2; Hardy, 1977: 46; Drew *et al.*, 1998: 604 (Synonym). Holotype male in BMNH.
Dacus vespoides – Bezzi 1909: 262 (Synonym).
Mellesis destillatoria Bezzi, 1916: 118 – Drew *et al.*, 1998: 604 (Synonym). Holotype female in MCSNM.
Mellesis eumenoides Bezzi, 1916: 119 – Drew *et al.*, 1998: 604 (Synonym). Lectotype male in BMNH.
Mellesis bioculata Bezzi, 1919: 437 – Drew *et al.*, 1998: 604 (Synonym). Lectotype male in MCSNM.
Callantra vespoides – Enderlein, 1920: 358; Hardy & Adachi, 1954: 169; Hardy, 1977: 46 (Synonym).
Dacus (Callantra) smieroides – Malloch, 1939b: 411; Drew, 1989: 248; White & Elson-Harris, 1994: 317.
Callantra destillatoria – Hardy, 1973: 9; Hardy, 1977: 45; Drew *et al.*, 1998: 604 (Synonym).
Callantra eumenoides – Hardy, 1973: 11; Hardy, 1977: 45; Drew *et al.*, 1998: 604 (Synonym).
Callantra bioculata – Hardy, 1974: 6; Hardy, 1977: 45; Drew *et al.*, 1998: 604 (Synonym).
Callantra longicornis – Hardy, 1977: 45.
Callantra unifasciatus Hardy, 1982: 184 – Drew *et al.*, 1998: 604 (Synonym). Holotype male in MZB.
Callantra variegata Wang, 1990: 73, 76 – Drew *et al.*, 1998: 605 (Synonym). Holotype male in IZAS.
Dacus (Callantra) variegata – Liang *et al.*, 1993: 139.
Dacus (Callantra) eumenoides – Kapoor, 1993: 83.
Dacus (Callantra) longicornis – Drew *et al.*, 1998: 604; 2007: 13; Norrbom *et al.*, 1998: 130.

Type material examined: See Drew *et al.* (1998) for an overview of type material which we have studied and the revised locations where these types are deposited.
Non-type material examined: BHUTAN: 1 male, Ganglephug, Tatapom, 9-13.ix.1990, coll. H.R. Fiejen; 1 male (2.x.2000), 1 male (16.x.2000), Tsirang; 1 male, Gelephu, Purana Bastey, (16.xi.2004), coll. Maha Prasard & Karma Namgyel; 2 males, Purana Bastey, (2.xi.2004), coll. Karma Namgyel & Maha Prasard. All specimens attracted to cue lure. **VIETNAM: Southern Vietnam** – 1 male (16.xi.2002), 1 male (23.xi.2002), Vietnam South, Ben Tre Province, Chau Thanh District, Tien Thuy Village, Site VS04, coll. SOFRI staff, attracted to cue lure. Specimens in QDPC.
Diagnosis: A large species; face fulvous with a pair of small irregularly oval black spots; postpronotal lobes yellow except posterodorsal corners fuscous; notopleura yellow; scutum dark red-brown without distinct dark patterns; lateral and medial postsutural yellow vittae absent; a narrow yellow triangle running along anterior margin of notopleural suture with base on notopleuron; mesopleural stripe narrow equal in width to notopleuron dorsally; scutellum yellow except for broad red-brown basal band; setae: *sc.* 2; *prsc.* absent; *ia.* 1; *p.sa.* 1; *a.sa.* 1; *mpl.* 1; *npl.* 2; *scp.* 4; legs with fore femora dark red-brown to fuscous, mid femora dark red-brown to fuscous except fulvous on basal ¼, hind femora dark fuscous; fore and mid tibiae dark red-brown to fuscous, hind tibiae dark fuscous; fore tarsi with all segments dark red-brown, mid and hind tarsi with basal segment fulvous and apical four segments red-brown; wing with cells bc and c fuscous, dense microtrichia over all of cell c and most of cell bc; a broad dark fuscous costal band overlapping R_{4+5} for its entire length and sometimes becoming darker at apex; anal streak indistinct but a broad pale fuscous area generally over cell cup and across wing margin towards cell dm; supernumerary lobe weak; abdomen strongly petiolate, abdominal terga III–V generally dark fuscous to black with a paler band often across posterior margin of tergum III, large orange-brown spots posterocentrally on terga IV and V with the spot on tergum V often expanded anteriorly into a medial

longitudinal orange-brown band, oval shining spots on tergum V red-brown to dark fuscous.
Attractant: Cue lure.
Female: As for male except no dense aggregation of microtrichia around A_1+CuA_2; pecten of cilia on abdominal tergum III absent; ovipositor basal segment fuscous, elliptical in transverse cross section and tapering posteriorly in dorsal view; apex of piercer needle-shaped.
Distribution:
Type locality: Indonesia (Java).
Previously recorded: Widespread across the region from southern Asia to South-East Asia (see Drew *et al.*, 1998).
Hosts: Recorded from four species in family *Cucurbitaceae* (see Drew *et al.*, 1998).
Remarks: *Dacus (Callantra) longicornis* Wiedemann has a widespread distribution across southern Asia and South-East Asia, and has been regularly misidentified with numerous synonyms being described (see above). It has been adequately described and illustrated by Drew *et al.* (1998) and is best distinguished by the combination of the following characters – red-brown scutum, cells bc and c fuscous, costal band overlapping R_{4+5}, face fulvous with a pair of black spots, anatergite fuscous, katatergite yellow, postpronotal lobes mostly yellow, mesopleural stripe narrow (slightly wider than notopleuron dorsally). It has minor pest status.

SUBGENUS *DIDACUS* Collart

Dacus (Didacus) ciliatus Loew

(Fig. 373)

Dacus ciliatus Loew, 1862: 7. Lectotype female in NRS.
Dacus brevistylus Bezzi, 1908b: 149 – Bezzi, 1914b: 154; Bezzi, 1916: 101. Munro, 1932: 152 (Synonym). Syntypes male and female in MCSNM.
Dacus (Bactrocera) cocciniae Premlata & Singh, 1987: 401. Drew *et al.*, 1998 and Norrbom *et al.*, 1998: 134 (Synonym). Holotype male in PAU.
Dacus (Afrodacus) cocciniae – Agarwal *et al.*, 1992: 20.
Dacus (Didacus) ciliatus – Drew *et al.*, 1998: 646; Norrbom *et al.*, 1998: 134; White, 2006: 81, Lectotype designation. For additional synonymy see White (2006).

Non-type material examined: PAKISTAN: 3 males and 3 females, Larkana, 16.vi.1963, bred from maggots in fruit of *Momordica charantia*; 1 male and 3 females, Sukkur, 20.v.1965, bred from maggots in fruit of *Luffa cylindrica* (in QDPC); 1 male and 1 female, Quetta, Baluchistan, 2.viii.1931, coll. Y.R. Rao, from *Momordica* fruits (in BMNH); 2 males and 2 females, Rawalpindi, 27.ix.1961, maggot in fruit of *Luffa acutangula* (in BMNH). **INDIA**: 2 females, Coimbatore, 26.vii.1946, coll. P.S. Nathan, bred from snake-gourd; 1 female, Coimbatore, 25.v.1931, C.V.S., in cucumber; 1 female, Coimbatore, 1936, Cherian, in *Coccinia indica* fruits; 1 female, S. India, vi.1950, coll. T.C. Lawrence & G. Beevor, ex *Cephalandra indica*; 1 male, Hessarghatta, S. India, v.1950, coll. G. Beevor & T.C. Lawrence, ex cucurbit; 1 male, Bareilly, Uttar Pradesh, vi.1950, coll. N.D. Waters, ex *Cucumis melo* var. *utilissimus* (in BMNH). **SRI LANKA**: **(formerly CEYLON)** – 1 male, Kandakudiye, HR/30, 2.x.1996, pumpkin flower (in QDPC).
Diagnosis: A small species; face fulvous with a pair of small circular black spots; postpronotal lobes and notopleura yellow; scutum red-brown with erratic blotched patterns due to sites of attachment of dorsoventral flight muscles; lateral and medial postsutural yellow vittae absent; a broad triangular marking along anterior margin of notopleural suture and which surrounds the notopleuron; broad mesopleural stripe reaching almost to anterior *npl.* seta dorsally; scutellum yellow except for narrow red-brown basal band; setae: *sc.* 2; *prsc.* absent; *ia.* 1; *p.sa.* 1; *a.sa.* absent; *mpl.* 1; *npl.* 2 (anterior one very weak); *scp.* 3–4; legs with all segments entirely fulvous, mid tibiae each with an apical black spur; wing with cells bc and c colourless, microtrichia in outer corner of cell c only; remainder of wings colourless except fuscous cell sc, narrow fuscous costal band confluent with R_{2+3} and remaining narrow around costal margin until it expands into a small spot just beyond apex of R_{4+5}; narrow fuscous anal streak ending before wing margin; supernumerary lobe only slightly developed; abdomen oval, abdominal terga III–V red-brown with large dark fuscous to black spots on anterior margin of tergum III midway between midline and lateral margins and tending fuscous in anterolateral corners of tergum IV in some specimens.
Attractant: No known record.
Female: As for male except no dense aggregation of microtrichia around A_1+CuA_2; supernumerary lobe weak; pecten of cilia on abdominal tergum III absent. Ovipositor basal segment red-brown, dorsoventrally flattened and tapering posteriorly in dorsal view; oviscape length 0.25 times length of tergum V.
Distribution:
Type locality: Africa ('Guinea' and Cape, South Africa).
Previously recorded: Type locality and the following locations – Africa (Senegal to South Africa); Egypt; Mauritius; Reunion; Iran; Pakistan; Bangladesh; India; Middle East including Oman, Saudi Arabia, UAE. (Drew *et al.*, 1998; White, 2006).
New records: Sri Lanka (Kandakudiye).
Hosts: Predominantly in the family *Cucurbitaceae*; occasionally in the families *Solanaceae, Leguminosae* and *Malvaceae* (Munro, 1984). A major pest of economic crops.
Remarks: *Dacus (Didacus) ciliatus* Loew is similar to *Dacus (Dacus) hainanus* Wang & Zhao and *Dacus (Didacus) keiseri* (Hering) in having a narrow costal band confluent with vein R_{2+3} and cells bc and c colourless. It differs from both in having a fuscous anal streak on the wing, from *D. hainanus* in lacking a medial yellow band on the scutum and from *D. keiseri* in having the scutum and abdomen red-brown. Asian specimens frequently show an indication of yellow on the lower part of the anatergite.

Dacus (Didacus) hainanus Wang & Zhao

(Fig. 374)

Dacus (Didacus) hainanus Wang & Zhao, 1989: 216, 219 – Liang *et al.*, 1993: 139; Norrbom *et al.*, 1998: 135. Holotype male in IZAS.
Dacus (Dacus) hainanus – Drew *et al.*, 1998: 641.

Type material examined: Holotype male, **CHINA**: Hainan I., Guangdong Province, 30.iii.1960, coll. Li Chang-qing, studied at IZAS by D.L. Hancock.

Non-type material examined: THAILAND: 1 female, nr Sai Yok, 500m, Kanchanaburi, 11.ix.1975, coll. H. Kurahashi (in BBM).

Diagnosis: A medium-sized species; face fulvous with a pair of large circular black facial spots; postpronotal lobes and notopleura yellow; scutum fuscous with irregular red-brown blotches representing sites of attachment of dorsoventral flight muscles (there are a medial and two lateral longitudinal narrow fuscous lines running full length of scutum); lateral postsutural yellow vittae absent; medial postsutural yellow vitta as a transverse band across posterior area of scutum; broad triangular marking along anterior margin of notopleural suture; broad mesopleural stripe reaching to anterior *npl.* seta dorsally, not extending onto katepisternum below; scutellum yellow except for a narrow dark basal band; setae: *sc.* 2; *prsc.* absent; *ia.* 1; *p.sa.* 1; *a.sa.* absent; *mpl.* 1; *npl.* 2; *scp.* 2; legs with all segments entirely fulvous, mid tibiae each with an apical black spur; wing with cells bc and c colourless, microtrichia in outer corner of cell c only; remainder of wings colourless except fuscous cell sc, narrow fuscous costal band confluent with R_{2+3} and expanding slightly across apex of R_{4+5}; anal streak absent; supernumerary lobe weak; abdomen oval, abdominal tergum III dark fuscous, terga IV and V red-brown except for anterolateral corners of both terga fuscous to dark fuscous and a narrow medial longitudinal fuscous line over tergum IV and onto anterior portion of tergum V. Ovipositor basal segment red-brown dorsoventrally compressed and tapering posteriorly in dorsal view, oviscape length 0.83 times length of tergum V. Apex of piercer needle-shaped.

Attractant: No known record.

Male: Not available for study.

Distribution:

Type locality: China (Hainan I.).

Previously recorded: Type locality and Thailand (Kanchanaburi) (Drew *et al.*, 1998).

Hosts: No known record.

Remarks: *Dacus (Didacus) hainanus* Wang & Zhao is similar to *Dacus (Didacus) keiseri* (Hering) and *Dacus (Didacus) ciliatus* Loew in having the costal band narrow and confluent with vein R_{2+3} and wing cells bc and c colourless. It differs in possessing a transverse yellow band across the posterior region of the scutum.

Dacus (Didacus) keiseri (Hering)

(Fig. 375)

Daculus keiseri Hering, 1956: 66. Holotype male in NMB.
Dacus (Polistomimetes) keiseri – Hardy, 1977: 56.
Bactrocera keiseri – White & Wang, 1992: 278.
Dacus (Didacus) keiseri – Drew *et al.*, 1998: 648; Norrbom *et al.*, 1998: 135.

Type material examined: Holotype male, **SRI LANKA: (formerly CEYLON)** – Hingurakgoda, N.C.P., 20.xii.[19]53, coll. F. Keiser (in NMB).

Non-type material examined: SRI LANKA: 1 male, Tricomali, 3.ix.1891, coll. Colonel Yerbury (in BMNH); 1 female, Put[talam] District, Kala-Oya, malaise trap, 1.viii.1975, coll. Y-M. Huang, E.L. Peyton, S. Kurunaratne, Mahinda (in USNM); 1 female, Ham[bantota] District, Palatupana, 3-6.ii.1975, coll. K.V. Krombein, P.B. Karunaratne, P. Fernando, E.G. Dabrera (in USNM); 4 females, Kiragolama, 20.iv.1995, ex Pasagilla (in BMNH); 1 male, Kalakaradawe, 19.ii.1996, ex Pasagilla, (in BMNH); 1 male, Sri Lanka, Kiralogama, 5.iii.1996, bred ex Diplocyclos palmatus; 1 male, Sri Lanka, Kalakaradawa, 8.iii.1996, bred ex *Diplocyclos palmatus* (in QDPC).

Diagnosis: A small species; face fulvous with a pair of small circular black spots; postpronotal lobes and notopleura yellow; scutum black with red-brown markings; lateral and medial postsutural yellow vittae absent; yellow triangular marking along anterior margin of notopleural suture; mesopleural stripe broad and reaching to postpronotal lobes dorsally; scutellum yellow; setae: *sc.* 2; *prsc.* absent; *ia.* 1; *p.sa.* 1; *a.sa.* absent; *mpl.* 1; *npl.* 2; *scp.* 4; legs fulvous; wing with cells bc and c colourless; a very narrow costal band broken in cell R_{2+3}, confluent with R_{2+3} and widening across apex of R_{4+5}; anal streak absent; supernumerary lobe weak; abdomen with terga mostly black except red-brown posterolaterally on tergum II and orange-brown medially from posterior margin of tergum III to tergum V, except for a narrow black medial band, pecten present on tergum III of males.

Attractant: No known record.

Female: No known record.

Distribution:

Type locality: Sri Lanka (Hingurakgoda).

Previously recorded: Type locality and other locations in Sri Lanka (Drew *et al.*, 1998).

Hosts: *Diplocyclos palmatus*, *Melothria maderaspatana* (family *Cucurbitaceae*) (Tsuruta *et al.*, 1997).

Remarks: *Dacus (Didacus) keiseri* (Hering) is similar to *Dacus (Didacus) hainanus* Wang & Zhao and *Dacus (Didacus) ciliatus* Loew in possessing a narrow costal band confluent with vein R_{2+3} and wing cells bc and c colourless. It differs from *D. hainanus* and resembles *D. ciliatus* in lacking a medial yellow vitta on the scutum. It differs from *D. ciliatus* in having a predominantly black scutum and black patterns on the abdomen, and in lacking a fuscous anal streak on the wing. In many respects it resembles the African taxon *Karphodacus* Munro, currently placed as a synonym of subgenus *Didacus*.

SUBGENUS *LEPTOXYDA* Macquart

Dacus (Leptoxyda) persicus Hendel

(Fig. 376)

Dacus (Leptoxyda) persicus Hendel, 1927a: 29. Lectotype female in BMNH; Drew *et al.*, 1998: 650; Norrbom *et al.*, 1998: 137.
Dacus ferrugineus – Becker, 1913: 641 [Misidentification].
Leptoxyda sp. nr *longistyla* Bezzi, 1913: 92.

Dacus longistylus Wiedemann, 1830 (Asian records) [misidentifications by various authors].
Leptoxyda persicus – Munro, 1984: 127 (Synonym).

Type material examined: Lectotype female and 1 male Paralectotype, **PAKISTAN**: P. Beluchistan, 25.ii.1901 (in BMNH).
Non-type material examined: PAKISTAN: 2 males and 4 females, Karachi, CIBC, FF22, ix.1963, maggot in fruit of *Calotropis procera* (in QDPC). **INDIA**: 2 males and 6 females, Karnataka, Bangalore, 916m, 9.x.1987, coll. G. Bhat, bred from *Calotropis gigantea* (in QDPC); 3 males and 6 females, Deesa, NW India, iii, iv and vii.1901, coll. C.G. Nurse, BM 1934-8 (in BMNH); 1 male and 2 females, Coimbatore, 8-22.xi.1945, coll P.S. Nathan (in BMNH); 2 males, Poona Ghauts, Bombay Presidency, 1888 (in BMNH); 1 female, Lyallpur, Punjab, 4.vii.1931 (in BMNH); 1 male, Hyderabad, 5.xi.1971, coll. A.C. Pont (in BMNH). **SRI LANKA**: 8 males and 2 females, Tricomali, 8.vii.1890, 1-2.x.1890, 11.xi.1891, 8.xii.1891, coll. Colonel Yerbury (in BMNH); 1 male and 1 female, Tricomali, 14.vi.1907, coll. Bainbrigge-Fletcher (in BMNH); 1 female, Suduganga, 7.ii.1922, coll. R. Senior-White, on *Calotropis procera* (in BMNH).
Diagnosis: A medium-sized species; face red-brown centrally, fulvous in antennal furrows and with a pair of small circular black spots; postpronotal lobes and notopleura yellow; scutum dull black with a fine silvery pubescence and with red-brown posterolaterally, along anterior margin and as a narrow medial longitudinal line from anterior margin to medial yellow vitta; two very narrow lateral longitudinal postsutural yellow vittae ending at or just before *ia.* seta; a moderately broad medial longitudinal postsutural yellow vitta; a triangular yellow marking along anterior margin of notopleural suture; mesopleural stripe of medium-width reaching midway between anterior margin of notopleuron and anterior *npl.* seta dorsally; scutellum yellow except for narrow black basal band; setae: *sc.* 2; *prsc.* absent; *ia.* 1; *p.sa.* 1; *a.sa.* absent; *mpl.* 1; *npl.* 2; *scp.* 2; legs with fore femora pale fuscous with dark fuscous on ventral surfaces, mid femora dark fuscous on apical ½ to ¾ and entire ventral surface with fulvous basally, hind femora dark fuscous on apical half and fulvous basally; tibiae pale fuscous; tarsi fulvous; mid tibiae each with an apical black spur; wing with cells bc and c colourless, microtrichia in outer corner of cell c only; remainder of wing colourless except pale fuscous cell sc, pale fuscous costal band confluent with R_{4+5} and with a slightly darker fuscous spot at apex, cell cup with a pale fuscous tint in base; supernumerary lobe weak; abdomen elongate oval, abdominal terga III and IV dark fuscous except for the posterolateral margins narrowly fulvous, tergum V fulvous except for the anterolateral corners dark fuscous and with a pair of large orange-brown shining spots, ovipositor basal segment elongate and narrow, conical in transverse section, orange-brown, oviscape length 1.75 times length of tergum V, apex of piercer broadly needle-shaped.
Attractant: No known record.
Male: As for female except abdominal sternum V with a moderately deep concavity on posterior margin and posterior lobe of surstylus moderately long and narrow.

Distribution:
Type locality: Hendel listed Persisch Belutschistan which referred to Belutschistan within Persia (Iran). The type locality is probably Baluchistan now within Pakistan.
Previously recorded: Type locality and the following locations – Pakistan; India; Sri Lanka; Middle East including Oman; UAE (Drew *et al.*, 1998; White, 2006).
Hosts: *Calotropis gigantea*, *Calotropis procera* (family *Asclepiadaceae*).
Remarks: *Dacus (Leptoxyda) persicus* Hendel is a distinct species with an extremely long ovipositor basal segment (oviscape) and is remotely similar to *Dacus (Mellesis) esakii* (Shiraki) (Part), *Dacus (Mellesis) icariiformis* (Enderlein) and *Dacus (Mellesis) vittatus* (Hardy) in having a broad costal band, medial and lateral postsutural vittae. It differs from these species in having short antennae, a pair of black facial spots, a black scutum and in the absence of *a.sa.* setae and spines on the fore femora.

Munro (1984) synonymised *D. (Leptoxyda) persicus* Hendel with *D. longistylus* Wiedemann but this does not appear to be justified. *Dacus persicus* is a smaller species with a shorter oviscape and lacks the brown apical spot on the costal band seen on the wing of *D. longistylus*. The latter species is restricted to Africa and Arabia; all Asian records belong to *D. persicus*.

SUBGENUS *MELLESIS* Bezzi

Dacus (Mellesis) ambonensis Drew & Hancock

(Fig. 377)

Dacus (Callantra) ambonensis Drew & Hancock, in Drew *et al.*, 1998: 572. Holotype male in BMNH.

Type material examined: Holotype male, **INDONESIA: MOLUCCAS (MALUKU)** – Waai (<20m), Ambon, 6.vii.1981, coll. J. Turner, attracted to cue lure (in BMNH). Paratypes: **INDONESIA: MOLUCCAS (MALUKU)** – 2 males, same data as holotype; 1 male, Paso (20-50m), Ambon, 7.vii.1981, coll. J. Turner; 1 male, Hila/Mamua (25-100m), Ambon, 5.viii.1981, coll. J. Turner. All specimens attracted to cue lure (in QDPC).
Diagnosis: A large species; face red-brown with a pair of medium-sized pear-shaped black spots; postpronotal lobes pale brown; notopleura yellow; scutum dull black except with dark red-brown on posterolateral corners and narrow lateral margins, inside postpronotal lobes and between postpronotal lobes and notopleura; lateral and medial postsutural yellow vittae absent; narrow triangular yellow markings along anterior margins of notopleural suture; narrow mesopleural stripe equal in width to notopleura dorsally; scutellum yellow except for a medium-width red-brown basal band; setae: *sc.* 2; *prsc.* absent; *ia.* 1; *p.sa.* 1; *a.sa.* 1; *mpl.* 1; *npl.* 2; *scp.* 4; legs with all femora and tibiae red-brown to pale fuscous except hind femora fulvous on basal third; tarsi fulvous; mid tibiae with an apical black spur; wing with cells bc and c fuscous, dense microtrichia covering all of cell c and most

of cell bc; remainder of wing colourless except fuscous cell sc, broad fuscous costal band just overlapping R_{4+5} and expanding apically to reach midway between R_{4+5} and M; broad pale fuscous anal streak ending before wing margin; supernumerary lobe weak in males (no females studied); abdomen elongate oval and petiolate, abdominal terga III–V mostly dark fuscous except for a slightly darker narrow medial longitudinal vitta from posterior region of tergum III to end just before posterior margin of tergum V and large orange-brown to pale red-brown shining spots on tergum V.
Attractant: Cue lure.
Female: No known record.
Distribution:
Type locality: Indonesia (Moluccas = Maluku – Ambon I., Waai).
Previously recorded: Type locality and other locations in Indonesia – Ambon I. (Drew *et al.*, 1998).
Hosts: No known record.
Remarks: *Dacus (Mellesis) ambonensis* Drew & Hancock is similar to *Dacus (Mellesis) nanggalae* Drew & Hancock in possessing a broad costal band overlapping vein R_{4+5}, abdominal terga III–V with a narrow medial longitudinal dark band and no lateral dark markings. It differs from *D. nanggalae* in having the anatergite and katatergite mostly red-brown, the costal band pale coloured and indistinct below vein R_{4+5} and a distinct broad anal streak.

Dacus (Mellesis) atrimarginatus White
(Fig. 378)

Dacus (Callantra) atrimarginatus White, in Drew *et al.*, 1998: 576. Holotype male in BBM.
Material examined: Holotype male, **THAILAND**: near Sai Yok, 500 m, Kanchanaburi, 9.ix.1975, coll. R. Kano (in BBM).
Diagnosis: A small species; face fulvous with a large black spot in each antennal furrow and joined by a transverse band in groove just above oral margin; postpronotal lobes and notopleura yellow; scutum black except for a narrow red-brown area inside postpronotal lobe and a pale medial red-brown line just anterior to scutellum (this may be a very reduced discoloured medial yellow vitta); lateral postsutural yellow vittae absent; medial postsutural yellow vitta apparently absent; triangular marking along anterior side of notopleural suture; mesopleural stripe of medium-width, reaching to posterior notopleural seta dorsally; scutellum yellow; setae: *sc.* 2; *prsc.* absent; *ia.* 1; *p.sa.* 1; *a.sa.* 1; *mpl.* 1; *npl.* 2; *scp.* 4 and very long; legs with femora red-brown except basal ⅔ hind femora fulvous; tibiae and tarsi pale fulvous; mid tibiae each with an apical spur; hind tibiae without a preapical bulge; wing with cells bc and c hyaline, cell c with microtrichia along entire anterior margin; remainder of wing hyaline except for fuscous narrow part of cell br, cell sc, broad costal band confluent with R_{4+5} and expanded into a shallow spot from just before apex of R_{2+3}, to overlap R_{4+5} at apex, no anal streak, supernumerary lobe weak in male (female not studied); abdomen elongate, clavate and petiolate, abdominal tergum III fuscous, terga IV and V fuscous basally and laterally and along midline, with large fulvous areas each side of midline, a pair of fulvous shining spots on tergum V.
Attractant: No known record.
Female: No known record.
Distribution:
Type locality: Thailand (Kanchanaburi).
Hosts: No known record.
Remarks: *Dacus (Mellesis) atrimarginatus* White is similar to *Dacus (Mellesis) crabroniformis* (Bezzi), *Dacus (Mellesis) maculipterus* White, *Dacus (Mellesis) ramanii* Drew & Hancock, *Dacus (Mellesis) satanas* (Hering) and *Dacus (Mellesis) sinensis* (Wang) in having the costal band confluent with or overlapping vein R_{4+5} and not expanded into a distinct spot at apex of wing, cells bc and c colourless and lacking lateral and medial postsutural yellow vittae. It differs from *D. crabroniformis*, *D. ramanii*, *D. satanas* and *D. sinensis* in having abdominal terga IV and V with pale areas posteriorly either side of a medial longitudinal dark band, and from *D. maculipterus* in having pale-coloured legs.

Dacus (Mellesis) bannatus (Wang)
(Fig. 379)

Callantra bannata Wang, 1990: 69, 75. Holotype male in IZAS.
Dacus (Callantra) bannata – Liang *et al.*, 1993: 139.
Dacus (Callantra) bannatus – Drew *et al.*, 1998: 578; Norrbom *et al.*, 1998: 130.

Type material examined: Holotype male, **CHINA**: Xishuangbanna, Yunnan, 650 m, 16.iii.1957, coll. Wang Shuyong, studied at IZAS by D.L. Hancock.
Non-type material examined: VIETNAM: Northern Vietnam – 1 male, Vietnam North, Hoa Binh Province, Mai Chau District, Pa Co, 30.vi.2001, Site N010, attracted to cue lure (in QDPC).
Diagnosis: A medium-sized species; face with distinct black facial spots in antennal furrows and a black spot below base of each antenna all joined by black lines; postpronotal lobes dark fuscous to dull black; notopleura yellow; scutum dark fuscous to dull black with a red-brown medial longitudinal band over anterior ½; lateral postsutural yellow vittae absent; a broad triangular yellow medial longitudinal vitta; a yellow triangle extending along anterior margin of notopleural suture; mesopleural stripe equal in width to notopleural dorsally; scutellum yellow with a narrow dark basal band; setae: *sc.* 2; *prsc.* absent; *ia.* 1; *p.sa.* 1; *a.sa.* 1; *mpl.* 1; *npl.* 2; *scp.* 4; legs black except basal quarter of hind femora fulvous; wing with cells bc colourless and cell c fuscous, a broad dark fuscous costal band overlapping R_{4+5} with an irregular margin; anal streak absent; abdomen petiolate and club shaped, abdominal terga mostly black with pale red-brown across posterior margin of tergum I, red-brown spots posterolaterally on tergum II and red-brown across posterior margin of tergum V, a pair of oval black shining spots present on tergum V.
Attractant: Cue lure.
Female: No known record.

Distribution:
Type locality: China (Yunnan Province, Xishuangbanna).
New records: Vietnam (Northern Vietnam, Hoa Binh Province).
Hosts: No known record.
Remarks: *Dacus (Mellesis) bannatus* (Wang) is similar to *Dacus (Mellesis) infernus* (Hardy) in possessing a broad costal band overlapping vein R_{4+5}, anal streak absent and scutum black with a broad triangular postsutural yellow vitta. It differs from *D. infernus* in having the legs mostly black, costal band in wing irregular in outline (reaching vein M at *r-m* crossvein) and abdominal terga III–V mostly black.

Dacus (Mellesis) bispinosus (Wang)

(Fig. 380)

Callantra bispinosa Wang, 1990: 71, 76. Holotype female in IZAS.
Mellesis esakii Shiraki, 1939: 410 (partim, Taiwan specimen), [Misidentification].
Dacus (Callantra) bispinosus – Liang *et al.*, 1993: 139; Drew *et al.*, 1998: 579; Norrbom *et al.*, 1998: 130.

Type material examined: Holotype female, **CHINA**: Kunming, Yunnan, 12.iv.1941, studied at IZAS by D.L. Hancock.
Other type material examined: TAIWAN: colour slide, 1 male Paralectotype of *Dacus (Mellesis) esakii*, Taihoku, 13.v.1937, coll. S. Issiki (in QDPC).
Non-type material examined: CHINA: 1 female, Jiangsu Province, 26.ix.1916, studied at IZAS by D.L. Hancock.
Diagnosis: A large species; face with a pair of small black spots joined by a black band across oral margin; postpronotal lobes and notopleura yellow; scutum red-brown with a medial longitudinal lanceolate fuscous band, narrow lateral longitudinal fuscous lines and fuscous areas laterally between postpronotal lobe and notopleuron; lateral postsutural yellow vittae absent; medial postsutural yellow vitta present as a small oval yellow marking posterocentrally; yellow triangle extending along anterior margin of notopleural suture; mesopleural stripe equal in width to notopleuron dorsally; scutellum yellow with a broad medial longitudinal red-brown band; setae: *sc.* 2; *prsc.* absent; *ia.* 1; *p.sa.* 1; *a.sa.* absent; *mpl.* 1; *npl.* 2; *scp.* 4; legs with fore femora with two preapical ventral spines, all femora and tibiae mostly fuscous; all tarsal segments fulvous; wing with a broad dark fuscous costal band overlapping R_{4+5} except for a small area just distal to *r-m* crossvein; anal streak absent; abdominal terga mostly red-brown with narrow transverse black bands across anterior margins of terga IV and V, a pair of red-brown oval shining spots on tergum V.
Attractant: No known record.
Distribution:
Type locality: China (Yunnan Province, Kunming).
Previously recorded: Type locality and China (Jiangsu Province); Taiwan (Taihoku) (Drew *et al.*, 1998).
Hosts: No known record.
Remarks: *Dacus (Mellesis) bispinosus* Wang is similar to *Dacus (Mellesis) polistiformis* (Senior-White) and *Dacus (Mellesis) esakii* (Skiraki) in characters listed under the comments for *D. polistiformis*. It differs from *D. polistiformis* in having an irregularly margined costal band narrowing in centre of wing distal to *r-m* crossvein and a small medial yellow spot on the scutum (in some specimens). It differs from *D. esakii* in having a narrow mesopleural stripe and a red-brown medial band on the scutellum. It is possible that *D. bispinosus* is a synonym of *D. esakii* but more material needs to be studied before this can be resolved. The specimen listed above as a Paralectotype of *D. esakii* is actually a specimen of *D. bispinosus*.

Dacus (Mellesis) calirayae Drew & Hancock

(Fig. 381)

Dacus (Callantra) calirayae Drew & Hancock, in Drew *et al.*, 1998: 580. Holotype male in BMNH.
Type material examined: Holotype male, **PHILIPPINES**: Lake Caliraya, Laguna, Luzon, 1.v.1983, coll. R.A.I. Drew, attracted to cue lure (in BMNH).
Paratypes: **PHILIPPINES**: 1 male, Guimaras I., Jordan, National Mango Research and Development Institute, 13–27.xii.1994, coll. I.M. White & H. Bignayan (in BMNH).
Diagnosis: A medium-sized species; face fulvous with two medium-sized circular black spots connected by a narrow pale fuscous line along oral margin; postpronotal lobes and notopleura yellow; scutum red-brown with large irregularly shaped dull black markings; lateral postsutural yellow vittae absent; medial postsutural yellow vitta present as a broad triangular mark in centre of scutum; a broad triangular marking along anterior margin of notopleural suture; broad mesopleural stripe reaching almost to postpronotal lobe dorsally; scutellum yellow except for narrow fuscous basal band; setae: *sc.* 2; *prsc.* absent; *ia.* 1; *p.sa.* 1; *a.sa.* 1; *mpl.* 1; *npl.* 2; *scp.* 2; legs with femora dark red-brown to pale fuscous except dark fuscous to black on outer and ventral surfaces of fore femora, on inner surfaces of mid femora and around apical ⅓ of hind femora; tibiae fuscous to dark fuscous; tarsi fulvous with apical four segments darker; mid tibiae each with an apical black spur; wing with cells bc and c fuscous, dense microtrichia covering all of cells bc and c; remainder of wing colourless except fuscous cell sc, broad fuscous costal band confluent with R_{4+5} and widening markedly at apex of wing; narrow fuscous to pale fuscous anal streak present; supernumerary lobe weak in male (female not studied); abdomen elongate oval, abdominal terga III–V orange-brown except for a narrow medial longitudinal black band and two broad lateral longitudinal dark fuscous to black bands over all three terga and joined along anterior margin of tergum III.
Attractant: Cue lure.
Female: No known record.
Distribution:
Type locality: Philippines (Luzon I., Lake Caliraya).
Previously recorded: Type locality and Philippines (Guimaras I.) (Drew *et al.*, 1998).
Hosts: No known record.
Remarks: *Dacus (Mellesis) calirayae* Drew & Hancock is similar to *Dacus (Mellesis) discophorus* (Hering), *Dacus*

(Mellesis) lagunae Drew & Hancock, *Dacus (Mellesis) leongi* Drew & Hancock, *Dacus (Mellesis) nummularius* (Bezzi) and *Dacus (Mellesis) subsessilis* (Bezzi) in possessing a broad costal band confluent with vein R_{4+5} and expanding distinctly at apex and possessing an anal streak. Further it is similar to *D. discophorus* and *D. nummularius* in having a medial longitudinal postsutural yellow vitta. *D. calirayae* differs from *D. lagunae*, *D. leongi* and *D. subsessilis* in having the basic colour of the scutum red-brown and possessing a medial postsutural vitta. It differs from *D. discophorus* and *D. nummularius* in having the medial postsutural vitta broad and triangular in shape.

Dacus (Mellesis) conopsoides de Meijere

(Fig. 382)

Dacus conopsoides de Meijere, 1911: 378. Lectotype male in BMNH.
Callantra conopsoides – Hardy, 1977: 45.
Dacus (Callantra) conopsoides. – Drew *et al.*, 1998: 582; Norrbom *et al.*, 1998: 130.
Dacus (Callantra) bimaculatus Lin, Wang & Zeng, in Lin *et al.*, 2011a: 620. Holotype female in HEIQ. **New Synonym**.

Type material examined: INDONESIA: JAVA – Lectotype male and 1 female Paralectotype, Depok, Java, ii.1908, coll. E. Jacobsen (in BMNH); 1 male Paralectotype, Depok, Java, x.1907 (in ZMAN).
Other type material studied: CHINA: A study of the colour photographs and descriptions of *Dacus (Callantra) bimaculatus* Lin, Wang & Zeng in the original publication.
Non-type material examined: INDONESIA: JAVA – 1 male, Tengger Highlands, E. Java, 1100m, v.1938, coll. Nongkodjarjar (in BMNH).
Diagnosis: A large species; face fulvous with medium-sized oval black spots in antennal furrows which are connected by a black line across oral margin; postpronotal lobes and notopleura yellow; scutum red-brown with irregular fuscous longitudinal bands; no lateral yellow postsutural vittae; a narrow medial longitudinal postsutural yellow vitta from line of notopleural suture to about ⅔ distance to the scutellum; triangular marking extended along anterior side of notopleural suture; a narrow mesopleural stripe equal in width to notopleuron dorsally and continued as a spot on the katepisternum; scutellum yellow except for narrow fuscous basal band; setae: *sc.* 2; *prsc.* absent; *ia.* 1; *p.sa.* 1; *a.sa.* 1; *mpl.* 1; *npl.* 2; *scp.* absent (only fine hairs present); legs with all femora red-brown except basal ⅕ of mid femora fulvous, fore femora with 2–4 (2 on right leg, 4 on left) stout ventral spines, tibiae red-brown, mid tibia with a black apical spur, hind tibia with a pre-apical microtrichose hump; wing with cells bc and c fuscous, dense microtrichia covering cells bc and c (non-type male with only the apical half of bc with microtrichia); remainder of wings hyaline except fuscous cell sc and basal sections of cell br and a broad costal band that extends evenly over R_{4+5} to about half-way to M, supernumerary lobe weak; abdomen elongate and petiolate, abdominal tergum III entirely red-brown, terga IV and V orange-brown with broad basal and lateral areas of red-brown and a narrow dark red-brown midline; a pair of orange-brown shining spots on tergum V.
Attractant: No known record.
Female: As for male except pecten of cilia on abdominal tergum III absent; no hump on hind tibiae; oviscape short (not visible in dorsal view).
Distribution:
Type locality: Indonesia (Java – West Java, Depok).
Previously recorded: Type locality and Indonesia (Java – East Java, Tengger Highlands).
New records: China (Hainan Province) due to **New Synonymy** of *D. bimaculatus*.
Hosts: No known record.
Remarks: *Dacus (Mellesis) conopsoides* de Meijere is similar to *Dacus (Mellesis) wallacei* White in having the costal band overlapping vein R_{4+5} and of uniform width, scutum basically red-brown and femora mostly red-brown with 2–4 ventral spines on fore femora. It differs from *D. wallacei* in possessing yellow on the anterior half of the anatergite (not entirely red-brown), a narrow medial longitudinal postsutural yellow vitta on the scutum and abdominal terga IV and V basically red-brown with orange-brown areas posteriorly either side of a narrow medial longitudinal red-brown band.

Dacus bimaculatus Lin, Wang & Zeng was separated from *D. conopsoides* on having the scutellum largely red-brown, fore femora with one stout ventral spine, mid and hind femora with fulvous around basal ends and two small red-brown spots on abdominal tergum III. These characters are not significant in defining species, e.g. in *Dacus* species the scutellum yellow patterns often change to darker colours after death of the specimen, the number of spines on the fore femora vary within a species as described for *D. conopsoides* de Meijere and *D. polistiformis* (Senior-White) by Drew *et al.* (1998), the extent of fulvous colouration around the bases of the femora usually exhibits intraspecific variation and the presence of pale spots on abdominal tergum III has no meaning at the species level. Consequently, we hereby make *D. bimaculatus* a **New Synonym** of *D. conopsoides*. This example demonstrates the problems that can arise by describing species from one specimen only.

Dacus (Mellesis) crabroniformis (Bezzi)

(Fig. 383)

Monacrostichus crabroniformis Bezzi, 1914b: 153. Holotype male in BMNH.
Mellesis crabroniformis – Bezzi, 1916: 117.
Callantra crabroniformis – Hardy, 1977: 45.
Dacus (Callantra) crabroniformis – Kapoor, 1993: 83; Drew *et al.*, 1998: 584; Norrbom *et al.*, 1998: 130.

Type material examined: Holotype male, **INDIA**: Shevaroy [Hills], Yercaud, 21.iV–4.v.[19]13, 4500 ft (in BMNH).
Diagnosis: A small species; face largely shiny black from just above oral margin to lower edge of antennal sockets, bordered laterally by fulvous; postpronotal lobes yellow

(except narrowly black along inner margin); notopleura yellow; scutum black; no lateral or medial postsutural yellow vittae; a wedge-shaped mark along anterior margin of notopleural suture; narrow mesopleural stripe barely wider than notopleuron dorsally and extended onto katepisternum as a small spot; scutellum yellow except for a broad black band over basal quarter; setae: *sc.* 2; *prsc.* absent; *ia.* 1; *p.sa.* 1; *a.sa.* 1; *mpl.* 1; *npl.* 2; *scp.* 4; legs with all femora red-brown except basal ⅔ hind femora fulvous; tibiae fulvous except apical ½ of hind tibiae darker; tarsi fulvous with segments 3–5 tending darker; mid tibiae each with an apical black spur; hind tibiae lacking preapical microtrichose pad; wing with cells bc and c hyaline, microtrichia in outer ½ of cell c only; remainder of wing hyaline except for narrow part of cell br and cell sc which are fuscous, a dark fuscous costal band confluent with R_{4+5} and then expanding into a shallow spot (just before apex of R_{2+3}) which extends ¼ of way to vein M; no anal streak; no dense aggregation of microtrichia around A_1+CuA_2; no supernumerary lobe in male (female not studied); abdomen elongate, clavate and petiolate, pecten of cilia on abdominal tergum III absent, abdominal tergum III black, tergum IV black except for fulvous apical areas each side of midline, tergum V black basally and laterally, fulvous apically, with red-brown shining spots.

Attractant: No known record.

Female: No known record.

Distribution:

Type locality: Southern India (Tamil Nadu, Yercaud).

Hosts: No known record.

Remarks: *Dacus (Mellesis) crabroniformis* (Bezzi) is similar to *Dacus (Mellesis) satanas* (Hering) in the characters listed in the comments on *D. satanas*. It differs in possessing red-brown femora and a dark inner margin to the postpronotal lobes.

Dacus (Mellesis) discipennis Walker, new combination

(Fig. 384)

Dacus discipennis Walker, 1861b: 294 – Hardy, 1959: 170. Holotype male in BMNH.
Dacus (Daculus) discipennis – Hardy & Adachi, 1954: 153.
Bactrocera (Tetradacus) discipennis – Norrbom *et al.*, 1998: 101.

Type material examined: Holotype male, **INDONESIA: MOLUCCAS (MALUKU)** – with the following seven labels – 'discipennis'; '68.4'; 'Bac'; 'Moluccas, Bachan, W.W. Saunders, BM 1868-4'; 'BMNH(E) No. 533011'; 'Dacus (Daculus) discipennis (Walker) det. 1954 D.E. Hardy'; 'Holotype male of Dacus discipennis Walker, 1861, Indonesia: Maluku Batchian. A.R. Wallace' (in BMNH).

Diagnosis: A large species; face fulvous with a pair of medium-sized oval black spots; postpronotal lobes and notopleura yellow; a lateral yellow band connecting the postpronotal lobe and notopleuron; scutum dull black with red-brown around margins; lateral postsutural yellow vittae present; medial postsutural yellow vitta absent; no yellow spot anterior to notopleural suture; mesopleural stripe equal in width to notopleuron dorsally; scutellum yellow; wing with cells bc and c dark fuscous, dense microtrichia covering all of both cells; a broad fuscous costal band almost confluent with M, a diffuse fuscous area across posterior of wing from cell cup to supernumerary lobe; abdominal terga III–V pale fuscous with orange-brown around posterior margin of tergum V.

Description: male.

Head: Height 2.2 mm. Frons – length 1.4 times breadth; red-brown without dark markings; orbital setae black: 1 *s.or.*, 3 *i.or.*; lunule red-brown. Ocellar triangle black. Vertex red-brown. Face fulvous with a pair of medium to large oval black spots; length 0.8 mm. Occiput red-brown, fulvous along eye margins. Antennae with all segments red-brown; length of segments: 0.3 mm; 0.54 mm; 1.2 mm; all three segments extremely long with the combined length greater than vertical length of head as for the subgenus *Callantra*.

Thorax: Scutum dull black with red-brown areas as follows: below and behind lateral postsutural vittae, around notopleural suture, inside postpronotal lobes and across anterior margin of scutum. Pleural areas fuscous to dark fuscous with red-brown below postpronotal lobes. Yellow markings as follows: postpronotal lobes; notopleura; mesopleural stripe equal in width to notopleuron dorsally, not extending to katepisternum below, anterior margin concave; a broad lateral yellow band joining postpronotal lobe and notopleuron; two moderately broad parallel-sided lateral postsutural vittae ending just before *ia.* seta. Medial longitudinal postsutural yellow vitta absent. Scutellum yellow except for narrow black basal band. Setae: *sc.* 2; *prsc.* absent; *ia.* 1; *p.sa.* 1; *a.sa.* absent; *mpl.* 1; *npl.* 2; *scp.* 4.

Legs: All segments entirely fulvous.

Wings: Length 8.8 mm; cells bc and c dark fuscous; dense microtrichia covering all of both cells; wings with small colourless areas through centre and with a broad fuscous costal band almost confluent with M, a large diffuse fuscous area across posterior region of wing from cell cup to supernumerary lobe; a dense aggregation of microtrichia around A_1+CuA_2; supernumerary lobe weak.

Abdomen: Elongate oval; terga fused; pecten present on tergum III. Tergum I fuscous; tergum II dark orange-brown anteriorly to fulvous posteriorly; terga III–V entirely pale fuscous except orange-brown around posterior margin of tergum V and between the shining spots. A pair of oval fuscous shining spots on tergum V. Posterior lobe of surstylus short, sternum V with a medium-sized concavity on posterior margin.

Attractant: No known record.

Female: No known record.

Distribution:

Type locality: Indonesia (Moluccas = Maluku, Bachan I.).

Hosts: No known record.

Remarks: *Dacus (Mellesis) discipennis* Walker has been placed in genus *Bactrocera* by previous workers on the basis of the abdominal terga believed to be free. However, careful examination of the Holotype male has revealed that the abdominal terga are fused thus placing this species in the genus *Dacus*. Further, the antennal segments are very long which is a major character for the subgenus *Mellesis* Bezzi. *Dacus discipennis* is unique and readily diagnosed in having the wing almost entirely infuscated, broad parallel-sided lateral postsutural vittae, a lateral yellow band joining the postpronotal lobe and notopleuron and all abdominal terga pale coloured without dark markings. As noted by previous workers and summarized by Norrbom *et al.* (1998), in the original description Walker (1861b) probably misstated the sex of the only specimen he used for the description. His description refers to a female while the only specimen in BMNH, which we have used for this revision, is a male with labels stating the correct original collection and locality data and Walker's Holotype label. Further, Walker did not describe any genitalia characters that would confirm the sex of the single specimen that he studied. This specimen must be the type.

Dacus (Mellesis) discophorus (Hering)

(Fig. 385)

Callantra discophora Hering, 1956: 64. Holotype male in NMB.
Dacus (Callantra) discophorus – Drew *et al.*, 1998: 586; Norrbom *et al.*, 1998: 130.

Type material examined: Holotype male, **SRI LANKA**: **(formerly CEYLON)** – Teldeniya, C.P., 16.xi.[19]53, coll. F. Keiser (in NMB).
Non-type material examined: SRI LANKA: 5 males, 1 female, Negombo, 19 & 27.xii.1995, ex *Dregea* (in BMNH); 2 males, Sri Lanka, Negombo, 27.xii.1995, ex *Dregea* (in QDPC). **INDIA**: 1 male, Karnataka, Chethalli (970m), 15.x.1986, host *Ceiba* sp. [error; bred from vine growing on *Ceiba* (S. Ramani, personal communication)] (in QDPC); 1 male, Bombay Presidency, Poona Ghats, 1888 (in BMNH); 3 males, Kallar, 6.v.1925, C.K.S., coll. at clove oil (in BMNH); 2 males, Kerala State, Walayar Forests, 700ft, x.1959, coll. P. Susai Nathan (in RNH); 1 male, Anamalar Hills, Chinchona, 1000m, xi.1959, coll. P. Susai Nathan (in ZMAN).
Diagnosis: A large species; face fulvous with two very large rectangular black spots covering antennal furrows; postpronotal lobes brownish (narrowly yellow along ventral and posterior margins); notopleura yellow; scutum pale fuscous with a mottled appearance from sites of connection of dorsoventral flight muscles and no dark markings; lateral postsutural yellow vittae absent; a short narrow medial longitudinal yellow postsutural vitta (often absent); broad triangular markings along anterior margins of notopleural suture; mesopleural stripe of medium-width, reaching midway between anterior margin of notopleuron and anterior *npl.* seta dorsally; scutellum yellow except for narrow pale fuscous basal band; setae: *sc.* 2; *prsc.* absent; *ia.* 1; *p.sa.* 1; *a.sa.* 1; *mpl.* 1; *npl.* 2; *scp.* 4; legs with fore and mid femora and tibiae pale fuscous except for a narrow fulvous ring around base of mid femora; fore and mid tarsi fulvous, hind legs missing; mid tibiae with an apical black spur; wing with cells bc and c fuscous, dense microtrichia covering all of cell c; remainder of wings colourless except fuscous cell sc, broad fuscous costal band confluent with R_{4+5} and expanding into a large round dark fuscous apical spot that overlaps M, pale fuscous within cell cup, supernumerary lobe weak in male; abdomen elongate, clavate and petiolate, abdominal terga III–V fuscous tending darker along anterior margin of tergum III and a dark fuscous to black medial longitudinal band, a pair of oval fuscous shining spots on tergum V. Posterior lobe of surstylus short, sternum V with a slight concavity on posterior margin, in males.
Attractant: Clove oil; cue lure (weak response).
Distribution:
Type locality: Sri Lanka (Central Province,Teldeniya).
Previously recorded: Type locality and Southern India (Drew *et al.*, 1998).
Hosts: Recorded from *Wattakaka volubilis* (family *Asclepiadaceae*) (Tsuruta *et al.*, 1997).
Remarks: *Dacus (Mellesis) discophorus* (Hering) is similar to *Dacus (Mellesis) calirayae* Drew & Hancock, *Dacus (Mellesis) lagunae* Drew & Hancock, *Dacus (Mellesis) leongi* Drew & Hancock, *Dacus (Mellesis) nummularius* (Bezzi) and *Dacus (Mellesis) subsessilis* (Bezzi) in possessing a broad costal band confluent with vein R_{4+5} and expanding distinctly at apex. It differs from *D. lagunae*, *D. leongi* and *D. subsessilis* in having the ground colour of the scutum red-brown and with a medial longitudinal postsutural yellow vitta. *D. discophorus* differs from *D. calirayae* and *D. nummularius* in having brown postpronotal lobes, a narrower mesopleural stripe (ending midway between anterior margin of notopleuron and anterior *npl.* seta), abdominal terga III–V fuscous without dark lateral margins and with a medial longitudinal dark fuscous to black band over terga I–V.

Dacus (Mellesis) discretus Drew & Romig, new species

(Fig. 386)

Material examined: Holotype male, **VIETNAM: Southern Vietnam** – Vietnam South, Tien Giang Province, Long Binh Dien Village, 7.ix.1999, Site S006, attracted to cue lure, deposited in BMNH.
Paratype: **VIETNAM: Southern Vietnam** – 1 male, Vietnam South, Tien Giang Province, Dao Thanh Village, 5.vii.1999, Site S005, attracted to cue lure. Specimen in QDPC.
Diagnosis: A large species; face red-brown with a pair of medium-sized oval black spots; postpronotal lobes yellow (red-brown anterodorsally); notopleura yellow; scutum basically red-brown with longitudinal irregularly shaped pale fuscous markings; lateral and medial postsutural yellow vittae absent; a large yellow triangle along anterior margin of notopleural suture with base adjacent to notopleuron; mesopleural stripe reaching midway between anterior margin of notopleuron and anterior *npl.* seta dorsally; scutellum yellow with a broad red-brown basal band; wing with cells bc and c dark fuscous, dense microtrichia covering all of cell c and most of cell bc; a broad dark fuscous costal band overlapping R_{4+5} throughout and expanding around apex of wing over M

as a pale tint; a narrow fuscous anal streak; abdominal terga III–V fuscous to dark fuscous with a medium-width medial longitudinal black band over all three terga, large orange-brown spots posterocentrally on tergum IV and orange-brown around posterior margin of tergum V.

Description: male.

Head: Height 1.96 mm. Frons – length 1.46 times breadth; red-brown with fuscous on anteromedial hump; orbital setae black: 1 *s.or.*, 2 *i.or.*; lunule red-brown. Ocellar triangle black. Vertex red-brown. Face red-brown with a pair of medium-sized oval black spots; length 0.39 mm. Genae red-brown, small dark fuscous subocular spot present; black seta present. Occiput red-brown, fulvous along eye margins; occipital row with 3–4 small black setae. Antennae with all segments entirely fuscous; length of segments: 0.57 mm; 0.66 mm; 1.08 mm.

Thorax: Scutum red-brown with a pair of irregularly shaped longitudinal pale fuscous markings. Pleural areas red-brown with dark fuscous anterior to mesopleural stripe. Yellow markings as follows: postpronotal lobes (anterodorsal corners red-brown); notopleura; mesopleural stripe reaching midway between anterior margin of notopleuron and anterior *npl.* seta dorsally, continuing to katepisternum as a small transverse spot, anterior margin straight; a large triangular marking anterior to notopleural suture with base adjacent to notopleuron. Anatergite (posterior apex dark fuscous); anterior ¾ katatergite (remainder red-brown). Lateral and medial postsutural yellow vittae absent. Postnotum red-brown. Scutellum yellow except for a broad red-brown basal band. Setae: *sc.* 2; *prsc.* absent; *ia.* 1; *p.sa.* 1; *a.sa.* 1; *mpl.* 1; *npl.* 2; *scp.* 4.

Legs: Fore and mid femora fuscous, hind femora fulvous with apical ⅓ fuscous; fore tibiae fuscous, mid tibiae dark red-brown tending fuscous basally, hind tibiae dark fuscous; fore tarsi with basal segment fulvous and apical four segments fuscous, mid and hind tarsi entirely fulvous; mid tibiae each with an apical black spur.

Wings: Length 6.5 mm; cells bc and c dark fuscous; dense microtrichia covering all of cell c and most cell bc; remainder of wings colourless except a broad dark fuscous costal band extending well over R_{4+5} and broadening around apex of wing and over M as a pale tint, a narrow fuscous anal streak; dense aggregation of microtrichia around A_1+CuA_2; supernumerary lobe weak.

Abdomen: Elongate oval and petiolate; terga fused; pecten present on tergum III. Tergum I and sterna I and II longer than wide. Tergum I fuscous, with a narrow transverse fulvous band across posterior margin, a black medial longitudinal band and narrow dark fuscous lateral margins; tergum II fulvous with a narrow transverse dark fuscous band across anterior margin and extending along lateral margins and a narrow medial longitudinal black band; tergum III fuscous to dark fuscous with a narrow- to medium-width medial longitudinal black band; terga IV and V fuscous with a narrow- to medium-width medial longitudinal black band, large orange-brown spots posterocentrally on tergum IV and orange-brown along posterior margin of tergum V. Posterior lobe of surstylus elongate (of medium-width), sternum V with a slight concavity on posterior margin. Abdominal sterna dark fuscous to black.

Attractant: Cue lure.

Female: No known record.

Distribution: Vietnam (Southern Vietnam, Tien Giang Province).

Hosts: No known record.

Remarks: *Dacus (Mellesis) discretus* n.sp. is similar to *Dacus (Mellesis) ooii* Drew & Hancock and *Dacus (Mellesis) pictus* (Hardy) in possessing fuscous cells bc and c with dense microtrichia over all of cell c and most of cell bc, a broad dark fuscous costal band overlapping R_{4+5} throughout and abdominal terga II–V with a similar colour pattern consisting of a broad medial longitudinal dark band over all four terga, lateral longitudinal dark bands connected by transverse dark bands over anterior margins of terga II and III. It differs from *D. ooii* in having a red-brown scutum and pale red-brown femora, from *D. pictus* in having a distinct anal streak in the wing and from both species in having a fuscous face with a pair of small black facial spots.

Dacus (Mellesis) donggaliae Drew & Romig, new species

(Fig. 387)

Material examined: Holotype male, **INDONESIA: SULAWESI** – Sulawesi Tengah, Donggala, Nupabomba, 23.ix.2006, coll. Rusdin, attracted to cue lure in forest, deposited in BMNH.

Diagnosis: A large species; face fulvous with a pair of medium-sized subquadrate black spots; postpronotal lobes yellow (anterodorsal corners fuscous); notopleura yellow; scutum entirely black; lateral and medial postsutural yellow vittae absent; a large yellow triangle along anterior margin of notopleural suture; mesopleural stripe equal in width to notopleuron dorsally; scutellum yellow; wing with cells bc and c dark fuscous, dense microtrichia covering all of cell c and outer ½ of cell bc; a broad dark fuscous costal band extending well over R_{4+5} and remaining of uniform width throughout; a broad dark fuscous anal streak; abdominal terga III–V fuscous with dark fuscous along anterior margin of tergum III and as narrow lateral margins on terga III–V.

Description: male.

Head: Height 2.1 mm. Frons – length 1.5 times breadth; red-brown with fuscous around orbital setae; orbital setae black: 1 *s.or.*, 2 *i.or.*; lunule fuscous. Ocellar triangle black. Vertex black. Face fulvous with a pair of medium-sized subquadrate black spots; length 0.54 mm. Genae fulvous, small pale fuscous subocular spot present; black seta present. Occiput red-brown, fulvous along eye margins; occipital row with a large number of strong black setae. Antennae with segments 1 and 2 red-brown, segment 3 red-brown with fuscous on apex and outer surface; length of segments: 0.6 mm; 0.6 mm; 1.1 mm.

Thorax: Scutum uniformly black without pale markings. Pleural areas black with red-brown below postpronotal lobes. Yellow markings as follows: postpronotal lobes (anterodorsal corners fuscous); notopleura; narrow mesopleural stripe equal in width to notopleuron dorsally, extending to katepisternum as a small transverse spot, anterior margin straight; a large yellow triangle along anterior margin of notopleural suture with base adjacent to notopleuron. Anatergite (posterior apex black); anterior ⅔ katatergite (remainder black); lateral and medial postsutural yellow vittae absent. Postnotum black. Scutellum yellow with a broad black basal band. Setae: *sc.* 2; *prsc.* absent; *ia.* 1; *p.sa.* 1; *a. sa.* 1; *mpl.* 1; *npl.* 2; *scp.* 4.

Legs: Fore and mid femora fuscous, hind femora fulvous with fuscous on apical ⅓; all tibiae entirely fuscous; all tarsal segments entirely fulvous; mid tibiae each with an apical black spur.

Wings: Length 6.7 mm; cells bc and c dark fuscous; dense microtrichia covering all of cell c and outer ½ of cell bc; remainder of wings colourless except dark fuscous cell sc, broad dark fuscous costal band extending well over R_{4+5} and remaining of uniform width throughout, broad dark fuscous anal streak; a dense aggregation of microtrichia around A_1+CuA_2; supernumerary lobe weak.

Abdomen: Elongate oval; terga fused; pecten present on tergum III. Tergum I and sterna I and II longer than wide. Tergum I fuscous with narrow dark fuscous lateral margins; tergum II dark fuscous anteriorly and fulvous posteriorly; terga III–V fuscous with a narrow transverse dark fuscous band across anterior margin of tergum III and widening over lateral margins and narrow dark fuscous lateral margins on terga IV and V. A pair of oval orange-brown shining spots on tergum V. Posterior lobe of surstylus short, sternum V with a slight concavity on posterior margin. Abdominal sterna dark fuscous to black.

Attractant: Cue lure.

Female: No known record.

Distribution: Indonesia (Sulawesi – Central Sulawesi, Donggala).

Hosts: No known record.

Remarks: *Dacus (Mellesis) donggaliae* n.sp. is similar to *Dacus (Mellesis) formosanus* (Tseng & Chu) in possessing a black scutum, postpronotal lobes mostly yellow, notopleura and scutellum yellow, a narrow mesopleural stripe and broad costal band of relatively uniform width and overlapping R_{4+5}. However, it differs from *D. formosanus* in having an anal streak in the wing, the abdomen elongate-oval but not petiolate, paler coloured femora and abdominal terga III–V with narrow lateral dark fuscous margins (not broad dark fuscous to black margins).

Dacus (Mellesis) dorjii Drew & Romig

(Fig. 388)

Dacus (Mellesis) dorjii Drew & Romig, in Drew *et al.*, 2007: 14. Holotype male in BMNH.

Type material examined: Holotype male, **BHUTAN**: Lumitshawa, 17.v.2000, attracted to cue lure, coll. C. Dorji (in BMNH).

Paratypes: 1 male, same data as holotype; 3 males (24.v.2000), 2 males (10.viii.2000), Lumitshawa, coll. C. Dorji; 4 males, Lumitshawa, 10.v.2000, coll. B.S. Fletcher; 1 male, Rimchu, 26.iv.2000, coll. Clarke & Fletcher; 1 male, Rimchu trap No. 1, 4.v.2000, coll. B.S. Fletcher; 1 male, Rimchu trap No. 2, 4.v.2000, coll. B.S. Fletcher; 6 males, Tsirang, Darachu, alt. 1500 m, 30.v.2001, coll. Brian Fletcher; 3 males, 16 km from Namling, alt. 1600 m, 9.vi.2001, coll. Brian Fletcher; 5 males, 14 km from Namling, alt. 1800 m, 7.vi.2001, coll. Brian Fletcher; 18 males, 11 km from Namling, alt. 2000 m, 9.vi.2001, coll. Brian Fletcher; 27 males, 7 km from Namling, alt. 2200 m, 9.vi.2001, coll. Brian Fletcher; 3 males, 19 km from Sengor, 6.vi.2001, alt. 2600 m, coll. Brian Fletcher; 2 males, 25 km from Sengor, alt. 2400 m, 7.vi.2001, coll. Brian Fletcher. All specimens attracted to cue lure (in ANIC; BMNH; NPPC; QMIC; QDPC; ZRC).

Non-type material examined: VIETNAM: Northern Vietnam – 2 males, Northern Vietnam, Lao Cai Province, Sapa 6 Village, 12.v.2000, attracted to cue lure, Site N017. Specimens in QDPC.

Diagnosis: A medium-sized species; face mostly black with oral and lateral margins fulvous; postpronotal lobes black; notopleura yellow; scutum black with red-brown behind *ia.* setae; lateral and medial postsutural yellow vittae absent; a yellow triangle along anterior margin of notopleural suture; mesopleural stripe equal in width to notopleuron dorsally; scutellum yellow except for narrow black basal band; setae: *sc.* 2; *prsc.* absent; *ia.* 1; *p.sa.* 1; *a.sa.* absent (occasionally one weak seta present); *mpl.* 1; *npl.* 2; *scp.* 2; legs with all femora entirely black except small areas of dark red-brown on apices of mid and hind femora; all tibiae dark fuscous to black; fore tarsi with basal segment fuscous and apical four segments slightly darker fuscous, mid and hind tarsi with basal segment fulvous and apical four segments red-brown; mid tibiae each with an apical black spur; wing with cells bc and c fuscous, dense microtrichia covering all of cell c and outer ½ of cell bc; remainder of wings colourless except dark fuscous cell sc, broad dark fuscous costal band overlapping R_{4+5} throughout; anal streak absent, supernumerary lobe weak; abdomen elongate oval and petiolate, all abdominal terga mostly black except for a narrow red-brown transverse band across intersegmental line between terga I and II and orange-brown spots posterocentrally on terga IV and V.

Attractant: Cue lure.

Female: No known record.

Distribution:

Type locality: Bhutan, Lumitshawa.

Previously recorded: Type locality and other locations in Bhutan (Drew *et al.*, 2007).

New records: Vietnam (Northern Vietnam – Lao Cai Province, Sapa).

Hosts: No known record.

Remarks: *Dacus (Mellesis) dorjii* Drew & Romig is similar to *Dacus (Mellesis) siamensis* Drew & Hancock, *Dacus (Mellesis) tenebrosus* Drew & Hancock and *Dacus (Mellesis) vijaysegarani* Drew & Hancock in the general colour patterns of the body, legs and wings. It differs from these species in possessing a black face, entirely yellow notopleura, the anatergite and katatergite entirely black and in lacking *a.sa.* setae. *D. dorjii* also differs from *D. tenebrosus* in having a narrow mesopleural stripe equal in width to the notopleuron and from *D. siamensis* in having distinct orange-brown spots posterocentrally on terga IV and V.

Dacus (Mellesis) esakii (Shiraki)

(Fig. 389)

Mellesis esakii Shiraki, 1939: 410. Lectotype male location unknown (possibly NTU).
Callantra ihai Shiraki, 1968: 4 – Drew *et al*., 1998: 588 (Synonym). Holotype female in NIAS.
Callantra esakii – Ito, 1983: 6.
Dacus (Callantra) esakii – Drew *et al*., 1998: 588, Lectotype designation; Norrbom *et al*., 1998: 130.

Type material examined: Lectotype male, **JAPAN**: Hikosan (Buzen), Kyushu, 20.viii.1933, coll. Teiso Esaki; Paralectotype female, Ikiriki (Nagasaki), 28.ix.1920, coll. Ishii. Colour slides of these types in QDPC.
Diagnosis: A large species; face fulvous with a black band across oral margin; postpronotal lobe yellow (sometimes tending brownish); notopleura yellow; scutum red-brown to dark red-brown; two narrow short lateral longitudinal vittae ending at level of *a.sa.* setae; medial postsutural yellow vittae present as a small marking in centre of scutum; broad triangular markings along anterior margins of notopleural suture with bases at notopleura; broad mesopleural stripe reaching to postpronotal lobe dorsally; scutellum yellow (tending fuscous in some specimens); setae: *sc.* 2; *prsc.* absent; *ia.* 1; *p.sa.* 1; *a.sa.* absent; *mpl.* 1; *npl.* 2; *scp.* absent; legs with spines on ventral surfaces of fore femora, femora pale fuscous except fulvous at bases of mid and hind femora; tibiae paler fuscous; tarsi pale fuscous except basal segments fulvous; mid tibiae with an apical black spur; wing with cells bc and c fuscous, remainder of wing colourless except for fuscous to dark fuscous cell sc; broad costal band overlapping R_{4+5} with an erratic margin, mostly fuscous with irregularly shaped dark fuscous areas apically and below R_{4+5} near junction of that vein with *r-m* crossvein, a pale fuscous tint in cell cup; supernumerary lobe weak; abdomen elongate, clavate and petiolate, abdominal terga III–V generally fuscous tending red-brown centrally and with a narrow transverse dark fuscous to black band across anterior margin of tergum IV, dark fuscous to black markings anterolaterally on tergum V (these are triangular in shape and widest on lateral margins of tergum), and a narrow medial longitudinal dark fuscous to black band on tergum V.
Attractant: No known record.
Distribution:
Type locality: Japan (Kyushu I., Hikosan).
Previously recorded: Type locality and Japan (Okinawa I.) (Drew *et al*., 1998).
Hosts: Follicles of *Pycnostelma paniculatum* (= *chinense*) (family *Asclepiadaceae*) (Shiraki, 1939; Ito, 1983).
Remarks: This description is taken from colour slide photographs of the male Lectotype and female Paralectotype of *Dacus (Mellesis) esakii* (Shiraki). Some characters are not visible on the photographs. *Dacus esakii* is similar to *Dacus (Mellesis) bispinosus* (Wang) and *Dacus (Mellesis) polistiformis* (Senior-White) in characters listed under the comments for *D. polistiformis*. It differs from these species in characters discussed under the comments of each.

In the original description, Shiraki (1939) did not designate a holotype. The male from Hikosan, Japan, was illustrated by the author and subsequently designated as the Lectotype by Drew *et al.* (1998). The remaining specimen, a female from Ikiriki, Japan, became a Paralectotype. The third specimen mentioned by Shiraki in the original description, a male from Taihoku, Taiwan, is a specimen of *D. bispinosus*.

In the two specimens from Japan, pins placed centrally through the scutum have obliterated the small central yellow spot.

Dacus (Mellesis) feijeni White

(Fig. 390)

Dacus (Callantra) feijeni White, in Drew *et al*., 1998: 590. Holotype male in BMNH.
Dacus (Mellesis) feijeni – Drew *et al*., 2007: 15.

Non-type material examined: Since the publication of Drew *et al.* (1998), the following specimens have been studied – **BHUTAN**: 2 males, Bhutan, Tashi, Lingmethang, 17.xi.1991, coll. H.R. Feijen; 1 male, Rimchu trap No. 1, 17.v.2000, coll. C. Dorji. Specimens attracted to cue lure (in QDPC).
Diagnosis: A large species; face fulvous with a pair of large circular to oval black spots; postpronotal lobes and notopleura yellow; scutum dark red-brown; lateral and medial postsutural yellow vittae absent; narrow mesopleural stripe equal in width to notopleuron dorsally; a yellow triangle along anterior margin of notopleural suture with base at notopleuron; scutellum yellow except for broad red-brown basal band; setae: *sc.* 2; *prsc.* absent; *ia.* 1; *p.sa.* 1; *a.sa.* 1; *mpl.* 1; *npl.* 2; *scp.* 4; legs with fore and mid femora entirely red-brown, hind femora fulvous tending fuscous to dark fuscous on apical ⅓; fore tibiae fuscous, mid tibiae red-brown tending fuscous apically, hind tibiae dark red-brown to fuscous; tarsi with basal segment fulvous and apical four segments fulvous to red-brown; wing with cells bc and c pale fuscous, dense microtrichia covering all of cell c and outer corner of cell bc; a broad dark fuscous costal band confluent with R_{4+5} basally then distinctly overlapping this vein from the region of *r-m* crossvein to the apex of the wing; anal streak absent; supernumerary lobe weak; abdomen elongate oval and petiolate, terga III–V red-brown with a 'T' pattern consisting of a broad fuscous to dark fuscous band across anterior margin of tergum III expanding to cover lateral margins, a narrow medial longitudinal dark fuscous band overall three terga, diffuse dark fuscous laterally on tergum IV and on anterolateral corners on tergum V. A pair of red-brown shining spots on tergum V. Posterior lobe of surstylus long, sternum V with a slight concavity on posterior margin.
Attractant: Cue lure.
Female: No known record.
Distribution:
Type locality: Bhutan (Manchungan).
Previously recorded: Type locality and the following locations in Bhutan (Namling; Rimchu; Tashi) (Drew *et al*., 1998; Drew *et al*., 2007).
Hosts: No known record.
Remarks: *Dacus (Mellesis) feijeni* White was adequately described and illustrated in Drew *et al. (*1998). It is best distinguished by possessing the combination of dark

red-brown scutum, yellow postpronotal lobes, red-brown femora and tibiae and abdominal terga III–V red-brown to fuscous.

Dacus (Mellesis) fletcheri Drew

(Fig. 391)

Dacus (Mellesis) fletcheri Drew, in Drew *et al.*, 2007: 16. Holotype male in BMNH.

Type material examined: Holotype male, **BHUTAN**: 19 km from Sengor, alt. 2600 m, 6.vi.2001, coll. Brian Fletcher, attracted to cue lure (in BMNH).

Paratypes: 10 males, same data as holotype; 8 males, 9 km from Sengor, alt. 2800 m, 9.vi.2001, coll. Brian Fletcher; 8 males, 25 km from Sengor, alt. 2400 m, 7.vi.2001, coll. Brian Fletcher; 1 male, 7 km from Namling, alt. 2200 m, 9.vi.2001, coll. Brian Fletcher; 1 male, 11 km from Namling, alt. 2000 m, 9.vi.2001, coll. Brian Fletcher; 1 male, Suntalay, Tsirang, 11.xi.2004, coll. Karma Duckpar and Karma Yangzom; 1 male, Lumitshawa, 10.v.2000, coll. Brian Fletcher; 1 male, Lumitshawa, 31.v.2000, coll. C. Dorji; 1 male, 2 km from Lumitshawa, 26.iv.2000, coll. Clarke & Fletcher. All specimens attracted to cue lure (in ANIC; BMNH; NPPC; QMIC; QDPC; ZRC).

Diagnosis: A large species; face with a pair of medium-sized oval black spots and a distinct black spot medially below bases of antennae; postpronotal lobes dark fuscous to black; notopleura yellow; scutum black; lateral and medial postsutural yellow vittae absent; a distinct yellow triangle along anterior margin of notopleural suture; narrow mesopleural stripe equal in width to notopleuron dorsally; scutellum yellow; setae: *sc.* 2; *prsc.* absent; *ia.* 1; *p.sa.* 1; *a.sa.* 1; *mpl.* 1; *npl.* 2; *scp.* 2; legs with femora entirely black except for small apical red-brown areas; fore tibiae dark fuscous to fuscous, mid and hind tibiae red-brown to fuscous; fore tarsi with all segments fuscous to dark fuscous; mid and hind tarsi with basal segment fulvous and apical four segments pale fuscous; mid tibiae each with an apical black spur; wing with cells bc and c dark fuscous, dense microtrichia covering all of both cells; remainder of wings colourless except for a broad dark fuscous costal band confluent with vein M at basal area of cell dm and distinctly overlapping R_{4+5} to apex of wing; anal streak absent; supernumerary lobe weak; abdomen elongate oval and petiolate, abdominal terga II–V black with distinct orange-brown spots posteromedially and increasing in size from terga II–IV and a medial longitudinal orange-brown area on tergum V which expands across posterior margin of tergum, terga II–IV with large protuberances posterocentrally along the midline best viewed in lateral view of the abdomen. A pair of oval shining spots on tergum V which vary in colour from orange-brown to black. Posterior lobe of surstylus long, sternum V with a slight concavity on posterior margin.

Attractant: Cue lure.

Female: No known record.

Distribution:

Type locality: Bhutan (Sengor).

Previously recorded: Type locality and other locations in Bhutan (Drew *et al.*, 2007).

Hosts: No known record.

Remarks: *Dacus (Mellesis) fletcheri* Drew is similar to *Dacus (Mellesis) siamensis* Drew & Hancock, *Dacus (Mellesis) tenebrosus* Drew & Hancock and *Dacus (Mellesis) vijaysegarani* Drew & Hancock in the general colour patterns of the body, wings and legs. It is distinct from these species in possessing large posterocentral orange-brown markings on abdominal terga II, III and IV and orange-brown medially on tergum V, large protuberances posteromedially on the midline of terga II, III and IV and the abdominal shape elongate oval, not distinctly club-shaped.

Dacus (Mellesis) formosanus (Tseng & Chu)

(Fig. 392)

Callantra formosana Tseng & Chu, 1983: 119, 122. Holotype male in MEA.

Dacus (Callantra) formosanus – Drew *et al.*, 1998: 591; Norrbom *et al.*, 1998: 130.

Non-type material examined: TAIWAN: 1 male, Jiauliping, Chiayi, Itsien, 1.ix-1.x.1977, coll. Y.H. Tseng, attracted to cue lure; 1 male, Wasna (900m), 6.x.1978, coll. Y.H. Tseng, attracted to cue lure, identified *Callantra formosana* by Y.H. Tseng; 1 male, Nanfeng, 27.xi.1978, coll. Lena Lee Wen (in QDPC).

Diagnosis: A large species; face fuscous with a large dark fuscous to black spot dorsocentrally below bases of antennae and two large irregularly oval black spots in antennal furrows; postpronotal lobes yellow (except inner posterior corner black); notopleura yellow; scutum black without pale markings; lateral and medial postsutural yellow vittae absent; narrow triangular markings along anterior margins of notopleural suture; narrow mesopleural stripe equal in width to notopleura dorsally; scutellum yellow except for narrow black basal band; setae: *sc.* 2; *prsc.* absent; *ia.* 1; *p.sa.* 1; *a.sa.* 1; *mpl.* 1; *npl.* 2; *scp.* 4; legs with all femora dark fulvous tending dark fuscous to black on basal ½ of fore and hind femora and basal ⅓; of mid femora; tibiae and tarsi entirely fulvous; mid tibiae with an apical black spur; wing with cells bc and c dark fuscous, dense microtrichia covering both cells; remainder of wings colourless or with a pale fuscous tint, except for dark fuscous cell sc, broad dark fuscous costal band confluent (or almost so) with M from base to *r-m* crossvein and overlapping R_{4+5} with an irregular line (pattern) from *r-m* crossvein to wing apex; anal streak absent; supernumerary lobe weak in male (female not studied); abdomen elongate, clavate and petiolate, abdominal tergum III dark fuscous to dull black except for a small red-brown area posterocentrally, terga IV and V dark fuscous to dull black except for a large red-brown area posterocentrally on each tergum. A pair of red-brown to fuscous shining spots on tergum V. Posterior lobe of surstylus short, sternum V with a slight concavity on posterior margin.

Attractant: Cue lure.

Female: No known record.

Distribution:

Type locality: Taiwan (Paulei, Kaoshiung Hsien).

Previously recorded: Type locality and other locations in Taiwan (Drew *et al.*, 1998).
Hosts: No known record.
Remarks: Three species of subgenus *Mellesis* have been recorded from Taiwan, viz. *Dacus (Mellesis) nummularius* (Bezzi), *Dacus (Mellesis) bispinosus* (Wang) and *Dacus (Mellesis) formosanus* (Tseng & Chu). *Dacus formosanus* is similar to *Dacus (Mellesis) trimacula* (Wang) in having abdominal terga III–V fuscous to black without a medial longitudinal dark vitta, the costal band broad and almost confluent with vein M basally, the anal streak absent, the anatergite fuscous and katatergite yellow. It differs from *D. trimacula* in having the scutum black and abdominal terga I and II mostly black. The posterior surstylus lobe of *D. formosanus* is short, not as illustrated in Fig. 7 of Tseng and Chu (1983).

Dacus (Mellesis) icariiformis (Enderlein)

(Fig. 393)

Callantra icariiformis Enderlein, 1920: 358 – Hardy, 1977: 45. Lectotype male in MZB.
Dacus (Callantra) icariiformis – Kapoor, 1993: 83; Drew *et al.*, 1998: 593, Lectotype designation; Norrbom *et al.*, 1998: 130.

Type material examined: Lectotype male, **INDIA**: Sikkim, [Bingham] in MFN. 1 male paralectotype with the following six labels – 'Sikkim, F. Muller, Per Lt Col C.T. Bingham'; 'Resembes a Hymenopteran (Polistes)'; 'Burma: Lt. – Col. Bingham, B.M. 1896-281'; 'TOPOTYPICAL, Dacus (Callantra) icariiformis (Enderlein), det. I.M. White, 1993, P.T.O.'; 'N B. FEMSPINE on male only – others with both sexes known both have femspine'; 'as type male but lacks fore-femoral spines'; the first two labels appear to be in the handwriting of Enderlein and the latter three in that of I.M. White; specimen in BMNH.
Diagnosis: A large species; face fulvous with a small elongate black spot in each antennal furrow and a narrow trace of a black line across oral margin; postpronotal lobes yellow (trace of yellow spot lateral to postpronotal lobe); notopleura yellow; scutum red-brown with dark fuscous patterns; narrow lateral postsutural yellow vittae incurved anteriorly and ending posteriorly at intra-alar seta; medial yellow vitta as an ill-defined spot; yellow triangle along notopleural suture; mesopleural stripe barely wider than notopleuron dorsally and tapered to a point at the katepisternum; scutellum yellow with narrow fuscous basal band; setae: *sc.* 2; *prsc.* absent; *ia.* 1; *p.sa.* 1; *a.sa.* 1; *mpl.* 1; *npl.* 2; *scp.* reduced to fine pale hairs; legs with femora red-brown with fuscous on ventral surfaces, hind femora pale over basal quarter, fore femora of male Lectotype with 1 or 2 stout ventral spines (absent in female); tibiae fulvous; tarsi fulvous; mid tibiae each with an apical black spur; hind tibiae without preapical pad; wing with cells bc and c fuscous, without microtrichia over cells bc and c; remainder of wings hyaline except fuscous cell sc, broad fuscous costal band which extends below R_{4+5} (about half-way to M) and not expanding into an apical spot; no anal streak; supernumerary lobe weak in both sexes; abdomen elongate, clavate and petiolate, abdominal tergum III red-brown, terga IV and V with fulvous spots in posterior ⅔ each side of a red-brown midline. A pair of oval fulvous shining spots on tergum V.
Attractant: No known record.
Female: As for male but lacking spines on fore femora and with a pair of fine orbital setae present.
Distribution:
Type locality: North-East India (Sikkim).
Hosts: No known record.
Remarks: *Dacus (Mellesis) icariiformis* (Enderlein) is similar to *Dacus (Mellesis) vittatus* (Hardy) in possessing a broad costal band (confluent with or overlapping vein R_{4+5}), *a.sa.* setae and lateral and medial postsutural yellow vittae. It differs from *D. vittatus* in having the costal band overlapping vein R_{4+5} throughout and the medial postsutural vitta short and posterior to the level of the notopleural suture. It is also very similar to *Dacus (Mellesis) murphyi* White, differing in having the costal band extending well below vein R_{4+5} throughout its length and not expanding apically.

In the original description, Enderlein (1920) had three specimens before him, 2 males and 1 female. As noted by Drew *et al.* (1998) the female is actually a specimen of *Dacus (Callantra) longicornis* Wiedemann from Salween, Burma, Jan. 1898, coll. Bingham. We regard the male in MFN as the Lectotype (designated by Drew *et al.*, 1998) and the male in BMNH as a Paralectotype. This specimen in BMNH was incorrectly listed as a female by Drew *et al.* (1998).

Dacus (Mellesis) indecorus (Hardy)

(Fig. 394)

Callantra indecora Hardy, 1974: 7 – Hardy, 1977: 45. Holotype male in BBM; paratype male in UZMH.
Dacus (Callantra) indecorus – Drew *et al.*, 1998: 595; Norrbom *et al.*, 1998: 130.

Type material examined: Holotype male, **PHILIPPINES**: **LUZON ISLAND** – Mayoyao, Ifugao, Mt Prov., Luzon, 1200–1500 m, 3.ix.1966, coll. H.M. Torrevillas (in BBM).
Diagnosis: A medium-sized species; face fulvous with two large facial spots that curve in at their lower ends, but separated across mouth opening by about their width; postpronotal lobes and notopleura yellow; scutum red-brown with a short lateral black stripe adjacent to the *a.sa.* seta and an irregular longitudinal black band on each side of the midline and which expands laterally behind each postpronotal lobe; lateral and medial postsutural yellow vittae absent; triangular yellow marking along anterior margin of notopleural suture; broad mesopleural stripe reaching anterior *npl.* seta dorsally; scutellum yellow; setae: *sc.* 2; *prsc.* absent; *ia.* 1; *p.sa.* 1; *a.sa.* 1; *mpl.* 1 (socket only in type); *npl.* 2; *scp.* 4; legs with femora red-brown except basal ⅔ of hind femora fulvous; tibiae red-brown; tarsi fulvous; mid tibiae each with an apical black spur; hind tibiae with a preapical microtrichose bulge; wing with cells bc and c with a pale fuscous tint, dense microtrichia covering both cells bc and c; remainder of wings hyaline

except for fuscous cell sc, fuscous costal band overlapping R_{2+3} but not expanded at apex and only crossing R_{4+5} at its extremity; anal streak narrow and only extending outside of cell cup at base of cup extension; supernumerary lobe weak; abdomen elongate, clavate and petiolate, abdominal terga II and III red-brown with a dark fuscous basal band and medial longitudinal band and lateral bands on tergum III, terga IV and V fulvous with large sub-rectangular red-brown lateral marks and a fuscous medial longitudinal band. A pair of fulvous shining spots on tergum V. Posterior lobe of surstylus of medium development, sternum V with a shallow concavity on posterior margin.
Attractant: No known record.
Female: No known record.
Distribution:
Type locality: Philippines (Luzon I., Mayoyas).
Hosts: No known record.
Remarks: *Dacus (Mellesis) indecorus* (Hardy) is similar to *Dacus (Mellesis) pedunculatus* (Bezzi) in having a broad costal band overlapping vein R_{2+3} but not confluent with or overlapping vein R_{4+5}, cells bc and c possessing a pale fuscous tint, scutum red-brown with dark patterns and similar abdominal colour patterns. It differs from *D. pedunculatus* in having two distinct dark spots on the face, cell bc completely covered with microtrichia, the anal streak present and in lacking dark markings between the eye margins and antennal bases.

Dacus (Mellesis) infernus (Hardy)

(Fig. 395)

Callantra inferna Hardy, 1973: 13 – Hardy 1977: 45. Holotype female in BBM.
Dacus (Callantra) infernus – Drew *et al*., 1998: 597; Norrbom *et al*., 1998: 130.

Type material examined: Holotype female, **THAILAND**: Tak Province, area on W side of Ping R., opp. Tak, 110 m, 8.vii.1969, coll. J.J.S. Burton (in BBM).
Diagnosis: A medium-sized species; face shallow, equal in height to a very large lunule (0.3 mm), fulvous, with a large slightly elongate black spot in each antennal furrow and a small black spot below antennal sockets; postpronotal lobes and notopleura yellow; scutum dull black with three ill-defined red-brown areas between the postpronotal lobes and a small area behind each yellow triangular sutural mark; lateral postsutural yellow vittae absent; medial postsutural yellow vitta from notopleural suture anteriorly almost to posterior margin of scutum and which is narrow in anterior half (part obscured by pin) then abruptly widening to form a triangle with base about half as wide as scutellum; triangular mark along anterior side of notopleural suture; narrow mesopleural stripe equal in width to notopleuron dorsally; scutellum yellow; setae: *sc*. 2; *prsc*. absent; *ia*. 1; *p.sa*. 1; *a.sa*. 1; *mpl*. 1; *npl*. 2; *scp*. 4; legs with femora red-brown except for ill-defined fulvous areas on basal ⅓ of fore and mid femora and basal ⅔ of hind femora; tibiae dark red-brown to fuscous; tarsi with basal segment pale fulvous; mid tibiae each with an apical black spur; wing with cells bc and c pale fuscous with dense microtrichia throughout; remainder of wings hyaline except for fuscous cell sc, basal part of cell br, broad costal band extending below R_{4+5} (almost half way to M apically); no anal streak; abdomen elongate, clavate and petiolate, abdominal terga III and IV mostly fuscous, except for small orange-brown areas posterocentrally, tergum V orange-brown with sub-rectangular fuscous areas anterolaterally and a broad midline. A pair of oval orange-brown shining spots on tergumV. Basal segment of ovipositor cylindrical in transverse cross section and tapering posteriorly in dorsal view. Ratio of length of oviscape to length of tergum V, 1:1.
Attractant: No known record.
Male: No known record.
Distribution:
Type locality: Northern Thailand (Tak Province, Ping River).
Hosts: No known record.
Remarks: *Dacus (Mellesis) infernus* (Hardy) is similar to *Dacus (Mellesis) bannatus* (Wang) in possessing a broad costal band overlapping vein R_{4+5}, anal streak absent and scutum black with a broad triangular postsutural yellow vitta. It differs from *D. bannatus* in having the postpronotal lobes yellow, costal band approximately even in width, legs mostly red-brown and abdominal terga III–V each with a pair of oval orange-brown spots posterocentrally each side of a dark midline.

Dacus (Mellesis) insulosus Drew & Hancock

(Fig. 396)

Dacus (Callantra) insulosus Drew & Hancock, in Drew *et al*., 1998: 599. Holotype female in BMNH.

Type material examined: Holotype female, **INDIA (ANDAMAN ISLANDS)**: Port Blair, 25.viii.1993, coll. Veena Kumani, S1 No. 36, bred ex Asclepiadaceae (in BMNH).
Paratypes: **INDIA: ANDAMAN ISLANDS** – 3 females, same data as holotype (in ANIC; QDPC).
Diagnosis: A large species; face red-brown with two small oval black spots in antennal furrows and a diffuse fuscous to dark fuscous spot dorsocentrally below bases of antennae; postpronotal lobes and notopleura yellow; scutum red-brown without dark markings (three specimens have some erratic dark fuscous patterns but these may be due to damage in preservation); lateral and medial postsutural yellow vittae absent; a narrow triangle along anterior margin of notopleural suture; mesopleural stripe of medium-width reaching midway between anterior margin of notopleuron and anterior *npl*. seta dorsally; scutellum yellow except for a very narrow dark basal band; setae: *sc*. 2; *prsc*. absent; *ia*. 1; *p.sa*. 1; *a.sa*. 1; *mpl*. 1; *npl*. 2; *scp*. 4; legs with femora red-brown except for dark fuscous apicoventrally on fore femora and along ventral margins of mid and hind femora; fore and mid tibiae red-brown, hind tibiae fuscous; tarsi with basal segments fulvous and apical four segments darker; mid tibiae

each with an apical black spur; wing with cells bc and c dark fuscous, dense microtrichia covering all of cell c and outer half of cell bc; remainder of wing colourless except for a broad dark fuscous costal band overlapping R_{4+5} and not expanding at wing apex (in two specimens there is an elongate colourless spot within the costal band at the apex); anal streak absent but with a pale fuscous tint in base of cell cup; supernumerary lobe weak; abdomen elongate-oval and petiolate, all abdominal terga red-brown except fuscous to black markings as follows: anterocentrally on tergum II, narrow transverse bands across anterior margins of terga III–V, narrow lateral margins over terga II–V. A pair of oval red-brown shining spots on tergum V.
Attractant: No known record.
Male: No known record.
Distribution:
Type locality: India (Andaman Is, Port Blair).
Hosts: Recorded from an undetermined species of family *Asclepiadaceae*.
Remarks: *Dacus (Mellesis) insulosus* Drew & Hancock is similar to *Dacus (Callantra) longicornis* Wiedemann and *Dacus (Mellesis) pullus* (Hardy) in having the costal band overlapping vein R_{4+5} throughout and postpronotal lobes yellow. It differs from *D. longicornis* in having an overall paler abdomen and the anatergite yellow. *Dacus insulosus* differs from *D. pullus* in having a red-brown scutum (not black) and abdominal terga mostly red-brown.

Dacus (Mellesis) lagunae Drew & Hancock

(Fig. 397)

Dacus (Callantra) lagunae Drew & Hancock, in Drew *et al.*, 1998: 600. Holotype male in BMNH.

Type material examined: Holotype male, **PHILIPPINES: LUZON ISLAND** – Mt Makiling (300 m), 15.vi.1983, coll. B. Ramos, attracted to cue lure (in BMNH).
Paratypes: **PHILIPPINES**: **LUZON ISLAND** – 1 male, Mt Makiling, 8.vi.1983, coll. B. Ramos; 1 male, Mt Makiling, Los Banos, 16.vi.1983, coll. A.M. Arida; 6 males, Mt Makiling (100 m), 8.vi.1983, coll. B. Ramos; 12 males, Mt Makiling (300 m), (8.vi.1983), (15.vi.1983), coll. B. Ramos; 5 males, Mt Makiling (400 m), (8.vi.1983), (15.vi.1983), coll. B. Ramos; 5 males, Mt Makiling (500 m), 9.v.1983, coll. B. Ramos. All specimens attracted to cue lure (in ANIC; BMNH; QDPC).
Diagnosis: A large species; face fulvous with two medium-sized circular black spots usually joined by a black band along oral margin; postpronotal lobes and notopleura yellow; scutum dull black with dark red-brown erratically along anterior and posterior margins, narrowly along lateral margins and (in some specimens only) between postpronotal lobes and notopleura; lateral and medial postsutural yellow vittae absent; broad triangular markings with bases at notopleura and running along anterior margins of notopleural suture; broad mesopleural stripe reaching almost to postpronotal lobe dorsally; scutellum yellow except for a narrow black basal band; setae: *sc.* 2; *prsc.* absent; *ia.* 1; *p.sa.* 1; *a.sa.* 1; *mpl.* 1; *npl.* 2; *scp.* 3–4; legs with femora red-brown except black along ventral margins of fore and mid femora and apical ½ of ventral margins of hind femora; fore tibiae fuscous, mid tibiae pale fuscous and hind tibiae dark fuscous; tarsi with basal segment fulvous and apical four segments pale fuscous; mid tibiae with an apical black spur; wing with cells bc and c fuscous, dense microtrichia covering both cells bc and c; remainder of wings with a pale fuscous tint across membrane except for fuscous cell sc, broad fuscous costal band confluent with R_{4+5} and overlapping this vein apically; a narrow pale fuscous anal streak ending before wing margin; supernumerary lobe weak; abdomen elongate, clavate and petiolate, abdominal terga III–V fulvous except for a broad medial and two broad lateral longitudinal dark fuscous to black bands joined along anterior margin of tergum III. A pair of oval fulvous shining spots on tergum V. Posterior lobe of surstylus short, sternum V with a slight concavity on posterior margin.
Attractant: Cue lure.
Female: No known record.
Distribution:
Type locality: Philippines (Luzon I., Mt Makiling).
Hosts: No known record.
Remarks: *Dacus (Mellesis) lagunae* Drew & Hancock is similar to *Dacus (Mellesis) calirayae* Drew & Hancock, *Dacus (Mellesis) discophorus* (Hering), *Dacus (Mellesis) leongi* Drew & Hancock, *Dacus (Mellesis) nummularius* (Bezzi) and *Dacus (Mellesis) subsessilis* (Bezzi) in possessing a broad costal band confluent with vein R_{4+5} and expanding markedly at apex. It differs from *D. calirayae*, *D. discophorus* and *D. nummularius* in having the scutum ground colour black and lacking a medial longitudinal postsutural yellow vitta. *Dacus lagunae* differs from *D. leongi* in having a broader mesopleural stripe reaching almost to postpronotal lobe dorsally, and face with two medium-sized circular black spots joined by a black band along oral margin. It differs from *D. subsessilis* in the same facial spot pattern, the wing with a distinct apical spot (not a diffuse pattern around the apical and hind margins) and with the broad lateral dark margins on abdominal terga III–V joined along base of tergum III.

Dacus (Mellesis) leongi Drew & Hancock

(Fig. 398)

Dacus (Callantra) leongi Drew & Hancock, in Drew *et al.*, 1998: 602. Holotype male in BMNH.

Type material examined: Holotype male, **EAST MALAYSIA: SARAWAK** – Sg. Sui, coastal area, Sarawak, 26.viii.1992, coll. S. Leong, attracted to cue lure (in BMNH).
Paratypes: **EAST MALAYSIA: SARAWAK** – 1 male, same data as holotype; 19 males, Sg. Sui, nr. Kuching, Sarawak, (16.iii.1992), (23.iii.1992), (2.iv.1992), (12.vi.1992), coll. S.

Leong; 14 males, Sg. Sui, Sarawak, 26.iii.1992, coll. S. Leong; 17 males, Sg. Sui, coastal area, nr. Kuching, (12. vi.1992), (18.vi.1992), (18.ix.1992), (25.ii.1993), (5. iii.1993), coll. S. Leong; 2 males, Tarat Agric Station, 16.x.1992, coll. S. Leong. All specimens attracted to cue lure (in ANIC; BMNH; QDPC).
Diagnosis: A medium-sized species; face fulvous with large oval black spots pointed dorsally; postpronotal lobes and notopleura yellow; scutum black without pale markings; lateral and medial postsutural yellow vittae absent; a broad band along anterior margin of notopleural suture; broad mesopleural stripe reaching to anterior *npl.* seta dorsally; scutellum yellow except for narrow black basal band; setae: *sc.* 2; *prsc.* absent; *ia.* 1; *p.sa.* 1; *a.sa.* 1; *mpl.* 1; *npl.* 2; *scp.* 4; legs with fore femora red-brown except dark fuscous to black on outer and ventral surfaces, mid femora fuscous to dark fuscous except red-brown at each end, hind femora fulvous tending dark fuscous to black on apical ⅓; fore tibiae dark fuscous, mid tibiae fuscous, hind tibiae dark fuscous to black; tarsi fulvous with apical four segments darker; mid tibiae each with an apical black spur; wing with cells bc and c fuscous, dense microtrichia covering all of both cells; remainder of wings colourless except for dark fuscous cell sc, broad dark fuscous costal band confluent with R_{4+5} and markedly overlapping this vein apically; narrow fuscous to pale fuscous anal streak present; supernumerary lobe weak; abdomen elongate, clavate and petiolate, abdominal tergum III dark fuscous to black (occasionally with two small orange-brown spots posterocentrally); terga IV and V orange-brown except for a moderately broad medial longitudinal dark fuscous to black band ending before posterior margin of tergum V and two broad lateral longitudinal dark fuscous to black bands over both terga. A pair of oval orange-brown shining spots on tergum V. Posterior lobe of surstylus short, sternum V with a slight concavity on posterior margin.
Attractant: Cue lure.
Female: No known record.
Distribution:
Type locality: East Malaysia (Sarawak, Sg Sui).
Previously recorded: Type locality and East Malaysia – Sarawak, Tarat Agriculture Station (Drew *et al.*, 1998).
Hosts: No known record.
Remarks: *Dacus (Mellesis) leongi* Drew & Hancock is similar to *Dacus (Mellesis) calirayae* Drew & Hancock, *Dacus (Mellesis) discophorus* (Hering), *Dacus (Mellesis) lagunae* Drew & Hancock, *Dacus (Mellesis) nummularius* (Bezzi) and *Dacus (Mellesis) subsessilis* (Bezzi) in possessing a broad costal band confluent with vein R_{4+5} and expanding markedly at apex. *Dacus leongi* differs from *D. calirayae*, *D. discophorus* and *D. nummularius* in having the scutum ground colour black and lacking a medial longitudinal postsutural yellow vitta. It differs from *D. lagunae* in having a narrower mesopleural stripe reaching to anterior *npl.* seta and by having two large oval black facial spots not joined along oral margin. It differs from *D. subsessilis* in having a distinct apical spot on the costal band (not a diffuse pattern around apical and hind margins) and by having abdominal tergum III mostly dark fuscous to black.

Dacus (Mellesis) maculipterus White

(Fig. 399)

Dacus (Callantra) maculipterus White, in Drew *et al.*, 1998: 607. Holotype male in BBM.

Type material examined: Holotype male, **THAILAND**: near Ban Sap Bon, 3 km E of Sara Buri, 4.x.1975, coll. R. Kano (in BBM).
Diagnosis: A medium-sized species; face yellow with a large elongate black spot in each antennal furrow joined across oral margin by a broad fuscous line; postpronotal lobes yellow (fulvous posteriorly but may be due to shrinkage of internal tissues); notopleura yellow; scutum black, lateral and medial postsutural yellow vittae absent; a yellow triangular marking along anterior side of notopleural suture; narrow mesopleural stripe only as wide as notopleuron dorsally and just reaching katepisternum ventrally; scutellum yellow except for narrow dark basal band; setae: *sc.* 2; *prsc.* absent; *ia.* 1; *p.sa.* 1; *a.sa.* 1; *mpl.* 1; *npl.* 2; *scp.* 4 (well developed); legs with femora dark red-brown to fuscous except basal quarter of hind femora fulvous; tibiae dark red-brown to fuscous, mid tibiae each with an apical black spur, hind tibiae each with a prominent preapical microtrichose bulge; tarsi with first segment fulvous, remainder fuscous; wing with cells bc and c hyaline, most of cell c with microtrichia at least along whole anterior region; remainder of wing hyaline except fuscous narrow part of cell br, fuscous cell sc, a broad fuscous costal band confluent with R_{4+5} and widening to overlap this vein at a point just before apex of vein R_{2+3}; no anal streak; supernumerary lobe weak; abdomen elongate, clavate and petiolate, abdominal tergum III black, terga IV and V black anteriorly, laterally and along midline with orange-brown posteriorly either side of midline. A pair of fuscous shining spots on tergum V. Posterior lobe of surstylus of medium development, sternum V straight on posterior margin.
Attractant: No known record.
Female: No known record.
Distribution:
Type locality: Central Thailand (Ban Sap Bon, near Saraburi).
Hosts: No known record.
Remarks: *Dacus (Mellesis) maculipterus* White is similar to *Dacus (Mellesis) atrimarginatus* White, *Dacus (Mellesis) crabroniformis* (Bezzi), *Dacus (Mellesis) ramanii* Drew & Hancock, *Dacus (Mellesis) satanas* (Hering) and *Dacus (Mellesis) sinensis* (Wang) in characters discussed under the comments on *D. atrimarginatus*. It differs from *D. atrimarginatus* in having dark legs and from *D. crabroniformis*, *D. ramanii*, *D. satanas* and *D. sinensis* in having abdominal terga IV and V with pale areas posteriorly either side of a medial longitudinal dark band.

Dacus (Mellesis) melanopectus Drew & Romig, new species

(Fig. 400)

Material examined: Holotype male, **INDONESIA: SULAWESI** – Sulawesi Tengah, Donggala, Nupabomba,

23.ix.2006, coll. Lyar, attracted to methyl eugenol in forest, deposited in BMNH.

Diagnosis: A medium-sized species; face entirely black; postpronotal lobes yellow (outer and inner margins black); notopleura yellow; scutum uniformly black; lateral and medial postsutural vittae absent; a large yellow triangle along anterior margin of notopleural suture with base adjacent to notopleuron; mesopleural stripe of medium-width reaching midway between anterior margin of notopleuron and postpronotal lobe dorsally; scutellum yellow with a broad black basal band; wing with cells bc and c fuscous, dense microtrichia covering all of cell c and outer ½ of cell bc; a broad dark fuscous costal band overlapping R_{4+5} at the apex of the wing where it continues around the apex to the hind margin and paling off from fuscous to a pale fuscous tint; anal streak absent; abdominal terga III–V entirely black.

Description: male.

Head: Height 1.47 mm. Frons – length 1.5 times breadth; fulvous with large areas of dark fuscous on anteromedial hump and around bases of orbital setae; orbital setae black: 1 *s.or.*, 2 *i.or.*; lunule dark fuscous. Ocellar triangle black. Vertex black. Face entirely black; length 0.39 mm. Genae dark fuscous, black subocular spot present; black seta present. Occiput black, yellow along eye margins; occipital row with a large number of small black setae. Antennae with segments 1 and 2 red-brown; segment 3 absent; length of segments: 0.44 mm; 0.44 mm.

Thorax: Scutum entirely black without pale markings. Pleural areas entirely black. Yellow markings as follows: postpronotal lobes (outer and inner margins black); notopleura; mesopleural stripe reaching midway between anterior margin of notopleuron and anterior *npl.* seta dorsally, extending to katepisternum as a transverse spot, anterior margin convex; a large triangular marking along anterior margin of notopleural suture with base adjacent to notopleuron. Anatergite (posterior apex black); anterior ⅔ katatergite (remainder black). Lateral and medial postsutural yellow vittae absent. Postnotum black. Scutellum yellow with a broad black basal band. Setae: *sc.* 2; *prsc.* absent; *ia.* 1; *p.sa.* 1; *a.sa.* 1; *mpl.* 1; *npl.* 2; *scp.* 4.

Legs: Fore and mid femora entirely black except for dark red-brown apices, hind femora fulvous with black on apical ⅓; all tibiae entirely dark fuscous; all tarsal segments entirely fulvous; mid tibiae each with an apical black spur.

Wings: Length 4.9 mm; cells bc and c fuscous; dense microtrichia covering all of cell c and outer ½ of cell bc; remainder of wings colourless except for dark fuscous cell sc, a broad dark fuscous costal band confluent with R_{4+5} basally and overlapping this vein apically where it extends around apex of wing and to hind margin where it pales to fuscous and then a pale fuscous tint; anal streak absent; a small area of dense microtrichia around A_1+CuA_2; supernumerary lobe weak.

Abdomen: Elongate oval and petiolate; terga fused; pecten present on tergum III. Tergum I and sterna I and II longer than wide. Tergum I black with a narrow dark red-brown band across posterior margin; tergum II black with dark red-brown spots posterolaterally; terga III–V entirely black. A pair of oval black shining spots on tergum V. Posterior lobe of surstylus long, sternum V with a slight concavity on posterior margin.

Attractant: Methyl eugenol.

Female: No known record.

Distribution: Indonesia (Sulawesi – Central Sulawesi, Donggala).

Hosts: No known record.

Remarks: *Dacus (Mellesis) melanopectus* n.sp. is similar to *Dacus (Mellesis) feijeni* White in possessing a similar wing colour pattern in the costal band, anal streak absent, cells bc and c fuscous and femora with dark patterns. It differs from *D. feijeni* in having a black face, black scutum, costal band expanding to M at apex of wing and abdominal terga III–V entirely black. *Dacus (Mellesis) dorjii* Drew and Romig is also similar in general body and wing colour patterns but differs from *D. melanopectus* in having black postpronotal lobes, the costal band overlapping R_{4+5} throughout and large orange-brown areas posterocentrally on abdominal terga IV and V.

Dacus (Mellesis) murphyi White

(Fig. 401)

Dacus (Callantra) murphyi White, in Drew *et al.*, 1998: 609. Holotype female in NUS.

Type material examined: Holotype female, **SINGAPORE**: Bukit Timah Forest, 1.v.1979, coll. D.H. Murphy (in NUS).

Diagnosis: A large species; face fulvous with a medium-sized black spot in each antennal furrow connected by a fuscous line across the oral margin; postpronotal lobes yellow (a spot inside each postpronotal lobe); notopleura yellow; scutum red-brown with irregular fuscous marks; a pair of elongate narrow lateral postsutural yellow vittae ending behind *ia.* seta; a short medial postsutural yellow vitta tapering anteriorly to end at line of notopleural suture; a yellow wedge along anterior margin of notopleural suture; mesopleural stripe of medium-width not reaching anterior notopleural seta dorsally and tapered to a point above katepisternum; scutellum yellow except for dark basal band; setae: *sc.* 2; *prsc.* absent; *ia.* 1; *p.sa.* 1; *a.sa.* 1; *mpl.* 1; *npl.* 2; *scp.* 4 (in type, central pair with one behind the other); legs with femora pale red-brown except basal ⅕ of mid and basal ⅐ of hind fulvous; tibiae red-brown, mid tibiae each with an apical black spur; basal tarsal segment fulvous, remainder slightly darker; wing with cells bc and c fuscous, cell c with microtrichia, cell bc without microtrichia; remainder of wings hyaline except for fuscous cell sc, fuscous costal band just extending over R_{4+5} basally and gradually widening further beyond apex of R_1 but not into a distinct spot; no anal streak; supernumerary lobe weak; abdomen elongate, clavate and petiolate, abdominal tergum III red-brown, tergum IV red-brown basally, laterally and along midline, leaving a pair of fulvous spots posteriorly each side of midline, tergum V as for tergum IV but with a pair of dark fuscous shining spots, in lateral view abdomen 'humped' though not as pronounced as in *Dacus solomonensis* Malloch, terga III–V with yellow hairs, ovipositor apex protruded in type and appears to be laterally compressed.

Attractant: No known record.

Male: No known record.
Distribution:
Type locality: Singapore (Bukit Timah Forest).
Hosts: No known record.
Remarks: *Dacus (Mellesis) murphyi* White is very similar to *Dacus (Mellesis) icariiformis* (Enderlein). It differs in having the costal band extending only slightly below vein R_{4+5} either side of *r-m* crossvein.

Dacus (Mellesis) nanggalae Drew & Hancock

(Fig. 402)

Dacus (Callantra) nanggalae Drew & Hancock, in Drew *et al*., 1998: 610. Holotype male in BMNH.

Type material examined: Holotype male, **INDONESIA: SULAWESI** – Palopo-Rantepao Rd, 16–23 km from Palopo, Sulawesi, (250–500 m), 24.vii.1981, coll. J. Turner, Nanggala III Forest, attracted to cue lure (in BMNH).
Diagnosis: A medium-sized species; face red-brown with a pair of large oval black spots; postpronotal lobes and notopleura yellow; scutum black except dark red-brown between postpronotal lobe and notopleuron and inside postpronotal lobes; lateral and medial postsutural yellow vittae absent; broad band along anterior margin of notopleural suture; broad mesopleural stripe reaching anterior *npl*. seta dorsally; scutellum yellow except for a broad black basal band; setae: *sc*. 2; *prsc*. absent; *ia*. 1; *p.sa*. 1; *a.sa*. 1; *mpl*. 1; *npl*. 2; *scp*. 4; legs with femora red-brown except for a dark fuscous spot ventrally on apical ⅓ of hind femora; all tibiae red-brown; tarsi fulvous; mid tibiae each with an apical black spur; wing with cells bc and c fuscous with dense microtrichia covering all of cell c and outer ⅔ of cell bc; remainder of wings colourless except fuscous to dark fuscous cell sc, broad fuscous to dark fuscous costal band overlapping R_{4+5} but not expanding apically; anal streak absent but with fuscous in basal area of cell cup; supernumerary lobe weak in male (female not studied); abdomen elongate oval, all terga red-brown except for a diffuse dark fuscous area across anterior half of tergum II but not reaching lateral margins and a narrow medial longitudinal fuscous to black band over terga III–V. A pair of oval red-brown shining spots on tergum V. Posterior lobe of surstylus short, sternum V with a slight concavity on posterior margin.
Attractant: Cue lure.
Female: No known record.
Distribution:
Type locality: Indonesia (Sulawesi – South Sulawesi, Palopo).
Hosts: No known record.
Remarks: *Dacus (Mellesis) nanggalae* Drew & Hancock is similar to *Dacus (Mellesis) ambonensis* Drew & Hancock in having a costal band overlapping vein R_{4+5}, a narrow medial longitudinal dark vitta and no lateral longitudinal dark bands on abdominal terga III–V. It differs from *D. ambonensis* in having the anatergite and katatergite mostly yellow, costal band dark fuscous and distinct below vein R_{4+5} for its entire length and anal streak absent with only colour in base of cell cup.

Dacus (Mellesis) nummularius (Bezzi)

(Fig. 403)

Mellesis nummularia Bezzi, 1916: 115 – Bezzi, 1919: 441. Lectotype male in MCSNM.
Mellesis apicalis Shiraki, 1933: 117 – Drew *et al*., 1998: 612 (Synonym).
Syntypes male and female possibly in NTU.
Callantra nummularia – Hardy, 1969b: 479, Lectotype designation; Hardy, 1974: 9; Hardy, 1977: 46.
Callantra apicalis – Hardy, 1977: 45.
Dacus (Callantra) apicalis – Kapoor, 1993: 83.
Dacus (Callantra) nummularius – Drew *et al*., 1998: 612; Norrbom *et al*., 1998: 130.

Non-type material examined: PHILIPPINES: LUZON ISLAND – 1 female, Lipa, Batangas, Luzon, xii.[19]52, 868, coll. C.B. Matabasco (in BBM). **PENINSULAR MALAYSIA**: 1 male, TN 3, 14.vii.1986, coll. Yusoff Salleh, attracted to cue lure; 1 male, TS 10, Bentong, 9.ii.1987, coll. Bahari, attracted to cue lure (in QDPC).
Diagnosis: A medium-sized species; face fulvous with two very large oval black spots filling most of antennal furrows; postpronotal lobes yellow, notopleura yellow (ventral margins narrowly fuscous); scutum dull black centrally, on lateral margins between postpronotal lobe and notopleuron and along anterior margins of yellow triangles adjacent to notopleural suture (remainder of scutum red-brown to fuscous); lateral postsutural yellow vittae absent; a short medial longitudinal yellow vitta rounded posteriorly and pointed anteriorly; broad triangular markings with bases at notopleura and running along anterior margins of notopleural suture; broad mesopleural stripe reaching anterior notopleural seta dorsally; scutellum yellow except for a narrow dark fuscous basal band; setae: *sc*. 2; *prsc*. absent in one specimen, 1 weak and 1 strong in the other specimen; *ia*. 1; *p.sa*. 1; *a.sa*. 1; *mpl*. 1; *npl*. 2; *scp*. 4; legs with fore and mid femora fuscous to dark fuscous except for small fulvous areas basally, hind femora fulvous; all tibiae fuscous to dark fuscous; tarsi with basal segment fulvous, apical four segments fuscous; mid tibiae with an apical black spur; wings with cells bc and c fuscous, dense microtrichia covering all of cell c; remainder of wing colourless except for dark fuscous cell sc, broad dark fuscous costal band confluent with R_{4+5} and expanding into a large round spot in wing apex (a small colourless spot occurs in front of R_{4+5} where the expansion into the apical dark fuscous spot begins); narrow dark fuscous anal streak ending before wing margin; supernumerary lobe weak in both sexes; abdomen elongate, clavate and petiolate, abdominal terga III–V orange-brown except for dark fuscous along anterior margin of tergum III, a broad medial longitudinal dark fuscous band and two broad lateral longitudinal fuscous to dark fuscous bands over all three terga. A pair of oval orange-brown shining spots on tergum V. Posterior lobe of surstylus short, sternum V with a deep concavity on posterior margin.
Attractant: Cue lure.
Distribution:
Type locality: Philippines (Luzon I., Mt Banahao).

Previously recorded: Type locality and the following locations – Taiwan, Peninsular Malaysia (Drew *et al.*, 1998). Records from Myanmar (Burma) and North-East India (Hardy, 1977) refer to *Bactrocera brachycera* (Bezzi) or a similar undescribed species.
Hosts: No known record.
Remarks: *Dacus (Mellesis) nummularius* (Bezzi) is similar to *Dacus (Mellesis) calirayae* Drew & Hancock, *Dacus (Mellesis) discophorus* (Hering), *Dacus (Mellesis) lagunae* Drew & Hancock, *Dacus (Mellesis) leongi* (Drew & Hancock) and *Dacus (Mellesis) subsessilis* (Bezzi) in possessing a broad costal band confluent with vein R_{4+5} and expanding distinctly at apex. It differs from *D. lagunae*, *D. leongi* and *D. subsessilis* in having the ground colour of the scutum red-brown (not black) with a medial longitudinal postsutural yellow vitta. *Dacus nummularius* differs from *D. calirayae* in having a short narrow medial longitudinal postsutural vitta and from *D. discophorus* in possessing a broad mesopleural stripe reaching the anterior notopleural bristle. It is superficially similar to *Bactrocera (Sinodacus) infesta* (Enderlein) and *Bactrocera (Sinodacus) hochii* (Zia) but differs in the obvious generic and subgeneric characters.

Dacus (Mellesis) ooii Drew & Hancock

(Fig. 404)

Dacus (Callantra) ooii Drew & Hancock, in Drew *et al.*, 1998: 614. Holotype male in BMNH.

Type material examined: Holotype male, **PENINSULAR MALAYSIA**: Baha Pahat, Johore, vii.1988, coll. C.S. Ooii (in BMNH).
Paratypes: **PENINSULAR MALAYSIA**: 3 males, Ulu Langat, near Kuala Lumpur, 18.i.1989, coll. Ooi Chee Seng, attracted to cue lure; 1 male, TS Gombak/Kuantan Road, 12.ii.1986, coll. Mahadzir, attracted to cue lure; 2 males, RSI Pontian, 14.vii.1986, coll. Salleh; 1 male, RSI, 8.ix.1986, attracted to cue lure; 1 male, Sungai Baging, 16.v.1986, coll. R. Samat Wahid, attracted to cue lure; 1 male, R S Telong, 16.vi.1986, coll. Fauzi; 1 male, Kemaman, 8.ix.1986, coll. Raja Azid; 1 male, TN 1, 23.ii.1987, coll. Y. Salleh, attracted to cue lure; 2 males, Gombak/Kuantan Road, 9.iii.1987, coll. Antony, attracted to cue lure; 1 male, Sungai Baging, 22.vi.1987, coll. A. Samat Wahid, attracted to cue lure (in ANIC; BMNH; QDPC).
Non-type material examined: INDONESIA: 2 males, Santong, Lombok, (1200-1300m), 2.vi.1981, coll. J. Turner, attracted to cue lure; 1 male, Suranadi, Lombok (200m), 31.v.1981, coll. J. Turner, attracted to cue lure. **THAILAND**: 1 male, Surat Thani, Chumporn Hor. Res. Centre, Sawi, Chumporn, 2.viii.1991, coll. Pichate, attracted to cue lure (in QDPC).
Diagnosis: A large species; face fulvous with a pair of large irregularly oval black spots; postpronotal lobes and notopleura yellow; scutum and pleura black except red-brown below postpronotal lobes and on posterolateral corners of scutum and dark fuscous inside postpronotal lobes; lateral and medial postsutural yellow vittae absent; broad band along anterior margin of notopleural suture; broad mesopleural stripe reaching to or almost to anterior *npl.* seta; scutellum yellow except for narrow black basal band; setae: *sc.* 2; *prsc.* absent; *ia.* 1; *p.sa.* 1; *a.sa.* 1; *mpl.* 1; *npl.* 2; *scp.* 4; legs with fore femora with outer surfaces and apical ½ of inner surfaces variable fuscous to black, remainder of inner surfaces red-brown, mid femora pale fuscous to black (except small areas of red-brown around basal and apical points), hind femora with apical ⅓ to ½ dark fuscous to black (remainder fulvous); all tibiae dark fuscous; tarsi with basal segment fulvous, apical four segments darker; mid tibiae each with an apical black spur; wing with cells bc and c fuscous, dense microtrichia covering all of cell c and outer ½ of cell bc; remainder of wing colourless except dark fuscous cell sc, broad dark fuscous costal band overlapping R_{4+5} from base of cell br and gradually widening to apex of wing (in a few specimens it reaches but does not overlap M, when reaching M it is usually pale fuscous in that area); a narrow fuscous anal streak ending well before wing margin; supernumerary lobe weak in male (female not studied); abdomen elongate, clavate and petiolate, abdominal terga III–V orange-brown except for a broad dark fuscous to black medial longitudinal vitta and two broad variable fuscous to dark fuscous lateral longitudinal vittae over all three terga and joining over all or most of tergum III. A pair of oval orange-brown shining spots on tergum V. Posterior lobe of surstylus short, sternum V with a slight concavity on posterior margin.
Attractant: Cue lure.
Female: No known record.
Distribution:
Type locality: Peninsular Malaysia (Jahore, Baha Pahat).
Previously recorded: Type locality and in locations Peninsular Malaysia; Southern Thailand; Indonesia (Lombok) (Drew *et al.*, 1998).
Hosts: No known record.
Remarks: *Dacus (Mellesis) ooii* Drew & Hancock is similar to *Dacus (Mellesis) pictus* (Hardy) in possessing a costal band overlapping R_{4+5} and of approximately uniform width (usually not touching or overlapping vein M except at base), abdominal terga II–V with a broad medial longitudinal dark band and terga III–V with broad lateral longitudinal dark bands. It differs from *D. pictus* in having a black scutum, costal band widening slightly at apex and a distinct anal streak in the wing.

Dacus (Mellesis) pedunculatus (Bezzi)

(Fig. 405)

Mellesis pedunculata Bezzi, 1916: 115 – Bezzi, 1919: 439. Lectotype male in MCSNM.
Callantra pedunculata – Hardy 1969b: 479, Lectotype designation; Hardy, 1970: 72; Hardy, 1974: 10; Hardy, 1977: 46.
Dacus (Callantra) pedunculatus – Drew *et al.*, 1998: 616; Norrbom *et al.*, 1998: 130.

Non-type material examined: PHILIPPINES: LUZON ISLAND – 1 male, Mayoyao, Ifugao, Mt. Prov., Luzon, 1200-1500m, 14.ix.1966, coll. H.M. Torrevillas (in BBM); 1 female, Tawi Tawi, Tarawakan, north of Batu batu, 23.x.1961, Noona Dan Exp. 61-62 (in BBM).

Diagnosis: A medium-sized species; face with a deep black line across oral margin that expands vertically into a spot in each antennal furrow; postpronotal lobes and notopleura yellow; scutum red-brown with black bands in lateral areas and along dorsocentral lines; lateral and medial postsutural yellow vittae absent; triangular marking along anterior side of notopleural suture; narrow mesopleural stripe only a little wider than notopleuron dorsally; scutellum yellow except for narrow fuscous basal band; setae: *sc*. 2; *prsc*. absent; *ia*. 1; *p.sa*. 1; *a.sa*. 1; *mpl*. 1; *npl*. 2; *scp*. 4–6; legs with femora predominantly red-brown except hind femora fulvous on basal quarter, mid femora damaged but appears pale basally; tibiae fulvous tending to fuscous on hind tibiae; tarsi fulvous (mid tarsi missing); mid tibiae each with a black apical spur, hind tibiae with a preapical microtrichose bulge; wing with cell c with a pale fuscous tint, cell bc pale fulvous, cell c with dense microtrichia throughout, cell bc with microtrichia in outer apical area; remainder of wing hyaline except for pale fuscous in narrow part of cell br, fuscous cell sc and fuscous costal band that almost reaches R_{4+5}, only crossing vein R_{4+5} as a shallow expansion beyond the end of vein R_{2+3}; no anal streak; supernumerary lobe weak in both sexes; abdomen elongate, clavate and petiolate, abdominal tergum III fuscous with a trace of pale spots either side of centre on hind margin, tergum IV fuscous laterally and on midline with narrow borders of fulvous either side of midline, tergum V fulvous including shining spots, fuscous anterolaterally and on midline. Posterior lobe of surstylus short, sternum V with a slight concavity on posterior margin.

Attractant: No known record.

Female: As for male except hind femora fulvous on basal half, mid femora entirely red-brown, pecten of cilia on abdominal tergum III absent; no dense aggregation of microtrichia around A_1+CuA_2; no microtrichose bulge on hind tibiae. Wing length 5.6 mm. Oviscape round in cross section; short, not visible in dorsal view.

Distribution:

Type locality: Philippines (Luzon I., Mt Makiling).

Previously recorded: Type locality and the Philippines (Tawi Tawi I.). Record of Mindanao (Drew *et al*., 1998) unconfirmed.

Hosts: No known record.

Remarks: *Dacus (Mellesis) pedunculatus* (Bezzi) is similar to *Dacus (Mellesis) indecorus* (Hardy) in characters given under the remarks for *D. indecorus*. It differs in having a broad dark band across the oral margin, a dark marking each side between the antennal base and eye margin, cell bc not entirely covered with microtrichia and the anal streak absent.

Dacus (Mellesis) pictus (Hardy)
(Fig. 406)

Callantra picta Hardy, 1970: 72. Holotype male in UZMC (Hardy, 1974 incorrectly stated that the holotype was a female).

Dacus (Callantra) pictus – Drew *et al*., 1998: 618; Norrbom *et al*., 1998: 131.

Dacus (Callantra) haikouensis Wang & Chen, 2002: 27. **New Synonym**.

Type material examined: Holotype male, **PHILIPPINES**: Uring Uring, Palawan, 26.ix.1961, in Malaise trap (in UZMC).

Other type material examined: CHINA: HAINAN PROVINCE – Holotype and 1 paratype of *Dacus (Callantra) haikouensis* Wang & Chen, Haikou, Hainan Province, 18. vii.1960, coll. Li, S.F. Photographs of types received from Chen Xiao-Lin (in IZAS).

Diagnosis: A large species; face yellow with large black spots that extend dorsally as traces of reddish colour to meet beneath antenna bases; postpronotal lobes and notopleura yellow; scutum mostly red-brown; no lateral postsutural yellow vittae; some trace of possible medial yellow vitta (area obscured anteriorly by pin); a dark fuscous stripe dorsocentrally behind notopleural suture; mesopleural stripe reaching anterior notopleural seta dorsally; scutellum yellow except for narrow dark basal band; setae: *sc*. 2; *prsc*. absent; *ia*. 1; *p.sa*. 1; *a.sa*. 1; *mpl*. 1; *npl*. 2; *scp*. 4; legs with femora red-brown (tending darker along ventral margins) and width basal ⅔ of hind femora fulvous; tibiae red-brown to fuscous; tarsi pale fulvous; mid tibiae each with a black apical spur, hind tibiae with preapical microtrichose hump; wing with cell bc pale fuscous, cell c fuscous, microtrichia in apical half of cell bc and throughout cell c; remainder of wing hyaline except for fuscous cell sc, narrow basal part of cell br, broad fuscous costal band overlapping R_{4+5} and expanding mid way to M apically; anal streak absent with a tint within cup; supernumerary lobe weak; abdomen elongate, clavate and petiolate, abdominal tergum III fuscous with a trace of fulvous posteriorly each side of a dark midline, tergum IV fulvous with broad fuscous marks laterally and fuscous midline, terga V fulvous, sub-rectangular fuscous marks anterolaterally and fuscous midline. A pair of fulvous shining spots on tergum V. Posterior lobe of surstylus of medium development, sternum V with a shallow concavity on posterior margin.

Attractant: No known record.

Female: No known record.

Distribution:

Type locality: Philippines (Palawan I., Uring Uring).

New records: China (Hainan Province, Haikou), due to **New Synonymy** of *D. haikouensis*.

Hosts: No known record.

Remarks: *Dacus (Mellesis) pictus* (Hardy) is similar to *Dacus (Mellesis) ooii* (Drew & Hancock) in possessing a costal band that overlaps vein R_{4+5} and of approximately uniform width (not touching or overlapping vein M except at base), abdominal terga II–V with a broad medial longitudinal dark band and terga III–V with broad lateral longitudinal dark bands. It differs from *D. ooii* in having the ground colour of the scutum red-brown, the costal band not widening at apex and the anal streak in the wing absent (reduced to a pale tint within cup).

After studying the types of *Dacus (Callantra) haikouensis* Wang & Chen and *D. pictus*, we can find no significant characters to separate them and are therefore making *D. haikouensis* a **New Synonym** of *D. pictus*.

Dacus (Mellesis) polistiformis (Senior-White)

(Fig. 407)

Mellesis polistiformis Senior-White, 1922: 156. Holotype male in BMNH.
Callantra munroi Zaka-ur-Rab, 1961: 538 – Drew *et al.*, 1998: 620 (Synonym). Holotype female in MUA.
Callantra nepalensis Hardy, 1964: 149 – Hardy, 1977: 46. Drew *et al.*, 1998: 60 (Synonym). Holotype female in BMNH.
Callantra polistiformis – Hardy, 1977: 46.
Callantra quadristriata Munro, 1984: 155 – Drew *et al.*, 1998: 620 (Synonym). Holotype male in SANC.
Dacus (Callantra) nepalensis – Liang *et al.*, 1993: 139.
Dacus (Callantra) munroi – Kapoor, 1993: 83.
Dacus (Callantra) quadristriata – Kapoor, 1993: 83.
Dacus (Callantra) polistiformis – Drew *et al.*, 1998: 620; Norrbom *et al.*, 1998: 131.

Type material examined: Holotype male, **INDIA**: Sukna, 500 ft (26°47'N, 88°22'E), West Bengal, near Darjeeling, 4.xii.1919 (in BMNH).
Other type material examined: NEPAL: Holotype female of *D. nepalensis*, evergreen shrubs in rocky ravine above E shore of River Arun, c. 2000 ft, 25.xii.1961, Brit. Mus. East Nepal Expedition, 1961–62, coll. R.L. Coe, BM 1962-177 (in BMNH).
Non-type material examined: CHINA: one female, Xishuangbanna, Yunnan Province (in IZAS). **INDIA**: one female NE India, coll. T. Jermyn, BM 1949-53 (in BMNH).
Diagnosis: A large species, face fulvous with a narrow black line across oral margin and joining traces of a narrow fuscous line in each antennal furrow; postpronotal lobes yellow (sometimes [not type] with a yellow spot adjacent to postpronotal lobe); notopleura yellow; scutum red-brown; lateral and medial postsutural yellow vittae absent; a narrow wedge-shaped band along anterior margin of notopleural suture; narrow mesopleural stripe; scutellum yellow except for narrow red-brown basal band; setae pale: *sc.* 2; *prsc.* absent; *ia.* 1; *p.sa.* 1; *a.sa.* absent; *mpl.* 1; *npl.* 2; *scp.* not differentiated from general vestiture; legs with femora red-brown except basal ⅕ of mid and hind femora indistinctly fulvous, fore femora with 2 or 3 prominent ventral spines (2 in Holotype); tibiae fulvous to red-brown; mid tibiae each with a black apical spur; hind tibiae with preapical microtrichose bulge; wing with cells bc and c pale fuscous, dense microtrichia in cell c and apical ½ of cell bc; remainder of wing hyaline except for pale fuscous costal band that overlaps R_{4+5} throughout and usually (including Holotype) abruptly darkened to fuscous beyond end of vein R_{2+3}; no anal streak; supernumerary lobe weak in both sexes; abdomen elongate, clavate and petiolate, all terga red-brown except for fuscous transverse markings across bases of terga II, IV and V. A pair of red-brown shining spots on tergum V. Posterior lobe of surstylus short, sternum V with a slight concavity on posterior margin.
Attractant: No known record.
Distribution:
Type locality: Northern India (West Bengal, near Darjeeling, Sukna).
Previously recorded: Type locality and other locations in northern India; East Nepal; China (Yunnan Province) (Drew *et al.*, 1998).
Hosts: *Oxystelma esculentum* (*Asclepiadaceae*). Munro (1984) noted that the holotype of *C. quadristriata* was bred from a pod of a wild plant (possibly an asclepiad) on *Tamarix*.
Remarks: *Dacus (Mellesis) polistiformis* (Senior-White) is similar to *Dacus (Mellesis) bispinosus* (Wang) and *Dacus (Mellesis) esakii* (Shiraki) in having a broad costal band not distinctly expanded at the apex and strong subapical spines on the fore femora. It differs from these species in having a costal band overlapping vein R_{4+5} and of uniform width throughout its length.

Dacus (Mellesis) pullus (Hardy)

(Fig. 408)

Callantra pullus Hardy, 1982: 181. Holotype male in MZB.
Dacus (Callantra) pullus – Drew *et al.*, 1998: 621; Norrbom *et al.*, 1998: 131.

Type material examined: Holotype male, **INDONESIA: SULAEWSI** – Sadaunta, 65 km SE of Palu, Sulawesi, 650 m, viii.1975, coll. D.E. & A.T. Hardy (in MZB).
Diagnosis: A large species; face with two large round slightly elongate black spots; postpronotal lobes yellow; notopleura yellow; scutum black with red-brown posteriorly; lateral and medial postsutural yellow vittae absent; broad triangular markings along anterior margins of notopleural suture, mesopleural stripe narrow; scutellum yellow; setae: *sc.* 2; *prsc.* absent; *ia.* 1; *p.sa.* 1; *a.sa.* 1; *mpl.* 1; *npl.* 2; legs with femora red-brown with a dark fuscous to black streak on basal half of ventral surfaces; all tibiae red-brown with hind tibiae darker; no spines on fore femora; wing with cells bc and c fuscous, broad fuscous costal band almost confluent with M and overlapping this vein at apex; anal streak present; abdomen with terga III–V mostly black except fulvous along posterior margin of tergum V and on the oval shining spots on this tergum. A pair of fulvous oval shining spots on tergum V.
Attractant: No known record.
Female: No known record.
Distribution:
Type locality: Indonesia (Sulawesi – Central Sulawesi, Sadaunta).
Hosts: No known record.
Remarks: *Dacus (Mellesis) pullus* (Hardy) is similar to *Dacus (Mellesis) insulosus* Drew & Hancock and *Dacus (Callantra) longicornis* Wiedemann in having the costal band overlapping vein R_{4+5} throughout and yellow postpronotal lobes. It differs from both species in having a black scutum, abdominal terga III–V mostly black and the

costal band almost reaching vein M in centre of wing and overlapping this vein at its apex.

Dacus (Mellesis) ramanii Drew & Hancock

(Fig. 409)

Dacus (Callantra) ramanii Drew & Hancock, in Drew *et al.*, 1998: 623. Holotype male in BMNH.

Type material examined: Holotype male, **INDIA**: 37 km E of Kodaikanal (850 m), 2.vi.1992, coll. D.L. Hancock, attracted to cue lure (in BMNH).
Paratypes: **INDIA**: 1 male, Bangalore (916 m), Kanataka, 24.ii.1988, coll. G. Bhat, attracted to cue lure (in QDPC). **SRI LANKA**: 1 male, Ham[bantota] District, Palatupana Tank (10–20 m), 27–29.ix.1977, coll. K.V. Krombein, P.B. Karunaratne, T. Wijesinhe, M. Jayaweera (in USNM); 1 male, Tabbowa, 24.ix.1994, [Cue lure] trap; 1 male, Tabbowa, 22.x.1994, C14; 1 male, Tabbowa, 7.x.1994, C16; 2 males, Tabbowa, 28.ii.1995, C16 (all Tabbowa specimens additionally labelled Sri Lanka NPQS Project Sp. R. No. 33) (in BMNH).
Non-type material examined: SRI LANKA: 1 male, Sri Lanka, Tabbowa, 24.ix.1994; 1 male, Sri Lanka, Kalpitiya, C14, 5.iv.1995 (in QDPC).
Diagnosis: A small species; face glossy black; postpronotal lobes yellow (except anterior and ventral margins black); notopleura yellow; scutum shining black without pale markings; lateral and medial postsutural yellow vittae absent; narrow triangle along anterior margin of notopleural suture; broad mesopleural stripe reaching postpronotal lobe dorsally; scutellum yellow except for narrow black basal band; setae: *sc.* 2; *prsc.* absent; *ia.* 1; *p.sa.* 1; *a.sa.* 1; *mpl.* 1; *npl.* 2; *scp.* 2; legs with fore and mid femora glossy black, hind femora fulvous except apical ⅓ glossy black; fore tibiae fuscous, mid tibiae fulvous tending fuscous basally, hind tibiae dark fuscous; tarsi entirely fulvous; mid tibiae each with an apical black spur; wing with cells bc and c colourless or with a very pale tint, microtrichia over most of cell c only; remainder of wings colourless except dark fuscous cell sc, broad dark fuscous costal band expanding along *r-m* crossvein and then widening apically to overlap R_{4+5} in an irregular pattern; anal streak absent; supernumerary lobe weak in male (female not studied); abdomen elongate oval, all abdominal terga shining black without pale markings. A pair of oval black shining spots on tergum V. Posterior lobe of surstylus slightly elongated, sternum V with a slight concavity on posterior margin.
Attractant: Cue lure.
Female: No known record.
Distribution:
Type locality: Southern India (Tamil Nadu, 37 km E of Kodaikanal).
Previously recorded: Type locality and locations in Southern India and Sri Lanka (See Drew *et al.*, 1998).
Hosts: No known record.
Remarks: *Dacus (Mellesis) ramanii* Drew & Hancock is similar to *Dacus (Mellesis) crabroniformis* (Bezzi), *Dacus (Mellesis) satanas* (Hering) and *Dacus (Mellesis) sinensis* (Wang) in having the costal band confluent with or overlapping vein R_{4+5}, cells bc and c colourless, lateral and medial postsutural vittae absent and abdominal terga mostly black. It differs from these species in having a different costal band shape, which is expanded medially over *r-m* crossvein.

Dacus (Mellesis) satanas (Hering)

(Fig. 410)

Callantra satanas Hering, 1939: 166 – Hardy, 1977: 46. Holotype male in MNHNP.
Callantra ziae Wang, 1990: 71, 75. Holotype male in IZAS.
Dacus (Callantra) ziae – Liang *et al.*, 1993: 139 (Synonym).
Dacus (Callantra) satanas – Drew *et al.*, 1998: 624; Norrbom *et al.*, 1998: 131.

Type material examined: Holotype male, **VIETNAM**: **Northern Vietnam** – Tonkin, Hoa Binh, coll. A. de Cooman, 1927 (in MNHNP).
Other type material examined: CHINA: Types of *Dacus (Callantra) ziae* (Wang) – Holotype male, Xiaomengyang, 850m, 2.iv.1957; one female paratype, Menghua, 1200m, 28.vi.1958; one male paratype, Daimenglong, 7.iv.1958; all types from Yunnan Province, studied in IZAS by D.L. Hancock.
Diagnosis: A small species; face black; postpronotal lobes and notopleura yellow; scutum mostly black; lateral and medial postsutural yellow vittae absent; triangular markings along anterior margins of notopleural suture; mesopleural stripe of medium-width; scutellum yellow; setae: *sc.* 2; *prsc.* absent; *ia.* 1; *p.sa.* 1; *a.sa.* 1 (weak); *mpl.* 1; *npl.* 2; *scp.* absent; legs with femora black except fulvous around apices and basal quarter of hind femora; fore and hind tibiae dark fuscous and mid tibiae fuscous; wing with cells bc and c colourless; broad fuscous costal band confluent with R_{4+5} and overlapping this vein at apex; anal streak absent; cell cup with apical lobe elongate, microtrichia present but indistinct around A_1+CuA_2; abdomen with pecten present on tergum III in Holotype male of *D. ziae* but apparently absent in the male Paratype, terga predominantly black except with dark red-brown along hind margins of terga IV and V.
Attractant: No known record.
Distribution:
Type locality: Northern Vietnam (Hoa Binh Province).
Previously recorded: Type locality and China (Yunnan Province).
Hosts: No known record.
Remarks: *Dacus (Mellesis) satanas* (Hering) is similar to *Dacus (Mellesis) crabroniformis* (Bezzi) in having a black face, wing with the costal band confluent with vein R_{4+5} except at apex, cells bc and c colourless, scutum and abdominal terga mostly black. It differs from *D. crabroniformis* in having all femora black except the basal ¼ of hind femora fulvous (in *D. crabroniformis* the femora are mostly red-brown) and the postpronotal lobes entirely yellow.

Dacus (Mellesis) siamensis Drew & Hancock

(Fig. 411)

Dacus (Callantra) siamensis Drew & Hancock, in Drew *et al*., 1998: 626. Holotype male in BMNH.
Dacus (Mellesis) siamensis – Drew *et al*., 2007: 18.

Type material examined: Holotype male, **THAILAND**: Doi Luang Natl Pk, 50 km S of Chiang Rai, Thailand, 18.i.1994, D.L. Hancock and Charuwan Kong Krong (in BMNH).
Paratypes: **THAILAND**: Central Region: 9 males, Khao Kheaw, Pakphli, Nakhonnayok, 18.x.1991, 8.vii.1991, 26.xi.1991, Montree, cue lure; 27 males, Wang Jum Pee, Pakchong, Nakhonratsima, 18.x.1991, 25.x.1991, 1.xi.1991, 15.xi.1991, 29.xi.1991, 6.xii.1991, 13.xii.1991, 20.xii.1991, 27.xii.1991, 3.i.1992, 24.i.1992, 31.i.1992, 10.iv.1992, 10.vii.1992, Montree, cue lure; Chiang Mai: 10 males, Mae Hae, 10.i.1992, 30.i.1992, 18.ii.1992, Chantanee, cue lure; 1 male, NRDC, 27.ix.1991, Chantanee, cue lure; Chiang Rai: 1 male, Chiang Sang, 10.vi.1991, Charuwan, cue lure; 65 males, Khun Khon, 9.i.1992, 15.i.1992, 23.i.1992, 30.i.1992, 4.ii.1992, 6.ii.1992, 13.ii.1992, 27.ii.1992, Charuwan, cue lure; 4 males, Phan, 8.i.1992, 15.i.1992, Charuwan, cue lure; 5 males, Pha Yao, 22.i.1992, Charuwan (in ANIC; BMNH; QDPC).
Non-type material examined: BHUTAN: 1 male, Suntalay, 22.v.2000, coll. S.W. & S. Dorji; 1 male, Suntalay, 5.vi.2000, coll. S. Wangdi, T.D. & S.D.; 1 male, Suntalay, 5.vi.2000; 1 male, Purana Bastey, 23.xi.2004, coll. Karma Namgyel & Maha Prasard. All specimens attracted to cue lure. **VIETNAM: Northern Vietnam** – 1 male (15.xi.1999), 1 male (13.v.2000), Vietnam North, Son La Province, Moc Chau Dist., Tra Long, 15.xi.1999, Site N007, attracted to cue lure; 1 male (1.xi.1999), 1 male (15.v.2001), Vietnam North, Son La Province, Moc Chau Dist., Co Do, Site N006, attracted to cue lure; 1 male, Vietnam North, Son La Province, Moc Chau Dist., Moc Chau house garden, 15.xi.1999, Site N008, attracted to cue lure; 1 male, Vietnam North, Son La Province, Moc Chau Dist., Moc Chau town, 29.xi.1999, Site N009, attracted to cue lure; 1 male (23.i.2000), 1 male (30.x.2000), 1 male (30.vi.2001), Vietnam North, Hoa Binh Province, Mai Chau Dist., Pa Co, Site N010; 1 male, Vietnam, Hoa Binh P, Mai Chau Dist., Pa Co V, 15.iv.2002, Site VN10, FAO Site, coll. NIPP staff, attracted to cue lure; **Southern Vietnam** – 1 male, South Vietnam, Dong Nai Province, Long Khanh District, Nhan Nghia Village, 26.v.2002, Site VS06, attracted to cue lure. Specimens in QDPC.
Diagnosis: Face fulvous with a 'U'-shaped black pattern consisting of medium-sized oval black spots connected with a narrow black band across the dorsal margin of face; postpronotal lobes dark fuscous to black; notopleura yellow; scutum primarily dull black; lateral and medial postsutural yellow vittae absent; a yellow triangle along anterior margin of notopleural suture with base at notopleuron; mesopleural stripe slightly wider than notopleuron dorsally; scutellum yellow except for a narrow black basal band; setae: *sc*. 2; *prsc*. absent; *ia*. 1; *p.sa*. 1; *a.sa*. 1; *mpl*. 1; *npl*. 2; *scp*. 4; legs with all femora entirely black (tending paler on apices of mid femora); all tibiae entirely dark fuscous; fore tarsi with all segments fuscous, mid and hind tarsi with basal segments fulvous and apical four segments fuscous; wing with cells bc and c dark fuscous, dense microtrichia covering all of cell c and most of cell bc; a broad dark fuscous costal band overlapping R_{4+5}; anal streak absent, in some specimens there is a very pale fuscous tint throughout most of wing membrane; supernumerary lobe weak; abdomen elongate, club shaped and petiolate, all terga generally dark fuscous to black except fulvous along intersegmental line between terga I and II and, in some specimens, dark orange-brown posterocentrally on terga IV and V. A pair of oval black shining spots on tergum V. Posterior lobe of surstylus short, sternum V with a slight concavity on posterior margin.
Attractant: Cue lure.
Female: No known record.
Distribution:
Type locality: Thailand (Chiang Rai, Doi Luang National Park).
Previously recorded: Type locality and other locations in Thailand; Bhutan (Drew *et al*., 1998; Drew *et al*., 2007).
New records: Vietnam (Dong Nai, Hoa Binh, Son La Provinces).
Hosts: No known record.
Remarks: *Dacus (Mellesis) siamensis* Drew & Hancock has been adequately described and illustrated in Drew *et al.* (1998). It is readily distinguished by the following characters: scutum black, postpronotal lobes black, notopleura yellow, anatergite black, katatergite mostly yellow, face fulvous with an inverted 'U'-shaped black pattern, costal band overlapping R_{4+5}, cells bc and c dark fuscous, femora and tibiae mostly black, abdominal terga mostly black.

Dacus (Mellesis) sinensis (Wang)

(Fig. 412)

Callantra sinensis Wang, 1990: 70, 75. Holotype male in IZAS.
Dacus (Callantra) sinensis – Liang *et al*., 1993: 139; Drew *et al*., 1998: 628; Norrbom *et al*., 1998: 131.

Type material examined: Holotype male, Allotype female, **CHINA**: Daimenglong, Yunnan, 650 m, 9.iv.1958, coll. Zhang Yiran (in IZAS).
Paratypes: **CHINA**: 1 male, 1 female, Jinping, Yunnan, 370–420 m, 17–21.iv.1956, coll. Huang Keren; 1 female, Cheli, Yunnan, 580 m, 12.iii.1957, coll. Pu Fuji; 1 male, 1 female, Xishuangbanna, Yunnan, 540 m and 1050–1080 m, 15.iii.1957 and 29.v.1958, coll. Liu Daihua and Wang Shuyang, studied in IZAS by D.L. Hancock.
Diagnosis: A small species; face fulvous with two large black spots; postpronotal lobes yellow; notopleura yellow; scutum mostly black; lateral and medial postsutural yellow vittae absent; triangular markings along anterior margins of notopleural suture; mesopleural stripe of medium-width; scutellum yellow; setae: *sc*. 2; *prsc*. absent; *ia*. 1; *p.sa*. 1; *a.sa*. 1; *mpl*. 1; *npl*. 2; *scp*. absent; legs with femora and tibiae black except fulvous around basal ⅗ of hind femora; wing with cells bc and c colourless, broad fuscous costal band confluent with R_{4+5} and overlapping this vein at apex; anal streak absent; microtrichia absent around A_1+CuA_2, cell cup with apical lobe short; abdomen with

pecten absent on tergum III, abdominal terga mostly black with red-brown along posterior margins of terga IV and V either side of a medial black vitta.
Attractant: No known record.
Distribution:
Type locality: China (Yunnan Province, Daimenglong).
Previously recorded: Type locality and other locations in China (Yunnan Province) (Drew *et al.*, 1998).
Hosts: No known record.
Remarks: *Dacus (Mellesis) sinensis* (Wang) is similar to *Dacus (Mellesis) satanas* (Hering) and *Dacus (Mellesis) crabroniformis* (Bezzi) in having the costal band confluent with vein R_{4+5} except at wing apex, cells bc and c colourless, scutum and abdominal terga mostly black. It differs from these species in possessing two black spots on the face and black only on apical ⅓ of hind femora.

In the Dacinae, species that have the pecten absent on the male abdominal tergum III also have the dense microtrichia absent in the wing around vein A_1+CuA_2. Within the genus *Dacus* this is an uncommon set of characters. The fact that this state occurs in *D. crabroniformis*, *D. satanas* and *D. sinensis*, together with major similarities in leg, wing and body colour patterns, suggests that these are conspecific. However, they are being treated as separate species until more material can be obtained for study.

Dacus (Mellesis) subsessilis (Bezzi)

(Fig. 413)

Mellesis subsessilis Bezzi, 1919: 435. Holotype female in MCSNM.
Callantra subsessilis – Hardy, 1974: 12; Hardy, 1977: 46.
Dacus (Callantra) subsessilis – Drew *et al.*, 1998: 631; Norrbom *et al.*, 1998: 131.

Type material examined: Holotype female, **PHILIPPINES**: Panay, Culasi, vi.[19]18, coll. McGregor (in MCSNM).
Non-type material examined: PHILIPPINES: 1 female, Culion I., 6km W of Culion, 6.ii.1962, Malaise trap, coll. H. Holtmann (in BBM).
Diagnosis: A large species; face with a pair of broad oval black spots; postpronotal lobes and notopleura yellow; scutum black with red-brown over anterior section and around margins; lateral and medial postsutural yellow vittae absent; broad triangular marking along anterior margin of notopleural suture; mesopleural stripe broad; scutellum yellow with a narrow dark basal band; setae: *sc.* 2; *prsc.* absent; *ia.* 1; *p.sa.* 1; *a.sa.* 1; *mpl.* 1; *npl.* 2 *scp.* 4; legs mostly red-brown, tibiae paler and basal segment of all tarsi fulvous; no spines on fore femora; wing with cells bc and c fuscous; a broad fuscous costal band confluent with R_{4+5} and expanded into a diffuse apical spot that overlaps vein M; anal streak present; supernumerary lobe weak; abdomen broadly petiolate, abdominal terga III–V red-brown with a medial longitudinal black vitta (Bezzi 1919); Hardy (1974) recorded the abdomen as having the sides of the terga broadly black and a broad black medial longitudinal vitta over terga II–V.
Attractant: No known record.
Distribution:
Type locality: Philippines (Panay I., Culasi).
Previously recorded: Type locality and Philippines (Culion I.) (Drew *et al.*, 1998).
Hosts: No known record.
Remarks: *Dacus (Mellesis) subsessilis* (Bezzi) is similar to *Dacus (Mellesis) calirayae* Drew & Hancock, *Dacus (Mellesis) discophorus* (Hering), *Dacus (Mellesis) lagunae* Drew & Hancock, D*acus (Mellesis) leongi* Drew & Hancock and *Dacus (Mellesis) nummularius* (Bezzi) in possessing a broad costal band confluent with vein R_{4+5} and expanding markedly at apex. It differs from *D. calirayae*, *D. discophorus* and *D. nummularius* in having the scutum mostly black and lacking a medial postsutural yellow vitta. *Dacus subsessilis* differs from *D. lagunae* and *D. leongi* in having the apical expansion of the costal band as a diffuse infuscation around the apical and hind margins of the wing and red-brown over the anterior part of the scutum.

The specimen described by Hardy (1974) has been examined and found to differ as follows: wing with a fuscous tint, except for an ill-defined area centrally (not as distinct as indicated in Fig. 5 in Hardy (1974)); abdomen with tergum I slightly broader than long and sides of terga III–V broadly marked with dark fuscous, the whole abdomen with a continuous medial longitudinal broad black band (Fig. 413 is based on this description).

Dacus (Mellesis) tenebrosus Drew & Hancock

(Fig. 414)

Dacus (Callantra) tenebrosus Drew & Hancock, in Drew *et al.*, 1998: 632. Holotype male in BMNH.

Type material examined: Holotype male, **MALAYSIA**: TN9, 11.viii.1986, coll. Yusoff Salleh, attracted to cue lure (in BMNH).
Paratypes: **MALAYSIA**: 2 males, TN9, 28.vii.1986 and 8. ix.1986, coll. Yusoff Salleh, attracted to cue lure; 5 males, TN9, 25.viii.1986, coll. Yusoff Salleh; 2 males, TN14, 25.viii.1986 and 4.v.1987, coll. Ahmad Radzuan and Panjang Abd.; 1 male, TN15, 18.v.1987, coll. Ahmad Radzuan and Panjang Abd.; 1 male, TS 8, 1.vii.1986, coll. Mahadzir; 1 male, Sungai Baging, 28.vii.1986, coll. A. Samat Wahid. **THAILAND**: Central Region: 2 males, Wang Jum Pee, A. Pakchong, Nakhonratsima, (21.viii.1992), (28.viii.1992), coll. Montree, attracted to cue lure; 3 males, Phliu Hort. Station, A. Laemsingh, Chanthaburi, (9.vii.1992), (10.xii.1991), (4.ii.1992), coll. Montree, attracted to cue lure; Chiang Mai: 4 males, N.R.D.C., (2.vii.1991), (20. viii.1991), (11.ix.1991), (30.vii.1992), coll. Chantanee, attracted to cue lure (in ANIC; BMNH; DOAC; QDPC).
Diagnosis: A large species; face red-brown and usually with an inverted 'U'-shaped black pattern that is enlarged in the antennal furrows by connecting to the large oval facial spots (in some specimens there is a broad black band across the oral margin in addition to the normal marking); postpronotal lobes patchy fuscous to dark fuscous (usually paler along ventral margin); notopleura yellow (sometimes with patches of dark fuscous); scutum black with irregular fuscous markings on

anterior and posterior margins; lateral and medial postsutural yellow vittae absent; a yellow band along anterior margin of notopleural suture; mesopleural stripe reaching to anterior *npl.* seta dorsally; scutellum yellow except for broad black basal band; setae: *sc.* 2; *prsc.* absent; *ia.* 1; *p.sa.* 1; *a.sa.* 1; *mpl.* 1; *npl.* 2; *scp.* 2–4; legs with fore and hind femora black and mid femora dark fuscous except fulvous basally on hind femora and basally and apically, in some specimens, on mid femora; tibiae dark fuscous; tarsi with basal segment fulvous and apical four segments fuscous; mid tibiae each with an apical black spur; wing with cells bc and c dark fuscous, dense microtrichia covering all of cell c and along costal margin of cell bc; remainder of wing colourless except dark fuscous cell sc, broad dark fuscous costal band overlapping R_{4+5} (narrowly basal to *r-m* and more so apical to *r-m*) but not expanding apically, fuscous in cell cup but not forming a distinct anal streak; supernumerary lobe weak; abdomen elongate, clavate and petiolate, all abdominal terga black except orange-brown along intersegmental line between terga I and II either side of a narrow medial longitudinal black line, posterolaterally on terga II and IV either side of a broad medial longitudinal black band and narrowly margining the medial longitudinal dark fuscous band on anterior half of tergum V and then widening markedly over posterior region of this tergum (in some specimens there is a fuscous to dark fuscous band across anterior margin of tergum V, joining the broad black lateral margins). A pair of orange-brown shining spots on tergum V. Posterior lobe of surstylus short, sternum V with a slight concavity on posterior margin.

Attractant: Cue lure.

Female: No known record.

Distribution:

Type locality: Peninsular Malaysia (Transect North, TN9).

Previously recorded: Type locality and other locations in Peninsular Malaysia and Thailand (Drew *et al.*, 1998).

Hosts: No known record.

Remarks: *Dacus (Mellesis) tenebrosus* Drew & Hancock is similar to *Dacus (Mellesis) siamensis* Drew & Hancock and *Dacus (Mellesis) vijaysegarani* Drew & Hancock in possessing a broad costal band overlapping vein R_{4+5}, scutum, legs and abdominal terga mostly black, postpronotal lobes dark fuscous to black and face with an inverted 'U'-shaped dark pattern joined to a black band across oral margin. *Dacus tenebrosus* differs from *D. siamensis* and *D. vijaysegarani* in having the anatergite and katatergite mostly yellow and the scutellum with a broad black basal band.

Dacus (Mellesis) trimacula (Wang)

(Fig. 415)

Dacus trimacula Wang, 1990: 68, 74. Holotype male in IZAS.
Dacus (Callantra) trimacula – Liang *et al.*, 1993: 138; Drew *et al.*, 1998: 634; Norrbom *et al.*, 1998: 131.

Type material examined: Holotype male, **CHINA**: Kunming, Yunnan, 9.viii.1984, coll. Zhao Mingzhu, studied in IZAS by D.L. Hancock.

Paratypes: **CHINA**: 1 male, Yuanjiang, Yunnan, 9.viii.1984, coll. Zhao Mingzhu; 1 male, Luodian, Guizhou, 20.viii.1985, coll. Luo Luyi; 1 male, Chongan, Fujian, 24.iv.1940; 1 male, Jinan, Shandong, 22.viii.1945 (in IZAS); 1 male, Yunnan, ix.1984, attracted to cue lure (in APQS).

Diagnosis: A large species; face with three black spots, two in the antennal furrows and one dorsocentrally below antennal bases; postpronotal lobes yellow; notopleura fuscous; scutum red-brown tending fuscous posteriorly and with a narrow medial longitudinal dull black marking and two smaller dull black markings posterolaterally; lateral and medial postsutural yellow vittae absent; yellow triangular marking along anterior margin of notopleural suture; mesopleural stripe of medium-width; scutellum yellow with a narrow dark basal band; setae: *sc.* 2; *prsc.* absent; *ia.* 1; *p.sa.* 1; *a.sa.* 1; *mpl.* 1; *npl.* 2; *scp.* 4; legs with femora mostly red-brown to fuscous; tibiae and tarsi fulvous; no spines on fore femora; wing with cells bc and c fuscous; a broad fuscous costal band overlapping R_{4+5} for entire length but not expanding apically; anal streak absent; supernumerary lobe weak; abdominal terga III–V broadly dark fuscous to dull black laterally and fuscous medially, pecten present on tergum III.

Attractant: Cue lure.

Female: No known record.

Distribution:

Type locality: China (Yunnan Province, Kunming).

Previously recorded: Type locality and the following provinces in China – Guizhou, Fujian and Shandong) (Drew *et al.*, 1998).

Hosts: No known record.

Remarks: *Dacus (Mellesis) trimacula* (Wang) is similar to *Dacus (Mellesis) formosanus* (Tseng and Chu) in possessing a broad costal band almost confluent with vein M basally, the anal streak absent, abdominal terga III–V fuscous to black laterally without a medial longitudinal dark vitta, a fuscous anatergite and yellow katatergite. It differs from *D. formosanus* in having a red-brown scutum and abdominal terga III–V slightly paler.

Dacus (Mellesis) triquetrus Drew & Romig, new species

(Fig. 416)

Material examined: Holotype male, **PHILIPPINES: LUZON ISLAND** – Forestry 2, UPLB MNH, Philippines, attracted to cue lure, coll. Greg Quimio, deposited in BMNH.

Diagnosis: A large species; face fulvous with a pair of large circular black spots; postpronotal lobes and notopleura yellow; scutum red-brown with lateral and medial longitudinal black bands; lateral postsutural yellow vittae absent; medial postsutural yellow vitta present as a yellow triangular star-shaped pattern; a large triangular yellow band along anterior margin of notopleural suture with base adjacent to notopleuron; mesopleural stripe reaching to postpronotal lobe dorsally; scutellum yellow; wing with cells bc and c dark fuscous, dense microtrichia covering all of cell c and most of cell bc; a broad dark fuscous costal band confluent with R_{4+5} and overlapping this vein at apex; anal streak absent; abdominal terga III–V dark orange-brown with a dark 'T' pattern consisting of a broad transverse black band across anterior of tergum III and a broad medial longitudinal black band over all three terga, broad lateral longitudinal dull black bands over terga IV and V.

Description: male.

Head: Height 1.8 mm. Frons – length 1.3 times breadth; fulvous ventrally, fuscous dorsally; orbital setae: 1 strong black *s.or.*, 2 weak and pale *i.or.*; lunule fuscous. Ocellar triangle black. Vertex fuscous. Face fulvous with a pair of large circular black spots; length 0.52 mm. Genae fulvous, large fuscous to dark fuscous subocular spot present; dark seta present. Occiput fuscous, fulvous along eye margins; occipital row with a large number of short black setae. Antennae with segments 1 and 2 fuscous, segment 3 dark fuscous; length of segments: 0.6 mm; 0.6 mm; 1.2 mm.

Thorax: Scutum dark red-brown with longitudinal black bands and black triangular markings between postpronotal lobe and notopleural suture. Pleural areas dark fuscous with red-brown above mid coxae and below postpronotal lobes. Yellow markings as follows: postpronotal lobes; notopleura; broad mesopleural stripe reaching postpronotal lobe dorsally, continuing to katepisternum as a small yellow spot, anterior margin straight; anatergite (posterior apex black); anterior ⅔ katatergite (remainder black); a triangular medial postsutural yellow vitta with all three sides indented to give a three-cornered star-shaped appearance; a large yellow triangular marking along anterior margin of notopleural suture with base adjacent to notopleuron. Lateral postsutural yellow vittae absent. Postnotum black. Scutellum yellow except for narrow black basal band. Setae: *sc.* 2; *prsc.* absent; *ia.* 1; *p.sa.* 1; *a.sa.* 1; *mpl.* 1; *npl.* 2; *scp.* 4.

Legs: Fore and mid femora fulvous with a large dark fuscous marking along outer surfaces, hind femora fulvous with dark fuscous around apical ⅓; fore and mid tibiae fuscous, hind tibiae dark fuscous; all tarsi with basal segment fulvous and apical four segments red-brown; mid tibiae each with an apical black spur.

Wings: Length 6.4 mm; cells bc and c dark fuscous; dense microtrichia covering all of cell c and most of cell bc; remainder of wings colourless except dark fuscous cell sc, broad dark fuscous costal band confluent with R_{4+5} and overlapping this vein at apex, anal streak absent; a dense aggregation of microtrichia around A_1+CuA_2; supernumerary lobe weak.

Abdomen: Elongate oval and petiolate; terga fused; pecten present on tergum III. Tergum I and sterna I and II longer than wide. Tergum I dark fuscous to dull black; tergum II dark red-brown with a narrow medial longitudinal dull black band and lateral dull black margins which extend to meet the medial band in centre of tergum; terga III–V dark red-brown with a dark 'T' pattern consisting of a broad transverse dull black band across anterior of tergum III and a broad medial longitudinal dull black band over all three terga and broad lateral longitudinal dull black bands over terga IV and V. A pair of oval red-brown shining spots on tergum V. Posterior lobe of surstylus short, sternum V with a slight concavity on posterior margin. Abdominal sterna dark fuscous to black.

Attractant: Cue lure.

Female: No known record.

Distribution: Philippines (Luzon I., Los Banos).

Hosts: No known record.

Remarks: *Dacus (Mellesis) triquetrus* n.sp. is similar to *Dacus (Mellesis) calirayae* Drew & Hancock in possessing a red-brown scutum with dark markings, a distinct triangular yellow marking posteriorly in the midline of the scutum, fuscous cells bc and c, a broad fuscous costal band confluent with R_{4+5} and abdominal terga with a medial longitudinal black band over terga II–V, a black 'T' pattern over terga III–V and broad lateral black margins on terga II–V but differs from this species in having a broad mesopleural stripe reaching the postpronotal lobe dorsally, a slight expansion only of the costal band at its apex, the anal streak absent and the triangular yellow marking on the scutum indented on each margin to approximate three-cornered star.

Dacus (Mellesis) vijaysegarani Drew & Hancock

(Fig. 417)

Dacus (Callantra) vijaysegarani Drew & Hancock, in Drew *et al.*, 1998: 636. Holotype male in BMNH.

Type material examined: Holotype male, **MALAYSIA**: TS8, 12.i.1987, coll. Mahadzir, attracted to cue lure (in BMNH). Paratypes: **MALAYSIA**: 1 male, TN6, 29.xii.1986, coll. Y. Salleh; 2 males, TN7, 1.xii.1986 and 15.xii.1987, coll. Yusoff Salleh; 2 males, TN9, 6.x.1986 and 20.x.1986, coll. Yusoff Salleh; 1 male, 6.x.1986, coll. Ahmad Radzuan and Panjang Abdullah; 1 male, TN14, 18.v.1987, coll. Ahmad Radzuan and Panjang Abdullah; 1 male, TS8, 29.xii.1986; 2 males, same data as holotype; 4 males, TS8, 26.i.1987, coll. Bahari; 1 male, TS9, 9.ii.1987, coll. Bahari; 1 male, TS9, 12.ix.1987, coll. Mahadzir. **THAILAND**: 1 male, SK96KC, Ko Hong Hill [Songkhla district], 22.iii.1993, coll. Anuchit and Sieng. All specimens attracted to cue lure (in ANIC; BMNH; QDPC).

Non-type material examined: VIETNAM: Southern Vietnam – 1 male, South Vietnam, Dong Nai Province, Long Khanh District, Nhan Nghia Village, 26.v.2002, Site VS06, attracted to cue lure (in QDPC).

Diagnosis: A medium-sized species; face fulvous to red-brown with an inverted 'U'-shaped black pattern formed by a broad black band across dorsal margin just below antennal bases joining the large oval black spots in the antennal furrows and in some specimens a narrow black line across oral margin; postpronotal lobes dark fuscous to black; notopleura black; scutum black with a narrow dark red-brown area along posterior margin; lateral and medial postsutural yellow vittae absent; a very narrow marking along anterior margin of notopleural suture; a very short and narrow marking dorsally where the mesopleural stripe would normally occur (this usually runs from the ventral margin of the notopleuron to the level of the *mpl.* seta); scutellum yellow except for a very narrow dark basal band; setae: *sc.* 2; *prsc.* absent; *ia.* 1; *p.sa.* 1; *a.sa.* 1; *mpl.* 1; *npl.* 2; *scp.* 2–4; legs with femora and tibiae black except for a narrow dark

red-brown apex on mid femora; fore tarsi fuscous to dark fuscous, mid and hind tarsi fulvous; mid tibiae each with an apical black spur; wings with cells bc and c dark fuscous, dense microtrichia covering all of both cells; remainder of wing colourless except for dark fuscous cell sc, broad dark fuscous costal band overlapping R_{4+5} and of uniform width; anal streak absent; supernumerary lobe weak; abdomen elongate, clavate and petiolate, abdominal terga II–V black except for small orange-brown to red-brown areas postero-centrally on terga IV and V. A pair of dark fuscous to black shining spots on tergum V. Posterior lobe of surstylus short, sternum V with a slight concavity on posterior margin.
Attractant: Cue lure.
Female: No known record.
Distribution:
Type locality: Peninsular Malaysia (Transect South, TS8).
Previously recorded: Type locality and other locations in Peninsular Malaysia and southern Thailand (Drew *et al.*, 1998)
New records: Vietnam (Southern Vietnam, Dong Nai Province).
Hosts: No known record.
Remarks: *Dacus (Mellesis) vijaysegarani* Drew & Hancock is similar to *Dacus (Mellesis) siamensis* Drew & Hancock and *Dacus (Mellesis) tenebrosus* Drew & Hancock in possessing a broad costal band overlapping vein R_{4+5}, scutum, legs and abdominal terga mostly black, postpronotal lobes dark fuscous to black and face with an inverted 'U'-shaped dark pattern joined to a black band across oral margin. *Dacus vijaysegarani* differs from *D. tenebrosus* in having the anatergite black and from *D. siamensis* in having the notopleura black (not yellow).

Dacus (Mellesis) vittatus (Hardy)

(Fig. 418)

Callantra vittata Hardy, 1974: 13 – Hardy, 1977: 46. Holotype male in BBM.
Dacus (Callantra) vittatus – Drew *et al.*, 1998: 637; Norrbom *et al.*, 1998: 131.

Type material examined: Holotype male, **PHILIPPINES: MINDANAO ISLAND** – Agusan, San Francisco, 10 km SE, Mindanao, 14.xi.1959, coll. L.W. Quate (in BBM).
Diagnosis: A large species; face fulvous with an irregular broad transverse black band across oral margin and extending dorsally half-way to antenna bases; postpronotal lobes yellow (yellow spot lateral to postpronotal lobe); notopleura yellow; scutum red-brown with a pair of longitudinal black bands, a pair of dorsocentral black stripes inside lateral postsutural vittae and lateral black patches presuturally; two lateral postsutural yellow vittae reaching to *ia.* seta; medial postsutural yellow vitta from centre of prescutum to ⅔ of distance between suture and posterior margins of scutum; broad triangular yellow band along anterior margin of notopleural suture; mesopleural stripe reaching postpronotal lobe dorsally; scutellum yellow (base obscured by pin); setae: *sc.* 2; *prsc.* absent; *ia.* 1; *p.sa.* 1; *a.sa.* 1; *mpl.* 1; *npl.* 2; *scp.* 2; legs with femora red-brown except basal ¼ of mid femora fulvous; tibiae fuscous; tarsi with basal segments fulvous, remainder red-brown; mid tibiae with apical black spur; wing with cells bc and c fuscous, dense microtrichia covering cell c and outer ⅔ of bc; remainder of wing hyaline except for dark fuscous cell sc and fuscous costal band which is confluent with R_{4+5}, some trace of expansion of costal band apically but not into a distinct spot; no anal streak; supernumerary lobe weak; abdomen elongate and petiolate, pecten of cilia on abdominal tergum III absent; abdominal tergum III red-brown to fuscous, tergum IV fulvous with broad fuscous midline and tending to red-brown laterally, tergum V fulvous with fuscous midline and antero-lateral triangles. A pair of fulvous shining spots on tergum V. Sternum V with a slight concavity on posterior margin.
Attractant: No known record.
Female: No known record.
Distribution:
Type locality: Philippines (Mindanao I., Agusan).
Hosts: No known record.
Remarks: *Dacus (Mellesis) vittatus* (Hardy) belongs to a group of *Dacus (Mellesis)* species that has a broad costal band confluent with vein R_{4+5}, anal streak absent and no preapical spines on the fore femora. It differs from all others in the group in having an elongate medial and two well-developed lateral postsutural yellow vittae.

Dacus (Mellesis) wallacei White

(Fig. 419)

Dacus (Callantra) wallacei White, in Drew *et al.*, 1998: 639.
Callantra longicornis (Wiedemann) – Hardy & Adachi, 1954: 149 [Misidentification].

Type material examined: Holotype male, **INDONESIA: MOLUCCAS (MALUKU): AMBON ISLAND** – Waai, Ambon I., Moluccas, 10.iii.1966, coll. A.M.R. Wegner, California Academy of Sciences Acc 1972 (in BBM).
Paratype: **INDONESIA: MOLUCCAS (MALUKU): AMBON ISLAND** – 1 female, Ambon ii.[19]08; recorded by Hardy and Adachi (1954) as *Dacus longicornis* (in BBM).
Diagnosis: A large species; face fulvous with indistinct, slightly elongate small facial spots; postpronotal lobes yellow (not much paler that scutum); notopleura yellow; scutum red-brown; lateral and medial postsutural yellow vittae absent; narrow triangular yellow bands along anterior margin of notopleural suture; narrow mesopleural stripe equal in width to notopleuron and not extended onto katepisternum; scutellum yellow with diffuse red-brown basal band; setae: *sc.* 2; *prsc.* absent; *ia.* 1; *p.sa.* 1; *a.sa.* 1; *mpl.* 1; *npl.* 2; *scp.* not differentiated from general vestiture; legs with femora red-brown, fore femora with 2 (right leg) to 4 (left leg) stout ventral spines; tibiae fulvous, mid tibiae each with a black apical spur, hind tibiae with preapical microtrichose bulge; wing with cells bc and c fuscous, dense microtrichia over cell c and most of cell bc (obscured by dirt: may be lacking at base); remainder of wings with a faint tint except fuscous narrow part of cell br, cell sc fuscous, a broad fuscous costal band extended below

R_{4+5} approximately ⅔ towards vein M; no distinct anal streak; supernumerary lobe weak; abdomen elongate, clavate and petiolate, abdominal terga I–IV red-brown, tergum V red-brown with a fuscous midline (covered with mould but some sign of a transverse fuscous band across anterior margin visible on slide of female). A pair of red-brown shining spots on tergum V. Posterior lobe of surstylus short, sternum V with a slight concavity on posterior margin.
Attractant: No known record.
Female: As for male except pecten of cilia on abdominal tergum III absent, no preapical microtrichose bulge on hind tibia, no dense aggregation of microtrichia around A_1+CuA_2; piercer laterally compressed as in *Dacus petioliforma* (May); length 1.2 mm.
Distribution:
Type locality: Indonesia (Moluccas = Maluku – Ambon I., Waai).
Hosts: No known record.
Remarks: *Dacus (Mellesis) wallacei* White is similar to and different from *Dacus (Mellesis) conopsoides* de Meijere in characters listed under the comments for *D. conopsoides*. The female used for this description (in BBM) was placed previously in *Dacus (Callantra) longicornis* (Wiedemann) by Hardy and Adachi (1954) but *D. wallacei* differs from that species in having postpronotal lobes entirely yellow, spines on ventral surface of fore femora, a black spot between antennal base and eye margin on each side and abdominal terga of uniform pale colouration.

SUBGENUS *NEODACUS* Perkins

Dacus (Neodacus) ortholomatus Hardy
(Fig. 420)

Dacus (Asiadacus) ortholomatus Hardy 1982: 188. Holotype male in MZB.
Dacus (Dacus) ortholomatus – Drew *et al.*, 1998: 643.
Dacus (Didacus) ortholomatus – Norrbom *et al.*, 1998: 135.

Type material examined: Holotype male, **INDONESIA: SULAWESI** – Sadaunta, 65 km SE of Palu, Central Sulawesi, 650 m, viii.1975, coll. D.E. & A.T. Hardy (in MZB).
Diagnosis: A large species; face with a pair of elongate black spots in antennal furrows; postpronotal lobes and notopleura yellow; scutum red-brown with a medial longitudinal dark fuscous pattern; lateral postsutural yellow vittae absent; elongate triangular medial postsutural yellow vitta present; yellow triangular-shaped band along anterior margin of notopleural suture with base at notopleuron; narrow mesopleural stripe present; scutellum yellow; setae: *sc.* 2; *prsc.* absent; *ia.* 1; *p.sa.* 1; *a.sa.* absent; *mpl.* 1; *npl.* 2; legs fulvous tending pale fuscous on apical ½ of all femora; wing with cells bc and c fuscous; a broad fuscous costal band slightly overlapping R_{4+5} and expanding over apex of wing to vein M; a broad fuscous anal streak; supernumerary lobe weak; abdomen elongate-oval, terga fused, abdominal terga III–V fulvous medially tending red-brown laterally and with a narrow dark fuscous to black medial longitudinal vitta over all terga, pecten present on tergum III.
Attractant: No known record.
Female: No known record.
Distribution:
Type locality: Indonesia (Sulawesi – Central Sulawesi, Sadaunta).
Hosts: No known record.
Remarks: *Dacus (Neodacus) ortholomatus* Hardy was originally placed in subgenus *Asiadacus* by Hardy (1982). However, the terga are fused, which, together with the absence of *prsc.* and *a.sa.* setae, confirms that this species belongs in *Dacus (Dacus)*. It is similar to *D. (Mellesis) sphaeroidalis* (Bezzi) in characters listed under the comments to that species. *Dacus ortholomatus* differs from *D. sphaeroidalis* in having a large medial postsutural yellow vitta on the scutum, the costal band overlapping vein R_{4+5} and a medial longitudinal black vitta on abdominal terga III–V.

Dacus (Neodacus) santongae Drew & Hancock
(Fig. 421)

Dacus (Dacus) santongae Drew & Hancock, in Drew *et al.*, 1998: 644. Holotype male in BMNH.

Type material examined: Holotype male, **INDONESIA: LOMBOK** – Santong (450 m), Lombok, 2.vi.1981, coll. J. Turner, attracted to cue lure (in BMNH).
Non-type material examined: INDONESIA: LOMBOK – 3 males, Batu Belik, Lombok, Indonesia, 450m, Oct 4.1976, coll. Eko M; 3 males, Batu Belik, Lombok, 300m, 9.ix.1979, coll. E. Mulyono, attracted to cue lure. Specimens in QDPC.
Diagnosis: A medium-sized species; face fulvous with a pair of small circular dark fuscous spots; postpronotal lobes and notopleura yellow; scutum black except red-brown around lateral and posterolateral margins, between postpronotal lobe and notopleuron along anterior margins of yellow triangles on notopleural suture and across anterior margin; lateral postsutural yellow vittae absent; medial postsutural yellow vittae as a broad transverse triangular marking; a broad yellow triangle along anterior margin of notopleural suture; mesopleural stripe slightly wider than notopleuron dorsally; scutellum yellow except for a narrow black basal band; setae: *sc.* 2; *prsc.* absent; *ia.* 1; *p.sa.* 1; *a.sa.* 1; *mpl.* 1; *npl.* 2; *scp.* 4; legs with femora fulvous except for a large diffuse fuscous spot on outer surfaces of fore femora and fuscous on apical ⅓ of hind femora; tibiae and tarsi fulvous except hind tibiae fuscous, mid tibiae each with an apical black spur; wing with cells bc and c fuscous, dense microtrichia covering both cells; remainder of wings colourless except for fuscous cell sc, broad fuscous costal band confluent with R_{4+5} and expanding into a large distinct spot in apex of wing; anal streak absent; supernumerary lobe weak in male (female not studied); abdomen elongate oval, abdominal terga I–V red-brown except for a diffuse medial longitudinal dark fuscous spot on tergum I, diffuse dark fuscous across anterior margin of tergum II but not reaching lateral margins, broad lateral longitudinal fuscous bands over terga III–V and a narrow medial

longitudinal dark fuscous to black band over terga II–V. A pair of oval fuscous shining spots on tergum V. Posterior lobe of surstylus short and pointed, sternum V with a slight concavity on posterior margin.
Attractant: Cue lure.
Female: No known record.
Distribution:
Type locality: Indonesia (Lombok, Santong).
Hosts: No known record.
Remarks: *Dacus (Neodacus) santongae* Drew & Hancock is similar to *Dacus (Mellesis) calirayae* Drew & Hancock, *Dacus (Mellesis) discophorus* (Hering), and *Dacus (Mellesis) nummularius* (Bezzi) in possessing a large spot in the apex of the costal band that crosses vein M, a large triangular yellow mark anterior to the notopleural suture, a yellow medial postsutural mark on the scutum and dark patterns on abdominal terga III–V. It differs from these species in lacking a fuscous anal streak in the wing and in having the medial yellow mark on the scutum as a unique transverse triangular pattern.

Dacus (Neodacus) sphaeroidalis (Bezzi)

(Fig. 422)

Mellesis sphaeroidalis Bezzi, 1916: 115. Holotype male in BMNH.
Callantra sphaeroidalis – Hardy, 1973: 11; Hardy, 1977: 46.
Dacus (Callantra) sphaeroidalis – Liang *et al*., 1993: 139; Drew *et al*., 1998: 629; Norrbom *et al*., 1998: 131.
Sinodacus chonglui Chao & Lin, 1996: 131. **New Synonym.**

Type material examined: Holotype male, **INDIA**: Dehra Dun, 16.vii.1912, on grass (in BMNH).
Other type material examined: CHINA: Holotype male of *Sinodacus chonglui* with the following two labels – 'Sinodacus chonglui Chao'; '1984.8 plus Chinese writing' (in PQMAB).
Non-type material examined: PAKISTAN: 3 males, 3 females, Rawalpindi, (28.iii.1962), (1,3,12.iii.1962), bred from *Telosma cordata* (in QDPC). **THAILAND**: 1 male, Phan, Chiang Rai, 18.ix.1991, coll. Charuwan, attracted to cue lure; 1 male, Lam Poon, Chiang Mai, 22.viii.1991, coll. Chantanee, attracted to cue lure (in QDPC). **CHINA**: 1 male, Chongan, Fujian Province (in IZAS).
Diagnosis: A large species; face red-brown with two large elongate black spots filling almost the entire antennal furrows; postpronotal lobes and notopleura yellow; scutum red-brown without dark markings other than a mottled appearance due to sites of attachment of dorso-ventral flight muscles; lateral and medial postsutural yellow vittae absent; a broad triangular yellow marking with base on notopleuron and running along anterior margin of notopleural suture; broad mesopleural stripe reaching midway between anterior margin of notopleuron and postpronotal lobe; scutellum yellow except for a narrow black basal band; setae: *sc*. 2; *prsc*. absent; *ia*. 1; *p.sa*. 1; *a.sa*. absent; *mpl*. 1; posterior *npl*. 1; anterior *npl*. absent (or very weak) *scp*. 2–3; legs with fore and mid femora red-brown, hind femora fulvous except red-brown around apical ⅓, elongate black markings on apical ventral surfaces of all femora; fore and mid tibiae red-brown, hind tibiae red-brown basally and pale fuscous apically; tarsi fulvous tending red-brown on apical segments; wing with cells bc and c fuscous, dense microtrichia covering both cells bc and c; remainder of wings with a pale fuscous tint across membrane except for fuscous cell sc, broad fuscous costal band confluent with R_{4+5} and expanding apically into a large round dark fuscous spot which overlaps M; pale fuscous anal streak present; supernumerary lobe weak in both sexes; abdomen elongate-oval, petiolate but not as strongly as species in subgenus *D. (Mellesis)*, abdominal terga III–V red-brown except for a narrow transverse black band across anterior margin of tergum III but not reaching lateral margins, a small diffuse dull black spot anterocentrally on tergum IV, a transverse black band across anterior margin of tergum V which widens around lateral margins and with a narrow medial longitudinal black band that does not reach posterior margin, two paler posterolateral spots on tergum IV that are tending towards fulvous. A pair of oval fuscous to dark fuscous shining spots on tergum V. Posterior lobe of surstylus short, sternum V with a slight concavity on posterior margin.
Attractant: Cue lure.
Distribution:
Type locality: India (Dehra Dun).
Previously recorded: Type locality and the following locations – Pakistan; northern Thailand; China (Fujian Province); Vietnam (Norrbom *et al*., 1998).
New records: China (Guangxi Province, Luzhai) due to **New Synonymy** of *Sinodacus chonglui*.
Hosts: *Telosma cordata* (Chinese violet) (family *Asclepiadaceae*).
Remarks: *Dacus (Neodacus) sphaeroidalis* (Bezzi) and *D. (Neodacus) ortholomatus* Hardy are similar in having a broad costal band that expands apically to meet or overlap vein M and in lacking spines on the fore femora. They differ in the absence of both a postsutural medial yellow vitta on the scutum and a medial longitudinal black band on abdominal terga III–V in *D. sphaeroidalis*.

We have studied the male Holotype of *Sinodacus chonglui* Chao & Lin and can confirm that this species is a **New Synonym** of *D. sphaeroidalis*.

GENUS *ICHNEUMONOPSIS* Hardy

Ichneumonopsis burmensis Hardy

(Figs 423A, 423B)

Ichneumonopsis burmensis Hardy, 1973:133   1977: 75, Drew & Hancock, 1994b: 830; Norrbom *et al*., 1998: 161. Holotype male in BMNH.

Type material examined: Holotype male and Allotype female, **MYANMAR (formerly BURMA)**: Mt Victoria, Chin Hills (in BMNH).

Non-type material examined: INDIA: 1 male and 1 female, Meghalaya, Nongph – Forest, 25-28.iv.1980, A. Freidberg (in TAUI).

Diagnosis: A large species; face fulvous with a medial longitudinal dark fuscous to black band joining a narrow transverse dark band across oral margin, three circular black spots on each lateral margin; postpronotal lobes yellow; notopleura pale brown; a narrow lateral yellow band running from hind margin of postpronotal lobe to anterior end of lateral postsutural vitta; scutum pale orange-brown with a pattern of dark markings; lateral postsutural yellow vittae present; medial postsutural yellow vitta absent; no yellow spot anterior to notopleural suture; a broad mesopleural stripe reaching anterior *npl.* seta dorsally; scutellum yellow; wing with cells bc and c fuscous, dense microtrichia covering both cells bc and c; a broad fuscous costal band overlapping R_{4+5}; anal streak absent but with a tint in cell cup; wing with cell c elongate and longer than cell sc, cell bm triangular and narrowing to a point basally, lobe of cell cup shorter than vein A_1+CuA_2; abdomen shape elongate, petiolate and markedly narrow; abdominal terga III–V orange-brown with dark fuscous bands across bases of terga III and IV and laterally on tergum V.

Description: male (Fig. 423A).

Head: Height 2.4 mm. Frons – length 1.0 times breadth; fulvous with black around orbital setae, two black spots adjacent to lunule and a narrow medial black line; anteromedial hump reduced and bare; orbital setae black: 1 *s.or.*, *i.or.* absent or with a markedly reduced lower seta; lunule fulvous centrally, dark fuscous laterally. A black spot against each eye lateral to base of antenna. Ocellar triangle black. Vertex fulvous. Face fulvous with a medial longitudinal dark fuscous to black band joining a very narrow transverse dark fuscous to black band across oral margin, three rounded black spots on each lateral margin (lateral to the long narrow antennal furrows); length 1.1 mm. Genae fulvous, large dark fuscous to black subocular spot present; black seta present. Occiput fulvous with four fuscous spots dorsally and two narrow fuscous vertical lines ventrolaterally; occipital row with a large number of small dark setae. Antennae with segments 1 and 2 red-brown, segment 3 fuscous; a small black dorsal seta on segment 2; arista plumose and dark (paler basally); length of segments: 0.24 mm; 0.48 mm; 1.4 mm.

Thorax: Scutum pale orange-brown except pale brown on lateral margin behind notopleural suture; a medial and pair of sublateral narrow longitudinal fuscous to dark fuscous bands joining with a fuscous area across posterior margin of scutum. Pleural areas orange-brown with a small black area below postpronotal lobes, anterior and posterior to mesopleural stripe and on katepisternum. Yellow markings as follows: postpronotal lobes; broad mesopleural stripe, reaching anterior *npl.* seta dorsally, continuing to katepisternum as a very large transverse spot, anterior margin slightly convex; narrow lateral postsutural vitta ending at or just before *ia.* seta; a narrow lateral band running from hind margin of postpronotal lobe to anterior end of lateral postsutural vitta at the transverse suture; anatergite and katatergite (posterior apices dark fuscous). Notopleura pale brown. Postnotum orange-brown with narrow fuscous lateral and dorsal margins. Scutellum yellow except for narrow black basal band. Setae: *sc.* 2; *prsc.* absent; *ia.* 1 (very weak); *p.sa.* 1; *a.sa.* 1; *mpl.* 1; *npl.* 2; *scp.* 3–4.

Legs: Femora fulvous except apical ⅓ of mid and hind femora and base of hind femora pale fuscous; fore tibiae pale fuscous, mid and hind tibiae fuscous; tarsi with basal segment fulvous and apical four segments pale fuscous to fuscous; mid tibiae each with an apical black spur; four strong preapical spines on ventral surface of fore femora.

Wings: Length 8.3 mm; cells bc and c fuscous; dense microtrichia covering both cells; remainder of wings with a pale tint across membrane except for darker fuscous cell sc, broad fuscous costal band overlapping R_{4+5}, anal streak absent but with a slightly darker tint in cell cup; no dense aggregation of microtrichia around A_1+CuA_2; supernumerary lobe weak. Cell c elongate and 1.45 times longer than cell sc; cell bm triangular, narrowing to a point basally; lobe of cell cup slightly shorter than length of vein A_1+CuA_2.

Abdomen: Elongate, petiolate and markedly narrow; terga free; pecten absent on tergum III. Terga and sterna I and II longer than wide. Tergum I fulvous with a dark fuscous 'U'-shaped pattern, the base of which lies on the intersegmental line between terga I and II; tergum II orange-brown with a fuscous to dark fuscous inverted 'V'-shaped pattern, the base of which lies on the intersegmental line between terga I and II; terga III–V orange-brown to dark orange-brown except for broad dark fuscous bands across bases of terga III and IV and laterally on tergum V. Shining spots not present on tergum V. Posterior lobe of surstylus long, sternum V with a moderate concavity on posterior margin.

Attractant: No known record.

Female: (Fig. 423B) As for male except with an extra segment present on abdomen and with abdominal terga III–V slightly wider. Oviscape very long and narrow, fuscous, dorsoventrally compressed; ratio of length of oviscape to length of tergum V, 7.7:1; apex of piercer rounded, flattened and shovel-shaped, with minute preapical setae. Two rounded, rough-surfaced spermathecae.

Distribution:

Type locality: Myanmar (Mt Victoria).

Previously recorded: Type locality, North-East India (Meghalaya) and northwest Thailand (Mae Hong Son Province, Pangmapha District) (Drew & Hancock, 1994b; Kovac *et al.*, 2013).

Hosts: *Melocalamus compactiflorus* (Kurz) (family *Poaceae*) (Kovac *et al.*, 2013).

Remarks: *Ichneumonopsis burmensis* Hardy is the only known species in the genus *Ichneumonopsis*. It differs from all other known species of Dacinae in possessing a short plumose arista on the antennae, wing with vein R_{4+5} setose, cell bm narrow basally and female with two round spermathecae. Kovac *et al.*, (2013) placed this species in the Tribe Gastrozonini.

GENUS *MONACROSTICHUS* Bezzi

Monacrostichus citricola Bezzi

(Fig. 424)

Monacrostichus citricola Bezzi, 1914a: 323 – 1919: 442; Hardy, 1969b: 481, Lectotype designation [invalid – see explanation under 'Remarks' for *Bactrocera occipitalis* (Bezzi)]. Hardy, 1974: 61; Hardy, 1977: 60; Munro, 1984: 159; Drew & Hancock, 1994b: 833; Norrbom *et al.*, 1998: 168. Syntypes in MCSNM.

Non-type material examined: THAILAND: 14 males, 10 females, Amphoe Wang, Wi Set, Trang, 20.ix.1991, bred ex *Citrus hystrix*, Sieng *et al.*; 2 males, 3 females, Amphoe Sichon, Nakhon Si Thammarat, 10.x.1991, bred ex *Citrus reticulata*, Sieng *et al.*; 10 males, 8 females, Amphoe Nabon, Nakhon Si Thammarat, 22.xi.1991, bred ex *Citrus hystrix* and *Citrus aurantifolia*, Sieng *et al.*; 4 males, 2 females, PSU, Songkhla, 25.xi.1991, bred ex *Citrus hystrix*, Vanpen *et al.*; 3 males, 4 females, Muang, Trang, Songkhla, 13. xii.1991, bred ex *Citrus hystrix*, Anuchit; 1 male, Tak Bai, Narathiwat, Songkhla, 21.xii.1991, ex *Citrus hystrix*, Anuchit *et al.*; 2 males, 1 female, Amphoe Chauang, Nakhon Si Thammarat, 3.i.1992, bred ex *Citrus hystrix*, Sieng *et al.*; 3 males, 1 female, PSU, Songkhla, 13.i.1992, bred ex *Citrus aurantifolia*, Vanpen *et al.*; 1 female, Amphoe Wang, Wiset, Trang, Songkhla, 23.i.1992, bred ex *Citrus aurantifolia*, Sieng *et al.*; 1 male, PSU, Songkhla, 4.ii.1992, bred ex *Citrus hystrix*, Vanpen *et al.*; 1 female, Yala, Songkhla, 3.iv.1992, bred ex *Citrus hystrix*, Anuchit *et al.*; 3 males, 6 females, PSU, Songkhla, 9.vi.1992, bred ex *Citrus hystrix*, Vanpen *et al.* (in QDPC).

Diagnosis: A large species; face orange-brown with a 'Y'-shaped dark fuscous to black pattern centrally and black over each antennal furrow; a deep transverse furrow across face; postpronotal lobes with anterior ⅓ fuscous and posterior ⅔ yellow; notopleura fuscous; a yellow band running posteriorly from hind margin of postpronotal lobe; scutum black tending dark fuscous posteriorly; lateral postsutural yellow vittae present; medial postsutural yellow vitta present as a large heart-shaped spot; a yellow spot anterior to notopleural suture; a broad mesopleural stripe reaching postpronotal lobe dorsally; scutellum yellow; wing with cells bc and c dark fuscous, dense microtrichia covering both cells bc and c; a broad dark fuscous costal band overlapping R_{4+5}; anal streak absent; wing cell bm long and narrow, cell sc long and wide, veins R_{2+3} and M with distinctive curvatures; abdomen shape elongate, clavate and petiolate; abdominal terga III–V dark fuscous to black with orange-brown across posterior margins.

Description: male.

Head: Height 2.1 mm. Frons – length 1.13 times breadth; orange-brown with fuscous ventrally and dark fuscous to black around orbital setae; no small hairs on anteromedial hump; orbital setae black: 1 *s.or.*; lunule orange-brown. A dark fuscous to black spot against eye lateral to base of antenna. Ocellar triangle black. Vertex pale fuscous. Face with a deep transverse furrow across centre and linking antennal furrows, orange-brown except for a 'Y'-shaped dark fuscous to black pattern and black covering entire length of each antennal furrow; length 0.68 mm. Genae orange-brown, large dark fuscous to black subocular spot present; very small dark seta present. Occiput orange-brown centrally, fuscous to dark fuscous laterally, orange-brown along eye margins; with a large number of small dark postocular setae. Antennae with all segments dark fuscous; a small black dorsal seta on segment 2; arista black (fulvous basally); length of segments: 0.52 mm; 0.6 mm; 1.84 mm.

Thorax: Scutum black tending dark fuscous posteriorly, except brown behind lateral postsutural vittae and above notopleuron. A transverse furrow across scutum connecting both parts of the notopleural suture. Pleural areas black except dark fuscous below postpronotal lobe. Yellow markings as follows: postpronotal lobe (anterior ⅓ and ventral margin brown to fuscous); band running from posterior margin of postpronotal lobe halfway to notopleuron; broad mesopleural stripe, reaching postpronotal lobe dorsally, continuing to katepisternum as a transverse spot, anterior margin deeply concave; anatergite (posterior apex black); anterior ⅔ katatergite (remainder black); large elongate spot along anterior margin of notopleural suture above notopleuron; moderately broad and curved lateral postsutural vitta ending well behind *ia*. seta; a large heart-shaped spot posterocentrally on scutum. Notopleuron fuscous. Postnotum black. Scutellum yellow except for broad black basal band. Setae: *sc*. 2; *prsc*. absent; *ia*. 1; *p.sa*. 1; *a.sa*. 1; *mpl*. 1; *npl*. 2; *scp*. 2.

Legs: Fore and mid femora fuscous with black on lateral and ventral margins, hind femora dark fuscous with black on lateral and ventral margins; fore and mid tibiae fuscous, hind tibiae dark fuscous; tarsi with basal segment fulvous and apical four segments fuscous; mid tibiae each with a small apical black spur; four preapical spines on ventral surface of fore femora.

Wings: Length 7.6 mm; cells bc and c dark fuscous; dense microtrichia covering both cells; remainder of wings colourless except for dark fuscous cell sc, broad dark fuscous costal band overlapping R_{4+5}; anal streak absent; no dense aggregation of microtrichia around A_1+CuA_2; supernumerary lobe very weak. Cell bm long and narrow (approximately 4 times longer than wide); cell sc long and wide (approximately 8 times longer than wide and 1.8 times longer than cell c); veins R_{2+3} and M with distinctive pronounced curvatures; vein A_1+CuA_2 ending well before wing margin.

Abdomen: Elongate, clavate and petiolate; terga free; pecten absent on tergum III. Tergum I and sterna I and II wider than long. Tergum I black except orange-brown anterolaterally; tergum II dark fuscous except large orange-brown spots anterolaterally; terga III–V dark fuscous to black except for irregularly shaped orange-brown bands across posterior margins. Shining spots not present on tergum V. Posterior lobe of surstylus short, sternum V with a moderate concavity on posterior margin.

Attractant: No known record.

Female: As for male except for an extra segment present on abdomen; oviscape red-brown, dorsoventrally compressed and tapering posteriorly in dorsal view; ratio of length of oviscape to length of tergum VI, 1.7: 1; apex of piercer needle-shaped, with microscopic preapical setae; two convoluted spermathecae.

Distribution:

Type locality: Philippines (Luzon I., Los Banos).

Previously recorded: Type locality and Peninsular Malaysia (Norrbom *et al*., 1998) and other locations in southern Thailand (Drew & Hancock, 1994b).

Hosts: *Citrus aurantifolia* (lime), *Citrus hystrix* (wild lime), *Citrus reticulata* (mandarin), *Citrus limon* (lemon), *Citrus limetta* (lime) and *Citrus maxima* (pummelo) (family *Rutaceae*).

Remarks: *Monacrostichus citricola* Bezzi can be separated from *Monacrostichus malaysiae* Drew & Hancock in having a heart-shaped yellow spot posterocentrally on the scutum, yellow spots along the anterior margin of the notopleural suture and four strong preapical spines on the fore femora.

Monacrostichus malaysiae Drew & Hancock

(Fig. 425)

Monacrostichus malaysiae Drew & Hancock, 1994b: 835; Norrbom *et al*., 168. Holotype male in BMNH.

Type material examined: Holotype male, **PENINSULAR MALAYSIA**: University of Malaya garden, Kuala Lumpur, Malaysia, 18.ix.1974, B.C. Stone, bred ex *Citrus halimii* (in BMNH).

Paratypes: **PENINSULAR MALAYSIA**: 5 males, 6 females, same label data as holotype (in ANIC; BMNH; QDPC).

Diagnosis: A large species; face fulvous with fuscous below antennal bases and across oral margin and dark fuscous to black over each antennal furrow; a deep transverse furrow across face; postpronotal lobes with anterior ½ dark fuscous and posterior ½ yellow; notopleura dark fuscous; a yellow band running posteriorly from hind margin of postpronotal lobe; scutum black tending dark fuscous posteriorly; lateral postsutural yellow vittae present; medial postsutural yellow vitta absent; no yellow spot anterior to notopleural suture; a broad mesopleural stripe reaching postpronotal lobe dorsally; scutellum yellow; wing with cells bc and c fuscous, dense microtrichia covering both cells bc and c; a broad fuscous costal band overlapping R_{4+5}; anal streak absent; wing cell bm long and narrow, cell sc long and wide, veins R_{2+3} and M with distinctive curvatures; abdomen shape elongate, clavate and petiolate; abdominal terga III–IV black, tergum V black with dark red-brown each side of a dark fuscous to black medial longitudinal band.

Description: male.

Head: Height 2.0 mm. Frons – length 1.2 times breadth; fulvous with pale fuscous around orbital setae and dark fuscous ventrally; no small hairs on anteromedial hump; orbital setae black: 1 *s.or*.; lunule fulvous. A dark fuscous to black spot against eye lateral to base of antenna. Ocellar triangle black. Vertex pale fuscous to fuscous. Face with a deep transverse furrow across centre and linking antennal furrows, fulvous with pale fuscous to fuscous centrally below antennal bases and on oral margin and dark fuscous to black covering entire length of each antennal furrow; length 0.6 mm. Genae fulvous, large dark fuscous to black subocular spot present; seta absent. Occiput fuscous dorsally, red-brown ventrally, fulvous along eye margins; with a few short dark postocular setae. Antennae with all segments dark fuscous; a strong black dorsal seta on segment 2; arista entirely fulvous; length of segments: 0.4 mm; 0.6 mm; 1.72 mm.

Thorax: Scutum black tending dark fuscous posteriorly. A transverse furrow across scutum connecting both parts of the notopleural suture. Pleural areas fuscous to dark fuscous. Yellow markings as follows: posterior ½ of postpronotal lobe (anterior ½ dark fuscous); band running from posterior margin of postpronotal lobe approximately halfway to notopleuron; broad mesopleural stripe, reaching postpronotal lobe dorsally, continuing to katepisternum as a transverse spot, anterior margin deeply concave; anatergite (posterior apex dark fuscous); dorsocentral region of katatergite (remainder dark fuscous); medium-width lateral postsutural vitta ending well behind *ia*. seta. Notopleuron dark fuscous. Postnotum dark fuscous. Scutellum yellow except for broad dark fuscous basal band. Setae: *sc*. 2; *prsc*. absent; *ia*. 1; *p.sa*. 1; *a.sa*. 1; *mpl*. 1; *npl*. 2; *scp*. 2.

Legs: All segments fuscous except basal tarsal segment of each leg fulvous; mid tibiae each with an apical black spur; two strong preapical spines on ventral surface of fore femora.

Wings: Length 6.6 mm; cells bc and c fuscous; dense microtrichia covering both cells; remainder of wings with a pale tint across the membrane except for fuscous cell sc, broad fuscous costal band overlapping R_{4+5}; anal streak absent; no dense aggregation of microtrichia around A_1+CuA_2; supernumerary lobe very weak. Cell bm long and narrow (approximately 4 times longer than wide); cell sc long and wide (approximately 8 times longer than wide and 2 times longer than cell c); veins R_{2+3} and M with distinctive pronounced curvatures; vein A_1+CuA_2 ending well before wing margin.
Abdomen: Elongate, clavate and petiolate; terga free; pecten absent on tergum III. Tergum I and sterna I and II wider than long. Tergum I black; tergum II black except for large orange-brown anterolateral spots; terga III–V black except for dark red-brown each side of a dark fuscous to black medial longitudinal band on tergum V. Shining spots not present on tergum V. Posterior lobe of surstylus short, sternum V with a moderate concavity on posterior margin.

Attractant: No known record.
Female: As for male except for an extra abdominal segment present. Oviscape orange-brown, dorsoventrally compressed and tapering posteriorly in dorsal view; ratio of length of oviscape to length of tergum VI, 3.7: 1; apex of piercer needle-shaped, with microscopic preapical setae.
Distribution:
Type locality: Peninsular Malaysia (Kuala Lumpur, University of Malaya garden).
Hosts: *Citrus halimii* (kumquat) (family *Rutaceae*).
Remarks: *Monacrostichus malaysiae* Drew & Hancock differs from *Monacrostichus citricola* Bezzi in lacking yellow spots anterior to the notopleural suture, lacking a yellow mark or vitta in the centre of the scutum, and in having abdominal terga III and IV entirely or mostly dark, only two preapical spines on the fore femora, all femora without black markings and a pale fuscous tint across the wing membrane.

Pest Species in the Genera *Bactrocera* and *Dacus*

Since 1983, we have been directly involved in field survey work aimed at recording dacine species, their host plants and geographic distributions. These surveys have been supported by projects funded by ACIAR, AusAID, the SPC-UNDP Regional Fruit Fly Project in the South Pacific, the project to eradicate *Bactrocera papayae* from the Torres Strait Islands and North Queensland, and other smaller projects within Australia funded by State and Territory Governments.

The surveys have been based on male lure trapping using cue lure and methyl eugenol in Steiner-type traps, and fruit collections from rainforests and plantations.

For the male lure trapping some 500 sites were used across Indomalaya and Wallacea, and another 1500 covering Australasia and 21 Pacific Island countries. The traps were maintained at each site for at least 1 year and up to 10 years in some Pacific Island countries, and serviced on 2–3 weekly cycles.

The fruit surveys involved regular field collections of wild forest fruits and edible cultivated fruits. Approximately 23,000 fruit samples were collected over Indomalaya and Wallacea, 90,000 in Australia and 20,000 in the South Pacific Islands. The host records have been published for Indomalaya and Wallacea in Allwood *et al.* (1999), for Australia in Hancock *et al.* (2000) and the Pacific Island countries in Leblanc *et al.* (2012).

These extensive surveys have provided an opportunity to define the dacine fruit fly species that can be classified as having significant economic importance, map their geographic distributions and define their host plant ranges. This current volume provides detailed information on the pest species of Indomalaya and Wallacea, including new host records since Allwood *et al.* (1999). See Drew (1989) for descriptions of the South Pacific species, including Papua New Guinea and Australia.

To classify a fruit fly species as one of economic importance, we believe that it has had to be reared from host fruits in large numbers (ten or more flies per sample) and/or from 10% of samples taken for each fruit species. There are always some unusual or unexpected results in fruit surveys of the magnitude noted above. These are a biological phenomenon, not an error in sampling, and such records cannot be used to assess the economic importance of fruit fly species. Hancock *et al.* (2000) overlooked such records as 'a result of contamination of samples during field transportation, larvae sometimes moving from one sample to an adjacent one by tunnelling through paper bags'. Having been directly involved in large segments of the field collection programmes recorded above, we wish to record that this statement of Hancock *et al.* (2000) is misleading and incorrect. Dacine larvae only leave their host fruit to pupate, and we have never observed them moving between samples, either when held in the field or during transport to laboratories.

The following list in Table 1 gives the dacine species we consider to be of significant economic importance across Indomalaya, Wallacea and Australasia. Our estimate of their pest status, potential to enter and colonize new locations, is given in Table 2.

Table 1. Species of the genera *Bactrocera* and *Dacus* of economic importance in Indomalaya, Wallacea and Australasia.

Genus *Bactrocera*	Subgenus	Lure	SEA	SP	Other
albistrigata (de Meijere)	*Bactrocera*	CUE	♦		
aquilonis (May)	*Bactrocera*	CUE		♦	
atrisetosa (Perkins)	*Paratridacus*	–		♦	
bryoniae (Tryon)	*Bactrocera*	CUE		♦	
carambolae Drew & Hancock	*Bactrocera*	M.E.	♦		
caryeae (Kapoor)	*Bactrocera*	M.E.	♦		
caudata (Fabricius)	*Zeugodacus*	CUE	♦		
correcta (Bezzi)	*Bactrocera*	M.E.	♦		
cucumis (French)	*Austrodacus*	–		♦	
cucurbitae (Coquillett)	*Zeugodacus*	CUE	♦	♦	N. Pacific
curvifera (Walker)	*Bactrocera*	M.E.		♦	
curvipennis (Froggatt)	*Bactrocera*	CUE		♦	
decipiens (Drew)	*Paradacus*	–		♦	
diversa (Coquillett)	*Hemigymnodacus*	M.E. (weak)	♦		
dorsalis (Hendel)	*Bactrocera*	M.E.	♦	♦	N. Pacific
facialis (Coquillett)	*Bactrocera*	CUE		♦	
frauenfeldi (Schiner)	*Bactrocera*	CUE		♦	
invadens Drew, Tsuruta & White	*Bactrocera*	M.E.	♦		Africa
jarvisi (Tryon)	*Afrodacus*	CUE (weak), ZING		♦	
kandiensis Drew & Hancock	*Bactrocera*	M.E.	♦		
kirki (Froggatt)	*Bactrocera*	CUE		♦	
kraussi (Hardy)	*Bactrocera*	CUE		♦	
latifrons (Hendel)	*Bactrocera*	–	♦		Hawaii/Africa
melanotus (Coquillett)	*Bactrocera*	CUE		♦	
minax (Enderlein)	*Tetradacus*	–	♦		
musae (Tryon)	*Bactrocera*	M.E.		♦	
neohumeralis (Hardy)	*Bactrocera*	CUE		♦	
obliqua (Malloch)	*Bactrocera*	–		♦	
occipitalis (Bezzi)	*Bactrocera*	M.E.	♦		
papayae Drew & Hancock	*Bactrocera*	M.E.	♦	♦	
passiflorae (Froggatt)	*Bactrocera*	CUE		♦	
psidii (Froggatt)	*Bactrocera*	CUE		♦	
pyrifoliae Drew & Hancock	*Bactrocera*	–	♦		
scutellaris (Bezzi)	*Zeugodacus*	CUE	♦		
scutellata (Hendel)	*Zeugodacus*	CUE	♦		
strigifinis (Walker)	*Sinodacus*	CUE		♦	
tau (Walker)	*Zeugodacus*	CUE	♦		
trilineola Drew	*Bactrocera*	CUE		♦	
trivialis (Drew)	*Bactrocera*	CUE		♦	
tryoni (Froggatt)	*Bactrocera*	CUE		♦	
tsuneonis (Miyake)	*Tetradacus*	–	♦		
tuberculata (Bezzi)	*Bactrocera*	M.E.	♦		
umbrosa (Fabricius)	*Bactrocera*	M.E.	♦	♦	
xanthodes (Broun)	*Notodacus*	M.E.		♦	
zonata (Saunders)	*Bactrocera*	M.E.	♦		Egypt
Genus *Dacus*					
ciliatus Loew	*Didacus*	–	♦		Africa, Egypt
solomonensis Malloch	*Callantra*	CUE		♦	

SEA = South-East Asia; SP = South Pacific; M.E.= Methyl Eugenol; ZING = Zingerone (Vanillylacetone).

Table 2. The pest status and potential to enter and colonize new locations, for the pest species of the genera *Bactrocera* and *Dacus* in Indomalaya, Wallacea and Australasia.

Genus *Bactrocera*	Pest status	Entry potential	Colonization
albistrigata (de Meijere)	Medium	Low	High
aquilonis (May)	Minor	Low	Low
atrisetosa (Perkins)	Minor	Low	Low
bryoniae (Tryon)	Minor	Low	Low
carambolae Drew & Hancock	Major	High	High
caryeae (Kapoor)	Major	Medium	High
caudata (Fabricius)	Major	Low	Low
correcta (Bezzi)	Major	High	High
cucumis (French)	Major	Medium	High
cucurbitae (Coquillett)	Major	High	High
curvifera (Walker)	Minor	Low	Low
curvipennis (Froggatt)	Major	Low	High
decipiens (Drew)	Medium	Low	Low
diversa (Coquillett)	Major	Low	Low
dorsalis (Hendel)	Major	High	High
facialis (Coquillett)	Major	High	High
frauenfeldi (Schiner)	Major	High	High
invadens Drew, Tsuruta & White	Major	High	High
jarvisi (Tryon)	Major	High	High
kandiensis Drew & Hancock	Medium	Low	High
kirki (Froggatt)	Major	High	High
kraussi (Hardy)	Major	High	High
latifrons (Hendel)	Medium	High	Medium
melanotus (Coquillett)	Major	Medium	Medium
minax (Enderlein)	Major	Low	Low
musae (Tryon)	Major (bananas)	Low–Medium	Low
neohumeralis (Hardy)	Major	High	High
obliqua (Malloch)	Medium	Low	Low
occipitalis (Bezzi)	Major	High	High
papayae Drew & Hancock	Major	High	High
passiflorae (Froggatt)	Major	High	Medium–High
psidii (Froggatt)	Major	High	High
pyrifoliae Drew & Hancock	Medium	Low	Low
scutellaris (Bezzi)	Medium	Low	Low
scutellata (Hendel)	Medium	Low	Low
strigifinis (Walker)	Minor–Medium	Low	Low
tau (Walker)	Medium	Low	Low
trilineola Drew	Medium–Major	High	High
trivialis (Drew)	Medium	Medium	Medium
tryoni (Froggatt)	Major	High	High
tsuneonis (Miyake)	Major	Low	Low
tuberculata (Bezzi)	Medium	Low	Low
umbrosa (Fabricius)	Medium	Low	Low
xanthodes (Broun)	Major	High	Medium–High
zonata (Saunders)	Major	High	High
Genus *Dacus*			
ciliatus Loew	Major	Medium	Medium
solomonensis Malloch	Minor	Low	Low

Figures

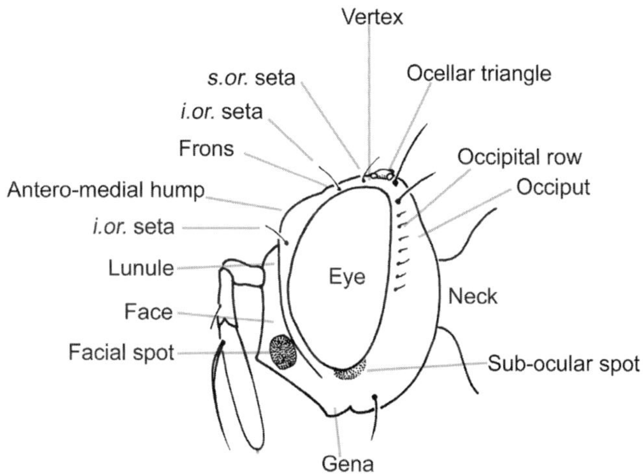

Fig. 1. Lateral view of Dacinae head showing key morphological features. **Abbreviations**: *i.or.* = inferior fronto-orbital seta; *s.or.* = superior fronto-orbital seta.

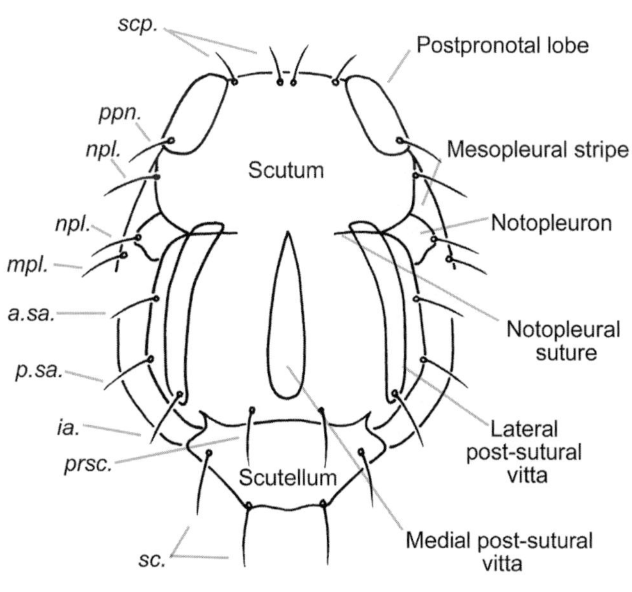

Fig. 2. Dorsal view of Dacinae thorax showing key morphological features. **Abbreviations**: *a.sa.* = anterior supra-alar seta; *ia.* = intra-alar seta; *mpl.* = mesopleural seta; *npl.* = notopleural seta; *ppn.* = postpronotal seta; *prsc.* = prescutellar seta; *p.sa.* = posterior supra-alar seta; *sc.* = scutellar seta; *scp.* = scapular seta.

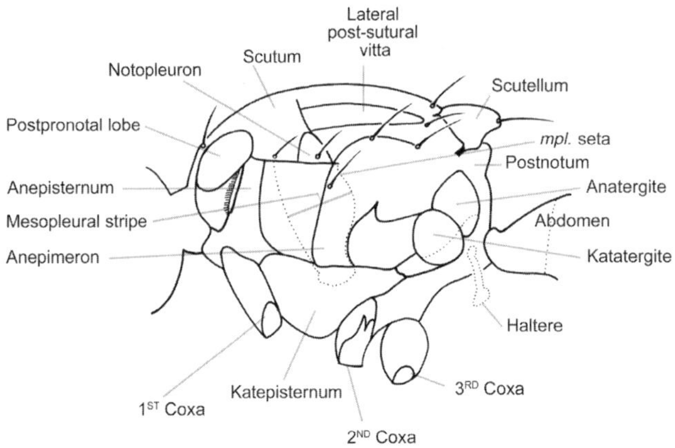

Fig. 3. Lateral view of Dacinae thorax showing key morphological features. **Abbreviation** *mpl.* = mesopleural seta.

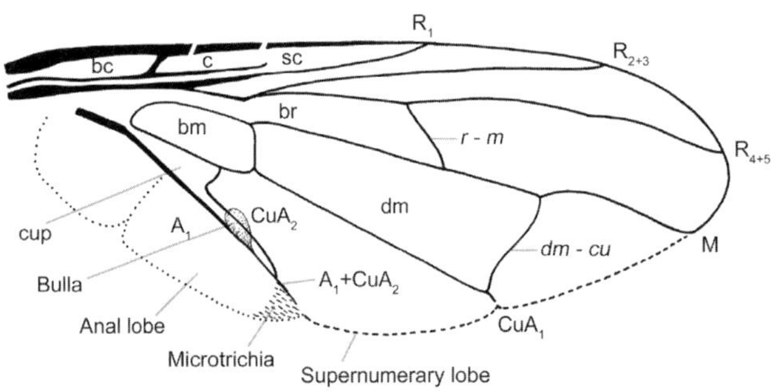

Fig. 4. Wing of Dacinae showing terminology of veins and cells. **Veins**: A_1 = branch of anal vein; CuA_1, CuA_2 = anterior branches of cubitus; M = media; R_1 = anterior branch of radius; R_{2+3}, R_{4+5} = combined posterior branches of radius; *dm-cu* = discal medial-cubital crossvein; *r-m* = radial-medial crossvein. **Cells**: bc = basal costal; c = costal; sc = subcostal; bm = basal medial; br = basal radial; cup = posterior cubital; dm = discal medial.

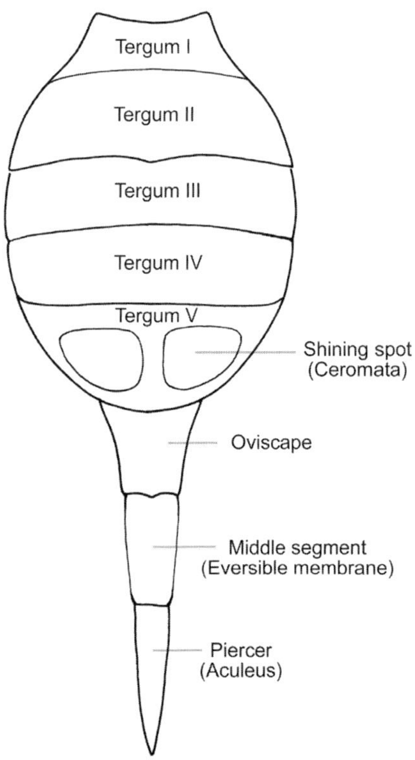

Fig. 5. Dorsal view of Dacinae abdomen and female ovipositor showing key morphological features.

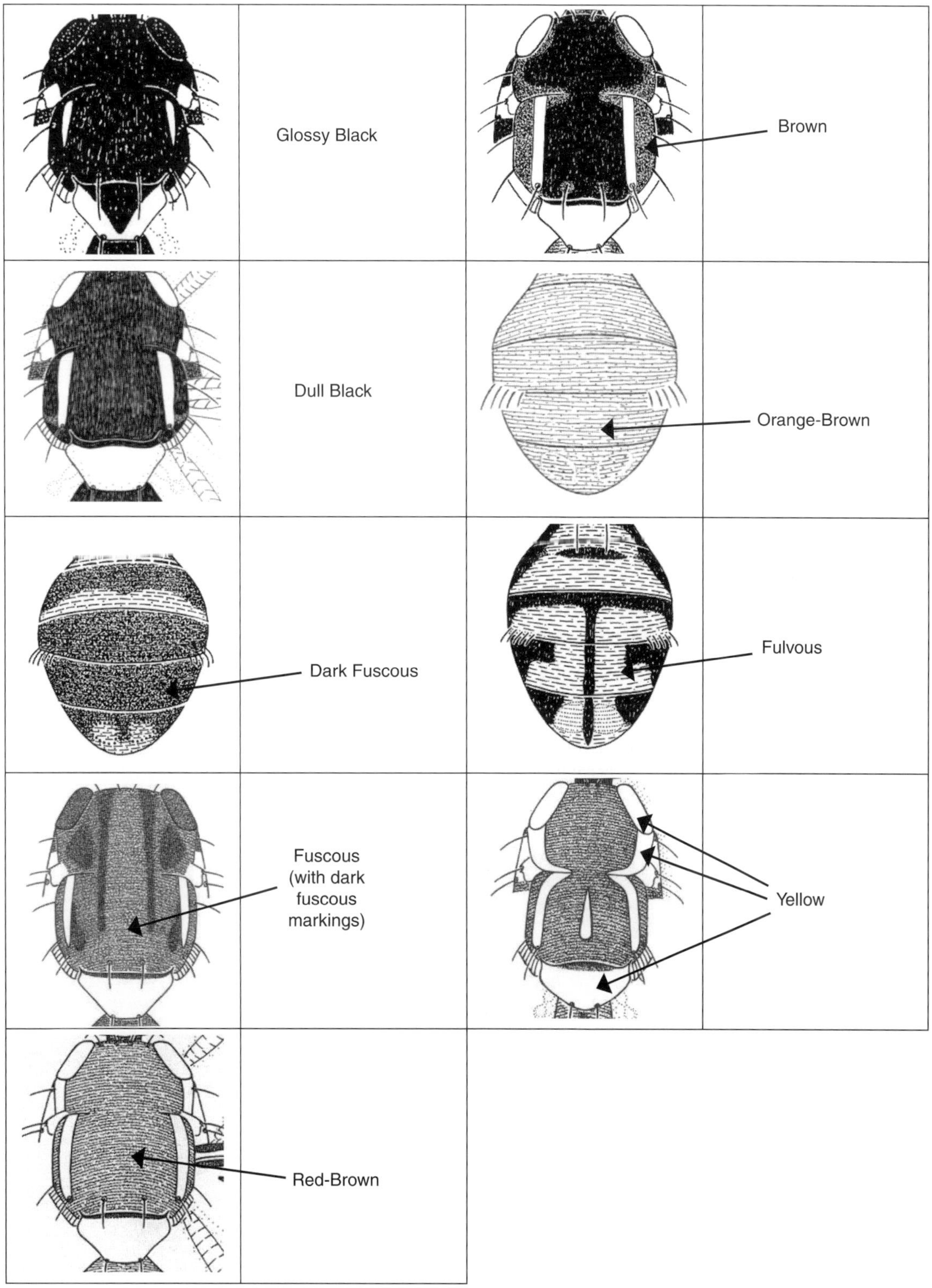

Fig. 6. Diagrammatic colours used in descriptions and illustrations. **Note**: fuscous: dark brown, approaching black (a mixture of black and red); fulvous: reddish-yellow (Nichols, 1989).

7

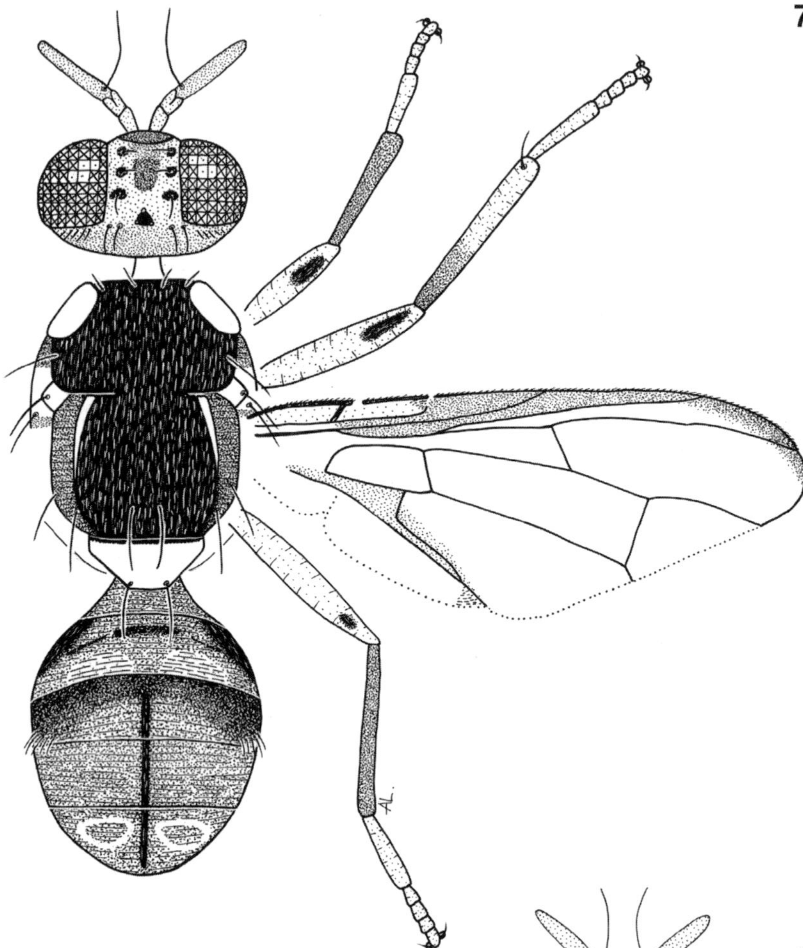

8

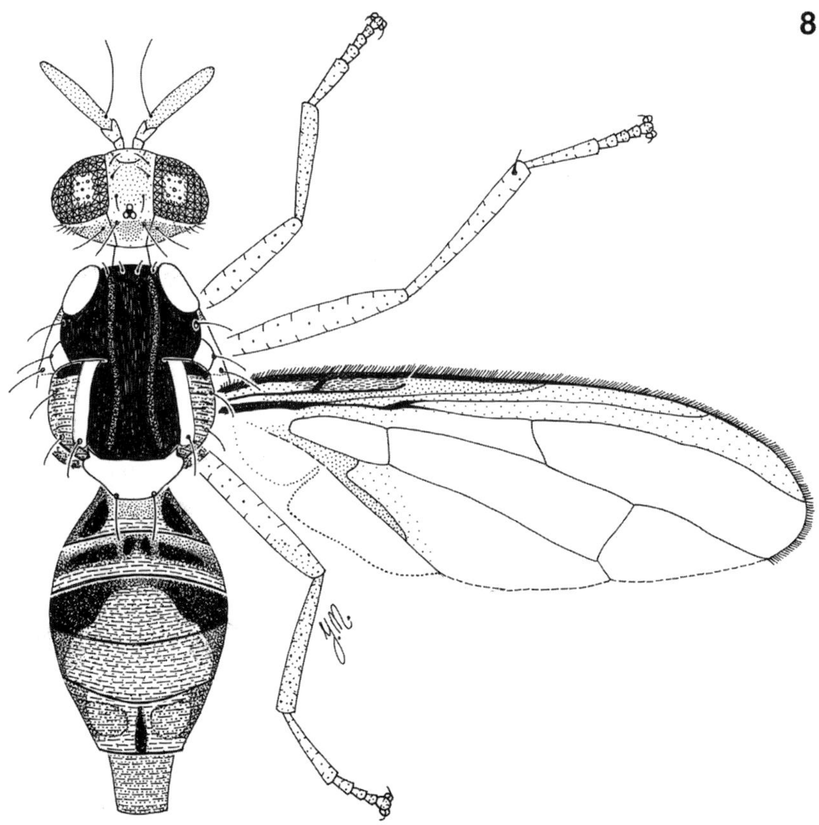

Fig. 7. *Bactrocera (Afrodacus) fastigata* Tsuruta & White, male. **Fig. 8.** *Bactrocera (Asiadacus) absoluta* (Walker), lectotype female.

9

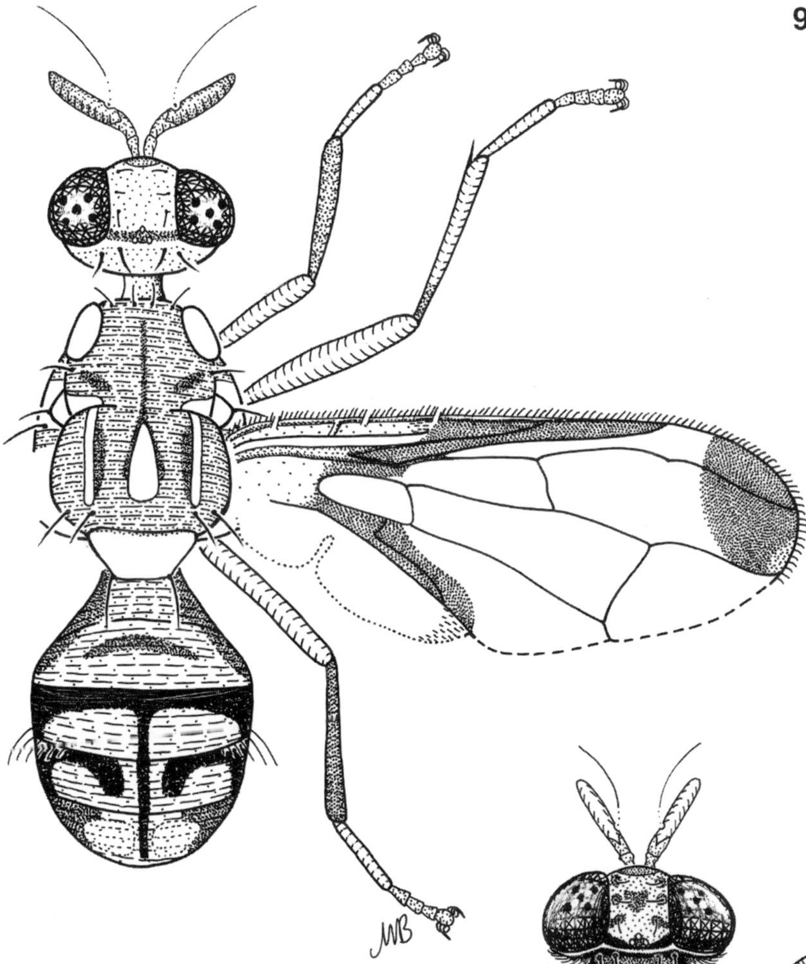

10

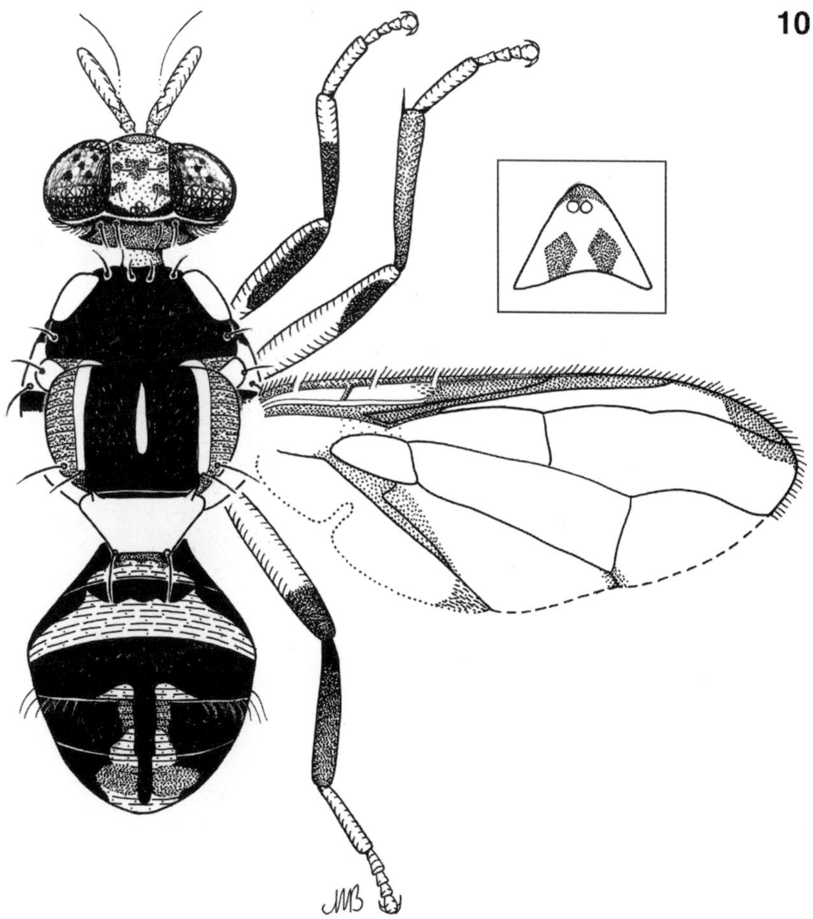

Fig. 9. *Bactrocera (Asiadacus) apicalis* (de Meijere), male. **Fig. 10.** *Bactrocera (Asiadacus) apiciflava* Yu, He & Chen, male and face showing facial spots.

11

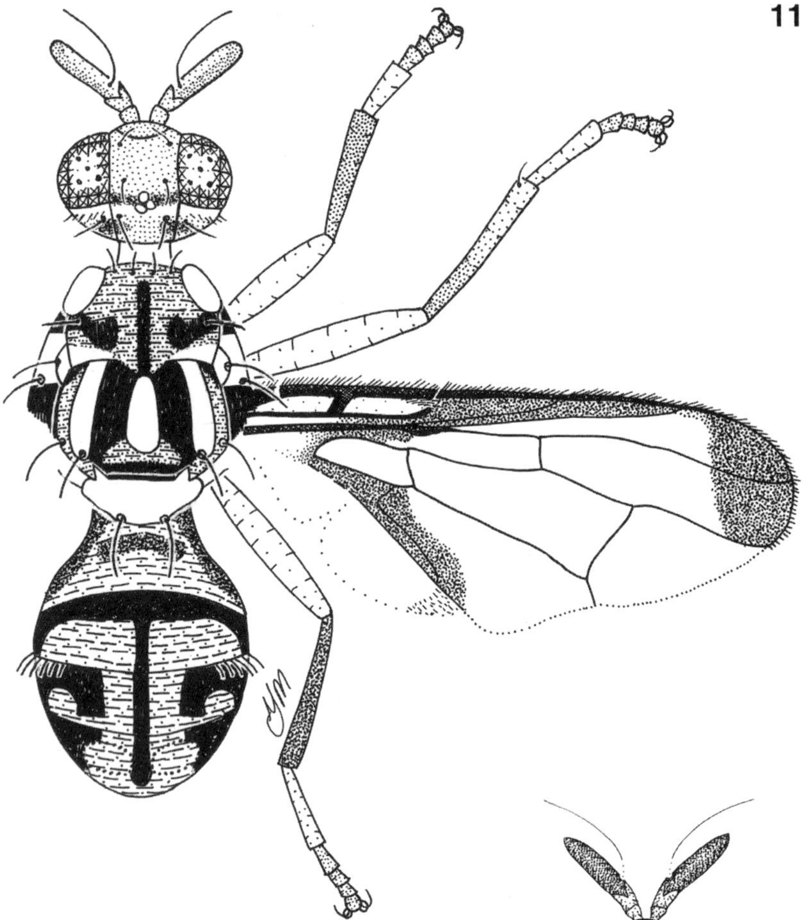

12

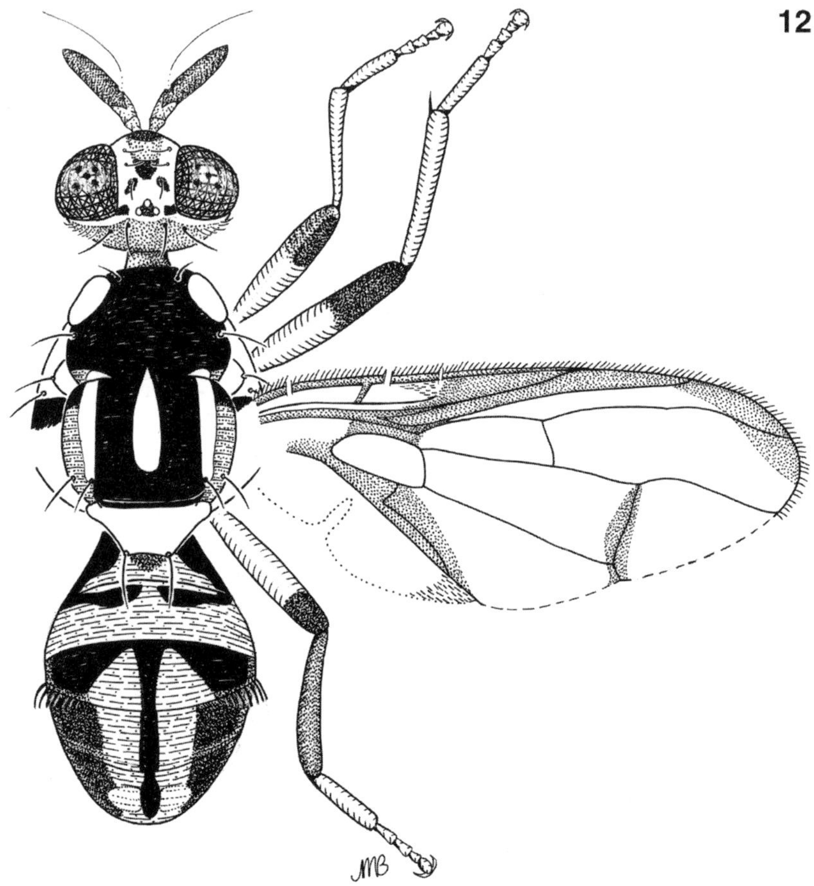

Fig. 11. *Bactrocera (Asiadacus) bakeri* (Bezzi), holotype male. **Fig. 12.** *Bactrocera (Asiadacus) careomacula* new species, holotype male.

13

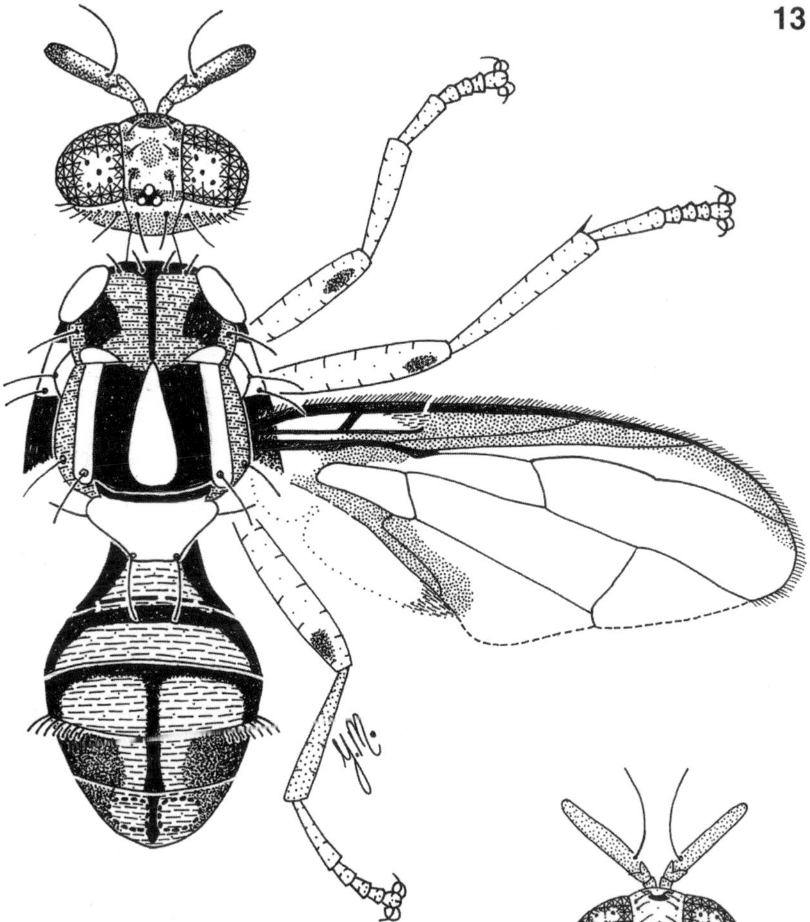

14

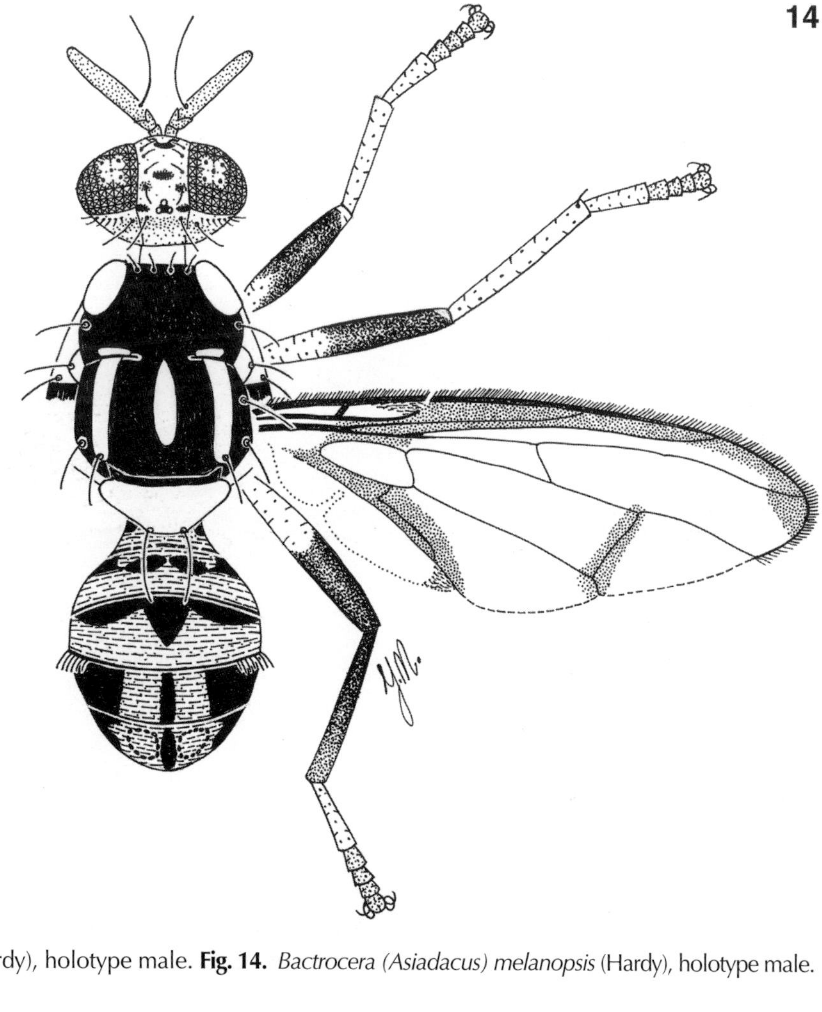

Fig. 13. *Bactrocera (Asiadacus) maculifacies* (Hardy), holotype male. **Fig. 14.** *Bactrocera (Asiadacus) melanopsis* (Hardy), holotype male.

15

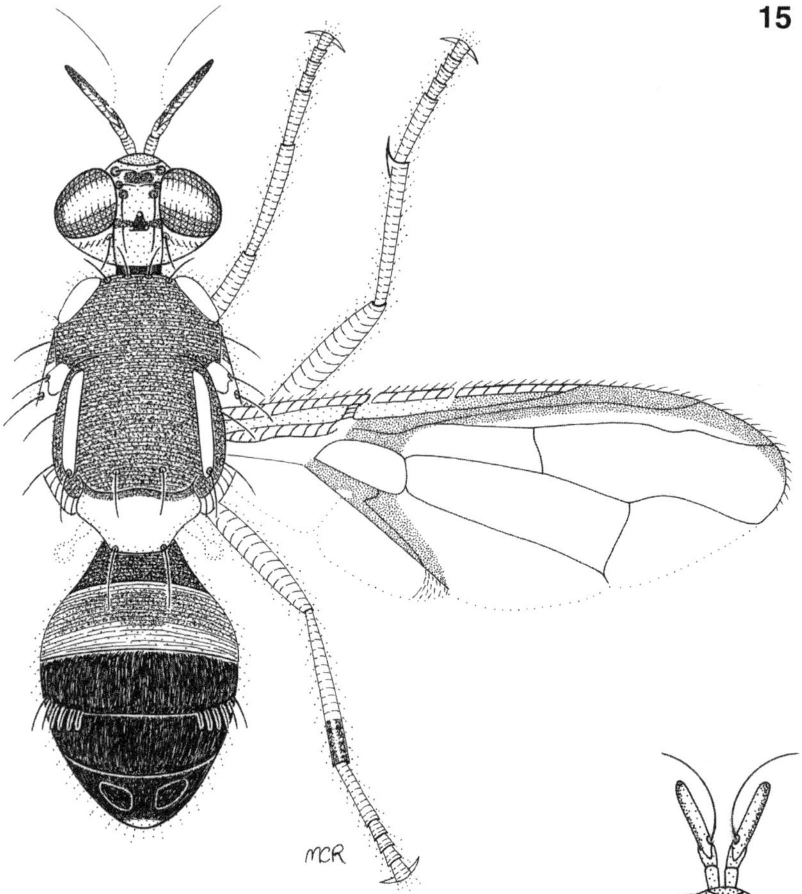

16

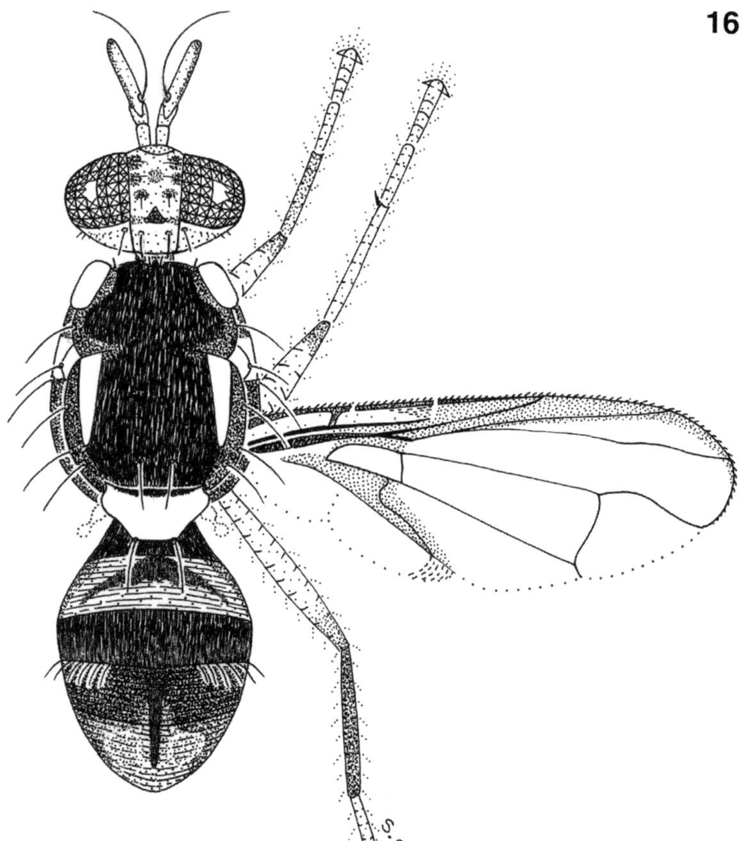

Fig. 15. *Bactrocera (Bactrocera) abdonigella* (Drew), holotype male. **Fig. 16.** *Bactrocera (Bactrocera) aemula* Drew, holotype male.

17

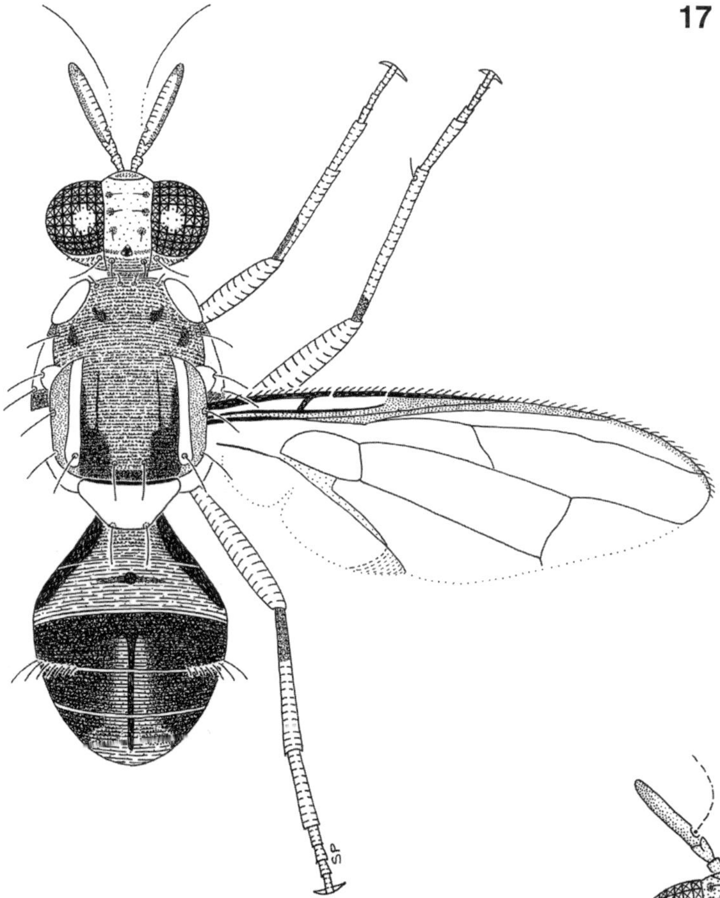

18

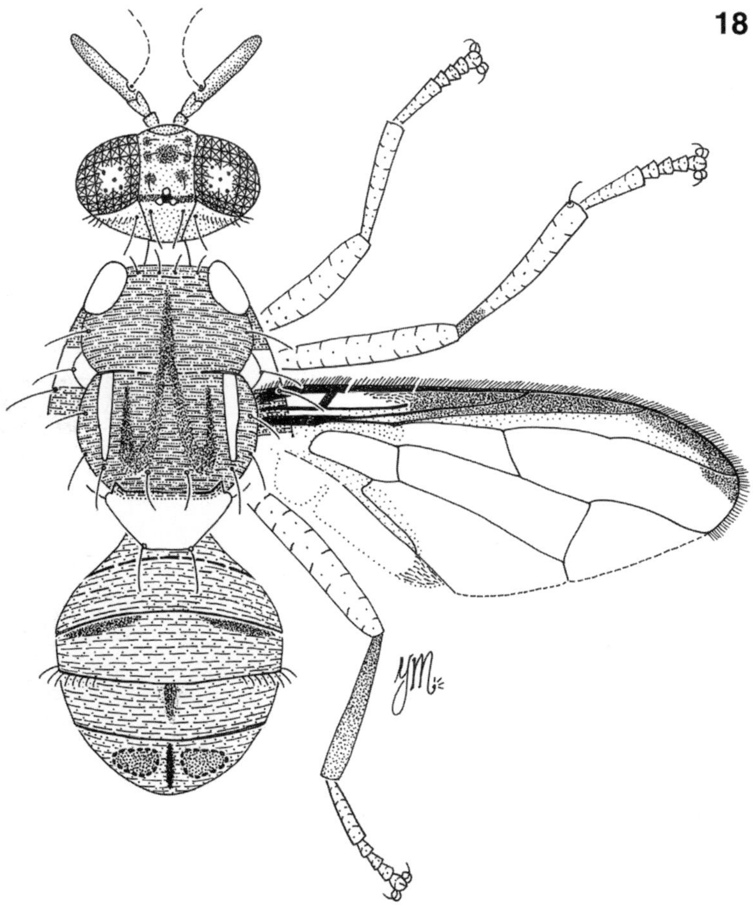

Fig. 17. *Bactrocera (Bactrocera) aethriobasis* (Hardy), male. **Fig. 18.** *Bactrocera (Bactrocera) affinibancroftii* new species, holotype male.

19

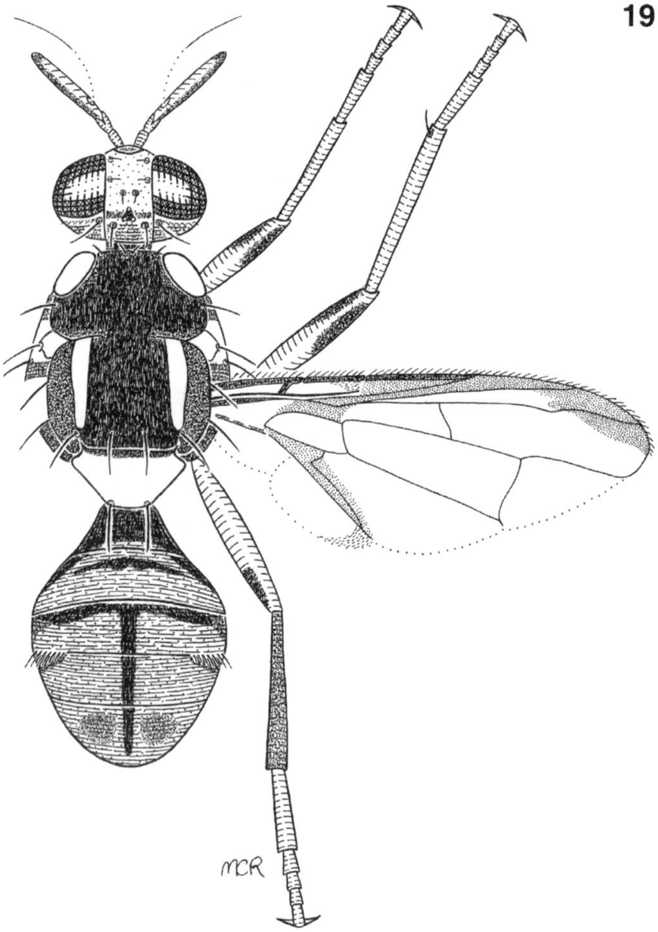

20A

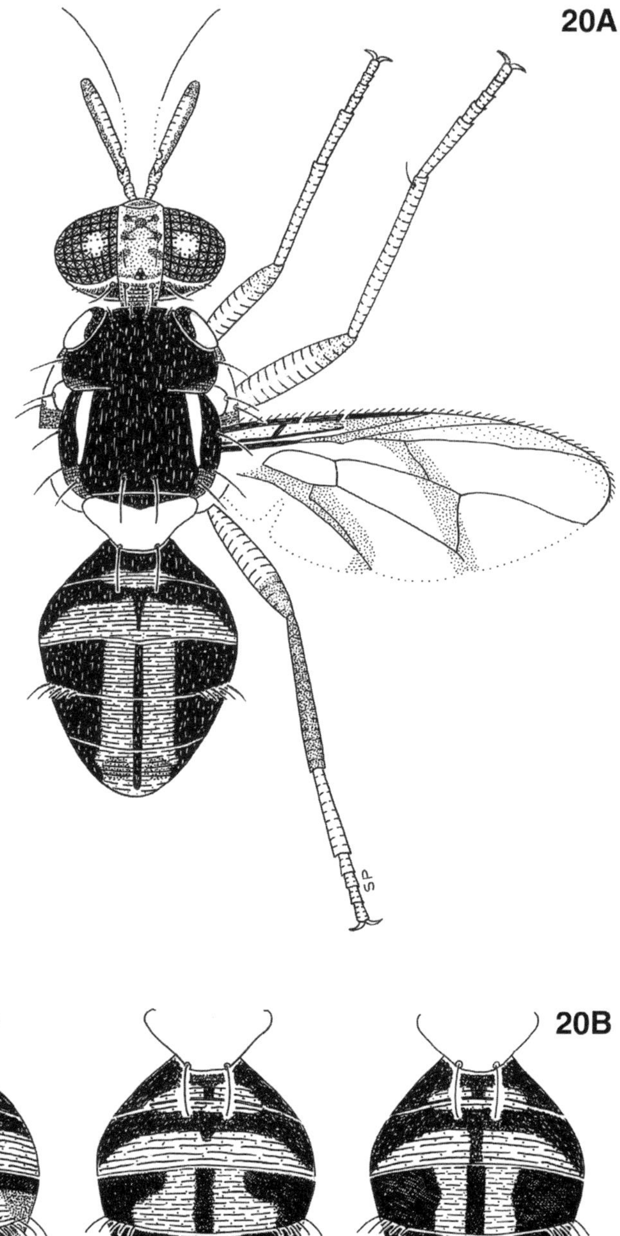

20B

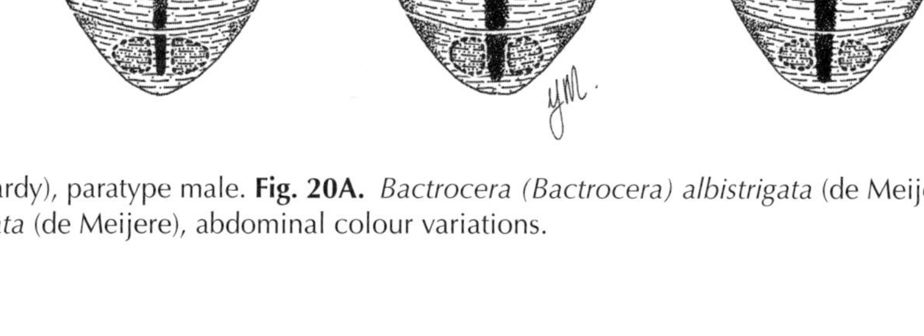

Fig. 19. *Bactrocera (Bactrocera) affinidorsalis* (Hardy), paratype male. **Fig. 20A.** *Bactrocera (Bactrocera) albistrigata* (de Meijere), male. **Fig. 20B.** *Bactrocera (Bactrocera) albistrigata* (de Meijere), abdominal colour variations.

21

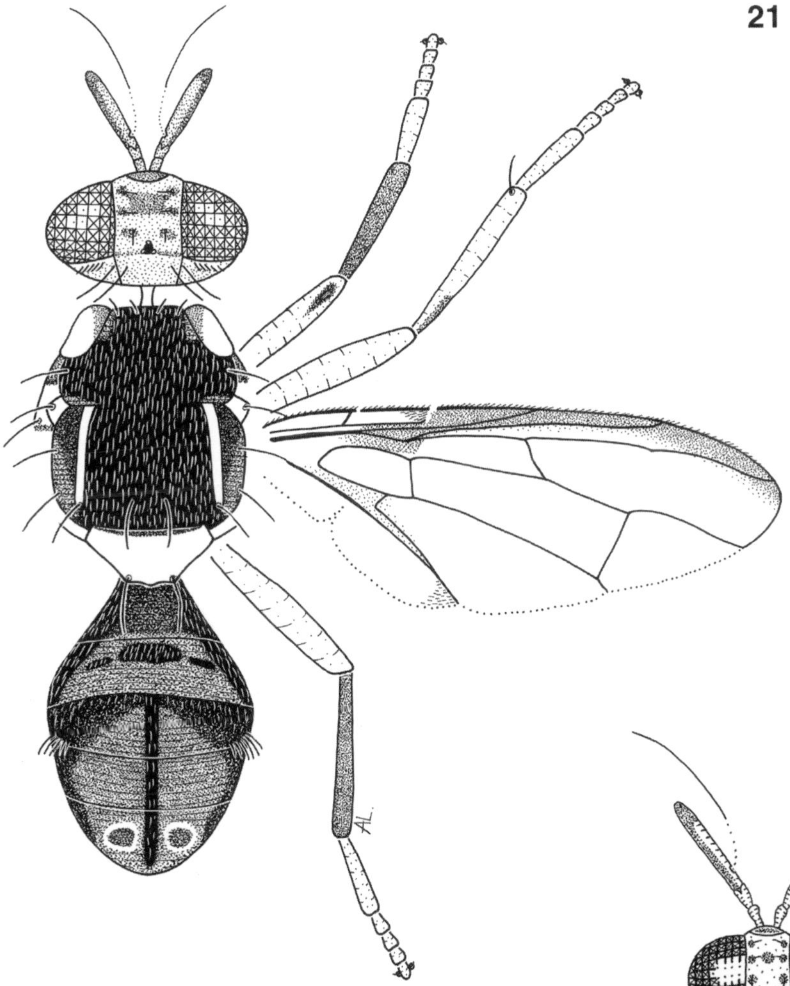

22

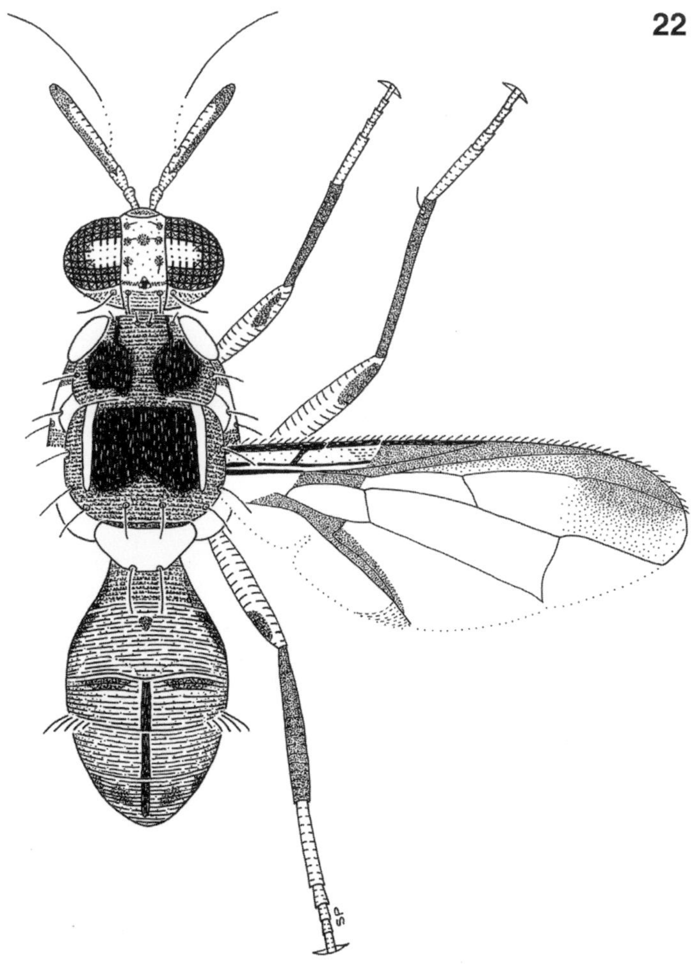

Fig. 21. *Bactrocera (Bactrocera) amarambalensis* Drew, holotype male. **Fig. 22.** *Bactrocera (Bactrocera) andamanensis* (Kapoor), male.

23

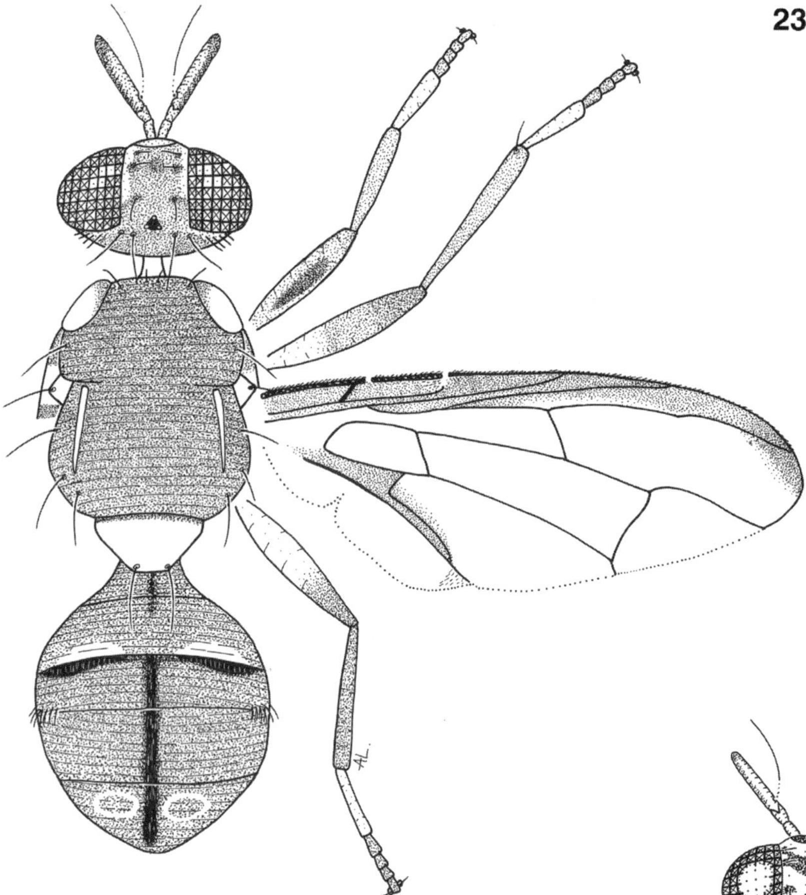

24

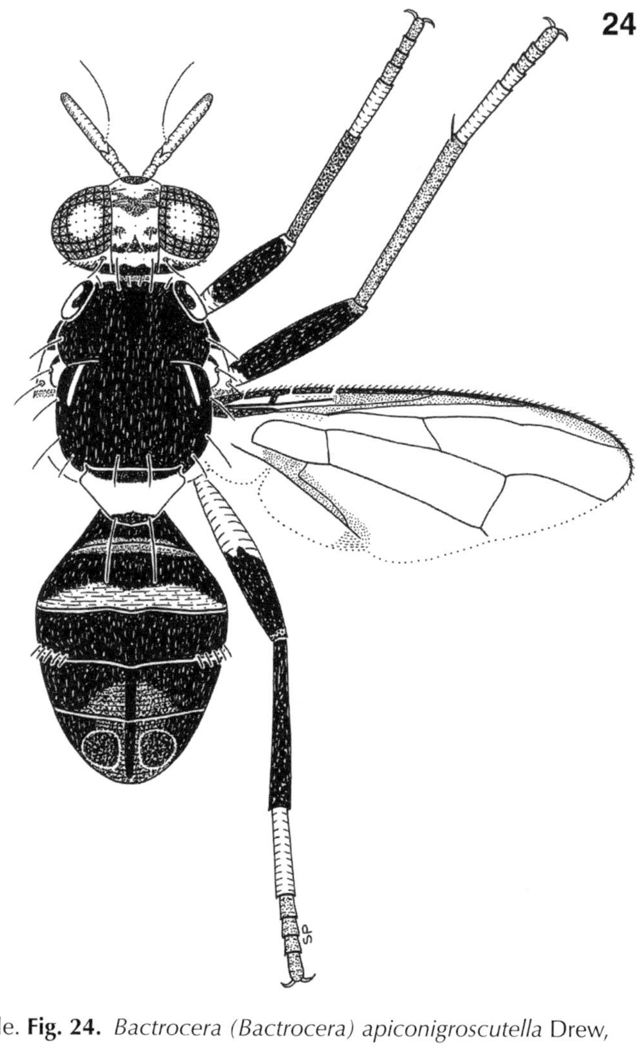

Fig. 23. *Bactrocera (Bactrocera) apicofuscans* White & Tsuruta, male. **Fig. 24.** *Bactrocera (Bactrocera) apiconigroscutella* Drew, holotype male.

25

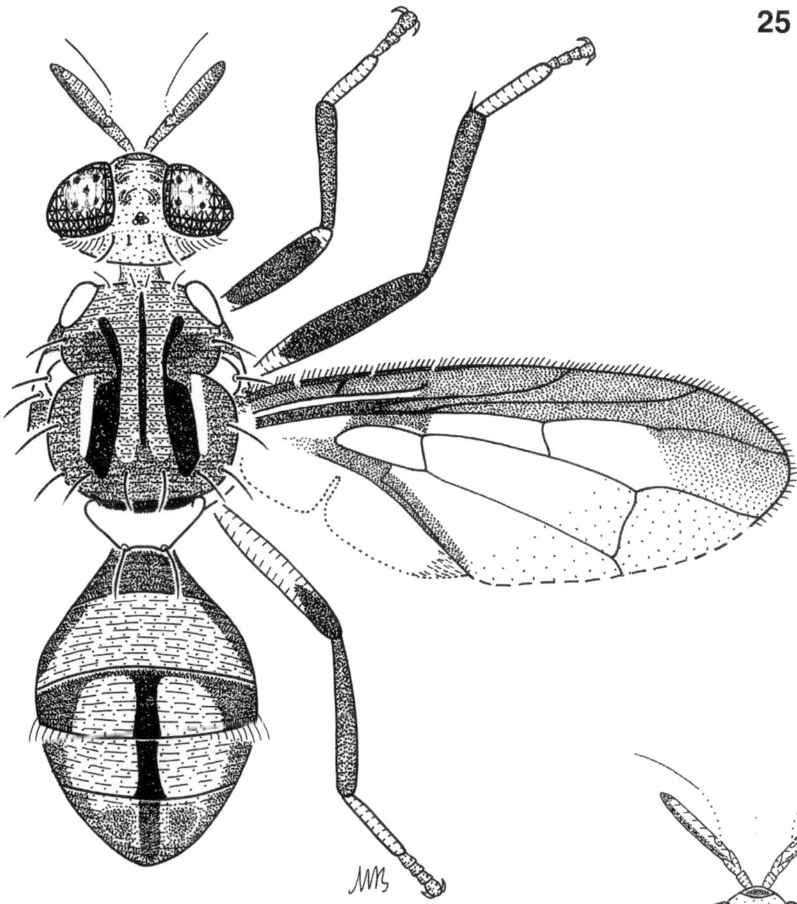

26

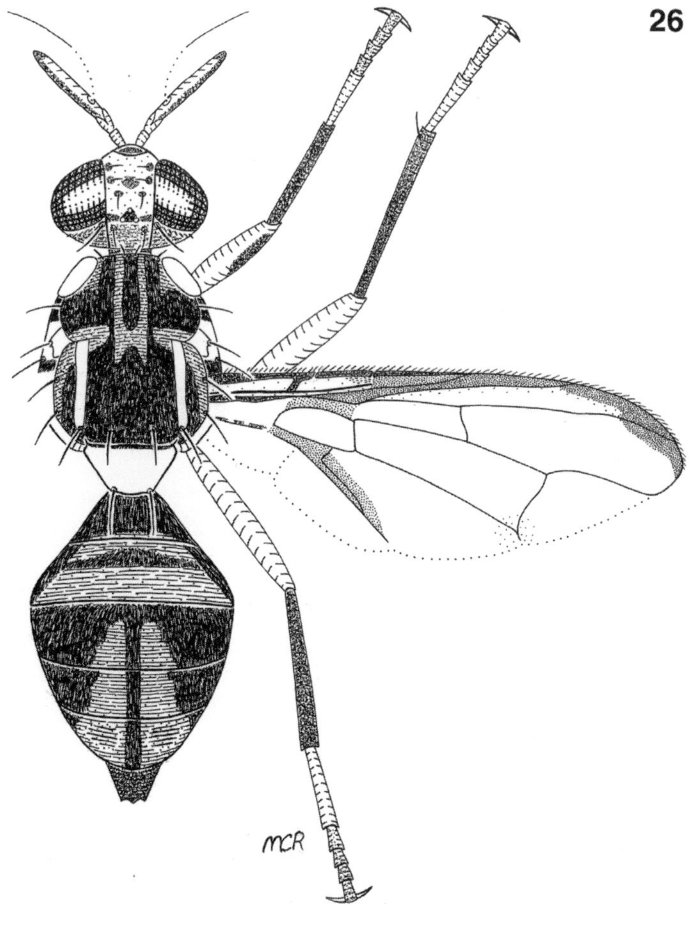

Fig. 25. *Bactrocera (Bactrocera) apicopicta* new species, holotype male. **Fig. 26.** *Bactrocera (Bactrocera) arecae* (Hardy & Adachi), paratype female.

27

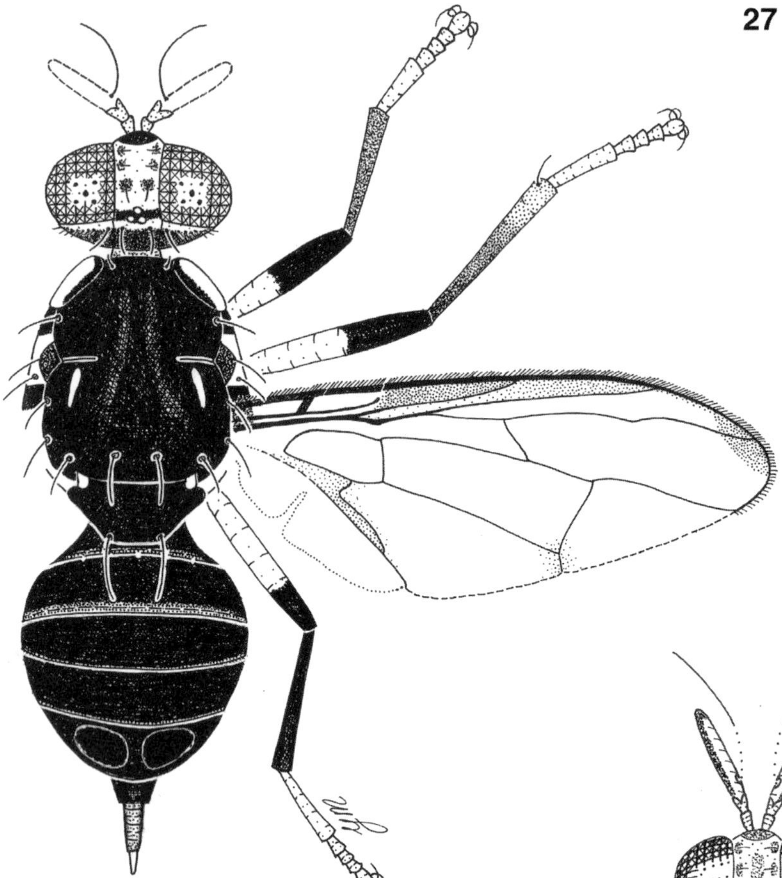

28

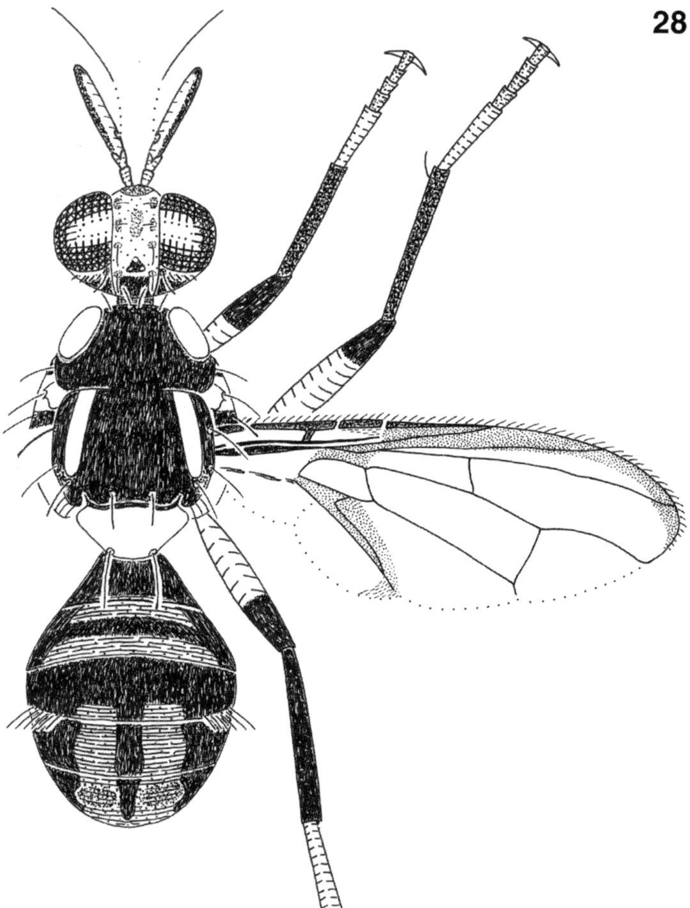

Fig. 27. *Bactrocera (Bactrocera) armillata* (Hering), paratype female. **Fig. 28.** *Bactrocera (Bactrocera) atrifemur* Drew & Hancock, holotype male.

29

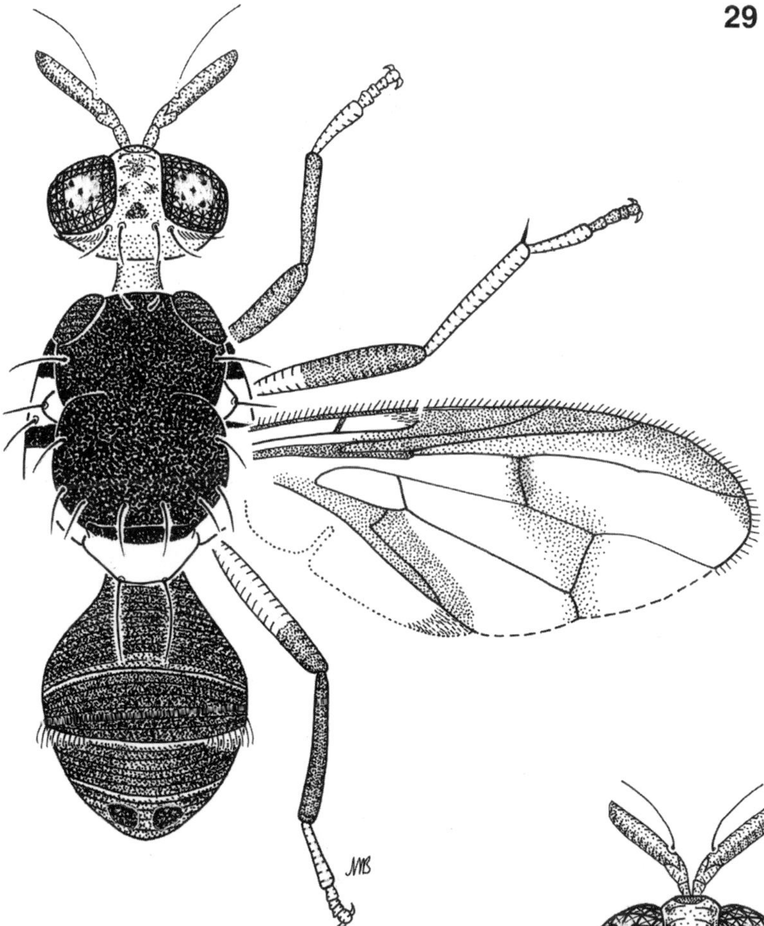

30

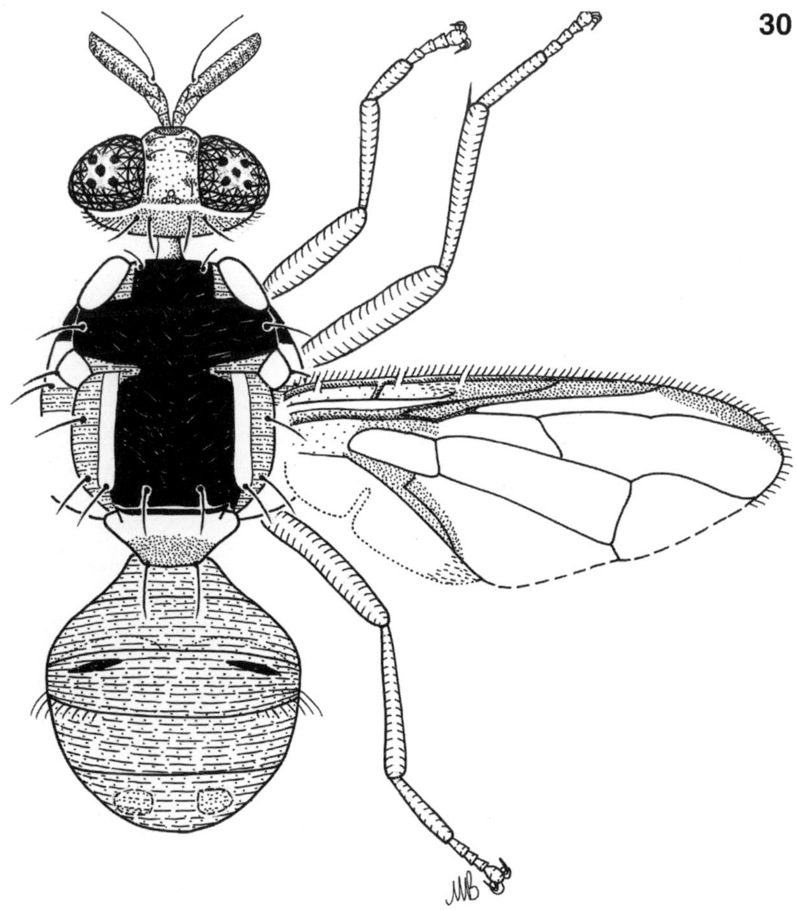

Fig. 29. *Bactrocera (Bactrocera) avittata* new species, holotype male. **Fig. 30.** *Bactrocera (Bactrocera) bancroftii* (Tryon), male.

31

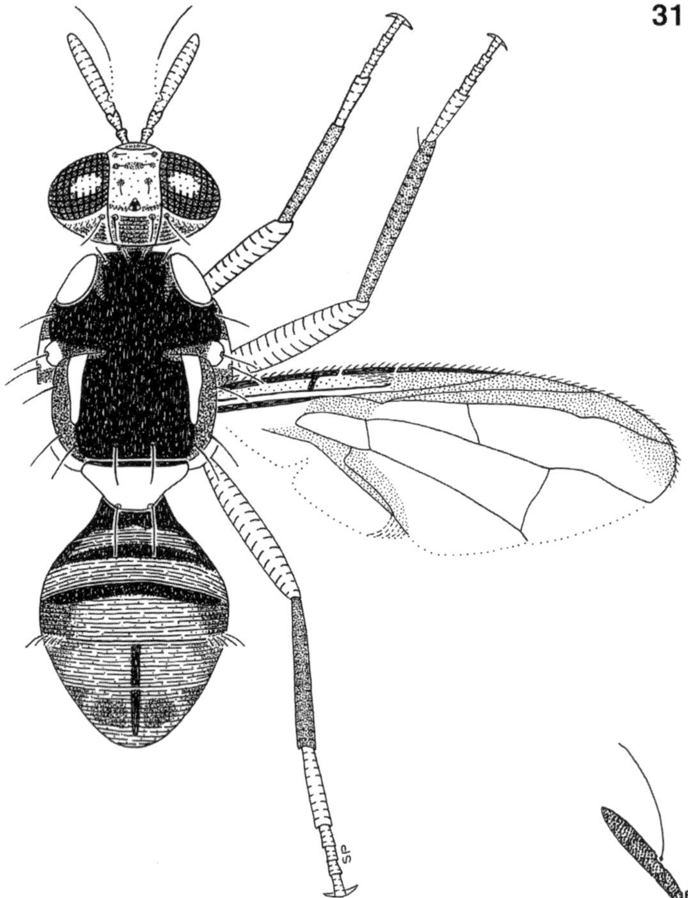

32

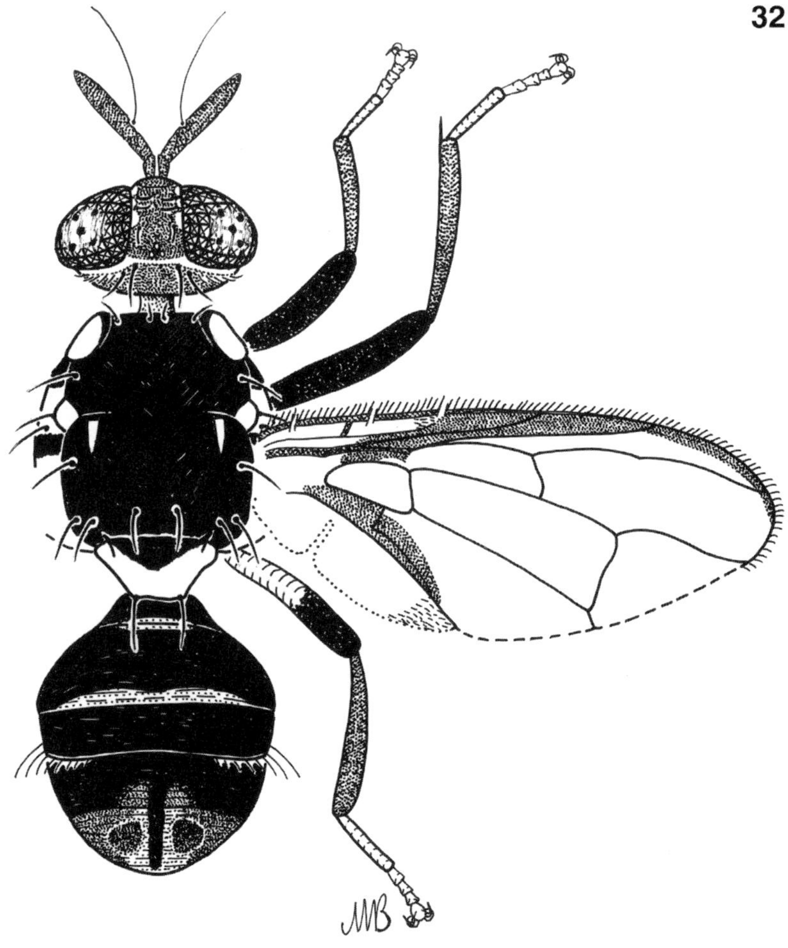

Fig. 31. *Bactrocera (Bactrocera) beckerae* (Hardy), male. **Fig. 32.** *Bactrocera (Bactrocera) bellisi* new species, holotype male.

33

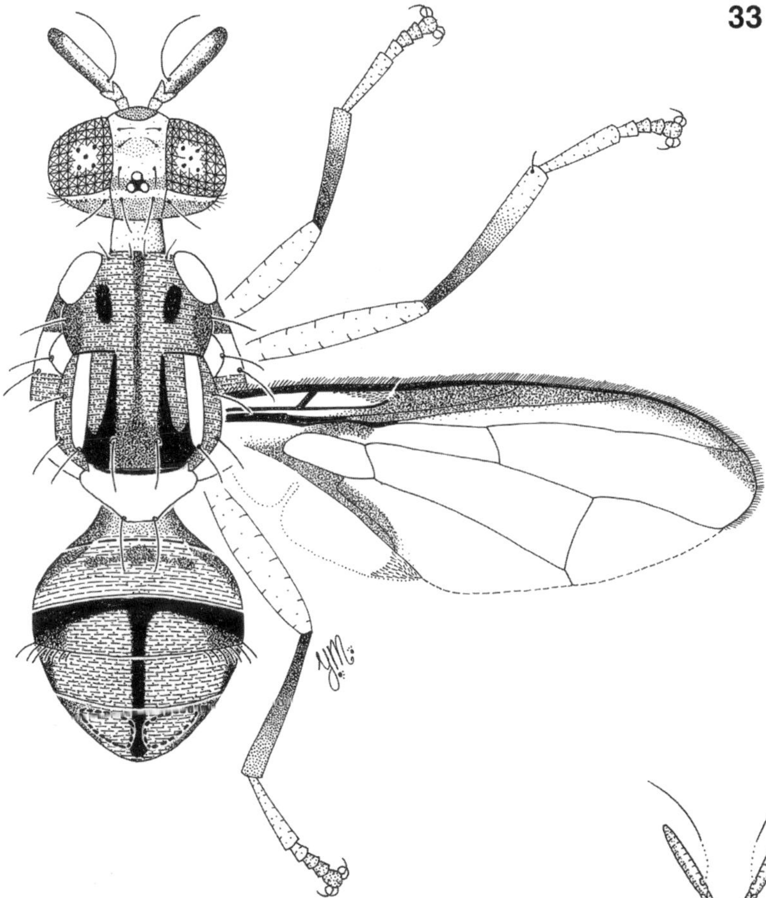

34

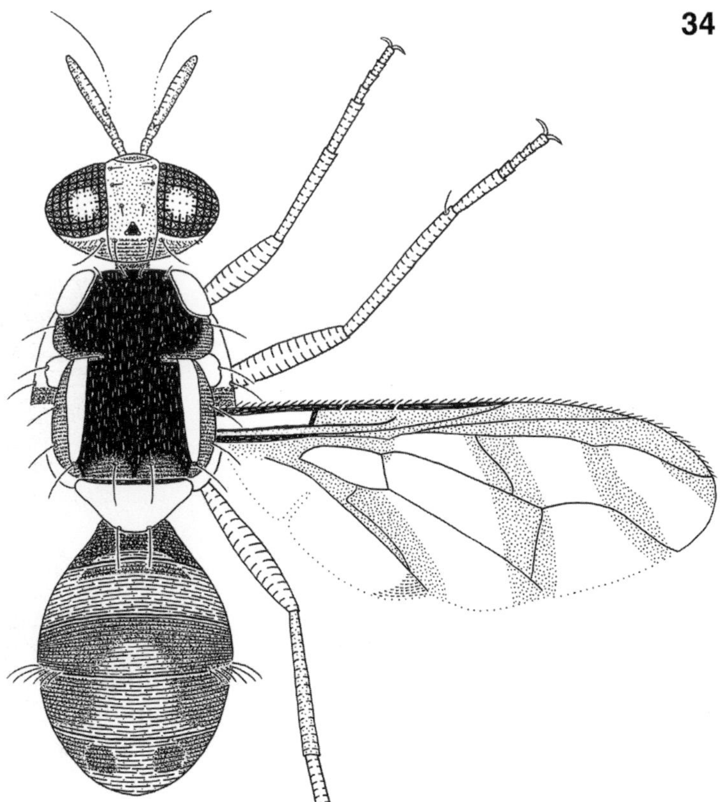

Fig. 33. *Bactrocera (Bactrocera) bhutaniae* new species, holotype male. **Fig. 34.** *Bactrocera (Bactrocera) bifasciata* (Hardy), male.

35

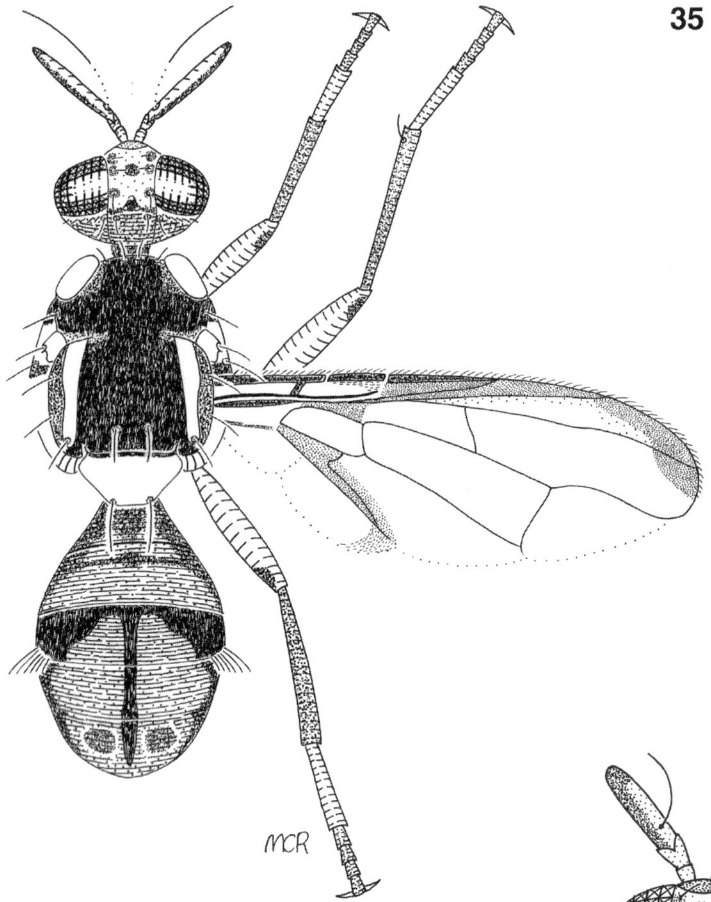

36

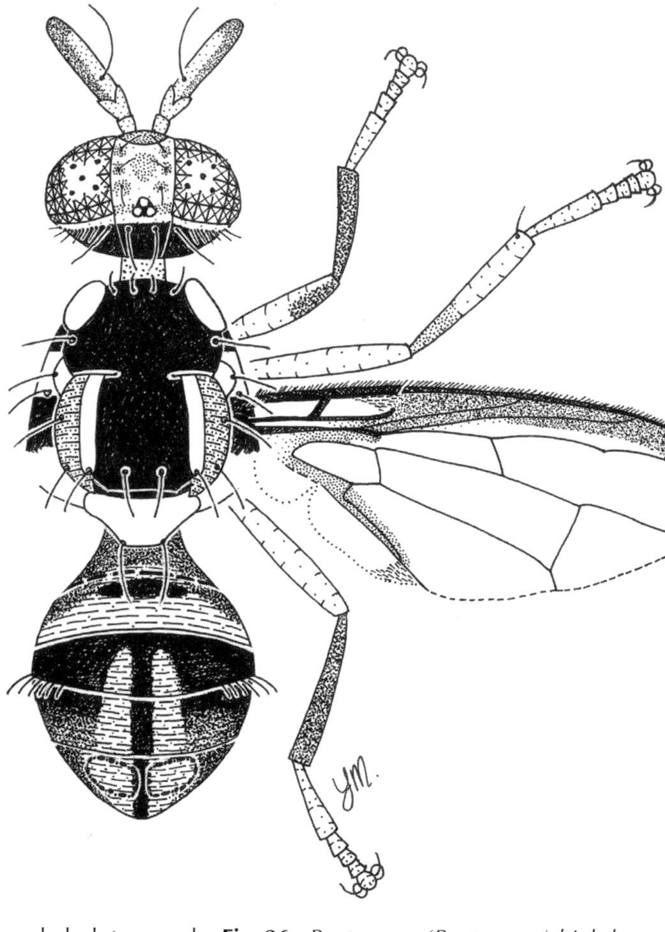

Fig. 35. *Bactrocera (Bactrocera) bimaculata* Drew & Hancock, holotype male. **Fig. 36.** *Bactrocera (Bactrocera) binhduongiae* new species, holotype male.

37

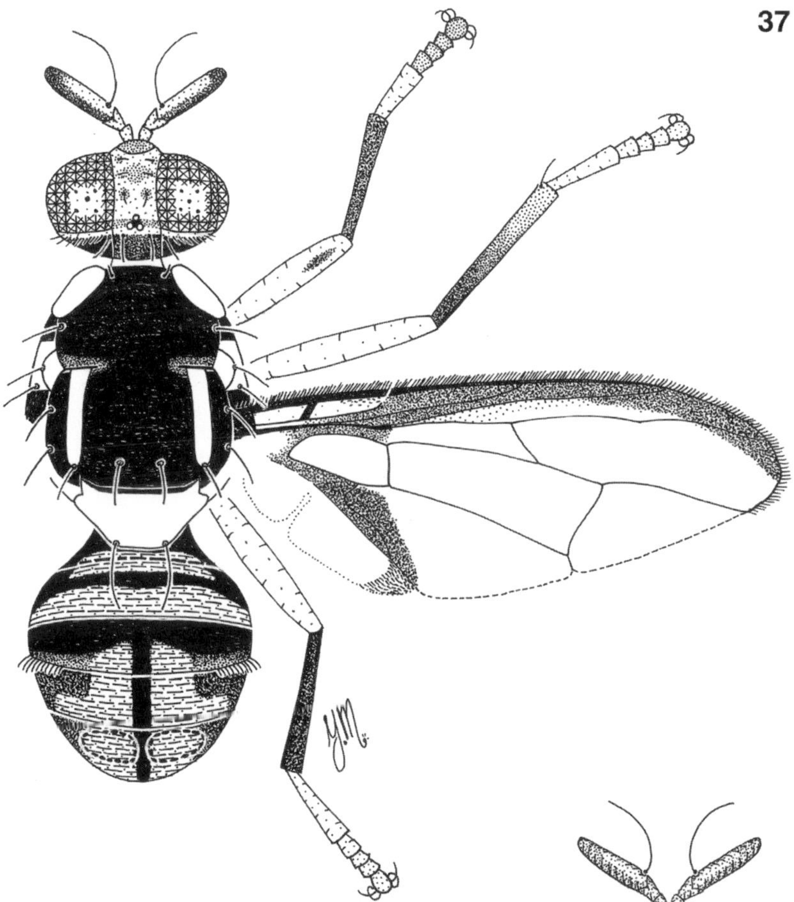

38

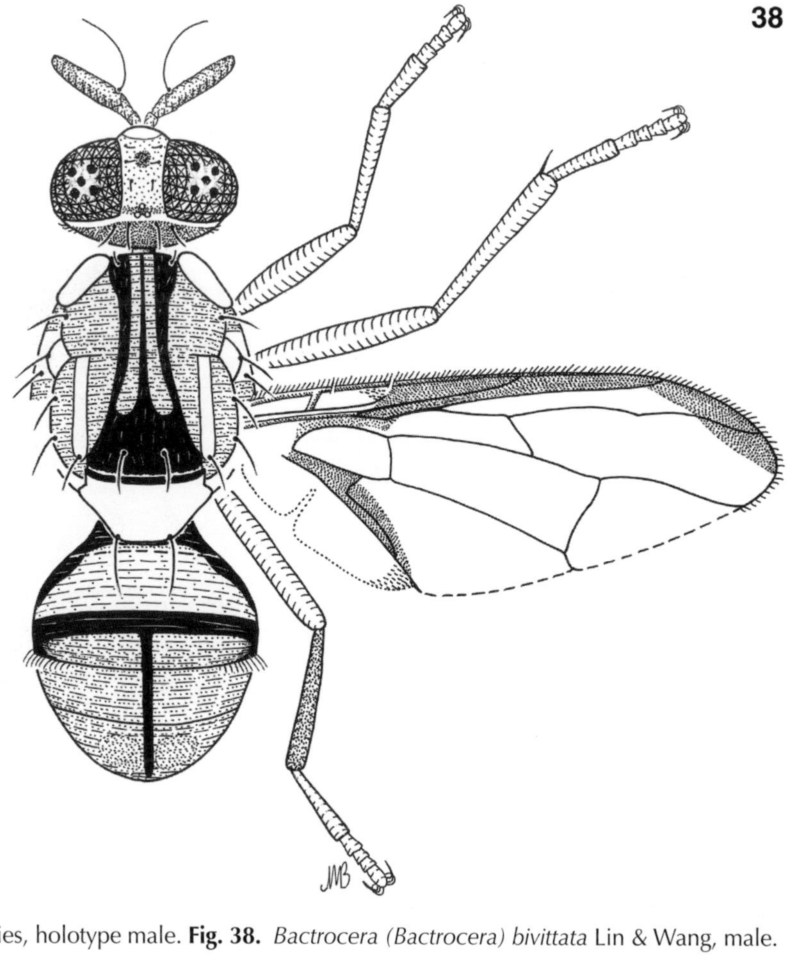

Fig. 37. *Bactrocera (Bactrocera) bitungiae* new species, holotype male. **Fig. 38.** *Bactrocera (Bactrocera) bivittata* Lin & Wang, male.

39

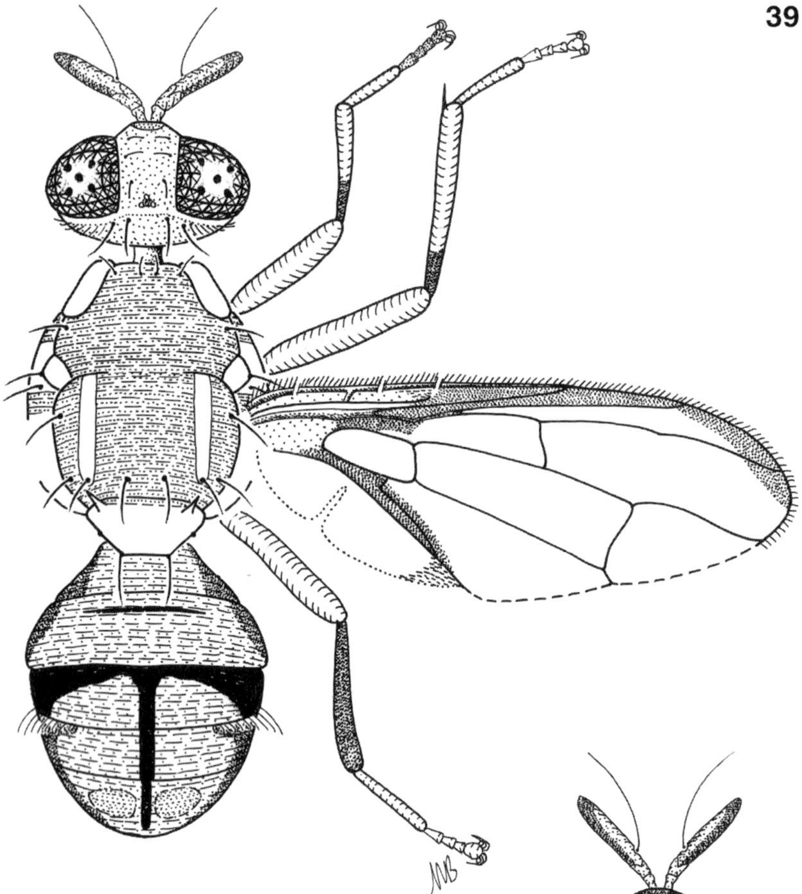

40

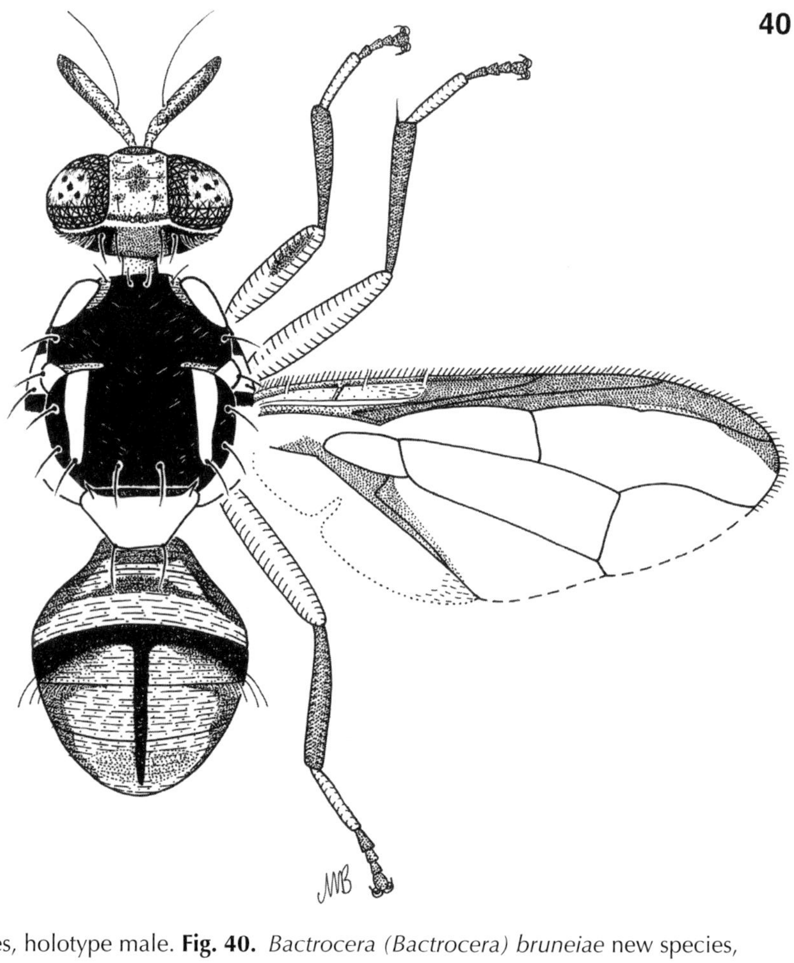

Fig. 39. *Bactrocera (Bactrocera) blairiae* new species, holotype male. **Fig. 40.** *Bactrocera (Bactrocera) bruneiae* new species, holotype male.

41

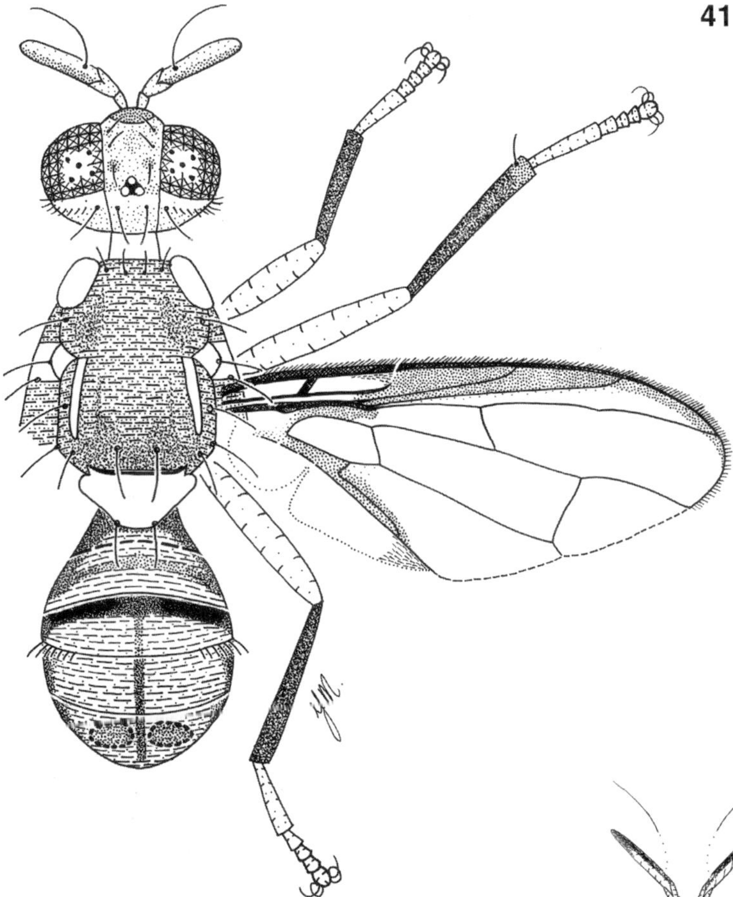

42

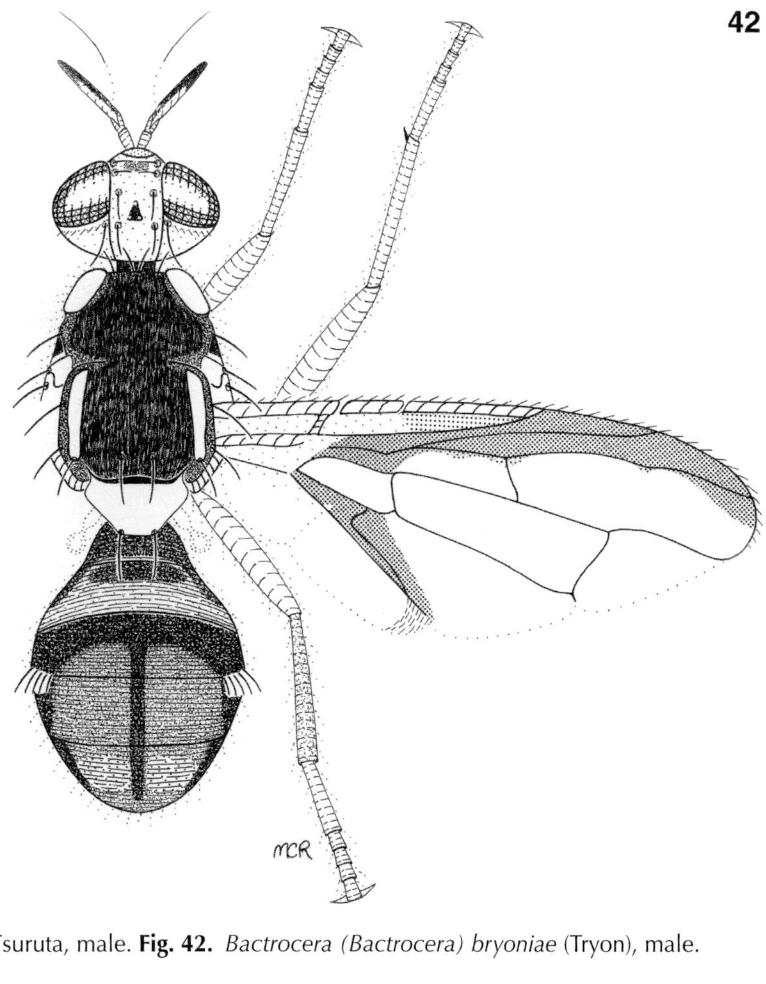

Fig. 41. *Bactrocera (Bactrocera) brunneola* White & Tsuruta, male. **Fig. 42.** *Bactrocera (Bactrocera) bryoniae* (Tryon), male.

43

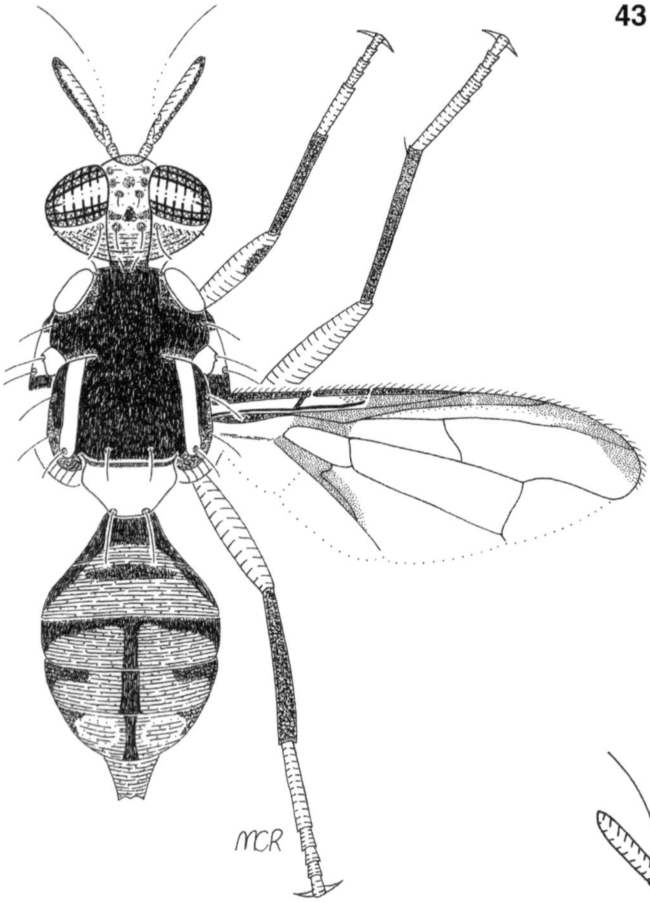

44

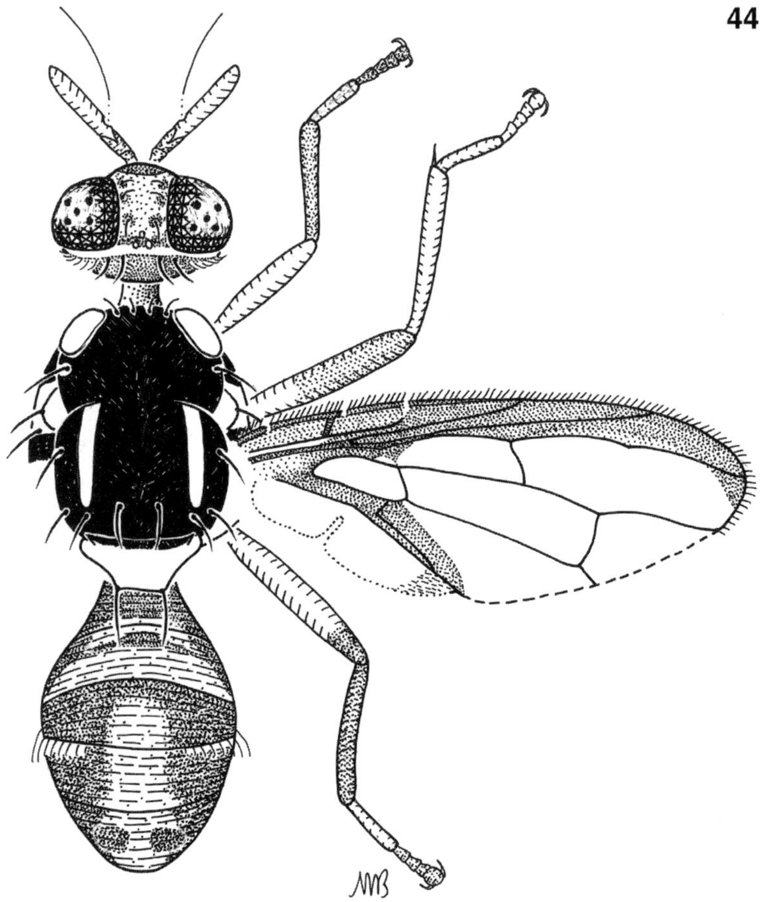

Fig. 43. *Bactrocera (Bactrocera) carambolae* Drew & Hancock, holotype female. **Fig. 44.** *Bactrocera (Bactrocera) careofascia* new species, holotype male.

45

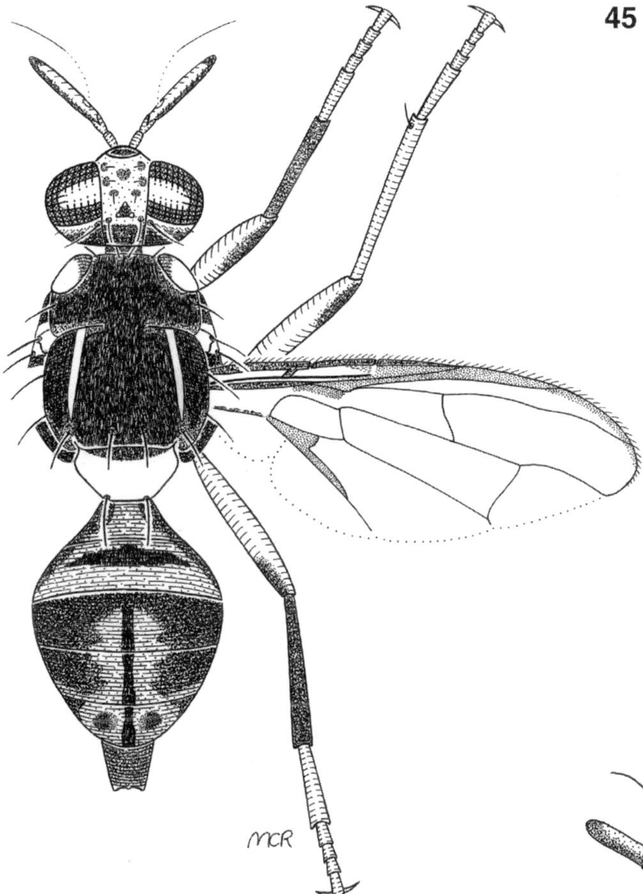

46

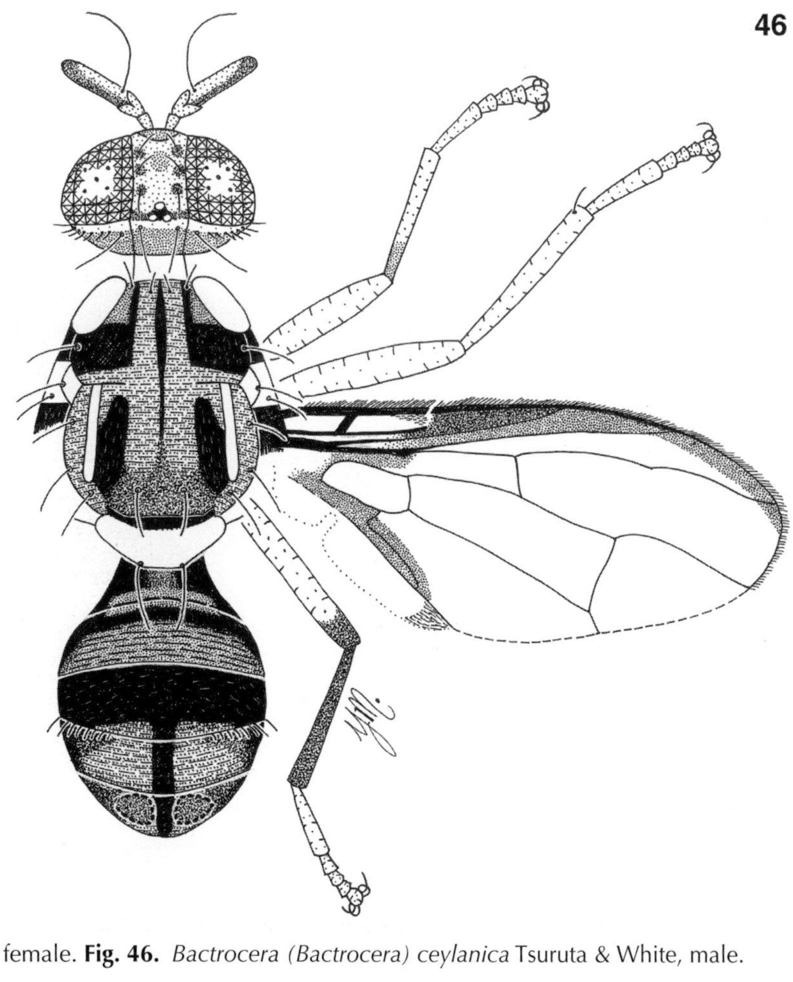

Fig. 45. *Bactrocera (Bactrocera) caryeae* (Kapoor), female. **Fig. 46.** *Bactrocera (Bactrocera) ceylanica* Tsuruta & White, male.

47

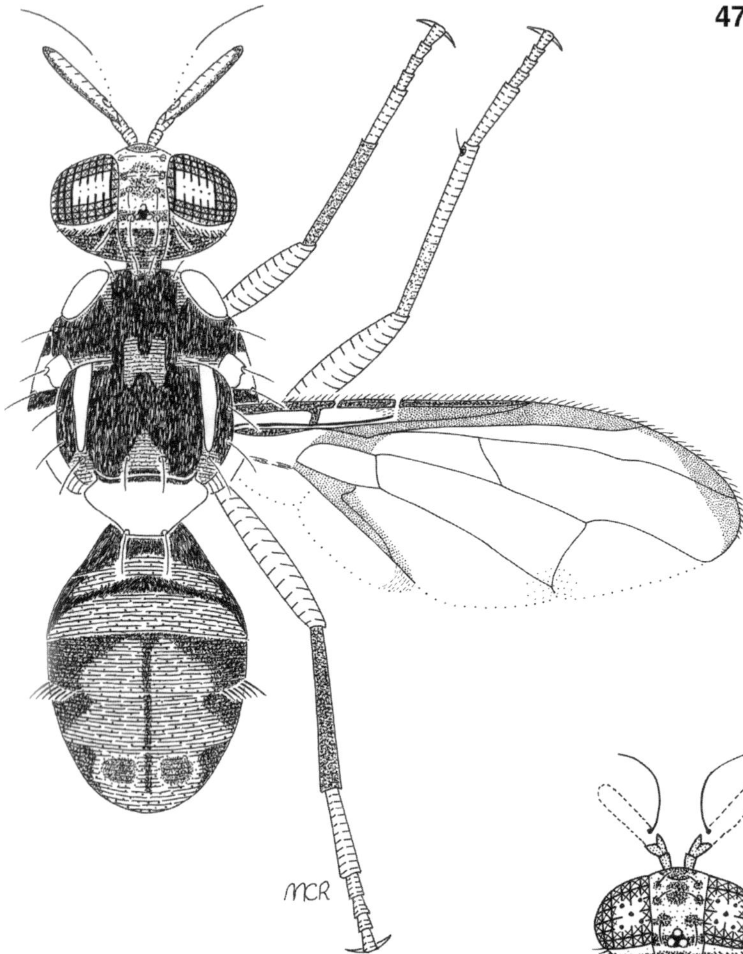

48

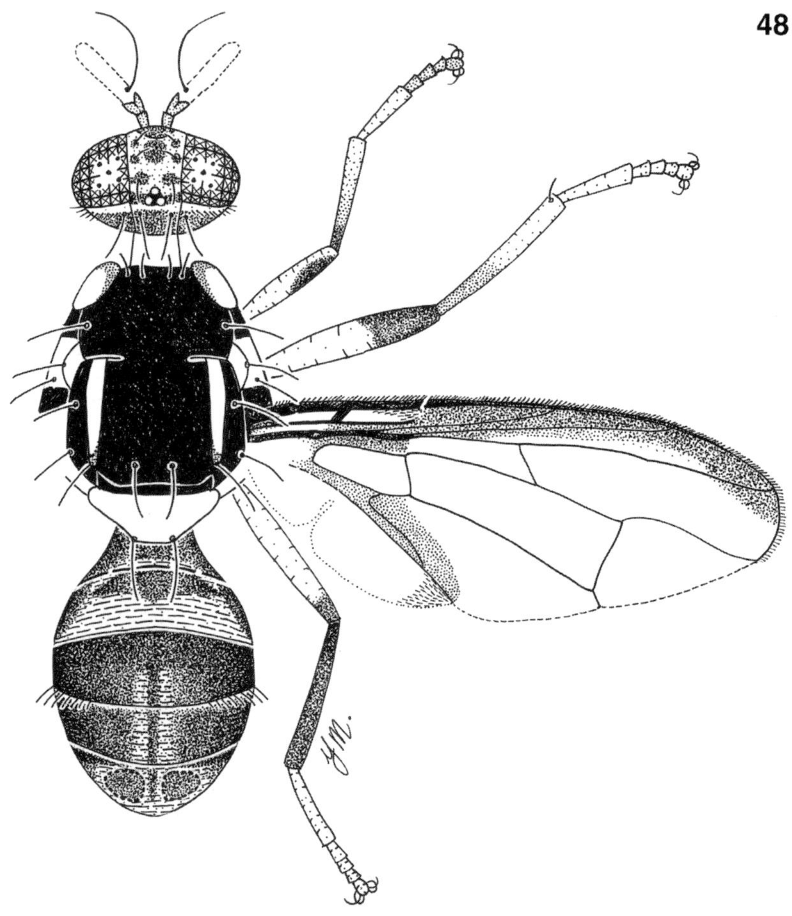

Fig. 47. *Bactrocera (Bactrocera) cibodasae* Drew & Hancock, holotype male. **Fig. 48.** *Bactrocera (Bactrocera) citima* (Hardy), holotype male.

49

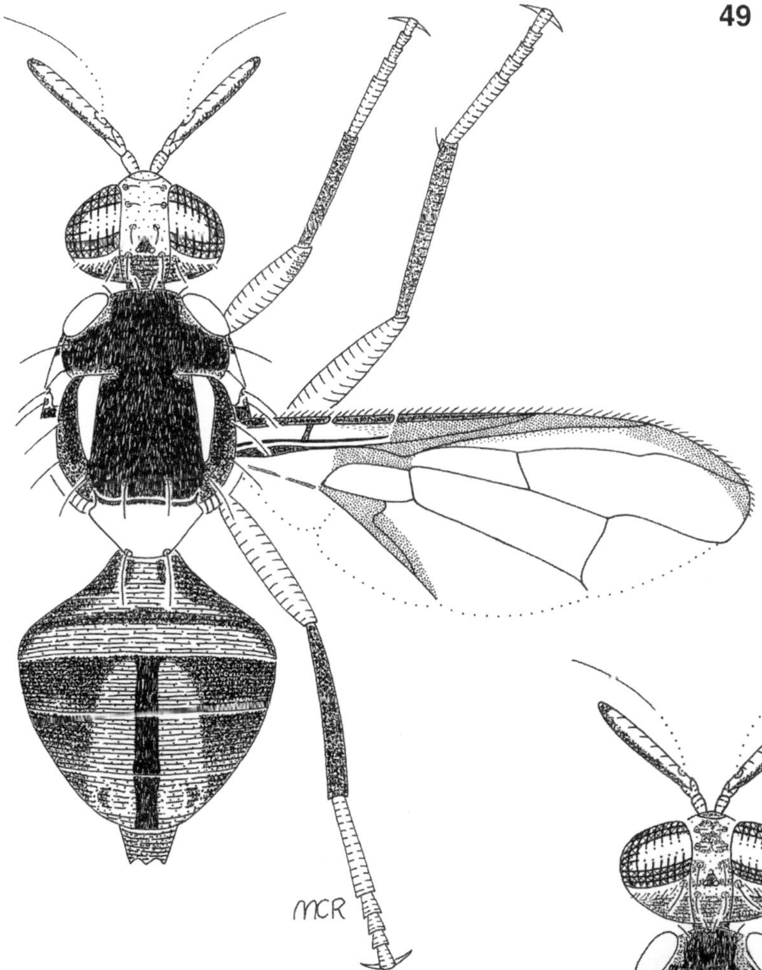

50

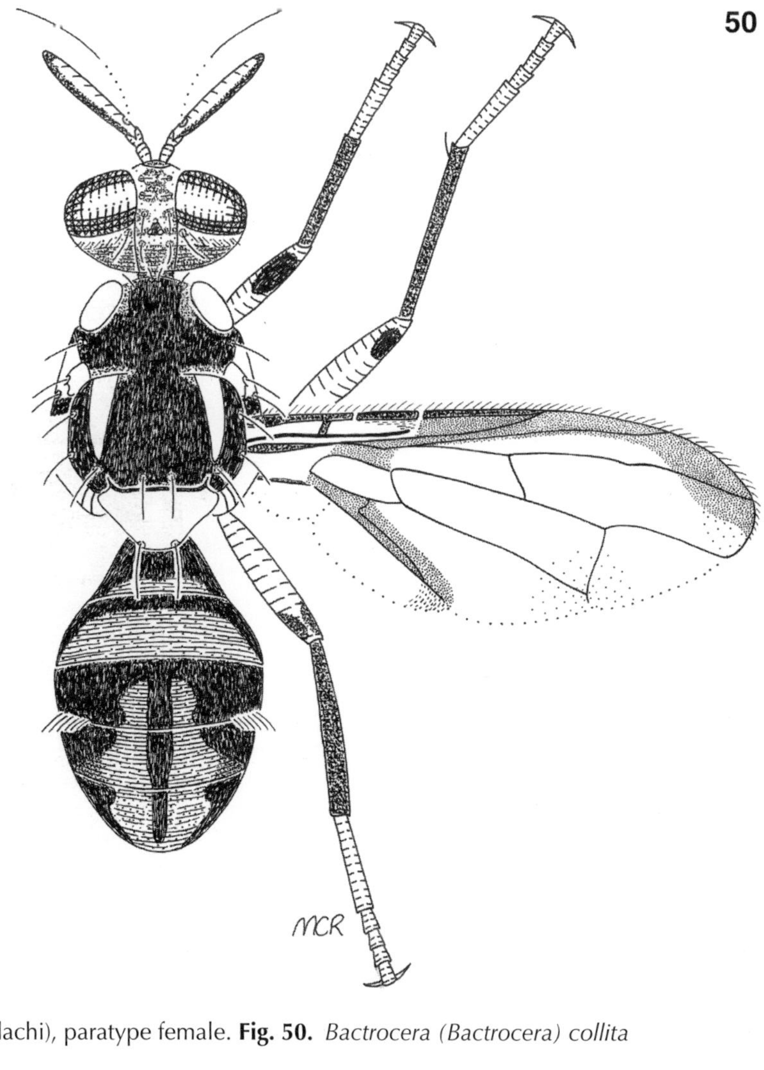

Fig. 49. *Bactrocera (Bactrocera) cognata* (Hardy & Adachi), paratype female. **Fig. 50.** *Bactrocera (Bactrocera) collita* Drew & Hancock, holotype male.

51

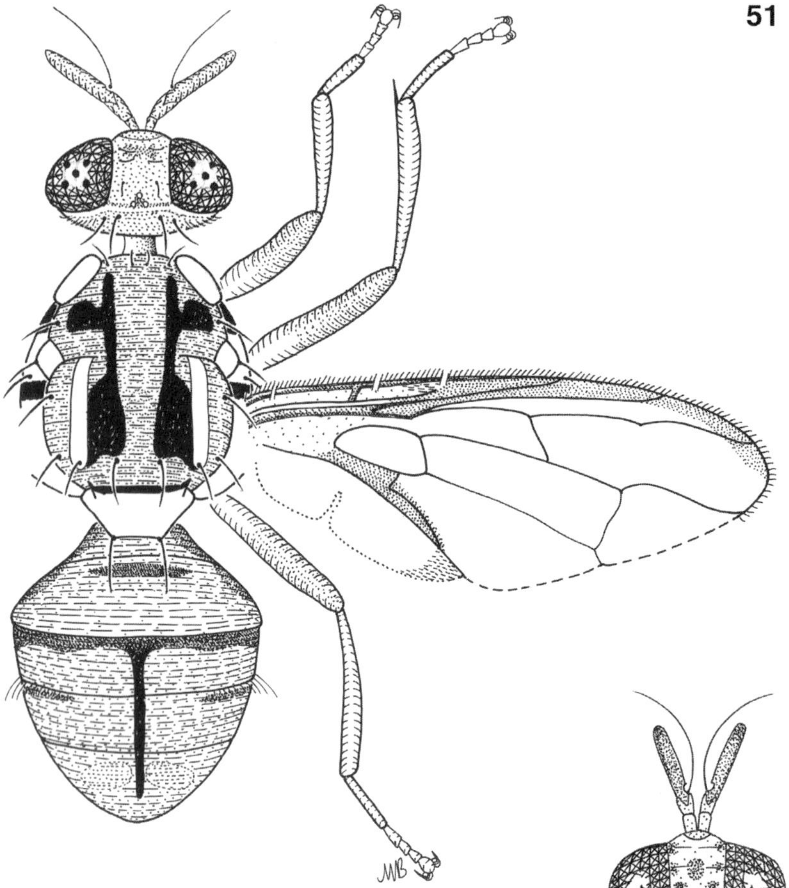

52

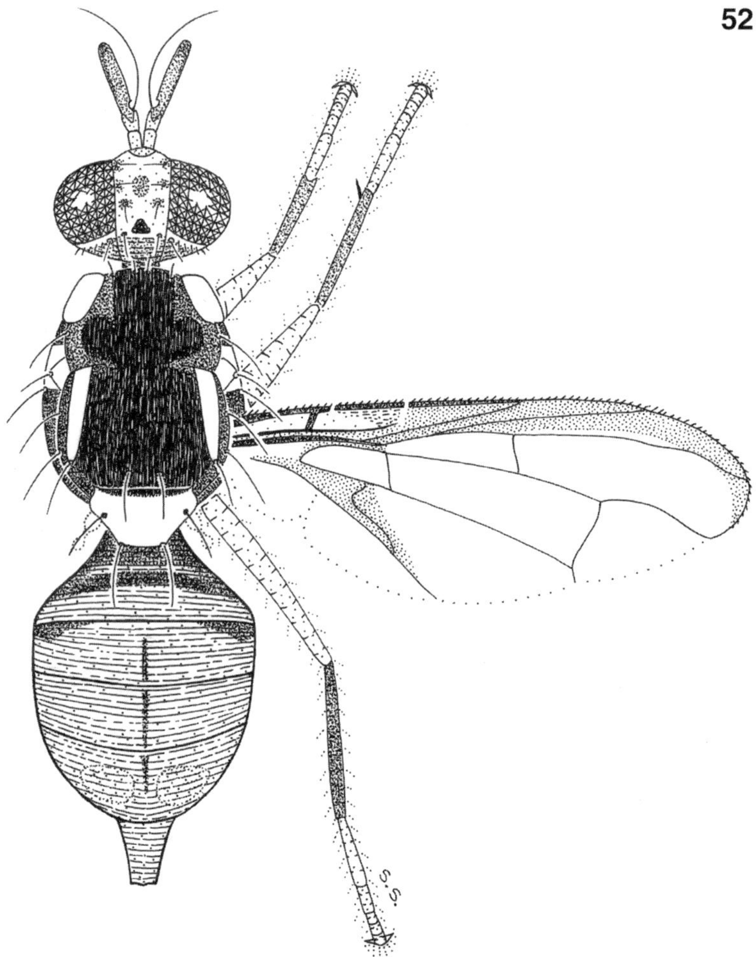

Fig. 51. *Bactrocera (Bactrocera) commensurata* new species, holotype male. **Fig. 52.** *Bactrocera (Bactrocera) contigua* Drew, holotype female.

53

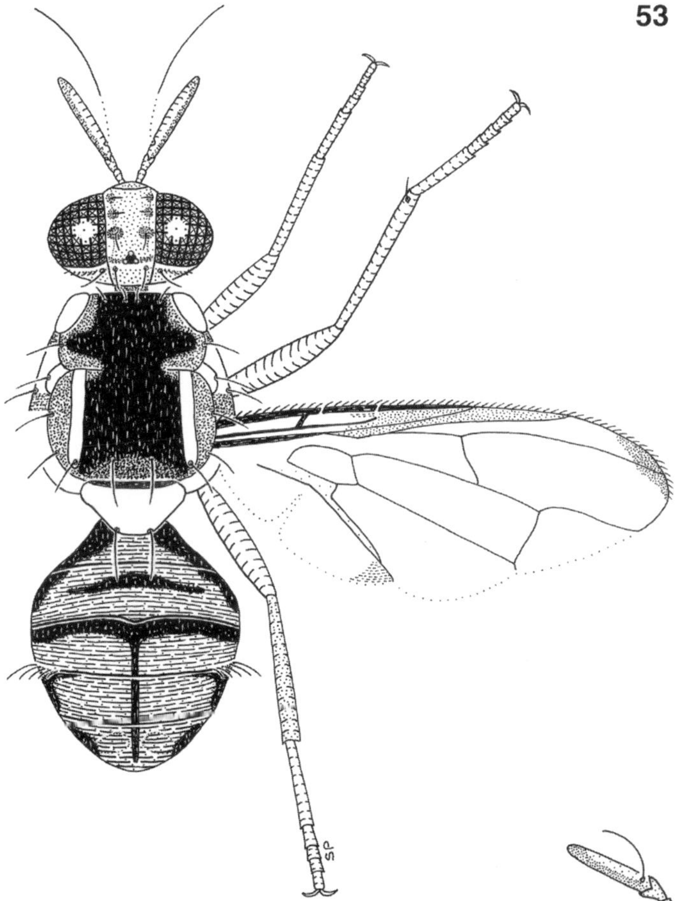

54

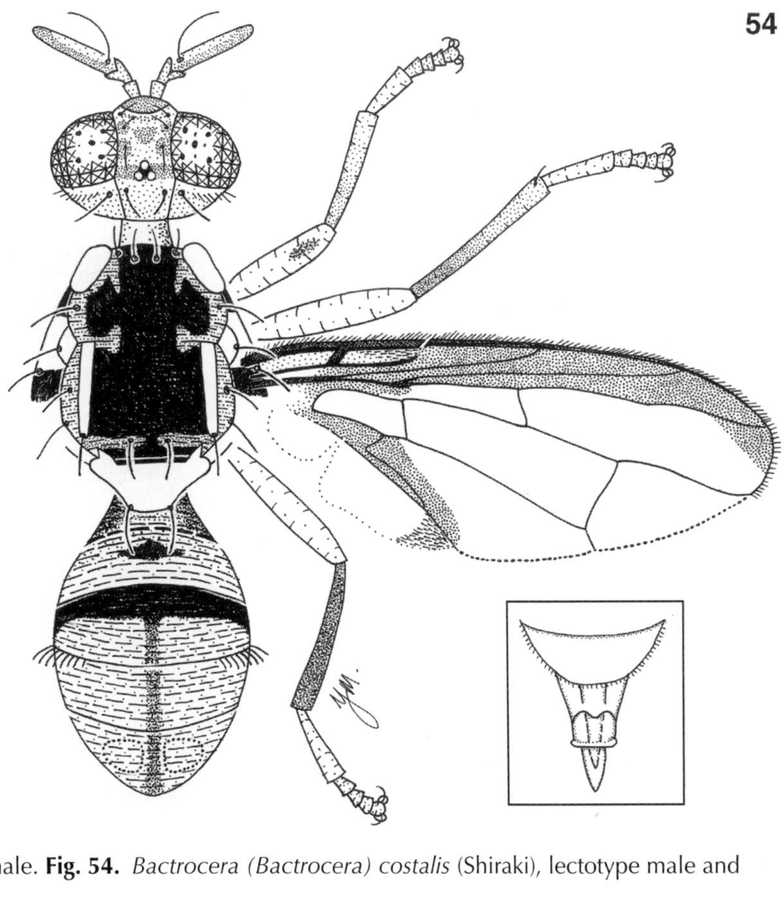

Fig. 53. *Bactrocera (Bactrocera) correcta* (Bezzi), male. **Fig. 54.** *Bactrocera (Bactrocera) costalis* (Shiraki), lectotype male and apex of piercer from female ovipositor.

55

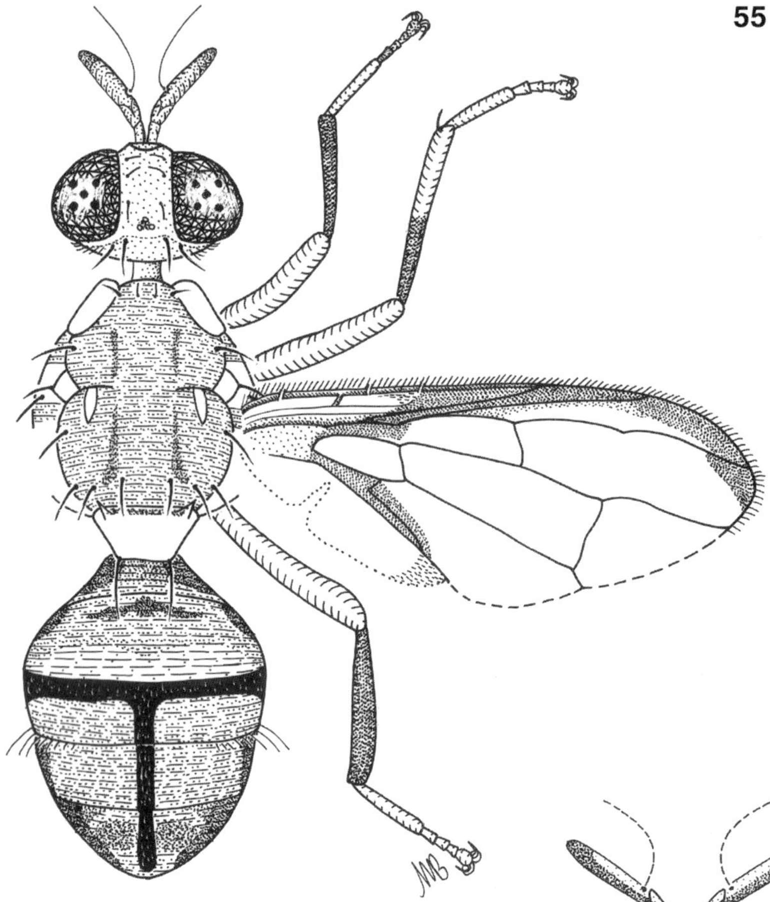

56A

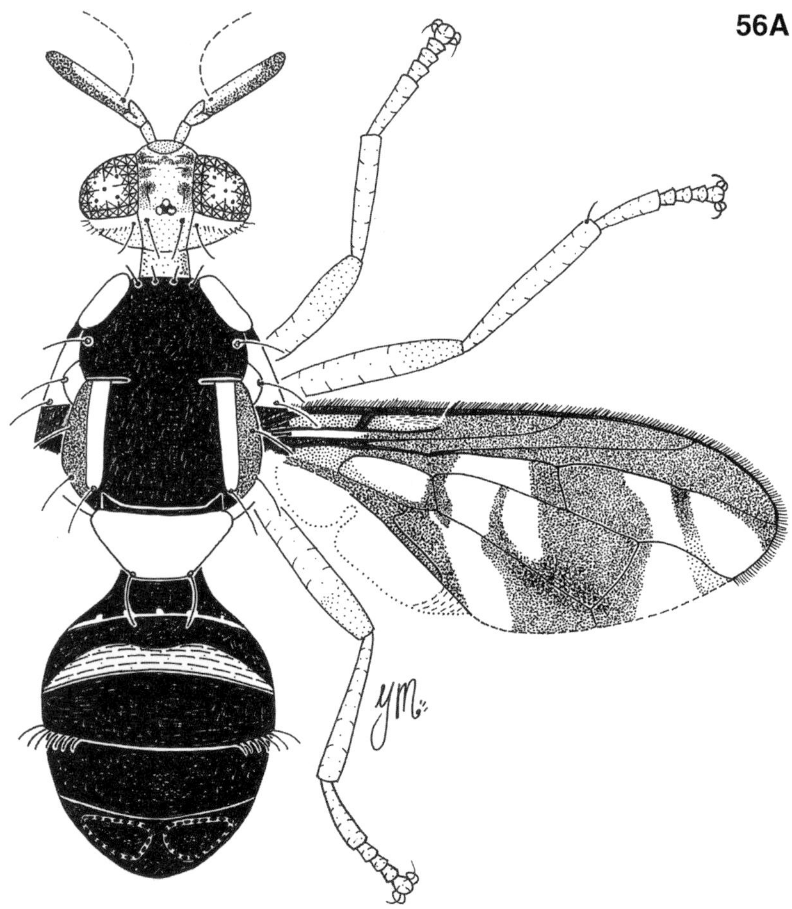

Fig. 55. *Bactrocera (Bactrocera) curtivitta* new species, holotype male. **Fig. 56A.** *Bactrocera (Bactrocera) curvifera* (Walker), male.

56B

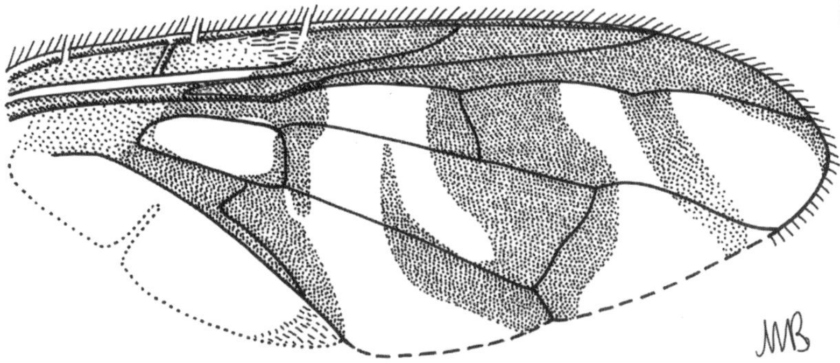

56C

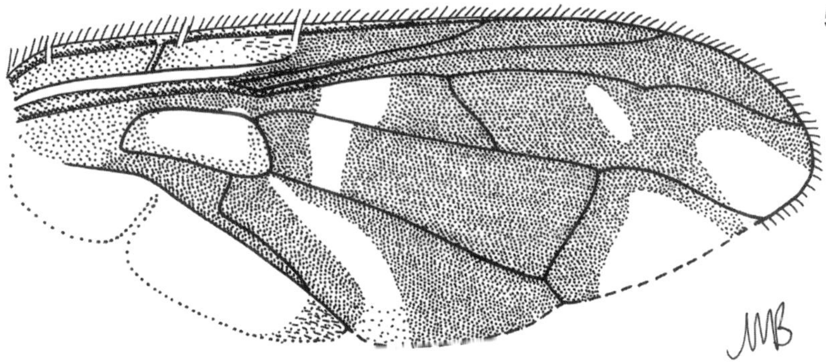

57A

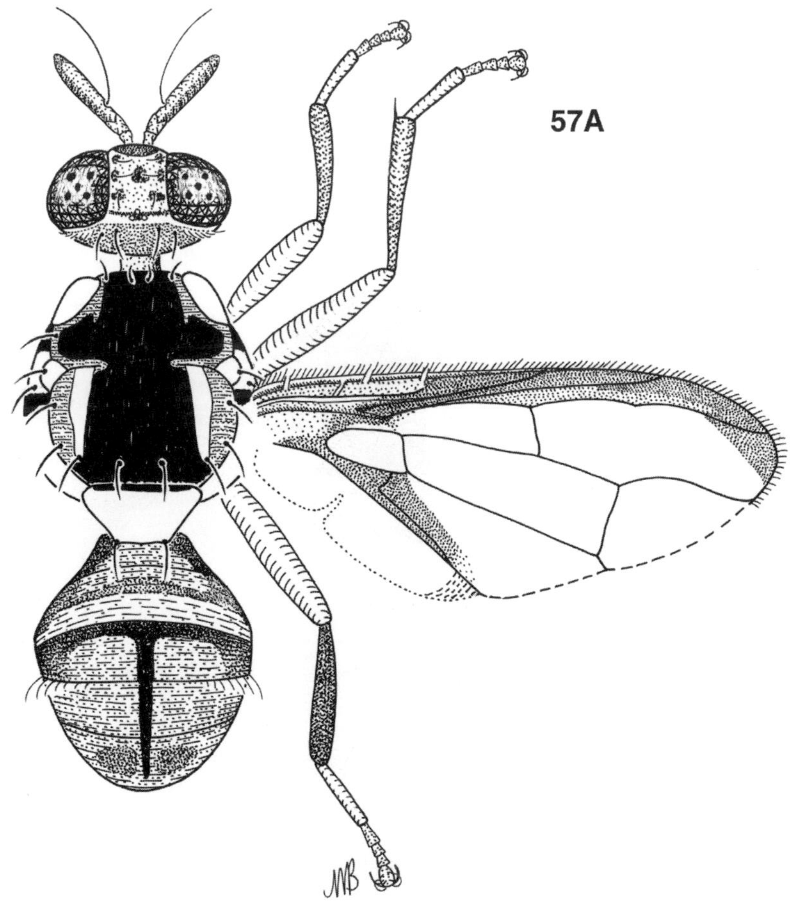

57B

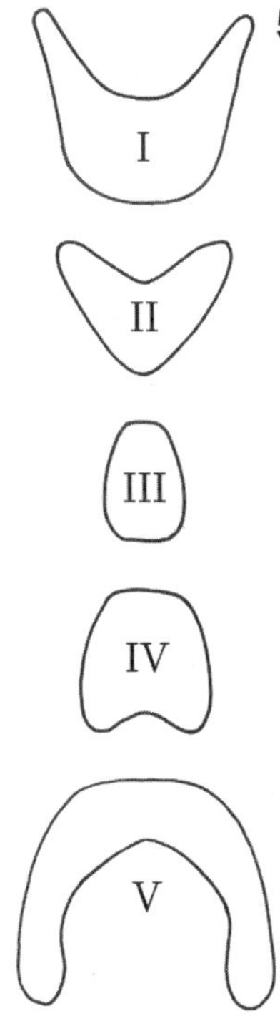

Fig. 56B. *Bactrocera (Bactrocera) curvifera* (Walker), wing of lectotype male. **Fig. 56C.** *Bactrocera (Bactrocera) speculifera* (Walker), wing of lectotype male. **Fig. 57A.** *Bactrocera (Bactrocera) curvosterna* new species, holotype male. **Fig. 57B.** *Bactrocera (Bactrocera) curvosterna* new species, holotype male abdominal sternites.

58

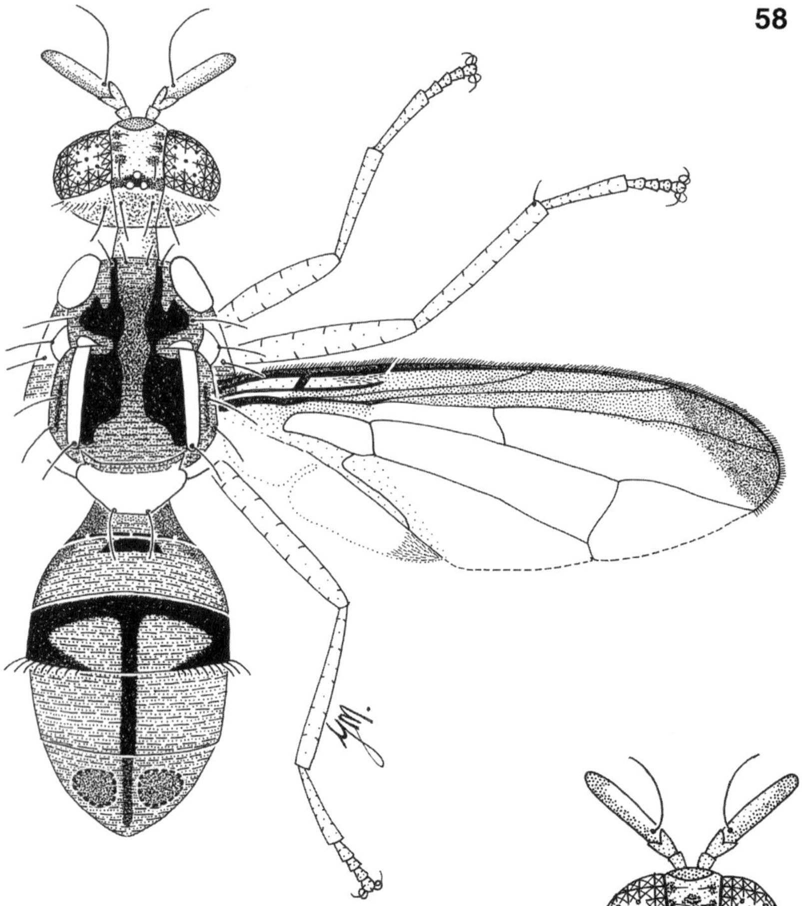

59

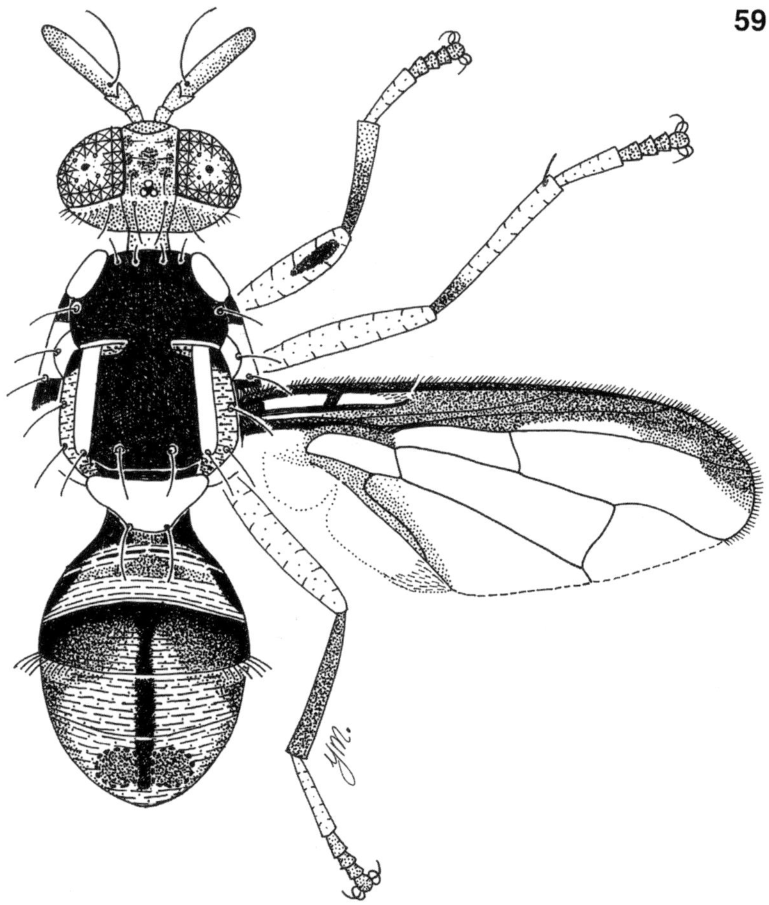

Fig. 58. *Bactrocera (Bactrocera) dispar* (Hardy), paratype male. **Fig. 59.** *Bactrocera (Bactrocera) dongnaiae* new species, holotype male.

60A

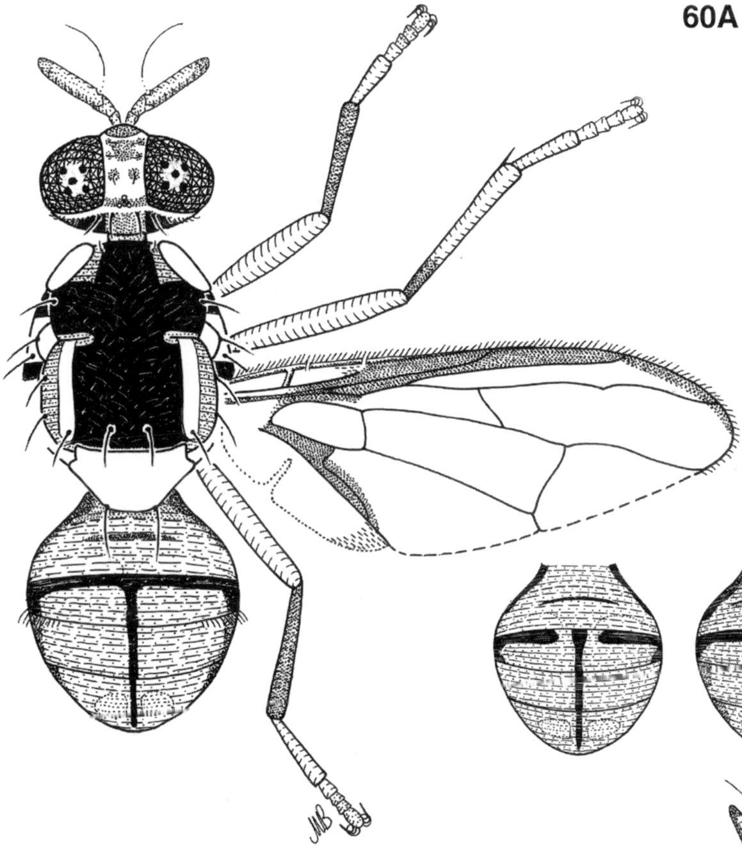

60B

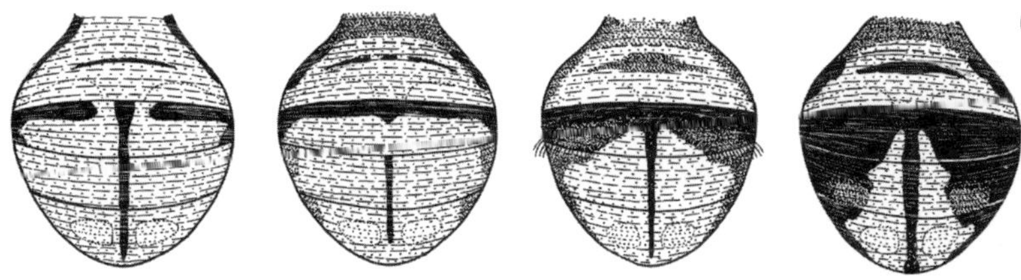

60C

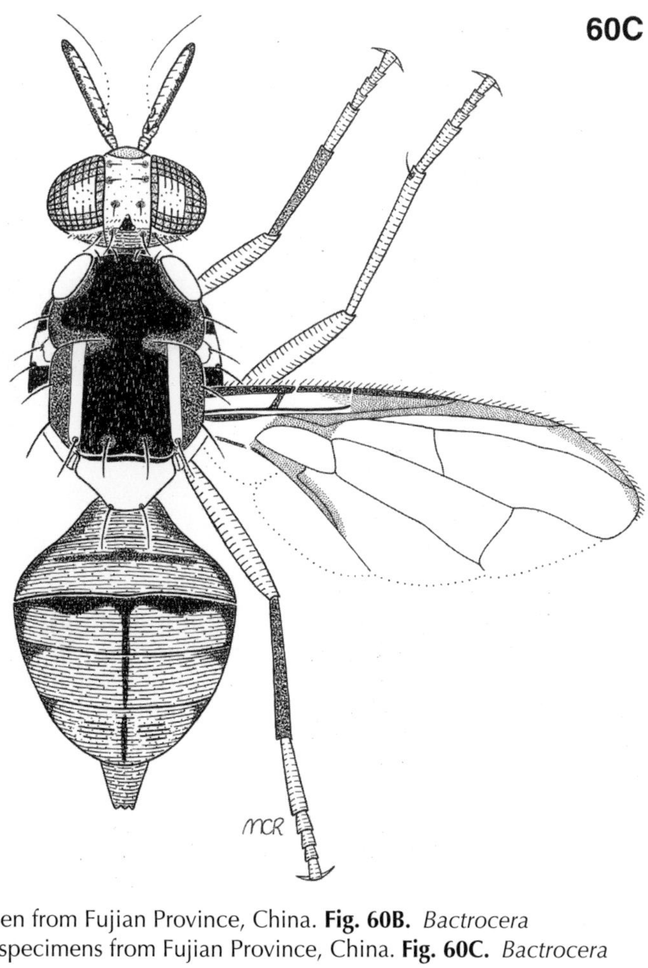

Fig. 60A. *Bactrocera (Bactrocera) dorsalis* (Hendel), male specimen from Fujian Province, China. **Fig. 60B.** *Bactrocera (Bactrocera) dorsalis* (Hendel), abdominal colour variations from specimens from Fujian Province, China. **Fig. 60C.** *Bactrocera (Bactrocera) dorsalis* (Hendel), female lectotype from Taiwan.

61

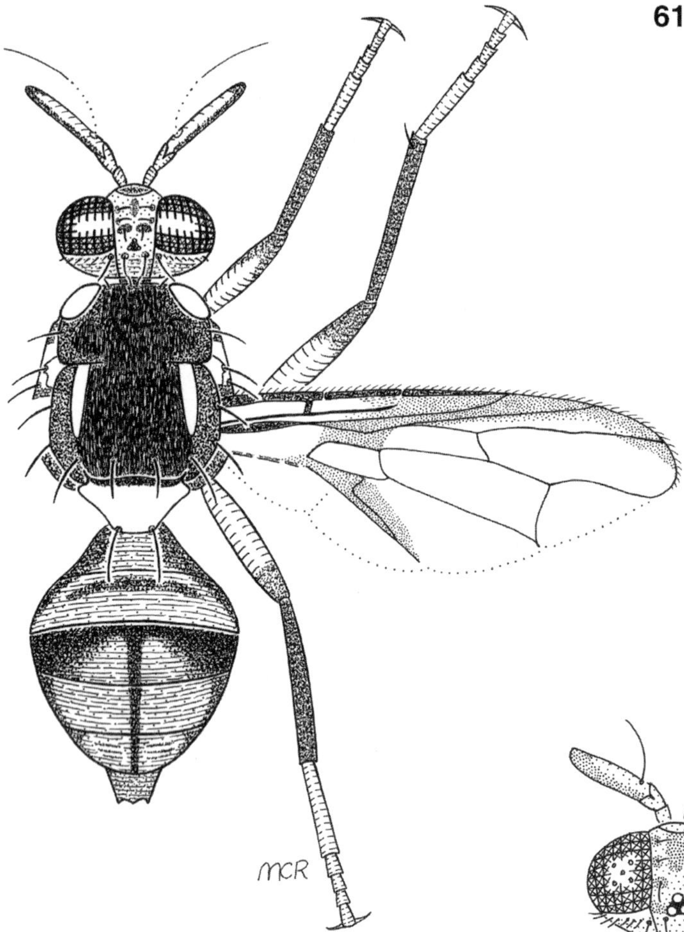

62

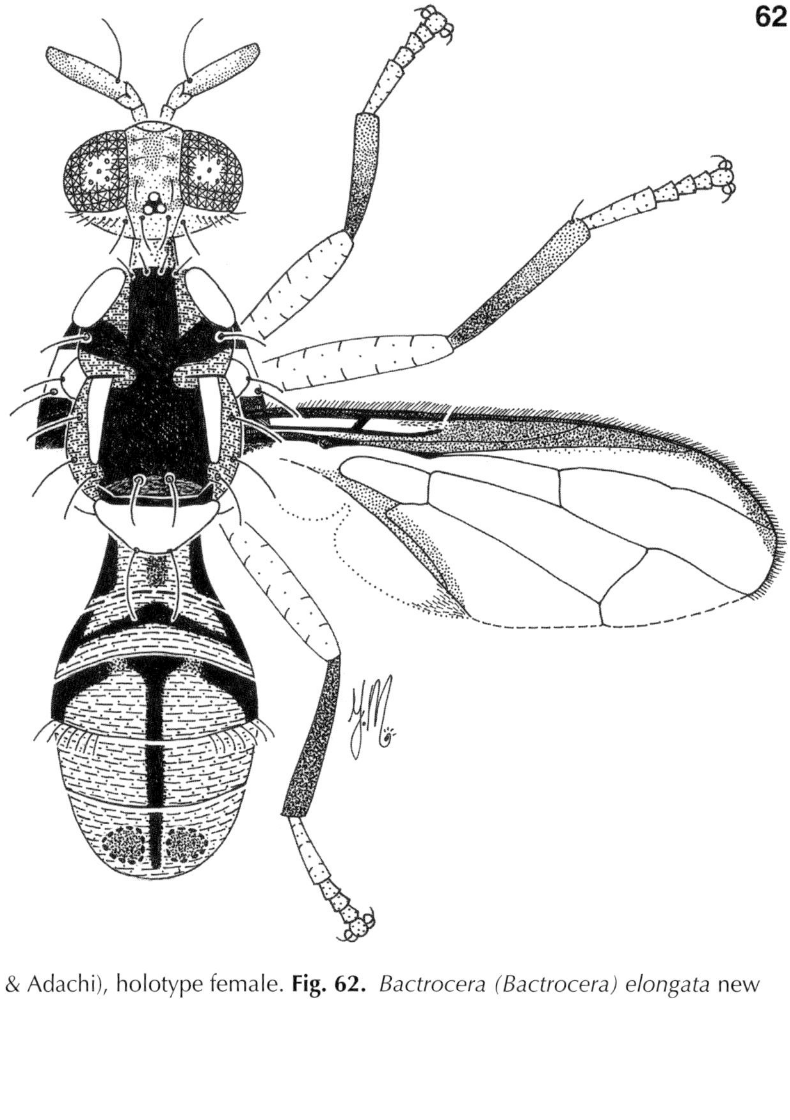

Fig. 61. *Bactrocera (Bactrocera) dorsaloides* (Hardy & Adachi), holotype female. **Fig. 62.** *Bactrocera (Bactrocera) elongata* new species, holotype male.

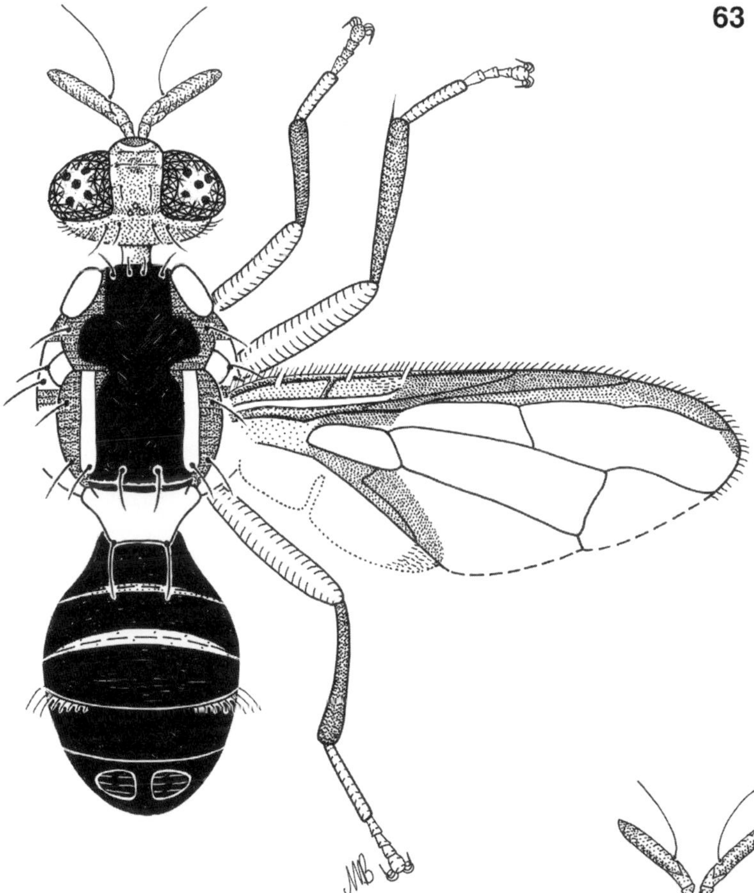

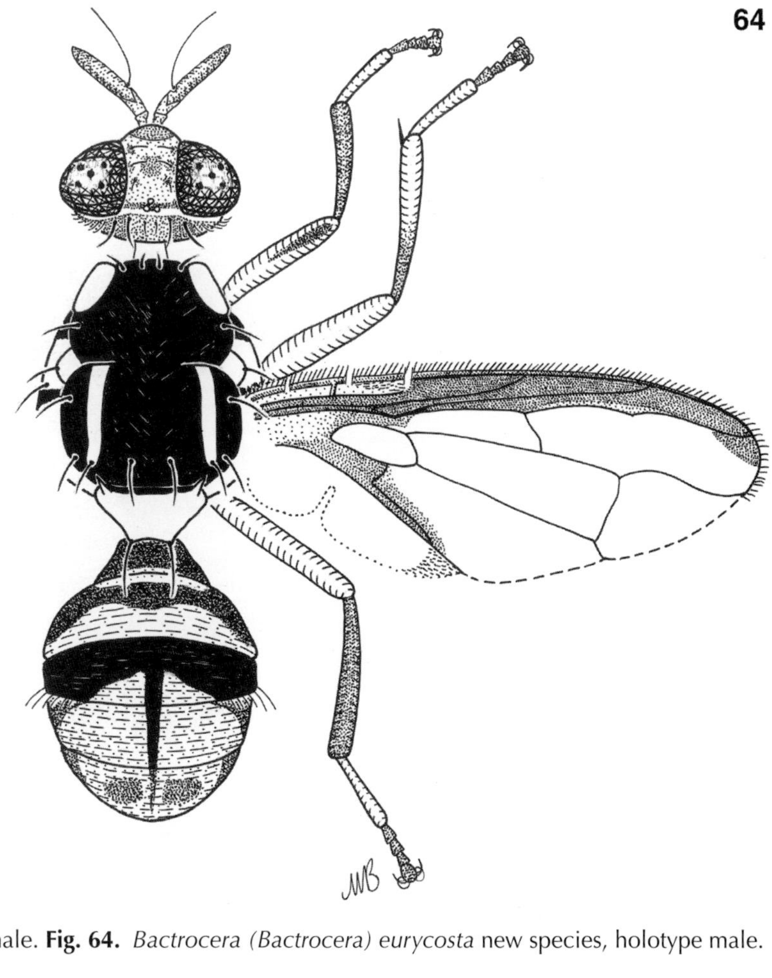

Fig. 63. *Bactrocera (Bactrocera) epicharis* (Hardy), male. **Fig. 64.** *Bactrocera (Bactrocera) eurycosta* new species, holotype male.

65

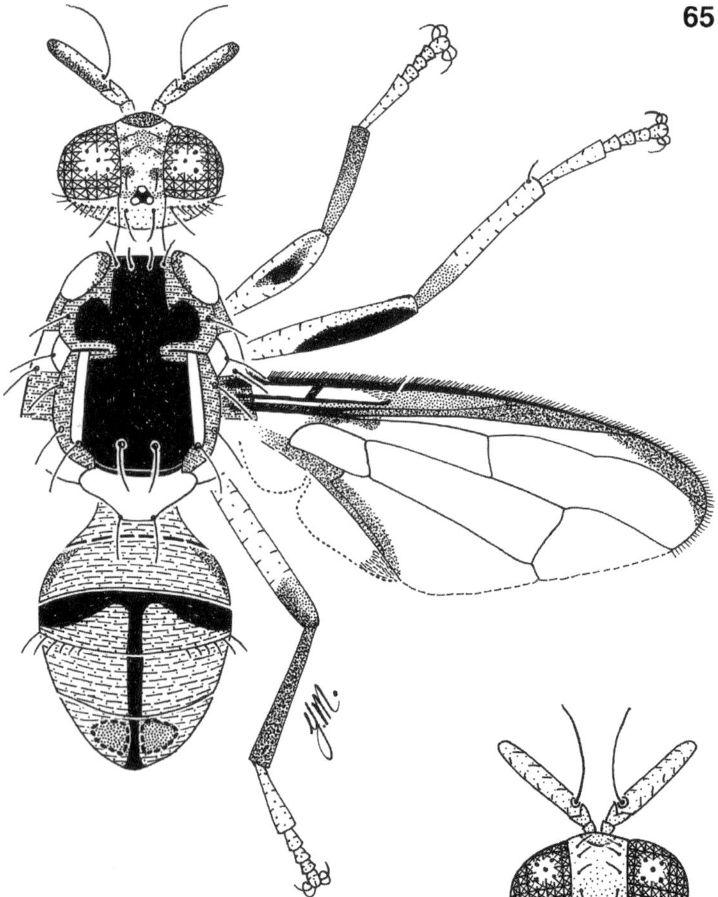

66

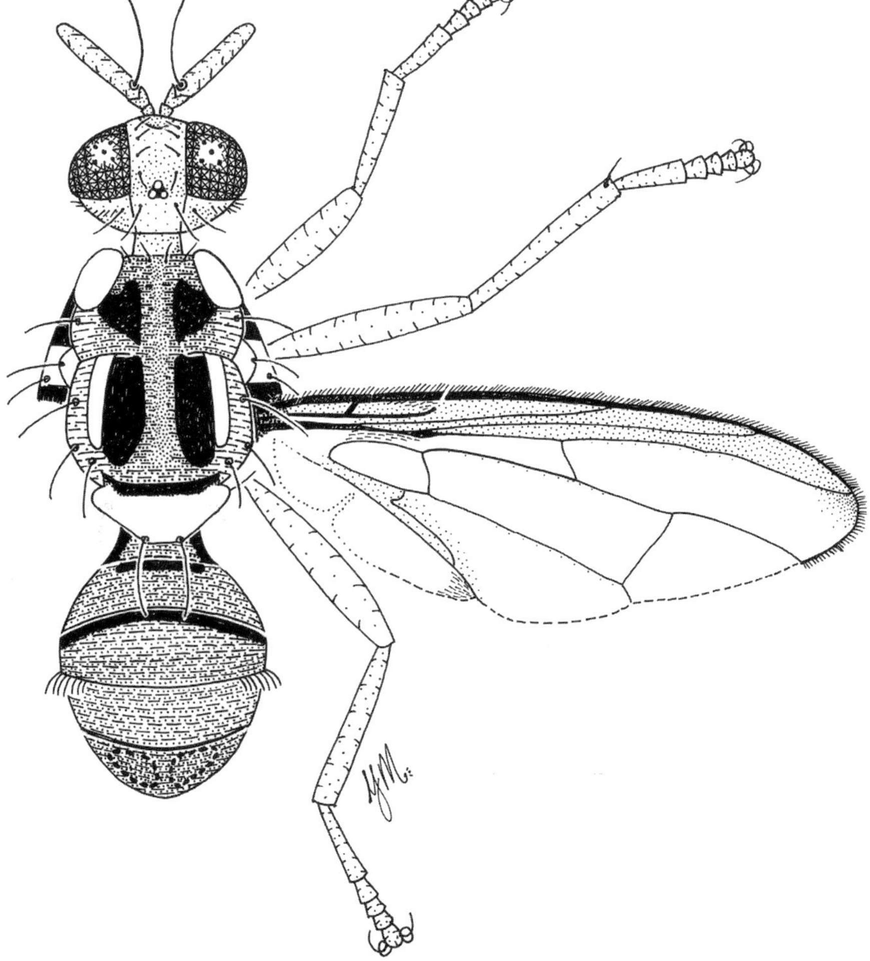

Fig. 65. *Bactrocera (Bactrocera) fernandoi* Tsuruta & White, male. **Fig. 66.** *Bactrocera (Bactrocera) flavipennis* (Hardy), holotype male.

67

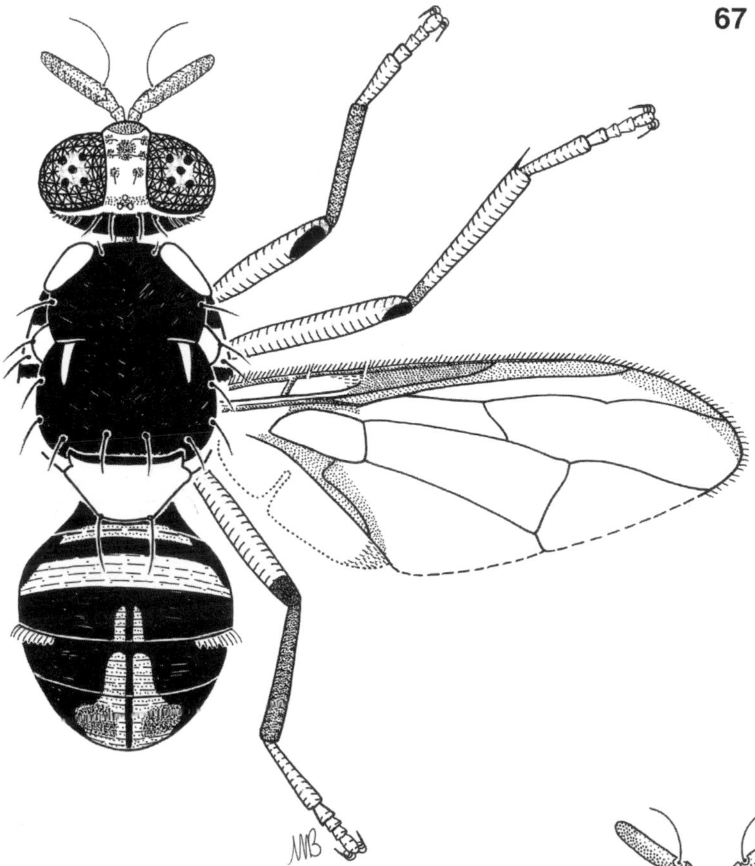

68

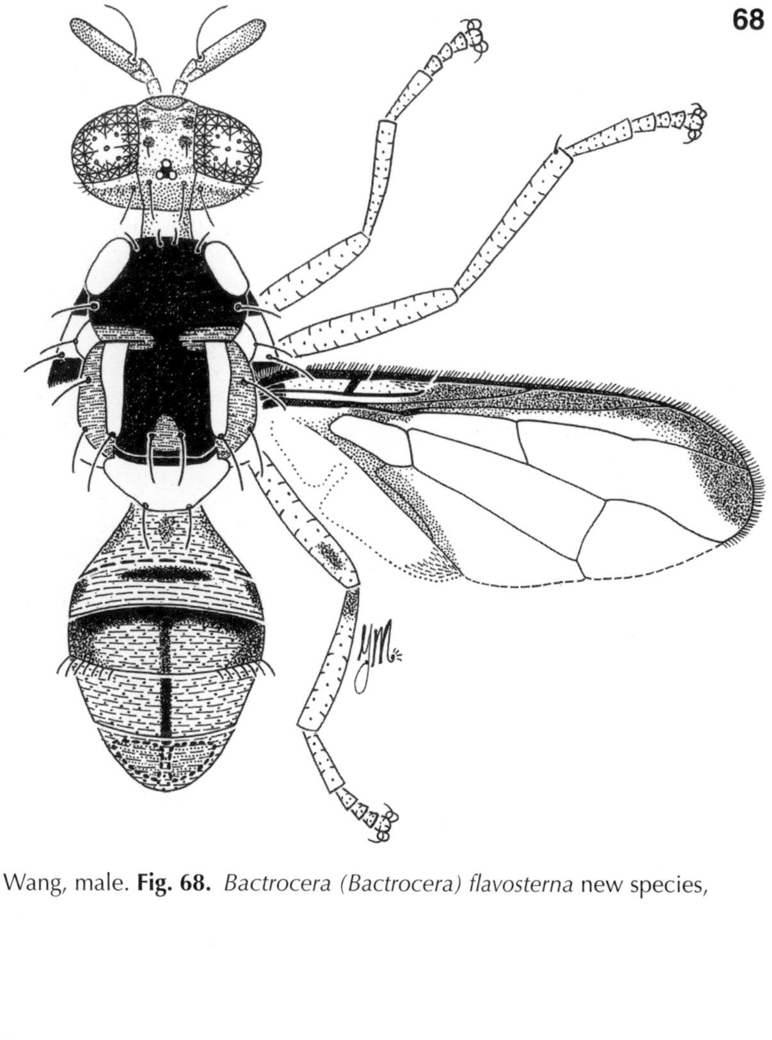

Fig. 67. *Bactrocera (Bactrocera) flavoscutellata* Lin & Wang, male. **Fig. 68.** *Bactrocera (Bactrocera) flavosterna* new species, holotype male.

69

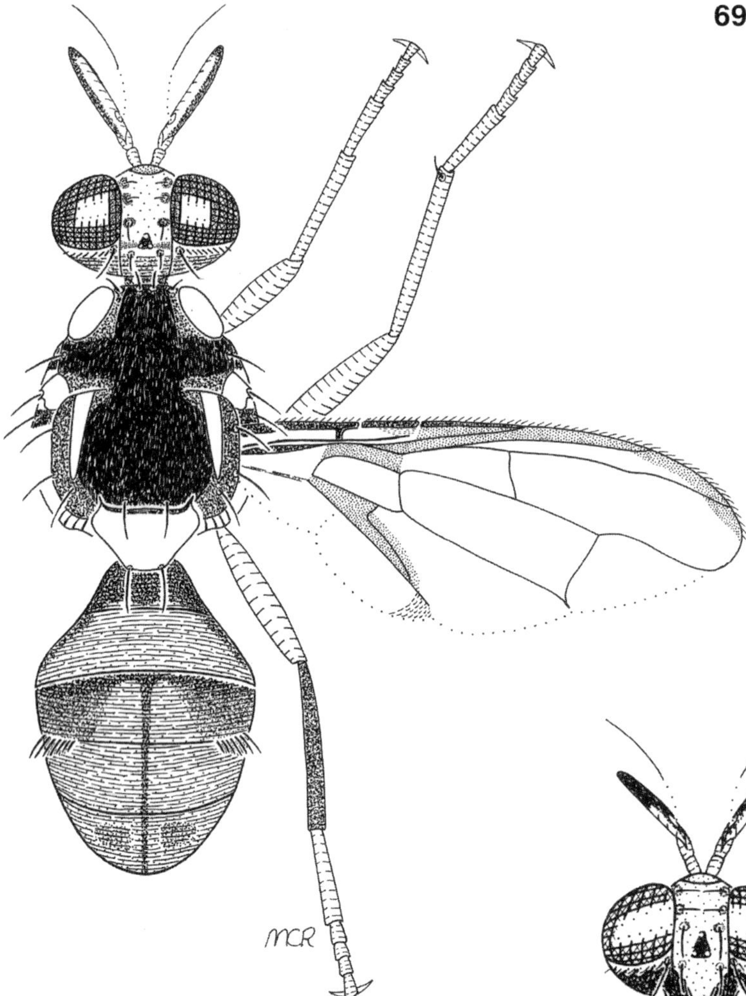

70

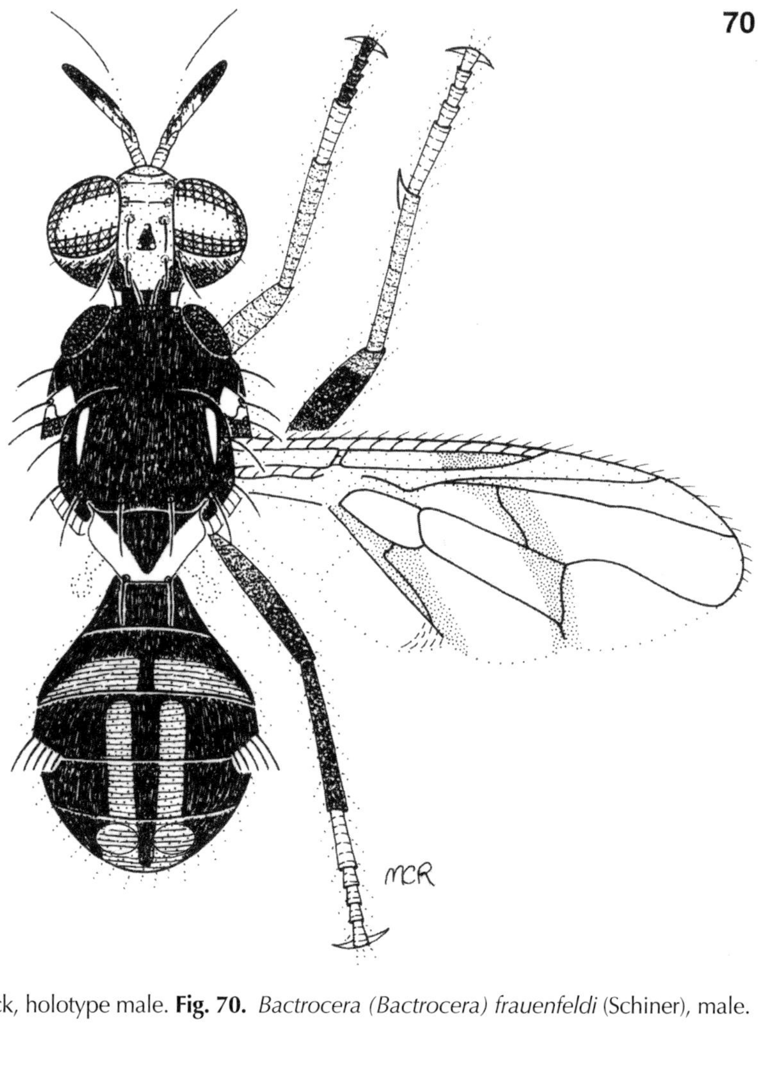

Fig. 69. *Bactrocera (Bactrocera) floresiae* Drew & Hancock, holotype male. **Fig. 70.** *Bactrocera (Bactrocera) frauenfeldi* (Schiner), male.

71A

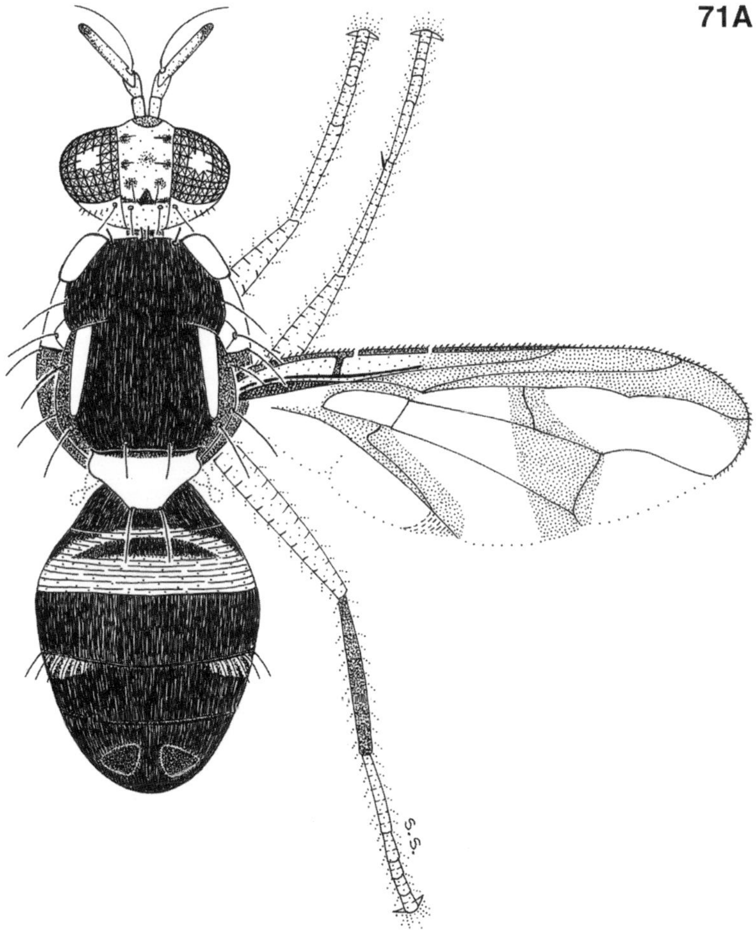

71B

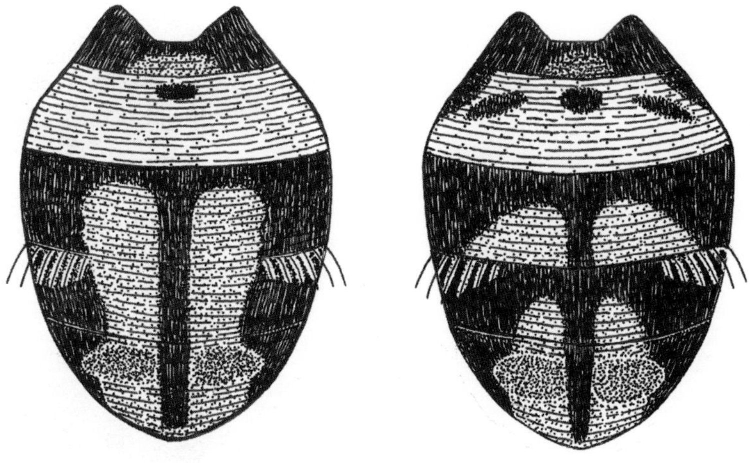

Fig. 71A. *Bactrocera (Bactrocera) fulvicauda* (Perkins), male. **Fig. 71B.** *Bactrocera (Bactrocera) fulvicauda* (Perkins), abdominal colour variations.

72

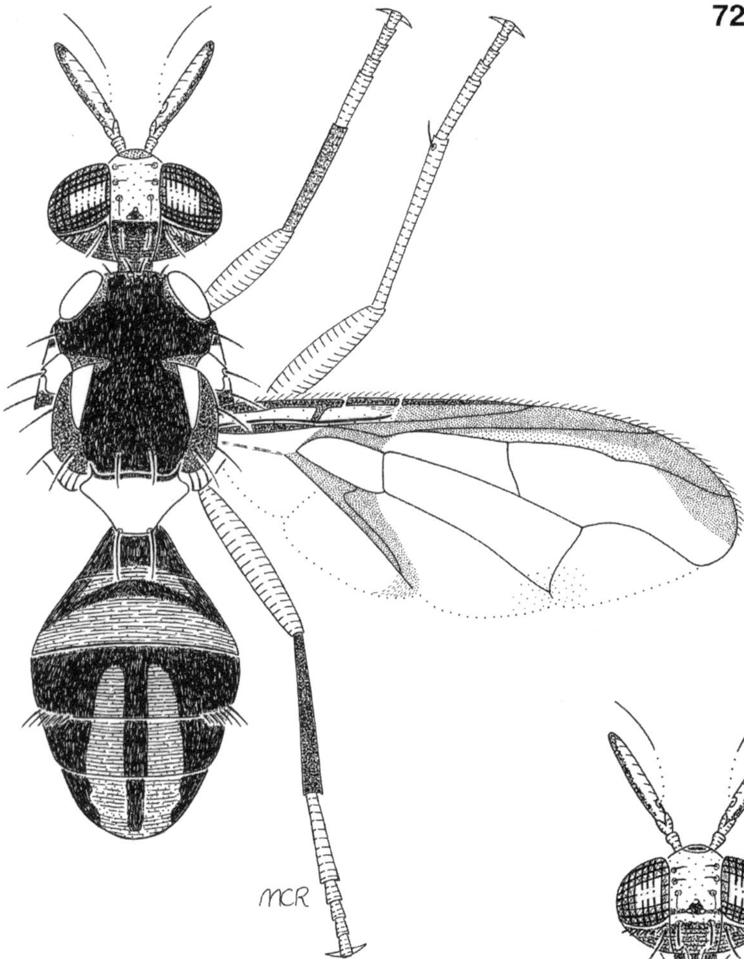

73

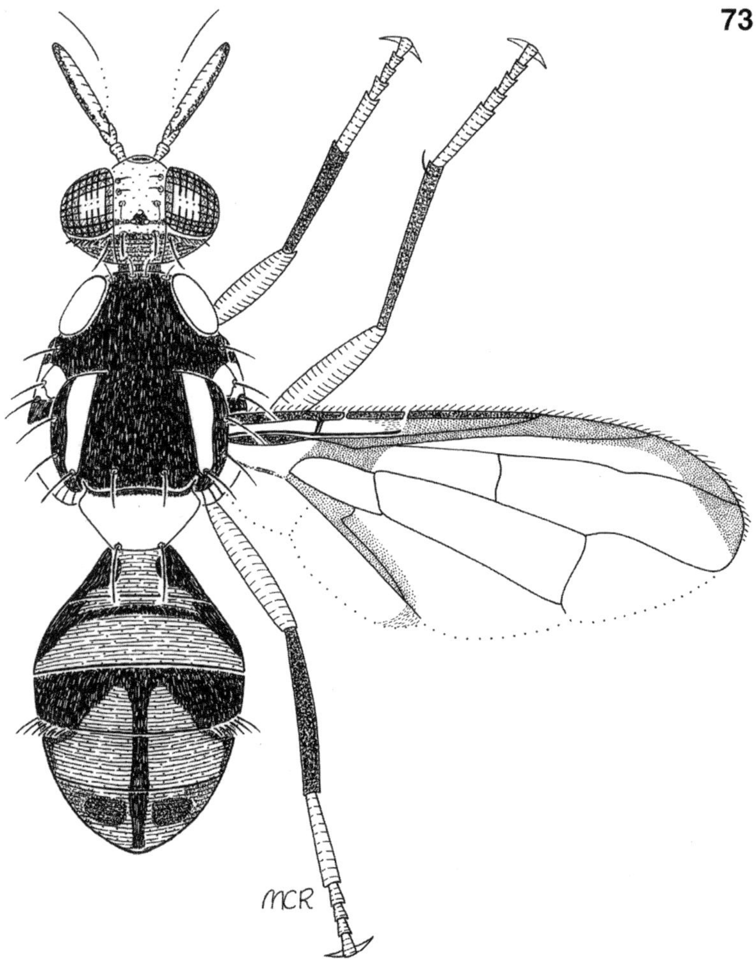

Fig. 72. *Bactrocera (Bactrocera) fulvifemur* Drew & Hancock, holotype male. **Fig. 73.** *Bactrocera (Bactrocera) fuscitibia* Drew & Hancock, holotype male.

74

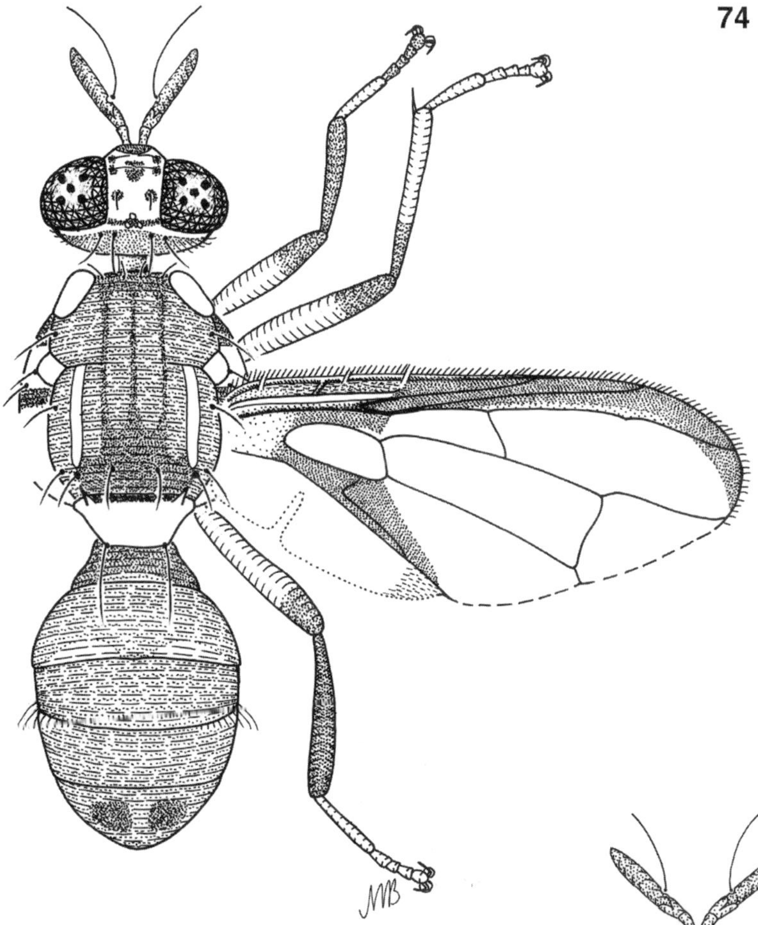

75

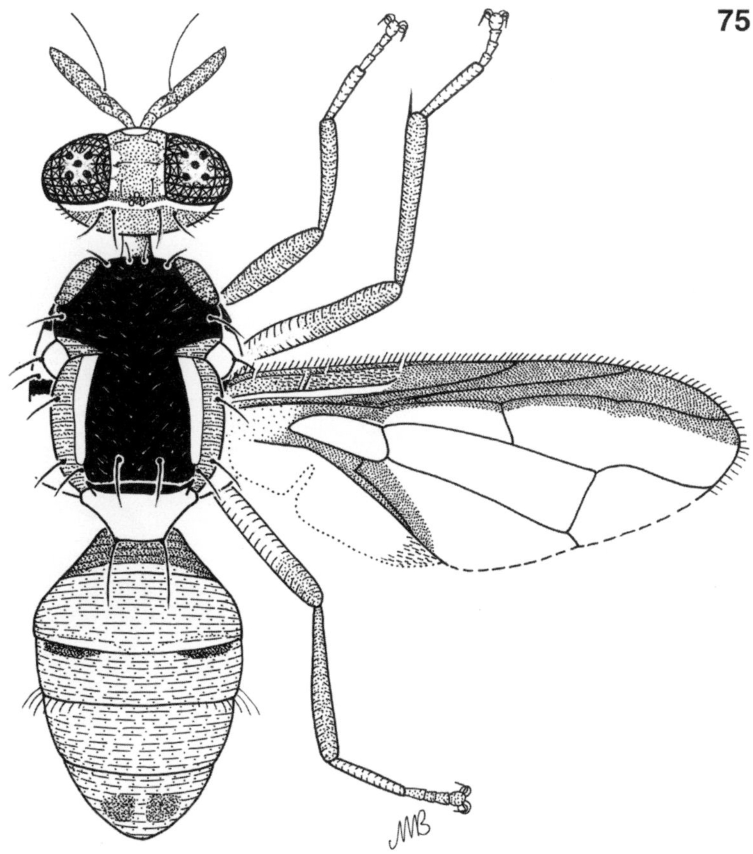

Fig. 74. *Bactrocera (Bactrocera) fuscoformosa* new species, holotype male. **Fig. 75.** *Bactrocera (Bactrocera) fuscolobata* new species, holotype male.

76

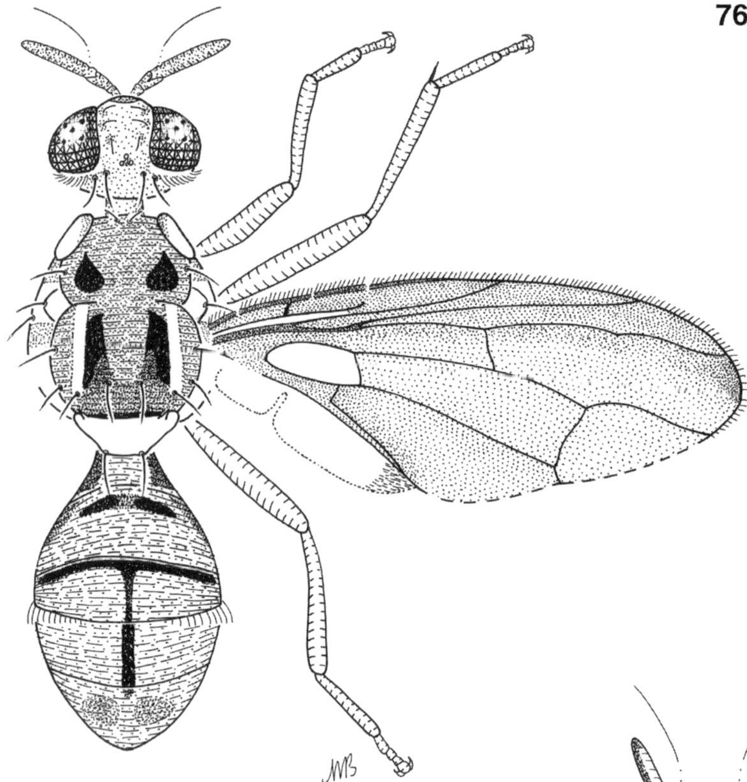

77

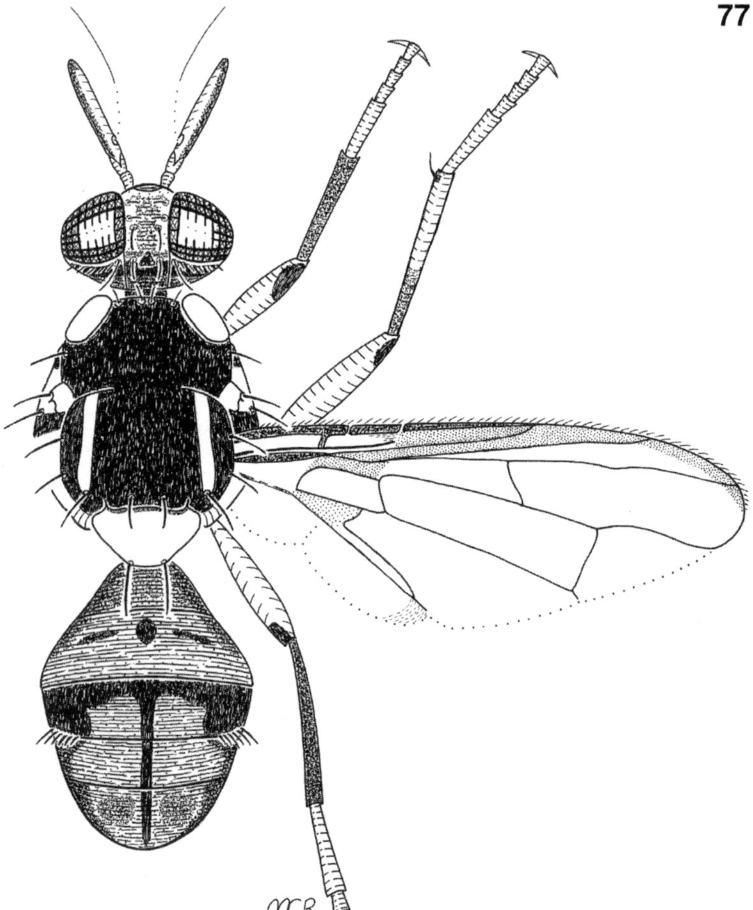

Fig. 76. *Bactrocera (Bactrocera) fuscoptera* new species, holotype male. **Fig. 77.** *Bactrocera (Bactrocera) gombokensis* Drew & Hancock, holotype male.

78

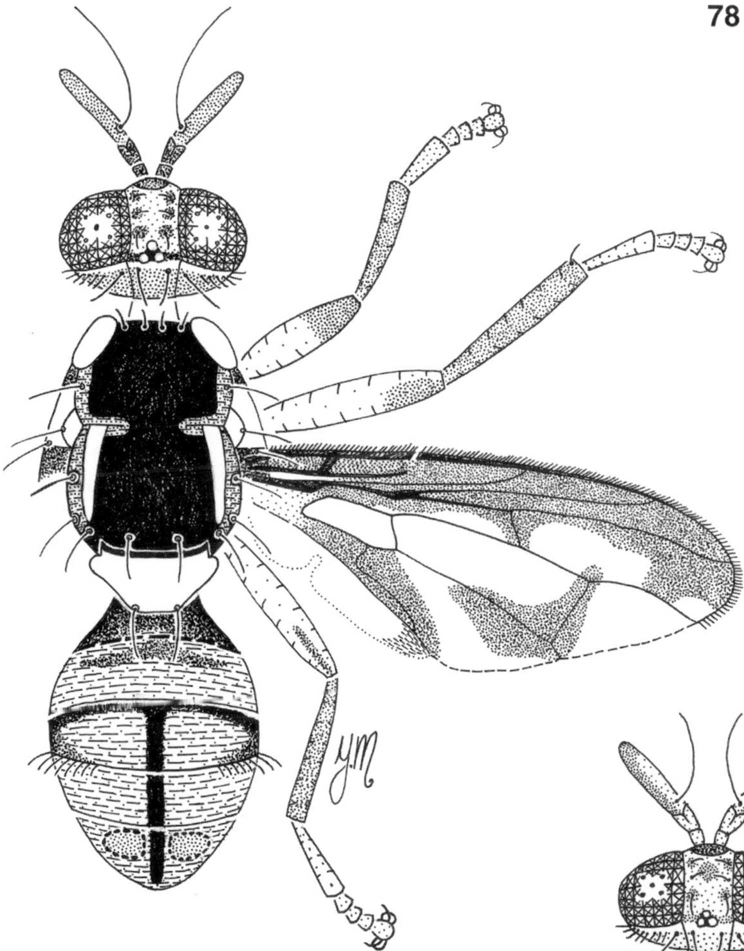

79

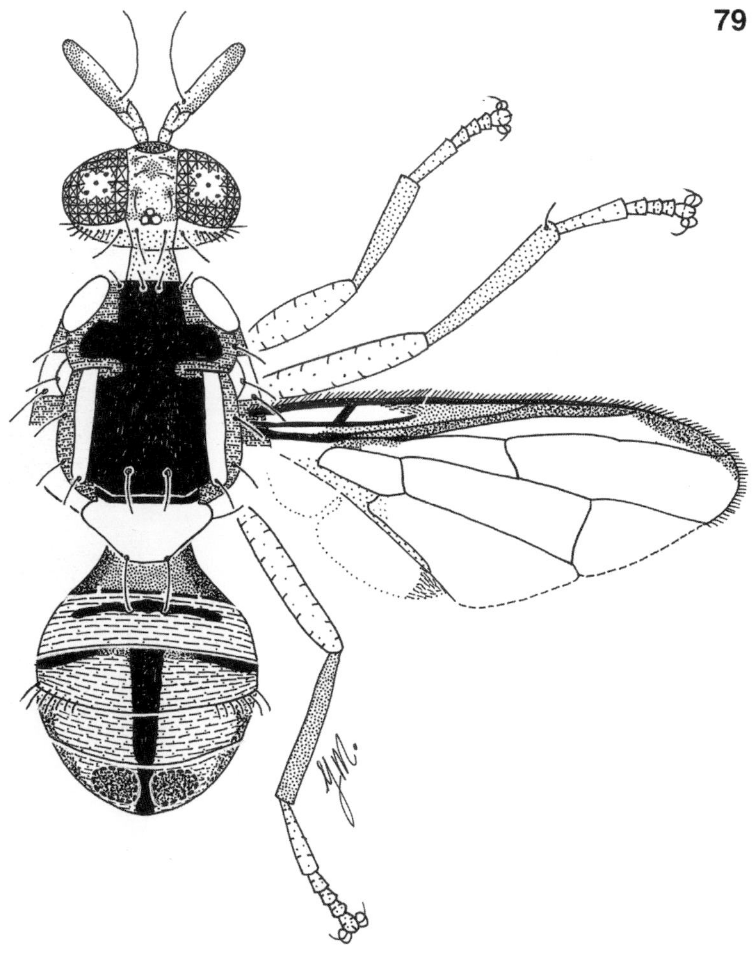

Fig. 78. *Bactrocera (Bactrocera) halmaherae* new species, holotype male. **Fig. 79.** *Bactrocera (Bactrocera) hantanae* Tsuruta & White, male.

80

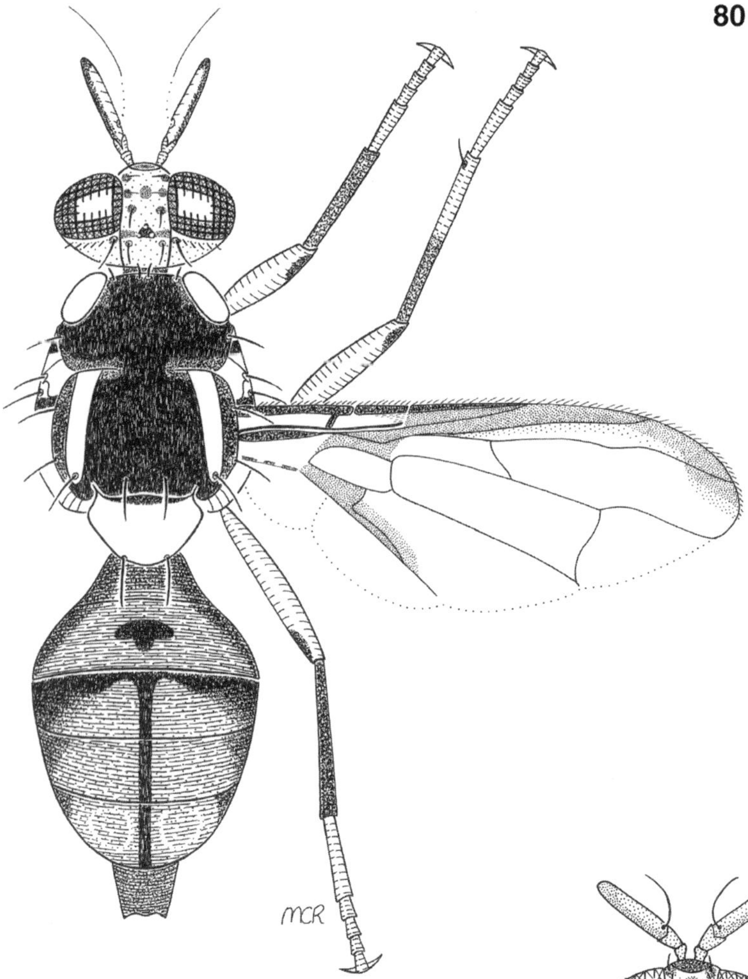

81

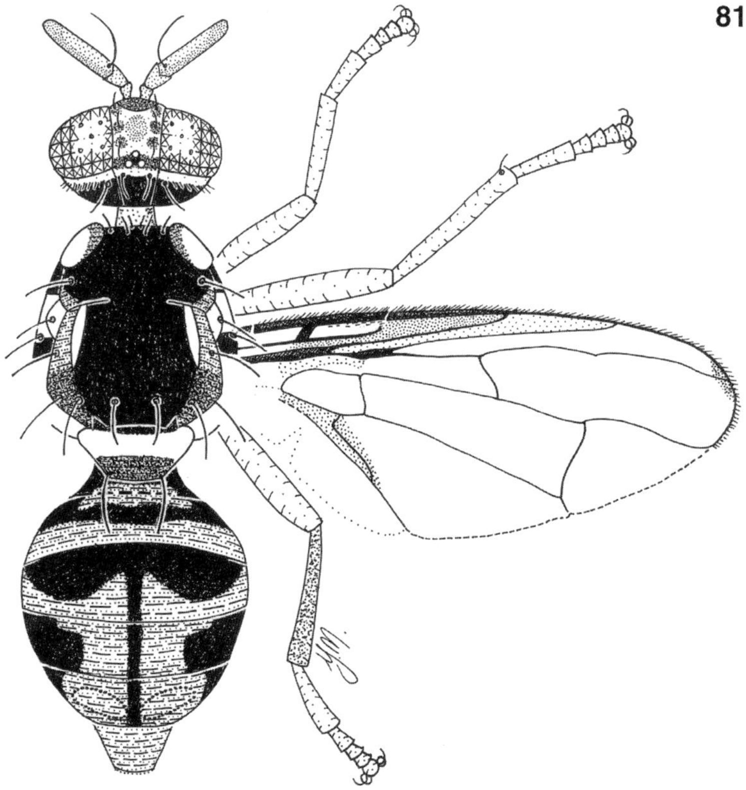

Fig. 80. *Bactrocera (Bactrocera) holtmanni* (Hardy), holotype female. **Fig. 81.** *Bactrocera (Bactrocera) hyalina* (Shiraki), holotype female.

82

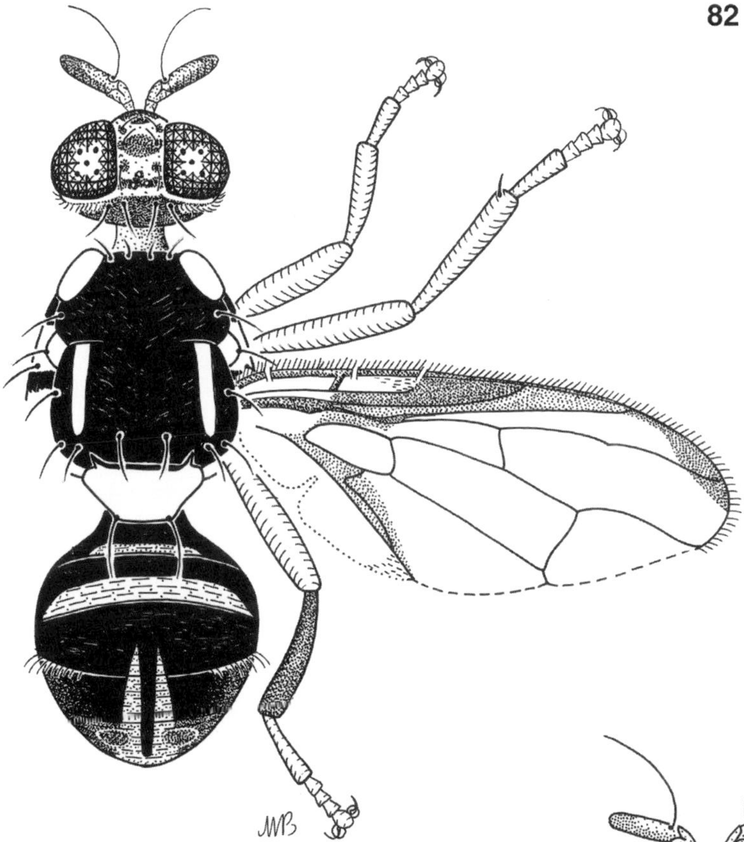

83

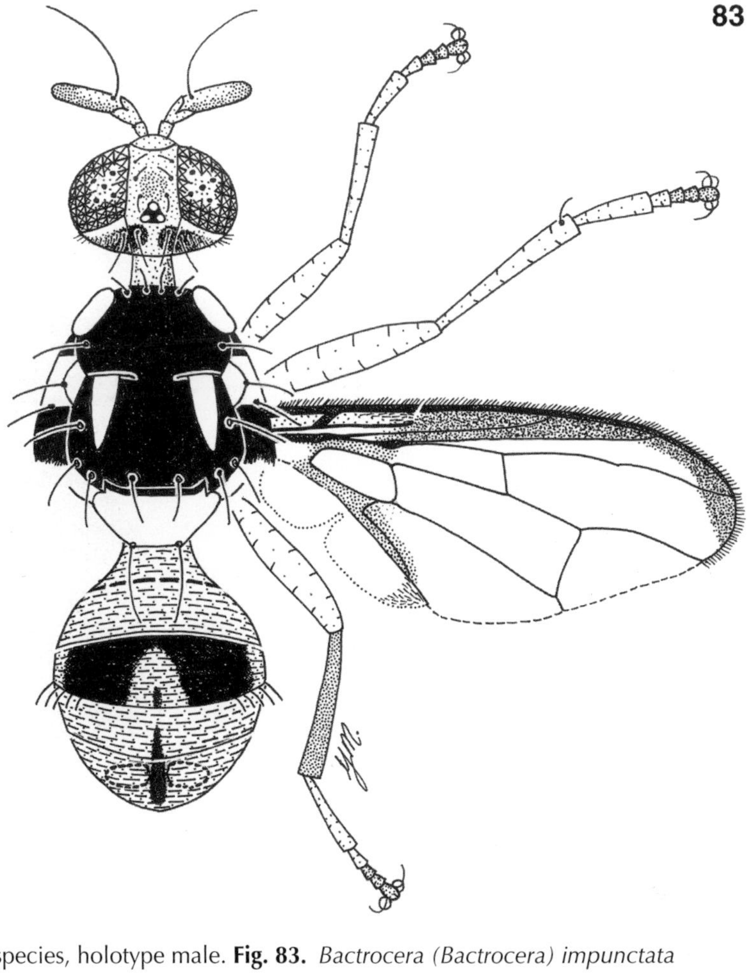

Fig. 82. *Bactrocera (Bactrocera) illusioscutellaris* new species, holotype male. **Fig. 83.** *Bactrocera (Bactrocera) impunctata* (de Meijere), male.

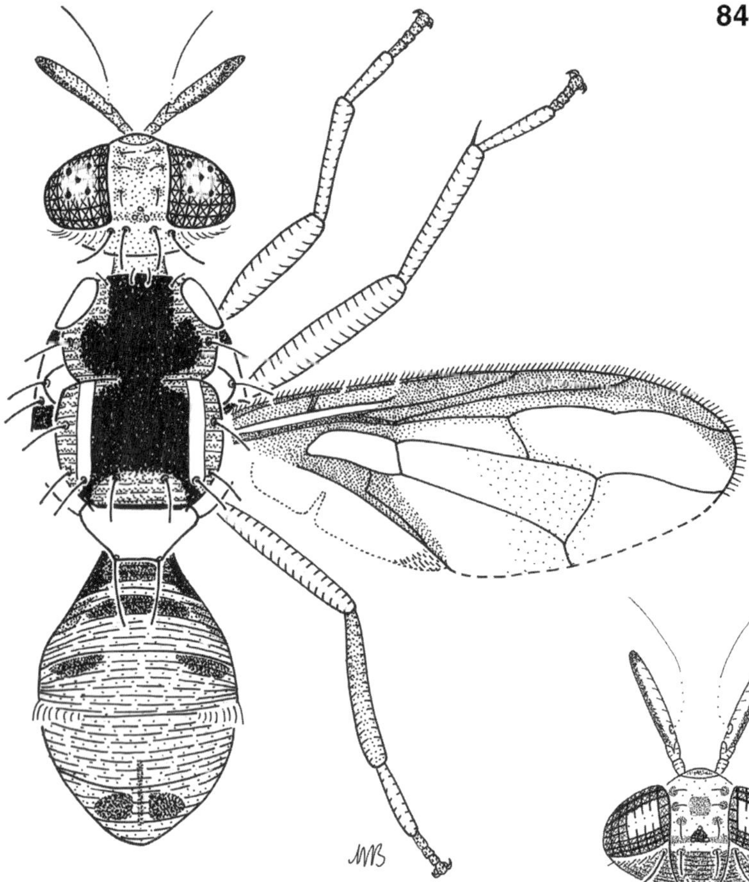

84

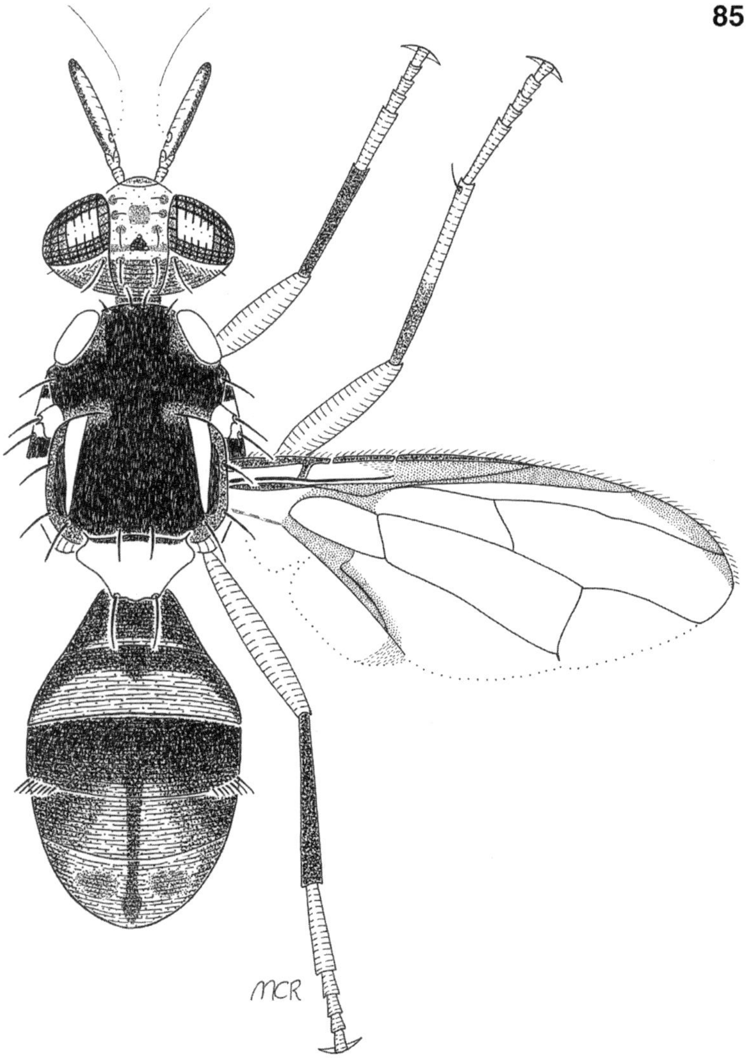

85

Fig. 84. *Bactrocera (Bactrocera) inconspicua* new species, holotype male. **Fig. 85.** *Bactrocera (Bactrocera) indonesiae* Drew & Hancock, holotype male.

86

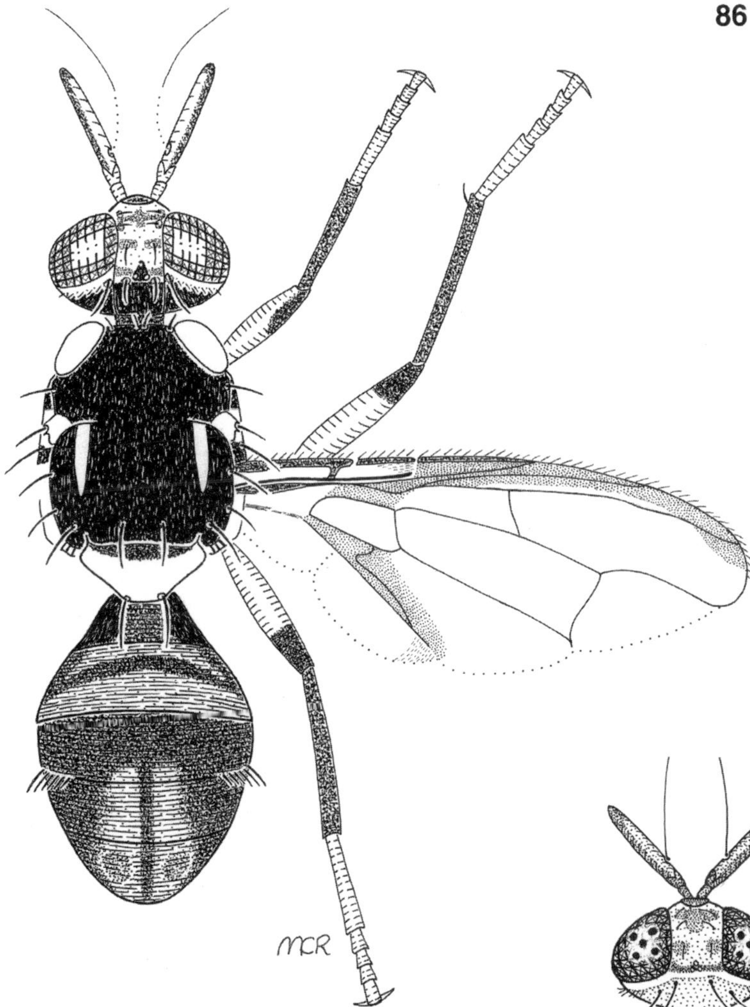

87A

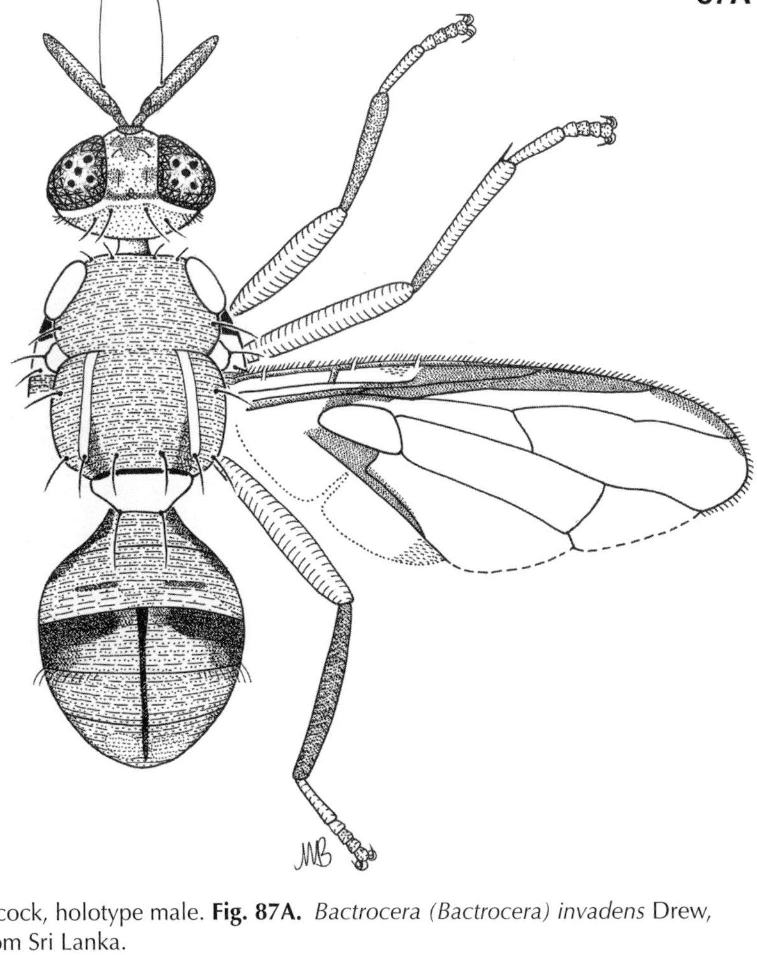

Fig. 86. *Bactrocera (Bactrocera) infulata* Drew & Hancock, holotype male. **Fig. 87A.** *Bactrocera (Bactrocera) invadens* Drew, Tsuruta & White, male, pale-colour form, specimen from Sri Lanka.

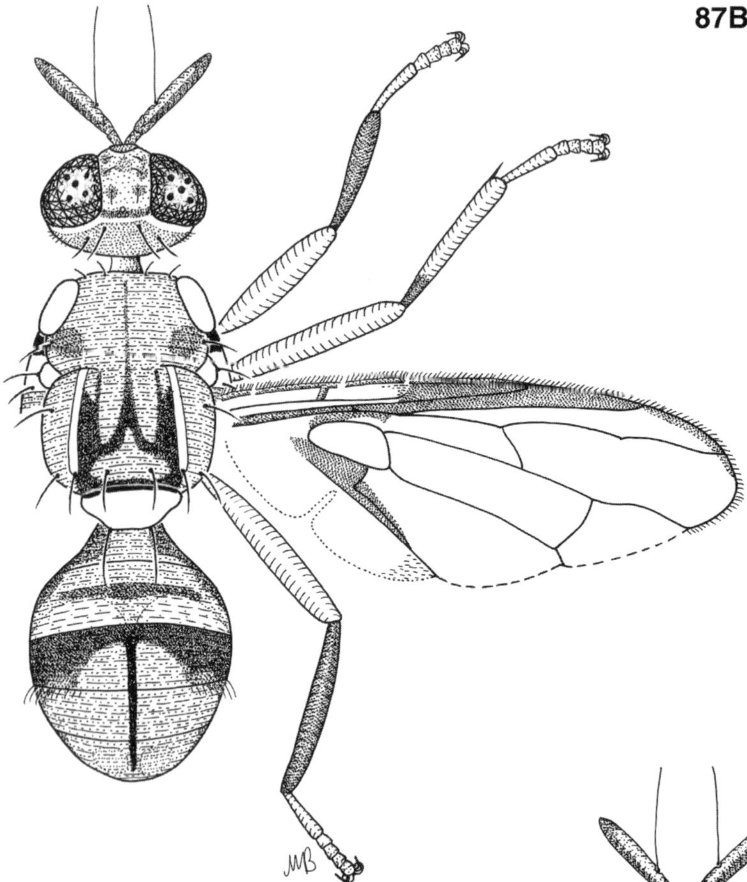

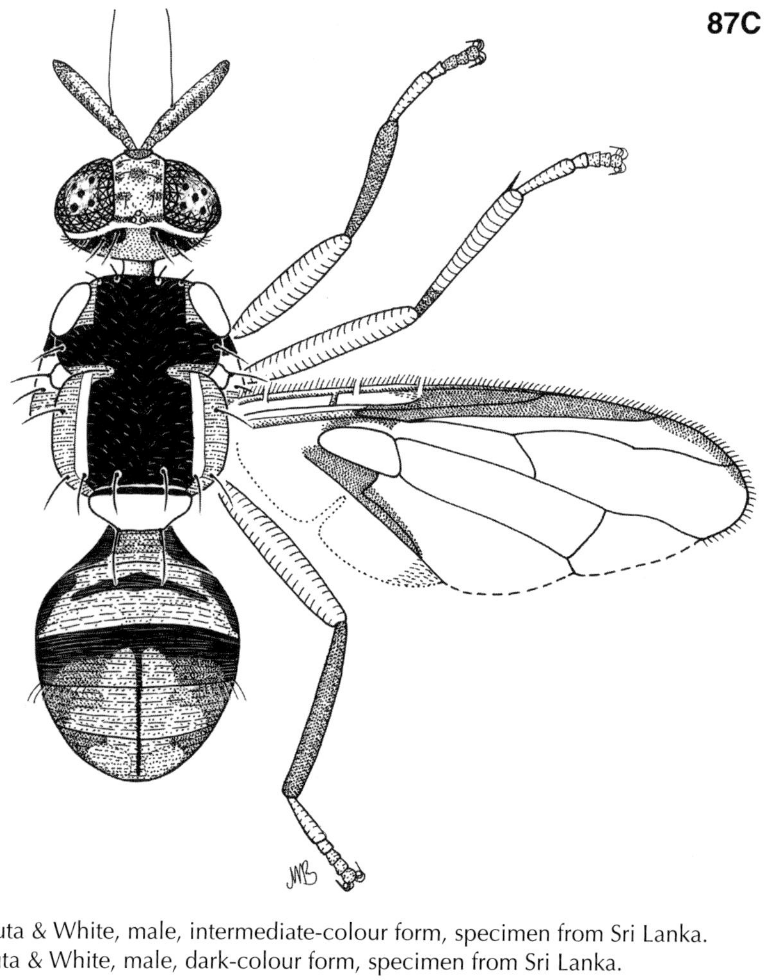

Fig. 87B. *Bactrocera (Bactrocera) invadens* Drew, Tsuruta & White, male, intermediate-colour form, specimen from Sri Lanka.
Fig. 87C. *Bactrocera (Bactrocera) invadens* Drew, Tsuruta & White, male, dark-colour form, specimen from Sri Lanka.

88

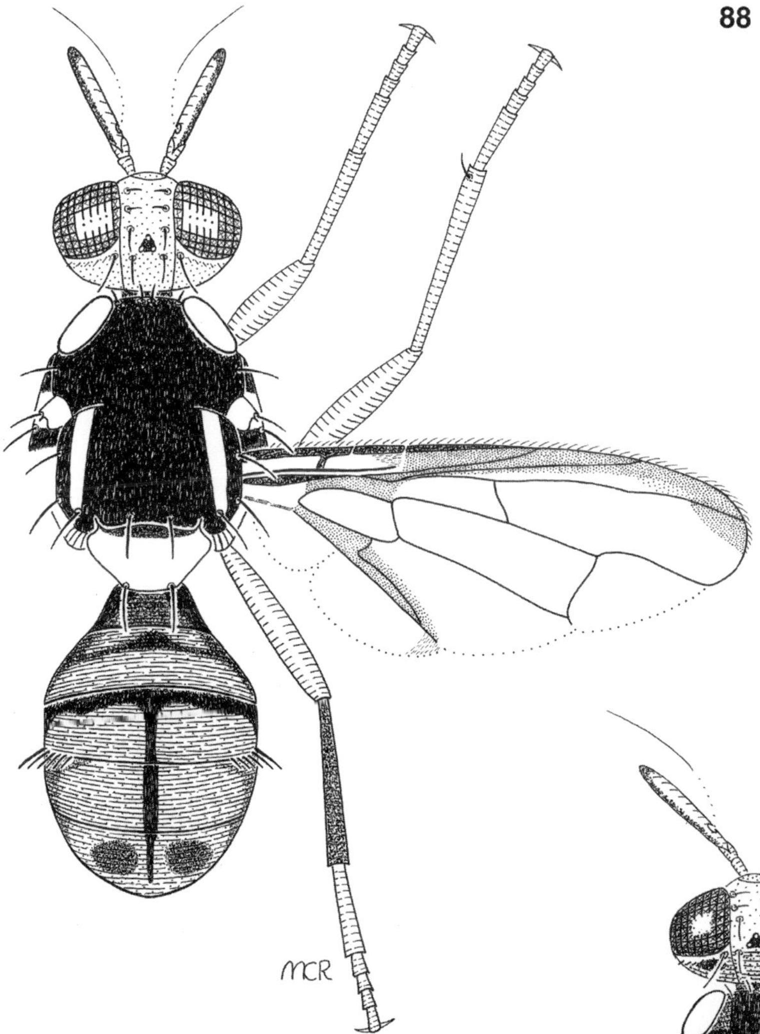

89

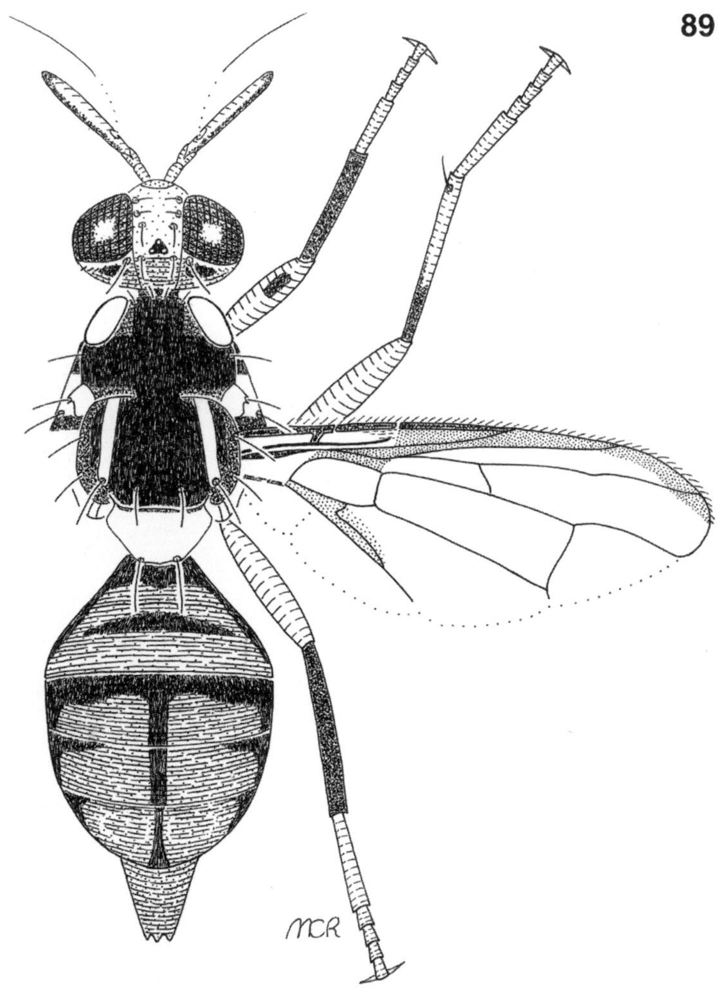

Fig. 88. *Bactrocera (Bactrocera) involuta* (Hardy), paratype male. **Fig. 89.** *Bactrocera (Bactrocera) irvingiae* Drew & Hancock, holotype female.

90

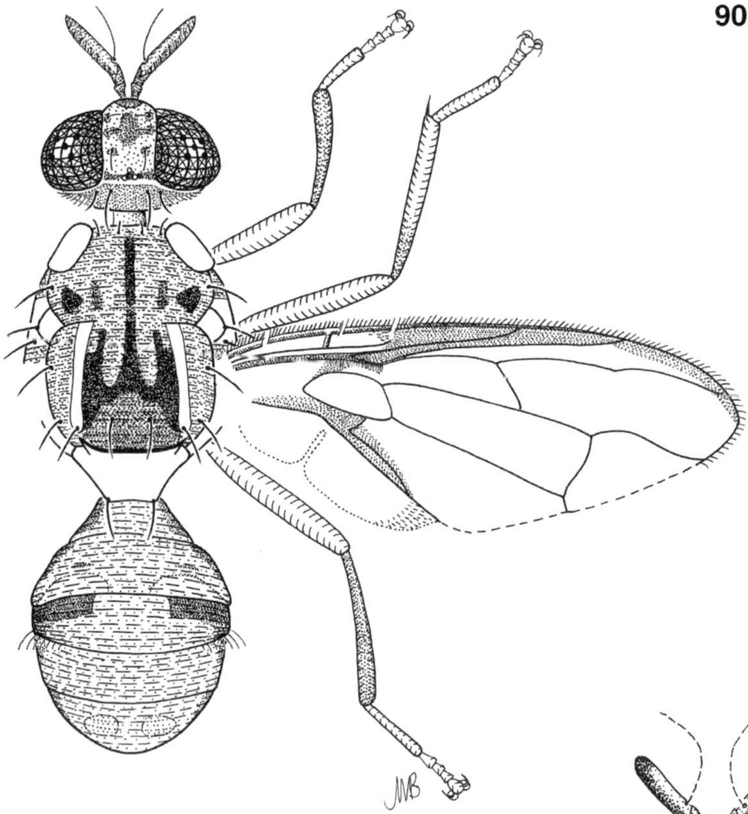

91

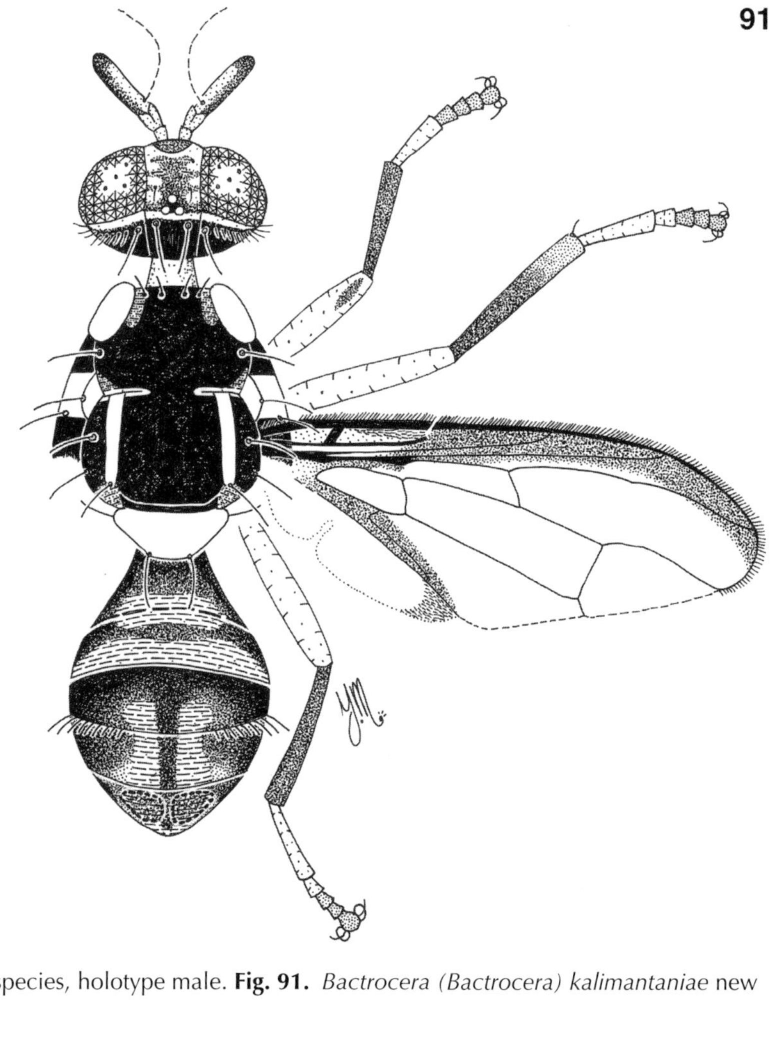

Fig. 90. *Bactrocera (Bactrocera) jaceobancroftii* new species, holotype male. **Fig. 91.** *Bactrocera (Bactrocera) kalimantaniae* new species, holotype male.

92

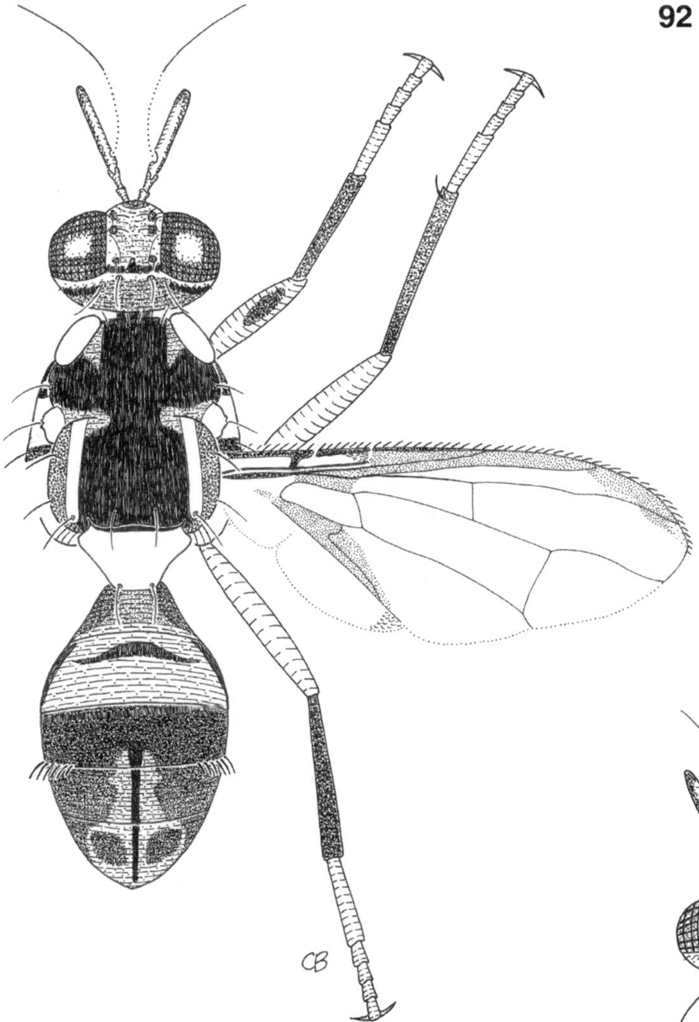

93

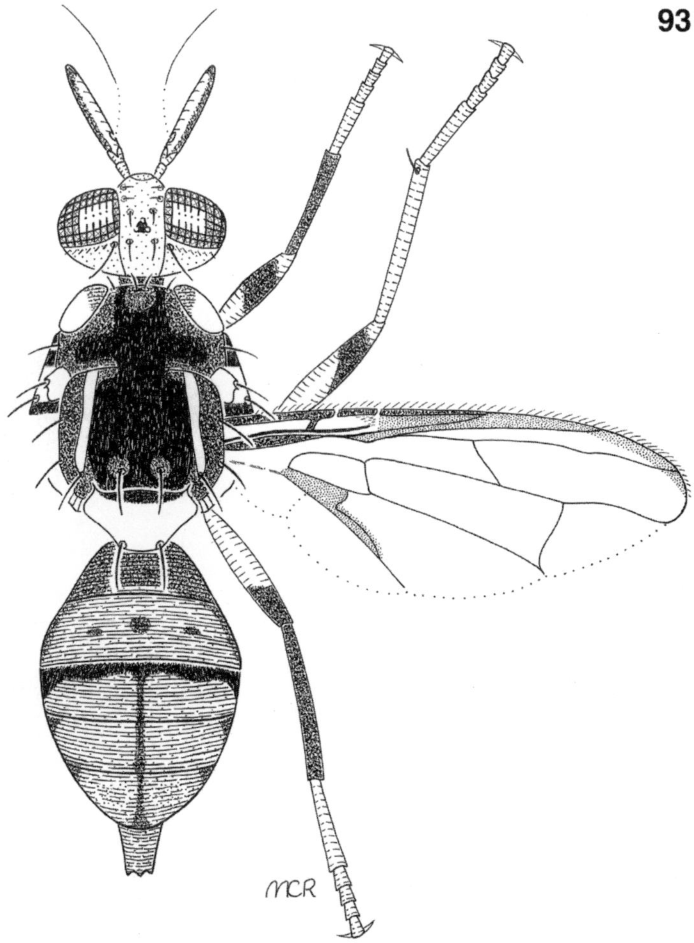

Fig. 92. *Bactrocera (Bactrocera) kanchanaburi* Drew & Hancock, holotype male. **Fig. 93.** *Bactrocera (Bactrocera) kandiensis* Drew & Hancock, holotype female.

94

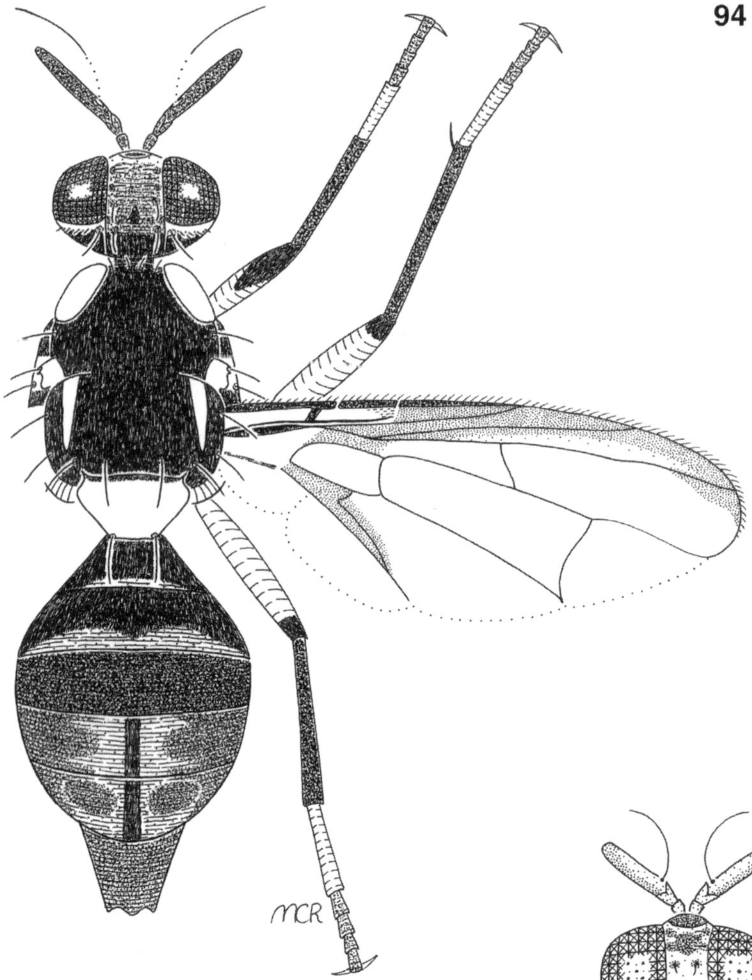

95

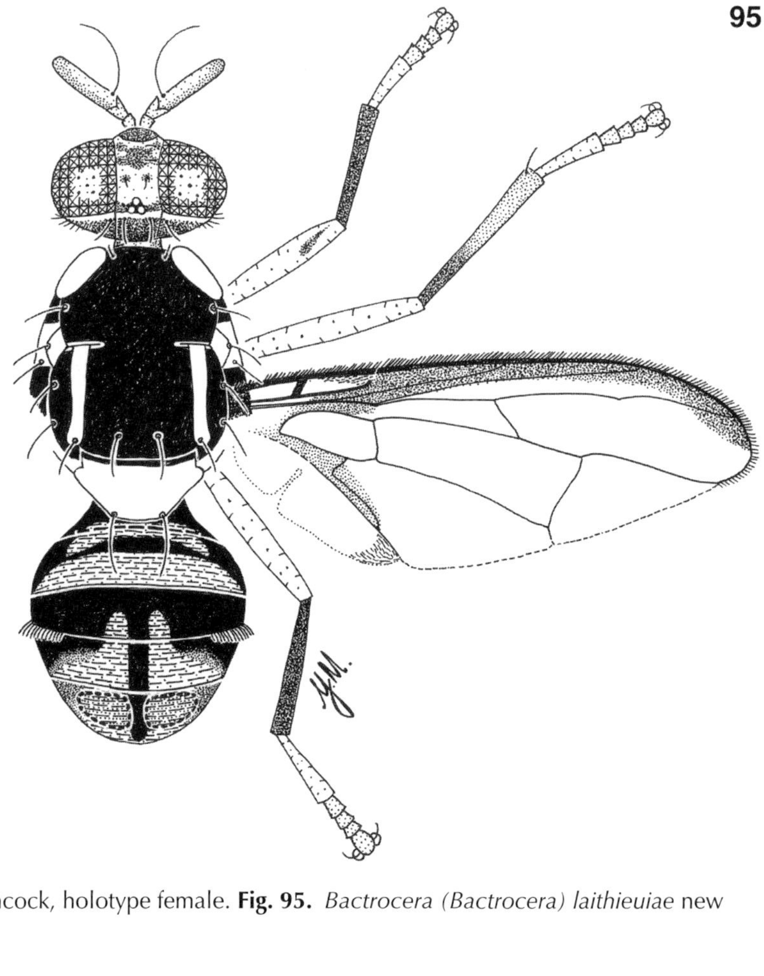

Fig. 94. *Bactrocera (Bactrocera) kinabalu* Drew & Hancock, holotype female. **Fig. 95.** *Bactrocera (Bactrocera) laithieuiae* new species, holotype male.

96A

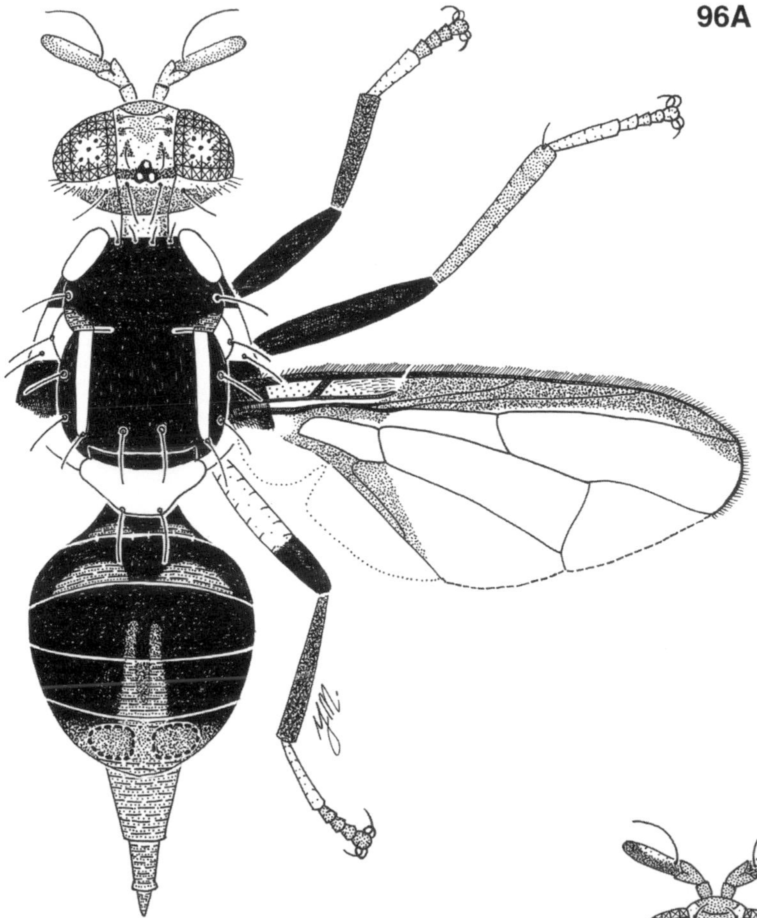

96B

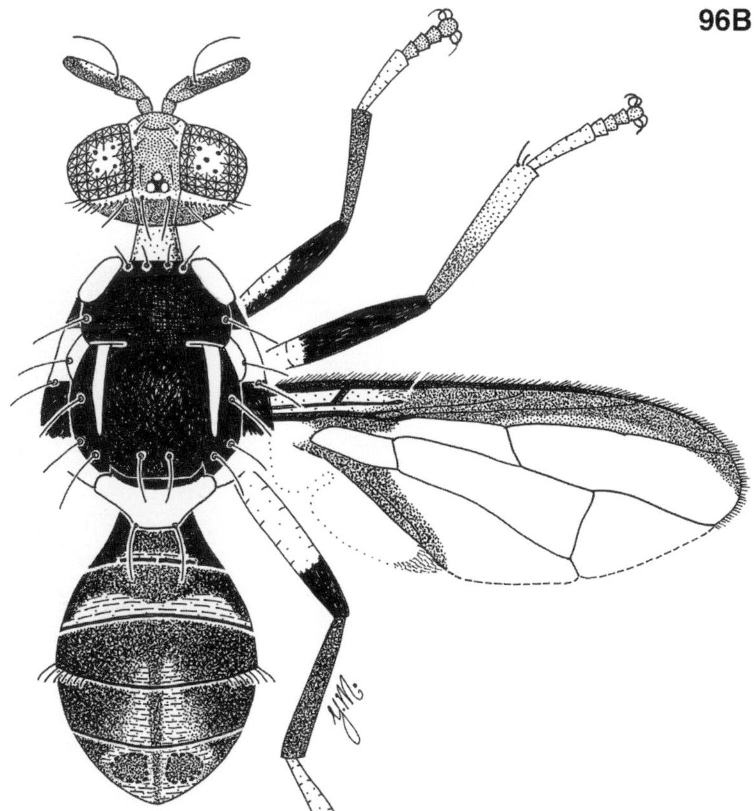

Fig. 96A. *Bactrocera (Bactrocera) lata* (Perkins), holotype female. **Fig. 96B.** *Bactrocera (Bactrocera) silvaticus* (Hardy), paratype male.

97

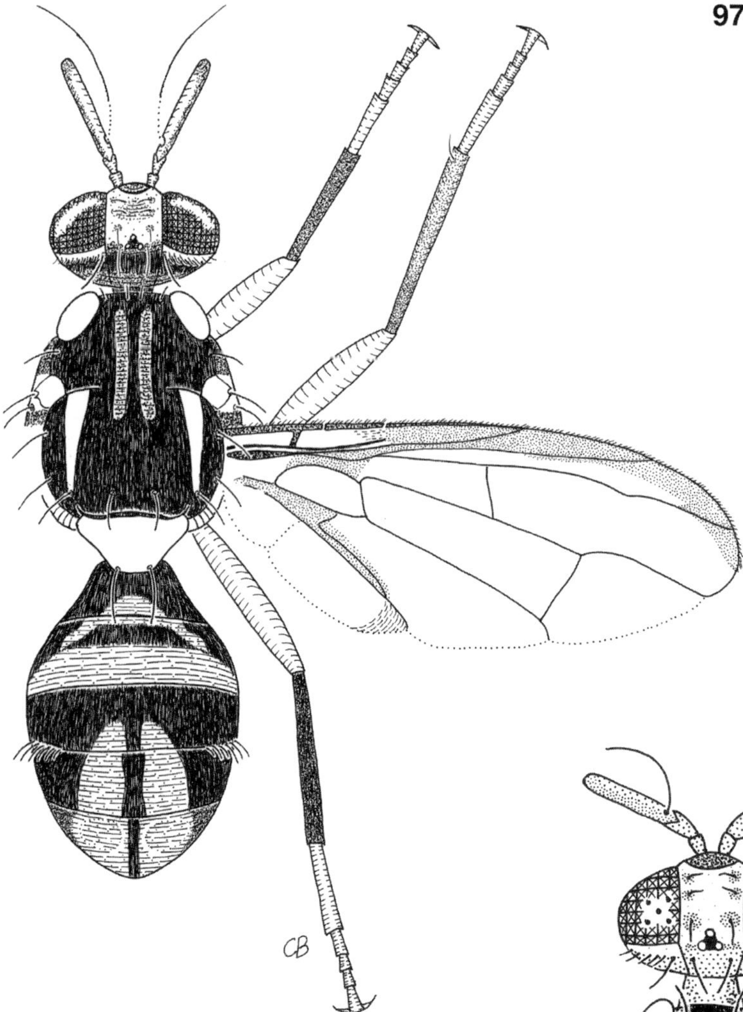

98

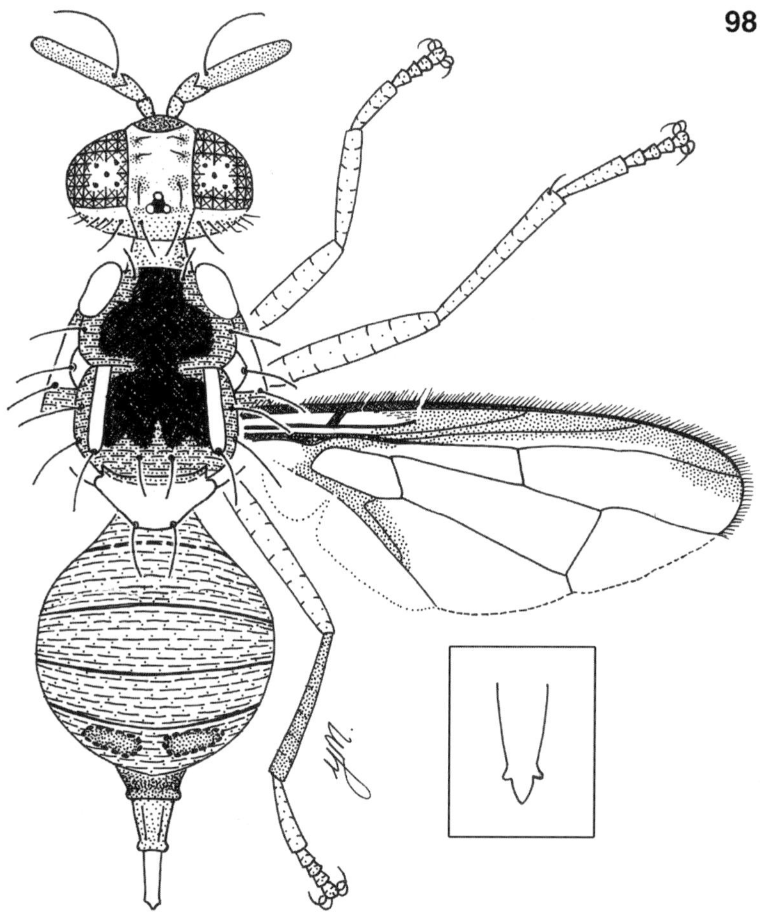

Fig. 97. *Bactrocera (Bactrocera) lateritaenia* Drew & Hancock, holotype male. **Fig. 98.** *Bactrocera (Bactrocera) latifrons* (Hendel), female and apex of piercer from female ovipositor.

99

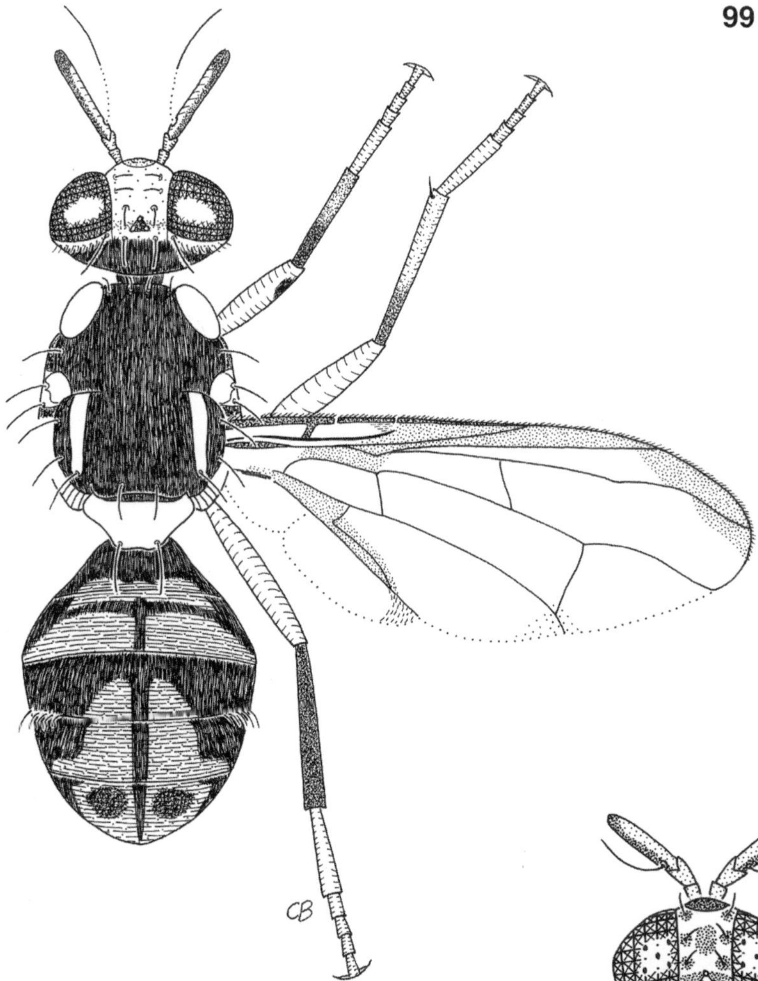

100

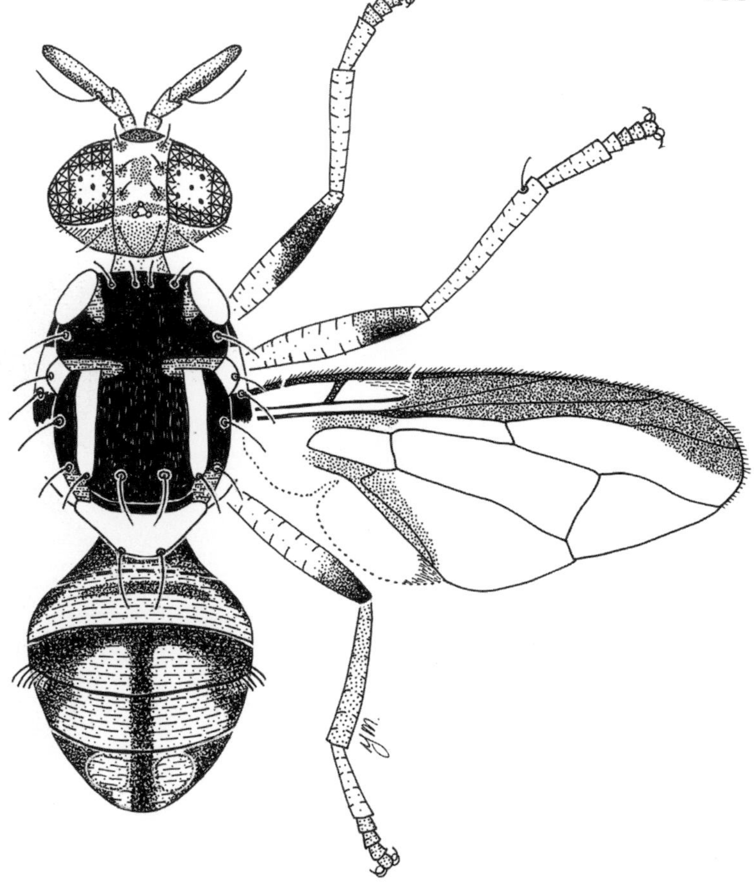

Fig. 99. *Bactrocera (Bactrocera) latilineola* Drew & Hancock, holotype male. **Fig. 100.** *Bactrocera (Bactrocera) limbifera* (Bezzi), lectotype male.

101

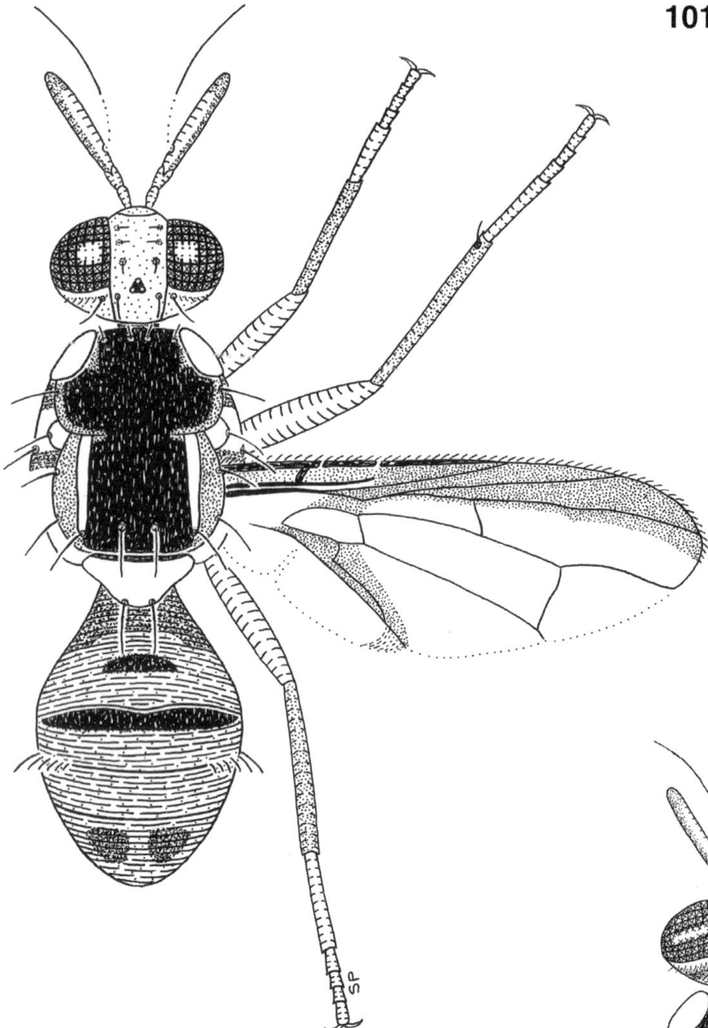

102

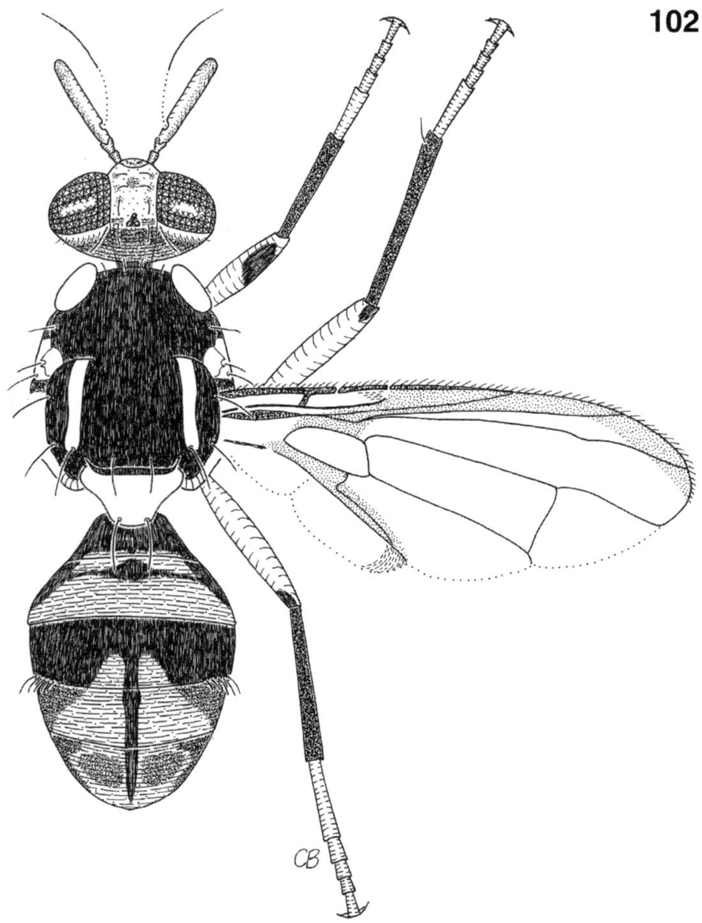

Fig. 101. *Bactrocera (Bactrocera) linduensis* new name, paratype male. **Fig. 102.** *Bactrocera (Bactrocera) lombokensis* Drew & Hancock, holotype male.

103

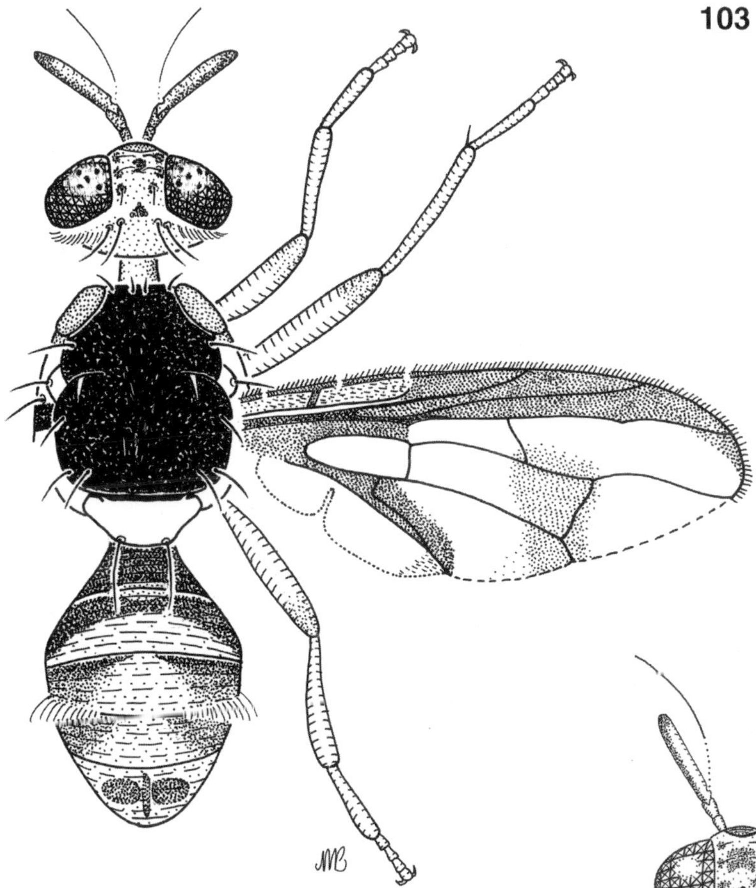

104

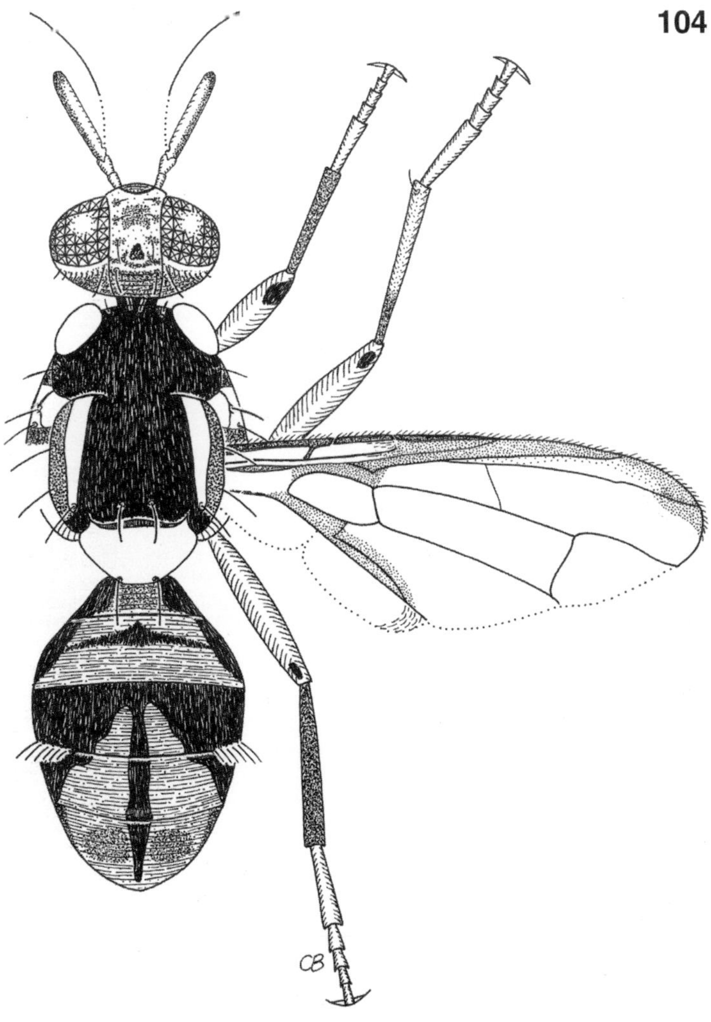

Fig. 103. *Bactrocera (Bactrocera) maculigera* Doleschall, male. **Fig. 104.** *Bactrocera (Bactrocera) makilingensis* Drew & Hancock, holotype male.

105

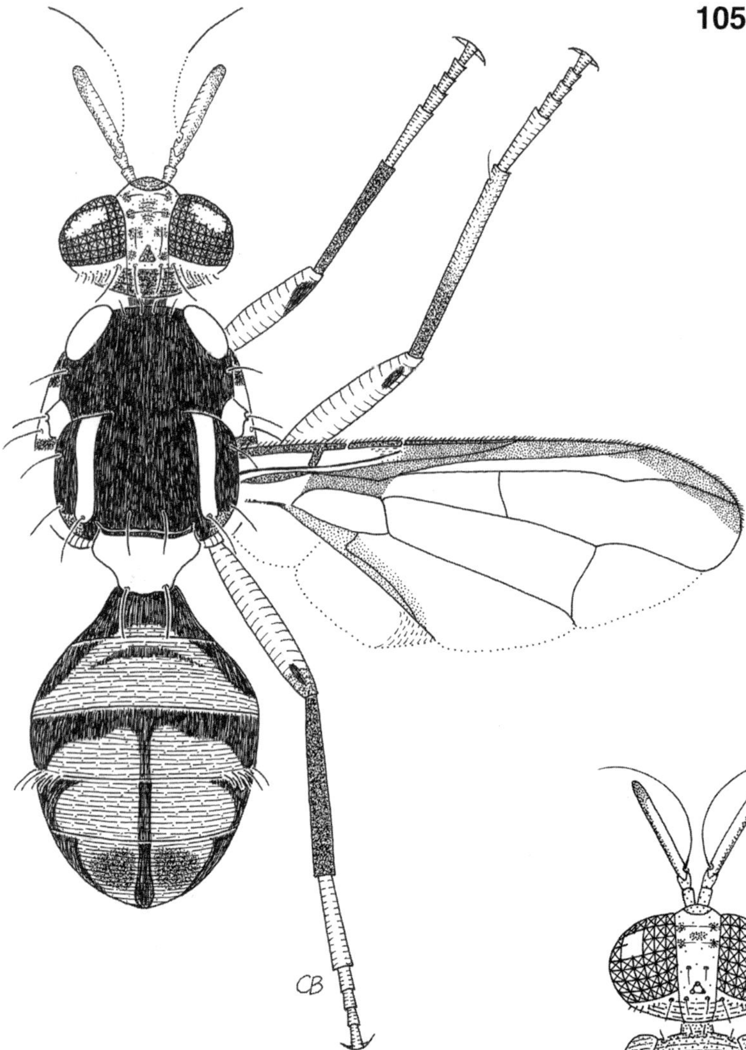

106

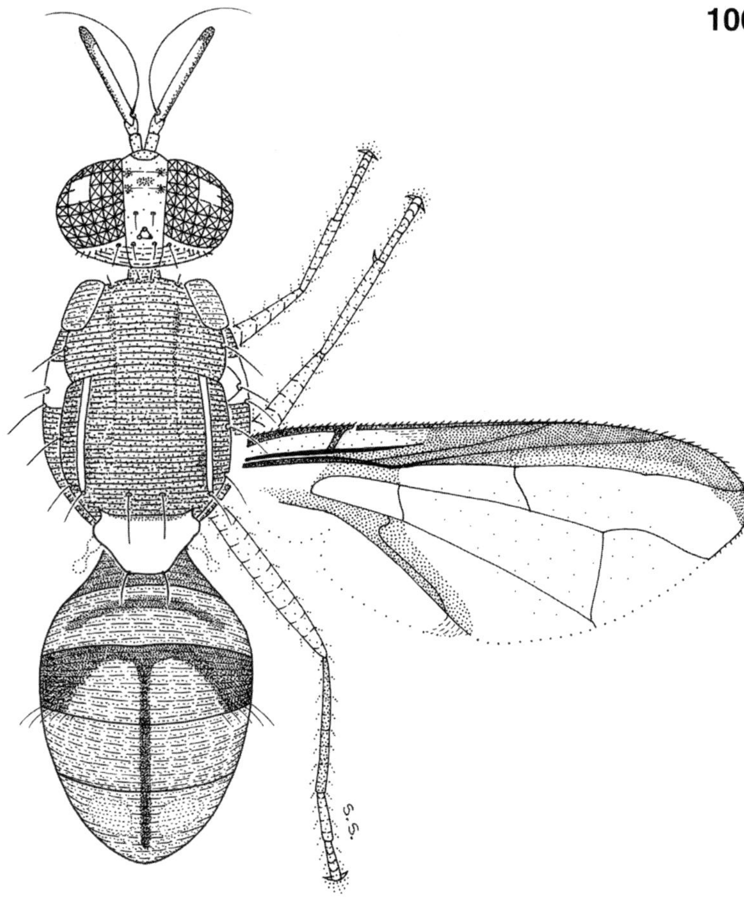

Fig. 105. *Bactrocera (Bactrocera) malaysiensis* Drew & Hancock, holotype male. **Fig. 106.** *Bactrocera (Bactrocera) mamaliae* new species, holotype male.

107

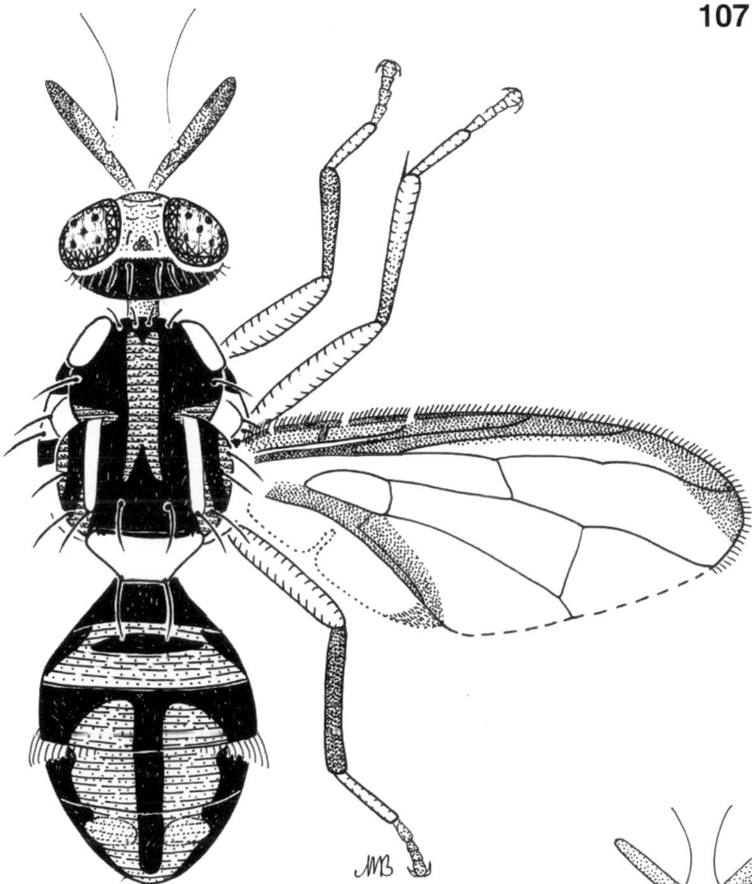

108

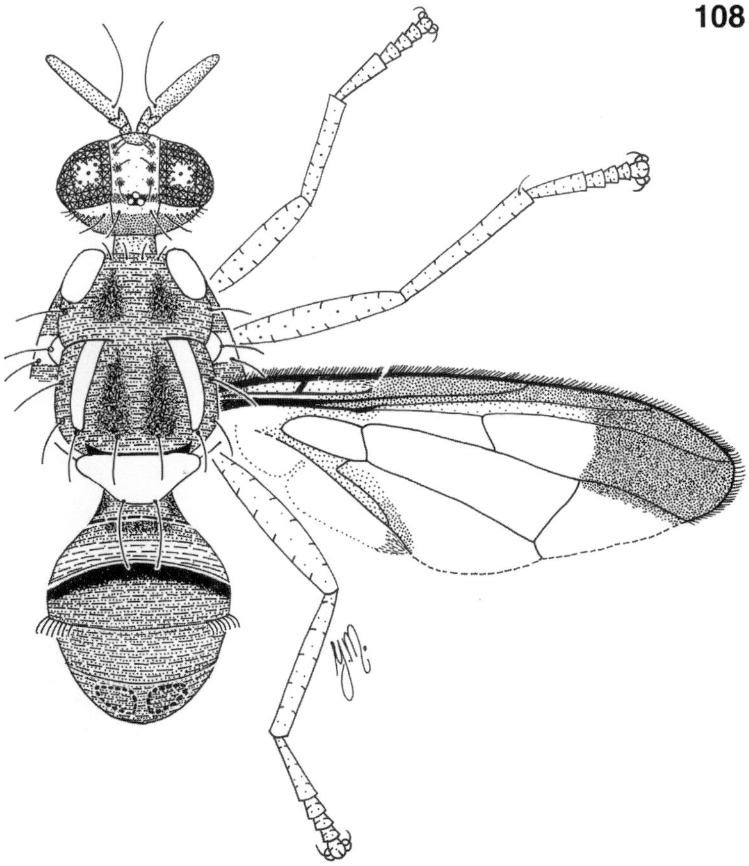

Fig. 107. *Bactrocera (Bactrocera) mediorufula* new species, holotype male. **Fig. 108.** *Bactrocera (Bactrocera) megaspilus* (Hardy), holotype male.

109

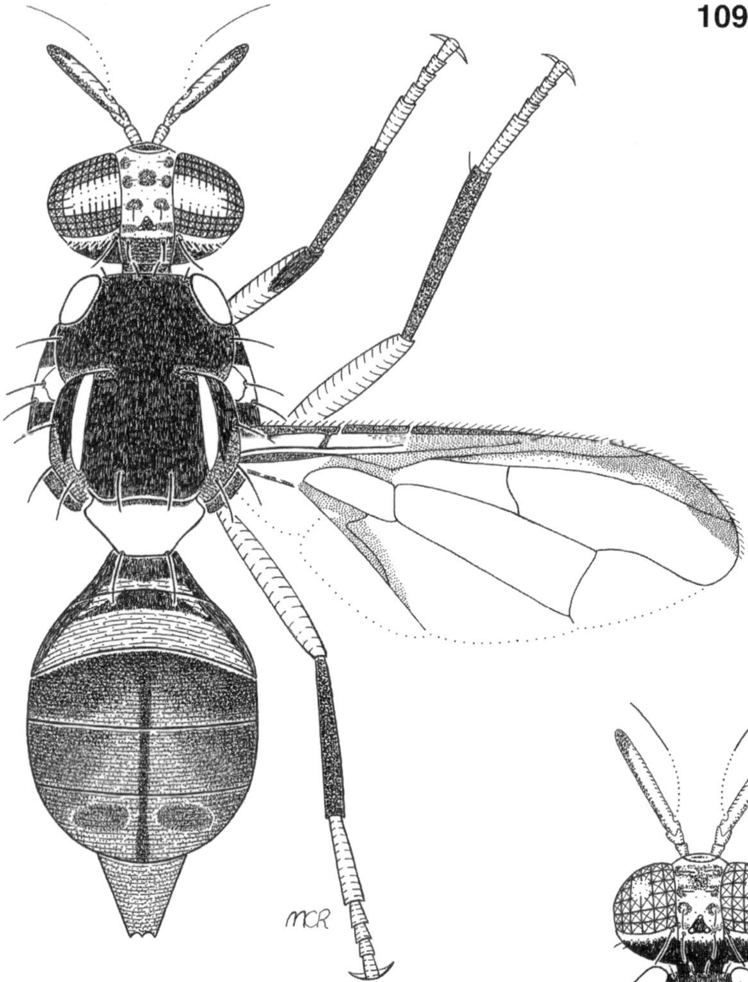

110

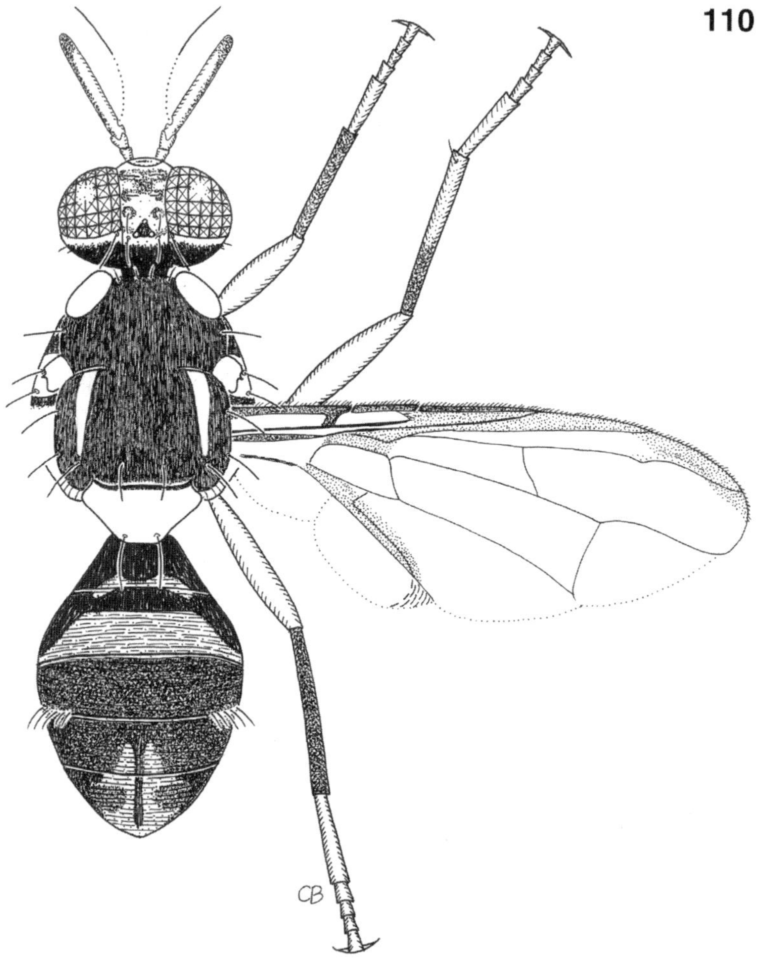

Fig. 109. *Bactrocera (Bactrocera) melastomatos* Drew & Hancock, holotype female. **Fig. 110.** *Bactrocera (Bactrocera) merapiensis* Drew & Hancock, holotype male.

111

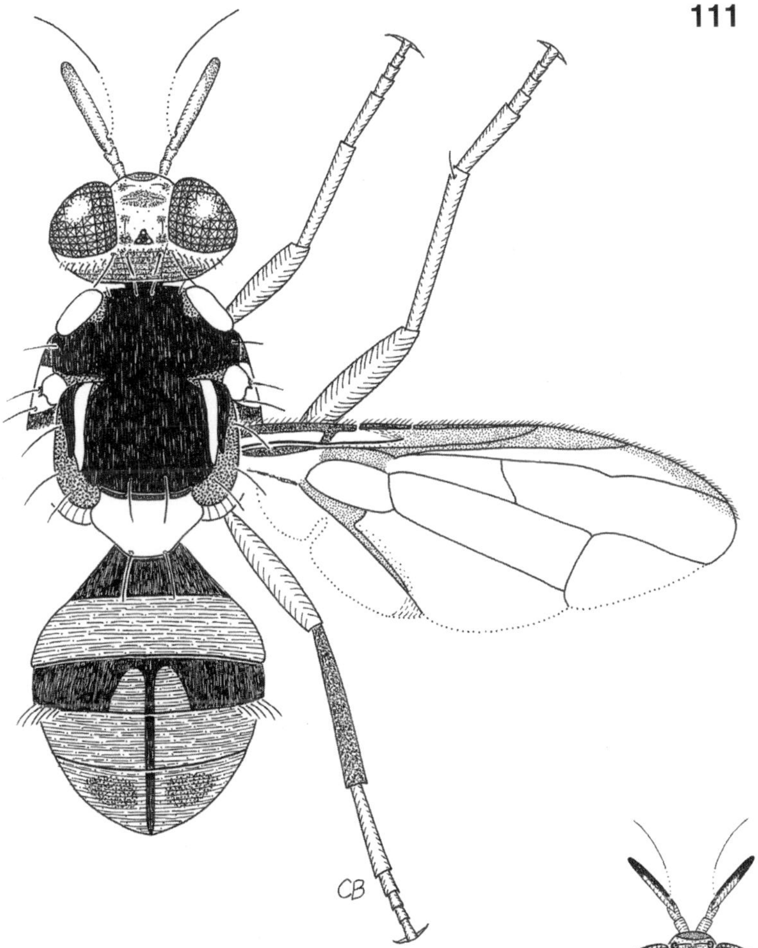

112

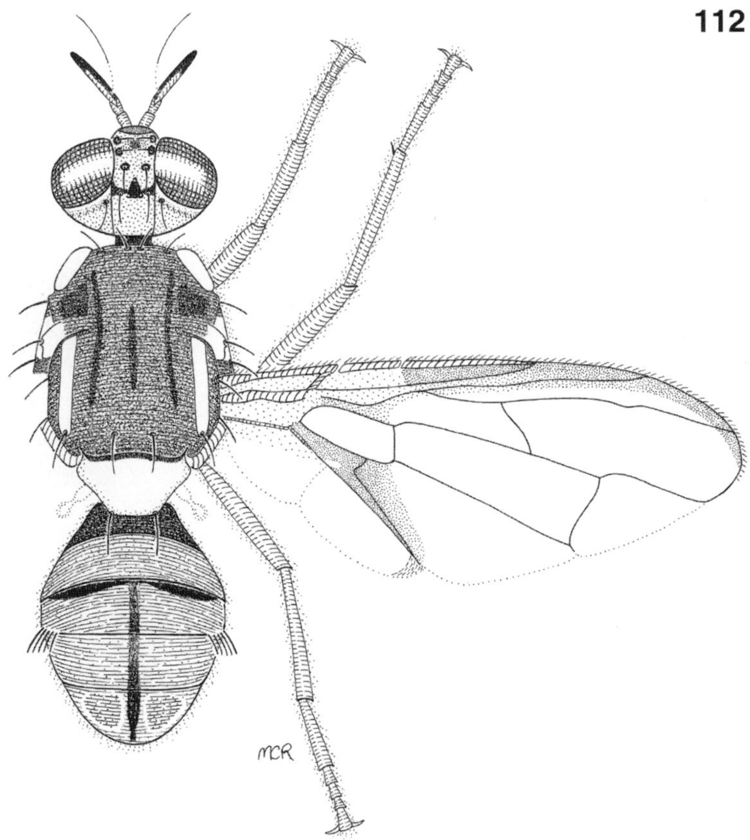

Fig. 111. *Bactrocera (Bactrocera) minuscula* Drew & Hancock, holotype male. **Fig. 112.** *Bactrocera (Bactrocera) moluccensis* (Perkins), male.

113

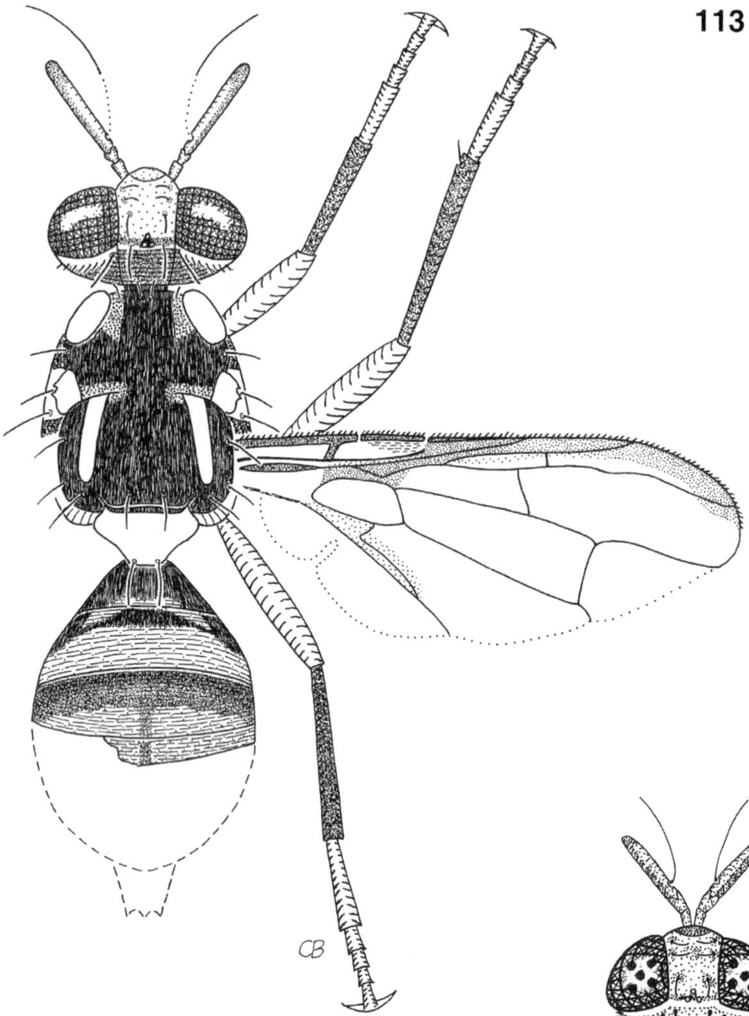

114

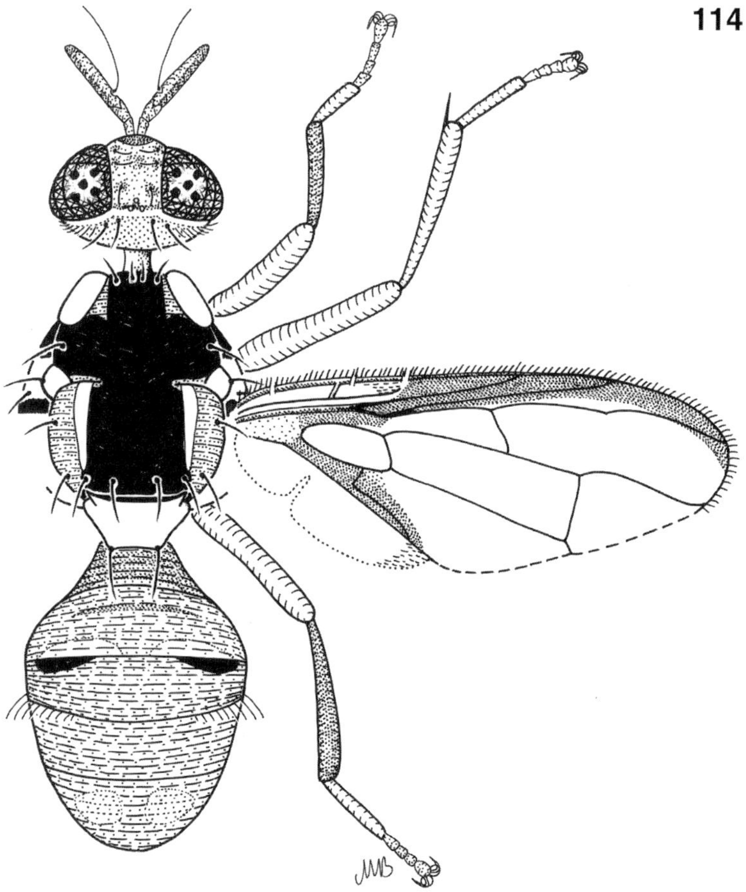

Fig. 113. *Bactrocera (Bactrocera) muiri* (Hardy & Adachi), holotype female. **Fig. 114.** *Bactrocera (Bactrocera) musae* (Tryon), male.

115A

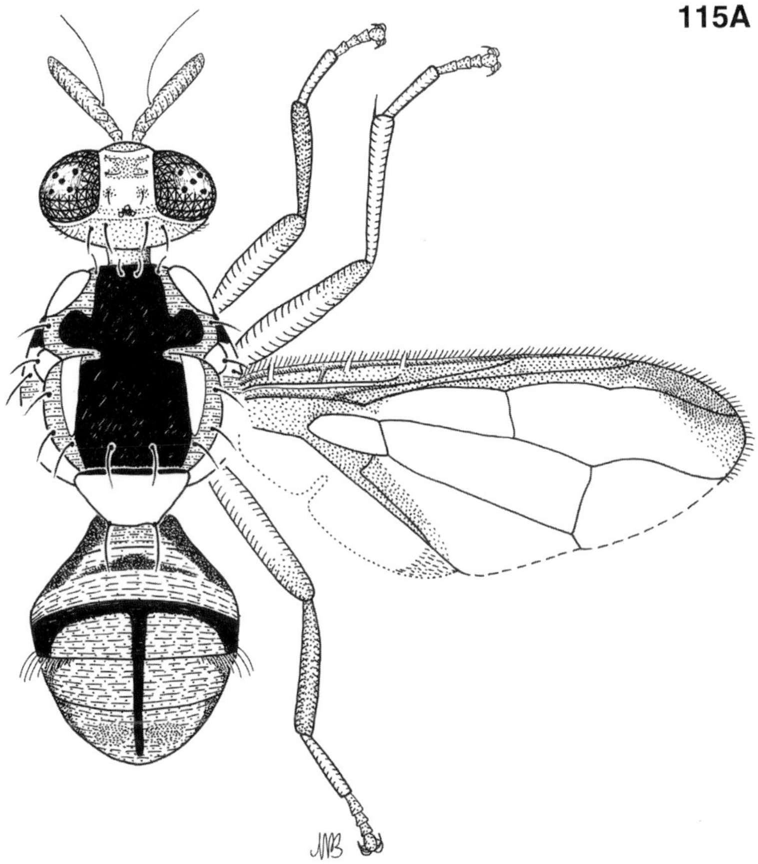

115B

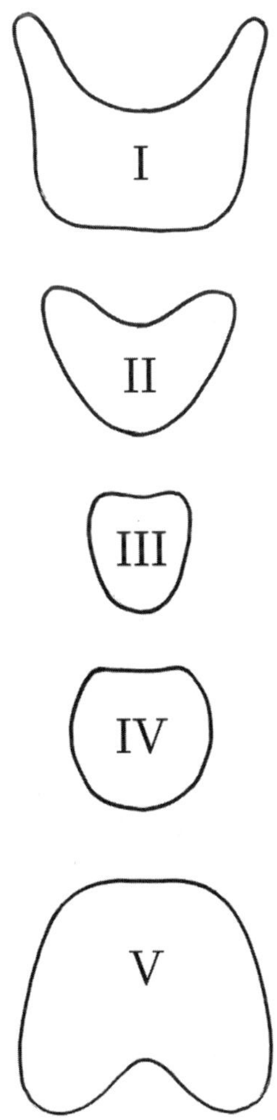

Fig. 115A. *Bactrocera (Bactrocera) nanoarcuata* new species, holotype male. **Fig. 115B.** *Bactrocera (Bactrocera) nanoarcuata* new species, holotype male abdominal sternites.

116

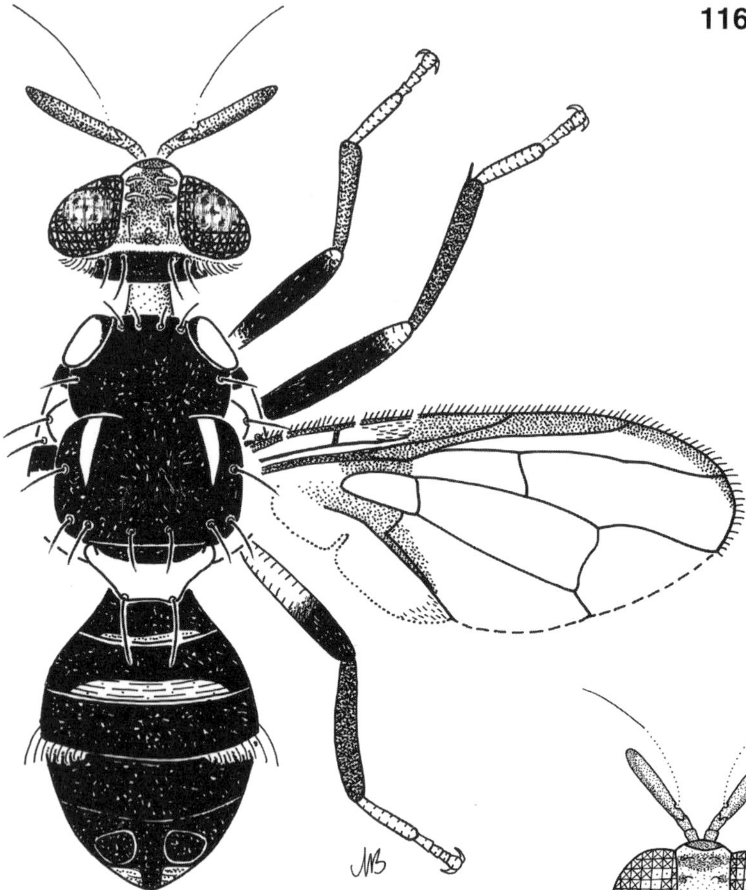

117

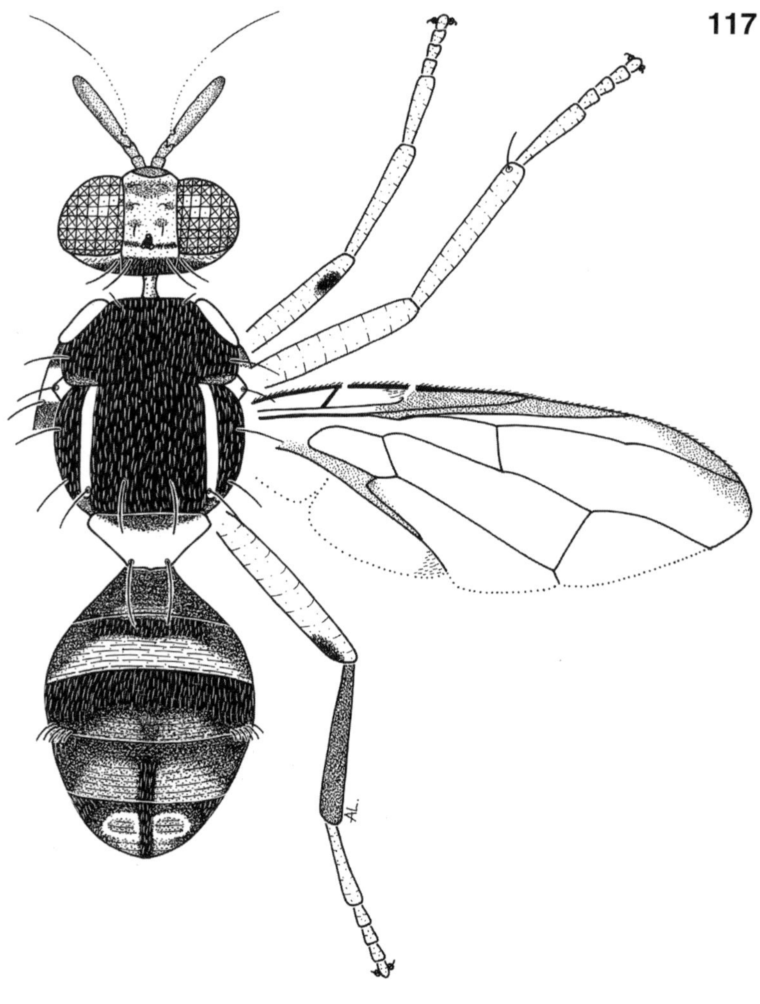

Fig. 116. *Bactrocera (Bactrocera) nationigrotibialis* new species, holotype male. **Fig. 117.** *Bactrocera (Bactrocera) neoarecae* Drew, holotype male.

118

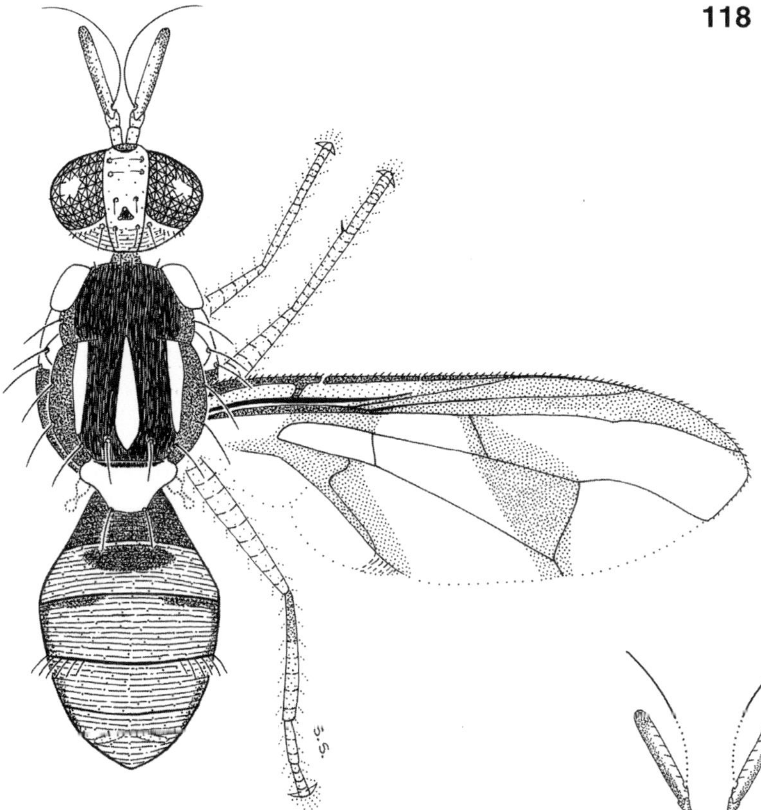

119

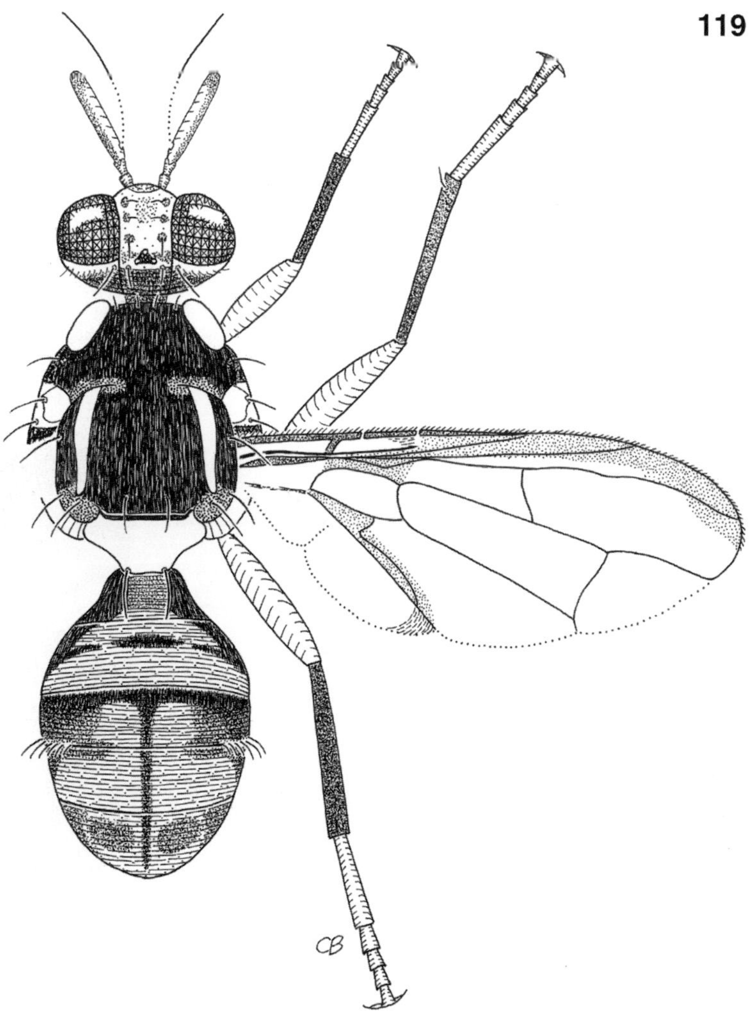

Fig. 118. *Bactrocera (Bactrocera) neocheesmanae* Drew, holotype male. **Fig. 119.** *Bactrocera (Bactrocera) neocognata* Drew & Hancock, holotype male.

120

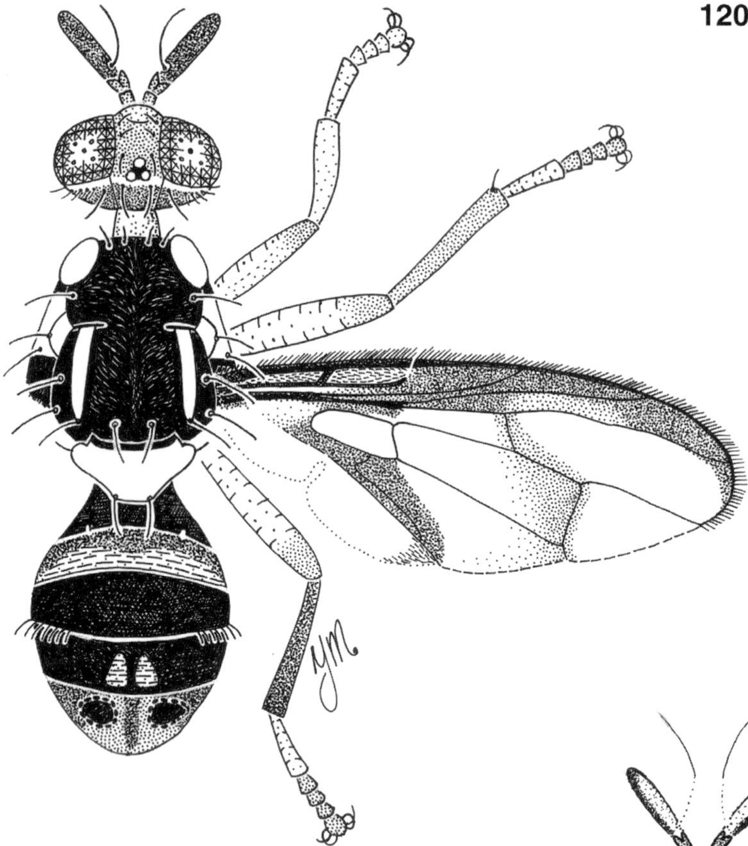

121

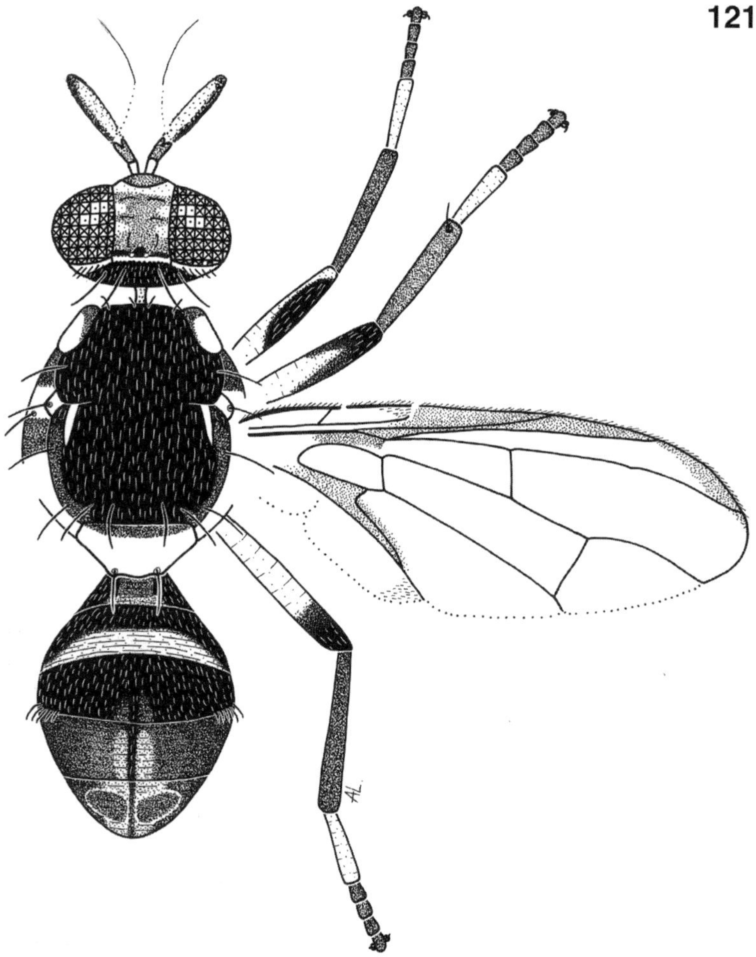

Fig. 120. *Bactrocera (Bactrocera) neofulvicauda* new species, holotype male. **Fig. 121.** *Bactrocera (Bactrocera) neonigrotibialis* Drew, holotype male.

122

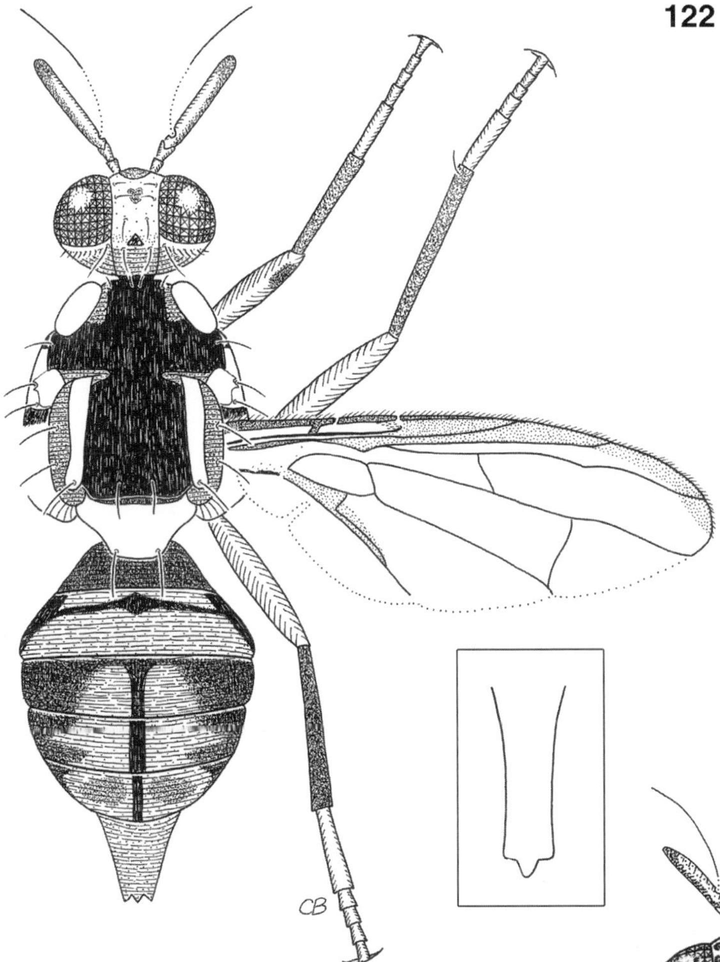

123

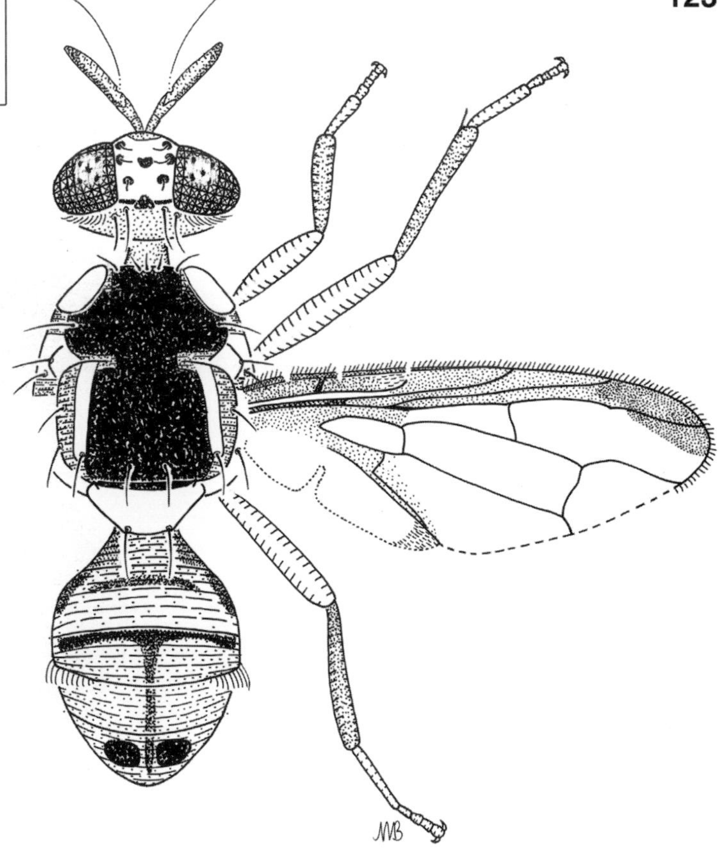

Fig. 122. *Bactrocera (Bactrocera) neopropinqua* Drew & Hancock, holotype female and apex of piercer from female ovipositor.
Fig. 123. *Bactrocera (Bactrocera) neoritsemai* new species, holotype male.

124

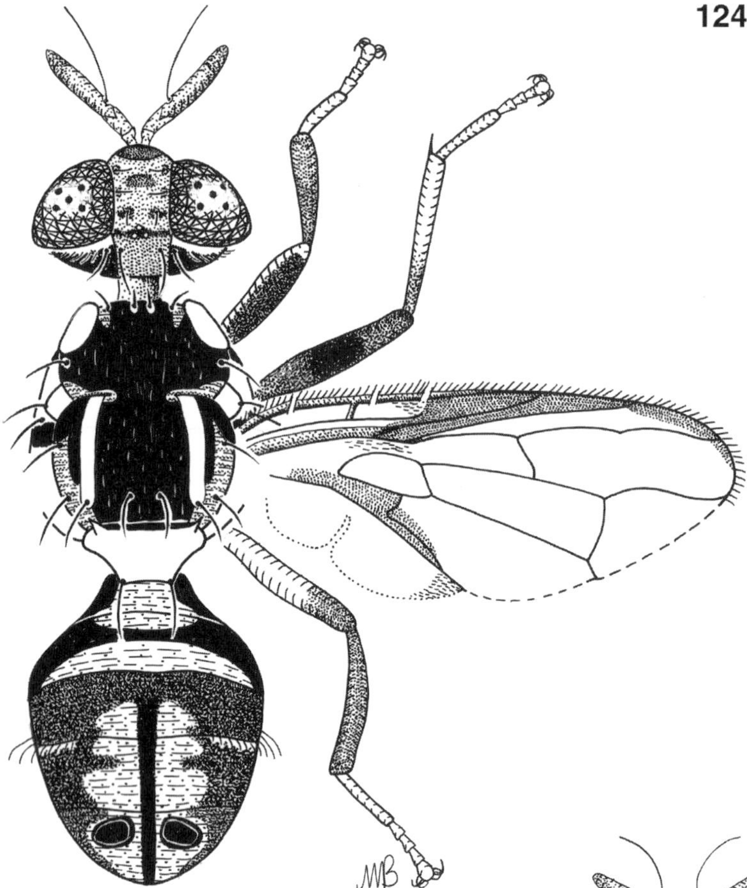

125

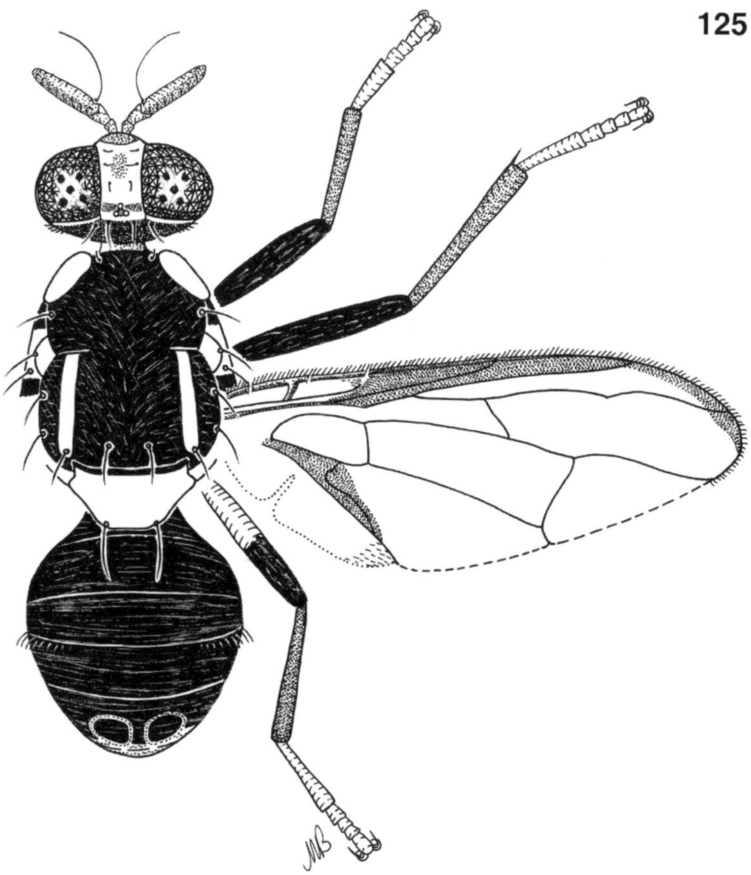

Fig. 124. *Bactrocera (Bactrocera) nigrifacia* Zhang, Ji & Chen, male. **Fig. 125.** *Bactrocera (Bactrocera) nigrifemorata* Lin & Wang, male.

126

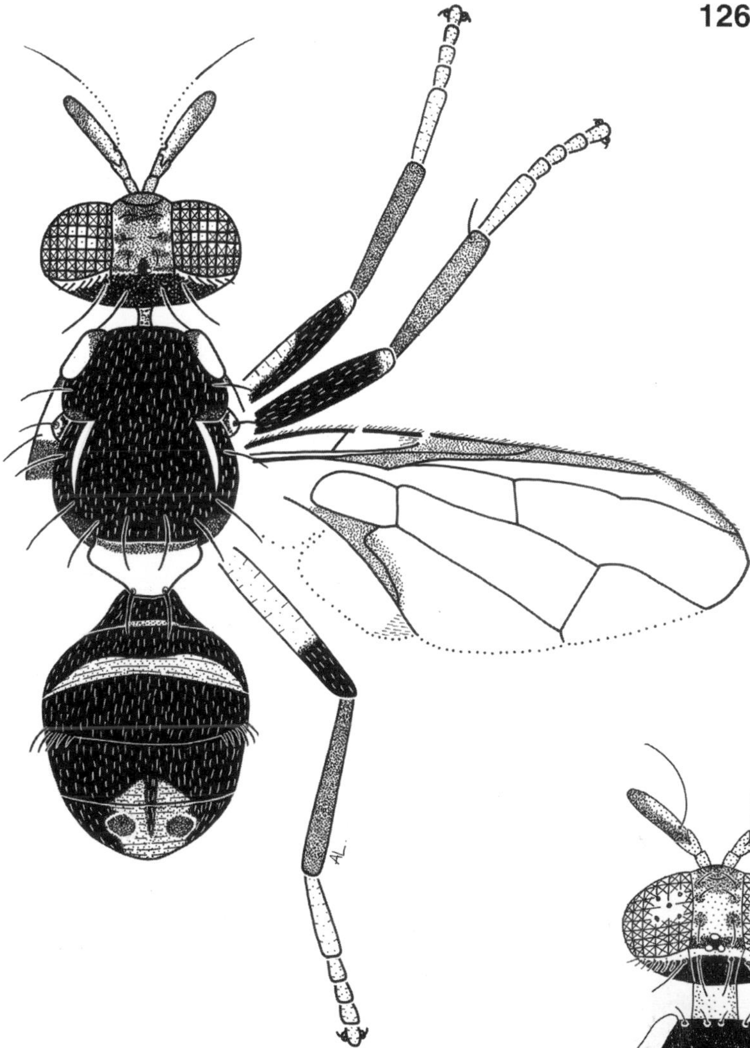

127A

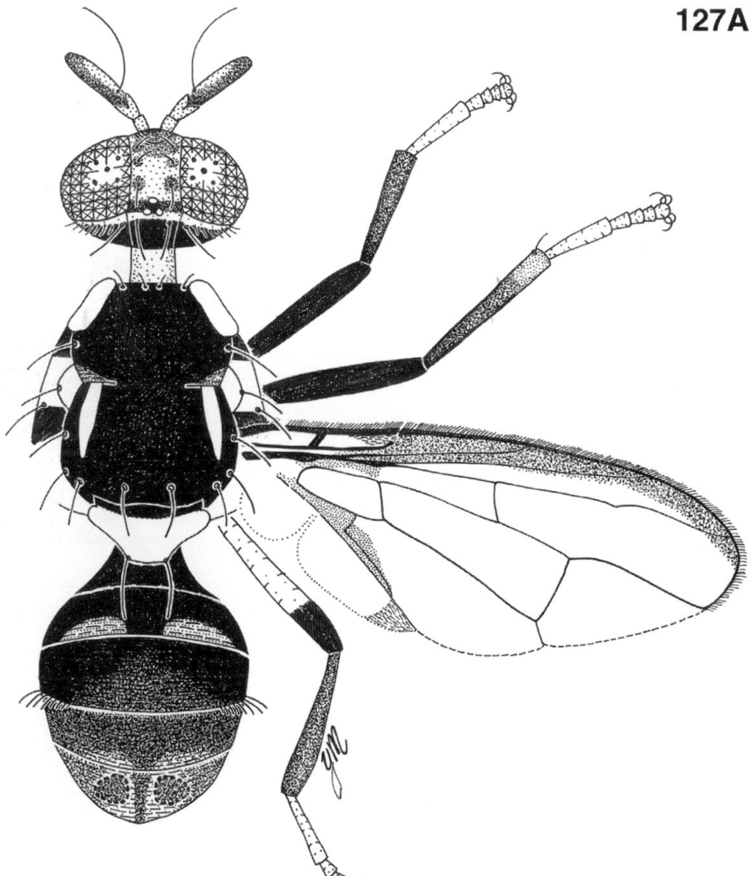

Fig. 126. *Bactrocera (Bactrocera) nigrofemoralis* White & Tsuruta, male. **Fig. 127A.** *Bactrocera (Bactrocera) nigrotibialis* (Perkins), lectotype male.

127B

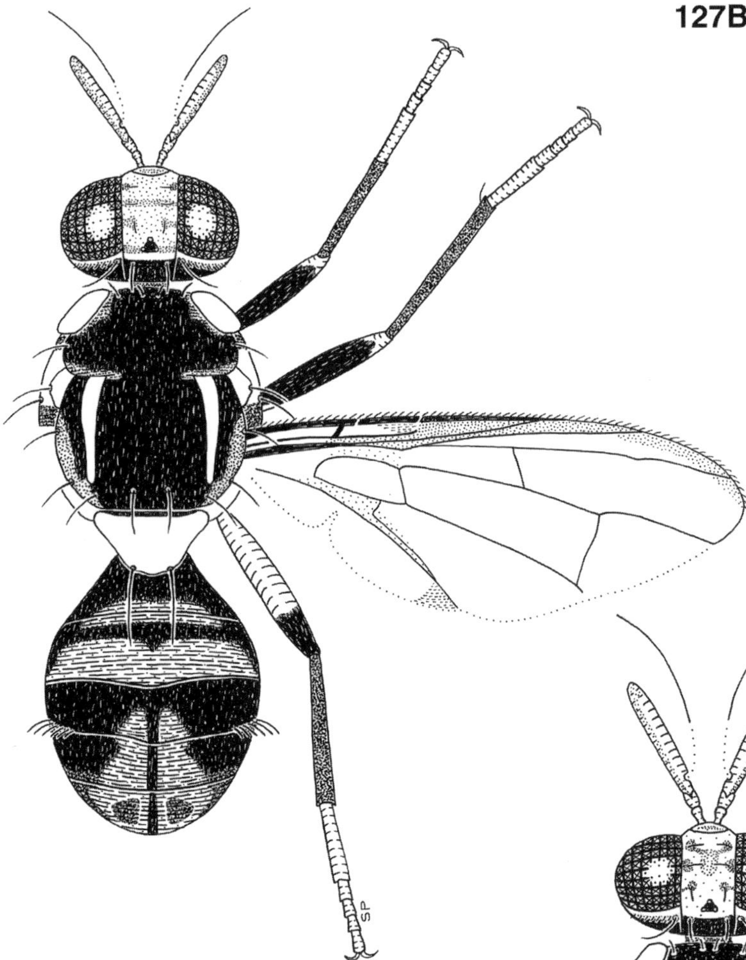

127C

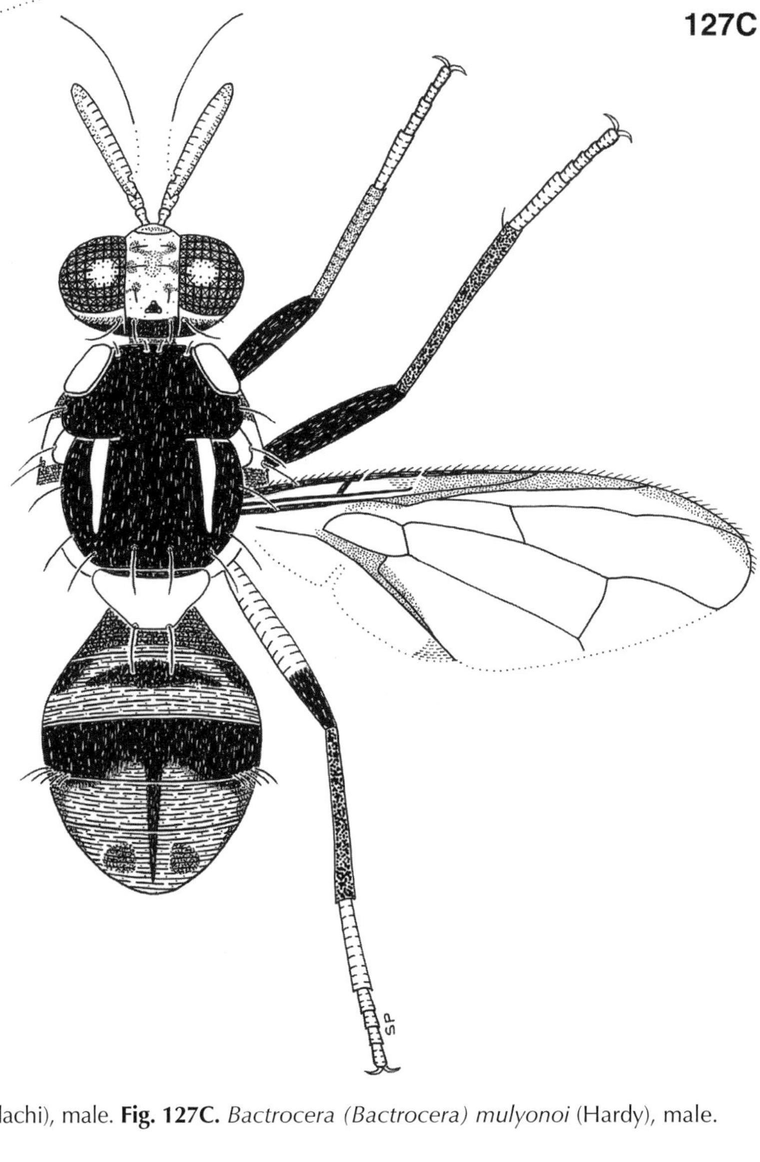

Fig. 127B. *Bactrocera (Bactrocera) luzonae* (Hardy & Adachi), male. **Fig. 127C.** *Bactrocera (Bactrocera) mulyonoi* (Hardy), male.

127D

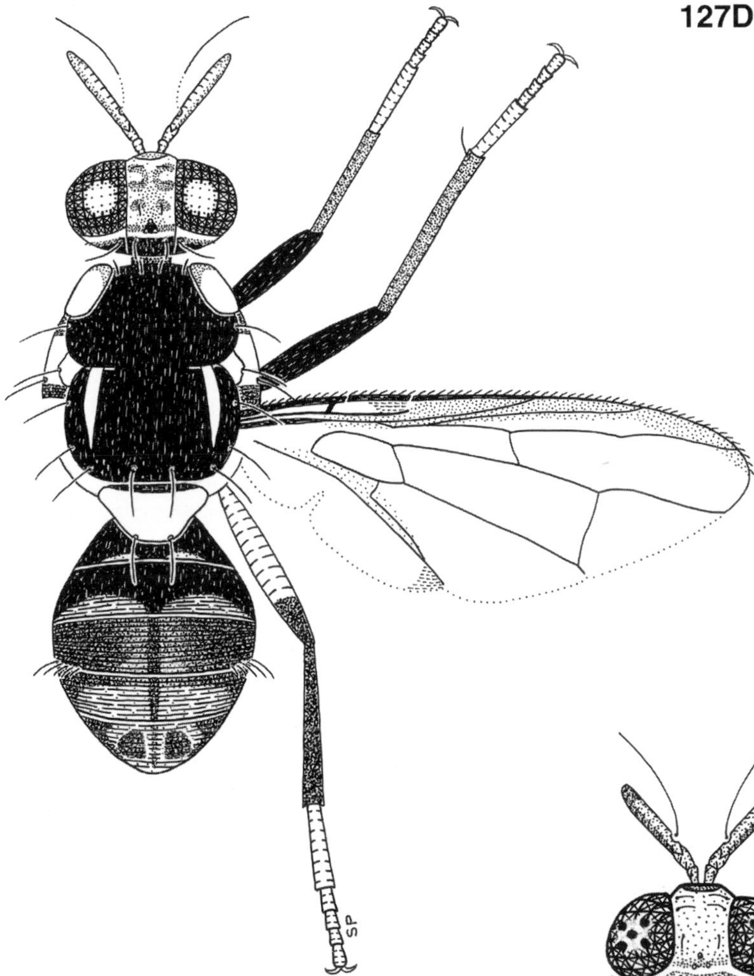

128

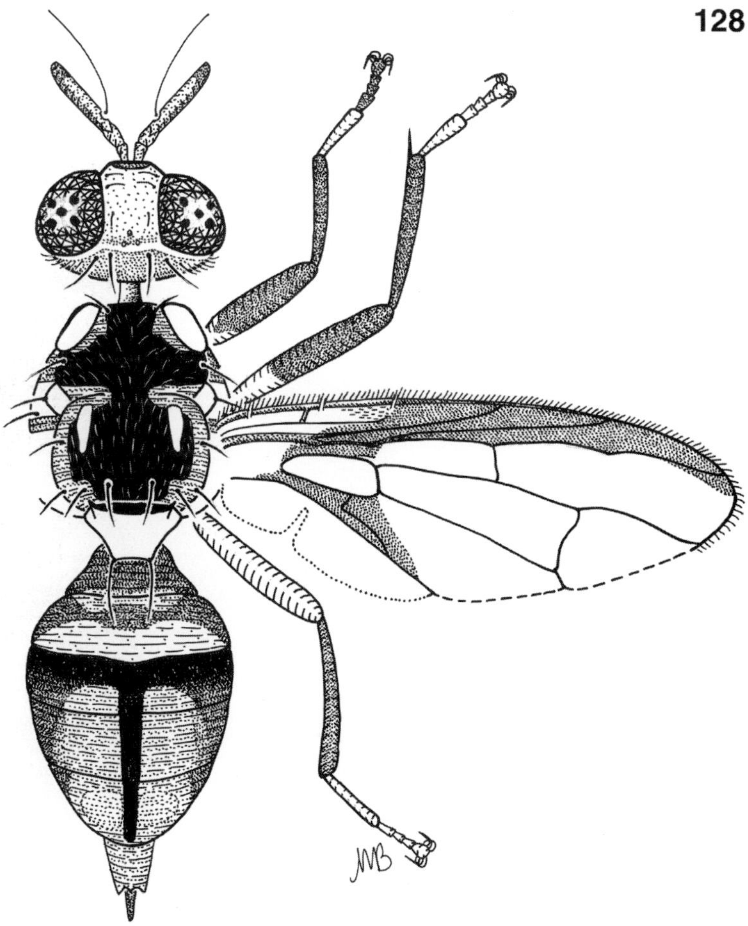

Fig. 127D. *Bactrocera (Bactrocera) sumatranus* (Hardy), paratype male. **Fig. 128.** *Bactrocera (Bactrocera) nothaphoebe* new species, holotype female.

129A

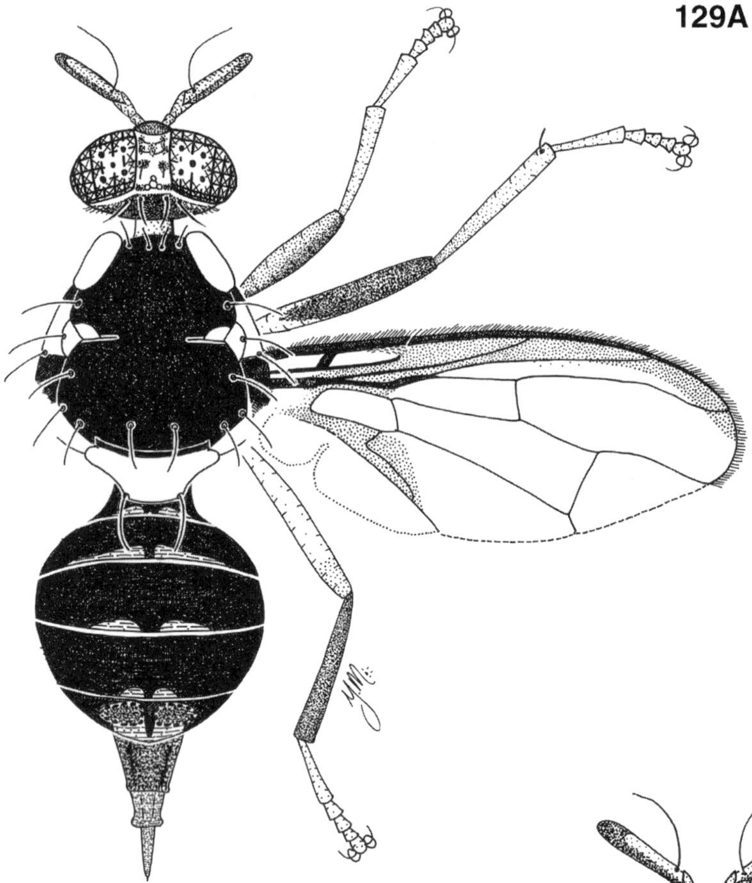

129B

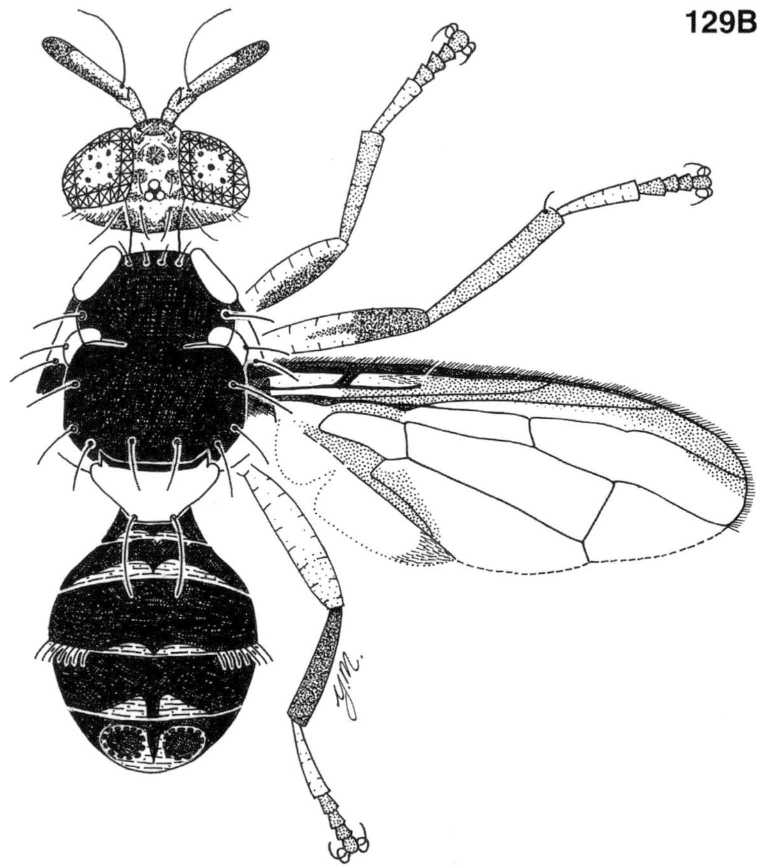

Fig. 129A. *Bactrocera (Bactrocera) obscurata* (de Meijere), holotype female. **Fig. 129B.** *Bactrocera (Bactrocera) heppneri* White, holotype male.

130

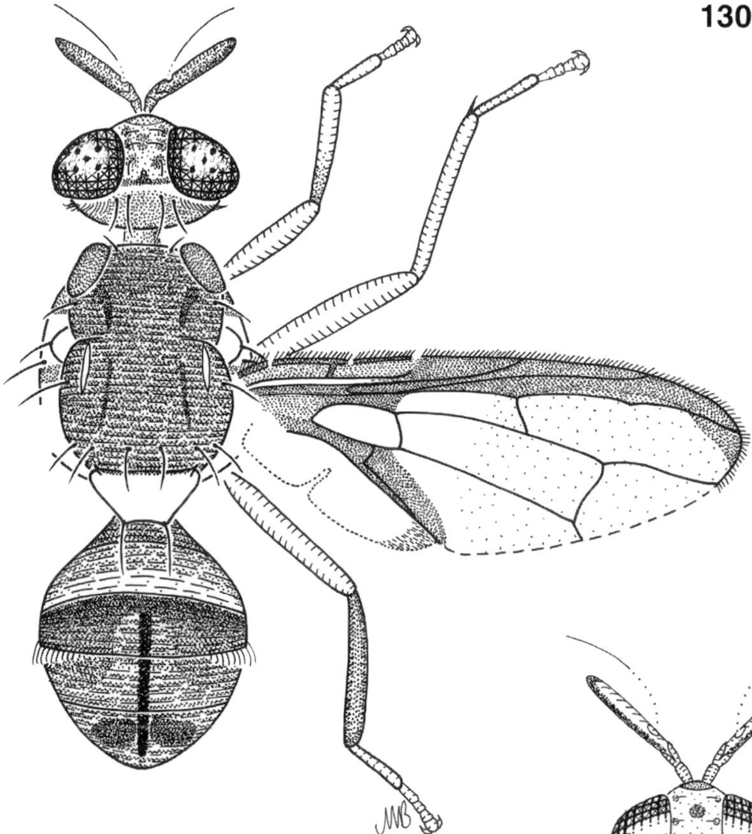

131

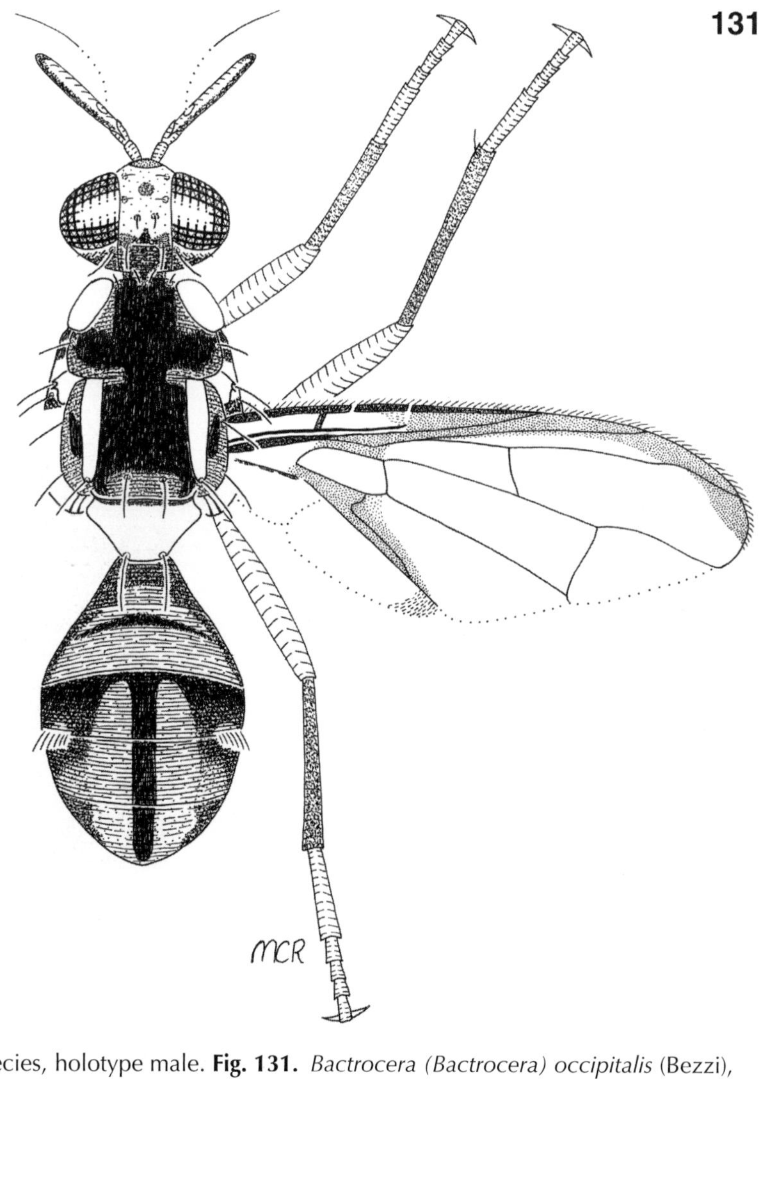

Fig. 130. *Bactrocera (Bactrocera) obscurivitta* new species, holotype male. **Fig. 131.** *Bactrocera (Bactrocera) occipitalis* (Bezzi), lectotype male.

132

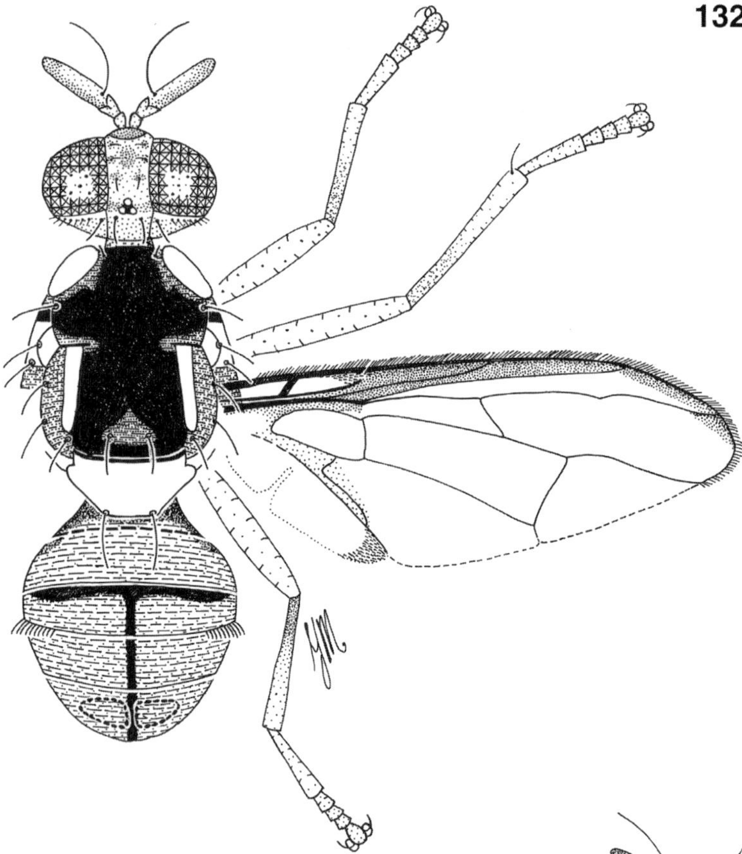

133

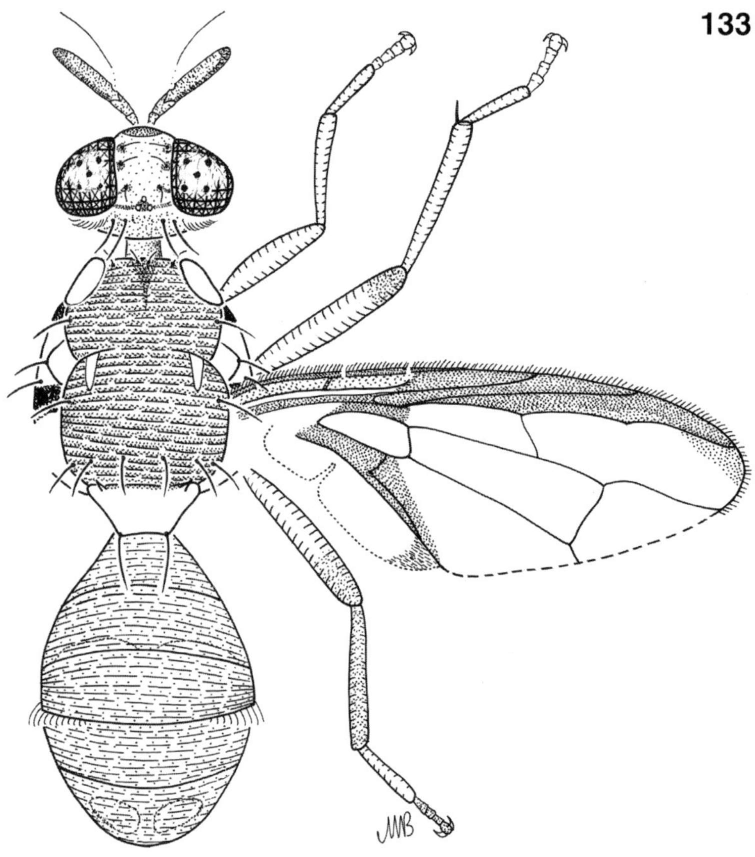

Fig. 132. *Bactrocera (Bactrocera) ochroma* new species, holotype male. **Fig. 133.** *Bactrocera (Bactrocera) ochroventer* new species, holotype male.

134

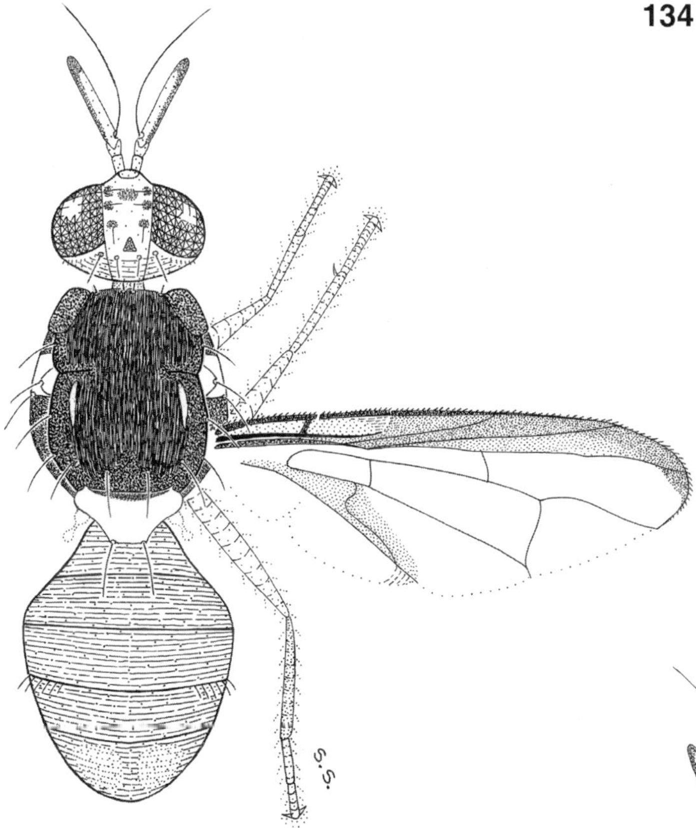

135

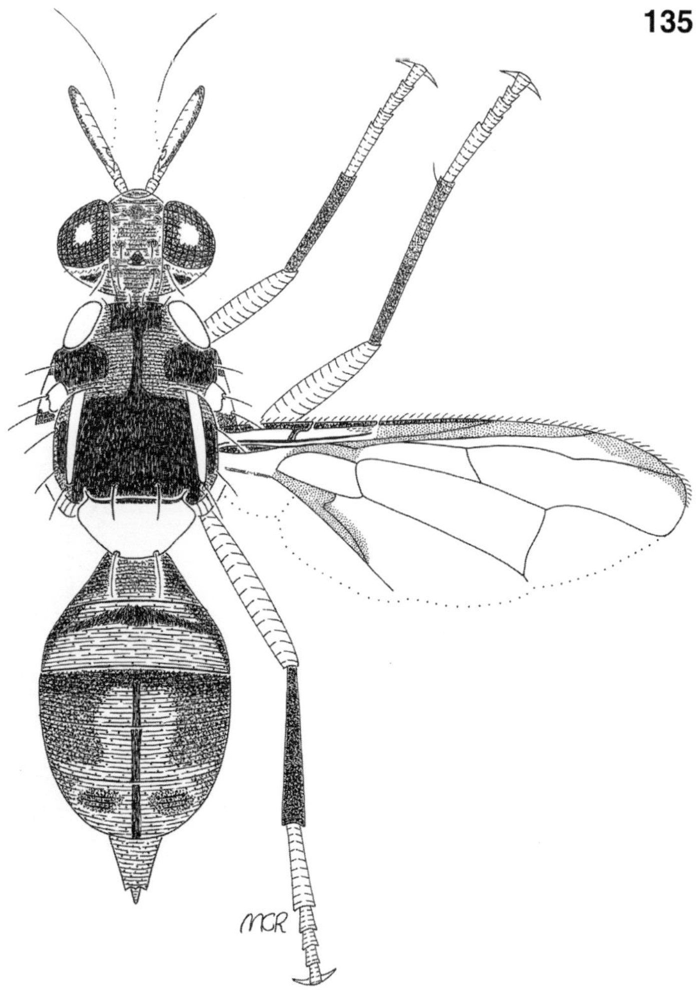

Fig. 134. *Bactrocera (Bactrocera) opacovitta* new species, holotype male. **Fig. 135.** *Bactrocera (Bactrocera) osbeckiae* Drew & Hancock, holotype female.

136

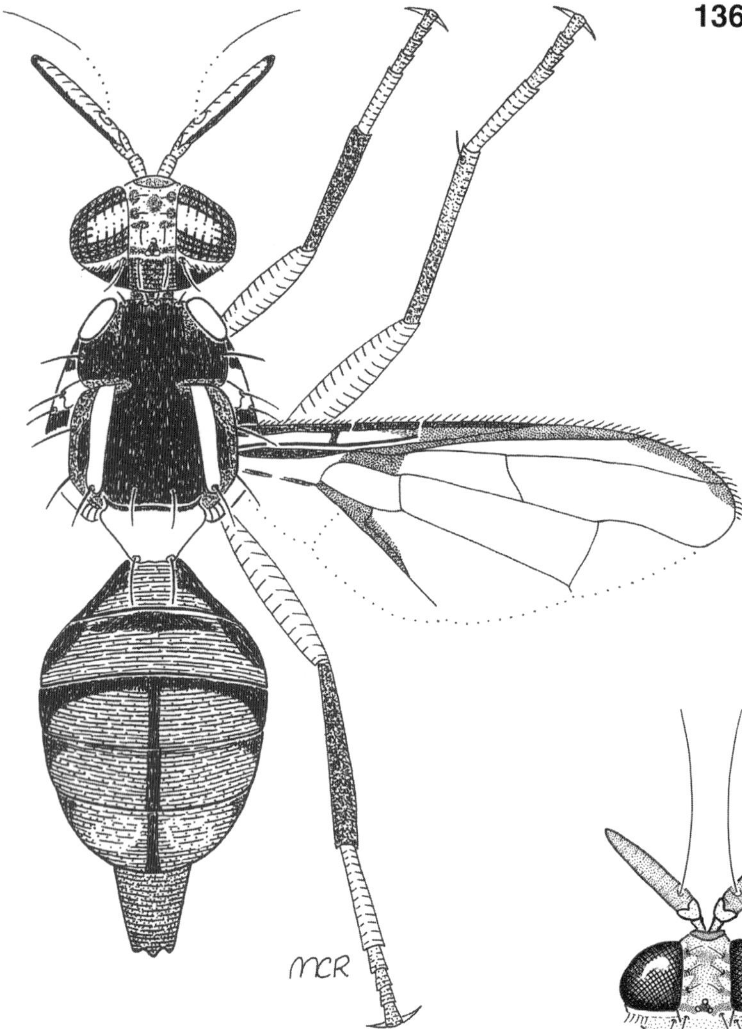

137

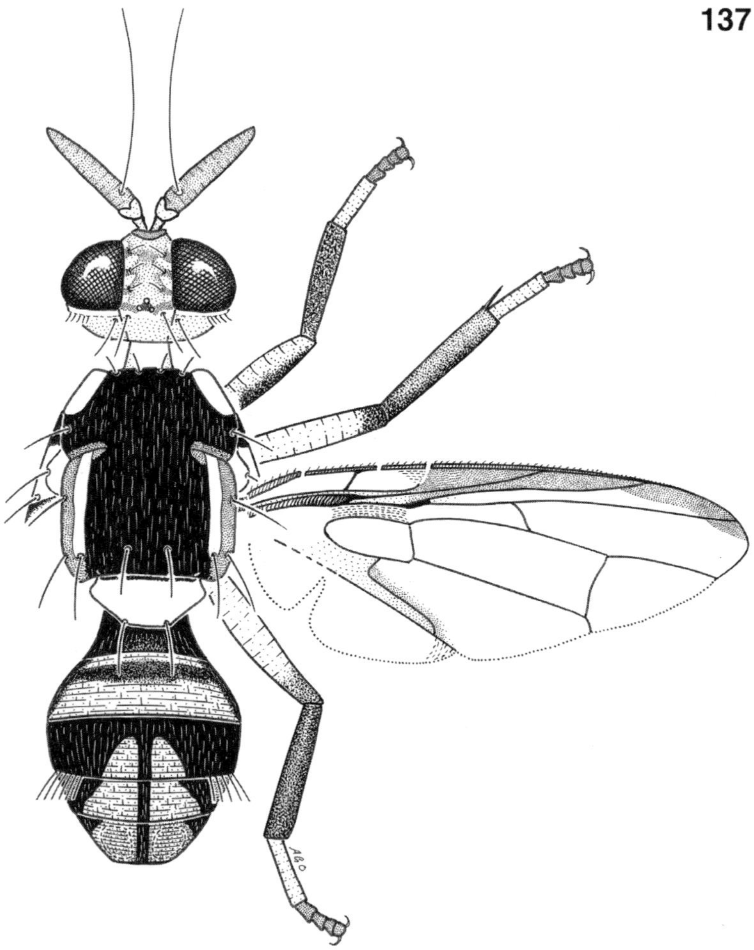

Fig. 136. *Bactrocera (Bactrocera) papayae* Drew & Hancock, holotype female. **Fig. 137.** *Bactrocera (Bactrocera) paraarecae* new species, holotype male.

138

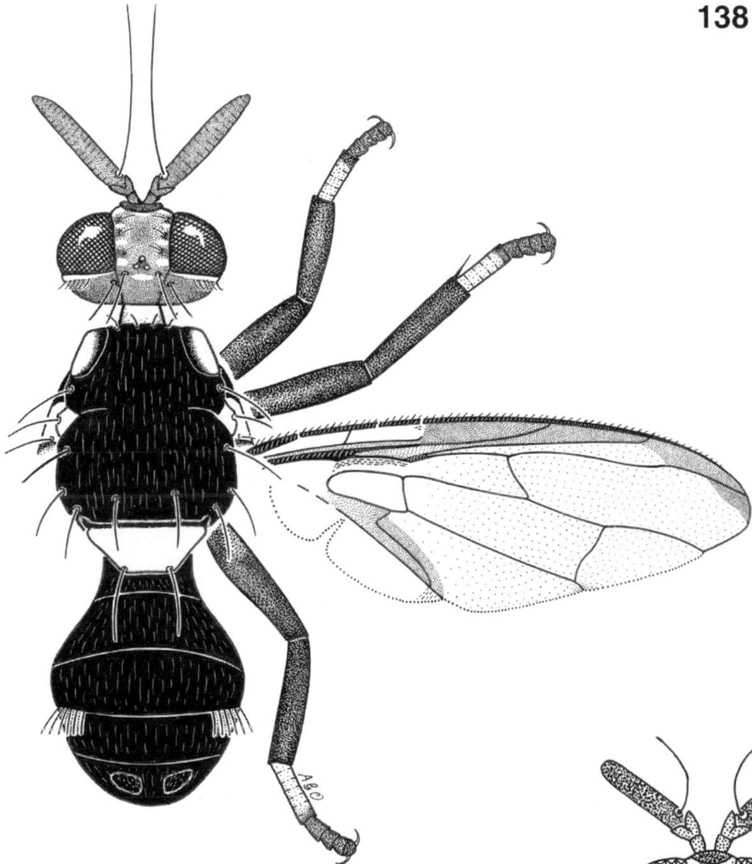

139

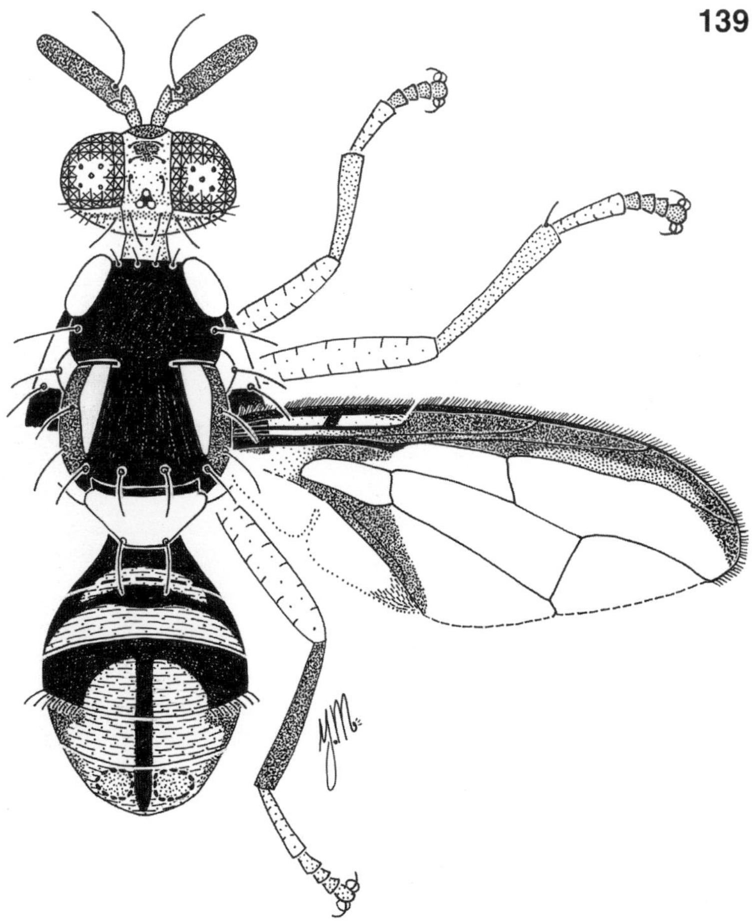

Fig. 138. *Bactrocera (Bactrocera) paradiospyri* Chen, Zhou & Li, male. **Fig. 139.** *Bactrocera (Bactrocera) paralatissima* new species, holotype male.

140

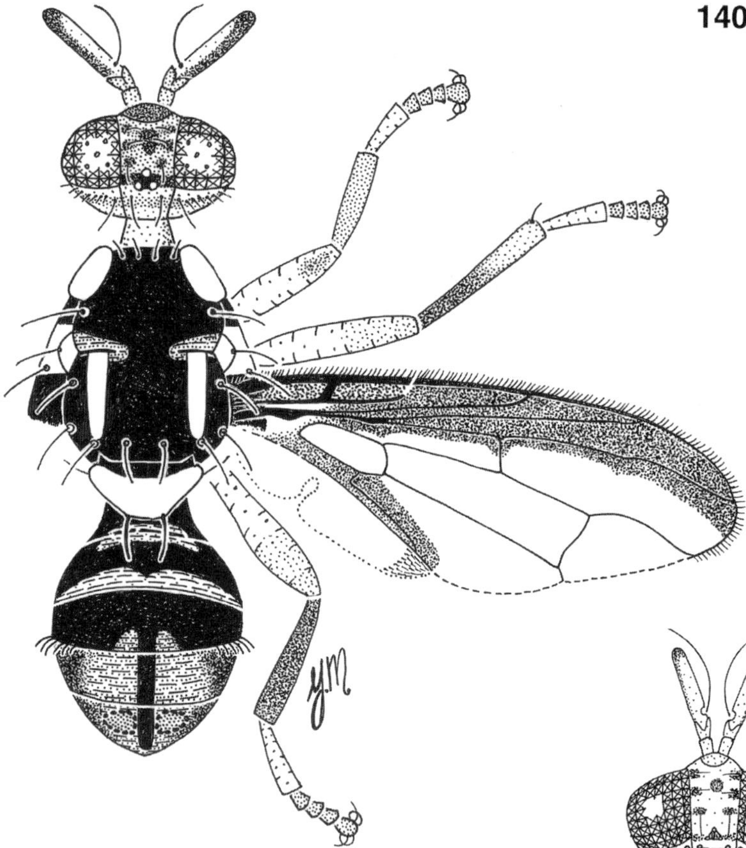

141

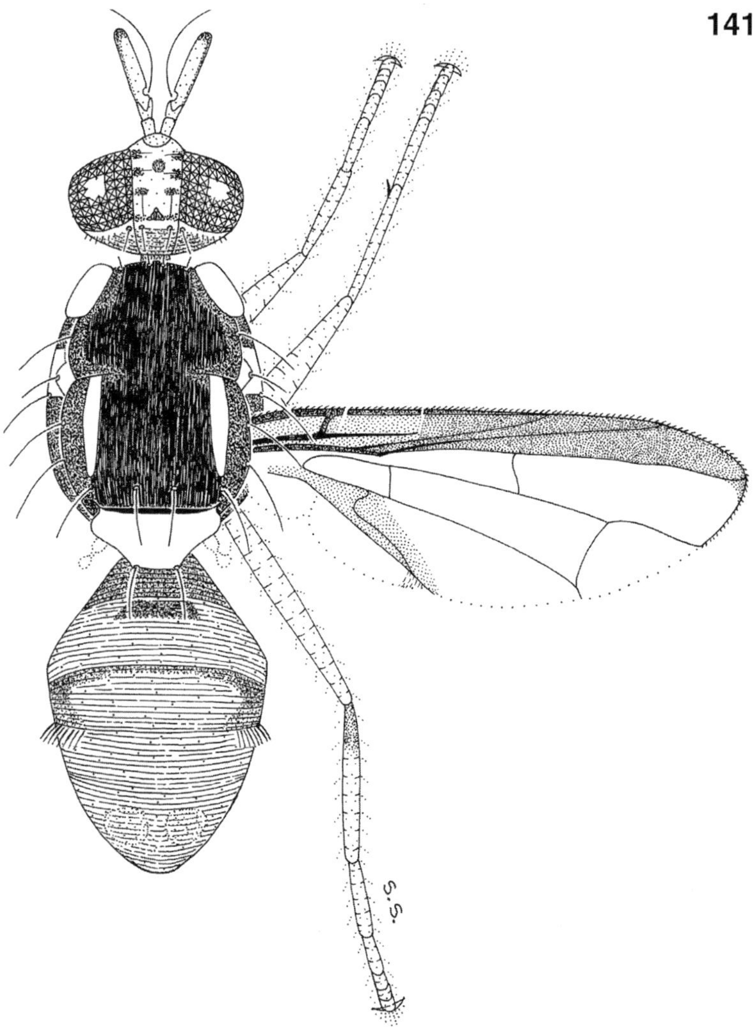

Fig. 140. *Bactrocera (Bactrocera) paralimbifera* new species, holotype male. **Fig. 141.** *Bactrocera (Bactrocera) paramusae* Drew, holotype male.

142

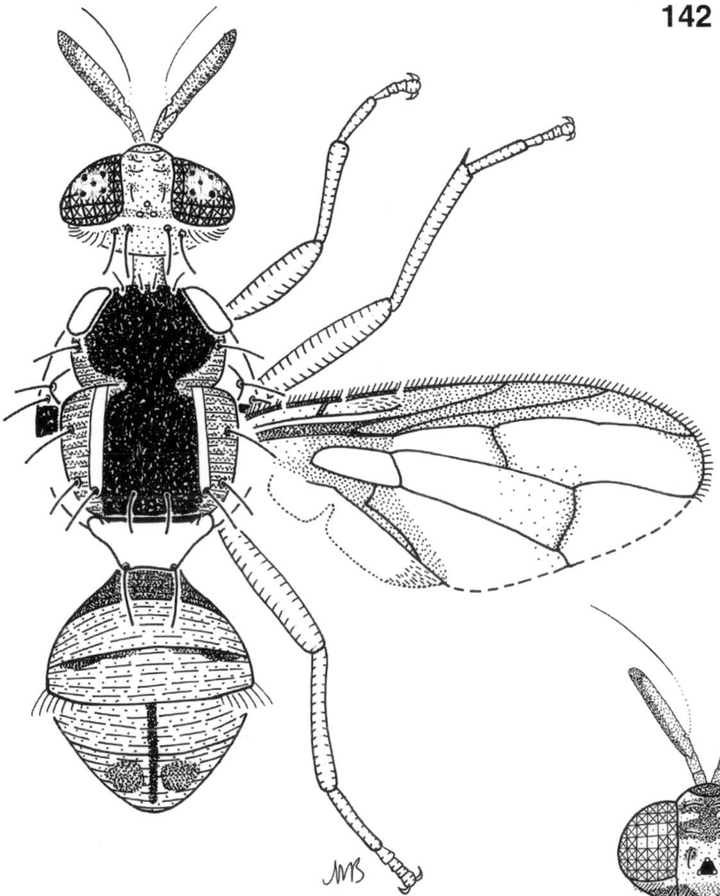

143

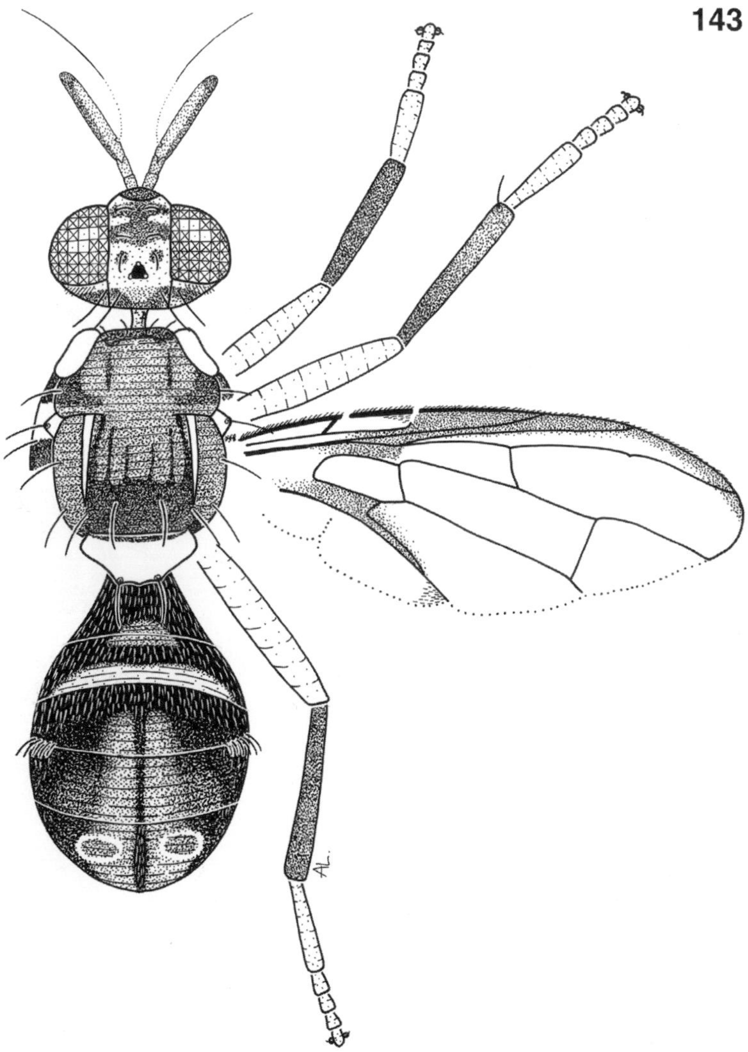

Fig. 142. *Bactrocera (Bactrocera) paranigrita* new species, holotype male. **Fig. 143.** *Bactrocera (Bactrocera) paraosbeckiae* Drew, holotype male.

144

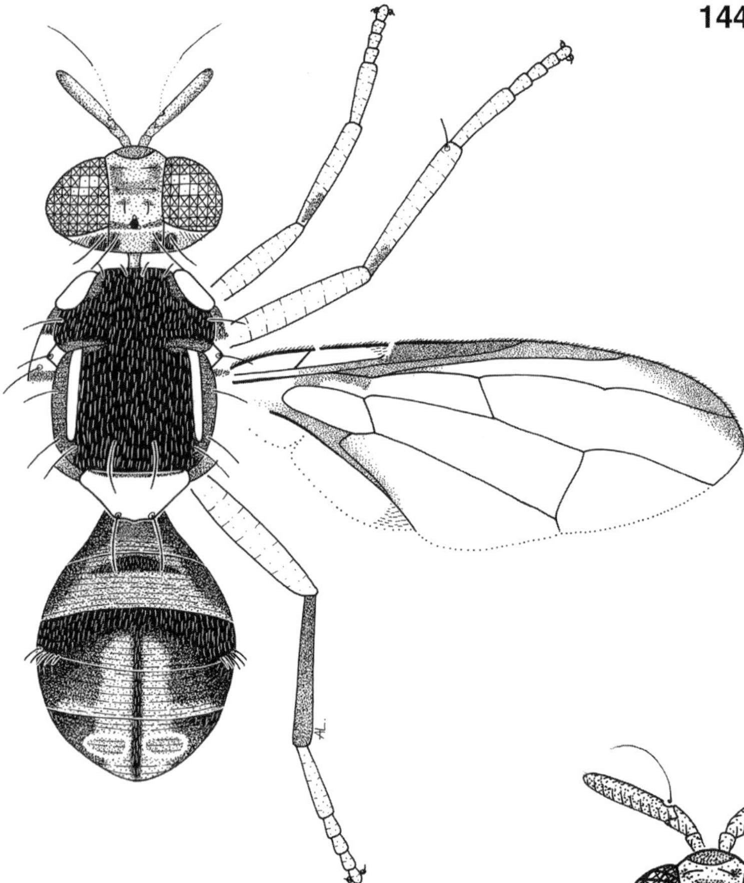

145

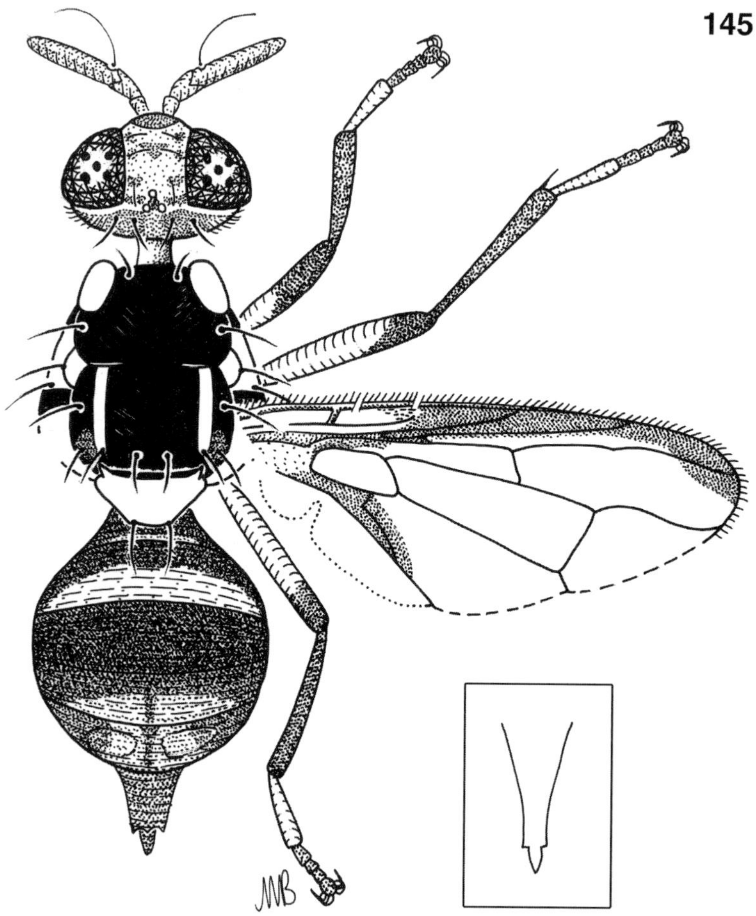

Fig. 144. *Bactrocera (Bactrocera) paraverbascifoliae* Drew, holotype male. **Fig. 145.** *Bactrocera (Bactrocera) parvula* (Hendel), syntype female and apex of piercer from female ovipositor.

146

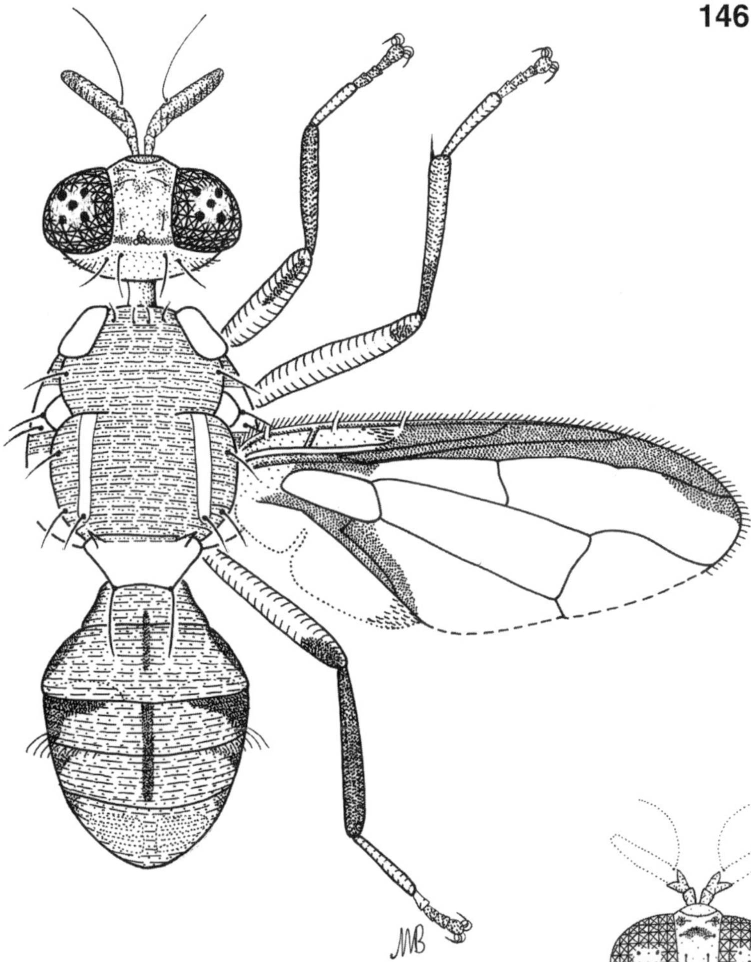

147

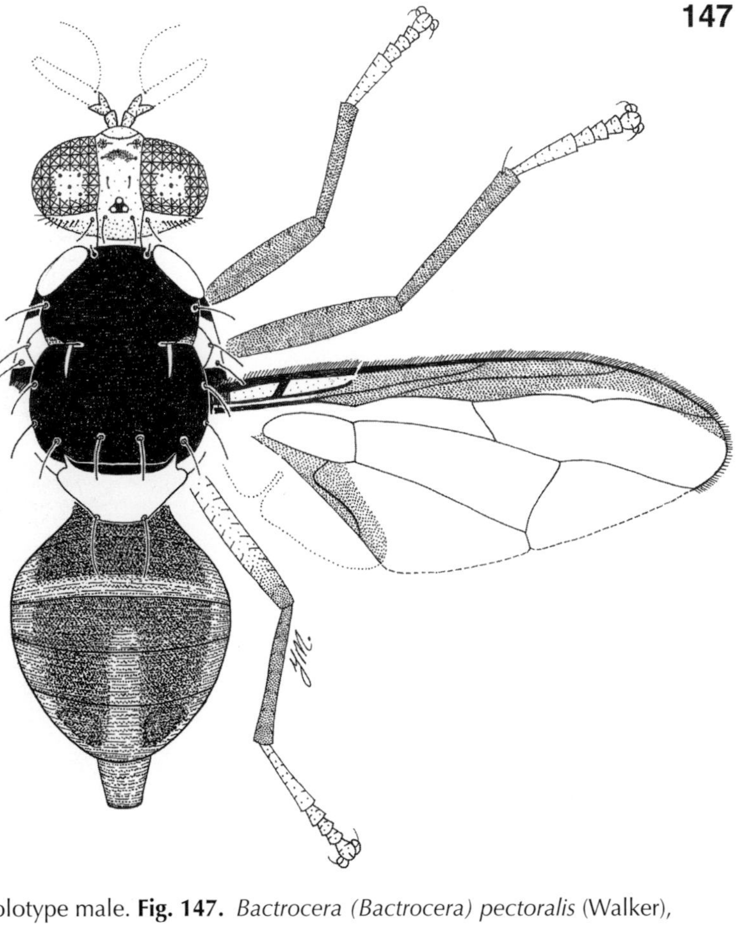

Fig. 146. *Bactrocera (Bactrocera) patula* new species, holotype male. **Fig. 147.** *Bactrocera (Bactrocera) pectoralis* (Walker), lectotype female.

148

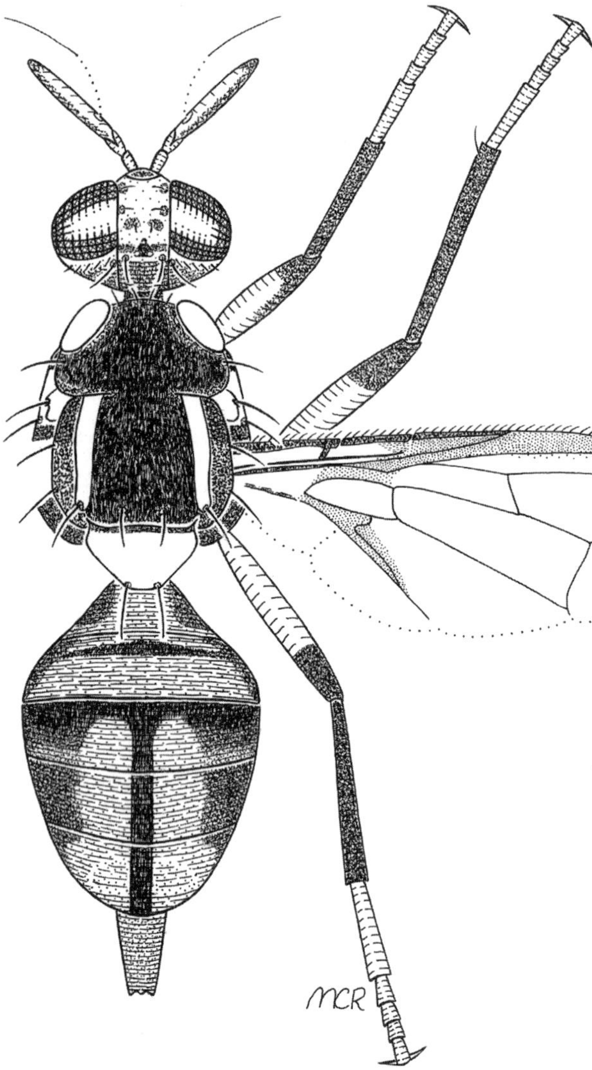

149

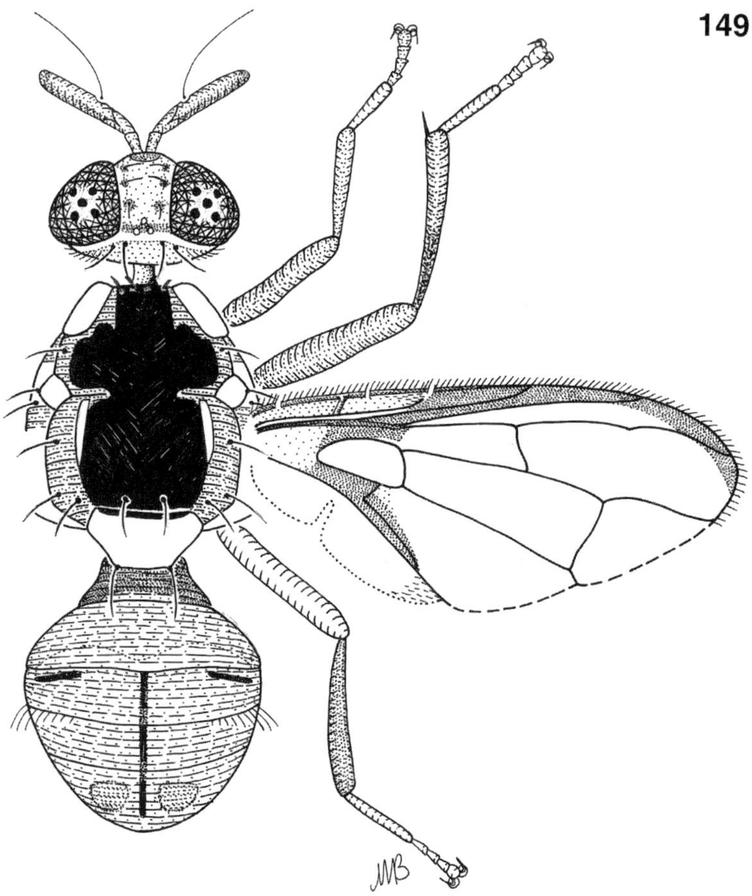

Fig. 148. *Bactrocera (Bactrocera) pedestris* (Bezzi), lectotype female. **Fig. 149.** *Bactrocera (Bactrocera) peneallwoodi* new species, holotype male.

150

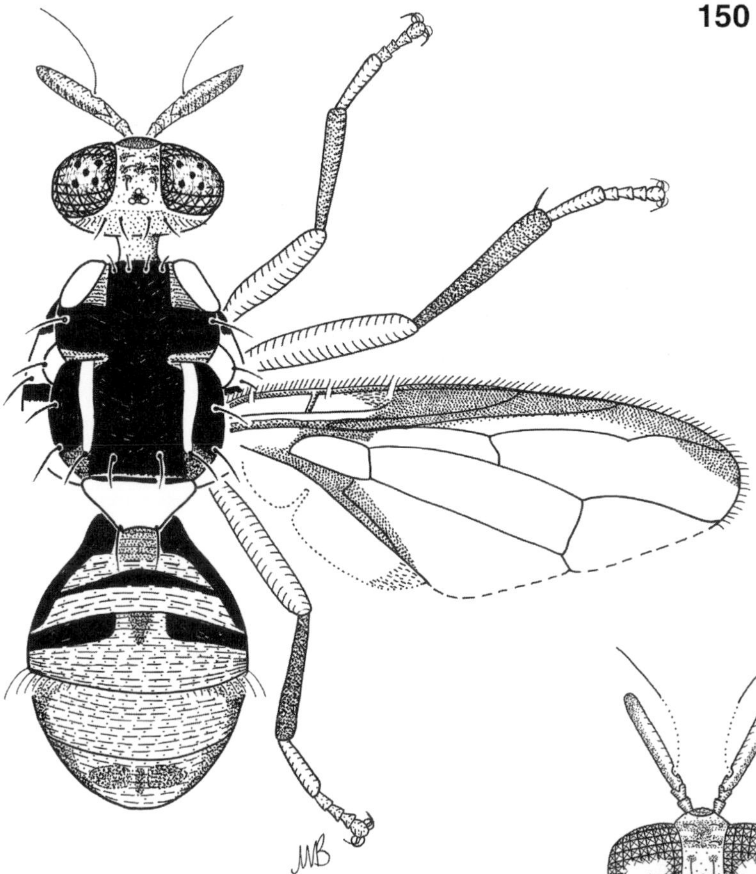

151

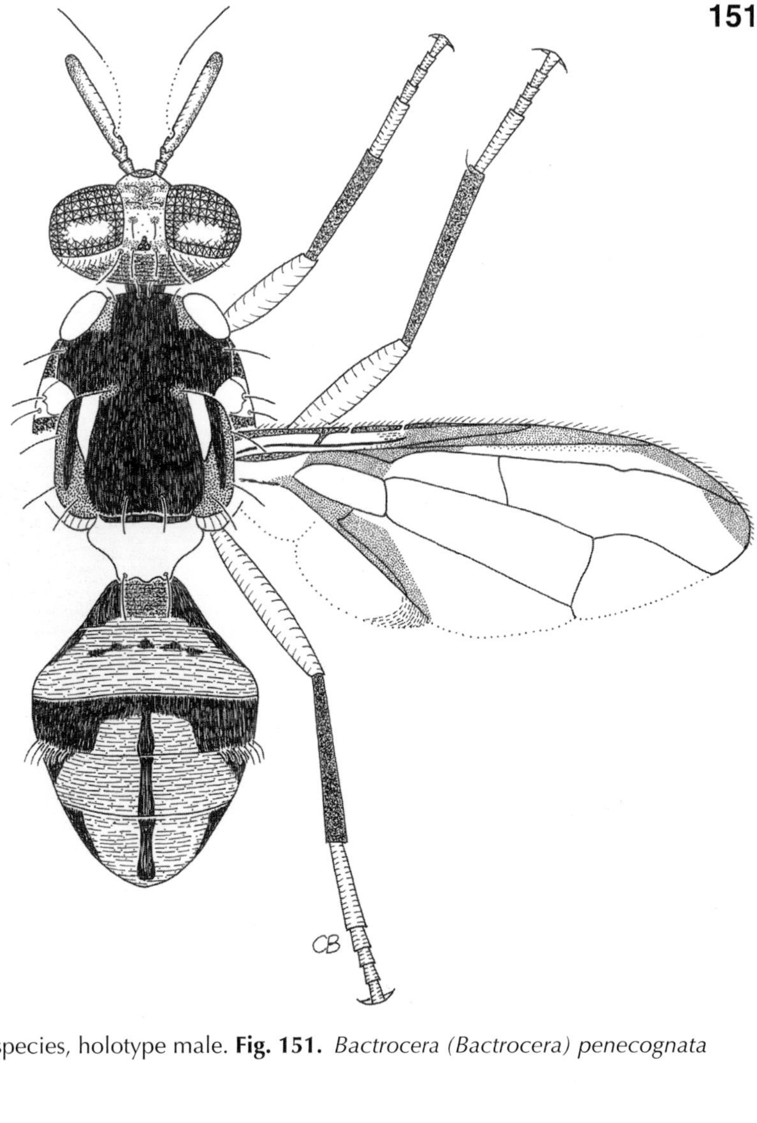

Fig. 150. *Bactrocera (Bactrocera) penebeckerae* new species, holotype male. **Fig. 151.** *Bactrocera (Bactrocera) penecognata* Drew & Hancock, holotype male.

152

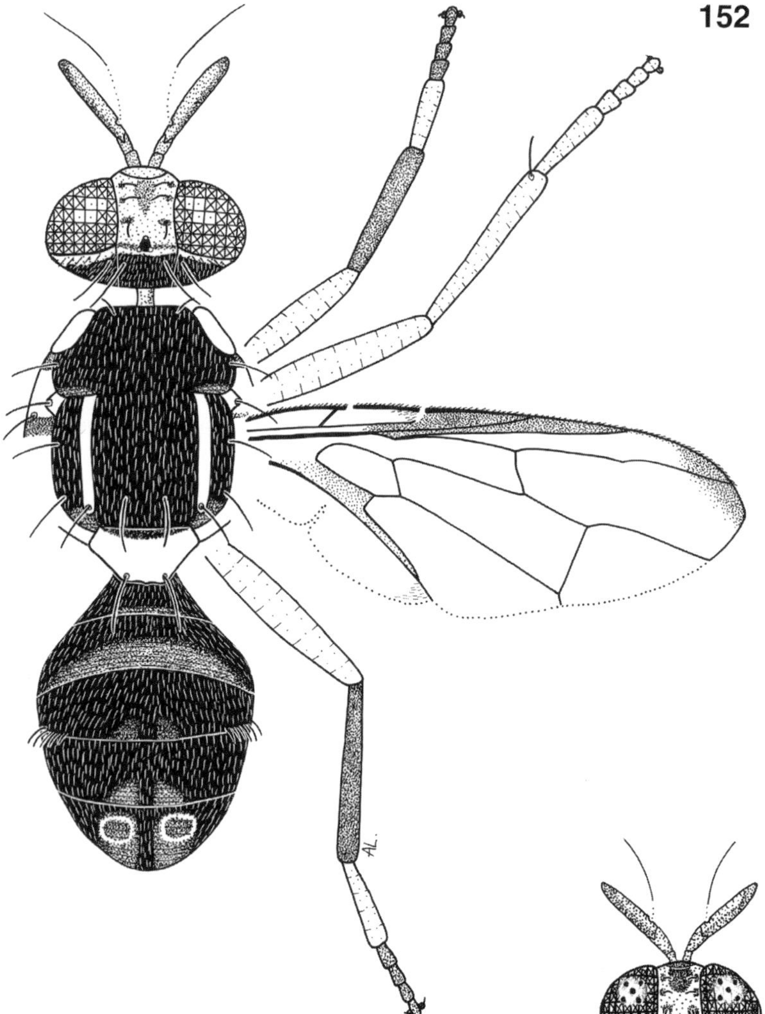

153

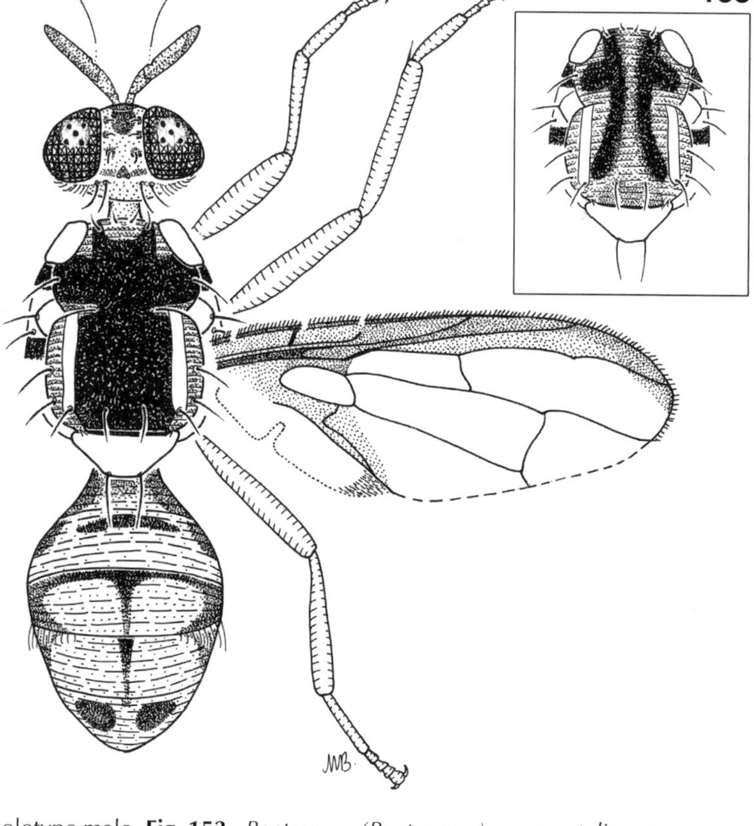

Fig. 152. *Bactrocera (Bactrocera) penecorrecta* Drew, holotype male. **Fig. 153.** *Bactrocera (Bactrocera) penecostalis* new species, holotype male and scutum colour variation.

154

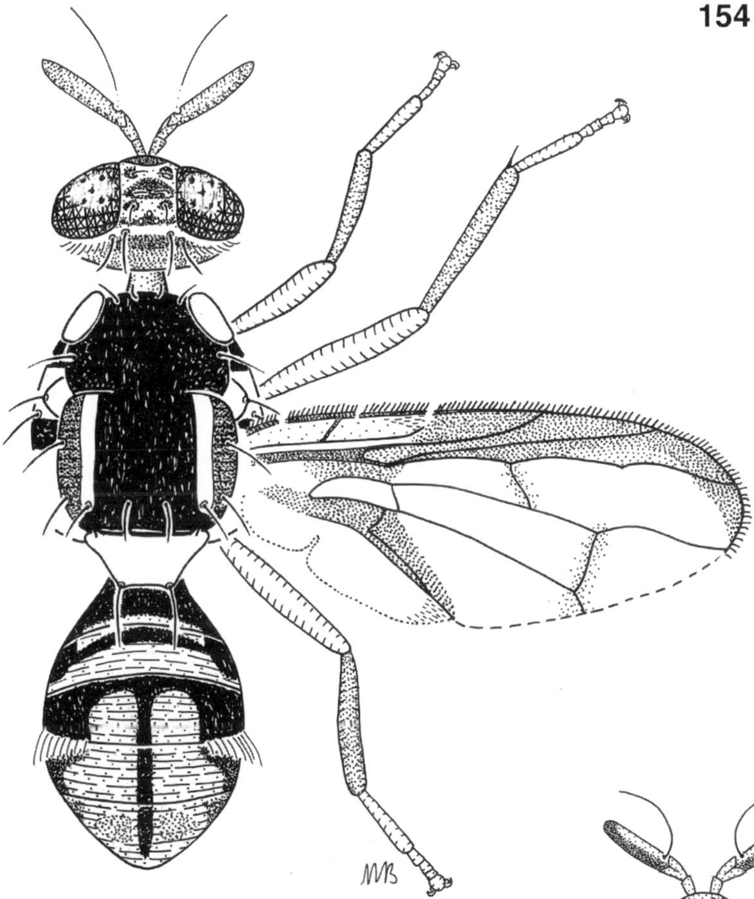

155

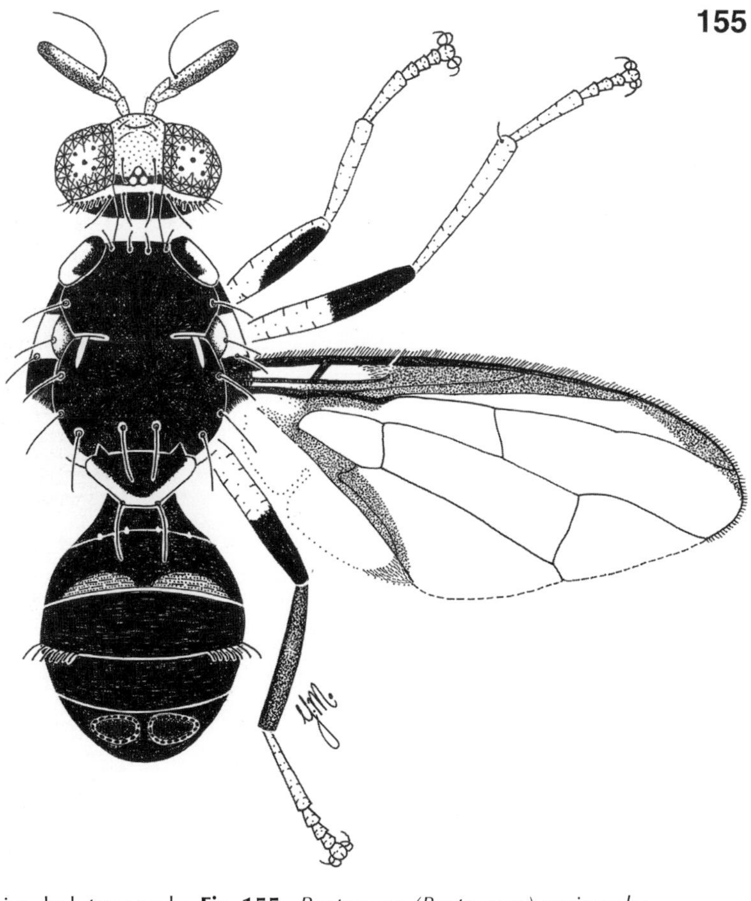

Fig. 154. *Bactrocera (Bactrocera) penephaea* new species, holotype male. **Fig. 155.** *Bactrocera (Bactrocera) perigrapha* White & Tsuruta, holotype male.

156

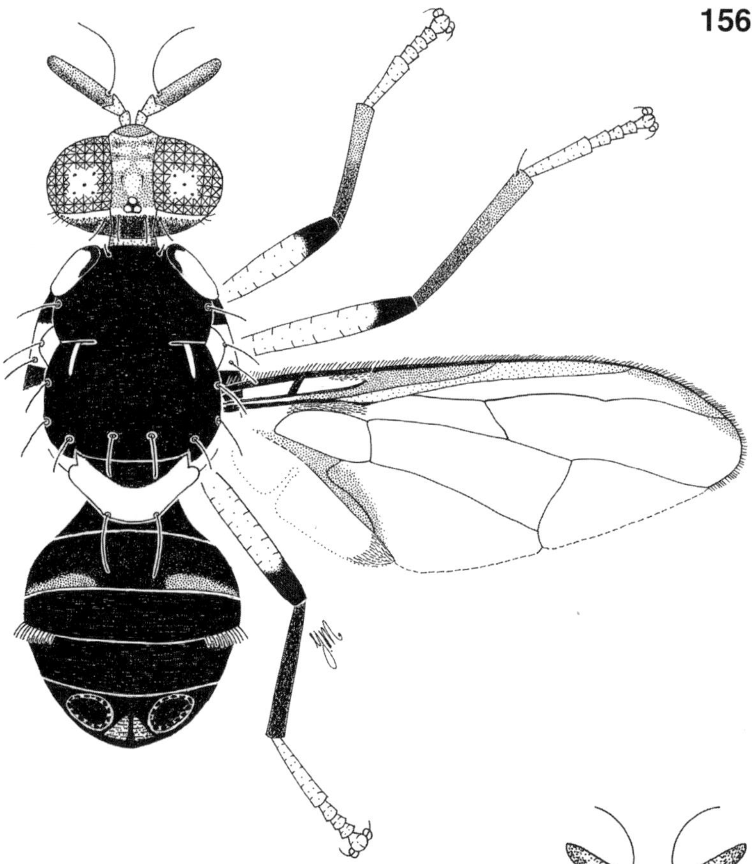

157

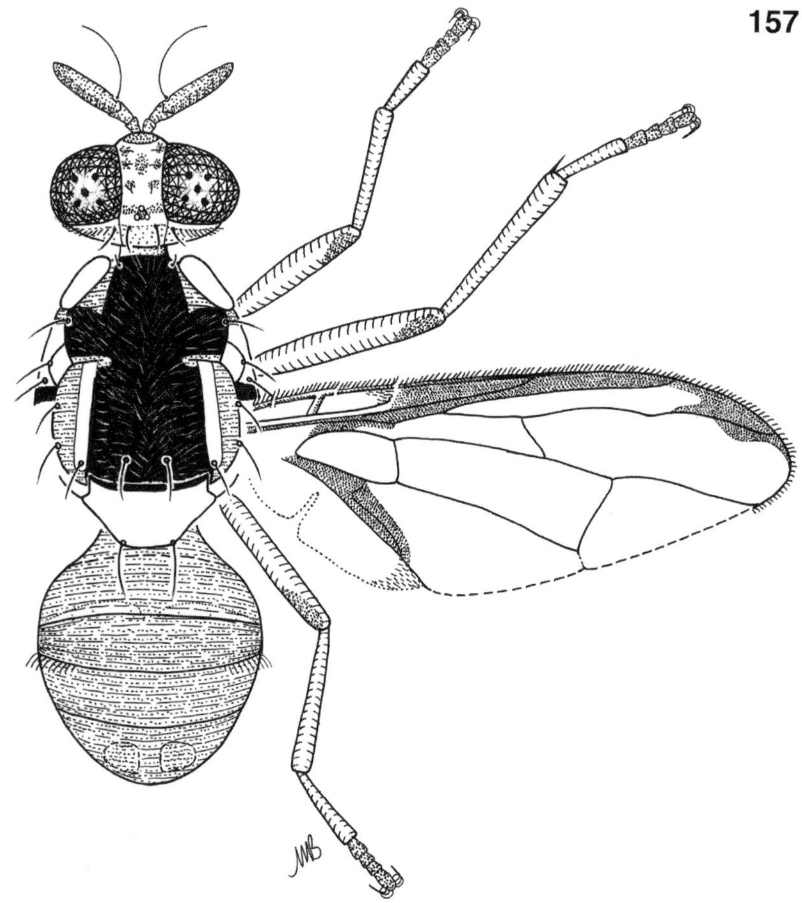

Fig. 156. *Bactrocera (Bactrocera) pernigra* (Ito), male. **Fig. 157.** *Bactrocera (Bactrocera) pictipennis* Lin & Zeng, male.

158

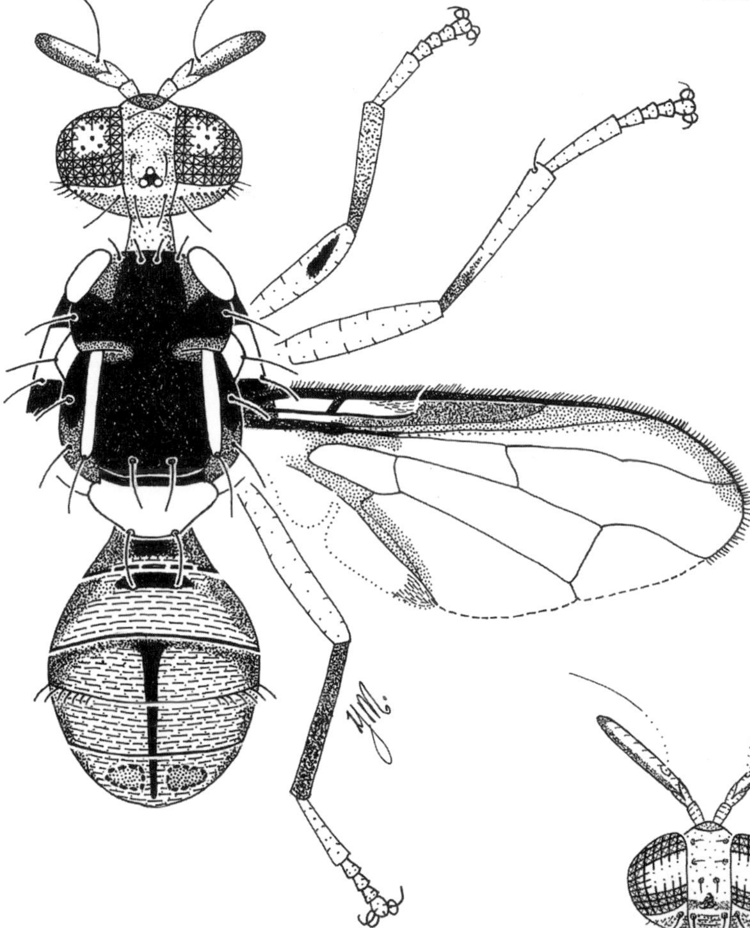

159

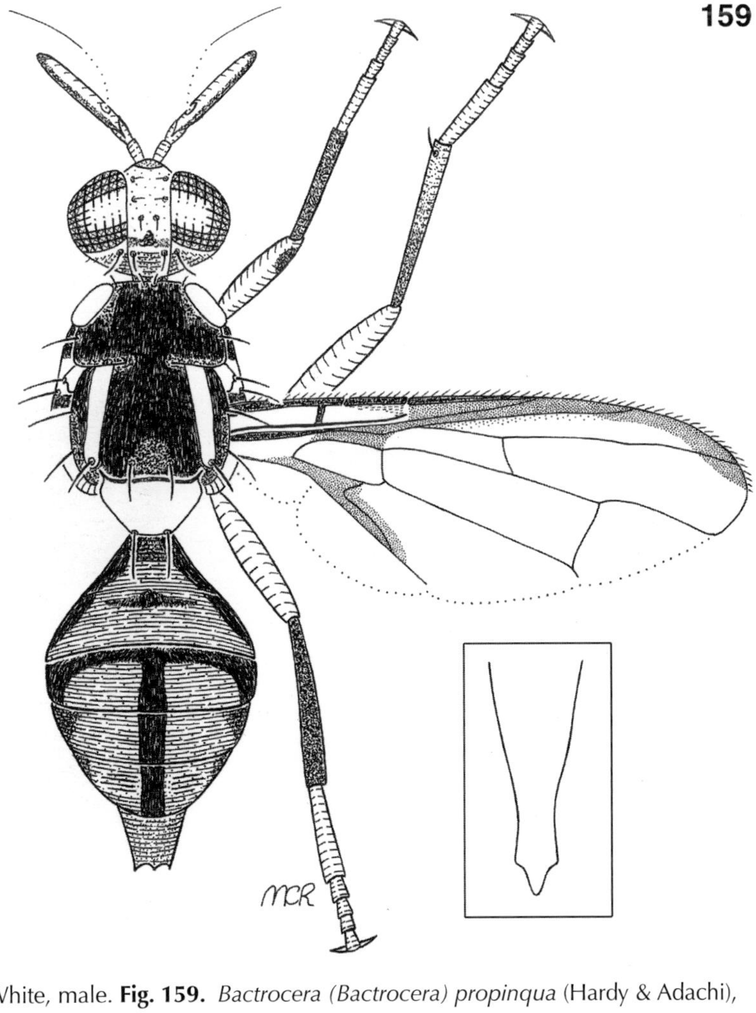

Fig. 158. *Bactrocera (Bactrocera) profunda* Tsuruta & White, male. **Fig. 159.** *Bactrocera (Bactrocera) propinqua* (Hardy & Adachi), holotype female and apex of piercer from female ovipositor.

160

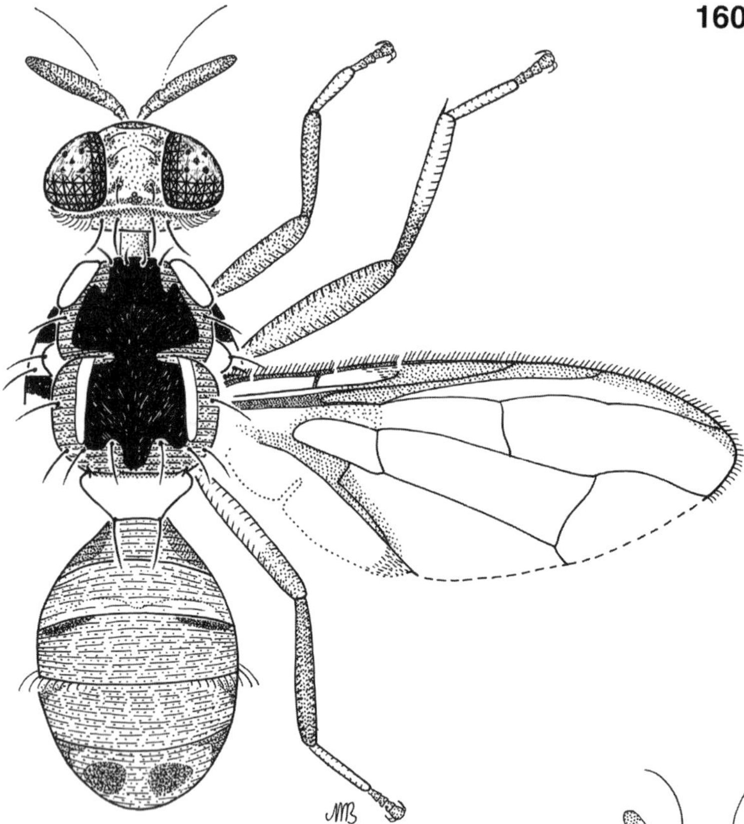

161

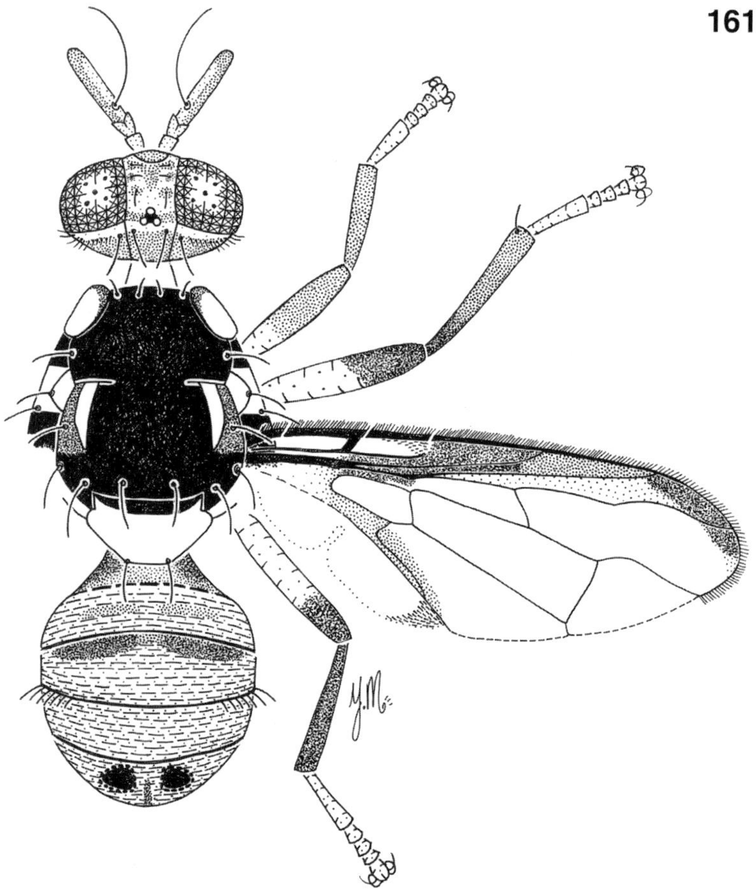

Fig. 160. *Bactrocera (Bactrocera) pruniae* new species, holotype male. **Fig. 161.** *Bactrocera (Bactrocera) pseudobeckerae* new species, holotype male.

162

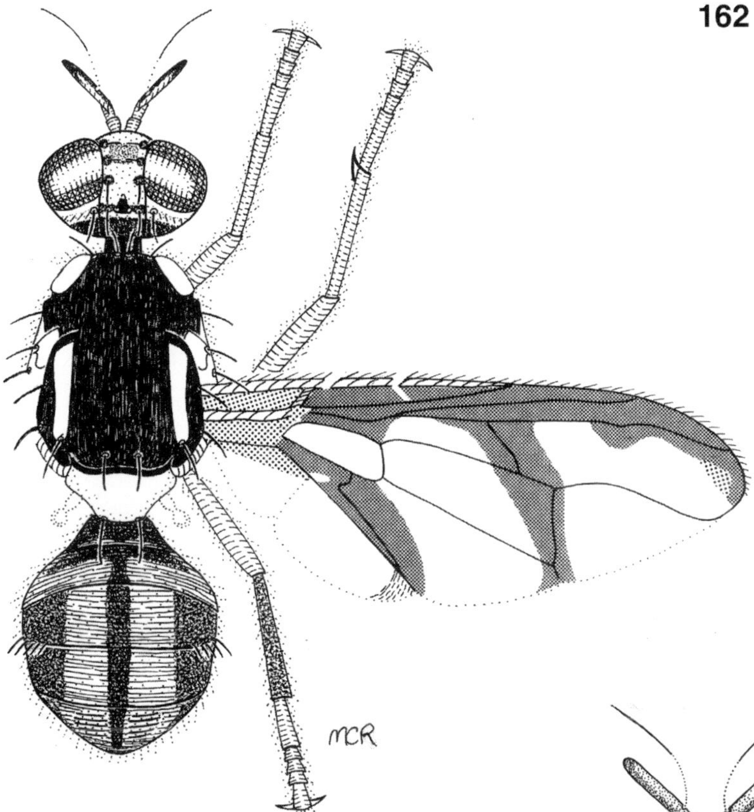

163

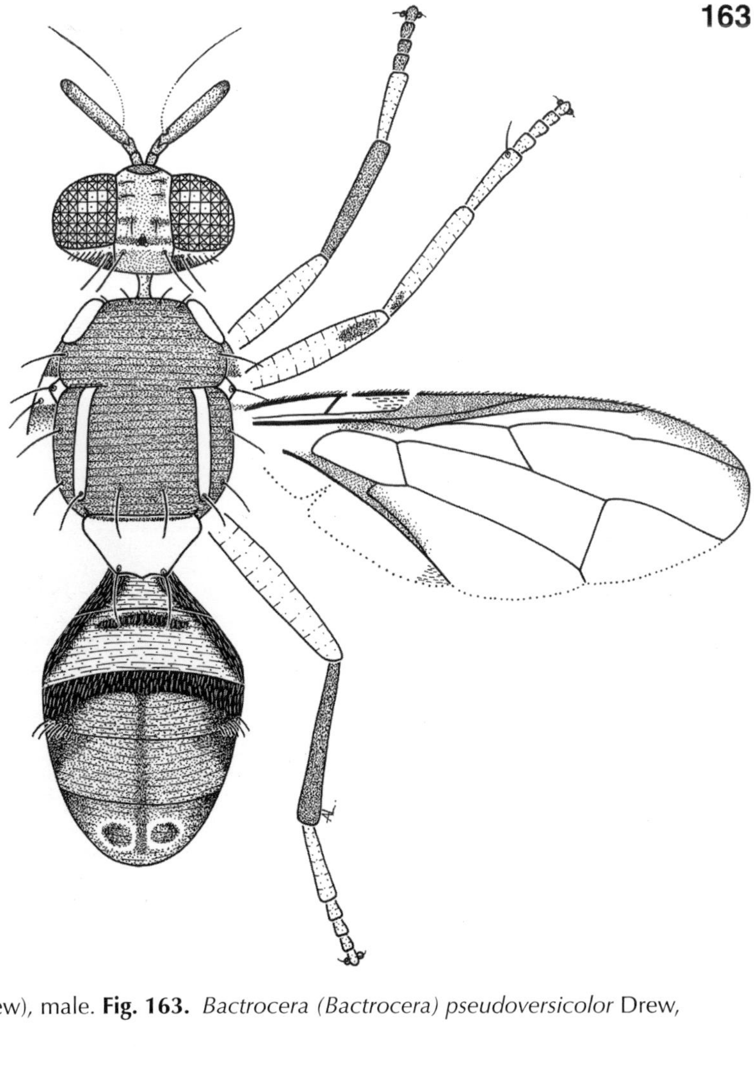

Fig. 162. *Bactrocera (Bactrocera) pseudodistincta* (Drew), male. **Fig. 163.** *Bactrocera (Bactrocera) pseudoversicolor* Drew, holotype male.

164

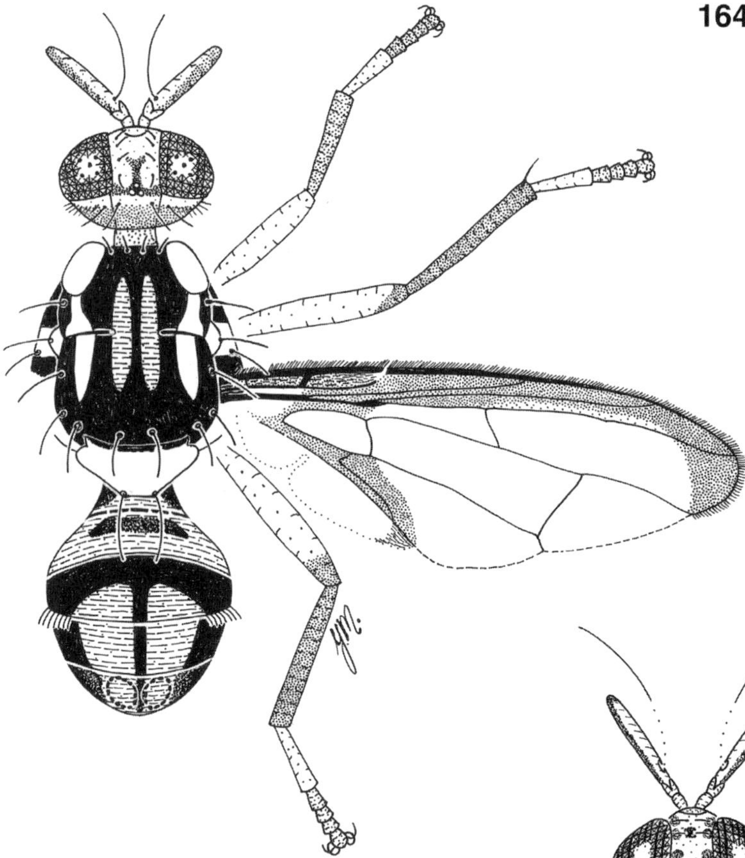

165

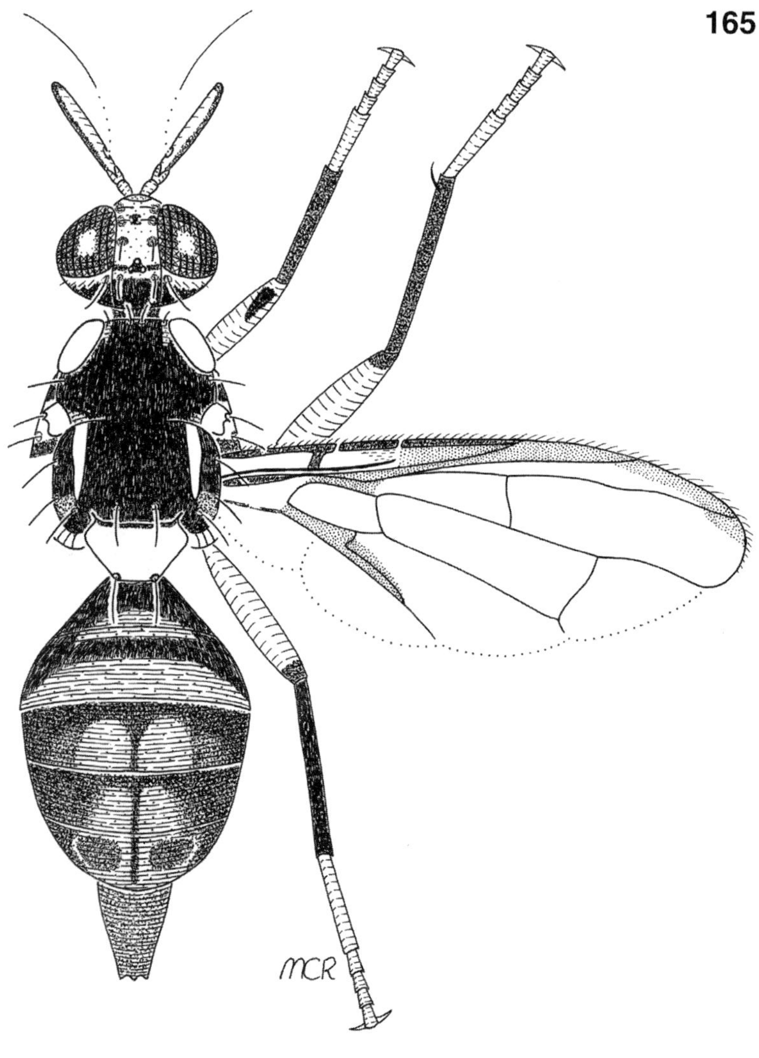

Fig. 164. *Bactrocera (Bactrocera) pusilla* (Hardy), holotype male. **Fig. 165.** *Bactrocera (Bactrocera) pyrifoliae* Drew & Hancock, holotype female.

166

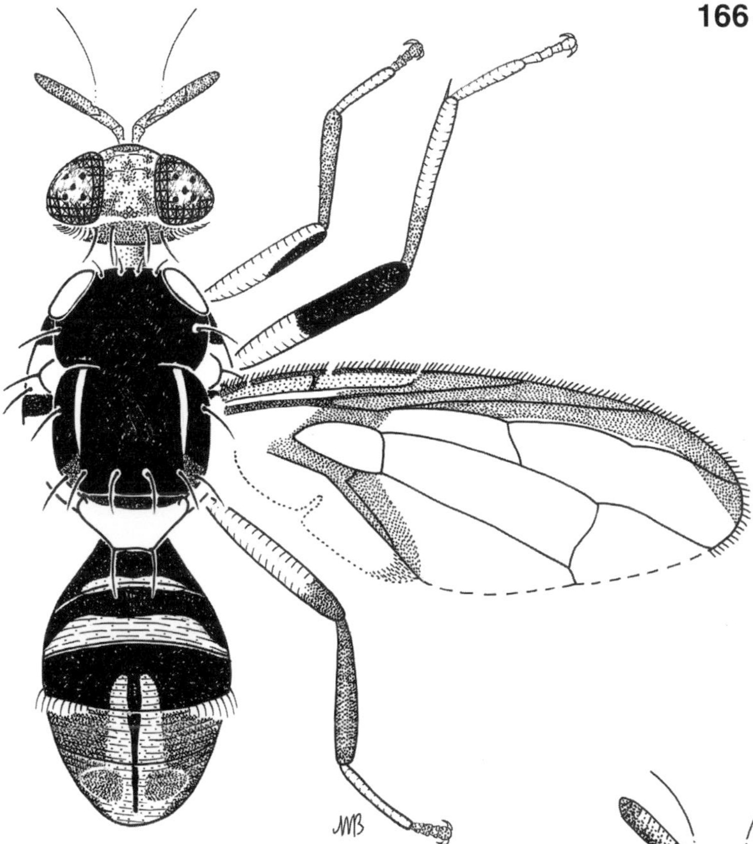

167

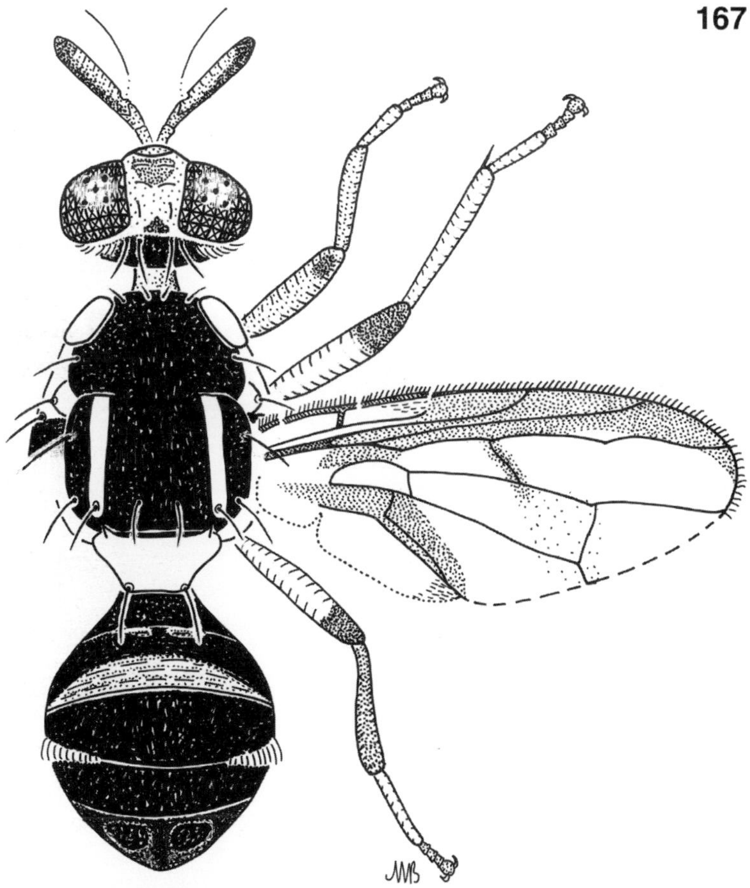

Fig. 166. *Bactrocera (Bactrocera) quasiinfulata* new species, holotype male. **Fig. 167.** *Bactrocera (Bactrocera) quasineonigrita* new species, holotype male.

168

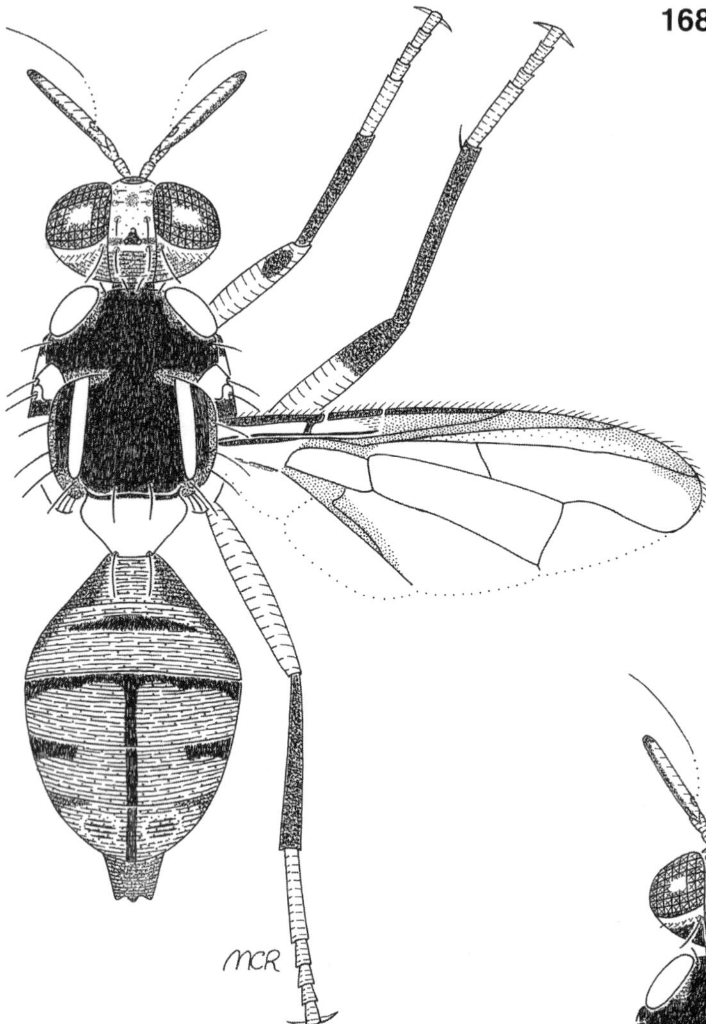

169

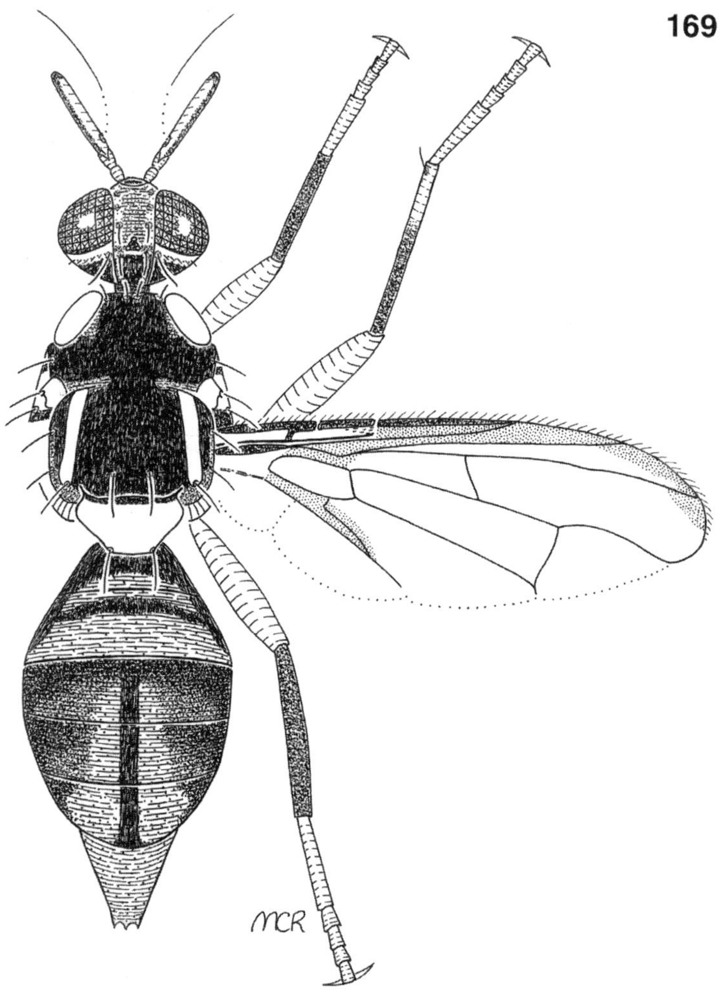

Fig. 168. *Bactrocera (Bactrocera) quasipropinqua* Drew & Hancock, holotype female. **Fig. 169.** *Bactrocera (Bactrocera) raiensis* Drew & Hancock, holotype female.

170

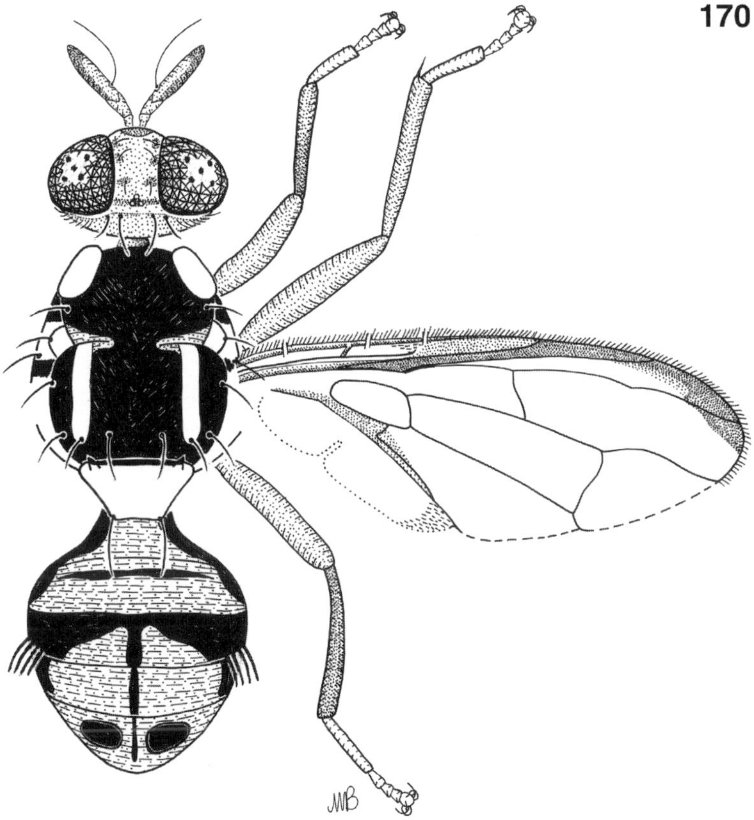

171

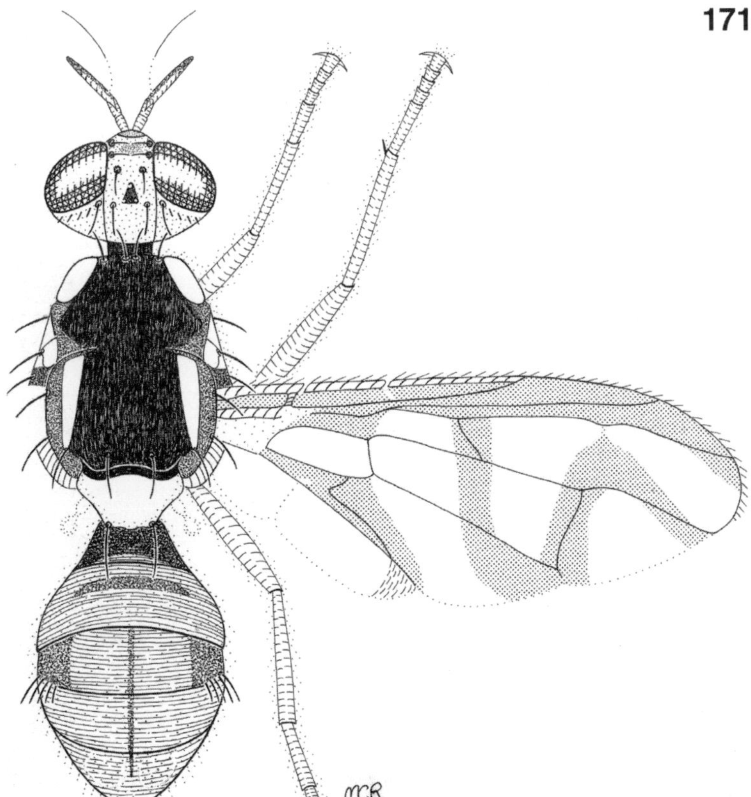

Fig. 170. *Bactrocera (Bactrocera) ranganathi* new species, holotype male. **Fig. 171.** *Bactrocera (Bactrocera) recurrens* (Hering), male.

172

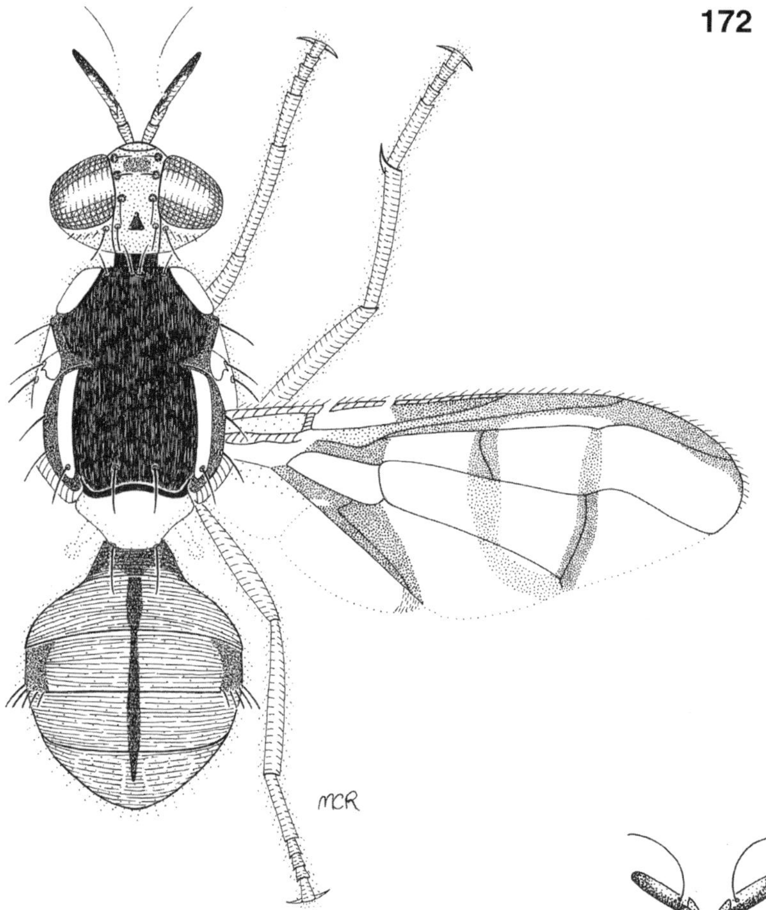

173A

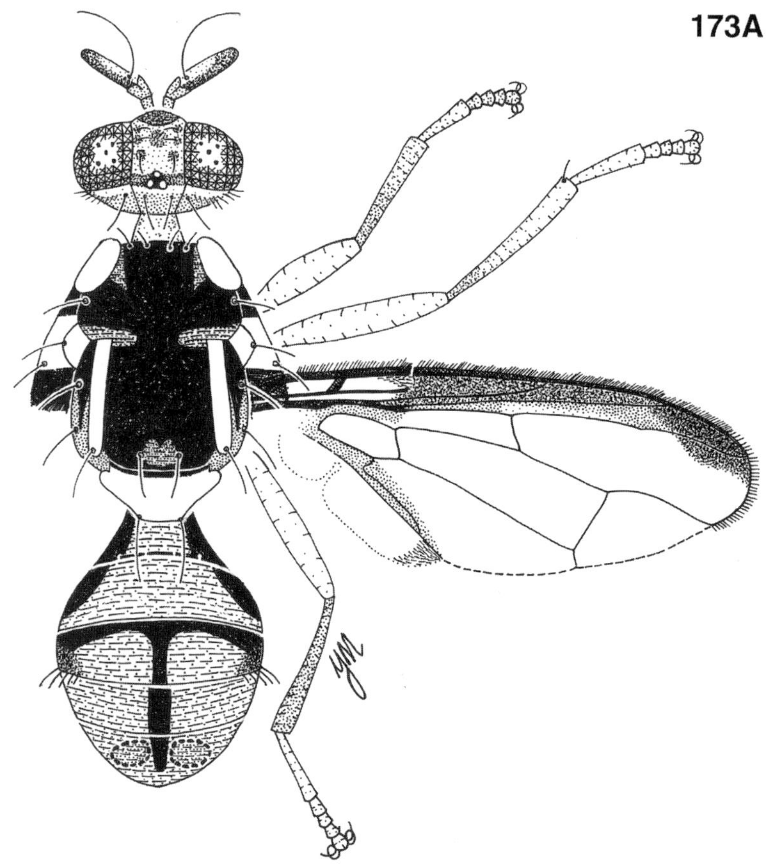

Fig. 172. *Bactrocera (Bactrocera) resima* (Drew), holotype male. **Fig. 173A.** *Bactrocera (Bactrocera) ritsemai* (Weyenbergh), male.

173B

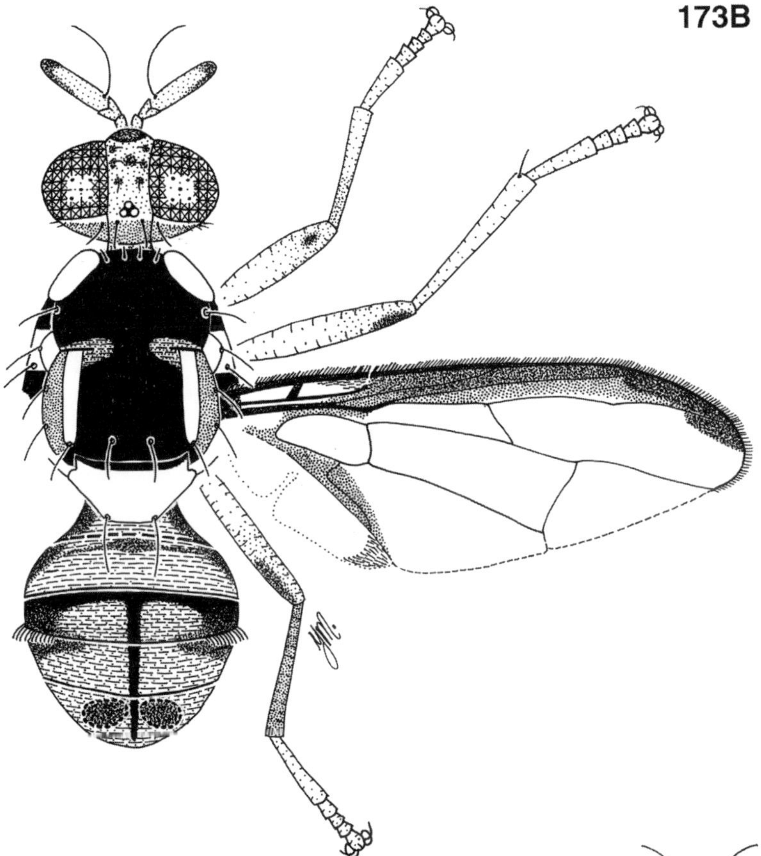

174

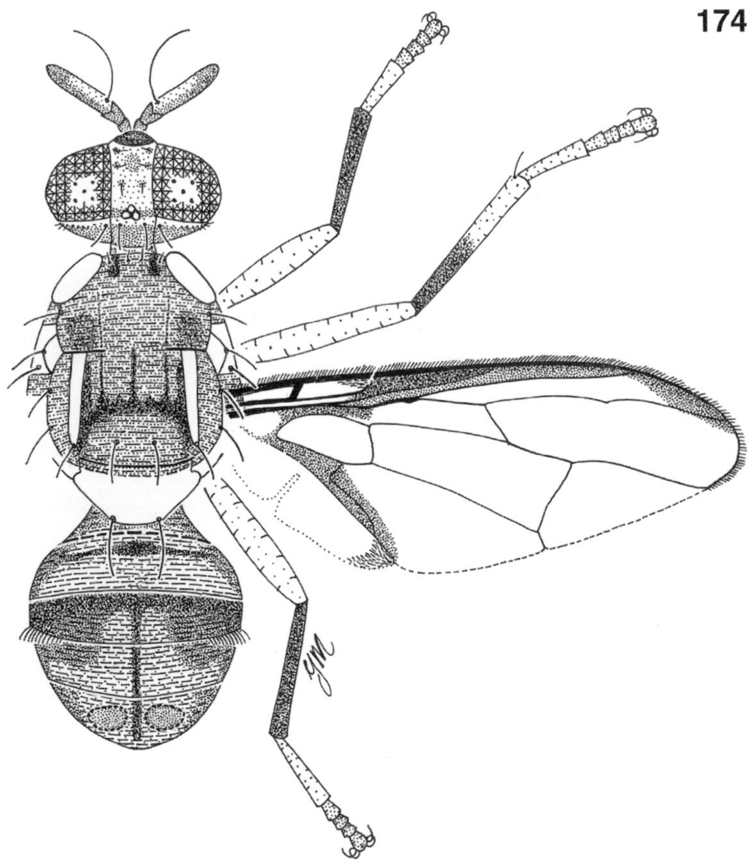

Fig. 173B. *Bactrocera (Bactrocera) rufula* (Hardy), holotype male. **Fig. 174.** *Bactrocera (Bactrocera) rubigina* (Wang & Zhao), male.

175

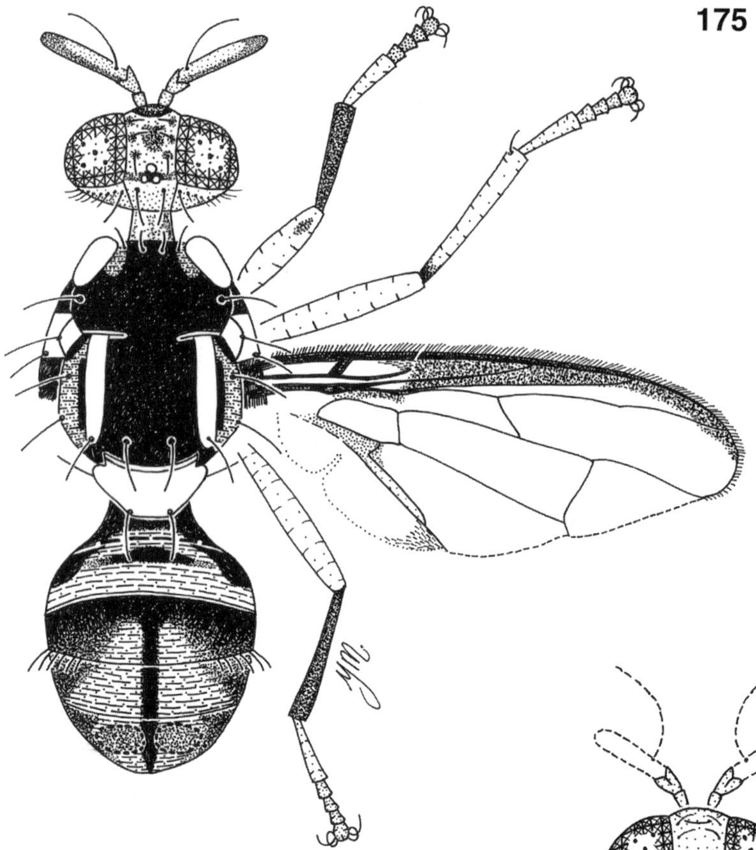

176

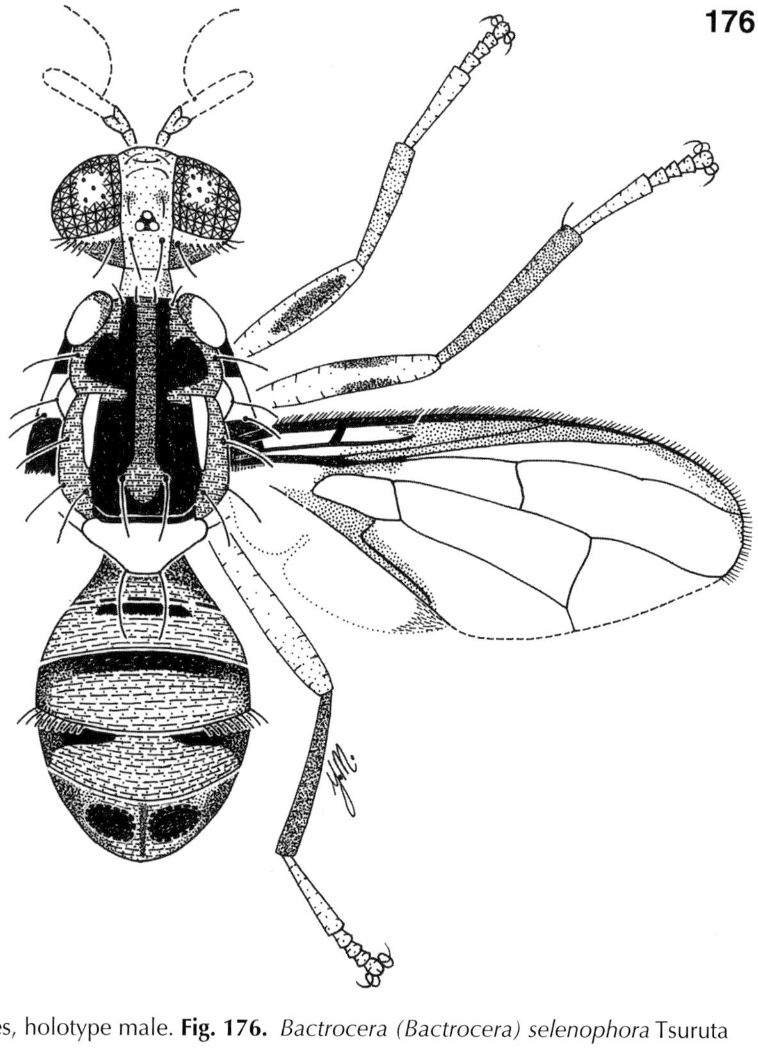

Fig. 175. *Bactrocera (Bactrocera) sapaensis* new species, holotype male. **Fig. 176.** *Bactrocera (Bactrocera) selenophora* Tsuruta & White, paratype male.

177

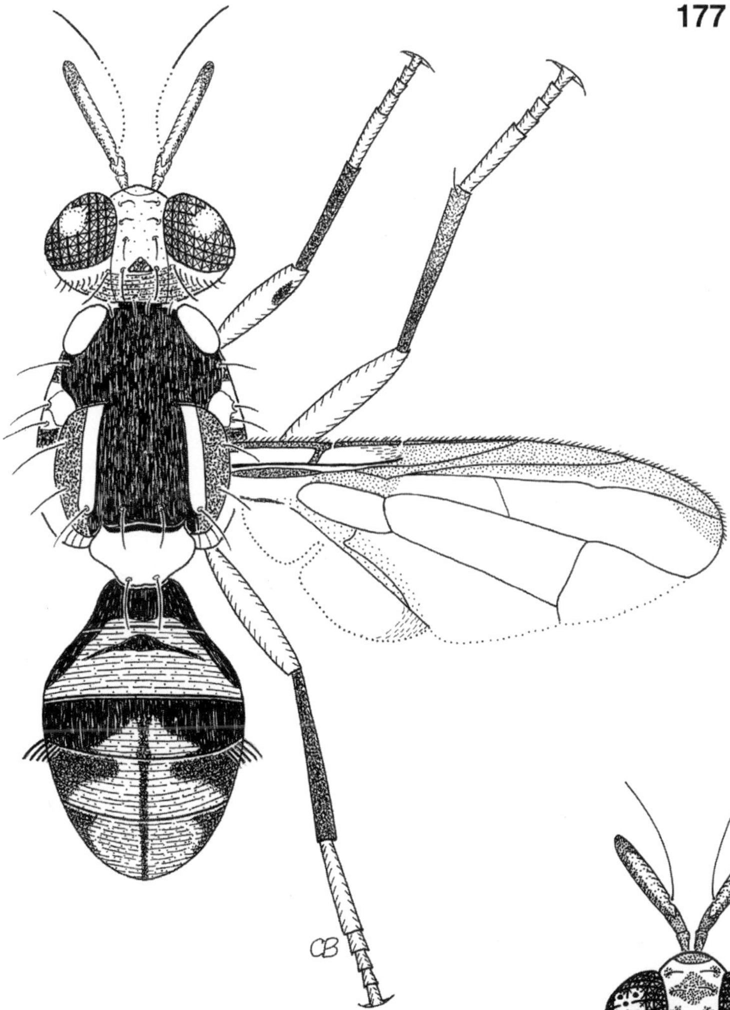

178

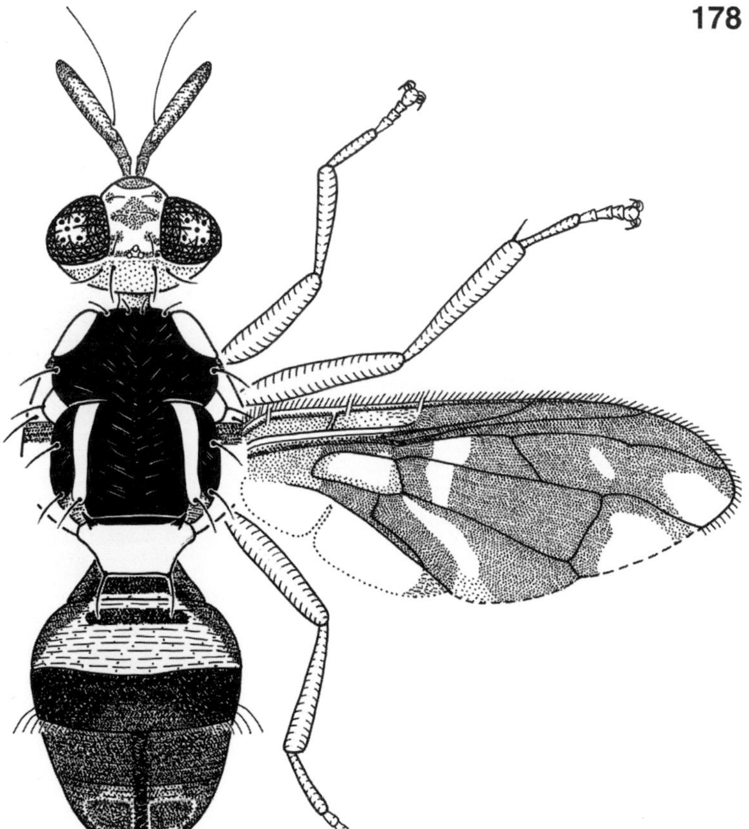

Fig. 177. *Bactrocera (Bactrocera) sembaliensis* Drew & Hancock, holotype male. **Fig. 178.** *Bactrocera (Bactrocera) speculifera* (Walker), lectotype male.

179

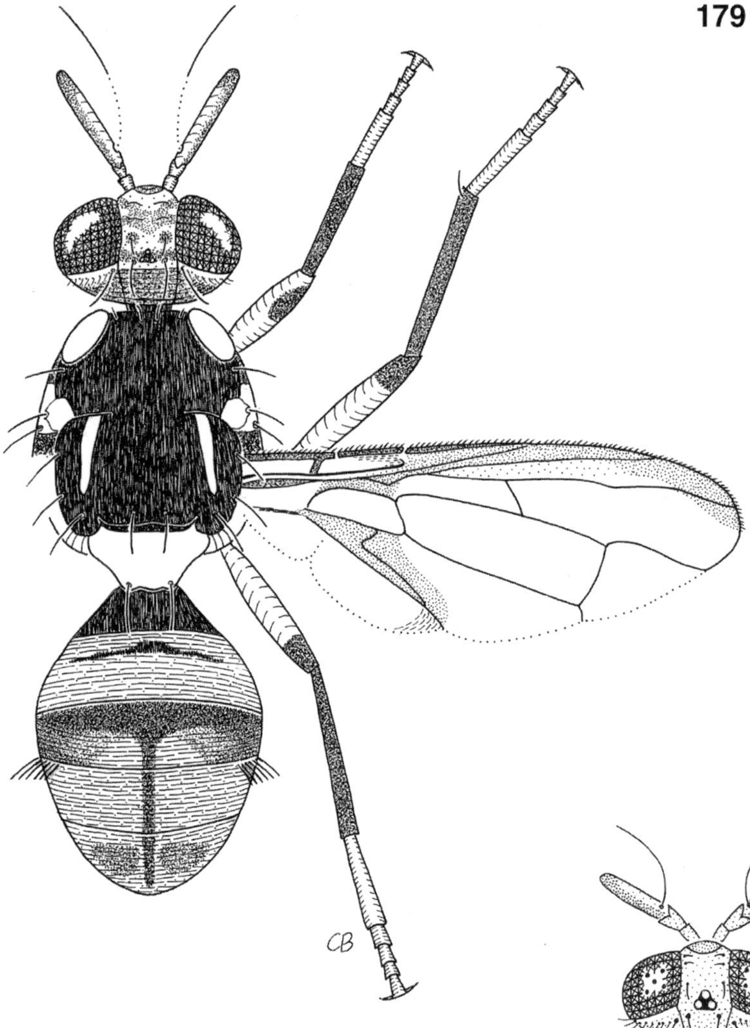

180

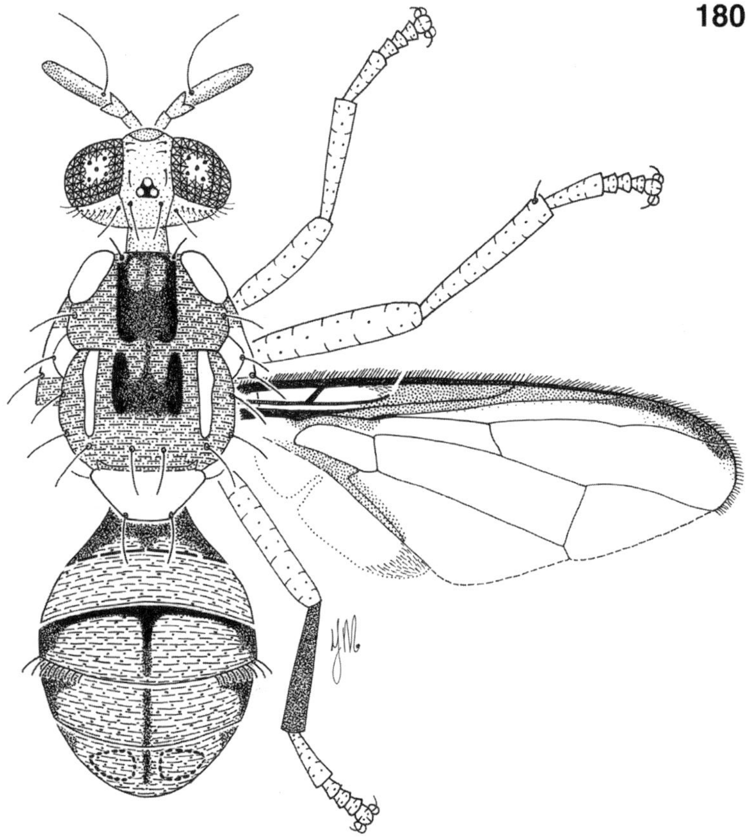

Fig. 179. *Bactrocera (Bactrocera) sulawesiae* Drew & Hancock, holotype male. **Fig. 180.** *Bactrocera (Bactrocera) suliae* new species, holotype male.

181

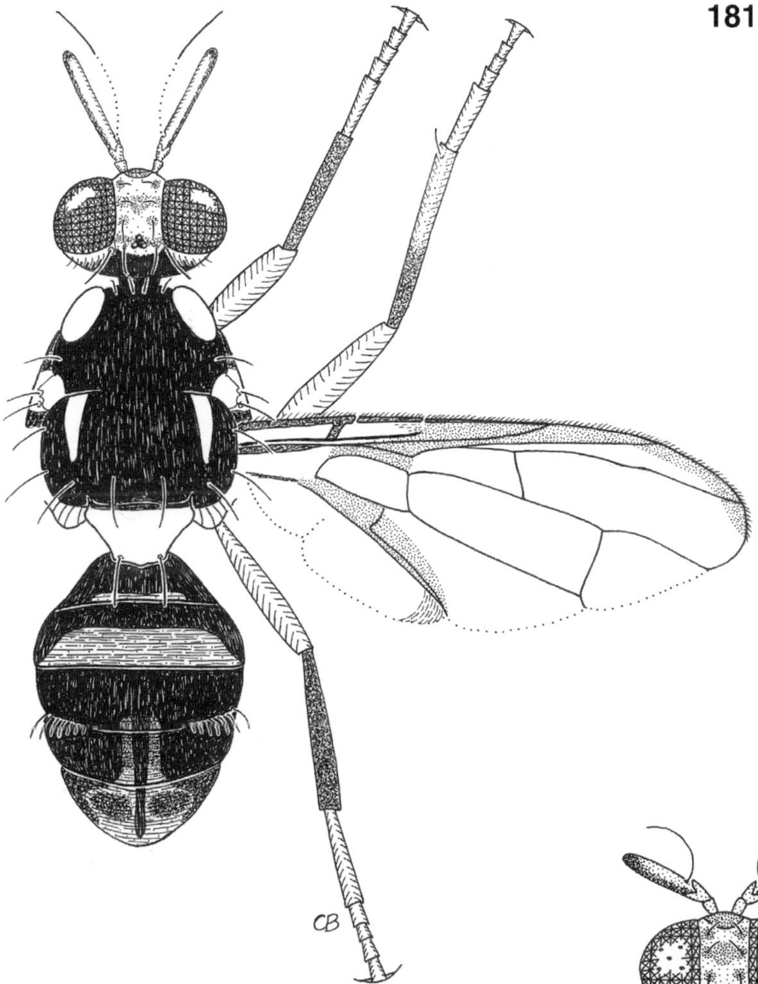

182

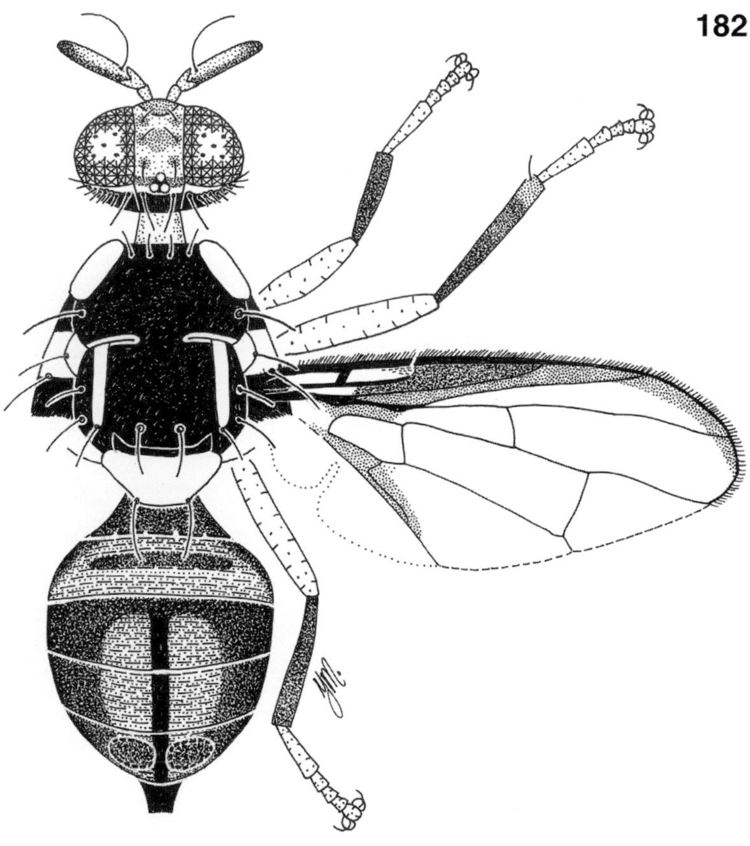

Fig. 181. *Bactrocera (Bactrocera) sumbawaensis* Drew & Hancock, holotype male. **Fig. 182.** *Bactrocera (Bactrocera) syzygii* White & Tsuruta, female.

183

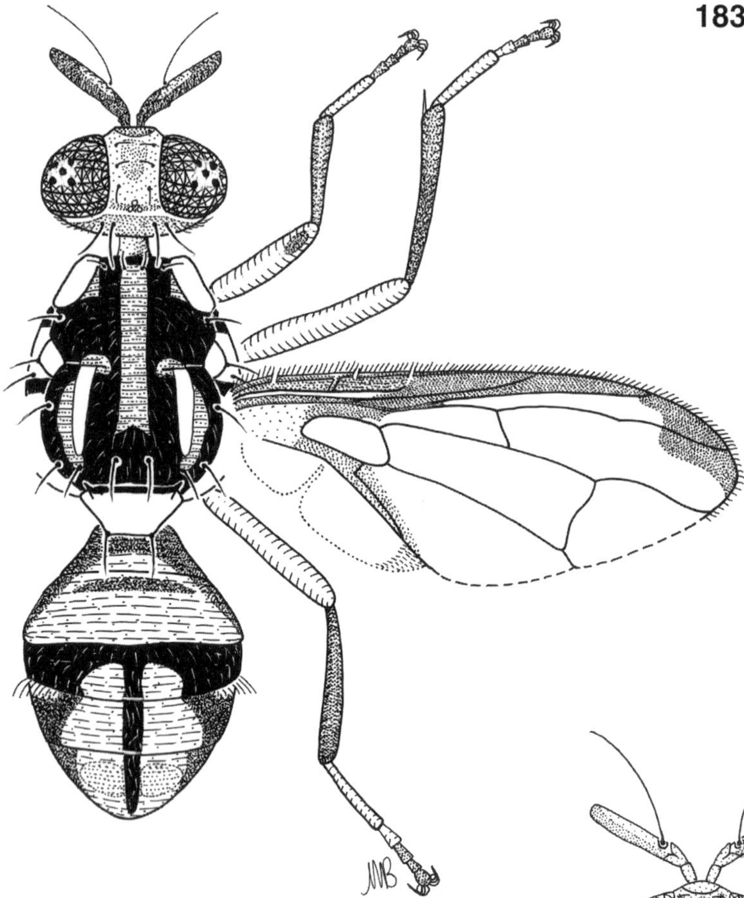

184

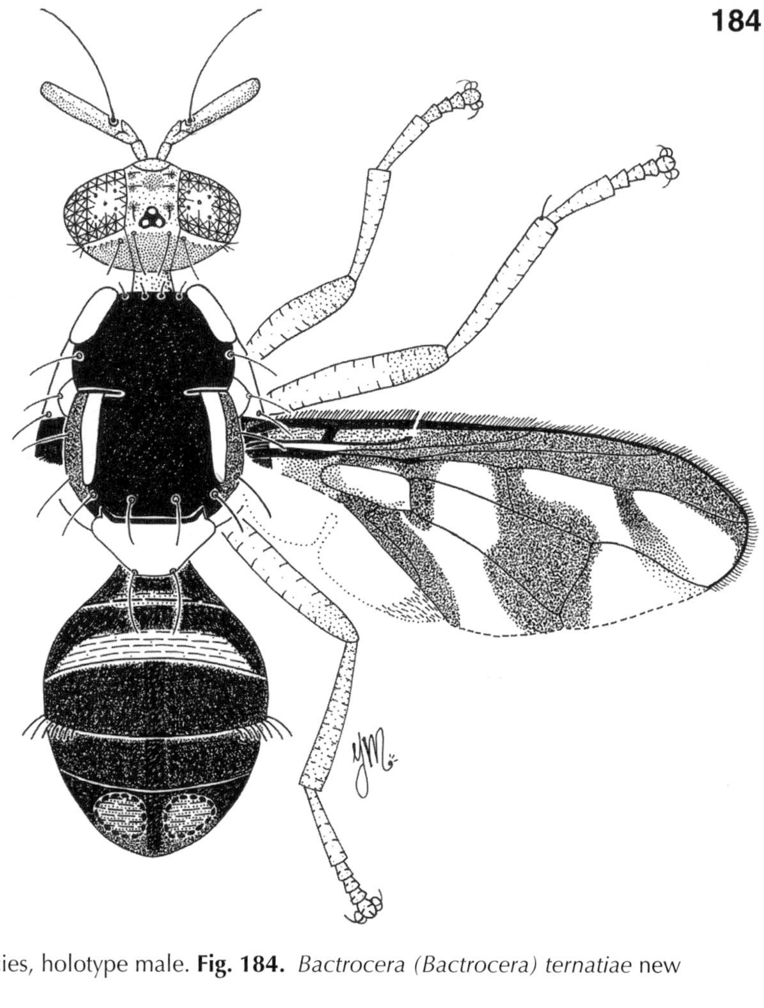

Fig. 183. *Bactrocera (Bactrocera) tapahensis* new species, holotype male. **Fig. 184.** *Bactrocera (Bactrocera) ternatiae* new species, holotype male.

185

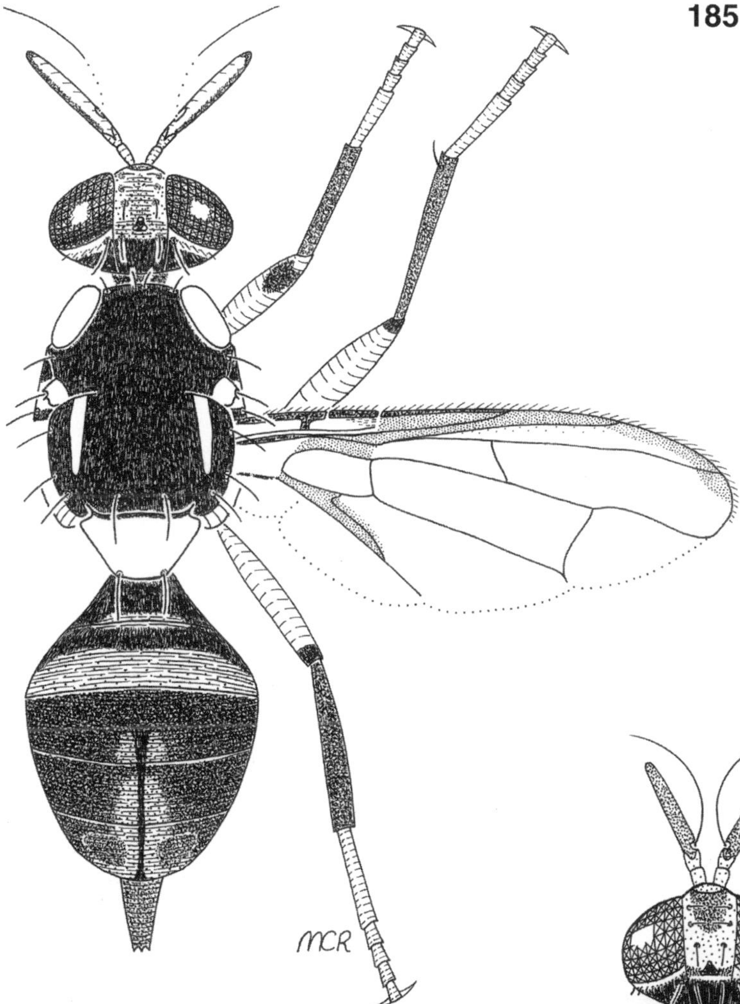

186

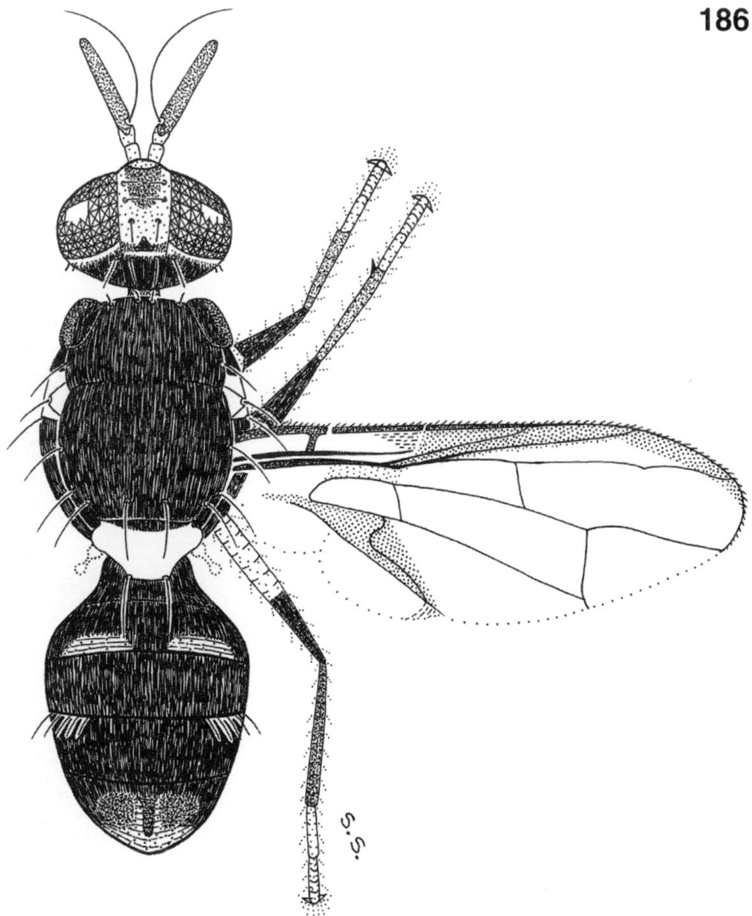

Fig. 185. *Bactrocera (Bactrocera) thailandica* Drew & Hancock, holotype female. **Fig. 186.** *Bactrocera (Bactrocera) thistletoni* Drew, holotype male.

187

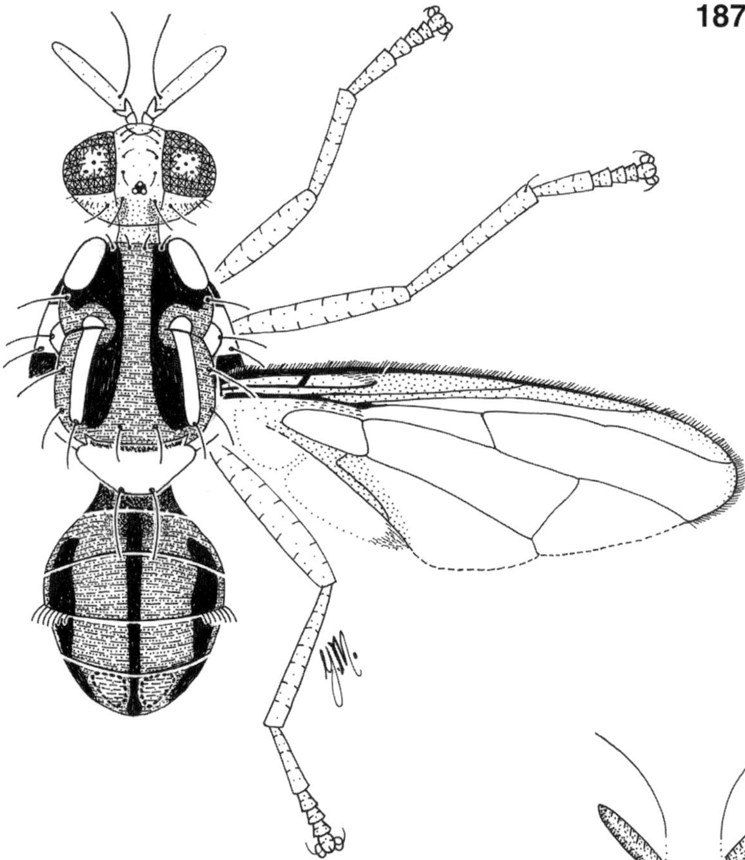

188

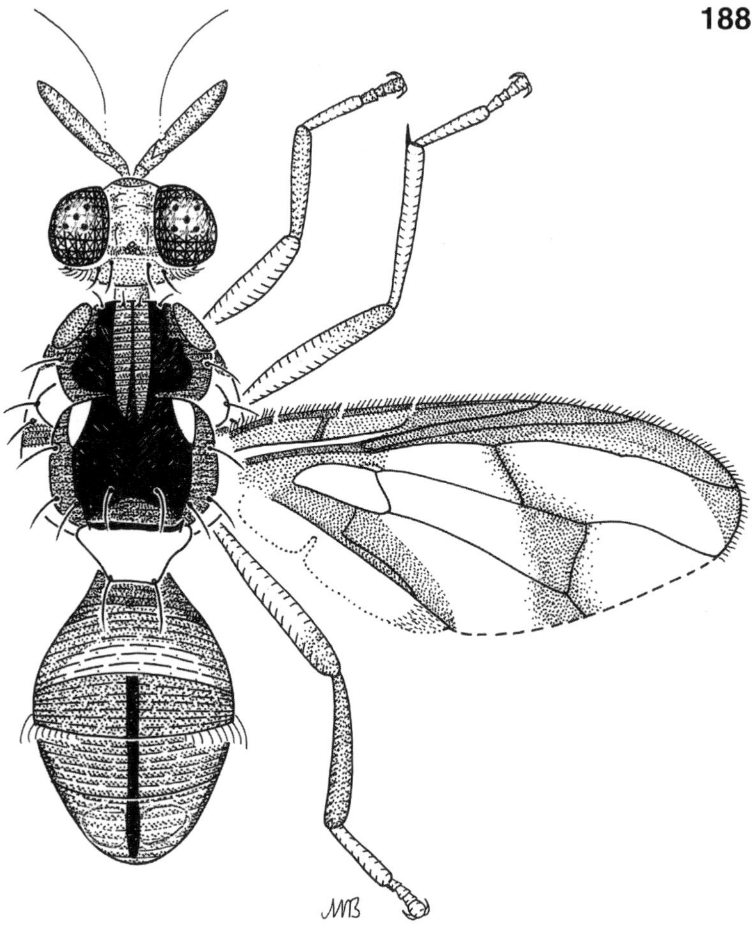

Fig. 187. *Bactrocera (Bactrocera) trifasciata* (Hardy), holotype male. **Fig. 188.** *Bactrocera (Bactrocera) truncata* new species, holotype male.

189

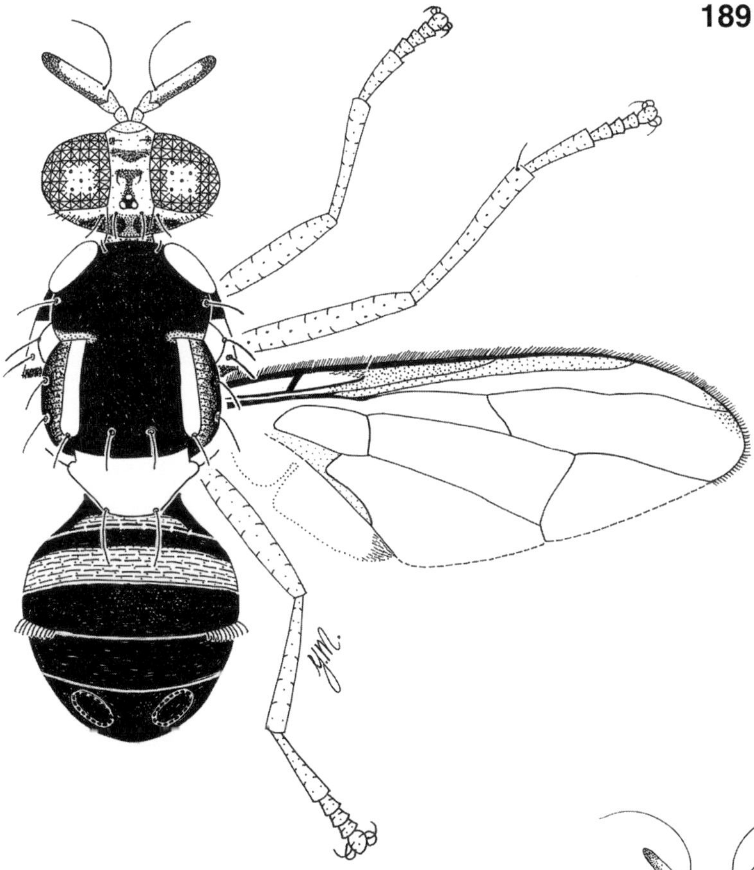

190A

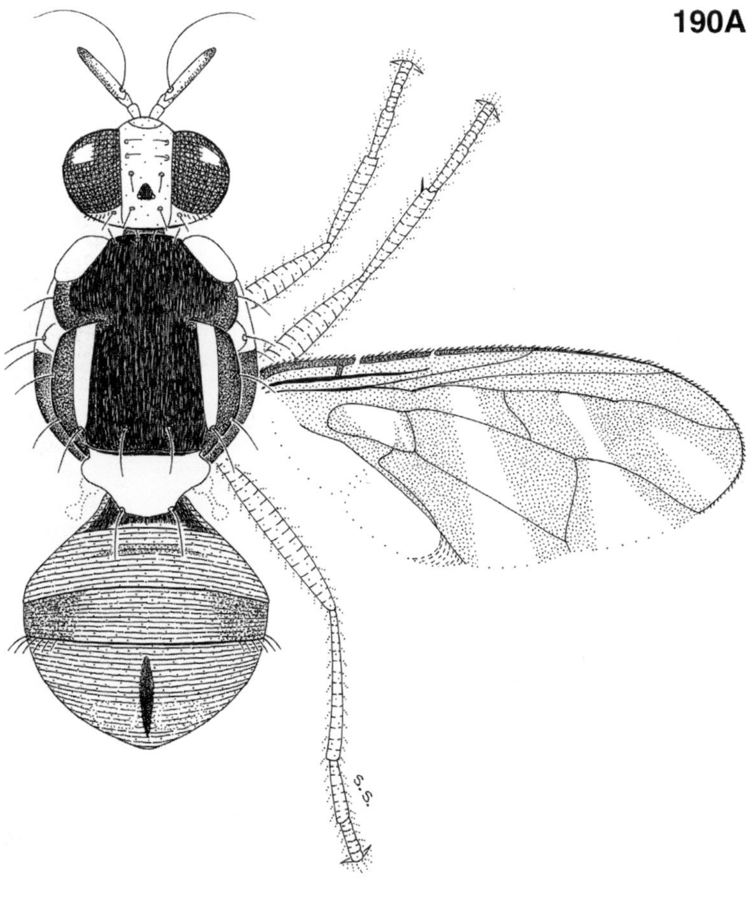

Fig. 189. *Bactrocera (Bactrocera) tuberculata* (Bezzi), holotype male. **Fig. 190A.** *Bactrocera (Bactrocera) umbrosa* (Fabricius), male.

190B

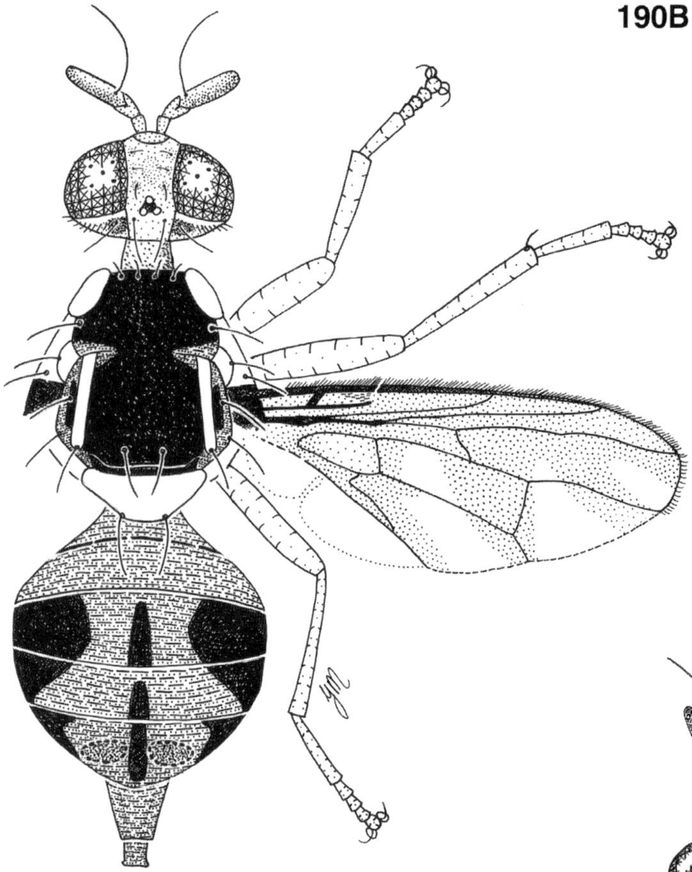

191

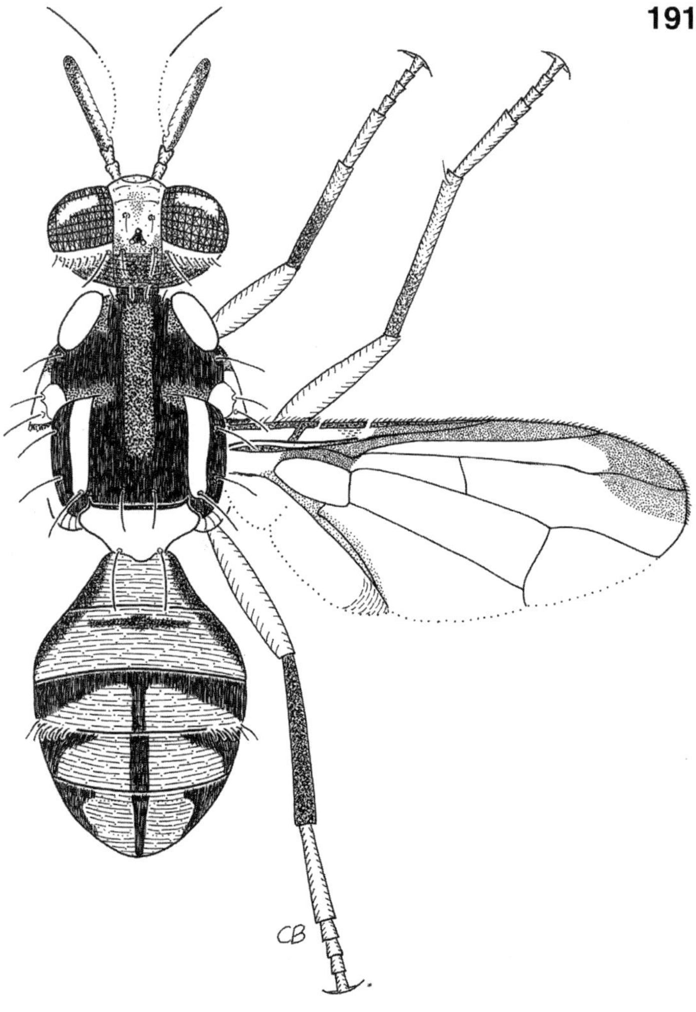

Fig. 190B. *Bactrocera (Bactrocera) lacerata* White & Evenhuis, holotype female. **Fig. 191.** *Bactrocera (Bactrocera) unimacula* Drew & Hancock, holotype male.

192

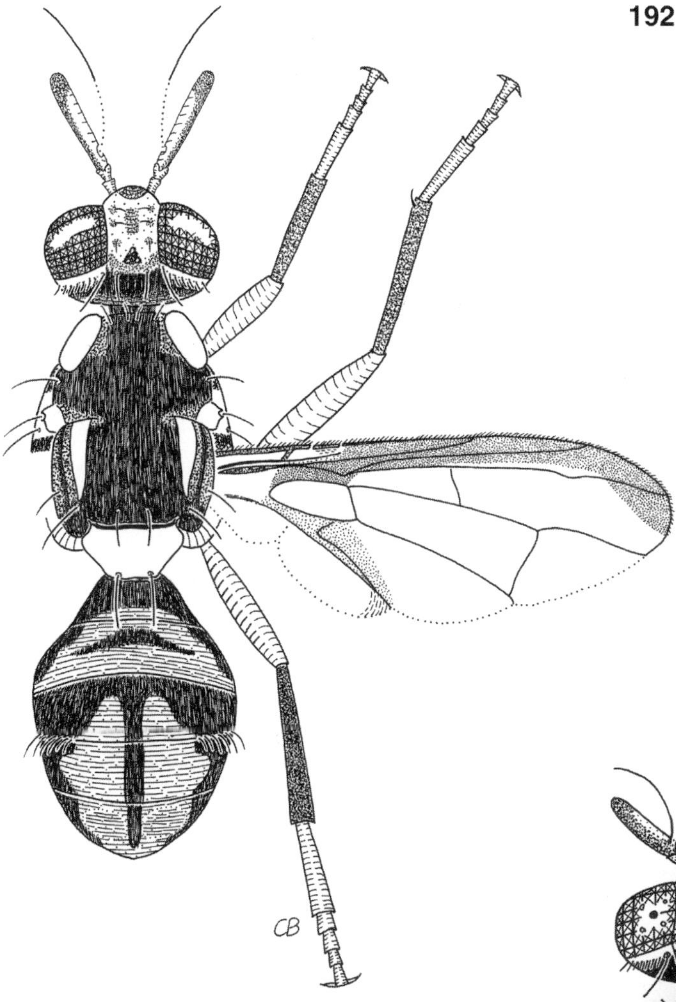

193

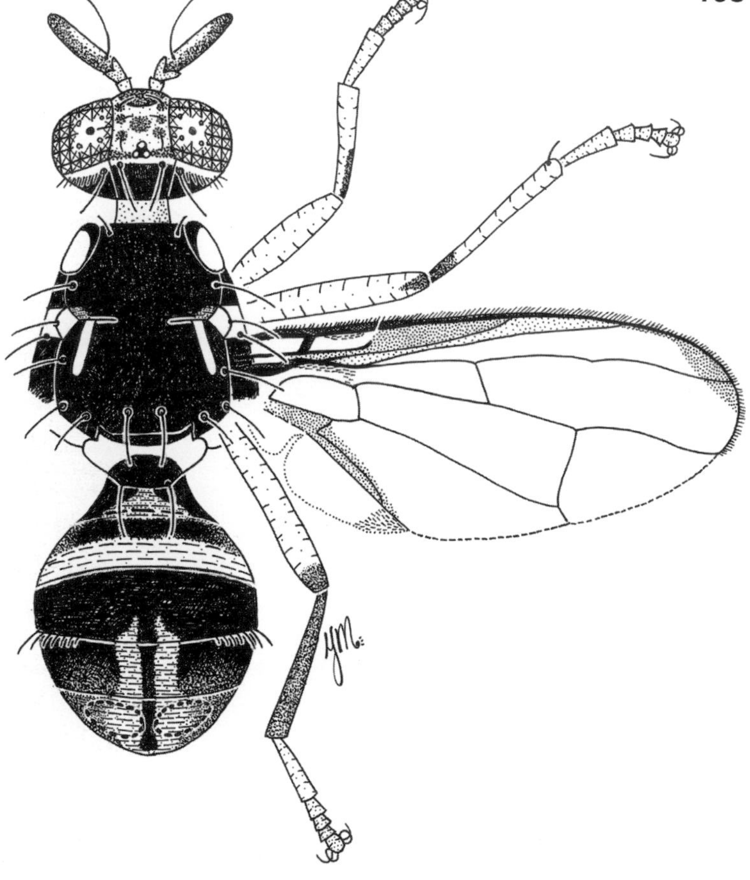

Fig. 192. *Bactrocera (Bactrocera) usitata* Drew & Hancock, holotype male. **Fig. 193.** *Bactrocera (Bactrocera) venefica* (Hering), holotype male.

194

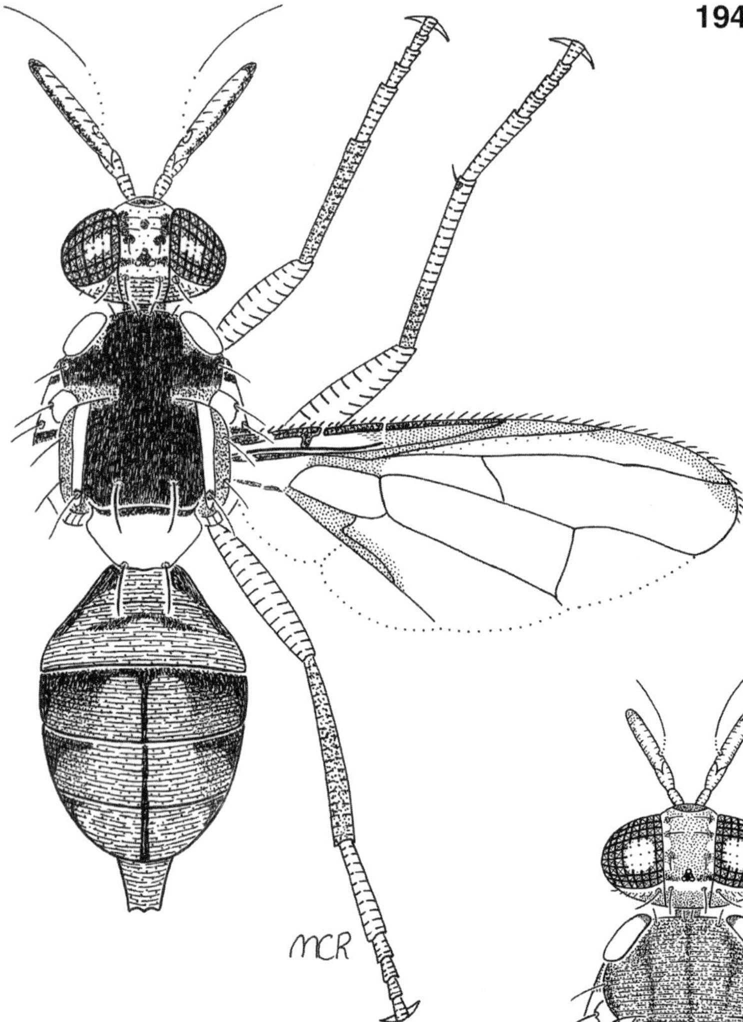

195

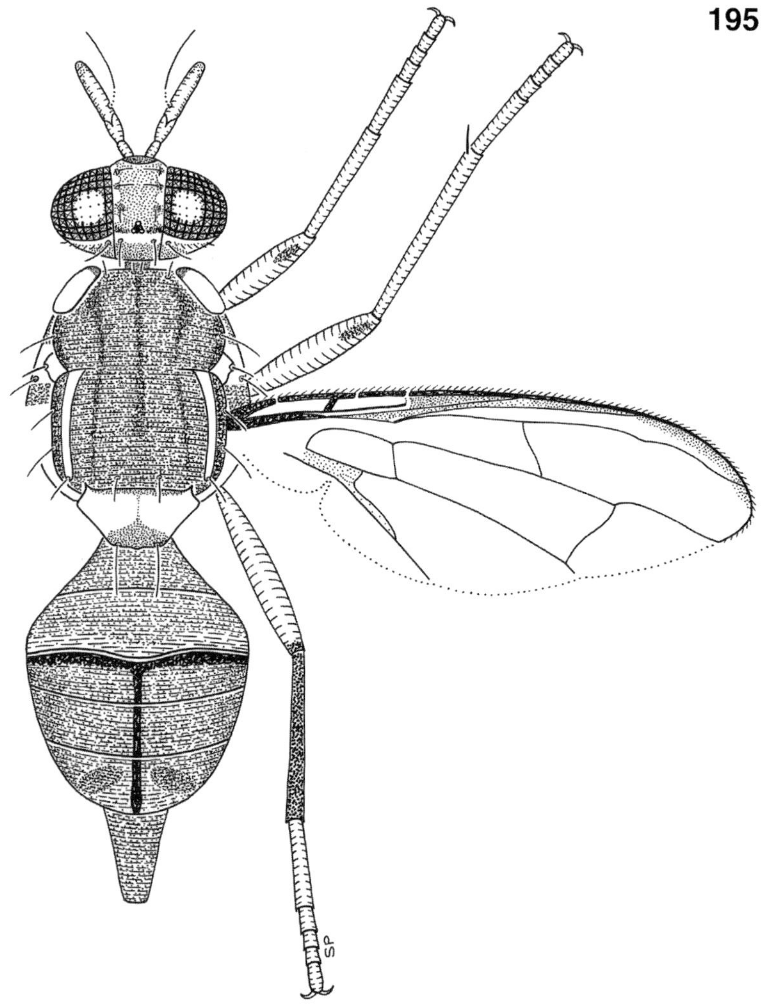

Fig. 194. *Bactrocera (Bactrocera) verbascifoliae* Drew & Hancock, holotype female. **Fig. 195.** *Bactrocera (Bactrocera) versicolor* (Bezzi), female.

196

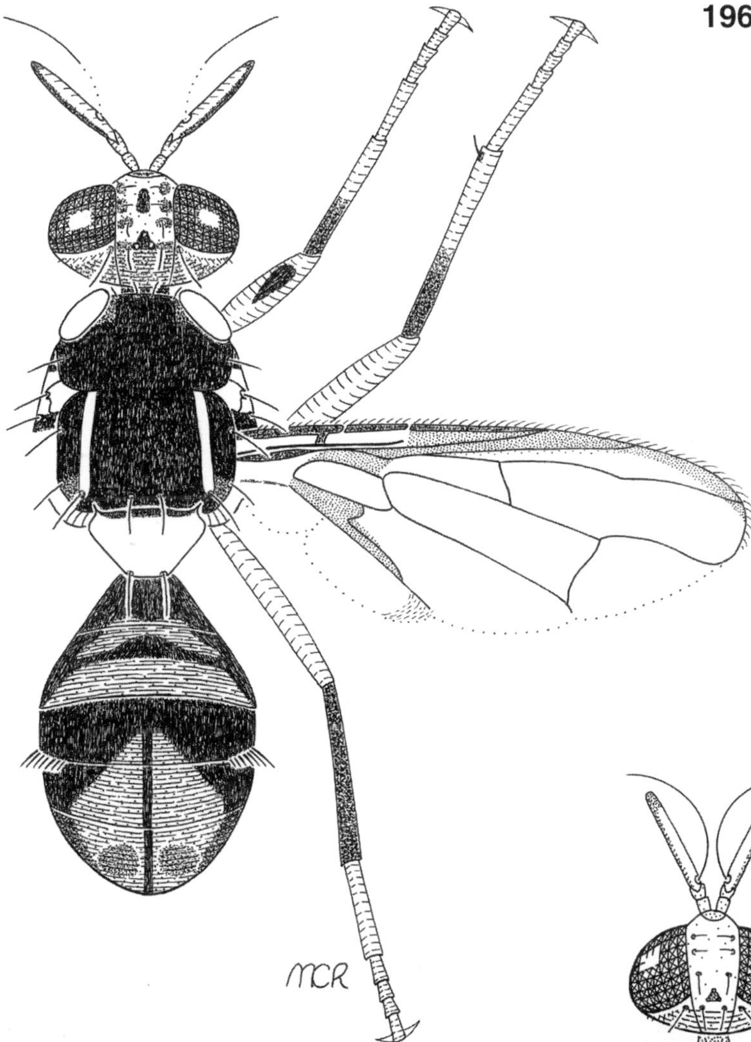

197

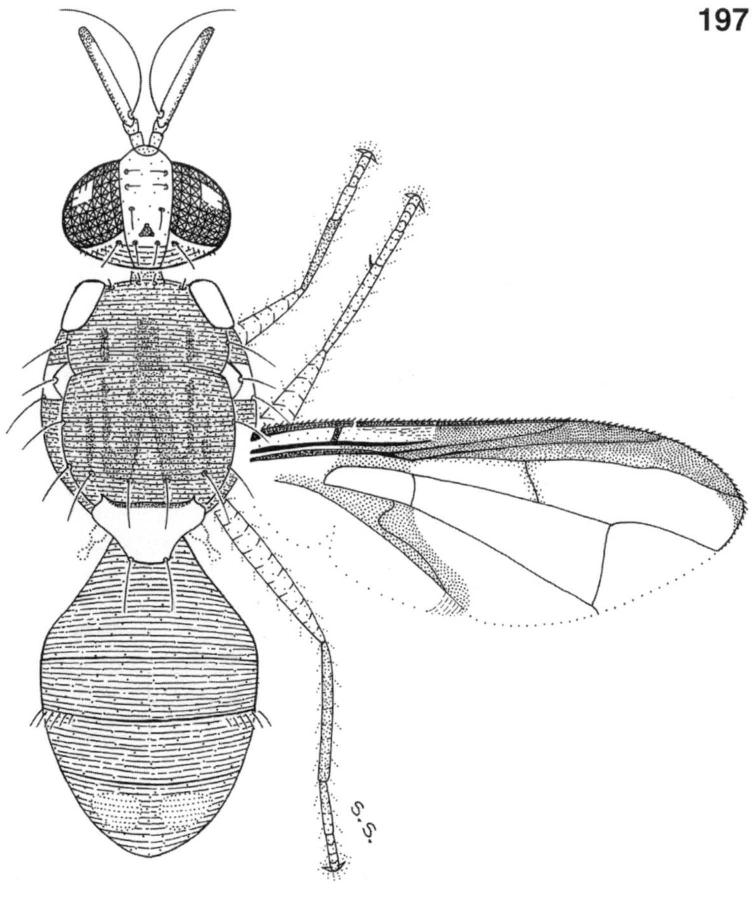

Fig. 196. *Bactrocera (Bactrocera) vishnu* Drew & Hancock, holotype male. **Fig. 197.** *Bactrocera (Bactrocera) waaiae* new species, holotype male.

198

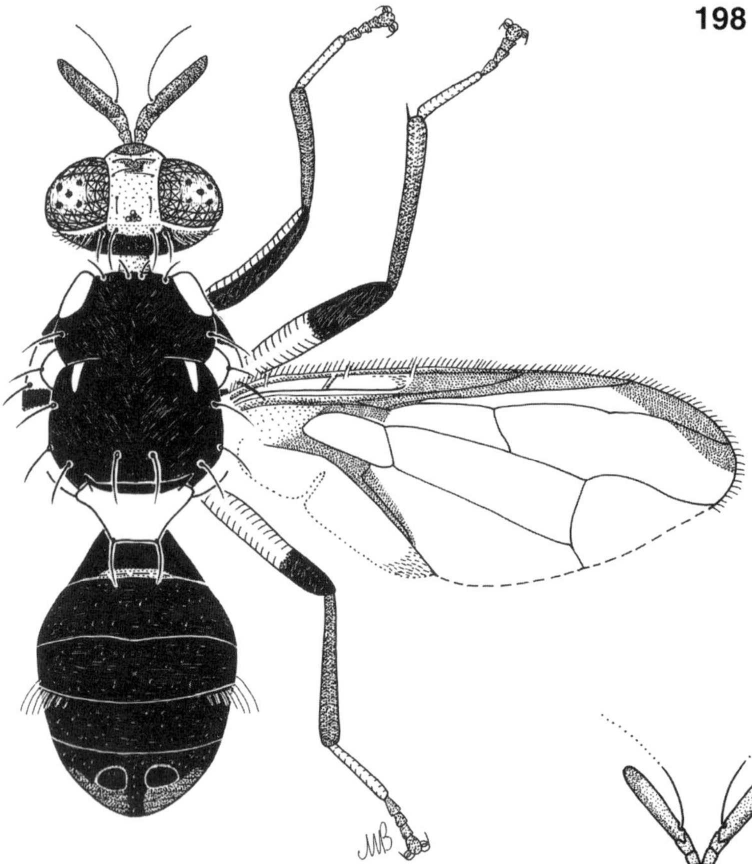

199A

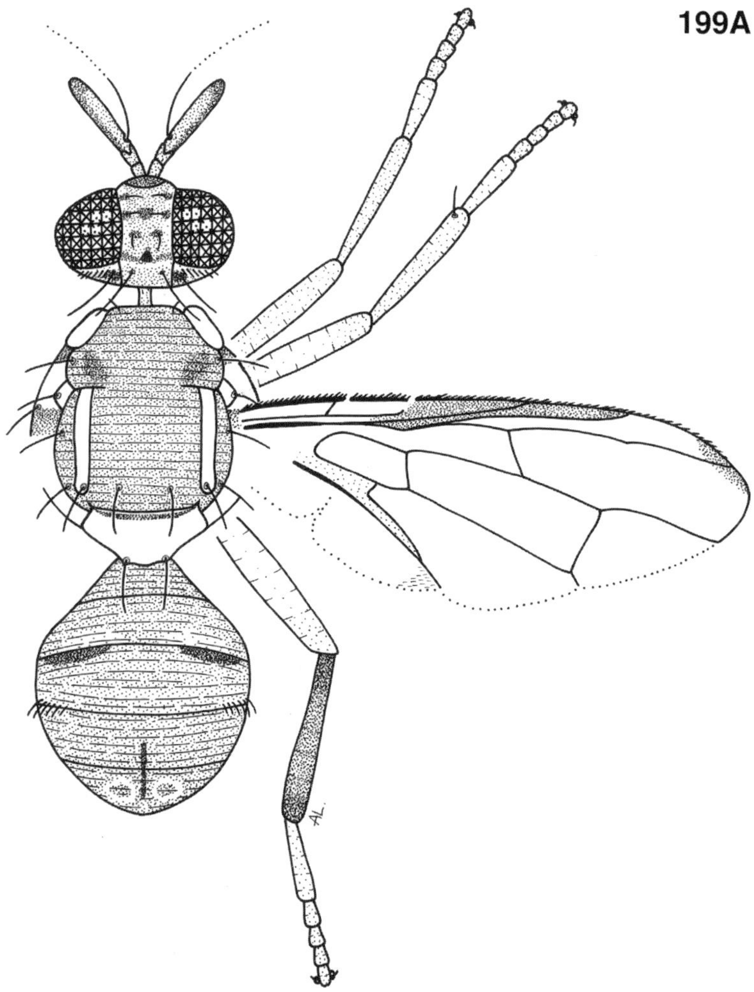

Fig. 198. *Bactrocera (Bactrocera) wuzhishana* Lin & Yang, male. **Fig. 199A.** *Bactrocera (Bactrocera) zonata* (Saunders), male.

199B

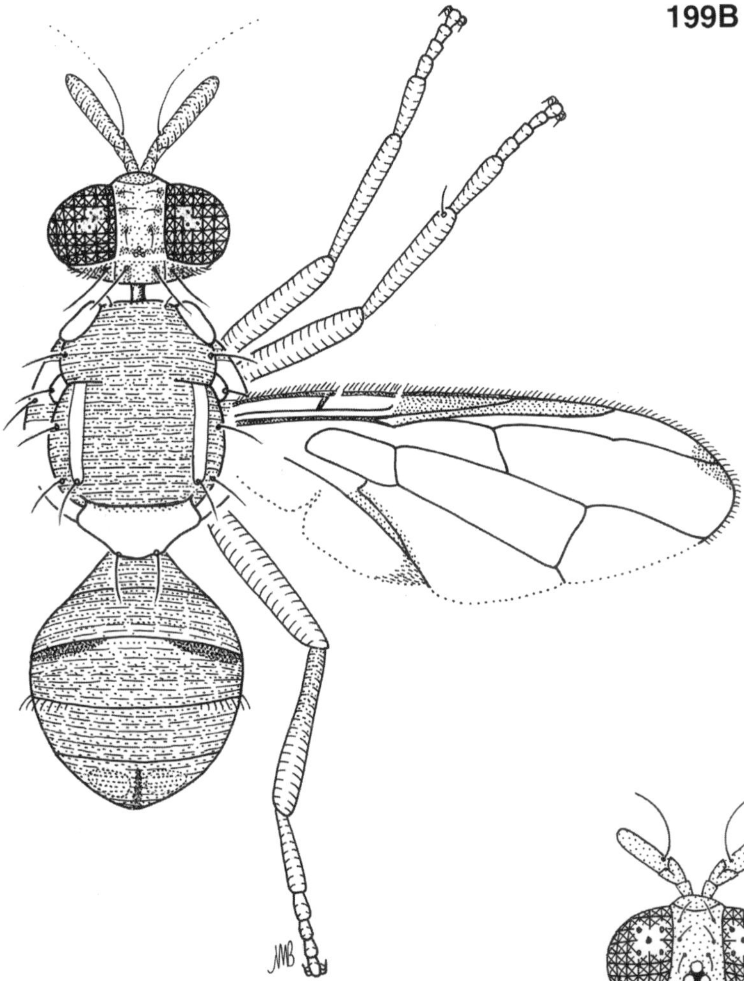

200

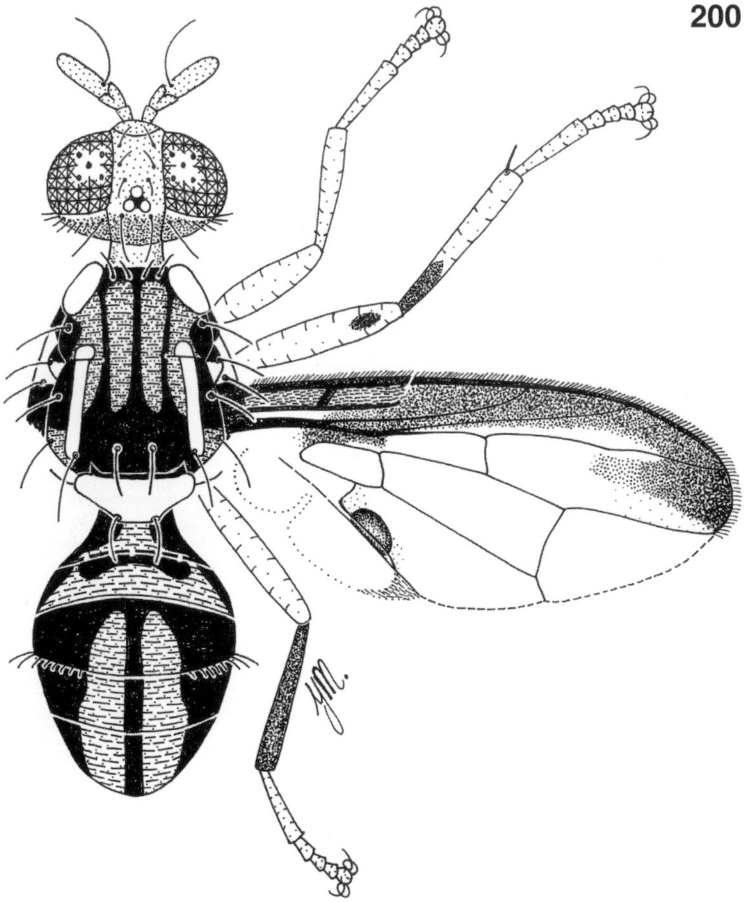

Fig. 199B. *Bactrocera (Bactrocera) affinis* (Hardy), paratype male. **Fig. 200.** *Bactrocera (Bulladacus) bullifera* (Hardy), holotype male.

201A

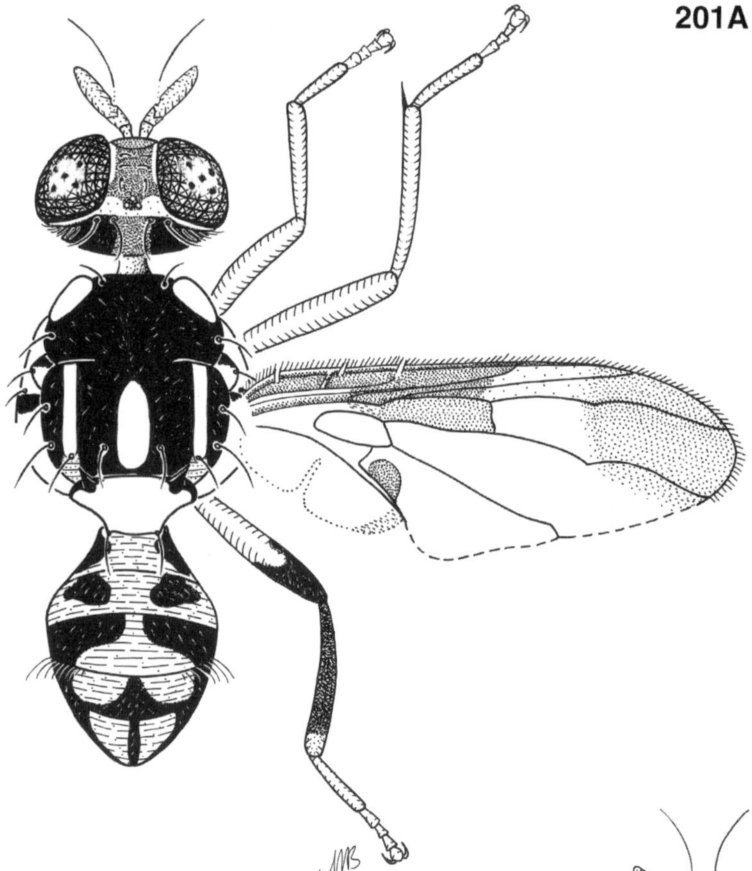

201B

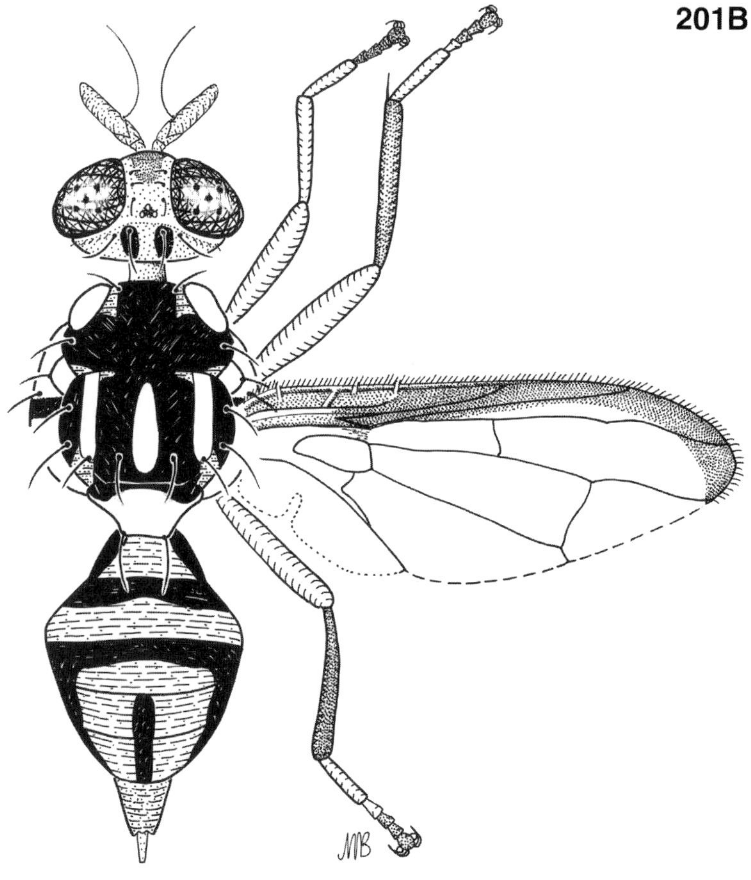

Fig. 201A. *Bactrocera (Bulladacus) captiva* new species, holotype male. **Fig. 201B.** *Bactrocera (Bulladacus) captiva* new species, paratype female.

202

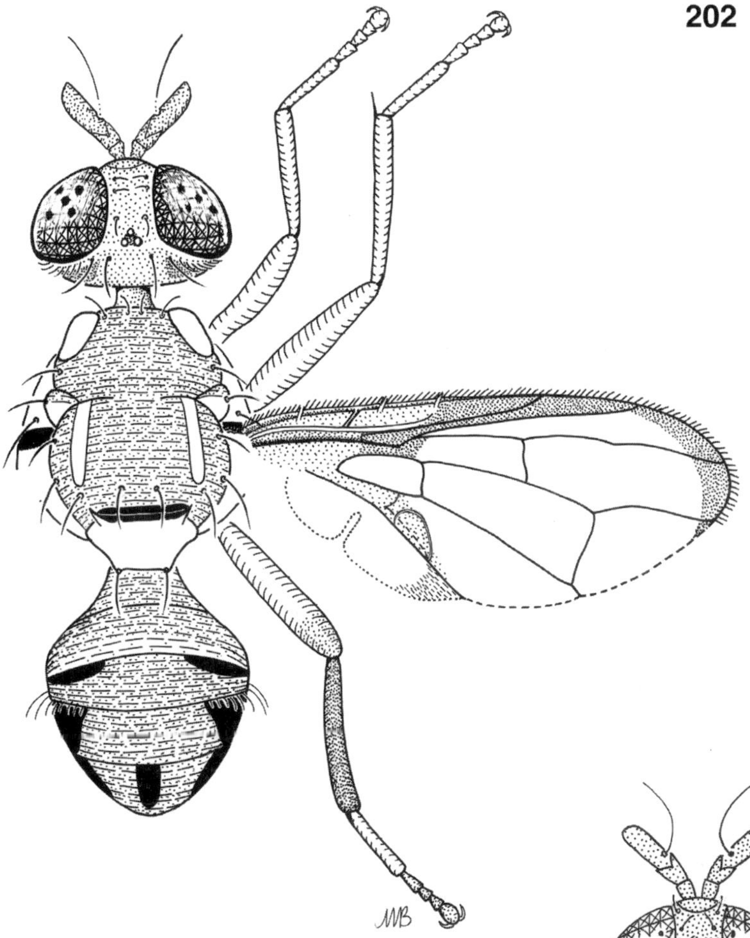

203

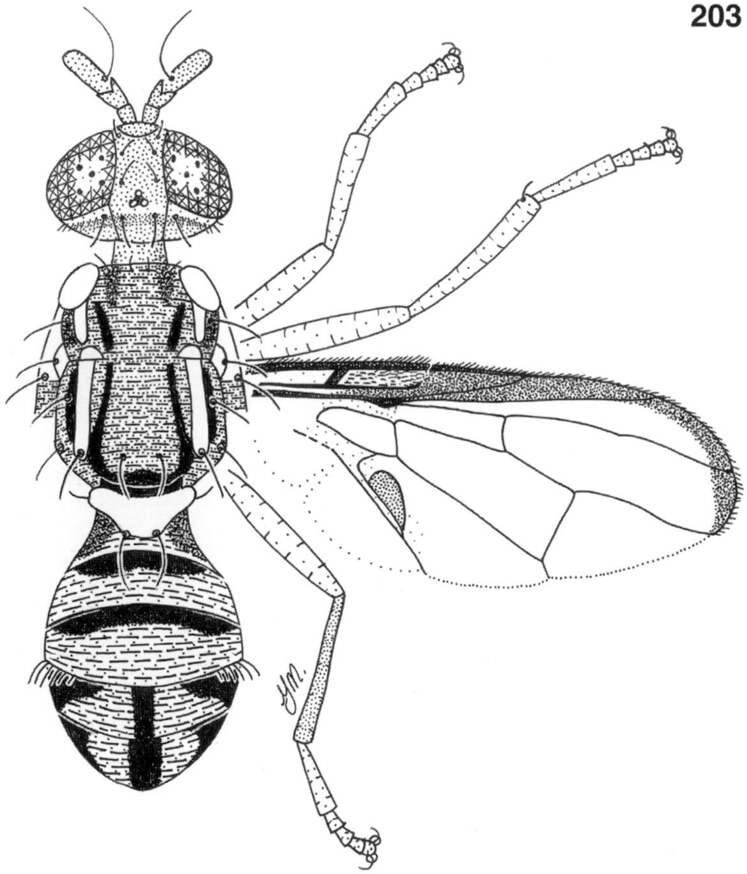

Fig. 202. *Bactrocera (Bulladacus) cinnabaria* new species, holotype male. **Fig. 203.** *Bactrocera (Bulladacus) mcgregori* (Bezzi), lectotype male.

204

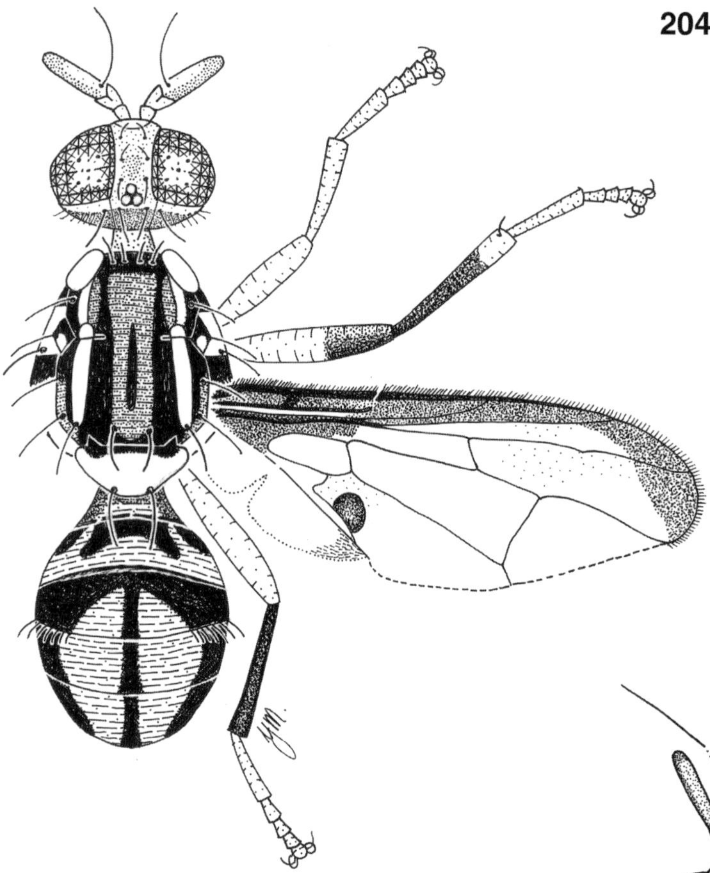

205

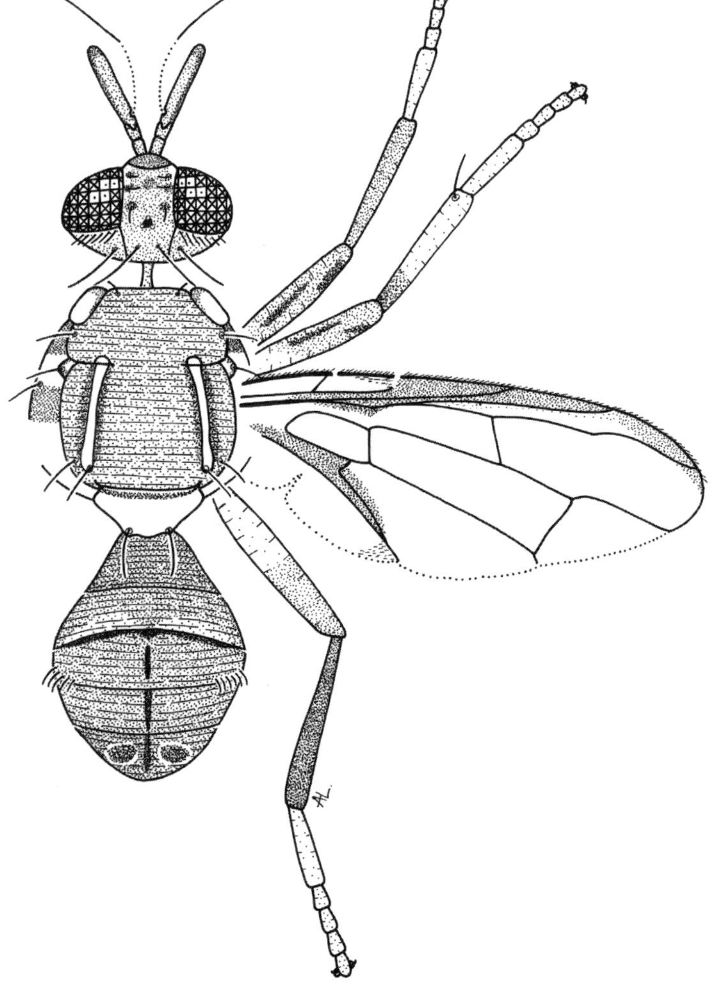

Fig. 204. *Bactrocera (Bulladacus) peterseni* (Hardy), holotype male. **Fig. 205.** *Bactrocera (Daculus) digressa* Radhakrishnan, male.

206

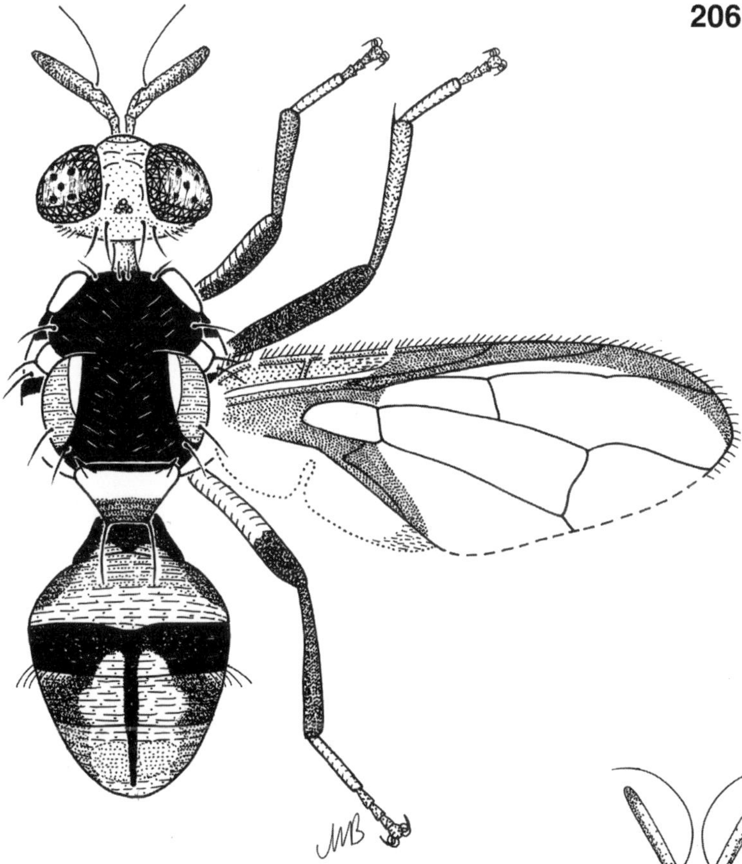

207

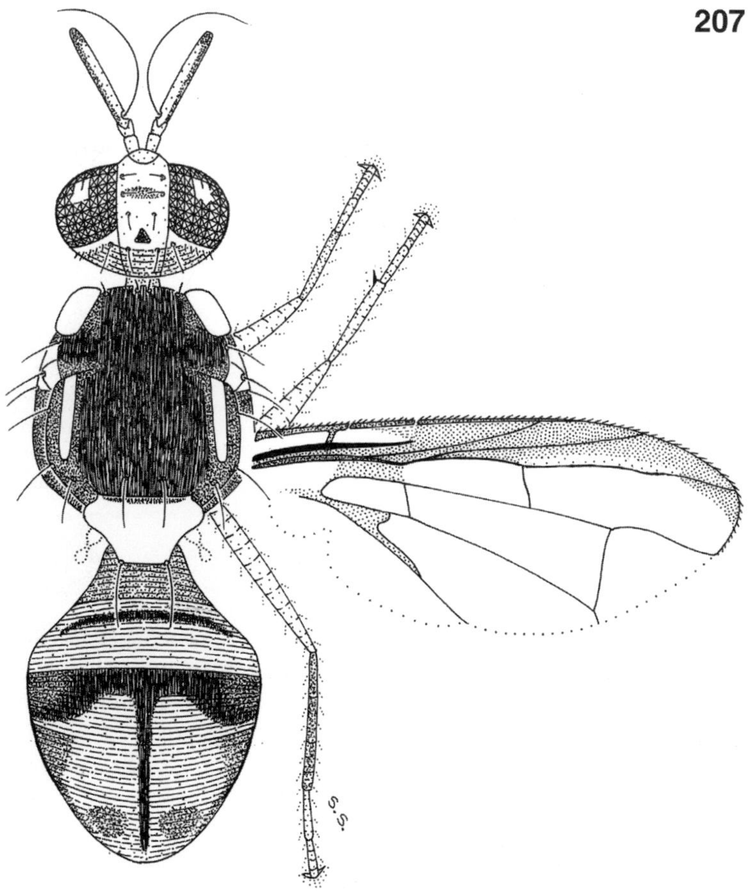

Fig. 206. *Bactrocera (Daculus) rutengiae* new species, holotype male. **Fig. 207.** *Bactrocera (Gymnodacus) calophylli* (Perkins & May), male.

208

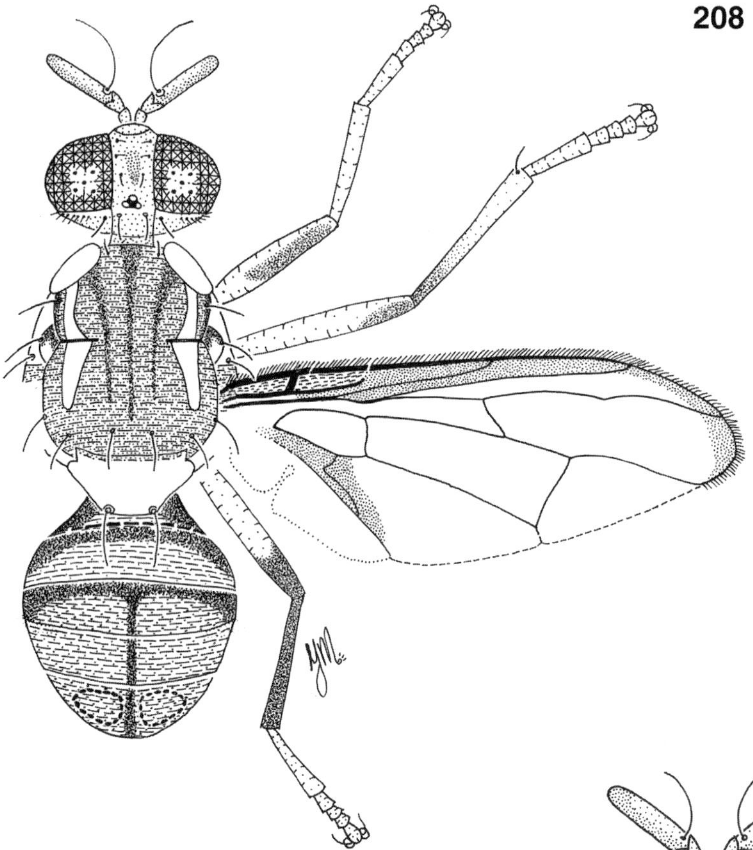

209

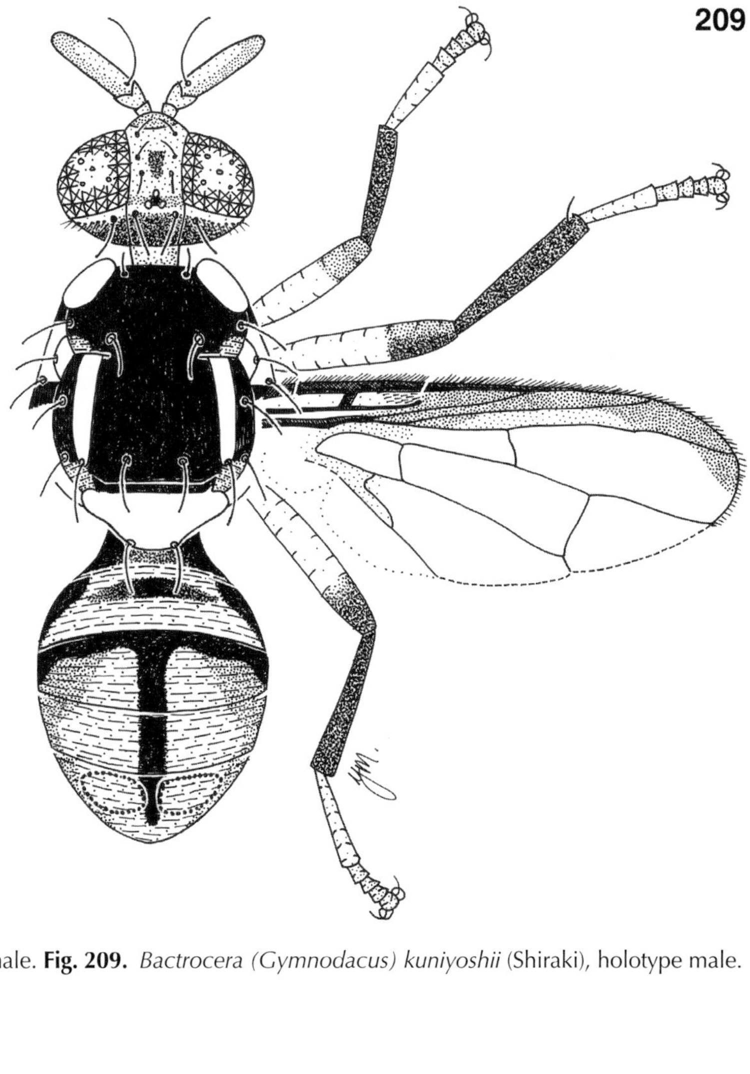

Fig. 208. *Bactrocera (Gymnodacus) continua* (Bezzi), male. **Fig. 209.** *Bactrocera (Gymnodacus) kuniyoshii* (Shiraki), holotype male.

210

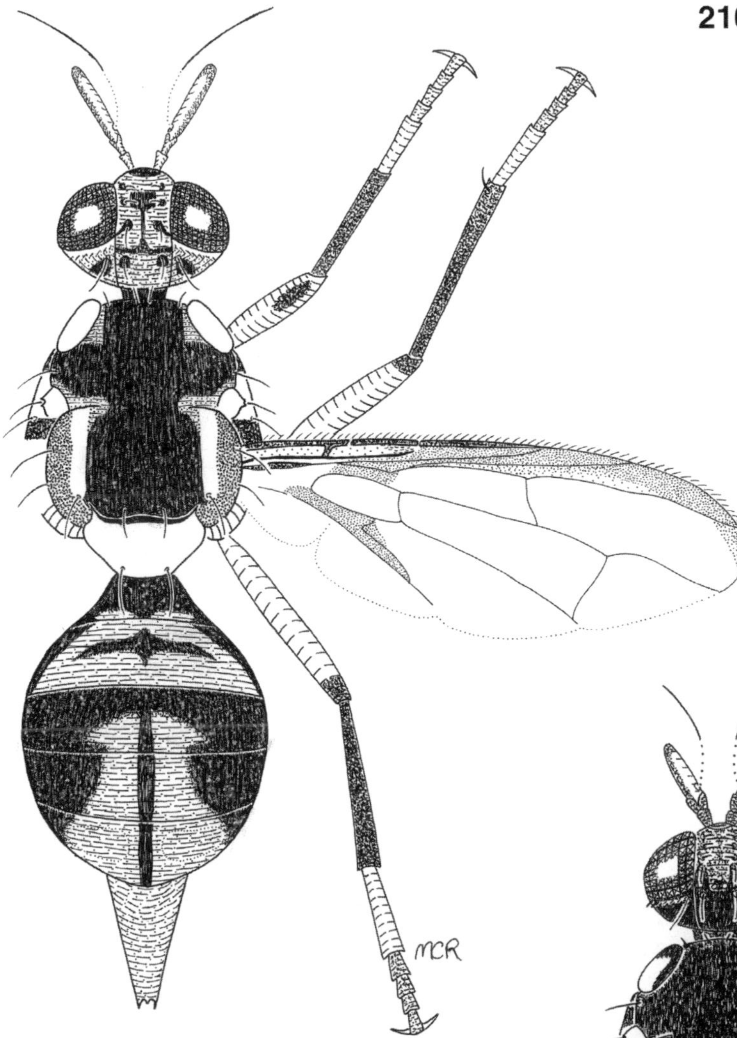

211A

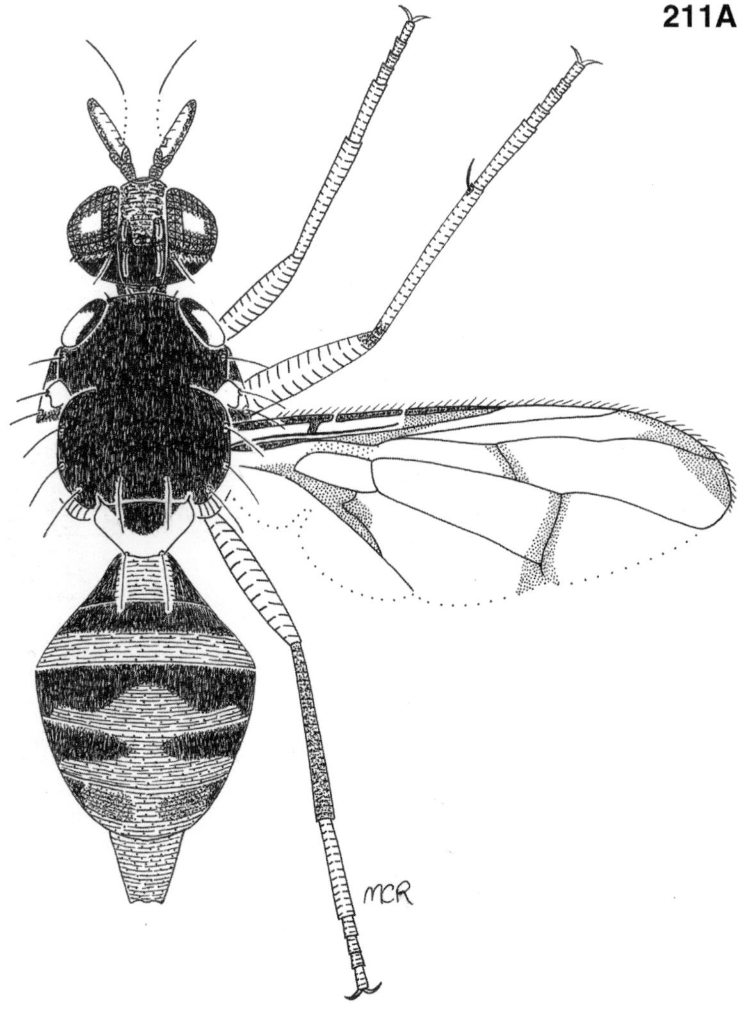

Fig. 210. *Bactrocera (Gymnodacus) symplocos* new species, holotype female. **Fig. 211A.** *Bactrocera (Gymnodacus) tillyardi* (Perkins), holotype female.

211B

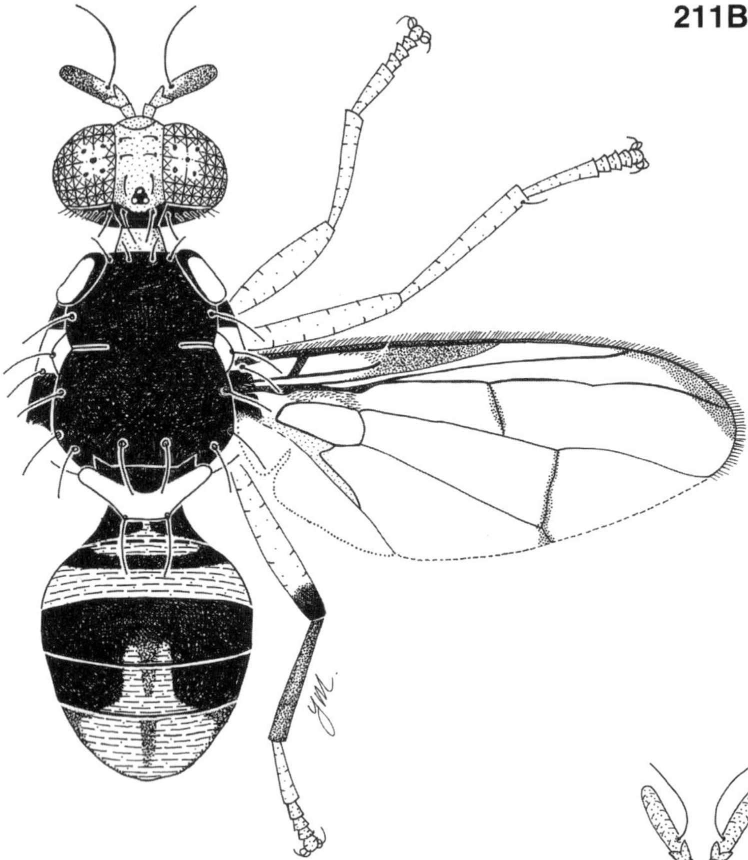

212

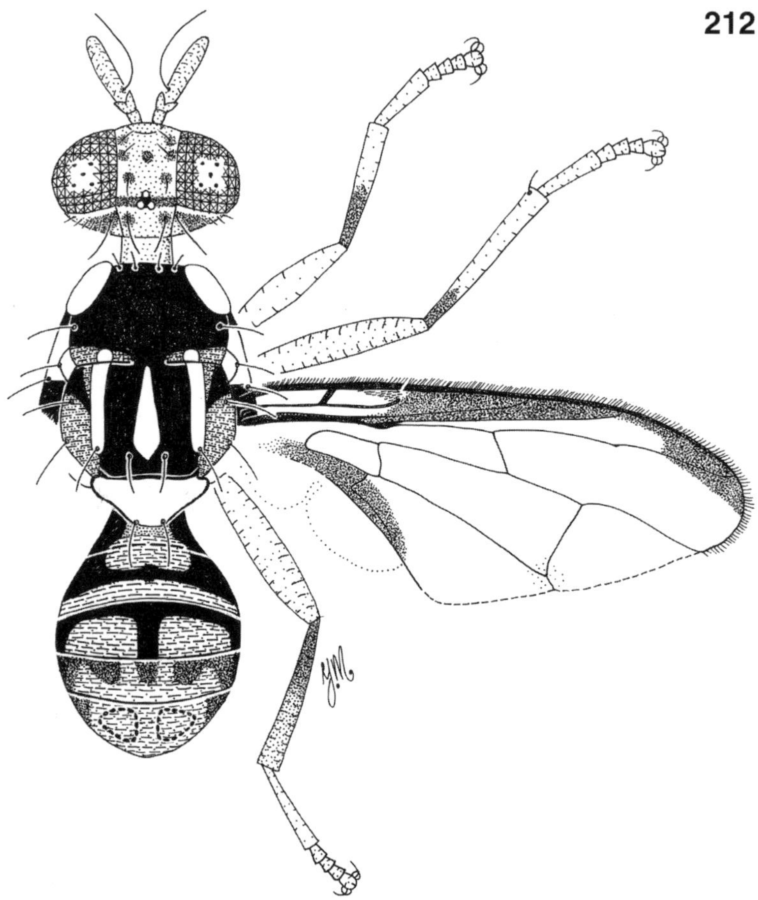

Fig. 211B. *Bactrocera (Gymnodacus) absona* (Hering), holotype male. **Fig. 212.** *Bactrocera (Hemigymnodacus) diversa* (Coquillett), male.

213A

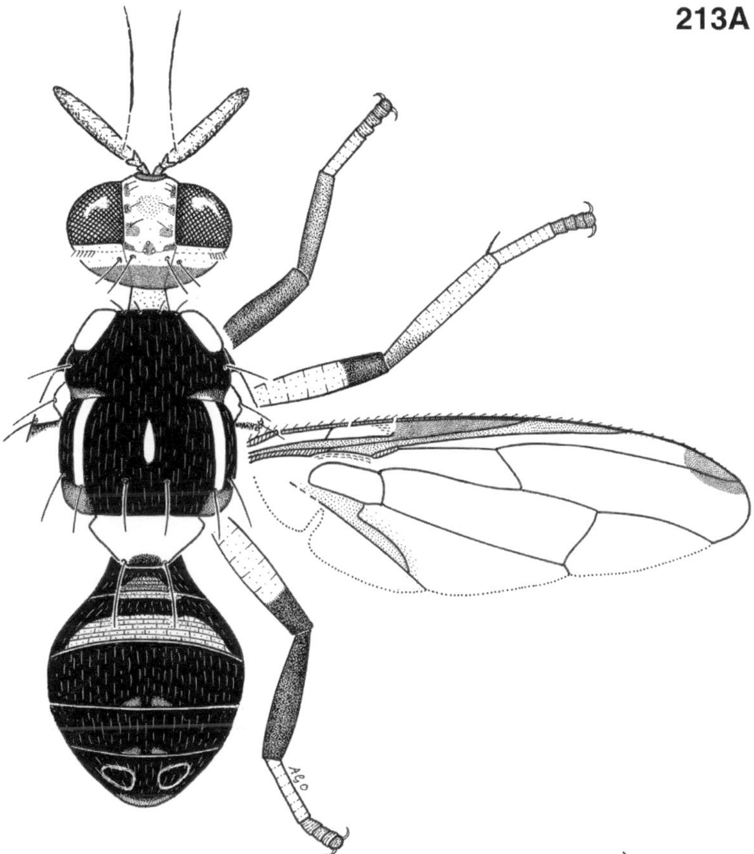

213B

Fig. 213A. *Bactrocera (Hemigymnodacus) mukiae* new species, holotype male. **Fig. 213B.** *Bactrocera (Hemigymnodacus) mukiae* new species, paratype female.

214

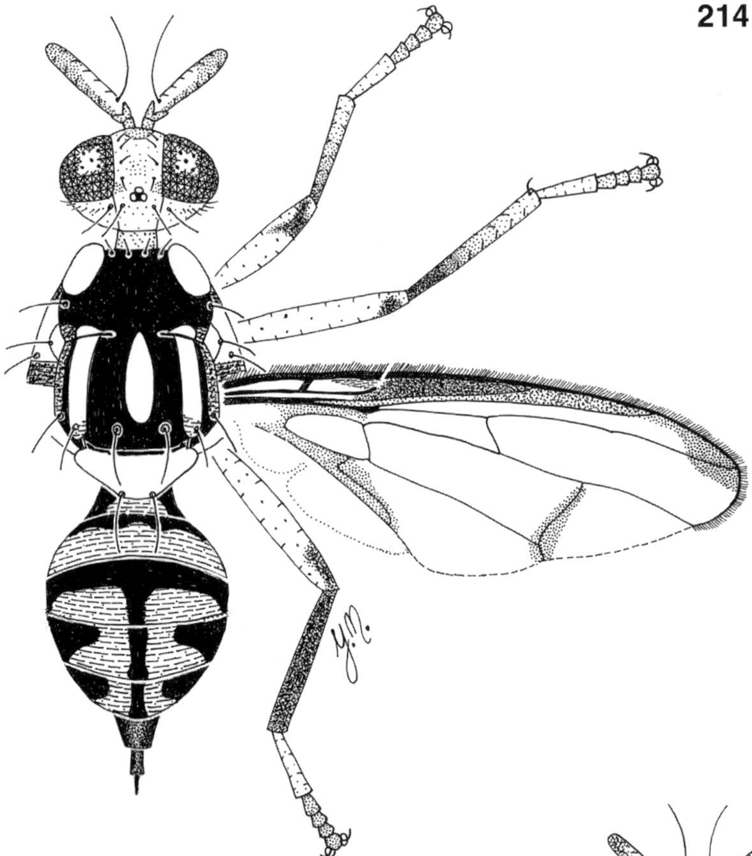

215

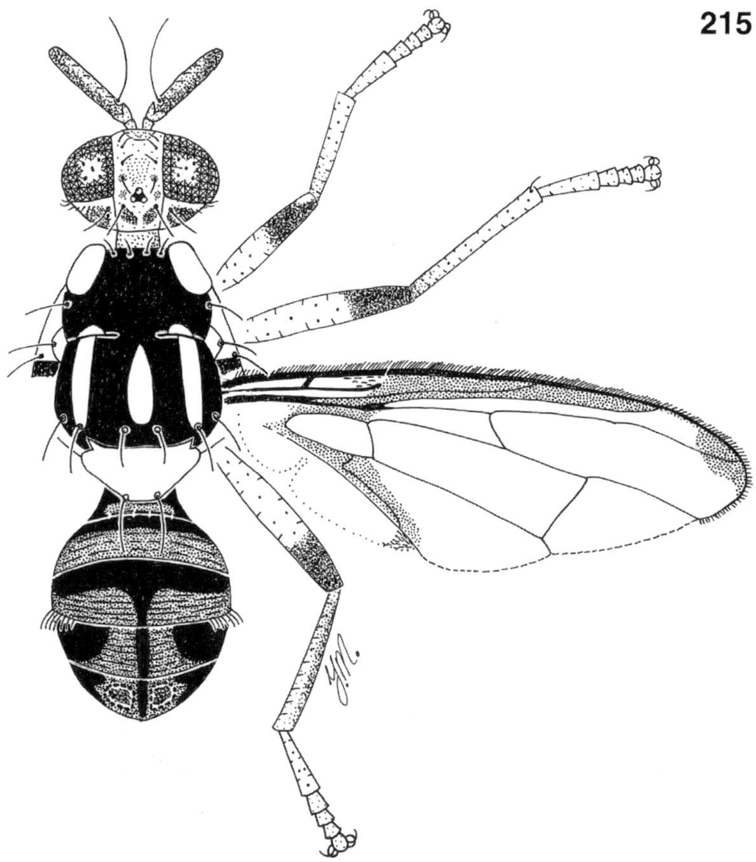

Fig. 214. *Bactrocera (Javadacus) javanensis* (Perkins), holotype female. **Fig. 215.** *Bactrocera (Javadacus) montana* (Hardy), holotype male.

216

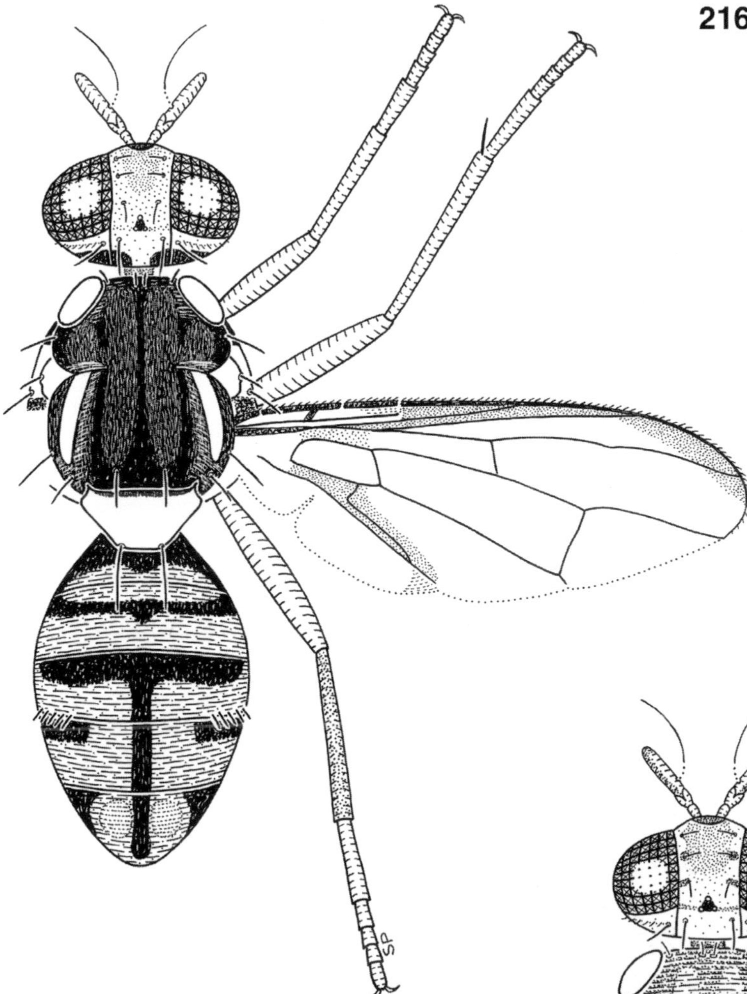

217

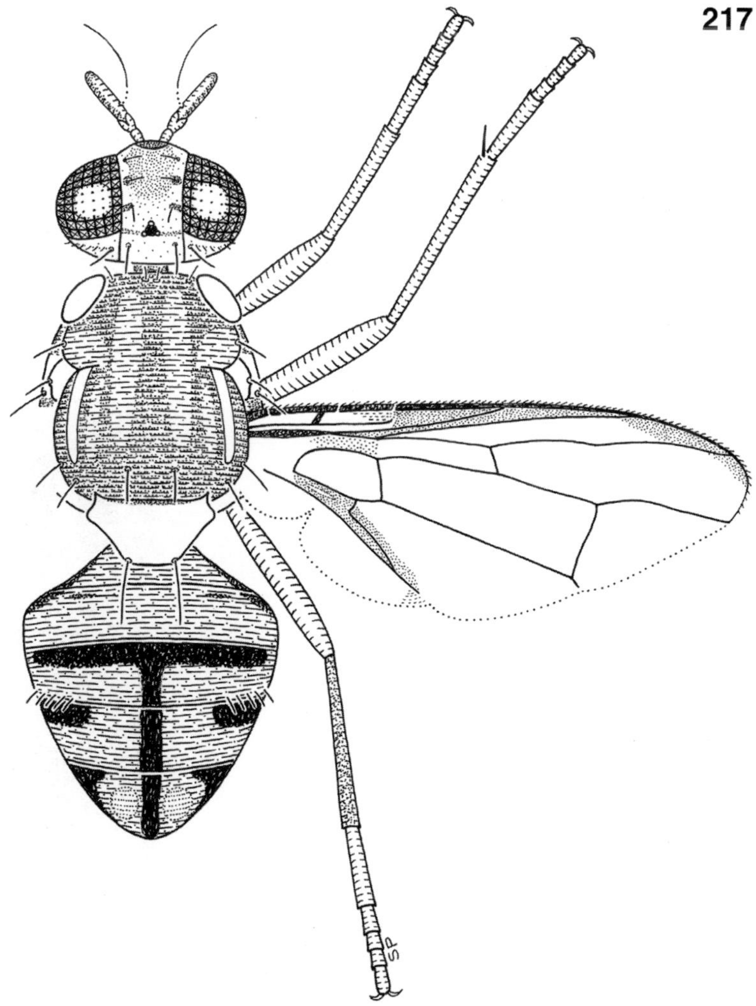

Fig. 216. *Bactrocera (Javadacus) nigrita* (Hardy), holotype male. **Fig. 217.** *Bactrocera (Javadacus) pallescentis* (Hardy), holotype male.

218

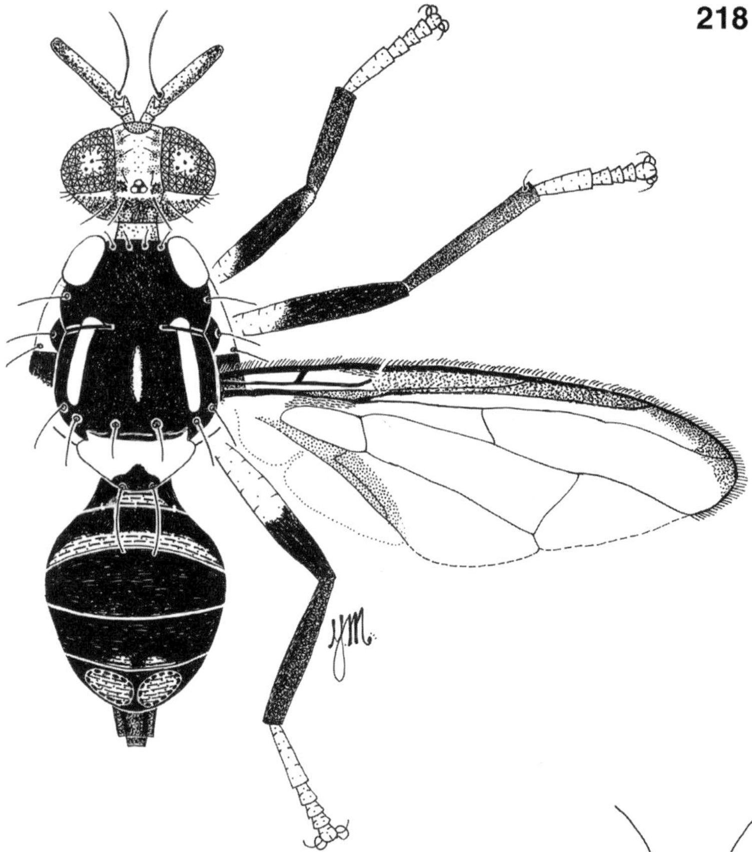

219

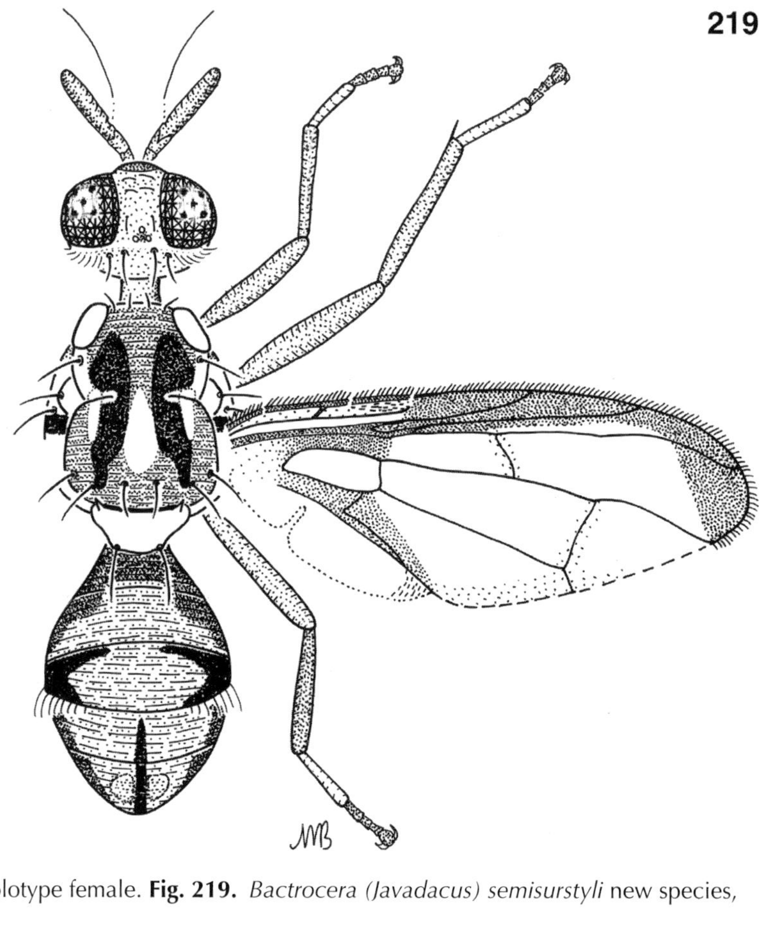

Fig. 218. *Bactrocera (Javadacus) scutellaria* (Bezzi), holotype female. **Fig. 219.** *Bactrocera (Javadacus) semisurstyli* new species, holotype male.

220

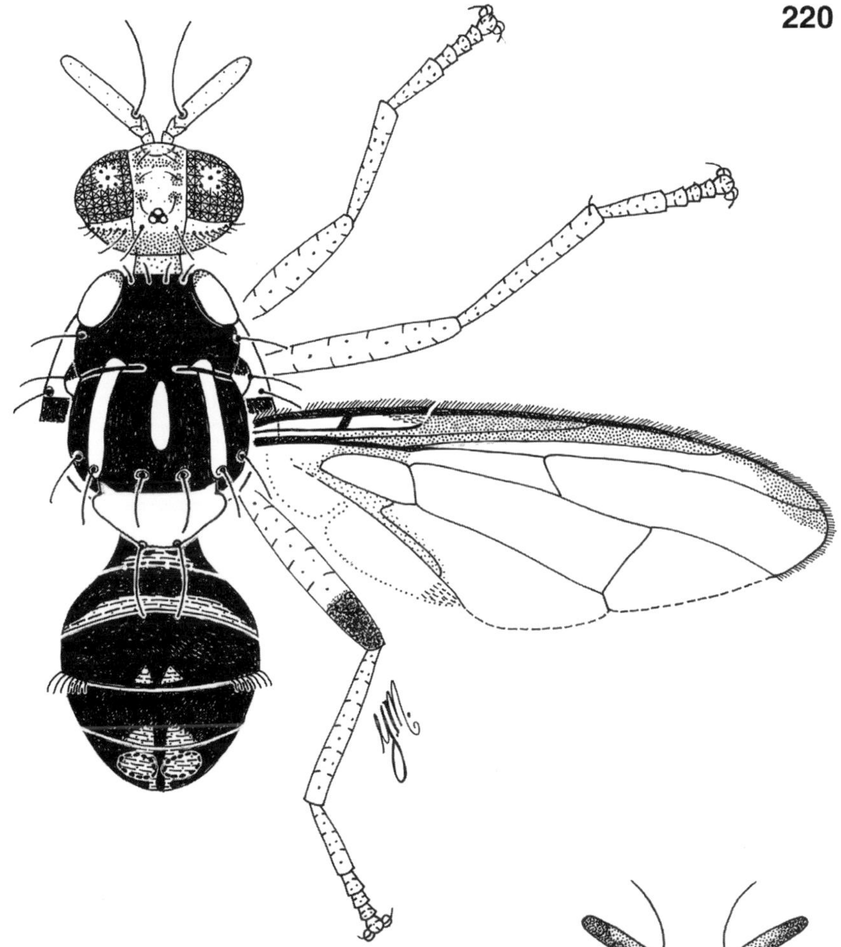

221A

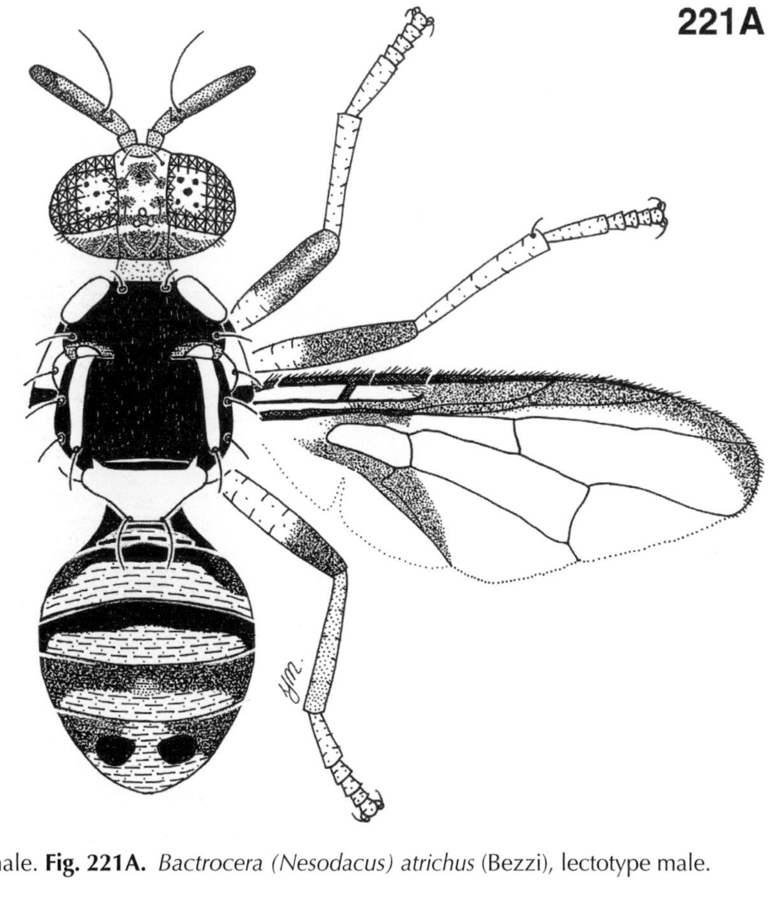

Fig. 220. *Bactrocera (Javadacus) trilineata* (Hardy), male. **Fig. 221A.** *Bactrocera (Nesodacus) atrichus* (Bezzi), lectotype male.

221B

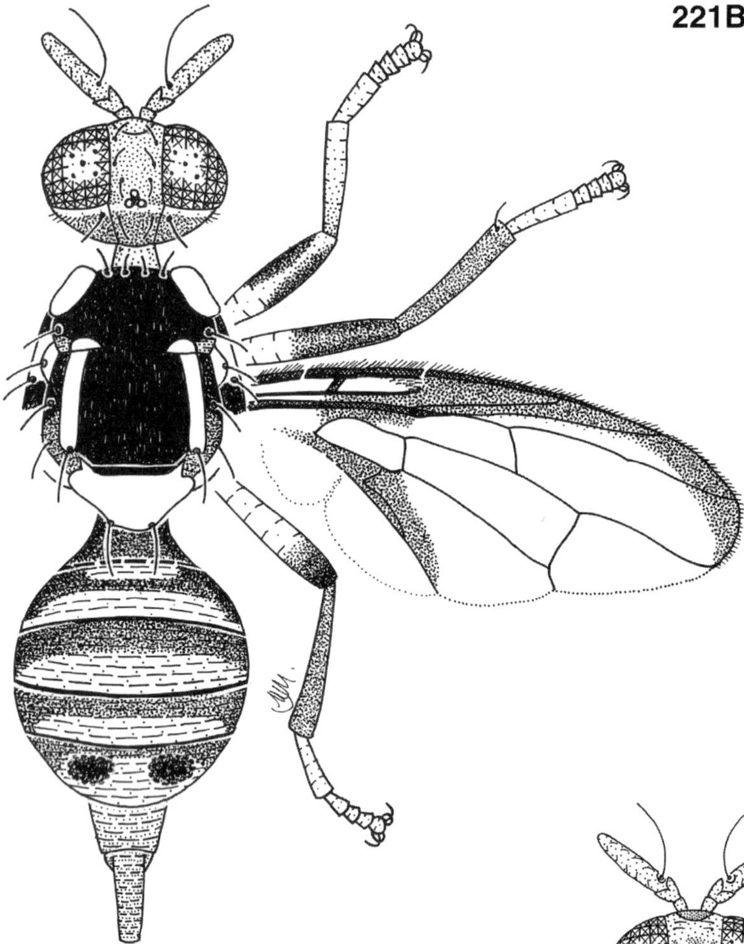

222

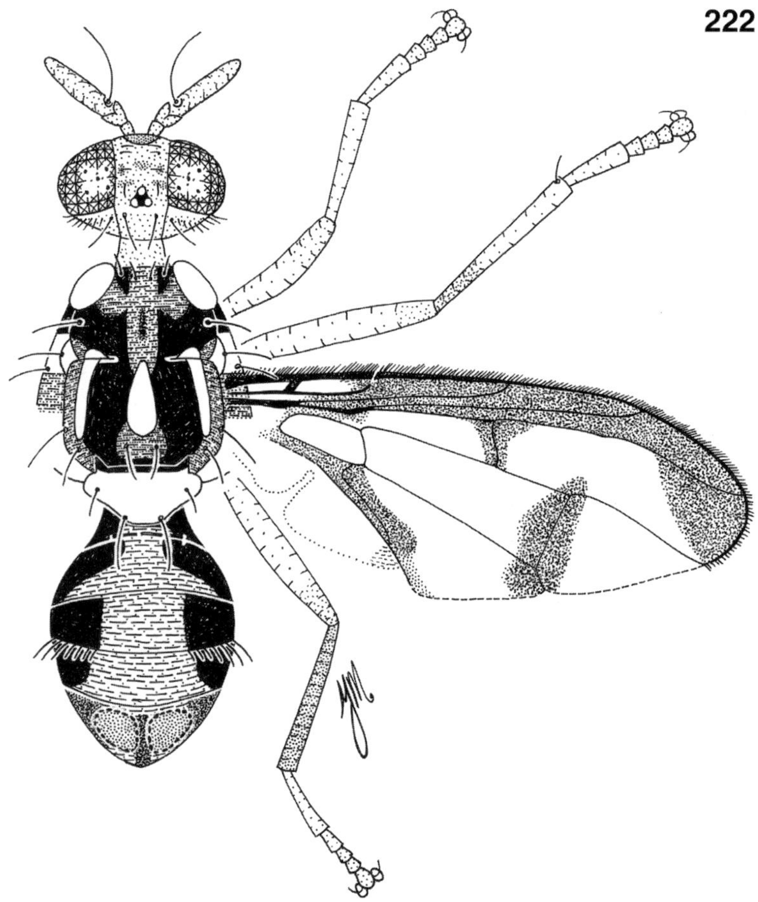

Fig. 221B. *Bactrocera (Nesodacus) davaoanus* (Bezzi), holotype female. **Fig. 222.** *Bactrocera (Papuodacus) complicata* White, male.

223

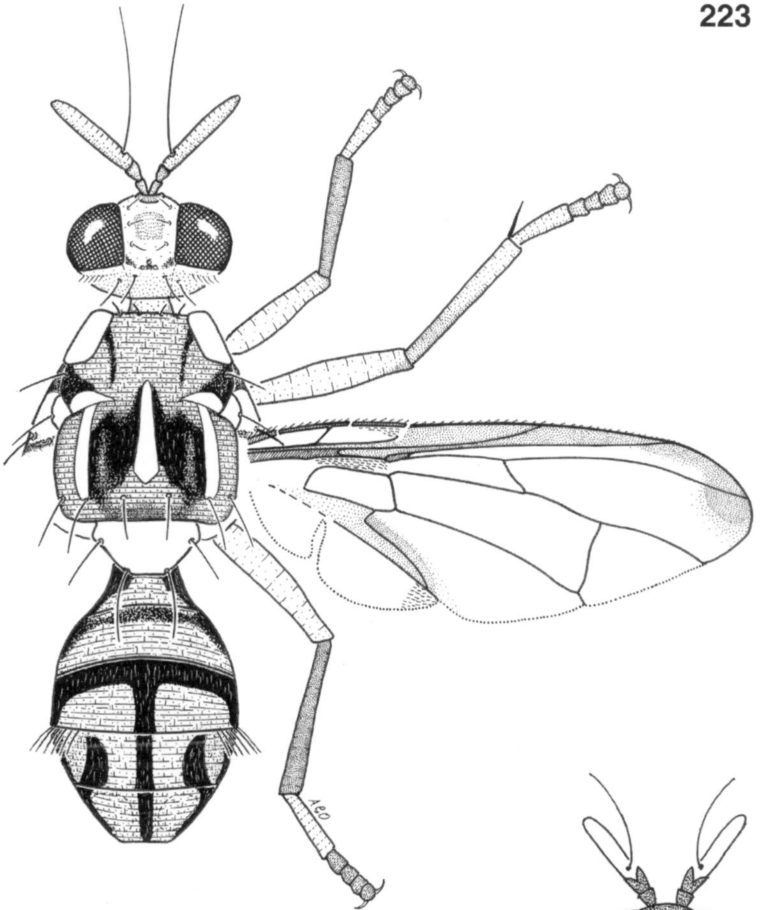

224

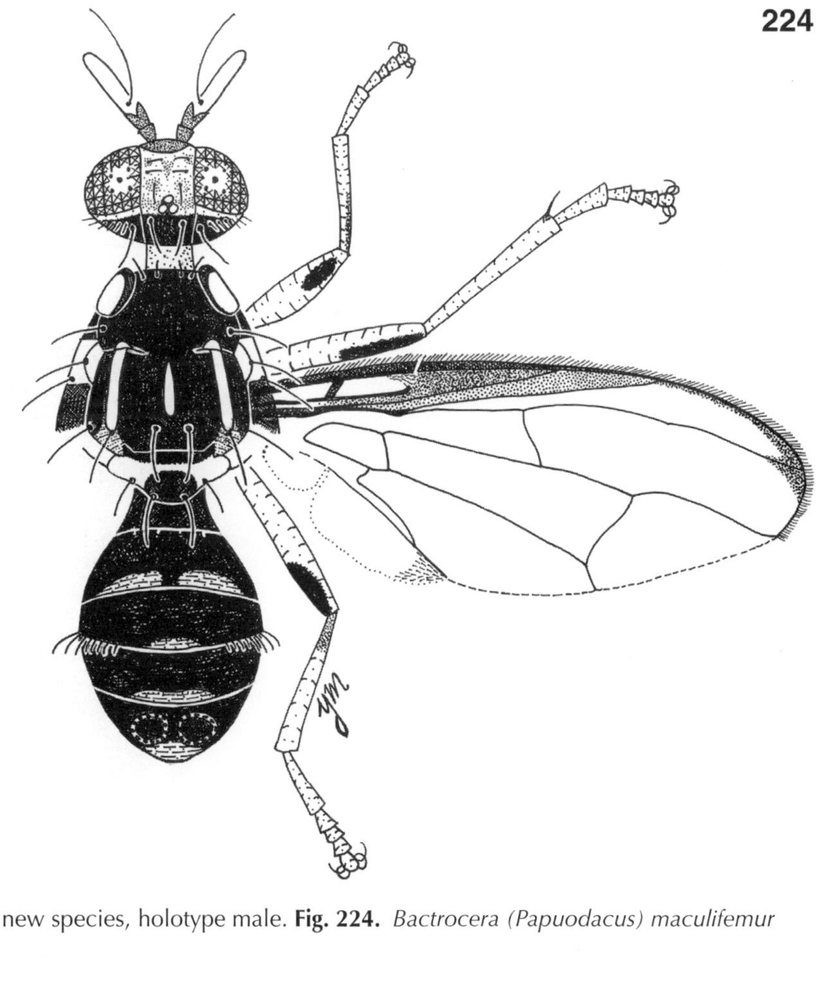

Fig. 223. *Bactrocera (Papuodacus) fereuncinata* new species, holotype male. **Fig. 224.** *Bactrocera (Papuodacus) maculifemur* (Hering), holotype male.

225

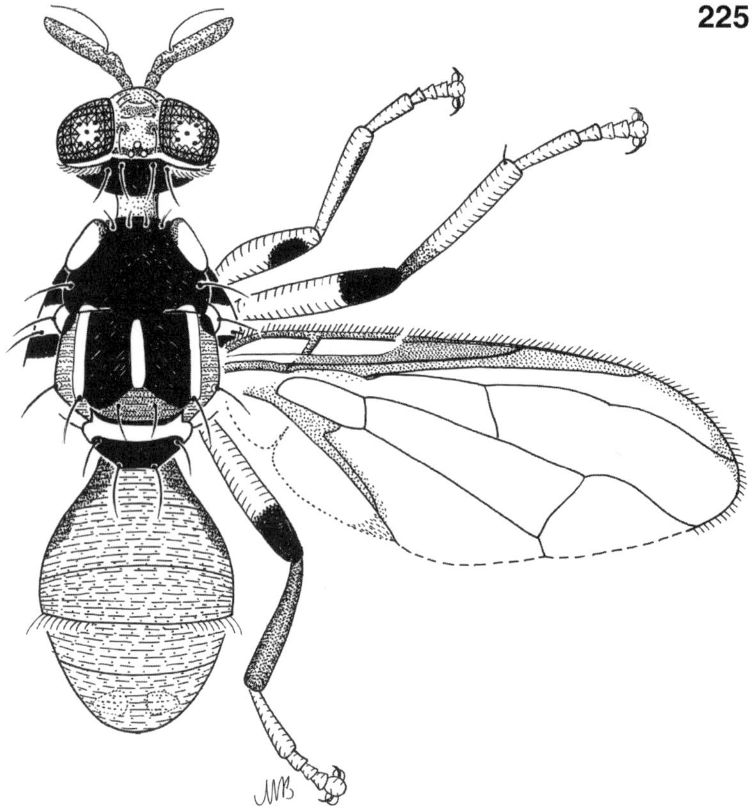

226

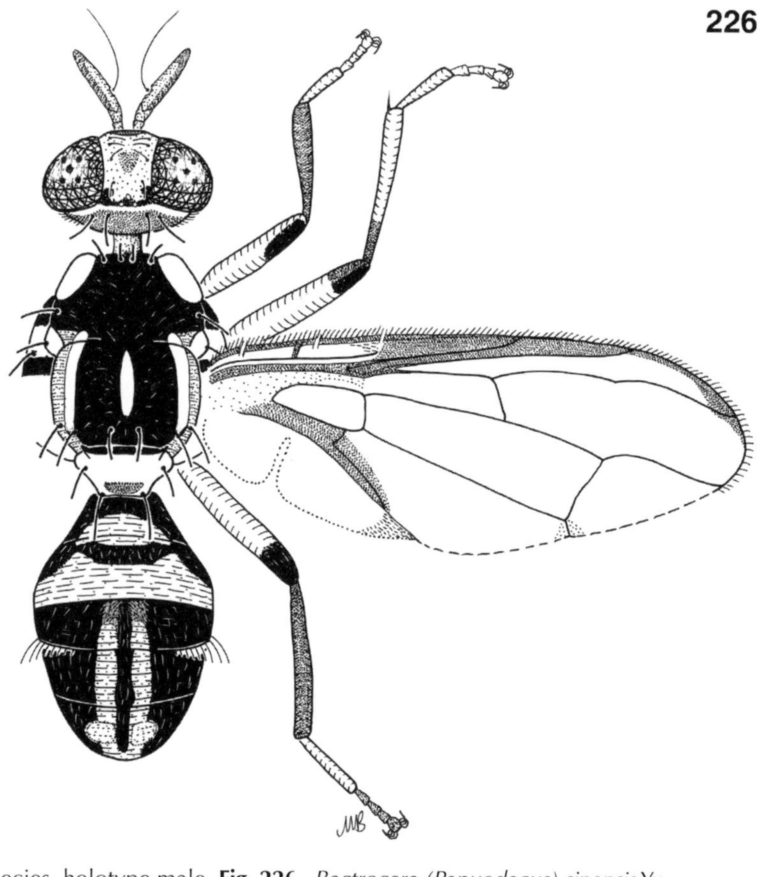

Fig. 225. *Bactrocera (Papuodacus) ochrosterna* new species, holotype male. **Fig. 226.** *Bactrocera (Papuodacus) sinensis* Yu, Bai & Chen, male.

227

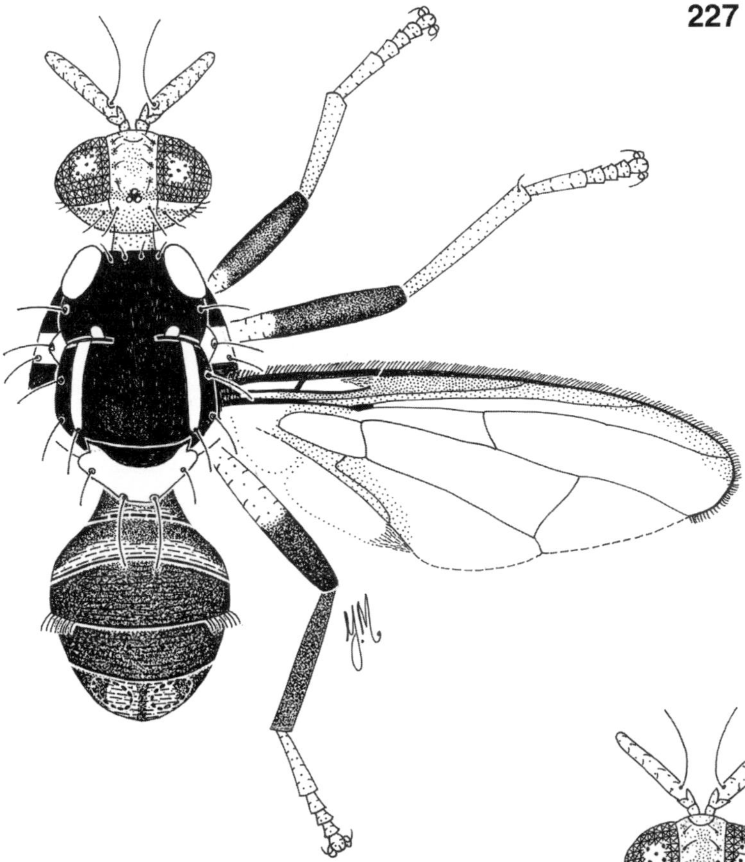

228

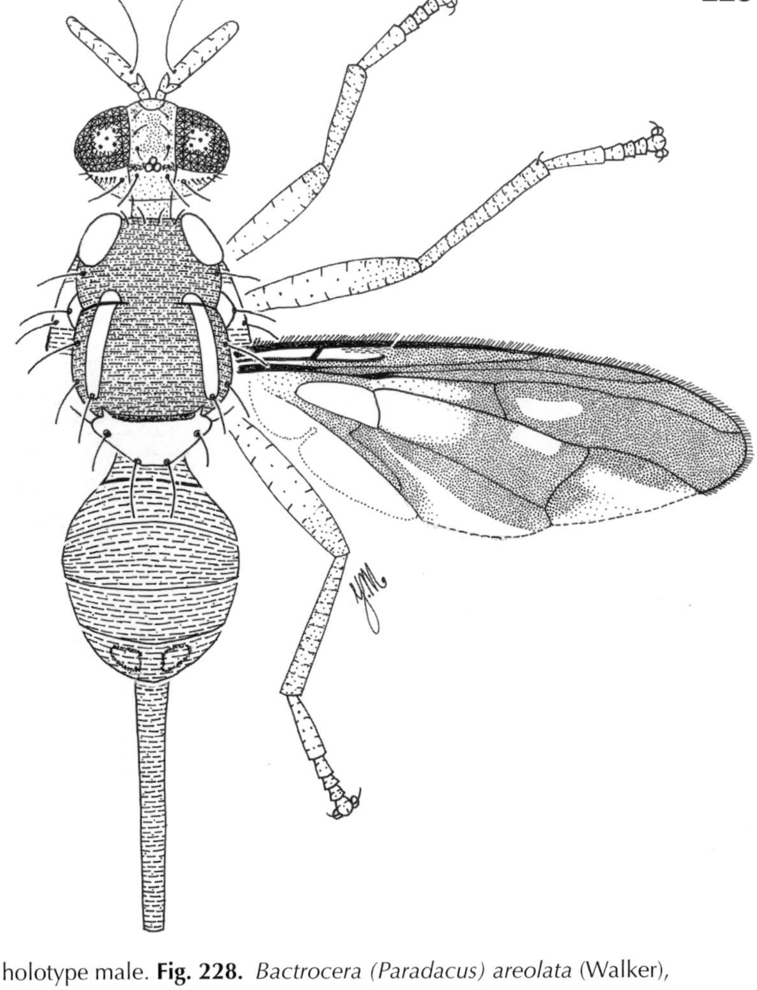

Fig. 227. *Bactrocera (Paradacus) angustifinis* (Hardy), holotype male. **Fig. 228.** *Bactrocera (Paradacus) areolata* (Walker), lectotype female.

229

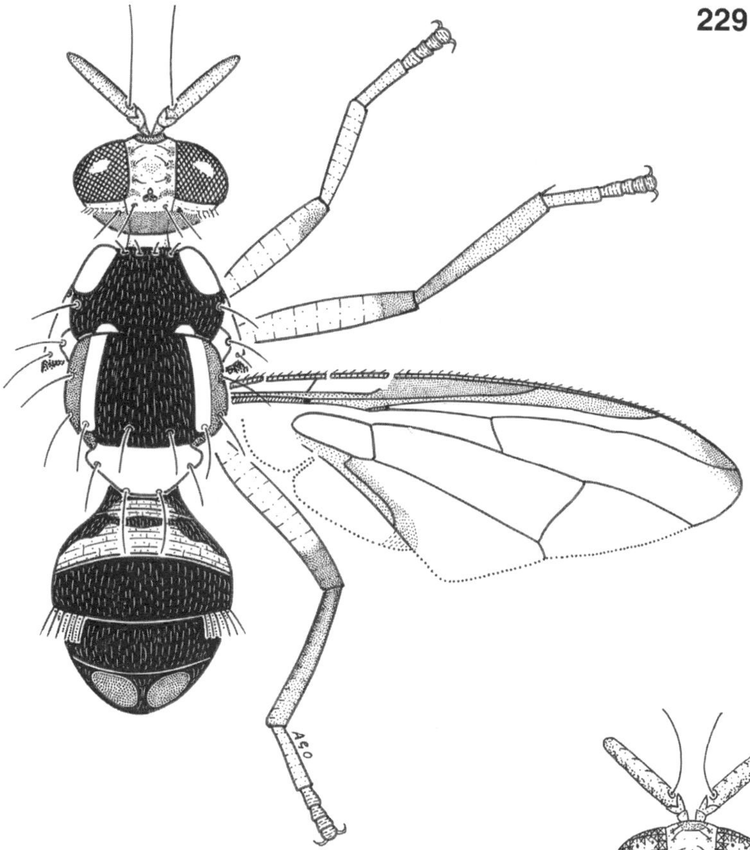

230

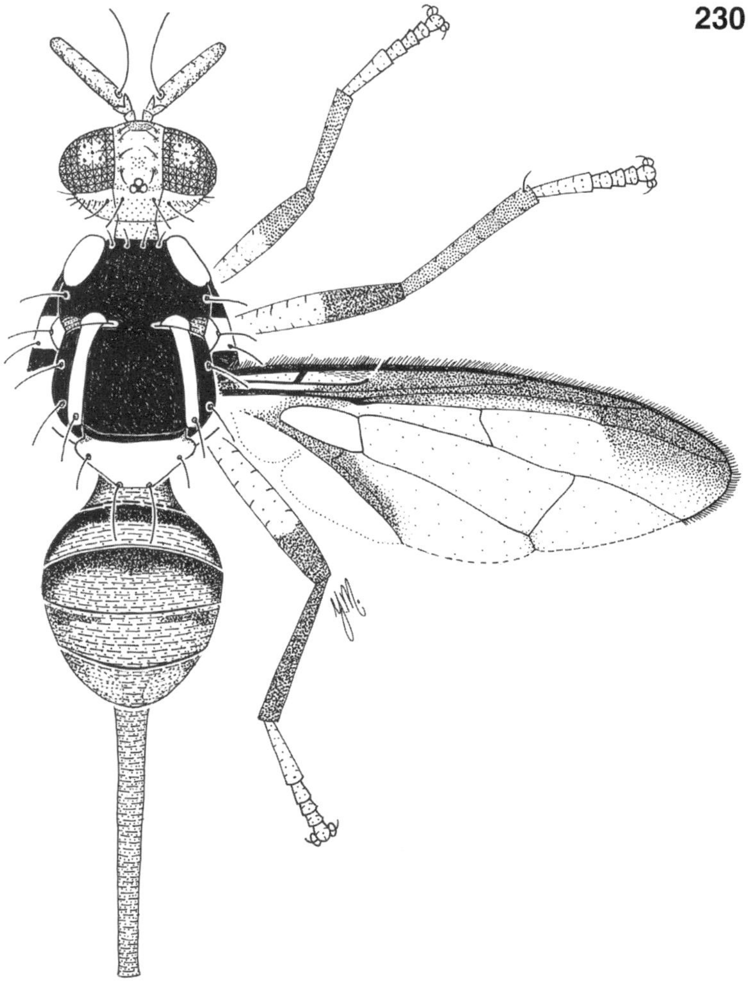

Fig. 229. *Bactrocera (Paradacus) duplicata* (Bezzi), syntype male. **Fig. 230.** *Bactrocera (Paradacus) fulvipes* (Perkins), female.

231

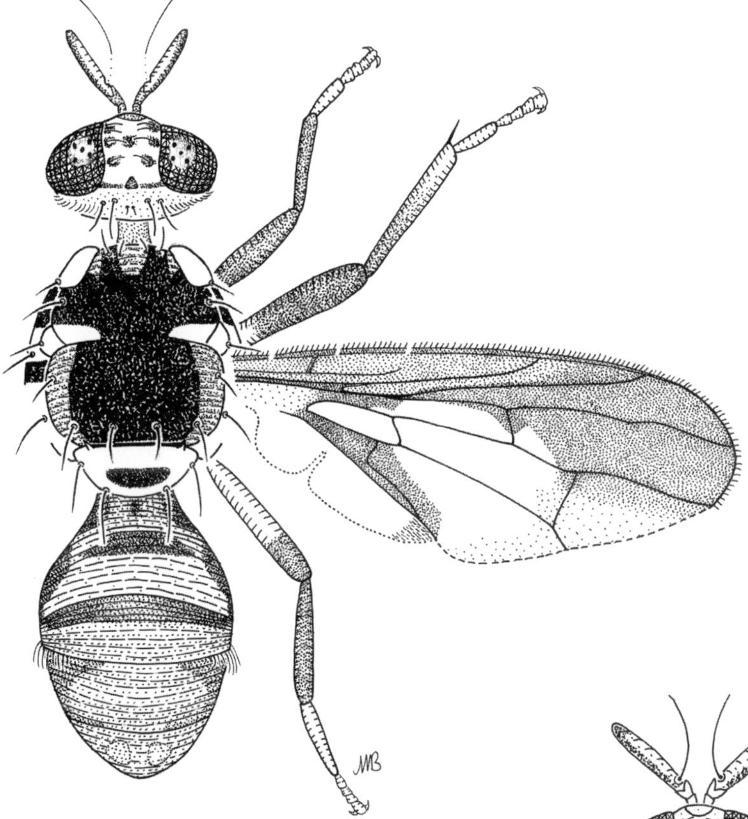

232

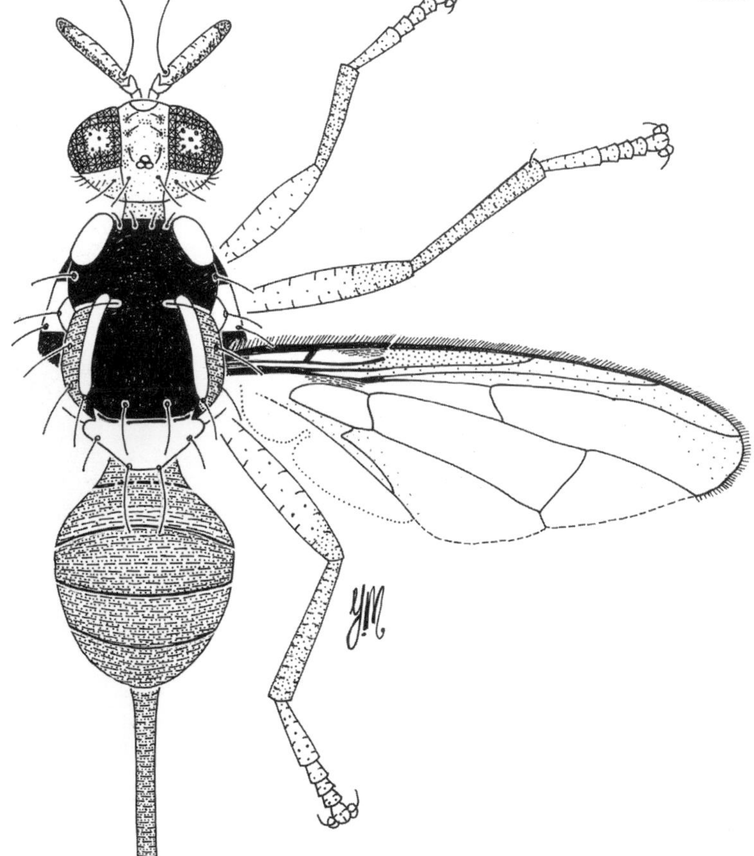

Fig. 231. *Bactrocera (Paradacus) hancocki* new species, holotype male. **Fig. 232.** *Bactrocera (Paradacus) urens* White, holotype female.

233A

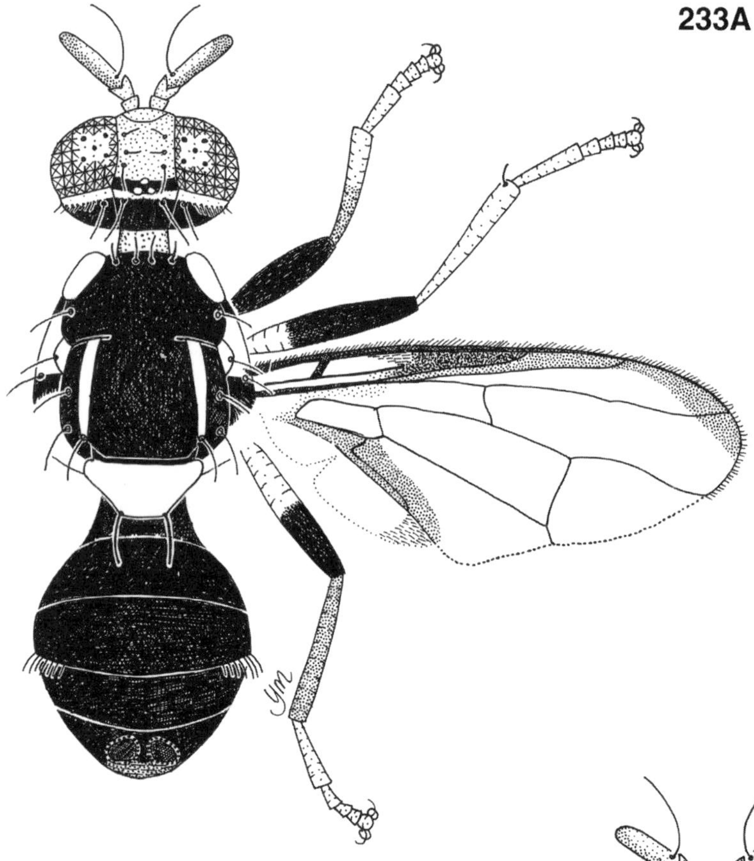

233B

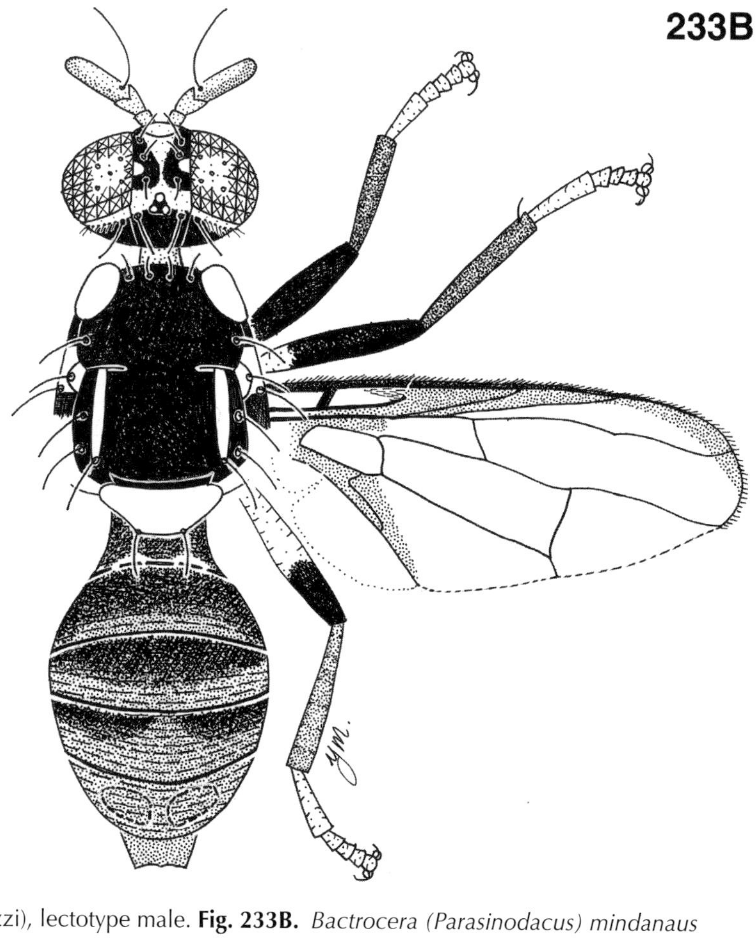

Fig. 233A. *Bactrocera (Parasinodacus) ablepharus* (Bezzi), lectotype male. **Fig. 233B.** *Bactrocera (Parasinodacus) mindanaus* (Bezzi), holotype female.

234

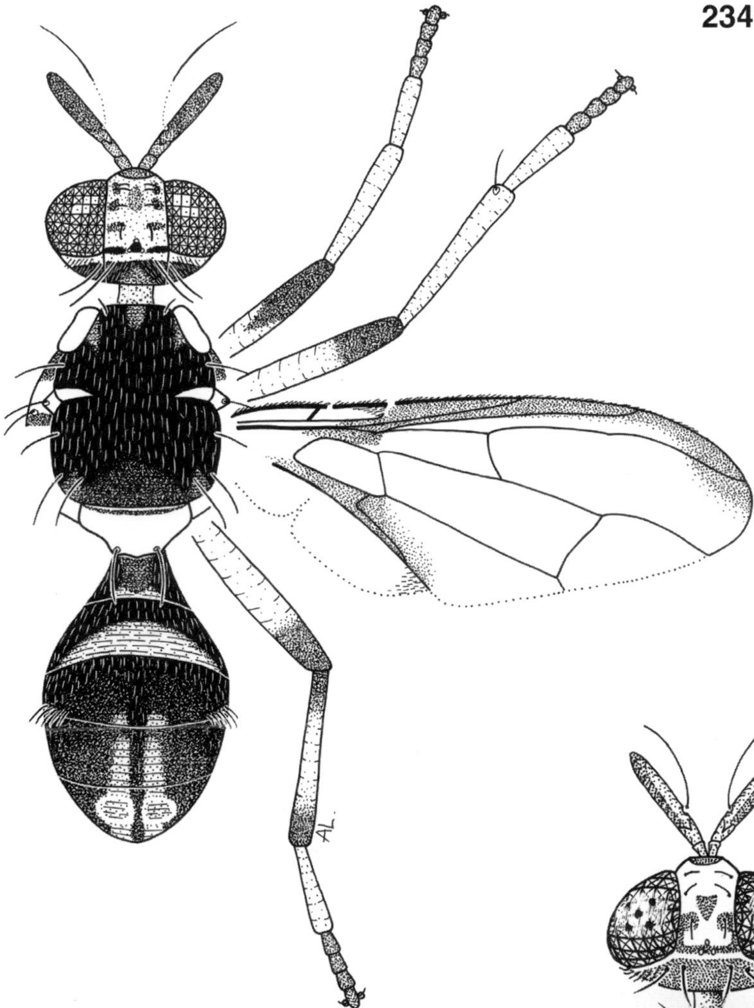

235

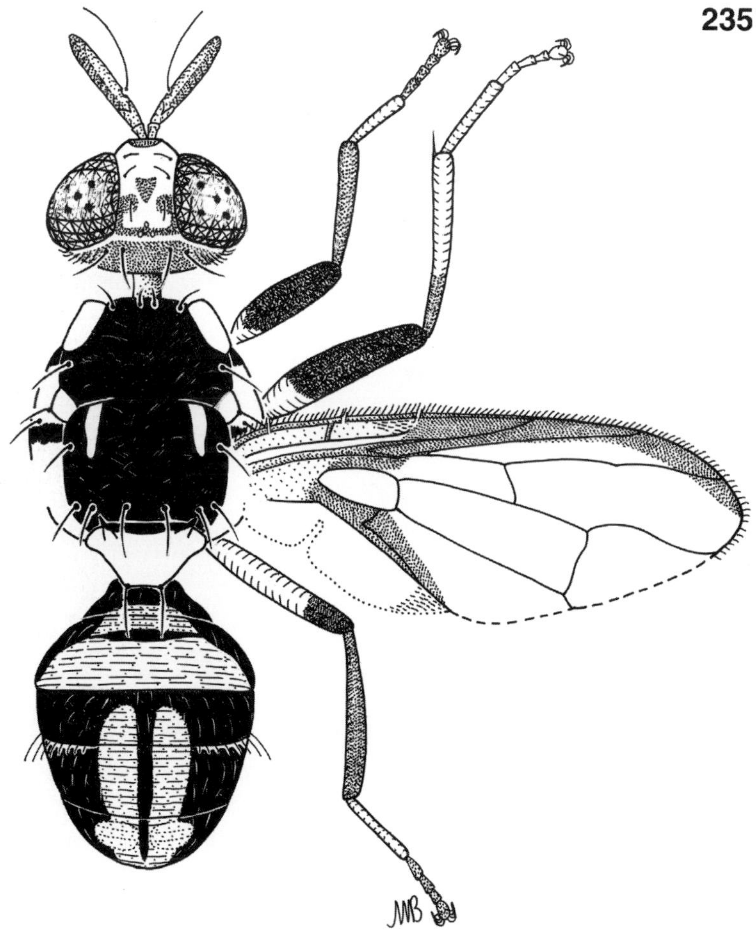

Fig. 234. *Bactrocera (Parasinodacus) binoyi* Drew, holotype male. **Fig. 235.** *Bactrocera (Parasinodacus) brevivitta* new species, holotype male.

236

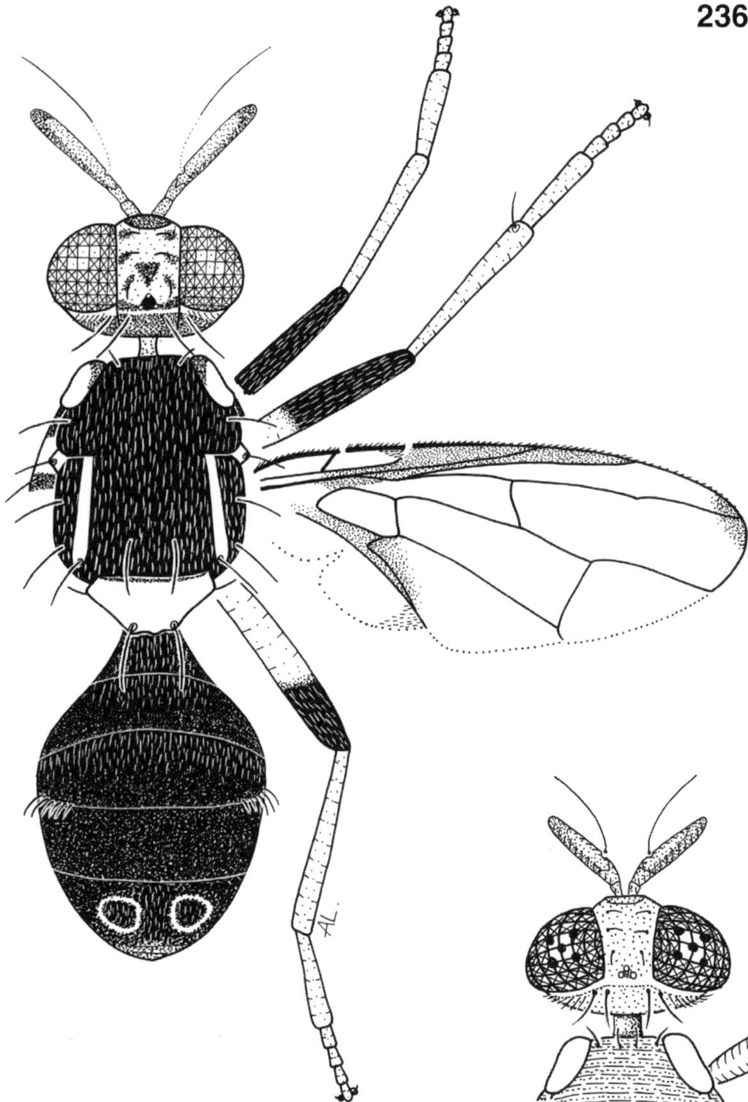

237

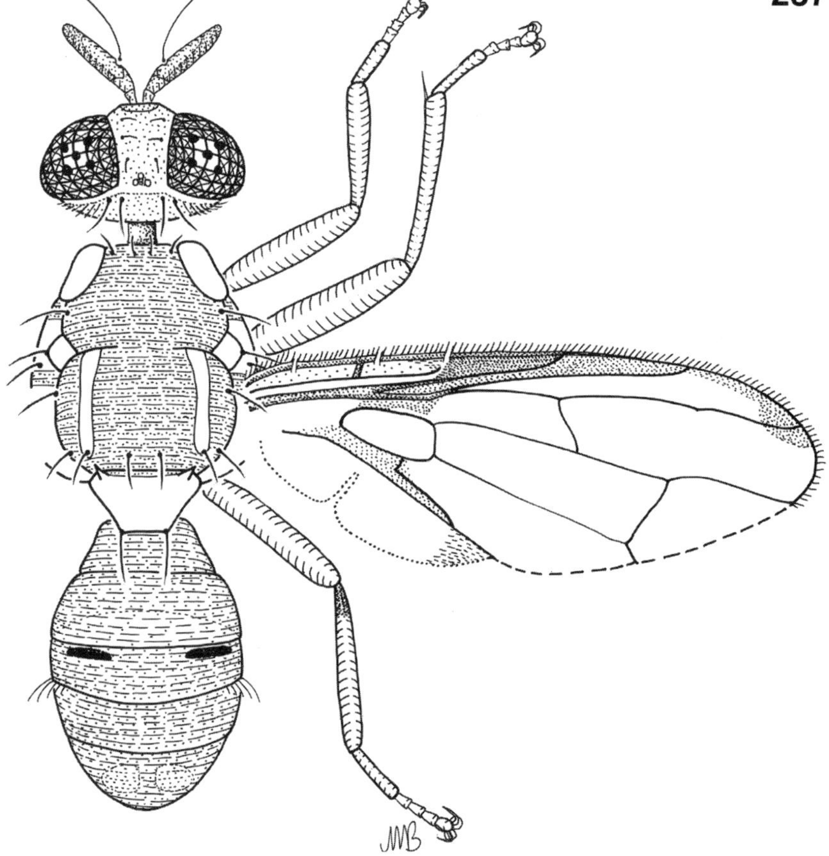

Fig. 236. *Bactrocera (Parasinodacus) cilifera* (Hendel), syntype male. **Fig. 237.** *Bactrocera (Parasinodacus) citrifusca* new species, holotype male.

238

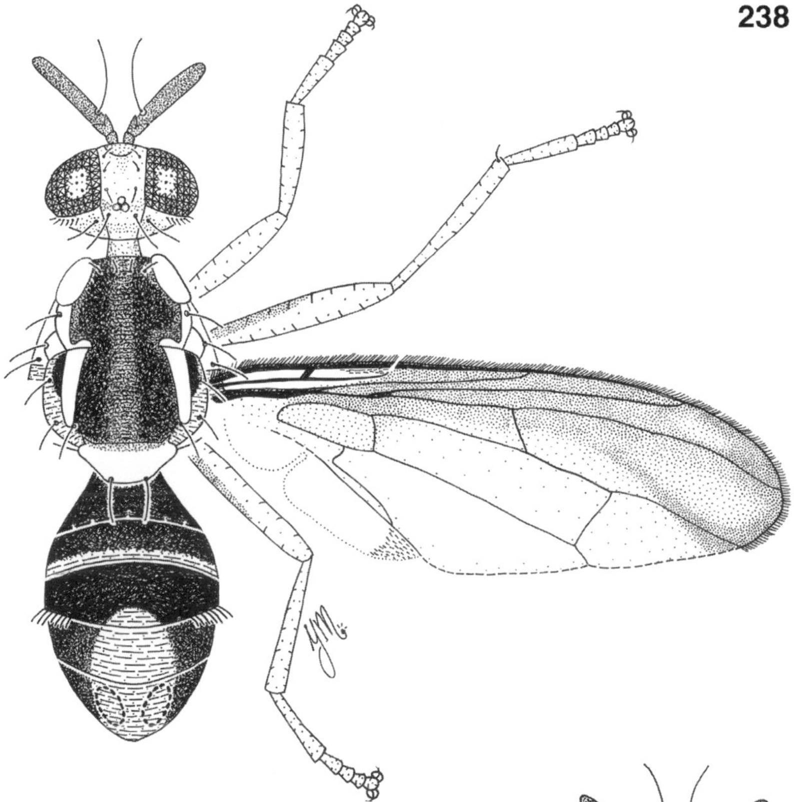

239

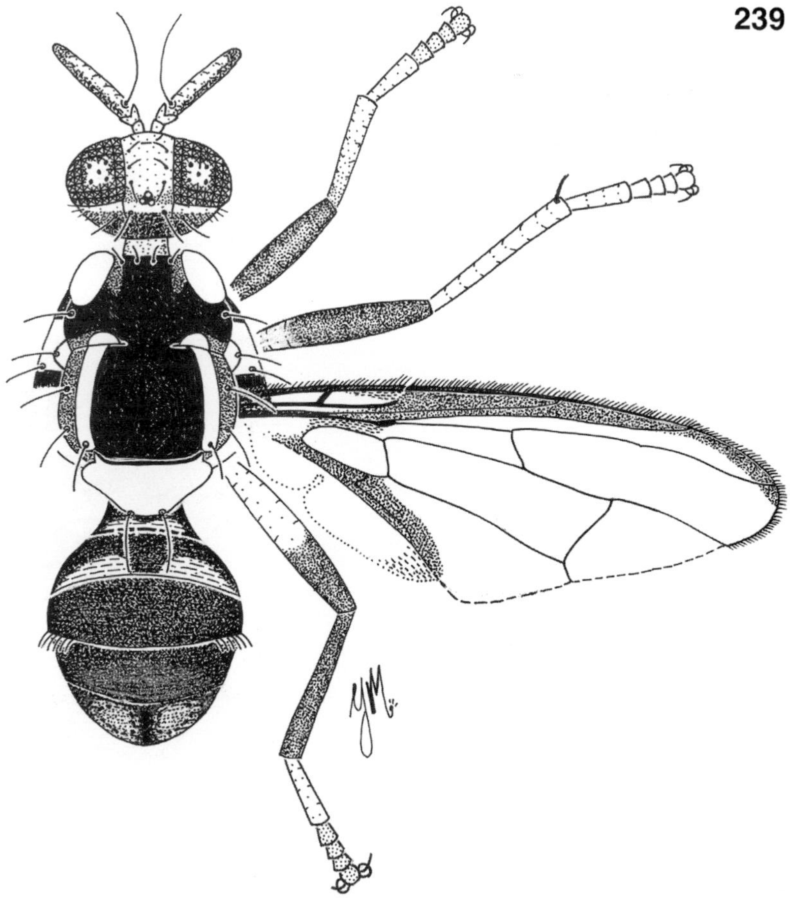

Fig. 238. *Bactrocera (Parasinodacus) eurylomata* (Hardy), holotype male. **Fig. 239.** *Bactrocera (Parasinodacus) incisa* (Walker), lectotype male.

240A

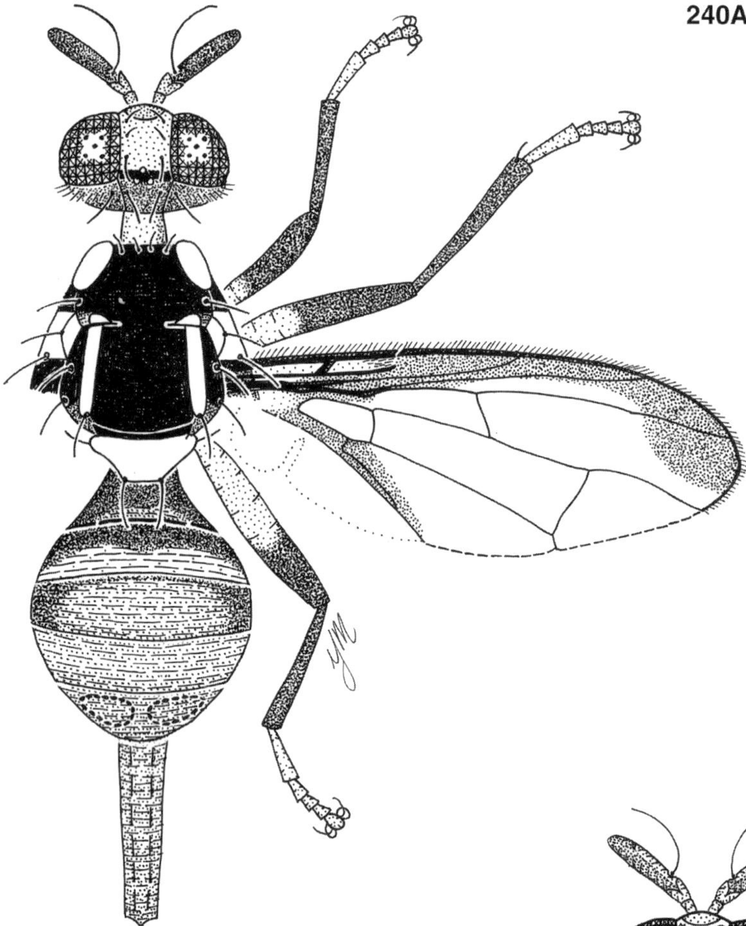

240B

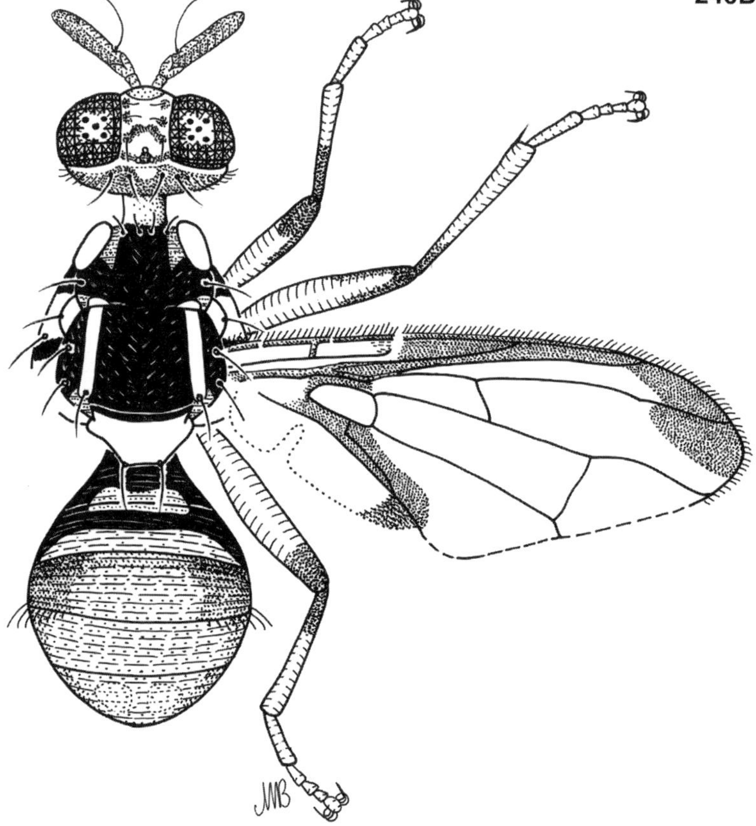

Fig. 240A. *Bactrocera (Parasinodacus) longicaudata* (Perkins), holotype female. **Fig. 240B.** *Bactrocera (Parasinodacus) longicaudata* (Perkins), male.

241

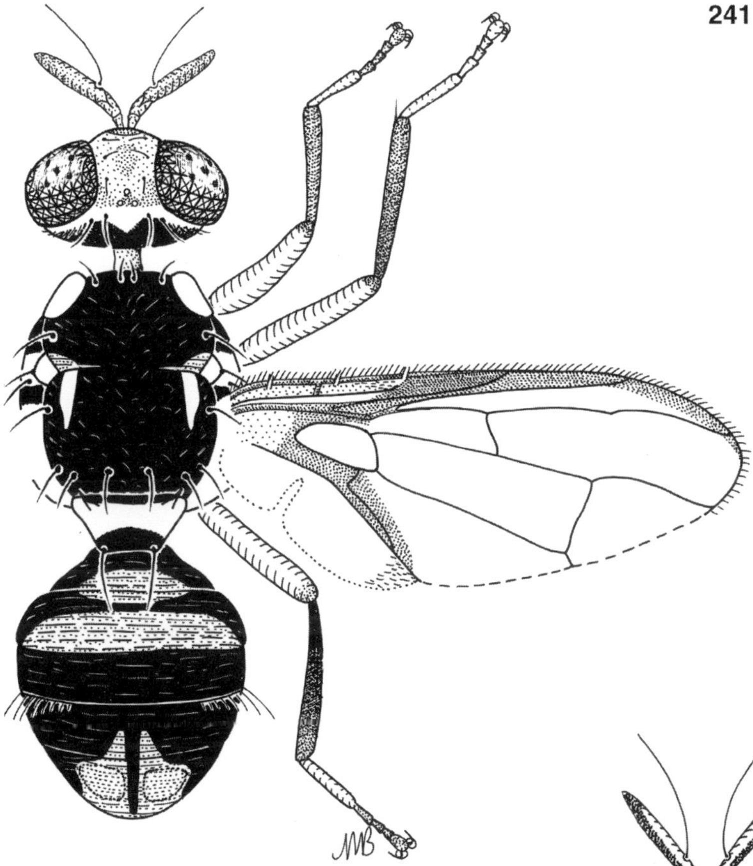

242

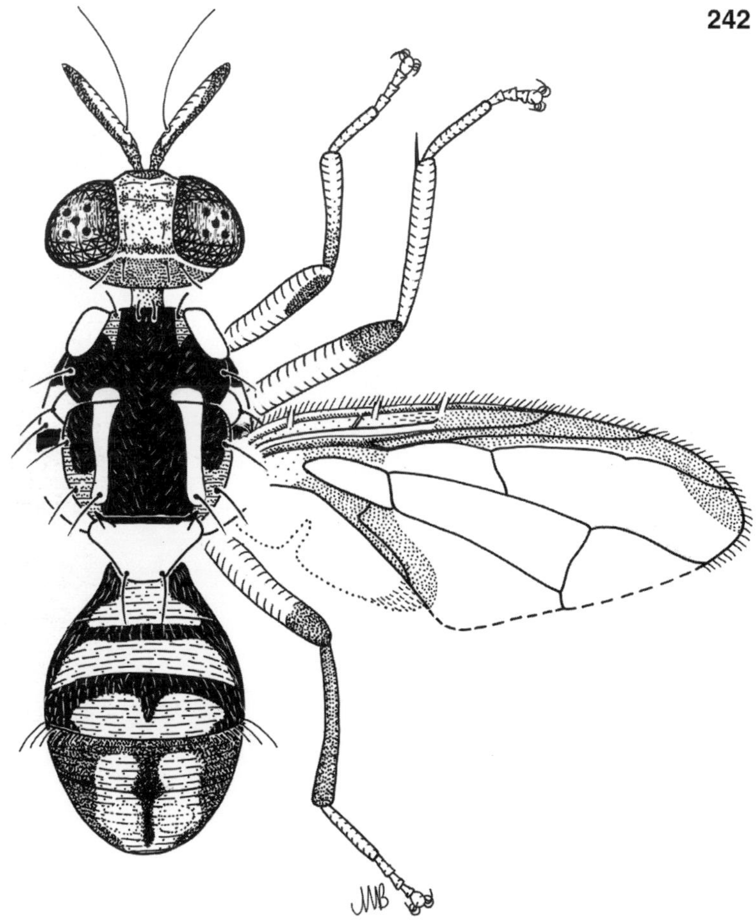

Fig. 241. *Bactrocera (Parasinodacus) pahangiae* new species, holotype male. **Fig. 242.** *Bactrocera (Parasinodacus) pantabanganiae* new species, holotype male.

243

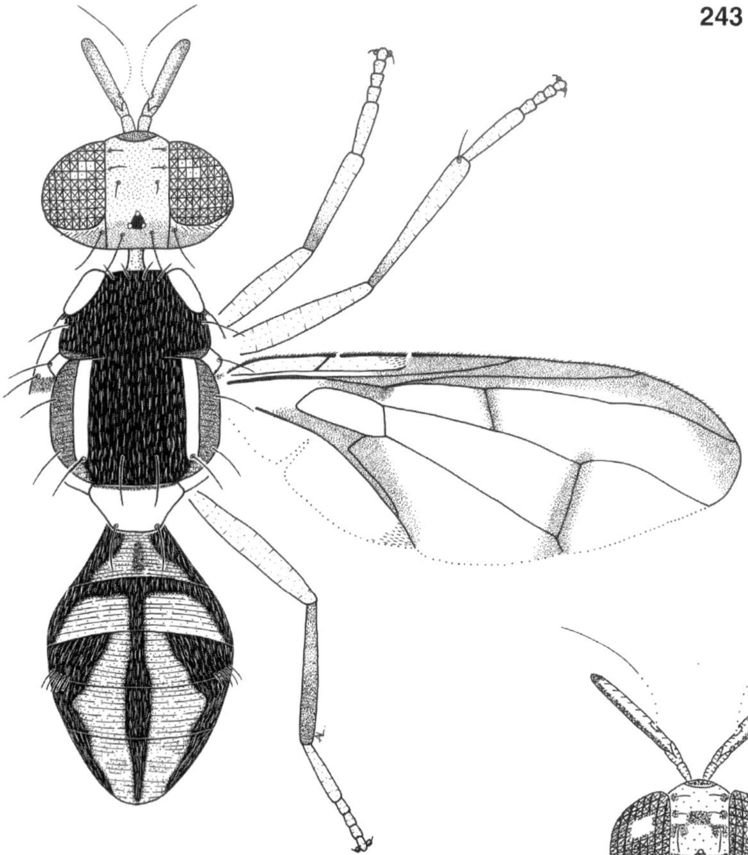

244

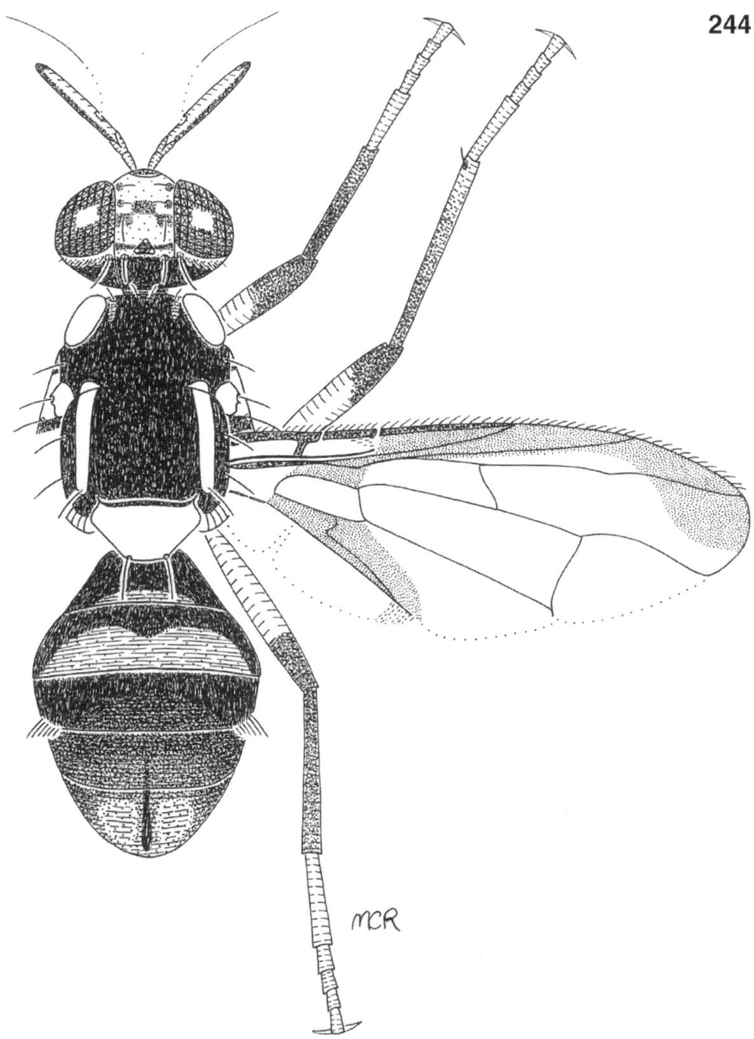

Fig. 243. *Bactrocera (Parasinodacus) pseudocucurbitae* White, holotype male. **Fig. 244.** *Bactrocera (Parasinodacus) vinnulus* (Hardy), male (illustration of paratype of *Bactrocera drewi* (Hardy)).

245

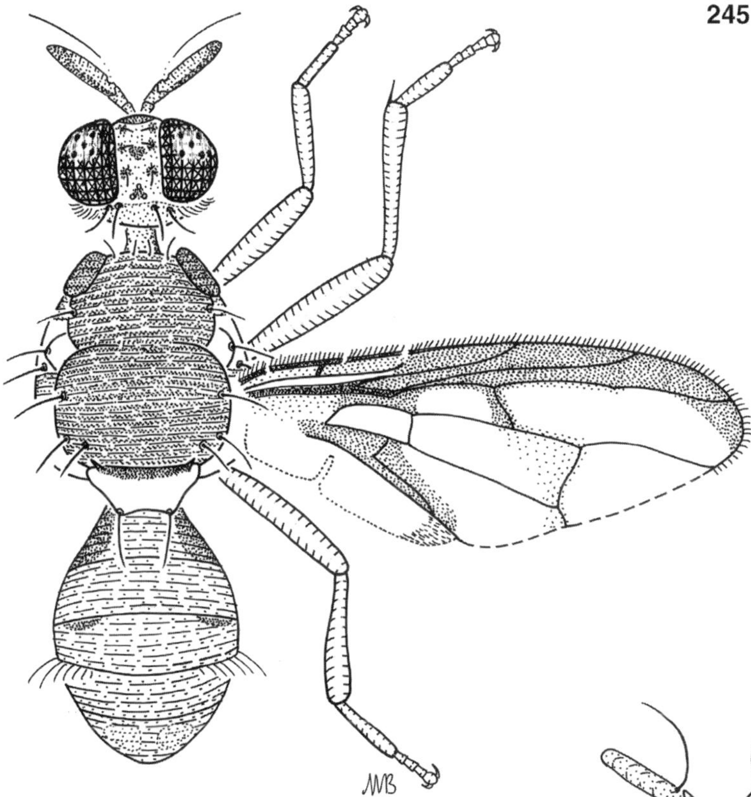

246

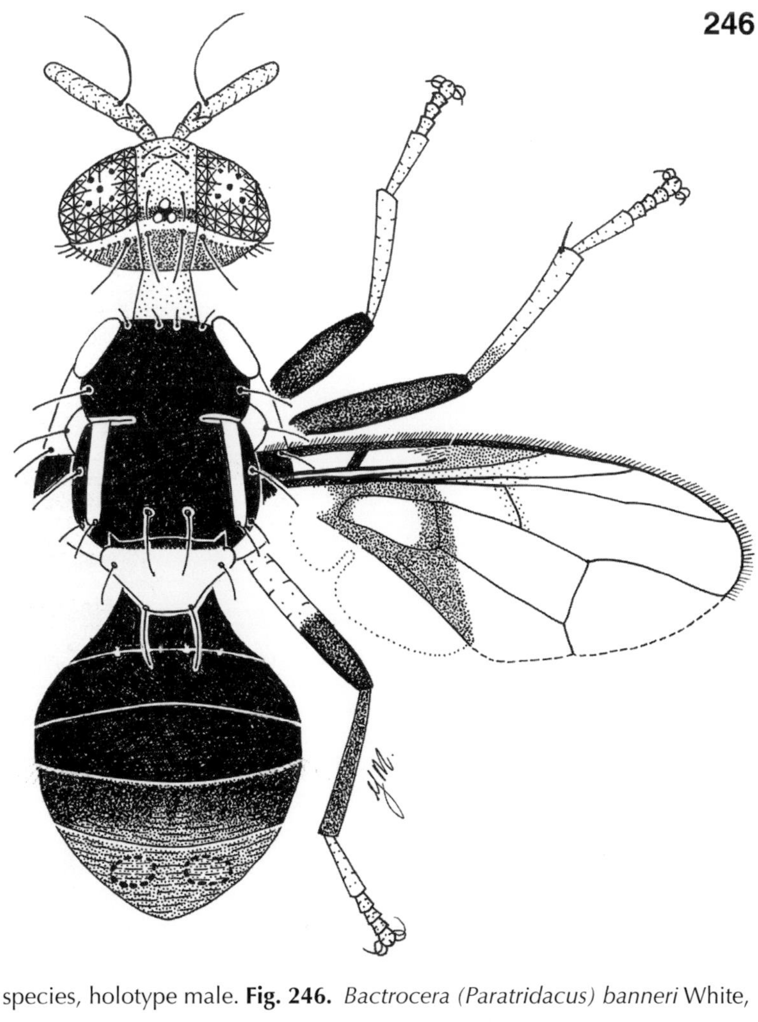

Fig. 245. *Bactrocera (Parasinodacus) waimitaliae* new species, holotype male. **Fig. 246.** *Bactrocera (Paratridacus) banneri* White, holotype male.

247

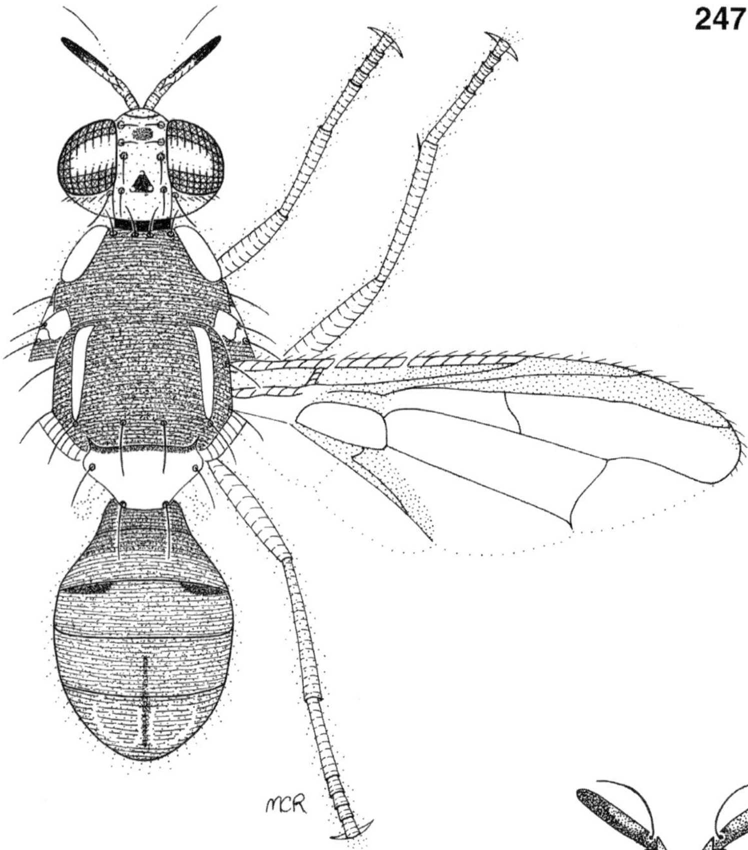

248

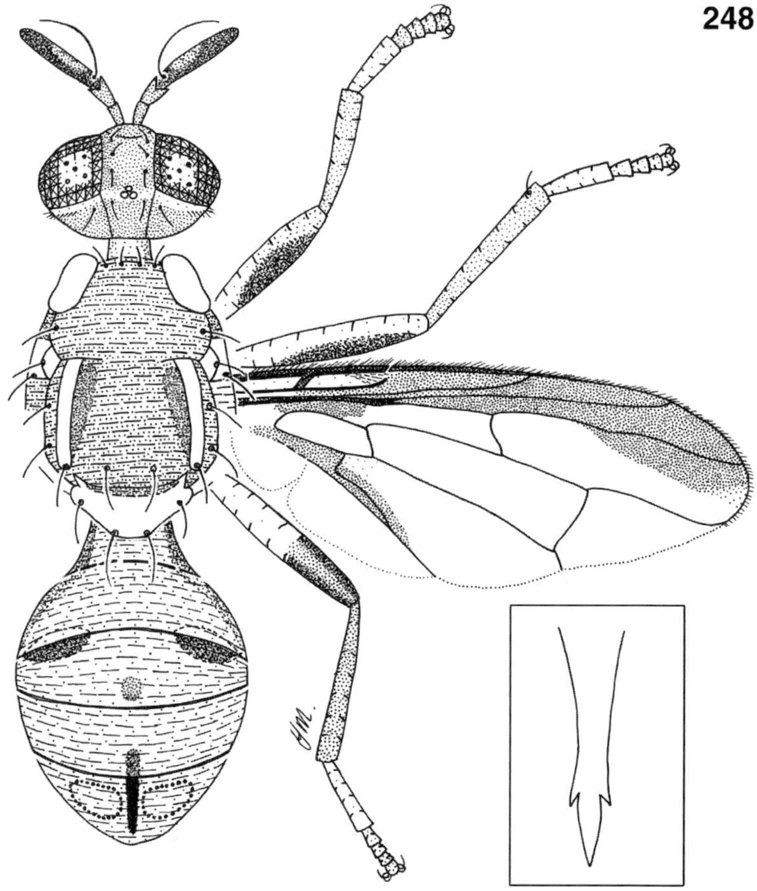

Fig. 247. *Bactrocera (Paratridacus) expandens* (Walker), male. **Fig. 248.** *Bactrocera (Paratridacus) garciniae* (Bezzi), syntype male and apex of piercer from female ovipositor.

249

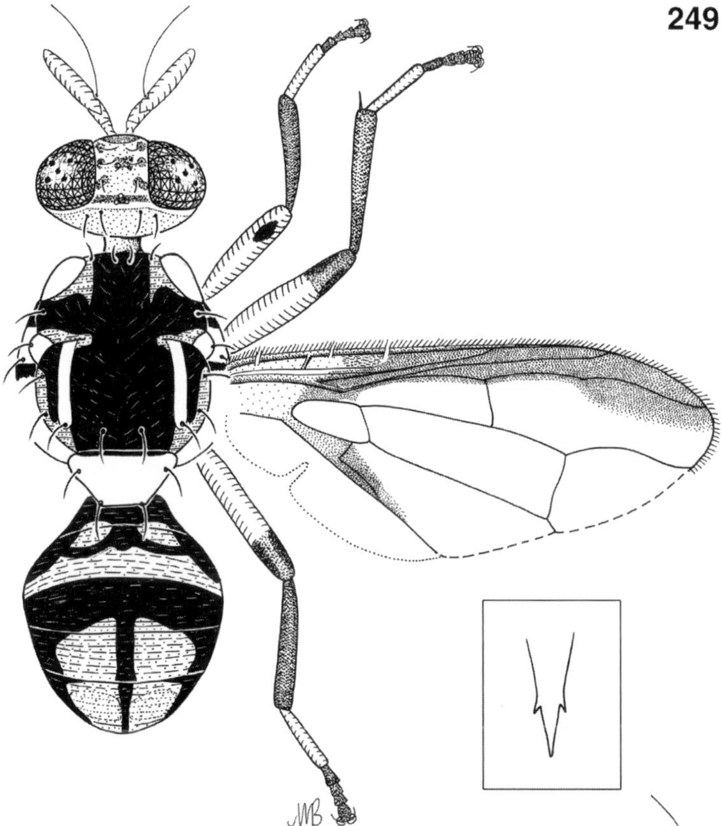

250A

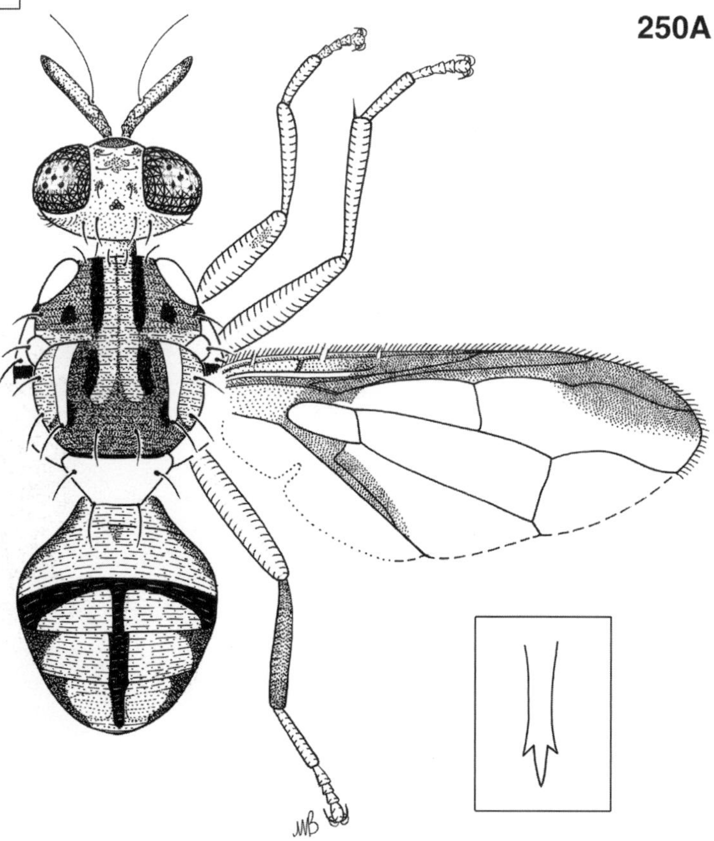

Fig. 249. *Bactrocera (Paratridacus) melanius* (Hardy & Adachi) new combination, holotype male and apex of piercer from female ovipositor. **Fig. 250A.** *Bactrocera (Paratridacus) yayeyamanus* (Matsumura) new combination, male and apex of piercer from female ovipositor.

250B

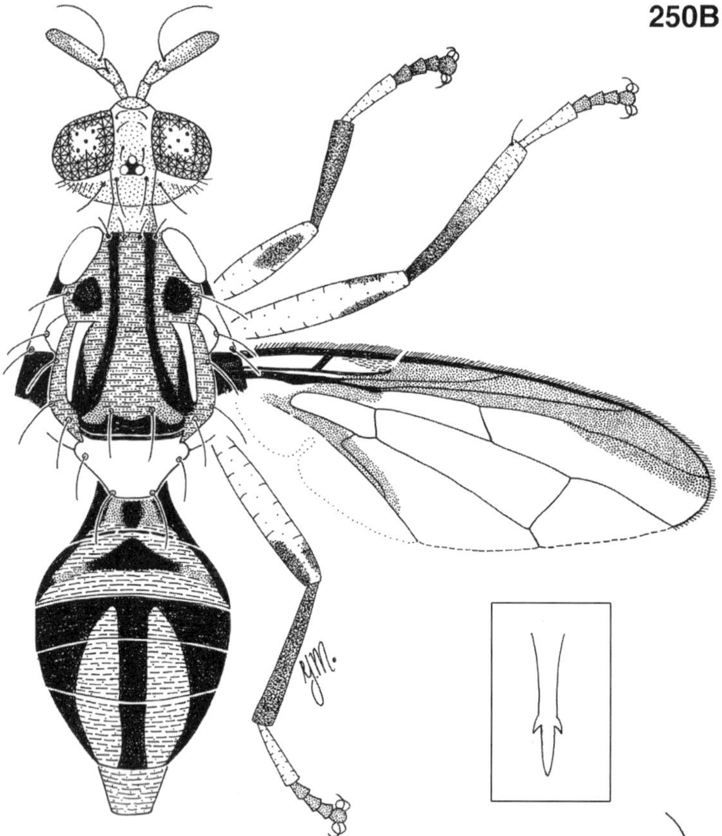

251

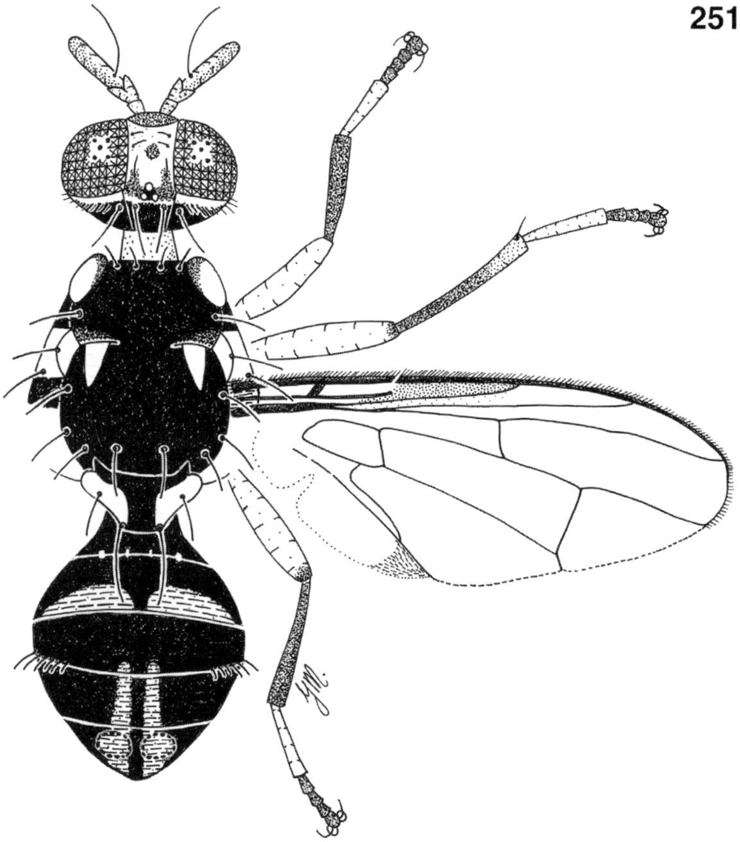

Fig. 250B. *Bactrocera (Paratridacus) katoi* (Hardy), paratype (allotype) female and apex of piercer from female ovipositor.
Fig. 251. *Bactrocera (Parazeugodacus) abbreviata* (Hardy), holotype male.

252

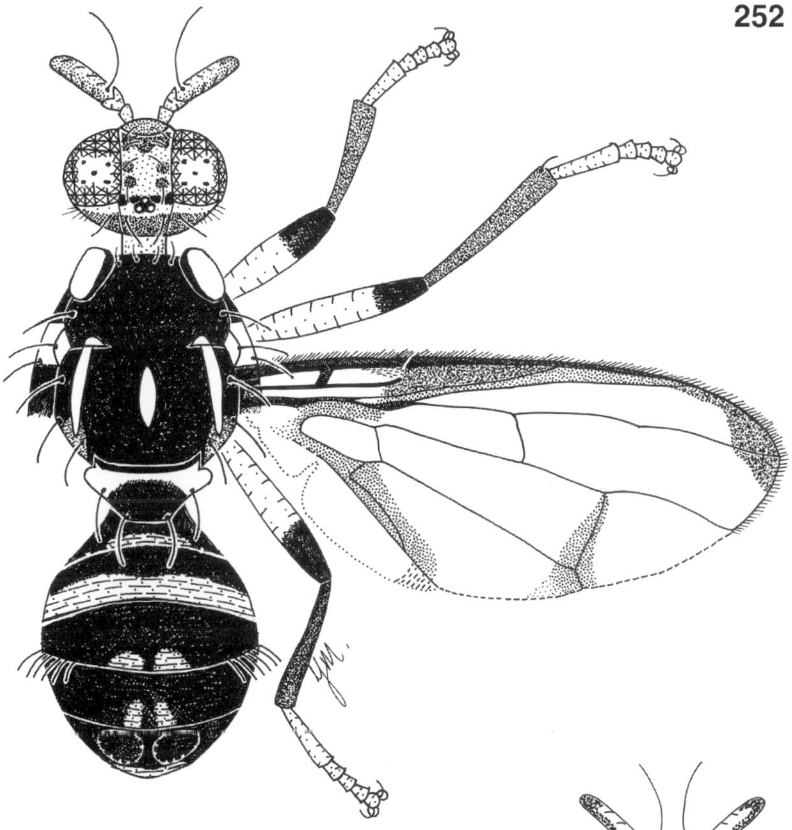

253

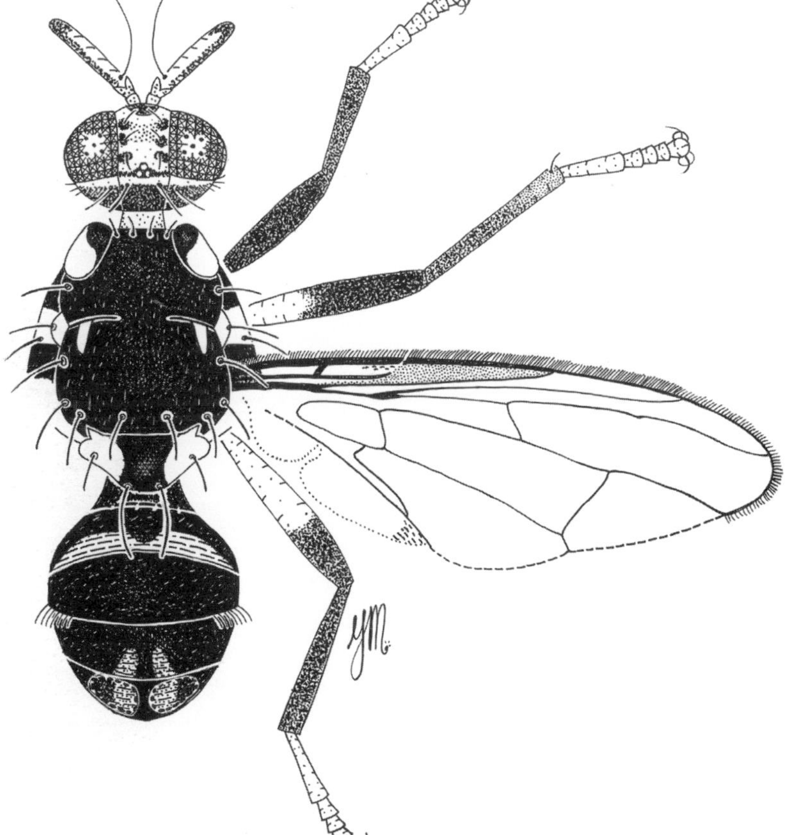

Fig. 252. *Bactrocera (Parazeugodacus) arisanica* (Shiraki), holotype male. **Fig. 253.** *Bactrocera (Parazeugodacus) bipustulata* (Bezzi), holotype male.

254

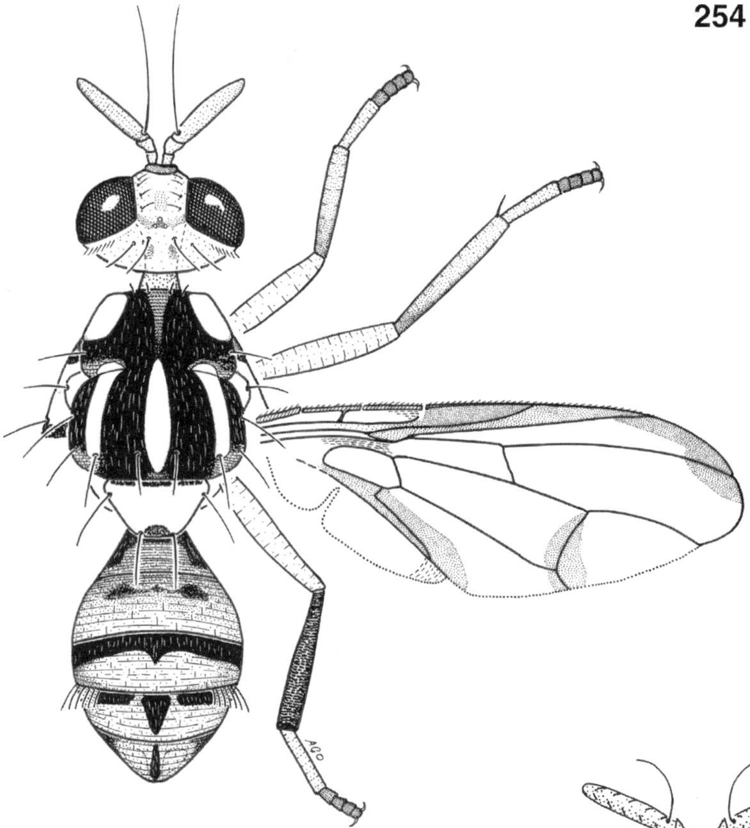

255

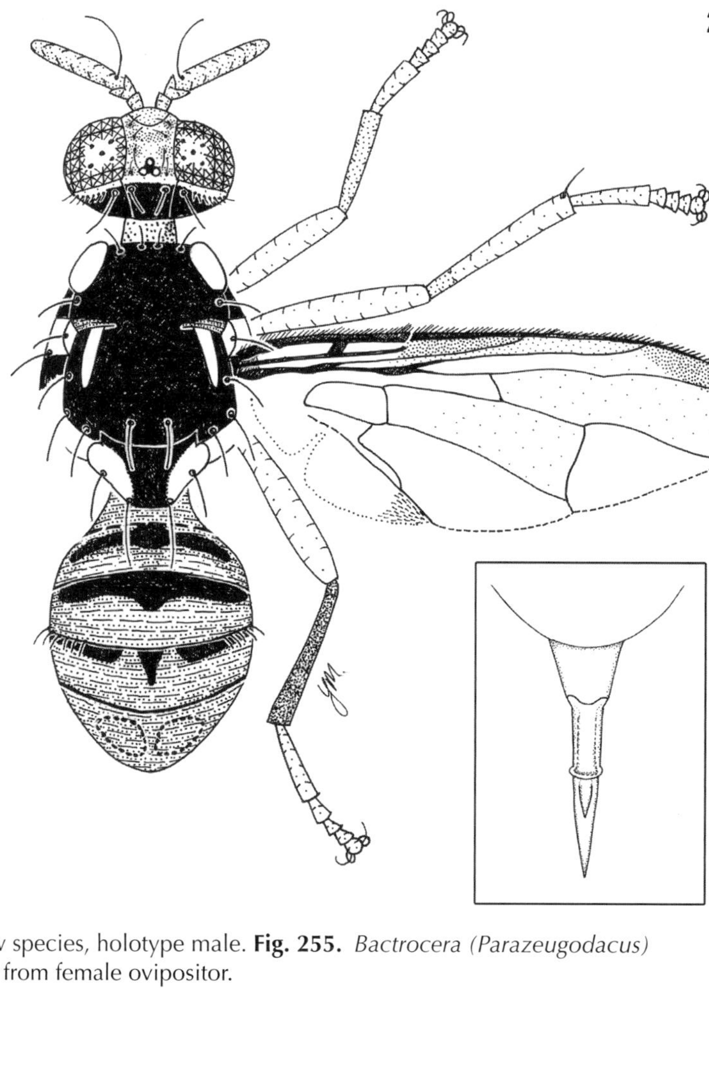

Fig. 254. *Bactrocera (Parazeugodacus) fulvosterna* new species, holotype male. **Fig. 255.** *Bactrocera (Parazeugodacus) matsumurai* (Shiraki), syntype male and apex of piercer from female ovipositor.

256A

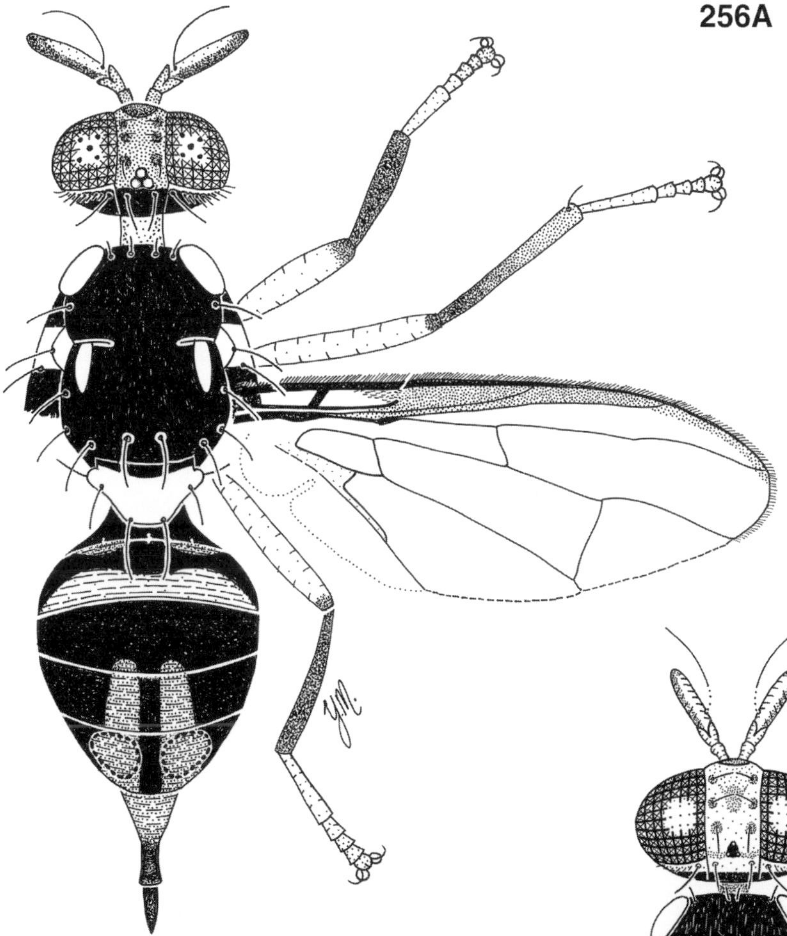

256B

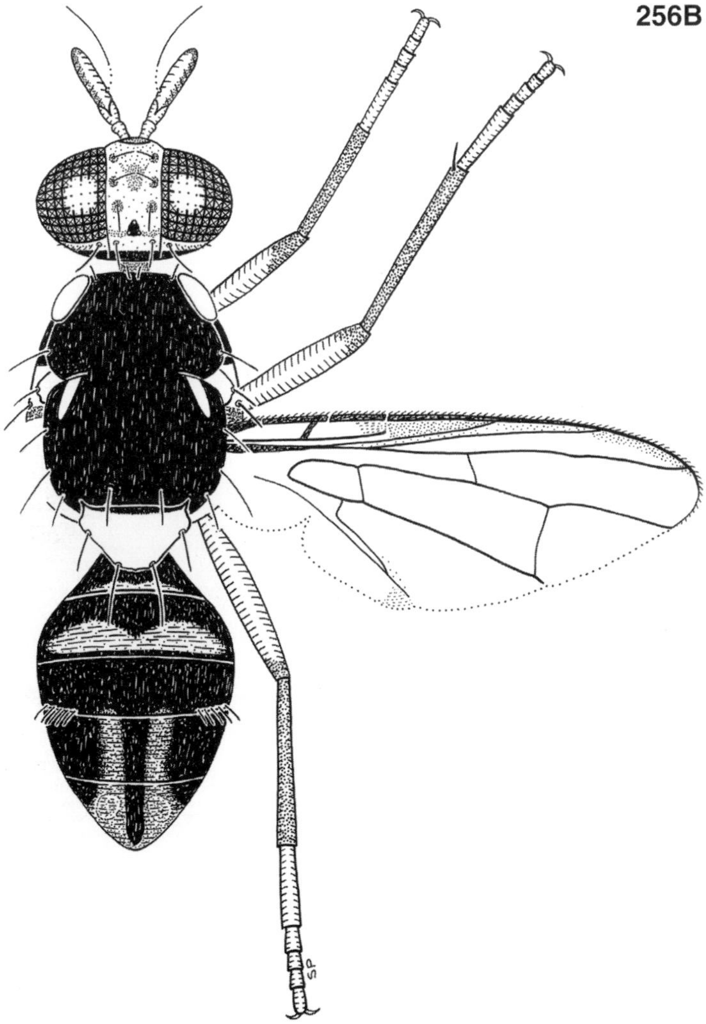

Fig. 256A. *Bactrocera (Parazeugodacus) pendleburyi* (Perkins), holotype female. **Fig. 256B.** *Bactrocera (Parazeugodacus) pendleburyi* (Perkins), male.

257

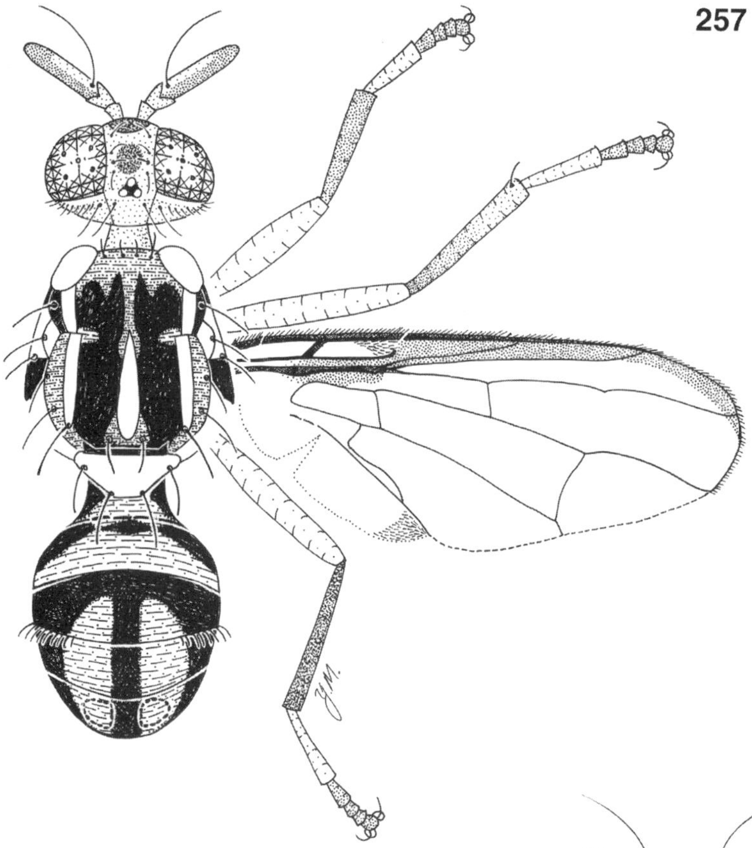

258

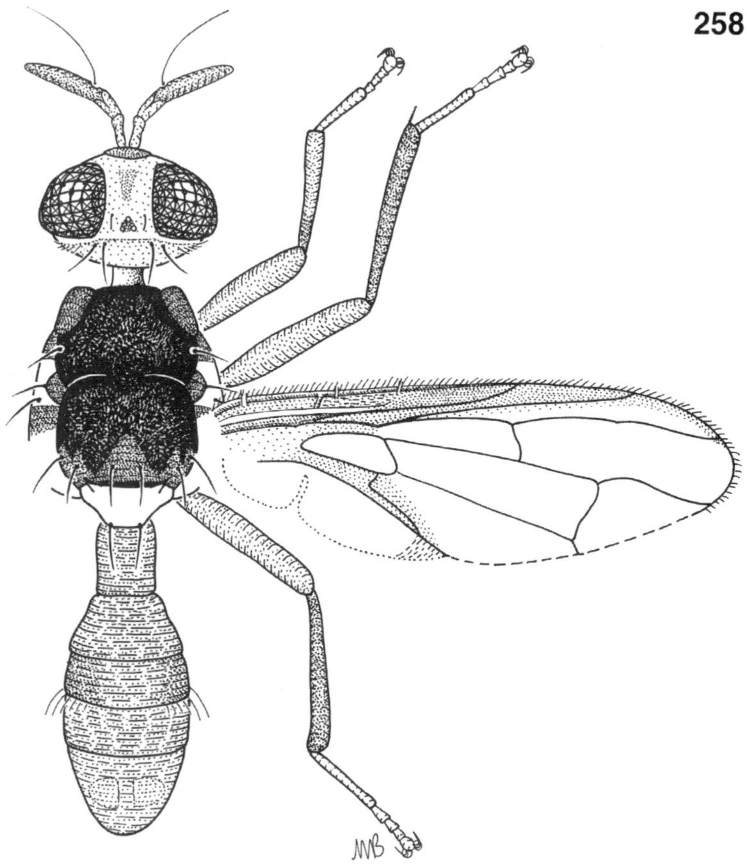

Fig. 257. *Bactrocera (Parazeugodacus) tetrachaeta* (Bezzi), lectotype male. **Fig. 258.** *Bactrocera (Semicallantra) incompta* new species, holotype male.

259

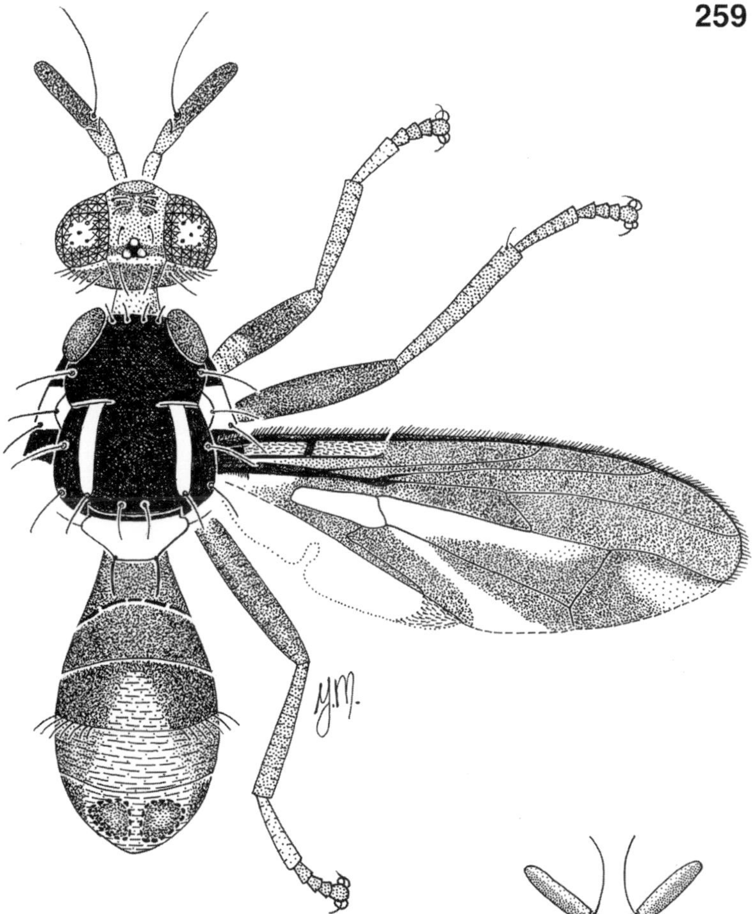

260

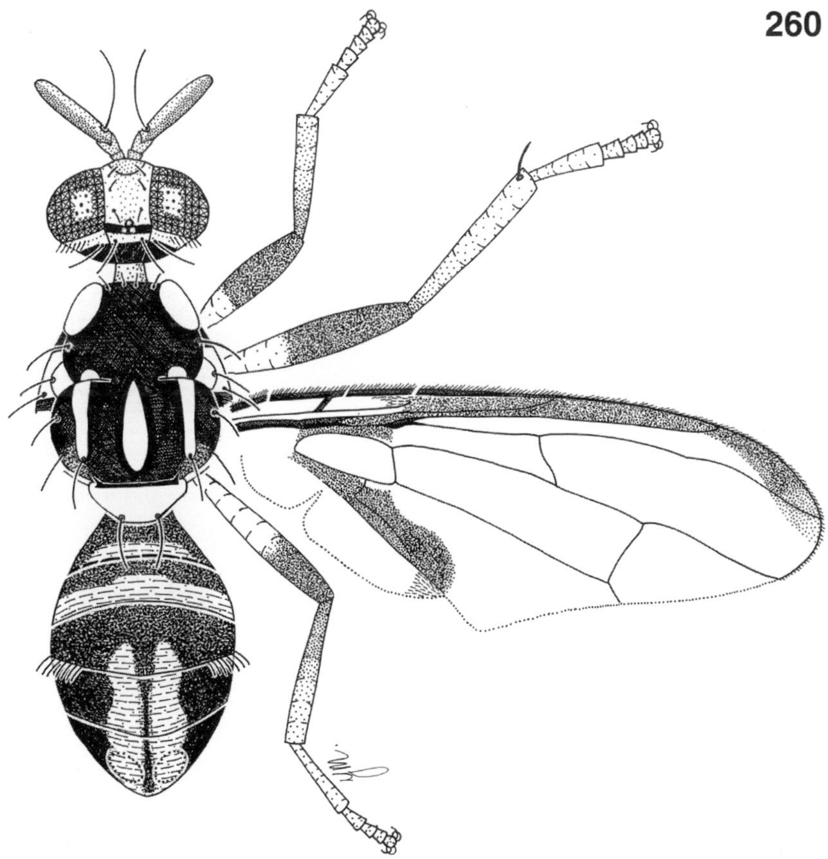

Fig. 259. *Bactrocera (Semicallantra) superba* new species, holotype male. **Fig. 260.** *Bactrocera (Sinodacus) bogorensis* (Hardy), holotype male.

261

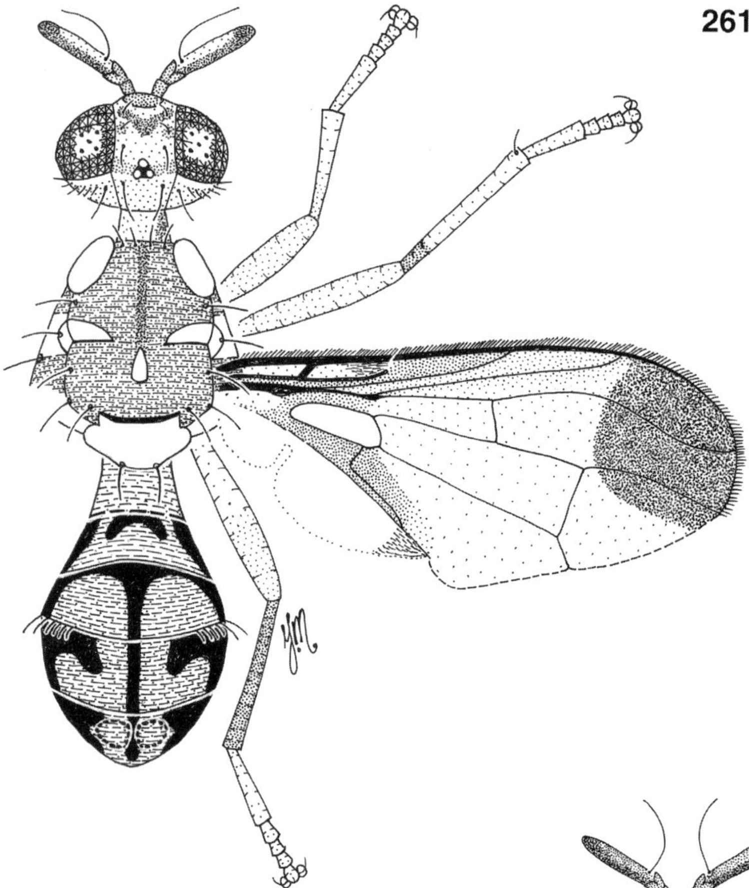

262

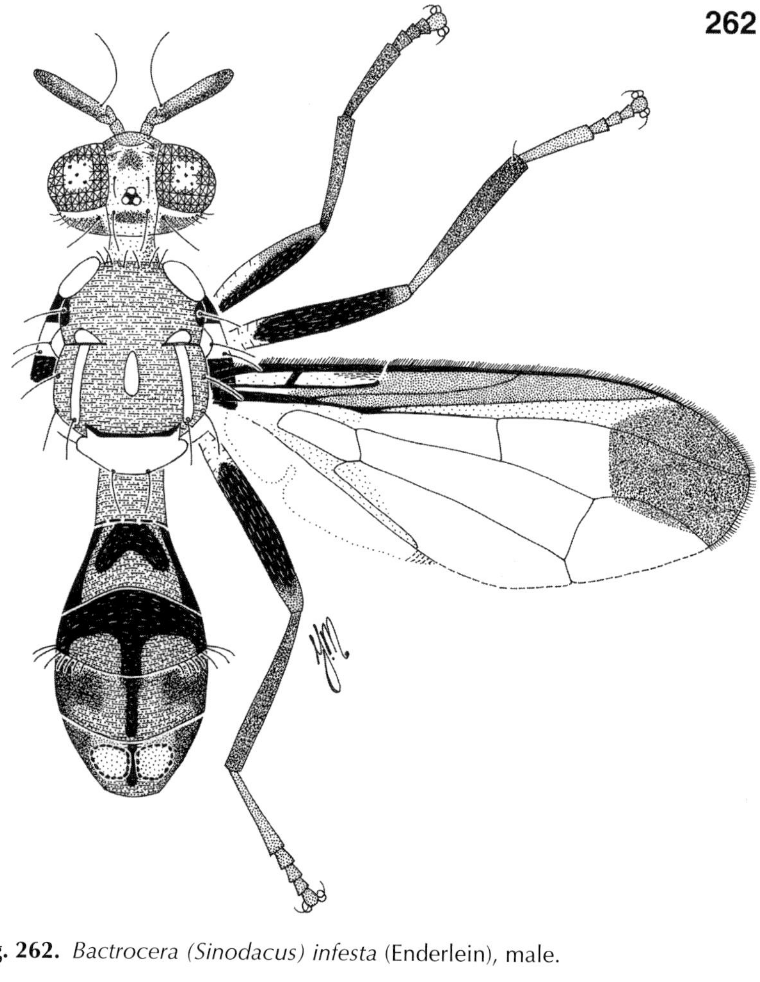

Fig. 261. *Bactrocera (Sinodacus) hochii* (Zia), male. **Fig. 262.** *Bactrocera (Sinodacus) infesta* (Enderlein), male.

263

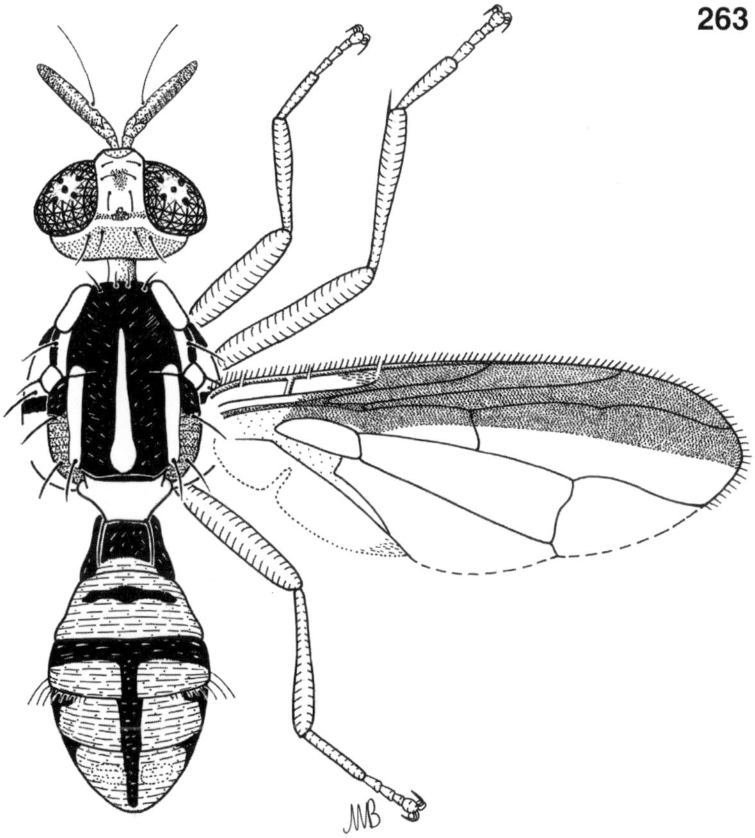

264

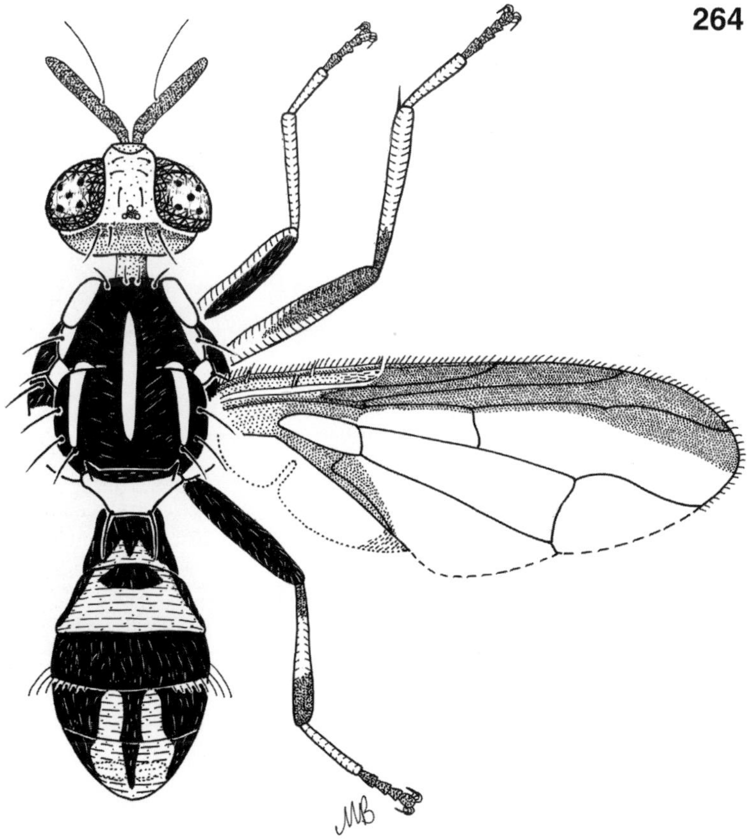

Fig. 263. *Bactrocera (Sinodacus) longivittata* Chua & Ooi, male. **Fig. 264.** *Bactrocera (Sinodacus) speciosa* new species, holotype male.

265

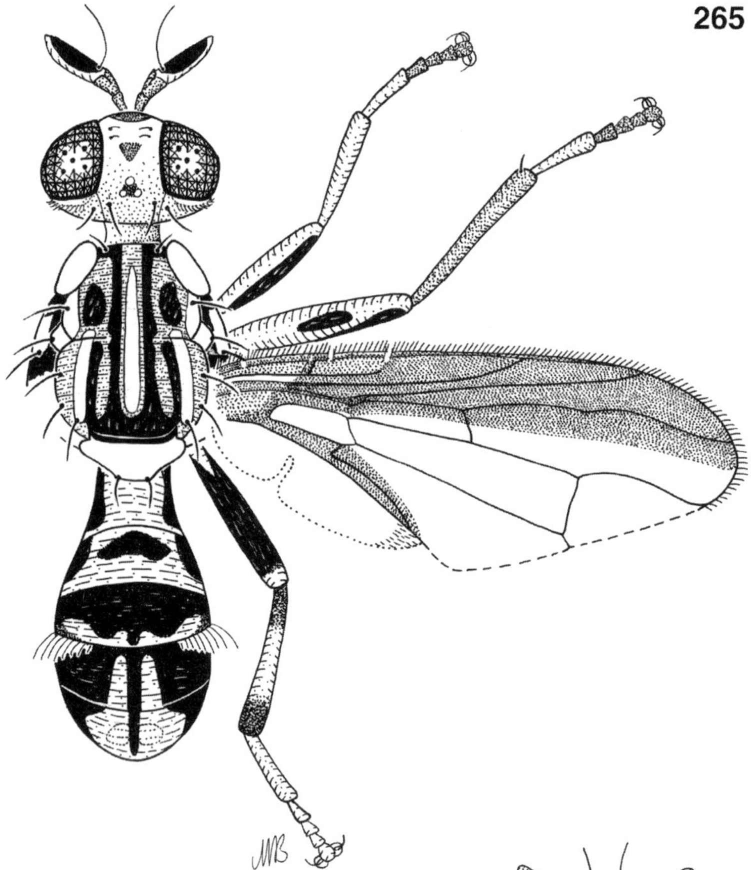

266

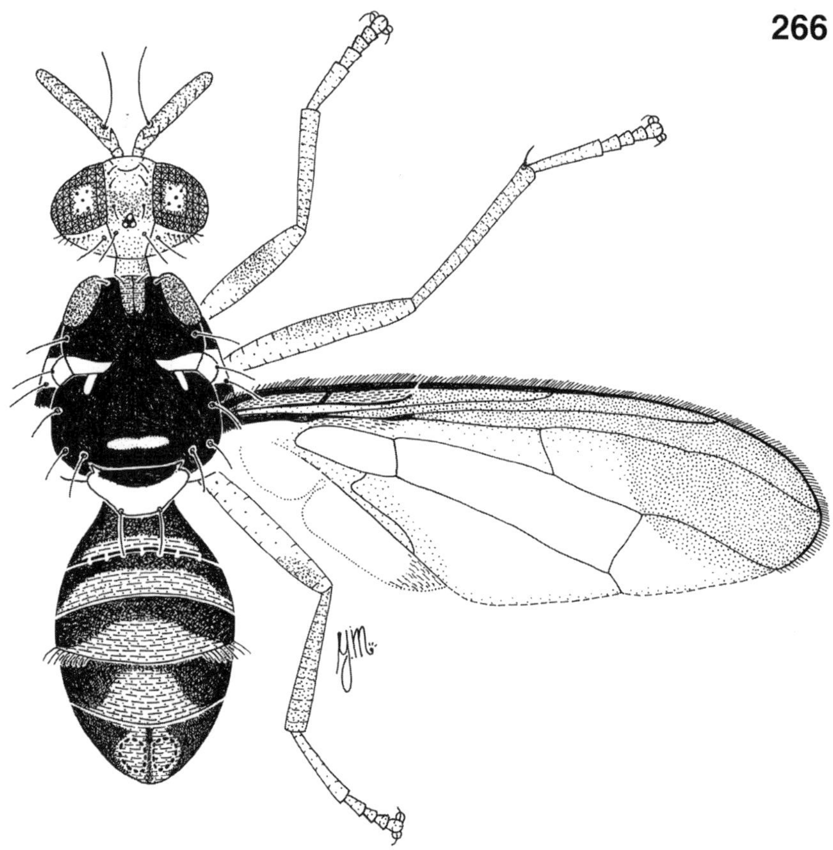

Fig. 265. *Bactrocera (Sinodacus) spectabilis* new species, holotype male. **Fig. 266.** *Bactrocera (Sinodacus) transversa* (Hardy), holotype male.

267

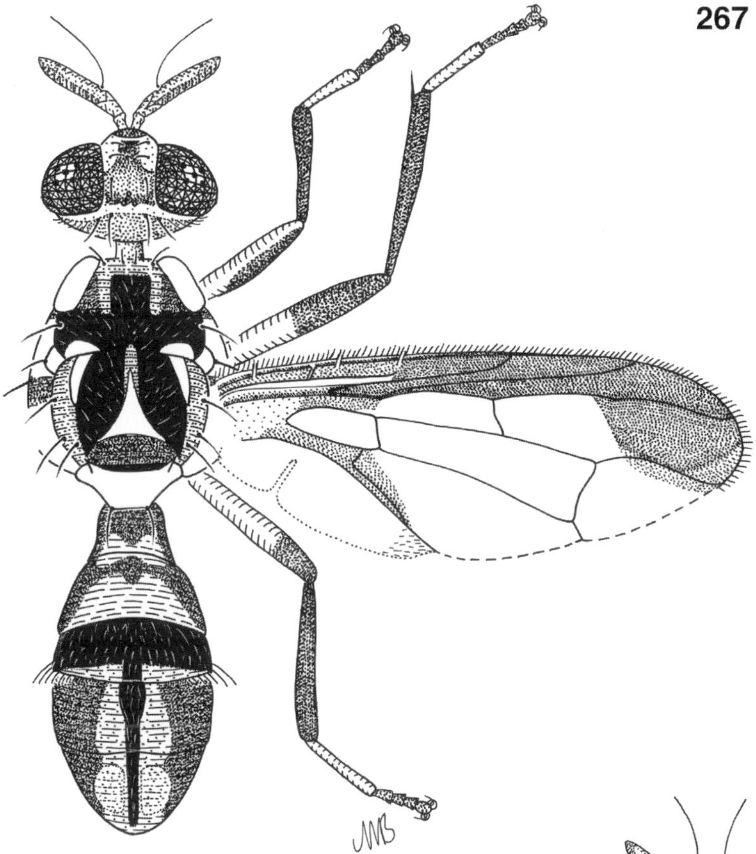

268

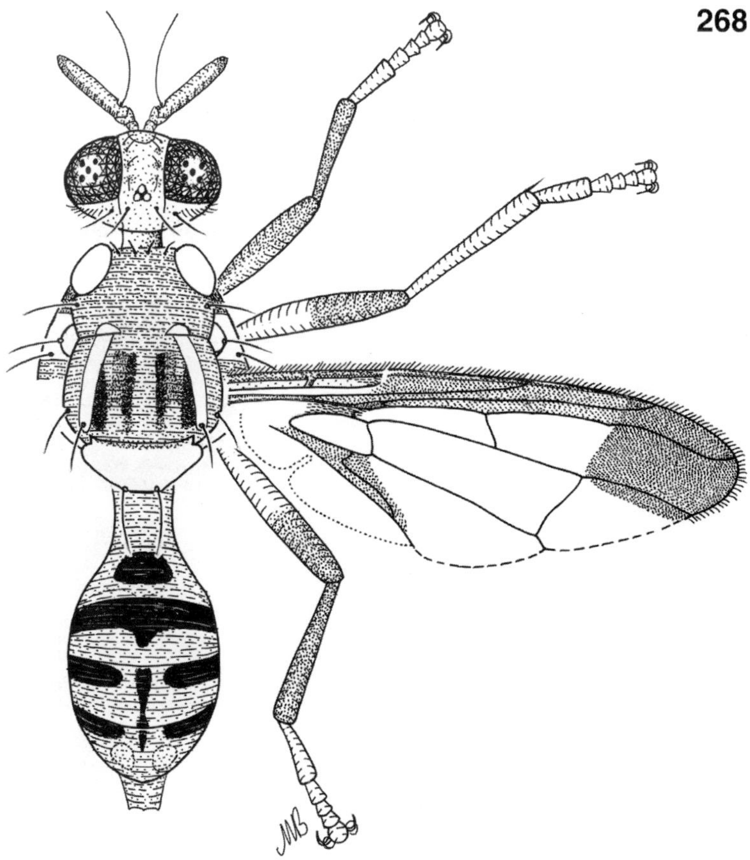

Fig. 267. *Bactrocera (Sinodacus) whitei* new species, holotype male. **Fig. 268.** *Bactrocera (Tetradacus) brachycera* (Bezzi), holotype female.

269

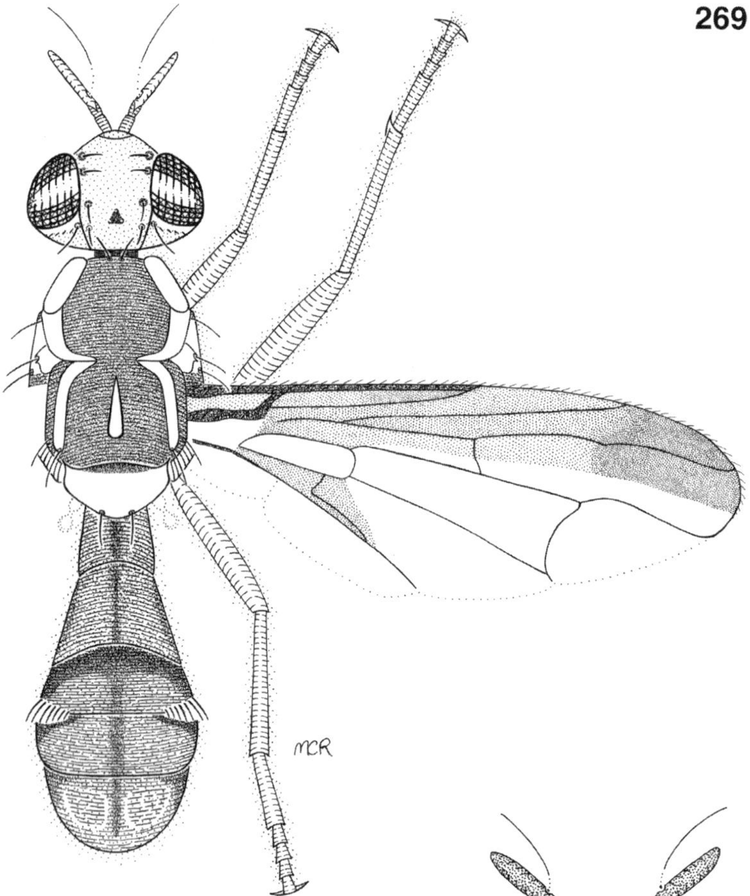

270

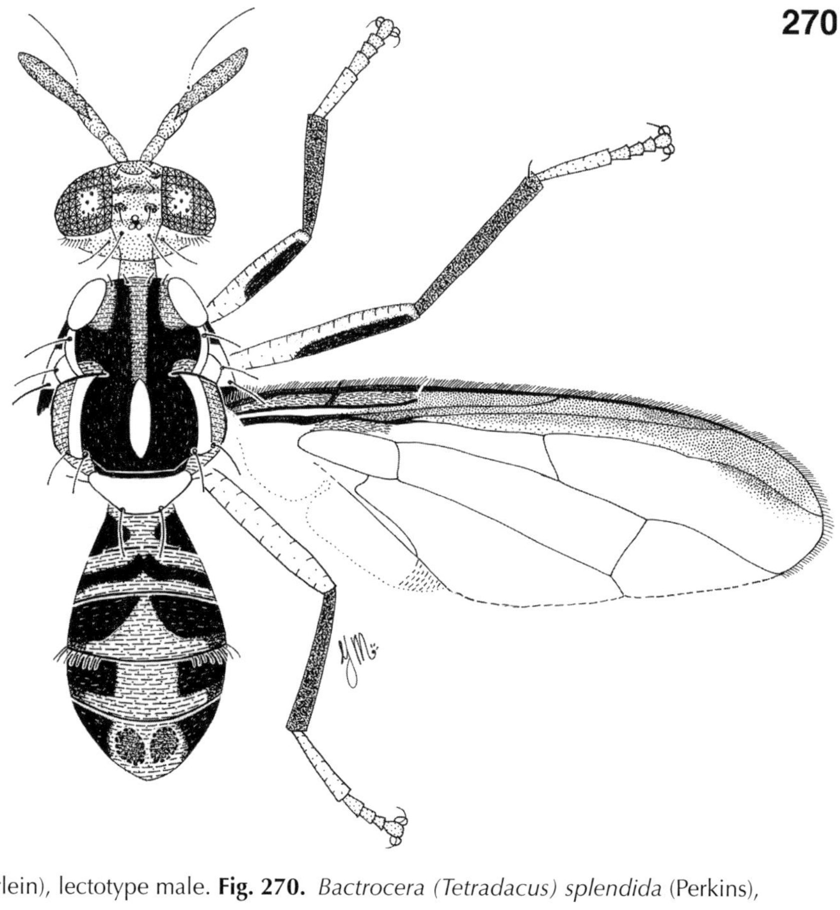

Fig. 269. *Bactrocera (Tetradacus) minax* (Enderlein), lectotype male. **Fig. 270.** *Bactrocera (Tetradacus) splendida* (Perkins), syntype male.

271

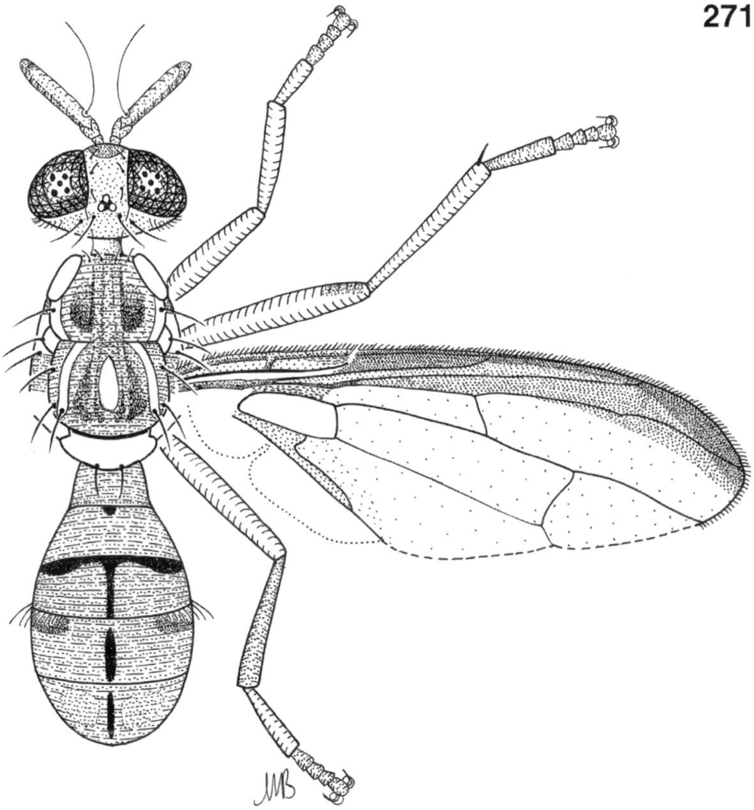

272

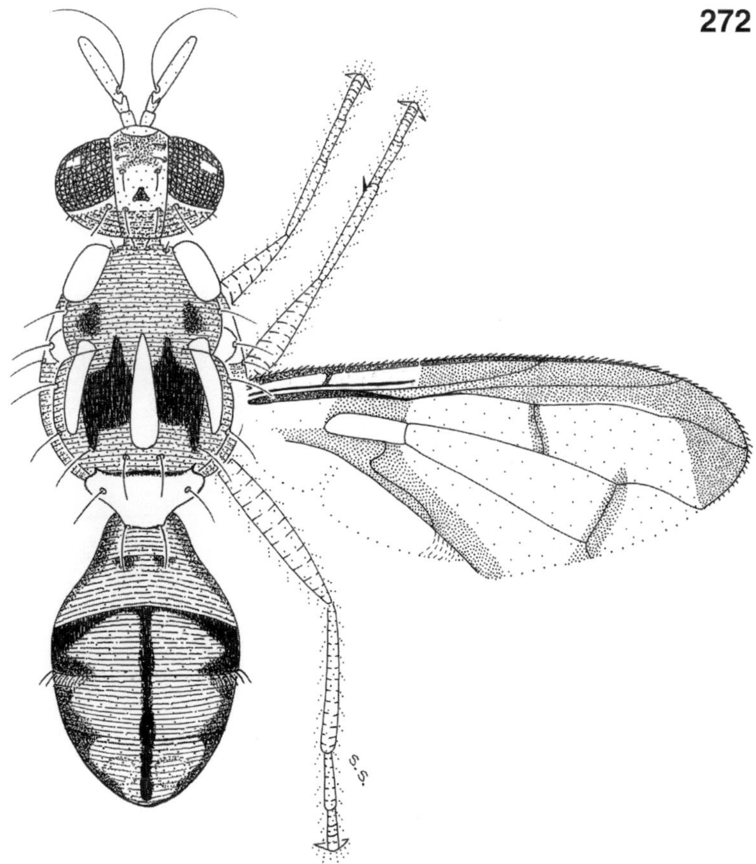

Fig. 271. *Bactrocera (Tetradacus) tsuneonis* (Miyake), male. **Fig. 272.** *Bactrocera (Zeugodacus) abdoangusta* (Drew), holotype male.

273

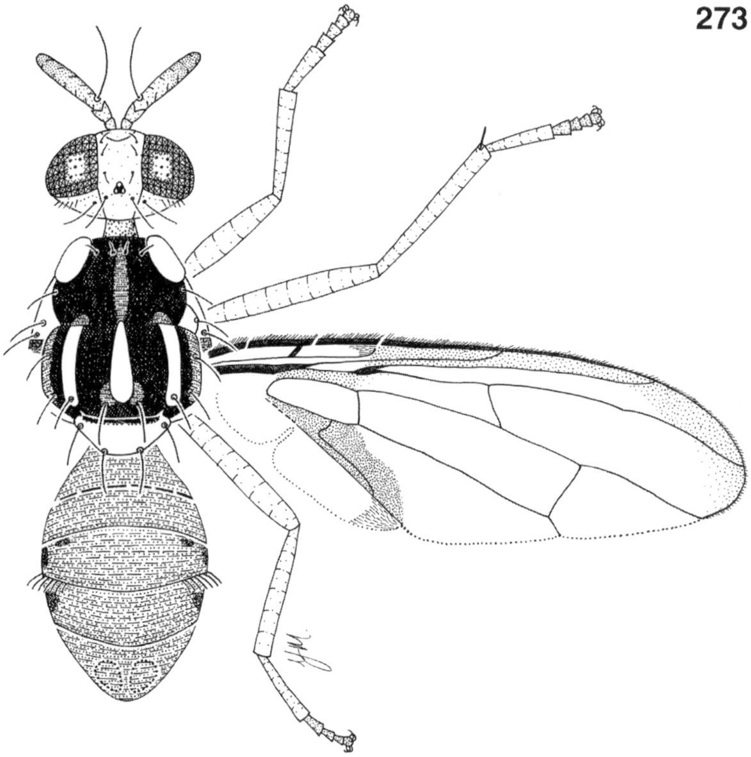

274

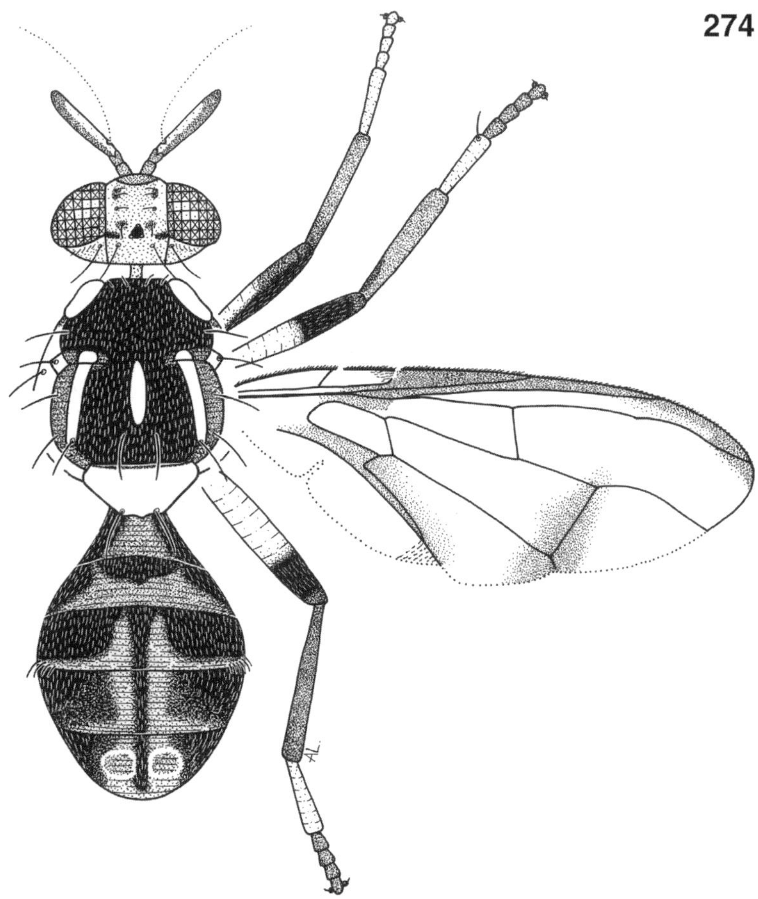

Fig. 273. *Bactrocera (Zeugodacus) abnormis* (Hardy), holotype male. **Fig. 274.** *Bactrocera (Zeugodacus) aithonota* new species, holotype male.

275

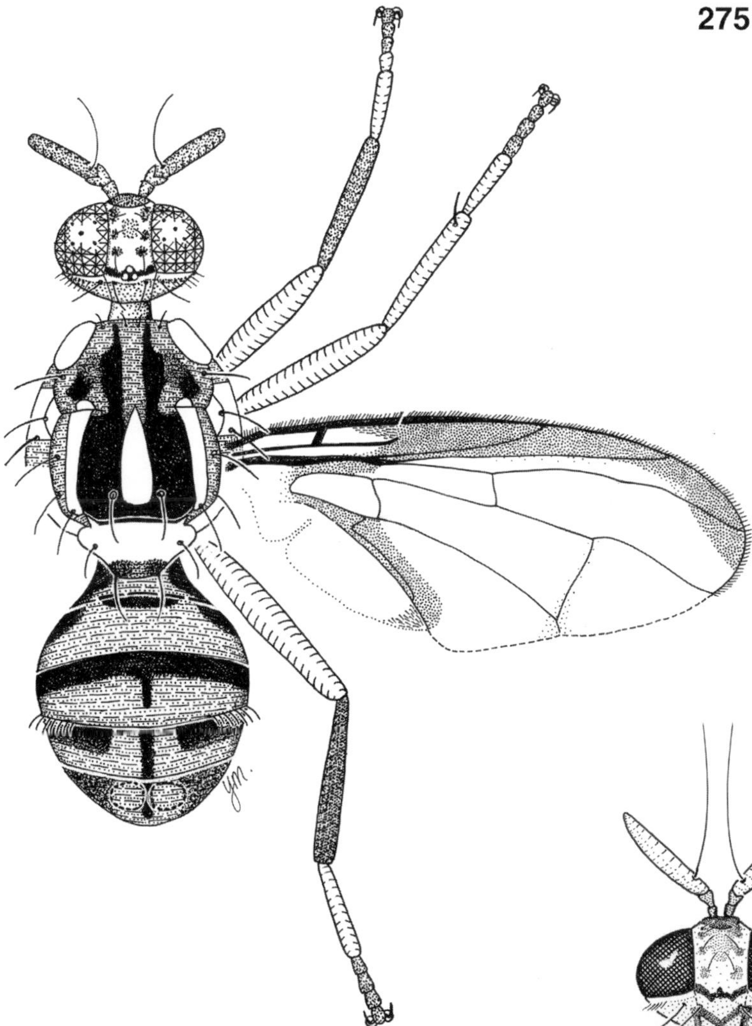

276

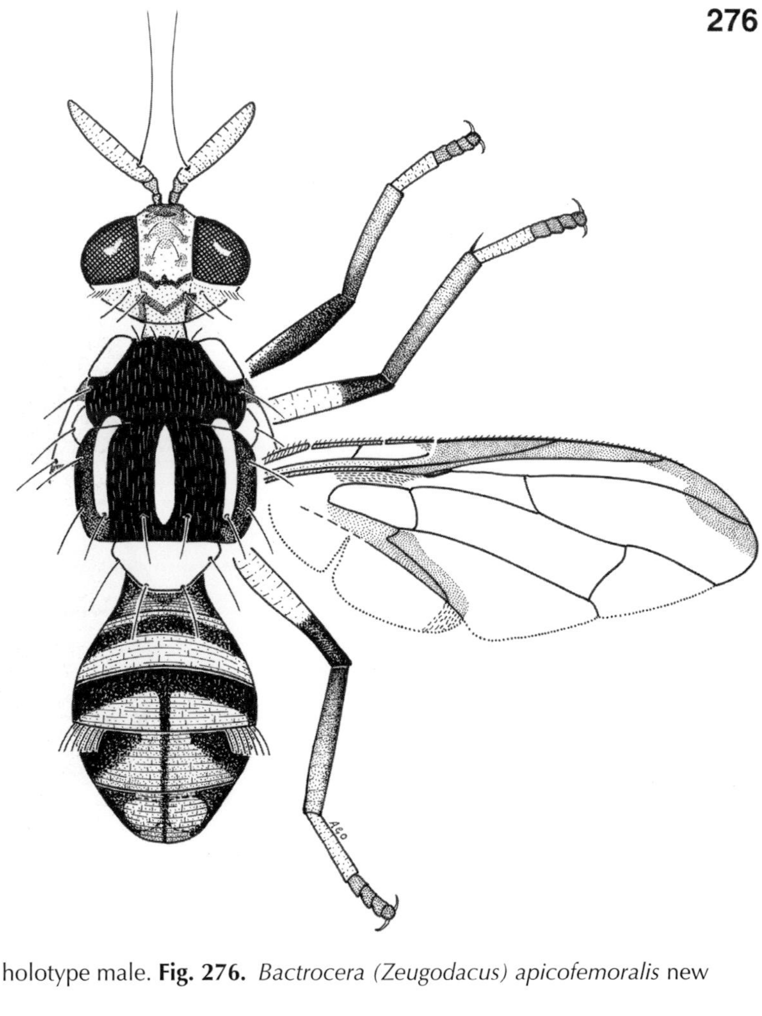

Fig. 275. *Bactrocera (Zeugodacus) ambigua* (Shiraki), holotype male. **Fig. 276.** *Bactrocera (Zeugodacus) apicofemoralis* new species, holotype male.

277

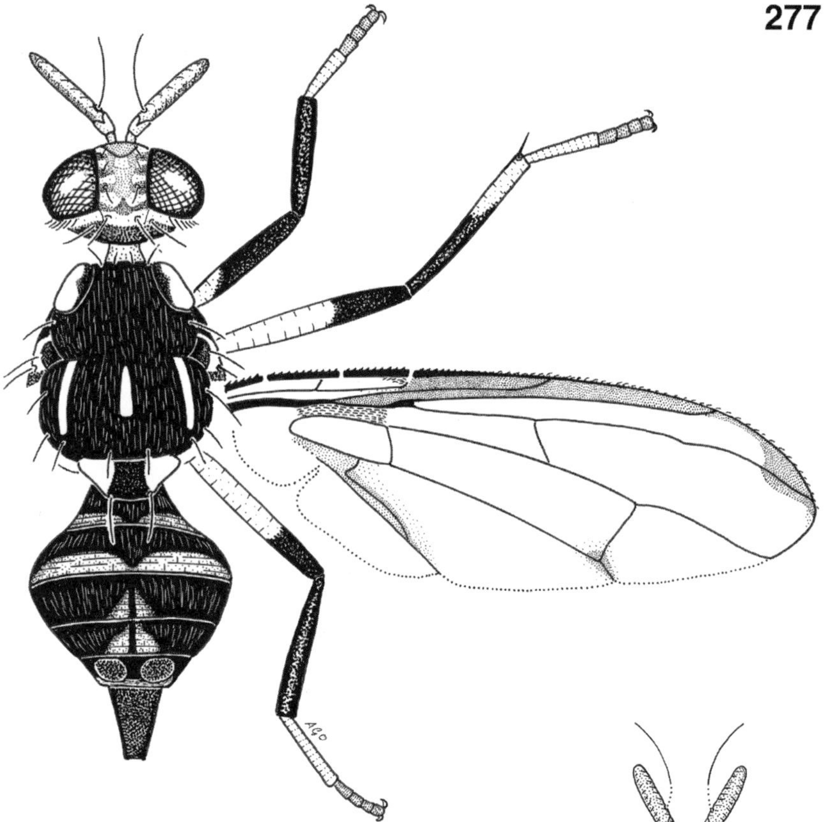

278

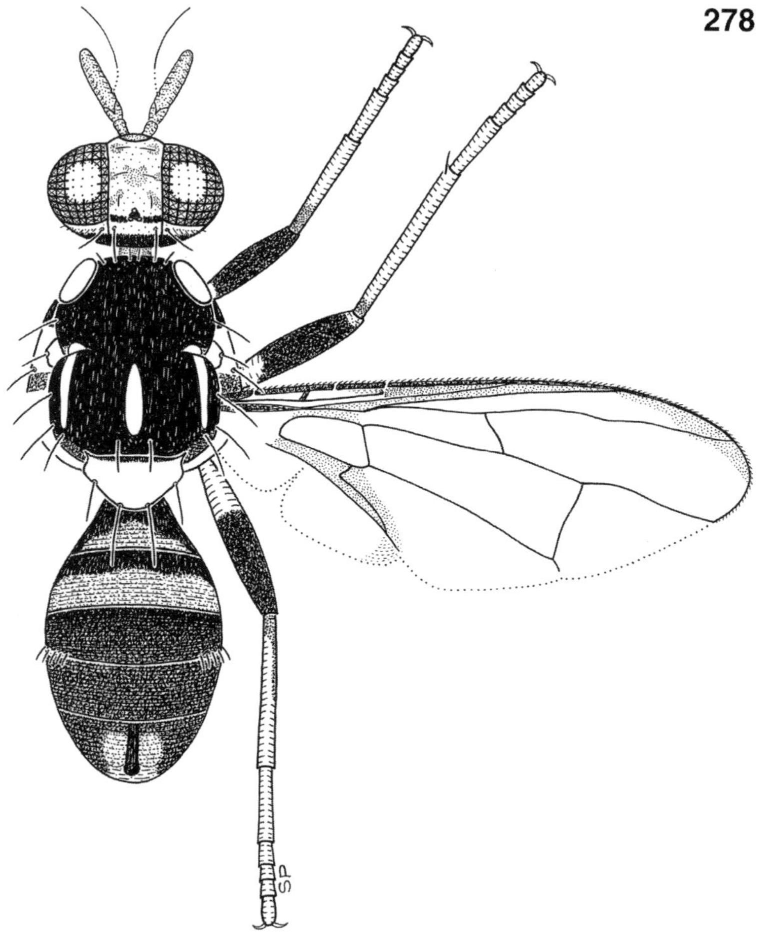

Fig. 277. *Bactrocera (Zeugodacus) assamensis* White, holotype female. **Fig. 278.** *Bactrocera (Zeugodacus) atrifacies* (Perkins), holotype male.

279

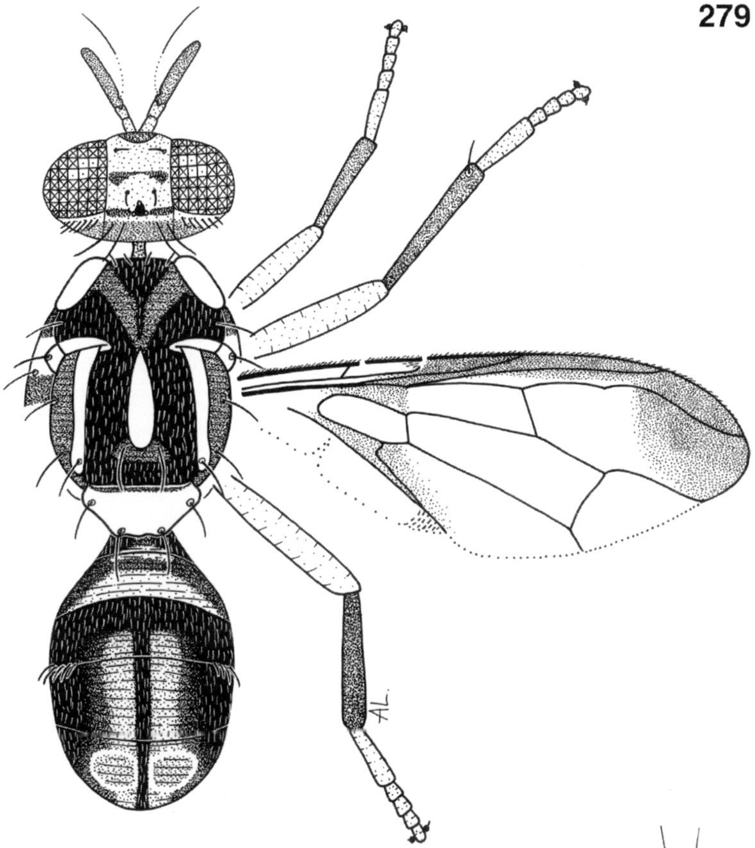

280

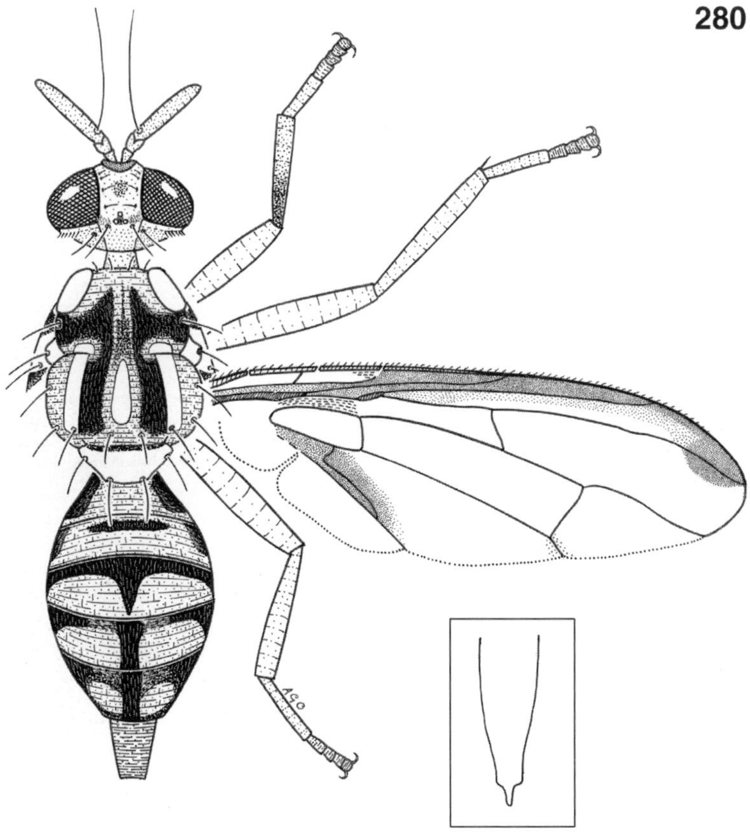

Fig. 279. *Bactrocera (Zeugodacus) baliensis* new species, holotype male. **Fig. 280.** *Bactrocera (Zeugodacus) bezziana* (Hering), holotype female and apex of piercer from female ovipositor.

281

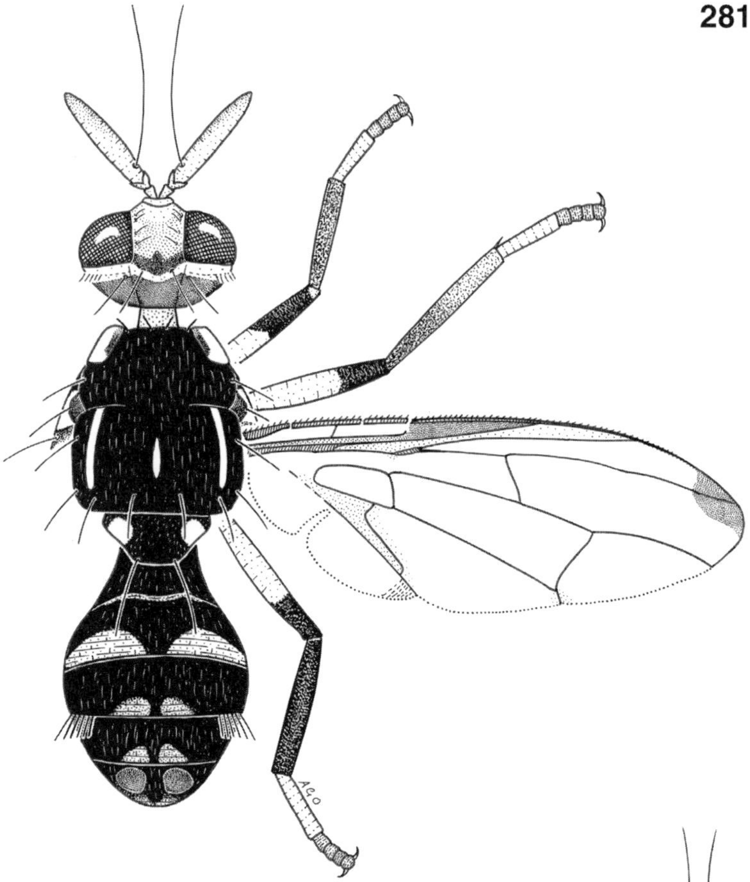

282

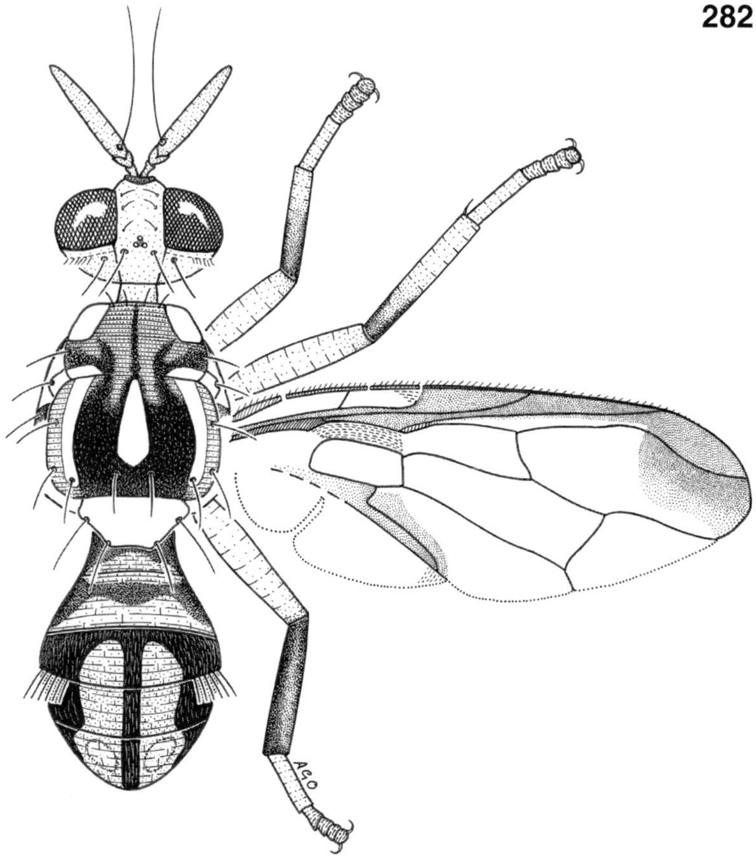

Fig. 281. *Bactrocera (Zeugodacus) biguttata* (Bezzi), male. **Fig. 282.** *Bactrocera (Zeugodacus) borongensis* new species, holotype male.

283

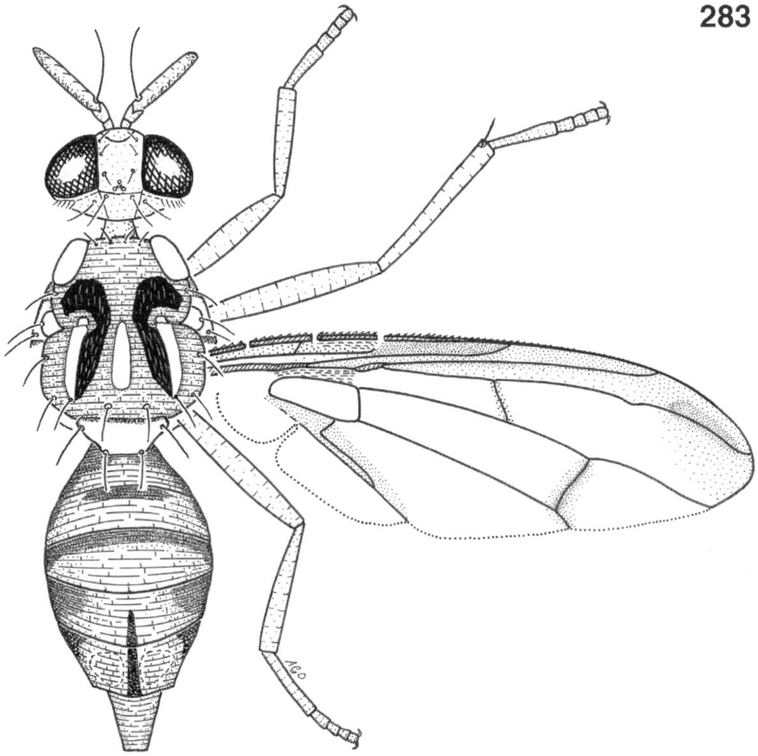

284

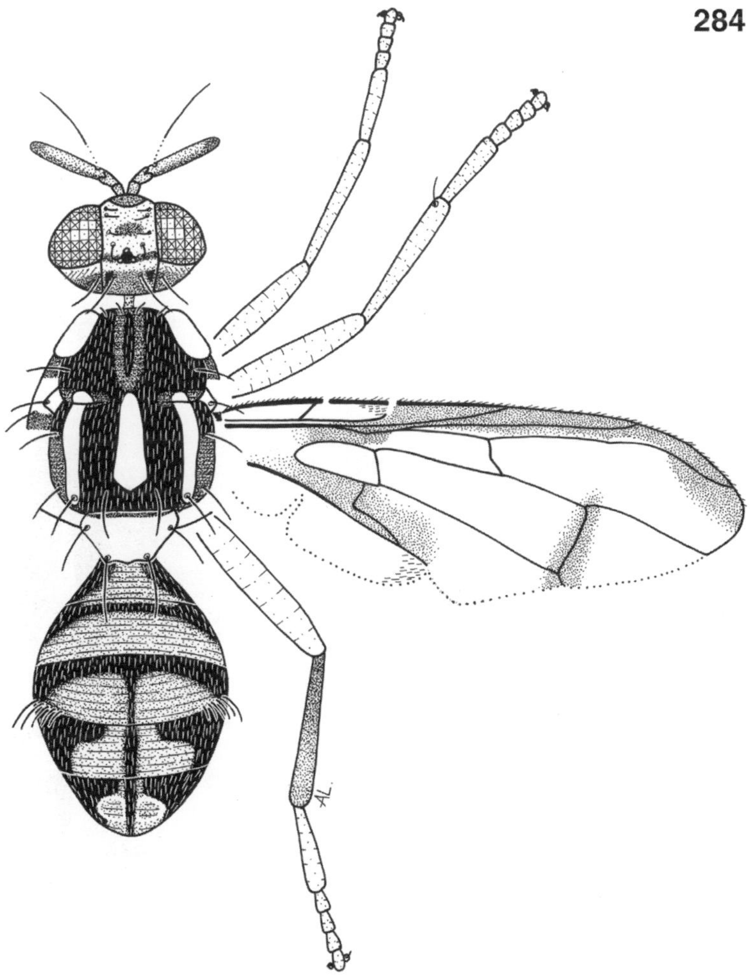

Fig. 283. *Bactrocera (Zeugodacus) buruensis* White, holotype female. **Fig. 284.** *Bactrocera (Zeugodacus) calumniata* (Hardy), holotype male.

285

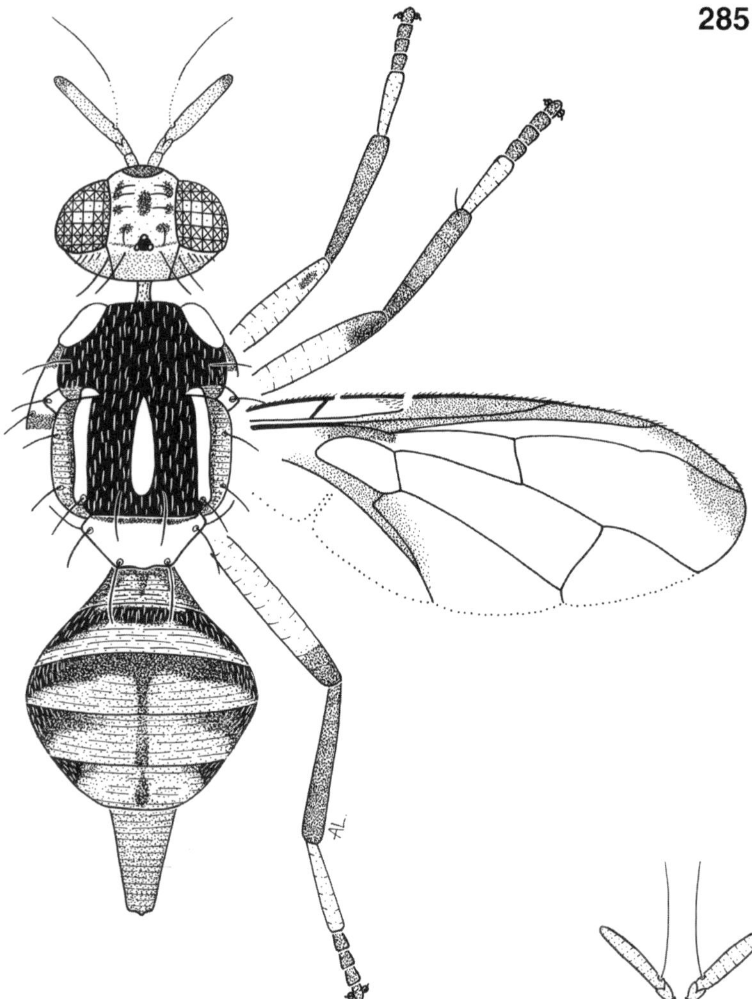

286

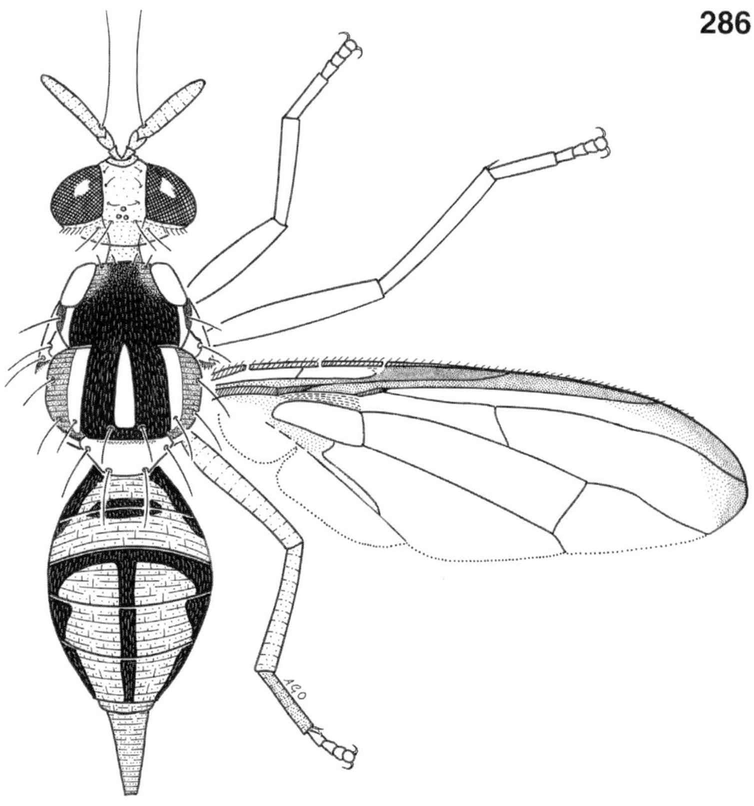

Fig. 285. *Bactrocera (Zeugodacus) caudata* (Fabricius), lectotype female. **Fig. 286.** *Bactrocera (Zeugodacus) connexa* (Hardy), holotype female.

287

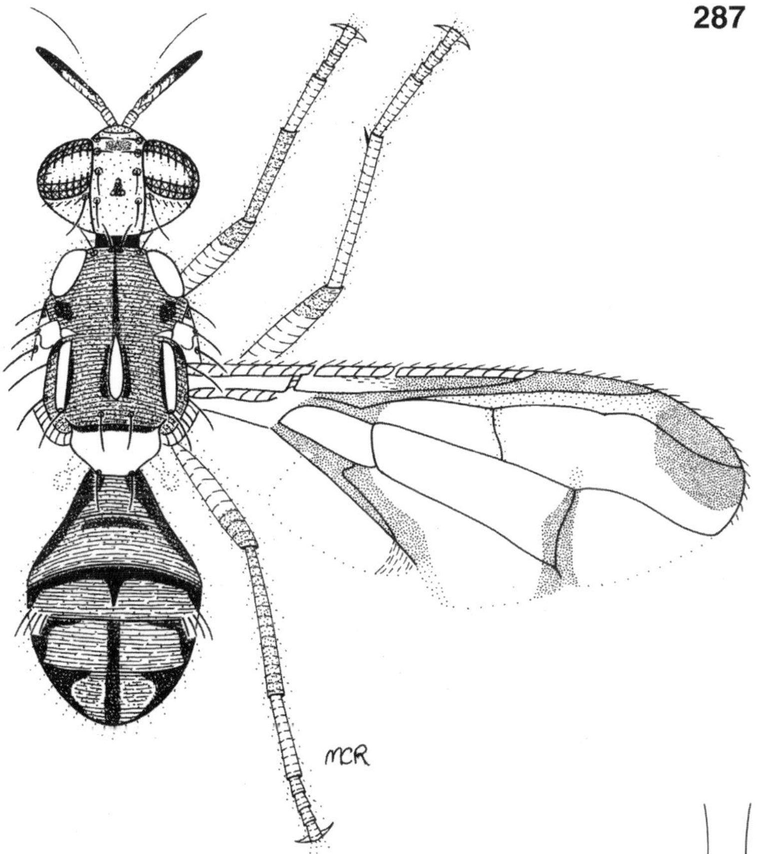

288

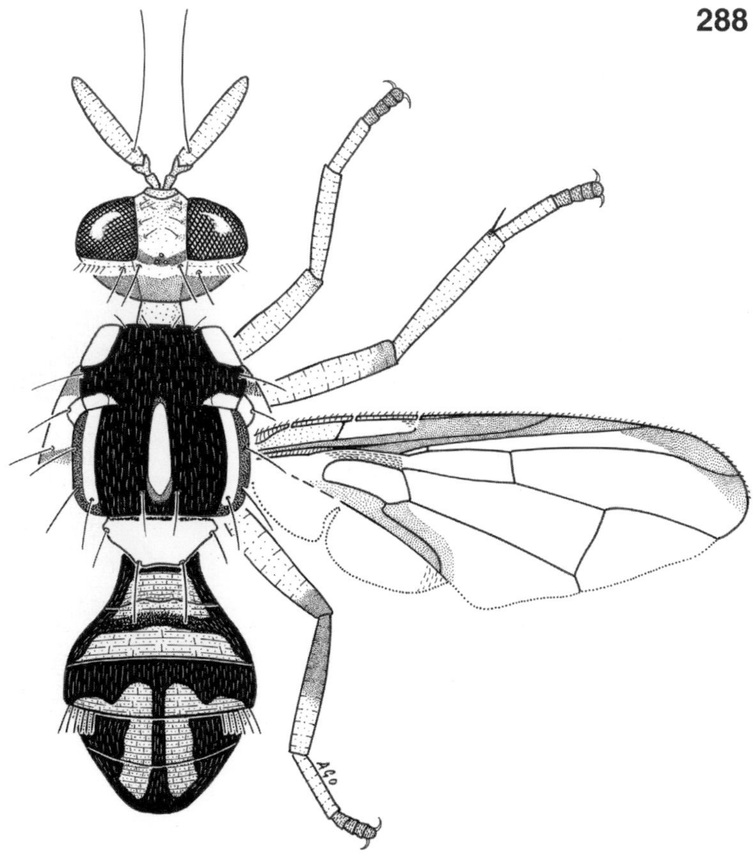

Fig. 287. *Bactrocera (Zeugodacus) cucurbitae* (Coquillett), male. **Fig. 288.** *Bactrocera (Zeugodacus) daclaciae* new species, holotype male.

289

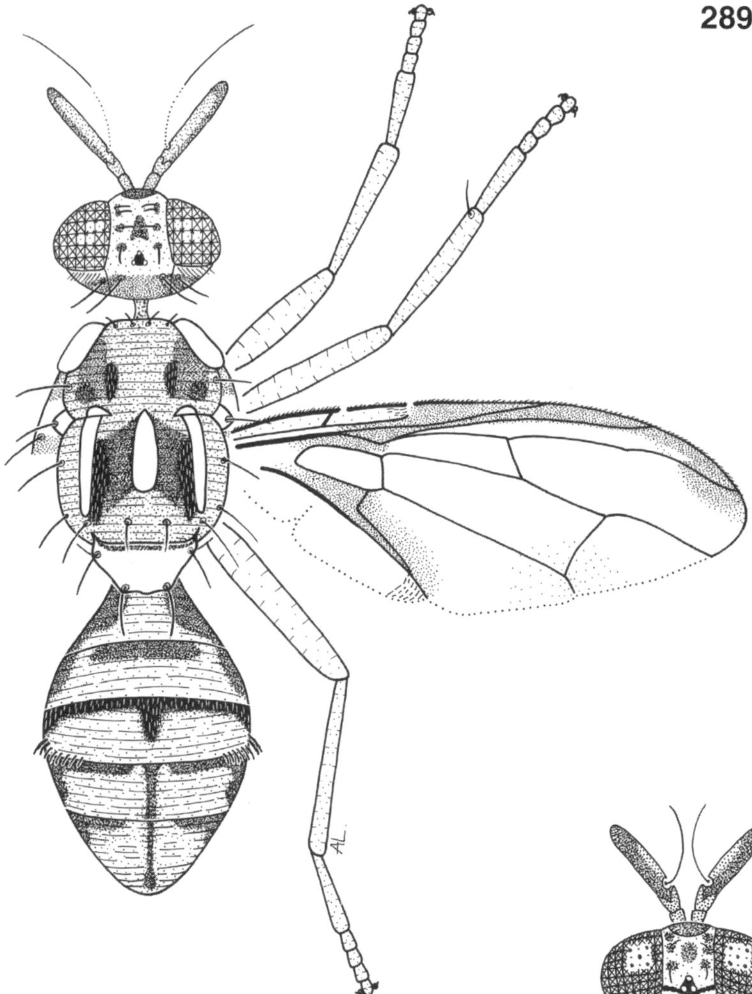

290

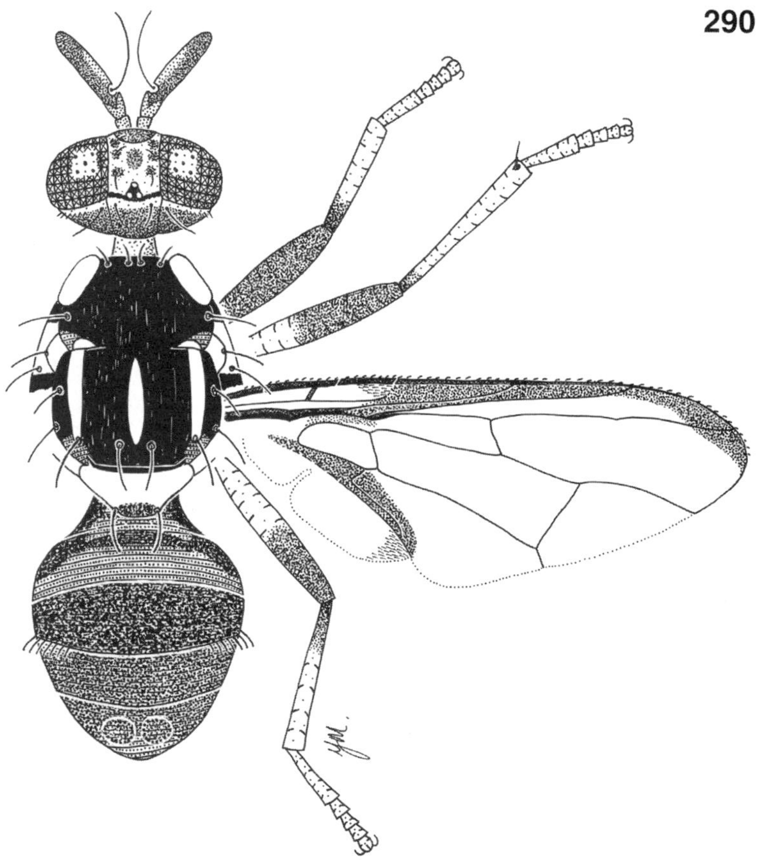

Fig. 289. *Bactrocera (Zeugodacus) depressa* (Shiraki), lectotype male. **Fig. 290.** *Bactrocera (Zeugodacus) diaphora* (Hendel), syntype male.

291

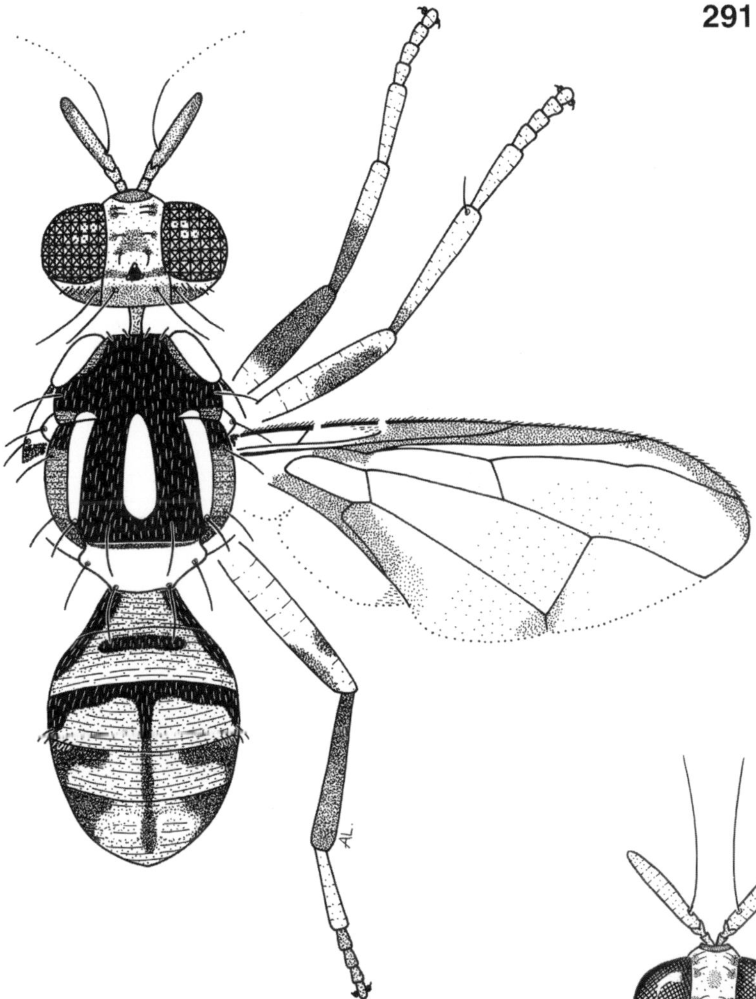

292

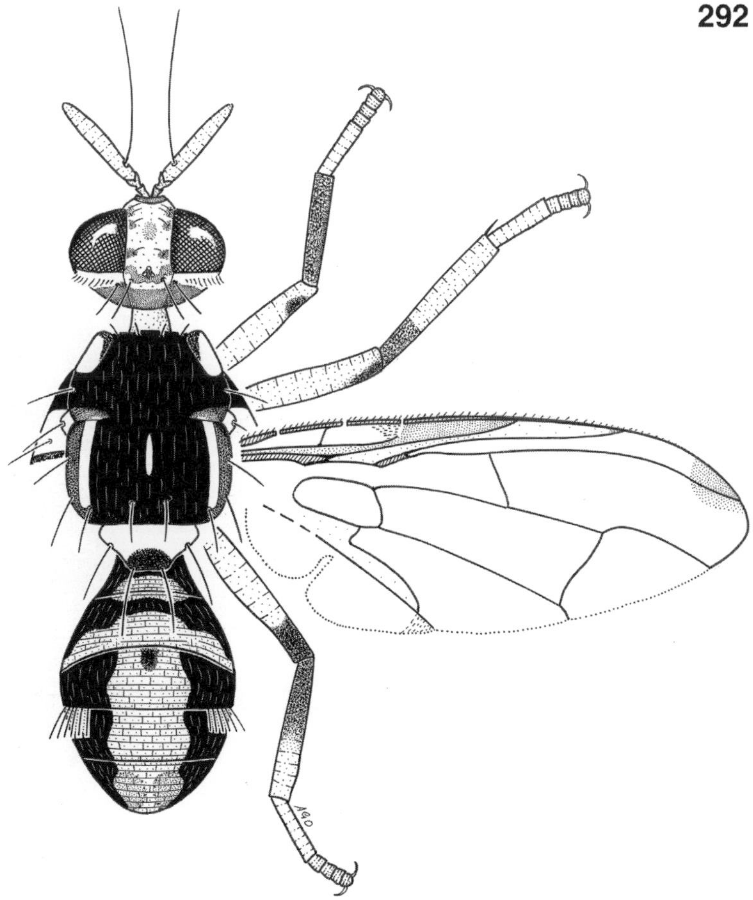

Fig. 291. *Bactrocera (Zeugodacus) diaphoropsis* (Hering), holotype male. **Fig. 292.** *Bactrocera (Zeugodacus) dorsirufa* new species, holotype male.

293

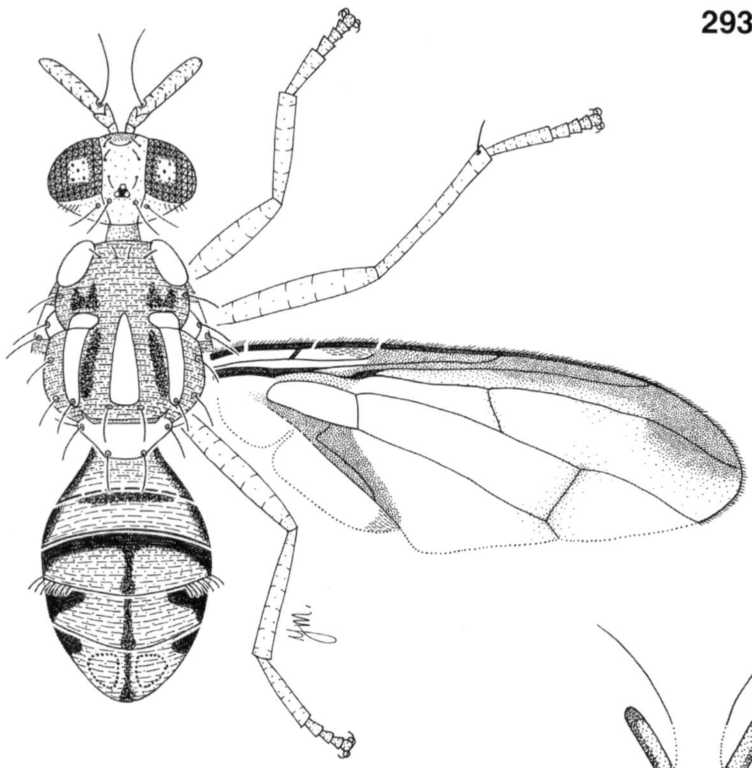

294

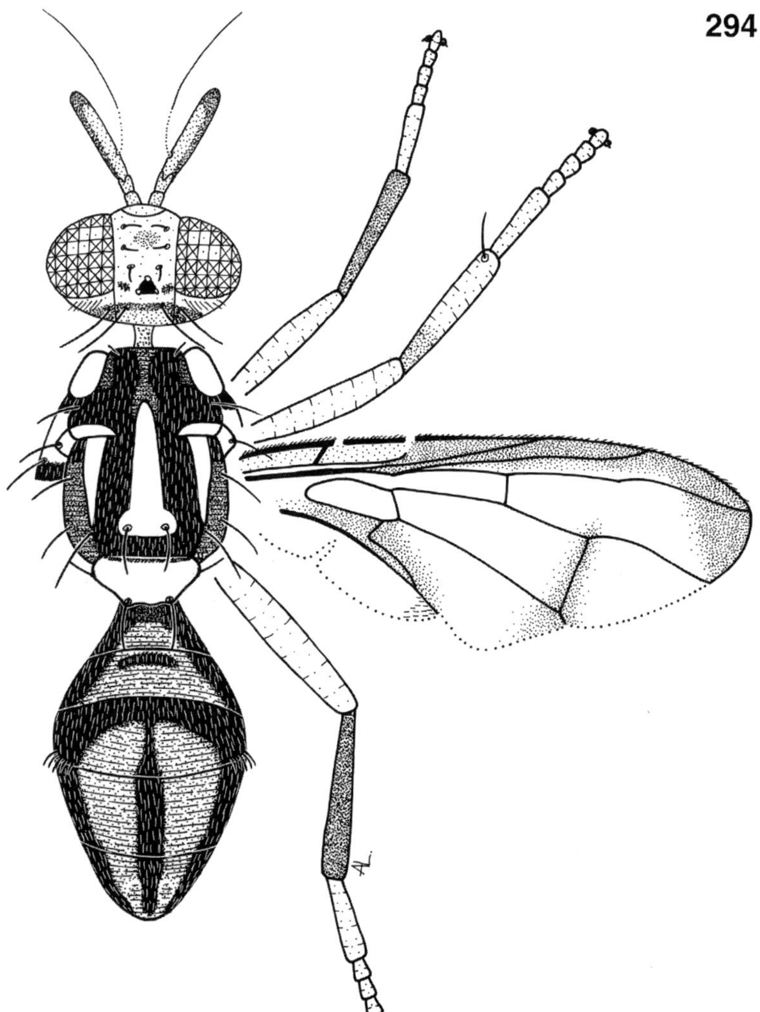

Fig. 293. *Bactrocera (Zeugodacus) dubiosa* (Hardy), holotype male. **Fig. 294.** *Bactrocera (Zeugodacus) elegantula* (Hardy), holotype male.

295

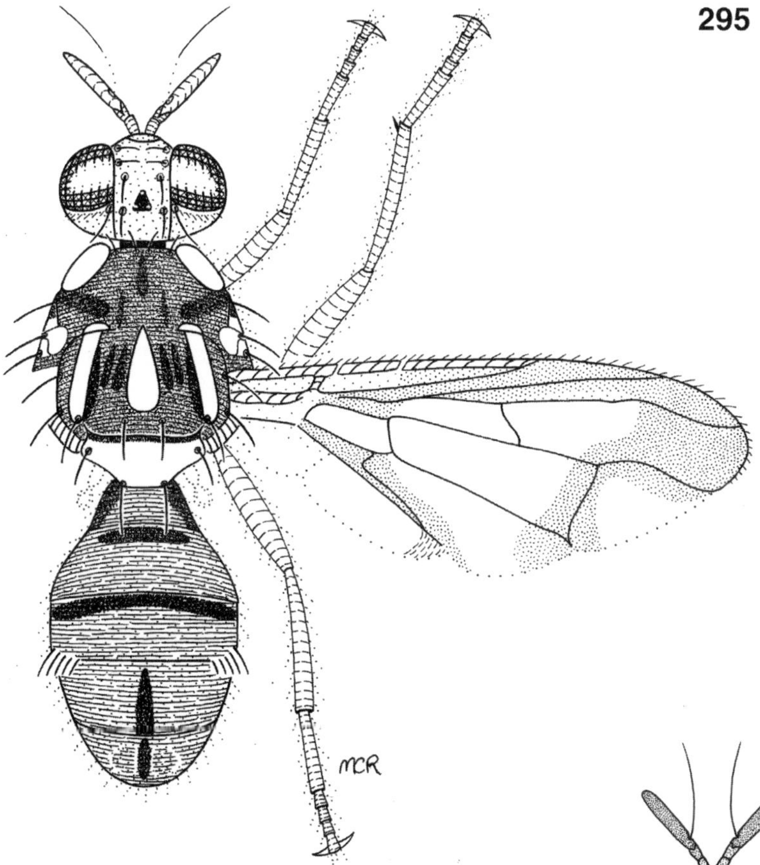

296

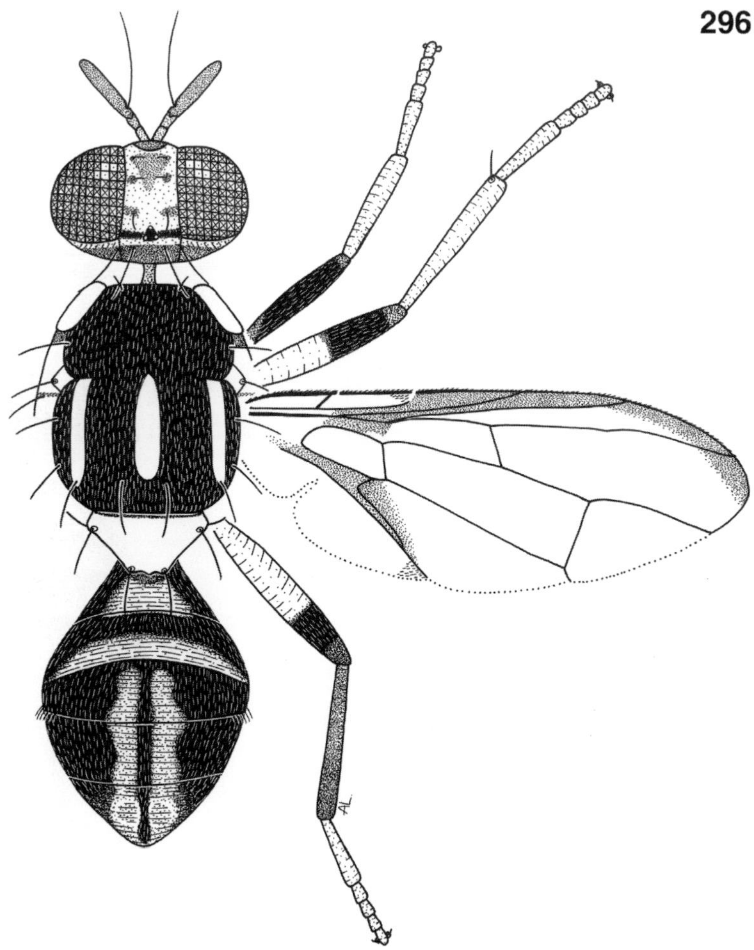

Fig. 295. *Bactrocera (Zeugodacus) emittens* (Walker), syntype male. **Fig. 296.** *Bactrocera (Zeugodacus) exornata* (Hering), male.

297

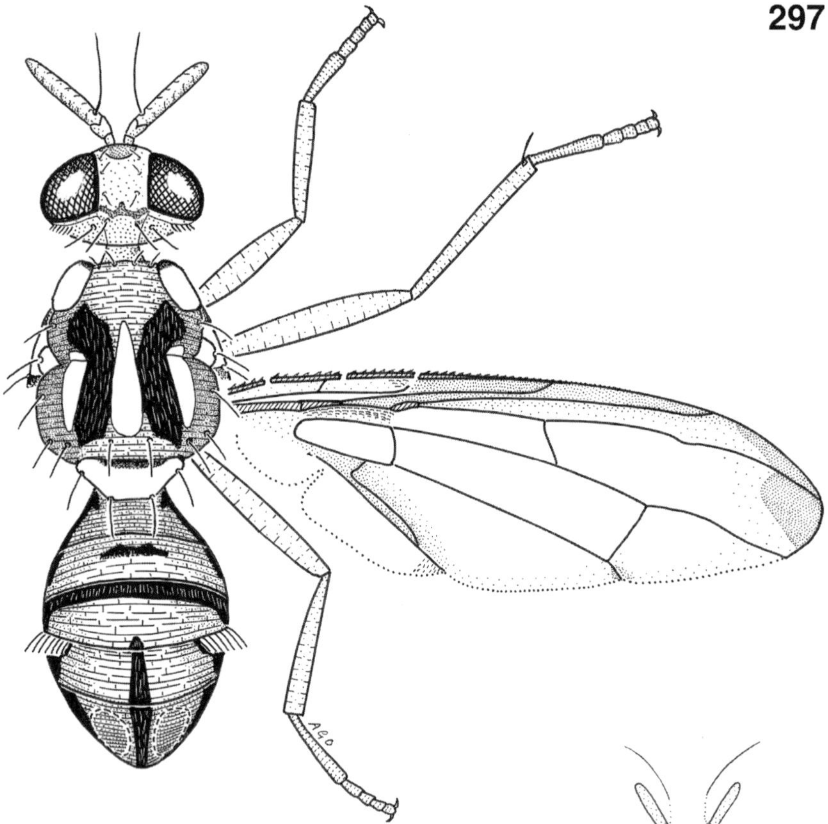

298

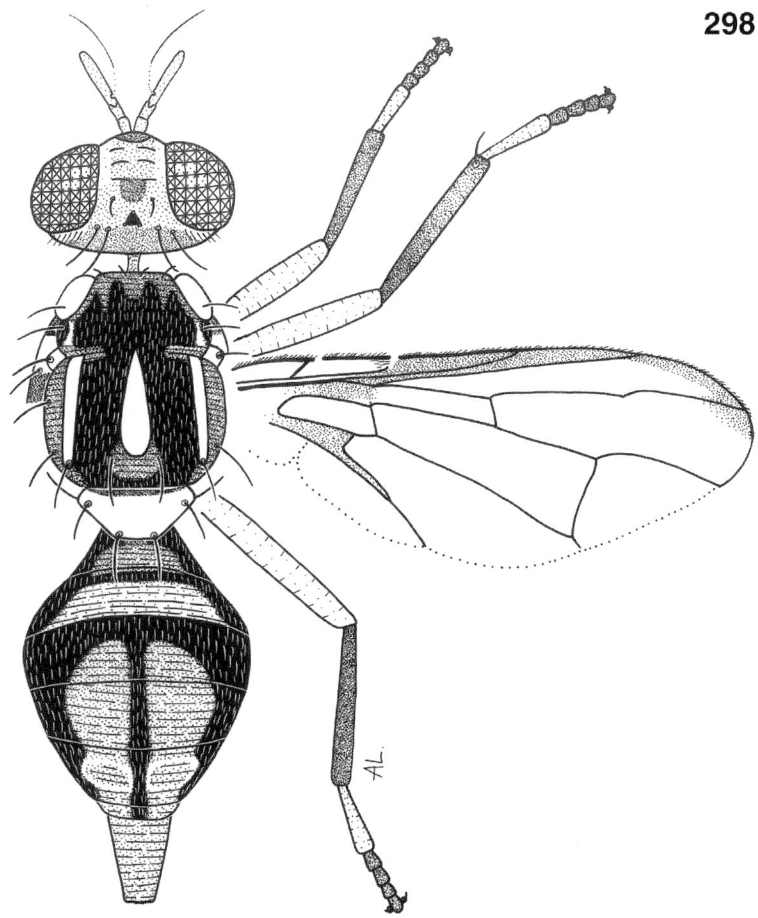

Fig. 297. *Bactrocera (Zeugodacus) flavipilosa* (Hardy), holotype male. **Fig. 298.** *Bactrocera (Zeugodacus) flavolateralis* new species, holotype female.

299

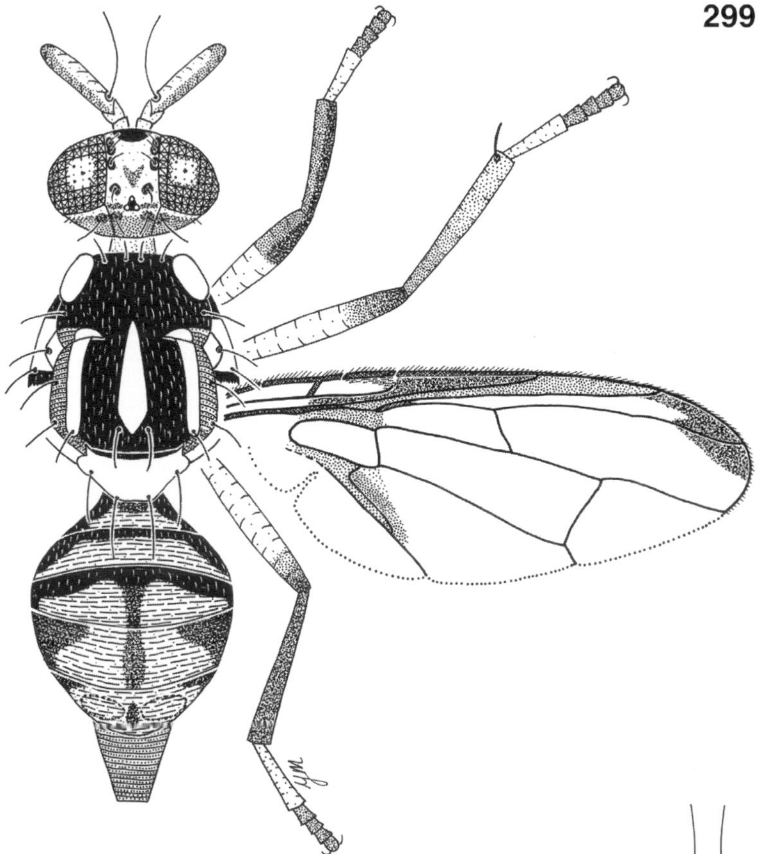

300

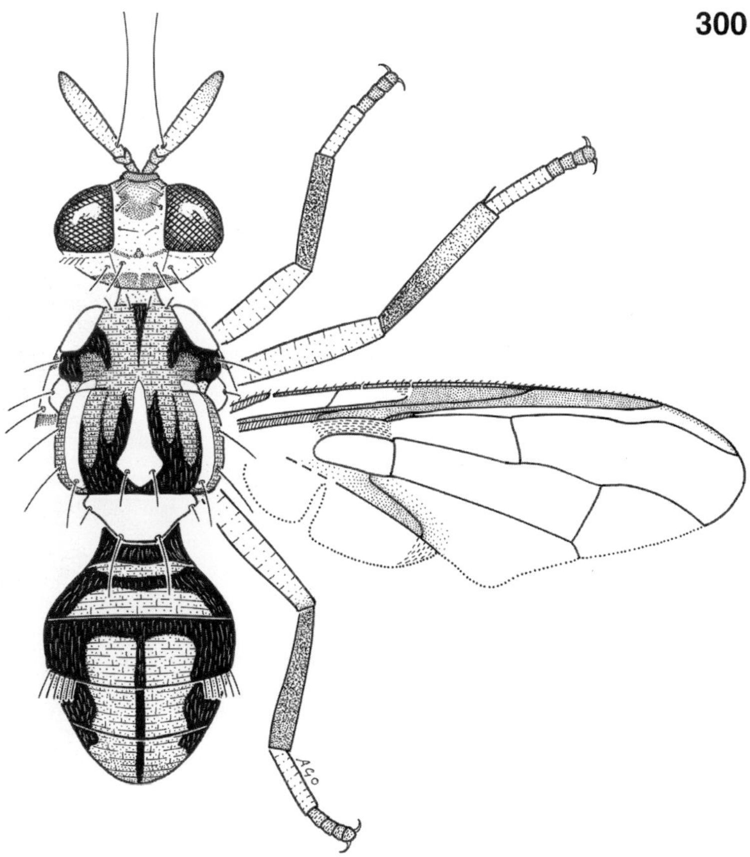

Fig. 299. *Bactrocera (Zeugodacus) flavopectoralis* (Hering), holotype female. **Fig. 300.** *Bactrocera (Zeugodacus) flavoverticalis* new species, holotype male.

301

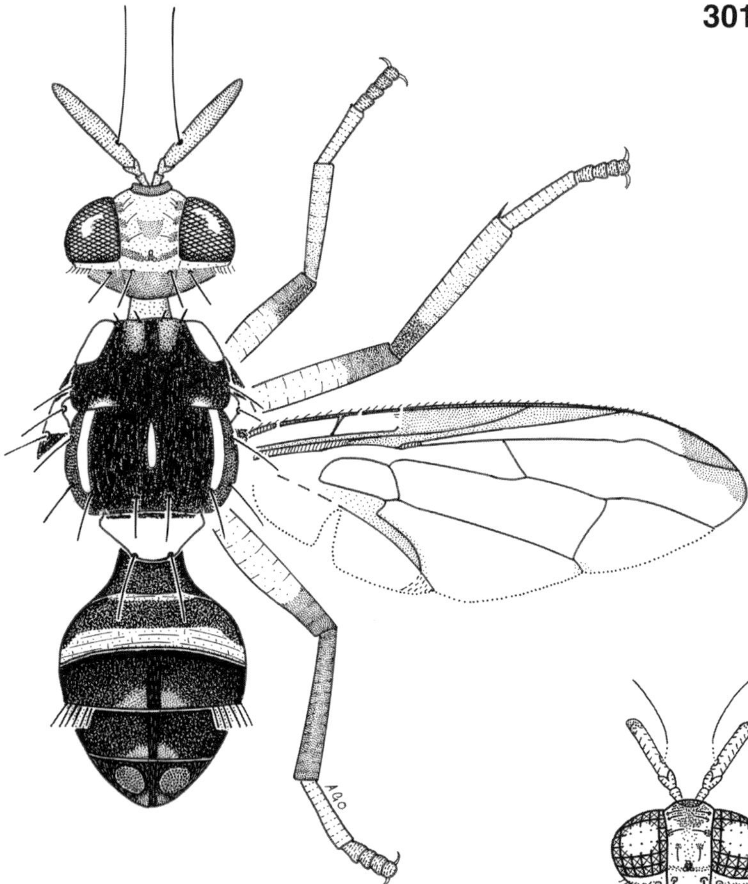

302

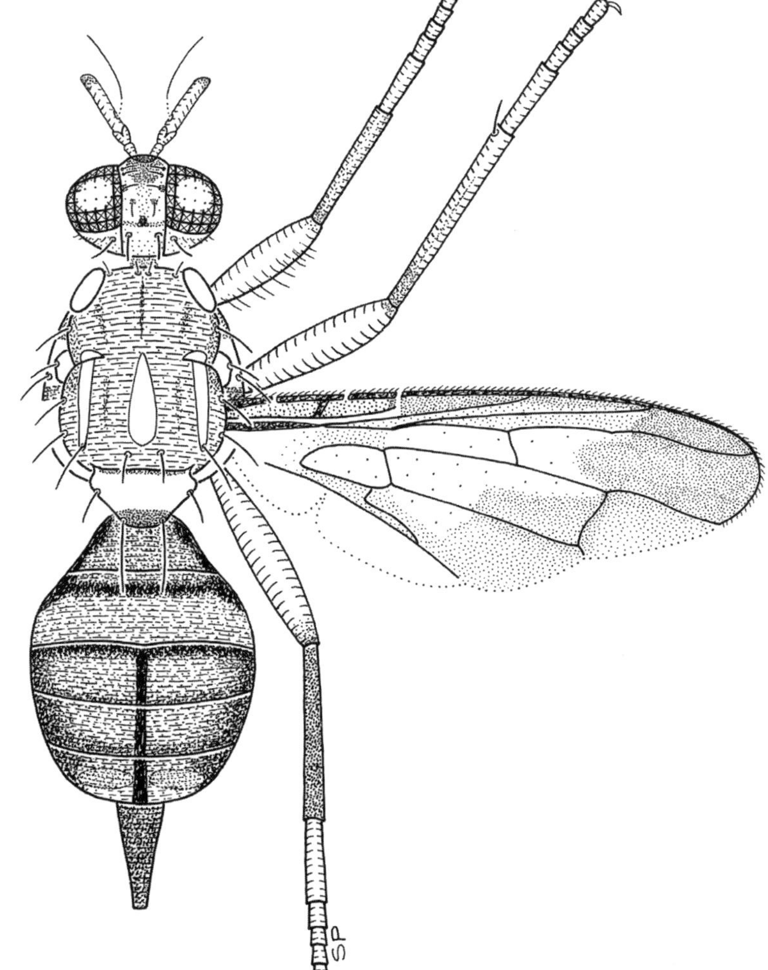

Fig. 301. *Bactrocera (Zeugodacus) freidbergi* White, holotype male. **Fig. 302.** *Bactrocera (Zeugodacus) fuscoalata* new species, holotype female.

303

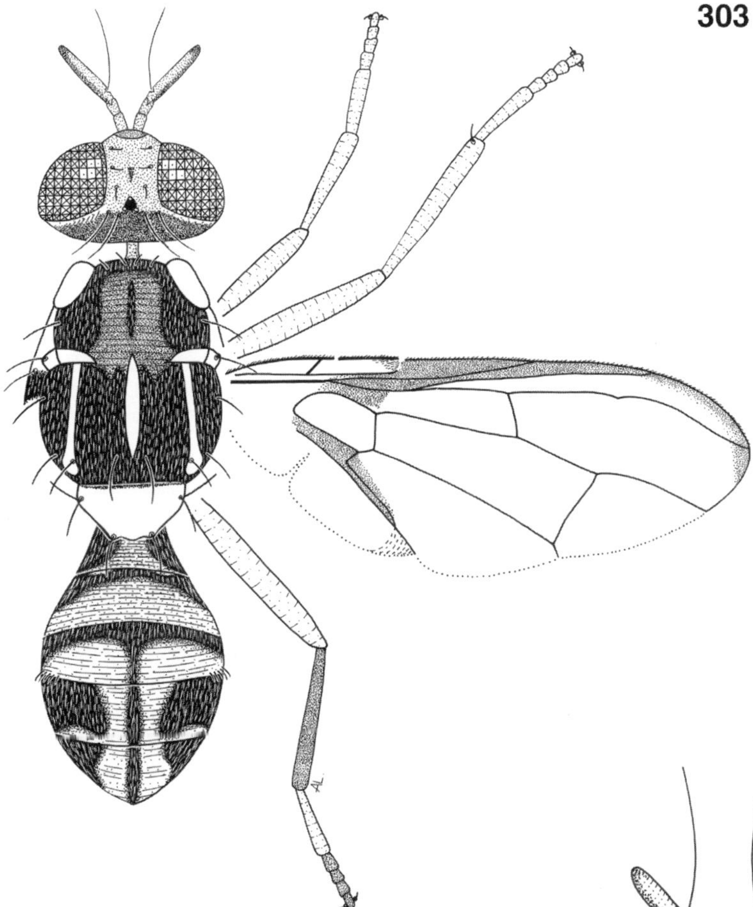

304

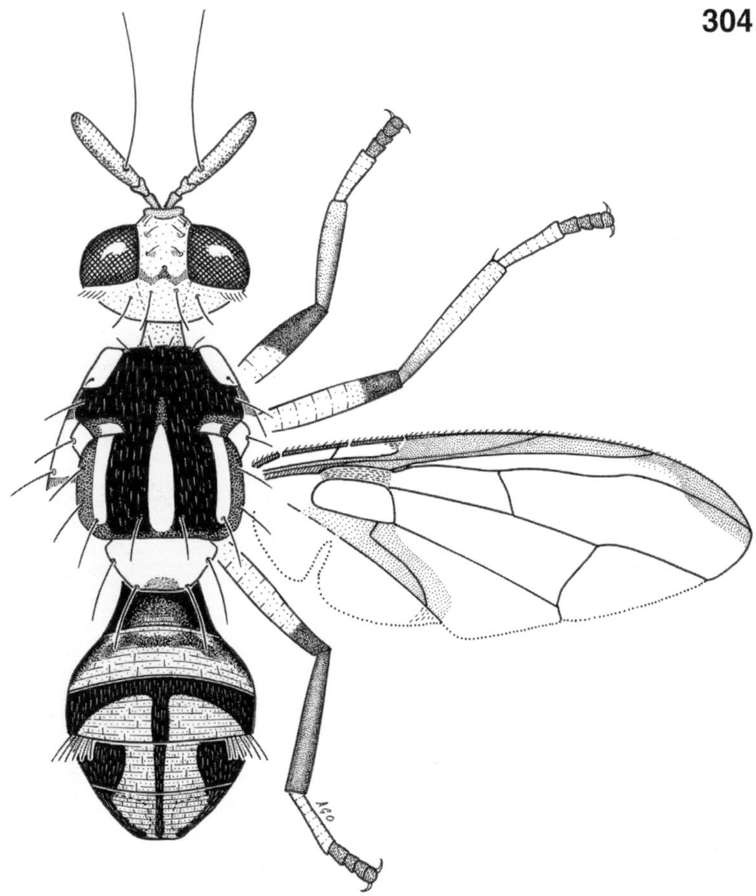

Fig. 303. *Bactrocera (Zeugodacus) gavisa* (Munro), holotype male. **Fig. 304.** *Bactrocera (Zeugodacus) hatyaiensis* new species, holotype male.

305A

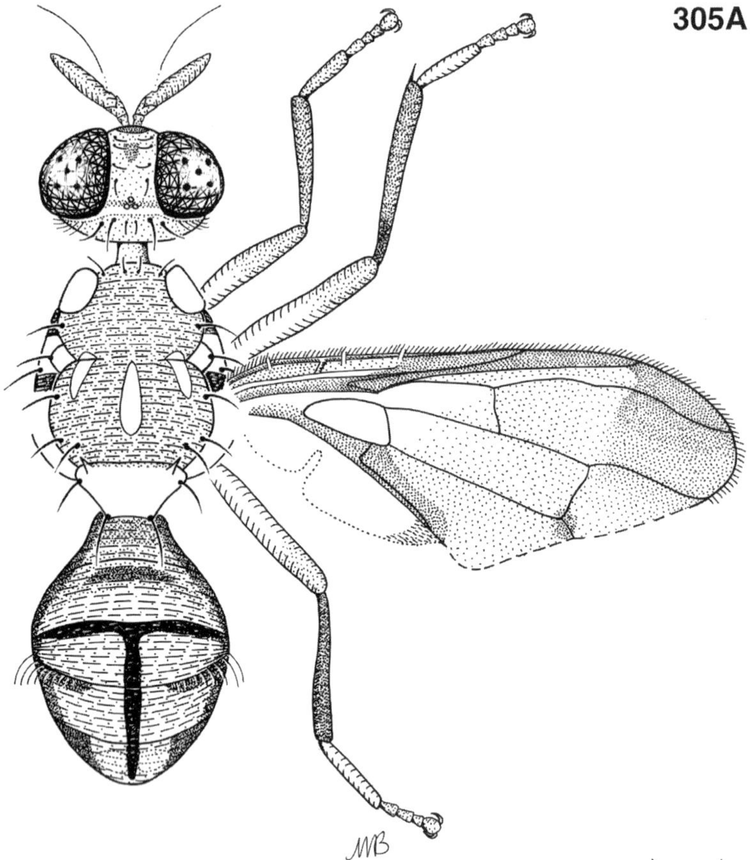

305B

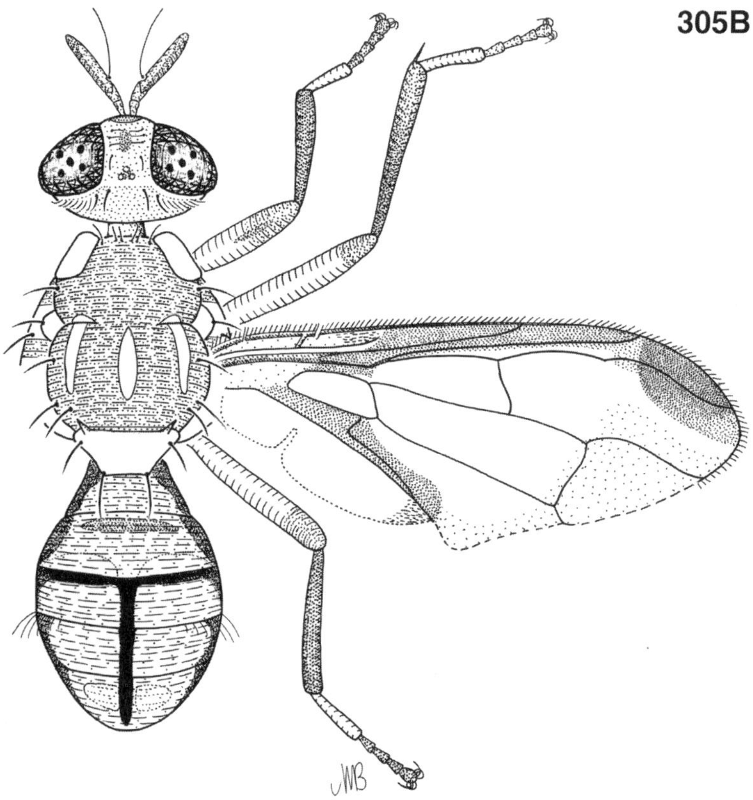

Fig. 305A. *Bactrocera (Zeugodacus) havelockiae* new species, holotype male. **Fig. 305B.** *Bactrocera (Zeugodacus) havelockiae* new species, paratype male showing variation in lateral postsutural vittae and wing colour patterns.

306A

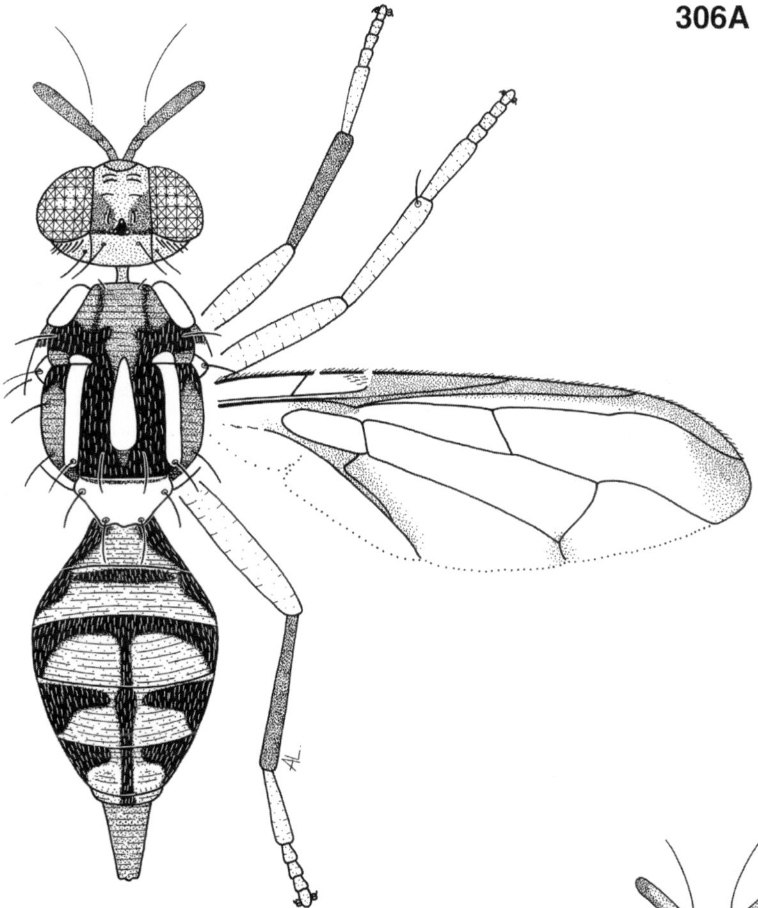

306B

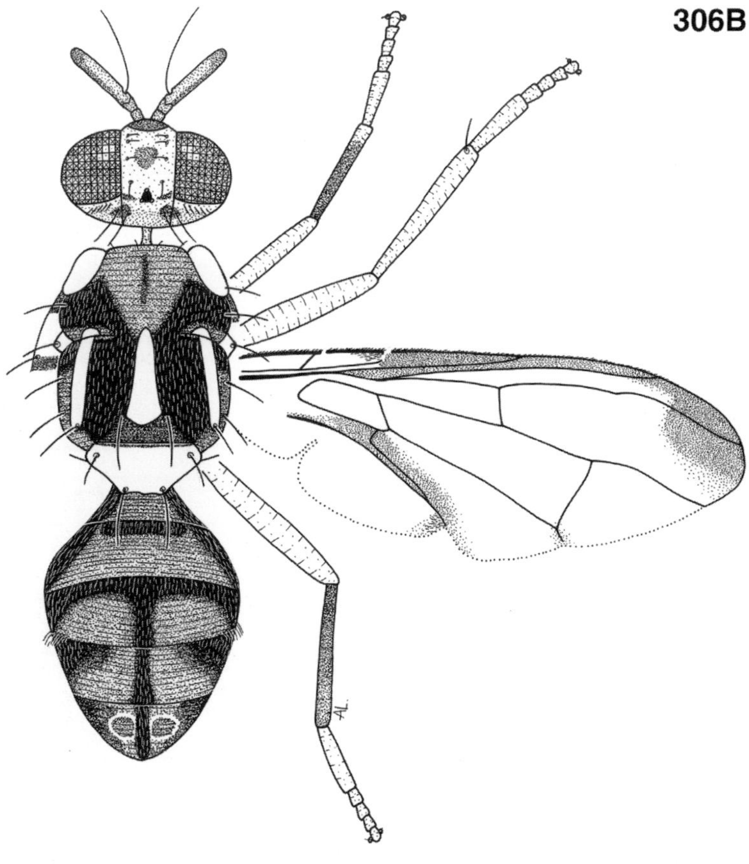

Fig. 306A. *Bactrocera (Zeugodacus) heinrichi* (Hering), holotype female. **Fig. 306B.** *Bactrocera (Zeugodacus) heinrichi* (Hering), male.

307

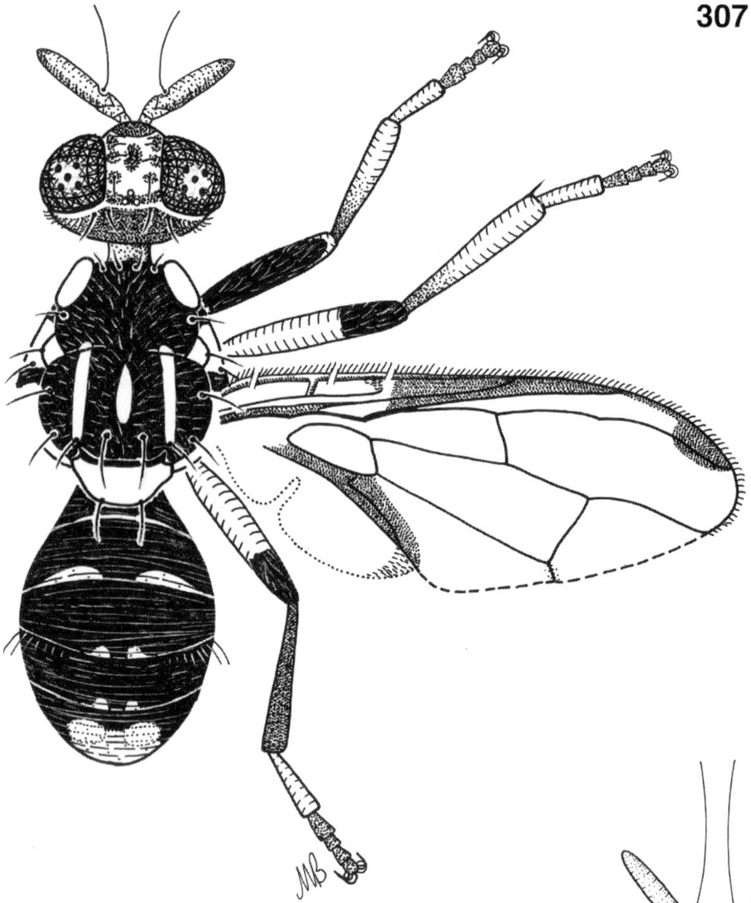

308

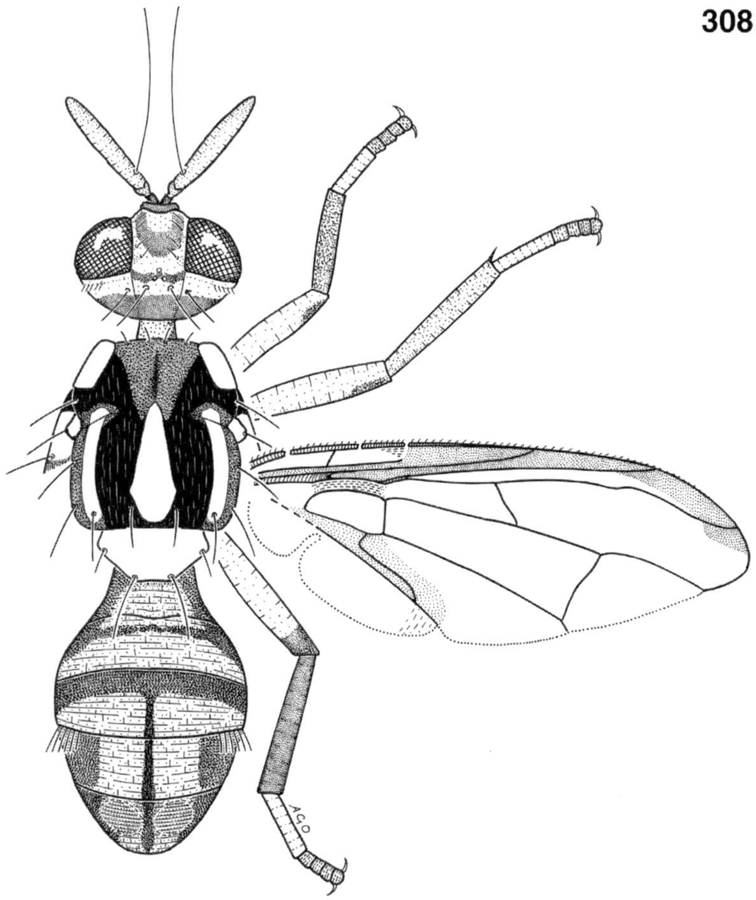

Fig. 307. *Bactrocera (Zeugodacus) hekouana* Yu, He & Yang, male. **Fig. 308.** *Bactrocera (Zeugodacus) hengsawadae* new species, holotype male.

309

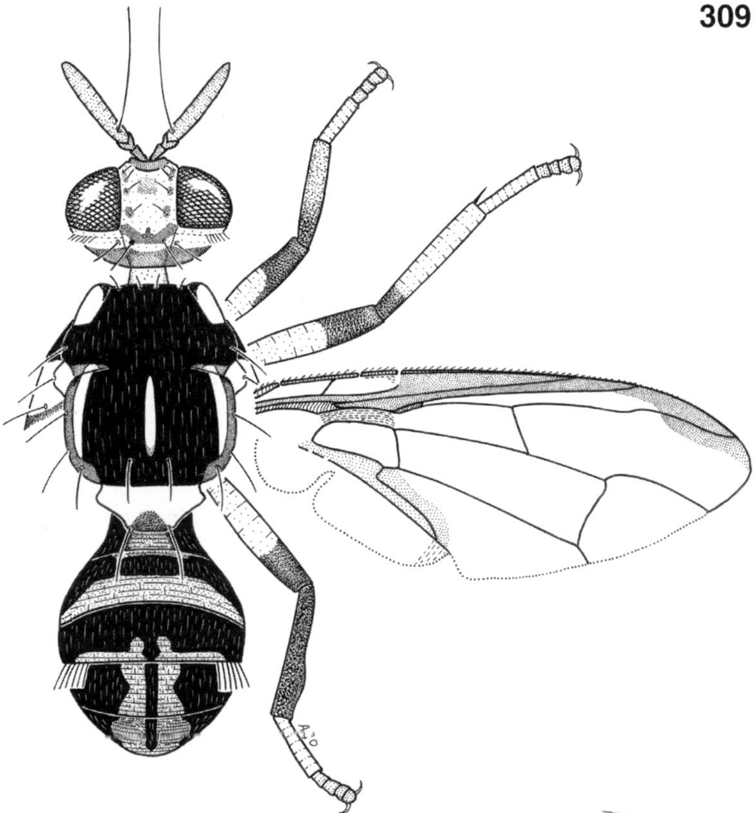

310

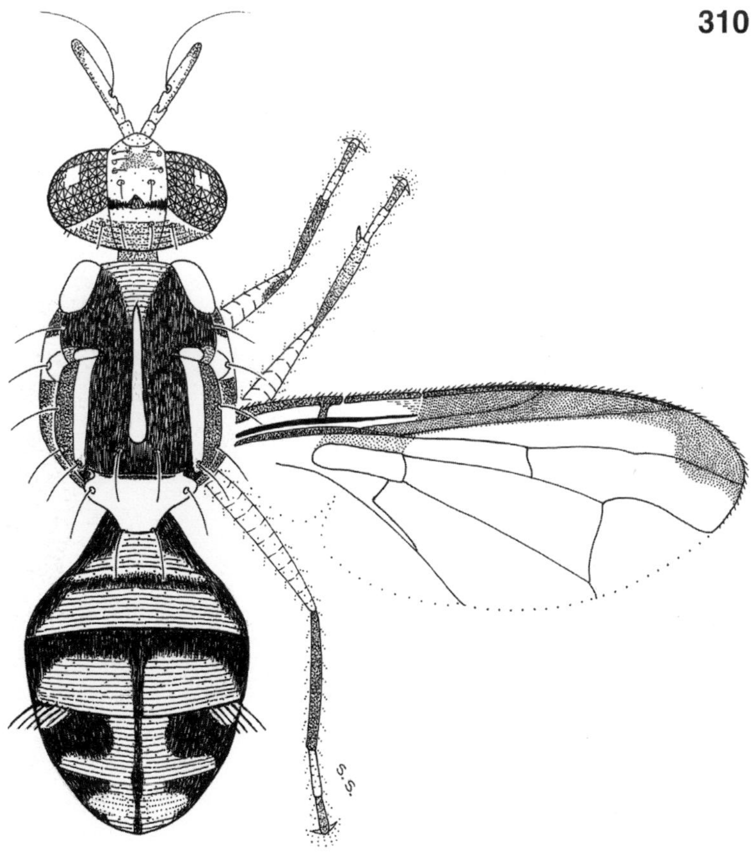

Fig. 309. *Bactrocera (Zeugodacus) hoabinhiae* new species, holotype male. **Fig. 310.** *Bactrocera (Zeugodacus) hodgsoniae* new species, holotype male.

311

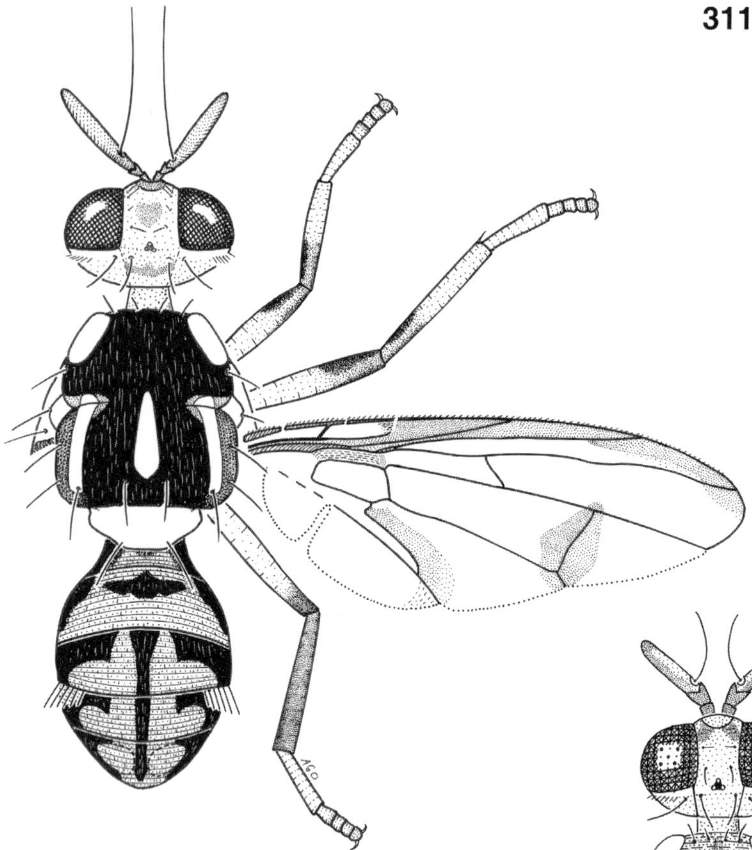

312

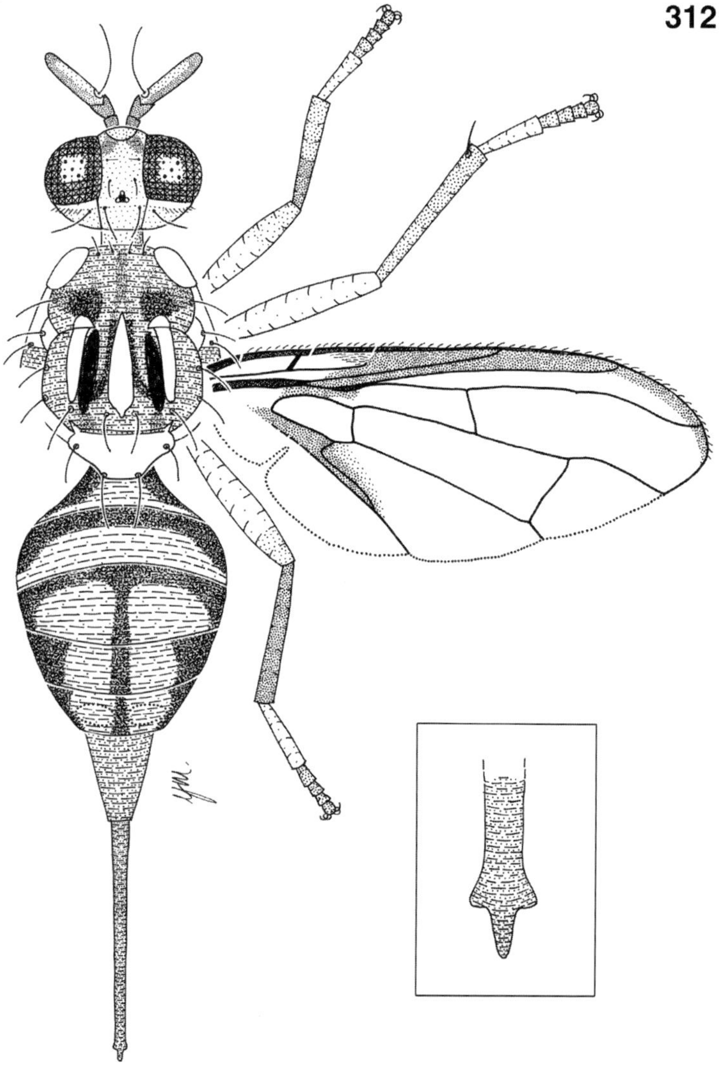

Fig. 311. *Bactrocera (Zeugodacus) hululangatiae* new species, holotype male. **Fig. 312.** *Bactrocera (Zeugodacus) indentus* (Hardy), holotype female and apex of piercer from female ovipositor.

313

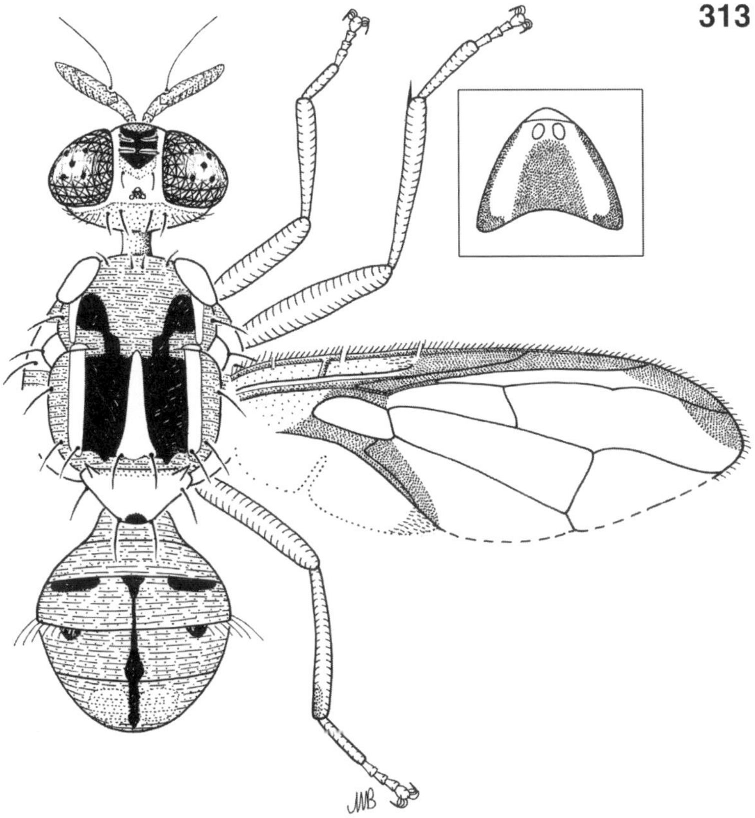

314

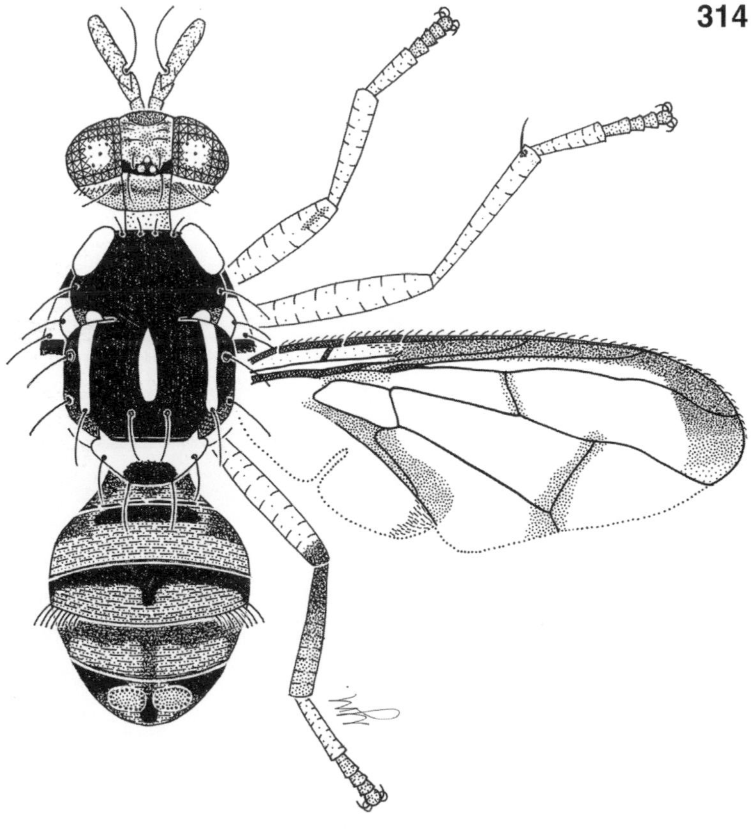

Fig. 313. *Bactrocera (Zeugodacus) iriomotiae* new species, holotype male and face showing unique colour pattern. **Fig. 314.** *Bactrocera (Zeugodacus) ishigakiensis* (Shiraki), holotype male.

315

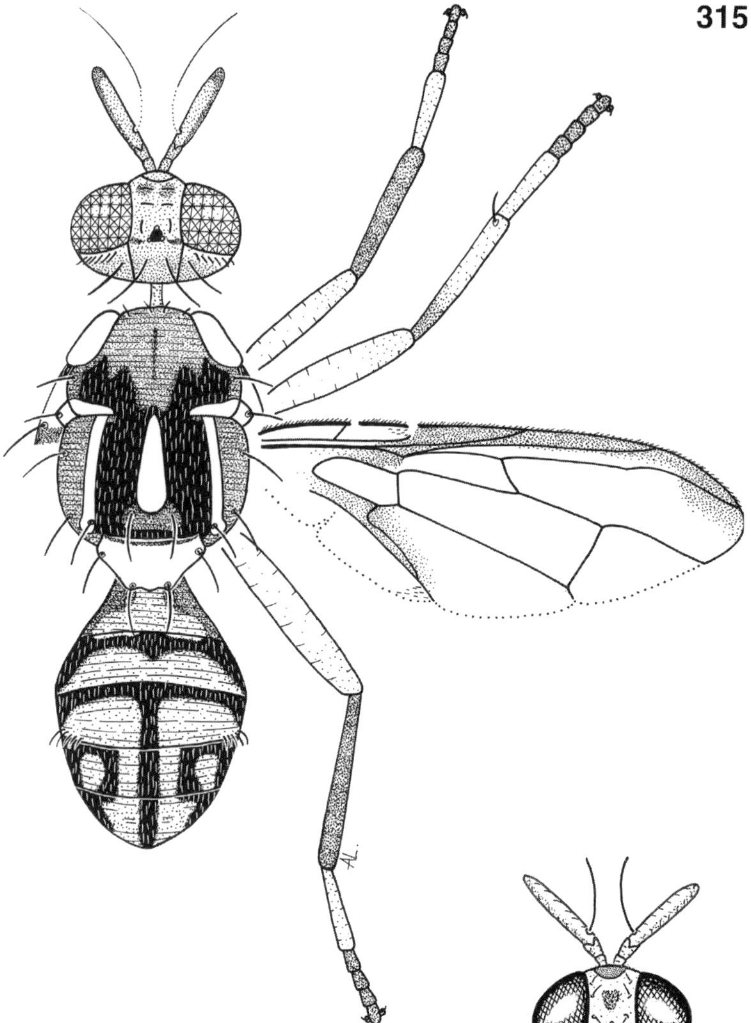

316

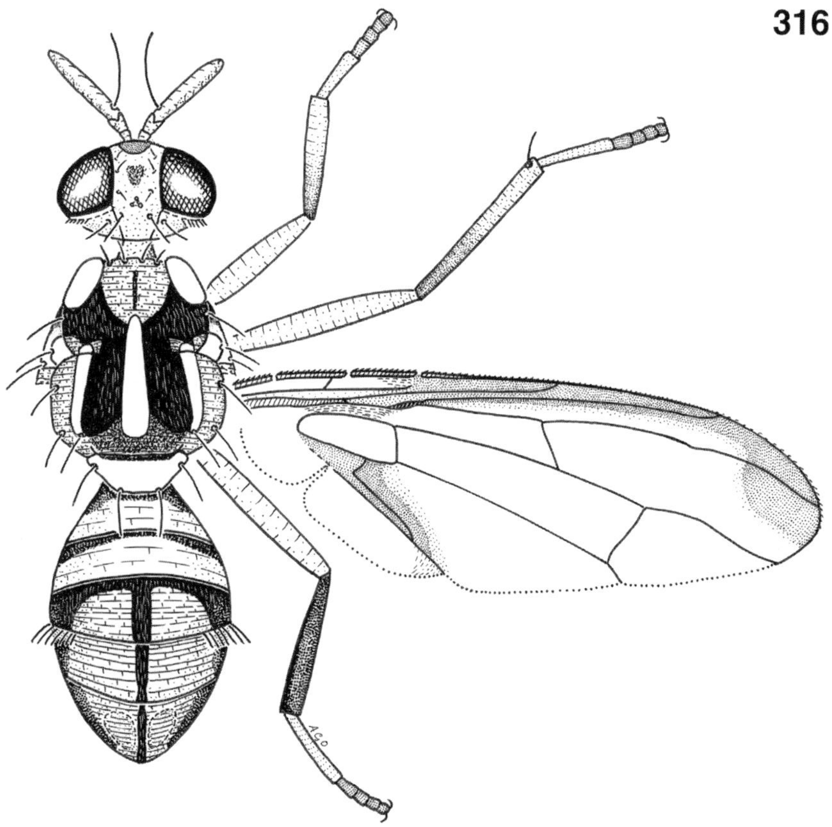

Fig. 315. *Bactrocera (Zeugodacus) isolata* (Hardy), holotype male. **Fig. 316.** *Bactrocera (Zeugodacus) javadica* Mahmood, holotype male.

317

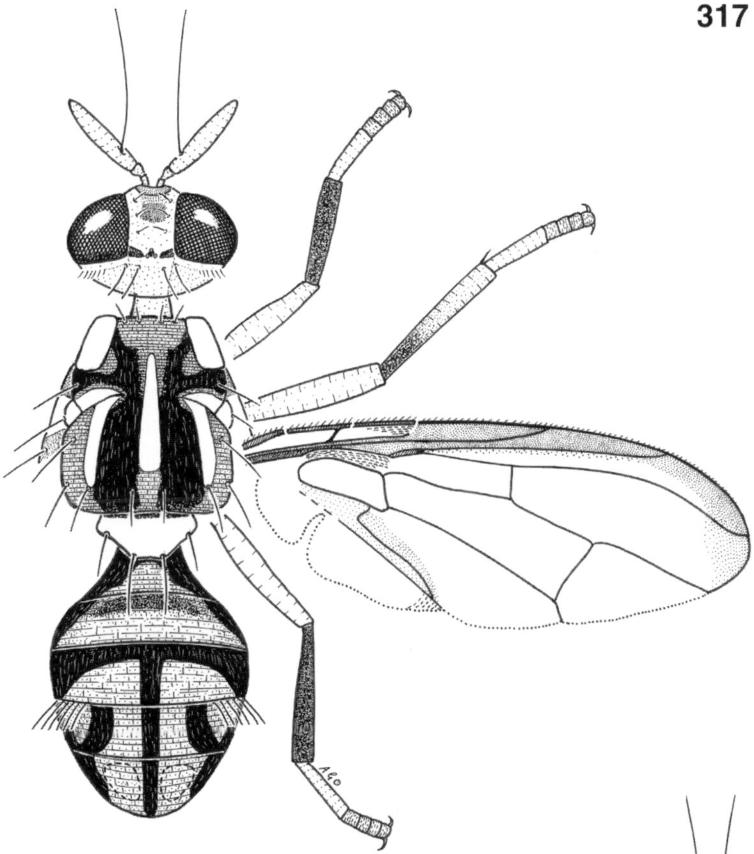

318

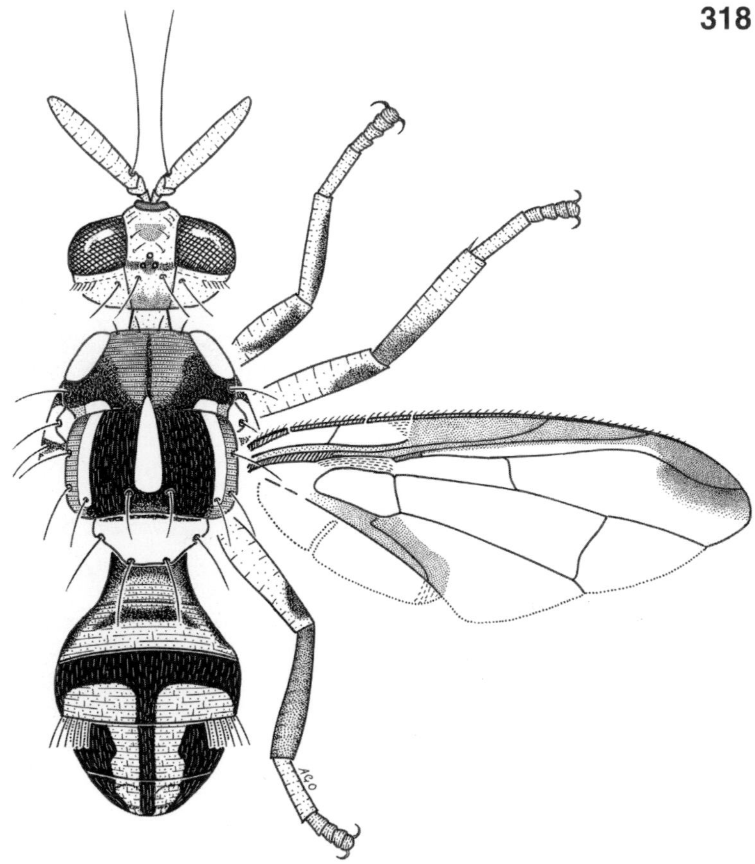

Fig. 317. *Bactrocera (Zeugodacus) juxtuncinata* new species, holotype male. **Fig. 318.** *Bactrocera (Zeugodacus) khaoyaiae* new species, holotype male.

319

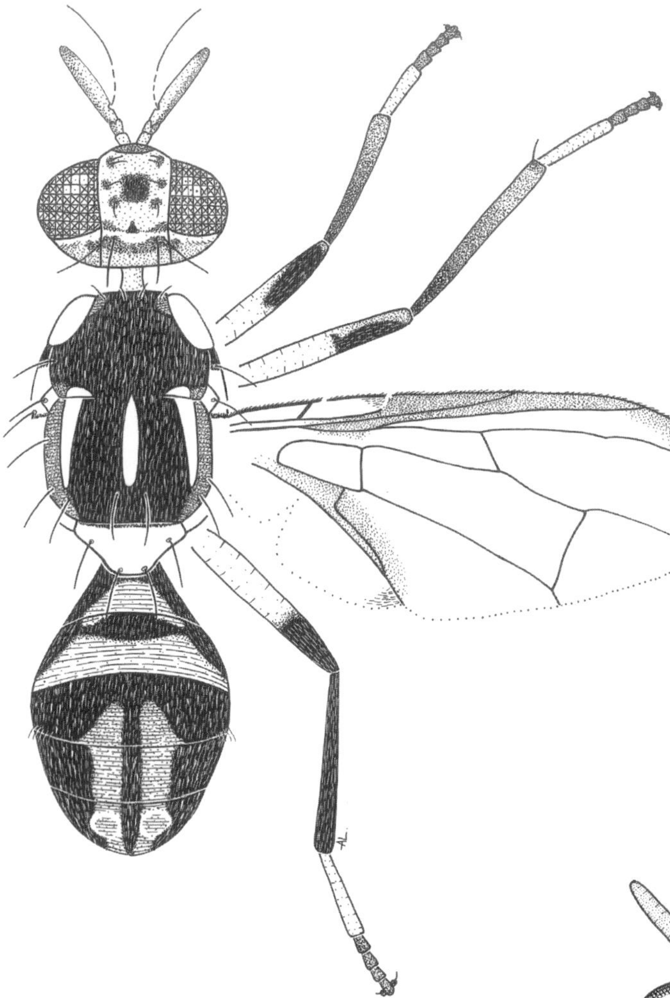

320

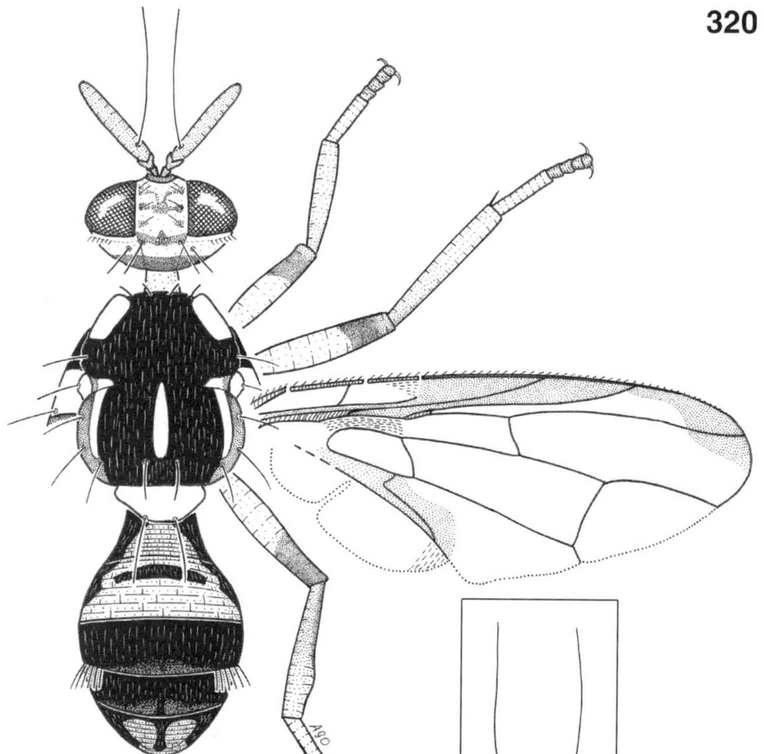

Fig. 319. *Bactrocera (Zeugodacus) laguniensis* new species, holotype male. **Fig. 320.** *Bactrocera (Zeugodacus) laocaiae* new species, holotype male and apex of piercer from female ovipositor.

321

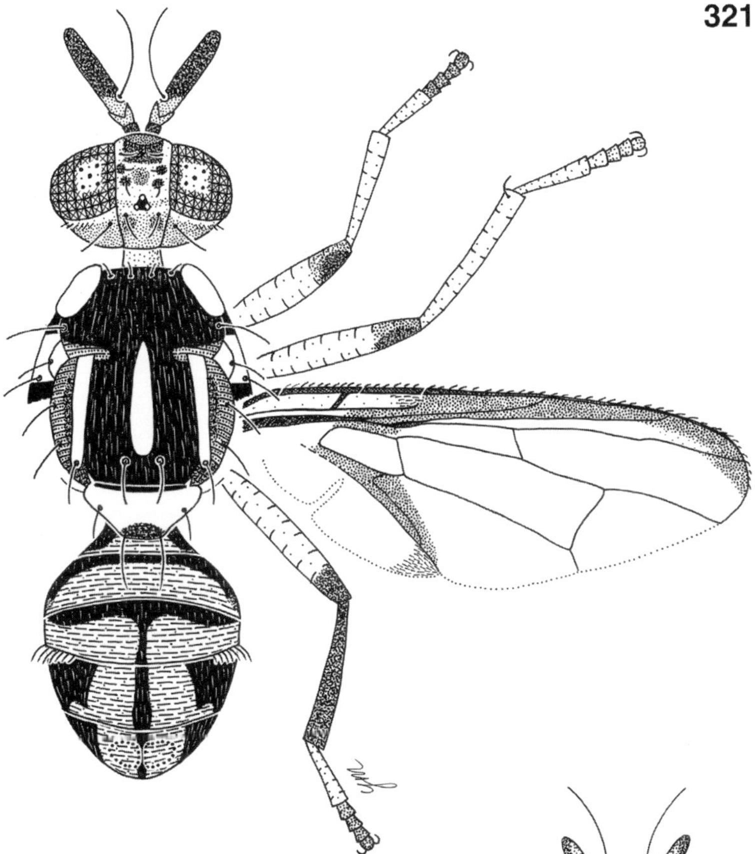

322

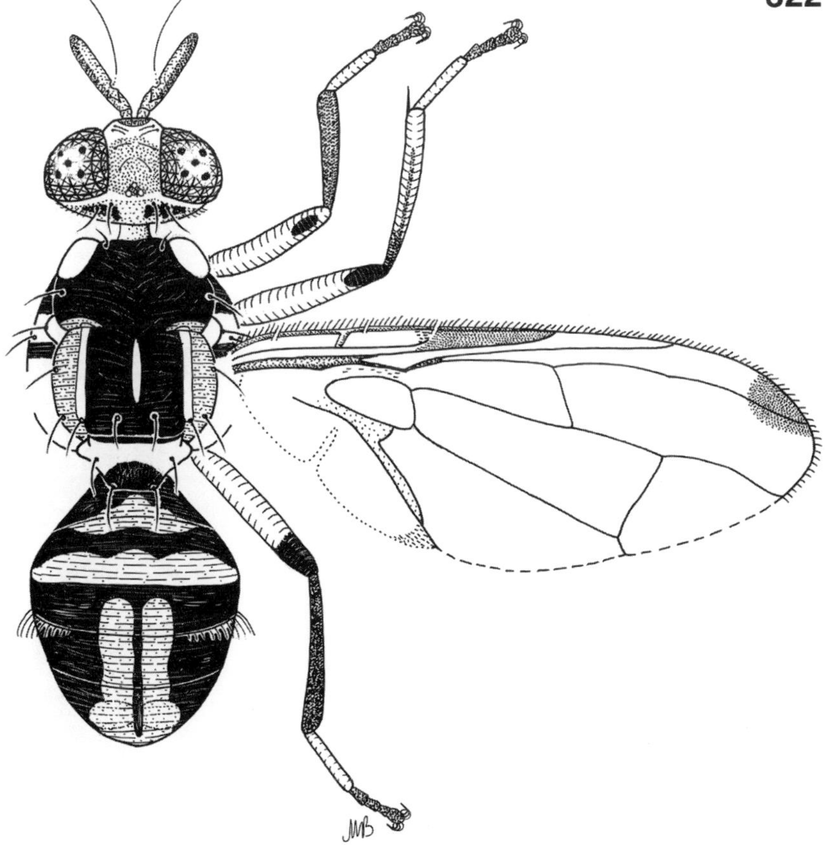

Fig. 321. *Bactrocera (Zeugodacus) lipsana* (Hendel), syntype male. **Fig. 322.** *Bactrocera (Zeugodacus) liquida* new species, holotype male.

323

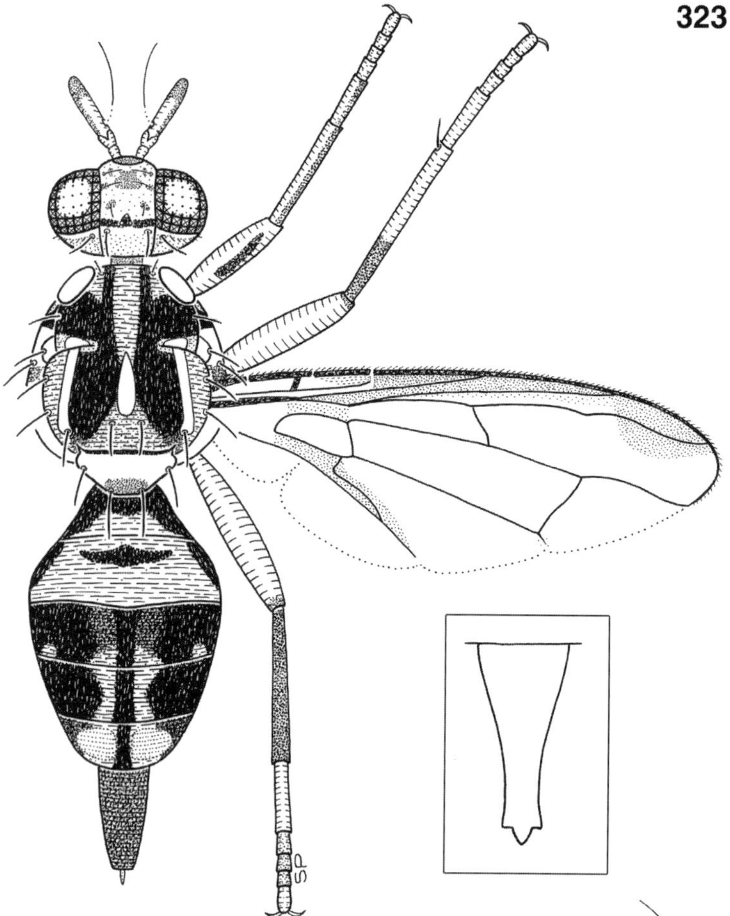

324

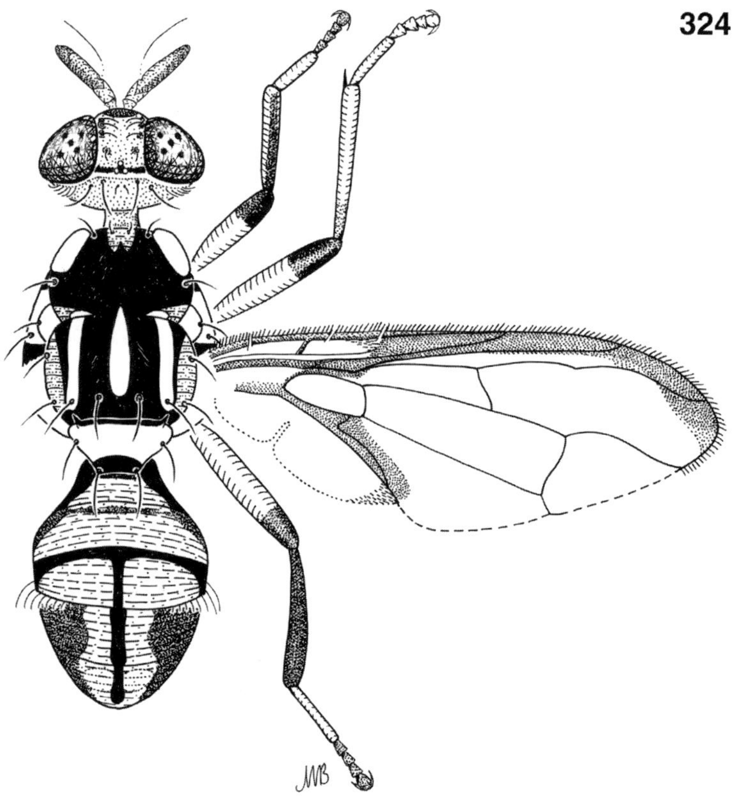

Fig. 323. *Bactrocera (Zeugodacus) macrophyllae* new species, holotype female and apex of piercer from female ovipositor.
Fig. 324. *Bactrocera (Zeugodacus) maculata* (Perkins), male.

325

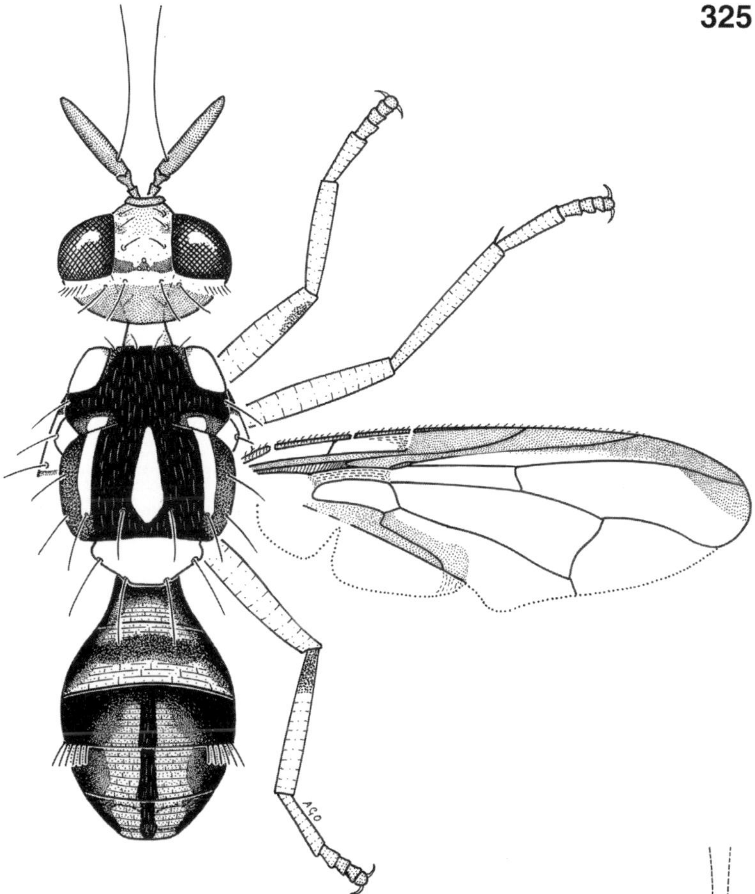

326

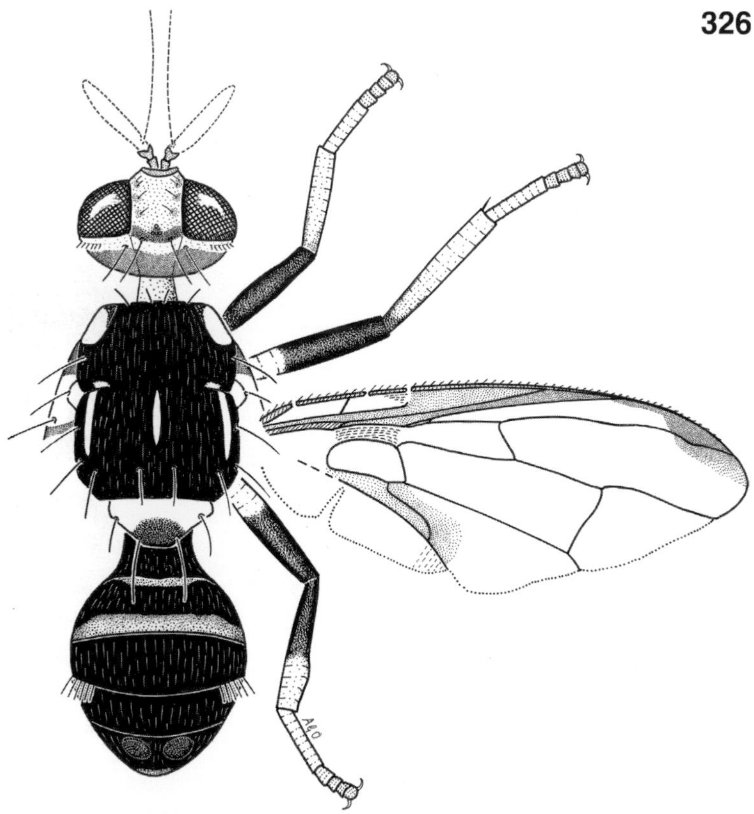

Fig. 325. *Bactrocera (Zeugodacus) melanofacies* new species, holotype male. **Fig. 326.** *Bactrocera (Zeugodacus) menglana* Yu, Liu & Yang, male.

327

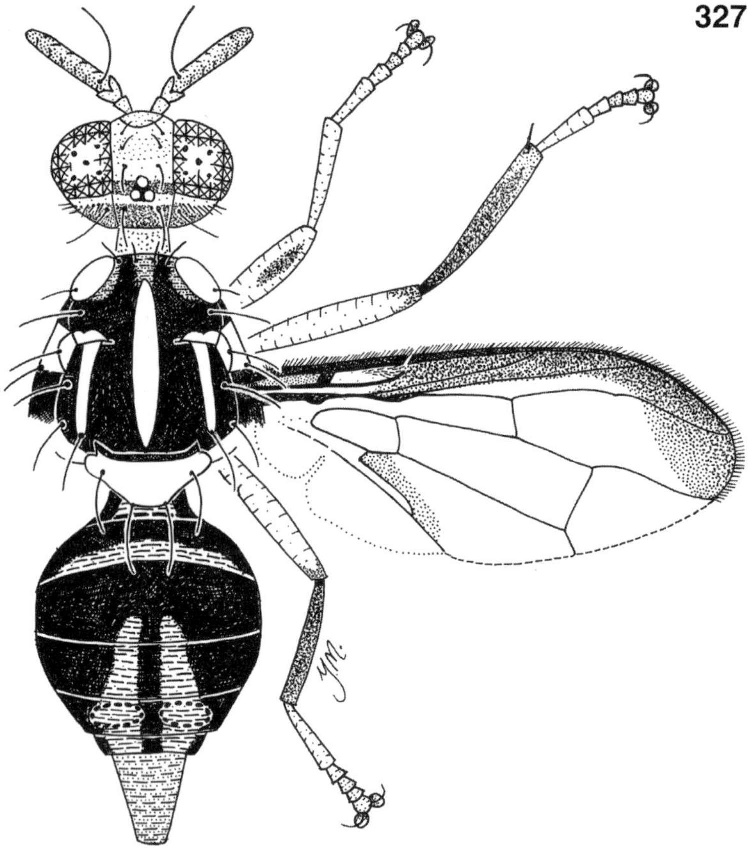

328

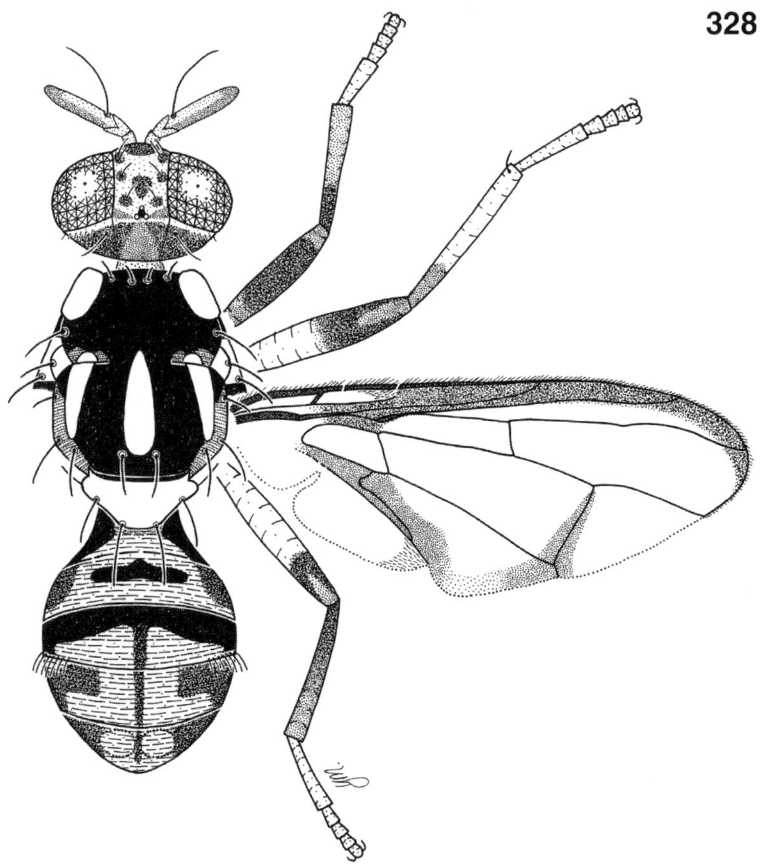

Fig. 327. *Bactrocera (Zeugodacus) minima* (Hering), holotype female. **Fig. 328.** *Bactrocera (Zeugodacus) munda* (Bezzi), syntype male.

329

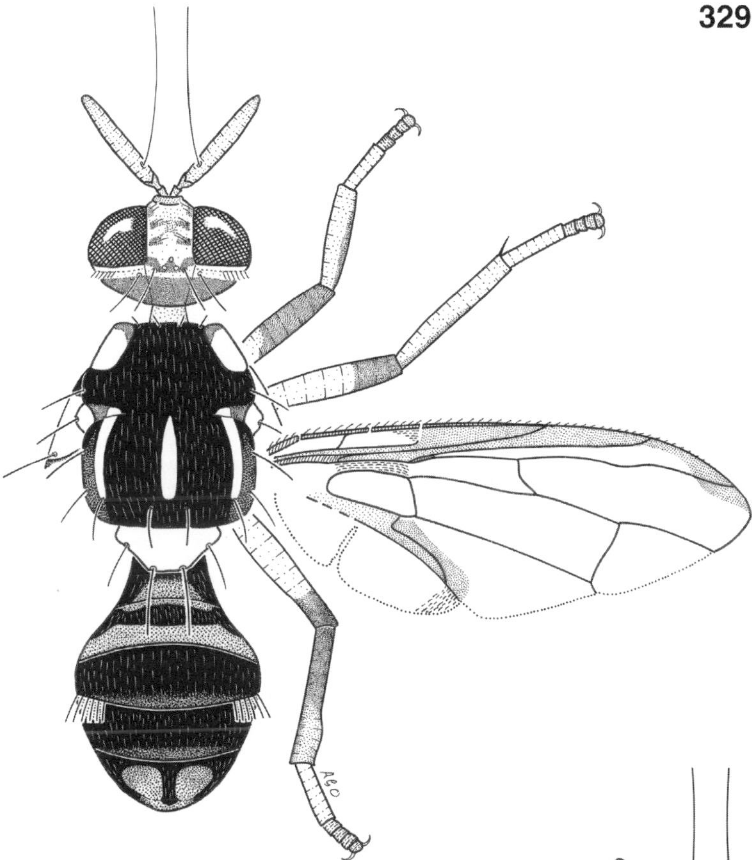

330

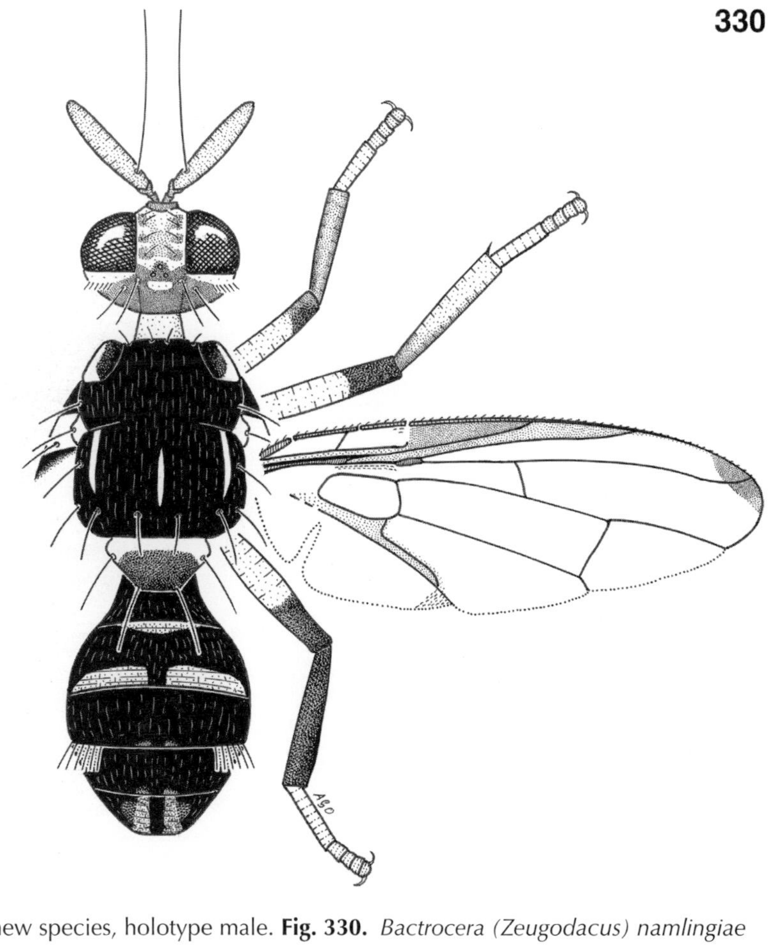

Fig. 329. *Bactrocera (Zeugodacus) nakhonnayokiae* new species, holotype male. **Fig. 330.** *Bactrocera (Zeugodacus) namlingiae* new species, holotype male.

331

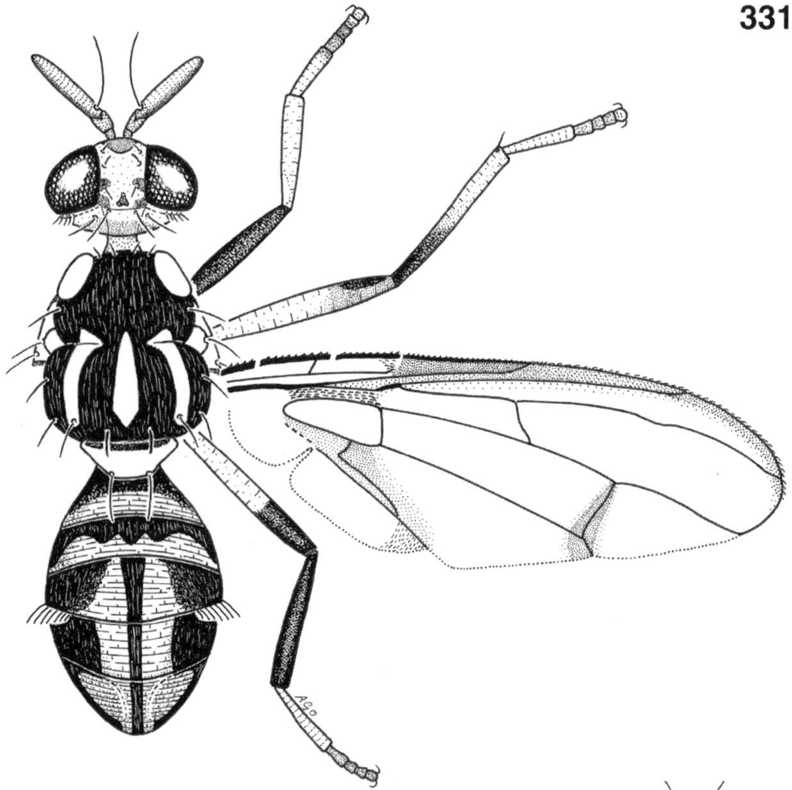

332

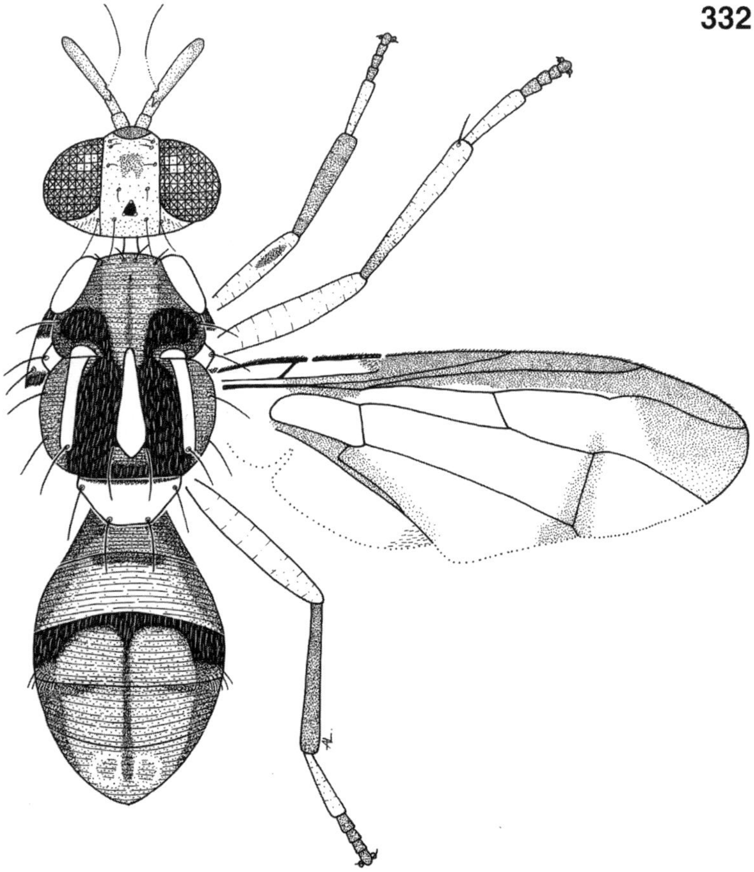

Fig. 331. *Bactrocera (Zeugodacus) neoelegantula* White, holotype male. **Fig. 332.** *Bactrocera (Zeugodacus) neoemittens* new species, holotype male.

333

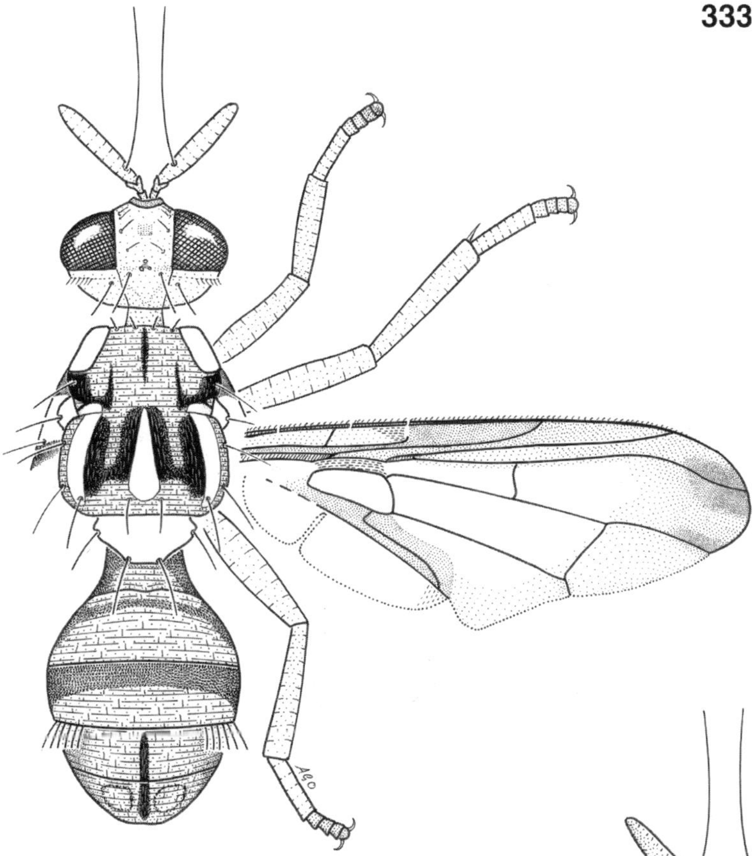

334

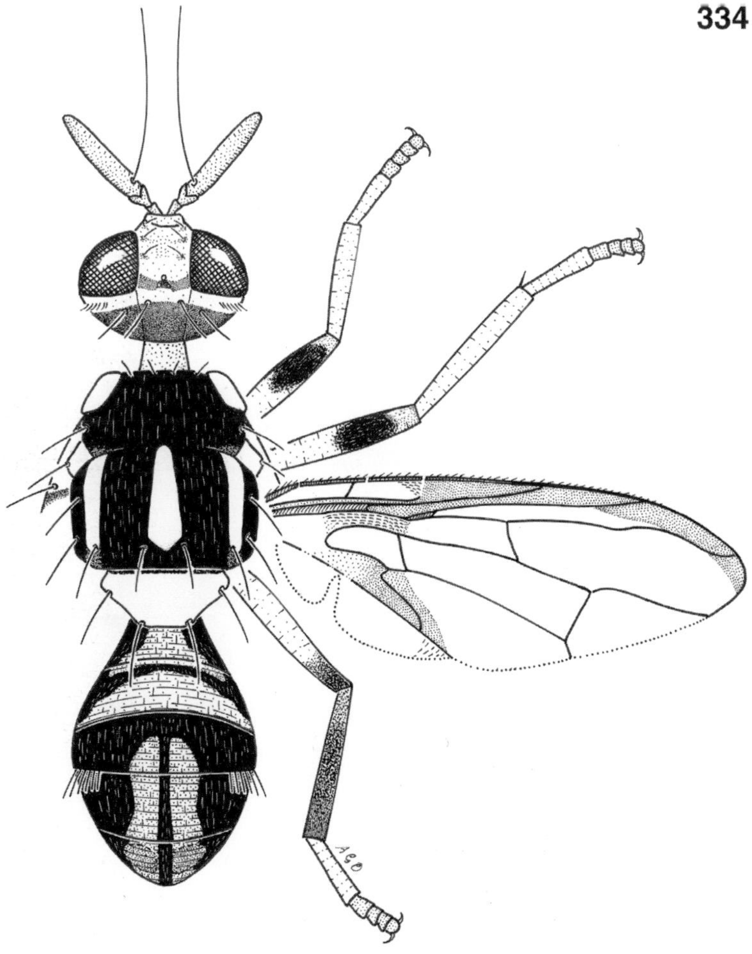

Fig. 333. *Bactrocera (Zeugodacus) neoflavipilosa* new species, holotype male. **Fig. 334.** *Bactrocera (Zeugodacus) neolipsana* new species, holotype male.

335

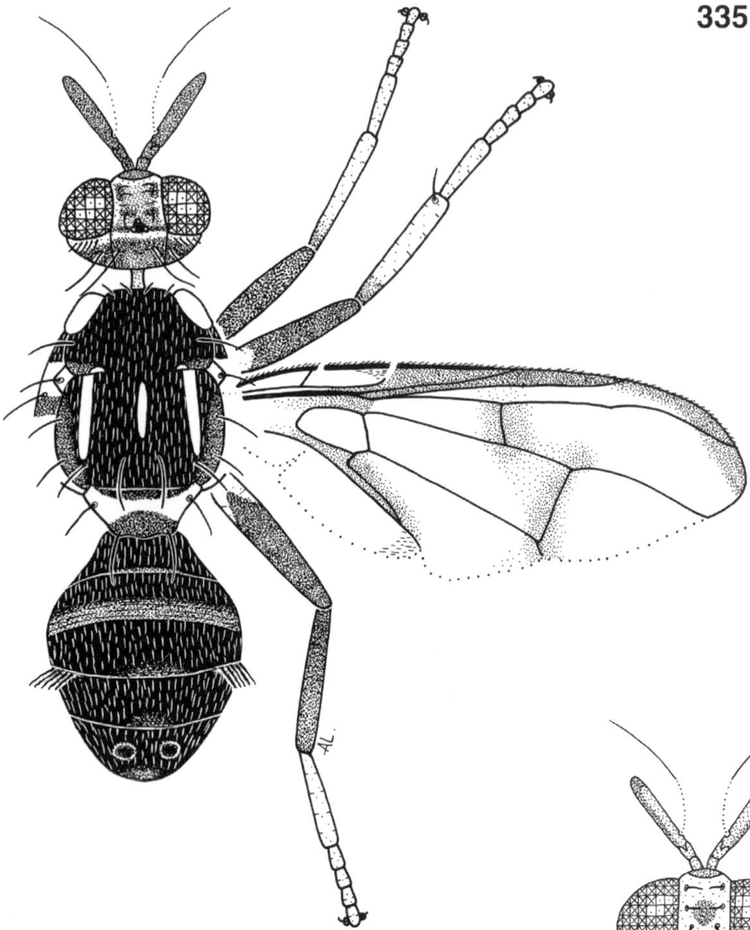

336

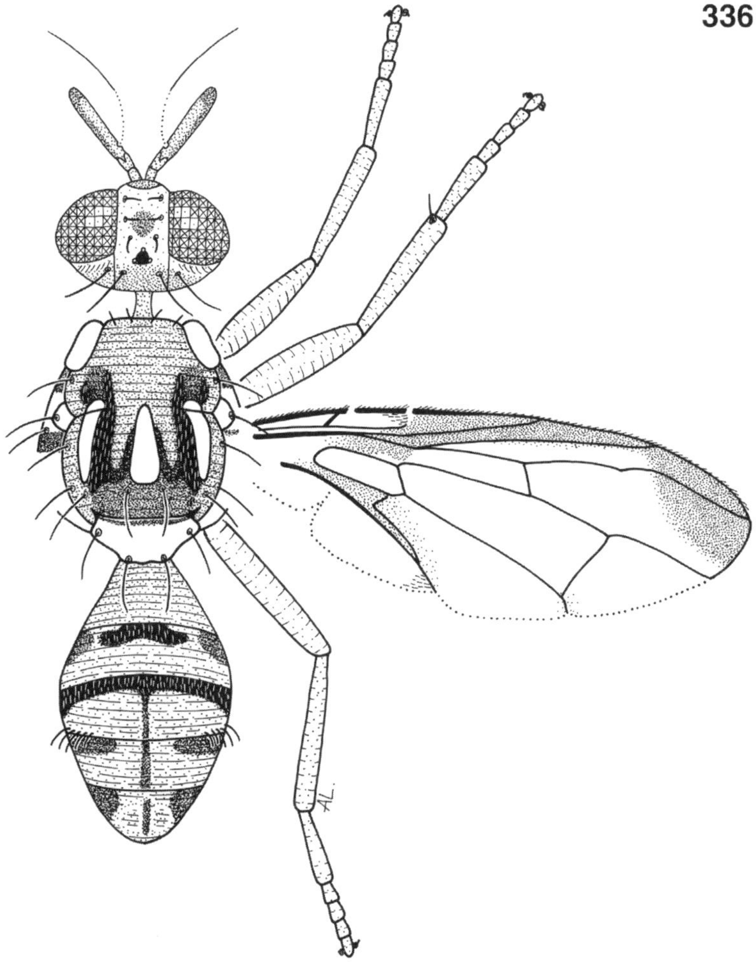

Fig. 335. *Bactrocera (Zeugodacus) nigrifacies* (Shiraki), lectotype male. **Fig. 336.** *Bactrocera (Zeugodacus) okunii* (Shiraki), lectotype male.

337

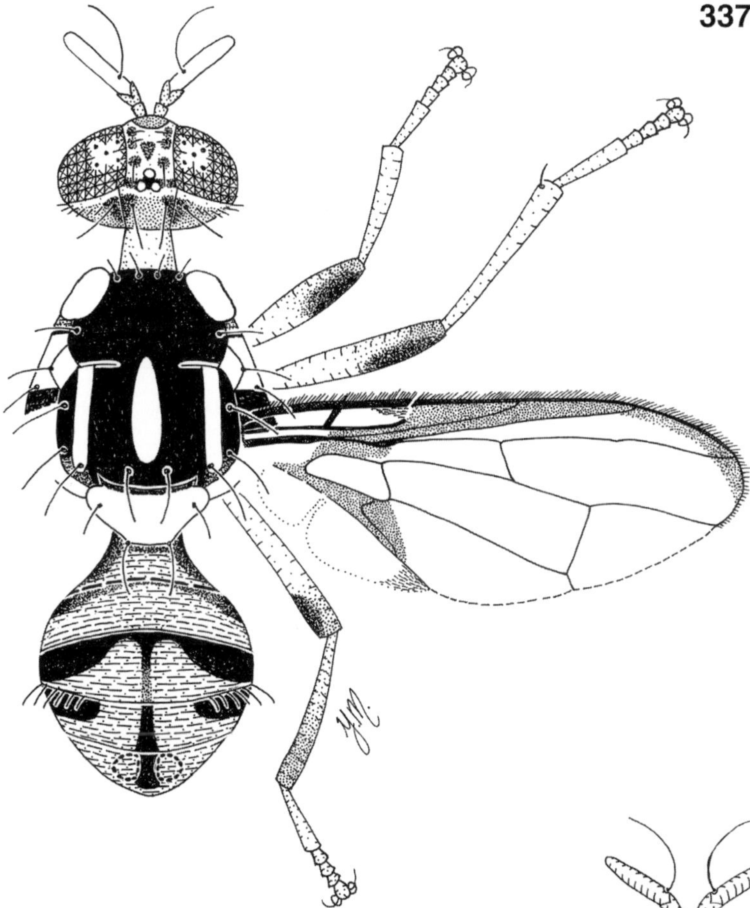

338

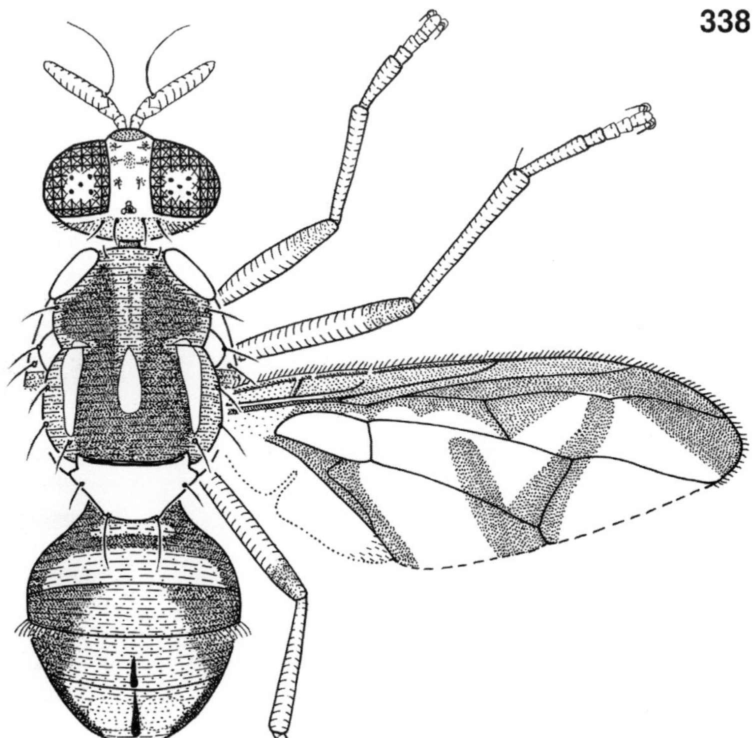

Fig. 337. *Bactrocera (Zeugodacus) pemalangiae* new species, holotype male. **Fig. 338.** *Bactrocera (Zeugodacus) perplexa* (Walker), male.

339

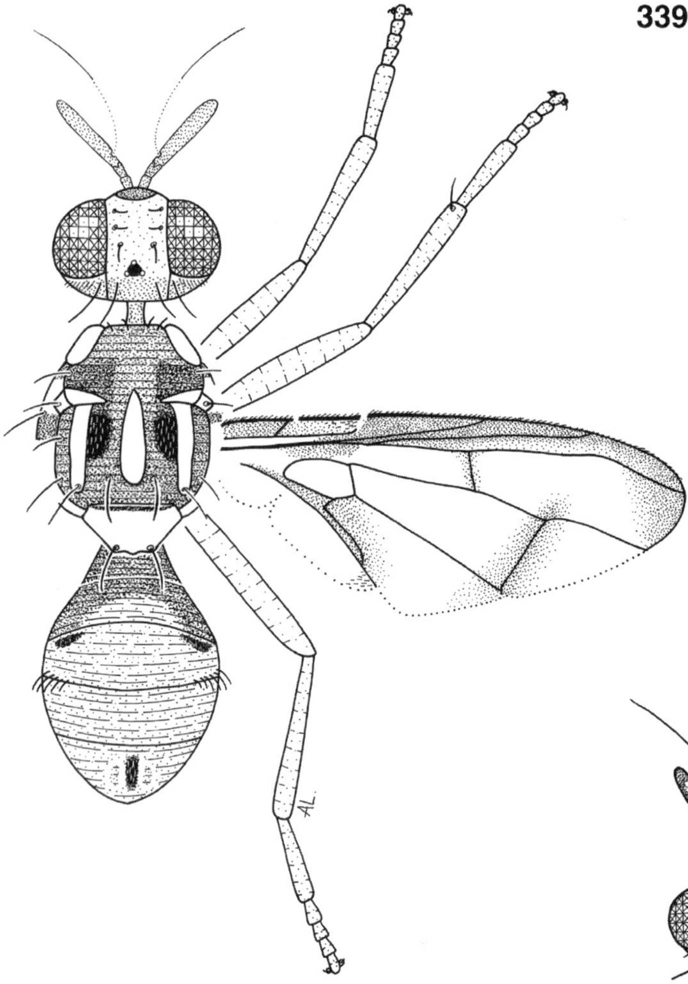

340

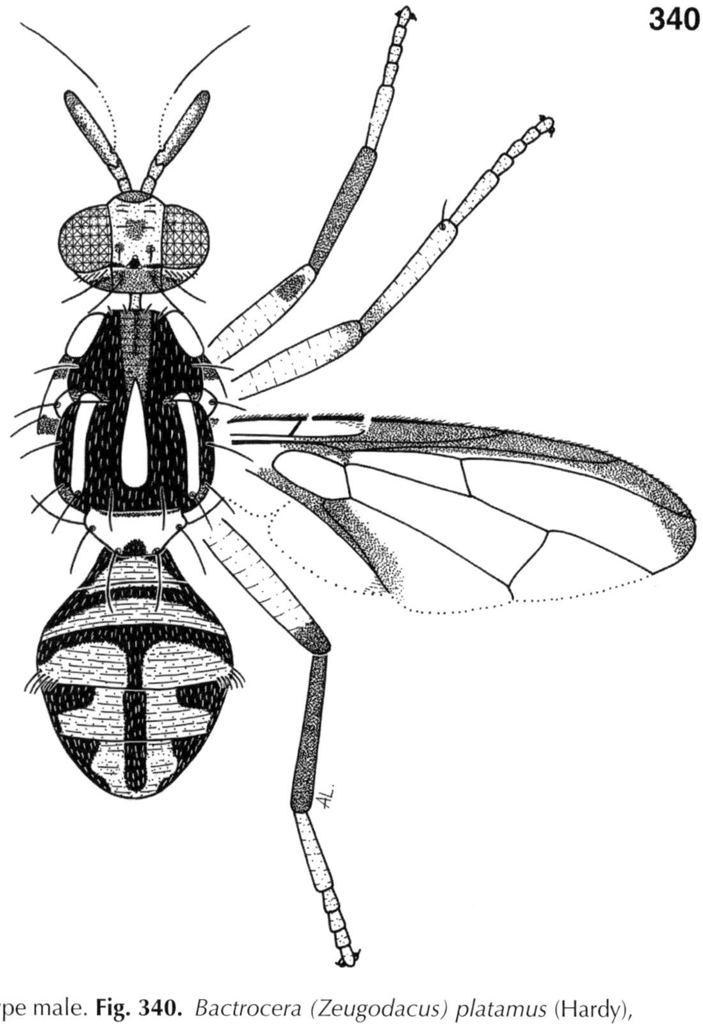

Fig. 339. *Bactrocera (Zeugodacus) persignata* (Hering), syntype male. **Fig. 340.** *Bactrocera (Zeugodacus) platamus* (Hardy), holotype male.

341

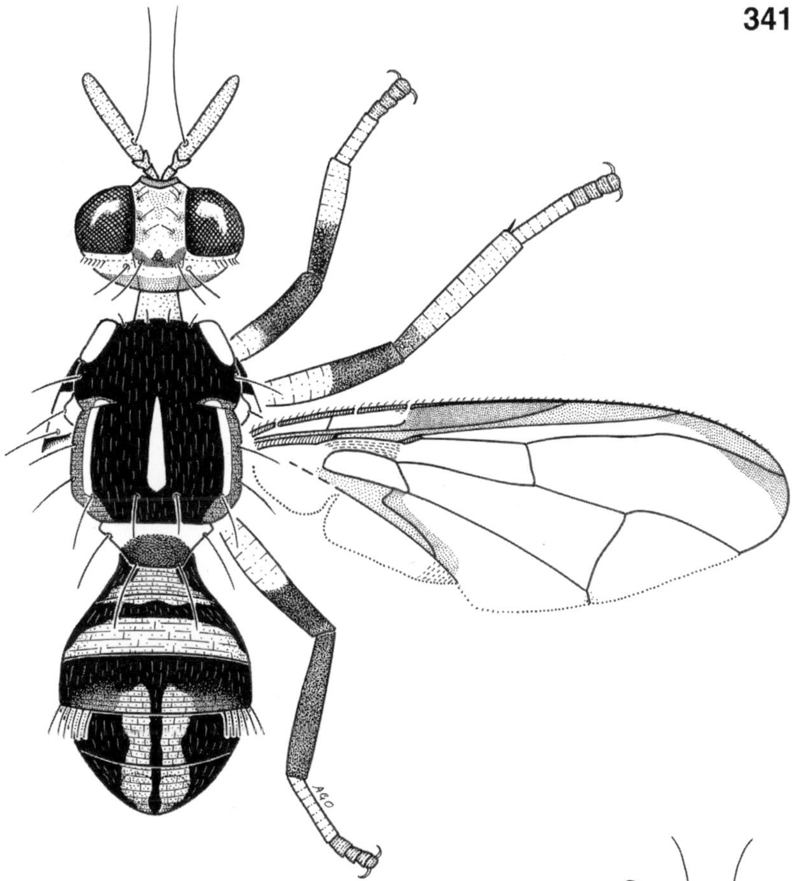

342A

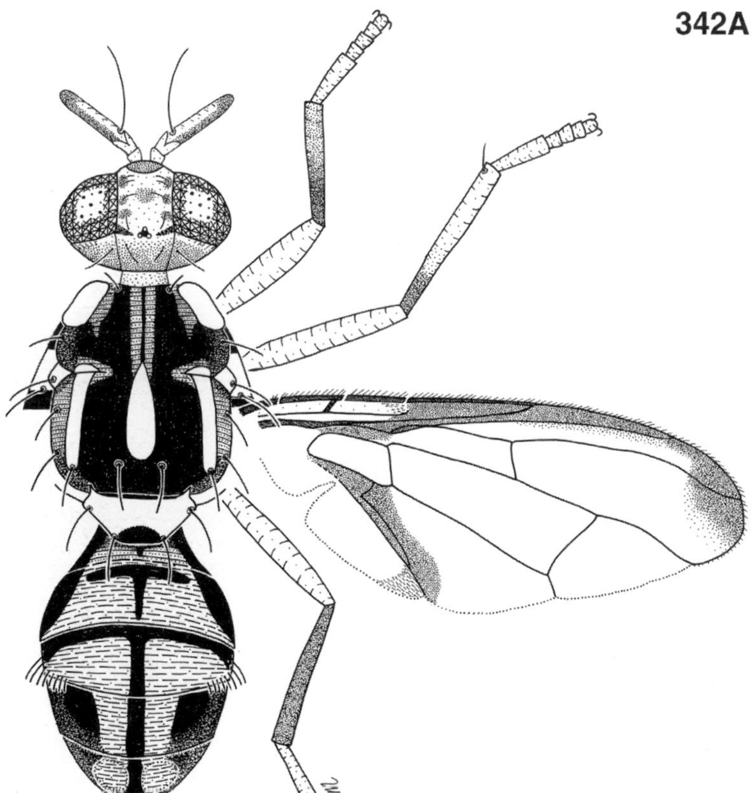

Fig. 341. *Bactrocera (Zeugodacus) proprescutellata* Zhang, Chen & Gao, male. **Fig. 342A.** *Bactrocera (Zeugodacus) pubescens* (Bezzi), lectotype male.

342B

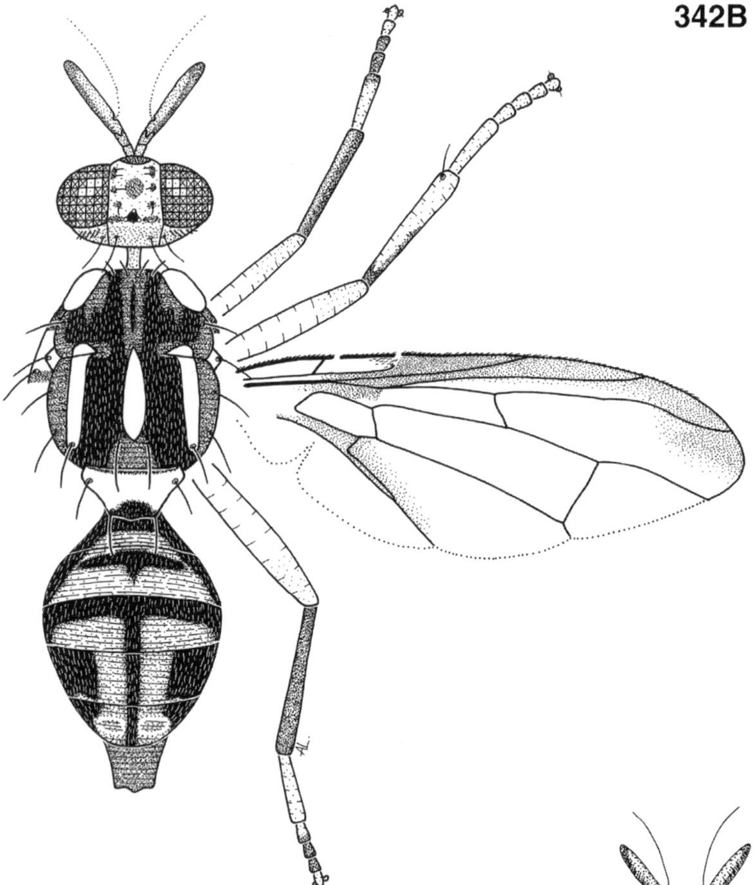

343

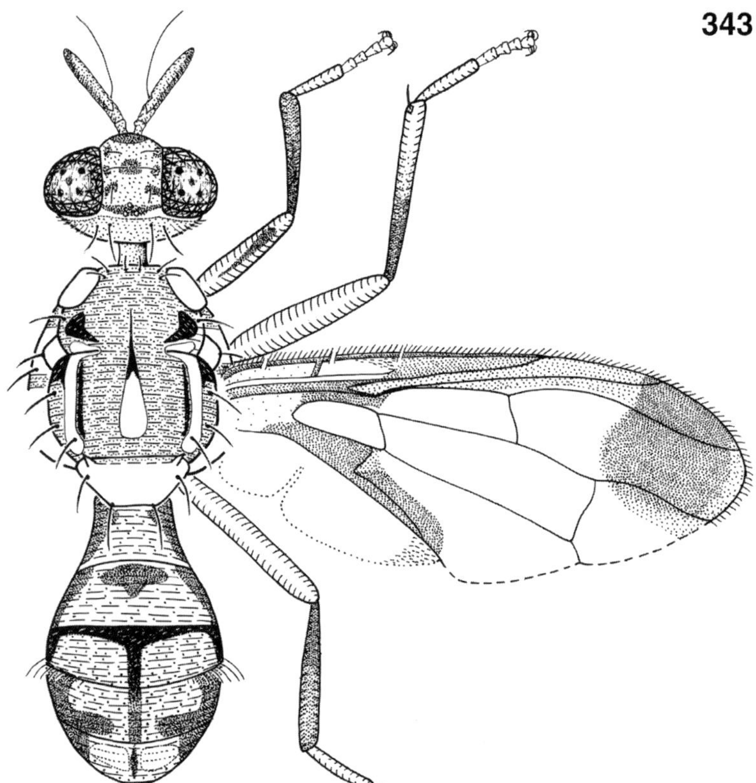

Fig. 342B. *Bactrocera (Zeugodacus) pubescens* (Bezzi), female. **Fig. 343.** *Bactrocera (Zeugodacus) quasiinfesta* new species, holotype male.

344

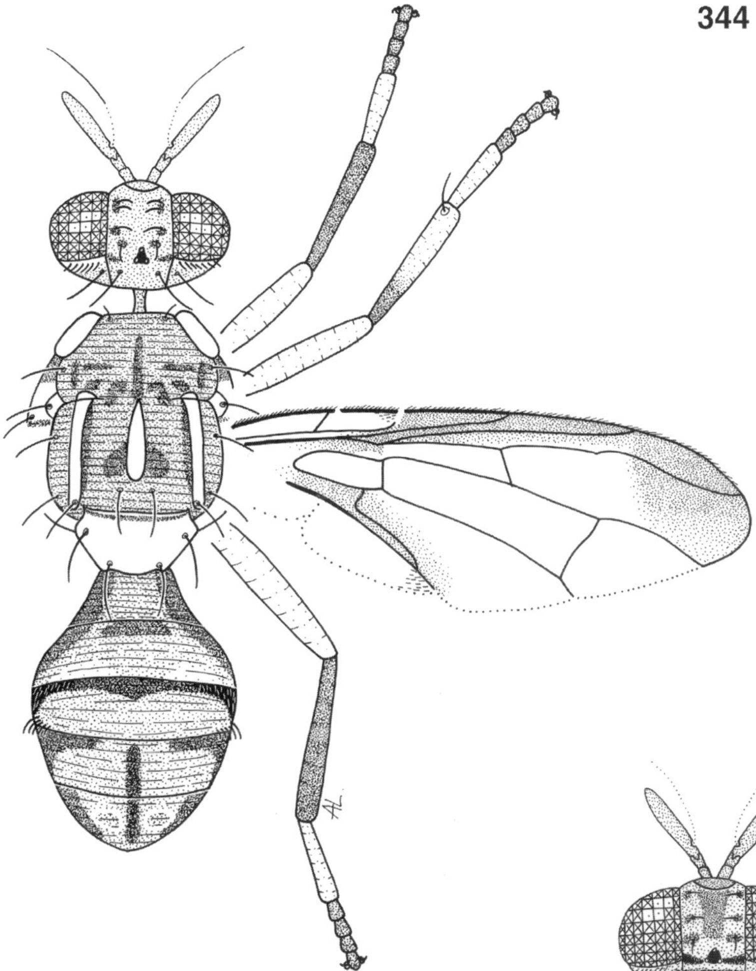

345

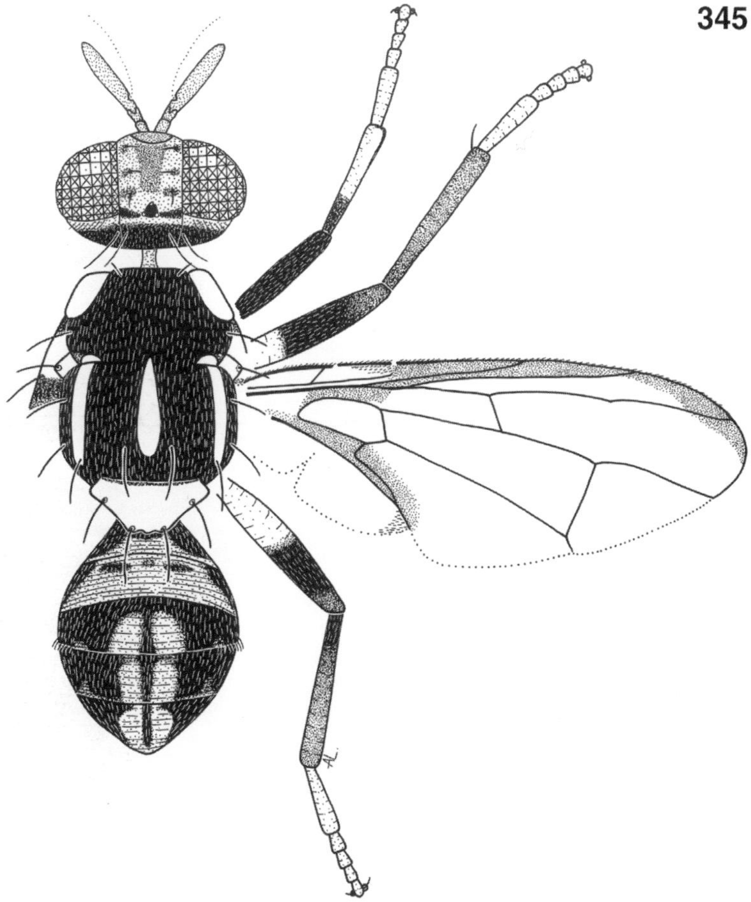

Fig. 344. *Bactrocera (Zeugodacus) rubella* (Hardy), holotype male. **Fig. 345.** *Bactrocera (Zeugodacus) sabahensis* new species, holotype male.

346

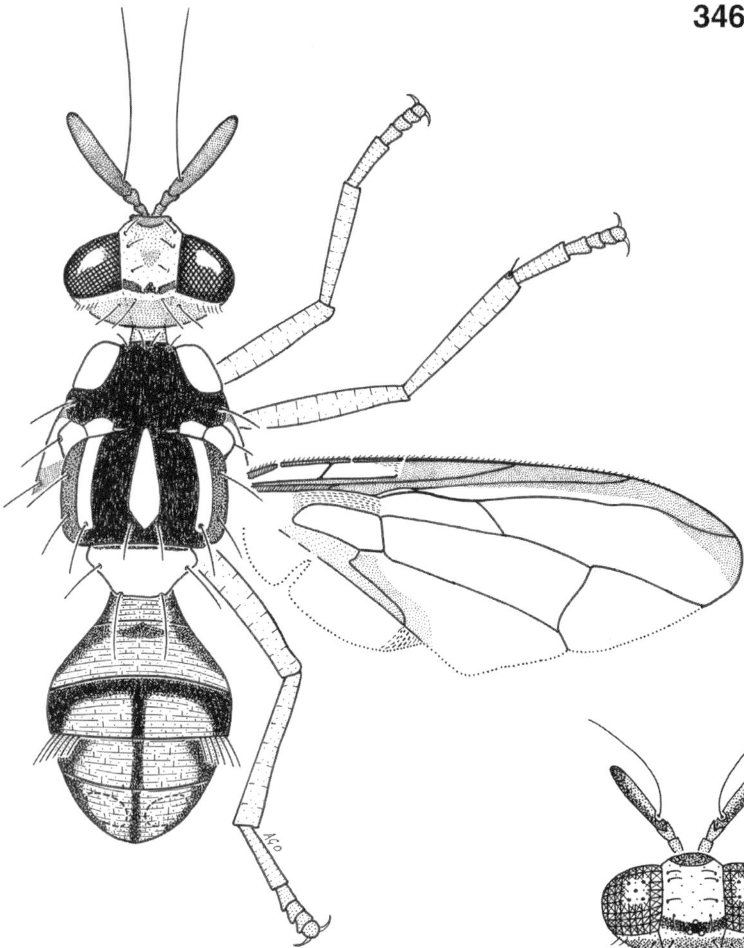

347

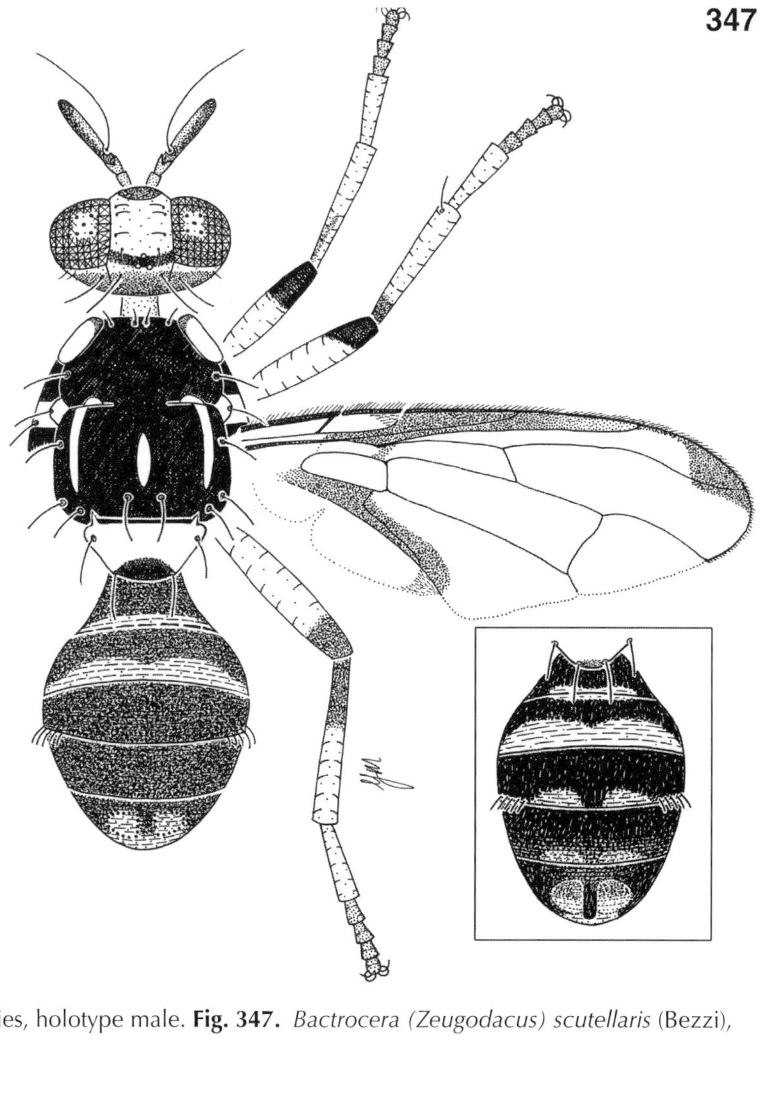

Fig. 346. *Bactrocera (Zeugodacus) sasaotiae* new species, holotype male. **Fig. 347.** *Bactrocera (Zeugodacus) scutellaris* (Bezzi), lectotype male and abdominal colour variations.

348A

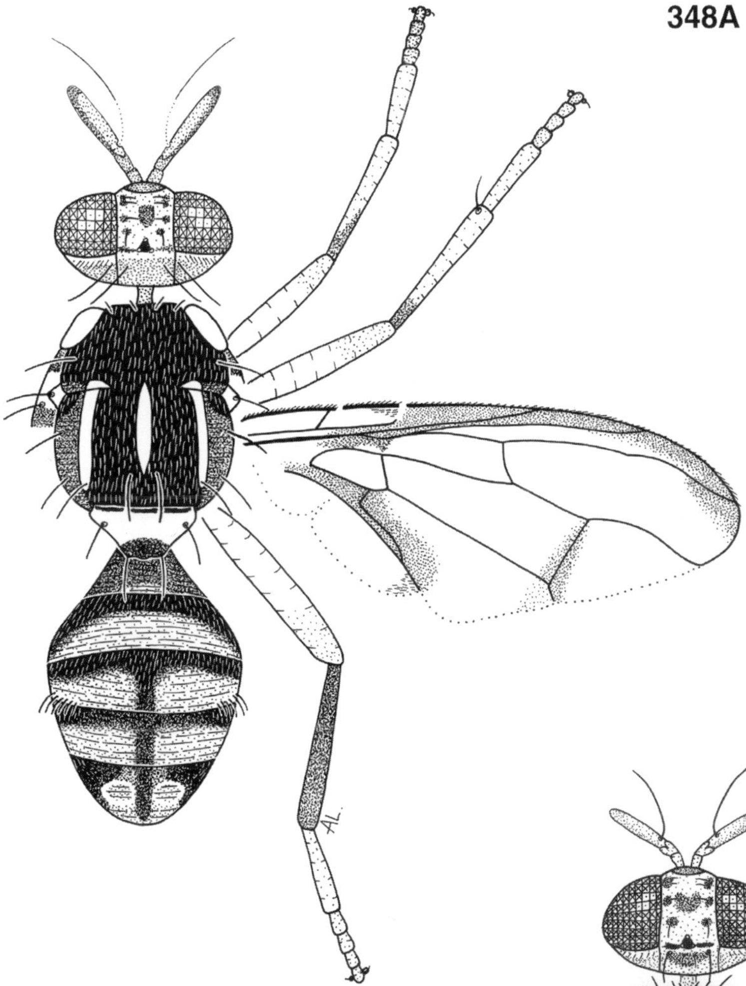

348B

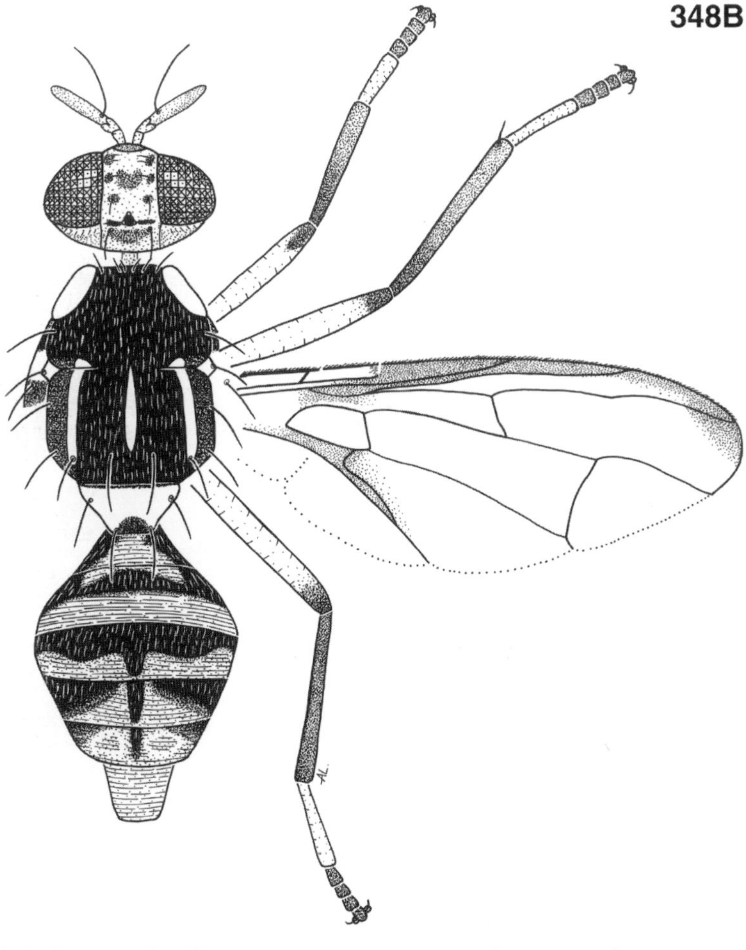

Fig. 348A. *Bactrocera (Zeugodacus) scutellata* (Hendel), holotype male. **Fig. 348B.** *Bactrocera (Zeugodacus) scutellata* (Hendel), female.

348C

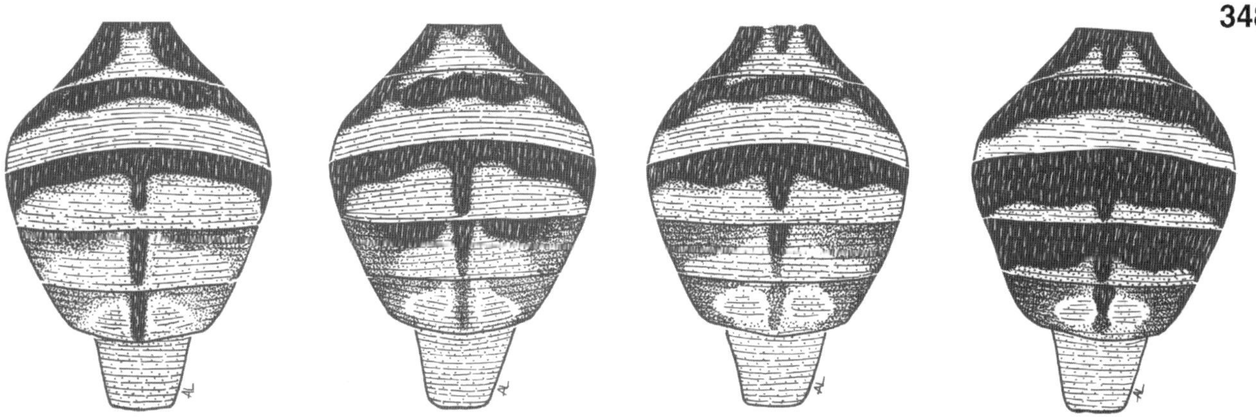

349

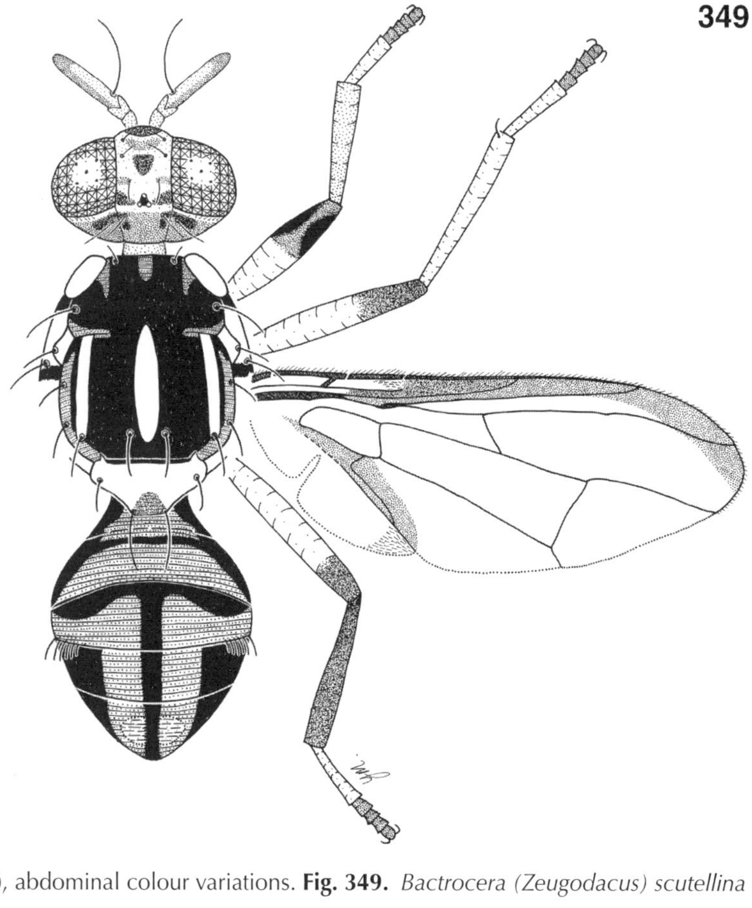

Fig. 348C. *Bactrocera (Zeugodacus) scutellata* (Hendel), abdominal colour variations. **Fig. 349.** *Bactrocera (Zeugodacus) scutellina* (Bezzi), syntype male.

350

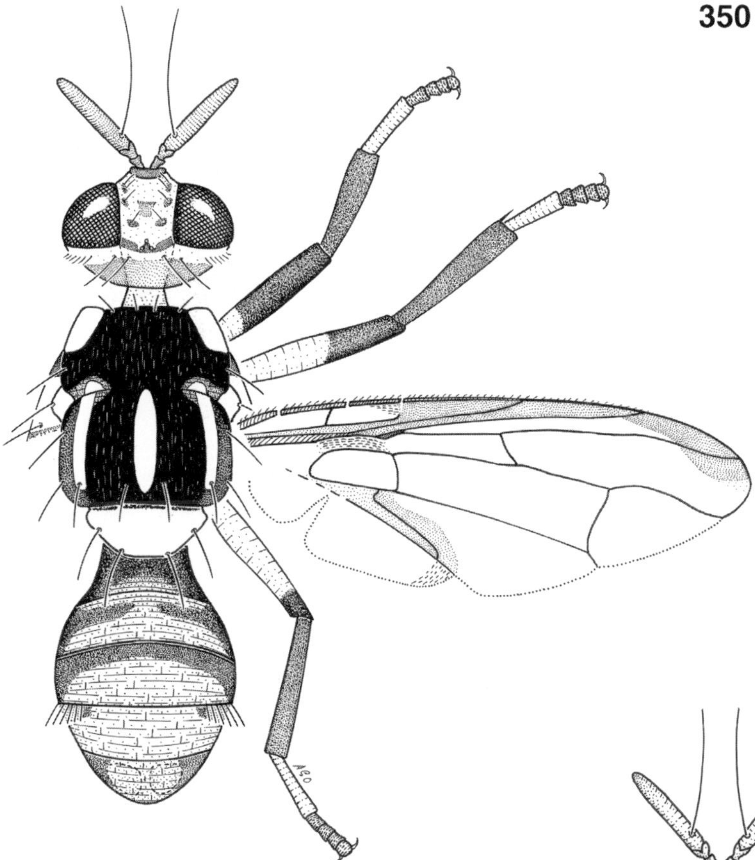

351A

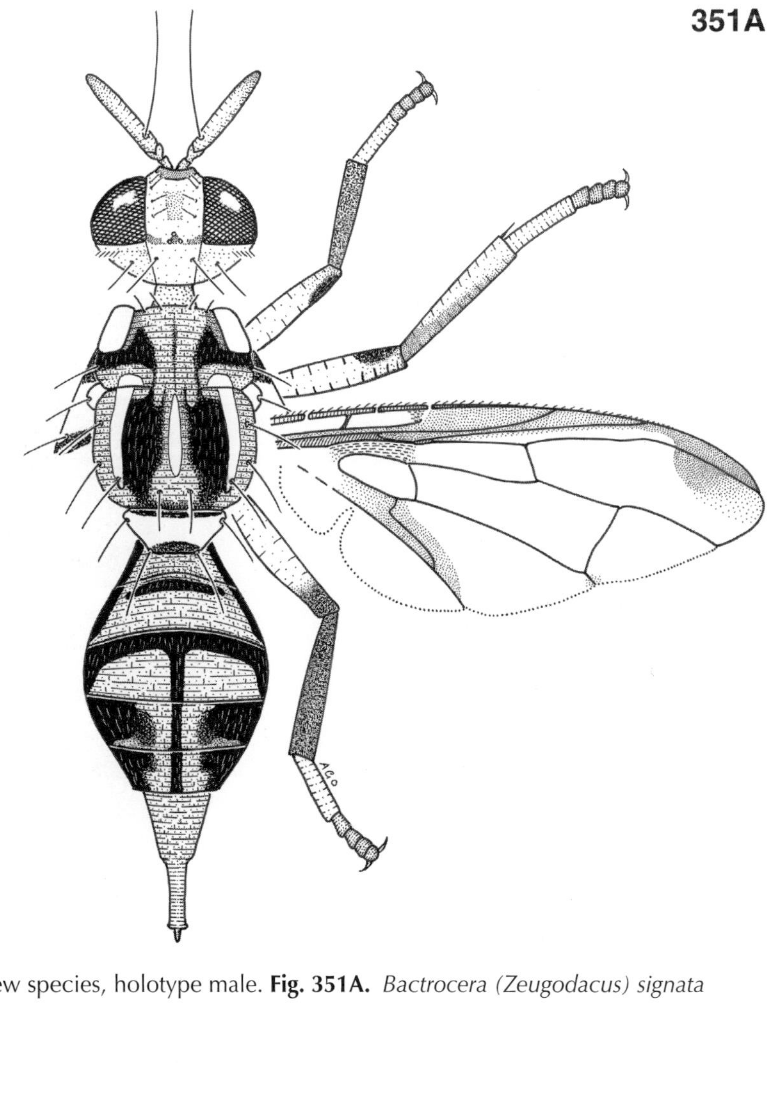

Fig. 350. *Bactrocera (Zeugodacus) semongokensis* new species, holotype male. **Fig. 351A.** *Bactrocera (Zeugodacus) signata* (Hering), syntype female.

351B

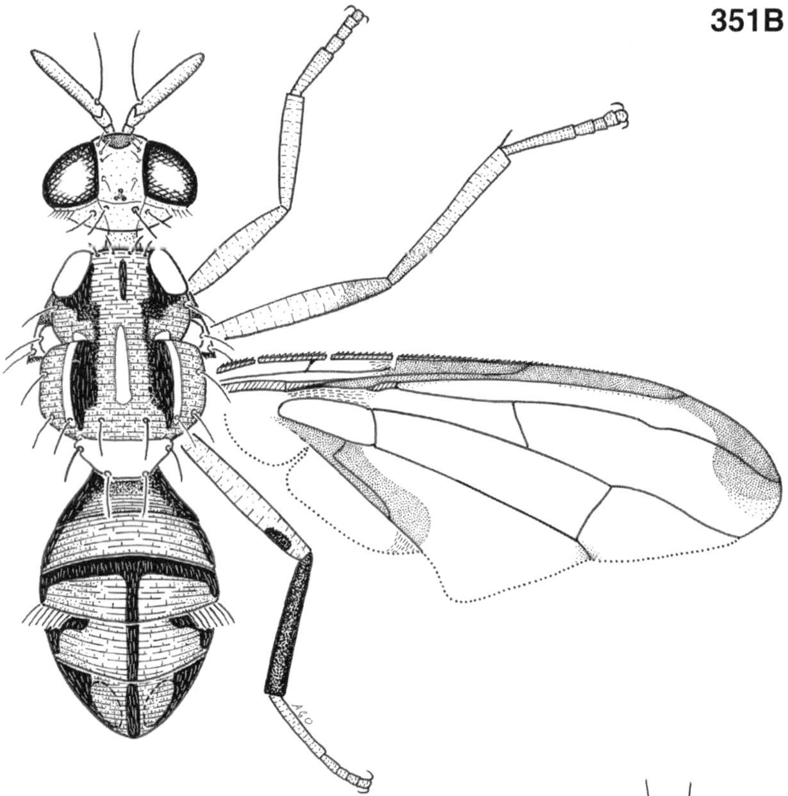

352

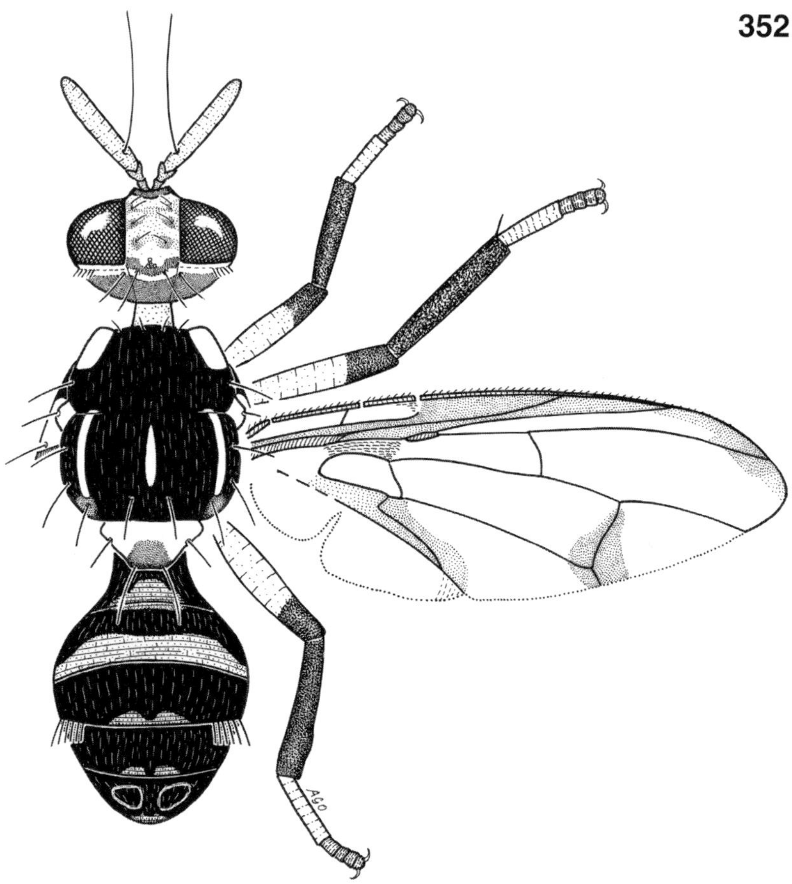

Fig. 351B. *Bactrocera (Zeugodacus) signata* (Hering), syntype male. **Fig. 352.** *Bactrocera (Zeugodacus) sonlaiae* new species, holotype male.

353

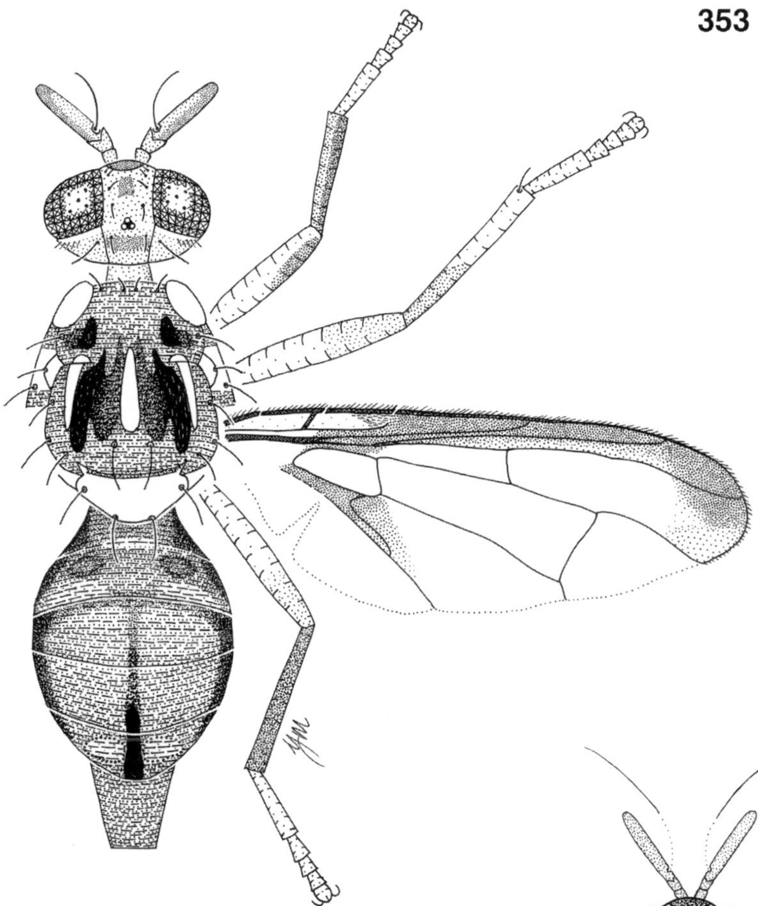

354A

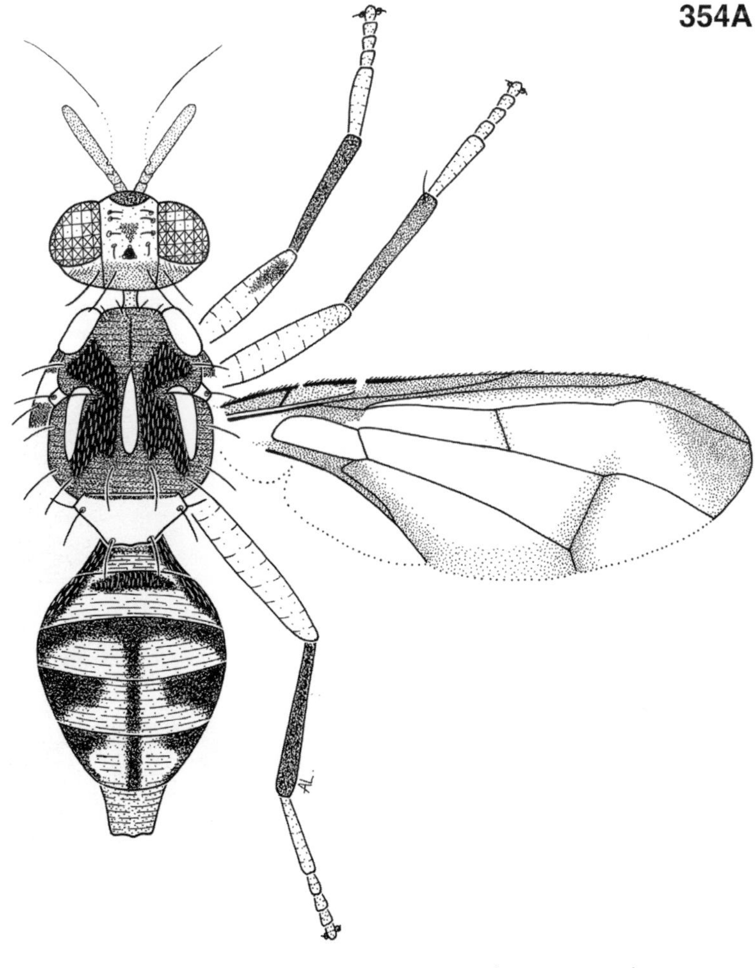

Fig. 353. *Bactrocera (Zeugodacus) sumbensis* (Hering), holotype female. **Fig. 354A.** *Bactrocera (Zeugodacus) synnephes* (Hendel), holotype female.

354B

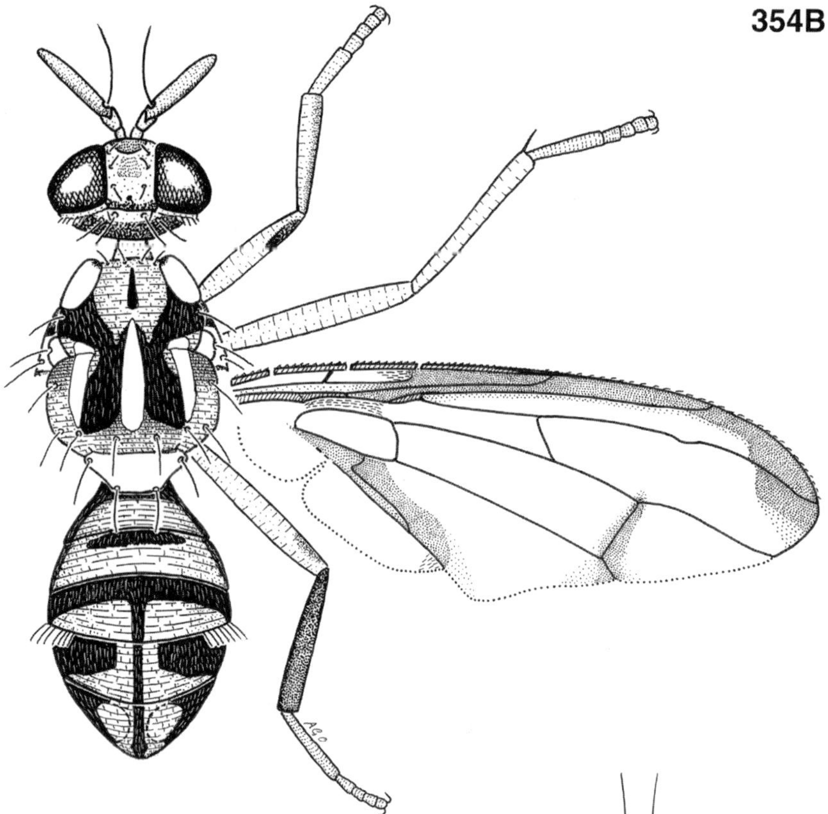

355

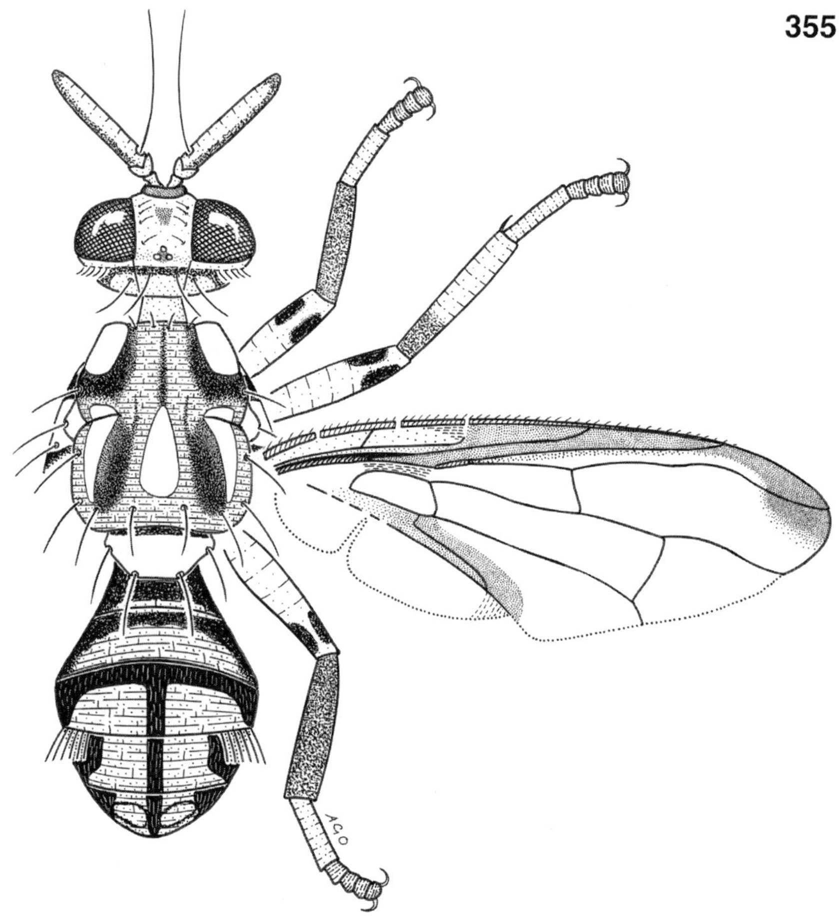

Fig. 354B. *Bactrocera (Zeugodacus) synnephes* (Hendel), male. **Fig. 355.** *Bactrocera (Zeugodacus) tapervitta* Mahmood, male.

356

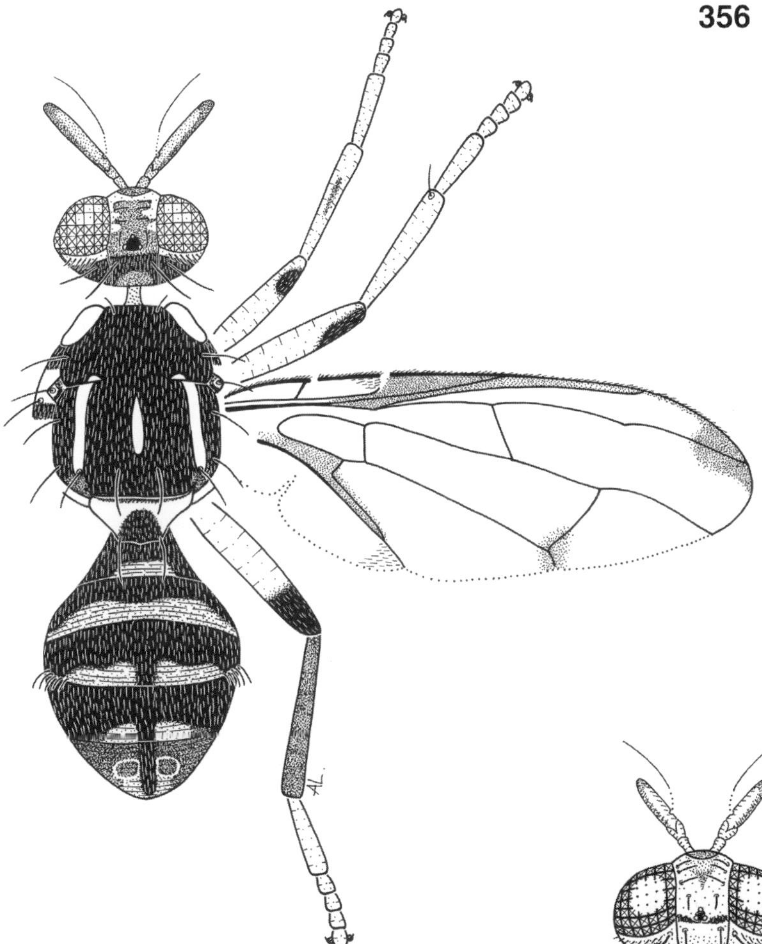

357A

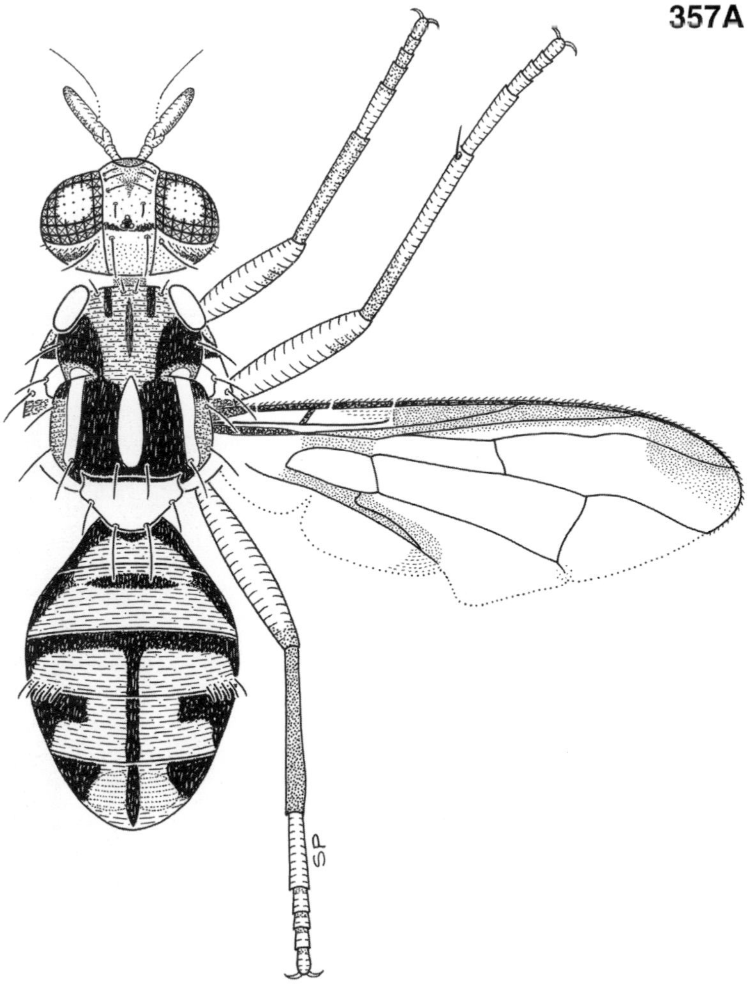

Fig. 356. *Bactrocera (Zeugodacus) tappanus* (Shiraki), holotype male. **Fig. 357A.** *Bactrocera (Zeugodacus) tau* (Walker), male.

357B

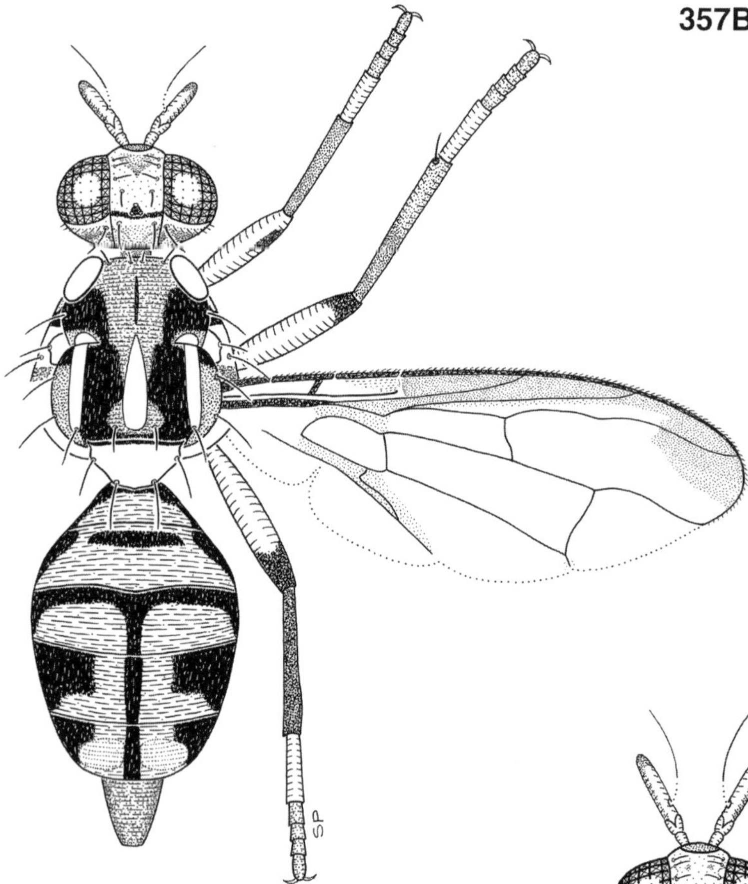

358

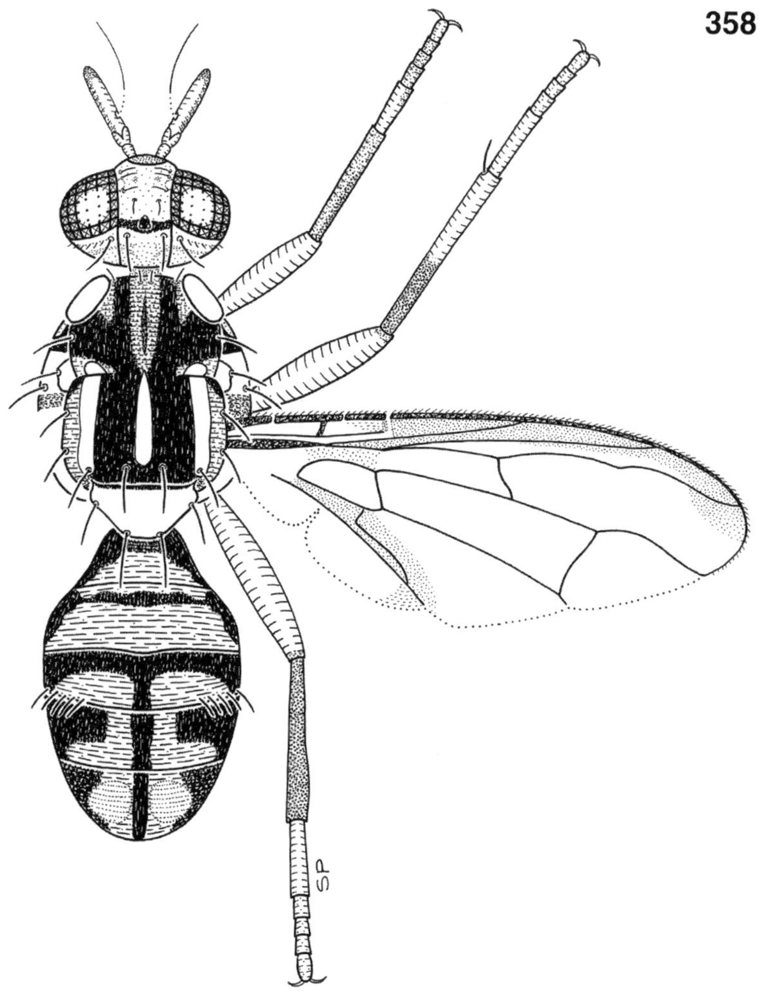

Fig. 357B. *Bactrocera (Zeugodacus) tau* (Walker), female. **Fig. 358.** *Bactrocera (Zeugodacus) tebeduiae* new species, holotype male.

359

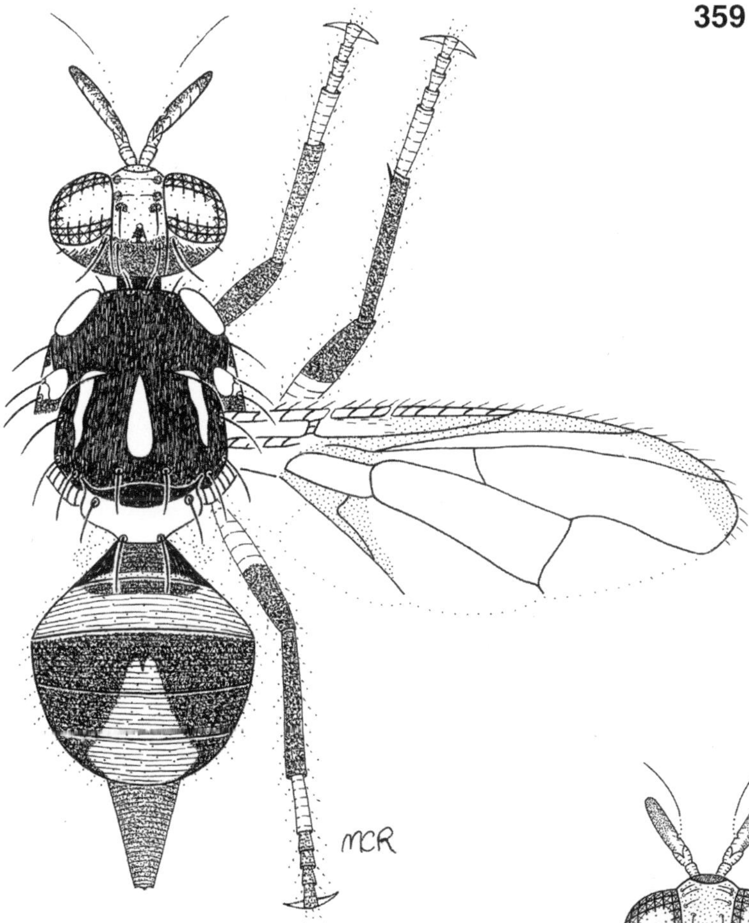

360

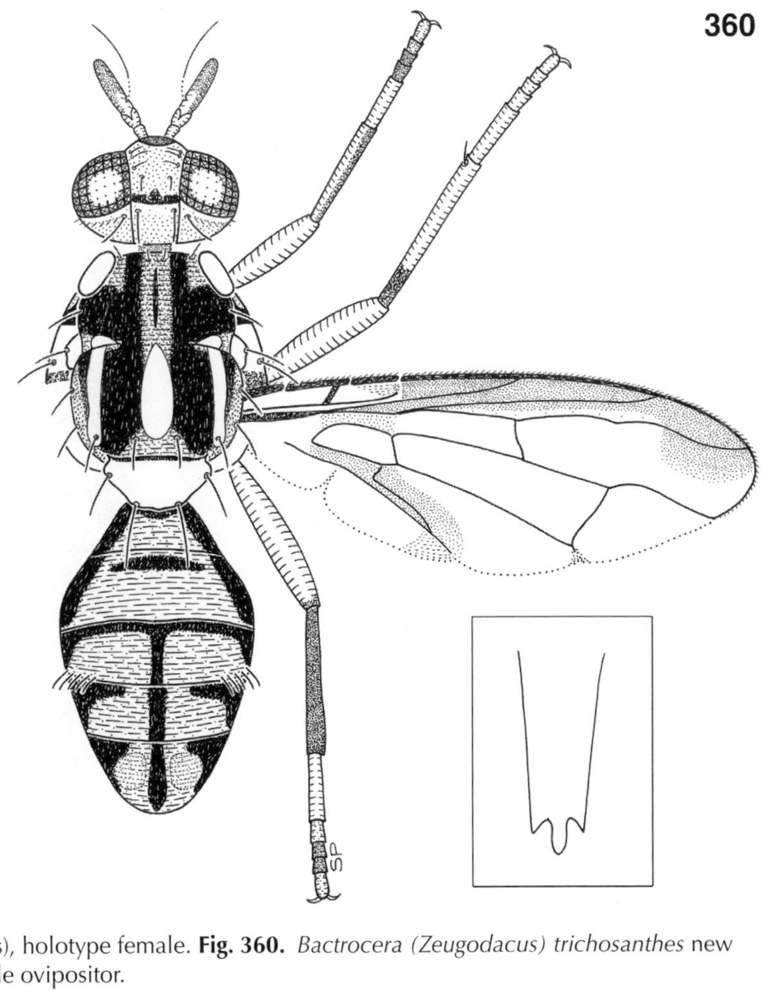

Fig. 359. *Bactrocera (Zeugodacus) timorensis* (Perkins), holotype female. **Fig. 360.** *Bactrocera (Zeugodacus) trichosanthes* new species, holotype male and apex of piercer from female ovipositor.

361

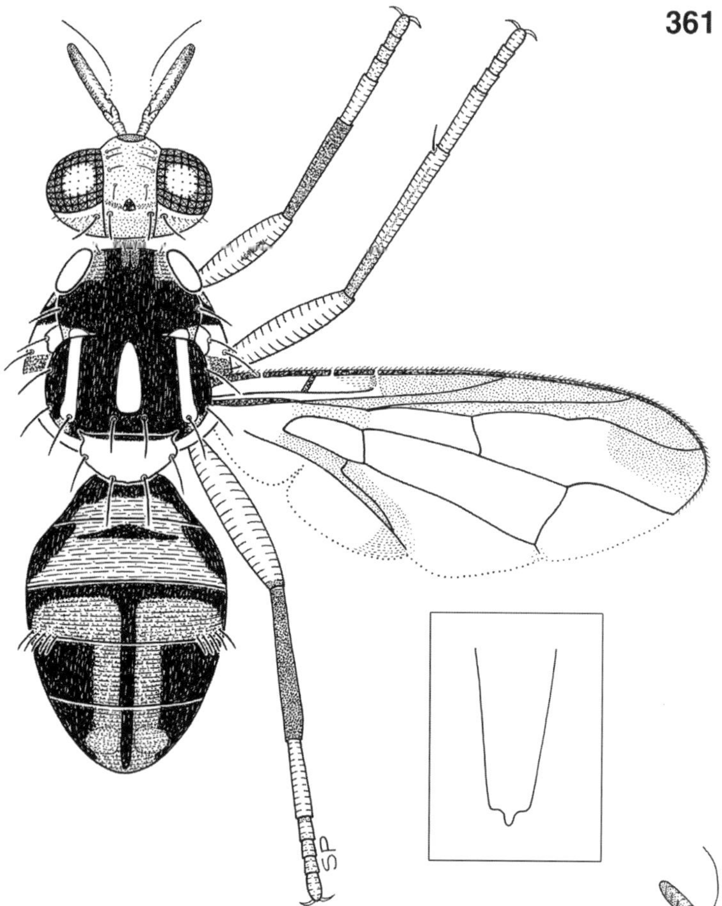

362

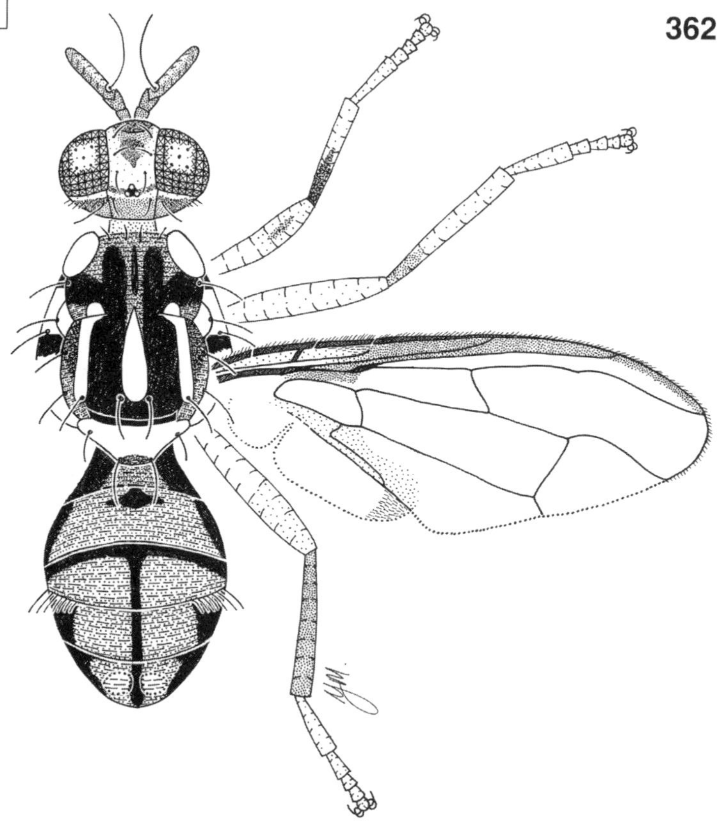

Fig. 361. *Bactrocera (Zeugodacus) tricuspidatae* new species, holotype male and apex of piercer from female ovipositor.
Fig. 362. *Bactrocera (Zeugodacus) trimaculata* (Hardy & Adachi), paratype male.

363

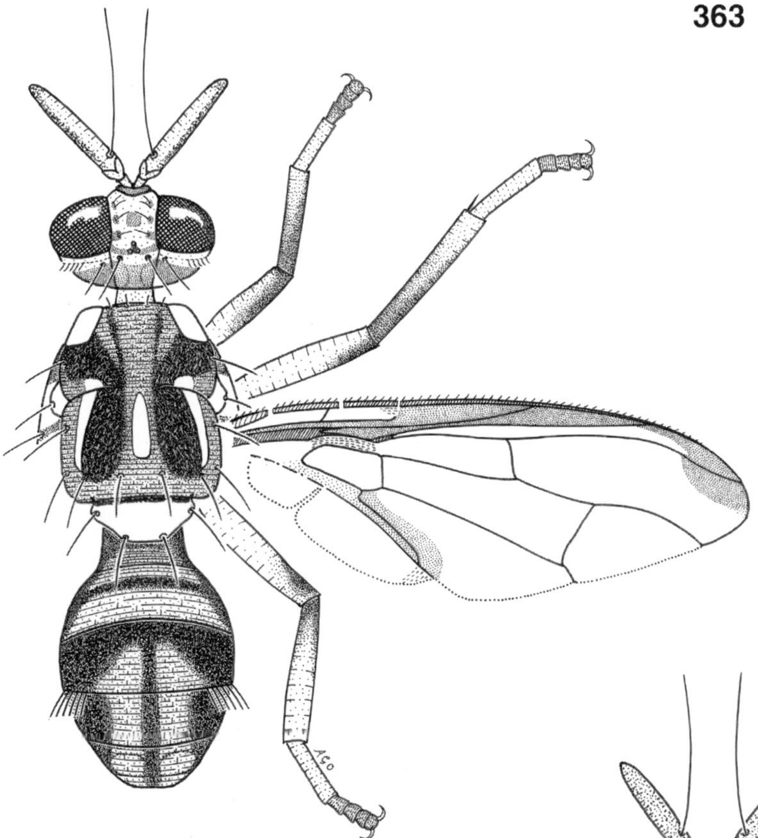

364

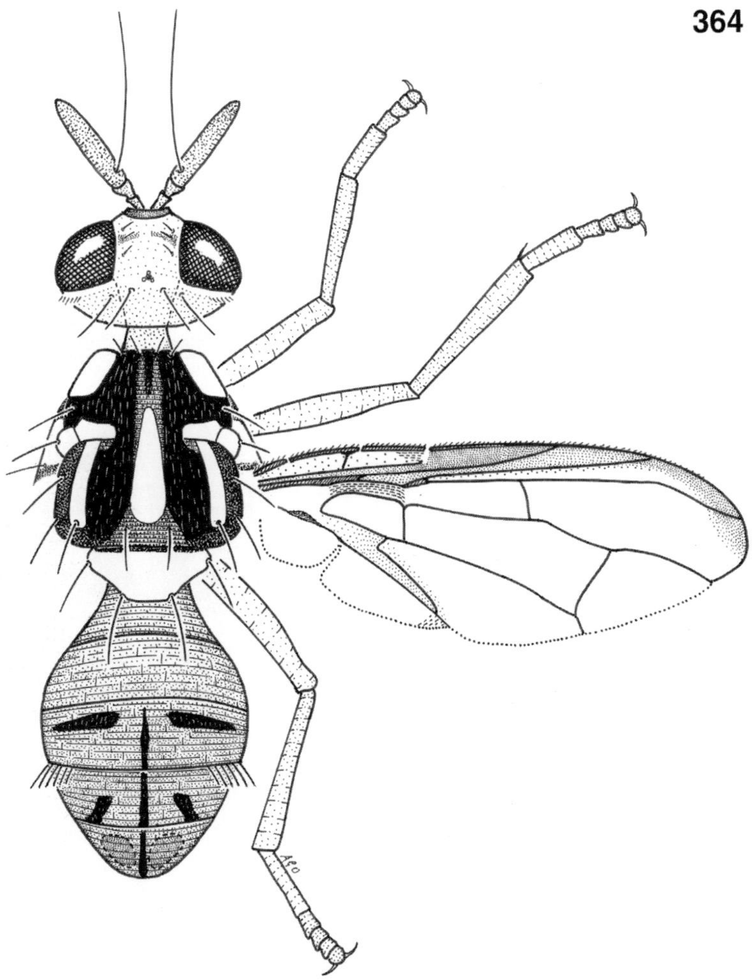

Fig. 363. *Bactrocera (Zeugodacus) trivandrumensis* new species, holotype male. **Fig. 364.** *Bactrocera (Zeugodacus) ujungpandangiae* new species, holotype male.

365

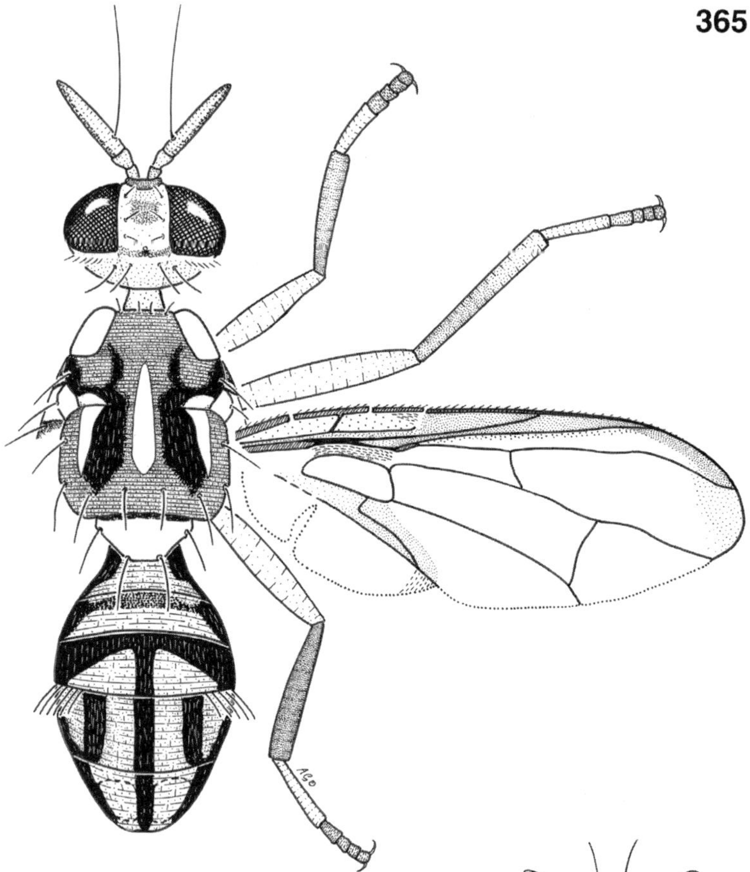

366

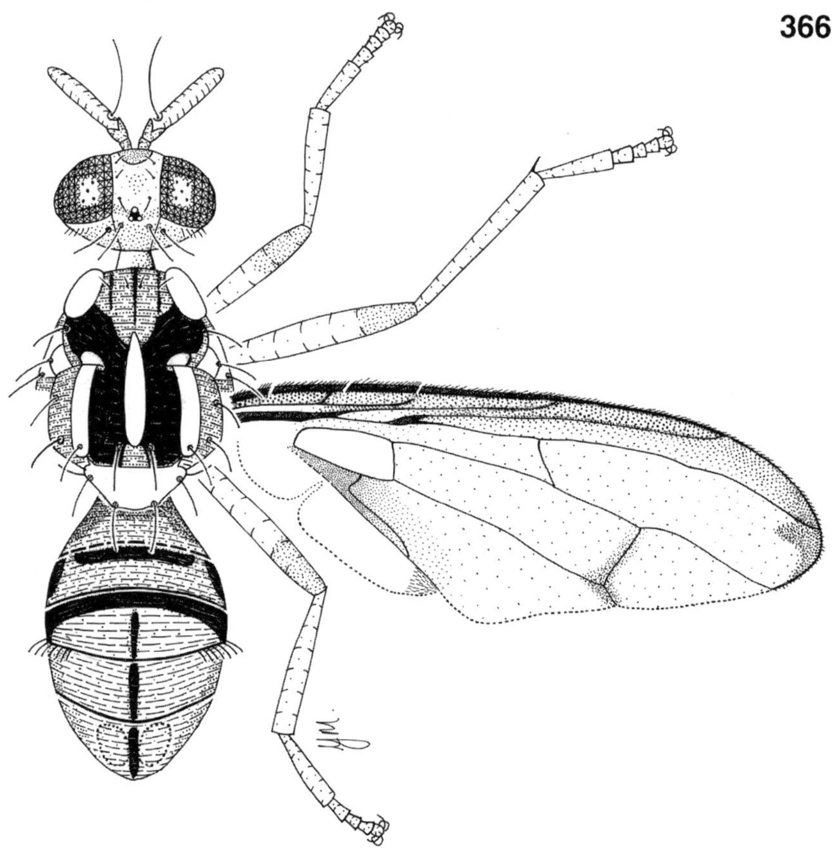

Fig. 365. *Bactrocera (Zeugodacus) uncinata* new species, holotype male **Fig. 366.** *Bactrocera (Zeugodacus) vargus* (Hardy), holotype male.

367

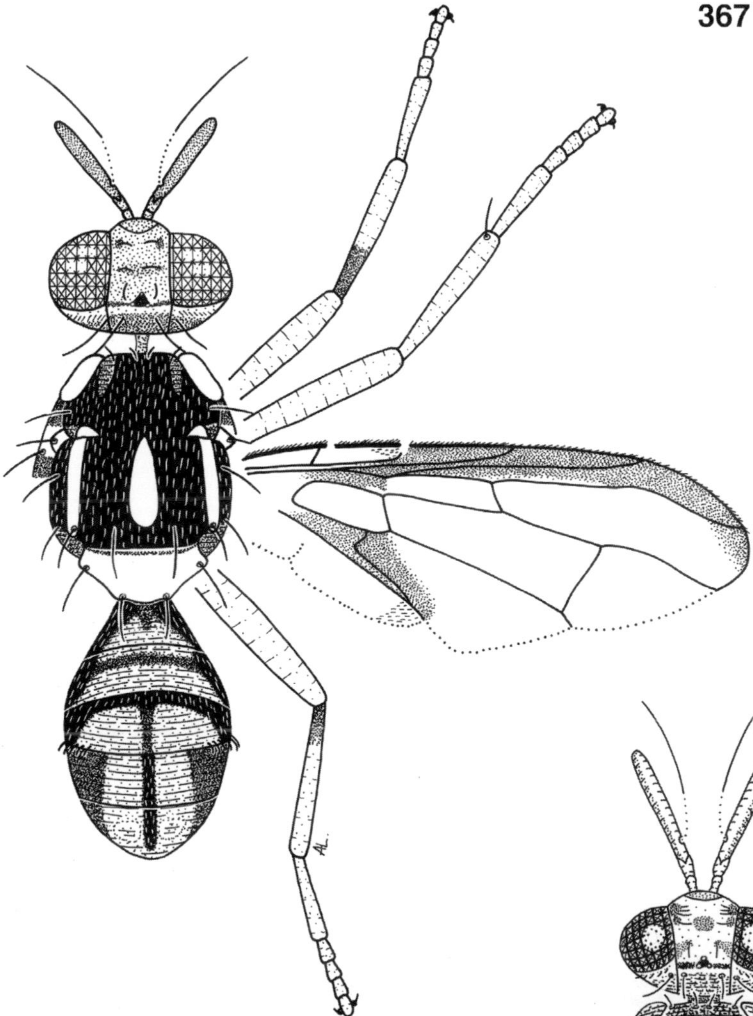

368

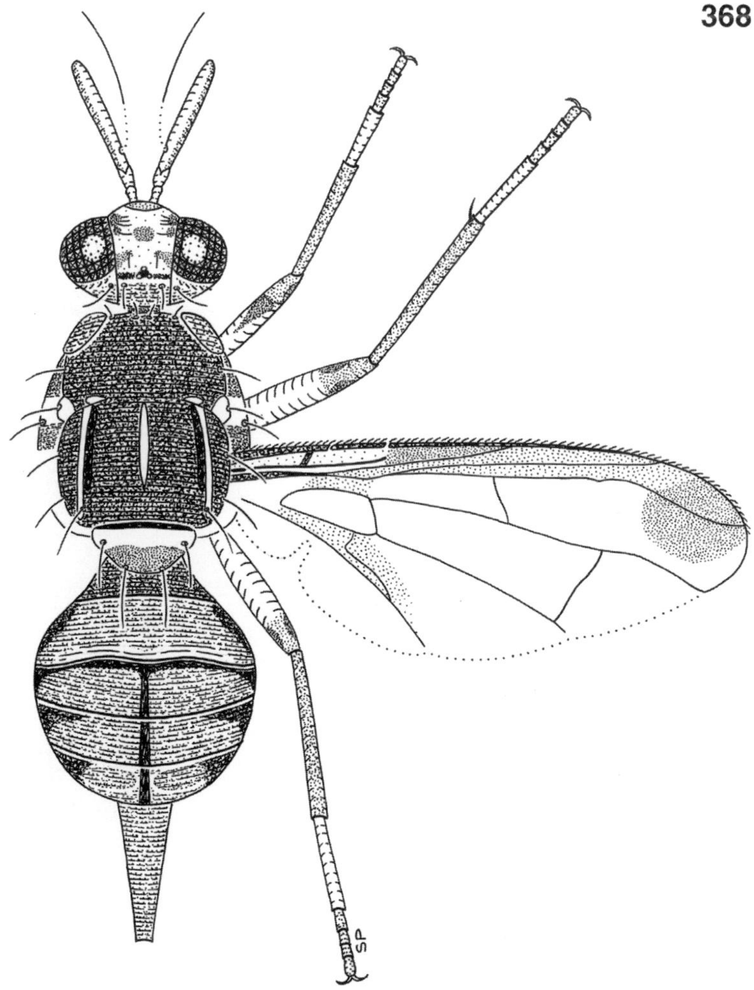

Fig. 367. *Bactrocera (Zeugodacus) vulta* (Hardy), holotype male. **Fig. 368.** *Bactrocera (Zeugodacus) watersi* (Hardy), female.

369

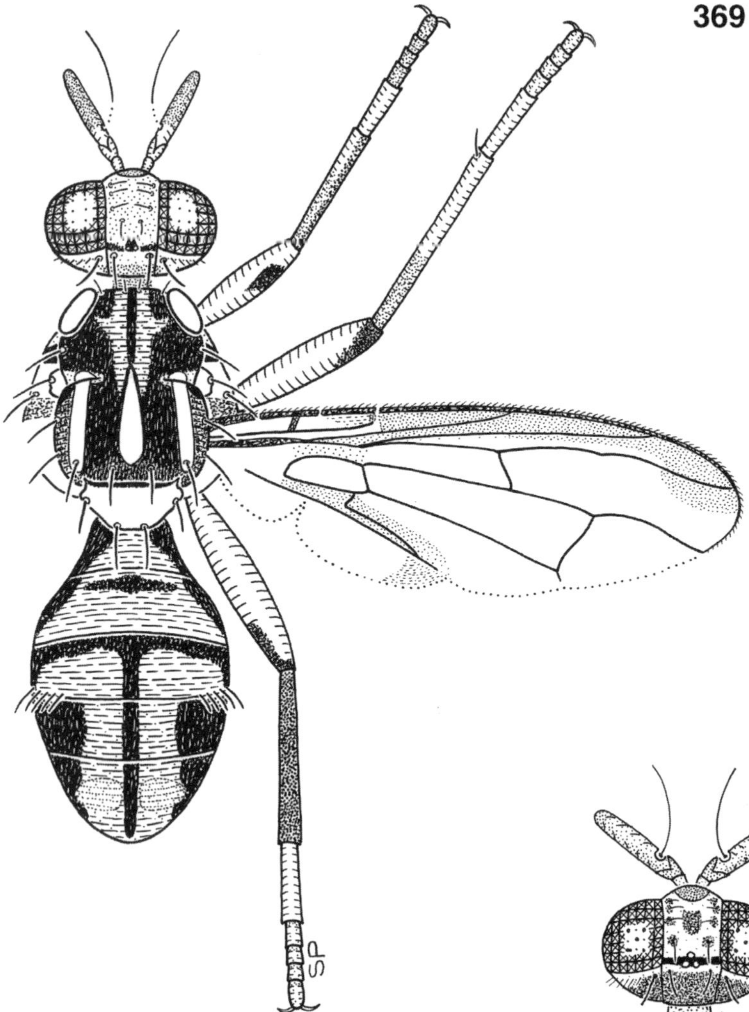

370

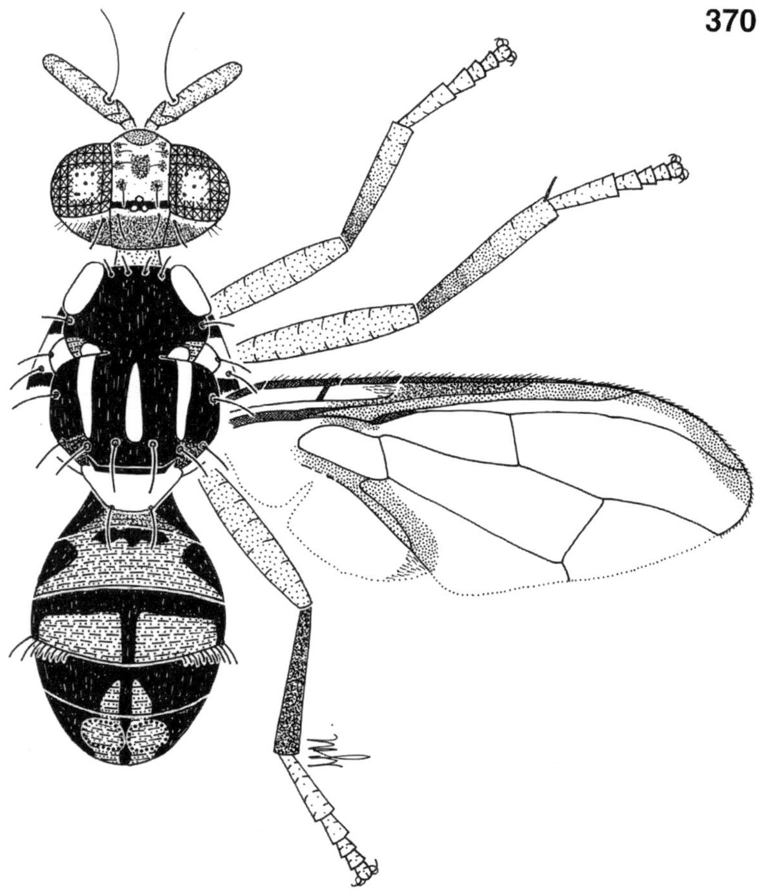

Fig. 369. *Bactrocera (Zeugodacus) yalaensis* new species, holotype male. **Fig. 370.** *Bactrocera (Zeugodacus) yoshimotoi* (Hardy), holotype male.

371

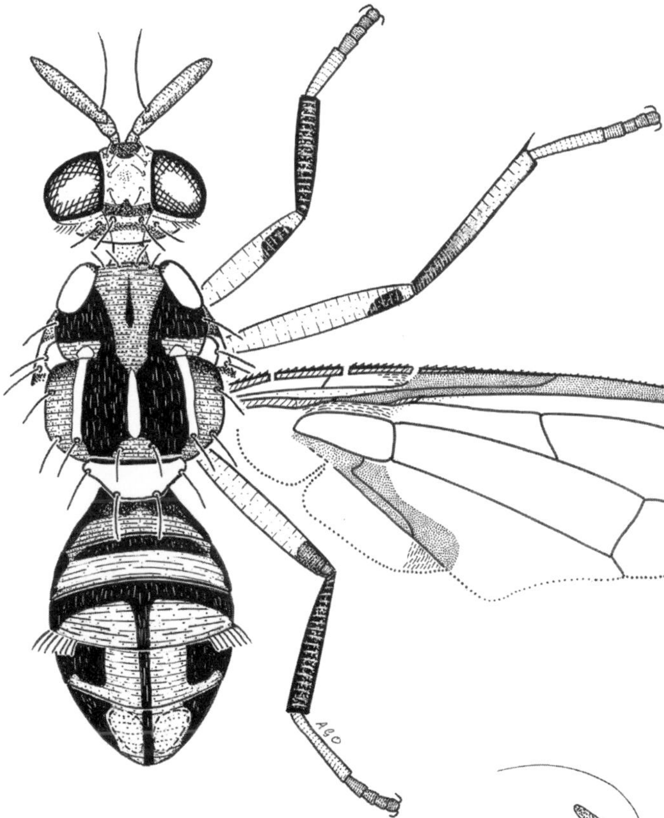

372

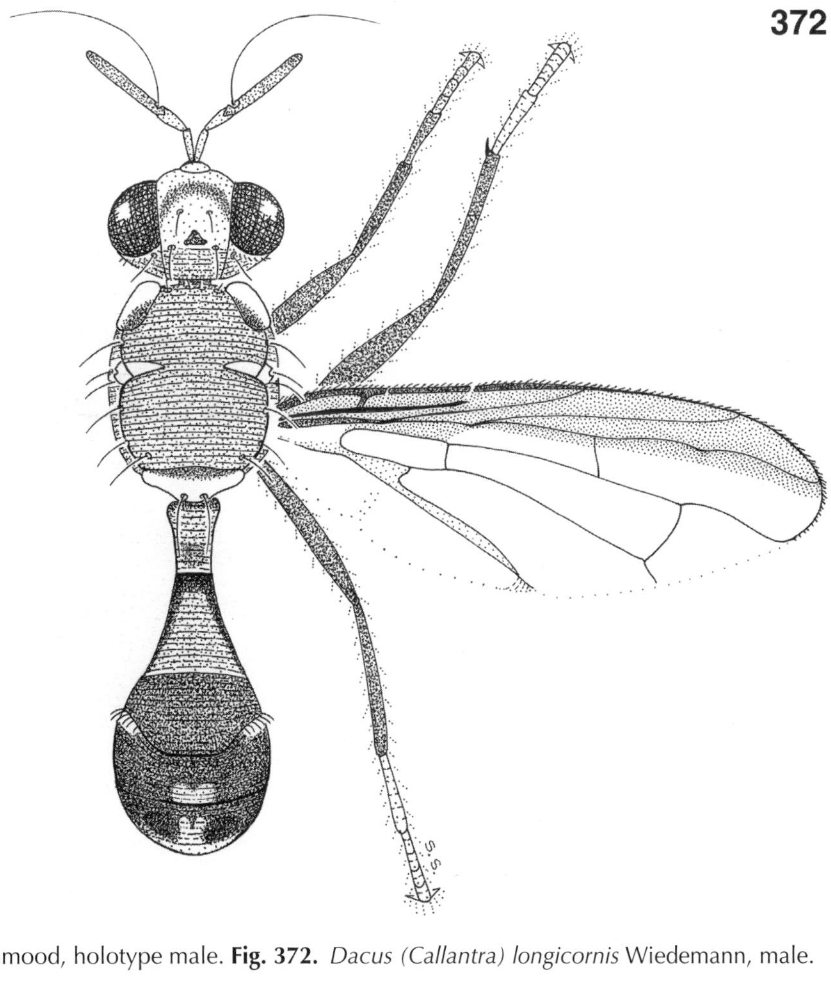

Fig. 371. *Bactrocera (Zeugodacus) zahadi* Mahmood, holotype male. **Fig. 372.** *Dacus (Callantra) longicornis* Wiedemann, male.

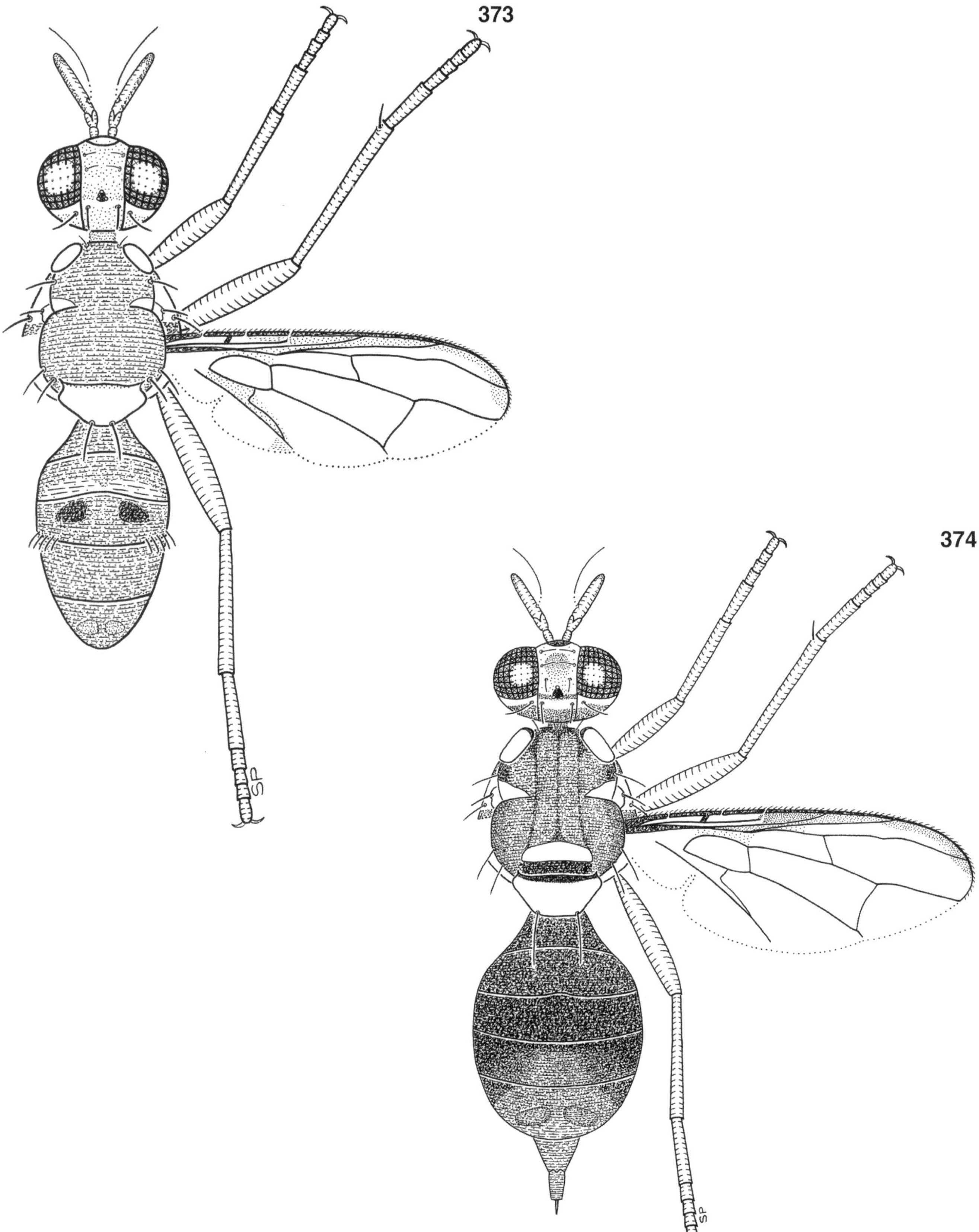

Fig. 373. *Dacus (Didacus) ciliatus* Loew, male. **Fig. 374.** *Dacus (Didacus) hainanus* Wang & Zhao, female.

375

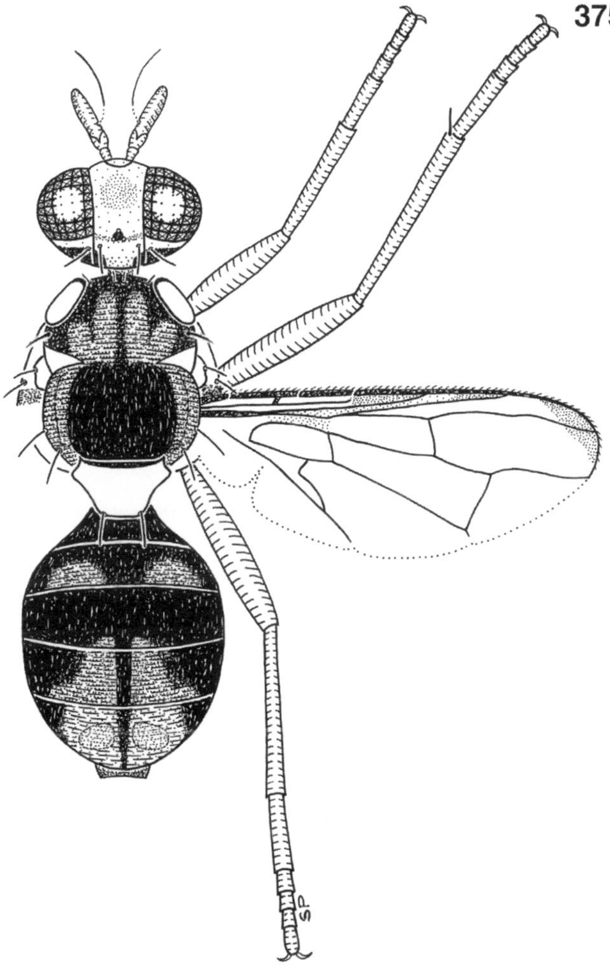

376

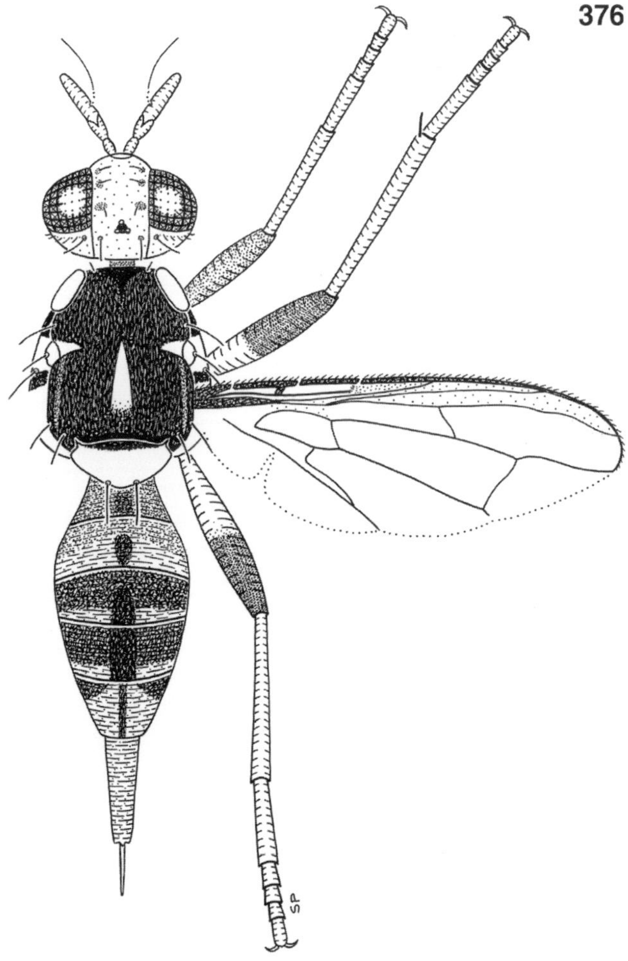

Fig. 375. *Dacus (Didacus) keiseri* (Hering), female. **Fig. 376.** *Dacus (Leptoxyda) persicus* Hendel, lectotype female.

377

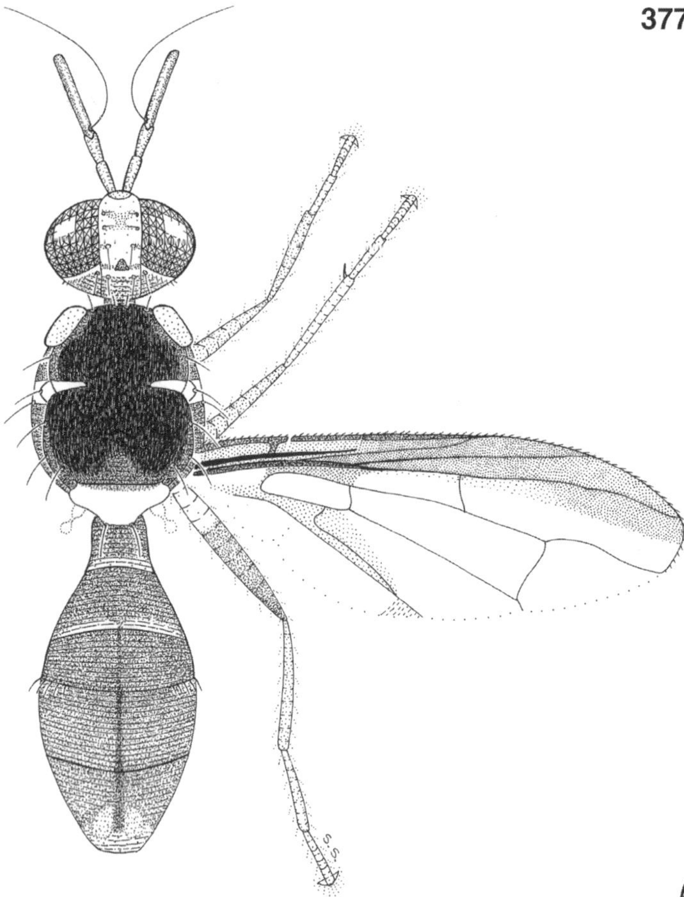

378

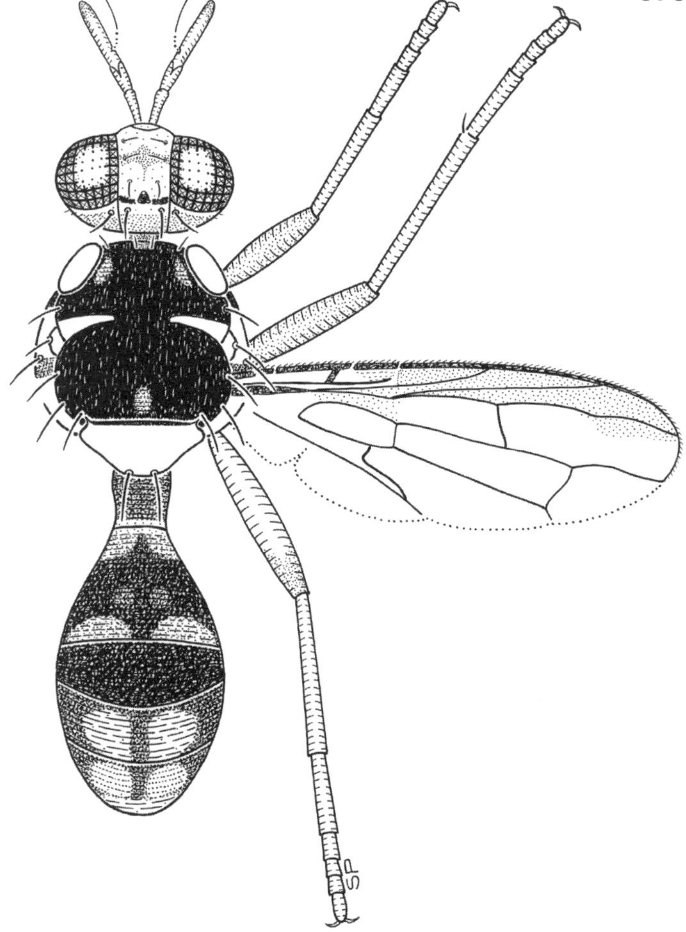

Fig. 377. *Dacus (Mellesis) ambonensis* Drew & Hancock, holotype male. **Fig. 378.** *Dacus (Mellesis) atrimarginatus* White, holotype male.

379

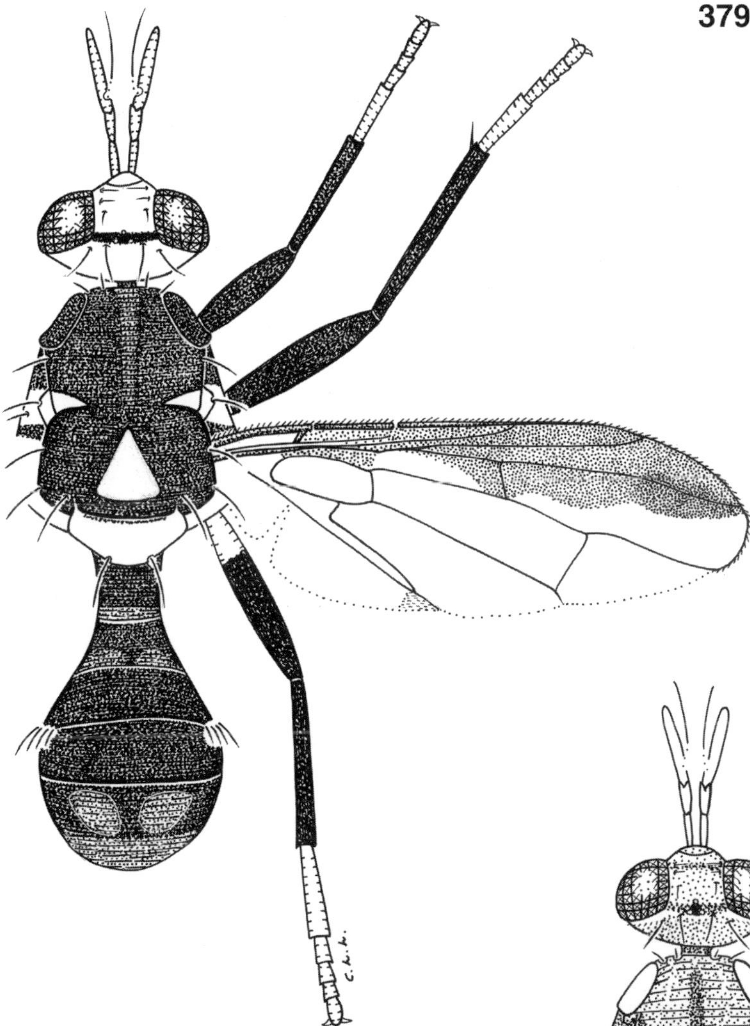

380

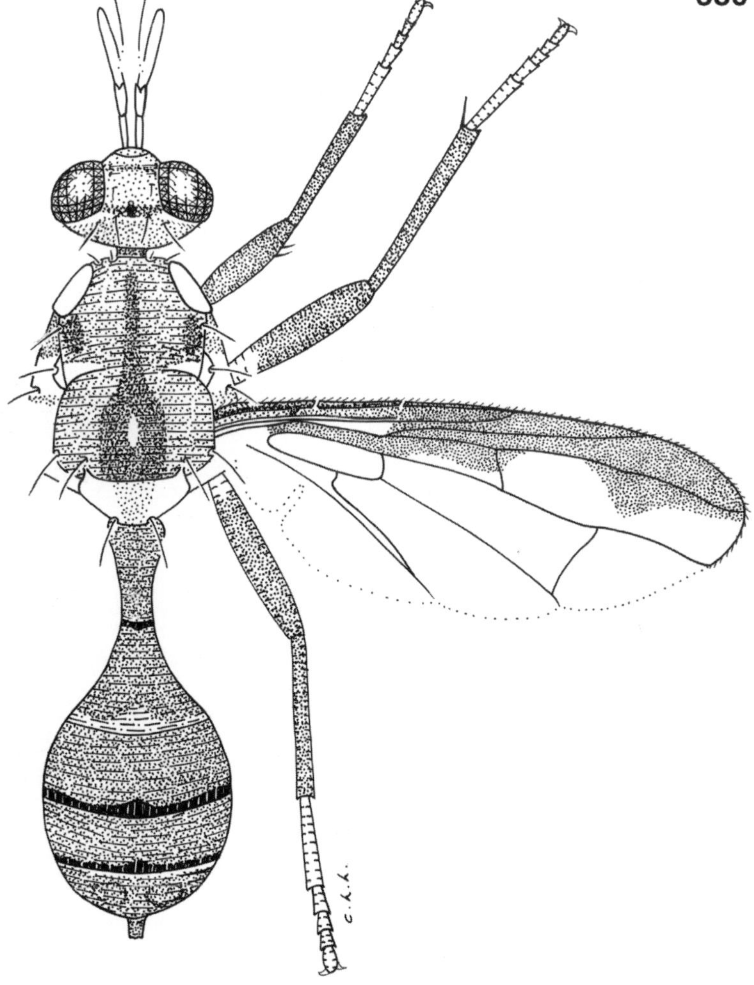

Fig. 379. *Dacus (Mellesis) bannatus* (Wang), holotype male. **Fig. 380.** *Dacus (Mellesis) bispinosus* (Wang), holotype female.

381

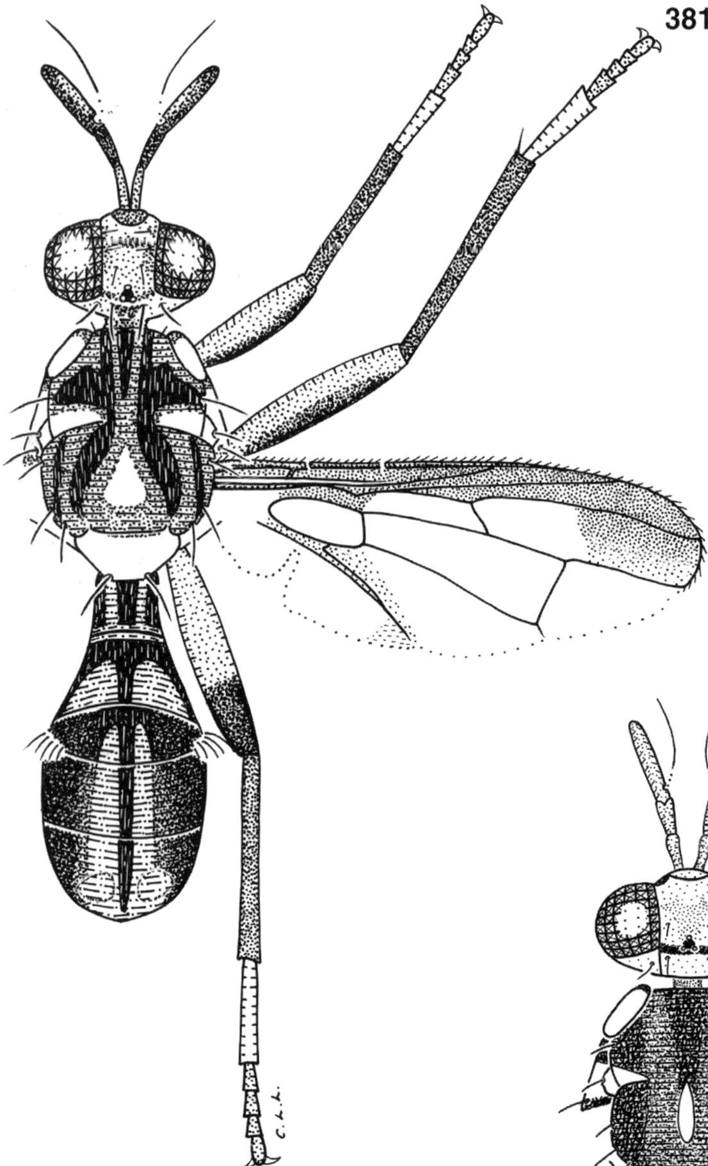

382

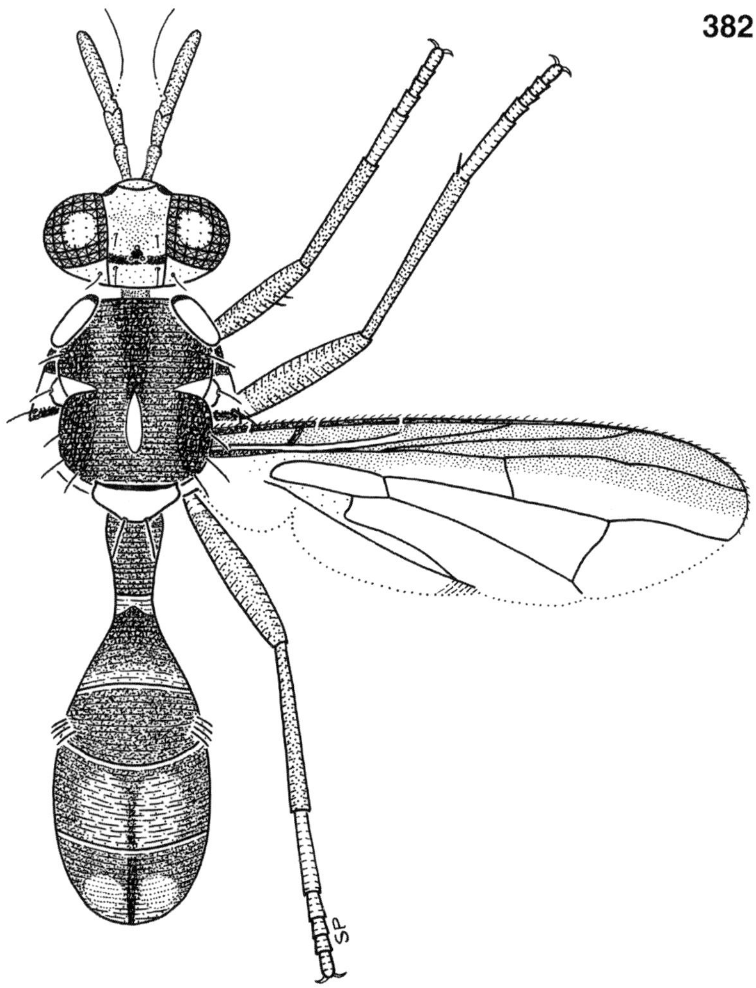

Fig. 381. *Dacus (Mellesis) calirayae* Drew & Hancock, holotype male. **Fig. 382.** *Dacus (Mellesis) conopsoides* de Meijere, lectotype male.

383

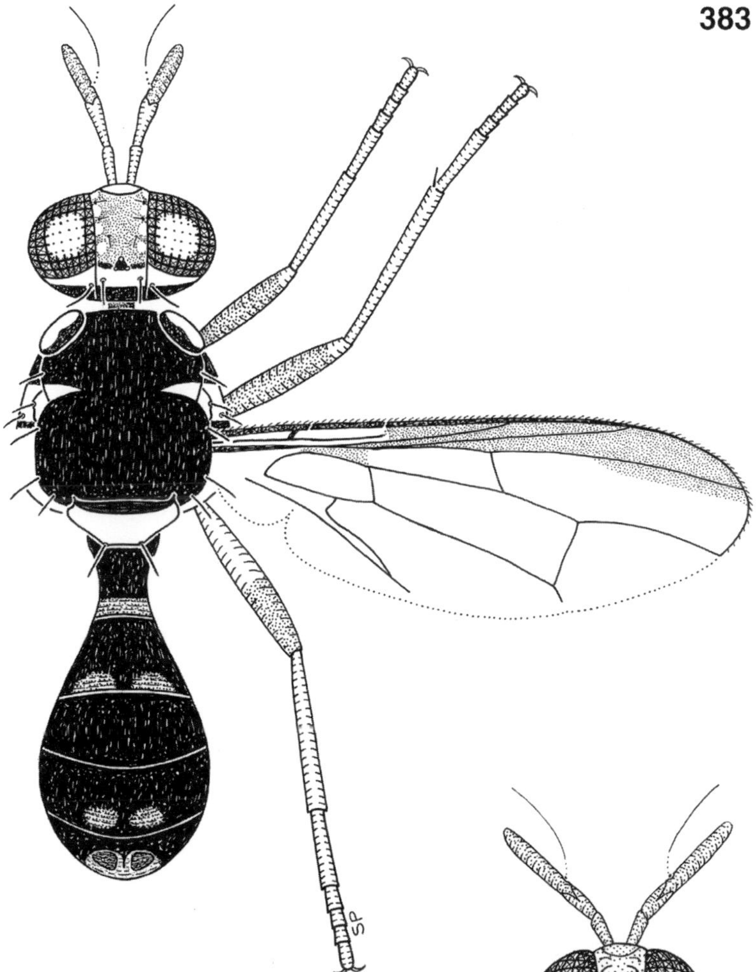

384

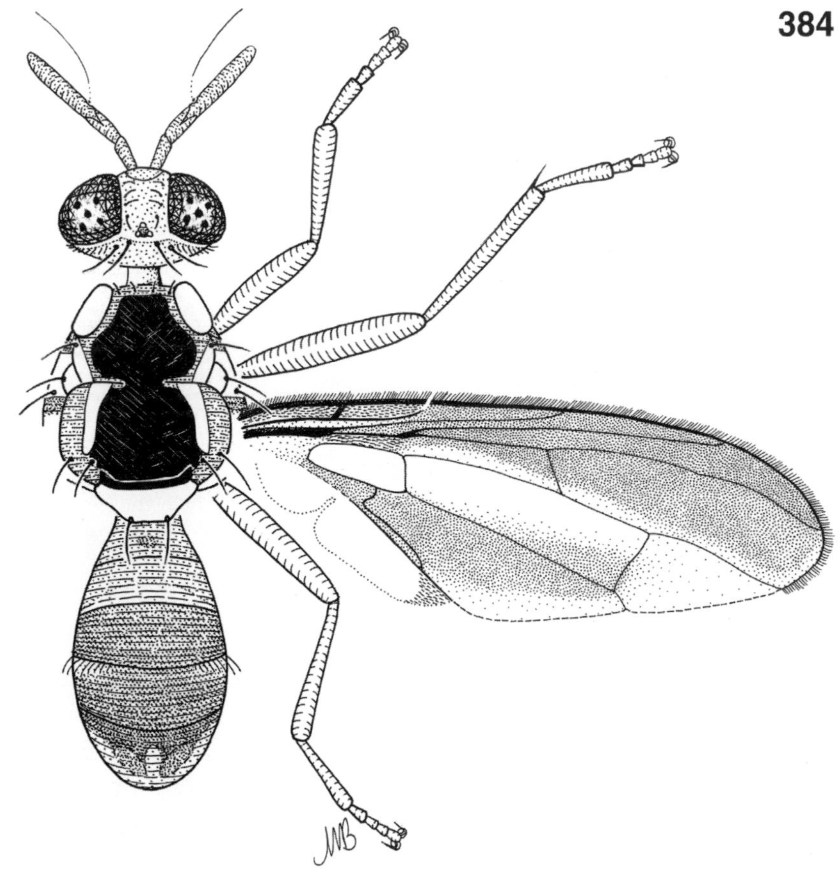

Fig. 383. *Dacus (Mellesis) crabroniformis* (Bezzi), holotype male. **Fig. 384.** *Dacus (Mellesis) discipennis* Walker new combination, holotype male.

385

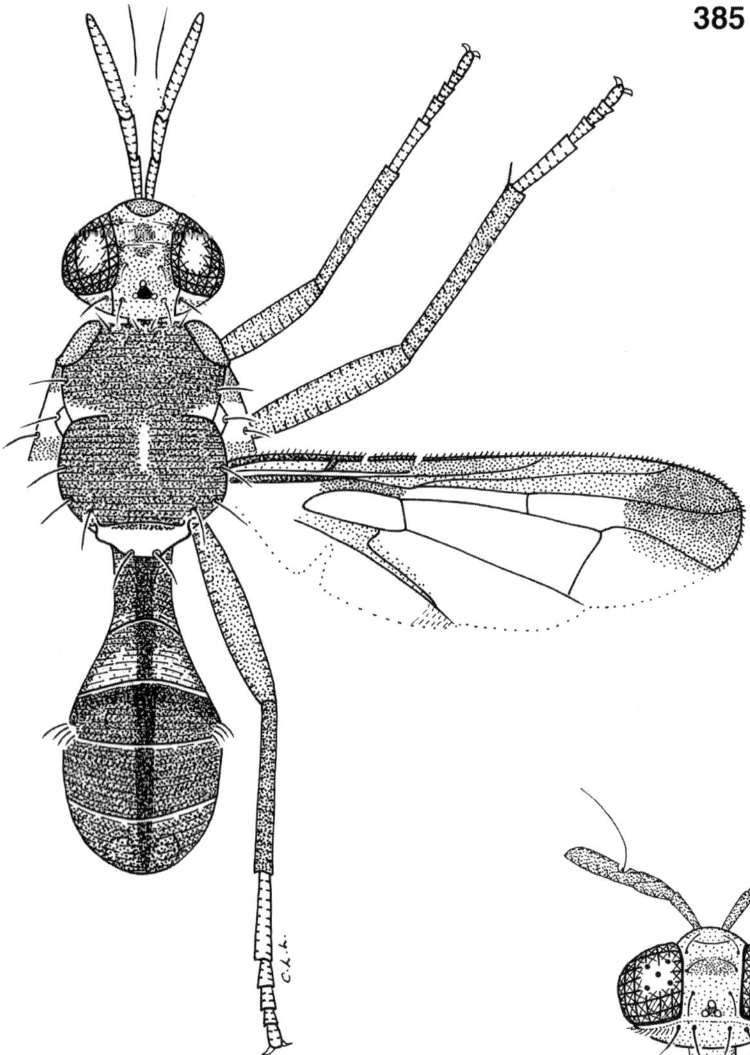

386

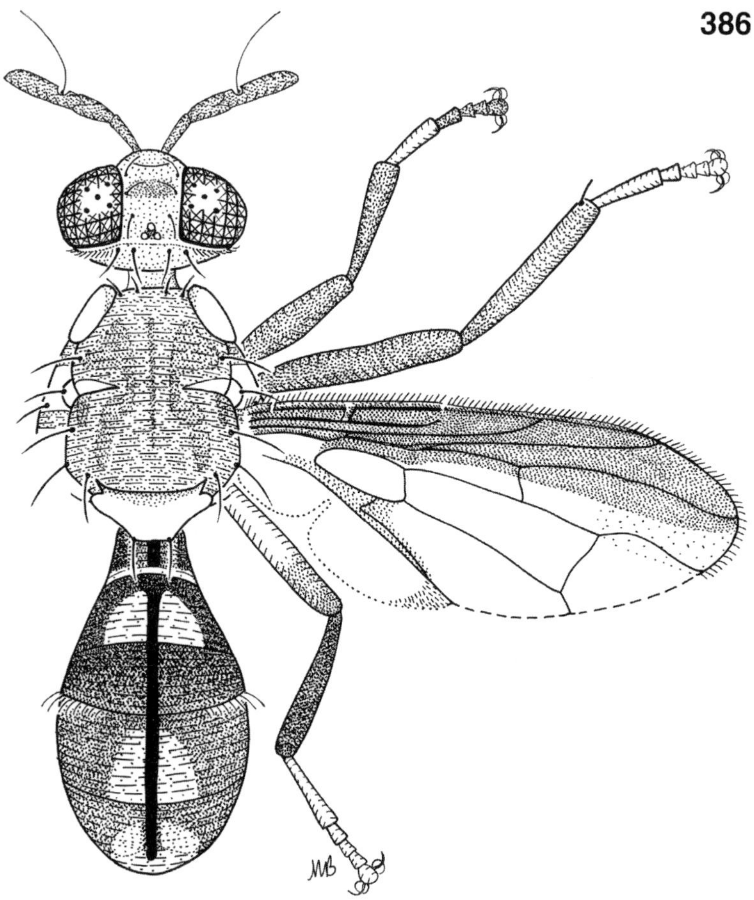

Fig. 385. *Dacus (Mellesis) discophorus* (Hering), holotype male. **Fig. 386.** *Dacus (Mellesis) discretus* new species, holotype male.

387

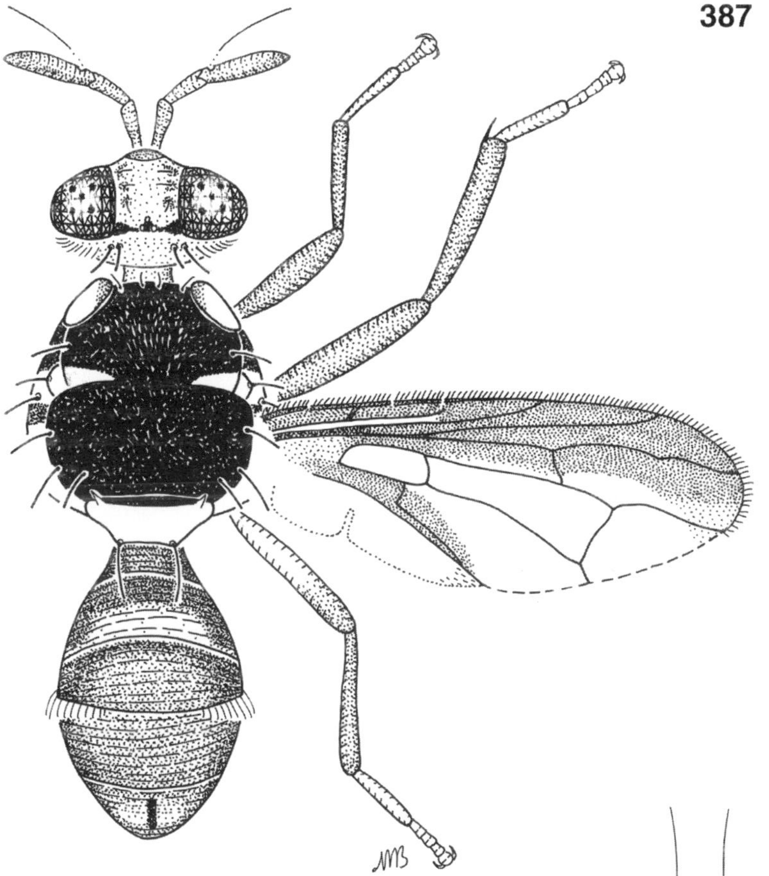

388

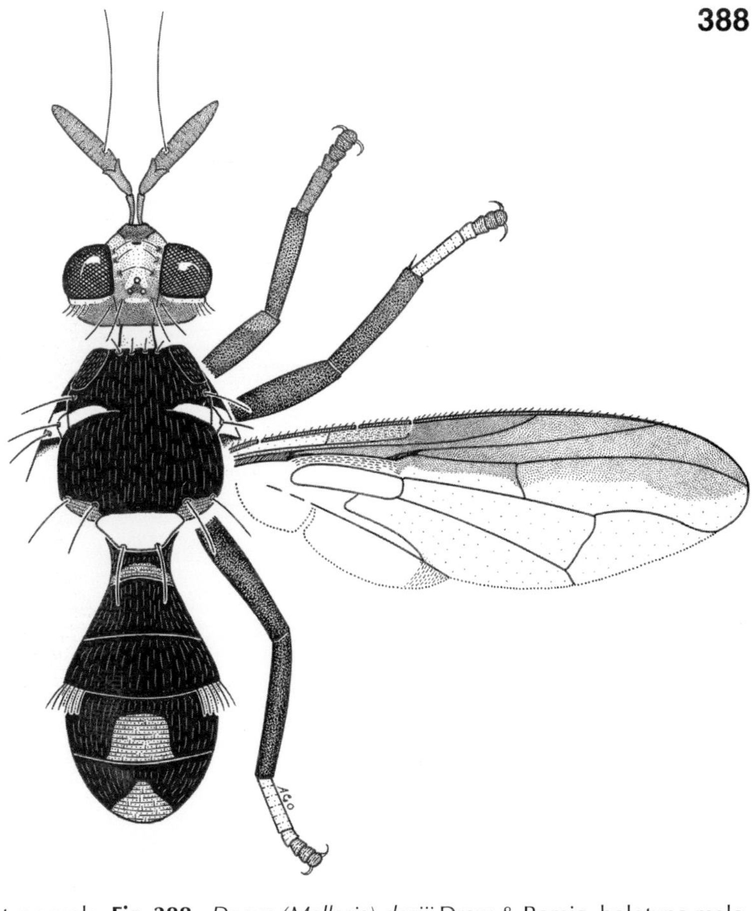

Fig. 387. *Dacus (Mellesis) donggaliae* new species, holotype male. **Fig. 388.** *Dacus (Mellesis) dorjii* Drew & Romig, holotype male.

389

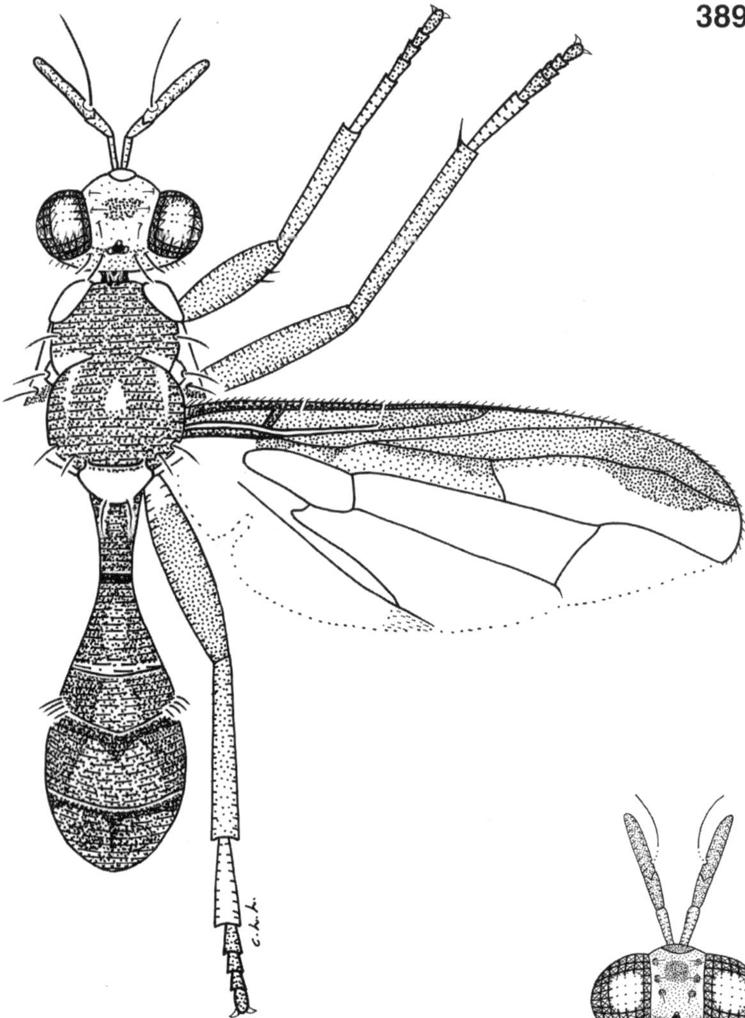

390

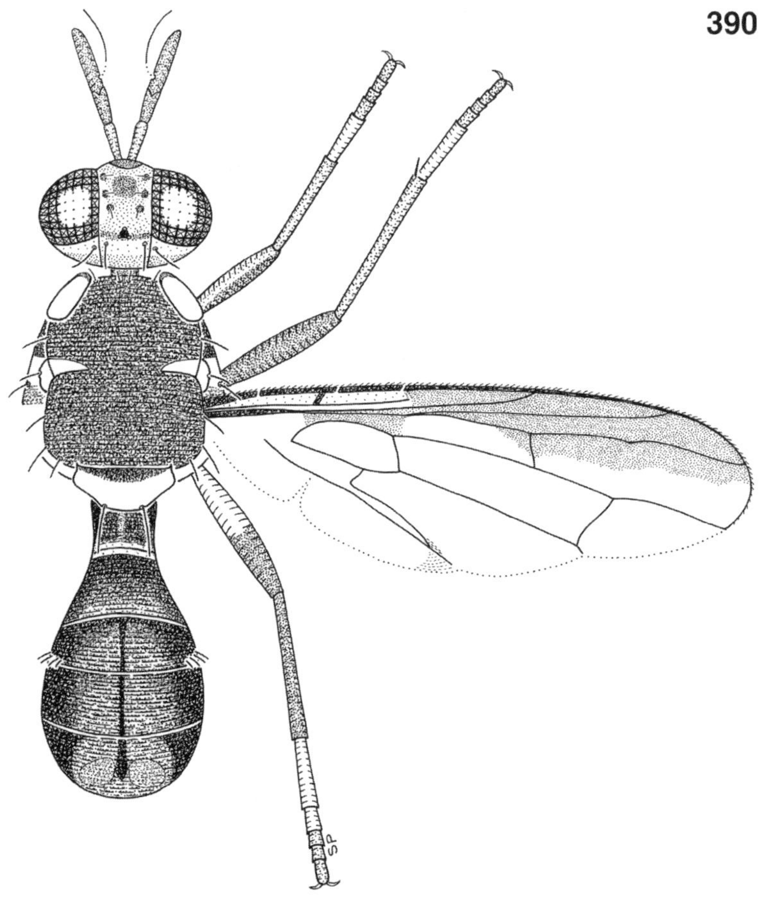

Fig. 389. *Dacus (Mellesis) esakii* (Shiraki), lectotype male. **Fig. 390.** *Dacus (Mellesis) feijeni* White, holotype male.

391

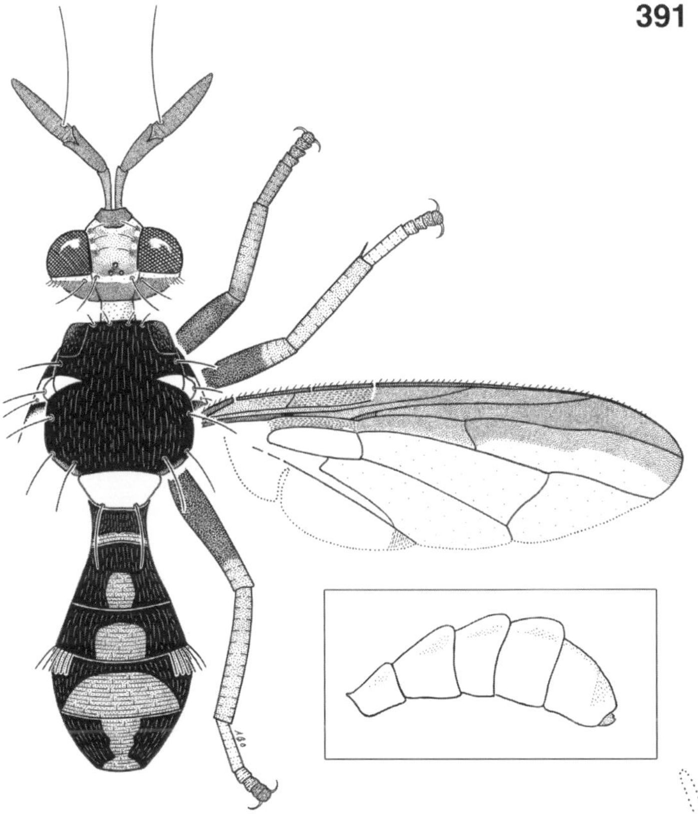

392

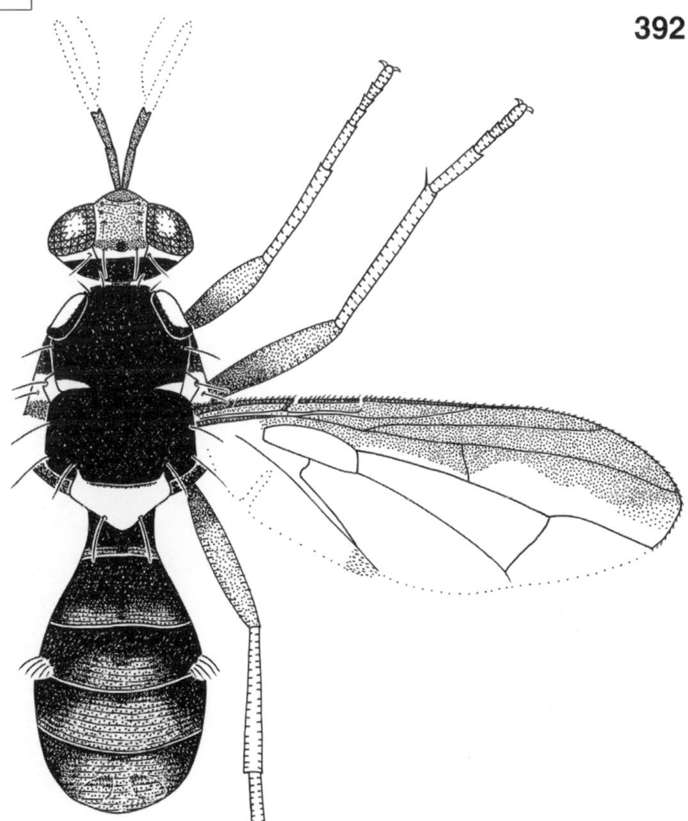

Fig. 391. *Dacus (Mellesis) fletcheri* Drew, holotype male and lateral view of abdomen. **Fig. 392.** *Dacus (Mellesis) formosanus* (Tseng & Chu), male.

393

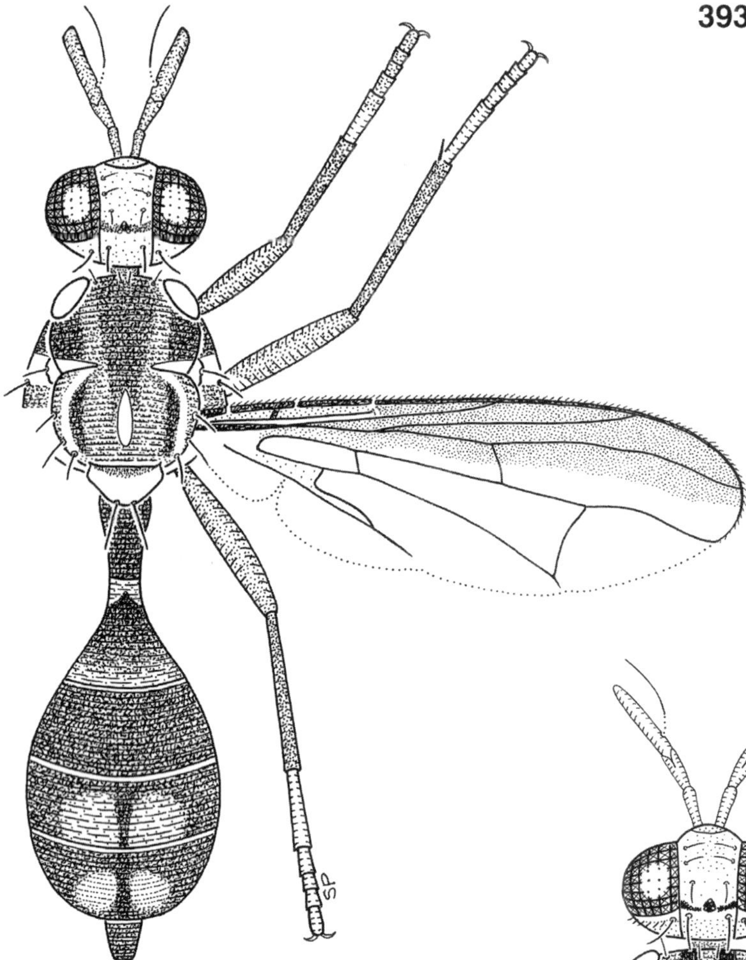

394

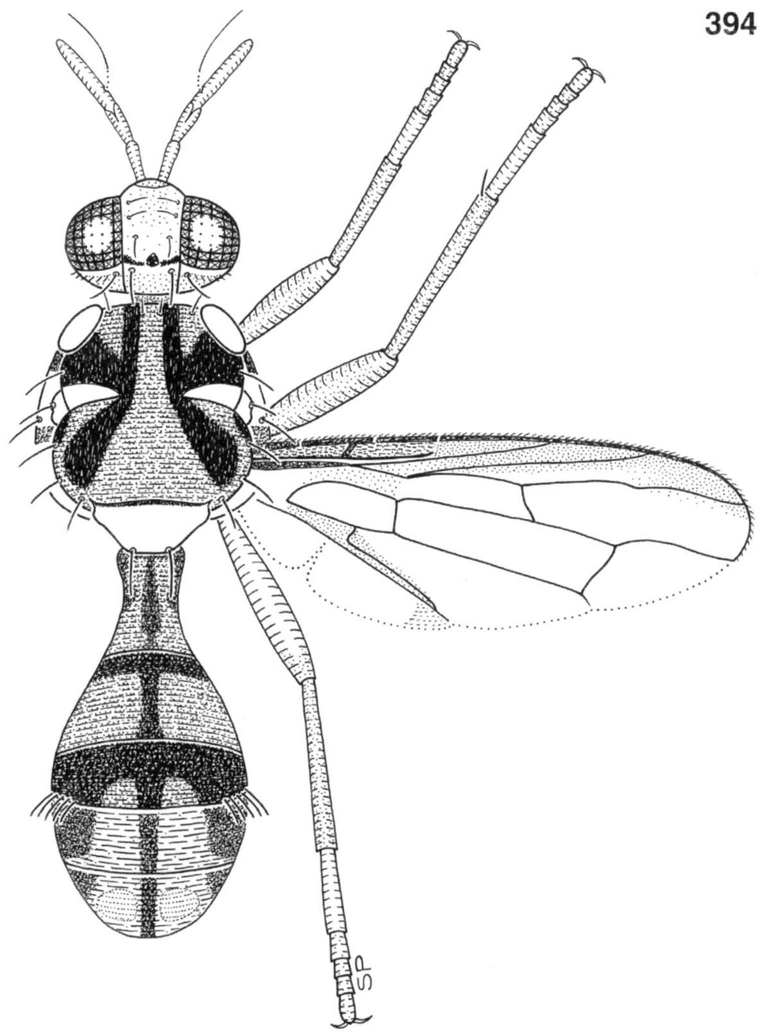

Fig. 393. *Dacus (Mellesis) icariiformis* (Enderlein), female. **Fig. 394.** *Dacus (Mellesis) indecorus* (Hardy), holotype male.

395

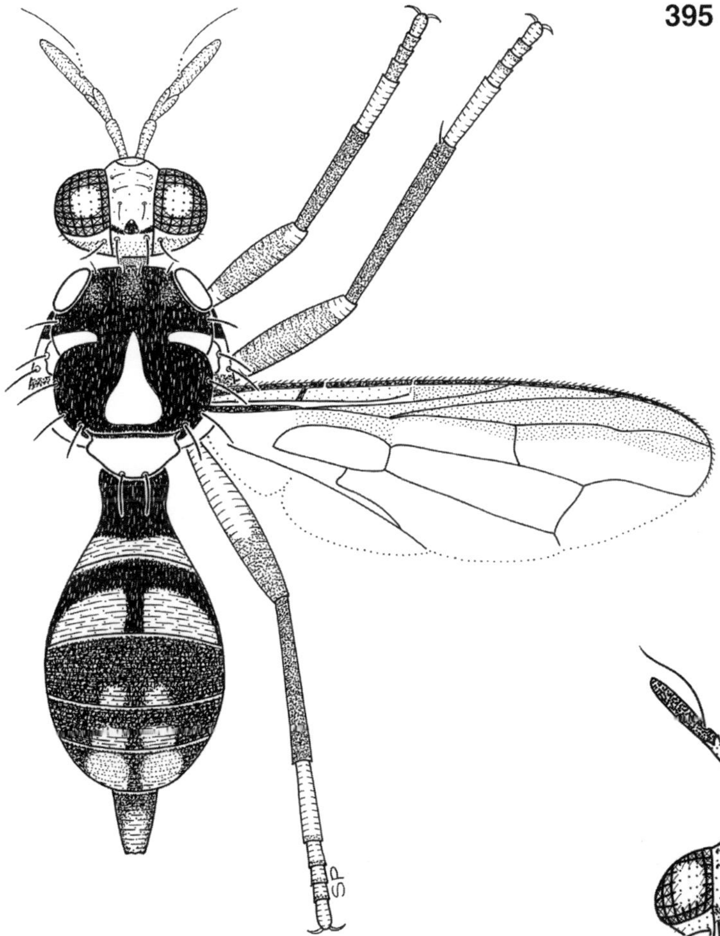

396

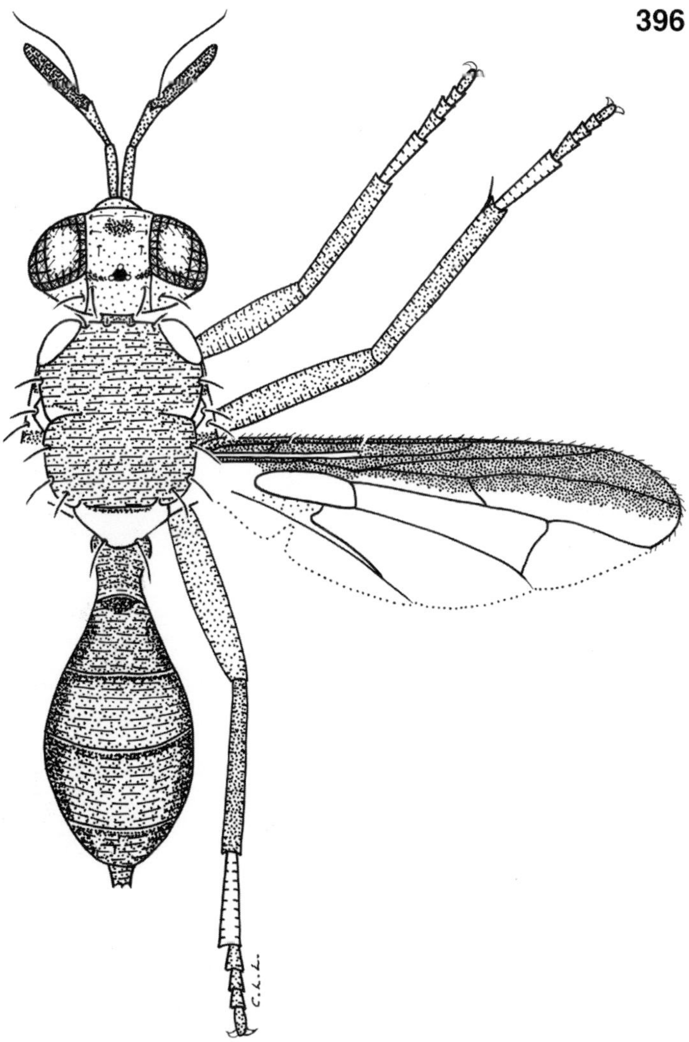

Fig. 395. *Dacus (Mellesis) infernus* (Hardy), holotype female. **Fig. 396.** *Dacus (Mellesis) insulosus* Drew & Hancock, holotype female.

397

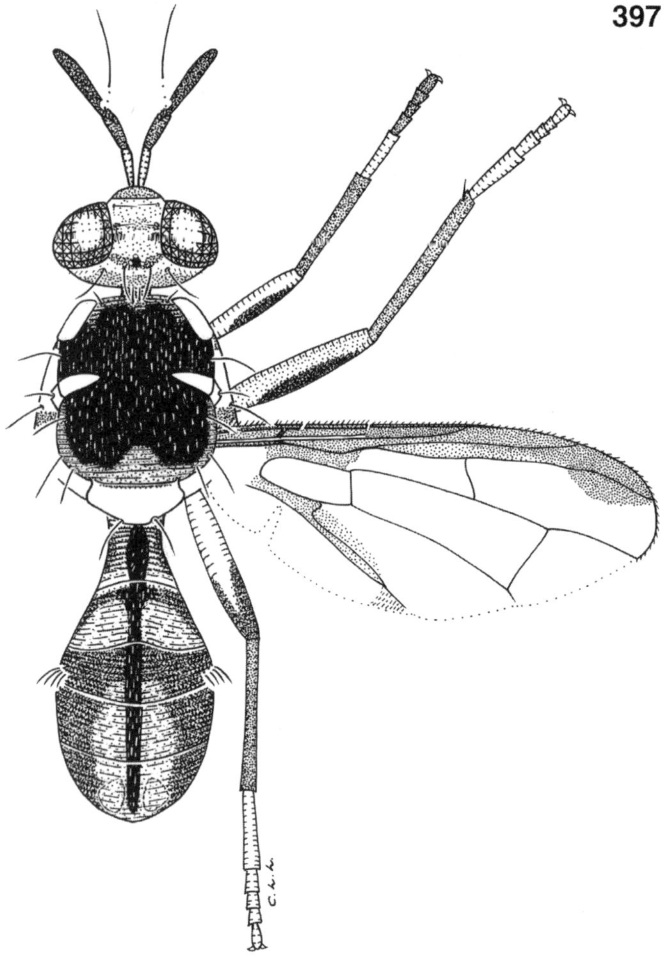

398

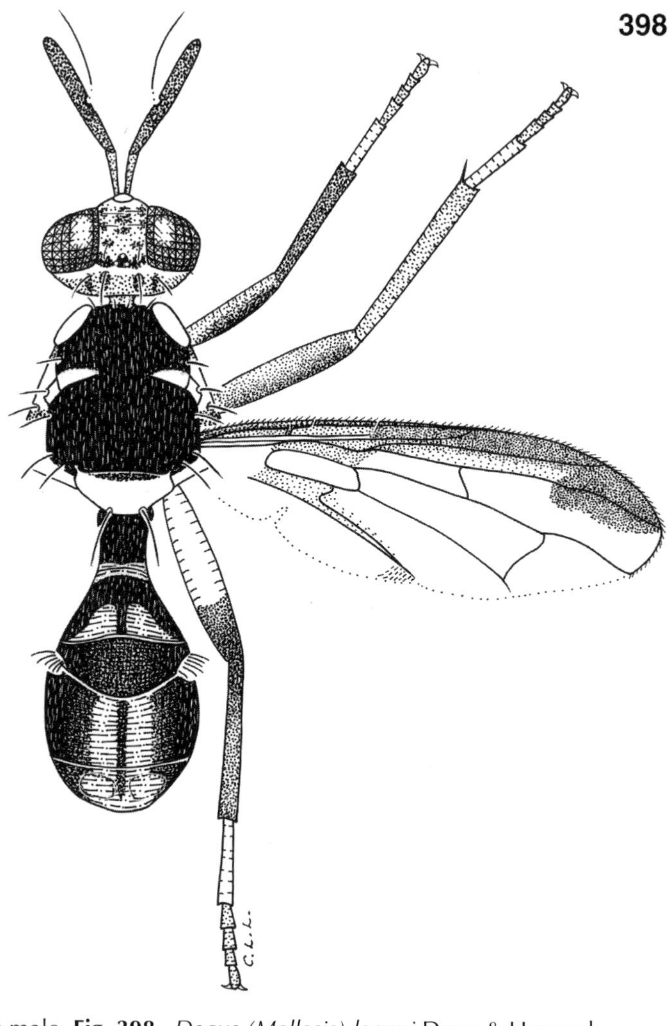

Fig. 397. *Dacus (Mellesis) lagunae* Drew & Hancock, holotype male. **Fig. 398.** *Dacus (Mellesis) leongi* Drew & Hancock, holotype male.

399

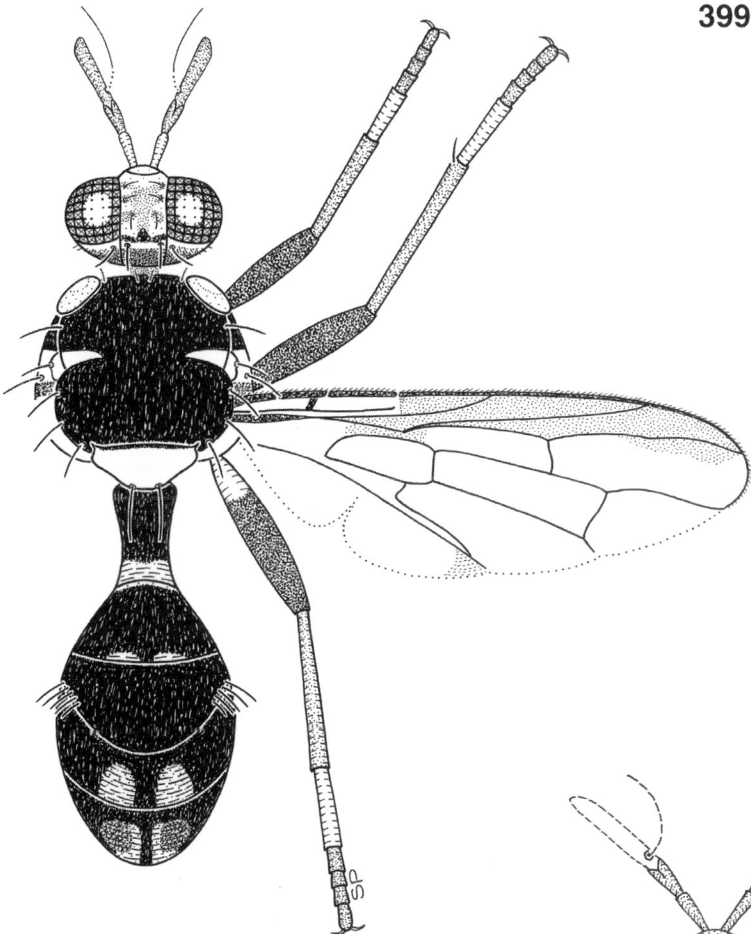

400

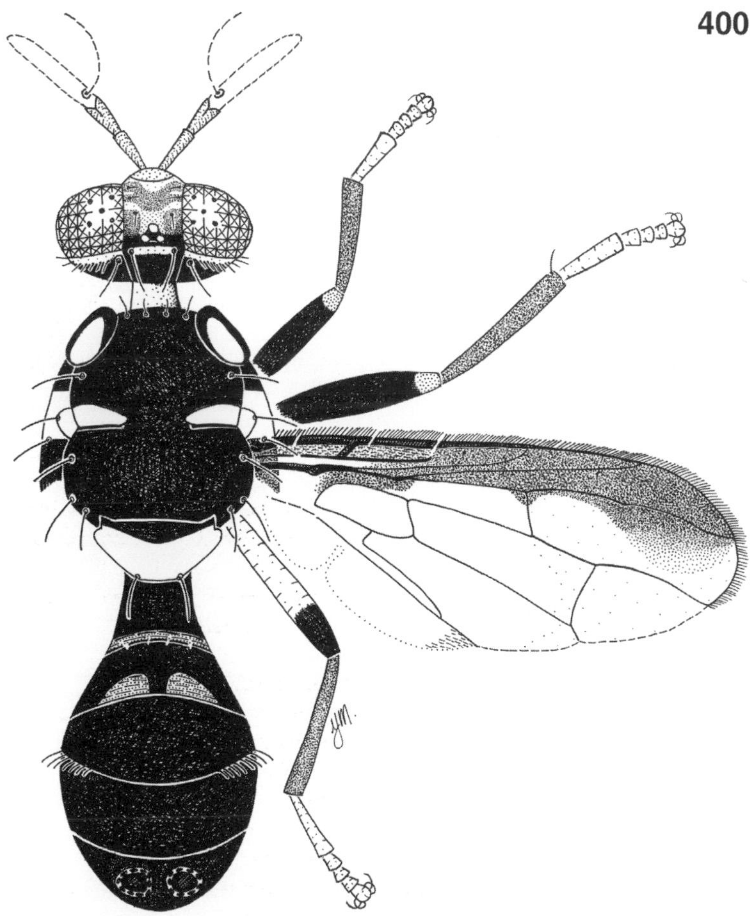

Fig. 399. *Dacus (Mellesis) maculipterus* White, holotype male. **Fig. 400.** *Dacus (Mellesis) melanopectus* new species, holotype male.

401

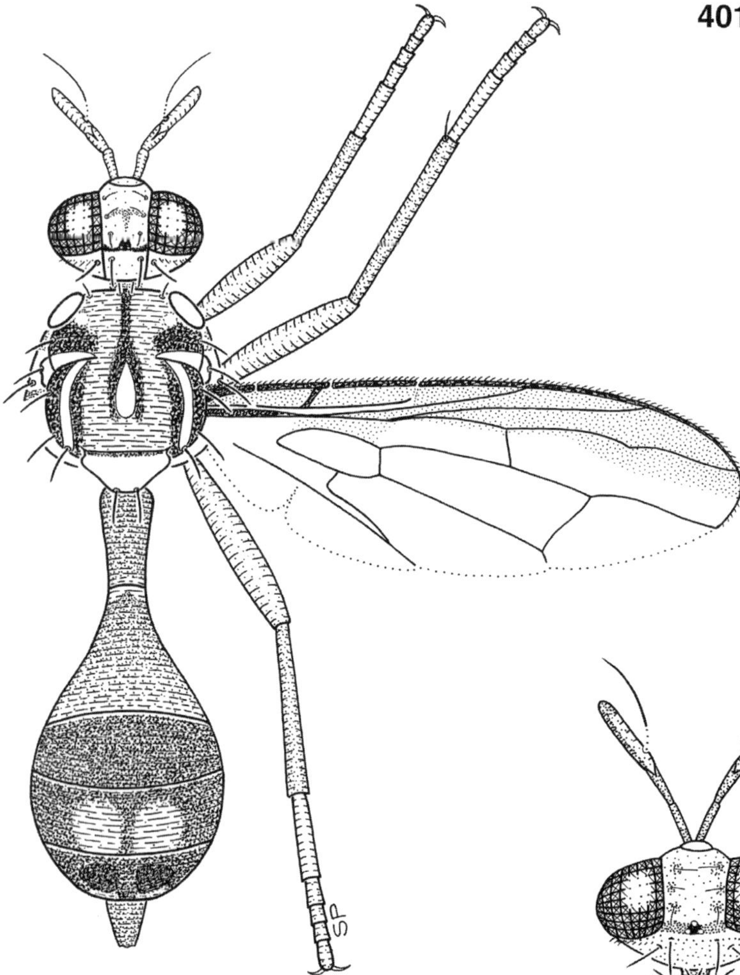

402

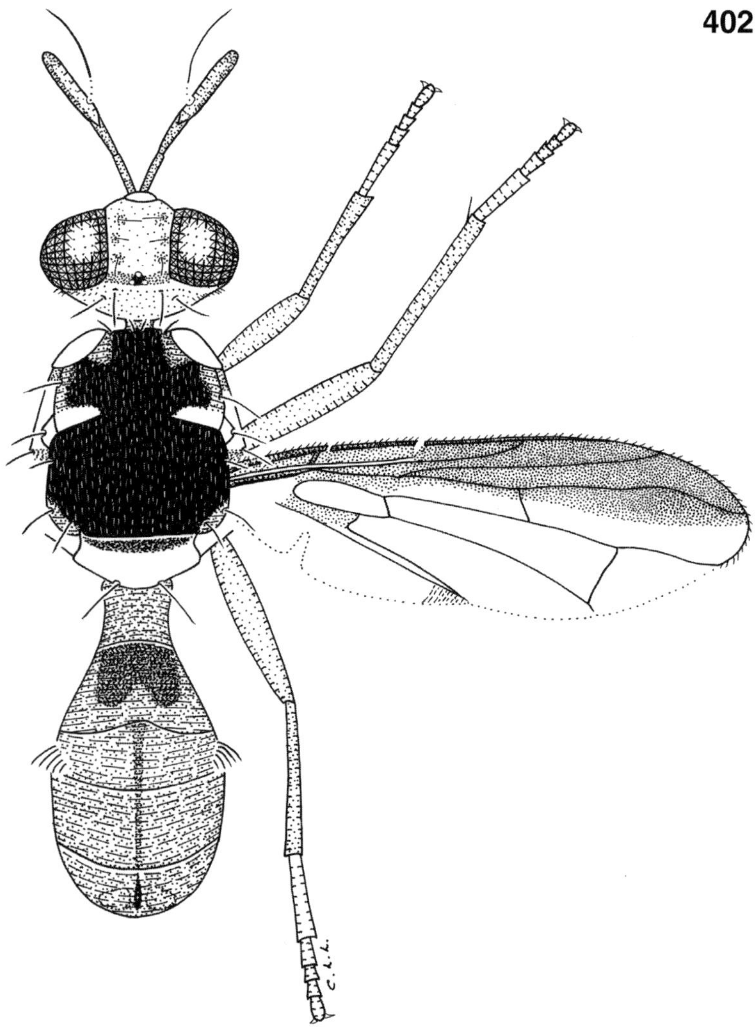

Fig. 401. *Dacus (Mellesis) murphyi* White, holotype female. **Fig. 402.** *Dacus (Mellesis) nanggalae* Drew & Hancock, holotype male.

403

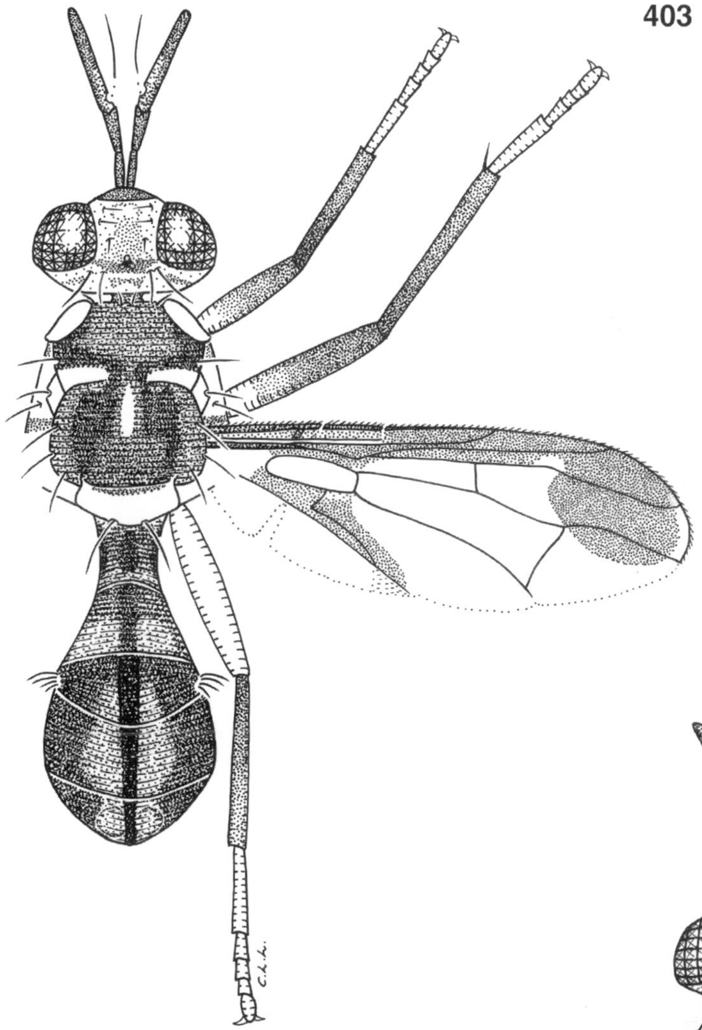

404

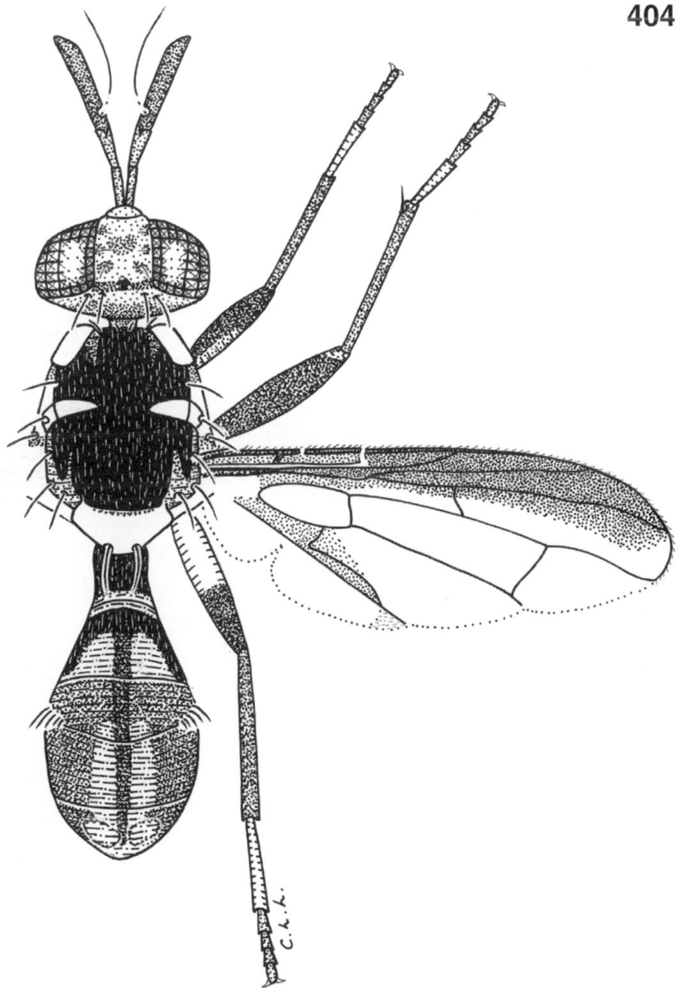

Fig. 403. *Dacus (Mellesis) nummularius* (Bezzi), male. **Fig. 404.** *Dacus (Mellesis) ooii* Drew & Hancock, holotype male.

405

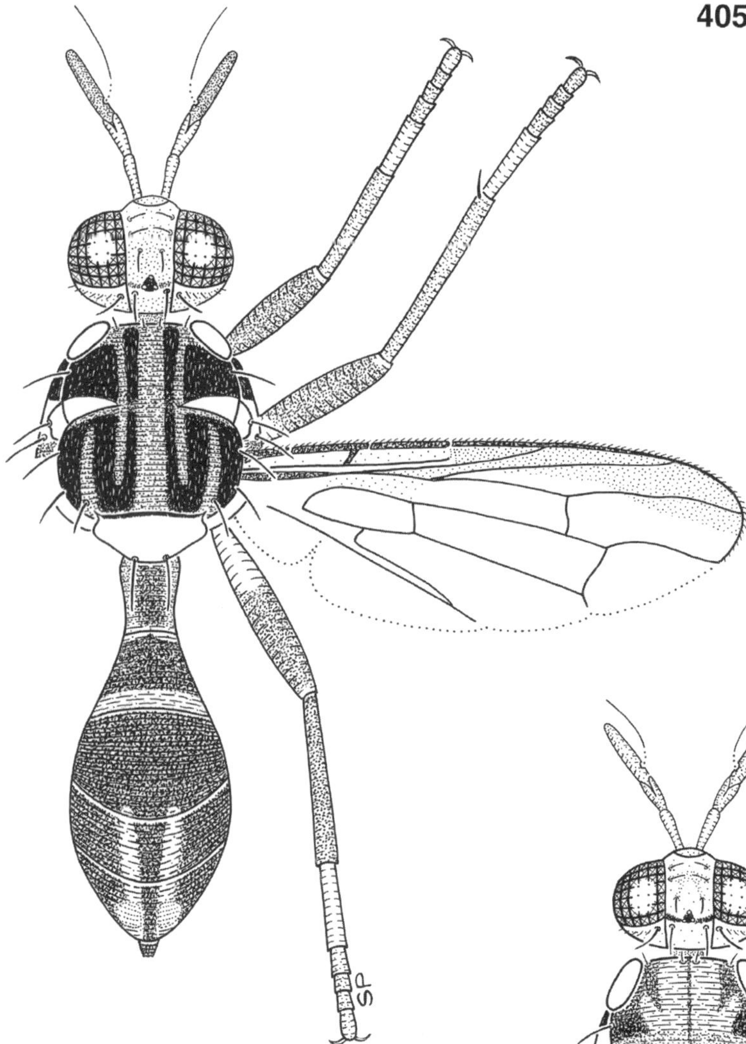

406

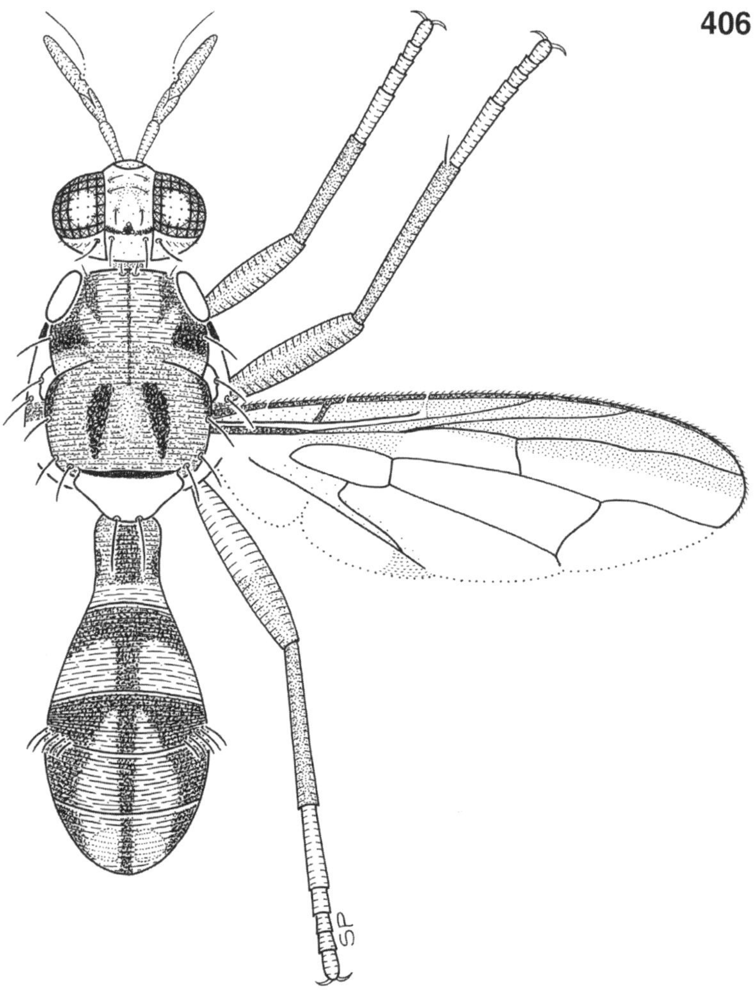

Fig. 405. *Dacus (Mellesis) pedunculatus* (Bezzi), female. **Fig. 406.** *Dacus (Mellesis) pictus* (Hardy), holotype male.

407

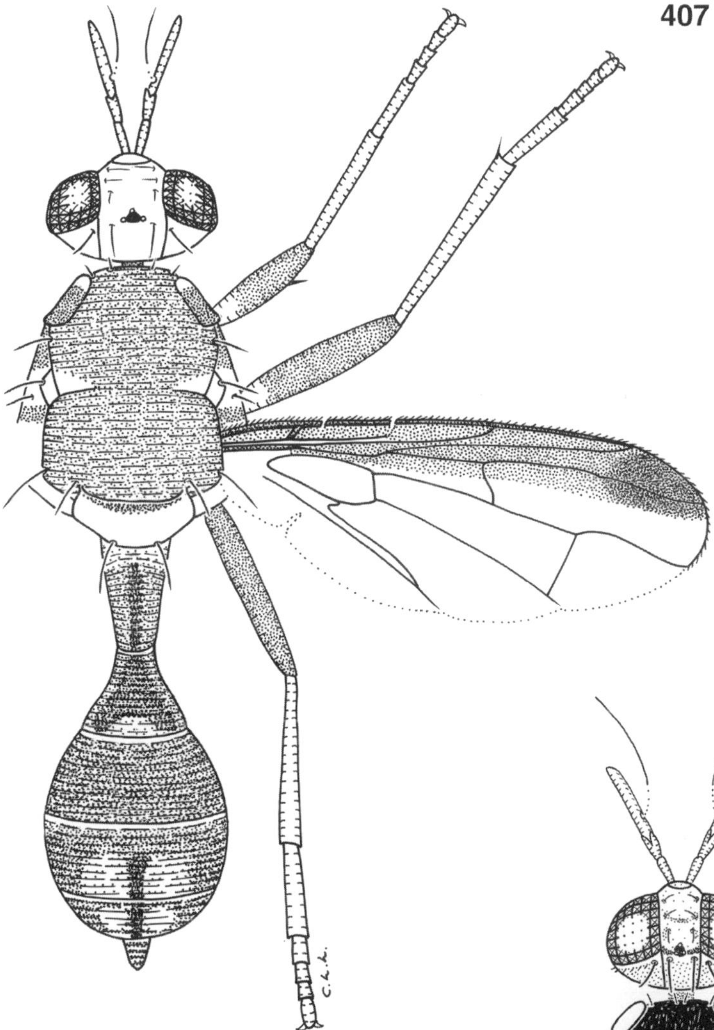

408

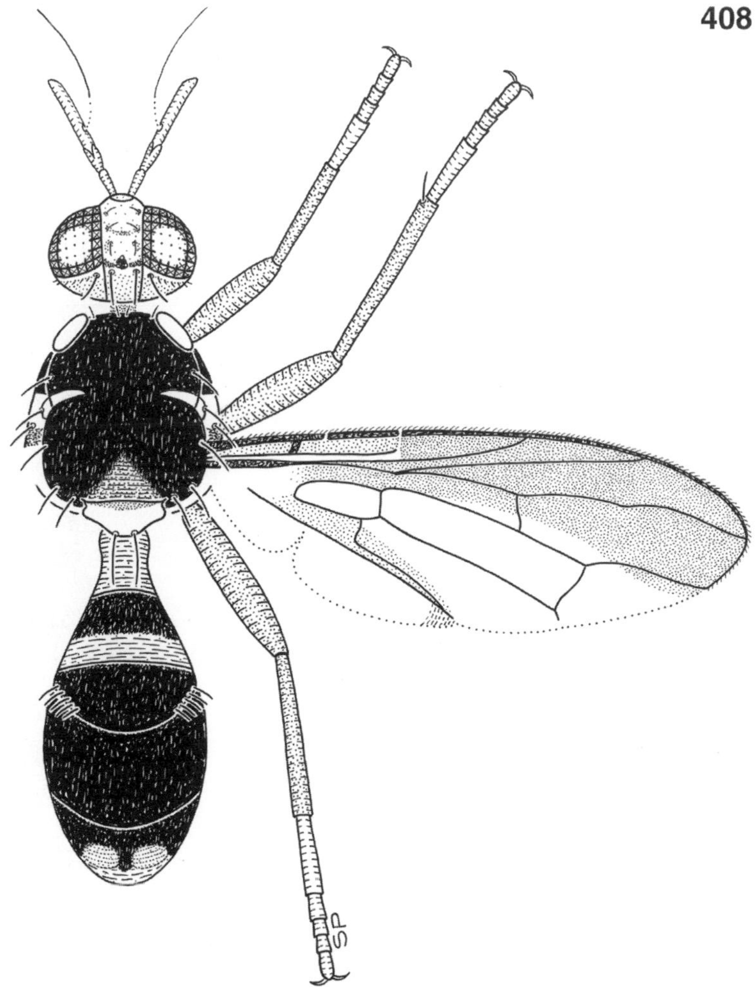

Fig. 407. *Dacus (Mellesis) polistiformis* (Senior-White), female. **Fig. 408.** *Dacus (Mellesis) pullus* (Hardy), holotype male.

409

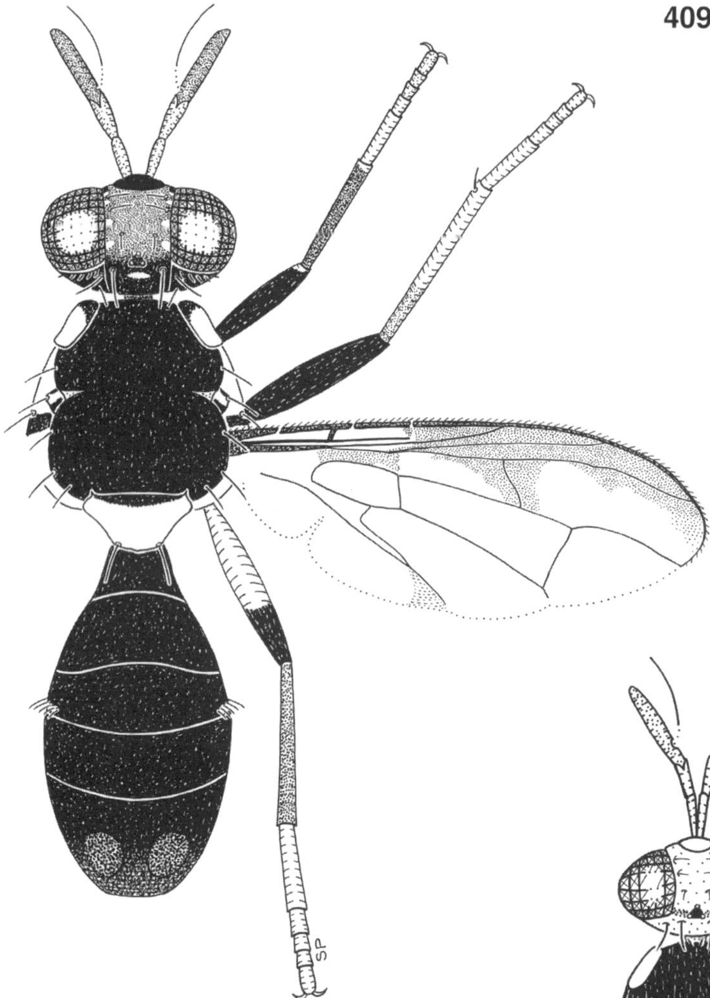

410

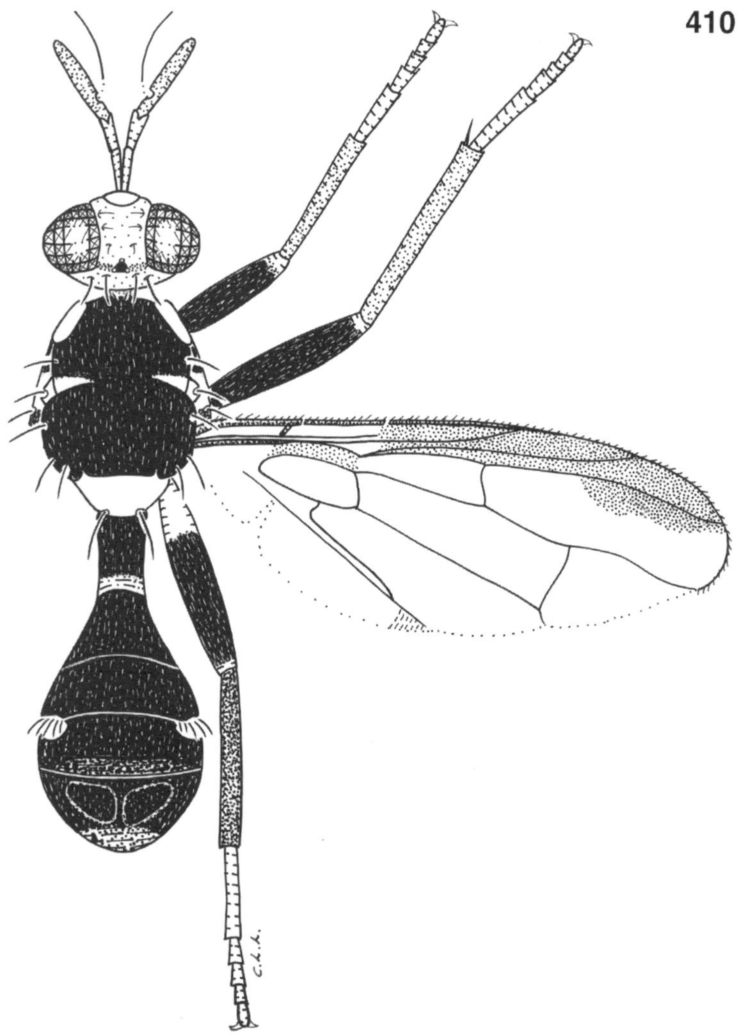

Fig. 409. *Dacus (Mellesis) ramanii* Drew & Hancock, holotype male. **Fig. 410.** *Dacus (Mellesis) satanas* (Hering), holotype male.

411

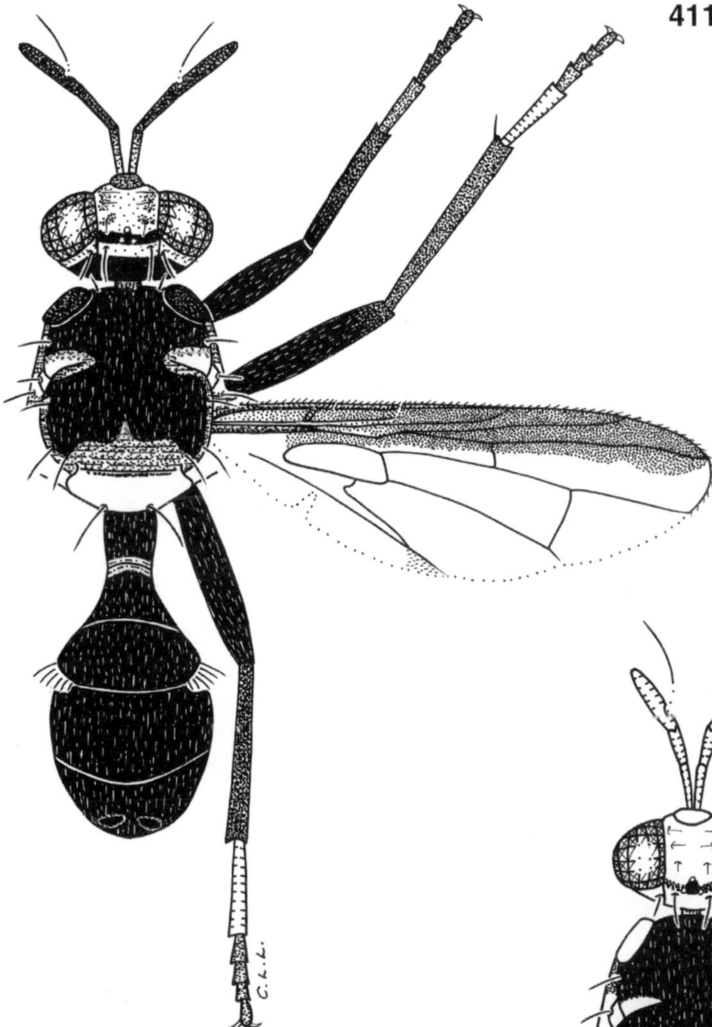

412

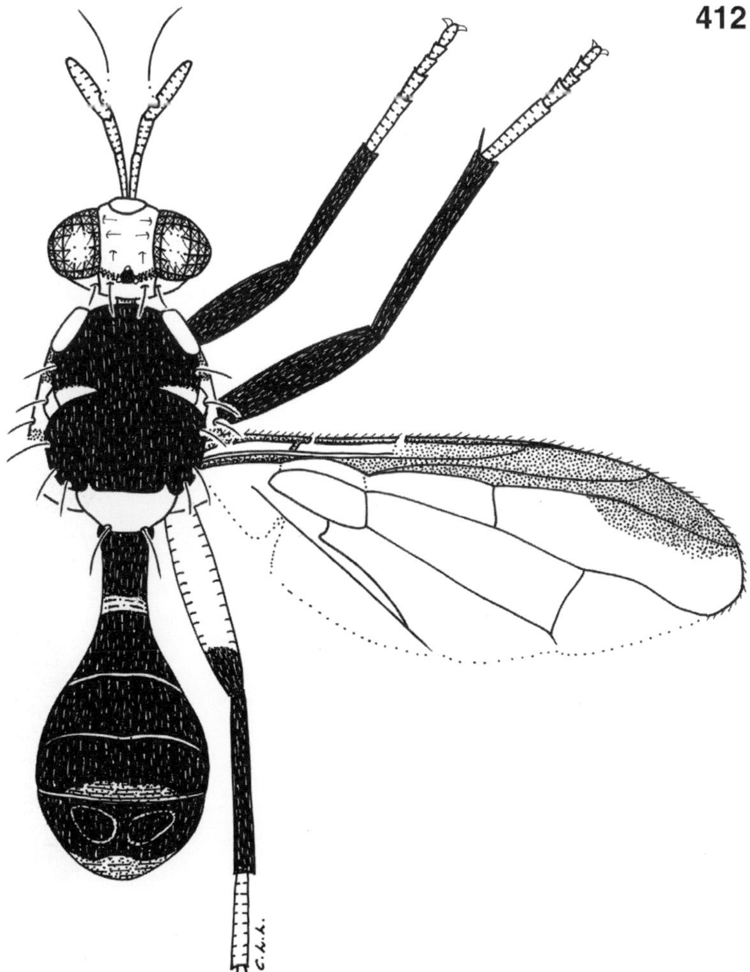

Fig. 411. *Dacus (Mellesis) siamensis* Drew & Hancock, holotype male. **Fig. 412.** *Dacus (Mellesis) sinensis* (Wang), holotype male.

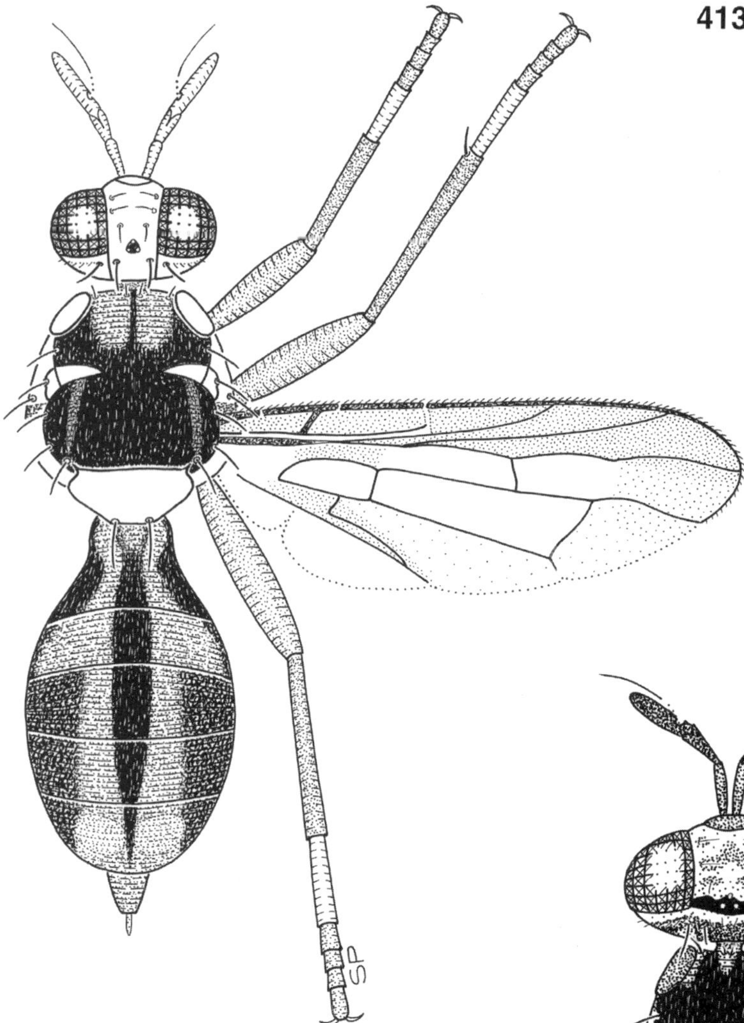

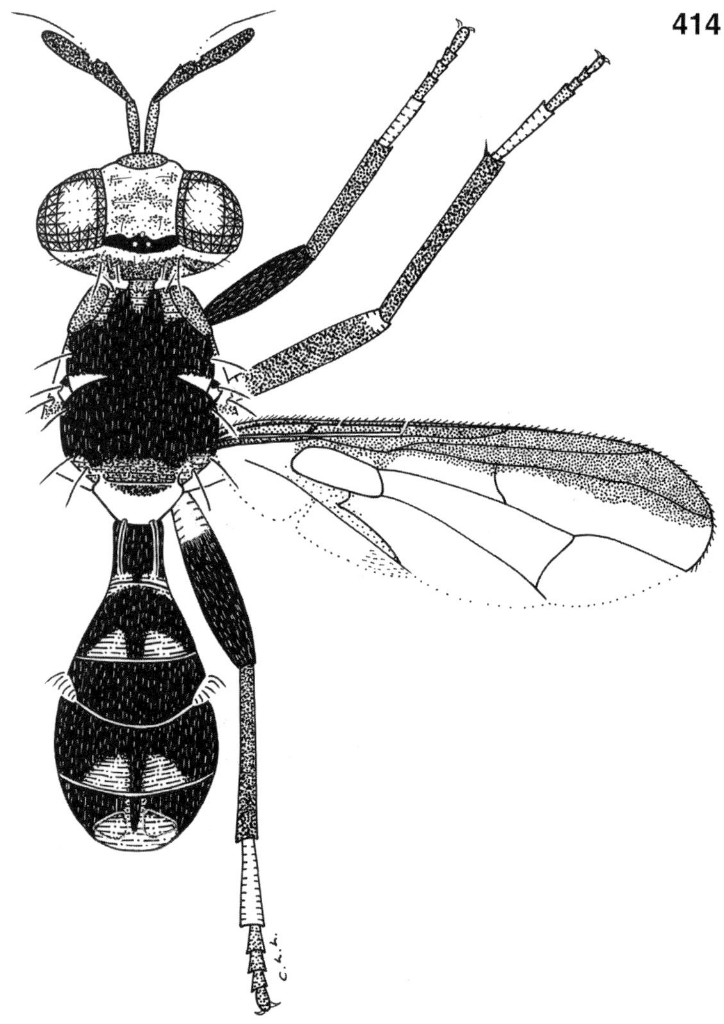

Fig. 413. *Dacus (Mellesis) subsessilis* (Bezzi), female. **Fig. 414.** *Dacus (Mellesis) tenebrosus* Drew & Hancock, holotype male.

415

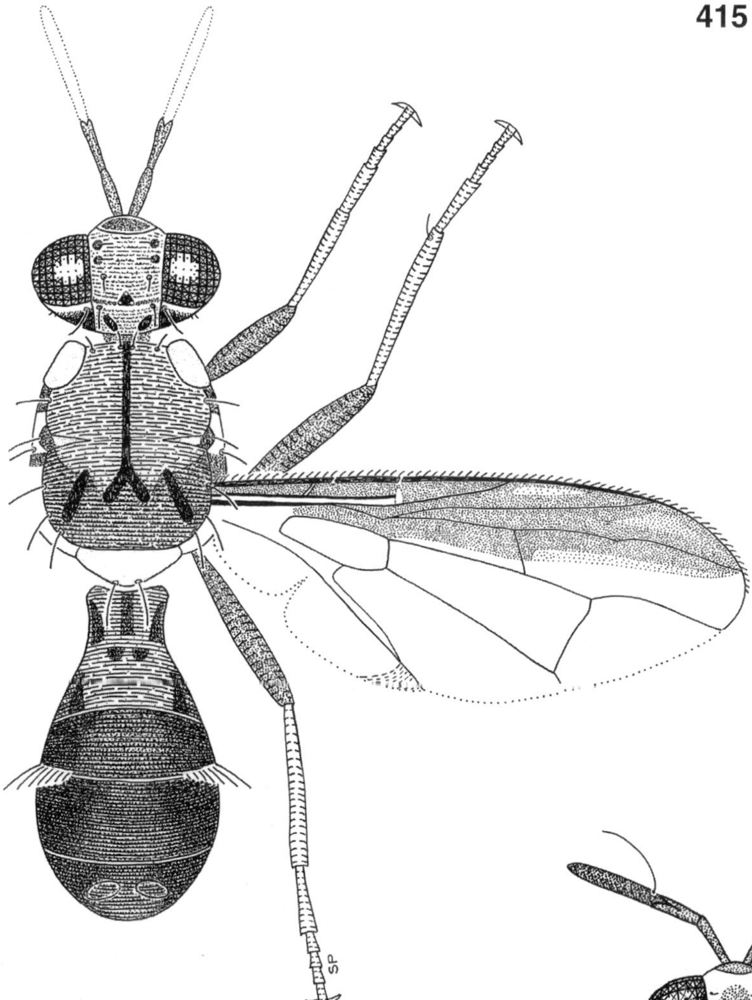

416

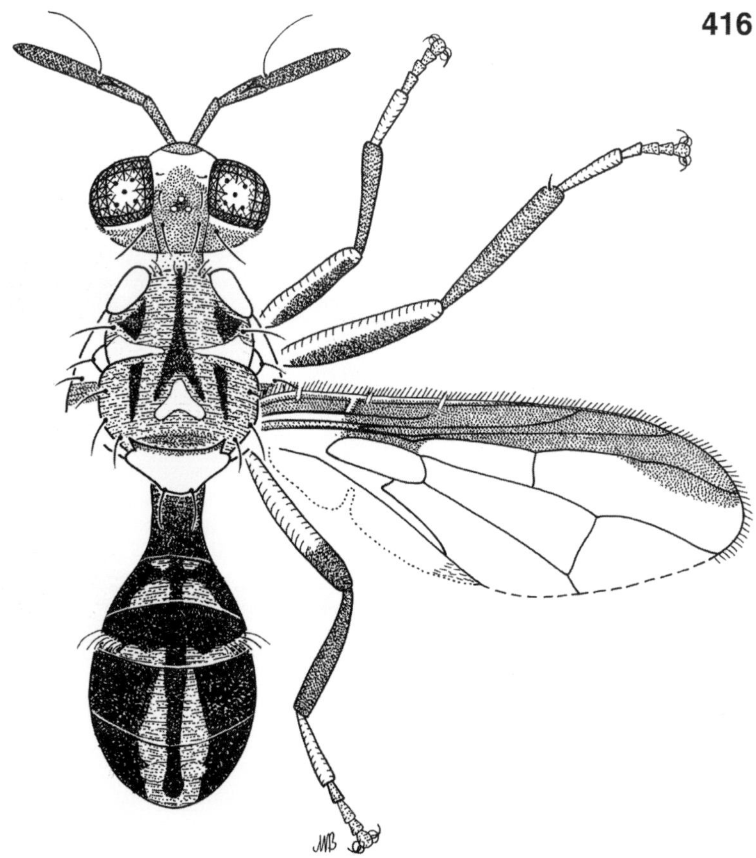

Fig. 415. *Dacus (Mellesis) trimacula* (Wang), holotype male. **Fig. 416.** *Dacus (Mellesis) triquetrus* new species, holotype male.

417

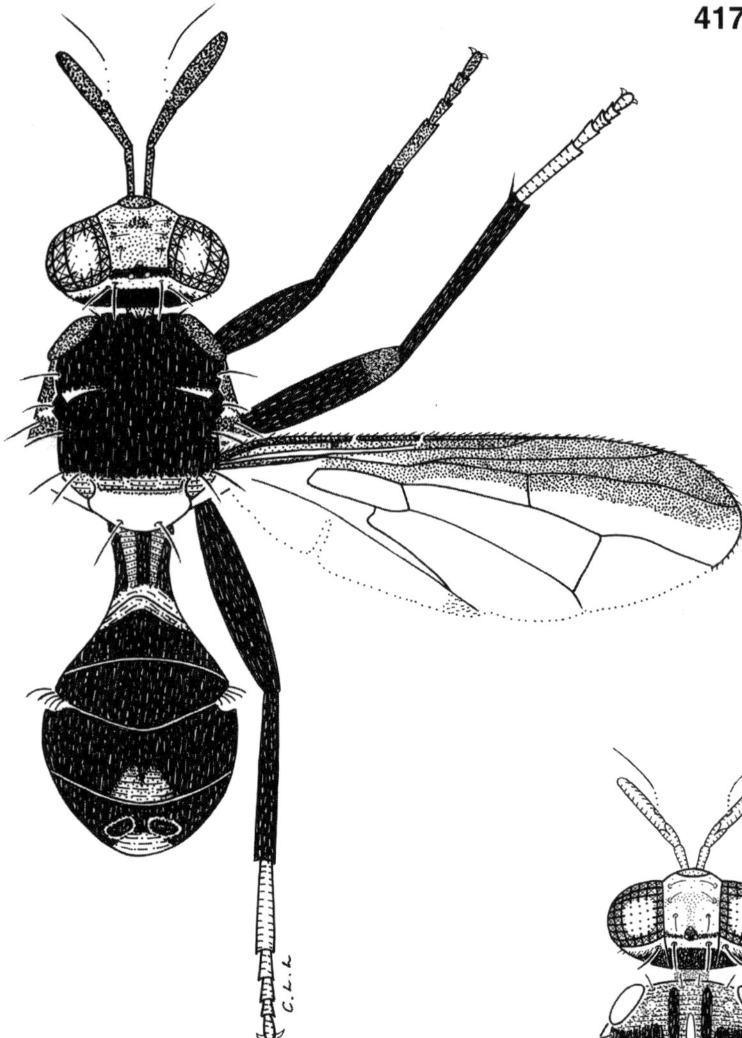

418

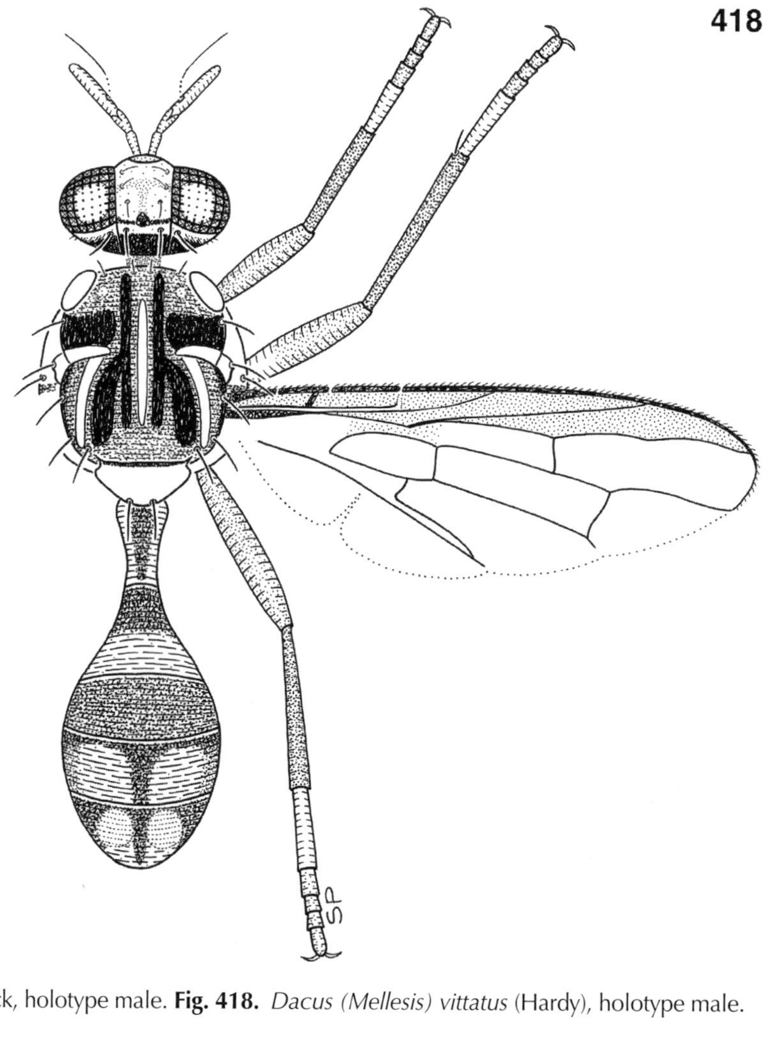

Fig. 417. *Dacus (Mellesis) vijaysegarani* Drew & Hancock, holotype male. **Fig. 418.** *Dacus (Mellesis) vittatus* (Hardy), holotype male.

419

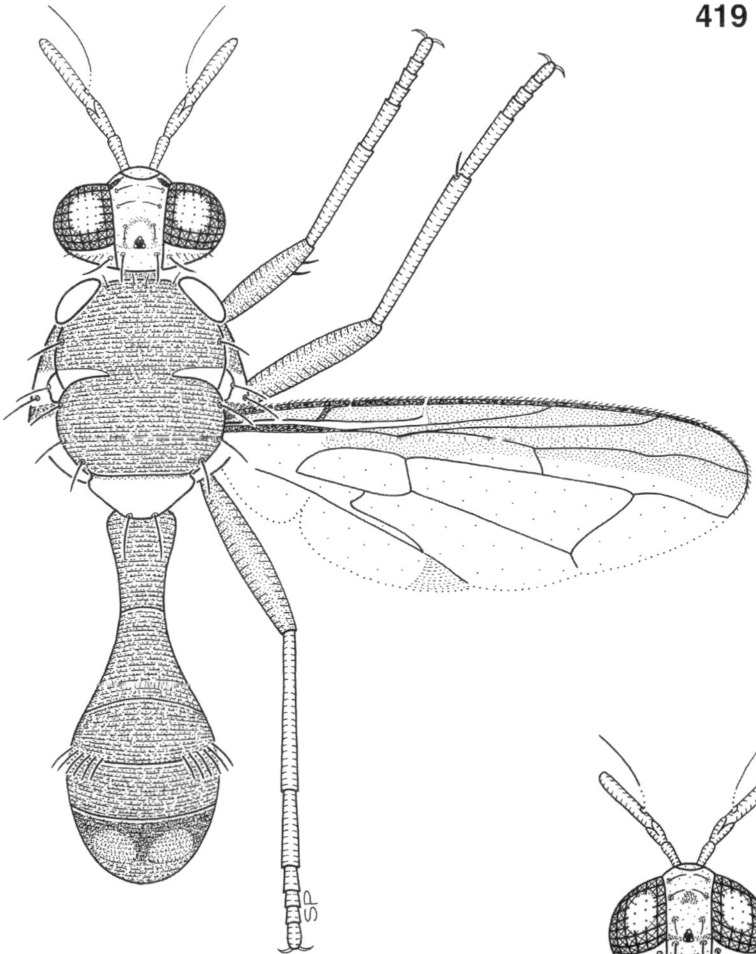

420

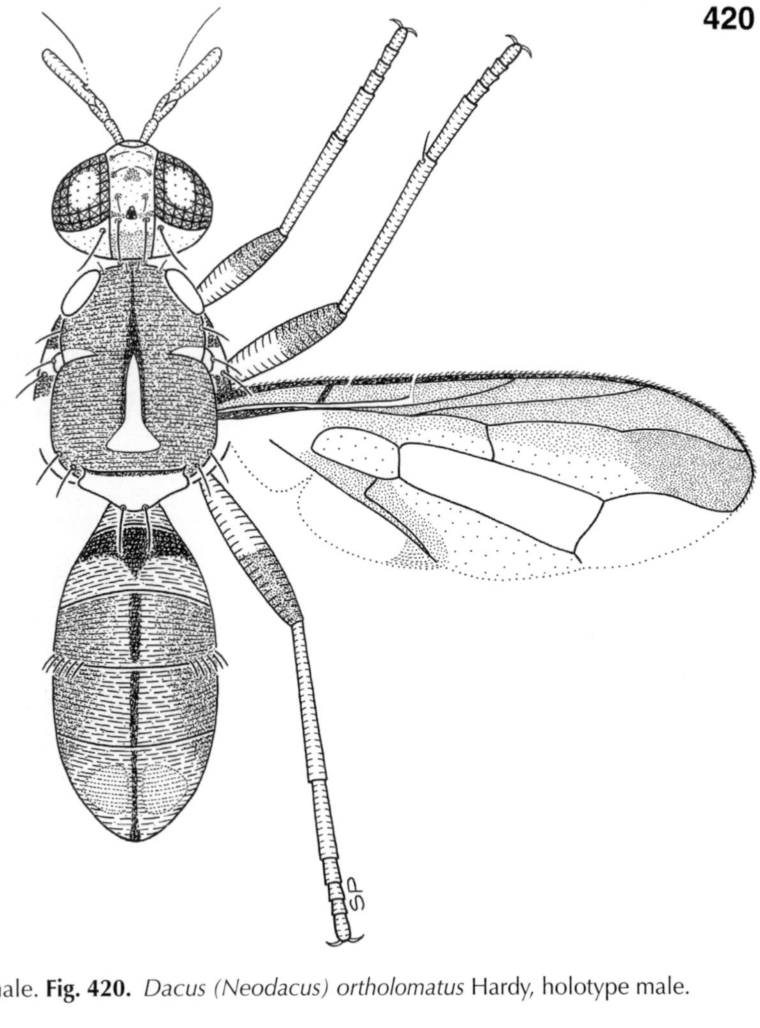

Fig. 419. *Dacus (Mellesis) wallacei* White, holotype male. **Fig. 420.** *Dacus (Neodacus) ortholomatus* Hardy, holotype male.

421

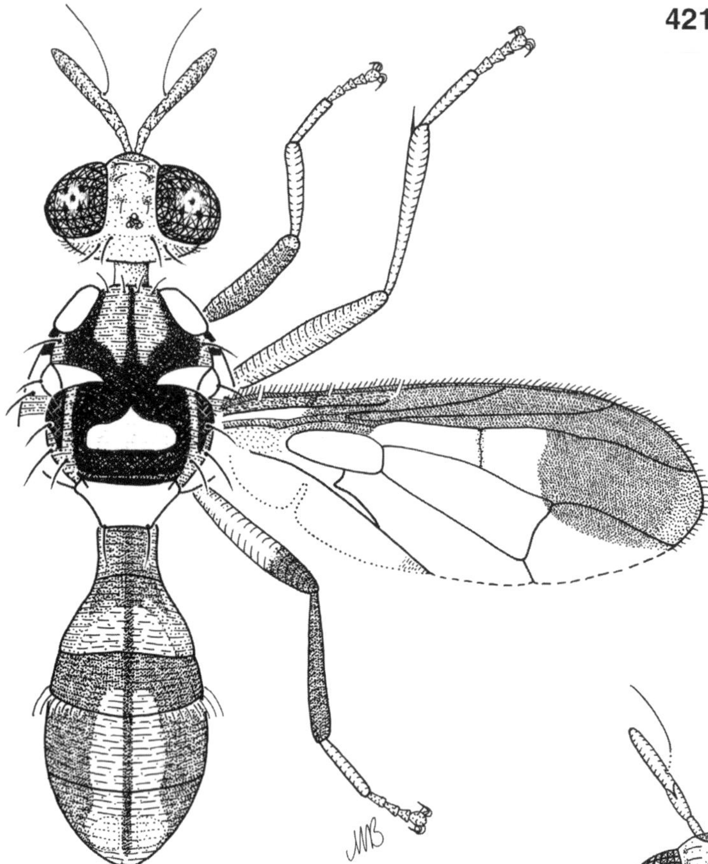

422

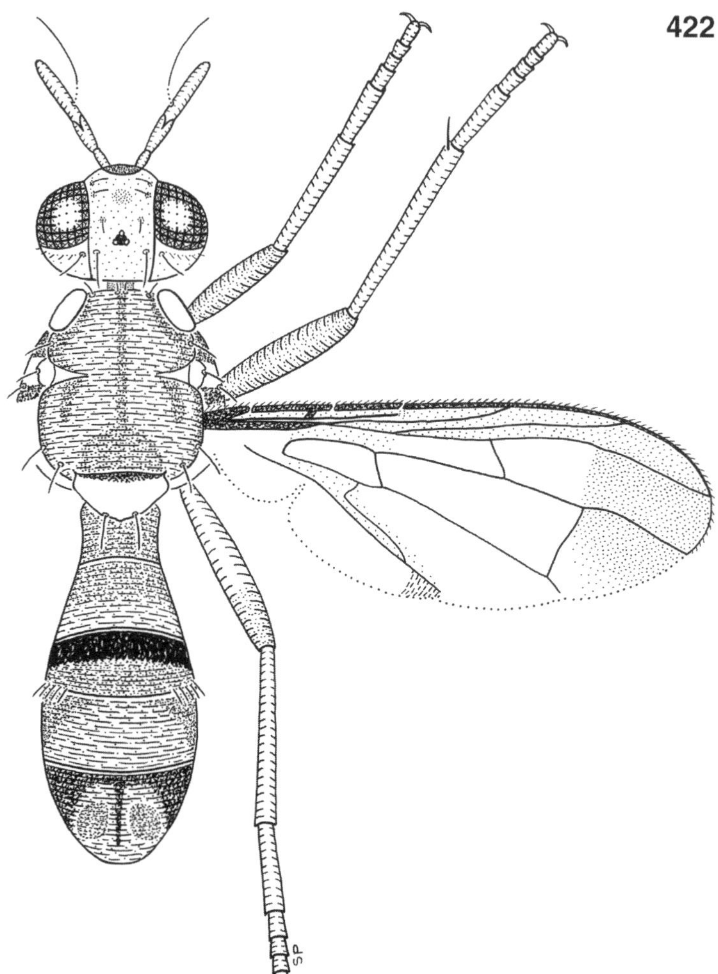

Fig. 421. *Dacus (Neodacus) santongae* Drew & Hancock, male. **Fig. 422.** *Dacus (Neodacus) sphaeroidalis* (Bezzi), male.

423A

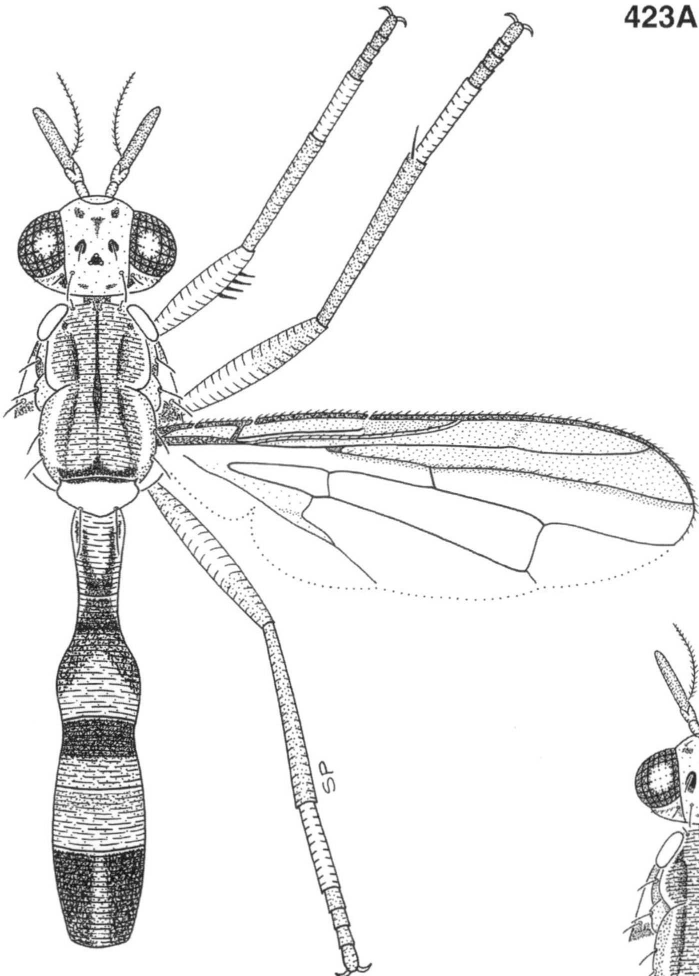

423B

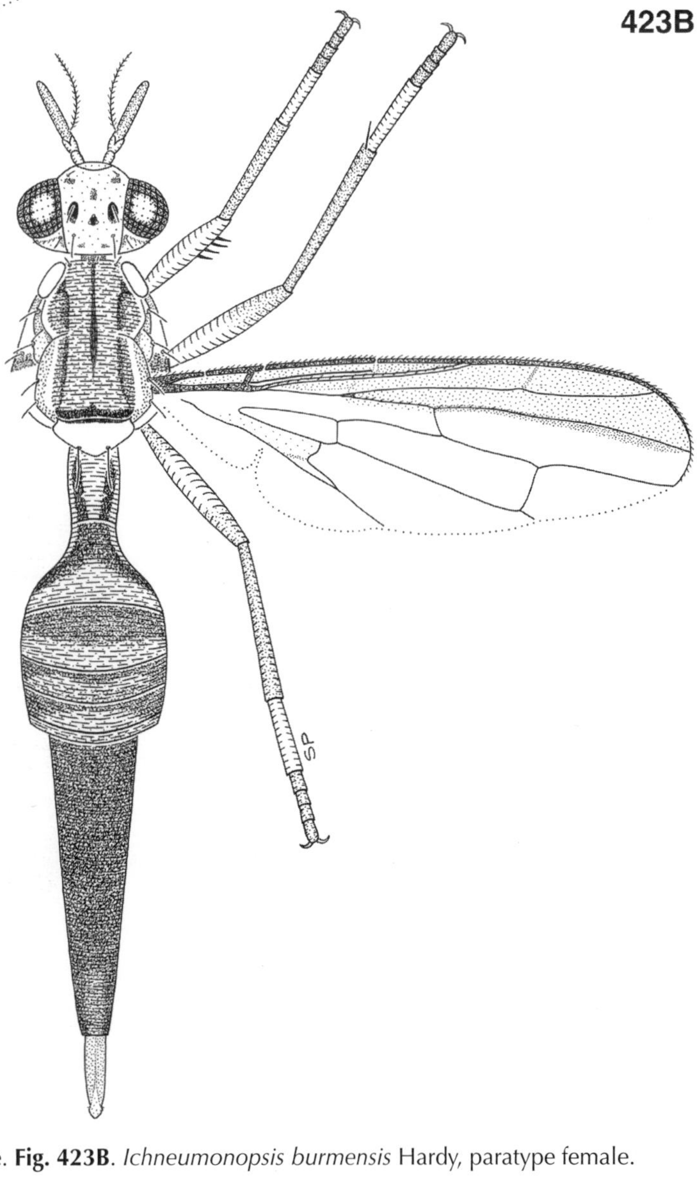

Fig. 423A. *Ichneumonopsis burmensis* Hardy, holotype male. **Fig. 423B**. *Ichneumonopsis burmensis* Hardy, paratype female.

424

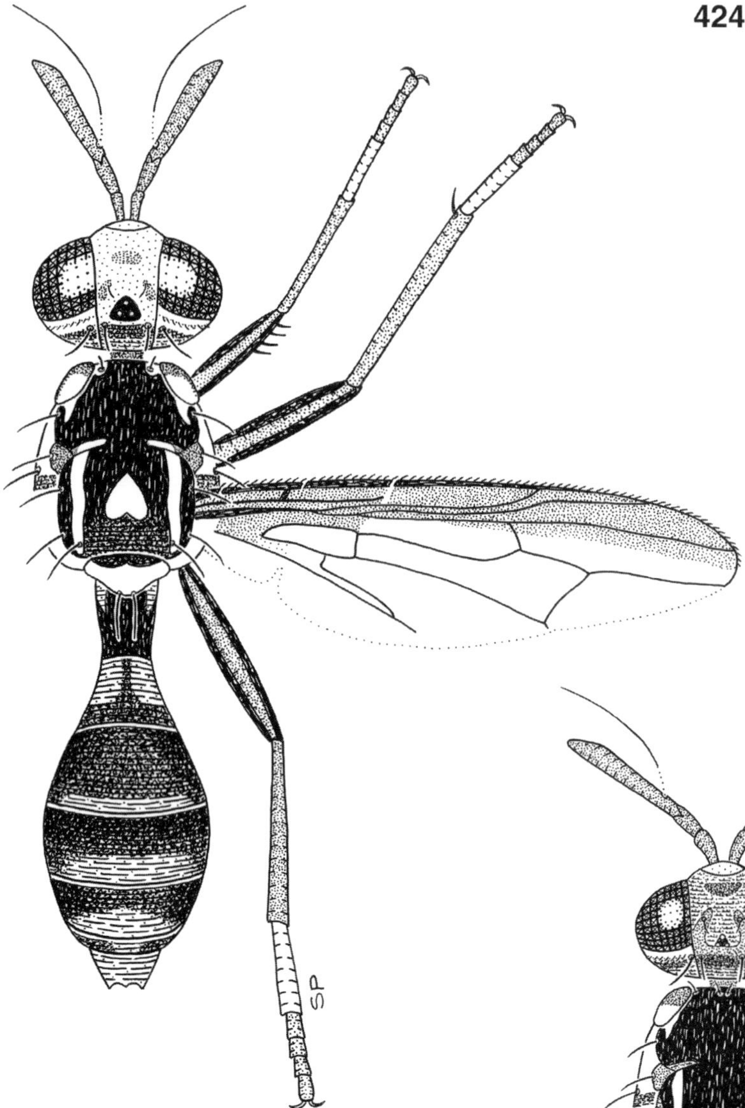

425

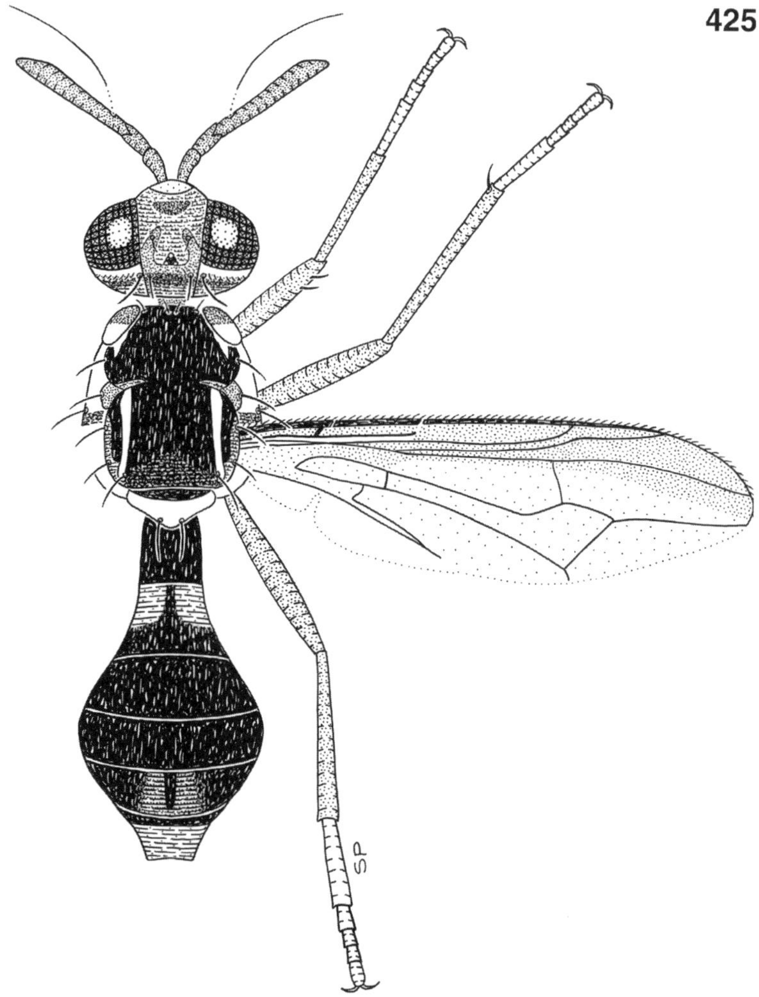

Fig. 424. *Monacrostichus citricola* Bezzi, female. **Fig. 425.** *Monacrostichus malaysiae* Drew & Hancock, paratype female.

References

Agarwal, M.L. and Kapoor, V.C. (1983) Two new species of *Dacus* Fabricius (Diptera: Tephritidae) from India. *Journal of Entomological Research* (New Delhi) 7, 169–172.

Agarwal, M.L. and Sueyoshi, M. (2005) Catalogue of Indian fruit flies (Diptera: Tephritidae). *Oriental Insects* 39, 371–433.

Agarwal, M.L., Grewal, J.S., Kapoor, V.C., Gupta, S.K. and Sharma, S.K. (1989) Three new species of fruit flies (Diptera: Tephritidae) from India. *Journal of Insect Science* 2, 88–94.

Agarwal, M.L., Grewal, J.S., Kapoor, V.C. and Rahman, O. (1992) A new species and nomenclatural status of some fruit-flies (Diptera: Tephritidae) from India. *Journal of Insect Science* 5, 20–22.

Allwood, A.J., Chinajariyawong, A., Drew, R.A.I., Hamacek, E.L., Hancock, D.L., Hengsawad, C., Jipanin, J.C., Jirasurat, M., Kong Krong, C., Kritsaneepaiboon, S., Leong, C.T.S. and Vijaysegaran, S. (1999) Host plant records for fruit flies (Diptera: Tephritidae) in South East Asia. *The Raffles Bulletin of Zoology Supplement* 7, 1–92.

Anon. (1999) *International Code of Zoological Nomenclature*, 4th edn. International Trust for Zoological Nomenclature, London, pp. 306.

Anwar Cheema, M. (1964) Taxonomic study of fruit flies of Lyallpur belonging to genus *Dacus*. *Pakistan Journal of Science* 16, 299–306.

Arakaki, N., Kuba, H. and Soemori, H. (1984) Mating behaviour of the Oriental fruit fly, *Dacus dorsalis* Hendel (Diptera: Tephritidae). *Applied Entomology and Zoology* 19, 42–51.

Baimai, V., Phinchongsakuldit, J. and Trinachartvanit, W. (1999) Metaphase karyotypes of fruit flies of Thailand (III): six members of the *Bactrocera dorsalis* complex. *Zoological Studies* 38, 110–118.

Baimai, V., Phinchongsakuldit, J. and Sumrandee, C. (2000) Cytological evidence for a complex of species within the taxon *Bactrocera tau* (Diptera: Tephritidae) in Thailand. *Biological Journal of the Linnean Society* 69, 399–409.

Becker, T. (1913) Persische Dipteren von den Expeditionen des Herrn N. Zerudny 1898 und 1901. *Annals of the Museum of St Petersburg* 17, 503–654.

Bezzi, M. (1908a) Diagnoses d'especes nouvelles de Diptères d'Afrique. *Annales de la Société Entomologique de Belgique (Liege)* 52, 374–388.

Bezzi, M. (1908b) Ditteri Eritrei. Part II. *Bollettino della Società Entomologica Italiana* 39, 3–199.

Bezzi, M. (1909) Le species dei generi *Ceratitis*, *Anastrepha* e *Dacus*. *Bollettino del Laboratorio di Zoologia Generale e Agraria della Regia Scuola Superiore d'Agricoltura, Portici* 3, 273–313.

Bezzi, M. (1913) Indian trypaneids (fruit-flies) in the collection of the Indian Museum, Calcutta. *Memoirs of the Indian Museum* 3, 53–175.

Bezzi, M. (1914a) Studies in Philippine Diptera, I. *Philippine Journal of Science (1913) (D)* 8, 305–332.

Bezzi, M. (1914b) Two new species of fruit flies from southern India. *Bulletin of Entomological Research* 5, 153–154.

Bezzi, M. (1916) On the fruit-flies of the genus *Dacus* (s.l.) occurring in India, Burma, and Ceylon. *Bulletin of Entomological Research* 7, 99–121.

Bezzi, M. (1919) Fruit flies of the genus *Dacus* sensu-latiore (Diptera) from the Philippine Islands. *Philippine Journal of Science* 15, 411–443.

Bezzi, M. (1922) Due nuovi Tripaneidi (Dipt.) infestanti frutti di *Olea* nell'Africa del Sud. *Bollettino del Laboratorio di Zoologia Generale e Agraria della Regia Scuola Superiore d'Agricoltura, Portici* 15, 293–301.

Bezzi, M. (1924) South African trypaneid Diptera in the collection of the South African Museum. *Annals of the South African Museum* 19, 449–577.

Bigot, J.M.F. (1890) New species of Indian Diptera. *Indian Museum Notes* 1, 191–192.

Chao, Y.S. (1987) The two species of fruit flies on oranges in China. *Technical Bulletin of Plant Quarantine Research* 5, 1–10.

Chao, Y.S. and Lin, X.L. (1993a) A new name of *Dacus (Zeugodacus) ater* (Diptera: Tephritidae). *Entomotaxonomia* 15, 77–78.

Chao, Y.S. and Lin, X.L. (1993b) Three new species of *Dacus* (Diptera: Tephritidae) from China. *Entomotaxonomia* 15, 137–143.

Chao, Y.S. and Lin, X.L. (1996) Notes on the genus *Sinodacus* Zia (Diptera: Tephritidae) with descriptions of six new species from China. *Entomotaxonomia* 18, 125–134.

Chen, S.H. (1940) Two new Dacinae from Szechwan. *Sinensia* 11, 131–135.

Chen, X.-L., Zhou, L., Wang, S., Li, Z. and Li, J. (2011) A new species and record of *Bactrocera* Macquart (Diptera, Tephritidae) from China. *Zootaxa* 3014, 59–64.

Chen, X.-L., Zhou, L., Han, H. and Bai, Y. (2012) New species and record of *Bactrocera* Macquart (Diptera: Tephritidae) from China and Laos. *Entomotaxonomia* 34, 351–355.

Chua, T.H. (2010) Fruit flies (Diptera: Tephritidae) from Malaysia and Brunei Darussalam: new species and records. *Florida Entomologist* 93, 483–488.

Chua, T.H. and Ooi, C.S. (1998) *Bactrocera (Sinodacus) longivittata*, a new species (Diptera: Tephritidae: Dacinae) from Peninsular Malaysia. *The Raffles Bulletin of Zoology* 46, 487–491.

Clarke, A.R., Armstrong, K.F., Carmichael, A.E., Milne, J.R., Raghu, S., Roderick, G.K. and Yeates, D.K. (2005) Invasive phytophagous pests arising through a recent tropical evolutionary radiation: the *Bactrocera dorsalis* complex of fruit flies. *Annual Review of Entomology* 50, 293–319.

Collart, A. (1935) Les Dacinae du Congo Belge (Diptera: Trypetidae). *Bulletin du Musée Royal d'Histoire Naturelle de Belgique* 11, 1–45.

Coquillett, D.W. (1899) A new trypetid from Hawaii. *Entomological News* 10, 129–130.

Coquillett, D.W. (1904) New Diptera from India and Australia. *Proceedings of the Entomological Society of Washington* 6, 137–140.

Cotes, E.C. (1893) Miscellaneous notes. *Indian Museum Notes* 3, 1–62.

Cruickshank, L., Jessup, A.J. and Cruickshank, D.J. (2001) Interspecific crosses of *Bactrocera tryoni* (Froggatt) and *Bactrocera jarvisi* (Tryon) (Diptera: Tephritidae) in the laboratory. *Australian Journal of Entomology* 40, 278–280.

David, K. and Ramani, S. (2011) An illustrated key to fruit flies (Diptera: Tephritidae) from Peninsular India and the Andaman and Nicobar Islands. *Zootaxa* 3021, 1–31.

de Meijere, J.C.H. (1911) Studien uber sudostasiatische Dipteren. VI. *Tijdschrift voor Entomologie* 54, 258–432.

de Meijere, J.C.H. (1914) Studien uber sudostasiatische Dipteren. IX (part). *Tijdschrift voor Entomologie* 57, 137–168.

de Meijere, J.C.H. (1914) Studien uber sudostasiatische Dipteren. IX (conclusion). *Tijdschrift voor Entomologie* 57, 169–275.

Doleschall, C.L. (1856) Eerste bijdrage tot de kennis der dipterologische fauna van Nederlandsch Indie. *Natuurkundig Tijdschrift voor Nederlandsch Indie* 10, 403–414.

Doleschall, C.L. (1858) Derde bijdrage tot de kennis der dipteren fauna van Nederlandsch Indie. *Natuurkundig Tijdschrift voor Nederlandsch Indie* 17, 73–128.

Drew, R.A.I. (1971) New species of Dacinae (Diptera: Trypetidae) from the South Pacific Area. *Queensland Journal of Agricultural Science* 28, 29–103.

Drew, R.A.I. (1972a) The generic and subgeneric classification of Dacini (Diptera: Tephritidae) from the South Pacific Area. *Journal of the Australian Entomological Society* 11, 1–22.

Drew, R.A.I. (1972b) Additions to the species of Dacini (Diptera: Tephritidae) from the South Pacific area with keys to species. *Journal of the Australian Entomological Society* 11, 185–231.

Drew, R.A.I. (1973) Revised descriptions of species of Dacini (Diptera: Tephritidae) from the South Pacific area. I. Genus *Callantra* and the *Dacus* group of subgenera of genus *Dacus*. *Queensland Department of Primary Industries, Division of Plant Industry Bulletin* No. 652, pp. 39.

Drew, R.A.I. (1974) Revised descriptions of species of Dacini (Diptera: Tephritidae) from the South Pacific area. II. The *Strumeta* group of subgenera of genus *Dacus*. *Queensland Department of Primary Industries, Division of Plant Industry Bulletin* No. 653, pp. 101.

Drew, R.A.I. (1979) The genus *Dacus* Fabricius (Diptera: Tephritidae) – two new species from northern Australia and a discussion of some subgenera. *Journal of the Australian Entomological Society* 18, 71–80.

Drew, R.A.I. (1982) Taxonomy. In: Drew, R.A.I., Hooper, G.H.S. and Bateman, M.A. (eds) *Economic Fruit Flies of the South Pacific Region*. Queensland Department of Primary Industries, Brisbane, Australia, pp. 1–97.

Drew, R.A.I. (1989) The tropical fruit flies (Diptera: Tephritidae: Dacinae) of the Australasian and Oceanian Regions. *Memoirs of the Queensland Museum* 26, 1–521.

Drew, R.A.I. (2004) Biogeography and speciation in the Dacini (Diptera: Tephritidae: Dacinae). In: Evenhuis, N.L. and Kaneshiro, K.Y. (eds) *D. Elmo Hardy Memorial Volume. Contributions to the Systematics and Evolution of Diptera. Bishop Museum Bulletin in Entomology* 12, 165–178.

Drew, R.A.I. and Hancock, D.L. (1994a) The *Bactrocera dorsalis* complex of fruit flies (Diptera: Tephritidae: Dacinae) in Asia. *Bulletin of Entomological Research* Suppl. 2, iii + pp. 68.

Drew, R.A.I. and Hancock, D.L. (1994b) Revision of the tropical fruit flies (Diptera: Tephritidae: Dacinae) of South-east Asia. I. *Ichneumonopsis* Hardy and *Monacrostichus* Bezzi. *Invertebrate Taxonomy* 8, 829–838.

Drew, R.A.I. and Hancock, D.L. (1995) New species, subgenus and records of *Bactrocera* Macquart from the South Pacific (Diptera: Tephritidae: Dacinae). *Journal of the Australian Entomological Society* 34, 7–11.

Drew, R.A.I. and Hancock, D.L. (2000) Synonymy, geographic distributions, lectotype designations and type depositories of some Australian and South Pacific Dacinae (Diptera: Tephritidae). *Australian Entomologist* 27, 27–30.

Drew, R.A.I. and Hardy, D.E. (1981) *Dacus (Bactrocera) opiliae*, a new sibling species of the *dorsalis* complex of fruit flies from northern Australia (Diptera: Tephritidae). *Journal of the Australian Entomological Society* 20, 131–137.

Drew, R.A.I. and Raghu, S. (2002) The fruit fly fauna (Diptera: Tephritidae: Dacinae) of the rainforest habitat of the Western Ghats, India. *The Raffles Bulletin of Zoology* 50, 327–352.

Drew, R.A.I. and Romig, M.C. (2001) The fruit fly fauna (Diptera: Tephritidae: Dacinae) of Bougainville, the Solomon Islands and Vanuatu. *Australian Journal of Entomology* 40, 113–150.

Drew, R.A.I., Hancock, D.L. and White, I.M. (1998) Revision of the tropical fruit flies (Diptera: Tephritidae: Dacinae) of South East Asia. II. *Dacus* Fabricius. *Invertebrate Taxonomy* 12, 567–654.

Drew, R.A.I., Tsuruta, K. and White, I.M. (2005) A new species of pest fruit fly (Diptera: Tephritidae: Dacinae) from Sri Lanka and Africa. *African Entomology* 13, 149–154.

Drew, R.A.I., Romig, M.C. and Dorji, C. (2007) Records of Dacine fruit flies and new species of *Dacus* (Diptera: Tephritidae) in Bhutan. *The Raffles Bulletin of Zoology* 55, 1–21.

Drew, R.A.I., Raghu, S. and Halcoop, P. (2008) Bridging the morphological and biological species concepts: studies on the *Bactrocera dorsalis* (Hendel) complex (Diptera: Tephritidae: Dacinae) in South-east Asia. *Biological Journal of the Linnean Society* 93, 217–226.

Drew, R.A.I., Ma Jing, Smith, S. and Hughes, J.M. (2011) The taxonomy and phylogenetic relationships of species in the *Bactrocera musae* complex of fruit flies (Diptera: Tephritidae: Dacinae) in Papua New Guinea. *The Raffles Bulletin of Zoology* 59, 145–162.

Enderlein, G. (1920) Zur Kenntnis tropischer Frucht-Bohrfliegen. *Zoologische Jahrbüecher. Abteilung für Systematik, Geographie und Biologie der Tiere* 43, 336–360.

Fabricius, J.C. (1794) *Entomologia systematica emendata et aucta. Secundum classes, ordines, genera, species, adiectis, synonymis, locis, observationibus, descriptionibus.* Tome 4, C.G. Proft, Hafniae [= Copenhagen], pp. 472.

Fabricius, J.C. (1805) *Systema antliatorum secundum ordines, genera, species, adiectis, synonymis, locis, observationibus, descriptionibus.* Reichard, Brunsvigae [= Brunswick], pp. 373.

Fletcher, M.T. and Kitching, W. (1995) Chemistry of fruit flies. *Chemical Reviews* 95, 789–828.

Froggatt, W.W. (1909) Part III. Fruit flies. A general account of the flies belonging to the family Trypetidae, that damage sound fruit, with descriptions of the different species (some described as new) and their habits, range, and suggestions for destroying them. In: Official report on fruit fly and other pests in various countries 1907–1908. Report on parasitic and injurious insects. New South Wales Department of Agriculture, Sydney, Australia, pp. 73–115. Reprinted in 1910 in *Farmers Bulletin of New South Wales Department of Agriculture* 24, 1–56.

Gmelin, J.F. (1790) Caroli a Linné, Systema naturae per regna tria naturae secundum classes, ordines, genera, species; cum caracteribus, differentiis, synonymis, locis. *Edition* 13, 2225–3020.

Hancock, D.L. and Drew, R.A.I. (2006) A revised classification of subgenera and species groups in *Dacus* Fabricius (Diptera: Tephritidae). *Instrumenta Biodiversitatis* VII, 167–205.

Hancock, D.L., Hamacek, E.L., Lloyd, A.C. and Elson-Harris, M.M. (2000) The distribution and host plants of fruit flies (Diptera: Tephritidae) in Australia. Queensland Department of Primary Industries Information Series QI99067, 1–75.

Hardy, D.E. (1951) The Krauss collection of Australian fruit flies (Tephritidae – Diptera). *Pacific Science* 5, 115–189.

Hardy, D.E. (1954) The *Dacus* subgenera *Neodacus* and *Gymnodacus* of the world (Diptera, Tephritidae). *Proceedings of the Entomological Society of Washington* 56, 5–23.

Hardy, D.E. (1955a) The *Dacus (Afrodacus)* Bezzi of the world (Tephritidae, Diptera). *Journal of the Kansas Entomological Society* 28, 3–15.

Hardy, D.E. (1955b) A reclassification of the Dacini (Tephritidae-Diptera). *Annals of the Entomological Society of America* 48, 425–437.

Hardy, D.E. (1959) The Walker types of fruit flies (Tephritidae-Diptera) in the British Museum collection. *Bulletin of the British Museum (Natural History) Entomology* 8, 159–242.

Hardy, D.E. (1964) Diptera from Nepal. The fruit flies (Diptera: Tephritidae). *Bulletin of the British Museum (Natural History) Entomology* 15, 147–169.

Hardy, D.E. (1968) The fruit fly types in the Naturhistorisches Museum, Wien (Tephritidae-Diptera). *Annalen des Naturhistorischen Museum in Wien* 72, 107–155.

Hardy, D.E. (1969a) Taxonomy and distribution of the oriental fruit fly and related species (Tephritidae-Diptera). *Proceedings of the Hawaiian Entomological Society* 20, 395–428.

Hardy, D.E. (1969b) Lectotype designations for fruit flies (Diptera: Tephritidae). *Pacific Insects* 11, 477–481.

Hardy, D.E. (1970) Tephritidae (Diptera) collected by the Noona Dan Expedition in the Philippine and Bismarck islands. *Entomologiske Meddelelser* 38, 71–136.

Hardy, D.E. (1973) The fruit flies (Tephritidae-Diptera) of Thailand and bordering countries. *Pacific Insects Monograph* 31, pp. 353.

Hardy, D.E. (1974) The fruit flies of the Philippines (Diptera: Tephritidae). *Pacific Insects Monograph* 32, pp. 266.

Hardy, D.E. (1977) Family Tephritidae (Trypetidae, Trupaneidae). In: Delfinado, M.D. and Hardy, D.E. (eds) *A Catalog of the Diptera of the Oriental Region, Vol. III, Suborder Cyclorrhapha (excluding Division Aschiza).* University of Hawaii Press, Honolulu, Hawaii, pp. 44–134.

Hardy, D.E. (1981) Clarification of status of *Strumeta persignata* Hering (Diptera: Tephritidae). *Proceedings of the Hawaiian Entomological Society* 23, 355–357.

Hardy, D.E. (1982) The Dacini of Sulawesi (Diptera: Tephritidae). *Treubia* 28, 173–241.

Hardy, D.E. (1983) The fruit flies of the genus *Dacus* Fabricius of Java, Sumatra and Lombok, Indonesia (Diptera: Tephritidae). *Treubia* 29, 1–45.

Hardy, D.E. (1984) *Dacus (Zeugodacus) persignatus* (Hering). Correct name for *goughi* Hardy. *Treubia* 30, 81.

Hardy, D.E. and Adachi, M.S. (1954) Studies in the fruit flies of the Philippine Islands, Indonesia and Malaya, Part 1. Dacini (Tephritidae-Diptera). *Pacific Science* 8, 147–204.

Hardy, D.E. and Adachi, M.S. (1956) Insects of Micronesia, Diptera: Tephritidae. *Bishop Museum Bulletin* 14, 1–28.

Hardy, D.E. and Foote, R.H. (1989) Family Tephritidae. In: Evenhuis, N.L. (ed.) *Catalog of the Diptera of the Australasian and Oceanian Regions*. Bishop Museum Special Publication 86, Bishop Museum Press (Honolulu) and E.J. Brill (Leiden, The Netherlands), pp. 502–531.

Hendel, F.G. (1912) H. Sauter's Formosa-Ausbeute. Genus *Dacus*, Fabricius (1805) (Diptera). *Supplementa Entomologica* 1, 13–24.

Hendel, F.G. (1915) H. Sauter's Formosa-Ausbeute. Tephritinae. *Annales Musei Nationalis Hungarici* 13, 424–467.

Hendel, F.G. (1927a) 49. Trypetidae. In: Lindner, E. (ed.) *Die Fliegen der palaearktischen Region, Vol. 5.* Stuttgart, pp. 1–221.

Hendel, F.G. (1927b) Einige neue Bohrfliegen (Trypetidae) aus dem Hamburger Museum. *Wiener Entomologische Zeitung* 44, 58–65.

Hendel, F.G. (1928) Neue oder weniger bekannte Bohrfliegen (Trypetidae) meist aus dem Deutschen Entomologischen Institut Berlin-Dahlem. *Entomologische Mitteilungen* 17, 341–370.

Hendel, F.G. (1934) Schwedisch-chinesische wissenschaftliche Expedition nach den nordwestlichen Provinzen Chinas, unter Leitung von Dr. Sven Hedin und Prof. Su Ping-chang. Insekten gesammelt vom schwedischen Arzt der Expedition Dr. David Hummel 1927–1930. 13. Diptera. – 5. Muscaria holometopa. *Arkiv foer Zoologi* 25, 1–18.

Hering, E.M. (1938) Entomological results from the Swedish Expedition 1934 to Burma and British India. Diptera: Fam. Trypetidae. (23. Beitrag zur Kenntnis der Trypetidae). *Arkiv foer Zoologi* 30A(25), pp. 56.

Hering, E.M. (1939) Neue Trypetiden der Erde. (25. Beitrag zur Kenntnis der Trypetidae). *Verhandlungen. Siebenter Internationaler Kongress fuer Entomologie, Berlin* 1, 165–190.

Hering, E.M. (1941a) Fruchtfliegen von Neu-Guinea (Dipt.). I. *Annales Musei Nationalis Hungarici. Pars Zoologica* 34, 45–53.

Hering, E.M. (1941b) Fruchtfliegen von Neu-Guinea (Dipt.). II. (38. Beitrag zur Kenntnis der Trypetidae). *Annales Musei Nationalis Hungarici. Pars Zoologica* 34, 54–65.

Hering, E.M. (1941c) Entomological results from the Swedish Expedition 1934 to Burma and British India. Diptera: Trypetidae. (30. Beitrag zur Kenntnis der Trypetidae). *Arkiv foer Zoologi* 33B(11), pp. 7.

Hering, E.M. (1941d) Neue Dacinae und Trypetinae des Zoologischen Museums der Universitat Berlin. *Siruna Seva* 3, 1–25.

Hering, E.M. (1941e) Dipteren von den Kleinen Sunda-Inseln. Aus der Ausbeute der Sunda-Expedition Rensch. II. Trypetidae. (34. Beitrag zur Kenntnis der Trypetidae). *Arbeiten ueber Morphologische und Taxonomische Entomologie aus Berlin-Dahlem (Berlin-Dahlem)* 8, 24–45.

Hering, E.M. (1942) Neue Gattungen und Arten von Fruchtfliegen aus dem Zoologischen Museum der Universitat Berlin (Diptera). (41. Beitrag zur Kenntnis der Trypetidae). *Mitteilungen aus dem Zoologischen Museum in Berlin* 25, 274–291.

Hering, E.M. (1952a) Trypetidae (Diptera) von Sumba und Flores. (46. Beitrag zur Kenntis der Trypetidae). *Verhandlungen der Naturforschenden Gesellschaft in Basel (Basel)* 63, 41–48.

Hering, E.M. (1952b) Fruchtfliegen (Trypetidae) von Indonesien (Diptera). (47. Beitrag zur Kenntnis der Trypetidae). *Treubia* 21, 263–290.

Hering, E.M. (1953) Trypetidae (Diptera) von Sumba. (49. Beitrag zur Kenntis der Trypetidae). *Verhandlungen der Naturforschenden Gesellschaft in Basel (Basel)* 64, 74–80.

Hering, E.M. (1956) Trypetidae (Diptera) von Ceylon. (53. Beitrag zur Kenntnis der Trypetidae). *Verharndlungen der Naturforschenden Gesellschaft in Basel (Basel)* 67, 62–74.

Ito, S. (1983) Lieferung 1. In: *Die Japanischen Bohrfliegen.* Maruzen Co., Ltd, Osaka, Japan, pp. 1–48.

Kapoor, V.C. (1971) Four new species of fruit flies (Tephritidae) from India. *Oriental Insects* 5, 477–482.

Kapoor, V.C. (1993) *Indian Fruit Flies (Insecta: Diptera: Tephritidae).* International Science Publisher, New York, pp. 228.

Kapoor, V.C. and Katiyar, K.N. (1969) New record of *Melanodacus* Perkins, a subgenus of *Dacus* Fabricius, with description of its new species (Dacinae: Tephritidae). *Bulletin of Entomology* 10, 123–125.

Kapoor, V.C. and Katiyar, K.N. (1970) New record of *Paradacus* (Perkins), a subgenus of *Dacus* Fabricius, with description of its new species (Dacinae: Tephritidae). *The Entomologist* 103, 252–254.

Kapoor, V.C., Hardy, D.E., Agarwal, M.L. and Grewal, J.S. (1980) *Fruit Fly Systematics of the Indian Subcontinent.* Export India Publications, Jullender, India, pp. 113.

Kovac, D., Freidberg, A. and Steck, G.J. (2013) Biology and description of the third instar larva and puparium of *Ichneumonopsis burmensis* Hardy (Diptera: Tephritidae: Dacinae: Gastrozonini), a bamboo-breeding fruit fly from the Oriental Region. *The Raffles Bulletin of Zoology* 61, 117–132.

Kuba, H. and Sokei, Y. (1988) The production of pheromone clouds by spraying in the melon fly, *Dacus cucurbitae* Coquillett (Diptera: Tephritidae). *Journal of Ethology* 6, 105–110.

Leblanc, L., Tora Vueti, E., Drew, R.A.I. and Allwood, A.J. (2012) Host plant records for fruit flies (Diptera: Tephritidae: Dacini) in the Pacific Islands. *Proceedings of the Hawaiian Entomological Society* 44, 11–53.

Liang, G.-Q., Hancock, D.L., Xu, W. and Liang, F. (1993) Notes on the Dacinae of southern China (Diptera: Tephritidae). *Journal of the Australian Entomological Society* 32, 137–140.

Lin, M.-G., Wang, X.-J., Li, W.-D., Xu, W. and Chen, X.-L. (2005) Taxonomic revision of the genus *Bactrocera* Macquart from Hainan, with descriptions of two new species (Diptera, Tephritidae, Dacinae). *Acta Zootaxonomica Sinica* 30, 842–847.

Lin, M.-G., Yang, Z.-J., Wang, X.-J., Li, J.-Y. and Li, W.-D. (2006) A taxonomic study of the subfamily Dacinae (Diptera: Tephritidae) from Hainan, China. *Acta Entomologica Sinica* 49, 310–314.

Lin, M.-G., Wang, X.-J. and Zeng, L. (2011a) A new species of the genus *Dacus* Fabricius (Diptera, Tephritidae, Dacinae) from Hainan, China. *Acta Zootaxonomica Sinica* 36, 620–621.

Lin, M.-G., Wang, X.-J. and Zeng, L. (2011b) Three new species of the genus *Bactrocera* Macquart (Diptera, Tephritidae, Dacinae) from Hainan, China. *Acta Zootaxonomica Sinica* 36, 896–900.

Liu, X.-F., Wang, X.-J. and Ye, H. (2009) One newly recorded species of the genus *Bactrocera* Macquart (Diptera, Tephritidae) from China. *Acta Zootaxonomica Sinica* 34, 661–663.

Loew, H. (1862) Bidrag till kännedomen om Afrikas Diptera. *Öfversigt af Finska Vetenskaps-Akademiens Forhandlingar, Stockholm* 19, 3–14.

Macquart, J.P.M. (1835) Histoire naturelle des Insectes. Dipteres. Tome deuxieme. Ouvrage accompagne de planches, IV + pp. 710. In: Roret, N.E. (ed.) *Nouvelles suites a Buffon, formant, avec les oeuvres de cet auteur, un cours complet d'histoire naturelle. Collection accompagnee de planches*. Roret, Paris. 82 vols. + 11 atlases.

Macquart, J.P.M. (1843) Dipteres exotiques nouveaux ou peu connus [2(3)]. *Memoires de la Societe Royale des Sciences, de L'Agriculture et des Arts de Lille* 1842, 162–460.

Mahmood, K. (1999) Taxonomy of the *Bactrocera (Zeugodacus) tau* (Tephritidae: Diptera) complex in Asia. *Pakistan Journal of Zoology* 31, 219–235.

Mahmood, K. (2005) A new species of fruit flies of genus *Bactrocera* (Dacinae, Tephritidae: Diptera) from Pakistan. *Pakistan Journal of Zoology* 37, 161–162.

Mahmood, K. & Hasan, S.A. (2005) Revision of subfamily Dacinae (fruit flies) (Tephritidae: Diptera) of South-East Asia. *Pakistan Journal of Zoology* Suppl. 7, 1–65.

Malloch, J.R. (1931) Trypetidae. In: *Insects of Samoa and other Samoan Terrestrial Arthropoda. Part VI. Diptera. Fasc. 7.*, pp. 253–266.

Malloch, J.R. (1938) Trypetidae of the Mangarevan Expedition. *Occasional Papers of the Bernice Pauahi Bishop Museum* 14, 111–116.

Malloch, J.R. (1939a) Solomon Islands Trypetidae. XVI. *Annals and Magazine of Natural History* 11, 228–278.

Malloch, J.R. (1939b) The Diptera of the Territory of New Guinea. XI. Family Trypetidae. *Proceedings of the Linnean Society of New South Wales* 64, 409–465.

Matsumura, S. (1916) *Thousand insects of Japan. Additamenta.* In: *Vol. 2 (Diptera).* Keisel–sha, Tokyo, pp. 413–424.

May, A.W.S. (1951) New genera and species of Dacinae (Tephritidae, Diptera) from Queensland. *Queensland Journal of Agricultural Science* 8, 5–13.

May, A.W.S. (1953) Queensland host records for the Dacinae (fam. Trypetidae). *Queensland Journal of Agricultural Science* 10, 36–79.

May, A.W.S. (1963) An investigation of fruit flies (Trypetidae: Diptera) in Queensland. 1. Introduction, species, pest status and distribution. *Queensland Journal of Agricultural and Animal Sciences* 20, 1–82.

McAlpine, J.F. (1981) Morphology and terminology – adults. In: McAlpine, J.F., Peterson, B.V., Shewell, G.E., Teskey, H.J., Vockeroth, J.R. and Wood, D.M. (eds) *Manual of Nearctic Diptera.* Monograph No. 27, Vol. 1. Ottawa, Canadian Government Publishing Centre, pp. 9–63.

Merz, B. (2011) Order Diptera, family Tephritidae, additions and the description of a new species. *Arthropod Fauna of the UAE* 4, 769–779.

Miyake, T. (1919) Studies on the fruit flies of Japan. *Bulletin of the Imperial Central Agriculture Experiment Station Japan* 2, 85–165.

Munro, H.K. (1932) Notes on *Dacus ciliatus* Lw and certain related species (Dipt. Trypetidae). *Stylops* 1, 151–158.

Munro, H.K. (1935) Records of Indian Trypetidae (Diptera) with descriptions of some apparently new species. *Records of the Indian Museum* 37, 15–27.

Munro, H.K. (1938) New genera of African Trypetidae (Diptera). *Proceedings of the Royal Society of London, Series B. Taxonomy* 7, 117–120.

Munro, H.K. (1984) A taxonomic treatise on the Dacidae (Tephritoidea, Diptera) of Africa. *Entomology Memoirs, Department of Agriculture and Water Supply, Republic of South Africa* 61, 1–313.

Nichols, S.W. (ed.) (1989) *The Torre-Bueno Glossary of Entomology*, revised edition. The New York Entomological Society, New York, New York, pp. 840.

Norrbom, A.L., Carroll, L.E., Thompson, F.C., White, I.M. and Freidberg, A. (1998) Systematic database of names. In: Thompson, F.C. (ed.) *Fruit Fly Expert Identification System and Systematic Information Database. Myia* 9, 65–251.

Perkins, F.A. (1937) Studies in Australian and Oriental Trypaneidae. Part I. New genera of Dacinae. *Proceedings of the Royal Society of Queensland* 48, 51–60.

Perkins, F.A. (1938a) Studies in Oriental and Australian Trypaneidae. – Part II. Adraminae and Dacinae from India, Ceylon, Malaya, Sumatra, Java, Borneo, Philippine Islands, and Formosa. *Proceedings of the Royal Society of Queensland* 49, 120–144.

Perkins, F.A. (1938b) Results of the Oxford University Expedition to Sarawak (Borneo), 1932. Diptera Trypaneidae. *Annals and Magazine of Natural History* 11, 401–409.

Perkins, F.A. (1939) Studies in Oriental and Australian Trypetidae. Part 3: Adraminae and Dacinae from New Guinea, Celebes, Aru Island, and Pacific Islands. *Papers of the University of Queensland Department of Biology* 1, pp. 35.

Perkins, F.A. and May, A.W.S. (1949) Studies in Australian and Oriental Trypetidae. *Papers of the University of Queensland Department of Biology* 2, 1–21.

Philip, A. (1950) Description of one new species of *Strumeta* Walker (Trypetidae: Diptera) from Burma and a record of one far-eastern species of the genus from India. *Indian Journal of Entomology* 10, 31–32.

Premlata and Singh, A. (1987) A new species of genus *Dacus* Fabricius (Tephritidae: Diptera) from India. *Journal of the Bombay Natural History Society* 84, 401–404.

Radhakrishnan, C. (1999) A new species of *Bactrocera* Macquart (Diptera: Tephritidae: Dacinae) from Southern India. *Records of the Zoological Survey of India* 97, 1–4.

Ranganath, H.R. and Veenakumari, K. (1995) Notes on the Dacine fruit flies (Diptera: Tephritidae) of Andaman and Nicobar Islands. *The Raffles Bulletin of Zoology* 43, 235–238.

Ranganath, H.R. and Veenakumari, K. (1996) Some new records of fruit flies (Diptera – Tephritidae) from the Andaman and Nicobar Islands. *Entomon* 21, 95–97.

Saunders, W.W. (1842) Descriptions of four new dipterous insects from central and northern India. *Transactions of the Entomological Society of London* 3, 59–61.

Schiner, J.R. (1868) Zoologischer Theil, 2. Band, 1. Abtheilung, B. Diptera. In: *Reise der Österreichischen Fregatte Novara um die Erde*. B.K. Gerold's Sohn, Vienna, pp. 388.

Schutze, M.K., Jessup, A. and Clarke, A.R. (2012) Wing shape as a potential discriminator of morphologically similar pest taxa within the *Bactrocera dorsalis* species complex (Diptera: Tephritidae). *Bulletin of Entomological Research* 102, 103–111.

Scopoli, J.A. (1763) Entomologia carniolica exhibens insecta carnioliae indigena et distributa in ordines, genera, species, varietates. *Methodo Linnaeana.* Johannis Thomae Trattner, Vindobonae [= Vienna], pp. 452.

Senior-White, R.A. (1922) Notes on Indian Diptera 1. Diptera from the Khasia Hills. 2. Tabanidae in the collection of the forest zoologist. 3. New species of Diptera from the Indian Region. *Memoirs of the Department of Agriculture in India, Entomological Series* 7, 83–170.

Shiraki, T. (1933) A systematic study of Trypetidae in the Japanese Empire. *Memoirs of the Faculty of Science and Agriculture, Taihoku Imperial University* 8 (Entomol. 2), pp. 509.

Shiraki, T. (1939) An interesting fruit-fly. *Zoological Magazine, Tokyo* 51, 410–411.

Shiraki, T. (1968) Fruit flies of the Ryukyu Islands (Diptera: Tephritidae). *U.S. National Museum Bulletin* 263, pp. 104.

SPC (2001) Fruit Flies in Papua New Guinea. Pest Advisory Leaflet No. 37, SPC, Noumea, New Caledonia.

Speiser, P.G.E. (1924) Eine Ubersicht uber die Dipteren-fauna Deutsch-Ostafrikas. In: *Beitrage aus der Tierkunde*. Grafe and Unzer, Konigsburg, Germany, pp. 90–156.

Tryon, H. (1927) Queensland fruit flies (Trypetidae), Series I. *Proceedings of the Royal Society of Queensland* 38, 176–224.

Tseng, Y.H. and Chu, Y.I. (1982) A new fruit fly from Taiwan (Diptera: Tephritidae). *Chinese Journal of Entomology* 2, 85–90.

Tseng, Y.H. and Chu, Y.I. (1983) A new fruit fly *Callantra formosana* from Taiwan (Tephritidae, Diptera). *Chinese Journal of Entomology* 3, 119–122.

Tseng, Y.H., Chen, C.C. and Chu, Y.I. (1992) The fruit flies. Genus *Dacus* Fabricius of Taiwan (Diptera: Tephritidae). *Journal of Taiwan Museum* 45, 15–91.

Tsuruta, K. and White, I.M. (2001) Eleven new species of the genus *Bactrocera* Macquart (Diptera: Tephritidae) from Sri Lanka. *Entomological Science* 4, 69–87.

Tsuruta, K., White, I.M., Bandara, H.M.J., Rajapakse, H., Sundaraperuma, S.A.H., Kahawatta, S.B.M.U.C. and Rajapakse, G.B.J.P. (1997) A preliminary note on the hosts of fruit flies of the tribe Dacini (Diptera, Tephritidae) in Sri Lanka. *Esakia* 37, 149–160.

Walker, F. (1849) List of the specimens of dipterous insects in the collection of the British Museum. Part IV. British Museum (Natural History), London, 689–1172.

Walker, F. (1856a) Catalogue of the dipterous insects collected at Singapore and Malacca by Mr. A.R. Wallace, with descriptions of new species. *Journal of Proceedings of the Linnean Society. Zoology. London* 1, 4–39.

Walker, F. (1856b) Catalogue of the dipterous insects collected at Sarawak, Borneo by Mr. A.R. Wallace, with descriptions of new species. *Journal of Proceedings of the Linnean Society. Zoology. London* 1, 105–136.

Walker, F. (1859) Catalogue of the dipterous insects collected in the Aru Islands by Mr. A.R. Wallace, with descriptions of new species. *Journal of Proceedings of the Linnean Society. Zoology. London* 3, 111–131.

Walker, F. (1860a) Catalogue of the dipterous insects collected at Makessar in Celebes, by Mr. A.R. Wallace, with descriptions of new species [conclusion]. *Journal of Proceedings of the Linnean Society. Zoology. London* 4, 90–172.

Walker, F. (1860b) Catalogue of the dipterous insects collected in Amboyna by Mr. A.R. Wallace, with descriptions of new species. *Journal of Proceedings of the Linnean Society. Zoology. London* 5, 144–168.

Walker, F. (1861a) Catalogue of the dipterous insects collected at Manado in Celebes, and in Tond, by Mr. A.R. Wallace, with descriptions of new species. *Journal of Proceedings of the Linnean Society. Zoology. London* 5, 258–270.

Walker, F. (1861b) Catalogue of the dipterous insects collected in Batchian, Kaisaa and Makian, and at Tidon in Celebes, by Mr. A.R. Wallace, with descriptions of new species. *Journal of Proceedings of the Linnean Society. Zoology. London* 5, 270–303.

Walker, F. (1861c) Catalogue of the dipterous insects collected at Gilolo, Ternate, and Ceram, by Mr. A.R. Wallace, with descriptions of new species. *Journal of Proceedings of the Linnean Society. Zoology. London* 6, 4–23.

Walker, F. (1861d) Characters of undescribed Diptera in the collection of W.W. Saunders, Esq., F.R.S. &c. *Transactions of the Entomological Society of London* 5, 297–334.

Walker, F. (1864) Catalogue of the dipterous insects collected in Waigiou, Mysol, and North Ceram, by Mr. A.R. Wallace, with descriptions of new species. *Journal of Proceedings of the Linnean Society. Zoology. London* 7, 202–238.

Walker, F. (1865) Descriptions of new species of the dipterous insects of New Guinea. *Journal of Proceedings of the Linnean Society. Zoology. London* 8, 109–130.

Walker, F. (1866) Synopsis of the Diptera of the Eastern Archipelago discovered by Mr Wallace, and noticed in the 'Journal of the Linnean Society'. *Journal of Proceedings of the Linnean Society. Zoology. London* 9, 1–30.

Wang, X.-J. (1988) A study on Chinese *Sinodacus* Zia with description of new species (Diptera: Tephritidae). *Sinozoologia* 6, 292–296.

Wang, X.-J. (1990) Notes on six new species of the genus *Callantra* from China (Diptera: Tephritidae). *Acta Zootaxonomica Sinica* 15, 67–76.

Wang, X.-J. (1996) The fruit flies (Diptera: Tephritidae) of the East Asian region. *Acta Zootaxonomica Sinica* 21, 1–338.

Wang, X.-J. and Chen, X.-L. (2002) A revision of the genus *Dacus* Fabricius from China. *Acta Zootaxonomica Sinica* 27, 631–636.

Wang, X.-J. and Zhao, M.-Z. (1989) Notes on the genus *Dacus* Fabricius in China with descriptions of five new species (Diptera: Tephritidae). *Acta Zootaxonomica Sinica* 14, 209–219.

Wang, X.-J., Xiao, S., Chen, X.-L., Long, R. and Zhang, C.-L. (2008) Two new species of the genus *Bactrocera* Macquart (Diptera, Tephritidae) from Yunnan, China. *Acta Zootaxonomica Sinica* 33, 73–76.

Weyenbergh Jr, H. (1869) Deux dipteres nouveaux de l'Archipel des Indes Orientales: *Dacus ritsemae*, *Lucilia leonardi. Archives Neerlandaises des Sciences Exactes et Naturelles* 4, 359–362.

White, I.M. (2006) Taxonomy of the Dacina (Diptera: Tephritidae) of Africa and the Middle East. *African Entomology* Memoir 2, pp. 156.

White, I.M. and Elson-Harris, M.M. (1992) *Fruit Flies of Economic Significance; their Identification and Bionomics.* CAB International, Wallingford, UK, pp. 601.

White, I.M. and Evenhuis, N.L. (1999) New species and records of Indo-Australasian Dacini (Diptera: Tephritidae). *The Raffles Bulletin of Zoology* 47, 487–540.

White, I.M. and Hancock, D.L. (1997) CABIKEY to the Dacini (Diptera: Tephritidae) of the Asia-Pacific-Australasian Regions. CAB International, Wallingford, UK, CD-ROM.

White, I.M. and Liquido, N.J. (1995) *Chaetodacus latifrons* Hendel, 1915 (currently *Bactrocera latifrons*; Insecta, Diptera): Proposed precedence of the specific name over that of *Dacus parvulus* Hendel, 1912. *Bulletin of Zoological Nomenclature* 52, 250–253.

White, I.M. and Wang, X.-J. (1992) Taxonomic notes on some dacine (Diptera: Tephritidae) fruit flies associated with citrus, olives, and cucurbits. *Bulletin of Entomological Research* 82, 275–279.

White, I.M., Headrick, D.H., Norrbom, A.L. and Carroll, L.E. (2000) Glossary. In: Aluja, M. and Norrbom, A.L. (eds) *Fruit Flies (Tephritidae): Phylogeny and Evolution of Behavior.* CRC Press, Boca Raton, Florida, pp. 881–924.

Wiedemann, C.R.W. (1819) Beschreibung neuer Zweiflugler aus Ostindien und Afrika. *Zoological Magazine (Wiedemann's)* 1, 1–39.

Wiedemann, C.R.W. (1830) *Aussereuropaische Zweiflugelige Insekten. Vol. 2.* Schulz, Ha mm, Germany, xii + pp. 684.

Yong, H.S. (1994) The gnemon fruit fly. *Nature Malaysiana* 19, 37–40.

Yu, H., Bai, Y.-H., He, W.-Z., Liu, Z.-S. and Yang, D. (2011) Three new species of the subgenus *Zeugodacus* from Yunnan, China (Diptera, Tephritidae). *Acta Zootaxonomica Sinica* 36, 315–320.

Yu, H., Bai, Y.-H., He, W.-Z. and Chen, N.-Z. (2011) Two new species and one new record species of the genus *Bactrocera* from Yunnan, China (Diptera, Tephritidae). *Acta Zootaxonomica Sinica* 36, 604–609.

Zaka-ur-Rab, M. (1961) A new species of the genus *Callantra* Walker from India (Diptera: Trypetidae). *Journal of the Bombay Natural History Society* 58, 538–541.

Zhang, N.-N., Chen, J.-H. and Gao, S.-D. (2011) A new species of subgenus *Zeugodacus* Hendel (Diptera, Tephritidae) from China. *Acta Zootaxonomica Sinica* 36, 321–324.

Zhang, N.-N., Ji, Q.-E. and Chen, J.-H. (2011) Three new species and one new record of genus *Bactrocera* Macquart (Diptera, Tephritidae) from Yunnan, China. *Acta Zootaxonomica Sinica* 36, 598–603.

Zhang, N.-N., Ji, Q.-E. and Chen, J.-H. (2012) A new species and of the genus *Bactrocera* Macquart from China (Diptera, Tephritidae). *Acta Zootaxonomica Sinica* 37, 206–208.

Zia, Y. (1936) New Diptera Trypetidae from Hainan. *Chinese Journal of Zoology* 2, 157–161.

Zia, Y. (1937) Study on the Trypetidae or fruit-flies of China. *Sinensia* 8, 103–226.

Zia, Y. (1939) Notes on Trypetidae collected from South China. *Sinensia* 10, 1–19.

Zia, Y. (1955) On three new species of trypetid flies from South China and Viet-Nam. *Acta Zoologica Sinica* 7, 63–68.

Zia, Y. (1963) Two new species of Trypetidae from Yunnan. *Acta Zoologica Sinica* 15, 457–460.

Taxonomic Index

abbreviata (Hardy) 19, **243**, 245, 247, 248, **546** (Fig. 251)
abdoangusta (Drew) 6, 20, **263**, 264, 350, **557** (Fig. 272)
abdonigella (Drew) 17, **36**, 37, 44, 60, **418** (Fig. 15)
aberrans ssp. *nigritus* Hardy **209**, 210, **527** (Fig. 216)
aberrans ssp. *pallescentis* Hardy **210**, 211, **527** (Fig. 217)
ablepharus (Bezzi) 19, **225**, 226, 228, 231, 235, **536** (Fig. 233A)
ablepharus var. *mindanaus* Bezzi, Synonym **225**, 226, **356** (Fig. 233B)
abnormis (Hardy) 20, **264**, 279, 306, 330, 364, **558** (Fig. 273)
absoluta (Walker) 17, **29**, 30, 33, **414** (Fig. 8)
absonus Hering, **New Synonym** 1, **204**, 205, **524** (Fig. 211B)
aculeus Hardy, **New Synonym** 1, **337**, 338
adustus Wang & Zhao, **New Synonym** 1, **273**, 274
aemula Drew 17, **37**, 66, 154, 162, **418** (Fig. 16)
aethriobasis (Hardy) 17, **37**, 38, **419** (Fig. 17)
affinibancroftii new species 1, 17, **38**, 39, 178, **419** (Fig. 18)
affinidorsalis (Hardy) 17, 23, **39**, 40, 57, 94, 110, 115, 116, 167, 175, 187, **420** (Fig. 19)
affinis Hardy 17, **40**, **517** (Fig. 199B)
***Afrodacus* Bezzi, subgenus 12**, 17, 29
aithonota new species 1, 20, **265**, **558** (Fig. 274)
albistrigata (de Meijere) 17, **41**, 42, 86, 409, 410, **420** (Figs 20A, 20B)
amarambalensis Drew 17, 23, **42**, 43, 46, 125, **421** (Fig. 21)
ambigua (Shiraki) 2, 20, **265**, 266, 343, **559** (Fig. 275)
ambonensis Drew & Hancock 21, **376**, 377, 391, **614** (Fig. 377)
amoyensis Froggatt, Nomen Nudum **108**
andamanensis (Kapoor) 17, **43**, **421** (Fig. 22)
angustifinis (Hardy) 19, **219**, 220, **533** (Fig. 227)
antennalis Shiraki, Synonym **108**
apicalis (de Meijere) 2, 17, **30**, 32, 36, 253, 255, 272, **415** (Fig. 9)
apicalis Shiraki, Synonym **391**
apiciflava Yu, He & Chen 17, 20, **32**, 33, 34, **415** (Fig. 10)
apicinigra Yu, Bai & Yang, **New Synonym** 1, **316**, 318
apicofemoralis new species 1, 20, **266**, 267, 278, **559** (Fig. 276)
apicofuscans White & Tsuruta 17, **44**, 60, 100, 164, **422** (Fig. 23)
apiconigroscutella Drew 17, 24, **44**, 45, 157, **422** (Fig. 24)
apicopicta new species 1, 17, **45**, **423** (Fig. 25)
aptatus Hardy, **New Synonym** 1, **240**, 241
aquilus Drew 14
arecae (Hardy & Adachi) 17, 23, 26, 43, **46**, 47, 63, 76, 103, 125, **423** (Fig. 26)
areolata (Walker) 19, **220**, 221, **533** (Fig. 228)
arisanica (Shiraki) 2, 19, **244**, 245, **547** (Fig. 252)
armatus Fabricius 11
armillata (Hering) 17, **46**, 47, 189, 273, **424** (Fig. 27)
asatoi Shiraki, Synonym **94**, 95
ascitus Hardy, **New Synonym** 1, **251**, 252
***Asiadacus* Perkins, subgenus 12**, 17, 29–36
assamensis White 2, 20, 24, **268**, 269, 273, 295, 311, 353, **560** (Fig. 277)
ater Chen, Synonym **283**
atrichus (Bezzi) 13, 19, **214**, 215, 226, **529** (Fig. 221A)
atrichus var. *davaoanus* Bezzi, Synonym **214**, 215, **530** (Fig. 221B)
atrifacies (Perkins) 2, 20, 24, **269**, 271, **560** (Fig. 278)
atrifemur Drew & Hancock 17, 23, **47**, 103, **424** (Fig. 28)
atrimarginatus White 21, **377**, 389, **614** (Fig. 378)
aureus Tseng & Chu, Synonym **279**
avittata new species 1, 17, **47**, 48, **425** (Fig. 29)

***Bactrocera* Macquart, genus** 1, **11**, 17, 29–372
***Bactrocera* Macquart, subgenus 12**, 17, 36–195
bakeri (Bezzi) 12, 17, 30, 32, **33**, 34, 35, 36, **416** (Fig. 11)
baliensis new species 1, 20, **271**, 272, 351, **561** (Fig. 279)
bancroftii (Tryon) 17, 39, **48**, 49, 102, 159, **425** (Fig. 30)
bangaloriensis Agarwal & Kapoor, Synonym **69**
bannatus (Wang) 21, **377**, 378, 387, **615** (Fig. 379)
banneri White 19, **237**, 238, **543** (Fig. 246)
beckerae (Hardy) 17, 37, **49**, 50, 113, 114, 153, **426** (Fig. 31)
bellisi new species 1, 17, 24, **50**, 51, **426** (Fig. 32)
bezziana (Hering) 2, 20, 24, **272**, 336, **561** (Fig. 280)
bezzianus f. *signata* Hering, Synonym **345**
bezzii Miyake, Synonym **342**
bhutaniae new species 1, 17, 39, **51**, 53, 57, 72, 100, 172, **427** (Fig. 33)

bifasciata (Hardy) 17, **53**, 54, 181, **427** (Fig. 34)
biguttata (Bezzi) 1, 20, 24, 47, 269, **273**, 274, 327, 353, **562** (Fig. 281)
biguttulus Bezzi 12
bimaculata Drew & Hancock 17, 23, **54**, 88, 101, **428** (Fig. 35)
bimaculatus Lin, Wang & Zeng, **New Synonym** 1, **379**
binhduongiae new species 1, 17, 23, **54**, 55, 103, **428** (Fig. 36)
binoyi Drew 19, **226**, 237, **537** (Fig. 234)
bioculata Bezzi, Synonym **373**
bipustulata (Bezzi) 19, 243, 244, **245**, 247, 248, **547** (Fig. 253)
bispinosus (Wang) 21, **378**, 384, 386, 394, **615** (Fig. 380)
bitungiae new species 1, 17, 23, **55**, 56, 88, 138, 160, 161, **429** (Fig. 37)
bivittata Lin & Wang 17, 23, 53, **56**, 57, 100, 172, **429** (Fig. 38)
blairiae new species 1, 17, **57**, 58, 68, **430** (Fig. 39)
bogorensis (Hardy) 1, 19, **251**, 252, 256, 327, **551** (Fig. 260)
boninensis Hardy & Adachi, Synonym **246**, 247
borongensis new species 1, 20, 24, **274**, **562** (Fig. 282)
brachycera (Bezzi) 19, **259**, 260, 392, **555** (Fig. 268)
brevistylus Bezzi, Synonym **374**
brevivitta new species 1, 19, **226**, 227, 229, 233, **537** (Fig. 235)
bruneiae new species 1, 17, **58**, 59, 71, 81, 171, **430** (Fig. 40)
brunneola White & Tsuruta 17, 39, 44, **59**, 60, 64, 72, 100, 178, **431** (Fig. 41)
bryoniae (Tryon) 6, 17, 50, **60**, 61, 80, 111, 113, 409, 410, **431** (Fig. 42)
***Bulladacus* Drew & Hancock, subgenus** **12**, 19, 195–199
bullifera (Hardy) 19, **195**, 196, 197, 199, **517** (Fig. 200)
burmensis Hardy 11, 21, **403**, 404, **637** (Figs 423A, 423B)
buruensis White 20, **274**, 275, 291, 326, **563** (Fig. 283)

calirayae Drew & Hancock 21, **378**, 379, 381, 388, 389, 392, 397, 399, 402, **616** (Fig. 381)
***Callantra* Walker, subgenus** **15**, 21, 373–374
calophylli (Perkins & May) 19, **201**, 202, 203, **521** (Fig. 207)
calumniata (Hardy) 20, 266, **275**, 276, 307, 325, 350, 351, **563** (Fig. 284)
captiva new species 1, 19, **196**, 197, **518** (Figs 201A, 201B)
carambolae Drew & Hancock 17, 23, **61**, 62, 77, 160, 409, 410, **432** (Fig. 43)
careofascia new species 1, 17, **62**, 63, **432** (Fig. 44)
careomacula new species 1, 17, **34**, **416** (Fig. 12)
caryeae (Kapoor) 17, 23, 46, **63**, 104, 125, 144, 409, 410, **433** (Fig. 45)
caudata (Fabricius) 15, 20, 207, 267, 271, **277**, 278, 317, 409, 410, **564** (Fig. 285)
caudatus var. *maculatus* Perkins **316**, 317, 318
caudatus var. *nubilus* Hendel, Synonym **353**, 356
ceylanica Tsuruta & White 17, **64**, **433** (Fig. 46)
cheni Chao, Synonym **262**
cheni Zhang, Ji & Chen, **New Synonym** 1, **113**
chonglui Chao & Lin, **New Synonym** 2, **402**
chrysotoxus Hendel, Synonym **288**, 289
cibodasae Drew & Hancock 17, 23, **64**, 65, 85, 98, 119, 120, 142, **434** (Fig. 47)
ciliatus Loew 15, 21, **374**, 375, 409, 410, **612** (Fig. 373)
cilifera (Hendel) 2, 19, 222, **227**, 228, 231, 233, 235, **538** (Fig. 236)
cinnabaria new species 1, 19, **197**, **519** (Fig. 202)
citima (Hardy) 2, 17, **65**, 66, 109, 150, 151, **434** (Fig. 48)
citri Chen, Synonym **260**
citricola Bezzi 12, 21, **405**, 406, 407, **638** (Fig. 424)
citrifusca new species 1, 19, **228**, 229, **538** (Fig. 237)
citronellae Kapoor & Katiyar, Synonym **205**
cocciniae Premlata & Singh, Synonym **374**
cognata (Hardy & Adachi) 17, 23, **66**, 108, 126, **435** (Fig. 49)
collita Drew & Hancock 17, 23, **66**, 67, 78, 105, 115, 176, 182, **435** (Fig. 50)
commensurata new species 1, 17, 58, **67**, 68, **436** (Fig. 51)
complicata White 19, **215**, 216, 219, **530** (Fig. 222)
conformis Doleschall, Synonym **142**
conformis Walker, Synonym **186**
connexa (Hardy) 20, 264, **278**, 279, 306, 330, 364, **564** (Fig. 286)
conopsoides de Meijere 1, 21, **379**, 401, **616** (Fig. 382)
contigua Drew 17, **68**, 69, 154, 162, **436** (Fig. 52)
continua (Bezzi) 19, **201**, 202, **522** (Fig. 208)
correcta (Bezzi) 17, **69**, 155, 186, 195, 409, 410, **437** (Fig. 53)
costalis (Shiraki) 3, 17, 50, 59, **70**, 71, 81, 113, 129, 156, 171, **437** (Fig. 54)
crabroniformis (Bezzi) 15, 21, 377, **379**, 380, 389, 395, 397, **617** (Fig. 383)
cucurbitae (Coquillett) 20, **279**, 281, 287, 409, 410, **565** (Fig. 287)
curtivitta new species 1, 17, **71**, 72, **438** (Fig. 55)
curvifera (Walker) 12, 17, **72**, 73, 176, 409, 410, **438** (Fig. 56A), **439** (Fig. 56B)
curvosterna new species 1, 17, 63, **73**, 74, 123, **439** (Figs 57A, 57B)

daclaciae new species 1, 20, **281**, **565** (Fig. 288)
***Daculus* Speiser, subgenus** **12**, 13, 19, 199–201
***Dacus* Fabricius, genus** 1, **11**, 12, 15, 21, 373–402
davaoanus Bezzi **214**, 215, **530** (Fig. 221B)
deceptus Hardy, **New Synonym** 2, **248**, 249
depressa (Shiraki) 3, 20, **282**, 283, 329, 348, **566** (Fig. 289)
destillatoria Bezzi, Synonym **373**
dianensis Wang & Zhao, **New Synonym** 2, **30**, 32
diaphora (Hendel) 2, 20, 24, 271, **283**, 285, **566** (Fig. 290)
diaphoropsis (Hering) 20, **285**, 286, 321, 328, **567** (Fig. 291)
***Didacus* Collart, subgenus** **15**, 21, 374–375
diffusus Walker, Synonym **186**
digressa Radhakrishnan 19, **199**, 200, 201, 211, **520** (Fig. 205)
discipennis Walker, new combination 3, 21, **380**, 381, **617** (Fig. 384)
discophorus (Hering) 21, 378, **381**, 388, 389, 392, 397, 402, **618** (Fig. 385)
discretus new species 1, 21, **381**, 382, **618** (Fig. 386)
dispar (Hardy) 17, **74**, 75, 83, 89, 118, 140, **440** (Fig. 58)
diversa (Coquillett) 2, 13, 19, **205**, 207, 409, 410, **524** (Fig. 212)
donggaliae new species 1, 21, **382**, 383, **619** (Fig. 387)
dongnaiae new species 1, 17, 23, 54, **75**, 76, 88, 89, **440** (Fig. 59)
dorjii Drew & Romig 21, **383**, **619** (Fig. 388)
dorsalis (Hendel) 2, 17, 23, 69, **76**, 77, 78, 100, 121, 140, 143, 169, 190, 409, 410, **441** (Figs 60A, 60B, 60C)
dorsalis var. *occipitalis* – Hardy & Adachi **137**
dorsaloides (Hardy & Adachi) 17, 23, 67, **78**, 105, 115, 176, 182, **442** (Fig. 61)
dorsirufa new species 1, 20, 24, **286**, 315, 341, **567** (Fig. 292)

drewi Hardy, **New Synonym** 2, **235**, 236
dubiosa (Hardy) 20, **286**, 287, **568** (Fig. 293)
duplicata (Bezzi) 19, **221**, 222, **534** (Fig. 229)
dutti Kapoor, Synonym **69**

elegantis Tseng, Chen & Chu, **New Synonym** 2, **353**, 356
elegantula (Hardy) 20, 265, **287**, 288, 305, **568** (Fig. 294)
elongata new species 1, 17, **78**, 79, **442** (Fig. 62)
emittens (Walker) 20, 45, 276, 287, **288**, 289, 296, 325, 366, **569** (Fig. 295)
enigmaticus Hardy, **Homonym** 1, **112**, 113
epicharis (Hardy) 17, 50, 61, **79**, 80, 113, **443** (Fig. 63)
esakii (Shiraki) 21, 376, 378, **384**, 394, **620** (Fig. 389)
eumenoides Bezzi, Synonym **373**
eurycosta new species 1, 17, 59, 71, **80**, 81, 171, **443** (Fig. 64)
eurylomata (Hardy) 19, 226, **229**, 230, **539** (Fig. 238)
exornata (Hering) 20, **289**, 290, 362, **569** (Fig. 296)
expandens (Walker) 19, **238**, 239, 241, 243, **544** (Fig. 247)
expandens ssp. *melanius* Hardy & Adachi, Synonym 3, **240**, **545** (Fig. 249)

fasciatipennis Doleschall, Synonym **186**
fascipennis Wiedemann, Synonym **186**
fastigata Tsuruta & White 17, **29**, **414** (Fig. 7)
feijeni White 21, **384**, 390, **620** (Fig. 390)
fereuncinata new species 1, 19, **216**, 217, 219, 310, **531** (Fig. 223)
fernandoi Tsuruta & White 17, 23, **81**, 82, 92, 113, 152, **444** (Fig. 65)
ferruginea Fabricius **76**
ferrugineus – Becker, Misidentification **375**
ferrugineus var. *dorsalis* Hendel **76**
ferrugineus incisus Bezzi, Misidentification **63**
ferrugineus var. *limbiferus* Bezzi **110**
ferrugineus var. *mangiferae* Cotes, Synonym **194**
ferrugineus var. *obscurata* de Meijere **135**
ferrugineus var. *occipitalis* Bezzi **137**
ferrugineus var. *okinawanus* Shiraki, Synonym **76**
ferrugineus var. *pedestris* Bezzi **151**
ferrugineus var. *versicolor* Bezzi **190**
flavipennis (Hardy) 17, 75, **82**, 83, 89, 116, 118, 140, 141, 147, 150, 291, 326, **444** (Fig. 66)
flavipilosa (Hardy) 20, 83, 275, 281, **290**, 291, 326, **570** (Fig. 297)
flavolateralis new species 1, 20, 223, **291**, 292, 293, **570** (Fig. 298)
flavopectoralis (Hering) 20, 24, **292**, 293, 297, **571** (Fig. 299)
flavoscutellata Lin & Wang 17, **83**, 84, 153, **445** (Fig. 67)
flavosterna new species 1, 17, 23, **84**, 85, **445** (Fig. 68)
flavoverticalis new species 1, 20, **293**, 294, **571** (Fig. 300)
flavus Tseng, Chen & Chu, **New Synonym** 2, **353**, 357
fletcheri Drew 21, **385**, **621** (Fig. 391)
floresiae Drew & Hancock 17, 23, **85**, 98, 120, **446** (Fig. 69)
formosanus (Tseng & Chu) 21, 383, **385**, 386, **621** (Fig. 392)
fraserensis Chua, **New Synonym** 2, **340**, 341, 342
frauenfeldi (Schiner) 17, 41, 42, **86**, 409, 410, **446** (Fig. 70)
freidbergi White 20, 24, 269, **294**, 295, 353, **572** (Fig. 301)
frenchi Froggatt, Synonym **186**
fulvicauda (Perkins) 6, 17, **86**, 87, 114, 127, 167, 181, 185, **447** (Figs 71A, 71B)
fulvifemur Drew & Hancock 17, 23, 66, **87**, 88, 108, 119, 126, 154, 188, **448** (Fig. 72)
fulvipes (Perkins) 13, 19, 221, **222**, 223, 224, 232, **534** (Fig. 230)
fulvosterna new species 1, 19, **245**, 246, 249, **548** (Fig. 254)
fuscans Wang, Synonym **259**
fuscitibia Drew & Hancock 17, 23, 54, 56, **88**, 101, 106, 138, 160, **448** (Fig. 73)
fuscoalata new species 1, 20, **295**, 296, **572** (Fig. 302)
fuscoformosa new species 1, 17, **89**, 116, **449** (Fig. 74)
fuscolobata new species 1, 18, **89**, 90, **449** (Fig. 75)
fuscoptera new species 1, 18, **90**, 91, **450** (Fig. 76)

gansuica Chen, Han & Zhou, **New Synonym** 2, **268**, 269
garciniae (Bezzi) 3, 19, 238, **239**, 241, **544** (Fig. 248)
gavisa (Munro) 20, **296**, **573** (Fig. 303)
gnetum Drew & Hancock 12
gombokensis Drew & Hancock 18, 23, 82, **91**, 92, 113, 152, **450** (Fig. 77)
goughi Hardy, Synonym **331**, 332
guangxianus Chao & Lin, **New Synonym** 2, **283**, 285
***Gymnodacus* Munro, subgenus** **13**, 19, 201–205

hageni de Meijere, Synonym 272, 300, **353**, 356
haikouensis Wang & Chen, **New Synonym** 2, **393**, 394
hainanus Chao & Lin, Synonym **252**, 254
hainanus Wang & Zhao 21, **374**, 375, **612** (Fig. 374)
halmaherae new species 1, 18, **92**, 93, **451** (Fig. 78)
hancocki new species 1, 19, 222, **223**, **535** (Fig. 231)
hantanae Tsuruta & White 18, 23, 78, **93**, 101, 168, 179, **451** (Fig. 79)
hardyi Zhang, Ji & Chen, **New Synonym** 2, **181**, 182
hatyaiensis new species 1, 20, 293, **296**, 297, **573** (Fig. 304)
havelockiae new species 1, 20, **297**, 298, 337, **574** (Figs 305A, 305B)
heinrichi (Hering) 2, 20, 24, **298**, 299, 300, **575** (Figs 306A, 306B)
hekouana Yu, He & Yang 20, 24, **300**, 301, **576** (Fig. 307)
***Hemigymnodacus* Hardy, subgenus** **13**, 19, 205–208
hengsawadae new species 1, 20, 36, 252, **301**, 302, 312, 339, **576** (Fig. 308)
heppneri White, **New Synonym** 2, **136**, **480** (Fig. 129B)
hoabinhiae new species 1, 20, 24, 208, 285, 301, **302**, 303, 314, 371, **577** (Fig. 309)
hochii (Zia) 2, 14, **252**, 253, 254, 255, 259, 392, **552** (Fig. 261)
hodgsoniae new species 1, 20, 24, **303**, 304, **577** (Fig. 310)
holtmanni (Hardy) 18, 23, **94**, 167, 175, **452** (Fig. 80)
hsui Tseng, Chen & Chu, **New Synonym** 2, **94**, 95
hululangatiae new species 1, 20, 265, 288, **304**, 305, 324, 350, **578** (Fig. 311)
hyalina (Shiraki) 2, 18, **94**, 95, 189, 247, **452** (Fig. 81)

icariiformis (Enderlein) 21, 376, **386**, 391, **622** (Fig. 393)
icelus (Hardy) 19, **239**, 240
***Ichneumonopsis* Hardy, genus** 1, **11**, 21, 403–404
ihai Shiraki, Synonym **384**
illusioscutellaris new species 1, 18, **95**, 96, **453** (Fig. 82)

implexus Perkins, Nomen Nudum **330**
impunctata (de Meijere) 18, **96**, 97, 174, **453** (Fig. 83)
incisa (Walker) 14, 19, 220, 226, **230**, 231, 234, 236, **539** (Fig. 239)
incompta new species 1, 19, **249**, 250, **550** (Fig. 258)
inconspicua new species 1, 18, **97**, 98, 185, **454** (Fig. 84)
indecorus (Hardy) 21, **386**, 387, 393, **622** (Fig. 394)
indentus (Hardy) 20, **305**, **578** (Fig. 312)
indonesiae Drew & Hancock 18, 23, **98**, 120, **454** (Fig. 85)
infernus (Hardy) 21, 378, **387**, **623** (Fig. 395)
infesta (Enderlein) 19, 253, **254**, 255, 259, 272, 392, **552** (Fig. 262)
infulata Drew & Hancock 18, 23, **98**, 99, 105, 165, 166, **455** (Fig. 86)
insulosus Drew & Hancock 21, **387**, 388, 394, **623** (Fig. 396)
invadens Drew, Tsuruta & White 18, 23, 26, 38, 53, 57, 58, 62, 78, **99**, 100, 102, 139, 172, 409, 410, **455** (Fig. 87A), **456** (Figs 87B, 87C)
involuta (Hardy) 18, 23, **100**, 101, 112, **457** (Fig. 88)
iriomotiae new species 1, 20, **306**, **579** (Fig. 313)
irvingiae Drew & Hancock 18, 23, 26, 78, 93, 94, **101**, 128, 168, 179, **457** (Fig. 89)
ishigakiensis (Shiraki) 20, 266, 276, **306**, 307, 325, 350, 351, **579** (Fig. 314)
isolata (Hardy) 20, 24, **307**, 308, 310, 365, **580** (Fig. 315)

jaceobancroftii new species 1, 18, **101**, 102, **458** (Fig. 90)
***Javadacus* Hardy, subgenus** **13**, 19, 29, 208–214
javadica Mahmood 20, 24, 274, **309**, **580** (Fig. 316)
javanensis (Perkins) 2, 19, **208**, 209, 212, 213, 214, **526** (Fig. 214)
jiannanus Chao & Lin, **New Synonym** 2, **252**, 254
jinghongensis Zhang, Ji & Chu, **New Synonym** 2, **65**, 66
jinreni Chao & Lin, **New Synonym** 2, **252**, 254
juxtuncinata new species 1, 20, 308, **309**, 310, 334, 365, **581** (Fig. 317)

kaghanae Mahmood 20, 24, **310**, 311
kalimantaniae new species 1, 23, **102**, 103, **458** (Fig. 91)
kanchanaburi Drew & Hancock 18, 23, 55, **103**, 128, 168, **459** (Fig. 92)
kandiensis Drew & Hancock 18, 23, 46, 63, 100, **104**, 125, 144, 409, 410, **459** (Fig. 93)
katoi Hardy, **New Synonym** 2, **241**, 243, **546** (Fig. 250B)
keiseri (Hering) 21, 374, **375**, **613** (Fig. 375)
khaoyaiae new species 1, 20, 24, **311**, 370, **581** (Fig. 318)
kinabalu Drew & Hancock 18, 23, 99, **104**, 105, 165, **460** (Fig. 94)
kuniyoshii (Shiraki) 19, 201, **202**, 203, 204, **522** (Fig. 209)

lacerata White & Evenhuis, **New Synonym** 2, **186**, 187, **512** (Fig. 190B)
lagunae Drew & Hancock 21, 379, 381, **388**, 389, 392, 397, **624** (Fig. 397)
laguniensis new species 1, 20, 267, 282, **312**, 339, **582** (Fig. 319)
laithieuiae new species 1, 18, 23, 56, 88, 89, **105**, 106, 138, 160, 161, **460** (Fig. 95)
laocaiae new species 1, 24, 208, 285, 301, 303, **313**, 314, 371, **582** (Fig. 320)
lata (Perkins) 2, 18, 24, **106**, 107, **461** (Fig. 96A)
lateritaenia Drew & Hancock 18, 23, 66, **108**, 126, 154, **462** (Fig. 97)
laterum Wang, **New Synonym** 2, **252**, 254
latifae Anwar Cheema, **New Synonym** 2, **205**, 207
latifrons (Hendel) 18, 49, 66, **108**, 109, 149, 150, 151, 159, 161, 409, 410, **462** (Fig. 98)
latilineola Drew & Hancock 18, 23, 40, 57, **109**, 110, 115, 187, **463** (Fig. 99)
leongi Drew & Hancock 21, 379, 381, **388**, 389, 392, 397, **624** (Fig. 398)
***Leptoxyda* Macquart, subgenus** **15**, 21, 375–376
limbifer rufulus Hardy, **New Synonym** 2, **170**, **503** (Fig. 173B)
limbifera (Bezzi) 18, 50, **110**, 111, 135, **463** (Fig. 100)
linduensis new name 1, 18, 50, 90, **112**, 113, 146, **464** (Fig. 101)
lipsana (Hendel) 20, **314**, 344, **583** (Fig. 321)
liquida new species 1, 20, 24, **315**, **583** (Fig. 322)
lombokensis Drew & Hancock 1, 18, 23, 82, 92, **113**, 152, **464** (Fig. 102)
longicaudata (Perkins) 19, 221, 222, 223, 224, 225, 226, 230, **231**, 232, 234, 236, **540** (Figs 240A, 240B)
longicornis Macquart 11,12
longicornis Wiedemann 15, 21, **373**, 374, 386, 388, 394, 400, 401, **611** (Fig. 372)
longicornis (Wiedemann) – Hardy & Adachi, Misidentification **400**
longistyla Bezzi **375**
longistylus Wiedemann, Misidentification **376**
longivittata Chua & Ooi 19, 252, **255**, 256, **553** (Fig. 263)
lunulatus Tseng, Chen & Chu, **New Synonym** 2, **283**, 285
luzonae Hardy & Adachi, **New Synonym** 2, **132**, 134, **478** (Fig. 127B)

macrophyllae new species 1, 20, 24, **315**, 316, 361, 370, **584** (Fig. 323)
maculata (Perkins) 1, 20, **316**, 317, 318, **584** (Fig. 324)
maculifacies (Hardy) 17, 30, 34, **35**, 36, 302, **417** (Fig. 13)
maculifemur (Hering) 19, **217**, 218, 219, **531** (Fig. 224)
maculigera Doleschall 18, **113**, 114, 194, **465** (Fig. 103)
maculipennis Doleschall, Synonym **277**
maculipterus White 21, 377, **389**, **625** (Fig. 399)
maculosus Walker, Nomen Nudum **37**
makilingensis Drew & Hancock 18, 23, 67, 78, 105, **115**, 176, 182, **465** (Fig. 104)
malaisei Hering, Synonym **339**
malaysiae Drew & Hancock 21, **406**, 407, **638** (Fig. 425)
malaysiensis Drew & Hancock 18, 23, 40, 57, 110, **115**, 187, **466** (Fig. 105)
mamaliae new species 1, 18, **116**, **466** (Fig. 106)
matsumurai (Shiraki) 14, **246**, 247, 248, **548** (Fig. 255)
mcgregori (Bezzi) 19, 196, **197**, 198, **519** (Fig. 203)
mediorufula new species 1, 18, **116**, 117, 180, **467** (Fig. 107)
megaspilus (Hardy) 18, 43, 83, **117**, 118, 140, **467** (Fig. 108)
melanius (Hardy & Adachi), new combination 1, 2, 19, 238, **240**, 241, 242, **545** (Fig. 249)
melanofacies new species 1, 20, 294, **318**, 368, **585** (Fig. 325)
melanopectus new species 1, 21, **389**, 390, **625** (Fig. 400)
melanopsis (Hardy) 17, 30, 34, **36**, **417** (Fig. 14)

melastomatos Drew & Hancock 18, 23, 88, 105, **118**, 119, 188, **468** (Fig. 109)
***Mellesis* Bezzi, subgenus** **15**, 21, 376–401
menglana Yu, Liu & Yang 20, 24, **318**, 319, 322, 341, **585** (Fig. 326)
merapiensis Drew & Hancock 18, 23, 65, **119**, 142, 178, **468** (Fig. 110)
mesomelas Bezzi 13
minax (Enderlein) 13, 19, **260**, 261, 263, 409, 410, **556** (Fig. 269)
mindanaus Bezzi **225**, 226, **356** (Fig. 233B)
minima (Hering) 20, **319**, 320, **586** (Fig. 327)
minuscula Drew & Hancock 18, 23, 98, **119**, 120, **469** (Fig. 111)
modicus Hardy, **New Synonym** 2, **30**, 32
moluccensis (Perkins) 18, 44, 60, **120**, 121, **469** (Fig. 112)
***Monacrostichus* Bezzi, genus** 1, **12**, 21, 405–407
montana (Hardy) 13, 19, **209**, 212, 214, **526** (Fig. 215)
muiri (Hardy & Adachi) 18, 23, 85, **121**, 143, 160, 192, **470** (Fig. 113)
mukiae new species 1, 19, **207**, 208, 285, 303, 371, **525** (Figs 213A, 213B)
mulyonoi Hardy, **New Synonym** 2, **132**, 134, **478** (Fig. 127C)
munda (Bezzi) 20, 286, **320**, 321, 328, **586** (Fig. 328)
munroi Zaka-ur-Rab, Synonym **394**
murphyi White 21, 386, **390**, 391, **626** (Fig. 401)
musae (Tryon) 6, 18, 66, 69, **121**, 122, 154, 161, 162, 409, 410, **470** (Fig. 114)
musae var. *dorsopicta* Tryon, Synonym **121**

nadanus Chao & Lin, **New Synonym** 2, **30**, 32
nakhonnayokiae new species 1, 20, 24, **321**, 322, **587** (Fig. 329)
namlingiae new species 1, 20, 24, 218, **322**, 323, **587** (Fig. 330)
nanggalae Drew & Hancock 21, 377, **391**, **626** (Fig. 402)
nanoarcuata new species 1, 18, 63, 74, **122**, 123, **471** (Figs 115A, 115B)
nationigrotibialis new species 1, 18, 24, **123**, 124, 194, **472** (Fig. 116)
neoarecae Drew 18, 23, 46, 104, **124**, 125, **472** (Fig. 117)
neocheesmanae Drew 6, 18, **125**, **473** (Fig. 118)
neocognata Drew & Hancock 18, 23, 66, **125**, 126, **473** (Fig. 119)
***Neodacus* Perkins, subgenus** **15**, 21, 401–402
neoelegantula White 20, 265, 288, 305, **323**, 324, 350, **588** (Fig. 331)
neoemittens new species 1, 20, 266, 289, **324**, 325, **588** (Fig. 332)
neoflavipilosa new species 1, 20, 83, 291, **325**, 326, **589** (Fig. 333)
neofulvicauda new species 1, 18, 48, 87, **126**, 127, 185, **474** (Fig. 120)
neolipsana new species 1, 20, 314, **326**, 327, 344, **589** (Fig. 334)
neonigrotibialis Drew 18, 24, 124, **127**, 132, 194, **474** (Fig. 121)
neopropinqua Drew & Hancock 18, 23, 103, **128**, 168, 173, **475** (Fig. 122)
neoritsemai new species 1, 18, 61, 80, **128**, 129, **475** (Fig. 123)
nepalensis Hardy, Synonym **394**
***Nesodacus* Perkins, subgenus** **13**, 19, 214–215
newmani Perkins 15
nigrifacia Zhang, Ji &Chen 18, 24, **129**, 130, 131, 194, **476** (Fig. 124)
nigrifacies (Shiraki) 3, 20, **327**, 328, **590** (Fig. 335)
nigrifemorata Lin & Wang 18, 24, **130**, 131, **476** (Fig. 125)
nigrita (Hardy) 19, **209**, 210, 211, **527** (Fig. 216)
nigrofemoralis White & Tsuruta 18, 24, 45, 51, 130, **131**, 132, 145, **477** (Fig. 126)
nigrotibialis (Perkins) 2, 3, 18, 24, 45, 51, **132**, 134, **477** (Fig. 127A)
nothaphoebe new species 1, 18, **135**, **479** (Fig. 128)
nubilus ssp. *femoralis* Hendel, Synonym **353**
nubilus ssp. *heinrichi* Hering, Synonym **298**
nubilus ssp. *sumbensis* Hering **348**
nummularius (Bezzi) 21, 379, 381, 386, 388, 389, **391**, 392, 397, 402, **627** (Fig. 403)

obscurata (de Meijere) 2, 18, **135**, 136, 137, 151, 192, **480** (Fig. 129A)
obscurivitta new species 1, 18, **137**, 192, **481** (Fig. 130)
occipitalis (Bezzi) 18, 23, 47, 56, 76, 85, 88, 103, 106, 119, **137**, 138, 139, 160, 405, 409, 410, **481** (Fig. 131)
ochroma new species 1, 18, 23, 78, 100, **139**, 140, **482** (Fig. 132)
ochrosterna new species 1, 19, **218**, 219, **532** (Fig.225)
ochroventer new species 1, 18, 89, 116, **140**, **482** (Fig. 133)
okinawanus Shiraki, Synonym **76**
okunii (Shiraki) 3, 20, 24, 283, **328**, 329, 348, 358, 360, **590** (Fig. 336)
oleae Gmelin 12
ooii Drew & Hancock 21, 382, **392**, 393, **627** (Fig. 404)
opacovitta new species 1, 18, **140**, 141, **483** (Fig. 134)
ornatissimus Froggatt, Synonym **121**
ortholomatus Hardy 15, **401**, 402, **635** (Fig. 420)
osbeckiae Drew & Hancock 18, 23, 65, 85, 98, 119, **141**, 142, 178, **483** (Fig. 135)
oscinae Agarwal & Kapoor, Synonym **40**

pahangiae new species 1, 19, 227, 229, **232**, 233, **541** (Fig. 241)
pallescentis Drew [= *neopallescentis* Drew] 13
pallescentis (Hardy) 19, 200, **210**, 211, **527** (Fig. 217)
pantabanganiae new species 1, 19, **233**, 234, **541** (Fig. 242)
papayae Drew & Hancock 2, 18, 23, 62, 68, 77, 85, 121, **142**, 143, 160, 169, 190, 192, 409, 410, **484** (Fig. 136)
***Papuodacus* Drew, subgenus** **13**, 19, 215–219
paraarecae new species 1, 23, **143**, 144, **484** (Fig. 137)
***Paradacus* Perkins, subgenus** **13**, 19, 219–225
paradiospyri Chen, Zhou & Li 18, 24, 51, **144**, 145, **485** (Fig. 138)
paralatissima new species 1, 18, 63, 74, 123, **145**, **485** (Fig. 139)
paralimbifera new species 1, 18, 90, **145**, 146, **486** (Fig. 140)
paramusae Drew 6, 18, 50, 113, 141, **146**, 147, **486** (Fig. 141)
paranigrita new species 1, 18, **147**, 148, **487** (Fig. 142)
paraosbeckiae Drew 18, 39, 44, 60, 64, **148**, 178, 191, **487** (Fig. 143)
***Parasinodacus*, new subgenus** 1, **14**, 19, 225–237
paratappanus Tseng, Chen & Chu, **New Synonym** 2, **352**, 353
parater Chao & Lin, **New Synonym** 2, **269**, 271
***Paratridacus* Shiraki, subgenus** **14**, 19, 237–243
paratuberculatus Philip, Synonym **69**
paraverbascifoliae Drew 18, 23, 43, **148**, 149, 190, **488** (Fig. 144)

***Parazeugodacus* Shiraki, subgenus** **14**, 19, 243–249
parvifoliaceus Tseng, Chen & Chu, **New Synonym** 2, **244**, 245
parvula (Hendel) 2, 18, 66, 109, **149**, 150, **488** (Fig. 145)
patula new species 1, 18, 89, 116, **150**, **489** (Fig. 146)
pauleiensis Tseng, Chen & Chu, **New Synonym** 2, **265**, 266
pectoralis (Walker) 18, 66, 150, **151**, **489** (Fig. 147)
pedestris (Bezzi) 18, 23, 82, 92, 113, **151**, 152, **490** (Fig. 148)
pedestris var. *limbiferus* – Perkins **110**
pedunculatus (Bezzi) 21, 387, **392**, 393, **628** (Fig. 405)
pemalangiae new species 1, 20, 306, **329**, 330, **591** (Fig. 337)
pendleburyi (Perkins) 19, **247**, 248, **549** (Figs 256A, 256B)
peneallwoodi new species 1, 18, **152**, 153, **490** (Fig. 149)
penebeckerae new species 1, 18, **153**, 154, 162, **491** (Fig. 150)
penecognata Drew & Hancock 18, 23, 66, 108, 126, **154**, **491** (Fig. 151)
penecorrecta Drew 18, 70, **154**, 155, 186, **492** (Fig. 152)
penecostalis new species 1, 18, 61, 80, **155**, 156, **492** (Fig. 153)
penephaea new species 1, 18, **156**, **493** (Fig. 154)
perigrapha White & Tsuruta 18, 24, 84, 153, **157**, 158, **493** (Fig. 155)
pernigra (Ito) 2, 18, 24, 84, 153, **157**, 158, **494** (Fig. 156)
perplexa (Walker) 20, **330**, 331, **591** (Fig. 338)
persicae Bigot, Synonym **194**
persicus Hendel 21, **375**, 376, **613** (Fig. 376)
persignata (Hering) 20, **331**, 332, **592** (Fig. 339)
personatus Hardy, **New Synonym** 2, **366**, 367
peterseni (Hardy) 19, 196, **198**, 199, **520** (Fig. 204)
petersoni Hardy, Misspelling **198**
philippinensis Drew & Hancock, **New Synonym** 2, **142**, 143
pictipennis Lin & Zeng 18, **158**, 159, **494** (Fig. 157)
pictus (Hardy) 2, 21, 382, 392, **393**, 394, **628** (Fig. 406)
platamus (Hardy) 20, 317, **333**, 334, 339, **592** (Fig. 340)
polistiformis (Senior-White) 21, 378, 379, 384, **394**, **629** (Fig. 407)
poonensis Kapoor, Synonym **230**
profunda Tsuruta & White 18, 23, 85, 121, 143, **159**, 160, 192, **495** (Fig. 158)
propinqua (Hardy & Adachi) 18, 23, 62, **160**, **495** (Fig. 159)
proprediaphora Wang, Xiao & Chen, **New Synonym** 2, **283**, 285
proprescutellata Zhang, Chen & Gao 20, 24, **334**, 335, 343, 347, **593** (Fig. 341)
pruniae new species 1, 18, 159, **161**, **496** (Fig. 160)
pseudobeckerae new species 1, 18, 154, **161**, 162, **496** (Fig. 161)
pseudocucurbitae White 19, 228, 231, **234**, 235, **542** (Fig. 243)
pseudodistincta (Drew) 18, 87, **162**, 163, 185, **497** (Fig. 162)
pseudoscutellatus Tseng, Chen & Chu, **New Synonym** 2, **342**, 343
pseudoversicolor Drew 18, 38, 44, **163**, 164, 191, **497** (Fig. 163)
pubescens (Bezzi) 20, 24, **335**, 336, 347, **593** (Fig. 342A), **594** (Fig. 342B)
pullus (Hardy) 21, 388, **394**, **629** (Fig. 408)
pusaensis Kapoor & Katiyar, Synonym **340**
pusilla (Hardy) 18, 117, **164**, 180, **498** (Fig. 164)
pyrifoliae Drew & Hancock 18, 23, 105, **164**, 165, 182, 409, 410, **498** (Fig. 165)

qionganus Chao & Lin, **New Synonym** 2, **252**, 254
quadrifidus Hendel, Synonym **205**
quadristriata Munro, Synonym **394**
quasiinfesta new species 1, 20, 298, **336**, 337, **594** (Fig. 343)
quasiinfulata new species 1, 18, 23, **165**, 166, **499** (Fig. 166)
quasineonigrita new species 1, 18, **166**, 167, **499** (Fig. 167)
quasipropinqua Drew & Hancock 18, 23, 94, **167**, 175, **500** (Fig. 168)
quaternum Wang, **New Synonym** 2, **252**, 254

raiensis Drew & Hancock 18, 23, 78, 93, 101, 103, 128, **167**, 168, 179, **500** (Fig. 169)
ramanii Drew & Hancock 21, 377, 389, **395**, **630** (Fig. 409)
ranganathi new species 1, 18, 23, **168**, **501** (Fig. 170)
recurrens (Hering) 18, **169**, **501** (Fig. 171)
resima (Drew) 18, **169**, 170, **502** (Fig. 172)
ritsemai (Weyenbergh) 2, 3, 18, 50, 58, 59, 71, 81, 112, 129, **170**, 171, **502** (Fig. 173A)
rubella (Hardy) 1, 20, **337**, 338, 346, 360, **595** (Fig. 344)
rubigina (Wang & Zhao) 18, 38, 52, 53, 57, 58, 60, 64, 100, 102, **171**, 172, **503** (Fig. 174)
rubzovi Chao & Lin, **New Synonym** 2, **252**, 254
rufula Hardy 112, **170**, **503** (Fig. 173B)
ruiliensis Wang, Long & Zhang, **New Synonym** 2, **181**, 182
rutengiae new species 1, 19, **200**, 201, **521** (Fig. 206)

sabahensis new species 1, 20, 312, **338**, 339, **595** (Fig. 345)
santongae Drew & Hancock 15, 21, **401**, 402, **636** (Fig. 421)
sapaensis new species 1, 18, 23, 128, **172**, 173, **504** (Fig.175)
sasaotiae new species 1, 20, 294, **339**, **596** (Fig. 346)
satanas (Hering) 21, 377, 380, 389, **395**, 397, **630** (Fig. 410)
scutellaria (Bezzi) 19, 29, 208, 209, **211**, 212, 214, **528** (Fig. 218)
scutellaris (Bezzi) 2, 3, 20, 24, 311, 319, 322, 323, **339**, 340, 341, 342, 347, 409, 410, **596** (Fig. 347)
scutellata (Hendel) 2, 20, 24, 266, 335, **342**, 343, 409, 410, **597** (Figs 348A, 348B), **598** (Fig. 348C)
scutellatus Bezzi, Misidentification **335**
scutellina (Bezzi) 20, 315, 327, **343**, 344, **598** (Fig. 349)
selenophora Tsuruta & White 18, 97, **173**, 174, **504** (Fig. 176)
sembaliensis Drew & Hancock 18, 23, **174**, 175, **505** (Fig. 177)
***Semicallantra* Drew, subgenus** **14**, 19, 249–251
semifemoralis Tseng, Chen & Chu, **New Synonym** 2, **76**, 78
semisurstyli new species 1, 19, **212**, 213, **528** (Fig. 219)
semongokensis new species 1, 20, **344**, 345, **599** (Fig. 350)
siamensis Drew & Hancock 21, 383, 385, **396**, 398, 400, **631** (Fig. 411)
sicieni Chao & Lin, Synonym **283**, 285
signata (Hering) 2, 20, 24, 338, **345**, 346, 347, **599** (Fig. 351A), **600** (Fig. 351B)
silvaticus Hardy, **New Synonym** 2, **106**, 107, **461** (Fig. 96B)
sinensis (Wang) 21, 377, 389, 395, **396**, 397, **631** (Fig. 412)
sinensis Yu, Bai & Chen 19, **218**, 219, **532** (Fig. 226)
***Sinodacus* Zia, subgenus** **14**, 19, 251–259
smieroides Walker, Synonym 15, **373**
sonlaiae new species 1, 20, 24, 341, **347**, **600** (Fig. 352)
speciosa new species 1, 19, 255, **256**, **553** (Fig. 264)
spectabilis new species 1, 19, 255, **256**, 257, **554** (Fig. 265)
speculifera (Walker) 2, 6, 18, 72, 73, **175**, 176, **439** (Fig. 56C), **505** (Fig. 178)
sphaeroidalis (Bezzi) 2, 15, 401, **402**, **636** (Fig. 422)

splendida (Perkins) 19, **261**, 262, **556** (Fig. 270)
stenomus Wang & Zhao, **New Synonym** 2, **349**, 351
subsessilis (Bezzi) 21, 379, 381, 388, 389, 392, **397**, **632** (Fig. 413)
sulawesiae Drew & Hancock 18, 23, 105, **176**, **506** (Fig. 179)
suliae new species 1, 18, 39, 72, **176**, 178, **506** (Fig. 180)
sumatranus Hardy, **New Synonym** 2, **132**, 134, **479** (Fig. 127D)
sumbawaensis Drew & Hancock 18, 23, 65, 85, 98, 119, 142, **178**, **507** (Fig. 181)
sumbensis (Hering) 20, **348**, **601** (Fig. 353)
superba new species 1, 19, **250**, 251, **551** (Fig. 259)
symplocos new species 1, 19, **203**, 204, **523** (Fig. 210)
synnephes (Hendel) 2, 20, 24, 263, 264, 265, 266, 276, 288, 305, 307, 324, **348**, 349, 350, 351, **601** (Fig. 354A), **602** (Fig. 354B)
synnephes var. *dobaensis*, Synonym **348**
synnethes – Hardy, Misspelling **348**
syzygii White & Tsuruta 18, 23, 78, 93, 94, 101, 168, **178**, 179, **507** (Fig. 182)

tapahensis new species 1, 18, **179**, 180, **508** (Fig. 183)
tapervitta Mahmood 20, 24, **351**, 352, 363, **602** (Fig. 355)
tappanus (Shiraki) 2, 20, 24, 269, 295, **352**, 353, **603** (Fig. 356)
tau (Walker) 2, 20, 24, 25, 217, 272, 298, 300, 304, 309, 345, 347, 352, **353**, 354, 356, 357, 358, 360, 372, 409, 410, **603** (Fig. 357A), **604** (Fig. 357B)
tebeduiae new species 1, 20, 24, **357**, 358, **604** (Fig. 358)
tenebrosus Drew & Hancock 21, 383, 385, **397**, 398, 400, **632** (Fig. 414)
tenuifinis Hardy, **New Synonym** 2, **227**, 228
tenuivittatus Tseng, Chen & Chu, **New Synonym** 2, **157**, 158
terminifera (Walker) 3, 19, **224**
ternatiae new species 1, 18, **180**, 181, **508** (Fig. 184)
testacea Macquart [= *longistylus* Wiedemann] 15
tetrachaeta (Bezzi) 2, 19, **248**, 249, **550** (Fig. 257)
***Tetradacus* Miyake, subgenus** 13, **14**, 19, 259–263
thailandica Drew & Hancock 2, 18, 23, 65, 67, 78, 85, 98, 105, 115, 119, 142, 176, 178, **181**, 182, **509** (Fig. 185)
thistletoni Drew 18, 137, **183**, 192, **509** (Fig. 186)
tibialis Shiraki, Synonym **320**
tillyardi (Perkins) 1, 19, **204**, 205, **523** (Fig. 211A)
timorensis (Perkins) 20, 345, **358**, **605** (Fig. 359)
transtillum Hering, **New Synonym** 2, **208**
transversa (Hardy) 19, **257**, 258, 259, **554** (Fig. 266)
trichosanthes new species 1, 20, 24, 316, 338, **358**, 360, 361, **605** (Fig. 360)
tricuspidatae new species 1, 20, 24, **360**, 361, **606** (Fig. 361)
trifasciata (Hardy) 18, **183**, 184, **510** (Fig. 187)
trilineata (Hardy) 19, 29, 208, 209, 212, **213**, 214, **529** (Fig. 220)
trimacula (Wang) 21, 386, **398**, **633** (Fig. 415)
trimaculata (Hardy & Adachi) 20, 290, **361**, 362, **606** (Fig. 362)
triquetrus new species 1, 21, **398**, 399, **633** (Fig. 416)
trivandrumensis new species 1, 20, 24, 352, **362**, 363, **607** (Fig. 363)
trivittatus Matsumura, Synonym **342**
truncata new species 1, 18, **184**, 185, **510** (Fig. 188)
tryoni var. *Musa* Tryon, Nomen Nudum **121**
tsuneonis (Miyake) 13, 14, 19, 113, 260, 261, **262**, 263, 409, 410, **557** (Fig. 271)
tuberculata (Bezzi) 18, **185**, 186, 409, 410, **511** (Fig. 189)

ubiquitus Hardy, **New Synonym** 2, **349**, 351
ujungpandangiae new species 1, 20, 334, 339, **363**, 364, **607** (Fig. 364)
umbrosa (Fabricius) 2, 18, 54, 181, **186**, 187, 409, 410, **511** (Fig. 190A)
uncinata new species 1, 20, 217, 283, 308, 310, 348, **364**, 365, **608** (Fig. 365)
unifasciatus Hardy, Synonym **373**
unimacula Drew & Hancock 18, 23, 40, 110, 115, **187**, **512** (Fig. 191)
urens White 19, 221, 222, **224**, 232, **535** (Fig. 232)
usitata Drew & Hancock 18, 23, 66, 88, 108, 119, 126, 154, **187**, 188, **513** (Fig. 192)

vargus (Hardy) 20, **365**, 366, **608** (Fig. 366)
variabilis Lin & Wang, **New Synonym** 2, **76**, 78
variegata Wang, Synonym **373**
venefica (Hering) 19, 95, **188**, 189, **513** (Fig. 193)
verbascifoliae Drew & Hancock 19, 23, 43, 77, 78, 103, 143, 149, 169, **189**, 190, **514** (Fig. 194)
versicolor (Bezzi) 19, 38, 44, 100, 164, **190**, 191, **514** (Fig. 195)
vespoides Doleschall, Synonym **373**
vijaysegarani Drew & Hancock 21, 383, 385, 398, **399**, 400, **634** (Fig. 417)
vinnulus (Hardy) 2, 19, 226, 230, 231, 234, **235**, 236, **542** (Fig. 244)
vishnu Drew & Hancock 19, 23, 85, 121, 143, **191**, 192, **515** (Fig. 196)
vittatus (Hardy) 21, 376, 386, **400**, **634** (Fig. 418)
vulta (Hardy) 2, 20, 318, 330, **366**, 368, **609** (Fig. 367)

waaiae new species 1, 19, **192**, **515** (Fig. 197)
waimitaliae new species 1, 19, **236**, 237, **543** (Fig. 245)
wallacei White 21, 379, **400**, 401, **635** (Fig. 419)
watersi (Hardy) 20, 296, **368**, 369, **609** (Fig. 368)
whitei new species 1, 19, **258**, 259, **555** (Fig. 267)
wuzhishana Lin & Yang 19, 24, 124, **192**, 194, **516** (Fig. 198)

yalaensis new species 1, 20, 24, 312, **369**, 370, **610** (Fig. 369)
yayemanus – Hendel, Misspelling **241**
yayeyamanus (Matsumura), new combination 2, 3, 14, 238, **241**, 242, 243, **545** (Fig. 250A)
yercaudiae Drew, Synonym **199**
yilanensis Tseng, Chen & Chu, **New Synonym** 2, **76**, 78
yoshimotoi (Hardy) 20, 24, 208, 285, 301, 302, 303, 314, **370**, 371, **610** (Fig. 370)
yuiliensis Tseng & Chu, Synonym **279**

zahadi Mahmood 20, 24, 292, 293, **371**, 372, **611** (Fig. 371)
***Zeugodacus* Hendel, subgenus** **15**, 20, 263–372
ziae Wang, Synonym **395**
zonata (Saunders) 19, 40, 69, 100, 113, 186, **194**, 195, 200, 409, 410, **516** (Fig. 199A)